ENGINEERING METROLOGY

(Including Quality Management and Reliability Analysis)
A TEXTBOOK FOR ENGINEERING STUDENTS

By

Er. R.K. Jain

formerly, Member
Central Electricity Authority
New Delhi
Fellow, Metrology Society of India.
(Formerly, Lecturer, Mech. Engg. Deptt.
Regional Engineering College,
Kurukshetra (Haryana)

KHANNA PUBLISHERS®
Investing in Learning®

Operational Office :
4575/15, Onkar House, Opp. Happy School,
Ground Floor, Daryaganj, New Delhi 110 002
Phones : 011-45033819 • *Mob.* 09811541460, 9650669626
email : contactus@khannapublishers.in
website : khannapublishers.in

Published by :
Romesh Chander Khanna & Vineet Khanna
for **KHANNA PUBLISHERS**
2-B, Nath Market, Nai Sarak
Delhi- 110 006 (India)

Website : www.khannapublishers.in

Hologram & Description

To all readers of our books, to prevent yourself from being defrauded by piraters, please make sure that there is an Hologram on the cover of our books with the below specifications. If you find any book without **Hologram and Description**, please mail us at **contactus@khannapublishers.in**

Thanking you

ISBN No. 978-81-7409-153-6

First Edition : 1984
Twenty Third Edition : 2023
Third Reprint : 2026

***Price :* ₹ 879.00**

Preface to the Twentieth Edition

The subject of Engineering Metrology is gaining lot of popularity and recognition in industry with the advent of automatic and precision machines which depend for their operation on sophisticated instruments capable of giving accurate measurements. The present-day manufacturing industries demand engineers with sound background of metrology and statistical quality control techniques. It is hoped that this book will continue to serve as an inspiration to the engineers to take up the challenging jobs, the industries offer.

The text has been revised in line with latest practices and trends in the industry. The syllabii of all the engineering colleges have also been kept in view while revising this edition.

Numericals play an important role in clarifying the text and accordingly a large number of solved numericals have been added. In recent years Quality Management and Reliability Analysis have been gaining attention. Hence, keeping it in view new chapters on Quality Control and Quality Assurance ; Total Quality Management and I.S.O. Series and Reliability have been updated considerably as per latest practices.

Latest trends on metrological aspects, particularly the role of computers and microprocessors alongwith aspects on Calibration, Accreditation, Traceability, Globalisation of trade etc. have been covered thoroughly.

A new chapter on "*Evaluating Uncertainity in Measurement*" and "*Metrology Laboratories*" has been added.

I am thankful to all who have given their valuable comments, and suggestions for the improvement of the book. I am grateful to my wife Sunita Jain for her assistance in bringing up this edition.

Assistance taken from literature and journals 'NAPAN' published by Metrology Society of India is thankfully acknowledged.

I am also thankful to the Director General, Indian Standards Institution, for his permission to reproduce some figures, tables and extracts from Indian Standards Specifications. These standards are available from Indian Standards Institutions, Manak Bhawan, Bhadur Shah Zafar Marg, New Delhi and its branch offices at Ahmedabad, Banglore, Calcutta, Hyderabad, Kanpur and Patna.

1st July, 2007 —*R.K. Jain*

Preface to the First Edition

The most important single factor is achieving quality and reliability in service of any product is its dimensional control. Hence, to meet with the increasingly exacting demands of industry for mechanisms and assemblies, the functioning of which must meet the stringent design requirements, closer attention need be paid to the Science of Precision Measurements. The development of engineering production in the past has been inseparable from that of Metrology and for further improvement of the control of the process of manufacture, the means precision measurements have to be continuously developed. This is in short what the book has been written for.

It is hoped that this book will not only benefit students of undergraduate and post graduate course but also those in industry who wish to acquire knowledge of the fundamentals of this technique.

The knowledge of Metrology and its practical applications is of vital and increasing importance in the modern industry. This is also applicable to those techniques which are allied to Metrology and do not strictly form a part of it. It is for this reason that two chapters dealing with statistical Quality Control and Non-destructive Testing of Metals and Alloys have also been added in this book.

An earnest endeavour has been made to make the work as complete as possible covering all aspects of the subject in this book. Throughout the book, line sketches have been used that students can easily understand and reproduce them. To make the concept of the subject clearer, actual photographs of some instruments have also been added in addition to line sketches.

I am sincerely thankful to Prof. M.L. Mandal for going through to manuscript to provide suggestions and appreciation for improvement in the text. I am also thankful to my colleagues Sh. R.C. Jain, tool room engineer for his valuable suggestions and all those who helped me in the preparation of this book. My sincere thanks are also for the firms who have extended their co-operation in sending me the diagram and related information. Lastly, I am thankful to the publisher and the artist for their co-ordination.

Any suggestion or comments for the improvements of the book shall be highly appreciated.

New Delhi, 1971

—R.K. Jain

Contents

Pages

Chapter

1

Introduction

1.1. What is Metrology

The word metrology has a long tradition and is derived from the Greek word for measure. It is well-known saying that the knowledge about anything is complete only when it can be expressed in numbers and something is known about it. Thus for every kind of quantity measured, there must be a unit to measure it and express it in numbers of that unit. Further, in order that this unit is followed by all and not one who is taking the measurements, there must be a universal standard and the various units for various parameters of importance must be standardised. Most important parameter in metrology is the 'length' which can be measured in several forms and in several ways.

Measurements play a vital role in every field of investigation and present day scientific and technological progress has resulted from progress in the field of measurements. In general, measurements are made to increase our knowledge and understanding of the world with a view to lead a better life. Measurement science is vital for trade and commerce and is the basis of modern science and technology.

In metrology, which literally is the science of measurements we have to go one step ahead and bother ourselves about the correctness of measurement also. We have to see whether the result is given with the sufficient correctness and accuracy for the particular need or not. Thus we are primarily concerned with methods of measurement based on agreed units and standards. Metrology is thus concerned with the establishment, reproduction, conservation and transfer of units of measurements and their standards. The practice of metrology involves precise measurements requiring the use of apparatus and equipments (instruments and necessary adjuncts) to permit the degree of accuracy required to be obtained.

Metrology, the science of measurements, includes all aspects both theoretical and practical with reference to measurements, whatever their uncertainty, and in whatever fields of science or technology they occur. Thus metrology is also the science of measurement associated with the evaluation of its uncertainty. It is important to understand that only to measure is not the specifity of metrology but the core of metrology lies in the validation of the result, particularly by specifying its actual limitations. Metrology is not restricted only to standards of length and mass but other parameters in sectors of social concern, such as health, safety, and environment protection also. Making mastery of the science of measurement is a prerequisite for progress in science itself. Industrial manufacturing and many fields of life call for activity at a high scientific and technical level at which any progress achieved has to be accomplished by progress in metrology. The increasing automation of manufacture requires the highest level of accuracy. One must remember the famous saying that man's knowledge of nature, the universe, and how to adapt nature to his purpose, advances in step with his ability to measure precisely.

The metrologist has to understand the underlying principles to be able to design and develop new instruments and also to use the available instruments in the best way. Metrology is therefore also concerned with the methods, execution and estimation of accuracy of measurements ; the measuring instruments and the inspectors. Today's standard of precision and reliability are so high that man's basic instincts and senses are inadequate to cope with them. To this end, use has to be made of precision measuring instruments and various types of conventional and sophisticated gauges and comparators.

Thus, it can be said that metrology is mainly concerned with (*i*) establishing the units of measurements, reproducing these units in the form of standards, and ensuring the uniformity of measurements, (*ii*) developing methods of measurement, (*iii*) analysing the accuracy of methods of measurement, establishing uncertainty of measurement, researching into the causes of measuring errors, and eliminating these.

In the broader sense, metrology is not limited to length measurement but is also concerned with the industrial inspection and its various techniques. Due to big industrial revolution and great advancement, industrial inspection does not simply mean the fulfilling of the specifications laid down by the manufacturers. Rather inspection in real sense is concerned with the checking of a product at various stages in its manufacture right from the raw material form to the finished products and even assembled parts in the form of machine also. Inspection is carried out with gauges and the metrologist is intimately concerned with the design, manufacturing and testing of gauges of all kinds. Dynamic metrology is concerned with measuring small variations of continuous nature. The measurement science today has developed to electronically operated and controlled equipments, computer-aided systems for on-line monitoring, opto-mechanical, laser and fibre optics based instruments, etc.

As regards length measurement (dimensional inspection), we will be dealing with non-precision and precision linear measurements and study the various instruments used for this purpose. The standardisation of various units is also important and we will study what are the various standards for linear measurements and how attempts are made to preserve and maintain these standards. We will also see how the light wave standard helps us in doing away with material standards. For very precise measurements, methods based on light wave interference phenomena will also form a separate chapter.

Metrological activities start from establishment of measurement standards, appraisal of various physical parameters including dimensions, development of measuring instruments and techniques, and calibration of test and measurement equipments. All this is essential for correct operative measurement for quality and products and services delivered by the industry. Present day industry demands not only one time achievability, but aims for conformity involving such aspects as repeatability, reproducibility, interchangeability, of very many dimensions and characteristics and evidence thereof, for confidence of both producers and customers. This is possible by creation of standards and measurement techniques.

Due to mass production, it can be very easily realised that it is not possible to measure the various elements of a component by conventional methods. Thus other devices, *i.e.* gauges and comparators will be studied in detail. Further it is also not advisable to measure all the components in mass production if they are coming out of automatic machines. It will be seen that inspection of a few components out of a big lot is sufficient under the study of statistical quality control, the knowledge of which is very essential now-a-days.

As regards assembly and fitting of various components, some system of limits and fits has to be followed throughout and we will study the Indian Standard for 'Limits and Fits'. For assembled products in form of machine, it is essential that the relative movements of various parts of machine take place in a desired way. For this purpose study of machine tool alignment tests is very essential.

Many times in actual production, angle measurement presents a big problem and its thorough understanding and various techniques involved in it and circular division form an important part of metrology. Sometimes we come across various measurements which are really quite typical and problems of this type can be easily solved with the help of some trigonometrical relations. Such measurements will be studied in the chapter of Miscellaneous Measurements.

It may be emphasised here that man has to handle various instruments and sense of 'feel' plays very important role. In order that all people get same readings for a component by the same instrument, the instrument should be designed in such a way that always constant pressure is applied between the component and the instrument. Also instrument must be held such that the sense of 'feel' present in hand is free to give correct decision. However, in Universal Machines, an attempt is made to eliminate human errors due to different senses of touch and feel.

Good machines and their proper functioning calls for very good finished surfaces and thus study of the surface finish and various methods to estimate it quantitatively is very essential. A most commonly used tool in the hand of metrologist is 'dial-indicator' which will be studied in detail separately. Testing of gauges and dynamic measurements also deserve full attention.

We will also be dealing with the screw thread measurement and gear measurement and gauges for screw threads, as these are most popular parts one comes across in a workshop and machines.

A chapter is also devoted to non-destructive testing of metals and alloys to make one familiar with those methods which do not destroy the material and at the same time check all the desired properties and examine the internal structure of materials for homogeneity.

There is a great awareness about quality and industries are adopting the approach of Total Quality Management. Chapters have thus been devoted to Quality Assurance Programmes and Total Quality Management with reference to ISO 9000. A chapter is also devoted to Machine Vision Systems.

In broader sense, metrology (the science and art of precision measurement, testing and evaluation) is the mother science for technological development. The advancements in industry depend, to a great extent, on the quality and reliability of dimensional accuracy and precision measurement of other physical characteristics.

1.1.1. Legal Metrology. Legal Metrology is that part of metrology which treats units of measurement, methods of measurement and the measuring instruments, in relation to the statutory, technical and legal requirements. It assures security and appropriate accuracy of measurements. Lack of legislation regarding various measures will lead to great uncertainty.

Legal metrology is directed by a national organisation, *viz.* National Service of Legal Metrology whose object is to resolve problems of legal metrology in a particular country. Its functions are to ensure the conservation of national standards and to guarantee their accuracy by comparison with international standards ; and also to impart proper accuracy to the secondary standards of the country by comparison with international standards.

The contemporary organisation of metrology includes a number of international organisations *viz.* (*a*) The International Organisation of Weights and Measures : and (*b*) National Service of Legal Metrology whose ultimate object is to maintain uniformity of measurements throughout the world.

The activities of the service of Legal Metrology are : control (testing, verification, standardisation) of measuring instruments ; testing of prototypes/models of measuring instruments ; examination of a measuring instrument to verify its conformity to the statutory requirements, etc.

Legal metrology has application in :

(*i*) Commercial transactions (net quantity)

(*ii*) Industrial measurements (proper control on accuracy of measurement, so as to ensure interchangeability with a view to promoting mass production.

(*iii*) Measurements needed for ensuring public health and human safety.

A national law relating to legal metrology covers the following points :

(*i*) Legal units of measurements. In 1976, Parliament enacted comprehensive law, the Standards of Weights and Measures Act 1976, to establish the International System of Units (SI), to regulate inter-State trade or commerce in weights and measures, and to provide for other matters important from the view-point of consumer protection.

(*ii*) physical presentation of legal units ;

(*iii*) hierarchy of standards—their maintenance and custody ;

National Standards (Echelon-I)
Reference Standards (Echelon-II)
Secondary Standards (Echelon-III A)
Working Standards (Echelon-III B)

(*iv*) specifications or technical regulations of measuring instruments as regards their metrological and technical requirements ;

(*v*) metrological control on measuring instruments ; (approval of model, initial verification, periodical verification ; verification after repairs, inspection of the use of measuring instruments)

(*vi*) metrological control on pre-packed goods ;

(*vii*) control on manufacture, repair and sale of measuring instruments ;

(*viii*) organisation/service concerned with legal metrology ;

(*ix*) levy and collection of fees ;

(*x*) penalties for contraventions ;

(*xi*) training of personnel.

1.1.2. Deterministic Metrology. This is a new philosophy in which part measurement is replaced by process measurement. In the deterministic metrology, full advantage is taken of the deterministic nature of production machines (machines under automatic control are totally deterministic in performance) and all of the manufacturing sub-systems are optimised to maintain deterministic performance within acceptable quality levels. In this science, the system processes are monitored by temperature, pressure, flow, force, vibration, accoustic "finger printing" sensors, these sensors being fast and non-intrusive. The new technique such as 3D error compensation by CNC (Computer Numerical Control) systems and expert systems are applied, leading to fully adaptive control. This technology is used for very high precision manufacturing machinery and control systems to achieve microtechnology and nanotechnology accuracies.

1.2. Need of Inspection

In order to determine the fitness of anything made, man has always used inspection. But industrial inspection is of recent origin and has scientific approach behind it. It came into being because of mass production which involved interchangeability of parts. In old craft, same craftsman used to be producer as well as assembler. Separate inspections were not required. If any component part did not fit properly at the time of assembly, the craftsman would make the necessary adjustments in either of the mating parts so that each assembly functioned properly. So actually speaking, no two parts will be alike and there was practically no reason why they should be.

Now new production techniques have been developed and parts are being manufactured in large scale due to low-cost methods of mass production. So hand-fit methods cannot serve the purpose any more. When large number of components of same part are being produced, then any part would be required to fit properly into any other mating component part. This required specialisation of men and machines for the performance of certain operations. It has, therefore, been considered necessary to divorce the worker from all round crafts work and to supplant hand-fit methods with interchangeable manufacture.

The modern production techniques require that production of complete article be broken up into various component parts so that the production of each component part becomes an independent process. The various parts to be assembled together in assembly shop come from various shops. Rather some parts are manufactured in other factories also and then assembled at one place. So it is very essential that parts must be so fabricated that the satisfactory mating of any pair chosen at random is possible. In order that this may be possible, the dimensions of the component part must be confined within the prescribed limits which are such as to permit the assembly with a predetermined fit. Thus industrial inspection assumed its importance due to necessity of suitable mating of various components manufactured separately. It may be appreciated that when large quantities of work-pieces are manufactured on the basis of interchangeability, it is not necessary to actually measure the important features and much time could be saved by using gauges which determine whether or not a particular feature is within the prescribed limits. The methods of gauging, therefore, determine the dimensional accuracy of a feature, without reference to its actual size.

The purpose of dimensional control is however not to strive for the exact size as it is impossible to produce all the parts of exactly same size due to so many inherent and random sources of errors in machines and men. The principal aim is to control and restrict the variations within the prescribed limits. Since we are interested in producing the parts such that assembly meets the prescribed work standard, we must not aim at accuracy beyond the set limits which, otherwise is likely to lead to wastage of time and uneconomical results.

Lastly, inspection led to development of precision inspection instruments which caused the transition from crude machines to better designed and precision machines. It had also led to improvements in metallurgy and raw material manufacturing due to demands of high accuracy and precision. Inspection has also introduced a spirit of competition and led to production of quality products in volume by eliminating tooling bottle-necks and better processing techniques.

1.2.1. Measuring Means. The means of measurements could be classified as follows :

(*i*) Standards (Reference masters or setting standards)—These are used to reproduce one or several definite values of a given quantity.

(*ii*) Fixed gauges—These are used to check the dimensions, form, and position of product features.

(*iii*) Measuring instruments—These are used to determine the values of the measured quantity.

1.3. Physical Measurement

Measurement is a complex of operations carried out by means of measuring instruments to determine the numerical value of the size which describes the object of measurement.

A physical measurement could be defined as the act of deriving quantitative information about a *physical object* or action by *comparison* with a *reference*. It will be noted from this definition that there are three important elements of a measurement, *viz*, (*i*) measurand, *i.e.* the physical quantity or property like length, angle etc. being measured, (*ii*) comparison or comparator, *i.e.* the means of comparing measurand with some reference to render a judgement and (*iii*) reference, *i.e.* the physical quantity or property to which quantitative comparisons are made. All these three terms would be very clear from the following example of a direct measurement using a calibrated fixed reference. Say, a mechanic has to measure the length of a surface table (measurand). For this, first he lays his rule (reference) alongside the table ; he then carefully aligns the zero end of his rule with one end of the table ; and finally he compares the length of table (measurand) with the graduation on his rule (reference) by eye (comparator).

Such examples of direct measurement using a calibrated fixed reference are plenty in the field of Metrology. For example, measurements of length using vernier calliper, micrometer screw gauge, measurement of angle using bevel protractor etc. Many measurements are made using inferential methods, *e.g.* measurement of pitch diameter or threads using three wire method. Such measurements are also frequently encountered when it is not possible to measure a measurand directly. Such methods are dependent on direct relationship (mathematical or otherwise) between the measurand and the actual measurement being made.

In modern measuring instruments which employ electrical principles, the measured parameter is sensed and converted into electrical quantity such as voltage (there being a direct relationship between the two), which is measured by a null measurement using a calibrated variable reference. The variable reference (by potentiometric voltmeter) is adjusted until an indicator shows that no voltage differential exists between the reference and the unknown voltage. Under such a condition of null, the reference voltage must be equal to the unknown voltage. The purpose of making measurement could be either to quantify the measurand (information), or, provide a historical profile to be referred to at other times (record) or use the information to produce action such as to produce correct size on machine tool (control). The measurement system to be employed will depend on the objective of the measurement to be fulfilled. A modern measurement system utilising electrical techniques responds to the measurand, effects a comparison with the reference and provides information based on that comparison.

However, the direct measurement system using a calibrated fixed reference is more conventional in the field of Metrology. Of course there are several other methods of measurements which are also used in this field.

Principle of measurement. It is the physical phenomenon utilised in the measurement.

Method of measurement. It is the way the measuring principles and measuring means are used.

Nominal size (Basic Size). It is the size on which the limits of size are based and which is assigned to a part in accordance with its function.

True size. It is the theoretical size of a dimension, which is free from any errors of measurement.

Actual Size. It is the value of size obtained through measurement with the permissible measuring error.

Exact size. It is the value of size obtained with the highest metrological accuracy attainable in practice.

Approximate Size. It is the value of size obtained with an error exceeding the permissible error of measurement and requiring refinement.

Error of measurement. It is the difference between the true value of the size being measured and the value found by measurement. Error pertains to a measurement and not to an instrument.

Correction. It is the amount which should be algebraically added to the indicated value to obtain the actual value of the size being measured. The correction is numerically equal to the error, but opposite in sign.

Correctness of measurement. It is quantitative characteristic showing how close to zero are the systematic errors of measurement results.

Reliability of measurement. It is a qualitative characteristic which implies confidence in the measured results depending on whether or not the frequency distribution characteristics of their deviations from the true values of the corresponding quantities are known.

Verification. It is the process of testing an instrument for the purpose of assessing the indication errors and determining whether setting standards or measuring instruments meet the prescribed specifications.

Calibration. It is the process of determining the values of the quantity being measured corresponding to a pre-established arbitrary scale.

1.4. Measuring Instruments

Measuring Instruments are measuring devices that transform the measured quantity or a related quantity into an indication or information.

Measuring instruments can either indicate directly the value of the measured quantity or only indicate its equality to a known measure of the same quantity (*e.g.* equal arm balance, or null detecting galvanometer). They may also indicate the value of the small difference between the measured quantity and the measure having a value very near to it (comparator).

Measuring instruments usually utilise a measuring sequence in which the measured quantity is transformed into a quantity perceptible to the observer (length, angle, sound, luminous contrast).

Measuring instruments may be used in conjunction with separate material measures (*e.g.* balances using standard masses to compare unknown mass), or they may contain internal parts to reproduce the unit (like graduated rules, a precision thread, etc.)

1.4.1. Measuring range. It is the range of values of the measured quantity for which the error obtained from a single measurement under normal conditions of use does not exceed the maximum permissible error.

The measuring range is limited by the maximum capacity and the minimum capacity.

Maximum capacity is the upper limit of the measuring range and is dictated by the design considerations or by safety requirements or both.

Minimum capacity is the lower limit of the measuring range. It is usually dictated by accuracy requirements. For small values of the measured quantity in the vicinity of zero, the relative error can be considerable even if the absolute error is small.

The measuring range may or may not coincide with the range of scale indication.

1.4.2. Sensitivity. It is the quotient of the increase in observed variable (indicated by pointer and scale) and the corresponding increase in the measured quantity.

It is also equal to the length of any scale division divided by the value of that division expressed in terms of the measured quantity.

The sensitivity may be constant or variable along the scale. In the first case, we get linear transmission and in the second case, we get a non-linear transmission.

1.4.3. Scale Interval. It is the difference between two successive scale marks in units of the measured quantity. (In the case of numerical indication, it is the difference between two consecutive numbers).

The scale interval is an important parameter that determines the ability of the instrument to give accurate indication of the value of the measured quantity.

The scale spacing, or the length of scale interval, should be convenient, for estimation of fractions.

1.4.4. Discrimination. It is the ability of the measuring instrument to react to small changes of the measured quantity.

1.4.5. Hysteresis. It is the difference between the indications of a measuring instrument when the same value of the measured quantity is reached by increasing or by decreasing that quantity.

The phenomenon of hysteresis is due to the presence of dry friction as well as to the properties of elastic elements. It results in the loading and unloading curves of the instrument being separated by a difference called the hysteresis error. It also results in the pointer not returning completely to zero when the load is removed.

Hysteresis is particularly noted in instruments having elastic elements. The phenomenon of hysteresis in materials is due mainly to the presence of internal stresses. It can be reduced considerably by proper heat treatment.

1.4.6. Response time. It is the time which elapses after a sudden change in the measured quantity until the instrument gives an indication differing from the true value by an amount less than a given permissible error.

The curve showing the change of indication of an instrument due to sudden change of measured quantity can take different forms according to the relation between capacitances that have to be filled, inertia elements and damping elements.

When inertia elements are small enough to be negligible, we get first order response which is due to filling the capacitances in the system through finite channels. The curve of change of indication with time in that case is an exponential curve. (Refer Fig. 1.1)

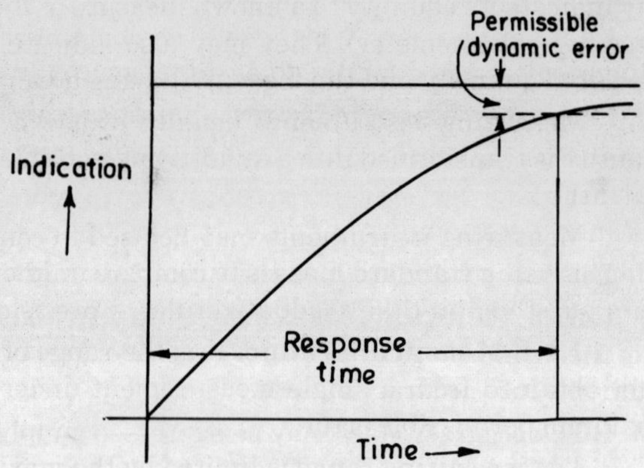

Fig. 1.1. First order response of an instrument.

If the inertia forces are not negligible, we get second order response. There are three possi-bilities of response (Refer Fig. 1.2.) according to the ratio of damping and inertia forces as follows :

— *overdamped system*—where the final indication is approached exponentially from one side.

— *under-damped system*—where the pointer approaches the position corresponding to final reading, passes it and makes a number of oscillations around it before it stops.

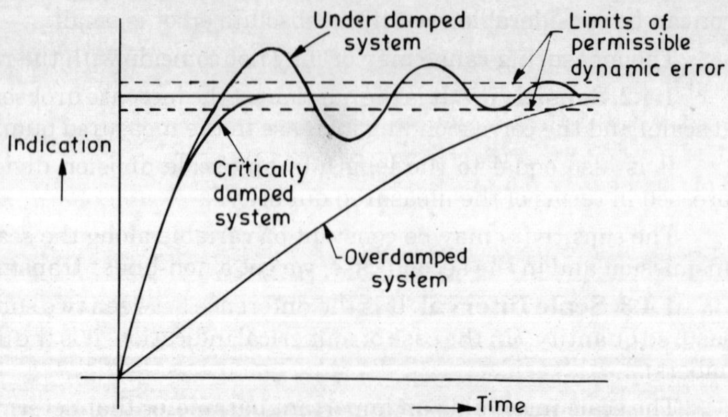

Fig. 1.2. Second order response of an instrument.

— *critically damped system*—where the pointer motion is aperiodic but quicker than in the case of overdamped system.

In all these cases the response time is determined by the inter-section of one (or two) lines surrounding the final indication line at a distance equal to the permissible value of dynamic error with the response curve of the instrument.

1.4.7. Repeatability. It is the ability of the measuring instrument to give the same value every time the measurement of a given quantity is repeated.

Any measurement process effected using a given instrument and method of measurement is subject to a large number of sources of variation like environmental changes, variability in operator performance and in instrument parameters. The repeatability is charactersied by the dispersion of indications when the same quantity is repeatedly measured. The dispersion is described by two limiting values or by the standard deviation.

The conditions under which repeatability is tested have to be specified.

1.4.8. Bias. It is the characteristic of a measure or a measuring instrument to give indications of the value of a measured quantity whose average differs from the true value of that quantity.

Bias error is due to the algebraic summation of all the systematic errors affecting the indication of the instrument. Sources of bias are maladjustment of the instrument, permanent set, non-linearly errors, errors of material measures etc.

1.4.9. Inaccuracy. It is the total error of a measure or measuring instrument under specified conditions of use and including bias and repeatability errors.

Inaccuracy is specified by two limiting values obtained by adding and subtracting to the bias error the limiting value of the repeatability error.

If the known systematic errors are corrected, the remaining inaccuracy is due to the random errors and the residual systematic errors that also have a random character.

This inaccuracy is called the "uncertainty of measurement".

1.4.10. Accuracy Class. Measuring instruments are classified into accuracy classes according to their metrological properties. There are two methods used in practice for classifying instruments into accuracy classes.

1. The accuracy class may be expressed simply by a class ordinal number that gives an idea but no direct indication of the accuracy (*e.g.* block gauges of accuracy class, 0, 1, 2 etc.).

2. The accuracy class is expressed by a number stating the maximum permissible inaccuracy as %age of the highest indication given by the instrument (*e.g.* an instrument of accuracy class 0.2 and maximum capacity 0—100 will have a maximum permissible inaccuracy of ± 0.2 at any point in the scale of instrument.

1.4.11. Precision and Accuracy. Both these terms are associated with the measuring process. Precision is defined as the repeatability of a measuring process, while accuracy is the agreement of the result of a measurement with the true value of the measured quantity. In most measurements it is the precision which is of greater importance. The chief concern is with comparing the dimension of measurement relative to each other, it being assumed that the scale used for measurement is a standard and accepted one. This would be clear from the example given below.

If a carpenter had to cut a board to fit the shelf into two projections in the wall, it does not matter whether his scale is accurate or not, provided he uses the same scale for the measurement of board and the distance between projections in the wall. Here the precision with which he measures two is of importance.

Now supposing he had to order for the board from market then it is necessary that the scale used by him and the one in market are in agreement with each other. One way to achieve this is

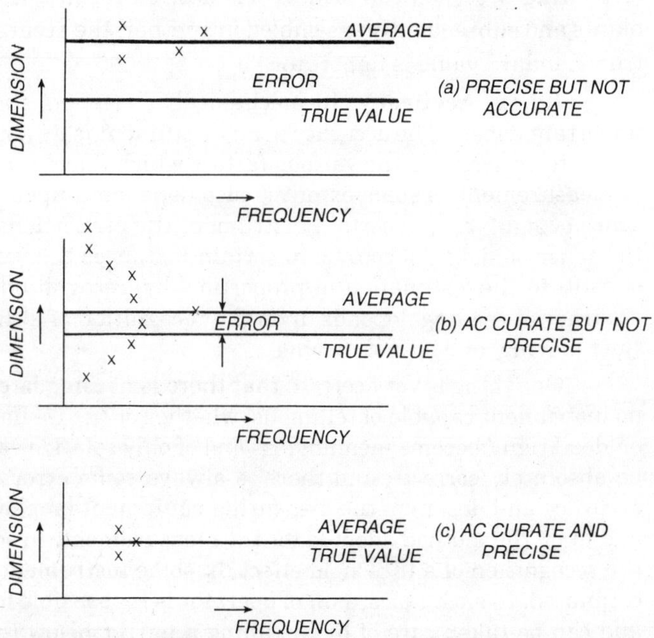

Fig. 1.3

Illustrating distinction between accuracy and precision by measuring a component several times and plotting the readings, by three instruments : (a) precise but not accurate, (b) accurate but not precise, (c) accurate and precise.

that both use the accurate scales in accordance with the standard scales. In this case, accuracy of the scale is important and it should be manufactured such that its units are in accordance with the standard units set.

The distinction between the precision and accuracy will become clear by the following example (shown in Fig. 1.3), in which several measurements are made on a component by different types of instruments and results plotted.

From Fig. 1.3, it will be obvious that precision is concerned with a process or a set of measurements, and not a single measurement. In any set of measurements, the individual measurements are scattered about the mean, and the precision tells us to how well the various measurements performed by same instrument on the same component agree with each other. It will be appreciated that poor repeatability is a sure sign of poor accuracy. Good repeatability of the instrument is a necessary but not a sufficient condition of good accuracy. Accuracy can be found by taking root mean square of repeatability and systematic error *i.e.*

$$\text{Accuracy} = \sqrt{(\text{repeatability})^2 + (\text{systematic error})^2}.$$

Error is the difference between the mean of set of readings on same component and the true value. Less is the error, more accurate is the instrument. Since the true value is never known, uncertainty creeps in, and the magnitude of error must be estimated by other means. The estimate of uncertainty of a measuring process can be made by taking care of systematic and constant errors, and other contributions to the uncertainty due to scatter of the results about the mean.

So wherever great precision is required in manufacture of mating components, they are manufactured in a single plant, where measurements are taken with same standards and internal measuring precision can achieve the desired results. If they are to be manufactured in different plants and subsequently assembled in another, the accuracy of the measurement of two plants with true standard value is important.

1.4.12. Accuracy. In mechanical inspection, the accuracy of measurement is the most important aspect. The accuracy of an instrument is its ability to give correct results. It is, therefore, better to understand the various factors which affect it and which are affected by it. The accuracy of measurement to some extent is also dependent upon the sense of hearing or sense of touch or sense of sight, *e.g.*, in certain instrument the proportions of sub-divisions have to be estimated by the sense of sight ; of course, in certain instances the vernier device may be employed in order to substitute the 'estimation of proportion' by 'recognition of coincidence'. In certain instruments, accuracy of reading depends upon the recognition of a threshold effect, *i.e.* whether the pointer is "just moving, or 'just not moving'.

One thing is very certain that there is nothing like absolute or perfect accuracy and there is no instrument capable of telling us, whether or not we have got it. The phrases like 'dead accurate' or 'dead right' become meaningless and of only relative value. In other words, no measurement can be absolutely correct ; and there is always some error, the amount of which depends upon the accuracy and design of the measuring equipment employed and the skill of the operator using it, and upon the method adopted for the measurement. In some instruments, accuracy depends upon the recognition of a threshold effect. In some instruments, proportions of sub-divisions have to be estimated. In such cases, skill of operator is responsible for accuracy, Parallax is also very common and can be taken care of by installing a mirror below the pointer. How method of measurement affect accuracy would be realised in angle measurement by sine bar, *i.e.* large errors may occur when sine bar is intended to be used for measuring large angles. Apparatus and methods should be designed so that errors in the final results are small compared with errors in actual measurements made. The equipment chosen for a particular measurement must bear some relation to the desired accuracy in the result, and as a general rule, an instrument which can be read to the next decimal place beyond that required in the measurement should be used, *i.e.*, if a measurement is desired to an accuracy of 0.01 mm, then instrument with accuracy of 0.001 mm should be used for this purpose.

When attempts are made to achieve higher accuracy in the measuring instruments, they become increasingly sensitive. But an instrument can't be more accurate than is permitted by degree of sensitiveness, the sensitivity being defined as the ratio of the change of instrument indication to the change of quantity being measured. It may be realized that the degree of sensitivity of an instrument is not necessarily the same all over the range of its readings. Another important consideration for achieving higher accuracy is that the readings obtained for a given quantity should be same all the time, *i.e.* in other words the readings should be consistent. A highly sensitive instrument is not necessarily consistent in its readings and, clearly, an inconsistent instrument can't be accurate to a degree better than its inconsistency. It may also be remembered that the range of measurement usually decreases as the magnification increases and the instrument may be more affected by temperature variations and be more dependent upon skill in use.

Thus, it is true to say that a highly accurate instrument possesses both greater sensitivity and consistency. But at the same time an instrument which is sensitive and consistent need not necessarily be accurate because the standard from which its scale is calibrated may be wrong. (It is, of course, presupposed that there always exists an instrument whose accuracy is greater than the one with which we are concerned). In such an instrument, the errors will be constant at any given reading and therefore, it would be quite possible to calibrate it.

It is very obvious that higher accuracy can be achieved by incorporating the magnifying devices in the instrument, and these magnifying devices carry with them their own inaccuracies, *e.g.*, in an optical system the lens system may distort the ray in a variety of ways and the success of the system depends upon the fidelity with which the lens system can produce the magnified images etc. In mechanical system the errors are introduced due to bending of levers, backlash at the pivots, inertia of the moving parts, errors of the threads of screws etc. Probably the wrong geometric design may also introduce errors. By taking many precautions we can make these errors extremely small, but, the smaller we try to make them, the greater becomes the complication of our task, and with this increased complication, the greater the number of possible sources of error which we must take care of. Thus the greater the accuracy is aimed at, greater the number of sources of errors to be investigated and controlled. As regards the instrumental errors, they can be kept as small as possible. The constant or knowable sources of errors can be determined by the aid of superior instruments and the instrument may be calibrated accordingly. The variable or unknowable sources of errors make the true value lie within plus or minus departure from the observed value and can't be tied down more closely than that. However, an accurate measuring instrument should fulfil the following requirements :

(*i*) It should possess the requisite and constant accuracy.

(*ii*) As far as possible, the errors should be capable of elimination by adjustment contained within the instrument itself.

(*iii*) Every important source of inaccuracy should be known.

(*iv*) When an error can't be eliminated, it should be made as small as possible.

(*v*) When an error can't be eliminated, it should be capable of measurement by the instrument itself and the instrument calibrated accordingly.

(*vi*) As far as possible the principle of similarity must be followed, *i.e.* the quantity to be measured must be similar to that used for calibrating the instrument. Further the measuring operations performed on the standard and on the unknown must be as identical as feasible and under the same physical conditions (environment temperature, etc. and using the same procedures in all respects in both the cases of calibration and measurement).

In some instruments, accuracy is expressed as percentage of full scale deflection, *i.e.* percentage of maximum reading of instrument. Thus at lower readings in the range, accuracy may be very poor. The range of such instruments should be selected properly so that the measured value is at about 70-90% of the full range.

Accuracy in measurement is essential at all stages of product development from research to development and design, production, testing and evaluation, quality assurance, standardisation, on-line control, operational performance appraisal, reliability estimation, etc.

The last word in connection with accuracy is that the accuracy at which we aim, that is to say, the trouble we take to avoid errors in manufacture and in measuring those errors during inspection must depend upon the job itself and on the nature of what is required, *i.e.* we should make ourselves very sure whether we want that much accuracy and that cost to achieve it will be compensated by the purpose for which it is desired.

1.4.13. Accuracy and Cost. The basic objective of metrology should be to provide the accuracy required at the most economical cost. The accuracy of measuring system includes elements such as :

(*a*) Calibration Standards,

(*b*) Workpiece being measured,

(*c*) Measuring Instruments

(*d*) Person or Inspector carrying out the measurement, and

(*e*) Environmental influences.

The above arrangement and analysis of the five basic Metrology elements can be composed into the acronym SWIPE for convenient reference :

S = Standard, W = Workpiece, I = Instrument, P = Person and E = Environment.

Higher accuracy can be achieved only if, all the sources of errors due to the above five elements in the measuring system be analysed and steps taken to eliminate them. An attempt is made here to summarise the various factors affecting these five elements.

1. *Standard.*It may be affected by ambient influences (thermal expansion), stability with time, elastic properties, geometric compatibility, and position of use.

2. *Workpiece*, itself may be affected by ambient influences, cleanliness, surface condition, elastic properties, geometric truth, arrangement of supporting it, provision of defining datum etc.

3. *Instrument* may be affected by hysteresis, backlash, friction, zero drift error, deformation in handling or use of heavy workpieces, inadequate amplification, errors in amplification device, calibration errors, standard errors, correctness of geometrical relationship of workpiece and standard, proper functioning of contact pressure control, mechanical parts (slides, ways, or moving elements) working efficiently, and repeatability adequacy etc.

4. *Personal* errors can be many and mainly due to improper training in use and handling, skill, sense of precision and accuracy appreciation, proper selection of instrument, attitude towards and realisation of personal accuracy achievements, etc.

5. *Environment* exerts a great influence. It may be affected by temperature ; thermal expansion effects due to heat radiation from light, heating of components by sunlight and people, temperature equalisation of work, instrument and standard ; surroundings ; vibrations ; lighting ; pressure gradients (affect optical measuring systems) etc.

The design of measuring systems involves proper analysis of cost-to-accuracy consideration and the general characteristics of cost and accuracy appears to be as shown in Fig. 1.4.

It will be clear from the graph that the cost rises exponentially with accuracy. If the measured quantity relates to a tolerance (*i.e.* the permissible variation in the measured quantity), the accuracy objective should be 10% or slightly less of the tolerance. In a few cases, because of technological limitations, the accuracy may be 20% of the tolerance ; because demanding too high accuracy may tend to make the measurement unreliable. In practice, the desired ratio of accuracy to tolerance is decided by considering the factors such as the cost of measurement versus quality and the reliability criterion of the product.

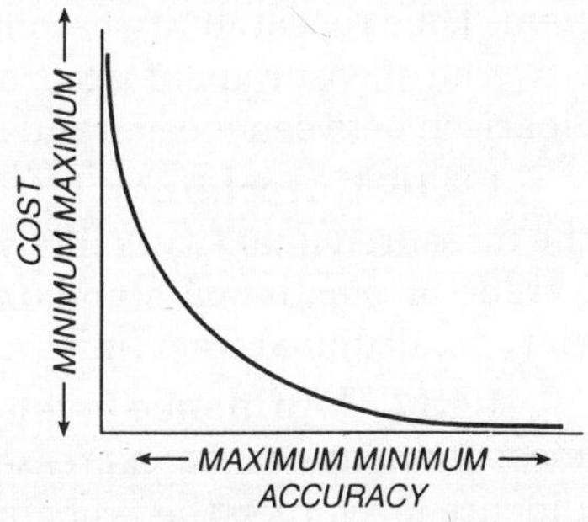

Fig. 1.4

1.4.14. Magnification. The human limitations or incapability to read instruments places limit on sensitiveness of instruments. Magnification (or amplification) of the signal from measuring instrument can make it better readable. The magnification is possible on mechanical, pneumatic, optical, electrical principles or combinations of these.

Magnification by mechanical means is the simplest and costwise most economical method. The various methods of mechanical magnification are based on the principles of lever, wedge, gear train etc.

In the case of magnification by wedge method, magnification is equal to tan θ, where θ is the angle of wedge.

Mechanical magnification method is usually grouped with other type of magnification methods to combine their merits.

Optical magnification is based on the principle of reflection by tilting a mirror, or on projection technique. In the case of reflection by a mirror, the angular magnification is 2 because the reflected ray is tilted by twice the angle of tilt of mirror. With 2-mirror system four-fold magnification is obtained.

In the case of optical lever magnification, various features of the test specimen are established on projector screen using reference lines as datum. Very high magnifications are possible in such systems.

Pneumatic magnification method is ideally suited for internal measurement. It offers better reliability and stability. Very high magnifications (upto 30,000 : 1) are possible.

Electrical magnification methods have the advantages of better control on the amount of magnification, quick response, large range of linearity, etc. Electrical magnification is based on sensing change in inductance or capacitance, which is measured by a wheatstone bridge.

Electronic method of magnification is more reliable and accurate. Electronic methods are ideally suited for processing of signals, *viz.* amplification, filtering, validation of signal, sensing for high and low limits, autocalibration, remote control telemetry etc.

1.4.15. Repeatability. Repeatability is the most important factory in any measuring system as it is the characteristics of the measuring system whereby repeated trials of identical inputs of measured value produce the same indicated output from the system.

It is defined as the ability of a measuring system to reproduce output readings when the same measurand value is applied to it consecutively, under the same conditions, and in the same direction. It could be expressed either as the maximum difference between output readings, or as "within ... per cent of full scale output."

Repeatability is the only characteristic error which cannot be calibrated out of the measuring system. Thus repeatability becomes the limiting factor in the calibration process, thereby limiting the overall measurement accuracy. In effect, repeatability is the minimum uncertainty in the comparison between measurand and reference.

1.4.16. Uncertainty. It is the range about the measured value within which the true value of the measured quantity is likely to lie at the stated level of confidence. It can be calculated when the trade or population standard deviation is known or it can be estimated from the standard deviation calculated from finite number of observation having normal distribution.

1.4.17. Confidence levels. It is the measure of a degree of reliability with which the results of measurement can be expressed. Thus if u be uncertainty in a measured quantity x at 98% confidence level, then the probability for true value to lie between $x + u$ and $x - u$ is 98%. Thus on measuring this quantity a large number of times, then 98% of the values will lie in between $x + u$ and $x - u$.

1.4.18. Calibration. The calibration of any measuring system is very important to get meaningful results. In case where the sensing system and measuring system are different, then it is imperative to calibrate the system as an integrated whole in order to take into account the error producing properties of each component. Calibration is usually carried out by making adjustments such that readout device produces zero output for zero-measurand input, and similarly it should display an output equivalent to the known measurand input near the full-scale input value.

It is important that any measuring system calibration should be performed under environmental conditions that are as close as possible to those conditions under which actual measurements are to be made.

It is also important that the reference measured input should be known to a much greater degree of accuracy—usually the calibration standard for the system should be at least one order of magnitude more accurate than the desired measurement system accuracy, *i.e.* accuracy ratio of 10 : 1.

1.4.19. Calibration vs. Certification. Calibration is the process of checking the dimensions and tolerances of a gauge, or the accuracy of a measurement instrument by comparing it to a like instrument/gauge that has been certified as a standard of known accuracy. Calibration is done by detecting and adjusting any discrepancies in the instrument's accuracy to bring it within acceptable limits. Calibration is done over a period of time, according to the usage of the instrument and the materials of its parts. The dimensions and tolerances of the instrument/gauge are checked to determine whether it has departed from the previously accepted certified condition. If departure is within limits, corrections are made. If deterioration is to a point that requirements can't be met any more then the instrument/gauge can be downgraded and used as a rough check, or it may be reworked and recertified, or be scrapped. If a gauge is used frequently, it will require more maintenance and more frequent calibration.

Certification is performed prior to use of instrument/gauge and later to reverify if it has been reworked so that it again meets its requirements. Certification is given by a comparison to a reference standard whose calibration is traceable to an accepted national standard. Further such reference standards must have been certified and calibrated as master not more than six months prior to use.

1.4.20. Sensitivity and Readability. The terms "sensitivity" and "readability" are often confused with accuracy and precision. Sensitivity and readability are primarily associated with equipment while precision and accuracy are associated with the measuring process. It is not necessary that the most sensitive or the most readable equipment will give most precise or the most accurate results.

Sensitivity refers to the ability of a measuring device to detect small differences in a quantity being measured. It may so happen that high sensitivity instrument may lead to drifts due to thermal or other effects, so that its indications may be less repeatable or less precise than those of the instrument of lower sensitivity.

Readability refers to the susceptibility of a measuring device to having its indications converted to a meaningful number. A micrometer instrument can be made more readable by using verniers. Very finely spaced lines may make a scale more readable when a microscope is used, but for the unaided eye, the readability is poor.

1.4.21. Uncertainty in Measurement. Whenever a value of physical quantity is determined through a measurement process, some errors are inherent in the process of measurement and it is only the best estimate value of the physical quantity obtained from the given experimental data. Thus quantifying a measurable quantity through any measurement process is meaningful only if the value of the quantity measured with a proper unit of measurement is accompanied by an overall uncertainty of measurement. It has two components arising due to random errors and systematic errors. Uncertainty in measurement could be defined as that part of the expression of the result of a measurement which states the range of values within which the true value, or if appropriate, the conventional true value is estimated to lie. In cases where there is adequate information based on a statistical distribution, the estimate may be associated with a specified probability. In other cases, an alternative form of numerical expression of the degree of confidence to be attached to the estimate may be given.

1.4.22. Random uncertainty and Systematic uncertainty. Random uncertainty is that part of uncertainty in assigning the value of a measured quantity which is due to random errors. The value of the random uncertainty is obtained on multiplication of a measure of the random errors which is normally the standard deviation by a certain factor '*t*'. The factor '*t*' depends upon the sample size and the confidence level. Systematic uncertainty is that part of uncertainty which is due to systematic errors and cannot be experimentally determined unless the equipment and environmental conditions are changed. It is obtained by suitable combination of all systematic errors arising due to different components of the measuring system.

It is necessary to understand difference between systematic uncertainty and correction.

The calibration certificate of an instrument gives a correspondence between its indication and the quantity it is most likely to measure. The difference between them is the correction which is to be invariably applied. However, there will be an element of doubt in the value of the correction so stated. This doubt is quantitatively expressed as accuracy or overall uncertainty in assigning the value to the correction stated and will be one component of the systematic uncertainty of that instrument. For example, in case of a metre bar, the distance between the zero and 1,000 mm graduation marks may be given as 1000.025 ± 0.005 mm. Then 0.025 mm is the correction and 0.005 mm is the component of systematic uncertainty of the metre bar.

1.4.23. Traceability. This is the concept of establishing a valid calibration of a measuring instrument or measurement standard by step-by-step comparison with better standards up to an accepted or specified standard. In general, the concept of traceability implies eventual reference to an appropriate national or international standard.

1.4.24. Fiducial Value. A prescribed value of a quantity to which reference is made, for example, in order to define the value of an error as a proportion of this prescribed value.

1.5. Selection of Instruments

The important characteristics to be considered in selection of an instrument are its measuring range, accuracy and precision. Usually accuracy is poor at the lower end of scale which should be avoided. In such a situation, where accurate measurement is required throughout full range, two instruments with different ranges may be used, one for lower range and other for full range.

The precision of instrument is very important feature since it should give repeatable readings which is possible with precise instrument. If an instrument is precise then accuracy can be taken care of by proper calibration of the instrument. It may be mentioned that accuracy is affected by systematic errors which can be accounted for.

Resolution or sensitivity, *i.e.* discrimination or relation between the movement of pointer and corresponding change in the measured quantity is also important aspect. It represents a smallest change in the measured quantity which produces a perceptible movement of the pointer.

1.6. Classification of Methods of Measurements

In precision measurements various methods of measurement are followed depending upon the accuracy required and the amount of permissible error.

There are numerous ways in which a quantity can be measured. Any method of measurement should be defined in such a detail and followed by such a standard practice that there is little scope for uncertainty. The nature of the procedure in some of the most common measurements is described below. Actual measurements may employ one or more combinations of the following.

(*i*) *Direct method of measurement.* In this method the value of a quantity is obtained directly by comparing the unknown with the standard. It involves, no mathematical calculations to arrive at the results, for example, measurement of length by a graduated scale. The method is not very accurate because it depends on human insensitiveness in making judgement.

(*ii*) *Indirect method of measurement.* In this method several parameters (to which the quantity to be measured is linked with) are measured directly and then the value is determined by mathematical relationship. For example, measurement of density by measuring mass and geometrical dimensions.

(*iii*) *Fundamental method of measurement.* Also known as the *absolute method of measurement*, it is based on the measurement of the base quantities used to define the quantity. For example, measuring a quantity directly in accordance with the definition of that quantity, or measuring a quantity indirectly by direct measurement of the quantities linked with the definition of the quantity to be measured.

(*iv*) *Comparison method of measurement.* This method involves comparison with either a known value of the same quantity or another quantity which is function of the quantity to be measured.

(*v*) *Substitution method of measurement.* In this method, the quantity to be measured is measured by direct comparison on an indicating device by replacing the measuring quantity with some other known quantity which produces same effect on the indicating device. For example, determination of mass by Borda method.

(*vi*) *Transposition method of measurement.* This is a method of measurement by direct comparison in which the value of the quantity to be measured is first balanced by an initial known value A of the same quantity ; next the value of the quantity to be measured is put in the place of that known value and is balanced again by a second known value B. When the balance indicating device gives the same indication in both cases, the value of the quantity to be measured is $\sqrt{AB}$. For example, determination of a mass by means of a balance and known weights, using the Gauss double weighing method.

(*vii*) *Differential or comparison method of measurement.* This method involves measuring the difference between the given quantity and a known master of near about the same value. For example, determination of diameter with master cylinder on a comparator.

(*viii*) *Coincidence method of measurement.* In this differential method of measurement the very small difference between the given quantity and the reference is determined by the observation of the coincidence of scale marks. For example, measurement on vernier caliper.

(*ix*) *Null method of measurement.* In this method the quantity to be measured is compared with a known source and the difference between these two is made zero.

(*x*) *Deflection method of measurement.* In this method, the value of the quantity is directly indicated by deflection of a pointer on a calibrated scale.

(*xi*) *Interpolation method of measurement.* In this method, the given quantity is compared with two or more known value of near about same value ensuring at least one smaller and one bigger than the quantity to be measured and the readings interpolated.

(*xii*) *Extrapolation method of measurement.* In this method, the given quantity is compared with two or more known smaller values and extrapolating the reading.

(*xiii*) *Complimentary method of measurement.* This is the method of measurement by comparison in which the value of the quantity to be measured is combined with a known value of the same quantity so adjusted that the sum of these two values is equal to predetermined comparison value. For example, determination of the volume of a solid by liquid displacement.

(*xiv*) *Composite method of measurement.* It involves the comparison of the actual contour of a component to be checked with its contours in maximum and minimum tolerable limits. This method provides for the checking of the cumulative errors of the interconnected elements of the component which are controlled through a combined tolerance. This method is most reliable to ensure inter-changeability and is usually effected through the use of composite "Go" gauges, for example, checking of the thread of a nut with a screw plug "GO" gauge.

(*xv*) *Element method.* In this method, the several related dimensions are gauged individually, *i.e.* each component element is checked separately. For example, in the case of thread, the pitch diameter, pitch, and flank angle are checked separately and then the virtual pitch diameter is calculated. It may be noted that value of virtual pitch diameter depends on the deviations of the above thread elements. The functioning of thread depends on virtual pitch diameter lying within the specified tolerable limits.

In case of composite method, all the three elements need not be checked separately and is thus useful for checking the product parts. Element method is used for checking tools and for detecting the causes of rejects in the product.

(*xvi*) *Contact and contactless methods of measurements.* In contact methods of measurements, the measuring tip of the instrument actually touches the surface to be measured. In such cases, arrangements for constant contact pressure should be provided in order to prevent errors due to excess contact pressure. In contactless method of measurements, no contact is required. Such instruments include tool-maker's microscope and projection comparator, etc.

For every method of measurement a detailed definition of the equipment to be used, a sequential list of operations to be performed, the surrounding environmental conditions and descriptions of all factors influencing accuracy of measurement at the required level must be prepared and followed.

1.6.1. Classification of Measuring Instruments. According to the functions, the measuring instruments are classified as :

(1) Length measuring instruments.

(2) Angle measuring instruments.

(3) Instruments for checking the deviations from geometrical forms.

(4) Instruments for determining the quality of surface finish.

According to the accuracy of measurement, the measuring instruments are classified as follows :

(1) Most accurate instruments *e.g.*, light-interference instruments.

(2) Second group consists of less accurate instruments such as tool room microscopes, comparators, optimeters etc.

(3) The third group comprises still less accurate instruments *e.g.*, dial indicators, vernier calipers and rules with vernier scales.

1.6.2. Metrological characteristics of Measuring Instruments. Measuring instruments are usually specified by their metrological properties, such as range of measurement, scale graduation value, scale spacing, sensitivity and reading accuracy.

Range of Measurement. It indicates the size values between which measurements may be made on the given instrument.

Scale range. It is the difference between the values of the measured quantities corresponding to the terminal scale marks.

Instrument range. It is the capacity or total range of values which an instrument is capable of measuring. For example, a micrometer screw gauge with capacity of 25 to 50 mm has instrument range of 25 to 50 mm but scale range is 25 mm.

Scale Spacing. It is the distance between the axes of two adjacent graduations on the scale. Most instruments have a constant value of scale spacing throughout the scale. Such scales are said to be linear.

In case of non-linear scales, the scale spacing value is variable within the limits of the scale.

Scale Division Value. It is the measured value of the measured quantity corresponding to one division of the instrument, *e.g.*, for ordinary scale, the scale division value is 1 mm. As a rule, the scale division should not be smaller in value than the permissible indication error of an instrument.

Sensitivity (Amplication or gearing ratio). It is the ratio of the scale spacing to the division value. It could also be expressed as the ratio of the product of all the larger lever arms and the product of all the smaller lever arms. It is the property of a measuring instrument to respond to changes in the measured quantity.

Sensitivity Threshold. It is defined as the minimum measured value which may cause any movement whatsoever of the indicating hand. It is also called the discrimination or resolving power of an instrument and is the minimum change in the quantity being measured which produces a perceptible movement of the index.

Reading Accuracy. It is the accuracy that may be attained in using a measuring instrument.

Reading Error. It is defined as the difference between the reading of the instrument and the actual value of the dimension being measured.

Accuracy of observation. It is the accuracy attainable in reading the scale of an instrument. It depends on the quality of the scale marks, the width or the pointer/index, the space between the pointer and the scale, the illumination of the scale, and the skill of the inspector. The width of scale mark is usually kept one-tenth of the scale spacing for accurate reading of indications.

Parallax. It is apparent change in the position of the index relative to the scale marks, when the scale is observed in a direction other than perpendicular to its plane.

Repeatability. It is the variation of indications in repeated measurements of the same dimension. The variations may be due to clearances, friction and distortions in the instrument's mechanism. Repeatability represents the reproducibility of the readings of an instrument when a series of measurements in carried out under fixed conditions of use.

Measuring force. It is the force produced by an instrument and acting upon the measured surface in the direction of measurement. It is usually developed by springs whose deformation and pressure change with the displacement of the instrument's measuring spindle.

1.7. The Measurement Problem

In practice, we come across four basic conditions to be controlled by tolerances, *viz*, (*a*) size, (*b*) form, (*c*) location and (*d*) conditions of assembly, operation, or function. Fig. 1.5 illustrates these. The difference between sizes *X* and *Y* determines the assembly condition. Size conditions are generally simple to specify and control ; the form and location conditions are more complex, especially where composite surfaces and cumulative tolerances are involved. It will be appreciated that all these four conditions inter-relate to define quality characteristics, and the problem of measuring quality characteristics to evaluate conformance to specifications becomes complex. Lack of true geometric perfection makes it difficult to define and control product quality characteristics. It may be noted that it is easy to define geometric form but difficult to produce. However, to maintain specific quality, the variation from perfect form must be defined and controlled. The geometric variations known as macro-errors concern straightness, flatness, parallelism, squareness, angular displacement, symmetry, concentricity, eccentricity, roundness. Lack of perfect rigidity caused due to material properties like expansion, stretching, springing, warping etc., also affect geometric form, size and location conditions.

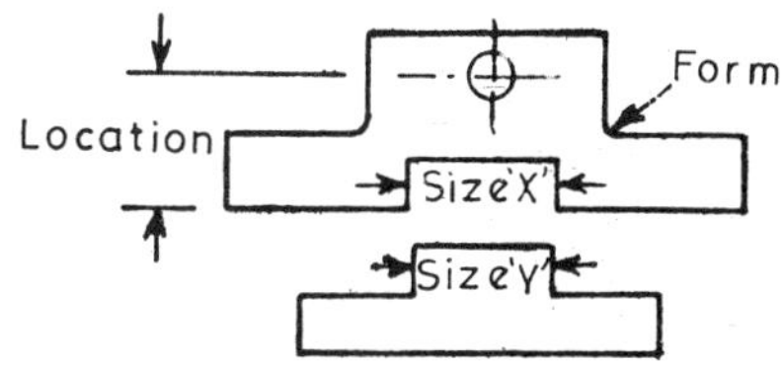

Fig. 1.5. Product conditions.

The inter-relationships of size, form and location conditions required to define quality characteristics, coupled with production variations due to geometric form and rigidity errors, lead to a variety of complex measurement problems involving sophisticated gauging method.

In some cases the configuration of parts is such that accurate measurement becomes difficult. In some cases, it is not possible for the standard gauge or tool to span the component. Such cases are : parts having not even a single common plane, irregular curved surfaces, odd number spline or gear, parts with phantom dimensions (*i.e.*, dimensions to be gauged have to be calculated or their dimensions are w.r.t. points in space) etc.

The measurement must be correct to a degree commensurate with the specified tolerances and the required functional service of the product. Following two factors need to be considered when evaluating the correctness of dimensional measurements : (*i*) gauging accuracy, (*ii*) the proper location of the measured dimension on the physical part. It must be remembered that the improperly located dimensions, can often have a greater effect on correctness of dimensional measurements than errors due to insufficient indicating accuracy.

The precision of the measurement can be affected by limitations in either of the basic requirements, *viz.* the accuracy of the instrument at the proper location of the gauging points which determine the dimension being measured on the physical part.

1.8. General Care of Metrological Equipment

The equipment (apparatus) used for precision measurements is designed to fine limits of accuracy and is easily liable to be damaged by even slight mishandling and such damage may not be noticeable. A great deal of careful handling is, therefore, required. As far as possible, the highly finished surfaces should not be touched by hand because the natural acids on the skin are likely to corrode the finished surface and also the temperature of body may upset the dimensions of the precision instruments. In order to overcome this many standard metrology laboratories recommend washing of hands thoroughly and coating them with a thin film of pure petroleum jelly before handling the instruments. Further very precise equipment like slip gauges is allowed to be handled only by using a piece of chamois leather or tongs made from a strip of "Perspex".

When the equipment is not in use, it should be protected from atmospheric corrosion. For this purpose, the highly finished surfaces are first wiped with a solvent to remove any finger marks and then coated with mixture of heated petroleum jelly and petrol. This mixture spreads much more easily and is applied with cloth or with fingers. Brushing is not recommended as it is liable to trap air which, with the moisture it contains, may cause rusting.

As the standard temperature for measurement is 20°C, for very precise measurements the instruments and workpieces should be allowed to attain this temperature before use and the handling should be as little as possible.

1.9. Objectives of Metrology

While the basic objective of a measurement is to provide the required accuracy at minimum cost, metrology would have further objective in a modern engineering plant with different shops like Tool Room, Machine Shop, Press Shop, Plastic Shop, Pressure Die Casting Shop, Electroplating and Painting Shop, and Assembly Shop, as also Research, Development and Engineering Department. In such an engineering organisation, the further objectives would be as follows :

(*a*) Thorough evaluation of newly developed products, to ensure that components designed are within the process and measuring instrument capabilities available in the plant.

(*b*) To determine the process capabilities and ensure that these are better than the relevant component tolerances.

(*c*) To determine the measuring instrument capabilities and ensure that these are adequate for their respective measurements.

(*d*) To minimise the cost of inspection by effective and efficient use of available facilities, and to reduce the cost of rejects and rework through application of Statistical Quality Control Techniques.

(*e*) *Standardisation of measuring methods.* This is achieved by laying down inspection methods for any product right at the time when production technology is prepared.

(*f*) *Maintenance of the accuracies of measurement.* This is achieved by periodical calibration of the metrological instruments used in the plant.

(*g*) Arbitration and solution of problems arising on the shop floor regarding methods of measurement.

(*h*) Preparation of designs for all gauges and special inspection fixtures.

1.10. Requirements of an Inspection Tool

The requirements of an ideal inspection tool are : It should be

(*a*) accurate
(*b*) require a minimum of operator skill
(*c*) inspect a specific type of error
(*d*) fast to use
(*e*) Self checking.

The degree of accuracy of calibration depends on the accuracy of the inspecting instruments. Devices which reduce dependence on operator skill contribute to both efficiency and accuracy. The requirement of speed is not for economic reasons but to avoid errors from changes in temperature where inspections become involved. A good inspection tool should be capable of being checked against itself. This feature increases the reliability.

1.11. Standardisation and Standardising Organisations

For overall higher economy, efficiency and productivity in a factory and country, it is essential that diversity be minimised and interchangeability among parts encouraged. All this is possible with standardisation. Standardisation is done at various levels, *viz.* International, National, Association, Company.

Realising the role of standardisation in the development of industry, organisations to handle the complexities of standardisation have been evolved in each of the chief industrial countries. In India, Bureau of Indian Standards (BIS) is responsible for evolving standards on metrological instruments, etc. There are several sectional committees, each dealing with various main branches of industry, in BIS. The detailed work of drawing up specifications is done by more specialised technical committees who prepare a draft standard based on practice in other countries and the needs of the country, and circulate it to relevant industries, government and service departments, research and teaching organisations, and others likely to be interested. Comments are invited both from producer and user to consider all aspects ; meetings help to discuss the matters in depth and final standards issued. The technical committees also keep on revising the existing standards from time to time.

The Bureau of Indian Standards is the National body for standardisation in India.

The functions of the Bureau are :

(*a*) Formulation, publication and promotion of Indian Standards ;

(*b*) Inspection of articles or process under Certification Scheme ;

(*c*) Establishment, maintenance and recognition of laboratories ;

(*d*) Formulate, implement and coordinate activities relating to quality maintenance and improvement in products and processes ;

(*e*) Promote harmonious development in standardization, quality systems and certification and matters connected therewith both within the country and at international level ;

(*f*) Provide information, documentation and other services to consumers and recognised consumer organisations on such terms and conditions as may be mutually agreed upon ;

(*g*) Give recognition to quality assurance systems in manufacturing or processing units on such terms and conditions as mutually agreed upon ;

(*h*) Bring out handbooks, guides and other special publications ; and for conformity to any other standard if so authorized.

Thus, the main functions of the Bureau can be grouped under standards formulation, certification marking and laboratory testing, promotional and international activities.

Bureau of Indian Standards has under the Mechanical Engineering Division Council, EDC, a separate Engineering Metrology Sectional Committee. This Committee was set up in 1958 and its main task is to formulate standards for the various aspects of dimensional metrological measuring instruments and accessories used in the mechanical engineering field. A large number of Indian Standards in the field of engineering metrology have been formulated.

In Great Britain, British Standards Institution plays similar role to BIS.

In Europe, the International Federation of National Standardising Association, known as I.S.A., co-ordinates the work of the continental countries. Before Second World War, U.K. and U.S.A. did not take any part in it, but after war, the countries like U.K., U.S.A. and Russia have taken part in its works. In 1946, the I.S.A., was re-formed as the International Organisation for Standardisation, I.S.O. In fact, for engineering matters, the foremost standards organisation at international level is I.S.O. The national standards organisation of individual countries are the members of I.S.O. The I.S.O. recommendations are used as basis for national and company standards. Lot of co-operative discussions in the field of standardisation have also been carried out in three countries—America, Britain and Canada known was ABC conference. The International Electro-technical Commission (IEC) deals with electrical engineering standards. Both ISO and IEC have published recommendations on some aspects of engineering metrology.

National Physical Laboratories (NPL) carry out lot of research work in various fields ; responsible for defining standards, and also issue certification marks for quality instruments.

1.11.1. International Organisation of Weights and Measures. It was established in 1975 under the "International Metre Convention" in Paris with the object of maintaining uniformity of measurements throughout the world. It comprises of :

(*a*) The General Conference of Weights and Measures.

(*b*) The International Committee of Weights and Measures.

(*c*) The International Organisation of Legal Metrology.

1.11.2. General Conference of Weights and Measures. Its objects are :

(*i*) To draw up and promote the decisions necessary for the propagation and perfection of an international system of units and standards of measurement.

(*ii*) To approve the results of new fundamental metrological determinations and the various scientific resolutions in the field of metrology which are of international interest.

1.11.3. International Committee of Weights and Measures. This Committee is placed under the authority of the General Conference of Weights and Measures and is responsible for promoting the decisions taken by the latter. Its objects are :

(*i*) To direct and supervise the work of the International Bureau of Weights and Measures.

(*ii*) To establish co-operation among national laboratories of metrology for executing the metrological work which the General Conference of Weights and Measures decides to execute jointly by the member states of the organisation.

(*iii*) To direct such work and co-ordinate the results and to look after the conservation of the International Standards.

1.11.4. Principal Global Organisations involved in Metrology

(*i*) **BIPM (Bureau International des Poids et Mesures).** It is created under the Metre convention for measurement standard activities. It provides leadership in ensuring collaboration on metrological matters and the maintenance of an efficient worldwide measurement system. It serves as the technical focal point to guarantee the equivalence of national standards. BIPM with its laboratories and offices at Serves act as a permanent international centre for Metrology under the supervision of the CIPM.

(*ii*) **ILAC (International Laboratory Accreditation Conference).** It is engaged in international laboratory accreditation and the standards writing bodies. It has demonstrated competence in calibration and testing.

(*iii*) **IEC (International Electrotechnical Commission).** A voluntary sector to prescribe standards.

(*iv*) **CIPM (Comite International des Poids et Mesures).** Most of the activities of CIPM are performed under the supervision of CIPM. Several (CCs) consultative committees have been set up by the CIPM.

(*v*) **CGPM (Conference Generale des Poids et Mesures).**

(*vi*) **ISO (International Organisation for Standardisation).** A voluntary sector to specify standards.

(*vii*) **NMI (National Metrology Institute).** A national laboratory responsible for the development and maintenance of measurement standards for the dissemination of the SI units, their multiples and sub multiples, and capable of making accurate measurements available to all users.

(*viii*) **International Organisation of Legal Metrology (OIML).** It was established in 1955 under the "International Convention of Legal Metrology" Paris to unify the metrological practices. Its objects are :

— To determine the general principles of Legal Metrology. Legal Metrology is concerned with the statutory technical and legal requirement of units of measurements, methods of measurements and measuring instruments with a view to assure public guarantee in respect of the security and the appropriate accuracy of measurements.

— To study with the object of unification, statutory and regulatory problems of legal metrology the solution of which is of international interest.

— To establish the draft of a model law and regulation on measuring instruments and their use.

— To prepare a plan for the physical organisation of a model service for the verification and control of measuring instruments and to establish the necessary and adequate characteristics and qualities which measuring instruments should possess in order that they may be approved by the member states and their use recommended on international basis.

This organisation comprises of the International Conference of Legal Metrology, the International Committee of Legal Metrology and the International Bureau of Legal Metrology.

OIML has made a number of international recommendations. They have also published a "Vocabulary of Legal Metrology-Fundamental Terms" the english translation of which is published in India by the Directorate of Weights and Measures, Ministry of Industry.

The functions of the Directorate of Weights and Measures are :

— to ensure the conservation of national standards and to guarantee their accuracy by comparison with international standards.

— to guarantee and impart proper accuracy to the secondary standards by comparison with national standards.

— to carry out scientific and technical work in all fields of metrology and methods of measurements.

— to take part in the work of other national organisations interested in metrology.

— to draw up draft laws relating to legal metrology and to promulgate the corresponding regulations.

— to regulate and advise on, supervise and control the manufacture and repair of measuring instruments.

— to inspect the use of instruments and the measurement operations when such use and such operations are covered under public guarantee.

— to detect frauds in measurement or sale of goods and to book offender for trials where necessary.

— to coordinate the activities of authorities exercising metrological supervision.

— cooperate with all to ensure respect for the regulations of legal metrology.

— to organise training in legal metrology

— to represent the country in international activities regarding legal metrology.

(*ix*) **National Service of Legal Metrology.** The National Service of Legal Metrology has following organisations to assist it in discharge of its duties :

— National Bureau of Legal Metrology. (It is the directing organisation)

— National Institute of Legal Metrology. (It is entrusted with the performance of scientific and research work)

— National Bureau of Verification.

There are Regional Bureau of Verification, Local Bureau of Verification, Mobile Bureau of Verification, and Verification Centres to assist National Bureau of Verification in ensuring appropriate accuracy of the standards, carrying out metrological supervision, verifying measuring instruments.

— Verification Agents (Authorised to exercise the functions of verification).

1.12. International System of Units (SI)

It is the system established in 1960 by the (CGPM) General Conference of Weights and Measures and abbreviated as SI (System International d'unites) in all languages. In India, we switched over to metric system of Weights and Measures conforming to SI units by an Act of Parliament No. 89, in 1956. This SI like traditional metric system, is based on decimal arithmetic. For each physical quantity, units of different sizes are formed by multiplying or dividing a single base value by powers of 10. Obviously this offers great advantage because the changes can be made very simply by adding zeros or shifting decimal point. In the metric system we have been following so far, this simplicity of a series of units linked by powers of 10 is limited to plain quantities like length, and this simplicity is lost when more complex units like energy etc. are encountered. For example energy is now represented by several units like kgm, H.P., kW etc. In contrast the SI provides only one basic unit for each physical quantity, and universality is thus achieved. This system in superior to other systems and also more convenient as it is coherent, rational and comprehensive.

The SI is a coherent system, in the sense that the product or quotient of any two unit quantities in the system is the unit of the resultant quantity, *e.g.* if unit of length is metre, then unit of area will be square metre and not acres or begas etc. It is rational system since it has absorbed in itself the rationalised MKSA system. It is also comprehensive because its seven base units cover all disciplines.

The seven base SI units established by the General Conference of Weights and Measures are given on next page and SI units having special names are given below.

SI UNITS HAVING SPECIAL NAMES

Physical quantity	*Name of unit*	*Unit symbol*
Force	newton	N = kg m/s^2
Work, energy, quantity of heat	joule	J = N m
Power	watt	W = J/s
Electric charge	coulomb	C = A.s
Electric potential	volt	V = W/A
Electric capacitance	farad	F = C/V
Electric resistance	ohm	Ω = V/A
Electric conductance	siemen	S = A/V
Magnetic flux	weber	Wb = V × s
Inductance	henry	H = V.s/A
Luminous flux	lumen	lm = cd.sr
Magnetic flux density	tesla	T = Wb/m^2
Illumination	lux	lx = lm/m^2
Frequency	hertz	Hz = cycles
Pressure	pascal	Pa = N/m^2

S. No.	*Physical Quantity*	*Name of unit*	*Unit symbol*	*Base of Definition*	*Definition*
1.	Length	metre	m	Wavelength of red light in Krypton 86	1,650,673.73 wavelengths in vacuo of the radiation corresponding to the transition between the energy levels 2*p* 10 and 5*d*5 of the Krypton 86 atom.
2.	Time	second	s	Cycles of radiation of cesium	The duration of 9,192,631,770 periods of the radiation corresponding to the transition between the two hyperfine levels of the ground state of the cesium-133 atom.
3.	Mass	kilogram	kg	Platinum-cylinder prototype	Mass of the International prototype which is in the custody of Bureau International des Poids et Mesures (BIPM) at Sevres, near Paris.
4.	Temperature	kelvin	K	Absolute zero and water	The fraction 1/273.16 of the thermodynamic temperature of the triple point of water.
5.	Electric current	ampere	A	Force between two conducting wires	The constant current which, if maintained in two parallel rectilinear conductors of infinite length, of negligible circular cross-section, and placed at a distance one metre apart in vacuum would produce between these conductors a force equal to 2×10^{-7} N/m length.
6.	Luminous intensity	candela	cd	Intensity of an area of platinum	The luminous intensity, in the perpendicular direction, of a surface of 1/600,000 square metre of a black body at the temperature of freezing platinum under a pressure of 101,325 newtons per square metre.
7.	Quantity of substance	mole	mol	Amount of atoms in carbon 12	The amount of substance of a system which contains as many elementary entities as there are atoms in 0.012 kilogram of carbon 12.

Other physical quantities are derived from these basic units. For example, volume is cubic metre (m^3), speed is metre per second (m/s), force is mass-metres per square second (kg m/sec^2). Realising that some derived units may be compex array of base units, some units have been given special names. They are given on page 24.

SI also recommends the use of supplementary units, the only authorised supplementary units being measures for plane and spherical angles. Though these could be expressed by base units but have been available as a convenience to users. Plane angles are represented by radians and solid angles by steradians.

Decimal multiples and sub-multiples of the SI units are formed by means of the prefixes given below, which represent the numerical factors shown :

Multiplication factor	*Prefix*	*SI Symbol*
10^{12}	tera	T
10^{9}	giga	G
16^{6}	mega	M
10^{3}	kilo	k
10^{2}	hecto*	h
10^{1}	deka	da
10^{-1}	deci*	d
10^{-2}	centi*	c
10^{-3}	milli	m
10^{-6}	micro	μ
10^{-9}	nano	n
10^{-12}	pico	p
10^{-15}	femto	f
10^{-18}	atto	a

*To be avoided where possible.

The two major advantages of coherency of SI units are (*i*) same system and unit of measurement are used regardless of industry, trade, or discipline ; and (*ii*) minimum of conversion factors are needed (other than powers of 10).

In connection with SI units, some rules of style, abbreviations, writing, and drafting practices are applied, some of which are described below :

(*a*) No dots, commas, etc., are used after SI symbols except at the end of sentences. For example 32 metres will be written as 32 m but when written as 32 m. is not correct.

(*b*) Plurals are never used in connection with SI unit symbols, for example 32 metres will be written as 32 m and not as 32 ms which would mean 32 milli-seconds.

(*c*) Decimal fractions are always started with 0. For example, half metre will be written as 0.5 m and not as .5 m.

(*d*) Multiplication or times sign is '.' This is used between the number to be multiplied and between unit symbols in derived units where two unit symbols adjoin for the purpose of clarity, *e.g.* unit of torque may be written as m.N (metre newton) which if written as mN can be misunderstood as millinewton.

(*e*) All symbols and prefixes are lowercase letters, except symbols derived from proper names, like W for watt, M, G and T for the largest three power-of-10 prefixes. All symbols should be used as they are to avoid any confusion.

(*f*) All units may be written either in full or using the agreed symbols. There is only one acceptable symbol in all languages for a metric unit. For example KILOGRAM be written as kg and not as kgm or kg, or kgr. etc.

(*g*) The product of two or more units is preferably indicated by a dot which can be dispensed with when there is no risk of confusion with another unit symbol *e.g.*, 63 N.m or 63 Nm is correct but not 63 mN.

(*h*) There is a mixture of capital and lower case letters in the symbols for the prefixes, but the full names of the prefixes commence with lower case letters only, *e.g.* 200 MW is written is 200 megawatts.

(*i*) No degree mark (°) is used with kelvin, the unit of temperature. A temperature interval can also be expressed in degree celsius.

(*j*) Double prefixes should not be used *e.g.* one kilo Megawatt may not be written as kMW, but as GW.

(*k*) Algebraic symbols representing quantities are written in *italics* while symbols for units are written in upright characters.

(*l*) The expression "per" in symbols of derived units is always indicated by a fraction line as m/s of $\frac{m}{s}$ but word "per" or letter "p" should not be used for this purpose.

(*m*) Always leave one space between numerical value and the symbol. For example, 92 N is commitment 92N is not.

(*n*) Only the numerator should be multiplied by powers of 10 in compound derived units and the denominator should always remain the base unit.

For example 0.032 m/s may be put as 32 mm/s and not as 0.000032 m/ms. (meter per millisecond).

(*o*) Numbers may be grouped in clusters of three in both directions from the decimal mark, and gap may be given for clarity and comma should not be used there.

e.g. 153297.3 m may be written as 153 297.3 m and not 153,297.3 m. A sequence of four figures is generally not broken in groups.

(*p*) Units with names of scientists should be capitalised when written in full.

(*q*) According to SI recommendations, litre is a special name given to cubic decimetre and the world litre should not be used for expressing results of high precision volume measurements.

(*r*) Some units which though strictly incompatible with SI units, have been allowed initially, like km/hr, rev/min.

(*s*) The appropriate integers, multiples and sub-multiples to which a unit is to be expressed is selected in such a manner that the numerical value to be expressed is between 0.1 and 1000.

(*t*) Figures are written in groups of three to the left and to the right of the decimal sign.

(*u*) Thousands are separated by a space and NOT by a comma *e.g.* 6000 000 kl is correct but 6,000,000 kl is not correct.

(*v*) In some countries, a point (dot) is used as the decimal marker, and in some countries a comma is used ; both are understood internationally.

Question. Correct the mistakes in following numbers and symbols :

1. mps
2. $\frac{1}{32\text{nd}}$, 1/32nd
3. 45mm .5, 45.5mms, 45.5m/m
4. 4.96w
5. $2\frac{1}{2}$ Hrs, 14.30 hours
6. Km, kms
7. 19 M
8. M., ms
9. Sq.m.
10. C-m, M-W
11. cms, cm., cent
12. cc, cu^3, C
13. m/m, mms, millim

14. Kg, kgr, kgs, kilo
15. gr, Gr, grs, grm
16. H
17. 962N
18. min., mins.,
19. ″, sec., secs
20. KW (electric consumption)
21. kph, k.p.h., km.p.h.
22. 396 70, 39,670
23. m/s/s
24. 96253.5297
25. $m.kg/s^2/J$
26. 1 m μm
27. μ kg
28. 9.6×10^4 N
29. 1523 Pa
30. 0.00036 m
31. 7.8×10^{-8} s
32. 100 kgs.

Sol. The correct notations are as under :

1. $\frac{m}{s}$, m/s, ms^{-1}
2. 1 : 32 or 1/32 or $\frac{1}{32}$.
3. 45.5 mm
4. .96 W
5. 14 h 30 min or 14 hours 30 minutes
6. km or kilometre
7. 19 m
8. m or metre
9. m^2 or square metre
10. cm, MW
11. cm or centimetre
12. cm^3 or cubic centimetre
13. mm or millimetre
14. kg or kilogram
15. g or gram
16. h or hour
17. 962 N
18. min or minute (time)
19. s or second (time)
20. kWh or kilowatt-hour

21. km/h
22. 39 670
23. m/s^2, ms^{-2}
24. 96 253.529 7
25. $m.kg/(s^2J)$ or $m.kg\ s^{-2}\ J^{-1}$
26. 1 nm
27. mg
28. 96 kN
29. 1.523 kPa
30. 0.36 mm
31. 78 ns
32. 100 kg

1.13. Role of National Physical Laboratory (NPL) in Metrology

NPL provides the basic backbone of organizational structure for metrology in India. NPL is the custodian of national measurement standards for physical measurement in the country. This premier national laboratory is established with the objective to strengthen and advance physics-oriented research for the overall development of science and technology in the country. NPL has the responsibility of physical measurements based on the International system (SI units) under the subordinate legislation of Weights and Measures Act 1956 (reissued in 1988 under the 1976 Act). NPL also has the statutory obligation to establish, maintain and update the national standards of measurement and calibration facilities for different parameters. NPL maintains seven SI base units, *viz.*, meter, kilogram, second, kelvin, ampere, candela, mole (mol) and the SI supplementary units radian (rad) and steradians (sr).

The main objectives of the NPL's work are as follows :

- To establish, maintain and improve continuously the national standards for measurements for the benefit of the country and also to realize the units based on International Systems (SI),
- To determine the major physical constants,
- To develop and evaluate measurement techniques,
- To assist industries, governmental and other agencies in their developmental task by precision measurements, calibration and development of devices and processes related to physics,
- Laboratory's commitment to good professional practice and quality of its calibration and testing in serving to industry, and
- Technical advisory services, information, extension, consultancy and training and collaboration with universities and other organization.

Under the Standards of Weights and Measures Rules 1988, the NPL has been further assigned the obligation to act as the custodian of the national standards of measurement (excepting those for ionizing radiations) compatible and traceable to the International System (SI units). The NPL fulfils this responsibility by developing, maintaining and updating of these standards through interaction and metrological intercomparison with other major standards laboratories in the NPL (UK), NIST (USA), WNIFTRI (Russia), PTB (Germany), IMGC (Italy), BIPM (France), ETL (Japan), NIM (China), CSIR (Australia), DSIR (New Zealand), NRC (Canada) and, directory of National Measurement Systems of Non-Aligned (NAM) Countries and other Developing Countries.

The NPL represents India in the Bureau of International Weights and Measures (BIPM), General Conference on Weights and Measures (CGPM) and International Committee for Weights and Measures (CIPM) besides several other international committees in the area of acoustics and vibration, time and frequency, vacuum and pressure, temperature, humidity and other parameters also.

The nodal work done by the NPL in metrication of the entire measurement system of the Country since 1956 forms the basis of legal metrology function in the country coordinated by the Directorate of Weights and Measures (DWM) Govt. of India. Measurement standards, calibration service, specialized testing back-up, technical advisory consultancy services and participation in the committees of the DWM form the bulk of NPL support for these important areas of consumer scientific knowledge, protection and improvement and securing of the living conditions.

The principal activities of the NPL in the field of metrology include the development of :

— Physicso-mechanical standards : length standards and dimensional metrology, mass, volume and viscosity, fluid flow measurement, standards of force and hardness, pressure/vacuum, temperature, optical radiation, ultraviolet radiation, infrared radiation, acoustics and ultrasonic.

— Electrical standards : dc measurement, Josephson voltage, ac standards and standards of LF and HF impedance, magnetic standards, HF, MW voltage, current power, HF, MW attenuation and impedance standards.

— Other standards : Time and frequency, NABL programme, piezoelectric accelerometer, humidity standards.

— Calibration and Testing Service Programme,

— Reference material standards, and

— Participation in international metrology activities.

1.14. Sources of Errors

In any measurement, there is always a degree of uncertainty resulting from measurement error, *i.e.* all measurements are inaccurate to some extent Measurement error is the difference between the indicated and actual values of the measurand. The error could be expressed either as an absolute error or on a relative scale, most commonly as a percentage of full scale. It is important to examine fully the errors in measurement systems that cause these uncertainties, the meaning and interpretations of these errors and methods of reducing or circumventing of errors. Each component of the measuring system has sources of errors that can contribute to measurement error.

Instrument or indication errors may be caused by defects in manufacture of adjustment of an instrument, imperfections in design, etc.

The error of measurement is the combined effect of component errors due to various causes. There may be errors due to method of location, environmental errors, errors due to the properties of object of measurement, *viz.* form deviation, surface roughness, rigidity, change in size due to ageing etc., observation errors.

The total error of measurement includes indication errors, errors of gauge blocks or setting standards, temperature change errors, and errors caused by the measuring force of the instrument.

During measurement several types of errors may arise such as static errors, instrument loading errors or dynamic errors, and these errors can be broadly classified into two categories *viz.* controllable errors and random errors.

Static Errors. These result from the physical nature of the various components of the measuring system as that system responds to a fixed measurand input. Static errors result from the intrinsic imperfections or limitations in the hardware and apparatus compared to ideal instruments. The environmental effect and other external influences on the properties of the apparatus also contribute to static erros. Other sources of static errors could be inexactness in the calibration of the system, displaying the output of the measuring system in a way that requires subjective interpretation by an observer. From above it could be concluded that static errors stem from three basic sources : reading error, characteristic error and environmental error. In the measurement of length of a surface table with a rule, these errors will be encountered when aligning the ends of the rule and surface table, and when estimating the length of the table. The static error divided by the measurement range (difference between the upper and lower limits of measurement) gives the measurement precision. *Reading error* describes such factors as parallax, interpolation, optical resolution (readability or output resolution). Reading errors apply exclusively to the readout device and have no direct relationship with other types of errors within the measuring system. Attempts have been made to reduce or eliminate the reading errors by relatively simple techniques. Where there is possibility of error due to parallax, the use of mirror behind the readout pointer or indicator virtually eliminates occurrence of this type of error. *Interpolation error* can be tackled by increasing the optical resolution by using a magnifier over the scale in the vicinity of the pointer. The use of digital readout devices is increasing tremendously for display purposes as it eliminates most of the subjective reading errors usually made by the observer. However, there exists a possibility of plus or minus one count error in digital readout devices also and its value can be effectively reduced by arranging full range to correspond to huge number of pulses so that one pulse has very negligible value. Digital counting devices are capable of counting each and every pulse, however short may be the duration, but it is only during start and at stop that one pulse is likely to be missed which can lead to error.

Environmental errors result from effect of surrounding temperature, pressure and humidity on measuring system. It can be reduced by controlling the atmosphere according to estipulated requirements. External influences like magnetic or electric fields, nuclear radiation, vibration or shock, periodic or random motion etc., also lead to errors. It is important to note that these factors affect both the measuring system and measurand, and usually the effects of these factors on each component are independent. Thus the environmental errors of each component of the measuring system make a separate contribution to the static errors. Due to this reason, the number of environmental variables and external influences that could affect the measurement should be minimised and where it is not possible to do so then their effect should be computed and taken into account.

Characteristic error is defined as the deviation of the output of the measuring system under constant environmental conditions from the theoretically predicted performance, or from nominal performance specifications. If the theoretical output is a straight line, then linearity, hysteresis, repeatability, and resolution errors are part of the characteristic error. Linearity errors, hysteresis and repeatability errors are present to some degree in each component of a measuring system. Other characteristic errors include gain errors and zero offset, often collectively called calibration errors.

Similar characteristic errors in each component of the measuring system tend to be additive. Thus, system linearity is usually the sum of the errors in individual components ; and as such the study of combination and accumulation of errors is very important and will be discussed later.

It has been found that the static erros introduced by the components of the measuring system are the cause of major concern. However, the loading errors and dynamic errors which are generally encountered in process measurements and not in the field of Metrology, will also be discussed in brief here to complete the subject.

Loading errors result from the change in the measurand itself when it is being measured, *i.e.* after the measuring system or instrument is connected for measurement. Instrument loading error is thus the difference between the value of the measurand before and after the measurement system is measured. One example of such an error could be the deformation of soft component under contact pressure of measuring instrument. The effects of instrument loading are unavoidable and must be determined specifically for each measurement and measurand. Such loading erros are often the single greatest uncertainty in a physical measurement. Therefore, measuring system should be selected such that its sensing element will minimise instrument loading error in the particular measurement involved. In a steady state measurement, the cumulative effect of static errors and instrument loading errors determines the accuracy of the measurement.

Dynamic error is caused by time variations in the measurand and results from the inability of a measuring system to respond faithfully to a time-varying measurand. Usually the dynamic response is limited by inertia, damping, friction or other physical constraints in the sensing or readout or display system. Dynamic error is characterised by the frequency and phase response (Bode criterion) of the system for the cyclic or periodic variations in the measurand input. For random or transient inputs, the dynamic error is described by the time constant of response time. In both the cases , it is essential that dynamic characteristics of the measuring system be known before putting the system to measure time varying inputs.

It is thus seen that different errors entering into any observation arise due to a variety of reasons. Many times it may not be possible to identify the source of errors. Therefore it is more fruitful to classify errors according to the effects they produce rather than on the basis of sources which produce them.

For statistical study and the study of accumulation of errors, errors are categorised as controllable errors and random errors.

(*a*) **Systematic or Controllable Errors.** Systematic error is just a euphemism for experimental mistakes. These are controllable in both their *magnitude* and *sense.* These can be determined and reduced, if attempts are made to analyse them. However, they can not be revealed by repeated observations. These errors either have a constant value or a value changing according to a definite law. These can be due to :

1. *Calibration Errors.* The actual length of standards such as slip gauges and engraved scales will vary from nominal value by small amount. Sometimes the instrument inertia, hysteresis effect do not let the instrument translate with complete fidelity. Often signal transmission errors such as drop in voltage along the wires between the transducer and the electric meter occur. For high order accuracy these variations have positive significance and to minimise such variations calibration curves must be used.

2. *Ambient Conditions.* Variations in the ambient conditions from internationally agreed standard value of 20°C, barometric pressure 760 mm of mercury, and 10 mm of mercury vapour pressure, can give rise to errors in the measured size of the component. Temperature is by far the most significant of these ambient conditions and due correction is needed to obtain error free results.

3. *Stylus Pressure.* Error induced due to stylus pressure is also appreciable. Whenever any component is measured under a definite stylus pressure both the deformation of the workpiece surface and deflection of the workpiece shape will occur.

4. *Avoidable Errors.* These errors include the errors due to parallax and the effect of misalignment of the workpiece centre. Instrument location errors such as placing a thermometer in sunlight when attempting to measure air temperature also belong to this category.

5. Experimental arrangement being different from that assumed in theory.

6. Incorrect theory *i.e.*, the presence of effects not taken into account.

(*b*) **Random Errors.** These occur randomly and the specific cases of such errors cannot be determined, but likely sources of this type of errors are small variations in the position of setting standard and workpiece, slight displacement of lever joints in the measuring joints in measuring instrument, transient fluctuation in the friction in the measuring instrument, and operator errors in reading scale and pointer type displays or in reading engraved scale positions.

Characteristics of random errors

The various characteristics of random errors are :

— These are due to large number of unpredictable and fluctuating causes that can not be controlled by the experimenter. Hence they are sometimes positive and sometimes negative and of variable magnitude. Accordingly they get revealed by repeated observations.

— These are caused by friction and play in the instrument's linkages, estimation of reading by judging fractional part of a scale division, by errors in positioning the measured object, etc.

— These are variable in magnitude and sign and are introduced by the very process of observation itself.

— The frequency of the occurrence of random errors depends on the occurrence probability for different values of random errors.

— Random errors show up as various indication values within the specified limits of error in a series of measurements of a given dimension.

— The probability of occurrence is equal for positive and negative errors of the same absolute value since random errors follow normal frequency distribution.

— Random errors of larger absolute value are rather than those of smaller values.

— The arithmetic mean of random errors in a given series of measurements approaches zero as the number of measurements increases.

— For each method of measurement, random errors do not exceed a certain definite value. Errors exceeding this value are regarded as gross errors (errors which greatly distort the results and need to be ignored).

— The most reliable value of the size being sought in a series of measurements is the arithmetic mean of the results obtained.

— The main characteristic of random errors, which is used to determine the maximum measuring error, is the standard deviation.

— The maximum error for a given method of measurement is determined as three times the standard deviation.

— The maximum error determines the spread of possible random error values.

— The standard deviation and the maximum error determine the accuracy of a single measurement in a given series.

From the above, it is clear that systematic errors are those which are repeated consistently with repetition of the experiment, whereas *Random Errors* are those which are accidental and whose magnitude and sign cannot be predicted from knowledge of measuring system and conditions of measurement.

Spurious errors and Dixon test. Errors due to operator mistakes or malfunction of instrument are called spurious errors and need to be ignored in the statistical analysis. Statistical outlier or Dixon test is applied to discard spurious readings. All good observations follow normal distribution and spurious reading will fall outside the normal distribution. In Dixon test, all observations are arranged in ascending order if spurious reading is suspected for high value, or descending order if spurious reading is suspected for low value.

If total readings upto suspected value X_n are n, then value of $\frac{X_n - X_{n-1}}{X_n - X_1}$ is calculated for n between 3 and 7, $\frac{X_n - X_{n-1}}{X_n - X_2}$ for n between 8 and 10, $\frac{X_n - X_{n-2}}{X_n - X_2}$ for n between 11 and 13, and

$\frac{X_n - X_{n-3}}{X_n - X_3}$ for n between 14 and 24. If this ratio is more than following critical value, then X_n should be discarded :

n	*Critical value*	n	*Critical value*	n	*Critical value*
3	0.941	11	0.576	19	0.462
4	0.765	12	0.546	20	0.450
5	0.620	13	0.571	21	0.440
6	0.560	14	0.546	22	0.430
7	0.507	15	0.525	23	0.421
8	0.554	16	0.507	24	0.413
9	0.512	17	0.490	25	0.406
10	0.477	18	0.475		

Statistical Treatment of Errors. Random and systematic errors are evaluated and studied by statistical procedures which make it possible to state from a limited group of data the most probable value of a quantity, the probable uncertainty of a single observation, and the probable limits of uncertainty of the best value that can be derived from the data. It may be noted that the object of the statistical methods ; based on laws of chance which operate only on random errors and not on systematic errors ; is to achieve consistency (precision) of value and not their accuracy (approach to the truth).

It is also important to note that in quality control of a product we must consider variations in the repeat measurement of a single part as well the variations in the single measurements of a large number of 'so-called' identical parts. The first is largely due to error in the instrument whereas in the second there is also a contribution caused by variations as a result of the manufacturing process. The first is the study of errors (dealt here) and the second is the subject of statistical quality control (dealt in chapter 18).

Let us first understand some terms used in statistical analysis as under :

Population of Measurement. An infinite number of independent measurements carried out for determination of a certain quantity constitute a population.

Sample of Measurements. In practice, only a finite number of measurements are carried out for determination of a certain quantity which constitute a sample.

Sample Mean. If $x_1, x_2, x_3, \ldots x_n$ be n measurements then sample mean $\bar{x}$ is

$$= \sum_{i=1}^{n} (x_i/n) = \frac{1}{n} \cdot \sum_{i=1}^{n} x_i$$

Sample Standard Deviation. Sample standard deviation 's' is defined as

$$s = \sqrt{\sum_{i=1}^{n} (x_i - \bar{x})^2/n}$$

$$\text{Variance of sample} \quad = s^2 = \frac{1}{n} \cdot \sum_{i=1}^{n} (x_i - \bar{x})^2.$$

Population Mean. The limiting value of sample mean as number of measurements tends to infinity is called population mean

$$\mu = \lim_{n \to \infty} \sum_{i=1}^{n} (x_i/n)$$

Population Standard Deviation. The limiting value of sample standard deviation as number of measurements tends to infinity is called population standard deviation

$$\sigma = \lim_{n \to \infty} \sqrt{\sum_{i=1}^{n} (x_i - \bar{x})^2 / n}$$

Estimate of Population Standard Deviation. An estimate S of the population standard deviation is obtained from sample standard deviation as

$S = s\sqrt{n/(n-1)}$, where s = sample standard deviation.

Random Uncertainty (Ur).

$$U_r = t \,.\, S/\sqrt{n} = t.\frac{s}{\sqrt{n-1}}$$

where t is Student's 't' factor and $S/\sqrt{n}$ is the standard error of the mean, assuming that measurements follow the Gaussian (Normal) distribution.

Systematic Uncertainty (US)

Contributions due to measuring instruments, operating conditions and inherent characteristics of the instrument are taken into consideration.

Uncertainty reported in the certificates of calibration for measurement standards and instruments normally follow Rectangular Distribution with semi-range α. Then variance may be taken a $\alpha^2/3$. Systematic uncertainty $Us = K\,\sigma_s$.

K is the value of student 't' for $n = \infty$ at the desired confidence level and value of $K = 0.675$ for CL of 50%, 1.00 for CL of 68.3%, 1.96 for CL of 95%, 2.58 for CL of 99% and 3.0 for CL of 99.7%.

Overall Uncertainty U

There is no universal agreement for combining the systematic and random uncertainties. One view is to add the two, another is to use the quadrature method while third is to report them separately.

Principles of Least Squares. Assessment of deviation of errors relative to some particular datum may be done with the help of the principle of least squares. The principle states that the most probable value of observed quantities is that which renders the sum of the squares of residual errors a minimum.

Let a number of repeated readings on a component be represented by $x_1, x_2, x_3, \ldots\ldots x_n$. It can now be shown by least squares principle that the most probable value of the series of observed results is the arithmetic mean $\Sigma\, x/n$ as follows :

Let the most probable value be assumed to be x'. Then the deviation of any particular value x from the most probable value x' is $(x - x')$. From the least square principle $\Sigma(x - x')^2$ should be minimum, *i.e.*.

$$\frac{d\Sigma(x - x')^2}{dx'} = -2\,\Sigma\,(x - x') = 0$$

Therefore, $\Sigma x - nx' = 0$ or $x' = \dfrac{\Sigma x}{n}$ = arithmetic mean.

Error distribution. Virtually all instrument errors are random in nature. Exceptions to these, such as non-linearity errors and other errors are called systematic errors.

Random errors have positive and negative values and their magnitudes are generally distributed in accordance with the Gaussian Distribution—the familiar bell-shaped curve shown in Fig. 1.6

$$P(x) = \frac{1}{\sigma\sqrt{2\pi}}\, e^{-x^2/2\sigma^2} \qquad \ldots(1.1)$$

$$P(x_1 < x < x_2) = \frac{1}{\sigma\sqrt{2\pi}} = \int_{x_1}^{x^2} e^{-x^2/2\sigma^2}\, dx \qquad \text{...(1.2)}$$

where $P(x)$ = probability density, x = error value

$P(x_1 < x < x_2)$ = probability that x (error value) lies within the interval x_1, x_2.

The curve (Fig. 1.6) and its mathematical expression [equation (1.1)] represents the probability distribution of the random errors. Since an error within the limits $-\infty$ to $+\infty$ is certain to occur, the area under the curve is numerically 1 ; representing a probability of 1. The probability that the error value lies between x_1 and x_2, $P(x_1 < x < x_2)$ is simply the area under the curve between these two points [equation (1.2)]. On a frequency basis, the area under the curve between error values x_1 and x_2 represents the percentage of all errors lying between these two values.

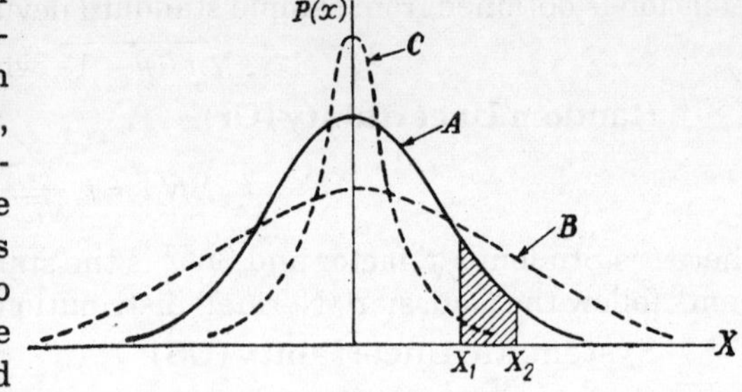

Fig. 1.6

Systematic errors are certain to occur and are, therefore, not treated statistically. If such errors are present in a system of random errors, they are simply added directly to the statistically combined random errors.

Errors Accumulation. The total static error of a measurement system can be measured in terms of root-mean-square (rms) of the component characteristic errors, if the following conditions are fulfilled :

Component characteristic errors are independent and of the same order of magnitude, and the distribution of errors is normal (Gaussian), *i.e.* we consider only the random errors ; and wherever possible this latter condition should be verified by experimental analysis.

The total static error of a measuring system, therefore,

$$= \sqrt{(LE_1 + LE_2 + ...) + RE^2 + CE^2 + EE_1{}^2 + EE_2{}^2 ......}$$

where LE = linearity errors of individual component ; RE = reading errors,

CE = characteristic errors (other than linearity) and EE = environmental errors.

Variance of Error Distribution

The basic measure of the random error distribution is the variance (σ^2) which indicates the spread or dispersion of the distribution function. Mathematically it is defined as :

$$\sigma^2 = \frac{1}{n}\sum_{i+1}^{n} x_i{}^2$$

where x_i = random error value and σ = standard deviation = RMS error.

If the variance is large, the error distribution curve (Fig. 1.6) is broad (curve B). Conversely, if the variance is small, the error distribution curve is quite narrow and peaked (curve C).

The square root of the variance or the "root mean square error" is also called the standard deviation (σ). It can be used to evaluate the most probable value of the measurement and is also quite useful in statistical analysis of Gaussian error distributions. For example, the mathematical nature of the Gaussian error distribution function is such that 68% of the errors represented by the distribution lie between the limits of $\pm$ 1σ, 94% between $\pm$ 2σ, and 99.7% between $\pm$ 3σ. Most of the instrument errors generally expressed are based on the 3σ limits. Thus, allowing for an error three times as great as the RMS static error gives a 99.7% probability that the measurement error is no greater.

1.15. Errors likely to Creep in Precision Measurements—Their Care

1.15.1. Effects of Environment—Temperature. Many applications specify tolerances in microns. It may be appreciated that 25 mm of steel will lengthen about 0.3 microns when its temperature is increased by 1°C. Obviously for precision measurements in terms of microns, error of 0.3 microns just by change of 1°C temperature is substantial. It should be understood that a piece of steel held in the hand absorbs heat very fast and upto 5°C rise in temperature can be expected in 5 minutes time, but it would take hours for it to cool to the laboratory temperature and obtain its original length. It would thus be realised that temperature has a great influence on accuracy of precision measurements. For such accuracy, it is essential that the gauge blocks and workpieces are handled with insulated forceps, or tweezers, with plastic pads or gloves. Usually a plastic shield is introduced in between the inspector and the machine so as not to influence temperature of machine surroundings due to presence by body radiant heat or hot breath or in any other way. In some interferometers, the machine is entirely enclosed in a box with transparent plastic and the operator manipulates the parts with long-handles, insulated forceps introduced through self-sealing rubber part holes. Prior to measurement, the parts, gauges and masters are all stored on the heat sink till they are stablised at the controlled room temperature. Heat sink consists of a slab of steel or aluminum of considerable mass with a clean-smooth surface.

To maintain such controlled temperature in air-conditioned gauging laboratory is essential. The cool air should be so deflected and diffused that the temperature in room will be same in any location but there should be no direct air currents to be noticed.

It may be mentioned that infra red rays from direct sunlight on the gauge would also tend to heat it even in air-conditioned room and, therefore, the sun rays should not enter the laboratory. Thus the stable atmospheric environment free from stray erratic temperature variations and soaking of gauges, masters and workpiece in heat sinks are essential for precise measurements. The temperature of the room should be measured precisely and recorded at regular intervals in the vicinity of gauges.

Internationally accepted temperature for measurement is 20°C and all instruments are calibrated at this temperature. In case workpieces and measuring instruments are of same material then there would be no need of maintaining temperature constant since both would expand equally. But different materials expand differently and hence the need of maintaining constant temperature.

Whenever gauges and workpieces have to be handled, as in the case of wringing of slip gauges, etc., these should be left on table for about an hour in order to enable them to attain uniform temperature of 20°C.

1.15.2. Effect of supports. In the case of long measuring bars, straight edges, these have been supported as a beam. The amount of their deflection due to supporting depends on the position of the supports. Slope and deflection at any point can be calculated by means of expressions derived from the theory of bending.

For a bar of length L, supported equidistant from the centre on supports by distance l apart, then for no slope at the ends, $l/L = 0.577$, and for minimum deflection of the beam, $l/L = 0.554$.

First condition is required for supporting standard bars and second condition is required in case of straight edges.

If a straight edge of rectangular section is quite true when laid on its side on a flat surface, it will not be straight when supported with its edge horizontal. The deflection is minimum when it is the same at the ends and the center. The necessary spacing of the supports is found by utilizing the mathematical relations for the bending of beam giving l/L equal to 0.554.

1.15.3. Effect of alignment. Abbe's principle of alignment should be followed in measurements to avoid cosine errors, sine errors, etc. According to this principle, the axis or line of measurement of the measured part should coincide with the measuring scale or the axis of measurement of the measuring instrument. Cosine error occurs in measuring the length of any part with a scale if the measuring scale is inclined to the true line of dimension being measured. In

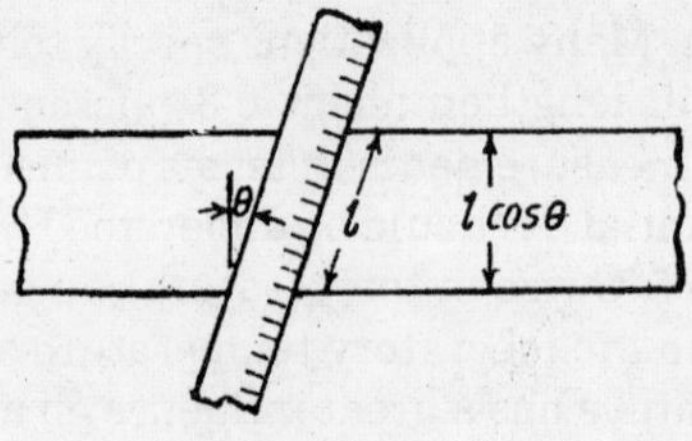

Fig. 1.7. (a) Cosine error.

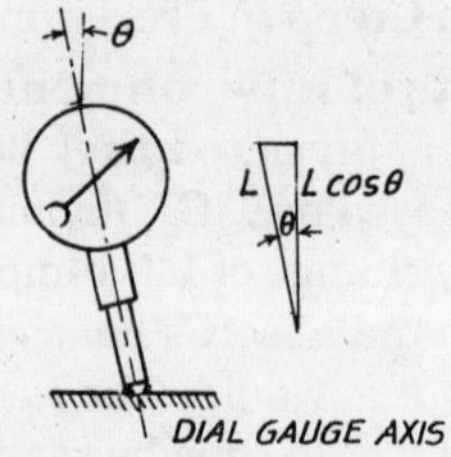

Fig. 1.7. (b) Errors due to non-alignment of plunger axis and line of measurement.

Fig. 1.7 (a) it may be seen that the length recorded is in excess of the true length by an amount $l(1-\cos\theta)$. However, this error is negligible in most cases since the angle θ is usually very small.

An alignment error of 2° over 1 metre introduces an error of approximately 0.6 mm.

Similarly error is introduced to dial indicator readings if the plunger axis does not coincide with the axis or line of measurement as in Fig. 1.7 (b).

If e = induced error and L = Change in indicator reading,
then $L\cos\theta$ = surface displacement and $e = L(1-\cos\theta)$.

To ensure correct displacement readings on the dial indicator, the plunger must, of course, be normal to the surface in both mutually perpendicular planes.

The measuring jaws of the vernier caliper are in effect extension of the scale markings on the instruments ; thus the jaws of an instrument in good condition are parallel and remain so at any measurement within the instrument range. The length l indicated on the scale corresponds to the displacement of the jaws from each other as in Fig. 1.8 (a). The effect of a bent scale beam is shown in Fig. 1.8 (b). The length l_1 between the extremes of the jaws has now become smaller than the length indicated on the scale. It is also evident that the error introduced is proportional to the length of the jaws and is at its minimum along the graduations line of the scale.

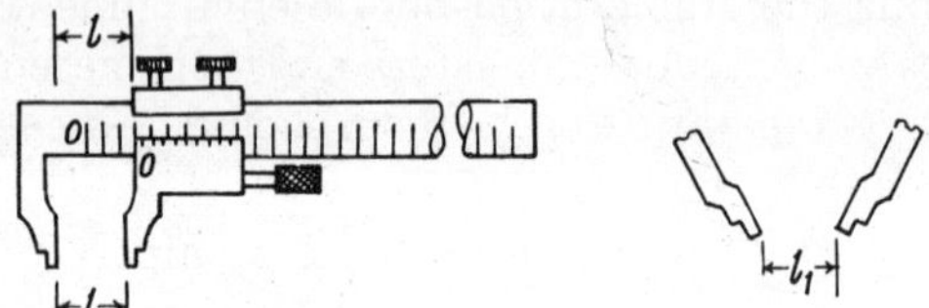

(a) (b)

Fig. 1.8. Error introduced due to bent jaws of the vernier caliper.

Similar errors occur in the case of an indicator fitted with a ball-end stylus arm. In this case, it is important for the arm to be set so that the direction of movement of the work is tangential to the arc long which the ball moves. In both the cases of Fig. 1.9 (a) and (b), the reading on the indicator will be $h\sec\theta$, where h is the distance moved by the work. It may be remarked that some lever type test indicators are designed for use with their styli inclined at about 30° to the line of measurement and in these cases, an error would occur if it were set with the stylus normal to the line of measurement.

The combined sine and cosine errors occur when measuring an end gauge in the horizontal comparator if the gauge is so supported that its axis is not parallel to the axis of the measuring anvils, or if its each end, though parallel to each other, are not square to the axis.

With reference to Fig. 1.10

$$L = M\cos\theta - d\sin\theta$$

Errors of the above nature can be avoided by using gauges with spherical ends. It may be noted that

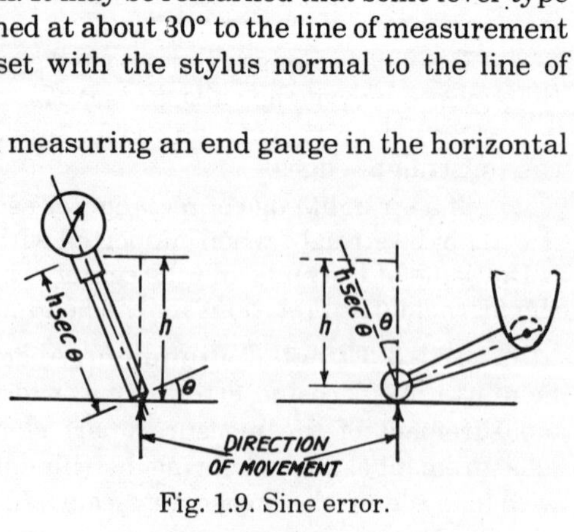

Fig. 1.9. Sine error.

such gauges need not be aligned accurately when used combination. Fig. 1.11 shows two gauges of this type with their axes out of line by an amount δ. The error in their combined length is very small, $(a + b)$.

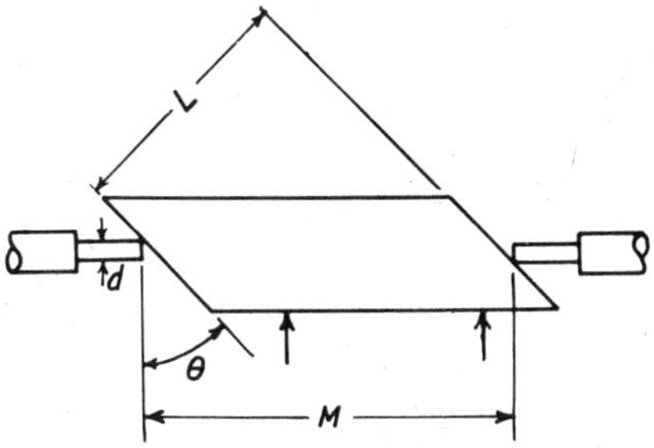

Fig. 1.10. Both sine and cosine errors.

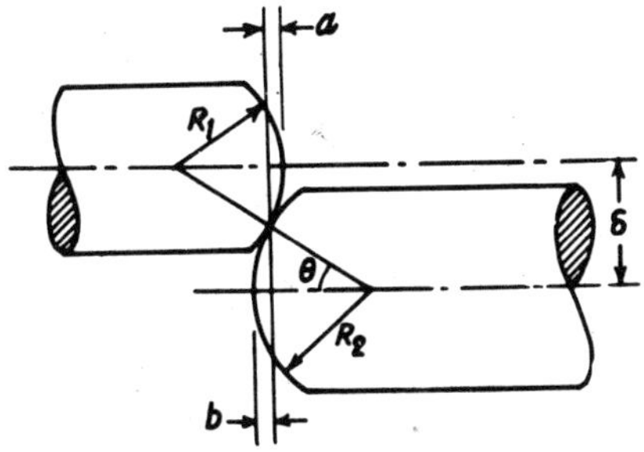

Fig. 1.11. Errors when spherical end gauges are not properly aligned.

$$a = R_1(1 - \cos\theta) = R_1\left(1 - \sqrt{1 - \sin^2\theta}\right)$$

$$R_1 = \left[1 - \sqrt{1 - \frac{\delta^2}{(R_1 + R_2)^2}}\right] = R_1\left[1 - \left(1 - \frac{\delta^2}{2(R_1 + R_2)^2}\right)\right] \text{ approx.}$$

$$= \frac{R_1\delta^2}{2(R_1 + R_2)^2}; \qquad \text{Similarly, } b = \frac{r_2\delta^2}{2(R_1 + R_2)^2}$$

Therefore,

$$(a + b) = -\frac{(R_1 + R_2)\delta^2}{2(R_1 + R_2)^2} = \frac{\delta^2}{2(R_1 + R_2)}$$

With 150 mm and 250 mm gauges out of line by 0.2 mm, the error in their combined length is only a tenth of micrometer. Longer gauges of course, result in a correspondingly smaller error, with standard special end pieces.

1.15.4. Dirt. Presence of dust in atmosphere may change reading by a fraction of micron. Where accuracy of the order of micron is desired, electrostatic precipitators need to be incorporated in the lab or in the air ducts in addition to air filters. Other sources of dust could be from clothing. The work-pieces and masters should, therefore, be cleaned by clean chamois or by a soft artists' brush. Dust inhibitors are also essential. The coated surfaces should be sprayed with a suitable filtered clean solvent. Gauges should never be touched with moist fingers.

1.15.5. Errors due to vibrations. For consistent and repetitive readings a gauge should be subjected to as little vibration as possible, or none. Vibrations can be eliminated by locating the lab properly away from vibration sources and slipping cork, felt or rubber pads under the gauge, and mounting the gauge pedestal or floor sections on tar mastic. Putting a gauge on a surface plate resting in turn on a heavy plate also reduces the effect of vibrations due to natural frequency of massive plate differing substantially from the immediate surroundings.

1.15.6. Metallurgical effects. The material for gauges should have been properly and naturally seasoned after heat treatment so that it attains stable dimensions ; otherwise measurement discrepancy may occur due to unstable, internal metallic structure.

The amount of surface roughness of gauge should be determined, because gauge almost invariably measures the peaks. The measurement may be disturbed by loosening up of tiny flakes of chrome on stainless steel. Tungsten carbides may have invisible pits which can influence readings.

Sometimes the size, area and weight of work-piece may create an error in obtaining a true dimension, for example, a 100 mm gauge block when tested in a horizontal position will give different readings than when placed vertically or suspended. Such errors would be noticeable on lengthy components above 450 mm in length.

1.15.7. Contact point penetration. Penetration is the bending, depressing, deforming and yielding of the surface of the workpiece under the pressure of the gauging contact. The penetration effect does occur and varies with the Young's modulus of the material being gauged. Tungsten carbide being rigid has nearly no penetration effect but brass yields readily. Penetration in the surface of a cylinder is found to be little greater than in a flat surface, which in turn is greater than in the bore of a ring.

For detecting size differences of less than 2-3 microns, gauging pressure is seldom trusted to human touch. Contact pressure in most precision gauges is controlled by spring action or by weights and counterweights. It may be appreciated that contact pressure (though it causes penetration errors) is essential in order to avoid any wavering of contact and ensuring certain contact. Vibration is the major cause of uncertain readings with light gauging pressure.

1.15.8. Errors due to deflection. Deflection of parts occurs due to contact gauging pressure. To avoid deflection errors, contact gauging pressure should be as small as possible, overhangs should be minimised, gauge frame should be made of rigid and adequate cross-section, and the gauge clamps or adjusting devices should be securely tightened. The deflection errors to some extent can be minimised by noting readings on comparator with master of same material as the workpiece. Sometimes free-rolling cylinders or balls are used under the work to avoid deflection errors by contact pressure. Jaws of precision calipers for internal diameter gauging are made of rigid material with cross-section areas great enough to reduce bending errors to negligible amounts.

In considering the total effect of stylus pressure, it is convenient to consider the two effects of deformation and deflection separately. The total error caused by the effect of stylus pressure may be obtained by adding the deformation and deflection values together. The deformation effect is shown in Fig. 1.12. The value of the deformation δ may be calculated from the relationship

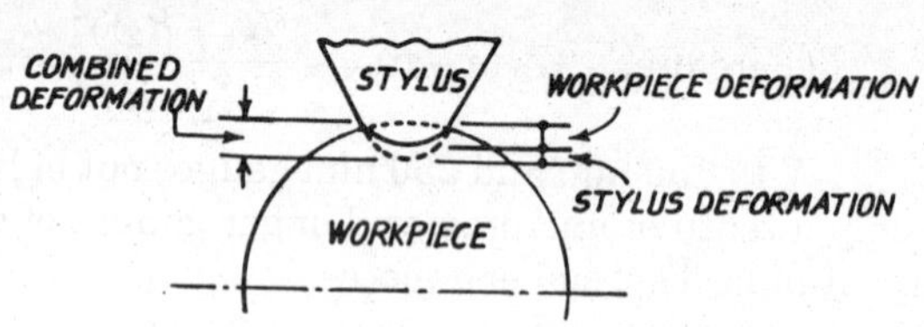

Fig. 1.12. Effect of contact pressure on measurement.

$$\delta = p^{2/3} B^{1/3} \lambda Y$$

where p = Stylus pressure, $B = \dfrac{1}{2R_1} + \dfrac{1}{2R_2}$

where R_1 = tip radius of stylus, R_2 = radius of workpiece

$$\lambda = \left(\frac{v_1 + v_2}{2}\right)^{2/3}$$

where v_1 = a constant dependent on the tip material

v_2 = a constant dependent upon the workpiece material

Y = a constant dependent upon the ratio of the geometry of the stylus to that of workpiece.

The deflection β of the hollow cylinder under the action of the stylus pressure has also been determined and may be calculated from the formula :

$$\beta = \frac{1.7856 R_m^2 p}{E L_e R_m t^3}$$

where p = stylus pressure, R_m = mean radius of workpiece, E = young's modulus of workpiece, L_e = effective length of workpiece, t = wall thickness of workpiece.

1.15.9. Errors due to looseness. Looseness leads to errors due to penetration and deflection. It results in inconsistent readings and inaccurate calibration. Looseness can be tested by setting the gauge contact on gauge anvil and zeroing the meter ; and then applying finger pressure or a light tab to each location, where looseness might be expected and noting the reading again. It is essential for this test that the gauge anvil should be clamped down securely to gauge frame. Metal chips or plain dirt between post and clamp can produce the pivoting effect. The gauge contact point should be tight in the spindle.

1.15.10. Errors due to wear in gauges. Wear on anvils and contacts is very troublesome. The internal instrument wear can also lead to measurement error, which can be checked by calibration. The extent of wear hollows in anvils and lack of parallelism can be readily measured with optical flats. The spherical contact can be checked by examination with a microscope. Wear can be minimised by keeping gauges, masters and workpieces clean and away from dirt. Carbide masters though wear resistant themselves, wear out hardened steel anvils. Cast-iron workpieces present the problem of sand particles. Gauge anvil and contact materials should be properly hardened. Chrome-plated parts have been found to withstand ten times more wear than unplated parts.

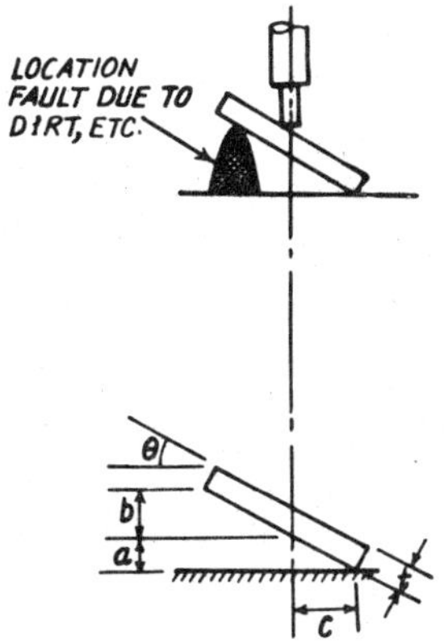

(*a*) Surface displacement

Reading $= a + b = c \tan\theta + t \sec\theta$

Error $= a + b - t = c \tan\theta + t(\sec\theta - 1)$

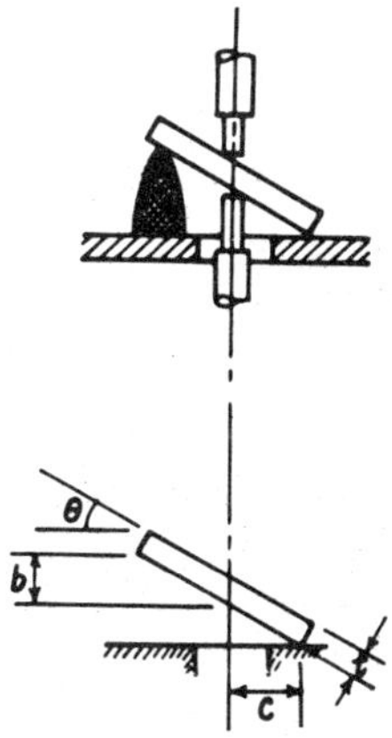

(*b*) Differential length measurement

Reading $= b$

Error $= b - t = t(\sec\theta - 1)$

Fig. 1.13. Errors due to faulty location.

1.15.11. Error due to location. Fig. 1.13 shows how imperfect location of surface causes errors and how differential measurement can reduce such errors.

1.15.12. The Parallax Effect. On most dials the indicating finger or pointer lies in a plane parallel to the scale but displaced a small distance away to allow free movement of the pointer. It is then essential to observe the pointer along a line normal to the scale, otherwise a reading error will occur. This effect is shown in Fig. 1.14, where a dial is shown observed from three positions

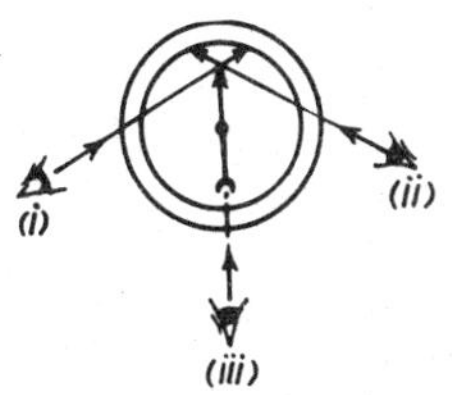

Fig. 1.14. Error due to parallax.

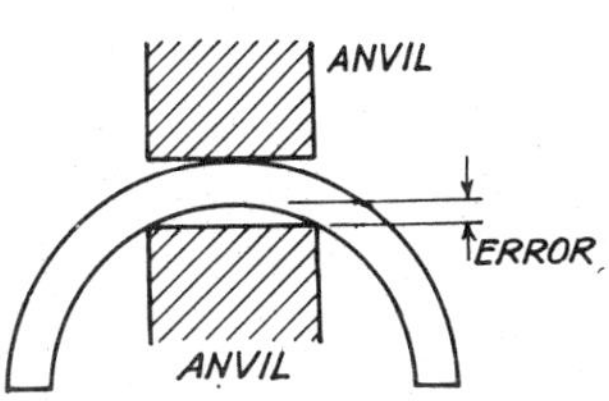

Fig. 1.15. Error due to poor contact.

where the pointer is set at zero on the scale. Observed from position (*i*) *i.e.* from the left, the pointer appears to indicate some value to the right of zero, and from position (*ii*) some value slightly to the left of zero, while only at position (*iii*) with the pointer coincide with zero on the scale. Rulers and micrometer thimbles are levelled to reduce this effect and on dials the indicators may be arranged to lie in the same plane as the scale thus completely eliminating parallax, or a silvered reflector may be incorporated on the scale so that the line between the eye and pointer is normal to the scale only when the pointer obscures its own image in the reflector.

1.15.13. Errors due to poor contact. Fig. 1.15 shows how the poor contact of working gauge or instrument with the workpiece causes an error. It would be appreciated that the gauge with wide areas of contact should not be used on parts with irregular or curved surfaces.

1.15.14. Error due to impression of measuring stylus. Every gauge stylus has relatively high weight. Fig. 1.16 shows how the impression of the gauge stylus indents the surface. Even a well supported, flat bar may be difficult to measure when the gauge indents the surface on contact. Optical or pneumatic instruments are best to avoid such error.

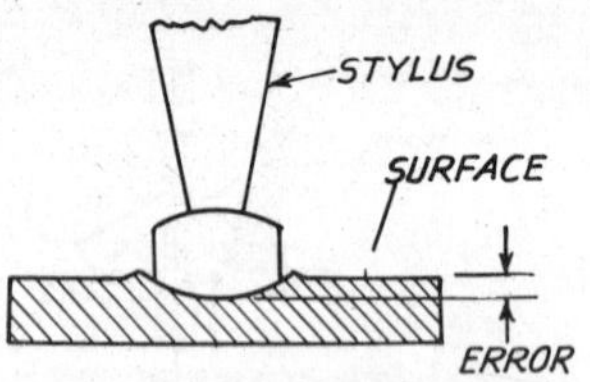

Fig. 1.16. Error due to impression of stylus on workpiece.

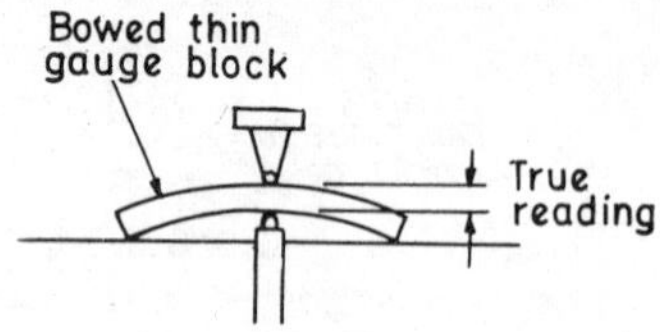

Fig. 1.17

Similarly distortions due to stylus pressure occur while measuring thin walled surface. Spring loaded gauges should be avoided and electronic/air gauges are best suited for such applications.

1.15.15. Importance of gauging set up to be square. Perpendicularity in gauging set ups should be ensured properly in precision measurements as an angular error of one minute in setting can cause 1 micron error in 100 mm. A comparator of gauge should be carefully checked for squareness from several different angles with the gauge clamped tight. Errors of perpendicularity may arise either due to gauge head not being set up perpendicular to workpiece or the reference anvil on which workpiece is placed may itself be tilted.

1.15.16. Gauging thin blocks. For gauging thin blocks, it is advisable to use dual gauging contact since these are liable to warp when not wrung together. Obviously enough if reading is taken with single gauging head with bowed thin gauge block placed on surface plate, the reading obtained will be much different, but two point gauging eliminates such problems.

1.16. Linearity

The output of a measuring system may be linear or non-linear. As long as the output is repeatable, it can always be calibrated, but for a non-linear system the construction of calibration curve becomes tedious, cumbersome and time-consuming. For a liner system, a simple two or three point calibration is sufficient and recalibration is also facilitated. Thus, a high degree of linearity, or the minimising of linearity error is very important characteristic of any measuring system. The more linear the measuring system is, the more readily it can be calibrated and the less uncertainty there will be about a particular output value indicated by the system.

As already indicated, characteristic error is defined as deviation of the output of a measuring system from the predicted performance, or from nominal performance specifications, under the condition of constant environment. According to this definition, the "theoretically predicted performance" or "nominal performance specification" is the norm of reference to which deviations are compared, and the basis of selecting a particular norm, particularly for linearity, is sometimes misunderstood. Linearity can be defined as the maximum deviation of the output of the measuring system from a specified straight line applied to a plot of data points on curve of measured values *versus* the measurand input value. As several lines can be used as linearity references, it becomes essential to clearly define the exact nature of reference straight line before interpreting or comparing the linearity specifications for a measuring system. The most common reference lines are :

(*i*) *Terminal line*—It is drawn from the origin to the data point at full scale output.

(*ii*) *End point line*—It is drawn between the end points of the data plot, usually without regard to the origin if the output is bipolar.

(*iii*) *Best fit line*—It is the line midway between the two closest parallel straight lines that enclose all the data points.

(*iv*) *Least square line*—It is the line for which the sum of the squares of the deviations of the data points from the line or curve being fit is minimised. This is the most favoured line and is truly best-fit curve in the sense that it comes as close as possible to each data point on a plot of output *versus* measurand input according to the least-square error criterion, *i.e.* to minimise the sum of the squares of the deviations of the data points from the curve being fit.

The least-squares line for a measuring system can be found from the established slope equations.

The slope m of the least square best-fit straight line for a collection of data points is given by :

$$m = \frac{\overline{x_i}\,\overline{y_i}}{\overline{x_i}^2} = \frac{\sum_{i=1}^{n} x_i y_i}{\sum_{i=1}^{n} x_i^2}$$

where x_i = input of measurand
y_i = output value (measured value)
n = No. of test data points.

Once the slope m is determined, it is possible to check the maximum deviation of linearity from the least squares line by the following relation :

$$\%\ \text{linearity} = \frac{y_{i\,max} - mx_i}{mx_{full\ range}} \times 100\%$$

where $y_{i\,max}$ = output value at point of maximum deviation from best-fit straight line
mx_i = point on best-fit straight line corresponding to maximum deviation point.

1.16.1. Application of the Least Squares Principle. The least squares principle has wide application in metrology for assessing the deviation of errors relative to some particular datum. In general the least squares principle states that the most probable value of observed quantities is that which renders the sum of the squares of residual errors a minimum.

1. *Series of Observed Values.* This situation occurs when a large number of readings are taken on the same dimension of a component.

Let the observed values be given by $x_1, x_2, x_3 \ldots x_n$. It can be shown by the least squares principle that the most probable value of the series of observed results is the arithmetic mean $\Sigma x/n$ as follows :

Let the most probable value be assumed to be X. Then the deviation of any particular observed value x from the most probable value X is $(x - X)$. From the least squares principle $\Sigma\,(x - X)^2$ should be a minimum, *i.e.*,

$$\frac{d\Sigma(x-X)^2}{dx} = 2\Sigma\,(x-X) = 0$$

$$\therefore \quad \Sigma x - nX = 0$$

$$X = \frac{\Sigma x}{n} = \text{arithmetic mean.}$$

2. *Series of Observed Values of Two Dependent Variables.* Such a problem arises in the measurement of straight edge when the variations (y) from a truly straight line are determined at a number of positions (x) along the straight edge.

Let the observed values be given by (x_1, y_1), (x_2, y_2) ...... (x_n, y_n).

The equation of the straight line is given by $X = ax + b$, where a is constant representing the slope of the line and b is constant representing the intercept on the y-axis. Referring to Fig. 1.18, let AB be any arbitrary straight line placed in relation to the observed points (x_1, y_1), (x_2, y_2), ... (x_n, y_n). Using this line to predict the value of y_1 from x_1, we get point P. Thus the difference of predicted value and the actual value will, therefore, be $y_1 - (ax_1 + b)$. What is required is to choose the best-fitting line with respect to the observed values by the method of least squares. This is done by choosing a line for which the sum of the squares of the deviations from it measured vertically is a minimum.

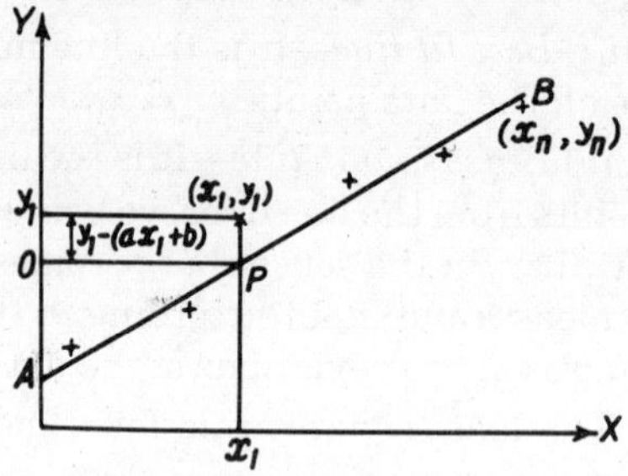

Fig. 1.18. Deviation of least-square line for two dependent variables.

Taking point (x_1, y_1), the square of the deviation of this point from the line is $[y_1 - (ax_1 + b)]^2$ and the sum of the squares of the deviation for all the points is given by :

$$E = \Sigma[y - (ax + b)]^2.$$

It is thus necessary to choose a and b to make E a minimum. This may be done by differentiating expression partially with respect to a and b and then equating each derivative to zero. Thus

$$\frac{dE}{da} = -2\,\Sigma x\,[y - (ax + b)] = 0 \qquad ...(1)$$

$$\frac{dE}{db} = -2\,\Sigma\,[y - (ax + b)] = 0 \qquad ...(2)$$

Equation (2) may be rearranged as follows :

$$\Sigma y = a\,\Sigma x + nb$$

$$\therefore \quad b = \frac{\Sigma y}{n} - \frac{a\Sigma x}{n} = \bar{y} - a\bar{x} \qquad ...(3)$$

where $\bar{y}$ = mean of $y_1, y_2, ..., y_n$ and $\bar{x}$ = mean of $x_1, x_2 ......, x_n$.

If these mean values are substituted in the expression $y = ax + b$ we obtain Eq. (3). The least squares line therefore passes through the mean of the points (x_1, y_1), (x_2, y_2), ..., (y_n, y_n).

Rearranging Eq. (1), we get $\Sigma xy = a\,\Sigma x^2 + b\,\Sigma x$

Substituting $b = \bar{y} - a\bar{x}$ in the above gives $a = \dfrac{(\Sigma xy - n\overline{xy})}{(\Sigma x^2 - n\bar{x}^{2})}$ ...(4)

If the origin of the straight line is taken through the mean point $\bar{x}, \bar{y}$ the Eq. (4) reduces to :

$$a = \frac{\Sigma x_m\, y_m}{\Sigma x_m^2}$$

where x_m = reading in x-axis relative to $\bar{x}$, *i.e.* $x - \bar{x}$

and y_m = reading in y-axis relative to $\bar{y}$, *i.e.* $y - \bar{y}$.

3. *Series of Observed Values of Three Dependent Variables.* Such a problem arises in the measurement of a surface table when the variations (z) from a truly flat plane are determined at a number of positions (x, y) on the surface plane.

Let the observed values be given by $(x_1, y_1, z_1), (x_2, y_2, z_2), ..., (x_n, y_n, z_n)$.

The equation of the flat plane is given by :

$$Z = ax + by + c.$$

Let any arbitrary plane be placed in relation to be observed points $(x_1, y_1, z_1), (x_2, y_2, z_2)$ ......, (x_n, y_n, z_n). Using this plane to predict the value of z_1 from (x_1, y_1), we get a point N. Thus the predicted value of $z_1 = Z_1 = ax_1 + by_1 + c$. The difference between the predicted value and the actual value will therefore be :

$$z_1 - (ax_1 + by_1 + c).$$

What is required is to choose the best-fitting plane with respect to the observed values by the method of least squares. This is done by choosing a plane for which the sum of the squares of the deviations from it measured vertically is a minimum.

Taking point (x_1, y_1, z_1), the square of the deviation of this point from the plane is $[z_1 - (ax_1 + by_1 + c)]^2$ and the sum of the squares of deviation for all the points is given by

$$E = \Sigma\,[z - (ax + by + c)]^2.$$

It is thus necessary to choose a, b and c to make E a minimum. This may be done by differentiating the expression for E, partially with respect to a, b and c and then equating each derivative to zero.

Thus
$$\frac{dE}{da} = 2\Sigma x\,[z - (ax + by + c)] = 0 \quad ...(5)$$

$$\frac{dE}{db} = 2\Sigma y\,[z - (ax + by + c)] = 0 \quad ...(6)$$

$$\frac{dE}{dc} = 2\Sigma z\,[z - (ax + by + c)] = 0 \quad ...(7)$$

Eqn. (7) may be rearranged as follows :

$$\Sigma z = a\,\Sigma x + b\,\Sigma y + nc$$

$$\therefore \quad c = \frac{\Sigma z}{n} - \frac{a\Sigma x}{n} - \frac{b\Sigma y}{n} = \bar{z} - a\bar{x} - b\bar{y} \quad ...(8)$$

where $\bar{x}$ = mean of x_1, x_2, ..., x_n ; $\bar{y}$ = mean of y_1, y_2, ..., y_n and $\bar{z}$ = mean of z_1, z_2 ......, z_n.

If these mean values are substituted in the expression $Z = ax + by + c$ we obtain desired equation of plane. The least squares plane therefore passes through the mean of the points $(x_1, y_1, z_1), (x_2, y_2, z_2)$, ..., (x_n, y_n, z_n). Eqns. (5) and (6) can now be rewritten as

$$\Sigma\, x = a\Sigma x^2 + b\Sigma xy + c\Sigma x \quad ...(9)$$

$$\Sigma yz = a\Sigma xy + b\Sigma y^2 + c\Sigma y \quad ...(10)$$

If the origin of the mean plane is taken through the mean point $\bar{x}, \bar{y}, \bar{z}$ then we may elimin‑‑e term c in expressions (9) and (10) and solve for a and b to give :

$$a = \frac{\Sigma y_m^2\, \Sigma x_m z_m - \Sigma x_m y_m\, \Sigma y_m z_m}{\Sigma x_m^2\, \Sigma y_m^2 - (\Sigma x_m y_m)^2}$$

$$b = \frac{\Sigma x_m^2\, \Sigma y_m z_m - \Sigma x_m x_m y_m\, \Sigma z_m}{\Sigma z_m^2\, \Sigma y_m^2 - (\Sigma x_m y_m)^2}$$

where x_m = readings in x-axis relative to $\bar{x}$, *i.e.*, $x - \bar{x}$
y_m = readings in y-axis relative to $\bar{y}$, *i.e.*, $y - \bar{y}$
z_m = reading in z-axis relative to $\bar{z}$, *i.e.*, $z - \bar{z}$.

1.17. Rigorous definitions of accuracy and precision. Say it is required to measure a dimension L. With highly accurate machine, we take several measurements and obtain a distribution (normal centred around mean ; strictly speaking, it is t-distribution which can be approximated with a normal distribution when the sample size or number of measurements is about 30 or greater). The difference δ between mean value and nominal dimension is called bias. Smaller the bias, the higher is the accuracy. Precision is measured as to how small is the dispersion ε from the mean value.

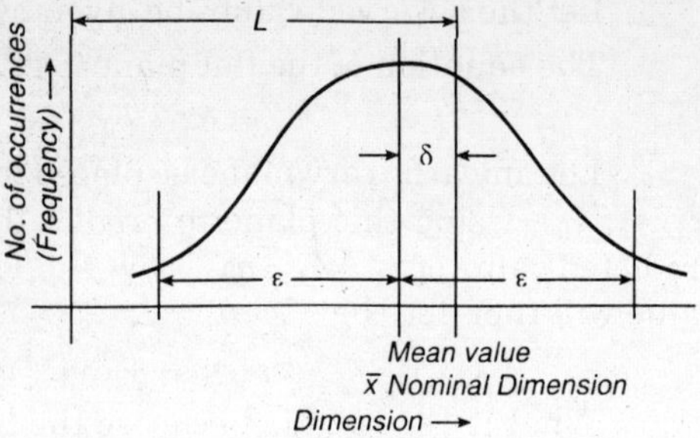

Fig. 1.19

When a bias exists, conditions are adjusted to compensate for δ to make $\bar{x}$ agree with L. Since compensation is done using sample mean $\bar{x}$ and not the population mean μ, there will be a slight error, and the corrected sample mean will still differ from L. The probabilistic range over which the difference $\delta = L - \bar{x}$ will be dispersed can then be used as a quantitative measure of accuracy and given as follows

$$\delta = t\left(\phi, \frac{\alpha}{2}\right)\lambda\sqrt{\frac{v}{n}}$$

where $t(\phi, \alpha/2)$ is the t-distribution value when the degree of freedom is $\phi = n - 1$
$1 - \alpha$ is the probability and can be determined from a mathematical table
n is sample size
α is the significance level
v is the unbiased variance defined as $v = \frac{1}{n-1}\sum_{i=1}^{n}(x_i - \bar{x})^2$.

If $1 - \alpha$ is taken as 0.95 (*i.e.* 95% probability), $n = 21$, $N = 4\ \mu m^2$, then
$t(20, 0.05/2) = 2.09$ (from a table)

and $$\delta \equiv \text{accuracy} = 2.09 \times \sqrt{\frac{4}{21}} = 0.912\ \mu\text{m}$$

Thus accuracy of such a process is 0.912 μm with 95% reliability ; *i.e.* $\mu - \bar{x}$ will be within ± 0.912 μm with a 0.95 probability. This can also be expressed as 60.912 μm (95%).

Precision $$\varepsilon = t\left(\phi, \frac{\alpha}{2}\right)\sqrt{v} = 2.09 \times \sqrt{4} = 4.18\ \mu\text{m}$$

This is expressed as '4.18 μm (95%)'.

It means that the finished dimension will be dispersed within ± 4.18 μm of mean value with a probability of 0.95.

Accuracy and precision can be combined to give a broader definition of precision *i.e.* error limit $\tau = \delta + \varepsilon$.

If accuracy and precision can be estimated with sufficient reliability, then precision, broadly defined, can be obtained from the variance of dispersion of δ and ε, *i.e.* $\tau = \sqrt{\delta^2 + \varepsilon^2}$

1.18. Principles of High-precision Measurements

(*i*) Temperature of measurements needs to be clearly defined. An object placed at 21°C may take more than 10 hours to attain room temperature of 20°C. This is so because temperature decreases exponentially. The process can be quickened by placing the object on a large mass (say, surface place which acts as a heat sink). Since heat also flows by radiation, careful attention must be paid to prevent radiation due to lighting, sunlight, human bodies, and electric equipment from affecting measurements.

If object's temperature is at different temperature, then it must be monitored at all times, and measurement must be corrected for difference between object's temperature and reference temperature by calculation.

(*ii*) Measurement force should be avoided (non-contact methods of measurement be preferred).

(*iii*) The precision of the measuring instrument must be five to ten times higher than the expected precision of the measured object.

(*iv*) Be aware of the instrument's characteristics like undirectional approach to avoid backlash effects, completing measurements quickly to avoid drift etc.

(*v*) Influence of work-affected layers on surface. The surface layer affected by machining needs closer look for precision measurements. The presence of an affected layer will influence residual stress (cause of changes with age), surface processability, reflectance, electrical conductivity, corrosion resistance, and abrasion resistance.

Work affected layer must consist of four or five layers as follows : Around 1 μm thick outermost layer consisting of a nearly amorphous structure, followed by 10 μm fibrous layer which has undergone severe plastic deformation and which resembles rolled steel in grain structure. Below that is the plastically deformed layer consisting of crushed grains. Then comes the transgranular slippage layer, in which the grains still retain their form but some internal slippage exists. Depth of affected layer can be measured by :

(*i*) Ultrasonic microscope method, (*ii*) Corrosion method, (*iii*) microscopic analysis, (*iv*) X-ray diffraction method, (*v*) hardness method, and (*vi*) recrystallisation method.

1.19. Why Precision Measurements are required and Why Struggle for ever increasing Accuracy ?

The need for reliable data was recognised very early in the developmental science. Fig. 1.20 shows how the experiments have helped in establishing of theoretical postulates, and the experiments were possible only with the developments in precision measurements. The logical consequences of preconceived views could be demonstrated by carrying out experiments. Thus precision measurements form the background of advanced technology engineering.

Advancements in Reference Standards

Quantitative measurements require some sort of stable measurement standard if the results of measurements in different places and at different times are to be usefully compared. Intimately connected with the definition of units of measurements is the practical task of the physical realization of these units, since a paper definition serves no useful purpose if it cannot be realized with the required precision.

It was only since science reached the state where engineering began to rely on any significant extent upon the discoveries of physics, that accurate standards of measurement became necessary. Towards the middle of the nineteenth century the need for word-wide agreement on the principal units of measurement was felt and this led to the signing in 1875 of the Convention. The founding of the major national metrological laboratories took place soon after. In 1875 the only standards deposited were those of mass and length. The latter very quickly required the establishment of

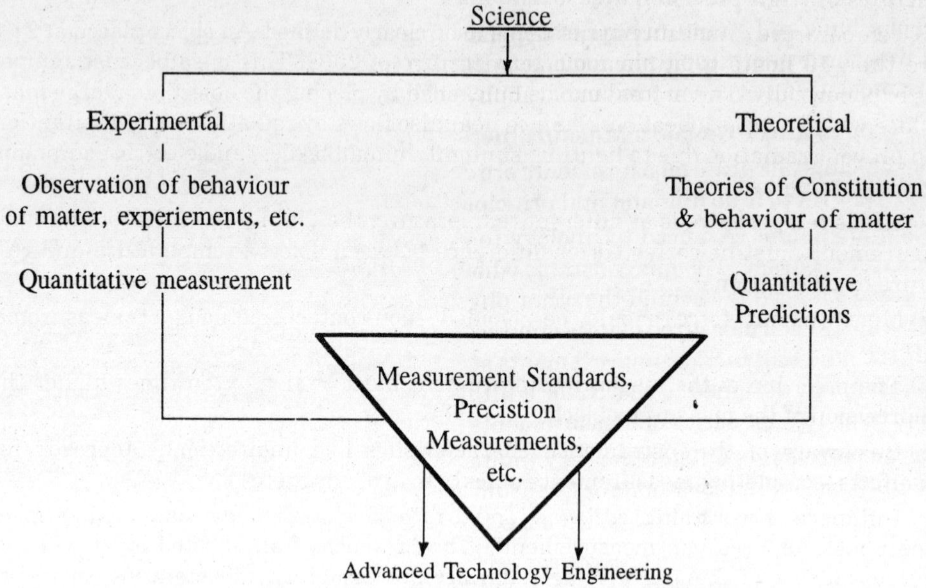

Fig. 1.20. The background of Advanced Technology Engineering.

accurate standards of temperature, since thermometers were needed to measure the thermal expansion of the meter bars.

During the second half of the nineteenth century some of the leading physicists devoted considerable time and effort to the problem of accurate measurement and the establishment of a consistent system of units. James Clerk Maxwell in 1870 stated, : "If we wish to obtain standards of length, time, and mass which shall be absolutely permanent, we must seek them not in the dimensions or the motion of the mass of our planet, but in the wavelength, the period of vibration, and the absolute mass of these imperishable unalterable, and perfectly similar molecules." Indeed, as physics progressed it became possible to envisage practical units based upon atomic phenomena.

The establishment of useful units based upon atomic and, in due course, quantum phenomena, has not turned out to be easy. In order to establish an atomic or quantum-based base unit at least three conditions must be satisfied. First, one requires a good theoretical understanding of the quantitative behaviour of the atomic or quantum phenomenon in question and, second, a practical way of realizing the phenomenon must exist that allows an explicit theoretical description to be written down of the process employed, and third, a method is required of making a sufficiently accurate link between the other base units and the proposed atomic or quantum-based unit. It is often the third of these that turns out to be the most difficult. Nevertheless, it is in the attempt to use the fundamental physical constants as the bases of measurement standards that the role of precise measurement in physics became evident. In order to check that we have an adequate theoretical understanding of atomic or quantum phenomena we need to be able to check quantitative predictions made by theory. This can sometimes require the most accurate and precise measurements possible, to distinguish between competing theories or interpretations of theory.

The bases of measurement standards, the fundamental physical constants are useful only in so far as physical phenomena exist that can be described explicitly in terms of the fundamental constants, and which can at the same time, be measured in terms of the macroscopic base units. The extent to which this can be done is an indication of the extent to which modern atomic and quantum physics can be used to produce much of today's advanced technology. The practical application of atomic and quantum physics depends upon our ability to make the precise measurements that link these microscopic phenomena to the macroscopic world. An essential part of this is the accurate measurement of the fundamental physical constants in terms of the base units.

Quantum Physics and Precision Measurements

Almost all high accurate measurements are dependent upon quantum mechanical properties of system. Further there is no fundamental principle that forbids accurate classical measurements.

The application of advanced technology to classical measurements can lead to significant improvements in accuracy. The gas constant, which together with the gravitational constant has in the past always lagged far behind the other physical constants in the accuracy with which it is known, has recently been measured using a spherical acoustic resonator with an accuracy approaching 1 part in 10^6, and still further improvements seem possible. Some weighings are carried out as part of an experiment designed to search for a fifth force, that have exhibited standard deviations of about 5 parts in 10^{12} for series of measurements lasting a few days.

Such measurements outside the very limited immediate field of application are of interest because one finds that when an atomic or quantum phenomenon is used as the basis of a particular technology, there appears a demand for all aspects of that technology to increase in precision.

1.19.1. Nanometrology. In order to connect monomode optical fibres together it is necessary to have mechanical connectors that can mate with axial errors of fractions of a micrometer. Such dimensional tolerances are characteristic of the demands now being placed upon length or dimension metrology by the optoelectronics industries.

The increase in the density of packing of circuit elements on silicon wafers has resulted in the dimensions of individual elements falling to tens of nanometers. In order to make such devices it is now necessary to be able to identity and reproduce the position that objects should have on a silicon wafer to this accuracy. Thus on a part of a silicon wafer roughly 10 cm square it is necessary to be able to locate and identify any point with a precision of 10 or 12 nm. The requirement in any particular production line may indeed be only for relative precision, but as soon as the process is transferred elsewhere, either to another factory or even to another part of the same factory, the position of a point on the wafer must be expressed in some unit. While in the short term a wholly local unit can be maintained, based upon the properties of a particular machine, in the long term it is necessary to measure in meters or nanometers. National laboratories are now finding that it is by no means a simple exercise to measure and identify the position of points on a 10 cm square with an accuracy of 10 nm or better. This is the level of accuracy and precision required of those national laboratories that are concerned with the testing of the commercial master grids that are used in the fabrication of masks for large-scale manufacture of integrated circuits.

Another example of extreme precision in manufacturing for the optoelectronic industries is the production of small aspheric lenses whose performance is diffraction limited, but whose dimensions are small. The production of the master mould for such lenses is possible with high-precision machining. Diamond machining, elastic-emission machining, and other processes now can produce surfaces smooth at the nanometer level. Intimately linked to the manufacture of objects having such properties is the measurement and testing of them. New optical devices now exist to measure surface profiles and at least one of them can observe large-scale surface structure at the level of tens of picometers.

The need for rapid inspection of complex objects manufactured to high precision by the new generations of numerically controlled machines has led to the development of coordinate-measuring machines. These machines are designed to measure the spatial coordinates of any point within a volume of up to 1 m^3 with an accuracy of about 1 μm. The calibration of such machines and the design of test objects for them is now a major activity of many national laboratories. The new "nanometrology" will require the development of new, more accurate techniques for surface profile, roundness, and diameter measurement. In these latter domains, existing optical or mechanical-contact methods for identifying the position of a surface are unlikely to be adequate. In order to identify the position of the surface of an object at the nanometer level it will be necessary first to define more carefully what it is meant by a surface and second to apply new methods, perhaps based upon the new tunnelling devices, to indicate contact.

Even if we can define what we mean by a surface and have devices that can detect its proximity, it is still necessary to be able to servo-control tools, work pieces and different parts of the device to maintain the required relative positions of all the component parts. The existing displacement detectors will have to be pushed to their limit. Capacitance devices are, of course, sensitive detectors of position but are difficult to make linear over a long range, and greater care is needed to avoid forces being applied to the object being displaced. Silicon cell detectors are not very sensitive at the level of fractions of a nanometer and, in any case, it can be difficult to supply enough light to obtain a sufficient signal-to-noise ratio without developing too much heat. Tunneling devices appear to be the most sensitive but have an infinitesimal range. Silicon X-ray interferometers are perhaps equally sensitive but are very difficult to adjust. Thus in the areas of length metrology, the advanced technology will require the ultimate in precision measurement. It will certainly have a great influence on the practical realization of the definition of the meter and the activities of the national laboratories in this field.

In order to detect relative movements, between different parts of an optical interferometer, of parts in 10^{18} to 10^{21}, great advances have been made and still need to be made in servo-control, position detection, and in ways of agumenting the signal-to-noise ratio in interferometers.

Why precise measurements and greater accuracy

We need to be able to make precise measurements for two reasons :

— to test fundamental physical theories of matter and the universe ;

— to provide the basis for the application of these tested physical theories to practical life through advanced technology.

We should continually strive for greater accuracy because advanced technology by its very nature is continually searching for greater efficiency and improved performance in all of its products. This inevitably results in continually increasing demands being placed on all of the factors contributing to efficiency and performance—one of these factors is measurement capability.

Needless to say, improved measurement accuracy leads to improved efficiency in manufacturing high-technology products. It must be obvious that being able to make a measurement to a given accuracy in only one tenth of the time will improve the efficiency of the process that relies upon that measurement. Such an increase in speed can only come about, however, by a greatly increased knowledge of the factors that previously limited the accuracy of the measurement, for example, improvement in signal-to-noise ratio that would probably have been acquired in the measurement system.

For international trade to be feasible, world-wide uniformity in measurement standards is essential. International trade in high-technology products, communication, and navigation networks, the exchange of scientific information, as well as a multitude of pure and applied scientific and technological projects carried out on an international basis, are all highly dependent upon traceability of measurements of national, and hence, to international measurement standards.

1.20. Establishing Calibration System

Market forces demand industry to ensure environment for supply of quality products to customers. Accurate measurements play a crucial role at each stage in the development and production of quality products. The aim of establishing calibration system is to develop internal metrology and calibration programmes to assure that the test and measuring instruments are maintained within desired accuracy. Systematic periodic checking of test and measuring instruments is thus becoming increasingly important to the development and maintenance of market competitiveness. This process of periodic checking against measuring equipments/standards of higher accuracy is termed as 'Calibration'.

An essential condition for establishment of calibration system is that the supplier shall control, calibrate and maintain inspection measuring and test equipment whether owned by the supplier, on loan, or provided by the purchaser, to demonstrate the conformance of product to the specified requirements, and further the equipments shall be used in a manner which ensures that measurement uncertainty is known and is consistent with the required measurement capability.

Important conditions for this purpose are :

(*i*) To ensure that the measuring equipment is capable of the desired accuracy and precision.

(*ii*) The calibration system should provide the means to directly or indirectly trace assigned values of measurement equipment to values in terms of nationally recognised standards.

(*iii*) Calibration should be carried out using measurement systems having adequate accuracy, stability and range to completely verify the performances of the calibrated item within its specified uncertainty limits.

(*iv*) The calibration procedure may be a document prepared internally, or by another agency, or by the manufacturer or a composite of the three. It is essential to designate and use documented procedures for all calibrations performed.

The contents for the calibration procedure are :

— Introductory description of the unit to be calibrated

— Preliminary actions prior to starting the calibration

— Measurement method used to verify the performance

— At table giving the specification tolerance of different parameters that can be measured by the unit and tolerance limits calculated.

— Listing of instructions to be performed.

(*v*) All test and measuring instruments should be securely and durably labelled, coded or otherwise identified to indicate its calibration status, *viz.* date of last calibration, date due for calibration, and the authenticity of calibration.

(*vi*) Access to adjustable controls/devices on measuring equipment whose setting affects the performance, should be sealed after calibration in order to prevent tampering by unauthorised personnel. Seals should be such that tampering is clearly apparent.

(*vii*) Test and Measuring equipment should be calibrated at periodic intervals established with the intent of achieving and maintaining a desired level of accuracy and quality.

(*viii*) Records should be maintained for all test and measuring equipment included in the calibration system.

(*ix*) It must be ensured that the environmental conditions (temperature, relative humidity, cleanliness, vibration, electromagnetic interference and voltage regulation of electric power) are suitable for the calibration and are maintained.

(*x*) Personnel performing calibrations of test and measuring instruments should have training in the area of job assignment, basic knowledge of metrology and calibration concepts, ability to follow instructions regarding maintenance and use of measurement equipment and standards.

(*xi*) The calibration control system should be subject to periodic audits conducted at a frequency and to the degree that assure compliance with all elements of the system procedures and documented requirements.

1.21. Geometry of Form or Shape

The measurement of geometrical values in modern engineering has become a factor of vital importance in the process of production due to higher and higher requirements for the accuracy of machined surfaces. Any machine is assembly of parts arranged in a definite relationship of each other. The surfaces of the various parts joining each other are called mating parts and their dimensions are called mating dimensions. The nature of the association required between the mating parts is determined by the conditions of operation of the machine and is called fit (*i.e.* more or less freedom in their relative movement of tightness in a fixed association). In the modern industry, mass production techniques are employed which aim at achieving strictly interchangeable assembly. Interchangeability concerns not only the size and form of parts, their position and surface finish, but the properties of the material of which they are made as well. In this book we shall be concerned with dimensional interchangeability only. (A strictly interchangeable part is one which can be assembled without selection, adjustment, or supplementary operations and which will function after assembly as stipulated by the manufacturing specifications, whereas an incompletely interchangeable part may require selection or adjustment, but will not require additional fitting or machining operation.)

In any machine, the following elements are to be inspected accurately for its functioning :

(1) Dimensional accuracy.

(2) Accuracy of geometrical form (which is usually a combination of flat, cylindrical, taper, spherical, and other surfaces) or surface macrogeometry.

(3) Accuracy of location of surfaces with reference to each other.

(4) Surface waviness.

(5) Surface roughness or surface microgeometry.

According to the form, the errors could be classified as :

(*a*) *For cylindrical form*

In a transverse section.

(1) Oval form (out-of-roundness) :

(2) Lobed form.

The degree of ovality (out-of-roundness) is judged by the difference between the major and minor axes in a single cross-section. The lobed form is one in which the cross-section contour is composed of arcs drawn from different centres. The amount of lobedness is determined as the difference between the diameter of the circle D into which the lobed cross-section is inscribed and the distance l between the parallel planes tangent to the part.

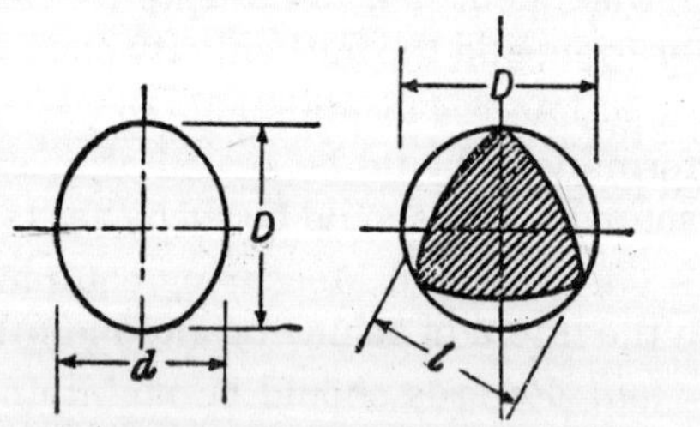

Ovality = $D - d$ Lobedness = $D - l$

Fig. 1.21

In a ligitudinal section :

(1) Barrel form ; (2) Bow (concave) form ;

(3) Curvature of the axis ; (4) Taper.

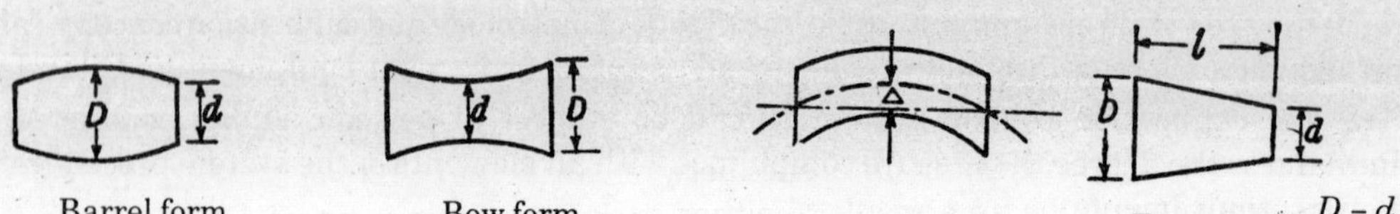

Barrel form
Barrelness = $D - d$

Bow form
Bowness = $D - d$

Curvature of axis = Δ

Taperness = $\dfrac{D-d}{l}$

Fig. 1.22

The barrel or bow forms are judged by the difference in diameters at the middle and end cross-sections. The curvature of the axis in determined by the deflection of the axis. The taper is evaluated by the difference in diameters over a given length.

(b) *For flat surface*

(1) Lack of straightness ; (2) Lack of flatness.

Lack of straightness in a flat surface is the distance between two parallel lines located in a plane P square with the tested surface *e.g.*, in Fig. 1.23 (*a*) plane P intersects the surface in the required direction and two lines are drawn to enclose the section profile of the testes surface. The lack of straightness in Fig. 1.23 (*a*) is represented by Δ.

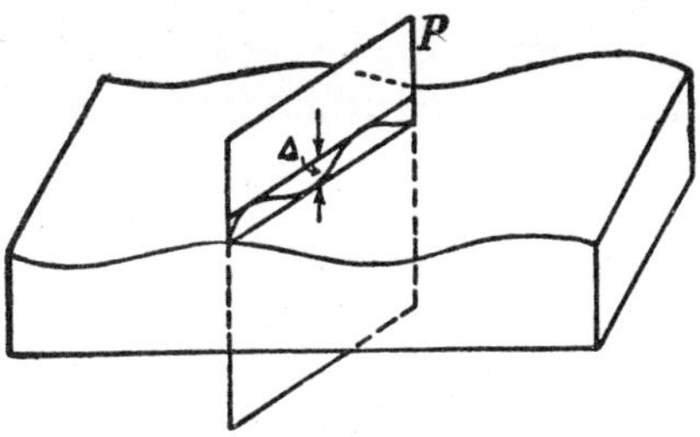

(*a*) Lack of Straightness

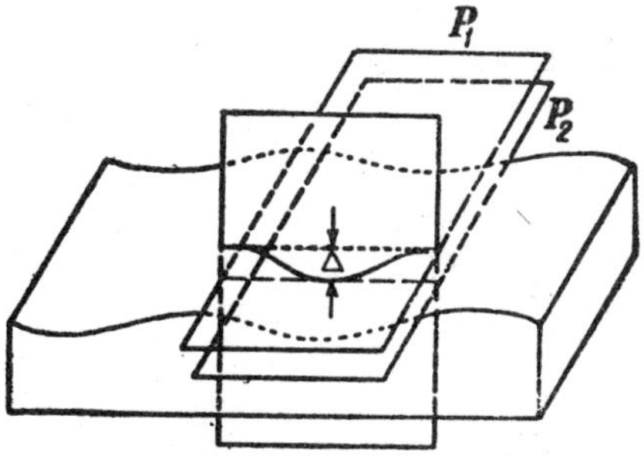

(*b*) Lack of Flatness

Fig. 1.23

Lack of flatness is expressed as the distance between the parallel planes P_1 and P_2 which enclose the tested surface between them. [Refer Fig. 1.23 (*b*)]

According to the relative location of surfaces, the errors could be classified as follows :

(1) Measurement ; (2) Radial run out ;

(3) Axial run out ; (4) Non-parallelism of axes ;

(5) Incorrect location of intersecting axes ; (6) Surfaces not parallel with each other ;

(7) Surfaces not square with each other.

Misalignment is the parallel deviation Δ of axes relative to one another. It is also the amount of cocking between two axes determined as the distance Δ between the centres of the holes on the tested length l. (Refer Fig. 1.24).

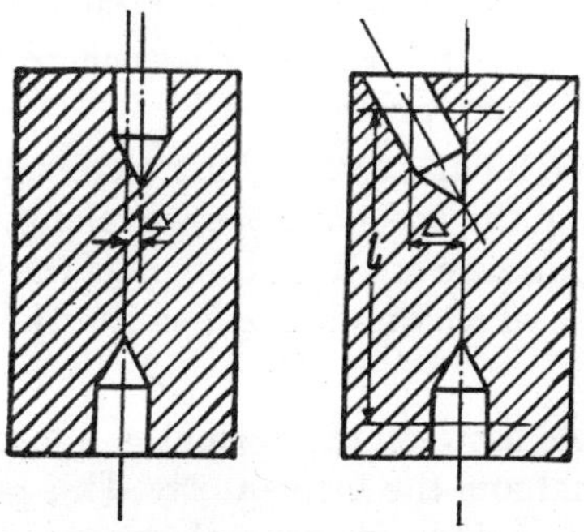

Fig. 1.24. Misalignment.

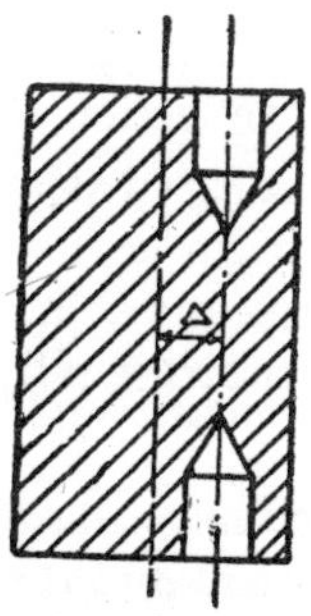

Fig. 1.25. Radial runout.

Radial runout is the maximum difference between the actual physical centre of the body and its axis of rotation. It may be as a result of either misalignment of axis, ovality, lobed form or curvature of axis.

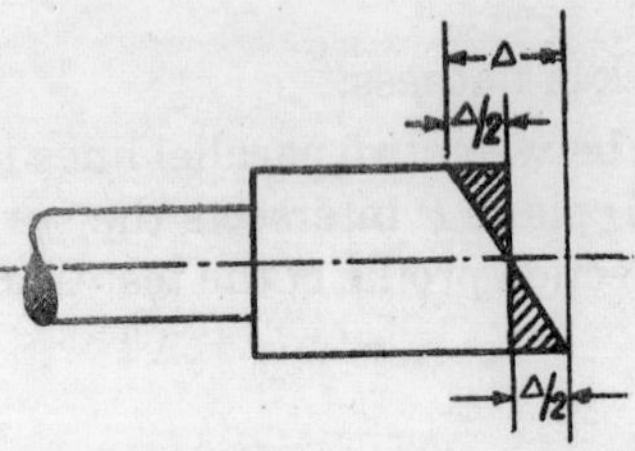

Fig. 1.26. Axial runout or slip.

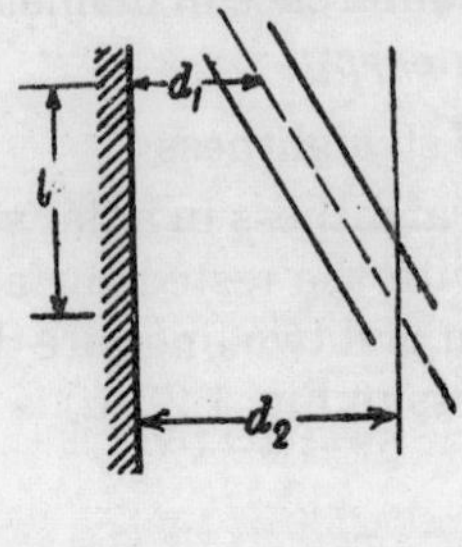

Fig. 1.27. Non-parallelism of axes.

Axial runout or slip is the maximum deviation Δ of an end surface from a plane square with axis of the part (Refer Fig. 1.26).

Non-parallelism of axes is the difference in the distance d_1 and d_2 between the axes measured in cross- section square with one axis and at the given distance l apart from each other. (Refer Fig. 1.27).

The amount by which surfaces are not square with each other is the deviation from a right angle of the angle between the surfaces.

The amount by which surfaces are not parallel with each other is the difference in distances between the surfaces measured at a given distance.

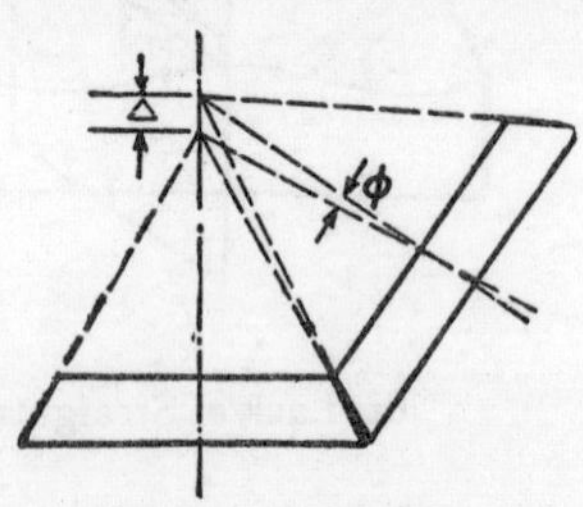

Incorrect location of intersecting axes
Fig. 1.28

Incorrect location of intersecting axes is determined by the angular deviation ϕ from the specified angle of intersection and the deviation Δ from the point of intersection.

1.22. Optical Principles

More and more precision measuring instruments employ optical principles and as such brief study of optical principles is considered essential.

The purpose of the optical system is to enable an image of the object being examined to be formed, generally to permit particular measurements to be made. The classes of measurements to which optics may be readily applied are those dealing with length, angle and irregular profiles.

The fundamental optical principles of importance in the design of precision measuring equipment are : (1) reflection, (2) refraction and (3) interference phenomena.

Light is propagated as an electromagnetic radiation having a very short wavelength, which varies according to the colour of light. At the violet end of the spectrum the wavelength is about 0.4 μm and at the red end about 0.7 μ m. White light is a mixture of all the colours of the spectrum. Rays of light travel in straight lines provided the medium in which they are transmitted is of uniform character.

When reflection occurs at a plane surface, the angle between the normal to the surface and the reflected ray is equal to the angle between the normal and the incident ray. This is shown in Fig. 1.29 (*a*). If the reflecting surface is tilted through an angle α, the normal also moves through this angle. Then, as shown in Fig. 129 (*b*), the angles of incidence of reflection each become equal to (θ + α) and the total angle between the incident and reflected ray becomes (θ + α). There is thus an increase of 2α in the total angle and a magnification of two is obtained. If, as shown at Fig. 1.29 (*c*), the reflected ray is again reflected from the tilted mirror by a second fixed mirror, the change in the angle of the final reflected ray becomes 4 α. In some instruments the second mirror is arranged to tilt through the same angle α, thereby giving a magnification of eight.

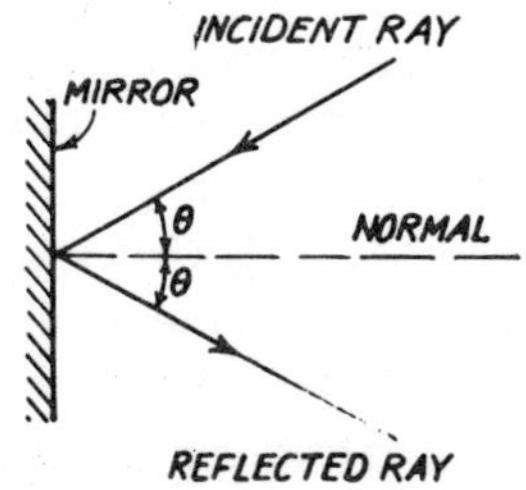

(*a*) Reflection at a plane surface.

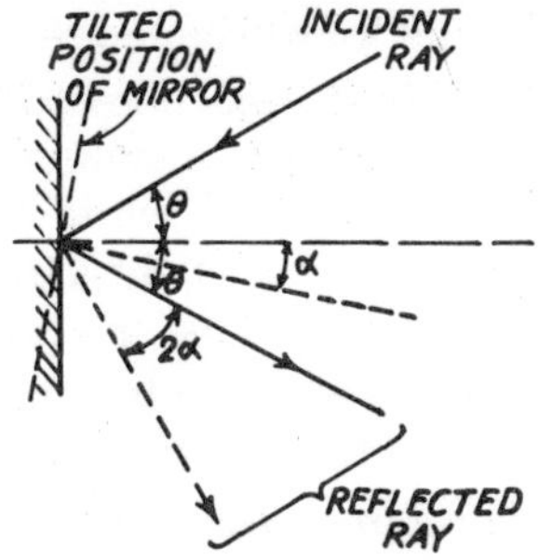

(*b*) Reflection from tilted mirror.

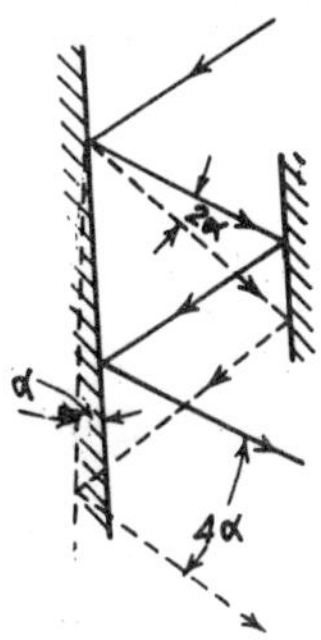

(*c*) Multi reflected.

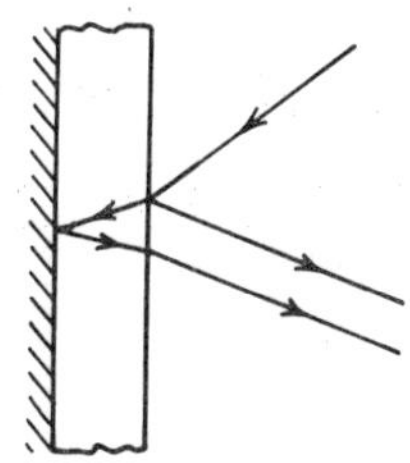

(*d*) Reflection from back silvered glass.

Fig. 1.29. Principles of light reflection from plane mirrors.

Most mirrors in everyday use consist of a piece of glass with the rear surface silvered and varnished over to protect it from damage.

Such mirrors are rearly suitable for use in metrological instruments because, as shown in Fig. 1.29 (*d*), there are two reflections, one from the front surface of the glass and one from the back silvered surface of the glass and one from the back silvered surface. Although the front reflection may be much weaker than that from the silvered surface, it almost coincides with the latter and makes it less distinct. Accurate mirrors for optical work consist of glass with the front surface coated with a thin deposit of a metal which has good resistance to corrosion. The reflecting surface is thus exposed and care must be taken in cleaning it.

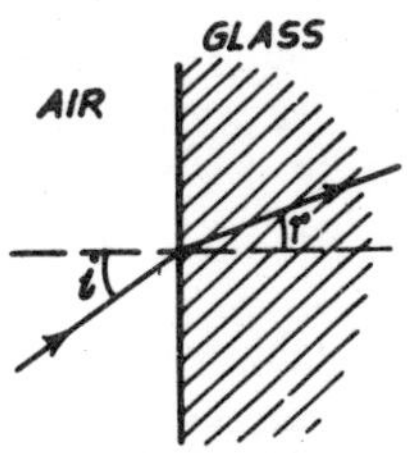

(*a*) Refraction through glass.

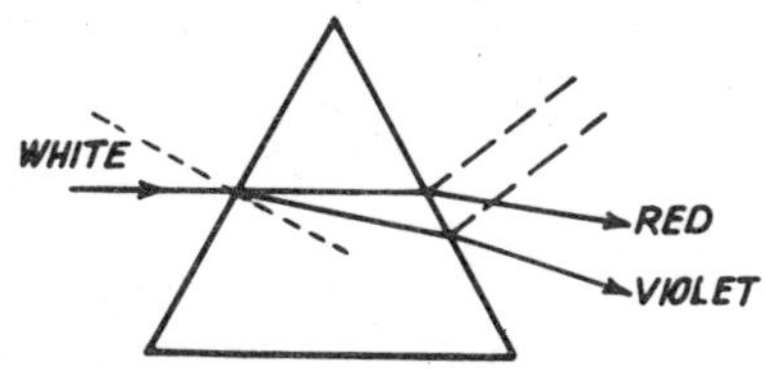

(*b*) Refraction through prism.

Fig. 1.30. Principles of light refraction.

When a ray of light passes from one transparent medium to another, it changes direction. Fig. 1.30 (*a*) shows a ray of light entering a block of glass. It will be seen that the ray is bent towards the normal to the surface separating the glass and air. A ray passing from glass to air is bent away from the normal. The amount of bending depends upon the angle of the incident ray and upon the relative densities of two media. It i denotes the angle of incidence and r the angle of refraction, then $\sin i/\sin r$ is a constant known as the refractive index.

The refractive index depends upon the wavelength of the light, being larger for the shorter wavelengths. This causes more bending of light at the violet end of the spectrum that at the red end and hence the separation of white light into its component colours. When white light passes through a parallel sided piece of glass, recombination of the colours occurs at exit ; but if through a prism, as shown in Fig. 1.30 (*b*), the colours remain separated and can be seen on a screen.

A number of different kinds of internal reflecting prisms may be employed in measuring instruments. The use of triangular prisms for deviation of 90°, 180° and 360° can be seen in Fig. 1.31.

One of the most important prisms of this class is the "Pentag" (pentagonal prism), a constant deviation prism used on some interferometers. It is also used as the basic feature of the optical square. An outline of one form of Pentag is shown in Fig. 1.31. The emergent ray is always at right angles to the incident ray, owing to the internal reflections undergone by the light ray. However, these internal reflections are the result of silvering the faces *SS* shown in Fig. 1.31, and are not due to the theory of the critical angle.

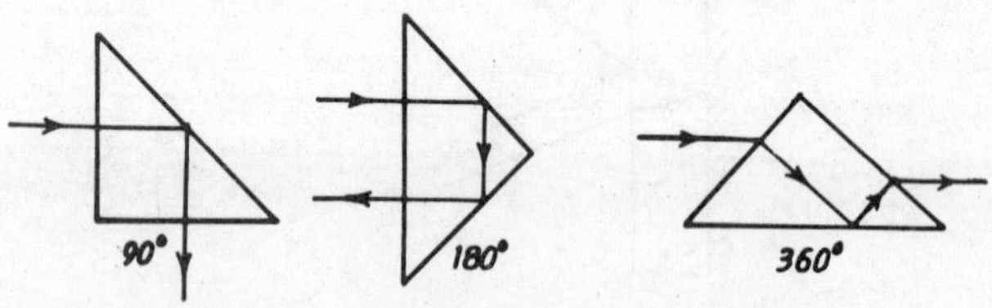

Fig. 1.31. Principles of internal reflecting prisms.

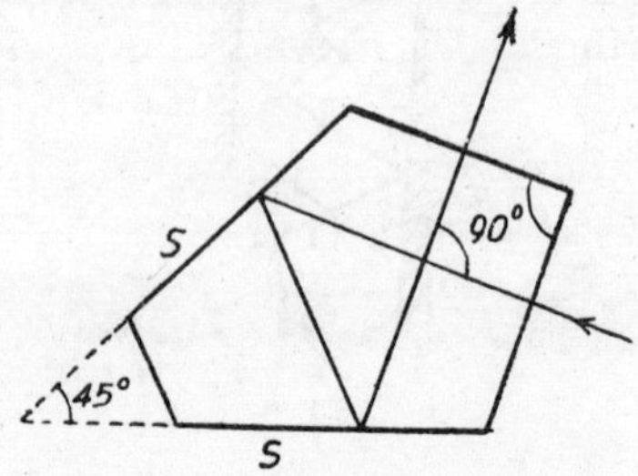

Fig. 1.32. Principle of "Pentag".

A lens can be regarded as an infinite number of parts of prisms having different angles. A convex lens causes a beam of light to converge, whereas a concave lens causes divergence. Fig. 1.33 shows a parallel beam of rays entering a convex lens, which brings the beam to a focus. From the concave lens, shown at Fig. 1.34 the parallel rays diverge and appear to come from a point behind the lens. Such a lens will not form an image on a screen, but only a virtual image which can be seen by looking through the lens.

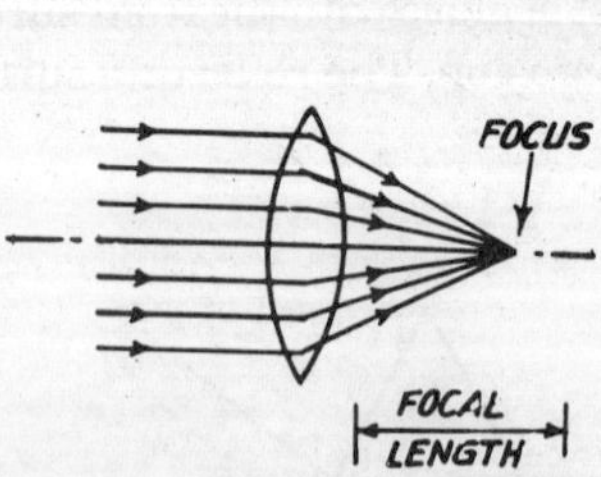

Fig. 1.33. Pinciples of light convergence.

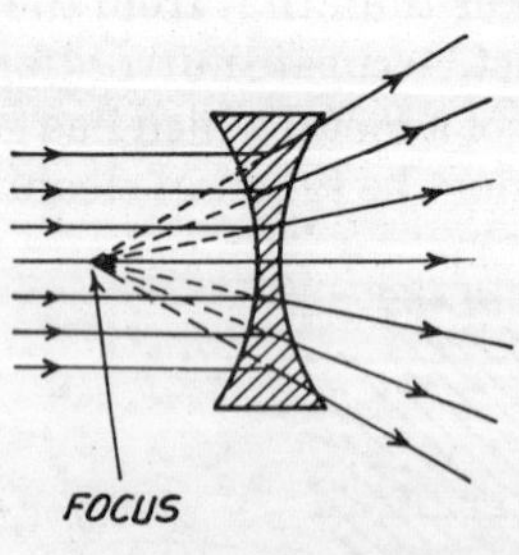

Fig. 1.34. Principles of light divergence.

The most important dimension associated with a lens is its focal length, that is, the distance from the lens to the point at which it brings parallel rays to a sharp focus. Knowing this distance the effect of a lens upon a beam of light can be determined. In Fig. 1.35 (*a*), rays of light from an object OO' pass through a convex lens and form an inverted image of the object at II'. The size and position of the image is found by the following construction. The line $O'L$ through the centre of the lens is drawn and continued beyond the lens. The line $O'L'$ is then drawn parallel to the axis of the lens and from L' a line $L'F$ produced is drawn through the focal point to meet the first line in I'.

This construction could be applied to every point on the object OO' and would form corresponding point on the image II'. The ratio of the size of image and object are given by the relation.

$$\frac{II'}{OO'} = \frac{IL}{OL}$$

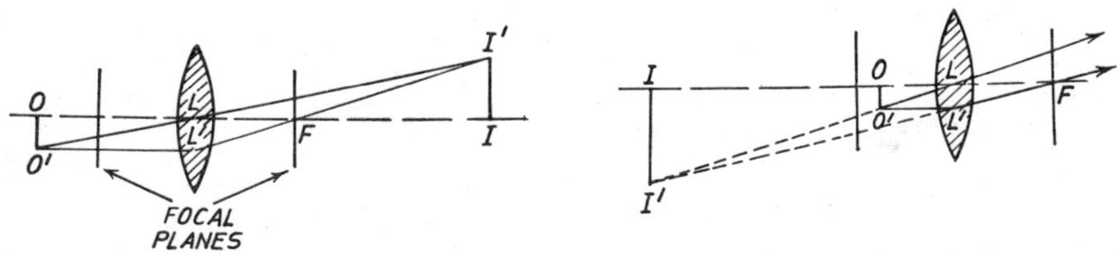

(a) (b)

Fig. 1.35. Principle of image formation by a convex lens.

If the object is brought nearer to the lens the image is larger and farther away. When the object is brought into the focal plane, rays from the lens become parallel and no image is formed. By bringing the object still near to the lens, a virtual, upright, image is formed as shown in Fig. 1.35 (*b*).

Simplified diagrams of the optical principles of some instruments are shown in Fig. 1.36. In all these, each lens is shown as single piece of glass, although, in practice, it may consist of several pieces. By combining into one unit several lenses made of glass having different refractive indices, various errors, such as the formation of coloured rings, can be reduced.

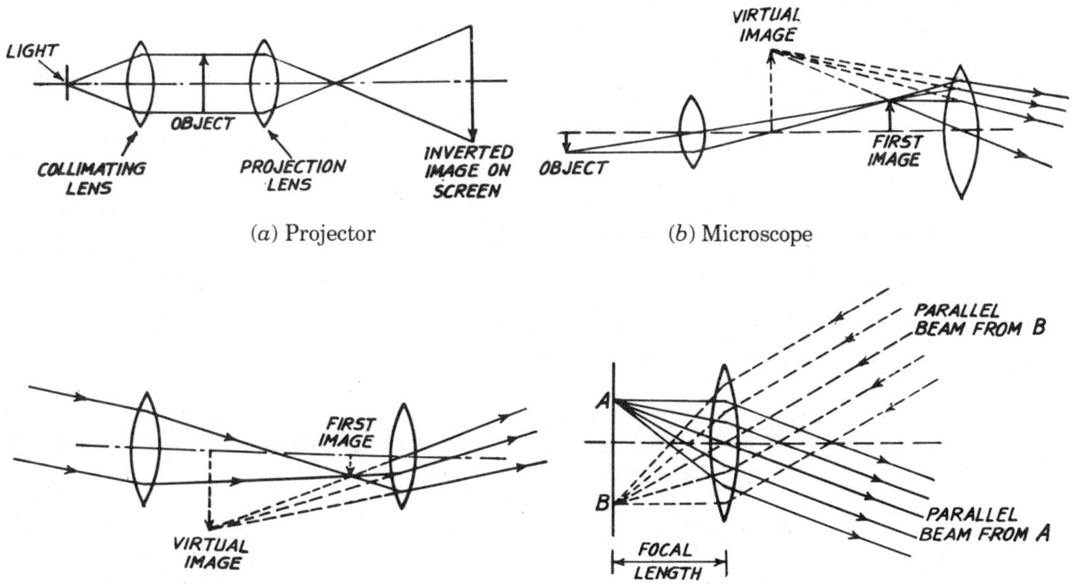

(*a*) Projector (*b*) Microscope

(*c*) Telescope (*d*) Collimator

Fig. 1.36. Principles of image formation.

The inherent disadvantages found in the mechanical pointer, as weight, bending properties, friction etc., may be overcome by using an *optical lever* shown in Fig. 1.37. The principle of this device involves the movement of a mirror (fixed to an appropriate feature of the instrument concerned) on which a beam of light is directed. The movement of the mirror is related to, and can be read off a scale, the image of which moves past a fixed fiducial line mounted in the eyepiece of a telescope. The beam of light forms wieghtless lever. An important advantage obtained when using optical lever is that the reflected beam deflects through twice the angle of rotation of the mirror, thus giving an automatic magnification.

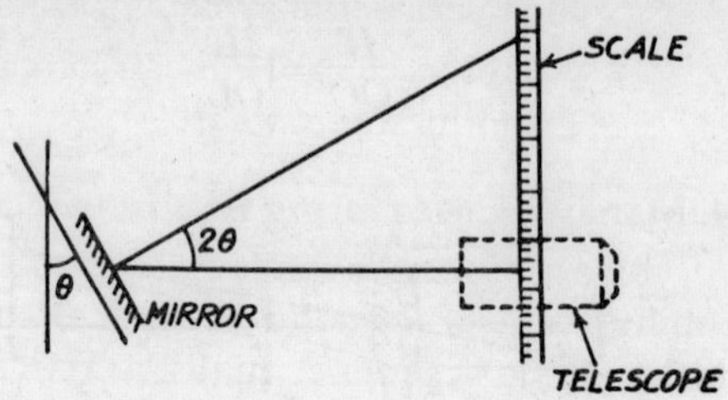

Fig. 1.37. Principle of an optical lever.

A screen upon which a spot of light is focused may also be used to avoid *parallax* errors (which result, with mechanical indicators, from different readings obtained when the observer's eye is moved to different positions). Alternatively, some instruments incorporate a fixed fiducial mark on the screen. An image of the scale moves over the screen, but only part of the scale is visible to the observer for a given position of the mirror.

1.23. Quality Improvement Programmes

Implementation of quality improvement programmes results into :

— dramatic reduction of defects
— improvement in overall product reliability
— dramatic decrease in the cost of quality
— increased customer satisfaction
— increased overall market share.

The philosophy behind quality improvement program is depicted by following :

— Quality is free since it pays for itself through reduced operating costs and increased sales through satisfied customers.
— Defect free products are possible and must be achieved. Acceptable Quality Levels (*AQL's*) are no longer good enough.
— Prevention is better than inspection *i.e.* quality must be built in and not inspected in. This is possible by eliminating operation and process variables.
— Everyone is a link in the quality chain, *i.e.* total employee involvement is essential, and teamwork is the basis of success.
— The concept of Total Business Quality (*i.e.* only a quality business can deliver a quality product) need to perculate throughout the plant. A total quality approach to the whole business is the best strategy.

The sequence of developments in regards of present status of metrology in regard to Quality is as under :

— Conventional inspection—sorting good from bad after manufacture. This is now a totally outmoded and inefficient philosophy and practice.
— Operator quality control—A technique in which operator is responsible for the quality he produces by provision of measurement/gauging equipment with the aim of determining trends and thus preventing defects from occurring.
— Accuracy capability studies on production machinery—An implicit acceptance, if not full recognition of deterministic performace of production machines leading to selection of "accuracy capable" machines and probable use of statistical sampling techniques (*SQC*).
— Post process measurement, analysis, and feedback control—It is frequently based on strictly sequential measurement of parts (from which patterns of performance can more readily be recognised).

— In process measurement and control—In it the speed of response of the closed-loop error feed back servo-system is at its highest.

— Adaptive control—It can compensate for relatively rapid trends/parameter changes like grinding wheel wear.

1.24. Managing Measuring Equipment

Measurement is the process of assigning a value to a given physical quantity. Quality is the capability of a product (or a service) to meet stated and implied needs. Measurements have significant role in product quality through the testing of products against a relevant specification, after completion of manufacture. In many cases tests are carried out during manufacture, on the partially assembled product or on key components, to ensure that the manufacturing process is going as planned and to enable any necessary process adjustments to be made.

Measurements have definite role in achieving product quality, leading to industrial competitiveness, consumer protection and related benefits. Measurements also have a vital part to play in enhancing quality in the more global sense of quality of life. The accuracy of an item of measuring equipment is established by testing that equipment in the appropriate way, using a more accurate measuring instrument. The accuracy of the measurement standard concerned may in turn be established, in a similar way, by calibration of the standard against a higher, more accurate standard, and so on, through an unbroken chain of calibrations, stretching back to the national measurement standards held at NPL.(measurement traceability). The national measurement standards held at NPL form the anchor for traceability chains serving different levels and sectors throughout the country. Traceability of measurement is essential, if accuracy of measurement is to be meaningfully assigned, and if measurements made on different instruments or at different times and locations are to be validly combined or intercompared.

Credibility of a measurement result deserves good attention. For example, a laboratory may possess accurate equipment, properly calibrated, but credibility will be lost if measurements are carried out with the equipment at the wrong temperature, or if measurement staff is insufficiently skilled, supervised or briefed in their tasks. Even where a measurement has been carried out satisfactorily, its usefulness may be undermined due to the result being reported or recorded in an ambiguous or misleading way. Only when there exists authenticated measurement traceability and when the measuring equipment is under proper control over all other aspects influencing credibility, can there be proper measurement assurance, giving overall confidence in a measuring equipment's capability to produce correct measurement results.

The management of measuring equipment requires the establishment of a system to demonstrate one's capability to provide reliable measurement results. The elements of the system demand an orderly approach to acquisition, calibration, maintenance, use and periodic recalibration of all measuring equipment. All equipment used in the measurement of characteristics of the product which affects its ability to comply with the specified requirement must be controlled, calibrated and maintained. Measuring equipment shall have the accuracy, stability, range and resolution required for the intended use. It should be ensured that the risk of measuring equipment producing results having unacceptable errors remains within acceptable bounds.

It is essential to establish and maintain an effective documented system for the managing, qualification and use of measuring equipment including measurement standards. The system must provide for the prevention of errors outside the specified limits of permissible error, by prompt detection of deficiencies and by timely action for their correction.

In performing measurements and in stating and making use of the results, it is required to take into account all significant known uncertainties in the measurement process including those that are attributable to measuring equipment (including measurement standards) and those contributed by personnel, procedures and environment.

Any item of measuring equipment that has suffered damage ; that has been overloaded or mishandled ; that shows any malfunction ; whose proper functioning is subject to doubt ; that has exceeded its designated qualification interval ; the integrity of whose seal has been violated ; must be removed from service by segregation, prominent labeling or marking. Such equipment should not be returned to service until it has been repaired and again qualified.

Access to adjustable devices on measuring equipment, whose setting affects the performance, has to be sealed or otherwise safeguarded immediately following the calibration and qualification in order to prevent tampering by unauthorized personnel. Seals must be designed in such a way that tampering is clearly apparent.

A system for receiving, handling, transporting, storing and despatching the measuring equipment, must be established and maintained in order to prevent abuse, misuse, damage and changes in dimensional and functional characteristics.

1.25. Measurements and Traceability

Measurement is necessarily a science of comparison. For a measurement result to be meaningful, it has to be traceable to national standards. The SI system of units consisting of 28 units (7 base, 2 supplementary and 19 derived units) are the basis of all modern measurements. The SI is a set of definitions. National laboratories are mandated to conduct experiments to realise and maintain the national standards of measurements. Every measurement has to have a linkage to the national standard. Traceability is a system of transferring the SI unit from the point of definition to the user. The use of hierarchical calibration system for achieving traceability is a necessary condition for credible measurements and it should be coupled with an appropriate degree of measurement quality assurance programme. The credibility of measurement results can be taken care of by use of educated and trained personnel, employing suitable measurement procedures, ensuring calibration status and maintenance of measuring equipment, and maintaining records for repeatability of measurements.

Accreditation (third party assurance about technical competence of a calibration laboratory) is done through an assessment process for laboratories. The overall programme of calibration of equipment needs to be designed and operated so as to ensure that all measurements made in a laboratory are traceable to national standards. The regulatory requirements for measurement traceability as prescribed must be strictly followed. The necessary condition for achieving traceability is that all the test and measuring equipment are calibrated by a laboratory whose standards have linkage to national standards.

Competent calibration and traceability of measurements are essential. The purpose of calibration is to limit the maximum error of the metrological characteristics of calibrated standards or instruments that enters into uncertainty levels of subsequent measurements.

1.26. Quality, Standardisation, Testing and Metrology (QSTM)

Global trade depends on quality and metrology to establish confidence between customers and supplier. Global trade is open to all and in order to take part effectively in today's global economy, QSTM activities are essential to improve productivity and ensure social and economic development. The basic concerns for QSTM programme are to ensure competitiveness, consumer protection (health and safety), and the protection of the environment. Competitiveness is possible by taking care of work culture, productivity, quality and environment. To meet customers requirements the new challenge is the production of high quality products, conformity to international standards, testing certification and high accuracy of measurements. Metrological practices have to take care of the increasing demand for international uniformity of measurements and the extension of normalisation and quality control based on metrology, the extension and globalisation of trade and to reduce all kinds of barriers, the continuous advent of new technologies and products calling for major reduction of uncertainties in measurements, the extension of metrology to new physical quantities, and the reinforcement of the presence of accurate metrology in fields of social importance such as environmental control and medical care.

The national measurement system provides a coherent formal system which ensures that measurements can be made on a consistent basis throughout the country. The contribution of International Standards Organisation in establishing standards and guides, and their transparent implementation by countries with the help of independent accrediation organisation has helped building the confidence between customer and supplier across international borders.

Quality Council of India has been established in India to implement accrediation system in line with international practice and to oversee and regulate conformity assessment process.

Conformity assessment organisation like Quality Council of India and Accrediation Boards establish rules and regulations to be followed by companies in the business.

SOLVED PROBLEMS

Q. 1.1. What is the difference between accuracy and uncertainty, and precision and accuracy ?

Ans. Every part has some true dimension and the purpose of measurements is to ascertain that true dimension. But for numerous reasons, the true dimension can never actually be determined—it can only be approximated. Accuracy of an instrument is closeness to the truth ; *e.g.* if an instrument of accuracy ± 0.01 mm indicates the dimension of a part is 20 mm, it means that true dimension could lie between 19.99 and 20.01 mm. Thus saying accuracy of ± 0.01 mm means that the measurement can be inaccurate by ± 0.01 mm or there is uncertainty about the true value to the extent of ± 0.01 mm. It would be noted that though the terms accuracy and uncertainty are complementary terms but they are used synonymously.

There exists a lot of confusion between the terms precision and accuracy. It is important to note that while the accuracy is concerned with closeness to true value, precision has no such connotation and it represents the degree of repetitiveness, and refers to a group of measurements rather than a single measurement. If an instrument is not precise, it will give different (widely varying) results for same dimension when measured again and again. Such an instrument thus is considered non-trustworthy. The first and fundamental requirement of any good instrument to be effective is that it should have adequate repeatability or precision. If an instrument is sensitive, it is likely to give different readings due to random influences like fluctuating temperature, presence/absence of dust, variation of gauging force. Thus if the measurements obtained are random in nature, and do not show any increasing or decreasing trend, then they can be characterised by two parameters, *viz.* average and standard deviation. The standard deviation, *i.e.* measure of spread represents the precision, and the closeness of average value to the true value refers to the accuracy of the measurement. Various influences can affect the precision and various sources of errors can affect the accuracy of measurement and, therefore, the measurements should be taken with great care and every effort made to eliminate fixed errors. The random errors can be minimised by defining and maintaining the changing conditions like temperature, gauging force etc. Therefore, many (minimum three) readings should be taken to get accurate results.

Q. 1.2. What is the relationship between sensitivity and range ? What is the disadvantage of very sensitive instruments ?

Ans. Sensitivity refers to minimum change in value that the instrument can reliably indicate. Obviously if an instrument is sensitive to read 0.01 mm and considering the limitation of length of the scale which may have about 100 divisions, each division representing 0.01 mm, the range can be 0.01 × 100, *i.e.* 1 mm. If a still more sensitive instrument (say 10 times more sensitive) is used then range has to be reduced to one-tenth. In other words, range and sensitivity have inverse relationship. It may be appreciated here that there is no advantage in using an instrument more sensitive than requirement ; in fact it is likely to be a disadvantage.

More the sensitive a gauge is, more susceptible it will be to extraneous influences, particularly vibration. It thus becomes very difficult for the operator to get reliable reading, and in fact it becomes a source of irritation to him. Another serious disadvantage is that the operator starts placing too much reliance in the value indicated by that instrument, *e.g.* though the operator may be interested in a dimension upto ± 0.01 mm but if he takes reading with instrument of ± 0.001 mm sensitivity, he is tempted to feel that reading exact dimension is as given upto third digit which infact may be meaningless unless the geometry of the part is good and the conditions of the instrument justify faith in the last digit.

Q. 1.3 One manufacturer of a co-ordinate measuring machine has indicated its accuracy as :

3 σ (three sigma)	**accuracy : ± 0.003 mm**
2 σ	**accuracy : ± 0.002 mm**
1 σ	**accuracy : ± 0.001 mm.**

What do you understand by this statement ?

Ans. More the complex a machine is, more are the possibilities of errors that effect the accuracy of machine. The errors in such a machine could be of the following nature : errors in straightness and squareness of ways, truè running of bearings, deflection of parts loaded in machine, temperature gradients, errors in scales and reading heads and other random errors. Thus the accuracy of such a machine cannot be quoted simply as a plus or minus value but it has to be quoted in statistical terms, *i.e.* state it as a high or low probability of being correct. The statement given above means that if 1,000 readings are taken with this machine then 99.7% or 997 of them will be within ± 0.003 mm of the true value, 95.4% or 954 will be within ± 0.002 mm of the true value and about 68% within ± 0.001 mm of the true value. In effect, with highly complex machines, this is the only meaningful way to express the accuracy.

Q. 1.4. Define the following terms so as to bring out the difference between them clearly :

(*i*) (*a*) **Quantity (Parameter)** (*b*) **Base Quantity** (*c*) **Influence quantity** (*d*) **Derived Quantity** (*e*) **True value of quantity.** (*f*) **Conventional true value of quantity.**

(*ii*) **Method of** (*a*) **direct measurement,** (*b*) **indirect measurement,** (*c*) **non-contact measurement,** (*d*) **fundamental measurement,** (*e*) **measurement by comparison,** (*f*) **differential measurement,** (*g*) **null measurement.**

Ans. (*i*) (*a*) *Quantity (Parameter).* It is the attribute of a phenomenon of a body which is capable of being distinguished qualitatively and determined quantitatively, *e.g.* length, temperature, pressure etc.

(*b*) *Base quantity.* It is the designation of quantities which in a system of quantities are accepted as being independent of one another and by means of which the quantities derived from them in that system may be expressed by equations of definition.

(*c*) *Influence quantity.* It is the quantity which is not directly of interest for measurement but is measured to represent some other quantity (of interest) as it influences the value of the quantity to be measured.

(*d*) *Derived quantity.* It is the quantity defined in a system of quantity as being a function of the base quantities of that system.

(e) *True value of a quantity.* It is the value which characterises a quantity determined perfectly in the conditions which exist at the moment when the value is observed. It is an ideal concept and, in general it can't be determined.

(f) *Conventional true value of a quantity.* It is the value approaching the true value of quantity and for all practical purposes may be considered as true value.

(ii) (a) Method of direct measurement is one in which the value of a quantity to be measured is obtained directly without the necessity of carrying out supplementary calculations based on a functional dependence of the quantity to be measured in relation to other quantities actually measured. For example, measurement of length by vernier caliper or by micrometer screw gauge.

(b) Method of indirect measurement is one in which the value of a quantity is obtained from measurements carried out by direct method of other quantities, connected with the quantity to be measured by a known relationship. For example, measurement of angle by sine bar and slip gauges.

(c) Method of non-contact measurement is one in which the sensing element is not placed in contact with object whose characteristics are to be measured ; *e.g.* measurement of diameter by pneumatic comparator.

(d) Method of fundamental measurement is also known as method of absolute measurement which is based on the measurements of base quantities entering into the definition of the quantity.

(e) Method of measurement by comparison is one which is based on comparison of the value of a quantity to be measured with a known value of the same quantity or a known value of another quantity which is function of the quantity to be measured.

(f) Method of differential measurement is based on the comparison of the quantity to be measured with a known quantity of the same kind but slightly different from that to be measured and measuring the difference between the values of these two quantities. Generally this is the most accurate method as it does not depend on the zero error of instrument, accuracy of calibration throughout the range.

(g) Method of null measurement is one in which the known quantity is so adjusted that difference between it and the quantity to be measured becomes zero. In this method, the difference between these two qualities is fed to an amplifier which drives a servomotor which in turn changes the value of the known quantity and also drives a pointer. At balanced condition, the pointer indicates the exact value of the quantity to be measured.

Q. 1.5. In gauging, the aim is only to ensure whether a component lies within tolerable limits, irrespective of its actual size. Such a method is called measurement by attributes. In measurement by variable systems, actual size is measured with a precise measuring instrument. Compare the characteristics of two systems.

Ans. Table below gives comparison of characteristics of two methods of measurements.

Characteristic	*Measurement by Attributes*	*Variable measuring system*
Type of instrument required	Fixed gauges or comparators	Precise measuring instruments
Cost of measuring instrument	low	high
Overall cost of measurement	low	high
Result of measurement	Good or bad, accepted or rejected	actual size precisely
Information value per observation	limited	can be used for many purposes
Skill of operator	unskilled	skilled
Speed of use	fast	slow

Q. 1.6. What is the difference between single sample test and multi sample test in connection with minimisation of the sources of errors in results of measurements ?

Ans. In measurement system, attempts are always made so that the results of measurement approach the true value, *i.e.* the errors in measurement which are inevitable in the system of measurement should be minimised. One of the methods to minimise errors is to take number of readings for the same parameter or attribute, and take the mean value. If the various observations are made again and again by the same person using same instrument then the results obtained are called to have been collected in a single sample test. However, if same parameter or attribute is measured a number of times by using alternate test conditions (like different operators, different instruments and procedures etc.) then the procedure is called multi-sample test. Multi-sample tests though being more costlier than single-sample test are more reliable as these eliminate controllable errors like personal errors, instrument zero error etc. However, in most of the cases, a number of readings under single sample test with statistical technique are found to be adequate.

Q. 1.7. Define the following terms :

(*i*) **Error and correction,** (*ii*) **Systematic and random errors,** (*iii*) **Parasitic and illegitimate errors,** (*iv*) **Absolute and relative errors,** (*v*) **Standard deviation** (σ) **of a single measurement in a series of measurements and** σ **of the arithmetic mean of a series of measurements,** (*vi*) **Uncertainty of measurement and inaccuracy of measurement.**

Ans. (*i*) *Error and correction.* While error is the disagreement between the result of measurement and the actual value of the quantity measured ; the correction is the value which should be added algebraically to the measured value to obtain correct result.

(*ii*) *Systematic and random errors.* Systematic error remains constant in absolute value and sign when a certain parameter is measured several times under the same conditions or it may vary according to a specified law if the conditions change. Systematic errors are, therefore, regularly repetitive in nature and result from improper conditions or procedures that are consistent in action. The causes of these errors may be known or unknown. Error from most of the known sources can be controlled. The random errors, unlike systematic errors, vary in an unpredictable manner both in absolute value and sign when a certain parameter is measured several times under the same conditions. These are inherent in the measuring system and as such can't be eliminated but the result of measurement can be corrected.

(*iii*) Parasitic error results from incorrect execution of measurement. The illegitimate error results due to blunders, errors in computation and offsets the final results so much that it can be detected.

(*iv*) Absolute error is the algebraic difference between the result of measurement and the value of comparison. If the value of the comparison is the conventionally true value of the quantity measured, it is known as true absolute error. If a series of measurement are made then the algebraic difference between one of the results of measurements and the arithmetical mean is known as apparent absolute error. Relative error is the quotient of the absolute error and the value of comparison (which may be true value, conventional true value or arithmetic mean of a series of measurements) used for the calculation of that absolute error.

(*v*) Standard deviation or root mean square deviation of a single measurement in a series of measurements is the parameter characterising the dispersion of the results obtained in series of n measurements of the same value of a quantity measured, and is expressed by the formula

$$\sigma = \sqrt{\sum_{i=1}^{n} \frac{(x_i - \bar{x})}{n-1}}$$

where x_i = ith result of measurement, and $\bar{x}$ = arithmetic mean of n results.

Standard deviation of the arithmetic mean of a series of measurements (σ_r) is the parameter characterising the dispersion of the arithmetic mean of a series of independent measurements of the same value of a quantity measured and expressed by

$$\sigma_r = \frac{\sigma}{\sqrt{N-1}}$$

where σ = R.M.S. deviation of a single measurement of a series,

N = No. of series of measurements.

(*vi*) Uncertainty of measurement represents the dispersion of the result of measurement defined by the limits of the error. Inaccuracy of measurement is expressed by all sorts of limiting errors of measurement including all systematic errors as well as limiting random errors.

Q. 1.8. Describe the following types of errors and how they can be taken care of ?

(*i*) **Environmental error**

(*ii*) **Error due to deformation of stylus and workpiece**

(*iii*) **Error due to application of measuring force**

(*iv*) **Parallax error**

(*v*) **Error due to non-alignment of centre lines of the workpiece and the measuring instrument**

(*vi*) **Error due to backlash and friction.**

Ans. (*i*) Environmental error occurs if the measurement is carried out at conditions other than standard ambient conditions of 20°C temperature and 760 mm of Hg barometric pressure, at which instruments are calibrated, but these can be easily accounted for. Temperature of instrument of workpiece may deviate from standard value of 20° either due to change in ambient temperature itself or due to handling. Error in measurement at different temperatures (say, instrument as t_1 from 20°C and workpiece at t_w from 20°C for workpiece of length 'l' is = $l\ (\alpha_w t_w - (\alpha_i t_i)$ where, α_w and α_i are the coefficients of thermal linear expansion of workpiece and instrument respectively.

(*ii*) In stylus type instruments, the contact pressure between stylus and workpiece can deform both. Further deflection may also be there depending on the method of supporting workpiece and the flexural rigidity of both workpiece and instrument. Scarr and Spencer gave the following formula to determine the value of deformation :

$$\delta = KP^{2/3}\left(\frac{1}{2R_1}+\frac{1}{2R_2}\right)^{1/3}\left(\frac{V_1+V_2}{2}\right)^{2/3}$$

where K = constant dependent on the ratio of geometry of the stylus and workpiece

P = stylus contact pressure

R_1 and R_2 are the tip radii of stylus and workpiece respectively

V_1 and V_2 are the constants depending on the materials of stylus and workpiece stylus and workpiece.

(*iii*) Application of measurement force deforms the workpiece as a result of which its dimension is altered. This type of error while comparing an unknown dimension with standard piece can be determined by the relation.

$$\text{Error} = L\left(\frac{F}{A_w E_w} - \frac{F}{A_s E_s}\right)$$

Where L = length to be measured, and F = measuring force

A_w and A_s are the cross-sections of workpiece and standard respectively

E_w and E_s are the values of modulus of elasticity of materials of workpiece and standard.

(*iv*) Parallax error occurs when the scale and pointer are not in the same plane, *i.e.* they are separated from each other and the reading is not taken normal to the plane of the scale. Referring to Fig. 1.38, say, pointer is located at distance l from scale and observer's eye is at distance L from scale. SN is normal to scale and if observer's eye is not along this line but at E (*i.e.* at angle θ with normal to scale) then while he views reading S at the scale, he sees reading corresponding to projection of point A (on the pointer) on the scale and thus PA becomes the error, and this error is equal to

$$e = \frac{l}{L} \times NE = \frac{l}{L} \times L \tan \theta = l \tan \theta$$

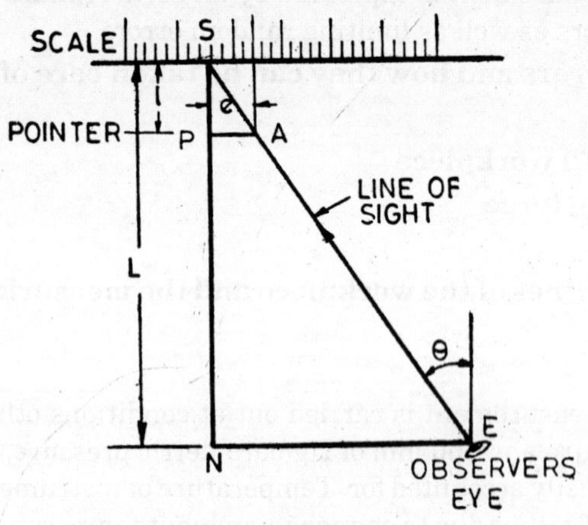

Fig. 1.38. Parallax error.

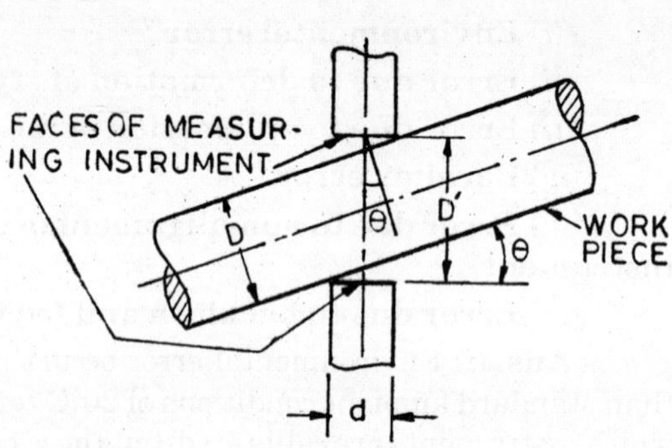

Fig. 1.39. Misalignment error on floating carriage type measuring instrument.

For least error, l should be minimum possible. Value of θ can be reduced to zero by placing mirror behind the pointer which ensures normal reading of scale.

(*v*) Many times the centre lines of workpiece and measuring instrument may be misaligned which also introduces error in measurement. Fig. 1.39 shows a simple commonly encountered case. In this case, workpiece is inclined at angle θ w.r.t. normal line between faces of measuring instrument and as such diameter D is measured as D'.

$$D' = D \sec \theta + 2\left(\frac{d}{2} \tan \theta\right) = D \sec \theta + d \tan \theta$$

$$\text{Error} = D' - D = D (\sec \theta - 1) + d \tan \theta.$$

(*vi*) Error on account of backlash and friction can be detected and corrected by observing reading first by increasing the instrument span from servo to desired value, and then decreasing the span from wide open to the desired value.

Q. 1.9. Discuss the characteristics of various materials for fabrication of different parts of measuring instruments.

Ans. Best possible materials should be used in manufacturing precision measuring instruments in order to achieve best results. Each part of the measuring instrument has to perform different duty and as such it requires different characteristics from materials. It, therefore, becomes very essential to understand the functions and duties of each part of the instrument, and also the properties and characteristics of various materials available in order to select the best choice.

The most commonly used materials for measuring instruments are described below.

Grey cast iron being easier to cast and having high compressive strength and easy to machine, is used for surface plates. Malleable cast iron being tougher and more resistant to bending and twisting is used for housings of measuring machines. Mild steel is very suitable for headed screws with rolled threads, rivets etc. Because of its low residual magnetism it is also used for elements of magnetic circuits.

Carbon steel with 0.1 to 0.14% carbon is used for light duty gears ; with 0.4% for spindles, studs, bolts ; 0.5% for keys, shafts and parts requiring moderate wear resistance ; and with 0.75 to 1.13% for springs of various types. Carbon and carbon manganese free cutting steel is used for shaft, pivot pins, heat treated bolts etc. which require good machinability.

Low alloy steel containing 0.15% C and 0.65% Cr is used for roler bearings and various parts of measuring instruments. Some of the low alloy steels show useful properties like hardening, case hardening, creep resistance after suitable heat treatment and alloy steel containing 0.25% C, 5% Cr and 0.55% Mo is used for parts requiring high surface hardness and wear resistance. Many chrome and chrome-nickel steels are often used for small precision parts. For example such a steel with 0.22% C, 13% Cr and 0.28% S is used for bolts, nuts, screws and other precision instrument parts. Nitriding steels find wide applications where very accurate and wear prone parts are required. Stainless steel with 12 to 14% Cr is used for corrosion resistant parts.

Silver steel is used for parts subjected to wear and where long life is not obtainable with free cutting tool steel.

Oil hardening tool steel is used for slip and angle gauges.

Austenitic stainless steel being acid proof is used for membrane diaphragms. It is also used for manufacturing small springs.

Tool steel with composition of 1.1% C, 2% tungsten and 1% chromium is used for gauges.

Aluminium and its alloys find wide application for parts (like housings, indicating needles, discs, etc.) where lightness, corrosion resistance with adequate strength is the criterion. Aluminium casting alloys (with silicion which promotes die castability) are used for frames, housing, levers etc.

Brasses are highly resistant to corrosion and machined easily. These also make good bearing materials. With 28 to 35% zinc, it is used for parts to be stamped, and deep drawn. Leaded wrought brass sheets are used for lightly loaded components and are machined very easily.

Phosphor bronze is used springs, hair springs, membranes exposed to corrosive surroundings.

Nickel due to its hardness and quality of resisting corrosion is sometimes used for shafts, gears, pivot pins etc. in corrosive media.

Magensium alloys are used for light and lightly loaded components.

Zinc alloys are used for intricate parts as these can be easily die cast to obtain high accuracy and smooth finish.

Q. 1.10. (*a*) **How the error in a quantity dependent on several other quantities can be determined by the method of Binomial approximations due to errors in individual quantities.**

(*b*) **In determining an end result it is directly dependent on multiple of three measurements of three dimensions of 20 mm, 30 mm and 40 mm and each of them can be measured to an accuracy of not more than 0.1 mm. Determine the possible percentage error in the end result.**

Ans. (*a*) If a quantity Q is dependent on three other quantities q_1, q_2 and q_3 related such that

$$Q = K(q_1)^{n_1} \times (q_2)^{n_2} \times (q_2)^{n_3} \qquad (K = \text{constant})$$

then according to Binomial approximation, the errors in quantity Q will be equal to $n_1 \times$ error in $q_1 + n_2 \times$ error in $q_2 + n_3 \times$ error in q_3. It can be proved as follows :

If δq_1 is a small change in q_1, δq_2 is a small change in q_2, δq_3 is a small change in q_3, and similarly if δQ is change in Q as a result of above three, then

$$Q + \delta Q = K(q_1 + \delta\, q_1)n_2 \times (q_2 + \delta q_2)\, n_2 \times (q_3 + \delta q_3)\, n_3$$

or
$$Q + \delta Q = Kq_2^{\,n_1} \times q_2^{\,n_2} \times q_3^{\,n_3} \times \left(1 + \frac{\delta q_1}{q_1}\right)^{n_1} \times \left(1 + \frac{\delta q_2}{q_2}\right)^{n_2} \times \left(1 + \frac{\delta q_3}{q_3}\right)^{n_3}$$

or
$$1 + \frac{\delta Q}{Q} = \left(1 + \frac{\delta q_1}{q_1}\right)^{n_1} \left(1 + \frac{\delta q_2}{q_2}\right)^{n_2} \left(1 + \frac{\delta q_3}{q_3}\right)^{n_3}$$

If δq_1, δq_2 and δq_3 are sufficiently small, then Binomial expansion gives

$$1 + \frac{\delta Q}{Q} = 1 + n_2\left(\frac{\delta q_1}{q_1}\right) + n_2\left(\frac{\delta q_2}{q_2}\right) + n_3\left(\frac{\delta q_3}{q_3}\right)$$

or
$$\frac{\delta Q}{Q} = n_1\left(\frac{\delta q_1}{q_1}\right) = n_2\left(\frac{\delta q_2}{q_2}\right) = n_3\left(\frac{\delta q_3}{q_3}\right).$$

(*b*) In this problem

$$\frac{\delta q_1}{q_1} = \pm\frac{0.1}{20},\ \frac{\delta q_2}{q_2} = \pm\frac{0.1}{30},\ \frac{\delta q_3}{q_3} = \pm\frac{0.1}{40}$$

$$\therefore \quad \frac{\delta Q}{Q} = \pm\left(\frac{0.1}{20} + \frac{0.1}{30} + \frac{0.1}{40}\right) \times 100\% = \pm\,(0.5 + 0.33 + 0.25) = \pm\,1.08\%.$$

Q. 1.11. (*a*) **How the error in a quantity which is function of other variables can be determined by method of partial differential due to individual errors in the variables.**

(*b*) **The radius of curvature (R) of a concave surface is determined by noting two dimensions d and h such that** $R = \dfrac{d^2}{8h} + \dfrac{h}{2}$.

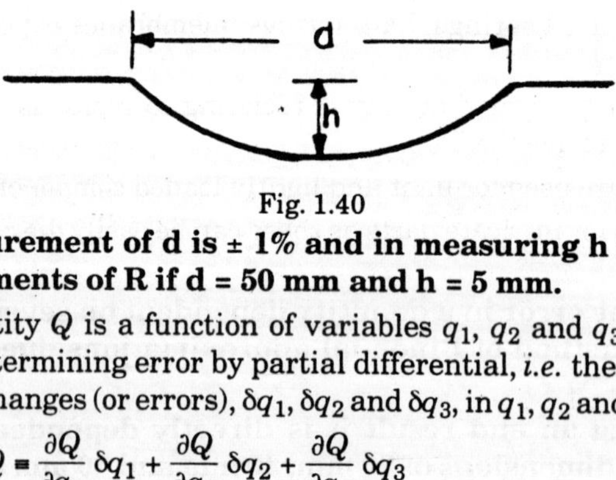

Fig. 1.40

If error in measurement of d is ± 1% and in measuring h is ± 0.5%, determine the total error in measurements of R if d = 50 mm and h = 5 mm.

Ans. (*a*) If a quantity Q is a function of variables q_1, q_2 and q_3, *i.e.* $Q = f(q_1, q_2, q_3)$ then according to method of determining error by partial differential, *i.e.* the total change in Q (or error δQ in Q) due to the small changes (or errors), δq_1, δq_2 and δq_3, in q_1, q_2 and q_3 is given by the relation.

$$\delta Q = \frac{\partial Q}{\partial q_1}\delta q_1 + \frac{\partial Q}{\partial q_2}\delta q_2 + \frac{\partial Q}{\partial q_3}\delta q_3$$

where $\dfrac{\partial Q}{\partial q_1}$ is the partial derivative, *i.e.* the rate of change of Q w.r.t. of q_1, keeping q_2 and q_3 constant.

Similarly, $\dfrac{\partial Q}{\partial Q_2}$ is the rate of change of Q w.r.t. q_2, keeping q_1 and q_3 constant and so on.

(*b*) Radius $\quad R = \dfrac{d^2}{8h} + \dfrac{h}{2}$

$$\therefore \qquad \frac{\partial R}{\partial d} = \frac{1}{8h} \times (2d) + 0 = \frac{2d}{8h} = \frac{d}{4h}$$

and
$$\frac{\partial R}{\partial h} = \frac{d^2}{8}(-1 \times h^{-2}) + \frac{1}{2} = -\frac{d^2}{8h^2} + \frac{1}{2}$$

$$\delta R = \frac{d}{4h} \times \delta d + \left(-\frac{d^2}{8h^2} + \frac{1}{2}\right) \times \delta h$$

$$= \frac{d}{4h}\left(\pm \frac{d}{100}\right) + \left(-\frac{d^2}{8h^2} + \frac{1}{2}\right)\left(\pm \frac{h}{200}\right) = \pm\left(\frac{d^2}{400h} - \frac{d^2}{1600h} + \frac{h}{400}\right)$$

$$= \pm\left(\frac{50 \times 50}{400 \times 5} + \frac{50 \times 50}{1600 \times 5} + \frac{5}{400}\right) = \pm (1.25 + 0.31 + 0.01) = \pm 1.57\%.$$

Q. 1.12. (*a*) **How the error in a quantity dependent on several other quantities can be determined by the method of logarithmic differentiation due to errors in individual quantities ?**

(*b*) **Determine the resultant error in angle θ determined by a sine bar by measuring height of slip gauge and the length between rollers of sine bar such that** $\sin \theta = \frac{h}{l}$.

Ans. (*a*) If a quantity Q is dependent on three other variables q_1, q_2 and q_3 related such that $Q = K\, q_1^{n_1}\, q_2^{n_2}\, q_3^{n_3}$ (K = constant) then $\log_e Q = \log_e K + n_1 \log_e q_1 + n_2 \log_e q_2 + n_3 \log_e q_3$ and differentiating it,

$$\frac{dQ}{Q} = n_1 \frac{dq_1}{q_1} + n_2 \frac{dq_2}{q_2} + n_3 \frac{dq_3}{q_3}$$

If small finite changes δq_1, δq_2 and δq_3 occur, then

$$\frac{\delta Q}{Q} = n_1 \frac{\delta q_1}{q_1} + n_2 \frac{\delta q_2}{q_2} + n_3 \frac{\delta q_3}{q_3}$$

which gives same result as in problem 1.11 above.

(*b*) $\sin \theta = \frac{h}{l}$, taking logs

$$\log_e (\sin \theta) = \log_e h - \log_e l$$

Differentiating, we have $\frac{\cos \theta}{\sin \theta} d\theta = \frac{dh}{h} - \frac{dl}{l}$

$$d\theta = \left(\frac{dh}{h} - \frac{dl}{l}\right) \tan \theta$$

or
$$= \left(\frac{dh}{h} - \frac{dl}{l}\right) \frac{h}{\sqrt{l^2 - h^2}} = \left(dh - h\frac{dl}{l}\right) \Big/ \sqrt{l^2 - h^2}$$

For small finite errors δh and δl, the error in θ is $\delta\theta = +\left(\delta h + h\frac{\delta l}{l}\right) \Big/ \sqrt{l^2 - h^2}$.

Q. 1.13. A dial indicator has a scale from zero to 1 mm over an arc of 270° and 100 divisions on scale ; the radius of scale line being 30 mm. During a calibration test, the following values were observed :

Calibration length (mm)	Scale value (mm)	Calibration length (mm)	Scale value (mm)
0.00	0.00	0.60	0.59
0.10	0.10	0.70	0.70
0.20	0.19	0.80	0.81
0.30	0.29	0.90	0.91
0.40	0.41	1.00	1.00
0.50	0.50		

(*a*) **Determine the sensitivity.**

(*b*) **Determine the maximum error as (*i*) percentage of scale value, (*ii*) percentage of full scale (fiducial) value.**

(*c*) **Determine whether the indicator conforms to the maker's specification of accuracy of within ± 1.0% of full scale deflection.**

Ans. (*a*) Sensitivity $= \dfrac{\text{Range}}{\text{No. of divisions}} = \dfrac{1}{100} = 0.01$ mm

(*b*) (*i*) From third column, maximum error as percentage of scale value is + 5%, – 2.5% ;

(*ii*) and from fourth column, maximum error as percentage of full scale value = ± 1%

(*c*) Yes

Calibration length lc (mm)	*Scale value ls (mm)*	*Error as % of scale value* $\frac{lc - ls}{lc}$ %	*Error as % of full scale value* $\frac{lc - ls}{lc_{max}}$ %
0.00	0.00	0	0
0.10	0.10	0	0
0.20	0.19	+ 5	+ 1
0.30	0.29	+ 3.3	+ 1
0.40	0.41	– 2.5	– 1
0.50	0.50	0	0
0.60	0.59	+ 1.7	+ 1
0.70	0.70	0	0
0.80	0.81	– 1.25	– 1
0.90	0.91	–1.1	– 1
1.00	1.00	0	0

Q. 1.14. A parameter 'R' is a function of 3 independent parameters 'A', 'B' and 'C'. The five readings of parameters A, B and C respectively are

A—100.01, 100, 99.98, 99.99 and 100.0.2 ;

B—20.01, 20.02, 19.98, 19.99, 20.00 ;

C—5.00, 5.01, 5.02, 4.98, 4.99.

The values of uncertainty in calibration of A, B and C are stated as 0.02%, 0.01%, and 0.005% respectively. Determine random uncertainty, systematic uncertainty and overall uncertainty in determination of value of 'R' for 95% confidence level. Take value of t for sample of 5 and 95% confidence level as 2.78 and value of K as 1.96.

Ans. For parameter *A*, mean $= \dfrac{100.01 + 100 + 99.98 + 99.99 + 100.02}{5}$

Sample standard deviation $S_A = \sqrt{\dfrac{.01^2 + 0^2 + .02^2 + 0.1^2 + 0.2^2}{5}} = .01\sqrt{\dfrac{10}{5}} = .01 \times 1.41$

Random uncertainty in parameter A

$$\mu_{rA} = t\frac{S_A}{\sqrt{n-1}} = \frac{2.78 \times .01 \times 1.41}{\sqrt{5-1}} = 1.96 \times 10^{-2}$$

Similarly U_{rB} and U_{rC} are also 1.96×10^{-2}

Random uncertainty of $R = 10^{-2}\sqrt{1.96^2 + 1.96^2 + 1.96^2} \times R$

and percentage value $= 3.45 \times 10^{-2}\%$

Systematic uncertainty

$$U_s = K\sqrt{\frac{0.2^2 + 0.01^2 + 0.005^2}{3}} = 1.96 \times 0.01\sqrt{\frac{4+1+.25}{3}}$$

$$= 1.96 \times 0.01\sqrt{1.75} = 1.96 \times 1.32 \times 0.01 \cong 2.59 \times 10^{-2}$$

Overall uncertainty $= \sqrt{U_R^2 + U_S^2} = 10^{-2}\sqrt{3.45^2 + 2.59^2} = 4.3 \times 10^{-2}\ \%$.

Q. 1.15. Give examples for following methods of measurement :

(*a*) **Direct** (*b*) **Indirect**
(*c*) **Fundamental** (*d*) **Comparison**
(*e*) **Direct comparison** (*f*) **Substitution**
(*g*) **Transposition** (*h*) **Differential**
(*i*) **Coincidence** (*j*) **Null**
(*k*) **Deflection** (*l*) **Interpolation**
(*m*) **Extrapolation** (*n*) **Complementary**
(*o*) **Resonance.**

Ans. (*a*) Direct method of measurement—Measurement of a mass on equal-arm balance ; and measurement of length by means of graduated scale.

(*b*) Indirect method of measurement—Measurement of speed on the basis of measurement of distance and time, and measurement of density of a body on the basis of measurements of its mass and dimensions.

(*c*) Fundamental method of measurement—Measurement of pressure with the aid of U-tube manometer by which the pressure is deducted from measurements of density, acceleration due to gravity and length, which in turn are measured based on the base quantities of length, mass and time.

(*d*) Comparison method of measurement—Measurement of pressure by Bourdon-tube gauge, and measurement of volume of liquid by means of a material of capacity.

(*e*) Direct comparison method of measurement—Measurement of length by means of a graduated rule ; and measurement of volume of liquid by means of a material measure of capacity.

(*f*) Substitution method of measurement—Determination of temperature with a Beckmann thermometer by reproducing the same indication by means of a known temperature.

(*g*) Transposition method of measurement—Determination of a mass by means of a balance and known weights, using the Gauss double-weighing method.

(*h*) Differential method of measurement—Determination of length by a comparator.

(*i*) Coincidence method of measurement—Measurement of length by vernier caliper.

(*j*) Null method of measurement—Measurement of an electrical resistance by means of wheatstone bridge and null indicator.

(*k*) Deflection method of measurement—Measurement of mass by spring balance ; and Measurement of length by dial indicator.

(*l*) Interpolation method of measurement—Linear interpolation between scale marks on a scale.

(*m*) Extrapolation method of measurement—establishment on the IPTS (International Practical Temperature Scale) of temperature above the highest established fixed point.

(*n*) Complementary method of measurement—Determination of the volume of a solid by liquid displacement.

(*o*) Resonance method of measurement—Measurement of frequencies by means of a sonometer or a vibrating-reed frequency meter.

Q. 1.16. Differentiate between indicating instrument, recording instrument, totalising instrument, integrating instrument, analogue/digital instruments, moving scale/moving index measuring instruments.

Ans. Indicating instrument is a measuring which displays the value of a measured or a related value. It is intended to give, by means of a single unique observation, the value of the measured quantity at the time of that observation.

A recording instrument provides a record of the value of measurand. The record may be permanent or semi-permanent from which the values of measured quantity can be obtained.

A totalising instrument is one which determines the value of a measurand by summation of partial values of the measurand obtained simultaneously or consecutively from one or more sources. For example, electrical power-summation meter ; discontinuous dispenser of liquid fuel with one or more measuring capacities.

An integrating instrument is one which determines the value of a measurand by integrating a quantity with respect to another quantity. For example, planimeter, continuous-belt-weighing machine.

Analogue instrument is one in which the output or display is a continuous function of the corresponding value of the measurand (pointer type of instrument) but a digital instrument has no pointer and it converts the quantity to be measured into coded discrete signals and provides a digitised output.

In moving scale measuring instrument, indications are given by the position of a moving scale in relation to a fixed index ; and in a moving index measuring instrument, indications are given by position of a moving index in relation to a fixed scale.

Q. 1.17. What is the difference between differential method of measurement and null method of measurement ?

Ans. Differential method of measurement is the method of measurement in which the measurand is compared with a quantity of the same kind,of known value only slightly different from the value of the measurand, and in which the difference between the two values is measured, *e.g.*, measurement using comparators.

Null method of measurement is the method of measurement in which the value of the measurand is determined by balancing, adjusting one or more quantities, of known values, to which the measurand has a known relationship at balance.

Q. 1.18. What is the difference between correction and correction factor ?

Ans. Correction is a value which when added algebraically to the uncorrected result of measurement, compensates for an assumed systematic error. Correction factor is the numerical factor by which the uncorrected result of a measurement is multiplied to compensate for an assumed systematic error.

Q. 1.19. What is the difference between the repeatability of measurements and the reproducibility of measurements ?

Ans. Repeatability of measurements is the quantitative expression of the closeness of the agreement between the results of successive measurements of the same measurand carried out subject to the conditions that no change is made in the method of measurement, observer, measuring instrument, location, conditions of use, and that the experiment is repeated over a short period of time. It may be expressed quantitatively in terms of the dispersion of the results.

Reproducibility of measurements is the quantitative measure of the closeness of the agreement between the results of measurements of the same measurand, where the individual measurements are carried out by changing the method of measurement, observer, measuring instrument, location, conditions of use, time etc. It may also be expressed quantitatively in terms of the dispersion of the results.

Q. 1.20. What is the difference between relative error, random error and systematic error ?

Ans. Relative error is the absolute error of measurement divided by the conventional true value of the measurand.

Random error is a component of the error of measurement which, in the course of a number of measurements of the same measurand, varies in an unpredictable manner. This error can't be corrected.

Systematic error is a component of the error of measurement which, in the course of a number of measurements of the same measurand, remains constant or varies in a predictable way.

Q. 1.21. Explain the difference between random errors and systematic errors.

Ans. Certain errors are inherent in any process of measurement. Errors which vary in an unpredictable manner both in magnitude and sign, following a normal distribution about a mean value, are referred to as random errors. Such errors may be due to uncontrollable environmental condition, personal judgement of the observer, and inherent instability of the measuring instrument. The normal distribution becomes clear when large number of readings are taken under controlled conditions. However for smaller number of observations, statistical results based on normal distribution need to be corrected by means of student's '*t*' factor.

Errors due to system which can't be reduced by taking large number of observations are referred to as systematic errors. These occur due to inability in detection of measuring system, errors in standards, constant bias, etc. These may either be constant in magnitude or variable (but only in one direction). These (constant) do not change with time and can be taken care of by zero setting. Those errors which vary with the magnitude of measured quantity can be taken care of by calibration. Variable errors are function of time (because of ageing effect.)

Q. 1.22. Fill in the blanks ?

1. .... is the field of knowledge dealing with the problems of measurements.

2. The operations conducted with a view to verify that the measuring instrument conforms to a standard is called .... instrument.

3. Partial examination of a measuring instrument whose final verification is done at the place of installation is called .... examination.

4. Set of operations carried out with regard to verifying the statutory provisions like inscriptions, dimensions, locations of stamps, etc. is known as .... examination.

5. The examination of a measuring instrument with the object of ensuring that the stamp is valid, at the instrument has not been subjected to any modification after verification, and that its errors do not exceed the maximum errors permitted in use is known as .... examination.

6. A set of experimental operations carried out with the object of determining the value of a quantity is called ....

7. The physical phenomenon which is the foundation of measurement is called the .... of measurement.

8. The sequence of operations necessary for the execution of a measurement is called the .... of measurement.

9. The set of operations having the object of ensuring that the measuring instrument satisfies the statutory provisions in relation to its metrological qualities is called .... examination.

10. The operation consisting of obtaining the indication of the measuring instrument is called ....

11. The value of the measured quantity obtained by a measurement is called .... of measurement.

12. Standards reduce diversity to a reasonable level and provide .... parts.

13. .... is the factor by which the indication of the instrument is multiplied to obtain the result of measurement.

14. The susceptibility of a measuring device to having its indications converted to a meaningful number is called ....

15. .... implies the ease with which observations can be made accurately.

16. The output signal from many measuring instruments need to be amplified so as to be .... In optical magnification, if the mirror is tilted by angle θ, the reflected lights gets displaced by angle ....

17. .... designates the degree of agreement of the measured size with its true magnitude as expressed in standard units of measurements.

18. .... expresses the degree of repeatability of the measuring process.

19. The capability of a measuring instrument to detect and to faithfully indicate even small variations of the measured dimension from nominal size of part is called ....

20. .... is the designation of the instrument's capability for consistent indications and is often expressed as the maximum drift over the specific time.

21. The .... of dimensional measurements may be regarded as a function of the occurring errors and of their assessment.

22. The maximum limit of size of an external dimension or the maximum limit of size of an internal dimension is referred to as ....

23. .... errors are primarily, yet not exclusively, associated with the human element and are considered as resulting from inaccurate scale reading, improper specimen staging, mistakes in recording, etc.

24. .... errors can neither be predicted nor eliminated, their probable magnitude and relative frequency are evaluated statistically for establishing the probable precision of the measuring process.

25. The concept of .... as a measure of the range of scatter in the measured values from a mean is applied for qualifying the precision of the instrument.

26. .... errors are those associated, with the measuring instrument's capabilities, environment and other extraneous conditions, which are generally controllable or, at least, measurable.

27. The verification of the measuring instrument's indicting accuracy, both in its original state and during usage at appropriate intervals, is accomplished by means of properly devised and implemented .... processes.

28. When the process requirements of the gauging are adequately assured, it is the .... of the instrument which will primarily determine the reliability of the measurements.

29. As a general rule, the measuring instrument's consistent indicating accuracy should be .... than the smallest graduation of the readout element.

30. .... is a comparison process whereby an unknown quantity is compared with a known quantity or standard.

31. The process of comparing one measuring device against a higher-order standard of greater accuracy is known as ....

32. The concept of .... has to be followed to ensure that all measurements are consistent.

33. .... refers to the minimum change in value that the instrument can reliability indicate.

34. The functions of .... are to control the dimensions of a product within the prescribed limitations and to segregate/reject a product that is outside those limitations.

35. Gauging by .... through the use of fixed gauges is a rapid method of determining whether or not a product is within prescribed limits.

36. Gauging by .... provides a means of evaluating the size of each part, establishing its relationship to limits of size, and determining to what extent limits of size are exceeded by rejected parts.

37. Gauging by attributes is accomplished by use of .... and gauging by variables by the use of ....

38. An attribute of a phenomenon, body or substance, which may be distinguished qualitatively and determined quantitatively is called ....

39. The quantity conventionally accepted as independent of other quantities in a system is called .... quantity but a quantity which is a function of base quantities of a system is called .... quantity.

40. The set of operations having the object of determining the value of a quantity is called ....

41. The set of theoretical and practical operations, in general terms, involved in the performance of measurements according to a given principle is called ....

42. An estimate characterising the range of values within which the trial value of a measurand lies is called .... measurement.

43. The portion of the scale over which the instrument purports to comply with specified limits of error is called .... range.

44. The concept of establishing a valid calibration of a measuring instrument or measurement standard, by step-by-step comparison with better standards up to an accepted or specified standard is called ....

45. The smallest change in the quantity measured which produces a perceptible movement of the index is called ....

46. The process of determining the characteristic relation between the values of the physical quantity applied to the instrument and the corresponding positions of the index is called ....

47. The property which characterises the ability of a measuring instrument to respond to small changes of the quantity measured is called ....

48. A curve which expresses the correspondence between the values of the quantity measured and the values indicated by the instrument is called ....

49. The algebraic difference between the upper and lower values specified as limiting range of operation of a measuring instrument is called ....

50. The reciprocal of gain where the gain is less than unity is called ....

51. Instruments in which the index is not in the same plane as the scale are subject to .... errors.

52. The progressive reduction or suppression of the free oscillations of a measuring instrument is called ....

53. The property of a measuring instrument whereby it gives different indications, or responses, for the same value of the measured quantity, according to whether that value has been

reached by a continuously increasing change or by a continuously decreasing change of that quantity is called ....

54. That part of the instrument error whereby a change of the indication, or response, departs from proportionality to the corresponding change of the value of the measured quantity over a defined range is called ....

55. The error which may result from the measurement of the value of a quantity, by a process which converts the quantity into a digital indication, or response is called ....

56. The quality which characterises the ability of a measuring instrument to give indentical indications for repeated applications of the same value of the measured quantity under stated conditions of use is called ....

57. Determination of a mass by means of a balance and known masses using the Borda substitution method is the example of .... measurement.

58. A method of measurement of a quantity in accordance with a definition of the unit of that quantity is called .... method of measurement.

59. A method of measurement in which the measurand is compared directly with a quantity of the same kind having a known value is called .... method of measurement.

60. A method of measurement in which the measurand is replaced by a quantity of the same kind, of known value, and chosen so that the effects on the indicating device are the same is called .... method of measurement.

61. All the operations for the purpose of determining the values of the errors of a measuring instrument are called ....

62. Human error of observation and judgement can be avoided by switching over to ....

Q. 1.23. Provide a single word for the following ?

(*a*) Science of precision measurement of length and angles and other quantities which can be expressed in terms of linear or angular units.

(*b*) Method of measurement in which the result is good or bad, accepted or rejected.

(*c*) Smallest change in the measured quantity which produces a perceptible movement of the index.

(*d*) Whether the pointer in instrument is just moving or just not moving.

(*e*) Extent to which the instrument repeats its results when making repeat measurements on the same unit.

(*f*) The measure of the scatter of measurement values of a given dimension by same instrument.

(*g*) Technique of measuring small variations of a continuous nature.

(*h*) The extent to which the average of a series of repeat measurements made by the instrument on a single dimension differs from the true value.

(*i*) Algebraic difference between the upper and lower values specified as limiting the range of operation of a measuring instrument.

(*j*) The property which characteristics the ability of a measuring instrument to maintain constant certain specified metrological properties.

(*k*) A reading error which is produced when the index being at a certain distance from the surface of the scale, the observation of the reading is not made in the appropriate direction.

(*l*) The mathematical function expressing the relationship between a stimulus and the corresponding response over a defined range of conditions.

(*m*) The relationship of the change of the response to the corresponding change of the stimulus.

(*n*) The error of a measuring instrument when used under reference conditions.

(*o*) A curve which expresses the correspondence between the values of the quantity measured and the values indicated by the instrument.

(*p*) The range through which a stimulus may be varied without indicating a change in response.

(*q*) The error which results from the dispersion of the indications of a measuring instrument under stated conditions of use.

(*r*) The discrepancy between the result of the measurement and the true value of the quantity measured.

(*s*) An error which varies in an unpredictable manner (both in magnitude and sign), when a large number of measurements of the same value of a quantity are made under effectively identical conditions.

(*t*) All the operations for the purpose of determining the values of the efforts of a measuring instrument.

(*u*) The operations having the object of determining the value of quantity.

(*v*) The value of a quantity conventionally accepted as having a numerical value equal to 1.

(*w*) A device which responds to the presence of particular quantity without necessarily measuring the value of that quantity.

(*x*) The divergence of the value of a quantity from a standard or reference value.

(*y*) Algebraic sum of the systematic errors affecting the indications of a measuring instrument under defined conditions of use.

(*z*) The interval between the lower and upper range limits.

Q. 1.24. Unit of measurement is the value of a quantity conventionally approved as having a numerical value equal to one.

Match the following two parts in connection with unit of measurement.

	Part A	*Part B*
1.	The conventional symbol designating the unit of measurement.	(*a*) Supplementary units
2.	The unit of measurement, the use of which is imposed or allowed by a law of the state	(*b*) Coherent system of units of measurement
3.	Unit of measurement of one of the base quantities.	(*c*) Coherent units
4.	Unit of measurement of a derived quantity	(*d*) Legal unit of measurement
5.	Unit of measurement which is expressed with the help of base units by a formula, the numerical coefficient of which is equal to 1.	(*e*) System of units of measurement
6.	A set of base and derived units, relating to a particular system of quantities like CGS, MKS, SI, etc.	(*f*) International system of units
7.	Measures for plane and spheroidal angles.	(*g*) Symbol of unit of measurement
8.	Unit of measurement which is multiple (sub-multiple) of the unit and which is formed in accordance with the grading principle approved for the corresponding unit.	(*h*) Unit of measurement outside the system
9.	A system of units of measurement composed of a set of base units and coherent derived units.	(*i*) Coherent unit of measurement

10. Unit of measurement which does not belong to the system of units of measurement envisaged.	(*j*) Base unit of measurement
11. System of units established by the General Conference of Weights and Measures.	(*k*) Multiple of unit of measurement
12. Same system and unit of measurement are used regardless of industry, trade or discipline.	(*l*) Derived unit of measurement

Q. 1.25. Quantity is the attribute of a phenomenon or of a body which is capable of being distinguished qualitatively and determined quantitatively, e.g. diameter of a shaft. Match the following two parts in relation to "quantity"

Part A	*Part B*
1. A quantity subjected to a measurement.	(*a*) scale of a quantity
2. A quantity which is not the subject of measurement but which influences the values of the quantity to be measured.	(*b*) Local value of a quantity
3. The value which characterises a quantity determined perfectly in the conditions which exist at the moment when that value is observed.	(*c*) Derived quantity
4. Pure number in the expression of a value of the specified quantity.	(*d*) Value of a specified quantity
5. The Value of a quantity measured at a place or a specified point in space.	(*e*) Base quantity
6. A set which comprises of a given group of base quantity and the corresponding derived quantities.	(*f*) Dimension of a quantity
7. The designation of quantities which in a system of quantities are accepted as being independent of one another and by means of which the quantities derived from them in that system may be expressed by equation of definitions.	(*g*) Influence quantity
8. The quantity defined in a system of quantities as being a function of the base quantities of that system.	(*h*) Numerical value of a quantity
9. The expression which represents a quantity of a system, as a product of the powers of the base quantities of that system, with a numerical coefficient equal to 1.	(*i*) Dimensionless quantity
10. A quantity which does not depend on any of the base quantities of the given system of quantities (like angle)	(*j*) System of a quantity
11. The set of values of quantity, determined in accordance with prescriptions and accepted by convention	(*k*) True value of quantity

12. The quantity expressed as the product of a number and the unit of measurement.	(*l*) Quantity to be measured

Q. 1.26. Verification involves a set of operations carried out with the objective of establishing and conforming that the measuring instrument fully satisfies the statutory provisions. It includes examination and stamping.

Match the following two parts concerning verification.

Part A	*Part B*
1. Verification of a homogeneous batch of measuring instruments on the basis of the results of examination of one or more specimens from that batch.	(*a*) Renewal of verification
2. Verification of a new measuring instrument not verified earlier.	(*b*) Full verification
3. Verifications after initial verification (like statutory periodical verification after repairs or before expiry of the period of validity, etc.)	(*c*) Periodical verification
4. Extension of the validity of verification of measuring instrument without a fresh examination (in case where that instrument has remained unused after verification).	(*d*) Initial verification
5. Subsequent verification of a measuring instrument for which a full examination of the instrument as for initial verification is necessary.	(*e*) Obligatory verification
6. Subsequent verification of a measuring instrument for which simplified examination is permitted.	(*f*) Exceptional verification
7. Verification carried out periodically at intervals of time and according to the methods established by regulations.	(*g*) Optional verification
8. Verification of a measuring instrument carried out for exceptional reasons.	(*h*) Verification by sampling
9. Verification to be carried out so that the use of measuring instrument is not prohibited.	(*i*) Loss of validity of verification
10. Non compulsory verification of a measuring instrument.	(*j*) Presentation for verification
11. Set of operations (administrative or physical) to be carried out by a person who presents a measuring instrument for verification.	(*k*) Simplified verification
12. Cancellation of the fact of validity of verification of a measuring instrument when it no longer conforms to the statutory provisions.	(*l*) Subsequent verification

Q. 1.27. Match the following two parts in connection with the methods of measurement.

	Part A	*Part B*
1.	Method of measurement in which the value of a quantity is obtained from measurements carried out by direct method of measurement of other quantities, connected with the quantity to be measured by a known relationship.	(*a*) Method of measurement without contact
2.	Method of measurement based on the measurements of based quantities entering into the definition of the quantity.	(*b*) Method of measurement by extrapolation
3.	Method of measurement based on comparison of the value of a quantity to be measured with a known value of the same quantity or a known value of another quantity, which is a function of the quantity to be measured.	(*c*) Method of differential measurement
4.	Method of measurement by direct comparison in which the value of a quantity to be measured is replaced by a known value of the same quantity, so selected that the effects produced in the indicating device by these two values are the same.	(*d*) Method of measurement by deflection
5.	Method of measurement by comparison, based on the comparison of the quantity to be measured with a quantity of the same kind, with a value known to be slightly different from that of the quantity to be measured, and the measurement of the difference between the values of these two quantities.	(*e*) Method of measurement by interpolation
6.	Method of differential measurement in which the difference between the value of the quantity to be measured and the known value of the same quantity with which it is compared, is brought to zero.	(*f*) Method of fundamental measurement
7.	Method of differential measurement in which a very small difference between the value of the quantity to be measured and the known value of the same kind with which it is compared, is determined by an observation of the coincidence of certain lines, or signals.	(*g*) Method of measurement by substitution
8.	Method of measurement by comparison in which the value of the quantity measured is determined by the deflection of an indicating device.	(*h*) Method of measurement by complement

9. Method of measurement in which the value of the quantity to be measured is compared by a known value of the same quantity, selected in such a manner that the sum of these two values is equal to a certain value of comparison fixed in advance.	(*i*) Method of measurement by coincidence
10. Method of measurement consisting of determining the value of the quantity measured on the basis of the law of correspondence and known values of the same quantity, the value to be determined lying between two known values.	(*j*) Method of measurement by comparison
11. Method of measurement consisting of determining the value of the quantity measured on the basis of the law of correspondence and known values of the same quantity, the value to be determined lying outside the known values.	(*k*) Method of null measurement
12. Method of measurement in which the sensor is not placed in contact with the object whose characteristics are being determined.	(*l*) Method of indirect measurement.

Q. 1.28. Match the parts A and B in regard to standards of measurements ?

Part A	*Part B*
1. Working standards used for setting diameter measuring machines.	(*a*) Interferometric
2. Slip gauge placed between the three balls mounted in a chuck on the tailstock and a single ball mounted on the measuring unit which is mounted on the main body of the machine by two steel strips.	(*b*) Reference discs
3. The technique by which the value of any measurement can be related in discrete steps and with known error to the value of the concerned national standard.	(*c*) Points of Bessel
4. If l is the length of bar, its supports placed 0.577 l apart so that ends of bar lie in horizontal plane.	(*d*) Precision polygons
5. If l is the length of bar, its supports at 0.54 l apart to give minimum deflection.	(*e*) Length standard
6. Concerned with angular standard.	(*f*) Slip gauges
7. Distance between the centres of engraved lines.	(*g*) Supplementary SI unit
8. Herts, newton, watt, joule. etc.	(*h*) Unit of time, second
9. One of the methods for calibration of slip gauges.	(*i*) Derived SI unit
10. End standards.	(*j*) Eden rolt comparator

11. The duration of 9192631770 period of the radiation corresponding to the transition between the two hyper fine levels of the ground state of the cesium 133 atom.	(*k*) Airy points
12. Radian and steradian.	(*l*) Traceability

Q. 1.29. Match the following two parts in connection with measuring instruments.

Part A	*Part B*
1. A measuring device which serves to transform and provides an output quantity having a given established relationship to the input quantity.	(*a*) Scale
2. The element of a measuring instrument to which a measurand is directly applied.	(*b*) analogue measuring instrument
3. The fixed or movable part of an indicating device (*e.g.* a pointer, luminous spot, liquid surface, etc.) whose position with reference to the scale marks enables an indicated value to be observed.	(*c*) adjustment
4. An ordered set of scale marks, together with any associated numbering, forming a part of an indicating device.	(*d*) scale interval
5. A device or substance which indicates the presence of a particular quantity without necessarily measuring the value of that quantity (like gold-leaf electroscope, temperature sensitive paint).	(*e*) indicating device
6. A measuring instrument in which the output or display is a continuous function of the value of the measurand	(*f*) dial
7. The set of components in a measuring instrument which displays the value of a measurand.	(*g*) expanded scale
8. The operation intended to bring a measuring instrument into a state of performance and freedom from bias suitable for its use.	(*h*) sensor
9. A scale of measuring instrument in which part of the scale range occupies a disproportionately large part of the scale length.	(*i*) scale division
10. The difference between the scale values corresponding to two successive scale marks.	(*j*) index
11. The part of a scale between any two successive scale marks.	(*k*) detector
12. The part of an indicating device, fixed or moving, which carries the scale (s).	(*l*) measuring transducer.

Q. 1.30. Match the parts A and B regarding types of errors of measuring instruments.

	Part A	Part B
1.	Difference between indication of a measuring instrument and true value of measurand.	(*a*) intrinsic error
2.	Error at a specified value chosen for checking instrument	(*b*) tracking error
3.	Initial error corresponding to zero value of the measurand	(*c*) datum error
4.	Error of a measuring instrument used under reference conditions	(*d*) bias error
5.	algebraic sum of the systematic errors affecting the indications of a measuring instrument under defined conditions of use	(*e*) relative error
6.	the error which results from the dispersion of the indications of a measuring instrument under stated conditions of use	(*f*) systematic error
7.	error of measuring instrument divided by a value specified for the instrument (like span, or upper limit of nominal range)	(*g*) zero error
8.	error arising from a lag in the response of a measuring instrument to a changing stimulus	(*h*) scatter/dispersion
9.	the slow variation with time of a metrological characteristic of a measuring instrument	(*i*) fiducial error
10.	absolute error divided by true value	(*j*) error
11.	the phenomenon exhibited by a measuring instrument whereby it gives different indications in a series of measurements of the same value of the quantity measured	(*k*) drift
12.	error which varies in a predictable way	(*l*) repeatability error

Q. 1.31. Match the following two parts in connection with the important characteristics of measuring instrument ?

	Part A	Part B
1.	change in the respone of a measuring instrument divided by the corresponding change in the stimulus	(*a*) dead band
2.	the range through which a stimulus can be varied without producing a change in the response of a measuring instrument	(*b*) drift
3.	the property of a measuring instrument whereby its response to a given stimulus depends on the sequence of preceding stimuli	(*c*) resolution
4.	the ability of a measuring instrument not to affect the value of the measurand	(*d*) sensitivity

5. ability to give indications approaching the true value of a measurand	(*e*) discrimination
6. slow variation with time of a metrological characteristic of a measuring instrument	(*f*) response time
7. a quantitative expression of the ability of an indicating device to distinguish meaningfully between closely adjacent value of the quantity indicated.	(*g*) accuracy
8. the ability of a measuring instrument to respond to small changes in the value of stimulus	(*h*) hysteresis
9. the modulus of the difference between the two limits of a nominal range of a measuring instrument	(*i*) instrument constant
10. the coefficient by which a direct indication must be multiplied to obtain the indication	(*j*) span
11. time interval between the instant an instrument is subjected to a specified abrupt change and the instant when the response reaches and remains within specified limits of its final steady value	(*k*) specified measuring range
12. the set values of a measuring instrument is intended to lie within specified limits.	(*l*) transparency

Q. 1.32. To which areas, legal metrology applies ?

Ans. Legal metrology applies mainly to the following areas :

(*i*) **Trade**—retail trade where consumer protection is of prime importance, and also wholesale and international trade.

(*ii*) **Health**—need to ensure credible measurement results from instruments and methods used in medical diagnosis and care.

(*iii*) **Safety**—speed of vehicles, light and noise levels at the place of work.

(*iv*) **Environmental protection**—need for accurate and credible measurements to tackle pollution.

(*v*) **Official controls**—sound metrological rules in the establishment of taxes and surveying, etc.

(*vi*) **Economic and social development**—to accurately determine the value of the products and services imported/exported through quantitative and qualitative measurements.

Q. 1.33. There is increasing demand for international uniformity of measurements. Elaborate this statement.

Ans. International uniformity of measurements is based on standards held by national metrological institutes and this being extended throughout countries by accredited laboratories and national calibration chains. To be practical several intercomparisons need to be organised for the same unit, and all results be linked to a single common reference. This leads to set up a hierarchical structure of the key comparisons. The attached uncertainty will have to take into account the uncertainty of transfer of the key comparison reference value by the link laboratories.

Q. 1.34. Traceability is a desired characteristic of every measurement. Define traceability.

Ans. Traceability is defined as the property of a result of a measurement or the value of a standard whereby it can be related to stated references, usually national or international standards, through an unbroken chain of comparisons all having stated uncertainties. Traceability is very closely linked with the concept of uncertainty.

A pyramid is often used to illustrate traceability ; the pyramid's apex showing the beginning of traceability chain with SI units maintained by the BIPM (Bureau International des Poids et Measures). The pyramid extends to national standards maintained by the national metrology institutes (NMI) in each country, to regional standards, to working standards, and ultimately to measurements made by end users.

Q. 1.35. Globalisation of trade calls for reduction of all kinds of barriers and mutual recognition agreements. Elaborate this statement.

Ans. It is not only enough to establish a certain level of equivalence between national standards of several countries, but same must be recognised by the partners in the world trade to enable global trade to take place smoothly. Thus there has to be mutual recognition of measurement standards and calibration certificates issued by National Metrology Institutes and the laboratories attached to them. NMIs should evolve some sort of quality system so that calibration certificates issued by other participating institutes can be recognised as part of global trade.

ANSWERS

Ans. to Q. 1.22.

1\. *Metrology* 2. *examination* 3. *preliminary* 4. *external administrative* 5. *supervisory* 6. *measurement* 7. *principle* 8. *process* 9. *metroloigcal* 10. *observation* 11. *result* 12. *interchangeable* 13. *constant* 14. *readability* 15. *readability* 16. *readable*, 2 θ 17. *accuracy* 18. *precision* 19. *sensitivity* 20. *stability* 21. *reliability* 22. *maximum material limit* 23. *Random* 24. *random* 25. *Standard deviation* 26. *Systematic* 27. *calibration* 28. *capability* 29. *better* 30. *Measurement* 31. *calibration* 32. *traceability* 33. *sensitivity* 34. *gauging* 35. *attributes* 36. *variables* 37. *limit gauges, indicating gauges* 38. *measurable quantity* 39. *base, derived* 40. *measurement* 41. *method of measurement* 42. *uncertainty* 43. *effective* 44. *traceability* 45. *discrimination* 46. *calibration* 47. *discrimination* 48. *calibration* 49. *span* 50. *attenuation* 51. *parallax* 52. *damping* 53. *hysteresis* 54. *non-linearily error* 55. *quantisation error* 56. *repeatability* 57. *substitution* 58. *definitive* 59. *direct-comparison* 60. *substitution* 61. *calibration* 62. *digital readouts.*

Ans. to 1.23.

(*a*) *Engineering Metrology* (*b*) *attribute* (*c*) *discrimination* (*d*) *threshold effect* (*e*) *precision* (*f*) *standard deviation* (*g*) *dynamic metrology* (*h*) *accuracy* (*i*) *span* (*j*) *stability* (*k*) *parallax error* (*l*) *transfer function* (*m*) *sensitivity* (*n*) *instrinsic error* (*o*) *calibration curve* (*p*) *dead band* (*q*) *repeatability error* (*r*) *error* (*s*) *random error* (*t*) *calibration* (*u*) *measurement* (*v*) *unit* (*w*) *detector* (*x*) *deviation* (*y*) *bias error* (*z*) *range*

Ans. to Q. 1.24. : 1—*g*, 2—*d*, 3—*j*, 4—*l*, 5—*i*, 6—*j*, 7—*a*, 8—*k*, 9—*b*, 10—*h*, 11—*f*, 12—*c*.

Ans. to Q. 1.25. : 1—*l*, 2—*g*, 3—*k*, 4—*h*, 5—*b*, 6—*j*, 7—*e*, 8—*c*, 9—*f*, 10—*i*, 11—*a*, 12—*d*.

Ans. to Q. 1.26. : 1—*h*, 2—*d*, 3—*l*, 4—*a*, 5—*b*, 6—*k*, 7—*c*, 8—*f*, 9—*e*, 10—*g*, 11—*j*, 12—*i*.

Ans. to Q. 1.27. : 1—*l*, 2—*f*, 3—*j*, 4—*g*, 5—*c*, 6—*k*, 7—*i*, 8—*d*, 9—*h*, 10—*c*, 11—*b*, 12—*a*.

Ans. to Q. 1.28. : 1—*b*, 2—*j*, 3—*l*, 4—*k*, 5—*c*, 6—*d*, 7—*e*, 8—*i*, 9—*a*, 10—*f*, 11—*h*, 12—*g*.

Ans. to Q. 1.29. : 1—*l*, 2—*h*, 3—*a*, 4—*a*, 5—*k*, —*b*, 7—*e*, 8—*c*, 9—*g*, *10*—*d*, 11—*i*, 12—*f*.

Ans. to Q. 1.30. : 1—*j*, 2—*c*, 3—*g*, 4—*a*, 5—*d*, 6—*l*, 7—*i*, 8—*b*, 9—*k*, *10*—*e*, 11—*h*, 12—*f*.

Ans. to Q. 1.31. : 1—*d*, 2—*a*, 3—*h*, 4—*l*, 5—*g*, 6—*b*, 7—*c*, 8—*e*, 9—*j*, 10—*i*, 11—*f*, 12—*k*.

TYPICAL EXAMINATION QUESTIONS

1. What is actually understood by measurement ? What are the fundamental methods of measurements ? Discuss generally the phases of most measurement systems.
2. Explain the following terms in mechanical measurements :
   (*a*) Calibration (*b*) Sensitivity
   (*c*) Precision (*d*) Standard length
   (*e*) Accuracy (*f*) Uncertainty.
3. Discuss the role of workshop engineer as regards precision measurements and inspection. How attempts are being made to eliminate the operator's sense of touch and feel in the instruments ?
4. Explain the following terms : Repeatability, accuracy, precision. Also discuss the relationship of accuracy and cost.
5. What are the various possible sources of errors in measurements ? What do you understand by systematic errors and random errors ? How the random errors are analysed ?
6. What is the role of standardisation and standardising organisations in an industrial country ?
7. What do you understand by cosine errors ?
8. What is the effect of temperature variation on the measurement ? What is the standard temperature for measurement ?
9. What is the difference between gauging and measurement ?
10. What is the role of metrology in international trade ?
11. With globalisation of trade ; standardisation, quality and traceability of measurements have attained high significance. Why ?
12. Range of measurement and precision of measurement have inverse relationship. Why ?
13. The prescribed methods of precision measurements have to be strictly followed with all precautions and there cannot be any compromise for them. Why ?
14. How the advancements in precision of measurements and fine production techniques are taking place side by side ?
15. How digital electronics is changing the methods of measurements ?
16. Why it is essential to be able to determine the degree of uncertainty in precise measurements and how same is expressed ?
17. Discuss the importance of calibration system in metrology laboratory.
18. What do you understand by accreditation ?

2

Linear Measurements

2.1. Introduction

The aim of the inspection of a component is to see whether it lies within the prescribed limits or not and as to whether it meets the workmanship demanded by service specifications. The critical examination of any component can be made by means of measuring instruments capable of dimensional control. Measuring instruments are designed either for line measurement (measuring distance between two lines, like steel rule), or end measurement (for measuring distance between two surfaces like micrometer screw gauge, etc). It is very difficult to use instruments designed for one system to be used for other system. The measuring instruments may be classified depending upon the accuracy that can be attained. The two categories are non-precision instruments and precision instruments.

2.2. Line Graduated Measuring Instruments

Line graduated measuring instruments incorporate graduation spacings representing known distances. These are used for direct measurements of specific distances within their capacity range. The sensitivity of the measurement is dependent primarily on the instrument's basic design (the least distance between the individual graduations). The measuring accuracy is affected by the original accuracy of the graduation, the level of resolution of the graduation lines and of the read-out members, excessive thickness or poor definition of the graduation lines, the instrument's design and the general workmanship exercised in its manufacture, the geometric deficiencies resulting from flatness and parallelism errors or caused by deflection. The precision of measurements made with line graduated instruments will also be a function of how truly are the actual distances to be measured associated with the corresponding instrument graduations. Observational errors may occur due to alignment deficiencies (due to improper coincidence of the distance boundaries with the selected scale graduations), and parallax errors when observing the scale and the object in a direction not perfectly normal to the surface being measured. However the observational errors can be reduced by means of auxiliary devices which improve the alignment of viewing and by keeping the measuring operation under control. But the instrument errors are inherent inaccuracies of the process and these must commensurate with the requirements of the particular measuring process.

Line graduated elements are used in a great variety of measuring instruments. Differences exist in various essential respects, such as the intended use, the designed sensivity, the accuracy of execution, the level of sophistication with regard to read-out and many others.

Various types of line graduated measuring instruments are :

(i) Line graduated rules and tapes—These are used for direct length comparison and they have no auxiliary devices. These are available in widely different degrees of accuracy to suit diverse requirements for plain length measuring tools.

(ii) Line graduated bar standards—These are usually made in modified "H", "U" or "X" cross-sectional forms, and the graduation lines are applied to that portion of the surface which lies in the neutral plane of the cross-section. It represents a very high level of accuracy. The essential

characteristic is the sustained accuracy of the graduation, and not the least distance between them. It requires an auxiliary device like microscope tube. Internals on bar may be subdivided by other equipment used in conjunction with the reference standard. Special precautions are required in supporting these reference bars in order to avoid the effect of sag.

Other devices include vernier caliper gauges, micrometers, line graduated angle measuring instruments, line graduated master scale (incorporating extra fine graduation lines ; increased resolution of the scale graduation for the purpose of fine incremental sub-division being accomplished through optical magnification of the master scale and the projection of the image on a screen which is equipped with a subdividing device), diffraction gratings, etc.

Ideally, all measurements or inspections should yield a result that is understood quantitatively through the use of numbers. A measurement is described more accurately if it includes a number that is referenced against a standard unit of measure. Quantitative measurements are described as variable data (in numbers) and offer more tangible information than qualitative assessment-(attributes data-pass or fail).

2.3. Steel Rule

It is also known as scale. It is a line measuring device. It is the simplest and most common measuring instrument used in inspection. It works on the basic measuring technique of comparing an unknown length to the one previously calibrated. It consists of a strip of hardened steel having line graduations etched or engraved at interval of fraction of a standard unit of length. Depending upon the interval at which the graduations are made, the scale can be manufactured in different sizes and styles. Various possible combinations of scales are found depending upon whether the scale is graduated either on both sides and both edges or on edges of only one side. The scale can be either 150 mm long or 300 mm or 600 mm or 1000 mm long. The scale need not be graduated uniformly throughout its length. In some part *e.g.* it may have 10 divisions of centimetre, in some portion 20 divisions, so that it can be used for all types of work and the particular range chosen depending upon the accuracy required.

Certain scales have got some attachments and special features with them to make their use more versatile *e.g.*, very small scales may be provided with a handle to use it easily. They may be made in folded form so that these can be kept in pocket also. Certain scales take into account the shrinkage of materials after cooling and are called shrink rules. These are used in foundry and pattern making shops. Scales can also be attached with an adjustable shoulder to make them suitable for depth measurement.

2.3.1. How to use scale. It is very necessary that for getting good results, certain techniques must be followed in the use of scale.

(1) Generally the scale is worn at the ends. Therefore, the end of the scale must never be set with the edge of the part to be measured. This is not only because end scale is usually worn but also it is very difficult to line up the end of the scale accurately with the edge of the part to be measured.

(2) The scale should never the laid flat on the part to be measured, because by doing so the graduations of the scale are not in direct contact with the surface of the part and it is difficult to read the correct dimension.

Some of the desirable qualities of the steel rule are : good quality spring steel, clearly engraved lines, graduations on two edges, and minimum thickness.

In using a rule the principle of *common datums* should be employed. This principle is shown in Fig. 2.1, the set-up

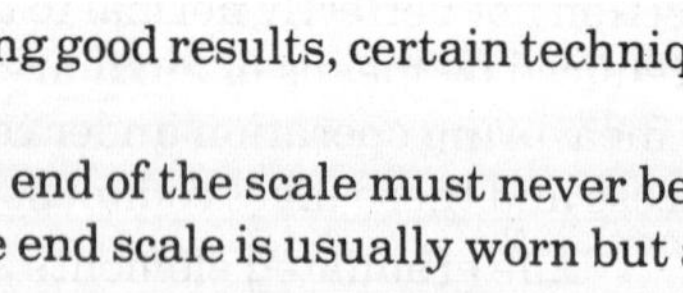

Fig. 2.1. Engineer's steel rule.

indicating the correct method of measuring the length of a component. A surface plate is used as a datum face and its purpose is to provide a common location or position from which the measurement can be made. It is important to note that both the rule and component are set perpendicular to the working surface of the surface plate, and the use of an angle plate simplifies the set up.

The degree of accuracy when measurements are made by a steel rule depends upon the quality of the rule, and the skill of the user in estimating part of a millimetre.

The engineer's rule, used for making direct measurements, depends upon visual alignment of a mark or surface on the work to be measured against the nearest division on its scale.

Various errors in measurement can be minimised by using a rule whose thickness is as small as possible.

It is important when making measurements with an engineer's rule to have the eye directly opposite and at 90° to the mark on the work, otherwise there will be an error—known as 'parallax'—which is the result of any sideways positioning of the direction of sighting.

In Fig. 2.2, point *P* represents the mark on the work whose position is required to be measured by means of a rule laid along-side it. The graduations of measurement are on the upper face of the rule, as indicated.

If the eye is placed along the sighting line *P* which is at 90° to the work surface, a true reading will be obtained at '*a*', for it is then directly opposite to *P*.

If, however, the eye is not on this sighting line, but displaced as at *B* or *C*, then an incorrect reading will be obtained.

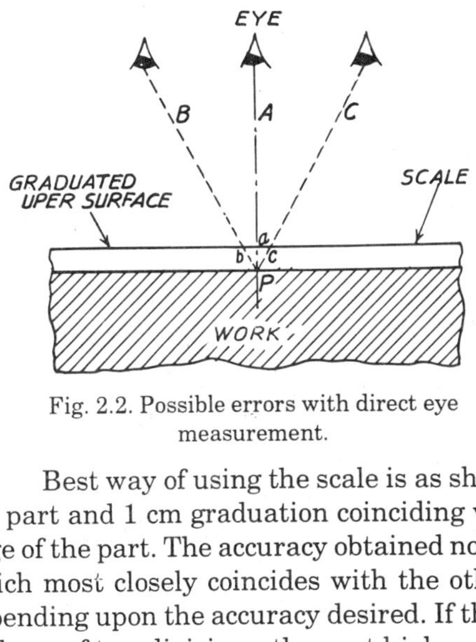

Fig. 2.2. Possible errors with direct eye measurement.

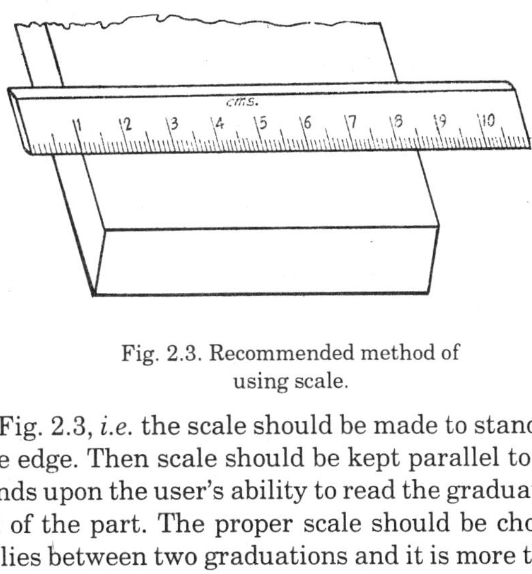

Fig. 2.3. Recommended method of using scale.

Best way of using the scale is as shown in Fig. 2.3, *i.e.* the scale should be made to stand on the part and 1 cm graduation coinciding with one edge. Then scale should be kept parallel to the edge of the part. The accuracy obtained now depends upon the user's ability to read the graduation which most closely coincides with the other end of the part. The proper scale should be chosen depending upon the accuracy desired. If the edge lies between two graduations and it is more than midway of two divisions, the next higher division must be read.

Sometimes it becomes necessary to use one end of scale in the measurement of certain parts, *e.g.* a circular bar having collar at one end. In that case a new scale must be used and end of scale held firmly against the collar. In order to obviate the difficulty of end being worn, sometimes a hook is provided at the end of the scale. The inner surface of the hook coincides with zero reading of scale. The hook is held against the flat portion of piece and readings taken. Such a scale is called-hook scale.

Since the edges of a scale are important in measurements, it should never be used for cleaning between parts or as a substitute for the screw driver. The sharpness of graduations must be maintained by the use of grease dissolving fluids.

2.3.2. Measuring from reference point to measured point. Any linear measurement can be broken down into two points, the reference point and the measured point, and we measure from reference point to the measured point by setting the instrument first on the reference point and then reading along it until we find the measured point.

There is always an uncertainty in measurement derived from an attempt to strictly and sufficiently define the geometric characteristics of a physical object. The act of measuring ensures the possibility of error in that assessment or description. Understanding the relationship between form and position ensures the proper use of a measurement system to assess and describe geometric relationships.

Measuring the diameter of a hole with a steel rule clearly establishes the above principle. The 1 cm mark on rule is the reference point. The rule is carefully set on one point of the circumference of the hole so that it can be pivotted about 1 cm mark. To locate the measured point, the rule is swung back and forth slightly along the arc of the hole opposite the stationary 1 cm mark—the 1 cm mark acting as the pivot or hinge-until the opposite point on the longest chord is located. Thus diameter is correctly measured and read.

The test of measurement is repetition, *i.e.* repeated readings that are the same. If same dimension is measured several times and different readings are obtained, then measurement is not being taken accurately, there being something wrong with the measuring instrument or the measuring procedure.

2.3.3. Folding Steel Rule. Folding steel rule has two or more graduated lengths of steel connected together by swivel joints, with square datum ends. It consists of legs of approximately equal length, connected end to end by permanent joints to form a straight line. Both extreme ends of the rule are flat and square to the rule edges.

The joints locate the legs in a straight line during service and also allow the rule to fold into a convenient length.

Both ends of the rule are sometimes fitted with tips within the graduated length. The tips are of the same width as the rule and fit flush with the rule end and edges. These tips are securely fastened to the outer face only of the appropriate legs.

Accuracy requirements. Flat ends of rule are square with the rule edges to within 0.05 mm over the width of the rule. Each joint ensures the alignment of the graduated edges of its two associated legs to within 0.25 mm. It also locates the positional continuity of the graduation to within 0.2 mm. Each individual leg of the rule is accurately graduated with a maximum permissible error between the centre lines of any two graduation lines on a single scale in accordance with permitted tolerances. The positional error of the centre line of the first and last 10 mm graduation lines relative to its nearest square end does not exceed 0.1 mm.

2.3.4. Graduated Rule Accessories. In the use of graduated rules for direct length measure-ment, the reliability of measurement is directly related with the proper positioning of the rule in relation to the workpiece. Several accessories are available to help improve the accuracy of the rule positioning.

(*i*) **Hooks.** These are used to provide a good alignment of the zero point on the scale with the border line of an object surface, and also to keep the rule in a position substantially normal to the edge of that surface. These may be of fixed length or extendable type, reversible and removable type. These are mounted on the ends of steel rules as shown in Fig. 2.4.

(*ii*) **Clamping Shafts.** These are used to hold short rule sections to permit direct measurements in cramped-locations, where the space confined within the walls of the feature does not provide access for regular size rules.

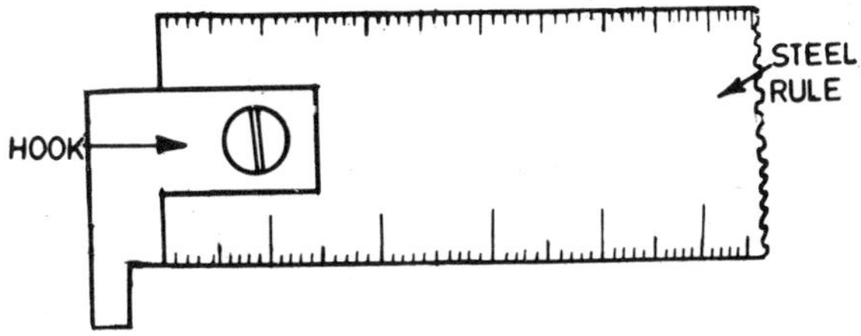

Fig. 2.4. Use of hooks to align the starting point of the rule graduations with the edge of the object.

(*iii*) **Parallel clamps.** These are used for aligning steel rules with the axis of cylindrical shafts.

(*iv*) **Foot rests.** These permit graduated steel rules to be held in a substantially vertical position with the starting point of the scale graduations on the support level of the foot block.

(*v*) **Square heads and centre finders.** These are used in layout work. Centre finders are used on steel rules to permit scribing centre lines on the face of round objects. Square head aligns rule normal to the edge of a straight sided object.

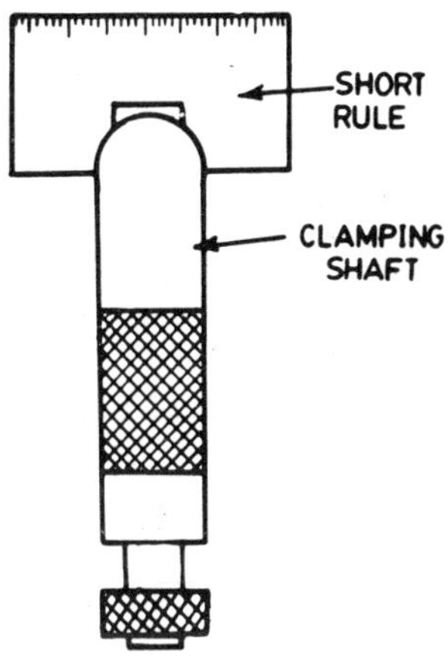

Fig. 2.5. Used for measuring in cramped areas.

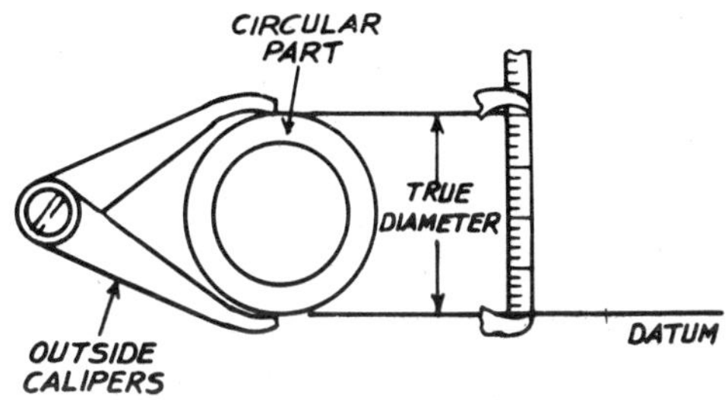

Fig. 2.6. Use of outside calipers.

2.4. Calipers

For the parts which can't be measured directly with the scale, assistance of calipers can be taken. Calipers thus act as accessories to scales. The caliper consists of two legs hinged at top, and the ends of legs span the part to be inspected. This span is maintained and transferred to the scale.

It would be noted that calipers easily sense diameter (*i.e.* maximum distance) and transfer the distance between the faces to the rule in such a way as to reduce sighting errors and increase the reading accuracy.

Calipers can be either spring type or firm-joint type. Again under spring calipers we can have outside and inside calipers and under firm-joint calipers we have outside, inside, transfer and hermophrodite calipers. In spring calipers, spring tension holds the caliper legs firmly against the adjusting nut. These are more accurate and permit accurate sense of touch in measuring.

Firm joint calipers work on the friction created at the junction of legs. These become loose after certain use. But they are easier to adjust and are particularly suitable for larger work.

The various types of calipers could thus be classified as firm joint calipers and spring calipers.

2.4.1. Firm Joint Calipers. These are the devices for comparing measurements against known dimensions. In the case of firm joint calipers, two legs (which are made from carbon and alloy steel containing not more than 0.05 per cent sulphur and 0.05 per cent phosphorus) and working ends are suitably hardened and tempered to a hardness of 400 to 500 HV and measuring faces are hardened to a hardness of 650 ± 50 HV exactly identical in shape with the contact points and equally distant from the fulcrum, the legs are joined together by a rivet. The legs are set correctly so that the working ends meet evenly and closely when brought together. In case the two legs are joined together by screw, nut and washer instead of rivet, then the threads of screw and nut should be full and true and the joints made such that these function freely with even tension without any undue play or stiffness. The component parts of the calipers should be free from seams, cracks, flaws and must have smooth bright finish.

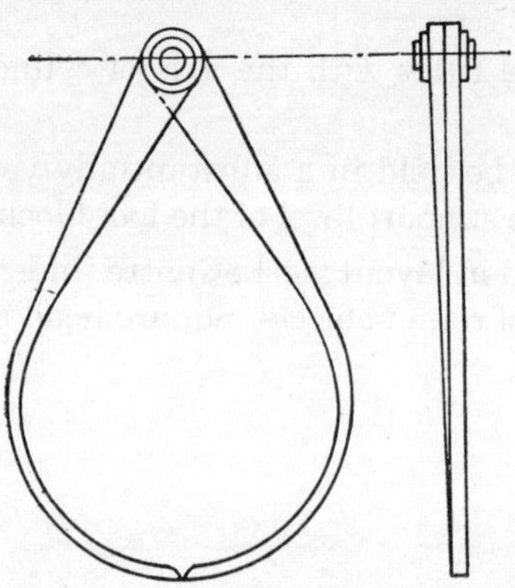

Fig. 2.7. Firm joint Outside Caliper.

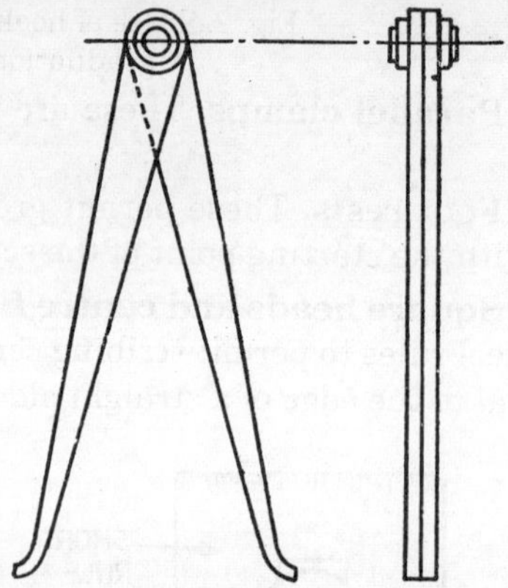

Fig. 2.8. Firm joint Inside Caliper.

The firm joint calipers can be designed for inside measurement as well as outside measurements depending upon the shape of the legs. The distance between the roller centre and the extreme working end of one of the legs is known as nominal size and these calipers are available in the nominal size of 100, 150, 200 and 300 mm. The capacity of the caliper is the maximum dimension which can be measured by it. The capacity of the caliper should not be less than its nominal size. The legs of these calipers are made of rectangular cross-section.

For assuring that the calipers function satisfactorily, it is essential to ensure that each caliper works smoothly and retains its size when adjusted to three test pieces of the sizes at extreme (lower and upper) ends and middle of the instrument range. The accurate use of calipers depends upon sense of feel of operator. For this purpose, caliper should be held gently, and held square to the work, and only light gauging pressure should be applied.

2.4.2. Firm Joint Divider. These are used for scribing arcs and circles. The distance between fulcrum roller centre and the extreme working end of one of legs is known as the nominal size and these are available in the size of 100, 200 and 300 mm. The working ends of the divider legs (upto 10 mm from extreme tips) are suitably heat treated in order to impart a hardness which will make it capable of scribing a clearly defined line across the face of a block (having hardness of 620 to 660 HV) and still the working ends should be sharp and visual examination should not reveal any sign of damage or failure. The working legs are made very sharp so that the trace on the test block does not show any sign of damage or failure. The fulcrum roller is hardened and tempered to a hardness of 625 HV and joint made such that the divider shall be capable of smoothly retaining its size. The rigidity of the leg and the proper functioning of the divider is judged by clamping securely one leg of the divider in a vice at a point about 13 mm from the fulcrum and opening the other leg slightly such that both the legs are in a horizontal plane. A test load of 4.5 kg is then gradually applied by suspension from the free leg as close to the working end as possible. On removal of the load, the joint should not be displaced and the legs should not show any signs of permanent set.

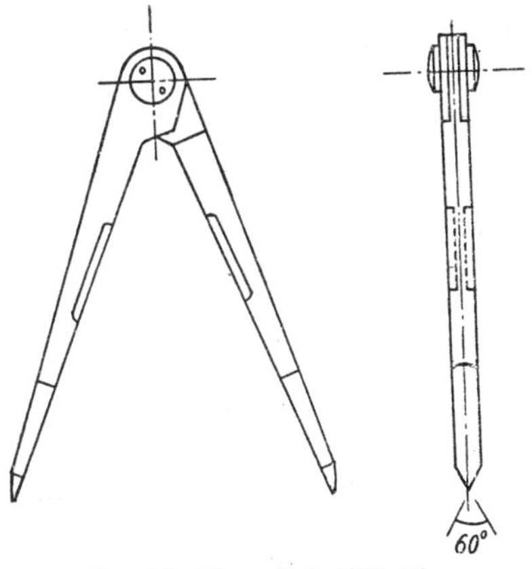

Fig. 2.9. Firm Joint Divider.

2.4.3. Spring Calipers. The legs of spring calipers are made from suitable alloy steel, measuring faces properly heat-treated and hardened to a hardness of 650 ± 50 HV. These should be free from seam, cracks, flaws etc. and all component parts should be smoothly finished. The spring which is made from carbon spring steel is properly hardened and tempered to a hardness of 470 to 520 HV and all components are manufactured and assembled in such a way as to achieve and maintain the smooth adjustment and efficient operation.

The working ends of the legs of each spring caliper should be identical in shape and have contact points equally distant from the fulcrum. The measuring ends of each pair are so shaped that all sizes within its capacity may be readily transferred from measuring scales or gauges. The cross-section of the legs may be either of rectangular or circular shape. The calipers are adjusted to a set dimension by means of either a knurled solid nut or a knurled quick-action release nut operating on a finely threaded adjusting screw, the latter one being preferred. The top portions of

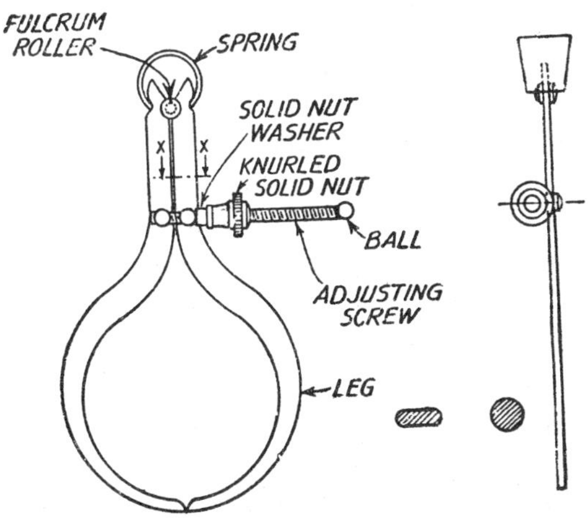

Fig. 2.10. Outside Spring Caliper.

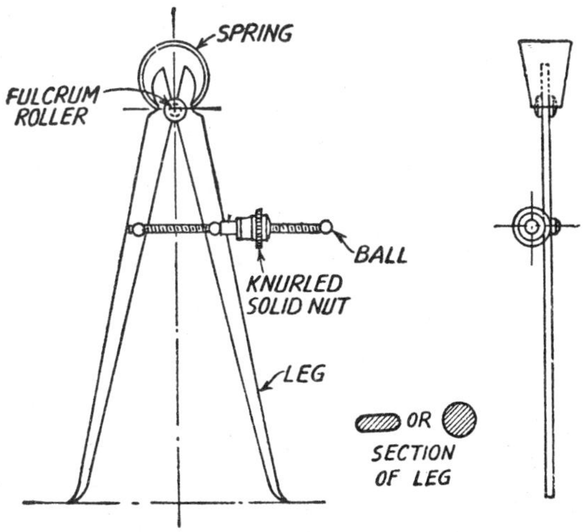

Fig. 2.11. Inside Spring Caliper.

the legs are located in a flanged fulcrum roller and held in position by a spring in order to maintain the alignment of the working ends. The spring provides sufficient tension to hold the legs rigid at all points of the adjustment. The friction between the adjusting nut and the leg is minimised by using a separate washer under the nut. One end of the adjusting screw is securely hinged to a leg of the caliper and a steel ball positively fixed to the free end of the screw for the purpose of retaining the adjusting nut. The inside and outside spring calipers are available in sizes of 75, 100, 150, 200, 250 and 300 mm. A good caliper should be able to retain its sizes when adjusted to three pieces near the lower and upper extremes of its range and in the middle.

According to other considerations, calipers could be classified as :

(*i*) Outside calipers, (*ii*) Inside calipers,
(*iii*) Transfer-calipers, (*iv*) Hermophrodite calipers.

2.4.4. Outside Calipers. These are designed to measure outside dimensions. The accuracy in caliper measurement depends upon the inspector's sense of feel. When a caliper has to be set from the scale to control the dimension of any component, then leg of the caliper must be held firmly against the end of the proper dimension by adjusting nut with the thumb and forefinger.

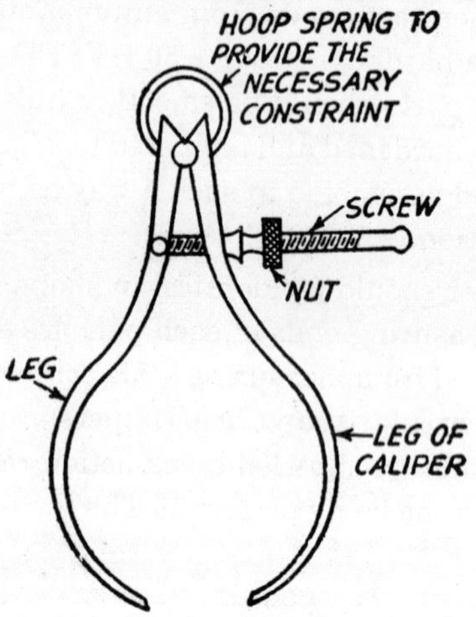

Fig. 2.12. Outside Spring Caliper.

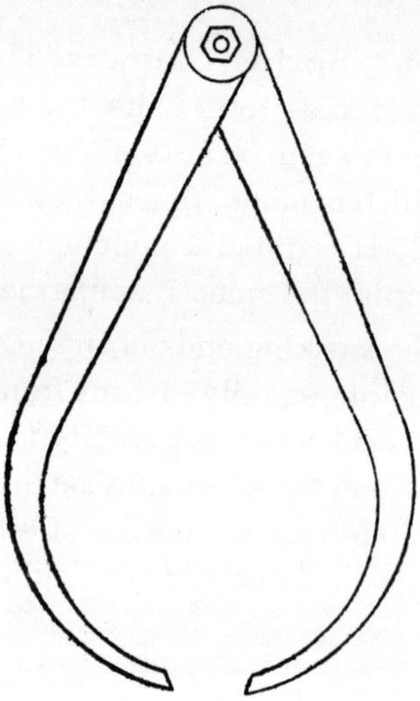

Fig. 2.13. Outside Firm Joint Caliper.

When caliper is used to measure the outside diameter of a cylinder, then it must be held in hand with its nut between fore-finger and thumb, and one end of leg touching with the surface of cylinder. The other leg is moved to and fro on the opposite surface so that other leg just passes through the cylinder. The caliper reading is transferred to the scale by holding one leg of the caliper against the scale firmly and the reading taken at the other end of leg.

For accurate settings, the distance between the outside calipers may be set by slip gauge or by micrometer anvils. Sometimes it may be required to transfer this dimension to caliper.

2.4.5. Inside Calipers. While using inside calipers, the use of hook scale to measure the distance between its legs is desirable. With ordinary scale, one leg of inside caliper must be coincided with a vertical brace and second leg adjusted accordingly. (Refer Fig. 2.15) It will be best if the legs of the inside caliper are adjusted with a micrometer.

2.4.6. Transfer Calipers. These are used for measuring recessed areas from which the legs of the calipers can't be removed directly but must be collapsed after the dimension has been measured. In the transfer calipers, an auxiliary arm is provided to preserve the original setting after the legs are collapsed.

2.4.7. Hermophrodite Calipers (Refer Fig. 2.14). It is also known as odd leg caliper and is a scribing tool consisting of one divider and one caliper leg. Thus it is used in layout work. It can have two types of legs, *viz.,* notched legs or curved legs

In order to avoid excessive wear of nut while closing the legs of spring calipers, the legs must be first collapsed.

The caliper must be held square to the work when checking (Refr Fig. 2.16. Also when measuring the outside diameter, a caliper should not be forced upon the work. Forcing the caliper upon the work results in springing of legs and the false readings.

For checking the inside diameters, one leg of caliper is held in contact with the bore while the other is rotated through arc to get the correct "feel", using no force.

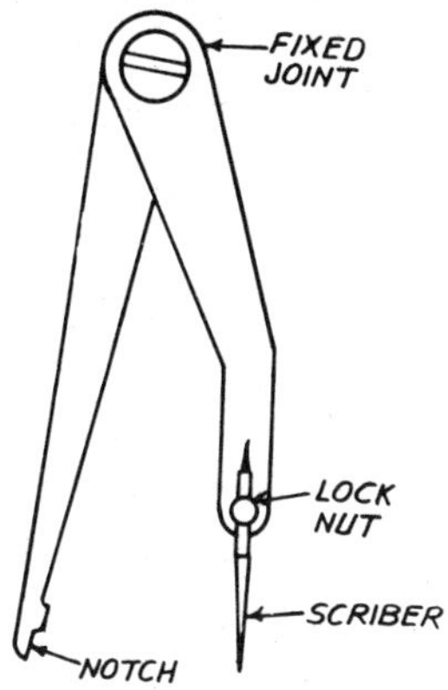

Fig. 2.14. Hermophrodite caliper.

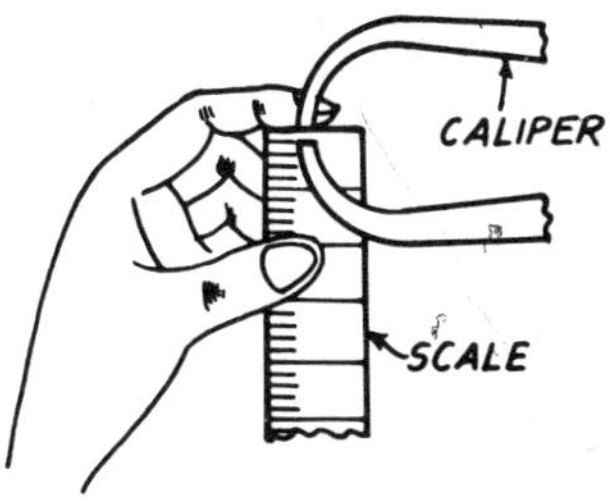

Fig. 2.15. Setting Caliper to Scale.

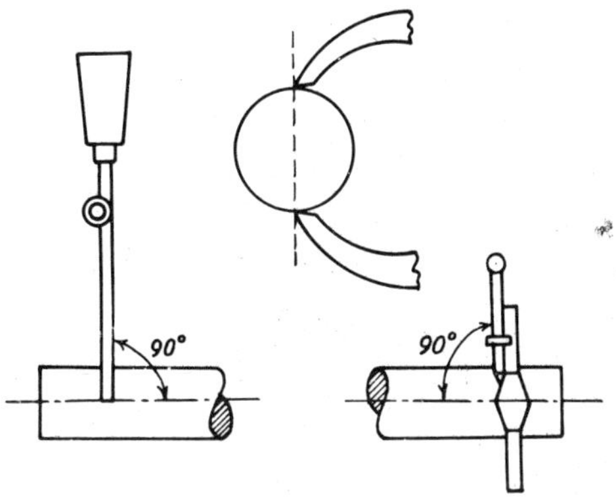

Fig. 2.16. Setting of caliper to a plug gauge (Note the perpendicularity in different planes).

2.4.8. Dial Calipers. (Refer Fig. 2.17) These are typical of all hand-held calipers. These provide a typical direct-reading capability of 0.02 mm. These units are quite flexible, offering a typical measurement range of 150 mm with accessories available for range extensions, as well as specific measurement accessories.

These use mechanical comparators. The translation of the linear form of the workpiece is accomplished by referencing the caliper jaws to a measurement scale by mechanical means. The workpiece variation is assessed by a gear-train and lever (mechanical) translation technique that

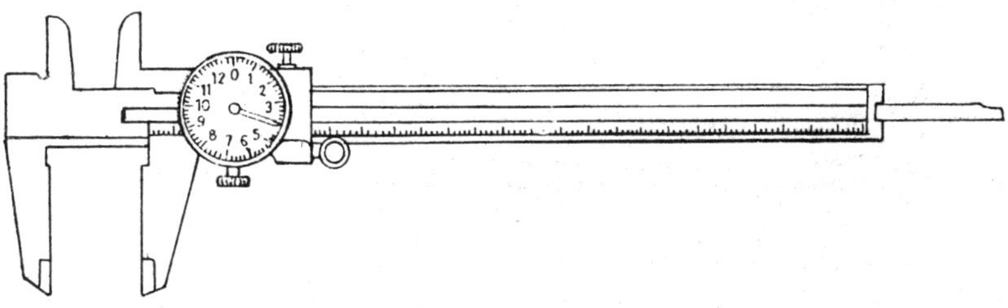

Fig. 2.17. Dial Calipers.

utilises principles of amplification. Calipers utilising electronic technology to assess and display dimensions are also available.

Many modern calipers are fitted with microprocessor-based direct reading devices. These are capable of direct hook up or interface with computerised storage and display unit ; which can chart and interpret assessed variation over time.

2.5. Surface Plates

The foundation of all geometric accuracy and indeed of all dimensional measurement is the flat plane, which is usually provided in the workshops and inspection laboratories by the surface plate.

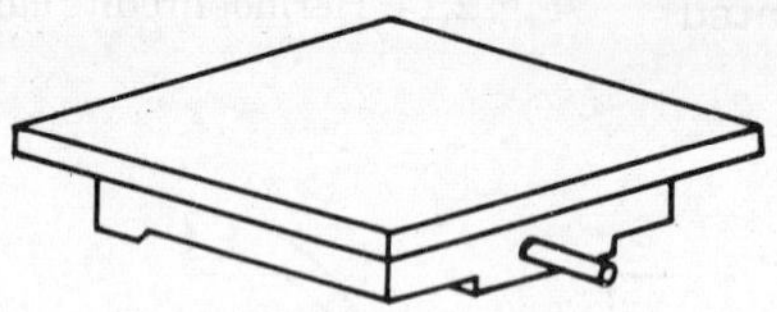

Fig. 2.18. Cast iron surface plate—The working surface.

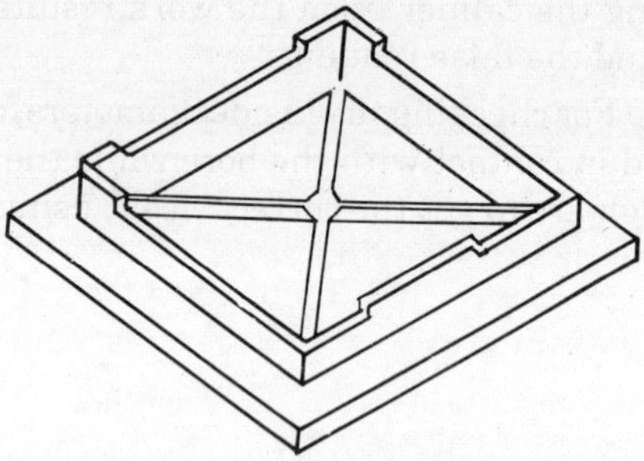

Fig. 2.19. Cast iron surface plate—the underside.

Figs. 2.18 and 2..19 show a typical cast iron surface plate and its working surface and the underneath. Most surface plates are rectangular, having 4 : 3 length : width ratio. Square or round plates are geometrically most correct for creating a flat plane by the three plate method.

The commonly used cast iron surface plates have special ribbing and structural formations under the surface so that these will be as light in weight as possible and yet will not warp or settle after the castings have aged, (Refer Fig. 2.20). In fact, the true plane of the surface is not machined and "scraped" until it has really aged and all internal stresses have subsided. During the last few years true, flat, smooth surfaces have been polished on granite blocks and granite surface plates have found easy popularity.

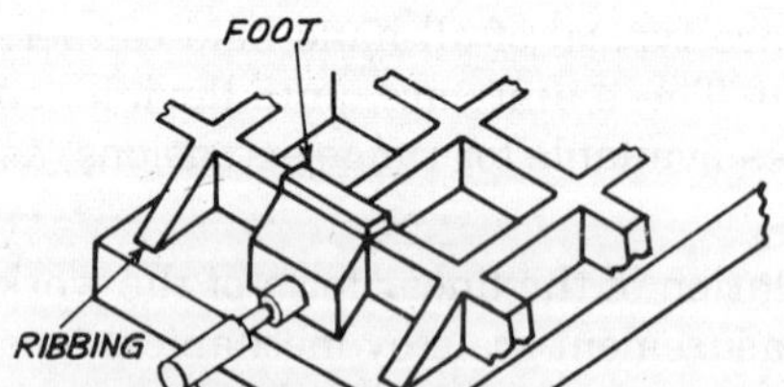

Fig. 2.20. Underside ribbing of a surface plate.

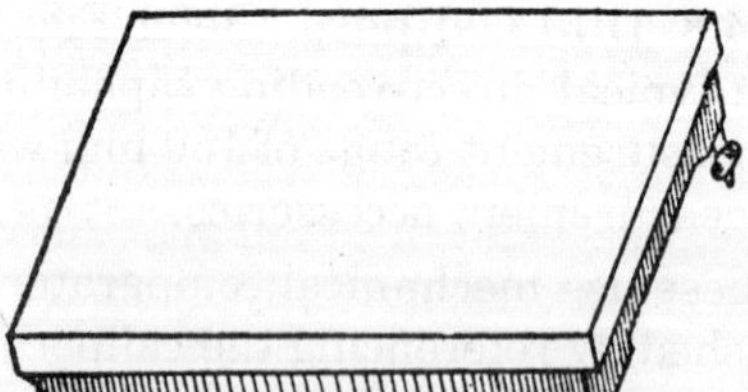

Fig. 2.21. Surface Plate.

The iron plate has the advantage of allowing a certain amount of wringing and surface or edges are not readily chipped out if something is dropped on it. The glass plate will shatter, or break and be destroyed if heavy pieces are dropped on it.

Surface plate (refer Fig. 2.21) forms the basis of measurement and is a must for metrology laboratory where inspection work is carried out. It provides a datum or reference surface for measurements. It is also used to check the flatness of another surface. It has got a flat surface and all other measuring instruments having flat base can be placed on the surface plate. The big surface plate generally rests on four levelling screws with the help of which its surface can always be kept horizontal. This is invariably made of cast iron, and in manufacture, is artificially aged to set free from all internal stresses. After this, its dimensions will never change with time. At the bottom it is strengthened by ribs. The top surface is scraped to true flatness. A wooden cover is always placed over it, when not in use. When in use the whole surface of the plate must be used and not the centre or corners in order to have uniform wear. Grease must be applied to top surface when not in use for long time. A regular check for flatness must be made after regular intervals.

It is well known fact that all precision measurements start from flat surfaces and that without the aid of an accurate plane surface it is impossible to do any precise mechanical work and that the surface plate, whose upper face is flat within close limits of accuracy, is of fundamental importance for precision measurements. The surface plate serves two important purposes. *viz.* it is used as datum plane for taking accurate linear and angular measurements, and with the aid of suitable marking compound, it serves for testing the degree of flatness of other surfaces. For very accurate work it is very essential that the form and dimensions of a surface plate must be such that distortion or deflection is reduced to a minimum even under comparatively heavy loads. For this reason it is customary to reinforce the plate proper by the provision of a frame and ribbing of considerable depth. The standard available sizes of the surface plates vary from 100 × 100 mm to 2000 × 1000 mm in about 13 ranges. Generally surface plates upto size of 1000 × 630 are provided with three supporting feet and above it, either 3 or 4 feet. If a surface has more than 3 feet, the supporting surface on which the feet rest ; a special stand or bench should be provided with adjustment to ensure that all the feet bear evenly. Moreover, it is also desirable that the bearing surfaces should be smoothly machined in order to permit slight movement with change in temperature. It is recommended to provide a projection of at least 25 mm on all sides of the surface plates measuring 250 × 250 or over. The undersides of these projections should be machined properly and reasonably flat to facilitate the clamping of work or measuring instruments. The threaded holes can also be provided in the plate without affecting the accuracy if the drilling and tapping operations are carried out before the surface is finished.

The four edges of precision plates are usually finished straight and square to each other within the accuracy of 0.08 mm per metre. The upper surfaces of the plates are finished by scraping operation. The idea of the accuracy of the flatness of the top surface can be had from the fact that the maximum permissible error from mean true plane (for a surface plate of diagonal 150 mm) is allowed between ± 0.002 to ± 0.005 mm ; and for surface plate of diagonal 2000 mm, maximum permissible error from mean true plane is allowed to vary between ± 0.02 to ± 0.04 mm. Further in addition to the conforming of above tolerances of whole of the top surface of the plate, bearing area of any local area from the mean true plane of that local area should not deviate by more than 0.01 mm per metre to 0.025 mm per meter. It is also very essential that the proportion of the area of the surface plate represented by high spots, on which another flat surface would rest, should be comparatively small and should not be more than 10 to 20%.

2.5.1. Material for Surface Plate. The surface plates should be made from a material which will provide a high degree of rigidity and freedom from warping and capable of taking a high finish and the surface must be resistant to wear. Most commonly used material for this purpose is either the plain or alloyed close-grained cast iron of good quality. After casting and rough machining and prior to finishing operation, the plates should be allowed to age either naturally by allowing a suitable period of time to elapse or by proper heat treatment in order to relieve the internal stresses.

The composition of most commonly used plain cast iron and alloy cast iron for surface plates are :

Plain Cast Iron		*Alloy Cast Iron*	
Total carbon	...3 to 3.5%	Total carbon	...2.8 to 3.2%
Combined carbon	...0.4 to 0.7%	Silicon	...0.8 to 1.5%
Manganese	...0.5 to 1.2%	Manganese	...0.6 to 1.0%
Silicon	...1.0 to 1.2%	Sulphur	...0.12% (max)
Sulphur	...0.15% (max)	Phosphorus	...0.33% (max)
Phosphorus	...1.2% (max)	Nickel	...1.4 to 1.6%
		Chromium	0.4 to 0.6%

The heat treatment of the surface plates made of above materials is carried out as follows :

The plates are placed in an annealing furnace and heated upto 450 to 500°C and maintained at this temperature for about 3 hours or longer depending on the mass of the casting. The castings are protected from the direct heat of the flames by means of suitable baffle plates and heating should be as uniform as possible throughout. In case of small castings, more uniform heating is obtained by packing the casting in iron fillings in boxes. Then plate is allowed to cool gradually in the furnace itself without opening the doors. It is preferable to hold the temperature at 400°C for about an hour during cooling.

Apart from cast iron, surface plates are also made of non-metallic substances, *viz.* granite and glass. Metallic objects like height gauges, squares etc. can be easily slid over their surface. Their advantages are : These are more rigid for same depth. Corrosion is virtually absent. A big problem with C.I. surface plate is that any damage to surface throws up a projecting burr but in case of granite and glass surface plates it causes indentation only.

2.5.2. Manufacture of cast iron surface plates. It is not uncommon for a cast iron plate to lose its straightness, etc. over a period of time if certain precautions are not taken during its manufacture and use. The composition of the iron is the first thing to be taken care of since the proper composition improves machinability, wear qualities and the quality of the scraped surface. It is also important to remember that stability of the iron depends mostly on slow, uniform cooling in the mould after casting, which in turn, depends on the moulds of proper design, and on allowing sufficient time for complete cooling before the casting is exposed to air.

The equilibrium of cast iron part can be altered during machining particularly when substantial amount of material (outer skin) is removed. Thus before the finish cut, the piece must be allowed to relax by release of the clamps, the parts then properly supported and re-clamped without bending. It is also observed that single-point planing produces least springing and machining distortion in piece (compared to machining by milling).

Large cast iron section can easily bend when subjected to a load and elastic deflections may occur. Deflection can be minimised by adopting proper design and supports.

Lastly, the geometric relationships are not affected by the ambient temperature if it is steady. Only temperature deviations cause problem, emphasing the need for maintaining temperature at constant value.

2.5.3. Sizes and Accuracy of Surface Plates. According to IS : 2285—1963, the cast iron surface plates are classified into two grades (grade I and grade II). The recommended size of the surface plates and the maximum permissible departure from flatness of bearing area are given below [Refer Fig. 2.22].

Nominal size $a \times b$ *mm*	*Minimum thickness of top (t)*	$m \times n$ *mm*	*h* *mm*	*s* *mm*	*Maximum permisible departure from flatness of bearing area in microns*	
					Grade I	*Grade II*
100 × 100	9	80 × 80	25	5	2.5	10
160 × 100	9	70 × 80	25	5	3.5	14
160 × 160	9	64 × 64	25	5	3.5	14
250 × 160	12	73 × 64	40	5	3.5	14
250 × 250	15	100 × 100	40	8	5.0	20
400 × 250	15	117 × 100	50	8	7.5	30
400 × 400	18	107 × 107	50	8	7.5	30
630 × 400	18	137 × 107	80	9	7.5	30
630 × 630	21	126 × 126	80	9	7.5	30
1000 × 630	24	175 × 126	125	12	10.0	40
1000 × 1000	30	200 × 200	160	15	12.5	50
1600 × 1000	30	280 × 200	200	18	17.5	70
2000 × 2000	36	300 × 200	250	18	20.0	85

In addition to the above provision, the departure from flatness should conform to the following requirements also for sizes 630 × 400 mm and larger :

Grade I. Maximum 5 micron over any local area of 300 × 300 mm of plate.

Grade II. Maximum 20 micron over any local area of 300 × 300 mm of plate.

The rigidity of grade I plates above 400 × 250 mm size should be such that when a concentrated load is applied at the centre of the plate (head should cover area of 100 cm^2 for smaller plates to 500 cm^2 for longer sizes of plates), the loaded area does not deflect below the remaining surface of the plate by more than 0.003 mm per 50 kg load.

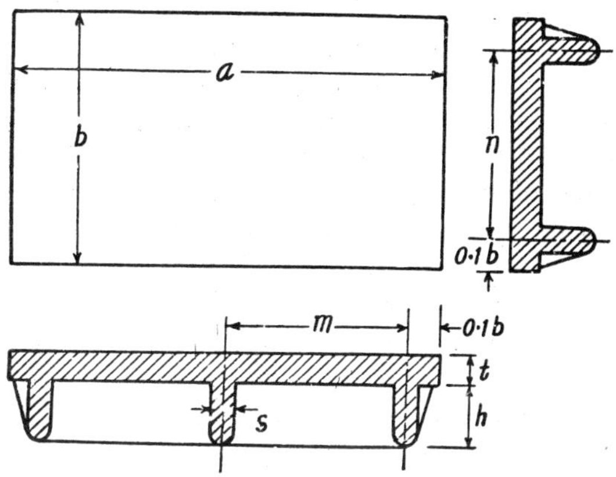

Fig. 2.22. Surface plate sizes as per BIS.

It is recommended that the top surface of grade I surface plate be finished by scraping process and the percentage of high spots on the top surface should not be less than 20% whereas the grade II plates should be finished by grinding operation and percentage of high spots should not be less than 10%. The percentage area of high spots should not be more than 50% in both the cases.

2.5.4. Principle of Flat Surface Generation. Principle of the three-plate technique of surface generation is commonly used for generating flat surface of surface plates. According to this principle, it is impossible to devise any combination of mating the three surfaces together unless they are flat.

1. Two plates can be scraped so that they mate together without being flat, as with plates 'A' and 'B'.

2. This out-of-flatness is shown up as soon as they are compared with plate 'C'.

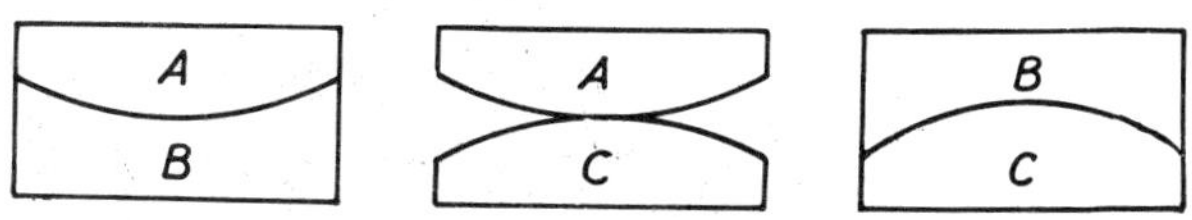

Fig. 2.23. Generation of a flat-surface.

3. Only if all three plates are perfectly flat can they be interchanged and still mate.

2.5.5. Scraping of Flat Surfaces. Cast iron surfaces are finished to a high degree of flatness by hand scraping method. This method requires considerable time and skill. The various steps followed for this purpose are described below :

(*i*) The surface is first cast, properly heat treated and allowed to age.

(*ii*) The surface is then rough machined and finish planed. After rough planing, castings are again aged either naturally or artificially and at least a month is allowed to elapse after the surface has been finish planed.

(*iii*) The surface is then scraped lightly once or twice all over by progressively removing the high spots in order to produce a smooth finish and an approximate flat surface before proceeding to correct any small irregularities that may be present.

(*iv*) The surface is then scraped by applying a suitable marking compound. The most commonly used compounds to show bearing marks are red lead, prussian blue and lampblack mixed with lard oil.

(*a*) For rough scraping, first the marking compound is applied fairly generously to flat plate, being thinned down as necessary to facilitate spreading. When an even coating has been obtained the work to be scraped is brought in contact with the flat plate and rubbed lightly thereon, or *vice versa*. The high spots thus indicated are removed by scraping and the process is repeated until fairly uniform spotting all over the surface is obtained.

(*b*) Then the compound in somewhat dryer form is applied to the work to be scraped. The flat surface is then rubbed on the work and thus the high spots produced by rubbing show up distinctly. The amount of compound applied is gradually reduced and in later stages petrol is also mixed to act as dryer. Finally only just enough compound is used to dull the surface without colouring it and the true high spots are thus shown and there is no risk of false bearing indications such as may occur if the surface plate is coloured.

(*c*) *Ornamenting Scraped Surfaces*. After the scraping operation, the surface plates are frequently finished by processes commonly known as frosting, snow-flaking and spotting in order to obliterate the somewhat irregular marks left by the scraping operation and to obtain pleasing appearance. In the frosting operation, no metal is removed and thus accuracy previously attained remains the same ; its effect being manifested primarily by the manner in which the light is reflected from the surface. The various patterns produced by frosting operation can be of the following

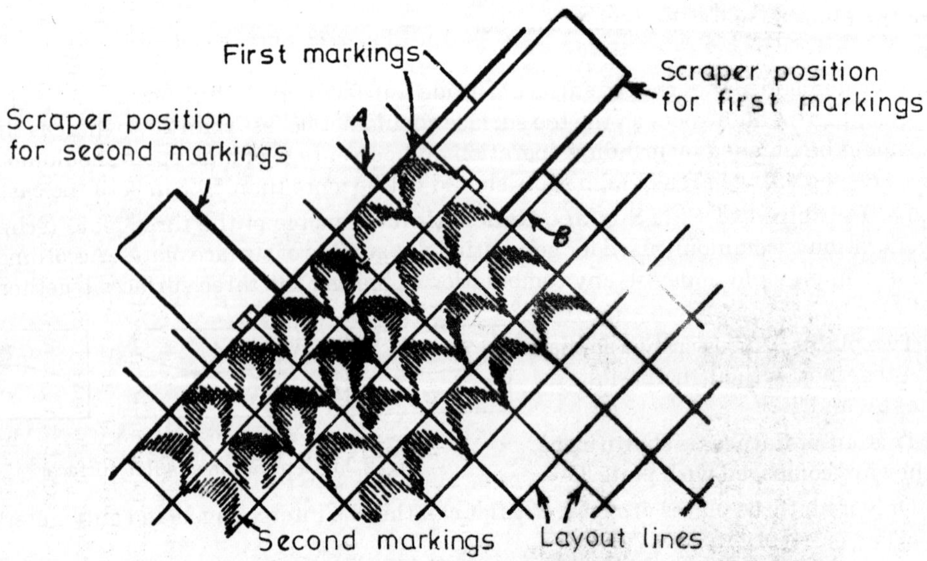

Fig. 2.24. Ornamenting scraped surface.

shapes : crescent shape, diamond frosting, evenly spaced crescent shaped marks, irregular checkered pattern. The method whereby evenly spaced crescent shaped marks can be readily produced on a surface plate is shown in Fig. 2.23.

Scraped surface of an iron surface plate is found to be free from wrap, waviness or chatter.

2.5.6. Originating Surface Plate. This is the method of producing an accurate surface plate without the aid of any reference plate or any measuring instruments. The method is based on the fact that the flat surfaces are self-checking and employs the principle known as "symmetrical distribution of errors".

For this method of originating a surface plate, it is necessary to have three suitable castings (say, 1, 2, 3) which have been aged and finish planed as accurately as possible. First of all the surfaces of all the castings are scraped to a smooth finish and the marks of planing tool completely removed. Prussian blue is then applied to plates 1 and 2, and each is roughly scraped using a wide, flat scraper blade (the cuts being long and deep) until a reasonably good bearing distribution is obtained. In the case of square plates, these should be successively turned 90° one to another to reveal "twist". The plate 3 is then roughly fitted to plate 1 by removing metal from plate 3 only and no metal being removed from plate 1. Thus plate 3 will be an approximate duplicate of plate 2, *i.e.* neither plate 2 nor 3 is perfectly flat but both have the same errors and the plate 1 is having errors reverse of 2 and 3. As the spots become progressively more uniform and evenly spaced, a narrower blade is used, the angle of cut is raised gradually from the horizontal, and the stroke is shortened. The most satisfactory scraping stroke is the "half noon", in which case the stroke starts from the valley, reaches its greatest intensity directly on the spot, and again fades to nothing as it passes to the other side.

On applying plates 2 and 3, the errors which were previously marked by the reverse deficiencies of No. 1 are now duplicated and accentuated. Now plates 2 and 3 are scraped to fit into one another, care being taken to remove as nearly as possible the same amount of metal from each during the process. Thus these two plates will be more accurate than No. 1. The next step is to scrap plate 1 to fit 2 and then fit plates 1 and 3 by scraping the same amount from each. In this way plates 1 and 3 are made more accurate than plate 2.

Plate 2 is then scraped to fit plate 1 after which 2 and 3 are scraped together removing the same amount from each. This cycle is repeated until all the plates will fit one another interchangeably thus affording positive indication that the plates are perfectly flat.

2.5.7. Finishing flat surfaces by hand lapping. The lapping process is employed for producing the flat surfaces required to possess great degree of accuracy and uniformity than can be obtained by scraping or grinding operation. The process of lapping is carried out by rubbing the work over the surface of a flat plate or lap made of either steel or copper where durability is the main consideration, and of cast iron for very precision work. Very fine particles of a suitable abrasive, which are either initially embedded in the surface of the lap or mixed with a suitable lubricant or distributed in a thin film between the lap and the work, gradually wear down the high spots on the work which initially bear on the lap. As the process is continued a close approximation to true plane is obtained. It will be appreciated that as the work is moved about over the surfaces of the lap the effect of errors in the flatness of the latter is distributed and as such the errors in the surface of the work will be small than those of the lap. Further, the more closely the surface of the lap approaches to a true plane, the greater will be the degree of accuracy ultimately obtainable in the work.

If a perfectly flat lap is used for lapping the flat surface plate, then many difficulties are encountered due to the presence of a film of oil, abrasive or air between the surfaces. The film of oil produces a light cushioning effect, thereby allowing one surface to rock upon the other to a limited extent. Trouble is also caused by uneven distribution of the abrasive which has a tendency to collect in the central portion of the plate. In order to overcome the above difficulties encountered when perfectly flat lap surfaces are employed, the surfaces of the laps are scored with small grooves of the form indicated to the right in Fig. 2.25 and spaced about 5 to 6 mm apart so as to produce a checkerboard pattern. These grooves allow the oil film to be pushed aside or sheared from the surfaces thereby permitting them to come into closer contact. The grooves also provide for ventilation so that the air pressure between the surfaces is partially relieved.

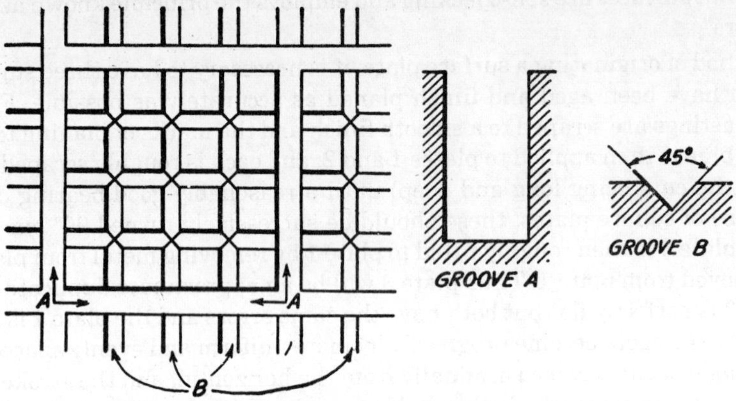

Fig. 2.25. Form of grooving for lapping plates.

The abrasive particles used for lapping purposes are of two types *viz.* charging and non-charging. The non-charging abrasives (such as Turkish lapping powder, Vienna lime etc.) when introduced between two surfaces of equal hardness remove equal amount of material from each by abrasion when the surfaces are rubbed together. Whereas in the case of charging abrasives *e.g.*, diamond dust, finely powdered emery and corundum as natural abrassives and SiC, Al_2O_3 as artificial abrasive) the laps of softer material are employed because the abrasives tend to become embedded in the surface of lap and the embedded particles projecting out of soft surface of the lap act as innumerable small cutting edges. For obtaining best results, it is essential that particles of abrasives be as nearly as possible uniform in size and as fine as possible in order not to produce the irregular scratches on the surface of the work and for greater accuracy and finish.

The lapping process can be either wet or dry. In the wet process the rate of metal removal is considerably higher whereas in the dry process a greater degree of accuracy is obtainable owing to more intimate contact between the work and the charged lap surface. In the wet process the abrasive is mixed with a suitable lubricant (soda water, paraffin, petrol, turpentine, machine oil and lard oil) and a thin film is applied over the lap. Under the pressure exerted by the work the abrasive particles tend to become embedded in the lap but there is always a surplus of oil and abrasive on the surface. In the case of dry process, first the lap is charged by pressing or rolling the abrasive into the surface. Though process is called dry, it is usual to apply a very slight amount of lubricant to the lap surface.

In the lapping operation, it is most essential that an irregular rotary movement be applied to the work in order to ensure, as far as possible, that every available part of the lay surface is utilised to the same extent. The work, therefore, should be pushed gently round, no two consecutive movements taking place over the same portion of the lap. If the work is only rubbed to and fro across the middle of the lapping plate, a depression would speedily be formed and satisfactory results cannot be expected. When irregular movement of the work takes place then it tends to distribute the effects of any inaccuracies that may have originally been present in the lap surface. The pressure exerted on the work should merely suffice to ensure constant contact with the lap surface. When parts of the highest precision are required, the question of relative temperature of the lapping plate must be taken into account The most common cause of temperature variation between work and lap is taken care of by following the practice of not holding the work with the hands while lapping. Generally a holder (handle) made of wood or other insulating material is made which fits over the work without exerting any distorting pressure on it. In order to avoid the inaccuracy due to the heat generated by the actual lapping operation, the rate of metal removal from the work must be very slow.

2.5.8. Testing Flatness of Surface Plate. Flatness is defined as the minimum distance between two planes containing all the irregularities of the surface under examination. While accurate testing and quantitative check requires sophisticated equipments, some simple techniques of testing flatness of surface plate are discussed first :

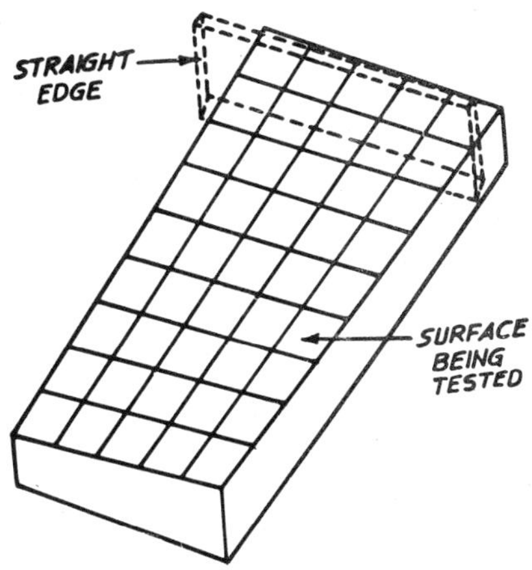

Fig. 2.26. Testing for flatness of a surface plate using a straight edge.

1. The surface to be checked is marked out with a soft lead pencil, so as not to damage the surface, into a series of parallel bands as shown in Fig. 2.26. It is then checked using a bevel edge straight edge and using a light box or feeler gauges, depending upon the size of the surface and the required accuracy of measurement. It is important to note that though the surface may appear flat if measured parallel to the sides, but additional checks need to be made across the diagonals to detect errors of flatness due to 'winding' of the surface.

2. The surface to be checked is first carefully cleaned and any burrs are removed with an oilstone. Engineers' blue is then smeared lightly over this surface. The blued surface is then inverted over an inspection grade surface plate and gently moved in a series of small circles as shown in Fig. 2.27 (*a*). This will transfer the blue to high spots of the scraped reference surface lightly and uniformly. Any attempt to apply the blue directly to the reference surface usually results in a poor distribution resulting in a false interpretation. The surface to be checked is now cleaned of the blue and brought once more into contact with the surface plate. Blue will now be transferred back to the high spots on the surface under test as shown in Fig. 2.27 (*b*).

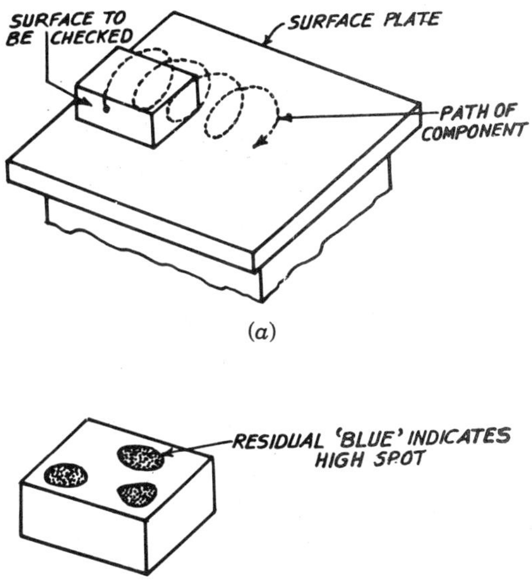

Fig. 2.27. Testing for flatness of surface plate.

3. The various methods of quantitative testing of surface plates are :

(*a*) *Using two-footed twist gauge.* Two-footed twist gauge is an electronic indicator, sensitive to one-tenth of micron. It is used for inspection of overall twist in surface plate. The base of the instrument is made to rest on the plate and the end feet rest on the diagonal corners of the plate.

The instrument is set so that its measuring probe registers zero or datum on a small central lapped pad on the gauge. Leaving the zero set, the twist gauge is swung 90° to the opposite diagonal corners.

The relative difference in heights, or "twist" is registered as a plus-or-minus reading against the lapped pad. It may be noted that even though the twist gauge sags appreciably, but this sag is constant in both the positions.

(b) *Using three-footed twist gauge.* It is used to detect local errors. It has two gauging lapped pads at the bottom corners of the single foot side. A small wedge shaped gauge having a slight taper of approximately 1 μm per 10 mm of length is pushed under the lapped flats (gauging pads) to calibrate change in height of the lapped flats. This twist gauge is self-proved upon being turned 90° on the surface plate and tested once again.

(c) *Surface gauge with a protruding arm having electronic probe.* This instrument is used to search more local errors. It employs an electronic indicator, a stand and a small lapped block. If the correct scraping procedure is followed then local errors will be minimum and errors may be of the nature of overall bow or twist.

(d) *Straddle surface gauge and a straight edge.* It is used to determine overall "bow". The straight edge has two support feet 0.554'l' apart and three indicating pads on the straddle surface gauge. The special straight edge is self-checked by reversal.

4. In testing the flatness of surface plate, it is the general practice to measure the actual deviation from the true plane at various points. The method of testing a flat surface with sensitive level or auto-collimator is described in detail in Chapter 7. Here we shall discuss the *Beam Comparator Method* which is very quick method and very accurate one also. This method is particularly convenient for testing the flatness of surface during manufacture.

This method is based on the principle of comparing the straightness of a succession of generators in the surface under test with that of a known reference straight edge. The comparison is carried out by the Beam Comparator, a sort of sensitive dial indicator which, in effect, measures the vertical displacement of the midpoint of any particular generator with respect to a theoretical straight line joining its ends. From these measurements, the idea of the apparent convexity or concavity of generator can be had, which may be used to determine the general departures, of the surface from mean true plane as explained later.

The body of the comparator consists of a robust but light beam having three supporting feet made of hardened steel and spherically ended, one fixed at the centre of beam but offset from the

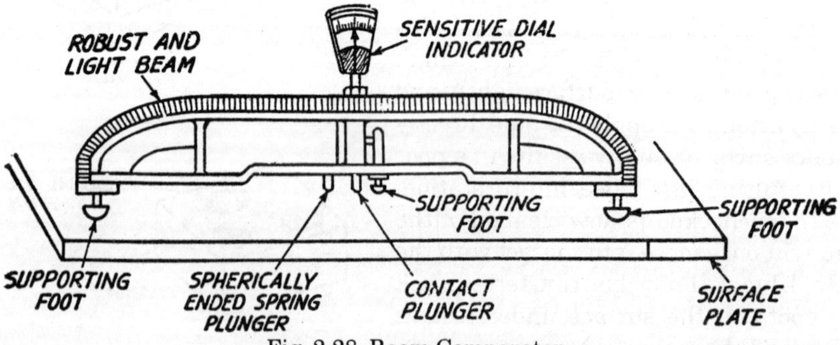

Fig. 2.28. Beam Comparator.

line joining the other two. The two supporting feet at the ends may be clamped to the beam in symmetrically disposed positions with respect to the central foot and at a distance apart depending upon the size of the surface under test. By having only three supporting feet, the beam is accidently liable to be rolled over on the side not supported by the offset foot. In order to avoid this, a fourth foot in the form of a spherically ended spring plunger is provided at a position symmetrically opposite the fixed offset foot. The fourth foot is normally arranged to be reasonably clear of the surface plate and comes into operation only if the beam tends to fall

A very sensitive dial indicator supported in the central web of the beam and the axis of its spherically ended contact plunger lies in the plane through the axis of the two adjustable feet.

For testing purposes a reference surface is employed whose straightness has already been checked by other methods and thus the errors of its straightness are known. The comparator is supported upon equal slip gauges or lapped pads on both the reference and the test surfaces so that the feet of the beam comparator do not accidently locate in the hollows of the scraped surface, and the readings are taken with respect to the high spots of the surface.

Method of Use. The method adopted for computing the results obtained from single generator is illustrated more clearly by taking an example of testing a surface plate of size 100 mm × 60 mm. Generally the flatness of the plate towards the edges is not bothered and thus a distance of about 5 cm from all the edges is not considered. The surface plate is marked as shown in Fig. 2.29 with a lead pencil. For marking purposes, chalk should never be used which gets deposited in the hollows of scraped portion and is harmful. The general and most commonly adopted practice of dividing the plate is as shown in Fig. 2.29 according to which eight generators (*AB, HF, DC ; AD, EG, BC ; AC, BD*) are tested. In addition, readings are also obtained for semi-diagonals of span *AX, BX, CX* and *DX*.

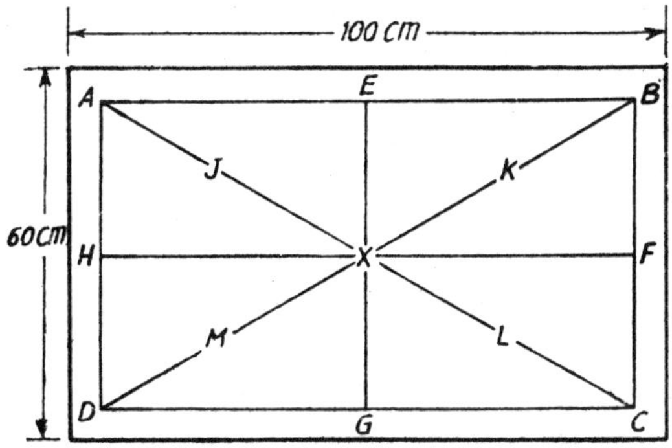

$AB = HF = DC = 90$ cm.
$AD = EG = BC = 50$ cm.
$AC = BD = 103$ cm.

Fig. 2.29. Scheme of test for a surface plate.

The span of the adjustable feet is set in turn for each of the different generators and observations are taken first on the reference surface, then on the surface plate and finally on the reference surface again.

Fig. 2.30 shows another way for calibrating surface plate flatness in the same manner. This method uses an autocollimator, reflecting target mirror.

In Table on next page for generator *AC* the total span between two adjustable feet was set as 103 cm (col. 2). The first reading was taken on the reference surface which was 5.1. (col. 2, Ist reading). Then reading was taken on the test plate which was noted as 5.7 divisions (col. 6). The reading was again taken on the reference surface and this time also it was 5.1 divisions (col. 3, 2nd reading). Thus the mean of Ist and 2nd readings on the reference surface is 5.1 divisions. The known error at the central region of the span on the reference surface was known to be hollow by 0.0002

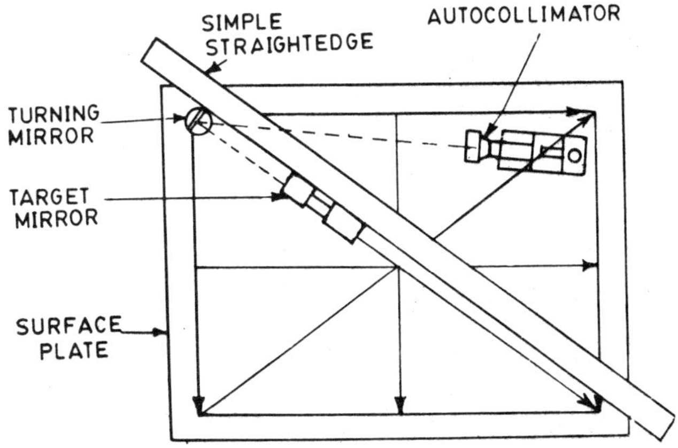

Fig. 2.30

mm. *i.e.*, equivalent to – 0.2 division on the indicator (col. 4), *i.e.* the indicator gave a reading of 0.0002 mm less than it would have done on a truly flat surface. Thus the indicator would be expected to read 5.3 divisions (col. 5) on a truly flat surface. As the reading on test plate is 5.7 divisions (col. 6), the mid-point *X* on generator *AC* is convex by (5.7 – 5.3) divisions (*i.e.* 0.0004 mm) in reference to the straight line joining points *A* and *C*.

The comparator is then turned until the end feet are stationed at *B* and *D*. In this case the reading on the indicator obtained was 4.5 divisions (col. 6). It may be noted that for this span the comparator has already been checked on the reference surface and thus the reading of reference surface will be same as for *AC*. This time the point *X* is (4.5 – 5.3 = – 0.8 division) 0.8 division below the straight line joining points *B* and *D*. It may be noted here that one setting and calibration of the comparator span served for checking the centre point of both the diagonals of the plate.

Similarly only one setting and calibration is required for generators *AB, HF, DC,* one for *AD, EG, BC* and one for semi-diagonals *AX, BX, CX* and *DX*. In the manner described above, the column (7) can be determined for all the generators.

The next step is now to correlate the results. For this first of all the departure of various points on the surface is derived from some arbitrary plane. For convenience, it is selected as that passing through the points *A, B* and *C* of the source plate to which zero errors are accordingly

Observations and Calculations relating to the test of 100 cm × 60 cm surface plate
(Division of the indicator = 0.001 mm)

1		2	3			4	5	6	7	8	9
Generator	*Central Point*	*Span (cm)*	*Reading on reference surface*			*Known error of reference surface*	*Corrected reference readings (3) – (4)*	*Reading on test plate*	*Convexity or concavity of Generator (6) – (5)*	*Basic reading at centre of span*	*Results*
			First reading	*Second reading*	*Mean*						
AC	X	103	5.1	5.1	5.1	– 0.2	5.3	5.7	+ 0.4	0	X + 0.4
BD	X	103	5.1	5.1	5.1	– 0.2	5.3	4.5	– 0.8	—	D + 2.4
AB	E	90	4.7	4.7	4.7	– 0.3	5.0	5.2	+ 0.2	+	E + 0.2
HF	X	90	4.7	4.7	4.7	– 0.3	5.0	4.8	– 0.2	+ 0.6	(X′ + 0.4)
DC	G	90	4.7	4.7	4.7	– 0.3	5.0	5.4	+ 0.4	+ 1.2	G + 1.6
AD	H	50	3.5	3.7	3.6	+ 0.1	3.5	3.7	+ 0.2	– 1.2	H + 1.4
EG	X	50	3.5	3.7	3.6	+ 0.1	3.5	3.0	– 0.5	+ 0.9	(X″ + 0.4)
BC	F	50	3.5	3.7	3.6	+ 0.1	3.5	3.3	– 0.2	0	F – 0.2
AX	J	51.5	5.4	5.4	5.4	0.0	5.4	5.6	+ 0.2	+ 0.2	J – 0.4
BX	K	51.5	5.4	5.4	5.4	0.0	5.4	5.5	+ 0.1	+ 0.2	K + 0.3
CX	L	51.5	5.4	5.4	5.4	0.0	5.4	5.2	– 0.2	– 0.2	L + 0.0
DX	M	51.5	5.4	5.4	5.4	0.0	5.4	5.0	+ 1.4	+ 1.4	M + 1.0

assigned, *i.e.* these points are treated as zero. From the table, it will be observed that the generator *AC* is convex by 0 0004 mm. As points *A* and *C* are treated as zero, the point *X* should be assigned a value of + 0.0004 or 0.4 division. The next step now is to assign a value to *D*. From the observations it is noted that generator *BD* is concave by 0.0008 mm. *i.e.*, point *X* is 0.8 divisions below the line joining points *B* and *D*. But it has already been seen in the generator *AC*, that point *X* with respect to datum plane *ABC* is 0.4 division above it. Since point *X* is common to both the diagonals, it is, therefore, necessary to adjust the direction of the generator *BD* about its end-point *B* until the mid-point *X* coincides with the assigned value of *X* at + 0.4 division. To do this, the mid-point *X* should be raised by 1.2 divisions and thus point *D* by 2.4 divisions. It is thus observed that for generator *AC*, the value of mid-point *X* remains same as + 0.4 because points *A* and *C* were assumed to be zero and, therefore, there was no basic reading at centre of span. Therefore, in column (8) for generator *AC*, value is put zero and the final result (col. 9) is that point *X* which is 0.4 division above datum plane ABC and it is put as $X + 0.4$. For generator BD, the method for finding the height of point D above the datum plane ABC has already been described above and the generalised formula does not hold for it and, therefore, entry in col. (8) is (–) and final result is that point D is 2.4 divisions above datum plane and thus entry in col. (9) is $D + 2.4$.

To obtain the values for E, F, G and H the following reasoning is applied. As the values assigned to A, B and C are zero, the values of E and F would also be zero if generators AB and BC were straight. Thus for generators AB and BC, the basic reading at centre of span is zero (col. 8) and the values of their mid-points E and F will be 0.2 division and – 0.2 division respectively from datum plane as obtained from col. 7. In col. 9 the entries for generators AB and BC would thus be $E + 0.2$ and $F - 0.2$. Similarly, as D is + 2.4 divisions and A and C are zero, then if generator AD and DC were straight, the values of H and G would be + 1.2 divisions (see col. 8 for generators AD and DC). It is helpful to note that the 'basic' value for the mid-point of any span is the mean of values already assigned to the end-points. Now as the mid-point H of generator AD is already convex by 0.2 division, then with reference to datum plane AB, it will be convex to 0.2 + 1.2 divisions, *i.e.* 1.4 divisions and thus entry in col. 9 against generator AD is $H + 1.4$. Thus col. 9 is algebraic sum of col. 7 and 8. In this way cols. 8 and 9 for all the generators can be computed.

The basic values for X'' and X' (col. 8) are obtained by making use of the results given in col. 9 for E, F, G and H. The results in col. 9 for X' and X'' are obtained by combining col. 7 with col. 8. It may be noted that X' and X'' are positions coincident with X but derived from spans HF and EG : whereas X was derived from span AC. The values of X', X'' should be same as of X and, therefore, it acts as a convenient check on the accuracy of observations.

Now the results of col. 9 are easily comprehend as shown in Fig. 2.31, *i.e.* the location of all points w.r.t. the datum plane ABC is marked on the points themselves.

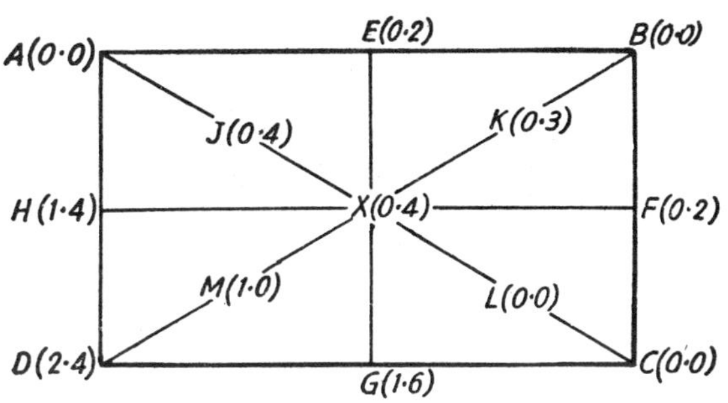

Fig. 2.31. Measurement errors referred to an arbitrary datum plane through the points $A = B = C = 0$.

It may be noted that the datum plane ABC was chosen quite arbitrarily and need not necessarily be the mean true plane. Mean true plane is actually such a plane which lies nearly in between the various points on the surface, *i.e.* exact location of various points is uniformly distributed on either side of the mean true plane and the various points are at minimum distance from it.

For determining such a plane, a visual inspection of the results plotted in Fig. 2.31 is usually sufficient *e.g.* in this case it will be noted that the surface lines within two planes [+ 2.4 and (0.2)] 2.6 divisions or 0.0026 mm apart. It will be observed that point D is most troublesome and is responsible for very much displacing apart the two planes under which all points lie. If the plane ABC be rotated about the line AC by an amount corresponding to 1.2 divisions at D and B, then the values of various points will be as shown in Fig. 2.32 and it seems to be mean true plane, of course, the exact mean true plane may never be determined and some error will always be there.

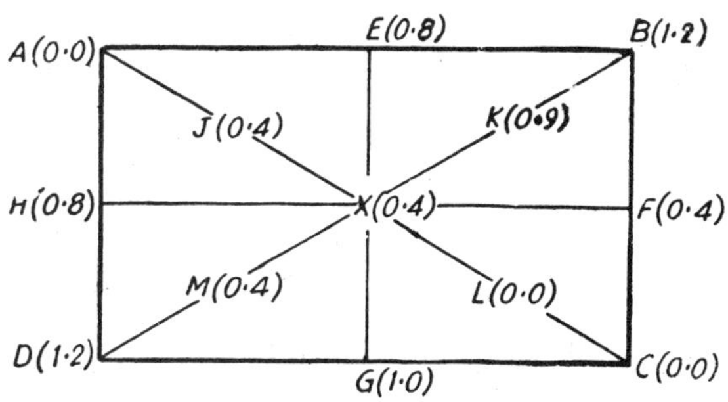

Fig. 2.32. Results referred to a datum plane selected to give best distribution of errors.

It is apparent from Fig. 2.32, that the surface lies within two planes separated by a distance of 0.0012 mm or the surface lies within ± 0.0006 mm from a mean true plane. (Please also refer Art. 7.2).

The working surface of surface plates can be explored with the variation gauge. The variation gauge (Refer Fig. 2.33) consists of a freely moving base with three co-planer pads for bearing on the surface being tested. It carries an indicator to indicate variations in deviation from flatness from adjacent areas of that surface through a fourth co-planer pad, which is itself mounted on an intermediate sprung block.

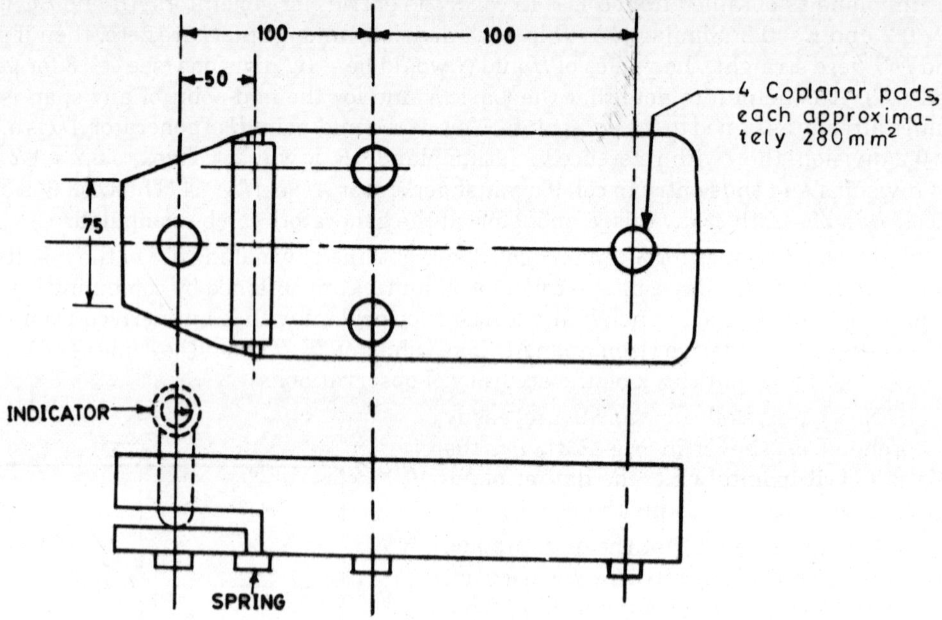

Fig. 2.33. Variation gauge.

2.5.9. Testing by comparison with a liquid surface. This method utilises the surface of a liquid as a basis of reference and is probably the most rapid method and can be conveniently applied to quite large surfaces with an accuracy of the order of 0.005 mm. The liquid is a weak soda solution which is contained in a trough made from a length of angle or channel iron and supported over the surface to be tested. In this method a special type of surface gauge is used to compare the flatness of the surface of the work with that of the liquid.

The instrument consists of a substantial base provided with a stiff column which carries a bracket holding a micrometer head in an overhanging vertical position with the spindle pointing downwards and acting like a height gauge. The gauge rests on the work with its base beneath the trough and the micrometer head over the liquid surface. The reading is noted down by screwing down the micrometer head until the conical tip of the spindle (having radius of about 0.5 mm) just makes contact with the liquid. This may be indicated electrically be connecting a dry cell and a galvanometer in series with the trough and the work. Care has to be taken to keep the tip of spindle perfectly dry.

This test is carried out by first levelling the surface such that the two readings obtained with the surface gauge in its extreme positions are the same. Readings are then taken with the gauge at intermediate points and any variation in these readings from those obtained in the end positions indicate errors in the work surface along the region beneath the trough. For large surfaces a number of troughs are set up parallel to each other and connected together at one end with rubber tubing so that their liquid surfaces are at the same level.

In a simple test shown in Fig. 2.34, two gauges are connected together by a stout rubber tubing. After setting the surface to be tested perfectly horizontal, one gauge is placed in the centre and the other gauge is moved from one place to other on the surface and for each position a reading is taken on each of the two micrometer heads. Difference in two readings as the other gauge is

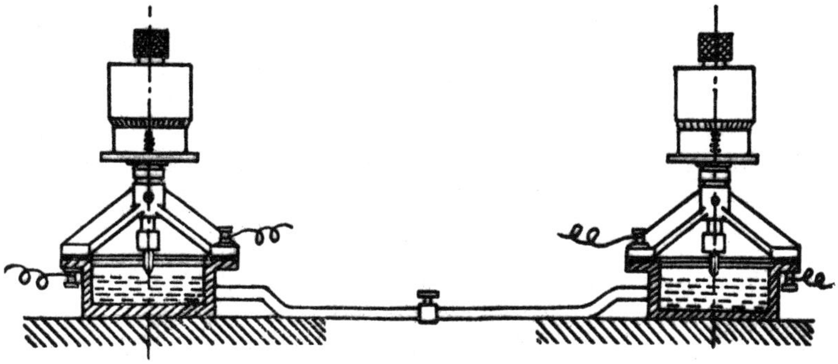

Fig. 2.34

moved from place to place indicates the errors in flatness at position tested. If the surface as a whole has not been set quite horizontal this will be indicated by the readings and be allowed for when plotting the results so as to give a diagram showing the errors in flatness of the surface from its mean true plane. The precaution to be observed is that interconnecting tubing must be free from air bubbles. In order to avoid the surging of liquid from one cup to another and for attaining better equilibrium, a stop cock may be provided in the tube which is shut while moving the second cup from one position to another. The work should be avoided from the effects of vibration which causes ripples on the surface of liquid.

2.5.10. Care and use of surface plates

1. The surface plate is a datum and needs to be protected against damage. When not in use, it should always be kept covered. During use, top should be frequently wiped clean from dust and other particles. When measurements are being made, a wiping cloth should be spread on the surface plate to receive small tools and gauge blocks.

2. Surface plate should be located in a stirred atmosphere under constant temperature control. Direct sunlight or sudden draughts should not be allowed to reach surface plate. Vertical gradient of temperature subjecting the top and underside of surface plate to different temperatures should be avoided.

3. The surface plate should be supported firmly and levelled. Stands should be located on a stable foundation.

4. The load on the surface plate should be distributed over the working surface, as far as possible.

5. Point engagement should not be made with scraped or machined surface plates. To avoid direct point engagement with the surface plate, contact should be made through an intermediate precision gauge block or a distance piece, preferably, not more than 10 mm tall.

6. Use should be made of the full available area of the working surface and not concentrated on any one area.

The variations in local flatness of the surface should be checked occasionally by using a variation gauge.

7. The excess metal due to burrs on the working surface of cast iron surface plate should be stoned away by confining attention to the burr. This operation should be followed by thorough cleaning from abrasive dust. If the surface plate is not required for some days, the surface should be coated with a corrosion preventive such as petroleum jelly.

8. Rusting can be reduced by frequently wiping the top when in use and by occasionally gently rubbing with another surface plate using a paste of paraffin and a little jeweller's rouge as a lubricant.

9. The wear on surface plate can be detected by rubbing it with a superior grade of surface plate and studying the rubbed appearance, by testing for deviation from straightness along lines on the surface plate, and/or using a variation gauge.

2.6. Tool Maker's Flats and High Precision Surface Plates

For the measurement work of the highest precision, small circular surface plates (commonly known as tool maker's flats) are used. The flats are usually made from an alloy steel, hardened, ground and lapped.

The selection of material for these is done from steel, cast iron or alloy cast iron on the basis of the following requirements :

(*i*) The requisite hardness with proper heat treatments.

(*ii*) A suitable surface for gauges to be wrung on to it.

(*iii*) Sufficient stability to maintain the final limits of accuracy.

The composition of cast iron should be such that after an appropriate heat treatment operation, the sorbo-pearlitic structure having a high hardness and wear resistance is formed.

The tool maker's flats are generally circular and made of solid steel free from inclusions (preferably from hammered and upset blanks) which after proper heat treatment gives a hardness of about 93 HRC or 850 HV. The sketch of the tool maker's flat is given in Fig. 2.35 and it shows various important dimensions also. It is, however, recommended that a shallow groove should be provided all around the periphery of larger flats (from 100 to 200 mm) to facilitate handling and also minimise the possibility of uneven hardening. The flats of size 50 to 75 mm generally have both faces finished as working surface whereas those above it have only one face finished as working surface.

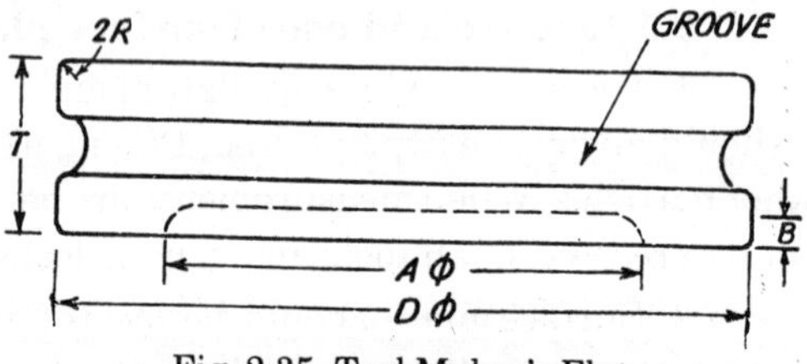

Fig. 2.35. Tool Maker's Flat.

D (*Nominal size*)	*T* (*min*)	*Recess*	
		Dia A	*Depth B*
50	12	—	—
75	15	—	—
100	18	80	3
125	25	100	4
150	30	120	6
200	35	160	6

2.7. High Precision Surface Plates

These are generally circular with adequate framing and ribbing underneath so that they are robust and the distortion is reduced to the minimum while in use. The plate is supported by three feet which are smoothly machined and ground. The top of the plate generally projects about 12 mm beyond the framing and is machined around the sides. The minimum thickness of top and minimum total depth are given below :

Dia (Nominal size) mm	*Min. thickness of top mm*	*Minimum total depth mm*
250	20	70
300	25	80
350	30	90

The working surfaces of the flats are generally finished by lapping and are free from noticeable scratches within the accuracy given below.

(The working surface should be between two parallel planes whose maximum distance is given here).

—	*Flat or plate dia mm*	*Marginal width which may be disregarded mm*	*Separation of limiting planes mm*
Flats	upto 150	2	0.0005
	upto 200	4	0.0005
Plates	upto 250	6	0.0008
	upto 300	8	0.0008
	upto 350	10	0.001

The unmachined parts are generally painted.

The base of each flat (other than flats finished with both faces as working faces) should be finished by grinding and should be parallel to the working face within 0.003 mm. Flats with both faces finished as working faces should have the two faces parallel to within 0.003 mm. In the case of plates the plane of feet should be parallel to the working face to within 0.01 mm.

2.8. Glass Surface Plates

Surface plates made of glass are now available in six sizes ranging from 150 × 150 mm to 600 × 900 mm. A range of circular glass surface plates are also available. These are supplied in a velvet lined box with lid. The accuracy of the glass surface plates is of the order of 0.004 to 0.008 mm from a mean true plane. The glass surface plates have the following advantages :

(*i*) These are light.

(*ii*) These have immunity from burring and atmospheric corrosion.

(*iii*) These maintain their accuracy for a longer period.

(*iv*) Their manufacture to a higher degree of accuracy is easier. However, these require careful handling.

2.9. Granite Surface Plates

Granite is widely used for surface plates for all types of shop, inspection and laboratory work. Table below shows a comparison of cast iron and granite surface plates.

Characteristic	*Cast Iron*	*Granite*
Wear Quality	Good	Good
Stability	Excellent	Excellent
Rigidity	Excellent	Excellent
Strength for given weight	High	Less
Strengthening by provision of ribbing	Yes	No
Effect of moisture	Rusts, but does not distort	Won't rust, but may distort
Fabrication, reconditioning	Lapped or Scraped	Must be lapped
Whether it will give up a bearing	Yes (advantage)	Not early (disadvantage)
Effect of burrs	raises up (disadvantage)	makes a hole (advantage)
Available shapes and sizes	Limited	Almost unlimited (can be sawed or shaped)
Versatility in use	Excellent—can be machined, drilled and tapped	Good, but needs inserts and fasteners
Effect of Temperature		
(*a*) Absorption	Fast to absorb heat and slow to dissipate	Slow to absorb heat and slow to dissipate
(*b*) Radiation	Reflects radiant heat (advantage)	Absorbs radiant heat (disadvantage)
(*c*) Temperature coefficient	Same as most of steels	unlike steel
Cost	Fairly expensive	Relatively inexpensive

2.10. Angle Plate

(Fig. 2.36). This is an accessory needed with surface plates for measurement purposes. The two surfaces of it are perpendicular to each other. It is available in various designs. Box squares are also sometimes used in place of angle plates. It is made of cast iron and its surfaces are scraped.

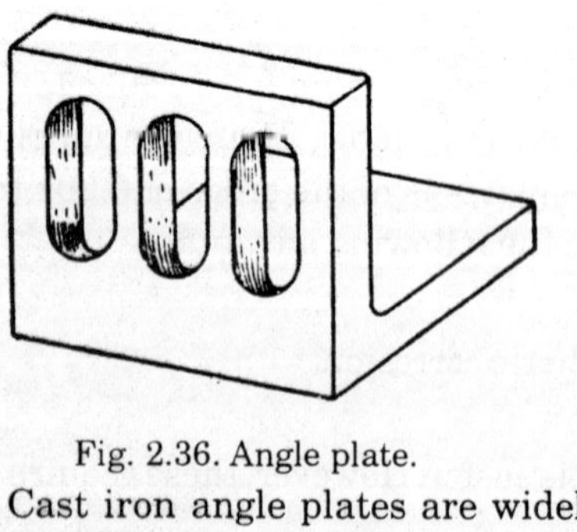

Fig. 2.36. Angle plate.

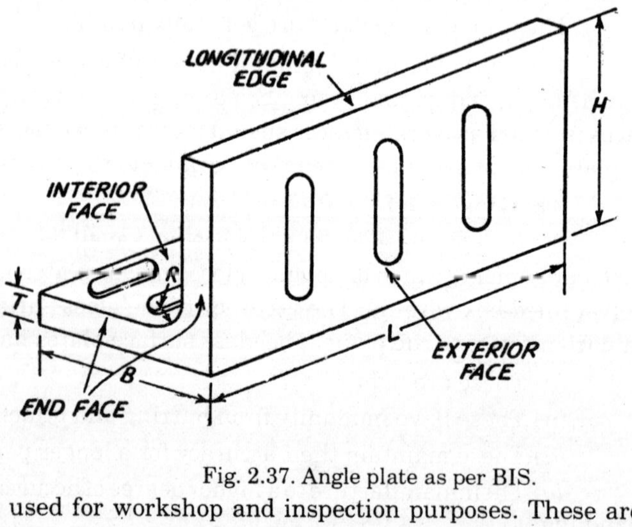

Fig. 2.37. Angle plate as per BIS.

Cast iron angle plates are widely used for workshop and inspection purposes. These are generally made from close grained cast iron and have a minimum hardness of 180 HB. After being cast and rough machined, the plates are given suitable heat treatment to relieve internal stresses before being finished. The heat treatment operation is carried out by placing the angle plates in an annealing furnace and heating up to a temperature of about 55°C and maintaining at this temperature for about 3 hours or more depending upon the mass of the casting. The castings are protected from the direct heat of the flames by suitable baffle plates and in smali plates more uniform heating is achieved by packing the casting in iron fillings in boxes. The castings are then cooled at a rate of 300°C per hour within the furnace itself without opening the doors. The material of the angle plates should be sound and free from blow holes and porous patches except for minor defects which can be repaired by plugging the material of the similar composition. Generally no sharp edges are allowed in the plates.

The angle plates (Refer Fig. 2.37) are available in two grades depending upon their accuracy, *i.e.* grades 1 and 2 The angle plates should be flat within the tolerances specified for each grade as specified in table below and the departure from flatness can be tolerated only in the nature of concavity and not convexity. It is the common practice to mark each angle plate legibly and permanently with the manufacturer's trade marks, the grade and other identification marks in such a way that the accuracy of the plates is not affected. Usually all the finished surfaces and edges of the plates are protected against climatic conditions by applying suitable corrosive preventive preparation on them.

Size No.	*L*	*B*	*H*	*T (minimum) For plates having no T-slots*		*R (Min) (for Grade 2 only)*
				Webbed	*Unwebbed*	
1	125	75	100	13	16	8
2	175	100	125	16	19	8
3	250	150	175	22	25	10
4	350	200	250	29	35	10
5	450	300	350	32	38	13
6	600	400	450	35	—	13
7	700	420	700	35	—	15
8	1000	600	1000	40	—	20
9	1500	900	1500	50	—	25
10	2000	900	2200	50	—	25

2.10.1. Standard Sizes of Angle Plates. The general view of the angle plate and its important dimensions for various size numbers as per IS : 2554—1963 are given in Fig. 2.37 and table above.

Grade 1 angle plates of size 3 to 6 when not finished on their interior faces are webbed. Grade 2 angle plates may or may not be webbed. When webs are provided, they are set in slightly from the ends of the plates.

It is recommended that grade 1 angle plates on all their exterior faces, edges and interior faces be finished by either grinding operation or by hand scraping and conform to the accuracies specified in tabular form below. In this case the area of high spots (bearing area) should not be less than 20%. For grade 2, all the exterior surfaces are finished by planing or milling operation and conform to the accuracies specified in the tabular form in Art. 2.10.2.

In the table given below by tolerance on flatness is meant the maximum permissible distance separating two imaginary planes within which the surface under consideration can just be enclosed. (For explanation Refer Art. 7.21).

The tolerance on squareness means the maximum permissible distance separating the two imaginary parallel planes (perpendicular to datum face) within which the surface under consideration can just be enclosed.

The tolerance on parallelism means the maximum permissible distance separating the two imaginary parallel planes (parallel to the datum face) within which the surface under consideration can just be enclosed. (Refer chapter 7 for these definitions.)

2.10.2. Tolerances for Grade 1 and Grade 2 Angle Plates (Refer Fig. 2.37)

Grade	*Size No.*	*Flatness of working faces (i.e. exterior and interior surfaces finished by any process) mm*	*Squareness of exterior faces over dimension H mm*	*Parallelism of opposite faces and edges over their total length mm*	*Squareness of end faces with respect to exterior faces as measured over dimension L mm*
1	1	0.005	0.010	0.013	0.013
	2	0.005	0.013	0.015	0.015
	3	0.008	0.015	0.018	0.018
	4	0.008	0.018	0.020	0.020
	5	0.010	0.018	0.020	0.020
	6	0.010	0.020	0.023	0.023
2	1	0.025	0.050	0.063	0.063
	2	0.025	0.050	0.063	0.063
	3	0.038	0.075	0.088	0.088
	4	0.038	0.075	0.088	0.088
	5	0.050	0.100	0.113	0.113
	6	0.050	0.100	0.113	0.113

2.10.3. Cast Iron Multipurpose Angle Plates

Brief Specifications. The multipurpose angle plate is made of close grained, high quality cast iron of adequate thickness to ensure rigidity. It has ten faces and all are accurately ground, and are 90 degrees true to one another. Casting strains are eliminated either by natural ageing or by artificial seasoning.

Accurately ground step is provided on the inside of angle plates.

Threaded holes are provided on the faces and sides which enable the angle plate to be used as a fixture.

The outer faces of the angle plates have recessed grooves.

Accuracy. (*a*) The working surfaces are square to accuracy of within 0.002 mm in 25 mm length.

(*b*) Straightness of the surfaces is within ± 0.001 mm in 25 mm length.

(*c*) The end faces are square to an accuracy of within 0.005 mm in 25 mm length.

2.10.4. Typical Adaptations of Surface Plates

(*i*) **Concentricity checking attachment** (Refer Fig. 2.38). This is an ideal attachment, which is to be used alongwith Cast Iron/Granite Base Inspection Bench Centres/T-slotted Surface Plates for checking concentricity of stepped and hollow shafts and spindles.

— V-Blocks/Supports are adjustable in all directions.

— Hardened and Ground Rollers per V-Block provides Straight line contacts and could be removed and replaced as they wear out.

— Soft brass rollers can also be supplied for inspection of super finishing surfaces on shaft/bearings.

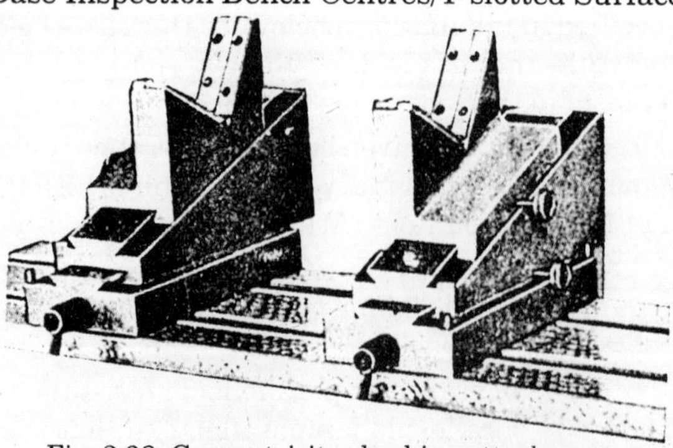

Fig. 2.38. Concentricity checking attachment.

(*ii*) **Clamping Pallets for CNC Machines** (Refer Fig. 2.39). Pallets allow easy, faster fixing and clamping of work pieces on machining centres. They can also be used as sub-plates to provide larger space for work setting and protect the surface of the machine tables.

Salient Features :

— Made from close grained Cast Iron Castings conforming to Grade FG-260 of IS-210 having hardness within 180 to 220 BHN duly stress relieved.

— Provided with hand-scrapped finish for top, bottom and side faces.

— Provided with T-slots of H-8 on top working faces in both the direction and also available with tapped holes at suitable pitch.

— Bottom faces relieved.

— Sides are hand-scrapped for fixing abutment plates.

— Counter bored holes provided for clamping with machine table.

— Flatness and parallelity within 10 microns over 300 mm.

Fig. 2.39

Standard Sizes

Overall size (in mm)	*Details of Tapped Hole*		*Details of T-Slots*	
	Size mm	*Pitch mm*	*Size mm*	*Pitch mm*
400 × 400 × 50	16	80	—	—
500 × 500 × 50	16	100	—	—
630 × 630 × 50	16	125	—	—
800 × 800 × 50	16	160	—	—
400 × 400 × 75	—	—	18	80
500 × 500 × 75	—	—	18	100
630 × 630 × 75	—	—	18/22	125
800 × 800 × 75	—	—	18/22	160

(*iii*) **T-slotted single angle sine tables and T-slotted compound sine tables.** Cast iron T-slotted sine tables (Refer Fig. 2.40) are extensively used with the aid of slip gauges for precise checking of tapers and also as a fixture for precision machining and grinding operation in machine shops, tool rooms and inspection shops and standard rooms.

Salient Features :

— Generally made as per IS-5339-70 (single angle) and IS-5943-70 (compound).

— Base and swivelling face are made from high duty close grained cast iron castings duly stress relieved, conforming to grade FG-200, of IS-210, having hardness within 170-220 HB.

— Working faces are offered in ground/hand scrapped finish so as to achieve requisite accuracies.

— Roller and slip gauge platform made from alloy steel hardened and ground to 55 to 60 HRC.

Standard Sizes

Table Dimn (mm)	*Base Dimm (mm)*	*C.D. of Rollers (mm)*
150 × 100	180 × 150	100
200 × 150	230 × 200	150
250 × 150	280 × 200	200
300 × 200	330 × 250	250
400 × 250	430 × 300	300

Fig. 2.40. T-slotted single angle sine tables.

Fig. 2.41. T-slotted compound sine tables.

(*iv*) **Modular Fixturing System with Hole Metrix/T-slot Type.** This clamping system consists of three basic units :

Pallets—Clamping Angle—Clamping Cube (Refer Fig. 2.42)

Basic components are made of stress relieved wear resistance cast iron having high damping properties. Supporting and clamping surfaces are precision hand-scrapped.

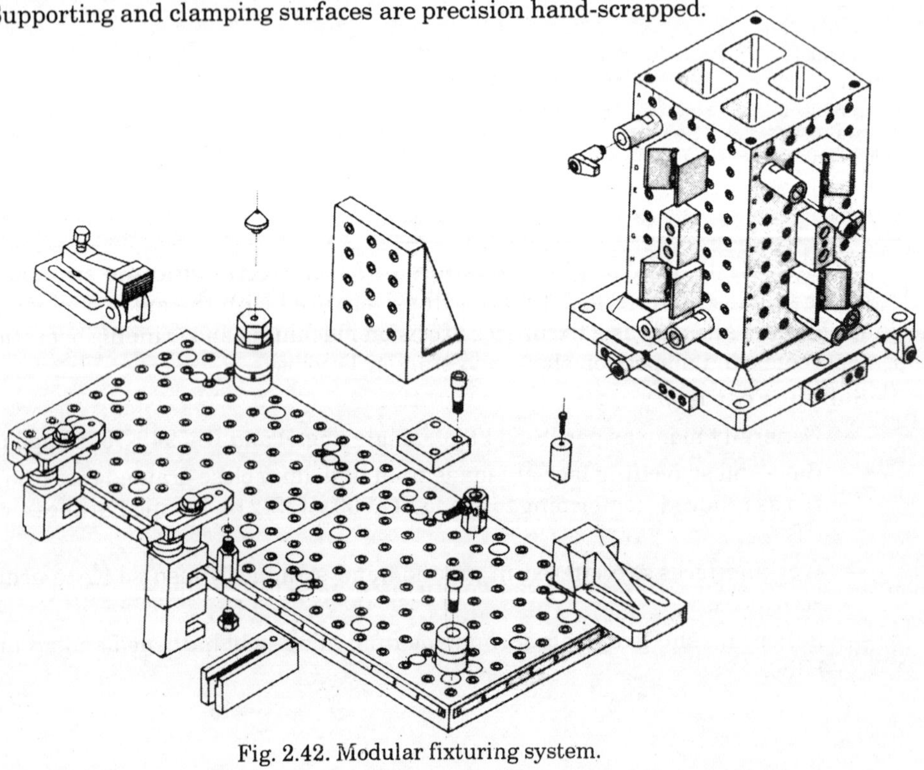

Fig. 2.42. Modular fixturing system.

A large variety of stoppers, locators, support blocks, V-blocks, tenons, isolators, threaded bolts, conical sockets, nut, washers, raiser blocks, strap clamps, etc. are provided for allowing easy positioning and clamping parts while machining thereby allowing heavy flexibility and cost reduction for fixturing.

(*v*) **Cast iron Tilting Tables** (Refer Fig. 2.43). Cast Iron tilting table is a useful accessory for setting work at an angle for milling or other operations. The table has machined T-slots along length and can swing upto 45 degree on either side. For setting purpose a degree scale, reading minimum one degree, is graduated on the end face of cylindrical base itself. This ensures better setting accuracy. The table is set manually to the required tilt and is locked in the set position by means of two locking clamps held through bolts.

Fig. 2.43. Cast Iron tilting table.

(*vi*) **Precision Sine Vice** (Refer Fig. 2.44). Hole and pin type, solid steel hardened, ground vice is mounted on hardened, ground rollers. The vice can be inclined in the desired position with the help of side strap clamp provided. The recess of 1 mm depth is made in base plate to measure small angles.

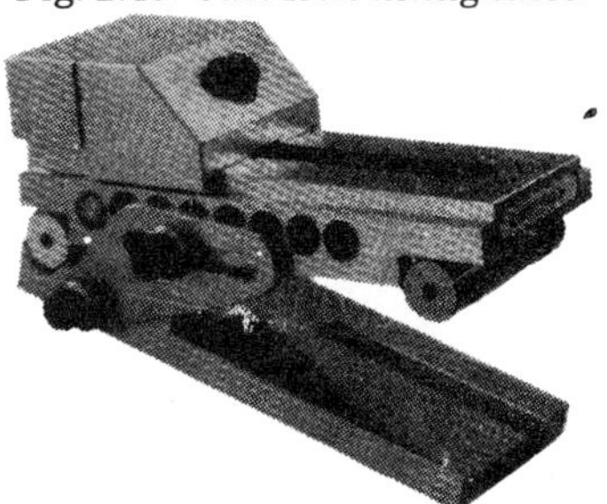

Fig. 2.44. Precision sine vice.

2.10.5. Method of testing flatness of plate by comparative method with the surface gauge. This method is employed when smaller plates upto size 1 are to be tested quickly. For this method, it is essential that a fairly large plate of good quality and known accuracy be available. The smaller plate to be tested is placed upon the larger one with a small tool maker's jack under each of its three feet. A sturdy surface gauge having a good flat base of ample size and fitted with 0.001 mm indicator is used as the comparator. The contact point of the indicator is provided with a swivelling pad about 20 mm in diameter in order to straddle over the scraping marks. The surface gauge is placed upon the larger plate with the indicator resting on the smaller one. The small plate is set parallel to the base in the regions above three supporting feet by using the surface gauge and the jacks. After that the surface gauge is moved about so as to pass the indicator over the surface of smaller plate and the variations in the indicator reading, if any, give a direct measure of the errors in flatness of the smaller plate.

2.10.6. Determination of the proportion of bearing area. In order to determine the proportion of the bearing area of a scraped plate its surface is first blued and rubbed with another plate so that the small bearing areas are brought up clearly into view. A small glass plate on which an area of 50 × 50 mm has been ruled into 400 smaller squares (2.5 mm × 2.5 mm) is then placed upon the surface. (This glass plate can be produced like lantern slides by photographing a chart drawn on paper). Even small square is then observed in turn and a note made of the estimated fraction of its area (in tenths) which is occupied by high spots on the surface underneath. The addition of all these fractions when divided by four gives the percentage of the bearing area over the region tested. In order to obtain a fair average figure the test is repeated at other positions on the surface also.

2.11. Cast Iron Surface Plates with Handles

Brief Specifications. Cast Iron Surface Plates with handles are made of close grained high quality iron.

Casting strains are eliminated either by natural ageing or by artificial seasoning.

Grade A : Handscraped and spotted.

Accuracy : ± 0.002 mm per 240 mm length.

Grade B : Ground handscraped.

Accuracy : ± 0.008 mm per 300 mm length.

2.12. Cast Iron Cubes

Brief Specification. The cast iron cubes are made of high quality closed grained cast iron, seasoned and accurately ground all over.

Accuracy. (*a*) The working surfaces are square to an accuracy of within 0.002 mm in 20 mm length.

(*b*) Straightness of the working surfaces is within ± 0.001 mm in 20 mm length.

(*c*) The end faces are square to an accuracy of within 0.004 mm in 20 mm length.

2.13. V-Block

These are widely used for workshop and inspection purposes for checking out the roundness of cylindrical workpieces and for marking centres accurately, etc. Generally the angle of V is 90°

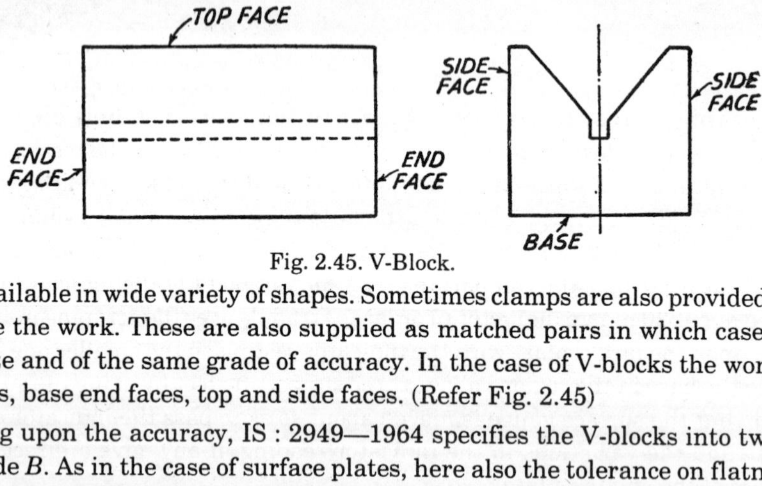

Fig. 2.45. V-Block.

and these are available in wide variety of shapes. Sometimes clamps are also provided which bridge the vee to secure the work. These are also supplied as matched pairs in which case two V-blocks are of similar size and of the same grade of accuracy. In the case of V-blocks the working surfaces are flanks of vees, base end faces, top and side faces. (Refer Fig. 2.45)

Depending upon the accuracy, IS : 2949—1964 specifies the V-blocks into two grades, *viz.* grade *A* and grade *B*. As in the case of surface plates, here also the tolerance on flatness is defined as the maximum permissible distance separating two imaginary parallel planes, within which the surface under consideration can just be enclosed. Similarly the tolerance on squareness is defined as the maximum permissible distance separating two imaginary parallel planes, (which are perpendicular to the datum face of the part in question) within which the surface under consideration can just be enclosed.

V-blocks are furnished by hand scraping process till they have a bearing area of not less than 20%. Any departure from flatness in case of vees should be a convexity and on other surfaces it should be a concavity. In a direction parallel to the axis of the vee the departure from flatness should be a concavity and not convexity.

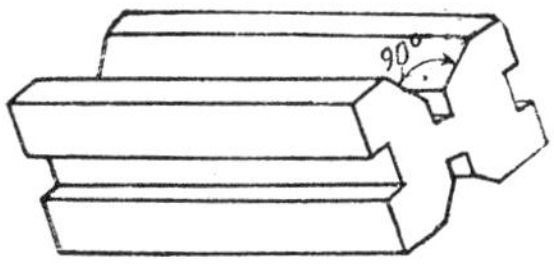

Fig. 2.46

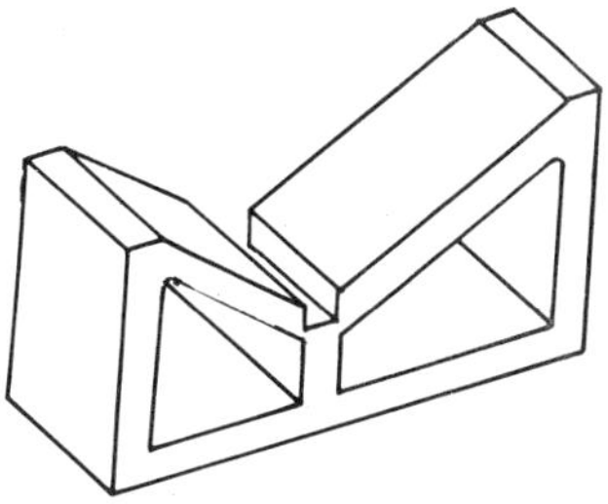

Fig. 2.47

Various forms of vee-blocks.

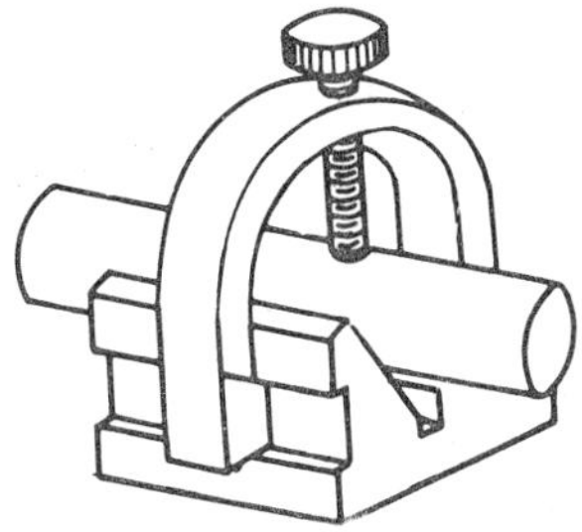

Fig. 2.48. Vee-Block with stirrup clamp.

Two types of V-blocks commonly used are shown in Fig. 2.46 and 2.47 Fig. 2.46 shows a V-block having two vees, one of them being deeper and wider than the other. Fig. 2.47 shows a larger vee block having one vee only. For special purposes such as checking triangle effects or taps and other three-fluted tools, 120-degree included angle vee-blocks are also available.

The major purpose of the vee-blocks is to hold cylindrical pieces, or move to the point, to establish precisely the centre line or axis of a cylindrical piece. In using a vee-block, it is very essential that the cylindrical piece should rest firmly on the sides of the vee and not on the edges of the vee as in Fig. 2.50.

Before using a vee-block, first check it visually, or with the finger nail, for dents, edge nicks, scratches and burrs which might either damage the surface plate or cause the vee-block to tilt imperceptibly from true flat or vertical.

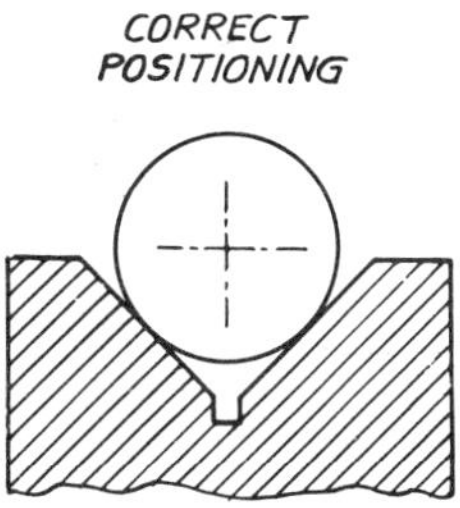

Fig. 2.49. Workpiece resting firmly on the side of the vee.

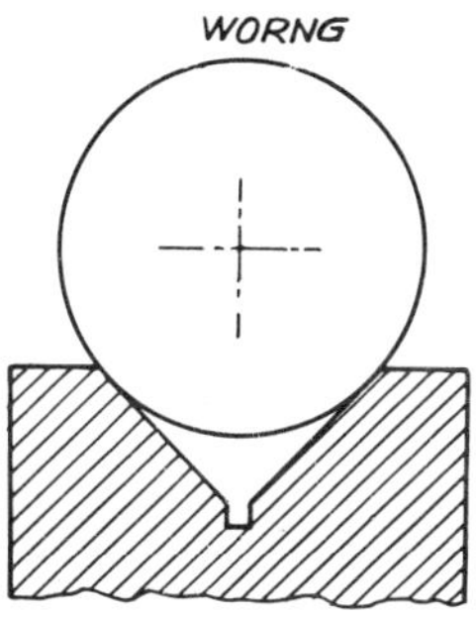

Fig. 2.50. Workpiece resting on the edge of vee.

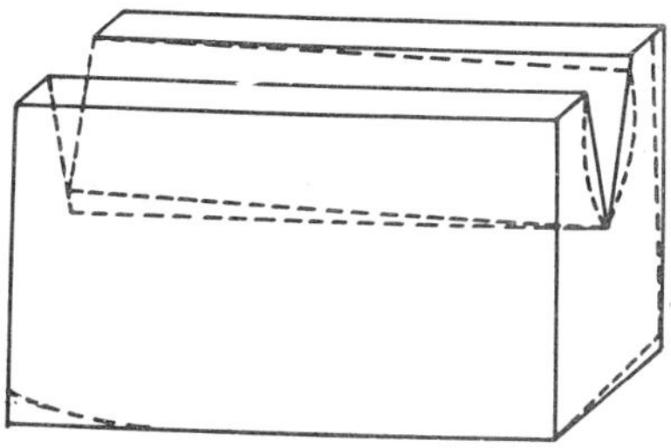

Fig. 2.51. Various inaccuracies in a vee-block

Using a vee-block correctly.

Any vee-block should be checked periodically for basic accuracy. If it has rusted, or worn, or warped a little ; or if it were made in-accurately ; its four sides and two ends may not be truly parallel or square to each other. The vee channel may be out of parallel. Constant use with the same size cylindrical workpieces wears hollows in the vee-sides. Some of possible types of inaccuracies discoverable in a vee-block are indicated by the dotted lines in Fig. 2.51 with a dial indicator.

2.13.1. Cast Iron Vee Block. *Brief Specifications.* The Vee Blocks are made of close grained high quality cast iron. They are accurately ground and numbered in pairs, so that the Vee-Grooves to the same numbers are always in alignment.

Accuracy. (*a*) Accurately ground to 90 degrees square.

(*b*) Straightness within ± 0.001 mm per 20 mm length.

(*c*) Symmetisation of Vs within 0.002 mm per 20 mm length.

2.14. Bench Centres

Bench centre consists of a rigid cast iron base made from close grained cast iron having minimum hardness of 180 HB (free from distortion, porosity and other casting defects). The base is provided with suitable *T*-solts for the attachment of the centre holders. The centre holders can be located in any desired position. The male or female centres are adjustable and can be locked in any position along the ground vees of the centre holder. These are used for checking eccentricity of rotary components between centres.

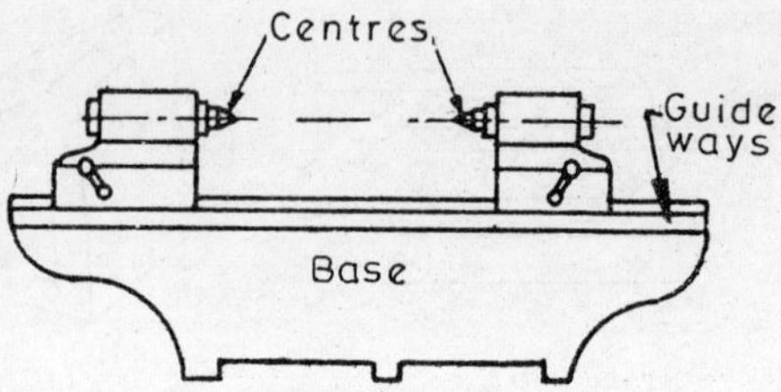

Fig. 2.52. Bench centre.

These are available with heights of centres 125, 160, 200, 250 and 300 mm. For easy loading and unloading, one of the centre is spring loaded. The distance between the centres may vary from 500 mm to 1500 mm depending on the height of the centres. The coaxiality of centres and parallelism of axis of centres with respect to guideways should be ensured within close tolerances.

The permissible deviations in parallelism of the axis of centres with respect to guideways, as per IS : 1980—1978 are 0.01 mm per 300 mm, and 0.015 mm/300 mm respectively leaning towards the free end of mandrel for centre heights of 125—160 mm and 200—300 mm respectively. The corresponding permissible errors in coaxiality of centres are 0.01 mm and 0.015 mm respectively over any length of 300 mm.

2.15. Straight Edges

These are used for checking the straightness and flatness of parts in conjunction with the surface plates and spirit levels. These may be made of steel or cast iron. Steel straight edges are available up to 2 m length and may be rectangular in section with bevelled edge. C.I. straight edges are made up to 3 m length and widely used for testing machine tool slide ways. They are heavily ribbed and bow-shaped (camel back construction) to prevent distortion. These are provided with feet for rest when they are idle to prevent distortion. Feet are placed at points of minimum deflection.

The straight edges are classified as follows : (Refer Fig. 2.53)

(*i*) Tool-maker's straight edge,

(*ii*) Wide-edge straight edge,

(*iii*) Angle straight edge,

(*iv*) Box straight edge.

Tool maker's straight edges have the highest accuracy and are available with various cross-sections, from one to four working edges and length from 75 mm to 175 mm. Straight edges with a single edge are used for checking straightness by sight test. The test consists of applying the straight edges along the full length of the surface (whose straightness is to be checked) against a bright back-ground. The absence of light between the straight edges and the surface indicates the straightness of element. Deviation of 1 or 2 microns from the straightness can be detected by the eye with some practice.

Flatness of a surface may be tested with single-edged straight edge by applying it in different directions at different places on the tested surface. The flatness of the surface is judged by the light showing through. A surface can also be tested by means of three and four edges type straight edges by applying a light coat of prussian blue on the working edges and then drawing the straight edge across the tested surface. Traces of the marking compound are rubbed in this way on the tested surface and due to irregularities in the surface it will be coated in spots with different density. All the high spots are painted more densely and the low spots are partly painted. The surface is then

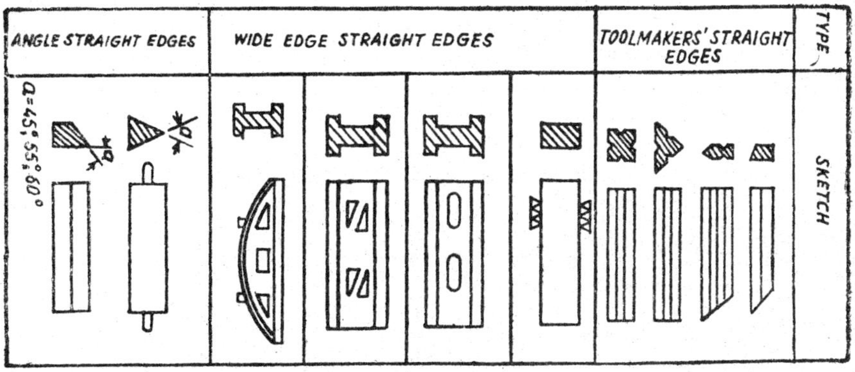

Fig. 2.53

scraped or ground until a uniform distribution of spots on the whole surface is obtained after subsequent tests.

Straight edges with a wide working edge are used for testing large surfaces or surfaces with large intermediate gaps or recesses. Such straight edges may have a length upto 0.6 metre, and have a form of a rigid beam or frame in order to possess ample of rigidity to maintain the straightness.

Angle straight edges are used for checking surfaces positioned at an angle in relation to each other. Such straight edges have a triangular or trapezoidal cross-section and are from 250 to 1000 mm in length and usually have handles at their ends to facilitate their application. Surfaces are usually tested with a marking compound as described above.

The recommended lengths of straight edges for bow shaped and I-section type are as follows. (Refer Figs. 2.54 and 2.55)

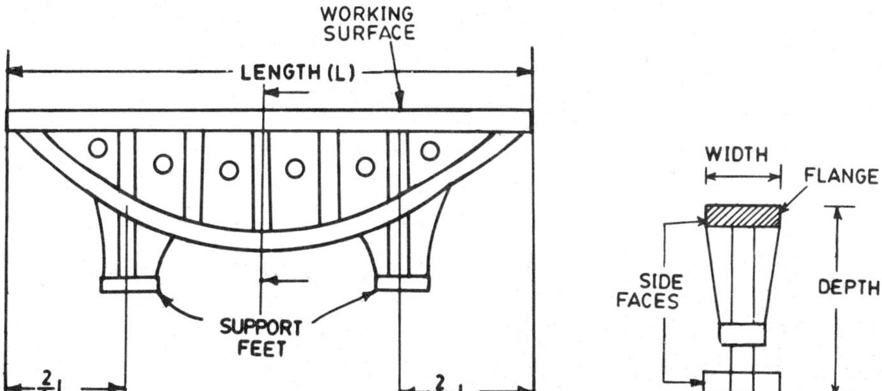

Fig. 2.54. Bow-shaped straight edge.

Lengths of bow-shaped straight edges (L) : 300, 500, 1000, 2000, 4000, 6000 and 8000 mm.

Lengths of I-section straight edge (L) : 300, 500, 1000, 2000, 3000, 4000 and 5000 mm

Bow-shaped straight edge has two feet of the same width as the working surface and these are so located as to allow the straight edge to be used in 3 positions *viz.* resting upon the feet ; with the working surface downward ; and lying on either side. The accuracy requirements specified for straight edge are applicable for all these three positions.

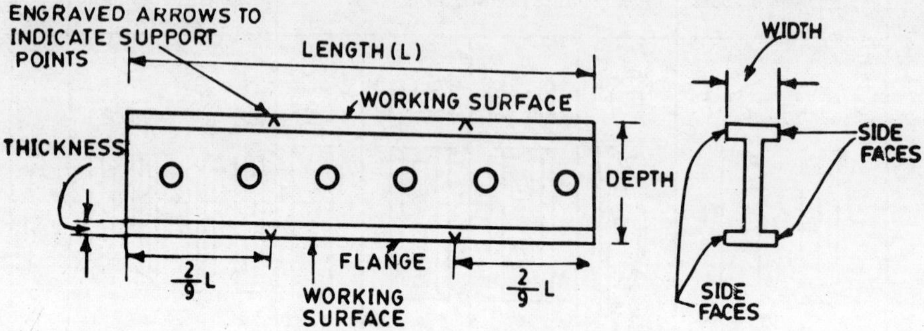

Fig. 2.55. I-section straight edge.

Usually two grades of accuracy (Grade A and Grade B) are specified. The working surface and supporting feet of grade A straight edges are finished by scraping, and the working surfaces of grade B and the side faces of all straight edge, are finished by smooth machining.

The proportion of bearing area of the working surface should not be less than 20% and 10% respectively for grades A and B ; High spots should be uniformly distributed and the percentage of bearing area should not be so high as to cause wringing. All sharp edges should be removed and the unmachined parts painted.

The bearing area of the working surfaces should be flat to 0.001 mm per 100 mm of working length for grade A and 0.002 mm per 100 mm of working length for grade B.

When a bow shaped straight edge rests on a flat base, the depths from working surface of the straight edge to this base, measured immediately over each support foot, shall be equal to each other to within the tolerances of 0.0001 mm per 100 mm of working length.

Any 300 mm length of each side face has to be flat to within 0.025 and 0.050 mm respectively for grades A and B respectively.

The maximum tolerance on flatness over the whole length of straight edges is 0.05 mm for grade A and 0.1 mm for grade B for lengths upto 4000 mm, and 0.1 mm and 0.2 mm (grades A and B) respectively for lengths above 4000 mm.

The working surface and side faces have to be square to within 0.008 mm per 25 mm and 0.015 mm per 25 mm for grades A and B respectively.

2.15.1. Accuracy. According to IS : 2220—1962, straight edges have been divided into two grades depending upon the accuracy of straightness (grade A and grade B). Grade A is intended to be used for inspection purposes, and grade B for workshop purposes during manufacturing. As it is generally more difficult to machine the ends of the straight edge accurately, it specifies two lengths, *i.e.* working length and total length. The total length is greater than the working length and the inspection is made without using the ends of the straight edge. This standard recommends rectangular cross-section for total lengths upto 180 mm and I-section beyond total length of 340 mm. In case of I-section used for bigger lengths, the side face is slightly relieved in order to reduce the weight. The straight edges beyond 340 mm total length are provided with drilled holes at uniform intervals in order to reduce the weight still further. Generally the straight edges are supported on two such points which results in the minimum natural deflection and these are at distance of 2/9 total length of the straight edge from each end. When the straight edge is supported at the points of best support, the error in the straightness of the working faces over its whole length should not exceed $(2 + 10L)$ microns for grade A and $(5 + 20L)$ microns for grade B ; where L is the length in metres. In addition, the error over any length of 300 mm should not exceed five microns for grade *A* and eight microns for grade *B*. Further the maximum error in parallelism should not exceed 1.5 times the tolerance on their straightness. Further when the straight edge is supported at its extreme points, then the natural deflection should not exceed 10 microns per metre of length.

It is also essential that the side faces of the straight edge be straight and parallel to each other. The maximum error that can be tolerated in the straightness of the side faces is ten times the error in straightness of working faces and the maximum tolerable error in the parallelism of the two side faces is 15 times the error in straightness of the working faces. Further, the side faces should be square also with the working faces to within one micron per millimetre of depth for grade *A* and two microns per millimetre of depth for grade *B* straight edges.

The accuracy of straightness of the working faces of a straight edge is tested by direct comparison with a master surface plate whose accuracy is already known. Two equal slip gauges are placed on the surface plate just below the points of best support and straight edge on its edge placed over the slip gauges. Then the gap between the lower working face of the straight-edge and the surface plate is measured at various points by fitting slip gauges of appropriate length, which gives an indication of the degree of parallelism along the length of straight edge.

2.15.2. Tool Maker's Straight Edges. These are intended for very accurate work and are available in the range of 15 to 300 mm length for tool room purposes. These have one working edge, which is bevelled and very slightly rounded off (commonly known as a knife-edge).

These are made of suitable cast or wrought alloy steel free from defects and hardened, and are suitably heat treated to remove the internal stresses resulting from the hardening process and to give stability. The hardness adjacent to working edges is of the order of 650 HV.

The faces adjoining the working edges or the straight edges are ground flat. The working face is very slightly rounded off to furnish the 'knife-edge'. The working edges are finished by

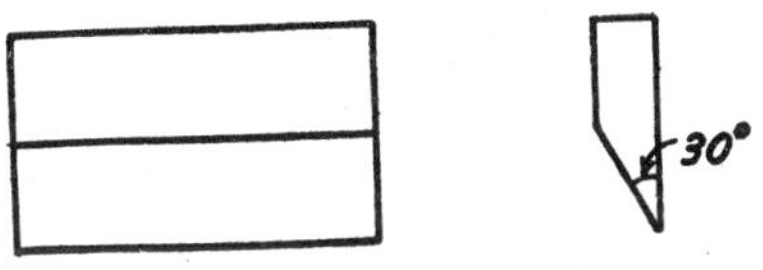

Fig. 2.56. Lengths 15 and 25 mm.

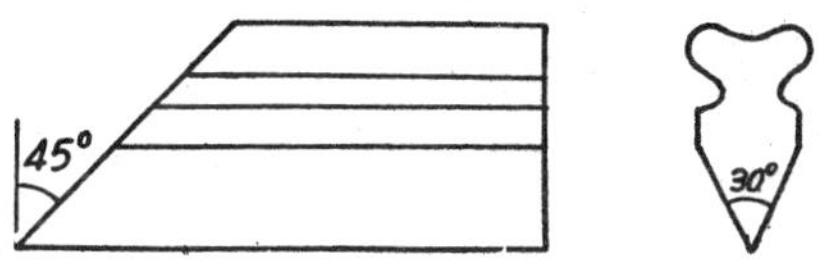

Fig. 2.57. Lengths 50 to 100 mm.

grinding and lapping and radius on the edge is made uniform from end to end. The sections of the straight edges for length 15 and 25 mm, 50 to 100 mm and 150 to 300 mm are shown below. The section of the straight edges for length between 25 and 50 mm can be either of shape given in Fig. 2.56 or Fig. 2.57. Similarly the sections of the straight edges for lengths between 100 mm or 150 mm can be either of shapes given in Fig. 2.57 or Fig. 2.58.

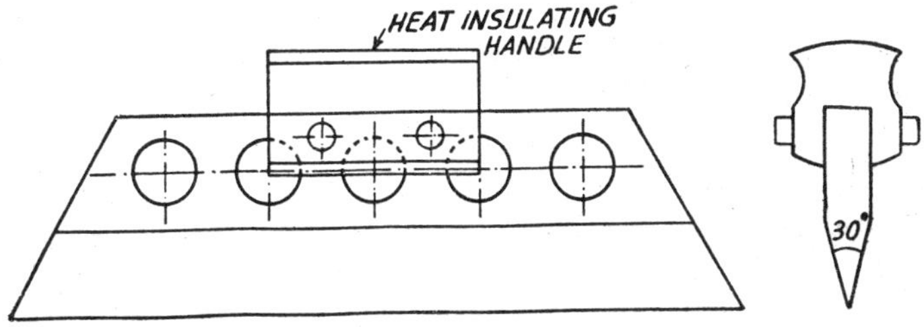

Fig. 2.58. Lengths 150 to 200 mm.

Recommended dimensions of straight edges are

Length *mm*	*Width* *mm*	*Thickness of flat portion* *mm*
15	15	3
25	25	6
50	30	8
75	35	8
100	40	10
150	40	10
200	50	12
250	50	12
300	60	12

The accuracy of straight edge should be such that when placed against a well-illuminated background on the tool maker's flat or plate no white light should be visible. This applies for angle upto 10° on either side of the normal.

For checking purposes, usually a flat back glass test plate is also provided with straight edges.

2.15.3. Steel or granite straight edges. These are available in lengths of 300, 500, 1000, 1500 and 2000 mm with corresponding depth and thickness varying from 40 to 125 mm and 6 to 13 mm for grade *A* steel straight edges, 30 to 80 mm and 5 to 12 mm for grade *B* steel straight edges, and from 75 to 150 mm and 25 to 50 mm for granite straight edges.

The handling slots are provided $\frac{2L}{9}$ away from each end as shown in Fig. 2.59. The side surfaces can be slightly relieved (I-section) to reduce weight.

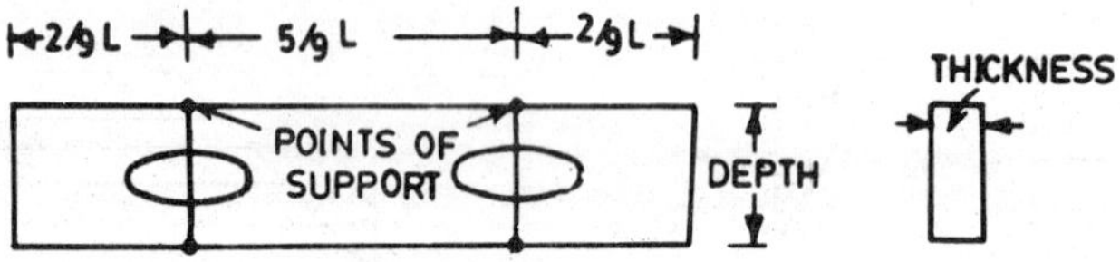

Fig. 2.59

Important accuracy requirements for these straight edges are :

(*i*) straightness tolerance of working face

(*ii*) tolerance on parallelism of working face

(*iii*) combined flatness and parallelism of side faces

(*iv*) squareness of side faces to working faces.

2.15.4. Generation of Straight Edge. A method commonly used for generation of straight edges requires two blanks clamped together side by side and the edges machined. These edges are then smeared with a marking compound and while still together, rubbed with a small flat block as shown in Fig. 2.60. This reveals any high spots, which are removed. The process of marking and correcting is continued until the edges are accurately matched. The plates are then separated and placed edge to edge as illustrated in Fig. 2.61. Marking compound is applied to one of the edges and the high spots located by relative sliding of the two plates. The widths of the gaps between the two edges are equal to twice the error in straightness as shown exaggerated in Fig. 2.62, which helps to locate the high spots more accurately. After matching in this position, they are again placed side by side and corrected for high spots. Testing edge to edge and side by side is repeated until the required accuracy of straightness is attained. A final check is made by reversing one of them end

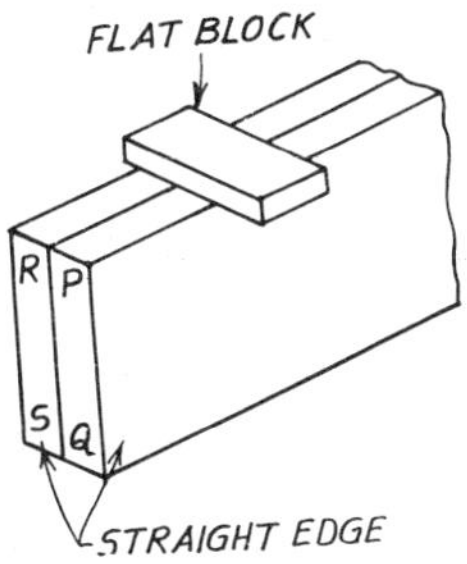

Fig. 2.60

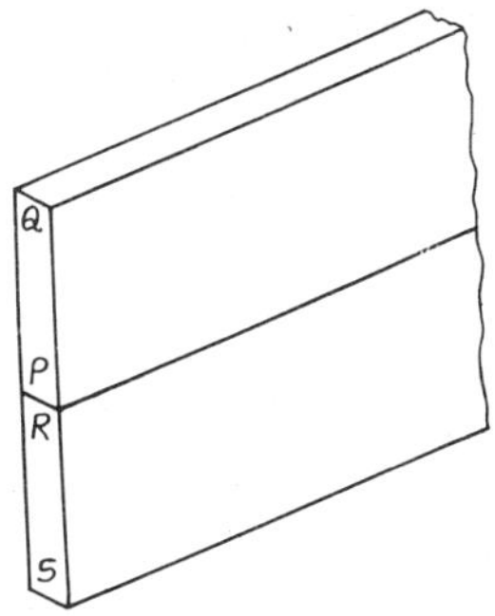

Fig. 2.61

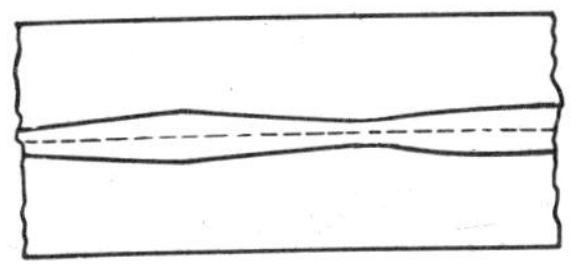

Fig. 2.62

Method of generating straight edges.

for end and subjecting it to be corrected by scraping. For hardened steel, a small stone or lap must be used, the latter being preferable during the concluding stages, when only small amounts of metal are to be removed.

2.15.5. Box Straight Edge. Fig. 2.63 shows a side view of box type straight edge and the method of creating a perfect box. The Important features of such a straight edge are : rigid cross section to minimise deflection in all directions, a symmetry and uniformity which minimises distortion when the straight edge is subjected to uneven temperature distribution ; a 90° form and straightness which are self-proving both in fabrication and in use.

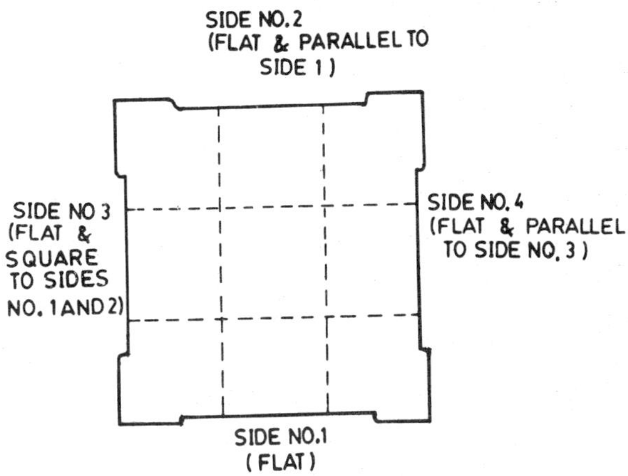

Fig. 2.63. Box straight edge.

For scraping the master straight edge, first side 1 is matched to the master surface plate (Refer Fig. 2.64). Side 2 is also matched and, by means of a thickness check, is scraped parallel to side 1. Inspection is done using master surface plate as a flat datum and an indicator stand with knife edge. Side 3 is scraped square to sides 1 and 2. The knife edge on the indicator stand is pressed against the lower area of side 3. Any out-of-squareness is revealed as half the difference of the indicator readings when the straight edge is turned over (18°), to side 2 and compared again. For highest accuracy and in order to avoid local flatness errors in the master surface plate, the straight edge should be turned end-of-end, as well as through 180° so that the position of the indicator on the surface plate is not altered.

Once 3 is completed, side 4 is scraped parallel to it, which means that side 4 is automatically square to side 1 and 2.

Thus both straightness and squareness are developed from the surface plate.

2.15.6. Using a Straight Edge. In fairly coarse measuring situations the deviation of a surface from the mean true plane can be approximately determined, as shown in Fig. 2.65. The irregular gap between the straight edge and the surface being tested can be determined by the use of feeler gauges as shown in Fig. 2.65.

Small components can be tested with a precision straight edge as shown in Fig. 2.66. The light box provides a means of uniform illumination that is independent of the angle of viewing. Any error of flatness in the component will leave a gap through which a strip of light will be visible between the straight edge and the component. With practice the error of straightness can be estimated by the 'quality' of the light seen.

The colour of light is white for gap greater than 0.002 mm and tinted for gap between 0.001 and 0.002 mm.

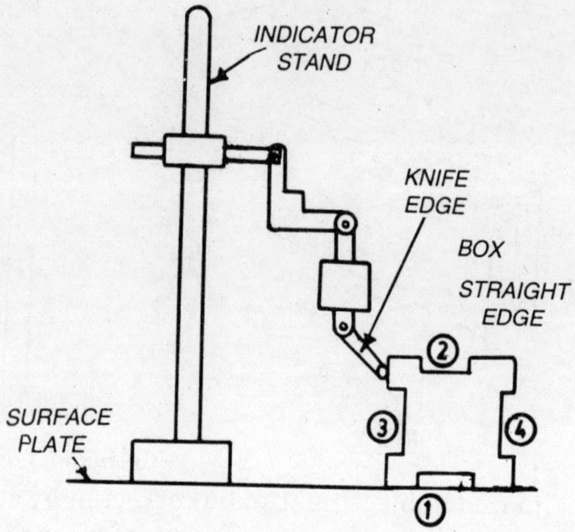

Fig. 2.64. Developing squareness using surface plate.

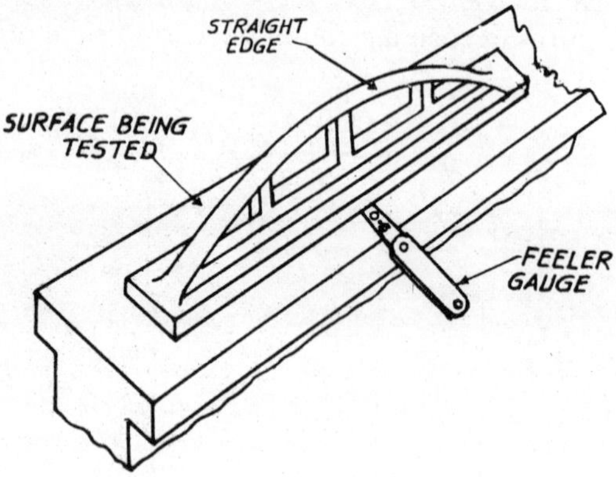

Fig. 2.65. Use of straight edge in conjunction with feeler gauges.

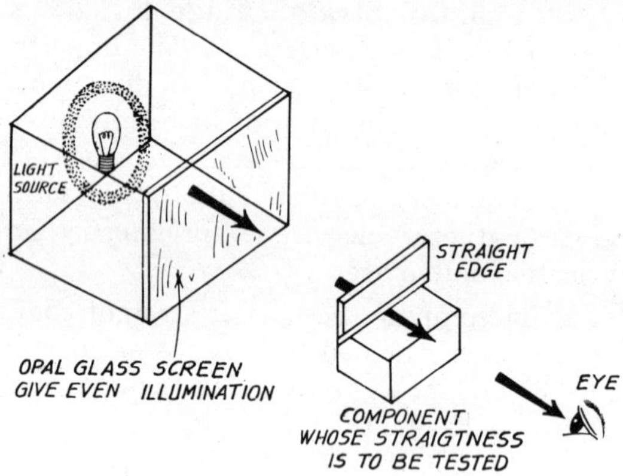

Fig. 2.66. Use of straight edge in conjunction with light box.

2.16. Spirit Levels

Spirit levels are used for measuring small angle or inclinations and also enable the position of a surface to be determined with respect to the horizontal.

A spirit level consists of a sealed glass tube, ground on its inside surface to a convex form with a large radius of curvature R. A scale is engraved on the glass at the top of the tube. The tube

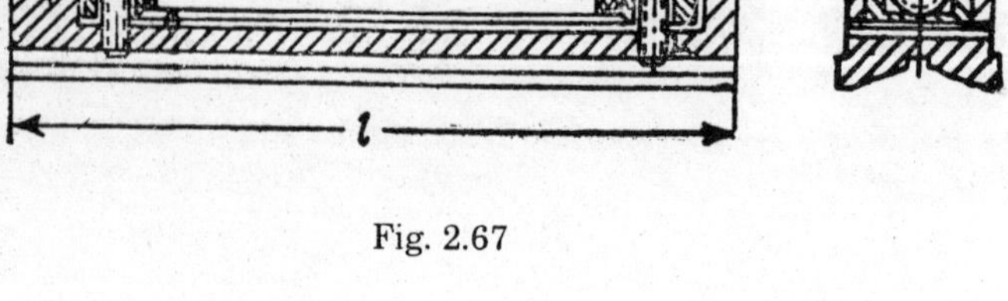

Fig. 2.67

Fig. 2.68. Frame spirit level.

is nearly filled with ether such that only a small volume remains at the top part of the tube which contains ether vapours in the form of a bubble.

The machinist's level consists of a body with a flat base surface and the glass tube (described above) mounted in the upper part of the body. (Refer Fig. 2.67)

For checking the vertical surfaces, a frame level shown in Fig. 2.68 is employed.

The side edges of the frame level are made strictly square with the base. A glass tube filled with ether is mounted in the base. For checking the vertical surfaces, the side edge of the frame level is placed into exact contact with the surface and reading of the bubble noted down. The position of the frame level in transverse direction is checked by another less accurate tube.

2.16.1. Principle. The glass tube is set in the base and adjusted in such a way that when the base is horizontal the bubble rests at the centre of the scale which is engraved on the glass. When the base of the level is moved out of the horizontal, the bubble tries to remain at the highest point of the tube and thus moves along the scale. The relations between the movement of bubble and other conditions involved are as follows :

In Fig. 2.69, B is the top of the tube radius and the position of the bubble when the base is at OA (horizontal). If the base is tilted through an angle α and base occupies position OA', the bubble will move a distance 1 to B', where angle $BOB' = \alpha$. If R is the radius of the tube then

$$\text{arc } l = R\alpha$$

If L is the length of the base and h is the difference in height between its ends, then for small value of h,

$$h = L\alpha$$

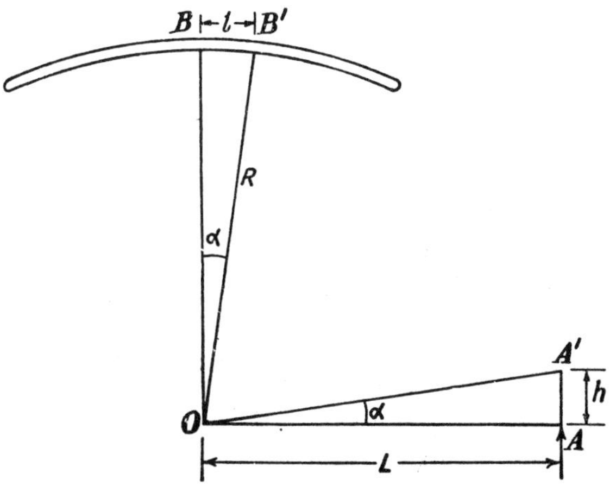

Fig. 2.69

$l = \dfrac{Rh}{L}$ and if α is taken in seconds, then $l = \dfrac{R\alpha}{206,265}$

(One radian equals 206, 265 seconds of an arc)

From above relation it is obvious that the sensitivity of the level increases as R increases.

The scale spacing or the distance between adjacent graduation is generally about 2 mm and thus for

$$R = 206 \text{ m.}$$

$$\alpha = \frac{2 \times 206,265}{206,000} \simeq 2 \text{ seconds.}$$

The inclination of 2″ causes bubble movement of 2 mm. This is the sensitive spirit level and is recommended for research laboratory.

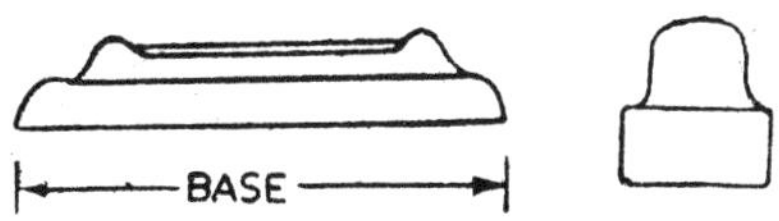

Fig. 2.70. Type 1 level.

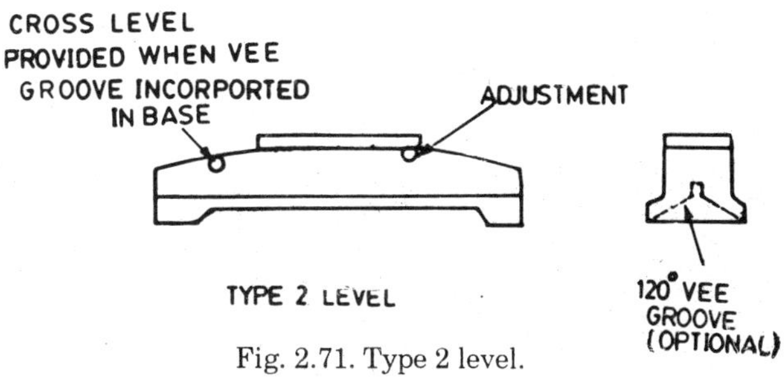

Fig. 2.71. Type 2 level.

For highly precise shop measurements, spirit levels with scale division value of 4 to 10 are employed. For ordinary purposes, scale division values of the order of 10 to 40 are sufficient.

It may be noted that spirit levels are very sensitive to variation in the temperature of their surroundings, since they change the tension of ether vapours in the tube. As such the spirit level should be used in controlled temperature conditions, or else precise spirit levels equipped with various devices that compensate for errors due to temperature changes be used.

2.16.2. Types of Spirit levels. According to *BS* 1958, three types of spirit levels are recommended.

Type 1. (Refer Fig. 2.70) This type of level has an unrelieved flat base of steel, hardened and lapped. The advantage of lapped base is that the effective length of the level can be varied by wringing two gauge blocks on the base at the desired distance apart. The base length of spirit level varies between 100 to 200 mm.

Type 2 (Refer Fig. 2.71) level is mounted in cast iron or steel body, (with base length 250-500 mm) having a base formed with feet bearing surfaces at the two ends (middle portion being relieved). The bearing surface may be plain or contain a longitudinal 120° vee groove for use on cylindrical surfaces, in which case a short cross-level or small circular level is provided.

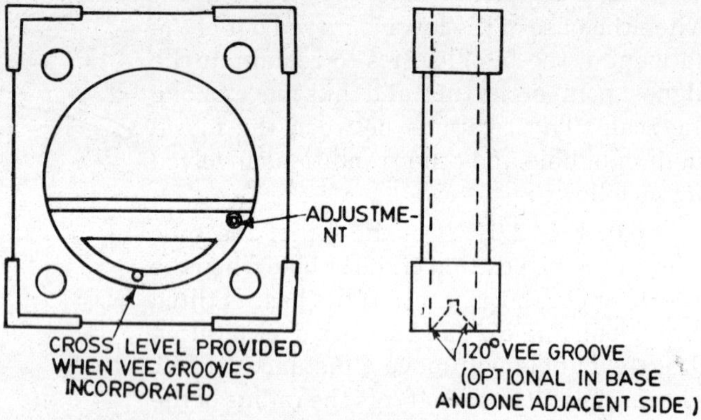

Fig. 2.72. Type 3 level.

Type 3 (Refer Fig. 2.72) level is a square block level about 200 mm square and made of cast iron. The four bearing surfaces are flat and may have the middle portion relieved. Alternatively, the base and one adjacent-surface may contain a longitudinal 120° vee groove for use on cylindrical surfaces in which case a short cross-level is provided.

2.16.3. Electrical Level. This is a type of precision level in which the usual spirit vial is replaced by a pendulum whose displacements are electrically amplified and shown on a separate meter.

2.16.4. Electronic Precision Levels with Digital Readout. For levelling and checking flat and cylindrical surfaces on machinery with high accuracy. Convenient manipulation. Clear and high contrast digital display assures good readability. Large measuring range permits checking of work even when still very much inclined. Therefore one electronic precision level will replace several levels of conventional design having different accuracies. Levelling of machines or surface plates is thus considerably simplified and accelerated.

Housing made of high quality, wear resistant cast iron. V-surface and flat contact surface precision ground. With fine setting of digital display. (Refer Fig. 2.73)

Typical Specification

Length	200 mm
Numerical division	0,01 mm/m
Measuring range	± 1,99 mm/m
Digital display	LCD
Stability of indication after	3 sec
Max. measuring error	≤ ± 5% of indicated measuring value
Repeatability	≤ 5 μm/m
Range of zero setting	± 1,75 mm/m

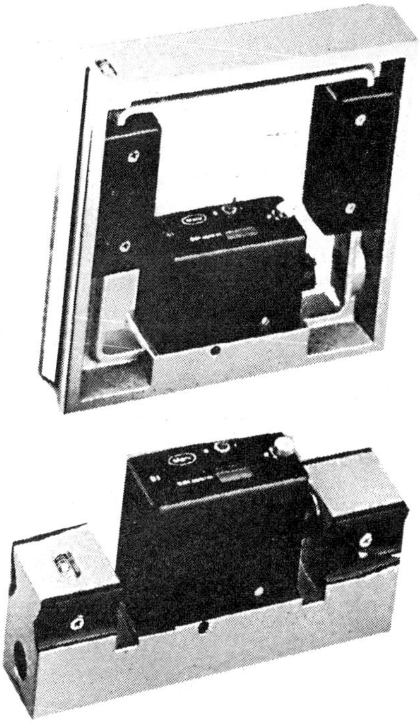

Fig. 2.73. Electronic precision level.

2.17. Combination Set

This is the most adaptable and commonly used non-precision instrument to be used in layout and inspection work (Fig. 2.74).

The combination set consists of scale, squaring-head, protractor and centre-head. It consists of a heavy scale, which is grooved all along its length. It is on this groove that sliding squaring head is fitted. One surface of the squaring head is always perpendicular to the scale and it can be adjusted at any place by a locking bolt and nut. The squaring head also contains a spirit level which is used to test the surfaces for parallelism. For laying out dovetails an included angle is also mounted on the scale. It can also slide to any position and be locked there. A scribing point is also inserted into the rear of the base for scribing purposes. The squaring head and scale can be used for height and depth measurements, inside and outside squaring operations.

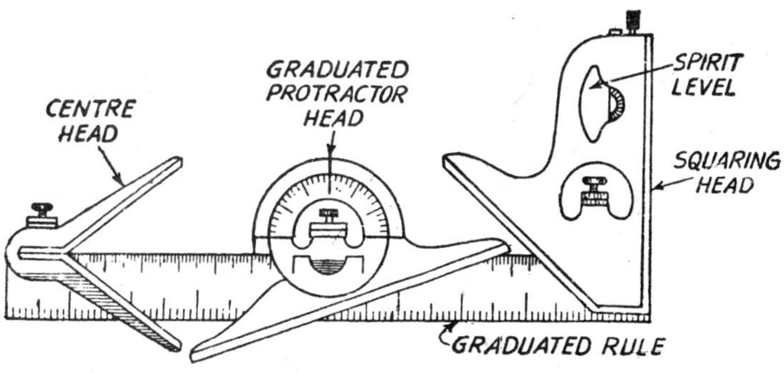

Fig. 2.74. Combination Set.

The centre head attachment is used with the scale to locate the centre of bar stock. For locating the centre, the part to be centred is held beneath the scale and against the sides of the centre head. Three lines are drawn along the scale by rotating the part every time. Thus three lines given three points of intersection. The middle point of the triangle is the centre of part.

The protractor is also capable of sliding along the scale. It contains a semi-circular disc graduated from 0 to 90° on either side of centre. For the purpose of levelling a surface or checking an inclined plane the protactor head also contains a spirit level. With the help of protractor, the correct angle of the work can be set on it and deviation of work from this angle can be checked by inserting a feeler gauge between the blade and the part to be inspected.

2.17.1. Combination Sets with square head, center head and nonreversible protractor head and blade. (Refer Fig. 2.75) With reversible lock bolts, scriber, spirit level in both square head and protractor head, direct reading double 180° protractor scale, and hardened steel, machine-divided blade. Also available with satin chrome blade and protractor head. Cast iron heads with black wrinkle finish.

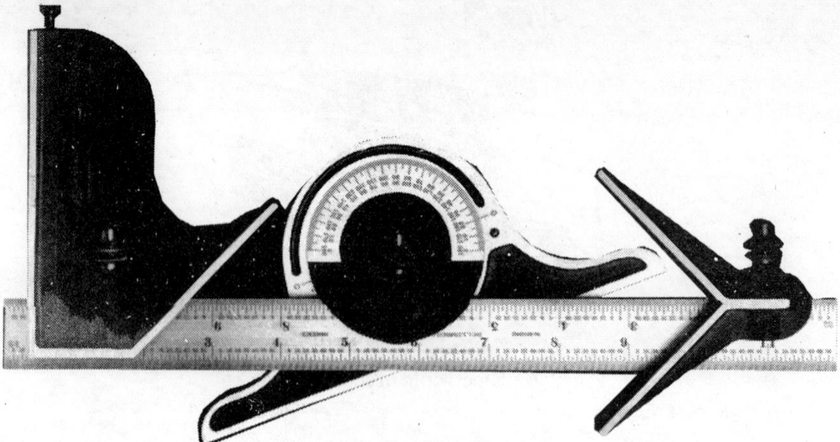

Fig. 2.75. Combination set.

2.18. Universal Surface Gauge

This is a most versatile instrument used in non-precision measurements. It is generally used with surface plate for layout work and inspection. It consists of a rigid base whose bottom surface is made perfectly flat. A spindle carrying a scriber in a universal clamp is attached to the base. The spindle can be inclined to the base in any position and clamped in place by tightening the spindle nut. After tightening it, slight inclination can be further given for finer adjustments by an adjusting screw. Actually the turning of the adjusting screw moves the whole spindle assembly up or down thereby tilting it very slightly. The scriber can be slid up or down the spindle and at the same time tipped back or forth in any position. The scriber can thus be set at any position desired. The base of the universal surface gauge is generally grooved so that it may be used on cylindrical

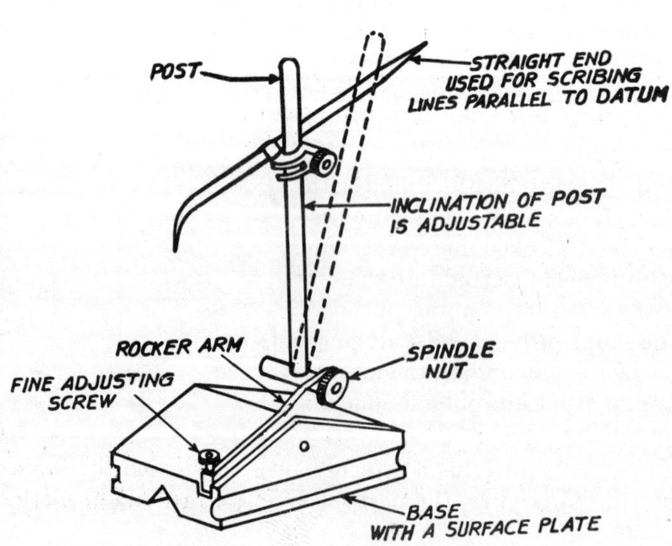

Fig. 2.76. Universal surface gauge.

surfaces also. Two pins are also placed in the base to serve as guides when scribing from the edge of the surface plate (Refer Fig. 2.76).

To set a universal surface gauge to a particular dimension, use can be made of a steel rule positioned at 90° to the datum with the help of angle plate and then lines can be scribed on the workpiece.

Many a time surface gauge is very successfully used for centring the work at lathe. It is capable of precision measurements also when used in conjunction with dial indicator.

2.19. Engineer's Square

Engineer's squares are of two types *viz* (*i*) with extra thick stock and (*ii*) without stock. The one without stock is less rigid and, therefore, requires thicker blade.

IS : 2103—1972 (specification for engineer's squares) prescribes three grades of accuracy (*viz*., grade *A*, and grade *B*, grade *C*) for engineer's squares suitable for workshop and inspection purposes. *A* square is designated by the length of the blade (measured from the outer working face of the stock to the end of the blade) and the grade of accuracy. In case of *A* grade squares without stock, the edges of the blade are usually bevelled to an angle of 60° by proper lapping operation in order to obtain line contact while testing flat surfaces. There is no need of bevelled edges for testing the cylindrical surfaces as they themselves provide the necessary line contact with flat edges of blade. Grades *B* and *C* squares are normally supplied with flat edges.

The squares are made of good quality tool or alloy steel suitable hardened, stabilized and lapped or finally ground and should be free from cracks, flaws or pits. Sharp edges are removed ; squares made smooth and bright all over. The blades are rigidly and permanently fixed in the stock by suitable means (riveting etc., the rivet bent flush with both the faces of the stock) such that when the stock is held rigidly and a load of 250 gm is applied to the end of the blade, it should not deflect by more than (0.7 $\sqrt{\text{length of blade}}$) micron (μm). The stock or the standard sizes of blade length

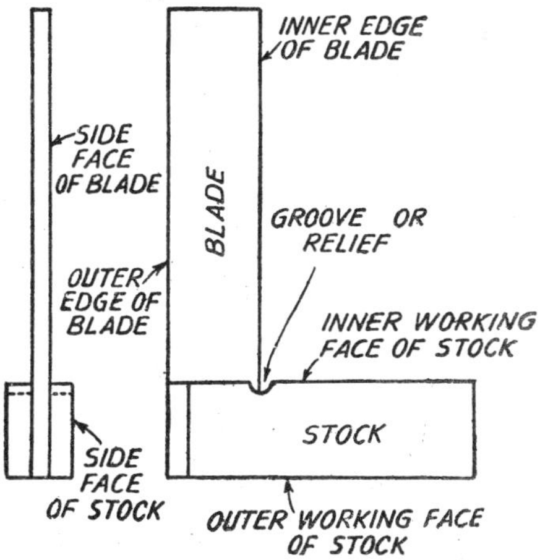

Fig. 2.77. Engineer's Square with Stock.

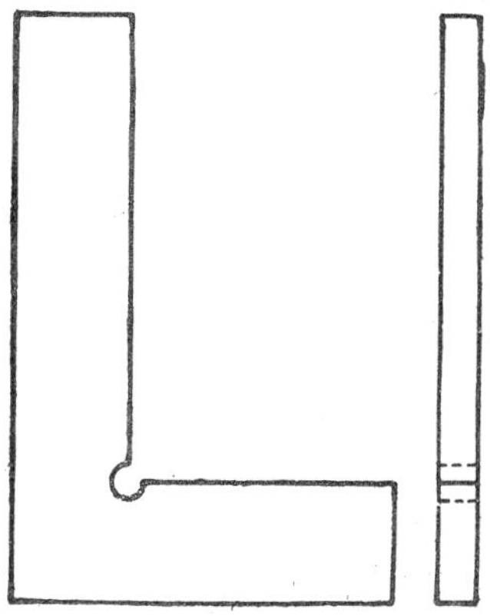

Fig. 2.78. Engineer's Square without Stock.

are : 50, 100, 150, 200, 300, 500, 700 and 1000 mm. The blade is generally slightly relieved across the width at the junction of the inner faces of the stock and blade.

The tolerable departure from straightness of each edge of blade and working faces of stock for three grades is given below.

(Length of blade or stock mm)		Deviation in microns		
Above	*Upto and including*	*Grade A*	*Grade B*	*Grade C*
1	150	1.00	1.25	2.50
150	300	1.00	1.50	3.00
300	1000	—	2.00	3.75

For good results, it is desirable that the edges of the blades and also the working faces of the stock should be parallel. The maximum tolerable error in the parallelism (*i.e.* the difference between the maximum or minimum width) between the edges of the blade or the working faces of stock is $\pm\left(2+\frac{L}{100}\right)$ microns for grade *A*, $\pm\left(5+\frac{L}{100}\right)$ microns for grade *B* and $\pm\left(10+\frac{L}{100}\right)$ micron for grade *C*, where L = length of blade or stock in mm. Also the error in the parallelism of the side faces of stock should not exceed thrice the parallelism error of the working faces of the stock specified above.

Length of stock (mm)		*Grade A*		*Grade B*		*Grade C*	
Above	*Upto and including*	*μm/mm*	*Angle*	*μm/mm*	*Angle*	*μm/mm*	*Angle*
0	100	0.001	0°03′	0.002	0°07′	0.003	0°10′
100	300	0.003	0°10′	0.005	0°17′	0.005	0°28′
300	750	—	—	0.007	0°24′	0.010	0°34′

The side faces of the stock should be truly square with the working face to within values given above. The outer edges of the blades should also be perfectly square with the outer working face of the stock and errors should not exceed those specified for parallelism of edges of blade and working faces of stock. Fig. 2.79 shows engineer's square with edges usually bevelled at 70°.

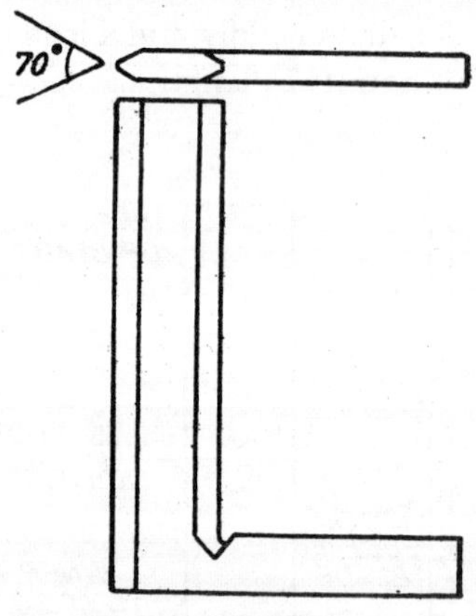

Fig. 2.79. Engineer's Square with bevelled edge.

2.20. Engineer's Parallels

Engineer's parallels are the devices of great value to tool room work and are used for setting apart the workpieces at desired distances in milling, grinding and shaper vices. These are also used for checking purpose, etc. on the face plates. These are available in three types, *viz.*

(*a*) Solid type

(*b*) Box type, and

(*c*) Adjustable type.

The solid type of engineer's parallels are made of suitable steel hardened to 550 HV and are available in two grades *A* and *B* depending upon their accuracy. These are generally supplied in pairs and the length of two pieces in the pairs should agree to within 0.5 mm. The thickness and the width of two members should also be equal to within ± 0.002 mm. All the faces should be straight to within ± 0.002 mm from the true mean plane. All the four sides of parallels are perfectly finished, ground and lapped and the sharp edges rounded off or bevelled. These are available in nine sizes varying from 5 × 10 × 100 (thickness × width × length) to 50 × 100 × 400 mm.

The box type of engineer's parallels are made of close grained cast iron and the surfaces free from porosity, burrs, cracks and grind chatters. These are available in 5 sizes varying from 100 × 100 × 100 (thickness × width × length) to 250 × 350 × 350 mm. In order to reduce the weight, these are made of hollow cross-section. Their dimensions are made accurate to within ± 0.02 mm. All the four working faces are finished ground and all sharp edges relieved to 0.5 mm.

The adjustable type of parallels are made of suitable close grained cast iron hardened up to about 180 HB and the surfaces free from porosity, burrs, cracks and grind chatters. All the four working faces are ground finished and sharp edges relieved. These consist of two pieces, one fixed member and other sliding member of the same shape and size as shown in Fig. 2.80. The sliding motion is obtained by a dovetail and the total width of wearing surfaces of the measuring member is made at least one-half of the total thickness of parallels. The two members can be held firmly in any desired position by means of a locking screw arrangement. For the two sliding members, their faces should be parallel to within 0.015 mm for all the positions.

These are available in seven sizes, all having a thickness of 8 mm and length from 40 to 100 mm. The width between the parallel faces varies from 10 to 13 mm for one having length of 40 mm and from 40 to 50 mm for the one having length of 100 mm.

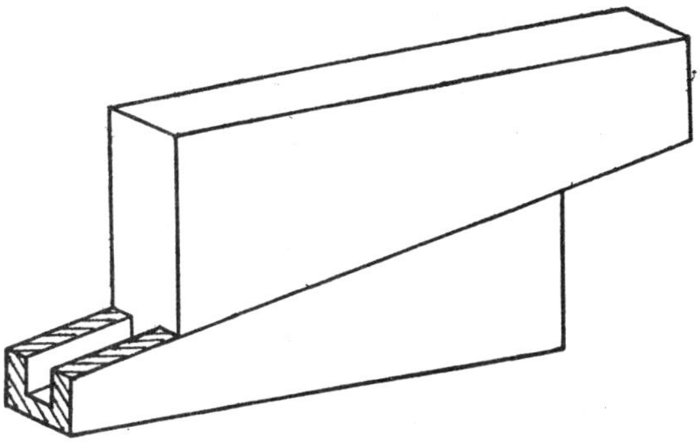

Fig. 2.80. Adjustable type of engineer's parallels.

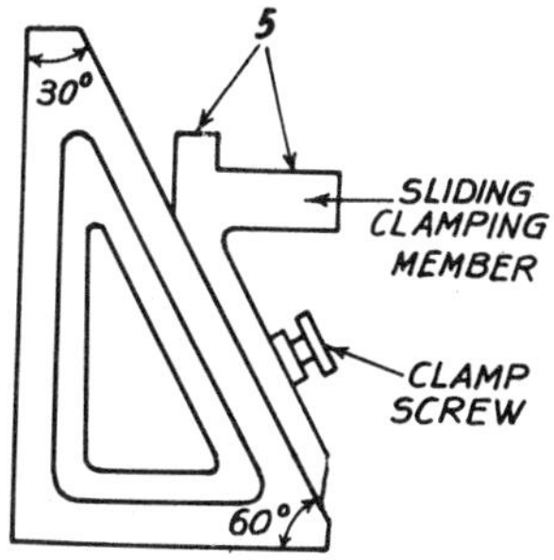

Fig. 2.81. Planer gauge.

2.21. Planer Gauge (Refer Fig. 2.81)

This adds to utility of measurements on surface plate. The planer gauge consists primarily of two triangular shaped blocks whose sloping sides can be clamped. A sliding clamping member having surfaces S and S parallel to datum base, can be slid along the hypotenuse of the triangle, and clamped in any position with clamp screw. Thus the surfaces S and S move vertically up down to the height and always being parallel to the surface plate on which planer gauge stands.

2.22. Radius Gauges

Wherever two surfaces meet, a radius is generally given to avoid stress concentration. The radius also plays an important role in case of heat treatment of parts. Thus the proper radius must be maintained and for checking it, radius gauges are employed. If the curved surface is external, *i.e.* male radius it is simply called a 'radius'. If curved surface is internal, *i.e.* female radius, it is called a 'fillet'. Radius gauges consist of sets of blades on each of which is stamped the corresponding radius. On one side we have external radius and on the other side internal radius, so that it may be suitable for fillets as well a radius. Radius gauges

Fig. 2.82. Fillet or Radius Gauges.

are used with proper light, placing the work to be gauged between light source and eye. It is only with the passage of light between gauge and work that the radius can be gauged properly.

Fillet or Radius Gauges. It contains external and internal radius on each leaf permitting both concave and convex surfaces to be measured. The leaves are specially shaped so that they can be used in any position at any angle to measure fillets and radius in corners or against shoulders (Refer Fig. 2.82).

2.23. Feeler Gauges

These are used to measure the width of the gap between two parallel flat faces *e.g.*, in gauging of the clearance between the piston and cylinder. A feeler gauge consists of a narrow strip of sheet steel made to a given thickness. The complete set consists of a number of gauging blades of different thicknesses assembled together. Their working depends entirely on the sense of feel. The feeler blade should neither be forced between the surfaces nor should it slide freely, rather the correct blade will give a characteristic 'gauge fit' type of feel. If necessary, two blades may be joined together for noting any dimension.

Feeler gauges generally comprise of a series of gauging blades of different grades and thicknesses, from 0.03 to 1 mm assembled in protective sheath. The blades are made of heat treated bright polished tool steel having a tensile strength of about 170 kgf/mm^2 for thickness up to 0.5 mm and about 70 kfg/mm^2 for thickness 0.5 mm and above. Generally the blades are available in overall length of 100 mm. These are 12 mm wide at heel and are tapered for the outer part of their length so that width at the tip is approximately 6 mm. These are hinged in the sheath on a screw and nut of such a design that the blades are removable. The nut is in the form of a bush, passing through both sides of the sheet and forms a hinge upon which the blades may be rotated. The sheath is so designed as to fully protect the blades when not in use.

Each blade is legibly and permanently marked with the nominal thickness and grade. These are available in two grades depending upon the accuracy. The various blades should be assembled in such a way that each thin blade is given the maximum protection by being inter-leaved between two thicker blades. The maximum variation in the thickness of a blade should not exceed 0.04 mm for blades upto and including 0.3 mm thick and 0.006 mm for blades over 0.3 mm in thickness.

The recommended set by Bureau of Indian Standards which has been so devised as to furnish sets of the greatest utility with a minimum number of blades is given below :

Set No.	*No. of blades in the set*	*Thickness of blades in mm*
1	8	0.03 to 0.1 in steps of 0.01
2	9	0.03, 0.03, 0.04, 0.04, 0.05, 0.05, 0.06, 0.07, 0.09.
3	16	0.03, 0.04, 0.05, 0.06, 0.07, 0.08, 0.09, 0.1, 0.15, 0.2, 0.25, 0.3, 0.35, 0.4, 0.45, 0.5
4	11	0.03, 0.04, 0.05, 0.06, 0.07, 0.1, 0.15, 0.2, 0.4, 0.5,
5	14	0.05, 0.06, 0.07, 0.08, 0.09, 0.1, 0.15, 0.2, 0.25, 0.3, 0.4, 0.5, 0.75, 1.0
6	11	0.05, 0.06, 0.07, 0.08, 0.09, 0.10, 0.15, 0.2, 0.4, 0.75, 1.0
7	11	0.5, 0.55, 0.6, 0.65, 0.7, 0.75, 0.8, 0.85, 0.9, 0.95, 1.0

2.24. Angle Gauges

These are similar to feeler gauges. A set of angle gauge consists of 18 blades with their ends cut at various angles from 2 to 45 degrees.

2.25. Engineer's Taper, Wire and Thickness Gauge (Fig. 2.83)

This is a very handy device which contains leaves for taper measurements, wire diameter measurements and thickness of small gaps.

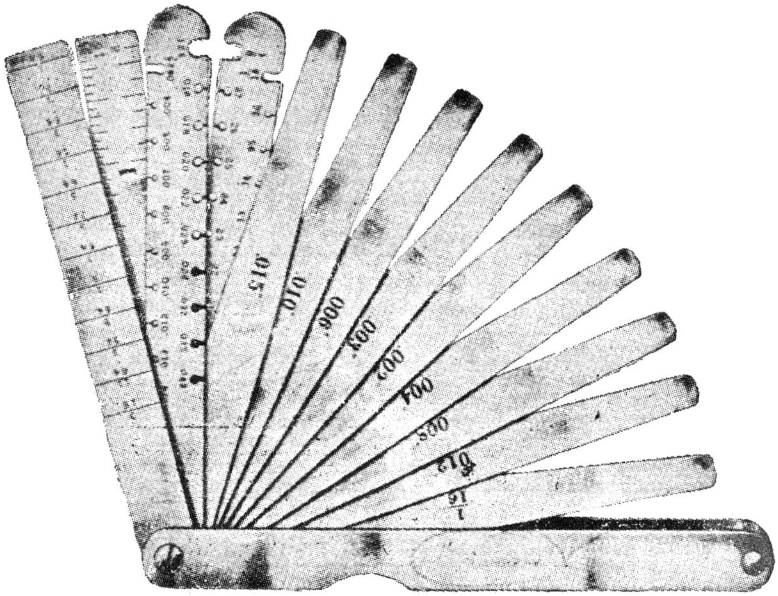

Fig. 2.83. Engineer's taper, wire and thickness gauge.

2.26. Pitch Screw Gauge

It is used to quickly determine the pitch of various threads by matching the teeth on the leaves with the teeth on the work (Refer Fig. 2.84).

Fig. 2.84. Pitch screw gauge.

2.27. Precision Linear Measurements

As modern production is concerned with interchangeable products, a great precision dimensional control is required in industry. For this to be achieved, importance of precision measuring instruments can be well visualised.

2.27.1. Characteristics of Precision Measuring Instruments. In order that precision measuring instruments function properly, they should possess the following characteristics : (*a*) high degree of sensitivity (*b*) high degree of accuracy and (*c*) minimum inertia in the moving parts of mechanism. Sensitivity of the instrument may be defined as the rate of displacement of the indicating device with respect to the quantity to be measured. Literally, sensitivity is defined with reference to the least change in the measured quantity which will cause an observable change in the instrument reading. In instrument design an attempt must be made that sensitivity remains constant throughout the range of measured quantity. The measure of accuracy of an instrument can be had by the amount of correction to be made to the instrument reading. The accuracy will be high, if calibration is proper and less wearing of parts takes place. The effect of inertia of moving parts is to make the instrument sluggish. All those instruments which depend on linkage and mechanical system, displacement of fluid, diaphragm etc. are subjected to disadvantages of inertia. Thus imperfect elasticity of the diaphragms and spring are causes of inertia in instruments. However instruments based on the principle of optics are entirely free from inertia effects.

In addition to these three characteristics, the measuring instruments must have freedom from variance. Variance is defined as the range of variance in instrument readings obtained from repeated measurements of same quantity. It is inherent in the instrument and depends upon the quality of instrument which in turn depends upon the quality of machines from which it is manufactured and quality of workmanship.

In order to achieve the above mentioned characteristics in the measuring instruments, the following principles must be observed.

2.27.2. Principle of Alignment. According to it the line of measurement (axis of scale) should be coincident with the line of the dimension being measured (the axis of the specimen whose dimension is being measured). The principle is known as Abbe principle of alignment. Abbe principle also states that if it is not possible to have the displacement measuring system in line with the functional point whose displacement is to be measured then either the slide-ways that transfer the displacement must be free of angular motion or angular motion data must be used to calculate the consequences of the offset. Abbe principle is followed in the case of micro-meter screw gauge. But in case of vernier caliper this is not followed. Therefore, there is possibility of error in vernier caliper. In vernier caliper, the specimen (whose dimension is to be measured) is placed between two jaws and the axis of this is quite far away from the axis of scale. Now let us say that due to force of tightening a slight curvature is brought in the vernier beam (scale), then obviously due to curvature of scale, it will not show the same reading as that of specimen since tightening pressure will vary from person to person ; reading obtained by different persons will be different. Hence generally the principle of alignment should be followed in the design of measuring instruments, as far as possible. (Also refer solved problem at the end of the chapter).

2.27.3. Principle of Kinematics. Here we are concerned with the theory of location and movement. Any rigid body in space can have six degrees of freedom of movement. These movements are linear along the three axes in space and three movements are rotational about the three axes. Thus in order to completely constrain a body from all movements, all the six degrees of freedom must be checked. For having one degree of freedom of movement, five degree of freedom must be constrained. Thus theoretically for a better design, the component parts should not have redundant locations or restraints.

The various ways to apply these constraints are conical holes, vee-grooves and flat surfaces. By conical hole the three linear movements are constrained. With V-groove two linear movements and two rotational movements are constrained while with a flat surface one linear movement and one angular movement is constrained. Using the proper combinations of these three methods, any degree of freedoms can be constrained.

Theoretically, the contact between the component parts of a kinematic design should take place at single point. But in that case they can't withstand much of impact, and robustness of the instrument is sacrificed. Hence these must be used very carefully. However to provide the necessary robustness kinematic principles can be modified slightly.

2.27.4. Measuring Contacts. Measuring faces are generally either flat or spherical at the position of contact.

In case of flat contacts, faces must be flat and parallel to each other. If not so then errors will be introduced in the measurement of a ball and flat piece of same size (Fig. 2.85). (Also refer solved problem at the end of the chapter). Secondly the work must be free to align itself relative to the measuring faces and normal operation of measurement should not upset the parallelism of the measuring faces. For flat pieces one measuring face must be spherical one and other flat one, while in case of circular specimen, both faces must be flat.

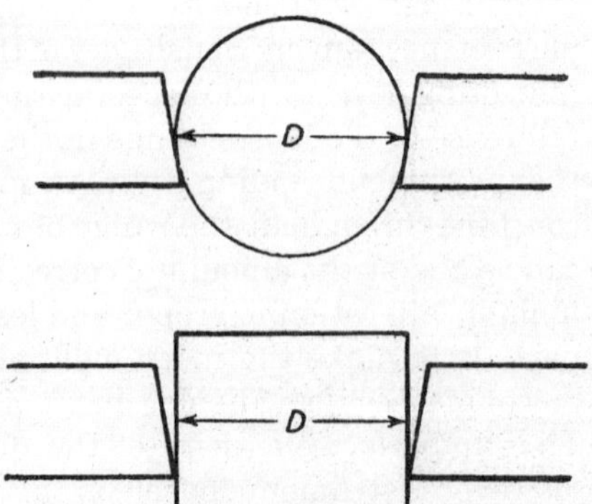

Fig. 2.85. Here the width D of rectangular block will appear larger than diameter D of the ball.

2.27.5. Selection of measuring instrument. Discrimination is a major factor in the selection of proper measuring instrument. One should consider the variable to be measured and the discrimination of the measuring device desired. Other points to be considered are the level of sensitivity or resolution that can be expected ; the accuracy and precision required of the measuring device and the conditions required to use this device properly.

Dealing with above factors successfully is important. It should be analysed as to what measuring devices, techniques and sequences would be most successful in optimising the inspection or measurement. How should the results be reported or displayed is also important consideration.

Lastly it should be ensured that what inspection system guarantees that this same measurement technique will yield comparable or repeatable results next time.

It must be remembered that the measuring devices possessed of a large measuring range offer a trade off in terms of discrimination. For instance, mechanical and electronic instruments having great amplification (and thus more discrete scales of discrimination), have less range.

Every instrument possesses some inherent amount of error or uncertainty, expressed as accuracy of the instrument. In assessing the performance of a measuring instrument, the performance expectations of the instrument must first be considered in terms of accuracy. Obviously, the choice of the proper measuring instrument for an inspection is based on a prior expectation of its accuracy. On the other hand precision of instrument refers to repeatability.

The number of graduations on an instrument or the degree to which it subdivides one centimeter is known as discrimination. Estimation of exact reading between two smallest graduations should not be attempted. If requirement is to measure exactly between two smallest divisions, then apparatus with fine discrimination (line vernier calipers or micrometer screw gauge) should be used.

2.28. Vernier Instruments

The principle of vernier is that when two scales or divisions slightly different in size are used, the difference between then can be utilised to enhance the accuracy of measurement. The vernier caliper essentially consists of two steel rules and these can slide along each other. One of the scales, *i.e.,* main scale is engraved on a solid L-shaped frame. On this scale cm graduations are divided into 20 parts so that one small division equals 0.05 cm. One end of the frame contains a fixed jaw which is shaped into a contact tip at its extremity.

The three elements of vernier caliper, *viz.* beam, fixed jaw, and sliding jaw permit substantial improvements in the commonly used measuring techniques over direct measurement with line graduated rules. The alignment of the distance boundaries with the corresponding graduations of the rule is ensured by means of the positive contact members (the jaws of the caliper gauges). The datum of the measurement can be made to coincide precisely with one of the boundaries of the distance to be measured. The movable jaw achieves positive contact with the object boundary at the opposite end of the distance to be measured. The closely observable correspondence of the reference marks on the slide with a particular scale value, significantly reduces the extent of read-out alignment errors.

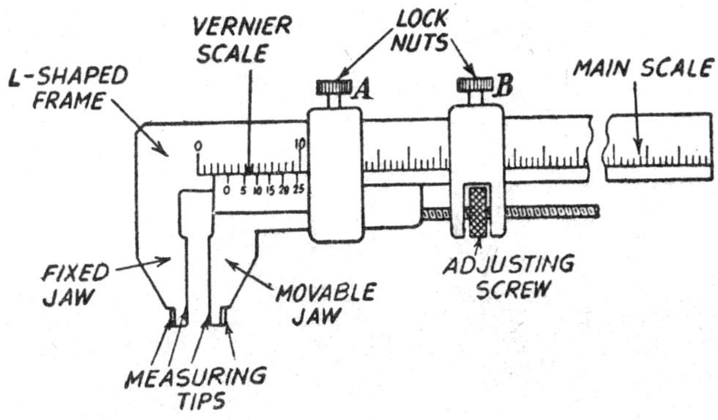

Fig. 2.86. Vernier caliper.

A sliding jaw which moves along the guiding surface provided by the main scale is coupled to a vernier scale. The sliding jaw at its left extremity contains another measuring tip. When two measuring tip surfaces are in contact with each other, scale shows zero reading. The finer adjustment of the movable jaw can be done by the adjusting screw (Fig. 2.86). First the whole movable jaw assembly is adjusted so that the two measuring tips just touch the part to be measured. Then lock nut *B* is tightened. Final adjustment depending upon the sense of correct feel is made by the adjusting screw. The movement of adjusting screw makes the part containing locking nut A and sliding jaw to move, as the adjusting screw rotates on a screw which is in a way fixed to the movable jaw. After final adjustment has been made, the locking nut *A* is also tightened and the reading is noted down. The measuring tips are so designed as to measure inside as well as outside dimensions.

2.28.1. Reading the Vernier Scale. For understanding the working of vernier scale let us assume that each small division of the main scale is 0.025 unit. Say, the vernier scale contains 25 divisions and these coincide exactly with 24 divisions of main scale. So now one vernier division is equal to 1/25 of 24 scale divisions, *i.e.*, 1/25 × 24 × 0.025 = 0.024 unit. Therefore, difference between one main scale small division and one vernier division (least count of the instrument) equals 0.025 – 0.024, *i.e.* 0.001 unit. *It means if the zero of main scale and zero of vernier coincide, then the first vernier division will read 0.001 unit less than the 1 small scale division.* Second vernier division will read 0.002 unit less than 2 small scale divisions and so on. Thus if zero vernier scale lies in between two small divisions on main scale its exact value can be judged by seeing as to which vernier division is coinciding with main scale division.

Thus to read a measurement from a vernier caliper, note the units, tenths and fortieths which the zero on the vernier has moved form the zero on the main scale. Note down the vernier division which coincides with a scale division and add to previous reading the number of thousands of a unit indicated by the vernier divisions *e.g.* reading in the scale shown in Fig. 2.87 is 3 units + 0.1 unit + 0.075 unit + 0.008 unit = 3.183 units. When using the vernier caliper for internal measurements the *width of the measuring jaws must be taken into account.* (Generally the width of measuring jaw is 10 mm for Metric System).

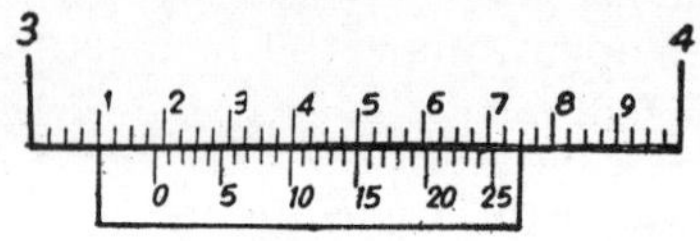

Fig. 2.87

2.28.2. Types of vernier calipers. According to IS : 3651—1974 (Specification for vernier caliper), three types of vernier calipers have been specified to meet the various needs of external and internal measurements upto 2000 mm with vernier accuracy of 0.02, 0.05 and 0.1 mm. The three types are called types *A*, *B*, *C* and have been shown in Figs. 2.88, 2.89 and 2.91 respectively. All the three types are made with only one scale on the front of the beam for direct reading. Type *A* has jaws on both sides for external and internal measurements, and also has a blade for depth measurements. Type *B* is provided with jaws on one side for external and internal measurements. Type *C* has jaws on both sides for making the measurements and for marking operations.

All parts of the vernier calipers are made of good quality steel and the measuring faces hardened to 650 H.V. minimum. The recommended measuring ranges (nominal sizes) of vernier calipers as per IS : 3651—1974 are 0—125, 0—200, 0—250, 0—300, 0—500, 0—750, 0—1000, 750—1500 and 750—2000 mm.

On type *A*, scale serves for both external and internal measurements, whereas in case of types *B* and *C*, the main scale serves for external measurements and for marking purposes also in type *C*, but on types *B* and *C* internal measurements are made by adding width of the internal measuring jaws to the reading on the scale. For this reason, the combined width for internal jaws is marked on the jaws in case of types *B* and *C* calipers. The combined width should be uniform throughout its length to within 0.01 mm.

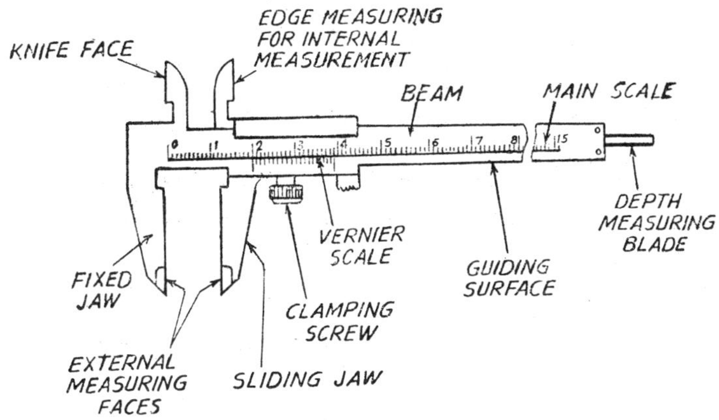

Fig. 2.88. Type A Vernier Caliper.

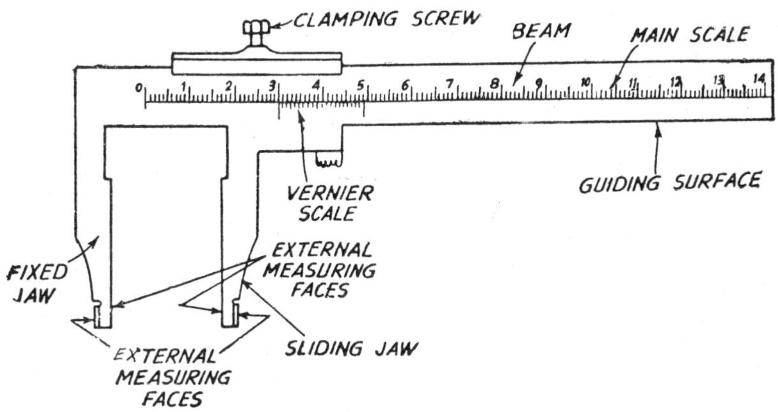

Fig. 2.89. Type B Vernier Caliper.

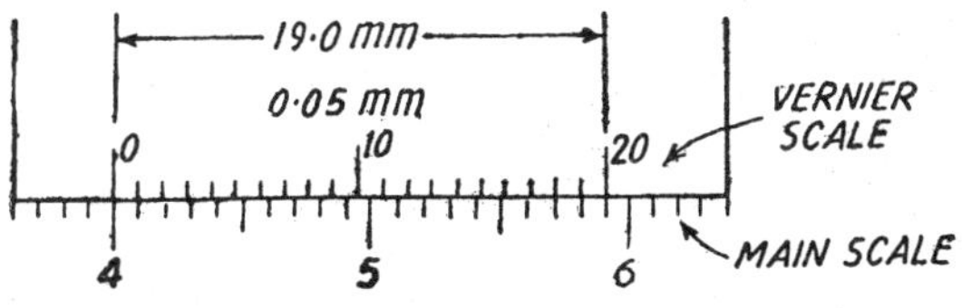

Fig. 2.90. Beam and Vernier for Vernier Calipers having least count of 0.05 mm.

Graduations on beam at every 1 mm and each 5 mm line extended and each 1 cm line is numbered. On vernier scale there are 20 divisions within a distance of 19 mm and 19 mm = 19 divisions of main scale.

In Fig. 2.92, Graduations on beam are at every 1/2 mm and every alternate mm lines are extended and numbered 2, 4, 6, 8. On vernier scale, there are 10 divisions within a distance of 9.5 mm and 9.5 mm = 19 divisions of main scale.

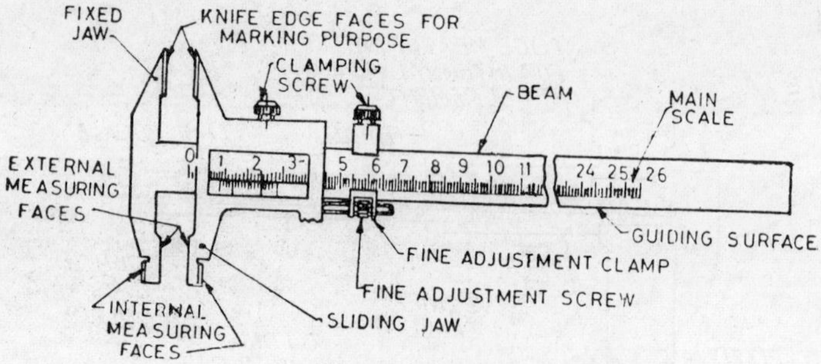

Fig. 2.91. Type C Vernier Caliper.

The beam for all the types is made flat throughout its length to within the tolerances of 0.05 mm for nominal lengths upto 300 mm, 0.08 mm from 900 to 1000 mm, and 0.15 mm for 1500 and 2000 mm sizes, and guiding surfaces of the beam are made straight to within 0.01 mm for measuring range of 200 mm and 0.01 mm every 200 mm measuring range of larger size. The measuring surfaces are given a fine ground finish. The portions of the jaws between the beam and the measuring faces are relieved. The fixed jaw is made an integral part of the beam and the sliding jaw is made a good sliding fit along with the beam and made to have seizure-free movement along the bar. A suitable locking arrangement is provided on the sliding jaw in order to effectively clamp it on the beam. When the sliding jaw is clamped to the beam at any position within the measuring range, the external measuring faces should remain square to the guiding surface of the beam to within 0.003 mm per 100 mm. The measuring surfaces of the fixed and sliding jaws should be coplanar to within 0.05 mm when the sliding jaw is clamped to the beam in zero position. The external measuring faces are lapped flat to within 0.005 mm. The bearing faces of the sliding jaw should preferably be relieved in order to prevent damage to the scale on the beam. Each of the internal measuring surface should be parallel to the corresponding external measuring surface to within 0.025 mm in case of type *B* and *C* calipers. The internal measuring surfaces are formed cylindrically with a radius not exceeding one-half of the their combined width.

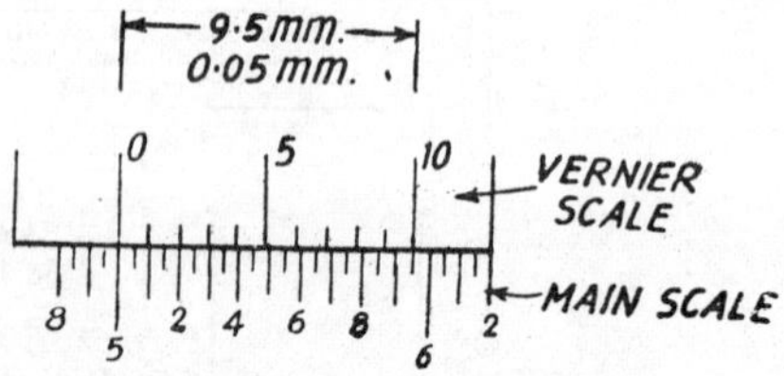

Fig. 2.92. Beam and Vernier for Vernier Caliper having least count of 0.05 mm.

2.28.3. Graduations. All graduations should be clearly engraved so that they are legible. Sometimes to facilitate reading, the surfaces of the beam and the vernier may be given a matt finish and the graduation lines filled with black pigment. For clarity sake, the length of visible portion of graduations on main scale on beam and vernier scale lines (dimension *h*) should be about 2-3 times the width of interval between adjacent lines (Refer Fig. 2.93) and the distance from the graduated face of the beam to the edge of the graduated bevelled face of the vernier (dimension *s*) in Fig. 2.93 should not exceed 0.1 mm.

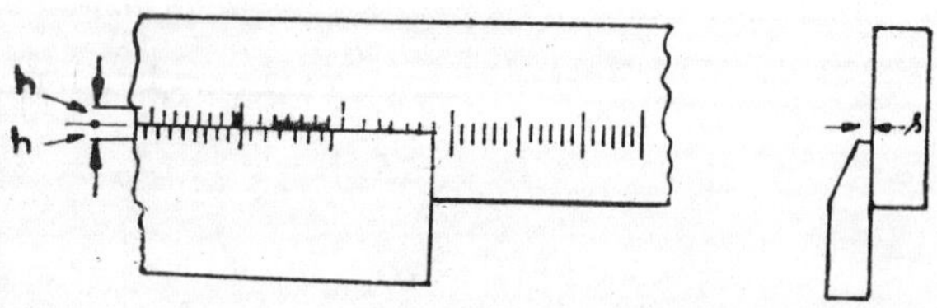

Fig. 2.93

The graduations on the beam and the vernier for least count 0.05 mm are illustrated in Figs. 2.90 and 2.92. In fact, various types of markings for each least count are possible.

The error in reading the vernier caliper should not exceed the values obtained by the following formulae :

Vernier with least count	*Permissible error in reading*
0.1 mm	$\pm (75 + 0.05l_1)$ μm
0.05 mm	$\pm (50 + 0.05l_1)$ μm
0.02 mm	$\pm (20 + 0.02l_1)$ μm

where l_1 = upper limit of the measuring range in mm.

The error in reading is found by placing slip gauges at right angles to the longitudinal direction of the measuring faces ; the readings being taken at three different points along the length of the jaws and same pressure applied to the sliding jaw each time. (The error measured in this way will include errors in the flatness and parallelism of the measuring jaws). The check should be repeated at a number of points distributed over the range of measurement and disposed in such a way that each measurement calls into play a different vernier graduation line.

2.28.4. Errors in measurements with vernier calipers. Errors are usually made in measurements with vernier calipers from manipulation of vernier caliper and its jaws on the workpiece. For instance, in measuring an outside diameter, one should be sure that the caliper bar and the plane of the caliper jaws are truly perpendicular to the workpiece's longitudinal centre line ; *i.e.* one should be sure that the caliper is not canted, tilted, or twisted. It happens because the relatively long, extending main bar of the average vernier calipers so readily tips in one direction or the other.

The accuracy of the measurement with vernier calipers to a great extent depends upon the condition of the jaws of the caliper. The accuracy and the natural wear, and warping of vernier caliper jaws should be tested frequently by closing them together tightly or setting them to the 0.0 point of the main and vernier scales. In this position, the caliper is held against a light source. If there is wear, spring or warp, a knock-kneed condition as shown in Fig. 2.94 (*a*) will be observed. If measurement error on this account is expected to be greater than 0.005 mm the instrument should not be used and sent for repair.

(*a*) (*b*) (*c*) (*d*)

Fig. 2.94. Various jaw conditions which result in inaccurate caliper measurements.

When the sliding jaw frame has become worn or warped so that it does not slide squarely and snugly on the main caliper beam, then jaws would appear as shown in Fig. 2.94 (*b*).

Where a vernier caliper is used mostly for measuring inside diameters, the jaws may become bowlegged as in Fig. 2.94 (*c*) or its outside edges worn down as in Fig. 2.94 (*d*).

2.28.5. Care in use of vernier calipers. These should not be treated or used as a wrench or hammer because these are not rugged instruments. They should be set down gently, preferably in the box and not dropped or tossed aside. They must be wiped free from grit, chips and oil. These should be brought to the workpiece.

The workpiece should not be clamped in the caliper jaws and waived in air.

2.28.6. Precautions in the Use of Vernier Caliper. No play should be there between the sliding jaw on scale, otherwise the accuracy of the vernier caliper will be lost. If play exists then the gib at the back of jaw assembly must be bent so that gib holds the jaw against the frame and play is removed.

Usually the tips of measuring jaws are worn and that must be taken into account. Most of the errors usually result from manipulation of the vernier caliper and its jaws on the workpiece.

In measuring an outside diameter it should be insured that the caliper bar and the plane of the caliper jaws are truly perpendicular to the workpiece's longitudinal centre line. It should be ensured that the caliper is not canted, tilted or twisted.

The stationary caliper jaw of the vernier caliper should be used as the reference point and measured point is obtained by advancing or withdrawing the sliding jaw.

In general, the vernier caliper should be gripped near or opposite the jaws ; one hand for the stationary jaw and the other hand generally supporting the sliding jaw. The instrument should not be held by the over-hanging "tail" formed by the projecting main bar of the caliper.

The accuracy in measurement primarily depends on two senses, *viz.*, sense of sight and sense of touch (feel). The short-comings of imperfect vision can however be overcome by the use of corrective eye-glass and magnifying glass. But sense of touch is an important factor in measurements. Sense of touch varies from person to person and can be developed with practice and proper handling of tools. One very important thing to note here is that sense of touch is most prominent in the finger-tips, therefore, the measuring instrument must always be properly balanced in hand and held lightly in such a way that only fingers handle the moving and adjusting screws etc. If tool be held by force, then sense of feel is reduced.

Vernier caliper must always be held at short leg of main scale and jaws never pulled.

2.28.7. Digital Caliper

Advantages

Easy readings of digital display with large 4, 7 mm high characters with decimal point, minus sign.

Zero setting at any position within measuring range by conveniently placed button.

One-hand operation only two function buttons.

No misreadings, due to extremly high measuring speed.

Max. speed 1.5 m/s.

Low power consumption. Battery has life of 1 year continuous operation.

Data exit. For connection of printers and computers for evaluation and processing of measuring results. Data acquisition is initiated by button on slide or by foot switch.

Positive locking screw for slide. This is of advantage for many applications.

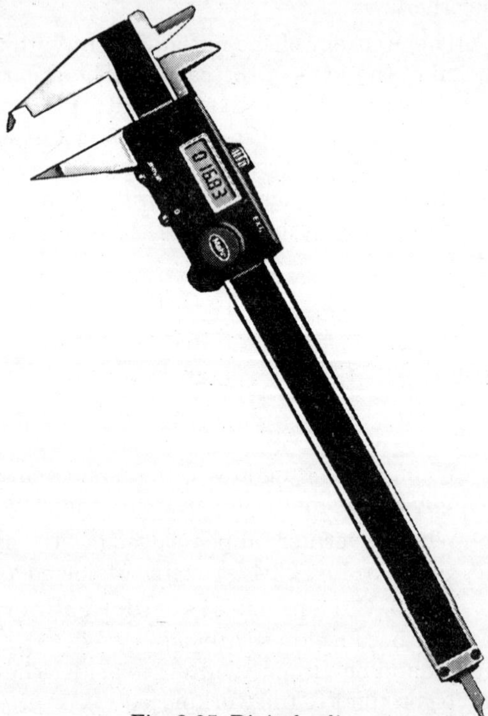

Fig. 2.95. Digital caliper.

2.28.8. Applications of Digital Caliper

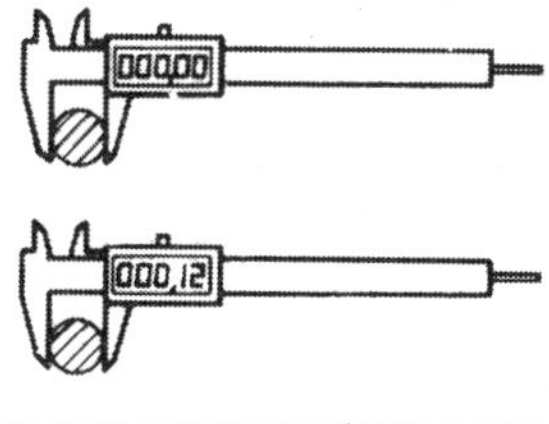

Measuring of deviations
Set nominal size, press zero button, perform comparison measurement.

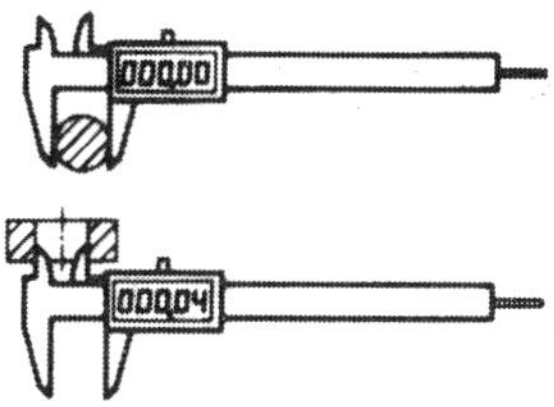

Measuring of clearance
e.g. between shaft and bore
Measure shaft, press zero button, measure bore and read clearance.

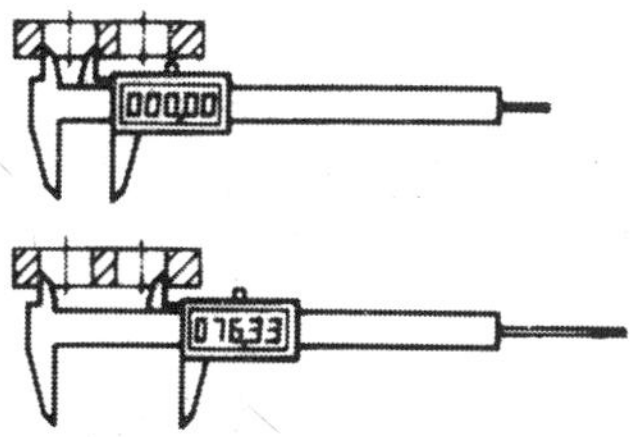

Measuring of center distances
of bores with equal diameter
Measure one bore, press zero button, measure largest distance and read center distance.

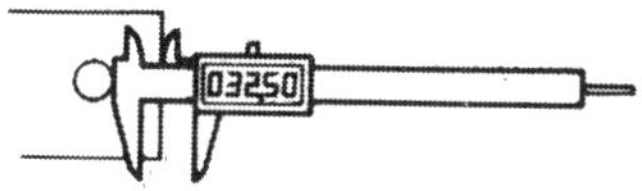

Measuring of distances
e.g. of a bolt from an edge.

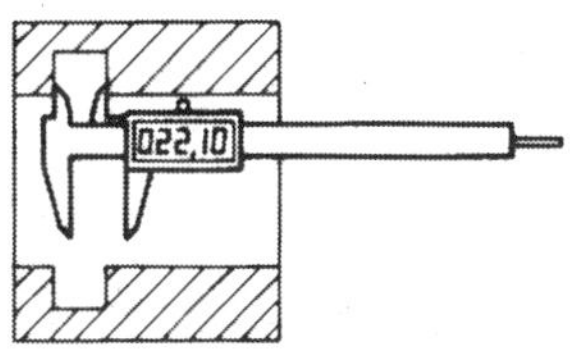

Measuring of positions with difficult access
e.g. width of recesses in bores. When locking screw cannot be reached, take measurement, press zero button, close measuring jaws and read results on closed caliper.

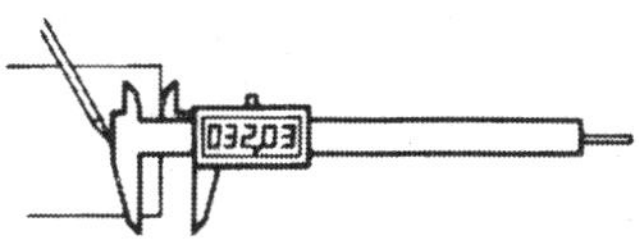

Scribing of components

Fig. 2.96. Applications of digital caliper.

2.28.9. Electronic Digital Caliper. Electronic Digital Caliper is the most advanced electronic digital measuring instrument for fast, accurate and reliable measurements and yet can be used by operators—(semi skilled and non-technical) on the shopfloor. It measures outside and inside diameters, depth and steps etc. In it is incorporated the capacitive measurement system and with the revolutionary new chip permitting interface with computer print-out facilities. The

Fig. 2.97. Electronic digital caliper.

powerful large scale integration (LSI) micro chip gives full scale computer functions to measure, monitor, and memorise. The clear easily read LCD reads upto 0.01 mm.

Salient features of the Caliper are :

— Error free accurate and fast reading.

— True inch/metric conversion.

— Floating zero allows comparative measurements.

— Display memory facilitates :

— 'HOLD' function freezes display reading.

— No vernier scale—rack or pinion, and there is no backlash.

— Unaffected by oil or dust and ideally suited for shopfloor use.

— Low power consumption (one button cell of 1.5 V) for one year continuous use.

The frame, jaws and depth rod are of hardened stainless steel precision ground.

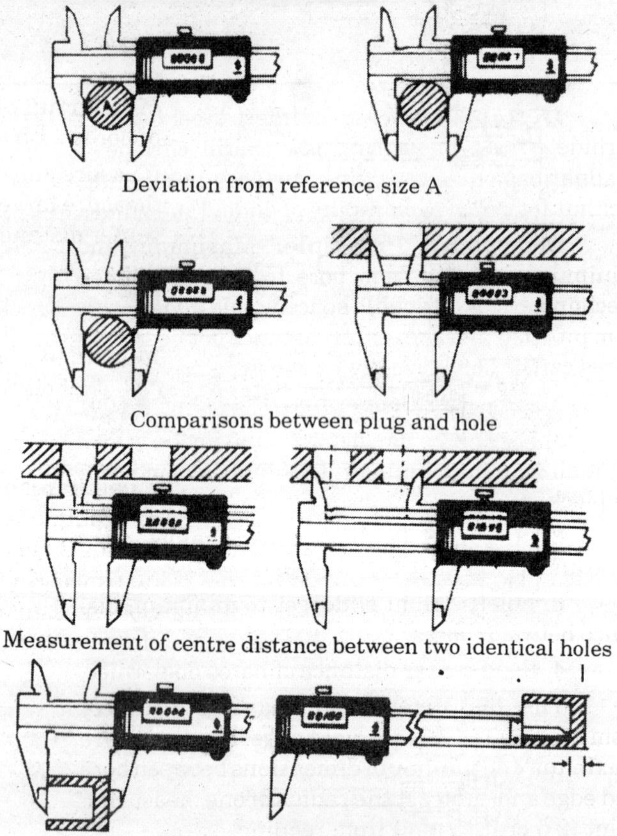

Fig. 2.98. Typical Application of Electronic Digital Caliper.

2.29. Distance Gauge (Refer Fig. 2.99)

Advantages. Center distances can be measured without complicated calculations and mathematical errors. Double spaced divisions provide high accuracy and easy 0.02 mm readings. Readings are not influenced by personal feel of the operator as measuring pressure acts in opposite directions when minimum and maximum distances are measured.

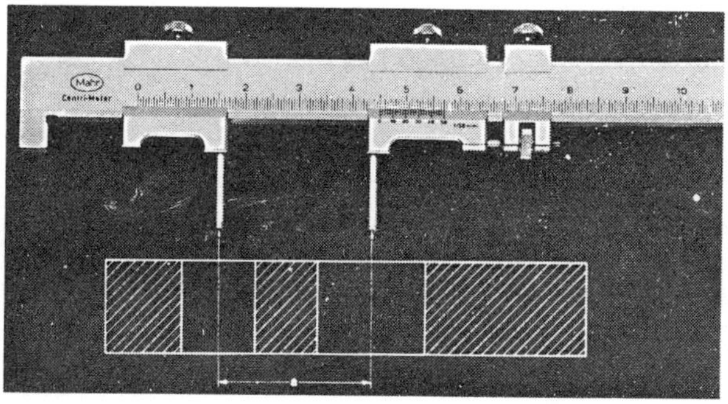

Fig. 2.99. Distance gauge.

Design Features. Stainless steel beam hardened throughout with 2 slides each with one carbide tipped measuring jaw. Satin chrome finished vernier and main scale with jet black graduations. Raised sliding surfaces to eliminate wear and marring of scales. Non-removeable locknut for positive clamping of slide. Fine adjustment.

Measuring Principle. Maximum and minimum distance from bore to bore are added mechanically. The double spaced scale divides this sum into half and the center distance between both bores can be read directly on the vernier.

Measuring Procedure (Refer Fig. 2.100)

1. Move left slide against stop and lock. Set right slide to maximum dimension between bores and lock.

2. Release left slide, move against right slide and lock.

3. Release right slide, set to minimum distance between bores.

4. Read center distance of bores on vernier.

This instrument also measures distances from bore center to workpiece edge. Here measure maximum and minimum dimensions between bore and edge and subtract the radius of one measuring point (2.5 or 0.75 mm) from reading.

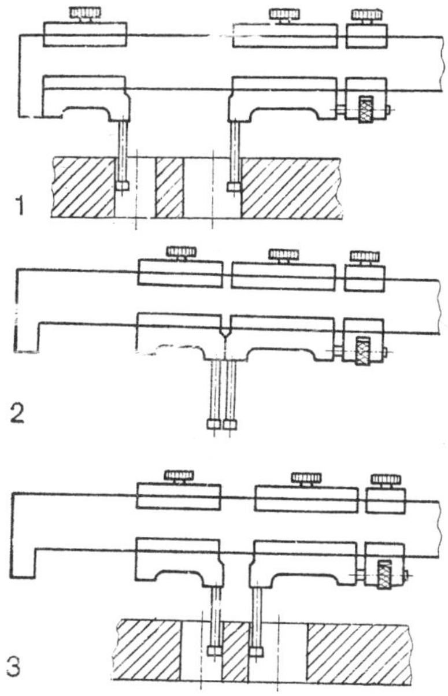

Fig. 2.100. Measuring principle for distance gauge.

2.30. Vernier Height Gauge

[Fig. 2.101] This is also a sort of vernier caliper, equipped with a special base block and other attachments which make the instrument suitable for height measurements. Along with the sliding jaw assembly, arrangement is provided to carry a removable clamp. The upper and lower surfaces of the measuring jaws are parallel to the base, so that it can be used for measurements over or under a surface. The vernier height gauge is mainly used in the inspection of parts and layout work. With a scribing attachment in place of measuring jaw, this can be used to scribe lines at certain distance above surface. However dial indicators can also be attached in the clamp and many useful measurements made as it exactly gives the indication when the dial tip is just touching the surface. For all these measurements, use of surface plates as datum surface is very essential.

2.30.1. Specification of Vernier Height Gauges. For specifying the vernier height gauge, one has to specify clearly the range of measurement, the type of scales desired, and any particular requirements in regard to the type of vernier desired. Generally, all the parts of the height gauges are made of good quality steel, or stainless steel also in certain cases. At the time of fabrication, the blanks for all height gauges are suitably heat-treated so as to remove any stresses resulting from possible cold working. The underside of the base and the faces of the measuring jaw and scriber are hardened to about 700 HV (58 RHC) for good quality steel and up to 650 HV (56 RHC) for stainless steel. In selecting a suitable high carbon stainless steel to give a hardness of not less than 650 HV, a balance should be drawn between hardening qualities and resistance to staining, because a steel which gives a high hardness number is likely to have low acid resisting qualities. The various parts and the nomenclature of the height gauge are illustrated in Fig. 2.102.

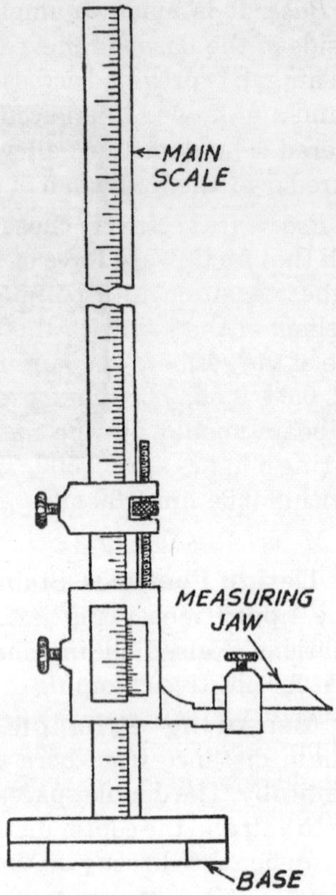

Fig. 2.101. Vernier Height Gauge.

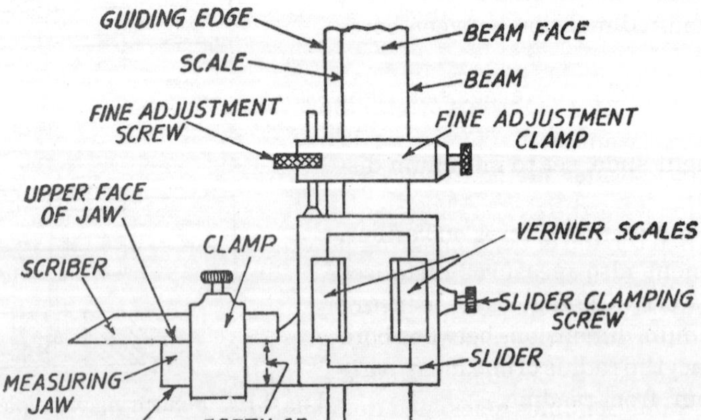

Fig. 2.102. Nomenclature for Vernier Height Gauge.

2.30.2. Description of parts and their requirements

Base. It is made to ample proportions to ensure rigidity and stability of the gauge. The underside of the base is relieved leaving a surface round the outside edge of at least 7 mm width and an air gap is provided across the surface to connect the relieved part with the outside. The base is ground and lapped to an accuracy of 0.005 mm as measured over the total span of the surface considered. The departure allowed from the true flatness is only in the nature of concavity, as measured over the total span of the surface concerned.

Beam. Its section is chosen such that it ensures rigidity during the use. The rigidity should be such that an upward force of 1 kg applied at the tip of the scriber when the measuring jaw is in its highest position in a plane containing the measuring jaw, shall not cause a change in the parallelism of the jaw with the base of more than 0.005 mm/10 mm of projection of the working surface of the scriber. The guiding edge of the beam should be perfectly flat within the tolerances of 0.02, 0.04, 0.06, 0.08 mm for measuring range of 250, 500, 750, 1000 mm respectively. The faces of the beam should also be flat within the tolerances of 0.04, 0.06, 0.10, 0.12 mm for vernier measuring heights of 250, 500, 750, 1000 mm respectively. These faces should also be square to the base within 0.04 mm/100 mm.

Measuring Jaw and Scriber. The clear projection of the measuring jaw from the edge of the beam should be at least equal to the projection of the base from the beam. For all positions of the slider, the upper and the lower gauging surfaces of the measuring jaw should be flat and parallel to the base to within 0.008 mm. The actual depth of the measuring jaw should agree with its nominal depth to within 0.013 mm. The nominal depth of the measuring jaw is always clearly marked on it. The measuring faces of the scriber should be flat and parallel to within 0.005 mm. The scriber should preferably be of the same nominal depth as the measuring jaws so that it can be reversed and should confirm to this depth to within 0.013 mm. The actual depth of the scriber is also clearly marked on it.

The projection of the scriber beyond the jaw should be at least 25 mm. Vernier height gauges may also have an offset scriber and the scale on the beam so positioned that when the scriber is coplanar with the base, the vernier is at zero position.

Graduations. The following requirements should be fulfilled in respect of graduations on scales. All graduations on the scale and the vernier should be clearly engraved and the thickness of graduation both on scale and vernier should be identical and should be in between 0.05 mm and 0.1 mm. The visible length of the shortest graduations should be about 2 to 3 times the width of the interval between the adjacent lines. The perpendicular distance between the graduations on scale and the graduations on vernier should in no case be more than 0.01 mm.

In most of the cases the graduations on the beam give the reading of heights from the underside of the base to the top of the measuring jaw, but in some spherical types of height gauges using offset scribers, the scriber can be positioned co-planar with the underside of the base and in that position the vernier reads zero. Generally provision is also made for adjustments of the vernier on the slider in order to reset the zero position if wear or any other cause makes this necessary.

For easy reading, it is recommended that the surfaces of the beam and vernier should have dull finish and the graduation lines blackened in. Sometimes a magnifying lens is also provided to facilitate taking the readings.

Slider. It has good sliding fit along the full working length of the beam. A suitable fitting is incorporated to give a fine adjustment of the slider and suitable clamp provided so that the slider could be effectively clamped to the beam after the final adjustment has been made.

2.30.3. Precautions. The height gauges are generally kept in their cases when not in use. Every care should be taken, particularly in case of long height gauges, to avoid its heating by warmth from the hands. The springing of the measuring jaw should be always avoided.

2.30.4. Tests for accuracy. For testing the accuracy of the vernier height gauge, the following conditions shall be kept in mind :

(*i*) The standard reference temperature of 20°C shall be maintained in the room where the test is going to be conducted.

(*ii*) The various tests shall be carried out by resting the height gauge on a precision surface plate.

(*iii*) The test measurements shall be made at a distance of 25 mm from the end of the jaw.

(*iv*) The standard gauges shall be used for testing purposes.

(*v*) The slider shall be firmly clamped.

(*vi*) The errors in the reading at any portion within the measuring range of height gauge should not exceed ± 0.02 mm for measuring range of height gauge of 250 mm, ± 0.04 mm for above 250 mm and upto 750 mm, and ± 0.05 mm for range of more than 750 mm.

The most important tests on vernier height gauge are straightness and squareness and the parallelism of the working faces of the beam, measuring jaw and scriber ; and the accuracy of the scale readings. However, the various recommended tests are described below :

(*a*) *Hardness of working faces.* It is assessed by any of the conventional methods. However, the scratch test is most commonly used. This test is performed by a hand scriber consisting of a small round Swiss file, the end of which is ground to a smooth hemisphere of about 0.5 mm radius, and is mounted on a wooden handle. The scriber is first drawn on a standard test piece of hardness 700 HV and then on the face under test with the same force. The resulting marks are compared and estimate of hardness of tested surface made.

(*b*) *Flatness of base.* This test is performed by blueing a known flat surface and then rubbing it against the surface to be tested. Afterwards the visual test which is sensitive to well within 0.003 mm may be employed. In this test a 'knife-edge' straight-edge is placed against the surface to be tested against a source of light and the colour of the light as visible against the illuminated background is noted which gives the measure of the flatness.

(*c*) *Rigidity of beam.* The change in parallelism between the surface of the scriber and the base, when a load of 1 kgf is applied to the tip of the scriber with the measuring jaw in its highest position, in a plane containing the measuring jaw, may be checked by a suitably mounted indicator and a precision square block as shown is Fig. 2.103 (*a*).

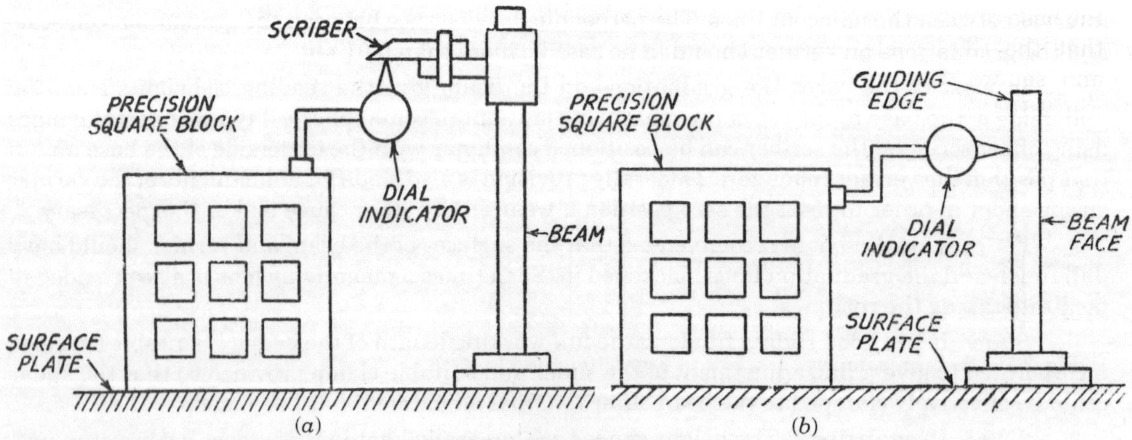

Fig. 2.103. Accuracy Tests.

(*d*) *Flatness and squareness of beam.* The test for the flatness and squareness of the beam is carried out with the aid of precision square block (whose tolerances of the squareness are known) and a precision surface plate. The square block is placed on the surface plate and height gauge also. The dial indicator is made to rest on the square block and its reading against the guiding edge of the beam is noted in several positions and thus the flatness and squareness assessed. [Refer Fig. 2.103 (*b*)]

(*e*) *Flatness of the working surface of the measuring jaw and scriber*. This is checked either by applying a 'knife-edge' straight edge or by the optical flat if the finish of the surface is very good.

(*f*) *Parallelism of the measuring jaw and scriber with the base*. This test is carried out in a similar manner as (*e*). The indicator is suitably mounted on the stand and indicator base directly made to rest either on surface plate or on the precision parallel block as the case may be and readings noted both on the measuring jaw and scriber and the parallelism assessed.

(*g*) *Thickness of graduations*. It is checked with a microscope fitted with an eye-piece graticule scale engraved at 0.03 mm interval when a graduation line thickness may be read directly against the graticule scale.

(*h*) *Accuracy*. The accuracy of reading of vernier height gauge is checked by making setting on end-bar and slip gauge combinations of known size. The combination and the gauge are made to stand side by side on a precision surface plate and the scriber is brought in position for the test and the reading on the scale is noted down. The size of the standard combination used should be chosen so as to cover a number of points over the range of the instrument and at the same time to provide a check on the accuracy of the vernier scale.

(*i*) The perpendicular distance from the graduated face of the beam to the graduated bevelled face of the vernier is checked with a 0.1 mm feeler gauge.

2.30.5. Master Dial Indicator Vernier Caliper. It combines the best features of the Vernier Caliper and the Dial Comparator Gauge in one instrument and greatly reduces the problem of operator's sense of feel. This is ideally suited for both external and internal measurements.

2.31. Vernier Depth Gauge

For measuring the depth of holes, recesses and distances from a plane surface to a projection, the vernier depth gauge is employed. In vernier depth gauge, the graduated scale can slide through the base and vernier scale remains fixed.

2.31.1. Errors in use of vernier depth gauge. For use of vernier depth gauge, its base or anvil is rested on or against a reference surface and the scaled beam or tongue is pushed beyond the base to contact the measured point. Errors are made due to manipulation. It must be ensured that the reference surface on which the depth gauge base is rested is satisfactorily true, flat and square. The gauge, though true and square, can be imperceptibly tipped or canted. [Refer Fig. 2.104]

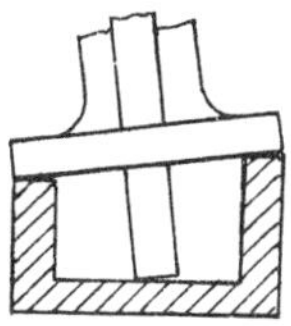

Fig. 2.104. Gauge is imperceptibly tipped

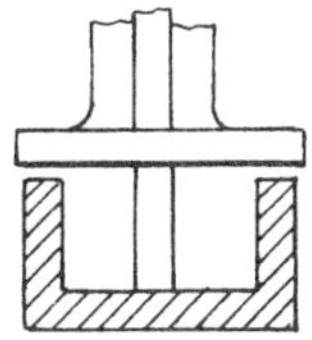

Fig. 2.105. Base is lifted when measured point is slided

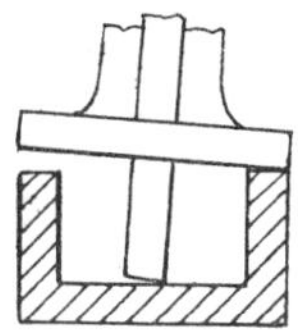

Fig. 2.106. Point pressure causes the base a trifle

Fig. 2.105 illustrates how the base or anvil of depth gauge which may be at first firmly rested on the reference surface, but is lifted due to applying high pressure in sliding the beam or tongue against the measured point. Sometimes measuring pressure may cause the base a trifle as shown in Fig. 2.106.

To avoid above errors, it is advisable to press the anvil firmly on the reference surface and keep several kilograms of hand pressure on it. Then in manipulating the gauge beam to measure depth, apply only standard, light measuring pressure, like making a light dot on paper with a pencil.

2.31.2. Combination Depth and Angle Gauge. It can be used as a protractor also for measuring angles in addition to being depth gauge. The rule can be set to any of the angles by swinging the rule so that line on the turret coincides with the corresponding graduation on the head. The scale can be locked in any position.

2.32. Digital Length Gauges

These are designed for universal application and are also ideally suited for the measurement of travel on small-sized measuring apparatus. These operate without wear and are highly accurate and maintenance free. The length measuring standard is a glass scale with graduations. The scale is directly connected to the plunger and carries a grating which is photoelectrically scanned. The measured value is obtained by electronic evaluation of pulses which are generated upon activation of the plunger.

2.32.1. Electronic Digital Machine Scales. These electronic linear measuring scales can be attached to machine or fixture slides or tables and these accurately and easily measure their movements. Large, right-sized LCD figures in a rugged display housing read to 0.01 mm. Accurate, linear glass scale is protected within the hardened, stainless steel bar.

Fig. 2.107. Electronic Digital Machine Scales.

Assembly Features

— Slide has convenient mounting holes, bar has adjustable mounting brackets with mounting holes

— Either bar or slide can be mounted as fixed or moveable. Fixed slide and moveable bar are preferred because a stationary slide can be mounted in a good, readable position without cable travel consideration.

Action Features with Four-Button Control

— Scale will automatically turn on with any movement and will turn off after one-half hour of nonuse

— ZERO set button will zero the reading at any position

— The "±" button designates a plus or minus indication

— PRINT button (on cable) sends readings to printers, data collectors, computers, etc.

— Two 3-volt batteries with one-year life under normal use

— Two output jacks conveniently located on either side of the slide, allow data transmission to peripherals for analysis, data collection, and hard copy documentation.

2.33. Master Height Gauge

This is basically a stack of slip gauge blocks in a robust housing which provides a traceable reference standard over considerable vertical lengths. Riser blocks extend the measuring capacity by upto 600 mm. They are particularly useful in setting other height gauges and providing a high precision standard.

It comprises a vertical rectangular column with sliding attachment. With digital readout system it can be used as a height standard for measuring hole location, step heights, groove widths and pitches, etc. With dial test indicator or lever had attached, the unit can be used for checking squareness and parallelism.

2.34. Universal Height Micrometer (Refer Fig. 2.108)

Universal height micrometers are used for setting height gauges, especially indicator stands, height calipers and lay-out devices and also bore gauges and inside micrometers.

Rugged construction. Three point contact surface with carbide pads adds stability and improves accuracy.

Height adjustable column with chromium plated incremental measuring surfaces which are stacked in such a manner that the contact surfaces of each step form a common plane. Full scale micrometer head controls height adjustment.

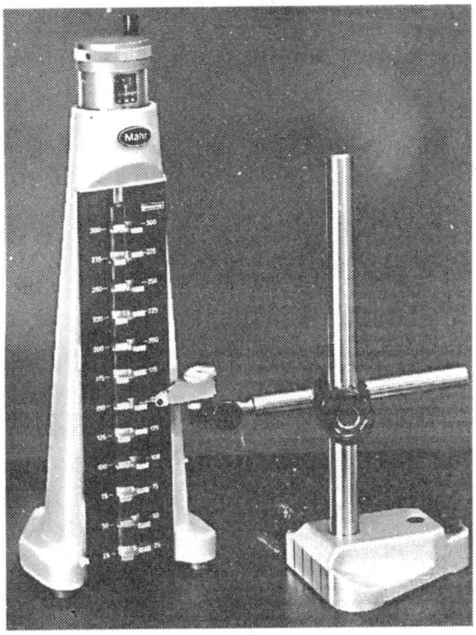

Fig. 2.108

Advantages

Quick and easy set-up by parallax-free reading of micrometer head.

White figures on black background and widely spaced graduations facilitate readings.

Versatility. Dual usage of each step. One measuring surface available from above, one from below.

Maximum accuracy by gauge block like surface finish of the measuring surfaces which form an integral part of the column. This adds stability and impact resistance to the instrument.

Maximum wear and corrosion resistance. Column has three rows of measuring surfaces which can be rotated for wear compensation. All measuring surfaces are chromium plated.

Easy access to the measuring surfaces by 12 mm throat depth and access from three sides, which also facilitates maintenance considerably.

2.35. Electronic Digital Readout Height Gauge

Vernier height gauges and height masters rely to some extent on the skill of the operator and are also relatively slow. A big advantage of the digital height gauge is that it provides an immediate unambiguous digital readout of the measurement value. It is possible to hold this value in memory and use as a datum for further readings, or be compared with given tolerances, etc. It is also possible to provide a binary coded digit (BCD) output to enable the results to be further statistically analysed and for a printout to be provided. These are provided with heavier steel or granite bases with air-bearing floatation systems. Thus these provide a cushion of air between the base of the stand and the surface plate which reduces, both the effort required to move the gauge and the possibility of damage to the surface plate. The instrument can be zeroed at any position after which it will display positive and negative dimensions with reference to the datum. There is very fine adjustment facility and the slide can be securely clamped for marking out.

2.35.1. Digital Height Gauge With Statistical Production Control. Today's modern production techniques reflect a growing need for Statistical Process Control (SPC). Systems capable of predicting or anticipating "out of tolerance" are thus of great value. The improvement in quality, the reduction of scrap and the increase of efficiency in production capacity are all closely related.

The digital height gauge with capability to provide SPC is best suited in this direction. Such a gauge has following features.

— storage of upto 400 dimensions
— data retention on failure of electric power
— automatic entry of data as component is measured
— entry of data via the custom touch key-pad
— calculation of capability indices for quick and easy assessment of machine/plant or process capability
— easy identification of component features out of tolerance
— calculation of mean, standard deviation and range
— assessment of 2-sigma, 3-signal, and 4-sigma confidence intervals
— comprehensive print-out of statistical data including histograms
— each sequence of measurement is recorded in numerical order and can be recalled subsequently and printed out in the format required.

2.36. Universal Measuring Instrument

Universal Measuring Instrument enables even an unskilled operator to quickly measure height, diameters, centre distances, squareness, grooves, concentricity, runout, parallelism, straightness, flatness, axial trueness, as well as the setting and checking of machine tools, instruments, toolings and fixtures and an almost unlimited range of additional applications with ease and confidence to a high degree of accuracy.

Features. Apart from direct digital reading and floating zero, it has automatic computational capabilities which saves time dramatically and eleminates possibilities of operator's error due to separate calculations.

Outstanding features

— Ease of reading digital display.
— Twin digital display in the front and side for convenience of operator.
— True instantaneous inch/metric conversion at any point without re-zeroing.
— Stores extra reference datum in memory for complex incremental measurements.
— Built-in microprocessor establishes centres of internal and external diameters, grooves, etc. automatically.
— Computes diameters of bores and shafts by automatically sensing highest and lowest points.
— On the same basis directly indicates axial trueness and face runout.
— High accuracy.
— Squareness testing facilitated.

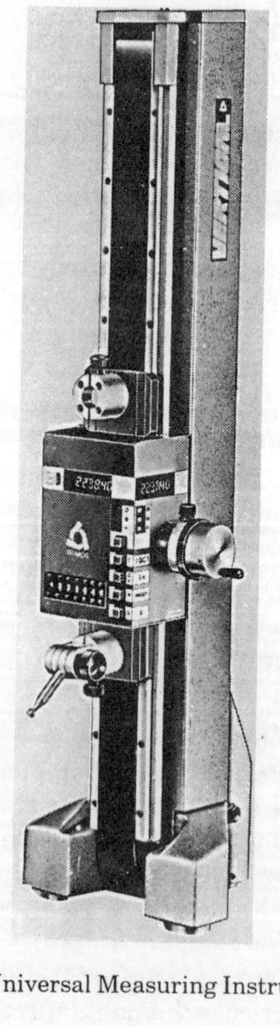

Fig. 2.109. Universal Measuring Instrument.

— Saves inspection time.

— Built-in microswitches in handwheel automatically calculate and compensate for the probe diameter/constant during the measuring process.

— Measuring system made of patented steel scale which has the same co-efficient of thermal expansion as steel components to be measured.

— Designed such that the servicing and repairs can be done at the customer's place, if required.

— All the accessories are designed to balance with the counter weight—No adjustment for the weight of any accessory is required when changed.

— Measuring speed upto 1 meter/second is permissible for error free reading.

— Electronic interface available to feed out-put informations to standard IEEE and IEC interface devices such as calculators, computers, recorders, plotters and machines.

— Fully tropicalised to withstand temperature upto 50°C and 95% relative humidity.

Operations Simplified

Only one push button to control all Measuring Programmes. Only one single push button key is sufficient for the selection of the correct measuring programme. All that operator has to see is whether the surface being measured is plane or cylindrical. Five push button keys are conveniently located near readout and their functions are as follows :

Push Button Controls (Refer Fig. 2.110)

A — Push button for selection of measuring programme (plane or cylindrical surfaces). Corresponding signal light indicates whether it is diameter, centreline or height measurements, in case of cylindrical surfaces maximum and minimum values are stored. Centreline distance appears when measuring pressure is removed.

B — Push button for input of probe constant while measuring pressure is applied.

C — Push button for changing one reference datum (*O*) to the other (*o*) or vice versa. Both the datums are stored in memory at a time (*e.g.* for incremental measurements).

D — Push button for adding or substracting preselected value. Useful for changeover of probes and probe holders.

E — Push button for zero setting of digital display at any position of the measuring head.

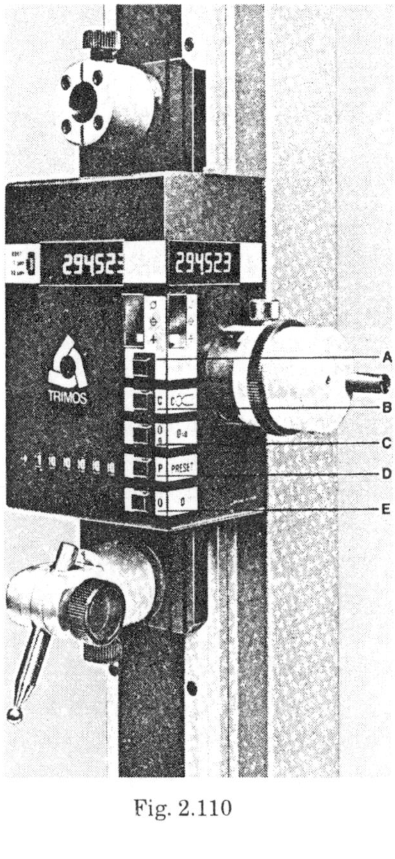

Fig. 2.110

Carefully Designed and Constructed for Easy Servicing (Refer Fig. 2.111)

A — Robust cast iron base with hammer-tone finish lacquer coating. Houses valve with air distribution to the 3 supporting pads. Fitted with snap action connector for compressed air and press button for air cushion.

B — 3 supporting pads with tungsten carbide coating except the top surface which can be lapped to restore squareness if necessary. Resists rust and does not wear-out. Guarantees well balanced air cushion movement.

C — Completely independent compact power supply unit fitted with four screws to the base and can be easily replaced. Provides continuous uninterrupted power to the electronic unit. Takes care of voltage fluctuations from—18% to + 20%. Sockets for mains cable, fuse and socket for printer and interface.

D — Measuring column with hardened stainless steel guide rail and precision lapped guide surfaces.

E — Unbreakable steel scale has same co-efficient of expansion as most of the steel components being checked hence temperature variations does not affect measuring accuracy. It also permits faster measuring speed (1 meter/second). Non-contacting system unaffected by friction, wear or back-lash. Insensitive to electrical interferences and oil film.

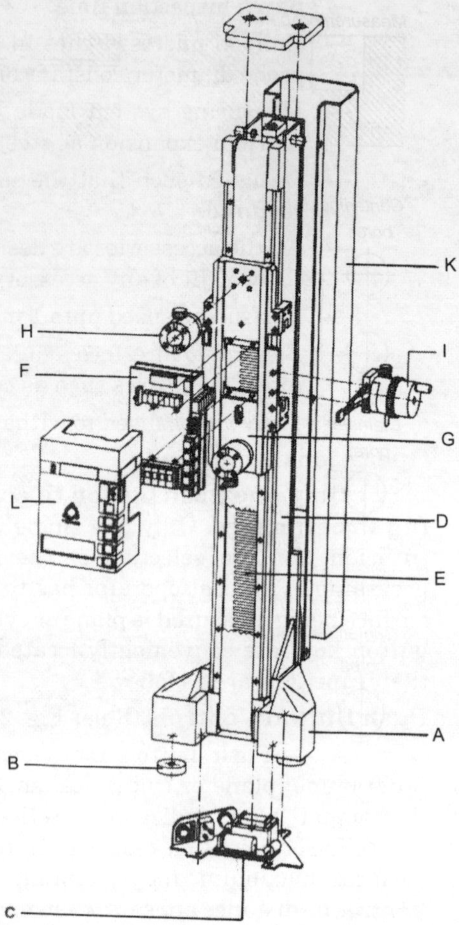

Fig. 2.111

F — Plug in type electronic display unit is completely independent and easily interchangeable. Read out is provided in the front as well as in the side for easy viewing. Fully tropicalised to withstand temperature upto 50°C and relative humidity 95%.

G — Measuring slide moves freely on precision ball bearings and its movement is controlled by steel tape and counter weight. (No rack and pinion movement or backlash).

H — Two probe holders to fix various types of probes with heavy duty locating shaft (dia 20 mm) for extra rigidity.

I — Ultra-smooth multi-functional handwheel controls rapid movement, provides unlimited fine adjustment movement, embodies preloaded ring for constant measuring pressure, locks measuring slide in desired position and also triggers built-in micro-switches to compensate automatically for probe constant through electronic unit.

K — Aluminium cover with hammertone finish lacquer coating.

L — Aluminium protection cover for electronic display unit.

Fig. 2.112 shows some typical applications of universal measuring instrument.

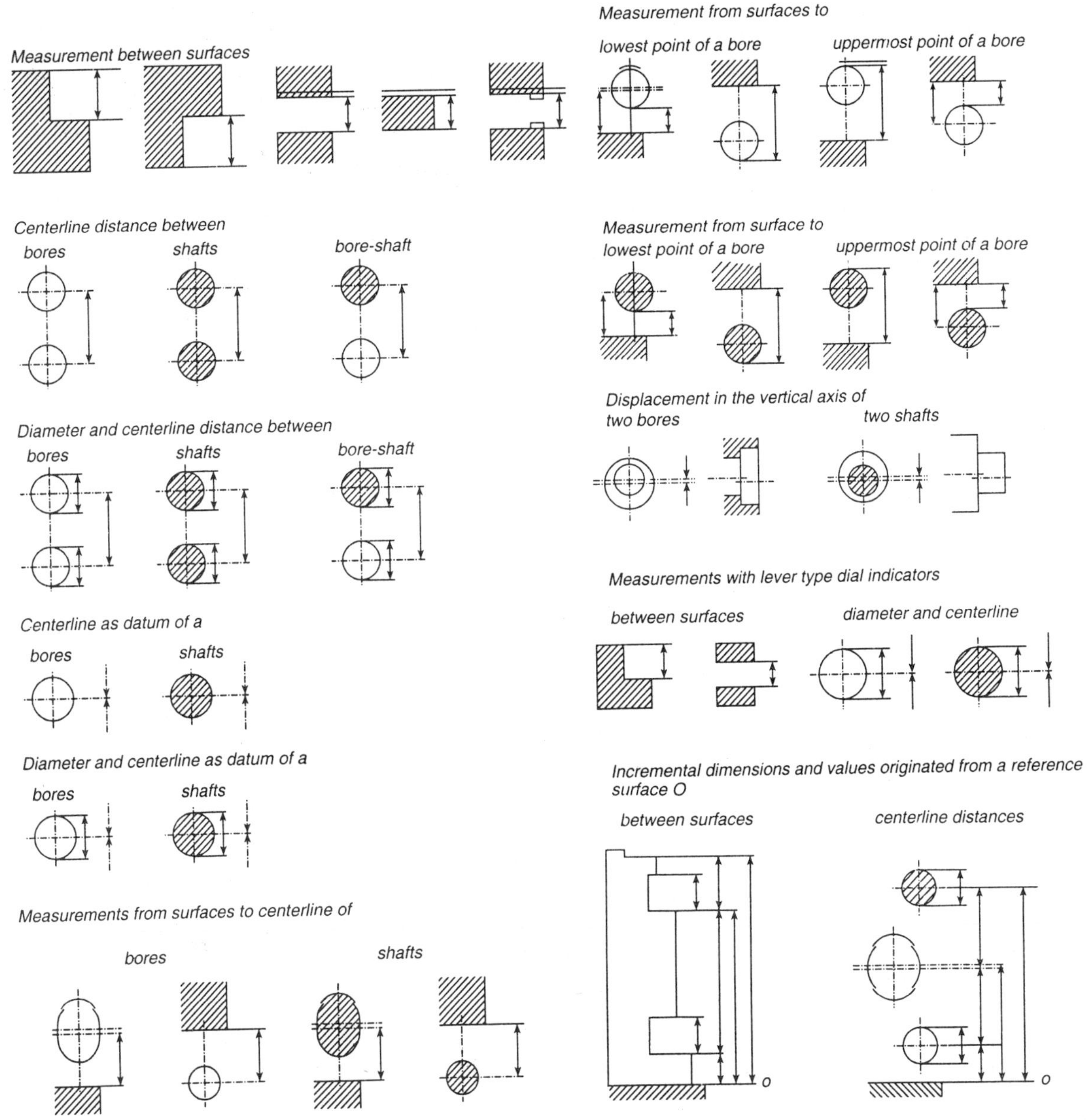

Fig. 2.112. Applications of Universal Measuring Instrument.

2.37. Micrometers (Fig. 2.113)

The micrometer screw gauge essentially consists of an accurate screw having about 10 or 20 threads per cm and revolves in a fixed nut. The end of the screw forms one measuring tip and the other measuring tip is constituted by a stationary anvil in the base of the frame. The screw is threaded for certain length and is plain afterwards. The plain portion is called sleeve and its end is the measuring surface. The spindle is advanced or retracted by turning a thimble connected to the spindle. The spindle is a slide fit over the barrel and barrel is the fixed part attached with the frame. The barrel is graduated in unit of 0.05 cm. *i.e.* 20 divisions per cm, which is the lead of the screw for one complete revolution. The thimble has got 25 divisions around its periphery on circular portion. Thus it sub-divides each revolution of the screw in 25 equal parts, *i.e.* each division corresponds to 0.002 cm.

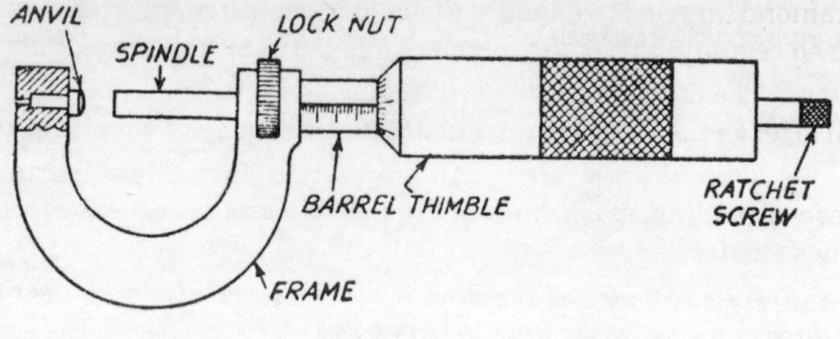

Fig. 2.113

A lock nut is provided for locking a dimension by preventing motion of the spindle.

Ratchet stop is provided at the end of the thimble cap to maintain sufficient and uniform measuring pressure so that standard conditions of measurement are attained. Ratchet stop consists of an overriding clutch held by a weak spring. When the spindle is brought into contact with the work at the correct measuring pressure, the clutch starts slipping and no further movement of the spindle takes place by the rotation of ratchet. In the backward movement it is positive due to shape of ratchet.

2.37.1. Description of various parts and their specifications

Frame. The frame of micrometer is so shaped as to permit measurements of the cylinder of diameter equal to the measuring range of micrometer, and the stiffness of the frame should be such that a test load of 1 kg weight does not alter the distance between them by more than 1.5 μm for range 0 to 25 mm, 2 μm for range 25 to 50 mm and so on. It should be neatly and evenly black enamelled or treated by other means to prevent corrosion and rusting. The frame is generally made of steel, cast steel, malleable cast iron or light alloy. The use of light alloys is recommended, for sizes over 300 mm in order to facilitate the handling of large micrometers. But it has to be understood that coefficient of expansion of light alloys is appreciably greater than that of steel or iron and thus the temperature changes during handling of the micrometer frame may result in appreciable changes in the zero reading. It is, therefore desirable in such cases that initially micrometer be checked against a standard length bar or setting gauge and appropriate correction made to the reading. It is desirable that the frame of the micrometers be provided with conveniently placed finger grips of heat insulating material.

Anvil and Spindle. The fixed anvil of micrometer screw gauge should protrude at least 3 mm from the frame in order to permit the attachment of measuring wire support. The measuring faces of these are hardened to about 800 HV (62 HRC approx.) and aged. In some cases, faces may also be tipped with tungsten carbide or some other suitable material. The anvil should be accurately ground and lapped with its measuring face flat and parallel to the measuring face of the spindle. The diameter of the anvil should be equal to the diameter of spindle within 0.04 mm and axes of both in exact alignment.

The front parallel portion of the spindle be a good free-turning fit in its bush, without perceptible shake. The spindle and screw are initially lubricated with a thin, light non-corrosive oil and in that condition the spindle should run freely and smoothly throughout the length of its travel.

There should be no perceptible backlash between the spindle screw and nut. There should be full engagement of nut and the micrometer screw when the micrometer is at its maximum reading. The design of the spindle clamp should be such that it effectively locks the spindle without altering the distance between the measuring faces by more than 0.003 mm. It is generally desirable that diamond knurled spindle lock-nut be provided with micrometer.

Ratchet Driver. The micrometer should be provided with a ratchet of friction stop sufficiently diamond knurled to enable satisfactory operation. The torsional moment of the ratchet drive should be so regulated that the force exerted between the measuring faces is between 0.5 tc 1 kgf.

The material used for thimble, barrel, ratchet and all other locking and clamping devices for all sizes of micrometers should be suitable quality wear resistant steel.

Thimble and barrel. All graduation lines on the barrel should be clearly engraved and for ease of reading the surface of the thimble and barrel should have a dull finish and the graduation lines should be blackened.

Adjusting nut. Micrometer screw gauges are generally provided with a friction or adjusting nut to compensate for wear between the screw portion of the spindle and nut. They are also provided with means of adjustment to compensate for wear on the measuring faces and for adjusting the zero setting. The adjustments are carried out by suitable spanners and keys which are provided with micrometer for this purpose. The means of adjustment should be such that after resetting, the parts are secured and the original accuracy of the instrument is not impaired.

2.37.2. Important Terms. The various important terms used in connection with the micrometer screw gauges are defined below :

Backlash. It is the lack of motion or lost motion of the spindle when the rotation of thumble is changed in direction.

Measuring range. It is the total travel of the measuring spindle for a given micrometer.

Total error. It corresponds to the maximum difference of ordinates of the cumulative error.

Cumulative error. It is the deviation of measurement from the nominal dimension determined at any optional point of the measuring range. It includes the effect of all possible individual errors such as errors of the thread, errors of measuring faces etc. It can be determined by some tests with slip gauges.

2.37.3. Indian Standard Specifications. As per IS : 2967—1964 (Specification for external micrometers) ; the micrometer screw should be of (M 10 × 1.5). It is recommended that the threads of the screw and nut be truncated so as to confine contact to the flanks of the thread.

The diameter of the graduated surface of the barrel should not be less than 10 mm. The angle on the bevel (angle α) at the graduated end of the thimble should not exceed 20° as measured from the barrel. The thimble should be sufficiently diamond knurled to facilitate ease of rotation. The distance from the barrel to the reading end of graduations (dimension B in Fig. 2.114) should not exceed 0.5 mm and should remain substantially constant for appropriately spaced measurements within the range of thimble movement.

The thimble should be graduated with 50 divisions, each representing 0.01 mm and numbered 0, 5, 10, 15...45. The barrel should bear longitudinal reference line parallel to the axis of the spindle. The graduations of barrel have to be provided in two parts, *viz.*, one above the reference line and other below. The graduations above the reference line are graduated at 1 mm intervals and the first and every fifth graduation are marked long and numbered 0, 5, 10, 15, 20 and 25.

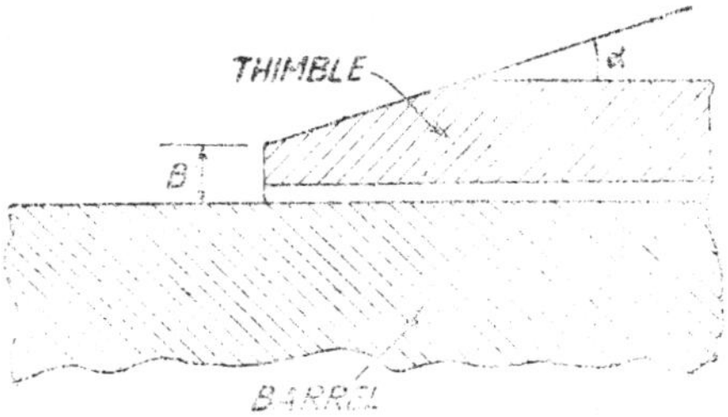

Fig. 2.114. Details of Barrel and thimble of screw gauge.

The lower graduations are graduated at 1 mm intervals but each graduation shall be placed at the middle of the two successive upper graduations to be read 0.5 mm. These lines are not numbered and distinguished from both of these by their greater length. The thickness of the graduations should be between 0.15 to 0.20 mm.

When tested at 20°C, the total error should not exceed the following values :

for grade 1, total error = $\left(4 + \frac{L}{100}\right)$ μm and for grade 2, total error = $\left(10 + \frac{L}{50}\right)$ μm

where L = upper limit of the measuring range in mm.

The micrometer must be so adjusted that the cumulative error at the lower and upper limit of the measuring range does not exceed half the total error.

2.37.4. Adjusting for wear of threads, wear of measuring surfaces, spindle locking arrangement and ratchet stop mechanism. In the external micrometer shown in Fig. 2.115, an abutment or anvil is provided which occupies a fixed position in relation to the main nut, in which the spindle moves. The anvil and the main nut are carried by a bow-shaped frame. The main

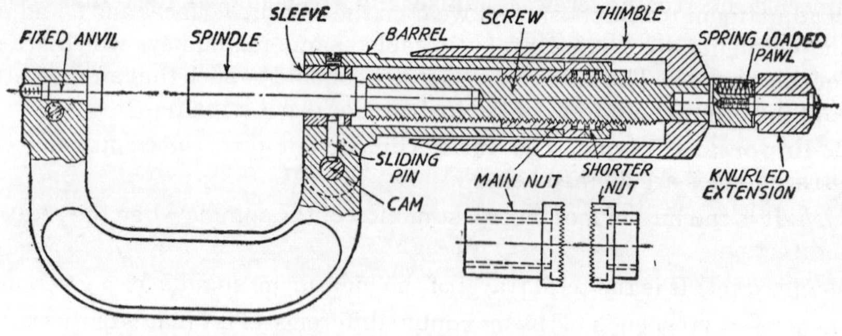

Fig. 2.115. Part-sectional view of an external micrometer showing wear adjustment, spindle locking arrangement and the mechanism of ratchet stop (Micrometer Range 25 to 50 mm).

nut is fitted tightly into the sleeve or barrel which is integral with the frame. The thimble is permanently secured to the screw and is knurled on the outer surface. The spindle having the measuring face at its end is tightly fitted into the screw. The spindle is thus rotated by rotating the thimble. It will be observed that the thimble extends over the end of the barrel so that threaded portions of the screw and the main nut are all the time completely enclosed. The nut and the thread portion of the screw are so proportioned as to ensure full length engagement in all positions of spindle. The fixed index line is marked upon the barrel and the angular graduations around the left hand chamfered end of the thimble. It will be noted that in addition to the main nut there is a shorter nut also by its side with which the screw also engages. The adjacent faces of the main nut and the shorter nut are provided with small V-shaped teeth which form a clutch and prevent the shorter nut from rotating with the screw. In the recess between these two nuts is housed a light coil spring, the tendency of which is to force the two nuts apart and in this way any slight backlash between the threads in the nuts and the screw is automatically taken up.

Adjustment for wear of threads. In addition to forming an abutment for the coil spring, the shorter nut also provides for adjustment in the event of wear on the threads. When the two nuts are originally assembled on the screw, the clutch teeth are meshed in a certain position and marks are provided on the outer surfaces to facilitate replacement in the event of the instrument being dismantled. The wear on the threads is compensated by withdrawing the screw and turning the smaller nut through one, two or more clutch tooth spaces from its original angular position and then bringing it in mesh with the main nut. In this way the threads on the smaller nut are slightly displaced. At the time of reassembling, care must be taken to see that the clutch teeth on the two nuts are meshing properly and that the small nut is held in position until the screw has entered the main nut.

In another most commonly used method of compensating for wear on the thread, provision of an external tapered thread is made on one end of the main nut. The tapered thread is engaged by an adjusting nut having a parallel thread. When sufficient wear occurs between the two threaded portions, the adjusting screw is advanced so as to contact the bore of the main nut slightly and thus making it take up the clearance between the threads of the nut and the screw.

2.37.5. Adjustment for wear of measuring surfaces. It will be appreciated that wear is bound to occur between the measuring surfaces of fixed anvil and the spindle and that this wear will result in errors being introduced. One method of compensating for such wear (shown in Fig. 2.115) is to accommodate the fixed anvil in a reamed hole in the frame and provide a thread extension which passes through a tapped hole of slightly smaller diameter than the reamed locating hole. Thus on rotating the anvil by means of a slot at the threaded end, it can be advanced to compensate for the wear which has taken place. After adjustment the anvil is locked by a screw in a transverse hole in the frame. The head of the screw is radiused to correspond with the radius of the anvil shank thereby obtaining a secure locking action.

In another design the main nut instead of being press fit inside the barrel is threaded into it. The adjustment for taking up wear in such a case can be easily made by screwing the main nut farther into the barrel and thus advancing the spindle without changing the reading on the thimble.

In yet another design shown in Fig. 2.116 the adjustment for wear of the measuring faces is carried out by rotating the thimble relative to the screw. Thus in this method, first the two

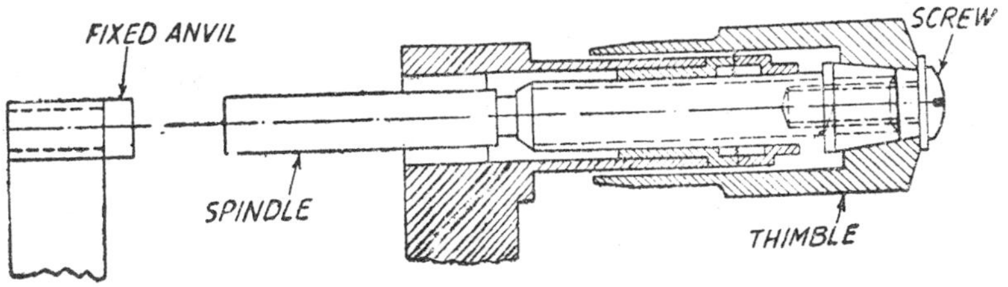

Fig. 2.116. A typical method of adjusting for wear of the measuring faces.

measuring surfaces are brought into contact with each other and the thimble is turned until the zero line marked on it coincides with the index of datum line on the sleeve. Referring to Fig. 2.116, it will be seen that a plain extension is provided at the right-hand end of the spindle beyond the threaded portion on which is mounted a sleeve having a slight external taper. The bore of the thimble is also finished to a corresponding taper. The thimble and the spindle at the tapered surface are held in secure engagement by a screw at the end of the thimble. The thimble is released from the spindle by slackening this screw and pulling thimble out. It is then turned independently of the spindle to obtain the desired setting and the screw is again tightened to secure thimble to the spindle.

In another design which is reverse of this method, a sleeve is fitted over the barrel and is retained in position by friction. The datum line is marked on this sleeve instead of marking on the barrel. This sleeve can be turned in relation to the barrel by means of a small spanner in order to bring the datum line opposite to the zero line on the thimble and thus compensate for wear on measuring faces.

2.37.6. Spindle locking arrangement. In order that any setting of the micrometer be definitely retained, it is desirable to have features of locking the spindle in any position temporarily. One way of achieving this is shown in Fig. 2.115, in which a sliding pin is located in a hole in the

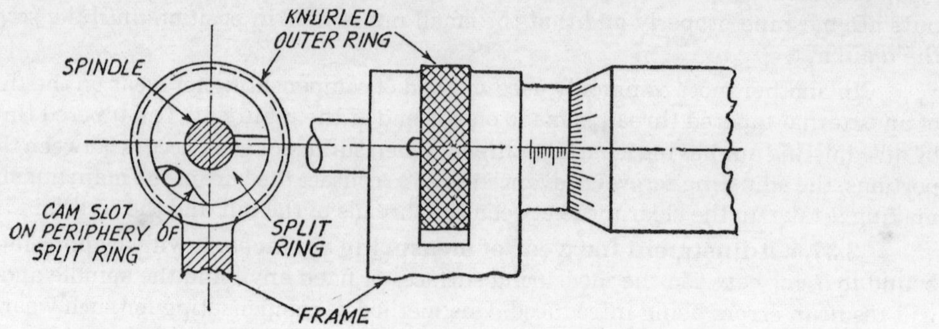

Fig. 2.117. Micrometer spindle locking arrangement.

frame and is pressed against the side of the spindle by a cam. The cam is fitted with dished and knurled heads on both sides of the frame. The shape of the cam is such that when a slight turning movement is imparted to it by gripping the dished and knurled heads on both sides of the frame between finger and thumb, the pin is forced against the spindle and a rigid lock is obtained.

In another design shown in Fig. 2.117, the micrometer is slotted to receive a split ring which surrounds the spindle and is prevented from rotating. The split ring is surrounded by a knurled outer ring. A cam slot is provided on the periphery of the split ring in which a roller moves up and down.

When the outer ring is turned in one direction the roller rides up and cam surface thereby tending to close the split ring and securely clamping the spindle. The grip is released on turning the outer ring in opposite direction.

2.37.7. Ratchet stop mechanism. The object of the ratchet stop is to ensure that a certain maximum torque on the spindle is not exceeded and the sense of the feel of operator is eliminated and consistent readings are obtained. By providing this arrangement, when the spindle has engaged the work with a certain pressure, further rotation causes the ratchet merely to slip, no additional movement being imparted to the spindle. The ratchet stop mechanism is incorporated in the knurled extension provided at the end of the thimble. The mechanism is shown in Fig. 2.115 and is illustrated below. The knurled extension is free to rotate on its retaining stud. The inner face of this is provided with fine ratchet teeth, and through these teeth and the spring loaded pawl (shown in Fig. 2.115) the movement is transmitted to the spindle. As soon as the resistance to the motion of the latter reaches a certain value, the pawl is forced back against the pressure of the spring and ratchet slips.

2.37.8. Reading a micrometer. In order to make it possible to read upto 0.0001 inch in micrometer screw gauge a vernier scale is generally made on the barrel. The vernier scale has 10 straight lines on barrel and these coincide with exact 9 divisions on the thimble. Thus one small division in thimble is further sub-divided into 10 parts and for taking the reading one has to see which of the vernier scale division coincides with a division of the thimble. Accordingly the reading for the given arrangement in Fig. 2.118 will be

on main barrel : 0.120″

on thimble : 0.014″

on vernier scale : 0.0002″

∴ Total reading = 0.1342″.

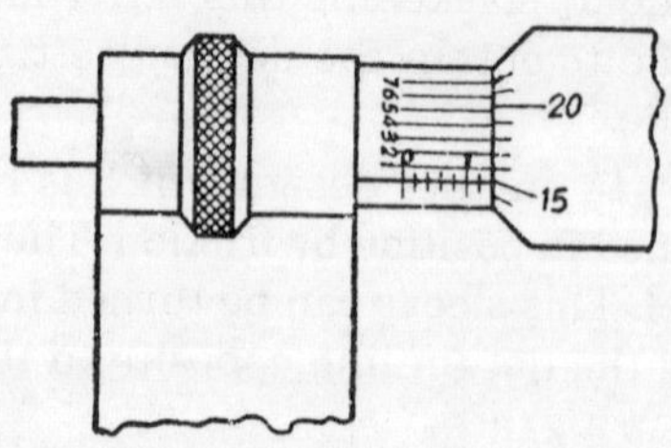

Fig. 2.118. Reading a micrometer.

Before taking readings, anvil and spindle must be brought together carefully and the initial reading noted down. Its calibration must be checked by using standard gauge blocks.

In metric micrometers, the pitch of the screw thread is 0.5 mm so that one revolution of the screw moves it axially by 0.5 mm. Main scale on barrel has least divisions of 0.5 mm. The thimble has 50 divisions on its circumference.

$\therefore$ One division on the thimble $= \frac{0.5}{50}$ mm = 0.01 mm.

If vernier scale is also incorporated then sub-divisions on thimble can be estimated upto an accuracy of 0.001 mm.

Reading of micrometer in Fig. 2.119 is 3.5 mm on barrel and 7 divisions on thimble

$$= 3.5 + 7 \times 0.01 = 3.5 + 0.07 = 3.57 \text{ mm}.$$

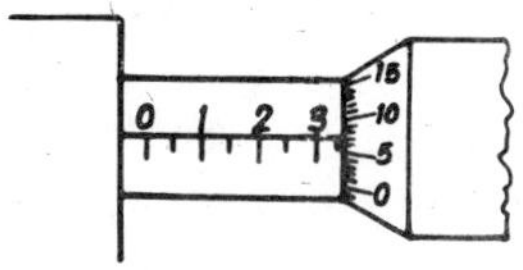

Fig. 2.119. Reading a micrometer.

2.37.9. Cleaning the micrometer. Micrometer screw gauge should be wiped free from oil, dirt, dust and grit. When micrometer feels gummy and dust ridden and the thimble fails to turn freely, it should never be bodily dunked in kerosene or solvent because just soaking the assembled micrometer fails to float the dirt away. Further it must be remembered that the apparent stickiness of the micrometer may not be due to grit and gum but to a damaged thread or to a warped and sprung frame or spindle.

Everytime the micrometer is used, the measuring surface, the anvil and spindle should be cleaned. Screw the spindle lightly but firmly down on to a clean piece of paper held between spindle and anvil. Pull the piece of paper out from between the measuring surface. Then unscrew the spindle a few turns and blow out any fuzz or particles of papers that may have clung to sharp edges of anvil and spindle.

2.37.10. Precautions in using micrometer. In order to get good results out of the use of micrometer screw gauge, the inspection of parts must be made as follows : Micrometer should be cleaned of any dust and spindle should move freely.

The part whose dimension is to be measured must be held in left hand and the micrometer in right hand. The way for holding the micrometer is to place the small finger and adjoining finger in the U-shaped frame. The forefinger and thumb are placed near the thimble to rotate it and the middle finger supports the micrometer holding it firmly.

Then the micrometer dimension is set slightly larger than the size of the part and the part is slide over the contact surfaces of micrometer gently. After it, the thimble is turned till the measuring tip just touches the part and the final movement given by ratchet so that uniform measuring pressure is applied. In case of circular parts, the micrometer must be moved carefully over representative arc so as to note maximum dimension only. Then the micrometer reading is taken.

The micrometers are available in various sizes and ranges, and the corresponding micrometer should be chosen depending upon the dimension.

Errors in reading may occur due to lack of flatness of anvils, lock of parallelism of the anvils at part of the scale or throughout, inaccurate setting of zero reading, etc. Various tests to ensure these conditions should be carried out from time to time.

2.37.11. Tests. The cumulative errors in the micrometer are determined by means of slip guages of Grade I accuracy. The tests are carried out over the entire measuring range, preferably in steps of 0.5 mm, as well as near the maximum and minimum points of the curve of errors in steps of 0.1 mm over a length of 0.5 mm towards the left and right of these points. If the cumulative errors are plotted for various lengths on a graph paper then the total error is the distance between the ordinates of the highest and lowest points determined. The total error of the micrometer is not

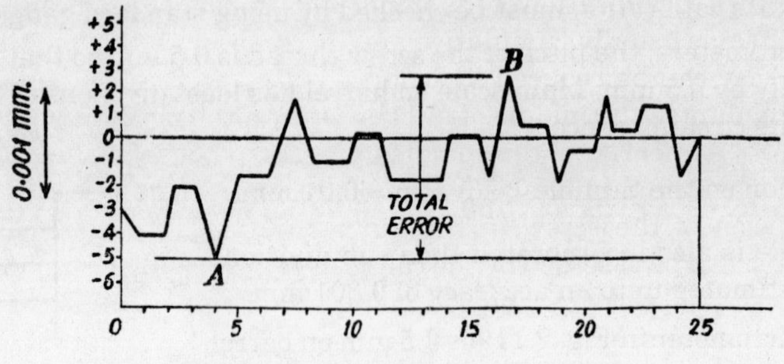

Fig. 2.120. Cumulative Error Curve.

considered as positive or negative deviation from the zero point, but as maximum difference of ordinates of the curve of cumulative errors.

Enlargement of curve A Enlargement of curve B

Fig. 2.121. Error curve enlargement.

Alternative method of calibrating micrometer screws. The accuracy of the readings of a micrometer is usually checked by taking readings on a series of slip gauges. The sizes of these gauges are chosen in such a way as to test the micrometer not only at complete turns of its thimble but also at intermediate positions. This is required as a check on the accuracy of the graduations round the thimble. The following series of readings serve for testing both progressive and periodic errors throughout its range : 2.5, 5.1, 7.7, 10.3, 12.9, 15.0, 17.6, 20.2, 22.8 and 25.0 mm. This series gives readings which work round the thimble twice over and so provide a double check on any periodic error which may be present.

Accuracy of Measuring Faces. Deviations of flatness of the measuring faces are tested with an optical flat moved in such a manner that the number of interference fringes are reduced to a minimum or closed curves are produced (Refer Chapter 6). The number of interference fringes observed on each of the two measuring faces should not exceed 2 for practically all the ranges of grade I micrometer.

The errors in parallelism of the measuring faces of micrometers of various ranges upto 100 mm are tested by observation of interference fringes produced between a flat parallel plate and the measuring faces. The sum of the interference fringes observed on the two measuring faces should not exceed 6, 8 and 10 for ranges 0 to 25, 25 to 75 and 75 to 100 mm in respect of grade I micrometers. Another suitable method for this test is by using auto-collimator.

Bending. The degree of bending of the frame is tested by suspending the frame directly at the spindle and loading with a weight of about 5 to 10 kg directly at the anvil. When the load is removed then the frame should return to its original position. The difference between the measurements of micrometer in unloaded and loaded conditions should not exceed 1.5 μm per kg for micrometer of range 0 to 25 mm. The corresponding values for other ranges are 2, 2.5, 3 μm for 25 to 50, 50 to 75, 75 to 100 mm range.

2.37.12. Accuracy with micrometer measurements. The accuracy of a micrometer depends on (*i*) the degree of calibration of the spindle movement, which is affected by the lead errors of the screw. This effect is usually cumulative and increases with the length of the spindle travel. This type of inaccuracy can be reduced by balanced calibration, *i.e.* by adjusting the thimble to produce error-free reading in the middle of full travel, or of the most frequently used section of the spindle traverse ; (*ii*) the linearity of the spindle movement. For this, any fractional rotation of the screw should result in a proportional advance of the measuring spindle. Drunken threads, or stick-and slip condition of the screw in the nut adds to this inaccuracy. The accuracy of the micrometer indication can be assessed by the calibration process. The accuracy can be improved by manufacturing the spindle of well stabilised material, precisely grinding the screw thread after hardening, using lapped nuts, etc.

The errors recorded on calibration chart will comprise the combined effect of flatness and parallelism of the measuring surfaces, deflection of frame by the applied measuring force, process errors (caused by head transfer while holding the micrometer), reading errors, inadequate alignment or stability in mutual positioning of object and measuring tool, wear, etc.

Following precautions in use of micrometer screw gauge will be of great help in reducing errors :

— Employing plastic insulating grips on the frame to reduce heat transfer.

— Reading errors can be minimised by providing satin chrome finish to eliminate glare, distinct graduation lines applied on a bevelled thimble surface, to facilitate reading with a minimum of parallax error, use of larger diameter thimbles.

— Using a stand which rests on the bench to clamp the hand micrometer instead of holding it in hand, thereby improving alignment and holding stability.

— Applying uniform and minimum measuring force.

— Taking care of wear after some use.

2.38. Inside Micrometer Calipers (Fig. 2.122)

This micrometer caliper has no U-shape frame and spindle. The measuring tips are constituted by the jaws with contact surfaces which are hardened and ground to a radius. One of the jaws is held stationary at the end and second one moves by the movement of thimble. A locknut is provided to check the movement of the movable jaw. This facilitates the inspection of small internal dimensions.

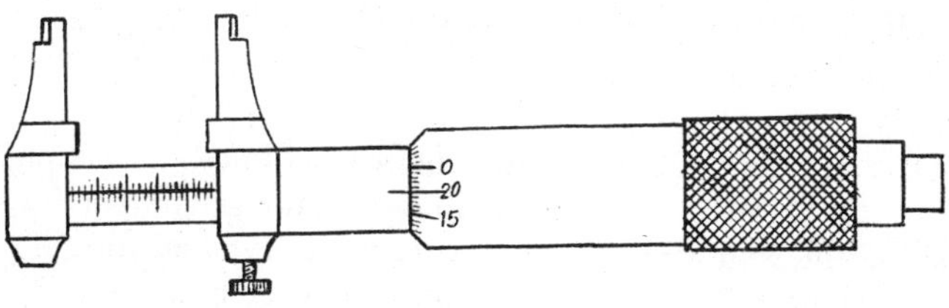

Fig. 2.122. Inside micrometer caliper.

This is not widely used as other types have various advantages over it. Its range is from 5 to 50 mm.

2.39. Inside Micrometer

For the measurement of larger internal dimensions, inside micrometer is employed. It consists of four parts : (*i*) Measuring head (micrometer unit), (*ii*) Extension rods, (*iii*) Spacing collars, (*iv*) Handle (Refer Fig. 2.124.) Using one range of extension rods, distance between the contacting surface is varied by rotating the thimble upto the extent of the screw length. Greater distances are obtainable by the use of different extension rods and spacing collars. The measuring tip on the thimble is adjustable one in order to compensate for the wear taking place. Spacing collars are used when those dimensions are needed which cant's be obtained by the use of extension rod and micrometer unit. For using this instrument first the diameter of bore is measured approximately by a scale. Extension rod is then chosen to the nearest one cm and inserted in the micrometer head. Then zero error is checked by taking the dimension on standard sizes specimen. The micrometer is then adjusted at a dimension slightly smaller than the diameter of bore. One micrometer head is then held firmly against the bore and other contact surface is adjusted by moving the thimble till correct feel is sensed. One contact surface must always be moved sideways up and down in order to ensure that diameter is being measured and not chord. The micrometer is then removed and reading taken. Dimension from the micrometer head is noted down and to this, lengths of extension rod and collar are added up.

By suitable base arrangement, inside micrometer can be converted into height gauge.

It is used for measuring the inside diameters of cylinders, rings etc. and also for measuring parallel surfaces and for setting calipers etc.

The desired range is obtained by assembling measuring rods and lapped gauges or spacing collars to the micrometer held.

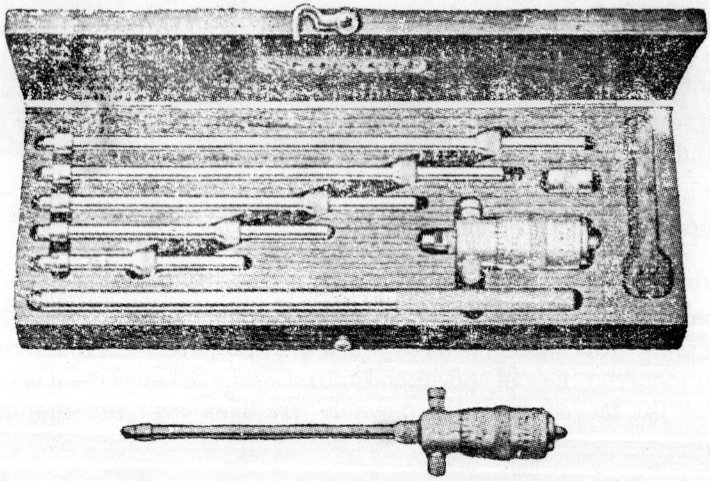

Fig. 2.123. Inside Micrometery (Commercially available inside micrometer set).

This form of internal micrometer is, in effect, merely an adjustable end measuring bar. As in the case of an ordinary external micrometer, there is a barrel and a thimble, but no frame and no spindle. The measuring points are at extreme ends and adjustment is effected by advancing or withdrawing the thimble along the barrel. A series of anvil bars (extension rods) of different lengths are provided in order to obtain a wide measuring range. When measurements are to be taken inside the bores of comparatively small diameters, a handle of sufficient length is provided which can be screwed into a radial hole in the barrel.

The description of various units of internal micrometer is given below :

Micrometer Unit. The body of the measuring head is made of suitable quality steel and its spindle and terminal measuring faces are made of high grade tool steel. The measuring face of the anvil is so designed as to have a hardness of about 800 HV (63.5 HRC). It is desirable that the measuring face be tipped with tungsten carbide or other suitable hard material or be faced with a deposit of hard chromium. The radius of curvature of the measuring face is made slightly less than one-half of the smallest measuring range of the micrometer. The micrometer screw has a pitch of 0.5 mm and the threads of the screw and nut are truncated so as to continue contact to the flanks

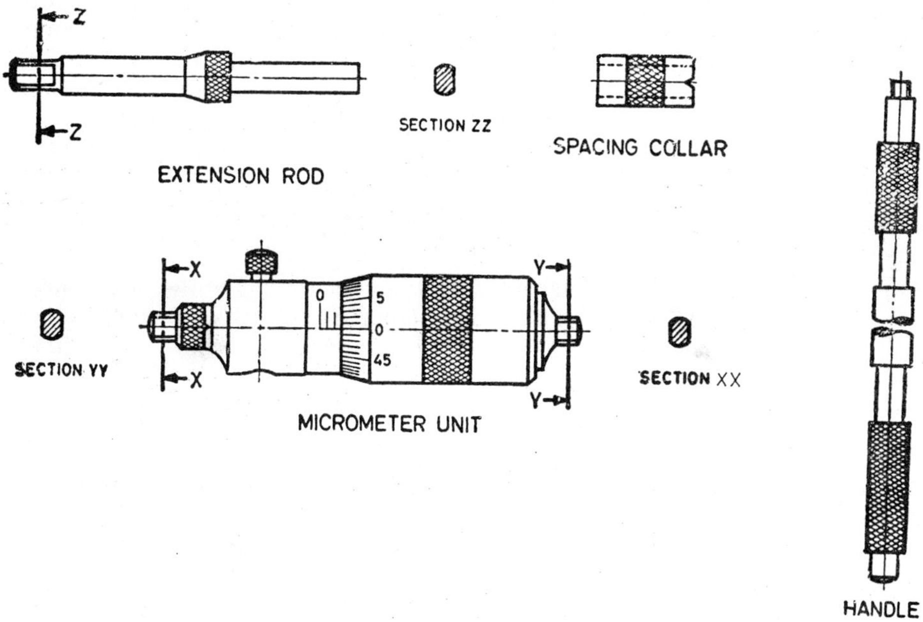

Nomenclature of Internal Micrometer
Fig. 2.124

of the thread. The screw should travel smoothly throughout the length of its travel. An adjusting nut is provided to arrange for the thimble to be sufficiently tight for the micrometer to retain its reading after being set. There should be no perceptible backlash between the spindle screw and nut. When the micrometer is at its maximum reading, there should be full engagement of the nut and micrometer screw. It is desirable that the travel of the micrometer be continued for at least one turn of the screw beyond each end of the graduated scale on the barrel. Generally means are also provided for compensation of wear between the screw and nut. The diameter of the graduated edge of the thimble is generally not less than 10 mm. For ease of reading, the surfaces of the thimble and barrel should have a dull finish and the graduation lines blackened.

Generally the micrometer is provided with means for adjusting the zero setting. The means of adjustment should be such that after resetting, the parts are secured, and the original accuracy of the instrument is not impaired. Suitable spanners and keys are provided for this adjustment.

The measuring head with its associated extension rods and collars should conform to the following limits of error when tested at standard reference temperature of 20°C.

Measuring range *mm*	*Maximum permissible error in length* *mm*
25—150	± 0.005
150—300	± 0.010
300—450	± 0.015
450—600	± 0.020

The error in traverse of the micrometer screw should not exceed 0.003 mm, whether errors are progressive or periodic in character.

Extension rods. These are provided in order to enable any measurement to be made throughout the range specified for the set, spacing collars being used whenever necessary. These are made of suitable material hardened upto 800 HV and the measuring faces are tipped with tungsten carbide or other hard material. The measuring faces and the contact face are finished by lapping operation. The extension rods may either incorporate their own measuring anvils or these may take the form of different pieces to which a separate common anvil piece may be fitted. When the extension rods have their own measuring faces, a suitable adjustment for zero setting is incorporated in it (*e.g.*, the attachment of the anvil by means of a friction-tight fine pitch thread) and there is no need of providing any such adjustment on the micrometer head. If the extension rods have separate anvil pieces for measuring faces then the adjustment for zero setting may be incorporated either in the anvil pieces or in the measuring head. The range of measurement for which an extension rod is applicable is always marked on it.

Spacing collars. These are intended for smaller adjustments in the range of measurement. The ends of the spacing collars are well finished by lapping and are flat and mutually parallel and square to the axis. No sharp edges are there and at least one face of the spacing collar is provided with dirt clearance grooves.

Handle. The micrometer set for the measuring range 25 to 50 mm is generally supplied with a suitable detachable handle so that micrometer head can be easily lowered into deep holes.

The micrometer assembly with any combination of extension rods and spacing collars should be very rigid. When the assembled unit is supported at its ends, then the deflection of mid-point under an applied load of 0.5 kg should not exceed 0.05 mm for an assembly 150 mm long with an increase of 0.05 mm for each increase in the length of 25 mm. The maximum deflection allowed for an assembly 1 m long is 1.75 mm.

2.39.1. Self-Centering Inside Micrometer. Measures through bores, blind bores, registers.

Consists of micrometer screw and screw-in, interchangeable measuring head.

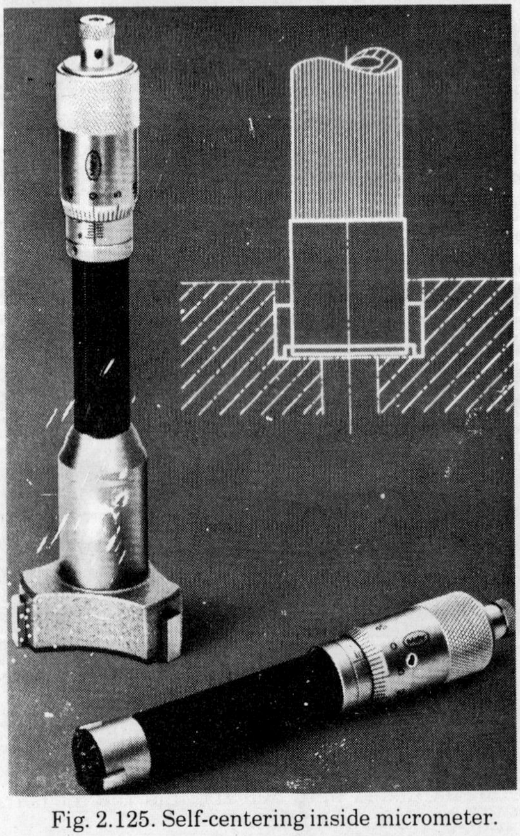

Fig. 2.125. Self-centering inside micrometer.

Components :

Micrometer Screw. Measuring spindle precision ground, hardened throughout. Ratchet stop. Heat insulator. Moving parts and scales satin chrome finished. Threaded for convenient mounting of measuring heads.

Measuring Head. Interchangeable. Three laterally protruding fingers, positioned at a 120° angle to each other, provide self-centering action and firm positioning. Positively no tilting or tipping. From 10 mm and up the fingers extend over the face of the head to allow measurements through the entire bore length. Individual measuring heads.

2.40. Stick Micrometers

Stick micrometers are designed for the measurement of longer internal lengths. These comprise of the following parts :

(*a*) A 150 mm or 300 mm micrometer unit fitted with a micrometer of 25 mm range and having rounded terminal faces.

(*b*) A series of extension rods which together with the micrometer unit, permit of a continuous range of measurements upto the maximum length required.

How the stick micrometer will look like is shown Fig. 2.126. Its accuracy is of the order of ± 0.005 mm throughout the range. It is essential that the radius of curvature of the terminal faces

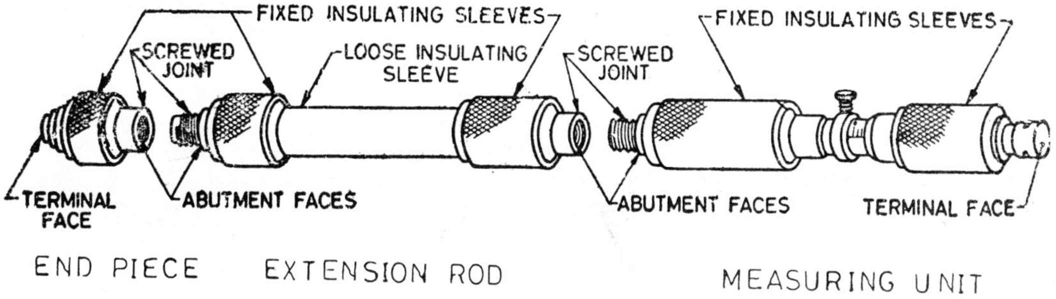

Fig. 2.126. Stick micrometer.

of the micrometer unit be slightly less than one-half the smallest measuring range of the micrometer unit, and abutment faces should have minimum radial width of 2 mm and properly lapped ; these should be parallel to each other and normal to the axis of the micrometer unit to within 0.005 mm across their diameters.

It may be noted that screwed joints are used for joining the end piece, extension rod and the measuring unit. The screw units generally have threads of 0.5 mm pitch. The extension rod is generally hollow and has a minimum external diameter of 14 mm. In the design of extension rod shown in Fig. 2.126, it will be noted that the contact faces come together by screwing and the pressure between the contact faces depends upon the heavy or light tightening of the screw. In order to overcome this difficulty, the extension rods of type shown in Fig. 2.127, are generally employed. This type consists of length standards which are mounted in protective bushing and when these extension rods are combined, the contact faces of the length standards come together under constant pressure and thus the measurement is not affected by heavy tightening of the screw. The accuracy of this instrument is of the order of ± 0.005 mm, *i.e.* the micrometer unit with terminal piece in position should record correctly throughout its range to within ± 0.005 mm. In this type of micrometer, there should be sufficient play between the external and internal threads of the joint to permit the abutment faces of the various parts of the micrometer to butt together solidly.

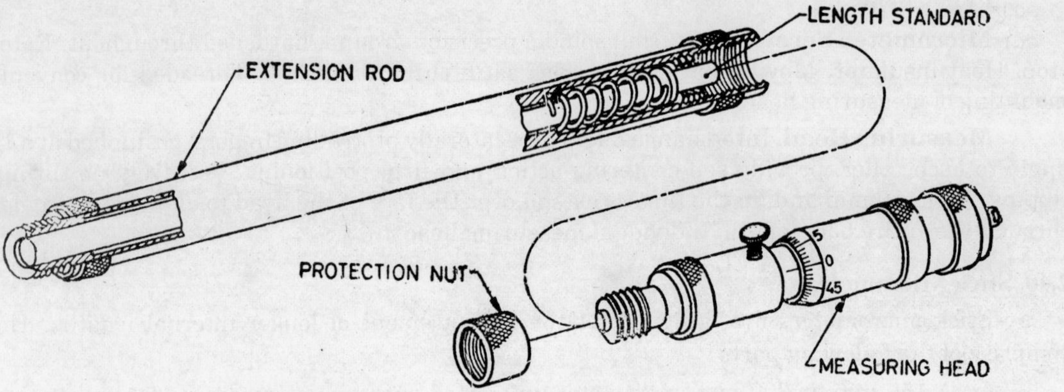

Fig. 2.127. Improved extension rods of stick micrometer.

2.41. Method of Testing Internal Micrometers

The accuracy of the internal micrometer reading is checked in two parts, *viz.*, checking the accuracy of the traverse of the measuring head and the checking of the accuracy of the overall lengths when the measuring head, set to zero, is associated with the various extension rods in turn.

The accuracy of the traverse of the measuring head is determined by clamping it in the Vee-block with its axis in line with a sensitive indicator and the feeler (with a flat contact face) of the indicator touching the rounded contact face of measuring head. The feeler is set initially such that it reads zero when the micrometer head is also reading zero and a slip gauge of convenient dimension has been introduced between the two faces. The micrometer head is then set at same reading and correspondingly the slip gauge size is reduced. Any error in the reading is revealed by a corresponding departure of the indicator pointer from its initial position. In this way the accuracy is tested at various readings of the micrometer head. In order to check both progressive and periodic errors, and in order that both the type of errors may be revealed, the reading in micrometer head of 0—25 mm range may be taken at following values 0, 2.4, 5.1, 7.7, 10.3, 12.5, 14.9, 17.6, 20.2, 22.8, 25.0 mm. The accuracy of the overall length is determined by using a vertical comparator. The indicator of the comparator is set at a height corresponding to that of the overall length to be measured, from the flat base. After this setting the internal micrometer is placed under it in the maximum position and the error noted down.

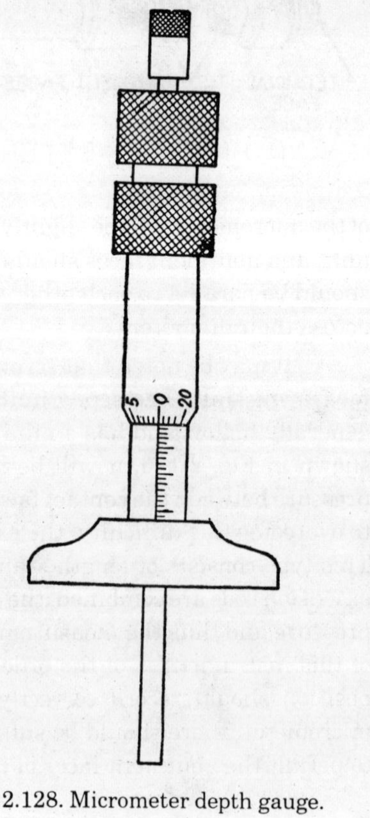

Fig. 2.128. Micrometer depth gauge.

2.42. Micrometer Depth Gauge (Fig. 2.128)

It is used for measuring the depths of holes, slots and recessed areas. It has got one shoulder which acts as reference surface and is held firmly and perpendicular to the centre line of the hole. Here also for larger ranges of measurements, extension rods are used. The screw of micrometer depth gauge has range of 20 mm or 25 mm. The length of the micrometer depth gauge varies from 0 to 225 mm. The rod is inserted through the top of the micrometer. The rod is marked after every 10 mm so that it could be clamped at any position. In using this instrument, first it must be ensured that the edge of the hole is free from burrs.

The scale here is calibrated in the reverse direction. The accuracy again depends upon the sense of touch.

2.43. Thread Micrometer Caliper

This is just like an ordinary micrometer with the difference that it is equipped with a special anvil and spindle. The anvil has an internal vee which fits over the thread. The anvil in this case is not fixed but is free to rotate. Thus vee of the anvil can accommodate itself to any rake range of thread. The spindle on the other hand has a ground conical shape. When the conical spindle is brought into contact with the vee of anvil, micrometer reads zero. Different set of anvils are provided for different types of threads and the contact points of the micrometer are so designed that some allowance for thread clearance is always made.

Thread micrometer is used for the measurement of the pitch diameter but the accuracy is influenced by the helix angle of the thread.

Screw thread micrometer caliper (Refer Fig. 2.129). It is used for accurate measurement of pitch diameter of screw threads. The micrometer has a pointed spindle and a double V-anvil, both correctly shaped to contact the screw thread of the work being gauged. It directly reads in terms of pitch diameter as the zero reading of the micrometer corresponds to the closed position of anvil and spindle when both are in perfect match with each other.

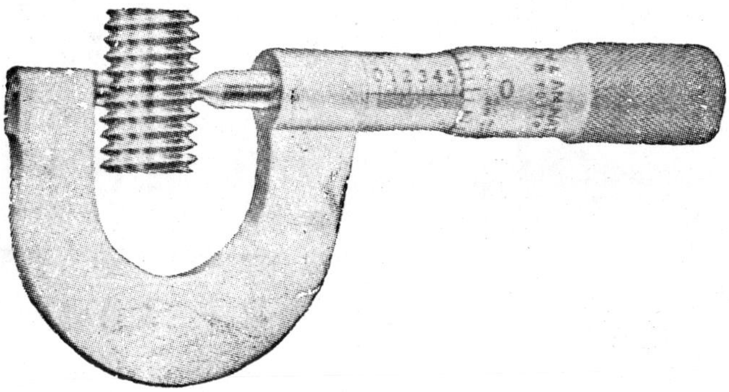

Fig. 2.129. Screw Thread Micrometer Caliper.

The angle of the V-anvil and the conical point at the end of the spindle corresponds to the included angle of the profile of the thread. The V-anvil is allowed to swivel in the micrometer frame so that it can accommodate itself to the helix angle of the thread. The extreme point of cone is rounded so that it will not bear on the root diameter at the bottom of the thread, and similarly clearance is provided at the bottom of the groove in the V-anvil so that it will not bear on the thread crest. The spindle point of such a micrometer can be applied to the thread of any pitch provided the form or included angle is always the same. The V-anvil is however limited in its capacity ; a number of different blocks being required to cover a full range.

2.44. Outside Micrometer Caliper (*Box type with Sliding Anvil*) (Refer Fig. 2.130)

It is designed for fast, easy and precision measuring of large work having dimensions above 250 mm. Generally six standards varying in length in step of 25 mm are furnished. It may be noted that it is impossible to check the zero readings by bringing the two measuring faces into contact. For this purpose a standard rod is employed. In order to increase the utility of instrument to cover wide range of measurements, standard rods of other dimensions are also provided which are fitted

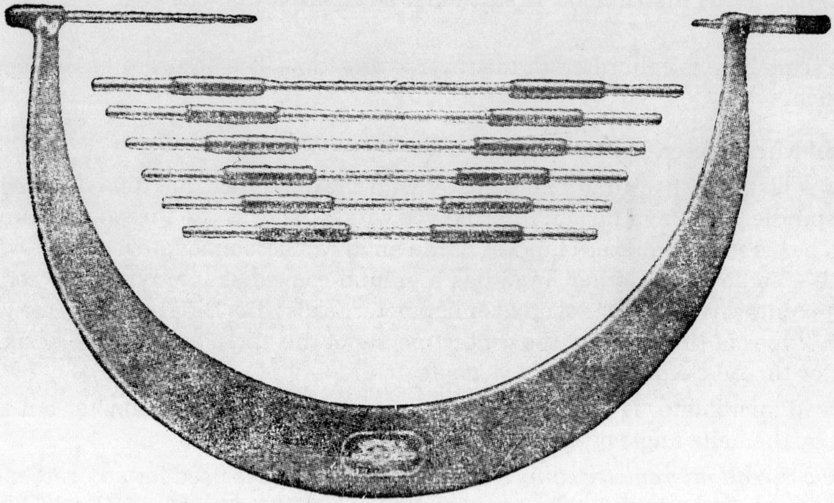

Fig. 2.130. Outside Micrometer Caliper.

in place of fixed anvil. All these rods are provided with ebonite pads in order to prevent the transfer of heat from operator's hand.

2.45. V-Anvil Micrometer Caliper (Refer Fig. 2.131)

Special features. Any out-of-roundness condition can be quickly checked in centreless grinding and machining operations.

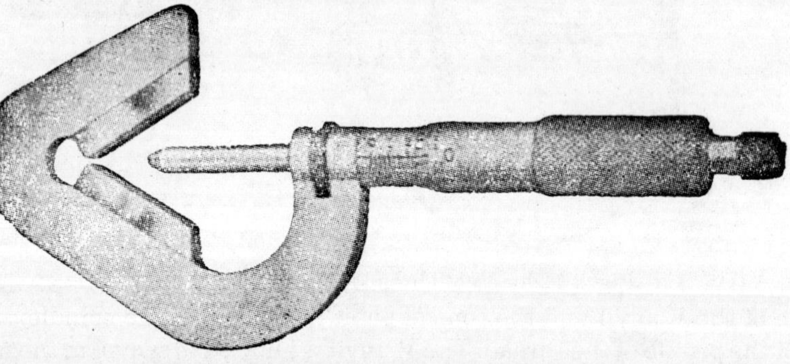

Fig. 2.131. V-Anvil Micrometer Caliper.

Direct reading eliminates use of special fixtures. It can be used for measuring odd-fluted taps, milling cutters, reamers etc.

In this micrometer, angle of Vee equals 60 degrees and the apex of the Vee coincides with axis of spindle. The zero reading of micrometer starts from a point where the two sides of Vee meet. It may now be seen that as the angle of Vee is 60°, the micrometer will measure a distance of $3d/2$ for a round piece of diameter d. It is thus simple matter to obtain the diameter from reading.

2.46. Blade Type Micrometer (Refer Fig. 2.132)

It is ideally suited for fast and accurate measurement of circular formed tools, diameter and depth of all types of narrow grooves, slots, keyways, recesses, etc. It has non-rotating spindle which advances to contact the work without rotation.

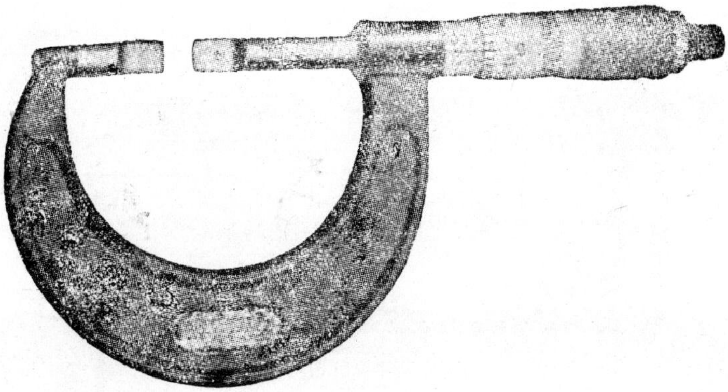

Fig. 2.132. Blade Type Micrometer.

2.47. Micrometer for Measuring thickness of Cylinder Walls

The ordinary micrometer can't be used for measuring the thickness of the wall of a tube, sleeve or bush because of the concavity of the internal surface. In the micrometers meant for this purpose, the fixed anvil is provided with a spherical measuring surface and the frame is cut away on the outside to permit of the anvil being introduced into tubes of diameters as small as 7.5 mm.

In another design shown in Fig. 2.133, the anvil is made of cylindrical form, its axis being perpendicular to the axis of the spindle.

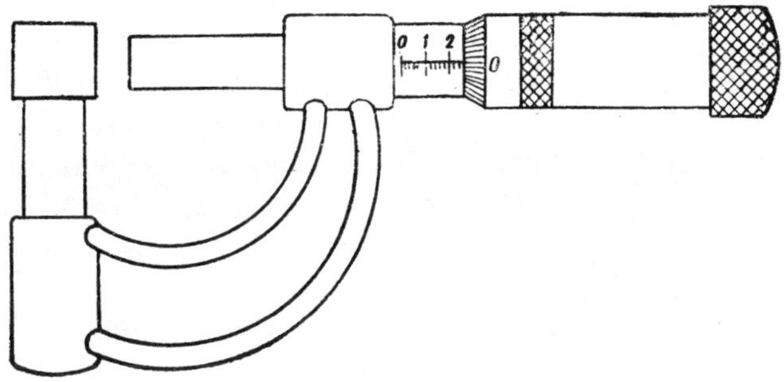

Fig. 2.133. Micrometer for thickness of cyliner walls.

2.48. Dial Micrometer Caliper

This is equipped with a movable anvil whose slight axial movement is indicated by a dial indicator, fixed into the frame. The arrangement is also provided to retract the anvil simply by pushing a button. Due to provision of indicator, the correct readings can be obtained as the indicator will show reading the moment some measuring pressure is felt. It can be very useful for statistical quality control work as variations in size can be noted by the dial indicator.

2.49. Bench Micrometer (Refer Fig. 2.134)

Special features. Super precision dial indicator for comparator measurements upto 1 micron (μm).

High precision micrometer head with direct readings in 25 mm units.

Anvil retractor device for repeated measurements.

Adjustable work-table.

Linear friction free motion transfer mechanism between anvil and indicator for high accuracy.

Anvil pressure adjustable and remains constant.

Fig. 2.134. Bench micrometer.

2.50. Taper-screw Operated Internal Micrometer (Refer Fig. 2.135)

In this micrometer, tapered screw threads are provided at the end of the main micrometer spindle, and the tapered threads engage with three similar feelers disposed 120° apart, their outer faces forming a perfect circle. These feelers slide in guides on the micrometer casing. The pitch of the screw cut on the taper serves as the micrometer measuring screw. The readings are taken in the usual way on the sleeve scale and the thimble. The taper screw is made of hardened and ground gauge quality steel. The shape of the thread is such that the feelers are supported at right angles to the taper thus minimising friction effects. At the end of the thimble is provided a ratchet stop mechanism which ensures constant measuring pressure between the feelers and the bore, and also eliminates the personal factors of sense of feel. When ratchet is revolved then slight vibrations are produced and the instrument is allowed to adjust itself in bore.

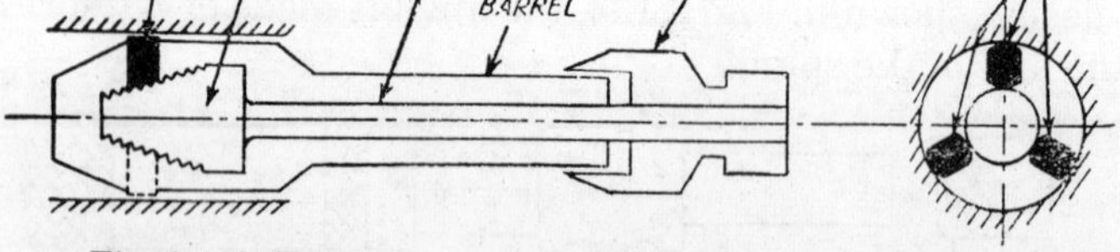

Fig. 2.135. Taper-screw operated internal micrometer.

The internal measuring instrument has no moving parts such as bearings and pinions which are liable to wear with constant use. It provides large contact surfaces for the sliding members.

2.51. Groove Micrometers (Fig. 2.136)

These Micrometers are designed for measuring grooves, recesses and shoulders located inside a bore. They have standard (12.7 mm) dia discs. 6.35 mm dia discs are used to reach hard to get at locations inside a small bore.

All disc thicknesses are 0.75 mm and are hardened and lapped to minimize parallax and to achieve a higher degree of accuracy.

Fig. 2.136. Groove Micrometer.

These groove micrometers measure not only thickness and spacing of grooves, but also measure from an edge to a land, or from shoulder to groove. Micrometers are satin-chrome finished throughout.

Several applications are shown in Fig. 2.137.

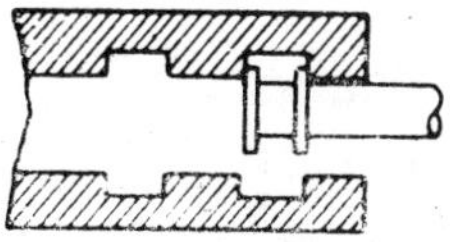

(*a*) Outside measurement

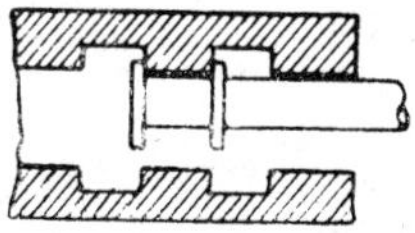

(*b*) Inside Measurement (add 1.5 mm to Reading)

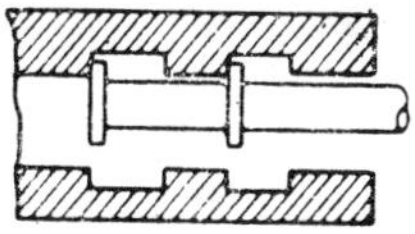

(*c*) Edge to Edge Measurement (add 0.75 mm to Reading)

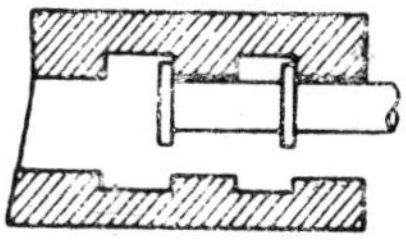

(*d*) Edge to Edge Measurement (add 0.75 mm to Reading)

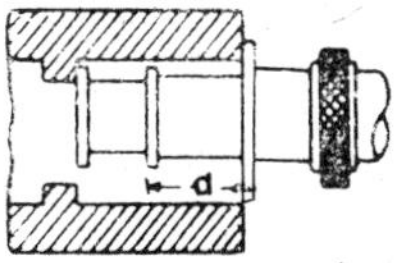

(*e*) Edge to edge measurement with locating disc. (Add 1.5 mm to reading)

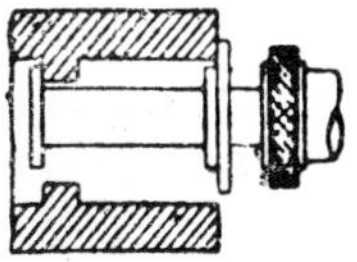

(*f*) Edge to edge measurement with locating disc. (Add 0.75 mm to reading)

Fig. 2.137. Typical applications of groove micrometers.

2.52. Digital Micrometers

Fig. 2.138 shows a digital micrometer available now-a-days. These offer direct reading to 0.001 mm and employ liquid display operating on a alkaline manganese battery.

Special Features. 1. Stainless steel spindles.

2. Every hundredth in metric (every thousandth in inches) is numbered for easy reading.

3. Measuring faces carbide tipped for a long tool life.

4. Friction thimble or ratchet stop for exact and repetitive readings.

5. Spindle thread hardened, ground and lapped.

6. Positive locking clamp ensures locking of spindle at any desired setting.

7. Large diameter pearl chrome thimbles for easy reading.

8. Clear, crisp black graduations and figures for easy reading against satin-chrome finish.

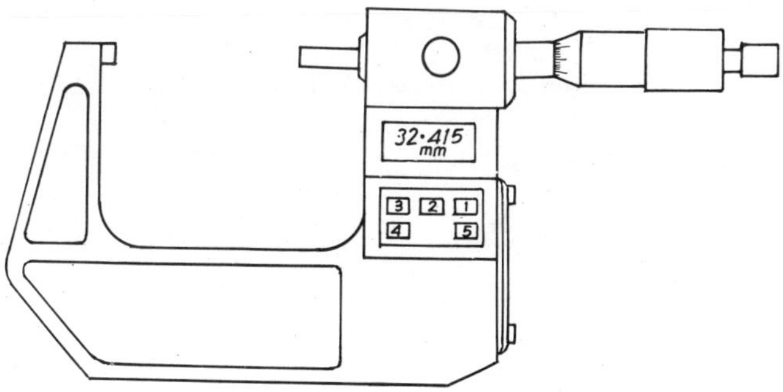

Fig. 2.138. Digital micrometer.

9. Anvil and spindle hardened and precision ground—with micro-lap finish on ends.

10. Quick and easy wear adjustment.

11. Operation is simple with push button controls for "Zero" reset and indication "hold"

Specifications

Thimble graduation	:	0.01 mm
Spindle screw thread	:	0.5 mm pitch

Feed error	:	± 0.001 mm (20°C)
Quantizing error	:	± 1 count max
Resolution	:	0.001 mm
Ranges	:	25, 50, 75 and 100 mm
Measuring force	:	6 ~ 10N (constant)
Ambient temperature	:	0 ~ 40°C (operation)—10 ~ 60°C (storage)
Weight	:	430 gm.

Control Push Buttons (Refer Fig. 2.138)

1. ON/OFF : Power ON or OFF.

2. IN/MM : To select the unit (in or mm) of measurement.

At power on, the unit of mm is present.

3. ZERO : To reset the display to zero at any desired position.

4. HOLD : To hold measurements taken until you push the button again. Then, the display shows up the current position of the spindle measured from predetermined zero point. The "hold" condition can also be released by the ZERO button.

5. ORIGIN : For LCD Micrometers range 25-50 mm (1-2″) and above, ORIGIN button sets minimum value for the micrometer depending on its size. Micrometer count then commences from this value.

Alarm indication showing excessive speed of thimble is provided and alarm can be re-set by the ZERO button.

2.52.1. Micrometer with Dial Comparator. Measures quickly and accurately diameters of cylindrical parts (shafts, bolts, shanks etc.), thicknesses and lengths. Very handy indicating micrometer for measuring precision parts produced in batches.

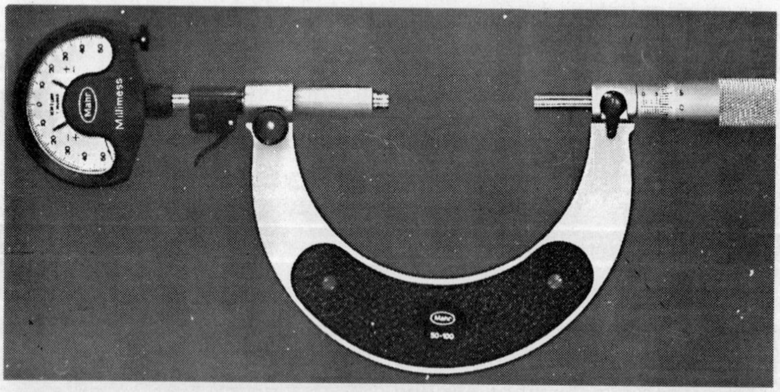

Fig. 2.139. Micrometer with dial comparator.

Rugged steel frame with heat insulators. Up to 100 mm one-piece construction of frame and spindle guide for highest stability. Precision ground stainless steel spindle, hardened throughout. Positive locking lever. Moveable anvil mounted in extra long guideway, retractable by trigger control.

Measuring faces carbide tipped. Dial Comparator interchangeable. Frame, moving parts and scale satin chrom finished.

2.52.2. Bench Micrometer with Dial Comparator (with interchangeable Anvils). Measures quickly and accurately diameters and lengths on workpieces of all types. Ideal for small precision parts. Rugged, heavy duty base. Micrometer screw incorporates precision ground, non-rotating sliding spindle. Extra large thimble. Moveable anvil mounted in extra long guideway,

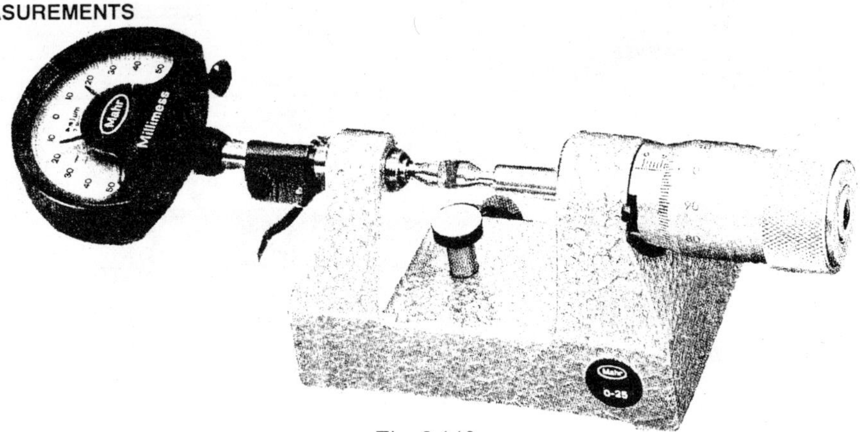

Fig. 2.140

retractable by trigger control. Positive locking lever. Comparator interchangeable. Spindle and moveable anvil with accurate tapered bores for quick and easy mounting of interchangeable anvils. Comes with workrest. Moving parts and scales satin chrome finished.

Range	0-25 mm
Reading of Micrometer screw	0.01 mm
Reading of Comparator	1 μm
Range of Comparator	± 50 μm
Spindle thread pitch	1 mm
Thimble dia.	30 mm
Throat depth	32 mm
Workrest dia.	14 mm
Workrest height adjustment	12 mm
Measuring pressure	7.5 N
Accuracy (deviations)	
Micrometer screw	≤ 2 μm
Comparator	≤ 1 μm
When setting with gauge block	≤ 1 μm

2.53. Differential Screw Micrometer

A very high degree of accuracy can be obtained in the micrometer screw gauges utilising the principle of differential screw on the operating spindle. In such a micrometer, the screw has two types of pitches, one smaller and one larger, instead of one uniform pitch as in conventional micrometer. Both the screws are right handed and the screws are so arranged that the rotation of the thimble member moves one forward and the other backward. If the larger screw has a pitch of 1.25 mm and smaller screw of 1.00 mm pitch, then the net result would be a total forward movement of 1.25 – 1.00 = 0.25 mm per revolution. Thus if thimble has 100 divisions and vernier scale be engraved on the sleeve, then the least count of the instrument $= 0.25 \times \frac{1}{100} \times \frac{1}{100} = 0.00025$ mm. It will be noted that this will have appreciably smaller total range of linear movement, although the main spindle's travel is larger. This is because the main spindle which is attached to the moving portion gets only differential movement.

In Fig. 2.141, it will be seen that larger screw is engaged with nut in the moving member which in turn ends in the main spindle.

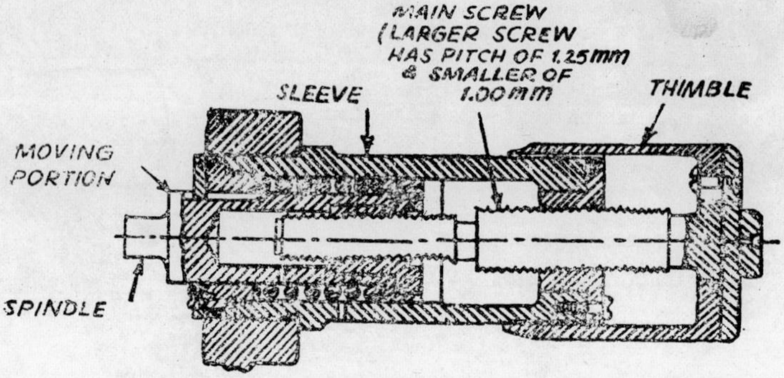

Fig. 2.141. Differential screw micrometer.

The larger screw and the smaller screw are integral parts. The screw is firmly attached with the thimble and this is rotated by rotation of the thimble.

Though, theoretically it is possible to read up to 0.0002 mm with this instrument, since the feel method is employed, the absolute accuracy would not necessarily by of this high order.

2.54. Graduated Scales

Graduated scales are accurately graduated at intervals of 1 mm. When used in conjunction with micrometer microscope, these can provide instrument and machine settings to an accuracy of the order of ± one-fourth of a micron. A microscope is fitted over the scale markings and have

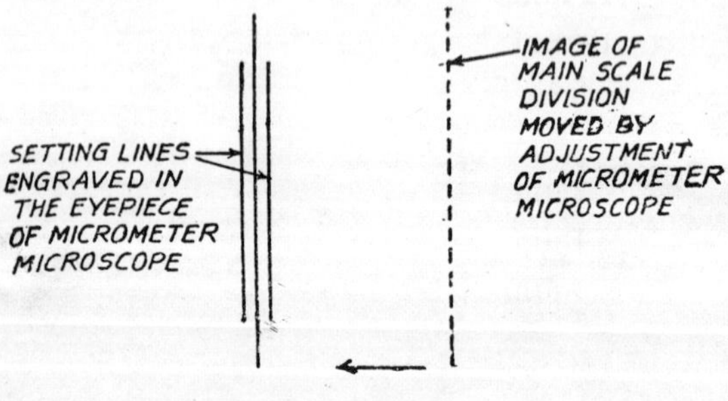

Fig. 2.142. Field of view of microscope.

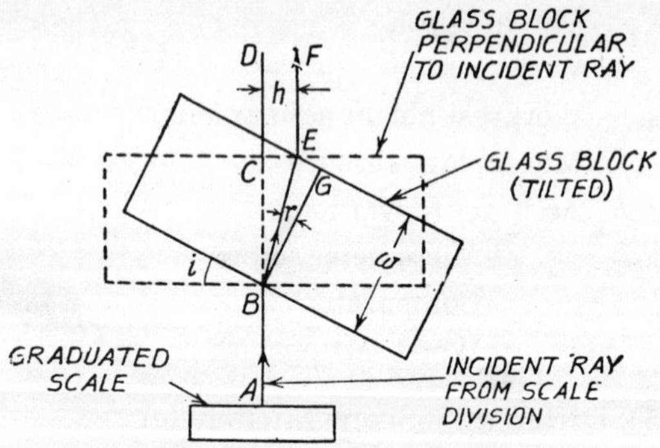

Fig. 2.143. Principle of operation.

micrometer provision to move the setting lines engraved on the microscope eye-piece. For observing intermediate readings, the micrometer is adjusted so that the two setting lines, symmetrically straddle the image of main scale division in the field of view of the microscope as shown in Fig. 2.142.

In other arrangement the optical arrangement is used (instead of moving the eye-piece table laterally by the micrometer) in which the image of the scale is moved optically as shown in Fig. 2.142.

A glass block is mounted between the eye-piece and the graduated scale. When the glass block is perpendicular to the optical axis, then scale division is sensed in the same position, there being no shift in the incident ray from the main scale and it travels straight along the line *ABCD*. (Refer Fig. 2.143) When the glass block is rotated by angle 'i', then the incident ray *AB* is refracted by angle r with reference to line *BG* and moves along *BE* and finally at other end again by refraction it travels parallel to original line *CD*, but now along *EF*, being displaced by distance h.

If w be the width of glass block, then, $h = CE = BE \sin (i - r)$

where i = angle of incidence of ray from scale

and r = angle of refraction of glass block

or $$h = \frac{BG}{\cos r} . \sin (i - r) = w \sin (i - r)/\cos r$$

For very small angles, $\cos r$ is nearly equal to 1 and $\sin i = i$.

If μ = refractive index of glass block, then

$$\mu = \frac{\sin i}{\sin r} = \frac{i}{r} \text{ or } r = \frac{i}{\mu}$$

$$\therefore \quad h = w \sin\left(i - \frac{i}{\mu}\right) = w\left(i - \frac{i}{\mu}\right) = wi\left(\frac{\mu - 1}{\mu}\right)$$

For a block of 5 mm thickness and $\mu = 1.5$, a rotation of 10 minutes causes displacement of 0.005 mm. Thus rotation of block can be easily calibrated in terms of linear displacement and high resolution over a long length can be obtained and the cumulative errors likely to occur with other methods of length measurement are avoided.

2.55. Other Instruments for Internal Measurements

2.55.1. Telescope Internal Gauge. It is used for the measurement of internal diameter of a hole during machining operation (Fig. 2.144). It consists of two spherical contacts carried on two rods. The two rods can slide within a tube and are forced apart by an internal spring. For taking the measurements the spring unit is compressed and inserted into the hole. The contacts then move out so as to press on the surface. They are then locked in position by means of a locking screw. The telescopic gauge is then taken out and the dimension across two tips is measured by some other external measuring instrument. The precaution to be taken in the use of telescopic gauges is that they should be inserted squarely into the bore, and centralised properly.

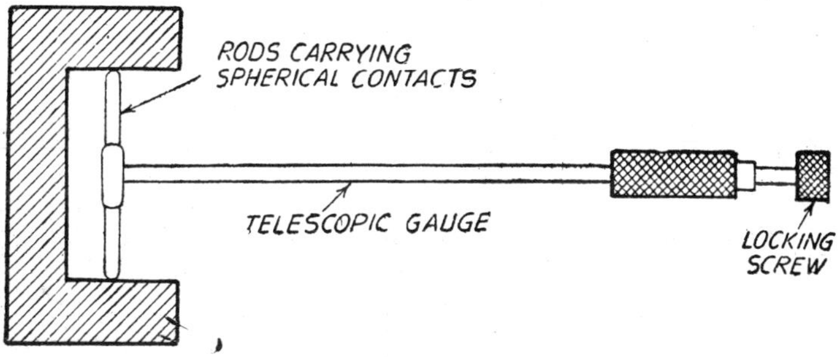

Fig. 2.144. Telescope internal gauge.

2.55.2. Hemispherical Gauge for Small Bores. The telescopic gauge described above can be used for big bores only. For the measurement of bores that are too small, hemispherical gauge is considered to be very useful. It consists of the spring steel arms having hemispherical ends, within which is a wedge shaped member attached to a spindle which can be moved endwise by a knurled head (Fig. 2.145). Thus the hemispheres can be expanded or contracted by the movement of the

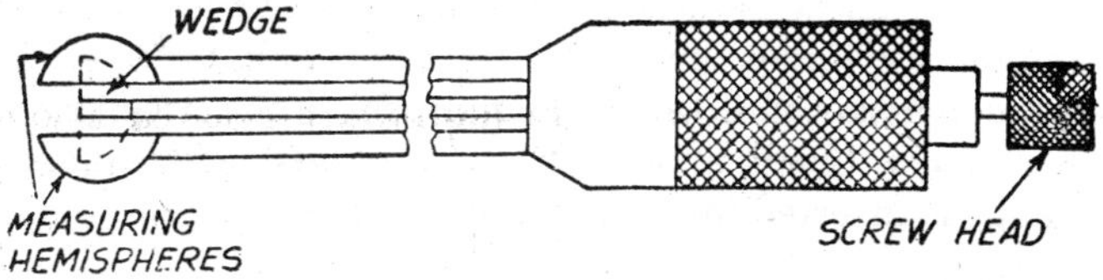

Fig. 2.145. Expanded position of hemispherical gauge.

wedge. Again this gauge is also operated by the sense of feel. When the hole to be measured is gauged in the above way, the gauge is taken out and the reading noted down by measuring over the hemispheres with an external micrometer. The gauges are available in four sizes to measure from 3 to 12 mm diameter.

2.55.3. Measuring diameter of deep bores. There are three units of gauges meant for internal measurement of deep bores. (1) calibrated rod, (2) the amplification unit with dial gauge, and (3) measuring head.

The calibrated rods are made in various sizes. Sometimes a number of short rods threaded at ends are also used in combination to get different desired lengths. The measuring head is placed in contact with the surface of hole and the movement of measuring head contacts is transmitted to the amplifying mechanism by the calibrated rods. These rods are located in tabular supports between the head and dial units. There is a spring to balance the weight of the rod and the tension of the spring can change the sensitivity of the operation.

2.55.4. Cylinder Gauges. In case of deep bores *i.e.,* in case of cylinders of internal combustion engines, steam engines etc. it is frequently necessary not only to measure the diameter at given depth but also to check the holes for (*i*) cylindrical shape, (*ii*) taper effect. The following instruments are mainly used for this purpose.

The cylinder gauge shown in Fig. 2.146 is ideal gauge for determining tapering, out-of-round or scored cylinders. Dial indicator shows instantly the condition of the cylinder to two thousandth of a cm. Adjustable rods are furnished to make it more versatile. The dial is mounted at right angles to the sled. The sled has two line contact points which are at all times in, alignment with the walls of the cylinder. Two hardened contact points, which independently cause the needle to travel over the dial, feature a unique double spring action making the gauge self-centering and absolutely non-collapsible. Handles of various forms and locking mechanism are also provided.

2.55.5. Keilpart Gauge. (Fig. 2.146). This is commercial gauge and its operation is just similar to the gauge described above. It consists of one fixed measuring head and one movable measuring head. The movement of the movable measuring rod is transmitted to dial indicator by push rod through a spring actuated hinged member. Thus the horizontal movement of the rod is transmitted into vertical direction and dial indicator gives indication of variation of size.

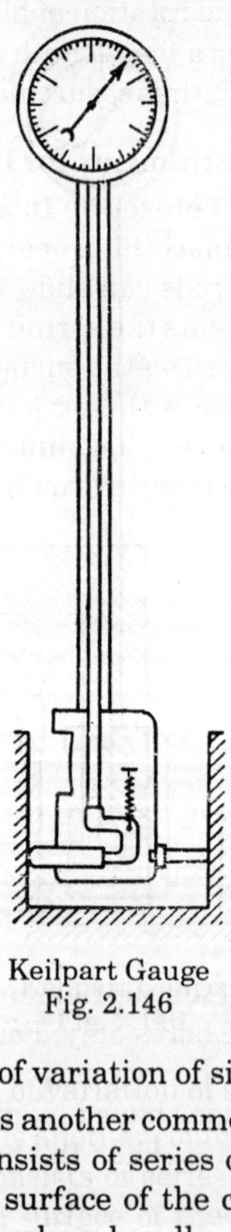

Keilpart Gauge
Fig. 2.146

2.55.6. Automobile Cylinder Bore Gauge. (Fig. 2.147). This is another commercially used gauge for estimating quickly the wear in automobile cylinders. It consists of series of steel balls arranged in a circular holder. These balls come in contact with the surface of the cylinder. For insertion of the gauge into the cylinder the circular holder is held by a long handle. A graduated tapered rod can pass through a pair of rectangular apertures in the circular holder. This rod bears upon the adjacent balls and by pressing upon them enables the effective diameter to be increased. In its application the gauge is inserted into the cylinder and the tapered rod is moved down so as to bring the balls into contact with the cylinder wall. From the graduation on the tapered rod, it is possible to ascertain the corresponding reading with reference to the calibration chart of the gauge. Although this type of gauge will not measure ovality or detect surface imperfections, it gives the mean diameter at the plane of the measurement. It can readily be used to indicate tapered cylinder bore.

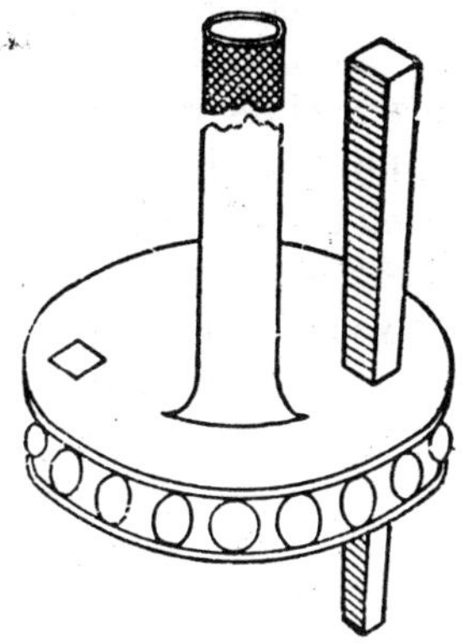

Fig. 2.147. Automobile cylinder bore gauge.

2.55.7. Ball Type Plug Gauge. This gauge is used for measurement of diameter of bores. This also detects the ovality and taper effect and also indicates the surface finish. This gauge is based on the principle of three equally spaced balls of same size which are moved outwards by a spring loaded cone as shown in Fig. 2.148. Thus when this gauge is inserted into any bore, the cone moves down under the action of the spring till all the three balls come in exact contact with the bore. The axial movement of the cone is transmitted to a dial indicator through a rod. The three balls move freely on the lapped surface of this plug gauge. The arrangement obviates the use of levers and cranks and their bearings as in the case of Keilpart gauge.

To measure the diameter of any bore first, the gauge is inserted into a ring gauge of nearly the same size and dial reading is set at zero by adjusting the bevel ring. The gauge is then inserted into the hole and the plus or minus reading noted.

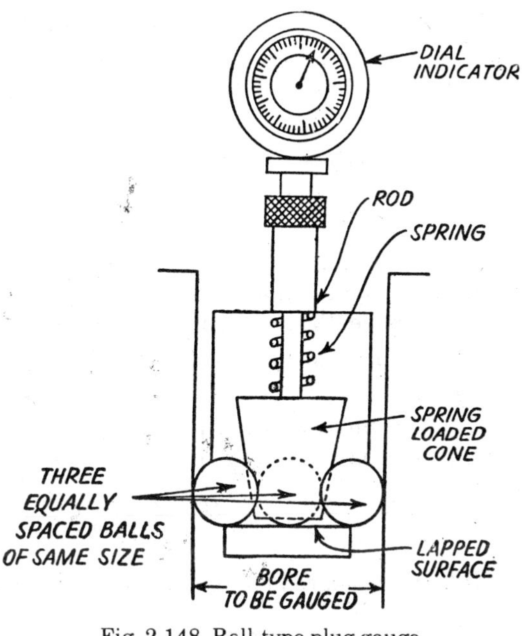

Fig. 2.148. Ball-type plug gauge.

2.55.8. Taper parallel gauges for measuring small hole diameter. Taper parallel gauges are used for the measurement of diameter of very small holes which are too small to permit the entry of an internal micrometer. One end of these gauges is tapered and the other edge is radiused. When it is desired to measure the diameter of small hole, a pair of gauges is introduced into the hole with the taper faces in contact. One gauge is then moved relative to the other in a direction parallel to the axis of the hole, the overall width of the pair being thus expanded until the two outer edges make simultaneous contact with the walls of the hole. A reading is then taken over that portion of the gauge assembly which projects from hole using an external micrometer.

2.55.9. Pneumatic hole gauging methods. Pneumatic method is now more widely being used for the precision hole gauging. It is discussed in detail in the chapter of comparators.

2.56. Pin Gauge for Larger Bores

The diameter of large bores which cannot be easily measured by micrometer or vernier etc., can be measured by means of a pin with spherical ends, the length of which is slightly less than the diameter to be measured (Refer Fig. 2.149). The pin is placed in the bore with its ends in contact with the bore surface, its axis thus being a chord of the circle. Keeping one end in contact with the bore surface, the other end is brought across to make contact at the other side as shown in Fig. 2.149. The two positions are shown by the full and dotted lines in Fig. 2.149. The distance between the two points of contact is measured and from this dimension, and the length of the pin gauge, the diameter of the bore is determined.

Referring to Fig. 2.150, B is the centre of the dotted circle along which the free end of the gauge moves. AB is the diameter D. BP_1 and BP_2 represent the two positions of bar of length 'L'. By the property of intersecting chords,

$$BP \times PA = P_1P \times PP_2 = l^2$$

Since $$BP^2 = BP_1{}^2 - P_1P^2$$

$$BP = \sqrt{L^2 - l^2}$$

and therefore $$PA = \frac{l^2}{\sqrt{L^2 - l^2}}$$

Diameter $$D = BP + PA = \sqrt{L^2 - l^2} + \frac{l^2}{\sqrt{L^2 - l^2}} = \frac{L^2}{\sqrt{L^2 - l^2}}$$

The expression can be simplified by applying Binomial Theorem. Thus,

$$L^2(L^2 - l^2)^{-1/2} = L\left(1 - \frac{l^2}{L^2}\right)^{-1/2} = L\left(1 + \frac{l^2}{2L^2}\right) \text{ approx.}$$

$$D = L + \frac{l^2}{2L} \text{ approximately.}$$

Since the approximation is permissible when the distance l is small in comparison with the length of the gauge, the length of pin gauge should be slightly smaller than D.

The approximate expression gives results which differ only slightly from the accurate expression.

It may be noted that it is a very accurate method in the sense that even with large errors in measurements, the error in final result will only be a small one.

$$D = \partial D \cong \frac{\partial D}{\partial l} . \delta l + \frac{\partial D}{\partial L} . \delta L$$

Error in measurement of diameter $\cong \dfrac{l}{L} . \delta l + \left(1 - \dfrac{l^2}{2L}\right) . \delta L$

It is seen that the error in the diameter is only a small fraction (l/L) of the error in the measurement of l.

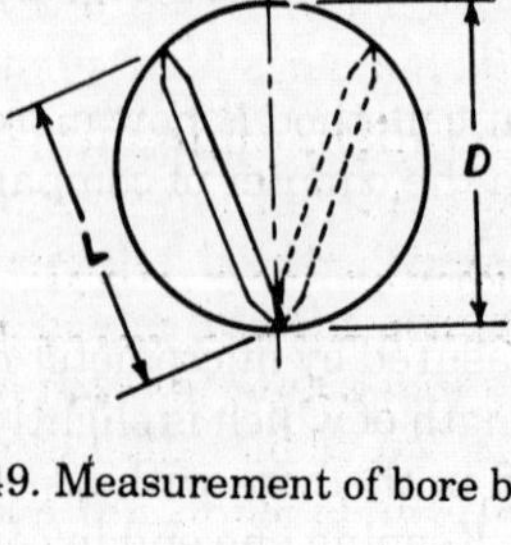

Fig. 2.149. Measurement of bore by pin gauge.

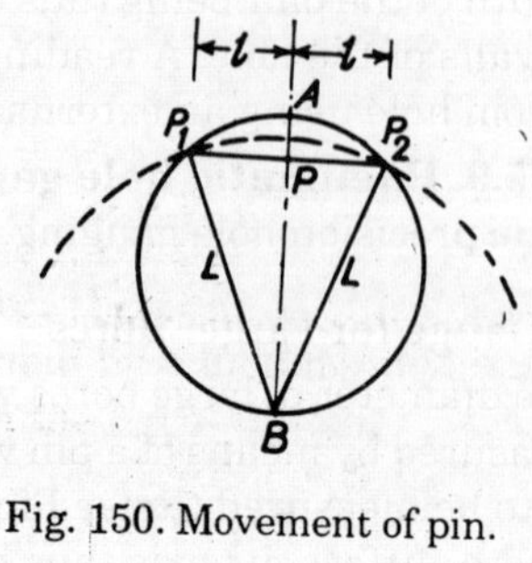

Fig. 150. Movement of pin.

In practice, readings are taken with two gauges of different lengths in order to see the effect upon the result. In addition reading along three different diameters are taken to check whether the bore is truly circular.

2.57. Four Balls Method for Measuring Diameter of Bore

A method of determining the diameter of bore which requires only simple apparatus is shown in Fig. 2.151. Three equal balls are spaced evenly within the bore, and a fourth ball, which need not be of the same size as the three others, is placed upon them.

The balls rest on a plane surface without having hollows in which one or more balls may rest. Referring to Fig. 2.151, height H is measured and from this, with diameter of the balls, the diameter of the ring is easily determined. Thus, in the triangle ABC shown in Fig. 2.151.

$$l^2 = \left(\frac{d_1 + d_2}{2}\right)^2 - \left(H - \frac{d_1 + d_2}{2}\right)^2$$

Then

$$= H(d_1 + d_2) - H^2.$$

$$D = 2l + d_1 = 2\sqrt{H(d_1 + d_2) - H^2} + d_1$$

If the fourth ball is of the same diameter, then

$$D = 2\sqrt{2Hd - H^2} + d_1.$$

The size of the balls is so chosen that the angle θ in Fig. 2.151 is between 30 and 60 degree. If this angle is large, small errors in measuring H will result in large errors in the diameter determined, and if it is small, heavy contact force between the balls will reduce the accuracy of the result. Readings are taken at both ends of the bore and at positions in between by standing the balls on three equal rollers placed with their axis vertical.

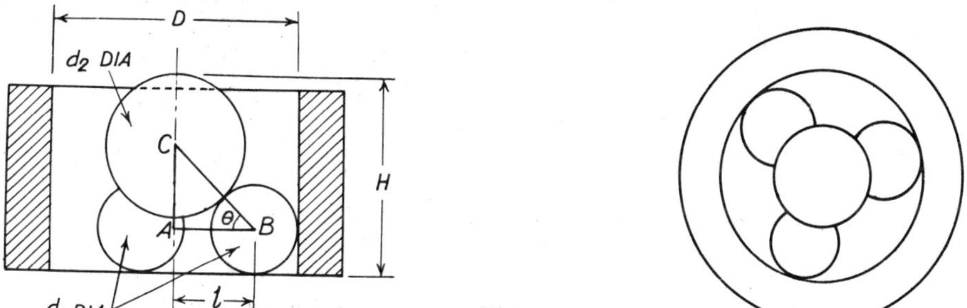

Fig. 2.151. Measurement of bore diameter.

This method can be applied to obtain the diameter of a recessed hole although in such cases the diameter may not be required to a very fine degree of accuracy and the simpler method shown in Fig. 2.152 is satisfactory. Referring to this figure, the dimension 'l' is given by

$$l^2 = \left(\frac{d_1 + d_2}{2}\right)^2 - \left(H_2 - H_1 + \frac{d_2 - d_1}{2}\right)^2$$

Then the diameter is $D = l + \dfrac{d_1 + d_2}{2}$.

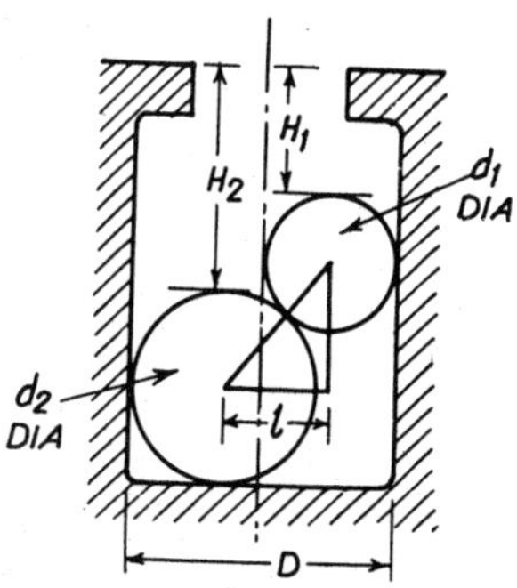

Fig. 2.152. A simple method of bore measurement.

2.58. Small Bore Diameter Measurement by a Ball on a Wire

The principle of this method is shown in Fig. 2.153. A piece of spring wire with a small ball near one end is fastened at its other end to a fixed base. The ring gauge or component to be measured is placed over the wire and supported on a movable carriage, the movement of which can be measured accurately. A fixed microscope is focused on to the top of the wire [Refer Fig. 2.153 (*a*)]. The carriage is moved along to bring the surface of the bore into contact with the ball on the wire and then given a small additional movement which deflects the wire until, when viewed through the microscope, the cross-line in the eyepiece is seen to be tangential to the circular end of the wire, as at (*b*) in Fig. 2.153. The gauge is then moved in the opposite direction to bring the other end of the diameter on to the ball and deflect the wire similarly as (*c*). Then the diameter of the bore is

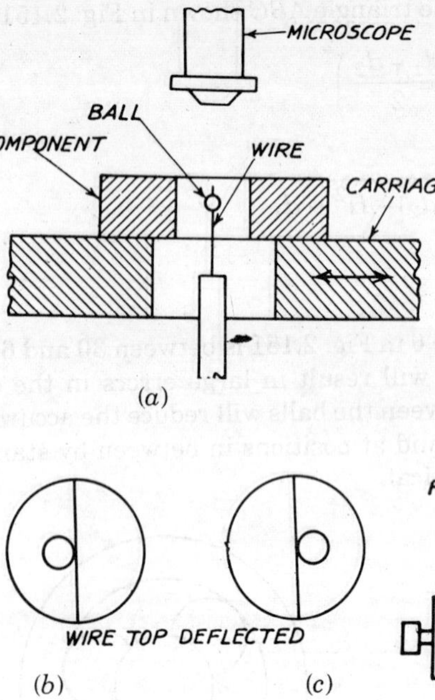

Fig. 2.153. Principle of small hole measurement by deflecting wire.

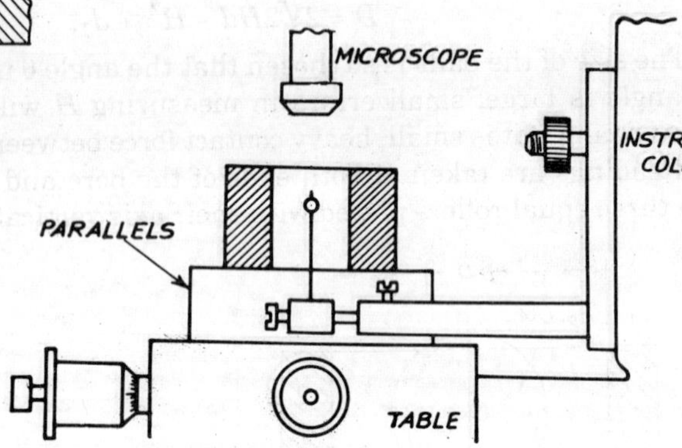

Fig. 2.154. Apparatus for measuring small holes on a toolroom microscope.

equal to the movement of the carriage plus the diameter of the ball and minus the sum of the two deflections. It is quite unnecessary to measure the ball diameter or to attempt to measure the deflections. By forming a gap of known size with slip gauges and finding the movement needed to deflect the wire into the same position as for the ring gauge, the amount to be added to the movement of the ring gauge is obtained. This is then a constant for the apparatus and can be used when measuring other holes. Thus if

Dimension of gap formed by the slips $= S$

Movement of carriage $= M$

Then, constant for the apparatus $= S - M = C$

This constant C is added to the movement of the carriage when carrying the ring gauge to obtain the diameter.

Apparatus for using this method is very easily made as an attachment for the Toolroom Microscope. A suitable arrangement is shown in Fig. 2.154 from which it will be seen that the wire is carried at the end of a rod projecting horizontally from a plate which can readily be fastened on to the column of the instrument. A small amount of adjustment is provided for the wire in order to align its top with the cross-wire in the microscope eyepiece. By using a small ball, holes of very small diameter can be measured by this method.

Care must be taken to set the component so that a diameter and not a chord of the circle is measured. This is done by bringing the surface of the hole into contact with the ball to deflect the wire and then, while viewing the wire top through the microscope, traversing the instrument table to and fro by the front micrometer and noting the position at which the wire is deflected the least.

Readings at several positions along the hole should be taken by varying the height of the component above the instrument table and at each position the component should be turned on its axis to obtain readings for at least three diameters.

This apparatus can, of course, be used for measuring the distance between parallel faces as in a gap gauge ; it is not restricted to hole measurements. The accuracy attainable is about ± 1 μm.

2.59. Measurement of Small Bore by a Pivoted Stylus and Autocollimator

This method is shown diagrammatically in Fig. 2.155. A ball ended stylus, slightly smaller than the hole to be measured, is set along a diameter of the component and swung into contact with the surface of the hole first at one side and then at the other. The mirrors shown in Fig. 2.155 are

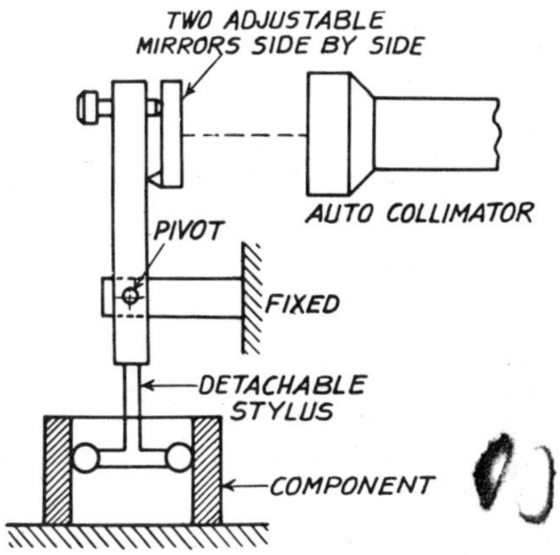

Fig. 2.155. Principle of small hole measurement by autocollimator and pivoted stylus.

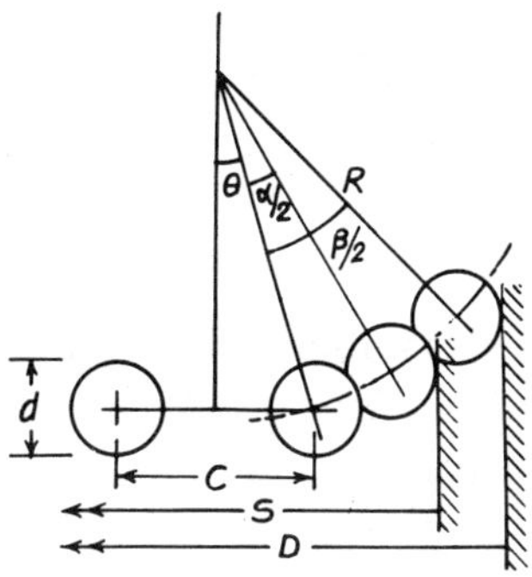

Fig. 2.156. Geometry of small hole measurement.

adjusted so that one of them gives a reflection into the autocollimator when the stylus is at one extreme position and the other when the stylus is swung over its other position. The readings of the autocollimator scale are compared with those obtained when a gap, built up with slip gauges to a dimension about equal to the diameter of the hole, is substituted for the component. The difference in angular tilt is then converted into difference in length by means of the formula below. Referring to Fig. 2.156, where S is the gap formed by the slip gauges and D is the diameter of the hole,

$$\frac{D}{2} - \frac{d}{2} = R \, . \sin\left(\theta + \frac{\beta}{2}\right) \qquad \ldots(1)$$

Subtracting,
$$\frac{S}{2} - \frac{d}{2} = R \, . \sin\left(\theta + \frac{\alpha}{2}\right) \qquad \ldots(2)$$

and
$$\frac{D - S}{2} = R \, . \sin\left(\theta + \frac{\beta}{2}\right) - R \, . \sin\left(\theta + \frac{\alpha}{2}\right)$$

$$D - S = 2R\left[\sin\left(\theta + \frac{\beta}{2}\right) - \sin\left(\theta + \frac{\alpha}{2}\right)\right] \qquad \ldots(3)$$

The value of θ is found from the relation $\sin \theta = C/2R$. This is then inserted in Eq. (2) above to obtain the value of the angle α. The difference in angular movement of the stylus, determined from the autocollimator readings, is the angle $(\beta - \alpha)$, which is then used in Eq. (3) to find the dimension D.

If a single ball is at the end of the stylus, the angle θ is zero and then

$$D - S = 2R\left(\sin\frac{\beta}{2} - \sin\frac{\alpha}{2}\right)$$

In Fig. 2.156, it has been assumed that the stylus is centraliy disposed in relation to the standard gap and to the hole being measured and this is a necessary condition for accuracy ; if it is not observed there is a slight cosine error in the result obtained.

An investigation of the possible error in the diameter resulting from errors in the measurements and with an instrument having a stylus with a single ball yields the following result.

Since $\quad D - S = 2R\left(\sin\frac{\beta}{2} - \sin\frac{\alpha}{2}\right)$

the error in $\quad D - S = \delta\,(D - S)$

$$\simeq 2\left(\sin\frac{\beta}{2} - \sin\frac{\alpha}{2}\right)\delta R + R\left(\cos\frac{\beta}{2}\,.\,\delta\beta - \cos\frac{\alpha}{2}\,.\,\delta\alpha\right)$$

If $R = 50$ mm $\pm$ 0.01 mm, $\alpha = 10° \pm 0.01'$ and $\beta = 9° \pm 0.1'$

then the error is found to be approximately 0.003 mm. By using an autocollimator capable of giving readings to a finer degree of accuracy, less error will occur, for the errors in angles α and β are the main source of inaccuracy in the final result.

2.60. Digital Measurement Equipment

Digital indication is better by far when an exact quantitative value is desired.

The important elements of any electronic digital readout system are the scale unit or transducer and the counter or digital readout unit.

The majority of linear digital readout systems comprise a glass scale and optical detector head. These are however, liable to be affected under workshop conditions by contaminants such as oil, swarf, and dirt. To overcome it some manufacturers have based their instruments on magnetic principles. These instruments also provide binary coded digit (BCD) output which can be directly fed to a computer or printer for further processing. In the case of instruments based on the principle of magnetic recording, the linear scale comprises a magnetic strip which is unaffected by workshop contaminants and can be of virtually any length.

Readout is provided by a range of digital display units and digital position readout units which vary in complexity from a simple unit with one or two axis display and thumb wheel reset to comprehensive three axis display multi-function capability. Functions such as datum point memory for each axis, touch sensor (a touch to a workpiece starts counting or holds the display), manual programming by key operation and automatic programming by play-back can be incorporated.

The advantages of such systems are ;

(*i*) Measuring element is free from errors caused by mechanical elements such as screws, gears etc.

(*ii*) Learning time is short. Amateurs measure as quickly and as well as professionals. Measurement takes half as long.

(*iii*) More accurate measurements, the spread of measurements and the mean error are reduced to about 1/3.

(*iv*) Excessive reading errors are eliminated.

(*v*) The clear readability of digital readout is advantageous for persons with impaired vision.

(*vi*) The display can be zeroed wherever desired.

(*vii*) BCD output makes the instrument computer compatible.

2.61. Slip Gauges

Slip gauges are often called Johannsen gauges also, as Johannsen originated them. These are rectangular blocks of steel having a cross-section of about 30 by 10 mm. These are first hardened to resist wear and carefully stabilised so that they are independent of any subsequent variation in size or shape. The longer gauges in the set and length bars are hardened only locally at their measuring ends. After being hardened, blocks are carefully finished on the measuring faces to such a fine degree of finish, flatness and accuracy that any two such faces when perfectly clean may be 'wrung' together. This is accomplished by pressing the faces into contact (keeping them perpendicular) and then imparting a small twisting motion whilst maintaining the contact pressure. The contact pressure is just sufficient in order to hold the two slip gauges in contact and no additional intentional pressure. It is found that phenomenon of wringing occurs due to molecular adhesion between a liquid film (whose thickness may be between 6 to 7 × 10^{-6} mm) and the mating surfaces. When two gauges are wrung together and the overall dimension of a pile made of two or more blocks so joined is exactly the sum of the constituent gauges. It is on the property of wringing units together for building up combinations that the success of system depends, since by combining gauges selected from a suitably arranged combination, almost any dimension may be built-up.

These may be used as reference standards for transferring the dimensions of the unit of length from the primary standard to gauge blocks of lower accuracy and for the verification and graduation of measuring apparatus, and length measures for the regulation and adjustment of indicating measuring apparatus and for direct measurement of linear dimensions of industrial component.

Slip gauges are classified according to their guaranteed accuracy :

AA for master slip gauges, *A* for reference purpose, and *B* for working slip gauges.

Johannsen *AA* slip gauges are accurate to plus or minus two microns per metre. Type *A* is guaranteed accurate up to plus or minus four microns per metre, while type '*B*' for plus or minus eight microns per metre. The guaranteed error is not divided for a block which is less than 25 mm ; such a slip gauge has same tolerance as 25 mm gauge. The workshop type, *i.e.*, *B* type gauges are finished on their measuring faces approximately to within 250 μm for flatness and parallelism. The corresponding figures for types *A* and *AA* are 125 and 75 μm respectively.

As regards grades or classes of slip gauges, these could also be designed in five grades as under :

Grade 2. This is the workshop grade. Typical uses include setting up machine tools, positioning milling cutters and checking mechanical widths.

Grade 1. Used for more precise work, such as that carried out in a good-class toolroom. Typical uses include setting up sine bars and sine tables, checking gap gauges and setting dial test indicators to zero.

Grade 0. This is more commonly known as the Inspection grade, and its use is confined to toolroom or machine shop inspection. This means that it is the Inspection Department only who have access to this grade of slips. In this way it is not possible for these slip gauges to be damaged or abused by the rough usage to be expected on the shop floor.

Grade 00. This grade would be kept in the Standard Room and would be kept for work of the highest precision only. A typical example would be the determination of any errors present in the workshop or Grade 2 slips, occasioned by rough or continual usage.

Calibration grade. This is a special grade, with the actual sizes of the slips stated or calibrated on a special chart supplied with the set. This chart must be consulted when making up a dimension, and because these slips are not made to specific or set tolerances, they are not as expensive as the Grade 00. It must be remembered that a slip gauge, like any other engineering component, cannot be made to an exact size. All slip gauges must have tolerances on the length, flatness and parallelism of measuring faces.

Except for the calibration grade, all slip gauge sets are manufactured to within specified limits ; the closer the limits the more expensive the slip gauges, but in the case of the calibration grade, greater tolerances on length are permissible. Because the actual lengths are known or recorded in the calibration chart, due allowance can be made when the slips are used.

Slip gauges are available in sets both in inch units and in metric units.

The five most usual sets available in inch units contain 81, 49, 41, 35 and 25 pieces respectively. *e.g.* in the 81-piece set, the slip gauges are arranged in the following order :

9 pieces from 0.1001″, to 0.1009″ in steps of 0.0001″.

49 pieces from 0.101″ to 0.149″ in steps of 0.001″

19 pieces from 0.050″ to 0.950″ in steps of 0.050″.

4 pieces of 1.000″, 2.000″, 3000″, 4.000″.

In metric units, sets of 103, 76, 48 and 31 pieces are available. Metric unit sets of 103 pieces are made up as follows :

49 pieces with a range of 1.01 mm to 1.49 mm in steps of 0.01 mm.

49 pieces with a range of 0.50 to 24.50 mm in steps of 0.50 mm.

4 pieces of 25, 50, 75 and 1000 mm respectively and 1 piece extra of 1.005 mm.

Slip gauge set of 56 slips is made up as shown below :

9 slips 1.001 to 1.009 in steps of 0.001 mm

9 slips 1.01 to 1.09 in steps of 0.01 mm

9 slips 1.0 to 1.9 in steps of 0.1 mm

25 slips 1 to 25 in steps of 1.0 mm

3 slips 25 to 75 in steps of 25 mm.

Slip of 1.0005 mm.

Many times to reduce wear on inspection workshop grades a pair of protector gauge blocks (1 to 1.5 or 2 mm length) are also supplied and these are wrung to end of slip gauge combination block. As these are made of tungsten carbide or similar wear resisting material these do not wear out and protect the slip gauges from wear. These are marked with letter '*P*' on one measuring face.

According to OIML, gauge blocks with rectangular section are length measures in the form of rectangular plates or rectangular parallelopiped blocks consisting of two opposite flat faces parallel to each other, called measuring faces, the surface quality of which is such that they have the property of wringing to the corresponding faces of other gauges blocks or the reference face of datum surfaces.

Gauge blocks are meant to be used as :

(*a*) reference standards for transferring the dimension of the unit of length from the primary standard to gauge blocks of lower accuracy and for the verification and graduation of measuring apparatus ;

(*b*) length measures for the regulation and adjustment of indicating measuring apparatus and for the direct measurement of linear dimensions of industrial components.

2.61.1. Basic Forms of Slip Gauges. Slip gauges with three basic forms are commonly found. These are rectangular, square with centre hole, and square without centre hole.

Rectangular form is the more widely used because rectangular blocks are less expensive to manufacture, and adapt themselves better to applications where space is restricted or excess weight is to be avoided.

For certain applications, square slip gauges, though expensive, are preferred. Due to their large surface area, they wear longer and adhere better to each other when wrung to high stacks.

Square blocks with centre holes are used to permit the use of tie rods as an added assurance against the wrung stocks falling apart while handling.

The greatest emphasis in the slip gauges is laid on the two gauging surfaces on opposite sides of the blocks. The length between measuring surfaces, flatness and surface conditions of measuring surfaces are the most important features of slip gauges which have to be controlled within desired limits.

2.61.2. Major Requirements for Slip Gauges. The essential purpose of slip gauges is to make available end standards of specific lengths by temporarily combining several individual elements—each representing a standard dimension—into a single gauge bar. The combination of single blocks must result in a bar of reasonable cohesion, whose actual dimension truly represents, within specific limits, the nominal dimension sought for a particular application.

This objective is accomplished by making slip gauges in dimensions (arithmetically determined sizes) needed to achieve any combination of sizes within the designed range and graduation of the set ; ensuring the accuracy of individual blocks to be within accepted tolerance limits ; attaching the individual blocks so closely to each other that the resulting bar will have a length equal to the added sizes of the individual blocks ; further the attachment of the elements to each other should be firm enough to permit a reasonable amount of handling as a unit, and yet when taken apart, the individual pieces should be reusable without any harm to their original size or other essential properties.

Slip gauges are used to provide end standards of specific length by temporarily combining several individual elements—each representing a standard dimension—into a single gauge bar. The success of system depends on formation of a bar of reasonable cohesion between individual elements and its actual dimension truly representing within specific limits, the desired nominal dimension. For achieving this, the individual gauges must be available in dimensions needed to achieve any combination with minimum number of gauges. The accuracy of individual blocks must be within accepted tolerance limits. The individual blocks must be attached so closely to each other that length of built-up combination is equal to the added size of the individual blocks of the assembly. This is achieved by wringing the slip gauges. Further the attachment of individual gauges must be firm enough to permit a reasonable amount of handling as a unit. It should be possible to detach all individual elements so that these are re-usable without any harm to their original size or other essential properties.

For this purpose, control of geometry of form, *i.e.*, flatness and parallelism of faces, squareness of the gauging surfaces ; surface condition, appearance, homogeneity of gauging surfaces, etc. is very essential.

The accuracy of gauges can be affected by the dimensional instability of material, or by wear in use, or damage during storage and handling. These can be taken care of by periodic calibration. The health of slip gauges can be very easily checked by checking its wringing quality by wringing the gauge to be tested with an optical flat and inspecting the wrung surface for colouration (which indicates an air gap and its area should not be more than 5 per cent of the entire wrung surface).

2.61.3. Indian Standard on Slip Gauge. According to accuracy, the slip gauges can be graded into three categories, *i.e.* Grade 0, Grade I and Grade II. As the high degree of finish and flatness of the slip gauges makes it possible for a number of gauges to be wrung together to form a composite size, it is essential that one should know more about the accuracy of length of gauges, their flatness, parallelism and hardness. Generally the slip gauges are made from high grade steel with coefficient of thermal expansion $(11.5 \pm 1.5) \times 10^{-6}$ per degree Celsius between 10°C and 30°C. These are hardened throughout and suitably heat-treated in order to stabilise their dimensions or are provided with hard measuring faces like carbide faces. The hardness of the slip gauges is always more than 800 HV. The measuring faces are free from any defects and finished by high grade lapping and have good wringing surfaces. All sharp edges are chamfered or rounded.

According to IS : 2984—1966, the size of a slip gauge is defined as the distance l between two plane measuring faces, one being constituted by the surface of an auxiliary body with which one of the slip gauge faces is wrung and the other by the exposed face to the slip gauge. In the absence of true parallelism, the gauge size counts as the vertical distance from the centre of the exposed face to the surface of the auxiliary body. The definition is conditional on the assumptions that the measuring faces of the slip gauge and of the auxiliary body are of the same material and of identical finish and condition, and that the contacting surface of the slip gauge and auxiliary body have been cleaned as thoroughly as possible with the agents generally used for this purpose and are free from any special preparations tending to promote adhesion.

The same definition applies analogously to the overall size of an assembly of slip gauges when wrung together.

An accepted method of measurement which avoids any stresses tending to modify the length of the slip gauges (zero measuring force) is based on the interference principle. The desired size (normal size) of a slip gauge is denoted by l_0.

Generally two sets of slip gauges (normal set and specials set) are used.

The normal set is made up of the following blocks :

(All dimensions in mm)

Range	*Step*	*Pieces*
1.001 to 1.009	0.001	9
1.01 to 10.9	0.01	9
1.1 to 1.9	0.1	9
1 to 9	1	9
10 to 90	10	9
		Total 45

The special set is made of the following blocks :

Range	*Step*	*Pieces*
1.001 to 1.009	0.001	9
1.01 to 1.49	0.01	49
0.5 to 9.5	0.5	19
10 to 90	10	9
1.0005	—	1
		Total 85

The most commonly used dimensions of the slip gauges which are always rectangular in cross-section, *i.e.* apex angles of these rectangles or angles of lateral faces are right angles correct to within (± 10 minutes of arc) are :

Normal Size	*Cross-sectional area (w × d) in mm (Refer Fig. 2.157)*
Upto 10 mm	$30^{+0.0}_{-0.3} \times 9^{-0.05}_{-0.3}$
Above 10 mm	$35^{+0.0}_{-0.3} \times 9^{-0.05}_{-0.3}$

The measuring faces are finished to high quality (only insignificant scratches not affecting the wringing properties and not influencing the conditions of interferometric measurements are permitted). The edges and the angles of the measuring faces are rounded over a breadth not exceeding 0.3 mm, or chamfered (the chamfer being merged with the measuring face by a gentle rounding) such that both the chamfer and rounding do not exceed a breadth of 0.3 mm.

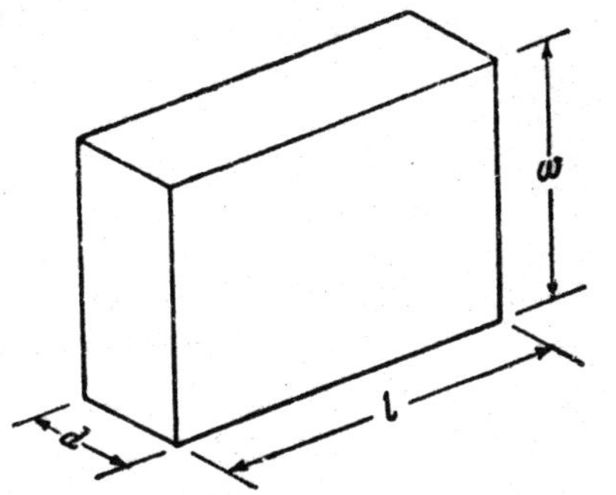

Fig. 2.157. Rectangular slip gauge.

2.61.3.1. Standard Terminology. The other terminology in connection with slip gauges as per IS : 2984—1984 is as follows :

Size l_b at an Optical Point on the Measuring Face. [l_b (l_{b1}, l_{b2})] (Refer Fig. 2.158). It is the vertical distance between some optional point on measuring face *A* and the plane surface of a body *B* made of the same material and having the same surface finish and condition, such body being wrung with the other measuring face *A* of the slip gauge.

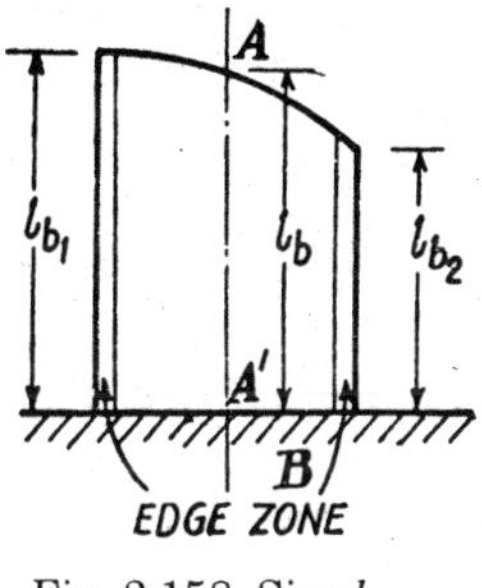

Fig. 2.158. Size l_b.

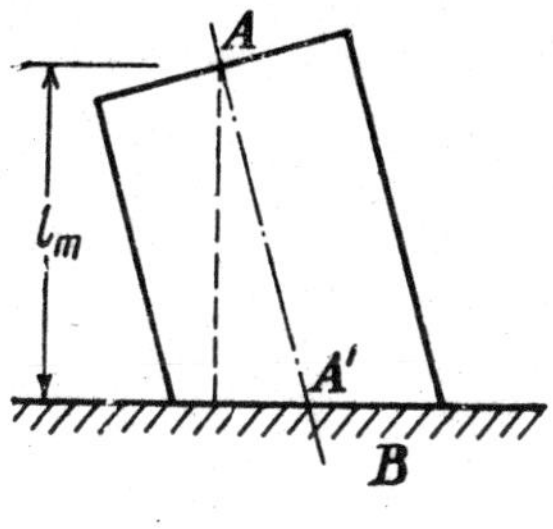

Fig. 2.159. Mean size l_m.

Means Size, l_m (Refer Fig. 2.159). It is the vertical distance between the mid-point of measuring face *A* and the plane surface of a body *B* made of the same material and having the same surface finish and condition, such body being wrung with the other measuring face *A'* of the gauge.

Gauge Error, f_b (f_{b1}, f_{b2}) (Refer Fig. 2.160). It is the amount by which the size l_b at some optional point on the measuring face varies from the desired size l_o.

$$f_b = l_b \sim l_o$$

Error f_m in Mean Size (Refer Fig. 2.160). It is the amount by which the mean size *lm* varies from the desired size l_o :

$$f_m = l_m \sim l_o$$

Measuring Face Error : f_f (f_{f1}, f_{f2}) (Refer Fig. 2.160). It is the amount by which the size l_b at some optional point on the measuring face varies from the mean size l_m :

$$f_f = l_b \sim l_m = f_b \sim f_m.$$

(**Note.** Symbol "~" denotes : subtract either of the quantities so that the result is positive value.)

Squareness Error (S). It is amount by which the side face deviates from a true right angled position relative to each measuring face. The error must be determined for all the four side faces and in relation to each of the measuring faces (S_1 to S_8).

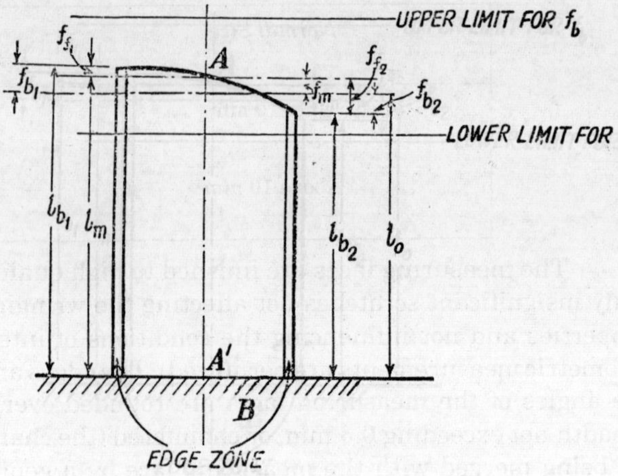

Fig. 2.160. Gauge error, error in Mean Size, and measuring face error.

Wringing and Enforced Adhesion. The term 'wringing' refers to the conditions of intimate and complete contact and of permanent adhesion between measuring faces which is brought about by wringing together the surfaces in question without application of pressure, assuming that the surfaces have been thoroughly cleaned and exhibit a good standard of flatness and smoothness. The wrung gauge can be handled as a unit, without the need for clamping all the pieces together.

It is believed that the phenomenon of wringing is due to molecular adhesion between a liquid film and the mating surfaces of the flat surfaces. In fact, the success of precision measurement by slip gauges depends on the phenomenon of wringing. It has been found that the gap between two wrung flat pieces is of the order of 0.00635 microns from which it would be very clear that the film thickness in two wrung flat pieces contributes no effect. It should be remembered that slip gauges are wrung together by hand through a combined blinding and twisting motion. First the gauge is oscillated slightly with very light pressure over other gauge so as to detect presence of any foreign particles between the surfaces. One gauge is then placed perpendicular to other using standard gauging pressure and rotary motion is then applied until the blocks are lined up. In this way air is expelled from between the gauge faces causing the two blocks to adhere. This adherence is caused partly by molecular attraction and partly by atmospehric pressure (Refer Fig. 2.162). Similarly for separating two wrung slip gauges, combined sliding and twisting motion should be used and no attempt should be made to separate them by direct pull because considerable load would have to be applied which may damage the slip gauges. The overall thickness of the wrung gauges, for most practical purposes, is the sum of the individual sizes in combination. If during wringing, there is slightest feeling of roughness the process should be stopped and the surfaces examined for contamination.

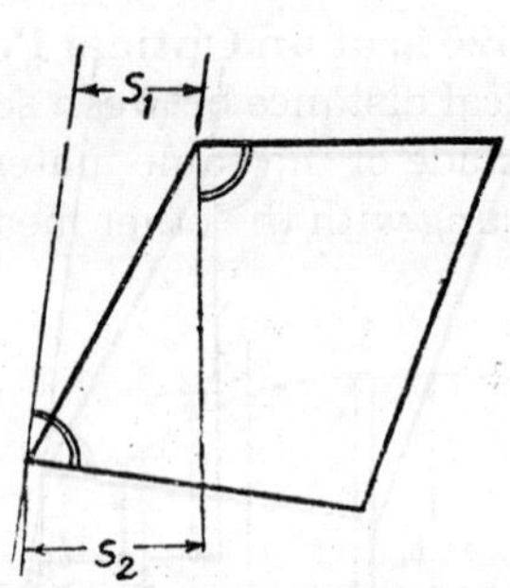

Fig. 2.161. Squareness Error.

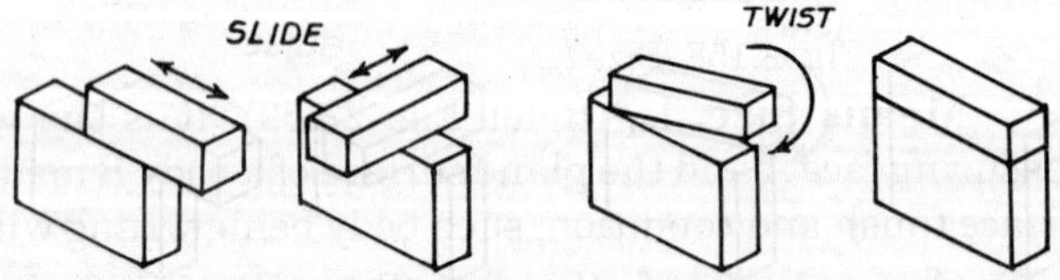

Fig. 2.162. Wringing of slip gauges.

Wringing is thus defined as the property of measuring faces of a gauge block of adhering, by sliding or pressing the gauge against the measuring faces of other gauge blocks or the reference faces of datum surfaces, without the use of any extraneous means.

The 'enforced adhesion' refers to the condition of contact over the entire surface and of permanent adhesion, which, owing to the lack of flatness or smoothness, can only be brought about by the application of pressure. It may be noted that the use of any extraneous agent to promote adhesion is not correct and it is sometimes specified that surfaces which are to be wrung together should be absolutely dry. On the other hand, a truly clean surface, though difficult to attain may not wring satisfactorily. The usual practice is to use silicone or filtered kerosene as lubricant, apply a thin coat of same and wipe it as thin as possible. There is as much danger from too little lubricant as from too much.

Measuring Faces. The top measuring face is that face on which the size is marked. If such marking is on a side face, then the right hand measuring face is the one which forms the right-hand boundary of the surface on which the size characters are marked assuming the characters are upright to the observer.

2.61.4. Permissible Errors. The various permissible errors for slip gauges of different grades are as given below :

	Grade 0	*Grade I*	*Grade II*
Permissible Gauge Errors f_b	± (0.1 + 0.002 l) μm	± (0.2 + 0.005 l) μm	± (0.5 + 0.01 l) μm
Permissible measuring face error f_f	± (0.1 + 0.00014 l) μm	± (0.16 + 0.00025 l) μm	± (0.25 + 0.00045 l) μm
Permissible squareness Error S	± (50 + 0.1 l) μm	± (60 + 0.12 l) mm	± (75 + 0.2 l) μm
Permissible variation of side face from flatness	± (20 + 0.02 l) μm	± (20 + 0.02 l) μm	± (20 + 0.02 l) μm
Permissible variation of side face from true parallelism	(40 + 0.04 l) μm	(40 + 0.04 l) μm	(40 + 0.04 l) μm
Permissible variation of side face from true squareness with each other	± 100 μm over 35 mm ± 90 μm over 30 mm ± 60 μm over 20 mm	Same as per Grade 0	Same as per Grade 0
Permissible variation in width and depth	+ 0 mm – 3.0 mm	+ 0 mm – 0.3 mm	+ 0 mm – 0.3 mm
Surface finish of measuring faces	Satisfy the condition for wringing and enforced adhesion	Same as for Grade 0	Same as for Grade 0
Edge Zone	An edge zone of l mm in width is disregarded on all surfaces	Same remarks as for Grade 0	

l is the length of slip gauge in mm.

For testing purposes, slip gauges upto 5 mm nominal size are wrung with a body having plane and parallel faces and a thickness of not more than 12 mm. In case of longer slip gauges the permissible errors apply with the gauge placed horizontally and supported at the optimum points of 0.2113 l from ends of slip gauge.

2.61.5. Grades of Slip Gauges. IS : 2984—1966, Bureau of Indian Standards for Slip Gauges specifies three grades of slip gauges : grade 0, Grade I and Grade II. Some examples of the maximum permissible errors in the mean length of gauges at 20°C are given below :

BIS Specification for Slip Gauges

Nominal Size mm	*Grade 0 micron*	*Grade I micron*	*Grade II micron*
10	0.02	0.10	0.35
20	0.04	0.15	0.45
50	0.10	0.30	0.75
100	0.20	0.50	1.20

Grade II is intended for use in workshops during the actual production of components, tools and gauges ; Grade I is of higher accuracy and intended for use in inspection departments ; and grade 0 is for use in laboratories and standards room. This latter grade serves as standards for periodically checking the accuracy of grade I and II gauges. The actual sizes of gauges in grade 0 sets should be used only in conjunction with the certificate of sizes obtained from a recognised testing authority such as the National Physical Laboratory.

2.61.6. Sets of Gauges. The recommended sets in the metric units are M112, M105, M87, M50, M33 and M27. The normal set of M112 is made up of blocks as given below :

Normal Set of M112 Gauges

Range mm	*Steps mm*	*Pieces*
1.001 to 1.009	0.0001	9
1.010 to 1.490	0.010	49
0.50 to 24.50	0.050	49
25, 50, 75, 100	25	4
1.0005	—	1
		Total 112

2.61.7. Protector Blocks. These are two additional 2 mm blocks with a letter *P* on measuring faces and are provided with high-grade sets of gauge blocks. These are accommodated at each end of a combination so that all the wear occurs on them. These are made from tungsten carbide or other suitable material and take all the wear due to rubbing on surface plates.

2.61.8. Accumulative error of slip gauges. Though the error by combination of two gauges in thickness due to separation is negligible (of the order of 6.35×10^{-6} mm), it is advisable to select least number of gauges for a given size when combining slip gauge blocks.

2.61.9. Selecting slip gauges for required dimension. Much time is wasted if the slip gauge combinations be made by hit and trial. Therefore standard procedure given below must be followed.

Let us say that the dimension to be arranged is 58.975 mm.

Always start with the last decimal place *e.g.,* here it is 0.005 mm and for this 1.005 mm slip gauge is selected.

Now dimension left is 50.975—1.005 = 57.970 mm.

Take second decimal place ; and for it select 1.47 mm slip gauge.

Therefore, the remainder is 57.970 – 1.47 = 56.500 mm.

[**Note :** One could have selected 1.07 mm piece also, but that way we would have been left with 56.900 and for it we need another 1.4 mm piece. Our aim should be to choose minimum number of slip gauges for a given dimension.]

Next for 56.500 mm, we choose 6.500 mm piece and finally 50.000 mm piece.

Thus, we have 50.000 + 6.550 + 1.47 + 1.005 = 58.975 mm.

All these four slip gauges are wrung properly to get required dimension.

When the dimension to be measured is unknown and is to be found using slip gauges, then first it must be determined to the nearest 0.1 mm size with a caliper etc. Let us say that size lies between 47.0 and 47.4 mm and we have to estimate it correct upto 3 decimal places of mm. It is now obvious that we require 1.005 mm for 0.005 mm and 1.01 to 1.09 mm for .01 to 0.09 mm. Thus 2 mm must be taken off from 47.3 mm to account for these three decimal figures. Thus first make combinations of 47.3 – 2.0 = 45.3 mm *i.e.* 1.3 + 19.0 + 25.00 mm pieces.

Next try with these and 1.005 and all the rest, *i.e.* 1.01 to 1.09 mm pieces. Then starting from 1.09 pieces one of these will give large dimension and next one small. Then decide whether 1.005 mm piece is needed or not by taking it out and adjusting the dimension near that value. The sense of 'feel' is very important in the use of slip gauge and depends upon the skill of the operator.

Since during the combination operation the gauges are in contact with body temperature, their dimension actually increases. Therefore, heat transferred from the body must first be rejected to surrounding by leaving the set at room temperature for some time. Other possible errors due to temperature can be due to heated parts or room temperature being higher or lower than 20°C. These can be avoided easily by necessary precautions.

2.61.10. Ceramic Slip Gauges (Refer Fig. 2.163)

Salient Features :

Corrosion Resistant

— Unaffected by water, acids and alkalies.

— Simple maintenance with no requirement of any anti-corrosion treatment.

— No adverse effects due to fingering while using.

Superior Wringability

— The uniform and close grain structure enables the blocks rigidly wrung together and easy to operate.

Dents and burrs are not easily produced on gauge surface.

Fig. 2.163

Resistant to Impact

— Zirconia ceramic blocks are hard and highly tough to withstand the knocks and drops that occur during use and will not chip or fracture easily.

Resistant to Wear

— About 10 times life or even more compared to steel slip gauges. The natural stability and durability due to its low friction co-efficient maintains their geometry long.

Thermal Expansion

— Co-efficient of thermal expansion of ceramic is 10×10^{-6}/°C is very close to that of steel.

— The lowest thermal conductivity makes these gauge blocks easy to use at different temperatures.

2.61.11. Long Series Slip Gauges and Length Bars

Long Series Slip Gauges (Ref. Fig. 2.164). Long Series Slip Gauges are made from high quality steel having cross section (35 mm × 9 mm) with holes for clamping 2 slips together. They are manufactured to DIN-861 of 1988 specifications in different grades. They are available in the following grades : '0', 'I' and 'II". These are available from 100 mm to 500 mm.

Fig. 2.164. Long series slip gauge.

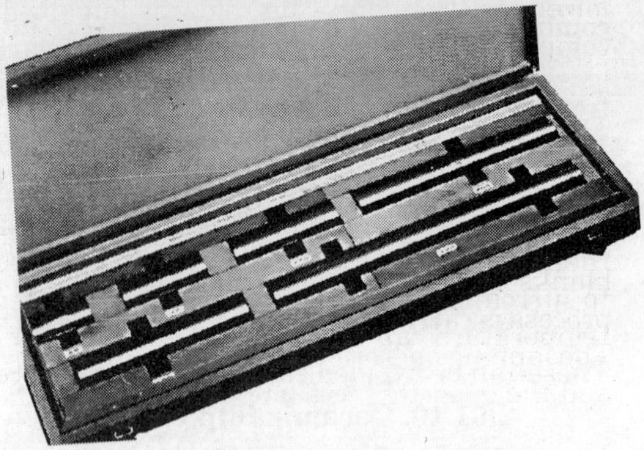

Fig. 2.165. Length bars.

Length Bars. Length Bars are made from High Carbon High Chromium Steel, ensuring that the gauge faces are hardened to 64 RC (800 HV). The bars have a round section of 30 mm for greater stability, and ease of handling. Both ends are threaded and recessed, and precision lapped to meet with the stipulated requirements of finish, flatness, parallelism and gauge length. Length Bars are available upto 500 mm in the following grades : Calibration ('0'), Inspection ('I') and Workshop ('II'). With 8 pieces set, length bars can be combined upto 2000 mm by using M6 stud.

2.61.12. Adjustable Slip Gauges (Refer Fig. 2.166). Some firms like Pitter have developed adjustable slip gauges with little refinement over Engineer parallels desrcibed in Art 2.2.13. These consist of two hardened, ground and lapped blocks having one tapered surface. The tapered surfaces are finished to wringing quality and it is possible to slide two tapered surfaces along the length of each other thereby obtaining different sizes within a very small range. The upper block has a rack fixed on bottom tapered surface, and lower block has receptacles on top tapered surface to receive the teeth in upper block. A scale with divisions corresponding to rack tooth in upper block is marked in lower block and a reference point is marked on upper block, so that by keeping upper block with its teeth placed at suitable distance, any dimension within the range can be obtained between the outer faces of two biocks.

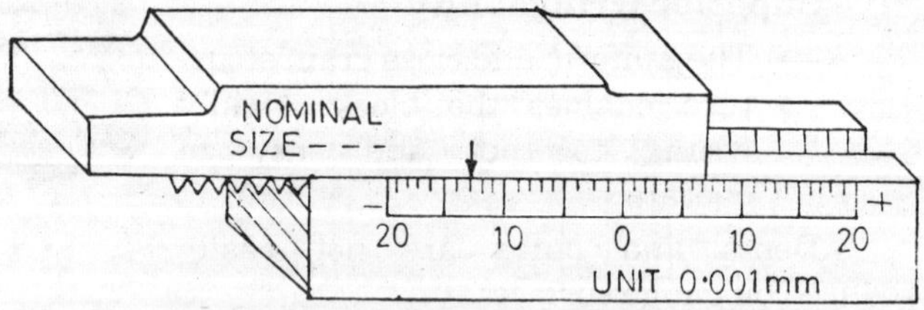

Fig. 2.166. Adjustable slip gauges.

2.61.13. Manufacture of Slip Gauges. The following sequence of operations is followed in order that slip gauges possess certain desirable qualities.

(*i*) Marking the approximate size by preliminary operations.

(*ii*) A special form of heat treatment to make the blocks hard and wear resistant.

(*iii*) An artificial and natural seasoning process to ensure stabilising for the whole life of the blocks. Stabilising is generally carried out by heating and cooling the gauges successively, after rough grinding. The successive temperatures used in the four stages of stabilising are 40°, 70°, 130° and 200°C, the gauges being heated in sand and cooled slowly at each stage.

(*iv*) A final grinding process to reduce the block to the approximate required dimension.

(*v*) A final-lapping operation to reduce the blocks to exact size and impart a beautiful finish to the surface. A special lapping machine is used for lapping. Lapping is done in a room maintained at 20°C and controlled humidity of 50%. The blocks are held in a moving spider between upper and lower cast iron laps. The spider is imparted both rotating and reciprocating motion to produce a complicated path of travel for each block and wear is distributed all over uniformly.

(*vi*) Comparison of finished gauges with grand master sets.

2.61.14. Manufacture and Generation. Various manufacturing methods have been devised for the production of slip gauges. The method employed by the National Physical laboratory, utilises special type of magnetic chuck on which eight similar steel blanks are mounted and spot ground on each face. A preliminary lapping operation is carried out on this chuck, by which all the blanks become parallel to about 0.0002 mm and within about 0.002 mm of size. The final lapping process is carried out on solid steel chuck on which the blocks are wrung as shown in Fig. 2.167 (*a*). The lapping operation is continued until they all lie in one true plane. The blanks are next reversed and the opposite faces lapped again until they also lie in one true plane. After lapping the blanks in this position, half of the blanks are interchanged and turned end for end as shown in Fig. 2.167 (*b*). Further lapping produces a very high degree of parallelism and equality of size between the eight blanks. In between, the eight blanks are wrung together and their overall length compared with an equivalent size standard gauge. In this way, the individual errors are reduced to one-eighth of the usual error for a single block. By this method, flat and parallel blanks are produced which are of equal size to within 0.002 mm.

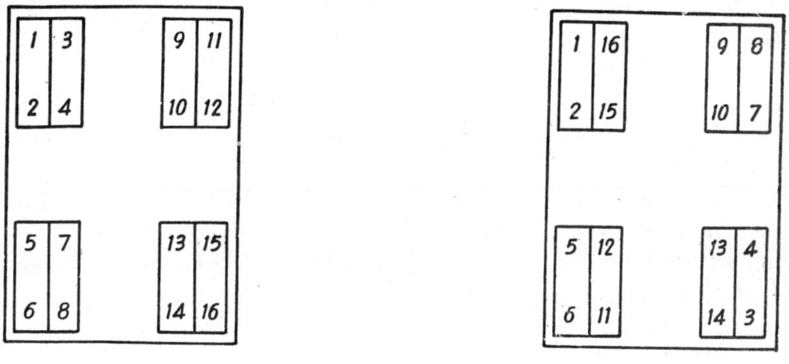

(*a*) First Arrangement (*b*) Second Arrangement

N.P.L. Arrangement of manufacturing slip gauges on lapping chuck

Fig. 2.167

Generation. The whole set of gauges is generated by this lapping technique by taking simultaneously a number of pieces which are of same size starting from the longest member. Hence eight 100 mm gauges are first lapped simultaneously, the process being stopped at intervals and any three of them are wrung together and compared with the 300 mm standard. The lapping is continued in stages until equality is obtained. Each one of the 100 mm gauge is thus correctly adjusted to size. One of the 100 mm gauge is again used as standard as we had used 300 mm gauge and 100 mm gauges in turn are used to generate the next lower series till the set is complete.

2.61.15. Mechanical Lapping of Slip Gauges. This method of finishing slip gauges to size involves the mechanical lapping of both faces simultaneously. The lapping machine employed for this purpose consists of three lapping plates which are of annular form. These plates are prepared

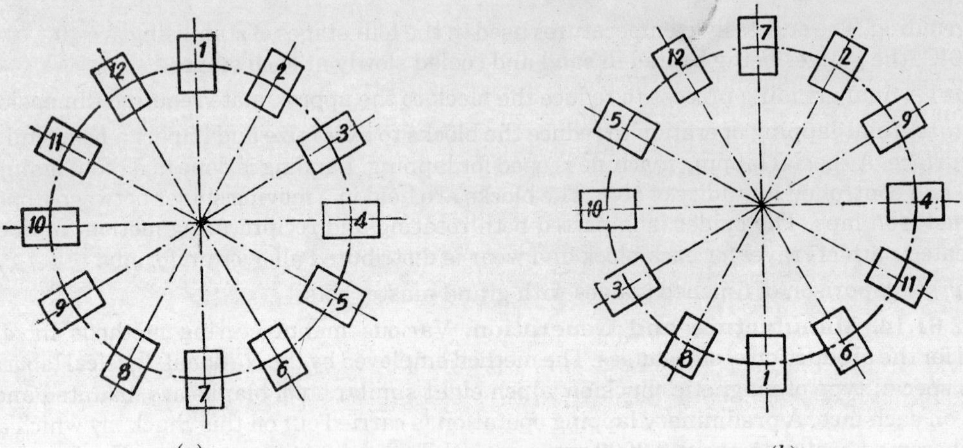

Fig. 2.168. Mechanical lapping operation of slip gauages.

by lapping operation keeping the plates together in pairs until they fit together interchangeably. In the machine, these lapping plates are disposed one above the other. The lower lap (lapping plate) is rigidly secured to the base, the upper most lap is so mounted that it is free to tilt in any direction, although restrained from rotating. Between these two laps is disposed a third flat steel plate (spider) in which a number of holes which are slightly larger in diameter than the distance across the corners of blocks, are provided for the reception of the gauges to be lapped. The block can thus move freely and the surfaces which project on their side of the plate can bear evenly on the faces of the upper and lower laps. The holes in the spider are disposed in a circle. The spider is imparted a planetary motion in addition to the rotary motion so that each block is caused to slide over every portion of the surface of both laps as the operation proceeds. The ratio between the rotary and planetary motions is so selected that the spider and the gauge blocks do not follow the same paths around the surfaces of the laps during successive turns. This is done in order to evenly distribute the wear on the laps. The gauge blocks before lapping operation are made flat and parallel by some other operations and are 0.025 mm over size. Initially the upper lap, which is free to take up any angular position, rests only on the three high points of the circle of gauge blocks. As the blocks are carried round between the laps, the area on which the upper lap bears is soon enlarged and finally the upper surfaces of all the blocks would be flat and lie in one plane and the lower surfaces of the blocks would be flat and would lie in another plane. It may be noted that there is no guarantee that the upper and lower planes of the gauge blocks are parallel. In order to achieve this, the alternate blocks are changed through 180 degrees to a position diametrically opposite to that which it previously occupied. In Fig. 2.168 (*a*), 12 blocks 1, 2, 3, ..., 12 are shown disposed in a circle. Actually 24 blocks are lapped simultaneously. Supposing, block No. 1 is the highest of the set and No. 7 is the lowest, and if the blocks are now disposed as shown in Fig. 2.168 (*b*) then as lapping operation is continued further, the upper lap will initially bear only on No. 1, 2 and 12 and thus the greatest amount of material will be removed from these. By repeated lapping and interchanging in this manner, a high degree of parallelism can be obtained.

2.61.16. American Method of Manufacturing slip gauges. By this method the gauges are produced to meet the requirements of the US National Bureau of Standards, for size, flatness, parallelism and surface finish. Other requirements taken care of are uniform hardness, stability (to prevent shrinkage or growth), good wringing characteristics, and wear and corrosion resistance. The gauges are made from steel containing 1.45 per cent chromium, 0.35 per cent manganese and approximately 1 per cent carbon. Tungsten carbide is used for the protective gauges.

The blocks, after being separated from the bar by an abrasive cut-off machine, are rough ground, except on the end gauging faces. About 0.25 mm of stock is left on all the faces, except the gauging faces, for final grinding. The blocks are then heat treated and brought to a temperature of 840°C and quenched at 50°C. Next follows tempering by immersion for one hour in an oil bath maintained at 105°C followed by cooling in still air at room temperature. The minimum hardness value at this stage is 65 Rockwell *C*.

The gauge blocks are then finished ground, with the exception of the end gauging faces. Stabilisation is achieved by alternately cooling and heating the blocks a number of times. In this process the gauges are placed in pans containing a wax lubricant (to prevent oxidation) and the pans placed in a dry-ice refrigerating unit, which cools them to a temperature between (– 73 to – 85°C). The length of the gauge determines the time taken, which varies between six and ten hours. The blocks are then removed and allowed to attain room temperature. To relieve transformation stresses, the gauges are tempered by heating to a temperature of 110°C for three house in a gas-fired furnace. The cooling tempering cycle is repeated about ten times, to increase the martensitic microstructure from approximately, 93 per cent (after heat treatment) to nearly 100 per cent. About 10 per cent of each batch of blocks is subjected to a Bureau of Standards approved stability test. The blocks concerned are immersed for twenty-four hours in a boiling 0.5 – 1 per cent potassium dichromate aqueous solution. The dimensions of the blocks must not change more than 0.05 micrometer per 25 mm, otherwise they are rejected.

Batches of gauge blocks are finished processed in quantities of sixteen, or multiple of this number, for any one size. The end gauging faces are ground to size plus 0.025 mm $^{+0.008}_{-0.000}$. The finish ground blocks are magnetically inspected for cracks while the blocks remain mounted on the magnetic chucks of the surface grinding machine. A magnetic comparator is also used for detecting internal flaws and for checking hardness. The blocks after being demagnetised, are now ready for lapping.

The end gauging faces of the blocks are lapped in four stages, namely, rough lapping, semi-finish lapping, and first and second "polishing" operations. Sixteen blocks are lapped at a time, on Norton lapping machines. The upper lap floats but does not rotate. The lower lap plate is driven by a variable-speed motor, the work being held in a spider in the usual way. Variable-speed drives enable various sized blocks to be lapped and the different operations to be performed. Roughening speeds are from 60 to 90 rev./min. whilst 30 rev. min. is the speed for finishing.

Lapping plates constructed from Mechanite cast iron are normalised, stabilised and available in sets of three. Their flatness is checked by means of optical flats. The work holding spiders are made of steel plates, each having four concentric rows of sixteen rectangular openings as shown in Fig. 2.169.

About 0.028 mm of stock is removed from the two end gauging faces of each block in rough lapping in twelve runs, each of thirty seconds duration. This brings the gauges to within 0.00045 mm of the required size. The surface finish at this stage is about 0.0625 μm. r.m.s. After each thirty second lapping period the positions of the gauge in the spider are changed to ensure uniform size

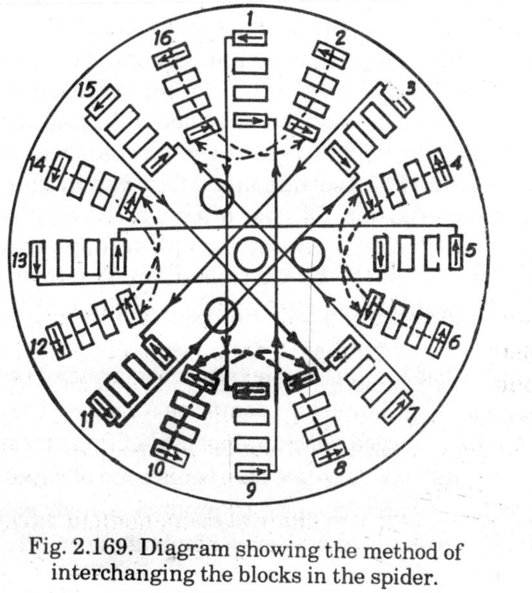

Fig. 2.169. Diagram showing the method of interchanging the blocks in the spider.

reduction. Fig. 2.169 shows the method of interchanging the blocks. The blocks are moved to other radial rows at subsequent changes.

At the end of each fourth lapping period the blocks are removed from the spider and cooled to room temperature by being placed on a steel block for about thirty minutes per 25 mm of length. An electric comparator, with 0.002 mm graduations is used to determine the lengths of the blocks.

In semi-finish lapping, 0.003 mm of stock is removed, bringing the blocks to within 0.0018 mm of the finished size, with a surface finish of about 0.0375 μm r.m.s. The procedure resembles that previously discussed for rough lapping. During the first "polishing" operation about 0.001 μm of stock is removed, to bring the blocks to within 0.0007 mm of size with a surface finish of 0.025 μm r.m.s. At this stage the size flatness and parallelism of each block are carefully checked. The second "polishing" operation, using a very fine abrasive, brings the blocks to size.

The National Bureau of Standards permit the following plus or minus tolerances per 25 mm of block length ; 0.00005 mm for grade *AA*, 0.0001 mm for grade *A* and 0.0002 mm for grade *B*. Flatness and parallelism of the measuring surfaces must be maintained within 0.00008 mm, and the surface finish specified is 0.0225 μm r.m.s.

The size, flatness and parallelism of the block gauges by comparison with "working masters", which, in turn, are recalibrated weekly by comparing them with "grand master". One set of six of the latter is calibrated by the National Bureau of standard every three months.

Final inspection is carried out by means of electric comparators and various size checks are made. Optical flats are used for checking both flatness and parallelism.

2.61.17. German (Zeiss) Method of manufacturing slip gauges. By this method, gauges are made to the highest quality (grade 0) and are held to limits of ± (0.1 + 0.002 × nominal length) microns. For example, a gauge of 10 mm nominal length is required to be finished to ± 0.12 microns. Variations from the mean size of a gauge resulting from face errors must be held to 0.10 micron over the range of sizes upto 200 mm nominal length. These gauge blocks are normally used as master or reference sets.

The gauges are manufactured from steel containing carbon 1%, chromium 1.8% and manganese 0.4 per cent. The cross-sectional dimensions of gauges of length upto 10 mm, and exceeding 10 mm are 9 by 30 mm and 9 by 35 mm respectively. Blanks are sawn from ground steel bars 0.3 to 0.4 oversize leaving 0.4 to 0.6 mm on the overall length of each block. The blocks are then hardened to Rockwell 65C to ensure dimensional stability. First, the gauges are heated to 900°C, and quenched in oil for hardening. Second, an artificial ageing treatment is undertaken. This involves holding the blanks at 120°C for ninety hours, and then at – 90°C for eight hours. Thin blanks are then surface ground, leaving about 0.07 mm on the length. Following this the side faces are ground, using angle blocks to ensure squareness within 0.07 mm of the finished sizes. Longer gauges are first ground on the side faces, followed by the measuring faces to give same finishing allowance. Demagnetisation follows, after which the blocks undergo a preliminary lapping operation. All lapping processes are carried out in fully air conditioned workrooms to permit both temperature and humidity to be controlled. In the room used for the preliminary lapping operation the temperature is held at slightly less than 20°C to provide compensation for heat generated in the gauges during the process.

The laps are made from close-grained cast iron, 450 mm diameter by 75 mm thick ; weighing about 75 kg. Three laps comprise a set, used to condition one another. At the preliminary stage the laps are charged with a compound of fine emery, with a mixture of oil and petrol. The work carrier plates have twenty-two apertures each of which holds two thin blanks. Longer blanks are mounted with a thickening piece to produce a near square section, to ensure greater stability, with less tendency to tip. Six or seven hours are normally required, for the preliminary operation. The blanks are changed around during the process to equalise the action, and to bring each block to the overall length within 0.003 mm of the nominal size for block gauges up to 10 mm, with a slightly larger allowance on longer blanks. Dimensional checks are carried out with the aid of Zeiss optimeter comparators.

Following preliminary lapping the gauge blocks are marked, by etching giving the nominal size, the grade of accuracy, etc. All the edges of the blanks are then bevelled 0.2 mm, at 45°.

Finish-lapping operations are carried out with corresponding position-changing sequences for the blanks in the work-carrier holes. The laps employed are specially heat treated to produce extremely hard iron phosphides in the matrix and cause the ferrite particles to break down slightly, leaving the phosphides in the matrix and cause the ferrite particles to break down slightly, leaving the phosphides embedded in the lap surface. No abrasive is used. Petrol is applied to clean the lap surfaces when required. The first lapping operation is continued until the blanks are within 0.0008 to 0.0009 mm of the nominal dimension.

For the final lapping stage, the gauges are transferred in trays lined with lint-free cloth to a room where the temperature is controlled within ± 0.5°C. Several hours elapse and further work is done to ensure that the temperature of the blocks is in equilibrium. The blocks are then ready for combined sizing and polishing operations. By varying the load on the upper lap by means of compression springs (to increase the almost non-existent cutting action) the gauges are reduced in length, leaving 0.0002 mm, on sizes up to 50 mm long, and slightly more for the longer blocks. The spring pressure is released an lapping continues to give a final polish to the measuring faces with a very slight amount of metal removal. The surfaces are finished within 0.00003 to 0.00004 mm. Frequent checks are made during the final lapping operation. When the gauges have been completed they are stored in the inspection department for several days to allow the temperature to reach equilibrium, after which the gauges are compared with approved masters. Records are maintained, of the difference in the length from the nominal dimension, and a month later the gauges are again checked. Blocks which show no measurable change on this occasion are accepted for service.

2.61.18. Calibration. Due to handling in the laboratory or inspection room, slip gauges are liable to show signs of wear appreciable period of use, and, therefore, they should be checked or recalibrated at regular intervals.

Workshop and inspection grade gauges are calibrated by direct comparison against the calibration grade gauges in a comparator. A variety of comparators are available which use mechanical, optical, electrical and pneumatic means of amplification. Eden-Rolt Millionth comparator and Brook-level comparator have been specifically designed for slip gauges.

2.61.19. Eden-Rolt Millionth Comparator. This comparator is capable of calibrating gauges upto 25 mm. It achieves a very high magnification of about 20,000 partly by mechanical and partly by optical means. Each division of its scale represents 0.0002 mm so that by estimation it is

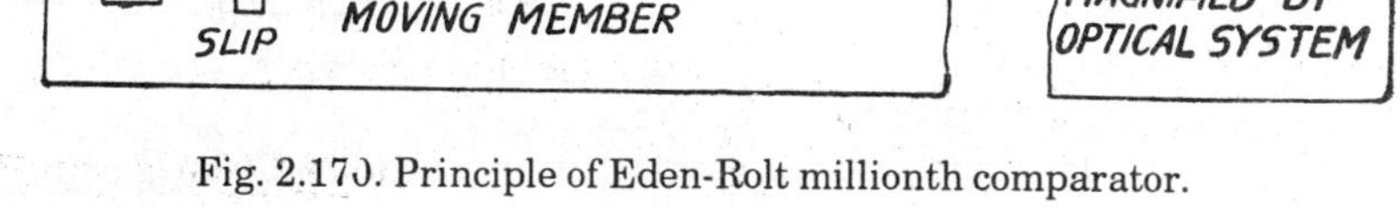

Fig. 2.170. Principle of Eden-Rolt millionth comparator.

possible to read 0.00002 mm. The measuring contacts consist of a steel ball on one side and a group of three balls in triangular formation on the other which eliminate many difficulties that would arise if plain parallel faces were used. The principle of the instrument is shown in Fig. 2.170.

The design of this comparator is extremely simple. In the mechanical amplifying system, a strip of spring steel is attached to the fixed block and a similar strip is fastened to the movable block. The moving member operates a lever at the end of which is fixed a light metal ring or cursor, with a piece of spider's web across it. Thus when a slip gauge is inserted between the contact faces of the instrument, the movable member is displaced, causing the light lever to twist. The magnification depends upon the relation between the length of the lever and the distance between the leaf springs. As the strip can be made quite small, of the order of 0.5 mm, a mechanical leverage of the order of 300 to 400 is obtained with a fairly short lever (about 200 mm long). The particular advantage of the spring-steel hinges is the almost entire elimination of friction and consequent wear.

The layout of the optical system is shown in Fig. 2.171 which further provides an amplification of about 60 : 1. A spider's web is inserted near the end of the lever and movements of this line (in the spider web) are further magnified by an optical shadow-projection apparatus. The light rays from a light source pass through a condenser, a prism and are further deflected towards a mirror (placed at a height of about 750 mm above the comparator) through the spider web and a lens. From the mirror the rays are reflected on a scale. The measurements are recorded from the position of a line image of the spider's web relative to the scale on the base of the machine.

To measure the difference in size between a slip gauge and a corresponding slip gauge of known size, the adjustable anvil of the comparator is unclamped and withdrawn sufficiently to enable the standard gauge to be inserted between the measuring faces. The anvil is then pressed forward so as to grip the gauge between the faces, at the same time bringing the image of the crosswire approximately to the middle of the scale and the reading is noted. Having clamped the anvil in this position, the trigger handles are pinched together so as to release the gauge, and the other gauge is inserted and its reading noted. It is usual to take three readings on each gauge, one at the centre and one each at either end. The difference between the two gauges is given by the difference between their mean readings.

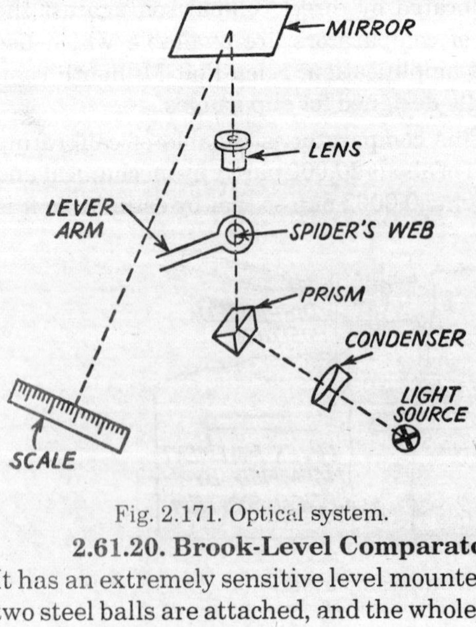

Fig. 2.171. Optical system.

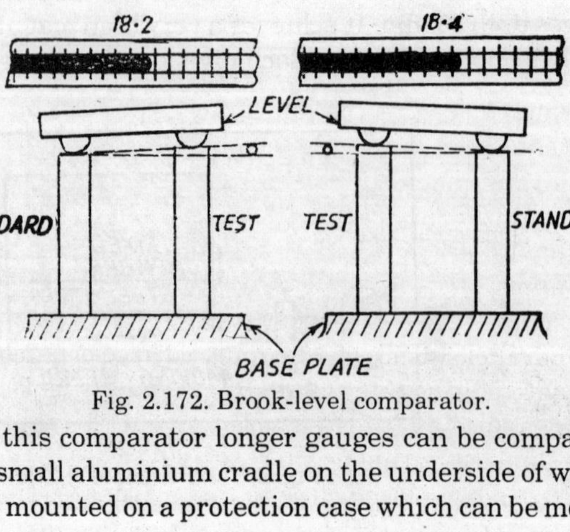

Fig. 2.172. Brook-level comparator.

2.61.20. Brook-Level Comparator. On this comparator longer gauges can be compared. It has an extremely sensitive level mounted on a small aluminium cradle on the underside of which two steel balls are attached, and the whole unit is mounted on a protection case which can be moved up and down a fixed pillar. The gauges to be compared are wrung on a circular steel platen in such a way that the two balls bridge the gauges. The platen can be rotated through 180° to reverse the position of the gauges under the level.

The position of the end of the bubble is read off a scale which is carried in the level housing and is graduated so that one division represents a difference in height of the balls of 0.00025 mm. The level is raised while the platen is rotated through 180°, then lowered again and a second reading taken. One-half of the difference between the readings obtained is equal to the difference in the length between the gauges.

In Fig. 2.172, if the first reading is 18.2, the standard being on the left hand side, and the second is 18.4 on metric scale, the difference in length is : $\frac{18.2 - 18.4}{2} = 0.1 \times 0.00025 = 0.000025$ mm, the gauge being shorter than the standard.

The principle of measurement is illustrated in Fig. 2.172, which shows the two opposite measuring positions.

2.61.21. Calibration by Interferometry. In engineering applications, wavelengths of light are used for extremely accurate measurements of surface flatness, slip gauges and other end gauges. These light waves have wavelengths of the order of 0.000375 to 0.000675 mm and by estimation of the widths of light interference bands, measurements to within 0.000025 or 0.00005 mm are possible.

Interference methods are suitable for determining the absolute size of calibration and reference grade gauges. For this the gauges must satisfy the following basic requirements :

(*a*) The end faces must be flat and parallel to each other ;

(*b*) They must have a high degree of surface finish ;

(*c*) The actual size must agree with its nominal size to within a very small tolerance ; and

(*d*) The edges must be properly rounded off and deburred.

Various types of interferometers have been designed from time to time over the last 50 years for the specific purpose of inspecting block gauges. Most prominent among these is the NPL-Hilger gauge interferometer. The interferometer is suitable for measuring sets of slip gauges up to 100 mm.

2.61.22. NPL-Hilger Gauge Interferometer. An important feature of the design of this interferometer is that batches of upto 20 gauges of differing sizes can be set up for measurement in one operation. The optical scheme is shown diagrammatically in Fig. 2.173.

Light from a cadmium or mercury-198 discharge lamp *A* is concentrated on a narrow slit *B* situated at the focus of a lens *C*. The parallel beam from *C* is directed into a constant deviation prism *D*, which disperses the light into its several constituent monochromatic rays. By rotating the prism, any one colour can be selected and directed vertically downwards on the optical flat *E*, which is supported above and at a slight inclination to the upper surface of the gauge *F* wrung on to a horizontal base plate. The rays are deflated from two parallel surfaces—the upper surface of gauge and face of base late—and after retraversing the prism and colimating lens are brought to a focus on a small reflecting prism *H* situated close to the slip *B*. An eye placed close to this prism sees two

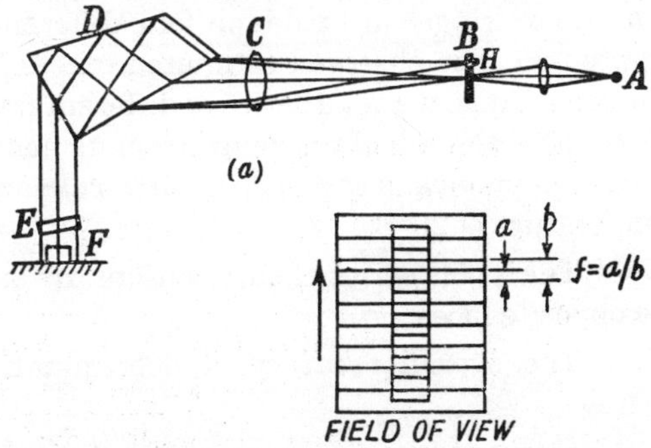

Fig. 2.173. Optical arrangement of the N.P.L. Hilger gauge interferometer.

sets of interference fringes which are formed between the lower semi-reflecting surface of the optical flat on the one hand, and the faces of the gauge and the base plate on the other. By suitably inclining the flat, the fringes can be seen as in Fig. 2.173 (*b*).

Theory. The two sets of fringes seen in the field of view are equispaced parallel to each other and in general, are displaced parallel to each other relatively by fraction a/b of the fringe-spacing as shown in Fig. 2.173 (*b*).

The height h of the gauge above the base plate is then equal to a certain whole number n plus the fraction part a/b of the half-wavelength of the light used. If the fringes are observed in several different radiations of wave-length λ_1, λ_2, λ_3, etc. and the corresponding fractions are $\frac{a_1}{b_1}$ (f_1), $\frac{a_2}{b_2}$ (f_2), $\frac{a_3}{b_3}$ (f_3), etc. then, $h = \frac{1}{2}\lambda_1(n_1 + f_1) = \frac{1}{2}\lambda_2(n_2 + f_2) = \frac{1}{2}\lambda_3(n_3 + f_3)$, etc.

where n_1, n_2, n_3 etc. are unknowns.

If the nominal length of the gauge is H, we have :

$$H = \frac{1}{2}\lambda_1(N_1 + F_1) = \frac{1}{2}\lambda_2(N_2 + F_2) = \frac{1}{2}\lambda_3(N_3 + F_3), \text{ etc.}$$

The error of the gauge is then given as :

$$h - H = \frac{1}{2}\lambda_1[(n_1 - N_1) + (f_1 - F_1)] = \frac{1}{2}\lambda_2[(n_2 - N_2) + (f_2 - F_2)]$$

$$= \frac{1}{2}\lambda_3[(n_3 - N_3) + (f_3 - F_3)]$$

The difference between the fractions in each case is known and it is also known that the whole numbers $(n_1 - N_1)$, etc. are only a few integers, since the errors of gauges do not usually exceed a few half-wavelengths.

The exact number of half-wavelengths to be associated with the known fractions to give the error of the gauge is found from a slide rule, on which various half-wavelengths are set out to scale from a common zero over a range of few units. (Also refer Ch. 6).

2.61.23. Determination of Length. The length of a slip gauge is defined as the perpendicular distance between the centre of the top of the gauge and the surface of an auxiliary base plate upon which the gauge has been wrung. This definition takes into account the thickness of one wringing film which is always associated with the gauge in its use.

For interferometric determination of its length, the gauge is brought to the centre of the field of view and an estimation is made of the fraction a expressed as a fraction of the fringe separation b [Fig. 2.173 (*b*)] at the centre of the gauge surface. Observations of the fractional displacements of two sets of fringes are made in four different wavelengths of either cadmium or mercury-198 or krypton-86. The observed estimated excess fractions are compared with the calculated nominal fractions. Thus, we get a series of different fractions corresponding to the uncorrected difference of the observed length of the gauge from its nominal size. The error in the length of the gauge is then evaluated by means of a special slide rule and is suitably corrected to give error at the standard temperature of 20°C.

The whole series of observations are subsequently repeated with the gauge wrung down on the opposite faces.

The overall accuracy of determination of length is one part in a million for lengths upto 25 mm.

Corrections. The measured lengths as derived interferometrically in ambient air require corrections on account of :

(*a*) Difference in temperature of the gauge from the reference temperature of 20°C ;

(*b*) Difference between conditions of ambient air and standards air ;

(*c*) Phase change due to the difference in finish between the exposed surface of gauge and top surface of base plate ; and

(*d*) Obliquity errors due to the inclination of incident light with the normal to the surface of base plate.

The temperature correction is calculated on the basis of an average value for the co-efficient of thermal expansion of hardened steel of 12 parts in a million per 1°C and is applied as $12\,l\,(20 - t) \times 10^{-6}$, where l is the observed length of the gauge at the value corresponding to 20°C, and t = room temperature.

The wavelength correction comprises of two parts, one for barometric pressure and air temperature, and the other for vapour pressure. This is given by the formula :

$$\frac{\lambda_a}{\lambda_b} = 1 + \left[A - B\frac{h}{1 + \alpha t} + C(f - 10) \right] \times 10^{-6}$$

where λ_a = wavelength in actual atmospheric conditions denoted by t, h and f ;

λ_b = wavelength in standard atmospheric conditions denoted by 20°C, 760 mm and 10 mm ;

t = temperature in °C ;

h = barometric height in mm of mercury ;

f = vapour pressure in mm of mercury ;

α = 0.003661°C ; and

A, B, and C = constants.

The total value of the wavelength correction is obtained by adding algebraically the value of the two terms and applying the results as a correction to the length of the gauge. Thus, if $+x$ and $+y$ are the values of the terms for a particular set of conditions t, h and f, then the correction to a gauge of nominal length N is given by

$$N(+x + y) \times 10^{-6}$$

The phase correction is zero, if the material and surface finish of the gauges and the base plate are the same. In case they are different, then it can be determined by taking five gauges from the set and finding the optical length of each by wringing on a base plate. The optical length of the combination is also measured on the same base plate, and from the length, the optical length of the uppermost gauge is subtracted. The resulting difference is the sum of the practical length of the other gauges. This is compared with the sum of the optical lengths of the same gauges and the average value of phase corrections is calculated.

For this interferometer, the viewing aperature is a rectangular hole of dimensions l and h, and obliquity correction for this is given by :

$$\text{Obliquity correction} = +N\left(\frac{S^2}{8F^2} + \frac{l^2 + h^2}{24F^2}\right)$$

where S is the distance between the centres of the illuminating and viewing apertures situated in the focal plane of the collimating lens of focal length F, and N is the nominal length of the gauge being measured.

The total correction factor is the algebraic sum of all the individual corrections. The value of the correction to be actually applied to the length of a gauge is derived by finding the product of the length and the total correction factor, taking due account of the sign of the correction.

2.61.24. On line calibration of Slip Gauges. In calibration of slip gauges by hand, other than time consuming process, disadvantage is that the accuracy of determination depends on the subjective judgement of the operator. In manual calibration, the operator will have tendency of play safe and thus high grade slip gauges may be classed as low grade which can be costly mistake. Automatic calibrator utilises microprocessor technology and provides an individual calibration and grading, and print-out of every gauge checked. It may take as small as 10 seconds to calibrate a gauge. Grading is done by measuring its length, parallelism between measuring faces and flatness of each face. Since the length tolerance on highest grade gauge is ± 50 mm, it is imperative that machine be capable of measuring 10% of this value, *i.e.* ± 5 nm. The handling by machine should not produce any scratch because a scratch of 1-2 micron width would affect the accuracy of slip gauge and lead to its rejection. The machine should be able to measure gauges of all sizes. The machine has to be installed in a tightly temperature controlled room, totally free of dust and totally vibration free foundation. Such a machine is totally enclosed and is fitted with hinged covers for access to the working zone and for loading and unloading of gauges. Operation is performed from a control station and all measuring cycles are performed automatically.

Each set of gauges is loaded into a cassette with a previously calibrated master gauge of the same nominal size and calibrated by interferometry. The cassettes are then loaded on to one end of a storage rack and a walking-beam mechanism brings them in turn to the measuring position. At measuring head, the master gauge is measured first and other gauges in sequence. Measuring heads consist of specially designed high-sensitivity, capacitive transducer to minimise risk of damage to the measuring faces. The gauges being measured are precisely positioned between two measuring heads, each of which incorporates fine probes to take measurements at the corners of each face. Extra probes are also included to cater for differences in cross-sectional area over the range of gauges which can be measured.

The probes comprise steel rods inserted into holes in the measuring head but are insulated from it by a thin film of epoxy resin. The measuring head acts as a guard ring to eliminate fringing of the electrical field between the gauge of the probes. The raise and lower measuring head also incorporates three gauge support rods on which the slip gauge rests while being measured. The heads are moved apart during gauge insertion. The lower measuring head is fixed and upper one is mounted on a ram which is raised/lowered hydraulically to adjust the separation of the measuring heads to suit the nominal size of the gauge being measured.

Each step of the control sequence is initiated only when a signal from a sensor has indicated completion of the previous action. The sequence is as follows : automatic loading of cassettes from the storage rack on to the beam ; adjustment of ram height to give the correct gap between the gauge and the upper measuring head, the placement, in turn, of each gauge at the measuring position ; and unloading the cassette from the beam after all the gauges have been measured.

2.61.25. Care and Use of Slip Gauges for Workshop and Inspection Purposes. All the surfaces are protected against climatic conditions by being covered with a high grade lanotin preparation or petroleum jelly or other suitable anti-corrosive preparation. The joint between the lid and the case is made with fillet and groove, to prevent the ingress of dust. The slip gauges are kept in a suitable case in which there is a separate compartment for each gauge. The case is designed such that the gauge can't become displaced. The sizes of the gauges are marked on the case, immediately adjacent to the appropriate compartment. The gauges and their case should be protected from dust and dirt. When the gauges are not in use they should be kept only in their case which should be kept closed. The gauges should be used only in air-conditioned rooms free from dust and maintained at constant temperature. Every care should be taken to protect the gauges from getting magnetised otherwise they will attract metallic dust. Gauge blocks should be handled using a piece of chamios leather or perspex tongs. In the case of new gauges, first the protective coating applied to it should be removed with petrol and finally gauges be wiped with a clean soft linen cloth. The wiping should preferably be done everytime before using the gauges.

During the actual use, the fingering of the lapped faces should be avoided as far as possible in order to preclude the risk of tarnishing, and the handling should be as minimum as possible to avoid transfer of heat from body to the gauges. Handling would also corrode high finish of gauges due to the natural acid in the skin. If handling is unavoidable, the hands should be washed and then coated with a film of pure petroleum jelly. If the gauges have been handled for some time, then they should be allowed to settle down to the prevailing temperature of the room. Actually both the work to be tested and the gauges wrung together should be allowed to settle down to the prevailing temperature of the room before doing any test. During actual use the slip gauges with their working surfaces should never be placed on the surface plates etc. After use, the gauges are wiped and finger marks removed, and replaced in their proper compartment in the case. For making combination, the proper gauges should be taken out of the case and case closed. Before starting the operation of wringing, the faces of the gauges should be wiped clean from the dust. The standard method of wringing the gauges together is first to bring the faces of the gauges into contact at right angles to one another and then turn them through 90°. This method causes less rubbing of the surfaces and for this reason the wringing should not be done by sliding two gauges parallel to each other. Thin slip gauges should not be wrung together without using a rigid gauge block as a basis for one of them since there is a danger of deforming thin gauge blocks permanently. The gauges should never be left wrung together for an unnecessary length of time. This may lead to micro cold welding and slight pitting of the surface when they are separated. The gauges should always be slid apart and the wringing joint between them should never be broken. If the gauges are not to be used for long time, they should be coated with petroleum jelly which should be applied only with a clean piece of soft linen. A brush should never be used for this purpose, as otherwise the jelly may be aerated and the moisture in the bubbles so formed may cause rusting of the faces.

When any gauge is knocked or dropped, its edges are most likely to be damaged. Such burrs may be removed with care by drawing an Arkansas stone lightly across the damaged edge in a direction away from the measuring face of the gauge and finally gauges washed and cleaned. Gauges with scratched surface should be returned to the manufacturer for the true surface to be restored. If during wringing process, any sign of roughness or scratching is felt the process of wringing should be stopped immediately and faces examined for burrs or scratches.

2.62. Slip Gauge Accessories

Although slip gauges can be used for many purposes without the aid of any auxiliary equipment, but their field of application is considerably extended by a few simple accessories, *viz.*, holders, jaws, scribers and centre points etc. The use of holders becomes very important when it is desired to use the slip gauge combination of considerable overall length, in order to ensure more

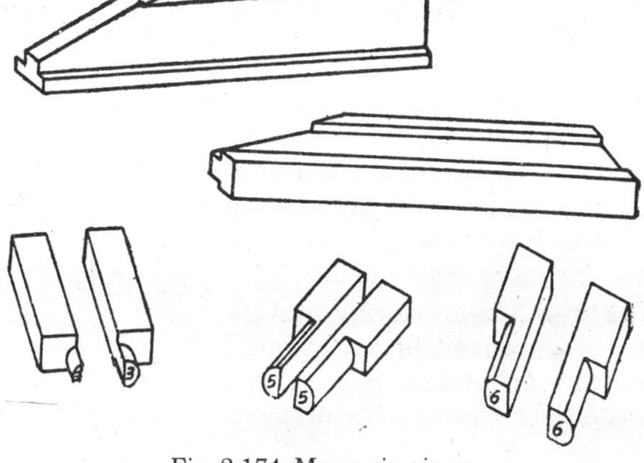

Fig. 2.174. Measuring jaws.

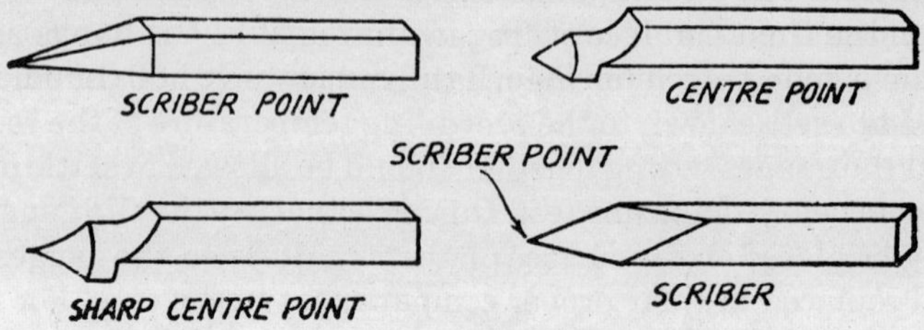

Fig. 2.175. Slip Gauge Accessories.

rigid assembly. The slip gauges after being selected are inserted in the holder (Refer Fig. 2.174) which comprises of two slotted side plates secured to end block whereby they may be lightly clamped in position. The gripping pressure is usually applied by means of a knurled head screw passing through a tapped hole in one of the end blocks ; however in case of the longer holders some form of rapid adjustment is desirable which is achieved by providing a sliding block which can be clamped to the side plates of the holder in any desired position by means of a knurled nut.

As already pointed, in order to increase the utility of slip gauges, it is essential that extensive use be made of slip gauge accessories. These in conjunction with slip gauges are used for the inspection of gauges and precision equipment ; for the measurement of plug and ring gauges ; and for marking out tools also. The beauty of these accessories is that these can be built up quickly and used as a temporary limit gauge.

The various accessories may be listed out as follows :

(*i*) Measuring jaws.

(*ii*) Scribing and centre point.

(*iii*) Holders and base.

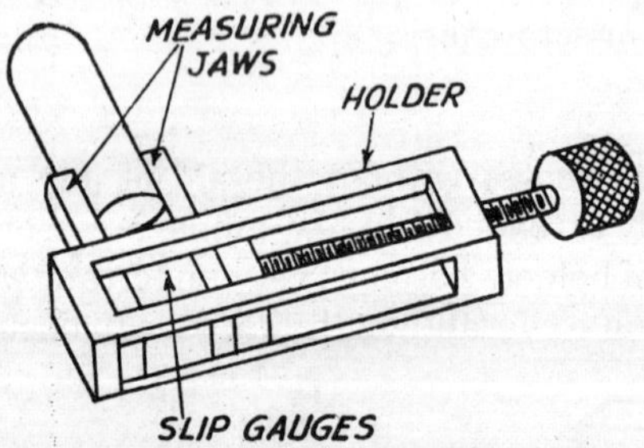

Fig. 2.176. Type A measuring jaws being used for external measurement.

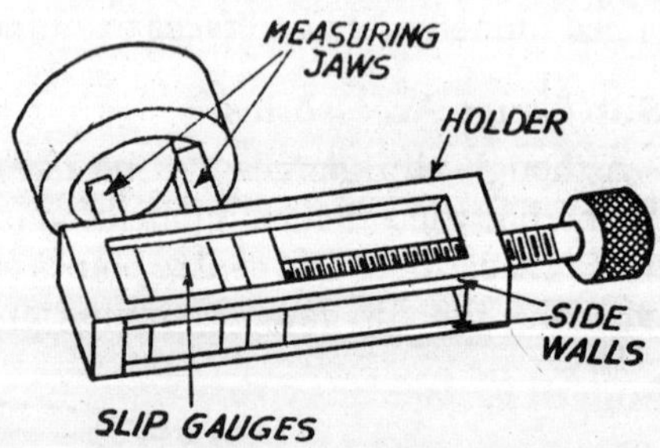

Fig. 2.177. Type A measuring jaws being used for internal measurement.

The slip gauge accessories are made from suitable steel free from defects and hardened uniformly upto a minimum hardness of 800 HV. The working surfaces of the measuring faces of measuring jaws and the working surface of the base are finished to obtain good wringing surface without the use of any extraneous agent.

2.62.1. Measuring jaws. The measuring jaws are always supplied in pairs. These are available in two types, *viz.*, for both external and internal measurements (Type A) and the other type intended for external measurement only (Type B).

Fig. 2.178. Type B measuring jaws intended for external use only.

In Fig. 2.179 the most important dimension is the size of measuring jaws. This nominal size is always engraved on the front end face of the jaw. In Fig. 2.179, the various other dimensions of the measuring jaw have been shown for the measuring jaw size of 5.0 mm. Other sizes available are 2.5 mm and 6.0 mm.

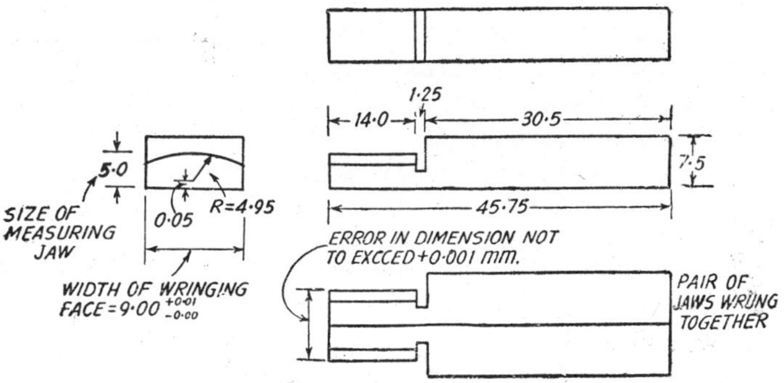

Fig. 2.179. All dimensions are in mm. Type A measuring Jaw.

It may be noted that in type *B* measuring jaw, no dimension is very important as it does not come into measurement. The only important item is the wringing face which should be flat to within 0.0002 mm. This limit of accuracy holds good for type *A* also. For type A jaw, the nominal size of the jaw should be accurate to within ± 0.0007 mm. When the pair of jaws are wrung together the error in overall dimension should not exceed + 0.0001 mm. The faces perpendicular to the wringing faces should be accurate to within 90° + 0.3′ from the wringing face.

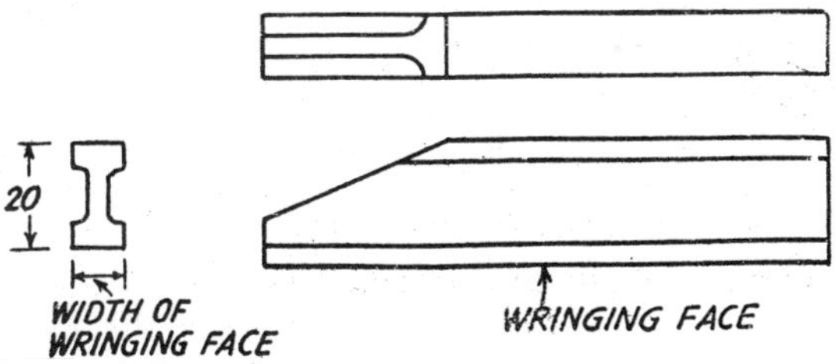

Fig. 2.180. Type B Measuring Jaw.

2.62.2. Scribing and Centre Points. These in conjunction with the holder and slip gauges constitute very accurate device for marking out purposes. The scribing and centre points are designed in such a way that when they are used in combination, the diameter of a final scribed circle does not differ from twice the size of the slip gauge combination inserted between the points by more than ± 0.05 mm. The scribing points and centre are shown in Fig. 2.181 and Fig. 2.182 respectively.

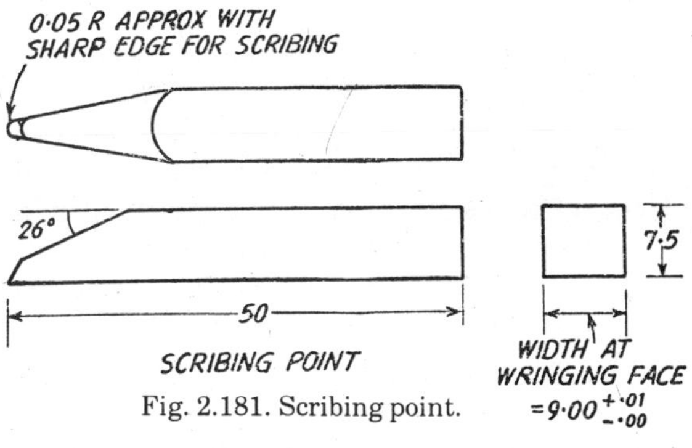

Fig. 2.181. Scribing point.

In this case too, the wringing face is very important and it should be flat to within 0.0002 mm. The faces perpendicular to wringing face should be accurate to within 9° + 0.3′ from the wringing face.

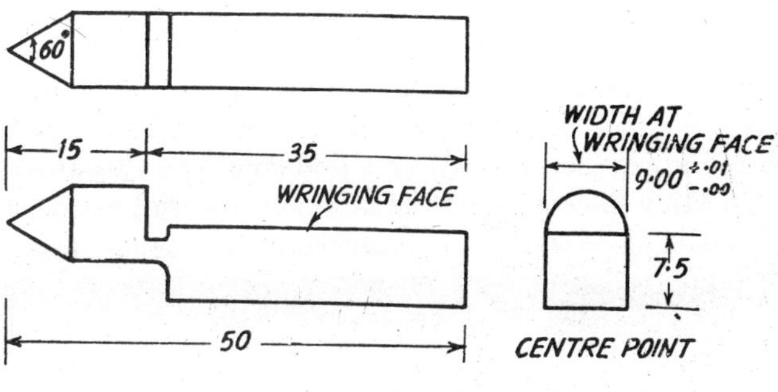

Fig. 2.182. Centre point.

2.62.3. Holder. The holders are available in three ranges, *viz.*, 0 to 60 mm, 50 to 110 mm, and 100 to 210 mm. These are made of suitable design for holding rigidly combination of slip gauges within their range. The surface upon which the gauges are wrung should be flat to within 0.0001 mm and parallel to the base.

2.62.4. Base (Refer Fig. 2.183). It is made of robust construction and designed such that it remains stable when used with the longest size holder. Suitable provision is made for the attachment of the holder with the base in a position, normal with respect to the wringing surface of the base. The wringing surface of the base should be flat to within 0.0002 mm and suitably relieved and an air vent provided. The sides of the base are suitably relieved so as to provide convenient finger grip. The surface upon which gauges are wrung should be flat within 0.0001 mm and parallel to the base within 0.0002 mm. The normal height of the base should be in agreement with the dimension marked on the base to within 0.0007 mm.

2.63. Reference Discs, Balls and Roller Gauges

Reference discs are working standards manufactured to desired tolerance and used for setting diameter measuring machines and other purposes.

In many measurements, precise balls and rollers manufactured to fine tolerances are used for inspection purposes (refer chapter 9). These are ground, lapped and calibrated to extreme limits of accuracy to permit inspection on comparators and for miscellaneous measurements.

2.64. Step Gauge (Refer Fig. 2.184)

A step gauge is a bar in which a series of tapped holes are provided to enable separation of the gauge head in nominal 30 mm increments upto 480 mm. This being an "end measure", great accuracy can be achieved and at the same time the inaccuracy, which can result from fastening a number of gauges together is avoided. The gauging surfaces of the step gauge are as close to the neutral axis as possible. It is always placed on exactly flat surface and great control of temperature exercised. If step gauge is placed on convex surfaces, effective length will increase, and if placed on concave surface it will decrease because measuring faces are slightly away from neutral bending plane for the purpose of rigidity of bar.

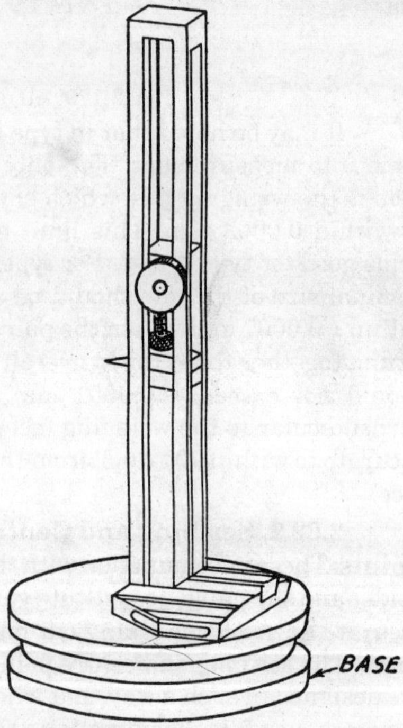

Fig. 2.183. Base.

Many times, a plexiglass cage is placed around measuring environment and a plexiglass cover is placed over step gauge, leaving only a small slot to permit movement of indicator. This is done for precise control of temperature in measuring zone as it is difficult to ensure constant temperature in big room. Both sides and also the bottom of the step gauge are lapped straight, parallel and square to 0.00025 mm. The gauging pads are also lapped square to the sides and bottom. During calibration, the step gauge is supported its full length on the measuring machine table to gain maximum surface contact. The large mass of measuring machine table (slow to change in temperature) acts as heat sink and lends stability to standards in contact with it. The idea of not supporting the step gauge on airy points is that temperature control becomes a problem which can be solved by having flat face and control its temperature through measuring machine.

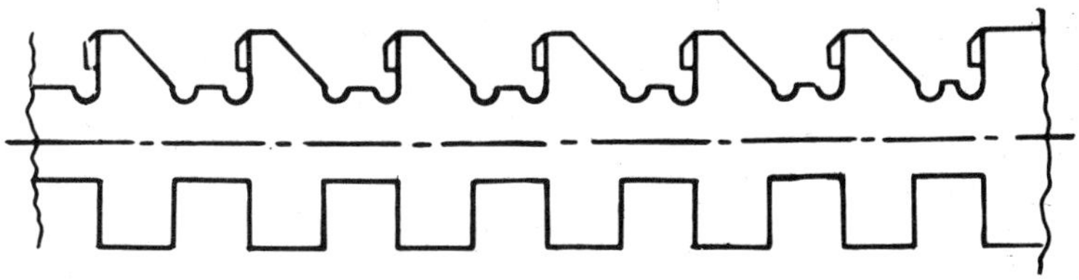

Fig. 2.184. Cross section of a step gauge.

The step gauge is calibrated by placing it and the end standard of appropriate length side by side. Two electronic gauge heads (reading upto 0.00001 mm) are fastened to a steel bar which in turn, is fixed to the spindle housing of measuring machine. The control of absolute temperature is highly essential for establishing total length but for subdivision, uniformity of temperature throughout is only important. The calibration of various steps is done by subdividing after establishing total length. For calibration of each step, several cross-checks are made.

An important thing to note about step gauge manufacturing is that if one of the steps is lapped short, then all the steps have to be reduced by same amount.

WORKED EXAMPLES

Q. 2.1. Scale of a measuring instrument is an ordered set of array of indicating marks, together with any associated numbering (forming a part of an indicating device) in relation to which the position of an index is observed.

Differentiate between scale mark, scale length, scale range, scale division, scale spacing, scale interval, scale numbering, and moving scale.

Ans. Scale mark is a line (mark) corresponding to one or more defined values of a measurand.

Scale length is the length of the line between the first and last scale marks and passing through the centres of all the shortest scale marks.

Scale range is the range of scale values between the extreme scale marks. It is normally stated in terms of its lower and upper limits, like 25 mm to 50 mm.

Scale division is the part of a scale between any two successive scale marks.

Scale spacing is the distance between any two successive scale marks measured along the same line as the scale length. It is expressed in units of length, regardless of the units of measurement marked on the scale.

Scale interval is the difference between the scale values corresponding to two successive scale marks. It is expressed in the units of measurement marked on the scale.

Scale numbering refers to the set of numbers, marked on a scale, which either correspond to the values of the measurand defined by the scale marks or merely indicate the numerical order of the scale marks.

Moving scale is used in some instruments in which the indications are given by the position of a moving scale in relation to a fixed index.

Q. 2.2. Differentiate between the following types of scales of measuring instruments : Linear scale, non-linear scale, suppressed zero scale, expanded scale.

Ans. Fig. 2.185 (*a*), (*b*), (*c*), (*d*) demonstrate the difference between four types of scale.

(*a*)

Linear or regular scale (Equal spacing throughout)

(*b*)

Non-linear scale (Here spacing is not linear but governed by other rule (Logarithmic scale)

(*c*)

Suppressed-zero scale (This scale has no zero but starts with 25 mm)

(*d*)

Expanded scale (Here range 5—10 occupies a disproportionately large part)

Fig. 2.185

Q. 2.3. What are the limitations of line graduated rules ?
What are the possible observational errors.
How can these be reduced.

Ans. The line graduated rules may have limitations like geometric deficiencies resulting from errors in flatness and parallelism or due to deflection ; (*b*) inaccuracies of scale graduations ; (*c*) excessive thickness or poor definition of the graduation lines.

The observational errors in length measurement by line graduated scales may be either due to alignment deficiencies (because of improper coincidence of the distance boundaries with selected scale graduations) or parallax errors (because of observing the scale and the object in a direction not perfectly normal to the surface to be measured). The observational errors can be reduced by use of auxiliary devices which improve the alignment of viewing and keep the measurement operation under control.

Q. 2.4. How does a caliper gauge permit substantial improvements over direct measurements with line graduated rules ?

Ans. The use of beam, fixed jaw and sliding jaw in a caliper gauge help improvements in linear measurements on following accounts.

(*a*) The datum of measurement gets coincided precisely with one of the boundaries of the object to be measured.

(*b*) The sliding jaw achieves positive contact with the object boundary at opposite end.

(*c*) The read-out alignment error is reduced due to closely observable correspondence of the reference marks on the slide with a particular scale value.

Q. 2.5. Enumerate the factors for (*a*) inherent accuracy and (*b*) process errors in the use of micrometer screw gauge.

Ans. (*a*) The inherent accuracy of a micrometer screw gauge may be affected by

(*i*) degree of calibration of the spindle movement.

(*ii*) linearity of the spindle movement.

(*iii*) flatness and parallelism of the measuring surfaces.

(*iv*) deflection of the frame.

(*b*) The process error of micrometer measurement can be caused by

(*i*) heat transfer during holding of the instrument.

(*ii*) reading errors.

(*iii*) inadequate alignment or stability in the mutual positioning of the object and gauge.

(*iv*) wear, etc.

Q. 2.6. (*a*) **A vernier scale consists of 25 divisions on 12 mm spacing and the main scale has 24 divisions on 12 mm. What is the least count.**

(*b*) **How you will compare the vernier caliper with a micrometer screw gauge ?**

Ans. (*a*) Least count of vernier scale.

$$= \text{main scale spacing} - \text{vernier scale spacing}$$

$$= \frac{12}{24} - \frac{12}{25} = 12\left(\frac{1}{24 \times 25}\right) = \frac{1}{50} = 0.02 \text{ mm}$$

(*b*) Single vernier caliper can be used over a much larger range of measurement but micrometer screw gauge can be used for a limited range. Four separate micrometers may thus have to be used for different readings in a span of 0—100 mm. Further same vernier caliper can measure both internal and external dimensions but separate micrometer screw gauges are required for this purpose.

The shape and construction of micrometer make it stiffer and more robust compared to vernier caliper.

Micrometer screw gauge is easier to record than a vernier scale.

The measuring screw of a micrometer is protected from dirt and accidental damage by the barrel, but the movable and main scales of vernier caliper are unprotected.

The accuracy of micrometer is more than that of vernier caliper.

Q. 2.7. Discuss the advantages and limitations of vernier calipers, micrometer screw gauges and dial indicators.

Ans.

Characteristic	*Vernier calipers*	*Micrometer screw gauge*	*Dial gauge*
Measuring Range	Large (upto 2 m)	Limited to 25 mm	upto 100 mm
Accuracy	0.02 mm	0.002 mm with vernier	0.001 mm
Adjustment for wear	No	Provided	No
Facility for "feel"	Lack of "feel"	Ratched or friction thimble provided to aid constant "feel"	Mechanism ensures constant "feel'.
Other advantages	Can be used for measuring external as well as internal dimensions	Wide variety of measuring tips available for different applications.	1. Easy to read 2. Quick in use if only comparison is required.
Other limitations	1. Point of measuring contact not in line with scale 2. Jaws can spring causing error	Separate instruments required for different ranges (in steps of 25 mm)	1. Usually used for comparison and not for measurement. 2. Can be easily damaged by mishandling. 3. Loss of accuracy because of number of moving parts. 4. Has to be rigidly supported.

Q. 2.8. The repetitive precision of measurements with a screw micrometer depends on the inherent accuracy of the micrometer caliper and the combined effect of process errors. Elaborate them.

Ans. The accuracy of micrometer caliper is governed by the degree of calibration of the spindle movement, and the linearity of the spindle movement.

The degree of calibration of the spindle movement is affected by the lead errors of the screw. Since its effect is cumulative, *i.e.* error increases with the length of the spindle travel, the aggregate effect of inaccuracies on this account can be reduced by "balanced calibration", *i.e.* by adjusting the thimble to produce error-free reading in the middle of the total, or of the most frequently used section of the spindle traverse.

The linearity of the spindle movement is dependent on the type of thread. Drunken thread, or stick-and-slip condition of the screw in the nut have adverse effect. It is required that any fractional rotation of the screw should result in a proportional advance of measuring spindle. The instrument accuracy is substantially improved by manufacturing the spindles of well stabilised material, precisely grinding the screw thread after hardening, using lapped nuts.

The flatness and parallelism of the measuring surfaces, and deflection of the frame also contribute to measuring errors.

The process errors of micrometer measurement can be caused by heat transfer while holding the instrument, reading errors, inadequate alignment or stability in the mutual positioning of object and measuring tool, wear, etc.

Heat transfer can be reduced by employing plastic insulating grips on the frame.

Reading errors are substantially reduced by (*i*) employing satin chrome finish to eliminate glare, (*ii*) providing distinct graduation lines on a beveled thimble surface to facilitate reading with a minimum of parallax error (*iii*) using large diameter thimbles.

Alignment and holding stability can be improved for measurements of small parts by using a stand which rests on the bench to clamp the hand micrometer instead of holding it with free hand.

The applied measuring force is usually limited by friction screw and ratchet. For precise control of the static measuring force indicator micrometers can be used.

Wear on measuring faces can be reduced by using carbide faced anvils and spindle tips.

Q. 2.9. Aluminium though light, is not used for micrometer frames. Why ?

Ans. Micrometer frames are made of steel forgings because it has practically the same coefficient of thermal expansion as most of the parts to be measured. With aluminium frame, the differential in the rate of thermal expansion would result in measuring errors.

Q. 2.10. Part A indicates the various types of micrometers and part B indicates various applications for which a particular type of micrometer from part A is suited. Match the parts A and B.

Part A	*Part B*
1. Disc type micrometer	(*a*) For measuring pitch diameters.
2. Blade type micrometer	(*b*) For measuring the wall thickness of tubes and other cylindrical bodies.
3. Micrometer with veeformed anvil and conical spindle tip	(*c*) For repeat accuracy of the measurement.
4. Pointed contact micrometer	(*d*) For measuring the distance of protruding features from a flat reference surface.

5. Ball contact micrometer	(*e*) To measure diameters at the bottom of narrow grooves.
6. Indicator micrometer	(*f*) For sensitive measurement of small parts.
7. Three-point contact micrometer	(*g*) For measuring diameter of deep bores.
8. Depth micrometer	(*h*) For measurements of thickness of closely spaced sections.
9. Bench micrometer	(*i*) For controlled displacement of slides.
10. Micrometer heads	(*j*) To measure inside recesses where regular tips would not penetrate.

Ans. 1 (*h*), 2 (*e*), 3 (*a*), 4 (*j*), 5 (*b*), 6 (*c*), 7 (*g*), 8 (*d*), 9 (*f*), 10 (*i*).

Q. 2.11. (*a*) **What do you understand by the term 'cosine error'. Give examples.**

(*b*) **Illustrate a case in which combined sine and cosine errors occur.**

Ans. (*a*) In measuring the length of any part with a scale, if the measuring scale is inclined to the true line of the dimension being measured, then there will be what is called 'cosine error'. For example in Fig. 2.186, it may be seen that the length recorded is in excess of the true length by an amount $l(1 - \cos\theta)$. However, in most cases, the angle θ is very small and thus the error is negligible.

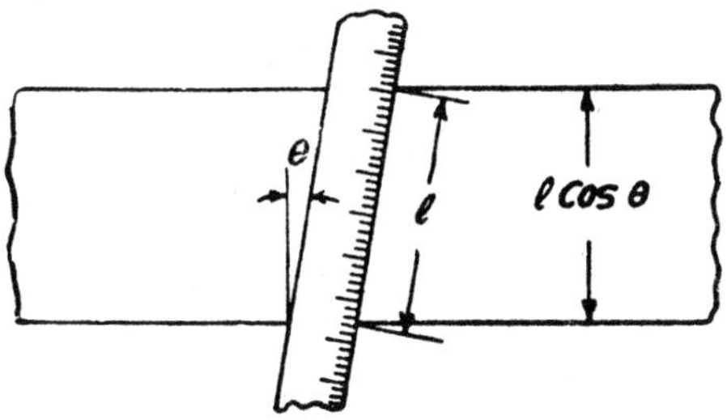

Fig. 2.186

Similar effect is observed when using dial gauges. It may so happen that the axis of plunger of dial gauge may not be along the direction of movement of work.

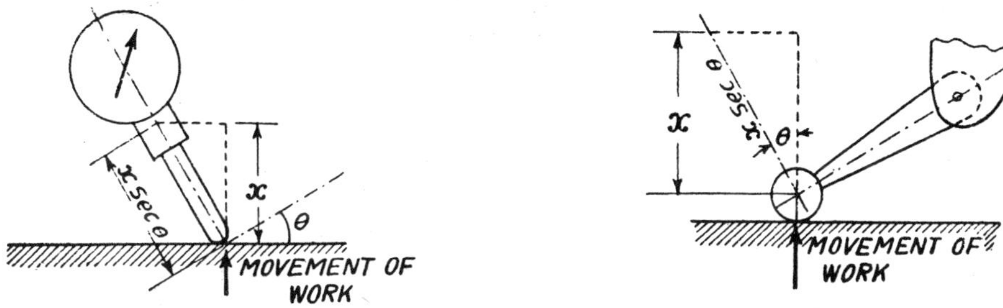

(*a*) (*b*)

Fig. 2.187

Similar is the case when an indicator is fitted with a ball-end stylus arm. In this case, it is important for the arm to be set so that the direction of movement of the work is tangential to the arc along which the ball moves. In both these case [Fig. 2.187 (*a*) and (*b*)] the reading on the indicator will be $x \sec\theta$, where x is the distance moved by the work. It may be noted that some lever-type test indicators are designed for use with their styli inclined at about 30° to the line of measurement, and in these cases in error would result if it were set with the stylus normal to the line of measurement.

(*b*) The combined sine and cosine errors occur when measuring an end gauge in the horizontal comparator if the gauge is not supported so that its axis is parallel to the axis of the measuring anvils, or if its ends, though parallel to each other, are not square with the axis. (Refer Fig.s. 2.188 and 2.189)

In such a case (Refer Fig. 2.188)

Length $L = M \cos \theta - d \sin \theta$

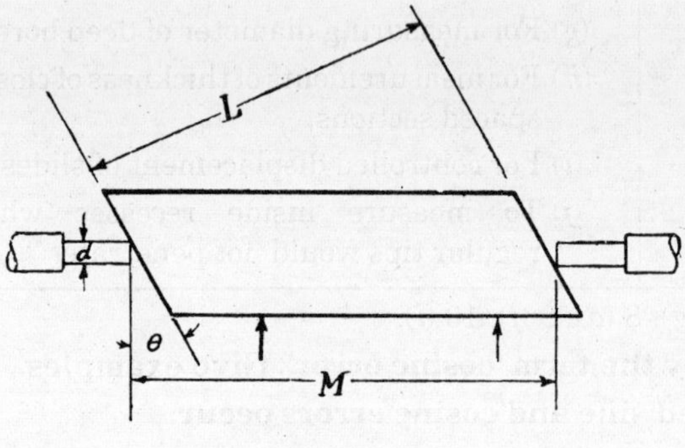

Fig. 2.188

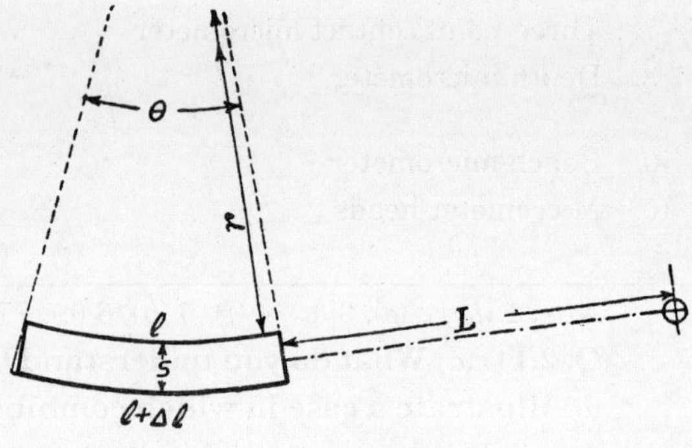

Fig. 2.189

Q. 2.12. (*a*) **Describe the feature of flexible strips used in amplifying mechanism in certain instruments. Also derive an expression for mechanical magnification.**

(*b*) **In an Edne-Rolt comparator, the mechanical lever is 150 mm long and is mounted on parallel strips 0.4 mm apart. Compute the overall magnification of the comparator, if the optical magnification of movement of the end of the lever is × 50.**

Ans. (*a*) Flexible strips are used to provide very low-friction pivots, parallel movements and other motions of high accuracy. In the Eden-Rolt comparator, the strips are used as part of the amplifying mechanism and not as supports for levers. The principle of operation is as follows.

Two flexible strips are used, one attached to moving member and other to the fixed one. Change in dimension of a component makes the moving members to be displaced, which in turn being attached to other strip effectively lengthens that strip by amount Δl. Since the strips are rigidly held at the ends, they are forced to bend into an arc by an amount depending on the lengthening of the strip Δl and the separation between them (s). Though the slight additional sideways movement of the inner end of the layer is added to the larger movement of the outer end by the bending of the strips but it is small by comparison.

If the strips are of length l and the outer one is displaced linearly by an amount Δl, then the outer arc becomes $l + \Delta l$ and the inner arc is still l. If the inner strip takes up a radius r in bending and the angle subtended is θ, then

$$\frac{l + \Delta l}{r + s} = \frac{l}{r} = \theta$$

$$\Delta l = s \times \theta \quad \text{or} \quad \theta = \frac{\Delta l}{s}$$

If radius of lever is L, then deflection at the end of lever $= L \,.\, \theta = L \,.\, \dfrac{\Delta l}{s}$

and the mechanical magnification $= \dfrac{\text{Deflection at end of lever}}{\text{Displacement of moving member}} = \dfrac{L \,.\, \frac{\Delta l}{s}}{\Delta l} = \dfrac{L}{s}$.

(*b*) In this numerical, $L = 150$ mm. and $s = 0.4$ mm.

$\therefore$ Mechanical magnification $= \dfrac{L}{s} = \dfrac{150}{0.4} = 375$

Since optical magnification is × 50.

$\therefore$ Overall magnification 375 × 50 = × **18750. Ans.**

Q. 2.13. Write short notes on :

(*i*) **Abbe principle of alignment and** (*ii*) **Kinematic slide.**

Ans. (*i*) All measuring instruments and the generating equipment, as far as possible, should conform to the Abbe principle of alignment, according to which the line of axis of measurement should coincide with the line of the scale or other dimensional reference. In certain cases where the different diameters have to be contended with and which cause variation in the position of the line of measurement, this principle may not be followed. This principle avoids error due to lack of straightness of slides.

(*ii*) Kinematic principle is one of the important principles which should be followed in the design of laboratory instruments. In metrology, we are often concerned with translational coupling, *i.e.* coupling with one degree of freedom ; in other words, to provide a kinematic arrangement which restricts all motions but allows only one of sliding or linear movement.

A simple form of kinematic slide consists of a cylinder sliding on to V-supports, with an arm (integral with the cylinder) ending in a point which is in contact with a plane arranged parallel to the cylinder axis. The device constrains the cylinder and prevents it from rotating on its V-supports.

Another scheme which reduces friction to a minimum is used in the N.P.L. diameter machine. The machine has two carriages, upper and lower. The parallel vees are machined in the lower carriage, and a vee and a flat in the upper carriage. On the lower member two balls are placed in each vee group and the upper carriage moves over these balls.

Q. 2.14. (*a*) **Explain a method to measure the diameter of a bore about 1 m in diameter precisely.**

(*b*) **Prove that the method adopted is reasonably accurate.**

Ans. (*a*) Obviously no vernier calipers or micrometers are available for this size and the ordinary scale will not measure the diameter accurately. A most commonly used method under such circumstances is using pin gauge (a bar with spherical ends) whose length is slightly less than the diameter to be measured and is exactly known.

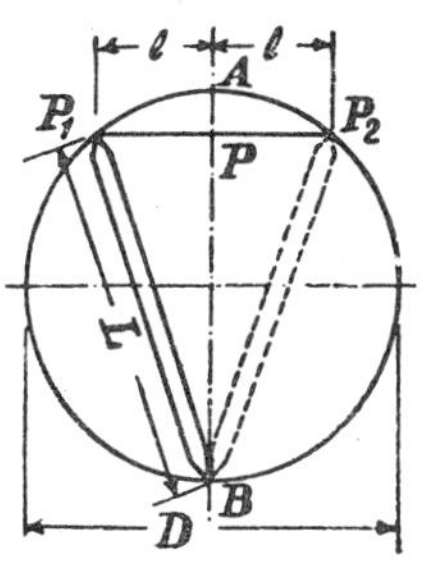

Fig. 2.190

This bar is placed in the bore as shown in Fig. 2.190. with one of its end resting at the bottom of the bore at point B and other end resting against the bore surface in one direction at point P_1. Keeping one end in contact with the base surface (point B), the other end is brought across to make contact at the other side (point P_2). The distance P_1P_2 is measured (it can be measured quite accurately being very small) and from this and the length of bar, the diameter of the bore can be easily computed.

In Fig. 2.190, by the properties of intersecting chords.

$$BP \times PA = P_1P \times PP_2 = l \times l = l^2$$

Since $\quad BP^2 = BP_1^{\ 2} - P_1P^2$

$\therefore \quad BP = \sqrt{L^2 - l^2}$

$\therefore \quad PA = \dfrac{l^2}{BP} = \dfrac{l^2}{\sqrt{L^2 - l^2}}$,

Diameter $\quad D = BP + PA$

$$= \sqrt{L^2 - l^2} + \frac{l^2}{\sqrt{L^2 - l^2}} = \frac{L^2}{\sqrt{L^2 - l^2}} = L^2(L^2 - l^2)^{-1/2}$$

$$= L\left(1-\frac{l^2}{L^2}\right)^{-\frac{1}{2}} = L\left(1+\frac{l^2}{2L^2}\right) \text{ approx. (By Binomial Theorem)}$$

(The above approximation is permissible because the distance l is usually small in comparison with L).

$$\therefore \qquad D = L + \frac{l^2}{2L}$$

In order to check the effect upon the results obtained, the above readings are taken with two length bars of different lengths. The circularity of the bore is checked by taking readings along three or four different diameters.

(*b*) Partially differentiating the equation for dia D, we get

$$\partial D = \frac{\partial D}{\partial l} \times \delta l + \frac{\partial D}{\partial L} \times \delta L = \frac{l}{L}\,\delta l + \left(1 - \frac{l^2}{2L^2}\right)\delta L$$

$$= \frac{l}{L}\,\delta l + \delta L - \frac{l^2}{2L^2}\,\delta L \qquad \left(\frac{l}{L} \text{ being too small, } \frac{l^2}{2L^2}\,\delta L \text{ can be neglected}\right)$$

$$= \frac{l}{L}\,\delta l + \delta L$$

where δl = error in estimating length l, δL = error in the calibration of pin.

Thus obviously error by this method will be mainly that in calibration of the pin.

Q. 2.15. Describe a method to determine the diameter of a bore by use of four balls and a height or depth gauge.

Ans. In this method the component whose bore is to be measured is placed on a surface plate and three balls of equal size are spaced evenly within the bore so that these rest on the surface plate and a fourth ball, which may be either of same size or of different diameter, is placed upon them. The height H is measured by the vernier height gauge.

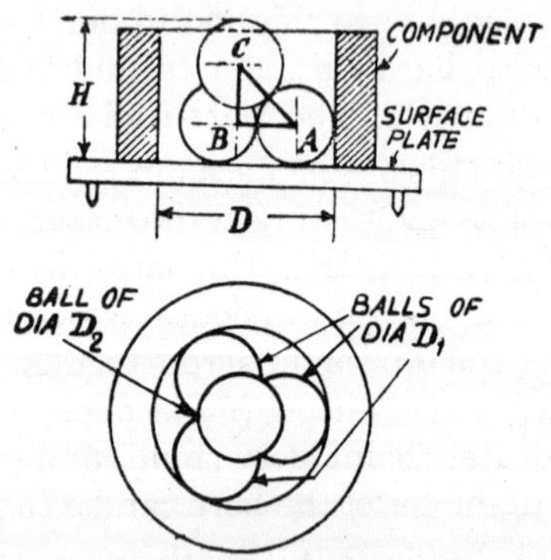

Fig. 2.191

Referring to Fig. 2.191,

$$AC^2 = AB^2 + BC^2$$

or
$$\left(\frac{D_1 + D_2}{2}\right)^2 = AB^2 + \left(H - \frac{D_1 + D_2}{2}\right)^2$$

$$\therefore \qquad AB^2 = \left(\frac{D_1 + D_2}{2}\right)^2 - \left(H - \frac{D_1 + D_2}{2}\right)^2$$

$$= \left(\frac{D_1 + D_2}{2}\right)^2 - H^2 - \left(\frac{D_1 + D_2}{2}\right)^2 + 2H \cdot \frac{D_1 + D_2}{2} = H(D_1 + D_2) - H^2$$

Now $\qquad D = 2AB + D_1 = 2\sqrt{H(D_1 + D_2) - H^2} + D_1.$

In this method ; it is important that the balls should be chosen of such size that the angle between lines AB and AC is between 30° and 60°. If this angle is large, then small errors in measuring H will result in larger errors in the diameter determined ; and if this is small, then heavy contact forces between the balls will reduce the accuracy. It is desirable to take the reading at both the ends and if possible, in the middle position also by placing the balls on equal slips or rollers.

Q. 2.16. (*a*) **Explain how a precise spirit-level vial is produced ?**

(*b*) **A vial is to be produced to a sensitivity of 5 seconds of arc for a bubble movement of 2.5 mm, calculate the radius to which it must be ground. (1 sec = 4.85 × 10^{-6} radian).**

(*c*) **What are the possible sources of errors in the use of spirit level.**

Ans. (*a*) Accurate level vials are not produced by bending but by internal lapping or grounding process. A lapping rod is bent to the required radius. A torque is applied to the lap and the tube is lapped by giving to and fro motion along the lapping rod as also the slight rotary motion. Radius of curvature to which tube has been ground is calculated by finding the radius at which the given bubble displacement corresponds to the given angle variation.

(*b*) Obviously 2.5 mm bubble movement at radius R substends 5 seconds.

$$R = \frac{2.5}{4 \times 4.85 \times 10^{-6}} = 1.0309 \times 10^{5} \text{ mm} = \mathbf{103.09\ m.}$$

(*c*) The errors in the use of spirit levels can be of two kinds :

(*i*) inherent in the level itself, and (*ii*) incorrect use.

Inherent errors are the constructional faults which include faults in the vial, such as the radius of curvature being non-uniform ; either the divisions, or the vial, or both being positioned incorrectly.

Incorrect use includes using it in conjunction with unsatisfactory workpieces, or neglecting to take precautions against temperature effects due to exposure to sun or simply carelessness.

Q. 1.17. Define the following terms relating to spirit levels : sensitivity, range, roll error, repeatability, and settling time.

Ans. Sensitivity. The angle of tilt which causes the bubble in the vial to move through one scale division. It is determined by the vial radius and the base length.

Range. The total angle of tilt through which a vial can be used.

Roll Error. The change in reading when the instrument is rotated about the axis of the vial.

Repeatability. The capability or measure of the ability of the level to repeat its readings under all ordinary methods of operation.

Settling Time. The time taken for a bubble to return to its steady position after the level has been disturbed. Usually, after the level has been disturbed to make the bubble move approximately one scale division, it should come to within one-tenth of a scale division of its steady position within 10 seconds.

Q. 2.18. What factors are to be considered in selecting the sensitivity of a level ?

Ans. The use of unnecessarily sensitive level will be troublesome and introduce difficulties. The changes in ambient temperature and any lack of rigidity will affect the readings considerably. Thus level of only appropriate sensitivity should be chosen for a given application.

Q. 2.19. How levels are calibrated ?

Ans. Levels are usually calibrated on a tilting table fitted with a pair of ball-ended feet at one end and a vertical micrometer head at the other. The tilting table should be of substantial construction and supported on a surface plate on a rigid foundation. For calibrating 5 second levels, the base length of the tilting table, *i.e.*, the normal distance between the axis of the micrometer and the centre line passing through the two feet at the other end, should not be less than 750 mm.

Q. 2.20. Describe the following in brief, clearly bringing out the difference between them :

(*a*) **Surface gauge ;** (*b*) **Surface plate,**

(*c*) **Tool maker's flat,** (*d*) **Optical flat,**

(*e*) **Profilometer.**

Ans. (*a*) **Surface gauge** also known as height transfer gauge is used to check the accuracy or parallelism of surfaces, and to transfer measurements in layout work by scribing them on a vertical surface.

(*b*) **Surface plate** is an accurately machined flat casting or lapped granite block upon which the parts to be checked and the surface checking instruments are placed for obtaining some measure of the accuracy of a surface or the condition of finish.

(*c*) **Tool maker's flat** is a small plate which is lapped to a greater degree of accuracy and is used for inspection of small parts with precision gauge blocks.

(*d*) **Optical flats** are flat lenses usually made from natural quartz with very accurately polished surfaces having light transmitting quality. These are used in connection with interferometric measurements (science of measuring with light waves) for testing of plane surfaces.

(*e*) **Profilometer** is the instrument used for measuring surface roughness. It measures the number of roughness peaks in unit traverse length above a preselected length by passing a fine tracing point over the surface.

Q. 2.21. Can surface plates be made from materials other than cast iron ? What are the advantages of making surface plates from those materials ?

Ans. Though surface plates are usually made out of cast iron, yet they can also be made of non-metallic substances like glass and granite.

Such surface plates have the advantages of more rigidity for same depth compared to cast iron surface plates, complete elimination of corrosion problems, ease in sliding metallic articles over them, and the damage to their surfaces does not throw up a projecting burr as in case of cast iron plate but only causes indentation.

Q. 2.22. What do you understand by cross-wind in cast iron surface plates ?

Ans. Cross-wind is a deviation from flatness, somewhat similar to twisting. This is an effect similar to holding one edge and twisting the opposite edge (Refer Fig. 2.192). In such a plate, for example, points P_1 *and* P_2 where crosswind is present will not be at same level w.r.t. a datum. It is possible that cross-wind may be confined to a particular area only in a surface plate and could be true for a portion of its surface.

Cross-wind in surface plates can be detected by mapping the contour of the plate.

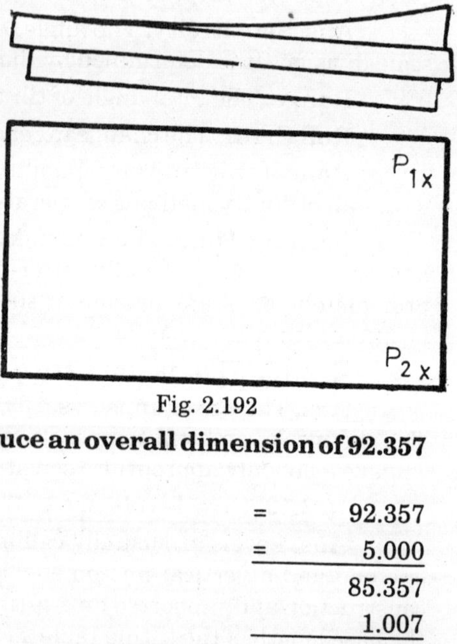

Fig. 2.192

Q. 2.23. List the slips to be wrung together to produce an overall dimension of 92.357 mm using two protection slips of 2.500 mm size.

Ans. Original dimension	=	92.357
Less two protector slips of 2.500 mm	=	5.000
		85.357
1 mm + 3rd place		1.007
		86.350
1 mm + 2nd place		1.050
		85.300
1 mm + 1st place		1.300
		84.000
To bring whole number 25, 50 or 75		9.000
		75.000

∴ Combination = 2 protector slips of 2.500 mm + 1.007, 1.050, 1.300, 9.000 and 75.000 mm.

Q. 2.24. (*a*) **What are the important points in wringing slip gauges together ?**

(*a*) **What is the function of protector.**

Ans. (*a*) The most important point to be remembered in method of wringing of two slip gauges together is to ensure that any moisture, oil, dust or similar foreign matter will be pushed away from gauging surfaces instead of being trapped between the gauges. Therefore, first the two surfaces should be slided transversely as shown in Fig. 2.193 (*a*) and then twisted together (Refer Fig. 2.193 (*b*)] to form the wrung pair [Fig. 2.193 (*c*)]

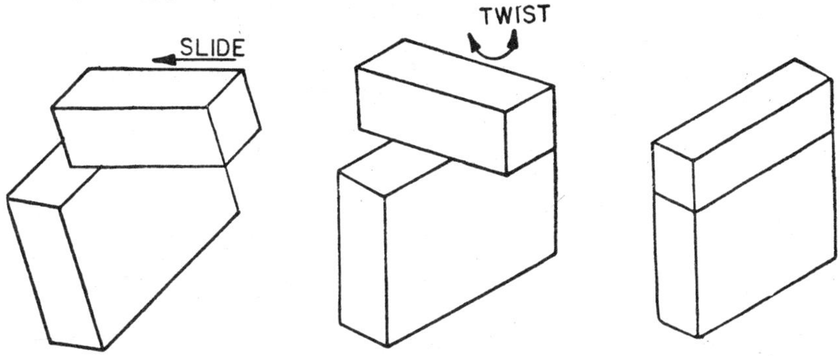

(*a*) (*b*) (*c*)

Fig. 2.193

(*b*) Usually, two protector slip gauges of 2.500 mm size are included in workshop and inspection gauges, but not in calibration and reference sets. The idea of these is to concentrate wear only on those two gauges because these are placed at top and bottom of other slip gauge combinations and hence only these two gauges come in contact with the surfaces to be gauged and get worn maximum. Other gauges are thus worn minimum and protector being of smaller size can be easily replaced.

Q. 2.25. It is proposed to measure the diameter of a large hole by a vernier caliper. What will be the disadvantage in doing so ?

Explain how the diameter of large holes could be determined accurately using

(*i*) **line standard,** (*ii*) **end standard.**

Ans. Measurement of diameter of a large hole by vernier caliper may not be suitable on account of following reasons :

(*a*) As the plane in which adjustment is made and the plane of measurement are not in direct line, any deflection can lead to wrong readings. This can be further aggravated by laxity of the fit of the sliding member on the stock.

The flexibility of the stock could lead to an inaccurate estimation of the 'feel'.

(*b*) There exists no provision to ensure constant measuring pressure between workpiece and measuring face of the instrument.

(*c*) The radiused jaws used for inside measurement lead to rapid wear and the length of the radiused portion limits its use to the mouths of holes.

(*d*) Accuracy is not more than 0.02 mm.

An accurate line standard for measuring diameter of larger holes is a 'stick' type inside micrometer which has rigidly assembled sticks and micrometer head having a travel of 25 mm and other fixed size end pieces.

Diameter of large holes could also be measured by using an end standard whose length is little less than the diameter. Its one end is kept in contact with one side of the bore and the other end is rocked to contact the opposite side of the bore at two points. For accurate results, the length of end standard should be selected such that rocking is small. If amount of rock is s and if L is the length of end standard, then diameter $= L + \frac{s^2}{8L}$.

Q. 2.26. The diameter of a large bore of about 1 m diameter is measured by using an accurately calibrated bar of length 999.00 mm having error in the measurement of not more than 0.01 mm. The bar is inserted in the hole and keeping one of its end at one place of the surface of bore, the other end is rocked to touch at two points and the amount of rock is measured as 80.0 mm marking blue on the surface. The accuracy of its measurement is of the order of 0.5 mm. Determine the diameter and likely accuracy of measurement. Also find whether poor accuracy of measurement of rock has any effect on accuracy of determining diameter.

Ans. Diameter of the bore $= \text{Length of bar } (L) + \frac{[\text{Swing } (s)]^2}{8 \times \text{Length of bar } (L)}$

$$D = 999.00 + \frac{(80.0)^2}{8 \times 999.00} = 999.00 + \frac{800}{990.00}$$

$$= 999.00 + 0.8008 = 999.8008 \text{ mm.}$$

Differentiating the equation $D = L + \frac{s^2}{8L}$ partially,

$$\frac{\partial D}{\partial L} \times L + \frac{\partial D}{\partial s} \times \delta s = \left(1 - \frac{s^2}{8L^2}\right) \delta L + \frac{s}{4L} \times \delta s$$

Since s is very small in comparison to L, $\frac{s^2}{8L^2}$ can be ignored

$$\therefore \qquad \delta D = \delta L + \frac{s}{4L} \delta s = 0.01 + \frac{80}{4 \times 999} \times 0.5 = 0.01 + 0.01 = 0.02 \text{ mm.}$$

It is obvious from above that poor accuracy of determining s has not much effect on overall determination of D, if s is small and L is very large.

Q. 2.27. How does optical micrometer work ?

Ans. Optical micrometer consists of a block of glass capable of being rotated about an axis perpendicular to a ray of light passing through it and a provision to measure the angle of rotation. When a ray of light AB falls perpendicular to block of glass, it passes straight as $ABCD$. When the block is tilted by angle i, then ray AB makes angle i with normal NN' to the block and is refracted along BE such that angle $N'BE$ $= r$ (angle of refraction) and finally emerges along EF parallel to CD but displaced from CD.

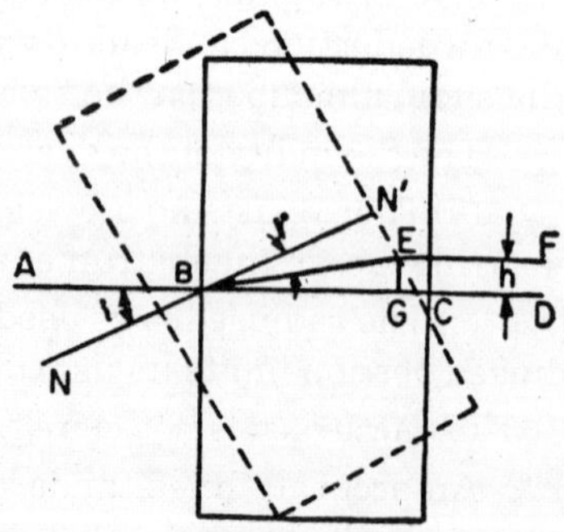

Fig. 2.194

If $\qquad \mu = \text{refractive index of glass} = \frac{\sin i}{\sin r}$

Displacement $\qquad h = DF = EG = BE \sin (i - r)$

$$= \frac{BN'}{\cos r} \sin (i - r) \approx w \sin (i - r)$$

For very small angles ; $\cos r = 1$ and $\sin i = i$

Since $\qquad \mu = \frac{\sin i}{\sin r} = \frac{i}{r}$

$$\therefore \qquad r = \frac{i}{\mu}$$

and $$h = w\left(i - \frac{i}{\mu}\right) = wi\left(\frac{\mu - 1}{\mu}\right)$$

For a block of about 5 mm width having refractive index of 1.5, the displacement caused due to rotation of 10 mm (0.003 radians) is 0.005 mm. It is thus possible to set very small displacement by rotating the glass block through relatively large angles.

Q. 2.28. What are the important design considerations in design of metrological instruments. Discuss them.

Ans. The following principles should, as far as possible, be applied in the design of metrological instruments.

(*i*) **Principle of alignment.** Also known as Abbe' principle, the line or axis of measurement should coincide with the line of the scale or other dimensional reference. If these two (*i.e.* measuring scale is offset from dimension being measured) are separated as in Vernier Caliper then any lack of straightness in the line of movement of the jaw will cause an error in measurement.

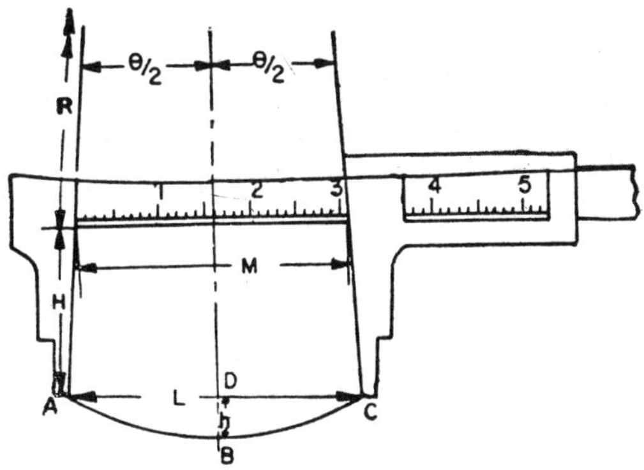

Fig. 2.195

In Fig. 2.195 if due to forcing of the jaws over the component being measured, the beam gets bent so as to subtend an angle θ, then the true dimension L of the component and distance M measured over scale will be different. Let H be the distance between measuring scale and line of component dimension being measured and R the curvature of the scale beam.

Then $$L = 2(R + H)\sin\frac{\theta}{2}$$

$$L = R\theta = 2R \,.\, \sin\frac{\theta}{2} \text{ (for small angles, } \theta/2 = \sin\theta/2)$$

$$L - M = 2H \,.\, \sin\frac{\theta}{2} = 2H\frac{L}{2(R + H)} = \frac{HL}{R + H}$$

By the property of intersecting chords of a circle $h[2(R + H) - h] = \dfrac{L^2}{4}$

or $$(R + H) = \frac{L^2}{8h} + \frac{h}{2} \approx \frac{L^2}{8h} \qquad \text{(neglecting } h/2 \text{ compared to } L^2/8h)$$

$$L - M = HL \times \frac{8h}{L^2} = \frac{8Hh}{L}.$$

(*ii*) **Kinematics.** It is important that components be so supported as to leave the desired degree of freedom, involving no redundant locations or restraints on the component parts. Where robustness is essential, then the principles of kinematics can't be fully applied. Redundant constraints should be avoided as these cause stresses and strains. For free and accurate movement, the theory of degrees of freedom and location must be followed instead of using more complex bearings in which the simple accuracy could be easily lost.

(*iii*) **Rigidity.** The instruments should be designed to be sufficiently rigid without being bulky. Parts subjected to flexure should be properly designed to take care of tolerance on the work for which the instrument is to be used. As a general rule, the forces applied in use should not cause more distortion than will result in a change of scale reading of about one-tenth of the smallest graduation.

(*iv*) **Amplification.** It is a general rule that larger is the amplification required, smaller will be the range of measurement and greater the need for care in using it. As far as possible, friction should be avoided by using flexible strips for amplification, or optical method should be employed.

(*v*) **Adjustment.** Adjustment for accurate setting to compensate for wear should be provided.

Q. 2.29. Calculate the radius of curvature (R) of the vial of a spirit level of 200 mm base length with one scale division indicating a change of level of 2.5 μm in 100 mm, if the length between scale divisions along the curvature is 3 mm. Also determine the sensitivity in terms of seconds of arc per scale division.

If the reading of a bubble position can be made to a fifth of a division, determine the discrimination.

Ans. If l be the movement of bubble corresponding to a tilt of base by angle α,

L = length of the base, h is the difference in height of base and points corresponding to tilt of angle α,

R = radius of vial,

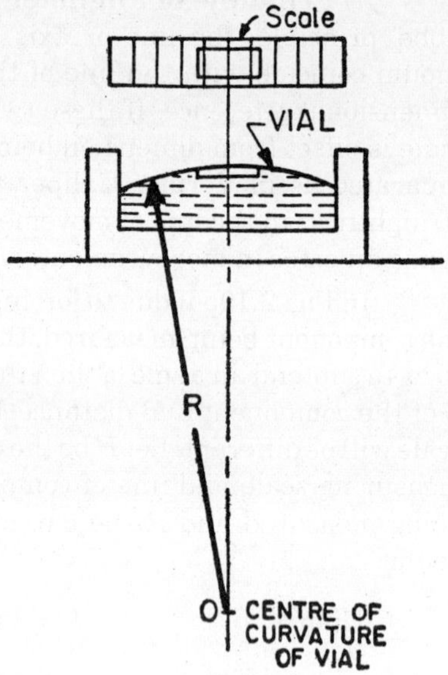

Fig. 2.196

then
$$\frac{l}{R} = \frac{h}{L}$$

Here
$$\frac{h}{L} = \text{gradient of the angle } \alpha,$$

and sensitivity of spirit level $= \dfrac{l}{\alpha} = \dfrac{l}{\left(\dfrac{h}{L}\right)}$

Here
$$L = 200$$
$$h = \frac{2.5\ \mu\text{m}}{100} \times 200 = 0.005 \text{ mm}$$
$$l = 3 \text{ mm}$$

$$\therefore \quad R = \frac{lL}{h} = \frac{3 \times 200}{0.005} \text{ mm} = \frac{3 \times 200 \times 1000}{5} = 120 \times 10^3 \text{ mm} = 120 \text{ m.}$$

Angle
$$\alpha = \frac{h}{L} = \frac{0.005}{200} \text{ radian} = \frac{0.005}{200} \times 206,265 \text{ sec.}$$

(1 radian being equal to 206, 265 sec)

$$\approx 2.15''.$$

$\therefore$ Sensitivity in terms of sec. of arc scale division = 2.15″.

Since h = 0.005 mm corresponding to scale division of 3 mm and if it can be estimated to 1/5th of division, then discrimination of spirit level.

$$= \frac{0.005}{5} = 0.001 \text{ mm.}$$

Q. 2.30. Fill in the blanks in regards to care and use of gauge blocks.

1. The greatest care should be exercised in protecting the gauges and their case from ...., ... and ....

2. When gauge blocks are not in use, they should always be kept in their .... and the case should be kept ....

3. Care should be taken that the gauges do not become ...., otherwise they will attract ferrous dust.

4. If the guages are new or have been covered with a protective coating after being last used, most of this coating may be removed with an appropriate ....

5. Gauges after washing with solvent should be wiped with a clean chamois leather or ....

6. Before using a gauge, it should always be .... first

7. It is undesirable to aim at removing all traces of grease, since a very slight film of grease is an aid to satisfactory ....

8. .... of the lapped faces should be avoided to preclude the risk of tarnishing

9. If the gauges have been handled for some time, they should be allowed to .... to attain room temperature before being used for test purposes.

10. A piece of work should .... be tested directly after machining and similarly large combinations of gauge blocks not be used .... after they have been wrung together

11. Gauges should not be held above the .... when being wrung together to avoid damage to other by accidental fall.

12. Before wringing gauges together, their faces should be .... free from dust and examined for ....

13. Damage to the gauges is most likely to occur on the ...., resulting from the gauge being knocked or ....

14. Slight burrs on gauges can be removed with care by drawing an Arkansas type stone across the damaged edge in a direction away from the measuring face of the gauge.

15. The gauges should never be left .... together for any length of time.

16. The wringing joint between gauges should never be broken, but should be removed by .... the gauges apart

17. Immediately after use, each gauge should be .... clean and be replaced in its proper .... in the case

18. It is particularly important to remove any .... from the measuring faces after use.

19. If the gauges are used infrequently, they should be .... with a suitable corrosion preventive before being put away

20. The .... should never be used for applying preparation on the faces of gauge as this may aerate the preparation and moisture in the air bubble so formed can cause .... of the faces.

Ans. 1. *Dust, dirt, moisture,* 2. *Case, closed,* 3. *Magnetised,* 4. *Solvent,* 5. *Soft linen cloth,* 6. *Wiped,* 7. *Wringing,* 8. *Fingering,* 9. *Settle,* 10. *Not, immediately,* 11. *Case,* 12. *Wiped, burrs,* 13. *Edges, dropped,* 14. *Lightly,* 15. *Wrung,* 16. *Sliding,* 17. *Wiped, Compartment,* 18. *Finger marks,* 19. *Coated,* 20. *Brush, rusting.*

TYPICAL EXAMINATION QUESTIONS

1. Explain the various instruments used for linear measurements and list them according to their accuracy. State the accuracy of measurement with rule, calipers and micrometers.
2. Sketch a micrometer and explain its working.
   Discuss the function and principle of ratchet drive in case of a micrometer.
   Explain why a micrometer provides a more positive reading than vernier caliper.
3. Discuss the relative advantage of vernier calipers and micrometer.

4. What tests will you carry out to fully check a 0 to 25 mm external and internal micrometer ? What error is expected in each case ?
5. Describe two methods of testing the parallelism of micrometer meausirng faces and also discuss the merits and limitations of each method.
6. How would you check the flatness and squareness with the axis of the spindle end of a 25—50 mm external micrometer ?
7. What are the basic principles that should be observed in the design of instruments and gauges ?
8. Write brief notes on the following, giving an example in each case :
   (*i*) Taylor's principle
   (*ii*) Cosine error
   (*iii*) Principle of alignment.
9. How slip gauges are manufactured ?
10. A block of steel is to be made 60 ± 0.02 square and approximately 40 mm thick. It is first milled on all faces and then ground. At milling time, about 0.1 mm is left as the stock to be removed during grinding. Describe how you would grind it on the four 40 mm wide face so that the error in squareness does not exceed one minute of arc at any corner. Explain how you would test it for squareness.
11. Name three materials commonly used for gauges. Discuss the merits of each. What is the stabilizing process for gauges ?
12. (*a*) Why adjustments are provided in the instruments ? Describe three different methods of adjustment and indicate in which instrument these are used.
    (*b*) Write short notes on :
    (*i*) crossed strip hinge
    (*ii*) optical micrometer
    (*iii*) rigidity, in connection with gauges and instruments.
13. Two end gauges are required to be 65.85 mm long and their faces accurately flat and parallel. Describe how you would make and test them, using 25 mm diameter bar already machined to 66 mm length and already fully heat treated.
14. An end gauge 30.800 mm long having accurately flat and parallel ends is to be measured using optical flat and master gauge of correct length. Explain how would you estimate any error in its length.
15. A circular end gauge of length 27.535 mm is to be measured by means of slip gauges and optical flats. Describe the method assuming the two ends to be :
    (*i*) parallel,
    (*ii*) not quite parallel.
    State the degree of accuracy.
16. Explain the various instruments most commonly used in inspection shops for various measurements involved.
17. Explain with neat sketches suitable methods of measuring or checking the following :
    (*a*) Diameter of a small hole upto 3 mm diameter.
    (*b*) Diameter of hole upto 50 mm diameter.
    (*c*) Diameter of holes greater than 50 mm.
18. Describe three different methods of measuring the diameter of a plain ring gauge approximately 45 mm in diameter, and also state the degree of accuracy you would expect in each case, and the advantage of each method.
19. The diameter of the bore in a plain ring was measured by laying it on a flat surface and placing 3 balls of 25 mm diameter equally spaced in it, with a 30 mm diameter ball resting on them. The distance from the surface to the top of the 30 mm ball was measured as 38.50 mm. Calculate diameter of the bore and state the possible sources of error in this method of measurement.

20. State the different methods of checking holes.

Bores of gun barrels are to be checked accurately and quickly for out-of roundness and taper defects. Discuss any method by which this can be done.

21. A bore of size 425.20 mm with a tolerance of ± 0.02 mm is to be checked by a pin gauge 425.00 ± 0.01 mm long. Determine the limits on the total movement of the free end.

22. One of the long edges of a piece of plate 500 × 125 × 5 mm thick is to be formed to a radius of 2.5 m. Describe the NPL method of generating the radius by grinding. Determine the suitable spacing for the pins and calculate the angular setting.

23. Describe clearly three methods of checking a try-square. What should be the accuracy of a good try-square ?

24. A 25 mm diameter hole is drilled and reamed 50 m deep in a 150 mm diameter shaft. The centre line of the hole is the plane of the circular cross-section and parallel to, but displaced 30 mm from the vertical centre line of the cross-section. Describe how would you check the accuracy of 30 mm dimension ?

25. Describe how a bored hole and 116.13 mm, with a tolerance of ± 0.01 mm. would be checked using slip gauges and accessories.

26. What do you understand by measuring pressure ? What is its effect ? How it is maintained constant during measurement ?

27. Why is it considered bad practice to lock a micrometer spindle and then use it as a gap gauge.

28. Give an example of typical application of length bars and accessories, indicating how a setting of 684.63 mm would be obtained.

29. The slip gauge set M 38 consists of the following :

Range	*Steps*	*Pieces*
1.005	—	1
1.01—1.09	0.01	9
1.1—1.9	0.1	9
1.0—9.0	1.0	9
10.0—100.0	10.0	10

Choose the suitable slip to give the following dimensions :

(*i*) 29.975 mm

(*ii*) 16.07 mm

(*iii*) 101.005 mm

30. A bore is to be made 500.10 mm in diameter, with a tolerance of ± 0.02 mm and for checking it, a pin gauge 500.00 ± 0.01 mm long is supplied. Determine the limits on the total movement of the free end.

31. The diameter of the bore of a ring was measured by laying it on a flat surface and then placing a ball of diameter d_1 touching one side of the ring and resting on a flat. The other ball of diameter d_2 is now placed resting on the ball of diameter d_1 and touching the other side of the gauge. If H is the measured height from the top of the ball of diameter d_2 (which also projects out the top of surface) to the bottom flat surface show that :

(*a*) the diameter D of the bore $= \dfrac{d_1 + d_2}{2} + \sqrt{H(d_1 + d_2) - H^2}$

(*b*) $(d_1 + d_2)$ should be greater than D

(*c*) for accurate measurement $(d_1 + d_2)$ should not be greater than 1.17 D.

32. In measuring the diameter (D) of a bore by four balls (3 balls of diameter D_1 and the fourth one of diameter D_2) placed over them, the height from the flat surface to the top of the ball of diameter D_2 is measured to be H. Show that $D = D_1 + 2\sqrt{H(D_1 + D_2) - H^2}$

Also show that for the success of the method, $(2D_1 + D_2)$ should be greater than D, and for accuracy of results, $D_1 + 0.414\, D_2$ should be less than $0.58\, D$.

3

Standards of Measurements

3.1. Introduction

Metrology concerns itself with the science of measurements and nearly all measurements in workshop involve measurement of dimensions. Production engineer is specially concerned with the measurement of the length and angle. Length is of fundamental importance as even angles can be measured by combination of linear measurements. Thus it is very essential that some standards be prescribed for the length. The importance of study of length has been realised long back and man has always sought a fixed and unvarying natural standard of length (fundamental unit). It is well known fact that without standards of fundamental units (length, mass, time), it would have not been possible for civilisation to exist.

Modern manufacturing technology is based on precise reliable dimensional measurements. Ultimately, all of these measurements are comparisons with standards developed and maintained by bureaus of standards throughout the world.

3.2. Standards

The two standard systems for linear measurement used throughout the world are English and Metric (yard and metre). Realising the importance and advantages of metric system, most of the countries are adopting metre as the fundamental unit of linear measurement. Length can be measured by using (*i*) line standard or (*ii*) end standard.

Search for a suitable unit of length has always been there and attempts made to keep that unit of length constant irrespective of the environmental conditions. Actually, previously the standards were made of materials which could change their size with temperature and other conditions. Thus a great care and attention was needed to maintain same conditions so that the fundamental unit remains same. Finally the natural and invariable unit of length was sought when it was found that the wavelength of monochomatic light never changed with other conditions. Due to it, the previously defined yard and metre could be very easily expressed in terms of the wavelength of light.

The various standards now known for linear measurements are (*i*) Line standard, (*ii*) End standard, and (*iii*) Wavelength standard.

3.3. Line Standard

According to it, yard or metre is defined as the distance between scribed lines on a bar of metal under certain conditions of temperature and support. These are legal standards and Act of Parliament authorises their use.

3.3.1. Yard. The Imperial Standard Yard is a bronze bar of one inch square cross-section and 38 inches long (Fig. 3.1). A round recess, one inch away from two ends is cut at both ends up to central plane of the bar. A gold plug $\frac{1''}{10}$ diameter having three lines engraved transversely, and two lines longitudinally is inserted into these holes so that the lines are in neutral plane. Yard is

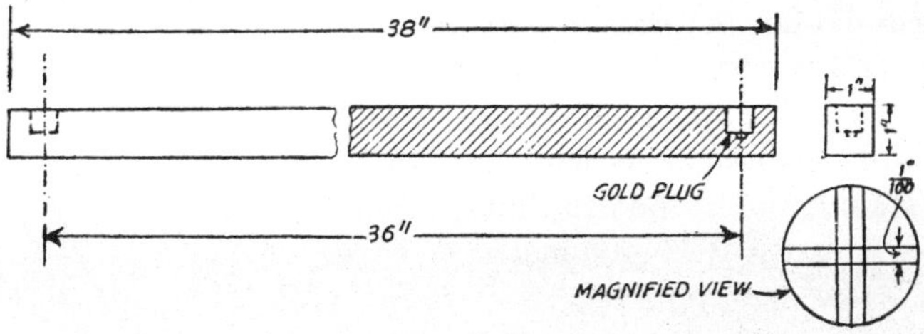

Fig. 3.1. Imperial Standard Yard.

then defined as the distance between the two central transverse lines of the plug at 62°F. Secondary standards were also made as copy of the above International yard for occasional comparisons. These standards are kept in safe custody. The purpose of keeping the gold plug lines at neutral axis is that due to bending of beam, neutral axis remains unaffected. Secondly the plug being in a well is protected from accidental damages.

In order to minimise slightest error in neutral axis due to support at ends, the supports must be placed such that the slope at ends is zero and the flat end faces of the bar are mutually parallel as shown in Fig. 3.2.

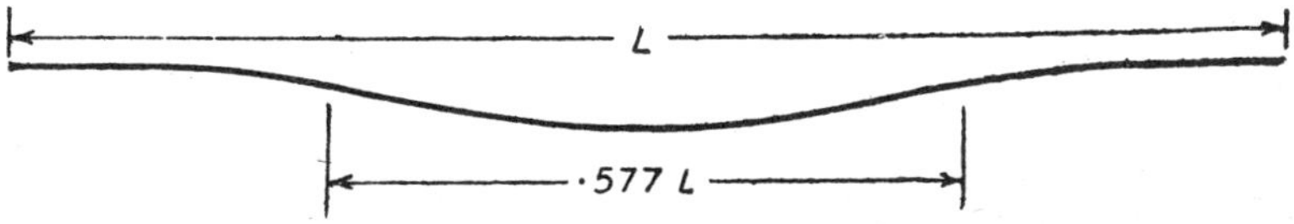

Fig. 3.2. Slope at ends is zero.

Sir G.B. Airy showed that this condition was obtained when the distance between the supports is $\frac{1}{\sqrt{n^2-1}} \times L$, where n = No. of supports, L = Length of bar. For a simply supported beam, the expression becomes $\frac{1}{\sqrt{n^2-1}} \times L = \frac{1}{\sqrt{3}} \times L = 0.577\ L$. These points of support are known as 'Airy' points.

Please note that in the case of straight edges, it is desirable that the deflection be minimum at the centre. For this condition, the supports should be 0.554 L apart.

3.3.2. Metre. This is the distance between the centre portions of two lines engraved on the polished surface of a bar of pure paltinum-iridium alloy (90% platinum and 10% iridium). It is inoxidisable and can have good polish required for ruling good quality of lines. The bar is kept at 0°C and under normal atmospheric pressure. It is supported by two rollers of at least 1 cm dia symmetrically situated in the same horizontal plane at a distance of 751 mm, so as to give minimum deflection (points of Bessel). It has a shape of winged section as shown in Fig. 3.3, having a web whose surface lines are on the neutral axis. The section chosen gives maximum rigidity and economy of costly material. Also the neutral axis lies at the top of the web and, therefore, whole of it can be graduated. Overall width and depth are 16 mm each. This reference is designated as International Prototype Metre—M in 1899. This standard is kept at BIPM at Sevres in Paris.

Earlier metre standard (before 1875 where the above described metre was adopted by International Bureau of Weights and Measures) used to be end standard which had great disadvantages associated with it and was found unsuitable. End standards are subjected to secular changes (*i.e.* changes in dimension over a period of time). Thus they are not suitable as fundamental standard. This is obvious from the fact that the standard yard described above was found to be decreasing in length when compared with metre line standard. The metre line being on platinum-iridium alloy was quite stable. Therefore, yard's relationship had to be defined in terms of metre as 1 yard = 0.91439841 metre or 1 metre = 39.37014 inches.

On International basis, this relation was modified as the American yard was longer by four parts in one million. Thus for stability of standards, International yard was adopted as 1 yard = 0.914400 metre, or 1 inch = 25.4 millimetres.

Secondly, it is very difficult to transfer accurately the line standard to end standard which is generally used for practical measurement.

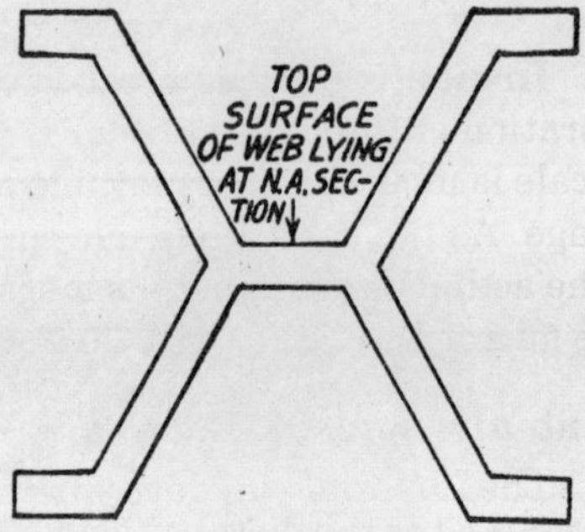

Fig. 3.3

3.3.3. The precision scale. Inspite of a natural standard, the need for physical standard remains. The obvious advantage of line standard over end standard is that the bar can be subdivided into very small increments further in the greater lengths (over one meter), fabrication of the end standard becomes increasingly difficult. The precision scale of reference standard caliber is usually of a Tresca or *H* section, designed to be supported and used horizontally. Fine lines are engraved

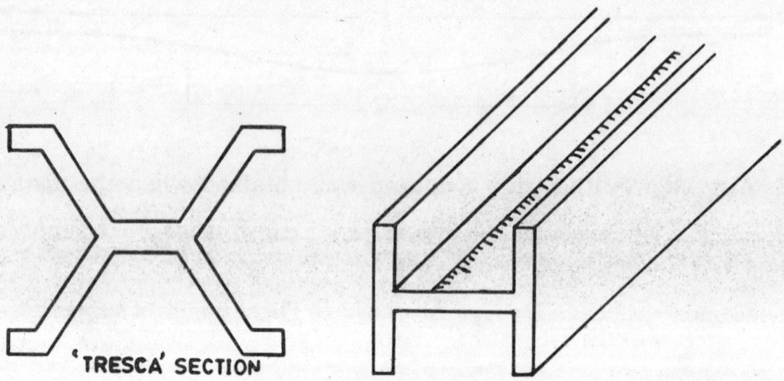

Fig. 3.4. Precision Scale.

90° to the longitudinal axis of the bar, on a highly polished, flat surface at the neutral bending plane of the bar. While it is the overall length between extreme end points, that is of interest to the bureaus, the precision scales are marked with lines at spacing of each 1 mm. The lines on scale are engraved by using a specially constructed ruling engine. The spacing of lines is governed by a precise, compensated lead screw, or alternately by reference to a master scale, using either an optical microscope or a photoelectric microscope. Lines are scribed with a sharp diamond cutter to width as small as 0.002 mm. The lines should be as fine as possible because these can be read by employing higher magnifications. Precautions are taken to produce lines to uniform thickness, parallelism and spacing. Vibrations are isolated by mounting the entire ruling engine on a massive concrete foundation which, in turn, is mounted on springs and uniform temperature of 20° is maintained. The factors which would influence the lines on precision scale to depart from their prescribed position are

— error of the standard or error in reading the standard
— deflection of the diamond in cutting
— uniformity of line width
— parallelism of lines
— variation in environmental temperature.

Inspection of the precision scale can be done by comparing with another reference scale. Calibration of the highest order is possible by using a Michelson type of interferometer, in which the scale is mounted on a movable carriage and the mirror is fixed to one end of the scale, or to the carriage. As the carriage is moved, the lines of the scale are picked up by photoelectric read out, but the actual displacement is measured in terms of the number of wavelengths of light movement of the mirror.

3.4. Sub-division of Standards

The Imperial standard yard and metre defined previously are just like master standards and can't be used for ordinary purposes. Thus depending upon the importance of standard, standards are sub-divided into four grades.

(*i*) **Primary Standards.** In order that standard unit of length, *i.e.* yard or metre does not change its value and it is strictly followed and precisely defined there should be one and only one material standard preserved under most careful conditions. This has no direct application to a measuring problem encountered in engineering. These are used after 10 to 20 years solely for comparison with secondary standards.

(*ii*) **Secondary Standards** are close copies of primary standards as regards both design, material and length. These are made, as far as possible, exactly similar to primary standards. Any error existing in these bars is recorded by comparison with primary standards after long intervals. These are kept at a number of places under great supervision and are used for comparison with tertiary standards whenever desired. This also acts as safeguard against the loss or destruction of primary standard.

(*iii*) **Tertiary Standards.** The primary and secondary standards exist as the ultimate controls for reference at the rare intervals. Tertiary standards are reference standards employed by N.P.L., and are the first standards to be used for reference in laboratories and workshops. These are also made as true copy of secondary standards and are kept as reference for comparison with working standards.

(*iv*) **Working Standards.** These are also line standards and similar in design to (*i*), (*ii*) and (*iii*), but being less in cost, are made of lower grades of materials. These are of general application in metrology laboratories.

Sometimes standards are also classified as (*a*) reference standards (used for reference purposes) (*b*) calibration standards (used for calibration of inspection and working standards), (*c*) inspection standards (used by inspectors), and (*d*) working standards (used by operators).

3.5. End Standards

For all practical measurements in workshop, we employ end standards *e.g.* slip gauges, gap gauges, end of micrometer anvils etc. Thus the importance of end standards (which are actually used in general measurement applications) arose. Length bars and slip gauges were then made which were equal in length to the legal line standard. The only difficulty realised with end standard was that of forming two accurately parallel surfaces at the end of a bar and to heat treat the ends so that they remained stable.

3.5.1. End Bars or Length Bars. These are used for the measurement of larger sizes of work. These consist of carbon steel round bar about 20 mm in diameter and made in sizes varying from 10 mm to 1200 mm. These are hardened only at ends upto 800 HV and supported at 'Airy' points so that end surfaces are parallel to each other. For bars above 150 mm size, the airy points are indicated by pairs of circumferential lines inscribed around the bars. Such bars are used for standardising the normally used one inch bars in combination. These are, therefore, generally not found in majority of engineering works but in standardising laboratories etc. These are also available in four grades of accuracy (*i*) reference, (*ii*) calibration, (*iii*) inspection, (*iv*) workshop.

Workshop and inspection grades have internal threaded ends and can be used in combinations with the help of a standard stud.

Slip gauges are also used as end standards of measurement in all metrology laboratories. These are made of high grade cast steel and hardened throughout. With the set of slip gauges, combination of slip gauge enables measurements to be made in the range of 0.0025 to 100 mm but in combination with length bars measurement range upto 1200 mm is possible.

3.5.2. Transfer from Line Standard to End Standard. It has already been clarified that the line standard of length is a highly inconvenient form for general measurement applications. In order to determine the position of the defining lines in the standard, a special microscope has to be employed. Since the line standard was defined first, and end standards being of real importance and more utility ; the end standards had to be produced of the highest accuracy in relation to the line standard. In end standards, the distance is defined between the working faces which are flat and mutually parallel. In order to transfer the line standard correctly to the ends of a bar, the use of an instrument called *Line-standard comparator* is made. It consists of two microscopes mounted about a yard apart over a table. A gauge, about 35 $\frac{1}{2}$ inches in length is produced with end faces flat and mutually parallel. Two $\frac{1}{2}$ inch blocks are taken and wrung at the ends of this gauge. The two $\frac{1}{2}$ inch blocks are engraved with a fine line on one surface approximately in the centre of the two end faces. Thus the distance between the centre lines is approximately 36 inches after wringing these $\frac{1}{2}$ inch blocks to the main 35 $\frac{1}{2}$ inches gauge.

The standard and the above blocks are mounted on the table. The microscopes have accurate micrometer screw controlled eyepieces. In eyepiece, there are cross wires to focus on the lines of the standard. The table is capable of being traveresed across so that either block may be brought under the microscope. The apparatus compares the positions of lines on the standard with lines on the gauge, and with micrometer eyepieces any small longitudinal variations between them can be determined.

Let the actual length $35\frac{1}{2}$ inches gauge be l. The distance between two lines on line standard is 36″. Let us ignore the effect of the wringing film between the surfaces in contact as it is always present in the use of end bars and gauges. The other possible erros are the misplacing of the line at the mid-position of the end faces of $\frac{1}{2}$ inch blocks and possible error in the length of $35\frac{1}{2}$ inches gauge. The two blocks at end are arranged in four ways. (Fig. 3.5 shows one position and other three positions are self-explanatory) and difference of readings between lines on standard and the line on gauges are noted every time. Let the difference be d_1, d_2, d_3 and d_4 respectively. Then for the successive positions of the $\frac{1}{2}$ inch blocks, we have

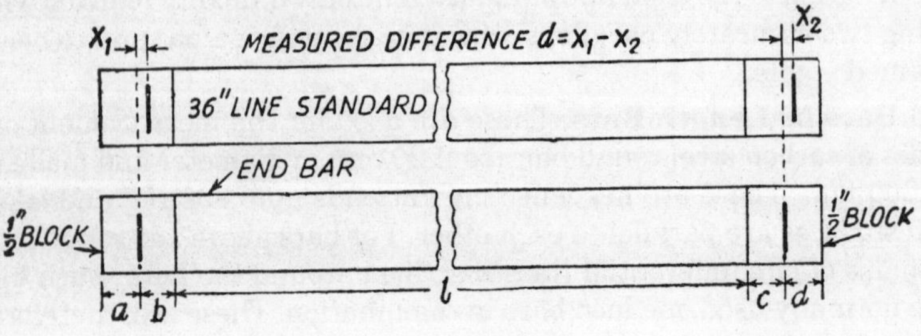

Fig. 3.5

$$l + b + c = 36 + d_1, \quad l + b + d = 36 + d_2, \quad l + a + c = 36 + d_3, \quad l + a + d = 36 + d_4.$$

Taking mean, $l + \frac{1}{2}(a + b + c + d) = 36 + \frac{\Sigma d}{4}$ ...(1)

In equation (1), it may be noted that the error due to the possible misplacing of the lines between the end faces of the $\frac{1}{2}$ inch blocks is eliminated.

Next 35 $\frac{1}{2}$ inches bar wrung with one of the $\frac{1}{2}$ inch blocks is compared with 36 inch end standard (to be calibrated) on a Brookes Level comparator and the deviation D_1 noted (Fig. 3.6). Then the other $\frac{1}{2}$ inch block is wrung and again it is compared with 36″ end standard (which is to be calibrated) and deviation D_2 noted down.

If L is the actual length of 36″ end standard, then

$$l + a + b = L + D_1, \quad l + c + d = L + D_2.$$

Taking the average

$$l + \frac{1}{2}(a + b + c + d) = L + \frac{\Sigma D}{2} \quad ...(2)$$

Combining equations (1) and (2),

$$L = 36 + \frac{\Sigma d}{4} - \frac{\Sigma D}{2}$$

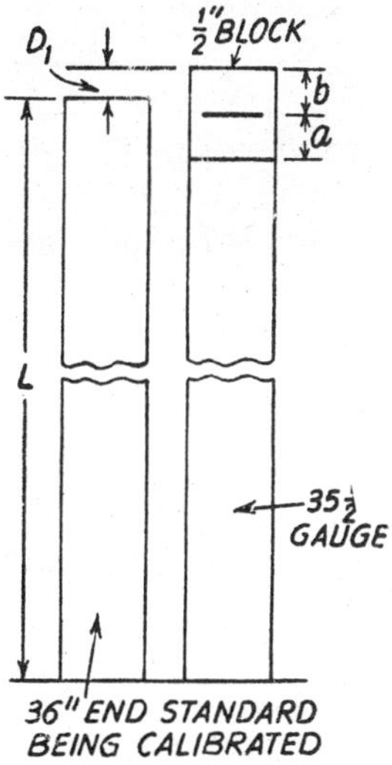

Fig. 3.6

Thus 36″ end standard has been calibrated and by this method the unknown erros in 35 $\frac{1}{2}$″ bar and $\frac{1}{2}$″ blocks are systematically eliminated. When these end standards are calibrated in this way, these are used as master gauges from which further sub-divisions are obtained *e.g.* two 18″ gauges may be prepared and when wrung in combination, their length is compared with that of 36″ end standard master gauge.

Again they are compared amongst themselves as the previous error is in both of them and need not be equally distributed. Thus individual errors of the 18″ gauges can also be isolated. A similar procedure is used to determine the length of any smaller gauge, in terms of 36″ master gauge.

3.5.3. Sub-divisions of the End Standard by Brookes Level Comparator. Brookes level comparator consists of a very accurate spirit level. The spirit level is supported on balls so that it makes only point contact on gauges. First the table on which gauges are to be compared is levelled properly by spirit level. Then the two gauges to be compared are wrung on the table and the spirit level is lowered on them. After it the position of bubble is noted down and table rotated by 180° so that position of two gauges is interchanged. Again spirit level is lowered down on gauges and the bubble reading noted (Fig. 3.7). If the two gauges are not of equal length then two readings will be different and effect of interchanging the position of the gauges will be to tilt the level through angle equal to twice the difference in the height of gauges divided by the spacing of level supports. The difference of gauges will, therefore, be equal to half of the above height difference. The bubble readings can directly be calibrated in terms of the height difference as the distance between the two balls is fixed. The advantage of turning the table by 180° is that the effect of table not being level initially is eliminated. Thus for other smaller gauges, they may be combined to form 36″ master gauge and then compared amongst themselves. This sub-division in similar way can be more accurately done by interferometry method to be discussed later.

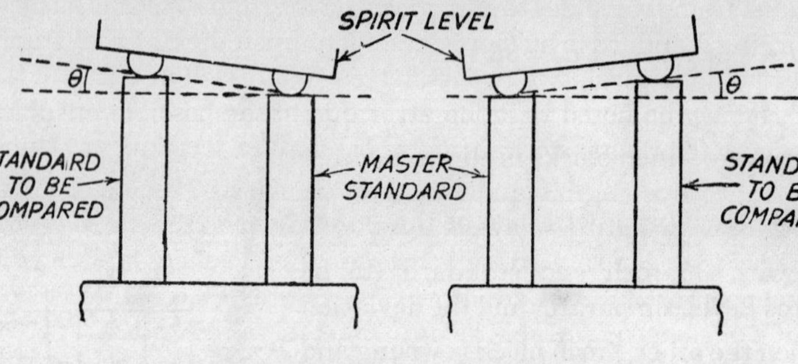

Fig. 3.7. Brookes level comparator.

Since we are interested here in extremely fine measurement, an important attention is required to be paid to keep the temperature of the room constant as most of the metals suffer slight changes in length when subjected to variations in temperature. It is, therefore, desirable that room must be kept at 20°C by thermostatic control and sufficient time must be allowed for the room and the gauges to attain steady temperature.

3.5.4. Comparison of an end gauge with a line standard by the displacement method. The set up for this is shown in Fig. 3.8. It utilises tool room microscope for small specimen and horizontal microscope for longer specimen.

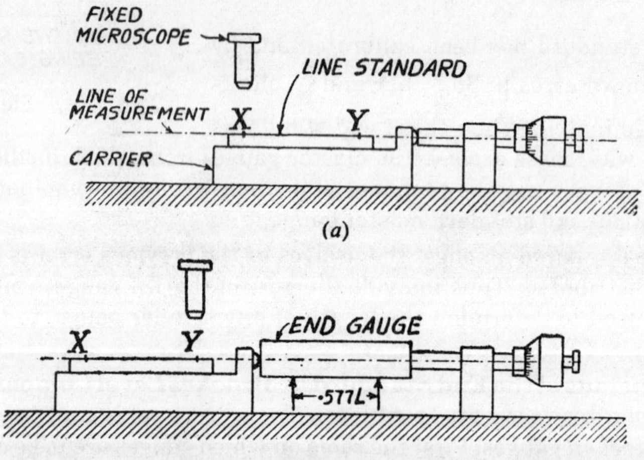

Fig. 3.8. Displacement method.

The line standard is placed on a carrier and adjusted to bring the line X under the cross-wires of the fixed microscope. The micrometer spindle is rotated so as to bring it in contact with the projection on the carrier and micrometer reading noted. The carrier is then moved along to bring the line Y under the microscope crosswires and the end gauge is inserted as shown in Fig. 3.8 (*b*) and the micrometer reading is noted again. The length of the end gauge is equal to the length of the line standard plus the difference between the micrometer readings.

3.6. Wavelength Standard

It is obvious from the method given above for comparing and verifying the sizes of gauges that a considerable difficulty has to be faced in their use. This is because the working standard to which one can refer is derived from a physical standard and in establishing the size of a working standard by above process, successive comparisons must be made. This leads to errors of unacceptable order of magnitude. Using wavelength of monochromatic light which is natural and invariable unit of length, the working standard is no more dependent upon the physical standard. Rather the definition of a standard of length relative to the metre is expressed in terms of the wavelength of the red radiation of cadmium. Thus for all practical purposes the use of phenomenon of the interference of light waves to provide working standard may be accepted as ultimate.

For some time light wavelength standard had to be objected because of the impossibility of producing pure monochromatic light as wavelength depends upon the amount of isotope impurity in the elements. But now with the rapid development in atomic energy industry, pure isotopes of natural elements have been produced. Cadmium 114, Krypton 86, and Mercury 198 are possible sources of radiation of wavelengths suitable as natural standard of length.

Since wavelength standard is not a physical one, it need not be preserved. This is reproducible standard of length, and the error of reproduction can be of the order of 1 part in 100 million.

Finally it was decided that Kr 86 is the most suitable element if used in a hot-cathode discharge lamp maintained at 68°K temperature. The orange radiation was selected for the measurement. According to this standard, metre is defined as 1650763.73 × wavelength of the radiation corresponding to the transition between the level : $2\,p_{10} - 5\,d_5$ of the Krypton 86 atom in vacuum.

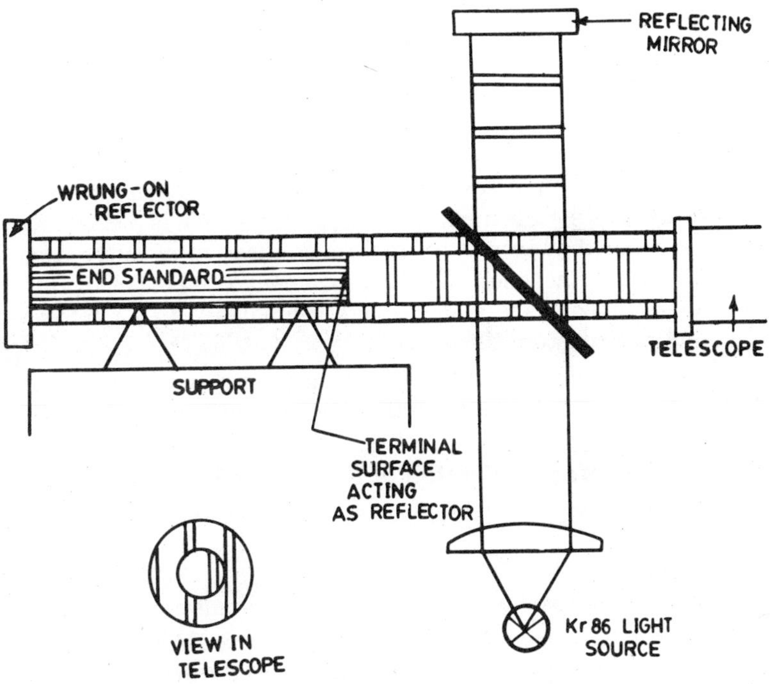

Fig. 3.9. No mechanical movement involved.

Some metrologists had predicted that the present length standard (krypton 86) will eventually be replaced by the laser (light amplification by stimulated emission of radiation). The great advantage of the gas laser compared to other light sources is its coherency and strongly deflective fringe pattern over long optical paths.

Today the gas laser is already established as a practical length-measuring tool in most of the material bureaus. The measurement and calibration of gauges using wavelength standard have been discussed in Chapter 6.

3.7. Relation of end standard and line standard to International (light waves) standard

The end standard can be directly compared to wavelengths of light by absolute means, without involving any mechanical movement or time lag (Refer Fig. 3.9). It would be noted that the terminal surfaces of end standard are used as optical reflectors in the interferometre arrangement shown in Fig. 3.9.

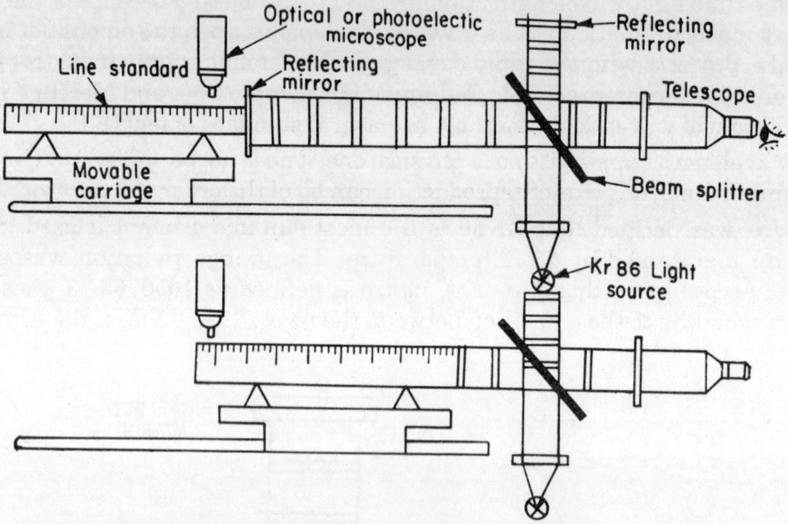

Fig. 3.10. Movement of carriage is required (not absolute method).

Fig. 3.10 shows the indirect comparison of line standard to the light wave standard. It involves movement of a carriage holding the mirror. The line on the scale is picked up optically or optoelectrically. It would be noted that in view of high magnification of microscope, the lines engraved on scale should be very fine ones. This arrangement (Refer Fig. 3.10) requires a fringe-counting interferometer, which takes a signal and collimates it to a photosensitive cell. Temperature has to be maintained constant throughout.

The error in straightness of travel of carriage, if any, has to be accounted for. It would thus be obvious that line standards can't be compared to the official light wave standard by absolute means.

3.7.1. New Trends for the Materialisation of the Meter. Significant advances in new sources, new lines, optical clocks, new techniques and tools have taken place for calibration of meter to accuracy of about 10^{-11}.

An *atomic meter* is defined as 1,650,763.73 wavelength of the $2p_{10} - 5\,d_5$ transition in krypton. The transfer to physical length measurement is achieved by interferometry.

In 1983, CGPM (*Conference Générale des Poids et Mesures*) fixed the SI numerical value of the speed of light in vacuum, and established a *light-related meter* as the distance travelled by light during a given time interval.

The set-up and the calibration of secondary references for the meter in the visible or nearby should ensure the traceability on the time reference, *i.e.* the calibration against the caesium clock.

International comparisons of the reference frequencies in iodine cells around 633 nm exhibit a standard relative deviation of about 3×10^{-11}.

Good progress has occurred in the set up of an optical clock, *i.e.* a physical system with an optical resonance, monitored using a ultra high resolution spectroscopic technique, that would be more stable and reproducible than the RF caesium clocks.

The iodine molecule presents promising characteristics to get references in an extended spectral range, to demonstrate the usefulness of new ultra high resolution (UHR) spectroscopic techniques. It has proved to be a very efficient wavelength reference tool for the transfer to calibration centers, which are using improved interferometric techniques to calibrate material references.

Solid state lasers having intrinsic low noise and long term stability are used to study new lines.

Frequency stabilisation of the laser sources used to materialise the meter in the optical domain generally requires two servo-loops. The first one is devoted to short term stabilization, using optical feedback on semiconductor lasers. The second one ensures the long term stability of the frequency on an atomic or molecular secondary reference.

3.8. Classification of Standards and Gauge Control Structure

The accuracy in a vast industrial complex can be maintained only if its standards are traceable to a single source, usually the National Standard of the country, which are further linked up with International Standards. Fig. 3.11 explains how the accuracy of measurement is transferred from National Standards to the working shop floor standards.

Since frequent use may impair the accuracy of any standard, direct comparisons with National Standards are less frequently performed. In such a case, a slightly lower order of standards, known as National Reference standards are used. Working standards are ordinarily used in calibration work. These are compared as frequently as necessary with National Reference Standards.

Classification of Standards

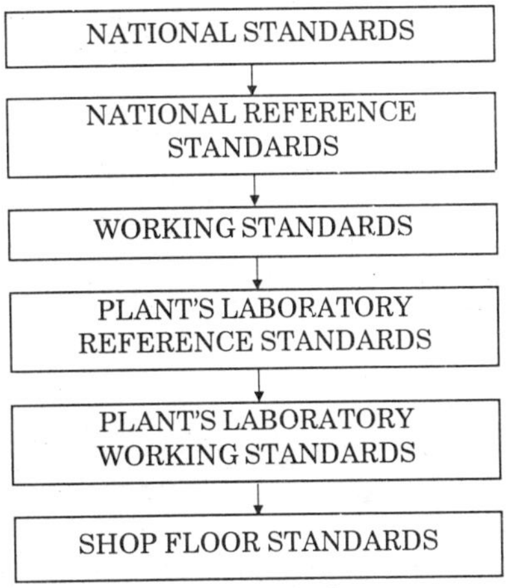

Fig. 3.11. Gauge control structure.

Clearly, there is degradation of accuracy in passing from the defining standards to the standards in use. The accuracy of a particular standard depends on a combination of the number of times it has been compared with a standard in a higher echelon, the recentness of such comparisons, the care with which it was done, and the stability of the particular standard itself.

3.9. Gauge Control Structure

There are three general classes of gauges which span the entire conformance-evaluation function. These are (*i*) working gauges used by process operators and set-up men (Shop floor standards), (*ii*) inspection gauges used by inspectors (Plant's laboratory standards), and (*iii*) master gauges used for calibrating working and inspection gauges (Plant's laboratory reference standards).

A series of two of three transfer measurements are involved in calibrating the working gauge against the primary gauge block standard. In this system two error factors are critical—transfer error and cumulative error. Every product measurement is subject to uncertainties due to size of the calibrating gauge, errors intrinsic to each gauge, and errors caused by operator technique and environment. This accumulated error must be kept to a practical minimum to reduce the respective probabilities of accepting non-conforming product and rejecting product which conforms to the specification. A common procedure for reducing these probabilities is the 10 to 1 rule illustrated in Fig. 3.12. Application of this rule means that measurements at the calibration stage must be 100 and often 1000 times as accurate as the product specification tolerance. As calibration and other

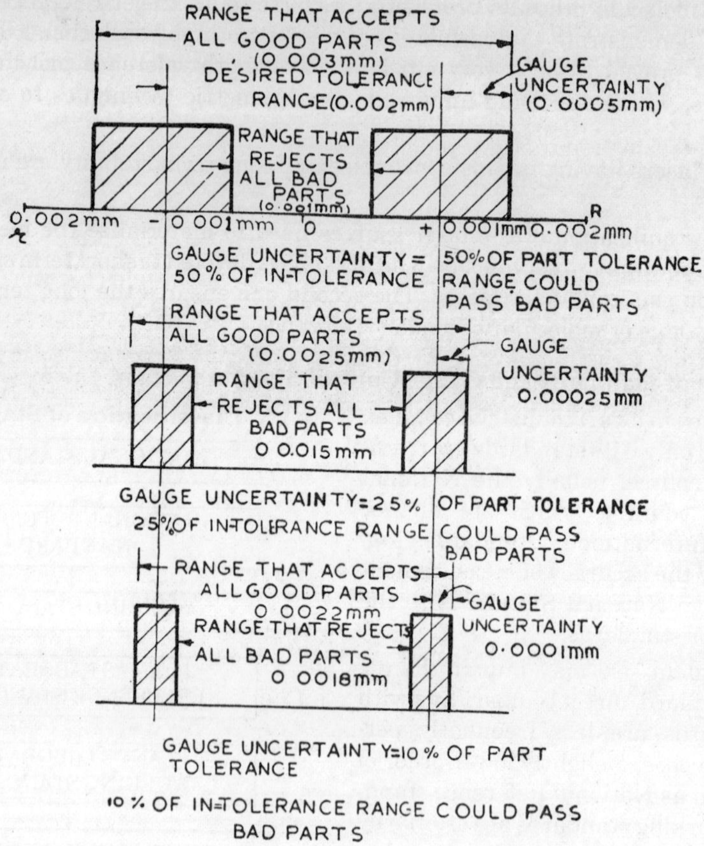

Fig. 3.12. Gauge uncertainty rules—50, 25 and 10%. For reliable measurement, each step in the gauging sequence, from work to gauge block calibration, should have ten times the accuracy of preceding step.

constant errors become a smaller fraction of the specification tolerance, the fractions of non-conforming product accepted and conforming product rejected decrease proportionately.

3.10. Conditions of Comparison

Determination of length and calibration must be made under specified conditions which need to be strictly maintained for accurate results.

The various factors to be rigidly maintained and which directly affect the accuracy of determination are :

(*i*) **Environment.** The following factors in environment need to be controlled within precise limits because each one of them influences the comparison of linear standards :

Temperature — Atmospheric pressure

Humidity — Vibration

It is very important that temperature in metrology laboratory is maintained at a constant value of 20°C and workpieces and gauges are allowed to soak to attain uniform temperature because non-uniform temperature within a piece may cause distortion. Maintenance of temperature at 20°C (internationally accepted value) is important because the rates of thermal expansion vary with different materials. If the coefficients of expansion were identical for all metals used, then it was possible to maintain temperature constant at some other value.

The ambient temperature may change by convection (effects of opening doors/windows, drafts from movement of personnel, non-uniform air currents, or stratification within room), conduction (flow of heat within a body or from one body to another like from body of person by handling of workpiece/standard), and radiation (lights, sunlight, personnel, electrical equipment, large masses above ambient brought into environment). Radiant energy effects are more troublesome because objects in direct path of (say) light will vary in temperature depending on their rates of absorption or reflectivity of that light and whether the body is white/black/dull/bright.

It must be remembered that a standard brought into controlled environment takes several hours to reach the ambient temperature.

Thermal Effects in Dimensional Metrology. In the field of close-tolerance work, thermal effect is the largest single source of error, large enough to make corrective action necessary if modern measurement systems and machine tools are to attain their potential accuracies.

It is observed that the usual efforts to correct for thermal error by applying "correction", or by air conditioning the working area do not always solve the problem. The specified accuracies of modern precision tools and gauges are attainable only if the thermal environment matches the requirements of each measurement system. If the thermal error index is more than 10% of the part tolerance, then it deserves attention.

(Thermal error index $TEI = NDE + UNDE + TVE$, where NDE = Nominal Differential Expansion = the difference between the nominal expansion of the part and the master, $UNDE$ = Uncertainty of Nominal Differential Expansion = Sum of the uncertainties of nominal expansion of the part and the master, and TVE = Temperature Variation Error = Estimate of the maximum possible measurement error induced solely by deviation of the environment from average condition and it is determined from the results of two drift checks ; one of the master and comparator, and the other of the part and the comparator). The thermal error index can be reduced by following corrective actions : making corrections for nominal differential expansion, establishing more accurate coefficients of expansion so as to increase the accuracy of the corrections, and minimising average temperature deviations from 20°C.

The errors resulting from temperature variation can be reduced by : improving procedures for soaking out workpieces and masters so that they are in thermal equilibrium with the environment, increasing rate of air flow and improving its distribution, increasing the frequency of temperature variation, decreasing the amplitude of temperature variation, redesigning masters and comparators so that their time constants and coefficients of expansion are in better balance with those of the parts to be measured.

For interferometric measurements, changes in atmospheric pressure affect the accuracy of measurement. Since air current may also affect the readings, such measurements are usually carried out in a vacuum or else compensations are made which include corrections for the refractive index of the air. When gauges are measured in a vacuum, corrections are required for the compression that occurs on return to normal atmospheric pressure. Various non-metallic substances are seriously affected dimensionally by humidity. Humidity can also corrode many ferrous materials. Thus humidity needs to be maintained at a constant value.

Sufficient amplitude of vibration might cause a shift of the standard's position during calibration. Even slightest vibration is frustrating for use of high magnification instruments. Thus it needs to be completely eliminated as far as possible.

(*ii*) **Material of the standard.** The parameters which affect accuracy of measurement in regard to material of standard are stability and coefficient of expansion.

Dimensional instability of a measuring standard due to residual stresses caused by machining or other processes results in bending and/or twisting, altering the effective length of standard. If heat treatment of gauge blocks is improper, then astonishing changes in length may be observed. It is generally observed that dimensional stability of surface hardened steel is more than that of through-hardened steel. The metallurgical composition and surface treatments also affect stability of material. For instance, nitrided and hardened "Nitralloy" are found to be among the ultra-stable materials. Thus maximum stability has to be ensured by proper selection of material, methods of fabrication and heat treatment. It is general observation that most of the metals expand when heated. In steel, the coefficient of expansion can be reduced by adding nickel. But no useful purpose will be served by having standard of high nickel steel because most components are made of steel and there will be different expansions, if temperatures are different. So attempts should be to have materials with equal coefficient of expansion and maintain the environmental temperature strictly at 20°C.

(*iii*) **Support of length standard.** Small gauges upto length of 75 mm do not undergo appreciable alteration in length with method of support. However a long standard will undergo a significant change in dimension even by its own weight. Even vertical supporting leads to reduction of size. It is, therefore, preferable to support long standards horizontally on the "Airy points" (supports 0.577 L apart).

(*iv*) **Geometry and Micro-geometry.** Gauges and standards must have good geometry. The calibration of end standard depends on the level of perfection attained in the flatness and parallelism of terminal surfaces. When measurements are made to fraction of micron, innumerable minute variables, become critical with 100 gm of pressure, tips of indicators and gauging surfaces squash and bulge to tenths of a micron. No surface can be claimed to be exactly perfect since with high magnification of X1,00,000 times, surface irregularities show up as valleys and hills. The term "microgeometry" best describes the special set of problems and conditions having to do with measurements at this level.

3.11. Environmental Control in Test Laboratory

To perform accurate and reliable measurements in test laboratories and to fulfill the requirements of compatibility it is absolutely essential that the environment in test laboratory be strictly controlled and maintained. Only then the measurements can be carried out as per standard specifications to meet customers requirements and also provide the basis for international multi-lateral recognition of product testing and conformance for trade in global market place. Environmental influences affect the accuracy of measurement and thus introduce uncertainty.

Various environmental factors to be controlled in test laboratories, their effect on measurement, the recommended values for such parameters and their monitoring, and suggestions to take care of their effect are discussed below :

(*i*) **Temperature.** As per international practices, temperature in test laboratory should be maintained at 20°C or 23°C or 25°C as specified. Temperature affects the linear dimensions of specimen and instruments and thus results have to be specified at the prescribed temperature. Temperature affects zero shift and scale error. Temperature variations can cause changes in thermal emfs and also change bias current. Temperature also affects relative humidity which has further influences on measurement results.

(*ii*) **Rate of Change of Temperature.** Normally it is maintained at 1.5°C per hour for stabilisation and 0.5° to 1°C per hour during measurements. The air conditioning system has to designed suitably to meet these requirements.

(*iii*) **Relative humidity.** It is usually maintained at 45 ± 10%. It may be understood that lower RH results in build up of static charge which can damage and invalidate measurement. Higher RH can cause leakage and erroneus measurement, affects corrosion resistance and reduces insulation resistance. Air conditioning system should be designed to take care of specified requirements of relative humidity.

(*iv*) **Barometric Pressure.** It is recommended to maintain a positive air pressure (10 to 20 N/m^2) in laboratory relative to outside area to eliminate possibility of dust entering in the laboratory. Electrostatic precipitators can be placed in corners to attract and collect dust particles, if any. The environment in laboratory should be stabilised.

(*v*) **Air velocity and air distribution.** Air velocity affects the temperature gradient. Air ingress in lab can be at a rate of 0.71 ± 0.28 m^3/mt. The air distribution is also important. Opening from air conditioning duct should be such that test area gets uniform cool area and the air at 1 to 1.5 m from floor should be stratified. Cool air must enter after passing through a filter of suitable size and the filter must be cleaned regularly. For cleaning lab, vacuum cleaner should be used. Floor should not be cleaned with bamboo which spreads dust. Entry to lab should be restricted and new equipment should not be cleaned in laboratory. Smoking, food and drinks should not be permitted inside laboratory.

(*vi*) **Dust particle Count.** Wet dust reduces insulation resistance and dry dust particles cause increase in surface resistivity. Double glass windows should be used and trap door employed so that only one door can open at a time. Filter in air conditioning dust should be of proper size so that particles greater than 50 micron find no entry in lab.

(*vii*) **Vibration and Shock.** Test laboratory should be designed so that it is adequately free from shock and vibration generated by central air conditioning plant, vehicular traffic and other sources. In general maximum acceleration at base of instrument should be restricted to 1 × 10 g displacement of 1 μm for vibration frequencies below 200 Hz. Vibrations and shock reduce measurement integrity, provoke wear and tear and accelerate breakdown. Unwanted currents can be generated by movement of cables. Vibrations also effect electromagnetic indicating devices.

(*viii*) **Acoustic noise.** Sensitive measurements are affected by acoustic noise which should be restricted below 45-60 dB.

(*ix*) **Illumination.** Lighting in lab also deserves proper attention. 450-700 lux level on working table with glare index 19 is preferred. To avoid local heating and temperature drift, bulbs should not be used. Fluorescent lighting is preferred with recessed fixtures which are interference free and fitted with line filters.

(*x*) **Electromagnetic interference.** RFI/EMI can result in noisy or erratic readings and sensitive test equipment may be damaged even. Permitted value is 100 microvolt per metre radiation field strength. Single point earthing, shielded cables are preferred to avoid EMI. Choke should not be used inside the test area. Doors should be sealed effectively. Measurements should be taken in Faraday cage.

(*xi*) **Stray magnetic fields.** These can also cause erroneous measurements. Low level voltage measurements are particularly vulnerable. Power transformers or other magnetic field sources should therefore be kept away from test system.

(*xii*) **Ionisation interference.** Sources like alpha particles can affect very low level current measurements.

3.12. Elastic Deformation in the Gauging of Standards

It has been observed that surface plates, large masters etc. compress, bend and twist from gravity or from other applied forces. Where very high accuracies of measurement are involved, elastic deformation (bending, squashing, compression and bulging) of gauging standards and workpieces can't be ignored, even with extremely light gauging force associated with high-magnification comparators. Fig. 3.13 shows the deformation (in exaggerated manner) likely to occur at four places during gauging process.

The deformation of various materials and shapes of bodies in contact can be calculated by the classical Hertzian formula

$$\Delta = K\sqrt{\frac{F^2}{D}}$$

where Δ = deformation, F = applied force, D = diameter of contact

K = a deformation variable depending on material, shape of bodies in contact, etc.

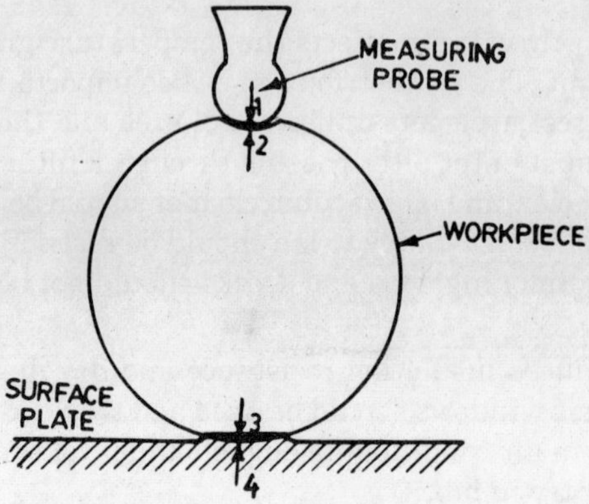

1. Deformation on the surface of measuring probe.
2. Deformation at the point on the workpiece where the force of the probe is applied.
3. Deformation at the area of workpiece resting on the gauging platform (surface plate).
4. Deformation of the surface plate in contact with the workpiece.

Fig. 3.13. Four points of elastic deformation during gauging.

It would be observed from this formula that deformation increases as two-third power of the measuring force. Thus even with a measuring force of 100 gms, differences in deformation amounting to tenths of a micron (μm) can be expected. Further, if the contact area on which measuring force acts is small, then deformation will be greater.

The deformation variable K, though dependent on so many variables, is considered for convenience for two variables, *viz*. geometry of contact area and the type of material in contact. The deformation is found to increase continuously in order for the following types of areas of contact : flat probe on flat, ball probe on flat, ball probe on cylinder, ball probe on ball, deformation due to surface roughness.

The quality of the surface also plays an important role in elastic deformation and precise measurements. Highly random errors occur as a result of probe falling alternately intc the peaks and valleys of surface roughness. Error also occurs due to the uncertain amount of local deformation, increasing according to the roughness of the finish.

It is obvious from above that variables in measurements should be eliminated as far as possible. All elements of comparison in the measuring set-up should be kept uniform. The comparison of unlike materials, shapes or type can cause unforeseen errors. If it is not possible to eliminate variables, either a large margin of uncertainty should be allowed, or a careful analysis of variables should be made and appropriate correction factor applied.

3.12.1. Type of Probe. Figs. 3.14 and 31.5 show two cases of gauging spherical and flat surfaces by spherical probe and flat anvil-type probe. It would be seen that flat anvil-type probe gives least deformation with gauge block but parallelism of probe to the opposing support surface

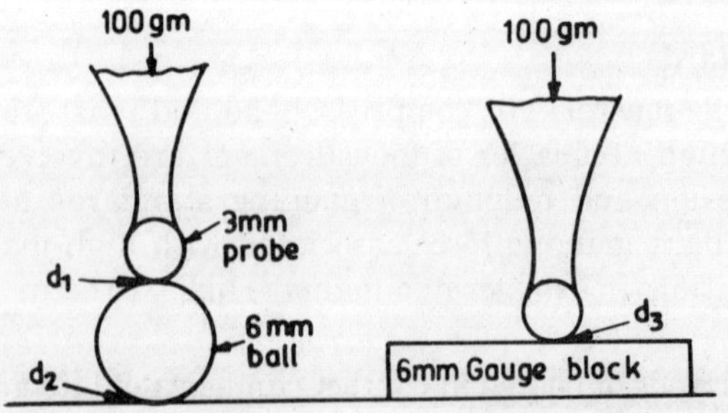

Total deformation $= d_1 + d_2 = 0.006$ mm

Total deformation $= d_3 + d_4 = 0.003$ mm

Difference in deformation = 0.006 – 0.003 = 0.003 mm

Fig. 3.14. Spherical probe.

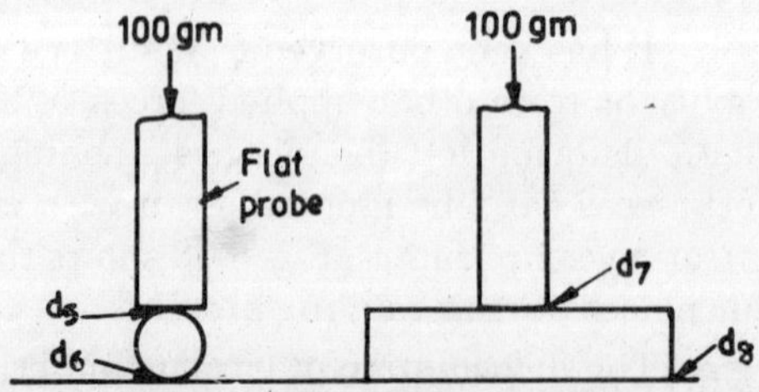

$d_5 + d_6 = 0.005$ mm $\quad d_7 + d_8 = 0$

Difference in deformation = 0.005 mm

Fig. 3.15. Flat probe.

can't be ensured. Errors in measurement may surpass those due to deformation. As a rule, for comparison of objects having plane surfaces, a large radius probe is preferred over one having a small radius because the danger of brivelling is reduced, wear of probe is minimised, and deformation is less. However small radius probe is preferred for internal measurement of small radii.

3.13. Measurement and Calibration Systems

For big factories it is desirable to establish and maintain an effective system for the control and calibration of measurement standards and measuring equipment employed. The calibration system should be designed to ensure that all measurement resources have the capability of making measurements within the specified limits. Such a system should provide for prevention of inaccuracy by prompt detection of deficiencies and timely action for their correction.

The calibration system should be periodically and systematically reviewed to ensure its continued effectiveness and the results should be recorded. All measurements should take into account all the errors and uncertainties in the measurement process that are attributable to the measurement standard or measuring equipment, as appropriate, those contributed by personnel, procedures and environment.

Documented calibration procedures should be available for reference and for the calibration of all measurement standards and measuring equipment used.

All measuring equipment and measurement standards should have records in regard to (*i*) the description of equipment and unique identification, (*ii*) the date on which each calibration performed, (*iii*) the results of calibration, (*iv*) the planned calibration interval, (*v*) the designated permissible limits of error, (*vi*) the reference to calibration procedures, (*vii*) environmental condition for calibration and corrections for reference condition, (*viii*) statement of cumulative effect of uncertainties on the data obtained in calibration, (*ix*) details of any maintenance (servicing, adjustment, repairs) or modifications that could affect the calibration status, (*x*) any limitations in use. All measurement standards and measuring equipment should be labelled, coded or otherwise identified to indicate their calibration status. Any limitation of calibration or restriction of use should be clearly indicated on the equipment.

Access to adjustable devices provided on measurement standard and measuring equipment should be sealed after fixing at the time of calibration. Seals should be so designed that tampering will destroy them.

The interval between periodic calibrations depends on the basis of stability, purpose and usage. Depending upon the results of preceding calibrations, intervals of calibration may be shortened, if necessary, to ensure continued accuracy.

The measurement standard or measuring equipment that has failed in operation or shows evidence of physical damage that may affect its accuracy, or that is suspected of being outside its designated limits, or which is outside its designated calibration period should be removed or identified to prevent their use. Details of such invalidation of calibration requiring corrective action should be recorded.

The system for handling, transporting and storing of measurement standards and measuring equipment to prevent their abuse, misuse, damage or change in dimensional/functional characteristics should be established and maintained.

All measurement standards and measuring equipment should be calibrated using measurement standards that are traceable to national or international measurement standards, except where they have been derived from acceptable values of natural physical constants or by the ratio type of self-calibration techniques.

All measurement standards used in the calibration system should be supported by certificates, reports or data sheets attesting to the date, accuracy, and conditions under which the results were obtained and are valid.

The cumulative effect of the errors in each successive stage of a calibration chain should be taken into account for each measurement standard or item of measuring equipment calibrated. Action should be taken when the total uncertainty is such that it significantly compromises the ability to make measurements within the required limits.

Measurement standards and measuring equipment should be calibrated and used in an environment controlled to the extent necessary to ensure valid measurements. Due consideration should be given to temperature, rate of change of temperature, humidity, lighting, vibration, dust control, cleanliness and other factors affecting measurement. Where pertinent, these factors should be continuously monitored and recorded, and when necessary, compensating corrections should be applied to measurement data. Records should contain both the original and the corrected data.

3.14. The Need and Function of Inspection

Inspection in industry is done to compare or determine the conformance of product to specifications. All inspections strive to isolate and assess the relevant detail or quality or characteristic of a product. Measurement is an important part of inspection. Practice and proficiency in efficiently securing accurate measurements are essential. Inspection details measure and assess relevant characteristics of a product in terms of requirements and specifications derived from a desire to meet expectations about the product's level of quality and performance.

Inspection ensures the ongoing maintenance of standards and requirements. This is done by continually measuring in order to control and reduce the variation that may reach alarming limit to threaten the integrity of specification.

Customer requirements, from a producer's point of view, balance a blend of quality and reliability within the product's performance with cost. Thus quality levels are customer-derived and it is for the inspection department to support these quality levels.

While the quality balance of acceptable versus unacceptable is the responsibility of inspector, he must also assess the causes for excessive variation.

Inspection department is responsible for ensuring measurable conformance and also for judging appearance.

According to the level of sophistication of inspection, it may be classified as : (*i*) manual, (*ii*) visual manual, (*iii*) test, (*iv*) mechanised or automatic.

According to the area in which the inspection is performed, it may be classified as (*i*) process (*ii*) batch, (*iii*) final, (*iv*) receiving, (*v*) tool and gauge.

3.15. Calibration Systems Requirements

The calibration and checking of measurement tools and instruments are critically important. The reliability, accuracy, and precision of any inspection measurement relies on the correlation between the measuring devices and the master or standard. We need exacting methods, procedures and equipment for verifying the various dimensional elements of the measuring instruments and gauges.

A calibration laboratory should be designed to furnish the following :

(*i*) Traceability of measurement

(*ii*) Proper environment

(*iii*) Appropriate measuring equipment

(*iv*) Standard measuring procedures consistent to the state of the art.

The standards established for calibrating the measuring instruments used in controlling product quality should have the capabilities for accuracy, stability and range required for the intended use. Environmental control, giving due consideration to temperature, humidity, vibration, cleanliness and other controllable factors affecting precision measurements is essential.

The calibration laboratory must maintain a set of standard procedures showing the equipment required to calibrate, control requirements, a step by step procedures on how to make the actual calibration and disposition of gauges.

Maintenance of records of calibration of all devices is essential. Each instrument and gauge should be given a specific serial number.

3.16. The Role of Metrology Laboratory

There is a need for exacting methods, procedures and equipment for verifying the various dimensional elements of the gauges. The accuracy of the measuring and test equipment can be controlled only by establishing and maintaining a calibration system. The calibration system available in metrology laboratory should be coordinated with the Inspection or Quality control systems and be designed to provide adequate accuracy in use of measuring and test equipment. The calibration system should also provide for the prevention of inaccuracy by ready detection of deficiencies and timely positive action for their correction. It is essential that the written description of the calibration system covering measuring and test equipment and measurement standards are properly documented and maintained. All measurements should be traceable to National Standards. For meeting all these objectives the need of an independent metrology laboratory in industry is justified.

The independent metrology laboratory is designed to furnish to the end user (*i*) traceability of measurement, (*ii*) proper environment, (*iii*) appropriate measuring equipment, and (*iv*) standard measuring procedures consistent to the state of the art.

The standards established for calibrating the measuring equipment used in controlling product quality should have the capabilities for accuracy, stability and range required for the intended use.

Environment controls for measuring and test equipment and gauges should be calibrated in an environment controlled to the extent necessary to assure continued measurements of required accuracy giving due consideration to temperature, humidity, vibration, cleanliness and other controllable factors affecting precision measurement.

The ideal environmental conditions in metrology laboratory for dimensional measurements of reference standards are :

Temperature	20°C ± 0.3°C
Temperature rate of change	Less than 0.5°C per hour
Relative humidity	35%—55% (constant within 2% around a regulated point)
Dust (particle count)	Less than 3.3×10^5 parts per m^3 over 1.0 micron, and less than 16.5×10^4 parts per m^3 over 0.5 microns.
Vibration	0.001 g (The instrument mount must provide such stability that readings can be consistently obtained to an accuracy of at least the value indicated for each unit)
Lighting (at bench level)	100 ft candles
Acoustic noise	50 db max.

The measuring equipment required in metrology laboratory depends on the requirements ; however the following minimum equipment must be provided.

Gauge blocks of grades AAA and AA
Gauge block measuring machines
Basic measuring machine
Thread measuring wires
Gear measuring wires
Internal measuring machines
Optical projector
Lead testing machine
Master thread setting plug gauges
Surface plates
Indicator tester
Master sequare
Digital height gauges
Co-ordinate measuring machine
Electronic amplifiers, etc.

The laboratory should maintain a set of standard procedures showing the equipment required to calibrate, control requirements, a step by step procedure on how to make the actual calibration and disposition of gauges.

The laboratory should furnish a test report for each gauge calibrated showing measurable or visual condition. The report should also tell the user the actual measurement or master size. The report should also reflect temperature, masters used, etc.

3.17. Metrology in Quality Assurance

Quality can be defined as the degree to which an item satisfies the given need it was designed to fulfill, *i.e.* its fitness for the purpose. As regards customer (actual user of product), quality product is one which satisfies the desired appearance and finish, the price and value for money, criterion of fitness for the purpose, and good after sales services. Good quality can be attained only when both the qualities of design and conformity are good.

Quality Assurance (QA) is concerned with all the functions and activities concerned with the attainment of the required quality. QA ensures improved quality and gives the guarantee to the user that the product manufactured will do what they are supposed to do, and be safer, more reliable and will last longer.

Quality control (QC) is the practical means of sustaining the product or service quality and ensuring that it meets the specified requirements. QC is defined as a system for programming and coordinating the efforts of various groups in an organisation to maintain or improve quality, at an economical level which allows for customer satisfaction. QC is an integral part of QA. Since QC is based on measurements so metrology is an important tool of QA.

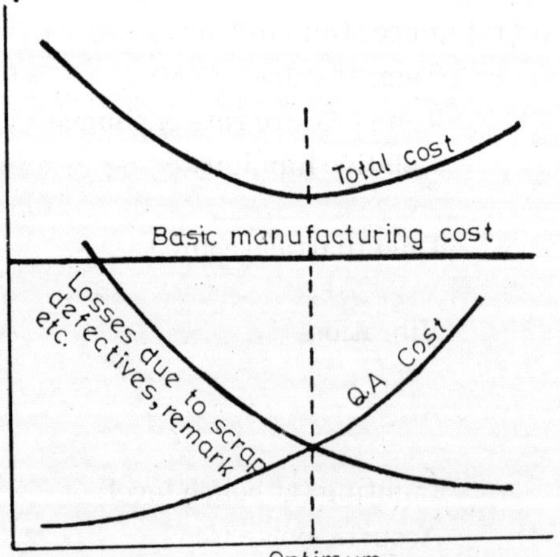

Fig. 3.16

On the manufacturing side, quality goods mean—less repairs, reduced replacement costs and reduced capital tie-up in spare stocks ; and reduced losses from rejects and subsequent re-work or scrapping of faulty items from despatch.

QA programme costs money but after an optimum level, the economic gains by QA are offset against the cost of QA. Referring to Fig. 3.16 it would be seen that upto the optimum point of minimum total costs a small amount spent on QA causes a comparatively big decrease in losses due to poor QA. Upto this optimum point, quality is cheap when the gains are considered. The optimum quality depends on how much effort (time and money) one is willing to place on QA. The area to the right of the optimum point where quality begins to become uneconomical is usually only ventured into when life support equipment is being considered. In these cases, perfection is desirable, which proves expensive.

QA depends not on QC alone but on the activities of an entire company. QA scheme begins with the proper identification of the customer's needs. These have to be converted into adequate designs, followed by the procurement of the necessary materials and conversion into the finished product.

The basic elements in QA are essentially—

(*i*) Specification/standards and design	— Involving the development of specification, standards and designs
(*ii*) Materials and component control	— Involving quality control and measurements
(*iv*) Manufacturing process control	—do—
(*iv*) Market and servicing	— Involving feedback and after sales service.

3.18. Setting up a QA System

Quality Management. A management representative, independent of other functions, should be responsible for resolving all matters pertaining to quality. Quality management should establish and maintain an overall programme of QA. The programme should cover all aspects of the company that affect quality and define all process functions that are not dependent on the particular product.

Various stages where quality management has to play role are :

Market analysis. A complete picture of the need the product is intended to fulfill must be obtained. All relevant features and characteristics and their associated criteria must be specified, defined and documented. Areas to be considered to give the complete picture are :

— maximum and minimum functional requirement

— design for need—there is no point in over-designing as it is expensive

— normal and worst case environmental conditions expected

— safety and reliability requirements

— cost limitation.

Design Development. Designers must be given all the necessary documentation covering the requirements for function, environment, reliability, etc. against which the product is to be designed. Quality has to be designed into a product. Reliability predictions are a useful tool during design as the various iterations can be compared for reliability and an idea of overall reliability can be obtained. Acceptance and rejection criteria for component parts and manufacturing/assembly processes should be specified. Tolerances on manufacturing and testing parameters should be sufficiently close and precise but should be no more stringent than necessary. Testing of prototypes to ensure compliance with the specifications and requirements should be carried out at the earliest as it is cheaper to correct it if failure is discovered earlier. Before drawing the design and specifications, the following questions should be answered :

— Does the design/specification include all customer requirements ?
— Does the design meet functional and operational requirements ?
— Is the design satisfactory for all expected environmental conditions ?
— Has data on similar designs been reviewed ?
— Have standard parts been utilised whenever possible ?
— Are drawing and specification tolerances achievable in production ?
— Does the design minimise installation problems ?
— Does the design minimise maintenance problems ?
— Has value analysis been utilised sufficiently ?
— Have necessary safety precautions been induced ?
— Has aesthetic image or appearance of the product been considered ?

During the development stage, manufacturing/production specifications, drawings, test specifications, etc., should be prepared in conjunction with other departments involved.

Purchasing, Receiving and Storing Goods. Items procured from outside should conform to the specified requirements. The material of component or part should be precisely defined. On receiving goods, these should be inspected, the amount and nature of inspection depending on the control exercised at source and objective evidence. A system should be established for the identification, preservation, segregation and handling of all material from the instant of receipt, and secure storage areas or stockroom provided.

Manufacturing / Production Control. Manufacturing should be carried out under controlled conditions. Ineffective, incomplete or intermittent control leads to unnecessary and costly defects. A system of identifying the state of progress of an item should be established and maintained. Process capability should be controlled and monitored on a continuous basis to detect and eliminate potential causes of non-conformance.

Inspection. Many inspections are normally performed during the manufacturing process, either for the purpose of process control or to ensure quality compliance of characteristics that can't be observed or measured at a later point. In addition, it is customary to perform a final inspection and test of the completed item to obtain an overall view of its conformance to requirements. A system for identifying the inspection status of the material during all stages of manufacture should be established. Proper inspection techniques need to be employed. All defects and failure data should be controlled and analysed.

Calibration / Control of Equipment. A system for controlling the calibration of measuring standard and equipment must be established and maintained. It must be ensured that all measurement resources have the capability of measuring within the required limits for meeting product specifications. It should provide for the prevention of inaccuracies through use of deficient equipment.

All measurement standards and measuring equipment should be calibrated using standards that are traceable to national or international standards. Measuring equipment should have a certificate of calibration. Intervals between calibrations should be decided.

Training of QA personnel, documentation/records and review procedures also need to be properly followed.

3.19. National Measurement System

All scientific investigations are based on the comprehension and exact observation of physical phenomenon. With increasing sophistication in technologies, the state of art in measurement technology attains higher precision, resulting in newer challenges to metrology—the science and skill of measurement (testing, calibration and related operations). Metrology, coupled with an effectively operating metrology service are of fundamental importance to the economy of every country. Accurate and reliable measurements are invariably required at all stages of productivity and product quality assurance, starting from conceptualisation to design, material management, standardisation and interchangeability of parts, production, quality control, test and evaluation, reliability determination and marketing services. Assurance of measurement quality is also vital for regulatory functions, consumer protection, public safety, scientific and technological research and development.

Measurement techniques of adequate capability and associated test and measuring equipments are used to carry out the complex measurements. The present day society is fast transforming into an information based society with an ever expanding and increasingly sophisticated industrial structure to support it, calling for still greater reliability in measurements. There is a need to underline the requirement for compatible and accurate measurements in the interest of international trade and for user satisfaction in domestic and foreign markets. This fact assumes further importance, since no economy can be truly self-sufficient and because self reliant economics of various nations require large scale import, export, absorption and adaptation of technology etc.

For attaining the objectives of reliable, adequately precise and accurate measurements, the measurement technology and the test and measuring equipment used are required to be made ultimately traceable to the international standards of measurement through regular and periodic calibration against the relevant national standards of measurement. Calibration activity thereby becomes an indispensable adjunct to the technology/product development cycle and an integral component of Quality Assurance practices.

The National Measurement System depends on types of industrial products, level of sophistication achieved in industries, the size and geographical conditions of the country, location of industries and the climatic conditions and application area requirements etc. It generally consist of two parts :

(*i*) The conceptual part : delineating the standards hierarchy with a view to achieve international compatibility.

TRACEABILITY OF ACCURACY IN MEASUREMENTS

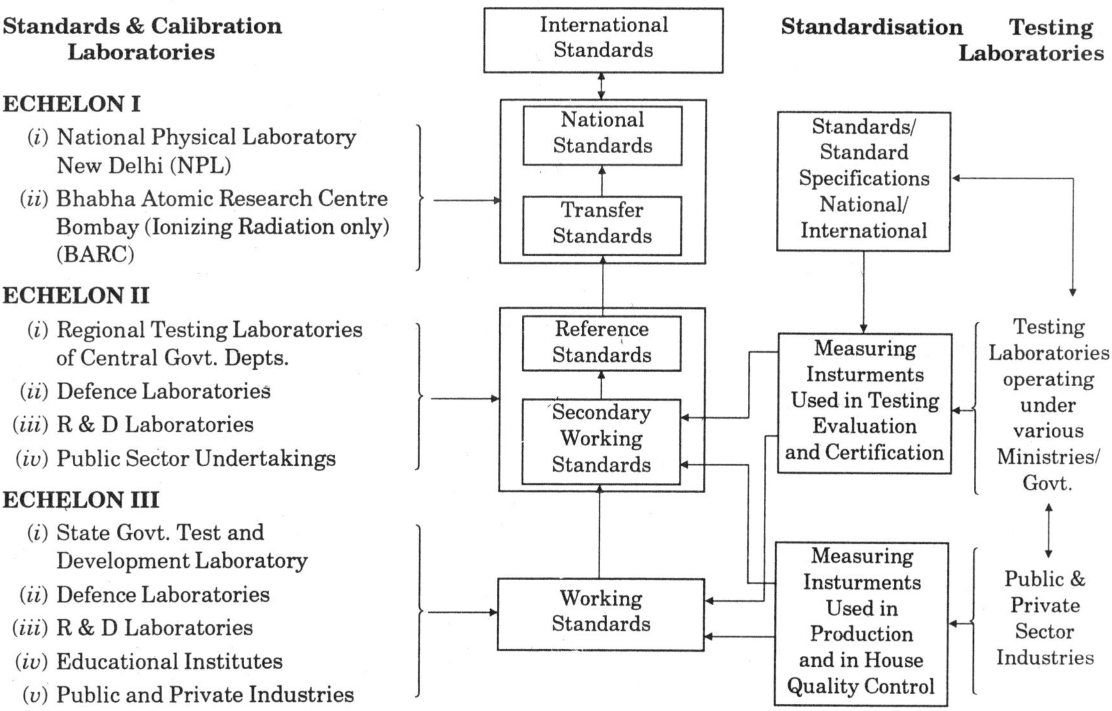

Fig. 3.17. National Measurement system.

(*ii*) The operation part : basically national in character and comprising of a set of operational elements, inputs, outputs, guidelines and details of the spectrum of activities etc.

The nodal point for National Measurement System in the country is National Physical Laboratory (NPL) at New Delhi, which has the statutory responsibility for :

(*i*) Development, maintenance and updating of national standards in respect of all physical parameters (except ionising radiation, for which BARC is the national standards laboratory)

(*ii*) Determination of major physical constants

(*iii*) Providing Apex level calibration facilities

(*iv*) Development and evaluation of measurement techniques.

The broad structure of the National Measurement System is explained in Fig. 3.17.

3.20. Calibration Service Programme

The Calibration Service Programme is implemented under the guidance of steering Committee for Calibration. It has the following objectives :

(*a*) To develop the Calibration hierarchy in order to coordinate the calibration activities of various laboratories and to ensure traceability of all measuring instruments to National Standards of Measurements by periodic calibration at appropriate level.

(*b*) To make recommendations on the level of accuracies for calibration laboratories at various echelons as well as at industrial level and to assist in identifying measurement results to be obtained from various calibration jobs at different laboratories in respect of the specified measurands.

(*c*) To lay down and provide assistance for development of appropriate calibration procedures and to ensure uniformity in reporting of measurement results carried at various calibration laboratories.

(*d*) To work out the criteria for assessment and accreditation of Echelon I, Echelon II and Echelon III laboratories.

(*e*) To assess the functioning of calibration laboratories in terms of the assessment criteria.

(*f*) To prepare guidance documents related to laboratory practices required to set up new calibration laboratories and their operation at industrial/user calibration level.

(*g*) To act as the National Calibration Service element in the operation of international quality systems in the country in order to promote industrial product quality in different areas of productivity.

(*h*) To establish information system and data base on measurement standards and calibration facilities and to arrange training programmes, seminars, symposia, exhibitions, audio visual programmes related to this activity.

(*i*) To promote and ensure compatibility in calibration through Measurement Assurance Programme and coordinate the intercomparison of standards amongst various laboratories using traveling standards.

(*j*) To work out guidelines for calibration charges at various levels to ensure uniformity.

(*k*) To assist in information of Acts and Status for National Coordination of Testing and Calibration Facilities.

3.21. Accreditation of Calibration Facilities

In the present day world markets which are becoming more and more competitive, the quality and affordable prices of products and services have become key determining factors in international business. A global compatibility in quality assurance procedures, standardisation and adequate control of production process have thus become essential. All these require adequate capabilities in measurements. As technology becomes more sophisticated and societies become more information oriented, measurements become more complex and increasingly important in every sphere. Accuracy of medical diagnosis, safety and environmental controls, quality assurance of consumer and industrial goods, all depend on precision measurements and comparison with standard materials and instruments, which is known as calibration. All industrialised nations have set up national measurement systems, consistent and compatible with international systems of units and quality through the mechanism of calibration. National Calibration system ensures that the measurements are made on a consistent basis throughout the country and are also in agreement with trading partners throughout the world.

Department of Science and Technology have launched "National Coordination of Testing and Calibration Facilities (NCTCF)" scheme. This envisages to continuously strengthen a network of accredited testing and calibration laboratories with uniform assessment procedures for compatibility in test and measurement data. National Physical Laboratory, New Delhi, being the custodian of national standards of measurements for physical parameters with international compatibility, has been entrusted to serve as the nodal organisation to implement this national scheme in the area of calibration. The accreditation of calibration laboratories is carried out and coordinated by a specially constituted cell at NPL under the Calibration Service Programme of the NCTCF scheme.

A calibration laboratory can get accreditation under this national scheme by fulfiling certain requirements and accepting the terms and conditions in line with the corresponding international practices.

3.22. Accreditation of Metrology (Calibration and Testing) Laboratories

The importance of test data in trade has been steadily growing in recent years. Most sophisticated products require testing for compliance with specifications and safety regulations before release into markets. Test documentation is very often the basis of trade. Third-party accreditation of testing and calibration laboratories is very common.

The following points need great consideration as regards the competence of calibration and testing laboratories :

— Organization and management
— Quality system, audit and review (including quality manual)
— Personnel
— Accommodation and environment
— Equipment and reference materials
— Measurement traceability and calibration
— Calibration and test methods
— Handling of calibration and test items
— Records
— Certificates and reports
— Sub-contracting of calibration and tests
— Outside support services and supplies
— Complaints

Laboratory Accreditation System, both for testing and calibration activities, are considered essential requirement throughout the industrial world for ensuring adequate reliability and confidence in the measured data throughout the life cycle of the industrial products and services. The accredited laboratory establishes confidence in the quality of laboratory's competence to provide testing and calibration services with commercial and technical integrity. The accredited laboratory should not get involved in such activities which might endanger the trust in its independence of judgement and integrity in relation to its calibration/testing.

Such a laboratory should be competent to carry out the test/calibration specified in its scope and provide supervision work by personnel who are familiar with the calibration/test methods and procedures, objective of the calibration/test and assessment of the results. The calibration/test laboratory should have adequate number of personnel having necessary education, training, technical knowledge and experience for their assigned functions. Layout of the laboratory premises should be suitable for performing the specific calibration/test jobs competently. The environment in the laboratory should be such that it does not invalidate or adversely affect the accuracy of calibration/test results.

The laboratory should be furnished with all necessary measuring and testing equipments required for accurate measurements. All the measuring and testing equipment of the laboratory should be properly maintained and a procedure of maintenance identified. The laboratory should be protected from excessive conditions, such as, heat, dust, vibrations, noise, moisture, steam, radiations and electromagnetic disturbance at all time. There should be sufficient space around the equipment to limit the risk of damage or danger and to allow the staff to perform the calibration/tests comfortably and smoothly. Record should be maintained for each major calibration test equipment significant to calibration/test performance.

The calibration/test laboratory should have documented instructions for the operation of all relevant equipments, for handling and preparation of items of the equipment and for calibration/test. The calibration laboratory should establish a quality system appropriate to the type, range and volume of calibration activities it undertakes. The laboratory quality manual and other related documents should include the policies and operational procedures established by the laboratory to meet the requirements of the criteria.

The laboratory should arrange for internal review and external audits of its activities periodically to ensure that its operation continues to comply with their requirements of the quality system.

3.23. Support or Background Services in Metrology

Instruments are required today in all fields of economy, may it be scientific research, quality systems, public health services, higher education, or industry. The smooth operation of instruments need support or background services. The management of instrumentation support services needs good attention. Standards, Measurement and Testing (SMT) Programme is corollary of this.

Today, quality is increasingly recognised as the only way to succeed in free market economies. The ISO 9000 family of standards and other quality programs such as Total Quality Commitment (TQC) determine what elements of production process are required to maintain quality regardless of the product manufactured. A key issue within these programs is product inspection and test. It is evident that quality cannot be assured without strict control and accurate measurements. Thus the support systems for metrology deserve all the attention. Fulfilling the measuring and control tasks requires : careful selection of the instruments used in the control process, deep knowledge of how to use them, training of personnel who carry out the task, and the regular control and calibration of the instruments.

For the repair and maintenance of the measuring etalons and special equipment special technology and a number of precision measuring and control instruments are used. For the smooth operation of these precision instruments well managed services like consultancy, repair and maintenance, calibration, etc. are needed. Proper knowledge of selection of most suitable instrument for given application, procurement problems, reliability of different supplies, repair and maintenance services, commissioning aspects and training to users deserve serious attention.

The prerequisites for a good after-sales service are : Well qualified multiply skilled experts. Technical specifications, drawings, service manuals, Special tools, tuning equipment and instruments for repair and calibration ; and lastly the stocking of spare parts needed for repair and maintenance service.

The preventive maintenance calls for checking the condition of instruments on a regular basis and perform the necessary adjustment on the spot, while the instrument is still capable of fulfilling its basic functions.

The maintenance and calibration of instruments is an essential activity of the support services. It has to be ensured that instruments while in use operate in satisfactory working condition. From time to time, the instruments must be calibrated through performing by comparison with a reference standard, which represents the best realisation of the unit of measurement at the place. Calibration plays an important role in the operation of the Quality Systems. The measuring and testing activity will be only useful if the results of measurement are reliable, *i.e.* if they are sufficiently accurate with a known degree of uncertainty. Calibration provides traceability, ensuring that the uncertainty level of every measurement made in the organization can be traced to known reference standards. A well organized calibration scheme gives the manufacturer the certainty that the features of the delivered products are within the specification and it gives the customer the assurance that he is buying exactly what he expects in terms of technical parameters and performance levels. Calibration of measuring and test equipment is essential to the reliability of test data. It is necessary to maintain a list of all equipment with associated details. The equipment register should also contain calibration details and the measurement standards against which the equipment has been calibrated. The calibration methods, procedures and references should be traceable to national or international measurement standards, and appropriate certificates certifying to this should be available on record.

Calibration is a precise activity requiring suitable procedures for all instruments and test equipment. Procedures should be as comprehensive and detailed as necessary for the type of the equipment to be calibrated. To carry out calibration service a calibration laboratory has to be established. To assure the correctness of its operation the calibration laboratory should follow international regulations, however, the best way to document this, is to obtain accreditation. Maintenance of proper environment in laboratory, availability of trained personel and adoption of proper procedures, and systems are equally important.

Quality control is not an isolated activity which can be carried out only by the inspection department of a firm. Well-planned quality management systems are needed. Support services are needed to support such activities.

At the same time support services can be provided for the independent laboratories and surrounding laboratories of different manufacturers. According to past experience, these services can be rendered by non-profit organisations or profit-oriented institutions in either centrally-planned or the market economies to meet the local requirements.

Effective management of the support services. The insight that human resource development plays critical role in industrial growth and competitiveness calls for greater emphasis being placed on building up efficient industrial support services. The instrumentation services form an essential part of these industrial support services.

WORKED EXAMPLES

Q. 3.1. Discuss the characteristics of line standards and end standards.

Ans. Line Standards. The characteristics of line standards are :

— Distance is measured between two lines.
— Scales can be accurately engraved but readings between two lines have to be estimated.
— A scale is quick and easy to use over a wide range.
— The scale markings are not subject to wear. However significant wear on leading end leads to "undersizing".
— A scale does not possess a "built in" datum which would allow easy scale alignment with the axis of measurement.
— Scales are subjected to the parallax effect.
— Scales are not convenient for close tolerance length measurement except in conjunction with microscopes.
— Line standards generally are not as accurate as end standards.

End Standards. These consist of set of standard blocks or bars and are used to build up a required length :

— Distance is measured between two surfaces.
— End standards are highly accurate and are well-suited to measurements of close tolerance. Dimensional tolerance as small as 0.005 mm can be obtained.
— These are time-consuming in use and prove only one dimension at a time.
— These are subjected to wear on their measuring faces.
— Groups of blocks can be "wrung" together to provide a given size.
— End standards have a "built in" datum because their measuring faces are flat and parallel and can be positively located on a datum surface.
— They are not subject to the parallax effect as their use depends on "feel".

The accuracy of both types is affected by temperature change. Care is needed in manufacture to ensure that change of shape with time, secular change, is minimal.

Q. 3.2. What do you understand by traceability ?

Ans. In order that items manufactured by various industries and manufacturing units in a country and also those manufactured in various countries be identical and interchangeable, it is essential that all follow standards of units and measurements traceable to some international standard. For purposes of accuracy, the standards followed in a vast industrial complex must be traceable to a single source, *i.e.* the national standards of the country which in turn must be according to International Standards like those of International Bureau of Weights and Measures. In the technique of traceability, the value of any measurement made in any industrial complex should be related in discrete steps and with known error to the value of the national standard. As the national standard cannot be used directly for calibration and standardisation of measurements in all industries, the process of traceability technique needs to be followed in stages, that is, national laboratories, standardising laboratories, etc. need to be established for country, states, and industries but all traceable to single source.

Q. 3.3. Three 100 mm end bars are measured on a level comparator by first wringing them together and comparing with a 300 mm bar and then inter comparing them. The 300 mm bar has a known error of + 42 μm and the three bars together measure 64 μm less than the 300 mm bar. Bar A is 18 μm longer than bar B and 23 μm longer than bar C.

Find the actual length of each bar.

Ans. Let the lengths of bars A, B and C be l_A, l_B and l_C mm respectively.

$$l_A, l_B + l_C = 300 \text{ mm} + 42 \ \mu\text{m} - 64 \ \mu\text{m}$$

$$= 300 \ \mu\text{m} - 22 \ \mu\text{m} \quad \text{...(1)}$$

$$l_A - l_B = 18 \ \mu\text{m} \quad \text{...(2)}$$

$$l_A - l_C = 23 \ \mu\text{m} \quad \text{...(3)}$$

Adding (1), (2) and (3), $3l_A = 300 \text{ mm} + 19 \ \mu\text{m}$

$\therefore$ $l_A = 100 \text{ mm} + 6.3 \ \mu\text{m}$

From (2), $l_B = 100 \text{ mm} - 11.7 \ \mu\text{m}$ **Ans.**

From (3), $l_C = 100 \text{ mm} + 16.7 \ \mu\text{m}$

Q. 3.4. A calibrated metre end bar has an actual length of 1000.0005 mm. It is to be used in the calibration of two bars, A and B each having a basic length of 500 mm.

When compared with the metre bar $L_A + L_B$ was found to be shorter by 0.003 mm. In comparing A with B it was found that A was 0.0006 mm longer than B. Find the actual length of A and B.

Ans. With reference to Fig. 3.18 and working first principles,

$$L - x_1 = L_A + L_B \text{ but } L_A = L_B + x_2.$$

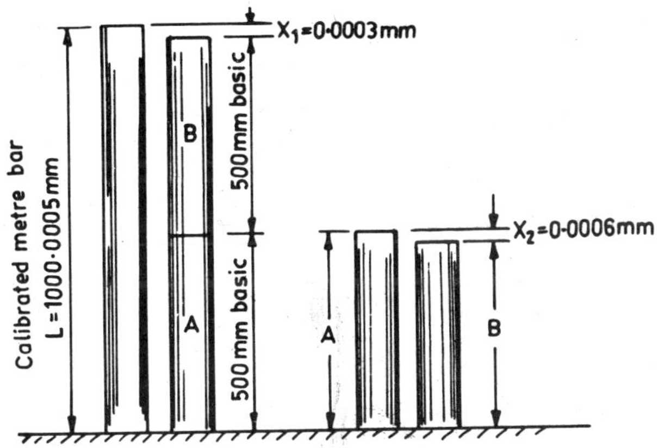

Fig. 3.18

Therefore, $L - x_1 = 2L_B + x_2$, or $2L_B = L - x_1 - x_2$

$$L_B = \frac{L - x_1 - x_2}{2} = \frac{1000.0005 - 0.0003 - 0.0006}{2} = \frac{999.9996}{2} \text{ mm}$$

i.e. $L_B = 499.9998$ mm

$L_A = L_B + 0.0006 = 499.9998 + 0.0006$ mm

or $L_A = 500.0004$ mm.

Q. 3.5. Enumerate the advantages of using wavelength standard as basic unit to define primary standards.

Ans. The advantages of wavelength standards are :

(*i*) It is not a material standard and hence is not influenced by effects of variation of environmental temperature, pressure, humidity and ageing.

(*ii*) It need not be stored under security and thus there is no fear of its being destroyed as in the case of metre and yard.

(*iii*) There are no problems of transferring this standard to other standards as in the case of metre and yard.

(*iv*) This standard is easily available to all standardising houses, laboratories and industries.

(*v*) It gives a unit length which can be produced consistently at all the times in all the circumstances at all the places. In other words, it is easily reproducible and thus identical standards are available with all.

(*vi*) It can be used for making comparative measurements of a much higher accuracy.

Q. 3.6. What do you understand by line and end measurements ? Discuss their relative characteristics.

Ans. Line measurement involves measurement of distance between two lines and end measurement involves measurement of distance between two surfaces. For example scale is a line measuring device. It is often very inconvenient to use line measuring device for measurement of end surfaces and similarly end measuring device for measuring distance between lines. The relative characteristics of both systems of measurements are discussed below.

	Line Standard	*End Standard*
Accuracy of measurement	Limited to ± 0.2 mm. For high accuracy, scales have to be used in conjunction with microscope	Highly accurate for measurement of close tolerances, upto ± 0.001 mm
Time of measurement	Quick and easy	Time consuming
Effect of use	Scale markings not subject to wear but end of scale is worn. Thus it may be difficult to assume zero of scale as datum	Measuring faces get worn out. To take care of this end pieces can be hardened, protecting type. Built-in datum is provided.
Other errors	Parallax errors can occur	Improper wringing of slip gauges may introduce errors. Change in laboratory temperature may lead to some errors
Manufacture and cost of equipment	Simple and low	Complex process and high.

Q. 3.7. (*a*) **What do you understand by airy point ? What are the conditions to achieve it ? For what length of bars, airy points are marked ? For what applications, airy points are useful ? What is the difference between airy points and supports to provide minimum central deflection ?**

(*b*) **Two slip gauges are joined together by wringing. How the two length bars are joined together and how are they (in combination) supported ?**

Ans. (*a*) Length bars above 125—200 mm length when used in horizontal position for measurement, are supported on two airy bands marked over them in order that the gauging faces at the ends are parallel with each other and perpendicular to the measuring plane. The airy points are achieved when the distance between the two supports is 0.577 × length of bar or distance of each support from the end of bar is 0.2115 × length of bar. Airy points are marked for lengths above

125—200 mm. Below 125 mm length the length bars can be used unsupported. Airy points are used for length standards. In the case of airy points, some sag is permitted at the centre of length bar and this sag is just sufficient to pull the ends of the bar square with the measuring plane. But in the case of points of minimum deflection, the central sag is minimum. When a length bar or straight edge is supported at points of minimum deflection, its ends sag so as to lift the centre and minimise the central deflection (or the deviation from the mean true plane). The condition of minimum deflection occurs when points of support are 0.554 × length of bar, or distance of support point from each end is 0.223 × length of bar.

(*b*) Wringing can't work in case of length bars because of their weight and length. Length bars are joined by screwed dowles and supported on two airy points for the combination and not at the airy point supports for each length bar.

Q. 3.8. (*a*) **Four length bars, A, B, C and D are to be calibrated using a calibrated bar whose length is nearly equal to sum of all the four length bars. Explain the procedure.**

(*b*) **Four length bars of basic length 100 mm are to be calibrated using a calibrated length bar of 400 mm whose actual length is 399.9992 mm. It was also found that lengths of bars B, C and D in comparison to A are + 0.0002 mm, + 0.0004 mm and – 0.0001 mm respectively and the length of all the four bars put together in comparison to standard calibrated bar is + 0.0003 mm longer. Determine the actual dimensions of all the four end bars.**

Ans. (*a*) Let the length of four bars be l_A, l_B, l_C and l_D respectively and that of master calibrated bar be l.

All the four bars are joined together and compared with calibrated bar. Let the error between the two be x.

$$\therefore \quad l_A + l_B + l_C + l_D = l + x \qquad \text{...(1)}$$

The three bars B, C and D are compared with A and let the difference observed be x_1, $-x_2$ and x_3 respectively.

$$\therefore \quad l_A + x_1 = l_B \qquad \text{...(2)}$$

$$l_A - x_2 = l_C \qquad \text{...(3)}$$

and

$$l_A + x_3 = l_D \qquad \text{...(4)}$$

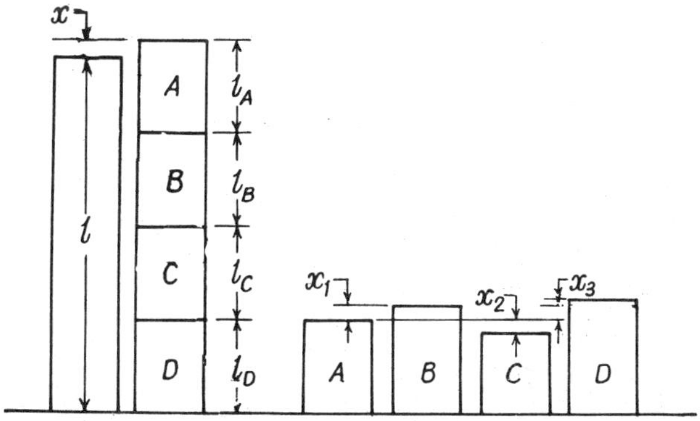

Fig. 3.19

Adding (1), (2), (3) and (4),

$$4l_A + l_B + l_C + l_D + x_1 - x_2 + x_3 = l + x + l_B + l_C + l_D$$

or $$4l_A = l + x - x_1 + x_2 - x_3$$

Thus $$l_A = \frac{l + x - x_1 + x_2 - x_3}{4}$$

Once l_A is known, then by equations (2), (3) and (4), l_B, l_C and l_D can be determined.

(*b*) $l = 399.9992$ mm

$x_1 = + 0.002$ mm, $x_2 = + 0.0004$ mm

$x_3 = - 0.0001$ mm, and $x = + 0.0003$ mm.

$$l_A = \frac{l + x - x_1 - x_2 + x_3}{4} = \frac{399.9992 + 0.0003 - 0.0002 - 0.0004 + 0.0001}{4}$$

$$= \frac{399.9990}{4} = 99.9997 \text{ mm}$$

$$l_B = l_A + x_1 = 99.9997 + 0.0002 = 99.9999 \text{ mm}$$

and $$l_C = 99.9997 + 0.0004 = 100.0001 \text{ mm}$$

$$l_D = 99.9997 - 0.001 = 99.9996 \text{ mm.}$$

Q. 3.9. Match the following parts in regard to standards of measurements.

A	B
1. A physical embodiment of unit.	(*a*) SI
2. A platinum iridium cylinder with diameter and height of 39 mm.	(*b*) Derived units
3. Coherent, rational and comprehensive system of units.	(*c*) tertiary standard
4. The amount of substance of a system which contains as many elementary entities as there are atoms in 0.012 kilogram of carbon-12.	(*d*) primary standard
5. Combination of two or more quantities.	(*e*) metre
6. Tera, giga, femto, atto, etc.	(*f*) yard
7. Only one material standard and preserved under most careful conditions.	(*g*) standard
8. Standards to be used for reference purposes in laboratories and workshops.	(*h*) wavelength
9. Distance between the two central transverse lines on the plugs.	(*i*) length
10. Distance between the axes of two lines engraved on the platinum iridium bar.	(*j*) kilogram
11. Standard based on the orange radiation of the isotope krypton-86.	(*k*) decimal multiples and sub-multiples
12. Non-material standard that can be reproduced to an accuracy of about one part in 10^9.	(*l*) mole—the unit of amount of substance.

Ans. 1—*g*, 2—*j*, 3—*a*, 4—*l*, 5—*b*, 6—*k*, 7—*d*, 8—*c*, 9—*f*, 10—*e*, 11—*i*, 12—*h*.

Q. 3.10. Match the parts A and B in relation to measurement standards.

A	B
1. A standard of a particular quantity which has the highest class of metrological qualities in a specified field.	(*a*) reference standard
2. A standard whose value is fixed by direct or indirect comparison with a primary standard.	(*b*) reference material
3. A standard, generally the best (of the highest metrological quality) available from which measurements made at that location are derived.	(*c*) measurement standard
4. A standard which, usually calibrated against a reference standard, is used routinely to calibrate or verify measuring instruments of lower accuracy.	(*d*) transfer standard
5. A standard used as an intermediary to compare standards, material measures, etc.	(*e*) group standard
6. A measuring device, sometimes of special construction, intended for transport between different locations or used for the comparison of values of a measured quantity at different locations. For example, a portable battery-operated caesium atomic frequency standard.	(*f*) traceability
7. A material/substance whose one or more of the properties are sufficiently well established to be used for the calibration of an apparatus.	(*g*) working standards
8. The concept of establishing a valid calibration of a measuring instrument of measurement standard, by step-by-step comparison with better standards upto an accepted or specified standard.	(*h*) conservation of standard
9. A material measure or system intended to define, realise, conserve or reproduce a unit of one or more known values of a quantity (or a multiple or sub-multiple of that unit) in order to transmit them to other measuring instruments by comparison.	(*i*) primary standard
10. A set of standards of specifically chosen values which individually or in suitable combination reproduce a series of values of a quantity over a given range.	(*j*) collective standard

11. A set of similar material measures or a group of measuring instruments of some pattern and the same metrological characteristics, fulfilling by their combined use, the role of the standard.	(*k*) travelling standard
12. All operations (regular calibration, storage under good conditions, care in use, etc.) necessary to preserve the metrological characteristics of a measurement standard within appropriate limits.	(*l*) secondary standard

Ans. 1—*i*, 2—*l*, 3—*a*, 4—*g*, 5—*d*, 6—*k*, 7—*b*, 8—*f*, 9—*c*, 10—*e*, 11—*j*, 12—*h*.

Q. 3.11. How one can ensure that products manufactured indigenously are comparable with their foreign counterparts ?

Ans. Three steps for this purpose are :

(*i*) Providing measurement facilities of high precision and accuracy for performance evaluation of indigenous products.

(*ii*) Ensure controlled laboratory environments satisfying national/international specifications.

(*iii*) The measuring equipments and systems should be calibrated with national standards which are traceable to international standards.

Q. 3.12. What are the advantages of documentary standards ?

Ans. Documentary standards offer advantages for both producer and user in the production of standardised goods. User is supplied with goods which are reliable and fit for the intended purpose and with components that can be interchanged. Store-keeping becomes easier, because there will be less variety in the range of equipment. Book-keeping becomes less burdensome, and goods and services become more easily available. Consumer enjoys better value for money spent. The producer enjoys following advantages : rationlisation of process of manufacturing ; improved control over processes, material and labour ; elimination of waste of materials and effort ; higher productivity, reduced cost ; less inventories ; increased sales and profits.

Glossaries of terms and definitions facilitate communication between people involved in such exchanges. Specifications define the ways in which value can be determined. Methods of sampling describe the most economic ways of selecting parts of materials or products in order to be able to make a representative evaluation. Methods of gauging, of testing and of analysis provide agreed yardsticks for measurement. Codes of practice describe the best procedures for installing, using and maintaining equipments. From a wider view it can be seen that standardisation is highly advantageous to a country as a whole. Documentary standards are thus essential for the organized development of the industries connected with the metrological tools. They also offer specific advantages to the users of all commodities by providing the base for accurate and reliable measurement and test results.

Trade enjoys minimisation of delays. There is workable basis for acceptance or rejection of goods, and for dealing with consequential disputes.

Q. 3.13. Describe the changing scene in international metrology resulting from the growing importance of traceable measurements in trade.

Ans. Today there is need to establish world-wide traceability of measurements to International System of Units. The international bodies like BIPM (Bureau International des Poids et Hesures) and CC (Consultative Committees) are working with the regional metrology organisations to establish world-wide equivalence of national measurement standards, taking fully into account the uncertainties of the standards from the different institutes. The double testing and calibration by both exporting and importing trade partners is a technical barrier. Thus there is a need for an agreement on international equivalence of national measurement standards which must be established beyond dispute.

Q. 3.14. How the international equivalence of national measurement standards can be organised ?

Ans. As per ISO 9000, each measurement result must be traceable to national or international measurement standards. An unbroken chain of comparisons must be established to standards which are considered to maintain a sufficiently accurate approximation to the SI value. It is necessary to build up a hierarchical system linking the working standards to generally agreed international standards as per Fig. 3.20.

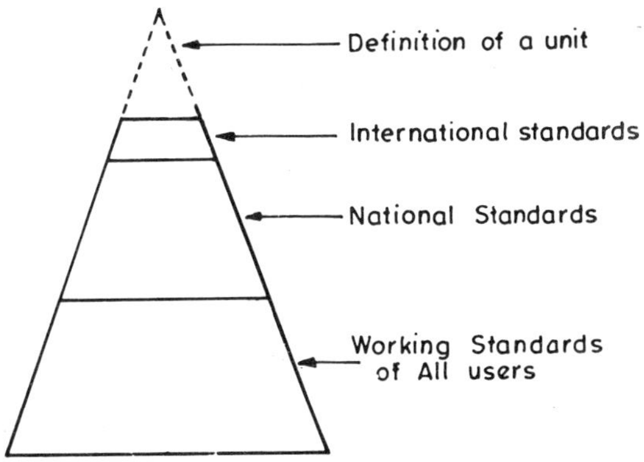

Fig. 3.20

Vertical traceability to users is usually ensured by intermediate links provided either by the laboratories of a verification service operating with the authority of law or, alternatively, by the accredited laboratories of a calibration service. The demand for international equivalence is gaining importance. General agreement is thus necessary on the required structures and procedures. The agreements for cooperation at the highest level of accuracy (indicated by arrows between the laboratories of the NMIs and by links with the BIPM, need to be made more formal. International recognition of the equivalence of national measurement standards is based on the use of agreed reference values derived from the results of so-called key comparisons between selected primary standards. The reference values, together with an associated uncertainty, are assumed at present to give the best approximation to the SI value referred to as international standards. Every national metrology institute should maintain primary standards, that is standards of the highest metrological quality by means of realization or reproduction of a unit's definition, only for those physical quantities which are essential for the national economy ; other quantities can be maintained by secondary standards, that is standards fixed by comparisons.

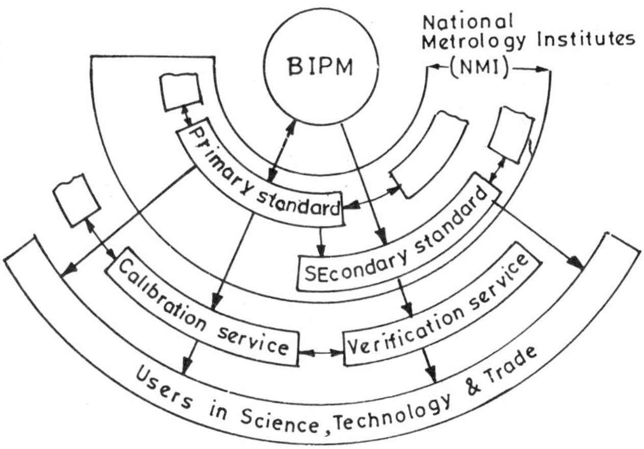

Fig. 3.21. Lines of traceability within world metrology.

TYPICAL EXAMINATION QUESTIONS

1. Distinguish between 'line standards' and 'end standards'. How are end standards derived from line standards ? Give examples of these two types of standards.
2. Explain the role of light wave standard in the future of precision measurements.
3. What is the difference between line standard and end standard ? How will you compare an end gauge with a line standard ?
4. An NPL type level comparator has a vial radius of 210 m, divisions 2.5 mm apart and contact feet at 25 mm centre distance. Calculate the difference in length of two gauges under comparison if the total bubble displacement is 6 divisions.
5. What is the standard temperature for measurement ? How measurements are affected by temperature variation ?
6. Distinguish clearly between the Airy points and the points for minimum deflection of a simply supported uniform beam.
7. (*a*) Explain clearly the difference between shop, inspection and master gauges.
   (*b*) Describe briefly the manufacturing of slip gauges.
8. Briefly describe the different types of standards for linear measurements.
9. What is meant by the term 'cosine error' ? Explain with examples.
10. Describe the precautions that should be taken to prevent corrosion of highly finished surfaces.
11. Discuss the possible effects upon accuracy of measurement due to (*i*) temperature variation, (*ii*) elastic deformation (*iii*) cosine errors.
12. What do you understand by accreditation of testing laboratories ? How it is done and what procedures are to be followed ?
13. Explain in detail the concept of traceability of accuracy in measurements.
14. What are the essentials of a metrology laboratory ?
15. Explain the procedure for accrediation of calibration facilities.
16. Discuss the role of metrology in quality assurance.

4

Limits, Fits and Gauges

4.1. Introduction

It is well known fact that no two things in the nature can be identical, they may be found to be closely similar. This is true of production of component parts in engineering also. We know that every process is a combination of three elements, *viz.*, man, machine and material. A change in any one of these will constitute a change in the process. All these three elements are subjected to inherent and characteristic variations. These variables result in the variation of size of components. If the process be under control, *i.e.* all the assignable causes of variation and controllable causes of variation have been removed or controlled, the size variation will be within reasonable limits. Thus when process is in control, most of the items will be on or about the mean value and other values will be distributed in a more or less symmetrical way about the mean value and remaining within expressible limits. It is, therefore, obvious that it is impossible to produce a part to an exact size and some allowance known as tolerance has to be allowed. The tolerance allowed also depends on the functional requirements which can't be sacrified. No difficulty arises when the tolerance allowed is sufficiently greater than the process variation (*i.e.* capability of machine). The designer should ensure that the tolerance on any component is neither restrictive nor permissive, *i.e.*, it should be as wide as the process demands and no wider.

Thus it can be said that due to inevitable inaccuracy of manufacturing methods it is not possible to make any part precisely to given dimension and it can only be made to lie between two limits, maximum and minimum. The difference between these two limits is called the permissive tolerance. When tolerance distribution is on only one side of basic size, it is referred to as unilateral tolerance. If permissible variation is both plus and minus, it is referred to as bilateral tolerance.

In brief it can be said that it is impossible to manufacture a part to a specified definite size—particularly if it is mass produced. Even if it is produced exactly to a specified definite size, it may not be possible to measure it accurately enough to prove it. Moreover considering the functional requirement, no purpose is served by producing any component to a specified definite size since some variation depending on function can be tolerated. The cost of attempting to produce parts to a specified definite size without variation would be too high which is uncalled for, thus justifying the need for tolerance.

Generally in engineering, any component manufactured is required to fit or match with some other component. The correct and prologed functioning of the two components in match depends upon the correct size relationships between the two, *i.e.*, the parts must fit with each other in a desired way : *e.g.* if a shaft is to rotate in a hole, there must be enough clearance between the shaft and hole to allow the oil film to be maintained for lubrication. If the clearance is too small, excessive force would be required in rotation of shaft. If clearance is too wide, there would be vibrations and rapid wear and ultimate failure. Thus desired clearance to meet requirement has to be provided. Similarly, if shaft is to be held tightly in the hole then there must be enough of interference between the shaft and hole so that forces of elastic compression grip them tightly and do not allow any relative movement between them.

The desired condition of match between shaft and hole could be obtained by specifying a definite size to one and varying the other. But actually, this is not possible in practice as it is impossible to make a perfect size and if at all, it is made, no means are there to measure it accurately and economically during machining. The variations in dimensions, however small, do exist because of the variations in tooling, machining, raw material and operators. If a machine is under control, *i.e.* no assignable causes of variation exist, then the resultant frequency distribution of dimension produced will be roughly in the form of normal curve, *i.e.* 99.7% parts will be within $\pm 3\sigma$ limits of mean setting. The value of σ depends upon the machine used to produce a component. If value of σ has to be reduced, then precision machines have to be used to produce the component having less variation in dimensions. It is thus important to note that the cost of production keeps on increasing tremendously for very precise tolerance [Fig. 4.1]. As the tolerance approaches zero, the task of achieving it becomes enormous and finally impossible. On the other end, fabrication cost decreases with the relaxing of tolerance, but a value is soon reached beyond which little manufacturing relief is gained. In general, tolerance *vs.* fabrication cost is hyperbolic curve. It is difficult to assign any typical numerical value to tolerance and fabrication cost because of the interdependence of tolerance upon the dimension. From above, it would be clear that magnitude to tolerance must be set so as to permit as low manufacturing cost as possible and still achieve the necessary functional performance.

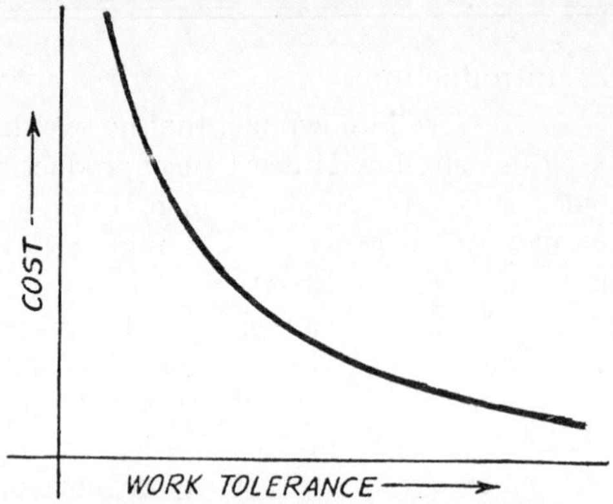

Fig. 4.1

In the early days the majority of components were actually mated together, their dimensions being adjusted until the required type of fit was obtained. But the interchangeable production and continuous assembly require some standard procedure to be followed. In order to obtain various fits, it is possible to vary the hole sizes and shaft size. Thus by variation of hole and shaft sizes, innumerable types of fits can be possible. All this has, therefore been standardised and the primary aim of any general system of standard limits and fits should be to provide guidance to the users in selecting basic functional clearance and interference for a given application or type of fit ; and in providing tolerances which provide a reasonable and economical balance between fits, consistency and cost.

For a system of limits and fits to be successful, following conditions must be fulfilled :

(*i*) It must be based on some standard so that everybody understands alike, and a given dimension has the same meaning at all places.

(*ii*) The range of sizes covered by the system should be sufficient for most purposes.

(*iii*) Number of limits and fits should be reduced and be adequate to cover all applications.

(*iv*) Each basic size of hole and shaft must have a range of tolerance values for each of the different fits.

(*v*) Both unilateral and bilateral methods of tolerance, and hole basis or shaft basis systems should be acceptable.

(*vi*) For any basic size it must be possible to select from a carefully designed range of fits the most suitable one for a given application.

(*vii*) The fundamental deviation required to give a particular fit must increase with the basic size so that a particular fit or same assembly characteristics are obtainable for smaller as well as bigger sized parts.

(*viii*) The system should cover work from high class tool and gauge work upto rough work.

4.2. Tolerances

As already indicated, tolerance can be defined as the magnitude of permissible variation of a dimension or other measured or control criterion from the specified value. Tolerances have to be allowed because of the inevitable human failings and machine limitations which prevent ideal achievements during fabrication. In order to maintain economic production and facilitate the assembly of components it is necessary to allow a limited deviation from the designed size. Due to its inevitability, tolerances constitute an engineering legality for deviation from the ideal value, and like any other legal matter, formulation of tolerances must also be given due consideration, and much thought and planning should go into it. The various factors affecting the choice of tolerances should be given due consideration, as the setting of tolerances is not an arbitrary matter. Though functional requirement is the primary consideration, *i.e.* the permitted deviation in size must permit the assembly to function correctly for its designed life, there are other factors like standardisation, manufacturing needs etc., which also influence the choice of tolerances and the primary consideration of functional requirement sometimes requires compromise with these factors.

The primary purpose of tolerances is to permit variation in dimensions without degradation of the performance beyond the limits established by the specification of the design. Where high performance is the criterion, there the functional requirements will be the dominating factor in setting tolerances. However, where the functional performance provides some latitude, then tolerance choice may be influenced and determined by factors like standardisation, methods of tooling, and available manufacturing equipment.

The numerical values of tolerances may range across the entire spectrum of measurements and if in all the cases, the functional requirement is taken as the only criterion to decide the value of tolerance then there may be serious disadvantages like—too many different tolerances which means excessive amounts of calculations and studies to establish tolerance values, excessive amount of special tooling, complications in inspection etc. If a limited number of standard tolerances are established, and the tolerances are chosen from these so that these are slightly closer than the function dictates, then we get the advantages of fewer varieties of tooling, few calculations and increased unit quantities because of repeated use of the same designs.

Functional and Non-functional Dimensions. Functional dimensions are those which have to be machined and fit with other mating components. Non-functional dimensions are those which need not be machined to a high degree of accuracy. These have no effect on the quality performance of the component or assembly.

4.2.1. Why tolerances are specified ? Ideal conditions would call for parts without any kind of dimensional variation, but in actual practice it is impossible due to following reasons :

(*i*) Variations in the properties of the material being machined introduce errors.

(*ii*) The production machines themselves have some inherent inaccuracies built into them and have the limitations to produce perfect parts.

(*iii*) It is impossible for an operator to make perfect settings. In setting up the machine, *i.e.* in adjusting the tools and workpiece on the machine, some errors are likely to creep in.

An attempt to entirely overcome these factors with a view to obtain ideal conditions would result into exhorbitant costs. The parts should, therefore, be made as inaccurate as tolerable to satisfy the functional requirements. Thus tolerances are specified to the dimensions of all manufactured parts and these should be just enough to do the intended job and no better. The tolerance is a compromise between accuracy required for proper functioning and the ability to economically produce this accuracy.

4.2.2. Design Considerations in the Selection of Engineering Tolerances. The establishment of correct engineering or manufacturing tolerances is a matter of serious concern because both too tight and too loose tolerances result in excessive cost. Tight tolerances result in unnecessary machining and inspection time, high rejects, and adversary relationship between design, manufacturing, and inspection. Loose tolerances result in assembly problems and poor product performance.

The designer for selection of tolerances must have broad knowledge of engineering metrology, must thoroughly understand how the part is to be measured. He should also be familiar with capability of a wide range of manufacturing processes. He must also have a clear understanding of the functional requirements of the piece part in assembly and operation with the other components of the system.

Reasons for tendency to select tight tolerances. The unexperienced designer in order to protect himself specifies significantly tighter tolerance than necessary for the following reasons : the creed for precision, fear of interference or excessive clearance between assembled parts, selection of tolerances from company or vender standards which tend to favour tight tolerances or selection from similar previous design which may have been established on unrealistic tolerances, practice of considering tight tolerances synonymous with good quality.

Optimum tolerance. A good designer must explore information and techniques available as guidance in the establishment of realistic and, where possible, optimum engineering tolerances.

It must be remembered that the cost of machining and finishing to tighter tolerances increases rapidly. One should have fairly good idea of relative cost of producing parts with different tolerances. The standard products available within the specified tolerances should be considered. One should also have good understanding of tolerances resulting from various production processes.

The understanding of statistical tools is of great help. It can be safely assumed that the most of the manufacturing processes produce parts which are normally distributed. For normal distribution, 95% of all production lies within ±2 standard deviations and 99.7% within ±3 standard deviations. It should be remembered that in a random assembly, probability of all large parts being assembled with large parts, or of all small parts likewise being assembled together, is extremely small. With this approach it is possible to use wider tolerances.

The use of quality control charts also enables one to be more realistic in selection of tolerances. Quality control chart is a powerful tool for the definition of the potential dimensional accuracy of manufacturing processes. Quality control charts are based on the assumption of normal distribution, random sampling, and independence between samples. Control charts are good guide for the design office in establishing of realistic specifications.

It is very important that the designer understands the capability of manufacturing process in producing parts to a given tolerance. Computer simulation techniques are used to provide understanding of the tolerance requirement of product of design. Computer simulation provides data on the probability of parts of a certain tolerance range interfering at assembly.

4.2.3. What is a basic dimension ? A basic dimension is the dimension, as worked out by purely design considerations. Since the ideal conditions of producing basic dimension, do not exist, the basic dimension can be treated as theoretical or nominal size, and it has only to be approximated. A study of function of machine parts would reveal that it is unnecessary to attain perfection because some variations in dimentions (however small) can be tolerated on sizes of various parts. It is, thus, general practice to specify a basic dimension and then indicate by tolerances as to how much variation in the basic dimension can be tolerated without affecting the functioning of the assembly into which this part will be used.

4.2.4. Different ways of expressing Tolerances. Tolerances are basically specified in two forms, *i.e.* unilateral and bilateral. In unilateral tolerances, the total tolerance as related to a basic dimension is in one direction only. This form of tolerance is usually indicated when the machining of mating parts is called for, as this greatly assists the operator. This form of tolerance is usually indicated when the machining of mating parts is called for, as this greatly assists the operator. The operator machines to the upper limit of shaft (lower limit for a hole) knowing fully well that he still has the whole tolerance left for machining before the parts are rejected. In case of bilateral tolerances, the total tolerance is specified on both sides (plus and minus) of the basic dimension. Bilateral tolerances usually have plus and minus tolerance of equal amount, but not necessarily always. This system permits operator to take full advantage of the limit system especially in positioning a hole.

The various methods of specifying unilateral tolerances are

(*i*) Specify as limiting dimensions :

Diameter of hole : 25.00 mm, 25.002 mm

Diameter of shaft : 24.999 mm, 24.997 mm

(*ii*) One limiting size may be specified with its tolerances :

Diameter of hole : 25.000 + 0.0002 mm, – 0.000 mm

Diameter of shaft : 24.999 + 0.000, 0.002 mm

(*iii*) The nominal size may be specified for both parts, with a notation showing both allowance and tolerance :

Diameter of hole : 25.000 + 0.0002, – 0.000 mm

Diameter of shaft : 25.000 – 0.0001, 0.003 mm

Bilateral tolerances may be expressed as :

Diameter of hole = $25.000^{+\,0.002}_{-\,0.000}$ mm,

Diameter of shaft = $25.000^{-\,0.001}_{-\,0.003}$ mm,

In the case of unilateral tolerances, the dimension is allowed to vary only in one direction. This system is used in machining processes like drilling in which case the dimensions are most likely to deviate in one direction only (in drilling, hole is always of over size rather than undersize). Bilateral tolerance system defines the theoretically desired size and the probable deviation permitted on either side of the basic size. This system is used in mass production techniques where machine setting is done for the basic size. Under such conditions, if tolerances are specified as unilateral, then these should be changed into bilateral tolerances by changing the basic size. The basic size should be mid way between upper and lower limits.

4.2.5. Specifying tolerances for given assemblies. The type of assembly, *i.e.* fit between two mating components is decided based on functional requirements. Based on it, tolerances on shaft and hole are decided using two approaches, *viz.* complete interchangeability, or statistical approach. In complete interchangeability approach, no risk is taken about even a single non-conforming assembly. If the fit between a shaft and hole is a clearance type, then for complete interchangeable approach (referring to Fig. 4.2 it will be seen that) tolerance on shaft = tolerance on hole = half the maximum clearance—half the minimum clearance.

The statistical approach bases the permissible tolerances on the normal distribution curve, considering that only 0.3% components would lie beyond $\pm\ 3\sigma$ limits. This approach, obviously,

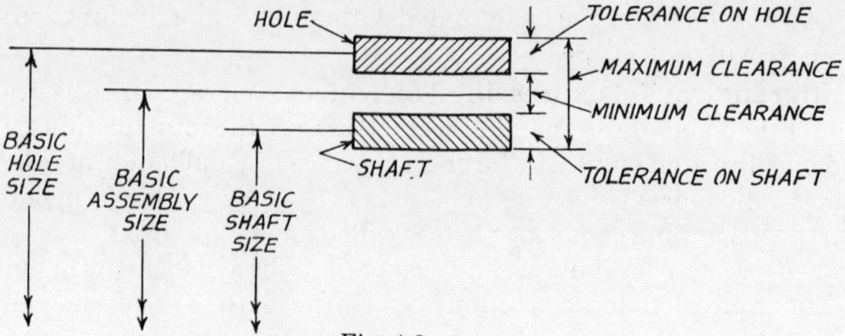

Fig. 4.2

allows wider tolerances compared to complete interchangeability approach. The statistical approach thus permits cheaper production methods but can be used when components are produced in bulk. If all the assignable causes of variation are fully controlled, then expected frequency curves of shafts and holes will be normal frequency curves, and accordingly the expected frequency of the clearances will also be the normal curve.

Mean clearance = Mean size of hole – Mean size of shaft

$$\sigma_{hole-shaft} = \sigma_{clearance} = \sqrt{\sigma^2_{hole} + \sigma^2_{shaft}}$$

Maximum spread of clearance – minimum clearance = $6\sigma_{clearance}$

or
$$\sigma_{clearance} = \frac{\text{Maximum clearance} - \text{Minimum clearance}}{6}$$

It is estimated that about 33% more tolerance is permitted by statistical approach in comparison to total interchangeability approach.

4.2.6. Tolerance Accumulation. If a part comprises of several steps, each step having some tolerance over its length, then overall tolerance on complete length will be sum of the tolerances on individual length.

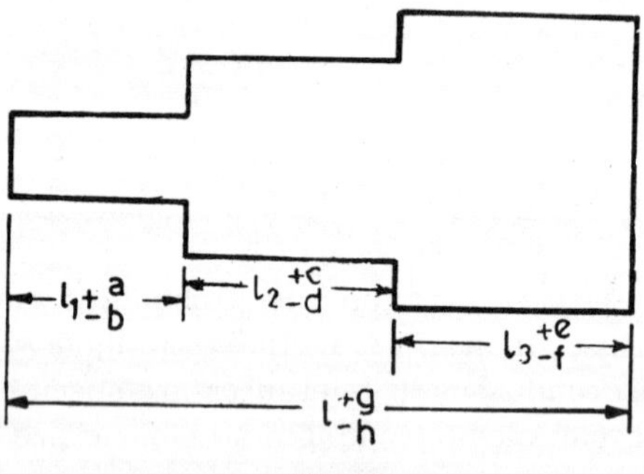

Fig. 4.3

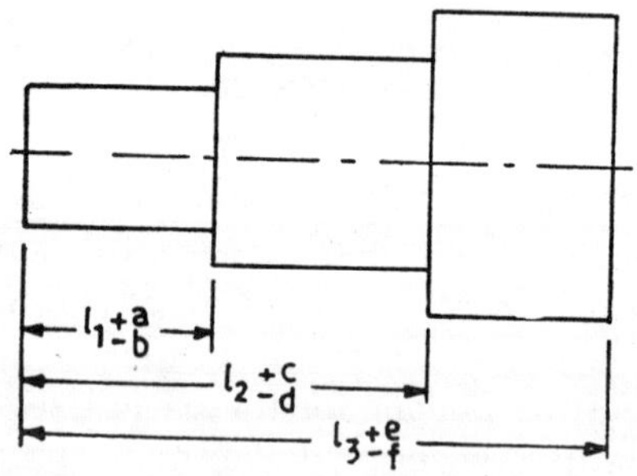

Fig. 4.4

For example, in Fig. 4.3, $l^{+g}_{-h} = l^{+a}_{1-b} + l^{+c}_{2-d} + l^{+e}_{3-f}$

and
$$g = a + c + e \text{ and } h = b + d + f$$

It would be appreciated that this practice of specifying tolerances would result in high tolerances on the overall length. The effect of accumulation of tolerances can be minimised by adopting progressive dimensioning from a datum as shown in Fig. 4.4.

4.2.7. Compound Tolerances. Compound tolerances are those which are derived by considering the effect of tolerances of more than one dimension. For example, in Fig. 4.5, the tolerances on dimension l are dependent on tolerances on L, h and θ. Thus compound tolerance of 'l' is the combined effect of these three tolerances. The minimum tolerance on 'l' will be corresponding to $L - b$, $\theta + \alpha$ and $h + c$.

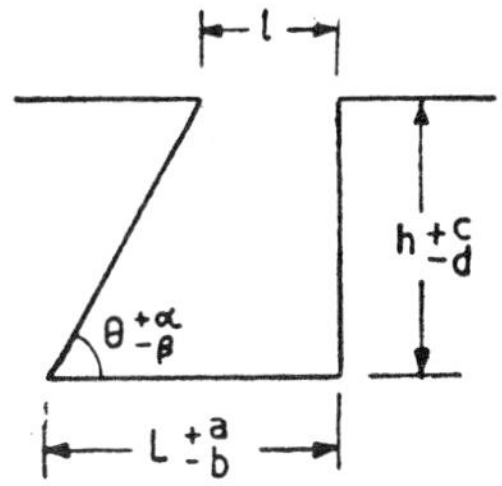

Fig. 4.5

4.3. Interchangeability

An interchangeable part is one which can be substituted for similar part manufactured to the same drawing.

In earlier times production used to be confined to small number of units and the same operator could adjust the mating components to obtain desired fit. With time the concept of manufacturing techniques kept on changing and today the same operator is no more responsible for manufacture and assembly too. With economic oriented approach, mass production techniques were inevitable, that led to breaking up of a complete process into several smaller activities and this led to specialisation. As a result various mating components will come from several shops, even a small component would undergo production on several machines. Under such conditions it becomes absolutely essential to have strict control over the dimensions of portions which have to match with other parts. Any one component selected at random should assemble correctly with any other mating component, that too selected at random. When a system of this kind is ensured it is known as interchangeable system. Interchangeability ensures increased output with reduced production cost.

In interchangeable system, every operator being concerned only with a limited portion of overall work, he can easily specialise himself in that work and give best results leading to superior quality. He need not waste his skill in fitting the components by hit and trial and assembly time is reduced considerably. In the case of big assemblies, several units to manufacture individual parts can be located in different parts of country depending on availability of specialised labour, raw material, power, water and other facilities and final assembly of all individual components manufactured in several units can be done at one place. The replacement of worn out or defective parts and repairs is rendered very easy and the cost of maintenance is very much reduced and shut down time also reduced to minimum.

Interchangeability is possible only when certain standards are strictly followed. Universal interchangeability (*i.e.* parts drawn from any two altogether different manufacturing sources for mating purposes) is desirable and for this it is essential that common standards be followed by all, and all standards used by various manufacturing units should be traceable to a single source, *i.e.* international standards. When all parts to be assembled are made in the same manufacturing unit, local standards may be followed (condition being known as local interchangeability) but for reasons of obtaining spares from any other source it is again desirable that these local standards be also traceable to international standards.

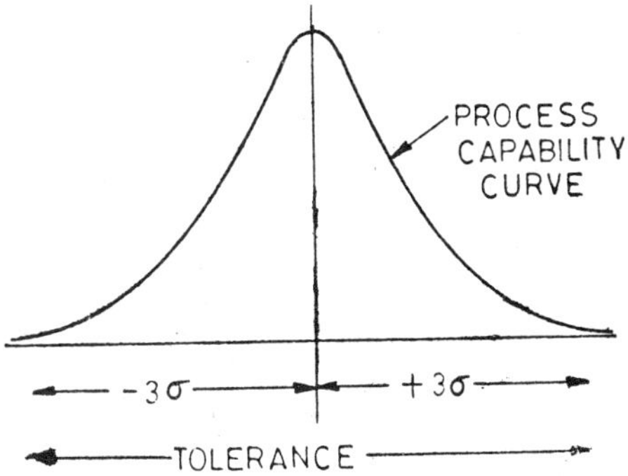

Fig. 4.6

The required fit in an assembly can be obtained in two ways, namely (*i*) universal or full interchangeability, and (*ii*) selective assembly. Full interchangeability means that any component will mat with any other mating component without classifying manufactured components in subgroup or without carrying out any minor alterations for mating purposes. This type of interchangeability is not must for interchangeable production and many times not feasible also as it requires machines capable of maintaining high process capability and very high accuracy, and very close supervision on production from time to time. (Process capability of a machine is defined as its $\pm 3\sigma$ spread of dimensions of components produced by it. If a plot is drawn of the actual dimensions of the similar components produced by a machine, it is found to follow natural law of distribution, *i.e.* having mean of all the components at central value with a spread of $\pm 3\sigma$ value, σ being known as standard deviation, and $\pm 3\sigma$ as the process capability of machine). For full interchangeable assembly it is essential that only such machines be selected for manufacturing whose process capability is equal to or less than the manufacturing tolerance allowed for that part. Only then every component will be within desired tolerance and capable of mating with any other mating component.

4.3.1. Selective Assembly. Today the consumer not only wants quality, precision and trouble free products but also he wants them at attractive prices. This has become possible only by adopting automatic gauging for selective assembly whereby parts manufactured to rather wide tolerances fit and function as though they were precisely manufactured in precision laboratory to very close tolerances. This is a concept which does away with old idea of inspection in which part is identified as 'good' or 'bad' ; good part being used for assembly and bad used to be scraped. In selective assembly the components produced by a machine are classified into several groups according to size. This is done both for hole and shaft and then the corresponding groups will match properly.

If some parts (shaft and holes) to be assembled are manufactured to normal tolerances of 0.01 mm (and both are within the curve of normal distribution), an automatic gauge can segregate them into ten different groups with a 0.001 mm limit for selective assembly of the individual parts.

Thus parts with tolerances of 0.0001 mm are obtained (due to segregation) and both the conditions of high quality and low cost can be served by selective assembly technique. However, it is very important that the two component parts to be fitted together must be kept within the normal distribution, *i.e.* the central or mean value should be at desired calculated value and the process capability of two machines producing shafts and holes must be identical otherwise for some components the mating components will not be available. If not so, then other techniques need to be adopted which are described below :

Fig. 4.7 shows a case in which the process capability of both shaft and hole producing machines is same but tolerances on parts are desired as one-tenth of process capability of machines. In such a case the parts are segregated by automatic inspection into ten groups and parts in shaft region S_1 are matched with parts in hole region H_1, S_2 with H_2 and so on. This results in matching of parts having tolerances 1/10th of machine capability. In this case as the process capability of both machines is same, equal number of parts are available in each segregated zone and no wastage will be there.

Fig. 4.8 shows another case in which the process capability of hole making machine is much wider than the tolerance of part but shaft making machine can produce component to the desired tolerance. In such a case the parts with hole are segregated into adequate number of groups depending on the desired tolerance.

$$\text{No. of groups} = \frac{\text{Process capability}}{\text{Tolerance desired}}$$

This curve when broken into several groups enables to determine the number of components likely to be produced in each group. The shafts are 'therefore' so produced that their setting needs to be changed as many number of times as there are groups formed for the holes and at the mean value

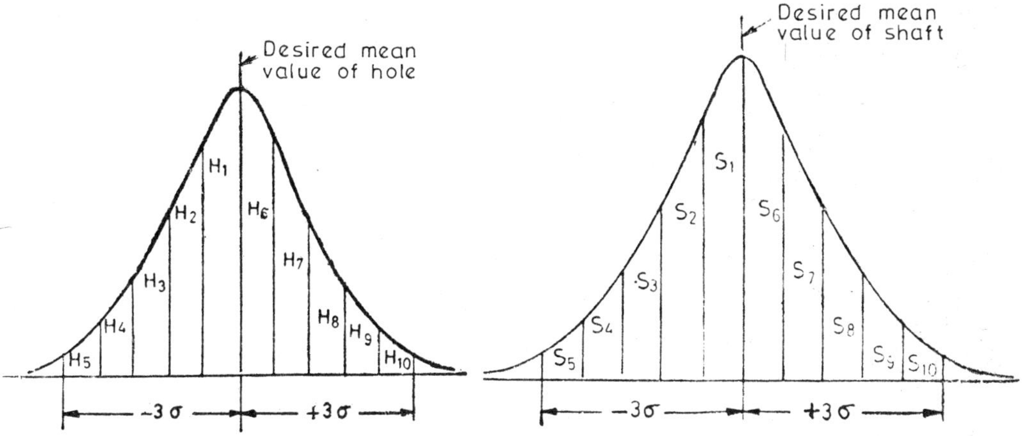

Process capability of hole making machine

Process capability of shaft making machine

Fig. 4.7

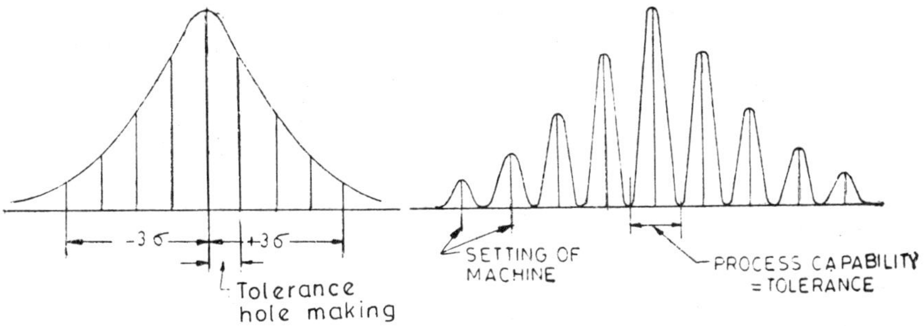

Shaft production (Area under curve dictates number of parts to be produced for each setting)

Fig. 4.8

of each sub-group. The number of shafts to be produced for each setting is determined by earlier curve for holes.

Another typical case can be when the process capability of both shaft producing and hole making machines are different and much wider than the permissible tolerances. In such a case the process having worst process capability is broken into adequate number of sub-groups as described before. The other process is then broken into adequate groups with their set points so adjusted that the number of components produced in this process corresponding to various groups made in earlier process will be same and as such no wastage will be there.

It will be noted that selective assembly technique can be followed for latter two cases also but is little complicated and the first case is more preferred one for selective assembly.

It would now be obvious that the selectively assembly is based on the natural distribution of components produced by a machine ; and proper means being available to ensure mean value and the dispersion of the process and the segregation of parts by automatic gauging. Thus machine produces parts of wider tolerance (cost being low) but for assembly, parts of one-tenth of tolerance of machine required for optimum performance are available. This is the best and cheapest method of assembling parts and is widely used in industry. Selective assembly is often followed in air craft, automobile industries and in ball and roller bearing unit as the tolerances desired in such industries are very narrow and not possible by any sophisticated machine at reasonable costs. The selective

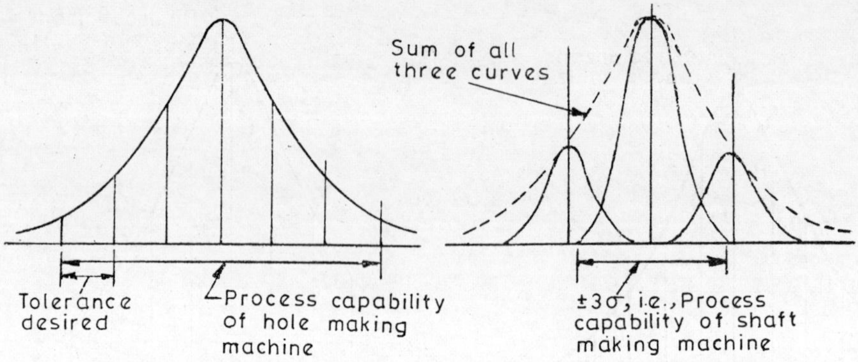

Fig. 4.9

assembly, however enables such tolerances to be arranged without their actually being produced as discussed above.

Limits of Size

There are three considerations in deciding the limits necessary for a particular dimension, *viz.* functional requirement (function of the component, what it is required to do), interchangeability (ease of replacement in the event of failure), and economics (minimisation of production time and cost). As already pointed earlier, the degree of tolerance thus calls for a compromise. To assist the designer in his choice of limits and fits, and to encourage uniformity throughout, a number of standards on limit and fit system have been published which must be strictly followed.

A limit system consists of a series of tolerances arranged to suit a specific range of sizes and functions, so that limits of size may be selected and given to mating components to ensure specific classes of fit.

4.4. Indian Standard (IS 919—1993) (Part I & Part II)

In India, we are following Indian Standard (IS 919—1993) for system of limits and fits. The system comprises suitable combinations of 20 grades of fundamental tolerances or in other words grades of standard tolerances or in other words grades of accuracy of manufacture ; and 28 types of standard deviations indicated by letter symbols for both holes and shafts (capital letters *A* to *ZC* for holes and small letters *a* to *zc* for shafts) in diameter steps upto 500 mm. The various 28 designations are represented by *A, B, C, CD, D, E, EF, F, FG, G, H, J, JS, K, M, N, P, R, S, T, U, V, X, Y, Z, ZA, ZB, ZC*. It may be noted that to avoid confusion, the following letters are not used : *I, L, O, Q, W*.

Innumerable fits ranging from extreme interference to those of extreme clearance may be obtained by a suitable combination of fundamental tolerances and fundamental deviations. All but very exceptional engineering requirements are covered for very coarse work to fine gauge manufacture.

A unilateral hole basis is recommended by BIS (Bureau of Indian Standards) but a full range of holes suitable for use on shaft basis (unilateral or bilateral) is also included. For the convenience, a basic size is ascribed to the part. The two maximum and minimum limits are defined by its deviation from the basic size.

Definitions. In the standard followed, certain conventions are used. In order to illustrate the definitions, simplified schematic diagram (Fig. 4.10) is conventionally used. Let us draw a shaft and hole system. It is assumed that shaft and hole touch at the bottom-most points and then complete tolerances will be considered on top portion. A zero line is defined corresponding to basic size, and all tolerances, etc. are considered with reference to this zero line. In the system of limits and fits we are simply interested in the tolerance on shafts and holes and not in their sizes

Therefore, for the conventional diagram the axis of shaft and hole is maintained very much below the diagram (Refer to the right-hand portion of Fig. 4.10) and we concentrate only on tolerances with reference to zero line as shown on extreme right side of Fig. 4.10. This representation should be clearly understood to avoid any confusion.

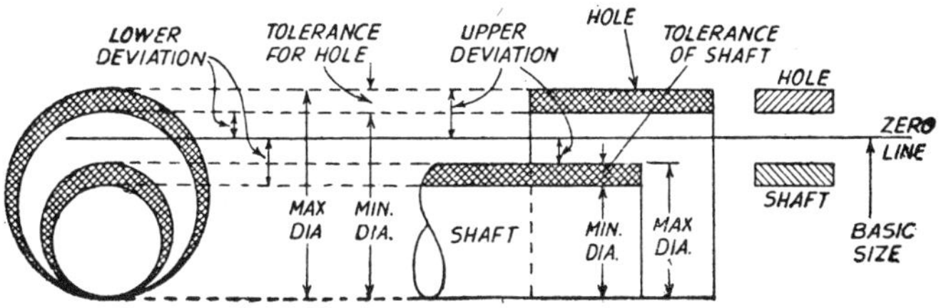

Fig. 4.10. A shaft and a hole system.

In Fig. 4.10, the tolerances etc., are shown too much exaggerated. It is assumed that the shaft touches at the bottom of hole and then other dimensions are considered. A line is assumed to be there in between the upper tolerance on shaft and lower tolerance on hole. This line is called zero line. This line refers to basic size from the axis of the shaft or hole. All tolerances and deviations are referred to from the datum line.

Thus it is very obvious from Fig. 4.10, that the setting of tolerance value alone is not sufficient to define particular limits, but the position of the tolerance zone relative to the basic size of the feature must also be specified.

Before we refer to definitions in connection with Fig. 4.10, let us define a few basic terms.

Nominal size. The nominal size of a dimension or part is the size by which it is referred to as a matter of convenience.

Basic size. It is got to be ascribed for the sake of convenience and is the size of a part in relation to which all limits of variation are determined. Often, basic size and nominal size of a part or dimension are used with the same sense. When used for fits between shaft and hole, same basic size is specified or both shaft and hole.

Zero Line. This is a line which represents the basic size so that the deviation from the basic size is zero. Fig. 4.11 illustrates the concept of basic size and zero line.

Actual size. This is the measured size.

Limits. These are two extreme permissible sizes for any dimension (high and low).

Tolerances. Tolerance on a dimension is the difference between the high and low limits of size. It is got to be allowed in order to cover the reasonable imperfection in workmanship and the inevitable inaccuracy of manufacturing processes, and varies with different grades of work. It can be unilateral or bilateral.

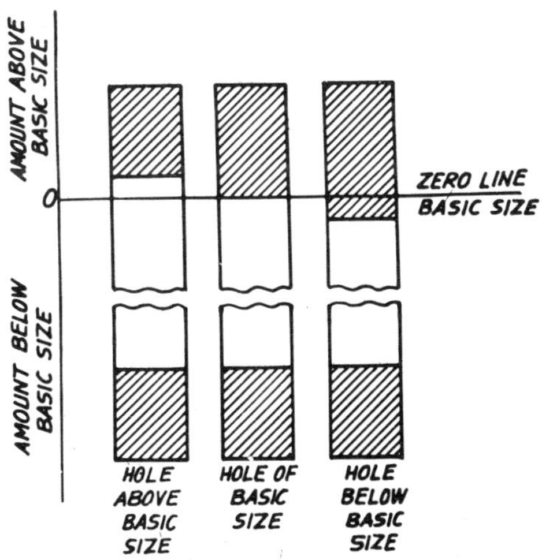

Fig. 4.11. Basic size and zero line.

Fig. 4.12 illustrates the concept of limits of size and tolerances.

Allowances. An international difference between the hole dimension and shaft dimension for any type of fit is called the allowance. Maximum allowance is obtained by subtracting the minimum shaft size from the largest hole size and the minimum allowance is the difference between the largest shaft and the smallest hole size. Thus allowance is positive for clearance fit and negative for interference fit.

Fits. When two parts are to be assembled, the relation resulting from the difference between their sizes before assembly is called a fit. Depending upon the actual limits of hole or shaft, the fit may be clearance fit, or a transition fit, or an interference fit.

In connection with Fig. 4.12 following may be noted :

Tolerance is considered on one side only. Actually it is representative of total tolerance. It is done for sake of clarity and simplification. In schematic representation of tolerances, no regard is given to show shafts and holes fully, but only their magnitude and relative position w.r.t. basic size are highlighted.

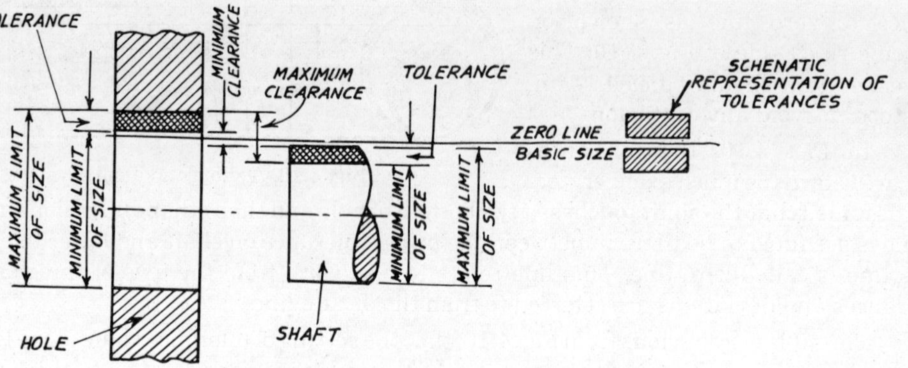

Fig. 4.12. Limits of size and tolerance.

(a) *Clearance fit.* In this type of fit, the largest permitted shaft diameter is smaller than the diameter of the smallest hole, so that the shaft can rotate or slide though with different degrees of freedom according to the purpose of the mating members.

(b) *Interference fit.* In this type of fit, the minimum permitted diameter of the shaft is larger than the maximum allowable diameter of the hole. In this case the shaft and the hole members are intended to be attached permanently and used as a solid component but according to the application of this combination, this type of fit can be varied. Thus, if during use, one of the two members is subjected to wear, it should be possible to drive or force the two members apart for replacement purposes. Examples of this type of fit are bearing bushes which are in an interference fit in their housing *e.g.* a small end in the connecting rod of an engine.

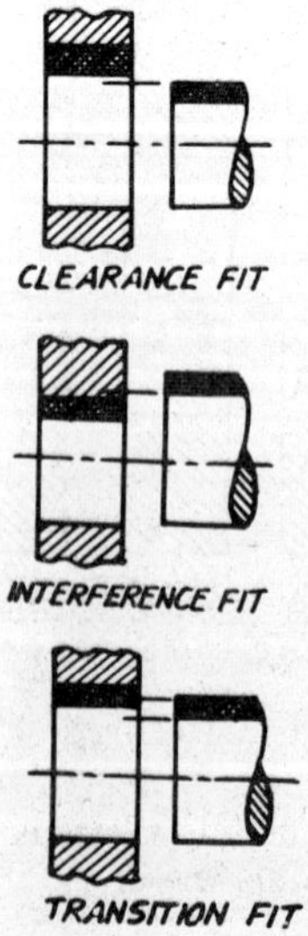

Fig. 4.13. Types of fit.

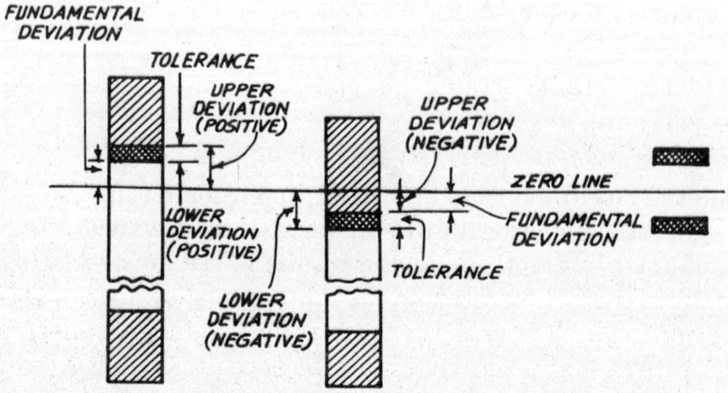

Fig. 4.14. Upper deviation, lower deviation and fundamental deviation for hole.

(c) *Transition fit.* In a fit of this type the diameter of the largest allowable hole is greater than that of the smallest shaft, but the smallest hole is smaller than the largest shaft, so that small positive or negative clearance between the shaft and hole members are employable. Location fits *e.g.*, spigot in mating holes, coupling rings and recesses are the examples of transition fit.

(Figure on right position in Fig. 4.13 shows schematic representation of tolerances).

Deviation is defined as the algebraic difference between a size (actual, maximum etc.) and the corresponding basic size.

Upper deviation is the algebraic difference between the maximum limit of size (of either hole or shaft) and the corresponding basic size. It is designated by letters ES for hole and es for shaft. It is a positive quantity when the maximum limit of size is less than the basic size and a negative quantity when the maximum limit of size is less than the basic size (Refer Figs. 4.14 and 4.15).

Lower deviation is the algebraic difference between maximum limit of size and the corresponding basic size. It is a positive quantity when the minimum limit of size is greater than the basic size and a negative quantity when the minimum limit of size is less than the basic size. It is designated by EI for a hole and ei for a shaft.

Fundamental deviation is that one of the two deviations which is conventionally chosen to define the position of the tolerance zone in relation to the zero line (Refer Fig. 4.16). This may be upper or lower deviation which is closest to the zero line (Refer Figs. 4.14 and 4.15). It fixes the position of zero line.

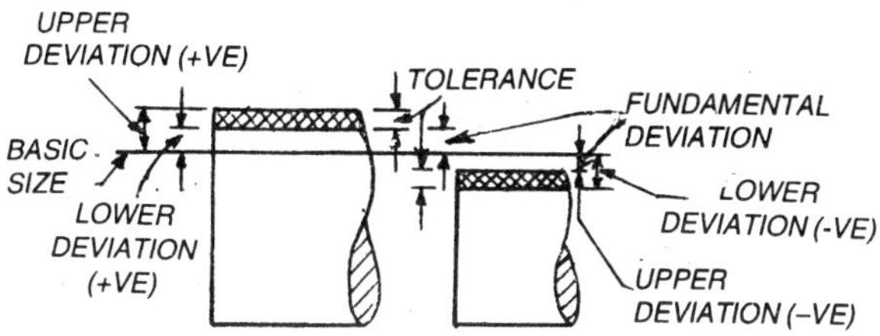

Fig. 4.15. Upper Deviation, Lower Deviation and Fundamental Deviation for shaft.

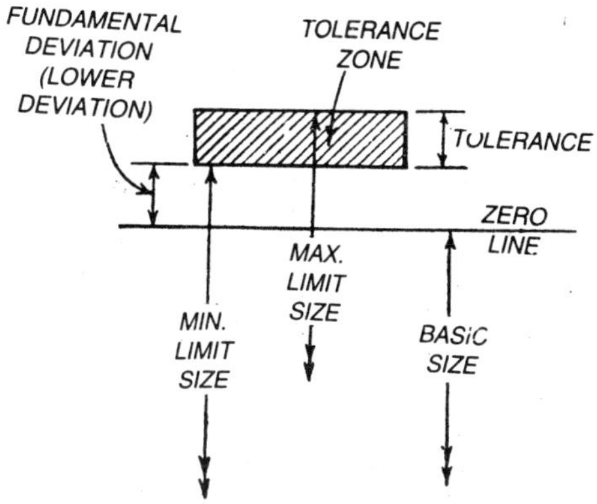

Fig. 4.16. Fundamental deviation.

Maximum Clearance. In a clearance fit, it refers to the difference between the minimum size of hole and the maximum size of shaft. (Refer Fig. 4.17).

Maximum Clearance. In the case of clearance or transition fit, it refers to the difference between the maximum size of hole and the minimum size of shaft. (Refer Fig. 4.17).

Minimum Interference. It is the difference between the maximum size of hole and the minimum size of shaft in an interference fit prior to assembly. (Refer Fig. 4.18).

Maximum Interference. In an interference or a transition fit it is the difference between the minimum size of hole and the maximum size of shaft prior to assembly. (Refer Fig. 4.18).

Allowance. The difference between the maximum shaft size and minimum hole is known as allowance. In a clearance fit, this is the minimum clearance and is *positive* allowance. In an interference fit, it is the maximum interference and is *negative* allowance. Positive and negative allowances are shown in Figs. 4.17 and 4.18 respectively.

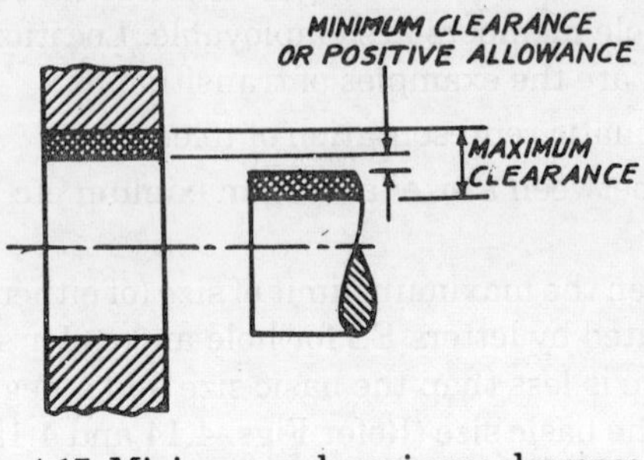

Fig. 4.17. Minimum and maximum clearance.

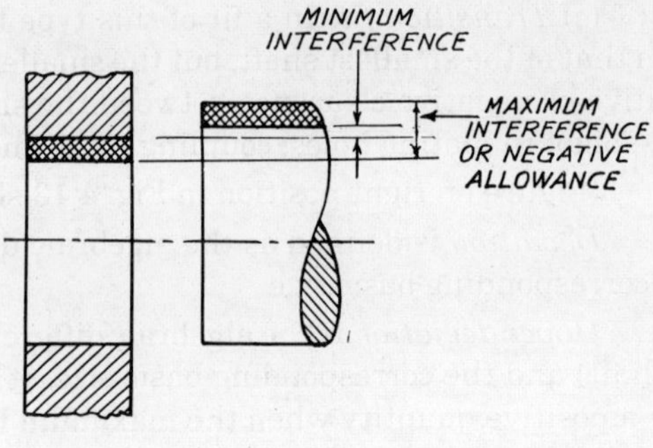

Fig. 4.18. Minimum and Maximum interference.

Shaft—This term is used to describe an external feature of a workpiece (including features which are not cylindrical).

Basic shaft is a shaft whose upper deviation is zero, *e.g.*, shaft '*h*'.

Hole—This term is used to describe an internal feature of a workpiece (including features which are not cylindrical).

Basic hole is one whose lower deviation is zero, *e.g.*, hole '*H*'.

For shafts '*a*' to '*h*', the deviation is below the zero line and for shafts '*j*' to '*zc*' it is above the zero line. For holes '*A*' to '*G*', lower deviation is above the zero line and for '*J*' to '*ZC*', it is below the zero line. Upper deviation for shafts is denoted by *es* and lower deviation by *ei*. For holes the corresponding deviations are denoted by *ES* and *EI* respectively. In the specifications, formulae are given to determine the fundamental deviation. For example, for shafts, the fundamental deviation (upper deviation *es* or lower deviation *ei*) is determined by means of the formulae given in the table 4.2. The other deviations may be derived directly using the absolute value of the tolerance *IT* by means of the algebraic relationship.

$$ei = es - IT \text{ and } es = ei + IT.$$

IT stands for "International Tolerance" grade.

The deviations given in the table 4.2 are those corresponding in principle to the limit close to the zero line, in other words, the upper deviation *es* for shafts *a* to *h* and lower deviation *ei* for shafts *j* to *zc*.

For holes also the deviations are derived from those of the corresponding shafts as follows. The general rule is that hole limits are identical with the shaft limits of the same symbol (letter and grade) but disposed on the other side of the zero line, *i.e.*, *EI* upper deviation of hole = *es* of the shaft of same letter symbol but of opposite sign.

Hole Basis System and Shaft Basis System are defined in Fig. 4.19 (*a*) and (*b*) for clearance fit, transition fit and interference fit. In hole basis system, different clearances and interferences

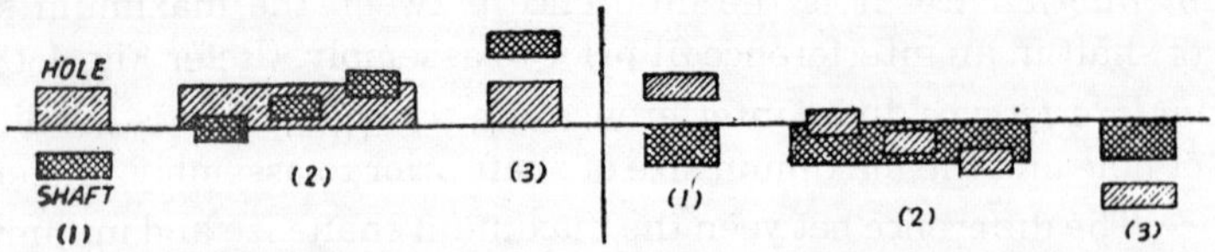

Hole Basis System
(*a*) Lower deviation (fundamental deviation) of hole is zero. (Hole H)

Shaft Basis System
(*b*) Upper deviation (fundamental deviation) of shaft is zero. (shaft *h*)

(1) Clearance fit (2) Transition fit (3) Interference fit
Fig. 4.19

are obtained by associating various shafts with a single hole whose lower deviation is zero (*H* hole). This is standard practice also as it is very convenient to make correct holes of fixed sizes. Since holes are produced by drilling, reaming, etc. their size is not easily adjustable because size of such tools are standard. On the other hand, size of shaft produced by turning and grinding can be easily varied. In shaft basis system upper deviation of shaft is zero, and different fits are obtained by varying the limits on the holes.

In I.S. specification, 20 grades of tolerances are designated : *IT*0, *IT*01, *IT1* to *IT*18. *IT*0 and *IT*01 are not in general use. These are known as standard tolerances and their numerical values have been determined in terms of the standard tolerance factor i, where i in microns is expressed as i (microns) = $0.45 \sqrt[3]{D} + 0.001D$.

(D is the geometric mean of the lower and upper diameters of a particular diameter step in which the diameter lies, D is in mm).

This formula has been empirically calculated on the basis of the former national standards of other countries and taking account of the fact that in most usual cases the tolerance varies more or less parabolically in terms of diameter for the same manufacturing conditions. This is so because of the difficulties encountered during manufacture and measuring of higher size diameters. The relative magnitude of each grade is given below for basic size upto and including 500 mm.

Table 4.1 (A)

Grades	*IT5*	*IT6*	*IT7*	*IT8*	*IT9*	*IT10*	*IT11*	*IT12*	*IT13*	*IT14*	*IT15*	*IT16*	*IT17*	*IT18*
Values	$7i$	$10i$	$16i$	$25i$	$40i$	$64i$	$100i$	$160i$	$250i$	$400i$	$640i$	$1000i$	$1600i$	$2500i$

The relative magnitudes of eighteen grades for sizes above 500 mm and upto 3150 mm are given below in Table 4.1 (B).

Table 4.1 (B)

IT1	*IT2*	*IT3*	*IT4*	*IT5*	*IT6*	*IT7*	*IT8*	*IT9*	*IT10*	*IT11*	*IT12*	*IT13*	*IT14*	*IT15*	*IT16*	*IT17*	*IT18*
$2I$	$2.7I$	$3.7I$	$5I$	$7I$	$10I$	$16I$	$25I$	$40I$	$64I$	$100I$	$160I$	$250I$	$400I$	$640I$	$1000I$	$1600I$	$2500I$

The two effects influencing the magnitude of tolerance are (*a*) size and (*b*) quality. The size effect is accommodated by the fundamental tolerance factor $i = 0.45 \sqrt[3]{D} + 0.001D$.

The standard tolerance factor I, in micrometer, is calculated as $I = 0.004D + 2.1$

The idea of various grades of tolerance can be had from the applications stated below for each :

Tolerance Grade	*Class of Work*
01, 0, 1	Gauge blocks.
2	High quality gauges, plug gauges.
3	Good quality gauges, Gap gauges.
4	Gauges. Precise fit produced by lapping.
5	Ball bearings. Machine lapping. Fine boring and grinding.
6	Grinding. Fine boring.
7	High quality turning, broaching, boring.
8	Centre-lathe turning and boring. Reaming. Capstan lathes in good condition.
9	Capstan or automatic lathes. Boring machines.

10 Milling, slotting, planing, rolling, extrusion.
11 Drilling, rough turning and boring. Precision tube drawing.
12 Light press work. Tube drawing.
13 Press work. Tube rolling.
14 Stamping.
15 Sand casting. Flame cutting.

It may be noted that the value of tolerance grade is not calculated on the basis of actual size but is taken as geometric means of upper and lower value of the range in which it falls.

The various diameter main steps specified by I.S.I. are 1—3, 3—6, 6—10, 10—18, 18—30, 30—50, 50—80, 80—120, 120—180, 180—250, 250—315, 315—400 and 400—500, 500—630, 630—800, 800—1000, 1000—1250, 1250—1600, 1600—2000, 2000—2500, 2500—3150 mm) D being taken as the average size (geometric mean) for a particular range of size to avoid a continuous variation of tolerance with size. In certain cases, intermediate steps between the above main stems are also used. Again the basic relationship between the tolerance and size is seen to be of the form $T = aD^n$; the additional term $0.001D$ being present to accommodate the uncertainties of measuring with increase in size, these being mainly due to difference in temperature and deformation of gauges and work which become noticeable on dimensions over about 75 mm.

The quality effect is accommodated by taking a multiple of fundamental tolerance units appropriate to the quality and consistant with the proposed or available method of production.

A fit is indicated by its basic size followed by symbols representing the limits of each of its two components, the hole being quoted first. For example, 25 H_8d_9. Its significance will be clear from the example given below.

Example 4.1. *Calculate the limits of tolerance and allowance for a 25 mm shaft and hcle pair designated* H_8d_9.

Solution. First of all find out, in which range of measurements 25 mm lies, from standard diameter steps specified. It is seen that 25 mm lies in the diameter steps 18 and 30 mm. Therefore, value of $D = \sqrt{18 \times 30} = 23.2$ mm. The value of fundamental tolerance factor

$$i = 0.45\sqrt[3]{D} + 0.001D = 0.45\sqrt[3]{23.2} + 0.023$$
$$= 0.45(2.850) + 0.023 = 1.285 + 0.023 = 1.308 = 1.3 \text{ microns.}$$

For hole quality 8, from table 4.1 the fundamental tolerance

$$= 25i = 25 \times 1.3 = 32.5 = 33 \text{ microns.}$$

For the 'H' hole, the fundamental deviation = 0.

Hence the hole limits are 25.000 mm, and 25 + 0.033 mm = 25.033 mm.

Hole tolerance = 25.033 – 25.000 = 0.033 mm.

For the quality '9' shaft, the fundamental tolerance

$$= 40i = 40 \times 1.3 = 52 \text{ microns.}$$

For 'd' shaft the fundamental deviation from the tables

$$= -16\,D^{0.44} = -16 \times (23.2)^{0.44} = -65 \text{ microns.}$$

∴ The shaft limits are 25.000 – 0.065 = 24.935 mm

and 25.000 – (0.065 + 0.052) = 24.883 mm.

Tolerance = 24.935 – 24.883 mm = 0.052 mm.

The disposition of these tolerances and deviations is shown in Fig. 4.20.

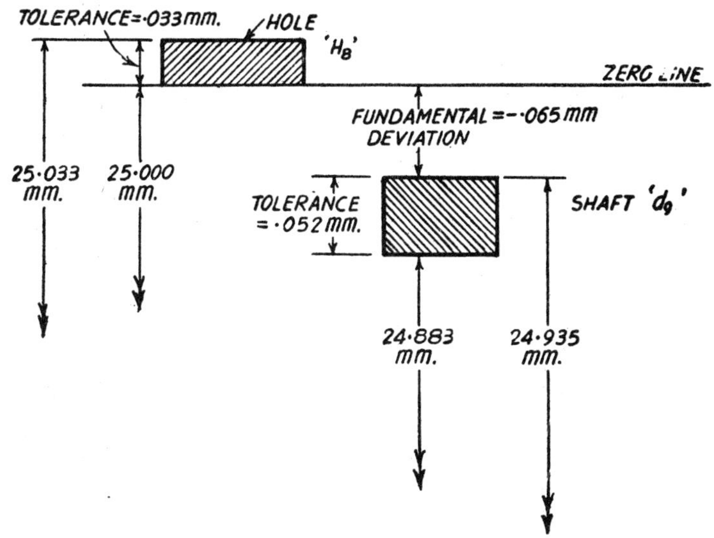

Fig. 4.20. Disposition of tolerances and deviations.

4.5. Terminology

The terminology as per IS : 919 (Part I) – 1993 is given below :

Tolerance of Parts. Due to the inevitable inaccuracy of manufacturing methods, it is not possible to make a part precisely to a given dimension and may only be made to lie between two limits, maximum and minimum, the difference between which is the permissible tolerance.

For the sake of convenience a basic size is ascribed to the part and each of the two limits is defined by its deviation from the size (the magnitude and sign of the deviation is obtained by subtracting the basic size from the limit in question).

Fits. When two parts are to be assembled, the relation resulting from the difference between their size before assembly is called fit.

Symbols for tolerances and deviations, and symbols for fits. In order to satisfy the usual requirements both for individual part and/or fits, the system provides for any given basic size, a range of tolerances together with a range of deviations defining the position of these tolerances with respect to the line of zero deviation—called the zero line.

The tolerance, the value of which is a function of the basic size is designated by a number symbol, called the grade.

The position of tolerance zone with respect to the zero line, which is a function of the basic size, is indicated by letter symbol (in some cases, two letters), a capital letter for holes and a small letter for shafts.

The tolerance size is thus defined by its basic value followed by a symbol composed of a letter (in some cases, two letters) and a numeral, *e.g.*, 50*g* 7.

A fit is indicated by the basic size common to both components, followed by symbols corresponding to each of the components, the hole being quoted first, *e.g.*, 50 *H* 8/*g* 7.

Here 50 is the basic size of shaft and hole, *H* is the position of tolerance zone w.r.t. zero line for hole, 8 is the tolerance size of the hole, *g* is the position of tolerance zone w.r.t. zero line for shaft, and 7 is the tolerance zone for shaft.

Size. A number expressing, in a particular unit, the numerical value of a length or of a linear dimension.

Actual Size. The size of a part as may be found by measurement.

Limits of Size. The two extreme permissible sizes between which the actual size is contained.

Maximum Limit of Size. The greater of the two is called the maximum limit.

Minimum Limit of Size. The smaller one of the two limits of size is called the minimum limit.

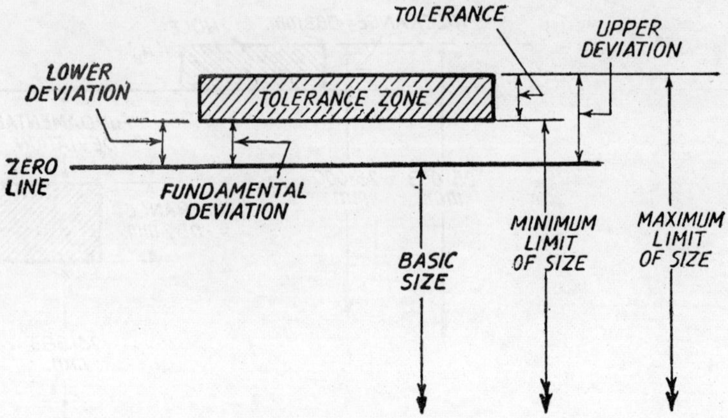

Fig. 4.21

Basic Size. (Normal size). The size with reference to which the limits of size are fixed. The limits of size are derived by the application of upper and lower deviations.

Deviation. The algebraic difference between a size (actual, maximum etc.) and the corresponding basic size. Symbols for shaft deviations are lower case letters (*es*, *ei*) and symbols for hole deviations are upper case letters (*ES*, *EI*).

Limit Deviations. Upper deviation and lower deviation.

Actual Deviation. The algebraic difference between the actual size and the corresponding basic size.

Upper Deviation. The algebraic difference between the maximum limit of size and the corresponding basic size. It is a positive quantity when the maximum limit of size is greater than the basic size and a negative quantity when the maximum limit of size is less than the basic size. It is designated by *ES* for a hole and *es* for a shaft.

Lower Deviation. The algebraic difference between the minimum limit of size and the corresponding basic size. It is a positive quantity when the minimum limit of size is greater than the basic size and a negative quantity when the minimum limit of size is less than the basic size. It is designated by *EI* for a hole and *ei* for a shaft.

Fundamental Deviation. It is that one of the two deviations which is conventionally chosen to define the position of the tolerance zone in relation to the zero line. It is the one nearest the zero line.

Zero Line. In a graphical representation of limits and fits, a straight line to which the deviations are referred. The zero line is the line of zero deviation and represents the basic size. By convention, when the zero line is drawn horizontally, positive deviations are shown above and the negative deviations below.

Tolerance is equal to the algebraic difference between the upper and lower deviations and has an absolute value without sign. In the context of this terminology for limits and fits, the difference between the maximum limit of size and minimum limit of size is called the tolerance.

For any basic size there are 28 different holes. These are obtained by providing a series of holes which are progressively oversize and a series of holes which are progressively undersize. The difference from basic size of the various holes is given by the fundamental deviation, and it is these differences in size which give the fit required. The 28 holes are designated by the Capital letters : *A, B, CD, C, D, E, EF, F, FG, G, H, JS, J, K, M, N, P, R, S, T, U, V, X, Y, Z, ZA, ZB, ZC.*

Limit System. A system of standardised tolerances and deviations.

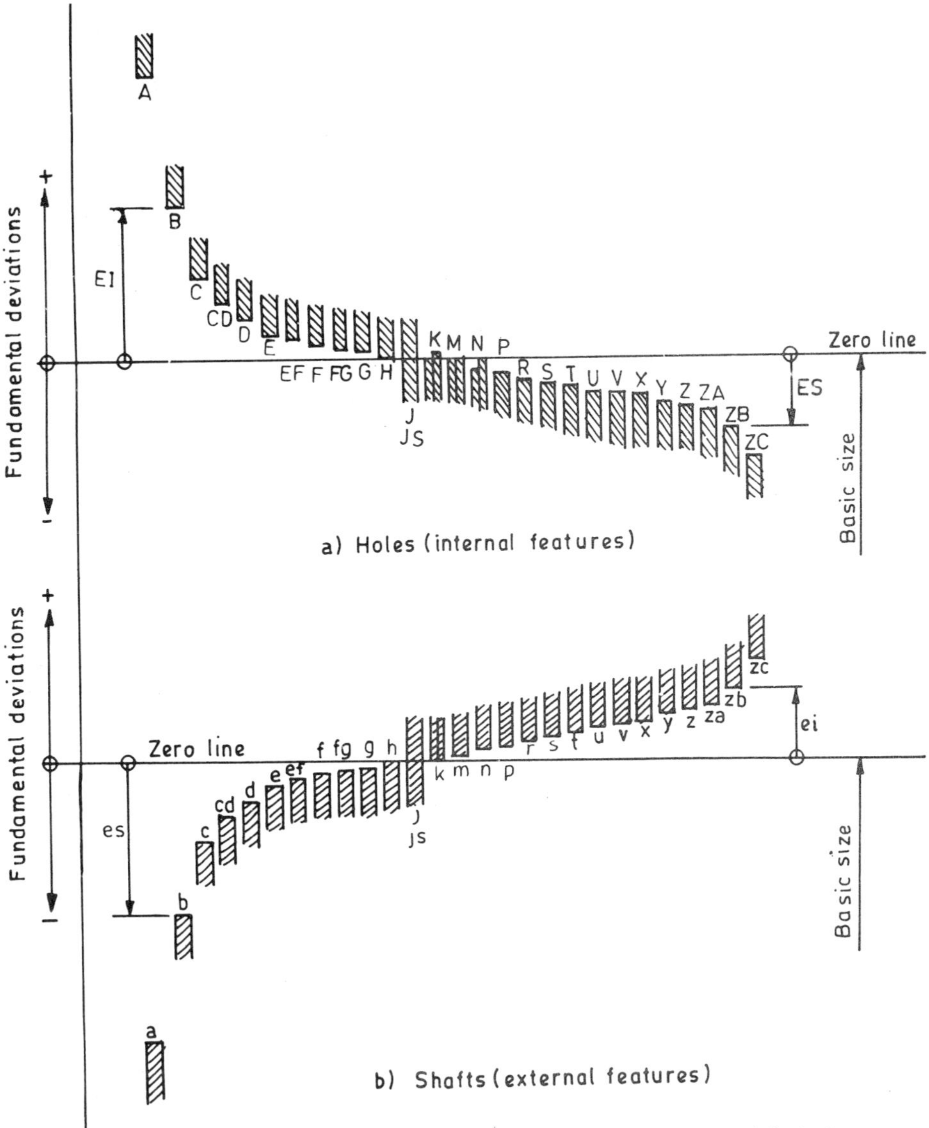

Fig. 4.22. Schematic representation of the position of fundamental deviations.

Each of the 28 holes has a choice of 20 tolerances which are designated : *IT*01, *IT*0, *IT*1, *IT*2 up to and including *IT*18. The tolerance grade decides the accuracy of manufacture. The seven finest grades (*IT*01 to *IT*05) cover sizes up to 500 mm and the eleven coarsest grades upto 3150 mm. The tolerance in each grade depends on the size of shaft/hole.

Similarly for shafts, for any given size there are 25 different shafts designated by small letters from *a* to *zc*. Also each shaft has 20 grades of tolerance grades which are designated as for the holes.

The setting of tolerance values is not by itself sufficient to define particular limit, the position of the tolerance zone relative to the basic size of the feature must also be specified. This is done by establishing fundamental deviations which are differences between the basic size and the nearest limit of tolerance. These fundamental deviations are obtained from empirical formulae given in Table 4.2. These are designated by capital letters for holes and small letters for shafts.

The fundamental deviations for holes A to H correspond exactly in value with those for shafts a to h but are in opposite directions. Hole A and shaft a have the largest fundamental deviations, hole being positive and shaft being negative, and the fundamental deviations for both H and h are zero. Thus the first eight designations represent a clearance fit system. The remaining groups JS to ZC (holes) and js and zc (shafts) do not correspond in their deviations in quite the same way, they are intended for use in interference and transition fits. The above facts are valid irrespective of any basic size.

Fig. 4.23 shows the deviations details for various shafts and holes.

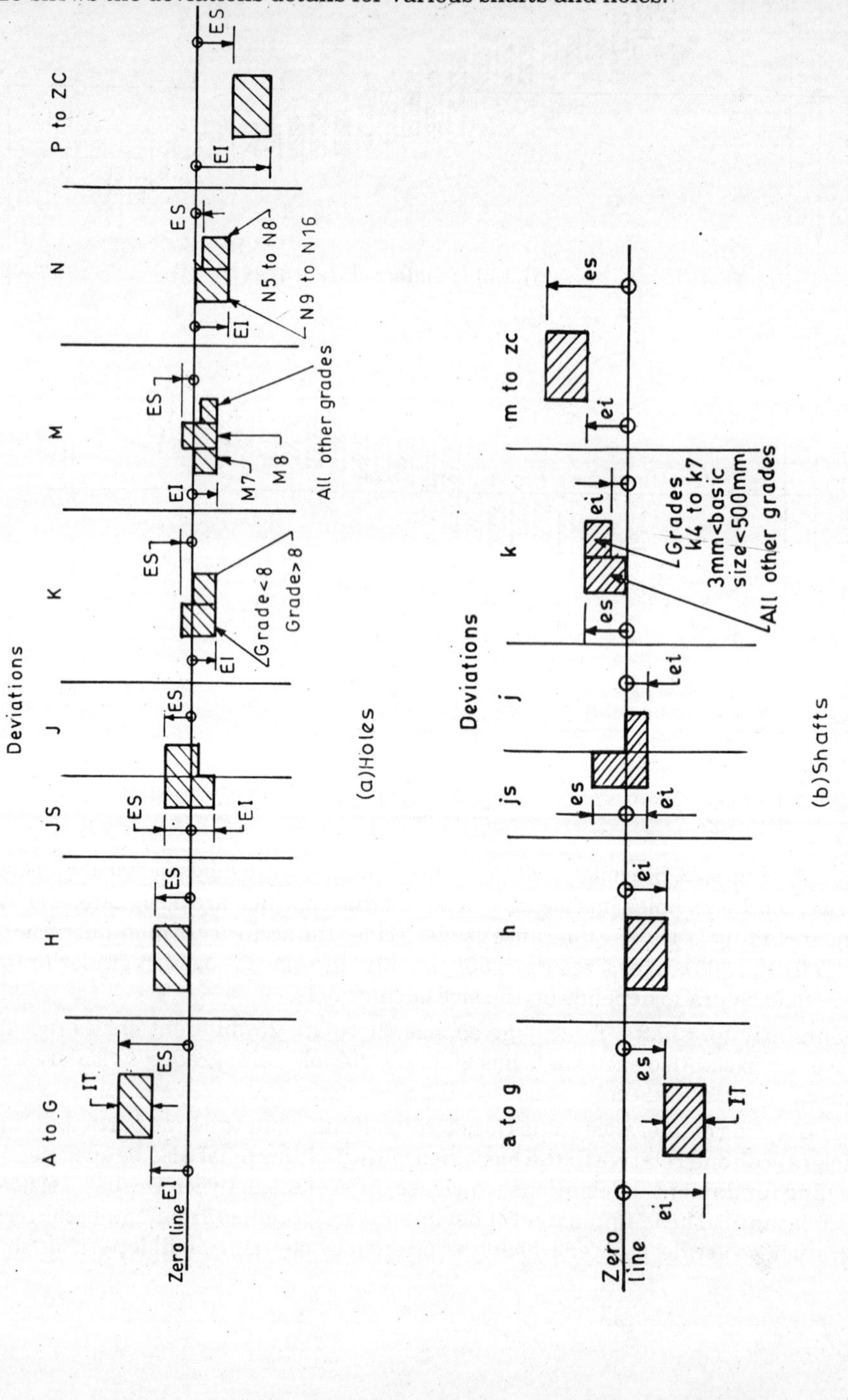

Fig. 4.23. Deviations for shafts and holes.
It may be noted that fundamental deviations for J/j, K/k, M/m and N/n are different than conventions for others.

The values of the fundamental deviations are functions not of the basic size but of the range of sizes in which the basic size falls.

Standard tolerance grades. A group of tolerance (*e.g. IT*6), considered as corresponding to the same level of accuracy for all basic sizes.

Tolerance Zone. In a graphical representation of tolerance, the zone bounded by the two limits of size of the part is called the tolerance zone. It is defined by its magnitude (*i.e.* tolerance) and by its position in relation to the zero line.

Grades of Tolerance. In a standardised system of limits and fits, group of tolerances are considered as corresponding to the same level of accuracy for all basic sizes. It designed by the letters IT followed by a number, *e.g.*, ITO1 ... 01 to 16.

Size Tolerance. The difference between the maximum limit of size and the minimum limit of size, i.e. the difference between the upper deviation and the lower deviation.

Standard Tolerance (IT). It is a tolerance belonging to any one of the standard grades.

Standard Tolerance factor (i, I). A factor which is a function of the basic size and which is common to the formulae defining the different grades of tolerances.

Shaft. A term used by convention to designate all external features of a part, including those which are not cylindrical.

Hole. A term used by convention to designate all internal features of a part, including those which are not cylindrical.

Basic Shaft. A shaft whose upper deviation is zero or where maximum limit of size is equal to basic size. It is *h*-shaft IS : 919.

Basic Hole. A hole whose lower deviation is zero or whose minimum limit of size is equal to basic size. It is *H* hole in IS : 919.

'Go' Limit. It refers to upper limit of a shaft and lower limit of a hole. Thus it corresponds to the maximum material condition.

'Not Go' Limit. It refers to the lower limit of a shaft and upper limit of a hole. Thus it corresponds to the minimum material condition.

Fit. The relationship existing between two parts, shaft and hole, which are to be assembled, with respect to the difference in their sizes before assembly is called fit.

Basic size of a Fit. It is that basic size which is common to the two parts of a fit.

Variation of Fit. This is the arithmetical sum of the tolerances of the two mating parts of fit.

Clearance. This is difference between the sizes of the hole and shaft, before assembly when this difference is positive.

Interference. This is the arithmetical difference between the sizes of the hole and shaft before assembly, when the difference is negative.

Minimum and Maximum Clearance. In a clearance fit, the maximum clearance is the difference between the minimum size of the hole and maximum size of the shaft.

In a transition or clearance fit, the maximum clearance is the difference between the maximum size of the hole and the minimum size of the shaft.

Minimum and Maximum Interference. In an interference fit, the minimum interference is the arithmetical difference between the maximum size of the hole and minimum size of the shaft before assembly.

In an interference or transition fit the maximum interference is the arithmetical difference between the minimum size of the hole and the maximum size of the shaft before assembly.

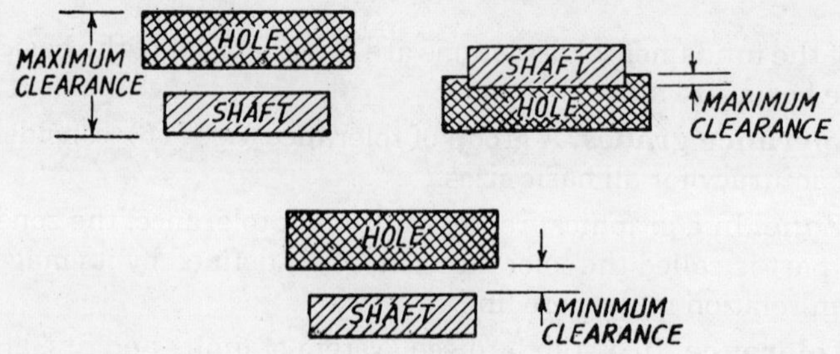

Showing disposition of maximum clearance and minimum clearance.
Fig. 4.24 (*a*)

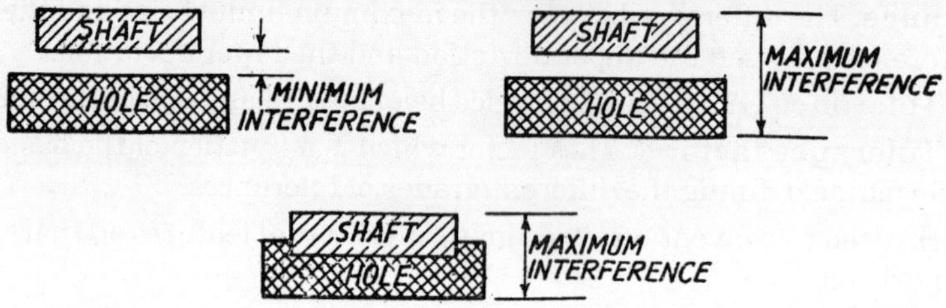

Showing disposition of maximum interference and minimum interference.
Fig. 4.24 (*b*)

Maximum Material Limit (MML). The designation applied to that of the two limits which corresponds to the maximum material size for the feature, i.e. upper limit of size for shaft, or the lower limit of size for a whole.

Least Material Limit (LML). The designation applied to that of the two limits which correspond to the minimum material size of the feature, i.e. the lower limit of size for an external feature (shaft), or the upper limit of size for an internal feature (hole).

Maximum and Minimum Metal Conditions. Maximum metal condition (MMC) corresponds to condition when a part has maximum amount of metal, *i.e.* corresponding to high tolerance of shaft and low tolerance of the hole. Similarly minimum metal conditions correspond to minimum size of shaft and maximum size of hole. MMC has special importance with regard to geometrical tolerancing as it critically affects the interchangeability of manufactured parts which are to be assembled together.

4.6. Tolerance and Deviations for Sizes Upto 500 mm

Formulae for Tolerances and Deviations. There are twenty grades of tolerances provided designations *IT*01, *IT*0, *IT*1, *IT*2, ... *IT*18. These are known as standard tolerances. For grades 5 to 16, the values of the tolerances are determined in terms of the standard tolerance factor 'i', where i (microns) = 045 $\sqrt[3]{D} + 0.001D$, (D = mean diameter in mm). For grade *IT*6, the value is 10 i and above it the tolerance magnitude is multiplied by 10 at each fifth step.

The values of standard tolerance corresponding to grades *IT*01, *IT*0, *IT*1 are :

Value in microns for *D* in millimetres	*IT*01	*IT*0	IT1
	0.3 + 0.008*D*	0.5 + 0.012*D*	0.8 + 0.020*D*

The values of *IT*2 to *IT*4 are regularly scaled approximately geometrically between the values *IT*1 and *IT*5. The values of *IT*5 to *IT*18 have already been given in Table 4.1 (A). From IT6 upwards, the standard tolerances are multiplied by a factor of 10 at each fifth step.

4.6.1. Fundamental Deviations of Shafts. For each letter symbol, the magnitude and sign for one of the two deviations, which is known as fundamental deviation (upper deviation *es*, or lower deviation *ei*) are determined by means of formulae given in the table 4.2. The other deviations can be derived directly using the absolute value of the standard tolerance *IT* by means of the following relationship :

$$ei = es - IT \text{ (for shafts with deviations } a \text{ to } h)$$
$$es = ei + IT \text{ (for shafts with deviation } k \text{ to } ze)$$

Table 4.2. Fromulae for Fundamental Deviations for Shafts for sizes upto 500 mm

Upper Deviation (es)		*Lower Deviation (ei)*	
Shaft Designation	*In microns (for D in mm)*	*Shaft Designation*	*In microns (for D in mm)*
a	$= -(265 + 1.3D)$ for $D \le 120$ and $= -3.5D$ for $D > 120$	j	No formula
b	$= -(140 + 0.85D)$ for $D \le 160$; $= -1.8D$ for $D > 160$	js	$ITI/2$
c	$= -52D^{0.2}$ for $D \le 40$; $= -(95 + 0.8D)$ for $D > 40$	$k4$ to $k7$	$= +0.6\sqrt[3]{D}$
cd	G.M. of values for c and d	k for grade ≤ 3 and ≥ 7	
d	$= -16D^{0.44}$	m	$= +(IT7 - IT6)$
e	$= -11D^{0.41}$	n	$= +5D^{0.34}$
ef	G.M. of values for e and f	p	$= +IT7 + 0$ to 5
f	$= -5.5D^{0.41}$	r	= geometric mean of values for p and s
fg	G.M. of values for f and g	s	$= IT8 + 1$ to 4 for $D \le 50$; $= +IT7$ to $+0.4D$ for $D > 50$
g	$= -2.5D^{0.34}$	t	$= +IT7 + 0.63D$
h	$= 0$	u	$= +IT7 + D$
		v	$= +IT7 + 1.25D$
		x	$= +IT7 + 1.6D$
		y	$= +IT7 + 2D$
		z	$= +IT7 + 2.5D$
		za	$= IT8 + 3 + 3.15D$
		zb	$= +IT9 + 4D$
		zc	$= +IT10 + 5D$

Same formulae also applicable for corresponding fundamental deviations for holes also. (Refer Table 4.2 A).

The ranges of shafts *cd*, *ef* and *fg* are only in the range of 0—10 mm.

Formulae for fundamental shaft deviations. Fundamental shaft deviation is upper deviation *es* for shaft *a* to *h* and lower deviation *ei* for shafts *j* to *zc*, as fundamental deviation is one closest to the zero line.

It will be observed from the table 4.2 that the value of the deviations is independent of the grade (even if the formulae include a term *IT*), except in the case of shafts *j* and *js*, for which strictly speaking there is no fundamental deviation.

4.6.2. Fundamental deviations for Holes. Deviations for holes are derived from those of the corresponding shafts.

The general rule is that hole limits are identical with the shaft limits of the same symbol (letter and grade) but disposed on the other side of the zero line

e.g. EI = upper deviation *es* of shaft of same letter symbol but of opposite sign.

Fundamental deviation for poles *A* to *H* are positive and *EI* and for holes *K* to *ZC* are negative and *ES*. For holes *A* to *H*

$$ES = EI + IT \text{ and for holes } K \text{ to } ZC,\ EI = ES - IT.$$

There are certain deviations in some types.

e.g. for hole of type *N* and 9, and coarser grades, $ES = 0$. For hole of *JS*, $ES = EI = \frac{IT}{2}$.

For holes of type *J, K, M* and *N*, and grades upto and inclusive 8 and holes of type *P* to *ZC* and grades upto and inclusive 7, *ES* = lower deviation *ei* of the shaft of the same letter symbol but one grade finer and of opposite sign increased in the difference between the tolerances of the two grades in question.

The most commonly used shafts and holes in the industries are indicated in the tables 4.3 and 4.4.

4.6.3. Limits and Fits for Sizes above 500 mm and upto 3150 mm. There are many practical difficulties in achieving interchangeable manufacture in large diameter, due mainly to the difficulty in ascertaining the error of measurement involved in gauging larger sizes. (The absolute accuracy of measurement can be more than ± 30 μm). In such cases, therefore, it is the general practice to introduce a safety zone to the extent of error of measurement (at least of minimum metal condition for ensuring minimum clearance or interference) ; particularly in case of smaller deviations. Because of the limitations, there is no use in having finer grades for larger diameters and only grades specified are *IT*6 to *IT*18.

The values of the standard tolerance factor *I* for various grades is

$$I(\text{in } \mu) = 0.004D + 2.1 \ (D \text{ is in mm})$$

Table 4.2 A. Formulae for Fundamental Deviations for Holes for Sizes upto 500 mm.

<table>
<tr><td colspan="3">All deviations except those given below :</td><td>General rule :
Hole limits are identical with the shaft limits of the same symbol (letter and grade) but disposed on the other side of the zero line.
EI = Upper deviation es of shaft of the same letter symbol but of opposite sign.</td></tr>
<tr><td rowspan="2"></td><td>N</td><td>9 and coarser grades</td><td>ES = 0</td></tr>
<tr><td>J, K, M and N</td><td>Up to Grade 8 inclusive</td><td rowspan="3">Special rule :
ES = Lower deviation ei of the shaft of the same letter symbol but one grade finer and of opposite sign increased by the difference between the tolerances of the two grades in question. i.e. $ei + IT_{n-1} = ES + IT_n$ and $ei - IT_n = ES - IT_{n-1}$.</td></tr>
<tr><td>For size above 3 mm</td><td>K, M and N</td><td>Up to grades</td></tr>
<tr><td>For sizes above 3 mm</td><td>P to ZC</td><td>Up to Grado 0 inclusive</td></tr>
</table>

Table 4.3. Most Commonly Used Holes

Letter Symbol →	*A*	*B*	*C*	*CD*	*D*	*E*	*EF*	*F*	*FG*	*G*	*H*	*J*	*JS*	*K*	*M*	*N*	*P*	*R*	*S*	*T*	*U*	*V*	*X*	*Y*	*Z*	*ZA*	*ZB*	*ZC*
Grade ↓ *0 & 01*											×		×															
1											×		×															
2											×		×															
3							×		×	×	×		×	×	×	×	×	×	×									
4							×	×	×	×	×		×	×	×	×	×	×	×									
5					×	×	×	×	×	×		×	×	×	×	×	×	×	×	×	×	×	×					
6				×	×	×	×	×	×	×	×	×	×	×	×	×	×	×	×	×	×	×	×	×	×	×		
7				×	×	×	×	×	×	×	×	×	×	×	×	×	×	×	×	×	×	×	×	×	×	×	×	×
8		×	×	×	×	×	×	×	×		×		×	×	×	×	×	×	×	×	×	×	×	×	×	×	×	×
9	×	×	×	×	×	×	×	×	×		×		×	×	×	×	×	×		×		×	×	×	×	×	×	×
10	×	×	×	×	×	×	×	×	×	×	×		×	×	×	×	×	×		×		×	×	×	×	×	×	×
11	×	×	×		×						×		×			×									×	×	×	×
12	×	×	×		×						×		×															
13	×	×	×		×						×		×															
14											×		×															
15											×		×															
16											×		×															
17											×		×															
18											×		×															

Notes : Holes *A* and *B* are provided only for basic sizes greater than 1 mm. Holes *K* in tolerance grades above *IT*8 are provided only for basic sizes less than or equal to 3 mm.

Holes *T*, *V* and *Y* are only provided for basic sizes greater than 24, 18 and 18 mm.

Holes *N* of tolerance grades above *IT*8 only are only provided for basic sizes greater than 1 mm. Tolerance grades *IT*14 to *IT*18 are only provided for basic sizes greater than 1 mm.

Table 4.4. Most Commonly Used Shafts

Letter Symbol → / Grade ↓	*a*	*b*	*c*	*cd*	*d*	*e*	*ef*	*f*	*fg*	*g*	*h*	*j*	*js*	*k*	*m*	*n*	*p*	*r*	*s*	*t*	*u*	*v*	*x*	*y*	*z*	*za*	*zb*	*zc*
0 & 10											×	×																
1											×		×															
2											×		×															
3							×	×	×	×	×		×	×	×	×	×	×	×									
4							×	×	×	×	×		×	×	×	×	×	×	×	×	×	×	×					
5				×		×	×	×	×	×	×		×	×	×	×	×	×	×									
6				×	×	×	×	×	×	×	×	×	×	×	×	×	×	×	×	×	×	×	×	×	×	×		
7				×	×	×	×	×	×	×	×	×	×	×	×	×	×	×	×	×	×	×	×	×	×	×	×	×
8			×	×	×	×	×	×	×	×	×	×	×	×	×	×	×	×	×	×	×	×	×	×	×	×	×	×
9	×	×	×	×	×	×	×	×	×	×	×	×		×	×	×	×	×	×		×		×	×	×	×	×	×
10	×	×	×	×	×	×	×	×	×	×	×		×	×			×	×	×				×	×	×	×	×	×
11	×	×	×		×						×		×	×		×									×	×	×	×
12	×	×	×		×								×	×													×	×
13	×	×			×						×		×	×														
14											×		×															
15											×		×															
16											×		×															
17											×		×															
18											×		×															

Notes : Shaft *a* and *b* are provided only for basic size greater than 1 mm.

Shafts *j*8 are provided only for basic sizes less than or equal to 3 mm.

Shafts *t*, *v* and *y* are provided for basic sizes greater than 24, 14 and 18 mm respectively. Tolerance grades *IT*14 to *IT*18 are only provided for basic sizes greater than 1 mm.

Table 4.5
Formulae for Shaft and Hole Deviations for sizes above 500 mm and upto 3150 mm

Shafts			*Holes*			*Formula for deviations in* μ
Type	*Fund Deviation*	*Sign*	*Type*	*Fund Deviation*	*Sign*	*(For D in mm)*
d	*s*	—	*D*	*EI*	+	$16D^{0.44}$
e	"	—	*E*	"	+	$11D^{0.41}$
f	"	—	*F*	"	+	$5.5D^{0.41}$
g	"	—	*G*	"	+	$2.5D^{0.34}$
h	"	No sign	*H*	"	No sign	0
js	*ei*	—	*JS*	*ES*	+	$0.5ITn$
k	"	+	*K*	"	—	0
m	"	+	*M*	"	—	$0.024D + 12.6$
n	"	+	*N*	"	—	$0.04D + 21$
p	"	+	*P*	"	—	$0.072D + 37.8$
r	"	+	*R*	"	—	Geometric mean of the values for *p* and *s* or *P* and *S*
s	"	+	*S*	"	—	$IT7 + 0.4D$
t	"	+	*T*	"	—	$IT7 + 0.63D$
u	"	+	*U*	"	—	$IT7 + D$

The values of the standard tolerances corresponding to various grades in terms of the tolerance factor for basic size from 500 to 3150 mm are given below :

Table 4.6

IT1	*IT2*	*IT3*	*IT4*	*IT5*	*IT6*	*IT7*	*IT8*	*IT9*	*IT10*	*IT11*	*IT12*	*IT13*	*IT14*	*IT15*	*IT16*	*IT17*	*IT18*
2*I*	2.7*I*	3.7*I*	5*I*	7*I*	10*I*	16*I*	25*I*	40*I*	64*I*	100*I*	160*I*	250*I*	400*I*	640*I*	1000*I*	1600*I*	2500*I*

It may be noted that here also the tolerance is multiplied by 10 at each fifth step from *IT*6 up to *IT*18.

Deviations. The magnitude and sign of one of the two deviations (the fundamental deviation, *i.e.* upper deviation *es* or *ES*, or lower deviation *ei* or *EI*) for various types of holes and shafts are determined by formulae given in table 4.5. The other deviations are derived by using the absolute value of *IT* tolerance by means of the following relationship :

$$ei = es - IT \qquad EI = ES - IT$$

or

$$es = ei + IT \qquad \text{or} \qquad ES = EI + IT$$

It may be noted that except for type *js* or *JS*, the value of deviation is independent of the given grade. (In case of *js* or *JS* it is 0.5 *ITn*, where *n* represents the grade).

Further it has been observed that symbols *G* and *g* for holes and shafts do not give practicable interchangeable fits, therefore, their use is not recommended.

The selection of the shafts and hole for larger diameters in industry is usually done from the table 4.8. It is also recommended that the shaft and holes in grades 6 and 7 be made in the same shop and measurements of fits in these grades be carried out with great care.

4.6.4. Converting a hole-based Fit into an Equivalent Shaft-based Fit.

For most general applications it is usual to recommend hole-basis system. However sometimes shaft-based system is also used for design of fits.

Where necessary, using the IS system, conversions may be readily made for either hole-based fits or shaft-based fits. For sizes upto and including 500 mm, conversion procedure for clearance fits on the hole-basis to identical shaft-basis is given below :

$$H7 - f6 = F7 - h3 \text{ or,}$$
$$H7 - d8 = D8 - h7 \text{ or,}$$
$$H11 - c11 = C11 - h11.$$

For transition and interference fits this conversion applies whether the tolerance grade for the shaft is the same as that for hole or, as is often the case, one grade finer. In the case of transition and interference fits with large total tolerances for the shaft it is usual to use the same grade of tolerance for the shaft as for the hole and this is done for transition fits *J*, *K*, *M*, *N* in grades above *IT* 8 and for interference fits *P* to *ZC* in grades above *IT*7.

In these cases, hole-basis may be converted directly into shaft basis fits, *e.g.*,

$$H10 - s10 \text{ is identical with } S - h10$$

When the total tolerance on a transition or interference fit is not very large, the practice of using a tolerance one grade finer for hole than on the shaft is followed.

Table 4.7. Equivalent Fits on the Hole-basis and shaft basis System

Clearance		*Transition*		*Interference*	
Hole Basis	*Shaft Basis*	*Hole Basis*	*Shaft Basis*	*Hole Basis*	*Shaft Basis*
*H*7—*c*8	*C*8—*h*8	*H*6—*j*5	*J*6—*h*5	*H*6—*n*5	*N*6—*h*5
*H*8—*c*9	*C*9—*h*8	*H*7—*j*6	*J*7—*h*6		
*H*11—*c*11	*C*11—*h*11	*H*8—*j*7	*J*8—*h*7	*H*6—*p*5	*P*6—*h*5
				*H*7—*p*6	*P*7—*h*6
*H*7—*d*8	*D*8—*h*7	*H*6—*k*5	*K*6—*h*5		
*H*8—*d*9	*D*9—*h*8	*H*7—*k*6	*K*7—*h*6	*H*6—*r*5	*R*6—*h*5
*H*11—*d*11	*D*11—*h*11	*K*8—*k*7	*K*8—*k*7	*H*7—*r*6	*R*7—*h*6
*H*6—*e*7	*E*7—*h*6	*H*6—*m*5	*M*6—*h*5	*H*6—*s*5	*S*6—*h*6
*H*7—*e*8	*E*8—*h*8	*H*7—*m*6	*M*7—*h*6	*H*7—*s*6	*S*7—*h*6
*H*6—*f*6	*F*6—*h*6	*H*8—*m*7	*M*8—*h*7	*H*8—*s*7	*S*8—*h*7
*H*7—*f*7	*F*7—*h*7	*H*7—*n*6	*N*7—*h*6	*H*6—*t*5	*T*6—*h*5
*H*8—*f*8	*F*8—*h*8	*H*8—*n*7	*N*8—*h*7	*H*7—*t*6	*T*7—*h*6
				*H*8—*t*7	*T*8—*h*7
*H*6—*g*5	*G*7—*h*5	*H*8—*p*7	*P*8—*h*7		
*H*7—*g*6	*G*7—*h*6			*H*6—*u*5	*U*6—*h*5
*H*8—*g*7	*G*8—*h*7	*H*8—*r*7	*R*8—*h*7	*H*7—*u*6	*U*7—*h*6
				*H*8—*u*7	*U*8—*h*7

To take account of this when converting from hole-basis to shaft-basis, value known as Δ (delta) is used to calculate the deviations for the holes. This represents the difference between the standard tolerance of the grade in question and that of the next finer grade where the values given in table 4.7 are used in following manner.

In converting the hole-basis, fit *H*7–*p*6 for a size of 25 mm to the shaft basis fit *P*7–*h*6, the upper limit (*ES*) for the hole

$$P7 = ei\ (p) + \Delta$$

Value of fundamental deviation for *p* for a 25 mm size (diameter step 18 – 25) can be calculated and is found to be – 22 microns. Also *IT*7 = 8 microns, hence *ES* (*p*7) = – 22 + 8, *i.e.* – 14 microns (0.014 mm). This rule applies to transition fits *J*, *K*, *M*, *N*, in grades upto and including *IT*7.

Based upon these principles. Table 4.7 which shows equivalent fits on the hole-basis and shaft-basis systems is given for convenience.

4.7. Guide for Selection of Fits (Refer IS : 2709—1982 and IS : 8841 : 1978)

It may be noted that as per IS : 919—1993, innumerable of combination of shafts and holes may be used for a fit. (There being 28 sizes of shafts and 28 sizes of holes, each having 20 grades of tolerances). In order to avoid confusion, some guidelines for selection of fits have to be followed. The first thing to be understood, is that it is normally easier to produce a shaft with a specified tolerance than a hole with the same tolerance due to the fixed character of hole producing tools. Therefore, it is the common practice to follow the system of hole basis (*i.e.* design size of hole is taken as basic size and allowance applied to the shaft). However, in certain cases when products are made from bright drawn bars, shaft basis system may also be employed and actually which system should be chosen depends upon many conditions, such as the nature of product, the manufacturing methods, the condition of raw material, etc.

On the hole basis system, we will first study in brief the important characteristics of various grades of the hole, unilateral from *H*5 to *H*11.

*H*5— This grade can be obtained by precision boring, fine internal grinding or honing.

*H*6— this grade too can be obtained by precision boring and honing or by hand reaming.

*H*7— This can be produced by grinding, broaching or careful reaming.

*H*8— This can be produced by boring or machine reaming.

*H*9— This grade is mostly used for non-circular fits and can be obtained by boring and reaming.

*H*10— This grade is not used for diameter fits. It is used for milled widths and unimportant parts of drilled holes.

*H*11— This grade too being very coarse is never used for fits. It is useful only for coarse drilled or punched holes.

We will now study about the **clearance fits.** This type of fit results in a positive clearance over the whole range of the tolerance. Shafts *a* to *h* produce a clearance fit with the basic hole *H*7.

Clearance Fits

Shafts a, b and c. These fits give a large clearance and are therefore, not widely used.

Shaft d. This fit is available in grade 8 to 11 and is generally suitable for loose running fits as in case of plummer block bearing and loose pulleys etc.

Shaft e. This fit is available in grades 7, 8 and 9, and best suited for general loose clearance fits such as properly lubricated bearings requiring appreciable clearance. In the finer grades, it may be used on large, high speed, heavily loaded bearings such as turbo-generator and large electric motor bearings etc., according to the working conditions.

Shaft f. This fit is available in grades 6, 7 and 8, and is recommended for use as a normal running fit. It is widely use for normal grease lubricated or oil lubricated bearings where no substantial temperature differences are encountered such as in gearbox shaft bearings, bearings of small electric motors and pumps, etc.

Shaft g. This fit is expensive to manufacture as the clearance is small and is, therefore, not recommended as a running fit except in precision equipment where shaft loadings are very light *e.g.*, in bearings for accurate link work, for piston and slide valves etc., and also for spigot and location fit.

Table 4.8 (A) Table for Selection of Holes for Sizes above 500 mm and upto 3150 mm

Letters	*A*	*B*	*C*	*D*	*E*	*F*	*G*	*H*	*JS*	*K*	*M*	*N*	*P*	*R*	*S*	*T*	*U*	*C*	*W*	*X*	*Y*	*Z*
	a	*b*	*c*	*d*	*e*	*f*	*g*	*h*	*js*	*k*	*m*	*n*	*p*	*r*	*s*	*t*	*u*	*v*	*w*	*x*	*y*	*z*
6				×	×	×	×	×	×	×	×	×	×	×	×	×	×					
7				×	×	×	×	×	×	×	×	×	×	×	×	×	×					
8				×	×	×	×	×	×	×	×	×	×	×	×	×	×					
9				×	×	×		×	×			×	×									
10				×	×			×	×													
11				×				×	×													
12				×				×	×													
13				×				×	×													
14								×	×													
15								×	×													
16								×	×													
17								×	×													
18								×	×													

Table 4.8 (B) Table for Selection of Shafts to Sizes above 500 mm and upto 3150 mm

Letters	A	B	C	D	E	F	G	H	JS	K	M	N	P	R	S	T	U	C	W	X	Y	Z
	a	*b*	*c*	*d*	*e*	*f*	*g*	*h*	*js*	*k*	*m*	*n*	*p*	*r*	*s*	*t*	*u*	*v*	*w*	*x*	*y*	*z*
1								×	×													
2								×	×													
3								×	×													
4								×	×													
5								×	×				×	×	×	×	×					
6					×	×	×	×	×	×	×	×	×	×	×	×	×					
7				×	×	×	×	×	×	×	×	×	×	×	×		×					
8				×	×	×	×	×	×	×												
9				×	×	×		×	×	ö												
10				×	×			×	×	ö												
11				×				×	×	ö												
12								×	×	ö												
13								×	×	ö												
14								×	×													
15								×	×													
16								×	×													
17								×	×													
18								×	×													

Shaft h. It exists in grades 5 to 11 and is widely used for non-running parts. It is useful for normal location and spigot fits. In finer grades, it is used as precision sliding fit.

Transition Fits. These may result in either a clearance fit or interference fit depending on the actual value of the individual tolerances of the mating components. Shafts j to n produce transition fit with the basic hole H 7.

Shaft j. This is available in grades 5 to 11 and is a transition fit averaging a slight clearance. It is recommended for location fits requiring less clearance than is given with the h shaft and where a slight interference is permissible *e.g.*, in coupling spigots and recesses, gear rings clamped to steel hubs, etc.

Shaft k. This is the true transition fit averaging virtually no clearance. Available in grades 5 to 11, it is best suited for location fits, where a slight interference can be tolerated as with the object of eliminating vibration.

Shaft m. This is available in grades 5, 6, and 7 and averages a slight interference, although appreciable force is required if extremes are encountered.

Shaft n. This transition fit gives clearances only on extreme sizes. Available in grades 5, 6 and 7, it is recommended for generally 'tight' assembly fits. The true interference fit would result with the finest grade shaft (*n*5) with the hole *H*6.

Interference Fits. These result in positive interference over the whole range of tolerance. This fit can be obtained by shafts *p* to *z* with the basic hole *H*7. The following factors have to be taken into consideration for determining the tolerances and grades for interference fits.

(*i*) It is necessary to check up the minimum interference from the point of view of torque transmission and longitudinal forces between the parts.

(*ii*) It also necessary to check up the minimum interference for taking into consideration the permissible stresses in the material and the forces required for assembling and dismantling.

(*iii*) The following also play an important part in deciding about the tolerancing of interference fits ; diameter, length of engagement, surface finish, material etc.

Shaft p. This gives a true interference fit with the holes of grade 6 or 7, but with holes of grade 8 a transition fit will be obtained. The amount of interference is not excessive but is sufficient to give non-ferrous parts a light press fit which can be dismantled when required. It is the standard press fit for steel, cast iron or brass-to-steel assemblies, *e.g.,* fixing of a bush on a gear. For light alloys (very elastic materials) the amount of interference is too small for getting satisfactory fits.

Shaft r. This combination gives a medium fit on ferrous parts and a light drive fit on non-ferrous parts which can be easily dismantled. An interference fit is obtained with the holes *H*8 with diameters over 100 mm ; with smaller diameters a transition fit results.

Shaft s. This is used for permanent or semi-permanent assemblies of steel and cast iron members, the amount of interference being sufficient to provide considerable gripping force. When using elastic materials, such as light alloys, it is equivalent to the *p* shaft fit in ferrous metals. Typical applications of this type of fit are collars pressed on to shafts, valve shaftings etc.

Shaft t, u, v. No general recommendations can be made about these fits as these involve an increasing degree of interference.

Shaft x, y and z. This fit gives a very large interference with the basic hole and is, therefore, not recommended for interference fits.

The most commonly used shafts and holes in the machines tool industry are as given below :

First preference, *a*11, *c*2, *d*9, *d*11, *e*8, *e*9, *f*6, *f*7, *f*8, *g*5, *g*6, *h*5, *h*6, *h*8, *h*11, *j*5, *j*6, *k*5, *m*5, *m*6, *n*6, *p*6, *r*6, *s*6.

Second preference, *b*9, *b*11, *c*8, *c*11, *d*8, *e*7, *h*7, *h*9, *h*10, *j*7, *k*7, *m*7, *n*7, *p*7, *r*7, *s*7, *t*6, *t*7, *u*7.

Holes

First preference, *A*11, *D*11, *E*11, *F*7, *F*8, *F*10, *G*7, *H*6, *H*7, *H*8, *H*11, *J*6, *K*6, *K*7, *M*6, *M*7, *N*6, *N*7, *P*9.

Second preference, *B*9, *B*11, *C*9, *C*11, *D*9, *D*10, *E*9, *G*6, *H*5, *H*9, *P*6, *R*7, *S*7.

The most commonly used types of fits are classified below (as per IS : 2709—1964) :

Table 4.9

Type of	Classes of fit	With holes				Remarks
		H6	H7	H8	H11	
CLEARANCE	Shaft a	—	—	—	a11	Loose clearance fit—not widely used
	Shaft b	—	—	—	—	—do—
	Shaft c	—	c8	*c9	c11	Slack running fit
	Shaft d	—	d8	*d9—d10	d11	Loose running fit
	Shaft e	e7	e8	*e8—e9	—	Easy running fit
	Shaft f	*f6	*f7	*f8	—	Normal running fit
	Shaft g	*g5	*g7	*g7	—	Close running fit or sliding fit. Also spigot and location fit.
	Shaft h	*h5	*h6	*h7—h8	*h11	Precision sliding fit. Also fine spigot and location fit.
TRANSITION	Shaft j	*j5	*j6	7	—	Push fit for very accurate location with easy assembly and dismantling.
	Shaft k	*k5	*k6	k7	—	Light keying fit (true transition fit) for keyed shaft, non- running locked pins etc.
	Shaft m	m5	*m6	m7	—	Medium keying fit.
	Shaft n	n5	*n6	n7	—	Heavy keying fit (for tight assembly of mating surface).
INTERFERENCE	Shaft p	p5	*p6	—	—	Light press fit with easy dismantling for non-ferrous parts. Standard press fit with easy dismantling for ferrous and non- ferrous parts assembly.
	Shaft r	r5	*r6	—	—	Medium drive fit with easy dismantling for ferrous parts assembly. Light drive fit with easy dismantling for non-ferrous parts.
	Shaft s	s5	*s6	s7	—	Heavy drive for ferrous parts, permanent or semipermanent assembled press fit and for non-ferrous parts.
	Shaft t	t5	t6	*t7	—	Force fit on ferrous parts for permanent assembly.
	Shaft u	u5	u6	*u7	—	Heavy force fit or shrink fit.
	Shaft v, x y and z	—	—	—	—	Very large interference fit. Not recommended for use.

4.8. The Newall System

This is the first standard evolved in Great Britain to standardise limits and fits and is still used to a certain extent although all the fits provided by this system can be obtained with approximately the same values by selection from B.S. 1916. The Newall system provides a range of clearance, transition and interference fits for sizes upto 12″. It is a hole basis system which stipulates two grades of holes, specified with bilateral tolerances, together with 6 grades of shaft tolerances.

It is the earliest of all the systems and is extremely simple. It specifies too few fits and those listed do not enforce to modern ideas as regards their basic deviations. Though this served a useful purpose in the past but is not considered suitable for modern production.

The Newall system is founded on the 'hole' basis. Thus in this system provision is made in the size of the hole for error in workmanship, and the variation to obtain the quality of fit required is allowed for on the size of the shaft which has to enter the hole.

It has two grades of bilateral holes (*A* and *B*), two grades of press fit (*F* and *D*), one transition fit (*P*), and three clearance fits (*Z, Y* and *X*). These are shown diagrammatically in Fig. 4.25. Hole *A* is of higher accuracy than *B* and can be produced by precision reamers.

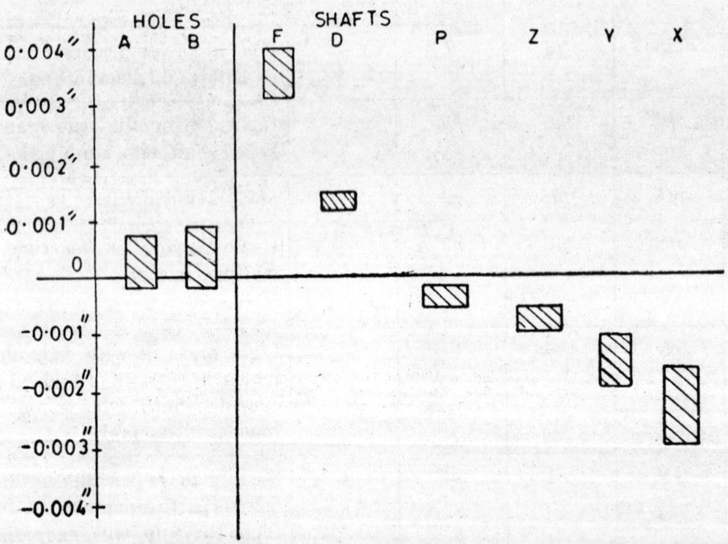

Fig. 4.25. Diagrammatic Representation of Newall Limits.
F = Force, *D* = Driving, *P* = Push, *Z* = Running (Close), *Y* = Running (Normal), *X* = Running (Free).

It specifies diameter steps as

$$0 - \frac{1''}{2}, \frac{9}{16} - 1'', 1\frac{1}{16} - 2'', 2\frac{1}{16} - 3'', 3\frac{1}{16} - 4'', 4\frac{1}{16} - 5''.$$

The limits are calculated from the following formulae.

$A : + 0.0006\sqrt{D}, - 0.0003\sqrt{D}$

$B : + 0.0008\sqrt{D}, - 0.0004\sqrt{D}$

$E : + 0.003D^{0.75}, + 0.0022D^{0.87}$ (D = geometric mean of diameter steps)

$D : = 0.0012D^{0.75}, + 0.0008D^{0.7}$ (D = G.M. of diameter steps)

$P : - 0.0002\sqrt{D}, 0.0006\sqrt{D}$

$Z : 0.0005 \sqrt{D}, 0.001 \sqrt{D}$

$Y : 0.001 \sqrt{D}, -0.0018 \sqrt{D}$

$X : -0.00125 \sqrt{D}, 0.0025 \sqrt{D}$.

The relative magnitudes of the tolerances in the various fits for approximately in the following ratio :

$$A : B : P : X : Y : Z : F : D$$
$$1 : 3 : 1 : 3 : 2 : 1 : \text{(special)}.$$

4.9. British Standard 1916

This was initially based on continental system (ISA system) which is detailed in ISA Bulletin 25. This standard specifies 16 grades of tolerances, (*ITI* to *IT* 16), each grade covering a range of sizes from 0.04 inch to nearly 20 inches. Like Indian standard, in British Standard also the tolerances in each grade vary over the range of sizes to produce the same standard of accuracy in each size.

Each grade of tolerance is built up from fundamental tolerances which are multiples of the fundamental tolerance unit i. For a diameter of size D,

$$i = (0.052 \times \sqrt[3]{D} + 0.001\, D) \times 0.001 \text{ inch}$$

or

$$i = (0.45 \times \sqrt[3]{D} = 0.001\, D) \times 0.001 \text{ mm}$$

(D is in inches and mm respectively and is the geometric mean of range in which it falls.)

For standard limits and fits specified, grades $IT6$ to $IT16$ only are used, and $IT6 = 10i$ and $IT16 = 1000 IT6$. $IT7 = IT6 \times 10^{0.2}$, $IT8 = IT7 \times 10^{0.2} = IT6 \times 10^{0.4}$ and so on.

It stipulates 21 types of holes and 21 types of shafts. Seeing the combination of various holes and shafts, and the tolerance grades, theoretically there will be around 4000 different fit combinations. However, in practice it is not reasonable to combine every hole with every shaft.

Having defined the tolerance zone, the next thing is the fundamental deviation, *i.e.* the position of the tolerance zone relative to the nominal size. The fundamental deviation (difference between nominal size and the nearest limit of tolerance) is designated by capital letters for holes, and small letters for shafts and its value is obtained from empirical formulae.

Deviations for H holes and for h shafts are zero. Deviations for A and a are largest, holes being positive and shafts negative. Deviations for holes designated A to H correspond directly in value with those for shafts a to h. The remaining groups J to Z and j to z do not correspond in their deviations in quite the same way.

Guide to the selection of fits from the very wide range of fits in B.S. 1916 is given in the second part of this standard issued by B.S.I. It specifies two systems based on either hole basis or shaft basis, and recommends hole basis wherever possible. The various advantages in using the hole basis for selection of fits are that the holes can be made by using conventional tools like drills, reamers, broaches, punches and so on ; standard sized holes can be gauged easily whereas non-standard holes have to be measured ; however there is no problem in manufacturing shafts of any size. Generally in the system using hole basis, hole H having fundamental deviation equal to zero is used allowing the designer a free choice of actual tolerance values. The hole H may have different tolerance grades, that is $IT5$ to $IT13$. Shafts with deviation a to h are used for clearance fits and other shafts are selected to give interference and clearance fits. The tolerances on interference fits are generally smaller than those permissible for clearance fits since the forces required to assemble them vary quite considerably with small changes to value of interference.

In the shaft basis system, the h series having zero deviation is chosen and variation in fits are obtained by making holes to tolerances in zones having various deviations. In this system, holes have to be made by using boring tools, internal grinders and similar methods.

BS : 1916 is largely similar to ASA, and recommends a primary selection of five fits which may cover about 95% of fits applications.

	Hole	*Shaft*	*Designation*
Primary	*H*8	*e*8	Loose clearance
	*H*8	*f*8	Average running
	*H*7	*g*6	Precision location
	*H*8	*h*7	Average location
	*H*7	*k*6	Push
	*H*7	*p*6	Press (ferrous parts)
Secondary	*H*8	*d*9	Coarse or loose
	*H*7	*f*7	Precision running
	*H*6	*g*5	Precision location
	*H*8	*j*7	Keying fit
	*H*7	*s*6	Press (non-ferrous parts)

4.10. American Standards Association Tolerance System (*AS A. B 4a*—1995)

It specifies eight fits, each shaft having one appropriate hole only (five fits using same hole). Holes are unilateral. It only describes the fits and does not refer them by symbols.

1. Heavy force shrunk fit	
2. Medium force fit	
3. Tight fit	With hole (1)
4. Wringing fit	
5. Sung fit	
6. Medium fit	With hole (2)
7. Free fit	With hole (3)
8. Loose fit	With hole (4)

It also indicates that fits 5 to 8 are intended for interchangeable manufacture and for remainder the selective assembly shall be used to achieve desired fits.

The diameter steps specified are $0, \frac{3}{16}, \frac{5}{16}, \frac{7}{16}, \frac{9}{16}, \frac{11}{16}, \frac{13}{16}, \frac{15}{16}, 1\frac{1}{16}, 1\frac{3}{16}, 1\frac{3}{8}, 1\frac{5}{8},$

$1\frac{7}{8}, 2\frac{1}{8}, 2\frac{3}{8}, 2\frac{3}{4}, 3\frac{1}{4}, 3\frac{3}{4}, 4\frac{1}{4}, 4\frac{3}{4}, 5\frac{1}{2}, 6\frac{1}{2}\, 7\frac{1}{2}, 8\frac{1}{2}.$

The formulae for calculating all limits are :

Hole	*Limits*	*Tolerance*	*Approx. Tolerance ratio*
1.	$+0, +0.0006\, D^{1/3}$	$0.0006\, D^{1/3}$	$1\frac{1}{2}$
2.	$+0, +0.0008\, D^{1/3}$	$0.0008\, D^{1/3}$	2
3.	$+0, +0.0013\, D^{1/3}$	$0.0013\, D^{1/3}$	3
4.	$+0, +0.0025\, D^{1/3}$	$0.0025\, D^{1/3}$	4

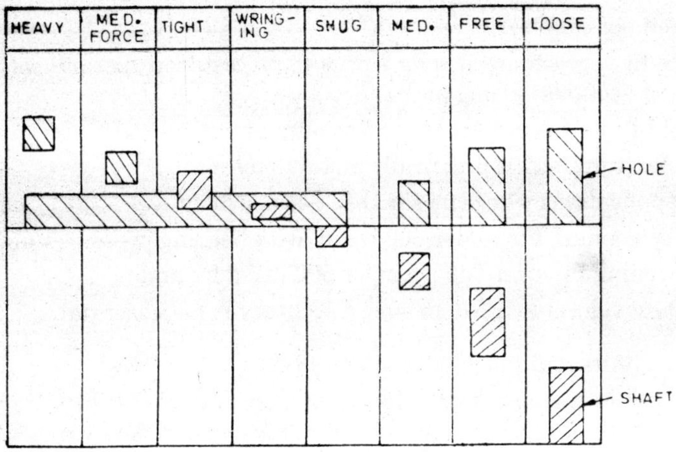

Fig. 4.26. Diagrammatic Representation of limits of ASA B. 4*a*.

Shafts	*Limits*	*Tolerance*	*Approx. tolerance ratio*
Heavy Force	$+ (0.001D + 0.0006D^{1/3}), + 0.001D$	$0.0006D^{1/3}$	$1\frac{1}{2}$
Medium Force	$+ (0.005D + 0.0006D^{1/3}), + 0.0005D$	$0.0006D^{1/3}$	$1\frac{1}{2}$
Tight	$+ (0.0025D + 0.0006D^{1/3}), + 0.00025D$	$0.0006D^{1/3}$	$1\frac{1}{2}$
Wringing	$+ 0.0004D^{1/3}, + 0$	$0.0004D^{1/3}$	1
Snug	$- 0, - (0.0009D^{2/3} + 0.0008D^{1/3})$	$0.0004D^{1/3}$	1
Medium	$- 0.0009\text{D}^{2/3}, - (0.0009D^{2/3} + 0.0008D^{1/3})$	$0.0008D^{1/3}$	2
Free	$- 0.0014D^{2/3}, - (0.0014D^{2/3} + 0.0013D^{1/3})$	$0.0013D^{1/3}$	3
Loose	$- 0.0025D^{2/3}, - (0.0025D^{2/3} + 0.0025D^{1/3})$	$0.0025D^{1/3}$	6

These limits are rounded off to nearest 0.0001 in, and D = arithmetic mean of dia. step extremeties :

Basic deviation formulae are :

Fit	*Formulae*	*Remarks*
Heavy Force	$0.001\ D - 0.0006\ D^{1/3}$	(This is negative below 7/16 in. nominal)
Medium Force	$0.0005\ D - 0.0006\ D^{1/3}$	(This is negative below $1\frac{3}{16}$ in. nominal)
Tight	$0.0025\ D - 0.0006\ D^{1/3}$	(This is negative below $3\frac{3}{4}$ in. nominal)
Snug	0	
Medium	$0.0009D^{2/3}$	
Free	$0.0014D^{2/3}$	
Loose	$0.0025D^{2/3}$	

According to this standard, heavy force and shrink fit are recommended for heavy shrink fits in steel holes, and not cast iron.

Medium force fit is recommended for shrink fit on cast iron ; permanently assembled parts ; shrink fits on medium sections or long parts.

Similarly tight fit for press fit more or less permanent.

Wringing fit for 'tunking' fit practically metal to metal.

Snug fit for applications where no shake is permissible.

Medium fit for running fits under 600 r.p.m. with bearing pressures upto 600 lb/sq. in.

Free fit for running fits over 600 r.p.m. or 600 lb/sq. in. and

Loose fits for considerable freedom where accuracy is not essential.

4.11. ISO System of Limits and Fits (ISO : 286-1 : 1998)

The present ISO Recommendation, ISO System of Limits and Fits, Part I, is based on the ISA System of Limits and Fits. It is the most comprehensive and detailed of all tolerance systems and the most modern one. It lists a full range of 28 holes and 28 shafts covering all possible requirements. It gives several qualities of tolerance for each grade of shaft and hole, and lists all the possible combinations and recommends for general use a few unilateral holes, an adequate range of shafts, the remainder being listed for special duty. It designates holes by capital letters and shafts by small letters. In numerical form, the toleranced size is defined by its basic value followed by a 'symbol' composed of a letter and a number *e.g.*, 50 $g7$; a fit is indicated by the basic size common to both components followed by symbol corresponding to each component, the hole being quoted first *e.g.*, 50 $H8/g7$ or $50H8 - g7$ or $50\ \dfrac{H8}{g7}$.

For the purpose of calculation of standard tolerances and fundamental deviations, it specifies 13 main diameter steps, and 22 intermediate steps, the results being computed on the basis of the geometric mean D of the extreme diameters of each step. The main diameter steps are above zero and upto 3, above 3 and upto 6 (represented as 3—6), 6—10, 10—18, 18—30, 30—50, 50—80, 80—120, 120—180, 180—250, 250—315, 315—400, 400— 500. The intermediate steps are used, in certain cases, when considered as necessary, for the deviations a to c and r to zc or A to C and R to ZC.

It provides 20 grades of tolerances (grades of manufacture) $IT01$, $IT0$, $IT1$, to $IT18$. The values of standard tolerances for grades 5 to 16 are determined from the tolerances factor i,

$$i = 0.45\sqrt[3]{D} + 0.001\,D \text{ microns } (D \text{ is in mm}).$$

The values of standard tolerances corresponding to grades 5 to 18 are as for IS : 919.

The standard tolerances corresponding to $IT01$, $IT0$ and $IT1$ are $0.3 + 0.008D$, $0.5 + 0.012D$, and $0.8 + 0.020D$ respectively. The values of $IT2$ to $IT4$ are scaled approximately geometrically between the values of $IT1$ and $IT5$.

The fundamental tolerances determine the dimensional difference between two limits, and the fits are determined by defining one of the limits, *i.e.*, the basic deviation or fundamental deviation. The other limits can be obtained by adding or subtracting the fundamental tolerance.

The formulae for fundamental shaft deviations are same as enumerated in IS : 919.

It specifies 28 grades of fits for both hole and shaft ranging from fits of extreme interference to fits of extreme clearance.

4.12. Plain Gauges

Gauges are inspection tools of rigid design, without a scale, which serve to check the dimensions of manufactured parts. Gauges do not indicate the actual value of the inspected dimension on the work. They can only be used for determining as to whether the inspected parts are made within the specified limits.

Plain gauges are used for checking plain (unthreaded) holes and shafts. Gauges are classified.

(1) According to their type :

(*a*) Standard and limit gauges,

(*b*) Limit gauges.

(2) According to their purposes :

(*a*) Workshop.

(*b*) Inspection.

(*c*) Reference or master gauges. These are better known as control gauges and comprise the setting plug gauges and wear check plug gauges and are mainly used for controlling the dimensions of ring gauges and for checking their wear during use.

(3) According to the form of the tested surface :

(*a*) Plug gauges for checking holes.

(*b*) Snap and ring gauges for checking shafts.

(4) According to their design :

(*a*) Single limit and double limit gauges,

(*b*) Single ended and double ended gauges,

(*c*) Fixed and adjustable gauges.

Advantage of fixed gauges

The various advantages of fixed gauges in comparison to comparator type gauges are :

(*i*) Fixed gauges are essentially free from errors due to drift and the original adjustment, non-linear response, effect of power supply variations and other extraneous factors which necessitate regular calibration and occasional correction on comparator type gauges.

(*ii*) These provide positive dimensional information.

(*iii*) These are portable and independent of power supply availability.

(*iv*) These involve no other auxiliary equipment and set ups.

(*v*) These can be designed to check combinations of several dimensions comprising lengths, diameters and angles.

(*vi*) These can be designed to inspect interrelated features, for size, location, form, alignment etc. so as to check the virtual size of a member (combined effect of all parameters with regard to the functional adequacy of the inspected features.)

(*vii*) These are particularly useful in the checking of part members whose meaningful geometric irregularities can't be readily detected by gauges which do not provide complete reverse replica of the part portions.

(*viii*) These provide uniform reference standards.

(*ix*) These are not expensive.

(*x*) Comparator have to be set from time to time using master fixed gauges.

4.12.1. Inspection Gauging of Plain Workpieces. As per IS : 3455—1971, in order to avoid any dispute, the manufacturer and purchaser should use gauges as follows :

The inspection department should normally use the same types of gauges as used in the workshop. However in order to attain same results by both workshop and inspection department, it is recommended that workshop should use new or only slightly worn gauges and inspection department should use gauges with their size nearer the permissible wear limits.

As regards purchaser, he can either use the gauges of manufacturer, after ensuring their accuracy, or use gauges having their sizes near the wear limit ; or use gauges having tolerance zones so disposed so as to ensure acceptance of sizes which are within the specified limits.

As the dimensions for workpieces and inspection instruments are specified at 20°C, all measurements should normally be carried out at 20°C. However if, both gauges and workpieces are of same material or materials having same coefficient of linear expansion then checking temperature may deviate from 20°C without deteriment to the result, provided same temperature is attained both by workpiece and gauges.

4.12.2. Types of Gauges. The gauges for cylindrical holes and internal screw-threads are customarily in the form of a cylindrical plug, plain or screwed, with a convenient handle. The gauging portion of the gauge may be either integral with the handle or the whole gauge may be machined from one piece of metal. Such gauges are usually called 'solid gauges'. Alternatively, the gauging portion and the handle may be separately manufactured and, for use, are engaged together to form a rigid assembly by means of a suitable locking device. The latter type is generally called as 'renewable end' type of gauges. The 'Go' and 'No Go' gauges may be in the form of separate single-ended gauge, or may be combined on one handle to form a double-ended gauge. In the case of plain plug gauges a form of convenient combined 'Go' and 'No Go' gauge is the progressive gauge, which is single-ended gauge with one gauging member having two diameters to the 'Go' and 'No Go' limits respectively. In the case of plain plug gauges of diameter larger than 75 mm, which are very heavy, it is common practice to use bar type (segmental type) of gauges, because the efficiency of the gauge would decrease rapidly if the sensitiveness of user is impaired by weight.

Another serious problem encountered in plug gauges is of inserting it into the bore of a component, because the axes of gauge and hole should be made coincident before entering which is a difficult task specially with heavy gauges. If this condition is not satisfied, gauges will have tendency to jam in the hole. In order to overcome these difficulties it is usual to provide a lead in chamber (Fig. 4.27) which minimises damages to holes and saves time.

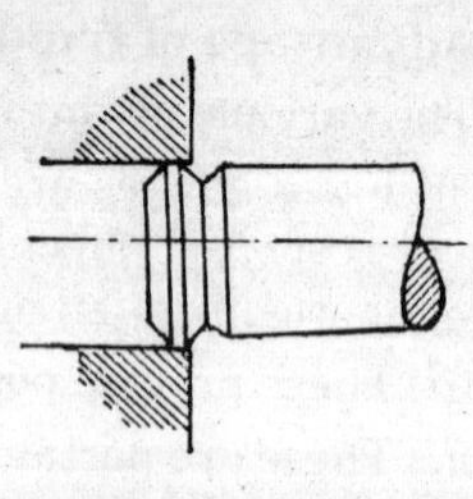

Fig. 4.27. Pilot plug gauge.

The gauges for shafts and external threads are either of 'ring' gauge form or of 'gap' gauge form. Ring gauges are generally of non-adjustable type and separate gauges are used for 'Go' and 'No Go' gauging. Gap gauges for plain work may be either of the solid, non-adjustable type or may be of the adjustable type. Gap gauges for screw threads are invariably of the adjustable type. Either the separate gauges may be used for 'Go' and 'No Go' gauging or a combined 'Go' and 'No Go' gauge may be used, and the adjustable gap gauges are generally of the latter form.

The entering end of the larger gauges, being liable to become burred if accidentally laid down heavily on metal surfaces, is provided with a guard extension at the entering end of gauges. All sharp edges on the gauging portions and handles should be removed and the larger and heavy plug gauges should be provided with venting holes.

The various type of limit plug gauges and limit snap gauges are shown below :

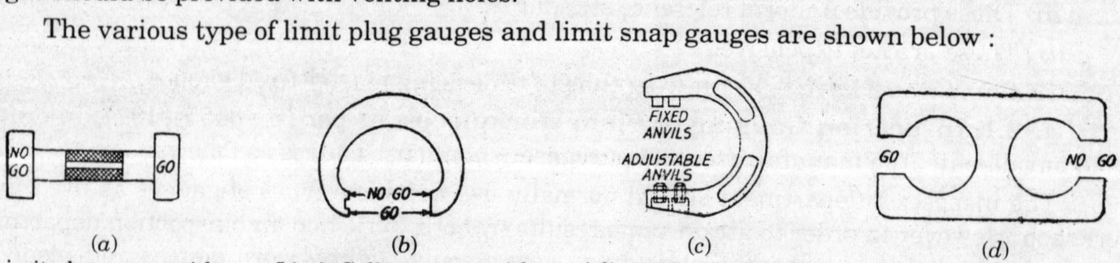

(a) Limit plug gauge with 'go' and 'no go' end

(b) Limit Caliper gauge with 'go' and 'no go' ends used for larger sizes

(c) Adjustable Snap Gauge

(d) Double ended solid snap gauge for smaller size

Fig. 4.28

4.12.3. Plain Plug Gauges. Generally the gauging members of the plain plug gauges are made of suitable wear-resisting steel and the handles can be made of any suitable steel *e.g.* handles may be made of light metal alloys for heavy plain plug gauges, or suitable non-metallic handles may be provided for smaller plain plug gauges. The gauging surface of plain plug gauges are normally hardened to not less than 750 H.V. and suitably stabilised and ground and lapped.

The plain plug gauges are normally of double ended type for sizes upto 63 mm and of single ended type for sizes above 63 mm.

The usual way of designating the plain plug gauges is by 'Go' and 'No Go' as applicable, the nominal size, the tolerance of the work-piece to be gauged and the number of the standard followed *e.g.*, a double ended 'Go' and 'No Go' plain plug gauge for gauging a bore of 10 mm with tolerance H7 and if designated according to Indian Standard (IS : 3484—1966) shall be designated as :

'Go and No Go plain plug 10 H7, IS : 3484'.

The various types of plain plug gauges in common use are shown below :

(*i*) 'Go' and 'No Go' plain plug gauges for sizes upto 10 mm, (solid type). (Refer Fig. 4.29)

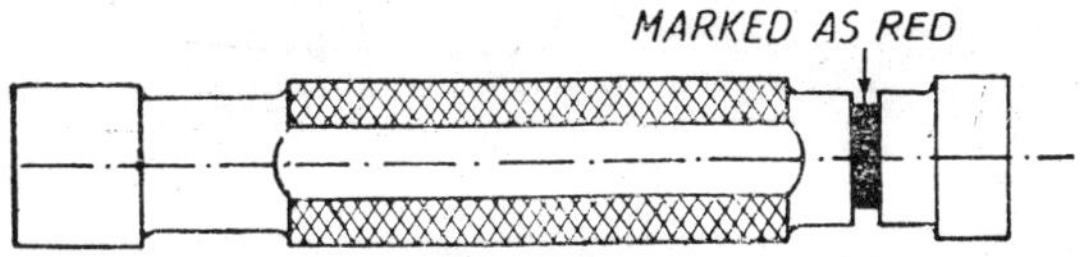

Fig. 4.29

(*ii*) 'Go' and 'No Go' plain plug gauges for sizes over 10 and upto 30 mm (Taper Inserted type). (Refer Fig. 4.30)

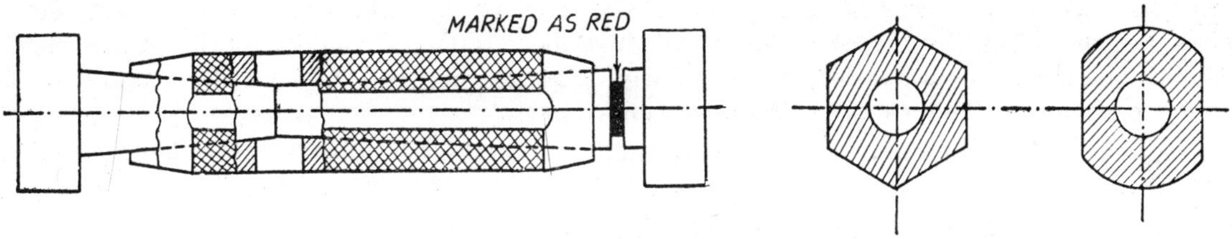

Fig. 4.30. Section of Handle.

(*iii*) 'Go' and 'No Go' plain plug gauges for sizes over 30 mm and upto 63 mm (Fastened type), (Refer Fig. 4.31).

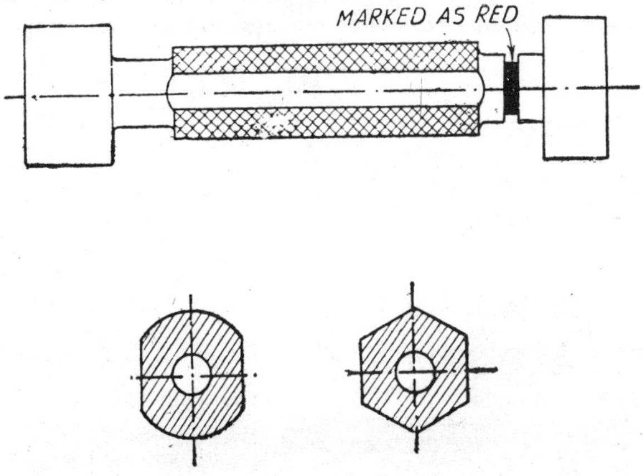

Fig. 4.31. Section of Handle.

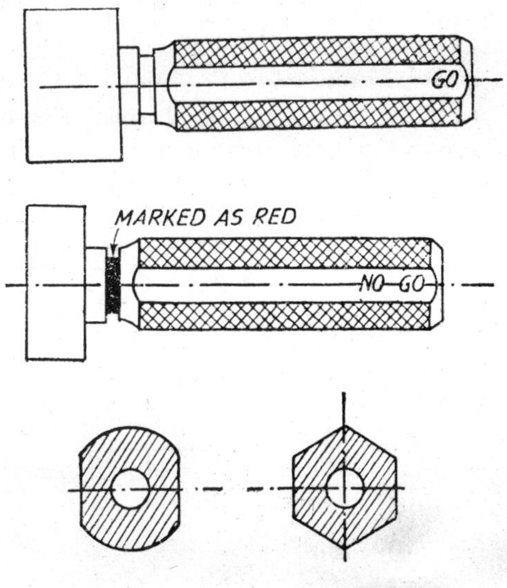

Fig. 4.32. Section of Handle.

(*iv*) 'Go' and 'No Go' plain plug gauges for sizes over 63 mm and upto 100 mm (Fastened type). (Refer Fig. 4.32)

(*v*) 'Go' and 'No Go' plain plug gauges for sizes over 100 mm and upto 250 mm (Flat type). (Refer Fig. 4.33). This is a shell form plug gauge. Each plug is relieved to reduce weight.

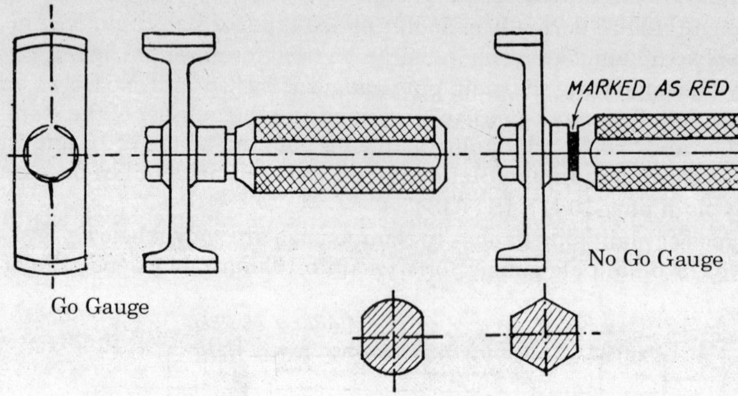

Fig. 4.33

For still further bigger holes and to restrict weight, use can be made of segmental cylindrical ended gauges.

Spherically ended rods are used for very large holes. It would be noted that with such types of gauges the full form of the gauge is lost and the errors of holes like ovality may not be detected. It is the general practice not to use cylindrical plugs above 100 mm diameter but to use cylindrically ended bars or spherically ended rods. Similarly Go gauges between the sizes of 100 and 200 mm diameter can take the form of a cylindrically ended bar.

The plain gauges are marked with the following on their handles for their identification :

(*i*) Nominal size.

(*ii*) Class of tolerance.

(*iii*) The word 'Go' on the 'Go' side.

(*iv*) The words 'No Go' on the 'No Go' side.

(*v*) The actual values of tolerances.

(*vi*) Manufacturer's name or trade mark.

The 'No Go' side is always painted with a red band. A typical example of the marking is shown in Fig. 4.34. It is usual practice to apply suitable anti-corrosive coating to the plug gauges

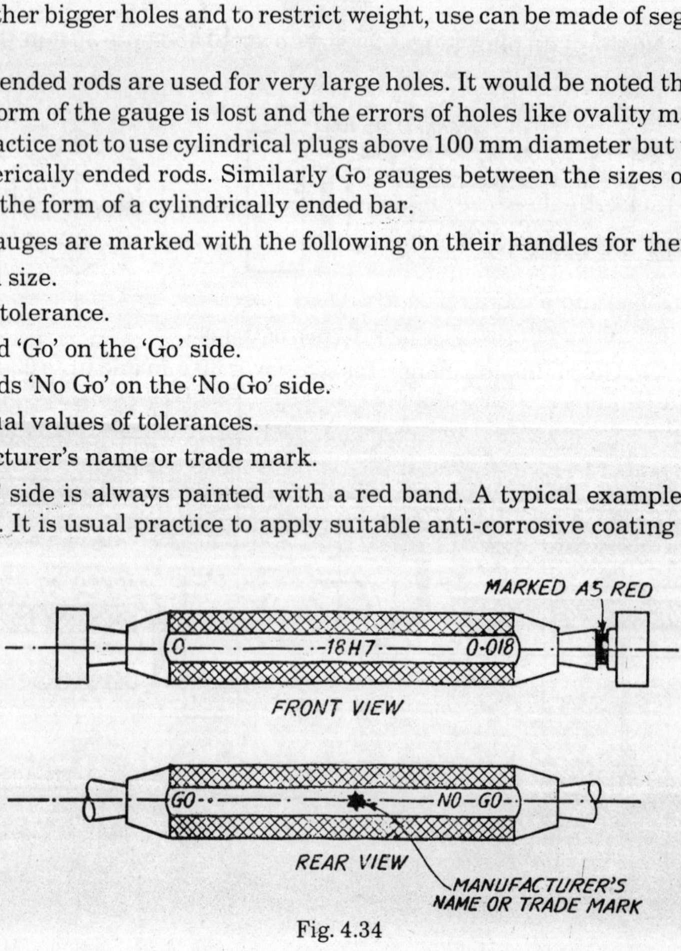

Fig. 4.34

in order to protect them against climatic conditions. In order to prevent damage in handling and transit, these are packed in suitable cases. It may be mentioned that gauges with the gauging portion integral with the handle are now becoming obsolete and gauges with renewable ends are gaining popularity (Refer Fig. 4.35) because of the following advantages :

(*i*) Worn or damaged end can be replaced conveniently.

(*ii*) In the event of scraping of gauge, handle can be used for other gauge.

(*iii*) To reduce the weight, handle can be made of plastic which also facilitates in handling the gauge, reduces cost and minimises risk of heat transfer.

For smaller through holes, another useful renewable and plug gauge is the progressive type of gauge in which both GO and NOT GO gauging members are provided on same side separated by a small distance (Refer Fig. 4.36). First Go portion is inserted in the hole which would be further obstructed by NOT GO portion if hole is of tolerable size.

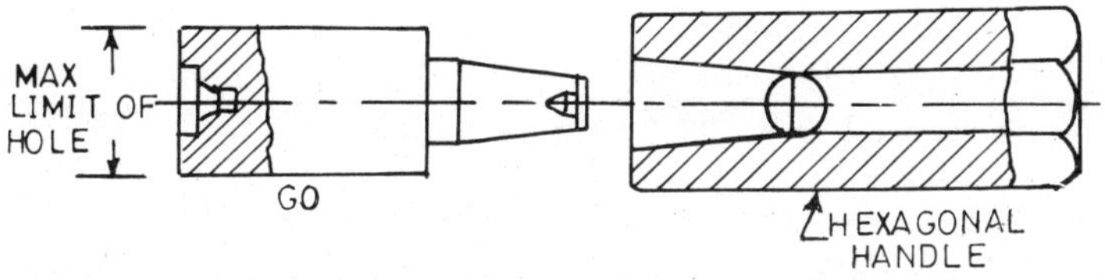

Renewable end type plug gauge—single ended
Fig. 4.35

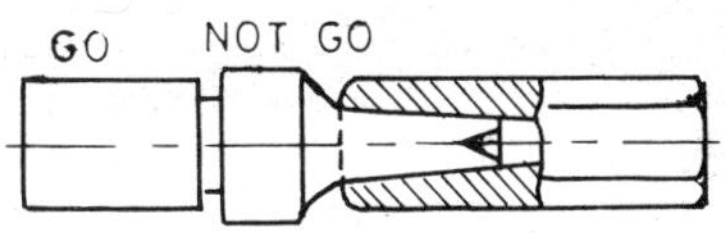

Progressive form of plug gauge
Fig. 4.36

4.12.4. Plain Ring Gauges. The plain ring gauges are made of suitable wear resisting steel and the gauging surfaces are hardened to a hardness of about 720 H.V. The gauging surfaces are first suitably stabilised using proper heat treatment process and then ground and lapped and other surfaces are finished smooth. These are protected against climatic conditions by applying a suitable anti-corrosive coating.

These are available in two designs, 'Go' and 'No Go'. These are designated by 'Go' and 'No Go' as may be applicable, the nominal size, the tolerance of the workpiece to be gauged, and the number of the standard followed.

The general shapes of Go' and 'No Go' gauges, for range from 3 to 70 mm in 10 steps, and from 70 to 250 mm in 17 steps are shown in Figs. 4.37 and 4.38 and Figs. 4.39 and 4.40 respectively.

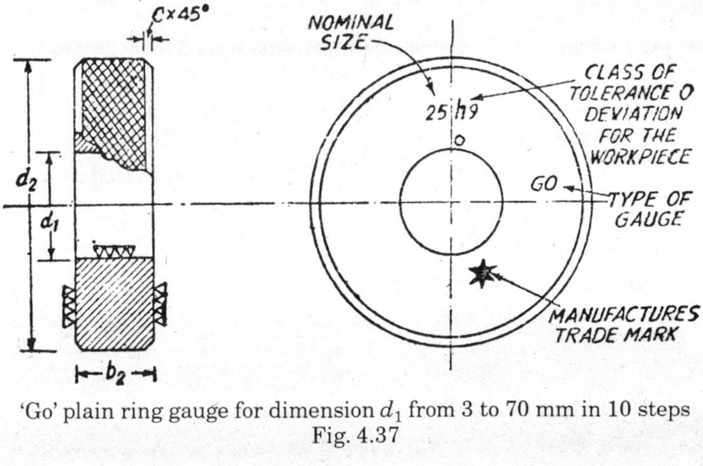

'Go' plain ring gauge for dimension d_1 from 3 to 70 mm in 10 steps
Fig. 4.37

For Fig. 4.37, dimension d_2 varies from 22 mm (for range of d_1 = 3—5 mm) to 112 mm (for range of d_1 = 60—70 mm). Correspondingly b varies from 5 to 22 mm and c from 0.4 to 1.6 mm.

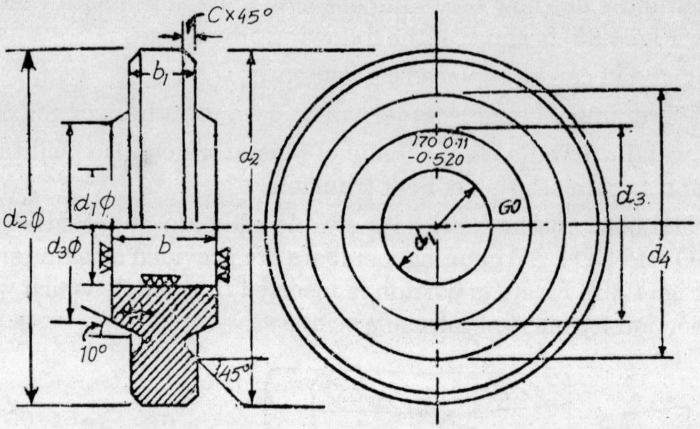

'Go' plain ring gauge for dimension d_1 from 70 to 250 mm in 17 steps.
Fig. 4.38

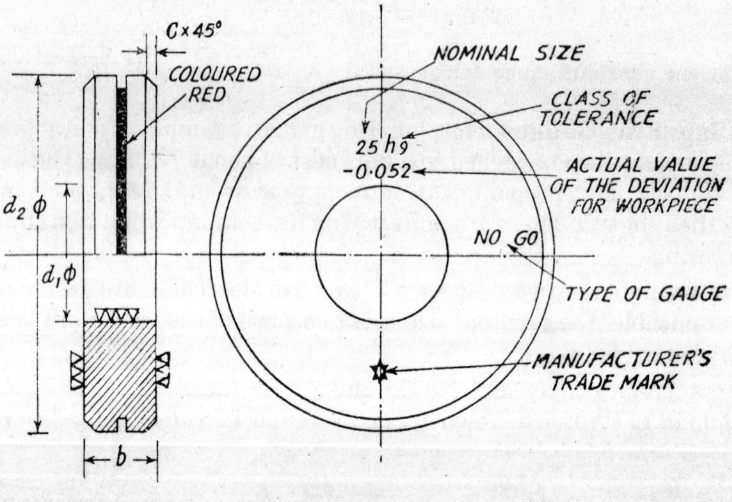

'No Go' plain ring gauge for dimension d_1 varying from 3 to 70 mm in 10 steps.
Fig. 4.39

For Fig. 4.38 dimension d_2 varies from 125 mm (for range of d_1 = 70—80 mm) to 355 mm (for range of d_1 = 236—250 mm). Corresponding to d_1 = 120 to 355, dimension d_2 varies from 100 mm to 280 mm ; b_1 from 12 to 28 mm ; c from 2.5 to 4 mm ; r from 3 to 6 mm ; t from 3 to 8 mm.

For Fig. 4.39 dimension d_2 varies from 22 mm for range d_1 = 3—5 mm, to 112 mm for range d_1 = 60—70 mm. Correspondingly b_1 varies from 3 to 8 mm, c from 0.4 to 1.6 mm.

For Fig. 4.40 dimension d_2 varies from 125 to 355 mm, d_3 from 100 to 280, d_4 from 113 to 335, b_1 from 10 to 25, b_2 from 7 to 16, t from 1.5 to 4, c from 1 to 3, and r from 3 to 6 mm.

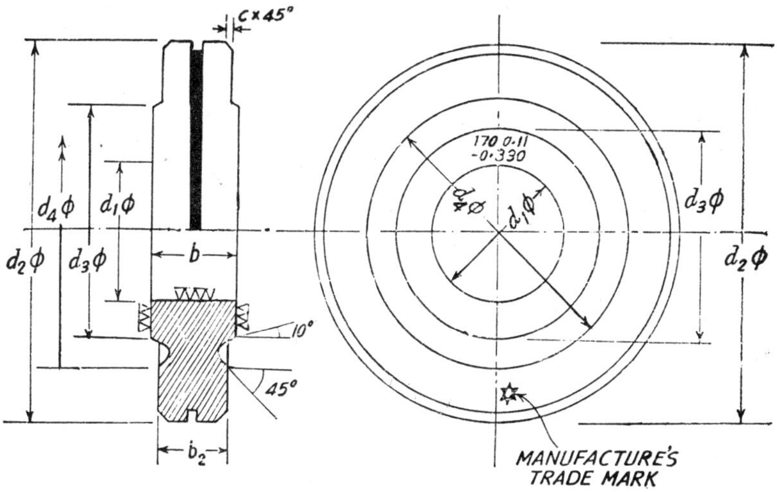

'No Go' plain ring gauge for dimension d_1 varying from 70 to 250 mm.
Fig. 4.40

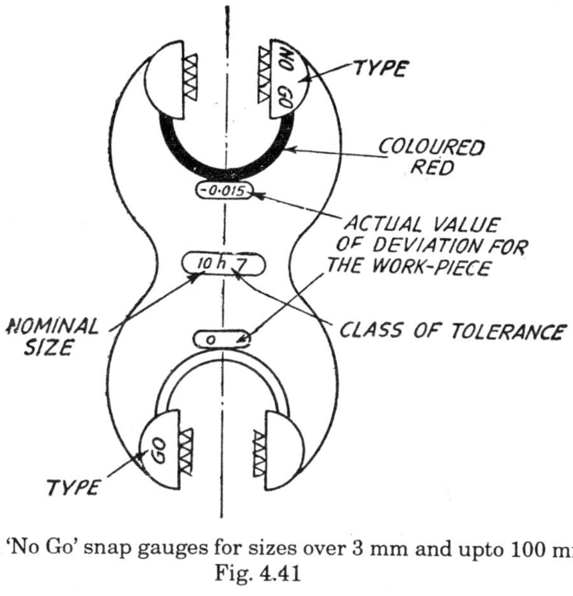

'Go' and 'No Go' snap gauges for sizes over 3 mm and upto 100 mm.
Fig. 4.41

4.12.5. Snap Gauges

4.12.5.1. Rib type snap gauges. Double-ended type snap gauges can be conveniently used for checking sizes in the range of 3 mm to 100 mm and single ended progressive type snap gauges are suitable for size range of 100 to 250 mm. The gauging surfaces of the snap gauges are hardened upto 720 HV and are suitably stabilised, ground and lapped. The other surfaces are finished smooth.

The way of marking is also shown.

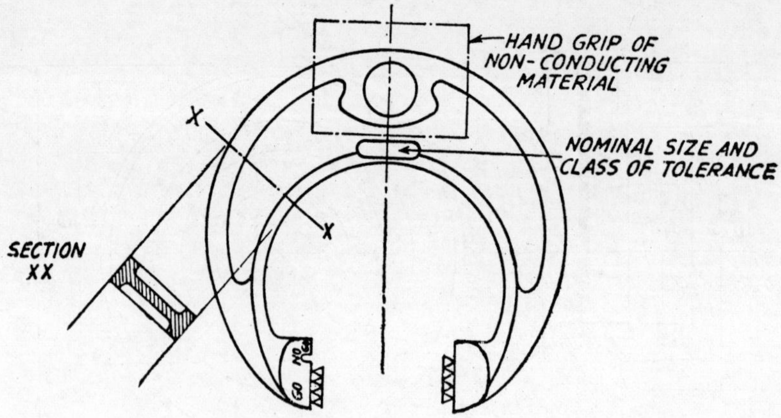

'Go' and 'No Go' snap gauges for sizes over 100 mm and upto 250 mm.

Fig. 4.42

4.12.5.2. Plate snap gauges. Double-ended type plate snap gauges are used for sizes in the range of 2 to 100 mm and single-ended progressive type in the size range of 100 to 250 mm. These

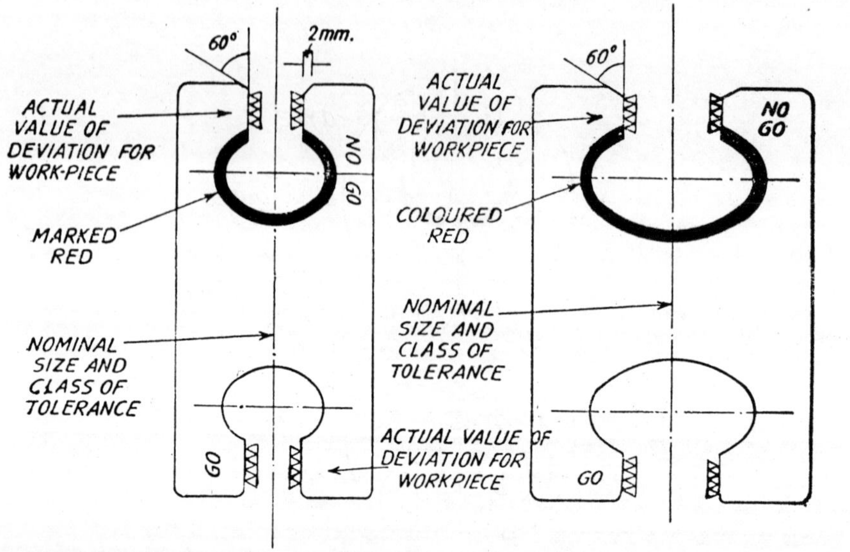

For sizes upto 10 mm For sizes above 10 mm

Fig. 4.43

are made of wear resistant steel of suitable quality. The gauging surfaces are properly hardened, stabilised, ground and lapped. Other surfaces are smooth finished. The gauges are reasonably flat and all sharp corners and edges are removed.

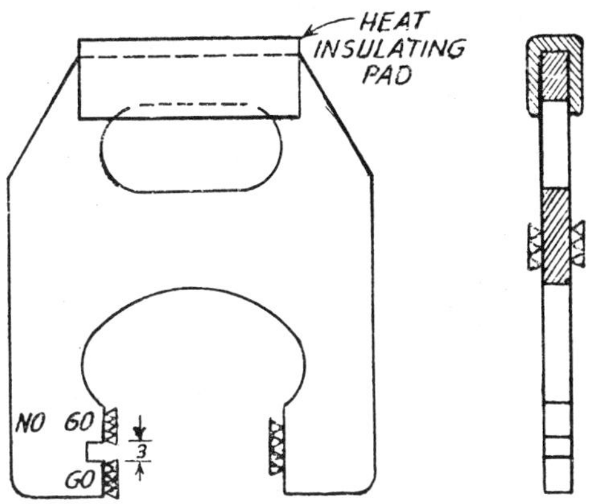

Plate snap gauges over 100 mm and upto 250 mm
Fig. 4.44

4.12.6. Adjustable Type Gap Gauges. In these gauges the gauging anvils are adjustable endwise in the horse-shoe frame. This type of gauge is set by means of slip gauges to any particular limit required. It is possible to set well made gauges to within about 0.002 mm of a desired size and thus their use enables fuller advantage to be taken of the manufacturing tolerance on the work than when solid gauges with an appreciable manufacturing tolerance of their own are employed. The adjustability also enables wear of the 'Go' anvils to be taken up at any time. If the anvil faces lose their flatness with use, they can be reground quite readily after removing them from the frame. The method of adjusting the anvils is usually by means of independent and finely threaded screws at the back end, and they are finally locked in position by some type of clamping screw.

For gap gauges the following conditions must be fulfilled :

(*i*) The frame should be of rigid design, and sufficiently strong to withstand workshop conditions.

(*ii*) In order to provide suitable supports for clamping during manufacturing and subsequent re-grinding, each side of the frame should have three finished coplanar faces, and these faces should be parallel.

(*iii*) The gauging anvils should have only a sliding and not a rotating movement for adjustment.

(*iv*) The anvils should have a sufficient length of bearing on their shanks to ensure parallel movement when being adjusted and also to obviate any tendency to tip when being locked or when in use.

(*v*) All adjustable parts should be close fitting.

(*vi*) The means of adjusting the gauging anvils should be simple.

(*vii*) The locking device for the anvils should be quite definite in its action.

(*viii*) There should be suitable provision for sealing the adjustment to prevent unauthorised readjustment.

(*ix*) The distance between the 'Go' and 'No Go' anvil should be sufficient, when the gauge is set for work of the largest diameter and largest tolerance, to enable the work to be in a free position when passing the 'Go' anvils and before meeting the 'No Go' anvils.

(*x*) The range over which the gauge can be adjusted should be marked on the frame.

4.12.7. Combined Limit Gauge. (Fig. 4.45). In the case of gauging of cylindrical holes, it is possible to combine both the 'Go' and 'No Go' dimensions of a plug gauge and thus a single gauge doing the work of checking both the upper and lower limits. It is achieved by employing a spherically ended gauge of the same diameter as the lower limiting dimension of the hole. Towards the outer edge of the spherical member, a spherical projection S is provided which is arranged in such a way that the distance from the surface S to the diametrically-opposite point on the spherical surface is equal to the maximum limiting dimension of the hole. For checking the hole for 'Go' position, the gauge is inserted into the hole with the handle parallel to the axis of the hole. For checking the hole for 'No Go' limit, the gauge is tilted so that the spherical projection S is normal to hole. The gauge in this position should not enter the hole.

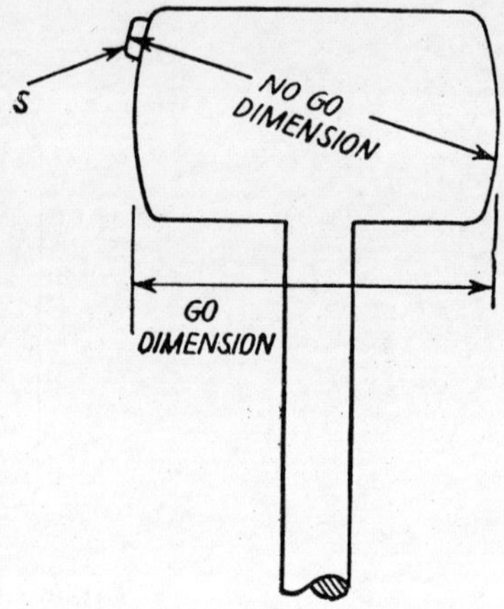

Fig. 4.45. Combined Limit Gauge.

Other types of gauges in common use employ pilot gauging methods. In pilot gauges, the possibility of the jamming of ordinary plug gauge in the hole when inserted obliquely, is eliminated. It is achieved by first proving a chamfer at the end of plug gauge and then machining a narrow ring or pilot after it.

4.12.8. Position Gauge. There is a wide variety of position gauges in common use and their design is based upon the shape of the work. Thus, practically a different position gauge is required for each design of work. The position gauges are employed for checking the position of some feature on the work in relation to another reference point or surface. Their design can be based either on the principle of sighting the gauge and work, or on the method of feel.

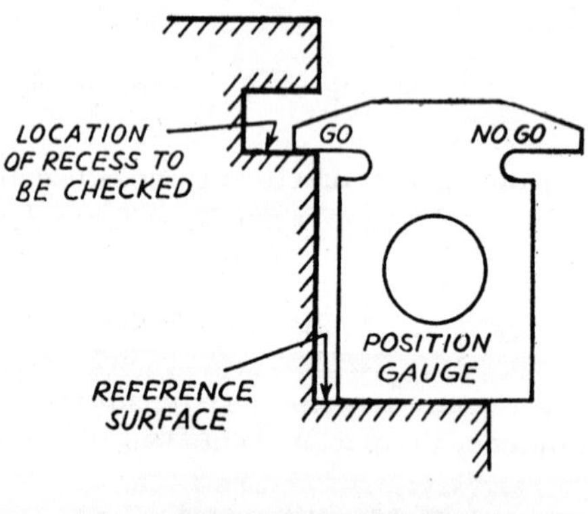

Fig. 4.46. Position gauge.

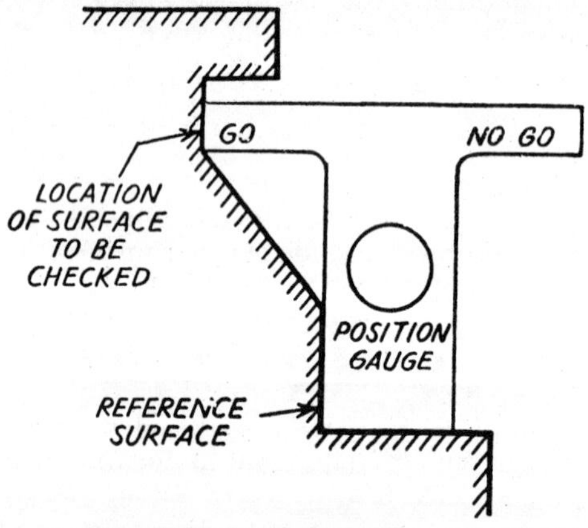

Fig. 4.47. Position gauge.

An simple gauge for checking the location of a recess in relation to a flat surface is shown in Fig. 4.46. Another simple design in which the location of surface parallel to the reference surface is to be located is shown in Fig. 4.47. It may be noted that no light will pass between the reference surface and gauge surface in contact with 'Go' side and light will pass with 'No Go' side.

4.12.9. Contour Gauges. These are employed for checking the dimensional accuracy and shape of irregular work. The contour gauges are made of similar profile as that of the work, *e.g.* radius gauges are employed for checking the shape of fillets. The radius gauge which is used for gauging both internal and external radii, contains a number of leaves having different radii profile and their radius being marked on the leaf. These leaves are hinged together at both the ends of a base member provided with a clearance section hole and whole assembly is in the form of a pocket-knife. Any leaf of desired radius can be slid outside and the rest remaining inside. These leaves are made of hardened steel plates.

Other types of simple contour gauges are those for thread pitches, cutting tool angles, form tools profiles and gear tooth profiles, etc.

4.12.10. Advantages of Fixed Gauges. The advantages of fixed gauges are :

(*i*) These are consistent in form and dimension and thus virtually free from errors due to drift of the original adjustment, non-linear response, and effect of power variation and other extraneous factors, which call for frequent calibration.

(*ii*) Except for careful handling, human errors are nil.

(*iii*) These provide unambiguous 'yes' or 'no' decision regarding acceptability of inspected part.

(*iv*) These do not require auxiliary equipment and setups.

(*v*) These are portable and independent of power supply requirements.

(*vi*) These can be designed to check combinations of several dimensions comprising lengths, diameters and angles. Other types of fixed gauges serve to inspect interrelated features for size, location, form, alignment etc. to check the combined effect of several parameters with regard to the functional adequacy of the inspected feature, by determining the virtual size of the component.

(*vii*) These provide uniform reference standards.

(*viii*) Fixed gauges are particularly useful in the checking of part members whose meaningful geometric irregularities cannot be readily detected by gauges which do not provide complete reverse replicas of the critical part portion.

(*ix*) Their moderate cost results in overall economy.

4.12.11. Miscellaneous Gauges. Gauges are used not only for shaft and hole checking but can be designed for all types of applications for checking purposes in production shop and also for gauge making and checking. Such gauges are usually complex.

As in slip gauges and angle gauges, all other gauges can also be of four types *viz.* master, reference, check and workshop gauges. Master gauge, which is ultimate reference, is used for controlling the reference gauges and can be found only in standards room. It is properly calibrated and its dimensional accuracy is beyond question. Reference gauge is used for checking check gauges and for arbitration purposes in case of doubt, and is reserved for reference purposes in an inspection department. Check gauge is used for testing the accuracy of workshop gauges. The general checking of component parts is carried out with workshop grade gauges.

4.12.12. Position Gauges. Position gauges are used to check the geometrical relationship of specific features, such as distance between two holes, distance of a hole from a reference surface, etc. Such gauges need not necessarily always be of solid form but can be designed with loose 'Go' and 'No Go' feeler pieces which need to be inserted between the workpiece and the gauge surfaces in contact after the workpiece is inserted in position. In some applications dial indicators are also used in conjunction with setting masters for contacting the work at the points to be gauged. Such arrangements result in decreased cost of gauges and the increased efficiency.

4.12.13. Receiver Gauges. Receiver gauges are designed to simultaneously check a number of features of workpiece. These are often used for checking components before assembly.

4.12.14. Profile Gauges. Profile gauges are used to check the form of the components. Profiles are difficult to be checked by limit gauges and it is usual practice to use fixed gauges mated to profile for checking profiles. There are two methods of tolerancing the form of profile (consisting of straight lines and curves.)

(*i*) To provide a tolerance zone within which the finished profile must lie. This method provides a uniform metal tolerance normal to the profile.

(*ii*) To use ordinates which are provided with individual tolerances. In this method the tolerance, normal to the surface, will vary with the form of the profile.

4.13. Gauge Design

To a greater or lesser extent, every gauge is a copy of the part which mates with the part for which the gauge is designed. For example, a bushing is made which is to mate with a shaft ; in this case, the shaft is the opposed (mating part) part. The bushing is checked by a plug gauge, which in so far as the form of its surface and its size is concerned, is a copy of the opposed part (shaft).

If a gauge is designed as an exact copy of the opposed part in so far as the dimension to be checked is concerned, it is called a 'Standard Gauge'.

In design of a gauge, simplicity should be the main aim as simple gauges can take measurements continuously and accurately.

4.13.1. Taylor's Principle. According to Taylor, 'Go' and 'No Go' gauges should be designed to check maximum and minimum material limits which are checked as below.

'Go' Limit. This designation is applied to that limit of the two limits of size which corresponds to the maximum material limit considerations, *i.e.* upper limit of a shaft and lower limit of a hole. The form of the 'Go' gauge should be such that it can check one feature of the component in one pass.

'No Go' Limit. This designation is applied to that limit of the two limits of size which corresponds to the minimum material condition, *i.e.* the lower limit of a shaft and the upper limit of a hole.

'No Go' gauge should check only one part or feature of the component at a time, so that specific discrepancies in shape or size can be detected. Thus a separate 'No Go' gauge is required for each different individual dimension.

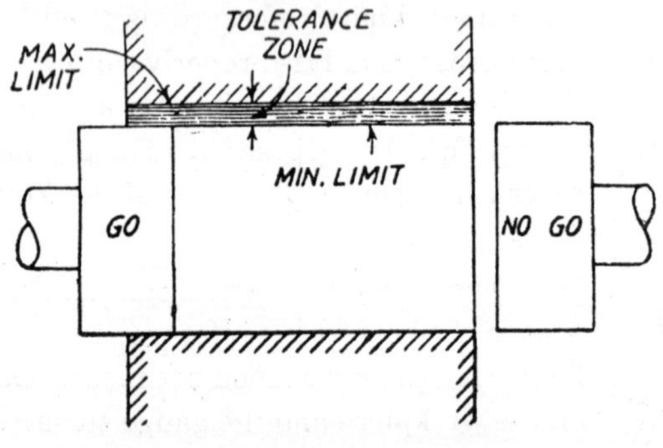

Fig. 4.48. Plug gauge.

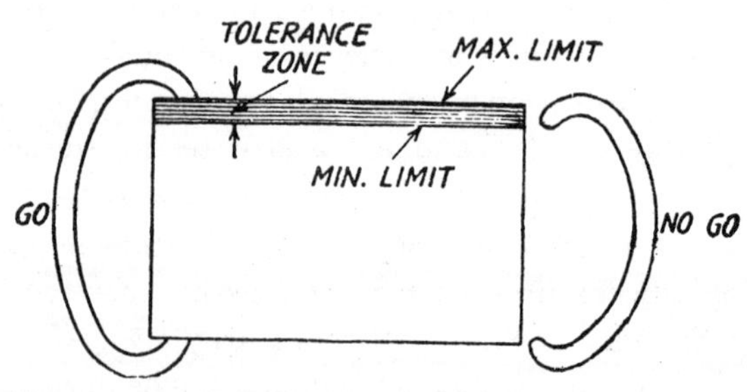

Fig. 4.49. Snap gauge.

The 'Go' plug gauge (Fig. 4.48) is the size of the minimum limit of the hole, while the 'No Go' plug gauge corresponds to the maximum limit.

The 'Go' snap gauge (Fig. 4.49) on the other hand, is of a size corresponding to the maximum limit of the shaft, while the 'No Go' snap gauge corresponds to the minimum limit. Gauging faces of a normal snap or gap gauge must be parallel and square to each other and the gauging points of contact with the work should be in the same plane. The difference in size between the 'Go' and 'No Go' snap gauges, as well as the difference in size between the 'GO' and 'No GO' plug gauges, is approximately equal to the tolerance of the tested hole or shaft in case of Standard Gauges, Rigidity and robustness of snap gauges are important features so that gauges function adequately and maintain size. Gauging diameters of components that are slightly larger than the gap setting can produce high wielding action which may lead to gauge distortion and wrong interpretation of reading. Therefore, larger gap gauges should, preferably, be forged in a deep I-section, ensuring maximum rigidity in the plane of gauge and sufficient rigidity in lateral direction.

Taylor's principle states that the 'Go' gauges should check all the possible elements of dimensions at a time (roundness, size, location etc.) and the 'No Go' gauge should check only one element of the dimension at a time.

To 'Go' plug gauge must be of corresponding section and preferably full length of hole so that straightness of hole can also be checked. Thus it not only controls diameter in any given section but also ensures bore alignability. However it cannot check the degree of ovality.

The 'No Go' plug gauge is relatively short and its function is dependent not only on the diameter but also on the circularity of the hole. Thus to some extent, variation of hole shape can be measured.

4.13.2. Wear Allowance Consideration on Gauge Maker's Tolerance. Since the gauge maker can't make absolutely accurately gauges, permissible deviation in accuracy must be assigned for gauge manufacture. Furthermore, the measuring surfaces of 'Go' gauges which constantly rub against the surfaces of parts in inspection, are consequently subjected to wear and lose their initial initial size. Thus due to wear, the size of 'Go' plug gauges is reduced, while that of 'Go' snap gauges is increased. It is of course desirable to prolong the service life of gauges and, therefore, a special allowance of metal, the wear allowance is added in a direction opposite to the wear. For this reason new 'Go' plug gauges are made with two positive deviations and 'Go' snap gauges with two negative deviations from the nominal size. (The nominal size on which limits of gauges are based are the limits of the parts to be checked.)

4.13.3. Important Points for Design. (1) The form of 'Go' gauges should exactly coincide with the form of the opposed (mating) parts.

(2) 'Go' gauges are complex gauges which enable several related dimensions to be checked simultaneously.

(3) In inspection, 'Go' gauges must always be put into conditions of maximum impassability.

(4) 'No Go' gauges are gauges for checking a single element of feature.

(5) In inspection, 'No Go' gauges must always be put into conditions of maximum passability.

4.13.4. Gauge maker's Tolerance. Keeping all above main points for gauge design in view there are three methods of giving tolerances on gauges (snap and plug gauges).

4.13.4.1. First system. Workshop and Inspection Gauges (Fig. 4.50). In this method workshop and inspection gauges are made separately and their tolerance zones are different. This was evolved many years ago in the development stage of limit gauges.

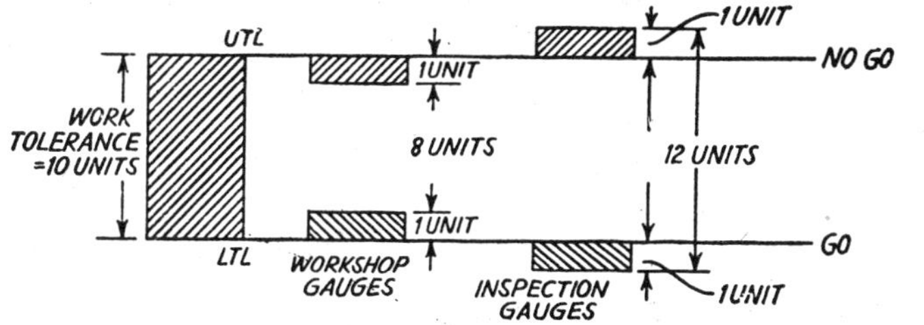

UTL—Upper Tolerance Limit
LTL—Lower Tolerance Limit

Fig. 4.50. Disposition of tolerances on workship and inspection gauges.

According to this system the tolerances on the workshop gauge are arranged to fall inside the work tolerance, while the inspection gauge tolerances fall outside the work tolerance. Further in workshop gauges, 'Go' gauge should eat away 10% of work tolerance and similarly the tolerance on 'No Go' gauge should be one-tenth of work tolerance, so if work tolerance is 10 units then only 8 units will be left as the difference between the minimum of 'No Go' and maximum of 'Go' ; the tolerance on 'Go' as well as 'No Go' gauges individually being 1 unit each.

In Inspection Gauges, gauges are kept beyond work tolerance by 10% of its value.

Disadvantages of Workshop and Inspection Gauges.

(1) Some of the components which are in work tolerance limits may be rejected under workshop gauges. So they are again checked by inspection gauges and may be accepted after that.

(2) Some components which are not in work tolerance limits may be accepted when tested by inspection gauges.

(3) The workshop and inspection gauges are to be made separately as their tolerance zones are different.

4.13.4.2. Second system (Revised Gauge Limits). Under this system the disadvantages of inspection gauges are reduced by reducing the tolerance zone of inspection gauge, and the workshop gauge tolerance remains the same.

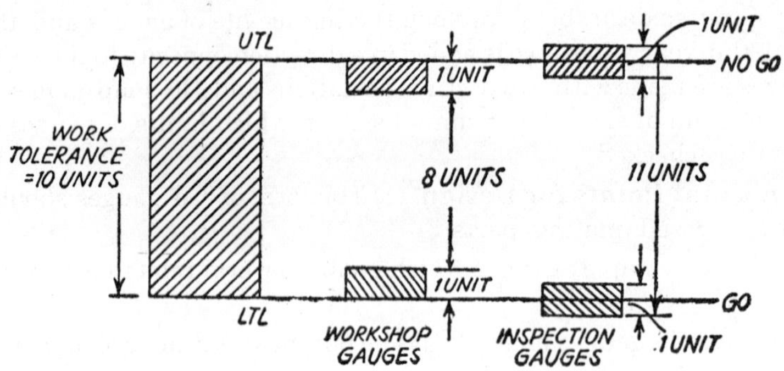

Fig. 4.51. Modified tolerances on inspection gauge.

For 'Go' and 'No Go' inspection gauges in this system, the 110% of the range of work tolerance is covered instead of 120% in the first system as shown in Fig. 4.51.

4.13.4.3. Third system (Present British System). In new system, following principles are followed along with Taylor's principle.

(*i*) Tolerance should be as wide as is consistent with satisfactory functioning, economical production and inspection.

(*ii*) No work should be accepted which lies outside the drawing pacified limits.

Thus modern system dispenses with workshop and inspection gauges and we give the same tolerance limits on workshop and inspection gauges and the same gauges can be used for both purposes.

The tolerance zone for the 'Go' gauges should be placed inside the work-limits and tolerance for the 'No Go' gauges outside the work-limits. Provision for wear of 'Go' gauges is made by introduction of a margin between the tolerance zone for the gauge and maximum metal limit of the work. Wear should not be permitted beyond the maximum metal limit of the work, when the limit is of critical importance. Its magnitude is one-tenth of the gauge tolerance. Thus when work tolerance is less than 0.09 mm there is no need of giving allowance for wear. If work tolerance is more than 0.09 mm then 10% gauge tolerance is given only on 'Go' gauge for wear.

The disposition of various tolerances and allowances on gauges according to this system is shown in Fig. 4.52 (*a*) and (*b*) and further clarified by the following example.

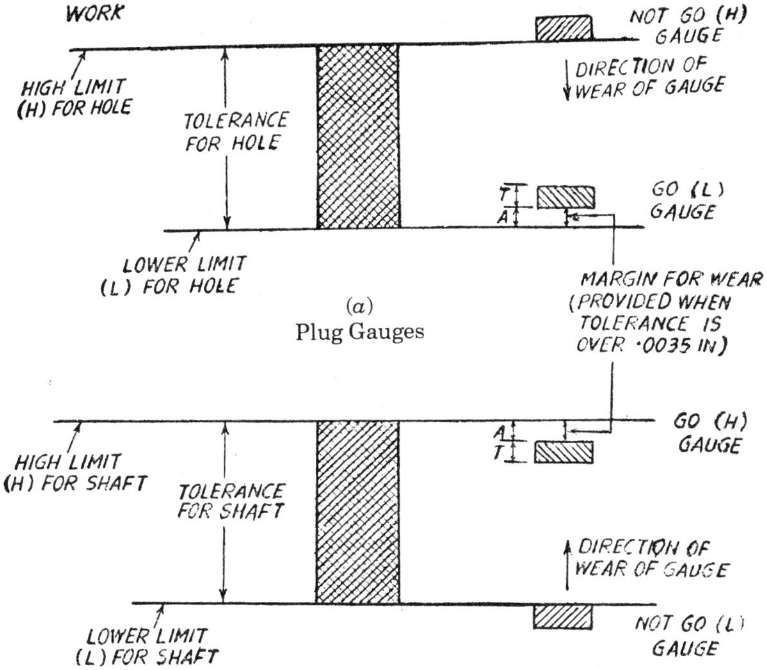

(a)
Plug Gauges

(b)
Gap and Ring Gauges

Fig. 4.52. Provision of margin for wear on GO gauges.

Example. *Calculate the dimensions of plug and ring gauges to control the production of 50 mm shaft and hole pair of H_7d_8 as per I.S. specification. The following assumptions may be made : 50 mm lies in diameter step of 30 and 50 mm and the upper deviation for 'd' shaft is given by— – $16D^{0.44}$ and lower deviation for hole H is zero. Tolerance factor i (microns) = $0.45\sqrt[3]{D} + 0.001D$ and IT6 = 10i and above IT6 grade the tolerance magnitude is multiplied by 10 at each fifth step.*

Solution. For calculation of tolerance, value of diameter is taken as the mean of range in which it lies.

$$D = \sqrt{30 \times 50} = \sqrt{1500} = 10\sqrt{15} = 38.68 \text{ mm} \simeq 38.7 \text{ mm.}$$

Value of tolerance unit $i = 0.45\sqrt[3]{D} + 0.001D$

$$= 0.45\sqrt[3]{38.7} + 0.001 \times 38.7 = 0.45(3.38) + 0.0387$$

$$= 1.521 + 0.0387 = 1.5597 \text{ microns} \simeq 0.00156 \text{ mm.}$$

Now hole is of type H and grade 7

$$IT7 = IT6 \times 10^{0.2} = 10i \times 10^{0.2} = 10i \times 1.585 = 15.85i$$

and for H hole, fundamental deviation = 0

and value of tolerance = 15.85 × 0.00156 = 0.02475 mm.

∴ For hole $H7$ disposition of work tolerance will be as shown in Fig. 4.53.

For shaft d_8

Tolerance $= IT8 = 10^{0.2} \times IT7$

$= 1.585 \times 15.85i = 25i$ (approx.)

$= 25 \times 0.00156 = 0.0391$ mm.

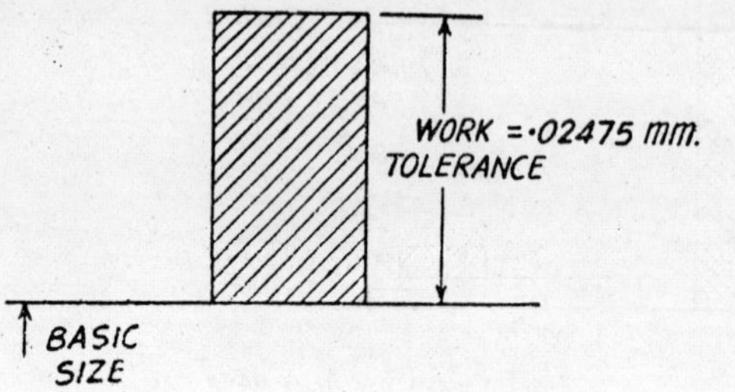

Fig. 4.53. Work tolerance for hole H_7.

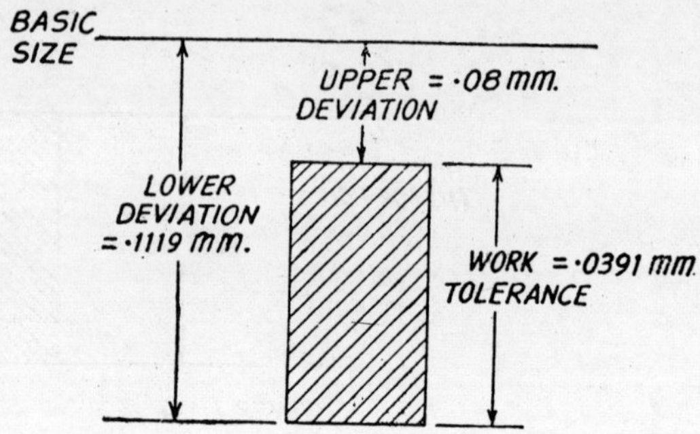

Fig. 4.54. Tolerance and deviations for shaft.

Fundamental deviation for '*d*' shaft

$$= -16D^{0.44} = -16 \times 38.7^{0.44}$$

$$= -16 \times 5 = -80 \text{ microns} = -0.08 \text{ mm.}$$

Upper deviation $= -0.08$ mm.

and Work tolerance $= -0.0391$ mm.

$\therefore$ Lower deviation $= -0.08 - 0.0391 = -0.1191$ mm.

Various tolerances and deviations for the shaft are shown in Fig. 4.54.

Thus the upper and lower deviations and tolerance zone for both shaft and hole have been determined.

Considering for Gauges

(1) *Plug gauges.* According to new system, 'Go' gauges are given 1/10th of work tolerance in the tolerance zone and 'No Go' gauges, outside it.

Since work is less than 0.09 mm.

$\therefore$ Effect of wear in 'go' gauges is not considered.

Work tolerance $= 0.02475$ mm.

$\therefore$ Gauge tolerance $= 0.002475$ mm.

$\therefore$ Limits for Go' gauges are $\left.\begin{array}{l} 50.000 \\ 50.0025 \end{array}\right\}$ mm.

and for 'No Go' gauge, dimensions are

$$\left.\begin{array}{r} 50 + 0.0248 = 50.0248 \\ 50.0248 + 0.0025 = 50.0273 \end{array}\right\} \text{mm.}$$

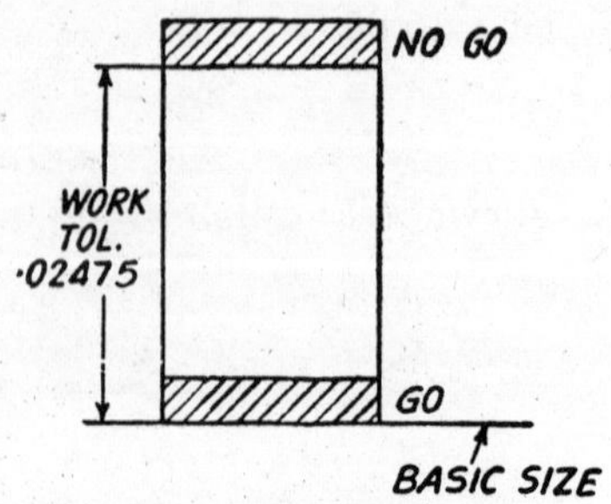

Fig. 4.55. Limits for plug gauge.

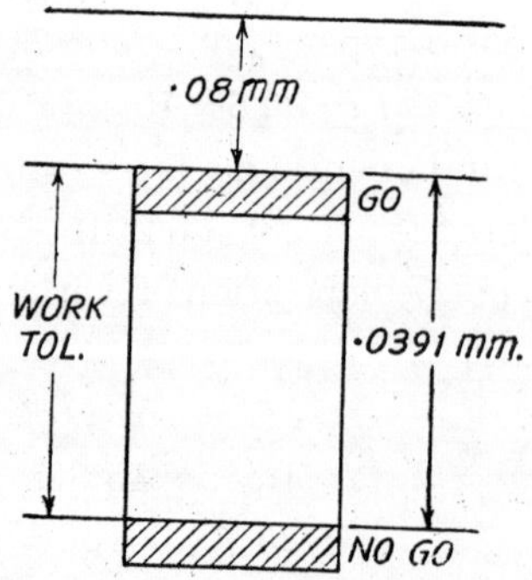

Fig. 4.56. Limits for ring gauge.

(2) *Ring gauges.* The various dimensions of Gap gauges are shown, in Fig. 4.56 and calculations made in a similar manner as for plug gauges.

Dimensions for 'Go' gauge are

$$50 - 0.08 = 42.92 \text{ mm.}$$

and $$49.92 - 0.0039 = 49.9161 \text{ mm.}$$

Dimensions for 'No Go' gauge are

$$50 - (0.38 + 0.391) = 49.8809 \text{ mm}$$

and $$49.8809 - 0.0039 = 49.8770 \text{ mm.}$$

Answers : Plug Gauge. Dimensions for

'Go' Gauge : $\left.\begin{array}{l}50.000\\50.0025\end{array}\right\}$ mm. *'No Go' gauge :* $\left.\begin{array}{l}50.0248\\50.0273\end{array}\right\}$ mm.

Ring Gauge : Dimensions for

'Go' gauge : $\left.\begin{array}{l}49.9161\\49.9200\end{array}\right\}$ mm. *'No Go' gauge :* $\left.\begin{array}{l}49.8770\\49.8809\end{array}\right\}$ mm.

4.14. Indian Standards Gauging Practice for Plain Workpieces

Generally inspection by limit gauges means an acceptable method for dimensional conformity to the specification of plain workpieces. The gauging of the workpieces is carried out both by the manufacturer as well as the purchaser. The tendency of the manufacturer is to pass all such pieces which are very close to the specified tolerance and the purchaser wants to be very much on the safe side. Moreover the purchaser does not rely on the manufacturer's gauges. The dispute requiring the checking of the conformity of the gauges of the manufacturer may be avoided by following the procedure given below.

4.14.1. Inspection by the Manufacturer. The best possible solution to avoid dispute is that the inspection department of the manufacturer uses the same types of gauges as those used in the workshop. Further the difference between the results obtained by the workshop and the inspection department could be avoided by recommending the workshop to use new or slightly worn gauges and the inspection department to use gauges having sizes nearer the permissible limit. Inspection should be carried out at 20°C.

4.14.2. Inspection by the Purchaser. There can be three possibilities : (*i*) The purchaser after ensuring the accuracy of the manufacturer's gauges may use the same ; (*ii*) he may use gauges in accordance with IS : 3455—1971 which have their sizes near the wear limit in order to avoid differences between the results obtained by the manufacturer and the inspector ; (*iii*) he may use any type of gauges provided he ensures that the workpieces, the sizes of which are within the specified limits will not be rejected. Inspection should be carried out at 20°C reference temperature only.

4.14.3. Taylor Principle. This principle is based on the use of a correct system of limit gauges to inspect shafts and holes. According to Taylor it is not adequate to use simple Go gauge on outer dimension only but the shape is an important factor, *i.e.* the Go gauge should be full form gauge and it should be constructed with reference to the geometrical form of the part being checked, in addition to its size. In other words, Go gauge should check all the dimensions of a workpiece in the maximum metal condition. As regards No Go gauges, Taylor was of the view that it was useless for Not Go gauge to be of full form, and each feature being dealt should be checked with a specific Not Go gauge. In other words, Not Go gauge shall check only one dimension of the workpiece at a time, for the minimum metal condition. Thus according to it, a hole should completely assemble with a 'Go' cylindrical plug gauge made to the specified 'Go' limit of the hole, having a length at least equal to the length of engagement of the hole and shaft. In addition, the hole is measured or gauged to check that its maximum diameter is not larger than the 'No Go' limit. The Taylor principle interprets the limit of size for gauging holes and shafts as follows :

4.14.3.1. For holes. If the Taylor principle is followed then the diameter of the largest perfect imaginary cylinder which can be inscribed within the hole so that it just contacts the highest points of the surface, shall not be a diameter smaller than the 'Go' limit of size. Further, the maximum diameter at any position in the hole should not exceed the 'No Go' limit of size.

4.14.3.2. For Shafts. If the Taylor principle is followed then the diameter of the smallest perfect imaginary cylinder which can be circumscribed about the shaft so that it just contacts the highest points of the surface, should not be a diameter larger than the 'Go' limit of size. Further the minimum diameter at any position on the shaft should not be less than the 'No Go' limit of the size.

It may be noted that the Taylor principle does not take care of the error of form, circularity or straightness etc., the tolerances for which should be specified separately.

Thus according to Taylor principle we require a plug ring gauge having exactly the 'Go' limit diameter and a length equal to the engagement length of the fit to be made for checking the 'Go' limit of the workpiece and this gauge must perfectly assemble with the workpiece to be inspected. The other gauge needed is the 'No Go' gauge which contacts the workpiece surface only in two diametrically opposite points and at those two points it should have exactly 'No Go' limit diameter. This gauge should not be able to pass over in the workpiece in any consecutive position in the various diametric directions on the workpiece length.

In certain applications, the Taylor principle cannot be strictly and blindly followed. The following deviations are allowed which basically do not deviate from the principle as such.

For 'Go' Limit. (*i*) In case the manufacturing process assures that the error of straightness will not affect the character of fit of the assembled workpieces, it is advisable to go for standard gauge blanks instead of using full form and length of engagement and make the gauge unnecessarily bulky and cumbersome and avoid the special gauge of exactly same working length for one workpiece.

(*ii*) If gauge happens to be too heavy, only segmental cylindrical bar could be used provided the manufacturing process ensures that errors of roundness will not have any effect on the character of fit of the assembled workpieces.

(*iii*) For shafts, particularly heavy ones ; it is generally not desirable to use full form ring gauges but only gap gauges. But for this purpose the manufacturing process used should take care of the error of roundness (especially lobing) and the error of straightness.

For 'No Go' Limit. The two-point checking devices are not feasible and practicable in actual practice because these are subjected to rapid wear etc. ; these can be safely replaced by small planes/cylindrical surfaces/spherical surfaces. For gauging very small holes and in cases where workpieces may be deformed to an oval by a two-point mechanical contact device, the 'No Go' gauge of full form may have to be used.

4.14.4. Allowable Deviation from the Taylor's Principle. In some applications, difficulties are experienced in conveniently using gauges if they are strictly based on Taylor's principle. Accordingly some deviations may be permitted.

According to Taylor's principle a Go gauge should be of full form having length equal to engagement length of fit, but this is not always necessary. For example, if it is known that the manufacturing process ensures that error of straightness of hole or shaft is so small that it would not affect the desired fit of assembly, then length of Go cylindrical plug or ring gauge may be less than the length of engagement.

For very big holes, the full form gauge may be too heavy and inconvenient to use. Therefore, segmental cylindrical bar or spherical gauge may be used if it can be assumed that the manufacturing process would not produce the error or roundness outside the permissible limits to affect the character of fit.

Similarly 'Go' gap gauge can be used in place of Go cylindrical ring gauge (which is often inconvenient for gauging shafts) provided manufacturing process can ensure the errors of roundness (lobing) and straightness of shaft within permissible limits. The straightness of long shafts of small diameter should be checked separately. Similarly, though Taylor's principle desired use of two-point checking device to check No Go limit, but same is not always necessary in following cases.

Since point contacts are subject to rapid wear, these can be replaced by small plane, cylindrical or spherical surfaces whenever appropriate. The two-point checking device is also found difficult to design and manufacturing for gauging very small holes and for such cases No go plug gauge of full cylindrical form can be used. However, possibility of accepting workpieces having diameters outside the No Go limit should be checked.

In certain cases the non-rigid workpiece may be deformed by 2-point mechanical contact device and in such cases No Go ring or plug gauges of full cylindrical form have to be used.

No Go gauges of full cylindrical form can also be used for thin-walled workpiece, which may be out of round due to heat treatment, but would become circular when such gauges are applied with force just sufficient to convert the elastic deformation into circularity.

4.14.5. Limit Gauges. The various types of limit gauges used for gauging internal diameters or holes are :

(1) *Full form cylindrical plug gauge.* The gauging surface is in the form of an external cylinder. Generally a small circumferential groove is cut near the leading end of the gauge and the remaining short cylindrical surface is slightly reduced in order to act as a pilot. The method of attaching gauge to the handle should be such as not to affect the size and form of the gauge by producing undesirable stresses (Refer Fig. 4.57)

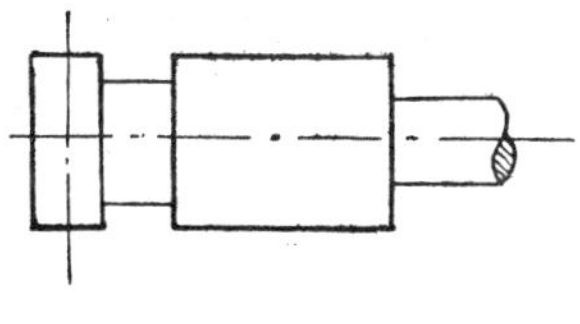

Fig. 4.57. Full form cylinderical plug gauge.

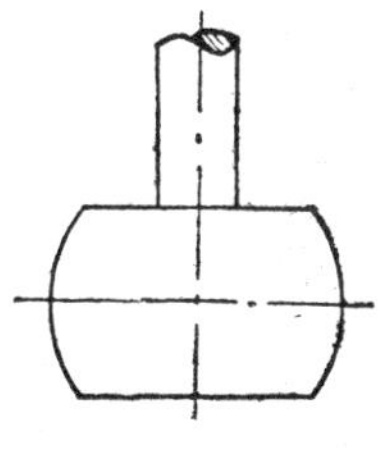

Fig. 4.58. Full form disc gauge.

(2) *Full form spherical plug or disc gauge.* This has gauging surface in the form of a sphere from which two equal segments are cut off by planes normal to the axis of the handle. (Refer Fig. 4.58).

(3) *Segmental cylindrical bar gauge.* It has gauging surface in one of the following two forms : (*i*) External cylindrical form from which two axial segments are made by lowering down surface at other places (Refer Fig. 4.59) or (*ii*) External cylindrical form in which segments are formed by removing remaining material (Refer Fig. 4.60).

(4) *Segmental spherical plug gauge.* It is similar to full form spherical plug or disc gauge but has two equal segments cut off by planes parallel to the axis of the handle in addition to the segments cut off by planes normal to the axis of the handle. (Refer Fig. 4.61).

(5) *Segmental cylindrical bar gauge with reduced measuring faces.* It is similar to segmental cylindrical bar gauge but has reduced measuring faces in a plane parallel to the axis of the handle. (Refer Fig. 4.62).

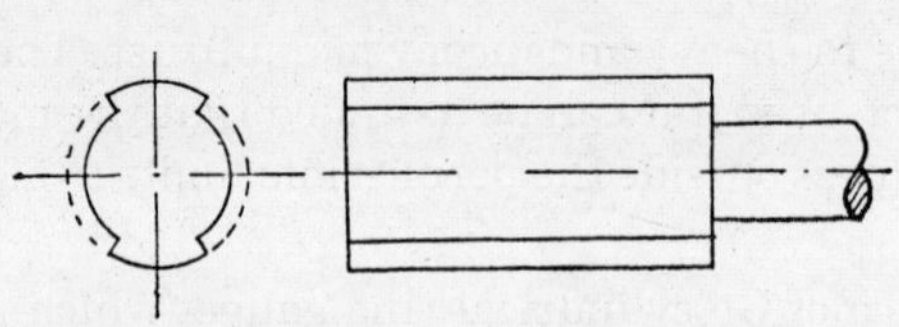

Fig. 4.59. Segmental cylindrical bar gauge.

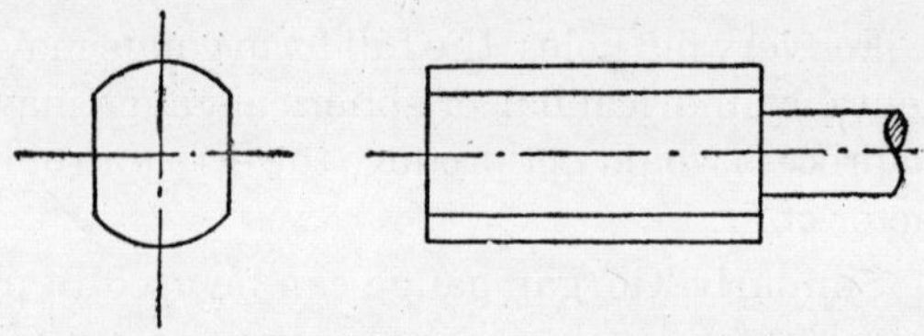

Fig. 4.60. Segmental cylindrical bar gauge.

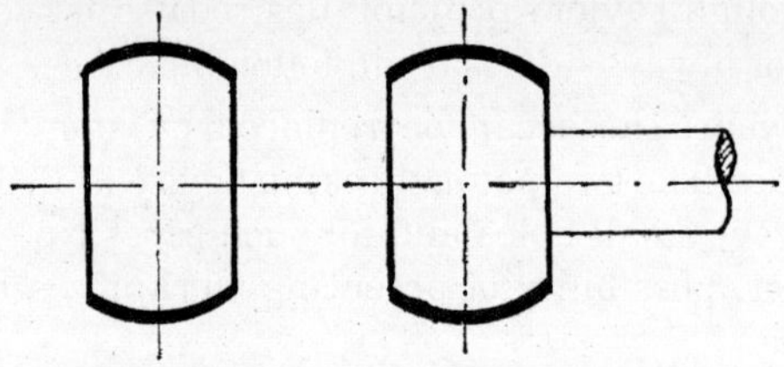

Fig. 4.61. Segmental spherical gauge.

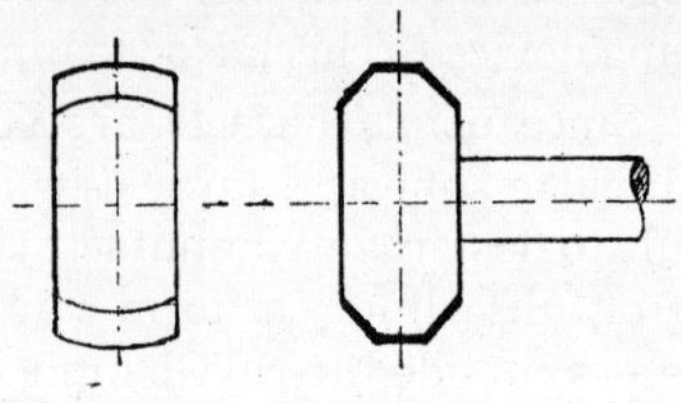

Fig. 4.62. Segmental cylindrical bar gauge.

(6) *Rod gauge with spherical ends.* It has spherical end surfaces which form part of one single sphere. (Refer Fig. 4.63).

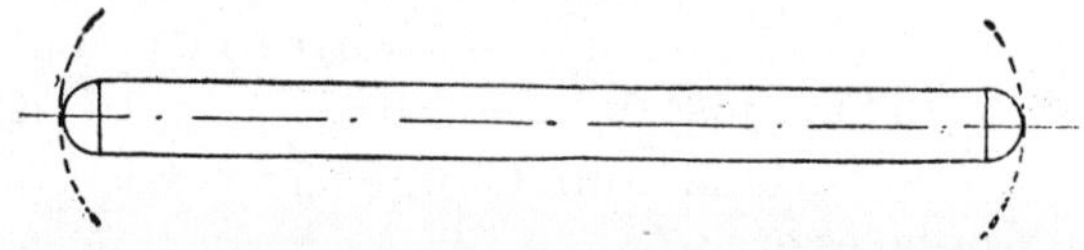

Fig. 4.63. Rod gauge with spherical ends.

The standard limit gauges used for gauging external diameters or shafts are :

(*i*) *Full form cylindrical ring gauge*, which has gauging surface, in the form of an internal cylinder and whose wall is thick enough to avoid deformation under normal conditions of use.

(*ii*) *Gap gauge.* It generally has one flat surface and one cylindrical surface, the axis of the two surfaces being parallel to the axis of the shaft being checked. The surfaces constituting the working size may both be flat or both cylindrical also.

In order to inspect and adjust the limit gauges, the following gauges are used :

(1) **Reference gauges.** These are either in the form of reference discs and intended to be used for setting gap gauges or in the form of cylindrical plug or ring gauges for calibrating gauges or indicating measuring instruments.

(2) **Block gauges.** These are standards of length having parallel plane end surfaces which are used for calibrating gauges of indicating measuring instruments.

4.15. Gauges for Tapers

A taper is tested by using taper plug and ring gauges. The important thing in testing a tapered job is to check the diameter at bigger end and the change of diameter per unit length. For

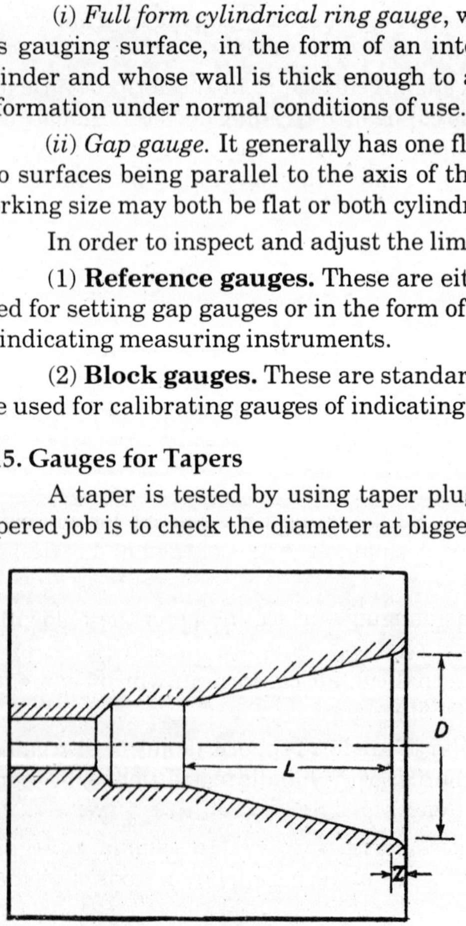

Fig. 4.64. Taper hole gauge.

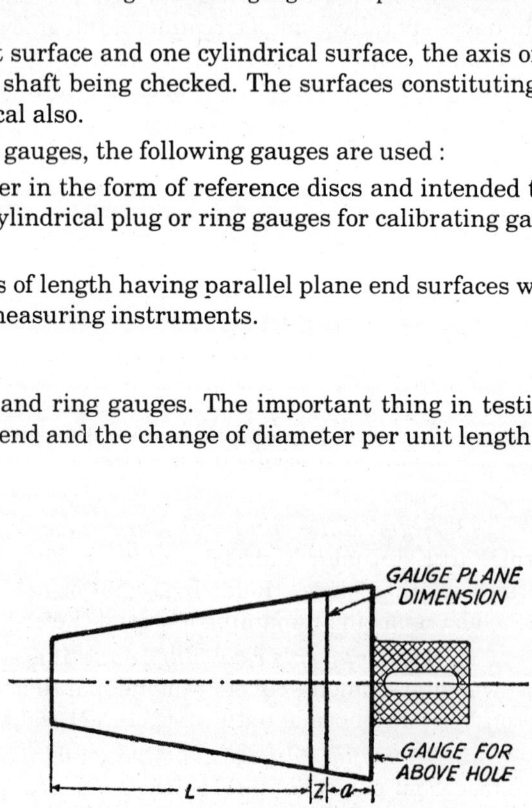

Fig. 4.65. Taper plug gauge.

testing the correctness of a taper, three light lines are drawn with persian blue about equidistant along the length on the (male portion) plug gauge or spindle to be tested and it is fitted in the gauge (female) and rotated once or twice. If persian blue marks do not rub off evenly, the taper is incorrect and setting must be adjusted until persian blue marks are rubbed equally all along its length.

4.15.1. Limit Gauges for Tapers. For checking the diameter of bigger end, a mark is drawn on the gauge to show the correct dimension for large end of taper. It is also known as gauge plane and taper actually starts from this plane. For bigger diameter (which is situated in gauge plane) the tolerance is ± *IT*6. Tolerance on smaller diameter will also be ± *IT*6 on basic size calculated from taper ratio.

Distance of bigger diameter from face depends upon the value of diameter itself *e.g.* for 200 mm diameter it is 2 mm and varies from 0.5 to 2 mm. In Fig. 4.65, it is represented by '*Z*' length for hole gauge.

On plug gauge, gauge plane is located at distance '*a*' from free end and it varies from 2 mm to 12 mm. Tolerance for plug gauge on basic size is + *IT*5.

The diameter is tested for size by noting how far the gauge enters the tapered hole or the tapered spindle enters the gauge.

Wear allowance on gauges is not considered because of high costs involved in manufacturing to that accuracy.

A double ended gauge in this instance is unnecessary because the difference in diameter between the high and low limit of the tolerance is converted into a lengthwise dimension and thus the hole will be within the tolerance, when on inserting the gauge, the large diameter of hole falls between the gauge plane and the other line at distance *Z* from it.

4.16. Plug and Ring Gauges for Self-holding Tapers

The most common use of self-holding tapers is in the machine spindle noses and tool shanks for the purpose of accurate alignment of tool and transmitting the torque. Tapers can be both internal as well as external and thus plug and ring gauges have to be employed to check them. The use of these gauges, of course, cannot help in arriving at the absolute deviation of any particular

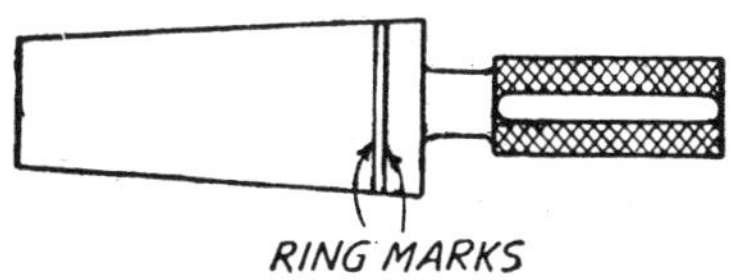

Fig. 4.66. Plug gauge, plain.

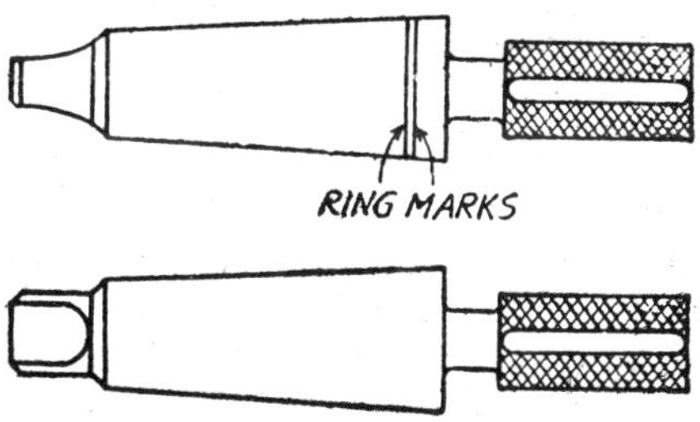

Fig. 4.67. Plug gauge, tanged.

dimension, but nevertheless it offers the easiest and quickest method of verifying inter-changeability of dimensions. The gauges are made from good quality steel and suitably heat treated for hardness (750 to 860 HV) and stabilisation. The gauges are generally well finished and free from crack, burrs, rust or other defects. The gauging surfaces are ground, lapped and covered with a suitable rust proof coating and packed in non-absorbent paper and packed in boxes for safe handling. Thus the plug and ring gauges for the self-holding tapers are of four types as shown in Figs. 4.66 to 4.69.

4.16.1. Plug Gauges, Plain. This is a full form 'Go' gauge representing a plain or tapped end shank. It is provided with one ring marked on the gauge plane and another ring to indicate the minimum depth of the internal taper.

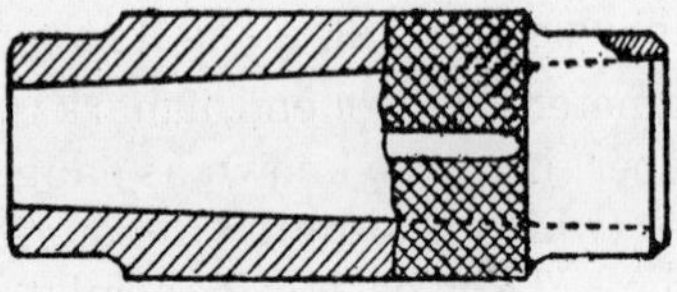

Fig. 4.68. Ring gauge, plain.

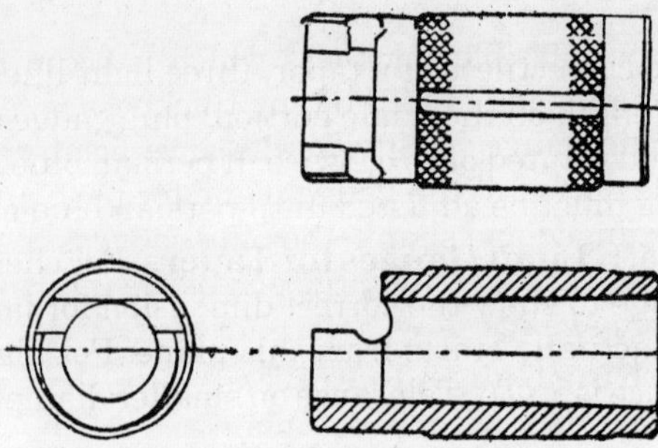

Fig. 4.69. Ring gauge, tanged.

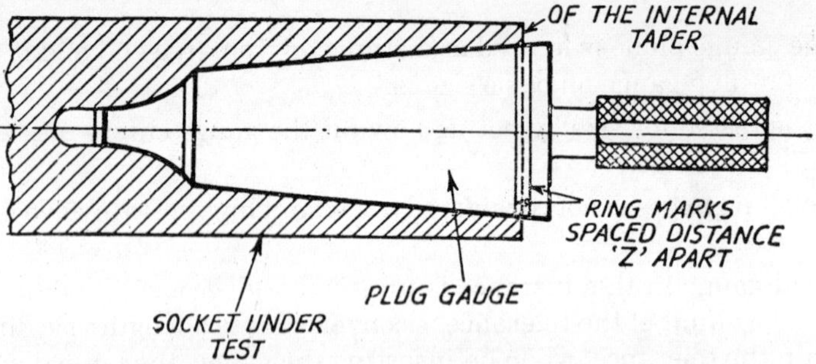

Fig. 4.70. Testing taper of socket with plain plug gauge.

As it is used for the verification of internal taper of corresponding size and thus the dimensions of the gauging portion represent those of a plain end shank of basic size. The distance between the two ring marks 'Z' corresponds to the permissible deviation of the gauge plane position for the particular taper.

For testing the internal taper, the plug gauge is inserted as far as it goes with light pressure. At the extreme point the face of large end of internal taper shall lie within the two ring marks given on the plug gauge.

4.16.2. Plug Gauge, Tanged. This is a full form 'Go' gauge representing the virtual size of the shank of basic dimension having a tang which is on the top limit of its thickness and eccentric by the maximum permissible amount. It is used for the verification of internal tapers in machine tools or sockets and to ensure that the tang slot will accept the tang. It is marked with a ring on the gauge plane and another ring to indicate the minimum depth of internal taper.

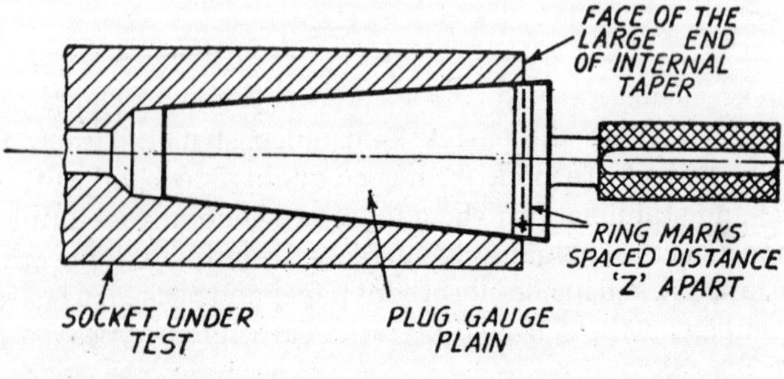

Fig. 4.71. Testing taper of socket with tanged plug gauge.

As it is intended to be used for the verification of internal taper of corresponding size, the plug gauge represents the size of the tanged end shank of the basic dimensions but having the tang of maximum permissible thickness which is eccentric by the maximum permissible amount. A further 0.01 mm is added to the tang thickness on either side to allow for the eccentrictiy of the tang of the plug gauge itself. The distance between the two ring marks '*Z*' corresponds to the permissible deviation of the gauge plane for particular taper. For testing the internal taper of a socket with a tanged end shank, the plug gauge is inserted as far as it goes with light pressure. At the extreme point, the face of large end of the internal taper should lie within the two ring marks given on the plug gauge.

4.16.3. Ring Gauge, Plain. This gauge represents an internal taper of basic size. It is used for verifying the taper of tapped or plain end shank.

As the ring gauge (plain) is intended to be used for verification of internal taper of plain end or tapped end shanks of corresponding size, the dimensions of its gauging portion represent an internal taper of basic size. For inspecting the external taper of a tapped or plain end shank, the

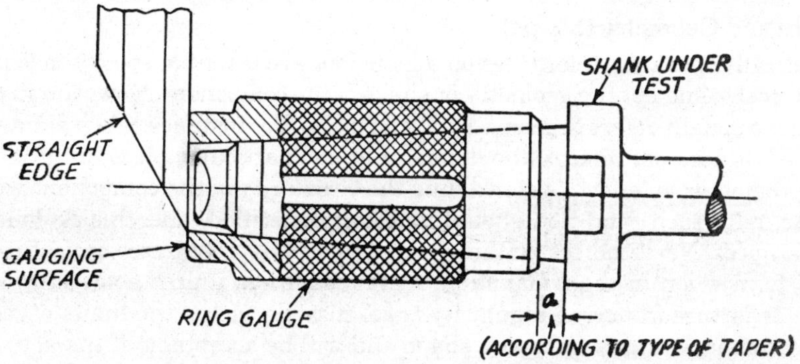

Fig. 4.72. Testing taper with ring gauge (plain).

ring gauge is inserted, as far as it goes with light pressure. At the extreme position, the small end of taper shank under test should lie flush or short of the face of the ring gauge on the small end. This can be verified with the help of a straight edge.

4.16.4. Ring gauge, Tanged. The gauge represents an internal taper of basic size. A limit step is provided at the small end of taper which verifies the length of shank from the gauge plane and also the combined effect of tang thickness and its offset. A ring gauge, tanged represents the basic dimensions of internal taper with a tang slot as it is used for the verification of the external taper of a tanged end shank of corresponding size. A step is provided to verify that the tang slot

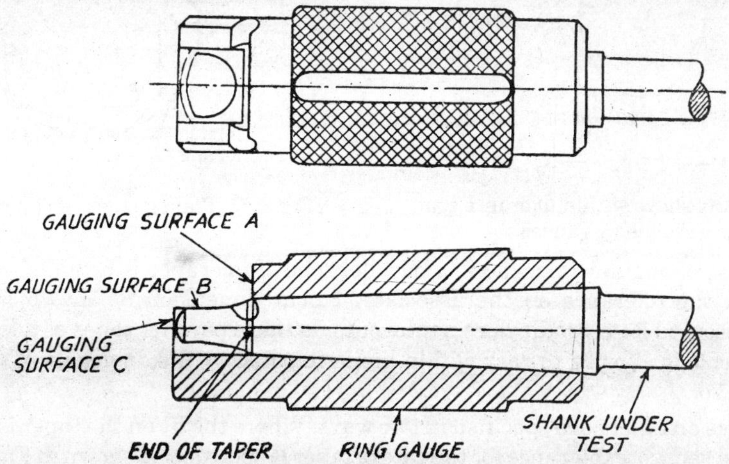

Fig. 4.73. Testing external taper with ring gauge (tanged).

will accept a shank even in the worst case. The step provided takes into account the maximum thickness of the tang which is eccentric by the maximum permissible amount and a further tolerance of 0.01 mm is allowed to offset the effect of variation in the manufacture or measurement of the ring gauge itself.

For testing the external taper of the tanged end shank, the ring gauge is inserted, as far as it goes with light pressure. At the extreme position, no part of the tang under test should extend beyond the surfaces *A, B* and *C* (Fig. 4.73). The shank surfaces may however, lie flush with these surfaces.

The plug and ring gauges are generally marked legibly and indeligible with the designation of taper manufacturer's name, initials or recognised trade mark and the year of manufacture. The gauges are designated by the (*i*) commonly used name of the gauges, (*ii*) designation of taper and (*iii*) number of the standard followed.

4.17. Tolerance and Geometry (Shape)

Theoretically speaking, tolerances on dimensions are meant to specify the extent of permissible variation of size and not the geometry or shape of the component. However, in actual practice, a given dimension usually refers to geometry or shape also. For instance, when diameter of a circular part is specified, it does not refer to the diameter at one particular place but at all the positions. Thus a single diameter, in effect, is controlling the circularity of the component without its being mentioned. Actually such condition should be clearly specified, and this is done by indicating positional tolerances. Nevertheless, within given dimensional tolerances on features these should be regular in form when in their maximum metal condition, unless something is stated to the contrary. The departure from true regularity is permissible within the limits of tolerance zone so established. A component may have any shape and will be acceptable if the actual profile of the feature lies wholly within the given tolerance zone. For example, for a shaft having diameter of 25 ± 0.02 mm, shown in Fig. 4.74, an oval shaft as long as within two circles of diameter 24.98 and 25.02 mm will be acceptable. (Refer Fig. 4.75, where even a circular or quadrilateral or any shaped part would be acceptable for a hexagonal shape, if the circle or quadrilateral or any shape lies within the tolerance zones).

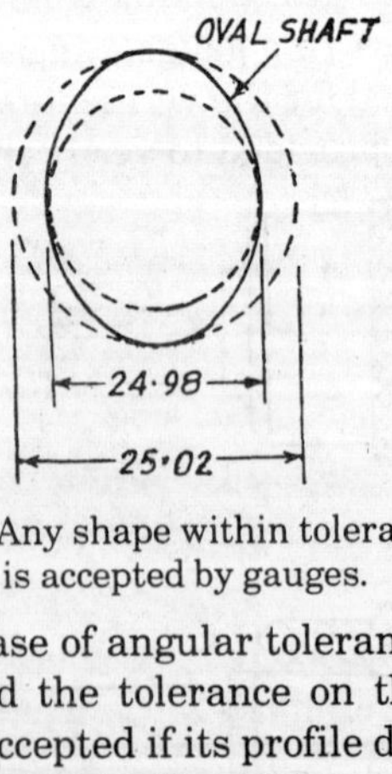

Fig. 4.74. Any shape within tolerance zone is accepted by gauges.

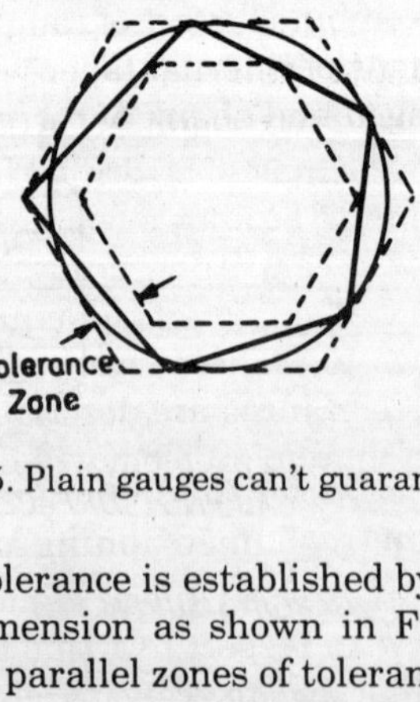

Fig. 4.75. Plain gauges can't guarantee shape.

In the case of angular tolerances, a parallel zone of tolerance is established by considering fixed angle and the tolerance on the associated linear dimension as shown in Fig. 4.76. The component is accepted if its profile does not lie outside these parallel zones of tolerance. It may be noted that angle θ is allowed to vary within limits imposed by the diagonals of tolerance zone as shown in Fig. 4.76.

Tolerances on tapers are specified in two ways. Where the fit on the taper is of great concern and critical, then separate tolerance for the rate of taper is specified as shown in Fig. 4.77. Tolerance

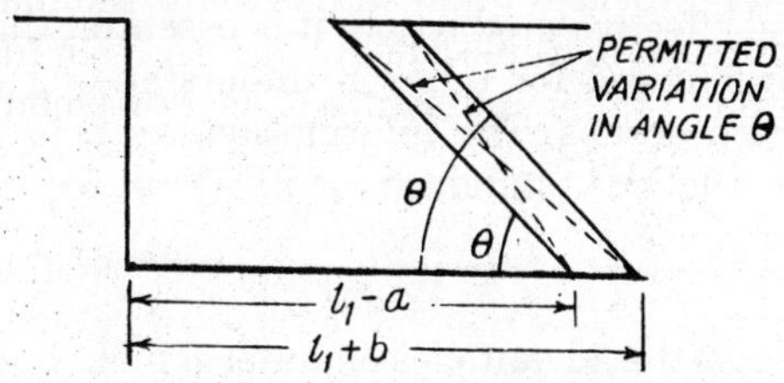

Fig. 4.76. Angular tolerance.

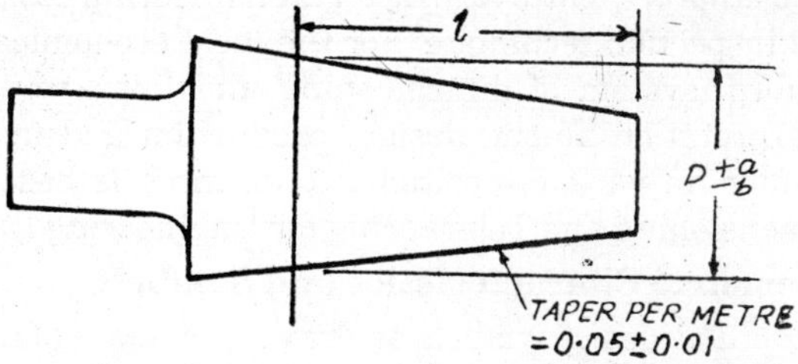

Fig. 4.77. Tolerance when fit on taper is critical.

on diameter is specified for a particular plane at a given length l but this tolerance on diameter is applicable only for this plane and not for diameters at other planes of taper. The diameter at other planes will be governed by the tolerance on the angle of taper.

However, if fit on taper is not critical then tolerance on diameter at a plane at a distance l is specified. Angle of taper is assumed to be same and two parallel zones corresponding to diameters $D - b$ and $D + a$. The taper surface can then take the profile corresponding to diagonals within the tolerance zones as shown in Fig. 4.78.

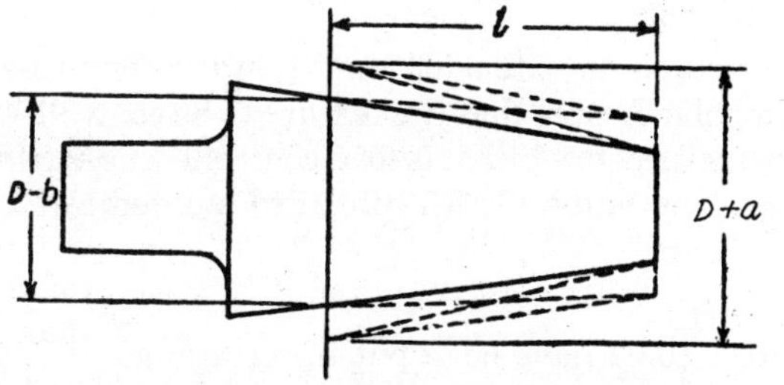

Fig. 4.78. Tolerance when fit on taper is not critical.

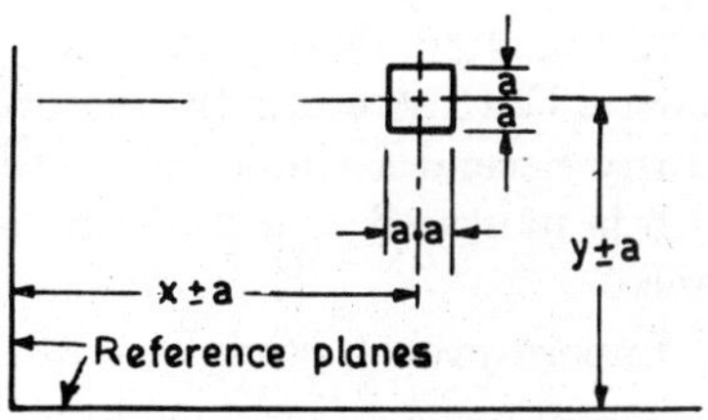

Fig. 4.79. Positional tolerances.

4.18. Positional tolerances

If attempts are made to specify positional tolerances in conventional way, then to specify a particular positional feature, one occasionally ends up with several tolerances. All this results in accumulation of tolerances and thus the desired positional tolerance would be too wide and not suitable for desired application. Some examples are given below to elaborate this point further. A particular hole has to be drilled in a plate. First the axis of hole will be defined and located. Some tolerance has to be allowed on this. Thus the centre of the hole itself can occupy any position within a square at the centre depending on the tolerances specified to locate the centre. Then some tolerance has to be specified for manufacturing the hole. Thus the hole obtained will be having cummulative effect of two tolerance. This problem can be obviated by specifying positional tolerances. (Some positional tolerances are described in next article) such as maximum metal condition (MMC) for the hole so that no problem will be experienced while carrying out assembly.

4.19. Geometric Dimensioning and Tolerancing

Tolerances on linear dimensions have been discussed and we will now concentrate on tolerances with respect to actual function and relationship among various components. Geometric dimensioning and tolerancing are means of specifying engineering design and drawing requirements with respect to actual 'function' and 'relationship' for part features ; (component portion of a part, *e.g.*, surface, hole, slot etc.) which can be most economically produced. Geometric tolerances are concerned with the accuracy of the relationship of one feature to another ; and these are specified separately. These represent the maximum permissible overall variation of form, or position of form, or position of a feature. In fact it is engineering design drawing language and a functional production and inspection technique. For the most economical and effective production, it is essential that a uniform system of dimensioning and tolerancing be followed for uniform understanding and interpretation among design, production and inspection groups. The information given in this chapter on various positional tolerances is based on the U.S. standard ANSI—Y 14.5—1973, Dimensioning and Tolerancing for Engineering Drawing.

Geometric Characteristics and Symbols

The use of symbols on drawing in place of notes has the advantages of uniform meaning, no misunderstanding, compact and quickly and easy drawn, international language requiring no interpretation etc.

The various geometric characteristics and symbols used and their definition are given on next pages. These can be broadly classified into three types.

Tolerance zone represents the zone within which the feature is to be contained. This zone may be a circle, cylinder, area between two parallel lines, etc. On drawings, the geometric tolerance is indicated in a rectangular frame, which is divided into compartments, as shown is Fig. 4.80.

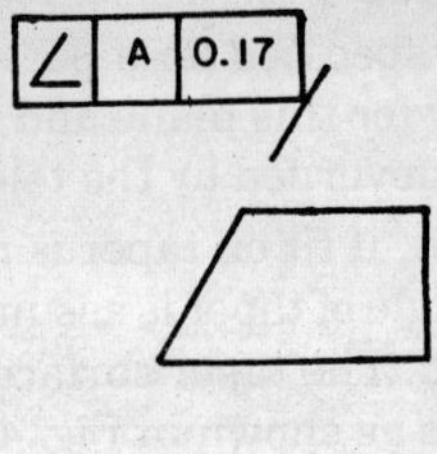

Fig. 4.80. Geometric tolerance.

In Fig. 4.80, feature control symbol is marked on an extension line from the feature, *i.e.* angularity. The first box in the feature control symbol indicates that the feature to be controlled is angularity. Second box indicates the references with respect to which it is to be controlled, *i.e.* surface *A* which is indicated in a rectangle at the bottom. The numerals 0.17 in third box indicate the tolerance.

Important features to be noted in Fig. 4.81 are :

(*i*) All basic dimensions like 14.6, 5.75, 20.4, 10.62 have been put into rectangular boxes instead of writing words BASIC against them.

(*ii*) Feature control symbol put on line *AB* extended indicates that surface *AB* should be perpendicular to datum *A* within 0.02, and the letter *B* in rectangular box below it indicates that the surface *AB* is perpendicular to datum *B* surface.

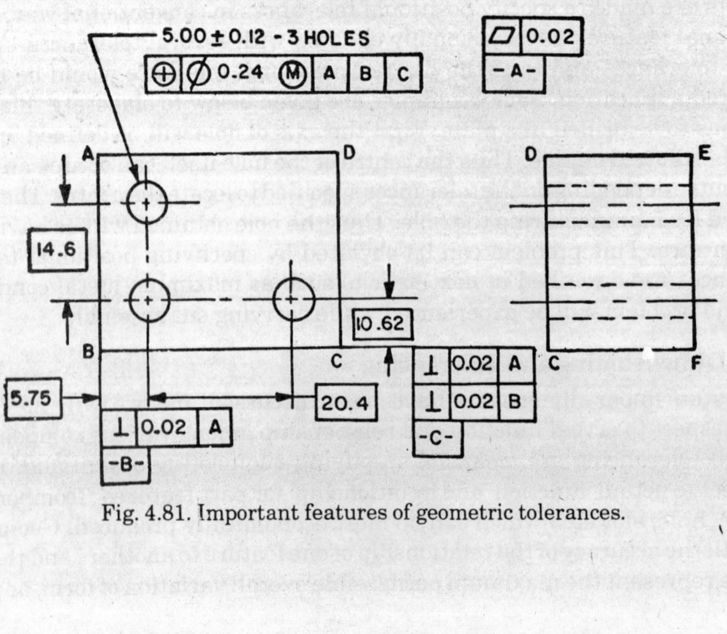

Fig. 4.81. Important features of geometric tolerances.

Broad classification (kind of feature)	*Geometric characteristics*	*Symbol*	*Definition*
(1)	(2)	(3)	(4)
Individual features A single surface, element, or size feature which relates to a perfect geometric counterpart of itself as the desired form.	Flatness	⏥	It is the condition of a surface having all elements in one plane.
	Straightness	—	It is a condition where an element of a surface or an axis is a straight line.
	Roundness (Circularity)	○	It is the condition on a surface of revolution (cylinder, cone, sphere) where all points of the surface intersected by any plane (*i*) perpendicular to a common axis (cylinder, cone) or (*ii*) passing through a common centre (sphere) are equidistant from the centre.
	Cylindricity	⌭	It is a condition of a surface of a revolution in which all points of the surface are equidistant from a common axis.
Individual or related feature A single surface or element feature whose perfect geometrical profile is described which may, or may not, relate to a datum.	Profile of a line	⌒	It is the condition permitting a uniform amount of profile variation, either unilaterally or bilaterally, along a line element of a feature.
	Profile of a surface	⌓	It is the condition permitting a uniform amount of profile variation, either unilaterally or bilaterally, on a surface.
Relation feature A single surface or element feature which relates to a datum, or datums in form and attitude (orientation).	Perpendicularity (squareness or normality)	⊥	It is the condition of a surface axis, or line which is 90° from a datum plane or datum axis.
	Angularity	∠	It is the condition of a surface, axis, or centre plane which is at a specified angle (other than 90°) from a datum plane or axis.
	Parallelism	∥	It is the condition of a surface, line or axis which is equidistant at all points from a datum plane or axis.
	Circular runout	↗	It is the composite control of circular elements of a surface independently at any circular measuring position as the part is rotated through 360°.
	Total runout	↗ Total	It is the simultaneous composite control of all elements of a surface at all circular and profile measuring positions as the part is rotated through 360°.

(1)	(2)	(3)	(4)
	Position	⌖	A position tolerance defines a zone within which the axis or centre plane of a feature is permitted to vary from true (theoretically exact) position.
	Concentricity	◎	It is a condition in which two or more features (cylinders, cones, spheres, hexagons etc.) in any combination have a common axis.
	Symmetry	⌯	It is a condition in which a feature (or features) is symmetrically disposed about the centre plane of a datum surface.

The detailed definitions of all these features and methods of their measurement have been explained in chapter 8.

Other symbols recommended are :

(1)	(2)	(3)
Maximum material condition MMC	Ⓜ	It corresponds to minimum hole size and maximum shaft size.
Regardless of feature size RFS	Ⓢ	This is the condition when the tolerance of form, runout or location must be met irrespective of where the feature lies within its size tolerance.
Projected tolerance zone	Ⓟ	It is a tolerance zone applied to a hole in which a pin, stud, screw or bolt etc. is to be inserted. It controls the perpendicularity of the hole to the extent of the projection from the hole. As it relates to the mating part clearance, it extends above the surface of the part to the functional length of the pin, screw etc. relative to its assembly with the mating part.
Datum target		It is specific datum point, line or area (identified on the drawing with this datum target symbol) used to establish datum points, lines, planes or areas for special functions or manufacturing and inspection repeatability.
Diameter (cylindrical) tolerance zero	∅	

(1)	(2)	(3)
Basic, exact dimension	X . X	A basic dimension is symbolised by boxing it *e.g.* 25.4. Basic dimension is a theoretical value used to describe the exact size, shape or location of a feature. It is used as a basis from which permissible variations are established by tolerances on other dimensions.
Datum identification symbol	—A—	It contains the datum reference letter in a drawn rectangular box.
Feature control symbol	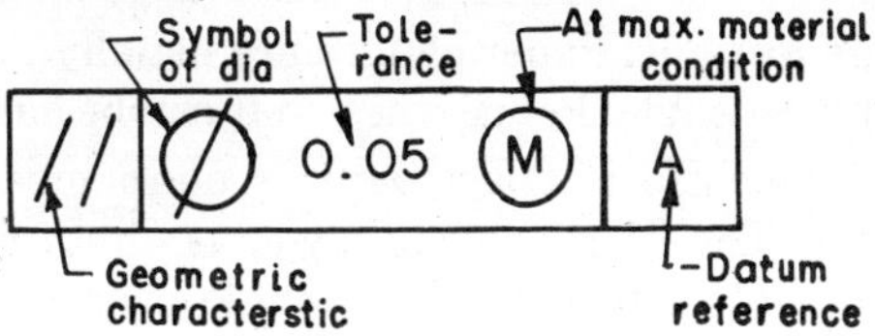	The feature control symbol consists of a frame containing the geometric characteristic symbol, datum references, tolerance and the MMC symbol, if applicable.

(*iii*) Similarly the feature control symbol on line *BC* extended indicates that this surface itself is datum *C* and that this surface *C* should be perpendicular to datum *A* within 0.02 and with datum *B* within 0.02.

(*iv*) Feature control symbol on hole indicates 3 holes of 5.00 ± 0.12 size which should be on true position within ϕ 0.24 at maximum material condition in relation to datums *A*, *B* and *C*.

(*v*) Feature control symbol over face *DE* indicates that it should be flat within 0.02 total with reference to datum *A*.

From this example, it will be appreciated that a lot of information can be put in a very easy manner on the drawing in symbolic form without requiring notes.

There are several rules to be followed in adopting these symbols and complete description of them is beyond the scope of this book. Those interested further may refer to the U.S. standard ANSIY 14.5—1973.

4.20. Paper Layout Gauging

Paper layout gauging is being commonly used since it is direct and inexpensive technique and paper gauges can be easily and quickly produced. It makes an immediate functional check of inspection results, permitting the adjustments possible with functional three-dimensional receiver gauging. Such gauges enable one to decide whether the part can be reworked and also enables evaluate tooling requirements and resetting required to produce an acceptable product. Unlike receiver gauges, paper gauges do not wear out and do not require storage space. Paper layout gauging is ideally suited where the datums are not properly specified and different inspectors obtain different results due to assumption of different datums like holes or edges, etc. Under such conditions, paper gauge gives best decision as to whether the part is functionally acceptable, regardless of the set-up used.

Paper gauge is not suited where all dimensions originate from a single specified datum reference frame. A part with concentricity requirements can be inspected by paper gauging but the procedure is more troublesome. Parts with independent patterns of axial features (holes, pins, etc.) related only to a primary datum, which would ordinarily be gauged with a feature relation receiver gauge, can be more effectively paper gauged.

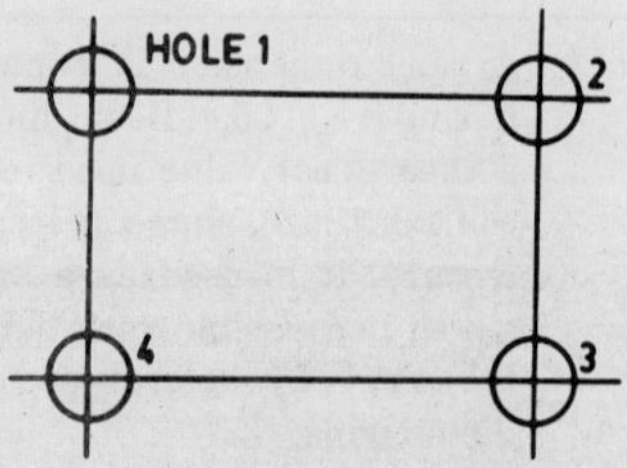

Fig. 4.82. Plot of positional tolerance.

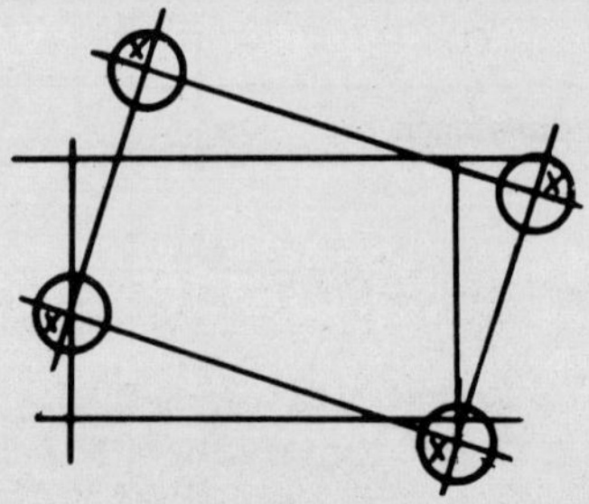

Fig. 4.83. Paper gauging mark X's represent measurement points of axes of holes.

Procedure to make paper gauges. Paper gauges are usually made on plastic materials like Myler which is more stable and reusable. Pencil lines on it can be easily erased.

Separate layouts are made for a pattern of part features, one of measured axis locations, and the other of positional tolerance zones. The tolerance zone layout is superimposed over the measured axis layout, and rotated and translated to determine if any single orientation allows all plotted axis points to fall within their respective tolerance zones. Tolerance zones and measured axis variations are scaled up so that they can be seen and accurately plotted and evaluated. Different scale factors are used for dimensions and tolerances but the layout format should resemble the geometry of the part. The measured locations of points at both ends of each feature axis are plotted in relation to the *X*- and *Y*-axes established by the datum specified on the inspection report. A grid plot is used to find actual deviations by counting the number of squares. The positional tolerance zone diameters taken from the inspection report are scaled up by the same factor selected for plotting the measured axis locations and drawn in with a compass. For example, Fig. 4.82 shows the plot of positional tolerance zones of four holes. The completed tolerance layout is placed over the inspection results layout and slightly shifted and/or rotated if necessary to determine if all sets of measured axis location points fall simultaneously within their respective tolerance zones, thus graphically demonstrating whether or not the part being gauged will assemble with its mating part.

SOLVED PROBLEMS

Problem 4.1. *What are the important rules for putting dimensions on drawings in respect of tolerances.*

Solution. The various important rules in this respect are :

(*i*) The dimensions should be shown at place where it can be measured directly. For example, no useful purpose will be served indicating dimensions from imaginary centre of the part.

(*ii*) Considering the interchangeability of parts, all important dimensions with reference to the locating surface should be clearly marked. For example, position of locating grooves on a part should be indicated from the main locating reference surface.

(*iii*) Contradictory additive dimensions which can affect the actual location and interchangeability should be avoided.

Problem 4.2. *Determine the dimension and tolerances on flange thickness T in Fig. 4.84.*

Solution. Dimension *T*

$$= 118.3^{+0.08}_{-0.09} - \left(25.2^{+0.01}_{-0.02} + 30.4 \pm 0.01 + 32.7 \pm 0.02 + 20.0^{+0.02}_{-0.03}\right)$$

$$= 10.0^{+0.08-(0.01+0.01+0.02+0.02)}_{-0.09-(-0.02-0.01-0.02-0.03)}$$

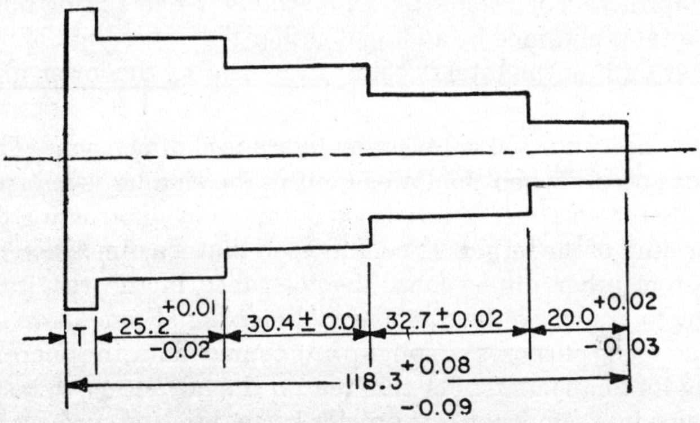

Fig. 4.84. All dimensions in mm.

$$= 10.0^{+0.08-0.06}_{-0.09+0.08} = 10.0^{+0.02}_{-0.01} \text{ mm.}$$

Problem 4.3. *Four parts with the following dimensions are to be assembled in random. What should be tolerance on the assembled length ?*

Part 1 = $25.32^{+0.02}_{-0.01}$ *mm* *Part 2* = *18.91 ± 0.03 mm*

Part 3 = $62.17^{+0.05}_{-0.07}$ *mm* *Part 4* = *46.25 ± 0.04 mm.*

Solution. Upper tolerance on the assembled length will be sum of all the individual upper tolerances.

i.e. = + 0.02 + 0.03 + 0.05 + 04 = 0.14 mm

and similarly lower tolerance on the assembled length will be the sum of all the individual lower tolerances,

i.e. = – 0.01 – 0.03 – 0.07 – 0.04 = – 0.15 mm

∴ Dimension of assembled length will be $152.65^{+0.14}_{-0.15}$ mm.

Problem 4.4. *Discuss several types of tolerances.*

Ans. As it is impossible to make any component exactly to given size due to inherent limitations in machine, man, and material, the desirable amount of tolerance based on design requirements is specified. The value of tolerance is usually a compromise between the functional requirements of the component and the cost of machining.

Tolerances on a component may be specified in two ways : (*i*) unilateral tolerance, (*ii*) bilateral tolerance. In the case of unilateral tolerance, the tolerance lies entirely on one side of the basic size,

e.g. $250^{+0.00}_{-0.01}$ mm

In the case of bilateral tolerance, the tolerance lies on both sides of the basic size, *e.g.*

$$25.0^{+0.00}_{-0.01} \text{ mm}$$

Unilateral tolerance is most commonly specified for interchangeable manufacture particularly where precision fits are required. The direction in which the unilateral tolerance is specified should be such that the variations in the basic size in this direction do not impair the functioning of the part. Bilateral tolerances are specified for mass production of jobs as these help in machine setting.

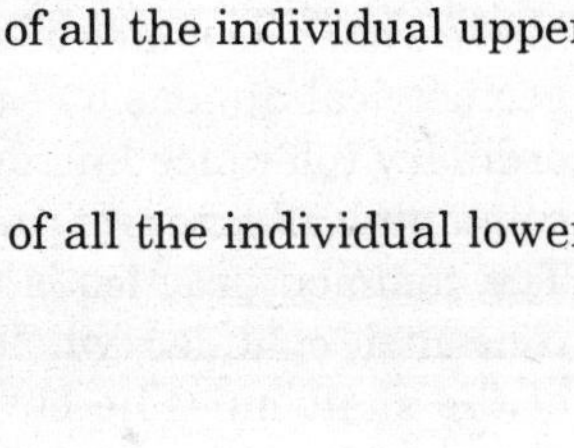

Fig. 4.85. Compound tolerance.

Addition and Subtraction of Toleranced Dimensions. When adding toleranced dimensions , the upper limit of the total is obtained by adding together the upper limit of individual dimensions and similarly, the lower limit of the total is obtained by adding the lower limits of the individual dimensions.

Particular care is needed when subtracting toleranced dimensions. The greatest deference between two dimensions occurs when the lower limit of the smaller is subtracted from the upper limit of the larger, while the smallest difference is obtained by subtracting the upper limit of the smaller from the lower limit of the larger. Thus, it is seen that if a dimension on a component is the resultant of a number of other dimensions, the tolerance in the resultant dimension is the arithmetical sum of the tolerance on the component dimension. From above it follows that if some dimensions for manufacturing purposes be split up into components, the tolerance on this dimension must be shared among its components. For this reason the advantage to be obtained by splitting up a functional dimension into components to provide better locating surfaces for ease, or inspection datums, must be weighed against the increased difficulty in producing the component owing to the necessarily reduced tolerances.

Compound Tolerances. A compound tolerance is one which is determined by the established tolerances on more than one dimension. For example, in Fig. 4.85, the combined effect of the tolerance on the 30.00 mm dimension and the tolerance on the 60 ° angle give a compound tolerance to dimension R. Compound tolerances should be avoided as far as possible.

Problem 4.5. *What do you understand by geometrical tolerances ?*

Ans. The tolerances provided in any standard refer only to the actual size or dimension of a feature such as a hole or shaft. In many instances, actual size alone is not of importance but other geometrical dimensions *e.g.* roundness and straightness of shaft, also matter. In such cases the ordinary tolerance laid down should cover such variation also ; and the problem becomes that of combining all errors of roundness, straightness and diameter within a single tolerance on diameter. The common practice is to consider the diameter tolerance to be represented by two imaginary concentric cylinders whole diameters are the upper and lower limits of size. The whole of the surface of the shaft must lie between these two limiting surfaces. In some instances depending on the functional requirement the tolerances on diameter, straightness, and roundness may be specified separately and can't be allowed to be mixed up. Similar considerations apply to tolerances of parallelism, squareness, concentricity and general location and positions. In certain cases the geometrical tolerances must be specific irrespective of variations in dimensions within tolerance : *e.g.* in case of precision parallels the faces should be parallel to high degree and the distance between the parallel faces is not that important.

Problem 4.6. *How the following are designated ?*

(*a*) *Standard tolerance grade* (*b*) *Position of tolerance zone*

(*c*) *Upper deviation* (*d*) *Lower deviation*

(*e*) *Tolerance class* (*f*) *Toleranced size*

(*g*) *Fit.*

Solution. (*a*) Standard tolerance grades are designated by the letters *IT* followed by one of the 20 numbers, *viz.* 01, 0, 1 to 18 ; *e.g. IT*8. The letters *IT* are omitted when the tolerance grade is associated with a letter representing a fundamental deviation *i.e. H*8.

(*b*) The position of the tolerance zone w.r.t. zero line is designated by letters *A* to *ZC* for hole and letters *a* to *zc* for shafts.

(*c*) Upper deviation is designated by letters *ES* for holes and *es* for shafts.

(*d*) Lower deviation is designated by letters *EI* for holes and *ei* for shafts.

(*e*) Tolerance class is designated by letter representing fundamental letter representing fundamental deviation followed by the number representing the standard tolerance grade. For example *H*8 (holes), *h*8 (shafts).

(*f*) Toleranced size is designated by the basic size in mm followed by the designation of the required tolerance class *e.g.*, 40*H*8, 80*g*7, $75^{-0.012}_{+0.025}$.

(*g*) A fit is designated by the common basic size in mm, followed by tolerance class symbols for hole and shaft.

For example, 60*H*8/*g*7 or $60\,\frac{H8}{g7}$.

Problem 4.7. *Fig. 4.86 shows a graphical representation of a shaft and a hole. Provide suitable words for various dimensions marked as A, B, C, D, etc.*

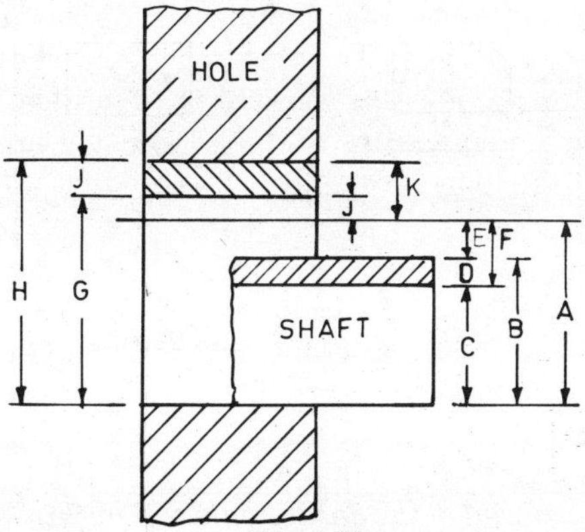

Fig. 4.86

Solution.

A — basic size
B — maximum limit of size for shaft
C — minimum limit of size for shaft
D — size tolerance for shaft
E — upper deviation of shaft (*es*)
F — lower deviation of shaft (*ei*)
G — minimum limit of size of hole
H — maximum limit of size of hole
I — size tolerance of hole
J — lower deviation of hole (*EI*)
K — upper deviation of hole (*ES*)

Problem 4.8. *Fig. 4.87 (a) shows deviations for shafts and figure (b) shows the deviations for holes. Mark the fundamental deviation, upper deviation and lower deviation on these.*

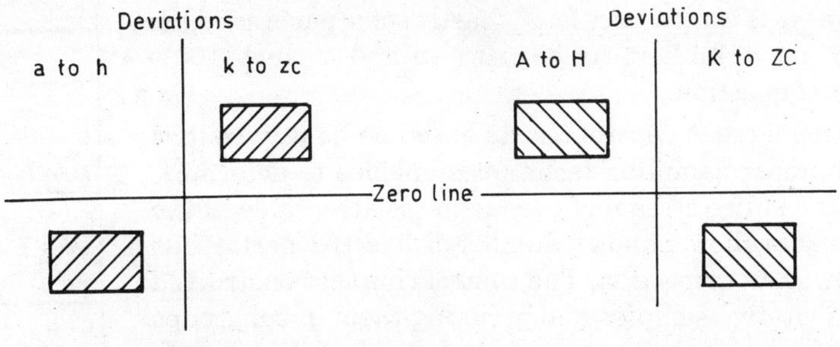

(*a*) Shafts (*b*) Holes

Fig. 4.87

Solution. Fundamental deviation for shafts *a* to *h* are closest to zero line (corresponding to upper deviation *es*), and – ve in value. For shafts *k* to *zc*, fundamentation deviation is again close to zero line and corresponding to lower deviation *ei* and is + ve.

For holes *A* to *H*, fundamental deviation is again close to zero line corresponding to lower deviation and + ve in sign, whereas for holes *k* to 2*c* fundamental deviation is close to zero line corresponding to upper deviation and – ve in sign. Thus Fig. 4.87 (*a*) and (*b*) get filled up as shown in Fig. 4.87 (*c*) and (*d*).

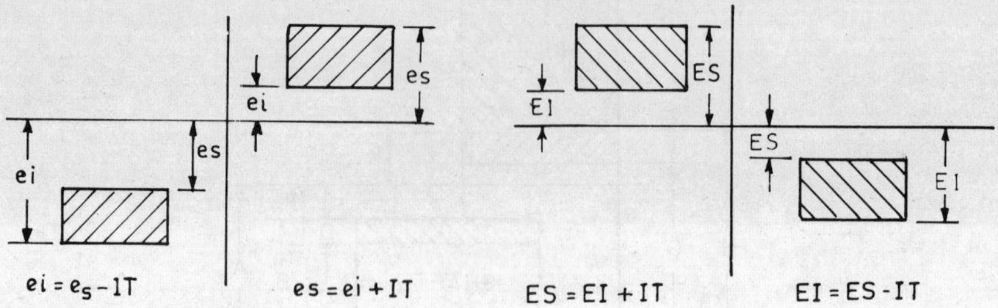

(*c*) (*d*)

Fig. 4.87

Problem 4.9. (*a*) *What per cent of the data should fall within ± 1σ on the normal distribution curve ? Within ± 2σ ? And within ± 3σ ?*

(*b*) *What per cent of the data would fall outside control limits naturally if ± 2σ were used as control limits ? ± 3σ were used as control limits ?*

Solution (*a*) Data within ± 1σ limits would be 68.27%

Within ± 2σ limits would be 95.45%

and within ± 3σ limits would be 99.73%.

(*b*) Data outside the ± 2σ limits would be 4.55% and outside ± 3σ limit would be 0.27%.

Problem 4.10. (*a*) *Why 100% inspection is not preferred generally ?*

(*b*) *What general statements can be made about the use of control chart for inspection and the effect of sample size on control chart construction ?*

Solution. The disadvantages of 100% inspection of each part are : It is slow, costly, results in boredom and fatigue to inspectors. Still is does not eliminate all the defective parts. Inspection gauges get worn out of adjustment more frequently. The risk of passing defective parts in 100% inspection technique is unknown and depends on unpredictable behaviour and mood of operator. On the other hand, in planned sampling procedures the risk of passing defective parts can be calculated. Lastly some parts are such that a test on it may result in its destruction and as such 100% inspection is out of question.

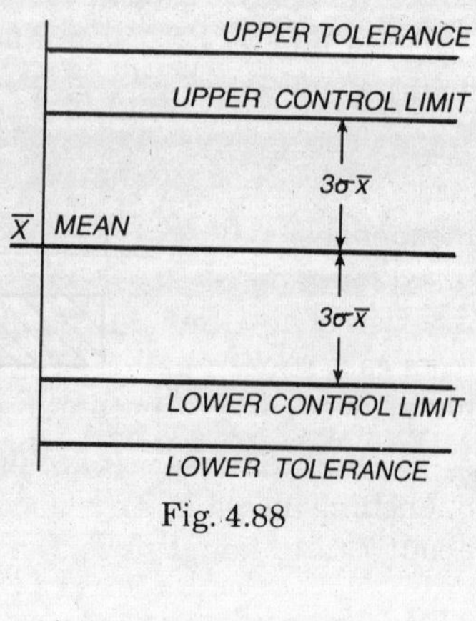

Fig. 4.88

(*b*) The technique of control charts based on quality control procedures and proper sampling techniques enables to determine whether or not the entire stream of production is acceptable, if the customer is willing to allow a known number of defective parts, thus doing away with 100% inspection. The control chart is constructed by taking representative samples and grouping them in subgroups of four or five and calculating the average and standard deviation of each sub-group, and then the average of averages of all sub-groups which is the central line of control chart. Then the average deviation $\overline{\sigma}$ is determined as $\overline{\sigma} = \frac{\Sigma\sigma}{N}$, where σ is standard deviation of each sub-group and N is the No. of sub-groups. Then the $\pm\ 3\sigma_{\bar{x}}$ limits on control chart will be $A_1\overline{\sigma}$, where the value of A_1 calculated by probability theory, dependent upon the number of pieces in sub-group = 3.76 for sample size of 2 ; 2.39 for sample size of 3 ; 1.88 for sample size of 4 and 1.60 for sample size of 5.

The tolerance limits are always placed outside the control limits. Beyond upper and lower tolerance limits, parts are unusable and between the upper tolerance and upper control limit or between the lower control limit and lower tolerance, these are usable but process is out of control. As long as parts are within upper and lower control limits, it can be said that not more than 3 out of 1000 will leave the process incorrectly made. Even if 5 to 7 consecutive points fall on one side of the mean, the process should be checked.

(For more details, one may refer chapter 19.)

Problem 4.11. *What is meant by nominal size and tolerance ?*

Solution. A nominal size is ascribed to a part for general identification purposes. Thus a shaft may have a nominal size of 60 mm, but for practical reasons this size cannot be manufactured without great cost. Hence certain tolerance or machining allowance must be added to it depending upon the intended application for which this part is to be used. Close tolerances are given when a part must fit properly with some other part, keeping in mind that the cost of a part goes up as the tolerance is decreased.

Problem 4.12. *Why hole-basis system of fit is generally employed ?*

Solution. It is used to produce holes by drilling or reaming operation and as such it is not possible to obtain any desired size of hole ; size being limited by availability of size of drills and reamers. It thus becomes convenient to fix the size of hole and vary the size of shaft for required fit. The shaft being generally turned and ground can be easily produced to any size.

However sometimes cold drawn steel shafts are employed and hole is made by boring, then shaft basis system of fit can be employed.

Problem 4.13. *What are the essential considerations in selection of materials for gauges, and what are the common materials used for gauges ? Explain various processes of manufacturing gauges.*

Solution. The best material for the gauges should fulfil most of the following requirements :

(*a*) *Hardness* to resist wear. This is the essential condition in selection of materials for gauges.

(*b*) *Stability* to preserve size and form.

(*c*) *Corrosion resistance.*

(*d*) *Machinability* for obtaining the required degree of accuracy.

(*e*) *Low coefficient of linear expansion* to avoid temperature effect.

A high grade finish (0.10 μm CLA value) is provided which reduces the initial wear on the working surface of the gap gauge.

The various materials commonly used are :

High Carbon Steel. Cast steel (0.8 to 1.0% carbon) either water or oil hardened is a relatively inexpensive and most commonly used material for gauges. It is available in the form of gauge plate (used for gauges) and silver steel (used for plug gauges). Oil hardening type is preferable if there is a risk of cracking during hardening. Plug gauges between 15 to 32 mm diameter may be made from cast steel, smaller sizes from silver steel, and bigger sizes from case-hardening steels which have lower carbon content and have to be hardened locally by carburising process.

Suitable heat treatment consists in hardening the steel by heating the steel to 730°C and quenching in water. To reduce the brittleness, a tempering temperature of 200°C is maintained for about 7 to 10 hours so as to achieve the stability also.

Mild Steel. Mild steel gauges are usually case hardened on the working surface. Suitable heat treatment consists in carburising the gauge to the necessary depth, either by heating in bath containing carbonaceous compounds or by means of a cyanide bath. The depth of case-hardening should be at least 0.75 mm to allow for final grinding and finishing.

After hardening, to relieve internal stresses and to promote stability, the gauge is held for about 5 hours at a temperature of 150° to 160°C, after which it is slowly cooled.

Mild steel is easily machined, stable and inexpensive. It is used for gauges of shapes which might cause cracking during hardening, if made of other steel.

Oil Hardened Steel. It is best suited where gauges are required in huge quantities. Contact faces can be surfaced with carbide to prolong gauge life.

Plated Gauges. The recent advances in chromium plating have led to its increasing use for gauges. It renders the surface of the gauge very hard coupled with resistance to abrasion and corrosion. In practice the gauge is first machined several thousands undersize, plated until it is a few thousands over size and then finally ground or lapped to the required size. The amount deposited on plug gauges is usually within the range of 0.2 to 0.3 mm on diameter.

Chromium plating has also proved a most useful method in reclaiming the worn gauges.

For gauging aluminium or other materials having an abrasive action, plug gauges having stellite ribs, inserted in the gauging surface are often used.

Cast Iron. It is often employed for the bodies or frames of large gauges, whose working faces are hard inserts of tool steel or cemented carbide. The bodies are made of good quality iron, of a close grain structure and free from blow holes. The iron is often stabilised before final machining in order to release casting stresses.

Glass gauges inspite of their good wearing qualities have not gained much popularity because if dropped or heavily knocked they may get damaged or definitely broken. Glass gauges have the advantage of eliminating the corrosive effects due to perspiration from hands. Their dimensions are not affected by temperature changes due to their low coefficient of expansion. Further, when a glass gauge is scratched or chipped, no burr is left on the gauge.

Invar containing 36% nickel has coefficient of expansion of less than 1×10^{-6} per °C but is unsuitable over a long period.

Elinvar is much more suitable and has an expansion coefficient of 8×10^{-6} per °C. Elinvar contains 42% nickel.

Problem 4.14. *Write brief notes on the following processes used in the manufacture of gauges :*

(a) Stabilisation, *(b) Spot Grinding, and* *(c) Lapping.*

Ans. (*a*) The process of *stabilisation* is applied to the gauges to prevent movement of the gauge—a phenomenon known as creep. In this process, the steel gauge is heated to 150°C and held at this temperature for about 5 hours and then the gauge is cooled slowly. The surface or skin of the gauge is grinded before stabilising. The process of stabilising does not relieve the internal stresses but it creates, in a very limited time the distortion, which otherwise would occur in a number of years. Thus while grinding before stabilisation, sufficient allowance should be provided both for finish grinding as well as this effect. It may be noted that the hardness of the gauge is reduced by a limited extent by this process.

(*b*) *Spot Grinding.* The process of spot grinding is used to produce very accurate flat surfaces. This process does not utilise machine slides and thus the accuracy of surface produced in independent a accuracy of machine slides. This process is used only for finishing operation and can produce accuracy of 0.002 mm for both flatness and parallelism. In this process, the movement of the work is controlled by hand and accordingly the work should not be very heavy and at the same time not too light also to be disturbed by the grinding force. The work is supported on an accurate surface plate rigidly secured in position, underneath a grinding wheel. The work is moved in all directions and ground on one surface till sparking completely stops. Maximum metal left to be removed by this process is around 0.05 mm and a cut not greater than 0.01 mm is taken, the finishing cut being 0.002 mm. It is also possible to support work on non-electric magnetic chuck whose two surfaces have been spot ground to be flat and parallel. After one surface has been ground the workpiece is turned over for grinding the opposite face. This process is called spot grinding because of the spotted appearance of the ground surface.

(c) *Lapping Process.* The process of lapping, which is not a polishing action but a cutting action is used after hardening. This can be used for both flat and cylindrical workpieces. It is used to remove irregularities on workpiece by a fine abrasive powder, suspended in a suitable liquid between two laps made of copper, lead for cast iron. It produces matt finished surface comprising of a number of minute scratches of very fine pitch.

Problem 4.15. *Write a short note on interchangeable manufacture and the 'selective assembly' of machine parts.*

Ans. Interchangeable manufacture consists in producing finished duplicate parts in lots to the desired accuracy, and in providing means to adhere to the limits of accuracy specified.

The interchangeable method of manufacture is especially applicable in the production of duplicate machine or other mechanical devices, particularly when the quantity is large. There are several degrees of interchangeability in machinery manufacture. Strictly speaking interchangeability consists in making the different parts of a mechanism so uniform in size and contour that each part of a certain model will fit any mating part of the same model regardless of the lot to which it belongs or when it was made. However, as often defined, interchangeability consists in making each part fit any mating part in a certain series ; this is the interchangeability exists only in the same series.

Selective assembly is sometimes termed interchangeability, but is merely assembly without fitting. It will be noted that the strict definition of interchangeability does not imply that the parts must always be assembled without hand work, although that is usually considered desirable.

Selective assembly consists is selecting by trial mating members of a mechanism that will give the desired fit at assembly, with little or no further machining. Companion parts made to extreme limits are not supposed to interchange. For instance, a shaft or pin of maximum size may not assemble with a mating part of minimum size, although the maximum shaft and minimum hole and also the minimum shaft and minimum hole must interchange. A practical example of this selective method of assembling is found in the production of ball bearings. The balls are sorted into groups, according to their size, to facilitate the assembly of any bearing with balls of uniform size. Nearly every so-called 'interchangeable' mechanism represents a combination of interchangeable and selective methods of quantity production.

The mater of clearances and tolerances, when manufacturing on an interchangeable basis, is quite different from the manufacturing on the basis of selective assembly. In interchangeable manufacture, the minimum clearances should be as small as the assembling of the parts and their proper operation under service conditions will permit. The maximum clearances should be as great as the functioning of the mechanism allows. The difference between the maximum and minimum clearance establishes the sum of the tolerances on the companion parts. When this allowable difference is smaller than normal manufacturing conditions will permit, then parts cannot be economically manufactured on an interchangeable basis. In situation of this type, the following method may be adopted. The tolerances are established which enable the parts to be manufactured economically and then sorted and assembled according to their size. This method is known as selective assembly manufacturing.

Problem 4.16. (*i*) *Define the following terms :*

(*a*) *Selective fit,* (*b*) *Push fit,* (*c*) *Driving fit,* (*d*) *Forced or Pressed fit, and* (*e*) *Shrinkage fit.*

(*ii*) *Write a brief note on design considerations of interference fits.*

Ans. (*i*) (*a*) *Selective fit.* Selective fit is required where the object is to make a shaft and hole with a finite fit and not a permissible range on fit. This fit is known as the ideal or 'selective fit'. It is customarily use for tight or interference fits where it is desired to avoid the extremes of maximum tightness or looseness.

The ideal selective fit for the tightest class of fit would stress the hole just to its elastic limit, thereby giving the maximum holding power without overstressing or distorting the grain structure. For low carbon steel this relation can be calculated and will be found to be 0.0001 d, where d is the basic size of hole.

(*b*) *Push fit.* Push fit also known as 'snug fit' represents the closest fit that permits assembling parts by had. In other words, anything closer would, in many cases, at least, require greater force that can be exerted by hand in assembling the parts. With a push fit, there should be no perceptible play between the mating parts.

(*c*) *Driving fit.* When a plug or shaft is made slightly larger than the hole into which it is to be inserted and the allowance in such that the parts can be assembled by driving, this is known as a driving fit. Such fits are employed when the parts are to remain in a fixed position relative to each other.

Before assembling parts with a driving fit, the bearing surfaces should be oiled. For assembling, a hydraulic press usually is preferable.

(*d*) *Forced or Pressed fit.* Forced or pressed fit is the term used when a pin, shaft or other cylindrical part is forced into a hole of slightly smaller diameter, ordinarily by the use of hydraulic press or some other type of press capable of exerting a considerable pressure. A force fit has a larger allowance than a driving fit, and therefore, requires greater pressure for assembling. Forced fits are restricted to parts of small and medium size. Crankpins, car wheel axels, and similar parts which must be held very securely, are given forced or pressed fits rather than driving fit.

(*e*) *Shrinkage fit.* A shrinkage fit is obtained by making the internal member slightly larger than the hole in the external member. In shrinkage fit, the pressure is not required for assembly but intead the external member is heated and expanded sufficiently to permit inserting the internal member easily. Then as the external part cools or is cooled by applying water or dry ice, it shrinks tightly around the internal part.

(*ii*) Interference fits are often preferred for costly splines, keyways, set screws or other fastening devices. They allow neat and tidy part geometries, and often they permit simpler designs with improved stress distribution and are economical. This type of fit is commonly used for holdings bushings in housings, for inserting dowel pins in holes, for attaching gears or sprockets to shafts, and for assembling bearings. But there are limitations to the use of this type of fit. The interference may not be sufficient, so that the two assembled parts may slip. Or the in interference may be too great so that the parts may fracture during assembly. Excessive interference can also make the parts too difficult to assemble even if they do not break.

If an interference fit cannot do the job, *press fit,* must be tried because of its simplicity and cheapness. But a press fit may impose too much strain on the parts and if it does, the designer has to go for a shrink fit. By calculating the axial force, torque to be transmitted and assembly forces etc., suitable interference fit can be designed.

Problem 4.17. (*a*) *State the Taylor's Principle for the design of 'limit gauge'.*

(*b*) *Write a short note on the various aspects foı deciding the limits on the limit gauges.*

(*c*) *Design the "general" type "Go" and "Not-Go" gauges for component having 25 H/f_8 fit.*

Being given with usual notations :

(*i*) *i* (*microns*) = $0.45 \sqrt[3]{D} + 0.001D$ (*D in mm*)

(*ii*) *the upper deviation for f shaft* = $-5.5.\ D^{0.41}$

(*iii*) *25 mm falls in the diameter step of 18 and 30.*

Take wear allowance as 10% of the gauge tolerance.

Also determine :

(*i*) *type of fit ;*

(*ii*) *allowance for the above fit.*

Ans. (*a*) The Taylor's Principle for limit gauges can be divided into the following two statements :

1. "GO" gauges should inspect all the features of a component at a time and should be able to control the maximum metal limit, or in other words the maximum metal limit of as many related dimensions as possible should be incorporated in the "GO" gauge.

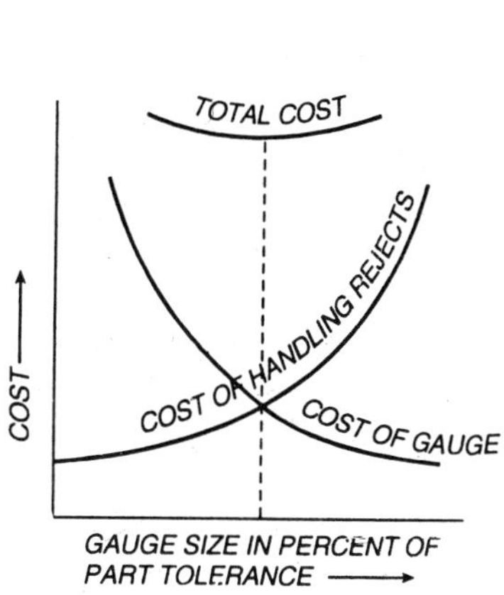

Fig. 4.89

2. "NOT GO" gauge should check only one element at a time for the minimum metal limit.

(*b*) Generally speaking the cost of a gauge is directly proportional to the accuracy to which it is manufactured. Extreme accuracy should, therefore, be maintained only where necessary.

It is not economical to attempt to make a fixed size gauge to an exact dimension. A gauge dimension may be restricted to a very small variation, but there must always be provision for some variation ; consequently the necessity for gauge tolerance.

Gauge wear is inevitable and this must be recognised and provided for in any practical gauging. The "GO" and to some extent the "NOT-GO" gauges will wear. To some extent wear is tolerable but excessive wear may reject parts which are above the minimum metal conditions. Therefore, a limit must be set on the amount of wear life which is economically feasible. Fig. 4.89 shows the method of determining the variation which gives the minimum cost.

$$D = \sqrt{18 \times 30} = 23.2 \text{ mm}$$

(*c*) $$i = 0.45 \sqrt[3]{23.2} + 0.001 \times 23.2 \text{ (microns)} = 0.45 \times 2.855 + 0.023$$

$$= 1.285 + 0.023 = 1.308 \text{ microns} \approx 0.00131 \text{ mm}$$

H_7 hole is the hole with no fundamental deviation and tolerance of IT_7.

Minimum size of hole = 25.00 mm.

$$IT_7 = 16i = 16 \times 0.00131 \approx 0.0209 \text{ mm}$$

Maximum size of hole = 25.0209 mm.

f_8 shaft is the shaft with fundamental deviation of $-5.5.D^{0.41}$ and tolerance of IT_8.

$$IT_8 = 25i = 25 \times 0.00131 = 0.0297 \text{ mm}$$

Fundamental deviation $= -5.5\,(23.2)^{0.41}$ microns $= -\dfrac{5.5. \times 3.645}{100}$ mm $= -0.0200$ mm

Gauge tolerance for hole gauging = 10% of 0.0209 = 0.0021 mm

Wear allowance on this gauge = 10% of 0.021 = 0.00021 mm $\simeq$ 0.0002 mm

Gauge tolerance for shaft gauging = 10% of 0.0297 = 0.00297 mm $\simeq$ 0.003 mm

Wear allowance on this gauge = 10% of 0.003 = 0.0003 mm

With reference to Fig. 4.90

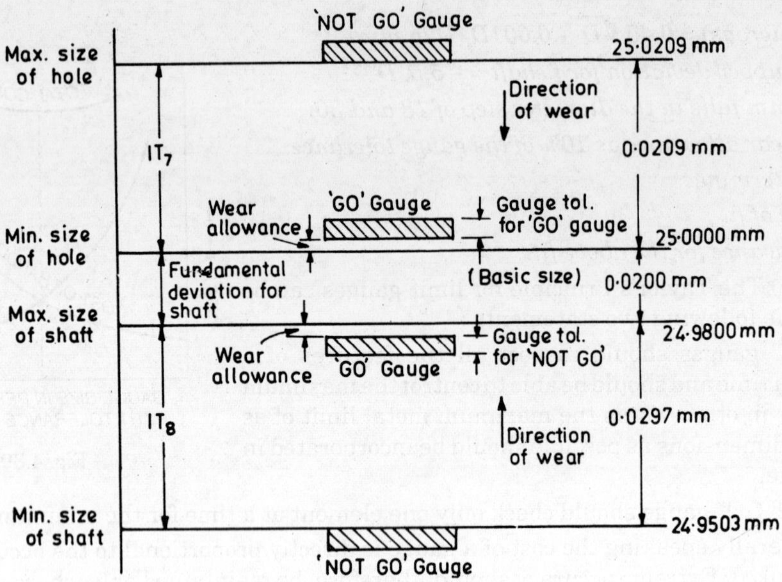

Fig. 4.90

For hole :

"GO" gauge $= 25^{+(0.0021 + 0.002)}_{+0.0002}$ "Not-Go" gauge $= 25.0209^{+0.0021}_{+0.0000}$ mm

i.e. $= 25^{+0.0023}_{+0.0002}$ *i.e.* $= 25^{+0.00230}_{+0.0209}$ mm

For Shaft :

The maximum size of shaft = 25 – 0.02 = 24.98 mm.

The minimum size of shaft = 25 – 0.02 – 0.0297 = 24.9503 mm

"GO" gauge $= 24.98^{-0.003}_{-(0.003 + 0.0003)}$ *i.e.* $= 24.98^{-0.003}_{-0.0033}$ mm

"NOT-GO" gauge $= 24.9503^{-0.0000}_{-0.003}$ mm

(*d*) (*i*) Fit for H_7/f_8 is the clearance fit because the minimum size of the hole is 25 mm while the maximum shaft size is 24.98 mm.

(*ii*) Allowance = minimum clearance between the mating members = 0.02 mm.

Problem 4.18. *Determine the dimensions and tolerances of shaft and hole having size of 30 H_7/h_8 fit. Also determine the allowance (i.e. minimum clearance) and maximum clearance.*

Solution. For hole H_7 and shaft h_8, the fundamental deviation is zero.

As 30 mm lies in the range of 18 and 30 mm, $D = \sqrt{18 \times 30} = 23.2$ mm

and the value of $i = 0.45\left(\sqrt[3]{D}\right) + 0.001\,D = 0.45\left(\sqrt[3]{23.2}\right) + 0.001 \times 23.2$ microns

$= 0.45 \times 2.855 + 0.023 = 1.308$ microns

Value of $IT7 = 16\,i = 16 \times 1.308$ microns

$= \dfrac{16 \times 1.308}{1000}$ mm $\approx 0.0020928 \approx 0.021$ mm

∴ Min. size of hole = 30.0000 mm and max. size of hole = 30.0021 mm

Max. size of shaft = 30.000 mm

and value of $IT8$ for shaft = $25\ i = 25 \times 1.308 = 32.7$ microns = 0.0033 mm

and min. size of shaft = 30.000 – 0.0033 = 29.9967 mm

∴ Allowance or minimum clearance

= Min. size of hole – Max. size of shaft = 30.0000 – 30.000 = 0

and Max. clearance = Max. size of hole – Min. size of shaft

= 30.0021 – 29.2967 = 0.0054 mm.

Problem 4.19. *What is the difference between unilateral and bilateral tolerances ? Why is unilateral tolerance preferred over bilateral tolerance ?*

Solution. In unilateral tolerance, the variation is made in only one direction from the nominal or basic dimension ; whereas in bilateral tolerance the variation is both over and under the nominal dimension. Examples of two respectively are :

$$60^{+0.02}_{-0.01} \text{ mm and } 60^{0.01}_{-0.01} \text{ mm.}$$

Referring to Fig. 4.91, it will be noted that in case or unilateral system of tolerancing, it is possible to change the tolerances (from 0.02 to 0.03 mm) for same allowance or type of fit, without changing the nominal size of shaft or hole. But in case of bilateral tolerancing system, for maintaining the same allowance, changing the tolerances is possible only by changing the nominal size of shaft, if nominal size of hole is maintained. Sometimes the nominal dimensions of both of

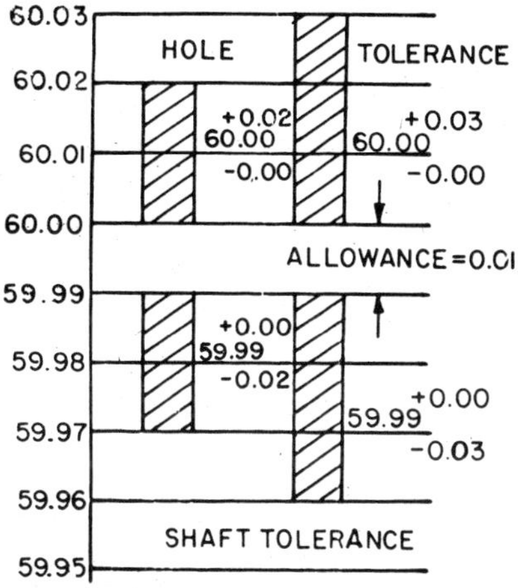

Unilateral tolerance

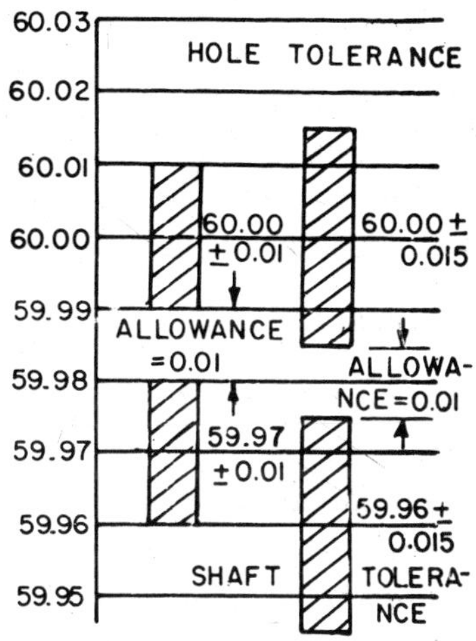

Bilateral tolerance

Fig. 4.91

the mating parts may have to be changed while changing the tolerances for attaining same type of fit. This is a big disadvantage of bilateral tolerancing system and for this reason, in mass production, where mating parts must be interchangeable, unilateral tolerances are customary.

Another advantage of unilateral system is that GO gauges for holes and NO GO gauges for shafts can be standardised.

Problem 4.20. (*a*) *What are the essential conditions for obtaining (i) interference fits and (ii) clearance fits ?*

(*b*) *It is possible to drill a 25 mm nominal hole to an accuracy of* $25^{+0.02}_{-0.02}$ *mm using standard drill and drilling machine available. A shaft is to be machined to obtain a clearance fit in above hole such that allowance should be 0.01 mm and maximum clearance should not be more than 0.08 mm. What should be the tolerance on the shaft ?*

(*c*) *Cold drawn shafts upto accuracy of ± 0.01 mm are available and an interference fit is to be designed for a 50 mm nominal size hole. Determine the tolerances for hole if maximum and minimum interferences are to be 0.01 mm and 0.07 mm respectively.*

Solution. (*a*) For an interference fit, the low limit of the shaft should be greater than the high limit of the hole and for clearance fit the low limit of hole should be greater than the high limit of the shaft.

(*b*) Max. size of hole is 25.02 mm and min. size of hole is 24.98 mm.

Since allowance is 0.01 mm, therefore, max. size of shaft

= Min. size of hole – Allowance = 24.98 – 0.01 = 24.97 mm.

Further the max. clearance specified is 0.08 mm.

∴ Lower size of shaft = Max. size of hole – Max. clearance = 25.02 – 0.08 = 24.94 mm.

∴ Tolerance on shaft = 24.97 – 24.94 = 0.03 mm.

and its dimension will be $24.95^{+0.02}_{-0.01}$ mm.

(*c*) Max. size of shaft = 50.01 mm

and min. size of shaft = 49.99 mm.

Since minimum interference = 0.01 mm

∴ Max. size of hole = Min. size of shaft – Max. interference

= 49.99 – 0.01 = 49.98 mm.

Also max. interference = 0.07 mm

∴ Min. size of hole = Max. size of shaft – Max. interference = 50.01 – 0.07 = 49.94 mm.

∴ Tolerance on hole = 49.98 – 49.94 = 0.04 mm

and size of hole should be = 49.96 ± 0.02 mm.

Problem 4.21. *A hole and shafting system has the following dimensions* $50\frac{H_8}{c_8}$. *The standard tolerance is given by* $i = 0.45\,(\sqrt[3]{D}) + 0.001D$

where D = *dia. (mm) of geometric mean of steps*

i = *standard tolerance, microns*

The multiplier for grade 8 is 25.

The fundamental deviation for shaft c, for D > 40 mm is given by – (95 + 0.8 D) microns

The diameter range lies between 50 to 80 mm.

Sketch the fit and show these upon the actual dimension of hole and shaft.

Solution. For calculation of tolerance, value of diameter $D = \sqrt{50 \times 80} = 10\sqrt{40} = 63.2$ mm.

Value of tolerance unit $i = 0.45\,(\sqrt[3]{D}) + 0.001D$

$= 0.45\,(\sqrt[3]{63.2}) + 0.001 \times 63.2 = 1.79 + 0.0632$

= 1.8532 microns = 0.00185 mm

Fundamental deviation for hole $H = 0$

∴ Min. size of hole = 50.0000 mm

For hole H_8, *i.e.*, IT_8, the value of tolerance = $25i$ = 25 × 0.00185 = 0.0462 mm

∴ Max. size of hole = 50.0462 mm.

Value of fundamental deviation for shaft

$$c = -(95 + 0.8\,D) \text{ microns} = -(95 + 0.8 \times 63.2) = -(95 + 50.56)$$
$$= -145.56 \text{ microns} = -0.14556 \text{ mm.}$$

$\therefore$ Max. size of shaft $= 50.00000 - 0.14556 = 49.85444$, say 49.85544 mm.

Value of tolerance for c_8 is $25\,i = 0.0462$ mm

$\therefore$ Min. size of shaft $= 49.8544 - 0.0462 = 49.8082$ mm.

Problem 4.22. *Determine the limit of size for a hole ϕ 130 N4, given that main step is 120—180 mm and intermediate step is 120 to 140 mm. Deviation N for standard tolerance grades from IT01 to IT8 is increased by $\Delta(IT_n - IT_{n-1})$. ES for deviation N is $-0.6\sqrt[3]{D}$.*

Solution. Basic size step is 120—180, but intermediate step being applicable for *A* to *C* and *R* to *ZC* is not applicable in this case.

$\therefore$ $D = \sqrt{120 \times 180} = 146.97$ mm

Standard tolerance $= IT4 = 5i$

$i = 0.45\sqrt[3]{D} + 0.001D \simeq 12\ \mu m$

Fundamental deviation, $ES = -0.6\sqrt[3]{146.97}$

$= \simeq -27\ \mu m$

$\Delta = IT_4 - IT_3 = 12 - 8 = 4\ \mu m$

$\therefore$ Upper deviation $= ES + D = -27 + 4 = -23\ \mu m$

Lower deviation $= ES - IT = -23 - 12 = -35\ \mu m$

$\therefore$ Limits of size are

Maximum $= 130 - 0.023 = 129.977$ mm

Minimum $= 130 - 0.035 = 129.965$ mm.

Problem 4.23. (*a*) *State Taylor's principle in design of limit gauges.*

(*b*) *What are the advantages and disadvantages of using ring gauge versus gap gauge for checking cylindrical work ?*

(*c*) *What are the advantages and disadvantages of using progressive type gauge versus double ended plug gauge for checking holes ?*

(*d*) *What is the guideline to identify GO and NO GO surfaces of a gauge ?*

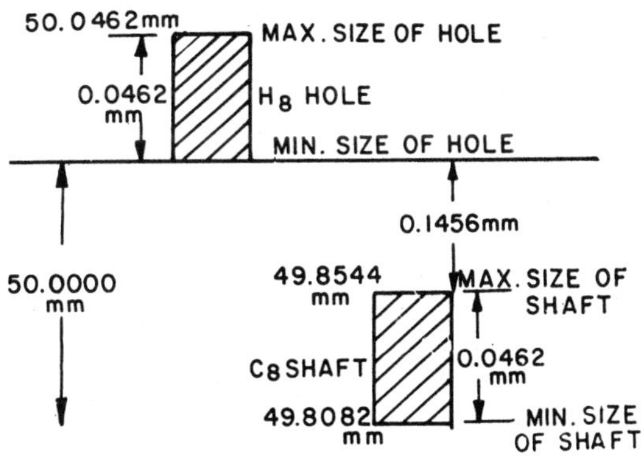

Fig. 4.92

Solution. (*a*) According to Taylor's principle for limit gauges, the GO gauge should be designed to check all or as many maximum metal limits of the part as possible in a single operation. In other words, Go gauge should be a full form gauge. If it is not done then holes of non-circular shape and not perfectly straight could be accepted. Further, NO GO gauges should be designed to check the minimum metal limit of only one dimension at a time. If NO GO gauge is made of full form, then elliptical holes could be accepted. On the other hand, NO GO gauge of bar type would detect major axis of elliptical hole and reject it.

(*b*) For cylindrical work, errors like lobing, raised surface imperfections can be detected by GO ring gauge but not by GO gap gauge. Further full form GO ring gauge can also check the straightness (bending and drunkenness) of shaft which is not possible with GO gap gauge. On the other hand, NO GO gap gauge can detect errors like barrelling, elliptical sections and other intended imperfections which is not possible with NO GO ring gauge. In fact NO GO ring gauge only can check the end of shaft and nothing else. Though gap gauge has the advantage of faster operation and its suitability for recesses, it is not suitable for GO gauge applications because of reasons stated above. In view of above and in conformity with Taylor's principle, it is now obvious that GO ring gauge and NO GO gap gauge should be used for cylindrical work for best results.

(*c*) In progressive type gauge, the GO gauge surface and NO GO gauge surface are provided in sequence on one side of the handle ; the NO GO surface being nearer to handle and GO surface on extreme. It has the advantage of faster operation for inspection as it need not be reversed for checking NO GO operation. However, it cannot be used for shallow or blind holes such as recesses where double ended gauges are used.

(*d*) The GO portion of a gauge is usually longer than NO GO portion of a gauge.

Problem 4.24. (*a*) *Discuss the different allowances that must be taken into account in the manufacture of a gauge.*

(*b*) *Determine the actual dimensions to be provided for a shaft and hole of 90 mm size for H_8e_9 type clearance fit. Size 90 mm falls in diameter steps of 80 and 100.*

Value of tolerance unit $i = 0.45\,(\sqrt[3]{D}) + 0.001D$.

Value of tolerances for IT8 and IT9 grades are 25i and 4i. Value of fundamental deviation for 'e' type shaft is $-\,11D^{0.41}$.

Also design the GO and NO GO gauges as per the present British system in which the same workshop and inspection gauges are used.

Solution. (*a*) The various allowances for gauges are :

(*i*) Gauge allowance—natural tolerance for machining allowances and is usually taken as 10% of the tolerance of the part.

(*ii*) Wear allowance—it is provided on GO gauge and not on NO GO gauges as latter have little wear compared to former. It is provided only when work tolerance is more than 0.0875 mm and is equal to 10% on gauge tolerance.

(*b*) For calculation of tolerance

$$D = \sqrt{80 \times 100} = 10\sqrt{80} = 89.5 \text{ mm}$$

Value of $\quad i = 0.45\,(\sqrt[3]{89.5}) + 0.001 \times 89.5$ microns $= 0.45 \times 4.46 + 0.0895$ microns

$= 2.0079 + 0.0895 = 2.0965$ microns $= 0.0020965$ mm.

$\therefore$ Value of tolerance on hole, corresponding to grate *IT*8 is

$$25i = 25 \times 0.0020965 = 0.0524 \text{ mm.}$$

As fundamental deviation for *H* type hole is zero

$\therefore$ Min. size of hole = 90.000 mm

and max. size of hole becomes

= 90.0524 mm

For 'e' type shaft, fundamental deviation

$= -\,11D^{0.41} = -\,11 \times 89.5^{0.41}$

$= -\,11 \times 6.23 = -\,68.53$ microns

$= -\,0.06853$ mm.

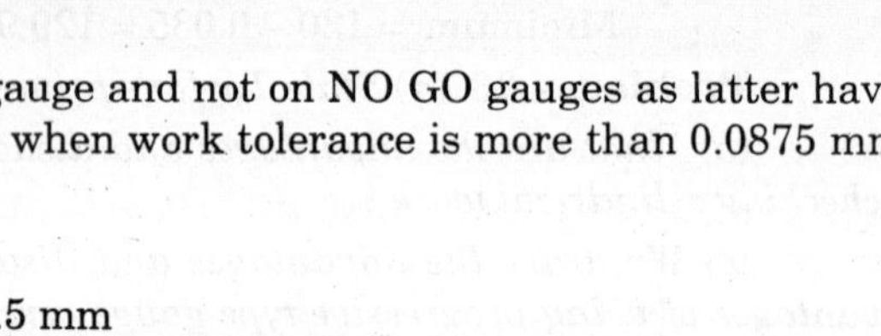

Fig. 4.93

$\therefore$ Max. size of shaft $= 90.000 - 0.06853 = 89.93147 \text{ mm} \simeq 89.9315 \text{ mm}$

Tolerance for $IT9$ grade is $40i$ $= 40 \times 0.0020965 = 0.08386$ mm.

$\therefore$ Min. size of shaft $= 89.9315 - 0.0839 = 89.8476$ mm

Gauge tolerance for hole = 10% of 0.0524 = 0.00524 mm ≈ 0.0052 mm

and wear allowance for hole = 0.000524 mm ≈ 0.0005 mm

Similarly gauge tolerance for shaft = 10% of 0.0839

= 0.00839 mm ≈ 0.0008 mm

With reference to Fig. 4.94

For hole

GO gauge $= 90^{(0.0005\,+\,0.0052)}_{+\,0.0005}$

$= 90^{+\,0.0057}_{+\,0.0005}$ mm.

NO GO gauge $= 90^{+\,0.0524\,+\,0.0052}_{0.0524}$

$= 90^{0.0576}_{+\,0.0524}$ mm

For shaft

GO gauge $= 89.9315^{-\,0.0008}_{-\,0.0008\,-\,0.0084}$

$= 89.9315^{-\,0.0008}_{-\,0.0092}$ mm

$= 90^{-\,0.0693}_{-\,0.0777}$ mm

NO GO gauge $= 89.8476^{-\,0.0000}_{-\,0.0084}$ mm

$= 90^{-\,0.1524}_{-\,0.1608}$ mm.

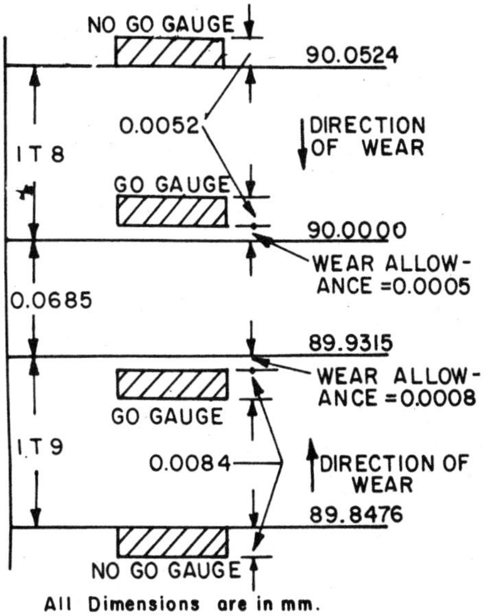

Fig. 4.94

Problem 4.25. *What do you understand by gauging force ?*

Solution. The gauging force is a function of so many variables such as material of the component and gauge, the elasticity of the material, gauging dimension, gauging conditions, the environment. A simple rule normally followed is as follows. If a GO gauge fails to assemble with the component *under its own weight* the component is quite definitely outside its maximum metal limit and needs reconsideration. Similarly, a component should be rejected only if the NOT NO gauge assembles freely under its own weight, and not merely because the NOT GO gauge has been forced to assemble.

Where such a criterion is used, the entry of the GO gauge should be facilitated by a chamfer as shown in Fig. 4.95 so as to ensure that cross cornering and jamming of the gauge do not prevent free assembly.

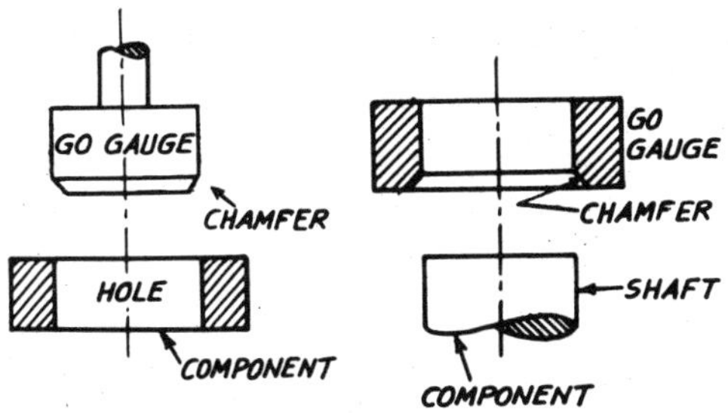

Fig. 4.95. Chamfering on GO gauges to avoid jamming etc.

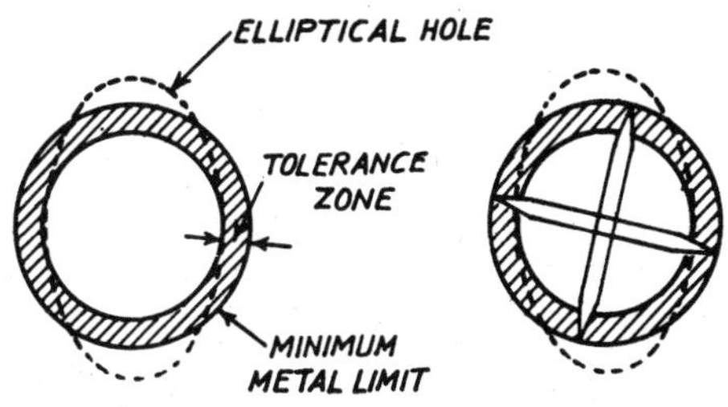

Fig. 4.96. Illustration of Taylor's Principle.

Problem 4.26. (*a*) *Explain Taylor's principle in detail.*

(*b*) *When does it become desirable to take deviations from Taylor's principle ?*

Ans. (*a*) Taylor's principle for designing the form of the gauges is of great importance in the design of limit gauges. For gauging a plain cylindrical hole, two plain cylindrical plug gauges, GO (diameter conforming to the maximum metal limit of the hole), and NOT GO (diameter conforming to the minimum metal limit) are employed for acceptance of hole, GO gauge should enter and NOT-GO should not. Since most methods of producing circular holes produce holes of reasonable geometrical accuracy, such gauges can be used with a fair degree of safety. In Fig. 4.96 (*a*), the two concentric circles represent the upper and lower limits of a circular hole. If the hole is truly circular and is within the tolcrance zone, gauges made to the size represented by the two circles will accept the component. However, if the hole departs from the true geometric shape, such as elliptical hole shown by dotted line, which for the most part lies outside the minimum metal limit, but enters the tolerance zone at two points, would be accepted by the cylindrical gauges described earlier because the GO gauge will enter the hole and NOT-GO will not because out of the two axes of the ellipse, one is smaller than the diameter of NOT-GO gauge. If the distortion of the hole is appreciable, it can be detected by rocking the GO gauge in the hole but this would give very little idea of the amount of variation or distortion. The trouble lies mainly with the failure of the NOT GO gauge to enter. The difficulty may be overcome by changing the form of the NOT GO gauge to that of a pin or bar gauge as shown in Fig. 4.96 (*b*). By offering the gauge to the hole in various positions, any lack of roundness may be detected. However, similar form of the GO gauge would not serve the purpose, as it is impossible for the outline to extend beyond the maximum metal limit and yet for the pin gauge to fail to reveal any error. This may arise when the hole is *lobed as* shown in Fig. 4.97.

A lobed figure is a constant diameter figure (at any two points) and offering the pin gauge in various positions will only show the constant diameter but will not reveal the lobed characteristic. But plug type full form GO gauge will not enter the hole because its entry will be obstructed by those portions of the hole which have exceeded the maximum metal limit.

Still another strong case advocating the full form of the GO gauge is where the diameter of the hole is constant but there is lack of straightness. In such a case the pin gauge will not indicate any lack of straightness but a plug of full form after entering the hole a little bit will not advance any further indicating lack of straightness.

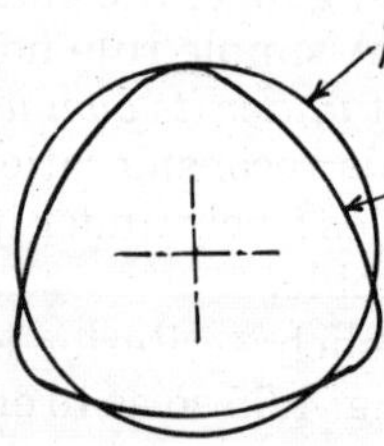

Fig. 4.97. Three-lobed shaft.

For these reasons it is obvious that the GO gauge must conform to the maximum metal limit over the whole profile if assembly is to be assured. It is also clear that in gauging the maximum metal limit of the component throughout the whole outline the GO gauge is, in reality gauging a whole family of common dimensions simultaneously. The principle is true, not only for one plane but for all planes along the axis of the hole, and may be extended to incorporate not only a measure of the roundness, but also of straightness of the hole in its maximum metal condition. For this reason, the GO gauge must conform to the mating surface entirely and is consequently described as "full form". Conversely, it is clear that on substituting a pin type NOT-GO gauge for the unsatisfactory cylindrical type, the effect is to reduce the area of contact and hence the number of *dimensions* gauged in their NOT-GO condition. The ideal case would be to gauge each individual dimension in its minimum metal condition separately. Practical difficulties arising from consideration of gauge wear makes it undesirable to employ pin gauges extensively for this purpose, and a useful compromise is to employ a bar type NOT-GO.

The foregoing principles were first enunciated by Taylor and, taken together are commonly known as *Taylor's Principles*. In general terms, Taylor's principles may be stated as follows :

1. The maximum metal limit of as many related dimensions as possible or convenient should be incorporated, in the GO gauge. Or stated a little differently, the GO gauge should be as far as possible, the geometrical equivalent of the mating part. It should be, in effect, a model of it.

Dimensionally the gauge cannot be the same as the mating component if a clearance or an interference fit is required. This will be appreciated if the GO gauge for a hole is considered.

(*i*) *Clearance fit.* In this case the shaft to assemble with the hole must always be smaller than the hole. The low limit of the hole is the high limit for the GO gauge, so that the gauge diameter must always be greater than the shaft by the amount of allowance provided for in the fit.

(*ii*) *Interference fit.* The shaft in this case must always be larger than the hole before assembly, a condition which is impossible for the GO plug gauge to meet.

2. *The minimum metal limits of the dimensions of the component should be checked by separate NOT-GO gauges.* This is because a NOT-GO gauge designed to check more than one dimension would fail to detect any dimension outside the minimum metal limit if one of the dimensions being checked is within the minimum metal limit as illustrated in Fig. 4.98.

(*b*) Some of the cases where departure from Taylor's principle are permitted are discussed below.

1. *Gauges for circular holes.* According to Taylor's principle, the GO gauge would be plug gauge having a minimum length equal to the length of the hole or the length of engagement of the associated component, whichever is shorter. The Not-Go gauge would be a pin gauge which could be used to check the minimum metal condition across any diameter at any position along the length of the hole as shown in Fig. 4.99. However, it is usually impracticable to make a plug of full length

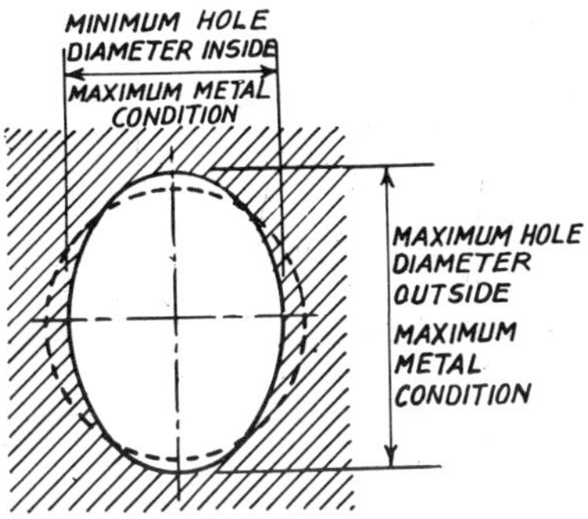

Fig. 4.98. Error not revealed by NOT-GO full form gauge.

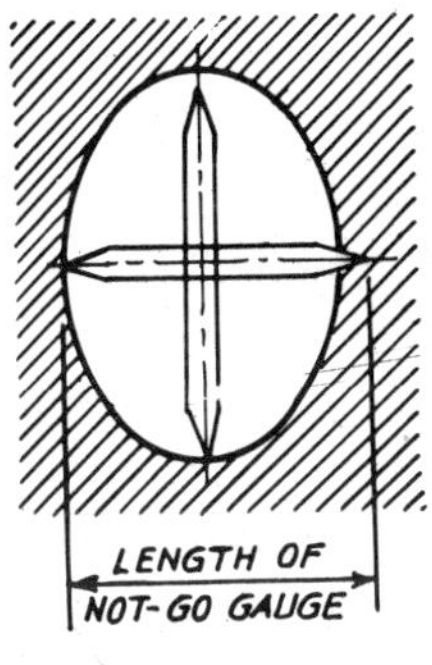

Fig. 4.99. Use of NOT-GO pin gauge for bores.

of the component. The usual GO plug gauge is often not the full length of the component, and the NOT-GO end is fully cylindrical. For instance many standards advocate combination of GO and NOT-GO gauges for gauging holes. In such a case, GO gauge is a plug of length not exceeding three to four diameter of the hole and the NOT-GO is a bar of some definite length. It violates the principle that GO gauge should be full form, *i.e.* length of the GO gauge should be equal to the length of hole and the NOT-GO gauge should be pin type approaching zero length.

In some cases, the plug is ground to a spherical diameter equal to the GO limit on the work, due allowance being made for the gauge-maker's tolerance and wear allowance.

A small circular area of the gauge is raised, by electroplating, an amount equal to the work tolerance. It is inserted into the hole by tilting the gauge forward so that the high portion clears the metal. This checks the low limit of the size. The gauge is then rocked back.

2. *Gauges for circular shafts.* For gauging circular shafts according to Taylor's principles, a ring gauge would be required for the GO gauge having a minimum length equal to the length of the shaft or the length of the engagement of the associated component, whichever is shorter. The NOT-GO gauge would be in the form of a gap gauge as shown in Fig. 4.100.

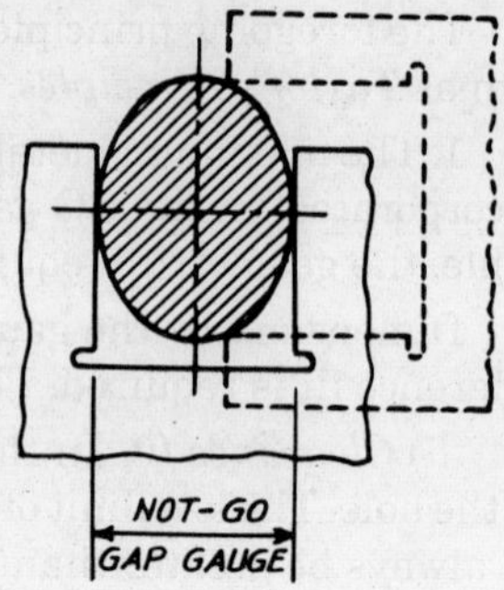

Fig. 4.100. Use of NOT-GO gap gauge for a shaft.

To gauge the maximum metal limit of all diameters simultaneously requires a ring gauge equal in diameter to the maximum metal limit of the shaft and theoretically equal in length. Again it is usually impracticable to make a ring gauge of length much exceeding two or three diameters. Further ring gauges for shafts are seldom used. While a gap gauge is ideal as a NOT-GO, it is imperfect as a GO. However, the present day permitted industrial practice is to use a combined unit of GO and NOT-GO gap gauges. This is convenient from the point of view of using the gauges and also it ensures that they are both kept together with less risk of misplacing one or the other.

From above it is obvious that practical considerations lead to variation from Taylor's principle. The departures of the above nature have been permitted due to accurate modern production machines and equipment available.

Problem 4.27. *A hole and shafting system has the dimension 60* $\dfrac{H-7}{m-6}$. *The standard tolerance is given by* $i = 0.45\sqrt[3]{D} + 0.0001D$

where D = *diameter (mm) of geometric mean step of 50 to 80 mm.*
i = *standard tolerance, micron*

The multipliers are

Grade	6	7
Multiplier	10	16

The fundamental deviation for fit 'm' is given by

$$FD = +(IT\text{-}7 - IT\text{-}6) \text{ for the relevant range of diameter.}$$

Find : (i) The Class of fit and state its application.

(ii) Sketch the fit and show the actual dimensions.

Solution. Nominal diameter of shaft and hole is 60 mm.

Since 60 mm diameter lies in the diameter steps of 50 and 80 mm.

$\therefore$ Value of D $= \sqrt{50 \times 80} = \sqrt{4000} = 63.2$ mm

The value of fundamental tolerance factor $i = 0.45\sqrt[3]{D} + 0.001D$

$= 0.45 \times \sqrt[3]{63.2} + 0.001 \times 63.2 = 0.45 \times 3.98 + 0.0632$ microns

$= 1.79 + 0.0632 = 1.85332$ microns $= 0.00185$ mm.

Hole is type H-7, and for H-Hole, fundamental deviation = 0

$\therefore$ Minimum size of hole = 60.0000 mm

For hole H-7, *i.e.* grade $IT7$, the value of tolerance $= 16i = 16 \times 0.00185 = 0.0296$ mm

$\therefore$ Maximum size of hole $= 60 + 0.0296 = 60.0296$ mm

Shaft is type m-6 : and value of fundamental deviation for shaft m is given by

$+ (IT7 - IT6) = 16i - 10i = 6i = 6 \times 0.00185 = 0.0111$ mm

$\therefore$ Minimum size of shaft

= Basic dimension + ($IT7 - IT6$) = 60.0000 + 0.0111 = 60.0111 mm

Value of tolerance for m-6 is $10\,i = 10 \times 0.00185 = 0.185$ mm.

$\therefore$ Maximum size of shaft = 60.0111 + 0.0185 = 60.0296 mm

Fig. 4.101 shows the fit and the actual dimensions.

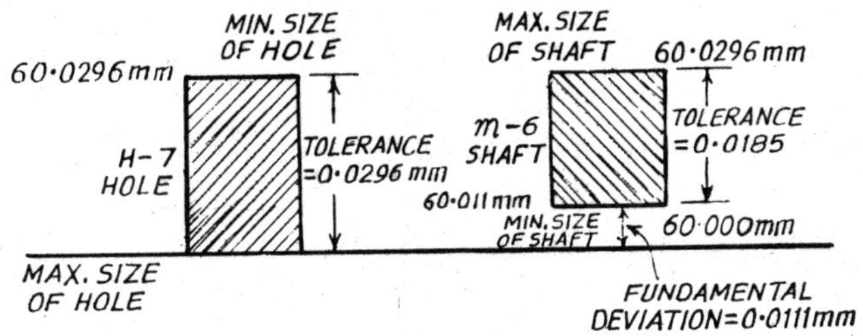

Fig. 4.101

It would be obvious from the disposition of dimensions that it is a transition fit. This type of fit is medium keying fit and finds application in medium tight assembly of mating surfaces as in the case of keyed shaft.

Problem 4.28. *Figure 4.102 given below, indicates a component with tolerances. Calculate the dimension 27 g8 as per IT 919 and determine the gap dimension G with tolerance.*

Given fundamental tolerance factor, $i = 0.45\sqrt[3]{D} + 0.001D$ *(microns)*

where D is the geometric mean of diameter step between 18 and 30 mm.

Grade of tolerance, $IT8 = 25i$ *(microns)*

Fundamental deviations, $g = 2.5\,D^{0.34}$ *(microns).*

Solution. Dimension 27 $g8$ has basic size of 27 mm but its tolerance values have to be calculated.

Fundamental tolerance factor $i = 0.45\sqrt[3]{D} + 0.001D$ microns

where D is the geometric mean of diameter steps, which in this case is $\sqrt{18 \times 30}$ since given

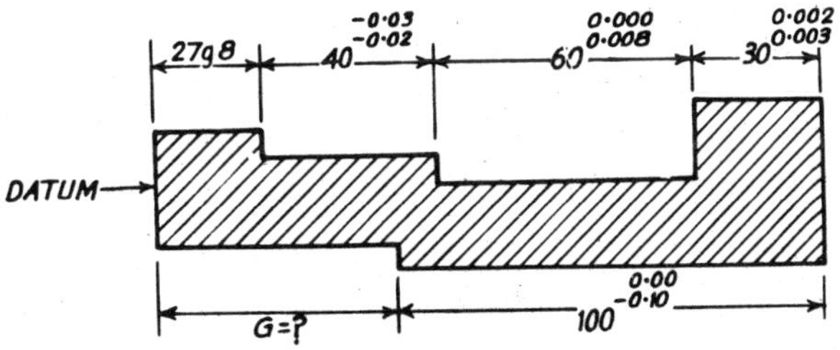

Fig. 4.102

dimension 27 mm lies in the diameter steps of 18 and 30 mm

$$D = \sqrt{18 \times 30} = 23.0 \text{ mm}$$

$$i = 0.45\sqrt[3]{D} + 0.001D$$

$$= 0.45\sqrt[3]{23.0} + 0.001 \times 23.0 \text{ microns} = 1.308 \text{ microns} \approx 0.0013 \text{ mm}$$

Tolerance of $IT8$ grade = $25i = 25 \times 0.0013 = 0.0325$ mm

Fundamental deviation for $g = 2.5 \times D^{0.34} = -0.5 \times 23.0^{0.34}$ microns

$= -0.5 \times 2.92 = -7.3$ microns $= -0.0073$ mm

$\therefore$ Maximum size $= 27.0000 - 0.0073 = 26.9927$ mm

and minimum size $= 26.9927 - 0.0325 = 26.9602$ mm

Thus tolerance on 27 mm size or dimension 27 *g* 8 is $27^{-0.0073}_{-0.0398} \approx 27^{-0.007}_{-0.04}$ mm

Basic dimension of gap $G = 27 + 40 + 60 + 30 - 100 = 57$ mm

Tolerances on G are as under

upper tolerance $= 0.00 - (0.007 - 0.03 + 0.000 + 0.002)$

$= 0.00 - (-0.035) = 0.035$ mm

lower tolerance $= 0.10 - (0.04 - 0.02 + 0.008 + 0.003)$

$= -0.10 - (-0.049) = -0.10 + 0.049 = 0.051$ mm

Thus gap dimension $G = 57^{-0.035}_{-0.051}$ mm.

Problem 4.29. *Figure 4.103 given below indicates a component with tolerances.*

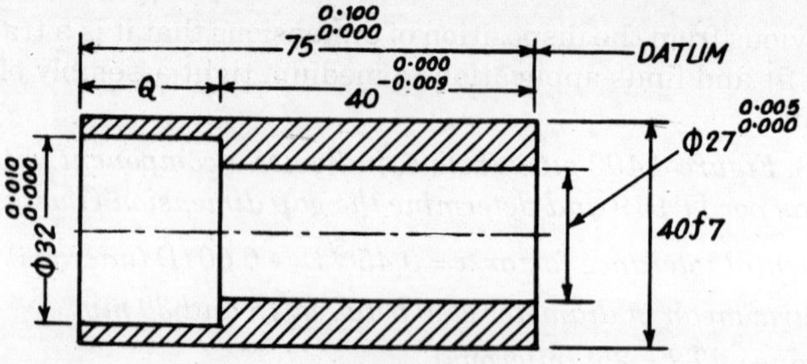

Fig. 4.103

Complete the dimensions as per IS : 919.

To what tolerance, can the dimension of the hole length Q be maintained ?

Given fundamental tolerance factor.

$i = 0.45 \sqrt[3]{D} + 0.001D$ *(microns), where D is the geometric mean of diameter steps between 30 and 50 mm.*

Grade of tolerance IT7 = 16 i (microns)

Fundamental Deviation, 'f' = $-5.5D^{0.41}$ (microns).

Solution. The first item to be determined is the dimension 40 *f* 7.

Its basic size is 40 mm.

Fundamental tolerance factor is

$$i = 4.05 \sqrt[3]{D} + 0.001D \text{ (microns)}$$

where $D = \sqrt{30 \times 50}$ mm (since diameter 40 mm lies in the steps of 30 and 50 mm)

$= 38.7$ mm

$$i = 0.45 \sqrt[3]{38.7} + 0.001 \times 38.7 = 0.45\,(3.38) + 0.0387$$

$= 1.521 + 0.0387 = 1.5597$ microns $= 0.00156$ mm

Thid dimension of shaft is of type *g* and grade 7.

Fundamental deviation of shaft $g = -5.5\, D^{0.41}$ microns

$= -5.5 \times 38.7^{0.41} = -5.5 \times 4.45$

$= -24.45$ microns $\simeq -0.0245$ mm

Tolerance of grade 7 is $16\,i = 16 \times 0.00156 = 0.02496$ mm

Since tolerance for shaft g is below the fundamental deviation

$$\therefore \text{ Dimension } 40\,f\,7 = 40^{-0.025}_{-0.0245-0.02496} = 40^{-0.025}_{-0.0494} \text{ mm}$$

Dimension of hole length Q can be maintained to

$$75^{0.100}_{0.000} \text{ mm} - 40^{0.00}_{-0.009} = 35^{0.100-0.00}_{0.000-(-0.009)}$$

$$= 35^{0.100}_{+0.009} \text{ mm.}$$

Problem 4.30. *Figure 4.104 below indicates a steel C-40 plug. The description of the components is to be made specific and precise for production on the shop-floor. The following information from the designer is available :*

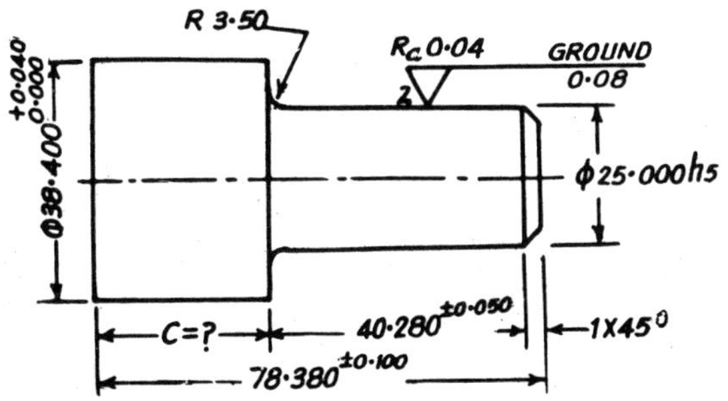

Fig. 4.104

Diameters are to be concentric and the concentricity error should not exceed 0.025 mm.

The squareness of the faces with the axis of rotation are to be within a permissible error limit of 0.02 mm. The following information is also given :

Grade of tolerance IT 5 = 7 i.

Standard tolerance unit, where D is the geometric mean of diameter steps between 18 mm and 30 mm.

(*i*) *Explain all the symbols used on the component drawing.*

(*ii*) *Calculate the tolerance on the smaller diameter.*

(*iii*) *Estimate the dimension on the closing link C with tolerance to indicate the accumulated error.*

(*iv*) *Sketch the component with full specifications.*

Solution. Let us first determine the dimension 25.000 h 5. The basic dimension of this shaft is 25.000 mm.

Fundamental deviation of shaft $h = 0$ and grade of tolerance $IT5 = 7i$

$$i = 0.45\sqrt[3]{D} + 0.001D, \quad (D = \sqrt{18 \times 30} = 23.21)$$

$$= 0.45\sqrt[3]{23.2} + 0.001 \times 23.2 = 0.45 \times 2.855 + 0.0223$$

$$= 1.285 + 0.023 = 1.308 \text{ microns}$$

$$\therefore \quad 7i = 7 \times 1.308 = 9.156 \text{ microns} = 0.0092 \text{ mm}$$

$$\therefore \quad \text{Dimension } 25\,h5 = 25^{+0.000}_{-0.0092} \text{ mm}$$

and $\quad$ tolerance on small diameter $= {}^{+0.000}_{-0.0092}$ mm

The symbols used on the drawing are :

R 3.50 – It stands for the provision of gradual change between two shafts of 25 and 38.4 mm diameters. A concave surface of radius 3.5 mm is provided as shown in Fig. 4.104.

This is an old method of representing surfaces roughness and other parameters on drawing. As per IS : 3073 – 1967 (old version), surface roughness is represented as under : (Refer Fig. 4.105)

Thus this symbol means that the surface on which Δ is marked shall be finished by grinding, with a machining allowance of 2 microns, and shall have a roughness not exceeding R_a = 5 μm on a sampling length of 0.08 mm.

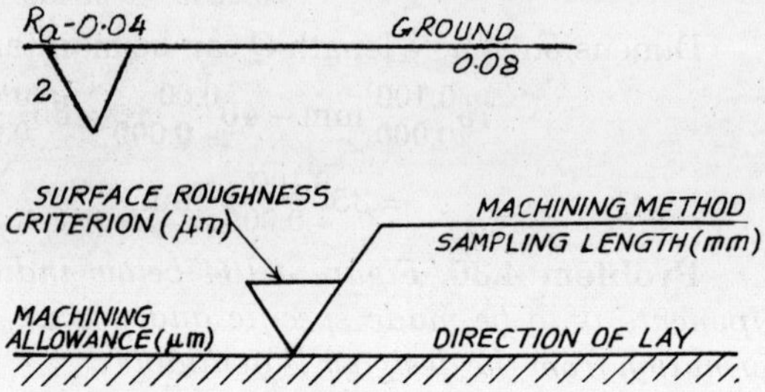

Fig. 4.105

Dimension on link C :

Chamber provided at the end of shaft of 25 mm diameter is 1 mm

$\therefore$ Dimension $\quad C = 78.380^{\pm\,0.100} - 40.280^{\pm\,0.050} - 1 = 37.100^{\pm\,0.050}$ mm

Component with full specifications :

It may be noted that vital information like concentricity error, squareness of the faces with the axis of rotation are not indicated on the drawing. These will be shown in symbolic form as shown in Fig. 4.106.

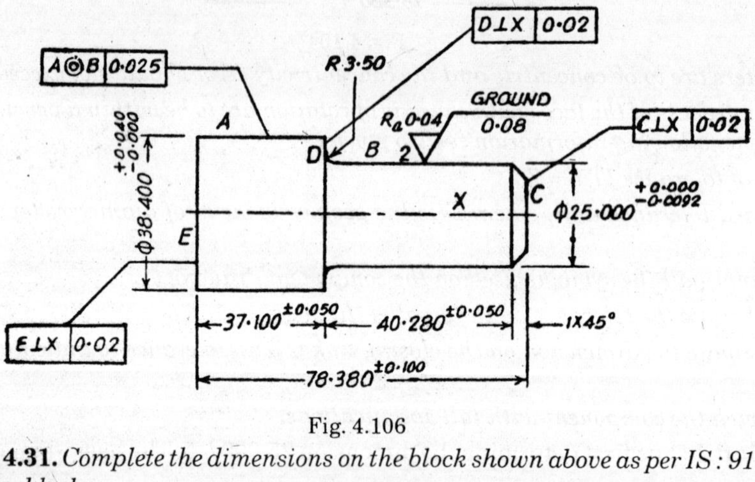

Fig. 4.106

Problem 4.31. *Complete the dimensions on the block shown above as per IS : 919 and explain the symbols on the block.*

Given fundamental tolerance factor, $i = 0.45\sqrt[3]{D} + 0.001D$ *(microns)*

where D is the geometric means of diameter steps.

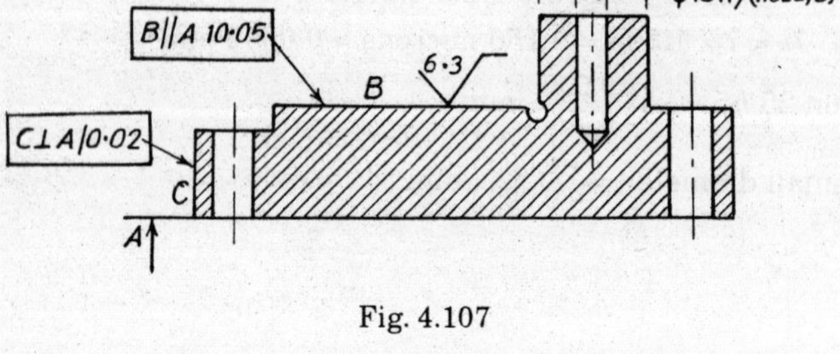

Fig. 4.107

Diameter steps for boss are between 30 mm and 50 mm and the diameter steps for hole are between 18 mm and 30 mm.

Grade of tolerance IT7 = 16i (microns).

Grade of tolerance IT8 = 25i (microns).

Solution. Two dimensions to be determined in this problem are 40 h 8 and 18 $H7$. Value of D for 40 mm basic hole is

$$D = \sqrt{30 \times 50} = 38.7 \text{ mm and } i = 0.45 \sqrt[3]{38.7} + 0.001 \times 38.7$$
$$= 0.45 \times 3.38 + 0.0387$$
$$= 1.521 + 0.0387 = 1.5597 \text{ microns} = 0.00156 \text{ mm.}$$

Fundamental deviation of shaft $h = 0$

and tolerance of grade 8 is = $25i = 25 \times 0.00156 = 0.039$ mm

$\therefore$ Dimension of 40 h 8 are $40^{+0.000}_{-0.039}$

Value of D for 18 mm is $\sqrt{18 \times 30} = 23.21$ mm

$$i = 0.45 \sqrt[3]{23.2} + 0.001 \times 23.2$$
$$= 0.45 \times 2.855 + 0.023 = 1.285 + 0.023$$
$$= 1.308 \text{ microns} = 0.001308 \text{ mm.}$$

Fundamental deviation of hole $H = 0$

Tolerance of grade 7 is $16\,i = 16 \times 0.001308 = 0.020928 \sim 0.0209$ mm

$\therefore$ Dimension of 18 $H7$ is $18^{+0.0209}_{+0.0000}$ mm

$\therefore$ Meaning of two symbols is $18^{+0.0209}_{+0.000}$ mm

Meaning of the symbols shown on block in Fig. 4.107 is

B ∥ A	0.05	Surface B is parallel to A and error of parallelity is 0.05 mm
C ⊥ A	0.02	Surface C is perpendicular to A and error of perpendicularity is 0.02 mm.

The symbol $^{6.3}\surd$ indicates that surface roughness height is 6.3 microns.

Problem 4.32. *Enumerate the advantages of fixed gauges.*

Ans. Fixed gauges offer many advantages, some of which are enumerated below :

(*i*) Consistency in form and dimension, no errors densto draft of original adjustment.

(*ii*) Provide positive dimensional information (supply unambiguous "yes" or "no" decision regarding acceptability of a part).

(*iii*) These are portable and do not need power supply.

(*iv*) Can be designed to check combinations of several dimensions comprising lengths, diameters, and angles. These can also be designed to inspect interrelated features for size, location, form, alignment, etc.

(*v*) These are used for assembly-wise functional inspection (provide complete reverse replicas of the critical part portion).

(*vi*) Provide uniform reference standards.

(*vii*) Provide most economical inspection.

Problem 4.33. *As the size of cylindrical plug gauges increases, their basic design also changes. Discuss some of the common basic designs.*

Ans. (*a*) For size from 0.75 mm to 1.9 mm, wire type plug gauge is used. (Refer Fig. 4.108).

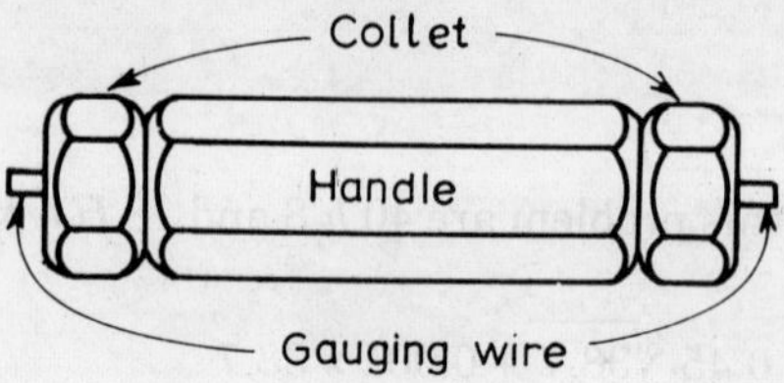

Fig. 4.108

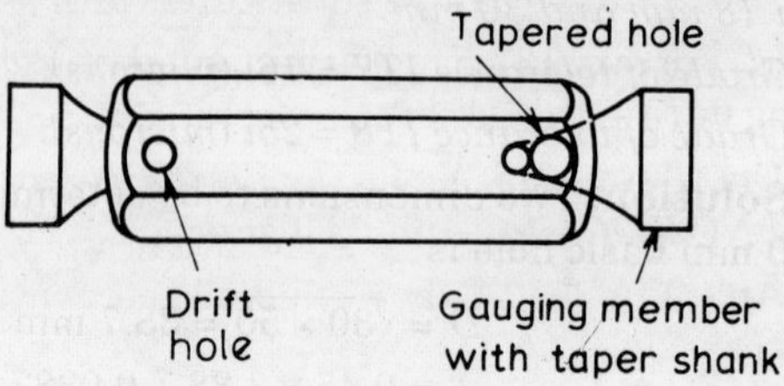

Fig. 4.109

It may be noted that wire type members of the gauge are held in a collet fitted at the end of a handle.

(*b*) For sizes, 1.5 mm to 37.5 mm, taper lock plug gauge design is adopted. The taper shank of gauging member is forced into the taper hole in the handle. Drift hole enables removal of gauging member.

(*c*) For sizes 19 mm to 200 mm, reversible or trilock plug gauge design is common.

In this arrangement the end faces of the handle have three radially arranged wedge shaped locking prongs which engage the corresponding grooves on the face of the gauging member. Both faces of the gauging members have similar grooves, permitting the plug to be reversed when worn. A screw secures the meshing wedge elements, causing them to stay in a rigidly locked position.

(*d*) For sizes 200 mm and upto 300 mm, angular plug gauge design is adopted. Here the main objective is to reduce weight and have high stability. Rim is made as gauging member and integral web is provided for rigidity. Centre of web is bored to reduce weight. Ball handles are inserted in holes.

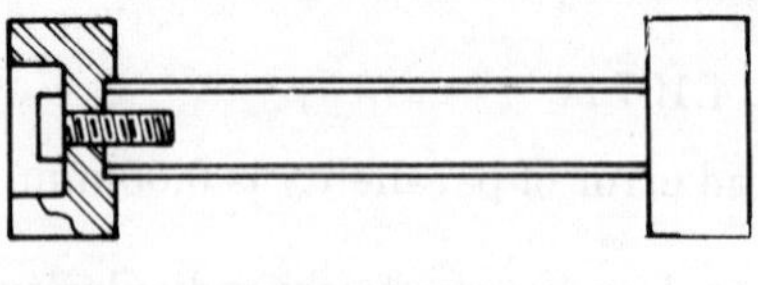

Fig. 4.110

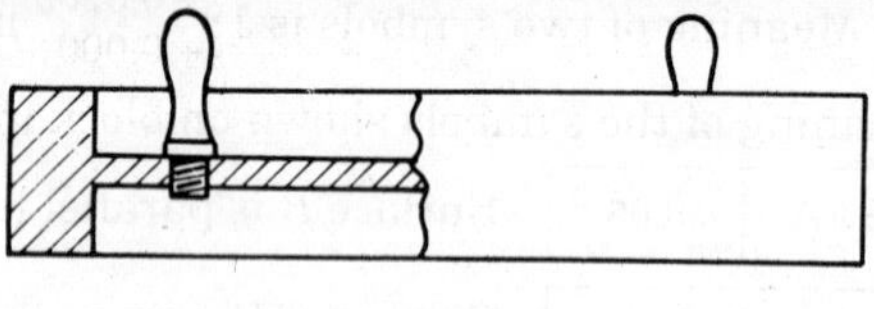

Fig. 4.111

Problem 4.34. *Discuss some of the basic design of cylindrical ring gauges for various sizes.*

Ans. (*a*) For sizes 1.5 mm to 12.5 mm, ring body is used with inserted bushing of hardened steel as gauging member. The gauge body is made of soft steel to ease manufacturing.

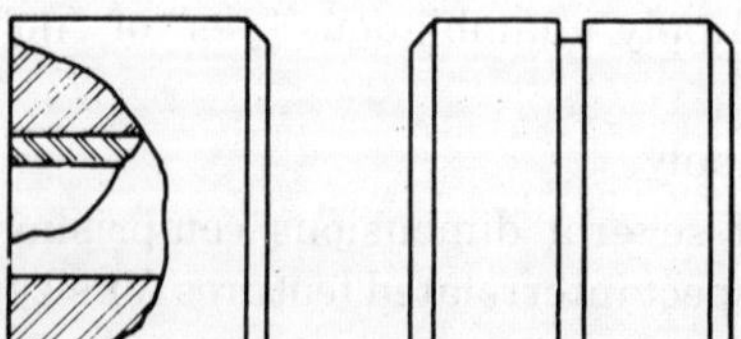

Fig. 4.112

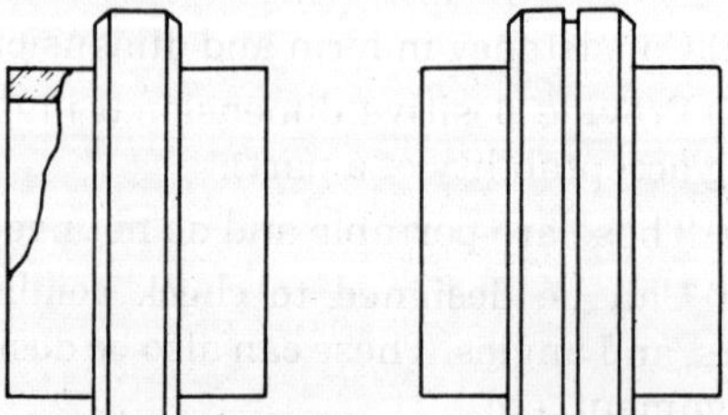

Fig. 4.113

(*b*) For sizes 12.5 mm to 37.5 mm solid ring gauge of plain cylindrical form made of hardened steel is used.

(*c*) For sizes 37.5 and 137.5 mm, solid ring gauge with flanged periphery for handling ease is used. Steel flange can be substituted with a neoprene ring.

(*d*) For size 137.5 to 300 mm, solid ring gauge with ball handles in flange are used to facilitate handling during the gauging process.

Problem 4.35. *Show the disposition of wear allowance and gauge maker tolerance on the GO and NOT GO plug and ring gauges.*

Ans. Fig. 4.114 shows the disposition of wear allowance and gauge maker clearances, in various cases.

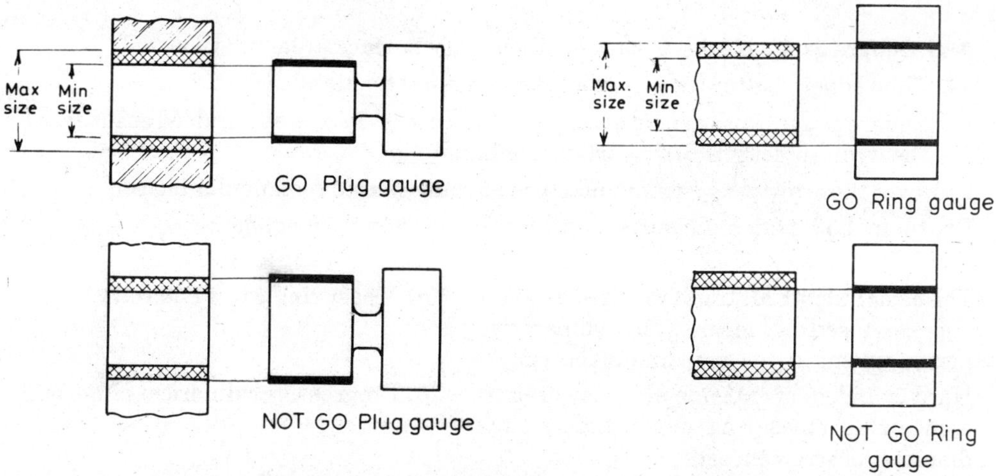

Dark area shows the wear allowance gauge maker tolerance for GO gauge and gauge maker tolerance for NOT GO gauge.

Fig. 4.114

Problem 4.36. *Enumerate the major requirements for gauge blocks.*

Ans. The major requirements for gauge blocks to serve the desired purpose are

— The individual elements of the set must be available in dimensions needed to achieve any combination of sizes.

— The accuracy of individual blocks must be within accepted tolerance limits.

— In the built-up combination, the individual blocks must be attached so closely to each other (a bar of reasonable cohesion) that the resulting bar will have a length equal to the added sizes of individual blocks of the assembly.

— The attachment of the elements to each other must be firm encugh to permit a reasonable amount of handling as a unit, and at the same time the individual pieces when taken apart must be reusable without any harm to their original size or other essential properties.

Problem 4.37. *Why the errors in flatness in case of gauge blocks have to be kept to minimum ?*

Ans. The errors in flatness have potential effect on the accuracy of length.

— Deficient flatness interferes with the exact measurement of the length.

— Lack of flatness can impede wringabililty.

— Deviations from flatness, by reducing the contacting area of the surface, can shorten the wear life of the gauge blocks.

Problem 4.38. *Why great importance is attributed to a high degree of smoothness of the gauging surfaces of the gauge blocks ?*

Ans. — A rough surface creates problem in dependable measurement of the gauge block length by contact instruments.

— for interferometric length measurements, a good reflecting surface is essential.

— high surface roughness can implede the dependability of dimension transfer.

— surface roughness reduces the wringability.

Problem 4.35. *What are the advantages of square gauge blocks over rectangular gauge blocks.*

Ans. Square gauge blocks, because of their large surface area, wear longer.

— They adhere better to each other when wrung the high stacks.

— Square gauges with centre holes permit the use of tie rods as an added assurance against the wrung stacks falling apart during handling.

However these are costly to manufacture as compared to rectangular blocks.

Problem 4.39. *Match the parts A and B in connection with gauges :*

Part A	*Part B*
1. The basic industrial standards used as reference and setting means for comparative measurement in precision manufacturing.	(*a*) Measuring wires and rolls
2. Hard cylinders of extreme accuracy used as a means of measuring angle, taper, pitch diameter of screw threads.	(*b*) Progressive cylindrical plug gauge
3. A gauge outside measuring surfaces of which are designed to verify the specified uniformity of holes.	(*c*) Spline plug gauge
4. A gauge the inside measuring surfaces of which are circular in form.	(*d*) Thread snap gauges
5. A complete internal gauge consisting of handle and gauging member in which the GO and NO GO gauging sections are combined in a single unit secured to one end of the handle.	(*e*) Taper ring gauge
6. A fixed gauge arranged with inside measuring surfaces for calipering diameters, lengths, thickness or width.	(*f*) Gauge Blocks
7. A complete internal gauge of single or double ended type. It consist of a handle and gauging member with suitable locking means.	(*g*) Flat plug gauge
8. An external gauge the internal contour of which conforms to the frustrum of a cone having diameter, taper and length suitable for the gauging of external taper dimensions.	(*h*) Ring gauge
9. These consist of conical points or serrated or threaded blades, plates, or cylinders which are adjustably mounted in a snap-gauge frame.	(*i*) Plain cylindrical plug gauge

10. A plug gauge having a series of projecting keys equally spaced about the perphery, when fit into the spline ways to be gauged.	(*j*) Wire type plug gauge
11. A plug gauge comprising a gauging member of straight cylindrical section throughout its length and which is held in a collet-type handle.	(*k*) Snap gauge
12. A plug gauge made in the form of a central axial segment of a plain cylindrical plug gauge.	(*l*) Plug gauge

Ans. 1. (*f*) 2. (*a*) 3. (*l*) 4. (*h*) 5. (*b*) 6. (*k*) 7. (*i*) 8. (*e*) 9. (*d*) 10. (*c*) 11. (*j*) 12. (*g*).

Problem 4.40. *Provide a single word for following statements :*

(i) A constant worst-case boundary generated by the collective effects of a size feature's specified material condition and the geometric tolerance for that material condition.

(ii) A composite tolerance control, capturing all errors of size, form, orientation, and location, used to specify the functional relationship of one or more feature of a part to a datum feature axis.

(iii) The process of inspecting and then adjusting a gauge or other measurement tool to meet a specific parameter.

(iv) The process of inspecting a gauge to verify that it meets a specific parameter.

(v) A device of defined geometric form and size used to assess the compliance of a feature or pattern of features of a workpiece with a dimensional specification and its accompanying tolerance.

(vi) A fixed—limit gauge used to verify of feature or a pattern of feature's compliance with virtual condition boundaries that are generated by the collective effect of the feature's maximum material condition and the applicable geometric tolerance at that size.

(vii) A fixed limit gauge that checks a workpiece at the maximum material condition.

(viii) A fixed limit gauge that checks a workpiece at cross section for violation of the least material condition.

Ans. (*i*) Virtual condition, (*ii*) Runout, (*iii*) Calibration, (*iv*) Certification, (*v*) Limit gauge, (*vi*) Functional gauge, (*vii*) Go gauge, (*viii*) NOT GO gauge.

Problem 4.41. Fill up blank choosing correct option from given choices.

(1) The essential characteristics of line graduated bar standards is the .... (least distance between graduations, sustained accuracy of the graduations).

(2) Vernier caliper gauge improves the .... (measuring sensitivity, accuracy, repeatability, resolution).

(3) Diffraction gratings need .... for providing desired measurement. (auxiliary master scale, optical magnifying device, photo electric sensor and electronic counter).

(4) Line graduated reference bars are usually made in the form of modified "H", "U" or "X" and the graduation lines are applied on .... (top surface, bottom surface, middle surface, neutral plane surface).

(5) Line graduated measuring bars represent a very high level of .... (accuracy, precision, sensitivity).

(6) During the measuring process the inspection bar and the rule to be calibrated must rest on a surface plate to avoid .... (temperature error, major deflections, parallax error).

(7) The essential element of a micrometer screw gauge is .... (linear and circumferential scales, screw frame).

(8) The ratchet mechanism in a micrometer screw gauge serves to .... (check wear out, ensure a uniform measuring force, eliminate play, use it as a snap gauge.

(9) For GO gauges toward increased material condition in the gauge, the plug diameter might be .... and ring bore diameter might be .... (smaller/larger), (larger/smaller).

(10) For NOT GO gauges toward the lesser material condition in the gauge, plug diameter might be .... and ring bore diameter might be (smaller/larger), (larger/smaller).

(11) Bilateral tolerances (spreading the gauge tolerance zone symmetrically in both sides of the gauge design limit) are applied in case of .... gauges. (GO/NOT GO).

(12) No wear allowances are needed for .... gauges. (GO/NOT GO).

(13) As a general rule, the wear allowance used in the design of GO limit gauges is equal to .... % of the total tolerance range assigned to the part dimension to a gauged. (1%, 5%, 15%).

(14) In case of 'GO' plug limit gauge, the gauge maker tolerances are added .... to the basic gauge size. (Unilaterally/bilaterally).

(15) In case of NOT GO plug limit gauge, the gauge maker tolerances are .... (Unilaterally subtracted from basic gauge size/bilaterally spreaded, straddling basic gauge size).

(16) For end standards (as in gauge blocks), the nominal dimension is the .... (largest boundary dimension/distance between the gauging surfaces).

(17) The American standard B 89. 1.9-1973 establishes .... accuracy grades for gauge blocks. (three/four/five).

(18) Flat contact points in dial gauges are preferred for measurements on .... (flat surfaces/convexly curved surfaces).

Ans. 1. sustained accuracy of the graduations, 2. measuring sensitivity, 3. photoelectric sensor and electronic counter, 4. neutral plane surface, 5. accuracy, 6. major deflections, 7. screw, 8. ensure a uniform measuring force, 9. smaller, 10. smaller, larger, 11. NOT GO, 12. NOT GO, 13. 5%, 14. unilaterally, 15. bilaterally spreaded, straddling basic size, 16. distance between the gauging surfaces, 17. four, 18. convexly curved surfaces.

Problem 4.42. State whether the following statements are true or false.

(1) In case of line graduated bar standards, high resolution is not the purpose because, it is intended as a reference standard with a limited number of distances.

(2) Diffraction gratings have graduation lines of known spacing but without assigned numerical values.

(3) In use of graduated rules for direct measurement, the proper positioning of the rule in relation to the workpiece, greatly affects the reliability of measurement. The accuracy of rule positioning can be greatly improved by various accessories like hooks, etc.).

(4) In use of line graduated rules, a big source of potential error is due to alignment of the distance boundaries with the corresponding graduations of the rule. This error is greatly reduced by means of positive contact members like the jaws of the caliper gauges.

(5) A locking screw is provided on vernier caliper gauges to enable quick and accurate reading.

(6) Depth gauge tongue provided on vernier calipers has same accuracy as for measurements between the jaws.

(7) The sleeve in a micrometer screw gauge provides a means of making zero adjustment.

(8) The contact surfaces on the spindle and on the anvail are always flat.

(9) Telescopic gauges are used to provide direct reading of bore dimensions.

(10) The GO limit gauge contains the reverse physical replica of the dimension to be measured at its maximum permissible material condition.

(11) The maximum material condition affects the assembly, by producing the minimum clearance in loose fits, and the maximum interference in press fits.

(12) The NOT GO limit gauge contains the reverse physical replica of the dimension to be measured its minimum acceptable material condition.

(13) If the NOT GO gauge assembles with the part being inspected, it is considered as a basis for acceptance.

(14) The phenomenon of the attachment of gauge blocks by wringing is attributed to the effect of molecular attraction.

(15) Gauge blocks are available in rectangular form.

(16) The essential purpose of gauge blocks is to make available end standards of specific lengths by temporarily combining several individual elements (arch representing a standard dimension) into a single gauge bar.

(17) Although gauge blocks, in their entirely, are precision products made to rigid specification, the greatest emphasis is laid on the two gauging surfaces on opposite sides of the blocks.

(18) Amplification mechanisms operating by gear trains permit a sturdy construction which is less sensitive to shocks and vibrations.

(19) The dial indicator accuracy is assessed by considering two categories of potential errors, *viz.* repetition of readings, and accuracy.

(20) The indicator gauges are always comparator type measuring instruments and require the use of a setting gauge for establishing the basic measuring position.

(21) Sector type indicator or sweep movement indicator fall under the category of high sensitivity.

(22) Air gauging is limited to a particular field of comparative measurements where the range of dimensional variations is relatively small.

(23) Air gauges are preferred for repetitive measurements and under conditions in which other sensitive measurements are not feasible.

(24) The response of back pressure pneumatic gauges is slower because of the compressibility of air.

(25) Obstruction of the narrow passages in the metering elements of the air gauge unit by condensed moisture or entrapped solids is the most frequent source of defective, erratic or inaccurate operation of air gauges.

(26) The zero setting of air gauges is commonly accomplished by means of a bleed valve and amplification adjustment is done by a precision valve of the control unit.

(27) The effect of the variations in the gauging position in pneumatic gauge reaches the control unit instantaneously.

(28) The flow type air gauges have a relatively quick response, but their use is essentially limited to applications where only visual observation of the indications is required.

(29) Air gauges utilising the changes in the flow characteristics of air are classified as belonging to the velocity system of pneumatic gauges.

(30) To create an effective obstruction, the solid surface facing the orifice must be located close enough to actually reduce the area of air escape.

(31) Air gauging is useful for comparative measurements for measuring wide variation of dimensions.

(32) The range of size variations which can be measured with air gauges is functionally limited.

(33) With open air jets, the magnitude of range is of the order of about 0.075 mm.

(34) With contact type sensing member in air gauge, the wider range of measurement (of the order of 0.1 mm) is possible.

(35) Direct air gauging with open jets is suitable for grooved, serrated or threaded surface areas as well.

(36) A one millimeter bore diameter nozzle in pneumatic gauge will consume air of the order of 0.3 cubic metre/hour.

(37) The air of pneumatic gauges must have minimum of humidity and be free of oil. (True)

(38) The most common cause of erratic or inaccurate operation of air gauge could be obstruction of the narrow passages in the metering elements of the air gauge by condensed moisture or entrapped solids.

(39) The back pressure type pneumatic gauges have a relatively quick response in comparison to flow type air gauges.

(40) In back pressure type air gauges, the response is slower because the compressibility of air contributes to the delayed transmission of the variations originated at the object surface.

(41) Two adjustments usually provided in all pneumatic gauges are

(*i*) Zero adjustment and

(*ii*) amplification or calibration adjustment.

(42) Zero setting in a pneumatic gauge is accomplished by a precision valve of the control unit.

(43) The purpose of amplification adjustment in pneumatic gauges is to obtain a range of gauge indications which, in terms of the scale graduations, are in correspondence with the size span of the setting gauges.

(44) The setting of pneumatic gauges is done using setting masters.

(45) Variable amplification in pneumatic gauges is utilised where the same gauge has to be used, in turn, for various measuring tasks requiring different measuring ranges.

(46) Air gauges having constant amplification can be set with single master gauge by zeroing on a specific mark on the scale.

(47) The advantages of velocity differential type air gauges are quick response, operation with a rather large clearance between nozzle and object surface (thereby reducing the wear of the gauging members) and low air consumption.

(48) Velocity differential type air gauges have magnification varying from 500 X to 5000 X.

(49) Back pressure type air gauges have magnifications varying from 5000 X to 40,000 X.

(50) For full calibration of pneumatic gauge, two masters of known sizes corresponding to total scale value are required.

(51) Although sensing by open-jet nozzle is the characteristics method used in pneumatic gauging, sensing with intermediate elements which operate by mechanical contact is also frequently applied.

(52) Air gauges permit the simultaneous measurement of several dimensions, even when closely spaced on the same object.

(53) The gauging force on the gauge spindle toward the specimen should be sufficient to assure positive contact and yet not be so high that the resulting elastic deformation could seriously affect the accuracy of gauging.

(54) Contact tips made of diamond, ground and polished to a flat form are preferred for gauge block calibration.

(55) Flat tip is preferred for measuring cylindrical or spherical parts.

(56) Gauge block comparator stands are made as light as possible since measurement forces involved are very low.

(57) Sine block is not a measuring instrument, but only a link in the angle measuring process.

Ans.

1. True	2. True	3. True	4. True	5. False
6. False	7. True	8. False	9. False	10. True
11. True	12. True	13. False	14. False	15. False
16. True	17. True	18. True	19. True	20. True
21. True	22. True	23. True	24. True	25. True
26. True	27. False	28. True	29. True	30. True
31. False	32. True	33. True	34. True	35. False
36. True	37. True	38. True	39. False	40. True
41. True	42. False	43. True	44. True	45. True
46. True	47. True	48. True	49. True	50. True
51. True	52. True	53. True	54. False	55. True
56. False	57. True.			

Correct Answers to False Questions

5. Locking screw enables to maintain a fixed jaw opening permitting repetitive readings of several identical parts.

6. Depth gauge torque is useful for rapid measurement of depth but the precision of measurement is not as much as between the jaws due to likely deflection and wear-out of torque surface.

8. In some micrometers, the contact tips may have forms like disk, blade, crew thread, pointed contact, ball contact, etc. for special applications.

9. The telescopic gauge is used to take the size of a bore and the same is then measured by an outside micrometer,

13. NOT GO gauge assembling indicates that the dimension being measured is incorrect and part needs rejection.

14. Wringing is effected due to combined effect of molecular attraction and adhesive action due to very thin film of moisture or oil.

15. Gauge blocks are available in rectangular, square with central hole, and round forms.

27. Because of long distance between gauge head and the control unit and because of compressibility of air, there may be lag of the order of one second.

31. Air gauging is limited to comparative measurements where the range of dimensional variation is relatively small.

35. It is not suitable for such applications.

39. Reverse is true.

42. It is achieved by a bleed valve.

54. For this purpose spherical form tip is preferred.

56. Gauge block comparator stands are made in generous dimensions, of heavily ribbed iron casting which is thoroughly stabilised, in order to provide high degree of rigidity.

TYPICAL EXAMINATION QUESTIONS

1. (*a*) Define the following :
   (*i*) Fits, (*ii*) Basic size,
   (*iii*) Fundamental deviation.
   (*b*) Determine the size of 'Go' and 'NO GO' gauges for components having 30 H_7/f_8 fit. Being given with usual notations. i (microns) = $0.45 \sqrt[3]{D} + 0.001\,D$ (D in mm)
   The upper deviation for f shaft = $-5.5.\ D^{0.41}$
   30 mm falls in the diameter step of 18 and 30.
2. Explain the terms 'wear allowance', 'Gauge Makers tolerance' and 'limit gauge'.
   Design and sketch working and plug snap 'GO' and 'NO GO' gauges for spindle $\frac{30.980 \text{ mm}}{31.960 \text{ mm}}$ and hole of $\frac{30.023 \text{ mm}}{31.000 \text{ mm}}$.
3. Explain why a GO gauge should be of full form and a NO GO gauge should check any one dimension of element or feature of the work.
4. (*i*) What is Taylor's principle as applied to the design of limit gauges ?
   Design a suitable 'GO' and 'NO GO' plug gauge for a bored hole $\frac{25.1}{25.0}$ mm diameter.
   (*ii*) Explain the terms :
   (*a*) Bilateral tolerance, (*b*) Basic hole system for dimensioning, and (*c*) Allowance.
   Explain the principle of 'NO GO' gauges.
5. (*i*) Explain Taylor's principle as applied to limit gauging. Why can't this be followed always in practice ?
   (*ii*) Discuss the disposition of tolerance on 'Workshop', 'Inspection' and 'General gauges'.
   (*iii*) How is wear allowance incorporated in the design of limit gauges ? Explain.
   (*iv*) Explain what is meant by :
   (*a*) Interchangeability part, (*b*) Universal interchangeability, and (*c*) Local interchangeability.
6. (*a*) Define the terms : 'tolerance', 'allowance', and 'limits'.
   Explain unilateral system of dimensioning and state its advantages.
   (*b*) Describe 'standard (basic) hole' and 'standard (basic) shaft' practices of fitting. What benefits are attributed to each ?
7. (*a*) Explain the term 'interchangeability'.
   (*b*) Describe briefly any system of limits and fits for engineering production.
   (*c*) Explain with neat sketch the basic hole system and unilateral tolerance. Discuss their suitability in comparison to other systems.
8. (*a*) Why is it necessary to give a tolerance on an engineering dimension ? Give an example of both the bilateral and unilateral tolerances.
   (*b*) State the hole used for a hole based system of limits and fits, and the shaft used for a shaft based system, as laid down in IS : 919.
   (*c*) Make a neat limit diagram showing the essential conditions for a clearance fit.
   (*d*) What is the importance of interchangeability with respect to an engineering assembly such as a motor cycle ?
9. A batch of pins has been turned to 20 mm ± 0.04 mm. Make a neat sketch, in good proportion, of a suitable limit gauge for checking the pins and insert the important dimensions.

10. (*a*) State three materials used for gauges.

(*b*) What is meant by wear allowance ?

(*c*) Does the GO gauge check the maximum or minimum metal condition ?

11. (*a*) How the effects and interactions between the 'Grade' machining and 'size' of job are correlated in computing actual tolerances according to I.S. system ?

(*b*) A turned shaft is to rotate in a reamed hole at low peripheral velocity. The designer has chosen grade *H*8 for tolerance. The diameter of the shaft is 50 mm and falls within size range of 50—80. The chosen dimension is shown as : 50 *H*8/*c*7.

Using IS : 919, calculate the actual dimensions of the hole and shaft giving upper and lower limits correct to third place.

12. (*a*) Explain Taylor's Principle.

(*b*) What are the consideration for deciding the limits on the limits gauges ?

(*c*) What is meant by universal and local interchangeability ?

13. (*a*) What are the Taylor principles for gauging finished workpieces ?

(*b*) A 60 mm, *m*6 shaft is to be checked by a GO NOT-GO snap gauge. Assume 5% wear allowance and 10% gauge maker's tolerance (% of the tolerance of the shaft). The fundamental deviation for *m* fit is (*IT*7 – *IT*6) where multiplies for grade 17 is 16 *IT*6 is 10.

Sketch the Workshop, Inspection and General gauges.

14. (*a*) Draw a neat sketch of 'progressive' type solid plug gauge and describe advantages and disadvantages of this type of gauging.

(*b*) Design "General", "Workshop" and "Inspection" type of GO and NOT-GO gauges for 50 mm *C* 7 hole. Given,

(*i*) Fundamental Deviation = $0.52D^{0.2}$; (*ii*) $i = 0.45(D)^{1/3} + 0.001D$

(*iii*) 50 mm lies in the diameter step 30 to 50 mm.

Take gauge tolerance as 10% of work tolerance.

Take wear allowance only if work tolerance exceeds 100 microns.

15. (*a*) Explain clearly what is meant by selective assembly. Given one practical example.

(*b*) How does a selective assembly differ from an interchangeable manufacture ?

(*c*) Name a limit system suitable if large number of shafts and holes are received from different sources.

16. (*a*) Explain clearly the following type of fits and how they can be achieved :

1. Selective 2. Push fit 3. Driving fit

4. Forced fit 5. Shrinkage fit.

(*b*) Which of the above are clearance, transition and interference fits ?

(*c*) Give a hole shaft combination for a push fit.

17. Indian Standard IS : 919 recommends the use of a Unilateral Hole Basis System for limits and fits for Indian industries.

(*a*) Explain the meaning of the following terms : (*i*) Hole Basis System, stating why it is recommended : (*ii*) Unilateral and Bilateral Limits.

(*b*) In a hole and shaft combination of 25 mm nominal size,

*H*7 hole limit are + 0.021 mm, + 0.000 mm

*e*8 shaft limits are – 0.040 mm, – 0.073 mm

State the values of (*i*) maximum and minimum clearance obtainable ; (*ii*) allowance, (*iii*) tolerance on the hole and the shaft ; (*iv*) type of fit.

18. In a modern engineering works, component parts are manufactured from dimensioned drawings. A feature of the method is that some sizes are shown as follows :

$\begin{matrix} 50.03 \\ 49.97 \end{matrix}$ when the normal size is 50 mm hole basis.

(*a*) Explain why this procedure is adopted and in your answer show that you understand the meaning of the following terms : limits, tolerance, fit, and transition fit.

(*b*) Explain clearly what is meant by selective assembly, when is it used, and how does it differ from interchangeable assembly ?

(*c*) When inspecting cylindrical work a ring gauge or a gap gauge may be employed. Explain the advantages of using both types.

19. (*a*) Indian Standard IS : 919 caters for either a 'hole-based' or a 'shaft-based' system of fits. Explain why a 'hole-based' system is generally preferred but quote one example where it would be desirable to use 'shaft-based' system in a standard limit system.

(*b*) What are the equivalent shaft-based system fits for the following hole based fits.

Clearance	:	$H8/g7$
Transition	:	$H8/r7$
Interference	:	$H8/u7$

(*c*) What type of fit you will recommended for permanent assemblies of machine components ? Give a suitable hole-shaft combination for the type of fit you have recommended on the hole-based system. What is the equivalent fit on a shaft-based system ?

20. (*a*) Explain clearly the terms :

(*i*) Limits, (*ii*) Allowance, (*iii*) Tolerance and (*iv*) Hole basis system of fit.

When one of the two systems of tolerance, unilateral or bilateral, is more advantages ? State reasons.

(*b*) What are limit gauges ? Sketch and explain any two types of limit gauges.

21. Discuss the terms 'interchangeability' and 'selective assembly'. Describe the different types of limit gauges commonly used.

22. Distinguish between a measuring instrument and gauge.

23. Briefly state the principal features of the Indian Standard system of fits. Explain how you would proceed to specify the tolerance on the mating components of clearance fit.

24. Explain Taylor's principle of limit gauging. Can this principle be strictly followed in practice ? Justify your answer. Determine the limit of dimensions of the gauge for checking holes of 30 ± 0.02 mm diameter and give a sketch of the limit gauge.

25. A plug gauge for a No. 3 Morse taper has a diameter of 28.25 mm at the large end, and is 52 mm long. The taper is 50.2 mm per metre on the diameter. List the checks that should be made on it and describe in detail how you would check the angle of taper.

26. A taper ring gauge has end diameters of approximately 60 mm and 45 mm and its length is 55 mm. Describe how would you measure its angle and end diameters. Derive any necessary formulae.

27. Sketch and describe the method of measuring a taper plug gauge and a taper ring gauge using rollers, balls and slip gauges.

A taper ring gauge is 62 mm long and has larger end diameter of 29.15 mm and included angle of $19° 25'$.

Describe method to check its dimensions and derive any formulae needed and estimate the accuracy of the results you would obtain.

28. A taper ring gauge has a mean dia of 17.5 mm and an axial length of 32.5 mm. Describe how you would find the angle of the taper. Show all necessary calculations.

29. What checks should be made for a complete inspection of a taper plug gauge ? Describe in detail a method of checking the angle of taper of such a gauge and include all necessary calculations.

5

Comparators

5.1. Introduction

A comparator works on relative measurements, *i.e.* to say, it gives only dimensional differences in relation to a basic dimension. So a comparator compares the unknown dimensions of a part with some standard or master setting which represents the basic size and dimensional variations from the master setting are amplified and measured. The advantages of comparators are that not much skill is required on the part of operator in its use. Further the calibration of instrument over full range is of no importance as comparison is done with a standard end length. Zero error of instrument also does not lead to any problem. Since range of indication is very small, being the deviation from set value, a high magnification resulting into great accuracy is possible. The comparators are generally used for linear measurements, and various comparators available differ principally in the method used for amplifying and recording the variations measured. According to the principles used for obtaining suitable degrees of magnification of the indicating device relative to the change in the dimension being measured, the various comparators may be classified as follows :

(1) Mechanical comparators
(2) Mechanical-optical comparators
(3) Electrical and Electronic comparators
(4) Pneumatic comparators
(5) Fluid displacement comparators
(6) Projection comparators
(7) Multi-check comparators
(8) Automatic gauging machines.

5.2. Characteristics of Comparators

Before we discuss the various types of comparators, let us first look into various fundamental requirements which every comparator must fulfil. These are as follows :

(*i*) The instrument must be of robust design and construction so as to withstand the effect of ordinary usage without impairing its measuring accuracy.

(*ii*) The indicating device must be such that readings are obtained in least possible time and for this, magnification system used should be such that the readings are dead beat. The system should be free from backlash, and wear effects and the inertia should be minimum possible.

(*iii*) Provision must be made for maximum compensation for temperature effects.

(*iv*) The scale must be linear and must have straight line characteristic.

(*v*) Indicator should be constant in it return to zero.

(*vi*) Instrument, though very sensitive, must withstand a reasonable ill usage without permanent harm.

(*vii*) Instrument must have the maximum versatility, *i.e.*, its design must be such that it can be used for a wide range of operations.

(*viii*) Measuring pressure should be low and constant.

5.3. Uses of Comparators

The various ways in which the comparators can be used are as follows :

(*i*) In mass production, where components are to be checked at a very fast rate.

(*ii*) As laboratory standards from which working or inspection gauges are set and correlated.

(*iii*) For inspecting newly purchased gauges.

(*iv*) Attached with some machines, comparators can be used as working gauges to prevent work spoilage and to maintain required tolerances at all stages of manufacturing.

(*v*) In selective assembly of parts, where parts are graded in three or more groups depending upon their tolerances.

5.4. Advantages and Disadvantages of Various Types of Comparators

Advantages of Mechanical Comparators

(*i*) These are usually cheaper in comparison to other devices of amplifying.

(*ii*) These do not require any external supply such as electricity or air and as such the variations in outside supplies do not affect the accuracy.

(*iii*) Usually the mechanical comparators have linear scale which is easily understood.

(*iv*) These are usually robust and compact and easy to handle.

(*v*) For ordinary workshop conditions, these are suitable and being portable can be issued from a store.

Disadvantages

(*i*) The mechanical comparators have got more moving parts than other types. Due to more moving parts, the friction is more and ultimately the accuracy is less.

(*ii*) Any slackness in moving parts reduces the accuracy considerably.

(*iii*) The mechanism has more inertia and this may cause the instruments to be sensitive to vibration.

(*iv*) The range of the instrument is limited as the pointer moves over a fixed scale.

(*v*) Error due to parallax is possible as the moving pointer moves over a fixed scale.

Advantages of Optical Comparators

(*i*) It has small number of moving parts and hence a higher accuracy.

(*ii*) In the optical comparators, the scale can be made to move past a datum line and thus have high range and no parallax errors.

(*iii*) It has very high magnification.

(*iv*) Optical lever is weightless.

Disadvantages

(*i*) As the instrument has high magnification, heat from the lamp, transformer etc. may cause the setting to drift.

(*ii*) An electrical supply is necessary.

(*iii*) The apparatus is usually large and expensive.

(*iv*) When the scale is projected on a screen, then it is essential to use the instrument in a dark room in order to take the readings easily.

(*v*) The instruments in which the scale is viewed through the eyepiece of a microscope are not convenient for continuous use.

Advantages of Electrical Comparators

(*i*) The electrical comparators have got small number of moving parts.

(*ii*) It is possible to have a very high magnification and the same instrument may have two or more magnifications. Thus the same instrument can be used for various ranges.

(*iii*) The mechanism carrying the pointer is very light and not sensitive to vibrations.

(*iv*) As the instrument is usually operated on A.C. supply, the cyclic vibration substantially reduces errors due to sliding friction.

(*v*) The measuring unit can be made very small and it is not necessary that the indicating instrument be close to the measuring unit, rather it can be remote also.

Disadvantages

(*i*) It requires an external agency to operate *i.e.*, the A.C. electrical supply. Thus the variations in voltage or frequency of electric supply may affect the accuracy.

(*ii*) Heating of coils in the measuring unit may cause zero drift and alter the calibration.

(*iii*) If only a fixed scale is used with a moving pointer then with high magnifications a very small range is obtained.

(*iv*) This is usually more expensive than mechanical instrument.

Advantages of Pneumatic Comparators

(*i*) The gauging member does not come into contact with the part to be measured and hence practically no wear takes place on the gauging member.

(*ii*) It has usually very small number of moving parts and in some cases none. Thus the accuracy is more due to less friction and less inertia.

(*iii*) Measuring pressure is very small and the jet of air helps in cleaning the dust, if any, from the part to be measured.

(*iv*) It is possible to have very high magnification.

(*v*) The indicating instrument can be remote from the measuring unit.

(*vi*) It is very suitable device for measuring diameter of holes where the diameter is small compared with the length.

(*vii*) It is probably the best method for determining the ovality and taperness of the circular bores.

Disadvantages

(*i*) It requires elaborate auxiliary equipment such as accurate pressure regulator.

(*ii*) The scale is generally not uniform.

(*iii*) When indicating device is the glass tube, then high magnification is necessary in order to avoid the meniscus errors.

(*iv*) The apparatus is not easily portable and is rather elaborate for many industrial applications.

(*v*) Different gauging heads are required for different dimensions.

5.5. Mechanical Comparators

In these comparators, magnification is obtained by mechanical linkages and other mechanical devices.

5.5.1. Systems of Displacement Amplification used in Mechanical Comparators

(*i*) **Rack and Pinion.** In it the measuring spindle integral with a rack, engages a pinion which amplifies the movement of plunger through a gear train. (Refer Fig. 5.1)

(*ii*) **Cam and gear train.** In this case the measuring spindle acts on a cam which transmits the motion to the amplifying gear train. (Refer Fig. 5.2)

(*iii*) **Lever with toothed sector.** In this case a lever with a toothed sector at its end engages a pinion in the hub of a crown gear sector which further meshes with a final pinion to produce indication. (Refer Fig. 5.3)

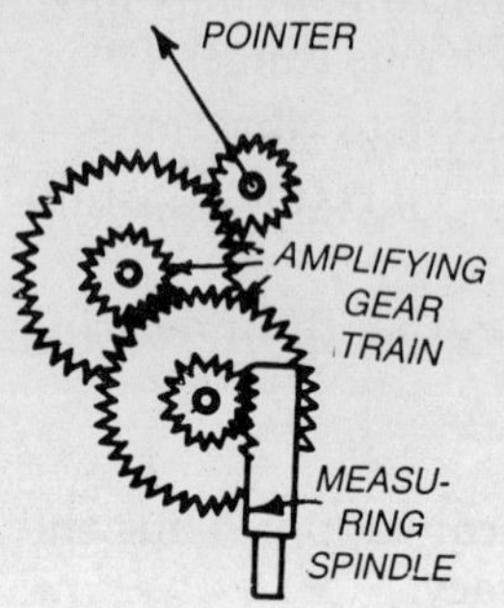

Fig. 5.1. Rack and pinion.

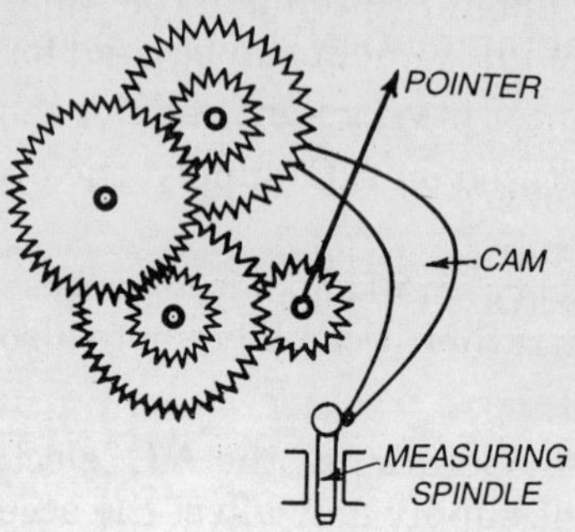

Fig. 5.2. Cam and gear train.

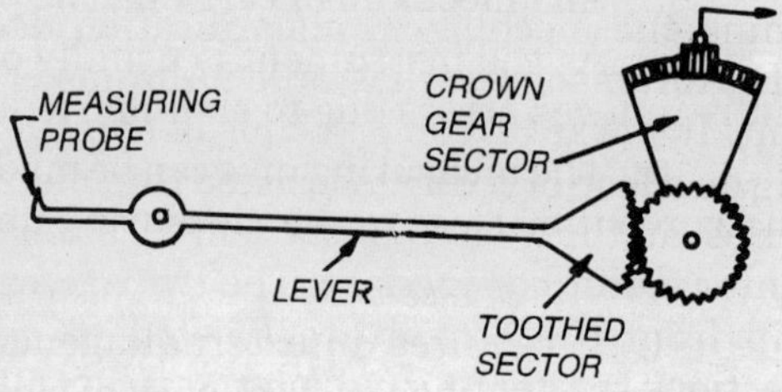

Fig. 5.3. Lever with toothed gear.

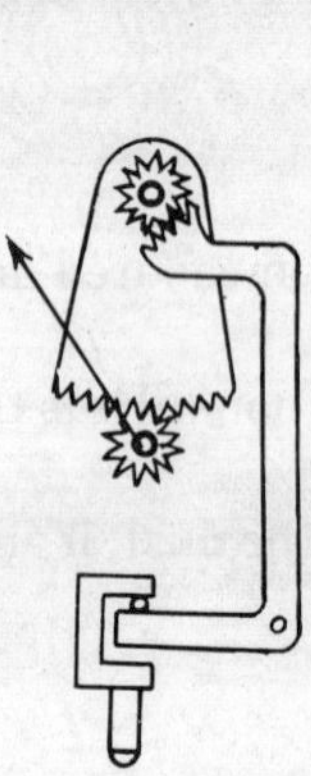
Fig. 5.4. Compound levers.

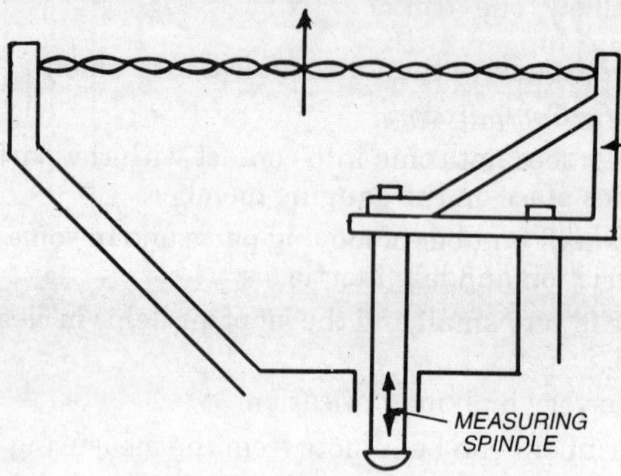

Fig. 5.5. Twisted taut strip.

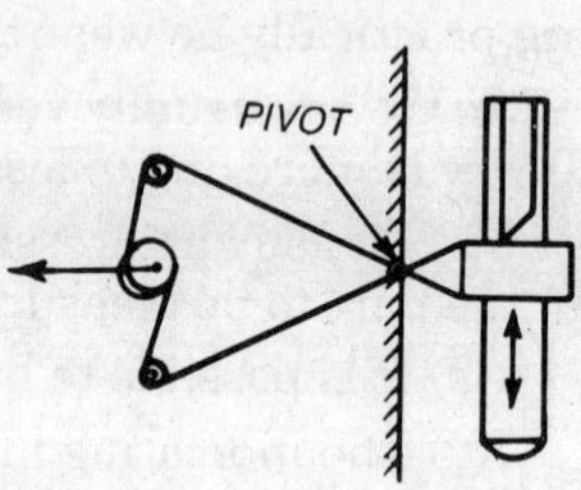

Fig. 5.6. Lever combined with band wound around drum.

(*iv*) **Compound Levers.** Here levers forming a couple with compound action are connected through segments and pinion to produce final pointer movement.

(*v*) **Twisted Taut Strip.** The movement of measuring spindle tilts the knee causing straining which further causes the twisted taut band to rotate proportionally. The motion of strip is displayed by the attached pointer.

(*vi*) **Lever combined with band wound around drum.** In this case, the movement of the measuring spindle tilts the hinged block, causing swing of the fork which induces rotation of the drum.

(*vii*) **Reeds combined with optical display.** In this case parallelogram reeds are used which transfer measuring spindle movement to a deflecting reed whose extension carries a target utilised in optical path.

(*viii*) **Tilting mirror projecting light spots.**

5.5.2. Dial Indicator. One of the most commonly used mechanical comparators is essentially of the same type as a dial indicator. It consists of a robust base whose surface is perfectly flat and a pillar carrying a bracket in which is incorporated a spindle and indicator. The linear movement of the spindle is magnified by means of a gear and pinion train into sizable rotation of the pointer on the dial scale. The indicator is set to zero by the use of slip gauges representing the basic size of the part. This is generally used for inspection of small precision-machined parts. This type of comparator can be used with various attachments so that it may be suitable for large number of works. With a V-block attachment it can be used for checking out-of-roundness of a cylindrical component.

5.5.3. The Johansson 'Mikrokator'. This comparator was made by C.F. Johansson Ltd. and therefore named so. It is shown diagrammatically in Fig. 5.7. This instrument uses the simplest and most ingenious method for obtaining the mechanical magnification designed by H. Abramson which is called Abramson Movement. It works on the principle of a button spinning on a loop of string. A twisted thin metal strip carries at the centre of its length a very light pointer made of thin glass. The two halves of the strip from the centre are twisted in opposite directions so that any pull on the strip will cause the centre to rotate. One end of the strip is fixed to the adjustable cantilever strip and the other end is anchored to the spring elbow, one arm of which is carried on the measuring plunger. As the measuring plunger moves either upwards or downwards, the elbow acts as a bell crank lever and causes twisted strip to change its length thus making it further twist or untwist. Thus the pointer at the centre of the twisted strip rotates by an amount proportional to the change in length of strip and hence proportional to the plunger movement.

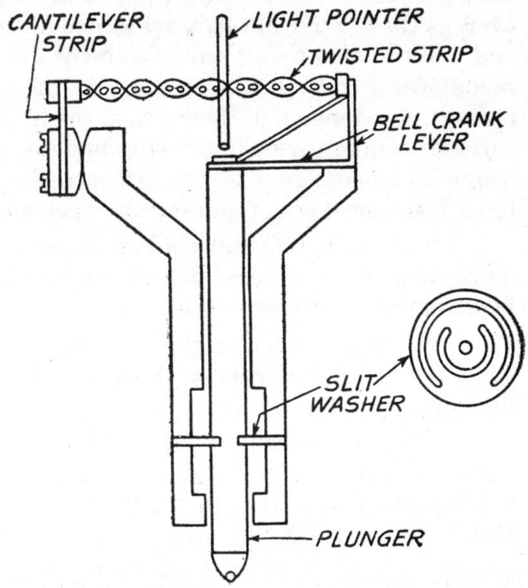

Fig. 5.7. Johansson Mikrocator.

The bell crank lever is formed of flexible strips with a diagonal which is relatively stiff. The length of cantilever can be varied to adjust the magnification of the instrument. Since the centre line of the strip is straight even when twisted, therefore, it is directly stretched by the tension applied to the strip. Thus in order to prevent excessive stress on the central portion, the strip is perforated along the centre line by perforations as shown in Fig. 5.7.

It can be shown that the amplification of instrument $\frac{d\theta}{dl} \propto \frac{l}{w^2 n}$ where θ is the twist of mid-point of strip with respect to the end, l is the length of twisted strip measured along its neutral axis and w is the width of twisted strip and n is the number of turns.

It is thus obvious that in order to increase the amplification of the instrument a very thin rectangular strip must be used. Further amplification can be adjusted by the cantilever strip which also provides anchorage. It increases or decreases effective length of strip. Final setting of the instrument amplification is made by a simple adjustment of the free length of cantilever strip.

A slit C washer as shown in Fig. 5.7 is used for the lower mounting of plunger. Thus this instrument has no mechanical points or slides at which wear can take place. The magnification of the instrument is of the order of × 5000.

5.5.4. Reed Type Mechanical Comparator. In mechanical comparator, the gauging head is usually a sensitive, high quality, dial indicator mounted on a base supported by a sturdy column. Fig. 5.8 shows the reed type mechanical comparator.

The reed mechanism is frictionless device for magnifying small motions of spindle. It consists of a fixed block A which is rigidly fastened to the gauge head case, and floating block B, which carries the gauging spindle and is connected horizontally to the fixed block by reeds C. A vertical reed is attached to each block with upper ends joined together. These vertical reeds are shown in the figure by letter D. Beyond this joint extends a pointer or target. A linear motion of the spindle moves the free block vertically causing the vertical reed on the floating block to slide past the vertical reed on the fixed block. However, as these vertical reeds are joined at the upper end, instead of slipping, the movement causes both reeds swing through an arc and as the target is merely an extension of the vertical reeds, it swings through a much wider arc. The amount of target swing is proportional to the distance the floating block has moved but of course very much magnified.

The scale may be calibrated by means of gauge block (slip gauges) to indicate any deviation from an initial setting.

Comparators using this type of linkage have sensitivities of the order of 0.25 micron per scale division.

The mechanical amplification is usually less than 100, but it is multiplied by the optical lens system. It is available in amplifications ranging from × 500 to × 1000.

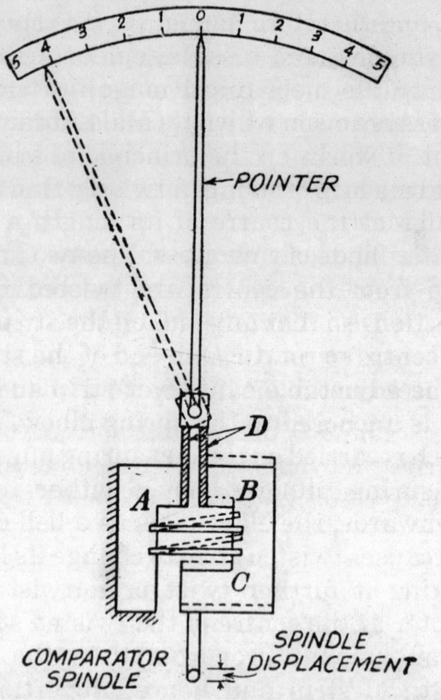

Fig. 5.8. Reed type mechanical comparator.

5.5.5. The Sigma Comparator. Fig. 5.9 shows the constructional details of the Sigma Mechanical Comparator.

The vertical beam is mounted on flat steel springs *A* connected to fixed members, which in turn are screwed to a backplate. The assembly provides a frictionless movement with a restraint from the springs.

The shank *B* at the base of the vertical beam is arranged to take a measuring contact, selecting from the available range.

The stop *C* is provided to restrict movement at the lower extremity of the scale.

Mounted on the fixed members, is the hinged assembly *D* carrying the forked arms *E*. This assembly incorporates a hardened fulcrum (provided with means for adjustment of controlling the ratio of transmitted motion) operative on the face of a jewelled insert on the flexible portion of the assembly.

The metal ribbon *F*, attached to the forked arms, passes around the spindle *G* causing it to rotate in specially designed miniature ball bearings. Damping action to the movement is affected by a metal disc, mounted on the spindle, rotating in a magnetic field between a permanent magnet and a steel plate. The indicating pointer *H* is secured to a boss on the disc.

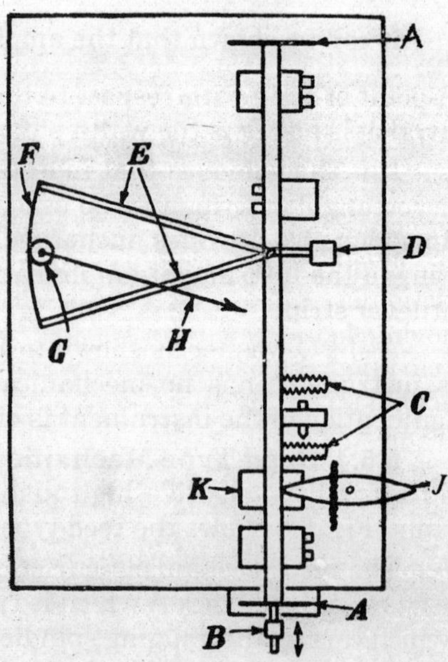

Fig. 5.9. Sigma mechanical comparator.

The trigger J (opposite K) is used to protect the measuring contact. At the upper end of the vertical beam, an adjusting screw is provided for final zero setting of the scale.

A new patented feature is shown at K. This is a magnetic counter-balance which serves to neutralise the positive "rate" of springs reaching on the measuring tip. In this way a constant pressure over the whole scale range is achieved.

The instrument is available with vertical capacities of 150 mm, 300 mm and 600 mm and magnifications of 500, 1000, 1500, 3000 and 5000. The scales are graduated in both English and Metric systems.

The least count which one division represents is of the order of 0.25 microns.

Advantage : It has got a bold scale and larger indicating pointer.

Disadvantages : (*i*) Due to motion of the parts, there is wear in the moving parts.

(*ii*) It is not as sensitive as optical or other types of comparators due to friction being present in the moving parts.

5.5.6.1. Diagrammatic sketch showing the movement of sigma comparator. The various movements in the 'sigma' comparator discussed in Art. 5.5.6, will be very clear with the help of Fig. 5.10 which shows the various movements in diagrammatic form.

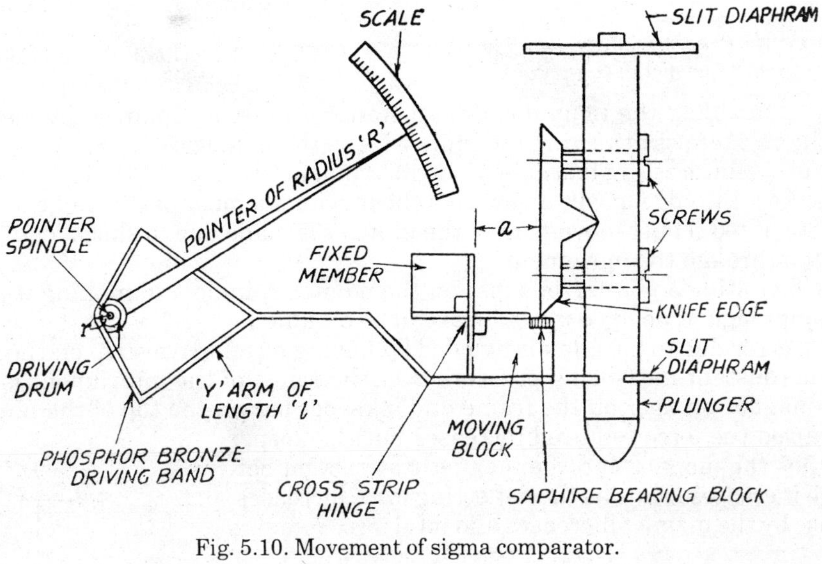

Fig. 5.10. Movement of sigma comparator.

The plunger in Fig. 5.10 is mounted on a pair of slit diaphragms in order to have frictionless linear movement. A knife edge is mounted on it and bears upon the face of the moving member of a cross strip hinge. For details of cross strip hinge refer Fig. 5.11. The cross strip hinge consists of the moving component and a fixed member which are connected by thin flexible strips alternately at right angle to each other. Thus if an external force is applied to the moving member ; it will pivot, as would a hinge, about the line of intersection of the strips.

To the moving member an arm of Y shape and having effective length l is attached. If the distance of the hinge from the knife edge be a then the magnification of the first stage is $\frac{l}{a}$.

A phosphor-bronze strip is attached to the two extremities of the Y arm and is passed around a radius r attached to the pointer spindle. If the length of pointer be R, then $\frac{R}{r}$ is the second stage of magnification. Thus total magnification is $\frac{l}{a} \times \frac{R}{r}$.

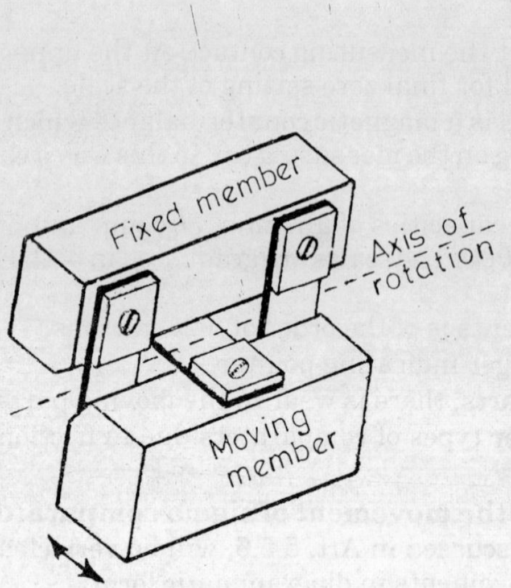

Fig. 5.11. The cross strip hinge used in sigma comparator.

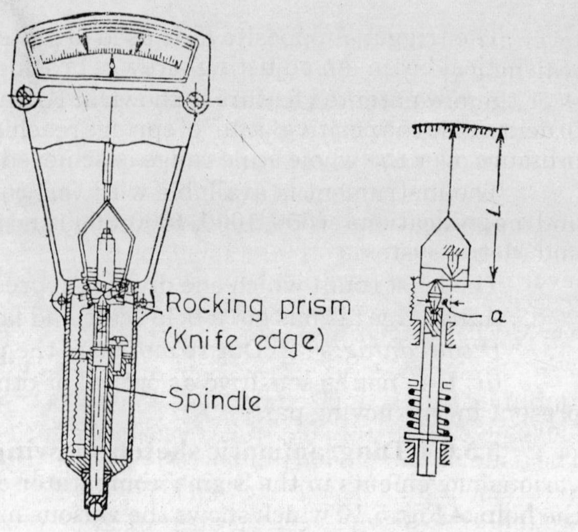

(*a*) Comparator (*b*) Movement used in comparator.

Fig. 5.12. Mechanical comparator using rocking prism.

In order to adjust the magnification, distance *a* must be changed by slackening and tightening the two screws attaching the knife edge to the plunger.

Some of the interesting features of the instrument are :

(1) As the knife edge moves away from the moving member of the hinge and is followed by it, therefore, if too robust movement of the plunger is made due to shock load, that will not be transmitted through the movement.

(2) By mounting a non-ferrous disc on the pointer spindle and making it move in field of a permanent magnet, dead beat readings can be obtained.

(3) The error due to parallax is avoided by having a reflective strip on the scale.

(4) The constant measuring pressure over the range of the instrument is obtained by the use of a magnet plunger on the frame and a keeper bar on the top of the plunger. As the plunger is raised the force required increases but the keeper bar approaches the magnet and the magnetic attraction between the two increases. Thus as the deflecting force increases, the assistance by the magnet increases and total force remains conotant.

5.5.7. Mechanical Comparator. The diagram of a type which is used for comparative measurement of external surfaces is shown in Fig. 5.12 (*a*). In this instrument the movement of the measuring tip attached at the end of the spindle is transmitted to the pointer through a mechanism shown in Fig. 5.12 (*b*). The upper end of the spindle bears against a rocking prism (knife-edge). There is a frame member having two V-slots offset to each other by distance *a*. The end of the spindle rests against the first V-slot and its movement is transmitted to this frame through a prism. A knife edge which is stationary relative to the body of the instrument enters the upper V-slot. The apex of the upper knife-edge is the centre for all the moving parts of the comparator. The distance *a* between the V-slots forms the shorter lever arm of the system, whereas, the longer

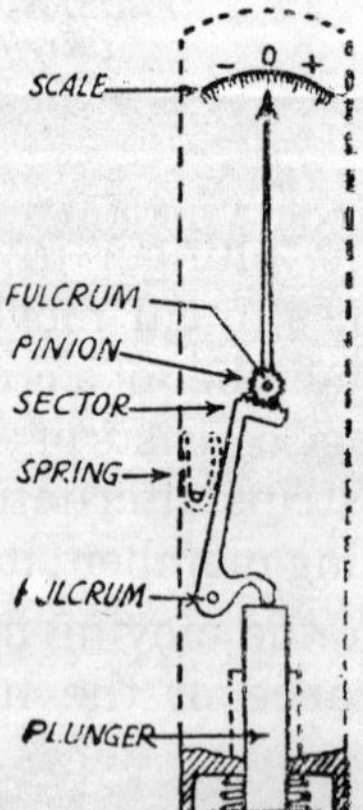

Fig. 5.13. Mechanical comparator using sector and pinion.

lever arm is the distance L from the centre of rotation of the system to the other end of hand, which moves along the comparator scale. Thus the magnification of the instrument is L/a and is of the order of 1000. The contact pressure is of the order of 300 to 400 grammes and is provided by a spring. The use of knife-edge pivots in the comparator movement excludes the influence of possible clearance in the pivots on the accuracy of this instrument.

Yet another type of mechanical comparator is shown in Fig. 5.13 in which the movement of the plunger (contact member) is transmitted to the pointer through an angular lever which ends in the form of a sector and pinion. It may be noted that a spring attached with the angular lever avoids the play in comparator.

5.6. Mechanical Optical Comparators

In mechanical optical comparators small displacements of the measuring plunger are amplified first by a mechanical system consisting of pivoted levers. The amplified mechanical movement is further amplified by a simple optical system involving the projection of an image. The usual arrangement employed is such that the mechanical system causes a plane reflector to tilt about an axis and the image of an index is projected on a scale on the inner surface of a ground-glass screen. Optical magnification provides high degree of measuring precision due to reduction of moving members and better wear resistance qualities. Optical magnification is also free from friction, bending, wear etc.

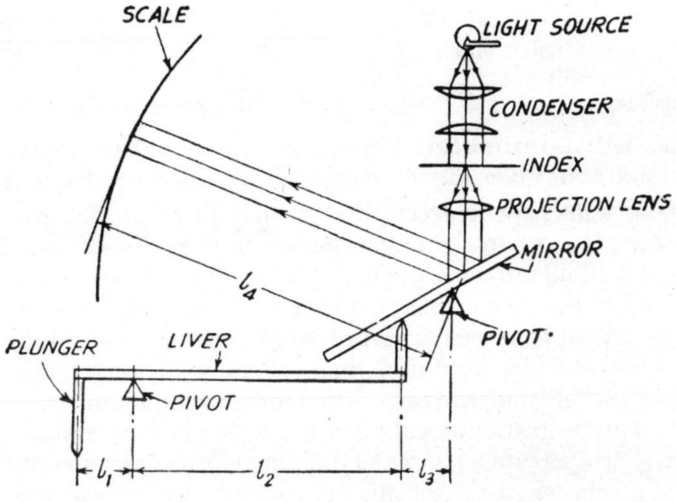

Fig. 5.14. Principle of Optical Comparator.

The whole system could be explained diagrammatically by Fig. 5.14, which gives very simple arrangement and explains the principle of above comparator.

In this system, Mechanical amplification = l_2/l_1

and Optical amplification = $l_4/l_3 \times 2$.

It is multiplied by 2, because if mirror is tilted by an angle $\delta\theta$, then image will be tilted by $2 \times \delta\theta$. Thus overall magnification of this system = $2 \times (l_2/l_1)\ (l_4/l_3)$.

Thus it is obvious that optical comparators are capable of giving a high degree of measuring precision owing to high magnification and the reduction of moving members to minimum. Further these possess better wear resistance qualities as the only wearing members are the plunger and its guide and the mirror pivot bearing. Another advantage of the optical comparators is that provision of an illuminated scale enables readings to be taken without regard to the room lighting conditions. The point of importance in optical comparator is that mirror used must be of front reflection type and not of normal back reflection type. In normal back reflection type there are two reflected images, one each from front and back. Thus the reflected image is not well defined one, as one bright and other blurred image are observed. If front reflection type of mirror is used, then it requires considerable care in its use to avoid damage to the reflecting surfaces.

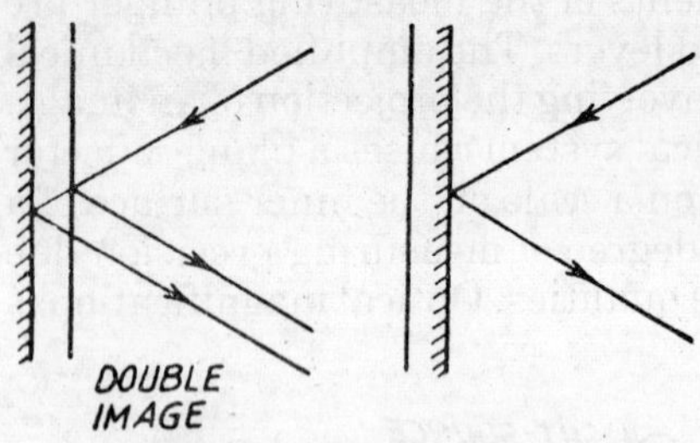

Back Reflection Type Mirror. Front Reflection Type Mirror.

Fig. 5.15

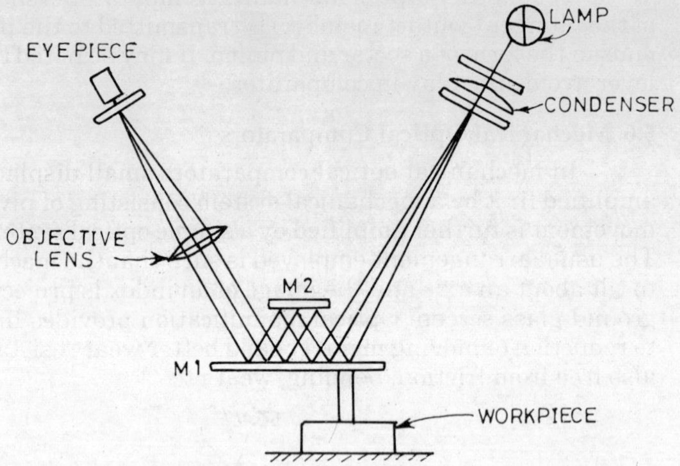

Fig. 5.16. The Optical arragnement of the Zeiss Ultra-optimeter.

5.6.1. Zeiss Ultra-optimeter. The optical system of this instrument involves double reflection of light and thus gives higher degree of magnification. (Refer Fig. 5.16)

A lamp sends light rays to green filter, which filters all but green light, which is less fatiguing to the eye. The green light then passes to a condenser which *via* an index mark projects it on to a movable mirror M_1, whence it is reflected to another fixed mirror M_2, and then back again to the first moveable mirror. The second objective lens brings the reflected beam from the first mirror to a focus at a transparent graticule containing a precise scale which is viewed by the eye-piece. The projected image of index line on the graticule can be adjusted by means of screw in order to set the zero. When correctly adjusted, the image of the index line is seen against that of the graticule scale. The special end of the contact plunger rests against the outer end of the first movable mirror so that any vertical movement of the plunger will tilt the mirror. The extreme sensitivity of this instrument necessitates special precautions in its operation to avoid temperature effects.

5.6.2. Zeiss Optotest Comparator (Refer Fig. 5.17). This is the commercial measuring instrument. This utilises a plunger tilted mirror, objective lens, prism and observing eye-piece to provide a high degree of magnification. The mirror is mounted on a knife-edge and it can be tilted about the fulcrum by any linear vertical movement of the contact plunger. A beam of light passes through a graticule suitably engraved with a linear scale and the movement of mirror causes this scale to move up or down past a translucent screen inside the observing hood of the instrument. The eye placed near the eye-piece views the image of a small scale engraved on glass after reflection from the plunger actuated mirror and the prism in the plan view in Fig. 5.17.

In the focal plane of the eye-piece, a fine reference line (Index) is provided and the system of lenses is so arranged that the image of the scale is projected in the same focal plane. Thus with movement of scale the image can be measured with reference to the fixed line. The division of the scale image opposite the index line indicates the amount of movement of contact plunger. The image of the scale and the index line could also be viewed through a projection system.

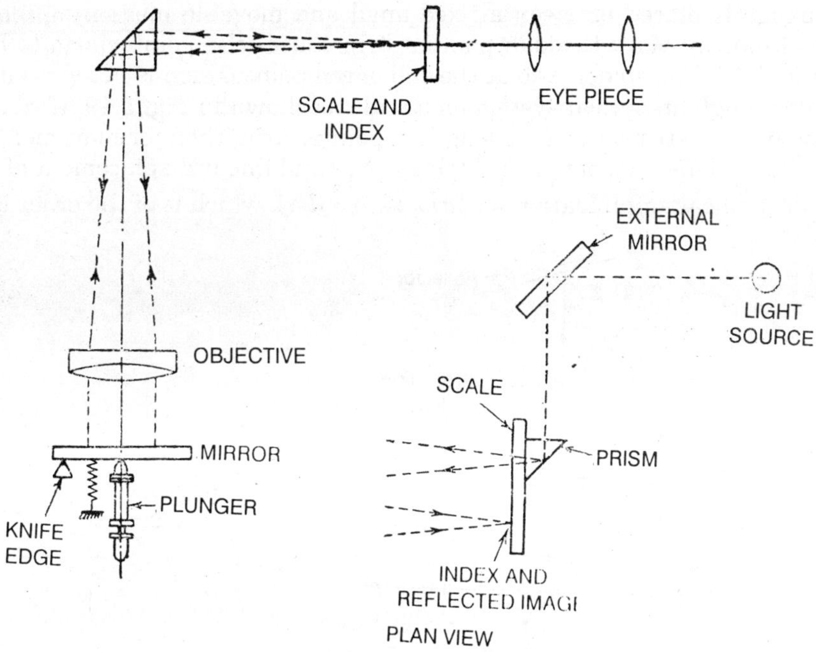

Fig. 5.17. Zeiss optotest optimeter.

The overall magnification of the comparator is given by $\frac{2f}{d}$ × Eye-piece magnification, where f is the focal length of the lens and d is the distance between the knife-edge and the plunger.

5.6.3. Newall OMS Horizontal Optimeter. This comparator is designed for very high magnification. It can measure very accurately with great precision in horizontal direction. It is very useful for measuring the internal and external dimensions of cylindrical gauges, screw gauges and thickness gauges.

The instrument consists of a cast iron base with three levelling screws and a spirit level for setting instrument horizontal. Two brackets are mounted on a transverse bar carried through the base. The brackets are kept in alignment by a keyway and a spline, and carry an optical measuring head and a tailstock respectively. The transverse bar is spiral grooved to allow easy and quick lengthwise adjustment of bracket. There are other attachments also on the base for various works. A vertically adjustable plunger with both titling and lengthwise movements is also housed in the base. Inter-changeable heavy and light duty contact arms are provided for internal measuring and a centre cradle for male cylindrical work.

5.6.4. Eden-Rolt Millionth Comparator. This comparator utilises both mechanical and optical magnifications and thus is highly sensitive one. Its design is simple and economical. The mechanical amplifying device utilises a parallel strip support, consisting of a fixed and a moving block connected together by parallel thin strips. The ends of fixed and moving strips through two pointer strips are connected to a pointer arm which carries at its end a ring across which is stretched a spider web. A

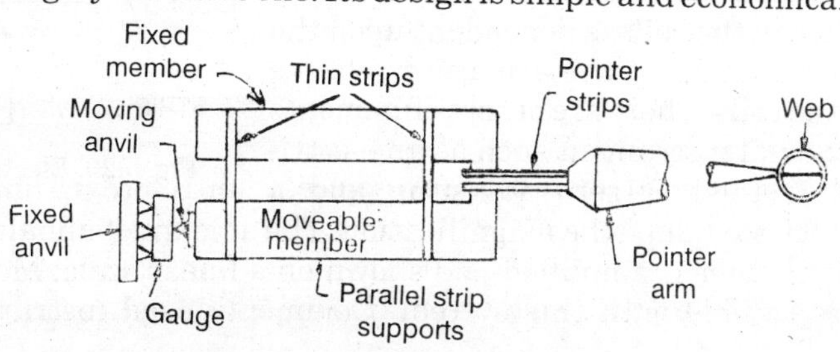

Fig. 5.18. Eden Rolt Millionth Comparator.

reference gauge is placed between a fixed anvil and movable measuring block. The slight movement of moving block through pointer strip causes the pointer arm to be deflected as shown in Fig. 5.18. The spider web at the end of the pointer arm is not viewed directly but it is projected through an optical system on a scale as shown in Fig. 5.19. If 'a' be the distance between the pointer strips and 'l' be length of pointer arm, then for movement 'δ' of movable member, angular deflection of pointer strips = (δ/a) and linear displacement of web = $(\delta/a) \times l$.

$\therefore$ Mechanical amplification = $(\delta/a) \times l \div \delta = (l/a)$, which is of the order of 400.

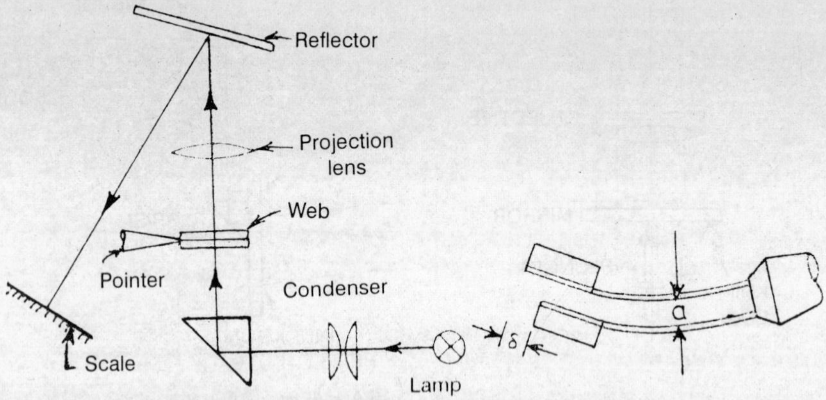

Fig. 5.19. Optical system of Eden Rolt comparator.

The optical magnification is of the order of 50, thus overall magnification is around 20,000.

5.7. Electrical and Electronic Comparators

These comparators depend for their operation on Wheatstone bridge circuit. In d.c. circuit, a change of balance of the electrical resistance in each arm of the bridge is caused by the displacement of an armature relative to the arm under the action of the measuring plunger. Once out of balance is caused in the bridge, it is measured by a galvanometer graduated to read in units of linear movement of plunger. This circuit is operated by battery. For the bridge to balance, the ratios of the resistances in two arms must be equal.

If alternating current is applied to the bridge, the inductance and capacitance of the arms must also be accounted for along with resistance. In actual measuring instruments, one pair of inductances is formed by a pair of coils in the measuring head of the instrument. The movement of the plunger displaces an armature, thus causing a variation in the inductance of a pair of coils forming one arm of a.c. bridge. The arm carries the armature (Fig. 5.20) and the inductance in the coils is dependent upon the displacement of the armature relative to the coils. There are other refinements in actual instrument such as an electrical method of zero adjusting and a switch to change the magnification. The amount of unbalance caused by movement of measuring plunger is amplified and shown on a linear scale. Magnifications of the order of × 30,000 are possible with this system. Commonly used instruments are Electrichek, Electricator,

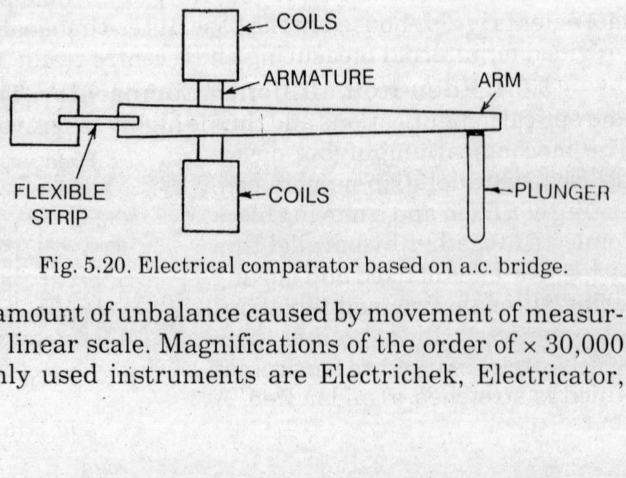

Fig. 5.20. Electrical comparator based on a.c. bridge.

Electrigage, Electrolimit and Electronic Measuring Equipment. A brief description of a few of these is given from 5.7.2 to 5.7.7.

5.7.1. Electrical Comparators. Electrical comparators are also known as electro-mechanical measuring systems as these employ an electro-mechanical device which converts a mechanical displacement into electrical signal.

Fig. 5.21 shows a block diagram of an electro-mechanical measuring system.

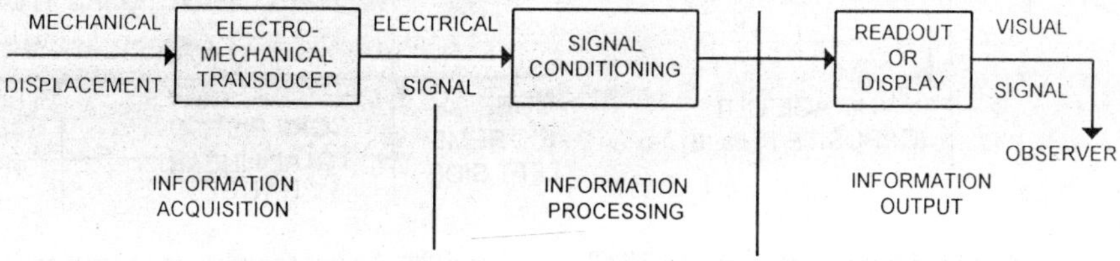

Fig. 5.21. Block diagram of an elementary electro-mechanical measuring system.

Linear variable differential transformer (LVDT) is the most popular electro-mechanical device used to convert mechanical displacement into electrical signal. It, in effect, is a transformer consisting of three symmetrically spaced coils carefully wound on an insulated bobbin. It works on mutual inductance principle and consists of a primary coil wound on an insulating form (bobbin) and two identical secondaries symmetrically spaced from the primary. AC carried excitation is applied to the primary and two secondaries are connected externally in a series opposition circuit. The lead wires exit through an opening in the outer shield, usually in the end-cover washers. A cylindrical shield of ferromagnetic material is spun over the metallic end-washer after the windings have been vacuum-impregnated with a potting compound suitable for the application environment. The finished transformer thereafter becomes quite impervious to humidity or ordinary magnetic influences. The device thus also becomes extremely rugged and reliable. There is a non-contacting magnetic core, made from a uniformly dense cylinder of nickel-iron alloy, carefully annealed to improve and homogenise its magnetic permeability, which moves in the centre of these coils wound on the insulating form and the motion of this core varies the mutual inductance of each secondary to the primary, which determines the voltage induced from the primary to each secondary. If the core is centered in the middle of the two secondary windings, then voltage induced in each secondary winding will be identical and 180° out-of-phase, and the net output will be zero. If the core is moved off middle position, then the mutual inductance of the primary with secondary will be greater than the other, and a differential voltage will appear across the secondaries in series. For off centre displacements within linear range of operation, the output is essentially a linear

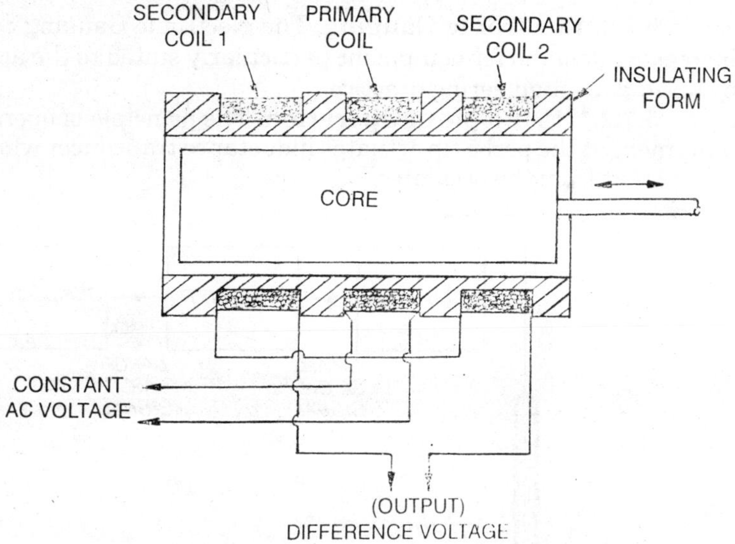

Fig. 5.22. Schematic Arrangement of LVDT (core-shown in the middle position).

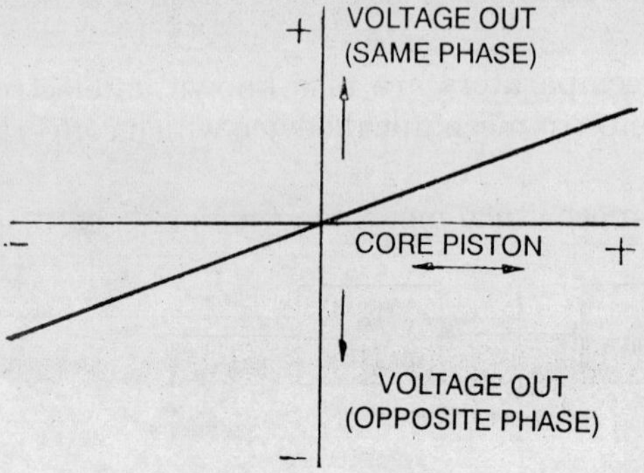

Fig. 5.23. Phase referenced output voltage of LVDT.

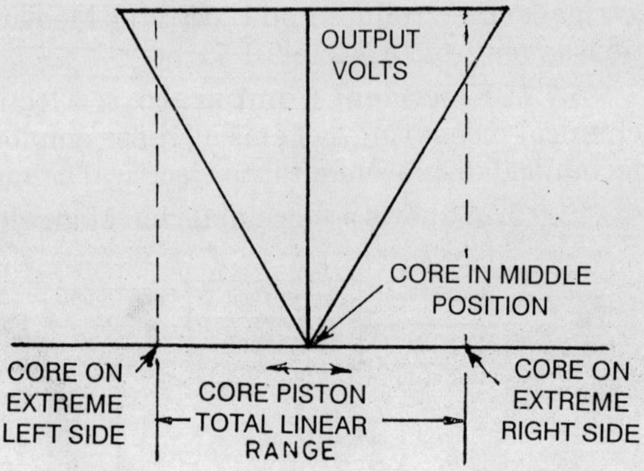

Fig. 5.24. Absolute value of output voltage of LVDT.

function of core displacement. The various unique features of LVDT are : Due to no physical contact between the core and coil its mechanical components do not wear out ; as a result the friction is absent and true infinite resolution with no hysteresis is obtained. The small core mass and the lack of friction enhance response capabilities for dynamic measurements and thus it becomes very suitable for taking measurements for on-line machining. Further as it is not affected by overload, its reliability is high.

5.7.2. Electronic Gauging. The electronic Gauging System is designed to fulfill the increasing demand for equipment particularly suited to the accuracy and versatility required by up-to-date engineering practice.

5.7.2.1. Principle of Operation. The principle of operation is shown in Fig. 5.25. The movement at the probe tip actuates inductance transducer which is supplied with an alternating current from the oscillator.

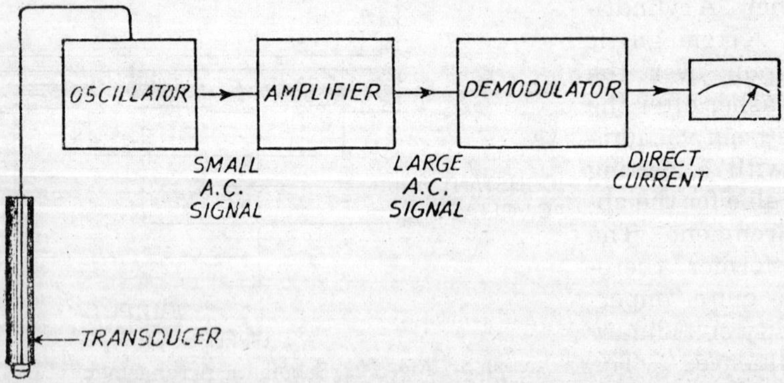

Fig. 5.25. Principle of operation in electronic gauging.

The transducer converts this movement into an electrical signal which is then amplified and fed *via* an oscillator to the demodulator. The current in D.C. form, then passes to the meter and the probe tip movement is displayed as a linear measurement.

Various measuring and control units can be incorporated which provide for an extremely wide range of single or multiple measurements to be made simultaneously.

Fig. 5.26 shows another design of electronic indicator capable of reading better than 1 micron. It consists of a gauging head and an amplifier. The mechanical movement of the contact probe alters the output or voltage of the gauge head proportional to displacement. The small alteration in voltage is amplified by means of a sensitive meter, calibrated in terms of units of length. Switching selection enables a convenient range of magnification. Such an instrument has many functional advantages like high magnification, range of magnifications, light

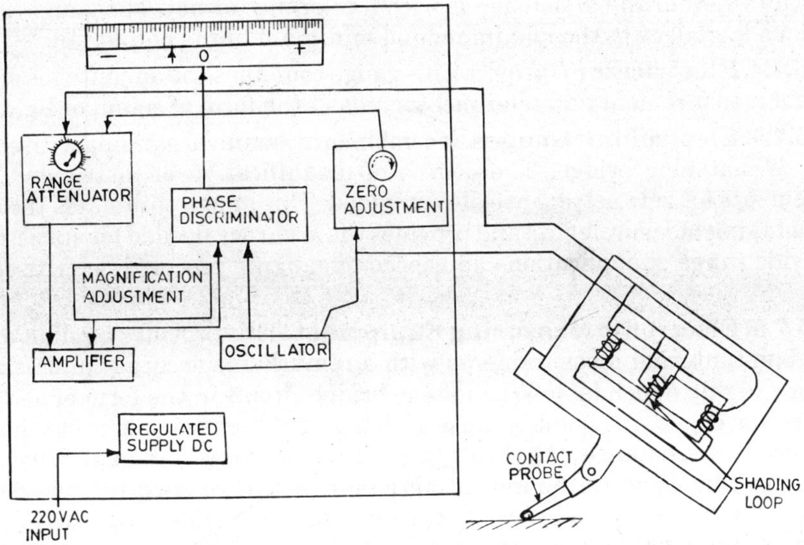

Fig. 5.26. Electronic indicator.

gauging force (40 grams or less), rugged construction, immediate response, unlimited applicability to inspection procedures.

5.7.2.2. Differential Unit. The differential gauge unit is used with two probes to show the combination of two readings and by selection of the appropriate probe input sockets, may add or subtract the readings obtained by the two probes. The sum, or difference of the two

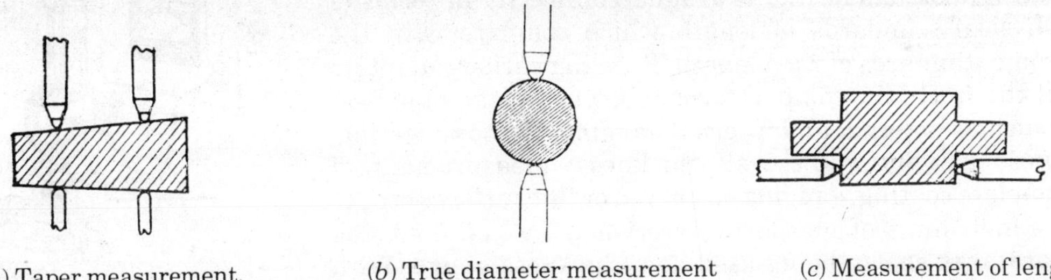

(*a*) Taper measurement. (*b*) True diameter measurement independent of datum. (*c*) Measurement of length irrespective of positional tolerances.

Fig. 5.27. Showing the measurement of various dimensions.

readings is amplified and displayed on the unit as a single scale reading. The magnification and output are exactly same as in the standard gauge unit, but an additional feature of this unit is a switch which enables one of the two probes to be isolated.

Differential units have many applications, including the measurement of tapers irrespective of changes in diameter of the component ; diameters, regardless of eccentricity and also for the normal measurement of both diameter and run-out as shown in Fig. 5.27.

5.7.3. Electricheck Gauges. This gauge simply shows whether the dimension of a component falls within a predetermined tolerance range without measuring the actual dimensions. It can also be used as an automatic check on machines in operation to stop the tool when the dimension is out of prescribed limits. It employs reed mechanism in its measuring head and reeds are caused to bend when the floating (movable) block is moved. The bending movement causes the extension arm to break an electric circuit. Electric contacts control signal to indicate whether the dimension is within tolerance or not. The gauge has first to be set by precision slip gauges to the maximum and minimum limits of tolerance.

5.7.4. Electricator Gauges. This gauge combines the advantages of both dial indicator for actual measurement and tolerance signals in the form of sound or light.

5.7.5. Electrolimit Gauges. It combines mechanical gauging with electrical magnification by Wheatstone bridge. The degree of magnification of instrument is adjustable and arrangements for zero setting are also provided. This gauge eliminates the errors due to sense of feel in taking measurements and provides the accuracy needed for statistical quality control. It has wide range of applications and speed of gauging process is increased very much by its use.

5.7.6. Electronic Measuring Equipment. This provides a reliable means for making external and internal measurements with a remarkable accuracy and ease. This equipment employs a highly dependable Wheatstone bridge circuit in the form of a strain gauge. Due to the deflection of gauging point because of change of dimension, a minute change is experienced in the electrical resistance of the bridge circuit. This creates voltage signals which are amplified and they in turn operate the indicating meter. It can be used for a variety of internal and external measurements and special attachments are provided to make it more versatile.

5.7.7. The Electronic Comparator. It is based on the principle of application of frequency modulation or radio oscillation. A special ratio oscillator is used which produces impulses of a definite value at a definite frequency. The frequency is controlled by the measuring head. When the measuring head is applied to the part to be measured the frequency of the instrument oscillator is modified because of the variation of the dimension from that of the preset standard. The modified frequency is annuled by suitable controls to restore it back to the basic value. The measuring instrument dial is graduated directly in terms of the physical standards of length which compare with the electronic standards of wavelength. By using various adaptors to suit the work, the comparator can be put to many applications such as external and internal gauging, flatness, testing, thickness gauging, tube wall thickness, measurement of electro-plate coating, gauging on curved or flat surfaces etc., to a very high limit of measuring precision. In Fig. 5.28, the comparator is shown being used as a thickness gauge. With high frequencies, it is extremely sensitive to such variables, as temperature, mechanical and electrical changes. Using different frequencies, the sensitivity of the instrument can be easily changed and it is achieved by providing a multi-position switch.

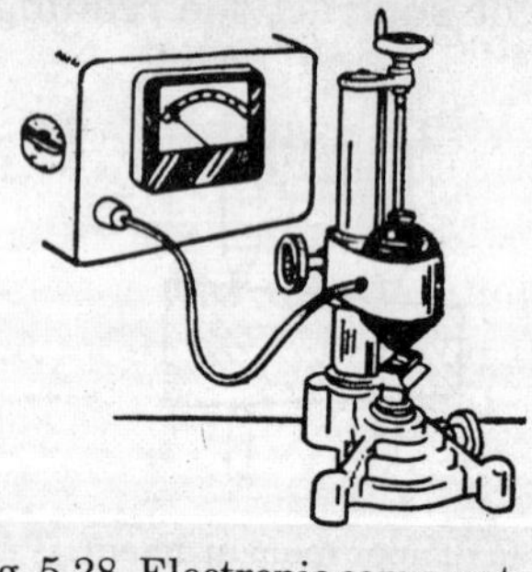

Fig. 5.28. Electronic comparator.

5.7.8. Digital Comparators (Special features of digital comparator shown in Fig. 5.29). Incremental measuring system with optoelectronic reading of a grid line glass scale. Clear and high contrast digital display. Large measuring range and high resolution.

Compact design with integrated but separately accessible case for batteries. Conveniently placed operating elements for the various measuring functions. Measuring spindle mounted in high precision ball bush guide, virtually friction free and without play. Screw-in cable release for raising measuring spindle. Fitted with dampened lowering motion of spindle. Contact point is fitted with ruby ball.

Highly versatile application. Can be combined with indicating hand measuring instruments. Compact design with short mounting shaft and measuring spindle. Therefore insensitive against overload.

Resolution selectable	1 μm
Measuring range	10 mm
Measuring accuracy over entire range	± 1 μm
Measuring pressure	1 N
Max. travel speed of spindle	≤ 0.5 m/s
Digital display	LCD, height of characters 8 mm
Power supply	Choice of batteries or net connection.

Fig. 5.29. Digital comparator.

Advantages

Digital read-out on large and clearly defined numerals.

Universal application by choice of various measuring functions such as reset, preset, tolerance indication, memory for maximum and minimum value.

Universal power supply by choice of batteries or net connection.

High measuring accuracy by incremental measuring system with optoelectronic reading of a grid line glass scale. Measuring system without mechanical components, therefore most accurate and free of wear.

Small dimensions. Digital Comparators with measuring range of 10 mm fit into practically all indicating hand measuring instruments.

Measuring spindle mounted in high precision ball bush guide virtually free of friction and insured against cocking. The next to zero friction not only insures maximum accuracy but also minimum hysteresis.

Corrosion-proof. Both mounting shank and measuring spindle are made of hardened stainless steel.

Processing of measuring results by data exit for the connection of computers *via* interface.

Measuring functions

A choice of various measuring functions is given by operating conveniently placed push buttons or switches on the front and the side of the housing. The digital comparator is therefore universally applicable.

Measuring function	Push button
Zero setting of digital display and clearing of functions max/.min., tolerance and store.	Reset
Storing of displayed measuring results. This does not influence the internal pick-up of measuring values which is continued.	Store

Setting of tolerance limits. The tolerance condition of any given part can be determined instantaneously by observing clear symbols for undersize, good and oversize.

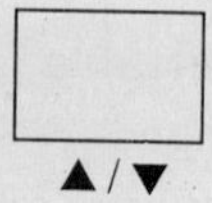

Memory for maximum and minimum value. Peak values for outside diameter checks. When performing this, simply move part under probe and any deviation from the preset nominal diameter is automatically entered into memory. Same applies to minimum value.

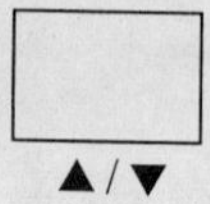

Direction of counting reversible for outside and inside measurements.

Data transfer from ASCII-data exit.

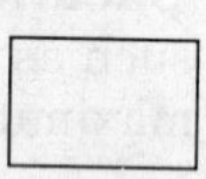

Data

Digital Preset. Basic measurements up to ± 999,999 mm can be preselected. They are automatically recognized at each measurement.

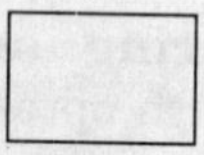

Preset

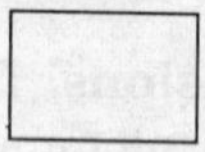

Cursor

5.7.9. Indicating Snap Gauge (Special features of snap gauge shown in Fig. 5.30). Measures cylindrical parts, such as shafts, bolts or spindles, thicknesses and lengths.

Rugged, forged steel frame with heat insulator. Measuring spindle mounted in extra long guideway, retracts by lever control. Anvil adjusts with thumb screw *via* worm and rack. Both measuring spindle and anvil made of hardened stainless steel, measuring faces carbide tipped. Adjustable centre stop for automatic alignment.

Advantages

Direct indication of measuring results. Actual dimension of the part can be determined with absolute accuracy.

Versatility. Each gauge covers a large range. Within its limits it is quickly and easily adjusted to any size and type of measuring application.

Maximum wear resistance by contact free positioning of measuring spindle plus carbide tipped measuring faces.

Constant pressure by built-in compression spring which eliminates interference from operator's personal feel.

Indicating Instruments. Any dial indicator/comparator with 8 mm mounting shank may be used.

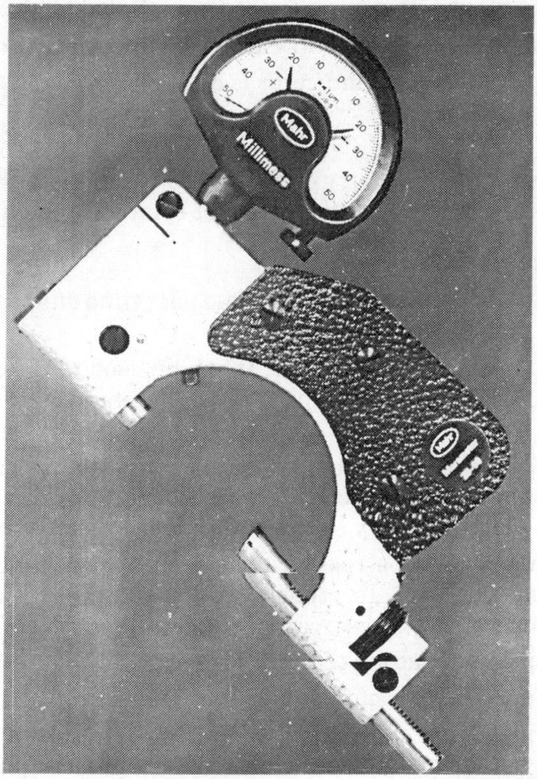

Fig. 5.30. Indicating snap gauge.

5.7.10. Instruments for checking and presetting tools Including a projector and measuring system :

Measuring range	:	X = 150 mm (dia. 300 mm) Z = 400 mm
Resolution	:	0.001 mm.

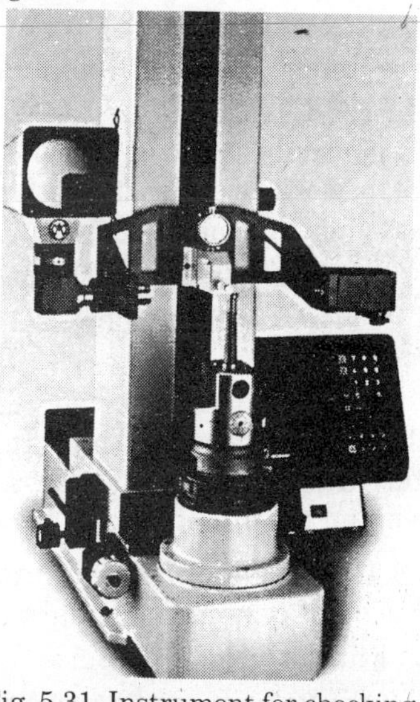

Fig. 5.31. Instrument for checking and presetting tools.

5.7.11. Possible sources of errors in Electronic Instruments. The following possible sources of errors from electronic instruments have a bearing on working accuracy :

(*i*) *Repeatability* or short run ability to repeat readings.

(*ii*) *Drift.* The indicator may be relied upon for a datum but wanders over a period of time.

(*iii*) *Temperature.* The changes in temperature, apart from the temperature of the workpiece, may alter the calibration of the instrument.

(*iv*) *Magnetism.* Some electronic instruments may be influenced by even the slightest residual magnetism.

(*v*) *Voltage fluctuation.* Electronic indicator may be subject to short-run wandering due to voltage fluctuation, specially in high magnification range.

(*vi*) *Loss of linearity*. Periodic adjustments for linearity may be required, but short-run changes stemming from unstable components can be dangerous in the inspection procedure.

(*vii*) *Degree of linearly*. Electronic indicator is comparator and can not be relied upon for wider ranges.

There are, in addition, sources of error in the environment, such as dirt, oil, air film, stability of the set-up, finish and geometry of the workpiece, deflection, angle of the probe tip to the work, vibration, temperature, parallax, etc. not related to the instrument itself.

5.8. Pneumatic Comparators

Air gauging has rapidly increased during some past time due to the following important characteristics :

(*a*) Very high amplifications are possible. It can be used to measure diameters, length, squareness, parallelism, concentricity, taper, centre distance between holes and other geometric conditions.

(*b*) As no physical contact is made either with the setting gauge or the part being measured, there is no loss of accuracy because of gauge wear. For this reason, air spindle and air snap gauges last very long. Also very soft parts which are easily scratched, can be gauged.

(*c*) Internal dimensions can be readily measured not only with respect to tolerance boundaries but also geometric form. In other words, while measuring a bore it can reveal complete story of size, taper, straightness, camber and bell mouth etc.

(*d*) It is independent of operator skill.

(*e*) High pressure air gauging can be done with cleansing of the parts which helps to eliminate errors due to dirt and foreign matter.

(*f*) Gauging pressures can be kept sufficiently low to prevent part deflection.

(In general, high pressure gauges are suitable for those parts in which tolerances are relatively large and low pressure air gauges are preferable for highly precise work.)

(*g*) Dimensional variations throughout the length of shaft or cylinder bore can be explored for out of roundness, taperness, concertricity, regularity and similar conditions.

(*h*) Not only it measures the actual size, but it can also be used to salvage oversized pieces for rework or to sort out for selective assembly, *i.e.*, it is suitable both for variable inspection (measurement of size) and attribute inspection (GO and NO GO) gauging and limits.

(*i*) The total life cost of the gauging heads in much less.

(*j*) It is accurate, flexible, reliable, universal and speedy device for inspecting parts in mass production.

(*k*) It is best suited for checking multiple dimensions and conditions on a part simultaneously in least possible time. It can be used for parts from 0.5 mm to 900 mm diameter having tolerance of 0.05 mm or less. It can be easily used for on line measurement of parts as they are being machined and take corrective actions.

5.8.1. Systems of Pneumatic Gauges. Based on the physical phenomena on which the operation of pneumatic gauges is based, these may be classified as :

(*i*) Flow or velocity type, (*ii*) Back pressure type.

Flow or velocity type pneumatic gauges operate by sensing and indicating the momentary rate of air flow. Flow could be sensed by a glass tube with tapered bore, mounted over a graduated scale. Inside the bore a float is lifted by the air flow.

Velocity of air in velocity type pneumatic gauges can also be sensed by sensing the velocity differential *i.e.*, differential pressure across a venturi chamber. Such systems have

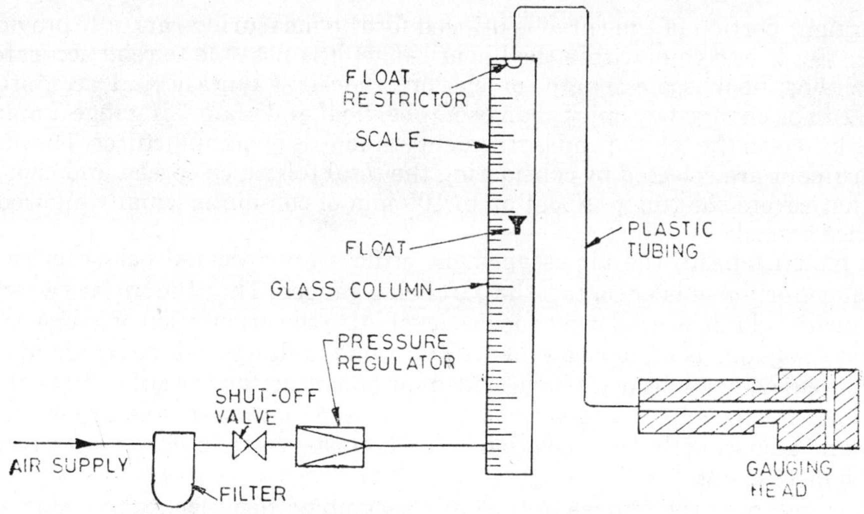

Fig. 5.32. Free flow air gauge.

quick response. These permit use of large clearance between nozzle and object surface, resulting in reduced wear of the gauging members. There is less air consumption. Magnification of the order of 500 to 5000 times is possible.

5.8.1.1. Free Flow Air Gauges (Flow or velocity type). In this case the compressed air after the filtering and pressure reducing unit flows through a tapered glass tube containing a small metal float and then through a plastic tube to the gauge head having two diametrically opposed orifices for air escapement into atmosphere (Refer Fig. 5.32). The position of the tube is dependent upon the amount of air flowing through the gauge head, which in turn is dependent upon the clearance between the bore to be measured and the gauge head. Fig. 5.34 shows a curve between the air flow and the clearance between the part and the orifice in gauge head.

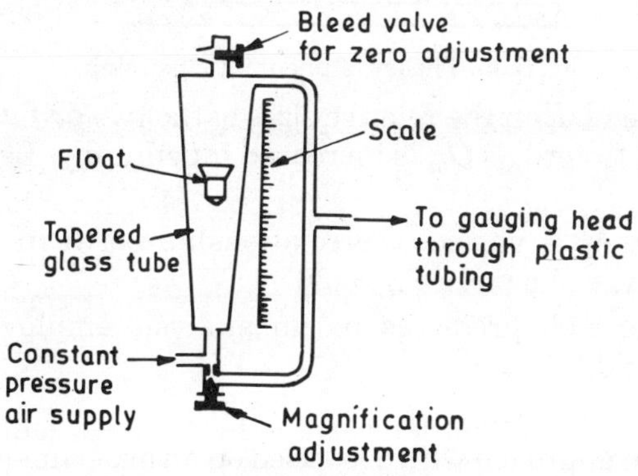

Fig. 5.33. Zero and magnification adjustment in flow type pneumatic comparator.

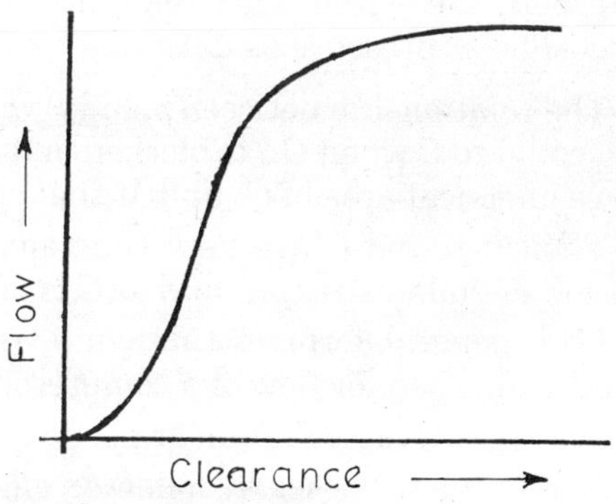

Fig. 5.34. Characteristic of air flow versus clearance of flow.

The flow velocity type pneumatic comparator with zero adjustment and magnification adjustment is shown in Fig. 5.33. Magnification can be changed by passing some of the air supply, using a screw at the inlet to the tapered glass tube. The float can be zeroed by a bleed valve installed at the top of the tube. Size is measured by the velocity of air in a tapered glass tube which is measured by the height of the float in tube.

The straight portion of the curve is utilized for the measuring range. It provides high amplification (10 : 1) and thus within the linear range, it is possible to read accurately upto microns depending upon scale length, or classify the sizes quickly and accurately. The amplification can be changed by quick change of tube, float and scale. Air gauge amplification and range are based on the tooling and instrument standards of manufacturer. The amplification and instrument are selected by considering the total tolerance spread and choosing the instrument that covers the range. About 50 to 100 mm of column is usually allowed for the actual tolerance spread.

In the gauging head, the air escapement orifices are recessed below its cylindrical surface so that the orifices never contact the part being gauged. Thus the surface wear will not affect the accuracy till it is worn down to orifice level. Also the orientation of gauge or the way operator holds the gauge is of no consequence and same readings will be obtained for given diameter. On the gauge, knobs are also provided for adjusting float position and calibration. Air gauge is set by placing masters for maximum and minimum tolerances on spindle alternatively and adjusting the float position for each master by turning the knurled knobs at the base of the instrument.

Free-flow column type gauges are usually assembled together side by side and thus multiple interrelated readings can be seen at a glance. This is the big advantage of air gauging that the multiple dimensions and conditions can be inspected with great ease, accuracy and speed.

Pneumatic circuits can be arranged to determine dimensional differences like taper (comprising the diameter of bore at different points along a part), bore centre distance and also to select parts to assemble to predetermined clearances or interference fits.

5.8.1.2. Back Pressure Gauges. The basic principle and the theory of pneumatic gauging in the back pressure gauges is described below. (Refer Fig. 5.35)

Air from a constant pressure source flows to the atmosphere through two orifices O_c and O_m in series.

P is the pressure upstream of the first orifice and p is the pressure between the two orifices, both measured with reference to the atmospheric pressure as datum.

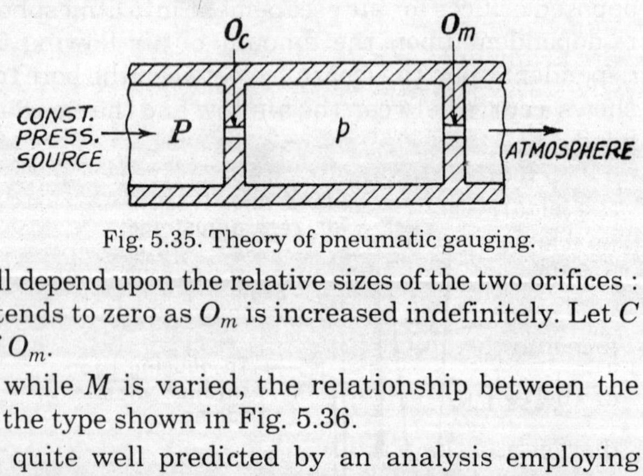

Fig. 5.35. Theory of pneumatic gauging.

The relationship between p and P will depend upon the relative sizes of the two orifices : p being equal to P when O_m is blocked and tends to zero as O_m is increased indefinitely. Let C be the geometrical area of O_c and M that of O_m.

Then if p and C are kept constant while M is varied, the relationship between the dimensionless quantities p/P and M/C is of the type shown in Fig. 5.36.

(The general form of this curve is quite well predicted by an analysis employing Bernouli's equation for flow of a compressible fluid.)

We are interested in linear portion of this curve.

For design purposes we follow an empirical approach which is based on an experimental study at N.P.L. (London) of the relationship between pressures and orifices areas.

The characteristics of p/P and M/C are determined experimentally for pressure P varying from 2 to 75 pounds per sq. inch (0.13 to 5 kg/cm^2) and inspection of any one of these shows that within the range 0.6p/P to 0.8p/P, the curve approximates to a straight line, the equation for which may be written as

$$\frac{p}{P} = A - b\,\frac{M}{C} \qquad \ldots(1)$$

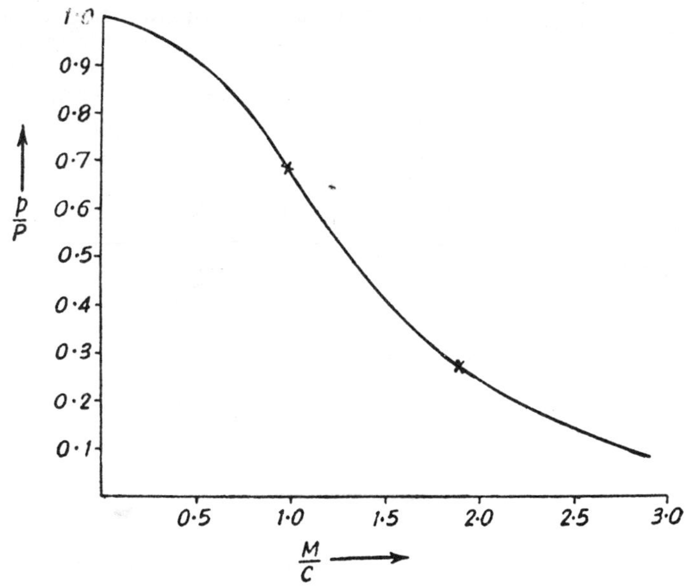

Fig. 5.36. Characteristic curve of pneumatic gauge.

Examination of the family of curves shows that constant A, the intercept on the p/P axis is closely constant over the range of pressures investigated and for practical purposes, the value of A = 1.10 can be adopted for any value of P likely to be used.

The slobe b of straight portion characteristics is however not independent of P, its numerical value decreases as P increases and the limiting values found in the investigation are as under :

b = 0.6 when P = 0.13 kg/cm^2

b = 0.4 when P = 5 kg/cm^2.

5.8.1.3. Area of escape orifice. When the clearance between the surface and the nozzle face is zero, no air escapes from the nozzle and the area of the escape orifice is zero.

When the clearance between the surface and the nozzle face is very large, the area of the escape orifice is $\frac{\pi}{4} d^2$, where d is the diameter of the nozzle.

Between these extremes, especially where the clearance is small and where air gauging can be employed, the area of the escape orifice is $\pi\, dl$, that is, the area of the curved surface of the cylinder shown in Fig. 5.38.

5.8.1.4. Range of linear measurement. The condition $0.6 < p/P < 0.8$ defines a section of the characteristics which experimental investigation has shown to be linear to within 1%.

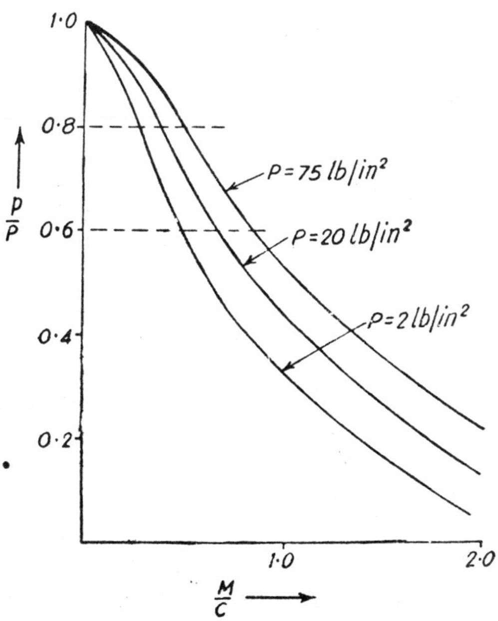

Fig. 5.37. Experimentally determined characteristics for different operating pressures.

i.e. to say, M_n is the minimum value of M for $p/P = 0.8$ and M_x is the maximum value of M for $p/P = 0.6$.

If a lower order of linearity is acceptable, the limits for p/P may be extended. Thus for $0.55 < p/P < 0.85$, the linearity is still within 2%.

From Fig. 5.36 for linear range, we have

$$b\frac{M_n}{C} - 1.10 - 0.8 = 0.3\ ;\ b\frac{M_x}{C} = 1.10 - 0.6 = 0.5$$

$$\frac{M_x}{M_n} = \frac{5}{3} \text{ and } \frac{M_x - M_n}{M_x + M_n} = \frac{1}{4}\ ;\ \text{Let}\left[M_a = \frac{1}{2}(M_a + M_n)\right]$$

Then $$\frac{M_x - M_n}{2M_a} = \frac{1}{4} \text{ and } M_x - M_n = \frac{1}{2}M_n \qquad ...(2)$$

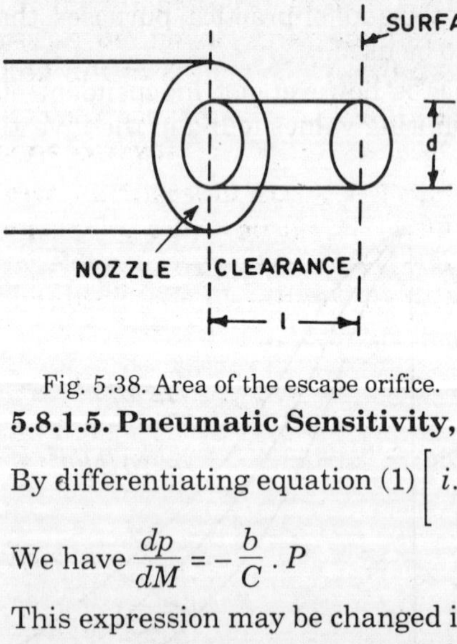

Fig. 5.38. Area of the escape orifice.

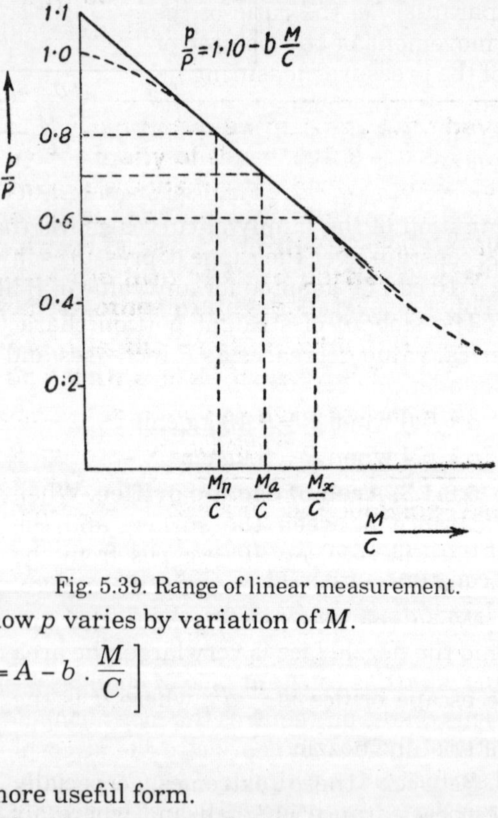

Fig. 5.39. Range of linear measurement.

5.8.1.5. Pneumatic Sensitivity, *i.e.* how p varies by variation of M.

By differentiating equation (1) $\left[i.e.\ \frac{p}{P} = A - b\,.\,\frac{M}{C}\right]$

We have $\frac{dp}{dM} = -\frac{b}{C}\,.\,P$

This expression may be changed into more useful form.

Since p will have a value $0.7\,P$ when M has its average value M_a.

$$0.7 = 1.10 - b\,.\,\frac{M_a}{C}\ ;\ i.e.\ \frac{b}{C} = \frac{0.40}{M_a}.$$

Hence $$\frac{dp}{dM} = -\frac{b}{C}P = -0.40\frac{P}{M_a} \qquad ...(3)$$

It is observed that sensitivity is directly dependent upon operating pressure P and inversely proportional to M_a.

5.8.1.6. Changing sensitivity (Magnification). A plot of pressure against escape orifice area for a number of different sizes of control orifice will show that sensitivity increases as the diameter of the control orifice decreases, *i.e.*, for small control orifices the change in pressure is greater for a given change in escape orifice area. Utilising this property, it is possible to set precisely the sensitivity (magnification) by incorporating a variable control orifice.

5.8.1.7. Overall Magnification. In a practical pneumatic measuring apparatus the area M will be associated with the measuring head and change in M will be the result of a change in the dimension which is being measured *e.g.* a change in the separation L between nozzle and surface (Fig. 5.40).

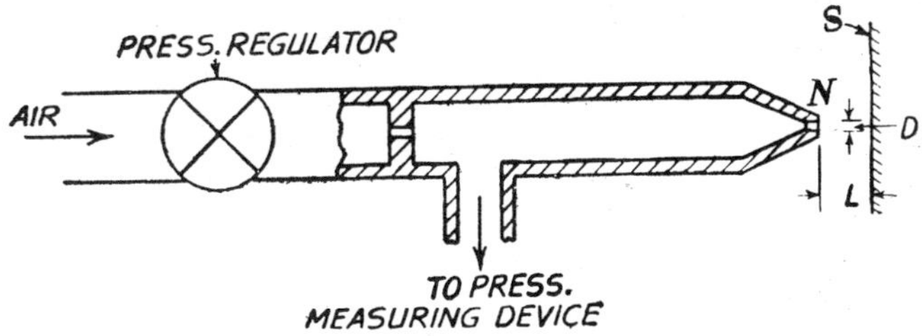

Fig. 5.40. Variation of M w.r.t. L.

The overall magnification of the apparatus, *i.e.* the ratio of the linear movement of the pointer or index of the pressure measuring instrument to the change in the dimension which produces it, will be proportion to the pneumatic sensitivity but will depend also on other two factors, *viz.*

(*i*) The way in which the area M changes as the dimension L changes, *i.e.* dM/dL.

(*ii*) The sensitivity of the pressure measuring instrument, *i.e.* dR/dp. (The rate of change of reading R of this instrument with respect to the varying pressure p).

Thus the overall magnification is given by ;

$$\frac{dR}{dL} = \frac{dR}{dp} \cdot \frac{dp}{dM} \cdot \frac{dM}{dL} \qquad ...(4)$$

where $\frac{dR}{dL}$ = Overall magnification, $\frac{dR}{dp}$ = Sensitivity of pressure measuring device,

$\frac{dp}{dM}$ = Pneumatic sensitivity, $\frac{dM}{dL}$ = Rate of change of M with L.

From equation (3), dp/dM is constant over linear range and thus overall magnification will be constant, provided :

(*i*) dR/dp is constant, *i.e.* the pressure measuring device is linear which is usually the case ;

(*ii*) dM/dL is constant ; *i.e.* M changes proportionately with L.

This condition requires the measuring head to be correctly designed. The final escapement of the air from the nozzle to the atmosphere is taken as being through an area of the curved surface of the cylinder of length L and diameter D, where L is the separation between the nozzle surface and the surface to be gauged and D is the internal diameter of the nozzle.

$$\therefore \qquad M = \pi DL \; ; \text{ and } \frac{dM}{dL} = \pi D = \text{constant.}$$

If L_a is average separation, *i.e.* corresponding to $p = 0.7\,P$. ; If $M_a = \pi DL_a$.

Now in equation $\frac{p}{P} = 1.10 - b \cdot \frac{M}{C}$; for $\frac{p}{P} = 0.7$, $M = M_a$,

$$b = (1.10 - 0.7) \times \frac{C}{M_a} = 0.4\,\frac{C}{M_a}$$

∴ Substituting the value of b in the following equation, $\frac{p}{P} = 1.10 - b \cdot \frac{M}{C}$

We have $$\frac{p}{P} = 1.10 - 0.4 \cdot \frac{C}{M_a} \times \frac{M}{C} = 1.10 - 0.4 \cdot \frac{M}{M_a}$$

Substituting the values of M and M_a, $\frac{p}{P} = 1.10 - 0.4 \frac{\pi DL}{\pi DL_a}$

or $$\boxed{\frac{p}{P} = 1.10 - 0.4 \frac{L}{L_a}}$$

This is the equation of the linear portion of the characteristics showing the relationship between $\frac{p}{P}$ and ratio of $\frac{L}{L_a}$. This equation shows that for given values of P and L_a, p varies directly with L.

Now equation (2) for the linear range may be written in terms of L, *i.e.* displacement of the surface from the nozzle.

$$\therefore \quad L_x - L_n = \frac{1}{2} L_a$$

i.e. the linear range of L is equal to half the average value of L_a.

Let λ be the length of scale of pressure measuring device corresponding to the pressure change from 0 to P (p may vary from 0 to P).

Then $$\frac{dR}{dp} = \frac{\lambda}{P}\text{; and } \frac{dp}{dM} = 0.40 \frac{P}{M_a}\text{; From equation (3) } \frac{dM}{dL} = \pi D$$

$$\therefore \quad \frac{dR}{dL} = \frac{dR}{dp} \cdot \frac{dp}{dM} \cdot \frac{dM}{dL} = \frac{\lambda}{P} \times 0.40 \frac{P}{M_a} \times \pi D = \frac{\lambda}{P} \times \frac{0.40P}{\pi DL_a} \times \pi D = 0.40 \frac{\lambda}{L_a}$$

Therefore, for design purposes, the equations needed to suffice the quantitative estimation of the performance of pneumatic systems are (for 1% linearity).

$$\frac{dR}{dL} = 0.04 \frac{\lambda}{L_a} \quad \text{...(5)}$$

$$L_x - L_n = \frac{1}{2} L_a \quad \text{...(6)}$$

$$P = 0.7\,P \text{ when } L = L_a$$

and λ = length of scale of pressure measuring device corresponding to the pressure change from 0 to P.

L_x = Maximum separation between nozzle and surface and L_n is the minimum separation.

(Refer solved problems at the end of this chapter for further clarity).

The main disadvantage of air plug gauges is its limited measuring range which is usually restricted to 0.10 mm.

5.8.1.8. Response Speed. For a back pressure system the speed of response is not as fast as for free-flow type, because some time is required for the pressure to build up. The speed of response becomes of concern when the gauging head is separated from indicating instrument by long distance.

A pneumatic measuring system will not correctly measure displacements of frequency greater than about 2 cycles/second, because of its slow speed of response. The response is considerably slower compared to the electrical system because of the following reasons.

Between the control orifice and the measuring head, there exists a closed volume associated with the measuring instrument used to measure the pressure p. A dimensional change, *e.g.*, a displacement of the surface alters the flow so changing the pressure p. The time needed to establish the new value of p depends on the total volume and on the rate of air flow into and out of it. The latter in turn depends upon the operating pressure P and the size of the control orifice and orifice in the measuring head. These orifices are related in size and determine the pneumatic sensitivity, the smaller orifices (corresponding to the higher magnification), having a more restrictive effect on the air flow and so slowing the response.

It has been examined theoretically and experimentally that response is slowed by using a large operating pressure P and a large volume, but by these high sensitivity (magnification) is obtained. In any practical pneumatic measuring system, the overall response will be influenced by dynamic characteristics of the pressure measuring device. Thus it follows that the use of a low operating pressure will not improve the overall response if a low pressure measuring device of slow response is used to measure the pressure changes. High sensitivity will inevitably be associated with slow response and the only factor left to the designer is the volume, which should be made as small as possible for quick response.

Since gauge is always located at some distance from the control unit, the effect of variations in the gauging position does not reach the control unit instantaneously, though the size variations of the object will promptly affect the air flow at nozzles. The time gap between the sensing and indication is known as response time which depends upon : Length of air line between the nozzles and indicator, (*ii*) type of gauge system, and (*iii*) the design of control unit. Response in case of flow type pneumatic gauges is relatively quick. Response in case of back pressure gauges is slow, the compressibility of air also contributing to the delayed transmission of the variations sensed at the nozzles.

Response time of back pressure type pneumatic gauges can be improved by utilising following devices :

(*i*) Using filled system pressure gauge, thereby reducing volume of air.

(*ii*) Restricting the unimpeded escape of air through the orifices when the gauging head is not in operation, by using a spring-charged cover sleeve around the gauge head.

(*iii*) Counteracting the unrestricted air escape by an auxiliary air supply relay whose operation automatically discontinues as soon as a specific back pressure develops during the actual gauging process.

(*iv*) Using a high speed relay to compensate for additions to the volume of the instrument system.

5.8.1.9. Zero Setting. It is accomplished by means of a bleed valve and consists in adjusting the indicating element of the gauge to that marking on the scale which was selected to signal coincident with the nominal limit size represented by a setting master.

5.8.1.10. Datum Control. If means be provided to change the pressure in the cavity (between control orifice and measuring orifice) using a variable bleed to atmosphere, a datum or zero can be provided which varies the pressure 'p' when the escape orifice area remains fixed. This addition to a circuit provides means of accommodating small differences which inevitably occur in the manufacture of gauge heads. Limited use of a datum control in the form of a bleed to atmosphere has an insignificant effect on linearity.

However, this system depends highly upon the pressure regulator to maintain the supply pressure within very close pressure limits. Thus the pressure regulator is a critical component in this circuit. This problem is overcome in differential back pressure circuit in

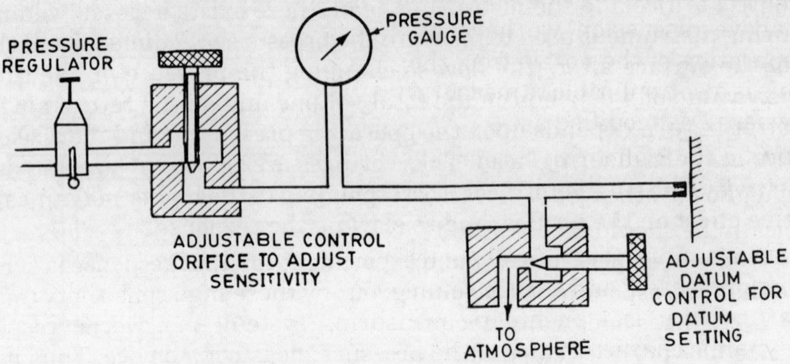

Fig. 5.41. Practical back pressure circuit.

which accuracy is maintained regardless of some variation in the regulator performance which controls supply pressure. Fig. 5.42 shows a practical flow-responsive system which is in common use. This system employs variable orifices for sensitivity and datum control.

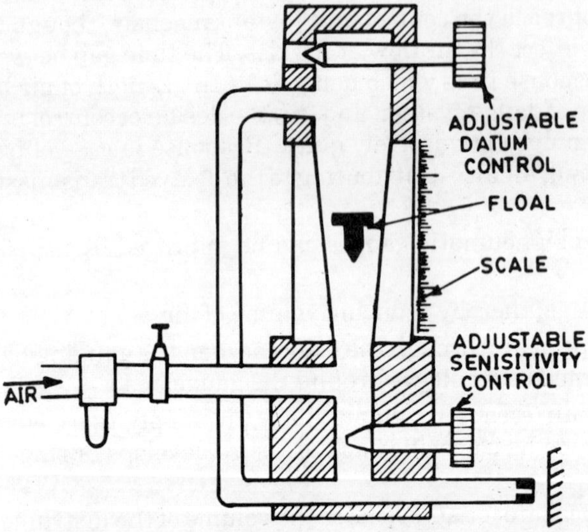

Fig. 5.42. Practical flow responsive system.

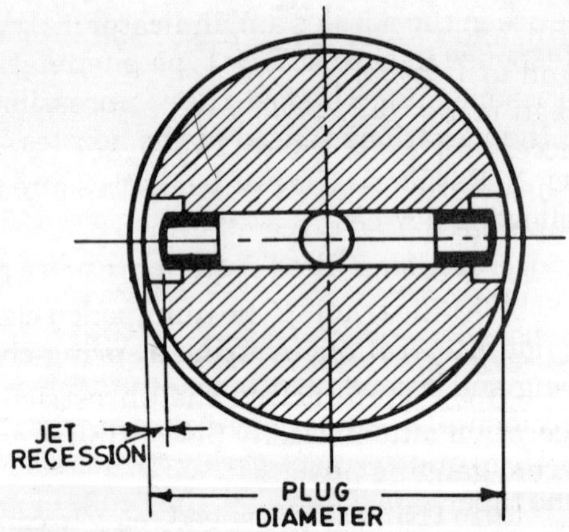

Fig. 5.43. Plug for pneumatic comparator.

5.8.1.11. Amplification Adjustment. This permits different range of gauge indications on same scale length and is carried out with a precision valve of the control unit. Both zero setting and amplification adjustment should be checked from time to time depending upon experience.

5.8.1.12. Jet Recession. It has already been seen that when the surface being measured is very close to the nozzle face, equal increments of change in clearance do not produce equal increments of pressure change. The system is not linear under these conditions.

Because of this the faces of the jets on air plug gauges are ground below the body diameter of the plug as shown Fig. 5.43. This grinding back is called jet recession and it is the means by which the non-linear portion at the high pressure (low flow), end of the pressure/clearance curves is avoided.

5.8.1.13. Measurement of Bore. [Refer Fig. 5.44 (*a*) and (*b*)]

In the above discussion we have assumed a jet separation of the nozzle from the flat surface. This is also applicable to measurement of bore, where L is modified as :

$L = L_1 + L_2$, L_1 = Radial gap for one jet and L_2 = Radial gap for other jet.

Theoretically L_1 should be equal to L_2, but even if $L_1 \neq L_2$ the back pressure would be closely equal to that corresponding to the condition $L_1 = L_2 = L/2$. This is a big advantage as operator need not be very meticulous about orientation of the measuring head in the bore, and thus the readings from operator to operator will be uniform as they do not depend upon a high degree of operator skill or sense of feel.

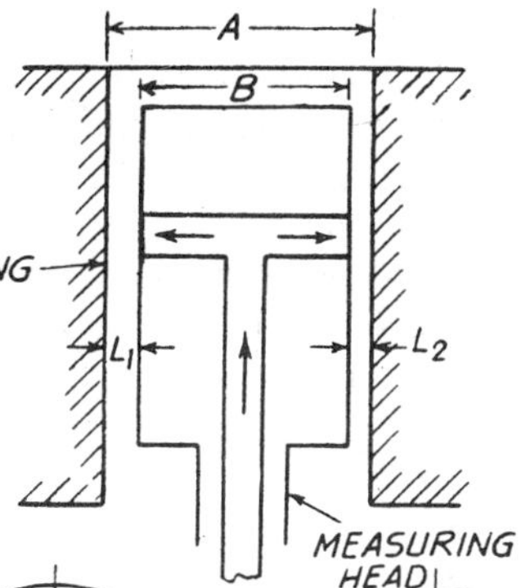

(*a*)

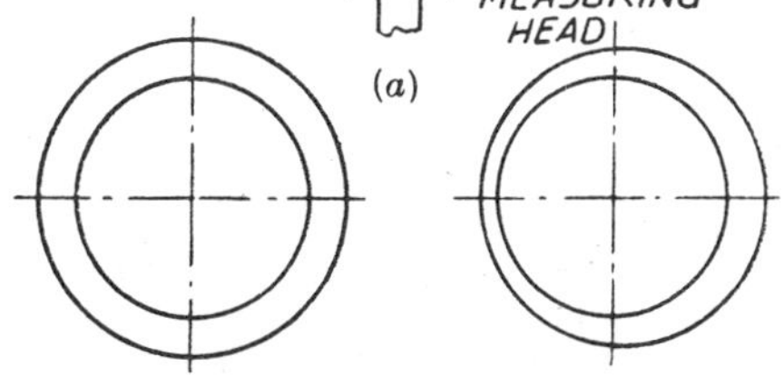

Symmetrical position. $L_1 = L_2$ $L = L_1 + L_2$

Non-symmetrical position. $L_2 \neq L_1$ $L = L_1 + L_2$

Fig. 5.44

5.8.1.14. Measuring Heads

Measuring heads fall under two categories, *viz.* direct head [Fig. 5.45 (*a*) and (*b*)], and indirect or contact head [Fig. 5.45 (*c*) and (*d*)]. Tapered nose type direct head [Fig. 5.45 (*a*)] is quite popular as it permits easy access in constricted measuring conditions. The ratio of land diameter (overall diameter at tapered small end) compared to jet diameter is twice in size. Bigger ratio would affect the escapement of air and the characteristics of the system.

Head at Fig. 5.45 (*b*) provides good protection to the nozzle due to incorporation of guard ring and escapement holes. In the case of indirect measuring heads, the jet is protected from accidental damage. The size of air escapement is controlled by a needle valve or flat plate [Fig. 5.45 (*c*) and (*d*) respectively] which move due to movement of measuring plunger. By changing the taper of needle valve, the range of measurement can be changed. Parabolic needle provides linear response. Fig. 5.45 (*e*) shows the plug gauge used for measurement of diameter, lobing, taper etc. The measuring side is made somewhat smaller than the bore so that it enters much more conveniently.

5.8.2. Solex Pneumatic Gauge. (Fig. 5.46). This instrument is produced commercially by Solex Air Gauges Ltd. This is generally designed for internal measurement, but with suitable measuring head it can be used for external gauging also.

It is obvious from the equation for sensitivity that in order that sensitivity (magnification) remains constant, the supply pressure P must be constant.

Thus it is necessary to have a simple pressure regulator which may control the pressure of air from the normal supply line, and if necessary the pressure might be reduced also. The arrangement used in Solex gauge is to pass the high pressure air after filtering, through a flow valve. There is a tank in which water is filled upto a certain level and a dip tube is immersed into it upto a depth corresponding to air pressure required. In Fig. 5.46, it is represented by H. Since air is sent at higher pressure than required one, some air will leak out from the dig tube and bubble out of water and the air moving towards control orifice will be at desired constant pressure H.

Now-a-days diaphragm type pressure regulators are readily available in the market and they are better for regulating the pressure than the above device. The air at reduced pressure then passes through the control orifice and escapes from the measuring jets. The back pressure in the circuit is indicated by the head of water displaced in the manometer tube. The tube is graduated linearly to show changes in pressure resulting from changes in internal diameter of the work measured. This instrument is capable of measuring to the accuracy of microns.

It is very obvious from Fig. 5.46 that the diameter being measured at any instant is corresponding to the portion against two jets. Now to find the concentricity (roundness of any job at any section), the workpiece may be revolved around measuring gauge. If no change in reading is there, then it is perfectly round hole. Similarly the diameter can be noted down at several places along the length of bore and thus tapering of hole is determined. This method is, therefore, best suited for measuring roundness and taperness of cylinder bores and gun barrel bores.

By having suitable measuring head this can be used for external gauging, and head in this case will be as

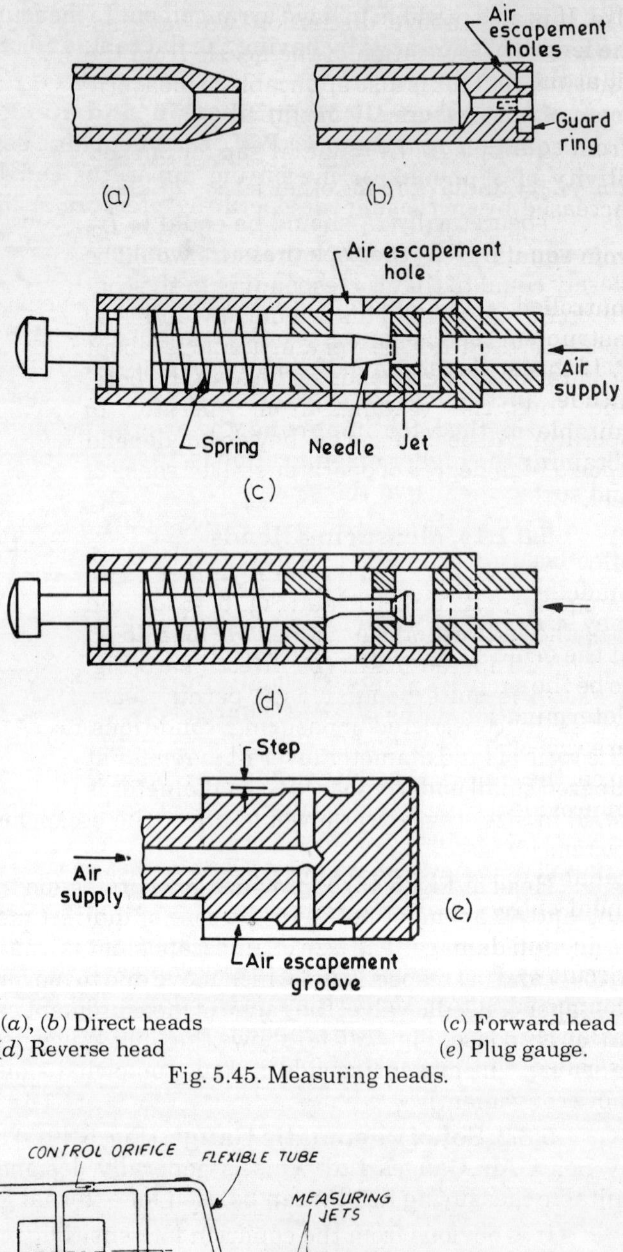

(*a*), (*b*) Direct heads (*c*) Forward head
(*d*) Reverse head (*e*) Plug gauge.

Fig. 5.45. Measuring heads.

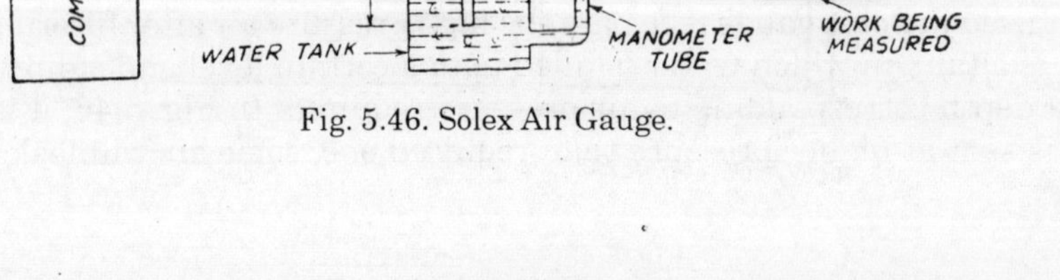

Fig. 5.46. Solex Air Gauge.

shown in Fig. 5.47. This can best reveal any lobing effect also. It is also possible to have arrangement to measure the length of slip gauge by having the flat table and one jet at the top.

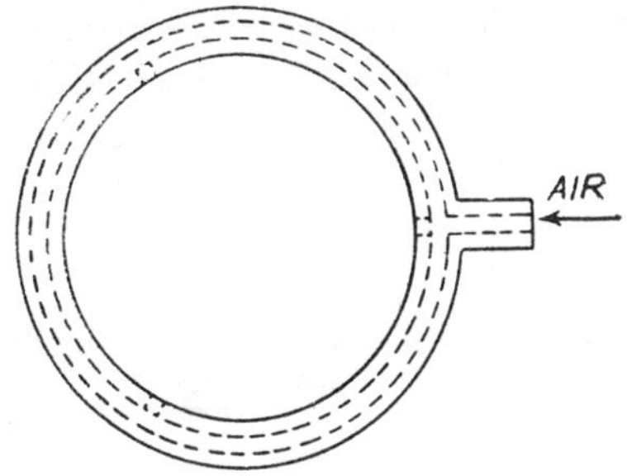

Three equally spaced measuring orifices (jets) reveal any lobbing effect also.
Fig. 5.47

5.8.2.1. Overall Magnification and Range. From equation $dp/dM = 0.40\ P/M_c$, the pneumatic sensitivity of a pneumatic measuring apparatus can be increased by increasing the operating pressure P, but from equation $\frac{dR}{dL} = 0.40 \frac{\lambda}{L_a}$ the overall magnification is controlled by the length of scale of pressure measuring instrument corresponding to pressure changes from 0 to P. If this scale length is to remain of convenient magnitude, increasing the operating pressure is not a suitable method for improving the overall magnification. The only effective method for obtaining the higher magnification is, therefore, to reduce average separation between nozzle and surface, which at the same time, of course, reduces the range of linear measurement.

5.8.2.2. Limitations of empirical approach. From the view-point of air flow, the effective area of an orifice is not usually identical with its geometrical area. If two orifices are made by producing holes of identical geometrical area in two thin discs, their effective areas may be appreciably different as a result of edge effects on the air flow arising at the peripheries of the orifices. Again the relationship between effective area and geometrical area is unlikely to be the same for air flow through an orifice and the jet of air from a nozzle. In the experimental determination of the p/P, M/C characteristics the value of M and C used were the geometrical areas of orifices O_m and O_c. Therefore due to effective area being different from the geometrical area, the empirical equation obtained by analysing these characteristics would not be expected to provide a completely accurate numerical forecast of performance.

Nevertheless, experience has shown that they do give a first approximation sufficiently reliable to permit the required performance to be obtained by a single-step corrective adjustment of the control orifice.

5.8.3. Differential Comparators. A later development brought out the balanced circuit type of air-gauge. In this equipment a differential pressure indicating mechanism, connected across the two air-paths and a built-in gauge zeroing valve is provided. Such a balanced circuit is shown schematically in Fig. 5.48. An air gauge based on this balanced circuit is called 'Differential Comparator'.

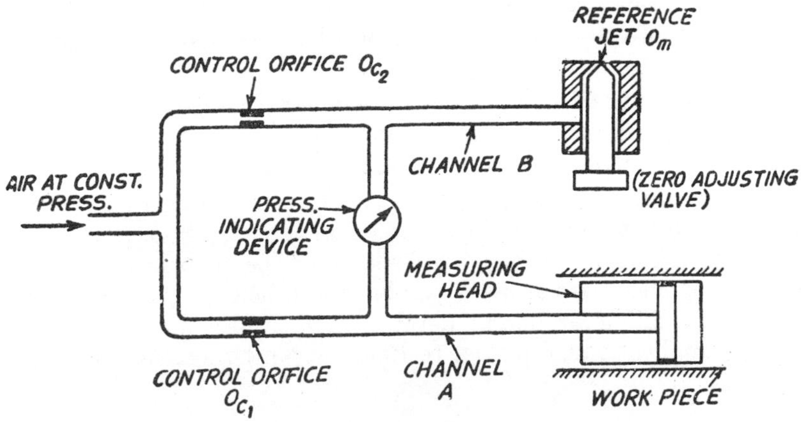

Fig. 5.48. Differential Circuit.

Compressed air from a suitable source, after passing through air-drier and filter is regulated for constant pressure by a pressure regulator. Now the air flows into two channels, each of which has a control orifice O_{c1} and O_{c2} (control orifices are also called master jets). From the control orifice O_{c1}, air flows to the measuring head where it meets further restriction of the workpiece or the master settings. The restriction of the workpiece builds up back pressure as explained earlier. At the same time, other half of the air is flowing through the other control orifice O_{c2} to the reference jet O_m. By closing or opening the valve of reference jet O_m, the pressure in the space between O_{c2} and O_m is regulated (adjusted) to match the back pressure from the measuring jets, which is sensed by the pressure indicating device fitted across the two channels as shown. At this adjustment of the reference jet, the pressure indicator would indicate equal pressure in the two channels and hence read zero on the scale. This zero setting (adjusting of reference jet O_m) is done with master workpiece whose dimension is exact nominal size.

Now the variation of the dimension at the measuring head would cause change of back-pressure in channel A. This pressure would be different from the mean pressure which has been already set in the channel B (by reference jet). Now the difference of pressure of the two channels would be indicated by the pressure indicating device which can be directly calibrated in terms of variation of dimension from the mean dimensions. Hence the instrument based on the measurement of differential pressure is called Differential Comparator.

If the dimension causes a decrease in gap L as compared to L_a, this in turn decreases M and hence increasing back pressure in channel A and *vice versa*. In these cases the pressure indicator would show readings on both sides of zero corresponding to ± deviation from master setting.

5.8.3.1. Advantages of Differential Circuit over Single Channel Circuit. (*i*) *Effect of change of operating pressure P.* The operating pressure may vary slightly from the designed value. It can be shown that error due to change of pressure would be 0.6 to 0.8 times the change in pressure in single channel in case of differential circuit the error would be 0.1 times the change in pressure.

(*ii*) Zero setting of master gauge is an extra advantage.

(*iii*) *Rectification for control orifice.* In a single channel system, the practical limitations may not give the perfectly correct and accurate dimension of the control orifice as designed. Sometimes it may go out of the useful range of the design and it may have to be discarded.

Therefore, in order to avoid the error of manufacture and also to take into account the fact that the geometrical area is different from the effective area, we need a needle valve so that area may be adjusted accordingly.

But in the differential circuit which incorporates a needle zero adjusting valve, the off-set of the actual size of the control orifice from the designed value can be nullified by adjusting this valve.

5.8.4. Non contact and Contact Type Pneumatic Gauging Elements. Non contact tooling is best suited for automatic gauging applications because of the advantages of no contact, clearing of oil or foreign particles from gauging area, etc.

In the case of non-contact air gauge tooling, only the air coming out of the air escapement orifice touches the part to be measured, the air flow rate depending on the cross-sectional area of the jet and the clearance between the jet and the part to be measured. It may have a single jet, two diametrically opposite jets or more evenly spaced jets. Single open-jet tooling can be used for checking outside diameter, height, depth, straightness, squareness, etc. and Fig. 5.49 shows a few of such applications. Dual jet techniques can be applied for determining true diameter, out or round, bell mouth, thickness, etc. The various gauges may be designed either for presenting the gauges to part or *vice versa*.

Many modern mechanical assemblies demand that holes should be closely controlled for straightness as well as diameter. An air plug gauge for gauging hole straightness is shown in Fig. 5.50.

In the contact tooling, a mechanical member is incorporated between the air escapement orifice and the part to be measured. The air flow from the jet changes due to displacement of this mechanical member when it contacts the part. The mechanical member could be a ball, lever, plunger or blade. A big advantage of contact type air gauge is that a much bigger measuring range (upto 2.5 mm) is possible *i.e.* it is suitable for wide range of gauging. Another advantage is that it eliminates surface roughness from size. It may be mentioned that open jet type method would be subject to error for rough surfaces because it measures a combination of size and surface finish ; further its range of measurement is limited. Ball jet spindle gives a point reading rather than the average over a small area and is best suited for gauging inside diameter of soft or porous parts and for rough bores. Leaf jet spindle can be used for checking laminated bores, blind holes in which keyways etc. do not permit the use of open jet spindles at extreme bottom of blind holes etc. Blade jet spindles are used for inspecting gun bores in which oil grooves, or slots do not permit the use of ball jet or leaf jet spindles.

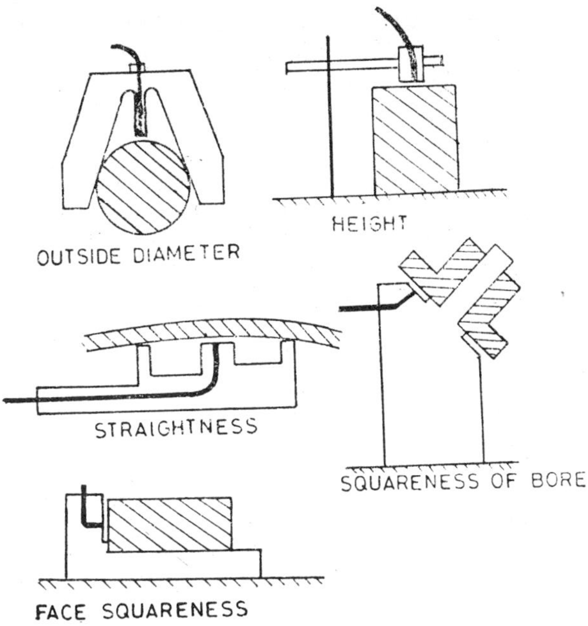

Fig. 5.49. Applications of single open jet tooling.

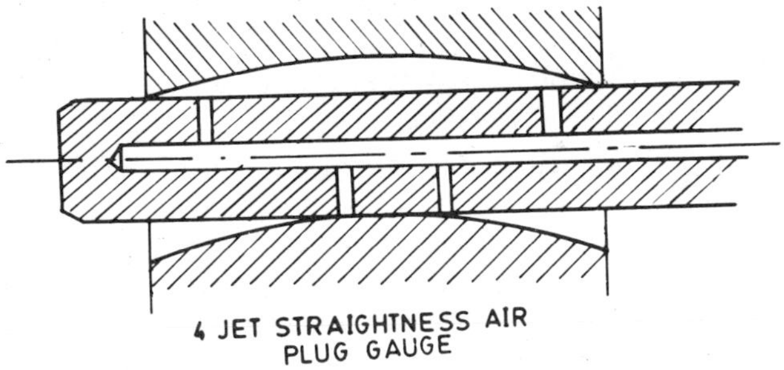

Fig. 5.50. Straightness measuring gauge.

Fig. 5.51 shows a small plunger type air gauging cartridge which is highly efficient size-sensing element for wide range of gauging, tooling, fixturing, and machine control applications. It essentially consists of a spring loaded plunger. The spring tries to keep the

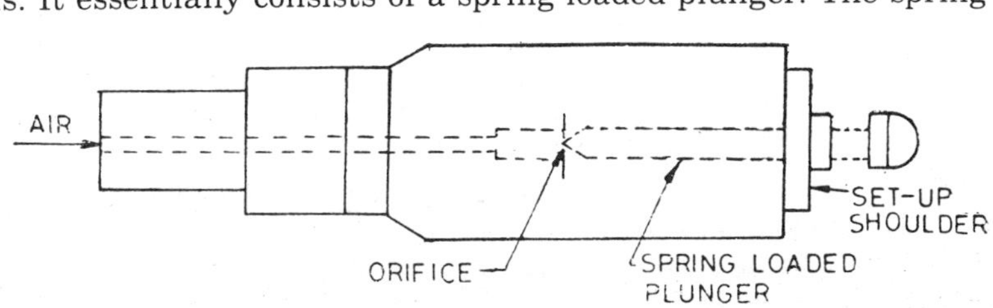

Fig. 5.51. Air Cartridge.

plunger outwards and when the part to be measured comes in contact with it or it comes in contact with the part it moves in and at the end restricts the orifice, thereby increasing the back pressure. The maximum and minimum limits of the plunger movement can be set with the help of masters. Such cartridges can be secured in gauging position on various types of fixtures and used for measurements like height, depth, flatness, concentricity, squareness, inside/outside diameter. etc.

Fig. 5.52 shows the same principle used in a test indicator which is very efficient for several applications. It has a tiny stylus capable of entering into small holes, keyways, slots

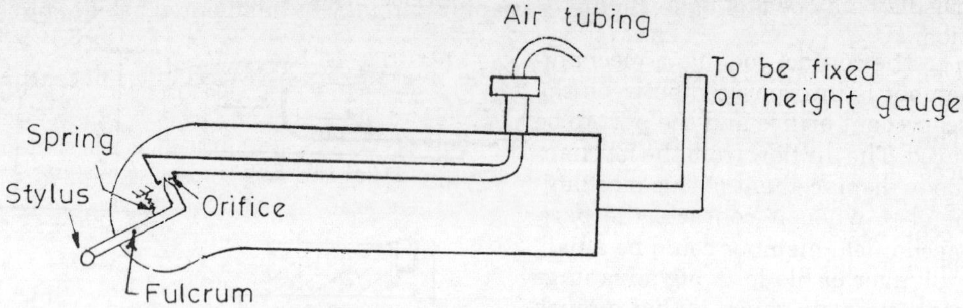

Fig. 5.52. Pneumatic test indicator.

etc. and its movement causes the tapered end to act as a precision valve stem to regulate the amount of air flowing through an orifice. It is free of hysteresis or lag or drag in indications when the stylus is moved in any direction across the workpiece.

The contact type adjustable spindle kits and adjustable air snap gauges available in market are found to be very suitable for handling new designs, altered dimensions and various other varied applications.

5.8.5. Indirect Pneumatic Gauging Devices. The open jet has the disadvantage of small measuring range. It can be overcome by using indirect pneumatic gauging devices by using a gauging cartridge. Such cartridge employs a contacting stylus, the inner end of which is tapered and forms the restriction in an escape orifice. The position of the stylus and consequently the position of the taper in the orifice causes changes in the area of the orifice. Changes in the rate of taper change the measuring range of the cartridge. Measuring ranges upto 3 mm can be obtained with this type of cartridge.

5.8.6. Air gauging with electronic sensors. Air gauging system operate on either low or high air pressure. While low pressure systems have greater sensitivity, quicker response time and minimal distortion when measuring flexible components, the high pressure systems are self cleaning type and have a large measuring range. Basically air gauging comprises air jet gauges—such as ring or plug, and air operated liquid columns for multi-dimensional measurement.

Now-a-days electronic flowmeters are used in place of air operated liquid columns. They have the advantage of measuring flow of air with the added benefit of electronic display. Such instruments can easily have 2 or 3-range selection to give an extra magnification factor. Tolerance limit lights can be incorporated to indicate whether parts are inside or outside manufacturing tolerances. Response time increases many folds.

The versatility of air gauging is enhanced by the wide range of measuring tools like 2 and 3 jet non-contact air plug gauges, setting rings and air jet ring gauges.

5.8.7. Multi-gauging Systems. Multi-gauging systems are used to measure a number of dimensions simultaneously. Parts to be gauged are compared with a setting master which simulates the component. The features gauged could be external/internal diameters, lengths, straightness, squareness, ovality, run-out of faces, etc. The measuring head gauging fixture is specifically designed to suit the component to be measured and may be completely special or it may be built using a series of modular elements. It contains the means for sensing the dimensional difference between the components and the master which may take the form of mechanical or electronic probes or air jets connected to the means for amplifying the difference.

The amplifying and display of readings may take the form of dial gauges or some form of electronic or air/electronic system.

Display may be analogue, digital or graphic and may be augmented by out of tolerance indication.

Using electronic differential methods, the relationship between different features can be related to a common datum.

The choice of system depends on number of factors like initial cost, dimensional tolerances of the features to be measured, complexity of the component, complexity of the features to be measured, number of features to be measured, speed of measurement required, skill or otherwise of the user.

The reasons affecting the choice of displays are given below :

(*i*) Pointer displays—best where a rapid check of run-out or concentricity is required.

(*ii*) Columns—natural choice where a considerable number of dimensions are involved.

These are fastest and most convenient form of displaying the readings for every dimension.

(*iii*) Digital—provides high accuracy over a longer measuring range and best for situations where it is required to work in drawing dimensions. Can be viewed without strain over longer distances.

(*iv*) Graphic on VDU—the most sophisticated display essential when statistical process control systems are employed.

Systems capable of dealing with very large number of inter-related dimensions can be developed.

Automatic inspection machines incorporate both automated loading hopper, magazine, pick and place robot and—automated segregation of inspected components. Automatic inspection is essential where the complexity of the component is such that manual methods can not achieve the desired levels of accuracy.

The results of inspection could be fed to electronic computer based system which may also control the machine operation. The use of such computer based processing also allows the results obtained to provide a wide range of control facilities including feedback for control of the manufacturing process.

Coordinate measuring machines accommodate multidimensional inspection by using a single point contact to take successive measurements over the component profile. The contact movement and processing of the dimensional information is under computer control, which can also provide similar facilities to those offered by multi-point gauging.

5.9. Fluid Displacement Comparators [Fig. 5.53]

This type of comparator has limited applications due to some of the disadvantages associated with it. But it is discussed here for the sake of the new and altogether different method of magnification. In this comparator, a fine bore capillary tube is arranged so that its lower end is placed in a chamber of relatively large cross-sectional area and containing a fluid of low viscosity. At the bottom of the chamber, a diaphragm is fitted, which gets deflected by pressure transmitted to it by the measuring plunger. Due to deflection of diaphragm, a small quantity of liquid rises up in the capillary tube. A scale is arranged at the side of tube and limit pointers are set relative to this to indicate the high and low limits of the dimension being measured. These limits are obtained by master setting.

$$\text{Magnification of this system} = \frac{A}{a} = \frac{\text{Area of cross-section of chamber}}{\text{Area of cross-section of capillary tube}}.$$

Fig. 5.54 shows the working of prestwich fluid gauge based on fluid displacement comparator principle. Instrument body comprises of a cone shape with a thin base containing a small button centrally. Any movement of the base causes a displacement of liquid up the tube to indicate actual dimension. The change in volume of liquid in conical portion due to deflection $(h_2 - h_1)$ can be taken as $\frac{1}{3}\pi\left(\frac{D}{2}\right)^2 h_2 - \frac{1}{3}\pi\left(\frac{D}{2}\right)^2 h_1$. This change in volume when pushed into pipe by length l will be equal to $\pi\left(\frac{d}{2}\right)^2 l$.

$$\therefore \quad (h_2 - h_1)\frac{1}{3}\pi\left(\frac{D}{2}\right)^2 = \pi\left(\frac{d}{2}\right)^2 l, \text{ and } \therefore \text{ Magnification} = \frac{l}{h_2 - h_1} = \frac{D^2}{3d^2}.$$

The magnification of this type of comparator is usually of the order of 1500.

Various disadvantages of the instrument are :

(1) It is too much dependent upon the temperature as the fluid changes its volume with change in temperature.

(2) The deflection of the diaphragm per unit measuring force is not constant.

(3) There is certain passivity in the instrument due to the characteristics of diaphragm and the viscosity of the fluid.

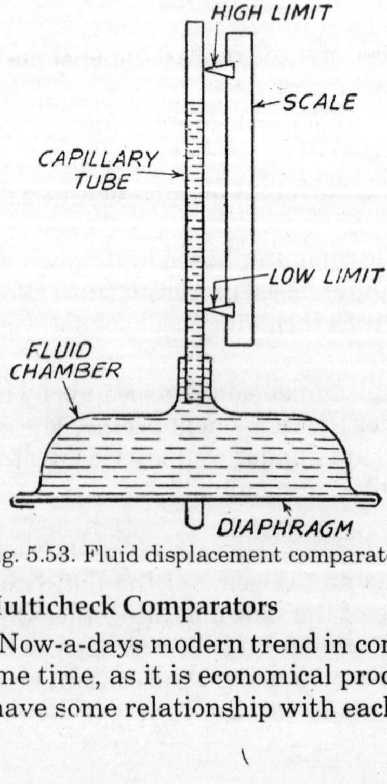

Fig. 5.53. Fluid displacement comparator.

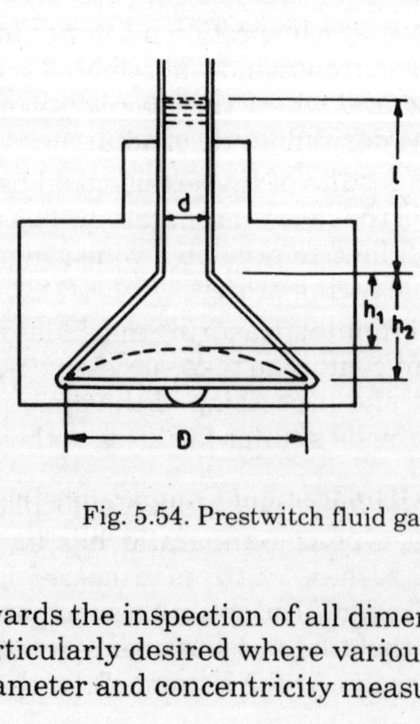

Fig. 5.54. Prestwitch fluid gauge.

5.10. Multicheck Comparators

Now-a-days modern trend in comparison is towards the inspection of all dimensions at the same time, as it is economical procedure. It is particularly desired where various dimensions have some relationship with each other, *e.g.*, diameter and concentricity measurement.

This job is done by multicheck comparators, which are of the following types and incorporate the following systems of amplification.

(*a*) Electrical, (*b*) Air, (*c*) Combination 'air-electric'.

5.10.1. Electric Multichecks. This is combination gauge incorporating about twenty or more electrical-check heads to measure simultaneously a number of dimensions of one part. For each dimension, there is one measuring head and signal lights for each dimension indicate whether it is within tolerance, undersize or oversize.

The instrument is used for high production checking and chances of any error in any dimension are less. As it will not be desirable to see all the lights for various dimensions, a master signal light is used to integrate all the individual lights, and inspector has to watch only one light. On seeing the master light warning individual signal lights are examined. This instrument is well suited for all sizes of works.

5.10.2. Air Multichecks. In this instrument, group of air comparators are set up to check a number of dimensions. In this type of instrument, initially some difficulty is experienced in setting the various comparators to check various dimensions and to arrange them in compact form. But once this is done things become very easy. In one setting all the dimensions can be checked and the time of inspection is reduced considerably. Mass inspection becomes possible and high quality can be ensured.

5.10.3. An air Electric Multicheck. This makes use of both types of comparators. It is very easy to use air comparator for checking diameter and concentricity, so for this purpose air comparator is used, while for other measurements electric-checks are used.

5.10.4. Automatic Gauging Machines. These machines incorporate comparator amplifying methods and are similar to multicheck devices. They eliminate manual inspection and parts are inspected for all the dimensions simultaneously and segregated and classified automatically.

5.11. Optical Projectors

It has become an indispensable instrument in modern metrological laboratories because of its versatility and ease in use. Using optical projectors it is possible to measure those components which are difficult to be measured by other methods due to typical size, material composition or dimensional characteristics. It displays the magnified images (which should be clear, sharp and dimensionally accurate) of the objects located by some holding device or staging fixture (ample working space being provided for the positioning of parts) on an appropriate viewing screen (located near the operator's normal eye level for his convenience). The magnified image serves as an aid to more precise determination (or comparison with chart gauge) of dimension, form, etc.

An optical projector of (*i*) a projector (having a light source, a condenser or collimating lens system to direct the light past the part and into the optical system), (*ii*) suitable work holding table which may be fixed or movable, (*iii*) projection optics including both mirrors and lenses (*iv*) screen where the image of workpiece is projected and where measurements or comparisons are made, (*v*) measuring devices. A good optical projector must have a precise optical system, and means for precise mechanical measurement.

While an optical projector displays a two dimensional projection of a part, it may be noted that all parts are not ideally suited to optical gauging. Operator's visual actuity and perceptiveness also matters in the use of optical projectors. The performance of optical projector also depends on its optical design, manufacture and assembly. All these factors directly influence the quality (clarity and definition) of image formed, the accuracy of the magnified image (freedom from distortion), and the contrast in the image (dark shadow with bright background). The quality, contrast and accuracy of image are also influenced by the aberrations in the optical systems. Actually the aberrations in the optical elements make the light rays passing through optical elements to deviate from the desire path.

The most common aberrations found in optical projectors are distorsion (non-uniform magnification from centre to edge of screen), field curvature (out of focus from centre to edge due to spherical shapes of the screen), chromatic aberration (dispersion of white light by optical system and inability to focus all the colours simultaneously in the plane of screen), lateral colour (appearance of distinct blue or red fringes at the edges of image caused by unequal magnification of these two colours in the image), image quality and contrast. The contrast of image is also influenced by viewing screen mirrors and mechanical mounting of the optical element to reduce flare, and stray light.

The image of test piece is produced on the screen by a beam of light ; the light source being a tungsten lamp, filament lamp or high pressure mercury or zenon arc lamp. A strong beam of light consisting of concentrated bundle of parallel light rays is produced by optical means. Beam should be of large enough diameter to provide coverage on the test piece and adequate illumination intensity at the screen. Two common arrangements of light sources used in optical projectors are (*i*) shadow and (*ii*) surface. The shadow image of the object may be produced either on horizontal screen or vertical screen. Surface illumination produces a reflected image of the face of the object. High intensity light is required for projecting surface characteristics. Using high intensity source which converges available light in a concentrated beam on a part, it is possible to project such dull and non-reflective materials as ceramic and carbon in the surface illumination system.

The objects to be tested are placed (staged) on the work-table. The space in which the parts can be staged is decided by the focal clearance or working distance of the optical system (*i.e.* the distance from the plane of the sharp focus to the nearest surface of the projection lens or projector housing). The sectional contour or maximum size of an object that can normally be projected to the screen is the contour lying in the focal plane, *i.e.* it is independent of the working distance of the projector. The working distance can however be increased optically by using a relay lens as a part of projection system.

The light beam after passing the object to be projected passes into the *projection system* comprising lenses and mirrors which must be held in accurate alignment on rigid supports. The lenses are used to obtain the desired magnification and mirrors to direct the beam of light on screen. It is, therefore, possible to change the magnification by changing the lenses. Using too high magnification is no good because the image produced at high magnification is not that sharply defined as with low magnification. The magnification used should be such which will show the smallest part tolerance limit clearly on the screen without strain to the operator. Screens are usually made of glass with the surface facing the operator ground to a very fine grain. The location of ground surface is of critical importance in the projection system because its location determines the position where accurate magnification is achieved. It is for this reason that the thickness of the glass must be held to close limits.

Projector *work-table* may be either stationary or moving type. Moving type may have facilities for measurement of movement, or be of non-measuring type. Some tables are also equipped with an angular adjustment for positioning to the helix of threads and worms. Stationary tables are used for projection inspection. These are equipped with keyways or T-slots for mounting fixture, and the inspection is co-ordinated with chart-gauge. Non-measuring moving tables usually have in and out movement parallel to the axis of the light beam for focusing purposes, and also provision for movements in other two planes. An optically flat glass is set into the top of the table and this glass surface, in addition to allowing passage of light beam, also serves a good surface for the positioning of a variety of flat parts. Measuring moving tables also have travel facilities in three planes—Focus, horizontal and vertical. Each direction of travel is carefully co-ordinated for true alignment to the optical axis. Micrometers in combination with dial indicators are fitted as measuring attachments for either two or all the three directions. For inspection of threads, worms and hobs, the helix is brought into parallel alignment with the optical axis of projector by having helix rotation provision on the table. Protactor rings mounted with the viewing screen (both mounted concentric with the optical axis of the projector) simplify the determination of angular measurement.

The following *measurement techniques* are employed on optical projectors, (*i*) measurement by comparison (possible on all projectors), (*ii*) measurement by movement (projector should have table travel and measuring facilities) and (*ii*) measurement by translation. It requires special tracer accessories.

In the measurement by comparison, the two-dimensional screen images are produced to the desired magnification by keeping the object in the focal plane of the optical system. A master layout of chart gauge is made to represent the high and low limits (double line outline) of the theoretically perfect part at the magnified size. There is usually a reference line for alignment of the part image to the chart, and then it is easier to compare. The contour characteristic such as irregular curves, chamfers, filler radii etc. can be all inspected in a single stretch without resorting to other inspection methods.

In the measurement by movement, the centre cross line on the screen provides optical reference.

Part containing such features (internal groove etc.) which cannot be projected by a light beam directly are gauged indirectly by a combination of tracing and optical projection. In such measurements (by translation), a tracer which is one-at-one pantograph is employed. One arm of the pantograph is stylus which traces freely over the part contour, and the other arm carries a follower (these may be of three types) which is projected by the light beam as it moves. Various types of followers used are (*i*) probe, (*ii*) dot, (*iii*) reticle gauge. These are shown in Fig. 5.55. It may be noted that probe follower is an exact duplicate of the stylus tracer in size and shape. In this case, the supporting stem restricts the shadow and so in some cases it is not preferred. Dot follower overcomes this difficulty. Dot follower is a glass recticle having an opaque dot of the same diameter as the stylus tracer. These two types of followers are used when the size of the part is less than the field of view of the projector at the given magnification. In their use, the contour layout of the part is made on the chart gauge (the contour layout may be a single profile or double line-profile showing the permissible limits of tolerance of part). As the stylus traces across the part having no error, the follower image (shadow) moves exactly along the contour on the screen. The reticle gauge follower, which is more versatile and not restricted by part size or magnification, is a glass reticle having an exact one-to-one actual size reproduction of the part-profile. Such followers are made by laying the contour of part on precision scribing machines at a magnified size, and then reducing photo-graphically one-to-one size by precision cameras. The chart used with reticle gauge follower has a circle (or two circles to represent tolerance spread) representing the diameter of the stylus tracer at the magnification being used. When a perfect part is traced, then the projected reticle gauge contour will move past the fixed circle on chart

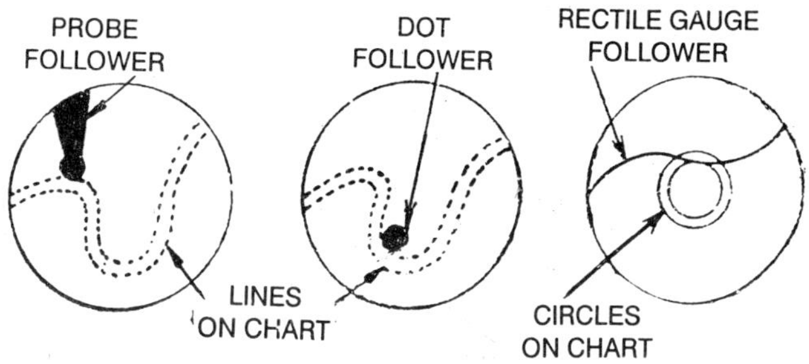

Fig. 5.55. Projector tracer followers.

gauge. If double circles are used then the projected reticle contour should fall between these circles at it passes across the screen, for the part to be within the given tolerance limits.

Chart gauges play an important role for the effective use of the optical projectors. These provide the point of reference for the operator. Mostly these are processed on glass or plastic, and may be either replacement type, or over-screen, or overlay type. Replacement chart gauge is used as a substitute for the standard viewing screen of a projector, and so it must have all the required physical characteristics (thickness, flatness, surface texture, etc.) of the original screen, and should be suitably positioned, and lines marked on the ground surface facing the operator. The overscreen chart gauges are suitably fastened in place over the screen and have the layout lines on the surface which is placed in contact with the viewing screen. These are easier to mount and replace and used where frequent changes of set up are required. Overlay chart gauges are held in hand and positioned by the operator on the part image, otherwise in all respect they are same as over screen chart gauges. Many types of standard overlay chart gauges are available for checking specific features like fillet raddi, small diameters, protactors for checking angles, etc.

5.11.1. Profile Projector

Features

— Medium-sized model that features high versatility and easy operation.

— Digital readout protractor screen facilitates angle measurement.

— Screen glass carries the chart clips.

— Easy-to-read digital XY-coordinate counters are located near the projection screen to minimise eye-movement.

— Floating-type table (X-direction only) for speedy measurement.

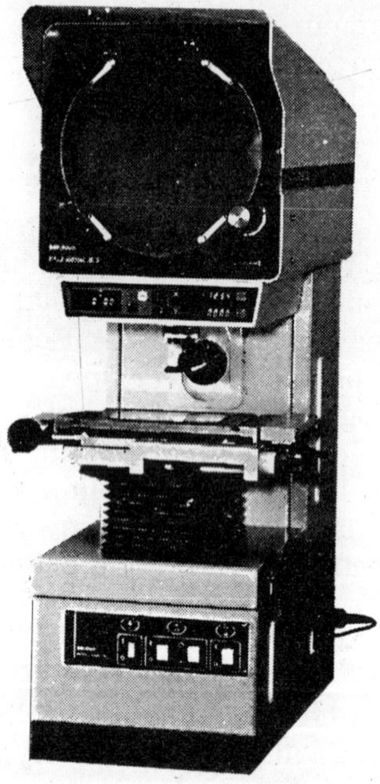

Fig. 5.56. Profile projector.

5.12. Recent Gauging Techniques

Some recent developments have taken place in other countries in the field of measurements. These are discussed here in brief of keep the students abreast of the latest innovations in the field of Metrology. These newer devices include ultrasonic gauging, nucleonic thickness gauges, and infrared measuring devices.

5.12.1. Ultrasonic Gauges. These employ ultrasonic waves (sound waves whose frequency is above the audible range) which are produced by a quartz crystal which converts the electrical energy applied to it by an oscillator into the sound waves or mechanical vibrations. For measurement purposes, frequency range of 0.5 to 30 MHz is employed. These waves are introduced into the component whose length/thickness is to be measured through a liquid couplant, and these get reflected back at the interface. The principle of operation is to measure the time taken by ultrasonic waves to make a sound trip through the component to be measured. The ultrasonic waves propagate in a fairly well-defined beam and continue to do so in the same material, but at the interface comprising of two different materials, these are reflected. The amount of energy reflected is dependent upon the area and the relative acoustic properties of the two materials making up the interface. It may be noted that incident face and the far side of the component also act as interfaces ; even internal discontinuities such as crack or flaw, shoulder or drilled hole also provide metallic grain interfaces to reflect ultrasonic waves. The velocity of the longitudinal sound waves is dependent upon the properties of material like density, Young's modulus and Poisson's ratio.

Two types of equipment *viz.* pulse echo and resonance type which have similar sound propagation characteristics but use different means of generating ultrasonic waves and the data readout, are available. In the pulse echo equipment, a short pulse is transmitted into the component and the number of multiple back reflections through the component are recorded on a CRT screen. First the pulse is passed on a standard component of known thickness and a relation is established between the number of echoes over a distance on CRT screen and the thickness of component. This provides calibration as the number of echoes and thickness which enables any unknown thickness to be determined to an accuracy of one to two per cent. A typical pattern of multiple echoes is shown in Fig. 5.57. It will be seen in this method that greater accuracy can be achieved by using longer total distance on CRT but there is limitation, and also the counting of greater number of back reflections becomes a tedious job.

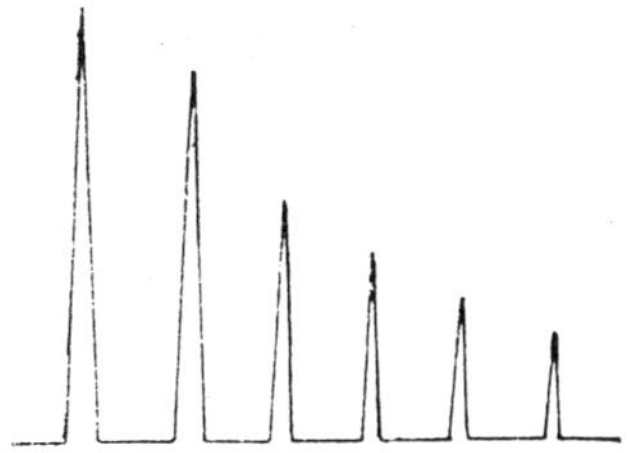

Fig. 5.57. Echoes obtained on ultrasonic gauge.

A refinement of this method, permitting higher accuracy to be obtained is by utilising the adjustable linear horizontal sweep on the CRT. The standard pattern of multiple echoes received on a standard piece of known thickness is super-imposed on CRT, and then with the unknown thickness the instrument sweep is adjusted so that the echoes from unknown thickness coincide with the standard pattern super-imposed on CRT. The thickness can then be directly read from the caliper scale used to adjust the sweep.

The resonant type of ultrasonic equipment works on the principle that a fundamental acoustic standing wave will be generated in the component whose thickness is to be measured, if the round trip distance, *i.e.* twice the thickness of component equals one wavelength. Harmonic resonances occur when the round trip distance equals whole-number multiples of a wavelength. The fundamental frequency, which is required to determine the thickness, can be known by finding the difference between two adjacent harmonic frequencies. The basic relationships are,

$$t = \lambda/2 \ (t = \text{thickness}, \lambda = \text{fundamental wavelength}) = \frac{V_L/2}{f_o} \left(\lambda = \frac{V_L}{f_o} \right)$$

where V_L = longitudinal velocity, and f_o = fundamental frequency.

Since $f_o = f_N - f_{N-1}$ (where f_N is harmonic frequency), $t = \dfrac{V_L/2}{f_N - f_{N-1}}$.

This instrument, therefore, requires the use of some devices by which the frequency of ultrasonic waves may be adjusted. It could be either a frequency-modulated continuous-wave generator, or an oscillator could be used to drive the ultrasonic transducer.

The presence of acoustic standing waves in the component, *i.e.* resonance condition, can be detected by either hearing an audible tone in the earphones or observing vertical pipe on the CRT depending on the method employed. The frequency information can then be translated into thickness and displayed on the read out system. Instruments are available that permit instantaneous and continuous readings as the transducer is moved over the part.

In the ultrasonic techniques, the transducer contacts the part to be measured from one side only through a thin film of couplant which acts as lubricant as well as excludes air. It is a portable device and has accuracy varying between 0.1 to 1% of nominal thickness depending on technique and system. Resonance method is generally used for measuring small thickness in the range of 0.25 to 50 mm, and the pulse echo method for thickness in the range of 2.5 to 250 mm.

5.12.2. Neucleonic Thickness Gauge. It is used to measure the thickness of sheet material in steel, rubber, paper, plastic industries. A radioisotope source is used to produce a penetrating radiation which is beamed through the part whose thickness is to be measured. The intensity of radiation as it passes through the part is reduced depending upon the part thickness and density. A radiation detector at the opposite end measures the amount of radiation available there and it is converted into a small current which provides continuous, non- contacting measurement of the material's thickness provided its density is constant. The reduction in intensity of beam as it passes through a material follows an exponential law as $I = I_o e^{-\mu\rho t}$ where I and I_o are the attenuated and incident radiation respectively, ρ is material density and t is its thickness, and μ is the mass absorption coefficient which is function of the particular radiation selected : greater being the penetrating power of radiation, lower the value of μ. The selection of proper radioisotope and this to be securely held is important. For thinner material upto 0.75 mm thickness, beta rays emitter such as krypton 85 etc. are used ; whereas for heavier and thicker materials upto 50 mm thickness, either gamma emitters (such as cesium 137, cobalt 60 etc.) or beta-excited X-rays may be used.

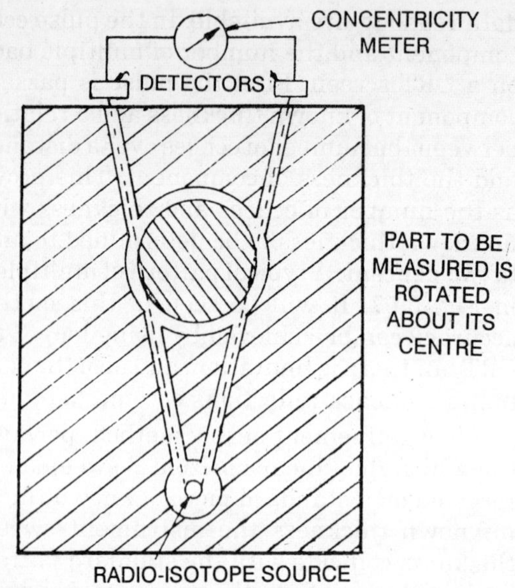

Fig. 5.58. Principle of nucleonic thickness gauge.

Fig. 5.58 shows how the concentricity can be measured by using two radiation paths through the component to be measured and comparing the emergent beams on a differential basis.

It is also possible to devise a set up to measure the change in wall thickness of pipe as shown in Fig. 5.59. It is observed that variations upto 2 to 5% for pipes upto 200 mm diameter with 6 mm thickness can be easily detected by this method.

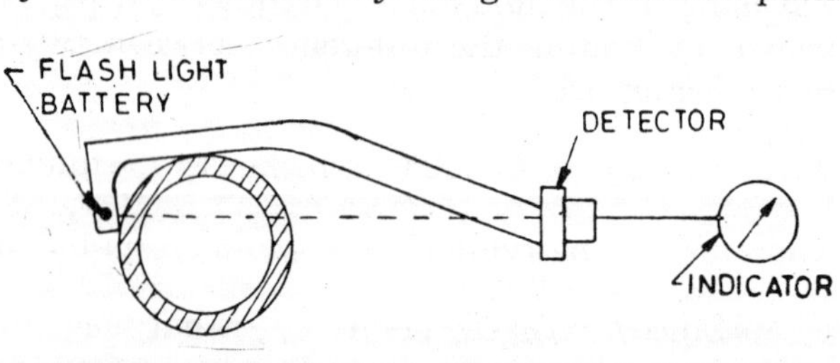

Fig. 5.59. Device to measure wall thickness of pipe.

It is important to note that these methods do not depend on external source of power supply. This factor alongwith the stability of radioisotope source makes these methods ideally suited for many industrial applications.

SOLVED PROBLEMS

Problem 5.1. *What are the characteristic advantages of mechanical indicators ?*

Solution. (*i*) **Long measuring range.** Mechanical indicators operating on the rack and pinion system have measuring ranges extending over several turns.

(*ii*) **Small overall size.** This property is of great help where space is confined or where several indicators have to be mounted at close distances to each other.

(*iii*) Positive contact and controlled measuring force.

(*iv*) **Rugged construction.** Ideally suited for operating machines where substantial vibrations are present. These are also less sensitive to inadvertently caused over-travel.

(*v*) **Independence of power supply.** No interference in mounting of instrument with power lines and inherent indication accuracy is unaffected by fluctuations in power supply.

(*vi*) **Stability and repeatability of indication.** Drift, usually a problem in electrical indicators is absent. Errors due to mechanical causes (binding, wear, play) can be detected and correction/compensation applied.

(*vii*) **Economical.** Initial cost is low. Can be easily maintained and repaired in plant at reasonable cost.

Problem 5.2. *What are the advantages and limitations of following types of measuring systems ?*

(*i*) *Capacitive systems* (*ii*) *Magnetic systems*

(*iii*) *Opto-electronic systems* (*iv*) *Inductive systems.*

Solution. (*i*) Capacitive systems have a low power consumption and are thus ideal for battery operated instruments.

(*ii*) Magnetic systems use a medium similar to a video tape and are more resistant to dust and liquids. These are thus suited for calipers used in rugged conditions.

(*iii*) Opto electronic systems are basically graduated glass scales with encoders and have a very high precision. These are difficult to package into compact instruments and are thus used in fixed displays.

(*iv*) Inductive systems (electromagnetic transducers) produce an electrical output proportional to the displacement of a movable core. Their repeatability is excellent but range is short.

Problem 5.3. (*a*) *Define the following terms in connection with pneumatic comparators.*

(*i*) *Overall magnification or sensitivity of the system.*

(*ii*) *Measuring head sensitivity.*

(*iii*) *Pneumatic sensitivity*

(*iv*) *Indicator sensitivity.*

How does each get affected ?

(*b*) *Discuss how the various parameters can be chosen to obtain :*

(*i*) *Maximum sensitivity.* (*ii*) *Maximum range.*

(*c*) *What are the advantages of differential type pneumatic comparators ?*

Solution. (*a*) (*i*) Overall magnification is the ratio of scale movement (dR) for a given change of dimension dL, *i.e.* overall magnification = dR/dL and it is the product of sensitivities of measuring head (dM/dL), pneumatic sensitivity (dp/dM), and the indicator sensitivity (dR/dp)

or
$$\frac{dR}{dL} = \frac{dR}{dp} \times \frac{dp}{dM} \times \frac{dM}{dL}.$$

(*ii*) Measuring head sensitivity (dM/dL) is the characteristic of measuring head.

Since $M = \pi\, dL$, D = measuring orifice size and L = separation or distance to be measured.

$\therefore$ $dM/dL = \pi D$ and hence it increases with increasing measuring orifice size.

(*iii*) Pneumatic sensitivity (dp/dM). It can be proved to be equal to $-\,b/C \times P = -\,0.40\,(P/M_a)$.

It can, therefore, be raised by raising the supply pressure or by decreasing the control orifice size C or by keeping value of M_a, *i.e.* mean separation area, small.

(*iv*) Indicator sensitivity (dR/dp) can be assumed to be constant if linear measuring device is used.

(*b*) From the equation $(dR/dL) = 0.40\ (\lambda/L_a)$ it is obvious that overall magnification and the range of measurement are inversely proportional and hence suitable compromise for a particular application is required.

(*c*) The advantages of differential type pneumatic comparator over ordinary pneumatic comparator are :

(*i*) the small variations in supply pressure are compensated for by the differential-pressure measurement.

(*ii*) the differential pressure can be zeroed, using the adjusting valve, corresponding to given mean size.

(*iii*) full range of scale of measuring device can be used.

Problem 5.4. *How do the measuring head sensitivity, pneumatic sensitivity and overall sensitivity vary with variation of diameter of measurement orifice.*

Solution. Measuring head sensitivity $= \dfrac{dM}{dL}$.

Since $M = \pi DL$ (D = dia. of measurement orifice), $\therefore \ \dfrac{dM}{dL} = \pi D.$

$\therefore$ Measuring head sensitivity varies directly in proportion of diameter of measuring orifice.

Pneumatic sensitivity $= \dfrac{dp}{dM}$.

Differentiating the equation $\dfrac{p}{P} = A - b\dfrac{M}{C}$ w.r.t. M, we have $\dfrac{dp}{dM} = -\dfrac{b}{C}\,.\,P.$

For average separation L_a, $M_a = \pi DLa$, and $\dfrac{p}{P} = 0.7$

$$\therefore \qquad 0.7 = 1.10 - b\,.\,\frac{M_a}{C}, \text{ or } \frac{b}{C} = \frac{0.40}{M_a}$$

$$\therefore \qquad \frac{dp}{dM} = -\frac{b}{C}P = 0.40\,\frac{P}{M_a} = -\,0.40\,\frac{P}{\pi DL_a}$$

Also for mid-point, *i.e.* $\dfrac{p}{P} = 0.7$

$$0.7 = 1.1 - b\,\frac{\pi DL_a}{\frac{\pi}{4}\times d^2} \quad (d = \text{diameter of control orifice})$$

or
$$L_a = -\,0.4\left(\frac{d^2}{4bD}\right) = \frac{K}{D}\,;\ K = -\,\frac{0.4d^2}{4b} = \text{constant}$$

$\therefore$ Equation $\dfrac{dp}{dM} = -\,0.40\,\dfrac{P}{\pi DL_a}$ becomes $\dfrac{dp}{dM} = -\,0.40\,\dfrac{P}{\pi D\dfrac{K}{D}} = -\,\dfrac{0.40P}{\pi K}$

i.e. pneumatic sensitivity is independent of D (measurement orifice diameter).

Since gauge sensitivity has nothing to do with D.

$\therefore$ Overall sensitivity is also proportional to diameter of measurement orifice.

Problem 5.5. *The operation of a pneumatic comparator is represented by the equation*

$$\frac{p}{P} = 1.10 - b\frac{M}{C} \text{ for } 0.6 > \frac{p}{P} > 0.8$$

where P = *supply pressure,* p = *pressure between measuring and control orifices*

b = *constant* = *0.5 for* P *of* $2\ kg/cm^2$; $M = \pi DL$

L = *separation between the nozzle surface and the surface to be gauged*

D = *measurement orifice diameter ;* d = *control orifice diameter*

C = *control orifice area* $\frac{\pi}{4}d^2$

The control orifice is 0.5 mm diameter, and the measuring orifice is 1 mm diameter hole.

Calculate (a) the range of linear measurement (b) measuring head, pneumatic gauge and overall sensitivities.

The back pressure gauge has a deflection of 20 mm for 0.01 kg/cm^2 *pressure change. Air supply pressure is constant at 2* kg/cm^2 *(gauge).*

Solution. (*a*) For $\frac{p}{P} = 0.6$, *i.e.* corresponding to maximum separation,

the equation $\frac{p}{P} = 1.10 - b\frac{M}{C}$ becomes $0.6 = 1.10 - 0.5 \times \dfrac{\pi D l_a}{\frac{\pi}{4}d^2}$

where L_x = maximum separation

$$0.6 = 1.10 - 0.5 \times \frac{\pi \times 1.0 \times L_a}{\frac{\pi}{4} \times (0.5)^2} \quad \text{or} \quad -0.5 = -0.5\left(\frac{L_x \times 4}{0.25}\right)$$

or $$L_x = \frac{1}{16} \text{ mm} = 0.062 \text{ mm}$$

For $$\frac{p}{P} = 0.8, \textit{ i.e. } \text{corresponding to minimum separation } L_n$$

$$0.8 = 1.10 - 0.5 \times \frac{\pi \times 1.0 \times L_n}{\frac{\pi}{4}(0.5)^2}$$

or $$-0.3 = -0.5 \times 16L_n \quad \text{or} \quad L_n = \frac{0.3}{0.5 \times 16} = 0.037 \text{ mm.}$$

$\therefore$ Linear range $= L_x - L_n = 0.062 - 0.037 = 0.025$ mm.

(*b*) Measuring head sensitivity $\frac{dM}{dL} = \pi D = 3.14 \times 1.0 \text{ mm} = 3.14$ mm.

Pneumatic sensitivity $\frac{dp}{dM} = 0.4\frac{P}{M_a}$

$$P = 2 \text{ kg/cm}^3 = 200 \text{ kg/mm}^2.$$

$$M_a = \pi DL_a = \pi D\frac{L_x + L_n}{2} = \pi \times 1.0 \times \frac{0.062 + 0.037}{2} = \pi \times 0.05 \text{ mm}^2$$

$$\frac{dp}{dM} = 0.4 \times \frac{200}{0.05\pi} = \frac{1604}{\pi} = 509 \text{ kg/mm}^4.$$

Gauge sensitivity = $\frac{dR}{dp}$, dR = 30 mm for dp of 0.01 kg/cm^2 = 1 kg/mm^2

$$\frac{dR}{dp} = \frac{20}{1} = 20 \text{ mm/kg/mm}^2$$

Overall Sensitivity $\frac{dR}{dL} = \frac{dR}{dp} = \frac{dp}{dM} \times \frac{dM}{dL} = 20 \times 509 \times 3.14 = 32{,}000.$

Problem 5.6. *Calculate the range and overall sensitivity of the pneumatic comparator in Problem 5.3, if the control orifice diameter is changed to 1.0 mm and all other factors remain same. What do you conclude from this problem ?*

Solution. For $\frac{p}{P} = 0.6,\ 0.6 = 1.10 - 0.5 \times \frac{\pi D L_a}{\frac{\pi}{4} \times d^2}$

$$= 1.10 - 0.5 \times \frac{\pi \times 1.10 \times L_a}{\frac{\pi}{4} \times 1^2}, \quad \text{or} \quad -0.5 = -0.5 \times 4\, L_x$$

and $\qquad L_x = \frac{1}{4}$ mm = 0.25 mm

Similarly for $\qquad \frac{p}{P} = 0.8,\ 0.8 = 1.10 - 0.5 \times \frac{\pi \times 1.0 \times L_n}{\frac{\pi}{4} \times 1^2}$

$$-0.3 = 0.5 \times 4\, L_n, \text{ and } L_n = \frac{3}{20} = 0.15 \text{ mm}$$

∴ Linear range = 0.25 – 0.15 = 0.1 mm

Measuring head sensitivity = πD = 3.14 mm

Pneumatic sensitivity $= 0.4 \times \frac{200}{\pi \times 1.0 \times (0.25 + 0.15)} = 122$ kg/mm^4

Gauge sensitivity is independent of size of control orifice.

∴ Overall sensitivity = 20 × 122 × 3.14 = 8000

From above problem, it is seen that linear range of measurement is increased by increasing size of control orifice, but pneumatic sensitivity and overall sensitivity are decreased.

Problem 5.7. *Calculate the range and overall sensitivity of the pneumatic comparator in problem 5.5, if the measuring orifice diameter is changed to 0.5 mm, all other factors remaining same.*

What do you conclude from this problem ?

Solution. For $\qquad \frac{p}{P} = 0.6,\ 0.6 = 1.10 - 0.5 \times \frac{\pi D L_x}{\frac{\pi}{4} d^2}$

or $\qquad -0.5 = -0.5 \times \frac{\pi \times 0.5 \times L_n}{\frac{\pi}{4} \times (0.5)^2}, \quad \text{or} \quad L_x = \frac{1}{8} = 0.125 \text{ mm}$

For $\qquad \frac{p}{P} = 0.8,\ 0.8 = 1.10 - 0.5 \times \frac{\pi \times 0.5 \times L_n}{\frac{\pi}{4} \times (0.25)^2}$

$$L_n = \frac{0.3}{0.5 \times 8} = 0.075 \text{ mm.}$$

$\therefore$ Linear range $= 0.125 - 0.075 = 0.05$ mm

Measuring head sensitivity $\frac{dM}{DL} = \pi D = 3.14 \times 0.5 = 1.57$ mm.

Pneumatic sensitivity $\frac{dp}{dM} = 0.4 \frac{P}{M_a}$

$$M_a = \pi \times D \times \frac{L_x + L_n}{2} = \pi \times 0.5 \times \frac{0.125 + 0.075}{2} = \frac{\pi}{2} \times \frac{0.2}{2} = \frac{0.314}{2} = 0.157$$

$$\therefore \quad \frac{dp}{dM} = 0.4 \times \frac{200}{0.157} \simeq 509 \text{ kg/mm}^4$$

Gauge sensitivity $= 20$ mm/kg/mm^2

$\therefore$ Overall sensitivity $= 20 \times 509 \times 1.57 = 16{,}000$.

From above example, it is concluded that reducing the measuring orifice diameter increases the measurement range but lowers down the overall sensitivity, whereas pneumatic sensitivity remains unaffected.

Problem 5.8. *Devise an air-jet sensing arrangement to determine :*

(*a*) *Ovality of shaft,* (*b*) *Ovality of hole,*

(*c*) *Ovality and other irregularities of hole,* (*d*) *3-point lobing of shaft,*

(*e*) *Centre distance of 2 holes,* (*f*) *Squareness of two faces.*

Solution. Fig. 5.60 (*a*) to (*f*) show all the above cases of measurements. In these figures symbol → indicates jet and its actual details are shown in Fig. 5.60 (*g*).

In Fig. 5.60 (*a*), the gauging head is first applied over an accurate setting plug and then over the shaft to be measured. The ovality is detected by rotating the shaft. [Refer Fig. 5.60 (*b*)]

Similarly the diameter of hole is compared with that of an accurate setting ring and ovality detected by rotating it. The range of the indicator for a given gauging head can be set by using a pair of standard plugs/rings.

Fig. 5.60 (*c*) shows a gauging head having 2 jets diametrically opposite connected together and forming one circuit and other 2 jets also diametrically opposite but at right angles to earlier pair of jets forming another circuit. The ovality of holes is checked by noting reading between difference of two circuits and other irregularities by rotating the hole.

In Fig. 5.60 (*d*) 3-point lobing of shaft is detected by rotating the shaft in gauging head having three jets spaced 120° apart.

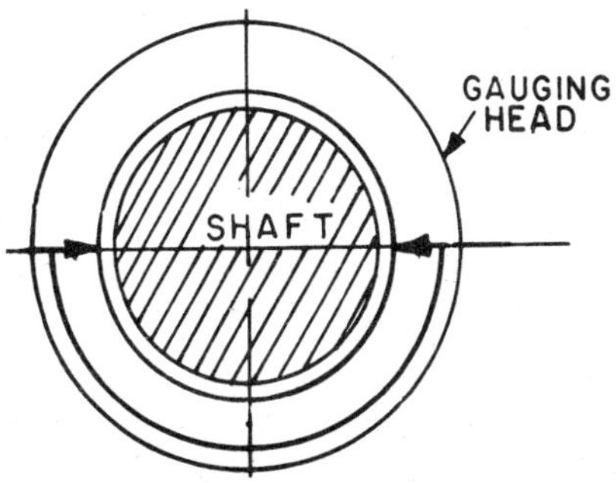

(*a*) Ovality of shaft is detected by rotating the shaft.

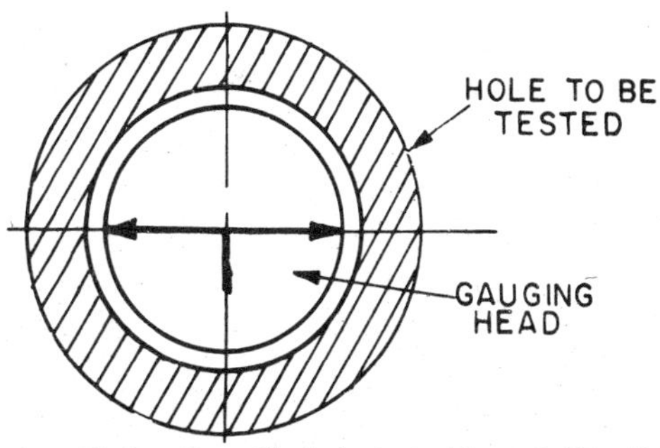

(*b*) Ovality of hole is tested by rotating it.

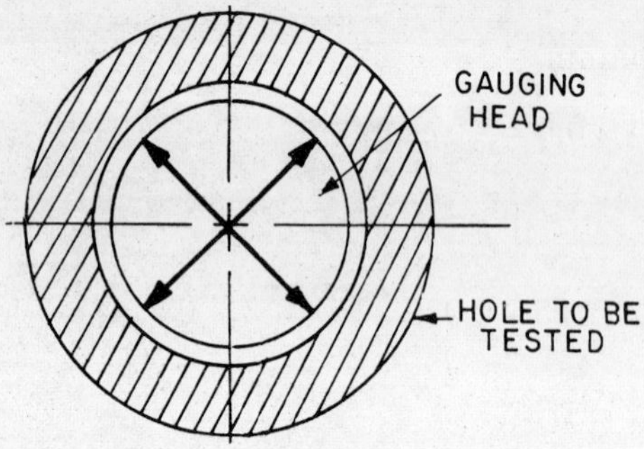

(c) Testing ovality of different indications on 2 circuits and other irregularities by rotation.

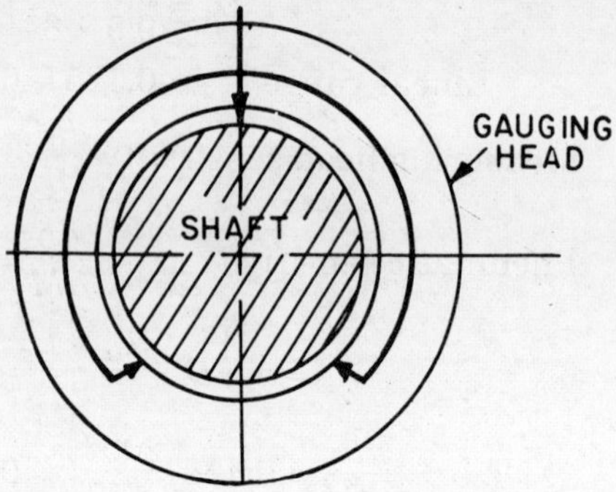

(d) Testing 3-point lobing by rotating the shaft.

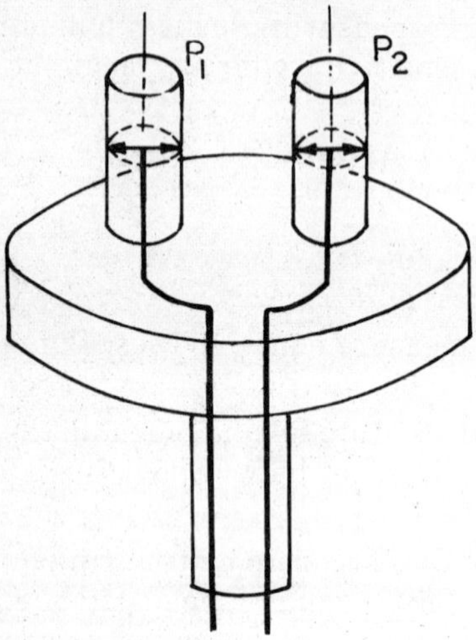

(e) Gauging head to test centre distance of 2 holes.

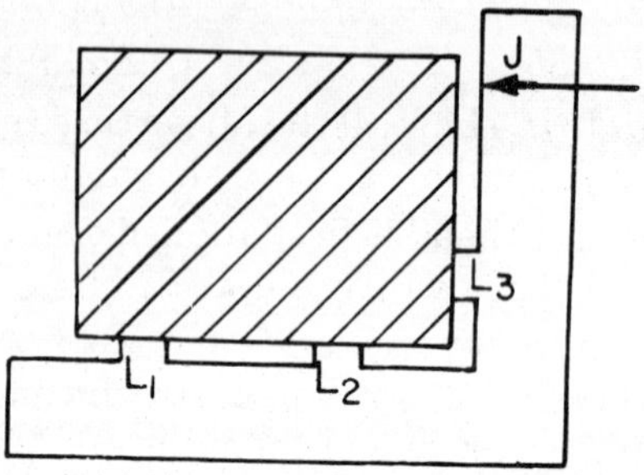

(f)

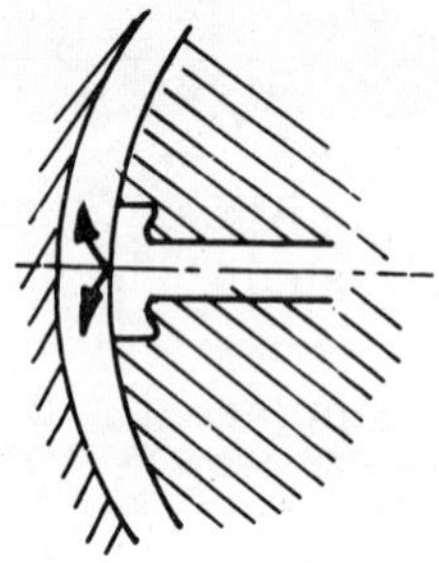

(g) Details of jet.

Fig. 5.60. Testing of squareness.

Fig. 5.60 (e) shows a double air-plug gauge with the plugs P_1 and P_2 at nominally the same centre distance as the holes to be measured. Measurements are made by comparison with standard reference.

Fig. 5.60 (f) shows a fixture for testing squareness. It locates specimen at points L_1, L_2 and L_3 and at J the air jet senses the distance of face depending on squareness. Readings are taken by comparison with standard reference.

Problem 5.9. (a) *Devise a set up with pneumatic jets to determine the error of angles on tapered components having small included angle and a reference vertical surface for location purposes. Determine the expression for such a comparator.*

(b) *If 50 mm movement of indication is obtained corresponding to a difference of 0.01 mm in diameter between the component and the master gauge used in setting the system at diameters d_1 and d_2, determine the angle errcr for a component producing a deflection of 100 mm on the comparator and the distance between jets along tapered surface being 25 mm.*

Solution. (*a*) The dark arrows on gauging heads show the location of measuring jets located distance l apart on tapered surface (Refer Fig. 5.61). Let θ be included angle on gauging head and θ' on the job. Let diameters on the job corresponding to location of jets be d_1 and d_2. When $d_2 - d_1$ is the same as on gauging head, then δh, *i.e.* magnified error on manometers corresponding to actual error $\delta(d_2 - d_1)$ is zero.

$$\text{Overall magnification or sensitivity } (S) = \frac{\delta h}{\delta(d_2 - d_1)}$$

Since value of θ is small, difference between the nominal diameter and the actual diameter $(\delta d_2 - \delta d_1)$ is very nearly the same as the radial difference $\delta d_2' - \delta d_1'$

$$\frac{\delta\theta}{2} = \frac{\delta(d_2' - d_1')}{2} \simeq \frac{\delta(d_2 - d_1)}{2l}$$

or
$$\delta\theta = \frac{\delta(d_2 - d_1)}{2} = \frac{\delta h}{\delta l} \text{ ;or } S = \frac{\delta h}{l \times \delta\theta}.$$

(*b*) Since 50 mm movement of manometer corresponds to $\delta(d_2 - d_1)$ or 0.01 mm.

$$\therefore \quad S = \frac{50}{0.01} = 5000.$$

$$\text{Angle error for 100 mm movement will be } \delta\theta = \frac{\delta h}{Sl} \; (l = 25 \text{ mm})$$

$$= \frac{110}{5000 \times 25} = \frac{1}{1250} \text{ radians} = \frac{1}{1250} \times \frac{180}{\pi} \times 60 \text{ minutes} = 2.75 \text{ minutes.}$$

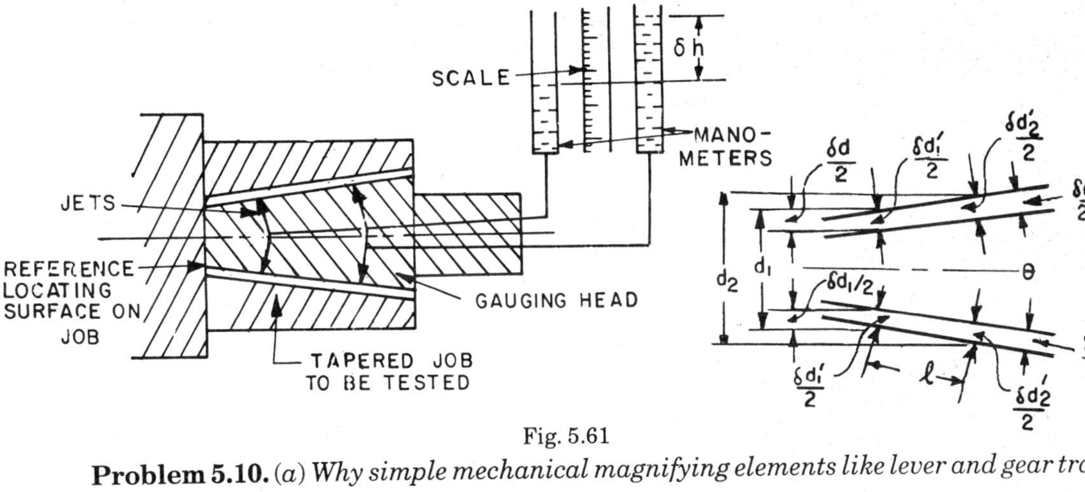

Fig. 5.61

Problem 5.10. (*a*) *Why simple mechanical magnifying elements like lever and gear train are avoided in comparators ?*

(*b*) *Describe three devices (elastic members) for amplifying linear movement in which the bearings and the disadvantages of friction are altogether eliminated.*

(*c*) *Explain by clear sketch the working of 'Reed' comparator utilising parallel strips (reeds).*

Solution. (*a*) Mechanical magnifying elements like lever and gear train suffer from the disadvantages of friction and backlash. The main difficulty with friction is its uncertainty, *i.e.* the value of the force or torque required to overcome initial static friction is not constant but varies within unknown zone depending on so many unknown factors (Refer Fig. 5.62). Both static as well as kinetic frictions create problems. In order to overcome static friction, a certain force or torque has to be applied before motion commences in either direction and this results in the dead zone of instrument. In other words, for a pointer movement *OP* corresponding to applied force *OC* (Refer Fig. 5.63), the input may be varied (reduced to *OA* or increased to *OB*) without changing pointer movement from *OP*. Obviously $AC = BC$ represents the torque to overcome static friction and results in what is known as dead zone of instrument.

Once the motion of pointer starts, *i.e.* the static friction is overcome, the kinetic friction acts and though of lower magnitude, but still opposes the motion and as a result pointer comes to rest at a smaller value than the actual value till some tapping is done or vibration introduced.

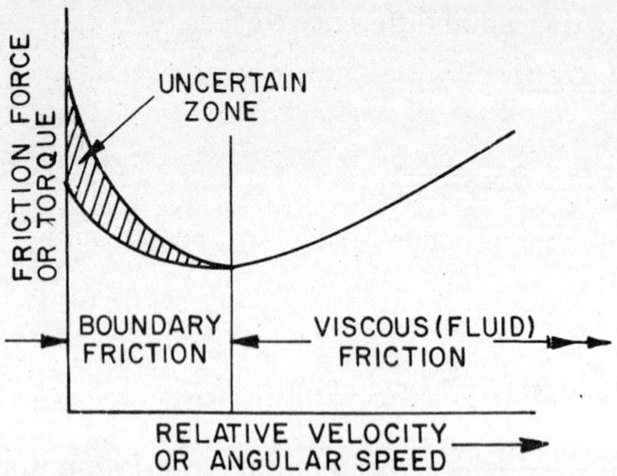

Fig. 5.62

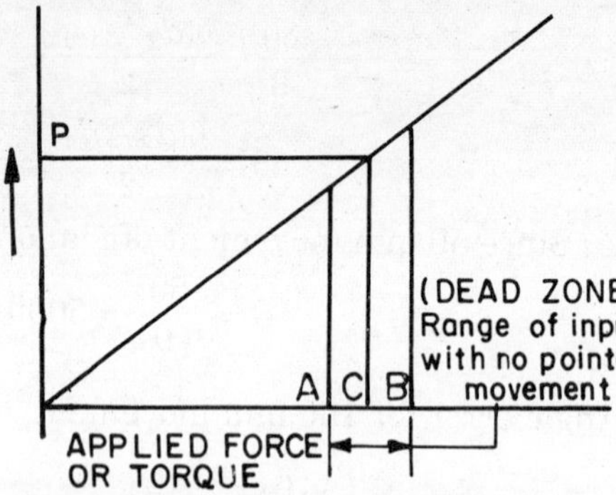

Fig. 5.63

Backlash in gear train can be taken care of by mounting torsion spring, otherwise the backlash may produce a difference of indicated value according to whether the value is approached by increasing or decreasing values.

(*b*) The three devices for amplification purposes commonly used in metrology work to eliminate friction altogether are :

(*i*) Pointer mounted on a thin rod secured at the ends [Fig. 5.64 (*a*)].

(*ii*) Parallel movement flexible strip hinges [Fig. 5.64 (*b*)].

(*iii*) Cross-strip hinge [Fig. 5.64 (*c*)].

Strip hinges are suitably only for small relative movements but these completely eliminate friction.

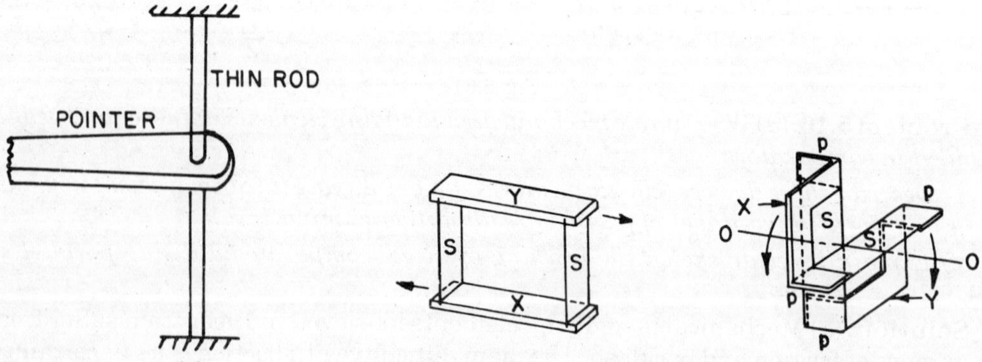

(*a*) (*b*) (*c*)

Fig. 5.64

In first arrangements [Fig. 5.64 (*a*)] ; the pointer is mounted on a thin slender rod fixed at both ends which acts as a torsion spring. Thus the pointer rotates with no bearing friction.

Fig. 5.64 (*b*) shows a parallel movement between two members *X* and *Y* which are joined with two parallel strips *S*. Fig. 5.64 (*c*) shows a cross-strip hinge for relative motion of *X* and *Y*, which take place approximately about the axis of intersection *OO*. It utilises two cross-strip hinges which are connected by strips parallel to *OO* secured at the points *P*. The hysteresis can be minimised by using suitable steel for strips.

(*c*) Fig. 5.65 shows the working of 'Reed' comparator.

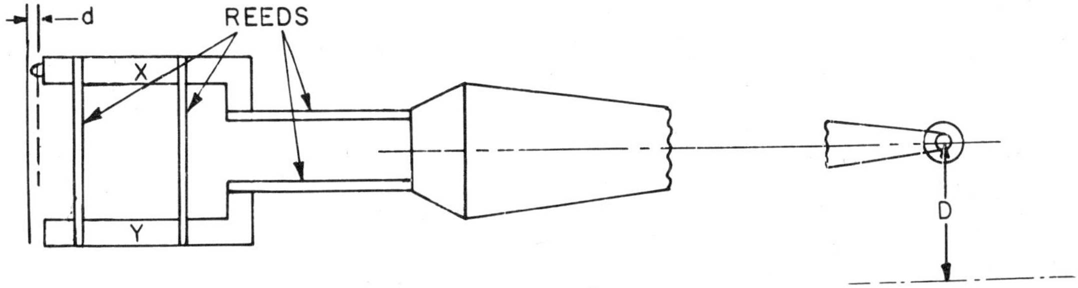

X—Fixed member, Amplification = $\frac{D}{d}$, *Y*–Moving member

Fig. 5.65. Reed Comparator.

Problem 5.11. (*a*) *For measuring diameter of cylindrical object by projection method, whether parallel or diverging or converging beam of light should be used and why ?*

(*b*) *For measuring screw threads by projection method, what type of light beams should be used and how they should be arranged ?*

Solution. (*a*) When measuring diameter of cylindrical object by projecting its image on a projector, it is essential that the true diameter be illuminated on screen which is possible only by using parallel light. If converging or diverging beam of light is used then instead of true diameter being illuminated, only the chord of the cylinder will be projected and true idea of diameter can't be had. It is also essential that light source should be as small as possible so that only a narrow angle of parallel rays is projected through the collimating lens. Screen of projector must be perpendicular to the final axis of the projected beam.

(*b*) For measuring screw threads by projection, the collimated beam must be raked to the mean helix angles of thread so as to project the highest points on the diameter in the axial plane. For moderate helix angles, the thread may be inclined to the parallel beam of light by the helix angle of the thread.

Problem 5.12. *Why the frequency of input supply to electrical transducer like linear variable differential transducer (LVDT) is made very high and not 50 cycles / sec.*

Solution. If frequency of input to LVTD is selected as 50 cycles/sec, there is possibility of picking up stray signals (noise) from other sources operating at mains frequency of 50 cycles/sec. However, with very high frequency, the stray signals at 50 cycles/sec from other sources will not affect the output of LVDT. Further with the use of high frequency supply to LVDT, the low frequency vibrations of the core do not have any effect on the measurement.

Problem 5.13. *Enumerate the characteristic advantages of mechanical indicators.*

Solution. The characteristics advantages of mechanical indicators, in general, are.

— These usually have long measuring range. The design permits measurements in small increments over quite substantial distances.

— Overall size of indicator is small (inclusive of sensing, amplifying and indicating elements).
— Possibility of selecting suitable instrument having desired contact force. Also flexibility of varying and adjustment the measuring force over substantial range.
— Usually rugged (their relative shock resistance under vibrations, and less sensitive to inadvertently caused over travel).
— No need of power supply and hence no effect of power supply fluctuations.
— Instrument errors due to mechanical causes can be detected and compensated.
— Initial cost is low and also maintenance is less. Repair cost is also low.

Problem 5.14. *The general form of the contract face in dial indicators is a spherical segment with around 5 mm radius. Explain why ?*

Solution. The spherical form of the faces provides essentially single point less sensitive to minor misalignment in the mutual position of the work and of the indicator.

Problem 5.15. *Enumerate the advantages of pneumatic gauging ?*

Solution.

— Being non-contact type due to provision of clearance between object surface and sensing member, ideally suited for soft as well as abrasive surfaces and where finish is not to be marred.
— Frictional forces are not encountered due to elimination of sliding members. Total elimination of wear and thus high useful life and original accuracy of gauge maintained for long time.
— Self-cleaning of size sensing area due to open air jet.
— Sensing members have small size and thus hard to reach surfaces can be gauged.
— Small clearances between air plug and bore wall, combined with effect of multiple jets, practically eliminate the effect of alignment and centerising errors and permit gauging by unskilled persons.
— Amplification from 250 to 20,000 times can be varied.
— Indicating part can be away from sensing point.
— Indication is continuous.
— It is possible to achieve average readings.
— Gauging system permits mounting of several indicators adjacent to each other for quick scanning of multiple dimensions.
— Initiating contacts (adjustable) can be used to indicate light or start/stop other devices.
— Initial cost is relatively low. Maintenance is less. Useful life of gauge is high.

Problem 5.16. *Explain the difference between flow type and back pressure type pneumatic comparators.*

Solution. Whenever pressurised air while escaping through an orifice of controlled size, is impeded in its free flow by impinging against a solid surface, the distance of that surface from the air escape opening will affect (*a*) air flow velocity and (*b*) create back pressure upstream of the orifice.

Within a specific dimensional range (dependent on design characteristics of the comparator), variations in flow and back pressure are almost proportional to the change in distance between orifice and the obstruction. Thus use can be made of either measuring flow or back pressure to estimate desired dimension.

The flow type gauge measures the quantity of air escaping from the measuring orifice. Measurement is possible either by a rotameter (flow type) or by a venturimeter (velocity type).

Back pressure type gauge measures the back pressure created by obstruction to flow of air and can be measured either by the level of liquid in graduated tube or by a Bourdon type pressure gauge.

Problem 5.17. *Under what conditions, the use of indirect gauging with mechanical contact elements is recommended over the use of direct air gauging with open jets ?*

Solution. In following cases, the direct air gauging will not give desired results and these can be gauged by indirect gauging technique.

(*i*) In case of rough surface where envelope size is to be measured in view of assembly requirements. On the other hand open jet air gauge will measure both crests and valleys differently and provide average dimension which is not desirable where the fit relationship between mating parts is the criterion.

(*ii*) In case of interrupted surfaces such as splines, gear teeth, threaded surfaces etc., contact type gauging is more meaningful.

(*iii*) In case of measurement at the edge of a hole, open jet can't be used because part of the surface against which the air stream must impinge is non-existant.

(*iv*) In case of porous substances (like powder metallurgy products where pores are undetected by unaided eye) false readings can be obtained with open jet.

(*v*) Gauge heads which combine the positive contact and a right angle transfer motion enable use contact type air gauging to be used for surface plate based measurements.

Problem 5.18. *It is possible to have either a single nozzle in an air plug or multi-nozzles. Discuss the applications for requirement of different nozzles.*

Solution. Single-nozzle type air plug is preferred to measure the straightness of a bore in relation to a datum which is established by the plug body. Single jet is also useful to inspect relative positions with regard to location or geometric relationship (like concentricity, location, squareness, flatness, straightness, depth, length, etc. Two-nozzle gauge plugs are used to check the diameter of hole for size as well as for consistency. These are used to check out-of-round, bell-mouth and taper also.

Three jet design is particularly sensitive to three lobed conditions of the bore, a form irregularity which passes undetected when relying on diametrical measurements only.

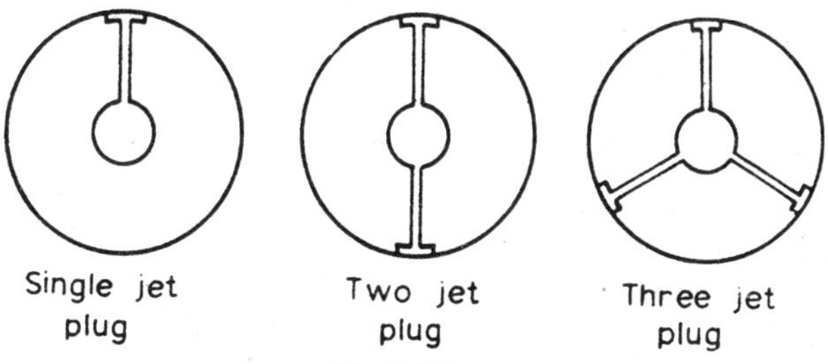

Fig. 5.66

The four-and six-nozzle designs of air plugs have an excellent averaging capability and are used for measuring bores whose mean size is considered to be desirable dimension.

Problem 5.19. *What are the essential requirements for comparative gauging ?*

Solution. For any comparative gauging, with any system of measuring instruments, following requirements must be satisfied.

(*i*) The variations in length must occur on a physical surface which is accessible to a mechanical sensing element.

(*ii*) The gauge must provide a datum plane which serves both the setting of the reference position of the sensing element and the locating of the object.

(*iii*) The direction of the size variations to be measured must coincide with the free movement of the sensing element whose displacement will represent the size variations being measured.

Problem 5.20. *Enumerate the advantages of electronic gauges.*

Solution. The various advantages of electronic gauges are :

(*i*) High resolution (sensitivity) *i.e.* these gauges react to a very small input due to practically frictionless suspension of the sensing member.

(*ii*) Immediate response and practically free from lag or hysteresis.

(*iii*) High repeatability due to elimination of backlash and play.

(*iv*) High calibration accuracy.

(*v*) High and variable amplification of signal.

(*vi*) Gauging force is light which greatly reduces the amount of deflections on the contacting surface.

(*vii*) Instantaneous response.

(*viii*) Gauge indications can be displayed by various means—meter, recorder, oscilloscope, digital meter, limit signals, etc.

(*ix*) Remote location of control unit and flexibility of adaptation.

(*x*) Signal, in addition to display, can be used for initiating alarm, stopping machine, or controlling various associated functions.

(*xi*) Operational convenience due to easy zero position adjustment and switching to different amplifications.

(*xii*) Readjustment by a single screw to restore original accuracy.

(*xiii*) High MTBF (mean time between failures) and low MTTR (Mean time to repair).

Problem 5.21. *List some of the commonly used electro-mechanical transducers used for electronic gauges.*

Solution. (*a*) Change of electrical capacitance-one metal plate is fixed and other (the reed) is free to move by the displacement of gauge contact tip.

(*b*) Strain gauges (variation in resistance of wire due to straining)

(*c*) Changes in electrical resistance caused by variations in the position of movable contact.

(*d*) Changes in self-inductance or mutual inductance.

(*e*) Voltage produced by bending a piezoelectric crystal.

(*f*) Eddy current probes.

(*g*) Electro mechanical transducer like linear variable differential transformer (LVDT).

Problem 5.22. *What are the functions of amplifier (control unit) of an electronic gauge ?*

Solution. While the basic sensor is installed in workshop, the control unit is located remotely and is connected to the gauge head by a cable. It performs the following functions :

(*a*) Amplify the relatively weak signal generated by transducer in response to the second mechanical displacement.

(*b*) Providing excitation current for the sensor gauge head.

(*c*) Demodulate the amplified signal for display purposes.

(*d*) Change of amplification, zeroing, gain adjustment for calibration.

(*e*) Signal display on a meter.

Problem 5.23. *For which cases the optical measurements are preferred ?*

Solution. For following applications, optical measurements are preferred :

(*i*) Measuring condition created by the particular geometric forms.

(*ii*) Insufficiently defined boundaries of object features.

(*iii*) Inaccessible location of the dimension to be measured.

(*iv*) Functional interrelationship of critical dimensions.

(*v*) Finish too sensitive for mechanical contact.

(*vi*) Desirability of visual observation for alignment of for contour comparison.

TYPICAL EXAMINATION QUESTIONS

1. What is a comparator ? Classify the different types of comparators.
   Describe mechanical comparator and clearly explain the magnification method adopted in it.
2. What are salient features of a comparator and how are they achieved in the 'Sigma comparator' ? Explain your answer with sketches.
3. Explain the difference between a comparator and measuring instrument. State the fields of application of comparators.
   Explain with a neat sketch any type of comparator you know and show clearly the advantages of magnification adopted.
4. Explain how a pneumatic comparator works and briefly enumerate the advantages of differential pneumatic comparators.
5. How does a comparator differ from a measuring machine ? Explain the principle and operation of optical comparators.
6. Discuss generally the use of comparators and the methods used for magnification.
   What is projection comparator ? Show a sketch to illustrate the principle and give the type of works for which this instrument is specially suitable.
7. What do you understand by damping as referred to instrument ? How the damping is achieved in various mechanical comparators ?
8. Why parallel light must be used when projecting cylindrical objects in a projector ?
9. In a prestwich fluid gauge, the diameter of cone is D and the diameter of tube is d. Show that its magnification approximately is given by $D^2/13d^2$.

6

Measurement by Light-Wave Interference

6.1. Introduction

For understanding the phenomenon associated with interferometry, let us first study the nature of light. According to Huygens Theory, light is considered as wave motion propagated in the ether. The light, therefore, can be considered as an electro-magnetic wave of sinusoidal form. The high point of the wave is called the crest and the low point is called the trough. The distance between two troughs or two crests is called the wavelength λ. In Fig. 6.1, light travels along the *OX*-axis and the time taken in travelling one wavelength λ is called the time period (T). The maximum disturbance of the wave is called the amplitude (A) and velocity of transmission is λ/T, $1/T$ being called the frequency. A ray of ordinary light can be considered as composed of an infinite number of wave lengths, the value of which determines the colour of light. The advantage and peculiar property of the monochromatic light source is that the above characteristics are virtually independent of any ambient conditions such as temperature and pressure etc. Its wavelength has precise value unlike the primary colours which have ill-defined wavelengths, and the monochromatic light such as from Mercury 198 or Krypton 86 are exactly reproducible also.

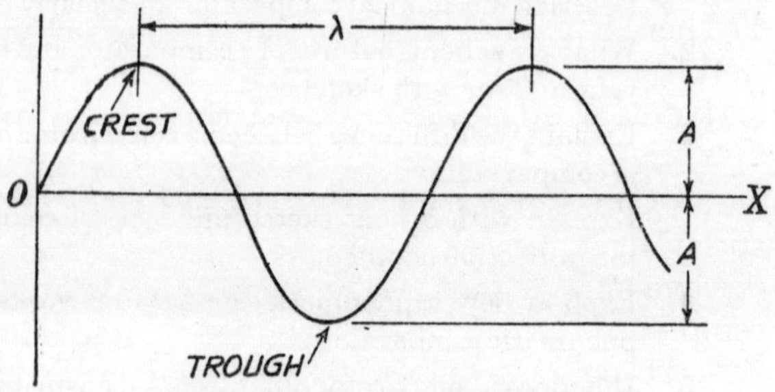

Fig. 6.1. Nature of light.

6.2. Interference of Two Rays

Let us consider the effect of combining two monochromatic rays of light of different intensities having the amplitudes A_1 and A_2. If it were possible to combine them in such a way that the crest of one coincided with the crest of other *i.e.*, they are in same phase, then the resultant effect is the production of maximum illumination. The resultant amplitude of the resulting wave will be equal to $A_1 + A_2$. (Refer Fig. 6.2)

Next if it is possibie to combine them in such a way that the crest of one and trough of other occur simultaneously *i.e.* they are out of phase by λ/2 or 180°, the resultant wave will have an amplitude which is the algebraic difference of A_1 and A_2 and is reduced to minimum (Fig. 6.3). If we have two rays of equal intensity then the resultant wave will have zero amplitude and complete interference will be produced. At this condition, no sensation of light is registered by the eye and the zero amplitude (intensity) of light produces darkness.

Thus for interference to occur, the two conditions are necessary, *i.e.* the light rays are obtained by division from a single source and the rays before being combined at the eye must travel paths whose lengths differ by an odd number of half wavelengths.

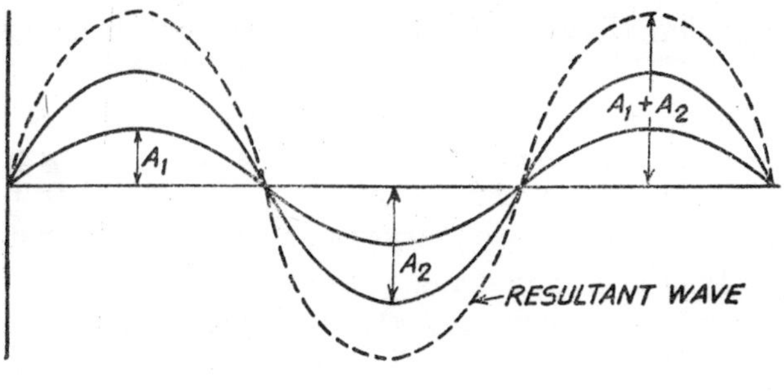

Fig. 6.2. Effect of combining two rays in same phase.

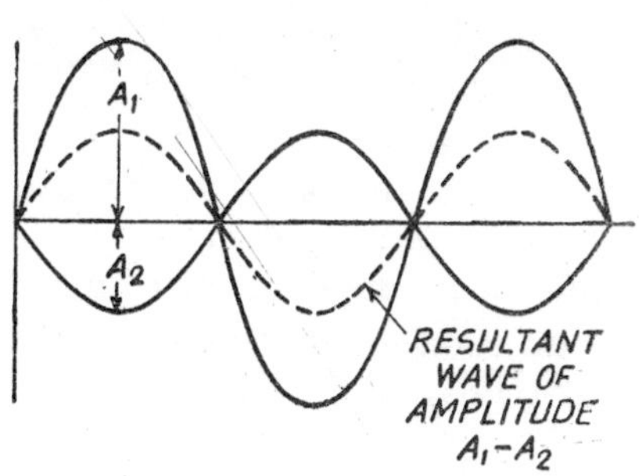

Fig. 6.3. Effect of Combining two rays out of phase by $\lambda/2$.

The interference of two monochromatic rays of light from two point sources X and Y can be demonstrated in the following way (Fig. 6.4). A screen is placed at certain distance from point sources X and Y, the plane of which is parallel to line joining X and Y. There is a point A on the screen which is equidistant from X and Y. The two rays of equal wavelength arrive at A in phase, and will combine to give the maximum illumination at this point.

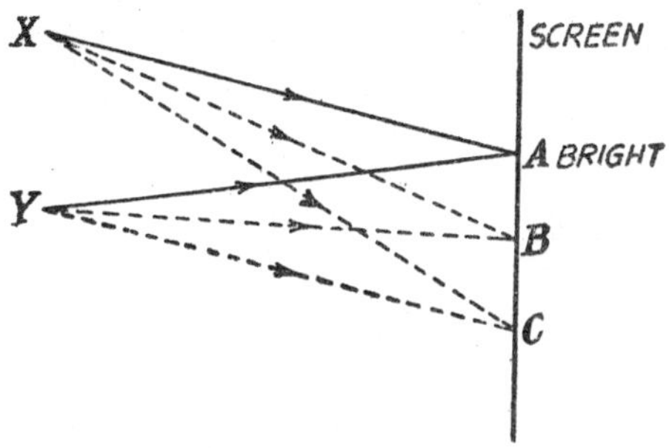

Fig. 6.4. Interference of two rays.

Considering another point B on the screen such that distance $XB—YB$ is equal to an odd number of wavelengths, then two rays will be out of phase and complete darkness will be observed at this point. Again, considering another point C, such that $XC—YC$ equals even number of half wavelengths, then no interference occurs and maximum illumination is observed at this point. Thus considering various points above and below A, alternate dark and bright bands will be observed. It may be stressed again that this phenomenon is possible only when two sources X and Y are from a single source of light.

Thus essential condition for interference of light wave is that light should be from coherent sources *i.e.* the two sources should continuously have light waves of same wavelength and equal intensity and maintain same phase difference. Further two sources should be very narrow and close to each other.

6.3. Light Sources for Interferometry

A wide variety of light sources is available for interferometry work but the selection of proper source for any application depends on the requirements or results to be obtained by interferometer, cost and convenience. For simple applications like testing of surface geometry, where the difference between interfering paths is of the order of a few wavelengths only, a tungsten lamp with a filter, transmitting only a narrow band of wavelength would be adequate.

However, sophisticated applications require the use of light sources such as mercury 198, cadmium, krypton, krypton 86, thallium, sodium, helium, neon and gas lasers. In these sources, the discharge lamp is charged with one particular element and contains means to vaporise them. The atoms of these elements are excited electrically so that they emit radiation at certain discrete wavelengths.

Characteristics of various light source are summarised below :

(*i*) *Mercury*. It is less expensive source having high intensity, and green line can be easily isolated with filters. Since natural mercury contains several isotopes, each isotopes emits light whose wavelength is very slightly different from each other. As a result, natural mercury light source radiates a mixture of wavelengths which can be treated as monochromatic only for short path difference.

(*ii*) *Mercury 198.* It is a pure isotope produced by neutron bombardment of gold. It is considered to be one of the best sources of very sharply defined wavelengths, and fringes are visible with path difference upto 500 mm. Light is emitted when mercury 198 is excited by microwave produced electric field. It is the international secondary standard of wavelength.

(*iii*) *Cadmium.* This is the only natural material producing a spectral line (red) almost completely symmetrical, having useful path difference of about 200 mm. Cadmium 114 is the official secondary international standard of length.

(*iv*) *Krypton.* It has the advantage of being easily excited, so used in some instruments. It is not as monochromatic as krypton 86 because natural krypton is a mixture of isotopes. It can be used upto path difference of 375 mm.

(*v*) *Krypton 86.* Krypton 86 lamp produces spectral lines of different wavelengths and, therefore, a fairly elaborate monochromator is required to separate them. Further its excitation takes place at very-very low temperatures, therefore, this lamp is used only in standardising laboratories. Next to laser, this enables the fringes to be observed with maximum path difference (about 800 mm.)

The orange-red line of krypton 86 isotope, produced under specified conditions, and at a temperature of 63.3°K (temperature of nitrogen triple point), is the new basic international standard of length-metre being defined as exactly 1,650,763.73 wavelengths of this source, measured in vacuum.

(*vi*) *Thallium.* As 95% of its light is emitted at one green wavelength, it can be used over a reasonable path difference without the use of any filter.

(*vii*) *Sodium.* It is used only in applications where interference path difference does not exceed a few hundred wavelengths. Usually yellow sodium light is used which contains two separate but closely spaced lines of equal intensity : and because of this the interference fringes wash out for higher path difference.

(*viii*) *Helium.* Orange line of helium is used where path difference is not great.

(*ix*) *Neon.* As conventional neon lamp has too many closely spaced lines (in red part of the spectrum) and not sharply defined, it does not find many applications. Neon in gas laser, however, has assumed a uniquely important role.

(*x*) *Gas lasers.* In metrology work gas lasers which produce highly monochromatic and intense light (1000 times more intense than others) are used to great advantage, enabling interference fringes to be observed with enormous path differences, upto 100 million wavelengths. (It may be noted that high-power, intermittently operating ruby laser is not of interest in metrology). Gas lasers are produced by exciting (by an electric discharge or a high-frequency field) a mixture of neon and helium stotred in a special discharge tube, and located inside a high efficiency Fabry-Perot etalon or a modified version of it. The wavelength at any given instant covers an extremely short portion of the spectrum, *i.e.*, highly monochromatic.

It has some limitations like high cost ; bigger size ; wavelength affected by temperature and vibration ; special optics required to spread the very small diameter ray having high degree of collimation over a larger area suitable for most instruments. Since the laser contains only a single wavelength, it can't be used for the Pitter-NPL gauge interferometer whose working for measuring length of gauge is based on 3 or 4 different wavelengths. However, it is best suited for length interferometry where fringes are counted, *i.e.* for fringe counting interferometers. (Refer A.C. laser interferometer described later in this chapter.)

Details of various wavelengths of cadmium and mercury in standard air *i.e.* air at 20°C, 760 mm mercury barometric pressure, 10 mm vapour pressure and containing 0.03% of CO by volume, are as given below :

Source of Radiation	*Wavelength (μm) = 0.001 mm*	*No. of half wave lengths per mm*
Cadmium		
Red	0.64485037	3106.31180
Green	0.50858483	3932.48064
Blue	0.47999360	4166.72222
Violet	0.46781743	4275.17205
Mercury-198 (for a cold cathode lamp)		
Yellow I	0.57006642	3455.81733
Yellow II	0.57696298	3466.42691
Green	0.54607826	3662.47871
Violet	0.43583611	4588.88090

6.4 Interferometry Applied to Flatness Testing

The essential equipment for measurement by light wave interference is a monochromatic light source and a set of optical flats. An optical flat is a circular piece of optical glass or fused quartz having its two plane faces flat and parallel and the surfaces are finished to an optical degree of flatness. Optical flats vary in size from 25 mm diameter to about 300 mm diameter. Quartz flats have high wearing qualities and have 200 to 400 per cent great wearing qualities than the optical glass. Quartz has minimum expansion due to heat and the thickness used insures freedom from distortion in use. The importance of optical flats is apparent in lightwave measurement because all inspection operations are performed with reference to the master surfaces of the optical flats. (For more details about optical flats, refer Art. 6.10).

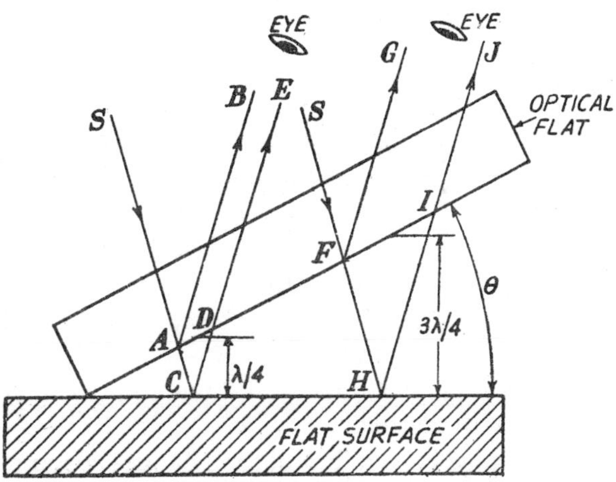

Fig. 6.5. Flatness testing by interferometry.

If an optical flat is placed upon another flat reflecting surface (without pressure) it will not form an intimate contact, but will lie at some angle θ making an inclined plane. If the optical flat be now illuminated by monochromatic source of light, the eye if placed in proper position, will observe a number of bands. These are produced by the interference of the light rays reflected from the lower surface of the top flat and the top surface of the lower flat through the very thin layer of air between the flats. In Fig. 6.5, the angle θ between the two flats is shown exaggerated too much. *S* is the source of monochromatic light. At point *A*, the wave of incident beam from *S* is partially reflected along *AB* and is partially transmitted across the air gap along *AC*. At *C*, again the ray is reflected along *CD* and passes out towards the eye along *CDE*. Thus the two reflected components, reflected at *A* and *C* are collected and recombined by the eye, having travelled paths whose lengths differ by an amount *ACD*.

If the path lengths of the two components differ by an odd number of half wavelengths, then condition for complete interference is achieved. If the surface is perfectly flat, then condition of complete interference is satisfied in a straight line across the surface as the surface at right-angles to the plane of the paper is parallel to the optical flat. Therefore, a straight dark line will be seen passing through point C. Consider another ray incident along path SFH. Again this ray is also splitted into two components. It is obvious that the path difference of the two component rays will keep on increasing along the surface due to angle θ. Thus if the path difference FHI be $3\lambda/2$ or the next odd number of half wavelengths, then interference will occur and a similar fringe will be seen. Next when path difference is $5\lambda/2$, again there will be another dark fringe. At the intermediate point between the points C and H, the path difference will be an even number of half wavelengths and the two components will be in phase producing a light band.

Thus, in case of a perfectly flat surface, we will have pattern of alternate light and dark straight lines on the surface, as shown in Fig. 6.6. Any deviation from this pattern will be a measure of the error in the flatness of the surface being inspected.

6.4.1. Checking the Flatness by above Method. Referring again to Fig. 6.5, if θ is very small, (which is generally the case, as it is made due to some particles between optical flat and surface being tested), then

$$AC = DC = \frac{\lambda}{4}, \quad \text{and } FH = HI = \frac{3\lambda}{4}$$

Fig. 6.6. Interference pattern on perfectly flat surface.

Therefore, the change in separation between the optical flat and the surface between two similar adjacent fringes is

difference between AC and $FH = \frac{3\lambda}{4} - \frac{\lambda}{4} = \frac{\lambda}{2}$.

Thus it is obvious that each adjacent fringe represents a change in elevation of the work surface relative to the optical flat of $\lambda/2$ and total change in elevation from point of contact to the outermost fringe will be $n \times \lambda/2$, if it contains n number of adjacent fringes.

Now there are four possible cases when the contact between the optical flat and the work surface occurs at one point only.

(*i*) If the surfaces are perfectly wrung together, then no air gap exists and no fringe pattern will be observable.

(*ii*) If angle θ is increased, then points C and H will be closer together and fringes are brought closer together.

(*iii*) If angle θ is reduced, then fringes spacing increases as the points C and H will occur at greater distances.

(*iv*) If θ is made too large, then fringes will be closely spaced as to be indistinguishable and no observable pattern will be visible.

The viewing angle, *i.e.* the angle at which the bands are viewed should be 0°, *i.e.* line of sight should be perpendicular to the reference surface of the optical flat. If viewing angle is 30°, then 3 fringes may be seen as 2.5, causing an error of nearly 15%. If it is 45°, 3 fringes mav be read as 1.75, causing error of nearly 40%.

In practice, on the surface to be tested,we have large number of hills and valleys and in these circumstances it is impossible that contact between the surface and the optical flat will occur at one point only. Rather the contact will be made at a large number of hills. If the optical flat is resting on any hill, then it will just behave as if it were placed on some spherically convex surface. In such a case, the contact is made at the central high point and in centre a bright circle will be visible. Around it, there will be concentric dark and bright circular fringes. As the distance from the centre increases, the separation between optical flat and surface keeps on increasing and the fringes become narrow and more closely spaced as shown in Fig. 6.7.

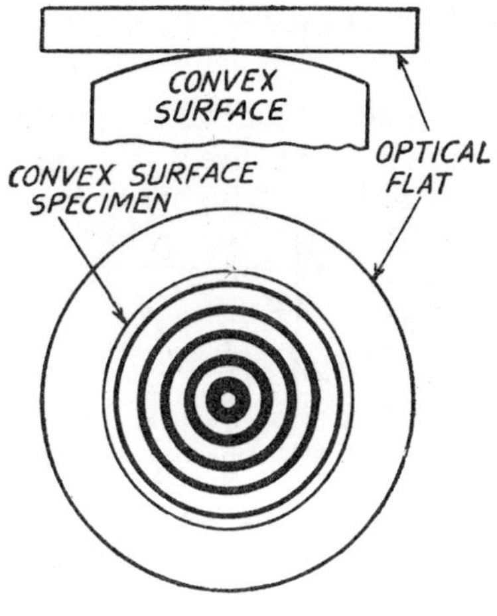

Fig. 6.7. Fringe pattern as observed through optical flat.

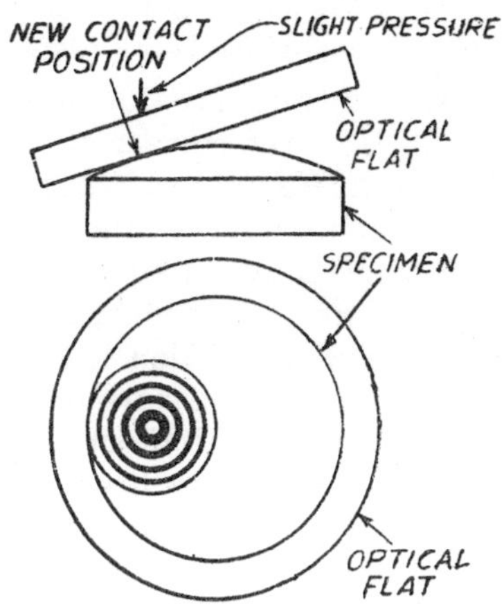

Fig. 6.8. Test for convex surface.

For testing any flat surface, we have to lower down the level of hills. Actually by naked eye, one can't differentiate between the hill and valley in case of fine finished surfaces. And the fringe pattern obtained by concave surface will be same as in case of convex surface.

To distinguish between the two conditions of convex and concave surfaces one edge of the optical flat is lightly pressed. In case of convex surface, by light pressure the optical flat will rock on a new high spot and the centre of the fringe pattern will move as shown in Fig. 6.8. Also the outer fringes will move closer together.

In case of spherically concave surface, the flat is resting on a line passing around the surface and on lightly pressing the edge of the optical flat, the edge line does not move as the pressure is varied. Rather, light pressure at the centre will cause the optical flat to be deflected and will become more nearly parallel to the concave surface, thus reducing the number of fringes observed.

Thus if by light pressure, the centre of fringes is displaced and the fringes are brought closer, it is convex (hill) surface and the level at that place must be lowered down to form a flat surface. If by light pressure the number of fringes is reduced and the fringes move apart, it is (valley) concave surface.

Sometimes surface might not be perfectly convex or concave. So the study of various patterns of fringes produced for different types of surfaces is very essential.

6.5. Surface Contour Test

For studying various surface contours, optical flat is kept at certain inclination over the surface to be tested. Normally daylight is used and spectrum of daylight has a wavelength of approximately 0.0005 mm. Thus each fringe interval corresponds to a change in elevation of the surface of 0.0025 mm. This clearly shows precision of the test. One more important point to be kept in mind is that the contour of each fringe lies on points of equal height relative to the surface of optical flat. Hence the fringe pattern as viewed through optical flat will represent the contour map of the surface under test and spacing of fringes will represent height intervals relative to the optical flat of $\lambda/2$.

In the study of the surface contours it is important to know as to where the optical flat is in contact with the surface being tested. The point or line of contact is usually where pressure is applied causing the air film to be squeezed thin at that place. In mono-chromatic light, the bands are sharper near the point of contact ; and in daylight the point of contact can be seen as the colour of the surface being tested.

Let *XX* be the line of contact in Fig. 6.9. Contour *BAB* shows that all the points on it are at equal height from the surface of the optical flat. Points *A* and *C* are at the centre of two contours *BAB* and *DCD*. It is obvious in Fig. 6.9 that edge at *B* is $\lambda/2$ higher or lower than *C*. The air gap will keep on increasing as we move away from *XX*. As *BAB* represents points at equal height, it means that *B* is actually higher than *C*. This means that edges of this surface are higher and central portion is lower, thus, it is concave surface. Hence the general rule is that if the bands curve around the point or line of contact, the surface is convex, and if the bands curve in opposite direction, the surface is concave (Fig. 6.10). The idea of the magnitude of convexity or concavity can be had by the curvature of the bands. If the curvature of bands is more it indicates more convexity and *vice versa* (Fig. 6.11).

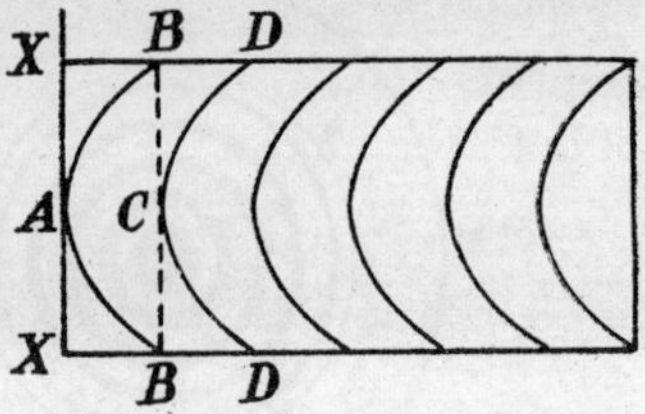

Fig. 6.9. Convex surface.

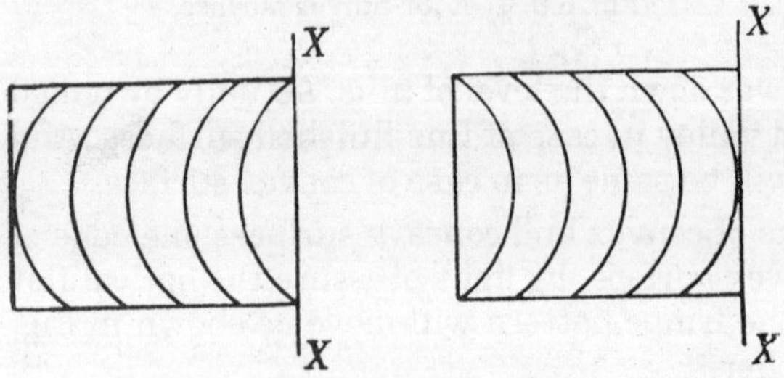

Concave surface   Convex surface

Fig. 6.10

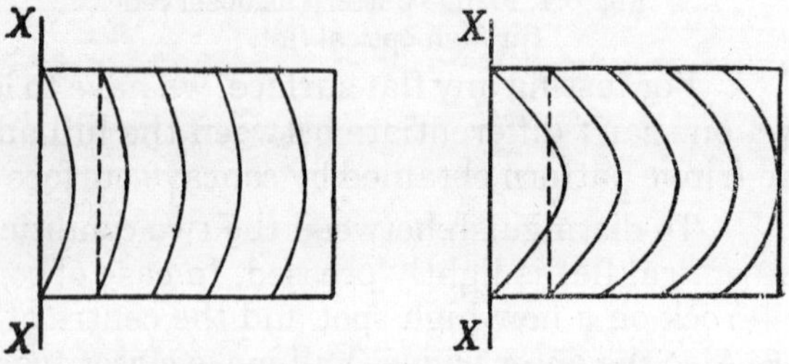

Less convex   More convex

Fig. 6.11

Let us consider any surface having a scratch. The interference dark bands is that case will be as shown in Fig. 6.10. Let *D* be the distance between two fringes and *d* be the distance due to scratch. Then depth of scratch = $\frac{d}{D} \times \frac{\lambda}{2}$. Thus it is possible to determine the depth of scratch also by this method.

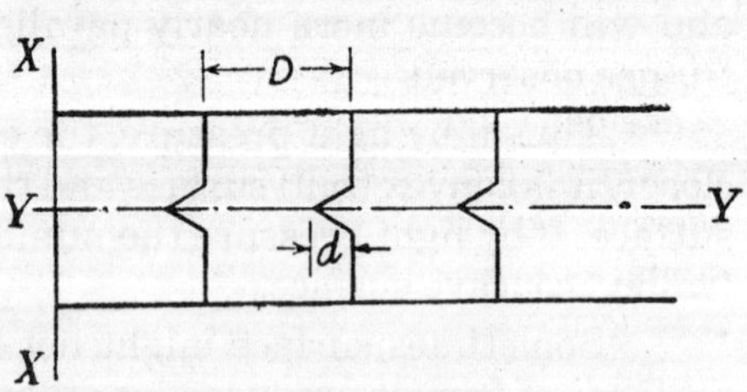

Fig. 6.12. *XX* is line of contact and scratch is along line *YY*.

Fig. 6.13 shows two high or low spots, which can be checked by pressure test, *i.e.* if by applying pressure at the point, the fringes move closer together, surface is valley. If fringes move apart, surface is hill.

Fig. 6.13. Surface with two high/low spots.

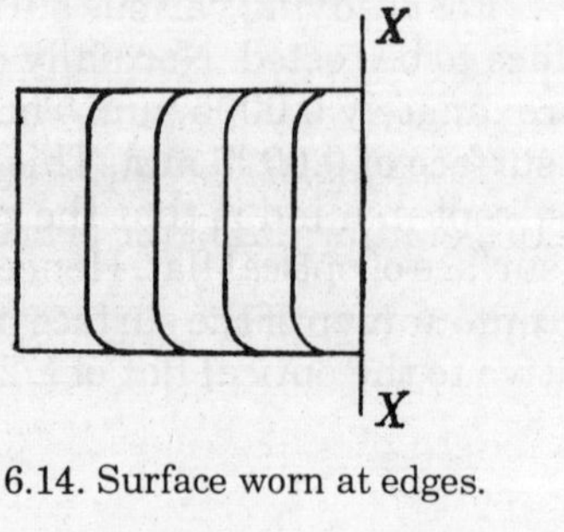

Fig. 6.14. Surface worn at edges.

Fig. 6.14 here shows that surface in the middle is flat and at ends, it has got worn away and has thus become convex. This type of surface is generally encountered in the workshop as the surface tends to round off from the edges due to lapping, wear etc.

Similarly other surfaces can also be predicted if these general rules are followed.

6.6. Testing the parallelism of any surface with reference to a standard optically flat surface by using optical flat

This method is also used for checking height of gauge blocks.

The two blocks, one whose surface is made optically flat and the other whose surface is to be checked are wrung on a flat reference plane surface. An optical flat is then placed on the top of both the blocks. In Fig. 6.15, A is the standard reference gauge and B is the gauge block

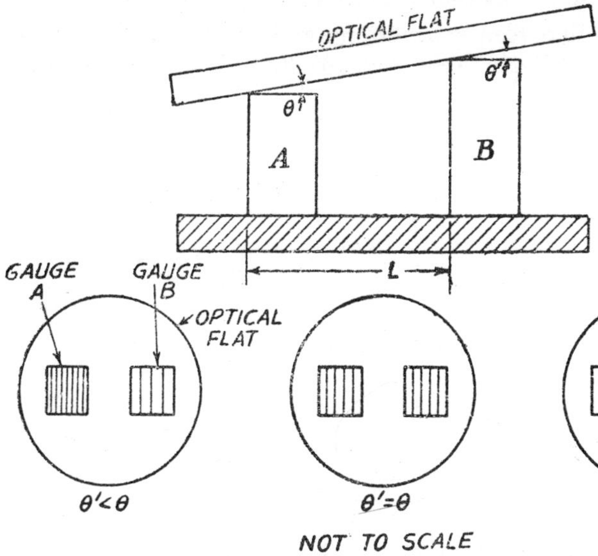

Fig. 6.15. Testing of parallelism.

under test. The distance L between the two gauges is measured very accurately. In case A and B are perfectly flat, but different in height ; then there will be equal wedge shaped air layers between the gauges A and B, and the optical flat and hence equal number of fringes will be seen on both the gauge surfaces. In case the surfaces are not parallel, the angles formed by the surfaces of the two gauges with the optical flat will be different. Let the angles be θ and θ' respectively. If the width of the gauge block l is same for both, then number of fringes per unit width will be different. These possibilities have been shown in Fig. 6.15.

Once it is assured that the surfaces of the two gauges are flat and parallel to each other, then it is possible to check the height of two gauges also. Let the number of fringes on the reference gauge block be N in the width of l units.

The difference in the height of blocks $h = \frac{L}{l} \times N \times \frac{\lambda}{2} = \frac{\lambda L N}{2l}$.

This is however a rough method and meant to explain principle only. For this purpose the use of interferometer is best.

6.7. Interferometers

Interferometers are optical instruments used for measuring flatness and determining the length of slip gauges by direct reference to the wavelength of light. The interferometer incorporates the extension of the application of the optical flat. It overcomes some of the disadvantages of the optical flats used in ordinary daylight or diffused source of monochromatic light by having some refined arrangements. In interferometers the lay of the optical flat can be controlled and the fringes can be oriented to the best advantage. Also there is arrangement to view the fringes directly from top and above the fringes thus avoiding any distortion due to incorrect viewing.

6.7.1. Michelson Interferometer. This is the oldest type of interferometer, which has subsequently been modified in several respects and lot of sophistication introduced. However, Michelson using this interferometer, established exact relationship between meter and red wavelengths of cadmium lamp ; so understanding of its working will be of interest to all.

The basic Michelson interferometer consists of a monochromatic light source, a beam splitter and two mirrors. It relies on the principle of constructive and destructive interference as one mirror remains fixed and the other is moved.

In schematic form, Michelson interferometer is shown in Fig. 6.16, which utilises monochromatic (or single wavelength) light from an extended source. This light falls on a beam

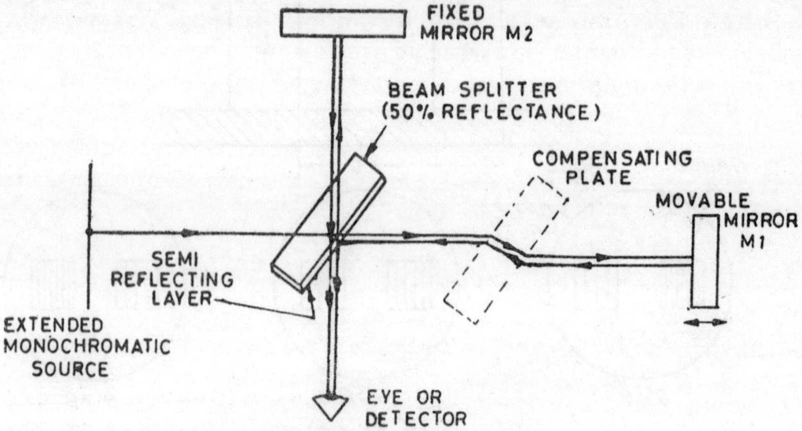

Fig. 6.16. Michelson Interferometer.

splitter (which is a plain parallel plate having a semi-transparent layer of silver at its back) which splits the light into two rays of equal intensity at right angles. One ray is transmitted to Mirror M_1 and other is reflected through beam splitter to Mirror M_2. From both these mirrors, the rays are reflected back and these reunite at the semi-reflecting surface from where they are transmitted to the eye as shown in Fig. 6.16.

Mirror M_2 is fixed and the reflected ray from M_1 serves as reference beam. Mirror M_1 is movable, *i.e.* it is attached to the object whose dimension is to be measured.

If both mirrors are at same distance from beam splitter, then light will arrive in phase and observer will see bright spot due to constructive interference. If movable mirror shifts by quarter wavelength, then beam will return to observer 180° out of phase and darkness will be observed due to destructive interference.

Each half wavelength of mirror travel produces a change in the measured optical path of one wavelength and the reflected beam from the moving mirror shifts through 360° phase change. When the reference beam reflected from the fixed mirror and the beam reflected from the moving mirror rejoin at the beam splitter, they alternately reinforce and cancel each other as the mirror moves. Thus each cycle of intensity at the eye represents $\lambda/2$ of mirror travel.

It may be noted that when monochromatic light source is used then fringes can be seen over a range of path difference that may vary from a few to a million wavelengths, depending on the source. However, when white light is used, then fringes can be seen only if both ray paths are exactly equal to a few wavelengths in total length in glass and air. The lengths themselves are not important, but only their differences affect fringe formation. So when white light source is used then a compensator plate is introduced in the path of mirror M_1 so that exactly the same amount of glass is introduced in each of the paths. (In the path of mirror M_2, the glass was coming due to rays passing through beam splitter back surface). The various sophistications which have undergone to improve the Michelson's basic apparatus are :

(*i*) Use of laser as the light source, which means that the measurements can be made over longer distances ; and also the beam laser compared to other monochromatic sources has exact and pure wavelength thus enabling highly accurate measurements.

(*ii*) Mirrors are replaced by cube-corner reflectors (retro-reflectors) which reflect light parallel to its angle of incidence regardless of retroreflector alignment accuracy.

(*iii*) Instead of observing the interference phenomenon by eye, photocells are employed which convert light-intensity variations in voltage pulses which are processed by electronic instruments to give the amount and direction of position change.

6.7.2. Single Frequency DC Interferometer System. It is much improved system over the Michelson simple interferometer. It uses a single frequency circular polarised laser beam. On reaching the polarising beam splitter, the beam splits into two components, the reflected beam being vertically polarised light and the transmitted beam being horizontally polarised light. These two beams referred to as reference arm and measurement arm respectively travel to their retroreflectors and are then reflected back towards the beam splitter.

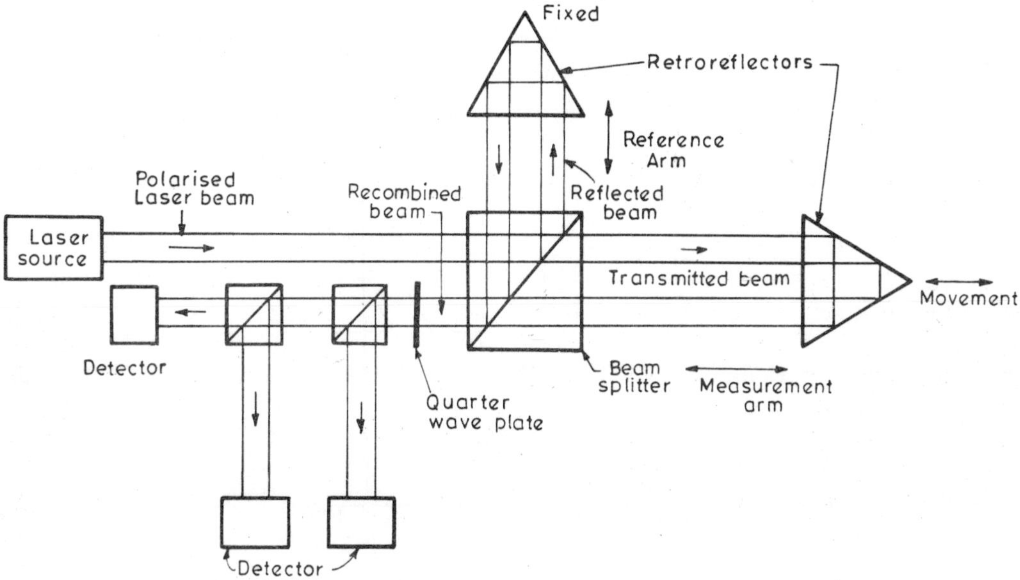

Fig. 6.17. Single frequency DC interferometer.

The recombined beam at beam splitter consists of two superimposed beams of different polarisation ; one component vertically polarised having travelled around reference arm and other component horizontally polarised having travelled around the measurement arm. These two beams being differently polarised do not interfere. The recombined beam then passes through a quarter waveplate which causes the two beams to interfere with one another to produce a beam of plane polarised light. The angular orientation of the plane of this polarised light depends on the phase difference between the light in the two returned beams.

The direction of plane of polarisation spin is dependent on the direction of movement of the moving retroreflector. The beam after quarter waveplate is split into three polarisation sensitive detectors. As the plane of polarised light spins, each detector produces a sinusoidal output wavefrom. The polarisation sensitivity of the detectors can be set so that their outputs have relative phases of 0°, 90°, and 180°. The outputs of three detectors can be used to distinguish the direction of movement and also the distance moved by the moving retroreflector attached to the surface whose displacement is to be measured.

For linear measurements (positional accuracy or velocity), the retroreflector is attached to the body moving along the linear axis. For angular measurement. For pitch and yaw), the angular beam splitter is placed in the path between the laser head and the angular reflector. In this way it is possible to measure flatness, straightness, rotatory axis calibration. Arrangements also need to be made for environmental compensation because the refractive index of the air varies with temperature, pressure and humidity. Heterodyne interferometer, an a.c. device avoids all the problems encountered in above d.c. device, *i.e.* effect of intensity level change of source, fringe contrast changes and d.c. level shifts which can cause fringe miscounting.

Interferometry is now an established and well developed technique for high accuracy and high resolution measurement.

6.7.3. Twyman-Green Specialisation of Michelson Interferometer. In the Michelson interferometer shown in Fig. 6.18, the rays actually describe a cone, giving rise to various types of fringe patterns which may be hard to interpret.

Twyman-Green modified Michelson interferometer utilises a pin-hole source diaphragm and collimating lenses. In this way, all rays are rendered parallel to the central rays and thus all rays describe the same path. All modern two-beam interferometers are based on this arrangement. The mirrors M_1 and M_2 are arranged perpendicular to the optical axis. If mirror M_1 is kept fixed, and M_2 is moved slowly exactly parallel to itself, the observer will note periodic changes in the intensity of the field being viewed, from bright to dark for every $\lambda/2$ movement of the mirror. In fact intensity variation is found to be sinusoidal. It may also be noted that if one of the mirrors is even slightly inclined to the optical axis then parallel fringes will be seen moving parallel to themselves by just one fringe for every $\lambda/2$ (half the wavelength of the light source used) mirror motion. Usually it is quite difficult to count such fringes by eye. However, photo detectors connected to high speed counters can do this job very accurately (accuracy of one part in million being obtainable). It is possible to calibrate the output of counter directly in terms of the linear movement of the mirror M_2, but several conditions must be met to achieve these results.

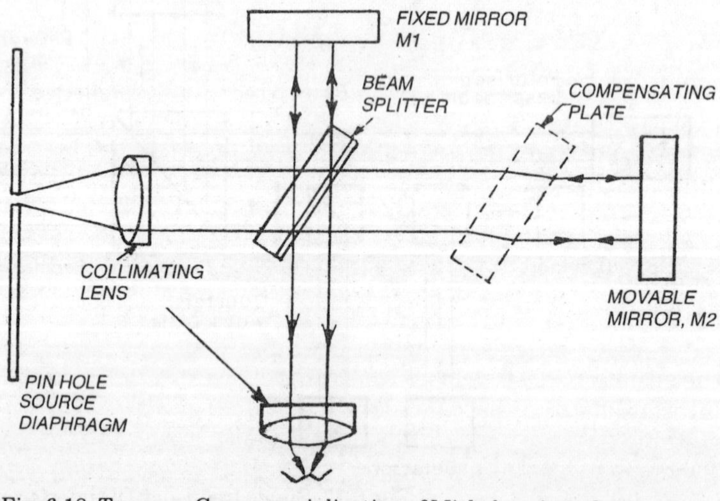

Fig. 6.18. Twyman-Green specialisation of Michelson interferometer.

A simple arrangement of fringe counting system based on Koster's prism is shown in Fig. 6.19. With the use of Koster's prism, the two interfering paths can be arranged parallel instead of at right angles. At big advantage of using Koster's prism is that if slight vibrations exist, then vibration tends to affect the arms equally and the annoying effect of vibration is nullified. In order to be able to count the fringes, the following must be taken care of :

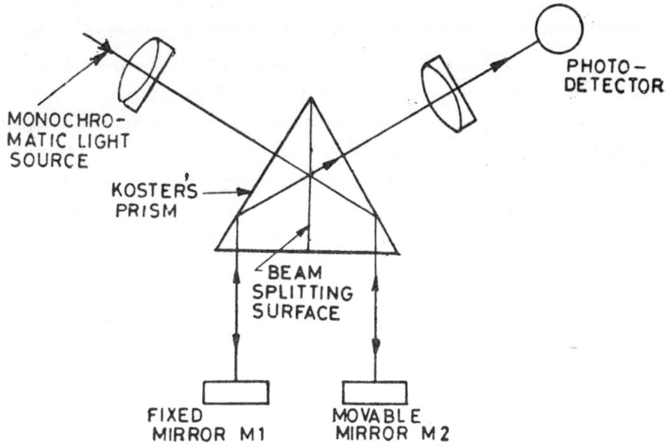

Fig. 6.19. Fringe counting Interferometer based on Koster's Prism.

(*i*) It has been indicated that mirror should travel exactly parallel to itself and no machines have ways sufficiently straight to maintain uniform fringe fields. The recent trend is to use corner-cube reflectors which are not at all sensitive to their own orientation and return the reflected ray exactly parallel to the indicident beam.

(*ii*) It is observed that the wavelength of light source is modified by the refractive index of air which is dependent on pressure, temperature and humidity of air (wavelength is fixed only in vacuum). The slight changes in wavelength may be immaterial in case of flatness or from measuring systems, but not in fringe counting and gauge block interferometers. So pressure, temperature and humidity should be measured and correction factors applied for. If optical paths are longer then the air currents between optical elements exert more and more influence ; and the system should, therefore, be properly shielded with insulating, and radiation reflecting enclosures.

(*iii*) It has already been indicated that the signal strength becomes poor if the path difference between the rays corresponding to two mirror systems is high. Thus it limits the range of movement of movable mirror because its movement means change in path length. It is found that using cooled mercury 198 lamps, speeds of 12.5 mm/sec. are possible when path lengths are nearly equal, but the traversing speed has to be reduced to 0.0025 mm/sec., when path difference is about 250 mm due to poor signal to noise ratio.

Two methods are possible to improve the situation. (*a*) Use laser source (A.C. laser interferometer described later). (*b*) Couple a suitably designed Fabry-Perot etalon with a Twyman-Green interferometer ; and then it is possible to attain 6 m path difference with good contrast. The use of Fabry-Perot etalon in shown in Fig. 6.20. The etalons are made up of two optical flats (flat to 1/20 fringe or better) and coated with a high efficiency semitransparent film on the two facing surfaces (as shown in Fig. 6.20). These flats are kept exactly parallel by means of a carefully designed spacer. In this interferometer

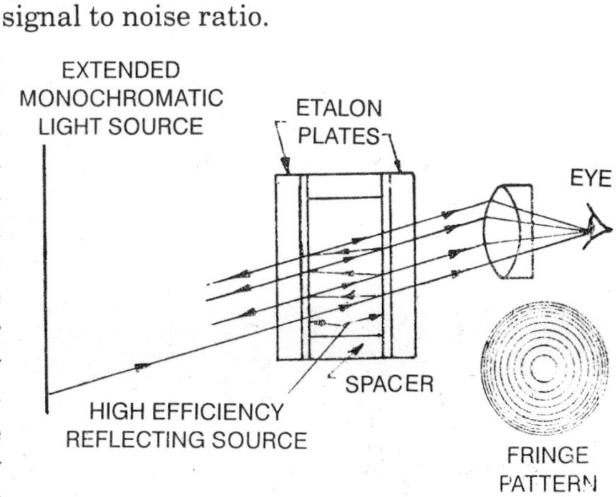

Fig. 6.20. Fabry-Perot Interferometer.

observer will see a series of very sharply defined bright circles resulting from interference between rays that are multiply reflected between the two working faces of the etalon.

(*iv*) These high sensitivity interferometers are highly susceptible to vibration. The presence of vibration may either wash out the fringes completely or may be very annoying for the observer. As already indicated, the use of Koster's prism can reduce the problem of vibration to great extent.

(*v*) In order to be able to count the fringes corresponding to both forward and backward movements, two detectors, one focussing the dark fringe and other focussing the adjoining bright fringe, must be arranged ; *i.e.* two detectors should be aligned with respect to the fringes, one located 90 fringe phase degree with respect to the other. If fringe position is to be detected to less than 1/4 fringe, then three photodetectors must be arranged to look at fringe sections 120° out of phase with each other.

6.7.4. N.P.L. Flatness Interferometer. This instrument, as the name suggests, is mainly used for checking the flatness of flat surfaces. This interferometer was designed by National Physical Laboratory and is commercially manufactured by Hilger and Watts Ltd. The flatness of any surface is judged by comparing with an optically flat surface which is generally the base plate of the instrument. This instrument essentially consists of a mercury vapour lamp (Fig. 6.21). As we are interested in having single monochromatic source of light, the radiations of the mercury lamp are passed through a green filter. The wavelength of the resulting monochromatic radiation is of the order or 0.0005 mm. This radiation is then brought to focus on pinhole in order to obtain an intense point source of light. A mirror is used in order to deflect the light beam through 90°. The pinhole is placed in the focal plane of a collimating

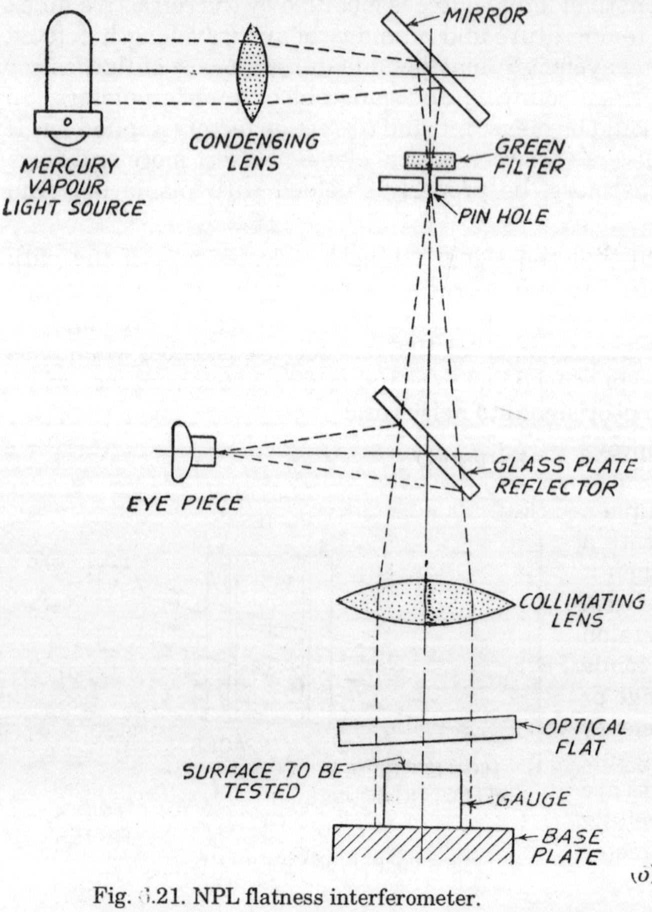

Fig. 6.21. NPL flatness interferometer.

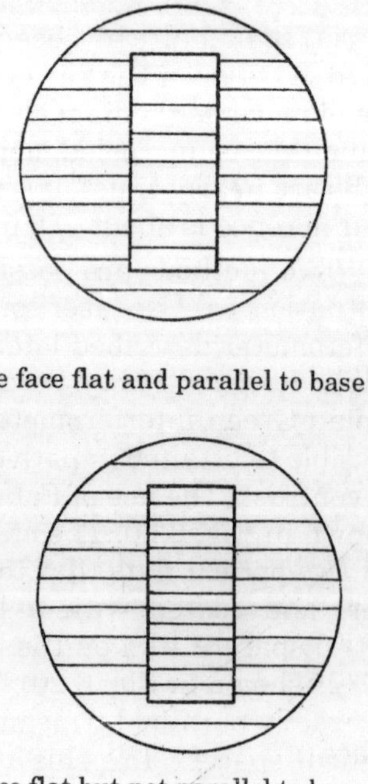

(*a*) Gauge face flat and parallel to base plate.

(*b*) Gauge face flat but not parallel to base plate.

Fig. 6.22

lens, thus the radiations out of the lens will be parallel beam of light. This beam is directed on the gauge to be tested *via* an optical flat. The fringes formed are viewed directly above by means of a thick glass plate semi-reflector set at 45° to the optical axis.

The gauge to be tested is wrung on the base plate whose surface is finished to a degree comparable to that of the highest quality gauge face. As the optical flat is placed above it in a little tilted position, interference fringes are formed ; one between rays reflected from the under surface of the optical flat and those reflected from the surface of the gauge, and the other between rays reflected from the under surface of the optical flat and those reflected from the base plate.

If the gauge face is flat and parallel to the base plate, then the optical flat being equally inclined on both the surfaces the fringe pattern from both the gauge face and the base plate will consist of straight, parallel and equally spaced fringes as shown in Fig. 6.22 (*a*). When the gauge is flat but not parallel to the base plate, then straight and parallel fringes of different pitch above the gauge face as compared with those of the base plate are seen [Fig. 6.22 (*b*)].

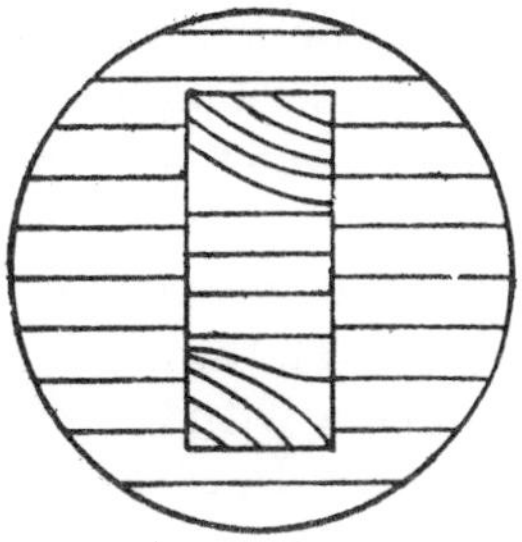

Fig. 6.23. Surface of gauge is inclined to base plate.

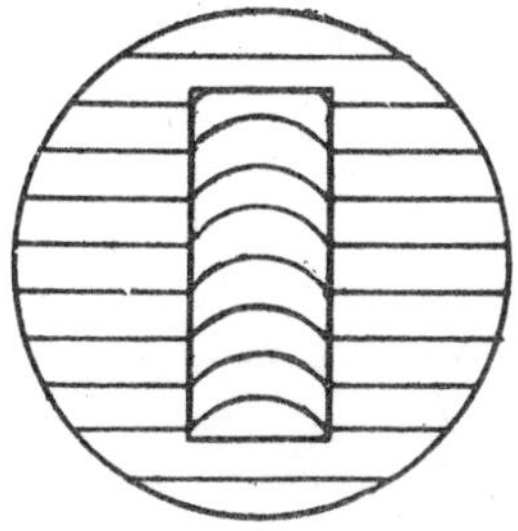

Fig. 6.24. Gauge surface convex/concave.

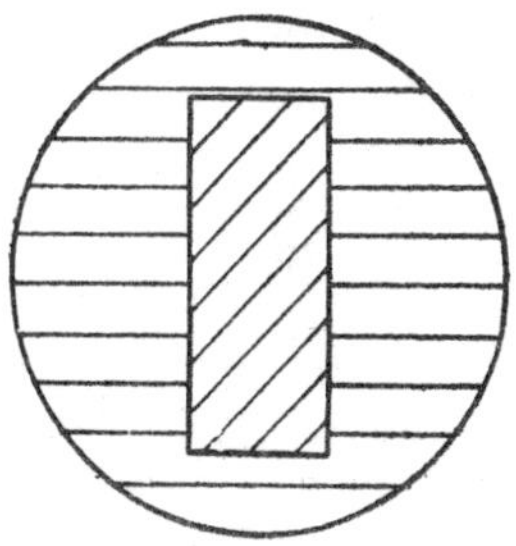

Fig. 6.25. Slight rounding off at corners.

In such case, to determine the amount of unparallelism, provision is made to rotate the base plate by 180° and this method is covered in detail after discussing the various other possible cases.

In case taper is present in some other direction, *i.e.* surface of the gauge is flat but its surface is inclined to the base plate at some other angle, then fringe pattern obtained is as shown in Fig. 6.23. Here the error is indicated by the amount by which the fringes are out of parallel with those on the base plate. When the gauge surface is convex or concave then fringe pattern as shown in Fig. 6.24 is obtained, *i.e.* the fringes on the gauge are curved lines.

Slight rounding off at the corners of an otherwise generally flat and parallel surface will give a fringe pattern as shown in Fig. 6.25 having closely curved lines at the ends, and straight and parallel fringes of equal pitch in the middle.

6.7.5. Quantitative Estimate of Error in Parallelism. The base plate on which the gauge is placed, is of rotary type and can be rotated at any angle. The underside of the gauge is lapped truly parallel with the working surface of the base plate. Now if a non-parallel gauge is viewed, the angle which it makes with the optical flat will be as shown in Fig. 6.26. In this position, face *AB* is nearly parallel with the

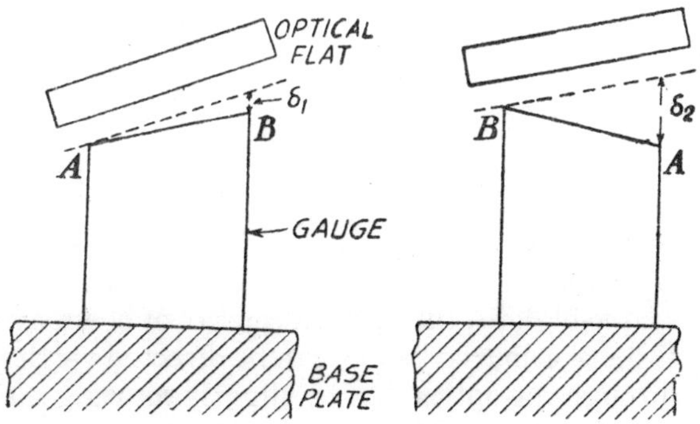

1st Position 2nd Position (Turned through 180°)

Fig. 6.26. Determining error in parallelism.

optical flat and distance δ_1 can be estimated by counting the number of fringes formed or the surface of gauge. If the table is turned through 180°, the surface *BA* will then be less parallel with the optical flat and a greater number of fringes will be observed. Let the number of fringes in first case be N_1 and in second case N_2.

In first position, the distance between the optical flat and gauge surface has increased by δ_1 over the length of the gauge and $\delta_1 = N_1 \times \lambda/2$.
where λ is the wavelength of light used. In second case, the corresponding distance $\delta_2 = N_2 \times \lambda/2$.

$\therefore$ Change in angular relationship is $\delta_2 - \delta_1$. But rotation through 180° causes doubling effect and, therefore, the error in parallelism is equal to $= \dfrac{\delta_2 - \delta_1}{2} = \dfrac{N_2 - N_1}{2} \times \dfrac{\lambda}{2} = \dfrac{\lambda}{2}(N_2 - N_1)$.

Thus knowing the value of λ and finding N_1 and N_2, error in parallelism can be determined quantitatively over length of the gauge.

6.7.6. The Pitter—N.P.L. Gauge Interferometer. This is also called the gauge length interferometer and used for determining actual dimensions or absolute length of the gauges. As the mechanical sub-division of end standards length tends to be laborious when small lengths are considered, and due to liability of error in that method, direct measurement interferometer based on the design of N.P.L. is most commonly used.

Since this involves very precision work, it is important to see that the physical conditions surrounding the measuring equipment are standardised and controlled. The standard conditions being temperature of 20°C, barometric pressure of 760 mm of mercury with water vapour at a pressure of 7 mm and containing 0.33% by volume of carbon dioxide. In case conditions are different, then correction factors have to be applied.

The diagrammatic arrangement of the instrument is shown in Fig. 6.27. *S* is suitable source of light and light form it is brought to focus on the illuminating aperture in the plate by passing it through a condensing lens. This plate is placed at the focal plane of a collimating

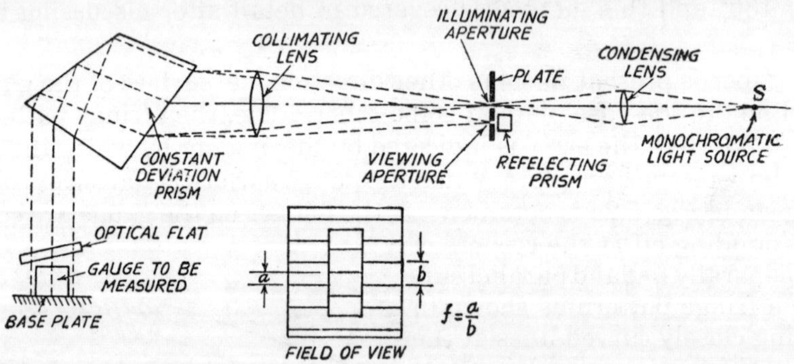

Fig. 6.27. Pitter—NPL gauge interferometer.

lens. Thus light from this plate acts as a point source of light and after collimating lens, rays of light in the form of parallel beam of light move into a constant deviation prism. The constant deviation prism disperses the light into its constituent colours. The beams of different colours are thus reflected downwards by the prism in slightly different directions. In case of cadmium source of light, the various colours of beam available are red, green, blue and violet. Any one of these coloured beams can be directed vertically downward on the gauge and the base plate through the optical flat by slightly rotating the constant deviation prism about a certain axis. The rays reflected at the gauge face and at the face of the base plate return along the same path approximately as the incident rays, but their axis is tilted slightly due to inclination of optical flat and thus brought to focus at some other point on the plate, where they are incident on a reflecting prism. Reflecting prism then reflects the rays into the eyepiece at normal to these rays.

The fringe pattern obtained is shown in the field of view in Fig. 6.27. Actually two interference systems are produced. One set of fringes is due to the upper surface of the gauge and the other is due to the base plate's reflecting surface. It is essential that the gauges being calibrated by this method possess a very high degree of flatness and parallelism. Only then the fringe pattern from the gauge and the base plate will consist of straight, parallel and equally spaced fringes of the same frequency. Generally two fringe patterns cannot be in phase and will be displaced as shown in Fig. 6.28. The amount of this displacement varies for each colour and, therefore, wavelength of light used. The displacement observed a, is expressed as a fraction of the fringe spacing b, *i.e.* 'f' = a/b. In order to determine the length of the gauge an estimation of f is made for each of the four radiations from the cadmium lamp.

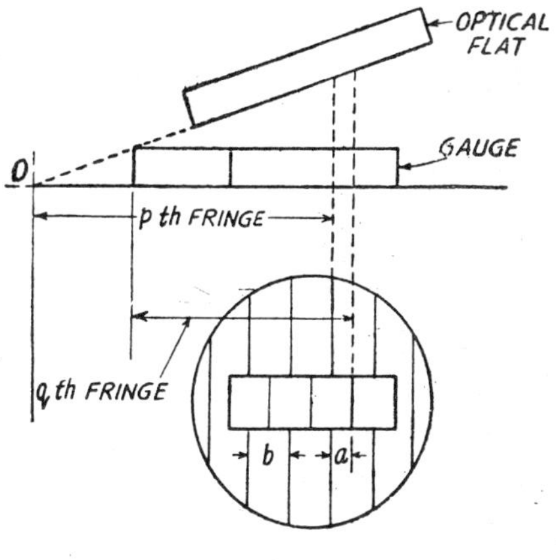

Fig. 6.28

As the two surfaces, *i.e.* surface of the gauge and of the base plate, are perfectly flat and parallel, the wedge shaped air film is of same angle in both the cases. Let the lower surface of the optical flat be produced to meet the base plate surface line produced at O. Let anywhere in the middle of the fringes system, a pair of fringes be chosen. Let pth be the fringe on base plate and qth on the gauge.

Then the length of the gauge $l = \dfrac{\lambda}{2}\left(p - q + \dfrac{a}{b}\right) = \dfrac{\lambda}{2}(p - q + f) = \dfrac{\lambda}{2}(N + f)$.

where f is the fraction a/b and N is the difference of the fringes p and q. Obviously the value of N will be very large and also complete number as p and q are full integers.

Before proceeding to estimate length upto fifth and sixth decimal place, it is essential that length of the gauge be determined accurately as far as possible by other methods.

Let the observed fractions for three different wavelengths be f_1, f_2 and f_3 and the corresponding integers, *i.e.* full number of fringes be N_1, N_2, N_3 respectively for lights of wavelengths λ_1, λ_2, λ_3, respectively.

Since the height of the gauge is same,

$$\therefore \quad \frac{\lambda_1}{2}(N_1 + f_1) = \frac{\lambda_2}{2}(N_2 + f_2) = \frac{\lambda_3}{2}(N_3 + f_3).$$

The nominal height of the gauge is known to us as told previously. For nominal height, the fractions will be different and they can also be calculated by knowing the wavelengths of lights.

e.g. for red light, wavelength is 25.348478 μ in.

and for green light, wavelength is 20.023055 μ in.

and for violet light, wavelength is 18.418017 μ in.

Thus if nominal length be 0.3″, then by calculation, the fractions will be 0.06, 0.46, 0.46 respectively. Let these fractions obtained by calculation be denoted by f_1', f_2', f_3' and the number of whole half wavelengths in nominal length be N_1', N_2', N_3' respectively.

$$\therefore \quad \frac{\lambda_1}{2}(N_1' + f_1') = \frac{\lambda_2}{2}(N_2' + f_2') = \frac{\lambda_3}{2}(N_3' + f_3')$$

Now the difference between the actual length of the gauge and the nominal length will be

$$\frac{\lambda_1}{2}(N_1 + f_1 - N_1' - f_1') = \frac{\lambda_2}{2}(N_2 + f_2 - N_2' - f_1') = \frac{\lambda_3}{2}(N_3 + f_3 - N_3' - f_3') \quad ...(1)$$

It is very obvious that in the above expression, $N_1 - N_1'$, $N_2 - N_2'$, $N_3 - N_3'$ will be a whole number as N, N' themselves are whole numbers and also it will be relatively small number.

∴ Equation (1) can be solved by hit and trial by substituting the values of $N_1 - N_1'$, $N_2 - N_2'$, $N_3 - N_3'$ as 1, 2, 3, etc. [$\lambda_1, \lambda_2, \lambda_3, f_1, f_2, f_3$; f_1', f_2', f_3', etc., in Eq. (1) are known]. Once this equation is solved, then length of gauge can be calculated. However this is a laborious job to solve this equation by hit and trial method. In order to make calculations easy, a special slide rule is made in which there are four scales and the wavelengths of red, green and violet radiations are set out to scale from a common zero. The lower scale has values corresponding to micro-inches. One cursor having one straight vertical line is also incorporated.

To solve equation (1) by this slide rule, since the values of fractions $f_1 - f_1'$, $f_2 - f_2'$, $f_3 - f_3'$ are known and it is known that values of $N_1 - N_1'$ etc., will be whole integer we try to locate the vertical line of cursor at such a place where the three fractions coincide approximately and the error between nominal value and actual value will be directly read from bottom line. (Also refer solved questions at the end of this chapter for further clarity).

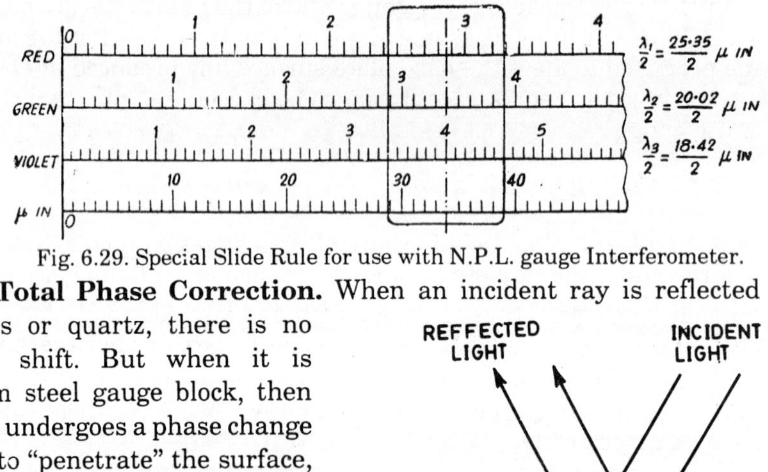

Fig. 6.29. Special Slide Rule for use with N.P.L. gauge Interferometer.

6.7.7. Total Phase Correction. When an incident ray is reflected from optically finished glass or quartz, there is no (zero) phase shift. But when it is reflected from steel gauge block, then incident light undergoes a phase change and appears to "penetrate" the surface, reflecting from a point below what might be called its mechanical surface (Refer Fig. 6.30). The correction factor varies from one material to another and it also varies with surface finish. In perfectly lapped steel, the phase change correction amounts to approximately 0.00002 mm.

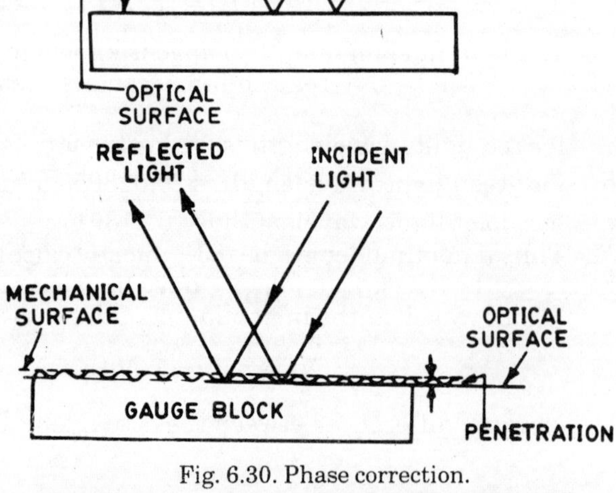

Fig. 6.30. Phase correction.

Depending on surface finish, the total phase change correction may amount to 0.00007 mm for gauge blocks.

Further wringing film thickness of the order of 0.000006 mm also needs to be accounted for. An improper wring may easily result in a value many times more than this amount. The thickness of film also depends on the type of materials in contact, their flatness and surface finish.

6.8. Laser Interferometers

This article is based on a similar article appearing in magazine "Machine Design" Vol. 47 No. 4. The measuring capacity in interferometers with lamp as source of light is limited because it is not possible to maintain the sharpness of interference fringes beyond certain distance due to the size of the lamp. Laser interferometer uses A.C. laser as the light source and thus enables the measurements to be made over longer distance because it is possible to maintain the quality of interference fringes over long distances when lamp is replaced by a laser source. It must be understood that white light emitted by a lamp is combination of waves at different frequencies but laser generates a continuous train of light waves, resulting into high coherence. Laser represents a source of intensely monochromatic optical energy, which can be collimated into a directional beam. Also laser beam wavelength is exact and pure for highly accurate measurements. It utilises the principles of both optical techniques and digital electronics ; and is a highly accurate and versatile measuring system that can cope with industrial environments. In case of AC laser interferometer (ACLI) position information is carried as phase deviation rather than as a signal amplitude deviation, thus giving a much improved signal to noise ratio over amplitude modulation, because the noise sources that affect signal amplitude have little effect on phase. In this way, ACLI is much more tolerant of environmental factors that attenuate the intensity of a laser beam, such as dust, smoke, air turbulance etc. It requires no warm-up time or standby power.

Thus ACLI has the following advantages : high repeatability and resolution of displacement measurement (0.1 μm), high accuracy, long-range optical path (60 m), easy installation, and no change in performance due to ageing or wear and tear. A single laser source can be used for as many as six simultaneous measurements in different axes. However, it is very much expensive ; since the basic instrument measures physical displacement in terms of wavelength instead of traditional units, conversion instrumentation is required for conventional read out. Highest possible accuracy is obtainable only by compensating changes in air pressure and temperature which affect wavelength of the laser beam.

Fig. 6.31 explains the operation of AC Interferometer.

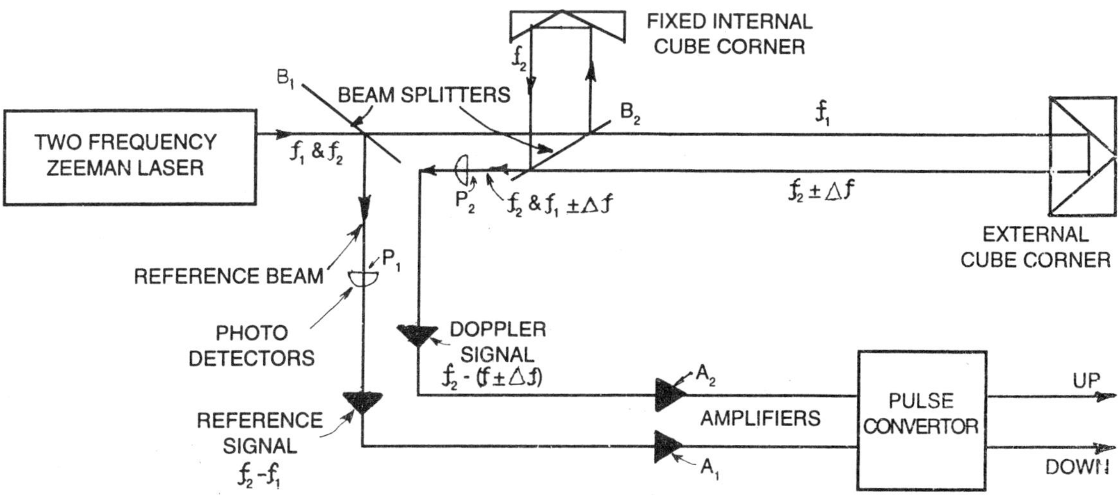

Fig. 6.31. AC interferometer.

It uses two frequency laser system, thus overcoming the shortcoming of d.c. laser interferometer. Whereas the d.c. system mixes out of phase light beams of the same frequency, the a.c. system mixes beams of two different frequencies thus permitting the distance information to be carried on a.c. waveform. Use is made of the fact that the AC amplifiers are insensitive to d.c. variation of a.c. inputs.

Two frequency Zeeman laser generates light of two slightly different frequencies with opposite circular polarisations. These beams get split up by beam splitter B_1 : one part travels towards B_2 and from there to external cube corner where the displacement is to be measured. It may be noted that mirror is not employed here like Michelson Interferometer, because mirror alignment is a critical procedure. This interferometer, instead, uses cube-corner reflectors (retroreflectors) which reflect light parallel to its angle of incidence regardless of retroreflector alignment accuracy. Beam splitter B_2 optically separates the frequency f_1 which alone is sent to the movable cube-corner reflector. The second frequency f_2 (optically separated) from B_2 is sent to a fixed reflector which then rejoins f_1 at the beam splitter B_2 to produce alternate light and dark interference flicker at about 2 Mega cycles per second. Now if the movable reflector (external cube corner) moves, then the returning beam frequency will be Doppler-shifted slightly up or down by Δf_1. Thus the light beams moving towards photo-detector P_2 have frequencies f_2 and $(f_1 \pm \Delta f_1)$ and P_2 changes these frequencies into electrical signal. (Photocells convert light-intensity variations into voltage pulses which can be processed by electronic instruments to give the amount and direction of position change).

Photo detector P_1 receives signal from beam splitter B_1 and changes the reference beam frequencies f_1 and f_2, into electrical signal. An A.C. amplifier A_1 separates frequency difference signal $f_2 - f_1$ and A_2 separates frequency difference signal $[(f_2 - (f_1 \pm \Delta f)]$. The pulse converter extracts Δf, one cycle per half wavelength of motion. The up-down pulses from the pulse converter are counted electronically and displayed in analog or digital form on the indicator. It may be noted that output in case of ACLI is in the form of pulses, whereas in d.c. systems, the output is in the form of a sinusoidal wave, the amplitude (intensity) of which depends upon laser aging, air turbulence or air pollutant and thus the change of amplitude leads to improper triggering and counting errors (Refer Fig. 6.32).

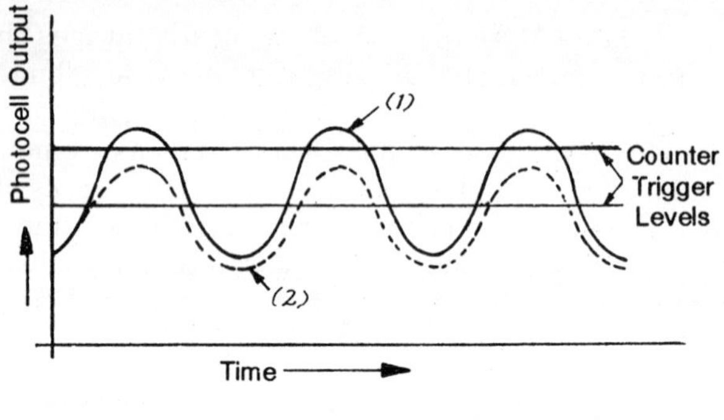

(1) Counter operating, if amplitude of wave is above counter trigger level.

(2) Counter disabled by small amplitude change of sinusoidal wave.

Fig. 6.32

6.8.1. Laser Interferometers and Automatic scale Measuring Interferometer. It is felt that the present length standard (krypton 86) may eventually be replaced by the laser (light amplification by stimulated emission of radiation) because the present light wave standard has a potential accuracy of determination of perhaps 1 or 2 parts in 10^9. The principle of the operation of laser-interferometer is demonstrated in Fig. 6.33 which has higher accuracies for measuring greater lengths.

Fig. 6.34 shows a cutaway view of a laser interferometer.

For measurement purposes, present techniques utilize the gas laser, and in particular, the helium—neon gas laser, married with a suitably designed interferometer. The great advantage of the gas laser compared to other light sources is its coherency and strongly defined

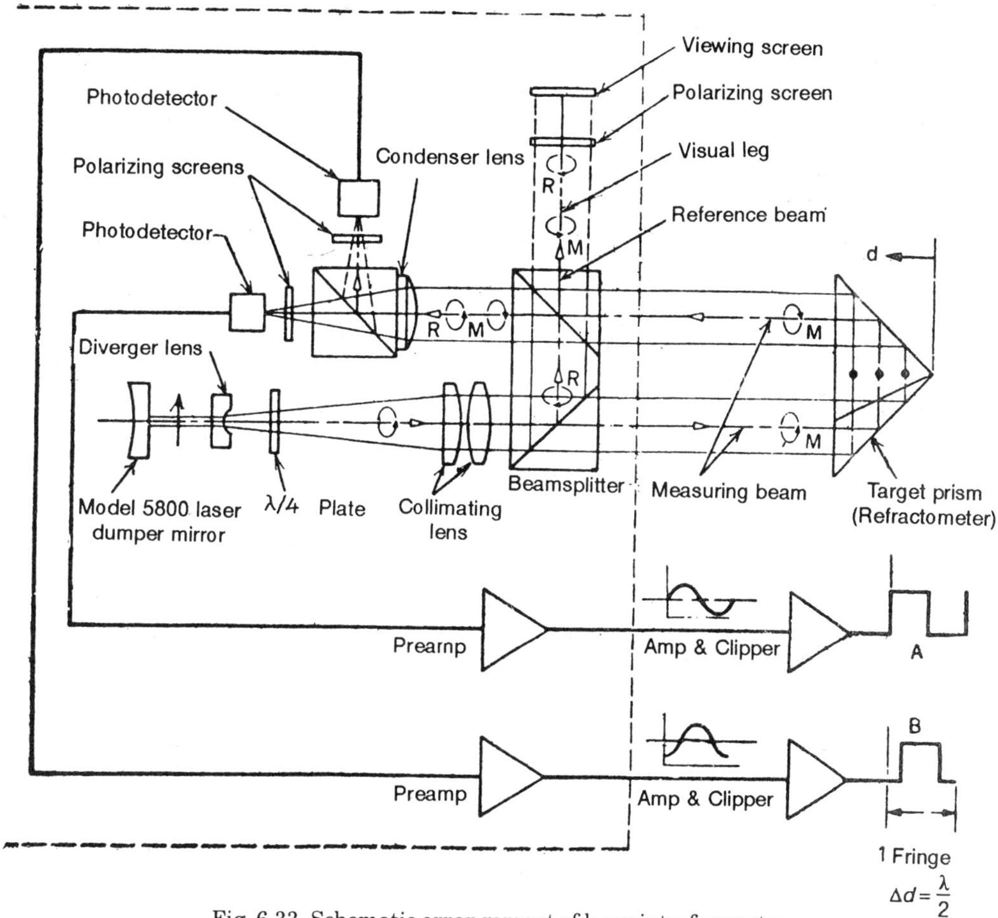

Fig. 6.33. Schematic arrangement of laser interferometer.

fringe pattern over long optical paths. The gas laser is already established as a practical length-measuring tool in most of the national bureaus, Fig. 6.35.

Precision scales up to 1 metre in length are verified automatically in terms of a laser wavelength determined by reference to the krypton 86 standard. The number of wavelengths in each scale interval is determined using a laser interferometer and electronic counting. The counts are punched onto paper tape which is processed by a computer to give a table of errors for the scale intervals.

The gas laser wavelength also drifts over a period of time and is dependent on (*i*) the conditions of the environment, (*ii*) on the adjustment of the reflecting mirror "chamber", which acts as an optical resonator for the helium—neon atoms. Latter can be adjusted very precisely by using piezoelectric or magnetostrictive mirror mounts. Laser standard is, therefore, periodically compared against the standard line of krypton-86.

6.8.2 Brief description of components involved

(*i*) **Two frequency laser source.** It is generally He—Ne type that generates stable coherent light beams of two frequencies, one polarised vertically and one horizontally relative to the plane of the mounting feet. The frequency stabilisation is based on Zeeman splitting of the atomic levels involved in the laser action. Laser oscillates at two slightly different frequencies by a cylindrical permanent magnet around the cavity. The two components of frequencies are distinguishable by their opposite circular polarisations. Beam containing both

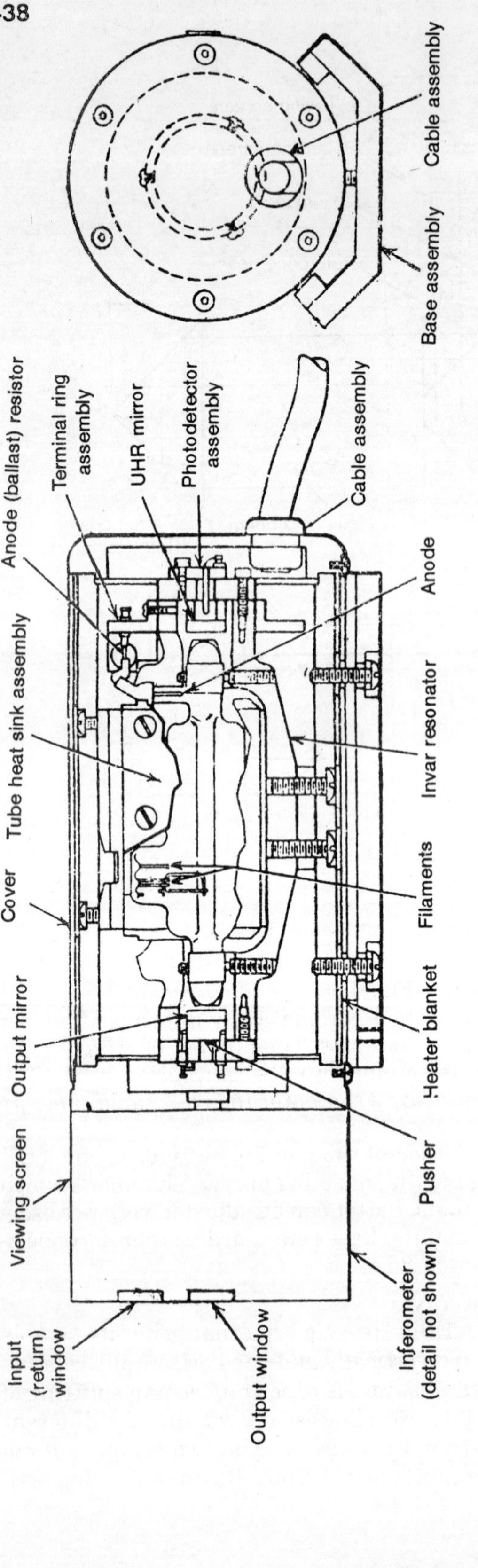

Fig. 6.34. Cutaway view of leser interferometer.

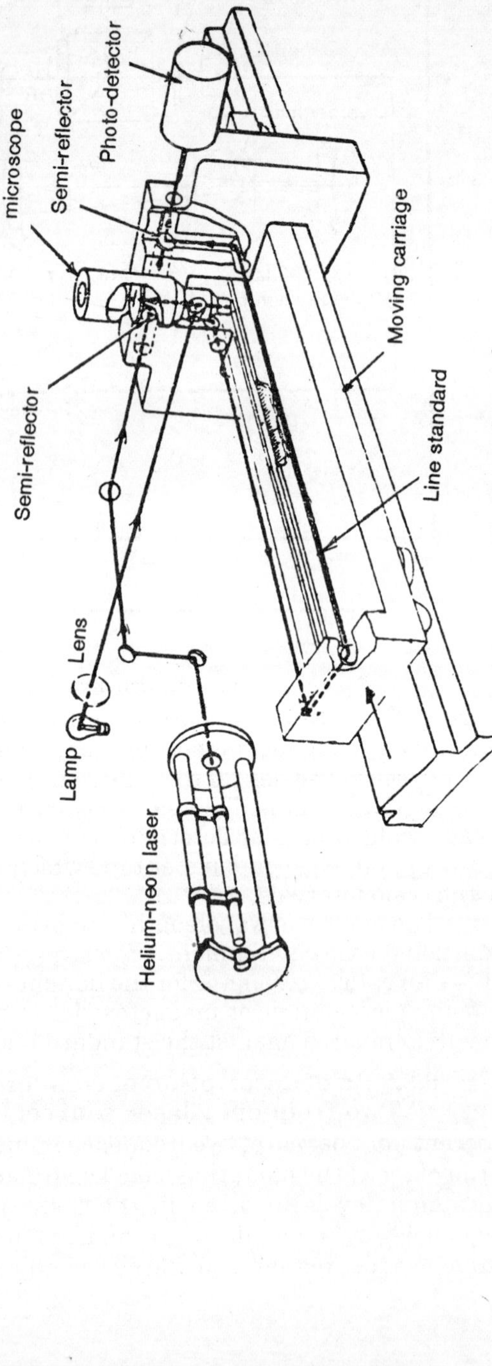

Fig. 6.35. Length measuring tool.

frequencies passes through a quarter wave and half wave plates which change the circular polarisations to linear perpendicular polarisations, one vertical and other horizontal. Thus the laser can be rotated by 90° about the beam axis (roll) without affecting transducer performance. If the laser source is deviated in roll from one of the four optimum positions, the photo-receiver signal will decrease. At a deviation of 45°, the signal will decrease to zero. The linearly polarised beam is expanded in a collimating telescope, after which most of the beam is transmitted through a 45° beam splitter and out of laser head.

(*ii*) **Optical Elements :**

(*a*) **Beam splitters.** These, (like partially silvered mirrors) divide the laser beam into separate beams along different axes. It is possible to adjust the splitted laser's output intensity by having a choice of beam splitter reflectivities. To avoid attenuation it is essential that the beam splitters must be oriented so that the reflected beam forms a right angle with the transmitted beam, so that these two beams are coplanar with one of the polarisation vectors of the input beam.

Fig. 6.36 illustrates the use of beam splitters to divide laser output equally along different axes.

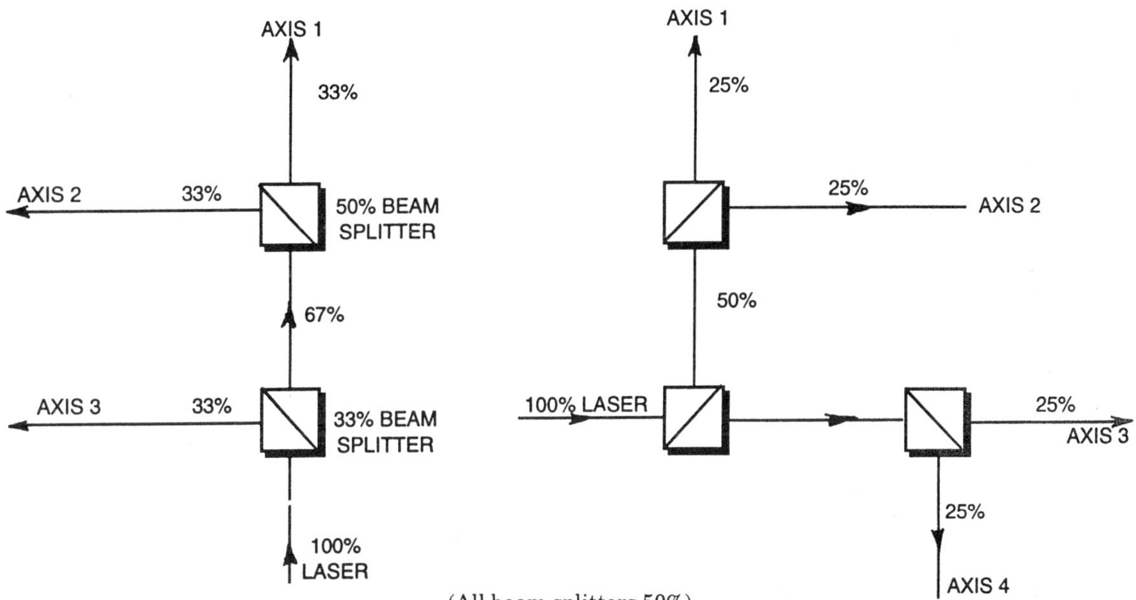

(All beam splitters 50%)

Fig. 6.36. Use of beam splitters to divide laser output equally.

(*b*) **Beam benders.** These are used to deflect the light beam around corners on its path from the laser to each axis. These are actually just flat mirrors, but having absolutely flat and very high reflectivity. Normally these are restricted to 90° beam deflections to avoid disturbing the polarising vectors.

(*c*) **Retro-reflectors.** These can be plane mirrors, roof prisms or cube-corners. Cube-corners are three mutually perpendicular plane mirrors, and in these devices the reflected beam is always parallel to the incidental beam. Each ACLI transducer axis needs at least two retro-reflectors as explained in Fig. 6.37. All ACLI measurements are made by sensing differential motion between two retroreflectors relative to an interferometer. Plane mirrors used as retroreflectors with the plane-mirror interferometer must be flat to within 0.06 micron per cm.

(*iii*) **Laser head's measurement receiver**. During a measurement, the laser beam is directed through optics in the measurement path and then returned to the laser head's measurement receiver which will detect part of the returning beam as $f_1 - f_2$ and a doppler shifted frequency component δf.

(*iv*) **Measurement display.** [Refer Fig. 6.37 (*a*)] It contains a microcomputer to compute and display results. The signals from reference receiver and measurement receiver located in the laser head are counted in two separate pulse counters and subtracted. Necessary calculations are made and the computed value is displayed. Other input signals for correction are temperature, coefficient of expansions, air velocity etc. which can also be displayed.

6.8.3. Various versions of ACLIs

(1) **Standard Interferometer** (Refer Fig. 6.37 (*a*)]. It is the least expensive and can be used whenever possible. The measurement retroreflector for this interferometer is a cube

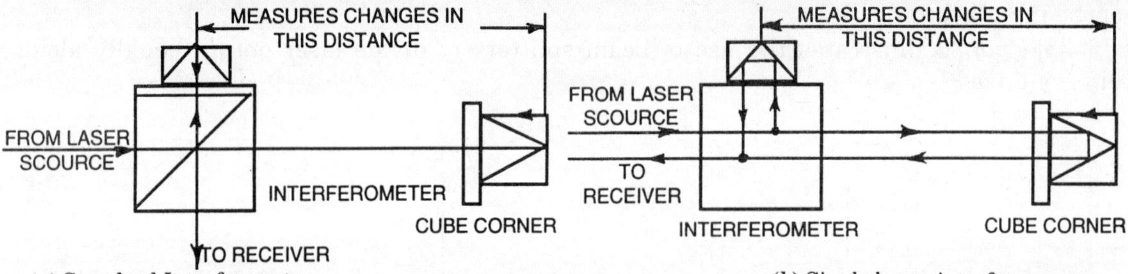

(*a*) Standard Interferometer. (*b*) Single beam interferometer.

Fig. 6.37

corner. Displacement is measured between the interferometer and the cube corner. Either one or both could move.

(*a*) **Single beam Interferometer** [Refer Fig. 6.37 (*b*)]. It has the outgoing and returning beam superimposed on each other, giving the appearance of only one beam travelling between the interferometer and the retroreflector.

This interferometer operates functionally the same way as the standard interferometer, but it is advantageous when space for optics and beam paths is limited. The interferometer and retro-reflector for this system are considerably smaller than the standard system.

6.8.4. Sources of Errors. The various sources of errors in the use of ACLIs are :

(*i*) uncertainty in the refractive index of air (velocity of light).

(*ii*) alignment (cosine error).

Avoiding Sources of Errors

(*i*) **Velocity of light (VOL) compensation.** It has been found that wavelength of laser changes by about one part per million for every :

(*a*) atmospheric pressure change of 2.5 mm of Hg ; *i.e.* if pressure varies by ± 20 mm of Hg, wavelength varies by about ± 7.5 ppm ; in other words, if axis of travel is 2 metres, this pressure change could cause an error of ± 15 microns.

(*b*) air temperature change of 1°C.

(*c*) relative-humidity change of 100%.

The best method to compensate for VOL changes is to take barometer, thermometer and hygro meter readings and apply correction by referring to compensation tables.

(*ii*) **Cosine error.** As the ACLI measures component travel in the direction of laser beam, and if two are not coincident, then error called cosine error will be introduced. To avoid it, the axis of measuring device should be aligned with the axis of travel or object.

(*iii*) **Beam Protection.** Though the ACLI is quite tolerant of the beam disturbance ; however the following types of disturbances should be avoided ;

(*a*) Beam must never be completely blocked by a solid object.

(*b*) Index of refraction of air path followed by a beam should be as uniform as possible over the entire optical path, *i.e.* no temperature or pressure gradient in air path.

6.9. Heterodyne Interferometry Technique

Simple d.c. fringe counting techniques suffer from problems of intensity level changes in source and also on account of motion of source or object. Fringe contrast changes and d.c. level shifts result in miscounting of the fringes. Heterodyne interferometer is an a.c. device and the problems of d.c. fringe counting techniques are overcome. In this type of interferometer, a Zeeman laser source emits two closely spaced orthogonal polarisation frequencies separated by around 1 MHz. A beam splitter placed in front of laser source separates off part of the signal

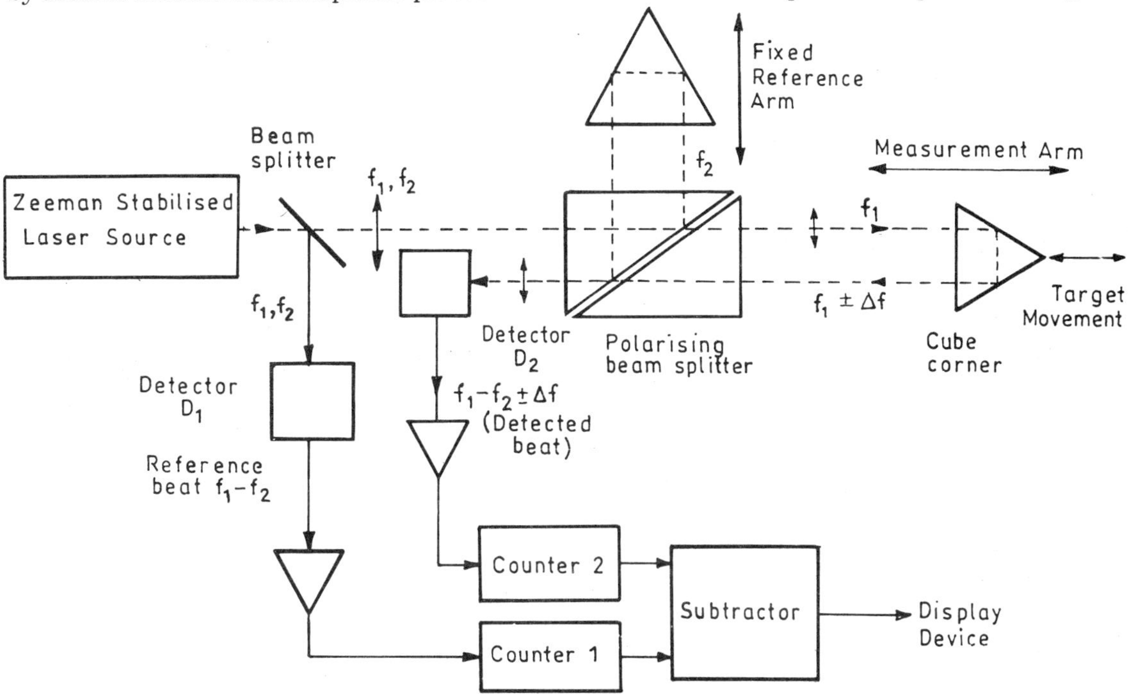

Fig. 6.38. Working principle of Heterodyne Interferometer.

from both polarisations which are mixed on detector D_1 to provide a reference beat $f_1 - f_2$. The transmitted component travels upto polarising beam splitter where it is splitted. Part of it travels to reference fixed arm and other to measurement arm connected with target movement. The two signals are recombined at the polarising beam splitter and detected by detector D_2. If target is stationary, the detected beam is $f_1 - f_2$. When it moves, then detected beat is $f_1 - f_2 \pm \Delta_f$. The reference and Doppler-shifted beats are counted by two independent counters and subtracted to give Δ_f. Integration of the count over time t measures $2d/\lambda$.

6.9.1. Dual-frequency Laser Interferometer. This instrument is used to measure displacement, high-precision measurement of lengths, angles, speeds and refractive indices as well as derived static and dynamic quantities. It operates on heterodyne principle. The two resonator modes (frequencies f_1 and f_2) are generated in a laser tube such that $f_1 - f_2$ = 640 MHz. These are controlled so that their maxima are symmetrical to the atomic transition. This permits a long reliable stability. The frequency stability of He-Ne laser is responsible for outstanding performance of the interferometer.

An amplitude beam splitter branches off part of the laser output create a reference beam, which an optical fibre cable relays to a photodetector 1. This detects the beat signal of the 640 MHz frequency difference produced by the heterodyning of the two modes. The other portion of the light serves as measuring beam. Via an interferometer arrangement it is directed to a movable measuring mirror and a stationary reference mirror, which reflects it on to a photo-detector 2. The two frequencies in the measuring beam are separated by a polarisation-sensitive beam splitter so that the measuring mirror receives light of frequency f_1 only, whereas the light that strikes the reference consists exclusively of frequency f_2. With the measuring mirror at rest, detector 2 also senses the laser differential frequency of $f_1 - f_2 = 640$ MHz. If the measuring mirror is being displaced at a speed v, the partial beam of frequency f_1 reflected by it is subjected to a Doppler shift df_1 ; where $df_1 = (2v)/\lambda_1$.

Accordingly, detector 2 now receives a measuring frequency of $f_1 - f_2 \pm df_1$ ($+ df_1$ or $- df_1$) depending on the direction of movement of the measuring mirror. The reference frequency $f_1 - f_2$ and the measuring frequency $f_1 - f_2 \pm df_1$ are compared with each other by an electronic counting chain. The result is the frequency shift $\pm df_1$ due to the Doppler effect, a measure of the wanted displacement of the measuring mirror. In a fast, non-hysteric comparator, the

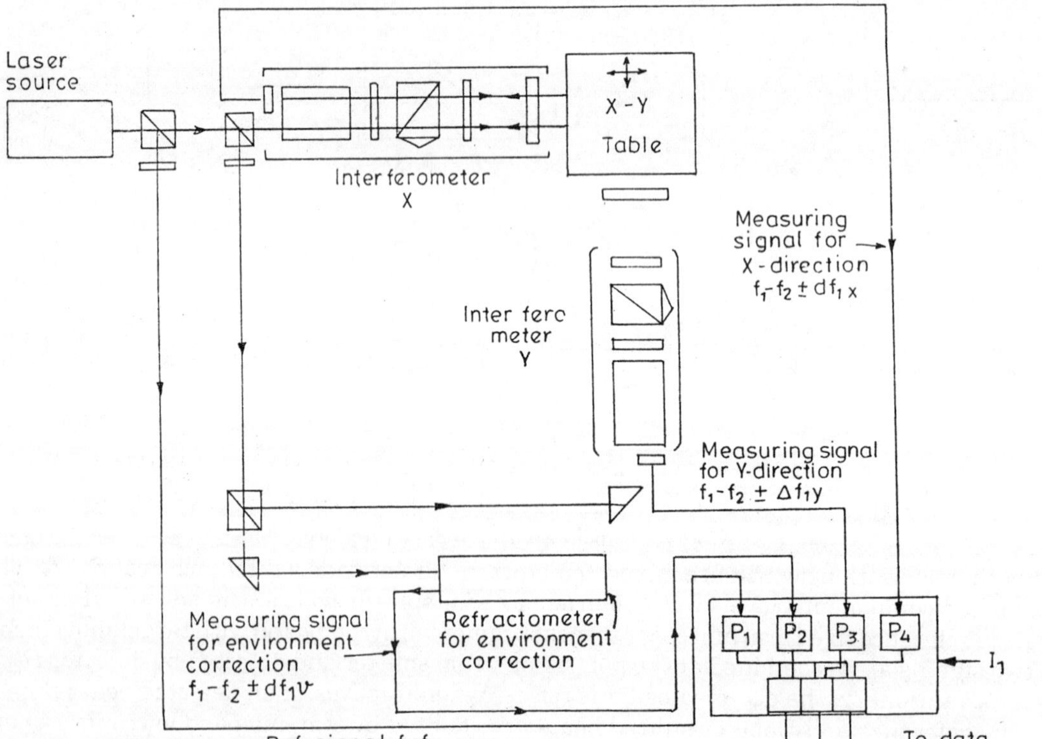

Fig. 6.39. Schematic of dual frequency laser interferometer.

P_1 = Photo detector for reference signal

P_2, P_3 and P_4 = photo detectors for measuring

I_1 = Basic Instrument signals with HF signal processing and interpolation facilities.

Doppler frequency df_1 is digitised and then fed to a counter, which registers the number of zero passages per unit time.

The forward and return movements of the measuring mirror can be distinguished by outcoupling the measuring signal $f_1 - f_2 + df_1$ at 'n' phase angles, via a delay line and feeding to 'n' mixers. The mixers are connected with the reference signal $f_1 - f_2$ (common feeding point for all mixers). Thus n Doppler frequencies get shifted in phase by π/n at the mixer outputs. They are symmetrical relative to zero. After comparison they are made available to low-frequency counting logic as TTL signals. The n phase angles and their tolerances are implemented by the geometry of the delay line.

This system can be used for both incremental displacement and angle measurements. Due to large counting range it is possible to attain a resolution of 2 nm in 10 m measuring range. Means are also provided to compensate for the influence of ambient temperature, material temperature, atmospheric pressure and atmospheric humidity fluctuations.

6.10. Optical Flats

Optical flats are cylindrical in form with the working surfaces flat and are of two types :

Type A—It has only one surface flat. The working surface of this type of flat is indicately by an arrow head on the cylindrical surface pointing towards the working surface.

Type B—It has both the surfaces flat and parallel to each other.

Type A are used for testing the flatness of precision measuring surfaces of flats, slip gauges, measuring tables, etc. Type B are used for testing measuring surfaces of micrometers, measuring anvils and similar length measuring devices for testing flatness and parallelism.

These are generally made of either fused quartz (whose co-efficient of linear expansion is not more than 0.6×10^{-6}/degree at standard temperature of 20°C) or borosilicate glass (whose co-efficient of linear expansion is not more than 3.6×10^{-6}/ degree at standard temperature of 20°C). This glass is clear and colourless and free from inclusions and defects like bubbles, internal strains and extraneous matter.

The cylindrical surfaces of the optical flats are finished by grinding and the working surfaces are finished by lapping and polishing process. The finishing is done to such an extent that straight and parallel interference fringes are produced when the optical flat is viewed under a monochromatic source of light by placing it in contact with a reference flat surface. Their edges are generally bevelled to 45° by a suitable grinding or polishing process so that the edges are free from any chips.

The working surfaces of each Type B flat are inclined to each other by about 5 minutes.

In order to avoid the optical flats from shocks, dust and moisture during transportation and storage, these are generally housed in a wooden box having seating lined with velvet.

Depending on their accuracy, *i.e.* tolerance on flatness for Type A and tolerances on flatness, parallelism and thickness for Type B, these are classified as :

(*i*) Reference Grade or Grade I, and

(*ii*) Working Grade or Grade II.

For type A optical flats, their diameter and grade are important and, therefore, these are usually designated by the above two ; whereas for type B flats, their thickness and grades are important and, therefore, their designation includes these two items.

As per IS : 5440—1969, the dimensions of optical flat of type A and grades I and II can be 25 × 10, 30 × 10, 50 × 15, 75 × 20, 100 × 25, 125 × 30, 160 × 35 (diameter in mm × thickness in mm) and the dimensions of type B optical flats of either grade can be 25.0 × 12.000,

25.0 × 12.125, 25.0 × 24.250, 25.0 × 12.375, 25.0 × 24.000, 25.0 × 24.125, 25.0 × 24.250 and 25.0 × 24.375 respectively. Further the tolerance on flatness should be 0.05 μm for type A, Grade optical flats of sizes upto 100 mm, and 0.10 mm for sizes 101—200 mm respectively ; and 0.01 μm for type A, Grade II opticalflats and sizes upto 100 mm, and 0.20 μm for sizes 101—200 mm respectively. For type B optical flats, the tolerances on flatness, parallelism and thickness should be 0.05 μm, 0.15 μm and 0.20 μm respectively for Grade I ; and 0.10 μm, 0.20 μm and 0.30 μm respectively for Grade II.

6.10.1. Testing of optical flats. As already indicated type A optical flats are tested for flatness of working face and type B for flatness of working faces, parallelism of working faces and thickness.

Flatness test. The flatness of the working surface of an optical flat is tested by comparing its flatness with a master flat of known flatness or by comparing the working surface of three separate flats of unknown flatness. The testing is usually carried out by observing the interference fringes using monochomatic light source, the incident rays from which may be either diffused or parallel.

In testing with the master flat whose flatness is known, the master flat is placed over the working surface of the flat under test (which itself is placed on some reference flat surface) in such as manner as to form a wedge shaped air film between the two flats as shown in Fig. 6.40.

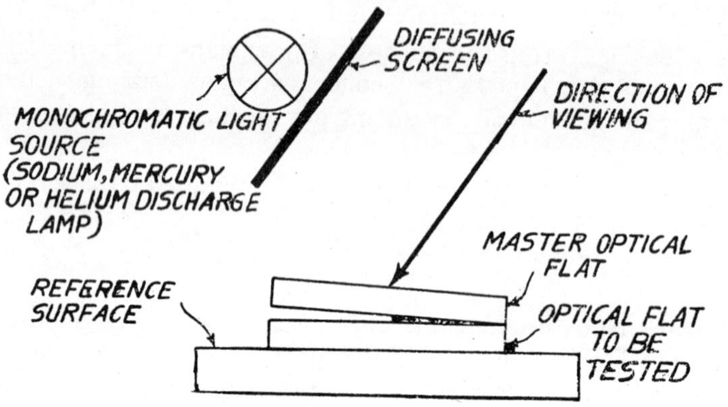

Fig. 6.40. Setup for flatness test.

If the optical flat under test has a perfectly flat surface then the fringes observed will be straight lines (dark and bright bands) parallel to each other. The degree of any deviation of flatness in the optical flat under test from the flatness of the master flat is given by the deviation of the fringes from straightness in any direction (the direction indicates the concavity or convexity).

The fringes are generally observed in a telescope which carries a micrometer eye-piece and in which the deviations of fringes are easily measured in terms of fringe width by means of the cross wires.

When the master optical flat of the known flatness is not available then the testing of flatness may be carried out by intercomparison with two other optical flats of unknown flatness. The three flats are examined in pairs as shown in Fig. 6.41 for the interference fringes, the testing procedure and arrangement being same as described above for testing with master flat. The deviations from straightness of fringes are noted for three cases, *i.e.* when flat to be tested is placed on unknown flat I, unknown flat I on unknown flat II, and unknown flat II on the flat to be tested (Refer Fig. 6.41).

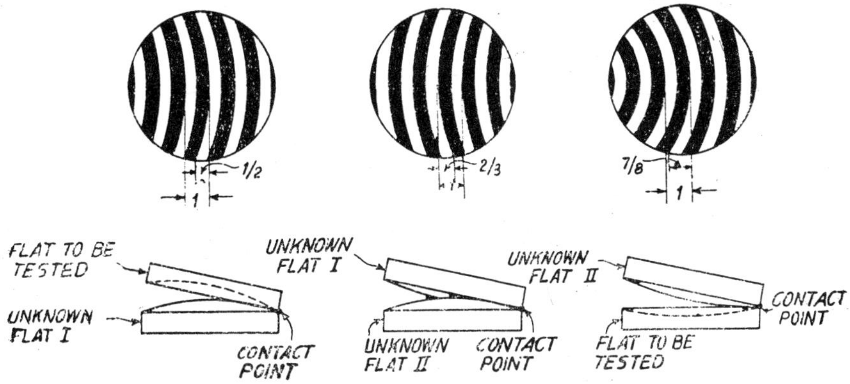

Fig. 6.41. Testing of optical flats.

Let us say that the deviations from straightness of fringes in three cases shown in Fig. 6.41 are $\frac{1}{2}$ fringe concave, $\frac{2}{3}$ rd fringe convex, and $\frac{7}{8}$ th fringe concave respectively. Then the flatness of the three optical flats may be determined by the following calculations :

Combined deviation from flatness between unknown flat I and flat to be tested $= -\frac{\lambda}{2} \times \frac{1}{2}$.

Combined deviation from flatness between unknown flat I and unknown flat II $= +\frac{\lambda}{2} \times \frac{2}{3}$

Combined deviation from flatness between unknown flat I and flat to be tested $= -\frac{\lambda}{2} \times \frac{7}{8}$

(where λ = Wavelength of monochromatic light used)

Let the deviation from flatness of the flat to be tested and of unknown flat I and II be D_T, D_I and D_{II} respectively.

$$D_T + D_I = -\frac{\lambda}{2} \times \frac{1}{2} \quad ...(1)$$

$$D_I + D_{II} = +\frac{\lambda}{2} \times \frac{2}{3} \quad ...(2)$$

$$D_{II} + D_T = -\frac{\lambda}{2} \times \frac{7}{8} \quad ...(3)$$

Substracting (2) from (1), we get $D_T - D_{II} = -\frac{7}{6} \times \frac{\lambda}{2}$ microns ...(4)

Adding (3) and (4), we get $2D_T = -\frac{49}{24} \times \frac{\lambda}{2}$ microns ; $\therefore\ D_T = -\frac{49}{48} \times \frac{\lambda}{2}$ microns.

Substituting the value of D_T in (1) and (3), the convexity or concavity of unknown flats I and II can be determined.

Parallelism Test

The parallelism of the working surfaces of type B optical flats can be tested by Fizeau Interferometer method.

In a modified arrangement of the Fizeau Interferometer method (Refer Fig. 6.42), light from a mercury vapour lamp is focused on to an opening in the eye-piece and is partly reflected by a beam splitter. The reflected light strikes the collimating lens which collimates it and throws it further on the flat under test. The flat under test is placed on a table provided with levelling arrangement. The level screws of the table are arranged such that the beam strikes

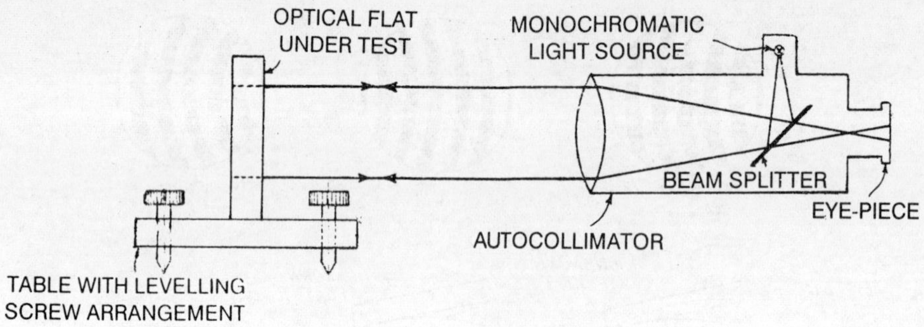

Fig. 6.42. Setup for parallelism test.

the surfaces of the flat perpendicularly. After reflection from the two surfaces of the flat under test, the beam retraces its path. Due to the interference of the light reflected at the two parallel surfaces of the optical flat, interference fringes are formed which can be viewed from the eyepiece of the autocollimator. These interference fringes help in determining the degree of flatness of the two surfaces of the flat. The difference of thickness of the flat at two points on the adjacent fringes is equal to $\frac{\lambda}{2} \times \frac{1}{\mu}$ microns, where μ is the refractive index of the substance of which the flat is made and λ is the wavelength of the light used. This gives the optical parallelism of the two surfaces of the flat. Since the flat is made of material of uniform refractive index, this parallelism also represents geometrical parallelism.

6.10.2. Care in the use of Optical Flat. Before using optical flat, it should be ensured that both flat and workpiece are clean and free from dirt, dust, oil or finger prints. This can be accomplished by swabbing the flat and workpiece with grain alcohol and polishing their surfaces with clean chamois or paper.

Optical flat should never be wrung on workpiece because it scratches too readily. It should only be rested carefully on the workpiece. It should be carefully and firmly pressed down evenly on workpiece with two fingers till interference bands become discernible. If the flat is pressed very close to workpiece, the air wedge between the flat and workpiece will be thin, and wider apart will be the interference bands.

Optical flat should never be slid over workpiece but lifted from it. If interference bands are not good, flat should be lifted and set down again, applying vertical finger pressure at various locations on the upper surface to obtain satisfactory bands. Sliding, creeping and wringing of flat over workpiece are extremely harmful and should be avoided. It can only be rocked and pressed.

If interference bands do not appear, it may be due to presence of dust or dirt between flat and workpiece or by lack of sufficient polish on the work to reflect the light.

6.11. Scales and Gratings

Scales are usually made of steel with lines (rulings) marked on them and spaced relatively far apart, so that some sort of interpolating device like vernier device is required to make accurate settings. In the case of gratings, the rules are so closely spaced as to produce a periodic pattern without blank gaps. It is impossible to make out anything with naked eye, and therefore, special readout systems such as photo-electric type are required for this purpose.

Metal scales are usually made from stainless steel which has the advantages of taking good finish, stability with time if properly treated, resistant to tranishing ; but has the disadvantages of lower thermal coefficient of expansion than steel or cast iron. Depending on the accuracy desired, the graduations on metallic scales may be produced by (*i*) cutting lines with a V-section milling cutter on milling machine, (*ii*) etching through a ruled resist or a photoetch resist copied from a glass master, or (*iii*) engraving directly into the polished surface with a sharp diamond.

Glass scales are also commonly used which can be easily polished, are stable, and are able to work in both transmitted and reflected light. Though ordinary glass has 15—30 per cent less thermal coefficient of expansion than steel, but glass scales, having same thermal coefficient of expansion as steel, have been produced. The graduations on glass scales can be produced (*i*) by etching through a ruled wax or photographic resist activated by contact or projection printing, (*ii*) by depositing material in the form of very thin lines having very high edge definition and highly resistant to wear, (*iii*) by depositing photographic emulsion on glass.

Most of the scales are produced by duplicating from a master scale on ruling engine. The master scale and the scale to be engraved are placed side by side, the master scale being on a fixed table and the other on a carriage. The carriage is successively positioned until master graduations are exactly in line with a reference in the form of a photoelectric microscope, and lines marked on the scale.

Master scales are produced on ruling engines having an accurate lead screw (with correction cams for errors), for successive indexing of the required interval. More sophisticated machines use interferometer feedback control which measures the positioning errors against the wavelength of a standard light source and takes corrective action till positioning error is zero. Thus intervals can be divided to a great accuracy. All this work is done in rooms having controlled atmosphere.

6.11.1. Readout Systems for Scales. Scales may be read by :

(*i*) naked eye, utilising vernier ;

(*ii*) aided by eye-piece or microscope ;

(*iii*) projection systems which are finding more and more applications as these reduce fatigue.

All scale readout systems utilise index mark, with reference to which scale reading is noted. Three types of visual scale line setting against an index are possible and are illustrated in Fig. 6.43. The type of index mark used plays a big role in the precision attainable. On sophisticated machines, the last type finds much application due to high setting precision available. It may be noted that in scales, the sharply defined edges are more critical and not the line widths.

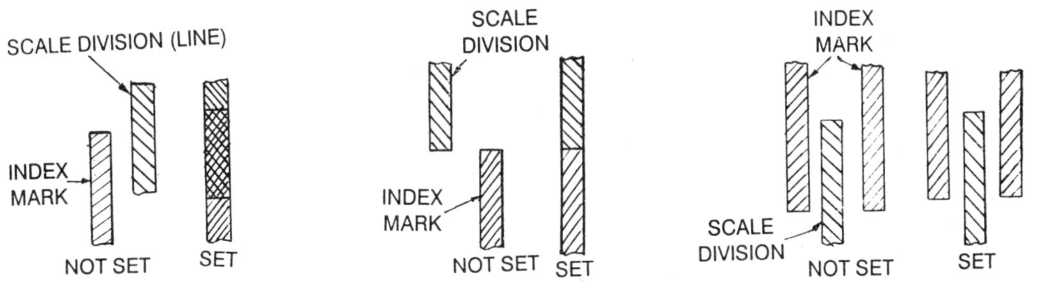

(*a*) Simple Superposition (*b*) Alignment of vernier setting (c) Symmetrical forming between two index Marks

Fig. 6.43. Types of visual scale line setting against an Index mark.

Two types of projection systems commonly used are :

(*i*) Index mark is moved mechanically until it frames the scale line, and the amount of movement required is read optically on the same screen, or externally.

(*ii*) Index line is kept fixed, and the image of the scale is moved with a suitable optical device such as a sliding lens or tilting glass plate, and the required motion to achieve settings are displayed optically on the screen.

The description of an optical projection system to read scale, employed in a universal length measuring machine is given below : (*This description is based on literature of VEB Carl Zeiss JENA.*)

Measuring Principle

In this machine, measurements are based upon direct comparison of the test specimen with built-in precision glass scales. The standard is made up of a number of aligned steel bars, which carry glass index marks at distances of 100 mm each. It combines the advantages of steel scales, *i.e.* the same coefficient of expansion as instrument and test piece, with the merits of glass scales, *viz.* excellent divisibility and parallax-free observation in transmitted light. As there is no graduation between the glass index marks on the steel bars, the intermediate values may be read in the microscope of the measuring carriage. Both the index glass marked and the graduation of the measuring carriage are imaged at a magnification of 75 × in the image plane of the microscope together with an index line by means of a special arrangement of three optically associated collimators. This arrangement compensates for any existing deviations in the guideways of the bed. The scales are located in the single focal length of the objective, and the measuring axis in the double focal length.

Description

A rugged bed carries the scale, the measuring unit and the test specimens. Measuring unit and holding devices for test specimens slide on two different guideways. This enables the measuring unit to be moved along the entire length of the test specimen and to take measurements at any optical plane of the latter. This arrangement also means that the guideways for the measuring carriage and for the scale are protected against premature wear. They are no longer overloaded by heavy test piece and cannot be damaged by awkward application of these test specimens.

The scale is rigidly connected to the bed over its entire length thus ensuring that any slight distortion that may occur, will not influence the measurement.

Tightly screwed down to the front guideway is the headstock with spindle, which is provided with a condenser for illuminating an index mark inserted into the steel scale and with a collimator lens which depicts the index.

Likewise attached to the front guideway of the bed is the two-element measuring carriage (lower and upper carriage). The lower one carries the reading microscope, with the aid of which readings down to 1 micron may be taken ; tenths of a micron being easily estimated. The upper carriage, which may be displaced relative to the lower one, can be provided with various measuring devices or with the measuring microscope.

A decimeter graduation at the front guideway of the measuring element enables the upper carriage to be roughly oriented relative to the lower one. With the aid of a clamp the lower carriage may be rigidly connected with the bed, the upper and lower carriage may also be interconnected by clamps. A slow motion drive permits of an exact setting of the lower carriage to the decimetre lines of the scale ; with a second slow-motion drive, orientation of the upper carriage relative to the test piece will be possible.

Mounted between clamping device and fine-setting of the upper carriage are spring contacts, which cause a green pilot lamp to light up, as soon as the upper carriage is pressed against the test piece (by moving the lower carriage) with a measuring pressure of about 300 g. When exceeding this pressure a red pilot lamp will light up.

Path of Rays. Refer Fig. 6.44.

The reading optics of the scales reposes on three different systems.

The optics of the first system, shown with less dense, dots, *e.g.* L_1, K_1, B_1 *MD*, F_1, M_0, M_1, M_2, M_3, M_3, M_4, M_5, M_6, is rigidly connected with the bed.

The optics of the second one (not marked by any dots) represents a definite part of the measuring equipment, which is fixed to the lower carriage and may be moved over the entire length of the three metre bed.

The third system, marked in more dense *e.g.* L_2, K_2, B_2, F_3, is a fixed part of the upper carriage onto which all measuring or aligning units may be clamped. Its range of movement on the lower carriage is 100 mm.

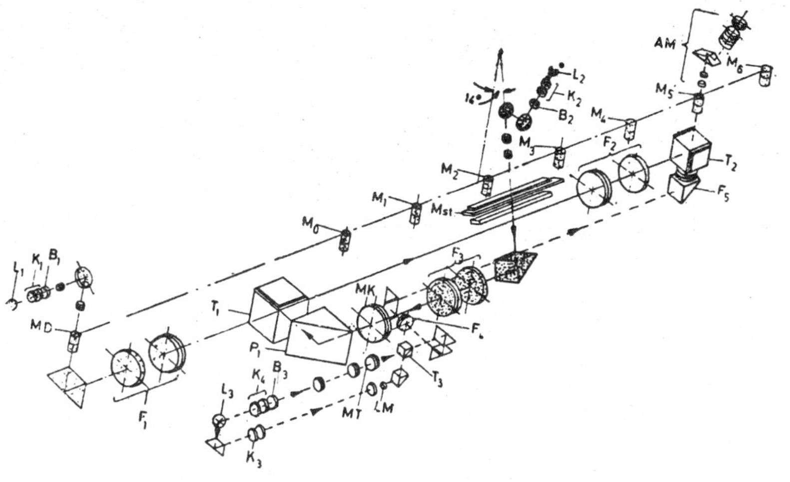

Fig. 6.44. Optics in universal length measuring machine.

By way of condenser K_1, the light of lamp L_1 of the headstock illuminates the double index mark *MD* which is imaged at infinity by the objective F_1. Passing through the beam splitting cube T_1, the beams arrive at objective F_2 and (depending on the position of the lower carriage on the bed) are then imaged in the image plane of the measuring mark M_0 upto M_{29} via beam-splitting cube T_2. In Fig. 6.45, the reading of microscope is set to measuring mark M_5. By looking into reading microscope *AM*, both image (*MD* and M_5) are set by the operator in such a way that double index mark *MD* symmetrically brackets the line of measuring mark M_5. This setting guarantees a constant, well-adjusted distance between scale *Mst* and the respective measuring marks M_1 upto M_{29}.

Via condenser K_2, the light L_2 is directed to the tenth-millimetre scale *Mst* which is imaged at infinity by objective F_3. Furthermore, the rays passing through measuring wedges *MK*, prism P_1 and beam splitting cube T_1, arrive at objective F_2, which images scale *Mst* in the plane of measuring mark M_5.

The light of lamp L_3, conducted to fluorescent mark *LM* by condenser K_3 is directed to objective F_4 via beam-splitting cube T_3, causing the flourescent mark to be imaged at infinity. By way of prisms, the light beams are conducted to objective F_5, which through beam-splitting cube T_2, images fluorescent mark *LM* in the plane of measuring mark M_5. Both, the image of fluorescent mark *LM* and of scale *Mst* are brought to coincidence with the aid of the microscope by displacing the image of the scale while turning measuring wedges *MK*. Micrometer scale *MT* is connected with measuring wedges *MK*, thus permitting the rotation of the measuring wedges to be read in microns. Illumination of micrometer scale MT is effected by filament lamp via condenser K_4, from where it is directed through beam-splitting cube T_3 to objective F_4.

It is likewise by way of prism that micrometer scale MT is directed to objective F_5, from where it is imaged through beam splitting cube T_2 in the plane of measuring mark M_5. This mark is viewed with the same reading microscope AM as used in the aforementioned paths of rays. Situated directly behind condensers K_1, K_2 and K_4 are diaphragms B_1 to B_3. They are intended to prevent irradiations in the three-partite field of vision of reading microscope AM. All scales and measuring marks (MD, $M5$ Mst, MT) are to be seen in one microscope (AM) and the measuring values may be taken at once (also refer Fig. 6.45).

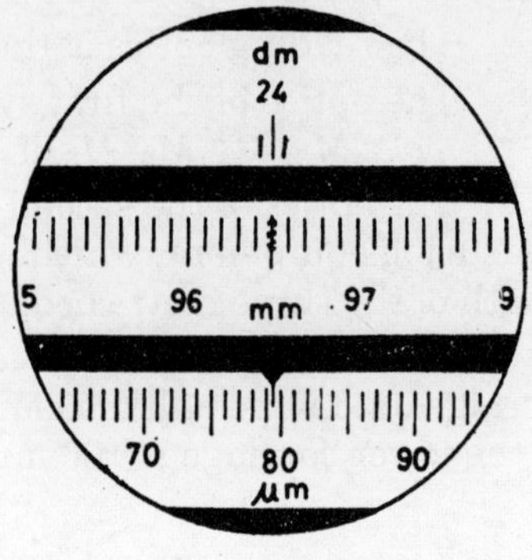

Fig. 6.45. Reading the scale.

Reading the measured value

The lower carriage of the measuring equipment is clamped in such a position that its index is in line with the decimeter mark on the metal scale. As soon as the scale line is symmetrically bracketed between the double index mark, the decimetre values are read in the upper scale of the visual field. Having done this, one scale division of the 100 mm scale is to be set symmetrically to the fluorescent thick line (to be seen in Fig. 6.45 as a dotted field) by optical deflection. The value thus received (96.5 in this case) represents the full millimetre and the tenths of a millimetre. Readings of the hundredths and thousandths of a millimetre are taken at the lower scale, ten thousandths may be estimated. (0.0794 mm in Fig. 6.45).

Total reading :

2400	reading of the higher scale
96.5	reading of the middle scale
0.079 (4)	reading of the lower scale
2496.579 (4) mm	

The optical schematic of another direct-reading optical gauge is shown in Fig. 6.46 in which the main scale is integral with the measuring spindle, and its graduations are imaged on two vernier scales and then both are projected superimposed on viewing screen. The zeroing glass has two lines engraved in it and serves a datum. The vernier scale can be slided in transverse direction and contains a line which moves up or down along incline to provide setting. For any reading, the vernier is so set that the horizontal line in it comes exactly in the centre of two index lines of zeroing glass. Then the reading can be noted from the image at ground glass screen.

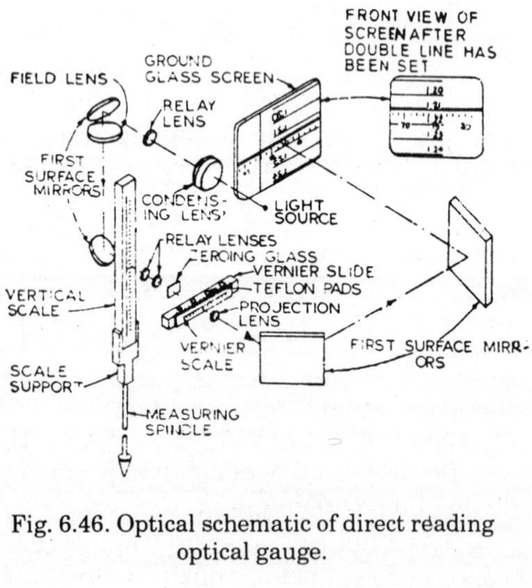

Fig. 6.46. Optical schematic of direct reading optical gauge.

In yet another system providing digital in-line scale readout (Fig. 6.47), the scale is imaged by means of relay lenses through dove prism onto plane of vernier scale which can be moved up and down by a knob. Combined image of main scale and vernier scale is projected by lens L onto a screen and viewed through the magnifier. To take reading, vernier slide with lens is moved until scale line is symmetric with index lines.

6.11.2. Gratings in Metrology. Gratings have brought about a great revolution in the methods of measurement. The visual readings of scales, micrometers, and measuring microscopes have been largely replaced by the use of transducers producing electrical signals suitable for operating counters, display units and servo devices.

Metrological transmission gratings consist of a regular succession of opaque lines separated by clear spaces of equal width. If a slit of width equal to one grating line is placed parallel to lines of grating and moved perpendicular to lines of grating over this grating, and

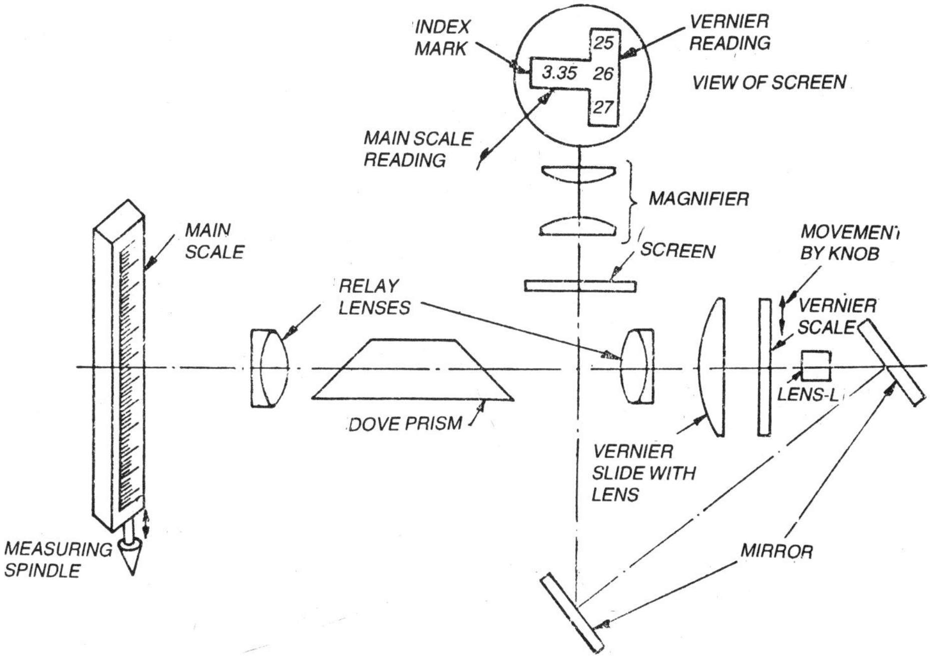

Fig. 6.47. Digital in-line scale readout device.

a photocell is placed at bottom of grating and the grating be illuminated by a normal incident light, then the photo-cell output will rise to a maximum when slit is exactly over the line and fall to zero when slit is exactly over the gap. One such cycle of maximum brightness and darkness sensed by the photocell is a measure of the traverse of one grating period. It is very interesting to note that if this slit be replaced by a short length of index grating exactly similar to the main grating, then output of photocell shall firstly be increased due to light passing through a multiple arrays of slits and secondly, the more important aspect, any local or periodic irregularities on any of the two gratings shall be averaged out or improved due to averaging effect of the multiple slits. In this way a measurement is made which is more accurate than the grating itself. This principle is utilised in manufacture of grating and not in measurement.

Moire Fringes. Moire is the name given to the patterns formed by the overlapping of two layers of fine fabrics. Moire fringes are observed when the index grating, instead of being kept parallel to the scale grating, is rotated slightly in its own plane. Under such condition, the lines of the two gratings intersect and the intersections are clearly visible as dark moire fringes running approximately at right angles to the grating lines. If the index grating is moved transversely, then the moire fringes move up or down and the position of the fringes repeats itself every time the grating has moved one grating period.

In fact when the index grating having same spacing as scale grating is moved transversely, then the fringes bisect at obtuse angle between the two gratings. Reading heads are used to measure the movement of the moire fringes. Various types of reading heads are :

(*i*) Spectroscopic, (*ii*) Normal incidence,
(*iii*) Mirror image, (*iv*) Phase modulated.

6.11.3. Formation of Moire' fringes. A metrological grating is a scale having a large number of equally spaced parallel lines. If two transparent gratings are put one over the other with their ruled surface as close together as possible and their ruling crossing at a small angle, a pattern of dark lines and fringes is formed in a direction perpendicular to that of the rulings, when the grating is observed against a uniformly illuminated background (Fig. 6.48). These fringes are called 'Moire', 'fringes'. The distance between successive moire' fringes is large as compared to the line spacing of the scale, if the angle of inclination of the two sets of lines is small. By keeping one of the grating at rest and moving the other slowly, the fringes are also found to move at a magnified rate. Thus a relatively coarse measurement of fringe displacement can give an accurate measure of the actual movement. The magnification is equal to the ratio of the distance between adjacent Moire' fringes(*s*) and the distance between adjacent lines on the grating (p_0). This method does not depend in any way on the wave nature of light. The distance between Moire fringes being much larger in comparison to spacings of grating, these can be easily counted by photocell.

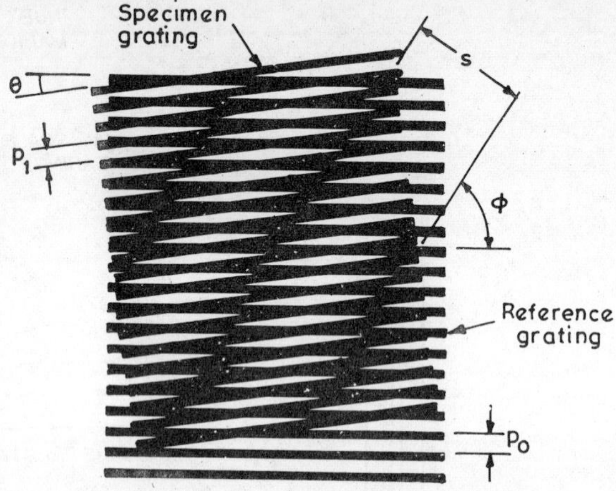

Fig. 6.48. The formation of Moire' fringes.

Fig. 6.48 demonstrates the formation of a Moire' pattern of this type. Let the line separation or pitch of the specimen grating be p_1 and the reference grating pitch be p_0, then it can be shown by geometry that the perpendicular spacing of the fringes resulting from the superimposition of the two is

$$s = p_0 p_1/[(p_0 \sin \theta)^2 + (p_0 \cos \theta - p_1)^2]^{1/2},$$

where θ is the angle between the two gratings.

The orientation of the fringes with respect to the reference grating is given by

$$\sin \phi = p_0 \sin \theta/[(p_0 \sin \theta)^2 + (p_0 \cos \theta - p_1)^2]$$

In most practical situations, the gratings are aligned with θ close to zero, then

$$s = p_0 p_1/(p_0 - p_1) \quad \text{or} \quad \frac{1}{s} = \frac{p_0 - p_1}{p_0 p_1} = \frac{1}{p_1} - \frac{1}{p_0}$$

$\frac{1}{s}$ is called **spatial frequency,** *i.e.* number of Moire' fringes per unit length, $\frac{1}{p_1}$ and $\frac{1}{p_0}$ are referred to as spatial frequency of specimen and reference gratings.

In other words, the Moire fringe frequency is the difference between the spatial frequencies of the gratings.

6.12. Moire Scales and Moire Fringes

In linear as well as rotary measurements, the grating (scales with a continuously repeating pattern of lines or grooves) play an important role. The gratings may have line spacings from 4 to 200 per mm, in lengths up 4 metre or so. These may operate by reflection (in which case the strips approximately equal in width are alternately reflecting and non-reflecting), or by transmitted light (in which case the strips approximately equal in width are alternately opaque and transmitting), and are usually employed with photoelectric readouts. As Moire scale is long and continuously ruled one having repeating pattern of lines, it is not necessary to separately indentify individual line or groove, *i.e.* have absolute measurement, but incremental measurement is best in this case. The position with Moire scale is sensed by the average of about 50 to 100 lines, and thus if a single line is out of place among 1000 accurate ones, the effect is reduced by a factor of 1000. In this scale, the response speed of the photodetectors (counting devices employed) limits the rate at which the positional information can be obtained.

Amplitude grating (also known as Ronchi rulings) is one which has plane lines, and the prismatic or phase grating is one which contains triangularly shaped, contiguous grooves. The latter type has 100 to 200 grooves per mm and its length is limited to 250 mm.

Moire fringes (a series of parallel bands, alternately dark and bright) can be observed by placing two similar continuous-type scales of gratings face to face, or nearly so, with their lines nearly parallel. The Moire fringes are readily visible to the naked eye. Distance between fringes depends on line spacing and the angle between the pair of gratings. For measurement purpose, one grating (index grating) is kept stationary and the other grating is moved over it in a direction perpendicular to the lines. When doing so, one can see the fringes moving at right angles to the motion of the scale and at a rate corresponding to exactly one fringe for a motion of one scale interval. As the Moire fringes are quite coarse particularly if scale with 40 lines per mm or coarser is used, relatively simple optics can be employed to watch Moire fringes.

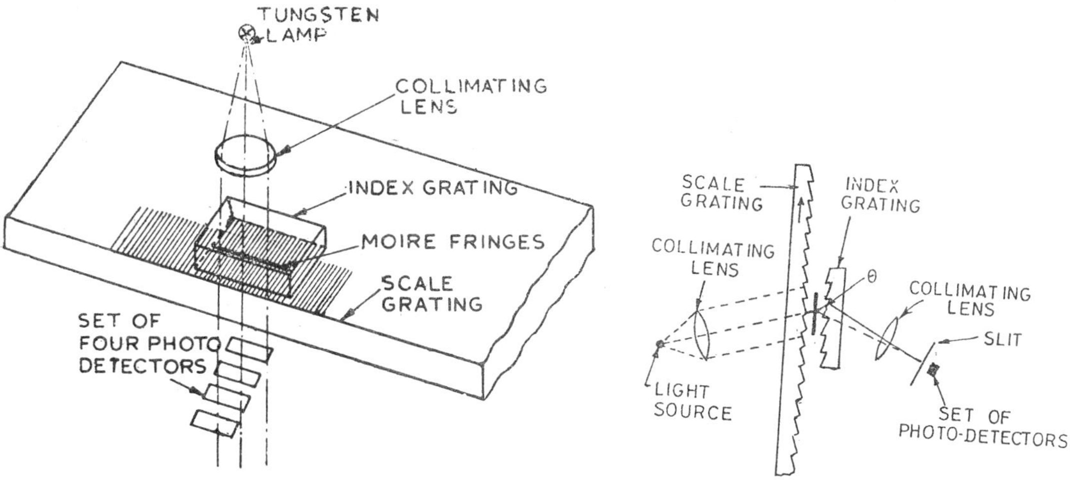

(*a*) Optical system for counting Moire fringes. (*b*) Using prismatic gratings.

Fig. 6.49

Fringes of high intensity (good contrast) can be obtained (which is essential condition for obtaining good electrical signals) only when the distance between the scale and index grating is minimum possible. The problem is not so severe for coarse gratings but it is found that for fine prismatic grating, distances (spacings) as low as 0.001 mm may be required for 200 lines per mm rulings. This condition can be achieved very easily by dispensing with the index grating altogether and arranging an optical system that images a scale back upon itself. This not only fulfils the above condition of very low spacing, but twice as many fringes are produced for a given scale motion, and further the fringe pattern is not affected by slight rotation about an axis normal to the scale.

As already indicated, for bi-directional fringe counting, two photocells spaced 90 fringe-phase degrees apart are adequate ; however, it is found that in practice better balance is obtainable with four detectors placed at 0, 90, 180 and 270 fringe phase degrees in which case counting intervals of one-quarter of the scale interval are provided.

When using prismatic grating, diffraction properties of such a grating must be considered for maximum efficiency ; the optimum arrangement being obtained for minimum deviation position [Refer Fig. 6.49 (*b*)]. Considering the central ray, it is split mainly into two approximately equal parts, one passing through straight undeviated, (zero order) and the other is diffracted by angle θ, depending on the grating spacing and wavelength of the light. It may be noted that scale grating and index grating are arranged with the grooves in opposition. With this arrangement, the first zero order ray on striking the index grating surface is diffracted in the first order direction whereas the first-order ray (*i.e.* the deviated one earlier) now goes through undeviated through index grating as the zero order. Thus both the rays are rendered parallel and these rays of equal intensity interfere with each other to form Moire fringes with nearly pure sinusoidal intensity distribution.

6.12.1. Photoelectric scale viewing system. The various advantages of photoelectric scale viewing systems are :

(*i*) Eyestrain is eliminated, as output is in the form of digital signals.

(*ii*) High precision obtainable (0.01 mm.)

(*iii*) Total carriage movement is measured directly, and intermediate drives such as lead screws, ball screws, rack-and-pinion devices, etc., are not required.

(*iv*) Setting can be done precisely and preset facilities exit for setting datum at any point relative to desired position.

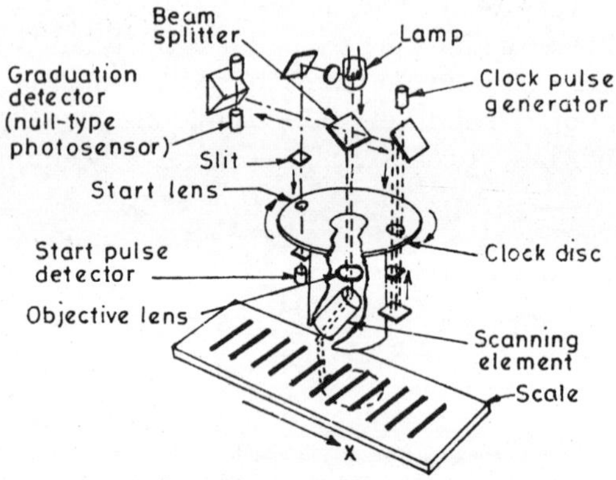

Fig. 6.50. Photoelectric scale viewing system.

In this system, two or three scales (especially designed for the purpose—having fine graduations, coarse tracks, lines of varying thickness along the scale etc., etc.) are mounted on the member whose movement is to be measured. The scales are so designed that when viewed through a rotating clock disc having alternate opaque and transmitting lines, then for every position of the scale, separate digital signals will be generated which can be interpolated and calibrated in terms of the increments of movement of length. Referring to Fig. 6.50, the position of the scale is continuously scanned by a rotating clock disc in front of an objective which images the line on to a pair of photodetectors. Alongwith the scanning, the pulses are generated in phase synchronism. The pulse duration corresponds to the least scale interval. The system also incorporates a device for starting the interval counting sequence once in each revolution. As a result of all this, a digital interpolation of the scale interval is obtained which is supplemented by information from the set of coded tracks (scales), and thus each scale interval is identified in a digital form.

6.12.2. Types of gratings used in Moire Fringe System of Measurements

Three types of gratings commonly used in Moire Fringe System of the measurement are

(*i*) Phase transmission grating, which is normally produced on selected glass blanks.

(*ii*) Metal reflecting grating, which is used for travel length greater than 1800 mm.

(*iii*) Line and space transmission grating. It includes the widely used 40 lines/mm variety which are reproduced photographically on glass.

The optical systems used also differ depending on the type of grating used.

Source of errors

The main sources of errors in measurement by moire fringe system are

(*i*) Grating error l_g. It can be taken as the calibration specification limit for the length and type of grating in use which is given for 99% confidence level.

(*ii*) Digital error l_d.

(*iii*) Analogue sub-division error l_a. It arises from tolerances in the d.c. level of signals from the reading head and in the analogue to digital converter stages. Its value is taken as 0.05 of grating pitch for glass transmission gratings and 0.08 for steel reflecting gratings.

(*iv*) Mechanical errors—l_m. Since each measurement involves two settings, it is necessary to add twice the square of first three errors to assess the accuracy of the measuring system by computing R.M.S. value.

Therefore, overall system error $l = \sqrt{(2l_g^2 + 2l_d^2 + 2l_a^2 + l_m^2)}$.

6.11.3. Leitz System. Fig. 6.51 shows in schematic form the arrangement adopted in an optical system using a linear transducer made by E-Leitz, Wetzlar. Light from a lamp passes through a condenser lens and red filter on to the scale. The image of the scale is deflected by the pentaprism through a beam splitting prism, which produces two sets of scale images, displaced relative to each other and polarised at 90°. Both images are reformed in an identical position on the scale surface by the action of a prism and objective lenses but are displaced to each other by 0.25 of the scale division width. Measuring errors cannot creep in since the light forming both images is the same.

When the scale is moved longitudinally, the images of the scale travel in the direction opposite to that the scale movement. The light beams comprising the scale images also pass through the scale to the second beam splitting prism, which separates the signals according to their direction of polarisation and reflects them through condenser lenses to the photocells P_1 and P_2.

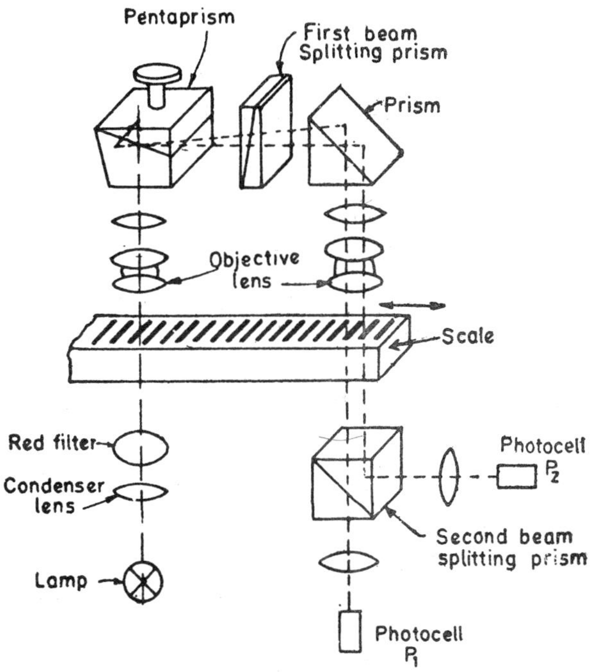

Fig. 6.51. Leitz system.

As the images traverse the scale markings, the intensity of light reaching the photo-cells is modulated, and the phases of the electrical signals produced cover 0.25 of a scale division width. Analogue signals corresponding to half the scale division width are thus obtained, and these signals can be divided electronically into 200 parts.

The intensity of light from the lamp is ensured at a constant value by sensing same by another photocell and varying the gain of the amplifier by a feedback circuit.

6.13. Contact-free Interferometric Inspection of Optical Components

To be able to measure precisely the curvature or flatness of optical components is essential for control of optical glass-working. The interferometric instrument developed for this purpose utilises the measuring ray path arranged vertically below the top of a vibration damped table. The interferogram is made visible via a video system. The entire interferometer arrangement is enclosed in a box in order to protect the fringe pattern from environmental influences. The test pieces to be tested are placed on circular knife-edges made of plastic materials or steel to ensure highly reproducible location and each part need not be adjusted separately. It utilises He-Ne laser emitting at 632.8 nm. The beam is expanded by a factor-4 Zoom collimator so as to produce a plane wave of 25 mm diameter. The beam can be further expanded to 100 mm diameter through a 4X beam a expander. A number of high-performance measuring objectives are needed to adapt to various kinds of test objects. The quality of the reference surface of the rear-most lens faces (which may have reflectance of 4%) determines the measuring accuracy of the overall system. Objectives can be changed simply by taking away one and placing the other in its position on the table top.

For focussing arrangements to move the measuring objective are made. Radius of test pieces to a measuring uncertainty of less than 1 μm can be measured. The visual evaluation of the fringe pattern is facilitated by overlaying a square grid on the VDU picture.

6.14. Precise, real-time interferometric testing of optical surfaces.

The visual method of testing interferometrically with the aid of test plates has been extensively used for testing the curvature of very small-diameter lenses with an accuracy of $\lambda/5$. By using contact-free, visual Fizeau interferometers, this error can be approximately halved. The precise testing of the surface geometries of diffraction-limited, high-performance optics demand still much higher accuracy of the order of $\lambda/50$.

Real time interferometeric testing using appropriate equipment permits testing of above order. Based on integrating micro computer and laser technology, it utilises a computer-linked Twyman-Green interferometer with real-time processing of the digitised interferogram data. The technique lies in analysing double-beam intensity distribution using phase sampling interferometry method in which the surface topology is established from the intensities for several values of the reference phase through the determination of the sample phases.

The Twyman-Green interferometer can be configured as a polarising interferometer with variable amplitude balancing between sample and reference waves. Such a polarising interferometric arrangement (shown in Fig. 6.52) has the following advantages.

(*i*) It permits testing of surfaces with wide varying reflectivities *i.e.* from *weakly* reflective surface without antireflection coating to highly reflective aluminised surfaces.

(*ii*) It avoids undesirable feedback of light reflected off the tested surface and the instrument optics.

(*iii*) It enables utilisation of the maximum of the available energy.

(*iv*) Polarisation permits phase variation to be effected with the necessary precision.

However the influences of stress birefringence of the optical media, the antireflection coats of the instrument opticals and fabrication inaccuracies of the polarisers should be suppressed, or compensated which is possible by high quality level of optical manufacture.

Special, diffraction—limited optics is used to serve as interferometric objectives. This permits testing of different curvature, and diameters of both convex and concave test surfaces.

The objectives focus near-ideal spherical waves into the centres of curvature of the surfaces to be tested. The deformation of these surfaces impresses itself on the waves reflected at these surfaces and can be visualised by interference with the standard wave reflected at the reference mirror.

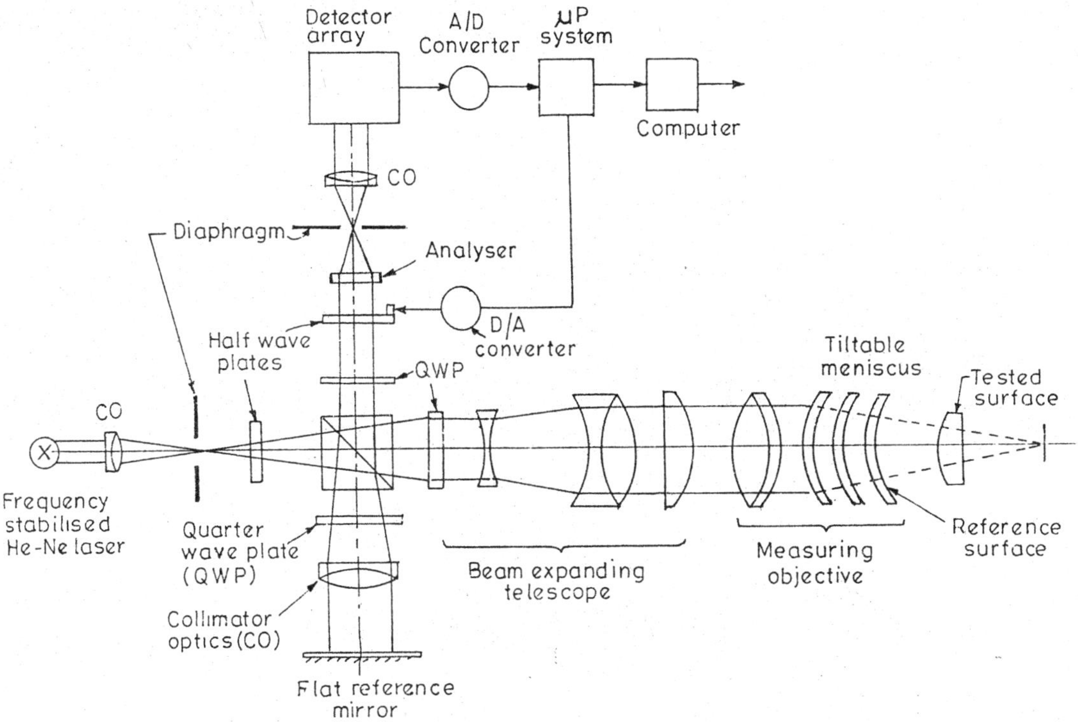

Fig. 6.52. Schematic Diagram of Twyman-Green Interferometer Arrangement.

The quasi-concentric meniscus front lenses of the interferometer—objectives must be specially designed to allow quick and easy conversion between Fizeau and Twyman—Green interferometer arrangements. In the Fizeau version, the surface of the meniscus that faces the test surface serves as reference surface and therefore has to be exactly centred. In the Twyman—Green version, however, the light reflected off that surface has to be coupled out of the sample beam. This requires the meniscus front lens to be tilted, which must not, however, produce any wave of aberration due to decentration.

With the Fizeau configuration requiring a unique reference surface definition, the rear surface of the meniscus front lens has to be tilted by 4 second for coupling out the disturbing light reflection. It is thus necessary to manufacture this meniscus deliberately as a decentred element with a corresponding wedge error. The interferometer objectives are conceived to keep the additional wave aberration caused by that wedge error within a negligible limit of $\lambda/250$. In the Twyman-green version, the front lens is tilted in such a manner that errors due to decentration, especially astigmatism, compensate each other. A special beam path is provided for the fast and precise alignment of the interferometer objectives and the surface to be tested.

For an exact measurement of the test surface, the instrument error can be determined by an absolute measurement of special precision standards made of Sitall, a vitroceram. This error is compensated by storing the same in microprocessor system and subtracting from the measurement of the test surface.

For the digital electronic analysis of the interferograms by means of the phase sampling method, a highly accurate polarising phase shifter, a special array of 32 × 32 detector elements and an 8-bit analog-to-digital converter are employed. Exact phase setting to an accuracy of 0.1 nanometer is enabled by a half-wave plate rotatable in azimuth and controllable digitally with an uncertainty of less than 1 second. Instrument control and measurement data processing are effected by a microcomputer with alphanumeric and graphic display.

Measurement data are the deviation of the tested surface from the mathematical best-fit sphere or plane as impressed on the test wave and measured at equidistant sample points. The grid sample point is established by the imaging of the test surface on the detector array. Adaptation of the test surface diameter to the array area is effected by means of a telescope of variable magnification. In this way, the deviations of any circular test surface from a best-fit sphere or plane in about 800 sample points can be determined in one measurement. The data can be displayed on CRT screen as a perspective profile, as contours, as orthogonal section through the test surface or as surface areas exceeding a specified tolerance limit. In addition to the maximum (peak-to-valley) surface deviation, the r.m.s. surface deformation can be determined.

With computer-assisted interferometers both a precise measurement of the surface geometry of lenses, mirrors and flat optical components and a wave front analysis of high-performance objectives can be accomplished.

6.15. Diffraction measurement technique

Small dimensions can be measured with high accuracy by diffraction measurement technique. When a laser beam of bigger size bends around a small object (say a wire), a diffraction pattern consisting of an array of dots extending in a line perpendicular to the wire is observed. A central undiffracted spot appears at the centre of the pattern projected on the screen. Separation of spots is inversely proportional to the diameter of the wire (D). The size of spots goes on decreasing from undiffracted spot. (Refer Fig. 6.53)

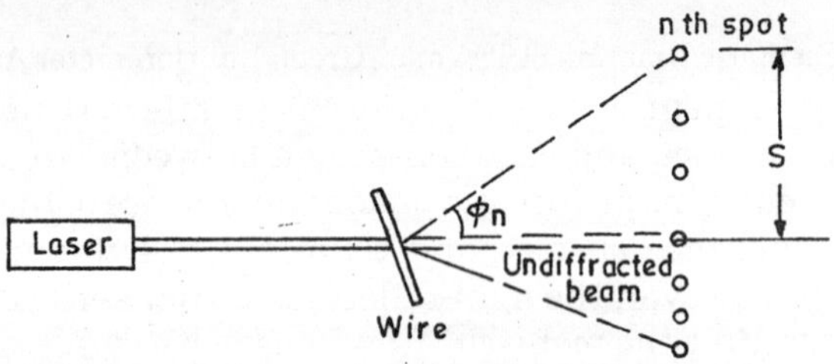

Fig. 6.53. Diffraction measurement technique.

Thus for nth spot, $\sin \phi_n = \dfrac{n\lambda}{D}$, λ = wavelength of laser beam.

Distance s for nth spot can be easily measured. Bright laser source is needed for success of this technique. An array of photodiodes or a TV vidicon detects the diffraction pattern. The signal from the detector is analysed to measure the separation of the maxima in the diffraction pattern and thus providing a direct readout of the wire diameter. It would be seen that this is a contactless method and the movement of wire along the length does not affect the diffraction pattern and thus wires can be measured while moving. The only limitation is that wire should continue to remain within the laser beam.

WORKED EXAMPLES

Problem 6.1. *What are the necessary conditions for interference of light waves ?*

Solution. The following conditions should be satisfied. To observe the phenomenon of sustained or continuous interference of light waves :

1. The two sources of light should be coherent, *i.e.*,

(*a*) The two sources should continuously emit waves of the same wavelength or frequency.

(b) For obtaining interference fringes, the amplitudes of the two interfering wave trains should be equal or very nearly equal.

(c) The two sets of wave-trains from the two sources should either have the same phase or a constant different in phase.

2. The two sources should be very narrow.

3. The emitting a set of interfering beams should be very close to each other.

Problem 6.2. *At what angle the bands should be viewed in interferometric measurement and what is the effect if it is too large.*

Solution. The line of sight for viewing the bands should be nearly perpendicular to the reference surface of the optical flat. If viewing angle varies by 5°, then no error is produced. However, when the viewing angle is bigger, then actual fringes will be read less and errors of around 15%, 40% and 100% may occur with viewing angles of 30°, 45° and 60° respectively.

Problem 6.3. *An object, placed 1200 mm from a convex lens forms an inverted image at a screen 800 mm beyond the lens. Determine the focal length of the lens.*

If the object is brought to a distance of 120 mm from the lens, then what will be the position and nature of the image formed.

Solution. From optical principles of formation of images using lenses,

$$\frac{1}{\text{Distance of object from lens } (l_0)} + \frac{1}{\text{Distance of image from lens } (l_i)} = \frac{1}{\text{Focal length of lens } (l_t)}$$

or $$\frac{1}{200} + \frac{1}{800} = \frac{1}{l_f}, \text{ or } \frac{4+1}{800} = \frac{1}{l_f}, \text{ or } l_f = \frac{800}{5} = 160 \text{ mm}$$

$\therefore$ Focal length of lens = 160 mm

When l_a = 120 mm, Then $\frac{1}{120} + \frac{1}{l_i} = \frac{1}{160}$

or $$\frac{1}{l_i} = \frac{1}{160} - \frac{1}{120} = \frac{3-4}{480} = -\frac{1}{480}, \text{ or } l_i = -480 \text{ mm.}$$

Distance of image = – 480 mm.

which means that the image is formed on the same side of the lens and obviously it will be erect magnified four times and virtual one.

Problem 6.4. (a) *Explain why monochromatic light is used for interferometry work and not the white light.*

(b) *On what factor the accuracy of laser interferometer mainly depends ? What is the biggest advantage of interferometer ?*

Solution. (a) As the white light contains a whole spectrum of wavelengths and since the pitch of the interference fringes will be different for each, the interference fringes formed will be mixture of all and it becomes very difficult to distinguish the various dark and light fringes. The whole pattern looks quite blurred and as the air gap between optical flat and the surface to be tested increases, it becomes absolutely impossible to distinguish the dark and light fringes at any one point.

In the case of monochromatic light, the spread of wavelength is very small and thus fringes are formed at considerable separations of optical flat and surface. The interference fringe pattern is much more clearly defined.

(b) The accuracy of laser interferometer mainly depends on the stability and reproducibility of wavelength of light from laser sources which is usually of the order of 0.1 part per million.

Interferometer has advantage of accuracy, portability and negligible wear or ageing.

Problem 6.5. *Fig. 6.54 (a) and (b) show the two patterns of fringes on non-flat surfaces. When pressure is applied towards left edge of optical flat, the new patterns of fringe for (a) is shown in Fig. 6.54 (c), and for (b) it remains same. But when pressure is applied in centre for (b) the new pattern of fringes as shown in Fig. 6.54 (d) is obtained. State which is convex surface and which is concave surface.*

Solution. Since fringes are closer towards the outer edges in both (*a*) and (*b*), both are spherical surfaces. Since application of pressure towards left edge of optical flat results in wide fringes below it, and too many fringes towards right, this surface is convex.

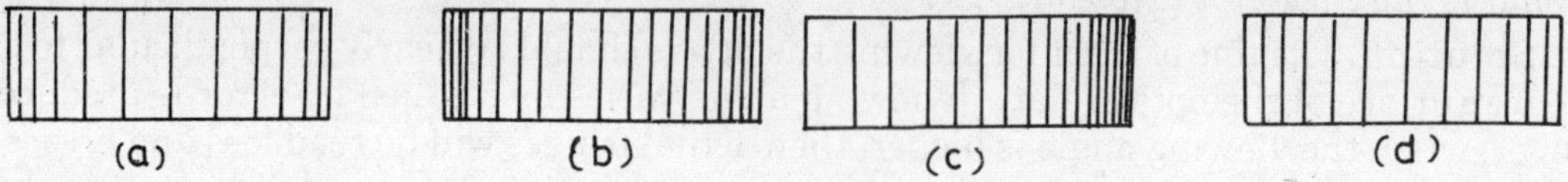

Fig. 6.54

Is case of surface at (*b*), since no change is observed by pressing optical flat at left edge, it means there is high point below it. Further since fringes in the centre and throughout become wider by applying pressure in the centre, this surface is concave.

Problem 6.6. *What do the following fringe contours obtained in checking gauge blocks by means of optical flat indicate.*

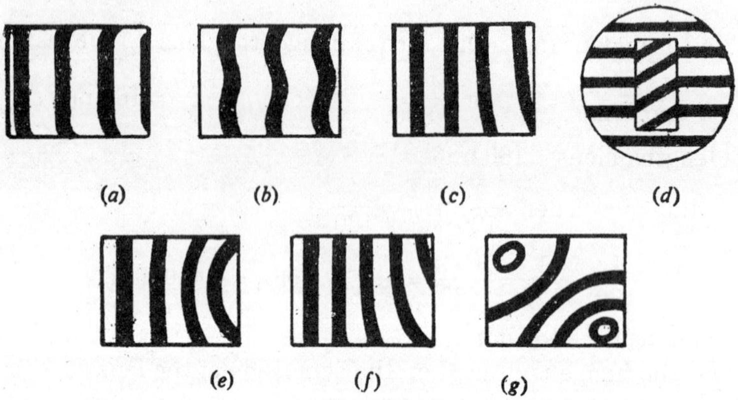

Fig. 6.55

Solution. The above seven figures indicate following characteristics of surface :

(*a*) Rounded edges or characteristic wear at edges on gauges.

(*b*) A depression in middle bottom region.

(*c*) Sloping area towards right bottom side.

(*d*) Gauge block face is not set parallel to that of the optical flat.

(*e*) Partly flat—changing or hump or hollow.

(*f*) Partly flat—falling off to or sloping hump.

(*g*) Two hills or two valleys or a hill and a valley.

Problem 6.7. *Fig. 6.56 (a), (b) and (c) show three patterns of fringes obtained while testing three flat gauges for their taper on a base plate with an optical interferometer.*

Determine the taper of gauge surface if half wavelength of light used is 0.5 μm. State the limitations of this method.

Solution. In case (*a*), since fringes on both are equal, gauge surface is both flat and parallel to the base. In case (*b*), gauge surface is flat but not parallel to the base in the direction *AB*.

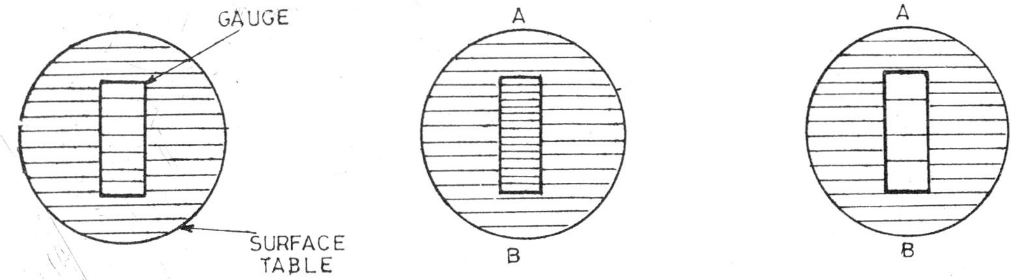

(*a*) Fringes on gauge and surface plate are straight, parallel and equifaced

(*b*) Fringes on gauge are narrower (16.5) than plate (6)

(*c*) Fringes on gauge are wider (3.5) than surface plate (6)

Fig. 6.56

Amount of taper = $(16.5 - 6)\frac{\lambda}{2} = 10.5 \times \frac{0.5}{2} = 2.625$ μm.

Similarly in case (*c*), amount of taper = $(6.3.5)\frac{\lambda}{2}$ $2.5 \times \frac{0.5}{2} = 2.625$ μm.

This method can be used upto gauge length of 25 mm, beyond which the fringe pattern on the base plate becomes difficult to observe. In such cases, fringes are observed by rotating the table alone (so as to rotate gauge and not optical flat) by 180° and difference of fringes on two cases gives degree of taper.

Problem 6.8. *Fig. 6.57 shows the fringe pattern on a gauge and the surface plate. Analyse the gauge surface.*

Solution. Since the fringes on gauge are parallel and equidistant, the surface is flat but it makes a compound angle with the base of surface plate. If the air wedge is increasing in the direction shown, then the direction on the latter is indicated by the arrow which is perpendicular to the fringe direction. Since fringe spacing on the gauge is less than the base fringe spacing, corner *A* is the lowest point on the gauge. Further as the surface is flat, corner *C*, is the highest point. Also the fringes along *DA* are less compared to fringes along *DC*, therefore, *D* is lower that *B*. The order of descending height, therefore, w.r.t. *C* is *B*, *D*, *A*.

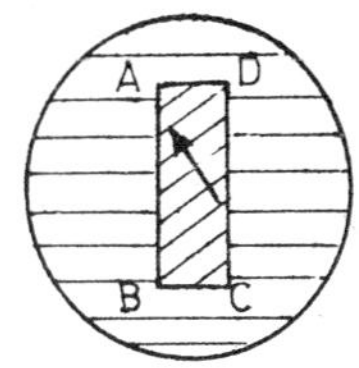

Fig. 6.57

Problem 6.9. (*a*) *Explain how flatness errors of lapped surfaces are measured with an optical flat.*

(*b*) *Sketch typical fringe patterns which might be observed on slip gauges which are (i) concave by 0.083 μm, and (ii) convex by 0.4 μm over their widths, using cadmium green light of wavelength λ = 0.5 μm. Show the fringe pattern that will be observed on the slip gauges by viewing through optical flats placed inclined on them.*

Solution. (*a*) When a lapped surface is observed by viewing through an optical flat placed on it in inclined position, a fringe pattern consisting of bright and dark bands is observed. The bands represent equal separation between the optical flat and the surface. Thus, on a concave surface, a given separation is nearer the apex of the air wedge at the centre of surface than at the sides, and as a result of this, the bands will be pointing towards the line where the optical flat and the lapped surface are in contact with each other (Fig. 6.57). On the other hand, for convex surface where the centre is higher than the sides. The bands will be pointing away from the line where the optical flat and the surface are in contact with each other. It may be remembered that the distance between one fringe and the next represents a change in separation by one-half wavelength of light.

(*b*) Pitch, *i.e.* distance between any two fringes (dark or bright) corresponds to $\lambda/2$, *i.e.* 0.5/2 = 0.25 μm slope of optical flat.

(*i*) When slip is concave by 0.083 μm, curvature is 1/3rd pitch. Since surface is concave, the crests of the bands will point towards intersection of optical flat and slip surface. [Refer Fig. 6.58 (*a*)].

(*ii*) When slip is convex by 0.4 μm, curvature is 1.6 times the fringe pitch with crests away from intersection. [Refer Fig. 6.58 (*b*)]

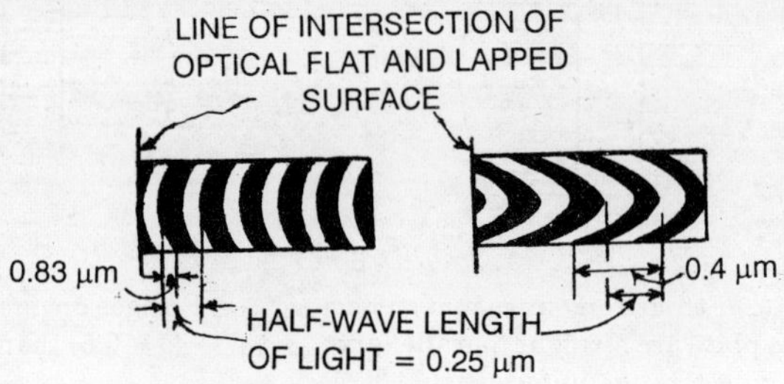

(*a*) Concave surface (*b*) Convex surface

Fig. 6.58

Problem 6.10. *Fig. 6.59 shows a set up in which a standard gauge S 25 mm high and 100 mm wide is wrung to a surface and is in close contact with a cylinder C of 20 mm diameter which in turn abouts an unknown gauge, U, 10 mm wide whose upper surface is not parallel. It is found that optical flat makes contact at point A on unknown gauge. Determine the height of other corners w.r.t. corner A. Take* $\lambda = 0.5\ \lambda m$.

Solution. Let h be the height of corner A above gauge S.

Height of edge FH w.r.t. EG is equal to

$$7 \times \frac{\lambda}{2} = 7 \times \frac{0.5}{2}\ \mu\text{m} = 1.75\ \mu\text{m}$$

$$\therefore \quad \frac{h}{30} = \frac{1.75}{10} \text{ or } h = +\ 5.25\ \mu\text{m}$$

Since there are 6 fringes on AB and 7 fringes on EF, B is higher than A by

$$(7-6)\frac{\lambda}{2} = 1 \times \frac{0.5}{2} = +\ 0.25\ \mu\text{m}.$$

Point C is lower than A by

$$(0-10) \times \frac{\lambda}{2} = -\ 10 \times \frac{0.5}{2} = -\ 2.50\ \mu\text{m}$$

Point D is higher than C by

$$(7-6)\frac{\lambda}{2} = 1 \times \frac{0.5}{2} = +\ 0.25\ \mu\text{m}$$

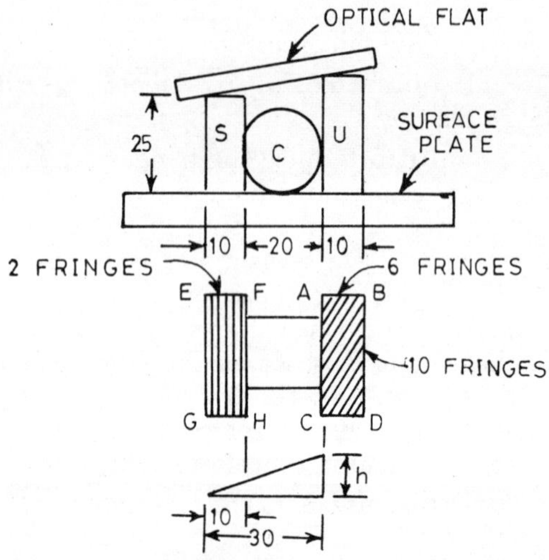

Fig. 6.59

$\therefore$ Height of D relative to $A = -\ 2.50 + 0.25 = -\ 2.25\ \mu$m.

Problem 6.11. *Explain the method of fractional coincidences in the measurement of end gauges by interferometry and explain why the unit of measurement is a half wavelength ?*

Solution. In the interferometres only the final fractier of half wavelength can be measured and it is not possible to count the whole number. By using several wavelengths, the combination of fractions which occurs for a given length can occur only for that length, within quite a wide range of length which could not be mistaken. With the help of special slide rules, it is possible to find the length of the error from the nominal length for a measured coincidence of fractions. The unit is a half wavelength because a change in length of the gauge by half-wavelength causes a path change of a whole wavelength and a shift of one interference fringe pitch.

Problem 6.12. (*a*) *What are the essential conditions to be observed while measuring absolute length of a gauge by an interferometer.*

(*b*) *What is the principle of absolute length measurement by interferometer ?*

(*c*) *Explain how an error in length can be determined when measurement is made using a number of scales. Illustrate with an example.*

Ans. The following are the essential conditions to be observed in measuring the absolute length by an interferometer :

(*i*) Gauge must be fully wrung to the platen and the gauging surfaces must be flat and parallel to each other.

(*ii*) The position of the reference surface must be adjusted so that the gauge and platen fringes are parallel and square to the sides of the gauge.

(*iii*) Temperature should be stabilised.

(*iv*) Corrections for temperature, humidity and atmospheric pressure must be made.

The measuring scale unit in absolute length interferometry is one-half wavelength $\lambda/2$ of the monochromatic light. Thus the length of a gauge can be expressed as a number of half wavelengths plus a fraction of one-half wavelength.

When the gauge is an exact number of half wavelength units long, then the fringes observed on the gauge will coincide with the fringes observed on the platen. Also the fringes on the gauge will be displaced relative to the platen fringes when the gauge length is any number of half wavelengths plus a fraction and the displacement of fringe on gauge with reference to fringes on platen will be proportional to this fraction.

Now if a gauge of a given nominal length be observed under 3 or 4 monochromatic lights, then the half wavelength fractions for each light can be calculated. However if the observed fractions differ from the calculated fractions, then an error in gauge length is indicated. The fractions observed, taken as a group, can be assumed to be unique to the length of the gauge under observation, *i.e.* no other gauge error can produce the same group of fractional values within the region of the expected error in gauge length. Since it is possible to determine error in length when measurement is made using a number of different scales, so actual length can be measured.

Let us illustrate it with an example. Let the nominal length of a gauge be 25 mm and the actual length be 26 mm, *i.e.* error is 1 mm. Let us use four measuring scales having 3, 4, 6 and 7 mm as the measuring unit.

Then scale fractions for nominal length will be

		Scale fraction
Scale 1	25/3 = 8.33	0.33
Scale 2	25/4 = 6.25	0.25
Scale 3	25/6 = 4.17	0.17
Scale 4	25/7 = 3.57	0.57

Scale fractions for actual length are :

		Observed fraction
Scale 1	26/3 = 8.67	0.67
Scale 2	26/4 = 6.50	0.50
Scale 3	26/6 = 4.33	0.33
Scale 4	26/7 = 3.71	0.71

The difference between the two group of fractions when added algebraically to the nominal length yields the actual length. The evaluation of the error from the calculated and observed fraction is shown in the following table.

Scale	*Calculated fraction*	*Observed fraction*	*Obs—Cal.* (3) – (2)	*Calculated error 4 × Scale unit*
(1)	(2)	(3)	(4)	(5)
1	0.33	0.67	0.34	0.34 × 3 or 1
2	0.25	0.50	0.25	0.25 × 4 or 1
3	0.17	0.33	0.16	0.16 × 6 or 1
4	0.57	0.71	0.14	0.14 × 7 or 1

It would be noted that the results in Col. 5 indicate 1 mm error.

Adding this error to the nominal length, the actual length is determined.

Problem 6.13. *The length of a 9 mm block gauge is measured on an absolute length interferometer. The results are as given below with cadmium-mercury discharge lamp*

Colour	*Radiation wavelength μm*	*Observed fraction*
Red λ_1	*0.6438*	*0.0*
Green λ_2	*0.5086*	*0.5*
Blue λ_3	*0.4800*	*0.3*
Violet λ_4	*0.4678*	*0.4*

Solution. If N is the no. of half wavelengths, and F = fractional half wavelength, and L = nominal length then for various wavelengths,

Red : $N_1 + F_1 = 2L/\lambda_1 = (9 \times 10^3 \times 2)/0.6438 = 17958.99\ \lambda_1$

Green : $N_2 + F_2 = 2L/\lambda_2 = (9 \times 10^3 \times 2)/0.5086 = 35391.27\ \lambda$

Blue : $N_3 + F_3 = 2L/\lambda_3 = (9 \times 10^3 \times 2)/0.4800 = 37500.00\ \lambda_3$

Violet : $N_4 + F_4 = 2L/\lambda_4 = (9 \times 10^3 \times 2)/0.4678 = 38477.98\ \lambda_4$

Let us draw table as shown below.

Colour	*Calculated fraction*	*Observed fraction*	*(3) – (2)*	*error (4)* $\times \frac{\lambda}{2}$
(1)	(2)	(3)	(4)	(5)
Red	0.99	0.0	– 0.99 or $\bar{1}.01$	$1.01 \times \frac{1}{2} \times 0.6438 = 0.325$
Green	0.27	0.5	+ 0.23	$1.23 \times \frac{1}{2} (0.5086) = 0.313$
Blue	0.00	0.3	+ 0.3	$1.3 \times \frac{1}{2} (0.48) = 0.312$
Violet	0.98	0.4	– 0.58 or $\bar{1}.42$	$1.42 \times \frac{1}{2} (0.4678) = 0.332$

Mean error = 0.32 ; ∴ actual length = 9 + 0.32 = 9.32 mm.

Problem 6.14. *Reproduced below are the nominal fractions for metric key values for cadmium radiation.*

Value (mm)	*Red*	*Green*	*Blue*	*Violet*
10.0	*0.118*	*0.806*	*0.222*	*0.720*
1.0	*0.312*	*0.481*	*0.722*	*0.172*
0.1	*0.631*	*0.248*	*0.672*	*0.517*
0.01	*0.063*	*0.325*	*0.666*	*0.752*
0.001	*0.106*	*0.932*	*0.167*	*0.275*

From the above table, compute the nominal fraction that will be observed with cadmium red radiation for a gauge length of 1.455 mm.

Solution. For 1.0 mm length, fraction $= 0.312 \times 1 = 0.312$

For 0.4 mm length, fraction $= 0.631 \times 4 = 2.524 = 0.524$

For 0.05 mm length, fraction $= 0.0663 \times 5 = 0.315$

For 0.005 mm length fraction $= 0.106 \times 5 = 0.530$

$\overline{0.681}$

$\therefore$ for 1.455 mm length, fraction = 0.681.

Problem 6.15. *Calculate the nominal fractions used in interferometer measurement using cadmium radiation for the following two gauge sizes, in the colours stated.*

10.00 mm gauge in cadmium red and green.

0.5 inch gauge in cadmium blue and violet.

What is the angle between the opticcl flat and the gauge surface when the pitch of the green fringes is 1 mm ?

The number of half-wavelengths per mm are as :

Red	*3106.31180*	*Green*	*3932.48064*
Blue	*4166.72222*	*Violet*	*4272.17205.*

Solution. The nominal fractions are derived by multiplying the number of half wavelength/mm by the gauge length. It may be noted that it is sufficient to use only the figures adjacent to the decimal point.

For 10.00 mm gauge,

Number of half-wavelengths of cadmium red. $= 3106.31180 \times 10 = 31063.1180.$

$\therefore$ Fraction only $= .1180 \approx 0.12$

Similarly fraction for cadmium green $= .48064 \times 10 = .80.$

For 0.5 inch gauge, *i.e.*, $0.5 \times 25.4 = 12.70$ mm gauge,

fraction in cadmium blue $= (6.7222 \times 12.7) \approx 0.37$; and fraction in cadmium violet

$= (2.172 \times 12.7) \approx 0.68.$

Problem 6.16. (*a*) *What are the advantages of light standards of wavelength ? How it became possible ?*

(*b*) *What is the advantage of using laser beam in interferometry ?*

(*c*) *What are the light sources and wavelengths employed in NPL interferometer ? And how the various wavelengths are obtained ?*

(*d*) *Fig. 6.60 shows the view obtained in testing a slip gauge wrung over a flat surface and viewing through an optical flat in the light of cadmium having wavelength of 0.644 μm. Determine the change in thickness of gauge along the length.*

Solution. (*a*) Light standard defines the length in terms of a standard which is not only constant, but also reproducible anywhere in the world. This is the major criterion for any standard. It does not depend on reference to some particular and possibly vulnerable piece of metal. It became possible because at a constant pressure and temperature, each pure colour of light from a vaporising element has a particular and constant wavelength, and with the advent of nuclear physics it was possible to obtain pure isotopes of various elements, serving as very pure monochromatic light sources.

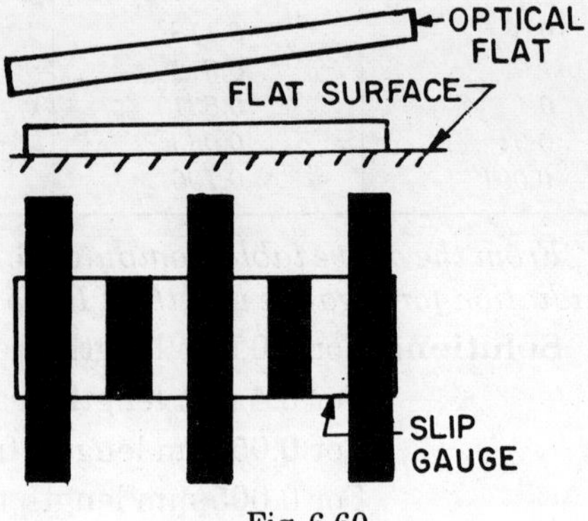

Fig. 6.60

(*b*) The laser provides a source of coherence and truly monochromatic light. Non-laser light is incoherent and does not exactly follow the sinusoidal wave, but is subject to small random variations. The property of coherence in laser beam enables it to be projected in a narrow pencil of beam (without any scatter).

(*c*) In NPL interferometer, two light sources having following wavelengths are used :

Mercury-198 isotope — Violet wavelength ; Green wavelength ;
2-yellow wavelengths ; — Cadmium hot-cathode discharge lamps
5-red wavelengths

The various wavelengths are obtained from the spectrum of the lamp by rotating the constant deviation prism to a suitable position.

(*d*) From interference fringe pattern shown in Fig. 6.60, it is observed that there are about $4\frac{1}{2}$ fringes along the length of the gauge and $2\frac{1}{2}$ fringes in the corresponding length of the flat surface. The separation between the top and bottom surfaces of the gauge is corresponding to $4\frac{1}{2} - 2\frac{1}{2}$, *i.e.* 2 fringes or 2/2 × wavelength of light *i.e.* $\frac{2}{2} \times 0.644\ \mu m \simeq 0.65\ \mu m$.

As the number of fringes on slip gauge is more compared to flat surface, the taper on gauge is opposite to the direction of inclination of optical flat.

Problem 6.17. *Estimate the difference in height between the slip gauge pile and the unknown specimen shown in Fig. 6.61. The fringes are observed using a cadmium light source having wavelength of 0.509 μm.*

Solution. (*a*) Over the distance *x*, *i.e.* separation of specimen and slip gauge, the number of fringes is nearly equal to $4\frac{3}{4}$.

∴ Change of separation of the optical flat and the base over distance *x*.

$$= \frac{4\frac{3}{4}}{2} \times \text{wavelength} = \frac{19}{8} \times 0.509$$

$$= \frac{9.671}{8} \approx 1.2\ \mu m.$$

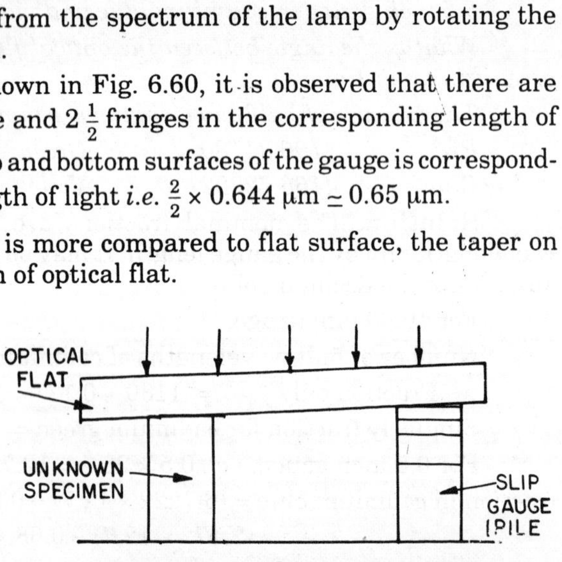

Fig. 6.61

(The result to an accuracy of 1 decimal place is quite adequate as the number of fringes can be estimated only roughly).

Problem 6.18. *The height of an unknown block (U) is to be determined in comparison to a standard specimen (S) of height 100 mm by interferometric methods. Fig. 6.62 shows the interference pattern on these two blocks S and U using red light of cadmium having wavelength of 0.64485037 microns. Determine the height of the four corners of unknown block (U). The width AB and EF of blocks is equal to 25 mm and distance between two blocks (i.e. BE) is also = 25 mm.*

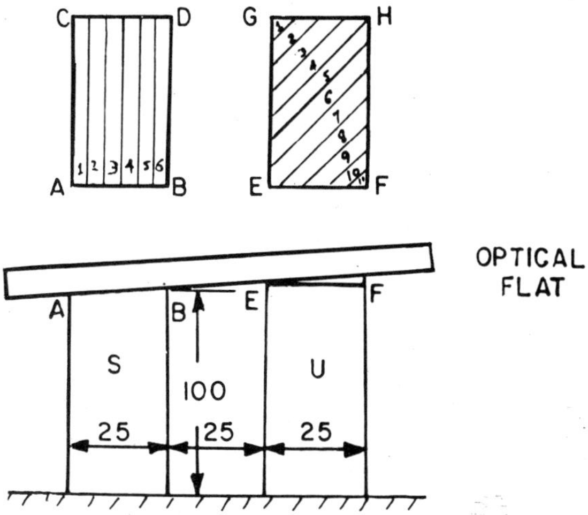

Fig. 6.62

Solution. From Fig. 6.62, it is noted that 6 equispaced straight fringes are observed over standard block S. Since face AB is perfectly flat and lapped one, this can allow slope of optical flat to be determined. It is obvious that optical flat is making contact with S at AC.

$$\text{Slope of optical flat} = 6 \times \frac{1}{2} \text{ over 25 mm} = 6 \times \frac{0.00064485037}{2} \text{ mm}$$

$$= 6 \times \frac{0.000645}{2} = 0.001835 \text{ mm/25 mm}$$

and it is in a direction perpendicular to the fringes.

Further observation of fringes above U indicates that optical flat is making contact at point G.

Height of point G = 100 + 0.001835 × 50/25 = 100.00367 mm

Point E is obviously lower than point G.

No. of fringes between G and E are 7

$$\therefore \text{ Height of corner } E = \text{height of } C - 7 \times \frac{\lambda}{2} = 100.00367 - 7 \times \frac{0.000645}{2}$$

$$= 100.00367 - 0.00226 = 100.00141 \text{ mm.}$$

As the optical flat is inclined at 0.001835 mm/25 mm, therefore, there will be air gap between points H and F w.r.t. G. Thus even if points H and F are at same distance as G, the heights of H and F will be equal to height of G + 0.0018350 = 100.0367 + 0.001835 = 100.005505 mm.

Depth of actual gap at H is corresponding to 4 fringes

$$\therefore \text{ Height of } H = 100.005505 - 4 \times \frac{0.000645}{2} = 100.005505 - 0.00129 = 100.004215 \text{ mm.}$$

Similarly, point F is 11 fringes away from G and, therefore, the height of point F

$$= 100.005505 - 11 \times \frac{0.000645}{2} = 100.005505 - 0.003542 = 100.002063 \text{ mm.}$$

Problem 6.19. *The parallelism and flatness of the faces of a micrometer are tested with a set of parallel-faced optical flats. Both micrometer faces are found to be flat within one fringe and the maximum and minimum number of parallelism fringes are 16 and 4 respectively, at* *rotational positions of the spindle 180° apart. What will be the number of fringes and their direction when the spindle is at 90° to the first position ?*

Solution. The error in squareness to the spindle axis of each face is given by half the sum and difference of the fringe numbers at the two positions. The angle of each face relative to the plane normal to the axis can be found.

Let $\lambda/2$ = height unit for 1 fringe and let side of plane be unity. Consider two planes *ABCD* and *EFGH* inclined to a reference plane (normal axis) to given angles 81 in opposition and 2λ in conjunction.

Individual values will obviously be 5λ and 3λ.

At 90° position, planes through the faces of micrometer meet at points *B* and *E* (and *J* being obtained by producing *DA* and *GH*).

$\therefore$ Triangles *FFC* and *JHA* are similar ; $\therefore$ $JH = \frac{3}{5}$ Units since $KD = 5\lambda$ and $HA = 3\lambda$

These planes intersect along *EJ* such that $\theta = \tan^{-1}\frac{3}{5} = 31°$

Obviously maximum angle between the plane passing through micrometer faces, in 90° position lies in plane normal to *EJ*.

Since angles of planes are very small, it can be assumed that angle = tan (angle) = sin (angle).

Now angle of the plane *ABCD* relative to reference plane is 5λ and *EFGH* is 3λ and the direction of these angles are perpendicular to each other.

Relative to the reference plane, angle of *ABCD* plane normal to *EJ*

= 51 cos 31° and the angle of *EFGH* in this plane = 3λ sin 30°

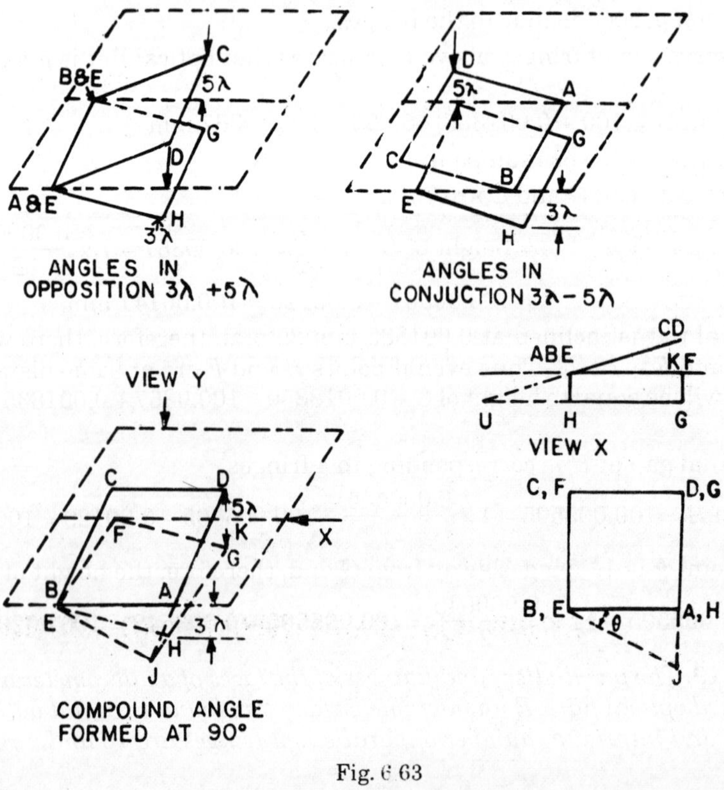

Fig. 6 63

∴ Angle between these planes = $5\lambda \cos 31° + 3\lambda \sin 31° = 5\lambda \times 0.85 + 3\lambda \times 0.52 = 5.9\lambda$

Fringes will lie parallel to *EJ*, *i.e.* at 59° to original fringes parallel to *EF*, and there will be 5.9 × 2, say 12 fringes over same area as previous 16 and 4 fringes.

Problem 6.20. (*a*) *Explain in brief the Moire fringes method of linear displacement measurement.*

(*b*) *Is it possible to use Moire fringe method for measuring angular displacement ?*

Solution. (*a*) The principle of Moire-Fringe method is that when two transparent plates, having lines ruled very close (of the order of 100 per mm) and parallel on them, are arranged so that the lines lie at a slight angle θ to each other, then the light-interference effects will cause light and dark fringes, *i.e.* fringes having nearly sinusoidal distribution of light intensity to be formed when light is directed through them (Refer Fig. 6.64). Now if one plate (of small size) moves relative to other (long and fixed one) longitudinally, then a corresponding but greatly increased movement of the fringes takes place transversely Photocells can be mounted to sense the fringes, *i.e.* distance between two sinusoidal light-intensity displacement signals, and it represents a definite distance in longitudinal direction. Relative displacement of the order of 0.0025 mm can be detected by this method. By mounting two transducers such that the voltage output curve of one is at 90° phase difference to other one, it is possible to determine the direction of motion also.

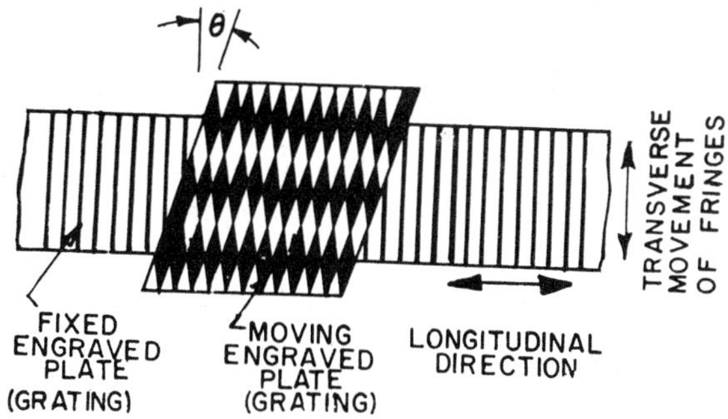

Fig. 6.64. Moire fringes.

(*b*) Moire-fringe method can be used to measure angular displacement by using radially ruled gratings.

Since one fringe pitch corresponds to a height change between optical flat and surface of one half-wavelength, therefore, angle for 1 mm pitch

$$= \frac{\lambda}{2} \text{ radian} = \frac{1}{3932} \text{ radian} = \frac{1}{3932 \times 5 \times 10^{-6}} \text{ sec.}$$

$$= 51 \text{ sec.} \qquad (1 \text{ second} = 5 \times 10^{-6} \text{ radian})$$

Problem 6.21. *A slip gauge of nominal length 10 mm was measured on a gauge length interferometer using the red, green and blue light of a cadmium lamp of which the wavelengths are :*

Red	$\lambda_1 = 0.6438\ \mu m.$
Green	$\lambda_2 = 0.5086\ \mu m.$
Blue	$\lambda_3 = 0.4800\ \mu m.$

The observed fractional displacements were respectively 0.8, 0.9 and 0.5. Determine the error in the length of the gauge.

Solution. The observed fractional displacement are $f_1 = 0.8, f_2 = 0.9, f_3 = 0.5$

Since the nominal length of the gauge = 10 mm.

∴ The numbers of half-wavelengths in 10 mm are

Red	31063.12
Green	39324.81
Blue	41667.22

$\therefore \quad N_1' = 31.063$ and $f_1' = 0.12 \sim 0.1$

$N_2' = 39.324$ and $f_2' = 0.81 \sim 0.8$

$N_3' = 41.667$ and $f_3' = 0.22 \sim 0.2$

The difference between the actual length of gauge and the nominal length

$$= \frac{\lambda_1}{2}(N_1 + f_1 - N_1' - f_1') = \frac{\lambda_2}{2}(N_2 + f_2 - N_2' - f_2') = \frac{\lambda_3}{2}(N_3 + f_3 - N_3' - f_3') \quad ...(1)$$

Since the nominal length and actual length are very close to each other.

$N_1 - N_1', N_2 - N_2', N_3 - N_3'$ are whole numbers and are relatively smaller.

Substituting the values, equation (1) becomes

$$\frac{0.6434}{2}(N_1 - N_1' + 0.8 - 0.1) = \frac{0.3219}{2}(N_2 - N_2' + 0.9 - 0.8) = \frac{0.4800}{2}(N_3 - N_3' + 0.5 - 0.2)$$

or $$0.3219\,(N_1 - N_1' + 0.7) = 0.2543\,(N_2 - N_2' + 0.1) = 0.24\,(N_3 - N_3' + 0.3)$$

By hit and trial, it will be seen that substituting

$$N_1 - N_1' = 1, N_2 - N_2' = 2 \text{ and } N_3 - N_3' = 2,$$

makes the three errors nearly equal, *i.e.* 0.3219 × 1.7, 0.2543 × 2.1, and 0.24 × 2.3 or 0.55, 0.54 and 0.55 μm respectively.

Thus the actual error can be taken as 0.55 μm. **Ans.**

Problem 6.22. *In an interferometry experiment, following patterns were observed with some typical specimen. Identify each of them.*

Solution. Since the fringes are straight and parallel in cases (*a*) and (*b*), both of them are flat. The difference in fringe pitch suggests that the optical flat is inclined at different angles on each one. The inclination of optical flat is more in case of (*b*) in comparison to (*a*).

The circular fringes with decreasing diameter at (*c*) indicate the surface to be spherical one and test of the convexity or concavity can be done by applying slight pressure at the centre of fringes. If fringes move towards the centre it is concave surface and if they move away from the centre then the surface is convex.

The elliptical pattern of fringes in (*d*) indicates that the surface is curved more in one direction than the other. The fringe pattern at (*e*) is a clear indication of smooth cylindrical surface.

The surface at (*f*) is having a ridge or valley in the middle and test of convexity or concavity can be done as suggested above in case (*c*).

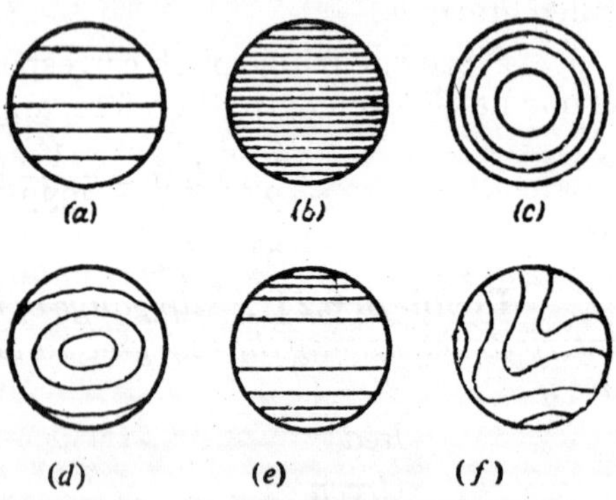

Fig. 6.65

Problem 6.23. *Optical instruments have been catalysts of scientific and technological progress. What is the importance of optical instruments ? List out various factors responsible for "renaissance" of optics.*

Solution. The developments in field of optics have gone hand in hand with achievements of the scientific-technological revolution in other fields. Optical principles have specific contribution to information capturing, measuring and machining processes. The optical principles have broad applications and optical instruments are known for their high accuracy and high reliability. The integration of precision optics, mechanics and electronics has been considerably promoted by the development of opto-electronic devices-links between information acquisition and information processing.

The integration of optical functions in systems endowed with artificial intelligence enables translation of optical principles into instrument design—the development of "instruments of refined sensory perception".

The factors responsible for renaissance of optics are

— an enormous increase in imaging performance due to computer-assisted optics design, correction and assembly.
— the availability of new optical media.
— the discovery and application of the laser
— the development of holography and coherent optics.
— the development of fibre-wave-guides.
— material structuring within micrometers.

TYPICAL EXAMINATION QUESTIONS

1. Explain the principle of measurement by light wave interference method.
   Sketch and interpret the different pattern of interference bands observed through optical flats for the following:
   (*a*) A perfectly flat surface.
   (*b*) A convex shape.
   (*c*) A concave shape.
   (*d*) A gauge block with edges rounded off.
2. Describe the use of optical flats and monochromatic light for dimensional comparison and testing flatness of surfaces.
   Sketch a set up for checking cylindrical plug gauge with reference to gauge blocks and optical flats. Explain how testing is done.
3. Explain the principle of an interferometer. How is it used in the absolute measurement of slip gauge dimensions ? Enumerate the various corrections to be applied to the measured value as estimated from the interferometer.
4. Give the basic principles of interferometry and show by means of diagrams the way in which interference fringes are produced. Sketch the optical arrangement of NPL Gauge Interferometer and explain how the relative dispositions of the interference bands are used to compute the length of a slip gauge ?
5. (*a*) Explain how the flatness and parallelism of the two faces of a slip gauge can be tested with the help of any method based on the principle of interference of light ?
   (*b*) The following data were obtained in an experiment for testing the dimension of a slip gauge with the help of a reference gauge and a pair or optical flats.
   Number of straight and equidistant fringes on each gauge = 10
   Width of each gauge = 20
   Gap between the two gauges placed on optical flats = 50 mm
   Wavelength of light used = 0.00005 mm
   Dimension of the reference gauge = 25 mm
   Determine the dimension of the test gauge.
6. What are the essential properties a light wave should have for interference work ? What are the common sources of light used for interferometry work ?
7. Explain the working of Michelson interferometer. How it has undergone modification in stages ? Also explain the working of C laser interferometer.
8. What do you understand by Moire Fringes and how these are utilised for accurate length measurement ? How can these be used for angular displacement ?
9. What is the fundamental difference between the length and flatness interferometers ?
10. At what angle to the optical flat, the fringe pattern should be observed and why ?

7

Straightness, Flatness, Squareness, Parallelism, Circularity and Rotation

Increasing demand for product reliability and efficiency has placed a corresponding emphasis on the geometric integrity of components and their assembly. In engineering applications, one often comes across the problems of measurement of geometrical parameters such as alignment, straightness, squareness, flatness, etc.

7.1. Straightness

At many places it is required that the surfaces must be perfectly straight, *e.g.*, in a lathe it is desired that tool must move in a straight path to generate perfect cylinder and it is possible only when the controlling guideways are themselves straight. Also straight line or plane is the basis of most methods of measurements. The quality of straightness in precision engineering is represented by straight edge.

The fundamental principle about straightness measurement is given by Bryan. According to Bryan principle, a straightness measuring system should be in line with the functional point at which straightness is to be measured. If this is not possible, either the slideways that transfer the measurement must be free of angular motion or angular motion data must be used to calculate the consequences of the offset.

7.1.1. Definition of straightness of a line in two planes. A line is said to be straight over a given length, if the variation of the distance of its points from two planes perpendicular to each other and parallel to the general direction of the line remains within the specified tolerance limits ; the reference planes being so chosen that their intersection is parallel to the straight line joining two points suitably located on the line to be tested and the two points being close to the ends of the lengths to be measured.

The tolerance on the straightness of a line is defined as the maximum deviation in relation to the reference straight line joining the two extremities of the line to be checked (Fig. 7.1).

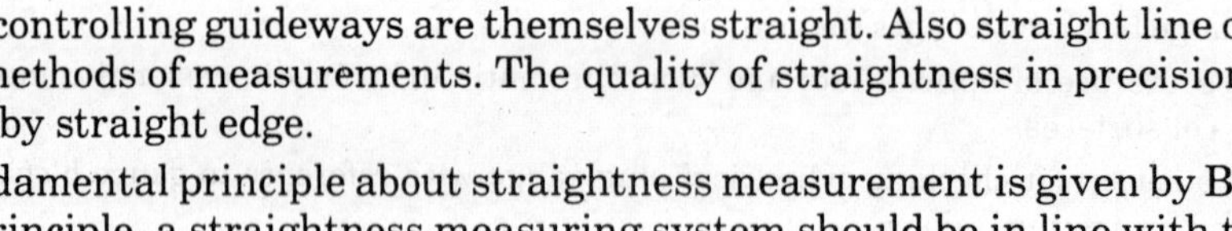

Fig. 7.1. Tolerance on straightness of a line.

It is the usual practice to state the range of measurement, *i.e.* the length to be checked ; and the position of the tolerance in relation to the reference straight line. In most cases, the parts very close to the ends, which most often have local errors of no great importance, may be neglected.

7.2. Straight Edge

A straight edge is a measuring tool which consists of a length of a steel of narrow and deep section in order to provide considerable resistance to bending in the plane of measurement without excessive weight. Precision straight edges are generally hardened and lapped on edge only to a small radius in cross-section, forming a blunt knife edge which is straight to a few thousandths of a millimetre. For checking the straightness of any surface, the straight edge is placed over the surface and two are viewed against the light, which clearly indicates the straightness. Usually the gap between the straight edge and surface will be negligible small for perfect straight surfaces and a quantitative measure of straightness could be had by observing the colour of light due to interference caused by diffraction of light while passing through the small gap. If the colour of light be red, it indicates a gap of 0.0012 to 0.0017 mm, while for blue light, gap is approximately 0.0075 mm.

A more accurate method of finding the straightness by straight edge is to place it on equal slip gauges at the correct point for minimum deflection and to measure the uniformity of space under the straight edge with slip gauges.

7.3. Test for Straightness by using Spirit Level and Auto-collimator

Tests for straightness can be carried out by using spirit level or auto-collimator. The straightness of any surface could be determined by either of these instrument by measuring the relative angular positions of number of adjacent sections of the surface to be tested. So first a straight line is drawn on the surface whose straightness is to be tested. Then it is divided into a number of sections, the length of each section being equal to the length of spirit level base or the plane reflector's base in case of auto-collimator. Generally the bases of the spirit level block or reflector are fitted with two feet so that only feet have line contact with the surface and whole of the surface of base does not touch the surface to be tested. This ensures that angular deviation obtained is between the specified two points. In this case length of each section must be equal to distance between the centre lines of two feet. The spirit level can be used only for the measurement of straightness of horizontal surfaces while auto-collimator method can be used on surfaces in any plane. In case of spirit level, the block is moved along the line on the surface to be tested in steps equal to the pitch distance between the centre lines of the feet and the angular variations of the direction of block are measured by the sensitive level on it. Angular variation can be correlated in terms of the difference of height between two points by knowing the least count of level and length of the base.

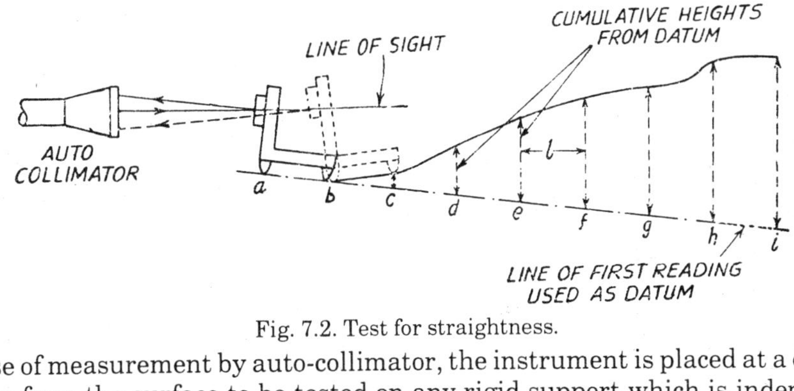

Fig. 7.2. Test for straightness.

In case of measurement by auto-collimator, the instrument is placed at a distance of 0.5 to 0.75 metre from the surface to be tested on any rigid support which is independent of the surface to be tested. The parallel beam from the instrument is projected along the length of the surface to be tested. A block fixed on two feet and fitted with a plane vertical reflector is

Position	Mean Reading of Auto-Collimator or spirit level (sec)	Difference from first reading (sec)	Rise or fall in interval length 'l' (mm)	Cumulative rise or fall (mm)	Adjustment to bring both ends to zero (mm)	Errors from straight line (mm)
(1)	(2)	(3)	(4)	(5)	(6)	(7)
a				0	0	0
$a-b$	θ_1	0	0	0	$-L/n$	$-L/n$
$b-c$	θ_2	$\theta_2-\theta_1$	$(\theta_2-\theta_1)l$	$(\theta_2-\theta_1)l$	$-2L/n$	$(\theta_2-\theta_1)l-2L/n$
$c-d$	θ_3	$\theta_3-\theta_1$	$(\theta_3-\theta_1)l$	$(\theta_2-\theta_1)l + (\theta_3-\theta_1)l$	$-3L/n$	$(\theta_2-\theta_1)l+(\theta_3-\theta_1)l - 3L/n$
...	...	...	...	...	...	...
...	...	...	...	...	...	...
...	...	...	...	...	...	...
$h-i$	θ_n	$\theta_n-\theta_1$	$(\theta_n-\theta_1)l$	$\Sigma(\theta_n-\theta_1)l = L$	$-L$	0

placed on the surface and the reflector face is facing the instrument. The reflector and the instrument are set such that the image of the cross wires of the collimator appears nearer the centre of the field and for the complete movement of reflector along the surface straight line, the image of cross-wires will appear in the field of eyepiece. The reflector is then moved to the other end of the surface in steps equal to the centre distance between the feet and the tilt of the reflector is noted down in second from the eyepiece. (Refer Fig. 7.2)

$$1 \text{ sec. of arc} = 0.000006 \text{ mm/mm}$$

Therefore, 1 sec. of arc will correspond to a rise or fall of $0.000006 \times \lambda$ mm, where λ is the distance between centres of feet in mm. The condition for initial and subsequent readings is shown in Fig. 7.2 in which the rise and fall of the surface is shown too much exhaggerated.

With the reflector set at $a-b$ (1st reading), the micrometer reading is noted and this line is treated as datum line. Successive readings at $b-c$, $c-d$, $d-e$ etc. are taken till the length of the surface to be tested has been stepped along. In order to eliminate any error in previous set of readings, the second set of readings could be taken by stepping the reflector in the reverse direction and mean of two taken. This mean reading represents the angular position of the reflector in seconds relative to the optical axis of auto-collimator. The results of measurement are tabulated as already explained in chapter 2. (Refer table above)

Column 1 gives the position of plane reflector at various places at intervals of 'l' *e.g.* $a-b$, $b-c$, $c-d$ ... etc., column 2 gives the mean reading of auto-collimator or spirit level in seconds. In column 3, difference of each reading from the first is given in order to treat first reading as datum. These differences are then converted into the corresponding linear rise or fall in column 4 by multiplying column 3 by 'l'. Column 5 gives the cumulative rise or fall, *i.e.*, the heights of the support feet of the reflector above the datum line drawn through their first position. It should be noted that the values in column 4 indicate the inclinations only and are not errors from the true datum. For this the values are added cumulatively with due regard for sign. Thus it leaves a final displacement equal to L at the end of the run which of course does not represent the magnitude of error of the surface, but is merely the deviation from a straight line produced from the plane of the first reading. In column 5 each figure represents a point, therefore, an additional zero is put at the top representing the height of point a.

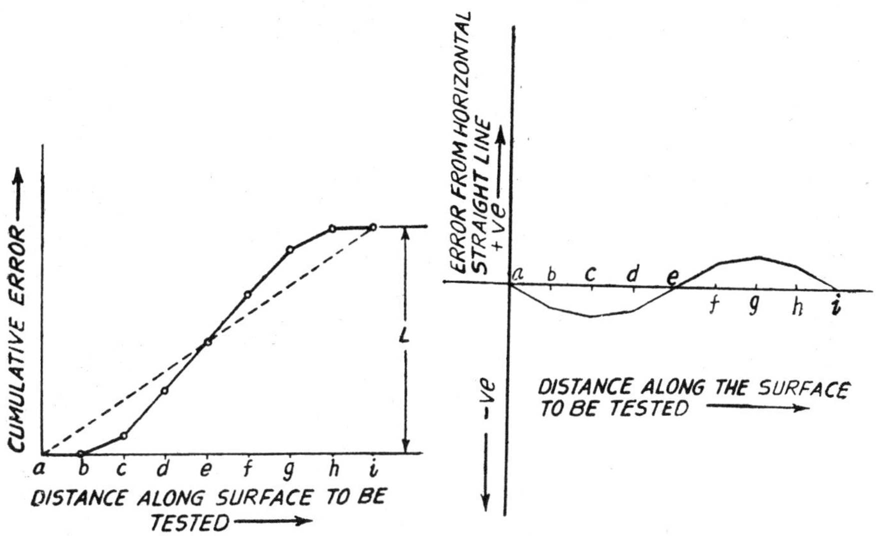

Fig. 7.3. Cumulative error ; and error w.r.t. straight edge.

The errors of any surface may be required relative to any mean plane. If it be assumed that mean plane is one joining the end points then whole of graph must be swung round until the end point is on the axis (Fig. 7.3). This is achieved by subtracting the length L proportionately from the readings in column 5. Thus if n readings be taken, then column 6 gives the adjustments $-L/n$, $-2L/n$... etc.,. to bring both ends to zero. Column 7 gives the difference of columns 5 and 6 and represents errors in the surface from a straight line joining the end points. This is as if a straight edge were laid along the surface profile to be tested and touching the end points of the surface when they are in a horizontal plane and the various readings in column 7 indicate the rise and fall relative to this straight edge.

7.4. Flatness Testing

Like straightness testing, the simplest form of flatness testing is possible by comparing the surface with an accurate surface. One of the surfaces is marked with Prussian blue and the other surface rubbed over it. The distribution of colour over the other surface gives a rough idea of high and low points on the surface. This method is usually suitable for small plates and not for large surfaces.

Mathematically, flatness error of a surface states that the departure from flatness is the minimum separation of a pair of parallel planes which will just contain all points on the surface. The deviation of a large surface such as surface table or machine table from the true plane may be determined by the use of either a spirit level or an auto-collimator. The principle of the method is same, whether the apparatus used is auto-collimator or spirit level.

According to IS : 2063—1962, a surface is deemed to be flat within a given range of measurement when the variation of the perpendicular distance of its points from a geometrical plane (this plane should be exterior to the surface to be tested) parallel to the general trajectory of the plane to be tested remains below a given value. The geometrical plane may be represented either by means of a surface plane or a family of straight lines obtained by the displacement of a straight edge or a spirit level or a light beam :

Flatness deviations (errors of flatness) are indicated as follows :

(i) ... μ or mm per metre when convexities are allowed as well as concavities ;

(ii) concave to ... μ or mm, when, between the ends, only concave surfaces are allowed ; and

(iii) convex to ... μ or mm, when, between the ends, only convex surfaces are allowed.

It is well known that a surface can be considered to be composed of an infinitely large number of lines. The surface will be truly flat only if all the lines are straight and they lie in the same plane. Let us study the case of rectangular table. From Fig. 7.4, it is obvious that all the generators (lines) are straight and parallel to the sides of the rectangle in both the perpendicular directions. Even then it is not truly flat, but concave and convex along two diagonals. Thus for the verification of a surface to be truly flat, it is essential to measure the straightness of diagonals in addition to the lines parallel to the sides. Thus the whole of the surface is divided by straight lines as shown in Fig. 7.5. The end lines *AB* and *AD* etc., are drawn away from the edges as the edges of the surface are not flat but get worn out by use and can fall off little in accuracy. The straightness of all these lines is determined and then those lines are related with each other in order to verify whether they lie in the same plane or not. In above setting of the lines, it should be kept in mind that the lines are whole multiples of the length of base of spirit level or reflector stand and the centre of the diagonals. I must be one of the contact points of the feet of the block.

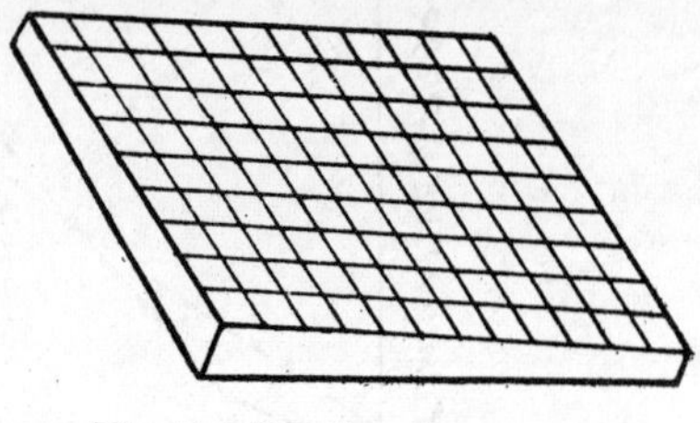

Fig. 7.4. Testing flatness.

7.4.1. Procedure for determining flatness (Fig. 7.5). The procedure for determining flatness is as follows :

(1) Carry out the straightness test already described on all the lines *AB*, *BC*, *AC* etc. and tabulate the readings upto the cumulative error column.

(2) Let a plane passing through the points *A*, *B* and *D* be assumed to be an arbitrary plane, relative to which the heights of all other points may be determined. For it, the ends of lines *AB*, *AD* and *BD* are corrected to zero and thus the height of points *A*, *B* and *D* are zero.

(3) The height of point *I* is determined relative to the arbitrary plane *ABD* = 000. As *I* is the mid-point of line *AC* also, all the points on *AC* can be fixed relative to the arbitrary plane by assuming *A* = 0 and correcting *I* on *AC* to coincide with the mid-point *I* on *BD*. In this way, all points on *AC* are corrected by amounts proportionate to the movement of its mid-point. *A* hint could be taken here that *C* is twice as far from *A* as the mid-point, the correction for *C* will be double that of *I*.

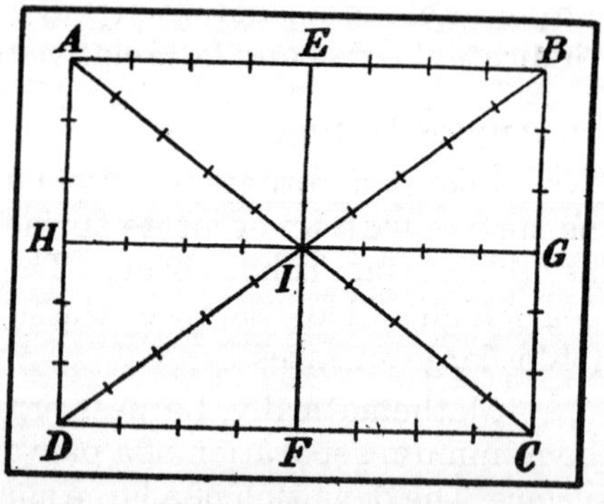

Fig. 7.5. Flatness testing procedure.

(4) Point *C* is now fixed relative to the arbitrary plane and points *B* and *D* are set at zero, all intermediate points on *BC* and *DC* can be corrected accordingly.

(5) The positions of *H* and *G*, *E* and *F* are known, so it is now possible to fit in lines *HG* and *EF*. This also provides a check on previous evaluation since the mid-point of these lines should coincide with the known position of mid-point *I*.

In this way, the height of all the points on the surface relative to the arbitrary plane *ABD* are known.

One thing to be noted here is that according to definition of flatness error, departure from flatness is determined by the *minimum* separation of a pair of parallel planes which will just contain all the points on the surface. Here it is possible to determine two points at either extreme of the reference plane to define the separation but the reference plane chosen may not be the best plane. Therefore, in order to determine the minimum separation some correction has to be made. The calculation for a final correction to determine the minimum separation of a pair of parallel planes which just contain all the points on the surface is made by graphical method as given below.

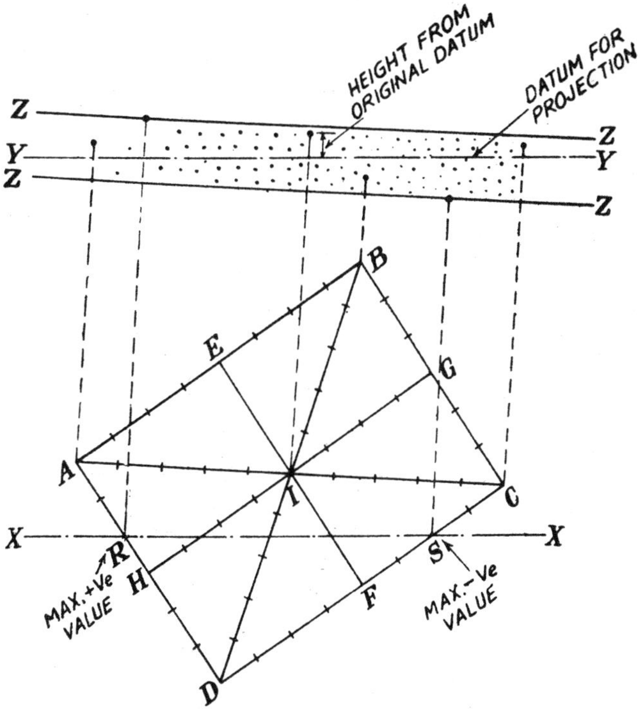

Fig. 7.6. Minimum separation of parallel planes containing all points on surface.

The various points on the surface have been determined with reference to *ABD* as reference plane as described previously. Two points on opposite sides having maximum positive and maximum negative values are selected and jointed together by a line *XX*. Let these points in Fig. 7.6 be *R* and *S*. Draw a line *YY* parallel to *XX* to represent the plane *ABD* as shown in Fig. 7.6. Set off to scale the height of all points relative to *YY* by taking projections from all the points on the surface.

In Fig. 7.6, projections from all points have not been shown for the sake of clarity. Next by inspection, draw a closest pair of parallel lines *ZZ* and *ZZ*, which will contain all of the points. It may be noted that one line will have two points on it, and the other line, one point only. The distance between these two lines is a measure of the error in flatness. Although it is not exact value but for practical purpose it gives sufficiently accurate results.

The optical flatness testing method for very flat and polished surface has already been discussed in the chapter of Interferometry.

7.5. Mathematical Treatment of Determination of Straightness and Flatness of Surfaces

For determining straightness, autocollimator is placed on a reference surface and a block containing reflector is traversed along a line on the surface whose straightness is to be determined and readings noted at regular intervals. Such observations can also be taken using precision level. These readings are first converted to give cumulative height difference from the starting and the best line through the points determined by above method can be found by applying in method of least squares, according to which the sum of the deviations of the points from the best line should be zero, and the sum of the square of the deviations from the best line should be minimum. For the first condition the centroid of the points $(\overline{X}, \overline{Y})$ must lie on this line. All the points are then retabulated taking points $\overline{X}$, $\overline{Y}$ as the zero reference. Then

$$x_1 = X_1 - \overline{X}, x_2 = X_2 - \overline{X}, \ldots\ldots x_n = X_n - \overline{X}$$

and $$y_1 = Y_1 - \overline{Y}, y_2 = Y_2 - \overline{X}, \ldots\ldots y_n = Y_n - \overline{Y}.$$

If A be the slope of the best line through the points $y = Ax$, then deviation of a point $(x_i, y_i) = \delta_i = y_i - Ax_i$. Summating all these deviations,

$$\sum_{i=1}^{n} \delta_i^2 = \sum_{i=1}^{n} (y_i - Ax_i)^2 \quad \ldots(1)$$

The slope of the best line through the observed points can be found by differentiating equation (1) w.r.t. A, and equating it to zero

i.e. $$\frac{d\Sigma\delta_i^2}{dA} = \sum_{i=1}^{n} x_i \left[2(Ax_i - y_i) \right] = 2A\sum_{i=1}^{n} x_i^2 - 2\sum_{i=1}^{n} x_i y_i = 0, \text{ or } A = \frac{\sum_{i=1}^{n} x_i y_i}{\sum_{i=1}^{n} x_i^2}$$

This slope of the best line through X, Y gives the mean true plane of the straight edge, from which errors of straightness can be determined.

For checking the flatness of a surface, first the equation of a mean plane through a number of points on the surface is calculated.

If surface to be measured is assumed to lie approximately on the X-Y plane, then the equation of the mean plane is $Z = A + BX + CY$.

The mean plane of a surface is one on which the sum of the deviations is zero and the sum of the squares of the deviations is a minimum. Accordingly the mean plane must pass through the centroid of the measured points $(\overline{X}, \overline{Y}, \overline{Z})$. By retabulating all points such that the centroid is made the origin of all the three axes, *i.e.* subtracting all values $\overline{X}$, $\overline{Y}$, and $\overline{Z}$, *i.e.* $x_1 = X_1 - \overline{X}$, $y = Y_1 - \overline{Y}$ and $z_1 = Z_1 - \overline{Z}$ etc.

The equation of plane then is $z = Bx + Cy$ and deviation from the mean plane at a point x_i, y_i, z_i is δ_i.

i.e. $$\delta_i = z_i - Bx_i - Cy_i \text{ ; or } \Sigma\,\delta_i^2 = \sum_{i=1}^{n} (z_i - Bx_i - Cy_i)^2$$

In order that $\Sigma\delta_i^2$ be minimum, $\dfrac{\partial\Sigma\delta_i^2}{\partial B}$ and $\dfrac{\partial\Sigma\delta_i^2}{\partial C}$ should be zero.

Now $$\frac{\partial\Sigma\delta_i^2}{\partial B} = \Sigma 2x_i (Bx_i + Cy_i - z_i) \text{ ; and } \frac{\partial\Sigma\delta_i^2}{\partial C} = 2B\Sigma\, x_i y_i + 2C\Sigma y_i^2 - 2\Sigma y_i z_i$$

It can now be shown that

$$C = \frac{\Sigma x_i^2\, \Sigma y_i z_i - \Sigma x_i y_i\, \Sigma x_i z_i}{\Sigma x_i^2\, \Sigma y_i^2 - (\Sigma x_i y_i)^2}, \quad \ldots(2) \quad \text{and } B = \frac{\Sigma y_i^2\, \Sigma x_i z_i - \Sigma x_i y_i\, \Sigma y_i z_i}{\Sigma x_i^2\, \Sigma y_i^2 - (\Sigma x_i y_i)^2} \quad \ldots(3)$$

Thus the true plane can be determined by marking off the surface in a uniform square grid and measuring the height of each intersection relative to a datum point and measuring the inclination of mean true plane by substituting in equations (2) and (3). The difference between the largest positive and largest-negative deviation gives the maximum error of the surface.

7.6. Laser Equipment for Alignment Testing

The use of laser permits alignment test to be carried over long distances compared to other conventional equipment like alignment telescope. The advantage of laser equipment is that it produces a real straight line that can be seen in space but the telescope provides only an imaginary line. Laser equipment for alignment testing is particularly suitable in applications like aircraft production, ship building, etc. where a number of components, spaced relatively long distances apart, have to be checked to a predetermined straight line. Other uses of laser equipment are testing of flatness of machined surfaces, checking squareness with the help of optical square, etc.

Laser equipment for alignment testing basically consists of a laser tube which produces a cylindrical beam of laster of about 10 mm diameter and an autoreflector (centring detector) which locates the centre of the cylindrical laser beam at any distance with a high degree of accuracy. Laser tube consists of a helium-neon plasma tube in a heavy aluminium co-axial cylindrical housing. Arrangements are made to support this housing on a stand. Sophisticated arrangements are carried out inside the tube to reduce transmission loss and a flat mirror is provided to provide high reflectivity. The laser beam comes out of the housing exactly from its centre (within 0.02 mm) and parallel to the housing within 10″ of arc. The alignment stability of such devices is of the order of 0.2″ of arc per hour.

Autoreflector consists of a detector head and a readout unit. The detector incorporates a number of photocells arranged to compare the intensity of laser beam in each half—vertically and horizontally. The panel meter (having two indicators) immediately detects any offset between the centroid of the beam and centre of the centering detector. The centering detector is housed on a stand which has two adjustments to translate the detector in its two orthogonal measuring directions perpendicular to the laser beam. The device thus can detect the alignment of flat surfaces perpendicular to a reference line of sight. The laser autoreflector projects the collimated laser beam to a flat target mirror and the perfect alignment of the mirror causes the beam to return directly superimposed on the projected beam. Misalignment of the mirror causes an angular offset of the return beam and it is detected by the two-axis angular position sensitive detector and data displayed on two panel meters.

Thus a reference straight line is established first and then using a second centering head, any number of intermediate parts can be aligned. Using optical square or pentaprism, which reflects the laser beam by exactly 90° within 1″ of arc it is possible to measure squareness of surface.

The surface plate can be checked for deviation from true planes by moving the laser to a number of positions and establishing as many intersecting lines as required. The level of surface can also be checked by using a levelling mirror.

7.7. Parallelism, Equidistance and Coincidence

7.7.1. Parallelism. First we shall be dealing with the parallelism of lines and planes, and then the parallel motion. Let us first understand what we mean by 'line' or 'axis' in metrology. An axis is represented by cylindrical surface of high precision of form, suitable surface finish and sufficient length. Under certain circumstances, test mandrels have to be used. These are fixed and centred on the end of the shaft or in the cylindrical or conical bore designed to take it. It may be understood that when a mandrel is employed to represent the axis

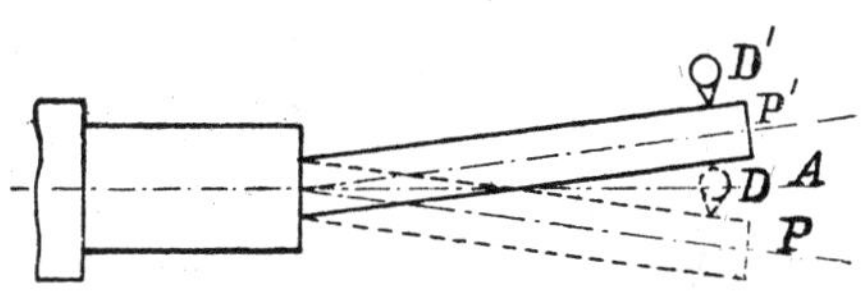

Fig. 7.7. Parallelism of lines.

of rotation of spindle etc. it is not necessary that mandrel might have been exactly centred and under such a case the axis of the mandrel on rotation may describe a hyperboloid, or a conical surface if the axis of the mandrel intersects that axis of rotation and gives two extreme positions P and P' lying in the plane of test. Under such circumstances, two readings should be taken, one in any position and the other after rotating the spindle by 180° (*e.g.*, two positions D and D' of dial indicator in Fig. 7.7), and their mean taken.

We shall now study the various cases of parallelism of lines and planes.

(*a*) **Parallelism of two axes.** This test is made in two planes, first plane being the one passing through one of the two axes and as near as possible to the second axis, and the other plane is perpendicular to the first one.

In this test, the dial indicator is held on a support with a base of suitable shape, so that it slides along a cylinder representing one of the two axes ; and the dial indicator is adjusted so that its feeler slides along the cylinder representing the second axis. The maximum deviation between the axes at any point may be determined be gently rocking the dial indicator in a direction perpendicular to the axes. In the same way the parallelism may be tested in the perpendicular plane.

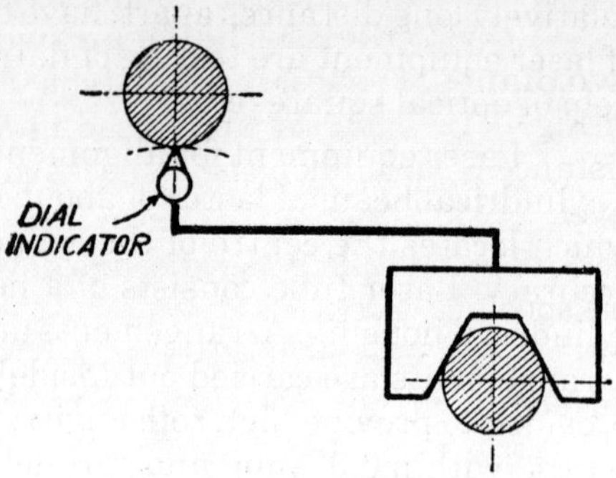

Fig. 7.8. Parallelism of two axe.

(*b*) **Parallelism of two planes.** Two planes are said to be parallel, when their distance from each other is measured anywhere on the surface and at least in two directions, and the maximum error (the difference between the maximum and the minimum dimensions obtained when measuring) over a specified length does not exceed an agreed value.

The test for parallelism of two planes is carried out in two directions (generally perpendicular to each other). The dial indicator, which is held on a support with a flat base, is moved in one plane over a given length, and the feeler is made to rest against the second plane ; and the deviations noted down (Fig. 7.9).

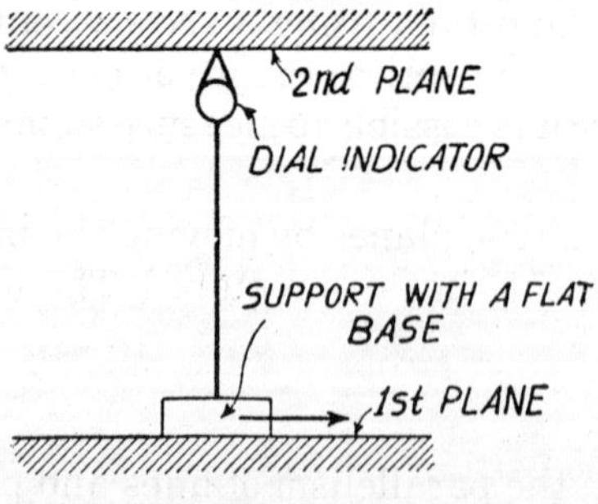

Fig. 7.9. Parallelism of two planes.

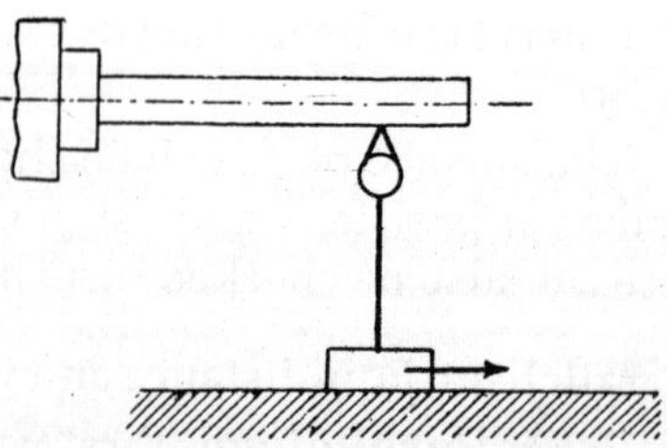

Fig. 7.10. Parallelism of an axis to a plane.

(*c*) **Parallelism of an axis to a plane.** An axis is said to be parallel to a plane, if the maximum difference between the several readings taken while measuring the distance of line from plane at a number of points, does not exceed a predetermined value. In testing, the feeler of the dial indicator is made to touch the surface of the cylinder representing the axis and the dial indicator (held on a support with a flat base) is moved along the plane by the specified amount (length over which the test is to be performed) (Fig. 7.10). At each point of measurement the shortest distance is found by slightly moving the indicator in a direction perpendicular to the axis.

(*d*) **Parallelism of an axis to the intersection of two planes.** The test set up is shown in Fig. 7.11 and this test is also conducted in two perpendicular planes of the greatest importance.

(*e*) **Parallelism of the intersection of two planes to a third plane.** This test is similar to the one given above. The feeler instead to touching the cylindrical surface touches the third plane to be tested. The feeler is set at right angles to the plane.

(*f*) **Parallelism between two straight line, each formed by the intersection of two planes.** The test set-up for checking the parallelism in two perpendicular planes is shown in Fig. 7.12. This set up is used in cases where the distance between two lines is small. If distance is large then the V-blocks are covered by a straight edge and check made by a spirit level.

The tolerance (permissible deviations) of parallelism of straight lines or plane surfaces are specified as :

0.015 mm on 300 mm (free end of spindle in upward direction only relative to the table surface).

If considered desirable, the tolerance of the form of feeler surface should also be specified.

7.7.2. Parallel Motion. The term 'parallel motion' refers to the position of the trajectory of a moving part of the machine in relation to a plane (support or slideway in case of machine tool), a straight line (axis, intersection of planes), and a trajectory of a point on another moving component of the machine. The tolerance on parallelism of movement is the permissible

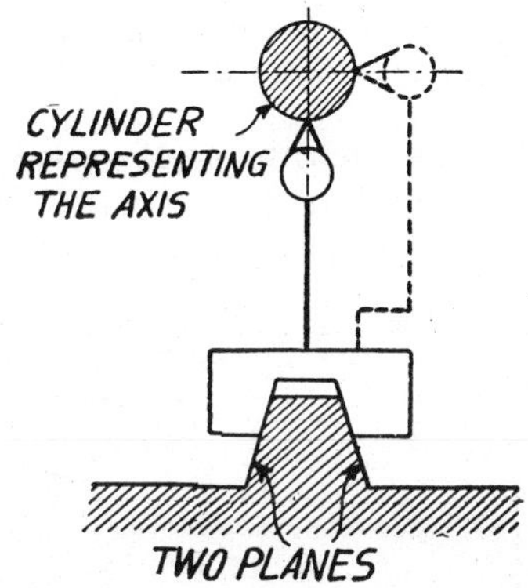

Fig. 7.11. Testing parallelism of an axis to intersection of two planes.

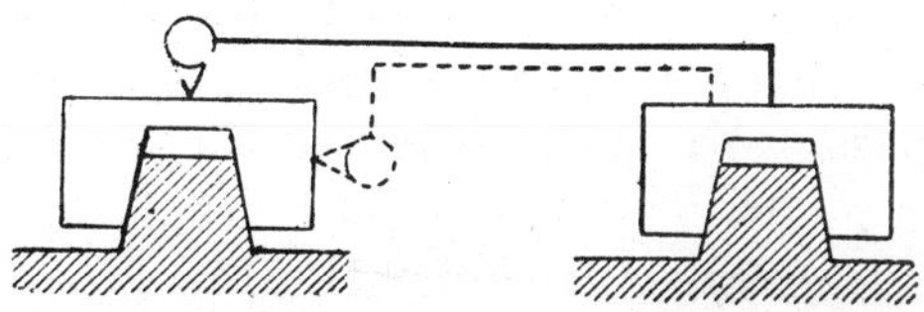

Fig. 7.12. Testing parallelism of two lines formed by interaction of two planes.

variation in the shortest distance between the trajectory of a given point on the moving part and a plane, a straight line or another trajectory within a stated length. As a general rule, unless all planes are of equal importance, the checks are made in two perpendicular planes selected as being those of most importance for the practical use of the machine.

The test for checking parallel motion in various cases are performed as below :

(*a*) **Parallelism between a trajectory and a plane**

(*i*) *When the plane is on the moving components itself.* In this case the dial indicator is attached to a fixed component of the machine and its feeler adjusted at right angles to the surface to be tested. The moving component is moved by the specified amount and the deviations noted down.

(*ii*) *When the plane is not on the moving component itself.* The set up for this test is shown in Fig. 7.14.

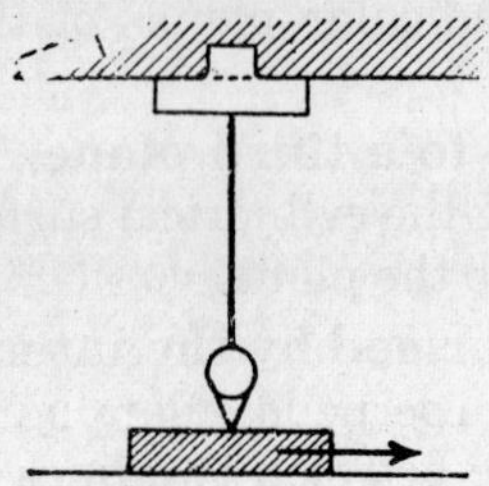

Fig. 7.13. Testing parallelism between a trajectiory and plane/on moving component.

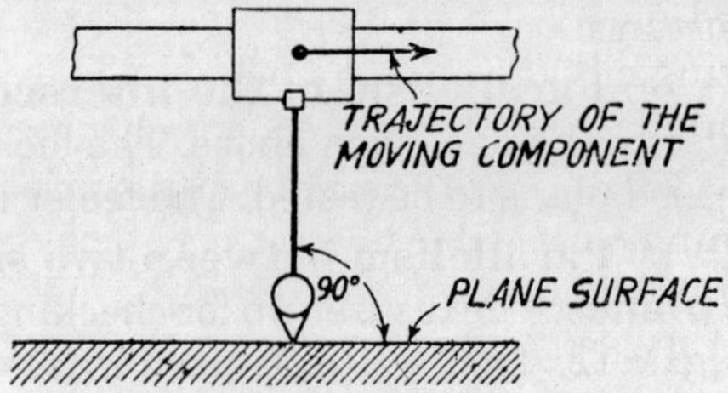

Fig. 7.14. Testing parallelism between a trajectory and a plane (not on moving component).

(*b*) **Parallelism of a trajectory to an axis.** The test set up is shown in Fig. 7.15.

(*c*) **Parallelism between two trajectories.** In this test, the dial indicator is attached to one of the moving components of the machine and its feeler made to rest against a given point on the other moving part. The two parts are moved together in the same direction by the same amount, and the variation in the reading noted (Fig. 7.16).

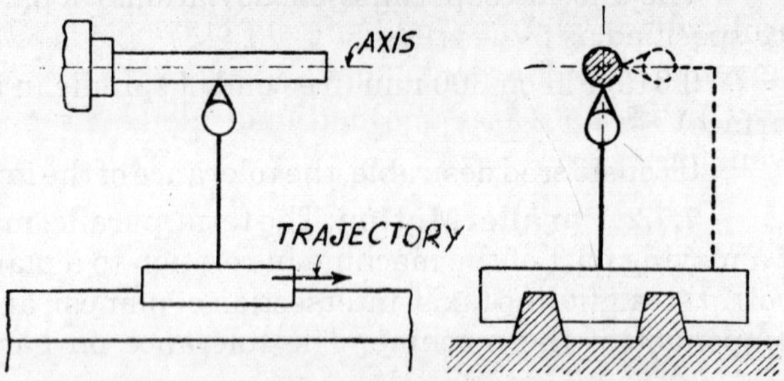

Fig. 7.15. Test set uf for testing parallelism of a trajectory to an axis.

7.7.3. Equidistance. It relates to the distance between the axes and a reference plane. There is equidistance when the plane passing through the axes is parallel to the reference plane. It may be different axes, or the same axis, occupying different positions after pivoting.

The method of measuring the equidistance is in effect identical with that of measuring parallelism between a plane passing through the axes and a reference plane. The test consists in first testing that the two axes are parallel to the plane and then testing that they are equidistant from this plane. For this purpose, a dial indicator is made to rest on the reference plane and with the same gauge, readings are noted on the cylindrical surfaces representing the two axes by sliding the dial gauge on the reference plane (Fig. 7.17). If the two cylinders are not identical, then the difference of radii of the tested sections should be taken into account.

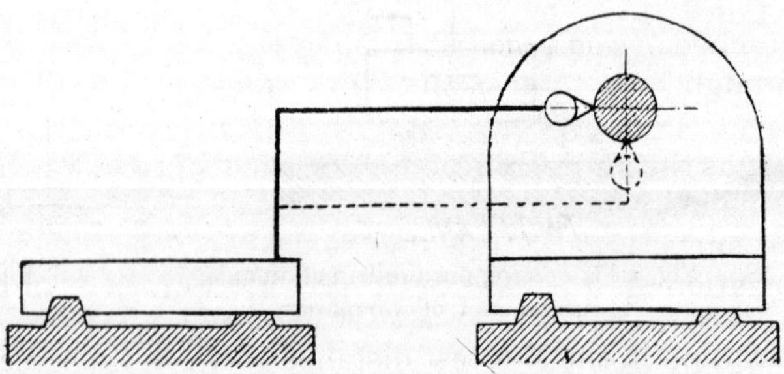

Fig. 7.16. Test set up for testing parallelism between two trajectories.

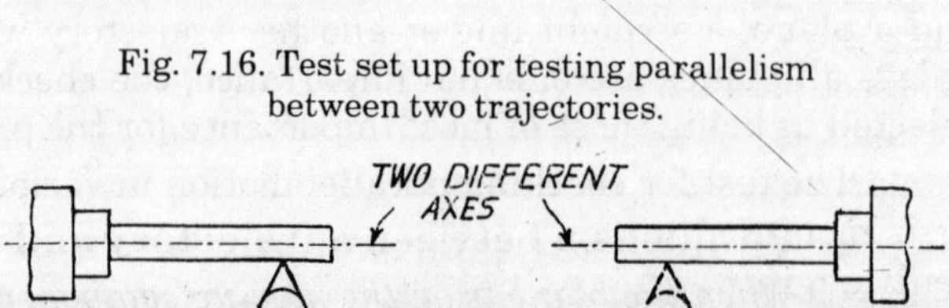

Fig. 7.17. Testing equidistance.

In the second case where the same axis occupies different positions after pivoting and where the plane of pivoting is not accessible for allowing movement of the instrument, an additional plane is constructed parallel to the plane of pivoting ; the setting and fixing of the auxiliary plane is carried out so that, when a level is placed on it, if possible, in two perpendicular directions, there is no deviation during the pivoting motion of the component. The equidistance of the pivoting axis is then tested both in the middle as well as extreme positions.

7.7.4. Coincidence or alignment. Many a times this word is used with more general meaning of parallelism. For metrology purposes this term refers to two axes merged in each other or where one axis extends beyond the other.

Two lines or axes are said to be coincident or in alignment when their distance apart at several points over a given length is measured and this distance does not exceed a given value. The distance measured may be located either on the actual lines or on their prolongation.

The tolerance on alignment of two axes is specified as under : When the direction of the permissible deviations in alignment of two axes (or two lines) does not matter, the tolerance is given as : Error of alignment of axis 1 to axis 2 .... μ or mm over a given length.

In special cases an additional indication depending on operating conditions may be given, for instance :

'axis 1 only higher than axis 2', or 'free end of axis 1 directed only outwards in relation to axis 2'.

In other cases, it may be useful to introduce beside the tolerance on coincidence, an additional tolerance, on parallelism between the two axes (Fig. 7.18). In this figure,

(*a*) 'Error of alignment of axis 1 to axis 2 : T μ or mm over a given length'.

(*b*) 'Error of parallelism between axis 1 and axis 2 : T μ or mm over given length ($T > T$)'.

Coincidence between two lines or axes is measured by the method given below. In case when both the axes whose coincidence is to be tested are fixed, then for one of the axes a mandrel is used to represent the axis (2nd axis in Fig. 7.19) ; and the dial indicator is attached to an arm and rotated through 360° around the other axis. (Fig. 7.19 shows the other axis, *i.e.*, first as the axis of rotation). The feeler of the instrument is adjusted to touch the surface of the mandrel and any variation in the reading represents twice the error of coincidence. Check is generally made into two planes because the first plane chosen may intersect both axes. When one of the axes is the axis of rotation, then arm carrying the dial indictor is mounted on a ring of sufficient length (in order to ensure that the reading is not affected by play in the ring) and rotating with a minimum amount of play (Fig. 7.19).

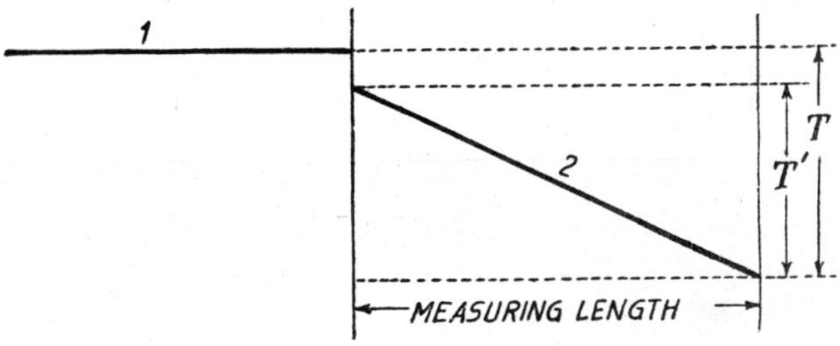

Fig. 7.18. Error of alignment of axis vis a vis error of parallelism of axes.

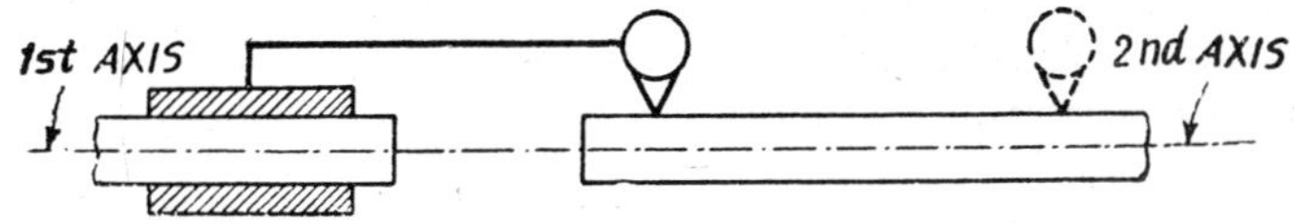

Fig. 7.19. Testing coincidence between two axes.

In this test, the dial indicator of very light weight should be used, as the direction of the measurement varies during rotation in relation to the direction of gravity, and if possible the sensitivity of the test apparatus to the force of gravity should be taken into account.

7.8. Squareness

First we shall be dealing with the squareness of straight lines and planes and then the perpendicularity of motion.

Two planes, two straight lines or a straight line and a plane are said to be perpendicular when the error of parallelism in relation to a standard square does not exceed a given value. The reference square may be a metrological square or a right angle level or may consists of kinematic planes or lines. The permissible errors are specified as 'Errors relating to the right angle : ± ... μ or mm on a given length' and if the error is determined in relation to another part of the machine then the permitted direction of error should also be specified, such as, 'Free end of spindle inclined only towards the support'.

We shall be considering the following cases for the measurement of squareness of lines and planes.

(*a*) **Squareness of an axis of rotation with a given plane**

For this test the dial indicator is mounted on an arm which is attached to the spindle representing the axis of rotation. The plunger of the dial indicator is adjusted parallel to the axis of rotation and made to touch the plane. As the spindle revolves, the dial gauge (or the end of plunger if revolving freely into air) describes a circumstances, the plane of which is perpendicular to the axis of rotation. When no testing plane is specified the dial gauge is rotated by 360° and the variation in the readings of instrument represents the deviation of parallelism between the plane of the circumstances and the plane to be tested. However, if planes are specified (*e.g.* planes 1 and 2) then the difference of the readings in the position of the dial gauge, 180° apart is noted for each of these planes (Fig. 7.20).

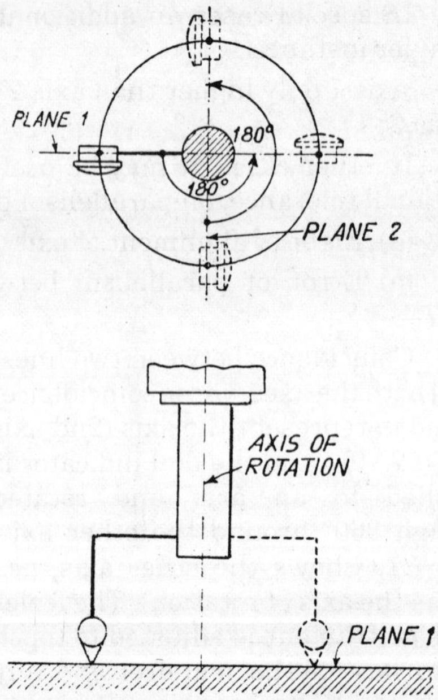

Fig. 7.20. Testing squareness.

The deviation is expressed in relation to the diameter of the circle of rotation of the instrument.

The effect of periodical axial slip of the spindle can be eliminated by repeating the above test after moving the dial gauge through 180° relative to the spindle and average of two sets taken. The effect of minimum axis play can be eliminated by means of a suitable axial pressure.

7.8.1. Checking of Perpendicularity of Motion. The term 'perpendicularity of motion' refers to the successive positions on the trajectory of a point on a moving part of the machine in relation to a plane (support or slide way), or a straight line (axis of intersection of two planes), or to the trajectory of a point on another moving point.

The checks for testing the perpendicularity of motion become the tests of parallelism by the use of a square suitable for the given conditions. As the tests of parallelism, have already been discussed in details, these are not being repeated. As in the tests for parallelism, here also, the moving part should be driven in the usual way so as to allow for the effects of play and errors in slideways.

7.8.2. Squareness Testing. The angle of 90° is probably the most important angle in engineering applications. It is assigned several names as square, normal, right angle and it is represented for most practical purposes by squares, rectangular blocks etc. Probably the achievement of modern sciences would have not reached the present state of advancement if right angle was unattainable to within a close degree of accuracy. Its importance is realised from the following applications. The cross slide of lathe must move exactly at 90° to the spindle axis in order to produce a flat face during facing operation. The spindle of depth micrometer must be square to the locating face in order to avoid any errors in measurement. The column and table of milling machine must be at 90° to each other.

For most of the purposes where high degree of accuracy is not desired, the workpiece can be tested against an engineer's square or square block. It simply shows whether the two surfaces of a workpiece are at right angle or not and judgement is purely based on eye. In order to know the amount of error and for checking squares and square blocks, the following methods can be used.

7.8.2.1. Indicator Method. This method is particularly suitable for checking the squareness of a block whose opposite faces are supposed to be parallel. It is assumed that the squareness of the block has already been assured to a reasonable accuracy by the use of square etc., as otherwise the full sensitivity of the method can't be obtained. The instrument for this purpose is designed by N.P.L. and is very suitable for checking squareness while manufacturing a square block. The instrument consists of parallel strip (framework) and a flat base. A knife edge and some form of indicator is mounted on the framework as shown in Fig. 7.21.

The other tests of squareness of lines and planes are given below in the tabular form.

Condition	*Test set up*	*Method in brief*
(*i*) Two planes (1 and 2) at 90° to each other	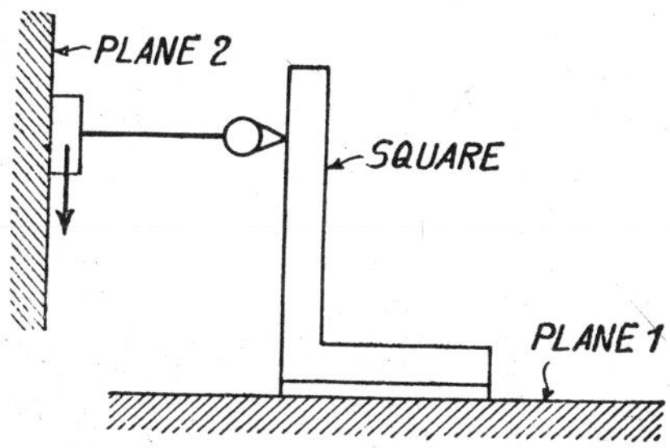	Squareness of two planes 1 and 2 is checked by placing the square on one plane and then checking the parallelism of 2nd plane with the free arm of the square by sliding the dial indicator (mounted on a base) along 2nd plane and its feeler moving against free arm of the square.
(*ii*) Two axes at 90° to each other (*a*) Both axes fixed	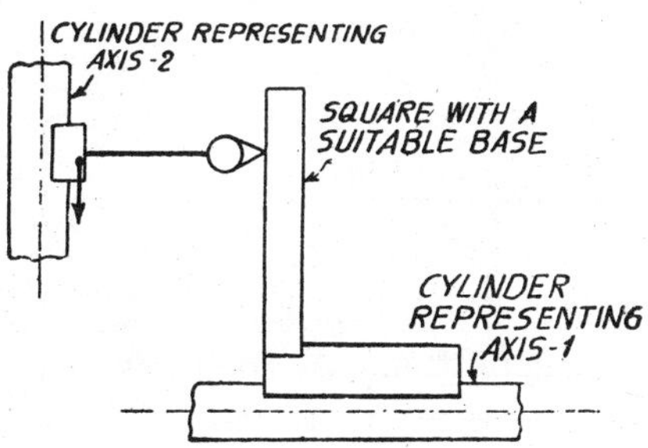	—do—

(*b*) One axis being axis of rotation and other fixed.	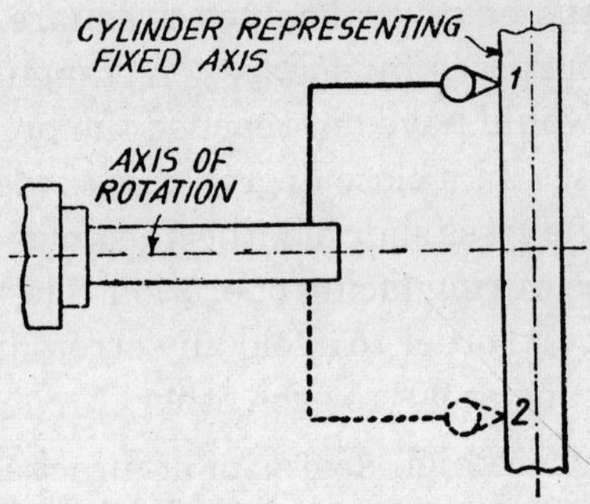	The dial gauge mounted on arm and fixed on the mandrel is brought into contact with the cylinder representing fixed axis at two points 1 and 2, 180° apart and deviation expressed in relation to distance between 1 and 2.
(*c*) Both the axes being axes of rotation.		The test is conducted in the same way as (*ii*)-(*b*) but the cylinder representing 2nd axis of rotation is brought into the mean position of the run out in the plane of measurement.
(*iii*) An axis at 90° to a plane. (*a*) axis is fixed.	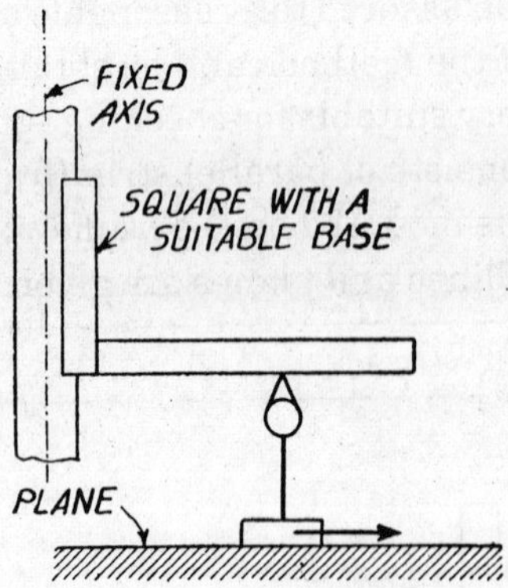	Test set up is self explanatory, but the test is carried out in two perpendicular directions.
(*b*) Axis being the axis of rotation.		This test has already been described.
(*iv*) An axis at 90° to the intersection of two planes. (*a*) Axis is fixed.	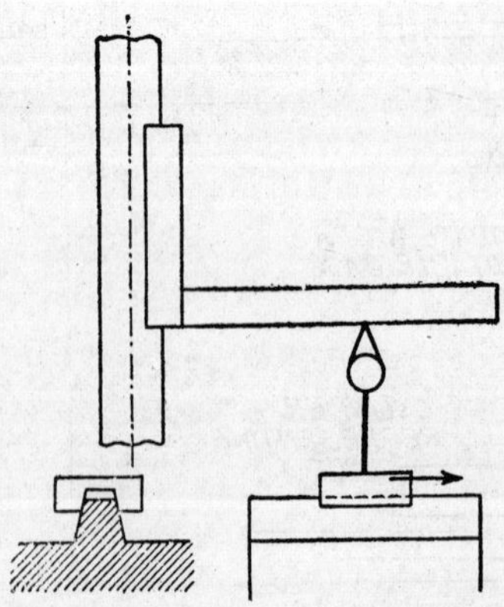	The test set up is self explanatory.

(*b*) Axis being the axis of rotation.	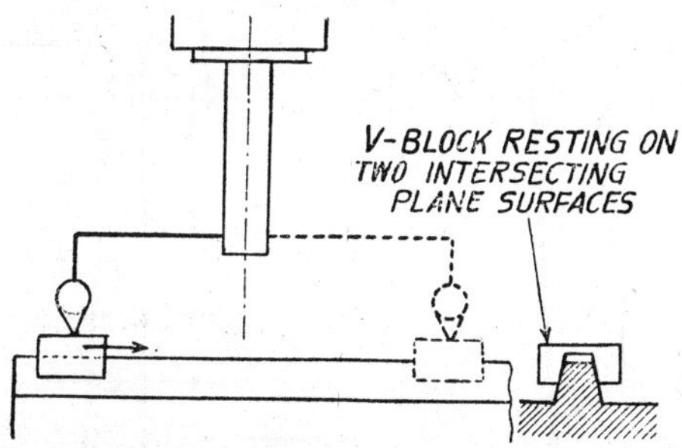	First reading is taken by making the feeler of the dial indicator to touch on a V-block resting on two intersecting plane surfaces. (The dial indicator is mounted on the spindle). The second reading is noted by rotating the spindle along with dial by 180° and moving the V-block so as to bring the feeler into contact with the same point on the block.
(*v*) Intersection of two planes is at 90° to another plane.	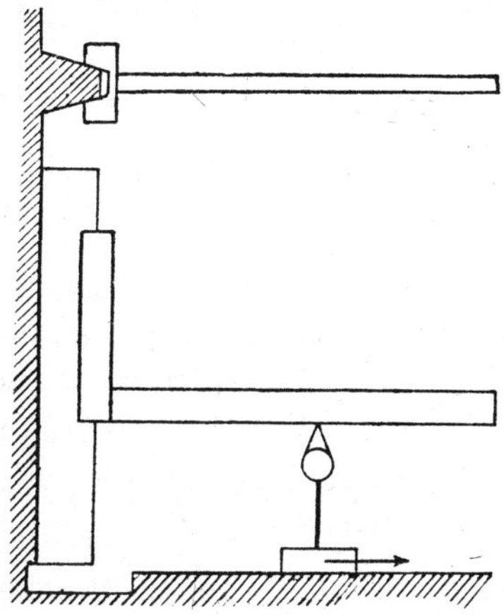	In this test, either the square or the dial indicator fitted with a suitable base is allowed to rest on the intersecting planes and the dial indicator is moved with its feeler resting against the arms of dial gauge. The test is made in two perpendicular planes.
(*vi*) Two straight lines, each formed by the intersection of two planes, are at 90° to each other.	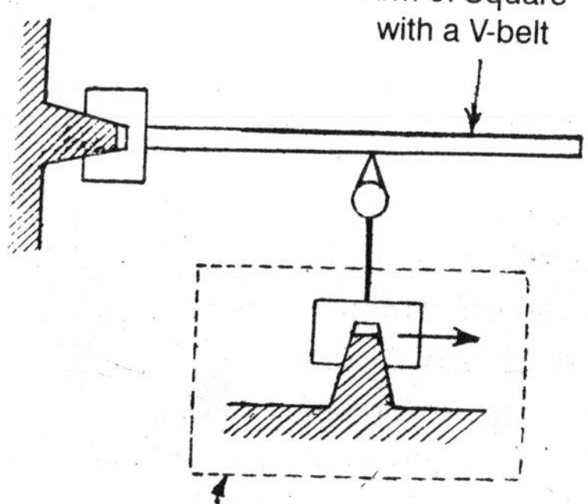	Test is self explanatory.

In this method, means are also available to produce and measure parallel surfaces by some form of grinder and a comparator respectively. The height of the indicator is adjusted such that it makes contact near the top of the side of the square block. The block is then placed against the knife edge as shown in Fig. 7.21 (*a*) and a reading is noted on the indicator. The block is then turned so that now the opposite side is facing the knife edge [Fig. 7.21 (*b*)] and again a reading is noted on the indicator. If two sides *AD* and *BC* are truly parallel then the two readings will be same for true right angle. In case the faces are not exactly at right angles,

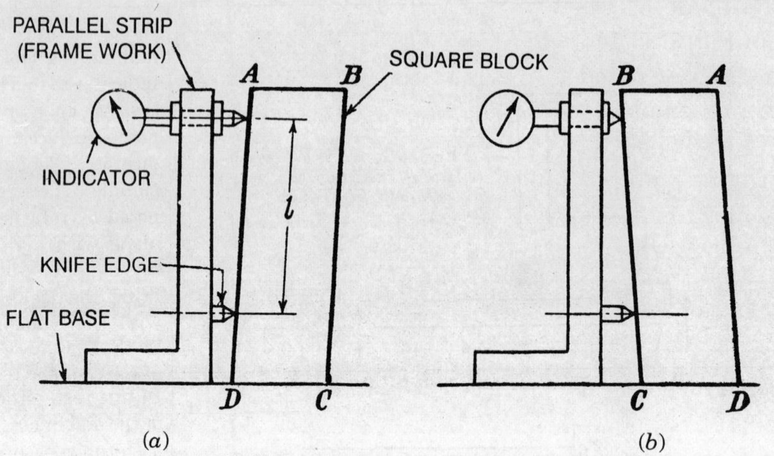

Fig. 7.21. Squareness testing by indicator method.

then the two readings will be equally above and below the reading for a true right angle. Thus the differences of two readings is double the error in squareness of work over the length 'l' between the two contact marks.

7.8.2.2. Correction for squareness Error. The purpose of determining the error in squareness of a workpiece is that it may be corrected, otherwise knowing the error will not be of much value. Let us assume that the two readings obtained as described previously are l_1 and l_2, and the distance between the contact points is l. Let the length of the square block AD be L. Then error in squareness $= \frac{l_1 - l_2}{2} \times \frac{L}{l}$. In other words, considering the same block in Fig. 7.22, point D is higher than A by amount $\frac{l_1 - l_2}{2} \times \frac{L}{l}$ and this much amount must be removed from D and nothing being removed from A. Similarly B must be brought down by same amount, no metal being removed at C. For this, block is set up on a surface grinder, face AD being uppermost and cut of the above said amount is taken across this face as shown in Fig. 7.22 (a). The block is then turned over as shown in Fig. 7.22 (b) and a cut taken to clean up face BC which will then be square with reference to AB and CD. The face AD can then be ground parallel to face BC and all the four faces will be square to each other.

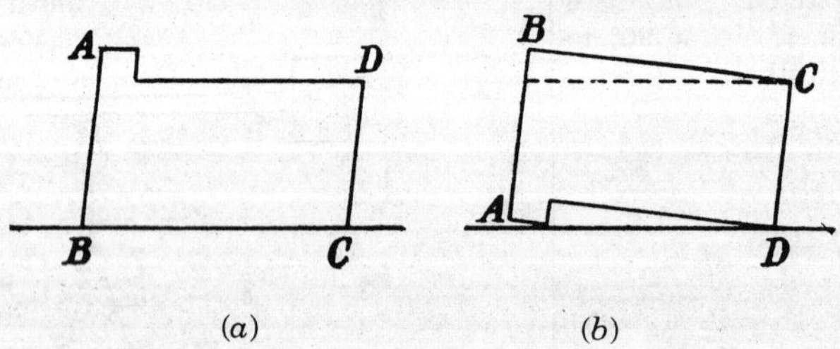

Fig. 7.22. Correction for squareness error.

The two readings taken by indicator could also be taken by using auto-collimator in which case a slip gauge is held against the surface to be tested and reading on auto-collimator taken. This thus becomes a still sensitive method.

7.8.2.3. Engineer's Square Tester. Its design is also devised by N.P.L. It consists of a bar or a tilting frame having a pair of hardened and ground steel cylinders of precisely the same size as shown in Fig. 7.23. A plane reflector is mounted upon the bar and the instrument is set up on a flat reference plane of suitable degree of accuracy.

The square is then placed on the same flat reference plane and the two cylinders just made to touch the blade of the square as shown in Fig. 7.23 and reading of auto-collimator noted down. The square as then placed on the other side as shown and again the bar so tilted such that the two cylinders just touch the blade of square and again reading in auto-collimator is noted down. Then half the difference in the two readings gives the angular error in the squareness of the square.

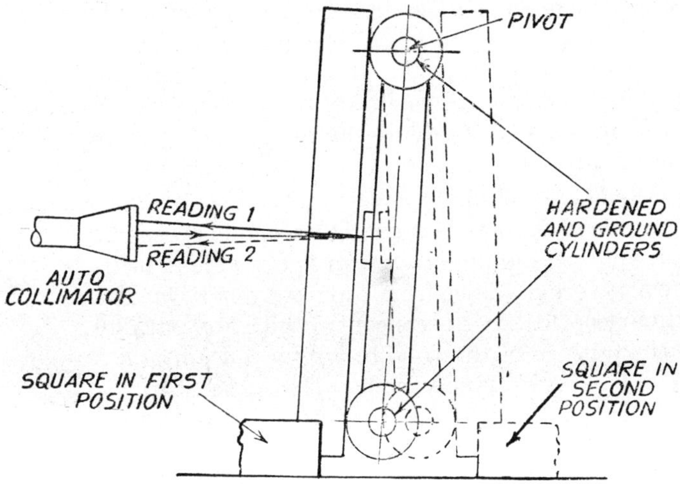

Fig. 7.23. Testing squareness with engineer's square tester.

In order to check whether two machined surfaces are exactly at right angles to each other or not, the use of optical square and the auto-collimator can also be made.

7.8.2.4. Optical Tests for Squareness. Squareness of any two machined surfaces can be easily checked by using the auto-collimator. The axis of the incident beam from the auto-collimator forms the measuring datum. An optical square is utilised for turning the incident bean through exactly 90°. In this test, it is assumed that the two surface faces are perfectly straight. A stainless steel mirror block with a flat base is used for the horizontal surface for aligning the collimator with the surface. A reading is thus taken in collimator at position *A* (Fig. 7.24). The mirror with base is then transferred to the vertical surface and the optical square placed in the angle as indicated. Another reading is then taken in position *B*. The two auto-collimator readings of the two mirror positions will indicate whether the machined surfaces are accurately at right angles, if not the reading will show the direction and amount of error.

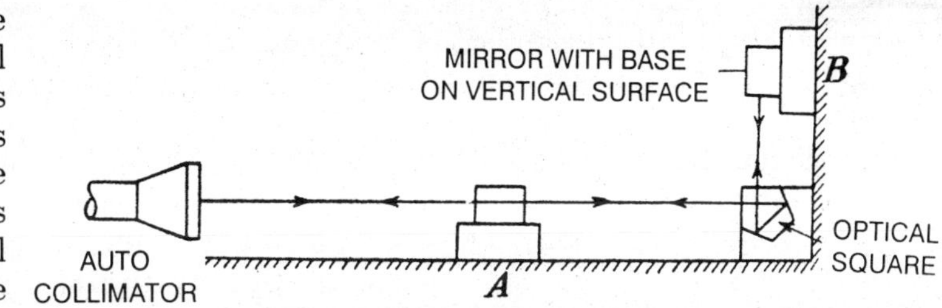

Fig. 7.24. Optical test for squareness.

7.8.3. Focusing telescope. Focusing telescope, also known as micro-alignment telescope, is an important and powerful optical instrument to check and ensure geometrical integrity of components and their assembly. The optical methods have the advantage of simplicity, non-contact measurements, versatility, cost effectiveness, and in some cases the only feasible solutions. Focusing telescope is based on the concepts of geometric optics and is simple and straight forward to use. It enables geometric problems to be solved from first principles. It is a portable instrument and requires simply power supply and thus can be conveniently used at site and in every area of the workshop/factory.

Micro-alignment telescope is used to measure deviation from straight line of sight to set and check alignment, squareness, straightness, flatness, parallelism, vertically and level. The focussing telescope is also used for achieving precise alignment settings on large engineering components and structures. The ability to move the focussing lenses with freedom from transverse movement or tilts is a critical element of the telescope design, determining the accuracy of the resultant line of sight. Horizontal and vertical displacement from a true line of sight are measured via a two-axis tilting plate micrometer coupled to graduated drums. (Refer Fig. 7.25 for schematic drawing of focusing telescope).

The Micro-Alignment Telescope is presently available to read directly to μm and is able to focus down to zero distance from the front objective. The primary optical axis is concentric with and parallel to the outside of the tube to within 6.4 μm and 3 seconds of arc respectively. The tube itself is cylindrical to within 5 μm. In practice one can readily achieve a setting accuracy of 50 μm at a distance of 30 meters and proportionally for longer and shorter distances down to 3 metres.

Alignment Telescope focuses from zero to infinity and incorporates an optical micrometer to measure deviations from a line of sight in two directions at right angles to each other.

The Micro-Alignment Telescope generates a straight line of sight which is the basic reference for all measurements. A prism is used to deviate the straight line to generate squareness, and a rotating prism generates flatness. The telescope is specially designed to facilitate autoreflection and autocollimation providing for squareness and angular measurement using reflection targets or polygons.

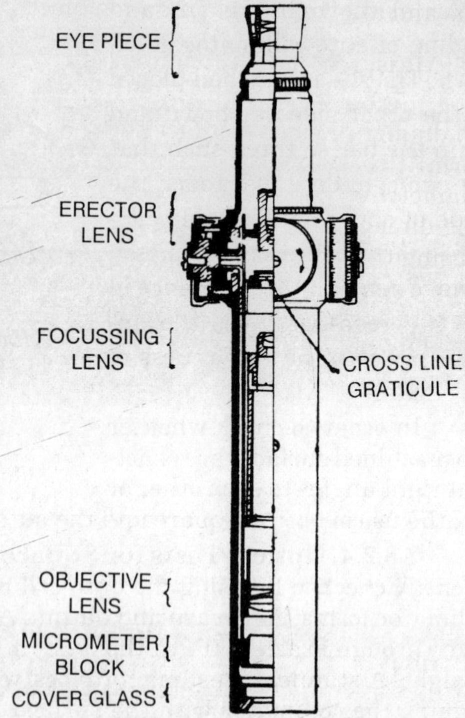

Fig. 7.25. Simplified general arrangement of the Micro-Alignment Telescope.

Focusing telescope becomes a powerful measurement tool to handle a vast range of applications when used in conjunction with a wide variety of accessories like mechanical mounting, targets and target holders, sweep optical square, optical squares, levels, beam dividing prisms etc. These accessories enable the focusing telescope to define straight lines of sight in related directions and to sweep out datum planes.

Mounting accessories are used for various requirements. Targets and target holders include mirror targets for auto-reflection and autocollimation. Sweep optical square is used to sweep out a reference plane at 90° to the telescope axis, from which errors of flatness can be measured. Optical squares are used to deviate the line of sight through 90 degrees to within 1 second of arc and to check squareness of axes. Optical squares are also used for setting out right angles lines of sight like checking that a machine column is square to the bed. Spherical mounts in conical seatings are used extensively to define a fixed point (the centre of the sphere) through which the telescope line of sight or target always passes irrespective of tilt. Telescope lamp house accessory or separate collimator unit is used to achieve angular setting and measurement from a datum. Well fixturing of the telescope is essential for most of the applications.

Applications of Focusing Telescope System

Typical applications include the measurement and setting of bearing alignment ; alignment and squareness of axles, spindles and bores ; straightness, flatness and squareness of bedways and slides ; alignment of engines with shafting, gearboxes and compressors ;

parallelism and squareness of rollers and conveyors ; squareness and alignment of assembly jigs and alignment to foundation blocks.

7.9. Measurement of Circularity

7.9.1. Introduction. In the assembly of circular parts, only the dimensional tolerances on diameter will not suffice the requirements, but it is the geometrical accuracy (accuracy of form) that needs closer attention. If cylindrical parts are measured from devices having diametrically opposite contacts such as micrometer etc., though they may be found to be within the dimensional tolerances and still may not be perfectly circular, which would be noticed during the assembly of parts. (Refer Fig. 7.26). Thus the method of circularity measurement has to be such which detects the various possible errors of circularity and this is always done by rotating a part and not measuring static diameter alone.

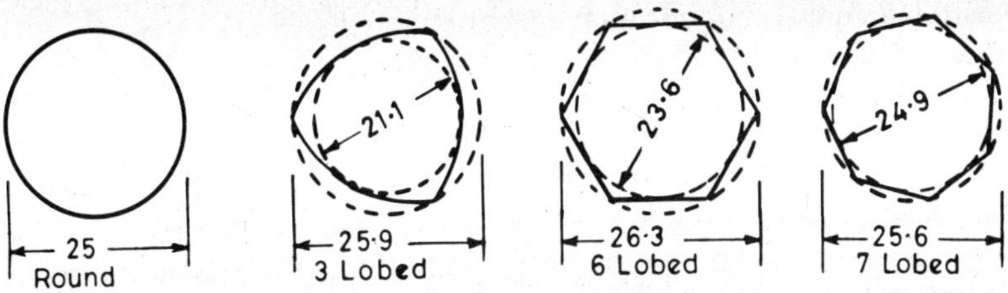

Fig. 7.26. Errors in circularity.

7.9.2. Sources of Out-of-Roundness. Several reasons when machining parts can be attributed to cause out-of-roundness. These are clamping distortion, spindle run-out, presence of dirt and chips on clamping surfaces, imbalance, heat and vibration. The characteristic roundness shape varies greatly depending on the method of generation.

The tolerance on roundness is critical and it should be much closer (usually five times) than that of the other dimensional tolerance which it effects.

7.9.3. Different types of Irregularities of a Circular Part. The errors of circularity at a cross-section can be of the following nature :

(*i*) *Ovality* (There is some difference between the major and minor axes.) [Refer Fig. 7.27].

(*ii*) *Lobing* (In this case the diameters at any two opposite points are constant, but still it is not circular form). [Refer Fig. 7.28]

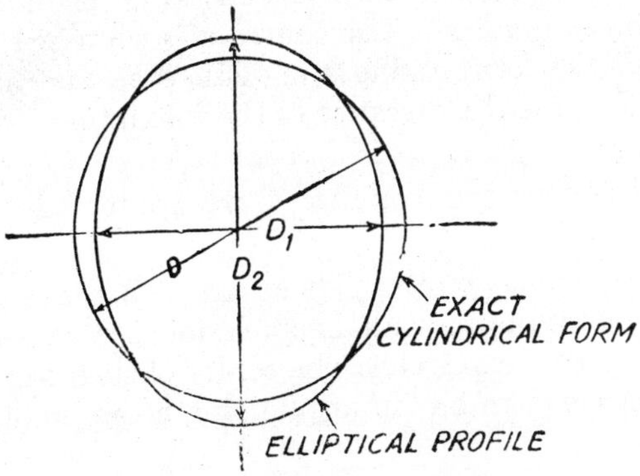

Fig. 7.27. Ovality, *i.e.* $D_1 \neq D_2 \neq D$.

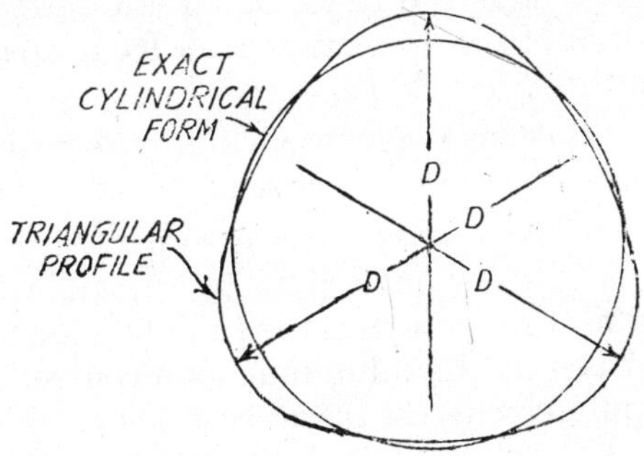

Fig. 7.28. Lobing : Diameters at places equal, but still not circular form.

(*iii*) *Irregularities of no specific form.* (Fig. 7.29)

7.9.4. Causes of out-of-roundness. Out-of-roundness of machined parts could be due to poor bearings in the lathe or grinding wheel spindle, or to deflections of the workpiece as the tool is brought to bear on it. Shafts, ground between centres can be out-of-round due to poor alignment of the centre or deflection of the shaft.

Lobing is particularly difficult to avoid in many machining processes. A round bar or ring-type part held in a 3-or-5-jaw chuk is compressed at the points of contact. Even if the part is turned or ground perfectly circular on the machine, when it is removed from the chuck the stress in the metal will be relieved, causing three or five lobes.

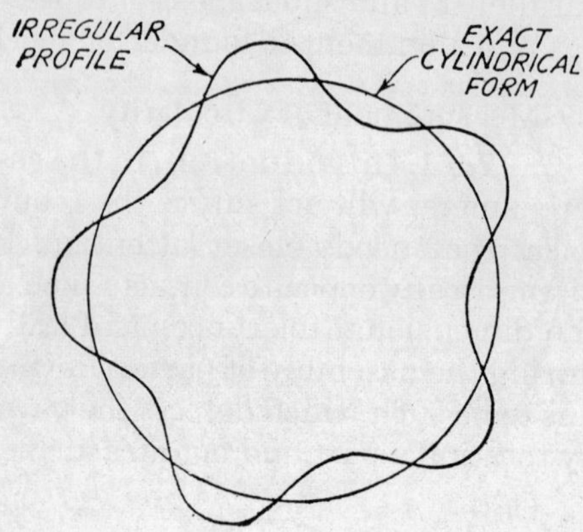

Fig. 7.29. Irregularities of no specific form.

7.9.5. Definition. The error of circularity is defined as the radial distance between the minimum circumscribing circle and the maximum inscribing circle, which contain the profile of the surface at a section perpendicular to the axis of rotation.

Polar graph (a graphical record, suitably magnified, of the displacement of the stylus of measuring elements, as either the stylus or the part rotates on the axis of the precision spindle) is interpreted by using concept of reference circle or a pair of circles which can be drawn/recorded or otherwise superimposed on the graph. The position of the reference circle and its centre is not arbitrary but is chosen to fulfil certain conditions.

Obviously, it is possible to draw a large number of circles for a given profile and infer different results. Accordingly four types of reference circles as described below have been standardised and the results should be specified indicating the reference circle adopted. Only in this can an unambiguous and repeatable value for the out-of-roundness of any particular profile be obtained.

7.9.6. Reference Circles. The four reference circles used in measurement of roundness are :

(*i*) *Least squares circle.* This circle represents the average of all the peaks and valleys, and can be defined mathematically as follows :

"The sum of the squares of a sufficient number of equally spaced radial ordinates, measured from the circle to the profile has minimum value". The centre of such a circle is referred to as the Least Squares Centre (L.S.C.). Out of roundness is defined as the radial distance of the maximum peak from this circle (P) plus the distance of the maximum valley from this circle (V) *i.e.* $P + V$.

It is very easy to calculate the position of the least squares circle by computer but very difficult by manual methods :

(*ii*) *Minimum zone or minimum radial separation circles.* These are two concentric circles that just enclose the profile and which have minimum radial separation. The value of the out-of-roundness is the radial distance between the two circles. The centre of such a circle is termed as the minimum zone centre. These circles can be found by using a template, or computed by necessary accessories.

(*iii*) *Maximum inscribed circle.* This is the largest circle that can be drawn inside the profile without cutting it. Its centre and radius can be found by trial-and-error using compasses, by template or computer. The out of roundness is specified as the height of the

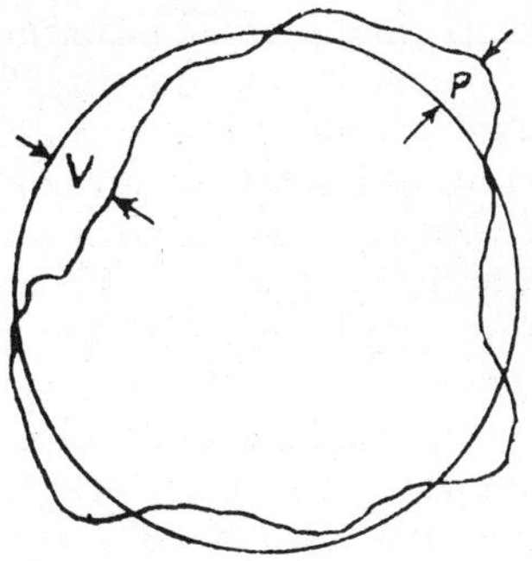

Fig. 7.30. Least Squares Reference Centre.

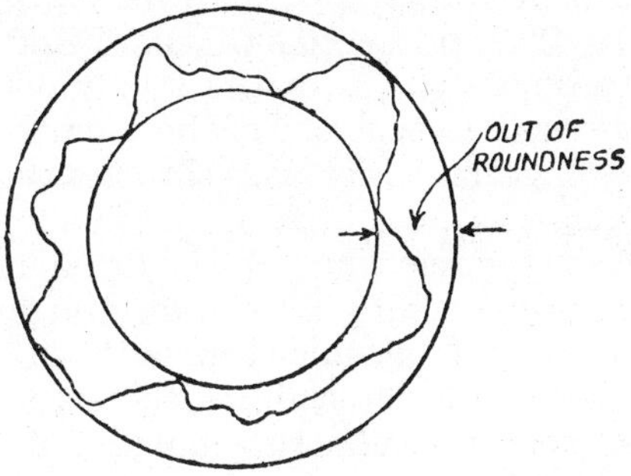

Fig. 7.31. Minimum radial separation circles.

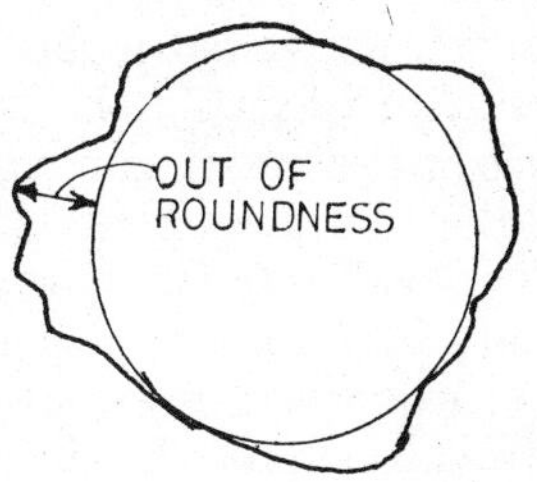

Fig. 7.32. Maximum inscribed circle.

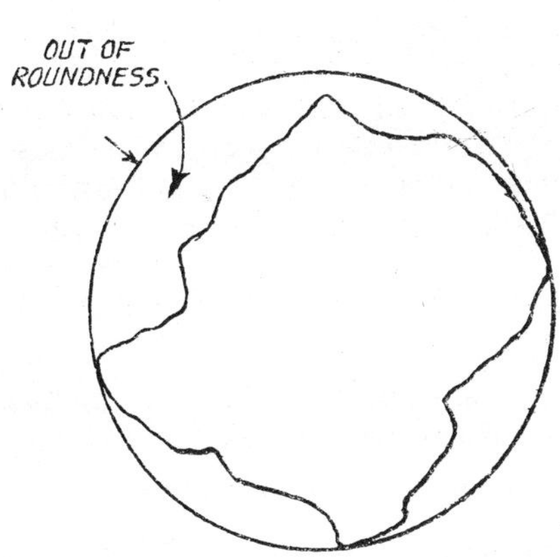

Fig. 7.33. Minimum circumscribed circle.

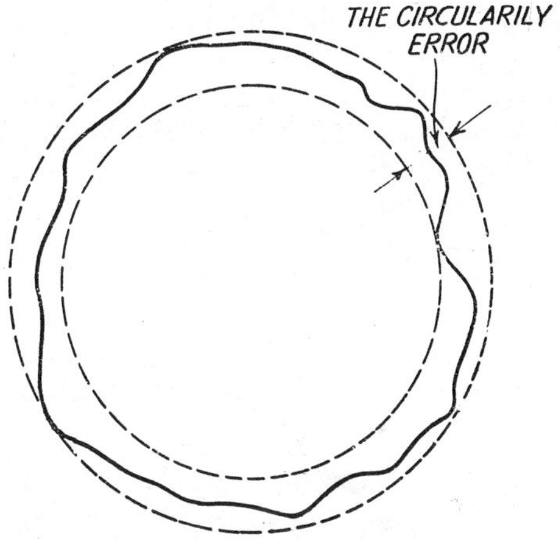

Fig. 7.34. Trace produced by a polar recording instrument.

largest peak above the circle. This value is nearly equivalent to $P + V$, since V is zero, there being no valleys inside the circle.

(*iv*)

Minimum circumscribed circle. This is the smallest circle that will completely enclose the profile without cutting it. Its centre and radius can be found in a similar manner to that of the inscribed circle. This out-of-roundness is the distance of the lowest valley from the circle, which is also equivalent to $P + V$, since P is zero, there being no peak outside the circle.

A template for rapid manual assessment of the graph is supplied with most instruments. It consists of some transparent material on which is engraved or printed a series of concentric circles matching the circular ordinates of the chart. To use it, the chart is placed under the template and then moved about to determine the position of reference circles. It can be used to find the circumscribed or inscribed reference circles and the minimum zone circles.

Error of circularity is determined by the modern circularity measuring instruments which make use of precise spindle on which the indicating means are carried and in which turn tables are provided for carrying the work. The radial variations discovered by the instruments are generally displayed in the form of a polar profile graph or a meter reading or both. From the polar graph, the departures from the ideal circle can be determined as "the difference in radius between two best fitting concentric circles, within which the profile is just contained." The radial distance between the minimum circumscribing circle and the maximum inscribing circle is the measure of the error of circularity.

In Fig. 7.34, is shown a trace produced by a polar recording instrument which is actually an amplified record of the displacement of the stylus of the measuring instrument. The procedure for making numerical assessment of the circularity error of this trace is to draw concentric circles on the polar graph which pass through the maximum and minimum points in such a way that the radial distance of the mininium circumscribing circle containing the trace or the maximum inscribing circle which can be fitted into the trace is minimum. In other words, many concentric circles touching maximum and minimum points respectively are drawn by hit and trial and the one set in which the radial distance between the outer and inner circles is minimum is considered for determining the error of circularity. This could also be interpreted as the smallest ideally circular ring gauge that can be passed over the workpiece or the largest ideally circular plug gauge that can be passed through the work-piece. The error, of course, can be expressed w.r.t. any of the four reference circles illustrated above.

7.9.7. Roundness and circularity. Often the terms roundness and circularity are used interchangeability. Roundness is defined as a condition of a surface of revolution (like cylinder, cone or sphere) where all points of the surface intersected by any plane perpendicular to a common axis in case of cylinder and cone (or passing through a common centre in case of sphere) are equidistant from the axis (or centre). Since the axis and centre do not exist physically, measurements have to be made with reference to surfaces of the figures of revolution only. Whatever is measured by referring to the surface of revolution is the circular contour. It may be understood that while roundness expresses a particular geometric form of a body of revolution in all the three dimensions, the circular contour is the characteristic form of the entire periphery of a plane figure. For measuring roundness, it is only the circularity of the contour which is determined.

7.9.8. Devices for measurement of roundness. The most commonly used devices for measurement of roundness are :

(1) Diametral. (2) Circumferential confining gauge—a shaft is confined in a ring gauge and rotated against a set indicator probe. (3) Rotating on centres. (4) V-Block.

Piece rotating against a set probe (*a*) of fixed angle, (*b*) of adjustable angle.

(5) Three-point probe (120° spacing).

(6) Accurate spindle.

(*a*) part fixed, exterior spindle with probe rotates, (*b*) probe fixed, part rotates with spindle.

7.9.8.1. Diametral Method. In this method, the measuring plungers are located 180° apart and the diameter is measured at several places. This method is suitable only when the specimen is elliptical or has an even number of lobes. Diametral check does not necessarily disclose effective size or roundness. This method is unreliable in determining roundness.

7.9.8.2. Circumferential Confining Gauge. Fig. 7.35 shows the principle of this method. It is useful for inspection of roundness in production. However, this method requires a separate highly accurate master for each size part to be measured. The clearance between part and gauge is critical to reliability. This technique does not allow for the measurement of other related geometric characteristics, such as concentricity, flatness of shoulders, etc. The values obtained are dependent on the shape of the specimen.

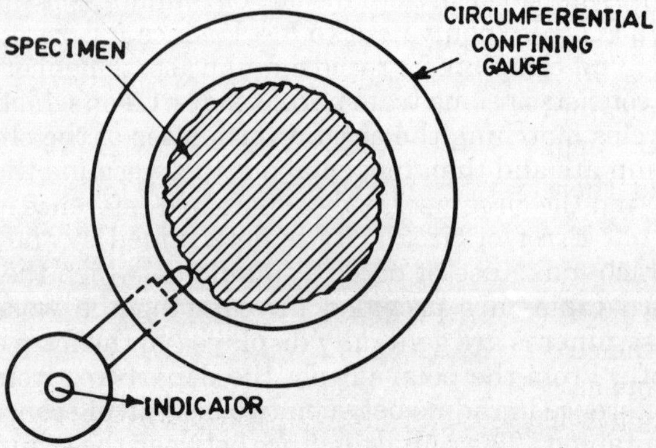

Fig. 7.35. Circumferential confining gauge.

7.9.8.3. Rotating on centres. (Refer Fig. 7.36). Some parts, (such as shafts) may be inspected for roundness while mounted on centres. In this case, reliability is dependent on many factors like angles of centres, alignment of centres, roundness and surface condition of the centres and centre holes, and runout of piece. Out-of-straightness of the part will cause a doubling runout effect and appear to be roundness error.

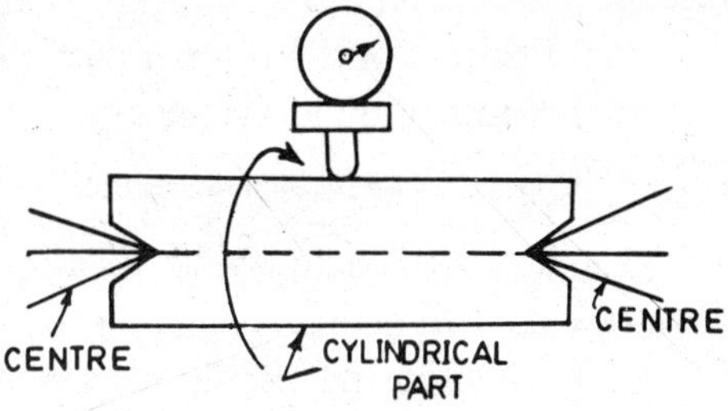

Fig. 7.36. Inspecting shaft for roundness.

Any or all of these factors may combine, creating a high degree of uncertainty as to exact nature of the error.

For workshop purpose, the V-block method is quite accurate as it is capable of indicating normal requirements of accuracy. However for very precise job where more reliable and more accurate results are desired, the second method is recommended which is quicker and also eliminates the effects of angle of the block and the number of lobes on part, but of course, is a very costly one.

7.9.8.4. Assessment using a V-block. The set up employed for assessing the circularity error (lobing) by using a V-block is shown in Fig. 7.37, *i.e.*, the vee-block is placed on a surface plate and the work to be checked is placed upon it. A sensitive dial indicator is firmly fixed in a stand and its feeler made to rest against the surface of the work. The work is rotated to measure the rise and fall of the work-piece. For determining the number of lobes on the work-piece, the work-piece is first tested in a 60° V-block and then in a 90° V-block. The number of lobes is then equal to the number of times the indicator pointer deflects during rotation of the workpiece through 360°. The idea of testing the work-piece in two V-blocks is that when an elliptically shaped part is rotated on a V-block is that when an elliptically shaped part is rotated on a V-block of angle 60°, no change in reading is indicated, whereas if the same part is rotated on a 90° angle Vee-block, two maximum and two minimum readings are indicated on the indicator.

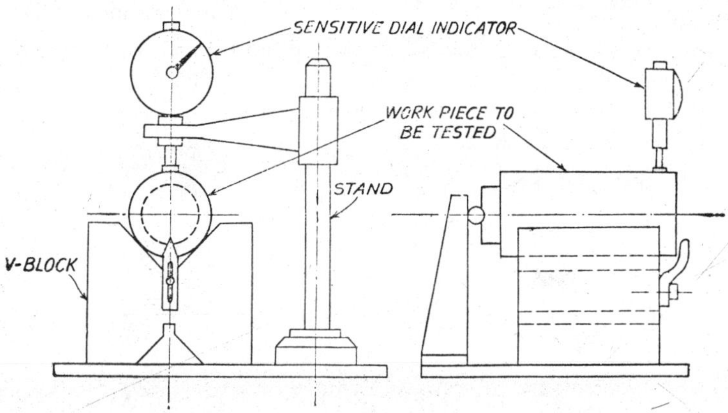

Fig. 7.37. Assessing circularity using a V-block.

The method of determination of the circularity error by V-block has certain limitations and, therefore, the following points should be born in mind.

(*a*) The error of circularity measured on a V-block is greatly affected by the following factors :

(*i*) Angle of V-block very much influences in the determination of circularity error, *i.e.* if the circularity error is say Δe, then it is possible that the indicator shows no variation, or same as Δe, or twice Δe, or thrice Δe, or some other value for each position of the instrument when V-blocks of different angles are used. This is because of the fact that as the angle of V changes the place where the work-piece rests also changes. Ultimately it will be noted that the same work-piece rests at higher place in V-Block of smaller angle and at lower place in V-block of larger angle and thus the indicator will show different readings for same work-piece kept in same position on different angle V-blocks.

(*ii*) Position of the instrument, *i.e.* whether measured from top or bottom.

(*iii*) Number of lobes on the rotating part (*e.g.*, elliptical, triangular, quadrilateral, pentagonal etc.)

(*b*) The instrument's position should be in the same vertical plane as the point of contact of the part with the V-block. If the error is measured at a point far from the V-block, the error of circularity will be influenced by the radial run out of the part.

(*c*) A leaf spring should always be kept below the indicator plunger and the surface of the part, otherwise readings are likely to be affected by minute undulations of the surface, such as surface roughness.

It is obvious that with different angle of V-blocks, dial indicator shows different readings for the same work-piece placed in the same position. If this problem is studied further by analysing an elliptical workpiece on different angle V-blocks, it will be found that some solution can be arrived at. Let us consider an elliptical work-piece whose major axis is 2Δ more than the minor axis as shown in Fig. 7.38.

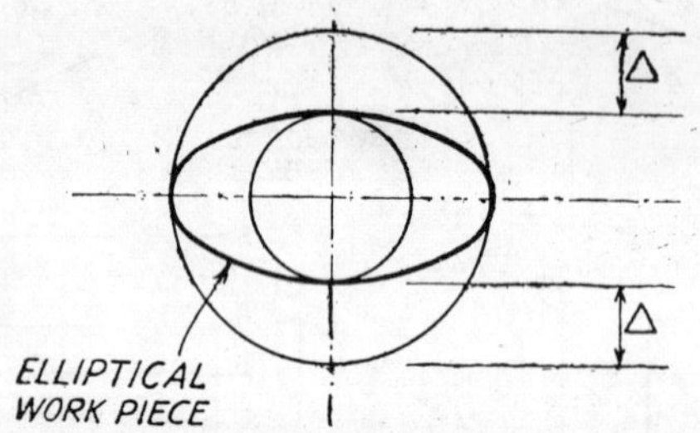

Fig. 7.38. Major axis 2Δ more than minor axis.

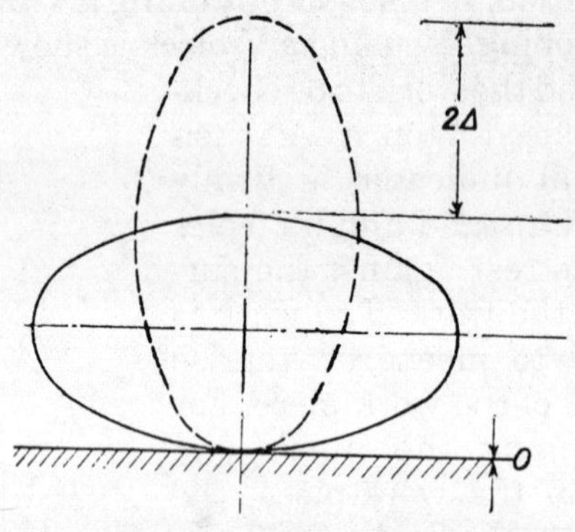

Fig. 7.39. 2Δ error revealed when elliptical object is placed on flat surface.

We will place this work-piece on different angle V-blocks first with major axis placed along the direction of dial movement and then with minor axis along the direction of dial-movement. This is so because the work-piece has two number of lobes.

It may be noted that it is possible to take dial readings both from top as well as bottom. Consider the case when work-piece is placed at flat surface in two positions as shown in Fig. 7.39. It will be noted that the error of circularity when measured from top = 2Δ and when measured from bottom = 0, whereas the actual error of circularity in the work-piece = Δ.

Similarly, if the work-piece is placed on a V-block of angle 120°, it will be found by calculation or otherwise that the error of circularity in this case when measured from top = 1.58Δ, whereas if measured from bottom = 0.42 Δ.

In the same way, it will be found that when work-piece is tested on the 108° V-block, the circularity error measurement when measured from

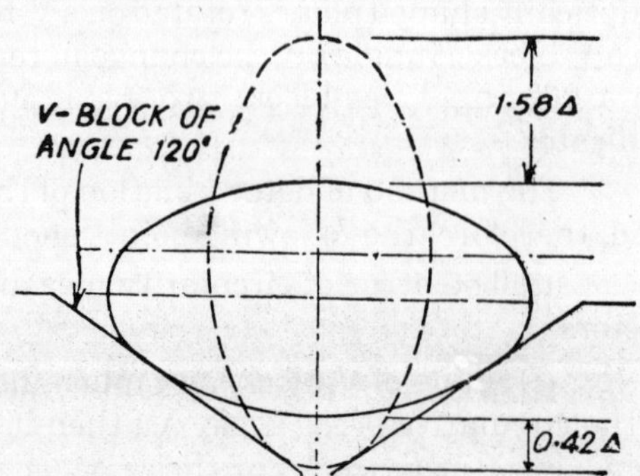

Fig. 7.40. Elliptical workpiece tested on V-block of 120°.

top = 1.38Δ and 0.62Δ if measured from bottom. The corresponding values for 90° and 60° V-block will be Δ, Δ and 0, 2Δ respectively.

Thus for an elliptical work-piece which has got 2 number of lobes, the ratio of circularity error measurement by dial indicator on different V-blocks and the actual error is as given below :

Angle of V-block	*Measurement from top*	*Measurement from bottom*
60°	0.00	2.00
90°	1.00	1.00
108°	1.38	0.62
120°	1.58	0.42

$$\text{Such values of the constant } K = \frac{\text{Measured value of error of circularity}}{\text{Actual value of error of circularity}}$$

can be determined for different shapes of work-pieces, *i.e.* having different number of lobes.

7.9.8.5. V-Block. (*a*) *Fixed Angle.* Depending on the number of lobes on a part, the following angles of V-blocks are recommended for measurement of correct roundness by V-block method.

Lobes	**Angle of V-block**
Three-point out of roundness	60°
Five-lobed part	108°
Seven-lobed part	128° 34′

(*b*) **Adjustable V-Block.** It is usually difficult to ascertain the number of lobes of a part and have large number of fixed angle V- blocks. V-block which can be adjusted to correct angle to show out-of-roundness is better choice.

V-Block method is limited in the determination of roundness of parts because it is suitable only when the number of lobes is known and is uniformly arranged, which is never the case.

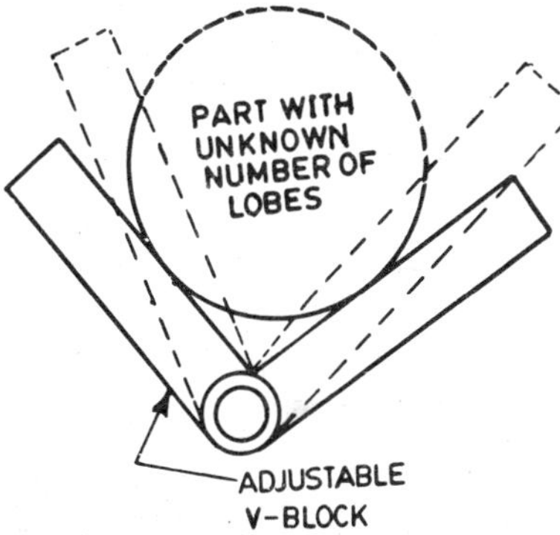

Fig. 7.41. Adjustable V-block.

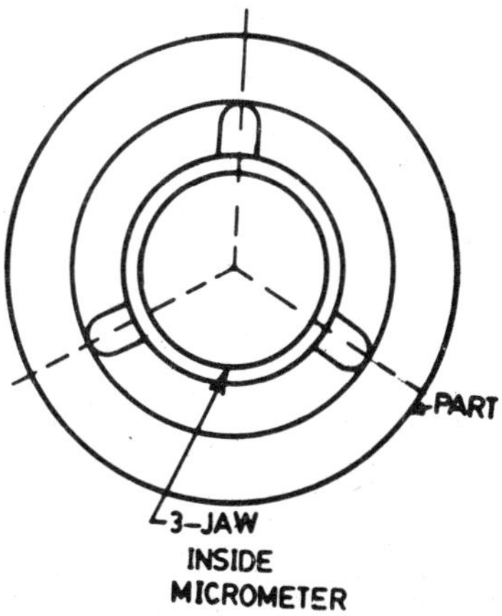

Fig. 7.42. Three point probe.

7.9.8.6. Three-Point Probe. (Refer Fig. 7.42). The three-probe with 120° spacing is very useful for determining effective size in cases of doubtful geometry of part. They perform like a 60° V-block. Like 60° V-block, it will show no error for 5 and 7 lobes, magnify the error for 3-lobed parts, show partial error for randomly spaced lobes.

7.9.8.7. Accurate spindle. None of the five methods discussed above are satisfactory for measuring roundness absolutely. Accurate spindle method is a method to provide a definitive value for calibration of roundness. There are innumerable types of spindle designs available for generation and inspection of round shapes. Various designs are : ball bearings, roller bearings, plain ball spindles, plain bearings (journal type), metered oil flow and fluid-bearing (gas bearing and oil-bearing). These spindles are available in a wide range of tolerances (down to a fraction of a micron).

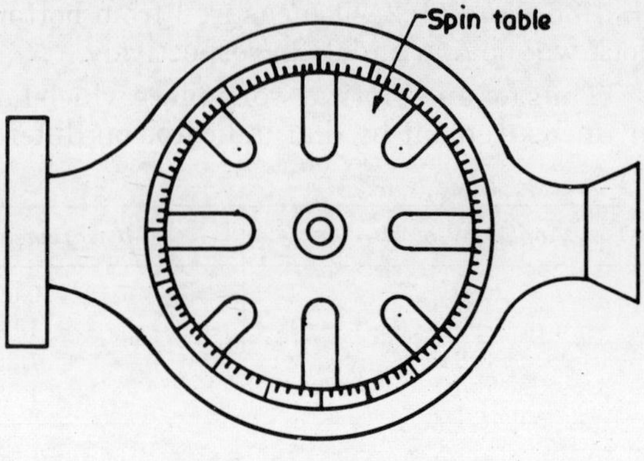

Fig. 7.43. Accurate spindle.

7.9.8.8. Spin Table. It is a highly precise spindle. It is a rotating table-type of spindle, useful for generating cylindrical, conical and spherical shapes when mounted on the jig grinder. It is designed for the generation of essentially round shapes. When the spin table is mounted on the Micro-Sine Table, tapers and conical shapes are ground and to an exact included angle. If the spin table is placed on the jig grinder so that its spindle is horizontal, either by mounting it on an angle iron or on the Sine Table.

The conventional design for this type of spindle includes the use of balls. For maximum rigidity and for minimising the danger of brivelling, the Spin Table instead uses precision rollers under stiff radial and axial preload. Trueness of rotation of the Spin Table is held to 0.0001 mm (Total Indicator Reading) and 0.0005 mm maximum axial deviation, including the effects of camming, flatness and parallelism. In rotation, these accurate spindles describe a perfect circle to which the part may be compared at all its radial ordinates.

Out-of-roundness is shown by the movement of high-magnification comparators. It often suffices to merely observe and note indicator readings. These are usually equipped with means of recording the (linear or polar) results. The polar recorder is an auxiliary device having several rotational speeds, synchronous to those of the measuring spindle. Deviations from true roundness registering on an indicator are traced on circular graph paper at magnifications upto 20,000 to provide a permanent physical record. By overlaying a special transparent template where there are concentric circles representing calibrated value (such as 1 line = 0.00025 mm) to the graph, deviations from roundness may be easily understood and analysed. This can be accomplished even without exact alignment of the part with the spindle.

7.9.9. Roundness Measurement by Plotting Polar Graph. The idea of the actual shape of the work-piece can be had by actually plotting the polar graph, which is explained later. Once the shape is known, the value of the constant can be noted from the table given on next page corresponding to that very shape and the actual error of circularity determined.

The procedure to be adopted for the measurement and drawing of the polar graph and therefrom the error of circularity is given below :

First of all 12 markings equally space at angles of 30° are made on the face of the work-piece to be measured. The work-piece after properly cleaning is then placed on the V-block against a rigid block with a steel ball in between as shown in Fig. 7.37. The dial indicator is placed just above the work-piece so that it is touching the work-piece nearly at the centre of the V-block. The work-piece is then rotated such that the marking on the work-piece is below the indicator plunger. A small strip (leaf spring) is fixed between the indicator plunger and the surface of the part. The readings of the dial indicator are noted down for all the markings. This is repeated three times to take the average value.

For plotting the polar graph, a suitable scale is chosen depending on the maximum value of the reading. Then a circle of diameter nearly 4 times the maximum reading of the dial indicator is drawn and is divided into 12 number of angular divisions as shown in Fig. 7.44. Inside the outer circle, another concentric circle of suitable diameter is drawn. Actually there are standard diameters drawn on polar graphs. Then the values of the indicator are plotted in radial direction taking the smaller circle as the reference circle in order that both the positive and negative readings are plotted within the prepared graph.

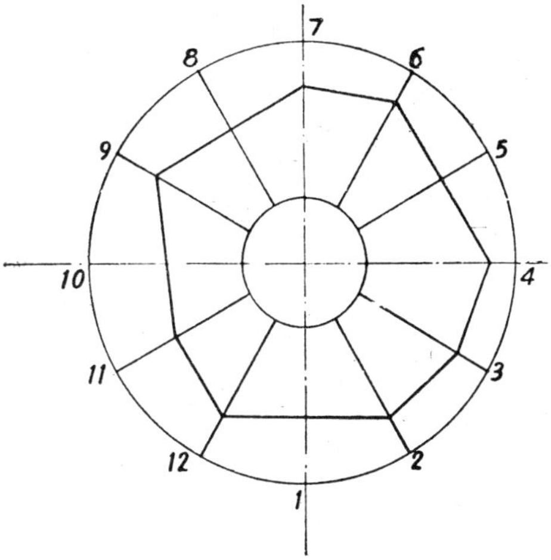

Fig. 7.44. Measuring roundness by plotting polar graph.

Table for determining the value of K different shapes of work-pieces

Shape of the Work-piece	*Angle of vee-block*							
	60°		*90°*		*108°*		*120°*	
	Measurement made from top	*Measurement made from bottom*	*Measurement made from top*	*Measurement made from bottom*	*Measurement made from top*	*Measurement made from bottom*	*Measurement made from top*	*Measurement made from bottom*
Elliptical	0.00*	2.00	1.00	1.00	1.38	0.62*	1.58	0.42
Equilateral Triangle	3.02	3.02	2.05	2.05	1.44	1.44	1.00	1.00
Quadrilateral	0.10*	2.00	0.415	2.40	1.01*	2.01	0.38	1.00
Pentagonal	0.5*	0.5*	2.07	2.07	2.24	2.24	2.03	2.03
Hexagonal	3.00	1.00	1.00	1.00	0.08*	2.00	0.16*	2.11
Heptagonal	0.10*	1.10*	0.02*	0.02*	1.40	1.40	2.04	2.04

*Out of experience it has been found out that the angles of vee-blocks corresponding to values of K indicated with aesterisk mark should not be used.

Individual points are then joined by straight lines, thus obtaining the actual profile of the measured work-piece. The error is measured as the radial distance between the minimum circumscribing circle and the maximum inscribing circle for the profile obtained. This is done by hit and trial method such that this distance is minimum.

The actual value of error of circularity is given by,

$$\text{Error} = \frac{\text{Measured error from polar graph}}{K}$$

where K is a constant, value of which depends upon shape of work-piece and angle of V-block.

7.9.10. Roundness Measuring Spindles. Roundness measuring spindles are of two kinds :

(*i*) Overhead spindle in which the part is fixed on a staging platform and the overhead spindle carrying the comparator rotates separately from the part (Refer Fig. 7.45).

It can determine roundness as well as camming (circular flatness). The concentricity can be checked by extending the indicator from the spindle and thus the range of this check is limited. Flatness and squareness can be inspected by physically sliding the workpiece past the indicator as is done on a surface plate. Height of the work-piece is limited by the location of overhead spindle.

Since the workpiece is stationary and separate from the spindle, the spindle is not affected by load of workpiece.

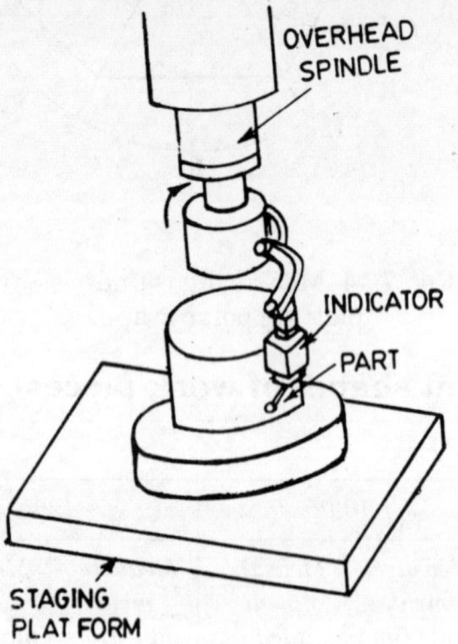

Fig. 7.45. Roundness measuring spindle-overhead spindle.

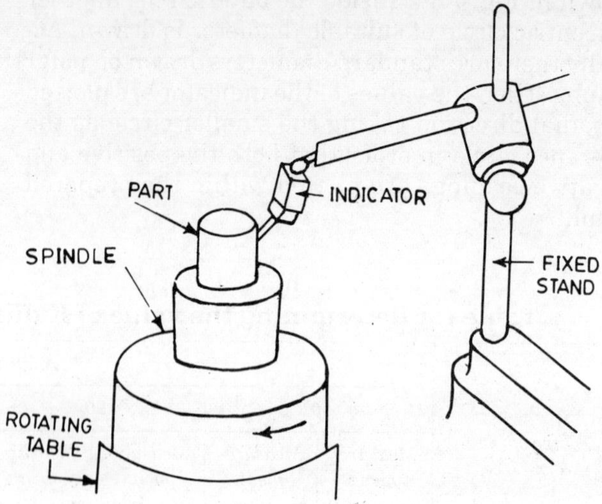

Fig. 7.46. Roundness measuring spindle-rotating table.

(*ii*) Rotating table—in which the spindle is integral with the table and rotates alongwith it. The part is placed over the spindle and rotates past a fixed comparator (Refer Fig. 7.46). It can determine more geometric characteristics without having to move the part.

The rotating table spindle can measure roundness, concentricity and camming (circular flatness). This design allows the inspection of concentricity over a great range along the part axis. Squareness inspection requires sliding the indicator and stand, or an auxiliary horizontal movement. The height of workpiece is no limitation. However, weight of workpiece may result in inaccuracy since it has to be supported on the rotating spindle. Spindle accuracy may be deteriorated due to eccentrically mounted load. There is also the danger of damaging the spindle during mounting of workpiece. With small parts such as bearings, ring gauges etc. no loss of accuracy may be there.

7.9.10.1. Measurement of camming, flatness and squareness. The overhead spindle can inspect camming. The inspection of squareness, taper and flatness is accomplished only by sliding the workpiece, which is not a good practice.

The rotating table can also inspect camming but squareness, taper and overall flatness are inspected by sliding indicator and stand, which is not good practice.

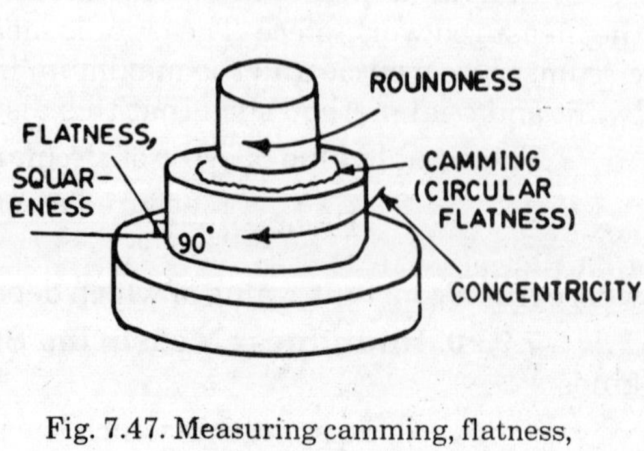

Fig. 7.47. Measuring camming, flatness, squareness, etc.

7.9.11. Features of Universal Measuring Machine Spindle. It is absolutely essential that any measuring machine employed for accurate hole location be fitted with a spindle that rotates true. The locating process is dependent on the use of an accurate spindle. If the spindle does not have a constant axis of rotation, it can not give correct readings for distance between two holes. Further, if the spindle does not rotate true, again it can't give actual centre-distance of two holes. If two holes to be measured are out-of-round, a true spindle will most accurately determine the correct centre line of the hole. Since Universal Measuring Machine measures hole location to less than one micron, its spindle must be accurate to a fraction of a micron.

Accuracy of a Universal Measuring Machine Spindle can be verified by comparison to a master glass hemisphere, (the latter's being around .025 micron) or a master round hole. A polar recording is made as the spindle is compared to the master. Accuracy of the hemisphere or master hole is self accuracy proved by turning it 90° and making a second polar chart. If this is not perfect, its error is shown displaced 90° on the second polar trace.

Spindle and Quill. The two main elements of Universal Measuring Machine are the spindle shaft and the quill. The tolerance of these parts as to size, straightness, and roundness is to a fraction of a micron. This accuracy is achieved by random motion of handlapping process. Between the spindle shaft and the quill inside are four hardened 3 mm diameter balls held in a brass retainer. To establish the correct preload, the size of the balls must be determined precisely. Uniformity of ball size is also prerequisite to rotational accuracy of the spindle.

Balls of 3 mm diameter are used as a thrust between a collar on the spindle, and a shoulder on the quill to enable spindle rotation free of axial camming.

The quill assembly slides vertically within the housing, guided by two lapped sleeves. Clearance is held to 1.3 micron, allowing enough space for oil film thickness.

Yoke. Yoke assembly is meant to elevate/lower the spindle. Spindle is elevated/lowered by a pinion which works against a rack set into the quill itself. The quill and yoke design must make it possible to attain a vertical quill movement which is straight, aligned with the spindle housing travel, and square to the table top.

Rotation of the Spindle. The spindle is rotated by hand through a knurled knob at the top of the spindle for normal measuring application like locating hole or exploring the geometry of a hole. When measuring roundness, the spindle is rotated at the rates or 1/2, 1, 2 or 4 revolutions per minute by power.

Spindle Housing Travel. The spindle housing is elevated/lowered by a large hand-wheel at the right of the column. It is primarily meant to accommodate larger workpieces. Its travel is accurate and straight.

Polar Recorder. It is fastened to the right of the machine. It has its rotational speed synchronous to that of spindle. It produces a permanent graphic record of roundness or other features of a part, such as concentricity, squareness and taper.

7.9.11.1. Characteristics that can be inspected by a Universal Measuring Machine. The following characteristics can be conveniently measured by a Universal Measuring Machine :

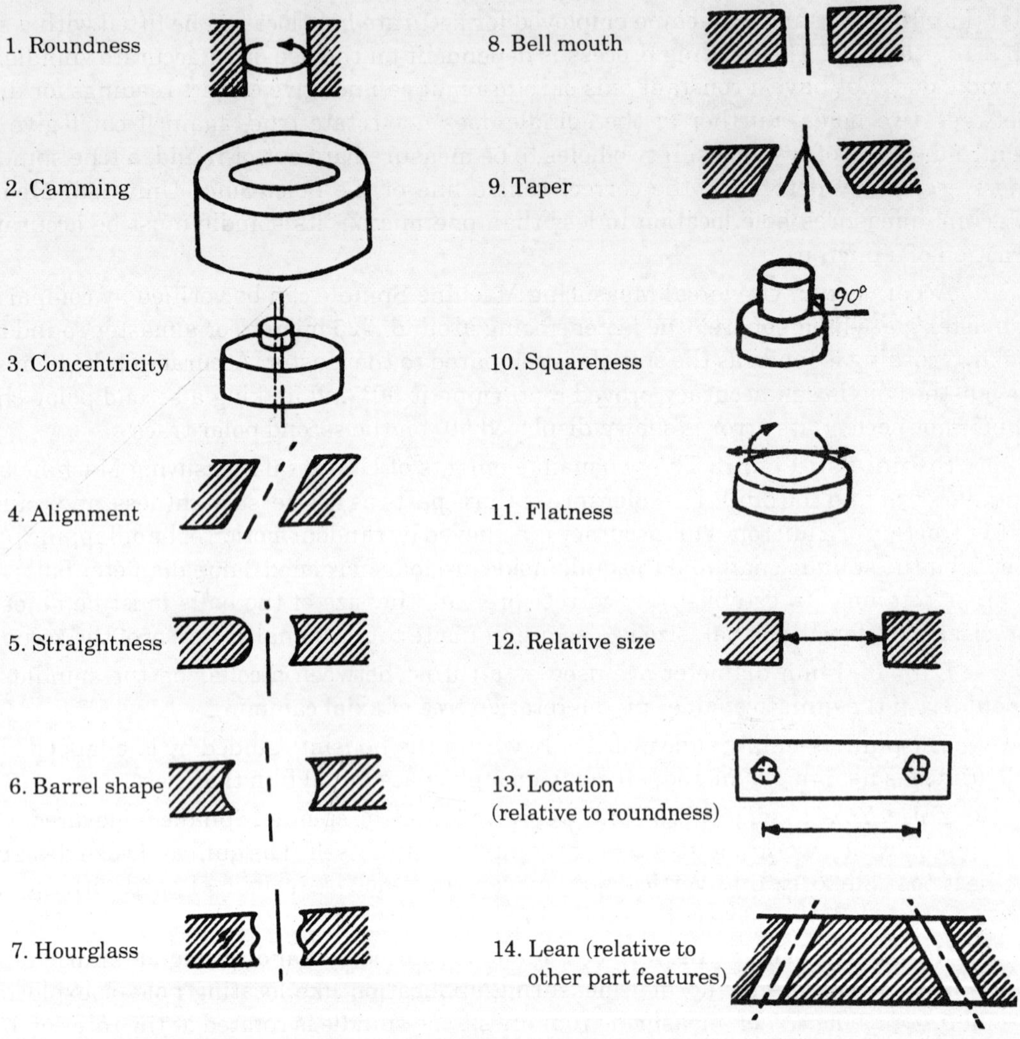

Fig. 7.48. Characteristics measureable by universal measuring machine.

All the above measurements are possible due to highly accurate spindle rotation, and all movement of quill and housing being straight and parallel and also mutually square to the table top and to the plane of travel of the base slideways.

7.9.12. Measurement of error of circularity by modern precision machine. In the precision measuring machine intended for this purpose, the effects of angles of V-block, the position of indicator and the shape of work-piece, *i.e.* the number of lobes are completely eliminated.

In this instrument, a truly running spindle mounted on the high precision ball bearings and carrying a micron indicator is employed. The accuracy of running of the spindle is of the order of 1 micron. The spindle is mounted vertically on a right stand and the workpiece is mounted on a co-ordinating table placed just below the spindle. The job is centered by means of the co-ordinating table such that the axis of the spindle coincides with the axis of the job. The remaining procedure for plotting the polar graph is same as with V-block, *i.e.* the 12 markings are made on the job and the spindle is slowly rotated starting from zero position and noting the indicator reading for all the positions.

In the very high precision measuring instrument such as Talyrond, which has high spindle running accuracy of less than 0.1 micron, a record of the exact profile of the job is made automatically and thus the waviness is also superimposed upon the profile of the job.

Modern roundness measuring instruments (machines) are designed to make the numerical assessment of departures from ideal roundness from one of the following two centres, *viz.*, (*i*) determination from minimum zone centre, *i.e.* a centre for which the radial difference has a minimum value ; (*ii*) determination from the least squares centre, *i.e.* a centre defined on a "least squares" basis (the sum of the squares of the errors is the minimum).

Most of the machines, employ air bearing (non-contact type bearing) for the rotating tables to ensure high accuracy of rotation. The other advantages of air bearings are (*i*) non-contact bearing prevents accuracy fall to be caused by wear, (*ii*) accuracy of rotation is not alliviated even if slight shock due to *e.g.*, striking of work-piece against the table, (*iii*) no temperature rise from long hours of use and no oiling or warming up is required, (*iv*) recording diagram having high reproducibility can be obtained in comparison with instruments employing mechanical bearing.

For the maintenance of accuracy of air bearing, fresh and clean air free from dust and moisture is very essential.

The workpiece to be tested is set on the rotating table and the gauge head is brought into contact with the workpiece. Rotating the workpiece, roundness is measured and the gauged value is recorded on the polar electric spark recorder synchronised with rotating table. The signal from the gauge head is amplified and applied to a polar co-ordinate recorder giving straight radial ordinates. The range of magnification extends from × 50 to × 10,000. The graph obtained shows the amount by which the periphery of the component being checked departs from a truly circular form.

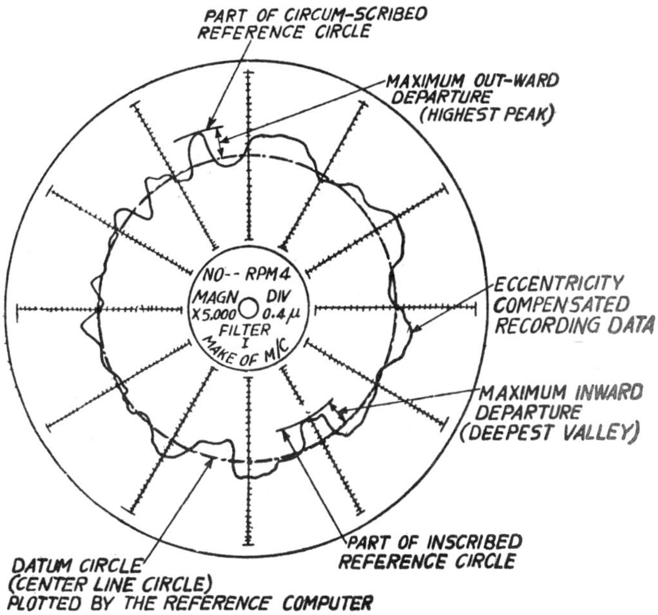

Fig. 7.49. Error of circularity measured in precision machines.

The sophisticated machines have the provision to check concentricity, roundness, alignment, squareness, parallelism and flatness. Usually tilting table is provided which can be used for alignment of the rotating shaft of air bearing to the work centre or for ensuring rectangularity between them when the measurement of concentricity or flatness is desired. The tilting can be easily set on the table. The table turns at constant speed of 4 RPM or 6 RPM which is specified on the recording chart.

The measured profile chart of the surface examined is plotted on circular chart paper printed with polar co-ordinates. (Refer Fig. 7.49) The departure from roundness can be assessed either by determining the centre of the two concentric circles having a minimum radial separation, within which the graph is contained, or by the "least squares" centre and circle from which the undulations are measured. The first method can be employed by using compasses or a transparent template with circular ordinates, under which the graph is floated until a "best fit" is obtained.

(The more sophisticated machines have the provision of automatic eccentricity compensation, which facilitates centering and assures distortion-free recording and makes judgement of recorded results very easy.)

The second method involves computation and automatic plotting. The x and y ordinates of the "least squares" centre for the reference circle are given by a meter reading on the reference computer.

7.9.13. Profile Filters. When all of the radial deviations of a circular cross-section are fully represented by a measured profile, the surface irregularities (of high frequency) can mask the form of the profile or lobing or widely spaced undulations (low frequency). Depending on the importance of requirements (whether low or high frequency irregularities), electronic filters are incorporated in the electronic unit to clarify the profile trace for visual assessment.

Filters could either suppress the closely spaced undulations to reveal the form of the profile more clearly, or suppress the widely spaced undulation and allow the magnifications to be increased to reveal the closely spaced irregularities.

Electronic devices also compute the least square reference circle of the profile. The analysis of departures from roundness can be speeded up by providing a visual display of workpiece profile and any of the four reference circles. It also includes a digital display of departures from the selected reference circles.

A single roundness graph is often of limited value because it can give only part of the information needed to determine the overall functioning of a part. However, several related roundness and straightness measurements can give a much more complete picture (like concentricity, alignment, squareness, parallelism, straightness and conicity).

7.9.14. Roundness Measuring Machines. While it is easiest to generate a circular cross-section and many objects we come across are circular, it is very difficult to check roundness.

By definition, roundness is the property of a surface of revolution, where all points on the surface are equidistant from the axis. Unfortunately this axis does not exist physically and measurements can only be made by sensing the large number of points on the surface.

The out-of-roundness of any profile can be specified only when some centre is found from which to make the measurements. Finding the centre from which to measure the variation in profile is an important part of roundness measurement.

It is important to distinguish between diameter and roundness. Two are measured quite differently and separately, by different methods, and instruments. While measurement of diameter is done statically, for measuring roundness, rotation is always necessary. Diameter, of course, is very much affected by out of roundness of a component. The simple measuring method with a dial gauge and V-block suffers form the limitation that results may vary according to the vee angle and spacing irregularities. It also gives a false impression when the undulations are regularly spaced. This method, thus has its limitations and the possibility of obtaining misleading results tends to offset the simplicity of the procedure. Modern engineering requires closer tolerances and more accurate measurement of roundness and tighter control over the geometric form of round and cylindrical surfaces. Roundness measuring instruments provide results of roundness measurements accurately, speedily and reliably overcoming demerits of workshop methods. These instruments are of two types ; *viz.*, the rotating pick up type in which the workpiece is stationary and the pick up revolved around at ; and the turntable type in which the workpiece is rotated and pick up is stationary. In the rotating pick up type instrument, precision spindle has to be designed to carry the comparatively light load of the pick up, which is easier to make. The weight of workpiece, being stationary, is not a limitation on measuring capacity.

In the case of turntable, the more commonly used type, since the pick up is not associated with the spindle, the instrument is more easily adapted to measurements associated with roundness, such as concentricity and alignment. Reposition the pick up has no effects on the reference axis. There is also more freedom in positioning the pick up to reach into slots or the undersides of the shoulders without having to use long or cranked stylus arms. It is also possible to measure straightness by providing straight vertical movement of the pick up, there being no need of modifying the spindle or its mounting.

The pick up converts the minute movements of the stylus into electrical signal, which is processed and amplified and fed to a polar recorder. A microcomputer is incorporated with integral visual display unit and system is controlled from a compact keyboards, which increases the system versatility, scope and speed of analysis.

Basically the final result is a roundness graph ; of course, it is not a magnified cross-section of the part. Results may be displayed graphically on a CRT and the out-of-roundness and other values shown digitally as well. The graph plotted uses a radial magnification of X2000 or so and the circumferential magnification is often less than one.

Usually several related roundness and straightness measurements are needed to give a much more complete picture about the overall functioning of the part. Extensive analysis of features like roundness, concentricity and cylindrically, can be done. System is programmed to assess the roundness of workpieces with respect to any of the four internationally recognised reference circles. A visual display of workpiece profile with the selected reference circle relative to the axis of the instrument spindle can be obtained. Workpieces can be assessed over a full circumference, and with undercut surface or an interrupted surface. With sufficient data the appropriate reference circle can be fitted to the profile.

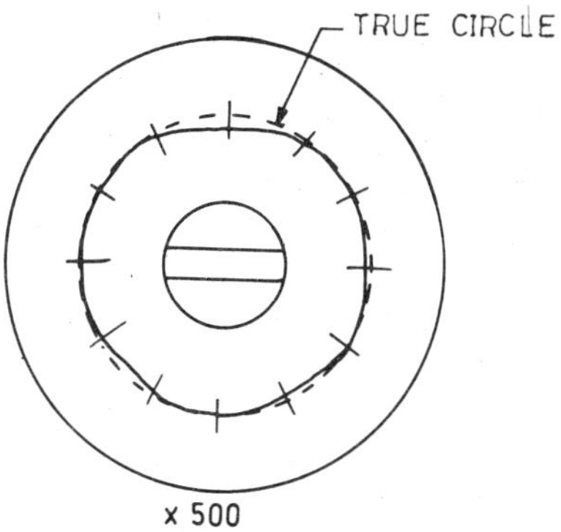

Fig. 7.50.

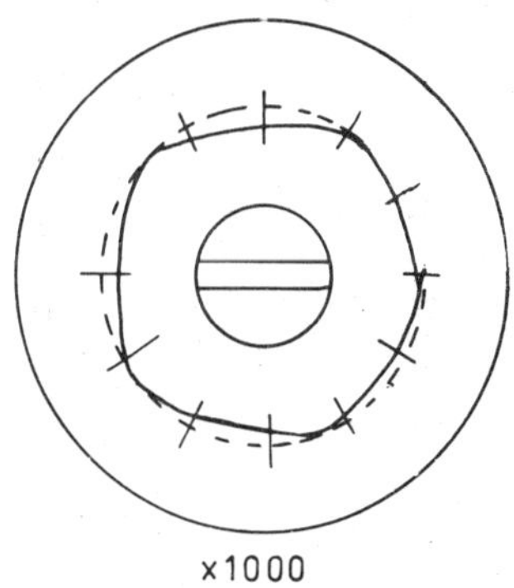

Fig. 7.51

Measurement of roundness at different magnification levels.

Measurements can be made with respect to the spindle axis of the measuring instrument, or with respect to an axis derived from the centre of selected measured plane on the component. The program also provides functions like auto-centering, auto-ranging, auto-calibration and concentricity. Figs. 7.50 to 7.53 show the effect on the graph of unequal magnification of same component using 4 magnifications. This part has 5 lobed parts. At low magnification the graph differs very slightly from a perfect circle. As magnification of radial distance is increased, the effect becomes obvious.

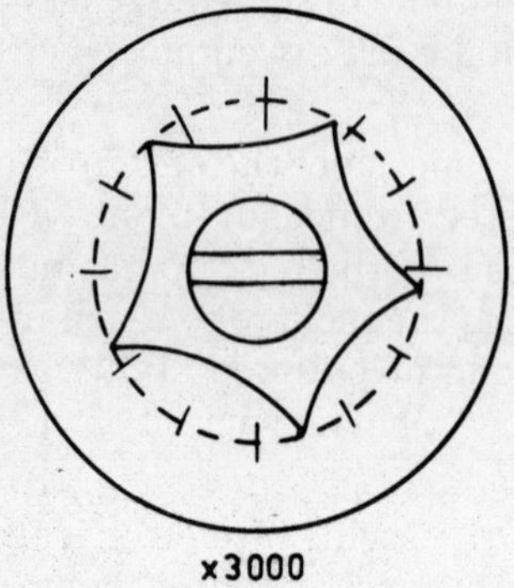

Fig. 7.52

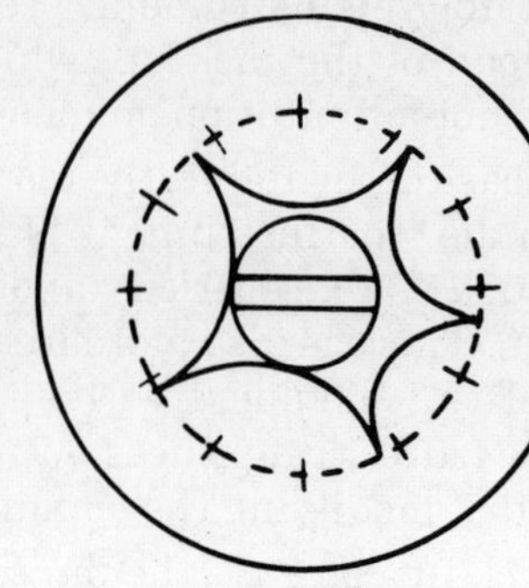

Fig. 7.53

Effect of magnification on roundness measurement.

Microprocessors are used to assess out-of-roundness. The desired data for out-of-roundness is presented on C.R.T. without the need of analysing the polar graph by the operator.

7.9.15. Modern Roundness Measuring Instruments. The modern roundness measuring instruments are based on use of microprocessor to provide measurements of roundness quickly and in a simple way. There is no need of assessing out-of-roundness. Machine can do centering automatically and calculate roundness and concentricity, straightness and provide visual and digital displays. There are two basic types of roundness measuring systems : (*i*) rotating worktable in which the workpiece rotates and the pick-up is stationary, and (*ii*) rotating spindle, in which the workpiece is stationary and the pick-up rotates. The bench top rotating worktable type of instrument is the most widely used in workshops and inspection areas by virtue of its lower cost and versatility. It consists of a rotary table of around 200 mm diameter spindle pick-up mounting, vertical magnifiers having magnifications of order of 100-10,000, and polar recorder. A computer is used to speed up calculations and provide the standard reference circles (least squares circle, minimum zone circle and maximum inscribed circle, and minimum circumscribed circle). Assessment of roundness can be done by templates. Computer can provide the least squares circle superimposed on the profile graph, meter display of the mean line average departure of the profile from the reference line, P, V and $P + V$, and the $x - y$ coordinates of the centre of the least square circle with respect to the axis of spindle rotation (Please note that the out-of-roundness is defined as the radial distance of the maximum peak (P) from the least square circle plus the distance of the maximum valley (V) from the least square circle, *i.e.* ($P + V$). It is possible, by the machine, to record a reference circle with the profile graph. Operator communication is via a small keyboard which can be held in hand. Instrument can be operated by untrained personnel. All the roundness analysis functions can be performed with the addition of harmonic and slope analysis. Cylindricity analysis can be done by collecting data from upto ten planes along a cylindrical component.

7.10. Tests for Checking Rotation

Tests for checking the rotation of a body are made for three conditions :

(*a*) Run-out. (*b*) Periodical axial slip. (*c*) Camming.

We shall now study the definition and meaning of these three conditions and then the ways to measure these.

7.10.1. Run-out. A round body can be said to be running true on its axis when a dial indicator with its feeler placed against the surface of the round body does not show any variation in reading. Run-out is caused when there is some deviation in the trajectories of points on a section of the surface is relation to the axis of rotation (errors of position) and the movement of the axis of rotation if bearing surfaces or bores are not exactly circular (errors of bearings). In other words, run-out is the resultant of the radial throw of the axis, of the out of roundness of the component, and of the errors of bearings.

Out of round is the error relative to the circular form of a component in a plane perpendicular to the axis at a given point of the latter. Its definition and the method of measurement have already been discussed in details. However, it may be again reminded that the value of out of round for a shaft is given by the difference between the diameter of the inscribed circle and the smallest measurable diameter of the shaft ; and for a hole, it is given by the difference between the diameter of the inscribed circle and the largest measurable diameter of the hole, each of them measured in a plane perpendicular to the axis.

Radial throw of an axis at a given point is the distance between the intersection of the geometrical axis and the rotating axis with a plane perpendicular to the axis of rotation at a given point (Fig. 7.54).

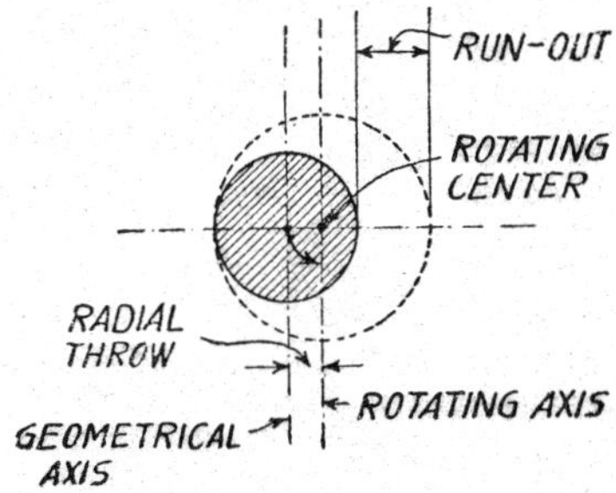

Fig. 7.54. Radial throw of an axis.

The geometrical and the rotating axes may be parallel to each other or inclined. They will be parallel if the radial throw is constant at any point of the axis. From the metrological point of view, the two axes are considered to be identical when the error of rotation of the cylindrical surface, at each measuring point, does not exceed the specified value.

Run-out (or out-of true running) of an exactly round component in a given section will be twice the radial throw of the axis in a given section.

7.10.1.1. Measurement of run-out. This is tested, for the external surface, by arranging the plunger of the dial gauge normal to the revolving surface and the readings of the dial noted as the spindle is slowly rotated by 360°, and for the internal surface by bringing the plunger into contact with mandrel fitted in the bore. The testing is usually carried out in both vertical as well horizontal planes.

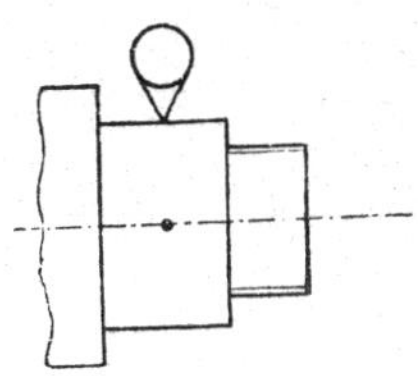
Test on external cylindrical surface.

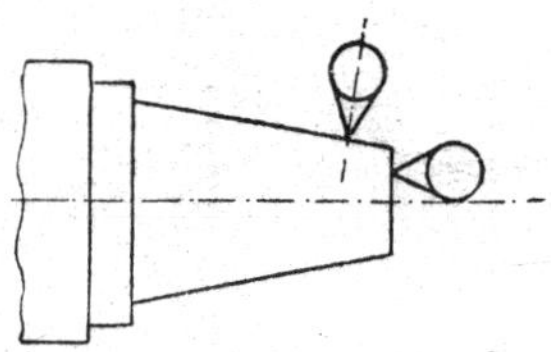
Test on external conical surface.

It is essential for this test that the spindle be rotated sufficiently before starting the test, in order to ensure that the lubrication film will not vary during the test and that the temperature attained may be considered as the normal running temperature of the machine. Further it may be noted that this test takes into account only errors in the bearings of spindle and does not give any information of the exact shape of the cylindrical or conical bore, or of actual position of the bore in relation to the axis of rotation.

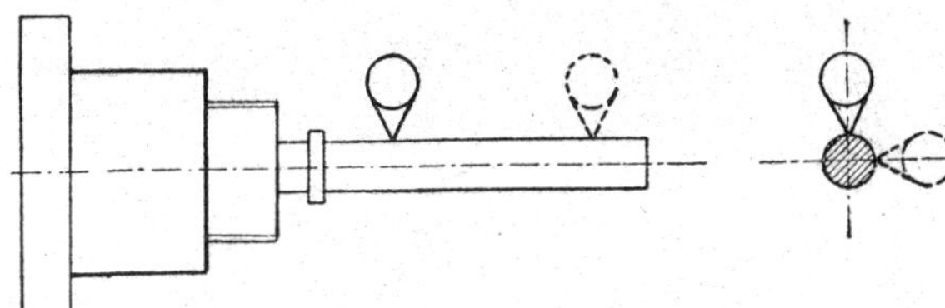
Fig. 7.55. Test on internal surface.

On the external conical surface, the feeler of the dial gauge should be set at right angles to the generation line. It may be noted that in this case if there is only axial movement in the spindle during its rotation then the diameter of the circle being checked will vary and as a result the run-out will appear greater than the actual. For this reason either the axial slip should be measured before hand and its effect on the measurement of run-out computed according to the angle of taper or else this test be applied only to tapered components having very small taper.

While performing this test on the internal surface with the help of mandrel, the following points should be remembered :

(*i*) Checking should be done at two sections, one near the housing and other at a specified distance apart. (This is necessary to check that the axis of the mandrel might not have crossed the axis of rotation in the plane of measurement).

(*ii*) Above checking to be repeated 4 items and average taken, by turning the mandrel, through 90° in relation to the spindle every time. (This test checks for any lack of accuracy in inserting the mandrel into the bore).

However, it may be noted that the testing of internal surface with test mandrel does not check the exact shape of the bore.

7.10.2. Periodical Axial Slip. Before we consider the definition and measurement of periodical axial slip, let us understand the meaning of minimum axial play. When a part rotates and if some axial movement (measured at rest at each of the several positions around the axis) along its axis is observed, it is known as axial play. And rotating part can be considered as fixed in its axial direction, when its axial slip remains within the tolerance zone. The minimum permissible axial slip defines the limit of the axial slip of the spindle during one slow rotation with a slight axial pressure.

Periodical axial slip represents the extent of reciprocating motion of a rotating component along its axis of rotation, when the influence of the minimum axial play is eliminated by applying axial pressure in a given direction.

The value of the axial slip is determined by applying a force along the axis of rotation and arranging the feeler of the dial gauge to touch the rotating surface a small distance away from the axis of rotation as shown in Fig. 7.56. Test is made for each of the two positions at 180° of the dial gauge and the algebraic mean of the readings taken. When testing the thrust bearing of the ball or roller type, the test should be made through at least two revolutions.

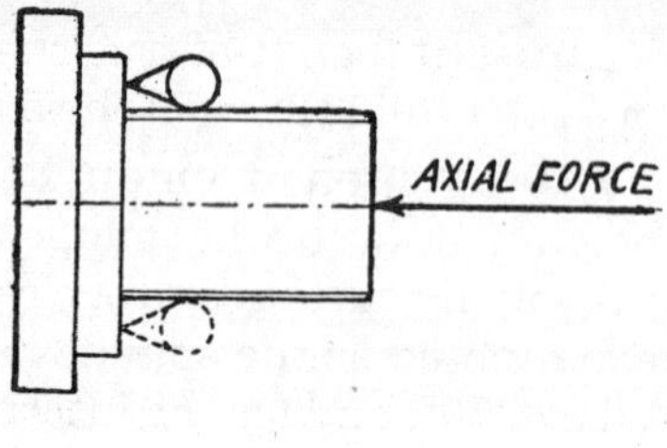

Fig. 7.56. Axial slip determination.

Periodical axial slip is determined by applying a slight pressure to the spindle in the direction of measurement and arranging the feeler of dial gauge along the centre of the front face, as far as possible along the axis of rotation. Thus it is essential that such a device be used which permits a force to be applied along the axis and at the same time permits a dial gauge to be placed on the same axis. For this purpose a dynamometer is used which applies force in the centre of spindle through its surface and also has opening in the middle for the dial gauge to be accommodated. In the case of a lead screw, the axial force is applied by movement of the slide when the nut is connected.

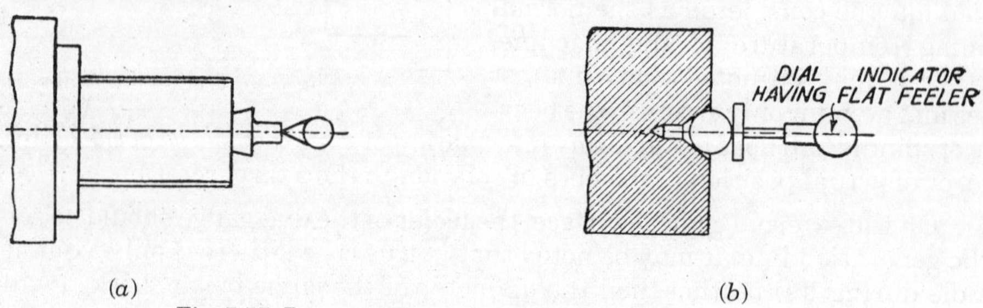

(*a*) (*b*)

Fig. 7.57. Determination of periodic axial slip.

If the spindle is hollow then a short mandrel, having a plane face perpendicular to the axis may be used [Fig. 7.57 (*a*)]. If the spindle has a centre, a steel ball as shown in Fig. 7.57 (*b*) may be inserted. In this case the feeler of the dial gauge should have flat surface.

7.10.3. Camming. The following definitions apply to the camming of a plane surface rotating around an axis :

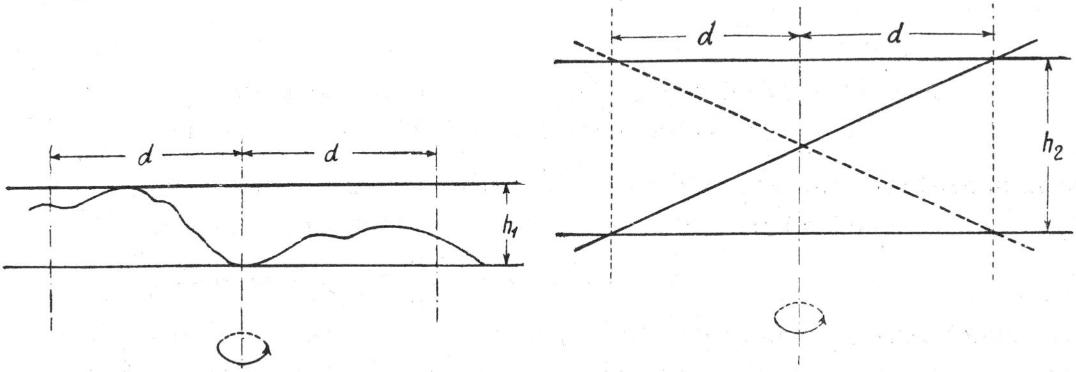

(*a*) Camming (when surface is not flat) (*b*) Camming at a distance.

(*a*) **Camming of a surface.** Camming is the defect of a plane surface which, when rotating around an axis, does not remain in a plane perpendicular to this axis. Camming is given by the distance *h* separating the two planes perpendicular to the axis between which the points of the surface are moving during the rotation.

(*b*) **Camming of the surface at a distance from the axis.** This is represented by the distance *h* separating two planes perpendicular to the axis between which moves the portion of surface delineated by a revolving cylinder, the diameter of which is 2*d*, and which has, as its axis of symmetry, the theoretical axis of rotation of the surface.

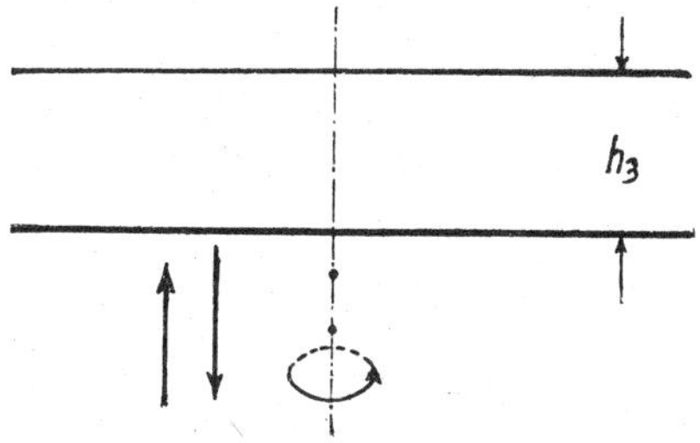

(*c*) Camming (with periodical axial displacement).
Fig. 7.58

Camming is the resultant of various defects of the surface and axis of rotation (h_1, h_2, h_3 in Fig. 7.58), for example, surface not flat [Fig. 7.58 (*a*)], surface and axis of rotation not perpendicular [Fig. 7.58 (*b*)], periodical axial displacement of the axis [Fig. 7.58 (*c*)].

[**Note.** When the plane concerned has a geometrical axis (part A, Fig. 7.59) which does not coincide with the axis of rotation, the resulting radial throw gives camming. This defect is due to the lack of perpendicularity of the face to the axis of rotation. In practice, the run-out does not have a great effect, as it affects camming only in the ratio 1 : 2. When the run-out is constant on all points of the axis (eccentricity), there is no camming.]

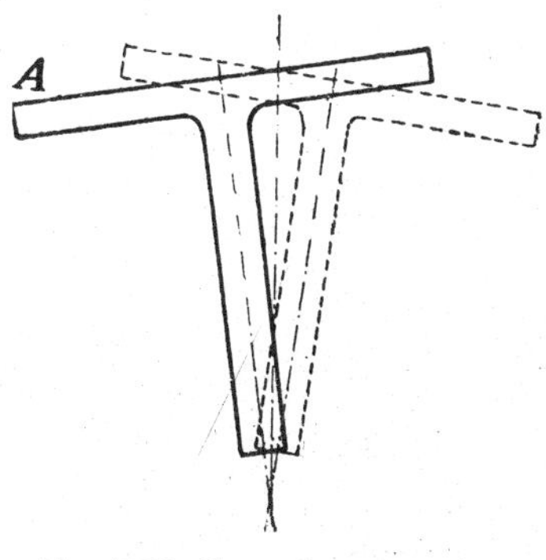

Fig. 7.59. Camming given by radial throw.

Thus in brief it could be said that camming relates to rotating plates and is measured in relation to a plane perpendicular to the rotating axis, and represents the deviation, at any point, of all the trajectories of all the points on a given circumference of the surface to be checked. It includes errors of shape of the face, the angle of the latter in relation to the axis of rotation, the radial throw and the periodical axial slip of the spindle. It may be noted that this does not include the minimum axial play of the rotating component.

The object of checking camming is to verify that all the points in the same circle on the front face are in the same plane perpendicular to the axis of rotation and that the axial position of this plane does not vary during the rotation of the spindle. Testing is always done on the points farthest from the axis on the circumference, because the camming increases for the points away from the axis of rotation. In order to analyse the origin of camming, the defects of the surface and of the axis should be measured separately. The axial slip should be measured in every event for measuring camming.

The testing is carried out by arranging the feeler of the dial gauge at a given distance L from the centre and at right angles to the face. Readings are taken by placing the plunger successively at a series of points spaced around the periphery (Fig. 7.59). The spindle is rotated continuously at slow speed and given a slight end pressure to eliminate the effect of play in the thrust bearings. The deviations between the maximum and minimum readings are noted for each position of the dial and the greatest deviation represents the camming.

7.11. Profile Measurements

Profile is that part of the contour of a component, which is observed as its boundary line in a functionally significant cross-sectional plane. The inspection and measurement of the profile can be accomplished in many different manners, depending on the size of the part, the required functional properties and level of accuracy of the profile, its form characteristics, etc. Various methods of profile measurement are described below in brief.

(*i*) Direct comparison with a washer having the inverse form of the normal contour. The gaps between specimen contour and washer can be observed by light transmission directly.

(*ii*) Limit gauges (in pairs for GO and NOT GO sizes) for the entire contour of the part. The sizes are determined by the maximum and minimum material conditions resulting from the compounded effects of the form and size tolerances.

(*iii*) Mounting indicators side by side in the plane which contains the profile to be inspected.

The indicators need to be located judiciously and must be set with the aid of a washer representing the nominal profile.

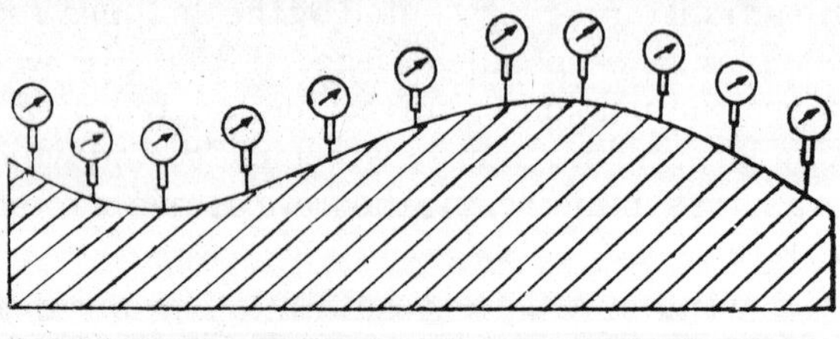

Fig. 7.60. Measurement of profile.

(*iv*) Measuring the coordinate positions of characteristic contour points. This is possibly by inspecting the size and form of features bounded by essentially straight lines.

(*v*) If a contour can be defined by polar coordinates as in case of cam, then functionally essential contour characteristics can be specified, laid out and inspected by individual points whose positions are defined by the means of dimensioning in a system of polar coordinates.

(*vi*) Magnification of part and comparison with a graticule carrying reference lines of normal form and size. The silhoutte image of the inspected object appears in the focal plane of a microscope where a transparent graticule with reference marks can be mounted.

(*vii*) Projection of magnified image on a special screen with charts having single or double contour lines of the object to be inspected.

(*viii*) Optical profile comparison by light sectioning. In this method, a narrow band of light is projected obliquely around the edge of the curved specimen, to illuminate a section which by means of optical correction, appears as a normal section in the field of view of the special microscope, to be observed concurrently with a washer drawing of the specimen.

WORKED EXAMPLES

Problem 7.1. *A measuring machine bed was tested for straightness using an autocollimator and reflector as shown in Fig. 7.61 and the following readings in minutes were obtained corresponding to various points a, b, c, d ...... etc. If one minute of arc increase in angle observed corresponds to a rise of 20 μm of the front of the reflector relative to the rear, construct a profile graph of the surface relative to the initial points a, b and state the maximum deviation from a line through the end points. Also determine the least square line and show it on the graph, and state the maximum deviation from it.*

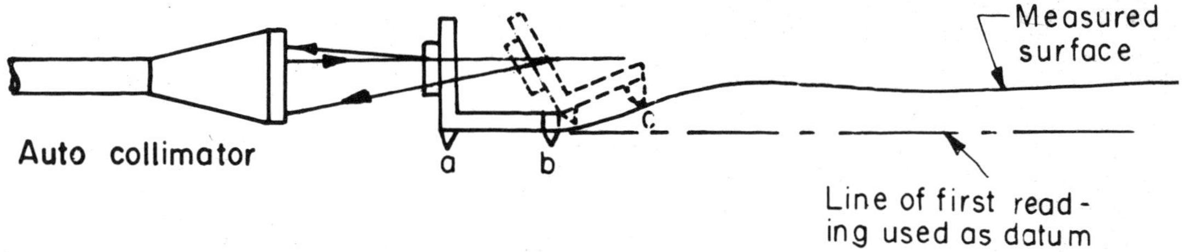

Fig. 7.61. Auto-collimator is firmly mounted relative to surface plate on which measuring machine bed is mounted.

Position	*Angle reading θ (minutes)*	*Position*	*Angle reading θ (minutes)*
a	—	*f—g*	*20.847*
a—b	20.70	*g—h*	*20.838*
b—c	*21.05*	*h— i*	*20.861*
c—d	*21.11*	*i—j*	*20.872*
d—e	*20.83*	*j—k*	*20.871*
e—f	*20.832*		

(With a vernier, the values of angle reading have been roughly estimated to third place of decimal).

Solution. Tabulating the results as per Table under article 7.3.

Position	*Angle reading (min.) θ*	*Difference from first reading δθ (min.)*	*Rise or fall in interval length (l) (μm)*	*Cumulative rise or fall w.r.t. line a-b extended (μm)*
a	—	—	—	—
a—b	20.70	0	0	0
b—c	21.05	+ 0.35	+ 7.0	+ 7.0
c— d	21.11	+ 0.41	+ 8.2	+ 15.2
d—e	20.83	+ 0.13	+ 2.6	+ 17.8
e—f	20.832	+ 0.132	+ 2.64	+ 20.44
f—g	20.847	+ 0.147	+ 2.94	+ 23.38
g—h	20.838	+ 0.138	+ 2.76	+ 26.14
h—i	20.861	+ 0.161	+ 3.22	+ 29.36
i—j	20.872	+ 0.172	+ 3.44	+ 32.80
j—k	20.871	+ 0.171	+ 3.42	+ 36.22
				+ 208.34

Plotting a curve between distance along the surface to be tested and the cumulative error (Fig. 7.61) and joining the end points, it is noted that w.r.t. this line the maximum deviation occurs at point d and is approximately equal to 3 μm.

The least squares line is one from which the sum of the squares of the errors is minimum and actually it is preferred to give the deviation from least squares line through the surface. Such a line can be defined by the relation, $y = mx + c$

where $m = \dfrac{\Sigma x_m y_m}{\Sigma (x_m)^2}$, $x_m = x - \bar{x}$ and $y_m = y - \bar{y}$.

Assuming distance $ab = bc = 1$ unit, $\bar{x} = 5$ units located at point f and $\bar{y}$

$$= \frac{\Sigma y}{N} = \frac{208.34}{10} = 20.834 \cong 20.83 \text{ μm}.$$

Reference	x_m	$y_m = y - y$	$x_m\, y_m$	x_m^2
a	– 5	– 20.83	+ 104.15	25
b	– 4	– 20.83	+ 83.42	16
c	– 3	133.83	+ 41.49	9
d	– 2	– 5.63	+ 11.26	4
e	– 1	– 3.03	+ 3.03	1
f	– 0	– 0.39	0	0
g	+ 1	+ 2.55	+ 2.55	1
h	+ 2	+ 6.69	+ 13.38	4
i	+ 3	+ 8.53	+ 25.59	9
j	+ 4	+ 11.97	+ 46.88	16
k	+ 5	+ 15.39	+ 76.95	25
			$\Sigma x_y\, y_m = 418.60$	$\Sigma x_m^2 = 110$

$$\therefore \quad m = \frac{418.6}{110}$$

$$= 3.805 \text{ μm/base unit.}$$

The least square line will, therefore, pass through the point k.

$$x = 10, y = \bar{y} + 5 \times m$$
$$= 20.83 + 5 \times 3.805$$
$$= 20.83 + 19.025$$
$$= 39.85 \text{ μm}.$$

From the graph, it is seen that the maximum deviation from the least squares line occurs at point b, the deviation being of the order of 6 μm approximately.

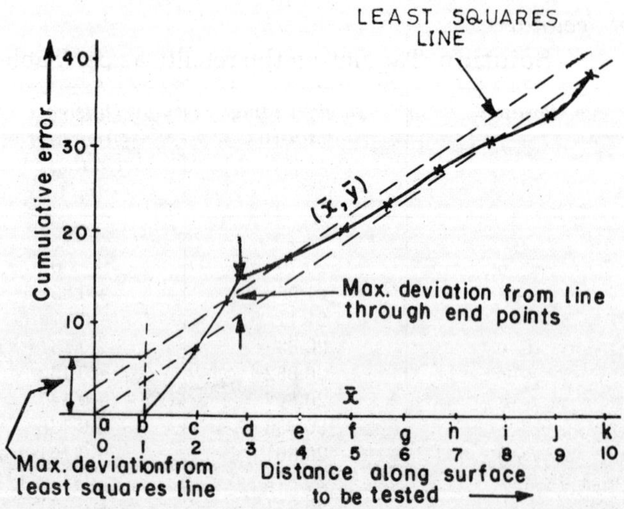

Fig. 7.62

Problem 7.2. *The straightness of a machine bed is measured with an autocollimator and a reflector. First reading is taken by placing the reflector at the left end and then the reflector is gradually moved towards the right. It is observed that the readings increase uniformly and that the total deviation* *is 30 seconds for the reflector movement of 750 mm from left end to right end. An increase in reading is also obtained when the leading end of the carriage is raised slightly. Calculate the amount and form of flatness error.*

Solution. It may be noted from Fig. 7.63 that increasing readings will be obtained when the surface is concave. In Fig. 7.63, *CD* is tangent at *C* to the surface of the machine bed. Let *O* be the centre of radius.

Now $OA = r \cos \theta/2$; and therefore $AB = OB - OA = r - r \cos \theta/2 = r(1 - \cos \theta/2)$

Also $OD = r \sec \theta/2$; and therefore

$$BD = OD - OB = r \sec \theta/2 - r = r(\sec \theta/2 - 1) = \frac{r(1 - \cos \theta/2)}{\cos \theta/2}$$

$$\frac{AB}{BD} = \frac{r(1 - \cos \theta/2)}{r(1 - \cos \theta/2)} \times \cos \theta/2 = \cos \theta/2$$

$$\cos \theta/2 = 1 - \frac{(\theta/2)^2}{2!} + \frac{(\theta/2)^4}{4!} \ldots$$

...(By binomial expansion)

since angle θ/2 is very small, the square and other power terms can be ignored ; ∴ $AB = BD$

since $AC = \frac{750}{2}$ mm and $\frac{\theta}{2} = \frac{30}{2}$ sec

$AD = AC \text{ arc } \theta/2 = 375 \times 15 \times 4.85 \times 10^{-6}$
$= 0.0273$ mm

[∴ 1 sec $= 4.85 \times 10^{-6}$ rad.]

AB = concavity at centre $= \frac{0.0273}{2}$

$= 0.0137$ mm. **Ans.**

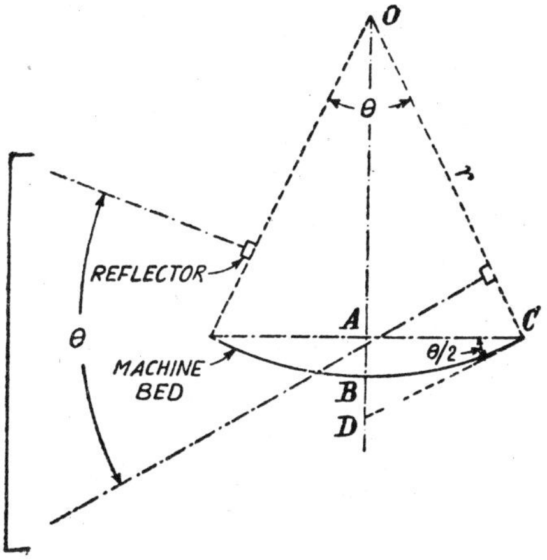

Fig. 7.63

Problem 7.3. *In a marking machine, the axis of the scale and the axis of the table (on which components are placed) are separated by 100 mm. A level is placed on the table and it is found that over a table traverse of 500 mm, the bubble shows a progressive change of reading of 50 seconds and it moves in the same direction as the table. Two transverse lines are to be marked on the upper surface of a component (50 mm in height) at a nominal separation of 200 mm. Ignore any other source of error, calculate what would be the true separation of these transverse lines marked on the component by this machine.*

Solution. Obviously this machine does not follow the principle of alignment. As the plane of measurement and plane of scale are separate, alignment errors are there.

Since the bubble movement is same as the table movement, it indicates that the leading end of the table rises relative to the first position.

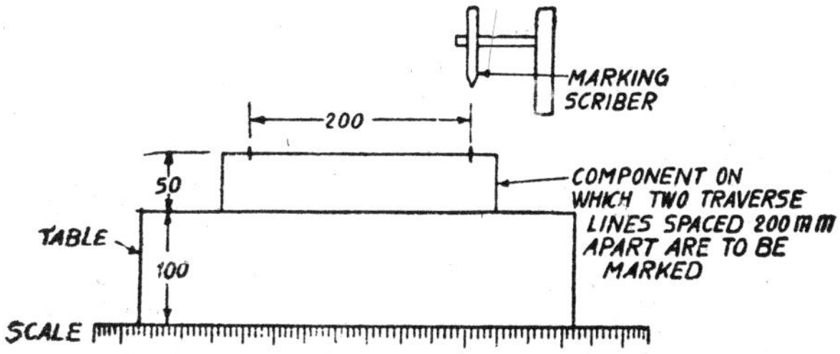

Fig. 7.64

It is given that the table movement of 500 mm produces a bubble movement corresponding to 50 seconds. Therefore, the deflection of bubble for a component on which two lines spaced 200 mm apart are to be marked will be equal to $\frac{200}{500} \times 50 = 20$ sec.

Now error in 200 mm length = arc of 200 sec on (100 + 50) mm distance

$$= 20 \times 150 \times 4.85 \times 10^{-6} \text{ mm } (\because \; 1 \text{ sec} = 4.85 \times 10^{-6} \text{ rad})$$
$$= 3000 \times 4.85 \times 10^{-6} \text{ mm} = 0.01455 \text{ mm}.$$

Since the top of the surface moves less than the bubble of table.

$\therefore$ Separation of lines will be 200 – 0.01455 = 199.98545 mm. **Ans.**

Problem 7.4. (*a*) *Describe in brief the reversal method used to test the squareness error of engineer's square. Explain why it is considered to be the simplest method.*

(*b*) *An engineer's square with a 225 mm blade is tested by the reversal method against a double-edged straight-edge set up approximately vertical. Slip gauges of length 5.000 mm are placed between the blade of the square at the bottom on each side in turn. With square on the right side, slip gauges of length 5.150 mm can be inserted at the top ; and with square on the left side slip gauge of 4.975 mm can be inserted at the top. Calculate the amount and direction of squareness error of the square over its 225 mm blade. If the straight edge is tapered by 0.005 mm over the 225 mm length and it is set up with its narrow end uppermost determine the amount and direction of squareness error of the square, and also determine the inclination of the centre line of the straight edge.*

Solution. (*a*) In the reversal method for testing the squareness of the engineer's square a double-edged straight edge is arranged vertical. The square is placed on either side of the straightedge, keeping its base at equal distance in both the cases. Then the slips are so adjusted which can be just inserted between the straight-edge and the square at the top on either side. If the straight edge is not exactly vertical then the tilt of the straight edge will add to the squareness error when the square is on one side and subtract from it on the other side. The mean of the apparent errors on both sides then gives the true squareness error.

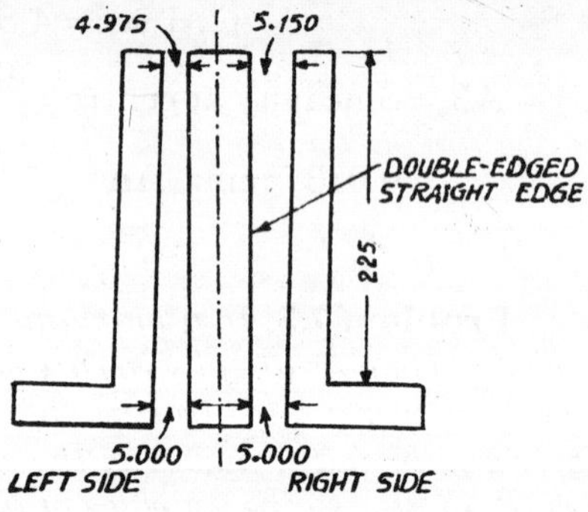

Fig. 7.65

A big advantage of this method is that the straight edge need not be set exactly vertical and need not be parallel also. If not parallel, the parallelism error must be determined exactly and then half the parallelism error applied as a correction in the correct sense.

(*b*) In solving this numerical, we well follow the notation that the positive error corresponds to excess at the tip of the square outside.

$\therefore$ Apparent error on right hand side = – (5.150 – 5.000) = 0.150 mm

(Obviously the reading at top is more, it means that tip of square is inclined inside.)

and apparent error on left hand side = + (5.000 – 4.975) = + 0.025 mm

$$\therefore \text{ Squareness error of square} = \frac{-0.150 + 0.025}{2} = \frac{-0.125}{2} = -0.0625 \text{ mm.}$$

i.e. the tip of the blade of square is inclined outside by 0.0625 mm over its length of 225 mm. **Ans.**

(*c*) Since the straight-edge itself is tapered by 0.0025 mm over 225 mm length with its smaller size portion on top, it means that the straight edge taper of 0.0025 mm on each side exaggerates square error. $\therefore$ Square error = – 0.0625 + 0.0025 = – 0.0600 mm. **Ans.**

Now the inclination of right hand of straight-edge = 0.150 – 0.060 = 0.090 mm

and the inclination of straight-edge centre line = 0.090 – 0.0025 = 0.0875 mm

and obviously the inclination is towards the left side. **Ans.**

Problem 7.5. (*a*) *Describe how a lobed surface is defined mathematically and show the dimensions of the geometric form of the cross-section.*

(*b*) *A cylindrical piece ground on centreless grinding machine when measured by micrometer screw gauge shows the readings to be 20.000 mm. It is found that it only-just fits into a truly circular ring gauge of 20.040 mm diameter and that it is uniformly three-lobed. Determine the dimensions of the geometric form of the cross-section.*

Solution. (*a*) Mathematically the contour of lobed surface can be expressed by two radii r_1 and r_2 (smaller radius and bigger radius respectively) whose centres are in succession, on the three apices of an equilateral triangle *ABC*, whose each side is equal to *s* (Fig. 7.66). Thus the diameter of such a component as measured between two parallel faces 180° apart (such as on a comparator or between a flat anvil and measuring plunger) will be $r_1 + r_2$, which will also be equal to $2r_1 + s$. Thus $r_2 = r_1 + s$. The size of the circumscribing circle or the diameter of the truly circular ring gauge which just fits in is equal to

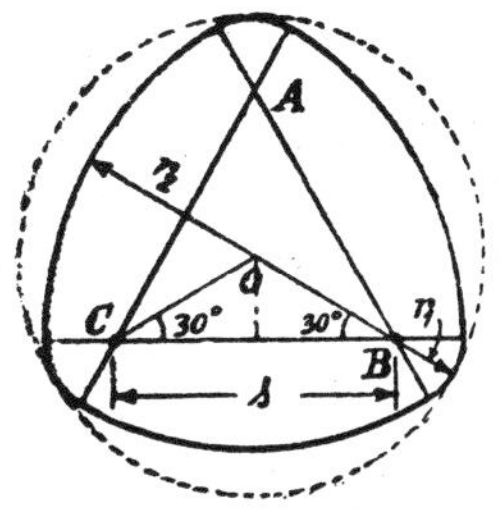

Fig. 7.66

$$2(OB + r_1) = 2\left(\frac{s}{2} \sec 30^\circ\right) + 2r_1 = s \sec 39^\circ + 2r_1.$$

Thus the lobed surface can be defined by r_1 and *s*.

(*b*) The size of apparent diameter as measured by micrometer screw

$$= 20.000 \text{ mm} = 2r_1 + s \qquad ...(i)$$

Diameter of circumscribing circle

$$= 20.04 \text{ mm} = 2r_1 + s \sec 30^\circ = 2r_1 + 1.55s \qquad ...(ii)$$

From equations (*i*) and (*ii*) $0.155s = 0.04$

and $$s = \frac{0.04}{0.155} = 0.258 \text{ mm.} \quad \textbf{Ans.}$$

and $$r_1 = \frac{20.000 - 0.258}{2} = \frac{19.742}{2} = 9.871 \text{ mm.} \quad \textbf{Ans.}$$

TYPICAL EXAMINATION QUESTIONS

1. Explain the various methods used for checking the squareness and parallelism part.
2. Describe a method to find out the flatness of a surface plate.
3. Explain how the straightness of surface is assessed in workshops.
4. How the accuracy of a straight-edged is specified, and how it can be checked ?
5. Describe the three methods of testing straight-edge one metre long, stating the advantage and sources of error in each.
6. A straight edge was tested be means of a spirit level having a vial radius of 60 m and standing on feet 100 mm apart. The readings taken were + 0.75, + 0.5, 0, – 2.5, + 2, 0, – 0.75 divisions on the vial, which are 2.5 mm apart. Plot the contour of the edge in micrometers and from the graph find the maximum deviation from the straight line joining the end points.
7. The departure from straightness of the rectangular section straight edge 1 metre long was determined by an autocollimator and reflecting block. The straight edge was supported at the points for minimum deflection and the centre distance of the feet to the block was 70 mm and the autocollimator readings (in minutes) were :

   + 0.6, + 0.9, – 0.2, 0, + 0.9, + 0.4, – 0.2, 0, – 0.4, + 1.1

Plot the graph in micrometers, and reduce the graph to show error from a straight line through the points of support.

8. Describe how the deviation from a true plane of a surface table can be found by using the surface of a liquid as a reference plane.
9. Explain how the flatness of a surface plate, approximately 1,000 × 700 mm in size, may be tested by means of a spirit level.
10. Describe the beam comparator and show how it is used to test the straightness of the surface of a long machine bed.
11. Describe two different methods of testing flatness of a surface, stating advantages and limitations of each.
12. Sketch and comment upon five typical interference patterns while testing a small plane surface by means of an optical flat.
13. How will you calibrate a given straight edge ? Indicate the various steps involved explaining the method of calculation. How will you use the calibrated straight edge to determine the straightness errors of other components ?
14. Explain how you will check the lobe error in a 20 mm diameter and 50 mm long cylinder.
15. What do you understand by cross wind in C.I. surface plates and how it is defected ?
16. Explain how a precision level can be used to determine the flatness and straightness of machine beds ? How the requirements are specified in choosing a level for this type of work ? How a precision level is more effective than an auto-collimator for this work ?

8

Angular Measurement and Circular Division

8.1. Introduction

Measurement of any angle precisely is one of the most important human activities. It is only because of precise angular measurement that ships and aeroplanes can navigate confidently without the help of the sight of the land. The relation of the stars and their approximate distances are also computed in astronomy by means of angular measuring devices. In workshop also, so often we come across problems involving angular measurements. This is rather indispensable in workshops for the manufacture of interchangeable parts, jigs, dies, etc.

As in linear measurement, where two kinds of standards (*viz.*, end standard and line standard) are used ; so also in angular measurement we have two types of tools. They are angle gauges corresponding to slip gauges and divided scales corresponding to line standard. The difference in the linear and angular measurement is that no absolute standard is required for angular measurement.

The angle is defined as the opening between two lines which meet at a point. If one is moved around a point in an arc, a complete circle can be formed and it is from this circle that units of angular measurement are derived. The basic unit in angular measurement is the right angle. It is defined as the angle between two lines which intersect so as to make the adjacent angles equal. Angle is also defined in terms of radian which is the angle subtended at the centre by an arc of a circle of length equal to its radius. For engineering purposes, the angular units are defined in sexagesimal system (degrees, minutes, seconds). If a circle be divided into 360 parts, each part is called a degree (°). So angle is such a thing which can be just generated very easily, requiring no absolute standard, and it is the precision with which the circle can be divided to get the correct measure of angle. Each degree is further divided into sixty parts called minutes (′), and each minute is further sub-divided into 60 parts called seconds (″).

In this chapter, we will be dealing with angular measurement and circular division which form an important part of Metrology. Angular measurement is generally concerned with the measurement of individual angles on gauges, tools as well as small angular changes and deflections etc. Circular division deals with the measurement and continuous division of a circle.

8.2. Vernier and Optical Bevel Protractor

It is probably the simplest instrument for measuring the angle between two faces of component. It consists of a base plate attached to the main body, and an adjustable blade which is attached to a circular plate containing vernier scale. The adjustable blade is capable of rotating freely about the centre of the main scale engraved on the body of the instrument and can be locked in any position. An acute angle attachment is provided at the top ; as shown in Fig. 8.1 for the purpose of measuring acute angles. The base of the base plate is made flat so that it could be laid flat upon the work and any type of angle measured. It is capable of measuring from 0 to 360°.

The vernier scale has 24 divisions coinciding with 23 main scale divisions. Thus the least count of the instrument is 5′. This instrument is most commonly used in workshops for angular measurements till more precision is required.

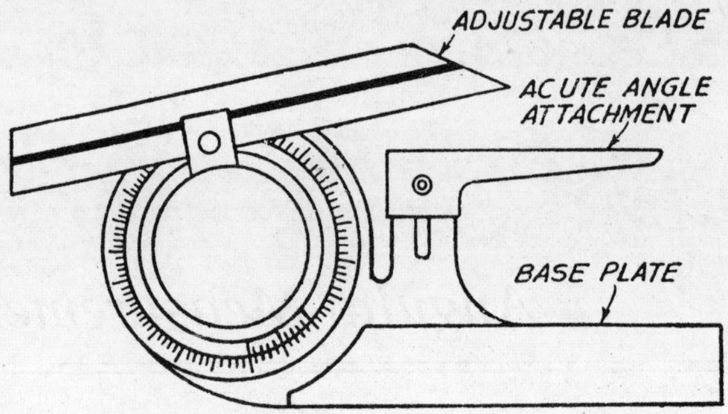

Fig. 8.1. Vernier and optical Bevel Protractor.

A recent development of the vernier bevel protractor is optical bevel protractor. In this instrument, a glass circle divided at 10′ intervals throughout the whole 360° is fitted inside the main body. A small microscope is fitted through which the circle graduations can be viewed. The adjustable blade is clamped to a rotating member which carries this microscope. With the aid of microscope it is possible to read by estimation to about 2′.

8.2.1. Universal Bevel Protractor. It is used for measuring and laying out of angles accurately and precisely within 5 minutes. The protractor dial is slotted to hold a blade which can be rotated with the dial to the required angle and also independently adjusted to any desired length. The blade can be locked in any position. (Refer Fig. 8.2)

8.2.2. Bevel Protractors as per Indian Standard Practice. The bevel protractors are of two types, *viz.*,

(*a*) Mechanical Bevel Protractor, and

(*b*) Optical Bevel Protractor.

The mechanical bevel protractors are further classified into four types ; *A*, *B*, *C* and *D*. In types *A* and *B*, the vernier is graduated to read to 5 minutes of arc whereas in case of type *C*, the scale is graduated to read in degrees and the bevel protractor is without vernier or fine adjustment device or acute angle attachment. The difference between types *A* and *B* is that type *A* is provided with fine adjustment device or acute angle attachment whereas type *B* is not. The scales of all the types are graduated either as a full circle marked 0—90—0—90 with one vernier or as semicircle marked 0—90—0 with two verniers 180° apart. Type *D* is graduated in degrees and is not provided with either vernier or fine adjustment device or acute angle attachment.

Fig. 8.2. Universal Bevel Protractor.

In the case of optical bevel protractor, it is possible to take readings upto approximately 2 minutes of arc. The provision is made for an internal circular scale which is graduated in divisions of 10 minutes of arc. Readings are taken against a fixed index line or vernier by means of an optical magnifying system which is internal with the instrument. The scale is graduated as a full circle marked 0—90—0—90. The zero positions correspond to the condition when the blade is parallel to the stock. Provision is also made for adjusting the focus of the system to

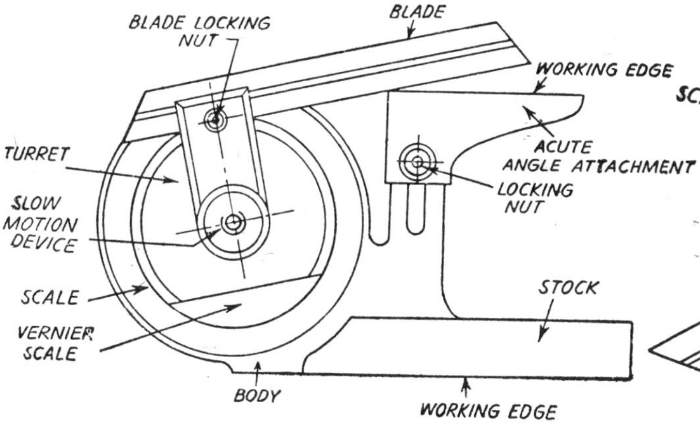

Fig. 8.3. Mechanical Bevel Protractor with Vernier and acute angle attachment.

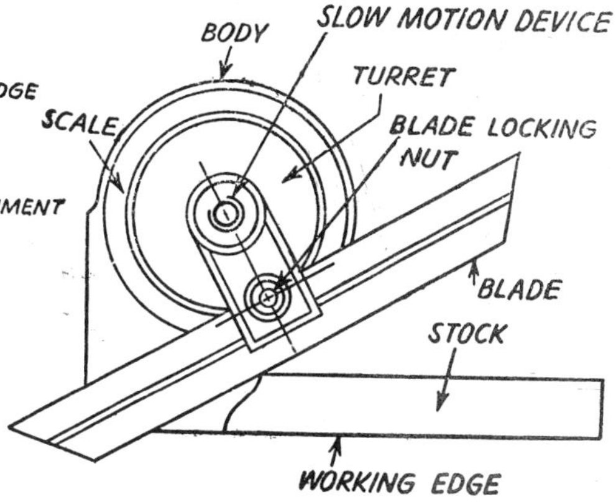

Fig. 8.4. Mechanical Bevel Protractor without Vernier and acute angle attachment.

accommodate normal variations in eye-sight. The scale and vernier are so arranged that they are always in focus in the optical system. The principle of optical scales has already been discussed in details in chapter 6.

8.2.3. General Description of Various Components of Bevel Protractors

Body. It is designed in such a way that its back is flat and there are no projections beyond its back so that when the bevel protractor is placed on its back on a surface plate there shall be no perceptible rock. The flatness of the working edge of the stock and body is tested by checking the squareness of blade with respect to stock when blade is set at 90°.

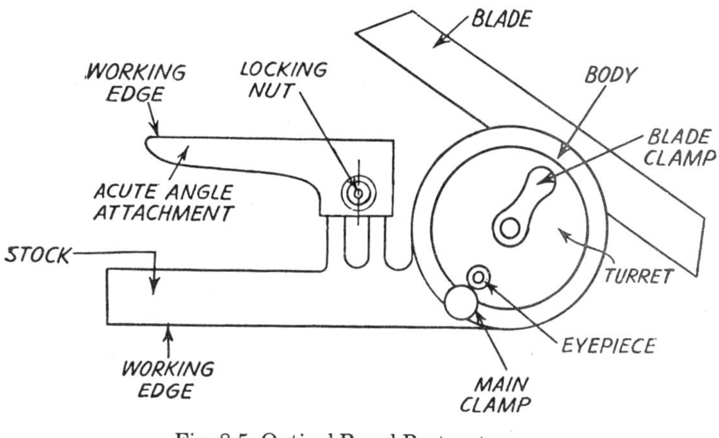

Fig. 8.5. Optical Bevel Protractor.

Stock. The working edge of the stock is about 90 mm in length and 7 mm thick. It is very essential that the working edge of the stock be perfectly straight and if at all departure is there, it should be in the form of concavity and of the order of 0.01 mm maximum over the whole span.

Blade. It can be moved along the turret throughout its length and can also be reversed. It is about 150 or 300 mm long, 3 mm wide and 2 mm thick and ends bevelled at angles of 45° and 60° within the accuracy of 5 minutes of arc. Its working edge should be straight upto 0.02 mm and parallel upto 0.03 mm over the entire length of 300 mm. It can be clamped in any position.

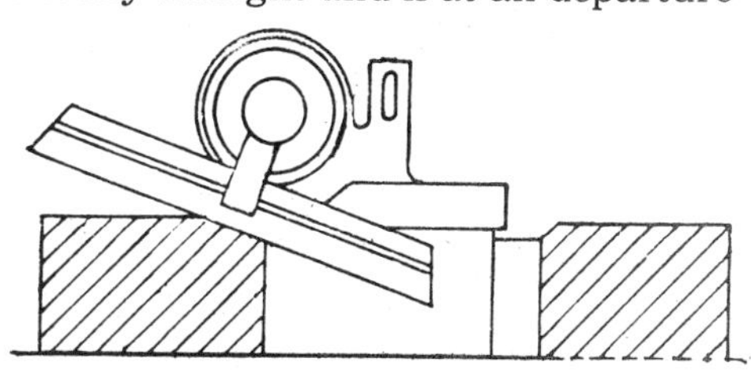

Fig. 8.6. Using bevel protractor for checking the inside bevelled face of a ground surface.

8.2.4. Acute angle attachment. It can be readily fitted into body and clamped in any position. Its working edge should be flat to within 0.005 mm and parallel to the working edge of the stock within 0.015 mm over the entire length of attachment.

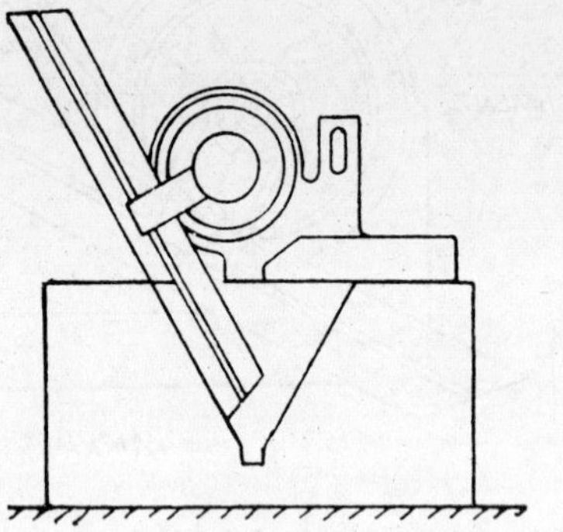

Fig. 8.7. Use of bevel protractor for checking of vee block.

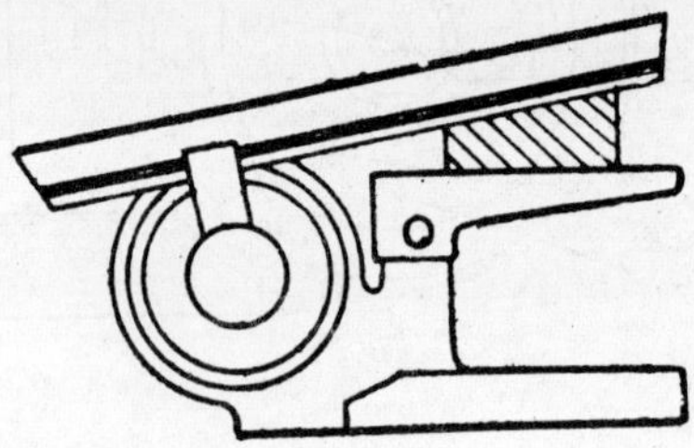

Fig. 8.8. Use of Vernier protractor for measuring acute angle.

The bevel protractors are tested for flatness, squareness, parallelism, straightness and angular intervals by suitable methods.

8.2.5. Use of Bevel Protractors. Figs. 8.6 to 8.8 show the use of bevel protractors for different applications.

8.3. Divided Circles

These correspond to the line standards in length measurement. The lines are ruled out into glass discs defining the angular positions.

8.4. Optical Dividing Head

Optical dividing head consists of a circular scale, suitably calibrated in degrees and fractions and fitted concentrically to the axis of the machine spindle. The scale is viewed through a microscope. It can measure directly upto 12″ of arc and readings can be estimated upto 3″ of arc. Fig. 8.9 shows the diagrammatic arrangement of the optical system employed in optical dividing head. It would be noted that the light passes through a system of lenses and illuminates the divided scale (mounted on the main spindle) thus projecting the readings on a ground glass screen. This arrangement thus dispenses with the focusing eyepiece and errors due to parallex are eliminated. The spindle on which the glass divided circle is mounted, runs in hardened and lapped plain bearings. The worm which is used only for rotation purposes is mounted in bearings independent of those for the main spindle. The worm wheel is driven by the worm, which is connected to a hand-wheel, and a lever is provided for engaging or disengaging the drive. If handle is turned against the locked spindle, it would simply override. In order to make a rough check on the required angle, an indexing ring graduated in degrees is provided at the end of the spindle.

8.6. Sine Principle and Sine Bars

The sine principle uses the ratio of the length of two sides of a right triangle in deriving a given angle. It may be noted that devices operating on sine principle are capable of "self generation". The measurement is usually limited to 45° from loss of accuracy point of view. The accuracy with which the sine principle can be put to use is dependent in practice, on some form of linear measurement. The sine bar in itself is not a complete measuring instrument. Another datum such as a surface plate is needed, as well as other auxiliary equipment, notably

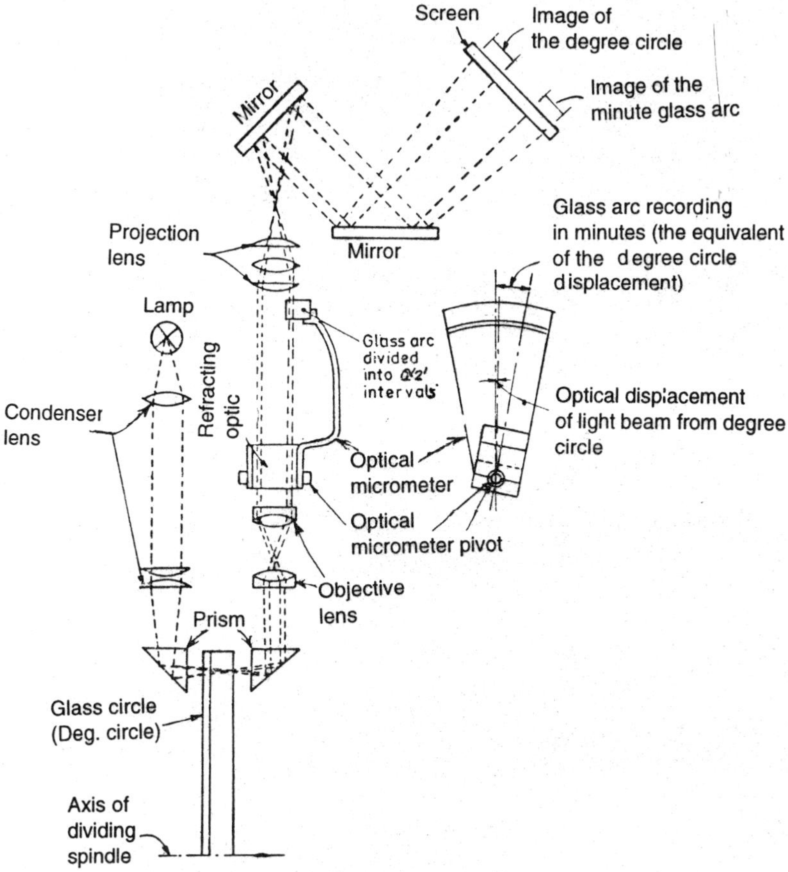

Fig. 8.9. Diagrammatic arrangement of the optical system employed in optical dividing head.

slip gauges, and indicating device to make measurements. Sine bars used in conjunction with slip gauges constitute a very good device for the precise measurement of angles. Sine bars are used either to measure angles very accurately or for locating any work to a given angle within very close limits. Sine bars are made from high carbon, high chromium, corrosion resistant steel, hardened, ground and stabilised. Two cylinders of equal diameter are attached at the ends. The axes of these two cylinders are mutually parallel to each other and also parallel to and at equal distance from the upper surface of the sine bar. The distance between the axes of the two cylinders is exactly 5 inches or 10 inches in British system, and 100, 200 and 300 mm in metric system. The above requirements are met and maintained by taking due care in

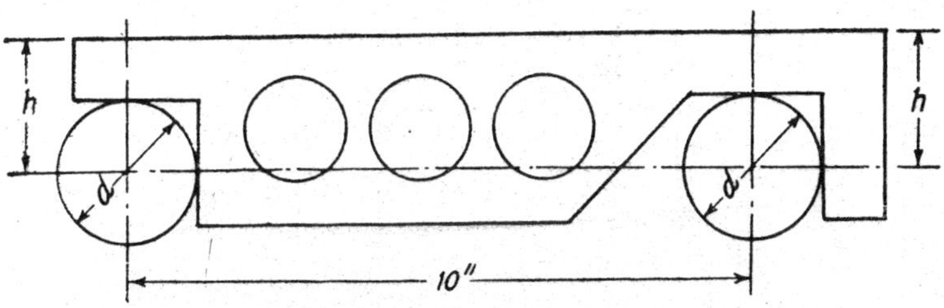

Fig. 8.10. Sine bar.

the manufacture of all parts. The various parts are hardened and stabilised before grinding and lapping. All the working surfaces and the cylindrical surfaces of the rollers are finished to surface finish of 0.2 μm *Ra* value or better. Depending upon the accuracy of the centre distance, sine bars are graded as of *A* grade or *B* grade. *B* grade of sine bars are guaranteed accurate upto 0.02 mm/m of length and *A* grade sine bars are more accurate and guaranteed upto 0.01 mm/m of length.

Although there are several forms of sine bars, but the one shown in Fig. 8.10 is most commonly used. Some holes are drilled in the body of the bar to reduce the weight and to facilitate handling.

However, from the point of view of ease in manufacturing to ensure an exact distance between the cylinders, the form shown in Fig. 8.11 is preferable. Also this type of sine bar can be set to a steep angle without the slip gauges fouling the underside of the bar. But this point is immaterial as the accuracy of setting appreciably decreases with steep angles and from the point of stability also we generally do not use sine bars for steep angles.

Before using any sine bar for angular measurement work, for accuracy and precision, it is better to ensure whether the following accuracy requirements and tolerances as specified by I.S. 5359—1969 for 100 mm sine bar are met with :

Characteristics	*Permissible Tolerance*
Flatness of upper and lower surfaces	0.001 mm
Parallelism of upper and lower surfaces w.r.t. datum surface when resting on it	0.001 mm
Flatness of side faces	0.005 mm
Squareness of side faces to upper surface	0.003/25 mm
Parallelism of side faces to the axes of rollers	0.01/25 mm
Flatness of end faces	0.003 mm
Squareness of end faces to the upper surface	0.003/25 mm
Parallelism of end faces to the axes of the rollers	0.01/25 mm
Straightness of individual rollers and freedom from lobing and uniformity in diameter	0.002 mm
Mean diameter of rollers	0.002 mm
Distance between the roller axes	± 0.003 mm
Roller axes :	
(*i*) In a common plane over the length of either roller	0.003 mm
(*ii*) Parallel to and equidistant from the upper surface over the length of either roller	0.003 mm
Flatness of the bearing surface of the setting foot	0.003 mm

Various types *i.e.*, different designs of sine bars required for different applications are shown in Fig. 8.10. Fig. 8.11 also shows the nomenclature of sine bars as recommended by IS : 5359—1969.

As already pointed, sine bar shown in Figs. 8.10 to 8.14 is more preferred form as the distance between the rollers in this case can be adjusted exactly.

This type of sine bar can be easily set on steep angle without the slip gauges fouling the underside of the sine bar ; of course, the accuracy of setting decreases with steep angles.

Sine bar shown in Fig. 8.12 is most commonly used form in which the rollers are so arranged that their outer surfaces on one side are level with the plane top surface of the sine bar.

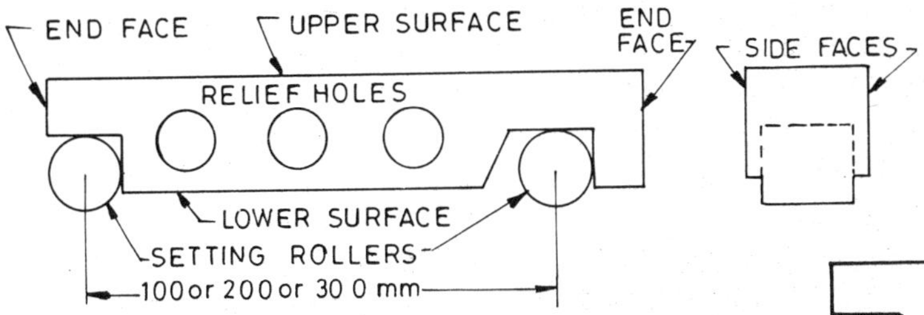

Fig. 8.11

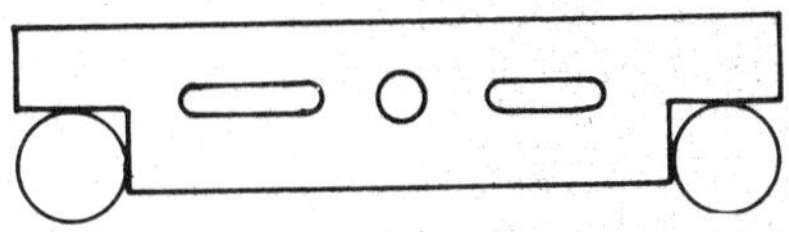

Fig. 8.12

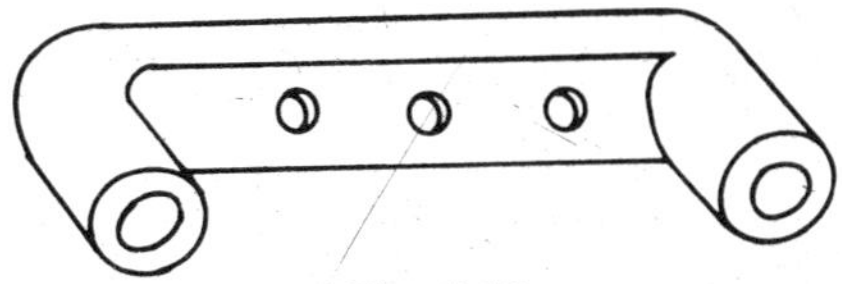

Fig. 8.13

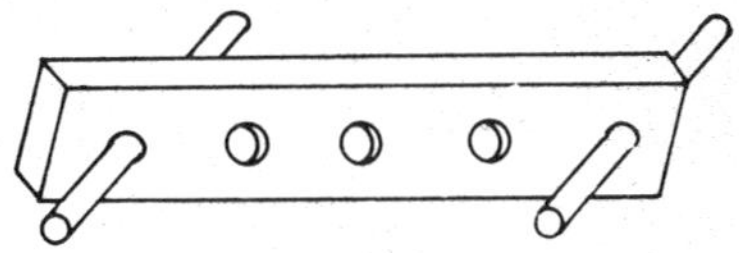

Fig. 8.14

Different types of sine bars.

The form of sine bar shown in Fig. 8.13 has hollow rollers whose outside diameter is equal to width of sine bar and is used where the width of the sine bar enters into calculation of work height.

The form of sine bar shown in Fig. 8.14 is used where ordinary type can't be used on the top surface due to interruption.

The accuracy of sine bar depends on its constructional features and on maintaining these.

These features are :

(*i*) The two rollers must have equal diameter and be true cylinders.

(*ii*) The rollers must be set parallel to each other and to the upper face.

(*iii*) The precise centre distance between the rollers must be known.

(*iv*) The upper face must have a high degree of flatness. The various characteristic tolerances have already been indicated above.

Fig. 8.15 shows the various types of sine bars as recommended by BS : 3064.

The accuracy requirements for sine as specified by British standard BS : 3064 are

(*i*) Flatness of whole of the upper surface of bar (and also of lower surface if it is working surface) : 0.0015, 0.002 and 0.003 mm respectively for 100, 200 and 300 mm bars.

(*ii*) Parallelism of lower and upper surfaces—Same tolerances as for (*i*) above.

(*iii*) Flatness of side faces of bars : 0.004, 0.005 and 0.006 mm for 100, 200 and 300 mm bars respectively.

(*iv*) Squareness of the side faces to the upper surface of bar : 0.0025 mm per 25 mm.

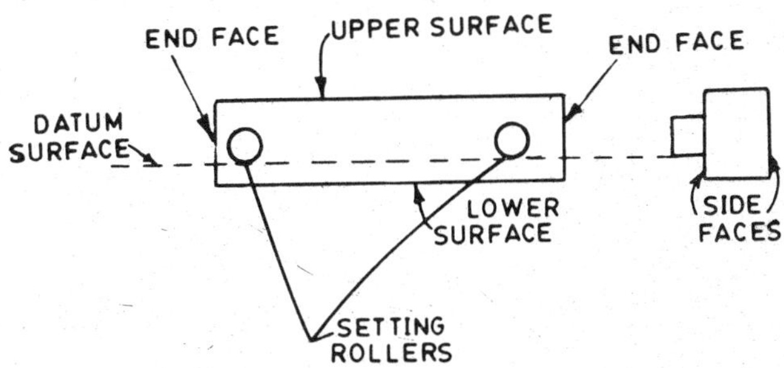

Type 1

(v) Squareness of the side faces to the axis of the hinge roller : 0.013 mm per 25 mm.

(vi) Flatness of end faces (if to be used as working surfaces) : within 0.0025 mm.

(vii) Squareness of end faces (if to be used as working surfaces) to the upper surface of the bar : within 0.0025 mm per 25 mm.

(viii) Parallelism of end faces to the axis of the hinge roller : within 0.013 mm per 25 mm.

(ix) Uniformity of radius and straightness over length of setting and hinge rollers : within 0.002 mm.

(x) Equal sizes of setting and hinge rollers—to be within 0.0025 mm.

(xi) Accuracy of distance between roller axes : 0.0025, 0.005 mm, 0.008 mm for 100, 200 and 300 mm bars.

(xii) *Additional requirement of rollers.* The roller axes should lie in a common plane within 0.002 mm over the length of either roller. Further the upper surface and lower surface (if working surface) should be parallel to the plane tangent to the lower surface of the rollers within 0.002 mm.

(xiii) Flatness of end and side faces—within 0.005 mm.

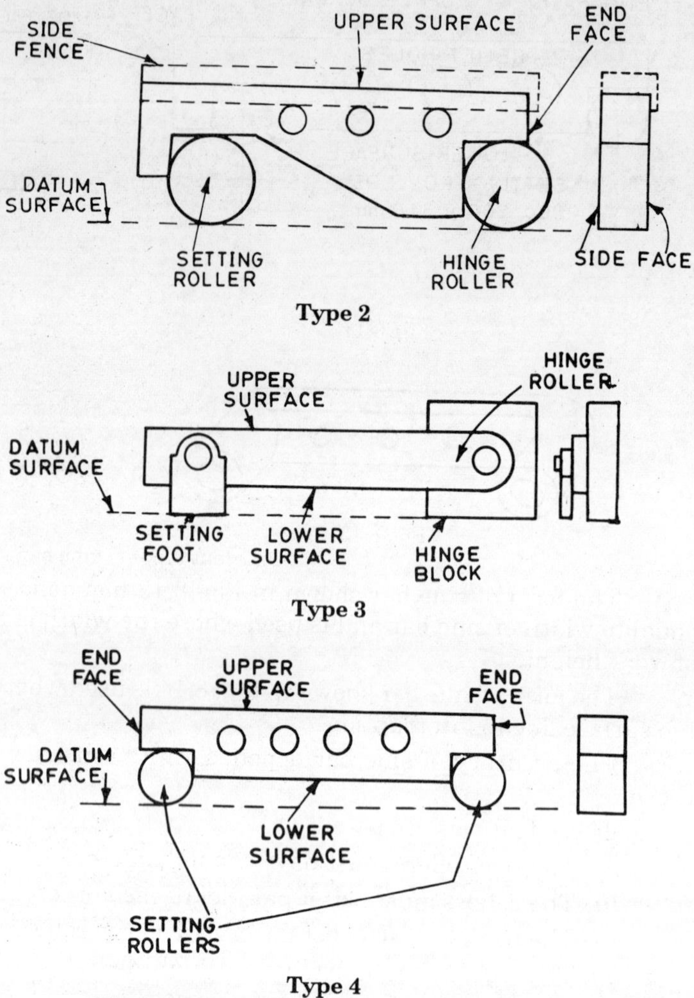

Fig. 8.15. Types of sine bars as per BS : 3064.

8.6.1. Use of Sine Bar. (1) *Measuring known angles or locating any work to a given angle.* For this purpose the surface plate is assumed to be having a perfectly flat surface, so that its surface could be treated as horizontal. One of the cylinders or rollers of sine bar is placed on the surface plate and other roller is placed on the slip gauges of height h. Let the sine bar be set at an angle θ. Then $\sin \theta = h/l$, where l is the distance between the center of the rollers. Thus knowing θ, h can be found out and any work could be set at this angle as the top face of sine bar is inclined at angle θ to the surface plate. The use of angle plates and clamps

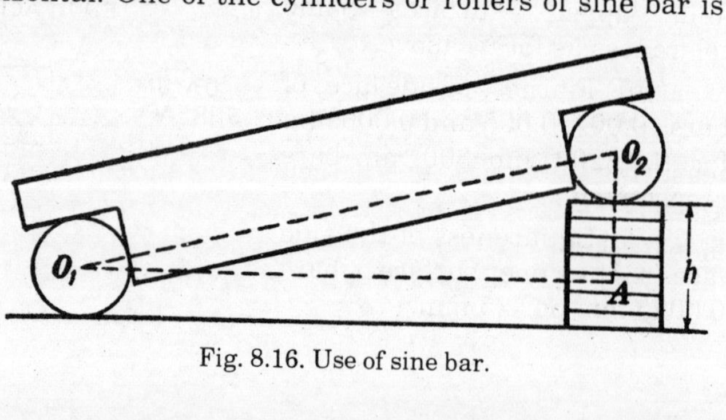

Fig. 8.16. Use of sine bar.

could also be made in case of heavy components. For better results, both the rollers could also be placed on slip gauges of height h_1 and h_2 respectively. Then $\sin \theta = (h_2 - h_1)/l$.

(2) *Checking of unknown angles.* Many a times, angle of a component to be checked is unknown. In such a case, it is necessary to first find the angle approximately with the help of a bevel protractor. Let the angle be θ. Then the sine bar is set at an angle θ and clamped to an angle plate. Next, the work is placed on sine bar and clamped to angle plate as shown in Fig. 8.17 and a dial indicator is set at one end of the work and moved to the other, and deviation is noted. Again slip gauges are so adjusted (according to this deviation) that dial indicator reads zero across work surface.

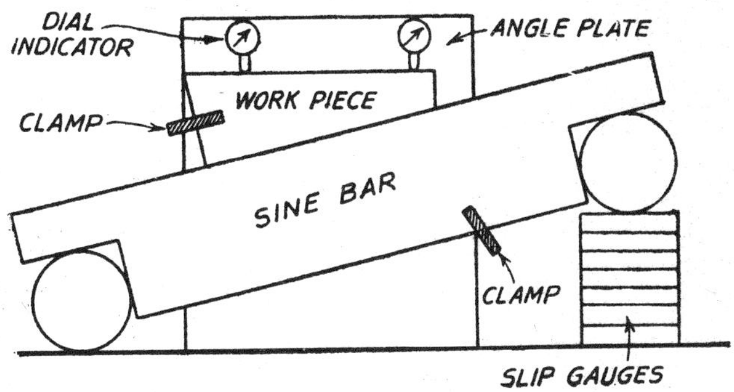

Fig. 8.17. Checking unknown angle using sine bar.

If deviation noted down by the dial indicator is δh over a length l' of work, then height of slip gauges by which it should be adjusted in equal to $= \delta h \times l/l'$.

(3) *Checking of unknown angles of heavy component.* In such cases where components are heavy and can't be mounted on the sine bar, then sine bar is mounted on the component as shown in Fig. 8.18. The height over the rollers can then be measured by a vernier height gauge ; using a dial test gauge mounted on the anvil of height gauge as the fiducial indicator to ensure constant measuring pressure. The anvil on height gauge is adjusted with probe of dial test gauge showing same reading for the topmost position of rollers of sine bar. Fig. 8.18 shows the use of height gauge for obtaining two readings for either of the roller of sine bar. The difference of the two readings of height gauge divided by the centre distance of sine bar gives the sine of the angle of the component to be measured. Where greater accuracy is required, the position of dial test gauge probe can be sensed by adjusting a pile of slip gauges till dial indicator indicates same reading over roller of sine bar and the slip gauges.

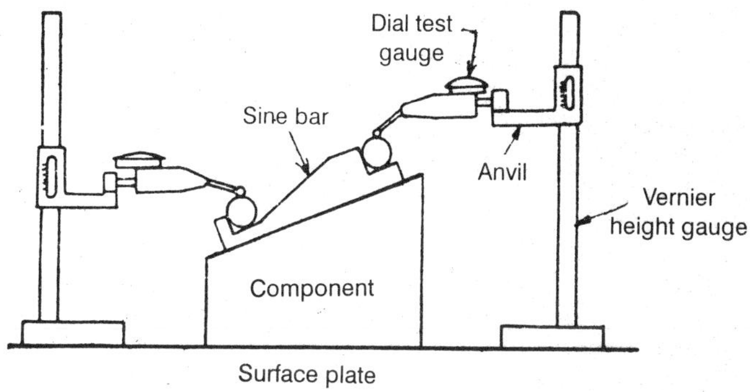

Fig. 8.18. Checking unknown angle of heavy component using sine bar.

8.6.2. Limitations of Sine Bars. The establishment of angle by the sine principle is essentially a length measuring process. Thus the accuracy, in practice, is limited by measurement of centre distance of two precision rollers. The geometrical condition involved in measuring the exact, effective centre distance existing between two rollers of the sine bar to a certainty of fraction of a μm is an infinitely complex problem. This fundamental limitation alone precludes the use of the sine bar as a primary standard of angle.

Devices operating on the sine principle are fairly reliable at angles less than 15°, but become increasingly inaccurate as the angle increases.

Sine bars inherently become increasingly impractical and inaccurate as the angle exceeds 45°.

The sine bars inherently become increasingly impractical and inaccurate as the angle exceeds 45° because of following reasons :

— The sine bar is physically clumsy to hold in position.

— The body of the sine bar obstructs the gauge block stack, even if relieved.

— Slight errors of the sine bar cause large angular errors.

— Long gauge stacks are not nearly as accurate as shorter gauge blocks.

— Temperature variation becomes more critical.

— A difference in deformation occurs at the point of roller contact to the support surface and to the gauge blocks, because at higher angles, the weight load is shifted more toward the fulcrum roller.

— The size of gauges, instruments or parts that a sine bar can inspect is limited, since it is not designed to support large or heavy objects.

8.6.3. Precautions in use of sine bars. (*i*) The sine bar should not be used for angle greater than 60° because any possible error in construction is accentuated at this limit. (Also refer Prob. 8.2).

(*ii*) A compound angle should not be formed by mis-aligning of workpiece with the sine bar. This can be avoided by attaching the sine bar and work against an angle plate.

(*iii*) Accuracy of sine bar should be ensured.

(*iv*) As far as possible longer sine bar should be used since many errors are reduced by using longer sine bars.

8.6.4. Sine Table. This is the development of the sine bar and the procedure of setting it at any angle is same as for sine bars. The sine table is the most convenient and accurate design for heavy workpiece. The table is quite rugged one and the weight of unit and workpiece is given fuller and safer support. The gauging platforms are self-contained and can be highly refined. The table may be safely swung to any angle from 0° to 90° by pivoting it about its hinged end. Two sets of sturdy non-influencing clamps are provided for supporting the table on both sides over the whole range of sine table. It may be noted that the table is a long level which bends and twists when put through various angles while supporting all sorts of shapes, sizes and weights of workpieces. The clamping mechanism of the sine table may also cause distortion, varying the angle set slightly. However the errors due to these are not generally large.

The sine table is capable of exceptional accuracy if user uses it properly, construction principles and all elements of sine table are correct, the gauge block stack is correct and at same temperature as sine table. The sine table should be elevated or lowered, using the fine adjustment feature to attain the desired "feel" between the gauge blocks and pins of both sides separately.

A further development of this is the compound sine table in which two sine tables having their axes of tilt set at right angles to each other are provided. These two tables are mounted on a common base and the table can be set at compound angle by resolving this compound angle into its individual angles in two planes at right angles to each other and setting each table accordingly.

The "double sine" principle employs gauge pins rather than gauging platform in both table and base of the sine table. This design allows angular settings to a full 90°, and minimises the errors normally inherent in a sine table at greater angles.

8.6.5. Sine Centre. Sine centre is basically a sine bar with block holding centres which can be adjusted and rigidly clamped in any position. These are used for inspection of conical objects (having male and female centres) between centres. These are used upto inclination of 60°. Rollers are clamped firmly to the body without any play. This is a very useful device for testing the conical work centered at each end. The principle of setting is same as of sine table.

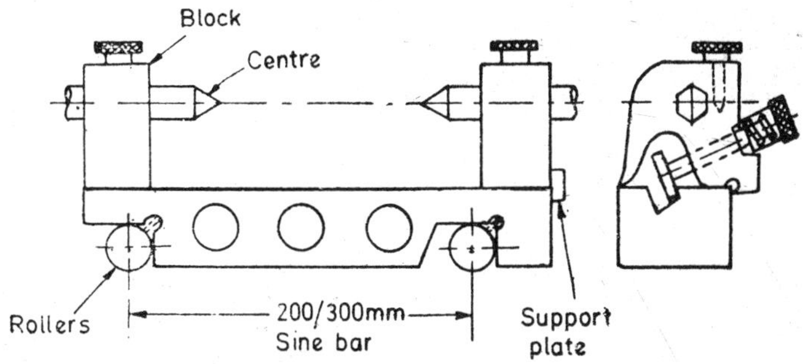

Fig. 8.19. Sine Centre.

8.7. Angle Gauges

The first set of combination of angle gauges was devised by Dr. Tomlinson of N.P.L. With thirteen separate gauges used in conjunction with one square block and one parallel straight-edge, it is possible to set up any angle to the nearest 3″. In the same way, as slip gauges are built up to give a linear dimension, the angle gauges can be build up to give a required angle. Angle gauges are made of hardened steel and seasoned carefully to ensure permanence of angular accuracy, and the measuring faces are lapped and polished to a high degree of accuracy and flatness like slip gauges. These gauges are about 3 inch (76.2 mm) long, 5/8 inch (15.87 mm) wide with their faces lapped to within 0.0002 mm and angle between the two ends to ± 2 seconds.

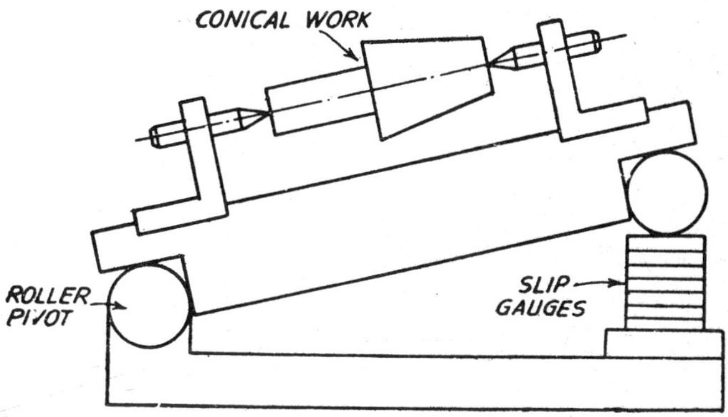

Fig. 8.20. Sine Centre.

The secret of this system in having any angle in step of 3″ is the adoption of a mathematical series of the values of the angles of various gauges of the set. The thirteen gauges can be divided into three series ; degrees, minutes and fractions of a minute. The gauges available in first series are of angle 1°, 3°, 9°, 27° and 41°. Second series comprises 1′, 3′, 9′ and 27′ angle gauges and this series has 0.05′, 0.1′, 0.3′ and 0.5′ (or 3″, 6″, 18″ and 30″) angle gauges. All these angle gauges in combination can be added or subtracted, thus, making a large number of combinations possible. There are two sets of gauges available, designated as *A* and *B*. The standard *A* contains all the above 13 gauges. Standard *B* contains only 12 gauges and does not have the 0.05′ angle gauge.

Direct combination enables computation of any angle upto 81° 40.9′ and angles larger than this can be made up with the help of the square block. However, an additional gauge of 9° can also be supplied with the set to obtain a full 90° angle without the use of the square. Fig. 8.21 illustrates how the gauges can be used in addition and subtraction. The procedure used for making various angles is as follows : *e.g.* say, we have to build up an angle of 57° 38′ 9″.

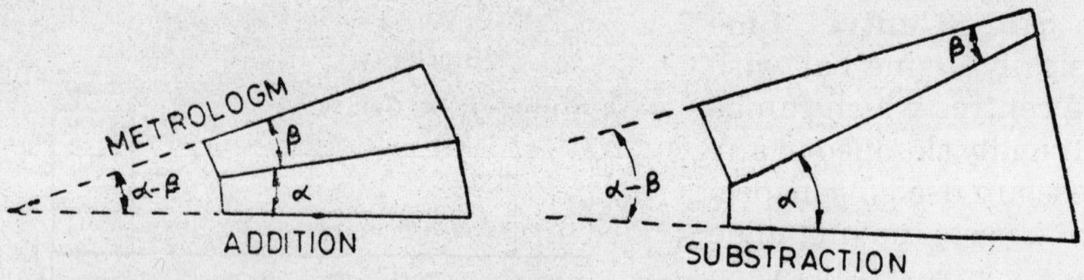

Fig. 8.21. Angle gauges can be used in addition and subtraction mode.

First we pay our attention towards degree only. So 57° could be built up as 41° + 27° – 9° + 1° – 3°.

Next if the minutes are less than 40′, they could be built up directly, otherwise number of degrees must be increased by 1° and the number of minutes necessary to correct the total is subtracted. Here now 34′ could be built as 27′ + 9′ – 3′ + 1′ and lastly 9″ is built up as 0.1′ + 0.05′.

It may be noted that each angle gauge is marked with engraved *V* which indicates the direction of included angle. When the angles of individual angle gauges are to be added up then the V_s of all angle gauges should be in line and when any angle is to be subtracted, its engraved *V* should be in other direction (Refer Fig. 8.22).

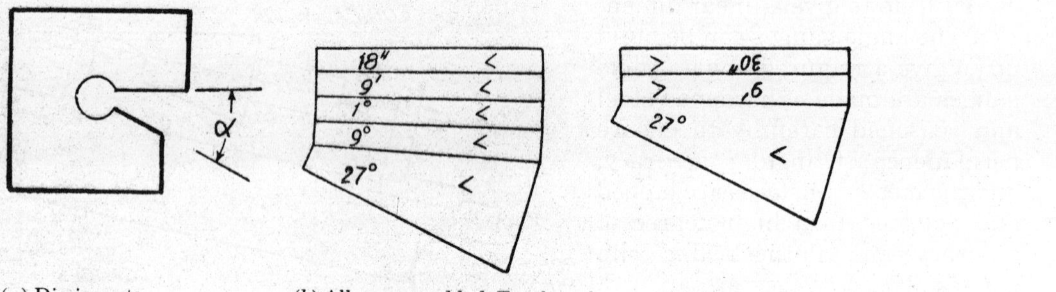

(*a*) Die insert

(*b*) All gauges added. Total angle = 37° 9′ 18″ (Not to scale)

(*c*) Angle set up = 27° minus 9′ 30″ = 26° 50′ 30″

Fig. 8.22

Thus it is seen that any angle could be made up but the block formed by the combination of a number of these gauges is rather bulky and, therefore, cannot the always directly applied to the work. But these gauges being used as reference and taking the aid of other angle measuring devices will be a good proposal at many places.

Angle gauge blocks seem to lack the requisites for use as primary standards because errors are easily compounded when angle blocks are wrung in combination. Further the absolute verification of angle blocks is usually dependent on some other primary standard.

8.7.1. Uses of angle gauges

(*i*) *Direct use of angle gauges to measure the angle in the die insert.*

To test the accuracy of the angle in the die insert, the insert is placed against an illuminated glass surface plate or in front of an inspection light box. The combination of angle gauges is so adjusted and the built-up combination of angle gauges carefully inserted in position so that no white light can be seen between the gauge faces and die faces. It may be noted that when all the engraved V_s on the angle gauges are in the same line, all angles are added up. In case some engraved V_s on angle gauges are on other side, those angles are subtracted.

(*ii*) *Use of angle gauges with square plate.*

As already indicated, the use of square plate increases the versatility of the application of angle gauges. Generally, the square plate has its 90° angles guaranteed to within 2 seconds of arc. Where very high degree of accuracy is required, the four corners of the square plate are numbered as *A*, *B*, *C* and *D*, and a test certificate is issued with each set of angle gauges, giving the measured angle of each corner. Fig. 8.23 shows a set up to test the angle of *V*-gauge whose included angle is 102°. The whole set up is placed against an illuminated glass surface plate. It may be noted that the use of slip gauges has to be made in order to facilitate the testing.

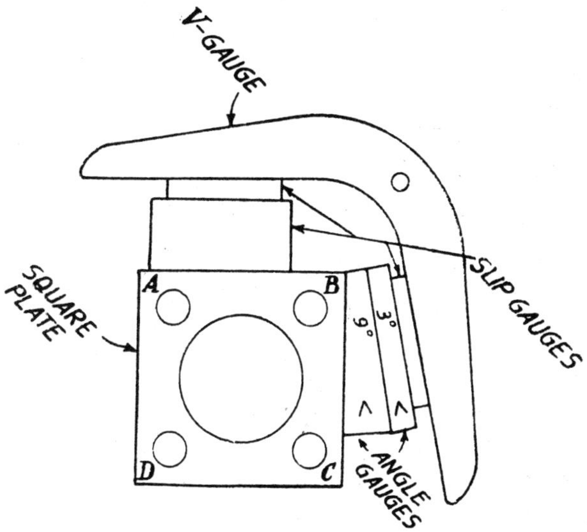

Fig. 8.23. Set-up for checking *V*-gauge having included angle 102°.

So far, we have used angle gauges to obtain a visual comparison of an angular dimension under test. It has also been realised that though it may be possible to obtain good results but it is difficult to give an estimate of the actual angular error. For very precise angular measurements, angle gauges are used in conjunction with angle dekkor described later in Art. 8.10.9.

8.7.2. Manufacture and Calibration of angle gauges. First the angle gauge is approximately machined to the desired shape, hardened by quenching, tempered and stabilised. The non-gauging sides of the block are then ground to obtain uniform thickness, and made flat and parallel by lapping operation. One gauging face is also lapped flat and made square to the lapped non-gauging sides. The simple method for manufacture of 27° and 9° angle gauges is described to explain the principle. This method requires the use of sine bar tilting table, slip gauges and magnetic chuck. The set up is shown in Fig. 8.24, in which the sine table is set at 27° and magnetic chuck fixed on tilting table surface, and the four 27° angle gauge blocks wrung side by side on to the magnetic chuck. Their upper surfaces are then surface ground parallel to the base of the table, thereby producing four blocks, of exactly equal angle which are again checked for accuracy of angle. In similar manner, 9° blanks are also finished and the three 9° blocks are wrung with 27° block in opposite direction so as to form a block with parallel sides, as shown in Fig. 8.25.

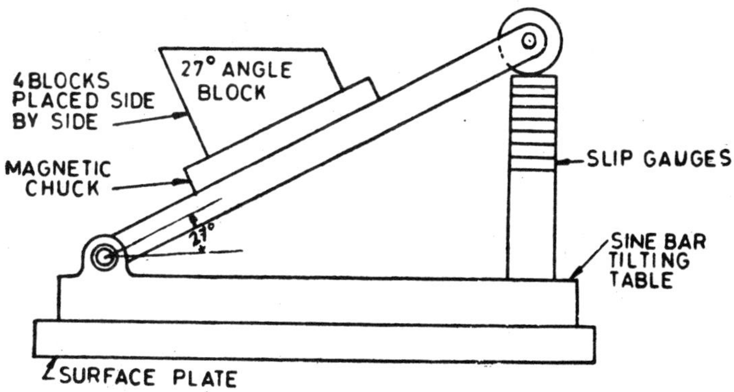

Fig. 8.24. Set-up for grinding 27° angle gauge blocks.

This combination of block is checked for error in parallelism. Thus error of each 9° block relative to 27° block will be one-third of the error noted in parallelism of composite block of Fig. 8.25. All the blocks of 9° are again lapped together till the error in parallelism is equal to the permissible value. In this way, all the gauges would be of equal angle but not necessarily of 9° and 27° exactly. The actual error in individual block from its true angle can be got rid of by the following procedure. Three blocks of 27° and one 9° block are wrung together to form a composite block of 90° and its squareness is tested by any of the method described in Art. 7.8.2 and the error noted down. On of the 27° angle block is then lapped to reduce its error to zero by making it similar to a standard angle gauge whose actual angle is known. In this process, the gauge being manufactured (lapped) is wrung with standard 27° angle gauge and error in parallelism noted and it is further lapped till error in parallelism becomes zero. Other 27° blocks are also corrected in same way individually. All the four 27° blocks are then placed on magnetic chuck and it is ensured that the outer faces of all are in one true plane by lapping them together carefully. Same procedure is followed for 9° blocks also. Finally 3 blocks of 27° and one block of 9° are again wrung to make a combination of 90° block which is tested for squareness by autocollimator method as described in Art. 7.8.2.3.

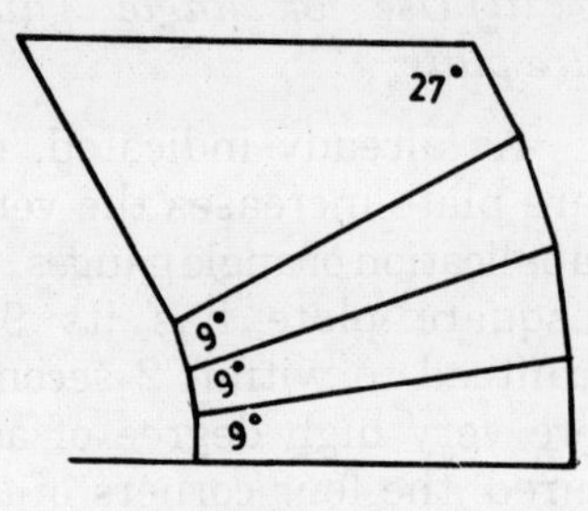

Fig. 8.25. Combination of 3 blocks of 9° and one block of 27° so as to form parallel set of blocks.

8.7.3. Calibration of angle gauges by interferometry. By this method, precision of the order of 0.1" of arc in angle measurement is possible. It is assumed that these gauges have high degree of flatness on their working surfaces. The gauge length interferometer is used for this purpose. The angle gauge is wrung on the platen carefully (Fig 8.26). On viewing through the eyepiece, two interference patterns are seen consisting of straight, parallel and equally spaced fringes. But due to angle between the two surfaces, the pitch of the two sets of fringes is different. Let there be p fringes above the platen and q fringes above the gauge on same length l.

Then $\theta = \theta_1 - \theta_2 = \lambda\,(p - q)/2l$

where λ is the wavelength of light used. There is however a limitation that angle upto only one minute can be measured directly by this method. Therefore, first the gauges of third series must be inspected and then others by combinations and adjusting optical flat to same angle.

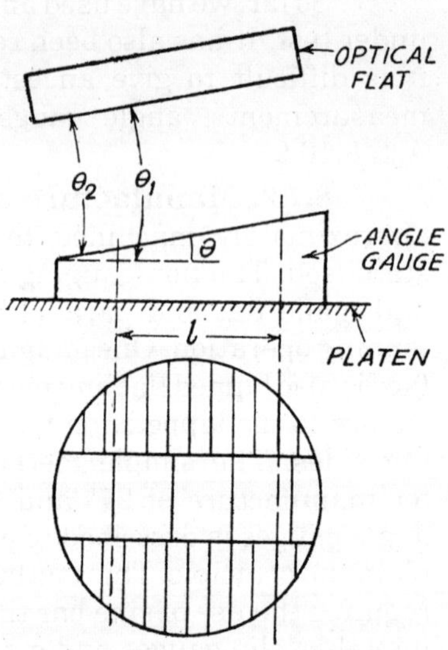

Fig. 8.26. Calibration of angle gauge by interferometry.

8.8. Spirit Level

It is generally thought that spirit level is used only for the static levelling of the machinery and other equipment. But calibrated spirit level is an angular measuring device of great precision. Spirit level is nothing but simply a glass tube, the bore of which is ground to a large radius. It is obvious that, if the liquid almost fills the tube, the bubble in liquid will always lie at the highest position in the tube. If the tube is tilted through a small angle the bubble will move along the radius of the tube through a certain distance depending on the angle of the tilt.

The sensitivity of a spirit level is expressed as the angle of tilt in seconds for which bubble will move by one division on the tube. One division is generally about 2.5 mm in length.

$$\text{Thus sensitivity} = \frac{\text{Angle in seconds}}{\text{1 division of tube}}.$$

Now if R is the radius of the tube, and l is the distance by which the graduations are separated, *i.e.* the length of one division, then the angle of tilt θ corresponding to 1 division movement of bubble will be given by $\theta = l/R$.

Generally the graduations are at 2.5 mm intervals and these represent a tilt of 10 seconds of arc, *i.e* sensitivity of level desired is 10 sec per 2.5 mm movement of bubble. Then 10 sec = 0.000485 radian = 2.5 mm/R or R = 25/0.000485 = 51500 mm, or R = 51.5 m approx.

Thus for the above sensitivity, radius of the tube or vial must be about 51.5 m and it is obvious now that sensitivity of the spirit level is governed solely by the radius of the tube and the base length of its mount.

Let the base length of any spirit level be about 250 mm, then the height h by which one end must be raised for 2.5 mm bubble movement is given by 0.0000485 = h/250

or $$h = 0.0121 \text{ mm}.$$

If the base length be reduced to 125 mm, then sensitivity is increased twice, and in this case each graduation represents 0.006 mm.

The accuracy of a spirit level depends upon the setting of the tube relative to the base. In all the higher sensitivity levels the tube is mounted kinematically in the body, one end of the tube resting on a cone which forms a part of the adjusting screw. Thus with the help of this fine pitch screw, it is capable of adjustment. Although it is now possible to adjust the tube such that bubble shows the same readings on a horizontal surface, even when the level is reversed. But in taking precise measurement, it should be assumed that some error exists and two readings must be taken along the same line by reversing the level. The mean of these two readings will indicate the true deflection of bubble.

8.8.1. Characteristic Elements of a Level. The characteristic elements of a level are the radius of curvature of the tube which decides its sensitiveness, and the distance between the two consecutive lines on its scale.

Sensitivity. It is defined as the displacement of the bubble for a tilt of 1 mm in 1 m or for 200 seconds of arc.

Constant of spirit level. It is the change in tilt, expressed in mm per m (or in seconds of arc), which produces a displacement of the bubble by one division.

Constant of level = length in mm of one division of scale/sensitivity

Accuracy of level. In order that spirit level be accurate, its base should be flat within prescribed limits. It is expressed by the movement of bubble by a division for a given change of angle. Usually it is 1 division for a change of angle of 0.05 mm per metre.

Errors. It could be due to error in the vial like radius of curvature being non-uniform, or vial or scale being positioned incorrectly. Errors could also creep in due to incorrect use of level. Temperature variation also influences readings and for this, repeated readings should be taken.

Now-a-days electronic levels have been developed in which a plate hangs from top of instrument and on horizontal plane it lies exactly midway between two fixed parallel plates. On even minute inclined surface the plate hung from top will be tilted and gap between this plate and other two plates will change. This change is detected by electrical circuits and is calibrated in terms of the angle of inclination. Some damping is also provided in movement of freely hanging plate so that it attains equilibrium position quickly and does not keep on oscillating like a pendulum.

8.9. Clinometers

A clinometer is a special case of the application of spirit level. In clinometer, the spirit level is mounted on a rotary member carried in a housing. One face of the housing forms the base of the instrument. On the housing, there is a circular scale. The angle of inclination of the rotary member carrying the level relative to its base can be measured by this circular scale. The clinometer is mainly used to determine the included angle of two adjacent faces of workpiece. Thus for this purpose, the instrument base is placed on one face and the rotary body adjusted till zero reading of the bubble is obtained. The angle of rotation is then noted on the circular scale against the index. A second reading is then taken in the similar manner on the second face of workpiece. The included angle between the faces is then the difference between the two readings.

Clinometers are also used for checking angular faces, and relief angles on large cutting tools and milling cutter inserts.

These can also be used for setting inclinable table on jig boring machines and angular work on grinding machines etc.

The most commonly used clinometer is of the Hilger and Watts type. The circular glass scale is totally enclosed and is divided from 0° to 360° at 10′ intervals. Subdivision of 10′ is possible by the use of an optical micrometer. A coarse scale figured every 10 degrees is provided outside the body for coarse work and approximate angular reading. In some instruments worm and quadrant arrangement is provided so that reading upto 1′ is possible.

In some clinometers, there is no bubble but a graduated circle is supported on accurate ball bearings and it is so designed that when released, it always takes up the position relative to the true vertical. The reading is taken against the circle to an accuracy of 1 second with the aid of vernier.

8.9.1. Precision Microptic Clinometer. These are used for measurement and checking of : angular faces, gauges, relief angles on large cutting tools, angle of milling cutter inserts, jigs and fixtures, levels of machine ways and bed plates, and for setting of inclinable tables on jig boring machines, and adjustable angle plates, angular work on grinding and lapping machines. With the appropriate accessories these can be used for measuring angular displacements of small parts, and setting out angles.

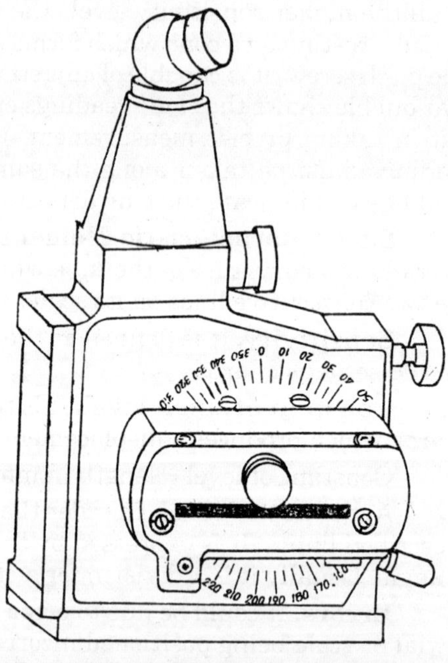

Fig. 8.27. Precision Microptic Clinometer.

The special features of precision microptic clinometer are direct reading over the range 0°—360°, optical reading system ; totally enclosed glass circles and easy-to-read scales ; main scale and micrometer scale visible simultaneously in the eyepiece external scale for rapid coarse setting, slow motion screw for fine setting, eyepiece rotatable to most convenient viewing position, and hardened ground steel base.

Precision Microptic Clinometer utilises bubble unit with a prismatic coincidence reader which presents both ends of the bubble as adjacent images in a split field of view. As the vial is levelled, the two half-images move into coincidence, making it very easy to see when the bubble is exactly centered, without reference to any graduations. [Refer Fig. 8.28].

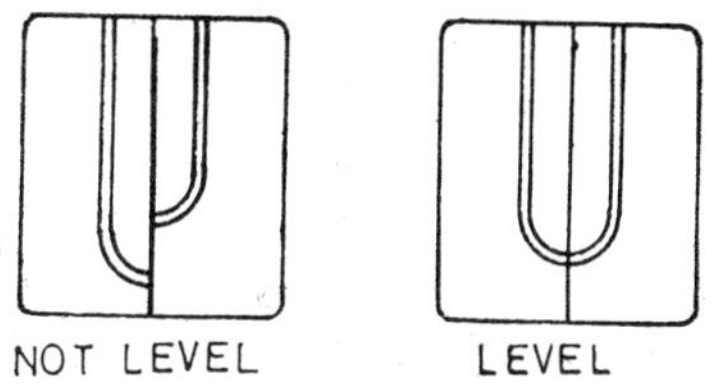

Fig. 8.28. Checking centering of bubble.

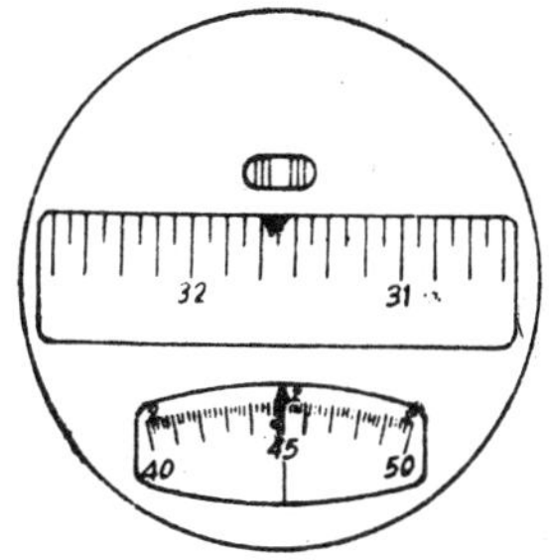

Fig. 8.29. Determining inclination of clinometer on scales.

To determine the inclination of the clinometer, the bubble unit is levelled and the scales read. On looking through the reader eyepiece, three apertures can be seen. The upper aperture contains two pairs of double lines and two single lines. To set the micrometer, the knob is turned until the single line is brought exactly central between the double lines. The scales can then be read, the required angle being the sum of the readings of the main scale and the micrometer scale. [Refer Fig. 8.29].

The double lines are imaged from one side of the circle and the single ones from a point diametrically opposite ; by using the double lines as an index for the single line, any residual centring error of the circle is cancelled out.

The scales are illuminated by an intergral low voltage lamp. The bubble unit is daylight illuminated, but is also provided with a lamp for alternative illumination.

A locating face on the back allows the instrument to be used horizontally with the accessory worktable or reflector unit.

The reference for inclination is the bubble vial. In order to measure the inclination of a surface, the vial—to which the circle is attached is turned—until it is approximately level ; then the slow motion screw is used for a final adjustment to centre the bubble.

To measure the angle between two surfaces, the clinometer is placed on each surface in turn and the difference in angle can be calculated.

The clinometer can be used as a precision setting tool to set a tool head or table at a specific angle. First the micrometer scale is set and then the glass scale is rotated to bring the relevant graduation to the index, using the slow motion screw for final adjustment. This sets the clinometer for the required angle. Then the work surface it tilted until the bubble is exactly centred. The work surface is thus set to the specified angle relative to a level plane.

8.9.3. Plain Index Centre. Index centres were actually meant for use on milling machines, but their application has now been extended to inspection work also. This is specially suited to those problems which involve the measurement of a large number of angular dimensions about a common centre. The work is set between the centres and correct angular dimensions are set directly from the indexing plate.

The index head of the centres mainly contains a crank, indexing plate, worm and worm wheel. The crank is attached to a worm which is in mesh with a wormwheel. Three types of indexing methods possible for obtaining a wide range of angles are simple, differential and compound indexing.

8.10. Optical Instruments for Angular Measurement

Autocollimator. This is an optical instrument used for the measurement of small angular differences. For small angular measurements, autocollimator provides a very sensitive and accurate approach. Auto-collimator is essentially an infinity telescope and a collimator combined into one instrument. The principle on which this instrument works is given below. O is a point source of light placed at the principal focus of a collimating lens in Fig. 8.30. The rays of light from O incident on the lens will now travel as a parallel beam of light. If this beam now strikes a plane reflector which is normal to the optical axis, it will be reflected back along its own path and focussed at the same point O. If the plane reflector be now tilted through a small angle θ, [Refer Fig. 8.31] then parallel beam will be deflected through twice this angle, and will be brought to focus at O' in the same plane at a distance x from O. Obviously $OO' = x = 2\theta . f$, where f is the focal length of the lens.

There are certain important points to appreciate here :

The position of the final image does not depend upon the distance of reflector from the lens, *i.e.* separation x is independent of the position of reflector from the lens. But if reflector

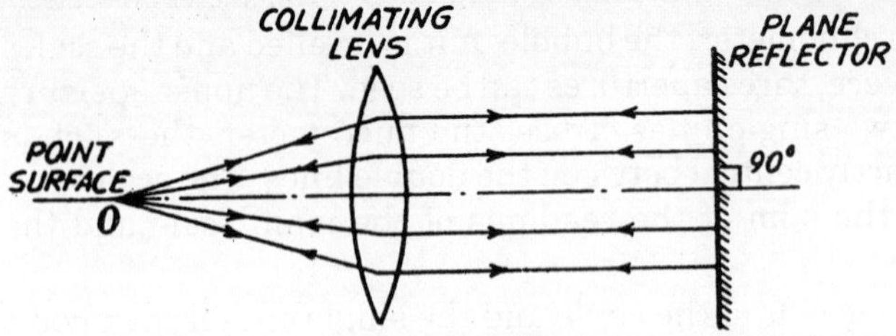

Fig. 8.30

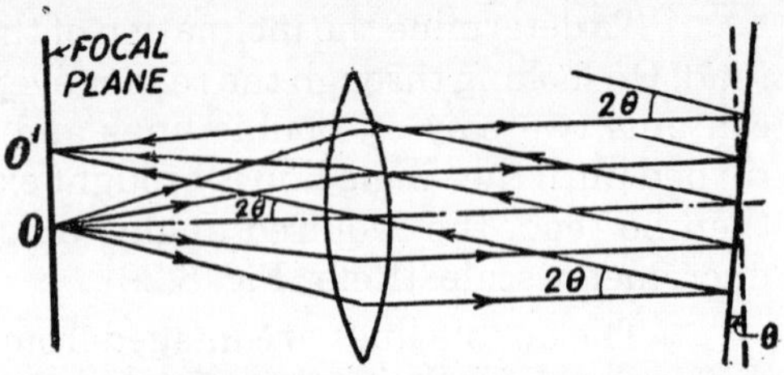

Fig. 8.31

Principle of working of autocallimator.

is moved too much back then reflected rays will completely miss the lens and no image will be formed. Thus for full range of readings of instrument to be used, the maximum remoteness of the reflector is limited.

For high sensitivity, *i.e.* for large value of x for a small angular deviation θ, a long focal length is required.

8.10.1. Principle of Autocollimator. A crossline "target" graticule is positioned at the focal plane of a telescope objective system with the intersection of the crossline on the optical axis, *i.e.* at the principal focus. When the target graticule is illuminated, rays of light diverging from the intersection point reach the objective via a beam splitter and are projected from the objective as parallel pencils of light. In this mode, the optical system is operating as a "collimator".

A flat reflector placed in front of the objective and exactly normal to the optical axis reflects the parallel pencils of light back along their original paths. They are then brought to focus in the plane of the target graticule and exactly coincident with its intersection. A

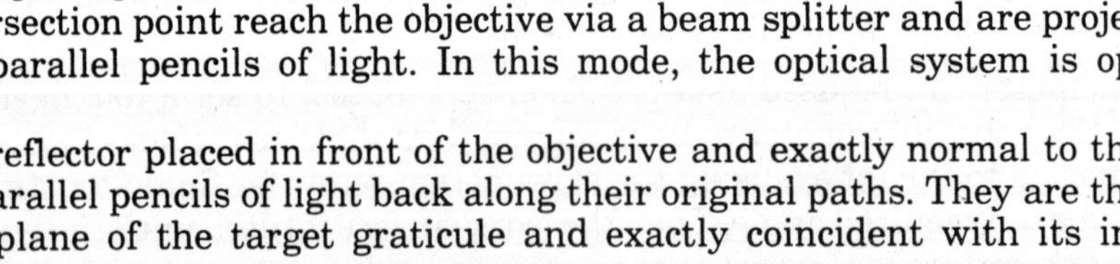

Fig. 8.32. Principle of Autocollimator.

← Reflected beam when reflector is square to beam

← ...... Reflected beam from tilted reflector

proportion of the returned light passes straight through the beam splitter and the return image of the target crossline is therefore visible through the eyepiece. In this mode, the optical system is operating as a telescope focused at infinity.

If the reflector is tilted through a small angle the reflected pencils of light will be deflected by twice the angle of tilt (principle of reflection) and will be brought to focus in the plane of the target graticule but linearly displaced from the actual target crosslines by an amount $2\theta \times f$.

Linear displacement of the graticule image in the plane of the eyepiece is therefore directly proportional to reflector tilt and can be measured by an eyepiece graticule, optical micrometer or electronic detector system, scaled directly in angular units. The autocollimator is set permanently at infinity focus and no device for focusing adjustment for distance is provided or desirable. It responds only to reflector tilt (not lateral displacement of the reflector). This is independent of separation between the reflector and the autocollimator, assuming no atmospheric disturbance and the use of a perfectly flat reflector.

Many factors govern the specification of an autocollimator, in particular its focal length and its effective aperture. The focal length determines basic sensitivity and angular measuring range. The longer the focal length the larger is the linear displacement for a given reflector tilt, but the maximum reflector tilt which can be accommodated is consequently reduced. Sensitivity is therefore traded against measuring range. The maximum separation between reflector and autocollimator, or "working distance", is governed by the effective aperture of the objective, and the angular measuring range of the instrument becomes reduced at long working distances. Increasing the maximum working distance by increasing the effective aperture then demands a larger reflector for satisfactory image contrast. Autocollimator design thus involves many conflicting criteria and for this reason a range of instruments is required to optimally cover every application.

Air currents in the optical path between the autocollimator and the target mirror cause fluctuations in the readings obtained. This effect is more pronounced as distance from autocollimator to target mirror increases. Further errors may also occur due to errors in flatness and reflectivity of the target mirror which should be of high quality.

When both the autocollimator and the target mirror gauge can remain fixed, extremely close readings may be taken and repeatability is excellent. When any of these has to be moved, great care is required.

8.10.2. Laser Interferometer. With laser interferometer it is possible to measure length to an accuracy of 1 part in 10^6 on a routine basis. With the help of two retro-reflectors, placed at a fixed distance, and a length measuring laser interferometer the change in angle can be measured to an accuracy of 0.1 second. The device uses the Sine principle. The line joining the poles of the retro-reflectors makes the hypotenuse of the right triangle. The change in the path difference of the reflected beam represents the side of the triangle opposite to the angle being measured. Such laser interferometer can be used to measure an angle upto ± 10 degrees with a resolution of 0.1 second.

The principle of operation is shown in Fig. 8.33.

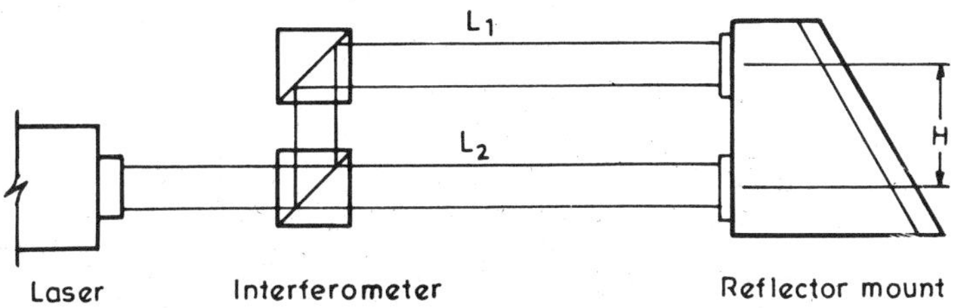

Fig. 8.33. Interferometric measurement of angle.

8.10.3. Photoelectric Microptic Autocollimator. Photoelectric setting makes measuring and checking by autocollimator far simpler and faster. Micrometer adjustment is provided for setting, but coincidence of setting graticule and target image is detected photoelectrically, and shown on a meter as a null reading. This provides a high degree of sensitivity and repeatability, also reducing eye fatigue to a minimum. The eyepiece is normally only used to assist in initial setting-up.

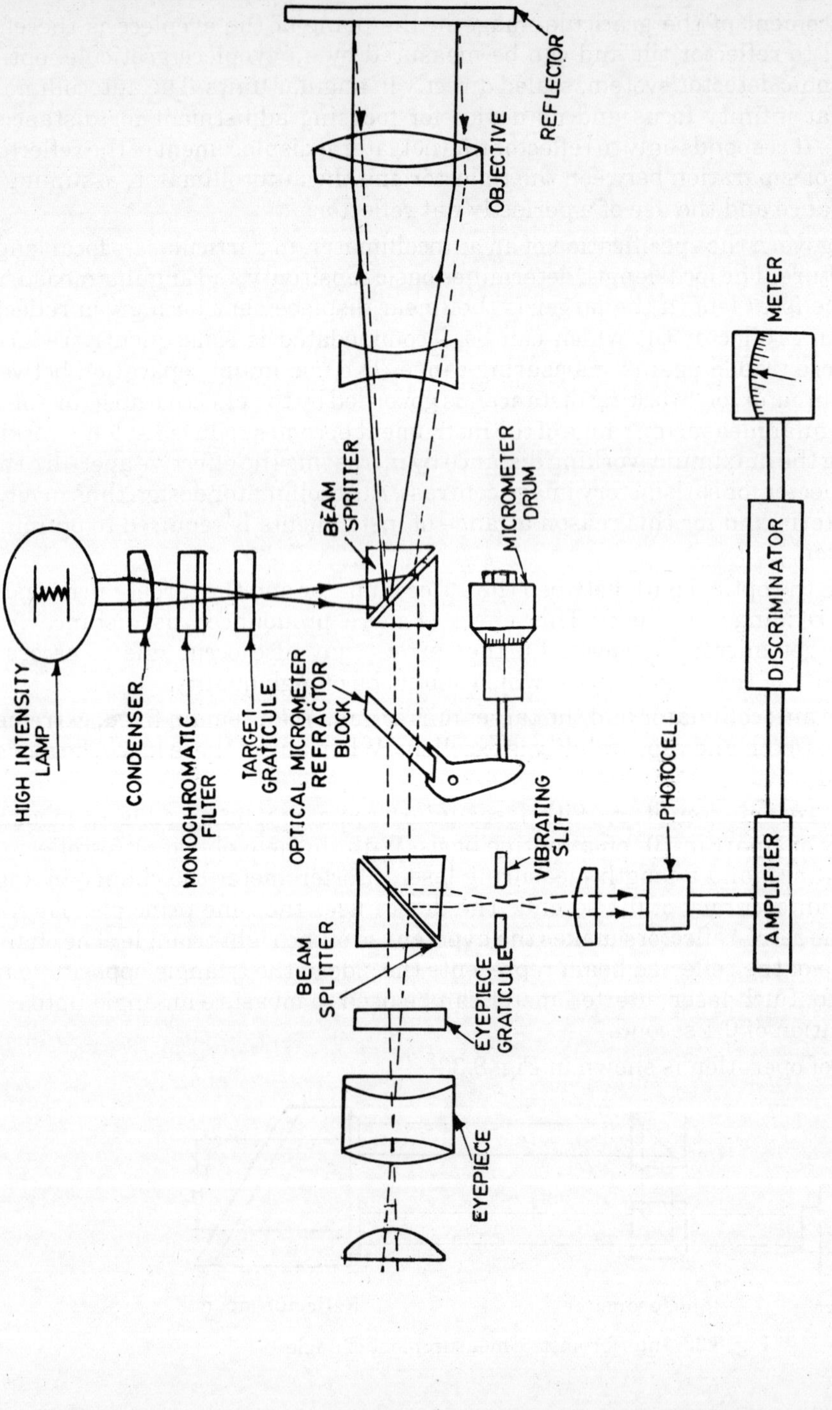

Fig. 8.34. Schematic diagram of Photo-electric autocollimator.

The photoelectric autocollimator is particularly suitable for calibrating polygons, for checking angular indexing and for checking small linear displacements.

It can be used as a visual autocollimator, if required, and is available with a dark field graticule as standard.

Fig. 8.34 shows a schematic of the operation of photoelectric autocollimator. It consists of a vibrating slit, a photoelectric detector, electronic amplifier for magnified viewing on a meter.

Operating Principle. (Refer Fig. 8.35). The photoelectric detecting unit consists of photocell and a vibrating slit which is attached to the micrometer screw. When the slit is positioned so that it vibrates about the reflected image, the intensity of light received by the photocell will vary. The output waveform is amplified and fed to a frequency discriminator and meter which, in effect, indicates the asymmetry of the waveform. When the slit is positioned so that it vibrates symmetrically about the image the meter indicates a null reading, and the angular displacement of the target mirror can be read from the micrometer.

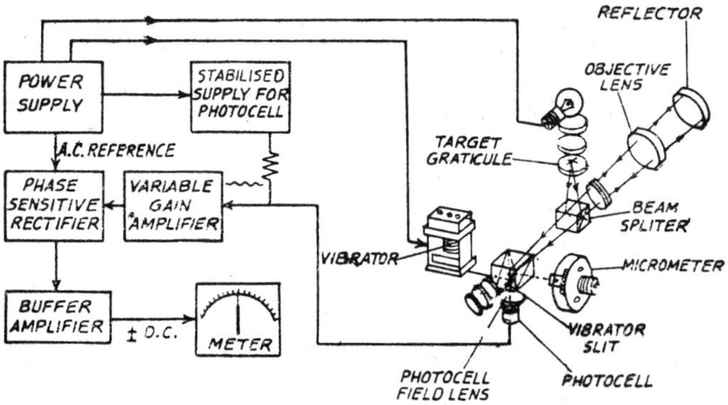

Fig. 8.35. Schematic diagram of Photoelectric Autocollimator.

The autocollimator, with suitable reflecting surfaces or optical gauges, is capable of calibrating straightness, flatness, squareness or division of the circle.

8.10.4. Automatic Position Sensing Autocollimators. Automatic position sensing autocollimators provide fully automatic setting and display. Angular displacement of the reflector is displayed on a digital readout—eliminating any micrometer reading for setting or measuring.

Automatic autocollimators can be used in cramped positions where it could be impossible to use a visual instrument, and no handling during measurement minimises the danger of accidental autocollimator movement.

Instruments measure in one plane only. For measuring in a second plane perpendicular to the first, the instrument is rotated through 90°. A dark field graticule is fitted as standard.

Accuracy is unaffected by normal mains fluctuations or lamp ageing.

Automatic autocollimators are ideal for the repetitive checking of production components, and for continuously monitoring angular displacement of slow moving parts.

Operating Principle : The vibrating slit image detecting system is similar to that described for the photo-electric autocollimator, but in addition, the slit is electrically biased across the field. The amplitude and polarity of the biasing signal is dependent upon the misalignment of the slit with the return image. By this arrangement the slit is biased to the position straddling the image, at which the misalignment is negligible. The current necessary to hold the slit biased in this position is used to provide a direct angular reading on a digital readout.

8.10.5. Autocollimator Accessories. The full versatility of autocollimators can be exploited by the addition of suitable accessories.

Levelling Base : It supports the autocollimator and enables it to be levelled to bring its axis parallel to the surface being measured. It incorporates spring-loaded clamps and a circular bubble level. Three pads are included for use under the foot screws. There is no necessity to remove the autocollimator from the base after use as the autocollimator case is constructed to take both items.

Surface Plate Stand : A multi-purpose stand of heavy duty construction for general bench use, comprises ground cast iron surface plate, column and bracket. The autocollimator clamping bracket has independent clamping and rotational adjustments, enabling the bracket to be turned without disturbing the height adjustment.

Steel Reflector : A reflector must be regarded as an integral part of any autocollimator system. Successful autocollimation requires a reflector of adequate flatness, reflectivity and diameter. The parallelism of the faces is such that negligible error is introduced when the unmounted reflector is back mounted.

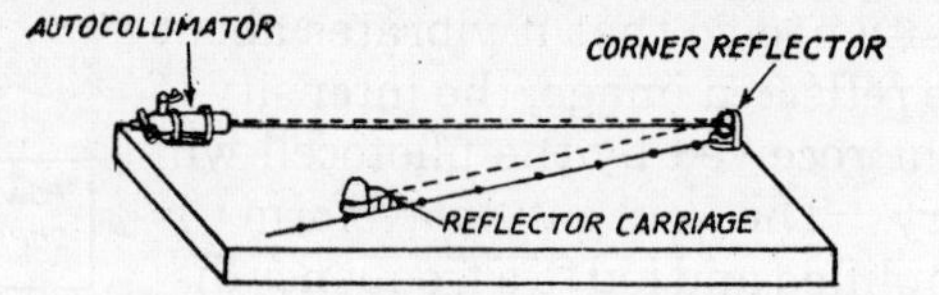

Fig. 8.36. Use of autocollimator accessories.

Mounted Glass Corner Reflector : This is used in conjunction with a reflector carriage and mounted reflector for calibrating a surface platé. It enables several calibration lines to be traversed without the autocollimator being moved, thereby saving setting-up time and making the subsequent correlation of reading easier.

Steel Cube Reflector : It can be used as a general purpose reflector and for providing a 90° angle standard in three planes, for setting or checking perpendiculars.

8.10.6. Microptic Autocollimator. (Fig. 8.37). In this, a pair of target wires take place of the point light sources as it is not convenient to visualise the reflected image of a point and then to measure the distance x precisely. So a pair of target wires placed in the focal plane of a collimating lens are illuminated from the back and their image projected which strikes a plane reflector and the reflection of the image is brought to a focus in the plane of the target wires. The field of view in the eyepiece in which the wires and their images are viewed simultaneously is shown in Fig. 8.37.

Setting wires are placed in the microscope unit and are adjustable by a micrometer until they straddle the reflected image. There is a scale in eyepiece which can directly read to the nearest 1/2 minute. The micrometer drum moves the setting wires at 1/2 minute per revolution and is divided into 60 equal parts. Thus with the aid of

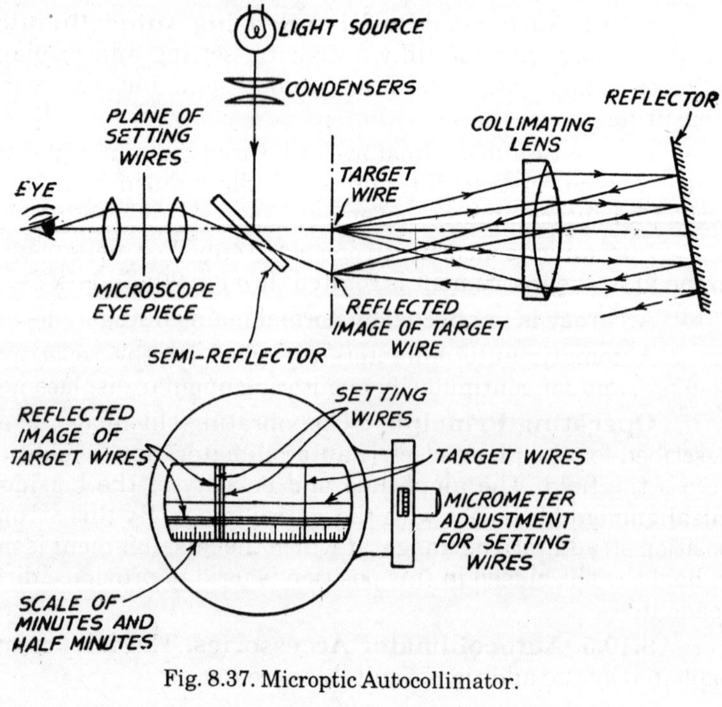

Fig. 8.37. Microptic Autocollimator.

micrometer it is possible to read the tilt of reflector to normal upto 1/2 sec. of arc. The two wires help to indicate tilt of reflector in two planes at right angles.

The instrument has generally a range of readings of 10 minutes of arc upto distances of 9 metres. It is used for a variety of purposes.

Autocollimator Applications. Autocollimators are applied to the measurement of straightness and flatness ; precise angular indexing in conjuction with polygons ; comparative measurement using master angles ; assessment of squareness and parallelism of components ; and the measurement of small linear dimensions.

Straightness is measured in conjunction with a reflector attached to a base having two co-planar locating pads at a known distance apart. The base is stepped in a straight line along the surface at intervals equal to the pitch of the locating pads and the angular change is recorded at each position. These readings are readily converted into changes in vertical height of the leading pad. A plot of the surface straightness can then be prepared from the data. Measurement of flatness is an extension of this method and involves a series of straightness measurements along straight line axes across the surface.

Polygons provide precise angular references for testing rotary tables and angular indexing systems. Supplied with a calibration of either ± 1 or ± 0.5 second of arc they can be used to directly measure functional errors in the readouts of such equipment. Standard polygons with 8, 9 or 12 sides allow measurement at 45, 40 and 30 degree intervals respectively. Used in combination, the measurement interval is reduced to the difference between the intervals of each polygon (*e.g.* a combination of 8 and 9-sided polygons allow measurement to 45 – 40 = 5 degrees). A great advantage of polygons is that they only require centering on the table to about 250 μm and levelling to 2 minutes of arc to give accuracies of + 1 second of arc used singly and ± 2 seconds of arc used in combination.

Angular comparison can be achieved by simultaneous viewing of a component and a master angle which can be either a combination of angle gauges or a master component. Two images are produced in the autocollimator and their separation is a direct reading of angular error. Squareness and parallelism measurements involve similar techniques of using reference squares or viewing two reflectors simultaneously.

8.10.7. Calibration of Angle Gauges using Autocollimator. Angle gauge is wrung on base platen. An optical flat is placed inclined at θ_2 to platen and θ_1 to angle gauge as shown

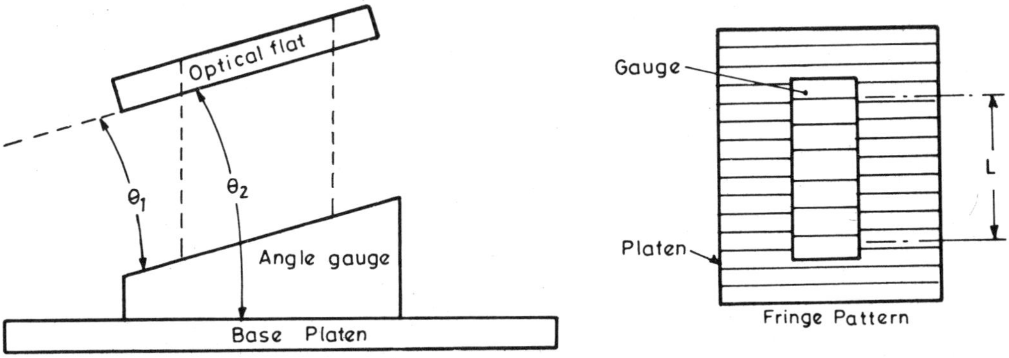

Fig. 8.38. Interferometric calibration of angle gauge.

in Fig. 8.38. Two sets of fringes are observed. The fringes are counted over length L for gauge and platen (say n_1 and n_2). The angle of gauge $\theta_2 - \theta_1 = \frac{\lambda}{2L}(n_2 - n_1)$, where λ is the wavelength of the light used. The values of n_2 and n_1 in general will not be an integral. This method can be used for angles upto 1 minute and the precision of 1 sec is possible.

Moire Technique

Goniometers. Fig. 8.39 shows in schematic form the working of Goniometer. It incorporates continuously rotating Moire gratings for high accuracy angular measurements. Two radial gratings rotating in unison are employed. These are scanned by two reading heads, one remaining stationary and the other moving with a rotatable work table *i.e.* through the angle to be measured. The phase difference between the two outputs from heads varies continuously as the second head attached to the rotary table is moved and provides an accurate measure of the angle by which the work table has been moved. The whole number of phase cycles is found by counting, while the fraction is measured with an accurate phasemeter.

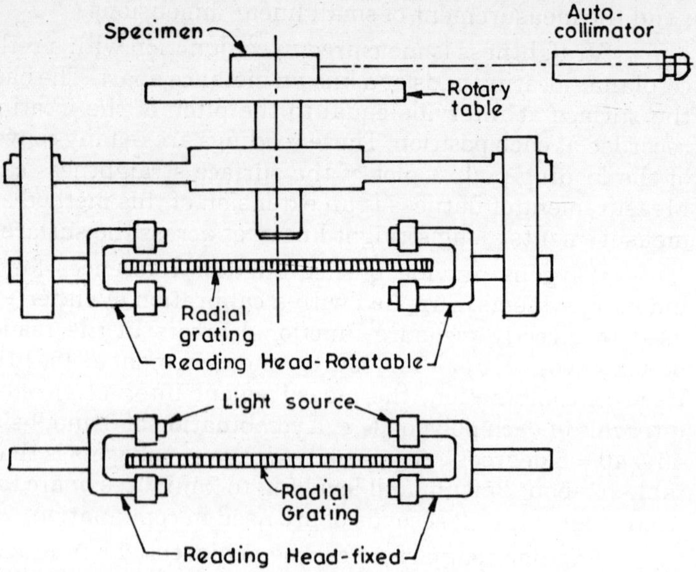

Fig. 8.39. Schematic diagram of Moire technique for angle measurement.

This system has the advantage that grating errors are almost entirely averaged to zero during the measurement time thus giving the instrument inherent accuracy, even with components manufactured with relatively poor accuracy.

With an angle goniometer fitted with two 32400 lines radial gratings, angle measurements to an accuracy of 0.1 second of arc are possible.

8.10.8. Angle Dekkor. This is also a type of an autocollimator. (Refer Fig. 8.40). It contains a small illuminated scale in the focal plane of the objective lens (collimating lens). This scale in normal position is outside the view of the microscope eyepiece as shown in Fig. 8.41 (*b*). The illuminated scale is projected as a parallel beam by the collimating lens which after striking a reflector below the instrument is refocused by the lens in the field of view of the eye-piece. In the field of view of microscope there is another datum scale [Fig. 8.41 (*c*)] fixed across the centre of screen and the reflected image of the illuminated scale is received at right angle to this fixed scale as shown in Fig. 8.41 (*a*) and

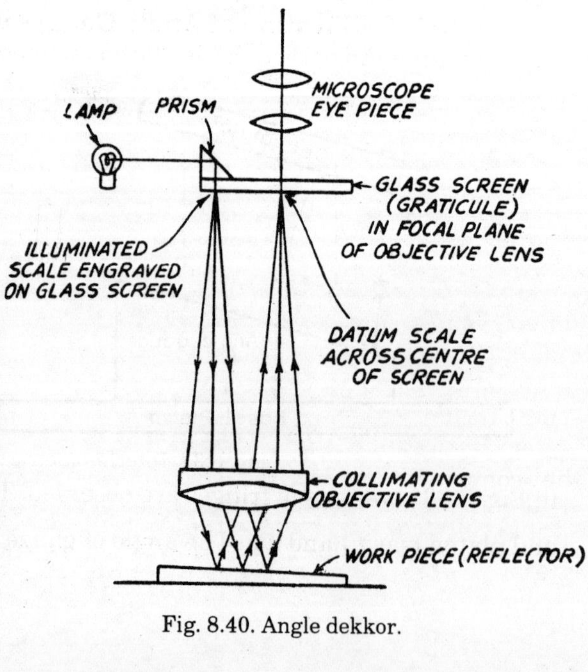

Fig. 8.40. Angle dekkor.

the two scales, in this position intersect each other. Thus the reading on the illuminated scale measures angular deviations from one axis at 90° to the optical axis and the reading on the fixed datum scale measures the deviation about an axis mutually perpendicular to the other two. In other words, changes in angular position of the reflector in two planes are indicated by changes in the point of intersection of the two scales. Readings from scale are read direct to 1′ without the use of a micrometer.

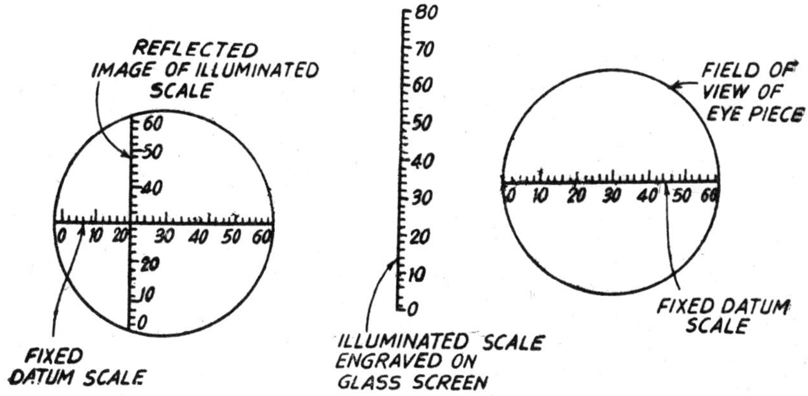

(a) (b) (c)

Fig. 8.41. View in the eyepiece of angle dekkor.

The whole of the optical system shown in Fig. 8.40 is enclosed in a tube which is mounted on an adjustable bracket. There is a lapped flat and reflective base on which all these things are placed. It is mostly used as a comparator. The instrument measures by comparing the readings obtained from a standard, a sine bar or combination of angle gauges with that from the work under test. Though this is not a precise instrument in comparison to autocollimator, it has wide field of application for general angular measurement, as angular variations are read direct without the operation of a micrometer.

8.10.9. Uses of angle dekkor in combination with angle gauges.

(*i*) *Measuring angle of a component.* It may be made clear that angle dekkor is capable of measuring small variations in angular setting, *i.e.* determining angular tilt. In operation the measuring principle is that of measurement by comparison ; the angle dekkor is set to give a fixed reading from a known angle (*i.e.* using known angular standards to obtain a zero reading). (Refer Fig. 8.43)

Thus first the angle gauge combination is set up to the nearest known angle of the component and the angle dekkor is set, (using special attachment and link), such that zero reading is obtained on the illuminated scale. The angle-gauge build up is then removed and replaced by the component under test, a straight-edge being used to ensure that there is no change in lateral positions. The new posi-

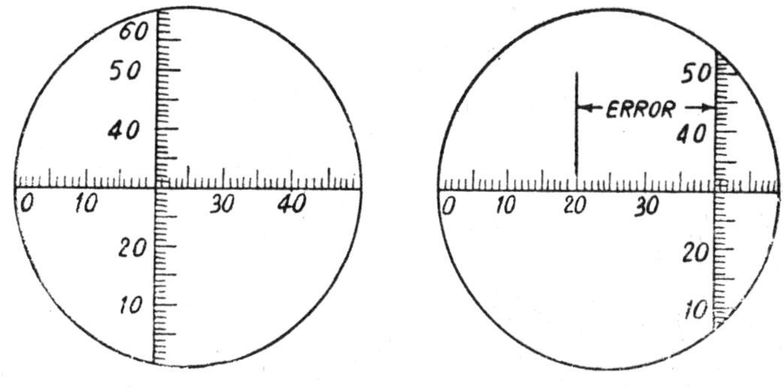

Zero-reading with angle gauge build-up

Reading with component in position error = 40 – 20 = 20 divisions = 20 minutes

Fig. 8.42

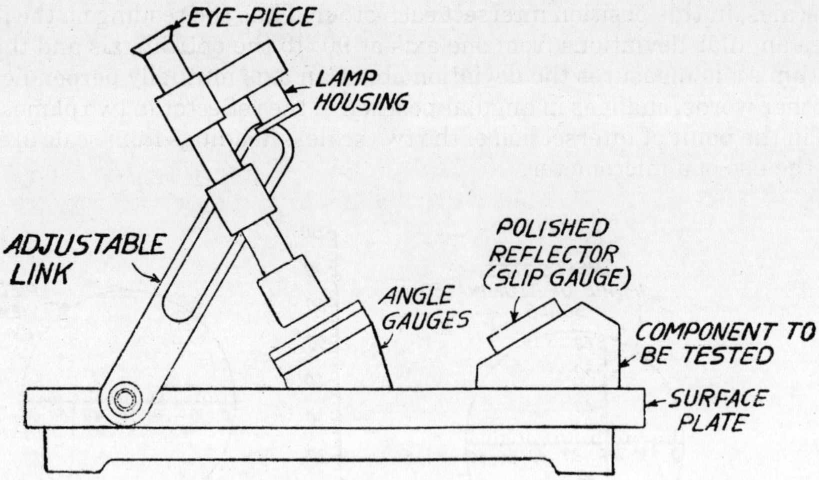

Fig. 8.43. Measuring angle of a component.

tion of the reflected scale with respect to the fixed scale gives the angular tilt of the component from the set angle. (Refer Fig. 8.42).

(*ii*) *To obtain precise angular setting for machining operations.* We will consider an example of milling a slot at a precise angle to a previously machined datum face. A parallel bar is used as a datum face, the component being securely clamped when in close contact with

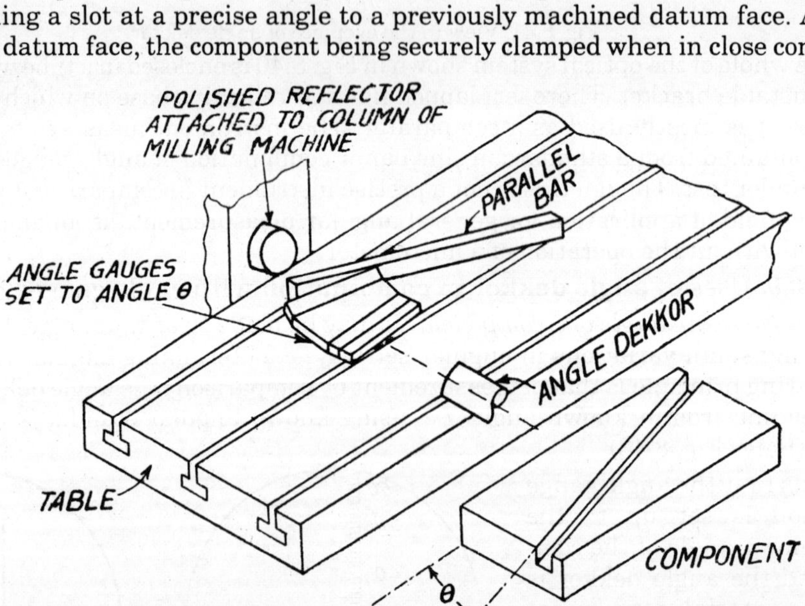

Fig. 8.44. Set up for milling angular slot.

it. The parallel bar is positioned on the table of milling machine with the aid of angle dekkor. The setting-up technique is illustrated in Fig. 8.44. It may be noted that a polished reflector is firmly attached to the column of the milling machine. With the aid of this surface as reference, the angle dekkor is set up such that zero reading is obtained ; in other words, the axis of the optical beam is truly at 90° to the table feed. Then build up the combination of angle gauges to the exact value θ, *i.e.* the inclination of the slot to be milled on the component. The angle gauges along with the parallel bar are placed on the table and adjusted in position such that the angle dekkor shows zero reading when viewing the flat surface of the angle gauge combination. It means that the angular inclination between the datum face of the parallel bar and the feed direction of the table is now θ°. The parallel bar is firmly clamped in this position, a check being made to ensure that no movement has taken place during clamping ; a few gentle taps will soon allow a zero reading on the angle dekkor to be regained. Finally, now the workpiece can be clamped on milling machine table, in close contact with this pre-set parallel bar.

(*iii*) *Checking the sloping angle of a V-block.* The set up for checking the sloping angle of V-block is illustrated in Fig. 8.45 (*a*). The principle consists of comparing the reading obtained from the polished slip gauge in close contact with the work-surface, and a zero reading obtained from the angle-gauge build-up.

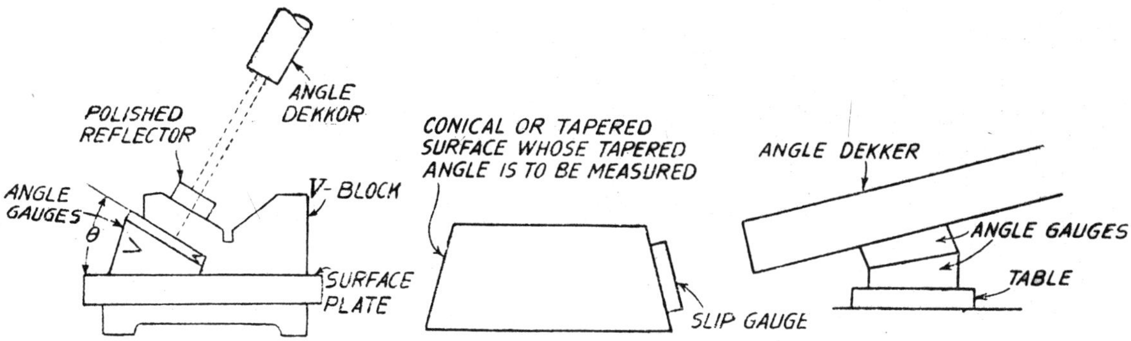

$$\theta = 90° - \frac{\text{angle of V - block}}{2}$$

(*a*) Checking V-block angle. (*b*) Measuring angle of cone.

Fig. 8.45

(*iv*) *To measure the angle of cone or taper gauge.* A simple set-up for this purpose is shown in Fig. 8.45 (*b*). The instrument is first set for the nominal angle of cone on a combination of angle gauges or on a sine bar set to the nominal angle. The cone is then placed in position with its base resting on the surface plate. A slip gauge or other parallel reflector is held against the conical surface as no reflection can be obtained from a curved surface. Any deviation from the set angle will be noted by the angle dekkor in its eye-piece and indicated by the shifting of image of illuminated scale, whose reading while setting with angle gauge is noted down before hand.

Angle dekkor and auto-collimator find a wide range of applications if used in conjunction with the constant deviation prism and Dowell Prism. It is, therefore, considered better to describe these prisms here.

8.10.10. Constant Deviation Prism. This enables both the projected and reflected beams to be turned through a right angle. It is, therefore, very suitable in alignment testing and checking of two surfaces at right angles. This is also called optical square. The special property of this prism is that it always reflects a ray of light through the same angle, irrespective of the angle of incidence. In the prism, the reflecting surfaces are disposed at an angle of 45°. As the ray is deflected through twice the angle between the reflectors, therefore, it will be reflected through 90°. Actually the rays bend as they move from rare to dense medium and similarly when coming out, but their effect is nullified and finally the beams are at right angles. The effect of bending of rays is not shown in Fig. 8.46.

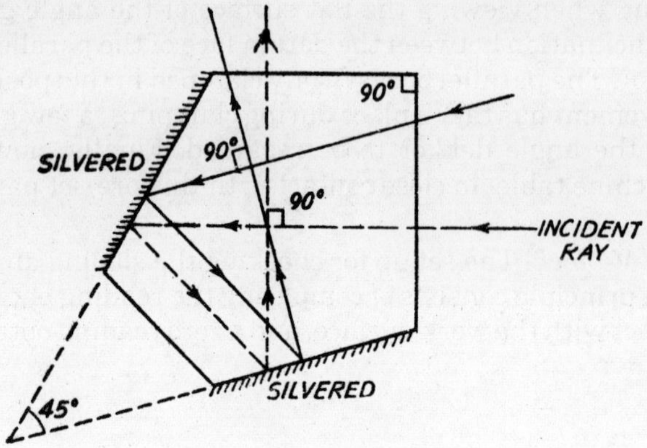

Fig. 8.46. Optical Square.

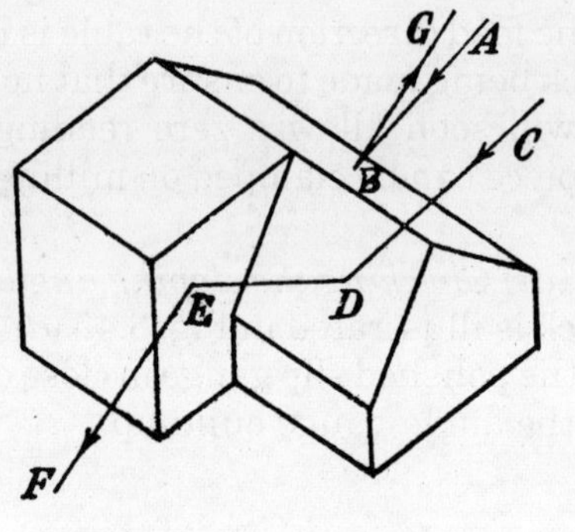

Fig. 8.47. Dowell Prism.

8.10.11. Dowell Prism. This is used to split up a beam of light into two beams which are exactly parallel but projected in opposite directions. The special feature of the prism is that its construction is such that the two projected beams are parallel under any conditions of light transmission, thus no setting of prism or angle dekkor is required. This prism is very suitable for testing the parallelism of two faces or surfaces. Since images from two surfaces will be seen simultaneously in eye-piece, the error can be measured directly. In Fig. 8.47, *AB* and *CD* are incident parallel rays and *BG* and *EF* are reflected parallel beams, but in opposite direction.

8.11. Circular Division

A circle can be divided into any number of parts. The accuracy with which it is performed is proven once the circle is "closed". The division of the circle is thus uniquely self proving. The level of accuracy depends only on the precision of available equipment and the techniques employed. The method of circular division is the only one inherently capable of deriving primary standards. No outside authority such as standards is necessary for verification of angle obtained by this method.

The level of accuracy depends only on the precision of available equipment and the techniques employed.

Angle is defined by sub-division of the circle and the divided circle is used extensively in practical applications for measuring and setting out angles.

Long back it was realised that for marking high-class instruments, precisely divided circular scales were required. For this purpose a circular dividing machine was also developed, but no means were available for checking the divisions. However, divided scales were made and used for circular division and these are still used in optical dividing heads and circular tables.

8.12. Rotary Tables

The rotary table is the best-known and most widely used circle dividing instrument. These fall under three categories :

(*i*) Optical type, employing a precision circular scale read by microscope as its angle measuring element. (Refer Fig. 8.48).

(*ii*) Mechanical, cam compensated (Refer Fig. 8.49). It employs worm-and-gear as the angle measuring element but the final accuracy is established by cam-compensation of gear error.

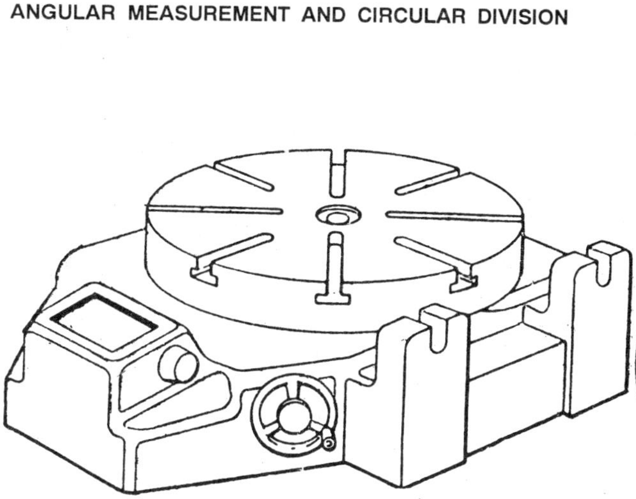

Fig. 8.48. Optical rotary table.

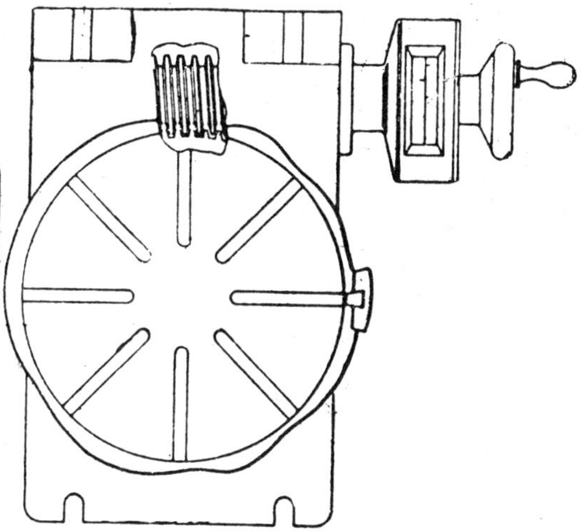

Fig. 8.49. Mechanical cam compensated rotary table.

(*iii*) Accurate worm-and-gear rotary table in which the gear is machined or lapped to final accuracy (Refer Fig. 8.50).

(*iv*) **Precision Index Table.** This type of table uses two face gears having serrated teeth. Both gears are identical in diameter, shape and number of teeth. This table is thus a serrated tooth table based on the principle of circle divider. When two face gears are brought into contact face to face (root to crest), then lock together and prevent any rotation unless they are disengaged, rotated and brought in contact. The indexing corresponds to the number of teeth through which the table is indexed. The serrated divider is limited to 360 teeth. Serrated teeth indexer, in 0.25° increment with accuracy of ± 0.1 second of arc, has been developed.

In **optical rotary** machines, the precision circular scale could be read by electrical (inductance) or by Moire fringe method. The number of lines which can be placed about the main circular scale of an optical rotary table depends on its diameter, the width of the lines and the magnification of the integral microscope. A graduated drum or other optical vernier arrangement enables settings to be made to finer increments of minutes and seconds of arc. In the "averaging" type, smaller increments of angle are achieved by the sine curve generated by electronic or optical pulsing. The errors of eccentric mounting of scale on the table are compensated by reading the circular scale simultaneously from opposite sides. The rotary motion is imparted by worm and gear (not made to any great accuracy). The measuring element in optical rotary table is free from wear. The table can be positioned from either direction. Repeatability is dependent on the rotary table's

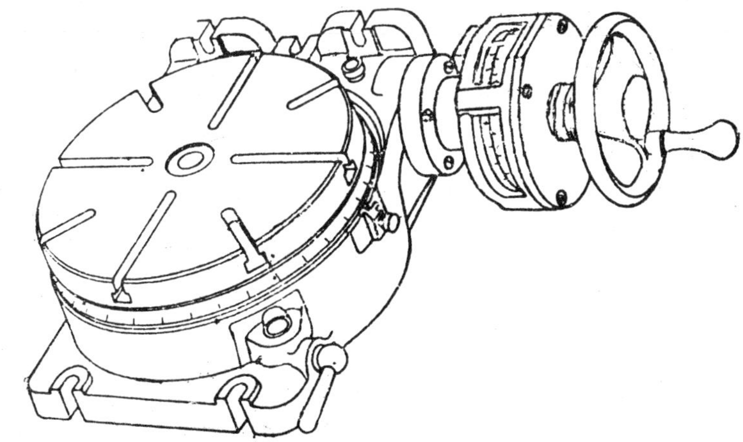

Fig. 8.50. Worn and gear rotary table.

ability to make very fine displacements, which in turn is dependent on the type of radial and axial bearing used, lubrication, and the perfection of the total worm and gear assembly.

In mechanical, cam-compensated rotary table, accuracy is dependent on the accuracy of the worm and gear. The final unresolved error are overcome by compensation, rather than by perfection of the gear. The shape of the circular cam attached to the rotating member is adjusted according to the calibrated error of the rotary table. A follower works against the cam and advances or retards the vernier dial or "zero" to match the error of the gear at that particular angle. With proper design, high repeatability can be obtained. This system offers reasonable accuracy at minimum cost. The periodic error, *i.e.* the error within one turn of the hand wheel, is not taken into consideration. This error is affected by factors like the thread form, drunkenness, pitch, line up, runout of worm, errors of eccentricity, tooth form, and tooth-spacing of the gear.

In accurate worm and gear rotary table, both the worm and the gear are made highly accurate. As such corrective mechanism is not required. Usually a unidirection approach is used for setting to take care of unavoidable presence of backlash. Thus rotary table has the advantages of trouble free operation, good repeatability, and a high potential of accuracy. The accurate worm and gear has one additional advantage for the generation and inspection of angles. It can be used together with an index plate to divide the circle, without the necessity of reading angle, and without loss of accuracy.

It may be noted that index can't be used with optical rotary table, because its worm-and-gear is not accurate.

It is also important to note that lack of true spindle rotation results in both angular error and centre distance error.

8.13. Dividing Heads and Circular Tables

Rotary tables are used for accurate circular indexing. These are normally designed to rotate in one plane but in some cases tilting action is also incorporated. Angular dimensions are either read from the common line measurement at the periphery of the table or by optical display units or digital readouts. Dividing heads are used for angular and linear measurements and for indexing. Clinometers, circular tables and dividing heads employ a variety of mechanical, optical and electronic techniques to divide a circular scale. Optical dividing head mainly consists of a glass scale which is mounted on the main spindle and the graduations of the scale are observed through a microscope fitted with an eye-piece graticule and mounted in the body of the instrument. This corresponds to the line standard in length measurement. This

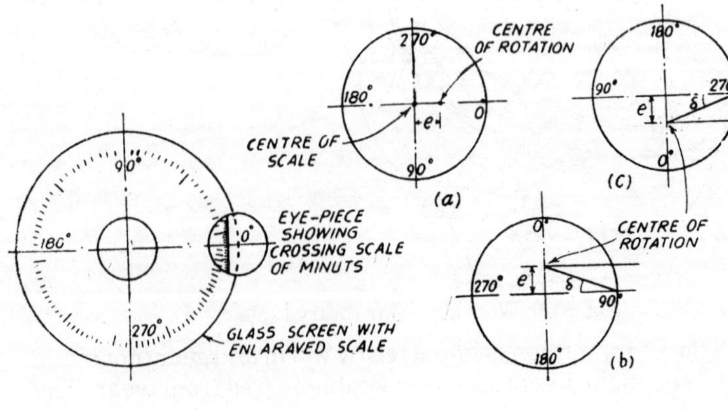

Fig. 8.51. Circular indexing.

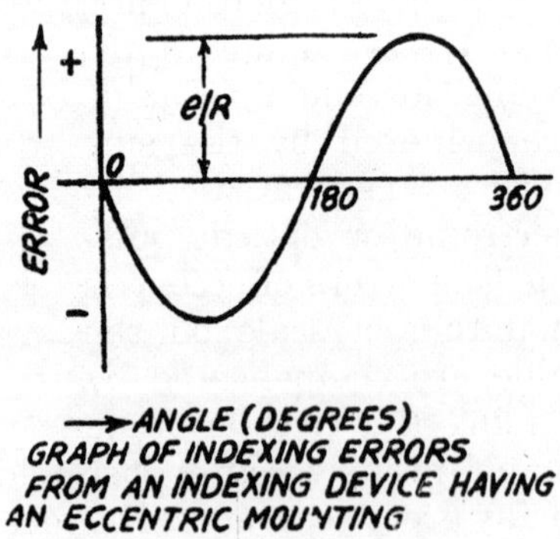

Fig. 8.52. Indexing error.

instrument can be used in inspection shops or on machine tools as it has a robust head and high standard of accuracy. This is very useful for rotating a workpiece through a given angle precisely. Optical dividing head thus has a big advantage over mechanical type *i.e* the possible inaccuracies due to wear of the worm and worm wheel mechanism are eliminated.

As readings are taken at only one point on the circle, the accurate centering of the glass circle with the axis of rotation is very important. The effects of eccentricity of mounting glass circle are very serious, though the glass is perfectly divided.

Let us assume that there is present an eccentricity of amount e between the centre of scale and centre of rotation. It is obvious that as the disc rotates, the centre of the disc describes a circle of radius e. In position (*a*) in Fig. 8.51, the disc will read correctly against the datum as centre of scale, centre of rotation, and fiducial mark lie in the same line. When disc is turned further and if we want 90° to be read against the centre of rotation, then it is obvious from Fig. 8.51 (*b*) that the disc must be turned through $(90 + \delta)°$, where $\delta = e/R$. (R is the radius at which readings are taken). Similarly to read 270°, an equal error of δ occurs and in this case Fig. 8.51 (*c*), the disc must rotate through $(270—\delta)°$ to read 270°.

Thus it can be visualised that error passes from 0 to $+ \delta$, through zero again to $- \delta$, and the graph of error will then be of sinusoidal nature as shown in Fig. 8.52. Obviously maximum error between any two readings is 2δ and this will occur when disc is rotated between the readings of 90° and 270°.

8.14. Precision Polygons

Multifaced polygons are used for calibration of rotary tables, dividing head. Any number of faces can be produced but the overall diameter should not exceed 300 mm in view of handling convenience. The minimum size of the reflecting face should be approx. 13 mm × 13 mm. Thus maximum number of faces can be 72, corresponding to an included angle of 5°. Steel polygons are made of high grade steel, hardened and subjected to a suitable heat treatment for stabilisation and the sharp corners are chamfered or rounded off suitably. The reflecting faces are lapped flat to within 0.0001 mm.

According to IS : 6987—1973, these are divided into three types, 1, 2 and 3 having working faces between 5—20, 21—40 and 41—72. The top and supporting surface are lightly lapped and legibily marked. In order to enable the result of the calibration to be recorded and subsequently applied in the case of polygon, the face of the polygon is identified. Each face is identified by engraving on the top surface, the nominal angles with the face and a datum face. The faces may be identified with series of number from 1 upwards if nominal angle is not a whole number of degrees. Around 4 holes (plain for type 1 and threaded for types 2 and 3) are provided for clamping purposes. The dimension of polygon between the top and supporting surfaces is 75, 150 and 300 mm for types 1, 2 and 3 respectively. The top and supporting surfaces are made flat and parallel to accuracies of 0.0025, 0.005 and 0.007 mm, 0.0035, 0.005 and 0.04 mm for types 1, 2 and 3 respectively and departure from flatness is permissible only in concave direction. All faces are made square to the bottom supporting surface to an accuracy of 2.5 μm over the thickness of the polygon. The error in angle between any working surface and the datum face and between any two adjacent working faces shall not be more than 10 seconds of arc.

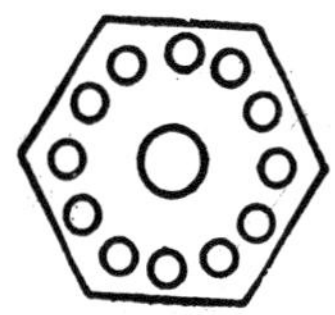

Fig. 8.53. Polygon.

In big polygons, a central hole is drilled alongwith small holes all around (Refer Fig. 8.53) for mounting purposes.

8.14.1. Criteria effecting polygon accuracy. For certainty of calibration, the primary consideration in accuracy is not how closely the mirror or reflecting faces agrees with the nominal angle, but how flat and square it is. The four major criteria to be fulfilled in selection of a polygon are : The reflecting surfaces (flats or mirrors) should be within 4 to 5 seconds of correct or nominal angle ; the reflecting surfaces should be flat to 0.0001 mm ; if individual mirrors are used, they should be mounted in such a way that there is no distortion ; and the mirror position should be permanent, *i.e.* there should be not possibility of alterations of mirror position from bumping or jarring, from poor mounting, or from secular change.

Accuracy of the polygon can be proved by first principles. It is calibrated by mounting it on a turntable. The angle segment being checked is contained between two rigidly fixed autocollimators nulled on the nominal angle.

8.14.2. Circular Division by Polygons. For normal metrology requirements, the most accurate method of circular division is possible by the use of polygons. This method is also used for calibrating the circular divided scales (circles of optical dividing equipment). A polygon consists of a hardened and suitabilised piece of steel having a number of optically flat and polished faces. The reflecting faces are lapped flat to within 0.0001 mm. Their faces are normal to equal divisions of a circle and are available in various ranges. The largest polygon has 72 faces at intervals of nominally°. However, normally a polygon having 12 sides at intervals of 30° is suitable for most of the work. The polygons generally have equiangular faces but for special purposes unequal angles can also be provided. These are generally about 15 mm thick and 75 mm across flats and have a central how with four smaller holes for mounting purposes. For handling convenience, maximum overall size should not exceed 300 mm. These are manufactured to an accuracy of 5 seconds maximum error between any two faces and errors present are known upto 1 second.

Polygons are used in conjunction with an auto-collimator. The polygon is mounted on the indexing plate to be calibrated or the circular scale to be divided. The auto-collimator is set to receive a reflection from any one of the faces of the polygon and the reading is noted down. The polygon is then rotated till the same reading is observed in auto-collimator and thus obviously the angle of the rotation will be equal to the angle between the polygon faces. If this angle is previously known, then it is possible to divide the circle as well as recalibrate it if it is already calibrated. The inherent accuracy of the method is due to the fact that no centring of the polygon is required because the variation in the distance of the reflector from the auto-collimator has no effect if the angular relationship between the reflector and the optical axis of the autocollimator is unaltered. (The eccentricity of the polygons can't affect the angles between its faces but can only affect the distance of succeeding faces from auto-collimator). However, in practice it is found that time spent for centering is worthwhile because it makes it possible to avoid any errors caused by variations of polygon form.

A correction factor (error of each polygon face from nominal) is furnished by the manufacturer. This factor is applied to the readings to find the actual error of the rotary table at the sectors calibrated.

8.14.3. Limitations of Polygons for calibrating rotary tables. The correction factor furnished by the manufacturers is applicable when the centre of the mirror is used in calibration but there is no means of reference to the exact centering. The calibration factor can change considerably if the target area used for calibrating the rotary table differs from that used when the polygon was originally calibrated. Each mirror has its own geometrical characteristics and reflectivity, and irregularity (wobble of rotary table, imperfect levelling, or reading the autocollimatory by eye rather than by photoelectric readout) may alter the supposed calibration value of the polygon.

Secondly, it is impossible to make a full comprehensive calibration of the rotary table with one polygon because of the limited sides it can have. Thus not all of the lines of the circular scale of an optical rotary table are inspected for error. The 13-sided polygon is often used because it gives a random check more likely to show cumulative errors of the total rotary table system. More than one polygon is usually employed (like 12-sided and 13-sided).

8.15. Angle Gauge Dividing Head

With the introduction of angle gauges, a new form of dividing head for exact rotation through any angle was devised by Dr. Tomlinson. It consists of a cast body having a central hole through which a large diameter spindle is mounted in precision lapped bearing. A lapped face plate is fitted on the front of the spindle and a lapped steel square is mounted behind the face plate. The body incorporates a worm and wheel mechanism and a clamping device on the spindle. The worm and wheel is very precise and smooth in action and plays no part in measurement. It is needed simply for the purpose of obtaining slow rotational motion. The spindle clamp is designed such that when it is operated, it does not change the setting.

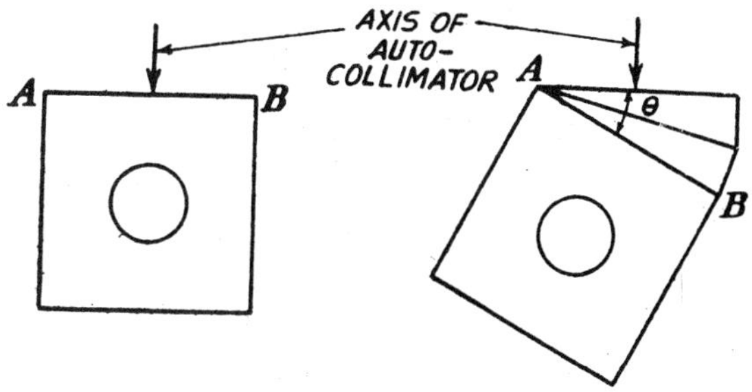

Fig. 8.54. Work is rotated through an angle θ in clockwise direction.

An auto-collimator is mounted on the dividing head and determines the fiducial position of gauge face. Its eye-piece is mounted on a micrometer slide and is capable of reading directly upto 1 second. For making angular measurements, the appropriate angle gauges are wrung to the faces of the square. First and initial setting is made on one face of square and the first setting is made on the work. The spindle is then rotated after the appropriate angle gauges are wrung, to such an extent that the same autocollimator reading is obtained. Thus the work is rotated exactly through the required angle. All the faces of square can be used in order to obtain the various angles with ease and convenience.

By this method it is possible to divide a complete circle into any number of equal or unequal parts. This is a very precises method, where accuracies of the order of 1 second are possible. But at the same time this is very slow and time consuming method.

8.16. Caliper Principle *(For circular division and calibration of polygons)*

The main advantage of angular measurement and circular dividing is that no arbitrary standard of angle is required but all angles can be obtained by uniform division of the circle. There may be some angles which are themselves not an exact fractional division of a circle, but can be considered as a combination of angles which are uniform divisions.

The fact behind the division of circles by caliper principle is that complete circular division of a circle must be equal to 360°. This method could be best illustrated by the following example for marking a circular disc into 90° divisions. In this method, we require the use of two microscopes. First two trial lines 1 and 2 are marked at proximately 180° and the two microscopes A and B are set against these marks as shown in Fig. 8.55. The disc is then rotated such that the mark 2 is exactly below the microscope

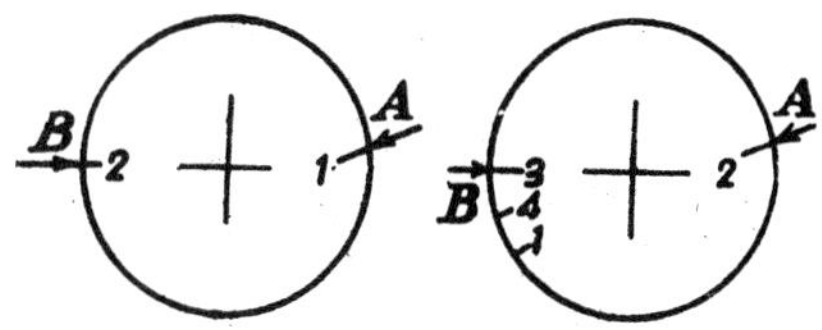

Fig. 8.55. Caliper principle.

A. The position below the microscope B is noted and let it be denoted by mark 3. If the division of the lines 1 and 2 was exactly 180°, the difference between 1 and 3 represents twice the error from 180° and this error can be measured. The space between lines 1 and 3 is divided by a line 4. Then lines 2 and 4 divide the circle into 180° divisions.

Further two trial lines are marked at approximately the 90° and 270° positions and also the microscopes are kept 90° apart approximately. By comparing the equality of the nominal 90° spacings and using the known errors of the 180° spacings, a full calibration of the lines is obtained. The four lines can thus be correctly engraved on the disc at 90° intervals by using the calibration of four lines and their known errors of setting.

8.17. Calibration of Polygons

Before using any precision polygon, it is usually calibrated by using two auto-collimators in order to know the errors. The two auto-collimators are set up by keeping angle between them equal to the nominal angle between adjacent faces of the polygon so as to give reflections from adjacent polygon faces as shown in Fig. 8.56 and are kept fixed in position. The polygon is mounted on a rotating table.

It is not necessary here that the auto-collimators be aimed at the centre of polygons as centring will have no effect with this device. Let the angle between the two polygon face AB and BC be P and the two auto-collimators be set at an angle R apart. If R_1 and R_2 be the readings on autocollimators T_1 and T_2.

Then $P + R_1 = R + R_2,$ or $P = R + (R_2 - R_1)$

If the complete polygon be considered in this way be rotating the table, and the readings R_1 and R_2 observed ; then adding the values of P, R, $(R_2 - R_1)$, we have $\Sigma P = \Sigma R + \Sigma(R_2 - R_1)$

But $\Sigma P = 360°$; $\therefore$ $360° = \Sigma R + W(R_2 - R_1)$.

If there be n number of sides of the polygon, then dividing throughout by n, we have

$$\frac{360°}{n} = \frac{\Sigma R}{n} + \frac{\Sigma(R_2 - R_1)}{n}$$

Here $360°/n$ is the nominal angle of the polygon, and as R, the angle between two auto-collimators is constant,

$$\frac{\Sigma R}{n} = R$$

$$\frac{360°}{n} = R + \frac{\Sigma(R_2 - R_1)}{n}$$

or
$$R = \frac{360°}{n} - \frac{\Sigma(R_2 - R_1)}{n}.$$

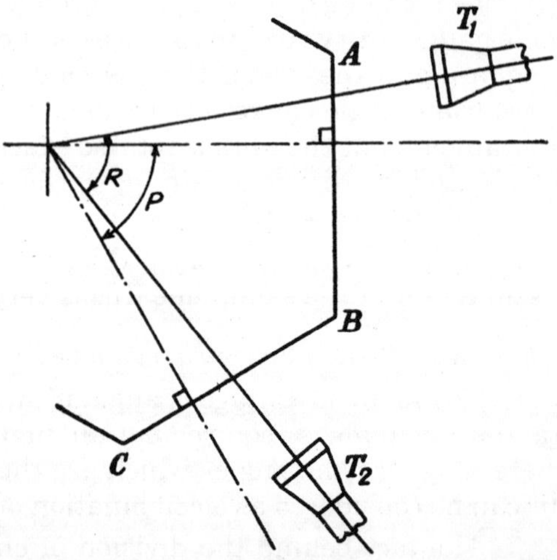

Fig. 8.56. Principel of calibration of polygons.

For a complete set of readings $\Sigma(R_2 - R_1)$ can be established and thus the value of R determined in above equation. Substituting back the value of R in the equation, $P = R + (R_2 - R_1)$ the value of P (angle between two polygon faces) can be determined and thus the whole of the polygon calibrated for angle between each set of faces. In this method, thus it is observed that calibration of polygon to a high degree of accuracy is possible without reference to any angular standard.

The same procedure could also be followed for the calibration of any graduated circular disc by using two microscopes set at angle R in place of auto-collimators, where R is nearly equal to the least spacing between the two marks on the disc.

Comparison Method

In this method two identical polygons and one rotatable table viewed by a single auto-collimator are used.

Two nominally identical polygons are mounted one above the other on a rotatable table. The auto-collimator is placed in such a position that it views the reflecting faces of both the polygon simultaneously.

All the exterior angles of the upper polygon are compared with the angle between any two faces of the lower polygon. By this method the accuracy of ± 3 seconds may be obtained. However, for a better accuracy, if each pair of faces of the upper polygon is compared with all the faces of the lower polygon and the angles are computed by the least square method, the accuracy of ± 1.0 second may be obtained for polygons upto 12 faces.

Calibration of Rotary Table

The calibration of a rotary table is performed by means of a precision polygon. The polygon is mounted concentrically on the table, preferably within 0.5 mm of the axis of rotation of the table. Although imperfect centering has a negligible effect on the calibration, it is advisable to spend some time in centering of the polygon in order to avoid the certain imperfections of the polygon form. The reflecting face of the polygon is read by a single, fixed auto-collimator. If the exterior angle of the polygon is θ, the table is set to read at θ, 2θ ... in turn as precisely as possible. The readings are then corrected with the known errors of the polygon. By this method the rotary table is calibrated at the intervals equivalent to the number of faces of the polygon only.

Calibration of Angle Gauges

The angle gauges are calibrated by using a rotary table and one autocollimator. The gauge is placed on the rotary table, which is carefully aligned. An autocollimator is placed radially to the rotary table, in a position so that the autocollimator views the working face of the angle gauge. The table is rotated in either direction so that the autocollimator is square to the face of the angle gauge. The reading of the rotary table is recorded. Now the rotary table is rotated so that the other face of the angle gauge is square to the autocollimator. This reading of the rotary table is also recorded. The difference of the two readings gives the included angle between two faces of the angle gauge. The accuracy of the measurement will primarily depend upon the accuracy and the resolution of the rotary table and the autocollimator. With precision table and autocollimator the accuracy of ± 1 second of arc may be obtained.

Comparison with Standard

The calibration of angle gauges may also be accomplished by direct comparison with standard angle gauges. In this method, a table and two autocollimators are used. The standard gauge is placed on the table, properly aligned. The autocollimators are placed radially to the table so that each collimator views one face each of the standard angle gauge. Readings of the reflected image of the cross wires of both the autocollimators are again recorded. The standard angle gauge is replaced by the gauge to be calibrated and the readings of the autocollimators are again recorded.

8.18. Calibration of Divided Circle

Calibration of divided circle, compared to linear line standard is simply and easy since it involves no arbitrary standard. It involves two micrometer microscopes fixed at some suitable angular relationship to each other. The divided circle is mounted on a rotary table and rotated equal to the nominal angular setting of the two microscopes and the difference between each successive position is noted as shown in Fig. 8.57.

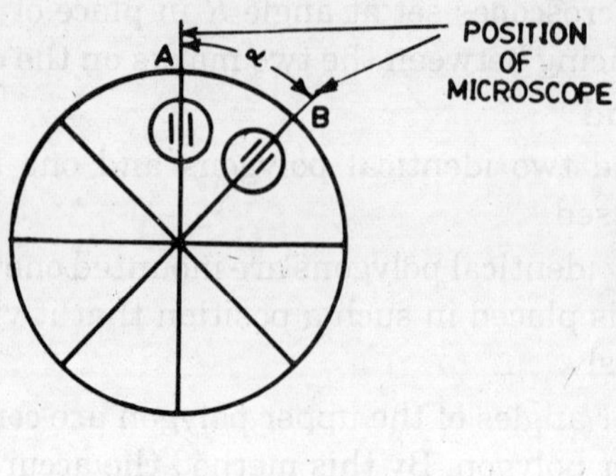

Fig. 8.57. Calibration of divided circle.

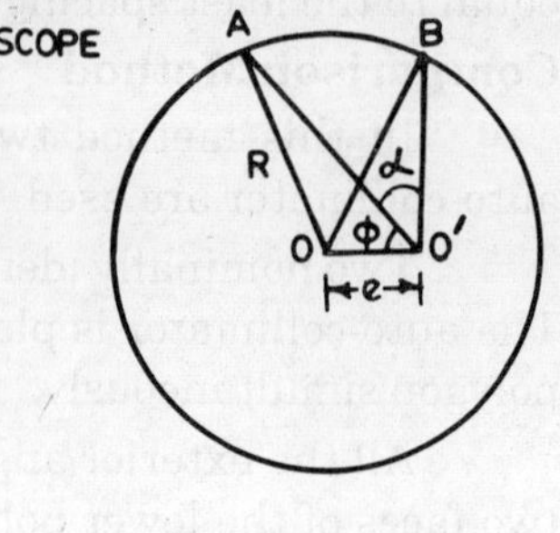

Fig. 8.58. Error due to eccentric mounting of a divided circle.

The calculations involved in this case are similar to those in the case of calibration of polygon.

The errors may occur due to positional error of engraved lines and due to eccentricity in the axis of rotation. The error due to eccentricity in the axis of rotation.

$$= \frac{e}{R} \{\sin (\phi + \alpha) - \sin \alpha\} \text{ radians}$$

where e = eccentricity

R = radius of divided circle

α = true rotation of circle

f = angle due to eccentric positioning.

This error can be eliminated by taking readings on two microscopes at 180° apart and obtaining the mean of the two readings.

8.19. Automatic Rotary Table Calibrator

It comprises of a serrated tooth indexer in 1/4° increments having 1440 teeth on 200 mm diameter. It has an accuracy of ± 0.1 second of arc at all 1440 indexed positions. The repeatability is of the order of 1/50th of a second. Such calibrator can be calibrated by self-proving (circle-closing) methods, utilising two calibrators. One is mounted concentrically on the table of the other. On the uppermost calibrator is placed a single mirror of optical quality, which is shielded from temperature variations within a protective housing. The mirror is used in conjunction with a monochromatic, photoelectric autocollimator sensitive enough to detect differences as small as 1/100th second and repeatable to about the same order of accuracy. The auto-collimator and mirror housing are close-coupled, with a minimum air gap to avoid variations which may arise from air currents passing between them. By rotating one of the

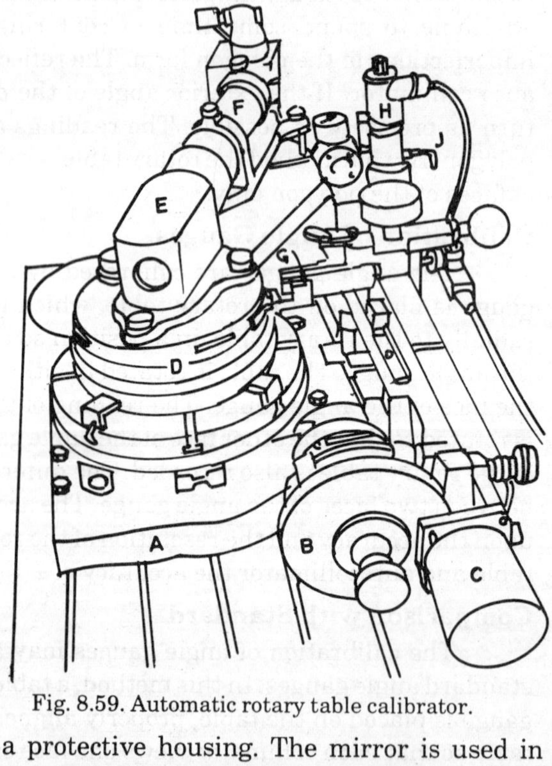

Fig. 8.59. Automatic rotary table calibrator.

calibrators clockwise and the other counterclockwise, each through the same nominal sector of arc, the mirror is returned essentially normal to the auto-collimator. The slight angular change in the attitude of the single mirror, as recorded by the auto-collimator, is representative of the difference between the particular angular segments of the calibrators. The rotation is continued till the circle is closed and readings noted for each position.

For calibration of the rotary table it is mounted on platform *A*. The automatic calibrator '*D*' is mounted on top of rotary table. (Refer Fig. 8.59). A single shielded reflecting mirror and autocollimator are used as described earlier for self calibration of calibrators. Mirror housing *E* is totally enclosed except for a slight gap to allow rotation between itself and the automatic position sensing autocollimator *F*. If the gap exceeds 3 mm, ordinary air currents would cause the collimator output to wander. A large gear *B* is mounted on a precise hub which concentrically locates it on the worm shaft. The rotary table is driven by a swing-away stepping motor *C* through gear *B* and worm shaft.

A small hydraulically actuated probe *G* keys the upper half of the calibrator to an approximate null while it is axially separated from the lower portion. The probe is retracted when the calibrator is lowered into engagement, allowing its teeth to mesh without outside influence. The hydraulic life thus disengages and engages the two halves of the calibrator during operation, and simultaneously feeds a plunger into by *V*-type indent in the upper member of the calibrator carrying the single mirror. Once the upper member is restrained, the lower member of the calibrator, being clamped to the rotary table, rotates integrally with the table as it is cranked to position. When the hydraulic lever is retracted, the upper member of the callibrator carrying the mirror floats down into mesh once again with its mate. The restraining plunger is also automatically retracted lest it influences the teeth as they engage. Errors of the rotary table are read directly on the autocollimator, requiring no "closing of the circle", nor use of calibration factors. Electric solenoid valve *H* supplies power to a hydraulic piston *J* mounted below, thus enabling the operation of the calibrator remotely from the programmed microswitch. The responses of the photoelectric autollimator are convered into an electric signal which is amplified and recorded on a linear graph. Calibration may be programmed for increments of 2°, 1°, 1/2° or 1/4°.

With the calibrator, the circle can be divided into only those parts which can be contained in the number 1440.

8.20. Taper Testing Machine

Application. For the determination of angle and form errors on outside and inside tapers and parts with angular surfaces and for form tests of cylindrical parts.

Design Features (Refer Fig. 8.60). On the rigid base which features hardened and precision lapped guide-ways the generously dimensioned measuring carriage is mounted on dust-proof high precision ball bearings. The wide sine table, provided with *T*-slots, supports an inclineable mounting table for the accommodation of the various fixtures required for the different taper checks. The sine table can easily be lifted by a convenient releasing lever which in turn facilitates the mounting of the parallel gauge blocks. The measuring head is height adjustable on the extra rigid column by means of a spindle. A swivel support arm is mounted on the sliding head and supports the lever type inductive gauging head to be used in conjunction with either the electronic comparator or the electric recorder. These instruments permit readings and evaluation of the measuring results with maximum accuracy.

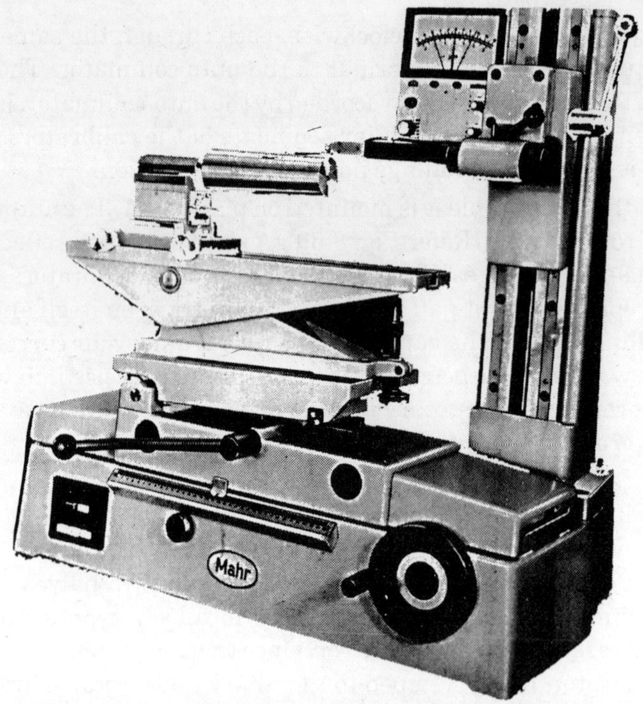

Fig. 8.60. Taper testing machine.

Operating Principle. The part to be checked (1) is mounted on the inclineable mounting table (2). The fixture is obliquely adjustable in reference to the axis of the part under test, thus permitting the perfect alignment of the part. At first at horizontal sine table (3) the upper crest line (4) of the taper is aligned parallel to the guide-ways of the measuring carriage (5) by inclining the mounting table. Parallelism of this crest line is checked by moving the measuring carriage back and forth (↔) while simultaneously gauging the crest line. If required, adjustments can be made by means of the adjustment screw (6) until the inductive gauging head (7) does not cause a deflection on the indicating instrument. The sine table is then swivelled around the entire taper angle by inserting a gauge block (8) which corresponds to the ideal angle α. At this point the lower taper crest line (9) is parallel to the guideways, provided the taper to be checked corresponds to the ideal value. This parallel alignment is checked by gauging the lower crest line. All taper angle and form errors are read on the indicator.

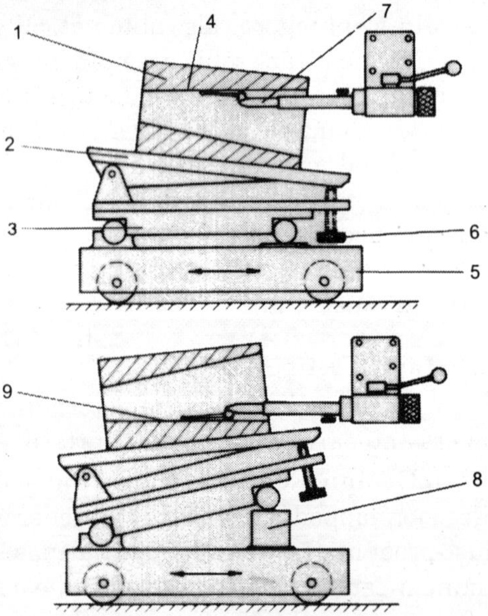

Fig. 8.61. Operating principle of taper testing machine.

By means of the said alignment procedure, all possible influences of mounting inaccuracies are eliminated. Therefore, the mounting surface of the part under test does not require precision machining nor does it require accurate centering in reference to the taper.

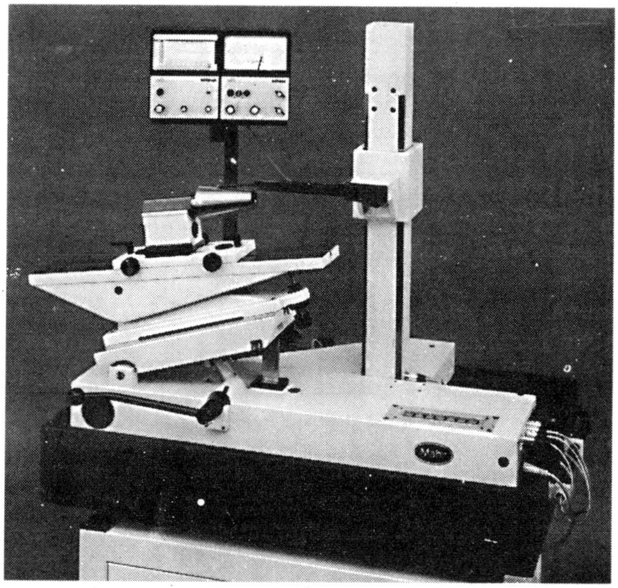

Fig. 8.62. Taper testing machine.

Features of taper testing machine (Fig. 8.62)

Max. taper angle	30°
Max. taper length (depending on taper angle)	275 mm
Range of taper gauges obtained with appropriate accessories	
for Morse Taper up to	MK6
for Steep Taper up to	size 60
for Metric Taper up to	size B 140/C 140
Guiding accuracy of measuring carriage	0,5 μm over 100 mm
	1 μm over 400 mm
Repeatability of measuring results	0,5 μm
Measuring inconsistency	= 1 μm
Hard granite base plate	1000 × 630 × 150 mm
Measuring table — mounting surface	110 × 650 mm
distance of sine rollers	300 mm
Travel of measuring carriage	400 mm
Height adjustment of measuring head	375 mm
Speeds of measuring carriage	20/5/2 mm/s
Speeds of measuring head	10/5/1 mm/s
Power supply	220 V, 50-60 Hz, 50 W
Air supply	5 bar, 1 m^3/h

Taper Testing Machine

For reference-free determination of angle-and form errors on outside-and inside tapers and parts with angular surfaces and for form tests on cylindrical parts.

The Taper Testing Machine consists principally of the following units :

Measuring Carriage

With wide guide surfaces and air cushion bearing for exact movement on the base plate and along the high grade guide edge of the measuring table. With motor-driven, vertically adjustable measuring head on sturdy column. Air filter and air regulating valve provide air for the air cushion bearings.

Hard Granite Base Plate. For accurate guiding of the measuring carriage upon an air cushion and for mounting the measuring table. Made of fine grain black granite (Diabase) of great hardness and stability.

Measuring Table. Consisting of lever operated sine table and inclinable mounting table with *T*-slot for fixing mounting devices for components to be tested. Self-locking safety device prevents damage of setting gauge block and mounting surface of gauge block at inadvertent slipping of sine table. With integrated switch board for controlling all movements of measuring carriage and measuring head as well as for the selection of testing speeds by push buttons.

Electronic Measuring and Recording Equipment. Consisting of inductive gauging system for contacting both crest lines without turning the stylus, electronic indicating instrument for reading the measuring results and also electrical recorder for evaluation of taper angle errors and taper form errors.

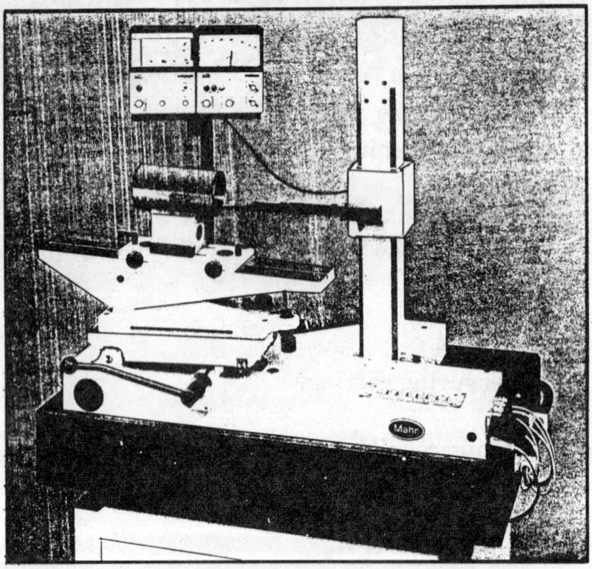

Fig. 8.63

Advantages

High measuring accuracy and repeatability of measuring results by friction-free movement of measuring carriage in wide air cushion bearing. Even heavy parts to be checked do not influence the accuracy of the measuring carriage.

Reference-free determination of angle- and form errors by eliminating all inaccuracies of mounting.

Long term accuracy by extremely stable base plate made of hard granite. No wear of guiding system owing to air cushion bearing.

Universal application. Many accessories allow testing of outside and inside tapers with various forms and sizes.

Indication of measuring results with selectable magnifications and with the possibility of recording diagrams of taper angle and taper form which are easy to evaluate.

Easy setting by indicating instrument with adjustable electronic limit switches. The setting movement of the motor-driven measuring head is automatically switched-off when electronic gauging system contacts the taper to be tested and is within the setting range.

Safety device for sine table. The lifting lever for the sine table is self-locking, this preventing inadvertent slipping of the sine table and excluding damage of setting gauge block and mounting surface for the gauage blocks. The inductive gauging system is equipped with a safety device to protect it from overload.

Conveniently placed switch board with push buttons for the selection of the various movements both of measuring carriage and measuring head and for the testing speed most suitable for the particular test.

WORKED EXAMPLES

Problem 8.1. *For precise measurement of length it is essential to have a primary standard of length such as the wavelength of a certain radiation or the international prototype metre, but no primary standard is essential for precise measurement of angle. Explain why ?*

Solution. Any angle can be traced or generated by accurately dividing a circle and thus the accuracy of the angular measurement depends on the ability to carry out the sub-division of a circle and not on the possession of a primary standard.

Problem 8.2. *What type of angle standards are available. Describe them in brief.*

Solution. Artefact angle standards are available either in the form of end standards with the angle defined between their adjacent faces ((*i*) polygons and (*ii*) Angle gauge blocks), or in the form of line standards with the lines engraved at regular intervals (Rotary tables and circular scales),.

Polygons are blocks of regular polygonal prism form having reflecting peripheral faces. The earliest polygon model was made out of steel with hardened, grounded and carefully lapped flat faces. Later, commercial models were also made out of glass with faces lapped optically flat and coated with aluminium for better reflectivity, and a protective layer. The faces are usually six (60°), eight (45°), nine (40°), twelve (30°) in number. The maximum number of flat faces that could be made are 72(5°). This is considered to be the maximum on a 300 mm diameter polygon.

An angle gauge block is a hardened and carefully lapped steel block of wedge shape. There is a specific included angle between two working faces. These are available in a number of sets. A set of 12 angle gauge blocks together with a square block is available in the following denomination.

1, 3, 9, 27, 41 degrees
1, 3, 9, 27 minutes
3, 9, 27 seconds.

With this set and a square block it is possible to generate any desired angle within 3 seconds by suitably wringing the gauges additively or subtractively.

Several types of rotary tables which can measure and generate angles by dividing circles are :

1. Optical rotary table with circular scaled graduations.
2. Mechanical cam compensated rotary table.
3. Accurate worm and gear rotary table.
4. Precision index table.

Optical, Mechanical and Worm and Gear Tables are fitted with circular scales (or Moire gratings) which are circular discs with graduations engraved on the farthest end of the circle (at the periphery). The size of the scale depends on the number of lines to be graduated for better resolution of the readings. Such a scale is subject to error if mounted eccentrically. The error can be eliminated by using two reading heads mounted 180° apart.

Problem 8.3. *Explain the testing of internal taper on the taper testing machine and describe what advantages the machine offers ?*

Solution. The workpiece with inside taper in example of Figs. 8.69 and 8.70 is mounted upon the swivelling table of the machine by the aid of mounting accessories. These are adjustable in a plane lateral to the axis of the taper, permitting accurate axial alignment of

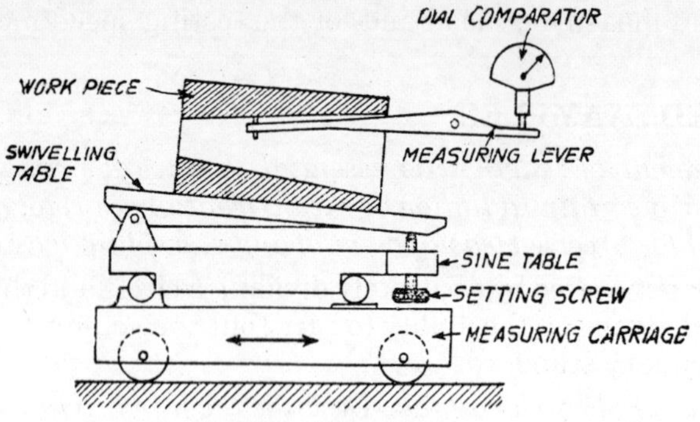

Fig. 8.69. Setting.

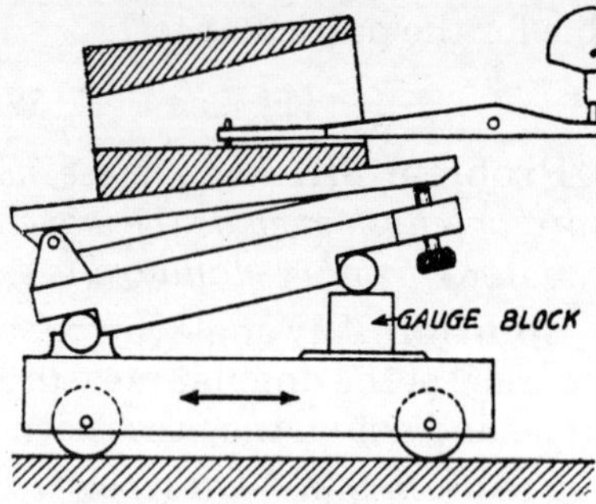

Fig. 8.70. Measuring.

the workpiece. At first, at horizontal sine table the upper crest line is adjusted parallel to the guide ways of the measuring carriage by inclining the swivelling table. Accurate parallel position of this upper crest line can be determined when moving the measuring carriage in longitudinal direction (↔) while contacting the taper with the measuring lever and by adjusting, if necessary, by means of the setting screws on the swivelling table.

After having aligned the workpiece the sine table is swivelled by the total taper angle θ using a corresponding gauge block. Now the lower crest line lies parallel with the guide way of the measuring carriage, provided that the taper under test is accurate. This parallel position is inspected by testing along the lower crest line of the taper with the measuring lever (↔). All errors of taper angle and taper form are indicated on the dial comparator. Perfect tapers indicate no deviation on the dial comparator.

Aligning of the taper under test (Fig 8.69) excludes completely all influences caused by inaccuracies of mounting. Therefore, the mounting surface of the workpiece does not have to be manufactured accurately or to be positioned centrically to the taper. Optimal measuring accuracy is thus obtained.

The various advantages offered by the taper testing machine are :

(*i*) Inspection of outside and inside tapers can be performed completely independent of the mounting accuracy of the workpiece. It is, therefore, possible to mount, for example, a workpiece with inside taper on its outside diameter, even if this is not accurately manufactured.

(*ii*) The rigid base supports accurately lapped and hardened guide ways for the hermetically sealed high precision ball bearings of the well dimensioned measuring carriage. The swivelling table is mounted upon the wide sine table and serves for mounting the various mounting accessories. The sine table can be lifted off by a well positioned lifting lever, thus facilitating the insertion of the gauge blocks. The measuring head with measuring lever is adjustable in height over 275 mm by a spindle on the rigid column of the machine.

(*iii*) Fast and accurate readings of the measuring result are obtained on the dial comparator of the measuring head (readings 1 μm or 0.5 μm). The interchangeable measuring lever is equipped with a ruby tipped measuring point. For obtaining an optimum of repeatability, the measuring lever is mounted in play-free, and friction-free high precision ball bearings. For recording angular and form deviations, an Inductive Recording Device can be used instead of the dial comparator.

(*iv*) For uniform motion of the measuring carriage, motor drive is employed with six measuring speeds to be selected. The measuring movement is introduced by well positioned control switch on the front of the machine base.

Problem 8.4. *Explain why it is preferred not to use a sine bar for generating angles larger than 45° if high accuracy is demanded.*

Solution. The slip gauge combination which is required to generate any particular angle θ, is given by

$$\sin \theta = \frac{h}{l}$$

in which, h = required slip gauge combination

l = spacing of the rollers.

The relationship between the angular setting accuracy ($d\theta$) and any error which may be present in the slip gauge combination (dh) or the spacing of the rollers (dl) can be determined by differentiating the equation $\sin \theta = h/l$.

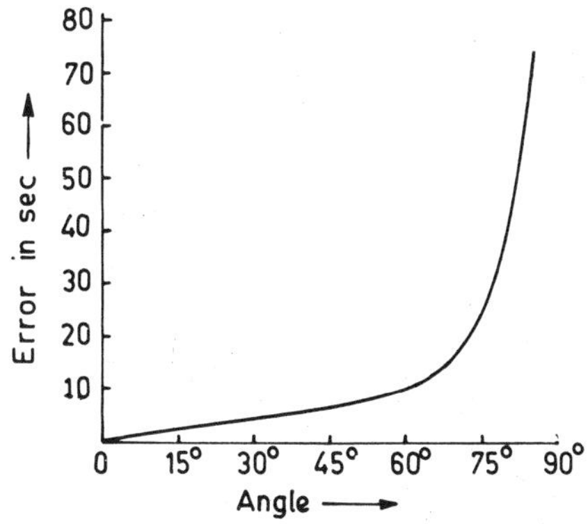

Fig. 8.71. Error with sine bar *vs.* angle to be measured.

Thus, $$\cos \theta \, d\theta = \frac{l\,dh - h\,dl}{l^2} = \frac{dh}{l} - \frac{h\,dl}{l^2}$$

$$= \frac{dh}{l} \sin \theta - \frac{dl}{l} \sin \theta$$

or $$d\theta = \tan \theta \left(\frac{dh}{h} - \frac{dl}{l} \right)$$

Obviously the effect and error in spacing of rollers or slip gauge is a function of the tangent of the angle θ.

Fig. 8.71 show the effect of an error of 0.00025 mm in the spacing of the rollers on a nominally 100 mm sine bar and error of + 0.0005 mm on the accumulated slip gauge blocks on the accuracy of angular setting. Below 45°, the effect is small. After that it increases rapidly.

Problem 8.5. *In the use of sine-bars, for getting accurate results it is important that the following may be checked. Diameter and uniformity of rollers, parallelism of rollers, roller centre distance, and parallelism of gauging face with rollers.*

In setting a sine bar of 125 mm length to an angle of 30° what errors will be introduced if :

(*i*) *The assumed 125 mm roller separation is actually 125 – 0.005 mm ?*

(*ii*) *The upper cylinder of the sine bar is 0.002 mm bigger than the actual size ?*

(*iii*) *Gauging face of the bar is out of parallel from rollers by + 0.002 mm.*

(*iv*) *The slip gauges used have an unsuspected error of 0.005 mm.*

(*v*) *All errors are present together.*

Solution. (*i*) In Fig 8.72, *AB* represents the position when centre distance is exactly 125 mm. When the centre distance is more than the angle to which bar is set then the angle set will be less than 30° as the bigger length *AC* has now to adjust itself on the slip gauges of length 125 sin 30°.

It means in Fig. 8.72, $DC = 0.005$ mm.

It may be noted that in the error triangle BCD, BD is an arc but may be considered as straight line perpendicular to AC, as the sizes of deviation are too small.

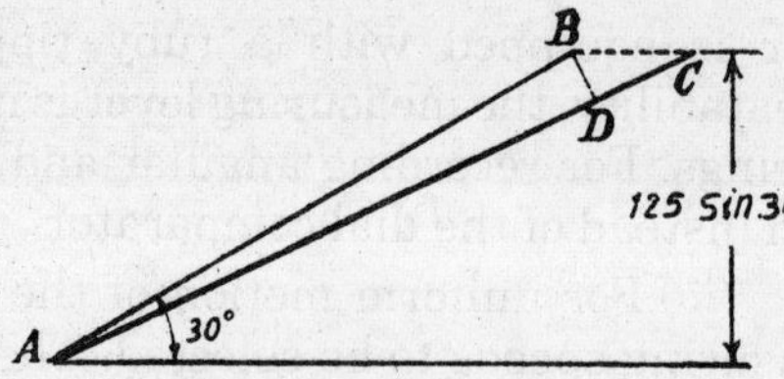

Fig. 8.72

Also $\angle BCD = 30°$

Now $BD = DC \tan \theta°$

$= 0.005 \times \tan 30° = 0.002775$ mm.

$\therefore$ Error angle $\angle BAD = \dfrac{0.002775}{125} = 0.000023$ radian $= 4.7$ seconds ≈ 5 seconds.

Thus, angle set $= 29° \, 59' \, 55''$

(*ii*) It may be noted that due to the upper roller being of bigger size, the angle set will be more than 30°. (Refer Fig. 8.73). The increase in angle is due to the increased dimension AB, and the component of CD along OB, *i.e.*, AD.

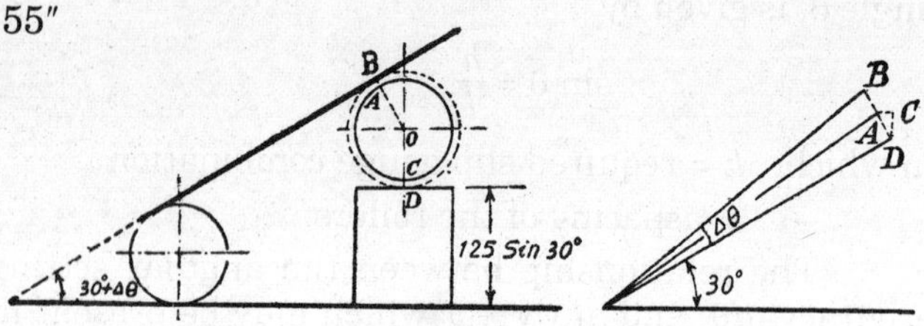

Fig. 8.73

$$\therefore \text{ Error angle } \quad \Delta\theta = \frac{AB + CD \sec 30°}{125} = \frac{0.001 + 0.0017 \times 1/0.866}{125}$$

$$= \frac{0.001 + 0.00197}{125} \text{ radian} = \frac{0.00297}{125} \times \frac{10^6}{4.85} \text{ sec}$$

$$= 4.45 \text{ sec} \approx 4 \text{ sec.}$$

Thus angle set $= 33° \, 0' \, 4''$.

(*iii*) Let us assume that the gauging face on upper side is out of parallelism by + 0.002 mm.

Again in this case angle set will be more by the amount $\dfrac{0.002}{125}$ rad.

$$= \frac{0.002}{125} \times \frac{10^6}{4.85} \text{ sec} = 3.3 \text{ sec} \approx 3 \text{ sec.}$$

Thus angle set $= 30° \, 0' \, 3''$.

(*iv*) In error triangle BCD, (Fig. 8.74)

BC = error in slip gauge = 0.005 mm

and $BD = BC \sec 30°$

$$= 0.005 \times \frac{1}{0.866} = 0.00575$$

and error angle $DAB = \dfrac{0.00575}{125} = 0.000046$ rad

$= 9.5'' \approx 9$ sec.

$\therefore$ Angle set $= 29° \, 59' \, 51''$.

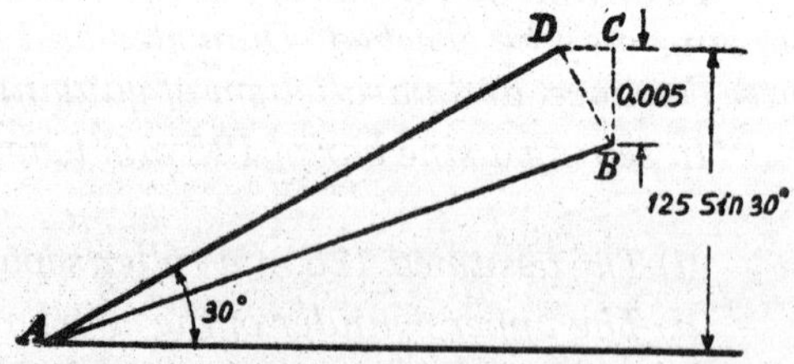

Fig. 8.74

(*v*) When all the above errors are present, error will be algebraic sum of above four cases, and angle set $= 29° \, 59' \, 53''$.

Problem 8.6. *In connection with spirit levels, describe the following :*

(*i*) *Method of reading changes in slope of level ;*

(ii) Characteristic elements of a level ;

(iii) Sensitivity of level ; *(iv) Constant of level ;*

(v) Relationship between sensitivity and constant of level ;

(vi) Accuracy of level ; *(vii) Precautions in its use.*

Solution. (*i*) Generally two types of arrangements are made to read changes in slope of level. In the case of levels fitted with a micrometer screw, the changes in slope are read on the division of the micrometer. In the case of levels fitted with graduation lines on the tube, the reading is taken directly by seeing the bubble and the graduation lines of the tube.

(*ii*) The characteristics element of a level are the radius of curvature of its tube and distance between the two consecutive lines on its scale.

(*iii*) Sensitivity is defined as the displacement of a bubble (or the micrometric device) for a tilt of 1 mm in 1 m or for 200 seconds of arc, and thus the sensitivity is equal to one thousandth of its radius of curvature expressed in metre.

(*iv*) Constant of a spirit level is also known as apparent sensitivity, and is the change in tilt, expressed in mm per m (or in seconds of arc), which produces a displacement of the bubble by one division.

(*v*) Sensitivity × Constant of level = Length in mm of one division of the scale.

(*vi*) Literally measure of accuracy can be judged from the statement ; the bubble should move through at least one division for change of angle not greater than 0.05 mm per m.

Further for accuracy purposes, the flatness of the base should also be specified.

(*vii*) *Precautions :* (*a*) Tests should be carried out in as short a time as possible and measurements should be repeated in the return direction in order to take account of temperature variations that may occur during test.

(*b*) The calibration should be carried out at regular intervals as the glass tube is likely to deform with time.

Problem 8.7. *What are the methods for calibration of angle standards ?*

Solution. The angle standards can be calibrated by :

(*i*) fundamental subdivision method

(*ii*) trigonometric generating methods

(*iii*) direct comparison with reference angle standards.

These can be further classified depending upon the technique used, instrument used, and the accuracy requirement.

Problem 8.8. *Explain the mechanical indexing equipment and optical dividing heads used in circular division.*

Solution. Mechanical indexing equipment of mechanical dividing heads is most commonly used on the milling machines for cutting teeth grooves. It consists of a single-start worm mating with a forty-tooth worm wheel and is used in conjunction with a series of index plates in the form of circular discs. It suffers from the disadvantage that the wear of the worm-wheel may lead to inaccuracies.

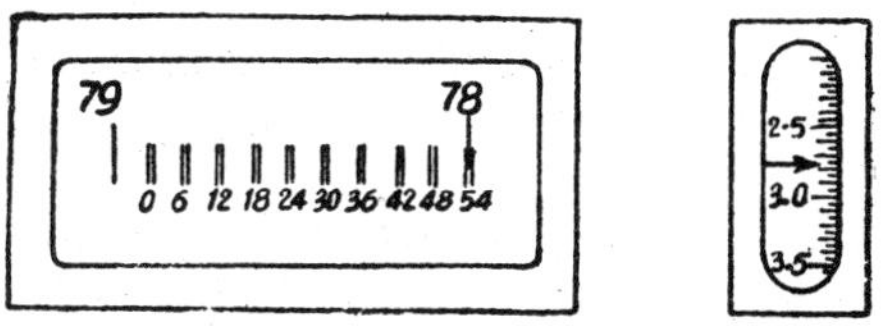

Screen reading = 78° 54′

Micrometer head reading = 2.75′ = 2′ 45″

∴ Total reading = 78° 56′ 45″

Fig. 8.75

Optical dividing heads incorporate a graduated scale engraved on glass circles (or metal in some instances). The scale is graduated in degrees and fraction thereof, and fitted concentrically to the axis of the machine spindle. The graduations are viewed by means of an optical system giving a suitable magnification. The graduated lines in glass circles are produced either by diamond ruling or by etching, or they may be reproduced photographically from a master. These reproduced photographically from a master. These are used mainly for setting and checking inclined work.

In the modern optical dividing heads, the readings are projected on a ground armour plate glass screen, thus dispensing with the conventional but less convenient focussing eye-piece. In this direct vision fixed focus principle, errors due to parallax are eliminated. The optical system comprises a system of lenses through which light passes and illuminates the divided scale (mounted on the main spindle) and projects the readings on a ground glass screen. On ground glass screen 10 division lines comprise two closely spaced lines as shown in Fig. 8.75. The scale can be moved linearly by a micrometer head having 120 divisions. One rotation of the micrometer head moves the scale by one division of the scale and thus 1 division of the micrometer head corresponds to 3 seconds. For noting down the reading the micrometer head is rotated till the degree mark comes exactly between the two lines on the ground scale (Refer Fig. 8.75).

Problem 8.9. (*a*) *What are the essential characteristics of a mirror for autocollimator ?*

(*b*) *What are the advantages of photoelectric autocollimator ?*

(*c*) *How an autocollimator is calibrated ?*

Solution. (*a*) In an autocollimator, the size and flatness of the reflecting mirror, as also its distance from objective play a very important role.

If the mirror is very small then rays will be sent back through different portions of the objective depending upon the location of centre of mirror with respect to the optical axis. This will cause the readings to vary, unless the objective is free of all aberrations, which is most unlikely.

Mirror should be flat to 1/4 fringe or better otherwise it will defocus the return image and present an undefined reference plane.

If light between the objective of autocollimator and the mirror is perfectly collimated, *i.e.* parallel rays are there ; then the distance between autocollimator and the mirror is immaterial. But in practice, perfect collimation is not possible due to so many limitations and in that case even a slight tilt of mirror will restrict this distance. When longer distances are required, a close attention must also be paid to adequate shielding of the air path from drafts, and gradients.

(*b*) The photoelectric autocollimator has the following advantages.

(*i*) These replace the judgement of the human eye with appropriate photoelectric systems.

(*ii*) Setting accuracy is increased (null setting sensitivity is better than 0.1 sec) and is constant for all operators.

(*iii*) Remote readings (digital or analog) are possible.

(*c*) The autocollimator can be calibrated by monitoring its output as the reference mirror is tilted through accuracy controlled angles. Mirror can be tilted accurately to 1 sec of arc by special sine bar fixture. It has a precisely defined axis at one end, and at the other end there is a well defined cylinder which can be raised or lowered with a special micrometer or with slip gauges.

Problem 8.10. *The centre of rotation of a glass circle 200 mm diameter lies 0.002 mm off the true centre of the circle in the direction of the 120° graduation. What will be the true angles turned through when the circle is rotated from a setting of 45° to setting of 180° and 250° respectively.*

For calculation purpose, take 1′ = 0.0003 rad. and 1″ = 0.000005 rad.

Solution. Whenever a divided circular scale in a circular dividing instrument is mounted eccentrically, sinusoidal division errors are introduced (irrespective of the random circle errors) which can be reduced by improving centring.

As at 120° position the centre of rotation and centre of circle lie in one line and $OC = 0.002$ mm, the rotation from 45° to 180° means that the actual angle turned is P_1CP_3, whereas the theoretically angle turned should have been $P_4\hat{O}P_3$ = 135°

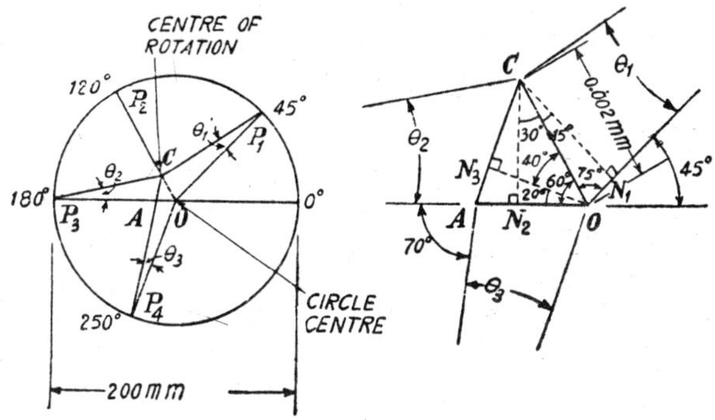

Fig. 8.76

Obviously, $$P_1\hat{C}P_3 = P_1\hat{O}P_3 + \theta_1 + \theta_2 \; ; \theta_1 \approx \frac{CN_1}{P_1N_1} = \frac{CN_1}{P_1C} \approx \frac{CN_1}{P_1O}$$

[∵ CN_1 is very small, and P_1N_1, and P_1C are relatively large ; and P_1N_1, P_1C and P_1O are nearly equal, therefore, this assumption will not introduce much errors]

$$CN_1 = OC \cos O\hat{C}N_1 = 0.02 \times \cos 15° = 0.00194 \text{ mm}$$

$$\therefore \quad \theta_2 = 0.00194/100 = 0.000019 \text{ rad.}$$

In the same way, values of θ_2 and θ_3 are found out.

$$CN_2 = OC \cos O\hat{C}N_2 = 0.002 \cos 30° = 0.00174 \text{ mm}$$

and $$\theta_2 = \frac{0.00174}{100} = 0.000017 \text{ rad.}$$

and $$ON_3 = OC \cos C\hat{O}N_3 = 0.002 \cos 40° = 0.00154 \text{ mm.}$$

and $$\theta_3 = 0.00154/100 = 0.000015 \text{ rad.}$$

Now actual angle between P_1 and P_3

$$= 135° + \theta_1 + \theta_2 = 135° + 0.000036 \text{ rad.} = 135° \; 0' \; 7.2''$$

Similarly actual angle between P_1 and P_2

$$= 250° + (\theta_1 + \theta_3) = 205° + 0.000034 \text{ rad.} = 205° \; 0' \; 6.8''$$

If rotation was made via 0°, then angle would be 225° minus 7.2″ and 155° minus 6.8″ respectively. **Ans.**

Problem 8.11. *It is desired to set a glass circle of 100 mm diameter exactly in the centre of an optical dividing head. For centering tests, a polygon and an auto-collimator are employed. Readings of the auto collimator at the four cardinal points shown the following errors, relative to the zero reading.*

0° set zero ; 40° + 45″ ; 180° + 35″, 270°—15″

With the help of above readings, locate the actual centre of rotation relative to the true centre of the circle.

Solution. The magnitude and the direction of the eccentricity error, *i.e.* the distance and location of line joining the circle centre and the centre of rotation can be found out by determining the resultant of the components of the eccentricity errors along 0°—180° diameter and along 90°—270° diameter.

In this case, setting at 0°, and moving via 90° to 180° setting, the angle turned through = 180° + 35″ ; and therefore the position of eccentricity error along 90°—180° diameter will be as shown in Fig. 8.77 (*a*). Similarly, setting at 90°, and moving via 180° to 270°, the angle turned through = 180°—60″ and position is as shown in Fig. 8.77 (*b*).

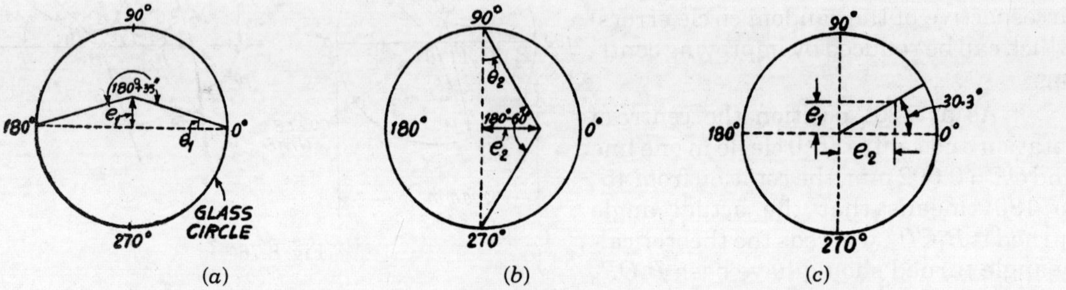

Fig. 8.77

Now obviously e_1 along 90° – 270° diameter = $r\theta_1$ = 17.5 × 4.85 × 10^{-6} rad

where r = radius of glass circle = 50 mm and θ_1 = angle shown in Fig. 8.77 (*a*).

$$\therefore \quad e_1 = 50 \times 17.5 \times 4.85 \times 10^{-6} \text{ mm} = 0.00425 \text{ mm}$$

Similarly e_2 along 0°—180° diameter = $r\theta_2$ = 50 × 60″/2 × 4.85 × 10^{-6} mm = 0.00728 mm

Actual centre is eccentric by the resultant of these and is equal to

$$= \sqrt{0.000425^2 + .00728^2} = 0.001\sqrt{(18.1 + 53)} = 0.001 \times \sqrt{71.1} = 0.0085 \text{ mm. } \textbf{Ans.}$$

It is along radius through angle of $\tan^{-1} \dfrac{0.00425}{0.00728} = 30.3°$. **Ans.**

Problem 8.12. *A six-sided precision polygon is calibrated by the calliper method using two auto-collimators R and S, set on adjacent faces of the polygon. R is to the left of S and the readings of R with respect to S always increase as the polygon rotates clockwise. The readings observed are given below :*

Polygon faces (degrees)		*Readings of autocollimator (seconds)*	
R	*S*	*R*	*S*
0	*60*	*26.1*	*18.3*
60	*120*	*24.6*	*17.0*
120	*180*	*22.7*	*16.4*
180	*240*	*23.9*	*18.1*
240	*300*	*24.5*	*19.2*
300	*0*	*25.8*	*20.9*

Complete the table to find the cumulative calibration errors and the true errors of the polygon.

Make calculations to nearest 0.1 sec.

Solution. Since the readings of *R* relative to *S* always increase, it shows an increase in angle interval between the faces. The difference of two readings is therefore found out by subtracting the reading of *S* from *R*. Further the mean of all differences must represent true 60°, as all angles add up to 360°.

Note how the zero for face *R* is inserted and the readings obtained in the various columns in the table given next page.

Table for Problem 8.12

Polygon faces (degrees) R	*Polygon faces (degrees)* S	*Auto-collimator Readings (seconds)* R	*Auto-collimator Readings (seconds)* S	*R—S*	*Differences of R—S from the mean of col. (3)*	*Cumulative*	*Correction*	*True Error*	*Face (degrees)*
(1)		(2)		(3)	(4)	(5)	(6)	(7)	(8)
—	—	—	—	—	—	0	0	0	0
0	60	26.1	18.3	7.4	+ 1.3	+ 1.3	+ 0.1	+ 1.4	60
60	120	24.6	17.0	6.6	+ 0.5	+ 1.8	+ 0.1	+ 1.9	120
120	180	22.7	16.4	6.3	+ 0.2	+ 2.0	+ 0.2	+ 2.2	180
180	240	23.9	18.1	5.8	– 0.3	+ 1.7	+ 0.2	+ 1.9	240
240	300	24.5	19.2	5.3	– 0.8	+ 0.9	+ 0.3	+ 1.2	300
300	0	25.8	20.9	4.9	– 1.2	– 1.2	+ 0.3	0	0
				36.3 Mean = 36.3/6 ≃ 6.1					

Problem 8.13. (*a*) *How does optical micrometer work ?*

(*b*) *Determine the displacement of a ray due to rotation of the glass block of an optical micrometer by 5 minutes, if the width of glass is 5 mm and its refractive index is 1.5. Assume 1 minute = 0.0003 radiation.*

Solution. (*a*) An optical micrometer consists mainly of a glass block capable of being rotated about an axis perpendicular to a ray of light passing through it, and a means to measure the angle of rotation, and a microscope to read the deflection (displacement) of light due to tilting of glass block. In Fig. 8.78, when ray *AB* falls perpendicular to face *XY*, it moves out as such along straight line *BCD*. When the glass block is rotated by angle *i* (shown in dotted), then the ray *AB* incident at angle *i* instead of moving along straight line *BCD* will be refracted along *BG* such that angle *GBF* = *r* (line *EBF* is ⊥ to *X'Y'*), where *r* is the angle of refraction such that refractive index $\mu = \dfrac{\sin i}{\sin r}$.

Displacement (*h*) of beam is corresponding to distance between *CD* and *GH*, *i.e.*, *JG*.

So $$h = JG = BG \sin (i - r) = \frac{BF}{\cos r} \sin (i - r)$$

For very small angle *r*, $\cos r \approx 1$

So that $$h = BF \sin (i - r)$$

Since $$\sin r = \frac{\sin i}{\mu}$$

and for very small angles of *i* and *r*

$$\sin r = r \text{ and } \sin i = i$$

and, therefore, $$r = \frac{i}{\mu}$$

and $$h = BF\left(i - \frac{i}{\mu}\right)$$

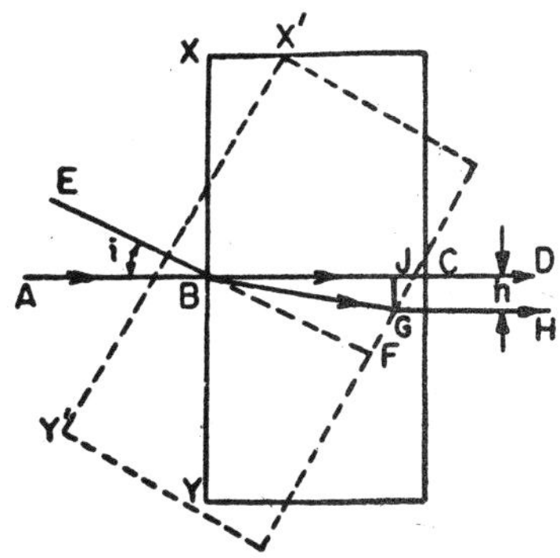

Fig. 8.78

Thus by measuring deflection (displacement) of image, it is possible to determine the angle of tilt.

(*b*) In this problem, width of glass BC = 5 mm, μ = 1.5

and $\qquad i = 5 \text{ min} \times 0.0003 = 0.0015$ radian

$$\therefore \qquad h = 5\left(0.0015 - \frac{0.0015}{1.5}\right) = 5\,(0.0015 - 0.0010) = 5\,(0.0005) = 0.0025 \text{ mm.}$$

Problem 8.14. *It is required to calibrate a twelve-sided reference polygon by mounting it on a rotary table and viewing adjacent faces with two autocollimators T_1 and T_2 positioned very nearly 30° apart as shown in Fig. 8.79. The following readings of the angular displacements R_1 and R_2 of the faces from the nominal positions were noted for each of the twelve faces.*

Faces →	*AB*	*BC*	*CD*	*DE*	*EF*	*FG*
R_1 *(sec)*	*+ 3.9*	*– 15.2*	*– 10.7*	*– 9.2*	*– 19.3*	*– 8.6*
Faces →	*BC*	*CD*	*DE*	*EE*	*FG*	*GH*
R_2 *(sec)*	*+ 6.5*	*+ 0.2*	*+ 0.4*	*+ 0.7*	*+ 9.2*	*0.0*
Faces →	*GH*	*HI*	*IJ*	*JK*	*KL*	*LM*
R_1 *(sec)*	*– 10.5*	*– 12.2*	*– 14.0*	*+ 0.2*	*– 9.1*	*– 7.1*
Faces →	*HI*	*IJ*	*JK*	*KL*	*LM*	*AB*
B_1 *(sec)*	*0.0*	*– 2.9*	*0.0*	*– 0.1*	*0.0*	*0.0*

Determine the error in the angle between each pair of adjacent faces.

Solution. Let the angle between the two polygon faces AB and BC be P and the two autocollimators T_1 and T_2 be set at an angle R apart.

$$\therefore \qquad P + R_1 = R + R_2$$

or $$P = R + (R_2 - R_1)$$

or $$\Sigma P = \Sigma R + \Sigma (R_2 - R_1)$$

or $$\frac{\Sigma P}{12} = \frac{\Sigma R}{12} + \frac{\Sigma (R_2 - R_1)}{12}$$

Since $\qquad \Sigma P = 360°$

and $$\frac{\Sigma R}{12} = R$$

$$\therefore \qquad \frac{360}{12} = R + \Sigma\,\frac{(R_2 - R_1)}{12}$$

or $$R = 30° - \frac{\Sigma (R_2 - R_1)}{12}$$

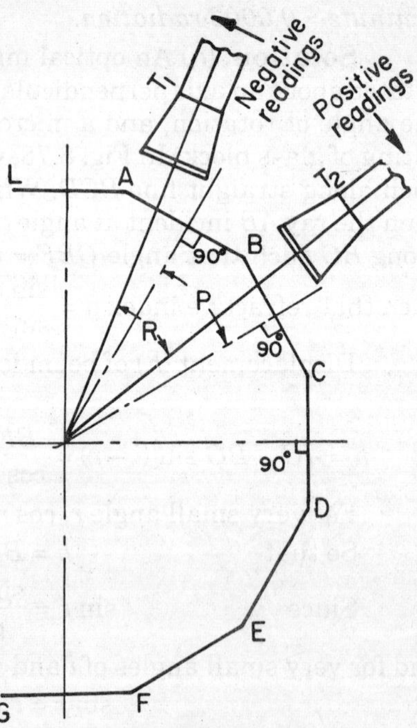

Fig. 8.79

Tabulating the above values to find ΣR_1 and ΣR_2.

Face	R_1 *(sec)*	*Face*	R_2 *(sec)*	*Errors in angle between two faces* $= -10.4'' + R_2 - R_1$	
AB	+ 3.9	*BC*	+ 6.5	– 7.8″	
BC	– 15.2	*CD*	+ 0.2		+ 5.0″
CD	– 10.7	*DE*	+ 0.4		+ 0.7″
DE	– 9.2	*EF*	– 0.7	– 1.9″	
EF	– 19.3	*FG*	+ 9.2		+ 18.1″
FG	– 8.6	*GH*	0.0	– 1.8″	
GH	– 10.5	*HI*	0.0		+ 0.01″
HI	– 12.2	*IJ*	– 2.9	– 1.1″	
IJ	– 14.0	*JK*	0.0		+ 3.6″
JK	+ 0.2	*KL*	– 0.1	– 10.7″	
KL	– 9.1	*LM*	0.0	–1.3″	
LM	– 7.1	*AB*	0.0	– 3.3″	
	$\Sigma R_1 = -111.8$		$\Sigma R_2 = +12.6$	– 27.9″	+ 27.5″

$$\therefore \text{ Angle } \quad R = 30° - \frac{\Sigma(R_2 - R_1)}{12} = 30° - \left(\frac{12.6 - (-111.8)}{12}\right) = 30° - \frac{124.4''}{12} = 30° - 10.4''$$

Angle $\quad P = R + R_2 - R_1$

or angle between faces *AB* and *BC* = 30° – 10.4″ + 6.5″ – 3.9″ = 30° – 7.8″

Similarly angle between faces *BC* and *CD* will be = 30° – 10.4″ + $R_2 - R_1$

= 30° – 10.4″ + 0.2″ – (– 15.2″) = 30° – 10.4″ + 15.4″ = 30° + 5.0″

Similarly, the values of errors between the other faces can be determined and are tabulated.

It will be observed that maximum and minimum errors are + 18.1″ and – 10.7″.

Problem 8.15. *An eight-sided precision polygon is to be calibrated by caliper method of calibration using two auto-collimators T_1 and T_2, set on adjacent faces of the polygon. T_1 is to the left of T_2 and readings on both autocollimators increase as the polygon rotates clockwise. The following readings were obtained.*

Polygon faces		*Readings (seconds)*	
		T_1	T_2
AB	BC	19.2	12.5
BC	CD	20.1	14.7
CD	DE	21.5	19.2
DE	EF	23.7	18.5
EF	FG	22.4	16.0
FG	GH	22.0	16.8
GH	HA	21.6	15.4
HA	AB	20.8	17.2

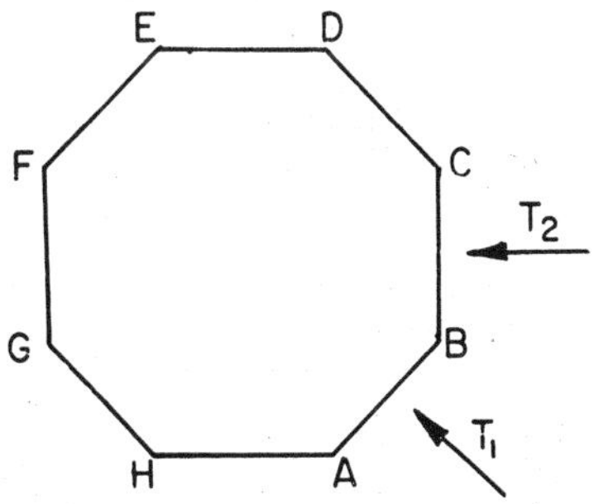

Fig. 8.80

Complete the table to find the cumulative calibration errors of the polygon and the true errors.

Solution. Since increase in reading T_1 relative to T_2 represents an increase in angle intervals between the faces, subtract the readings of T_2 from T_1 to obtain correct signs of difference. The mean of all differences must represent true 45° as all angles add up to 360°.

Readings		Difference $T_1 - T_2$	Diff. from mean	Cumulative error	Correction	True error	Face
T_1 (sec)	T_2 (sec)						
—	—	—	—	0	0	0	AB
19.2	12.5	6.7	+ 1.6	+ 1.6	– 0.1	+ 1.5	BC
20.1	14.7	5.4	+ 0.3	+ 1.9	– 0.1	+ 1.8	CD
21.5	19.2	2.3	– 2.8	– 0.9	– 0.1	– 1.0	DE
23.7	18.5	5.2	+ 0.1	– 0.8	– 0.1	– 0.9	EF
22.4	16.0	6.4	+ 1.3	+ 0.5	– 0.2	+ 0.3	FG
22.0	16.8	5.2	+ 0.1	+ 0.6	– 0.2	+ 0.4	GH
21.6	15.4	6.2	+ 1.1	+ 1.7	– 0.2	+ 1.5	HA
20.8	17.2	3.6	– 1.5	+ 0.2	– 0.2	0	AB
Sum = 41.0 ; Mean = 41.0/8 ≈ 5.1							

Cumulative error of + 0.2 is distributed proportionally (in approximation) to all readings so as to make error for face *AB* = 0.

In above table, true error of + 1.5 sec is for the angle between the normals to faces *AB* and *BC*, + 1.8 sec between normals to faces *BC* and *CD*, and so on.

Problem 8.16. *Two six-sided precision polygons, placed one above the other, are calibrated together by the pick-a-back method, i.e. fixing one face of one polygon against the auto-collimator and taking readings on other autocollimator on all faces of other polygon one by one and repeating similarly for all other faces of first polygon. Thus there will be 36 sets of readings which are tabulated below. Calculate the error from true 60° for each angle of each polygon*

All readings are in seconds

	UV	VW	WX	XY	YZ	ZU
AB	– 4.2	– 0.5	– 0.1	0.0	– 0.9	– 2.5
BC	– 2.1	– 0.1	+ 1.9	– 0.1	+ 1.0	– 1.1
CD	– 0.8	+ 1.7	+ 1.7	+ 0.6	+ 1.5	– 0.0
DE	– 1.9	+ 1.9	+ 0.5	– 1.2	– 0.1	– 1.5
EF	– 2.1	+ 0.7	– 1.1	– 1.5	+ 1.2	+ 2.7
FA	– 3.2	+ 0.9	+ 1.5	+ 0.2	– 1.1	+ 1.6

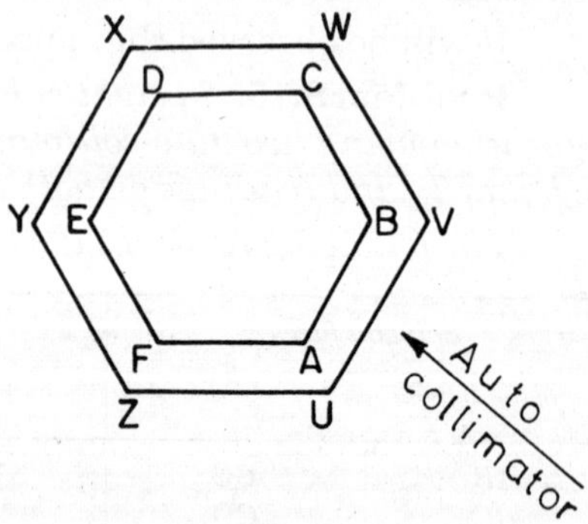

Solution. Six sets of readings have been taken for each face of polygon so that one face corresponds to all the six faces of other polygons in turn. Comparison of differences between readings on adjacent face for each polygon gives relative values of the angles on each polygon.

Let $\alpha_1, \alpha_2 \ldots\ldots \alpha_6$ be the angles for polygon *ABCDEF* and $\beta_1, \beta_2 \ldots\ldots \beta_6$ for other polygon. Then adding first row we have

$$(\alpha_1 - \beta_1) + (\alpha_1 - \beta_2) + \ldots (\alpha_1 - \beta_6) = -4.2 - 0.5 - 0.1 + 0.0 - 0.9 - 2.5.$$

or $$6\alpha_1 - (\beta_1 + \beta_2 + \ldots \beta_6) = 8.2'' ; \text{ or } 6\alpha_1 - 360° = 8.2°$$

or $$\alpha_1 = \frac{360° - 8.2''}{6} = 60° - 1.5''$$

$\therefore$ Error on angle $\alpha_1 = 1.5''$.

Similarly errors for $\alpha_2 \ldots\ldots \alpha_6$ can be determined by adding all rows and dividing by 6.

Adding entries in first column

$$(\alpha_1 - \beta_1) + (\alpha_1 - \beta_1) + (\alpha_3 - \beta_1) + \ldots\ldots = -4.2 - 2.1 - 0.8 - 1.9 - 2.1 - 3.2 = -14.3''$$

or $$(\alpha_1 + \alpha_2 + \ldots \alpha_6) - 6\beta_1 = -14.3'' ; \text{ or } 360° - 6\beta_1 = -14.3''$$

or $$\beta_1 = \frac{360° + 14.3}{6} = 60° + 2.4'' \quad \therefore \text{ Error in } \beta_1 = +2.4''$$

In other words, in case of determining error for β_1 the sign will have to be changed as obtained by adding up the readings of columns.

In this way table can be completed as follows :

	UV	*VW*	*WX*	*XY*	*YZ*	*ZU*	*Sum*	*Sum/6*	*Error*
AB	– 4.2	– 0.5	– 0.1	0.0	– 0.9	– 2.5	– 8.2	– 1.5	– 1.5
BC	– 2.1	– 0.1	+ 1.9	+ 0.1	+ 1.0	– 1.1	– 0.5	– 0.1	– 0.1
CD	– 0.8	+ 1.7	+ 1.7	+ 0.6	+ 1.5	+ 0.0	+ 4.7	+ 0.8	+ 0.8
DE	– 1.9	+ 1.9	+ 0.5	– 1.2	– 0.1	+ 1.5	+ 0.7	+ 0.1	+ 0.1
EF	– 2.1	+ 0.7	– 1.1	– 1.5	+ 1.2	+ 2.7	– 0.1	0.0	0.0
FA	– 3.2	+ 0.9	+ 1.5	– 0.2	– 1.1	+ 1.6	– 0.1	0.0	0.0
Sum	– 14.3	+ 4.6	+ 5.4	– 2.0	+ 1.6	+ 2.2	– 3.5		
Sum/6	– 2.4	+ 0.8	+ 0.9	– 0.3	+ 0.3	+ 0.4	– 0.6		
Error	+ 2.4	– 0.8	– 0.9	+ 0.3	– 0.3	– 0.4	+ 0.6		

Problem 8.17. *Name some examples of angle measurements by comparison to angular reference body.*

Solution. (*a*) Measurement with fixed reference taper limit plug and ring gauges.

(*b*) Use of polygon block for inspecting the accuracy of circular divisions by optical observation of the polygon's reflecting faces.

(*c*) Using combination of angle gauges for measurements of different sizes by comparative methods.

(*d*) Locating the part by means of sine blocks. Slip gauges and sine bar create a controlled slope to a datum plane.

(*e*) Measurement of taper by assembly of gauge rollers and balls. (Creating a virtual cone by combining the line contacts of two gauge rollers and the point contacts of two gauge balls.

Problem 8.18. *What do you understand by measurement of cone shaped parts by comparison to a virtual body or by attributes ?*

Solution. The taper angle of a regular cone can be determined by the interrelated position of a limited number of points which are contained in a common axial plane of that cone. In such cases, specific number of properly located physical points can represent the practical equivalent of a reference body. The term "virtual body" is used to designate a set of discrete points which are used to determine the taper angle of a cone.

Elements used to create virtual bodies are balls, rollers and gauge blocks (Refer Fig. 8.82).

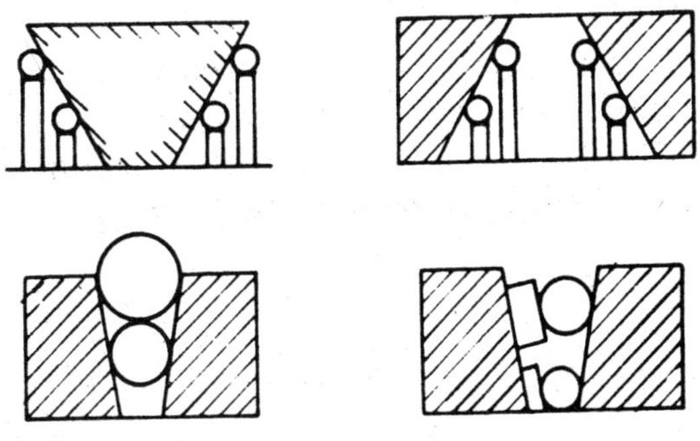

Fig. 8.82. Taper angle measurement by comparison to a virtual cone.

The relationship of the specific diameters being measured is considered a distinctive attribute of the taper angle, and therefore, sometimes designation "attribute gauging" is used for this method of measurement of cone angle.

Problem 8.19. *What are the reasons for using virtual bodies for measurement of cone angles etc. instead of solids as reference bodies ?*

Solution. Some of reasons are :

(*a*) Virtual bodies can be built up from standard elements, but solid gauges have to be manufactured in a wide variety for each particular angle and size group.

(*b*) Fixed gauges can't enter hidden areas, but virtual bodies permit the exploration of hidden area dimensions.

(*c*) Virtual bodies can be used for measuring meaningful linear dimensions which reflect angular or diametrical size variations of a cone shaped body.

Problem 8.20. *What are the two categories of measurement for which clinometer can be used ?*

Solution. Clinometer may be used for

(*i*) The measurement of an inclined surface in relation to the basic horizontal. (In this case the instrument is placed on the surface to be measured and its graduated disk is rotated to produce zero indication on the fiducial bubble. The scale value of the disk position will then be equal to the angle of incline).

(*ii*) Measurement of the relative position of two mutually inclined surfaces (The clinometer is placed on each of the surfaces in turn and readings are taken w.r.t. basic horizontal. The difference of two readings indicates the angular value of the relative decline).

TYPICAL EXAMINATION QUESTIONS

1. What are the different methods of measuring angles ? State the accuracy of each method suggested.
2. What do you understand by a marking "1 Div. = 0.005 mm in 100 mm" found on a precision spirit level. Also describe how the accuracy of the graduations on a spirit level can be checked ?
3. Sketch two forms of sine bars in general use. What are the important dimensions of sine bar and what precautions are necessary in its use ? List the possible sources of error in a sine bar and derive an expression for the error any in the angle set up resulting from :
   (*i*) an error in the centre distance between the rollers.
   (*ii*) error in size of the roller.
   (*iii*) setting the sine bar at angles more than 45°.
4. State the essential requirements for accuracy in the construction of a sine bar.
   Why is it that the use of a sine bar is not recommended for angles larger than 45° with the reference plane.
   Explain how a sine bar may be employed to determine the inclined angle of a taper plug gauge.
5. A 100 mm sine bar is to be set up to an angle of 32.5°. Determine the slip gauges needed. Estimate the error in the angle, if :
   (*i*) distance between rollers is not correct by ± 0.01 mm.
   (*ii*) If dia. of rollers is out by ± 0.005 mm.
   (*iii*) If error of parallelism between top surface and line joining centres of rollers is ± 0.02 mm.
   (*iv*) If all the three errors exist simultaneously.
6. Explain the importance of tolerance in taper shanks of twist drill. What methods would you recommend for the measurement of the accuracy of taper ? Explain with neat sketches.
7. How are angular measurement made ? Describe the instrument for angular measurement correct upto a few seconds.

8. Select the sizes of angle gauges required to build the following angles :

   22°, 11° 20′, 29° 54″, 31° 49′ 24″.

9. Describe three methods of precision angular measurement other than those using balls and rollers as accessories, and sketch a practical application of each.

10. Describe with a sketch the principle of working of an auto-collimator. What is meant by the term 'Flatness' of a surface and how it is determined with the help of an auto-collimator ? State some of the important uses of this instrument.

11. Draw the diagram of an instrument to check the departure from squareness of the adjacent edges, to an accuracy of ± 1 minute of arc, of a metal plate 75 mm × 90 mm × 12 mm thick. Describe briefly how the measurement is made.

12. It is intended to employ a precision spirit level to measure to an accuracy of ± 1 minute of arc the included angle of 10° 15′ between the two accurately flat ground faces of a large number of components measuring 250 mm × 25 mm and approximately 40 mm high. Describe the set up and select a suitable sensitive level and state the maximum permissible bubble variation.

13. An internal angle between two faces of a female form gauge 115° ± 10′. Show by a sketch how three 25 mm dia. discs could be used, with a micrometer depth gauge is check the angle. Also calculate the limits of the micrometer reading which will ensure that the angle is within the prescribed tolerance.

14. Two rectangular-section straight edges are to be fastened flat on to a machine table so as to be side by side and with a taper gauge approximately 16 mm wide between them. They are to be inclined to each other at an angle of 5° 23′ within a tolerance of ± 1 minute. Describe how this could be done, showing the necessary calculations.

15. A 125 mm diameter shaft has light parallel-side splines. Describe how you would check the width of the splines, the angular spacing, and their parallelism with the shaft axis.

16. (*a*) Describe the method of calibrating a polygon by means of two precision auto-collimators.

    (*b*) The following readings were obtained in a calibration test on a six-sided polygon :

Corner	1	2	3	4	5	6
Auto-collimator No. 1	6.2	4.4	7.0	5.7	7.1	7.9
Auto-collimator No. 2	14.9	17.8	19.6	16.1	19.8	22.1

    Determine the error in the angle at each corner.

17. Describe how you would check the accuracy of a circular table by means of an auto-collimator, a precision polygon, and a set of angle gauges ?

18. A 20 mm diameter hole is drilled and reamed at an angle of 75° to a flat surface. Describe how the angle could be checked by using two unequal rollers and calculate the values of the measurement which must be made, using suitable sizes of rollers.

19. State the accuracy to which readings can normally be taken on

    (*a*) 0.01 mm micrometer

    (*b*) an angle dekkor

    (*c*) a 20 sec level.

    Describe fully how would you check the accuracy of one of these instruments.

9

Miscellaneous Measurements

Introduction

Some parts are very difficult to be measured by actual tool and measuring instruments. A lot of time is likely to be wasted in thinking about the ways in which some miscellaneous measurements could be made unless some guiding principles are followed. Instead of using tools alone, some parts are best inspected and measured by mathematical calculations. Once the methods given below for some typical problems are visualised they can be applied to any problem being faced in the workshop.

9.1. Checking the angle of a Piece Tapered on One Side

This can be done by using two discs of different sizes, slip gauges and a dial indicator. The length of slip gauges required for the purpose is calculated as given later and the slip gauge combination of that length prepared. The two discs whose diameters are known and slip gauges are placed on a surface plate as shown in Fig. 9.1 and the discs are clamped in position against an angle plate by *C*-clamps. The work is then placed on top of discs and clamped against the angle plate by a *C*-clamp. If the angle of the piece is all right, then the top edge will be parallel to surface plate and the dial indicator will show no variation when traversed along its surface.

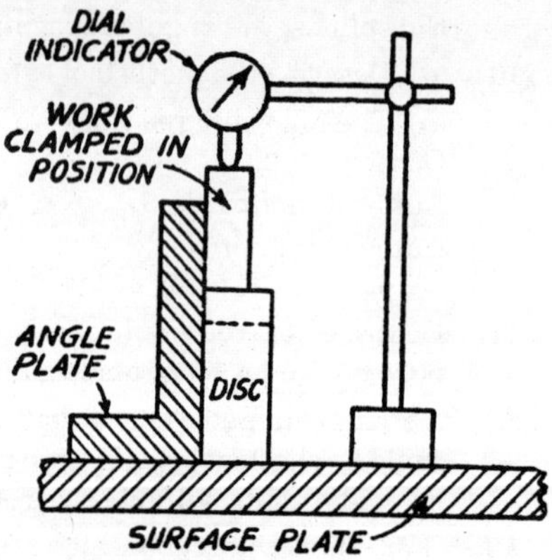

Fig. 9.1

In Fig. 9.2, O_1 and O_2 are the centres of the two discs. Line *XY* is drawn perpendicular to line joining O_1O_2. O_2A is horizontal line and O_2B is parallel to the tapered surface of the piece *i.e.* inclined at an angle α to the horizontal line. The angle $O_1O_2A = \alpha/2$.

Now in right angled triangle O_1O_2A,

$$\tan\frac{\alpha}{2} = \frac{O_1A}{AO_2} = \frac{(d_1 - d_2)/2}{(d_2 + d_2)/2 + l}$$

where l is the length of slip gauges and d_1 and d_2 are the diameters of two discs.

or $$\left(\frac{d_1 + d_2}{2} + l\right)\tan\frac{\alpha}{2} = \frac{d_1 - d_2}{2}$$

and $$l = \frac{(d_1 - d_2)/2}{\tan \alpha/2} - \frac{d_1 + d_2}{2}$$

Thus the length of slip gauges is adjusted by the above formula and two discs placed just in contact with end of slip gauges as shown in Fig. 9.2.

By this method, the straightness of the edges of the tapered piece can also be checked.

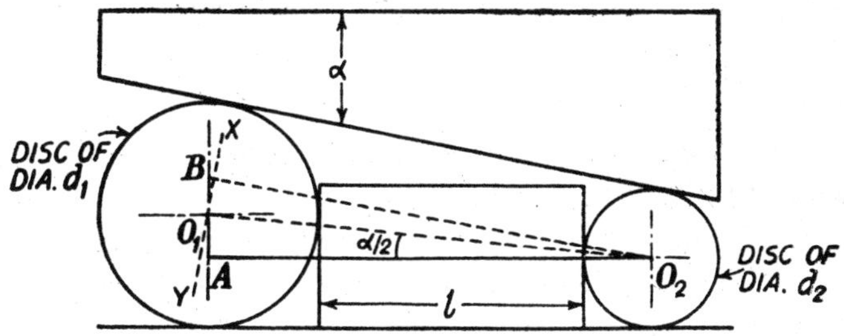

Fig. 9.2

9.2. To Check the Angle of a Transverse Recess whose Angle of Taper in Very Small

In Fig. 9.3, *XY* and *MN* are two tapered faces, inclined at angle *A*. For this measurement we require one sphere of certain diameter depending upon the distance *XM*, set of slip gauges and a micrometer depth gauge.

A slip gauge of certain length l_2 is chosen and inserted in tapered recess along with sphere as shown in Fig. 9.3. In this position, the depth H_2 is measured with the micrometer depth gauge. Then a slip gauge of bigger length, 'l_1' is selected and again inserted with the same sphere in the tapered recess. Again in this position depth *H* is measured with the depth gauge.

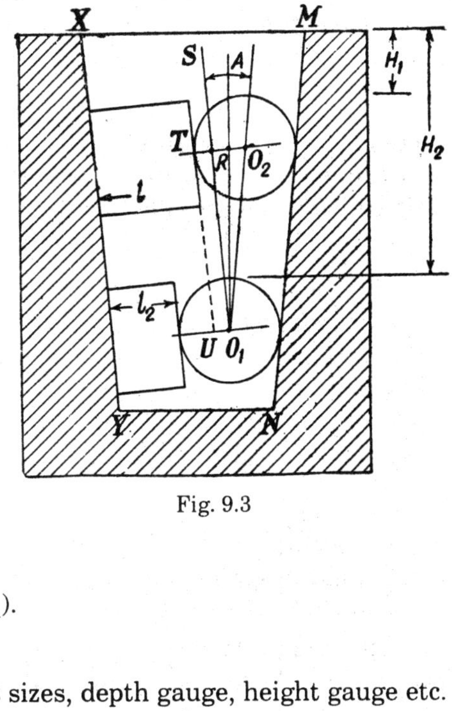

Fig. 9.3

In Fig. 9.3, O_1 and O_2 are centres of spheres in two different positions. Then O_1O_2 is parallel to the tapered surface *MN*. O_1S is drawn parallel to the other surface *XY* of recess and O_2R is drawn perpendicular to O_1S. Then angle O_2O_1R is equal to *A*.

Then $$\sin A = \frac{O_2R}{O_1O_2} = \frac{l_1 - l_2}{H_2 - H_1}$$

(O_1O_2 being very approximately equal tc $H_2 - H_1$).

9.3. To Check the Angle of a Tapered Hole

In this method, we require two balls of different sizes, depth gauge, height gauge etc. This method is particularly suitable for checking the angle of a tapered ring gauge.

First a small ball of radius R_1 is inserted in the hole in lower position and depth H_1 from upper surface of tapered hole to the top of the ball is measured. Small ball must be of such size as to be seated somewhere in between hole. Then bigger ball of radius R_2 is placed in the hole and distance H_2 *i.e.* between the top of ball and tapered hole is measured with height gauge. In Fig. 9.4 (*a*) O_1 and O_2 represent the centres of two balls. Draw O_1S parallel to the line *XY* of tapered hole in section and O_2S perpendicular to O_1S. Then $O_2\hat{O}_1S = A/2$; (where *A* is the angle of tapered hole.)

$$\sin \frac{A}{2} = \frac{O_2S}{O_1O_2} = \frac{R_2 - R_1}{H_1 + H_2 + R_1 - R_2}$$

Thus the angle of taper *A* can be found out. This method is also very suitable for measuring the relief clearance of dies etc. In case both the balls lie inside the groove, then formula is modified as shown in Fig. 9.4 (*b*).

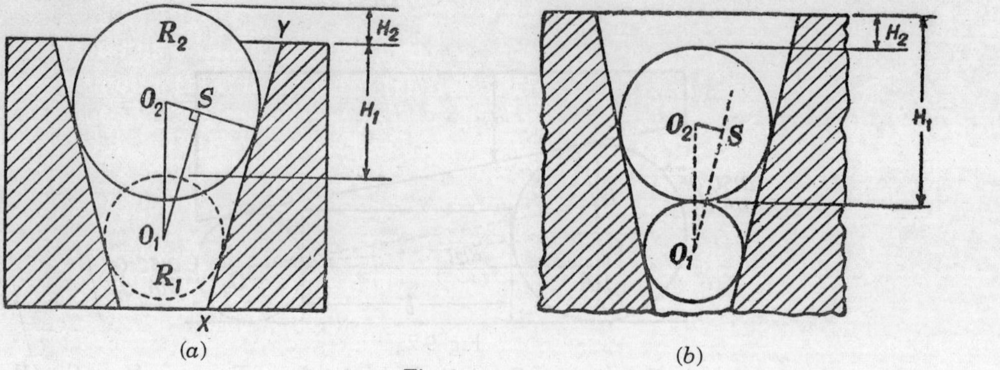

Fig. 9.4

$$\sin\frac{A}{2} = \frac{R_2 - R_1}{H_1 + R_1 - (H_2 + R_2)}$$

A word of caution in measurement of small taper measurement by balls should always be remembered. If ball is allowed to drop in tapered hole, it will wedge in the taper and may be several thousands of mm lower. The weight of the ball and the measuring pressure exerted over it can cause appreciable axial movement to give inaccurate readings.

9.4. To Determine the Included Angle of an Internal Dovetail

In dovetail the sloping sides act as guide and prevent the lifting of the female mating part during mating operation. The angle which the sloping face makes with an imaginary vertical centre plane is important in case of dovetail. For measuring this angle we require two pins of equal size, slip gauges and micrometers. First the two pins are placed touching both the sides of dovetail as shown in Fig. 9.5 and distance l_1 is measured across these pins with a micrometer. Then the pins are raised on two sets of equal slip gauge blocks. Here care should be taken that the pins do not extend above the top surface of dovetail. Again distance l_2 across the pins is measured. Let height of slip gauges be h.

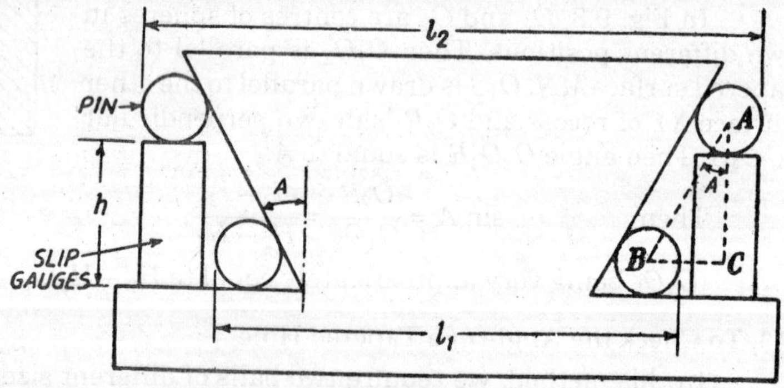

Fig. 9.5

Then $\tan A = \dfrac{BC}{AC} = \dfrac{\frac{l_2 - l_1}{2}}{h}$. Thus knowing l_2, l_1 and h, the angle A can be calculated.

This method is also suitable for measuring the angle of a taper plug gauge or any round or flat tapered work which can be placed on a surface.

9.5. To Measure in Internal Taper in Blind Hole

Let tapered angle = A.

For this method, we require two steel balls of equal size and slip gauges. First two balls of equal diameter are placed in each bottom corner of the taper shown in Fig. 9.6 and the space between the balls is filled with a correct combination of slip gauges. The length of slip gauges which just fits in the space between the balls must be noted very carefully and let it be l_1.

Next prepare a combination of slip gauges of convenient height 'h' and place it in the hole as support. Again place the two balls in upper position, each touching the tapered hole and the space between them again being filled up by suitable combination of slip gauges. Let the length of slip gauges in this case be l_2.

Then angle O_1O_2B equals half the taper angle, where O_1 and O_2 are the centres of the balls and O_1O_2B is a right angled triangle.

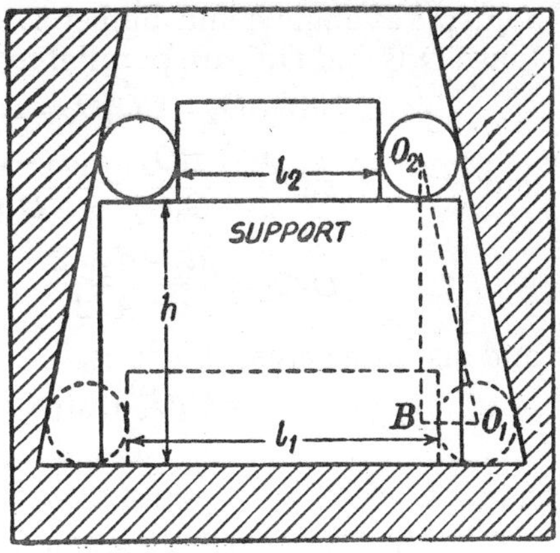

Fig. 9.6

$$\tan\frac{A}{2} = \frac{O_1B}{O_2B} = \frac{\frac{l_1 - l_2}{2}}{h}$$

$$A = 2\tan^{-1}\left(\frac{l_1 - l_2}{2h}\right)$$

Thus the internal tapered angle can be calculated.

In this method, h should be chosen as large as possible. The radius of rollers should be small so that contact between the rollers and the flanks of sloping sides is made in a plane near to the base and close to the top portion (mouth). The error in angle measurement can be decreased by increasing $l_1 - l_2$ and hence large h and small r.

9.6. To Determine the Intersecting Dimension 'l' of the V-shaped Slide

The method is similar to Art. 9.5.

Let A is the tapered angle of V-shaped slide,

then $\tan\frac{A}{2} = \frac{l_1 - l_2}{2h}$

Let r be the radius of the roller

Then $O_1C = r \sec A/2$, and $ED = r \tan A/2$

$$l = l_1 + 2[r + O_1C + ED]$$

$$= l_1 + 2\left[r + r\sec\frac{A}{2} + r\tan\frac{A}{2}\right]$$

$$= l_1 + 2r\left[1 + \sec\frac{A}{2} + \tan\frac{A}{2}\right].$$

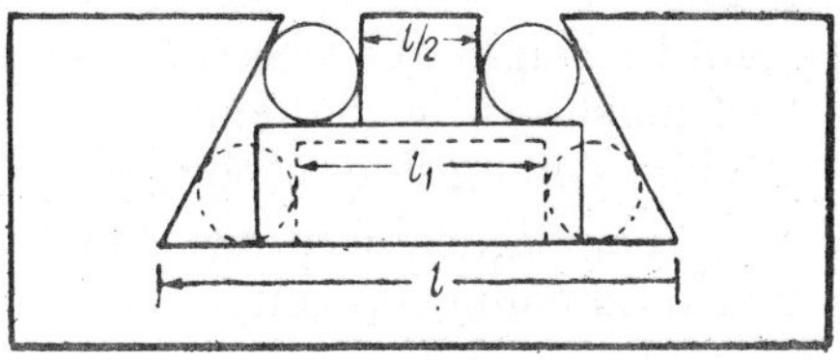

Fig. 9.7 (*a*)

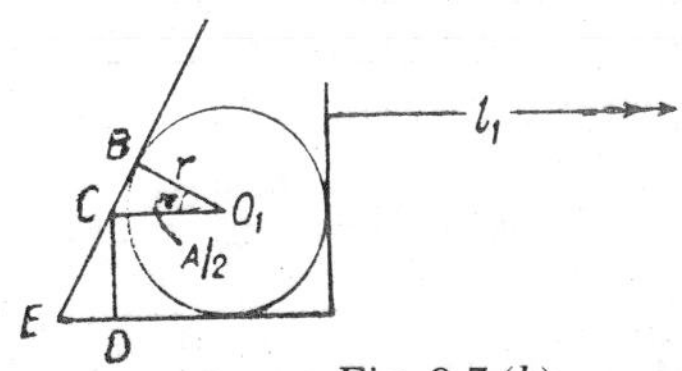

Fig. 9.7 (*b*)

9.7. To Set two Straight Edges at a Desired Angle to Close Limits (Fig. 9.8)

This method requires two discs of different diameters and a vernier caliper. This method is very suitable for setting the two straight edges of a taper gauge to check tapered components. In taper gauge, the two straight edges are pivoted at ends and slots at ends permit them to be adjusted to the desired angle. A source of light placed below the straight edges indicates clearly whether or not the tapered work being inspected fits perfectly against them.

This method can also be used to set any two machine parts at desired angle.

The diameters of the two discs are measured very carefully. Then the next important thing is to determine the distance between the two balls such that the two straight edges be adjusted at angle A and diameters of two discs be d_1 and d_2. O_1 and O_2 are centres of two discs. Lines O_1B and O_1C are parallel to two straight edges.

Then $\quad B\hat{O}_1O_2 = C\hat{O}_1O_2 = A/2$

and $\quad \sin\frac{A}{2} = \frac{BO_2}{O_1O_2} = \frac{(d_1 - d_2)/2}{O_1O_2}$

$\therefore \quad O_1O_2 = \frac{(d_1 - d_2)/2}{\sin A/2}$

and distance across the discs,

$$l = O_1O_2 + (d_1 + d_2)/2$$

$$= \frac{(d_1 - d_2)/2}{\sin A/2} + (d_1 + d_2)/2.$$

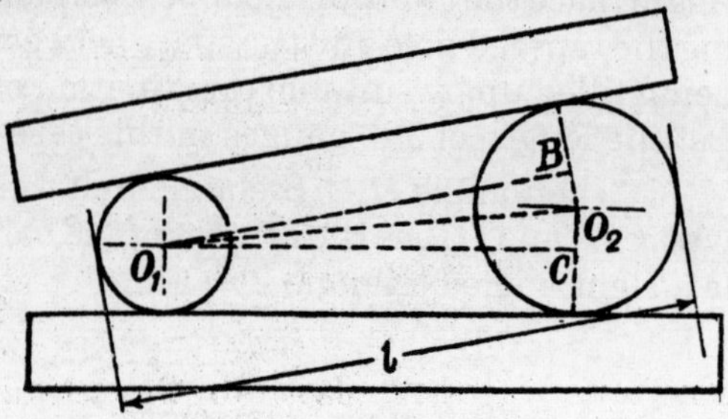

Fig. 9.8

Thus the two discs are adjusted exactly to above distance by tapping one disc slightly with a mallet and the distance checked by a vernier caliper.

9.8. To Measure the Angle of a V-groove

For this method we require two discs or pins of different sizes and a height gauge. The sizes of the two discs are determined exactly. First the small disc is placed in the groove and height of disc above the surface plate is measured with a height gauge. Next small disc is removed and bigger disc is placed in the groove and again its height above surface plate is measured. The discs must be of such diameter that they always touch only the sloping sides of the groove.

For higher accuracy, height h_1 and h_2 can be measured using slip gauges and sensitive indicator as shown in Fig. 9.9 (*a*).

In Fig. 9.9, O_1 and O_2 are the centres of the two balls. O_1B is drawn parallel to one of the inclined surfaces and $O_2B \perp O_1B$. Then $O_2\hat{O}_1B$ equals half the angle of V-groove.

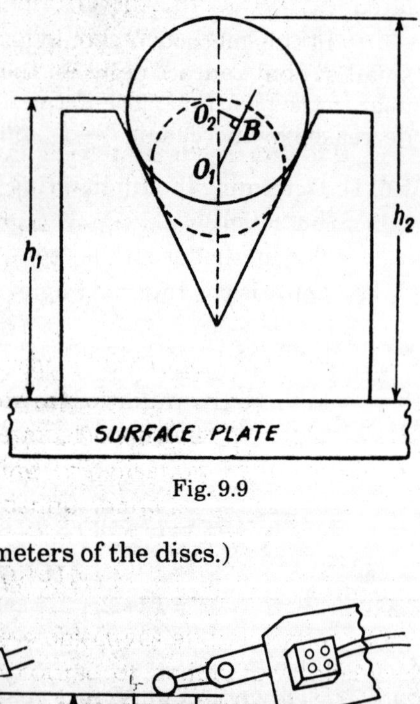

Fig. 9.9

Now $O_1B = \frac{d_2 - d_1}{2}$, (where d_1 and d_2 are the diameters of the discs.)

and $\quad O_1O_2 = \left(h_2 - \frac{d_2}{2}\right) - \left(h_1 - \frac{d_1}{2}\right)$

$$= h_2 - h_1 - \left(\frac{d_2 - d_1}{2}\right)$$

Let angle of V-groove = A,

Then

$$\sin\frac{A}{2} = \frac{O_2B}{O_1O_2} = \frac{\frac{d_2 - d_1}{2}}{h_2 - h_1 - \left(\frac{d_2 - d_1}{2}\right)}$$

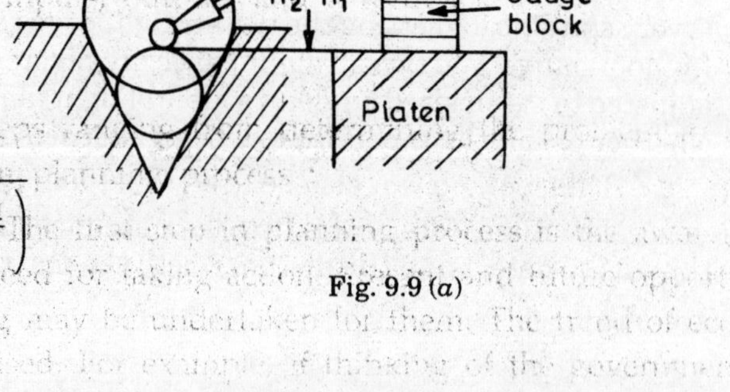

Fig. 9.9 (*a*)

This method is very suitable for measuring the angle of machine-tool slides, angle of profile and the angle of shallow grooves and threads.

In case of threads etc., $h_1 - h_2$ can be found out by using a dial indicator.

9.9. To Determine the Width of a V-groove

It is actually very difficult to measure the width of a *V*-groove directly with standard tools. This method requires the use of a steel ball and a vernier height gauge. By only one reading, *i.e.* height of ball above the base, one can determine the width of groove. Since reading can be taken without removing or changing the set up, this method can be used while machining or grinding operation is being carried out.

First the angle of *V*-groove is determined by the two balls method or it must be known exactly. A ball of diameter d is then placed in the groove and reading h_1 noted down (Fig. 9.10). Next the height of *V*-block is noted down and let it be h_2. Let angle of groove be A. Let width of groove be w.

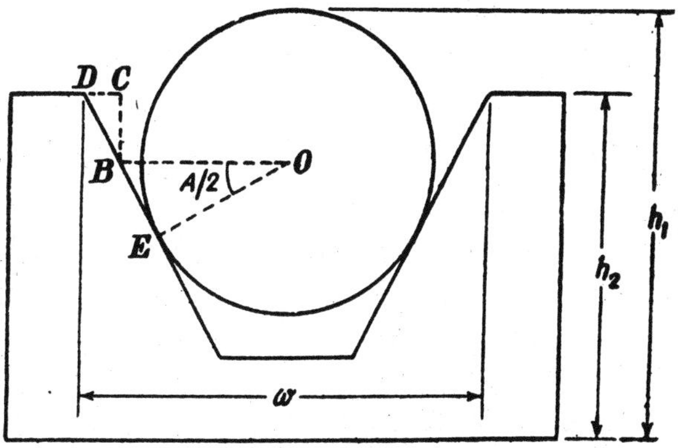

Fig. 9.10

O is the centre of the ball. Draw $OE \perp$ to one of the sloping sides and OB a horizontal line.

Then $BOE = \frac{A}{2}$, *i.e.* half the angle of groove.

$$\cos\frac{A}{2} = \frac{OE}{OB};\ OB = \frac{OE}{\cos A/2} = \frac{d/2}{\cos A/2}$$

In rt. $\angle$d ΔBCD, $D\hat{B}C = A/2$ and $BC = d/2 - (h_1 - h_2)$

$$DC = BC \tan A/2$$

$$= [d/2 - (h_1 - h_2)] \tan A/2$$

$\therefore$ Width of slot $= 2[OB + DC]$

$$= \frac{d}{\cos A/2} + [d - 2(h_1 - h_2)] \tan A/2$$

In case centre of the ball lies above the flat surface, then width of slot

$$= \frac{d}{\cos A/2} - [2(h_1 - h_2) - d] \tan A/2.$$

9.10. To Check Interior Angle of a Profile Gauge

(When one of the sides is at right angle to the reference surface.)

This method requires the use of two discs of different sizes and a Vernier height gauge. First a small disc of diameter d_1 is placed in the slot and the height of disc above some reference surface is determined (Fig. 9.11). Let it be h_1. Then the bigger disc is placed in the slot after removing the smaller one and again height h_2 with reference to same plane is noted down. The difference of height $(h_2 - h_1)$ could also be noted down by means of dial-indicator.

Let O_1 and O_2 be the centres of two discs. Through O_1 draw a line O_1C parallel to the vertical surface and O_1B parallel to the sloping surface. Through O_2 draw $O_2C \perp O_1C$ and $O_2B \perp O_1B$. Let angle of taper be A.

$$\therefore \quad C\hat{O}_1O_2 = \frac{A}{2} = O_2\hat{O}_1B$$

$$\tan\frac{A}{2} = \frac{O_2C}{O_1C} = \frac{\frac{d_2-d_1}{2}}{\left(h_2-\frac{d_2}{2}\right)-\left(h_1-\frac{d_1}{2}\right)}$$

$$= \frac{d_2-d_1}{2(h_2-h_1)-(d_2-d_1)}$$

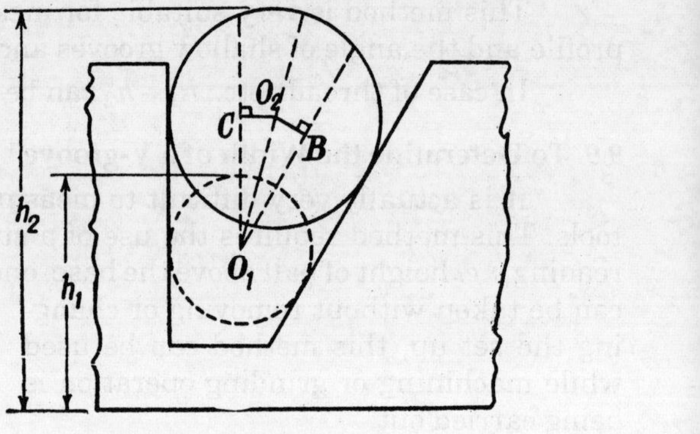

Fig. 9.11

By the same set-up, the angle of any inclined surface with vertical plane can be calculated by making the part to rest against angle plate.

9.11. To Measure the Angle of Small Work Located in a Difficult Position

This method requires the use of a gear-tooth Vernier caliper. This method is very useful in those cases where the work is small and enclosed by sides. All that has to be done in this method is to take readings at two different places. The two reading have to be noted down at one section, *i.e.* thickness of the piece t_1 and its depth from the top surface h_1. Similarly two readings t_2 and h_2 are to be noted down at other section with the help of gear tooth Vernier caliper.

Let the angle of the taper be A.

Then in right angled triangle BCD in Fig. 9.12,

$$B\hat{D}C = \frac{A}{2}$$

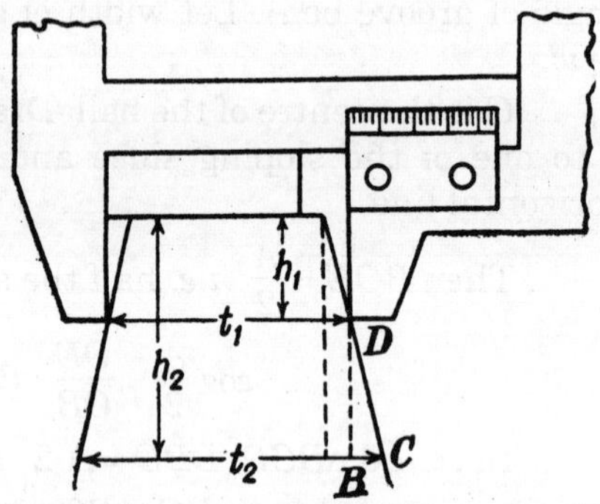

Fig. 9.12

and $\quad \tan\frac{A}{2} = \frac{BC}{BD} = \frac{\frac{t_2-t_1}{2}}{h_2-h_1}$, Thus the angle A can be calculated.

9.12. To Measure the Taper Angle of a Shallow Circular Groove

It is actually very difficult to measure the internal angle of a shallow circular groove by any method. The method proposed below requires the use of two small balls of equal size and two large balls of equal size and slip gauges. First the small balls are placed in the hole and the slip gauges of maximum length inserted in between the balls. The small balls are removed and large balls inserted and again space between the balls filled up by slip gauges of maximum length. Let us assume that the diameters of the balls are d_1 and d_2, and length of slip gauges in each case is l_1 and l_2 respectively. Let the angle of taper be A.

In Fig. 9.13, O_1 and O_2 represent the centres of two balls in two different positions.

In rt. ∠d ΔO_1O_2B, $O_2\hat{B}O_1 = 90°$ and $O_2\hat{O}_1B = \frac{A}{2}$

and $\quad \tan\frac{A}{2} = \frac{O_2B}{A_2B} = \frac{\frac{d_2-d_1}{2}}{\frac{l_1+d_1-(l_2+d_2)}{2}} \quad$ or $\quad \tan\frac{A}{2} = \frac{d_2-d_1}{l_1-l_2+d_1-d_2}$.

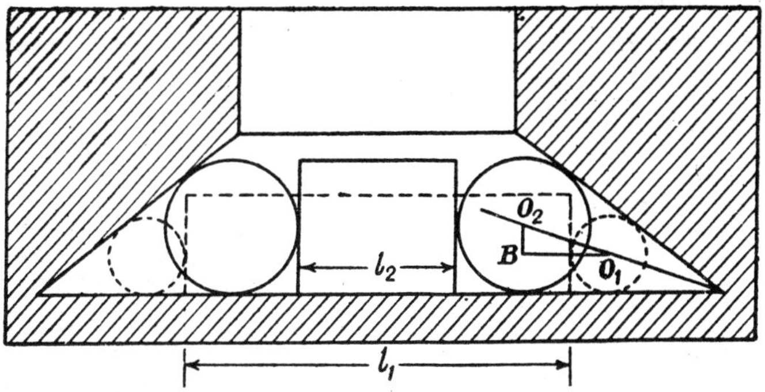

Fig. 9.13

9.13. To find out the Radius of Circle of any Job having a Portion of a Circle

This method requires the use of surface plate, Vernier caliper, C-clamp and two pins of equal size. This method could be best applied to jobs like cap of a bearing. The job is first clamped to surface plate with the help of C-clamp. It should be clamped in such a way that central position of the circular part is in contact with the surface plate as far as possible. Then two pins of equal diameter are placed on each side of the work as shown in Fig. 9.14 and the reading over the balls is taken with the Vernier caliper. Let the reading be l. Let the diameter of pins be d and radius of job R.

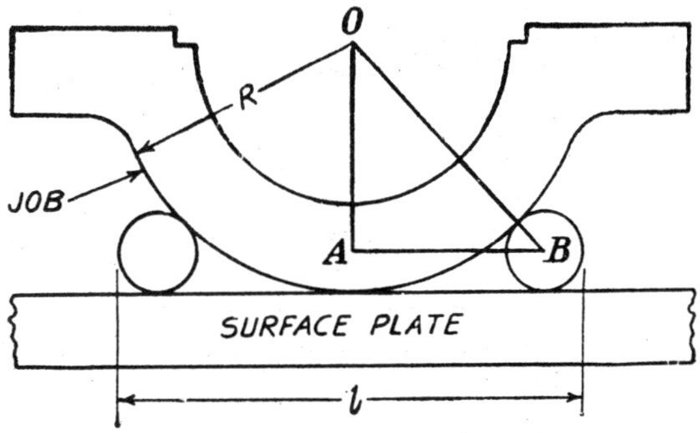

Fig. 9.14

Let 'O' be the assumed centre of the circle. Then in rt. ∠d ΔOAB, in Fig. 9.14,

$$OB^2 = OA^2 + AB^2$$

or $$\left(R + \frac{d}{2}\right)^2 = \left(R - \frac{d}{2}\right)^2 + \left(\frac{l-d}{2}\right)^2$$

or $$R^2 + \frac{d^2}{4} + Rd = R + \frac{d_2}{2} - Rd + \frac{(l-d)^2}{4}$$

or $$2Rd = \frac{1}{4}(l-d)^2$$

and $$R = \frac{(l-d)^2}{4 \times 2d} = \frac{(l-d)^2}{8d}.$$

9.14. To find out the Radius of a Concave Surface

(*i*) When the edges are well-defined.

(*ii*) When the edges are rounded up.

(*i*) *When the edges are well-defined.* This method is applicable to those parts which have large radius to curvature. This requires the use of a surface plate, angle plate, height gauge, depth micrometer, slip gauges and a C-clamp.

The part to be tested is kept on a surface plate and with the help of a depth micrometer the maximum depth of the cavity is determined. Let it be h. Next the part is kept in such a way that cavity is resting against an angle plate and the part is then clamped in this position. The hole is then measured from edge to edge with a height gauge having a sharp scribing arm. Let the maximum reading, *i.e.* diameter of the hole be d (Fig. 9.15).

Let O be the assumed centre of the cavity and R the radius of curvature.

Then in rt. ∠d ΔOAB,

$$OA^2 = AB^2 + BO^2$$

or
$$R^2 = \left(\frac{d}{2}\right)^2 + (R-h)^2$$

$$= \frac{d^2}{4} + R^2 + h^2 - 2Rh$$

or
$$2Rh = \left(\frac{d}{2}\right)^2 + h^2$$

$$R = \frac{\left(\frac{d}{2}\right)^2 + h^2}{2h} = \frac{d^2}{8h} + \frac{h}{2}.$$

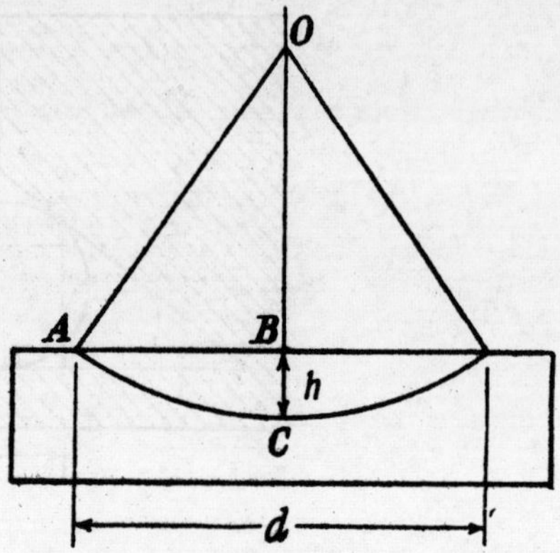

Fig. 9.15

(*ii*) *When the edges are rounded up*. When the edges of the cavity are rounded up, then the radius of curvature can be measured by a depth micrometer and slip gauges. The width of the depth micrometer base is measured with the help of slip gauges. Let it be d. Then it is placed in the cavity till it fully rests in the cavity, its frame touching all the sides of cavity (Fig. 9.16). The measuring tip is then lowered down till it touches the base. The reading is then noted on the thimble and let it be h. Now the case is similar to previous one and the radius of curvature R can be found out by the same formula.

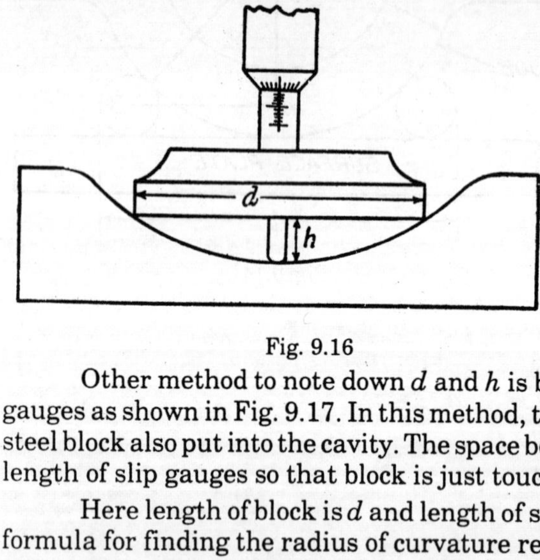

Fig. 9.16

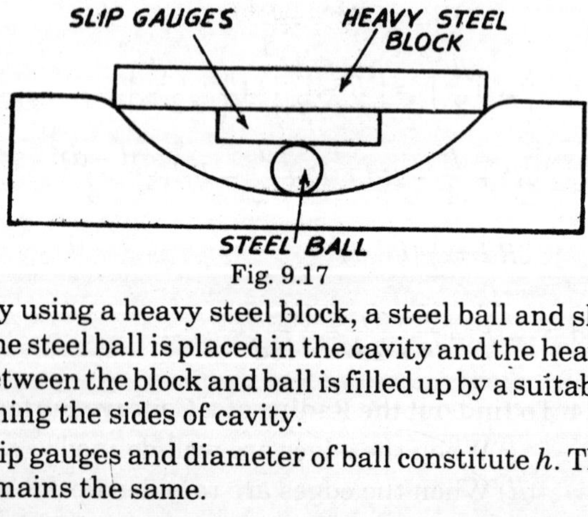

Fig. 9.17

Other method to note down d and h is by using a heavy steel block, a steel ball and slip gauges as shown in Fig. 9.17. In this method, the steel ball is placed in the cavity and the heavy steel block also put into the cavity. The space between the block and ball is filled up by a suitable length of slip gauges so that block is just touching the sides of cavity.

Here length of block is d and length of slip gauges and diameter of ball constitute h. The formula for finding the radius of curvature remains the same.

9.15. To Measure the Diameter of a Recessed Hole

It is impossible to measure the diameter of a recessed hole by micrometer or any of the conventional instruments. A lot of difficulty is generally experienced in measuring the diameters of the recessed hole. A simple method of measuring the diameter by two balls of different diameters is given below.

The two balls are placed upon each other as shown in Fig. 9.18, *i.e.* balls must be touching each other and also the sides of the walls. With the help of a depth micrometer the distance from top surface, to the top surface of both the balls is measured. Let these distances be h_1 and

h_2 and the diameter of the balls d_1 and d_2 respectively. O_1 and O_2 are the centres of the two balls.

In rt. ∠d ΔO_1AO_2, $O_1A^2 + AO_2^2 = O_1O_2^2$

or $$O_1A^2 = O_1O_2^2 - AO_2^2$$

$$= \left(\frac{d_1 + d_2}{2}\right)^2 - (h_1 + d_1 - h_2 - d_2)^2$$

$$\text{Diameter of hole} = \frac{d_1 + d_2}{2} + O_1A$$

$$= \frac{d_1 + d_2}{2} + \sqrt{\left(\frac{d_1 + d_2}{2}\right)^2 - (h_1 + d_1 - h_2 - d_2)^2}.$$

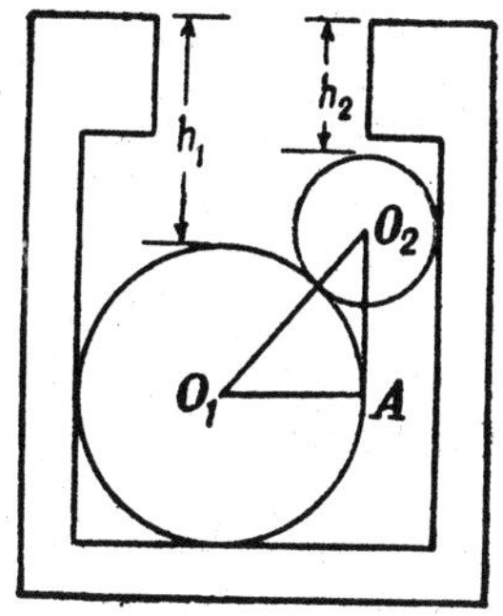

Fig. 9.18

9.16. To Measure the Diameter of a Large Hole Accurately

This method requires the use of two pins, C-clamps, slip gauges and a dial indicator. The work is clamped on the surface plate with C-clamps.

Two pins of equal diameter are placed against the inside edge of the circle and the intervening space between the pins is filled in with a combination of slip gauges. It is not necessary that combination of slip gauges and pins should fit into the space exactly but a slight amount of play can be tolerated and in that case the slip gauges and one pin can swing around the second pin. A dial indicator is clamped in such a position as to register arc of the pin and slip gauges when they swing from side to side as shown in Fig. 9.19.

Let the diameter of pins be d and length of slip gauges l and swing noted by dial indicator S.

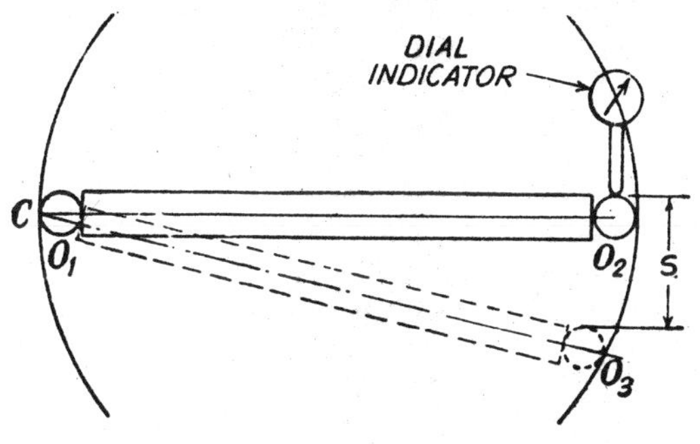

Fig. 9.19

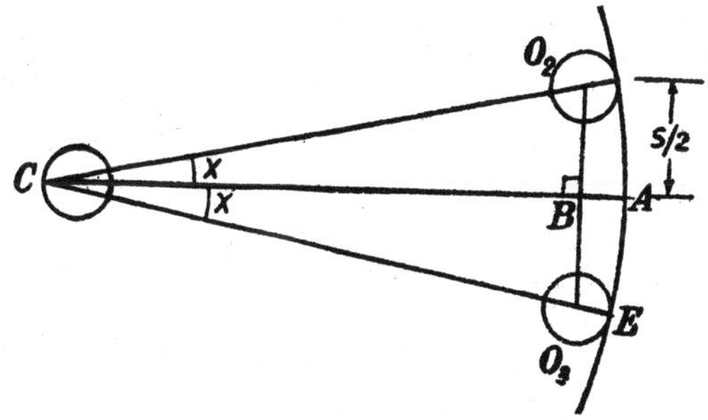

Fig. 9.20

O_1, O_2 and O_3 are the centres of three balls. Line O_1A divides $O_2\hat{O}_1O_3$. Point C lies on circle where O_2O_1 produced meets the circle.

Draw $O_2B \perp O_1A$ and join AE, where A lies on circle and divider of $O_2\hat{O}_1O_3$ (Fig. 9.20) and E lies on the circle where the ball with centre O_3 is touching it.

Now $$\sin X = \frac{O_2B}{CO_2} = \frac{S/2}{\left(1 + \frac{3d}{2}\right)}$$

Let diameter CA of circle be equal to D. Also $A\hat{E}C = 90°$

$\therefore$ in rt. ∠d ΔCEA, $\cos X = \dfrac{CE}{CA} = \dfrac{1 + 2d}{D}$; $\therefore$ $D = \dfrac{1 + 2d}{\cos X}$. Thus D can be calculated.

WORKED EXAMPLES

Problem 9.1. *(a) Explain the method of measuring the included angle and nominal values of diameter for a given tapered hole having small angle of taper with (i) two unequal balls of nearly same diameter, and (ii) two balls and two rollers.*

Also explain the precautions to be taken and the limitation in each case.

(b) In method (i) if balls' dimensions are measured to an accuracy of ± 0.001 mm and the height measurements to an accuracy of ± 0.01 mm, calculate the error likely to be introduced in the measurement of angle of taper.

(c) The nominal dimensions and tolerances for a No. 4. Morse taper ring gauge (in inches) are

Diameter of gauge plane $2.2310^{-0.0003}_{+0.000}$. *Taper per inch on diameter = 0.05193 (0 to 0.000025 fast).*

If two standard balls 1.0625 in. and 30 mm (1.1811 in.) in diameter are used to measure such a gauge, calculate values for depth gauge measurements from the gauge plane over the balls for

(i) Nominal values of diameter and taper.

(ii) Upper limit values for diameter and taper in combination.

What tolerance on component diameter is represented by a step of 0.060 in. on the gauge.

(In this method both inch and metric balls have been used because in one system it becomes difficult to find two balls on nearly same diameter ; and with slow tapers only small difference in size of balls is required.)

(d) A taper ring gauge is resting on a surface plate with its small end down. Its taper and diameter are measured using a 10.0 mm ball together with a 2.5 mm cylinder and a 15.0 mm ball with a 1.5 mm cylinder respectively. Heights measured over the balls with reference to the surface plate are respectively 18.00 mm and 46.05 mm. Calculate the value of the included angle of taper.

Solution. *(a) (i)* It may be noted from Fig. 9.21, that sometimes it may be necessary to put slip gauges of equal height below and above the gauge, if the ball falls outside the smaller bore (as in this case) or the bigger bore. Usually the balls of such sizes are selected that one rests in the cone near the bigger end and one near the smaller end. Sometimes it may become necessary to use balls from both inch and metric systems [Refer part (*c*) of this question]. The distances of the upper surfaces of the balls are measured from a datum surface, such as one end of the gauge while measuring with micrometer depth gauge, or the surface plate when measuring with a vernier depth gauge. On the basis of these measurements, the difference between the height of the centres of two balls can be determined.

$$\left[e.g.\ H_2 - H_1 - \left(\frac{D_1 - D_2}{2}\right) \text{ in this case.} \right]$$

If taper angle is θ, height of the taper gauge is H and the balls of diameter D_1 and D_2 are used.

then $$\sin\frac{\theta}{2} = \frac{O_1A}{O_1O_2} = \frac{(D_1 - D_2)/2}{(H_2 - H_1) - (D_1 - D_2)/2}$$

Diameters in the planes of the ball centres are

$$D_{B_1} = D_1 \sec\frac{\theta}{2} \qquad \text{and} \qquad D_{B_2} = D_2 \sec\frac{\theta}{2}$$

Diameters at the ends of the taper are

$$D_B = D_{B_1} + 2\left(\frac{D_1}{2} + H_1\right)\tan\frac{\theta}{2} = D_1 \sec\frac{\theta}{2}$$

$$+ (D_1 + 2H_1)\tan\frac{\theta}{2}$$

$$D_B = D_{B_2} - 2\left(H - H_2 - \frac{D_2}{2}\right)\tan\frac{\theta}{2}$$

$$= D_2 \sec\frac{\theta}{2} - (2H - H_2 - D_2)\tan\frac{\theta}{2}.$$

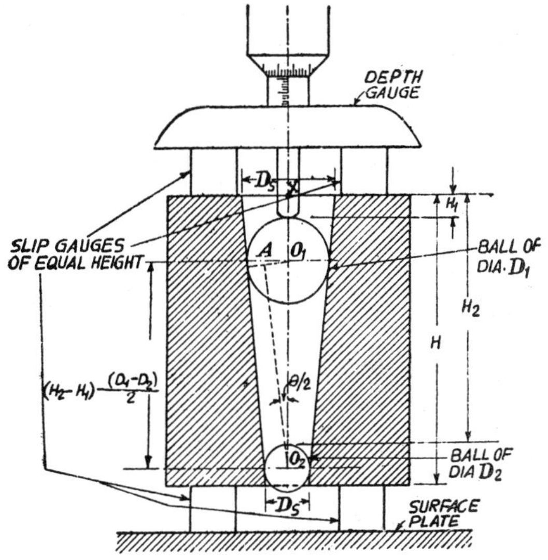

Fig. 9.21

For testing the uniformity of the taper, a third measurement is given by choosing a ball of such a size which gets seated midway along the taper.

In this method, practical precautions have to be taken in order to prevent the balls jamming in the gauge. The weight of the ball and the force applied to it by the micrometer can cause appreciable axial movement. For this purpose the gauge may be held vertically, with one finger through the small end and the ball gently lowered into position. As the angle of taper is small, if care is taken, relatively coarse height measurements give high accuracy of diameter measurement.

(*ii*) In the previous method, sometimes it so happens, particularly with slow tapers that two balls can't be found to fit in the taper gauge at suitable positions. This difficulty can be overcome by putting a cylinder between such ball and one side of the taper gauge. In this method too, there being point contact between balls, cylinders and gauge, the compression effect must be minimised.

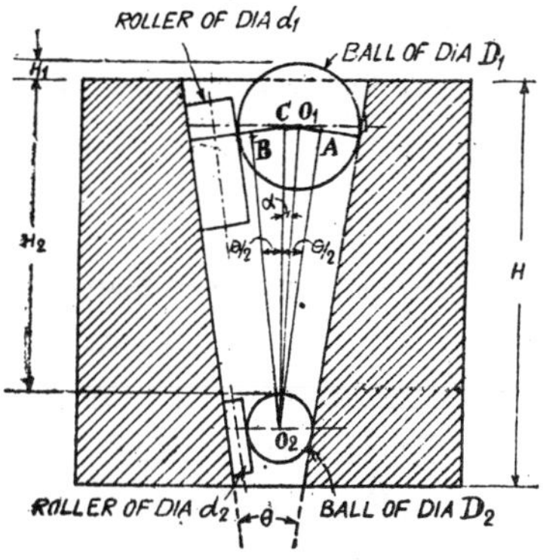

In Fig. 9.22, O_1 and O_2 are the centres of balls of diameters D_1 and D_2 ; and O_2A is parallel to the right tapered edge and O_2B is parallel to the left tapered edge and O_2C is drawn perpendicular to the surface plate or parallel to the axis of the tapered hole.

Fig. 9.22

In rt. ∠d ΔO_1BO_2,

$$\sin O_1\hat{O}_2B = \frac{O_1B}{O_1O_2}; \qquad \text{or } \sin\left(\frac{\theta}{2} + \alpha\right) = \frac{O_1B}{O_1O_2} \qquad ...(i)$$

Similarly in rt. ∠d ΔO_1AO_2,

$$\sin O_1O_2A = \frac{O_1A}{O_1O_2}; \quad \text{or } \sin\left(\frac{\theta}{2} - \alpha\right) = \frac{O_2A}{O_1O_2} \qquad ...(ii)$$

Expanding and adding (*i*) and (*ii*), $2\sin\frac{\theta}{2}\cos\alpha = \frac{O_1B + O_1A}{O_1O_2}$

$$\therefore \quad \sin\frac{\theta}{2} = \frac{O_1B + O_1A}{O_1O_2} \times \frac{\sec\alpha}{2} \qquad ...(iii)$$

$$O_1B = \left(\frac{D_1}{2} + d_1\right) - \left(\frac{D_2}{2} + d_2\right), \ O_1A = \frac{D_1}{2} - \frac{D_2}{2}$$

$$O_1O_2 = O_2C. \sec\alpha = \left(H_1 + H_2 - \frac{D_1}{2} + \frac{D_2}{2}\right) \sec\alpha$$

Substituting the values in Eq. (*iii*), we have

$$\sin\frac{\theta}{2} = \frac{\left(\frac{D_1}{2} + d_1 - \frac{D_2}{2} - d_2 + \frac{D_1}{2} - \frac{D_2}{2}\right)}{\left(H_1 + H_2 - \frac{D_1}{2} + \frac{D_2}{2}\right)\sec\alpha} \times \frac{\sec\alpha}{2} = \frac{D_1 - D_2 + d_1 - d_2}{2H_1 + 2H_2 - D_1 + D_2}$$

Diameters in the planes of the ball centres are

$$D_{B_1} = (D_1 - d_1) \sec\theta/2 \ ; D_{B_2} = (D_2 + d_2) \sec\theta/2$$

Diameters at smaller and bigger ends are

$$D_S = D_{B_2} - (2H - 2H_2 - D_2) \tan\theta/2 \ ; \text{ and } D_B = D_{B_1} + (D_1 - 2H_1) \tan\theta/2.$$

(*b*) In the case of taper measurement by two unequal balls, angle of taper is given by the relation

$$\sin\frac{\theta}{2} = \frac{\frac{D_1 - D_2}{2}}{(H_2 - H_1) - \left(\frac{D_1 - D_2}{2}\right)} = \frac{x}{y} \text{ (say)}$$

$\therefore$ The error in the angle is $\delta\theta \cong \frac{\partial\theta}{\partial y} . \delta x + \frac{\partial\theta}{\partial y} . \delta y$

$$\cong 2\sec\frac{\theta}{2} . \frac{1}{y} . \delta x - 2 . \sec\frac{\theta}{2}\frac{x}{y^2} . \delta y = \frac{2}{y\sqrt{y^2 - x^2}}(y . \delta x - x . \delta y)$$

Since x involves two measurements, *i.e.* the diameters of the balls, the error

$$\delta x = \pm 0.002 \text{ mm.}$$

Similarly $\delta y = 0.01 + 0.01 + 0.001 + 0.001 = 0.022.$

Substituting these values in equation for $\delta\theta$, the values of $\delta\theta$ can be calculated.

(*c*) For the No. 4 Morse taper ring, it is given that taper per inch diameter = 0.05193

$$\therefore \quad \tan\frac{\theta}{2} = \frac{0.01593}{2}, \text{ or } \frac{\theta}{2} = \tan^{-1}\frac{0.05193}{2} = 1° \ 29' \ 15''$$

Referring to Fig. 9.21, $O_1O_2 = AO_1 \operatorname{cosec}\frac{\theta}{2}$

and $$AO_1 = \frac{D_1 - D_2}{2} = \frac{1.1811 - 1.0625}{2} = \frac{0.1186}{2} = 0.0593 \text{ inch.}$$

$$\therefore \quad O_1O_2 = 0.0593 \operatorname{cosec} 1° \ 29' \ 15'' = 2.2845 \text{ inch}$$

Diameter of gauge through centre $O_1 = \frac{1.1811 \sec\theta}{2} = 1.1815$ inch

(*i*) Now $$H_1 = O_1X - \frac{D_1}{2} = \frac{(1.2310 - 1.1815)}{0.05193} - \frac{1.1815}{2}$$

$$= \frac{0.0495}{0.05193} - 0.5905 = 0.9531 - 0.5905 = 0.3626 \text{ inch. } \textbf{Ans.}$$

and $$H_2 = O_1X + O_1O_2 - \frac{D_2}{2} = 0.9531 + 2.2845 - \frac{1.0625}{2}$$

$$= 3.2376 - 0.5312 = 2.7064 \text{ inch. } \textbf{Ans.}$$

(*ii*) The variations due to tolerances are calculated as small differences instead of recalculating for altered initial values.

Effect of tolerance of 0.0003 on diameter $= \frac{0.00015 \cot \theta}{2}$ on height = 0.0058 (both balls) lower.

Similarly 0.000025 taper over 1 in. (increasing angle) or 0.0003 inch on taper over 12 inch = rise of 0.0058 at 12 inch from gauge plane.

At centre of 30 mm ball, O_1X from gauge plane, rise

$$= O_1X \times \frac{0.0058}{12} = \frac{0.9531}{12} \times 0.0058 = 0.0005 \text{ inch.}$$

At centre of 1.0625 inch ball, O_2X from gauge plane, rise $= 3.2376 \times \frac{0.0058}{12} = 0.0016$

∴ New depth of 30 mm ball = 0.3626 + 0.0058 – 0.0005 = 0.3679 inch. **Ans.**

And of 1.0625 in ball = 2.7064 + 0.0058 – 0.0016 = 2.7106 inch. **Ans.**

Diameter equivalent of 0.060 inch step = 0.060 × 0.05193 = 0.0031 inch. **Ans.**

(*d*) For this method, (Refer Fig. 9.22) $\sin \frac{\theta}{2} = \frac{D_1 - D_2 + d_1 - d_2}{2H_1 + 2H_2 - D_1 + D_2}$

In the numerical, $D_1 = 15$ mm, $D_2 = 10$ mm, $d_1 = 1.5$ mm, $d_2 = 2.5$ mm

$$H_1 + H_2 = 46.05 - 18.00 = 28.05 \text{ mm}$$

∴ Substituting the values, we have $\sin \frac{\theta}{2} = \frac{15 - 10 + 1.5 - 2.5}{2 \times 28.05 - 15 + 10} = \frac{4}{51.10}$

∴ $$\frac{\theta}{2} = 4° \, 30', \text{ and } \theta = 9°. \textbf{ Ans.}$$

Problem 9.2. (*a*) *Explain in brief the method of measuring a taper plug gauge by rollers, slip gauges and micrometer.*

(*b*) *What precautions must be taken in this method ?*

(*c*) *Derive an expression for the error likely to creep in the measurement of angle of taper by this method.*

(*d*) *Two taper plug gauges are being measured by this method. In both the cases, dimensions over rollers are measured at the height difference of 100 mm and these values are 80 mm and 8 mm respectively with a possible error of 0.005 mm. The angle of taper is 44° for the first and 4.5° for the second. Calculate the error likely to creep in. What conclusion you draw from this question ?*

Solution. (*a*) In this method, measurements are made over two equal rollers standing on slip gauges at each side of the taper plug gauges at two positions, one near the lower end and one near the upper end as shown in Fig. 9.23.

The various formulae used are $\tan \frac{\theta}{2} = \frac{L_1 - L_2}{2(H_1 - H_2)}$

The diameter in the planes through the rollers centres are

$$D_{R_1} = L_1 - D\,(1 + \sec \theta/2)$$

and $D_{R_2} = L_2 - D(1 + \sec \theta/2)$

Diameters at bigger and smaller ends are

$$D_S = D_{R_2} - 2(H_2 - h_2) \tan \theta/2$$

and $D_B = D_{R_1} + 2(h_1 - H_1) \tan \theta/2$

The uniformity of taper is checked by taking a third diametral measurement about mid-way along the gauge and comparing this with the value at this position calculated from the two other readings.

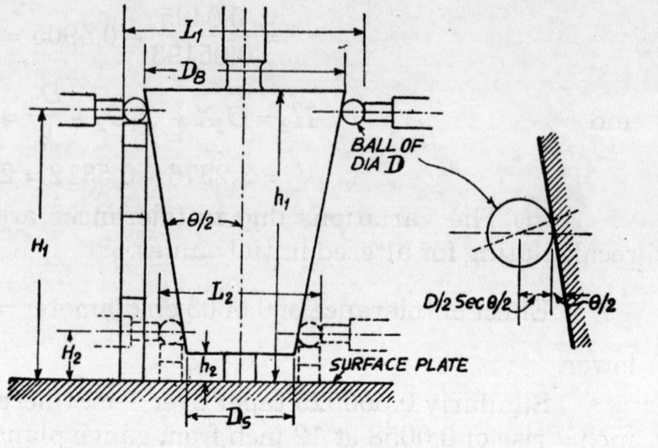

Fig. 9.23

(b) The following precautions are essential for obtaining the satisfactory results from this method :

(i) The plug gauge should be set upon its small ends, so that the force exerted by the micrometer used for the diametral measurement tends to wedge the rollers against the plug and the slip gauges ; if the gauge is set with its bigger end down, then the rollers have the tendency to ride up the taper.

(ii) The force exerted by micrometer must be kept to a minimum to avoid errors due to compression effect, there being a point contact.

(iii) The rollers should preferably be of about the same diameter as the micrometer anvils.

(c) Possible error in the angle of taper can be calculated as follows :

Let $L_1 - L_2 = L$, and $H_1 - H_2 = H$; Then $\tan \dfrac{\theta}{2} = \dfrac{L}{2H}$

$$\delta\theta \cong \frac{\partial\theta}{\partial L} \cdot \delta L + \frac{\delta\theta}{\partial H} \cdot \partial H \cong 2 \cdot \cos^2 \frac{\theta}{2} \cdot \frac{1}{2H} \cdot \delta L - 2 \cdot \cos^2 \frac{\theta}{2} \cdot \frac{L}{2H^2}\, \delta H$$

since $$\cos \frac{\theta}{2} = \frac{H}{\sqrt{\left(\frac{L}{2}\right)^2 + H_2}}, \;\therefore\; \delta\theta \cong \frac{2H^2}{\left(\frac{L}{2}\right)^2 + H^2}\left[\frac{1}{2H} \cdot \delta L - \frac{L}{2H^2} \cdot \delta H\right]$$

$$\cong \frac{1}{\left(\frac{L}{2}\right)^2 + H^2}\,[H \cdot \delta L - L \cdot \delta H]$$

Usually the error in H is very small, being that in slip gauges used.. If this be neglected, and the expression reduces to $\delta\theta \cong \dfrac{H}{(L/2)^2 + H^2} \cdot \delta L$.

(d) In first case, $$\delta\theta \cong \frac{H}{(L/2) + H^2}\,\delta L \cong \frac{100}{(80/2)^2 + (100)^2} \times 0.005$$

$$\cong \frac{100}{1600 + 10000} \times 0.005 \cong \frac{100}{11600} \times 0.005 \cong 0.0000431 \text{ radian} \cong 9''.$$

As the angle of taper = 44° ; Percentage error = $\frac{9''}{44°} \cong 0.006\%$

In second case, $\delta\theta \cong \frac{100}{(8/2)^2 + (100)^2} \cong \frac{100}{10016} \cong 10$ sec

and Percentage error $\cong \frac{10''}{4.5°} \cong 0.06\%$

From this question, it is clear that the error in the measured value of the angle of slow taper is larger than that of a quick taper.

Problem 9.3. (*a*) *Describe the method of measuring included angle by three discs method.*

(*b*) *Also derive expressions for the possible errors likely to occur in the measurement.*

Solution. (*a*) In this method, three hard steel discs of rollers of equal diameter (D) are arranged as shown in Fig. 9.24 (*a*). The distance H is measured by a depth micrometer gauge or a straight edge and slip gauges.

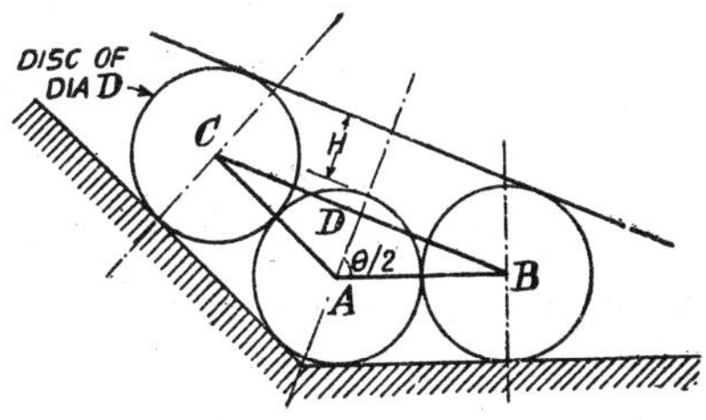

Fig. 9.24 (*a*)
For angles between 90° and 180°

In ΔABC, $AD = H$

$$\therefore \quad \cos\frac{\theta}{2} = \frac{AD}{AB} = \frac{H}{D} \qquad ...(i)$$

Though theoretically this method can be used for angles between 180 degrees and about 60 degrees, but for angles below about 90 degrees, it is preferable to measure the gap between the two outer discs (L) as shown in Fig. 9.24 (*b*), in which case

$$\sin\frac{\theta}{2} = \frac{L + D}{2D} \qquad ...(ii)$$

For angles less than 60° it may be preferable to use only two discs as shown in Fig. 9.24 (*c*) and the distance L is measured by slip gauges. Referring to Fig. 9.24 (*c*), obviously $BC = L$.

$$\therefore \quad \sin\theta = \frac{BC}{AB} = \frac{L}{D} \qquad ...(iii)$$

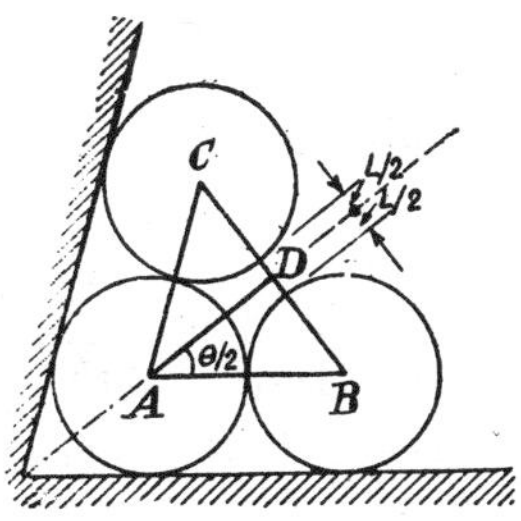

Fig. 9.24 (*b*)
For angles between 90° and 60°

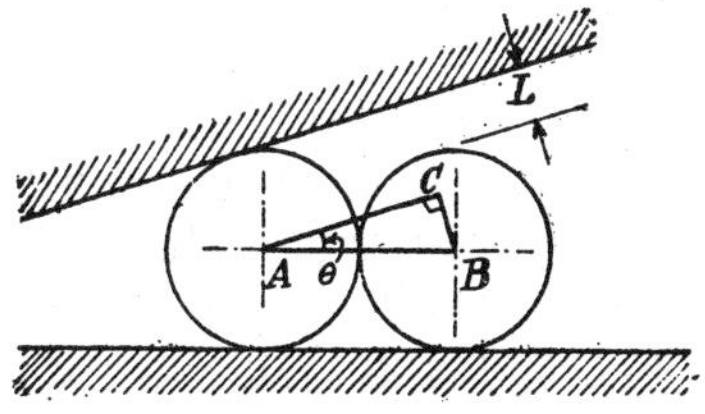

Fig. 9.24 (*c*)
For angles below 60°

(*b*) Possible errors in the measured angles are determined as follows :

For equation (*i*), $\cos\frac{\theta}{2} = \frac{H}{D}$;

$$\delta\theta \cong \frac{\partial\theta}{\partial D}\cdot\delta D + \frac{\partial\theta}{\partial H}\cdot\delta H \cong 2\,\text{cosec}\,\frac{\theta}{2}\cdot\frac{H}{D^2}\,\delta D - 2\,\text{cosec}\,\frac{\theta}{2}\cdot\frac{1}{D}\cdot\delta H = \frac{2\,\text{cosec}\,\theta/2}{D^2}(H.\,\delta D - \delta H)$$

If $D = 40$ mm with a possible error of 0.001 mm, and $H = 20$ mm with a possible error of 0.01 mm then $\cos\frac{\theta}{2} = 0.5$ and $\theta = 120°$; and $\delta\theta = \frac{2 \times \text{cosec } 60°}{40^2}(20 \times 0.001 + 40 \times 0.01)$

(Here + sign is used to take the worst case of all the errors being added up)

$$\delta\theta = \frac{2 \times 1.1547}{1600}(0.02 + 0.4) \cong \frac{2.309}{1600} \times 0.42 \cong 0.0058122 \text{ radian} \cong 20.1 \text{ min.}$$

It is obvious that unless measurements are made to still higher accuracy, errors in angle measured will be high. For equation (*ii*), $\sin\frac{\theta}{2} = \frac{L + D}{2D}$

$$\delta\theta = \frac{\partial\theta}{\partial L}.\delta L + \frac{\partial\theta}{\partial D}.\delta D \cong 2\sec\frac{\theta}{2}.\frac{1}{2D}.\delta L - 2\sec\frac{\theta}{2}.\frac{L}{2D^2}.\delta D \cong \frac{\sec\theta/2}{D^2}(D.\delta L - L.\delta D)$$

For equation (*iii*) $\sin\theta = \frac{L}{D}$; $\delta\theta \approx \frac{\partial\theta}{\partial L}.\delta L + \frac{\partial\theta}{2D}.\delta D \approx \frac{\sec\theta/2}{D^2}(D.\delta L - L.D)$.

Problem 9.4. *Describe the various methods used to measure the radius of a circular arc on a component, when the centre of the circle is not accessible and the radius is large in comparison with the dimensions of the component.*

Solution. One of the simplest method is such cases is to use surface plate and rollers. Two set-ups are possible as shown in Fig. 9.25 (*a*) and (*b*).

In the first method, the work is placed on a surface plate and two equal rollers of diameter (d) are placed in position as shown in Fig. 9.25 (*a*). The distance over the rollers (L) is measured by vernier gauge. Let R be the radius of circular arc.

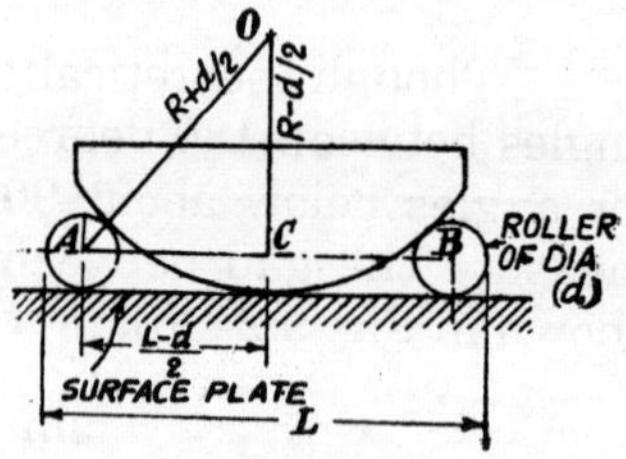

Fig. 9.25 (*a*)

In ΔOAC, $OA^2 = OC^2 + AC^2$

or $\left(R + \frac{d}{2}\right)^2 = \left(R - \frac{d}{2}\right)^2 + \left(\frac{L - d}{2}\right)^2$

Solving it, we get $R = \frac{(L - d)^2}{8d}$

If δL and δd be errors in measurements of L and d, then error δR in measurement of radius R is $\delta R = \frac{\partial R}{\partial L}\delta L + \frac{\partial R}{\partial d}\delta d = \frac{L - d}{4d}\delta L - \frac{(L^2 - d^2)}{8d^2}\delta d = (L - d)/8d^2\,[2d.\,\delta L - (L + d)\,\delta d]$.

In the other method [Fig. 9.25 (*b*)], two rollers of equal diameter (d) are set up against a straight edge and clamped on a surface plate with a certain standard end gauge of length (L) set accurately between them. The end gauge is then removed and the work is held against the opposite sides of the rollers. Slip gauges of appropriate dimension (l) are then inserted between the work and the straight edge. Let R be the radius of the work.

In Fig. 9.25 (*b*), DE = dimension of slip gauges inserted between work and the straight edge = l.

For circle AFB, AD is the chord, $\therefore\ AC \times CB = FC \times (CO + FO)$ ...(*i*)

$$AC = CB = \frac{L + d}{2},\ FE = DC = \frac{d}{2}\ DE = l \text{ and } FC = FE - DE = \frac{d}{2} - l$$

$$FC = CD + FD = \frac{d}{2} + \frac{d}{2} - l = d - l$$

$$CO = FO - FC = R + \frac{d}{2} - d + l$$

$$CO + FO = R + \frac{d}{2} + R + \frac{d}{2} - d + l = 2R + l$$

Substituting the values in equation (*i*)

$$\left(\frac{L+d}{2}\right)\left(\frac{L+d}{2}\right) = (d - l)\,(2R + l).$$

Knowing the values of L, d and l ; and substituting them in this equation, the value of R can be found out.

In such a test, uniformity of radius is checked by moving the work round to various positions and noting any variations in slip-gauges needed.

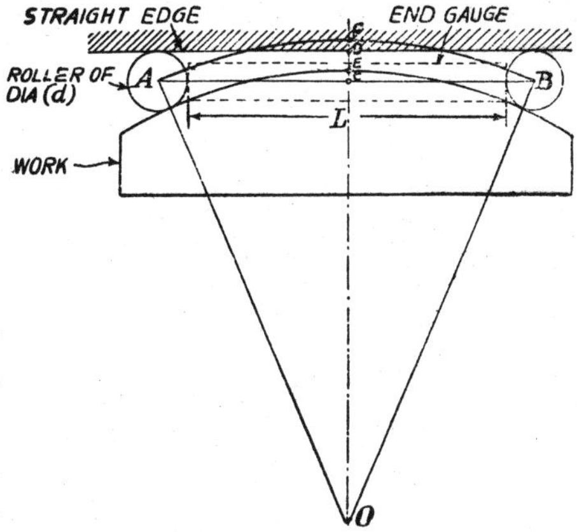

Fig. 9.25 (*b*)

Problem 9.5. *In Prob. 9.4, for first method, derive an expression for the possible error. Determine its value and the percentage error, if distance measured over the roller = 475 mm with a possible error of 0.05 mm and the rollers are 25 mm in diameter within 0.001 m.*

Assuming a case that the surface plate is not exactly true and as a result the point of contact of the workpiece with the surface plate is not on the tangent to the roller and that the height of the lowest point of the workpiece above the tangent is 0.002 mm, determine the calculated radius.

Solution. For the first method, $R = \dfrac{(L-d)^2}{8d}$.

Let errors in the measurement of L and d be δL and δd respectively, then error in

$$R \cong \delta R = \frac{\partial R}{\partial L}\,.\,\delta L + \frac{\partial R}{\partial d}\,.\,\delta d \cong \frac{L-d}{4d}\,.\,\delta L - \frac{(L^2 - d^2)}{8d^2}\,.\,\delta d \cong \frac{L-d}{8d^2}\,[2d.\,\delta L - (L+d)\,\delta d]$$

$L = 475$ mm, $d = 25$ mm and $\delta L = 0.05$ mm and $\delta d = 0.001$ mm

$$\delta R = \frac{450}{8 \times 25^2}\,[50 \times 0.05 + 500 \times 0.001]$$ [Errors are likely to add up ∴ use + sign]

$$\cong \frac{450}{8 \times 25 \times 25}\,[2.5 + 0.5] \cong \frac{18 \times 3}{8 \times 25} \cong 0.27 \text{ mm}$$

Calculated $$R = \frac{(475 - 25)^2}{8 \times 25} = \frac{450 \times 450}{8 \times 25} = 1012.5 \text{ mm}$$

∴ Percentage error $$= \frac{\delta R}{R} \times 100\% = \frac{0.027}{1012.5} = 0.0267\%$$

Due to surface plate being slightly lower at the point of contact by amount δS ; dimension OC in Fig. 9.25 (*a*) becomes.

$$= R - \frac{d}{2} + \delta S\ ;\ \delta S = 0.002 \text{ mm}\ ;\ \therefore\ R = \frac{(L-d)^2}{8(d - \delta S)} - \frac{\delta S}{2}$$

Inserting the values, we get $R = 1012.6$ mm, *i.e.* here is a difference of 0.1 mm. **Ans.**

Problem 9.6. *Describe three methods to determine the diameters of parallel bores using steel balls, dial indicator and slip gauge.*

Solution. (*i*) The job to be measured is assumed to have its one face perpendicular to the axis of job. The balls of known diameters d_1 and d_2 are kept one over the other in bore as shown in Fig. 9.26. The values of d_1 and d_2 should be greater than diameter of bore. The height

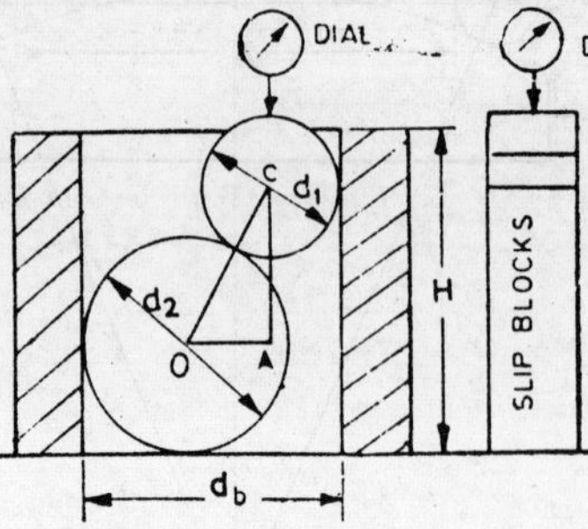

Fig. 9.26

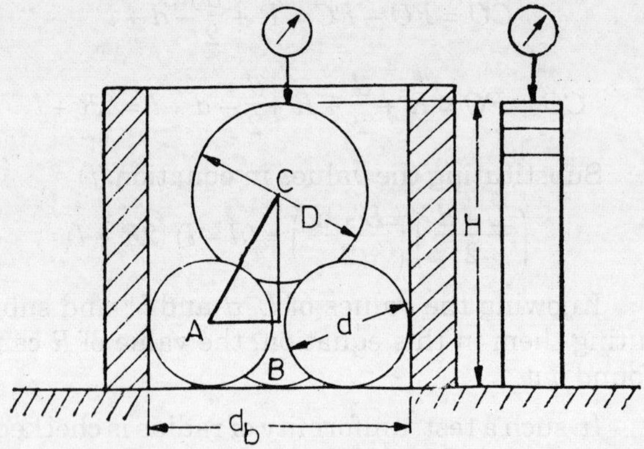

Fig. 9.27

of high point of upper ball from surface plate is measured with the help of slip gauges and dial indicator. The slip blocks are piled such that the dial indicator reading at slip blocks and highest point of ball is same. Knowing distance H, the diameter of bore can be calculated as follows :

$$\text{Diameter of bore} = \frac{d_1 + d_2}{2} + AO$$

$$AO^2 = OC^2 - CA^2 = \left(\frac{d_1 + d_2}{2}\right)^2 - \left(H - \frac{d_1 + d_2}{2}\right)^2 = -H^2 + (d_1 + d_2) \times H$$

and $$AO = \sqrt{(d_1 + d_2)H - H^2}\ ;\ \text{So}\ \ d_b = \frac{d_1 + d_2}{2} + \sqrt{H(d_1 + d_2) - H^2}.$$

(*ii*) The diameter of bore can also be found by using three equal balls of diameter (d) less than half of the bore diameter and a fourth ball of any diameter (D) but less than 0.75 of the bore diameter. Balls are placed as shown in Fig. 9.27. The height of highest point of upper ball is measured accurately with the help of slip gauge and dial mounted on a height gauge. It will be appreciated that bigger ball of diameter D resting over three balls of equal diameter d will always be located in the centre of bore, and therefore,

Diameter of bore $d_b = d + 2AB$

$$(AB)^2 = (AC)^2 - (CB)^2 = \left[\frac{D + d}{2}\right]^2 - \left[H - \frac{D + d}{2}\right]^2$$

$$AB = \sqrt{H(D + d) - H^2}\ ;\ \text{So}\ \ d_b = d + 2\sqrt{H(D + d) - H^2}.$$

(*iii*) Figs. 9.28 and 9.29 show another way for checking diameter of bore. Here 2 balls (or rollers are taken such that $(d_1 + d_2)$ is less than the diameter of bore. Suitable slip blocks are inserted between the balls (or rollers). Diameter of bore = $(d_1 + d_2)$ + slip inserted.

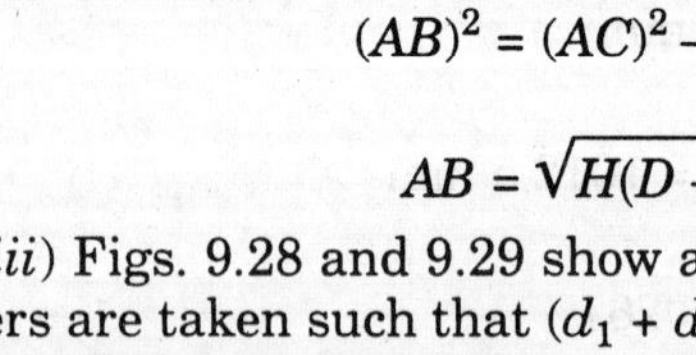

Fig. 9.28

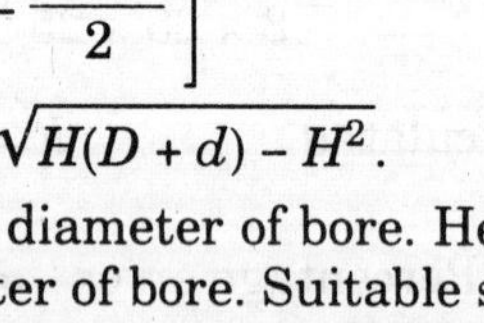

Fig. 9.29

This method is specially useful when :

(*a*) The thickness of job is less.

(*b*) Required combinations of balls are not available.

(*c*) End surface of job is not perpendicular to the axis of the job.

Problem 9.7. *Describe a method to measure radius on a part.*

Solution. Quite often it is required to check the radius of a gauge or template. The special methods are generally adopted according to the design of gauge of template and a suitable reference is chosen. For measuring big radius the special method using balls and rollers is the only way. Rollers are found more convenient to use.

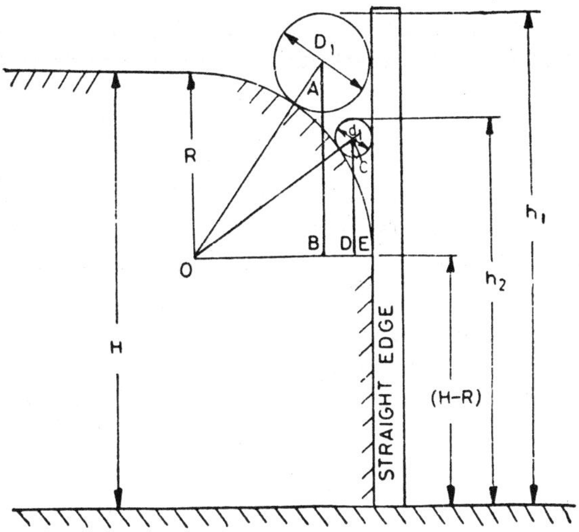

Fig. 9.30

Fig. 9.30 shows a small template whose radius is to be determined. It has been placed on a surface plate and an adjacent surface is touching the straight block. Height of template, H is measured by slip gauges and height gauge. Roller of known diameter is placed and allowed to rest in between the straight block and radius, and h_1 is measured over the roller.

In ΔAOB, $\qquad AO^2 = OB^2 + AB^2$

$$\left(R + \frac{D_1}{2}\right)^2 = \left(R - \frac{D_1}{2}\right)^2 + \left[\left(h_1 - \frac{D_1}{2}\right) - (H - R)\right]^2$$

From the above expression R can be calculated. For cross checking the results, other roller of known diameter is taken and R is again calculated.

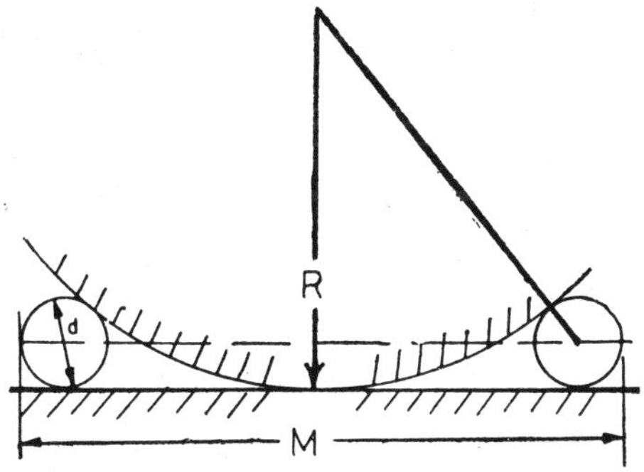

Fig. 9.31

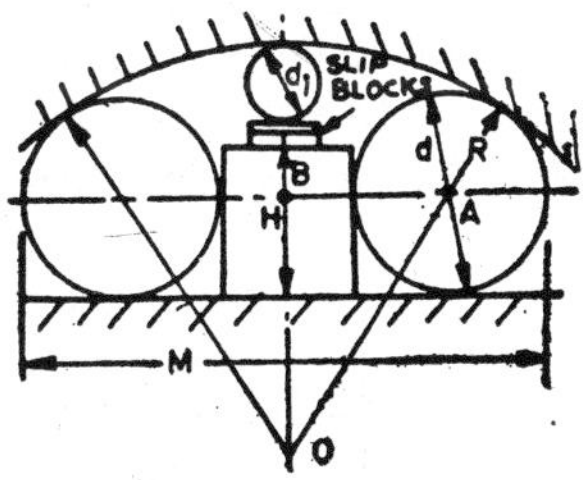

Fig. 9.32

(*ii*) External big radius of cylindrical form or spherical forms can be calculated using standard balls or rollers. Fig. 9.31 represents the way of checking external radius.

Two rollers of equal diameters are allowed to take position touching the surface plate and the curved surface. The size over these rollers is measured and radius R can be calculated as follows :

$$R^2 = \left(\frac{M - d}{2}\right)^2 + \left(R + \frac{d}{2}\right)^2$$

In above expression only R is unknown, which can be calculated easily.

(*ii*) Internal radius of segment can also be found by the special methods. Fig. 9.32 represents the method of placing the balls and slips. $OA^2 = OB^2 + BA^2$

$$\left(R - \frac{d}{2}\right)^2 = \left[R - (d_1 + S + H) + \frac{d}{2}\right]^2 + \left(\frac{M - d}{2}\right)^2$$

In above expression only R is unknown which can be calculated.

Problem 9.8. *Check smaller and larger diameters of tapered gauge within an accuracy of = 0.001 mm without using sophisticated measuring instruments. The edges have been made round by giving radius ; so the imaginary smaller and larger diameters have to be measured.*

Solution. The diameters can be checked by using one precision ball or roller of known diameter. For obtaining high accuracy it is essential that only one set-up be used to take readings. Set-up is shown in Fig. 9.33. Slip gauges, S_2 and S_1 are so adjusted that upper taper

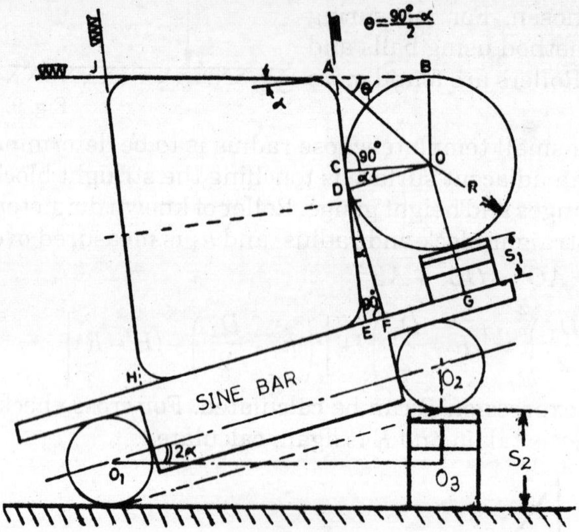

Fig. 9.33

portion of gauge is exactly horizontal and the highest point of the roller of radius R over slip gauge S_1 is in line with upper taper surface of gauge.

The following calculation will give required results

$$2\alpha = \sin^{-1}\frac{S_2}{\text{Distance between rollers of sine bar}}$$

So the angle a can be calculated.

The smaller diameter

$$AE = R\cot\theta + R\tan\alpha + (S_1 + R)\sec\alpha, \qquad \text{where } \theta = \frac{90 - \alpha}{2}$$

Knowing smaller diameter, the larger diameter can be known by measuring height of the job.

Problem 9.9. *Describe method to determine the depth (h), width (w) and angle (q) of the dovetail shown in Fig. 9.34.*

Solution. For determining depth h, place a mandrel over top surface AB and try suitable slip gauge combination that would just pass under the mandrel and bottom surface of dovetail. The slip gauge pile should be tried at several positions to obtain average value of h.

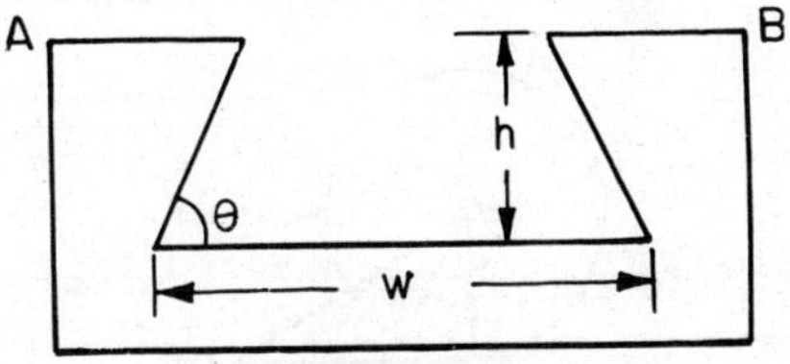

Fig. 9.34

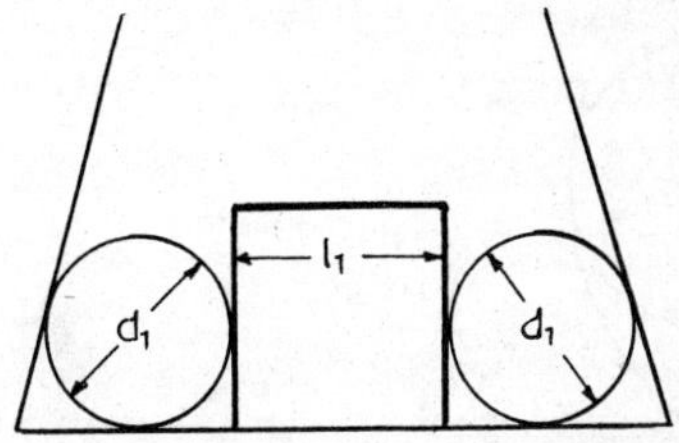

Fig. 9.35

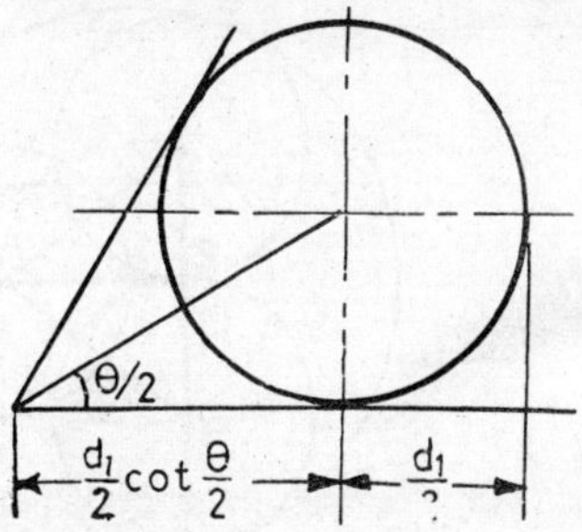

Fig. 9.36

To determine the values of w, θ, place two rollers of diameter d_1 in vees and insert suitable gauge pile of length l_1 in between the space. Similarly place other big rollers of diameter d_2 and determine length l_2 of slip gauge to fill up the remaining space.

From Figs. 9.35 and 9.36 it will be noted that

$$w = l_1 + 2\left[\frac{d_1}{2} + \frac{d_1}{2}\cot\frac{\theta}{2}\right] = l_1 + d_1\left(1 + \cot\frac{\theta}{2}\right)$$

Similarly $$w = l_2 + d_2\left(1 + \cot\frac{\theta}{2}\right); \text{ or } l_1 + d_1\left(1 + \cot\frac{\theta}{2}\right) = l_2 + d_2\left(1 + \cot\frac{\theta}{2}\right)$$

$$l_1 + d_1 + d_1\cot\frac{\theta}{2} = l_2 + d_2 + d_2\cot\frac{\theta}{2}; \text{ or } \cot\frac{\theta}{2}(d_2 - d_1) = l_1 - l_2 + d_1 - d_2$$

and $$\cot\frac{\theta}{2} = \frac{l_1 - l_2 + d_1 - d_2}{d_2 - d_1}; \text{ and } \theta = 2\cot^{-1}\left(\frac{l_1 - l_2 + d_1 - d_2}{d_2 - d_1}\right)$$

As the values l_1, l_2, d_1 and d_2 are known, the value of θ can be calculated.

Knowing the value of θ, value of w can be calculated from any one of the above equations.

Problem 9.10. *A cylindrical ring gauge, having its end face square to its axis stands on a lapped surface plate and its diameter is measured by placing three balls of 10 mm diameter in the gauge, two of them resting on the plate and the third making contact with other two and the bore of the gauge only. The latter ball lies 2.50 mm above the other two. Determine the diameter of the gauge.*

What is the limitation of this method ?

Solution. Let A, B and C be the centres of the three balls. Let C be the projection of centre C on a horizontal plane passing through the centres A and B.

The circumscribing circle of the triangle formed by joining points A, B and C' will be 10 mm smaller in diameter than the diameter of gauge.

$$AB = BC = CA = 10.0\ ;$$

$$EC = 10.0 \times \sqrt{3}/2 = 8.66 \text{ mm}$$

$$(EC')^2 = EC^2 - (2.5)^2 = 75 - 6.25 = 68.75$$

and $$EC = 8.3$$

To find radius of circle through ABC'

$$AO = OC'\ ; \text{ Let } EO = x$$

$$AO^2 = AE^2 + EO^2, \text{ or } (OC')^2 = AE^2 + EO^2$$

or $$(EC' - OE)^2 = AE^2 + x^2$$

or $$(8.3 - x)^2 = AE^2 + x^2 = 5^2 + x^2$$

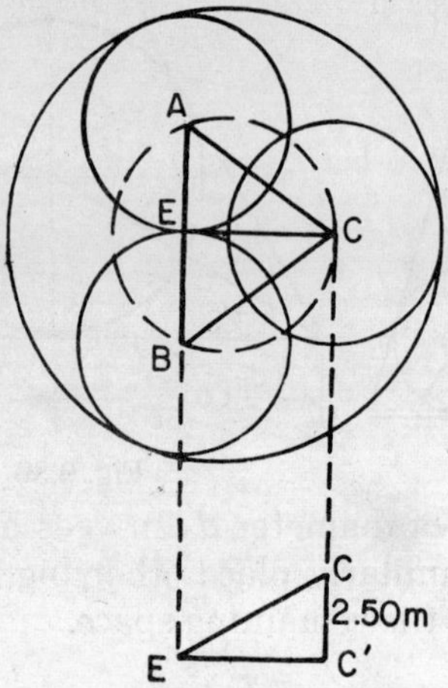

Plan of the set up.

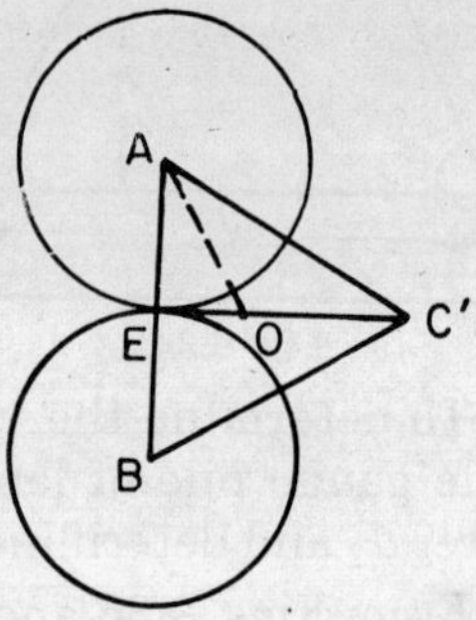

View through a horizontal plane passing through the centres A and B.

Fig. 9.37

$$\text{or} \quad 8.3^2 + x^2 - 2 \times 8.3\, x = 25 + x^2$$

$$\text{or} \quad 16.6x = 68.75 - 25 = 43.75$$

$$\text{and} \quad x = 43.75/16.6 = 2.635$$

$$\therefore \quad C'O = EC' - EO = 8.3 - 2.635 = 5.665$$

Thus is the radius of circle through ABC'

Diameter $= 11.33$ mm

Diameter of circle circumscribing balls or diameter of gauge

$= 11.33 + 10.0 = 21.33$ mm

The limitation of this method is that this does not measure a true diameter in one plane and variations in diameter along and round the cylinder will affect accuracy. This also does not detect the ovality. However if the height difference between the ball is quite small and the cylinder is fairly uniform in diameter, then quite good accuracy can be obtained and any error in height measurement will give only small errors in diameter.

Problem 9.11. *Obtain an expression for the measurement of the diameter of a cylindrical ring using two balls of equal diameter d_1 less than the radius of the gauge and another large ball of diameter d_2 which will rest on the two smaller balls and the ball of the gauge.*

Solution. Let A, B and C be the exact centres of three balls and C' the projection of C on a horizontal plane passing through centres A and B. Next we can locate a circumscribing circle which is concentric with the diameter to be measured and this passes through the centres of the lower balls and through another point which is perpendicular on horizontal plane through a point at a distance along the diameter of the larger ball from its point of contact, equal to the radius of the smaller balls.

Let E be such a point and E' its projection on the horizontal plane through A and B (Refer Fig. 9.38). $DC^2 = AC^2 - AD^2 = (d_1 + d_2)^2 - d_1{}^2$; or $DC = \sqrt{d_2{}^2 + 2d_1d_2}$

Let height difference $CC' = h$, which can be measured

$$\therefore \quad DC' = \sqrt{DC^2 - h^2} = \sqrt{d_2^2 + 2d_1d_2 - h_2}$$

E is a point on CG so that $EG = d_1$.

E' is the projection of E on DC', $CE = d_2 - d_1$ and

$$DE' = DC' + C'E' = \sqrt{d_2^2 + 2d_1d_2 - h^2} + d_2 - d_1$$

Let O be the centre of circle circumscribing ABE' and if $DO = x$ and $AO = EO$ = radius of circumscribing circle.

$$AO^2 = AD^2 + DO^2 \text{ , or } OE^2 = d_1^2 + x^2$$

or $\quad (DE - DO)^2 = d_1^2 + x^2$; or $(DE - x)^2 = d_1^2 + x^2$

from which x can be determined

$$DE' = \sqrt{d_2^2 + 2d_1d_2 - h^2} + d_2 - d_1$$

$\therefore$ Diameter of gauge $= 2\,(DE' - x + d_1)$.

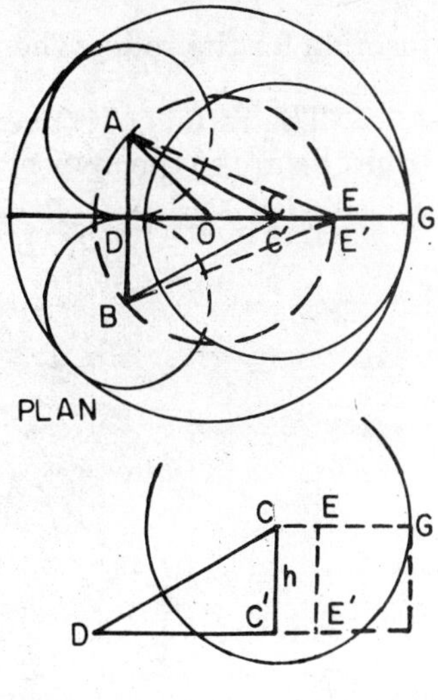

Fig. 9.38

Problem 9.12. *Describe a fixed roller type micrometer used to determine the radii of convex and concave surfaces. Derive the mathematical relationship.*

Solution. Fig. 9.39 shows a fixed roller type mircometer which is used to determine the radii of both convex and concave arcs. Let d be the diameter of rollers and l the distance between the centre lines of the roller.

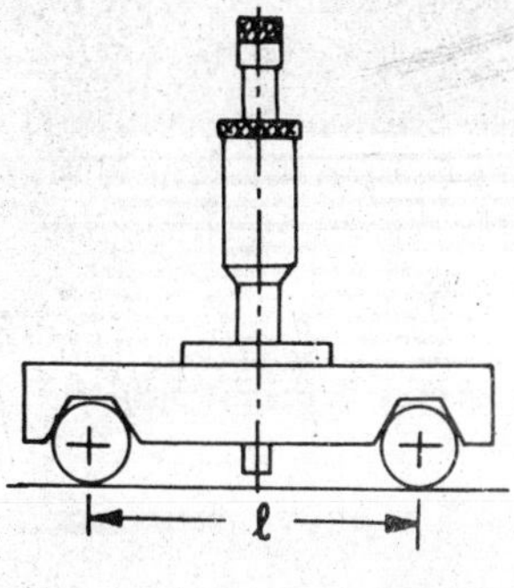

Fig. 9.39

The instrument is first placed on a true straight edge and a reading of the micrometer taken with the spindle end in contact with the straight edge. It is then placed on the arc and another reading taken on this. The difference between the two readings is the height of the arc above the tangent to the rollers, *i.e.* dimension h_2.

Then $h_1 = AB = d/2 - h_2$

In right ∠d ΔOO_1A, $OO_1^2 = O_1A_1^2 + OA^2$

or $$\left(R + \frac{d}{2}\right)^2 = \left(\frac{1}{2}\right)^2 + (R + h_1)^2 = \left(\frac{l}{2}\right)^2 + \left(R + \frac{d}{2} - h^2\right)^2$$

from which $$R = \frac{l^2}{8h_2} - \frac{(d - h_2)}{2}$$

Then possible error in the radius is given by

$$\delta R \cong \frac{\partial R}{\partial h^2} \cdot \delta h_2 + \frac{\partial R}{\partial l} \cdot \delta l + \frac{\partial R}{\partial d} \cdot \delta d$$

$$\cong \left(\frac{1}{2} - \frac{l^2}{8h_2^2}\right) \cdot \delta h_2 + \frac{l}{4h_2} \cdot \delta l - \frac{1}{2} \cdot \delta d.$$

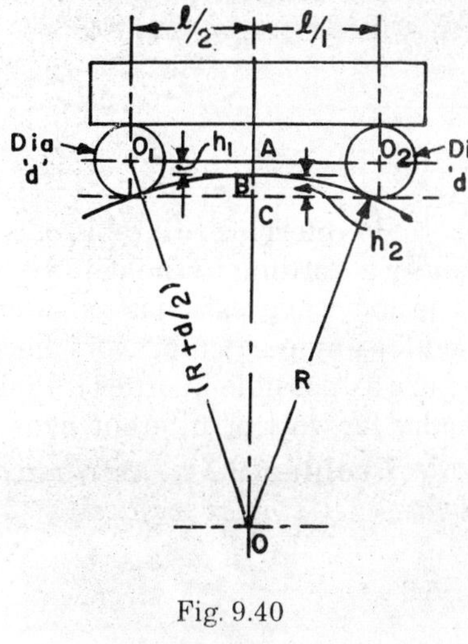

Fig. 9.40

Problem 9.13. *Describe a method to determine the radius of convex surface using V-plate. What are the limitations and disadvantages of this method ?*

Solution. When measuring concave arcs, a ball end is fitted over the spindle and the formula for the radius then becomes $R = \dfrac{l^2}{8h^2} + \dfrac{(d + h_2)}{2}$.

The radius of convex surface can be determined from a measurement of the distance from the arc to the apex of a Vee-piece resting against it [Refer Fig. 9.41 (*a*)]

In Fig. 9.41 (*a*), $\dfrac{R+h}{R} = \operatorname{cosec}\dfrac{\theta}{2}$, or $R = \dfrac{h}{\operatorname{cosec}\dfrac{\theta}{2} - 1}$

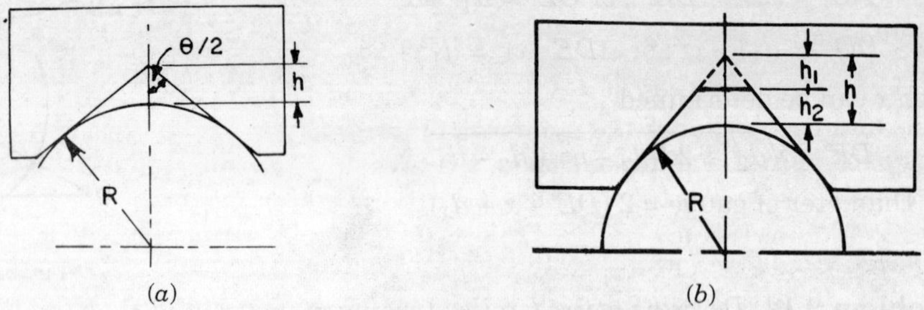

Fig. 9.41

The value of $1/(\operatorname{cosec}\theta/2 - 1)$ is usually marked on the instrument so that the radius can be easily calculated by multiplying the distance h by this constant.

As it is very difficult to measure distance from convex surface upto apex of the vee with sufficient accuracy, sometimes a flat is formed in the vee-angle and slip gauges are used to measure the distance between arc and the flat. The latter arrangement is shown in Fig. 9.41 (*b*). In this case,

$$R = \frac{h}{\operatorname{cosec}\frac{\theta}{2} - 1} = \frac{h_1}{\operatorname{cosec}\frac{\theta}{2} - 1} + \frac{h_2}{\operatorname{cosec}\frac{\theta}{2} - 1}$$

In above expression $\dfrac{h_1}{\operatorname{cosec}\dfrac{\theta}{2} - 1}$ is constant and only variable is h_1.

The possible error in the radius is

$$\delta R \approx \frac{\partial R}{\partial h} . \delta h + \frac{\partial R}{\partial \theta} . \delta\theta \cong \frac{\delta h}{\operatorname{cosec}\frac{\theta}{2} - 1} + \frac{h \operatorname{cosec}\frac{\theta}{2} . \cot\frac{\theta}{2}\,\delta\theta}{2\left(\operatorname{cosec}\frac{\theta}{2} - 1\right)^2}$$

By referring to a table of cosecants, it will be seen that the value of $(\operatorname{cosec}\theta/2 - 1)$ varies rapidly according to the value θ, and consequently the error in the radius obtained by means of the above expression is considerably influenced by the angle of the vee-piece. The instrument used for any particular workpiece should therefore be chosen so that it embraces as much of the arc as possible in order it make the angle small. This means that several instruments are needed for work of different sizes and this is the main disadvantages of this type of instrument.

Problem 9.14. *Describe a dynamical method to determine the radius of arc of (i) concave surfaces, (ii) convex surfaces.*

Solution. For concave surface, a small roller of radius r is placed on the arc and allowed to oscillate about its equilibrium position with S.H.M. In Fig. 9.42 roller is shown displaced from equilibrium position by angle θ and the restoring force F is then $= mg \sin \alpha$ (m = mass of roller).

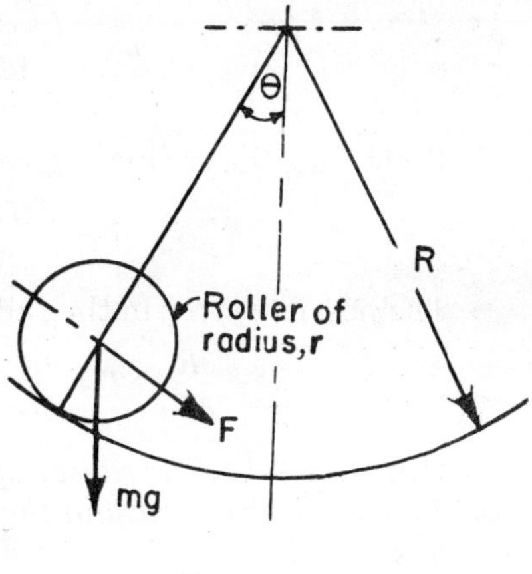

Fig. 9.42

Part of this force causes a linear acceleration f and part causes angular acceleration α, thus,

$$F = mf + \frac{l\alpha}{r} = mf + \frac{If}{r^2}, \ (I = M.I. \text{ or roller}, \ \alpha = f/r),$$

or $$mg. \sin \theta = mf + \frac{If}{r^2}$$

For small values of angle θ, $\sin \theta$ can be replaced by θ radians and for a cylinder the moment of inertia is $mr^2/2$. Substituting in the above equation, the acceleration is found to be $f = 2g\ \theta/3$.

This shows that the acceleration is proportional to the displacement and therefore the motion is simple harmonic. The linear displacement is $(R - r)\,\theta$; hence the time of one oscillation is

$$T = 2\pi \sqrt{\frac{\text{Displacement}}{\text{Acceleration}}} = 2\pi \sqrt{\frac{3(R-r)}{2g}}$$

From this, the radius is given by $R = \dfrac{T^2 g}{6\pi^2} + r = \dfrac{T^2 \times 9810}{6\pi^2} + r = 165.66T^2 + r$ millimetre (when T is in seconds and is obtained by noting time for 20 oscillations and finding T for one oscillation).

(*i*) In the case of convex surfaces, a rod of rectangular section is balanced horizontally on the arc and caused to oscillate. Referring to Fig. 9.43, when the bar oscillates from the horizontal position, the centre of gravity G moves to G', and the original point of contact of the bar A moves to the position A'. For small angles of oscillation, A' is almost vertically above A. It can, therefore be assumed that

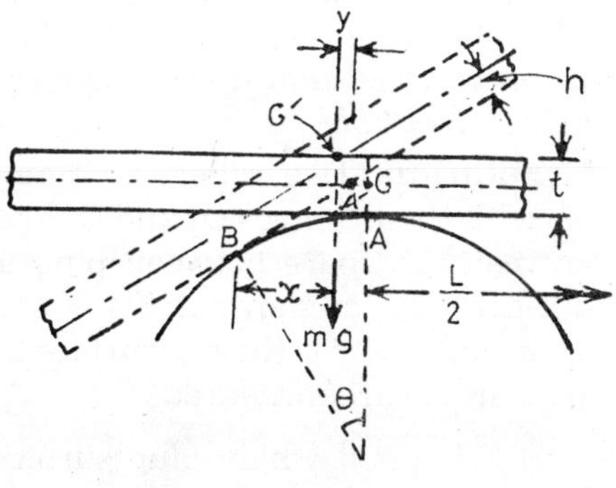

Fig. 9.43

$$y = h \,.\, \sin \theta, \ x \cong R \,.\, \sin \theta - h \,.\, \sin \theta$$

and the restoring torque $= mgx = mg\,(R - h) \sin \theta$

The moment of inertia of the rod about its lower surface is $m\,(k^2 + h^2)$. Since torque is equal to moment of inertia multiplied by angular acceleration, the angular acceleration is

$$\alpha = \frac{mg(R-h) \,.\, \sin \theta}{m(k^2 + h^2)} = \frac{g(R-h)\theta}{(k^2 + h^2)} \text{ – (for small values of } \theta)$$

This is proportional to the angular displacement, hence the motion is Simple Harmonic, and

$$T = 2\pi \sqrt{\frac{k^2 + h^2}{g(R-h)}}$$

For a rectangular bar of length L and thickness t (= $2h$),

$$k^2 = \frac{L^2 + t^2}{12} \; ; \quad \therefore \quad T = 2\pi \sqrt{\frac{L^2 + 4t}{12g\left(R + \frac{t}{2}\right)}}$$

From this, the radius is given by

$$R = \frac{\pi^2(L^2 + 4t^2)}{3gT^2} + \frac{t}{2} = 0.000335 \left(\frac{L^2 + 4t^2}{T^2}\right) + \frac{1}{2} t \text{ millimetres}$$

The possible error in the radius is given by

$$\delta R \cong \frac{\partial R}{\partial T} . \delta T + \frac{\partial R}{\partial L} . \delta L + \frac{\partial R}{\partial t} . \delta t \cong \frac{2\pi^2}{3gT^2}\left[-\frac{(L^2 + 4t^2)}{T} . \delta T + L.\delta L + 4t.\delta t \right] + \frac{1}{2}\delta t.$$

Problem 9.15. *The radius of an arc was measured by allowing a 20 mm diameter roller to oscillate to and fro on it and the time for 25 oscillations was noted at 56.25 sec. Determine the radius.*

Solution. When a roller is allowed to oscillate about its equilibrium position by placing it on an arc, then it executes S.H.M. and the radius of arc R, radius of roller r and time of one oscillation T are related by the relation.

$$R = \frac{T^2 g}{6\pi^2} + r \; ; \quad \therefore \quad T = \frac{56.25}{25} = 2.25 \text{ sec. } ; g = 9810 \text{ mm/sec}^2$$

$$\therefore \qquad R = \frac{(2.25)2 \times 9810}{6\pi^2} + \frac{20}{2} = 840 + 10 = 850 \text{ mm.}$$

Problem 9.16. *A solid cylindrical gauge of diameter D has an inclined hole of diameter d_1 passing transversely through its axis at angle θ to the axis. Devise a method to determine the height of the intersection of centre of hole from the plane end of the gauge. It can be assumed that a well-fitting plug for the hole is available.*

Solution. Two rollers of diameter d_1 and d_2 with their centres O_1 and O_2 are placed in between the space between plug and the cylindrical gauge as shown in Fig. 9.44 and the height h_1 and h_2 over the top surface from the bottom surface are measured.

If C is point where plug surface intersects with the outer surface of gauge then line CO_1O_2 bisects the angle θ.

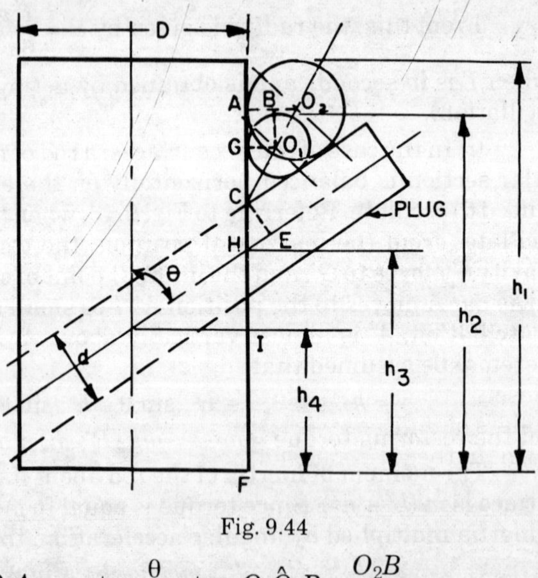

Fig. 9.44

O_2A is ⊥ to gauge surface and O_1B is ⊥ O_2A, $\therefore \quad \tan\frac{\theta}{2} = \tan O_2\hat{O}_1B = \frac{O_2B}{O_1B}$

$$= \frac{\frac{d_2 - d_1}{2}}{\left(h_1 - \frac{d_2}{2}\right) - \left(h_2 - \frac{d_1}{2}\right)}$$

$$h_3 = GF - (GC + CH)$$
$$= \left(h_2 - \frac{d_1}{2}\right) - \left(\frac{d_1}{2}\cot\frac{\theta}{2} + \frac{d}{2}\operatorname{cosec}\theta\right)$$

and $$h_4 = HF - HI = h_3 - \frac{D}{2}\cot\theta.$$

Problem 9.17. *Describe a method to determine the angular spacing of holes of same size located radially at angle q.*

Also derive the expressions for errors.

Solution. A plug which must be a good fit for the hole is inserted in each hole and rollers are placed in positions as shown in Fig. 9.45. The dimension L over the rollers is measured and from this, together with the diameters of the pins (P), rollers (d) and shaft (D), the angle q can be calculated as follows.

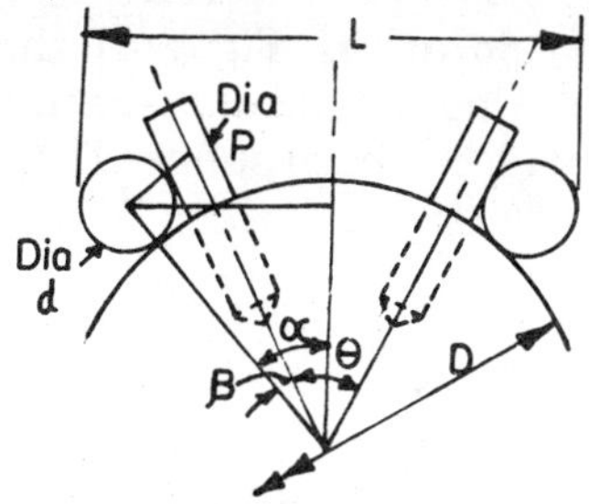

Fig. 9.45

Referring to the Fig. 9.45, $\theta = 2(\alpha - \beta)$

$$\sin\alpha = \frac{\frac{L}{2} - \frac{d}{2}}{\frac{D}{2} + \frac{d}{2}} = \frac{L - d}{D + d};\ \sin\beta = \frac{\frac{P}{2} + \frac{d}{2}}{\frac{D}{2} + \frac{d}{2}} = \frac{P + d}{D + d}$$

Assuming the centre lines of the holes to be truly radial, the accuracy to which the angle can be measured is given by the following expression. $\delta\theta = 2(\delta\alpha - \delta\beta)$

Now

$$\delta\alpha \cong \frac{\partial\alpha}{\partial L}\,.\,\delta L + \frac{\partial\alpha}{\partial d}\,.\,\delta d + \frac{\partial\alpha}{\partial D}\,.\,\delta D \cong \frac{\sec\alpha}{D + d}\,.\,\delta L - \sec\alpha\,.\,\frac{L + D}{(D + d)^2}\,.\,\delta d - \sec\alpha\,\frac{l - d}{(L + d)^2}\,.\,\delta D$$

$$\cong \frac{\sec\alpha}{(D + d)^2}\,[(D + d)\,.\,\delta L - (L + D)\,.\,\delta d - (L - d)\,.\,\delta D]$$

and $$\delta\beta \cong \frac{\sec\beta}{(D + d)^2}\,[(D + d)\,.\,\delta P + (D - P)\,.\,\delta d - (P + d)\,.\,\delta D].$$

Problem 9.18. *Describe an arrangement for measuring the angle of a hole drilled at an angle to a flat surface. Also derive an expression for the same and the error likely to be introduced on account of error in measuring distances and diameters of the rollers etc.*

Solution. The arrangement for the measurement of angle is shown in Fig. 9.46. A closely fitting plug is inserted into the hole and rigid block fixed on the horizontal surface. Two rollers of different diameters are required and measurements are made with these placed, in turn, in the position shown.

With reference to Fig. 9.46,

$$\tan\frac{\theta}{2} = \frac{\frac{1}{2}(D_1 - D_2)}{\left(l_1 - \frac{D_1}{2}\right) - \left(l_2 - \frac{D_2}{2}\right)}$$

$$\cot\frac{\theta}{2} = \frac{2(l_1 - l_2)}{D_1 - D_2} - 1$$

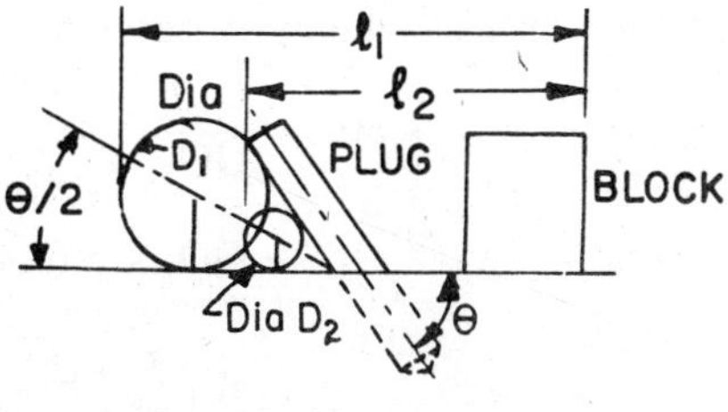

Fig. 9.46

The possible error in the angle q is quite small.

Thus substituting l for $l_1 - l_2$; and D for $(D_1 - D_2)$; $\cot \frac{\theta}{2} = \frac{2l}{D} - 1$

Then

$$\delta\theta \cong \frac{\partial\theta}{\partial l} . \delta l + \frac{\partial\theta}{\partial D} . \delta D \cong \frac{-2}{\operatorname{cosec} \frac{\theta}{2}} . \frac{2}{D} . \delta l + \frac{2}{\operatorname{cosec}^2 \frac{\theta}{2}} . \frac{2l}{D^2} . \delta D \cong \frac{4 \sin^2 \frac{\theta}{2}}{D_1} [l. \delta D - D. \delta l].$$

Problem 9.19. *Devise a method to measure the angular spacing of a spline shaft having splines with parallel faces.*

Solution. Fig. 9.47 shows a cross-section of the shaft and Fig. 9.48 the method of measuring the angular spacing. It is first necessary to measure the root diameter of the shaft and width of each spline. Then, referring to Fig. 9.48.

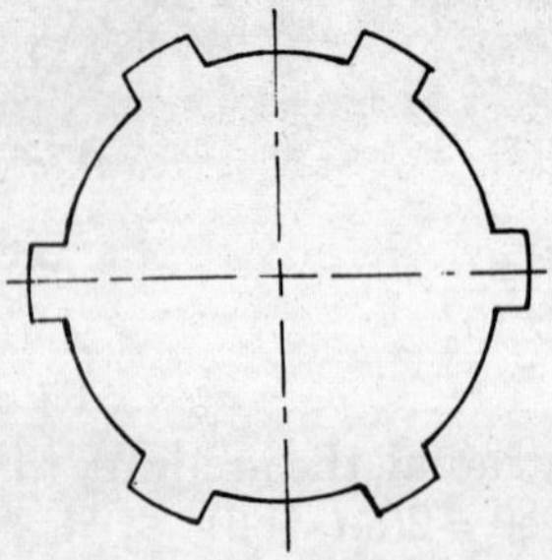

Fig. 9.47

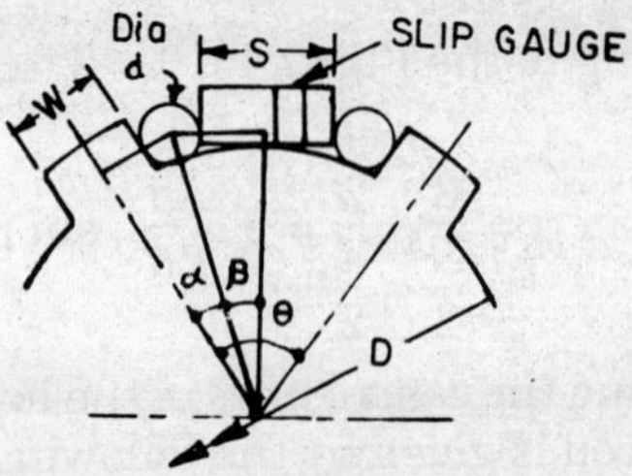

Fig. 9.48

$$\sin \alpha = \frac{\frac{1}{2}(d + W)}{\frac{1}{2}(D + d)} = \frac{W + d}{D + d} ; \sin \beta = \frac{\frac{1}{2}(S + D)}{\frac{1}{2}(D + d)} = \frac{S + d}{D + d} \text{ and angle } \theta = 2(\alpha + \beta).$$

Problem 9.20. *Describe a method for checking the angular spacing of a 3-splined shaft with the help of rollers.*

Solution. Fig. 9.49 shows the set up for measuring the accuracy of three-splined shaft. The two rollers of diameter 'd' are placed against spline faces and distance 'L' is measured over the rollers.

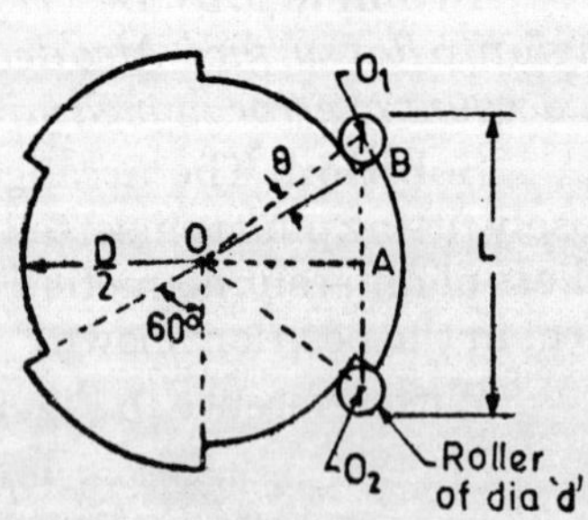

Fig. 9.49

Now $\quad L = 2(O_1A + d/2)$

$$\sin \theta = \frac{O_1B}{OO_1} = \frac{d/2}{D/2 + d/2}$$

Thus value of θ can be determined.

Also $\sin(\theta + 30°) = \dfrac{O_1A}{OO_1}$

and $\quad O_1A = OO_1 \sin(\theta + 30°) = \left(\frac{D}{2} + \frac{d}{2}\right) \sin(\theta + 30°)$

$$\therefore \quad L = 2\left[\left(\frac{D + d}{2}\right) \sin(\theta + 30°) + \frac{d}{2}\right]$$

Thus L is measured and also calculated from above formula. If angular spacing is in order, then the measured and calculated values should tally.

Problem 9.21. *Describe a suitable method to determine the diameter of small end of a taper gauge.*

Solution. Fig. 9.50 shows the arrangement to determine small end diameter of a taper plug gauge in which the gauge is mounted over a sine bar. The height of slip gauge pile below the sine bar is adjusted so that the upper surface of taper plug is horizontal. A roller having its diameter slightly less than the small end diameter is then placed at the end of the plug gauge and slip gauges of such length are inserted between the roller and end of plug gauge so that the dial reading over the gauge and roller is same.

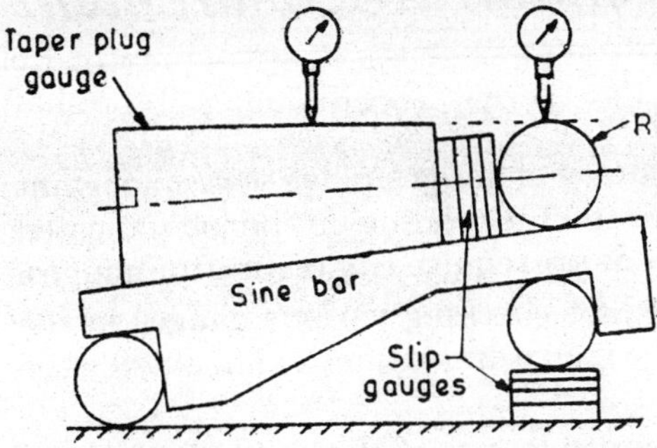

Fig. 9.50

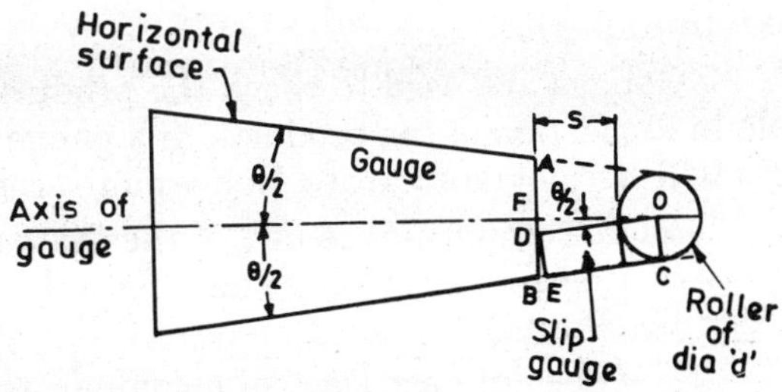

Fig. 9.51

Referring of Fig. 9.51, distance OF along the axis of taper gauge is $S + d/2$ (where S is the length of slip gauge and d is the diameter of roller).

$$FD = (S + d/2) \tan \theta/2 \ ; \quad \text{and} \quad BD = d/2 \ . \sec \theta/2$$

$\therefore$ Small end diameter $AB = 2\,(BD + DF) = (2S + d) \tan \theta/2 + d \ . \sec \theta/2.$

TYPICAL EXAMINATION QUESTIONS

1. State the different methods of checking tapers.
2. Taper bore of a socket is to be measured with the help of precision steel balls. Sketch a set-up to explain how you would measure it.
3. Sketch an instrument which can be used to determine the radius of arc on a component. Describe the method and derive the formula for obtaining the radius. Derive an expression for the error in radius caused by an error in the dimension measured by the instrument.
4. A convex arc on a component is approximately 150 mm. long and has a nominal radius of 250 mm. Describe three methods to measure the radius, indicating the relative advantage and the degree of accuracy obtained in each case.
5. The radius of an arc was measured by allowing 0.25 mm diameter roller to oscillate to and fro on it. Prove that the radius is equal to 866.1 mm if the time for 20 oscillations was 45.4 secs.
6. The radius of the convex arc on a segmental gauge was measured by using a surface plate, two rollers, and a micrometer. If the rollers were 25 mm in diameter and the measurement over them was 202.15 mm, calculate the radius. If the accuracy of measurements was ± 0.01 mm, determine the possible error in the calculated value of the radius.
8. The radius of a concave arc on a component was measured by fastening the component on to the table of a milling machine and noting the coordinates of three points on the micrometer dials, with a dial gauge in contact with the arc. The readings obtained were :

On long traverse	0.25	52.50	87.75 mm
On cross traverse	0.41	43.90	19.25 mm

Determine the radius.

10

Testing and Calibration of Gauges and Dynamic Measurement

10.1. Introduction

Gauges are used to check the products. The accuracy of gauges plays very important role in inspection of the products. If a gauge is not accurate, the reading on the job indicated by it will also not be accurate. Hence more accurate methods are required for testing the gauges.

Micro-comparator is one of the instruments used for checking various gauges in the industry. It ensures that the wear effects do not allow the gauge dimensions to fall outside the prescribed limits.

The absolute accuracy of a comparator is guaranteed by use of standard slip gauges, and slip gauges employed for setting working gauges are themselves checked by micro-comparator for wear effects. For very accurate and for absolute results light wave interference method is used.

The indirect method of testing gauges consists in using two comparators where relative difference from standard size is obtained. Comparators are available upto the accuracy of 75 millionth of a mm.

In addition to gauges, all linear and angular measuring instruments must be calibrated from time to time for best results. So we will first consider the calibration of linear and angular measuring instruments.

10.2. Calibration of Linear and Angular Measuring Instruments

Every measuring instrument must be provable, *i.e.* it must be caused to prove its ability reliably. The procedure for this is calibration. The variation in any observation on a product depends upon the variation in the product due to process of manufacture and variation due to measuring process, *i.e.* it can be expressed as : $\sigma_{observation} = \sigma_{process} + \sigma_{measurement}$.

In order to keep $\sigma_{observation}$ minimum so that product as a whole is reliable, $\sigma_{measurement}$ should be kept minimum which in other words means that measurement process or the instruments used should be precise with minimum of variation in the measured values. In order to maintain the precision accuracy of measuring device its periodical calibration is essential as from the moment an instrument is put into use it begins to deteriorate in accuracy and its precision. To a degree, this takes place even if the instrument is not being used. Regarding calibration it is said that in a plant where accuracy is not properly organised, 30% to 50% of the measuring equipment used do not give true results.

In order to maintain accuracy of measuring instruments, following procedure should be followed :

(*i*) Each instrument should be numbered. It serves the easy location of the instrument.

(*ii*) A card record should be established for each instrument.

Table 10.1

Calibration Card

Defects Found (if any)	*Errors / Defects After repairs (if any)*	*Details of repairs carried out*	*Remarks*	*Calibrated by*	*Place of use*	*Next calibration due on*	*Initial Lab. I/C*

Instrument .... Type and Class .... Inventroy No. ....

Table 10.2

Annual Calibration Programme For General Measurement Instruments

Empty—Indicates that calibration is due.

O—Indicates that calibration has been completed for all the instruments due for calibration in the month.

Months	*SHOPS*								
	I							*II*	
	Calipers, Micrometers, Protractors, dial gauges etc.	*Sine bars*	*Limit gauges*	*Slip gauges*	*Levels, surface plates etc.*	*Others*			
January	O		O						
February	O	O	O	O	O				
March	O		O						
April	O		O			O			
May	O		O	O					
June	O	O	O						
July	O		O	O					
August	O		O		O				
September	O		O	O					
October	O	O	O						
November	O		O	O					
December			O			O			

(*iii*) Checking interval should be established.

(*iv*) Some system should be adopted for providing adherence to the checking schedule.

(*v*) The record of the findings of the check should be maintained.

(*vi*) Record of checks should be further studied and analysed so as to improve upon the system.

It shall be preferable to have individual history card for each instrument. History card can be of type given in Table 10.1. These types of history cards can be prescribed using Kardex cabinets.

As regards checking interval, this mainly depends upon the frequency of use of the instruments, and the precision requirement of the measurements. For example in machine shop and tool room the measurements made are of more precise nature as compared to those made in casting and fabrication shops, so instruments used in these shops require more frequent calibration. It is always better to prepare annual calibration programme for instruments used in various shops, so as per the programme the shops send the instruments for calibration. The annual programme can be of the type as suggested in Table 10.2 which is for general measuring instruments such as vernier calipers, micrometers, protectors, limit gauges, slip gauges, dial gauges etc. For optical measuring instruments like universal microscopes, tool maker's microscopes etc. annual programme can be made for periodical calibration as well as for general preventive maintenance which besides general cleaning and lubrication will also consist of carrying visual checks like relative movements of moving parts, presence of corrosion, scratch marks, visibility and correct working of optical system etc. Such visual checks should be carried out more frequently. In Table 10.3 is shown such chart wherein the annual schedule for calibration and preventive maintenance of optical instruments is drawn. Table 10.4 also shows the history card to be made for limit gauges. For every gauge or gauge

Table 10.3
Annual Schedule for Preventive Maintenance and Calibration of Optical Metrological Instruments

Δ Indicates about preventive maintenance being due.

□ Indicates about calibration being due.

Instruments	*Months*											
	Jan.	*Feb.*	*Mar.*	*Apr.*	*May*	*June*	*July*	*Aug.*	*Sept.*	*Oct.*	*Nov.*	*Dec.*
Universal Microscope Sl. No. ...	Δ □	Δ	Δ	Δ □	Δ	Δ	Δ □	Δ	Δ	Δ □	Δ	Δ
Tool Makers Microscope Sl. No. ...	Δ □	Δ	Δ	Δ	Δ □	Δ	Δ	Δ	Δ □	Δ	Δ	Δ
Interferometer Sl. No. ...	Δ	□	Δ		Δ		Δ	□	Δ		Δ	
Shadow graph Sl. No. ...	Δ	Δ □	Δ	Δ	Δ	Δ □	Δ	Δ	Δ	Δ □	Δ	Δ

Table 10.4

GAUGE CARD SPECIFICATIONS

Gauge	*Major Dia and Tolerance*	*Effective Dia and Tolerance*			*Tolerance*		*Minor Dia and Tolerance*	
		For New Gauge	*Transfer to Insp.*	*Wear Limit*	*Pitch*	*Half Profile Angle*		
Go								Wire size ______
Not Go								Factor X ______

RESULTS OF PERIODIC CHECKING

Date	*DIMENSIONS OBTAINED*													*Remarks About Stability*		*Place of Use*	*Next Checking Due*	*Checked By*
	Go						*No Go*											
	Major Dia	*Effective Dia*	*Half Profile Angle*		*Pitch*	*Minor Dia*	*Major Dia*	*Effective Dia*	*Half Profile Angle*		*Pitch*	*Minor Dia*						
			Left	*Right*					*Left*	*Right*				*Go*	*Not Go*			

Type .... Size .... Card .... Instt. No ....

set an individual history card should be maintained. For gauges of any particular size, sometimes it becomes difficult for shop or inspection personnel to know whether the gauges are available or not.

For this purpose the charts could be maintained which indicate the maximum/minimum inventory stock quantity level for these gauges so that at any time when the quantity available in the stores goes down the minimum value, action should be taken for the procurement.

Here are described the methods of calibration of some of the important metrological instruments.

10.2.1. General Metrological Instruments. (*a*) *Vernier Calipers and other Vernier Instruments.*

The following table gives the allowable deviation in the parameters.

Parameters	*Permissible Error*	
	Least count 0.02 mm	*Least count 0.05 mm*
Zero error	0.02	0.05
Flatness of measuring jaws	0.003	0.004
Parallelness of measuring jaws	0.010	0.015
Error in readings	0.02	0.05
Spherical portion size of inside measuring jaws	± 0.02	± 0.03

The zero error is checked by bringing in contact the jaws, and the shift of zero of main scale is observed with respect to zero of vernier scale. The flatness of the measuring jaws is checked using a straight edge having sharp edge of class 1 accuracy. The straight edge is put over the surface and the light gap observed between the straight edge and the surface and compared with standard light gap formed between another straight edge and an optical flat. The parallelness is checked by inserting a slip gauge of any value between the jaws at various positions and determining the out of parallelness using slip gauges.

Error in readings along the entire range is also found out using slip gauges. In case of vernier calipers having spherical inside measuring jaws, the width of spherical portion is checked using a passmeter or dial type micrometric comparator.

(*b*) *Dial Gauges.* The dial indicators of least count 0.01 mm can be conveniently calibrated using passmeter or micrometer dial comparator of least count (L.C.) 0.002 mm. Here dial indicator is inserted in place of fixed right hand side flat jaw and the dial tip rests on the movable jaw. The dial indicators can also be calibrated using slip gauges and by fixing the dial gauge in a comparator stand.

The following table gives the permissible errors in the dial indicators :

Parameter	*Permissible Error*		
	L.C. 0.01 mm	*L.C. 0.001 mm*	*L.C. 0.002 mm*
Maximum error along entire range and in any one turn	0.02 mm along entire range	0.003 mm	0.004 mm
	0.006 if any one turn.		
Variation in readings along entire range.	0.02 mm.	0.003 mm	0.004 mm

(c) *Micrometers.* In case of micrometers the following are the main points to be checked :

(i) General appearance and relative movement of moving parts.

(ii) Checking initial zero setting for micrometers of size 25—50 mm or more.

(iii) Flatness of measuring surfaces.

(iv) Parallelness of measuring surfaces.

(v) Error in readings.

In general appearance, the micrometer is thoroughly checked for presence of scratches, dents etc. on measuring jaws, as well as, for corrosion marks, scratches, dents etc. on the surfaces of measuring drums, for proper working of ratchet system. The relative movement of moving parts is also checked which should be smooth. The working of lock system is also checked. The zero error of micrometer is checked and if it is found wrong it is adjusted easily for micrometers of size 25—50 mm and more. The size of the setting piece is checked on interferometer or any other comparator set to read upto 0.0001 mm. The permissible error allowed in its size 0.001 mm for micrometers upto size of 100 mm. The flatness error is checked by keeping optical flat on each jaw. The maximum permissible error is 0.0009 mm. The parallelness error is also checked using four optical flats of different width so that one complete turn of micrometer drum is made. The optical flat is set in such a way that total fringes on both sides are minimum. The permissible error in parallelity is 0.002 mm for micrometers upto 100 mm size and 0.004 mm for micrometers above 100 mm and upto 200 mm size.

The error in readings is checked using slip gauges so as to cover the entire range. The maximum permissible error is 0.004 mm for class I micrometers and 0.008 mm for class II micrometers.

10.2.2. Optical Measuring Instruments

Calibration of Universal Microscope. The calibration of the microscope is done as follows :

(i) *General visual checking.* Under this, levelling, movement of carriages, micrometric drums and vertical column, clear visuality of optical systems focusing etc. is checked. The measuring surfaces should be free from corrosion marks, scratches dents, etc.

(ii) *Straight movement of longitudinal and transverse tables.* The permissible error in these movements is 0.002 mm in horizontal plane and 0.005 mm in vertical plane.

This is checked using a rectangular straight edge and a dial indicator of least count 0.001 mm alongwith attachment for fixing dial indicator. The straight edge is put over the work table of microscope and dial tip is set at its top working face. The table is moved longitudinally and the dial tip deviation is the required error in vertical plane. For determining error in horizontal plane the dial tip is set horizontally touching the vertical side of straight edge and again table is moved and deviation is noted. Similarly, the straight edge is put in transverse direction over the table and deviations of movements in horizontal and vertical planes are noted.

(iii) *Relative perpendicular motion of the longitudinal and transverse carriage.* The permissible error allowed is 0.005 mm in 1100 mm length. This is done using a precision square of class 0 or class I accuracy having maximum error of ± 10″ in squareness. The movement of one of the carriages is set first coinciding with the shadow of one of the sides of square. Then the other side of square is brought in coincidence with transverse line of ocular microscope and the transverse table is moved. The deviation of the shadow and the square from the transverse line as observed through microscope is noted when the transverse carriage is moved.

(*iv*) *Perpendicular motion of ocular microscope column with respect to the work stage.* The permissible error in this motion is 0.060 mm over 100 mm movements. This is checked using a precision try square of class I accuracy and dial indicator 90.01 mm) with fixing attachment. The try square is put with its base over the work stage and the other side is held vertical with dial tip touching on it. The column is moved up and down and the deviation of dial readings is noted.

(*v*) *Checking reading of column rotation used for setting of helix angles.* The permissible error allowed in the entire range of column rotation is ± 5′. This is checked by using a clinometer of 1′ least count. The clinometer is put over column top and inclination is given which is further checked with clinometer.

(*vi*) *Change of ocular lines' position due to column rotation.* This is checked using a special control shaft where a right angled blade is provided at right angle to shaft diameter line. The ocular line is set coinciding with the blade and the column is titled and the shift in position of ocular line is noted which is the required error. The maximum error permissible is 0.010 mm in entire ± 12.5° rotation of column from centre position.

(*vii*) *Checking of main ocular microscope magnification.* This is checked using a special glass scale where graduation lines in millimetres and tenth of millimetres are marked very accurately. The permissible error in this case is 0.0005 mm.

(*viii*) *Change of job image due to minor shift of focusing ring.* For this a ground sheet is taken with a *X* mark engraved and it is focused by the microscope stage using reflecting light. The position of mark is observed through the ocular and then an optical flat of nearly 20—25 mm thickness is put over the mark and refocusing is done. In this position, the shift of mark from the original position is noted which should not exceed 0.005 mm.

(*ix*) *Parallelness of the longitudinal axis of millimetre scale to the movement of carriages.* For this test, the millimetre line is brought in contact with any of micron line and its position is observed. Now the carriage (longitudinal or transverse) is moved to its other extreme and the position of test millimetre line is noted. Its shift from the original position of test millimetre line is the required error which should not be more than half division of micron scale as seen visually.

(*x*) *Change of reading on micron scale by repeated 3—4 settings of the millimetre line in the centre of spiral double lines.* The error should not exceed 0.0005 mm.

(*xi*) *Checking of centres and centre stocks.*

(*a*) *Wear of centres.* The centres are set and checked for straightness which is observed while rotating the centres. The maximum error permissible is 0.01 mm.

(*b*) *Radial eccentricity of the centre sleeves (Fitted in stocks).* This is done using a dial indicator of least count 0.001 mm and measured by touching the dial tip on inside surface of sleeves where centres are fitted and the sleeve is rotated in the stock. The maximum variation in the reading of the dial indicator is the required error which should not exceed 0.002 mm.

(*c*) *Radial eccentricity of each centre when the sleeve is rotated.* In this case the dial tip is put over the centre surface with the centre fitted in the sleeve and the sleeve is rotated. The maximum deviation of dial readings is the required error which should not exceed 0.005 mm.

(*d*) *Coaxiality of both centres in horizontal and vertical plane.* For checking the coaxiality in horizontal plane both the centres are brought in field of view of main microscope and viewed for coaxiality. For determining coaxiality in vertical plane, a precision ground mandrel of 200 mm length is held between the centres. The dial indicator (L.C. 0.001 mm) is clamped with a suitable attachment to the microscope tube with its tip resting over the mandrel at one end. By moving the transverse carriage the maximum dial reading over one end is noted and the longitudinal carriage is moved by approximately 200 mm and again the maximum dial reading is noted at this end over the mandrel. The difference in these two readings is the required error in vertical plane which should not exceed 0.001 mm.

(*xii*) *Error in readings.* The error in readings is checked in two ways :

(*a*) Error in readings while measuring on flat glass measuring table.

(*b*) Error in readings while measuring in between centres using measuring knives.

(*a*) *Error in measurement on flat tables.* The permissible error in measurement is given as follows : Error = ± (1 + 1/10) microns ; where L is the length being measured in millimetres.

This error is determined with the help of a standard scale which is flat transparent glass scale on which graduations in millimetres are made from 0 to 200 mm with an accuracy of ± 0.0002 mm. Along the longitudinal movement of the measuring carriage, measurements are made in the range 0—200 mm in the intervals of every 5 mm. Along the transverse movement of measuring carriage, measurements are made in the range 0—50 mm in the interval of every 1 or 2 mm. Every time the difference between actual interval distance and the measured value with the microscope is made which is the error. The maximum error in the entire measurement range in both longitudinal and transverse movement should not exceed the value as calculated from the abore mentioned formula. From these data of errors the standard deviation of the error variation can also be found which should remain within 1/3 to 1/5 of the least count of the instrument, and should in no case exceed the value of least count of the instrument even in case of an old used instrument.

(*b*) *Error in measurement between centre using measuring knives.* The error in measurement should not exceed the values given in the following table :

Table

	Check	*Parameter to be Checked*	*Permissible error*
1.	Diameter of precision plug whose diameter is certified using interferometer.	Diameter	± 0.0015 mm
2.	Thread plug gauge of size around M 48 × 5 and M 90 × 5 with certified effective diameter pitch and half profile angle.	(1) Effective diameter (2) Pitch (3) Half profile angle	± 0.0025 mm ± 0.0025 mm ± 3′

For all the above measurements, measuring knives supplied in set alongwith the instrument are used.

Every parameter is measured three times and average of these three measurements is taken as the required parameter.

10.3. Measurement of Limit Gauges

Before making any dimensional measurement of limit gauges, preliminary inspection is made about in general and the non-gauging dimensions of the gauges checked. In addition, the flatness, hardness and straightness of the gauging surfaces are also checked. The checking of the general dimensions is very useful to find the reasons of wear of the gauge etc.

The hardness of the gauging surface may be usually measured on a hardness testing machine, preferably on the diamond pyramid machine. In certain types of gauges like plain and screw ring gauges, hardness is checked by means of a scratch test. A scratch made on gauging surface with the help of a suitable scriber is compared with the scratch put with same scriber on a mild steel piece. The comparison of the two gives the desired result. Same pressure should be applied in both the cases.

The flatness and straightness of the gauging surface may be checked accurately and quickly by applying a knife edge straight edge to the surface and viewing any gap between the two against an illuminated background. Under these conditions it is possible to see a gap of light as small as 0.008 mm. Substantial error of flatness may produce a gap of light exceeding 0.008 mm in magnitude and estimation of gap of such a size is not possible with any degree of accuracy by the above method. In this case slip gauges of different size should be introduced between the gauging surface and straight edge so as to reduce the gap of light to a value which can be accurately estimated. If this method is not convenient, the gauging surfaces should be set up horizontally on a surface plate, the flatness or the straightness being checked by means of a sensitive dial gauge mounted on a stand which is placed on the surface plate.

Having completed the preliminary inspection of gauges, the gauging dimensions are measured. This is described below for different gauges.

10.3.1. Parallel Gap Gauges. These may be measured directly by means of slip gauges and length bars depending upon whether the gap is smaller or larger than about 10 cm. The slip gauge or the length bar combination should be laid on its side on a surface plate with the end faces vertical. The gap gauge should be supported lightly at both ends and lowered over the slip gauges combination so that only its own weight allows is to move downwards, and the slip gauge combination should be gradually increased until the gap gauge just nips the slip gauge combination and yet moves freely over a combination which is 0.002 mm smaller in size (Refer Fig. 10.1).

The size of gap gauge is determined like this for both "GO" and "NOT GO" dimensions.

In order to detect errors in parallelism of the gauging surface of the gauge, it is important to present both sides of gauge to slip gauge combination, every time keeping one face in exact contact with slip gauges and noting the difference in other face of gauge surface.

Small gap gauges may be measured to an accuracy of ± 0.001 mm by above method but the accuracy of measurement of larger gap gauges is not so high, due to probable springing open of jaws when inserting the slip gauge combination and any other reason such as temperature difference between the gap gauge and the slip gauges. Other methods of testing parallelism of surfaces have been described in chapter 7.

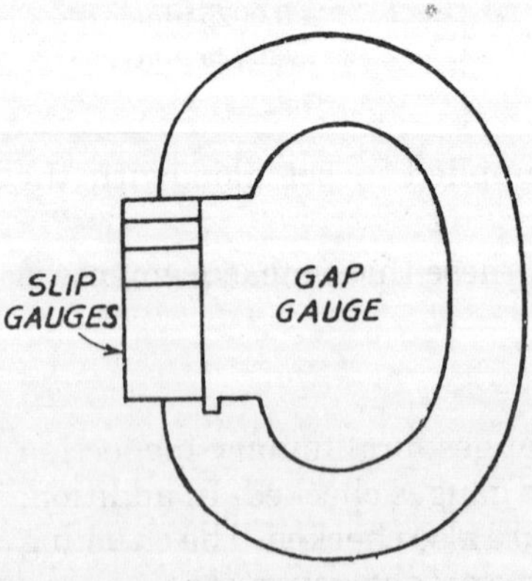

Fig. 10.1. Checking dimensions of gap gauge.

Fig. 10.2. Plain plug gauge.

10.3.2. Plain Plug Gauges. These may be measured by hand micrometer to an accuracy of ± 0.002 mm. For such measurements the micrometer should be used as a comparator, *i.e.* by taking successive readings on a standard known size and on the gauge and then adding or subtracting the difference between the readings in relation to size of the standard. The standard size should be quite near to the size of the plug gauge so that the travel of micrometer spindle during the measurement is small enough to minimise the effect of any progressive pitch error of the micrometer screw. The standard should preferably be a plain gauge and not a slip gauge combination to lessen the effect of any possible errors of parallelism of micrometer anvil and the spindle faces and any compression effects caused by measuring force on the gauge and standard. If the tolerance on the plug gauge is appreciably less than 0.025 mm. it is necessary to use more accurate method of measurement. In that case, bench micrometer, diameter measuring machine or high magnification dial comparator should be employed. The previous way of using the hand micrometer also applies to bench micrometer or the diameter measuring machine. High magnification dial comparators can only measure over a limited range of size at any one setting and the size of the standard has to be very approximately close to that of the gauge ; slip gauge combinations are, therefore, used for setting standards of comparators. The measuring forces of a dial comparator are very small and the errors caused by different compression effects on the flat surface of the slip gauges and gauge surface may be neglected. An advantage of the use of slip gauge standards for dial comparators is that size of the standard may be equal to one of the limits of the size of the plug gauge and hence the effect of any errors of the comparator scale graduations is minimised.

When measuring the plug gauge, sufficient readings should be taken to ascertain the presence of any ovality and taper of gauging surface. It may be necessary to check the gauge for "lobing" by rotating it in a suitable angle block, beneath dial gauge of high sensitivity or by some other suitable method.

A reading on standard should be taken after the plug gauge has been measured, to make sure that the instrument setting has not altered during measurements. This rule should always be followed for all types of measuring instruments.

10.3.3. Cylindrical Ring Gauges. Those gauges which are less than 1 cm. in diameter are difficult to be measured accurately and it may be best to measure the approximate size of such gauges by suitable plug and then to insert a tapered mandrel having mean size equal to the measured approximate size and having a taper of 0.0002 mm on the diameter per mm. of the length. The size of the ring gauge is then given by the size of the mandrel where it fits the ring gauge. The measurement of the ring gauge by tapered mandrels or plug gives the minimum size of the gauge and so the presence of any ovality is not checked. Care should be exercised when inserting a tapered mandrel in the gauge, to avoid any considerable wedging force on the gauge. Ovality can be checked by setting the gauge concentrically on the face plate of the dividing head and inspecting the bore of the gauge with a sensitive dial indicator as shown in Fig. 10.4.

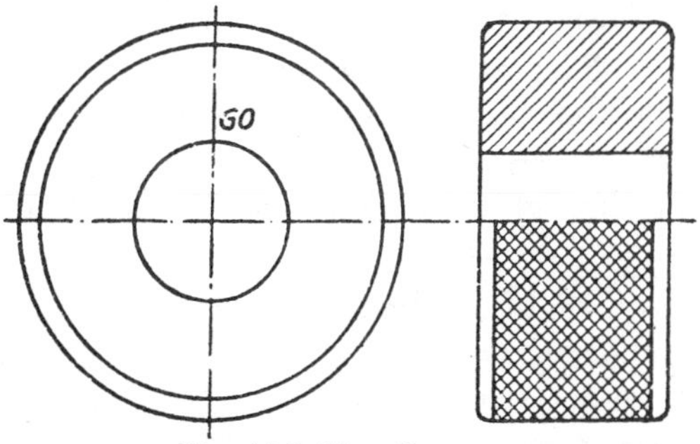

Fig. 10.3. Ring Gauge.

The size of the ring gauge having diameter more than 1 cm may be measured by using two small rollers or balls of equal size and slip gauges. The balls are placed diametrically opposite in ring gauge (Fig. 10.5) and a slip gauge combination is found which will be just nipped by the rollers while a combination 0.002 mm smaller will enter between the rollers by virtue of its own weight. The diameter of the ring gauge is then calculated by adding the mean size of the slip gauge combinations to the sum of the diameters of the rollers. By using rollers, the size of the gauge may be estimated to an accuracy of ± 0.001 mm for gauge upto 4 cm. in diameter, an increased accuracy can be obtained by the use of ball although more skill is required for that.

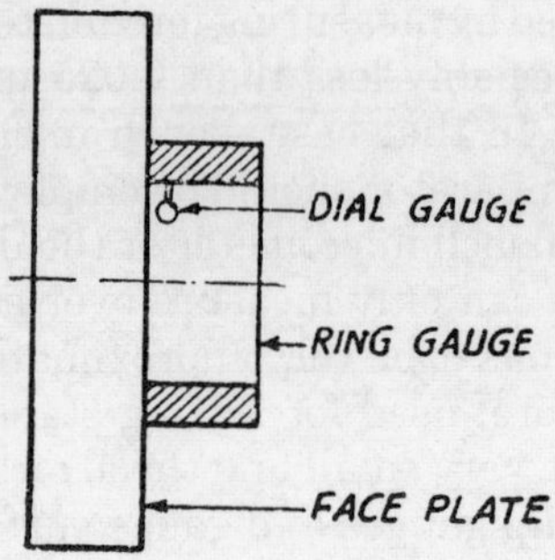

Fig. 10.4. Inspecting bore of a ring gauge.

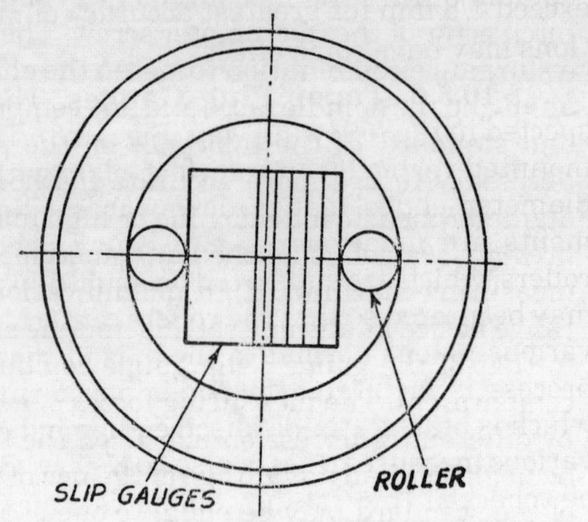

Fig. 10.5. Checking size of ring gauge.

Cylindrical ring gauges may be most conveniently measured on various types of pneumatic comparators ; the ovality and taper of the gauges may be quickly investigated : and accuracy of the order of ± 0.001 mm is obtained by using such comparators.

10.3.4. Form of Profile Gauges. These are most conveniently inspected by optical projection at a standard magnification and comparing the projected shadow image with standard projection diagram. The standard magnification most generally used is 50. Another standard magnification used is 100 but due to reduced intensity of illumination at the projector screen and consequent loss of the definition of illumination accuracy of measurement at this magnification, it is generally not more than 50.

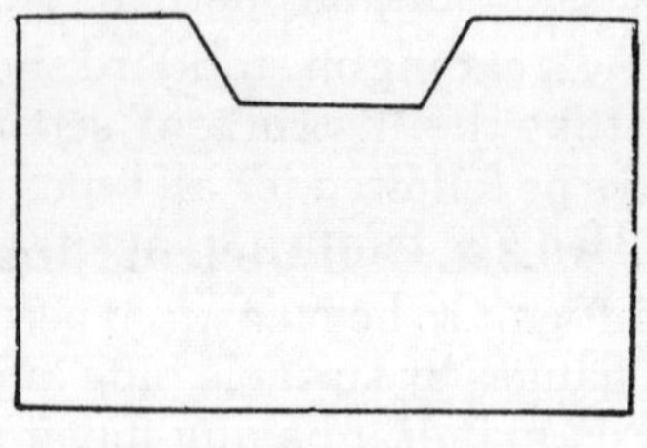

Fig. 10.6. Profile Gauge.

The magnified diagram with which the form is compared should be drawn on a material which has a good drawing surface and also is dimensionally stable.

Form gauges may be measured on Tool Maker's Microscope, especially when minute detailing of individual dimensions of the form makes the projection of enlarged projection diagram difficult.

10.3.5. Height and Depth Gauges. These may be measured by means of depth micrometers, bench micrometers or comparators. The 'gap of light' method is recommended for this. This gauge is placed on a surface plate in front of light and gauges are built up until the top of the slip gauge combination and the gauging surface of the gauge are at the same height above the surface plate. This is possible when no gap of light is visible below a knife edged straight edge resting simultaneously on the two (Fig. 10.7).

Alternatively a sensitive dial gauge mounted on a stand on the surface plate may be employed to measure the difference of the height of the gauge and the slip gauge combination. This difference should not

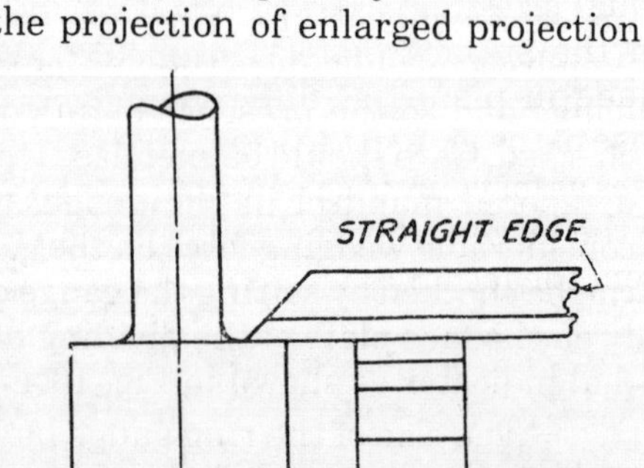

Fig. 10.7. Cheching height of depth gauge.

exceed 1/8 mm for greatest accuracy of measurement otherwise errors in dial gauge graduations may have some effect.

10.3.6. Taper Plug Gauges. These are inspected in the following way (Fig. 10.8). The gauge is mounted vertically on a surface plate with the small diameter end of gauge downwards, and measurements are made across two equal sized precision rollers, which are supported on equal slip piles which may be altered so that the rollers contact the taper in various planes normal to the axis of the taper. The process is facilitated by taper measuring machine which is having standardised equipment, and taking various measurements is also very easy. Ovality may be checked by taking diametral measurements around the circumference of the taper plug in various measuring planes, and the uniformity of the rate of taper is checked by taking measurements in three or more planes uniformly spaced and normal to the axis of the taper. The diameter of the small end of the gauge is then calculated from the formula :

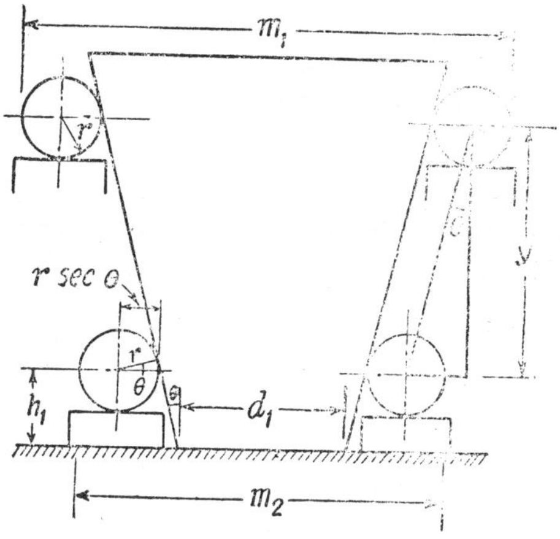

Fig. 10.8. Checking taper plug gauge.

$$\theta = \tan^{-1}\frac{(m_1 - m_2)}{2y}, \; d_1 = m_2 - 2r(\sec\theta - 1) - \frac{h_1}{y}(m_1 - m_2).$$

10.3.7. Taper Ring Gauges. Those gauges which are above 20 mm in diameter may be measured in a similar way as taper plug gauges (Fig. 10.9). The ring gauge is mounted on a surface plate with the axis of the ring vertical and the small diameter end uppermost. Measurements are made by means of slip gauge combinations between two precision balls of equal size supported on equal slip piles placed on the surface plate. The measurements are made in planes at different heights from the base of the gauge, the same criterion of fit being uses as for cylindrical ring gauges. The diameter of the large end of the gauge is then calculated from the formula

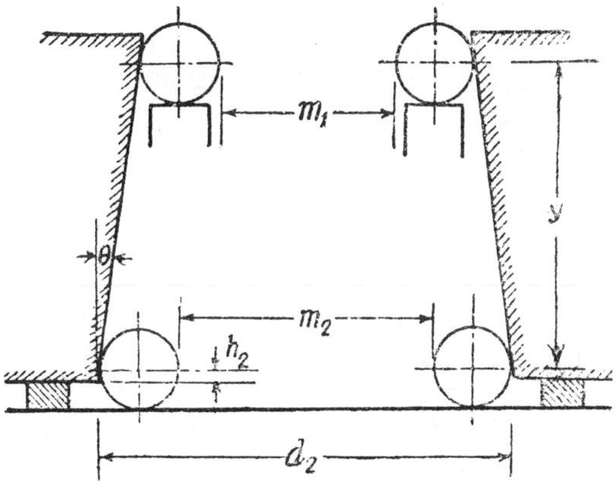

Fig. 10.9. Checking taper ring gauge.

$$\theta = \tan^{-1}\frac{m_2 - m_1}{2y}; \; d_2 = m_2 + 2r(\sec\theta + 1) + h_2\left(\frac{m_2 - m_1}{y}\right)$$

Taper ring gauges of sizes below 20 mm diameter are best measured by inserting precision balls of various sizes in the gauge and measuring the distance between the top of each ball and one of the end faces of the gauge (Fig. 10.10). The diameter in the plane of one end faces is then calculated from the formula

$$\theta = \sin^{-1}\frac{(r_1 - r_2)}{y}, \; y = (h_1 + h_2) - (r_1 + r_2), \; d_2 = 2r_2\sec\theta - 2h\tan\theta \quad (\text{where } h = h_2 - r_2).$$

10.3.8. Taper Plate Gauges. There are measured by the methods described for taper plug and ring gauges, depending whether the taper is external or internal. In the latter case it is more convenient to use precision rollers in place of balls.

10.3.9. Parallel Screw Plug Gauges. Various elements of a screw plug gauge are measured in the following way :

(*i*) *The major diameter* is measured on a diameter measuring machine which is standardised in a similar way as a bench micrometer for this purpose.

(*ii*) *The minor diameter* is obtained by making measurement on a diameter measuring machine between the flat ends of Vee-pieces placed in the thread groove on opposite sides of the gauge. These Vee-pieces have a radius at the apex which is smaller than the root radius of the thread so that during the minor diameter measurement they contact the root of the thread. The diameter measuring machine has arrangements for holding either gauge or a standard cylindrical plug. Two V-pieces hang freely and they can be placed across gauge in same plane and readings taken across them. First readings are taken on the cylindrical plug standard of known size with the Vee-pieces in position on both sides of the standard. The standard is then replaced by the gauge and a series of readings is taken on the minor diameter (Fig. 10.11). The minor diameter K of the gauge will be equal to the known diameter of the standard plus or minus the difference between the reading taken on the gauge and the standard respectively. The sizes K and S should be nearly equal for accurate results.

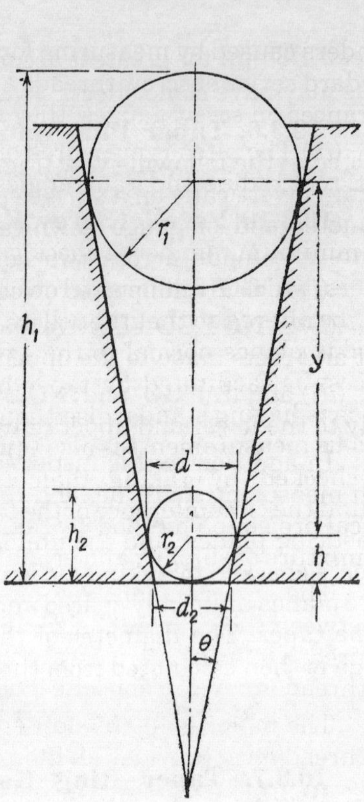

Fig. 10.10. Checking small taper ring gauge.

(*iii*) *The effective diameter* is obtained by noting the dimension measurements T under a pair of thread measuring cylinders placed in the thread groove on opposite sides of the gauge (Fig. 10.12). A diameter measuring machine is used for this purpose. Next readings are taken on a standard cylindrical plug of nearly same size as T with the thread measuring cylinders in position on both sides of the standard. The effective diameter is then calculated from the formula

$E = T + P$ + Compression correction—C ; $T = S \pm$ Difference between the readings taken over the gauge and the standard.

Where T is the measured diameter under the cylinders, P is a constant depending on the pitch and angle of the thread, and the diameter of the cylinders ; C is a correction for the rake of the thread and the compression correction is for the compression of the

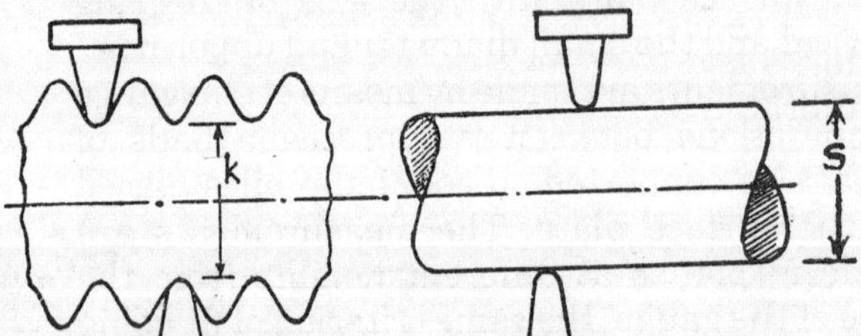

Fig. 10.11. Measurement of minor diameter of screw plug gauge.

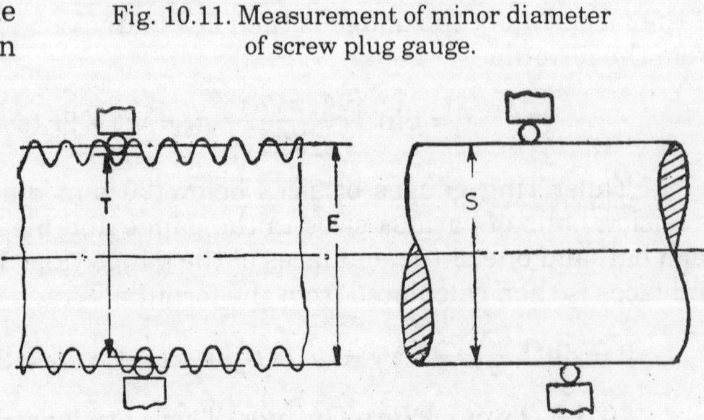

Fig. 10.12. Measurement of the effective diameter of screw plug gauge.

cylinders caused by measuring force of the machine. The rake and compression corrections for standard series of screw threads are usually very small and tend to cancel out, but to the small tolerances on screw gauges, they must not be neglected. (Also refer chapter 13).

(*iv*) The pitch is measured on a pitch measuring machine and the thread form, including crest, root radius and flank angles are examined by optical projection method.

10.3.10. Parallel Screw Ring Gauges. These are best inspected by plug check gauges. The minimum number of check gauges required for this purpose are :

(*i*) A 'GO' full form screw check gauge made to low limit of the screw ring gauge.

(*ii*) A 'NOT GO' effective diameter screw plug check gauge having threads cleared at the crest and roots made to the upper limit of screw ring gauge.

(*iii*) and (*iv*) 'GO' and 'NOT GO' minor diameter plain plug check gauges made respectively to the lower and upper minor diameter limits of the screw ring gauge.

In addition to this inspection, it is usual to measure the pitch of the ring gauges on a pitch measuring machine, and to inspect the thread form and measure the flank angles by 'optical projection method' by taking a cast of the inside of threads made from plaster of paris or a mixture of sulphur and graphite. (For small ring gauges a cast of dental wax may be used.)

The inspection of screw ring gauges by plug check gauges verifies that the ring gauges lie between their limits of size. If the actual effective diameters of the screw ring gauges are required, a suitable horizontal comparator is necessary, employing two styluses which seat in the thread groove on opposite sides of a gauge.

The major diameters of screw ring gauges are usually not measured but inspection of the thread form provides an indirect verification by checking of the depth of the thread.

10.3.11. Position Gauges. The design of position gauges varies so much that it is difficult to lay down any procedure for their measurement.

10.4. Checking of Slip Gauges

As a result of handling in the laboratory or inspection room, slip gauges are liable to exhibit signs of wear after appreciable period of use and should, therefore, be checked or recalibrated at regular intervals. This can conveniently be done with the aid of a millionth measuring machine such as the Newall interferometer.

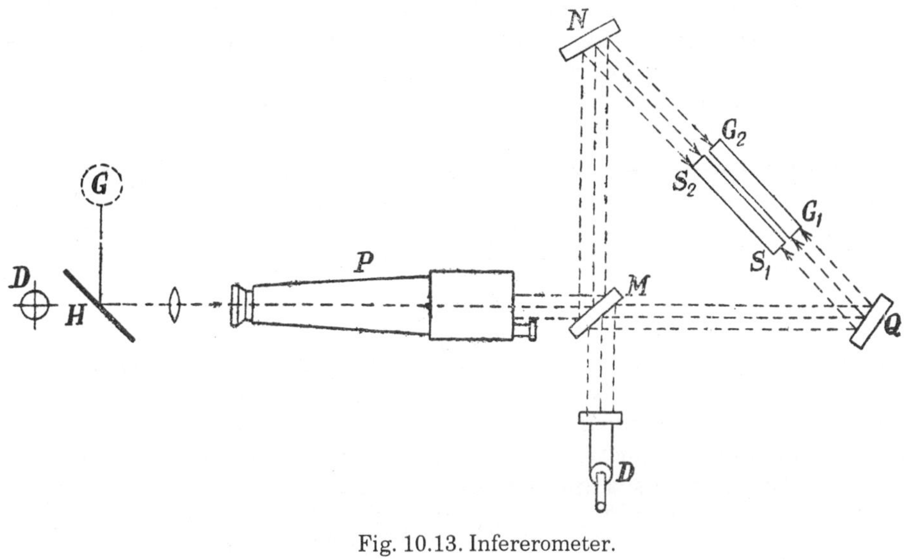

Fig. 10.13. Infererometer.

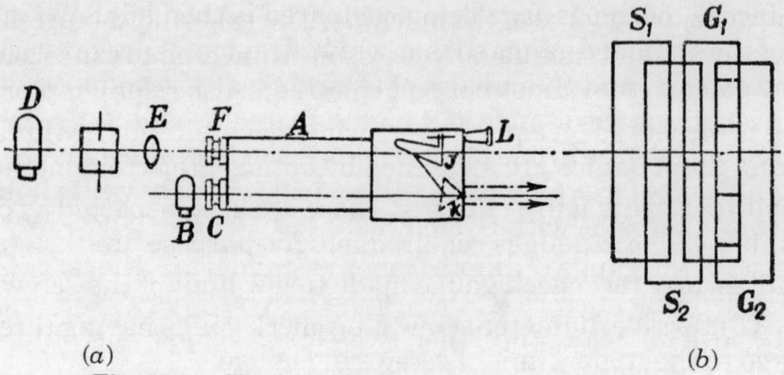

Fig. 10.14. Checking slip gauges with interferometer.

The gauge comparison and absolute measurement are done by an interferometer. Plan and elevation of this instrument are shown in Figs. 10.13 and 10.14. It consists essentially of a Higler interferometer arranged in such a way that both the ends of the gauge being tested are examined simultaneously in the same field of view. After simple adjustment, the gauges show interference fringes across the faces at both ends and from the relative positions of these differences in length can be measured to accuracy of 0.00002 mm. For all lengths to be tested by the instrument, the reference gauge and the gauges being tested are placed side-by-side and the optical system adjusted to give, with white light, a pattern of interference fringes across the face of the gauges. The illumination is then changed to different light of three or four different wavelengths, one after the other from the monochromatic source ; the position of the interference fringes being noted each time. The absolute length is then computed with the help of the special slide rule provided for this purpose.

10.4.1. Description of the Instrument for Checking Slip Gauges. A dual collimator A, with a white light source B and slit C provides white light ; and a monochromatic light source D, condenser E and slit F provide monochromatic light. An alternative monochromatic light source G may also be employed and introduced when required by the mirror H which can be swung in and out so that either light source D or G can be selected. The monochromatic light is reflected by the mirror J to the depression prism K. This mirror can be rotated by the micrometer L to select suitable wavelengths, which are engraved on the micrometer drum. The prism K can be swung in or out of the optical axis, so that white light or monochromatic light can be transmitted to the interferometer. This light is transmitted to the dividing mirror M and some of it is reflected to the mirror N from where it is turned towards the end faces of the gauges. The light is then reflected back along a similar path to the observing telescope P. Some of the light transmitted by the dividing mirror goes to the mirror Q from where it is turned to the opposite end faces of the gauge and reflected back to the observing telescope.

10.4.2. Comparison of the Lengths of two Slip Gauges. The two gauges are mounted side by side as shown in Fig. 10.13 and both ends of each gauge are observed through the telescope, but the mirrors are so arranged that the end faces of the gauges appear to overlap each other as shown in Fig. 10.14. For clearness, the end faces are shown displaced vertically in the illustration.

The standard gauge S is adjusted so that the end faces S_1 and S_2 are at an optical path distance from the dividing mirror M when illuminated by the white light source B. Under these conditions the interference fringes known as 'white light band' appear. These are produced only when the optical paths from the dividing mirror are equal. The exact point of equality is indicated by a prominent black band which has other bands, fringed with colour on either side, successive bands being of minimised intensity until invisible.

The end face G_1 of the gauage being compared is then adjusted into the same plane as the end face S_1 of the standard gauge so that white light bands are produced on the overlapping halves of the end faces S_1 and S_2 and also of G_1 and G_2 as shown in Fig. 10.14 (*b*). If the gauge G is of the same length as the standard S and the faces G_1 and S_1 are in the same plane then the faces G_1 and S_2 will also be in one plane. If this is so, the white light bands produced between the overlapping halves of faces G_1 and G_2 will coincide with the white light band formed by the overlapping halves of faces S_2 with G_1 and S_1.

If the length of the gauge G differs from the standard S, the face G_2 will be at a different distance from the dividing mirrors M, than is the face G_1. This will cause the white light bands to be displaced above or below the other two sets of bands, the difference in length being proportional to the number of bands and fraction of band displaced. Each band corresponds approximately to a difference in length of 0.0001 mm. A micrometer eye-piece is provided accurately measuring the fractions of bands, but it is preferable to estimate the fraction.

10.4.3. The Zeiss Slip Gauge Interferometer. This compact instrument is used for absolute comparative measurement of slip gauges. With it, the absolute length of a slip gauge can be determined by three different methods : (1) By comparison with a standard block. (2) By a combined measurement of three gauge blocks. (3) By measurement with a wrung-on gauges block. The Zeiss optical system, which employs three wavelengths of cadmium light is illustrated in Fig. 10.15 (*a*) which shows the set-up for the differential method of measurement. The light of cadmium lamp falls on prim P and is made parallel by a lens. This light is divided by the dividing plate T. One part penetrates the compensation plate K (compensating for the path of light) and this part from plate T is deflected by mirror S_1 and reaches the surfaces F_2 and F_4 of the gauge blocks. From there it is reflected and returns to T in the same way. The other part of the light deflected from the dividing plate falls on the surfaces F_1 and F_3 of the gauge blocks by way of the angle mirrors S_2 and S_3, and is then reflected back to T. Both parts coincide behind the dividing plate, that is to say those reflected by F_1 and F_4 and those reflected by F_2 and F_3. Interference bands on the surfaces of the slip gauges can be observed by means of a second lens. By tilting one of the slip gauge tables, they are then adjusted to the position shown in Fig. 10.15 (*b*). As one band system originates from the superimposition of the rays reflected from F_1 and F_4 and the other system from F_2 and F_3 the displacement of the two band system is a measure of the different between the gauge blocks. This displacement for all three bands is then measured by means of the measuring device. The measured bands of both systems are brought to coincidence and the fractions are read off directly from the scale. The corresponding linear measurement in microns is then determined by means of a special slide rule.

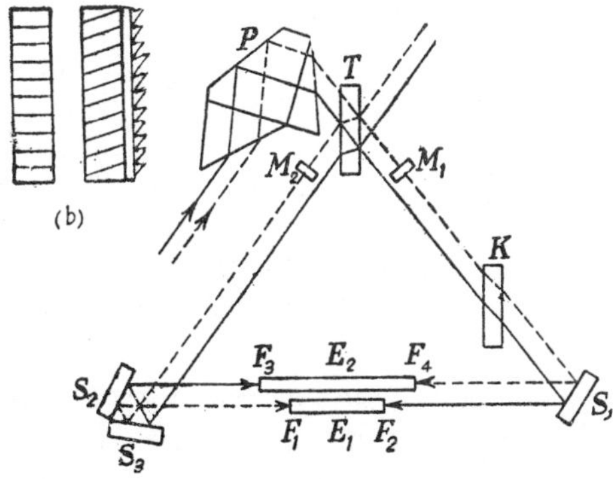

Fig. 10.15. Zeiss slip gauge interferometer.

10.5. Dynamic Measurement of Size, Form and Position

By the dynamic measurement we mean the measurement and control of dimensions of a component when it is being produced and without interrupting the process of production. This is utilised to achieve accuracy as well as time saving which is a must in the competitive period of today. The dynamic measurements also include the automatic gauging of component when they are coming in a flow line out of the position process. In high precision processes such as grinding operation, these techniques help to control the dimensions within close tolerances and thus avoiding the possibility of any rejection.

The various dynamic measurements are ; the shaft diameter measurements during precision machining or grinding operations, checking the internal bores when the process is going on, automatic gauging of components coming in a mass production line, checking the dimension of steel ingots or sheets coming out of the production machines etc. For these measurements various instruments and comparators used are discussed below.

10.5.1. Three point Shaft Gauging. The V-block type of gauge illustrated in Fig. 10.16 is used for shaft diameter measurement during precision machining. It operates on three-point principle whereby contact is made at points A and B of a chord and C also. In this method, the point C is on the plunger of a dial gauge and if this gauge is set to zero with a master gauge the dial readings show differences in relation to the master diameter setting. Thus during a grinding operation if the gauge is applied to the shaft at frequent intervals the reduction of work diameter can be followed.

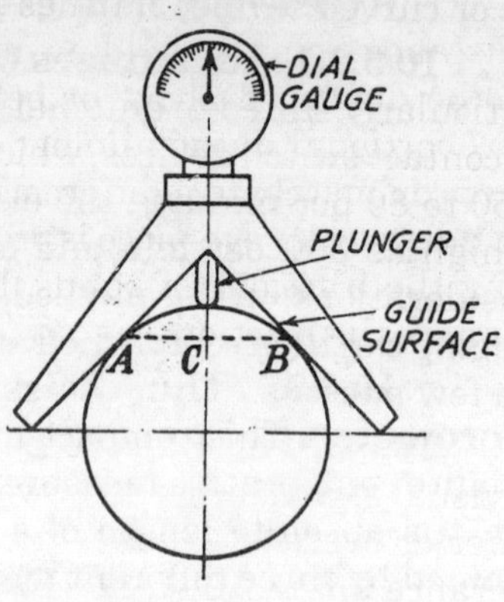

Fig. 10.16. Three point shaft gauging.

In the Churchill and Johnson shaft gauges, the contact faces of the vice member are lined with tungsten carbide so that the surfaces can be held continuously in contact with the shaft during sizing operation. The plunger contact consists of a diamond point. The effect of wear does not influence the accuracy of the readings, since the gauge is checked at regular intervals and if necessary, reset with the aid of the master gauge.

This gauge can be adopted for automatic sizing of shafts and other cylindrical parts during machining operations. The gauge travels along the direction of tool with the wheel head and when the correct diameter in reached, operates an electric control causing tool wheel head to withdraw. In this method, presetting is required to be very correct.

10.5.2. The Electronic Comparator. A range of high precision instruments for measuring purposes, now available, are based on the principle of application of frequency modulation in radio oscillator. A special radio oscillator produces pulses of a definite value at a basic frequency, this being controlled by the measuring head when the latter is applied to an engineering part that is to be measured. Any variation in the dimension from that of the present standard modifies the frequency of the instrument oscillator. Suitable control permits this frequency variation to be annulled and frequency restored toits basic value. The measuring dial is graduated directly in length units and by means of suitable switches on the instrument, different degrees of linear magnification upto those corresponding with wavelengths of light can be employed. Thus, in a typical bench micrometer, six ranges of metric values for full scale movement, from 1 to 100 microns are provided, while the work capacity between the measuring jaws is 50 mm. For bench comparator, the range is from 0.001 to 1.25 mm.

In regard to the measuring head, this is based upto the inductance coil and condenser (or capacitance) principle, which has the property of resonating at a given frequency, which can be readily calculated. Variation in the capacity between the measuring head fixed plate and the surface of the test piece, due to change of air gap, will affect the resonance frequency ; the actual variation being utilized to form a measure of the air gap dimension.

The electronic method involves no mechanical movements other than mechanical contact at the surface to be measured but it involves the use of valve circuits and electrical components and knowledge of their principles and adjustments. The upper limits of magnification, however, are much higher than for the mechanical comparator, and compared with the standards of light wavelengths.

There are many applications of this instrument to internal and external gauging and it is very suitably for very high limits measuring precision. The one typical use is in the measurements of fuel injection pump barrel diameters where the total tolerance is not more than 4 microns. In this application the variation in the capacity between the pilot head of the instrument and inside diameter cf the barrel causes the frequency to change and this variation can readily be translated into a length change. Other applications for static measurement are measurements of tube wall thickness variation, the thickness of electroplated coatings with flat or curved surface finishes etc.

10.5.2.1. Advantages of Automatic Gauging. The electronic comparator method is particularly suited to automatic gauging of mass produced components since unlike mechanical contact switch devices, which are subjected to time lags, limiting the role of gauging usually to 50 to 80 per minute ; the electronic type is capable of dealing with components at rates of as high as 500 per min and more. This limit being fixed only by the mechanical handling considerations. In this method, inertia and other mechanical errors do not occur and respective gauging can be accomplished at this high rate with magnification and the accuracy of the order of a few microns.

10.5.3. The Electrichek. This electric device utilises the reed mechanism to indicate by visual signals whether a work part dimension is within tolerances or not and whether it is under or over size. It does not, however, indicate the amount of variation under or over the tolerance limits. It is, therefore, useful either as an inspection gauge, or mounted directly on a precision grinding machine it will inform the operator when the part being ground is brought to size.

In regard to the tolerances and accuracy of this device it may be set for any tolerance gauge, from 0.001 to 1.5 mm according to the model of gauging head used, by the use of appropriate precision gauge blocks. Apart from these tolerances the electrichek gauge is aimed to have an accuracy of 0.002 mm.

10.5.3.1. Principle of Operation. This instrument employs a comparatively simple but highly accurate magnifying system representing a combination of the mechanical and optical principles. The mechanical part is so designed that there can be no wear effects in any of the moving members as springs are used instead of levers and bearings. The instrument is robust in design and available in a number of different models with magnification ranging from 500 to 20,000.

The principle of the reed mechanism is as follows (Refer Fig. 5.2). Essentially, it consists of two metal blocks, one being fixed and the other floating or movable. The two blocks are connected by two springs or reeds so that the moving block will move parallel to the fixed one for small movements. The moving blocks carries a plunger contact. The two blocks are connected at their upper ends by means of two alloy steel springs or reeds, the latter being attached at their lower ends to the blocks and rivetted together at their upper ends. A needle is attached to the junction of two reeds to form an extension unit.

The action of the mechanism is as follows : when the plunger contact and its movable block are moved upward the horizontal reeds deflect slightly, but vertical reed on the floating block tends to step apart the fixed block reed ; the result being that since the reeds are fixed at their upper ends, the combination swings through an arc, needle pointer follows the same movement but in a wider arc.

At both extreme ends of needle swings, it makes an electrical contact with fixed pointer which makes a circuit on one side with a terminal connected to a red signal light and on the other with a similar terminal connected to a green signal light. The actual position of those terminals, which fix the sweep of the needle, are controlled by two micrometer adjusting screws, one for each terminal. The sweep of the needle between contacts is proportional to the dimensional tolerance for which the gauge is set, the two terminals controlling the upper and lower limits of the tolerance zone. For setting an electrichek head to any given dimension, a master fixed size gauge or stock of precision gauge blocks equivalent to the minimum tolerance limit, it placed in gauging position on the anvil. By means of the adjusting knob on the spindle, the position of the fixed pointer is adjusted which controls the red signal light. Its movement may be viewed through the window in the gauge head case.

Thus, any work part smaller than the minimum master gauge will allow the needle to swing far enough to make the red light 'on', and shows that the part is under-size.

In the same way, the second fixed pointer is set with a master maximum limit gauge. When a work part is larger than that, the green light will be 'on'. The electrical circuit is designed in such a way that when the part is under-size, red light will be 'on' and when part is over-size, green light will be 'on'. If part-size is within the higher and lower tolerance zone, amber light will be 'on'.

The instrument possesses the advantage of portability since the three main units, namely, the gauging head, light unit and base assembly, can be used as a single instrument. Further, the gauge head and light unit may be used directly on a machine tool in order to control the size of work being machined.

10.5.4. Gauging thin strips. In connection with the manufacture of thin metal strips, it is necessary to check the thickness from time to time during the manufacturing process. Although the special designs of sheet measurement could be used for this purpose but it is undesirable to stop the metal rolls for such an object of measurement. Moreover, numerous individual measurements could be needed in view of the large quantities of metal strip normally made at each work.

To overcome these difficulties the continuous indicating and recording type of gauges have been developed which use mechanical, electro-magnetic, or electronic principle.

In some instances the strip from the rolling mills makes contact with the measuring device, while in other cases, notably the electronic and radio-isotope methods, the strips do not touch the measuring instrument. The pneumatic gauge also belongs to the non-contact type.

10.5.4.1. Mechanical Roller Thickness Gauge. It is based on the principle of passing the strip continuously, between a pair of rollers on the gauging device. The bottom roller spindle is fixed in location, but the upper roller is movable in a vertical direction and spring controlled so as to keep it in contact with the work. The vertical movements are magnified by suitable means and indicated by an instrument or recorded automatically on a graph paper.

A typical example of such an instrument is the Cambridge strip gauge which will indicate or record continuously the thickness of steel strip travelling at speeds upto 90 m per minute with an accuracy of 0.025 mm. The strip to be measured passes through two hardened steel rollers, of which the lower one's spindle is fixed to the base, whilst the upper one is pivoted on a crossed spring frictionless bearing. Variations in thickness cause the upper roller to move vertically, these movements being measured by means of an electric micrometer which operates on the following principle. The displacements to be measured is communicated to an armature situated between the poles of two fixed opposed electro-magnets. Though the windings of these magnets are passed alternating currents from equal windings of a transformer. The current through each magnet depends upon the air gap between it and the armature. Included in each magnet circuit is a full-wave rectifier, and the two rectifiers are so interconnected that any difference in the currents through the two magents is shown on a D.C. indicator or recorder calibrated in terms of the movement of the armature. The instrument is operated directly from the A.C. mains through a supply unit. The indicator has a clearly graduated scale from minus 0.05 mm to plus 0.05 mm with centre as zero.

The motor driven recorder is available for use with the gauging device. It provides records of the variations in thickness at three points, namely at the centre and two edges of the strip.

10.5.4.2. Pneumatic Thickness Gauge. A special non-contact type of pneumatic gauge, for continuous thickness measurement of rolled sheet metal, paper, rubber, plastic sheeting etc. is made by Solex (Gauges) Ltd.

The strip or sheet runs between two jets, placed in the centres of guide anvils (Fig. 10.17). The thickness variation causes change in the total flow through both jets, which are indicated in a manometer column having a suitable graduated scale giving direct readings of thickness changes. This gauge being of the comparator type, requires an initial setting to zero by means of a standard gauge. The readings show the exact size of the average surface thickness, irrespective of local irregularities. The speed of the strip through the gauge can be a relatively high one.

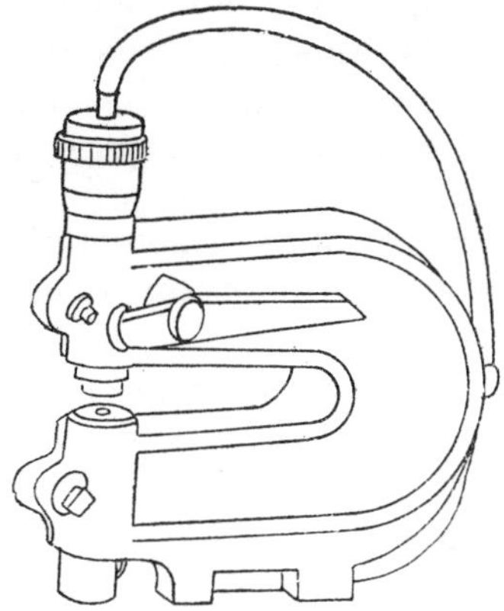

Fig. 10.17. Pneumatic thickness gauge.

10.5.4.3. Electrical Comparator. Usually *Electro-limit type* of electrical comparator is used for the purpose of comparison of dimensions with the standard dimension.

Its characteristics are :

(*i*) A great magnification power to magnify small movement of the plunger is obtainable.

(*ii*) Variation of the magnification power (in order to suit the instrument of various requirements, *i.e.* for fine work as well as high precision work) is possible.

(*iii*) With special attachments, it can gauge the sheets continuously and give the record on graph per (it can give the electroplating coating thickness also).

But it is very costly and very cumbersome instrument.

Operation. It operates on the principle of the change of inductance of a coil by the movement of a steel armature, which is moved by a plunger in contact with the surface to be checked.

Two of the inductances of the bridge are between armature and two are outside. These four inductances form a bridge. Initially the inductors are set so that no resultant current flows through them. When armature is moved, the balance of bridge is disturbed and a resulting current flows which is calibrated in terms of length of movement of the plunger.

The basic elements of instrument are :

(*i*) Electric gauge head employing two arms of bridge armature and feeler.

(*ii*) Milliammeter.

(*iii*) Power supply.

(*iv*) The base and case of the head.

Construction. The construction of the main head is as shown in Fig. 10.17. There is a plunger which rests on the surface to be checked. It rests against a diamond ball which rests against a steel armature, *C* and *D* are inductance elements and form two arms of a bridge.

The steel armature is held on one side by a spring. On other side, on both sides of it are two inductances *C* and *D*. Upon these there are two screws *B* and *B* which are used to change the magnification. Naturally when electromagnets are near the armature then it will be sensitive one and will have higher magnification and when it will be at distance from steel armature then magnification will be less. Thus by adjusting the screws B, magnification is changed.

Two such arms (inductances) are outside and these four form a bridge. When the armature position is changed by any irregularity in the surface the balance of bridge is disturbed and a current flows which is calibrated on milliammeter in terms of plunger movement.

Also screws B are used for initial adjustment of the bridge to give zero reading when there is no irregularly, *i.e.* for plane surface.

The accuracy of the instrument is of the order of 0.1 micron and thus it is very sensitive one.

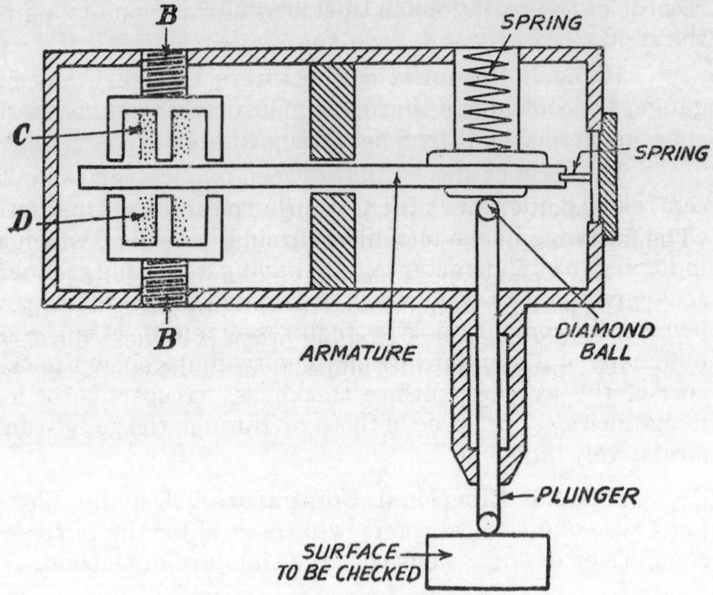

Fig. 10.18. Electrical comparator.

10.6. Automatic Inspection Machines

Automatic inspection machines are very economical for final inspection in high production manufacturing facilities. Arrangements for automatic transport of parts from one measuring station to other by electro-mechanical, hydraulic, or pneumatic mechanisms are provided. Part transportation system is designed for efficient and safe movement through machine. No human intervention is required to judge the quality. One or more measuring stations may be required to inspect part. Information received from the measuring station(s) is processed by the gauging amplifier and compared to a desired and predetermined standard. Good or bad components are identified automatically and bad components are discharged in a reject chute for collection of scrap parts. A logic system is incorporated that controls all of the machine movements and also keeps track of good and bad part conditions.

Most measurements are taken statically to simplify the inspection process. Direct process control of machining process is achieved by using electronic gauging techniques and monitoring the quality output just after a component has been machined and applying a proper compensation feedback if a part size is approaching an out-of-tolerance limit. If the part size is out-of-tolerance then machine is stopped before production of scrap. Electronic gauging machines are used because these provide output information that can be easily interfaced with statistical analysers, computers, CRTs and printers.

The development of microprocessor based systems results in following advantages :

— increased measuring capacity with reduction in physical size of equipment.

— availability of precise and clearer measurement information (displayed as a message on a screen).

— good parts are further segregated automatically to permit selective assembly.

— bad parts are segregated and discharged into reject chute to avoid any possibility of using them in a final assembly.
— automatic or semi-automatic mastering of the system is easily performed.
— self-diagnostic systems allow for easy and fast identification of faulty circuits reducing downtime and improving maintenance efficiency. Retooling is less expensive and faster because no hardware changes are required and the software program is able to achieve these.

10.6.1. Important Considerations for Design of Automatic Gauging Systems. The following points should be considered while designing automatic gauging systems :

— As far as possible, gauging probes should be designed to be actuated by an auxiliary probe, rather than the gauging probe connecting the part directly. This way the service life of gauging probe is enhanced considerably because now only auxiliary probe rams into slides against the parts. (Refer Fig. 10.19)
— As far as possible, the probes should be made retractable type to prevent possible wear or damage to the probes. The probe heads are thus held back till the part is in gauging position.

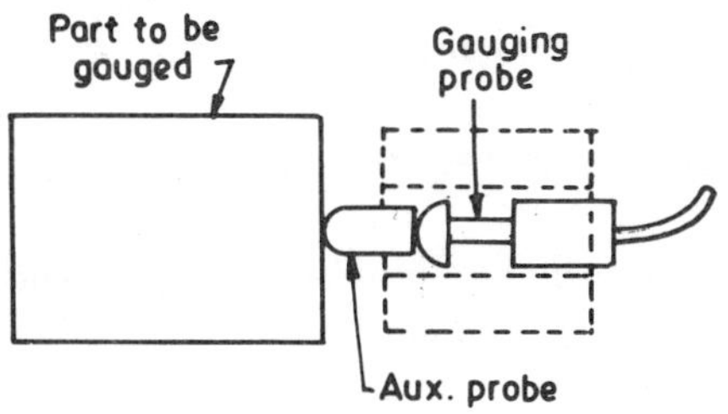

Fig. 10.19. Use of auxiliary probe to enhance life of gauging probe.

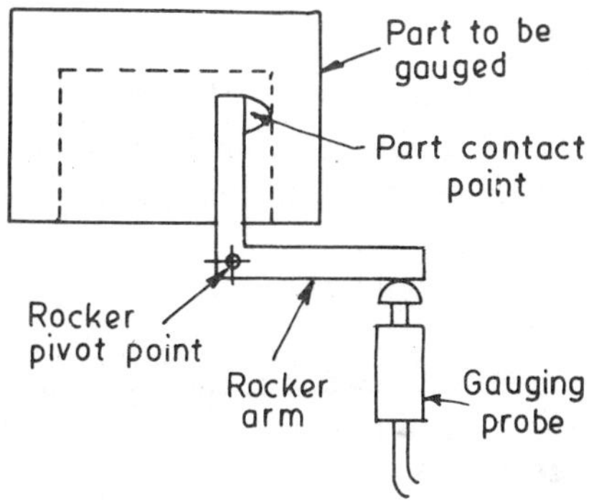

Fig. 10.20. Use of rocker arm to enable probe to reach difficulties.

— For greater accuracy, rocker-arm components could be used to transfer movement from the datum through one or more pivot points to the gauging probe. Such a system amplifies the motion, enables probe to reach difficult areas, and enables positioning of part on gauging station. (Refer Fig. 10.20)
— Parts should be designed so that all dimensional checks can be made from the same clamped position.
— The structure of the machine is designed to be completely separate for transportation system and separate for the gauging system. Special anti-vibration mounts should be used on each of the supporting leg of the gauging structure to insulate the gauging fixture from any possible vibrations from the rest of the machine and the surrounding environment.
— To avoid out-of-sequence mishaps, the sequence of part lifting, clamping between centres, the gauge fixtures moving down and up, etc. are linked by a single control mechanism, and also all movements of the gauge structures in each gauging station are cam operated and powered by a single electric motor reducer.
— To avoid damage to the gauging fixture, safety devices need to be incorporated. For instance, the gauging devices entering the part should be spring loaded into their

proper position instead of being positively pushed. If excessive force is needed anywhere, further operation should be stopped to avoid damage.

— Necessary safeguards are required to avoid wrong orientation of parts.

10.7. Measurement During Machining

If such devices are developed which measure the work during the process of operation, then much time can be saved in stopping the machine, taking out the part and resetting it. Automatic control over machining, inspection and assembly assures greater productivity, higher quality, and maximum use of machine capability. The comparators are the devices which serve this purpose and are usually set initially from an accurately dimensioned sample and then indicate any deviation of this work from the standard. It is essential that the instrument of this type must satisfy some or all of the following requirements :

(*i*) If possible, the gauging member should not come in contact with the work to be measured. The measuring and control device must be accurate, have high speed of response.

(*ii*) It should be unaffected by the variation of the machine, oil, dirt and moisture.

(*iii*) If sliding contact is used for measurements then there should be no damage done to the work by it.

(*iv*) The device must be easily capable of being attached to a machine or tool slide.

(*v*) The contact member must skip over grooves and keways.

(*vi*) It should be capable of checking the work for true cylindrical shape, taper, out or round and geometrical conditions like squareness, parallelism, concentricity and centre distance.

(*vii*) It should be simple to manipulate and capable of rapid resetting to a different diameter.

(*viii*) After resetting, the measuring contacts should be capable of assuming the right measuring position automatically.

10.7.1. Automatic Sizing. The concept of the automatic sizing is to produce the components accurately within the prescribed limits and relieve the operator of a considerable amount of responsibility once the machine is set up. Also the operation time is reduced considerably as the operator need not take the readings on the component several times during its machining and thus avoid the time of stopping the machine. Automatic sizing devices are generally applied to precision machine tools only *e.g.*, grinding and honing, etc. The various devices in use are described here in brief.

10.7.2. Mechanical Type of Automatic Sizing Instrument. It utilises a diamond-pointed finger in contact with the work. The diamond is held firmly by a lever which by a spring is constantly kept pressed towards the work. The movement of the diamond-pointed finger is amplified by a system of levers and finally transmitted to a long pointer which moves on a graduated scale. The whole assembly is mounted on a pillar secured to the table of the machine. As the grinding operation proceeds, the reduction of the work diameter allows the pointer to move under the action of a spring until it approaches a slot in its casing and in so doing intercepts a beam of light falling on a photo-electric cell. This at once stops the wheel-in-feed and the operator is informed of this condition by visual and audible indication. At the same time a solenoid is energised which operates the hydraulic transmission system and brings the machine to a standstill and recedes the wheel head to its standing position away from the work. This instrument is capable of producing workpieces upto the accuracy of 0.003 mm.

10.7.3. Electronic Sizing Device. In the mechanical system, since the sizing device which contacts the work operates under arduous conditions, therefore, they can't embody any delicate mechanical amplifications making the system less sensitive. Actually it is desirable that these devices should respond readily to the quick variations caused by out-of-roundness just before the work reaches final size. This is very easily achieved by employing an electronic device consisting of Wheatstone bridge circuit and supplied from a high-frequency generator. The bridge is made up from two fixed condensers C_1 and C_2 and the gauge itself, which contains two fixed plates P_1 and P_2 and intermediate movable plate P. When the movable plate P is exactly midway between the plates P_1 and P_2 (it is adjusted for the exact component size), the capacitance between it and the outer plates is exactly equal and the bridge is balanced.

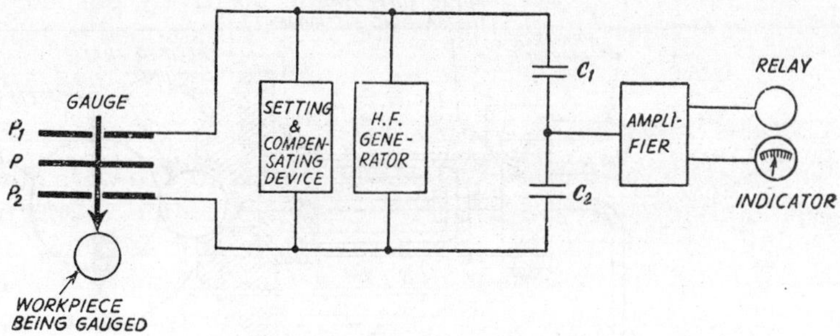

Fig. 10.21. Principle of Electronic Sizing Device.

Initially when the work-piece size is more, the plate P is nearer to P_1 and thus the bridge is out of balance and a current flows in the electronic amplifier which after being detected prevents a current flowing in the output stage of the amplifier. When the exact size is reached, the bridge is balanced and the current into the amplifier ceases and thus the maximum current is permitted in the output stage which operates a relay and causes the withdrawal of grinding wheel. The variations in temperature, humidity, ionisation of air, frequency and supply voltage act on both sides of bridge and consequently have no influence on the reading.

10.7.4. Pneumatic Sizing Device. The principle of the pneumatic sizing device is that if the air escapes from a nozzle, the pressure in the line is affected by the size and shape of the outlet through which air is escaping. If the air is supplied at constant pressure then the back pressure is a direct index of the size of the work-piece, provided the work-piece and the nozzle are arranged such that the size variation of work-piece changes the distance between nozzle and work-piece accordingly. The pressure change is used to change the level of mercury switch which in turn controls the feed movements of the machine.

Air-to-electric type transducers are also available which supply electrical impulse when the size being gauged reaches one of two present size limits. The electrical impulse may be utilised to light signal lamps or to actuate solenoids or relays. But utilising a number of such units in the same gauge circuit, a series of size classification can be carried out.

In Fig. 10.22, an air pump supplies the filtered air through a pressure regulating device to the nozzle mounted just behind the tungsten carbide tip and supported in a caliper frame. The tungsten carbide tip is loosely fitted into two gauging shoes and always remains in light contact with the work-piece. The shoes are mounted in the caliper frame which itself is supported through a pillar on to the frame of the machine. As grinding proceeds, the sizing head is moved forward by spring pressure, thereby reducing the gap between the work and the air outlet from the nozzle. The back pressure in the line thus keeps on building up and raises the mercury in the switch-box.

In this device, arrangements are made to have automatic different in-feeds of the grinding wheel depending upon the roughening operation and finishing operation. This is achieved by incorporating two electrodes in two tubes arranged at different levels. At the end of the roughening operation, just before the work reaches finished size the mercury touches the first-contact and a solenoid is energised through making of the contact. The energisation of the solenoid lowers down the speed of the in-feed of the grinding wheel. As the grinding proceeds further and when final size is reached another contact is made and another solenoid

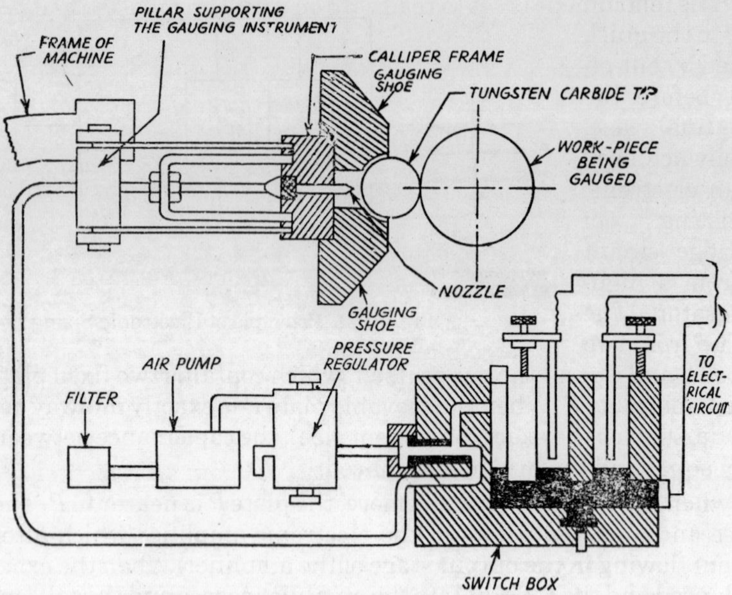

Fig. 10.22. Pneumatic Sizing Device.

energised which stops the machine and causes the grinding wheel to move away from the work. The position of the electrodes (contacts) in the mercury switch-box and the position of the carbide tip are adjustable.

10.7.5. In-process Measurement for Turning by a Mechanical Caliper. Mechanical caliper is the simplest device used for on-line measurement of workpieces being turned on

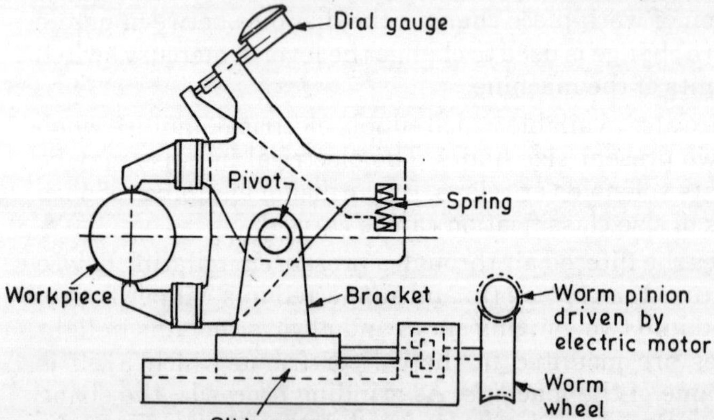

Fig. 10.23. In-process measurement using mechanical caliper.

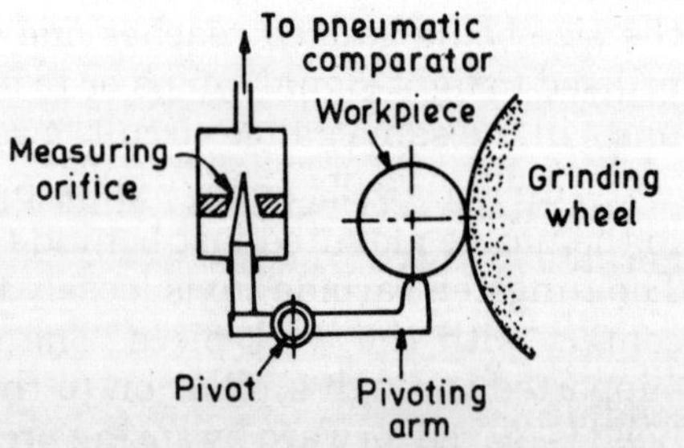

Fig. 10.24. Automatic sizing using pneumatic comparator.

a lathe. Fig. 10.23 shows such a device which is quite rugged and simple, has a wide range and can detect changes in size upto ± 0.01 mm. It uses scissors caliper jaws which are mechanically simple and trouble free. It may have amplification ratio of 1 : 1 or higher but the geometry of a similar triangle must be maintained throughout the measuring range for accurate readings. The caliper assembly is mounted on a bracket which can be moved transversely on the machine on slideway by an electric motor. The movement of jaws can be sensed at other arm by a dial gauge or displacement transducer like LVDT (Linear variable differential transformer) or capacitance type. The instrument is set to read zero on a rotating master gauge of the required mean diameter.

10.7.6. Use of Pneumatic Comparator for Automatic Sizing. Devices based on pneumatic comparator principle can be easily employed to measure diameter of work as it is being finished and also for setting the slides of machine tools precisely as shown in Fig. 10.24. A pivoting arm makes contact with the work-piece. As the size of workpiece is reduced, the conical tip at other end closes the measuring orifice forming part of the pneumatic comparator, thereby increasing the back pressure which is connected to the indicating device. Pressure switches can be connected across this back pressure and these can actuate the tripping devices to stop the in-feed mechanism of machine when desired size has been attained. The disadvantage of such a radial reading device is that its accuracy is affected by any changes in work axis due to thermal affects. Tip wear can be minimised by using hard materials but needs to be accounted. In the case of slide positioning, the tapered pivot is connected to machine slide.

10.8. Electronic Gauging

It has been general experience that the speed of gauging and the accuracy of measurement can be increased only by adopting the electronic techniques which offer numerous advantages. Most of the electric gauging systems are based on a displacement transducer (linear voltage differential transducer) which produces an e.m.f. linearly proportional to the displacement. The electrical impulse is displayed by an indicating instrument.

The automatic inspection and size control system have been developed based on the principles of electronic gauging. In these systems it is possible to set the desired dimension on a setter very precisely and whenever this set value is exceeded (when measuring any dimension) even by 0.000025 mm, a transistor-switch circuit is operated and signal given on the control panel. Further it is also possible to take the corrective action on the grinding machine during grinding operation or to stop the machine automatically if the dimension of the component falls in the reject limits. Any number of the circuits can be used in parallel, each operating at a different signal level to form a dimensional grading ladder.

The other advantages of electronic gauging systems are :

(*i*) Sensing devices are of smaller size permitting use of a simple, inexpensive measuring fixture.

(*ii*) Capacitive pick-ups do not make any contact and thus no wear problem on the measuring tip.

(*iii*) Accuracy of measurement is very high.

(*iv*) It is possible to carry out simple arithmatic operation on the electrical signals generated by transducers.

(*v*) It is possible to make datum setting adjustments from remote, which obviates the difficulties often experienced in positioning a gauging element accurately in a fixture by mechanical means.

(*vi*) Response time is very fast. Tolerance boundary circuits are available which operate in about 5 milli seconds or faster.

(*vii*) High resolution transducers having a linear measuring range of 2.5 mm may have a resolution of the order 0.00003 mm.

(*viii*) The same transducer can be used to give any one of a rang of magnifications by a selector switch that alters the gain of the amplifier.

(*ix*) Reliability of gauging is very high.

(*x*) Solid-state circuitry and plug-in modules used in this system offer least maintenance.

(*xi*) It is possible to use ancillary circuitry at later stages also which provides for control functions, digital and analogue recording and statistical analysis.

10.9. Contactless Three Dimensional Measurements by Laser Based System

This instrument consists of an optical head and a controller. The head uses a laser as the optical source and projects a laser spot onto the surface of the object to be measured. The position of the spot is sensed by a detector in the head and its distance from the head is

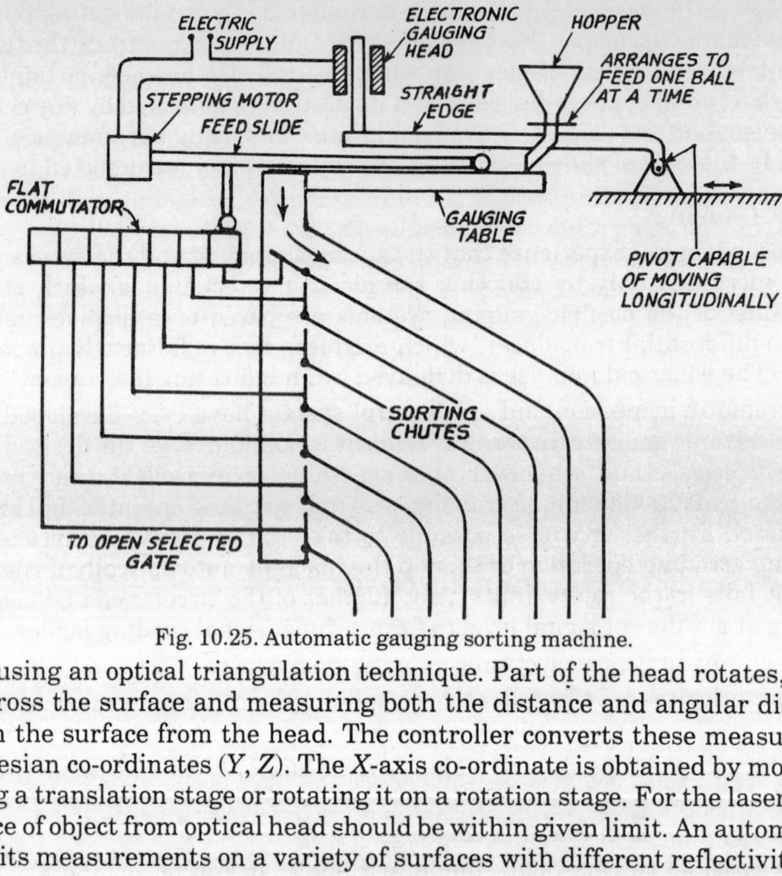

Fig. 10.25. Automatic gauging sorting machine.

determined, using an optical triangulation technique. Part of the head rotates, scanning the laser spot across the surface and measuring both the distance and angular displacement of each point on the surface from the head. The controller converts these measurements from polar to Cartesian co-ordinates (Y, Z). The X-axis co-ordinate is obtained by moving the head linearly along a translation stage or rotating it on a rotation stage. For the laser sport to be in focus, distance of object from optical head should be within given limit. An automatic exposure feature permits measurements on a variety of surfaces with different reflectivities.

10.10. Multi-Dimension Automatic Gauging and Sorting Machines

While for manual inspection, it can said that 100% inspection in likely to lead to operator fatigue and acceptance of faulty components. Same is not true for automatic inspection machines. These can carry out 100% inspection without fatigue and thus eliminating possibility of even a single defective component likely to be accepted when sampling inspection is done to avoid operator fatigue in carrying 100% inspection. With the present automatic inspection machines it is possible to check as many as 32 dimensions like lengths, diameters, taper, eccentricity, squareness, ovality, lobing, etc. simultaneously. The components are routed and moved automatically to different stations and depending on their dimensions, are graded for selective assembly. For simple components involving only a few dimensions, like ball and roller bearings, piston rings, etc., these can be fed via hopper to the gauging stations and sorted and carried into proper track to be collected in a container.

Fig. 10.25 shows schematically an automatic inspection and sorting machine for balls and rollers. It basically consists of a straight edge which is pivoted at one end and other end rests over the ball resting on gauging table. This straight gauge, in effect, works as a taper gauge and its inclination is sensed by an electronic gauging head fitted over it in which an extension rod fitted firmly to straight edge moves. Depending on the size of the ball being gauged, the straight edge will be inclined at a particular angle. The system is so desined as to have (seven gradings in this case) adequate gradings of dimensions of ball. Only one ball is fed over the gauging table from the hopper. If ball is of smallest size, it will keep straight edge in the desired inclination near the hoppe. If it is of bigger size then straight edge will be tilted more which is sensed by the gauging head and it passes a command to stopping motor to move the feed slide by one step. The ball under gravity moves over the gauging table to next position away from the hopper end. If its size is still bigger, then another command is given to stepping motor to move feed slide to next position and ball keeps on following it. Finally ball comes to a desired size position where straight edge makes the desired angle and ball size then corresponds to that position. An extension from feed slide moves over a flat commutator which in turn opens the corresponding gate of the storing chute and ball at the end of correct size checking flows from gauging table into the correct hopper via the sorting chutes. The straight edge can be moved longitudinally along with its pivot for size adjustments. It is possible to obtain size steps between the groups from 0.001 to 0.1 mm. A system of indicating lamps indicates the manner in which grading is progressing.

10.11. Electro-Optical Inspection

Electro-optical inspection techniques are basically contactless measuring techniques, are 100% reliable and very fast in operation.

These are used for

(*i*) Inspection of part dimensions,

(*ii*) Surface defects, and

(*iii*) Inspection of parts for proper assembly.

These techniques are generally applied in high volume and high cost parts. The basic advantages of electro-optical systems are speed, accuracy, non-contact operations, large range, and ability to see very small areas. Cleanliness is must for success of such a system. The pros and cons of various types of electro-optical measurement techniques need to be studied in depth before choosing a particular type for the given application.

The outside diameter is sensed by the scanning laser-beam shadow technique and profile imaging. This duplicates closely a shadowgraph or contour projector where the ground glass screen has been replaced by an image sensor. The imaging approach with the solid state diode array is considered to be best for piece-part measurements. Very high scan rates of the order of thousands of parts per hour with extremely high resolution and accuracies of the order of 0.05 microns are possible. Scanning photo-diode array imaging system has another advantage of operation with stroboscopic illumination which allows the part position to be frozen in space. Further it has no moving parts. Outer diameter can also be sensed by scanning laser beams and collimating optics. It is very useful for large range but where resolution is less than 10 microns.

Surfaces that can't be seen in profile like concave surfaces on combustion chamber, turbine blade, airofoils, gear teeth etc. can be measured by the triangulation technique. Fibre optic technique has also been tried for this purpose but not successfully. Fibre optic technique depends on the measurement of amount of light reflected from the part back upto the fibre. Since the amount of light reflected is not only a function of part dimension but also surface finish texture, colour, and form, so this method is not very successful for this application. However, the triangulating system does not rely on the amount of light but on the physical location of an image spot. While spot intensity and its distribution may vary but its location does not vary. Thus by sophisticated scanning and analysis of centroid distribution of shifts and location, dimensional differences in the range of 1 micron can be accurately picked out. Surface defects by electro-optical techniques can be measured by (*i*) imaging with scanning diode arrays, (*ii*) scanning laser beams, and (*iii*) fibre optics.

The image scanning approach is considered to have several advantages. This method is similar to process of visual inspection since the human eye, lens and retina are closely approximated by the sensor's lens and diode array. Its scan rate is faster than lasertype scanners. This technique has a high resolution (facility for magnifying a part of surface) which is very desirable for piecepart inspection. Such an arrangement is not possible in laser scanning and fibre optics. Another advantage of the photodiode array is that it can be strobed to freeze surface areas in two axes. Hardware used in compact, solid state and can stand off the part at a large distance.

Laser scanning approach can detect certain types of flaws better because of extremely directional property of laser.

Fibre optics is like a small light source and detector near the surface of the part. The light illuminates, and the detector sees light coming back from a very small area. However to cover large areas of the part, many multiplexed fibers have to be used. Further resolution is limited by fibre diameter and is impaired by dead zones between the fibres. The fibre system has to be extremely close to the part as there are no imaging elements. This method virtually precludes complicated surface flaw processing and analysis.

Diode array approach can digitise the size or location of flaws and thus furnishes a whole-part map. However in this method, the choice of lighting type and angle is very important in accentuating or diminishing the signals from certain types of conditions on the part surface. Utilisation of the diode array requires the intelligent use of light sources and hardware processing and microcomputer hardware and programming.

10.12. Some Recent Developments in Optical Measurements

Over the last 20 years a major impact on optical measurement has been made by developments in electro-optics, providing automation, greater acuity of setting and faster response times. Photo-electric detectors have progressed in parallel with active electronic devices, and radiation sources have received considerable attention.

The most basic form of photo-electric detector, the individual cell, acts as an on-off or grey scale device. More recent types, such as the Avalanche effect photo-diode and photo-transistor, have integral amplification which provides greatly improved low-light performance. Of particular significance to measurement applications are position detecting arrays which can be divided into those with analogue and those with digital structure.

The analogue (or continuous) arrays, available in either linear or quadrant form, operate as potentiometric devices to define the position of an illuminating spot of light. Digital arrays, on the other hand, are matrices of accurately spaced photo-sensitive elements or diodes (usually charge-injection or charge-coupled devices) on an integrated circuit chip. These are electronically scanned in sequence to generate video-type outputs. Available as complete image-sensing packages, with associated shift registers, they have typical element spacings 15—25 μm. Linear arrays with up to 2048 elements and two-dimensional arrays with up to 1024 × 1024 elements are currently on the market. They are being used in an increasing number of non-contact optical measuring applications (profile, form and position measurement, for example) and can be readily interfaced to digital image-processing systems.

Radiation sources have developed in a number of areas. Whilst filament and discharge sources have been refined, the most important developments are light-emitting diodes and lasers. Light-emitting diodes combine low power consumption with very high efficiency. Some of the more recent types have built-in optical elements which produce well defined beams. The laser, with its highly controlled frequency modes and coherent output, is used extensively for interferometry (particularly the He-Ne gas type) ; the laser distancer-measuring interferometer has become an industry standard. Gradual cost reduction has made the laser suitable for more general application where a very convenient, collimated, high-intensity source is required. Semiconductor lasers, or laser diodes, now give many of these advantages at even lower cost.

A laser beam projected directly onto a position detector is a method of alignment used in a number of commercially available systems. A major disadvantage of this type of "two-point" approach is the error caused by slight tilts of the laser support, this being magnified at increasing working distance. The projected beam is also prone to instability due, largely, to heat generated in the laser tube.

As position detecting arrays become more linear and more consistent in quality, possibilities for their use in automatic autocollimators become increasingly evident. Future instrument will undoubtedly feature semiconductor sources which closely match the spectral response of these detectors. The promise of high resolution coupled with wide range and small size is an attractive prospect. As a further step, micro-processor technology offers the means of compensating for detector non-linearities by storing measured errors in P.R.O.M. (programmable read-only memory). This will greatly enhance overall system accuracy.

Micro-computers are greatly extending the capabilities of optical measurement systems. Their facility to process and store large volumes of data enables measurement results to be presented in directly useable form for control and inspection purposes.

10.13. Precision Instruments Based on Laser

Development of laser enabled production of clear coherent light. The biggest advantage of coherent light is that whole of the energy appears to be emanating from a very small point. Further by a system of lenses, such a beam can be focused easily into either a parallel beam or on to a very small point. A typical helium-neon laser source produces a 1 to 2 mm diameter beam of pure red light having power of 1 mW. When such a beam is focused at a point, it has very high intensity.

Upto a great distance beam has no divergence but then it begins to expand at a rate of about 1 mm/m. The laser beam is visible on any screen and on virtually any surface and it can be observed easily from a distance and it remains in position for reference. Its centre can be easily judged to an accuracy of 1 mm over 2 m. It can be used for very accurate measurements of the order of 0.1 μm in 100 m.

10.14. Laser Telemetric System

Laser telemetric system is a non-contact gauge that measures with a collimated laser beam (Refer Fig. 10.26). It measure at the rate of 150 scans per second. It basically consists of three components, a transmitter, a receiver and processor electronics. The transmitter module produces a collimated parallel scanning laser beam moving at a high, constant, linear speed. The scanning beam appears as a red line. The receiver module collects and photoelectrically senses the laser light transmitted past the object being measured. The processor electronics takes the received signals to convert them to a convenient form and displays the dimension being gauged.

The transmitter contains a low-power helium-neon gas laser and its power supply, a specially designed collimating lens, a hysteresis synchronous motor, a mutli-faceted reflector prism, a synchronous pulse photodetector and a protective replaceable window.

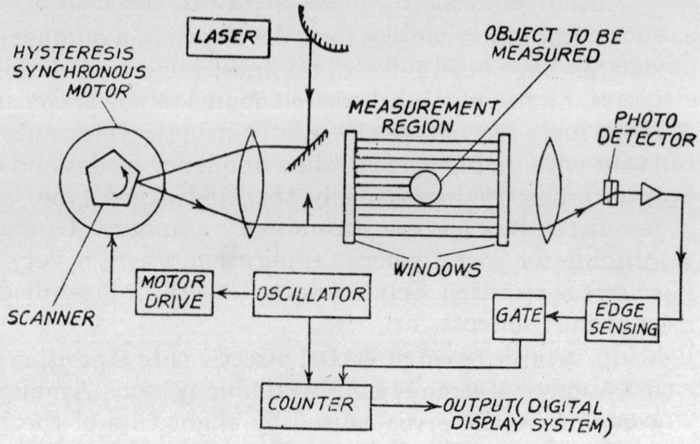

Fig. 10.26. Schematic diagram of laser telematic system.

The high speed of scanning permits on-line gauging and thus it is possible to detect changes in dimensions when components are moving or a continuous product such as in rolling process moving at very high speed. There is no need of waiting or product to cool for taking measurements. This system can also be applied on production machines and control them with closed feedback loops. Since the output of this system is available in digital form, it can run a process controller, limit alarms can be provided and output can be taken on digital printer. It is possible to write programs for the microprocessor to take care of smoke, dust and other airborne interference around the workpiece being measured.

10.15. Laser and LED Based Distance Measuring Instruments

These can measure distances from 1 to 2 m with an accuracy of the order of 0.1 to 1% of the measuring range. When the light emitted by laser or LED hits an object, it scatters and some of this scattered light is seen by a position sensitive detector or diode array. If the distance between the measuring head and the object changes, the angle at which the light enters the detector will also change. The angle of deviation is calibrated in terms of distance and output is provided as 0—20 mA. Such instruments are very reliable because there are no moving parts. Their response time is in milliseconds. The measuring system uses two distance meters placed at equal distance on either side of the object and a control unit to measure the thickness of an object. The distance meter is focused at the centre of the object.

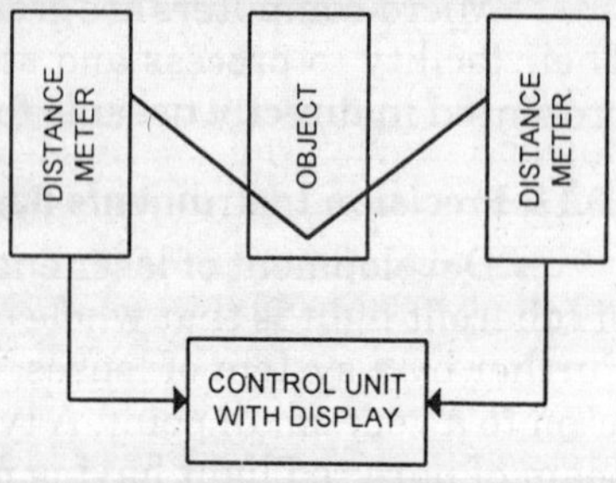

Fig. 10.27. Distance meter.

10.16. Intelligent Vision

Intelligent vision system is used to provide in-process monitoring and control functions. It is capable of performing on-line part recognition, classifying, inspecting and sorting. It can also make real time precision measurements.

An intelligent vision system does the repetitive and monotonous jobs continuously with 100% perfection, whereas operator being subject to fatigue and boredom under such conditions is liable to err. The inspection is done at the speed of electronic whereas operators take time. However such systems can do only whatever they are told to do. They can't differentiate between critical and non-critical factors. On the other hand, operators can exercise their judgement and readily adapt to changes in the environment or process.

In some systems, parts for sorting have to be presented suitably for controlling visual access. In other systems the parts can be presented in a random way.

Intelligent vision systems operate on the basis of threshold, or binary, video signals and as such are much tolerant to variations in lighting conditions and uniformity. In such systems, anything above an established brightness level (threshold) is considered white, and everything else in the scene is black. Sometimes problems may be encountered with low contrast scenes which are illuminated at levels near the threshold point ; because minor changes in lighting can cause marginal areas to be considered white one time and black another. Erroneous results are also likely from shadows across white regions and specular reflections from back regions.

Thus the most important consideration in such systems is the lighting arrangement. Most commonly used arrangements are structured lighting, polarised or coherent light sources, spectral pass or reject filters, stoboscopic lighting, etc. Fig. 10.28 shows in schematic form the working of a typical intelligent vision system. Camera (solid state type) converts the optical image into an electronic video input signal (analog or digital). The signals from one or more cameras are fed into video processing unit where image are accepted for processing only when the part position sensor calls for them and stores it for subsequent processing. Grayscale scenes are converted to 1-bit binary images. Image data processing systems extract two-dimensional spatial data from the scene. It is possible to derive third dimension information with the aid of structural lighting. Dimensional data like height, area and distance between holes or edges is extracted easily. Sorting and classification is possible by comparing the measurement with established tolerance limits. Some systems operate through object recognition. It recognises object by taking its multiple views and storing a set of orientation-independent features which represent a signature for that object. When an unknown part is presented to the system, signature comparisons are performed and best fit relationships are established. Results of image data processing are used to generate the decisions fed back to the system under control. Machine interface communicates with operators, controllers, robots, etc. A flexible intelligent vision system can be achieved through software programming particularly in decision generators and particularly in image data processing. Thus product changes can be handled by generating application program.

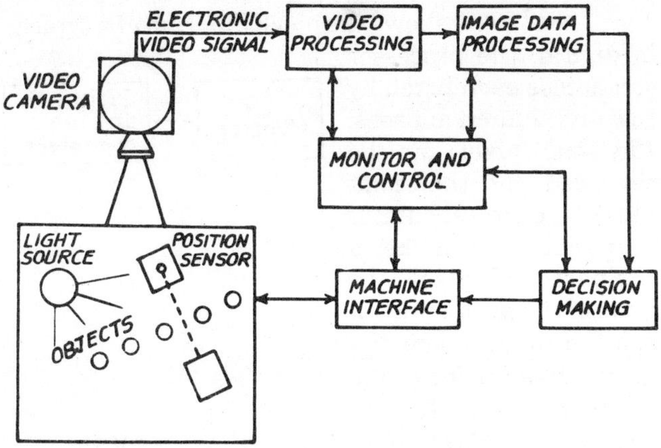

Fig. 10.28. Intelligent vision system.

10.16.1. Vision Methodology (Video Image Analysis) for on-line non-contact Inspection. Video metrology technique utilises a camera, video memory, and micro computer is used for inspecting on-line the several parts per minute with 100% reliability. The camera image is temporarily stored in a solid state memory device. Computer logic reviews subsequent stored image data and makes comparisons of picture contrast differences, by enhancing inspection areas with certain lighting and optical conditions.

Image cameras are of three types : Line scan, solid state matrix, and analog. Line scan image sensor has high resolution and high scan speed but it required intense lighting, constant part speed, pixel normalisation and high cost. Solid state matrix image sensor has excellent linearity and can't burn image. It also requires pixel normalisation, has unusable pixels, low resolution and high cost. Analog image sensor has high resolution, high contrast sensitivity, requires low lighting, has capability to preprocess and has low cost. But it has disadvantages of poor linearity, image drift with time and temperature, and image burn. Analog image sensor is more popular.

The video signal has to be transformed into a digital form because computer can only work with digital information.

The analog picture obtained by image sensor is diseched by rows and columns into finite segments, say 512 × 512. Each segment (pixel) has a known size and contrast value. Hardware to distinguish between any shades of gray, (around 256 contrast differences between white to black) is available.

Fig. 10.29 shows the block diagram of a typical optical inspection system. With the video in digital, the processor can access each pixel, by row and column address. Any one pixel can be reviewed by the computer for contrast data. It is possible to make a comparison, pixel by pixel of the part being inspected to an image of a good part in memory and take decision to accept/reject the part.

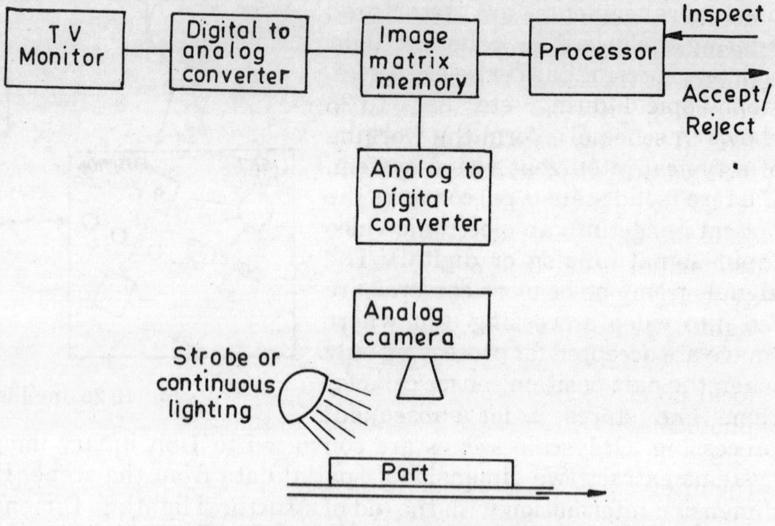

Fig. 10.29. Optical inspection system.

Nowadays "grey scale analysis technique" is used to cope with large good part variations and find defects with 100% reliability. In this technique, discrete areas, or windows, are formed around only the portions of the image to be inspected. The presence of bracket can be inspected by using high intensity lighting and looking for the presence of a dark shadow. If there is no bracket, no shadow will be cast. An easy analysis of bracket presence/absence would be to set a contrast threshold between the dark and bright pixel value area and have the computer count up the number of pixels which must meet this threshold value. Discrete area analysis is a powerful tool and can be used for inspections of presence/absence, correct part assembly, orientation, part integrity etc. However, such a system can be successful only if the part is precisely fixtured and presented to the inspection system in a repetitive fixtured position. However by execution of a simple set of preinspection instructions, computer can reorient the part in memory.

To increase the resolution, several cameras can be multiplexed, each having a portion of the total measurement problem.

10.17. Optical methods for fast, non-contact, on-line measurements

Optical techniques are used to measure linear dimensions—like length, diameter, gap and displacement. These systems are also used for automatic inspection. Optical techniques for this purpose employ linear diode arrays, lasers, and diffraction phenomena. An optical measurement system consists of a light source, optical components, photodetectors, and an electronics unit which converts the light into an electrical signal for control and/or display. System accuracy depends on the way in which the dimensional information carried by the light beam is coded and decoded.

A linear diode array consists of photodetector diodes regularly spaced at constant intervals along a straight line. It is thus optical gauging's analogy to the ruler. The object whose dimension is to be measured, is illuminated and its image formed on the linear diode array. The electronic devices are used to count the number of adjacent, illuminated diodes, the length being directly proportional to the number of diodes counted—of course, taking magnification of image into account. In such a system, the accuracy is dependent on the fidelity of the image and on the spacing of the diodes and the uniformity of their response.

Scanning laser beam. It employs a low-mass rotating mirror to sweep a single beam of laser light across a planar area. The laser beam is reflected from the rotating mirror at its axis of rotation which coincides with the focal point of a lens. The rays originating at the focal point of an ideal lens are parallel after emerging from the lens, thus the laser beam is continuously translated parallel to itself, generating a continuous plane of light.

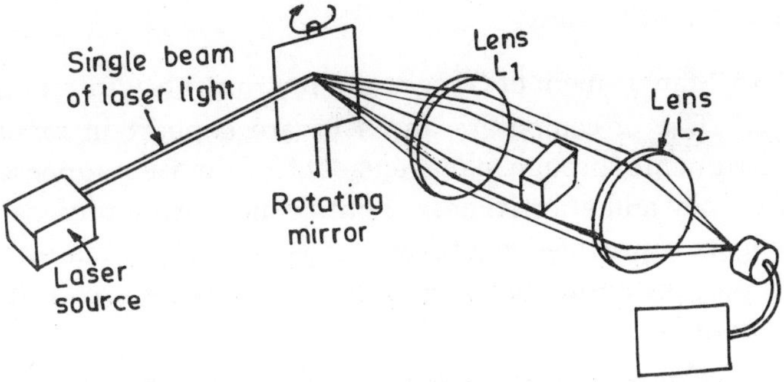

Fig. 10.30. Scanning Laser beam.

A second lens focusses the parallel rays of laser light onto a photodetector located at the focal point of the second lens. In the absence of any object in the planar light beam, the photodetector provides a continuous output signal. If any object is placed between the two lenses, it partially blocks the plane of light and the length of the object is measured by timing the intervals over which no laser light reaches the photodetector.

Diffraction and Symmetry

Diffraction effects are used for gauging objects like small-diameter wire and narrow gaps. A narrow beam of light incident on a small-diameter wire is spread due to diffraction, with the spread inversely proportional to the wire diameter. The diffraction beam can be focused on a screen using a suitable lens, resulting in the typical diffraction pattern of alternate light and dark bands. The linear distance between these bands is measured using a linear diode array.

At the focal plane of the lens, the angle of twist θ is equal to the wavelength of the light divided by the diameter of wire ($\theta = \lambda/d$). The distance in the focal plane from the lens axis to the first minimum is equal to the angle θ multiplied by the focal length of the lens (f). Thus the wire diameter can be calculated directly from the diffraction pattern measurement.

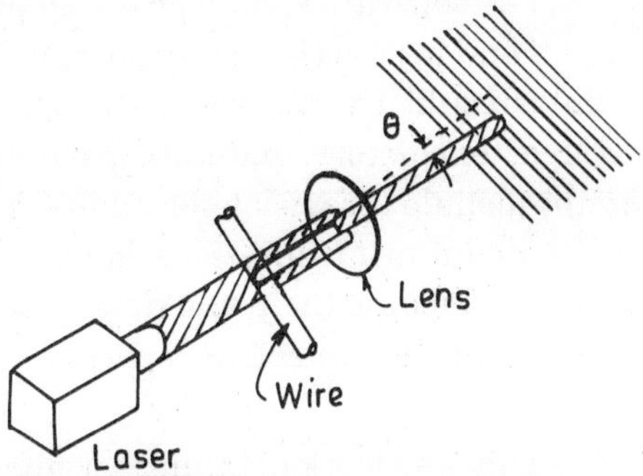

Fig. 10.31. Diffraction technique for gauging.

The displacement along a straight line can also be measured by the method Moire fringes produced by using two similar, transmission diffraction gratings placed nearly face to face and with the grating lines nearly parallel. The fringes are quite visible to the naked eye even though the grating-line spacing is too fine for visual discrimination. The distance between adjacent dark fringes depends on the grating-line spacing and on the angle between the lines on the two gratings. If one grating is moved relative to the other, the pattern of fringes moves perpendicular to the direction of relative motion. The ratio of fringes pattern displacement to grating displacement is a function of the angle between the two gratings (of the order of 1000).

Objects with rotational symmetry properties can be inspected by projecting its image onto a ring-shaped sensor made up of an appropriate number of multi-detector sensing array having same symmetry as object. Any deviation from symmetry, like a worn out gear tooth, causes an uneven illumination, which can be interpreted as detect by electronics unit as an unacceptable object.

10.18. Management of Gauges and Instruments with Computers

Use of computers and software support in management of gauges and measuring instruments through management of information support can result in upgradation of quality, efficiency and effectiveness of entire measuring process. It is important to ensure that such support is provided right from inspection of raw material stages to finished product, always keeping in mind that metrology applications are traceable to national and international standards.

System for surveillence and control of gauges, measuring instruments and other inspection equipment can be either in the form of acceptance inspection of new items or the periodic calibration of gauges and instruments. The flow charts for both the types of requirements need to be developed, clearly identifying where the support of software will be needed. A list of all the gauges and instruments to be calibrated is prepared specifying an ID number to each, also indicating the period of calibration. A list of all the measuring equipment available for calibration with full specification and capabilities is also prepared and stored. Similarly a list of personnel available for carrying out calibration work, with their specialisation etc. are also prepared.

The software used should be as user friendly as possible so that persons can use computers with minimum training. Separate work stations have to be developed for calibration of gauges, instruments and inspection equipment in various shops as per requirement. The entire information shall be made available to a central monitoring and development station.

The software should be developed in the form of modules, each module being complete in itself and interaction being possible between various modules as per requirements. Various modules could be : master data maintenance module (incoporating new item code entry, retrieve item history, maintain test specification, vendor master, etc.), purchase order and item receipt module (provision for addition and updation of orders and receipts, inventory control etc.). Scheduling module (for schedule generation and schedule adjustment to meet work load), inspection module (help and data entry, approval by supervisors etc.), analytical information module to generate various reports menus etc., and system maintenance module for maintaining security.

This way work planning becomes smooth as work schedules for calibration are prepared automatically by computer. Software provides on line help to inspector during calibration and can take decisions regarding acceptance or rejection on basis of evaluation of tolerances. Calibration history of any item can be conveniently retrieved. Several types of reports can be generated which help in taking important decisions in deciding overall view of stocks and inventory, pending items etc. It is possible to draw useful inferences like wear analysis trends to decide calibration intervals, locationwise analysis of rejections to identify need of improved instruments, etc.

TYPICAL EXAMINATION QUESTIONS

1. (*a*) Explain the role of dynamic measurements in modern production techniques and how it could cut down production cost.

   (*b*) Describe some instruments for dynamic measurements commonly used in industries.
2. Discuss the importance of gauging system in workshop and explain how the accuracy of the gauges is maintained and tested.
3. Describe with neat sketches the method to measure the absolute length of slip gauges.

11

Measurement of Surface Finish

11.1. Introduction

A great advancement in the present machine age has been possible due to the increased knowledge and constant improvement of the surface textures. The need of producing smoother and harder surfaces was realised as greater strength and bearing loads were demanded. It was also found that the functioning of machine parts, load carrying capacity, tool life, fatigue life, bearing corrosion, and wear qualities of any component of a machine have direct bearing with its surface texture. Therefore, these effects made the control of surface texture very important.

Failure due to fatigue always starts to occur at sharp corners where stress concentration occurs. The root of any surface irregularity acts as sharp corner and such part fails earlier. It has been also shown that the surface irregularity at the non-working surface also matters. Thus in order to increase the life of any part which is subjected to repeated reversals of stress, the working and non-working surfaces of that surface must be given very good finish.

Good bearing properties in any part are obtained when the surface has large number of irregularities, *i.e.* a large number of hills and valleys. If the surface is perfectly smooth then seizure would occur due to difficulty of maintaining the lubricating oil film. The hills in irregular surface reduce the metal to metal contact and valleys help to retain the film of lubricating oil. The rate of wear is proportional to the surface areas in contact and the load per unit area.

Thus it is seen that different requirements demand different types of surfaces. Therefore, it became essential to measure the surface texture quantitatively and methods were devised for this purpose. Also the greater demands of high surface finish resulted in refined processes like grinding, honing and lapping.

It may be appreciated that it is not possible to produce perfectly smooth surface. The manufactured surface always departs from the absolute perfection to some extent. The imperfections on the surface are in the form of succession of hills and valleys varying both in height and in spacing. The departures from a truly smooth surface may arise from a variety of causes and may be of several kinds. The texture or roughness (succession of minute irregularities) on surface is influenced by the machining process employed.

For instance, a surface machined by a single point tool, will have roughness which is uniformly spaced and directional. In the case of finishing process, the texture is irregular and non-directional. Thus while tool produces its characteristic roughness, there may be produced additional more openly spaced components of roughness. Complete roughness is the resultant of irregularities of various kinds. If the hills and valleys on a surface are very close, *i.e.* the wavelength is very small, then surface appears as rough. This is due to action of the cutting tool and is referred to as primary texture. If the hills and valleys on a surface are far apart, it is due to imperfection in the machine tool and is referred to as secondary texture or waviness.

It would be appreciated that the imperfections to be found on any surface take the form of peaks and valleys of different height and wavelength. This distinction between primary and secondary texture is due to difference in wavelength.

It is however impossible to specify any particular wavelength which could be treated as the dividing line between the primary and secondary texture. This depends on the requirements and class of the work desired.

A surface actually is quite complex and consists of many different wavelengths caused due to feed of the tool, cutting action, vibrations, imperfections in machine tools, etc.

11.2. Meaning of Surface Texture and Some Definitions

Usually the concept of surface roughness is sensory and it is capable of being understood in an intuitive way. Generally the surface roughness has been experienced and understood by the sense of light and touch etc.

Therefore, to define surface roughness in terms of measurable attributes presents great difficulties. The definition of surface roughness could probably be best understood on a causative basis and independent of intuition.

As we know that any material being machined by chip removal process can't be finished perfectly due to some departures from ideal conditions as envisaged by the designer. Due to conditions not being ideal, the surface produced will have some irregularities ; and these geometrical irregularities could be classified into four categories.

First Order. This includes the irregularities arising out of inaccuracies in the machine tool itself *e.g.* lack of straightness of guideways on which tool post is moving. This includes the surface irregularities arising due to deformation of work under the action of cutting forces and the weight of the material itself.

Second Order. Some irregularities are caused due to vibrations of any kind such as chatter marks and are included in second order.

Third Order. Even if the machine were prefect and completely free of vibrations, some irregularities are caused by machining itself due to characteristic of the process. This also includes the feed marks of the cutting tool.

Fourth Order. This includes the irregularities arising from the rupture of the material during the separation of the chip.

Further these irregularities of four orders can be grouped under two groups :

First group includes irregularities of considerable wave-length of a periodic character resulting from mechanical disturbances in the generating set-up. These errors are termed as macro-geometrical errors and include irregularities of first and second order and are mainly due to misalignment of centres, lack of straightness of guide-ways and non-linear feed motion. These errors are also referred to as *Waviness or Secondary Texture.*

Second group includes irregularities of small wavelength caused by the direct action of the cutting element on the material or by some other disturbance such as friction, wear, or corrosion. These errors are chiefly caused due to tool feed rate and due to tool chatter, *i.e.* it includes irregularities of third and fourth order and constitutes the micro geometrical errors. Errors in this group are referred to as *Roughness* or *Primary Texture.*

Thus any finished surface could be considered to be combination of two forms of wavelength (large wavelength for waviness and smaller wavelength for roughness) superimposed upon each other. One of the problems in measuring surface finish is to separate the waviness from the roughness.

The various terms used in connection with surface finish could now be shown on the diagram below (Fig. 11.1).

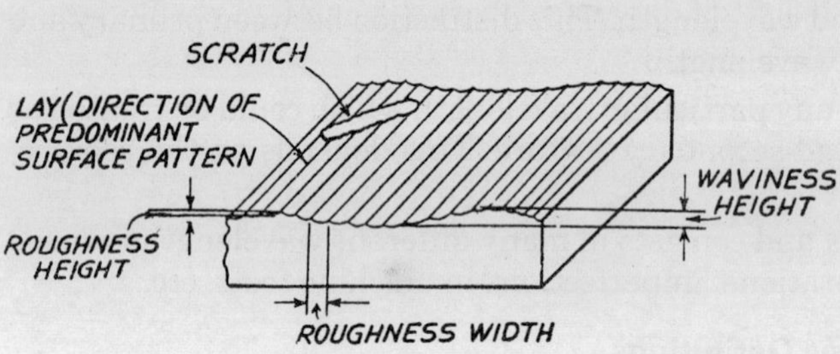

Fig. 11.1. Terms for surface roughness.

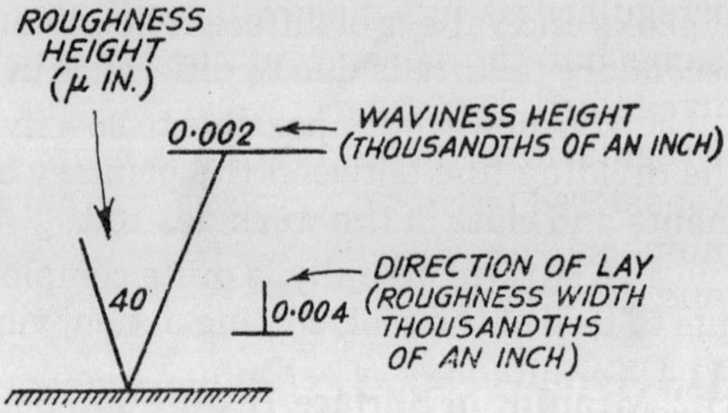

Fig. 11.2. Representation of surface finish.

These terms are represented systematically in Fig. 11.2 and this notation is followed universally. Roughness width may be sometimes omitted.

11.3. Surface Roughness

On any finished surface, imperfections are bound to be there and these take the form of a succession of hills and valleys which vary both in height and in spacing and result in a kind of texture which in appearance or feel is often characteristic of the machining process and its accompanying defects. The several kinds of departures are there on the surface and these are due to various causes. Roughness or texture in the form of a succession of minute irregularities is produced directly by the finishing process employed *e.g.* while using a single point tool. The roughness on the surface tends to be both uniformly spaced and directional ; whereas surfaces produced by straight and cylindrical grinding tend to have irregularly spaced but directional roughness and in the case of boring or lapping process the surface roughness tends to be irregular and non-directional in character. The characteristic roughness produced by the tool is not the only cause of roughness in case of machining operations, but the more openly spaced component or roughness are also produced from faults in the machining operation. In practice the complete roughness commonly represents a combination of irregularities of various kinds and magnitudes arising from several different causes and the individual effects of the separate contributing factors can't always be readily distinguished. Thus for the complete study of the surface roughness, it is essential that the measurement and analysis of all the component elements and an assessment of the effects of the resulting combined texture be made. All this being very difficult and tedious job, in practice all that is essential is that a practical method of assessment be followed, the result of which can be readily compared with a specified requirement of quality, preferably on the numerical basis. It is done by analysing the form of profile revealed by a plane section through the surface. The measurement of surface roughness poses a problem in three dimensional geometry, but for simplification purposes, it is better to reduce it into two dimensional geometry by confining individual measurement to the profiles of plane sections taken through the surface. The direction of measurement is usually perpendicular to the direction of the predominant surface markings or 'lay'. The shape of the hills and valleys which constitute surface roughness is best studied in a cross-section normal to the surface using a stylus instrument which plots a graph in which the minute irregularities of the surface are magnified to 50,000 times or even greater, whereas the length over which observations are made is magnified only by 50 to 300 times. Thus the above graph does not give a real representation of the surface nature, but a distorted view, but it gives very useful information. From this curve the height of the irregularities (total or average value), separation of the crests (separation) can be easily noted.

When the irregularities are comparatively uniform in shape and size, then the distance between the successive peaks is described as pitch or dominant spacing. It may be appreciated here that the surface roughness is concerned both with the size and the shape of the irregularities *e.g.*, in certain profile the height of departure from the nominal profile may be same but the spacing of the irregularities may be wider or closer, or the space of the irregularities may be of various forms. Thus realising that both size and shape (*i.e.* height of irregularities, their spacing and form) are important for specifying surface roughness, it is not considered possible to express the complete roughness characteristics by means of any single number. The normal practice is, therefore, to specify R_a value to define the quality or grade, and the manufacturing process which serves to produce the type of roughness.

11.4. Terminology as per Indian Standards

Real Surface : is the surface limiting the body and separating it from the surrounding surface.

Geometrical Surface : is the surface prescribed by the design or by the process of manufacture, neglecting the errors of form and surface roughness.

Effective Surface : is the close representation of real surface obtained by instrumental means.

Surface Texture : Repetitive or random deviations from the nominal surface which form the pattern of the surface. Surface texture includes roughness, waviness, lay and flaws.

Surface Roughness : It concerns all those irregularities which form surface relief and which are conventionally defined within the area where deviations of form and waviness are eliminated.

Primary Texture (Roughness) : It is caused due to the irregularities in the surface roughness which result from the inherent action of the production process. These are deemed to include transverse feed marks and the irregularities within them.

Secondary Texture (Waviness) : It results from the factors such as machine or work deflections, vibrations, chatter, heat treatment or warping strains. Waviness is the component of surface roughness upon which roughness is superimposed.

Flaws : Flaws are irregularities which occur at one place or at relatively infrequent or widely varying intervals in surface (like scratches, cracks, random blemishes, etc.).

Centre line : The line about which roughness is measured.

Lay. It is the direction of the 'predominant surface pattern' ordinarily determined by the method of production used.

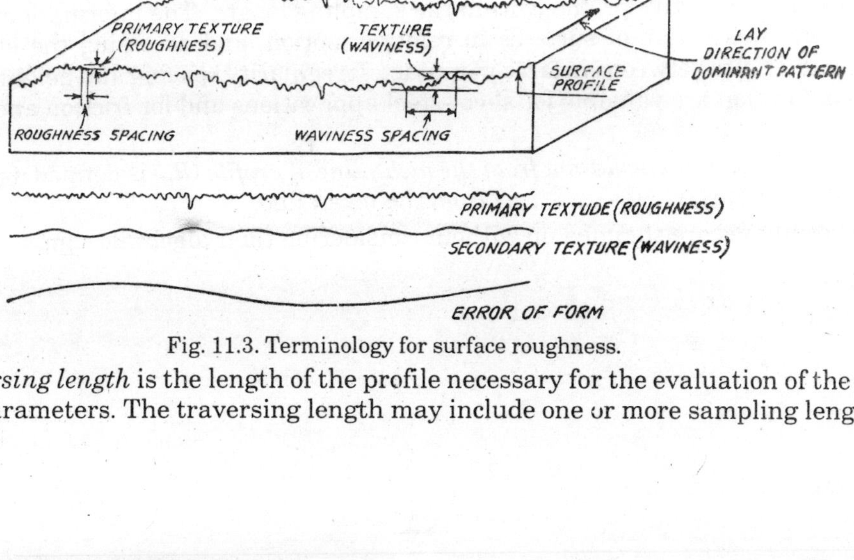

Fig. 11.3. Terminology for surface roughness.

Traversing length is the length of the profile necessary for the evaluation of the surface roughness parameters. The traversing length may include one or more sampling lengths.

Sampling length (l) is the length of profile necessary for the evaluation of the irregularities to be taken into account. This is also known as the 'cut-off' length in regard to the measuring instruments. It is measured in a direction parallel to the general direction of the profile. It is very difficult to specify any value for spacing (*i.e.* the length over which the surface profile is to be considered) which could be regarded as forming a definite line of demarcation between roughness and to much extent, it depends upon the manufacturing process employed. Similarly no theoretical value could be assigned to the maximum spacing for considering the waviness, as at the higher value the waviness tends to become merged in more general errors of geometric form. However, for majority of engineering work, value of 0.8 mm is generally considered to be quite satisfactory for instrument cut-off and upper limit of 25 mm is commonly accepted as suitable for most waviness measurements.

Mean line of the profile is the line having the form of the geometrical profile and dividing the effective profile so that within the sampling length the sum of the squares of distances (y_1, y_2, ... y_n) between effective points and the mean line is a minimum.

Centre line of profile is the line parallel to the general direction of the profile for which the areas embraced by the profile above and below the line are equal. When the waveform is repetitive, the mean line and the centre are equivalent.

It may be noted that true repetitiveness in any manufacturing process is impossible and as such some difference in mean line and centre line is bound to exist. But, however, in view of its insignificance in relation to other errors or measurement of surface geometry by index, the mean line and the centre line may be considered to be equivalent for practical purposes.

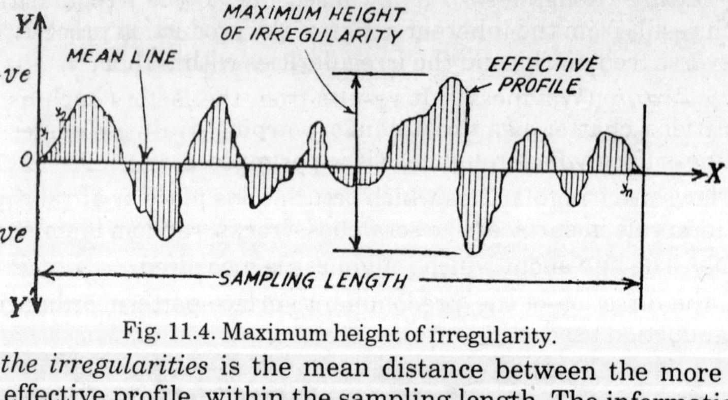

Fig. 11.4. Maximum height of irregularity.

Spacing of the irregularities is the mean distance between the more prominent irregularities of the effective profile, within the sampling length. The information is useful for measuring the wearing in of surfaces in relative motion, and assessing the electrical and thermal conductivity between surfaces in contact. Irregularity spacing and height parameters used in combination are valuable for sheet-steel applications and for friction and lubrication studies.

Arithmetical mean deviation from the mean line of profile (R_a) is defined as the average value of the ordinates (y_1, y_2, ... y_n) from the mean line.

The ordinates are summed up without considering their algebraic sign,

$$i.e.\ R_a = \frac{1}{n}\int_0^l |\,y\,|\,dx$$

$$\text{Approximately}: \ R_a = \frac{\sum_{0}^{n} |y_i|}{n}$$

where n is the no. of divisions over the sampling length l.

R_a readings serve well for surface finish control in most instances. It may be mentioned that R_a provides an average reading and several different surfaces can have same average, *i.e.* although R_a readings may be approximately same, the surfaces will function quite differently. Thus, more knowledge of the surface texture is required. For instance, measurement of peak heights may be needed in painting, plating and glass applications to control the pitting of gear teeth, improve seals between surfaces and increase the stiffness of press fits.

Maximum height of irregularity (R_{max}) is defined as the distance between two lines parallel to the mean line and touching the profile at highest points within the sampling length. (Refer Fig. 11.4)

Ten point height of irregularities (R_z) is defined as the average difference between the five highest peaks and the five deepest valleys within the sapling length measured from a line, parallel to the mean line and not crossing the profile. (Refer Fig. 11.5)

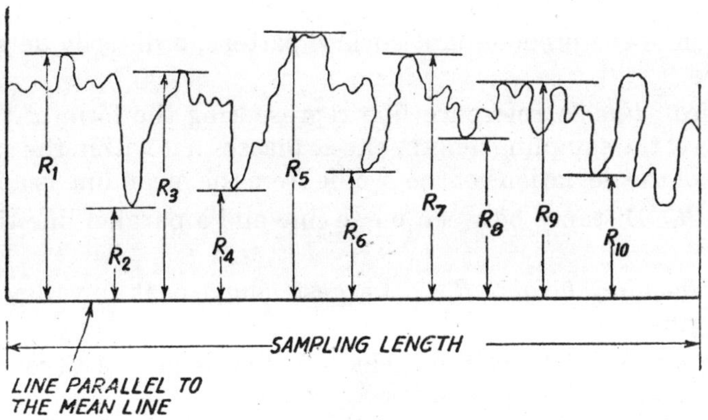

Fig. 11.5. Ten point height of irregularities.

$$\text{Thus,} \quad R_z = \frac{(R_1 + R_3 + R_5 + R_7 + R_9) - (R_2 + R_4 + R_6 + R_8 + R_{10})}{5}$$

11.4.1. Terms used to describe surface roughness

Arithmetic average roughness $R_a = \frac{1}{L}\int_0^L |h|\,dx$ over 2-20 consecutive sampling lengths.

Average peak-to-valley height R_z. This is the average of single peak-to-valley heights from five adjoining sampling lengths.

R_t *measurement*. It is the maximum peak to valley height within the assessment length. This measurement is valuable for analysing finish to provide guidance for planning subsequent metal-cutting operations. *Average wavelength* = $2\pi \times R_a$/Mean slope.

Bearing area. (Bearing area fraction). This is the fraction of surface at a given height above or below the mean line.

Centre line. A line representing the form of the geometrical profile and parallel to the general direction of the profile throughout the sampling length, such that the sums of the areas contained between it and those parts of the profile which lie on either side of it are equal.

Depth of surface smoothness. $R_p = \frac{1}{L}\int_0^L (h_{max} - h)\, dx$

R_p value indicates amount of material to be removed from a workpiece to obtain 50% bearing area.

Effective profile. The contour that results from the intersection of the effective surface by a plane conventionally defined with respect to the geometrical surface.

Effective surface. The close representation of real surface obtained by instrumental means.

General surface texture. The vertical distance between reference lines formed by truncating the 10 per cent highest peaks and lowest valleys of the surface texture. It takes into account initial wear and represents a measure of the maximum possible surface wear.

Geometrical profile. The contour that results from the inter-section of the geometrical surface by a plane conventionally defined with respect to this surface.

Geometrical surface. The surface determined by the design or by the process of manufacture, neglecting errors of form and surface roughness.

Land length. Length of intercept with the profile of a line drawn parallel to the profile mean line.

Lay. The direction of the predominant surface pattern, ordinarily determined by the production method used.

Least-squares mean line. A reference line representing the form of the geometrical profile within the limits of the sampling length, and so placed that within the sampling length the sum of the squares of the deviations of the profile from the mean line is a minimum.

Levelling depth, R_u. Distance between mean line and a parallel line through highest peaks.

Maximum peak-to-valley height. R_{max}. Largest single peak-to-valley height in five adjoining sampling lengths.

Mean apparent amplitude. Mean of the peak-to-valley heights of 13 consecutive 50 mm lengths.

Mean depth, R_m. Distance between mean line and a parallel line through the deepest valley.

Mean roughness step. Mean roughness wavelength or frequency. $A_R = \frac{1}{n}\sum_{i=1}^{n} h_i$ (n is within measuring length).

Mean waviness step. Mean waviness wavelength or frequency. $A_w = \frac{1}{n}\sum_{0=i}^{n} Li.$

Measuring traversing length. The length of the modified profile used for measurement of surface roughness parameters. It is usual for the measuring length to contain several sampling lengths.

Meter cut-off. In a profile instrument, the conventionally defined wavelength separating the transmitted component from the attenuated components of the effective profile.

Modified profit. The effective profile modified by such defined filter means as are used for suppressing those undulations of the real profile that are not or are not fully to be included in the measured parameters of the surface.

Peak roughness. $R_p = \frac{1}{L}\int_0^L (h_{max} - h)\, dx$

Peak-to-valley height. Separation of highest peak and lowest valley.

Real profile. The contour that results from the intersection of the real surface by a plane conventionally defined with respect to the geometrical surface.

Real Surface. The surface limiting the body, separating it from the surrounding space.

Recording traversing length. The maximum recording movement of the stylus along the surface.

Reference line. A line chosen by convention to serve for the quantitative evaluation of the roughness of the effective profile.

Root-mean-square roughness. $R_q = \sqrt{\frac{1}{L}\int_0^L h^2\, dx}$.

Bearing ratio (t_p) is the length of bearing surface and provides guidance for improving production. $t_p\%$ is the ratio at selected depth P (Refer Fig. 11.6), and

$$t_p\% = \frac{b_1 + b_2 + b_3 + \ldots\ldots b_n}{L} \times 100$$

This measurement helps improve production processes.

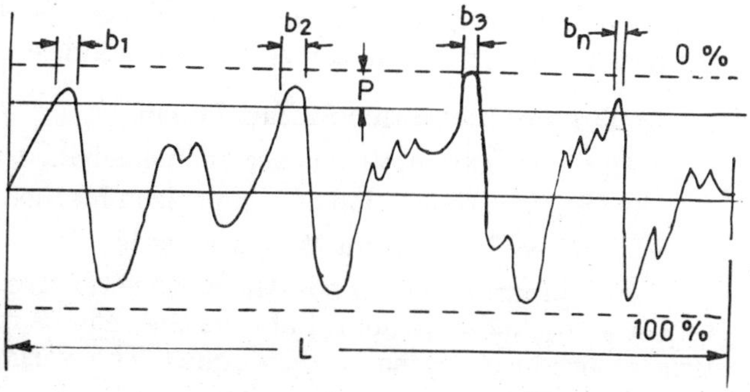

Fig. 11.6. Bearing ratio.

Roughness. The irregularities in the surface texture which are inherent in the production process, but excluding waviness and errors of form.

Roughness factor. Ratio of true to projected surface area.

Roughness width. $A_r = \frac{1}{n}\sum_{i=1}^{n} h_i$ (n within roughness sampling length.)

Run out length. Vertical projection on to the mean line of the last part of the traversing length which is not used for numerical evaluation.

Sampling length. The length of the effective profile selected for the evaluation of the surface roughness, without taking into account other types of irregularities.

Secondary texture. Irregularities outside the bandwidth of wavelengths of the primary texture.

Single peak to valley height. Separation of highest peak and lowest valley within a single sampling length.

Shape factor or smoothness index. Ratio of the separation of the mean line and a parallel line through the highest peak to the average roughness.

Skew will show whether porous, sintered and cast iron surfaces will yield a meaningful R_a measurement. Also, the criterion for a good bearing surface is that it should have a negative value of skew.

Start-up length. Length of the first part of the traversing length projected vertically on the mean line and not used for numerical evaluation. Transient oscillations shall fade out within the start-up length.

Surface texture. Irregularities which, recurring many times across the surface, tend to form on it a pattern or texture. Also known as primary texture.

Surface volume. The volume per unit area enclosed by the surface and a plane located by the highest summits.

Ten-point height. Separation of average of five highest peaks and five lowest valleys within a single sampling length.

Traversing length. The length of the effective profile necessary for the evaluation of surface roughness parameters for the surface inspected. The traversing length may include one or more sampling lengths.

True profile length. The length obtained by following all the undulations of the profile along a sampling length.

Waviness. That component of surface texture upon which roughness is superimposed. Waviness may result from such factors as machine or work deflections, vibrations, chatter, heat treatment or warping strains.

Waviness height. Separation of highest peak and lowest valley of waviness over a waviness sampling length, corrected for roughness.

11.5. Methods of Measuring Surface Finish

There are two methods used for measuring the finish of machined part :

(*i*) Surface Inspection of Comparison Methods.

(*ii*) Direct Instrument Measurements.

In comparative methods, the surface texture is assessed by observation of the surface. But these methods are not reliable as they can be misleading if comparison is not made with surfaces produced by same techniques. The various methods available under comparison method are :

(*i*) Touch Inspection, (*ii*) Visual Inspection, (*iii*) Scratch Inspection, (*iv*) Microscopic Inspection, (*v*) Surface Photographs, (*vi*) Micro-Interferometer, (*vii*) Wallace Surface Dynamometer and (*viii*) Reflected Light Intensity.

11.5.1. Touch Inspection. The main limitation of this method is that the degree of surface roughness can't be assessed. Also the minute flaws can't be detected. This method can simply tell which surface is more rough. In this method, the finger-tip is moved along the surface at a speed of about 25 mm per second and the irregularities as small as 0.01 mm can be easily detected. A modification of it is possible by using a table tennis ball, which is rubbed over the surface and vibrations from the ball transmitted to hand and surface roughness judged thereby.

11.5.2. Visual Inspection. Visual inspection by naked eye is always likely to be misleading particularly when surfaces having high degree of finish are inspected. The method is, therefore, limited to rougher surfaces and results vary from person to person. More accurate inspection can be done by using illuminated magnifiers.

11.5.3. Scratch Inspection. In this method, a softer material like lead babbit or plastic is rubbed over the surface to be inspected. By doing so it carries the impression of the scratches on the surfaces which can be easily visualised.

11.5.4. Microscopic Inspection. This is probably the best method for examining the surface finish but suffers due to limitation that only a small portion of the surface can be inspected at a time. Thus several readings are required to get an average value. In this method, a master finished surface is placed under the microscopic and compared with the surface under inspection. In another method a straight edge is placed on the surface to be inspected and a beam of light projected at about 60° to the work. Thus the shadows cast into the surface scratches are magnified and the surface irregularities can be studied.

11.5.5. Surface Photographs. In this method magnified photographs of the surface are taken with different types of illumination. In case we use vertical illumination, then defects like irregularities and scratches appear as dark spots and flat portion of the surface appears as bright area. In case of oblique illumination, reverse is the case. Photographs with different illumination are compared and the results assessed.

11.5.6. Micro Interferometer. In this method, an optical flat is placed on the surface to be inspected and illuminated by a monochromatic source of light. Interference bands are studied through a microscope. Defects, *i.e.* scratches in the surface appear as interference lines extending from the dark bands into the bright bands. The depth of the defect is measured in terms of the fraction of the interference band. This has been explained in detail in the chapter of 'Measurement by Light Wave Interference'.

11.5.7. Wallace Surface Dynamometer. This is a sort of friction meter and consists of a pendulum in which the testing shoes are clamped to a bearing surface and a predetermined spring pressure can be applied. In this method the pendulum is lifted to its initial starting position and allowed to swing over the surface to the tested. If the surface is smooth, then there will be less friction and pendulum swings for a longer period. Thus time of swing is a direct measure of surface finish.

11.5.8. Reflected Light Intensity. In this method a beam of light of know quantity is projected upon the surface. This light is reflected in several directions as beams of lesser intensity and the change in light intensity in different directions is measured by a photocell. The measured intensity changes are already calibrated by means of reading taken from surface of known roughness by some other suitable method.

11.5.9. Other Methods of Measuring Surface Roughness

Taper Sectioning—In this method, a section is cut through the surface to be examined at a shallow angle θ, thus magnifying height variations by a factor of cot θ, and the section is examined by optical microscope. It is very accurate method but is a destructive method.

Gloss measurement—There is a variation in the optical properties of different rough surfaces as judged subjectively by the human eye and various instruments have been developed to try and quantify this variation more objectively. In one instrument the reflection of a series of vertical bars in the surface is reflected at the specular angle. This angle is gradually decreased until the bars can no longer be distinguished. The roughness is inversely proportional to the cosine of the angle of extinction.

Diffraction techniques—Speckle is produced when a rough surface is illuminated with spatially coherent light and observed with an optical system of finite aperture. Light is received at each point in the image from several different points on the object because of the resolution limitations of the system. The path length of the light from each point on the surface depends on the height of the surface contour at that point. If this height varies significantly across the width of the point spread function, interference effects will appear as speckle. Surface roughness can be deduced from measurements of speckle pattern.

Other methods are tactile tests, pneumatic gauging, optical sections, thermal comparator, capacitance gauges etc.

11.6. Direct Instrument Measurements

These methods enable to determine a numerical value of the surface finish of any surface. Nearly all instruments used are stylus probe type of instruments. These operate on electrical principles. Further, these electrical instruments can be of two kinds. In first type, they operate on the carrier-modulating principle. The movements of the stylus exploring the surface are caused to modulate a high frequency carrier current. The second type includes those cperating on voltage-generating principle. In these the movements of the stylus are caused to generate a voltage signal. In both these types the output has to be amplified and the amplified output is used to operate a recording or indicating instrument.

The carrier modulated frequency type of instruments have the advantage that the signal fed to the recorder depends only upon the position of the stylus. While in the voltage generating type, when the oscillatory movement of the stylus stops, the output falls to zero no matter where the stylus may be.

Some pneumatic instruments are also used for measuring surface finish.

11.6.1. Stylus Probe Instruments. This type of instrument generally consists of the following units :

(*i*) A skid or shoe which is drawn slowly over the surface either by hand or by motor drive. The skid when moved over the surface, follows its general contours and provides a datum for the measurements. In case a skid is not used and only a probe is used then probe will trace the actual profile, but upward and downward movement of probe will be dependent upon the setting of the work under probe. But since the roughness of the surface does not depend upon the position of the work, it will be necessary to choose a datum from which the measurement is to be taken. A line touching the crest of the profile, that is, the envelope line which defines the macro-geometrical form is generally chosen as datum line and this is obtained by using a skid of such a size which can span a large number of surface undulations. A lot of work has been done on shape of skid, and different types of skids are available for different purposes by which the true macro-geometrical form may be obtained.

(*ii*) A stylus or probe which moves over the surface with the skid. The stylus for R_a measurement on new instrument can have a radius of 10 microns ± 30%. When in use, tip radius is allowed to vary ± 50% (5 to 10 μm). However for calibration purposes, stylus should have 10 μm nominal size radius ± 20%. The stylus should be cone shaped with a spherical tip. This records the micro-geometrical form of the surface. It moves vertically up and down relative to skid movement due to roughens of the surface. Generally it is desired that if the skid is moving up then the stylus must also be moving up. But when the pitch between the skid and stylus is comparable to that of half the wavelength of surface, then record will not be desirable one, *e.g.*, in the skid and stylus position shown in Fig. 11.7 skid is moving up, while stylus is giving downward indication.

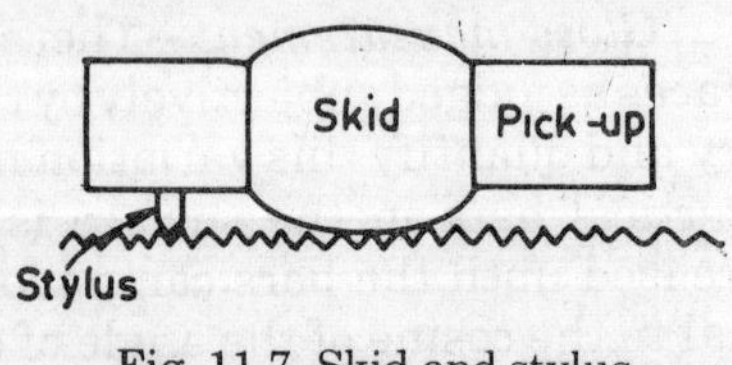

Fig. 11.7. Skid and stylus.

(*iii*) An amplifying device for magnifying the stylus movement and an indicator.

(*iv*) A recording device to produce a trace or record of the surface profile. Usually the vertical movement is magnified more in comparison to horizontal movement, thus the record will not give the actual picture of surface roughness but a distorted trace obtained.

(*v*) A means for analysing the trace is obtained. The analysis can be done separately or some automatic device may be incorporated in the instrument for analysis.

11.6.1.1. Profilometer. This instrument is most commonly used in U.S.A. for direct measurement of surface quality. This is a dynamic instrument similar in principle to a gramophone pick-up. A finely pointed stylus mounted in the pick-up unit is traversed across the surface either by hand or by motor drive. The instrument records the rectified output from the pick-up which is amplified further and operates an indicating device. Thus this records the average height of the surface roughness. In this instrument, roughness together with waviness and flaws comprises the irregularities found on the surface. An indication is obtained only when the pick-up is moving. This instrument is best suited for measuring surface finish of deep bores.

11.6.1.2. The Tomlinson Surface Meter. This instrument was designed by Dr. Tomlinson. This instrument uses mechanical-cum-optical means for magnification (Fig. 11.8).

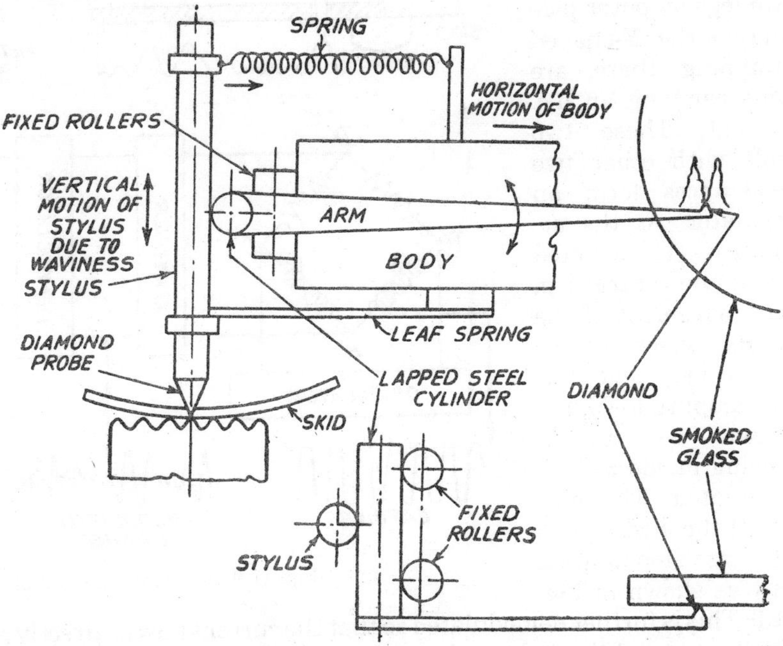

Fig. 11.8. Tomlinson Surface Meter.

The diamond stylus on the surface finish recorder is held by spring pressure against the surface of a lapped steel cylinder. The stylus is also attached to the body of the instrument by a leaf spring and its height is adjustable to enable the diamond to be positioned conveniently. The lapped cylinder is supported on one side by the stylus and on the other side by two fixed rollers as shown in Fig. 11.8. The stylus is restrained from all motions except the vertical one by the tensions in coil and leaf spring. The tensile forces in these two springs also keep the lapped steel cylinder in position between the stylus and a pair of fixed rollers. A light spring steel arm is attached to the horizontal lapped steel cylinder and it carries at its tip a diamond scriber which bears against a smoked glass.

When measuring surface finish, body is traversed across the surface by a screw rotated by a synchronous motor. Any vertical movement of the stylus caused by the surface irregularities, causes the horizontal lapped steel cylinder to roll. By its rolling, the light arm attached to its end provides a magnified movement on a smoked glass plate. This vertical movement coupled with the horizontal movement produces a trace on the glass magnified in vertical direction and there being no magnification in horizontal direction. The smoked glass trace is then, further projected at × 50 or × 100 magnification for examination. This instrument is comparatively cheap one and gives reliable results.

11.6.1.3. The Taylor-Hobson Talysurf. The Talysurf is an electronic instrument working on carrier modulating principle. This instrument also gives the same information as the previous instrument, but much more rapidly and accurately. This instrument as also the previous one records the static displacement of the stylus and is dynamic instrument like profilometer.

The measuring head of this instrument consists of a diamond stylus of about 0.002 mm tip radius and skid or shoe which is drawn across the surface by means of a motorised driving unit (gearbox), which provides three motorised speeds giving respectively × 20 and × 100 horizontal magnification and a speed suitable for average reading. A neutral position in which the pick-up can be traversed manually is also provided. In this case the arm carrying the stylus forms an armature which pivots about the centre piece of *E*-shaped stamping as shown in Fig. 11.9. On two legs of (outer pole pieces) the *E*-shaped stamping there are coils carrying an a.c. current. These two coils with other two resistances form an oscillator. As the armature is pivoted about the central leg, any movement of the stylus causes the air gap to vary and thus the amplitude of the original a.c. current flowing in the coils is modulated. The output of the bridge thus consists of modulation only as shown in Fig. 11.9. This is further demodulated so that the current now is directly proportional to the vertical displacement of the stylus only.

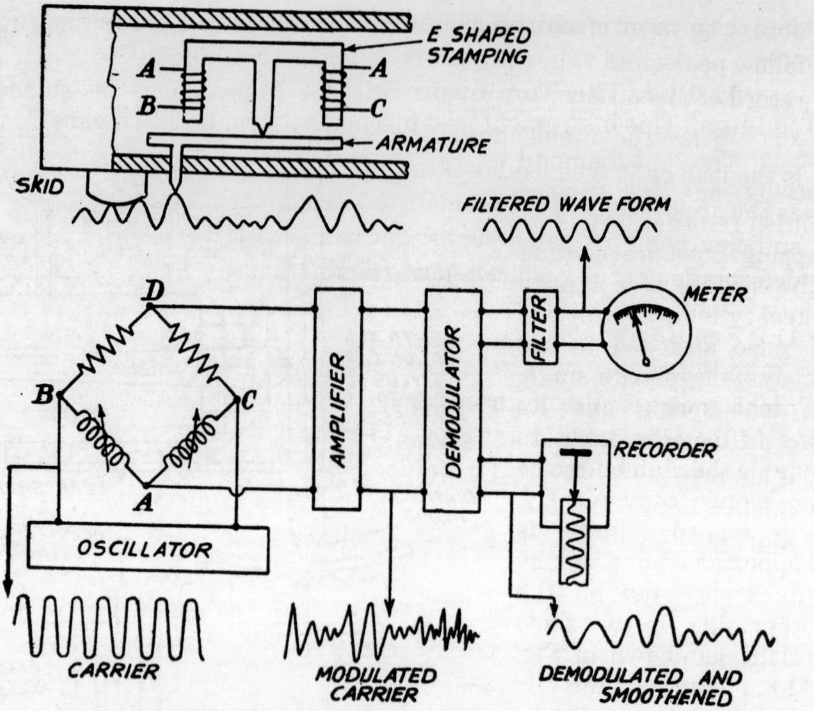

Fig. 11.9. Schematic Layout of Talysurf.

The demodulated output is caused to operate a pen recorder to produce a permanent record and a meter to give a numerical assessment directly. In recorder of this statement the marking medium is an electric discharge through a specially treated paper which blackens at the point of the stylus, so this has no distortion due to drag and the record strictly rectilinear one.

Now-a-days microprocessors have made available complete statistical multi-trace systems measuring several places over a given area and can provide standard deviations and average over area-type readings and define complete surface characterisation. These systems lend themselves to research applications where specialised programming can achieve auto-correlation, power spectrum analysis and peak curvature.

11.6.1.4. Stylus. Phonograph needles, though used in some cases are found to be too large and too heavily loaded. It also causes damage. Diamond styli are used universally. Some of them are cones of 90° include dangle and tip radius 4-12 μm. A popular stylus with truncated pyramid is shown in Fig. 11.10. The angle between the faces is 90°. The short edge is parallel to the direction of motion. Thus this stylus cannot resolve a wavelength shorter than 6 μm, and integrates over a narrow strip of surface 8 μm wide.

It may be noted that this pick up has finite dimensions, and it is constrained to move in a nearly vertical plane, relative to the moving pickup. Thus the stylus cannot record re-entrant features, an unimportant drawback for engineering investigations as re-entrant structures are absent on most machined surfaces. This stylus will fail to follow peaks and valley faithfully and produces a distorted record of the surface.

Since the dimensions of the stylus are finite, so also is the load on it. The load is of the order of 70 mg force. But as the area of contact is too small, the local pressure may be sufficiently high to cause significant local elastic downward deformation of the surface under examination.

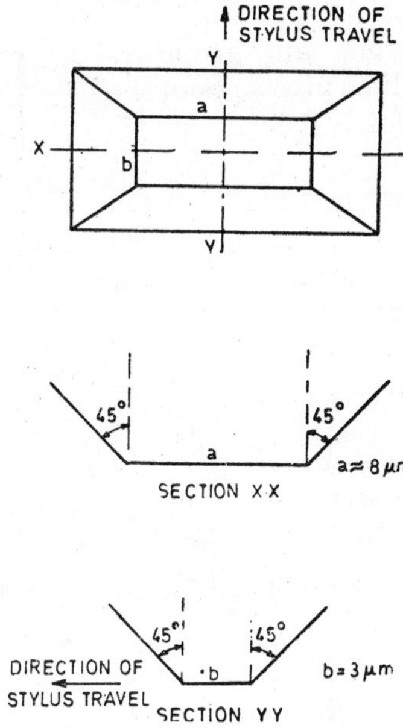

Fig. 11.10. Diamond stylus showing typical dimensions.

11.7. Replica Method

When the surface is not readily accessibly to the probe, then a replica of surface is made by dampening a piece of cellulose acetate film in acetone to dampen it and pressing it against the surface till it hardens. By this method, 80% fidelity can be obtained. Modern techniques using expoxy and other resins approach 100% fidelity.

11.8. Sample Length or Cut-off Length

Difficulties are generally experienced, when measuring the surface of irregular surfaces, in choosing the sample length. The nature of the surface many a times is such that the good indication of the surface can't be obtained when the selected length is different, *e.g.* in Fig. 11.11 different values would be obtained for various measures of surface irregularities if determined for lengths. L_1 and L_2 etc. This is because, the macro-geometrical irregularities affect each to a varying extent. Then the next problem is as to which type of irregularity should be allowed to influence the assessment of other. Sampling length should bear some relation to the type of profile. It is found that the required length can be related to the manufacturing process. In a surface there are, different kinds of irregularities having different frequencies and it is never clear which has got more importance. Thus the best thing would be to standardise some sampling length and that can be always associated with the surface roughness value. To standardise the sampling length is also not an easy task as the wavelengths vary to a considerable extent according to the way in which the surface is

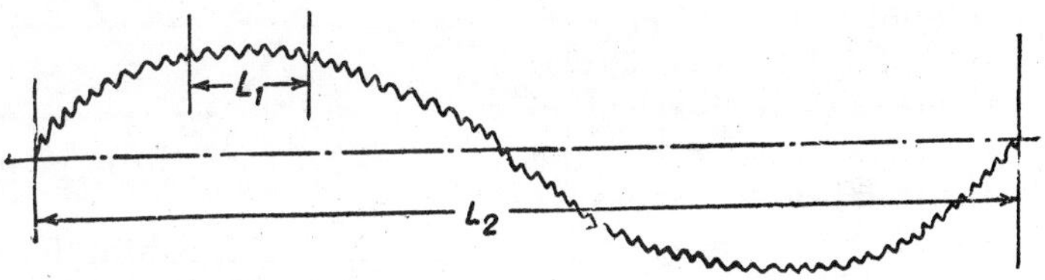

Fig. 11.11. Different results for different cut off lengths.

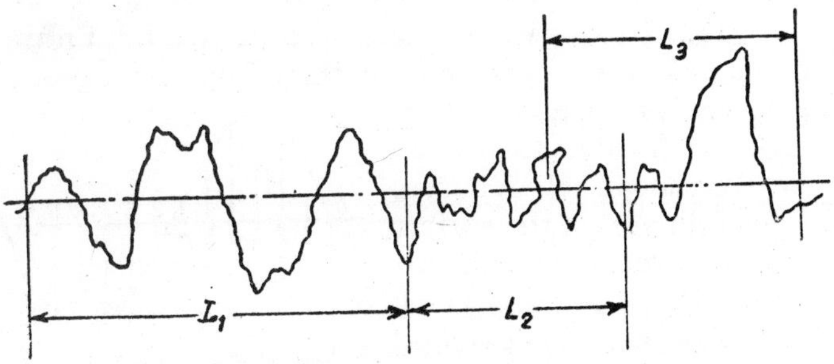

Fig. 11.12. Different positions of sampling length give different results.

produced. Further referring to Fig. 11.12 another problem is as to where the sample length must be chosen, *i.e.* position of sample length on surface will also give different results. This effect is taken into account by taking the readings at several places and taking the mean value. Sampling length is standardised for various operations considering all the above facts as follows :

Process	*Symbol*	*Sample length in mm*			
Milling	M	—	0.75	2.5	7.5
Turning	T	0.25	0.75	2.5	—
Grinding	G	0.25	0.75	2.5	—
Planing	P	—	—	2.5	—
Lapping	L_P	0.25	0.75	—	—
Polishing	P_o	—	—	2.5	—

11.9. Analysis of Surface Traces

A numerical assessment is assigned to indicate the degree of smoothness (roughness) in a number of ways. In practice three roughness measures have shown themselves to be particularly useful.

11.9.1. Maximum Peak to Valley Height of Roughness. This is obviously the most common measure of roughness but is not by any means a complete definition of roughness, *e.g.* in the two cases in Fig. 11.13 peak to valley height is same, but frequencies of irregularities are different and second surface is more rough in comparison to first one. But, since this is a relatively simple method of analysis, this will be a satisfactory measure where it is desired to control the cost of finishing for checking the rough machining. This method is also very advantageous in cases where the condition of surface is likely to exert an important influence on such properties as fatigue resistance and it is intended to clear the surface of the irregularities left by a previous operation.

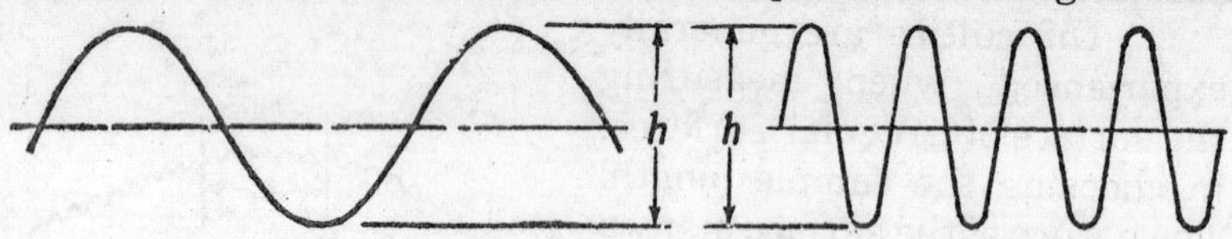

Fig. 11.13. Peak to valley height.

11.9.2. Root Mean Square Value (R.M.S. Value). This measure was in use previously and now-a-days superseded by Centre Line Average measure, as latter has the properties of being easily measured. R.M.S. value is defined as the square root of the mean of the squares of the ordinates of the surface measured from a mean line.

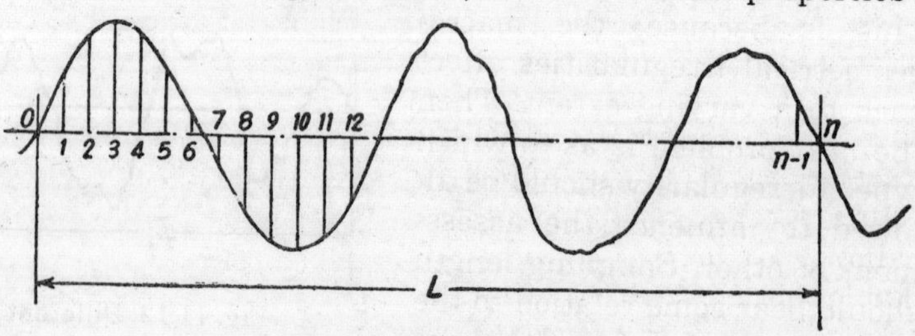

Fig. 11.14. R.M.S. value.

Referring to Fig. 11.14, the selected length L is divided into n equal parts. Ordinates are erected at the points 1, 2, 3, 4, ..., n, whose heights are (say) h_1, h_2, h_3, h_4, ... , h_n.

Then $$h_{r.m.s.} = \sqrt{\frac{h_1 + h_2^2 + h_3^2 + ... + h_n^2}{n}}.$$

11.9.3. Centre Line Average Method (C.L.A. Value). This is defined as the average height from a mean line of all ordinates of the surface regardless of the sign.

Thus referring to Fig. 11.14, C.L.A. $= \dfrac{h_1 + h_2 + h_3 + h_4 + \ldots\ldots h_n}{n}$.

But of find C.L.A. value like this will be a laborious job. Also by this method, spacing chosen may be such that important ordinates are likely to be missed.

Things can be much simplified by using a planimeter which can find out the area of any curve. Referring to Fig. 11.15, let us say that somehow or other the mean line is exactly known. Then C.L.A. value

$$= \frac{A_1 + A_2 + A_3 + \ldots\ldots}{L} = \frac{\Sigma A}{L}$$

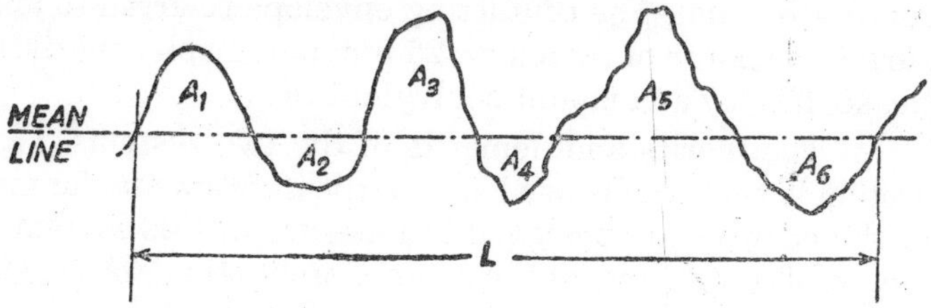

Fig. 11.15. C.L.A. value.

One has to take care of units carefully to find the C.L.A. value in micron.

11.9.3.1. How to Determine Mean Line. For it first the mean line is estimated by eye-judgement. Then total area above and below the assumed mean line are measured and a correction is applied to the assumed mean line to get the correct mean line. Error [ΣA (above) $- \Sigma A$ (below)]/L is applied to the assumed mean line to get the correct mean line. Thus in Fig. 11.16, X' Y' is assumed mean line.

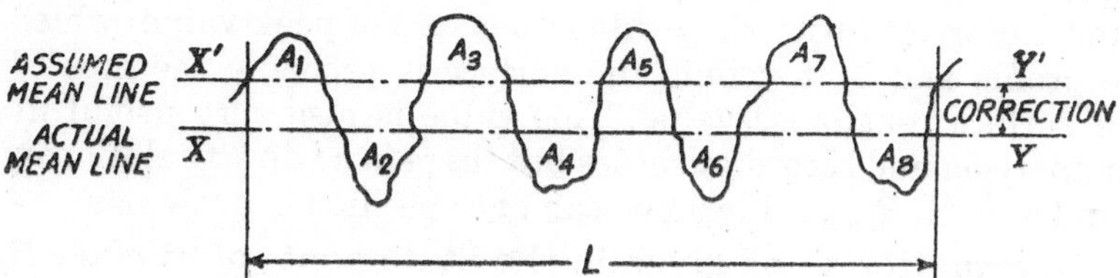

Fig. 11.16. Mean line.

Then correction $= \dfrac{(A_1 + A_3 + A_5 + A_7) - (A_2 + A_4 + A_6 + A_8)}{L}$ and this correction added to the assumed line algebraically gives the actual mean line. The C.L.A. value can be found out as described previously.

"Talysurf" has got built in arrangement for integrating the areas and the average value is directly given.

C.L.A. value does not give any idea regarding the greatest extent and the nature of the surface irregularities. It is likely to give identical values for surfaces of vastly different characteristics. So this is the main disadvantage of C.L.A. value. But when the characteristics of a surface have been found out to be satisfactory and such conditions are produced which do not allow the surface to change radically in manufacturing, then C.L.A. value provides a workable control.

11.10. Assessment of Surface Roughness as per Indian Standards

There are various methods of evaluating the surface roughness but the most prominent and commonly used methods are the M [Centre Line Average method (CLA) or the mean line] system and the E (Envelope system) system. The M system expresses the arithmetical average departure of the actual surface both above and below a mean line, within a specified sampling length, and in a plane substantially normal to the direction of surface. A similar system also based on the mean line expresses the departure of the actual surface as a root mean squared value. CLA system is more popular than RMS value method. To eliminate the effects of secondary texture the profile is split into a number of sampling lengths and a separate mean line computed for each length. In this way, the mean lines of adjacent sampling length may not coincide even. The *E* system expresses the arithmetical departure of a surface both above and below a 'mean' curve. This mean curve is developed from a contacting envelope by displacing it to a position, where the areas enclosed by the profile above and below the mean curve are equal. The contacting envelope referred to above is obtained by rolling across the surface a sphere of radius $r = 25$ mm and displacing the locus of centre for this circle towards the surface by an amount equivalent to *r*.

The merits and demerits of the two systems (*i.e.*, *M* and *E*) are given below. The *M* system is very widely used and all the instruments for measuring surface roughness available in the country are designed to measure in this system. It is more useful and a satisfactory means of controlling, at the point of production, the consistency of results from a process when the production parameters have been established.

This system has the limitation that it is unable to control the functional qualification of a surface whence associated with a machine process.

The *E* system, though, at present in its early stage of development is going to be widely accepted in future because of its better definition of peak measure and the fact that it is more easily applied to the surface finish instruments based on the interference principle.

M system. The surface roughness in this system is expressed by R_a value which is very useful for comparison of various surfaces obtained by similar operation ; of course, it does not indicate the limits of irregularity R_{max} value (measure of peak value) which is perceived by looking at the profile and by touching the surface is probably the most direct method of expressing the surface roughness value. This value is also very useful in identifying the surfaces for comparison and record purposes but there is no direct relationship between R_{max} and R_a value and as such R_{max} value alone can't be specified in drawings.

11.10.1. Information to be given in the Statements of Surface Roughness. The various roughness grade numbers *N* 1 to *N* 12 in 5 groups are specified as under by BIS. The relationship between the roughness grade numbers and the commonly used and generally accepted system of indicating surface roughness by symbol is given below.

Roughness grade number	*Roughness value R_a μm*	*Roughness symbol*
N 12	50	~
N 11 N 10	25 12.5	∇
N 9 N 8 N 7	6.3 3.2 1.6	∇∇
N 6 N 5 N 4	0.8 0.4 0.2	∇∇∇
N 3 N 2 N 1	0.1 0.05 0.0025	∇∇∇∇

11.10.2. The values of the surface roughness expected from various manufacturing processes are indicated below.

Manufacturing Process		R_a *value in μm*	
Casting	Sand casting	5	to 50
	Permanent mould casting	0.8	to 6.3
	Die-casting	0.8	to 3.2
	High pressure casting	0.32	to 2
Hot working	Hot rolling	2.5	to 50
	Forging	1.6	to 25
	Extrusion	0.16	to 5
	Flame cutting, sawing and chipping	6.3	to 100
Machining Operation	Radial cut-off sawing	1	to 6.3
	Hand grinding	3.6	to 25
	Disc grinding	1.6	to 25
	Filing	0.25	to 25
	Planing	1.6	to 50
	Shaping	1.6	to 25
	Drilling	1.6	to 20
	Turning and milling	0.32	to 25
	Boring	0.4	to 6.3
	Reaming	0.4	to 3.2
	Broaching	0.4	to 3.2
	Hobbing	0.4	to 3.2
Grinding and Super-finishing Operation	Surface grinding	0.063	to 5
	Cylindrical grinding	0.063	to 5
	Honing	0.025	to 0.4
	Lapping	0.012	to 0.16
	Polishing	0.04	to 0.16
	Burnishing	0.04	to 0.8
	Super-finishing	0.16	to 0.32

11.10.3. Lay. This is the direction of the predominant surface pattern and is determined by the production method used. The surface roughness is generally measured across the direction of the lay.

The various directions of lays depending upon the production process can be as shown in Fig. 11.17.

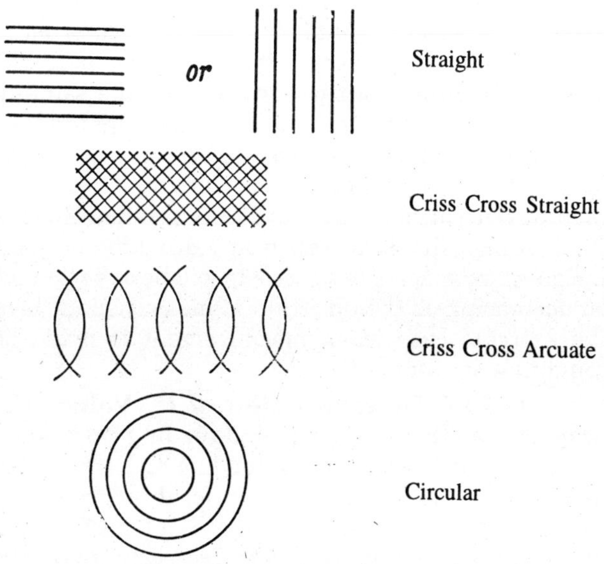

Fig. 11.17. Lay.

11.10.4. Preferred Values of R_a and R_z. Preferred values for arithmetical mean deviation R_a in μm are selected from :

0.025, 0.05, 0.1, 0.2, 0.4, 0.8, 1.6, 3.2, 6.3, 12.5 and 25 and the preferred values for ten point height of irregularities R_z in μm are selected from :

0.05, 0.1, 0.2, 0.4, 0.8, 1.6, 3.2, 6.3, 12.5, 25, 50 and 100.

11.10.5. Selection of Sampling Length. For measuring surface roughness, the value of the sampling length *l* is selected from the following series depending upon the process of manufacturing :

0.08, 0.25, 0.8, 2.5, 8, 10 and 25 mm.

	Process	*Suitable Sampling Length*						
Machined Surfaces	Milling	—	—	0.8	2.5	8	10	—
	Boring	—	—	0.8	2.5	8	10	—
	Turning	—	—	0.8	2.5	—	—	—
	Grinding	—	0.25	0.8	2.5	—	—	—
	Planing	—	—	—	2.5	8	10	25
	Reaming	—	—	0.8	2.5	—	—	—
	Broaching	—	—	0.8	—	—	—	—
	Diamond boring	—	0.25	0.8	—	—	—	—
	Diamond turning	—	0.25	0.8	—	—	—	—
	Honing	0.08	0.25	0.8	—	—	—	—
	Lapping	0.08	0.25	0.8	—	—	—	—
	Super-finishing	0.08	0.25	0.8	—	—	—	—
	Buffing	—	—	0.8	2.5	—	—	—
	Polishing	—	—	0.8	2.5	—	—	—
	Shaping	—	—	0.8	2.5	8	10	—
	Spark machining	—	—	0.8	—	—	—	—
Non-machined surfaces	Burnishing		—	0.8	2.5	—	—	—
	Drawing		—	0.8	2.5	—	—	—
	Extrusion		—	0.8	2.5	—	—	—
	Moulding		—	0.8	2.5	—	—	—
	Electron-polishing		—	0.8	2.5	—	—	—

It is usual practice to choose shorter value of the finer and larger value for the coaser grade of a given process when more than one values are given.

11.10.6. Measurement. It may be emphasised here again that the surface texture represents the combined effects of several causes, of which the roughness, waviness and the error of form are more predominant. If only the total height (*i.e.*, the difference between the highest and lowest points on the plot) is considered then idea of above three characteristics can be had by considering different sample lengths. If we are interested only in the measurement of roughness of the surface then the sampling length should be neither too big as to include the waviness, nor too small as to ignore the occasional deep scratches. The best thing is, therefore, to limit the measurement to a sufficiently short length of the surface and in order to take into consideration the considerable variation of roughness from place to place, several readings at various places are taken. When the surface roughness has a directional quality, the readings should be taken in a direction perpendicular to the direction of lay as this way the effective spacing of the crests will be at short distances. The effective spacing of crests goes on decreasing as the direction of measurement is reduced from 90° to direction of lay and in the extreme case when measurement is made parallel to the lay, a minimum result of roughness is obtained.

11.10.7. Determination of R_a Value. The R_a value can be determined either by graphical methods or by direct reading instruments. Whenever the surface texture has a directional quality, the direction in which the measurement is made is approximately at right angles to the 'lay'. It is always preferable to compute the R_a value by taking mean results from the measurement of several sampling lengths taken consecutively along the profile so that it gives better indication and is the correct representation of the whole of surface roughness.

Graphical Methods. In this method, it is presumed that the surface is nominally flat and that the record is produced in rectilinear co-ordinates in which a truly flat surface is represented as a straight line. First of all mean line of successive sampling lengths is determined. In case profile is not nominally straight, then mean line of appropriate curvature could be assumed ; for the record could be developed such that base line is straightened. For such cases, the profile records are made on plastic replicas by taking from curved surfaces and flattening out ; or such instruments may be used which operating on a curved surface under appropriate conditions produced records in which the nominally curved base line is automatically straightened.

The mean line is determined by first drawing any straight line parallel to the general course of the record over the sampling length. Then the sum of areas above this line and below it are computed and their difference divided by the sampling length gives the distance by which the above drawn line is to be shifted in order to obtain the central line. Then the arithmetical sum of all the areas above and below the central line divided by the sampling length and the vertical magnification of the record gives the R_a value in mm, if areas are computed in mm^2 and sampling length in mm. This value multiplied by 1000 gives R_a value in microns.

11.10.8. Some Notes on Stylus and Skid Method. It is obvious that the stylus which rests on the surface and traverses across it responds to the irregularities of the surface relative to a selected datum provided by a skid having a large radius of curvature in the direction of the traverse. The skid simply follows the general contour, riding over the crests of smaller irregularities without responding to them individually. This method is subjected to some limitations in giving exact response corresponding to the shape of the irregularities. It is firstly due to the impossibility of having a mathematically sharp point for the stylus, which at the most can be finished with a small but finite radius. The effect of this is that the full depth of the groove is reduced, the trace is smoothed out over the sharp corner and crests, and the trace offers no evidence of fine deep scratches in the actual surface. However, it may be appreciated that for an average reading the loss in the valley is fully offset by the gain resulting from the apparent rounding of the crests. Further detailed studies indicated that the surface irregularities being relatively shallow compared with their spacing, the errors due to finite radius of the stylus are not very serious and that a nominal radius of 0.003 mm for the tip of the stylus gives quite satisfactory results. Secondly, the error is also introduced due to the displaced position of the stylus in relation to the skid, because it is just possible that the stylus is touching a valley when the skid is at its lowest point and *vice-versa*. In order to reduce the chances of meeting extreme phase conditions, it is common practice to have two skids having a small radius and both touching the surface with the stylus acting midway between them.

Keeping the above in view, the stylus type instruments, employed for determining the surface roughness should comply with the following requirements :

Stylus. For graphical recording, nominal radius of the tip of stylus should be 0.003 mm. For determination of R_a value by direct reading instruments the nominal radius of the tip of stylus should be 0.01 mm. The variation in the tip radius of the stylus should be within 30% of its nominal dimension or 0.003 mm whichever is greater.

The shank may be either conical or pyramidal, and in either case the included angle of cone should not exceed 90°.

The maximum force between a stylus and the surface under test should not be more than 0.016 × (tip radius in microns)^2 grams.

Skid. If a single skid is employed its radius in the direction of traverse should not be less than 50 times the sampling length. In case of two skids, the radii-should not be less than 8 times the sampling length.

11.11. Roughness Comparison Specimens

The roughness comparison specimens are used for qualitative assessment of the surface roughness. In this qualitative assessment of the roughness of the finished product is evaluated by visual or tactile comparison (*i.e.* lightly moving a finger nail across the lay of the roughness) with roughness comparison specimen of known roughness value finished by similar machining process. The roughness comparison specimens are produced in a variety of materials (steel having a hardness of about 400 HB) by direct application of the production process or by copying techniques and are available in various grades of roughness. The copying techniques consist of (*a*) electro-forming positive replicas of master surfaces and (*b*) making positive replicas of master surfaces in plastic materials by coating or otherwise and having the feel and appearance of the natural machined metal surface. It is general practice to produce the specimen surfaces as replicas of master surfaces in a hard material like nickel alloy or metallised plastic.

The roughness comparison specimens are available either in flat form or segmented cylindrical form. The flat specimens have the size : length—25 mm, width—16 mm, thickness—1.6 to 10 mm. The specimen is rigidly supported on a base to prevent it from being warped or distorted during use. The segmented cylindrical specimens will have either convex surface or concave surface having a surface radius of about 125 mm.

These specimens have also proved to be very good devices for training the personnel to judge the R_a value of the feel and appearance of the surfaces. While these are being used for direct comparison purposes, it is important to note that

(*i*) two surfaces being compared should resemble each other in material, shape and method of finish ;

(*ii*) two surfaces must be closely alike in general character in the sense that the cutting technique and machine feeds used for making the specimen should resemble those used for the products ;

(*iii*) surfaces being compared should be clean and dry.

11.12. Other Methods of Interest Used for Evaluating Surface Roughness

11.12.1. Pneumatic Methods of Evaluating Surface Finish. This Method has been successfully used for evaluating the micro irregularities. This method is popularly known as air-leakage method. The instrument works on the following principle. The compressed air is made to pass through a self-aligning nozzle which moves on the surfaces to be tested. The gap between the nozzle tip of the instrument and the tested surface depends upon the height of the micro irregularities. The size of the gap, in turn, affects the air discharge through the nozzle. The air discharge is measured by a special variable-area air flowmeter, (a rotameter) and determines the height of the micro irregularities.

It will be observed that it may be very difficult to calibrate this instrument directly in terms of surface micro irregularities. It is, therefore, used to test the surfaces with instrument by the comparative method. Certified samples are used as reference for comparison. Under these circumstances, the instrument is best suited for inspection of parts in mass production.

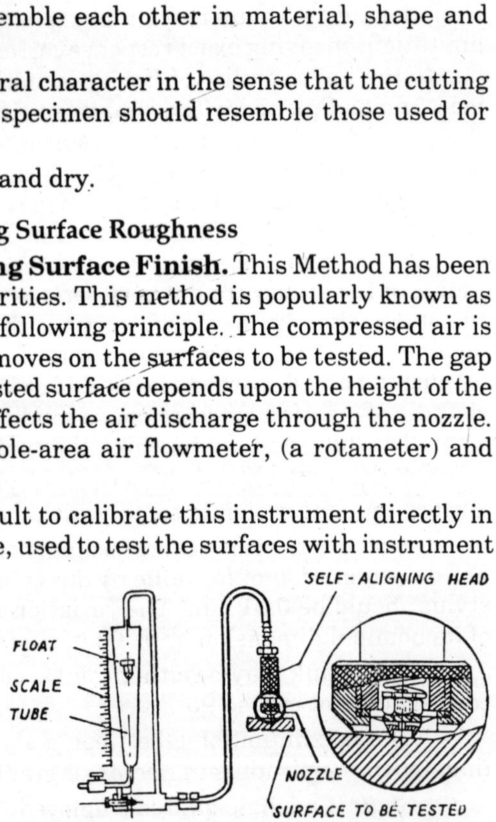

Fig. 11.18. Diagram of the instrument used for checking surface finish by air-leakage principle.

The diagram of the instrument and the design of the nozzle are shown in Fig. 11.18. Air is first passed through a tube in which there is a float which can freely move up and down in the tube depending upon the flow of air through tube. The air from this tube is led to the measuring nozzle of the self-aligning head through a flexible hose. The air leakage between the nozzle and the test surface will be same as in the tube. To allow the air to pass, the float is lifted by the air stream to a definite height depending upon the flow of air through the tube or the amount of air passing through the nozzle. The float height is read on the scale placed by the side of the tube and it indicates the quality of surface finish.

11.12.2. Tracer Type Profilogram. The principle of the tracer-type profilogram is shown in Fig. 11.19. The surface to be tested is placed on a working table. The table can move to and fro because there is a nut at the bottom of table, which fits into a lead screw driven by a motor. The stylus or tracer point which is made to rest on the surface to be tested, is pivoted with mirror as shown in the figure. The pivot point is fixed in space with the stand and its position can be changed initially. When the surface to be tested moves, the oscillations of the tracer are transmitted to the mirror. The beam of light (received from light source and through the lens and precision slit) strikes on the oscillating mirror and is reflected on to the revolving drum, upon which sensitised film is arranged. The drum is also rotated from the same lead-screw that traverses the table through bevel gears.

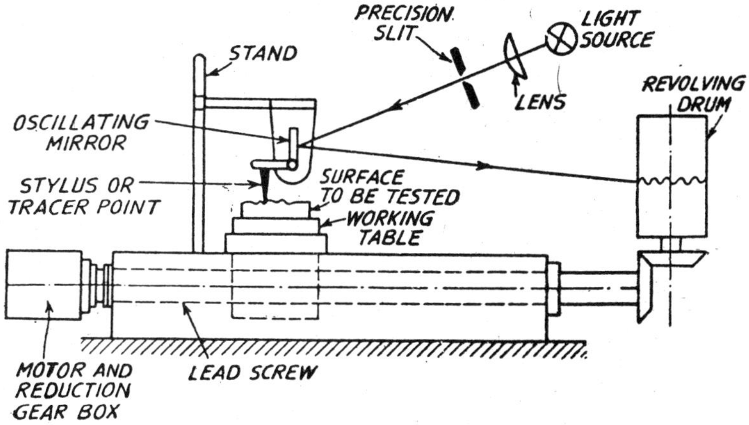

Fig. 11.19. Principle of the tracer-type profilogram.

The trace obtained on the drum is very much magnified one and can be further assessed by various methods.

11.12.3. Light Cross-section Method. The principle of this method is that when a thin film of light strikes the surface to be tested at an angle of 45°, a band of reflected light will appear on the surface. The edge of this band will reproduce the profile of the surface irregularities or so to say it will show an optical cross-section. This profile is magnified and observed by an optical device, arranged at an angle of 45° (perpendicular to the light source).

A double microscope working on the above principle is shown in Fig. 11.21, and is used to evaluate the surface finish by this method. The general view and optical system of the microscope is shown in Fig. 11.20. From a light source, a beam of light passes through the condenser and precision slit and is directed at an angle of 45° to the surface to be tested. The observing microscope having objective and eyepiece is also inclined at an angle of 45° with the tested surface. The field of view of the eyepiece contains a reading scale called an eyepiece micrometer, which is used to measure the height of surface irregularities.

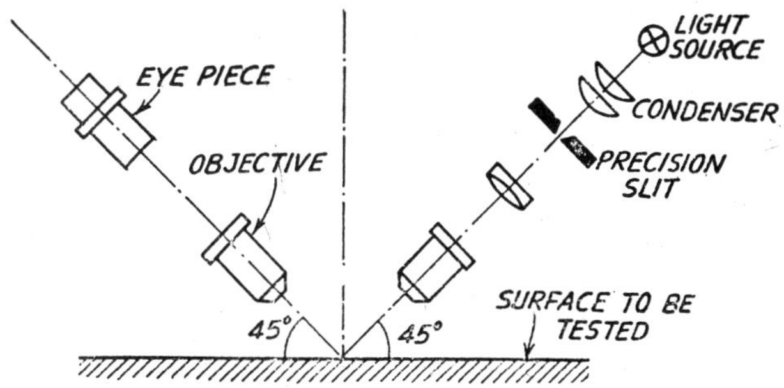

Fig. 11.20. Surface meter based on light cross-section.

11.13. Mechanical Roughness Indicator (MECRIN)

It is recent development from the tactile method. It is suitable for non-regular surfaces where the frictional properties measured compare with those provided by electronic stylus instrument, such as ground, honed, lapped, hand filed and emery clothed surfaces, etc. to approximately 4.5 μm R_a and on the finer grades of turning and milling to 0.8 μm R_a. On regular surfaces of rougher values, it is used as a comparator against surface roughness blocks to decide which is rougher. This instrument assesses the surface through its frictional properties and the average slope of the irregularities. It is a common experience that a thin metallic blade when pushed against a surface at a certain angle will either slide or buckle depending on (*a*) the surface roughness, and (*b*) the angle of attack, *i.e.* the angle of the blade to the surface. At angles below the critical angle the blade tip will simply slide along the surface but if the angle is successively increased, at a certain angle the blade just begins to buckle. This angle is the critical angle of attack. The smoother the surface the greater the critical angle and thus this angle is a function of the degree of roughness encountered.

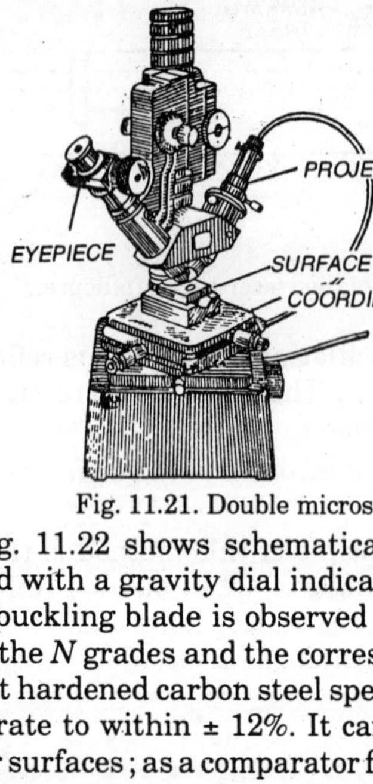

Fig. 11.21. Double microscope.

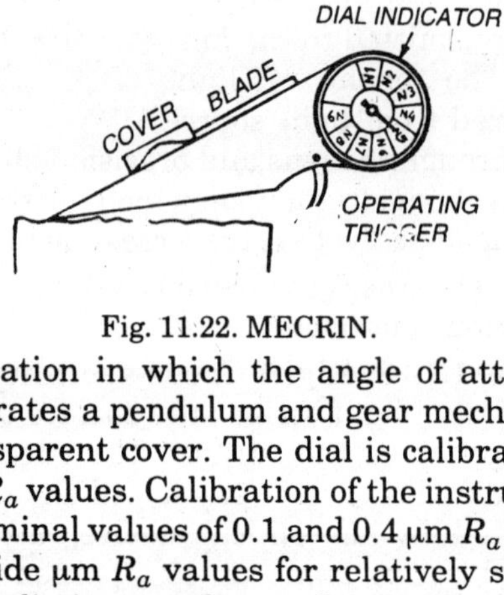

Fig. 11.22. MECRIN.

Fig. 11.22 shows schematically such an indication in which the angle of attack is measured with a gravity dial indicator which incorporates a pendulum and gear mechanism and the buckling blade is observed through the transparent cover. The dial is calibrated in terms of the N grades and the corresponding central R_a values. Calibration of the instrument is against hardened carbon steel specimens having nominal values of 0.1 and 0.4 μm R_a which are accurate to within ± 12%. It can be used to provide μm R_a values for relatively smooth irregular surfaces ; as a comparator for two surfaces as a limit gauge for roughness ; to establish the direction of the lay of a surface to determine the direction of maximum roughness if the surface has a definite lay ; to measure short lengths down to 1 mm which may be inaccessible to stylus instruments.

11.14. Recent Advances in Measurement and Evaluation of Surface Roughness

Now-a-days there is a growing awareness of the importance of surface metrology, because of the need to better understand the role played by the surface geometry of the part during its function. Basic research in diverse fields such as friction, wear, corrosion, lubrication, fatigue, strength, dynamic response of machine tools, etc. cannot be satisfactorily conducted without the knowledge of the surface geometry. The present system of specifying the surface finish by parameters like R_a, R_{max}, R_z, etc., specifies the geometry in the vertical direction only. Even though these parameters are adequate for control purposes in production, they are inadequate for analysing the functional behaviour of surfaces. For example, profiles with same R_a values can have different profile shapes differing in functional behaviour. They do not indicate the closeness or openness of the surface, texture, number of peaks or valleys contained in the surface, etc. This requires specification and measurement of new parameters of surface geometry which have direct influence on the function, rather than parameters like R_a, R_{max}, etc. The computation of these sophisticated parameters has been possible by the use of digital computation of surface profiles. Some of the parameters proposed are the Auto-correlation function, Amplitude-density function, cross correlation function, power spectral density etc. which give an indication of the surface in both amplitude as well as in frequency domain.

Q. 11.1. Calculate the CLA (R_a) value of a surface for which the sampling length was 0.8 mm.

The graph was drawn to a vertical magnification of 10,000 and a horizontal magnification of 100, and the areas above and below the datum line were :

Above :	**150**	**80**	**170**	**40 mm²**
Below :	**80**	**60**	**150**	**120 mm²**

Solution. The C.L.A. or R_a value is given by

$$\frac{\text{Sum of area (mm}^2)}{\text{Sampling length (mm)}} \times \frac{1000}{\text{Vertical magnification}} \times \frac{1}{\text{Horizontal magnification}}$$

Sum of areas $= 150 + 80 + 170 + 40 + 80 + 60 + 150 + 120 = 850 \text{ mm}^2$

$$\therefore \quad \text{C.L.A. or } R_a \text{ value} = \frac{850}{0.8} \times \frac{1000}{10{,}000} \times \frac{1}{100} = 1.06 \ \mu\text{m}.$$

Q. 11.2. In the measurement of surface roughness, heights of 20 successive peaks and troughs were measured from a datum and were 35, 25, 40, 22, 35, 18, 42, 25, 35, 22, 36, 18, 42, 22, 32, 21, 37, 18, 35, 20 microns.

If these measurements were obtained over a length of 20 mm, determine the C.L.A. (Ra) and R.M.S. value of the rough surface.

Solution. Approximate C.L.A. or R_a value $= \dfrac{\int_{i=1}^{1} h_i}{n}$

$$= \frac{35 + 25 + 40 + \ldots\ldots + 18 + 35 + 20}{20} = \frac{280}{20} = 29 \text{ microns}$$

Approximate R.M.S. value

$$= \sqrt{\frac{\int_1^n |h_1{}^2|}{n}} = \sqrt{\frac{(35)^2 + (25)^2 + \ldots\ldots(35)^2 + (20)^2}{20}} = \sqrt{\frac{182.32}{20}} = 31.18 \text{ microns.}$$

Q. 11.3. What is surface metrology ?

Ans. The science of measurement of surface parameters is known as surface metrology. This includes measurement of (*i*) surface geometry (*i.e.* form errors like waviness, straightness, flatness, cylindricity, sphericity etc.) and of (*ii*) surface irregularities known as the surface roughness or the surface finish.

Q. 11.4. Why surface finish is important in engineering applications ?

Ans. Surface finish has marked influence on function, lubrication, pressure holding ability, resistance to wear and corrosion, noise reduction, bearing area, friction etc. Surface finish affects, the overall result, particularly when the tolerances are close. If the limits are close, a rough surface after a little wearing may result in a slack fit. The effect is more pronounced when the surfaces are in relative motion or subjected to fatigue stresses.

Rough surfaces tend to collect and absorb moisture between peaks and valleys. Fine surfaces are more resistant to corrosion. Concentration of stresses at the peaks tends to fracture rough surfaces faster. On surfaces which are to be painted, the paint does not adhere if the surface is too fine.

Q. 11.5. (*a*) *Define Surface Roughness.*

Any surface produced by a chip removal process departs from the ideal form due to tool geometry, inaccuracies of movements in machine tool slides, deformation of workpiece due to cutting forces and due to its own weight, vibration of machine, method of machining and its characteristics, rupture of material during chip separation, etc.

(*b*) *Define the following irregularities on a surface texture produced by a machining process :*

(*i*) *Primary texture* (*ii*) *Secondary texture*

(*iii*) *Surface flaws,* (*iv*) *Errors of form.*

Ans. (*i*) **Primary Texture**. These are irregularities of comparatively shorter wavelengths and are usually produced because of cutting action *i.e.* from built up cutting edges, abrasives etc. These constitute the micro-geometrical errors and are referred to as surface roughness or primary texture.

(*ii*) **Secondary Texture**. These are irregularities of comparatively longer wavelengths than those of surface roughness and are periodic in character. These are caused mainly by feed rate, chatter or vibration of the machine. These constitute the macro-geometrical errors and are referred to as waviness or secondary texture.

It may be noted that waviness refers to longer waves present in the surface profile and roughness to the shorter ones.

(*iii*) **Surface Flaws.** Irregularities such as scratches, dents, blow holes, metallurgical or heat treatment defects, etc., which do not occur that frequently are referred to as surface flaws.

(*iv*) **Errors of Form.** In addition to primary texture and secondary texture, irregularities of wavelengths much longer than those of secondary texture or waviness also are present on a machined surface. These irregularities are called errors of form. These are caused by the deflection of the work piece or of the machine, inaccuracies of machine tool slide movements, heat treatment distortions etc.

Q. 11.6. How is surface roughness assessed ? Define M-system.

Ans. Surface roughness can be assessed by considering the variations of the ordinate heights of the points on the surface profile. A reference line, usually, M-line is chosen from which the ordinate heights of the points are measured and the surface roughness is evaluated with reference to this reference line.

In the M-System, the reference line chosen is the mean line of the profile, which is defined as a line following the general direction of the profile and dividing the surface profile in such a way that within a sampling length, the sum of the areas covered by the profile above and below the mean line are equal.

Q. 11.7. Surface roughness parameters could be classified into the following groups :

(*a*) **Amplitude parameters (based on ordinate heights of the irregularities**

(*b*) **Spacing parameters (provide indication of the nature of surface in horizontal direction)**

Discuss the various methods of amplitude and spacing measurements.

Solution. (*a*) Various amplitude parameters as per M-system are :

(*i*) **R_a—Centre Line Average (CLA) or Arithmetic Average (AA).** This is the most widely used parameter for specifying surface roughness. It is the arithmetic mean of the departures Y of the points on the profile from the mean line (Fig. 11.23). The mean line is first

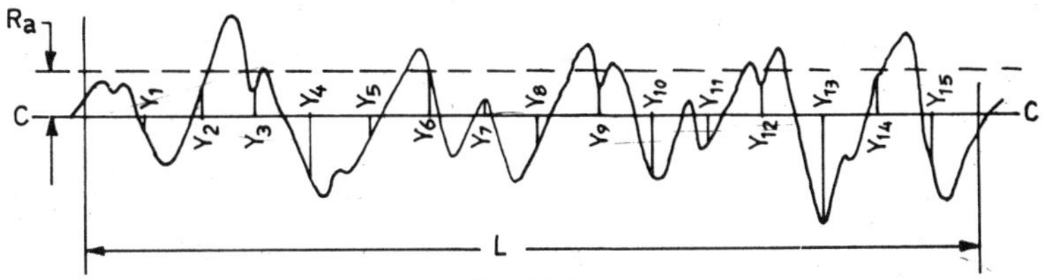

Fig. 11.23

determined and the ordinates of the points from the mean line are added without considering their sign (*i.e.* irrespective of whether the points are above or below the mean line).

$$R_a = \frac{1}{L}\int_0^L Y\,dx = \frac{1}{n}\sum_{i=1}^{n} Y_i$$

where n is the number of ordinates in the sampling length L and Y is the ordinate height. Ordinates are taken at equal intervals.

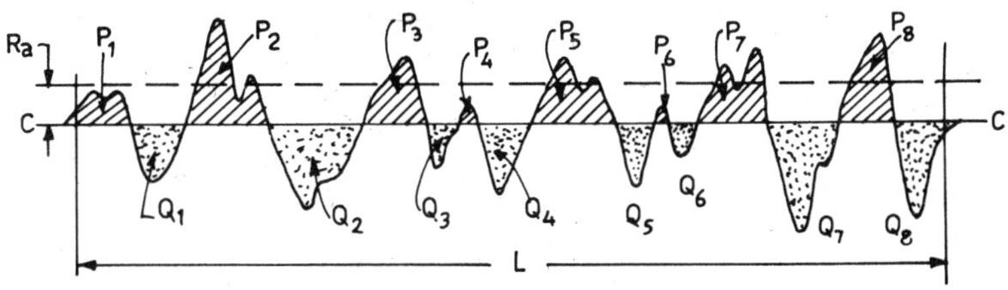

Fig. 11.24

The CLA value can also be calculated from the areas between the profile and the mean line (Fig. 11.24).

$$\text{CLA } (R_a) \text{ in } \mu\text{m} = \frac{(P_1 + P_2 + P_3 \,.......) + (Q_1 + Q_2 + Q_3 \,.......)}{L} \times \frac{1000}{M}$$

where P_1, P_2, P_3 ...... and Q_1, Q_2, Q_3 ...... are areas in mm^2

L is the sampling length in mm

M is the vertical magnification of the recorded profile.

The parameter R_a gives a fairly useful average value for all types of machined profiles with the exception of regular and periodic surfaces. Instrumentation used is easy and reliable. It has considerable sorting power and discriminating quality on regular and periodic profiles. However R_a does not reveal the actual depth of irregularities, and several surfaces with altogether different functional characteristics may have same value of R_a.

(*ii*) **Roughness Average (Geometric Mean) or Root Mean Square Average Height (RMS)—R_a**

It is defined as the average root mean square deviation Y of the profile from its mean line within the sampling length L.

Referring to (Fig. 11.23)

$$R_q \text{ (RMS)} = \sqrt{\frac{1}{L}\int_0^L Y^2\,dx} = \sqrt{\frac{1}{n}\sum_{i=1}^{n} Y_i^2}$$

where n is the number of ordinates in the sample length L.

(*iii*) **R_z—Ten Point Peak to Valley Average Height.** R_z is the average difference between the five highest peaks and five deepest valleys within the sampling length, the heights being measured from a line parallel to the mean line and not crossing the profile (Fig. 11.25).

$$R_z = \frac{(R_1 + R_3 + R_5 + R_7 + R_9) - (R_2 + R_4 + R_6 + R_8 + R_{10})}{5}$$

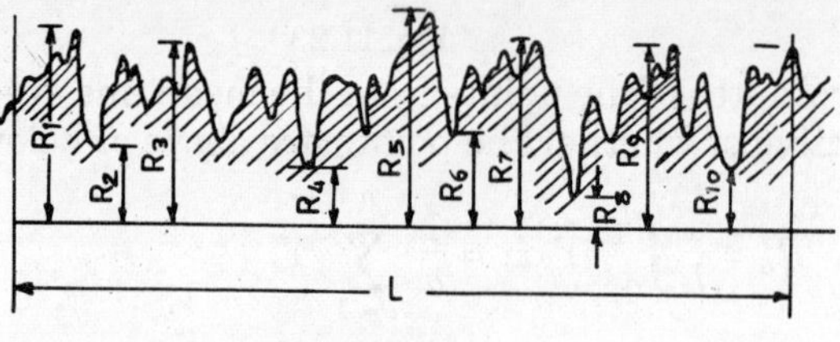

Fig. 11.25

(*iv*) **R_{max} —Maximum Peak to Valley Height in One Sampling Length.** The maximum peak to valley height is the distance between two lines parallel to the mean line, one line touching the profile at its highest point and the other at its lowest point. R_{max} is the maximum peak to valley height within the sampling length L. This parameter, however, fails to give a true picture of the surface as surfaces which are entirely different can have the same value of R_{max}. This parameter is used only when, for some technological or functional reasons, it is necessary to determine the maximum depth of irregularities and if they have to be removed in the subsequent machining operation.

(*v*) **R_t—Peak to Valley Height within the Assessment Length.** R_t is the maximum peak to valley height within the assessment length (Fig. 11.26). The disadvantage of R_t and R_{max} is that there can be a single predominant peak or valley which influences the value and will not give a true picture of the actual profile of the surface.

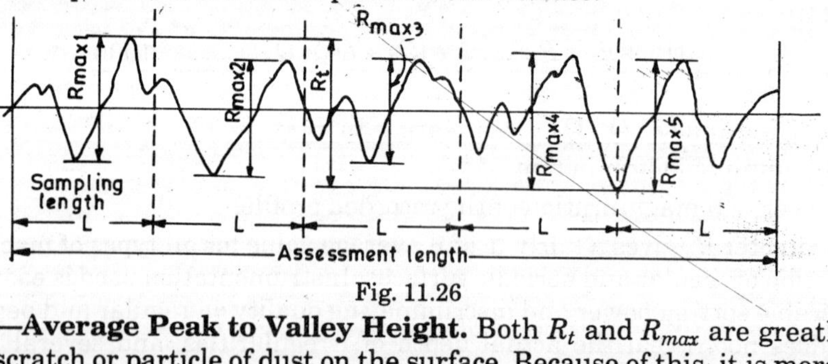

Fig. 11.26

(*vi*) **R_{tm}—Average Peak to Valley Height.** Both R_t and R_{max} are greatly affected by any spurious scratch or particle of dust on the surface. Because of this, it is more usual to use the average of the maximum peak to valley heights of 5 consecutive sampling lengths. This parameter is designated as R_{tm} (Fig. 11.26) which is the average of maximum peaks and valley heights for 5 consecutive sampling lengths.

$$R_{tm} = \frac{R_{\max 1} + R_{\max 2} + R_{\max 3} + R_{\max 4} + R_{\max 5}}{5} = \frac{1}{5}\sum_{i=1}^{n} R_{max\,i}$$

where R_{max1}, R_{max2}, ... are maximum peak to valley heights in each individual sampling lengths.

(*vii*) **R_p—Levelling Depth.** It is the maximum profile height from the mean line or the centre line, within the sampling length (Fig. 11.27).

$$R_p = \frac{1}{L}\int_0^L YdL$$

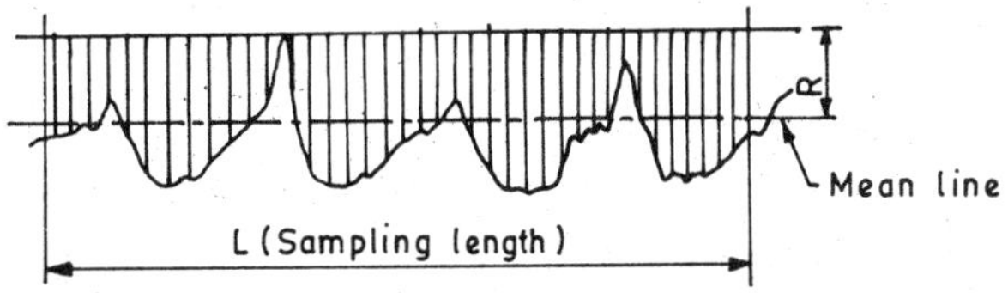

Fig. 11.27

This parameter describes the functional quality of a surface more closely than any other parameter. It gives information regarding the minimum initial volume of lubricant that could be contained between two sliding surfaces.

(*viii*) **R_{pm}—Average Levelling Depth.** This is the mean value of R_p determined over 5 sampling lengths (Fig. 11.28).

$$R_{pm} = \frac{R_{p1} + R_{p2} + R_{p3} + R_{p4} + R_{p5}}{5} = \frac{1}{5}\sum_{i=1}^{5} R_{pi}$$

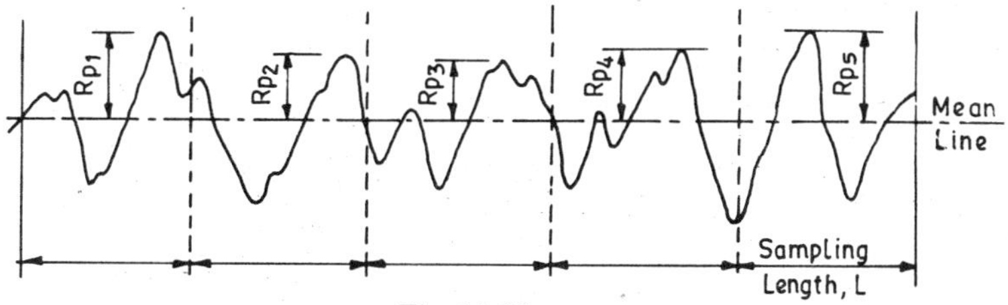

Fig. 11.28

Since it is an average value, it gives a more close representation of the surface.

(*b*) Various **spacing parameters** which give an indication of the nature of the surface in the horizontal direction are as follows :

(*i*) Δ_a	Average slope (R_a)
Δ_q	Average slope (*RMS*)
(*ii*) λ_a	Average wavelength (R_a)
λ_q	Average wavelength (*RMS*)
(*iii*) S_m	Mean spacing of profile irregularities
(*iv*) *HSC*	Total high spot count at the mean line within the assessment length.
(*v*) $t_p\%$	Bearing ratio at a selected depth below the 0% level of the profile. It gives an idea of the amount of contact that gets established when the surface contacts an ideally smooth surface.

	The bearing ratio t_p is the length of bearing surface (expressed as a percentage of length L) at a height above the lowest valley on the surface. This level can also be at some depth below the highest peak.
(*vi*) t_p (level)	Depth from the 0% level at which the given bearing ratio is obtained

All the above spacing parameters can be assessed from the profile after eliminating waviness from the profile.

While height parameters are important in assessment of surface roughness, equally important is index to the crest spacings. Autocorrelation function or power spectrum and the statistical equivalents in length to the amplitude probability density in height provide much valuable information about profile of surface.

TYPICAL EXAMINATION QUESTIONS

1. Discuss what you understand by the following terms in connection with surface finish measurement :
   (*i*) Waviness (*ii*) Lay
   (*iii*) Envelope method (*iv*) Crest line method.
2. Why is the assessment of surface texture important ? Describe two methods to obtain a numerical value of the texture from a given graphical record.
3. What is the difference between primary texture and secondary texture ? Describe an instrument that may be used to obtain a graphical record of primary texture.
4. (*a*) Describe an instrument for measuring surface finish.
   (*b*) Discuss the relative merits and demerits of the *M* (mean line) and the *E* (envelope) system of measurement of surface finish.
5. (*a*) What is meant by roughness and waviness of machined surfaces ?
   (*b*) Name the describe the methods used for measuring surface roughness.
6. How the Tomlinson surface recorder and Talysurf machine work ? What are their relative merits ?
7. How CLA Index Number is determined ? Explain why CLA Index Number alone is not sufficient to specify the surface texture required and to make the information complete, what else is to be specified ? How the required texture is specified on drawing ?
8. Describe various methods of measuring surface texture giving their relative advantages.
9. State the possible causes of each of the various types of irregularities found in surface texture. Show how surfaces having the same numerical assessment may have different properties and textures.
10. What are roughness comparison specimens and how they assess surface roughness ? What are the limitations ?
11. Describe briefly some recent advances in the field of surface texture measurement.
12. Which of the methods is recommended by IS : 3073—1967 for specifying the surface texture on machined parts ?
13. Calculate the R_a value of a surface for which the sampling length was 8.0 mm, the graph was drawn to a vertical magnification of 1000 and the areas above and below the datum line were

Above :	180	90	155	55 mm^2
Below :	70	90	170	150 mm^2

14. State the possible causes of each of the various types of irregularities found in surface texture. Show how surfaces having the same numerical assessment may have different properties and textures.

12

Dial Indicators and Indicating Gauges

12.1. Introduction

Dial indicators are instruments used for making and checking linear measurements. These require less skill in their use than other precision instruments such as plug gauges, gap gauges, micrometers and vernier scales etc. Dial indicators are smaller indicating devices containing a graduated dial scale and these depend upon gear and pinion or lever magnification system. These are generally used as comparators, *i.e.* compare a part to a master setting. Some indicators known as dial micrometers or dial thickness gauge are however, also used for actual (direct) measurements (without comparison). When dial indicator is used as essential part in the mechanism of any set-up for 'comparison' measurement purposes, it is referred to as dial gauge.

Most of the dial indicators take the form of a circular or semicircular scale upon which a pointer gives a direct indication of the movements of a contact arm or spindle. Small changes of dimension of the component in contact with the spindle are made to assume large proportions by employing some mechanical amplification and reading to be obtained on scale.

A dial indicator by itself is not of much use unless it is properly mounted and set before using for inspection purposes. By mounting a dial indicator on any suitable base and with various attachments, it can be used on thousands of special gauges which manufacturers themselves design to meet the requirements of the job.

12.2. Requirements of Good Dial Indicators

(1) The instrument should be capable of giving trouble-free and dependable measurements over a long period without attention. For this purpose, the design and construction must be robust enough.

(2) No oscillation must be present in the pointer when readings are being taken. For this, pointer movement must be damped properly.

(3) There should be some provision to avoid damage to the instrument due to larger movements of the plunger than the specified amount on the scale.

(4) The instrument is likely to be used in such a way that plunger may move sharply in either direction. So the instrument should be so constructed that by such movement, the instrument will not be damaged and its accuracy and setting will not be impaired. A finger lever or knob is usually provided for raising the plunger.

(5) The relative movement of pointer and scale should be such as to suggest the direction of movement of the measuring plunger *e.g.* an increase in the size of work must correspond to the upward movement of pointer in vertical scales and to right or clockwise movement in horizontal or circular scales.

(6) Accuracy of reading should be laid down to close limits for various sizes and ranges of instruments.

(7) Operating pressure required on a measuring head to obtain the zero reading should be of the order of 80 oz. and must remain reasonably constant over the whole range.

12.3. Advantages of Dial Indicators

(1) Ease of reading a dial indicator is the biggest advantage over other conventional types of linear measuring instruments, which require degree of skill and considerable practice before proficiency in their use can be attained.

(2) This is extremely useful for quality control in the inspection room and unskilled labour can be employed for its use. Once a dial indicator has been set up, the operator has only to refer to the work and observe the position of the pointer.

(3) It has made it unnecessary to depend upon the uncertain human ability to 'feel' or judge the variations in pressure when a conventional gauge is applied to the work, as the measuring pressure between the spindle and work automatically remains constant. This feature is specially important in the measurement of light materials. In the case of other gauges, the nerves of the inspector tell him, how a gauge feels on work and this can't be same in case of any two persons. Thus wherever we have part tolerances in thousands, other gauges are not so reliable.

(4) This is best suited for precision dimensional control, where tolerances are very small and by its use dimensions can be easily controlled in mass production.

(5) This is not subjected to problems of gauge wear and temperature variations etc. as in the case of conventional gauges.

(6) These are able to inspect dimensional variations which are impossible to be detected even by the conventional gauges.

(7) The speed, adaptability, and positive visibility of dial indicator gauges, in addition to their accuracy and economy place them in vanguard of precision measurement.

(8) It is possible to use it for many different measurements by means of a few simple attachments, thus making it more versatile.

(9) By its intelligent use, a large and costly collection of specially made gauges can be avoided.

(10) Conditions which influence the accuracy of a specified dimension and which cannot be detected by conventional gauges are readily discovered by dial indicators *e.g.* out of roundness, taper etc.

(11) Combination of dimensions having a certain relation to each other can be easily inspected by the proper design of multiple inspection gauges, thereby saving time and avoiding the necessity of using different inspectors and expensive 'set-ups'.

12.4. Classifications

Dial indicators of purely mechanical operation may be broadly classified into groups as follows :

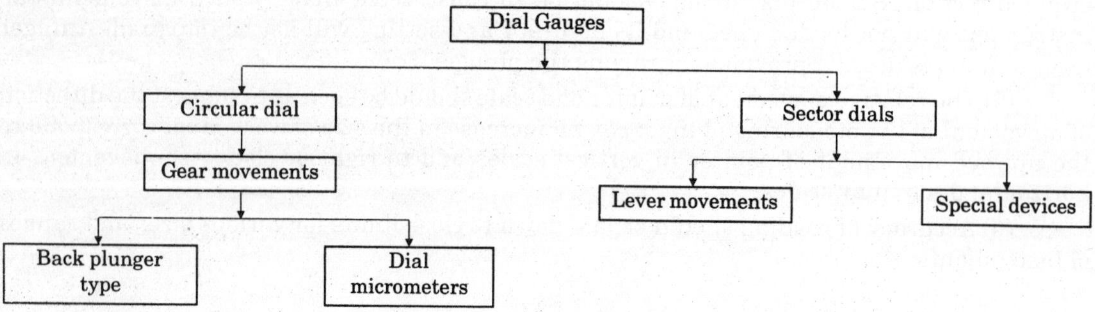

Simple lever system used in some of the gauges is shown in Fig. 12.1. *ABC* constitutes the simple form of magnifying lever, which is pivoted at *B* and magnification ratio is obviously *BC* : *AB*. A simple dial indicator as shown in Fig. 12.2 is constructed on this principle. The gauge is restricted to few degrees on either side of zero. But it is very robust and very useful within its range for comparison purposes.

Sometimes a compound lever arrangement as shown in Fig. 12.3 is used to give an increase in magnification. Here the magnification ratio is $\frac{ED}{DC} \times \frac{CB}{AB}$. However, this type of lever is more difficult to construct due to the difficulty of making fork joint *C*. It has a low order of accuracy and less range.

Most of the dial gauges used have circular scales, while some have sector shaped scales also. Sector scale is usual on those instruments intended for extremely accurate measurement over a limited range. The circular scale type is most commonly used in workshops. There are two main types of circular scale dial gauges. In one, we have a back plunger at right angles to the dial and other type has the plunger parallel to the plane of the dial. The back plunger type shown in Fig. 12.4 is not very widely used. We will be dealing here with the circular scale dial indicators.

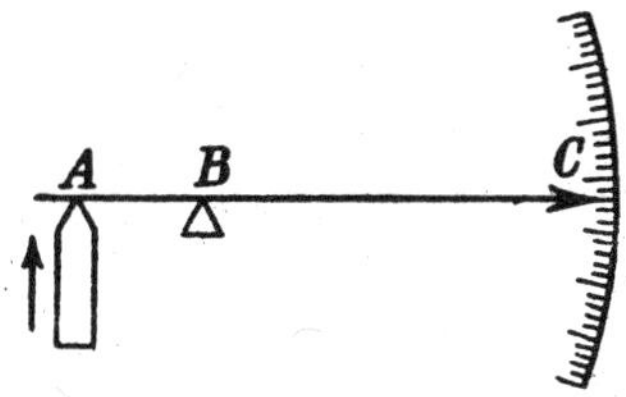

Fig. 12.1. Lever system of magnification.

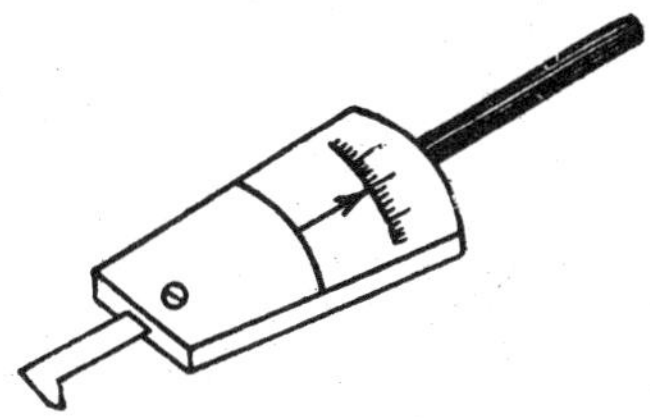
Fig. 12.2. Simple dial indicator.

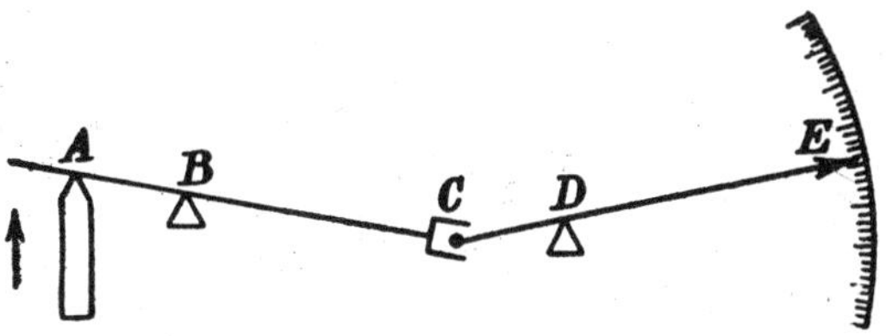

Fig. 12.3. Compound lever.

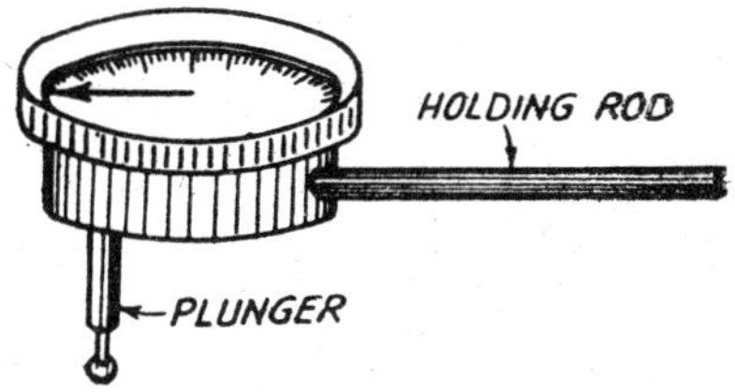

Fig. 12.4. Back plunger dial indicator.

12.5. Working Mechanism of Dial Indicators

The necessary high magnification ratio in this instrument is achieved by the use of gears and pinions arrangement in a small area, by arranging the gear train in a manner similar to a clock movement. Method of magnification used is shown in Fig. 12.5.

There is a plunger which is a perfect sliding fit in its own bearings. This carries a rack which accurately meshes with a pinion *A*. The rotation of the plunger about its own axis is prevented by a pin attached to it, which is located in a slot in a rack guide *G*. In order to keep the plunger in an extended or normal position a light coil spring *S* is employed. The spring exerts a pressure of approximately 6 oz. A small movement of the contact point causes the rack to turn the pinion *A* with which it is meshed. A large gear *B* is attached to the same spindle as pinion *A*. The gear *B* is further meshed with a pinion *C*, which thus magnifies the movement of pinion *A*. Attached to the second pinion *C* is another gear *D* which meshes with a

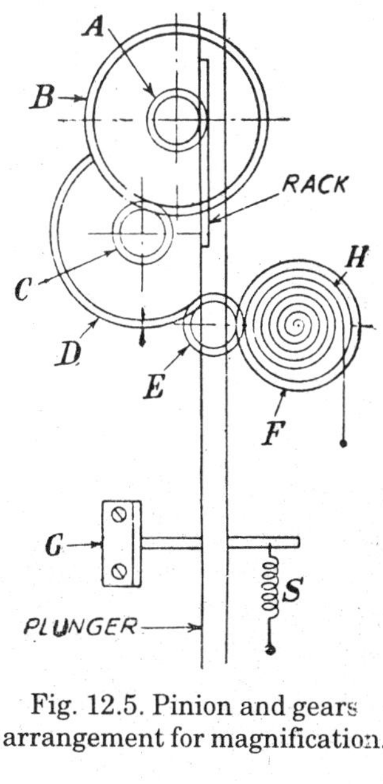

Fig. 12.5. Pinion and gears arrangement for magnification.

third pinion E mounted on the same spindle as the indicator pointer. The overall magnification of final pinion is thus = $\frac{T_D}{T_E} \times \frac{T_B}{T_C}$, where T_B, T_C etc. represent the number of teeth of gears B and C respectively. This magnification is further enlarged at the tip of the pointer by an amount dependent upon its length. The overall magnification for any dial gauge may be thus calculated by measuring the distance between division on the scale and dividing this dimension by the equivalent movement of the measuring plunger.

In order to take up the backlash, a light hair spring H is always fitted to the dial gauge gear trains by meshing a gear F to pinion E and attaching spring H with F. In order to prevent gear backlash, the gears are precision cut and the movement assembly resembles that of high-grade watch. In expensive instruments, jewelled bearings are fitted to all pivots After final assembly, each dial gauge is thoroughly tested and checked and properly calibrated by either a calibrated micrometer head or by the use of slip gauges.

If the pointer makes more than one revolution, a revolution counter is also incorporated which will correctly indicate the number of revolutions made by the main pointer and thus enable a complete reading to be made when the gauge is used for direct measurement.

12.6. Range

The accuracy attained in measurement is inversely proportional to the range of travel. With most of the applications of dial gauges, they are mostly used as comparators on finished work and as such are only called upon to work over a narrow range. So the range need not be very large in order to have high accuracy. Generally the range is of the order of 10.0 mm.

12.7. Description of Various Parts

The nomenclature of the various parts of a dial indicator is shown in Fig. 12.6 and a brief description of them is given below.

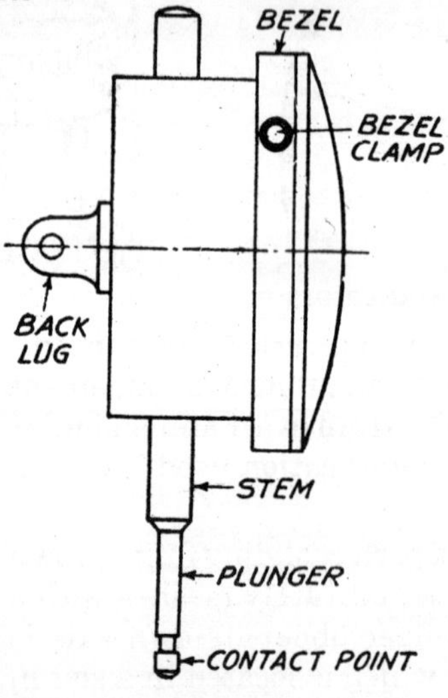

Fig. 12.6. Nomenclature of dial indicator.

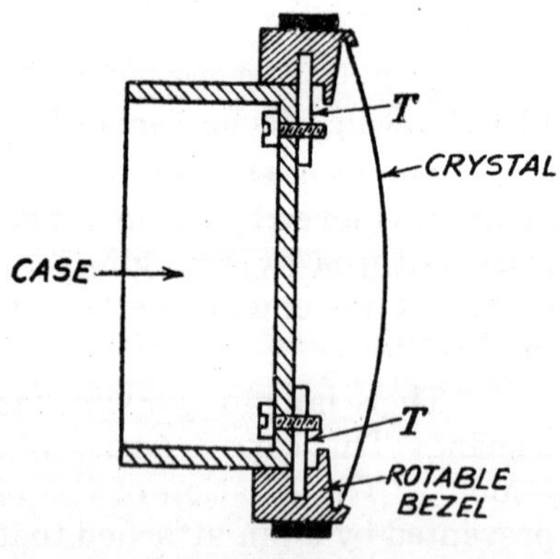

Fig. 12.7. Rotatable bezel.

12.7.1 Rotatable Bezels. Most dial indicators incorporate a rotatable bezel for making the final zero setting at any position throughout the full circle. A clamp for locking it in any position is also provided. The bezel must be free to rotate and must be exactly concentric as it carries the outer scale. Also when unclamped, it must be securely located on the case. This is achieved by an arrangement shown in Fig. 12.7. Three equidistant tongues (T) locate the rotatable bezel in a groove cut in the bezel wall, thus preventing it from being pulled away from the gauge. By careful fitting of the bezel surface on the case, free rotation of the bezel can be ensured.

12.7.2. Contact Points. In order to make the dial indicator more versatile in its use, arrangement is provided to change the contact points. The measuring plunger is tapped out and corresponding threads are provided at the end of contact points.

The contact points are generally made of hardened steel and blackened in order to avoid rusting. In order to prevent scratching of workpiece, the contact end of each point is polished. The general purpose contact point is of hardened spherical-formed shape. This is very satisfactory in all the applications, if care is taken to renew the ball when a flat is worn at the point of contact.

For measurement over a flat work surface, a flat surface contact anvil is generally preferred. The contact anvil's gauging surface is lapped flat and true in position and this must be parallel to the flat work surface. The above two contact points are most commonly used. For some special purposes, as for finding the exact depth of etching in printing blocks, a needle point can be used. In gauging of continuous moving stock, the rolling contact will be ideal. Similarly for thread inspection, conical points are used. Fig. 12.8 shows the various shapes of contact points.

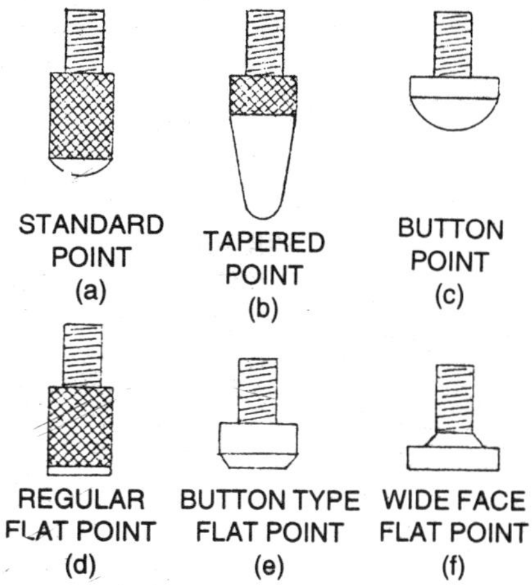

Fig. 12.8. Various shapes of contact points.

12.7.3. Tolerance Pointers. For rapid checking of parts, tolerance pointers can be attached to the dial face. By these pointers any tolerance, whether bilateral or unilateral, can be clearly indicated on the dial gauge. These are located in the centre and can be moved independent of each other by means of projecting knobs as shown in Fig. 12.9. This is very useful in statistical quality control. The parts below and above the minimum and maximum limits can thus be easily sorted out.

12.7.4. Special Stems. Where deep recesses are to be inspected, long-stem indicators are available.

12.7.5. Shock-Proof Cushioned Movement. Some dial indicators are designed to withstand sudden shocks due to rough usage and at the same time it reads correctly. In such indicators, before the impact reaches indicator movement, it is cushioned by a specially loaded attachment on the plunger. Thus the small and sensitive parts like gear, jewels, points etc. are protected.

12.7.6. Lever Attachments. Lever attachments are very useful and facilitate the work of machine settings. The right angle offset attachment shown in Fig. 12.10 enables internal measurements to be made. Also the awkward corners and recesses can be easily gauged by this attachment.

The attachment transfers motion at right angles to the indicator spindle directly. There is provision to avoid any errors due to wear or play in bearings etc.

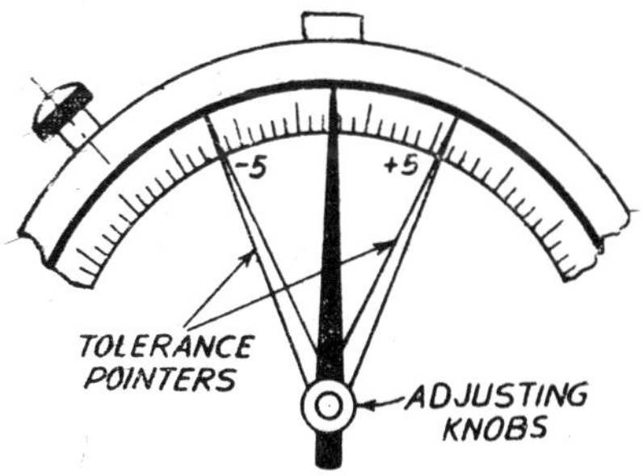

Fig. 12.9. Tolerance pointers.

12.8. General Care and Use of Dial Gauges

This instrument being precision one, must be handled very carefully. If plunger is allowed to strike the work with force, the teeth of rack and gears will get distorted and the pointer may slip on the spindle. This must always be fixed very rigidly to the stand etc. for attaining accurate results.

The following points need attention in use of dial gauges :

Since any fall or sharp blow can damage the contact points or upset the alignment of the bearings, these should be avoided. If after such an incident, the motion is not found to be smooth, the dial gauge should be checked.

The plunger of dial gauge is designed as a close fit in its bush and intended to work dry. Any oiling of plunger will only result in picking up of dust and sticking of plunger. Accordingly the plunger should never be oiled or greased.

Dial gauges should not be used as direct measuring instruments but be used only as comparators.

The stand used for dial gauge should be very rigid for accurate readings.

In the case of clamping of dial gauge by means of a back lug, the clamp should be quite tight ; otherwise false readings may be obtained by occurrence of slip. If the dial gauge is clamped by means of the steel stem, then too great a clamping force may cause the plunger to jam in the stem. The hole in the fixture must be true and smooth to avoid distortion, binding and locking of stem in it.

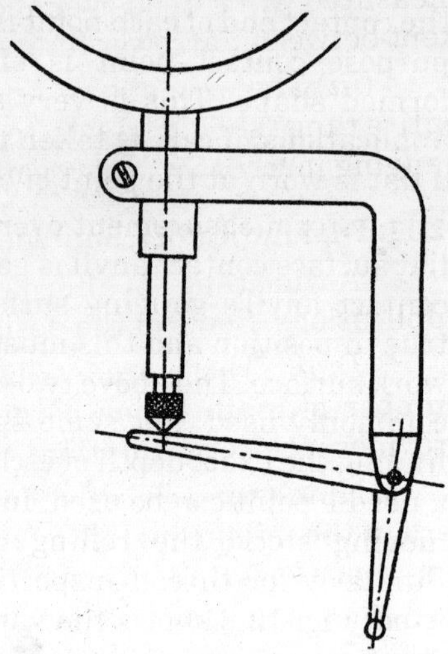

Fig. 12.10. Use of lever attachments.

The plunger lifting lever, when raised to allow work to be inserted under the contact point, must be released slowly. Sudden release with force will distort the teeth of rack.

If contact points of dial gauge are worn, these should be replaced.

The flat surface of flat contacts should be truly square to the plunger and parallel to the measuring surface of the fixture for accurate readings.

The dial gauge should be free from any trace of stickiness or backlash, particularly when it is used for detecting small errors in alignments, or for testing the accuracy of the centering of work mounted in a machine.

The magnification of dial gauge selected should be in relation to the accuracy required.

Tests for repeatability of reading, discrimination, and calibration should be carried out from time to time.

12.9. Accuracy of Dial Indicators

Two categories of potential errors are described by the American Gauge Design Standard in order to assess the accuracy of dial indicator :

(*i*) *Repetition of Readings.* Reading at any point should be reproducible through successive measurements of the spindle to ± one-fifth graduation.

(*ii*) *Accuracy (Calibration) or linearity of indications.* The dial indication should be accurate to within ± 1 graduation, at any point with 2.3 turns, *i.e.* the variation of the actual object size in relation to the setting master must be displayed by the pointer position on the dial with a maximum error of ± 1 graduation, at any position within 2.3 turns for effective measuring range. The calibration accuracy determines the confidence which can be placed in the measurement of size variations by the use of the instrument.

12.10. Guidelines in use of Dial Indicators to reduce the Effect of Calibration Errors

(*i*) Indicator should be set to a dimension which is approximately in the centre of the spread over which the actual object size is expected to vary.

(*ii*) Dial amplification should be so selected that the expected size variations to be measured will spread over about 1/3rd turn or less. Thus the effect of calibration errors can be kept between narrower limits than the total calibration tolerance range of the instrument.

(*iii*) The area of best linearity (obtainable from the actual calibration chart) within the entire range should be selected for the actual measurement, thereby reducing even further the gauging inaccuracy which would result from instrument calibration errors.

12.11. Dial Indicator Components, Accessories and Attachments

A wide range of alternative designs are available to fulfill the diversity of dial indicator applications. Some of the designs are described below :

Dials. These are available with 200, 100, 80, 60, 50, 40 and 20 graduations. Greater number of graduations permit larger measuring range and wider spacing of graduations provides a better resolution.

Numbering on dials could be either systematically arranged in both directions from the zero in the middle (balanced dials) or be in sequential order (continuous dials).

Mountings. A dial indicator could be mounted either by the stem or by the back. Stem (usually 8 mm diameter) mounting provides the advantage of permitting simple adjustment in the axial direction, as well as the rotation of the indicator around its axis for the convenient observation of the dial. Back mounting provides more positive clamping, and is preferred where design of fixture mount permits adjustment and also where the space is limited.

Contact points. The spindle end may vary in length of contact member and the form of the contact face. Six mm long contact member with spherical segment contact face (5 mm radius), which provides essential single point engagement with most object surface forms and makes measuring contact less sensitive to minor misalignment is most commonly used. Flat contact points are preferred for diameter measurements on convexly curved surfaces (balls, cylinders, cones etc.) to make the precise centering of the object less critical for obtaining the correct indicator reading. It should be remembered that flat contact surfaces are sensitive to alignment errors in a plane perpendicular to direction of measurement and thus care should be taken. Taper points are used where measurements are to be taken in the bottom of grooves, at the root of gear tooth, etc. Tips of contact points are made of suitable material depending on requirement of application.

Dial indicator accessories. These may include adjustable tolerance limit hands, lifting lever to raise the rack spindle while inserting the part into the measuring position, devices to protect the instrument spindle when it has to withstand adverse conditions of use or clamping.

Attachments for dial indicators. These include hole attachments for indicator measurements on internal surfaces which are not accessible to regular dial indicators. (This attachment consists of an equal arm lever, pivoting around a fulcrum pin mounted in a bracket, which can be attached to the stem of the dial indicator), right angle attachment for use when the indicator dial must be observed at right angles to the direction of measurement (bell crank lever having arms of equal effective length) ; adjustable mounting brackets used with dial indicators with rail-type backs, in order to accomplish a substantial range of adjustment by simply sliding the indicator in the guides of the mounting bracket and locking it in the desired position.

12.12. Test Type Indicators

Test type indicator is special type of indicator which is characterised by the direction of the sensed displacement. Test indicators are designed to sense and to measure displacements which occur in a direction essentially perpendicular to the shaft of the contact point.

The displacement of the contact point in a direction normal to the axis of the probe, causes the swing motion of a lever, the shorter arm of which is the shaft of the contact point and the longer arm ends in a toothed segment which engages a pinion. The latter is in the hub of a sector with geared periphery and functioning like a crown gear.

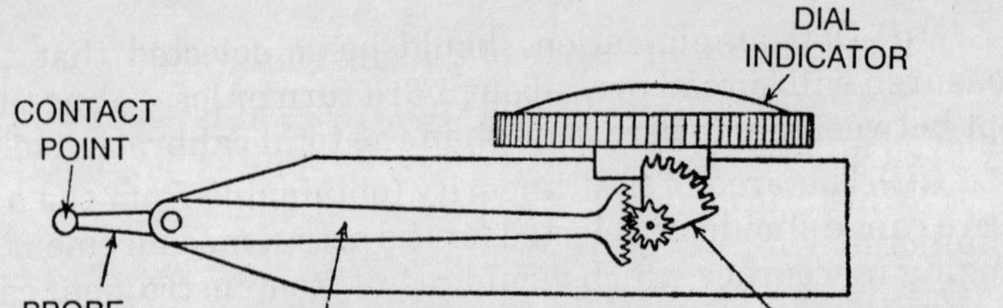

Fig. 12.11. Test-type Indicator.

This crown gear is meshing with a final pinion which is attached to the spindle of the indicator hand. The displacement of the probe is thus amplified by a transmission train comprising the main lever of unequal arm length, the sector and two pinions. The effective gauging range is usually less than a circle. Dial indicator portion could be either parallel or perpendicular to main lever of the gauge. The gauging force is provided by a flat spring acting on the lever.

The test type indicators permit a wide variety of uses in several difficult applications, such as contacting hard-to-reach surfaces, for measurements on internal surfaces, transfer of height dimensions, measuring along vertical surfaces, centering of rotating members, etc.

12.13. Test Indicators

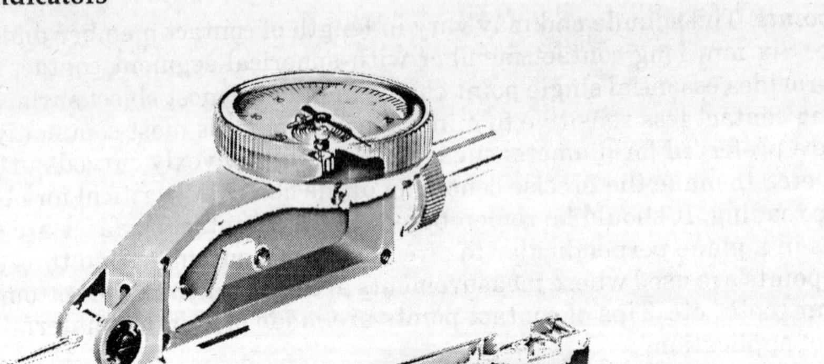

Fig. 12.12. Test indicator.

Design Features—Advantages

The Test Indicators are outstanding in regard to accuracy, dependability, performance and long life. Proven on actual jobs in test after test over a long period of time, they demonstrated their unique quality and durability. Even after millions of measurements they are still accurate and fully meet the speci⁻cations.

Listed below are some of the most outstanding features of these instruments :

Self-contained movement. This is a self-contained, separate unit which can be easily removed and replaced when in need of repair or for maintenance.

Maximum sensitivity and accuracy insured by jewel mounting of the movement in conjunction with precision ground gears and pinions.

Box type housing opens at the front only. One piece construction of side panels and bottom surface. The housing totally encloses the movement and provides maximum protection against shock and any kind of damages. The front end of the housing is designed in the shape of a fork and the axis of the stylus is mounted in between this fork, thus providing added stability and shock resistance. The precision alignment of the two mounts insures also maximum accuracy.

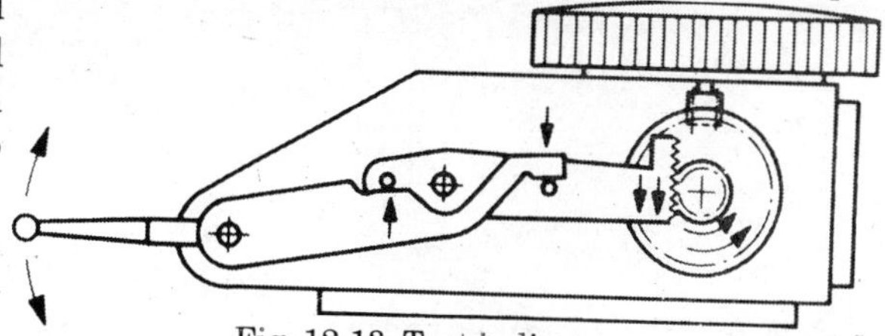

Fig. 12.13. Test indicator.

Corrosion-proof by satin chrome finish of the housing and stainless hardened steel for the stylus.

Easy to read by sharply defined contrasting numerals on the dial face.

Rigid mounting. Dovetail guideways at top and bottom of the housing permit easy movement of the mounting shanks. They can be clamped at any desired point by simply tightening a knurled clamp nut. No separate tool is required.

No Plastic parts are used on this instrument with the exception of the cover over the dial face.

Automatic reversing eliminates reversing lever. The indicator hand always moves clockwise which insures error-free readings.

12.14. Dial Gauges as Per Indian Standards

According to IS : 2092 dial gauges have been divided into two categories, *viz.*, type 1 and type 2 for general engineering purposes depending upon the movement of the plunger. These are manufactured in two grades, *viz*, grade *A* and grade *B*, with total plunger movement or lift of 3, 5 and 10 mm. Type 1 dial gauge has the plunger movement parallel to the plane of dial and type 2 has the plunger movement perpendicular to the plane of dial.

Dial gauge with lift 3 mm has 50 divisions instead of 100. Permissible errors for two grades of accuracy (grade *A* and grade *B*) are given

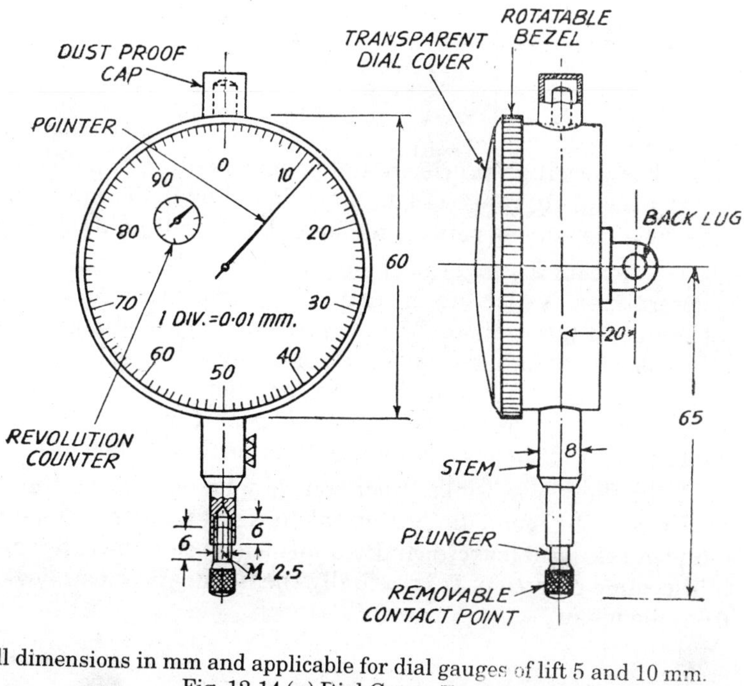

All dimensions in mm and applicable for dial gauges of lift 5 and 10 mm.

Fig. 12.14 (*a*) Dial Gauge

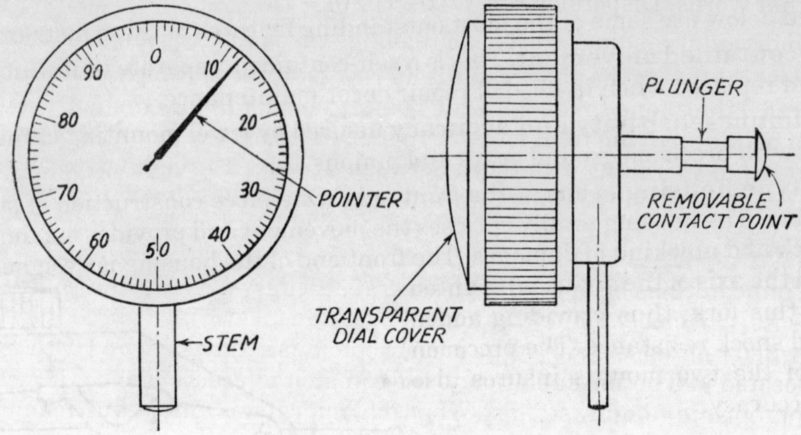

Fig. 12.14 (*b*) Dial Gauge Type 2

below, which prescribes the errors for part measuring range 0.1 mm and for the total measuring range of 1 mm, 2 mm, 3 mm, 5 mm, 10 mm. These errors are applicable only for the vertical movement type plunger. In addition the gauge should be capable of indicating small gradual changes of the order 0.025 mm in either direction of the plunger to within 0.003 mm without resort to tapping.

Permissible error for dial gauges

All dimensions in microns (1 micron = 0.001 mm)

Grade of Accuracy	*Permissible error (plunger moving outwards)*						*Dispersion*	*Reverse Interval*
	Part Measuring Range of 0.1 mm	*Total Measuring range of*						
		1 mm	*2 mm*	*3 mm*	*5 mm*	*10 mm*		
A	5	6	7	10	12	15	± 1	3
B	8	11	13	15	20	25	± 2	5

For very high precision dial gauges reading to 0.001 mm, the permissible deviation in a part measuring range of 0.05 mm should be 0.003 mm and the permissible deviation over the whole measuring range should not be more than 0.003 mm.

In actual testing, the error over the whole measuring range and part measuring range is determined by plotting the error curve. The error curve is drawn by obtaining readings on the gauge in intervals of 0.1 mm with the help of slip gauge blocks. The highest and lowest points on the error curve are further analysed by taking measurements at intervals of 0.01 mm for a range of 0.1 mm. The error for whole range is the difference between the highest and lowest points of error curve, whereas for part measuring range the error is the difference in the highest and the lowest point of that portion of the curve.

In the above table, dipersion may be defined as the difference between two or more readings of the same dimension taken with the same gauge under same conditions and it is found by taking at least ten individual measurements of the same work with plunger movement in the same direction. It is actually the quadratic mean deviation of the individual readings from the mean.

In other words, Dispersion $= \pm \sqrt{\Sigma(x - \bar{x})^2/(n-1)}$

where x is individual reading, n is No. of readings taken and $\bar{x}$ is mean of all the n readings.

The last column in the above table is for reverse interval which is the difference between two measured values (one taken with the pointer moving from a higher to a lower value and second with pointer moving from lower to higher value) on the same work and with the same gauge. It is found by taking difference between at least 10 readings at any position of the measuring range by moving plunger inwards and outwards.

12.15. Indicator Gauges

Dial indicator has been used with several auxiliary devices for a wide variety of length measurements. Obviously dial indicator can be used for carrying out measurements of length variations when reference plane of the measurement is established and both the indicator and object are located in a definite relation to the common reference plane. These are achieved by means of appropriate gauging set-ups. When dial indicator is united with members which serve the needed complementary functions, resulting in a single tool, it is known as indicator gauge. It must be remembered that indicator gauges are always comparator type measuring instruments and require the use of a setting gauge for establishing the basic measuring position.

In this way, it may be used as adjustable indicator snap gauge, bench comparator gauge, plate gauge, depth indicator gauge, shallow diameter gauge, internal indicator gauge, etc.

12.15.1. Passmeter. It is also known as dial-type adjustable snap gauge and is used for external measurements. It utilises the lever and gear movement for magnification purposes. This instrument is somewhat similar to the micrometer screw gauge and has very big scale in its frame on which a pointer moves. The adjustable anvil is adjusted along its axis only for setting up the instrument. The adjustment of the adjustable anvil is carried out by first removing the protecting cover and releasing the clamping screw. The setting is done to the zero graduation of scale with the aid of slip gauges, by holding the movable anvil against slip

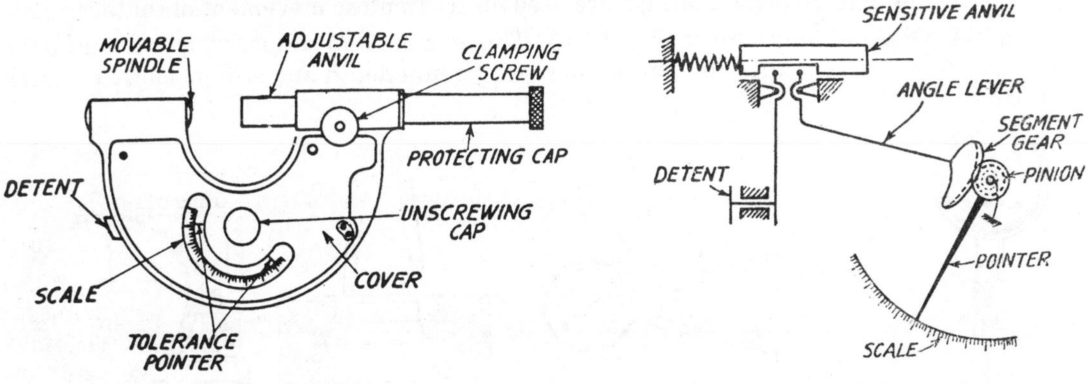

(*a*) Passmeter. (*b*) Movable anvil.

Fig. 12.15.

gauges with constant contact pressure exerted by spring. After setting up and reclamping, the adjustable anvil remains stationary during checking operation. The transmission of movable anvil to the pointer takes place through the angle lever and the gearing (segment gear and pinion) as shown in Fig. 12.15 (*b*). A detent is also provided which retracts the moving anvil to relieve the load when the instrument is not in use, and to protect it against shocks and damage. The detent is also used for retracting the movable anvil during measurement. The spring at the back of the movable anvil always exerts a constant contact pressure of the order of 700 ± 200 grammes. Two tolerance pointers are also provided on the scale which can be set in any position by unscrewing the cap on the cover of the frame and using a special setting wrench. This instrument is much accurate than a dial indicator even.

12.15.2. Internal Dial Gauges. These are designed for checking bore diameters by the comparative method. The principle of operation is shown in Fig. 12.16. The instrument basically consists of a hollow tube (body) into which is contained a lever pivoted about its intermediate support. One end of the lever is linked to the movable contact of the instrument (the instrument has three contacts, equally spaced along the circumference of the head, one being movable referred above and two fixed ones) and other end of the lever actuates the pointer. The three contacts (one movable and two fixed, not shown in the figure) bear against the internal surface of the part and properly centre the instrument in relation to the axis of the bore being checked. The contacts are interchangeable with extension rods in order to broaden the range of measurement. The smallest size, this instrument can check is of the order or 11 to 18 mm in diameter and largest size range is from 86 to 120 mm. The range of measurement in a single setting is about ± 0.06 and ± 0.22 mm with scale division value of 0.002 mm and 0.01 mm respectively.

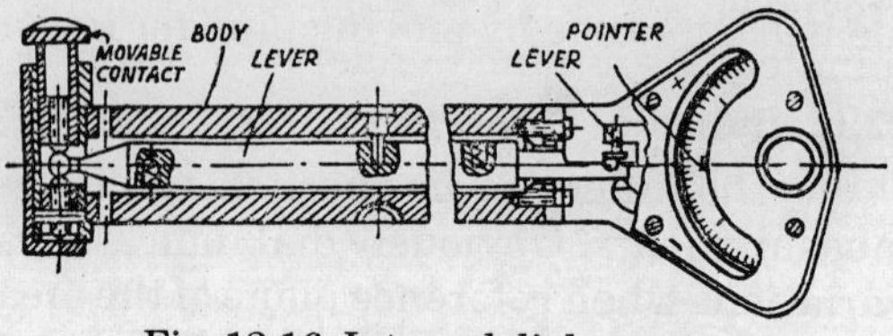

Fig. 12.16. Internal dial gauge.

12.16. Applications of Dial Gauges

12.16.1. Dial Test Indicators (Automatic Reversal Type). In the dial test indicators, finger probe can be moved by around 240° and on releasing returns back to initial setting. The rotary movement of finger probe is transmitted to dial mechanism through a cam and cam follower.

It uses fully jeweled movements.

Precision miniature pivot bearings are used for frictionless movement of contact point.

12.16.2. Dial Caliper Gauge (Fig. 12.18). The internal dial caliper gauges are used for quick inside diameter comparisons. These can reach 80 mm depth and still measure recesses for "O" rings etc.

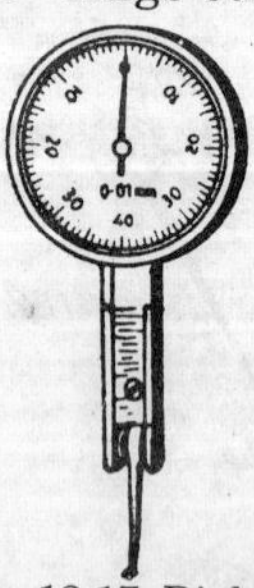
Fig. 12.17. Dial test indicator.

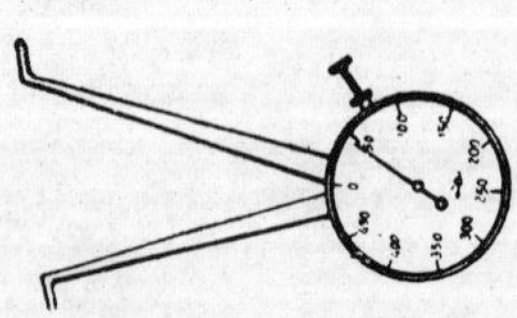
Fig. 12.18. Dial caliper gauge.

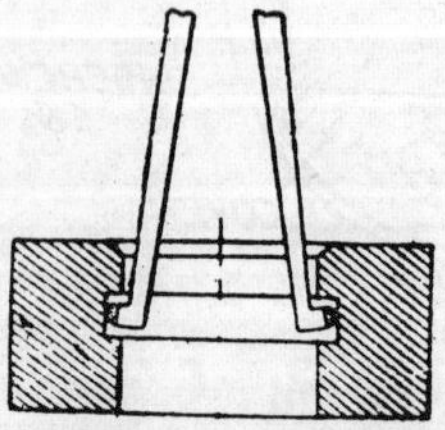
Fig. 12.19. Measuring recess diameter.

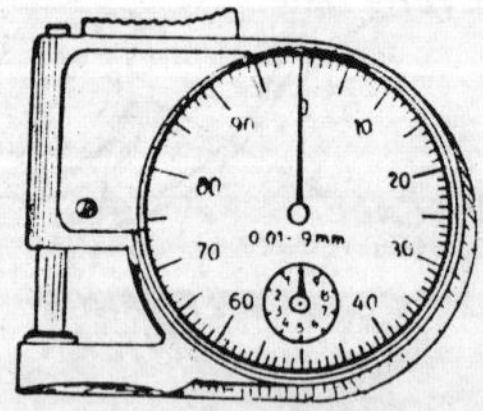
Fig. 12.20. Pocket thickness gauge.

The dial caliper gauge is a comparison gauge and should be used together with master ring or a micrometer for zero setting. The movable bezel dial must be adjusted to zero before gauging by means of a master of known size. Black oxide finish caliper arms have 1.59 mm steel ball on the measuring tips.

12.16.3. Pocket Thickness Gauges (Fig. 12.20). Pocket thickness gauge made for carrying convenience, makes precise measurement to 0.01 mm of the thickness of paper, leather, wire, plastic sheet metal etc.

12.16.4. Dial Pipe Gauge. (Refer Fig. 12.21) This is a specially designed dial thickness gauge for easy measurement of the wall thickness of pipes. Contact point of spindle is 2.5 mm diameter.

Measurable minimum inside diameter of pipe is 3.5 mm.

12.16.5. Dial Depth Gauge. (Refer Fig. 12.22) These are used for rapid, direct reading measurement of hole depths. The base is hardened steel, ground and microlapped for accuracy. Provision is made for wear adjustment. Spindle lifting lever for raising and lowering the spindle is maintained under spring pressure.

The dial indicator is attached to the base by a clamping screw. Measurements are automatically registered directly on dial face in 0.01 graduations.

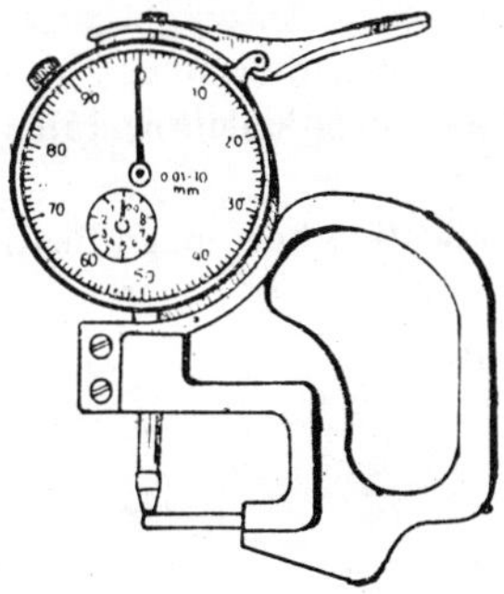

Fig. 12.21. Dial pipe gauge.

Fig. 12.22. Dial depth gauge.

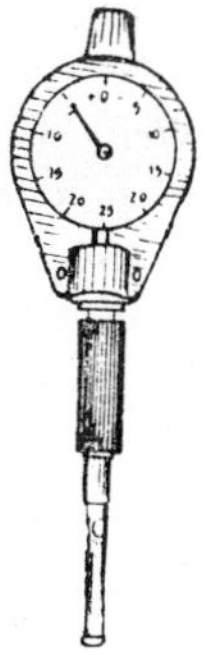

Fig. 12.23. Bore gauge.

12.16.6. Bore Gauges. The dial bore gauges shown in Fig. 12.23 are for miniature hole measurements.

The gauge is supplied with a set of split ball measuring contact points which are hard chrome-plated to retain original spheres.

Along with the measuring probes, setting rings are also provided to zero set the indicator whenever the probes are interchanged.

Actual ring size is engraved on the ring frames to the closest 0.001 mm value.

SOLVED PROBLEM

Q. 1. Enumerate the advantages and limitations of dial indicators.

Ans. Advantages. (*i*) It is ideal for shop use because it involves amplification by gear trains which permits a sturdy construction and is less sensitive to shocks and vibrations than other system of designs.

(*ii*) Long measuring range is possible.

(*iii*) These usually permit a reasonable amount of over-travel without injuring the mechanism.

(*iv*) Dial form of display with several turns of the hand, permits extended indicating range with a high resolution.

(*v*) Initial as well as maintenance cost are low.

Limitations. (*i*) Wear results in backlash and lagging response. Of course, hairsprings are provided to eliminate backlash and wear can be reduced by precisely machining and heat treating gears.

(ii) The effect of imperfect gear form, or play, results in cumulative error which is usually proportional to the length of the gauging travel. It is for this reason that dial gauges are not considered suitable for extended measuring range for precision applications.

(iii) The clearance of the measuring spindle in its guides can result in positional variations of the rack in relation to the meshing pinion. The precision dial indicators, therefore, employ cam in place of rack and pinion system in order to overcome the effect by looseness of measuring spindle guides.

(iv) Play in pivot bearing due to initial inaccuracies or wear affects the precise meshing of the gear teeth causing measuring errors. Accuracy of gear locations can be substantially improved by use of jeweled bearings.

PROBLEMS

1. Write an essay on the importance of dial indicators in workshops, inspection shops and its overall utility for production engineer.
2. Sketch a dial indicator and explain its working mechanism clearly indicating how magnification is achieved.
3. What are important characteristics of dial indicator ? Enumerate its uses.

13

Metrology of Screw Thread

13.1. Introduction

Screw thread has generally two functions to perform, *viz.*, transmission of power and motion, and to act as fastener. Second function is rather more important, so we will be more concerned with Vee-form of threads. The object of dimensional control in case of plain shaft and hole is to ensure a certain consistency of fit. In the case of threaded work, the object is to ensure mechanical strength which is dependent upon the amount of flank contact and not upon the fit.

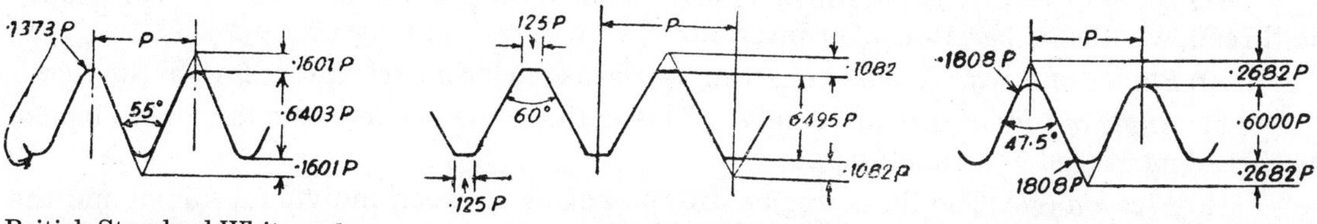

British Standard Whitworth American Standard British Association

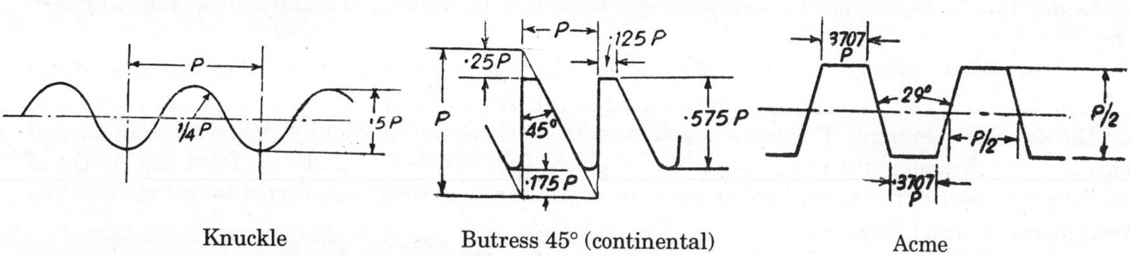

Knuckle Butress 45° (continental) Acme

The most commonly used thread-forms all over the world are shown in Fig. 13.1. These mainly consist of the British Whitworth and B.A. series, and in the Metric series the American Standard series. Unified thread which is an attempt to achieve interchangeability between threads used in England and America has been added recently. The ISO metric threads recommended for use in Indian Industry by BIS have been dealt with separately in details in Chapter 14.

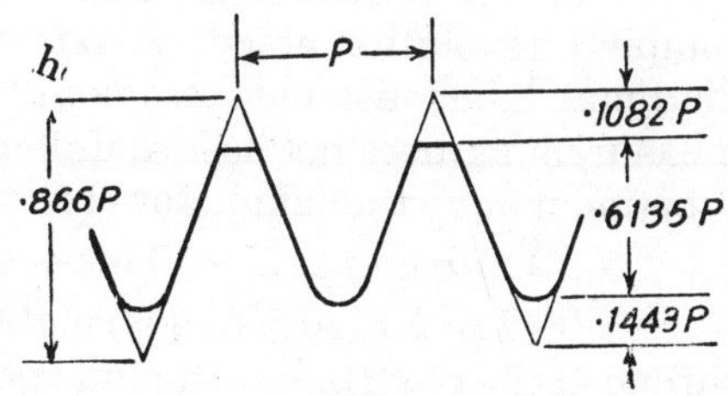

Unified

Fig. 13.1. [Not to scale]
Commonly used thread forms.

13.2. Screw Threads Terminology

(1) *Screw thread.* A screw thread is the helical ridge produced by forming a continuous helical groove of uniform section on the external or internal surface of a cylinder or cone. A screw thread formed on a cylinder is known as straight or parallel screw thread, while the one formed on a cone or frustrum of a cone is known as tapered screw thread.

(2) *External thread.* A thread formed on the outside of a workpiece is called external thread *e.g.,* on bolts or studs etc.

(3) *Internal thread.* A thread formed on the inside of a workpiece is called internal thread *e.g.* on a nut or female screw gauge.

(4) *Multiple-start screw thread.* This is produced by forming two or more helical grooves, equally spaced and similarly formed in an axial section on a cylinder. This gives a 'quick traverse' without sacrificing core strength.

(5) *Axis of a thread.* This is imaginary line running longitudinally through the centre of the screw.

(6) *Hand (Right or left hand threads).* Suppose a screw is held such that the observer is looking along the axis. If a point moves along the thread in clockwise direction and thus moves away from the observer, the thread is right hand ; and if it moves towards the observer, the thread is left hand.

(7) *Form of thread.* This is the shape of the contour of one complete thread as seen in axial section.

(8) *Crest of thread.* This is defined as the prominent part of thread, whether it be external or internal.

(9) *Root of thread.* This is defined as the bottom of the groove between the two flanks of the thread, whether it be external or internal.

(10) *Flanks of thread.* These are straight edges which connect the crest with the root.

(11) *Angle of thread (Included angle).* This is the angle between the flanks or slope of the thread measured in an axial plane.

(12) *Flank angle.* The flank angles are the angles between individual flanks and the perpendicular to the axis of the thread which passes through the vertex of the fundamental triangle. The flank angle of a symmetrical thread is commonly termed as the half-angle of thread.

(13) *Pitch.* The pitch of a thread is the distance, measured parallel to the axis of the thread, between corresponding points on adjacent thread forms in the same axial plane and on the same side of axis. The basic pitch is equal to the lead divided by the number of thread starts. On drawings of thread sections, the pitch is shown as the distance from the centre of one thread crest to the centre of the next, and this representation is correct for single start as well as multi-start threads.

(14) *Lead.* Lead is the axial distance moved by the threaded part, when it is given one complete revolution about its axis with respect to a fixed mating thread. It is necessary to distinguish between measurement of lead from measurement of pitch, as uniformity of pitch measurement does not assure uniformity of lead. Variations in either lead or pitch cause the functional or virtual diameter of thread to differ from the pitch diameter.

(15) *Thread per inch.* This is the reciprocal of the pitch in inches.

(16) *Lead angle.* On a straight thread, lead angle is the angle made by the helix of the thread at the pitch line with plane perpendicular to the axis. The angle is measured in an axial plane.

(17) *Helix angle.* On straight thread, the helix angle is the angle made by the helix of the thread at the pitch line with the axis. The angle is measured in an axial plane.

(18) *Depth of thread.* This is the distance from the crest or tip of the thread to the root of the thread measured perpendicular to the longitudinal axis or this could be defined as the distance measured radially between the major and minor cylinders.

(19) *Axial thickness.* This is the distance between the opposite faces of the same thread measured on the pitch cylinder in a direction parallel to the axis of thread.

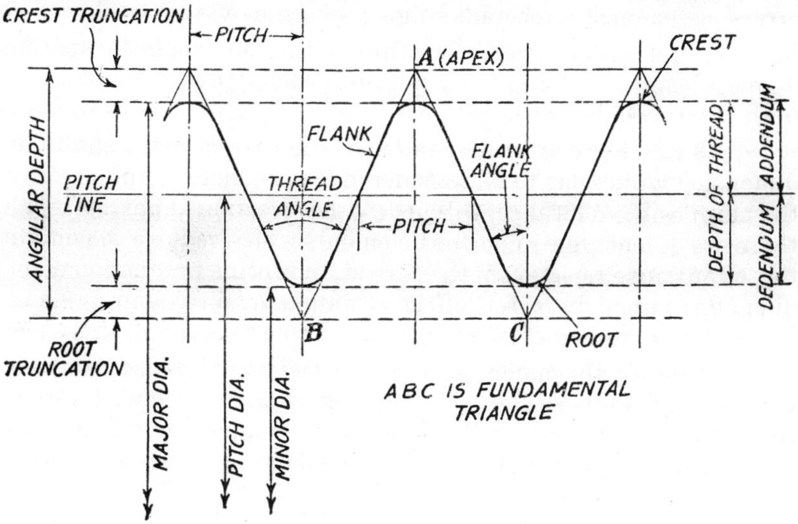

Fig. 13.2. Screw-thread terms relating to external thread.

(20) *Fundamental triangle.* This is found by extending the flanks and joining the points B and C. Thus in Fig. 13.2, triangle ABC is referred to as fundamental triangle. Here BC = pitch and the vertical height of the triangle is called the angular or theoretical depth. The point A is the apex of the triangle ABC.

(21) *Truncation.* A thread is sometimes truncated at the crest or at the root or at both crest and root. The truncation at the crest is the radial distance from the crest to the nearest apex of the fundamental triangle. Similarly the truncation at the root is the radial distance from the root to the nearest apex.

(22) *Addendum.* For an external thread, this is defined as the radial distance between the major and pitch cylinders. For an internal thread this is the radial distance between the minor and pitch cylinders.

(23) *Dedendum.* This is the radial distance between the pitch and minor cylinder for external thread, and for internal thread, this is the radial distance between the major and pitch cylinders.

(24) *Major diameter.* In case of a straight thread, this is the diameter of the major cylinder (imaginary cylinder, co-axial with the screw, which just touches the crests of an external thread or the root of an internal thread). It is often referred to as the outside diameter, crest diameter or full diameter of external threads.

(25) *Minor diameter.* In case of straight thread, this is the diameter of the minor cylinder (an imaginary cylinder, co- axial with the screw which just touches the roots of an external thread or the crest of an internal thread). It is often referred to as root diameter or cone diameter of external threads.

(26) *Effective diameter or pitch diameter.* In case of straight thread, this is the diameter of the pitch cylinder (the imaginary cylinder which is co-axial with the axis of the screw, and intersects the flank of the threads in such a way as to make the width of threads and width of the spaces between the threads equal). If the pitch cylinder be imagined as generated by a straight line parallel to the axis of screw, that straight line is then referred to as the pitch line. Along the pitch line, the widths of the threads and the widths of the spaces are equal on a perfect thread. This is the most important dimension at it decides the quality of the fit between the screw and the nut.

(27) *Functional (virtual) diameter.* For an external or internal thread, this is the pitch diameter of the enveloping thread of perfect pitch, lead and flank angles having full depth of engagement but clear at crests and roots. This is defined over a specified length of thread. This may be greater than the simple effective diameter by an amount due to errors in pitch and

angle of thread. The virtual diameter being the modified effective diameter by pitch and angle errors, is the most important single dimension of a screw thread gauge.

In the case of taper screw thread, the cone angle of taper for measurement of effective diameter, and whether pitch is measured along the axis or along the pitch cone generator also need to be specified.

13.2.1. Errors in Threads. In the case of plain shafts and holes, there is only one dimension which has to be considered (*i.e.* diameter), and errors on this dimension if exceed the permissible tolerance, will justify the rejection of part. While in the case of screw threads there are at least five important elements which require consideration and error in any one of these can cause rejection of the thread. In routine production all of these five elements (major diameter, minor diameter, effective diameter, pitch and angle of the thread form) must be checked and method of gauging must be able to cover all these elements.

Errors on the major and minor diameters will cause interference with the mating thread. Due to errors in these elements, the root section and wall thickness will be less, also the flank contact will be reduced and ultimately the component will be weak in strength. Errors on the effective diameter will also result in weakening of the assembly due to interference between the flanks.

Similarly pitch and angle errors are also not desirable as they cause a progressive tightening and interference on assembly. These two errors have a special significance as they can be precisely related to the effective diameter.

Now we will consider come errors in detail and define some terms.

13.2.2. Drunken Thread. This is the one having erratic pitch, in which the advance of the helix is irregular in one complete revolution of the thread.

Thread drunkenness is a particular case of a periodic pitch error recurring at intervals of one pitch. In such a thread, the pitch measured parallel to the thread axis will always be correct, the only error being that the thread is not cut to a true helix. If the screw thread be regarded as an inclined plane wound around a cylinder and if the thread be unwound from the cylinder, (*i.e.* development of the thread be taken) then the drunkenness can be visualised. The helix will be a curve in the case of drunken thread and not a straight line as shown in Fig. 13.3.

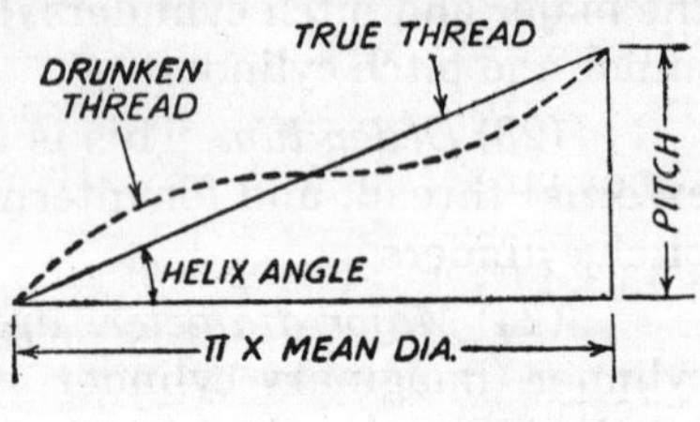

Fig. 13.3. Drunken thread.

It is very difficult to determine such errors and moreover they do not have any great effect on the working unless the thread is of very large size.

13.2.3. Pitch Errors in Screw Threads. Generally the threads are generated by a point cutting tool. In this case, for pitch to be correct, the ratio of the linear velocity of tool and angular velocity of the work must be correct and this ratio must be maintained constant, otherwise pitch errors will occur. If there is some error in pitch, then the total length of thread engaged will be either too great or too small, the total pitch error in overall length of the thread being called the cumulative pitch error. Various pitch errors can be classified as :

13.2.3.1. Progressive Pitch Error. This error occurs when the tool work velocity ratio is incorrect though it may be constant. It can also be caused due to pitch errors in the lead screw of the lathe or other generating machine.

The other possibility is by using an incorrect gear or an approximate gear train between work and lead screw *e.g.,* while metric threads are cut with an inch pitch lead screw and a translatory gear is not available. A graph between the cumulative pitch error and the length of thread is generally a straight line in case of progressive pitch error (Fig. 13.4).

13.2.3.2. Periodic Pitch Error. This repeats itself at regular intervals along the thread. In this case, successive portions of the thread are either longer or shorter than the mean. This type of error occurs when the tool work velocity ratio is not constant. This type of error also results when a thread is cut from a lead screw which lacks squareness in the abutment causing the leadscrew to move backward and forward once in each revolution. Thus the errors due to these cases are cyclic and pitch increases to a maximum, then reduces through normal value to a minimum and so on. The graph between the cumulative pitch error and length of threads for this error will, therefore, be of sinusoidal form.

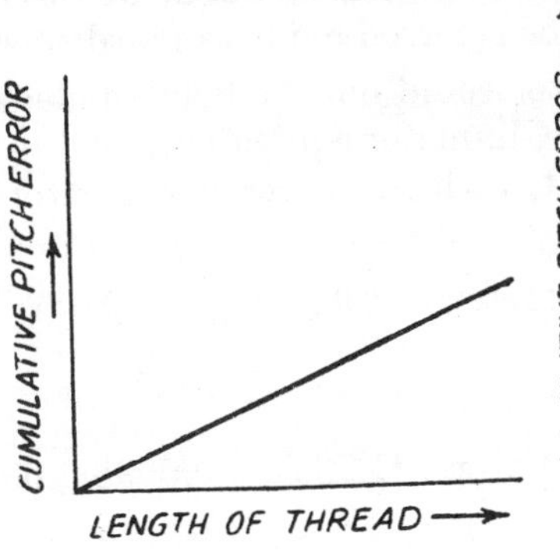

Fig. 13.4. Progressive Error.

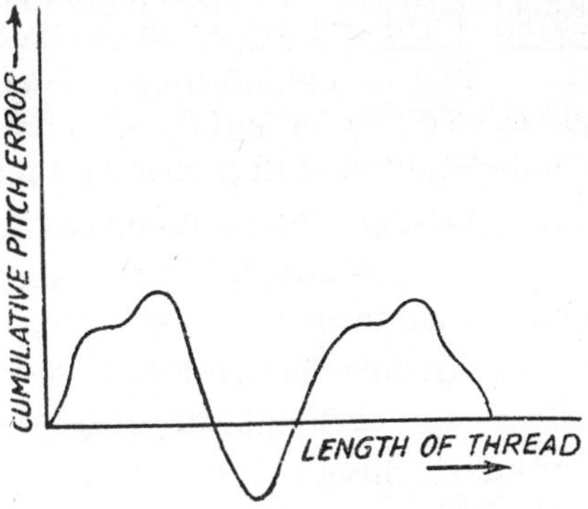

Fig. 13.5. Periodic Error.

13.2.3.3. Irregular Errors. These arise from disturbances in the machining set-up, variations in the cutting properties of material etc. Thus they have no specific causes and correspondingly no specific characteristics also. These errors could be summarised as follows :

Erratic Pitch. This is the irregular error in pitch and varies irregularly in magnitude over different lengths of thread.

Progressive Error. When the pitch of a screw is uniform, but is shorter or longer than its nominal value, it is said to have progressive error.

Periodic Error. If the errors vary in magnitude and recur at regular intervals, when measured from thread to thread along the screw are referred to as periodic errors.

13.3. Effect of Pitch Errors

An error in pitch virtually increases the effective diameter of a bolt or screw and decreases the effective diameter of a nut. The meaning of the virtual change in effective diameter is that if any screw is perfect except for pitch error, it will not screw easily into a perfect ring gauge of same nominal size until its effective diameter is reduced.

For Whitworth thread, if δp is the error in pitch then the virtual increase (decrease) in the effective diameter of the thread in case of bolt (nut) is given by the relation :

Virtual change in effective diameter = $1.921 \times \delta p$. Similarly errors in flank angles also require a corresponding reduction in the effective diameter if the screw is to fit a perfect ring gauge of the same nominal size.

It $\delta\theta_1$ and $\delta\theta_2$ are the errors in flank angles in degrees (regardless of sign), the corresponding virtual change (increase or decrease) in effective diameter of the thread in case of a bolt or nut is given by (for Withworth thread) $\delta E = 0.0105 \times p\ (\delta\theta_1 + \delta\theta_2)$, where p is the normal pitch.

Thus it means that the error in pitch and angle can be accounted for by a suitable alternation of the effective diameter. There is however a limit to the magnitude of these errors for accurate work. This is done by fixing a maximum limit for the equivalent of the combination of pitch and angle errors in terms of the effective diameter of the gauge. Thus for Whithworth thread, a maximum value of the effective diameter equivalent $[1.921 \times \delta p + 0.0105 \times p\,(\delta\theta_1 + \delta\theta_2)]$ is fixed and these values can be referred to in standard specifications.

The virtual effective diameter could now be defined clearly as the sum of the simple effective diameter and the effective diameter equivalent, as it is the effective diameter of the smallest nut of perfect form and pitch which the screw will enter *e.g.*, for Whitworth threads :

Virtual effective diameter

$$= \text{Simple effective diameter} + [1.921 \times \delta p + 0.0105p\,(\delta\theta_1 + \delta\theta_2)]$$

For Unified thread

Virtual effective dia

$$= \text{Simple effective diameter} + [1.73\,\delta p + 0.0098p\,(\delta\theta_1 + \delta\theta_2)]$$

For metric thread.

Virtual effective dia = Simple effective diameter + $[1.722\,\delta p + 0.0131p\,(\delta\theta_1 + \delta\theta_2)]$.

13.3.1. Mathematical Derivation for Effect of Pitch Errors. Let us imagine a perfect bolt, having some pitch error and it has to enter a nut of perfect form and pitch (Fig. 13.6). It will not be possible without experiencing a lot of strain, as the error has to be accommodated by the strain.

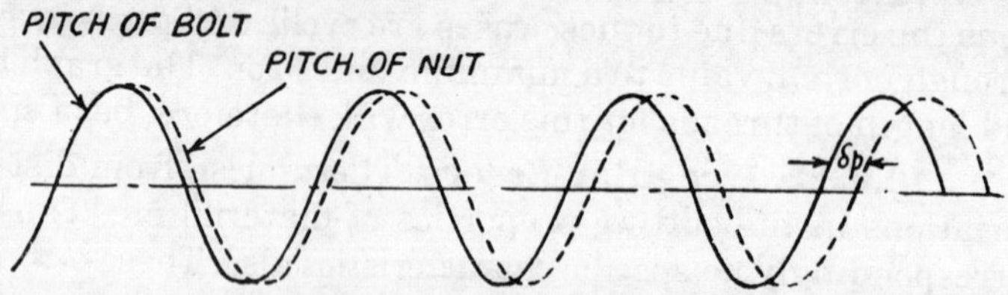

Fig. 13.6. Effect of pitch error.

The other way out is to increase the effective diameter of the nut. Let δp be the cumulative pitch error over the length of the engagement.

By increasing the effective diameter of the nut and retaining the same pitch, the two threads can assemble without interference as shown in Fig. 13.7.

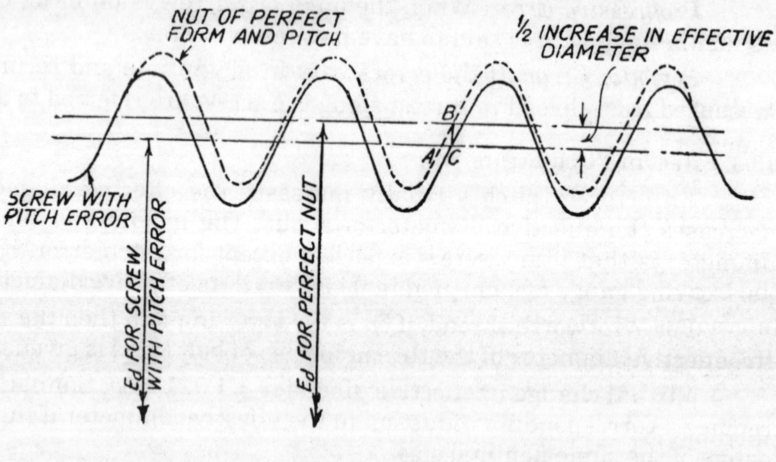

Fig. 13.7. Pitch error is accommodated by increasing virtual diameter.

It is assumed that the maximum pitch error over the length of engagement is equally distributed at each end of engagement. Increase in effective diameter will obviously be the vertical movement of flanks necessary to produce coincidence.

It may be mentioned here that effect of long or short pitch will be same, *i.e.* increase of the interference between the mating threads, so each will lead to increase in effective diameter of nut.

In ΔABC (Fig. 13.7) $\qquad A\hat{B}C = \theta$ = half the angle of thread

$\therefore \qquad \cot\theta = BC/AC = (\delta E_d/2)/(\theta_p/2)$, or $\delta E_d = \delta p \cot\theta$

$\therefore$ Increase in effective diameter $= \delta p \times \cot\theta$.

Since cot 55°/2 = 1.921 (for Whitworth), its effect is nearly doubled when the equivalent increase in effective diameter is calculated.

Similarly the effect of pitch error will be to reduce the effective diameter of the screw.

13.3.2. Angle Errors. Angle errors on threads may be either due to errors on one or both flanks. Any error in angle of thread results in interference between the bolt and nut and to accommodate it, the effective diameter of nut has to be increased. Thus like pitch errors, the angle errors also increase the virtual effective diameter of a bolt and decrease that of a nut. Assuming that one of the pairs is correct, it is possible to satisfactorily assemble the thread pairs by modifying the effective diameter. The effective diameter of an incorrect bolt must be decreased to permit a correct mating thread to mate and similarly the effective diameter of an incorrect nut must be increased.

If $\delta\theta_1 + \delta\theta_2$ be equivalent to the errors in the adjoining flank angles of any thread, then the corresponding correction = $Cp\ (\delta\theta_1 + \delta\theta_2)$

where C = 0.0100 for Unified thread = 0.0105 for Whitworth thread, = 0.0091 for British Association thread, = 0.0115 for ISO metric thread, p = basic pitch of thread,

and $\delta\theta_1 + \delta\theta_2$ = sum of errors in adjacent flank angles in degrees (regardless of signs of the errors).

13.3.3. Diametral errors. Errors of major, minor and pitch diameter and their mutual non-concentricity give rise to interference and strain in the joint. More force is required for fitting.

13.4. Measurement of Various Elements of Thread

The methods discussed hear are from the point of view of measurement of gauges, but they can obviously be applied to precise work, threading tools, taps and hobs etc. We will be dealing with the measurement of most important six elements, *i.e.* major, minor and effective diameters, pitch, angle and form of thread.

13.4.1. Measurement of major diameter. For the measurement of major diameter of external threads, a good quality hand micrometer is quite suitable. In taking readings, a light pressure must be used as the anvils make contact with the gauge at points only and otherwise the errors due to compression can be introduced. It is, however, also desirable to check the micrometer reading on a cylindrical standard of approximately the same size, so that the zero error etc., might not come into picture.

For greater accuracy and convenience, the major diameter is measured by bench micrometer. This instrument was designed by N.P.L. to estimate some deficiencies inherent in the normal hand micrometer. It uses constant measuring pressure and with this machine the error due to pitch error in the micrometer thread is avoided. In order that all measurements be made at the same pressure, a fiducial indicator is used in place of the fixed anvil. In this machine there is no provision for mounting the workpiece between the centres and it is to be held in hand. This is so, because, generally the centres of the workpiece are not true with its diameter. This machine is used as a comparator in order to avoid any pitch errors of micrometers, zero error setting etc. A calibrated setting cylinder is used as the setting standard.

The advantage of using cylinder as setting standard and not slip gauges etc., is that it gives greater similarity of contact at the anvils. The diameter of the setting cylinder must be nearly same as the major diameter. The cylinder is held and the reading of the micrometer is noted down. This is then replaced by threaded workpiece and again micrometer reading is noted for the same reading of fiducial indicator. Thus, if the size of cylinder is approaching that of major diameter, then for a given reading the micrometer thread is used over a short length of travel and any pitch errors it contains are virtully eliminated.

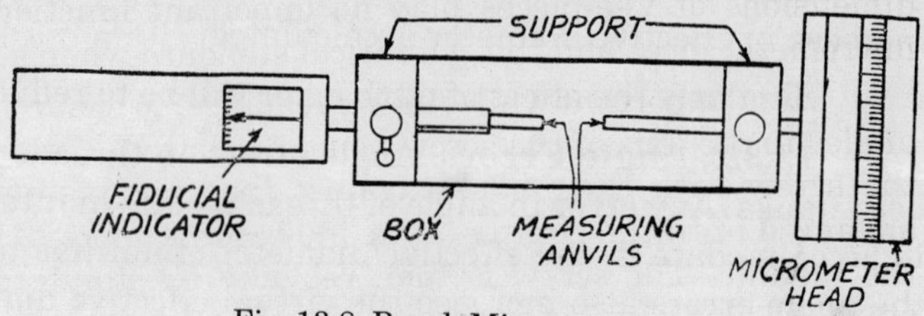

Fig. 13.8. Bench Micrometer.

If D_1 = diameter of setting cylinder.

R_1 = reading of micrometer on setting cylinder.

R_2 = micrometer reading on thread,

then major diameter = $D_1 + (R_2 - R_1)$.

In order to determine the amount of taper, the readings should be taken at various positions along the thread and to detect the ovality, two or three readings must be taken at one plane in angular positions.

13.4.2. Major diameter of internal threads. The measurement of the elements of an internal thread is more cumbersome. Since it is difficult to approach the elements of internal thread, an indirect approach is followed by making a cast of the thread. The main art thus lies in obtaining a perfect cast, because once good cast is available the various elements can be measured as for external threads.

Cast may be made by plaster of paris, dental wax, or sulphur. The part whose internal thread is to be measured is first cleaned and brushed with a fine oil. The part is then mounted between two wooden blocks whose upper surface lie about half-way up the ring. Cast material is then poured to a depth less than the radius of part to permit easy removal of cast without screwing it out. After the plaster is set, it should be taken out without rotating, but by pulling up the middle portion of the cast. It may be mentioned that taking out of sulphur cast is easier than the plaster. Oiling is not necessary in case of sulphur cast.

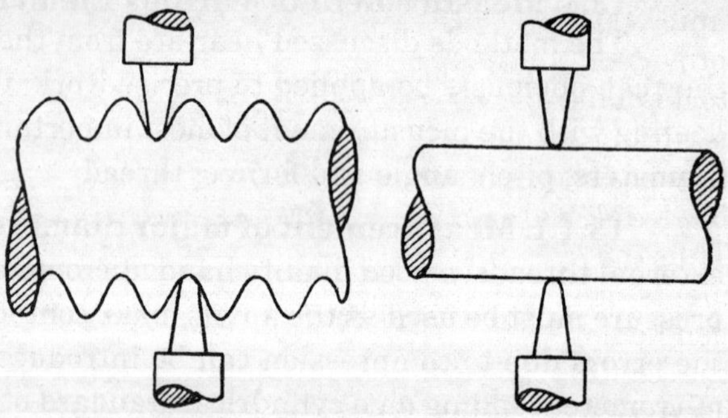
Fig. 13.9. Measurement of minor diameter.

13.4.3. Measurement of minor diameter. This is also measured by a comparative process using small Vee-pieces which make contact with a root of the thread. The Vee-pieces are available in several sizes having suitable radii at the edges. The included angle of Vee-pieces is less than the angle of the thread to be checked so that it can easily probe to the root of the thread. To measure the minor diameter by Vee-pieces is suitable for only Whitworth and B.A. threads which have a definite radius at the root of the thread. For other threads, the minor diameter is measured by the projector or microscope.

The measurement is carried out on a floating carriage diameter measuring machine in which the threaded work-piece is mounted between centres and a bench micrometer is constrained to move at right angles to the axis of the centre by a Vee-ball slide. The method of the application of Vee-pieces in the machine is shown diagrammatically in Fig. 13.9. The dimensions of Vee-pieces play no important function as they are interposed between the micrometer faces and the cylindrical standard when standard reading is taken.

It is important while taking readings, to ensure that the micrometer be located at right angles to the axis of the screw being measured. The selected Vees are placed on each side of the screw with their bases against the micrometer faced. The micrometer head is then advanced until the pointer of the indicator is opposite the zero mark, and note being made of the reading. The screw is then replaced by standard reference disc or a plain cylindrical standard plug gauge of approximately the core diameter of the screw to be measured and second reading of the micrometer is taken.

If reading on setting cylinder with Vee-pieces in position = R_1 and reading on thread = R_2 and diameter of setting cylinder = D_1, Then minor diameter = $D_1 + (R_2 - R_1)$.

Readings may be taken at various positions in order to determine the taper ovality.

Before proceeding to the measurement of effective diameter, the screw diameter measuring-machine is first described in brief here. The machine is shown in Fig. 13.10. Also refer Fig. 13.16 for schematic sketch. If consists of three main units. A base casting carries a pair of centres, on which the threaded work-piece is mounted. Another carriage is mounted on it and is exactly at 90° to it. On this is provided another carriage capable of moving towards the centres. On this carriage one head having a large thimble enabling reading upto 0.002 mm is provided. Just opposite to it is a fixed anvil which is spring loaded and its zero position is indicated by a fiducial indicator. Thus the micrometer elements are exactly perpendicular to the axis of the centres as the two carriages are located perpendicular to each other. On the fixed carriage the centres are supported in two brackets fitted on either end. The distance between the two centres can be adjusted depending upon the length of the threaded job. After job is fitted between the centres the second carriage is adjusted in correct position to take measurements and is located in position. The third carriage is then moved till the fiducial indicator is against the set point. The readings are noted from the thimble head. It is now obvious that the axis of the indicator and micrometer head spindle is same and is perpendicular to the line of two centres. The indicator is specially designed for this class of work and has only one index line, against which the pointer is always to be set. This ensures constant measuring pressure for all readings. Sufficient friction is provided by the conical pegs to restrain the movement of carriage along the line of centres. The upper carriage is free to float on balls and enables micrometer readings to be taken on a diameter without restraint. Squareness of the micrometer to the line of centres can be adjusted by rotating the pegs in the first carriage which is made eccentric in its mounting.

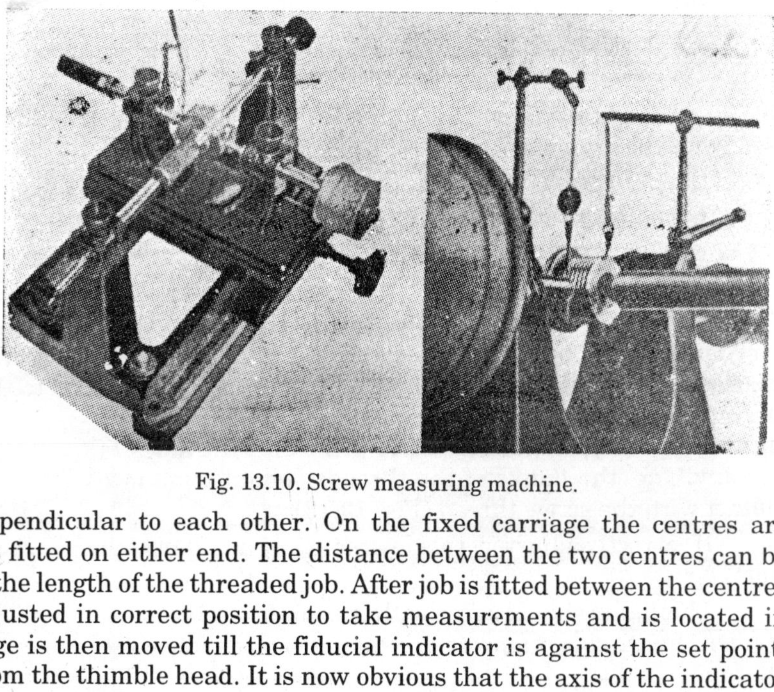

Fig. 13.10. Screw measuring machine.

Above the micrometer carriage, two supports are provided for supporting the wires and Vee-pieces for measurement of effective diameter etc.

13.4.4. Minor diameter of internal threads. Minor diameter of internal threads can be measured conveniently by the following methods :

(*i*) **Using taper parallels.** The taper parallels are pairs of wedges having radiused and parallel outer edges. The diameter across their outer edges can be changed by sliding them over each other as shown in Fig. 13.11. The taper parallels are inserted inside the thread and adjusted until firm contact is established with the minor diameter. The diameter over the outer edges is measured with a micrometer. This method is suitable for smaller diameter threads.

(*ii*) **Using rollers.** For threads bigger than 10 mm diameter, precision rollers are inserted inside the thread and proper slip gauge inserted between the rollers as shown in Fig.

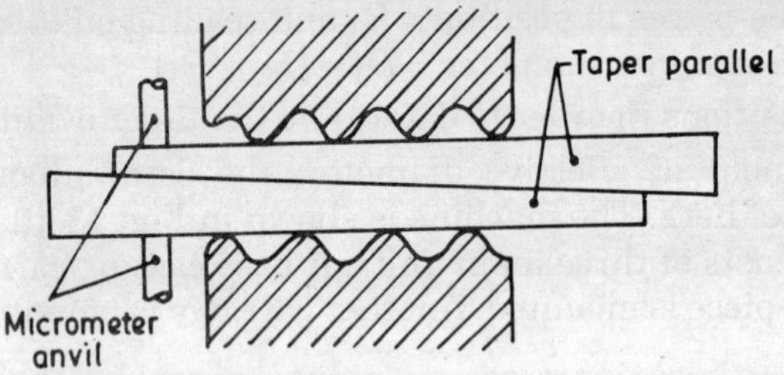

Fig. 13.11. Measuring minor diameter of internal thread using taper parallels.

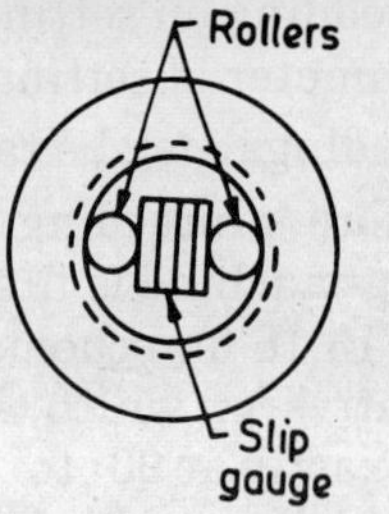

Fig. 13.12. Measurement of minor diameter of internal thread using rollers.

13.12, so that firm contact is obtained. The minor diameter is then the length of slip gauges plus twice the diameter of rollers.

13.4.5. Effective diameter measurements. The effective diameter or the pitch diameter can be measured by any one of the following methods :

(*i*) Micrometer method

(*ii*) One wire, two wire, or three wire (or rod) method.

13.4.5.1. Thread micrometer method. The thread micrometer resembles the ordinary micrometer, but it has special contacts to suit the end screw thread form that is to be checked. In this micrometer, the end of the spindle is pointed to the Vee-thread form with a corresponding Vee-recess in the fixed anvil. When measuring threads only, the angle of the point and the sides of Vee-anvil, *i.e.* the flanks of the threads should come into contact with the screw thread (Fig. 13.13).

If correctly adjusted, this micrometer gives the pitch diameter.

This value should agree with that obtained by measurement by outside diameter and pitch from the following relation :

Pitch dia $= D - 0.6403p$ (in case of Whitworth thread)

where $0.6403p$ = depth of thread, D = outside dia., p = pitch.

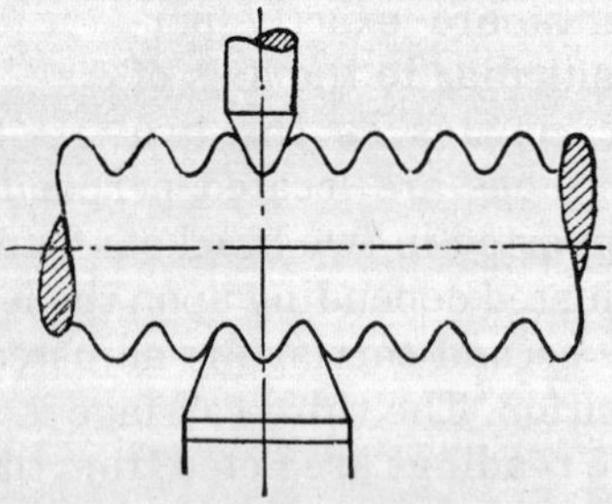

Fig. 13.13. Thread micrometer for effective diameter.

Limitations of Thread Micrometer : The micrometer must be set to a standard thread plug. If not done so in the first instance, there will be error due to helix angle of the thread being measured. When setting the instrument to a standard plug gauge it will be observed that the reading is not exactly zero, as previously inferred, when the spindle and anvil are brought together.

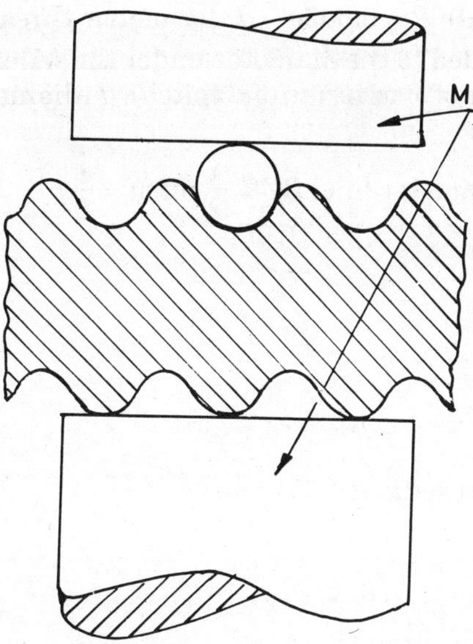

Fig. 13.14. One-wire method.

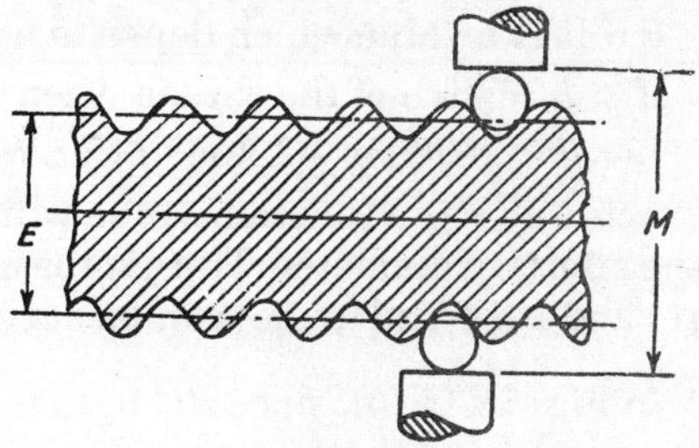

Fig. 13.15 (*a*)

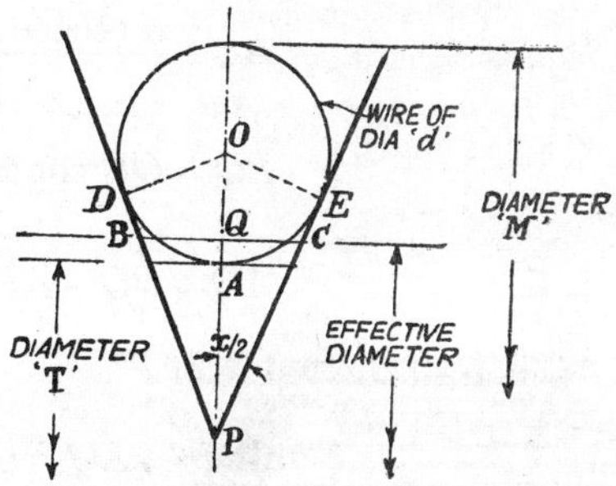

Fig. 13.15 (*b*). Two wire method.

For correct results it is necessary to use a separate thread micrometer for every size of screw thread to be gauged, otherwise there will always be a small amount of error inherent in thread micrometer.

A big advantage of thread micrometer is that this is the only method which shows the variation for the drunken thread.

13.4.5.2. One-wire method. In this method, one wire is placed between two threads at one side and on the other side the anvil of the measuring micrometer contacts the crests as shown in Fig. 13.14. First the micrometer reading is noted on a standard gauge whose dimension is nearly same as to be obtained by this method. Actual measurement over wire on one side and threads on other side = size of gauge ± difference in two micrometer readings.

This method is used for measuring effective diameter of counter pitch threads, and during manufacture of threads.

The difficulty with this method is that the micrometer axis may not remain exactly at right angles to the thread axis.

13.4.5.3. Two-wire method. The effective diameter of a screw thread may be ascertained by placing two wires or rods of identical diameter between the flanks of the thread, as shown in Fig. 13.15, and measuring the distance over the outside of these wires. The effective diameter E is then calculated as

$E = T + P$, where T = Dimension under the wires = $M - 2d$

M = dimension over the wires, d = diameter of each wire

The wires used are made of hardened steel to sustain the wear and tear in use. These are given a high degree of accuracy and finish by lapping to suit different pitches.

Dimension T can also be determined by placing wires over a standard cylinder of diameter greater than the diameter under the wires and noting the reading R_1 and then taking reading with wires over the gauge, say R_2. Then $T = S - (R_1 - R_2)$.

P = It is a value which depends upon the dia of wire and pitch of the thread.

If p = pitch of the thread, then

$P = 0.9605p - 1.1657d$ (for Whitworth thread), $P = 0.866p - d$ (for metric thread).

Actually P is a constant value which has to be added to the diameter under the wires to give the effective diameter. The expression for the value of P in terms of p (pitch), d (diameter of wire) and x (thread angle) can be derived as follows :

In Fig. 13.15 (b), since BC lies on the effective diameter line, $BC = \frac{1}{2}$ pitch $= \frac{1}{2}p$

$$OP = \frac{d \operatorname{cosec} x/2}{2}$$

$$PA = \frac{d (\operatorname{cosec} x/2 - 1)}{2}$$

$$PQ = QC \cot x/2 = \frac{p}{4} \cot x/2$$

$$AQ = PQ - AP = \frac{p \cot x/2}{4} - \frac{d (\operatorname{cosec} x/2 - 1)}{2}$$

AQ is half the value of P

$$\therefore \qquad P \text{ value} = 2AQ = \frac{p}{2} \cot \frac{x}{2} - d \left(\operatorname{cosec} \frac{x}{2} - 1 \right)$$

Two wire method can be carried out only on the diameter measuring machine described for measuring the minor diameter, because alignment is not possible by 2 wires and can be provided only by the floating carriage machine. In the case of three wire method, 2 wires on one side help in aligning the micrometer square to the thread while the third placed on the other side permits taking of readings.

A simplified diagram of this measuring machine is shown in Fig. 13.16. As already pointed out the machine ensures that the axis of the micrometer is maintained at 90° to the axis of the screw under test. In Fig. 13.16, the lower slide (wrongly indicated as lower side in Fig. 13.16) is capable of movement parallel with the axis of thread while the top slide moves at 90° to the thread axis.

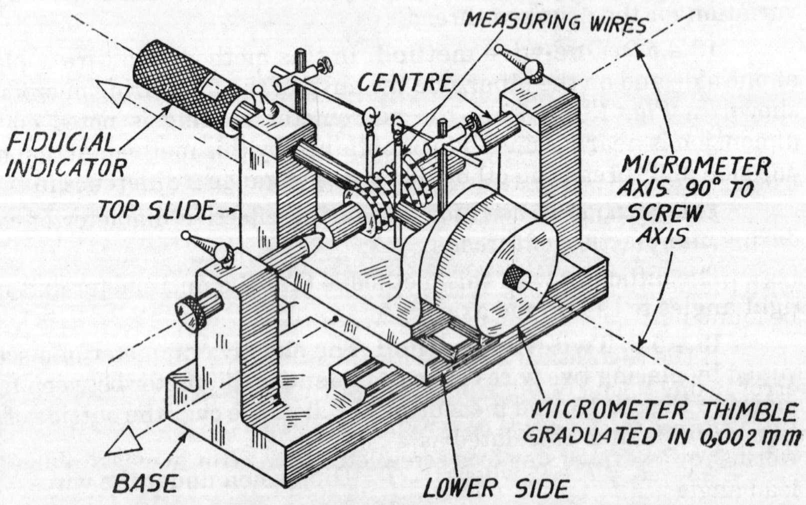

Fig. 13.16. Thread measuring machine.

13.4.5.4. Three wire method. This method of measuring the effective diameter is an accurate method. In this three wires or rods of known diameter are used : one on one side and two on the other side [Fig. 13.17 (a) and (b)]. This method ensures the alignment of micrometer anvil faced parallel to the thread axis. The wires may be either held in hand or hung from a stand so as to ensure freedom to the wires to adjust themselves under micrometer pressure.

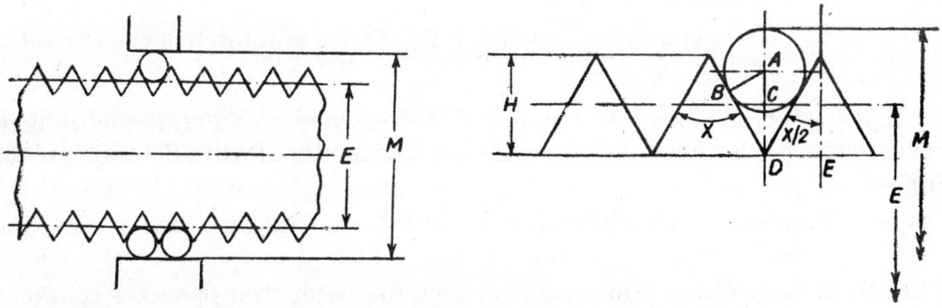

(a) (b)

Fig. 13.17. Three wire method of measuring effective diameter.

M = distance over wires, E = effective diameter, r = radius of the wires, d = diameter of wires, h = height of the centre or the wire or rod from the effective diameter, x = angle of thread.

From Fig. 13.17 (*b*), $AD = AB \text{ cosec } x/2 = r \text{ cosec } x/2$

$$H = DE \cot x/2 = p/2 \cot x/2$$

$$CD = H/2 = p/4 \cot x/2$$

$$h = AD - CD$$

$$r = \text{cosec } x/2 - p/4 \cot x/2$$

Distance over wires $= M = E + 2h + 2r$

$$= E + 2\,(r \text{ cosec } x/2 - p/4 \cot x/2) + 2r$$

$$= E + 2r\,(1 + \text{cosec } x/2) - p/2 \cot x/2$$

or $$M = E + d\,(1 + \text{cosec } x/2) - p/ \cot x/2$$

(*i*) *In case of Whitworth thread :*

$x = 55°$, depth of thread $= 0.64p$, so that, $E = D - 0.64p$ and $\text{cosec } x/2 = 2.1657$, $\cot x/2 = 1.921$

$$M = E + d\,(1 + \text{cosec } x/2) - p/2 \cot x/2 = D - 0.64p + d\,(1 + 2.1657) - p/2\,(1.921)$$

$$= D + 3.1657d - 1.6005p$$

$M = D + 3.1657d - 1.6p$, where D = outside dia.

(*ii*) *In case of metric threads* : Depth of thread $= 0.6495p$.

so, $E = D - 0.6495p$, $x = 60°$, $\text{cosec } x/2 = 2$; $\cot x/2 = 1.732$

$$M = D - 0.6495\,p + d\,(1 + 2) - p/2\,(1.732) = D + 3d - (0.6495 + 0.866)\,p = D + 3d - 1.5155p.$$

We can measure the value of M practically and then compare with the theoretical values with the help of formulae derived above. After finding correct value of M and knowing d, E can be found out.

If the theoretical and practical values of M (*i.e.* measured over wires) differ, then this error is due to one or more of the quantities appearing in the formula.

Effect of lead angle on measurement by 3-wire method. If the lead angle is large (as with worms ; quick traversing lead screw, etc.) then error in measurement is about 0.0125 mm when lead angle is 4.5° for 60° single thread series.

For lead angles above 4.5° compensation for rake and compression must also be considered.

There is no recommendation for B.S.W. threads.

Rake Correction in U.S. Standard :

$$E = M + \frac{\cot x/2}{2n} - x\left(1 + \operatorname{cosec}\frac{x}{2} + \frac{S^2}{2}\cos\frac{x}{2}\cot\frac{x}{2}\right)$$

where $x/2$ = half the included angle of threads, E = effective diameter, M = actually measured diameter over wires, n = number of threads/inch, d = diameter of wire, S = tangent of the helix angle in thread.

(Also refer Problem 13.5 at the end of this chapter for rake correction and compression correction).

13.4.6. Best Size Wire. This wire is of such diameter that it makes contact with the flanks of the thread on the effective diameter or pitch line. Actually effective diameter can be measured with any diameter wire which makes contact on the true flank of the thread, but the values so obtained will differ from those obtained with 'best size' wires if there is any error in angle or form of thread. It is recommended that for measuring the effective diameter, always the best size wire should be used and for this condition the wire touches the flank at mean diameter line within ± 1/5 of flank length (Refer Solved Problem 13.2). With best size wire, any error on the measured value of simple effective diameter due to error in thread form or angle is minimised.

It can be shown that size of best wire diameter = $d = \dfrac{p}{2\cos x/2}$

[Refer Solved Problem 13.1 at the end of this chapter]

With best size wire, P-value = $d\,(\operatorname{cosec} x/2 + 1) - d\cos x/2 \cot x/2$

$$= d\left(\frac{1 + \sin x/2 - \cos^2 x/2}{\sin x/2}\right) = d\,(1 + \sin x/2) = \frac{p}{2}\cdot\frac{1 + \sin x/2}{\cos x/2}.$$

13.4.7. Measurement of Effective Diameter of Tapered Threads. The measurement of the effective diameter of taper threads is *not made perpendicular to the axis,* but at an angle depending on the taper. The measurement is made at a given point or distance from the end of the thread, and in the three wire method, the single wire is placed at this point. The other two wires are placed in the two opposite grooves and care must be taken to ensure that the micrometer or *measuring anvils make contact with each of the three wires.*

The formula for the effective diameter of the taper thread is :

$$E = (M - d)\sec h + \frac{\cot x/2}{2n} - d\operatorname{cosec} x/2$$

where E = effective diameter, M = measurement over the wires, d = diameter of the wires, h = half the angle of taper, $x/2$ = half the included angle of the thread form, n = number of threads per inch.

13.4.8. Effective Diameter of Internal Threads (Fig. 13.18).

Thread comparator. In this case a pair of ball tips engage the flanks of the threads in the work and measure the effective diameter only.

The ball tip on the right is fixed at the end of a measuring jaw attached to a floating head in the sliding bracket (B). The floating head has extension in contact with the spindle of the dial indicator and the movement of floating head towards the indicator is constrained by a spring. (The set-up in Fig. 13.18 does not show the ball tips).

The instrument is set to a reference standard, with the dial pointer at zero. To use the gauge, the floating head is retracted to insert the ball tips in the internal threads of the work, and released to allow the tips to engage the flanks of the thread under the pressure of the spring. The dial indicator then shows the deviation from the nominal size to which the gauge is set. The instrument may be used on work in the machine, or on the working bench. The fixed head (A) carrying the left hand ball tip is adjusted by a fine screw to set the gauge to the reference standard. The reference standard is built up from slip gauges as shown in Fig. 13.19. The two end pieces have V-jaws of an angle of Vee corresponding to the thread, *i.e.* 60 degree or 55 degree.

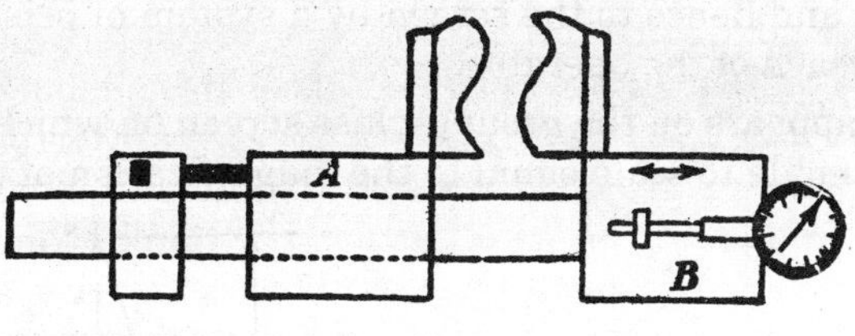

Fig. 13.18

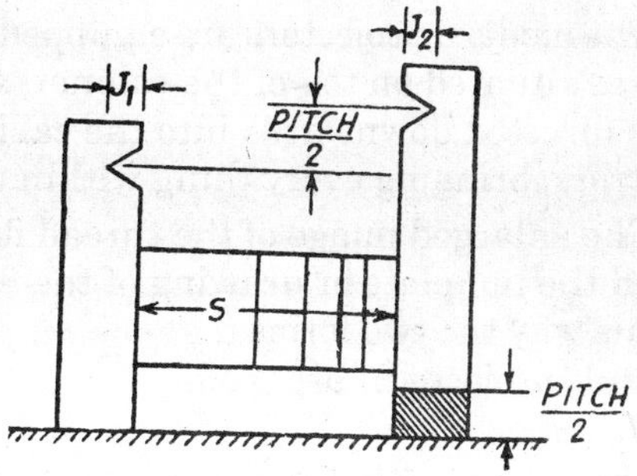

Fig. 13.19

The dimensions J_1 and J_2 are marked on the pieces, and are the depths from the face to the apex points of the Vees. Assuming the effective diameter and pitch of the thread to be known, the distance S is found from the formula

$$S = X + y - Z$$

where X = Mean effective diameter.

y = Depth of the thread from apex to the apex of the V-form.

The value of y (Fig. 13.20) depends on the included angle of the thread, and is equal to $0.9605p$ for 55° threads and $0.866p$ for 60° threads.

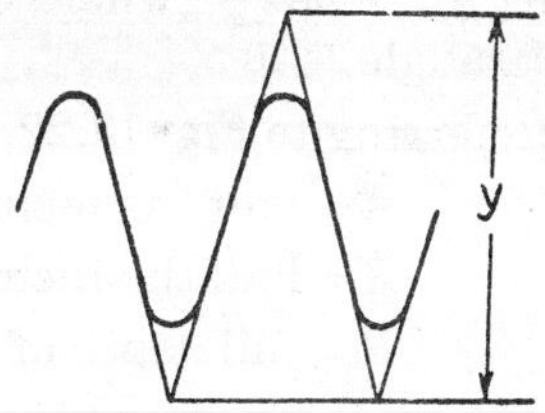

Fig. 13.20

$Z = J_1 + J_2$ *i.e.*, constants for the end gauge pieces.

The assembled slips are set in a holder with a slip equal to half the pitch, beneath one end piece to compensate for the helix angle.

The reference gauge thus assembled is ready for setting the comparator. Ball tips must be of a suitable size for the thread. The size is not critical provided the ball point fist the thread so as to bear on the flack near the mean pitch line.

For threads from 4 to 7 t.p.i., a ball of 0.095 inch dia is used, from 7 to 12 t.p.i., 0.060 inch diameter and from 12 to 20 t.p.i., 0.035 inch diameter balls are used. A pair of V-jaws, 55° or 60° covers all pitches from 4 to 20 t.p.i. The method of calculating the value of S from the effective diameter excluded the radius OY at the crest and root of the thread, (Fig. 13.21) as the form is considered to extend to the apex of the Vee. In some cases it may be necessary to accept the major diameter as it may be the basic dimension of the thread, and the form at the crest and root of the thread must then be taken into account.

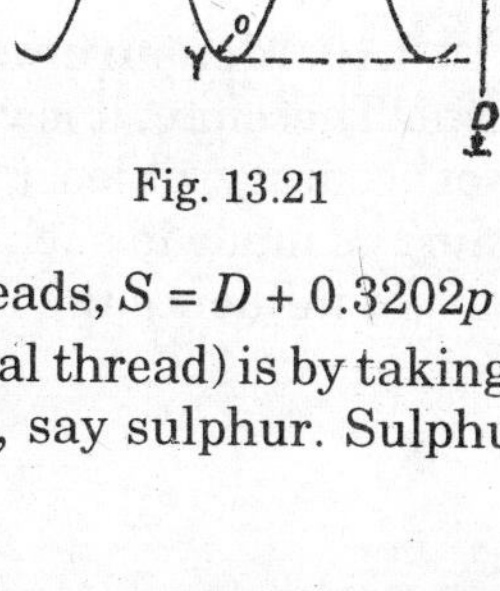

Fig. 13.21

For metric threads, $S = D + 0.2165p - Z$; For Whitworth threads, $S = D + 0.3202p - Z$.

(*iii*) Layman's method of finding the effective diameter (internal thread) is by taking the impression of threads with the help of wax or any other material, say sulphur. Sulphur is mostly used because it can be used many times.

13.4.9. Checking the "Thread Form" and "Angle" by Optical Projection of Thread. This method is applicable only to external threads because internal threads cannot be projected.

The standard type of projector is used, consisting of a projector lamp, a condenser lens or collimator, the projection lens and the screen.

The screw thread to be examined is placed in the parallel beam of light between the condenser lens and the projector lens.

The modern projectors are equipped with work holding fixtures, the projection lamp and the lenses situated on top of the cabinet, and the screen at the front. The light rays from the lens are directed downwards into the cabinet, and hence to the screen by a system of prisms and mirrors, bringing every thing within the reach of the operator.

The enlarged image of the thread form appears on the ground-glass screen on which is mounted the template or drawing of the form made to scale equal to the magnification of the lens. This way the two forms (*i.e.* ideal and projected) are compared.

One of the difficulties in projecting screw thread is the fact that form is specified on an axial plane. So we must consider the correction for it.

Referring to Fig. 13.22 (*i*), the normal pitch p is less than the axial pitch P and is given by the relation : $p = P \cos \phi$; where ϕ is the helix angle.

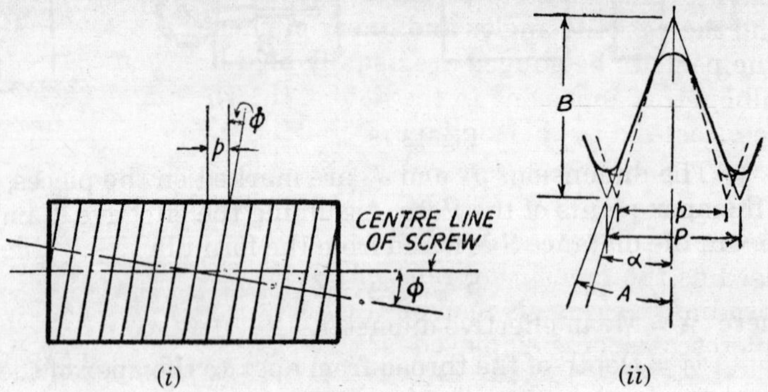

Fig. 13.22

Referring to Fig. 13.22, (*ii*),

If A = half the included angle of thread on the axial plane,

X = half the included angle of the thread on normal plane.

and B = full depth of thread to apex.

$$\tan A = \frac{0.5P}{B} \; ; \; \tan X = \frac{0.5P \cos \phi}{B} \left(\text{or } \tan X = \frac{0.5p}{B} \right)$$

or we can say : $\tan X = \tan A \cos \phi$.

Values of A and ϕ are known and $2X$ is the included angle of the thread projected on the screen.

On the projected thread form we find the included angle $2X$ and then compare it to the theoretically calculated value. $2X = 2 \tan^{-1} (\tan A \cos \phi)$.

The included angle can also be determined by two ball method (Refer Art. 9.8.3 and Prob. 13.7).

13.4.10. Measurement of Pitch. The accuracy of pitch in any form of thread is very important. Therefore, it is very important to be able to measure this element of thread to high degree of accuracy, at least double that of the effective diameter measurement. The measurement must be made in such a way that other features or dimensions *e.g.*, diameter and thread angle do not influence the result.

13.4.10.1. External Threads. 1. For less accurate methods, the Zeiss pitch or lead measuring instrument may be used. It utilises contact members having two ball points which are applied to the effective surface of the thread. These points are aligned parallel to the thread axis either by a thread pin at the back or a special back rest having a plane face parallel to the thread axis. The instrument is adjusted to zero before making a measurement, with the aid of a special micrometer gauge supplied for the purpose, or by using a standard plug gauge. Upon applying the instrument to the thread it registers the pitch deviation from the standard measurement. The scale of the indicator has a range of ± 0.1 mm and each division reads to 0.01 mm. The measuring accuracy of the indicator is ± 0.003 mm.

2. The pitch of external threads can be measured by using screw pitch or profile gauge. Such a gauge consists of series of thread forms with varying pitch. The one which coincides perfectly with the thread under test gives the pitch. The accuracy of measurement depends on the method of sighting used to judge the perfectness.

3. A more accurate method is the microscope method. Screw threads can be inspected and their profile angles and linear pitches checked with the aid of a goniometric microscope. The parts to be gauged are usually held between centres and illuminated from below, their silhouettes appearing in the field of the viewing eyepiece. Effective pitch diameters can also be measured by this method.

The method of measuring pitch is shown in Fig. 13.23. The microscope has two raticules that can be oriented to the slopes of the thread and the point of intersection of these is used as the measuring reference. The movement of the longitudinal carriage is read off the linear scale, the micrometer microscope being employed for this purpose.

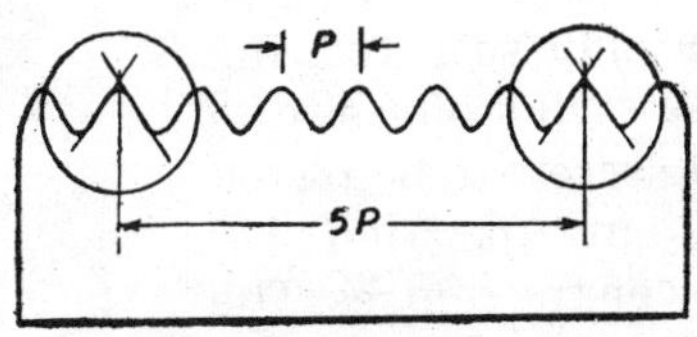

Fig. 13.23

The linear measuring accuracy is within 0.001 mm and for angles, it is 10 sec. of arc.

A comparatively simple method of testing the pitch of a screw thread with the Cooke tool room microscope fitted with its projection screen is as follows :

The screw to be checked is mounted in a cradle under the microscope objective and the necessary adjustments made to project the sharp enlarged image of the thread on the screen. The appropriate thread form on the microscope thread template is then brought into coincidence with the projected image, as shown in (Fig. 13.24) and a reading of the longitudinal table micrometer screw taken ; this can be done to an accuracy of 0.0025 mm. The table is then moved by means of the micrometer screw until the image of the next thread on the screen under inspection fills the template profile and the reading of the micrometer again taken. The difference between the readings gives the measured pitch of the screw. The procedure is repeated for each individual thread in order to find the separate pitch error, if any. Finally, the difference between the initial and last readings of the micrometer when divided by the number of threads that have been measured enables the "mean pitch" of the screw to be estimated.

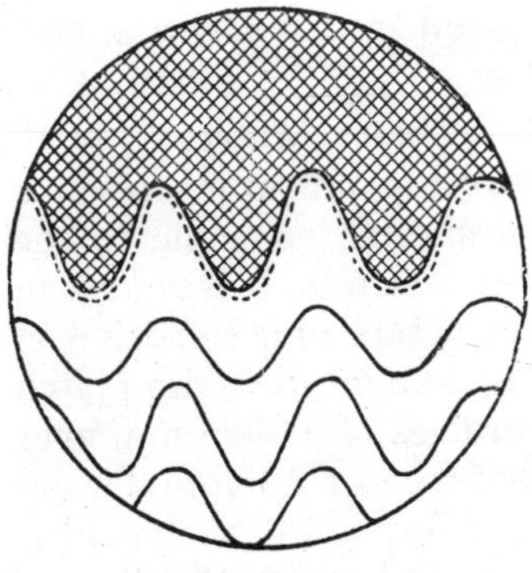
Fig. 13.24

For still more accurate purposes it is necessary to employ a special screw pitch measuring machine by which the actual pitch error of individual threads can be measured. The Pitter and Matrix are typical examples of pitch measuring machines.

The Pitter screw measuring machine employs various stylus points to suit screw threads that are to be checked. The screw under measurement is held stationary between centres on the machine. The indicator unit, carrying the stylus which bears on the flanks of each thread successively, is carried on a slide which is mounted on balls. The slide is actuated by means of a micrometer. The act of rotating the micrometer spindle causes the slide to move in relation to the fixed centres, *i.e.,* causes the indicator to move in relation to the work being measured. The stylus which is mounted on a leaf spring, falls in and out of each thread ; the pointer of the indicator reads zero (it is adjusted to read zero in the first groove) when this stylus is in a central position in each successive thread. The micrometer reading is taken each time the indicator reads zero ; these readings then shown the pitch error of each thread of the screw that is being measured. Special graduated discs are provided to fit the micrometer to suit all ordinary pitches whilst special can be provided for.

It may be mentioned that a small handwheel below the micrometer actuates screw for the purpose of moving the indicator in relation to the slide so as to bring the stylus opposite to the screw to be tested in any position between the centres. The total travel of the micrometer is 25 mm.

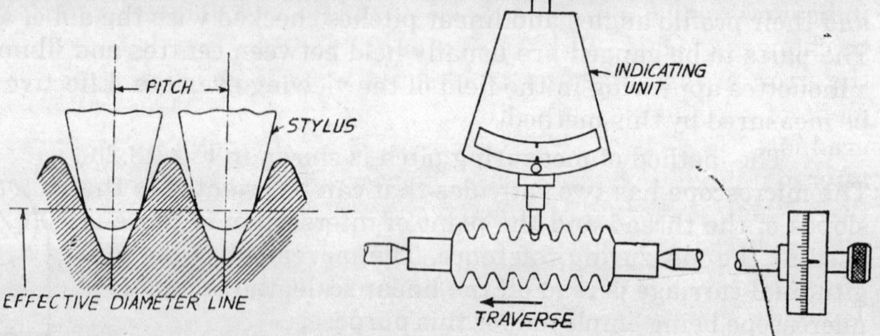

Fig. 13.25 (*a*) Fig. 13.25 (*b*)

Illustrating that for best results the stylus point should make contact on or near the effective diameter.

As the pitch of the micrometer screw is checked accurately when the machine is inspected and a curve of errors is provided, it is possible to attain a high standard of precision in measuring screws. The pitch errors are extremely small, being of the order of 0.002 mm for a thread. A test screw is also supplied with the machine and a chart of pitch error for this screw.

The Matrix pitch measuring machine operates on a similar principle to the Pitter machine. It is robust in construction and sensitive in measurement, revealing pitch accuracies of 0.0025 mm over a distance of 50 mm for all thread forms. In this machine (Refer Fig. 13.26) a micrometer head is provided on the headstock which is fixed to the base. The rotation of micrometer head produces movement of the longitudinal carriage along the bed of the base.

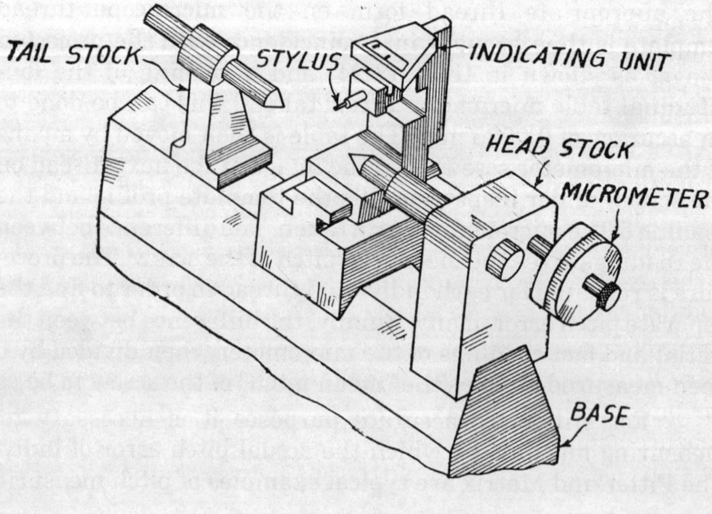

Fig. 13.26

Another carriage carrying the indicating and amplifying units comprising a radiused stylus and visual scale allowing a zero reading to be taken, and also capable of moving at 90° to the longitudinal carriage ; is mounted on the former carriage, and can be traversed longitudinally and locked in any position. A weight ensures a unidirectional thrust at all times. The micrometer screw of 40 t.p.i. has a 50 mm traverse and also has a compensator for any small residual pitch errors. In operation, the screw thread to be checked is placed between centres and the correct stylus mounted in the indicating head.

When the test screw is in position between the centres, and the correct stylus chosen *i.e.,* the one which makes contact at or near the effective diameter, the carriage carrying the indicating unit is traversed until the stylus is located in the first thread of the test screw and the indicator is coincident with the fiducial line ; the second carriage is then locked. The stylus, by virtue of an ingenious mounting device, is capable of free movement riding up and down the thread flanks on linear movement of the screw thread by rotation of micrometer head. The stylus is now traversed along the thread, pitch by pitch, reading being taken each time the indicator is set at zero. The micrometer can be fitted with a series of graduated dials that can be changed quickly. With the proper dial for the pitch that is to be measured the readings of the errors are obtained from the displacement of the lines on the disc which is graduated in (0.002 mm) divisions. It is after making this test, to turn to the first thread and repeat the readings, and the micrometer should read zero again.

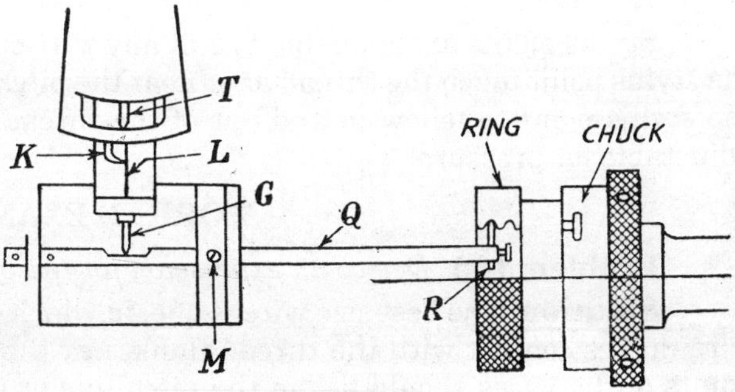

Fig. 13.27

G — Stylus
Q — Rod holding stylus *R*
L — Line for adjusting stylus *G*. This is arranged opposite to pointer *K*.
M — Locking screw for *Q*.
T — Indicator pointer which is arranged to read zero while taking micrometer reading.

13.4.10.2. For Internal Threads. The pitch of an internal thread can be measured on any of the standard pitch machines by using an adaptor. This adaptor carries a bar which can be inserted into the ring, the stylus being fitted to the bar end engaging with the thread in the usual manner. The ring gauge is mounted on a face plate or on the head stock of the machine, which will accommodate rings upto several cm. in diameter. For very large rings, a special set up on a surface plate is necessary, utilising an indicator and slip gauges.

13.4.10.3. Additional description of pitch measuring machines. To correct any errors in pitch of the micrometer screw a compensator bar is provided.

The instrument is checked periodically with a master reference screw which is placed between centres and measured for the pitch over full range of micrometer. In this case variation in the reading is taken to indicate errors in the micrometer screw, and the compensator bar modified accordingly.

The micrometer screw has 40 t.p.i. and with a graduated dial of 250 divisions numbered every 10 divisions, the instrument is read as an ordinary micrometer calibrated to 0.0001 inch. The micrometer dial may be replaced by any one of the five alternative dials to simplify the measurement of the threads of certain pitches. Each of the dials is marked with a number of divisions to suit a range of pitches as follows :

Dial	No. of div.	For measurement of
A	6 9	6, 12, 24, 48, 15, 30, 60 t.p.i. 4, 5, 9.18, 36 t.p.i.
B	7 11	7, 14, 28, 56, 5, 10, 20, 40 t.p.i. 11, 2 t.p.i.
C	8	4, 8, 16, 32 t.p.i.
D	13 19	13, 26 t.p.i. 19 t.p.i.
E	250 20	(Each 0.002 mm. numbered every fifth division) Pitch-multiples of 0.025 mm.

Dial C is for British association, metric or non-standard pitches. Dial E is for metric machines only. The provision of a dial marked to suit a particular pitch simplifies pitch measuring, a division on the dial is opposite the zero mark for nominal pitch of each thread. Any variation of the division from the zero may then be read directly to 0.0001″ on either side of the zero line.

Stylus points are available to suit any particular thread. Care should be taken to make the stylus point touch the thread at or near the pitch line. The stylus holder is pivoted to allow the stylus point to follow in and out of the threads, as the carriage is moved along, and is adjustable for pressure.

WORKED EXAMPLES

Problem 13.1. *Derive an expression for the best size wire.*

Solution. The best size wire is one, in which case the wire makes contact with the thread flank, *i.e.*, the contact points of the wires should be, on the pitch line or effective diameter. In other words, OP is perpendicular to the flank position of the thread. Let half the included angle of thread be x.

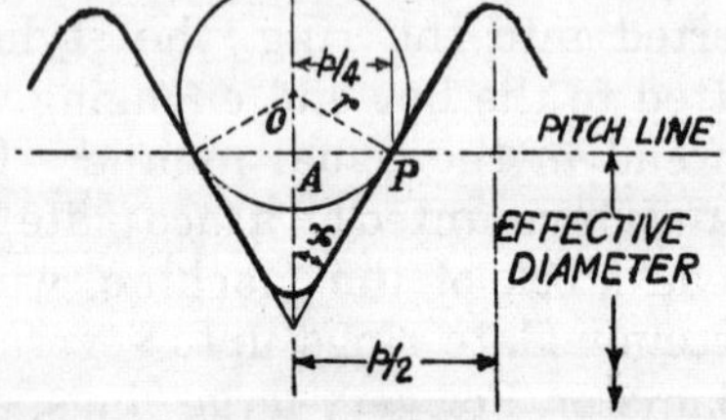

Fig. 13.28

Then in ΔOAP, $\sin P\hat{O}A = \dfrac{AP}{OP}$, or $\sin(90° - x) = \dfrac{AP}{OP}$

or $$OP = \frac{AP}{\sin(90° - x)} = \frac{AP}{\cos x} = AP \sec x$$

Since $AP = r$, and wire diameter $= 2r$.

$\therefore$ Best wire diameter $= d_b = 2r = 2AP \sec x$

As AP lies on the pitch line, $\therefore$ $AP = \dfrac{p}{4}$ (where p = pitch of the thread)

$$\therefore \quad d_b = \frac{2p}{4} \sec x = \frac{p}{2} \sec x.$$

Problem 13.2. *Derive an expression from first principles for the limits of diameter for 'best size' wires for measuring threads of BA form in terms of pitch.*

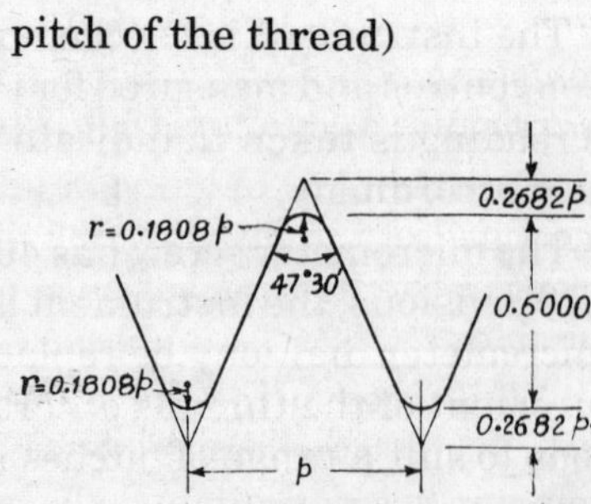

Fig. 13.29

Solution. Best wire size is : $d = \dfrac{p}{2} \sec \dfrac{x}{2}$

where x = included angle of the thread

$$= \frac{p}{2} \sec\left(\frac{47° \, 30'}{2}\right) = \frac{p}{2} \times \frac{1}{0.9150} = 0.5465p.$$

(*a*) Upper limit/lower limit : $\frac{1}{5}$ flank length $= \frac{1}{5} \times BF$.

(Refer Fig. 13.30 ; please note that point B could not be shown in Fig. 13.30. Actually B lies on line OF such that $AB \perp OF$. Point C lies on inter-section of line AD and OF).

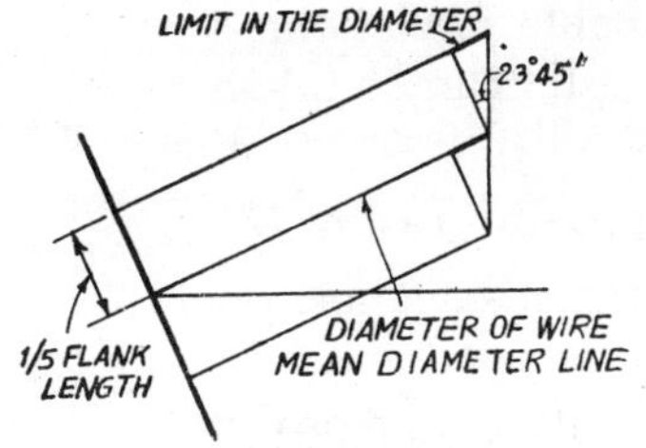

Fig. 13.30

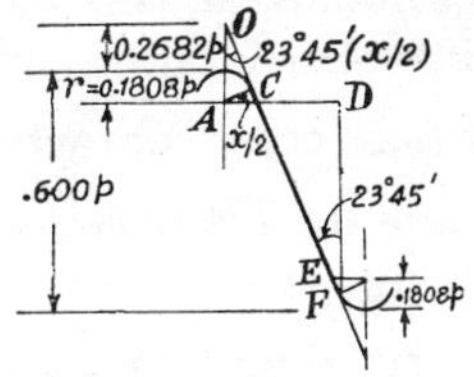

Fig. 13.31

$$BF = CE + BC + EF = CE + 2BC \qquad [\therefore \; BC = EF]$$

$$BC = (OA \sin x/2) \tan x/2$$

$$= [(0.1808p + 0.2682p). \sin 23° \, 45'] \times \tan 23° \, 45'$$

$$= [(0.4490p) \times 0.4015] \times 0.44 = 0.07924p.$$

$$CE = [0.6000p - 2 \times 0.1808p] \times \frac{1}{\cos 23° \, 45'}$$

$$= 0.2384p \times \frac{1}{0.9150} = 02719p$$

$\therefore \; BF = 0.2609p + 2 \times 0.0792p = 0.4293p.$

$\therefore$ 1/5 flank length $= 0.0859p$; Limit $= 0.0859p. \tan 23° \, 45' = 0.0378p$

Hence upper limit for best wire size $= 0.5465p + 0.0378p = 0.5843p$ and lower limit for best wire size $= 0.5465p - 0.0378p = 0.5087p$. **Ans.**

Problem 13.3. (*a*) *Generally three wires are used to measure a screw thread with a hand micrometer and two wires to measure with a floating carriage measuring machine. Why ? Also a screw thread is never placed on centres to measure its major diameter whereas for measuring effective or minor diameters, it is placed on the centres. Why ?*

(*b*) *A screw thread of size 10 mm × 10 threads per centimeter is measured for major and effective diameter in a screw measuring machine. The micrometer is standardised on a slip gauge and, due to mal-alignment in its plastic handle, the plug gauge can lie at an angle of 5 minutes to the line joining centres of the machine, which is itself satisfactorily parallel to the micrometer faces. What errors are likely to be introduced into both major and effective diameter measurements due to this condition ? The micrometer anvils are 4 mm in diameter.*

Solution. (*a*) In using hand micrometer the three wires are used because the two wires on one side help in aligning the micrometer square to the thread ; whereas the alignment is inherent in the floating carriage measuring machine and thus two wires can be used, one on either side.

While measuring the major diameter, the micrometer must be allowed to align itself in order to avoid "sine errors" which are likely to be introduced if the thread on centres and the micrometer are mal-aligned. Therefore, for measuring the major diameter, the screw thread should not be placed on the centres. In the measurement of effective diameter and minor diameter, vees are used and with the help of wires or vees only a negligible cosine error can be introduced by such mal-alignment.

(*b*) Both the measurements introduce either a secant or tangent or both the errors, but the error over wires is negligible for such a small angle.

For major diameter : For maximum length (l) of thread portion that can be accommodated within the micrometer faces of 5 mm diameter $= \frac{5}{\text{No. of threads/mm}} = \frac{5}{1} = 5$ pitches

Since, it will not be possible and practicable for all the 5 pitches to be accommodated, this being theoretical, l can be taken as 4.5 pitches = $9/2 \times 1 = 4.5$ mm.

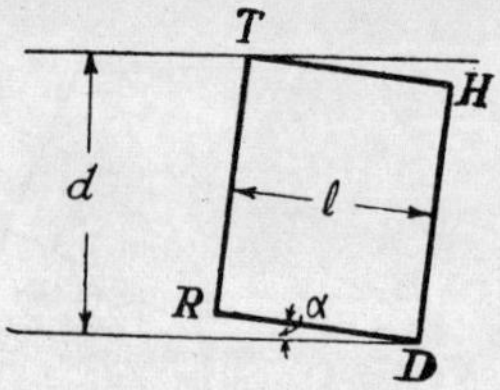

Fig. 13.32

Now one thread crest contacts at T and opposite crest at D.

$\therefore$ True diameter TR or $HD = d \sec \alpha - l \tan \alpha$ since $\sec 5' =$ 1.000001, it can taken as 1.

$\therefore$ TR, *i.e.*, true diameter $= d - 4.5 \times \tan 5' = d - 4.5 \times 0.00145$ $= d - 0.006525$

For effective diameter : $l = 1/2$ pitch = 0.5 mm,

$\therefore$ $0.5 \times \tan 5' = 0.5 \times 0{,}000145 = 0.00725$ mm

When opposing wires are at T and D, true diameter is larger than measured and when opposing wires are at H and R, true diameter is larger than measured.

Hence, gauge ends should be removed from handles for measurement.

Problem 13.4. (*a*) *A special form of buttress thread has a sharp crest and root, and an included angle of 50°. The leading flank is inclined at 5° and the trailing flank at 45° to a line perpendicular to the thread axis. If this thread is measured by the NPL method, show that the P value is given by*

$$P = 0.9105 \times \text{Pitch} - 1.1233 \times \text{Cylinder diameter.}$$

(*b*) *Two thread measuring needles having diameters of 0.03448 inch and 0.03449 inch respectively are used to check a screw of 16 t.p.i. BS. Whitworth threads having included angle of 60°. Determine whether these needles are smaller or bigger compared to the best size wires. Also find the P value for this set up.*

Solution. (*a*) In Fig. 13.33,

$$OD = d/2, \text{ and } DF = 1/2 \text{ (pitch)}, P/2 = OC - OB,\ OB = OA' - BA'$$

Now $\quad OA' = OA \,.\, \sin 70° = \left(\frac{d}{2 \sin 25°} \,.\, \sin 70°\right) = 1.112\, d.$

In ΔDEA, $\quad \frac{EA}{\sin 45°} = \frac{DE}{\sin 45°}\,;\ EA = DE \qquad ...(i)$

In ΔEFA, $\quad \frac{EF}{\sin 5°} = \frac{EA}{\sin 85°}\,;\ EA = EF.\frac{\sin 85°}{\sin 5°} \qquad ...(ii)$

Comparing (*i*) and (*ii*), we get $DE = 11.43\ GF$; Also $DF = \frac{\text{Pitch}}{2} = \frac{p}{2} = DE + EF$

Eliminating EF, we get $\frac{p}{2} = DE + \frac{DE}{11.43}$; or $DE = 0.4597p$

Now $\quad DE = EA + BA'\,;\ OB = 1.112d - 0.4597p$

or $\quad \frac{P}{2} = \frac{d}{2} - (1.112d - 0.4597p)$

Rearranging, we get $p = 0.9194p - 1.224d$.

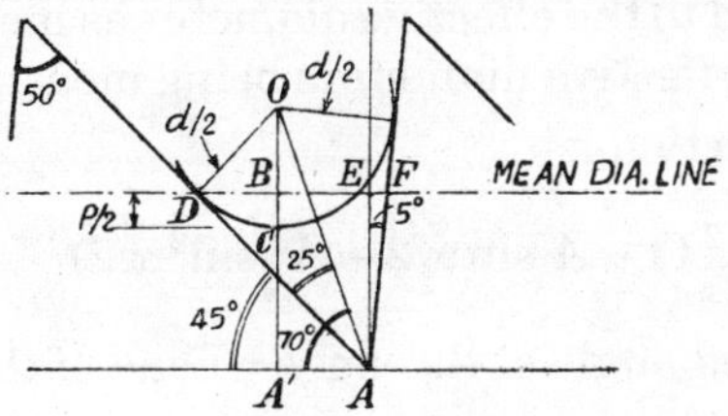

Fig. 13.33

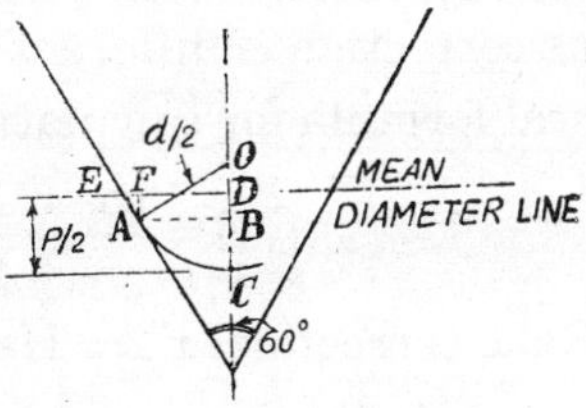

Fig. 13.34

(*b*) Pitch of the screw = 1/15 inch ; $p = 0.0625$ inch

$\therefore$ Best wire size diameter $= \dfrac{p}{2 \,.\, \cos \dfrac{x}{2}}$; $x = 60°$; or $x/2 = 30°$

$\therefore$ Diameter $= \dfrac{0.0625}{2 \times \cos 30°} = 0.02609$ inch.

Hence, the given needles are small.

We shall calculate $P/2$ for the given needles, and their sum is the required value.

For wire size 0.03448 inch ; $OA = OC = d/2 = 0.01724$

$$OB = OA \,.\, \sin 30° = 0.01724 \times \sin 30° = 0.008620$$

$$\therefore \quad BC = OC - OB = 0.01724 - 0.008620 = 0.00862$$

$$DB = AF = \frac{EF}{\tan 30°} = \frac{ED - AB}{\tan 30°} \text{ ; } ED = \text{Pitch}/4 = 0.0625/4 = 0.015625$$

$$AB = OA \,.\, \cos 30° = 0.01724 \times \cos 30° = 0.01493.$$

Putting the value, we get $DB = \dfrac{0.015625 - 0.014930}{\tan 30°} = \dfrac{0.000695}{\tan 30°} = 0.001204$

$$\therefore \quad \frac{P}{2} = DB + BC = 0.001204 + 0.00862 = 0.009824$$

For wire size 0.03449 inch : $OA = OC = d/2 = 0.017245$

$OB = OA \sin 30° = 0.017245 \times \sin 30° = 0.008623$; $BC = 0.017235 - 0.008623 = 0.008622$

$ED = 0.015625$; $AB = 0.017245 \cos 30° = 0.014931$

$$\therefore \quad DB = \frac{0.015625 - 0.014931}{\tan 30°} = \frac{0.000694}{\tan 30°} = 0.001202$$

$\therefore$ $P/2 = 0.008622 + 0.001202 = 0.009824$;

$\therefore$ P value $= 0.009824 + 0.009824 = 0.019648$ inch.

Problem 13.5. *What are the two corrections applied in the measurement of effective diameter by the method of wires.*

Solution. The two corrections applied are : (*i*) Rake correction, and (*ii*) Compression correction.

(*i*) *Rake correction.* The rake correction becomes necessary because in the determination of formula for effective diameter by three wires method, a plane axial section of the thread had been considered and it is assumed that the wire touches each flank of the thread in this plane. This assumption is true for angular grooves with zero helix angle, but not for screw threads which have a helix ; and in the later case wire lies parallel to the helix at the radius of the point of contact. The points of contact on opposite flanks will lie on opposite sides of the mean axial plane. As a result of this, the wire lies slightly farther from the thread axis than what has been assumed and a correction has to be applied to the effective diameter as measured and calculate. This correction is different for different effective diameters being measured.

A general formula for calculating rake correction is

$$C = \frac{\cos x/2 \cot x/2}{2\pi^2} \frac{l^2}{d} A^2 (1 + A \sin x/2 + A^2 \sin^2 x/2)$$

where C = Rake correction, $x/2$ = Half the included angle of thread, l = Lead of thread,

d = Diameter of wire, A = Constant $\cong \dfrac{d}{T+d}$

where T = Diameter under the wires.

This correction is always subtracted from the measured diameter.

(*ii*) *Compression correction.* As the micrometer exerts some force on the wires while measuring the effective diameter of threads, some degree of compression takes place and as a result the diameter observed is less. This correction is, therefore, added to the value of diameter obtained. This correction is more pronounced on fine threads and those whose included angle is small *e.g. B, A,* threads. For measuring forces upto about 350 gm. The correction is within 0.0025 mm for thread diameter down to about 3.5 mm and only 0.005 mm at 1 mm diameter. For larger threads, for the some measuring force, the compression is less and can be ignored.

Formula for determining compression correction is $= 0.001 \dfrac{E^{2/3}}{E^{1/3}}$ mm

where E = measuring force in newtons ; and E = effective diameter in mm.

Problem 13.6. *Draw a table giving value of best size wire and effective diameter for various values of pitch for metric thread. Also give formulae for maximum and minimum size of wire for various threads.*

Solution. See table below :

Formulae : Effective diameter = Size over wires – K, or $E = M - K$

where K = a constant = d (1 + cosec $x/2$) – $p/2$ cot $x/2$

where $x/2$ = Half of profile angle, and d = size of wire

The nearest available size of wire is used in determining the value of K and effective diameter in the table.

Thread Profile	*Pitch (p) in mm*	*Size of best wire*	*Value of K*
Metric Thread 60°	0.3	0.170	0.250
	0.35	0.201	0.300
	0.4	0.232	0.350
	0.45	0.260	0.390
	0.5	0.291	0.440
	0.7	0.402	0.600
	0.8	0.461	0.690
	1.0	0.572	0.850
	1.25	0.724	1.090

Thread Profile	*Pitch (p) in mm*	*Size of best wire*	*Value of K*
	1.5	0.866	1.299
	2.0	1.157	1.739
	2.5	1.441	2.158
B.S.P. and B.S.W. 55°	0.907 (28 TPI)	0.572	0.940
	1.337 (19 TPI)	0.724	1.008
	1.814 (14 TPI)	1.008	1.449
	2.309 (11 TPI)	1.157	1.444

Size of best wire $= \dfrac{p}{2 \cos x/2}$

The maximum and minimum size of wire for particular thread can be calculated by the expressions :

Min. wire dia. $= d_{min} = \dfrac{p \cos x/2 - 2h \sin x/2}{1 + \sin x/2}$; and Max. wire dia. $= d_{max} = \dfrac{p}{\cos x/2} - 2r$

for truncated threads $d_{max} = \dfrac{p \cos x/2 - 2h \sin x/2}{\cos^2 x/2}$;

where r = crest and root radius, and h = height of truncation.

Thread	*Angle*	*Size over wire, M*	*Best wire*	*Min. wire d_{min}*	*Max. wire d_{max}*
Whitworth	55°	$E + 3.1657\,d - 0.9604p$	0.5637p	$6p$	$0.853p$
Metric	60°	$E + 3d - 0.866p$	0.5773p	$0.505p$	$1.010p$
B.A.	$47\frac{1°}{2}$	$E + 3.4829d - 1.1263p$	0.5462p	$0.498p$	$0.730p$

Problem 13.7. *In measuring the thread angle of a standard metric thread the diameters measured over the wires of 0.750 and 1.500 are 19.50 and 21.75 mm respectively. Find the angle of the thread and the possible error in the measurement if the diameters of the wire are certified to an accuracy of within 0.001 mm and each measurement over the wires is having a possible error of 0.01 mm.*

Solution. In Fig. 13.35, O_1 and O_2 are centres of wires of diameters d_1 and d_2 respectively and O_2A and O_1C are to thread flank and $O_1B \perp O_2A$; ∴ Angle $O_2O_1B = x/2$

and $\sin \dfrac{x}{2} = \dfrac{O_2B}{O_1O_2} = \dfrac{O_2A - BA}{(h - d_2) - (h_1 - d_1)}$; or $\sin \dfrac{x}{2} = \dfrac{d_2 - d_1}{(h_2 - h_1) - (d_2 - d_1)}$

where d_2 and d_1 are wire diameters and h_2 and h_1 are the respective measurements over them.

$$\therefore \quad \sin \frac{x}{2} = \frac{1.500 - 0.750}{(21.75 - 19.50) - (0.750)} = \frac{0.750}{2.25 - 0.75} = \frac{0.75}{0.5} = 0.50$$

$$\therefore \quad x/2 = 30° \text{ ; and } x = 60°$$

If $\frac{1}{2}(h_2 - h_1 - d_2 + d_1) = c = 0.75$ and $\frac{1}{2}(d_1 - d_2) = r = 0.375$

then $\quad \sin \dfrac{x}{2} = \dfrac{r}{c}$

Since r and c both have errors, differentiating partially the error in angle x, we have

$$\delta(x) \approx \frac{2}{c\sqrt{c^2 - r^2}}(c \, . \, \delta r - r \, . \, \delta c) \text{ radians}$$

$$= \frac{2}{0.75\sqrt{(0.75)^2 - (0.375)^2}}(0.75 \times 0.001 - 0.375 \times 0.01)$$

$$= \frac{2}{0.75 \times 0.65}(0.0075 - 0.00375)$$

$$= \frac{2 \times 0.00375}{0.75 \times 0.65} = 0.0123 \text{ rad} = 42 \text{ min.}$$

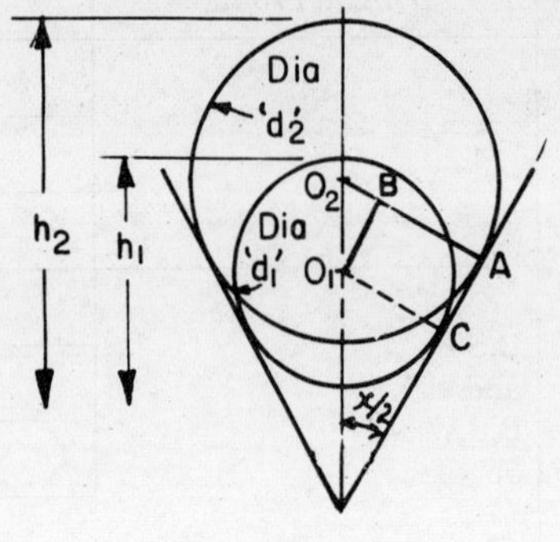

Fig. 13.35

Problem 13.8. *It is desired to measure the*

(a) Effective diameter of a 32 × 3.5 mm pitch metric plug screw gauge. For this purpose, following readings (average values) were noted.

Micrometer reading over the standard cylinder of 30.500 mm and the wires of 2.000 mm diameter as 13.3768 mm.

Micrometer reading over the gauge and wires as 12.248.

(b) Calculate the effective diameter considering the rake and compression corrections, assuming a measuring force of 2.5 newtons.

Solution. (*a*) Effective diameter = $T + P$

where T = dimensions under the wires = $S - (R_1 - R_2)$; S = diameter of standard = 30.500 mm

R_1 = reading over wires located above the cylinder = 13.3768 mm

R_2 = reading over wires located over the gauge = 12.2428 mm

$T = 30.500 - (13.3768 - 12.2428) = 29.3600$ mm

For metric threads P is a constant = $0.866p - d = 0.866 \times 3.5 - 2 = 1.0312$ mm

$\therefore \quad E = T + P = 29.3660 + 1.0312 = 30.3972$ mm.

(*b*) Rake correction : $C = \dfrac{\cos x/2 \cot x/2}{\pi^2} \dfrac{l^2}{d} \times A^2 \times (1 + A \sin x/2 + A^2 \sin x/2)$

$x/2$ = half the included angle of thread = 30°, l = lead = pitch (single start thread), d = diameter of curve = 2.000 mm

$$A = \text{constant } \frac{d}{T+d} = \frac{2.000}{29.3660 + 2.000} = 0.06376$$

$$\therefore \quad C = \frac{\cos 30° \cot 30°}{2\pi^2} \times \frac{(3.5)^2}{2.000}(0.06376)^2 \times (1 + 0.05376 \times \sin 30°)$$

$$+ [(0.06376)^2 \times \sin 30°]$$

$$= \frac{\sqrt{3}}{2} \times \frac{\sqrt{3}}{2\pi^2} \times \frac{(3.5)^2}{2.000} \times (0.06376)^2 \times [1 + 0.03188]$$

$$= 0.00195 \text{ mm} = 1.95 \text{ µm}$$

Compression correction $= 0.001 \dfrac{F^{2/3}}{E^{1/2}} = 0.001 \times \dfrac{(2.5)^{2/3}}{(30.3972)^{1/2}} \cong 0.0006 \text{ mm} = 0.6 \text{ µm}$

$\therefore$ Effective diameter = 30.3972 – Rake error + Compression error

= 30.3972 – 0.00195 + 0.0006 = 30.3959 mm.

Problem 13.9. *If in Problem 13.8, the pitch error is 0.006 mm and thread error angle on right hand and left hand flanks are 8 and 10 minutes of arc, then determine the virtual effective diameter.*

Solution. Virtual effective diameter = $E + 1.722\delta p + 0.0131 \times (8/60 + 10/60)$

$= 30.3972 + 1.722 \times 0.006 + 0.0131 \times 18/60 = 30.3972 + 0.01033 + 0.0039 = 30.4214$ mm.

TYPICAL EXAMINATION QUESTIONS

1. (*a*) The following elements of screw threads are to be measured :

(*i*) Outer diameter, (*ii*) Effective diameter, (*iii*) Core diameter, (*iv*) Pitch diameter, and (*v*) Thread profile.

Explain the methods with sketches wherever possible, you propose for the measurements.

(*b*) What is the 'Best Size' wire ?

2. What are the various types of pitch errors on thread components and what is the difference between them ? State the causes of each of them. How these errors effect the effective diameter of screw thread ?

3. What do you understand by drunken thread ? How it is caused ? Describe a method to test drunkenness of a 31 mm dia × 2 mm pitch component machined on centres.

4. How you would check the accuracy of pitch of a plug screw gauge ?

5. How the thread form can be checked and flank angle of internal thread measured ?

6. (*a*) Describe a pitch measuring machine.

(*b*) Explain how the various elements of internal threads are measured.

7. Describe a method to measure the effective pitch diameter of a screw plug gauge. Also explain how the errors in pitch and angle affect the virtual effective diameter ?

8. Show by means of a diagram the tolerance zones for I.S. metric bolts and nuts with no allowance.

9. (*a*) Name the three most important dimensions of a Vee thread which control the fitting of threads. Show in a sketch all the dimensions which are necessary to completely define thread.

(*b*) Describe in some detail the types of screw threads available with the ISO-system and explain the principle underlying the designation of these screw threads ?

(*c*) What does *M* 30 denote.

10. How will you proceed to check the thread angle or flank angles of a screw thread ? If you are using an optical projection for the purpose, explain the precautions you will observe in setting angle of the thread.

14

Gauges for ISO Metric Screw Threads

14.1. Introduction

The thread geometry can be measured accurately on certain measuring machines or by projecting its profile, but in normal production work these procedures cannot be adopted, so the use of gauges is made which takes the form of mating surfaces. By suitably designing the gauges it is possible to provide a limit gauging system which will control the complex dimension of the thread within the limits laid down by its specifications.

14.2. Classification of Thread Gauges

These can be mainly classified in two broad groups, *i.e.* one group comprising gauges used to check the product and the other used for reference. In first group are the *working gauges* which are used to check the product as it is being manufactured and the *inspection gauges* which are used to determine the acceptance or rejection of the product. In the second group are setting or checking gauges which are generally plug gauges with the help of which adjustable thread ring gauges, thread snap gauges and other thread comparators are set for checking size of master or basic gauges (threaded plug gauges representing the physical dimensions of the normal or basic size of the part).

In first group of gauges the working gauges are sometimes set to limits which are within the limits of inspection gauges. This practice ensures that any part which is passed as being within tolerance by the working gauges will also be passed by inspection gauges, thereby, reducing the possibility of a disagreement between the machine operator and the inspector in border line case ; a disagreement which often arises when working and inspection gauges are set to identical limits.

Working, inspection and master gauges differ in respect of the accuracy with which these are made. A working gauge is made to the widest tolerance and is, therefore, the least accurate, while master gauge has narrowest tolerance and is the most accurate. The accuracy of inspection gauges lies in between, depending upon the intended application.

14.3. Various Forms of Thread Gauges

14.3.1. Plug Screw Gauges. For gauging nuts or internal threads it is obvious that full form plug gauge made accurately to minimum dimensions of internal thread will ensure that all dimensions of the thread are not less than that minimum if it will assemble with the thread.

Major, minor and effective diameters will be checked and it will ensure that size of pitch, angle or form of the threads is not reducing below minimum. It is not possible to judge individual error of a particular component. Strictly speaking a plain NO GO gauge is required for minor diameter, NO GO gauge for effective diameter and NO GO gauge for major diameter is also required. The NO GO major diameter gauge has its major diameter on the corresponding upper limit of the work with its flank and minor diameter being well cleared off the work dimensions.

In ordinary production practice, NO GO effective diameter gauge is the only one used apart from full form GO gauge. This gauge is all that is necessary to ensure that the threads of the work are not thin even though the major and minor diameters are within the limits. Since tolerances on major and minor diameters are larger than those on effective diameter, these can be easily controlled.

Form of threads for NO GO effective diameter gauge is shown in Fig. 14.1. It may be noted that NO GO plug is truncated and the roots of the threads are cleared away. This is done with a view to avoid contact with the crest of the mating thread which, while being large on the effective diameter, may have a minor diameter on the low limit. The truncations also avoid contact with the roundings at the root of the nut. The effect of all these modifications from full form is to limit the NO GO end of the gauge to checking only the effective diameter.

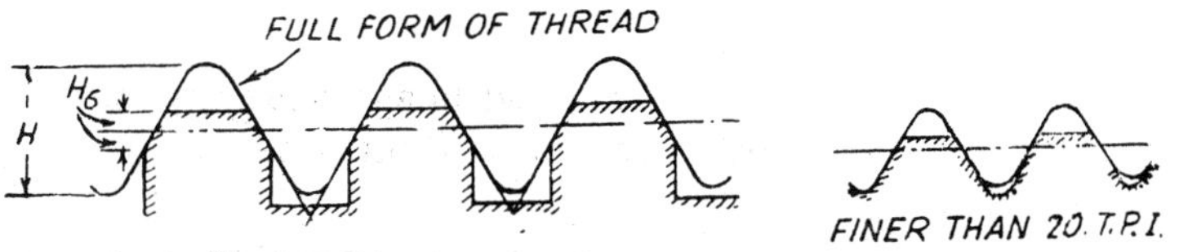

Fig. 14.1

The gauges are made up of special quality gauge steel, hardened and seasoned before the threads are grooved and finally tapped to dimension. Plug gauges are made in different designs according to the purpose for which they are required. Most widely used patterns are shown in Fig. 14.2 (*a*), (*b*), (*c*) and (*d*).

For gauging parallel external screw threads, minimum of following 3 gauges are recommended.

(*i*) a plain caliper gauge with gauging features to check the tolerance on the major diameter.

(*ii*) a screw caliper gauge with full form gauging features to check the maximum effective diameter, or a screw ring gauge.

(*iii*) A screw caliper gauge with special gauging features to operate slowly on the flanks of the thread, to check the minimum effective diameter.

For gauging parallel internal screw threads, following 3 gauges are recommended.

(*i*) GO and NO GO plug gauges to check the tolerance on the minor diameter.

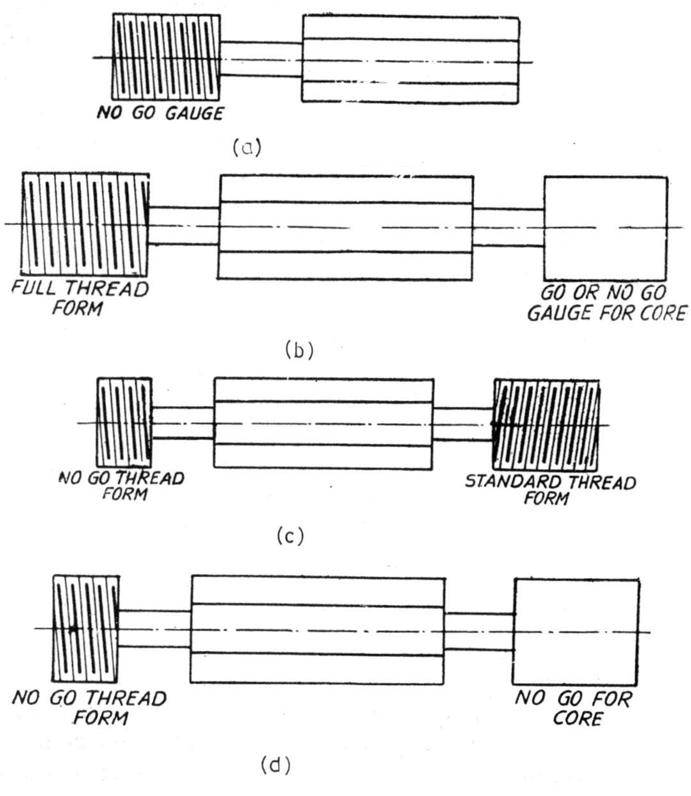

Fig. 14.2

(*ii*) a GO screw plug gauge to check the minimum effective diameter.

(*iii*) a NO GO screw plug gauge to check the maximum effective diameter.

14.3.1.1. Notes on Screw Gauges. Since the entering surface of these is liable in service to become burred if placed too heavily on metal surface, the British Standard Institute has, therefore recommended that there should be a guard extension at entry as indicated in Fig. 14.3. The extension is formed by reducing the diameter of end portion of the gauge.

To minimise the risk of damage, the feather edges of all screw gauges (both plug and ring) of GO and NO GO types should be removed to the extent of one pitch so that thread at the end of gauge commences abruptly at its full section or with slight taper at the end.

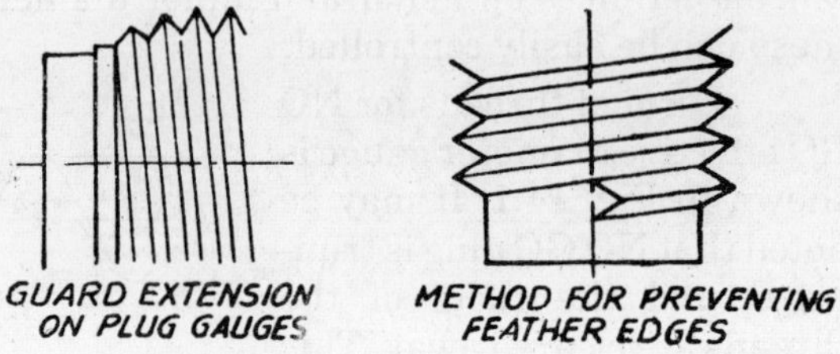

Fig. 14.3

Sometimes it is considered desirable to provide a dirt clearance groove in case of plug screw gauges. This groove is cut axial to the thread to a depth slightly below the root of thread, in a position where the thread commences its full section. The provision of such a groove on the gauge also prevents it from sticking. All sharp edges on gauging position as well as on handle should be removed.

In case of gauges above 100 mm diameter, lightening holes are advisable which enable the worker to handle them conveniently and also for venting purposes so that there is no risk of air being trapped in inner portion on blind hole.

14.3.2. Ring Screw Gauges. For production gauging of bolts the equivalent mating surface of the bolt threads is known as a ring gauge. As in case of plug gauges a system of limit gauges can be provided by full form NO GO and GO effective diameter ring gauges. The NO GO ring gauge is truncated on its minor diameter and cleared on its major diameter as shown in Fig. 14.4.

The first step in using a set of ring thread gauges is to read legend stamped on the rings. The legend should give size and pitch, for example, 7/16-20 NF means the workpiece should be of 7/16″ outside or major diameter having 20 threads per inch (National fine threads series). The pitch diameter upper limit is 0.4050″ which should also be stamped on the gauge. NO GO gauge member should also have same legend except diameter would be 0.4024″ (corresponding to lower limit). Sometimes class of fit also appears on the ring.

A thread ring cannot ordinarily be used to analyse the individual errors present in a screw. If the pitch diameter is over-size the ring will not engage : of course neither will it turn on an over-size major diameter. If the minor diameter is off (*i.e.*, the screw whose roots are filled with dirt, or made from worn die or lathe tool) the gauge will bind. Excessive screw thread lead errors, like taper will ordinarily be detected after a few turns of ring gauge. Where very fine threads are being gauged, care must be exercised not to force the threads of the ring gauge on screw.

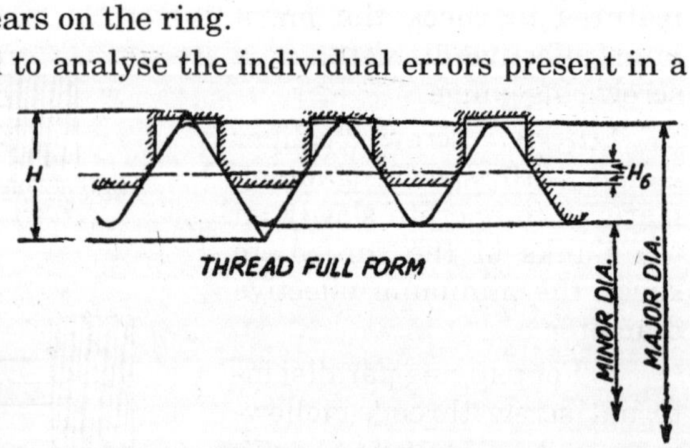

Fig. 14.4

14.3.3. Caliper Gauges. In principle these gauges are equivalent of gap gauges with thread form on anvils. In the metric system, external threads are gauged with a caliper type gauge with two sets of anvils. The front anvils have the full thread form cut on them and are set at such a distance apart as to accept the screws below the upper size limit, *i.e.*, they are the 'GO' anvils of the gauge. The width of these anvils is such as to test the required length of engagement of the screw so that they would reject a screw with a total pitch error over the length sufficient to raise the effective diameter to a value larger than anvils settings. If the screw passes through front GO anvils it reaches near NO GO anvils. There, NO GO anvils test the effective diameter, the tooth form on these anvils is cleared off at root and truncated at crest. They gauge one thread only as their function will not be influenced by any pitch error in the screw being engaged.

Now if a screw is cut of correct effective diameter but has a pitch error, it will not be able to pass through GO anvils. The only way available except by correcting pitch would be to reduce its effective diameter. Then it will pass through GO anvils as well as NO GO anvils and will be finally rejected.

14.3.4. Taylor's principle applied to screw thread gauges. It has already been indicated that effect of errors of pitch and flank angle is to bring about a virtual increase in the effective diameter in case of external threads and a decrease in effective diameter in case of internal threads. Accordingly Taylor's principle can be applied to threads also. It may be remembered that according to Taylor's principle, a GO gauge should check both geometric features and size and thus be of full form, whereas NO GO gauge should check only one dimension. In line with this, GO gauge is made of full form to the full length of thread to the maximum diameter of thread. If NO GO thread gauge is also made of full form, then the virtual reduction in effective diameter due to pitch errors may give misleading results.

To obviate this difficulty, and to account for Taylor's principle, NO GO gauges for the threads are made to check for major diameter and the effective diameter which is not influenced by the errors in pitch or form of the thread.

14.4. Gauge Wear

The wear takes place more rapidly in screw thread gauges than in plain gauges. It is not uncommon for a hardened ring thread gauge to wear from 0.025 mm to 0.075 mm during a few days of operation. The gauge tolerance has to be adjusted in such a way as to take into account the wear allowance. This reduces the work tolerance by an amount that must be allowed for wear in full form ring gauge. This allowance for wear is overcome in Caliper gauges to a great extent. The allowance recommended by B.S.I. for wear on generally used solid ring and plug gauges is 0.05 mm. Therefore, frequent checking of solid pattern gauges is necessary.

14.5. Tolerance for Screw Thread Gauges and Screw Threads

The screw threads combination is very much complex. There is no necessity of grading fits as in case of plain shaft and hole because there is a complicated interplay between pitch, thread angle and effective diameter. Although in an ideal case, a single class of fit would suffice for most of the purposes, the variables and conditions are such that three grades of accuracy are found desirable.

14.5.1. Effective diameter. The effective diameter tolerance must take into consideration the three variables (*i*) the major diameter ; (*ii*) the length of thread engagement ; (*iii*) the pitch. Furthermore the variables are related to the tolerance as indicated by following expression :

Tolerance on effective diameter : $= a\sqrt[3]{D} + b\sqrt{L} + c\sqrt{p}$

where D = major diameter, L = length of engagement, p = pitch, a, b, c are constants.

The value of these constants will depend upon prevailing accuracy of tools and machines used to produce the screw threads. Increase in accuracy will reduce the value of constants. The value of these constants according to British Standard Specifications for Withworth threads are

$a = 0.002$, $b = 0.003$ and $c = 0.005$ for medium fit

i.e. Tolerance on effective diameter is $= 0.002\sqrt[3]{D} + 0.003\sqrt{L} + 0.005\sqrt{p}$

The specifications provide for three grades of fits, (*a*) close, (*b*) medium, (*c*) free.

The basic formula gives effective diameter tolerance for medium fit. Close fit tolerances are two-thirds of free fit, and free fit tolerance is one and a half times that of medium fit. This results in tolerance increasing by 50% each time from close-medium-free fit.

Close fit. This applies to screw requiring a fine snug fit.

Medium fit. Suitable for better class of ordinary interchangeable screw threads.

Free fit. Suitable for bulk of screw threads of ordinary commercial quality.

Major Diameter. Its tolerance is given by expression :

Effective Dia. tolerance $= +\ 0.01\sqrt{p}$

This is applicable for all grades of fits (for external threads)

For internal threads no such tolerance is laid down.

Minor Diameters. Tolerance (external threads)

(*a*) *Close fit*—Effective diameter tolerance $= +\ 0.013\sqrt{p}$

(*b*) *Medium and free fit*—Effective diameter tolerance $= 0.02\sqrt{p}$

Tolerance (Internal Threads)

(*a*) 26 tpi and finer $= 0.2p + 0.004''$, (*b*) 22 tpi and 24 tpi $= 0.2p + 0.005''$, (*c*) 20 tpi and coarser $= 0.2p + 0.007''$.

The wide tolerance on the core diameter of internal threads permits of a tapping drill being used of ample size to prevent binding at the root of the tap. This follows a general practice and results in a flat crest to the root of the thread.

14.5.2. Example Determine the limiting dimensions for a bolt and nut 3/4 inch B.S.F. (Medium fit).

(*a*) Basic sizes : Full diameter $= 0.75''$,

Core diameter $= 0.75'' - 2 \times \text{Depth} = 0.75'' - 2 \times (0.64 \times 1/12) = 0.6434$ inch

Effective diameter $= 0.75 - 0.64p. = 0.75 - 0.64 \times 1/12 = 0.75 - 0.053 = 0.6967$ inch.

(*b*) Effective diameter tolerance : $= 0.002\sqrt[3]{D} + 0.003\sqrt{L} + 0.005\sqrt{p}$

$= 0.002\sqrt[3]{0.75} + 0.003\sqrt{0.75} + 0.00\sqrt{1/12} = 0.0018 + 0.0026 + 0.0014 = 0.0058$ inch.

(*c*) Major diameter tolerance for bolts (external threads)

$= \text{Effective Diameter tolerance} + 0.001\sqrt{p} = 0.0058 + 0.01\sqrt{1/12} = 0.0087''$

(*d*) Minor diameter tolerance

(*i*) For Bolts (external threads) and Medium fit

$= \text{Effective Diameter tolerance} + 0.02\sqrt{p}$

$= 0.0058 + 0.02\sqrt{1/12} = 0.0058 + 0.0058 = 0.0116''$

(*ii*) On Nut (internal threads) For 20 tpi and coarser

$= 0.2p + 0.007'' = 0.2 \times 1/2'' + 0.007 = 0.0166 + 0.007 = 0.0236''$

Hence limiting size for Bolt and Nut would be

	Effective Dia.			*Major Dia.*			*Minor Dia.*		
	Max.	*Min.*	*Tol.*	*Max.*	*Min.*	Tol.	*Max.*	*Min.*	*Tol.*
BOLT		0.6909	0.0058	0.750	0.7413	0.0087	0.6434	0.6318	0.0116
NUT	0.7025	0.6967	0.0058	—	—	—	0.667	0.6434	0.0236

14.6. Screw Thread Gauge Tolerance

N.P.L. London during First World War laid down tolerance for plug and ring screw gauges which remained practically unchanged till Second World War. These were published in 'Notes on Screw Gauges' and covered in three classes, *i.e.* Reference, Inspection and Production ; workshop gauges being within the work tolerance, and inspection and reference being outside.

The tolerances were graded solely in terms of diameter. The inspection and workshop tolerance on gauges upto 1.5 inches being 0.0006 inch on effective diameter and 0.002″ for 6″. Reference gauge tolerance was half of these values. This tolerance included virtual effect of pitch and angle errors.

In 1940, B.S.S. No. 919 was issued on screw gauge tolerance as was emergency. Specifications widened tolerance somewhat, rather more on smaller size than larger. At the same time fresh conception of effective diameter was introduced. In order to discourage manufacture of gauges with simple effective dia. on high limits so as to allow maximum pitch and angle error, the tolerance was divided into two separate elements :

For gauges upto 1.5″ dia., Tolerance on effective diameter = 0.0004″

Equivalent of pitch and angle tolerance on effective diameter = 0.0005″

The simple effective diameter tolerance on NO GO gauge was same but virtual tolerance was halved. This is quite logical because short length of thread is not liable to much pitch error.

The manufacturing tolerances on screw gauge are laid down in B.S.S. 919—'Screw Thread Gauge Tolerances' and in the earlier notes on screw gauges of N.P.L.

In connection with GO screw gauges, the latter publication contains tables of limits for both plug and ring gauges grouped under the headings of 'Reference', 'Inspection' and 'Workshop'.

The compensations for pitch and angle errors are also provided for by the application of corrections for each of these quantities. Effect of pitch error upon effective dia. of a screw thread is : $D_1 = E \times K$

where D_1 = Effective dia. equivalent of pitch error, E = max. cumulative pitch error along the thread, K = constant for the shape of thread = 1.92 for Whitworth = 2.27 for B.A. and = 1.73 for International metric threads.

The compensation for pitch error ΔE_1, used in this connection is the maximum error which is found between any two threads along the gauge.

The compensation for angle error, ΔE_2 is related to the pitch p as follows :

$$\Delta E_2 = p \times \Delta\theta/100$$

where $\Delta\theta$ is the angle error in degrees.

It is necessary that ΔE_2 and p should be expressed in the same measure, *i.e.* English or metric. The angle error in this formula is the arithmetical sum of the errors of the two flank angles irrespective of their signs.

If the general form of thread and the major and minor diameter are satisfactory, there still remains the following requirements on effective dia which must be met before the gauge can be accepted.

The items defined under (*a*) and (*b*) below must fall within the prescribed tolerances :

Plug gauges : (*a*) The measured effective diameter along the pitch line.

(*b*) The measured effective diameter + ΔE_1 + ΔE_2

Ring gauges : (*a*) The measured effective diameter along the pitch line.

(*b*) The measured effective diameter – ΔE_1 – ΔE_2.

It should be noted that the max. allowable pitch error, if no angle error is present, or the max. allowable angle error, if no pitch error is present, can readily be found from the formulae given, by evaluating ΔE_1 or ΔE_2 to the full effective diameter tolerance allowed on any given size of gauge.

14.7. Adjustable Thread Gauges

The fixed or solid pattern plug and ring screw thread gauges possess certain disadvantages that may be enumerated briefly as follows :

(1) *Gauge wear.* This wear takes place more rapidly in screw thread gauges than in plain gauges. It is not uncommon for hardened ring—thread gauge to wear from 0.02 mm to 0.03 mm during a few days use. When full form GO and effective diameter NOT GO ring gauges are used, manufacturing tolerance of work is reduced by amount that must be allowed for wear in the full form ring gauges.

(2) *Gauge making tolerances.* Solid gauges, being unadjustable for wear, their use reduces considerably the tolerance or limit of accuracy within which the work must actually be made, thus increasing the cost of manufacture.

(3) *Possible Inaccuracy.* Owing to wear effects, which in time may take up the full amount of tolerance allowed, there is always the risk of a gauge being used beyond its designed capacity with the result that the work checked with the gauge will be inaccurate ; frequent checking of screw thread gauges is necessary in order to obviate this effect.

The minimum effective diameter of a thread hole must necessarily be greater than the maximum effective diameter of the screw. The allowance recommended by the B.S.I. for this is 0.002 inch (0.05 mm) to allow for the wear on the solid ring and plug gauges generally used.

In order to overcome the disadvantages of solid gauge wear, inaccuracy and increased cost of production of parts manufactured for this type of gauge ; the adjustable pattern screw thread gauge was evolved which is now widely employed for precision gauging purposes.

Before describing a typical gauge of this class it may be mentioned that when adjustable gauges are employed, although wear can be compensated for but the B.S. allowance of 0.002″ for solid gauges is reduced by about 0.0005″, and the difference of 0.0005″ represents manufacturing tolerance when producing the same quality of work. The result of having this greater tolerance is to lower the manufacturing cost of the work ; further, whatever the tolerance may be, the more closely to size the limit gauges are made and maintained the more cheaply can the work be produced.

14.7.1. The Wickman adjustable thread gauge shown in Fig. 14.5 can be used for inspecting both right and left hand threads. It is adjustable to 0.002 mm and will test for out of roundness or eccentricity. It can be sealed after adjustment and is temper proof.

The principal feature of this gauge is the combined fine adjustment and lock as indicated in the upper part of Fig. 14.5. To set the gauge, the anvils are first adjusted by means of the adjusting screws until the setting of master gauges held between them is just sufficiently right to support the weight of the gauge. If the locking screws are tightened by means of a special key supplied then the ball will be slightly compressed ; their elasticity enables the setting gauge to be gripped or loosened alternately. An intermediate adjustment, where the setting gauges can just slide through the anvils is the desirable setting. After adjustment, a monogram lead seal can be fixed over the adjusting screws.

The frame of the gauge is of cast iron with specially selected characteristics to ensure that it will break rather than take a permanent set in the event of an accident. The anvils are of hardened tool steel and made sliding fit in their holder. Hardened abutments fixed solidly to the frame prevent the anvils from turning or tilting. The gauges are made to suit all standard screw thread forms.

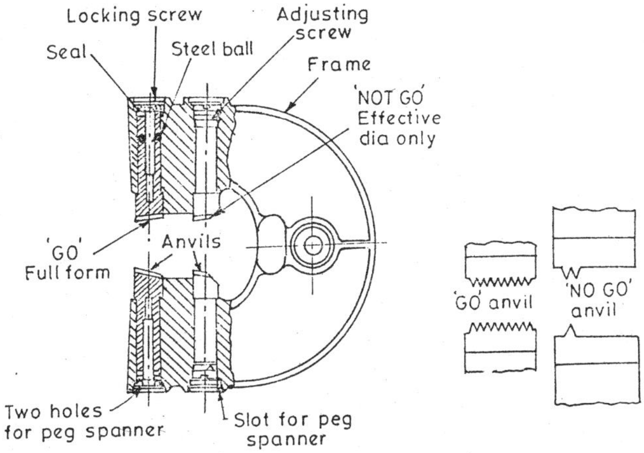

Fig. 14.5. Wickman Adjustable Screw Thread Limit Gauge.

The principle of the Wickman thread gauging system is to ensure truly interchangeable thread work of sound mechanical strength by controlling within the prescribed limits, the major, minor and effective diameters. As mentioned previously, the effective dia. is virtually increased by errors of pitch and angle so it must not be too large. For this reason the 'front' or 'GO' portion of the anvil is provided with a full form of thread of sufficient width to cover the engagement of the screw ; it is set to the nominal diameter. The rear anvil is of specially truncated form and consists of a few threads only, in order that contact may be made only on the central portion of the flanks of the thread and that no effects of pitch error are taken into account. In this way, the gauge certifies that the effective dia. is not too small. The threads on the anvils are relieved so that there is no interference with the helix of the thread being inspected ; thus inspection takes place on the axial line of the thread, where the true thread is located. As there is no interference with the helix angle, the same gauge can be used for both right and left hand threads.

14.7.2. Roller type Adjustable Thread Gauge. In this type of thread limit gauge, annular rollers replace the sectioned threads that are employed on the type of gauge described above. The wearing surfaces of the gauge are thus greatly increased and suitable form is, therefore, available for workshop and coarse tolerance inspection purposes.

The rollers, of which there are four, are not of helical thread but annular pattern of corrected form to compensate for helix angle interference. They are simple to adjust and can be sealed in position. In general, however, this type of gauge does not give the same accuracy as the Wickman Gauge.

It consists of a pair of full form GO rollers at the front having about three times as many threads as the rear NOT GO rollers. The latter are employed for the effective dia. and are truncated on the crest and root so that only flank contact is made.

14.7.3. Gauging of Taps. As the majority of internal threads in production work are tapped, it is necessary to employ a satisfactory limit for controlling the dimensions of taps. The Wickman gauge system for taps fulfills this requirement and comprises a pair of adjustable thread caliper gauges designed for four fluted taps. The four gauging positions of the pair of gauges are illustrated at (*a*) , (*b*), (*c*) and (*d*) in Fig. 14.6. The first gauge controls the form of the thread, pitch and effective dia. and ensures that neither the full nor the core dia. is too large [(*a*) and (*b*), (*c*) and (*d*)]. The second gauge ensures that full and core dia. are not too small.

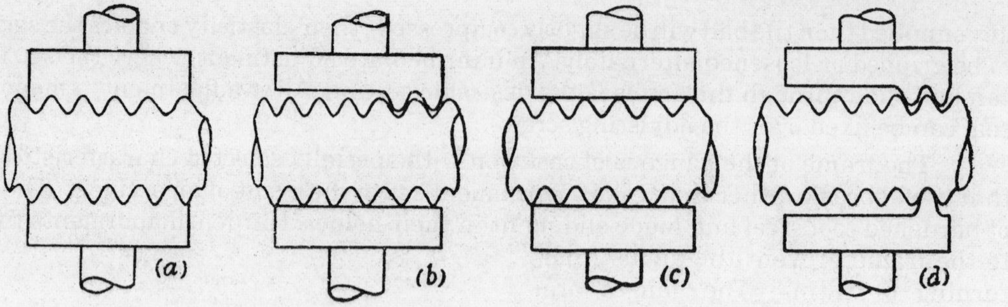

Fig. 14.6. Wickman Tapping Gauging System.

The full form GO front anvils of the first gauge are shown at (*a*) and the rear pair of NOT GO anvils for effective dia. at (*b*).

The NOT GO full dia. front anvils of the second gauge are shown at (*c*) and the NOT GO core dia. rear anvils at (*d*).

Tap gauges are set in similar manner to thread gauges except that on the full and core dia. an allowance is made for wear on tap threads. The anvils on tap gauges are, therefore, made to special thread form, *i.e.* some dimension from the tip is etched and must be taken into account when calculating dia. of setting plug gauge.

ISO Metric Threads and Gauges for them

Bureau of Indian Standards recommends the use of ISO (International Organisation for Standardisation) metric screw threads for use in Indian industries and thus it is essential that these threads be studied in details from all the angles. It may be noted that ISO metric thread is closely related to Unified thread.

14.8. Basic Profile

The basic profile of ISO metric screw thread is shown in Fig. 14.7. It is a theoretical profile and shows the basic sizes of the major, pitch, and minor diameters of the thread. The design profile of threads is shown in Fig. 14.9. The design profile of an external thread in its maximum material condition is the same as the basic profile except that the root of the thread (at the minor diameter) is rounded to a theoretical radius equal to $H/6$ = $0.14434P$ below the flat of width $P/4$. The design profile of an internal thread in its maximum material condition is the same as the basic profile. In practice, in order to avoid sharp corners at the root of threads (major diameter), the roots are rounded and cleared beyond a width of $P/8$ as shown in Fig. 14.9.

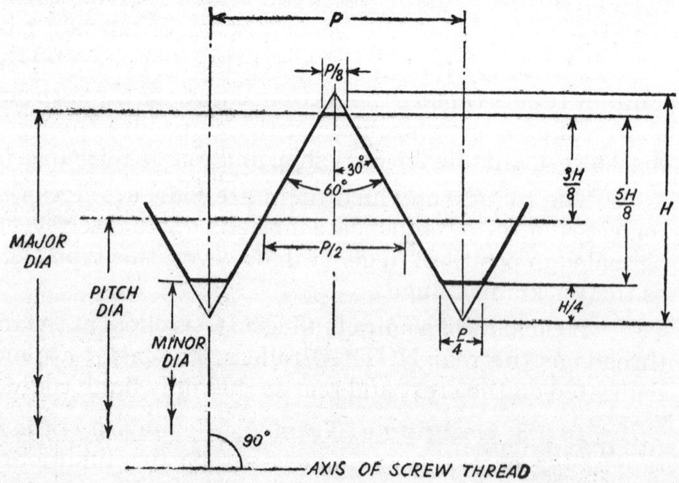

H = height of fundamental triangle = $\frac{\sqrt{3}}{2}P = 0.866025\,P$

Fig. 14.7

Notes on the design profile

Maximum depth of engagement (H_1). The design profiles of external and internal threads as shown in Fig. 14.9, are such that if a pair of screw threads of the same basic size (each in max. material condition) is assembled, contact between the two threads will be at the flanks over the radial depth of 5/8*H* which is the maximum depth of engagement.

Roots of external and internal threads. In order to provide resistance to fatigue and shock loads, the roots of the external threads are rounded off within the limiting profile and cleared beyond a flat of width *P*/4 at the minor diameter of the internal thread. The root of the internal thread is rounded off and cleared beyond a flat of width *P*/8 at the major diameter of the external thread.

The crests of the threads of GO screw ring and plug gauges are made flat and of *P*/4 and *P*/8 width respectively. They may not, therefore, be used to inspect the form of the rounding at the roots of external and internal threads. Optical methods may, however, be used to inspect them.

Crests of External Threads. The basic and design profiles of ISO metric external threads (Fig. 14.7 and 14.9) show the crest as flat but modern methods of manufacture result in large quantities of external threads with crests partially or even completely rounded. The departure from flat crest is not detrimental. While the limiting profiles within which the form of the crests of external threads should lie are represented by two full lines shown in Fig. 14.8, the minimum limiting profile shall, in practice, be taken as rounded at the crest as shown by the dotted arc. The actual form of any rounding of the crests of external threads is not checked by GO screw ring gauge, since the roots of the threads of these gauges are cleared to accept maximum threads with flat crests of width *P*/8.

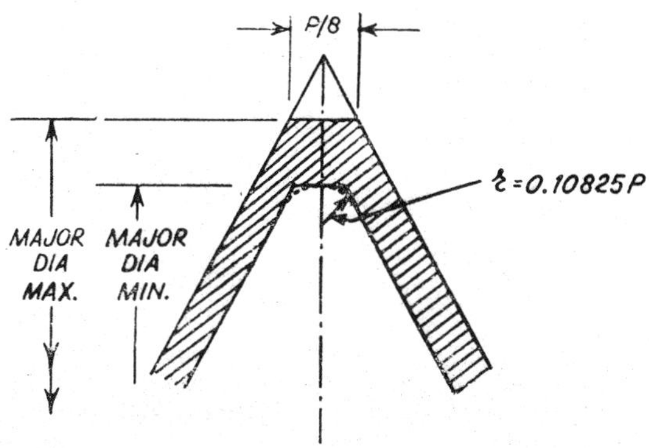

Limiting profiles at crests of external threads.
Fig. 14.8

In some cases, the external thread may be required to have rounded crest. This should be stated following the appropriate designation. The normal radius of such rounding is 0.10825*P*.

14.9. Diameter and Pitch Combinations

The diameter and pitch combinations for ISO Metric screw threads in the diameter range of 1 to 500 mm have been specified in the IS : 4218 (Part II)—1976. This standard specifies several pitches for various diameters. According to this standard there are two types of pitches, *viz.*, coarse series and fine series. For diameters, it specifies three choices, *i.e.* first choice, second choice and third choice, and recommends the use of diameters given in the first choice as far as possible for various applications. The diameters given in the second choice may be selected only if those given in the first choice do not meet the requirements, and similarly last choice is for 3rd category. For any particular diameter, the largest of the pitches associated with that diameter is called the coarse pitch (it has nothing to do with quality), the rest of the pitches being always finer than the coarse pitch are called fine pitches. For clarification, one entry of the table is explained here for diameters in the ranges of 17 and 25 mm *e.g.* here as far as possible 20 mm or 24 mm diameter should be selected. For diameter of 20 mm various pitches allowed are 1, 1.5, 2 and 2.5 mm. Of these 2.5 is, of course, coarse pitch, and the remaining 1, 1.5 and 2 are fine pitches. Second choice of diameters will be 18 or 22 mm ; and if all these do not suit then 17 or 25 mm may be selected. Table 1 of IS : 4218 (Part II)—1976 also puts some pitches in brackets which should be avoided as far as possible. The following diameters are indicated by IS in the 1st choice, 1, 1.2, 1.6, 2, 2.5, 3, 4, 5, 6, 8, 10, 12, 16, 20, 24, 30, 36, 42, 48, 56, 64, 72, 80, 90, 100, 110, 125, 140, 160, 180, 200, 220, 250 and 280.

Nominal Diameters			*Pitches*			
1st Choice	*2nd Choice*	*3rd Choice*	*Coarse*	*Fine*		
		17	—	—	1.5	1
	18		2.5	2	1.5	1
20			2.5	2	1.5	1
	22		2.5	2	1.5	1
24			3	2	1.5	1
		25	—	2	1.5	1

According to this standard, the pitch diameter combination of an ISO Metric screw thread is designated by the letter *M* followed by the values of the nominal thread diameter and the pitch, (expressed in mm) the two being separated by the sign '×' *e.g.*, '*M* 33 × 2 would mean a pitch diameter combination of thread size 33 mm and pitch 2 mm.

If there is no indication of pitch in the designation, it shall mean standard coarse pitch is implied *e.g.*, *M* 20 shall designate a pitch diameter combination of threads size 20 mm and pitch 2.5 mm.

The standard values of major diameters and the pitches for coarse and fine threads (ISO Metric screw threads) are given below :

Minor dia	*Coarse Threads*		*Minor dia*	*Coarse Thread*		*Fine Thread*	
	Pitch	*Minor dia*		*Pitch*	*Minor dia*	*Pitch*	*Minor dia*
1.6	0.35	1.22	8	1.25	6.65	1.00	6.92
2	0.4	1.57	10	1.5	8.38	1.25	8.65
2.5	0.45	2.01	12	1.75	10.11	1.25	10.65
3	0.5	2.46	16	2.0	13.83	1.50	14.38
4	0.7	3.24	20	2.5	17.29	1.50	18.38
5	0.8	4.13	24	3.0	20.75	2.00	21.83
6	1.0	4.92	30	3.5	26.21	2.00	27.83
			36	4.0	31.67	3.00	32.75

14.10. Design Profile

Basic dimensions for design profiles for ISO Metric screw threads are given below. Their exact values in tabular form for threads in the diameter range of 1 to 300 mm have been indicated in IS : 4218 (Part III)—1976. The standard symbols for denoting the various important quantities are :

D = major diameter of internal thread, D_1 = minor diameter of internal thread, D_2 = pitch diameter of internal thread, d = major diameter of external thread, d_2 = pitch diameter of external thread, d_3 = minor diameter of external thread, H = height of the fundamental triangle.

Dimension $H_1 = 5H/8$ = Maximum depth of engagement, Dimension $h_3 = 17\,H/24$ = Basic depth of external thread, P = pitch, R = root radius of external threads.

The values of D_2, d_2, D_1 and d_3 are given by following formulae.

$$D_2 = D - 2 \times 3\ H/8 = D - 0.64952\ P,\ d_2 = d - 2 \times 3\ H/8 = d - 0.64952\ P$$
$$D_1 = D_2 - 2\ (H/2 - H/4) = d - 2\ H_1 = d - 2 \times 5\ H/8 = d - 1.08253\ P$$
$$d_3 = d_2 - 2\ (H/2 - H/6) = d - 2 \times 17\ H/24 = d - 1.22687\ P.$$
$$H_1 = (D - D_1)/2 = 5\ H/8 = 0.54127\ P,\ h_3 = (d - d_3)/2 = 17\ H/25 = 0.61343\ P,$$
$$R = H/6 = 0.14434\ P$$

The tensile stress area is given by formula : Tensile stress area $= \dfrac{\pi}{4}\left(\dfrac{d_2 + d_3}{2}\right)^2$.

14.11. Tolerancing System

According to IS : 4218 (Part IV)—1967, the tolerances on ISO Metric screw threads are defined by tolerance grades and tolerance positions. There are 7 tolerance grades (*viz.*, grades 3, 4, 5, 6, 7, 8 and 9) in use, out of which 3, 4 and 5 are for precision threads, where little variation of fit is desired and where length of thread engagement is short), 6 for medium quality (*e.g.* commercial bolts and nuts) and 7, 8 and 9 for threads produced by less precision methods).

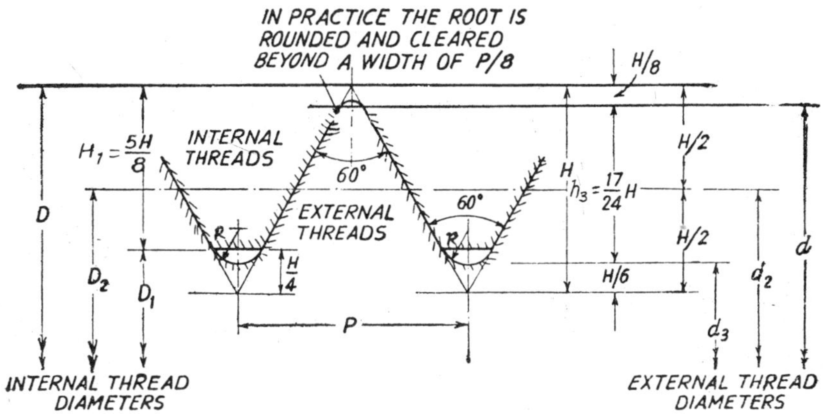

ISO Metric threads—Design Profiles of Internal and External threads.
Fig. 14.9

The various tolerance grades recommended for the crest and pitch diameters are as follows :

	Tolerance Grade
Minor diameter of internal threads	4, 5, 6, 7 and 8
Major diameter of external threads	4, 6 and 8
Pitch diameter of internal threads	4, 5, 6, 7 and 8
Pitch diameter of external threads	3, 4, 5, 6, 7, 8 and 9

The tolerance positions for external threads are designated by letters *e*, *g*, *h* and for internal threads by *G* and *H*, and their disposition are shown in Fig. 14.10.

A selection of grades and positions constitutes tolerance classes. Actually large number of tolerance classes are possible by combining various preferred tolerance grades and positions.

In order to guarantee sufficient overlap, the finished components are generally made to form the following fits : *H*/*g*, *H*/*h* or *G*/*h*

e.g., for *M* 6 threads, the usual fit is 6*H*/6*g* ; for *M* 12 × 1.25 threads, the usual fit is 5 *H*/4*h* and for *M* 20 × 2 threads, the usual fit is 7*G*/6*h*.

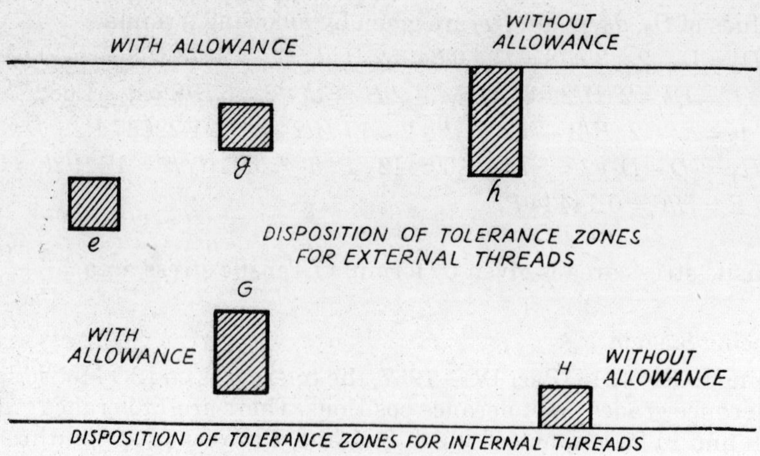

Fig. 14.10

The values of fundamental deviations for internal and external threads for various tolerance positions are calculated from the following formulae :

Lower deviation for tolerance position $G = +(15 + 11)$ μm,

Lower deviation for tolerance position $H = 0$

Upper deviation for tolerance position $e = -(50 + 11\,P)$ μm

Upper deviation for tolerance position $g = -(15 + 11)$ μm

Upper deviation for tolerance position $h = 0$, where P is the pitch in mm.

Value of tolerance Td on major diameter of external thread for grade 6,

$$Td(6) = 180\sqrt[3]{P^2} - \frac{3.15}{\sqrt{P}}\ \mu\text{m}$$

and for grade 4 its value = 0.63 Td (6), and for grade 8 its value = 1.6 Td (6).

The value of tolerance TD on minor diameter of internal thread for grade 6 is calculated from the formula :

$$TD\,(6) = 433\,P - 190P^{1.22}\ \mu\text{m for pitch 0.2 to 0.8 mm}$$

and $$= 230\,P^{0.7}\ \mu\text{m for pitches 1 mm and coarser.}$$

The values of the pitch diameter tolerances for external thread are calculated from the formula

$$Td_1\,(6) = 90_P{}^{0.1}\,d^{0.1}\ \mu\text{m}$$, where p = pitch in mm, and d = crest diameter in mm.

For other tolerance grades, the corresponding values will be

$$Td_1\,(3) = 0.5\,Td_1\,(6),\ Td_1\,(4) = 0.63\,Td_1\,(6),\ Td_1\,(5) = 0.8\,Td_1\,(6)$$

$$Td_1\,(7) = 1.25\,Td_1\,(6),\ Td_1\,(8) = 1.6\,Td_1\,(6),\ Td_1\,(9) = 2\,Td_1\,(6)$$

and the pitch diameter tolerance on internal threads for various grades is as given below :

$$TD_1\,(4) = 0.85\,Td_1\,(6),\ TD_1\,(5) = 1.06\,Td_1\,(6),\ GD_1\,(6) = 1.32\,Td_1\,(6)$$

$$TD_1\,(7) = 1.7\,Td_1\,(6),\ TD_1\,(8) = 2.12\,Td_1\,(6).$$

The value of length of thread engagement varies from $2.24\,Pd^{0.2}$ mm to $6.7d^{0.2}$ mm where P = pitch in mm, and d = crest diameter in mm.

A complete screw thread is designated by its (*i*) size and (*ii*) tolerance class.

Size is designated by letter '*M*' followed by diameter and pitch ; the two being separated by sign '×'.

The tolerance class is designated by pitch diameter tolerance followed by crest diameter tolerance.

Each class designation shall consist of (*i*) figure indicating tolerance grade and (*ii*) a letter indicating tolerance position.

e.g. *M*6— 5*g* 6*g*

- *M*6 — Thread size having crest diameter of 6 mm and fine pitch
- 5*g* — Class designation for crest diameter tolerance.
- 6*g* — Class designation for pitch diameter tolerance

In case the class designations for both pitch diameter and crest diameter are same then only one class may be indicated, *e.g.* $M12 \times 1.25 - 5\,H$.

Where no tolerances are specified, it is implied that classes are 6*H* and 6*g* (used for commercial bolts and nuts).

British Standard BS 3643, ISO Metric Screw Threads specifies three classes of fits, *viz.*, close, medium, and free. Medium class is appropriate for general engineering purposes, free class for application in which quick and easy assembly is desired, and close class is applied to threads requiring a closer fit than that normally obtained with the medium class of fit.

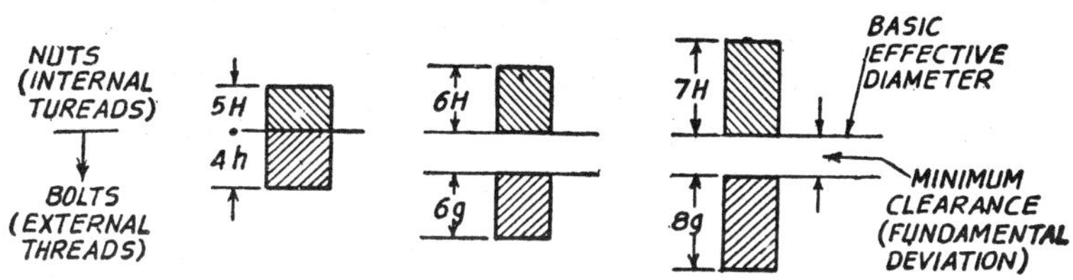

Close fit 5*H*/4*h* | Medium fit 6*H*/6*g* | Free fit 7*H*/8*g*

Fig. 14.11

The disposition of tolerance zones and fundamental deviation for the three types of fits is shown in Fig. 14.11.

14.12. Gauging Practice for ISO Metric Screw Threads

The screw threads have to be inspected both by the manufacturer and the purchaser and gauging practice should be such that it avoids any dispute between the two parties. In order to obviate the various difficulties, it is better practice if the purchaser inspects the thread workpieces by the manufacturer's gauges and of course his screw plugs may be checked by direct measurement. In the case of manufacturer, inspection is carried out both in the workshop and the inspection department. Here too, the better thing would be that both workshop and inspection department use same gauges. If not so, then for consistency of results the workshop should use new 'GO' gauges in which the pitch diameter lies in the inner half of the permissible wear limit and the 'NO GO' gauges should be such in which the pitch diameter lies in the outer half of the wear limit, whereas the inspection department should use new 'NO GO' gauges and only such 'GO' gauges in which the pitch diameter lies in the outer half of permissible wear limit. If the workpiece and gauges have different coefficients of expansion, then inspection should be carried out only at 20°C reference temperature.

14.12.1 Gauges for Workpiece External Threads and their Check Plugs and Setting Plugs

According to IS : 2334—1975, following gauges are prescribed.

(*a*) Solid GO screw ring gauge, (*b*) Adjustable GO screw ring gauge.

(*c*) Check plugs (GO and NO GO) for new solid GO screw ring gauge.

(*d*) Setting plug for adjustable GO screw gauge.

(*e*) Wear check plug for solid or adjustable GO screw ring gauge, (*f*), GO screw caliper gauge.

(*g*) Setting plug for GO screw caliper gauge, (*h*) NO GO screw caliper gauge.

(*i*) Setting plug for NO GO screw caliper gauge, (*k*) Solid NO GO screw ring gauge.

(*m*) Adjustable NO GO screw ring gauge.

(*n*) Check plugs (GO and NO GO) for new solid NO GO screw ring gauge.

(*o*) Setting plug for adjustable NO GO screw ring gauge.

(*p*) Wear check plug for solid or adjustable NO GO screw ring gauge.

(*q*) Plain gauge for major dia (ring or caliper gauges).

Gauges for Workpiece Internal Threads,

(*a*) GO Screw plug gauge, (*b*) NO GO screw plug gauge.

(*c*) Plain gauges for minor diameter (plug gauges).

It is not essential that all these gauges have to be used but GO and NO GO limits should be ensured by proper gauges for every piece.

14.12.2. Functions of Various Types of Gauges. The functions and the methods of use of the various gauges recommended above are described here.

(*i*) *Solid or adjustable GO screw ring gauge.* It is employed for checking the maximum material limit of the virtual pitch diameter, (the virtual pitch diameter is the actual pitch diameter as apparently decreased or reduced by errors of pitch and flank angle of thread) the maximum size of the minor diameter and the adequacy of the length of straight flank. The workpiece meets with specifications if the gauge can be screwed throughout over the work without using excessive force. The solid GO gauge has to be inspected periodically for wear with wear check plug. The maximum size of pitch diameter of the GO screw ring gauge is checked by NO GO check plug gauge. The adjustable GO screw ring gauge is set with the setting plug and controlled by the wear check plug.

(*ii*) *Check plug for new solid GO screw ring gauge.* It is used for checking the limit of the pitch diameter of new solid GO screw ring gauges. The checking is done by both GO and NO GO check plugs. The GO check plug gauge should screw through the corresponding solid GO screw gauge without the use of excessive force. It checks the profile of the new solid GO screw ring gauges. The NO GO check plug should not enter by more than one turn of thread on either side of the new solid screw ring gauge.

(*iii*) *Wear check plug for solid and adjustable GO ring gauge.* It checks the maximum wear limit of the pitch diameter of the GO screw ring gauge and should not enter by more than 1 turn when screwed from either side of the GO screw ring gauge.

(*iv*) *Setting plug for adjustable GO screw ring gauge.* It is used for setting the adjustable GO screw ring gauge to the specified pitch diameters. It has got complete flank threads and the screw ring gauge is adjusted for a snug fit on this. Further checking is done by NO GO screw check plug.

(*v*) *GO screw caliper gauge.* It checks the maximum limit of virtual pitch diameter in an axial plane taking into account the errors of pitch and flank angle. This gauge violates Taylor's principle in the respect that it does not detect errors of form, departures from roundness and periodic errors of pitch.

(*vi*) *Setting plug for GO screw caliper gauges.* The GO screw caliper gauges should just pass over the setting plug gauge under its own weight or under a fixed working load, otherwise the GO screw caliper gauge should be readjusted

(*vii*) *NO GO screw caliper gauge.* It checks the minimum limit of pitch diameter. The checking of the work by this gauge, *i.e.*, its not passing by more than two threads over the work should be carried in at least for three positions evenly distributed over the circumference.

(*viii*) *Setting plug for NO GO screw caliper gauge.* NO GO screw caliper gauge is set by this gauge.

(*ix*) *Solid or adjustable NO GO screw ring gauge.* This is intended to check the minimum pitch diameter limit of the actual pitch diameter. Its use is recommended for thin walled parts which would get deformed by checking with a NO GO screw caliper gauge. The solid NO GO screw ring gauge is made to a GO check plug and should be periodically inspected by the wear check plug and by the NO GO check plug for complete assurance. The adjustable NO GO screw ring gauge is set by a setting plug provided for this purpose and further checked by a NO GO check plug.

(*x*) *Screw check plugs for new sold NO GO screw rings gauge.* These check plugs (GO and NO GO) are used to check the limits of the pitch diameter of the new solid NO GO screw ring gauge.

(*xi*) *Wear check plug for solid and adjustable NO GO screw ring gauge.* It is used for checking the maximum wear limit of the solid NO GO screw ring gauge.

(*xii*) *Setting plug for adjustable NO GO screw ring gauge.* It has a portion with complete flanks and the screw ring gauge is adjusted for a snug fit on this portion and further checked by the NO GO screw check plug.

(*xiii*) *Gauges for major diameter.* The plain GO and NO GO caliper gauges or ring gauges are provided for checking the major diameter of the workpiece threads. In most of the cases for checking GO limit, the ring gauges are preferred ; however, if the workpieces have been manufactured by methods which do not involve the risk of errors of circularity, the use of caliper gauges may be made.

In the case of internal thread gauging :

(*i*) *GO* screw plug ensures the minimum limit of the virtual pitch diameter and the major diameter and also the adequacy of the straight flanks of thread. It should be periodically checked for wear by actual measuring.

(*ii*) *NO GO screw gauge.* It checks the maximum limit of the pitch diameter.

(*iii*) *Gauges for minor diameter.* GO and NO GO plain cylindrical plugs are used for checking the minor diameter of the workpiece thread. The use of spherical ended gauges and rod gauges is not permissible.

14.12.3. Tolerance Zones for Pitch Diameter of Screw Gauges

Gauges for External Thread. A GO or NO GO screw ring gauge that has errors of pitch and/or flank angle, has a virtual pitch diameter which is smaller than the simple pitch diameter. It can only be screwed on to a perfect GO screw check plug if the simple pitch diameter of the latter is also smaller by a certain amount (the sum of the diametral equivalents of the pitch errors and flank angle errors of the screw thread concerned) than the simple pitch diameter of the screw ring gauge. If the GO screw check plug also has errors of pitch and/or flank angle, opposite in sign to those of the gauge, its simple pitch diameter must be smaller by yet another amount to enable the screw ring gauge to screw on to the check plug. (The purchaser and the manufacturer shall agree whether the screw ring gauge should be measured or inspected by screw check plugs).

Gauges for Internal Threads. The simple pitch diameter of screw ring gauges may be measured directly, rather than inspected by screw check plugs. In order to ensure as far as practicable that a screw ring gauge accepted as satisfactory by direct measurement is also acceptable as satisfactory by the GO screw check plug, it is necessary for the simple pitch diameter tolerance zone of the GO check plug to be displaced downwards by an amount m in relation to the simple pitch diameter tolerance zone of the screw ring gauge (Fig. 14.12). The value of m given in Table corresponds to the sum of the diametral equivalents of the average errors to be expected in the pitch and flank angles of the check plug and the screw ring gauge in relation to the specified tolerances on these elements.

14.12.3.1. External Thread. The dispersion of the magnitude to tolerances on the gauges for external threads is shown in Fig. 14.12.

In the diagram shown in Fig. 14.12, the various notations are as follows and their values are given in the Table below Fig. 14.12.

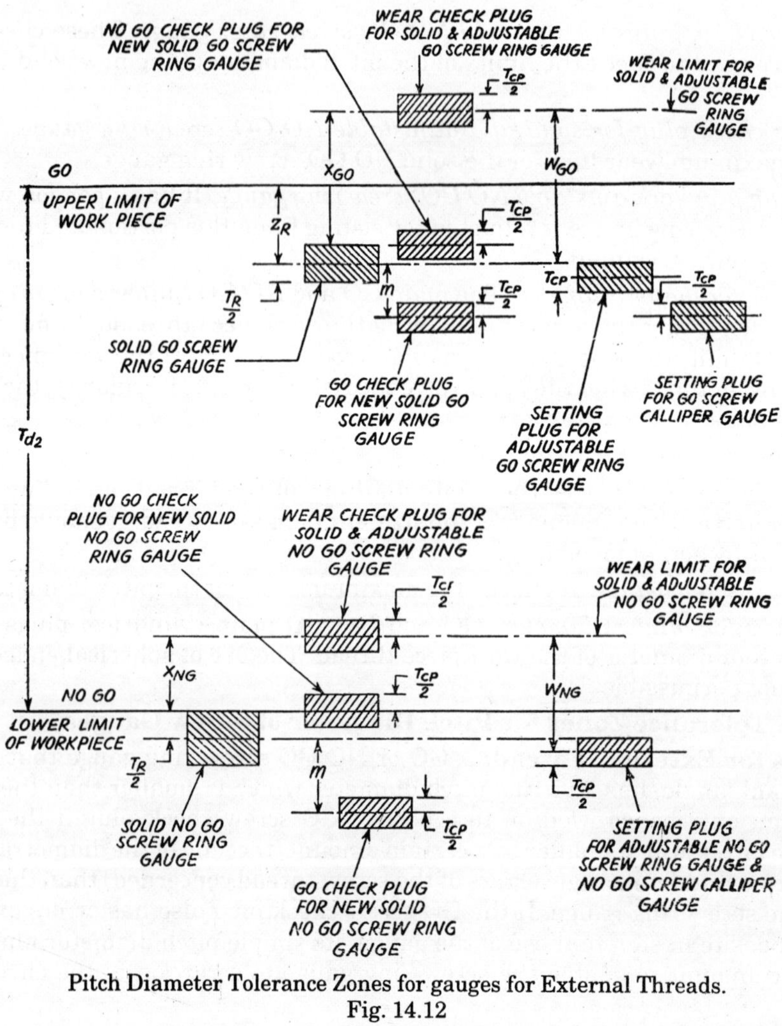

Pitch Diameter Tolerance Zones for gauges for External Threads.
Fig. 14.12

T_{d2} = tolerance for the pitch diameter of the workpiece.

T_R = tolerance for the pitch diameter of GO and NO GO screw ring gauges.

T_{PL} = tolerance for the pitch dia of GO and NO GO screw check plugs, wear check plugs and setting plugs.

T_{CP} = tolerance for the pitch dia of GO and NO GO screw check plugs.

W_{GO} = average amount available for the permissible wear of the GO screw ring. (The amount W_{GO} permits economical use of GO screw ring gauges.

W_{NG} = average amount available for the permissible wear of the NO GO screw gauge.

Z_R = distance from the middle of the tolerance zone T_R of the GO screw ring gauge to the upper limit of the workpiece.

m = distance of the middle of the tolerance zones T_R of the screw ring gauge and T_{CP} of the GO check plug gauge.

X_{GO} = distance between the wear limit of the solid GO screw ring gauge and the maximum size of the solid GO screw ring gauge in the upper limit pitch diameter of the workpiece.

X_{NG} = distance between the wear limit of the solid NO GO screw ring gauge and the max. size of the solid NO GO screw ring gauge at the lower limit pitch dia. of the workpiece.

Tolerance and Permissible Wear for the Pitch Diameter Values in μm (microns)

T_{d2} and T_{D2}		T_R	T_{PL}	TCP	m	Z_{PL}	W_{GO}		Z_R	W_{NG}	
Above	Upto						GO screw Ring Gauge	GO screw Plug Gauge		NO GO screw Ring Gauge	NO GO screw Plug Gauge
24	50	8	6	6	10	0	10	8	– 4	7	6
50	80	10	7	7	12	2	12	9.5	– 2	9	7.5
80	125	14	9	8	15	6	16	12.5	2	12	9.5
125	200	18	11	9	18	12	21	17.5	8	15	11.5
200	315	23	14	12	22	16	25.5	21	12	19.5	15
315	500	30	18	15	27	24	33	27	20	25	19
500	670	38	22	18	33	32	41	35	28	33	23

The minus value of Z_R means that Z_R is situated outside the tolerance T_{d2}.

14.12.3.2. Gauge for Internal Threads. The disposition and the magnitude of tolerances of internal threads are shown in Fig. 14.13.

T_{D2} = tolerance for pitch dia. of the workpiece.

T_{PL} = tolerance for the pitch dia. of GO and NO GO screw plugs.

W_{GO} = average amount available for the permissible wear of the GO screw plugs.

W_{NG} = average amount available for the permissible wear on NO GO screw plug.

X_{GO} = distance between the wear limit of the solid GO screw plug and minimum size of the solid GO screw plug at the pitch dia.

X_{NG} = distance between the wear limit of the solid NO GO screw plug and the minimum size of the solid NO GO screw plug.

Z_{PL} = distance from the middle of the tolerance zone T_{PL} of the GO screw plug to the lower limit of the workpiece.

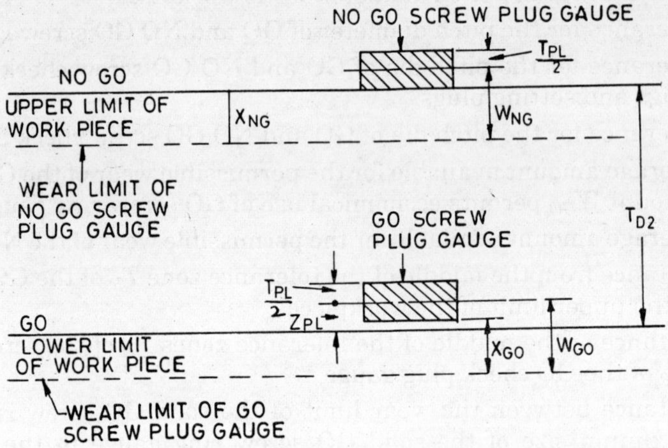

Fig. 14.13. Pitch diameter tolerance for gauges for internal threads.

14.12.4. Tolerance Zones for Plain Gauges for Screw Threads. The disposition and magnitude of the tolerance for plain gauges for the major diameter of external threads are shown in Fig. 14.14.

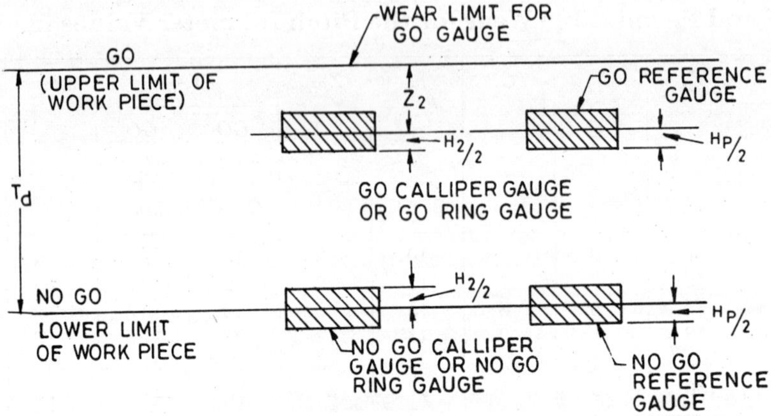

Fig. 14.14

H_2 = tolerance for plain caliper gauge and plain ring gauges.

Z_2 = distance from the middle of tolerance zone H_2 of the GO caliper gauge or the GO ring gauge to the upper workpiece limit.

Tolerances for plain gauges for major diameter of external threads
(All dimensions in micrometers)

D_T—Tolerance for the major diameter of the work piece external thread		$\frac{H_2}{2}$	Z_2	$\frac{H_p}{2}$
Over	*Upto*			
36	85	4	8	1
85	140	5	20	1.5
140	335	8	38	2
335	850	14	54	3
850	950	21	60	4

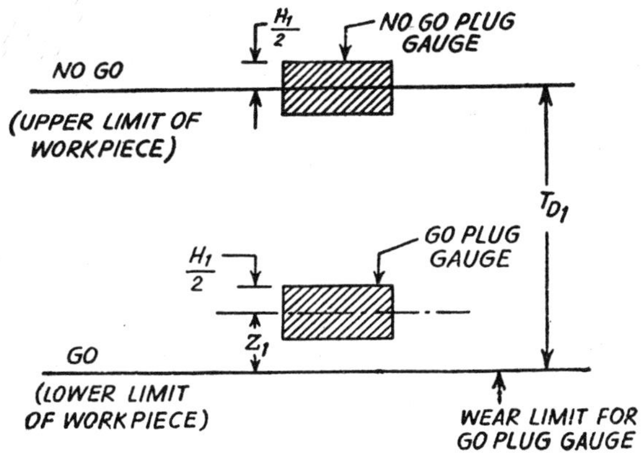

Fig. 14.15

The disposition and magnitude of the tolerances for the plain gauges for minor diameter of internal threads is shown in Fig. 14.15.

T_{D1} = tolerance for the minor dia. of workpiece internal threads.

H_1 = tolerance for plain plugs.

Z_1 = distance from the middle of the tolerance zone H_1 of the GO plug of the lowerpiece limit.

Tolerance for plain gauges for minor diameter of internal threads
(All dimensions in micrometers)

T_{D1}		$H_1/2$	Z_1
Over	*Upto*		
38	100	4	9
100	180	5	22
180	375	8	38
375	710	13	52
710	1250	23	65

Gauging Taper Screw Threads. External taper threads are usually checked by using a plain taper ring gauge and a taper screw ring gauge. Internal taper threads are checked by using a plain taper plug gauge and a taper screw plug gauge.

14.12. Threads Pitch Gauges for ISO Metric Screw Threads

According to IS : 4211—1967, the gauges which help to identify the thread pitch of ISO Metric screw threads in the pitch range 0.25 to 6.0 mm are made in the form of 24 blades (made of suitable tool steel sheet and 0.5 mm thick), 23 for checking the pitches and one having an ISO profile of 60°. All these blades are assembled in a protective sheath. All the blades are suitably hardened and tempered to a hardness of 540 HV and properly finished and sharp edges removed. These are suitably hinged in the sheath with screw and nut arrangement, and can be easily removed and rotated about.

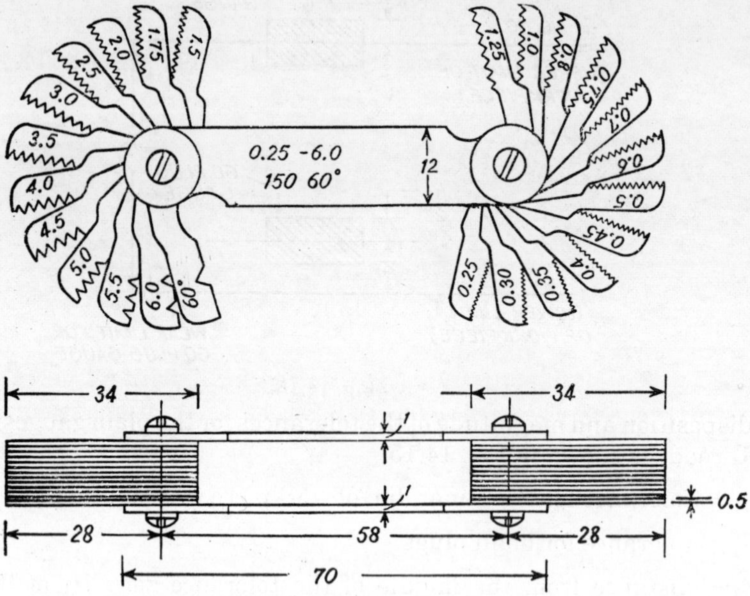

Fig. 14.16

Pitch	*Half angle tolerance in minutes*	*Tolerance on each consecutive pitch*	*Pitch*	*Half Angle tolerance in minutes*	*Tolerance on each consecutive pitch*
0.25	± 80		1.75	± 35	
0.3	± 80		2.0	± 30	
0.35	± 80		2.5	± 30	
0.4	± 80		3.0	± 30	± 0.02
0.45	± 75	± 0.015	3.5	± 30	
0.5	± 70		4.0	± 25	
0.6	± 60		4.5	± 25	
0.7	± 55		5.0	± 25	
0.75	± 50		5.5	± 25	
0.8	± 50		6.0	± 25	
1.0	± 40				
1.25	± 40				
1.5	± 35				

14.13. Indicating Thread Comparator

It is used for extra fast measuring the diameter of outside threads and plain cylindrical parts.

The base carries two holders each equipped with a mounting shank for the accommodation of measuring rollers. Top holder mounted virtually free of friction and without play on a parallel spring system. Bottom holder, which has a particularly large adjustment range, adjusts with a knurled screw for coarse setting. Adjustable stop provides automatic center alignment. Measuring pressure is variable.

Range 2-50 mm

Advantages

Direct indication of measuring results. The actual dimension of the part can be determined with absolute accuracy.

Versatility. Comparator covers a large range. Within its limits it is quickly and easily set to any desired diameter. The measuring rollers are furnished with parallel grooves with thread profile which allows the use of the same roller for each diameter within the range of the comparator provided it has the same pitch.

Fast measurements by placing part between rollers, moving it against stop and taking reading. Works particularly well with electric comparators.

Processing of Measuring Results. For testing components produced in batches the instrument may be used with electronic comparators in conjunction with computers allowing evaluation and documentation of measuring results, print out of measuring values, determination of work tolerances, statistical evaluation and trend observation.

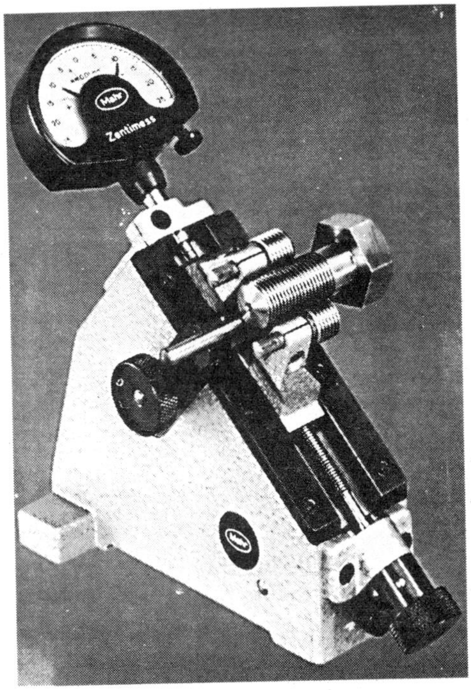

Fig. 14.17

Minimum wear of measuring rollers is ensured by use of hardened rollers in conjunction with rolling friction when the part is positioned.

Constant pressure by built-in compression spring which eliminates interference from operator's personal feel.

TYPICAL EXAMINATION QUESTIONS

1. Distinguish between simple and virtual effective diameter. Calculate the simple effective diameter of a M 33 × 2 mm ISO metric plug gauge and the diameter of the best size wires to use in measuring.
2. Describe the limit gauging system for threads which will take into account all the elements of the thread.
3. Describe the Wickman type gauge, for gauging thread elements.
4. Give a list of measuring instruments for the measurement of screw thread elements.

   Screw manufactured by an automatic screw machine are to be checked. Discuss any method you suggest for checking threads quickly with respect to outer diameter, root diameter and pitch.
5. How will you check the pitch of a screw ring gauge ? Explain with the help of a sketch the principle of the instrument used, clearly incorporating therein the mechanism of stylus movement. What will be the expected accuracy of this instrument ?

   Give a method of checking the pitch diameter of an indicating type of instrument.
6. It will be noted that the standard tolerance tables do not specify tolerances for pitch and flank angle. Explain how errors in these elements are controlled ?
7. What are the three most important dimension of V-form screw threads ?

   Explain Wickman type of screw thread gauge.
8. Describe briefly the devices you would adopt for checking (*i*) the diameter of the head of round headed screw, and (*ii*) the various elements of threads of screws produced in an automatic screw machine.
9. How do the angle and pitch errors of a screw thread affect its virtual effective diameter. Derive suitable analytical expression in support of your answer.

   Explain how a thread caliper gauge serves the purpose of checking the screw thread in shop production.

15

Measurement and Testing of Gears

15.1. Introduction

Gears are mainly used for transmission of power and motion. In order that the rotary motion of the driven shaft be perfectly uniform relative to the driving shaft, it is essential that both gears be of perfect geometrical form and be perfectly mounted on perfect shafts running in perfect bearings. It is thus obvious that a big factor which decides the accuracy of gearing is the precision with which gears are manufactured. For closer control over the accuracy of manufacture, precision measurement of gears plays a vital role. Fortunately, developments in manufacture of gear and its inspection techniques have gone hand-in-hand. For measurement of gears, which is rather complex field, it has been possible to follow general rule of metrology, *i.e.* the accuracy of measurement should be better than the component tolerance by a factor of 10.

The most commonly used forms of gear teeth are (*i*) involute and (*ii*) cycloidal. The involute tooth is derived from the trace of the point on a straight line, which rolls without slipping around a circle, which is the base circle, or it could be defined as the locus of a point on a piece of string which is unwounded from a stationary cylinder. The cycloidal tooth is derived from the curve which is the locus of a point on a circle rolling on the pitch circle of the gear. Here the addendum of the tooth is trace of the point on a circle rolling outside of pitch circle and this is an epicycloidal curve whereas the dedendum portion of the tooth is the trace of the point on a circle rolling on inside of the pitch circle of the gear and is a hypocycloidal curve. Involute gears are also called straight tooth or spur gears and are mainly used for general purpose in precision engineering due to some advantages. The cycloidal gears are not generally used in modern engineering, but used for some crude purposes where heavy and impact loads come on the machine. These along with bevel and spiral gears will not be considered here. The two great advangates associated with involute for the flank curve of gear teeth are ;

(*i*) Variations in the centre distance between two gears have no effect on the velocity ratio between a pair of involute gears.

(*ii*) The involute rack has straight teeth. Thus the complex involute form on gear can be generated from a simple cutter.

(*iii*) The involute system has a standard pressure angle which is either 20° or 14.5° whereas on the cycloidal system the pressure angle varies from zero at pitch line to a maximum at the tips of teeth.

(*iv*) In the cycloidal system for achieving correct meshing the gears must be operated on centre that will maintain theoretical pitch circles in exact contact which is not necessary in involute system.

Various types of gears commonly used are :

Spur Gear : It is a cylindrical gear whose tooth traces are straight lines.

Helical Gear : It is a cylindrical gear whose tooth traces are straight helices.

Spiral Gear : A gear whose tooth traces are curved lines.

Straight Bevel Gear : A gear whose tooth traces are straight line generators of a cone. It is conical in form operating on intersecting axes usually at angles.

Worm Gear Pair : The worm and mating worm wheel have their axes non-parallel and non-intersecting.

15.2. Involute Curve

An involute curve is defined as the locus of a point on straight line which rolls around a circle without slipping. It could also be defined in another way as the locus of a point on a piece of string which in unwound from a stationery cylinder.

Thus it is obvious that in an involute curve the length of the generator ($G_1 R_1$) will always be equal to the arc length (GR_1) of the base circle from the point of tangency to the origin of involute at G (Fig. 15.1).

Similarly generator $G_2 R_2$ = arc GR_2.

It is also clear that the tangent to the involute at any point will be perpendicular to the generator at that point. This condition fulfils the requirements of laws of gearing.

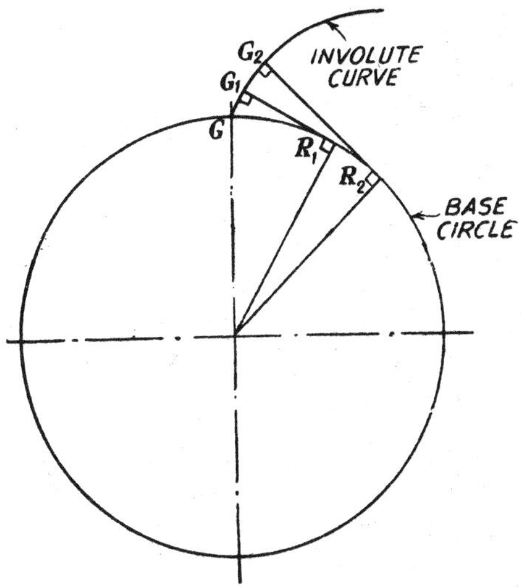

Fig. 15.1

Further, will also be noticed that the shape of the involute curve is entirely dependent upon the diameter of base circle from which the involute is generated. The curvature of the involute goes on decreasing as the base circle diameter goes on increasing and finally involute becomes straight line when base circle diameter is infinity.

15.3. Terminology of Gear Tooth

A gear tooth is formed by portions of a pair of opposed involutes. Most of the terms used in connection with gear teeth are explained in Fig. 15.2.

Base Circle. It is the circle from which involute form is generated. Only the base circle on a gear is fixed and unalterable.

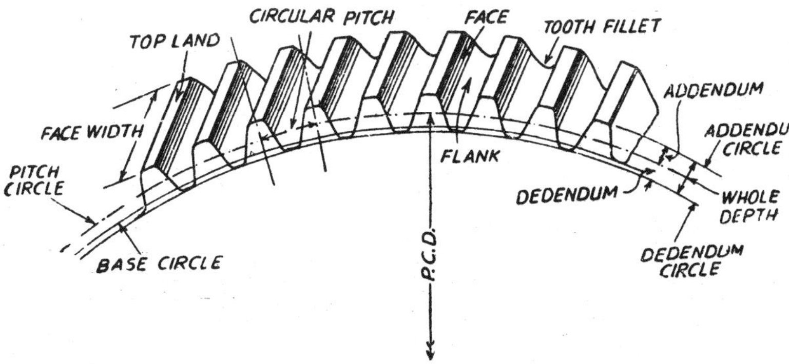

Fig. 15.2

Pitch Circle. It is an imaginary circle most useful in calculations. It may be noted that an infinite number of pitch circles can be chosen, each associated with its own pressure angle.

Pitch Circle Diameter (P.C.D.). It is the diameter of a circle which by pure rolling action would produce the same motion as the toothed gear wheel. This is the most important diameter in gears.

Module. It is defined as the length of the pitch circle diameter per tooth. Thus if P.C.D. of gear be D and number of teeth N, then module $(m) = D/N$. It is generally expressed in mm.

Diametral Pitch. It is expressed as the number of teeth per inch of the P.C.D.

Circular Pitch (C.P.). It is the arc distance measured around the pitch circle from the flank of one tooth to a similar flank in the next tooth. $\therefore$ C.P. $= \pi D/N = \pi m$

Addendum. This is the radial distance from the pitch circle to the tip of the tooth. Its value is equal to one module.

Clearance. This is the radial distance from the tip of a tooth to the bottom of a mating tooth space when the teeth are symmetrically engaged. Its standard value is 0.157 m.

Dedendum. This is the radial distance from the pitch circle to the bottom of the tooth space.

$$\text{Dedendum} = \text{Addendum} + \text{Clearance} = m + 0.157\,m = 1.153\,m.$$

Blank Diameter. This is the diameter of the blank from which gear is out. It is equal to P.C.D. plus twice the addenda.

$$\text{Blank diameter} = \text{P.C.D.} + 2m = mN + 2m = m\,(N + 2).$$

Tooth Thickness. This is the arc distance measured along the pitch circle from its intercept with one flank to its intercept with the other flank of the same tooth.

Normally tooth thickness = C.P./2 $= \pi m/2$

But thickness is usually reduced by certain amount to allow for some amount of backlash and also owing to addendum correction.

Face of Tooth. It is that part of the tooth surface which is above the pitch surface.

Flank of Tooth. It is that part of the tooth surface which is lying below the pitch surface.

Line of Action and Pressure Angle. The teeth of a pair of gears in mesh, contact each other along the common tangent to their base circles as shown in Fig. 15.3. This path is referred to as line of action. As this is the common generator to both the involutes, the load or pressure between the gears is transmitted along this line. The angle between the line of action and the common tangent to the pitch circles is therefore known as pressure angle ϕ. The standard values of ϕ are 14.5° and 20°.

In Fig. 15.3, $\dfrac{OA}{OP} = \cos\phi = \dfrac{R_b}{R_p} = \dfrac{D_b}{D}$

$\therefore$ Dia. of base circle D_b = P.C.D. $\times \cos\phi$.

Base Pitch. It is the distance measured around the base circle from the origin of the involute on the tooth to the origin of a similar involute on the next tooth.

$$\text{Base Pitch} = \frac{\text{Base circumference}}{\text{No. of teeth}}$$

$$= \frac{\pi \times \text{Dia. of base circle}}{N}$$

$$= \frac{\pi \times D \cos\phi}{N} = \pi m \cos\phi$$

Involute Function. It is found from the fundamental principle of the involute, that it is the locus of the end of a thread (imaginary) unwound from the base circle.

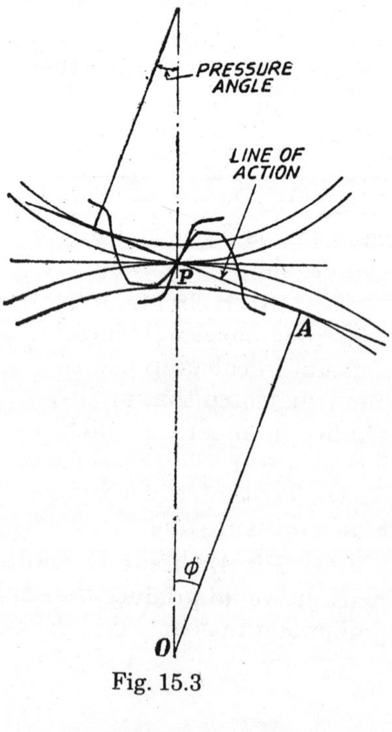

Fig. 15.3

Mathematically its value is Involute function $\delta = \tan \phi - \phi$

where ϕ is the pressure angle.

The relationship between the involute function and the pressure angle can be derived as follows :

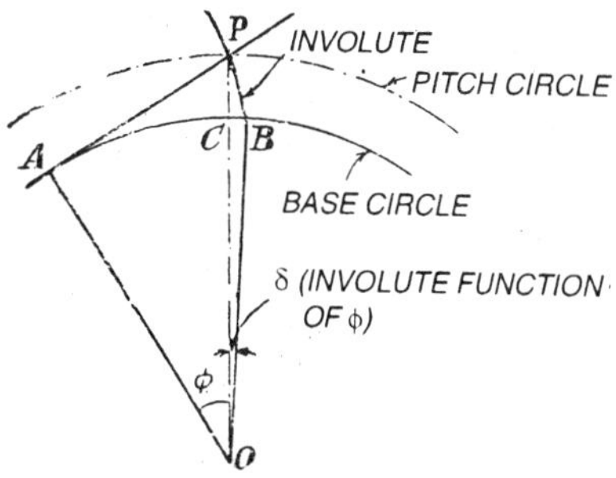

Fig. 15.4

In Fig. 15.4, OA = base circle radius = R_b

OP = pitch circle radius = R_p

and BP = involute profile of gear tooth.

AP is tangent to base circle at A,

$A\hat{O}C = \phi$ = pressure angle

Now $OA = OP \cos \phi$, or $R_b = R_p \cos \phi$

$C\hat{O}B$ = Involute function of ϕ.

By definition of involute, length AP = arc AB

and $\tan \phi = \dfrac{AP}{OA} = \dfrac{AP}{R_b} = \dfrac{\text{arc } AB}{R_b}$, Also $\phi + \delta = \dfrac{\text{arc } AB}{R_b}$

$\therefore \quad \phi + \delta = \tan \phi$, or $\delta = \tan \phi - \phi$.

Helix Angle : It is the acute angle between the tangent to the helix and axis of the cylinder on which teeth are cut.

Lead Angle : It is the acute angle between the tangent to the helix and plane perpendicular to the axis of cylinder (Refer Fig. 15.5).

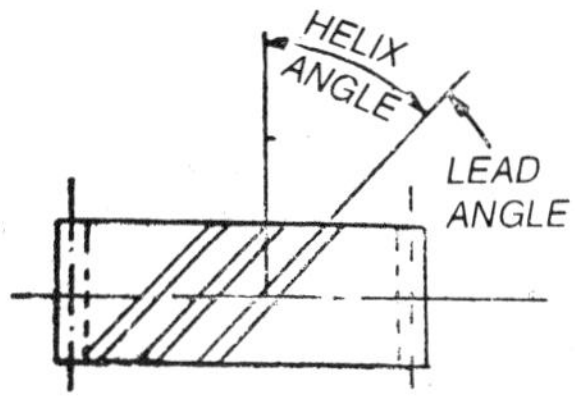

Fig. 15.5. Illustration of Helix and lead angle.

Back Lash : The distance through which a gear can be rotated to bring its non-working flank in contact with the teeth of mating gear. (Refer Fig. 15.6).

Basic Tooth Proportions for Involute Spur Gears

	Pressure Angles	
	20°	*14.5°*
Addendum	m	m
Dedendum	1.25 m	1.157 m
Teeth Depth	2.25 m	2.157 m
Circular teeth thickness	$\pi m/2$	$pm/2$
Fillet radius	0.3 m	0.157 m
Clearance	0.25 m	0.157 m

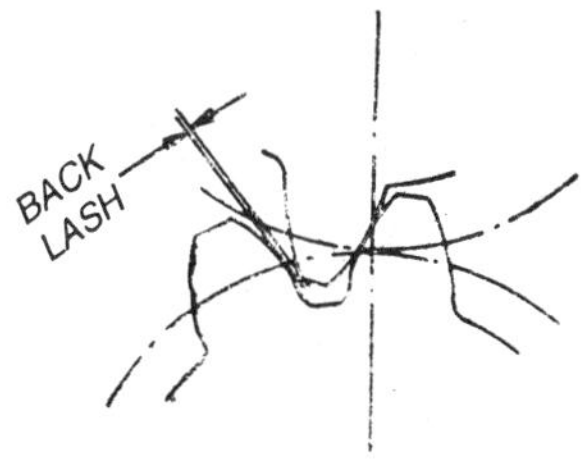

Fig. 15.6. Illustration of Backlash.

Some Important Relationships between Various Elements of Gears :

To find	*Having*	*Formula*
(*a*) *Spur Gears*		
Module (m)	No. of teeth (N) and pitch diameter (D)	$m = D/N$
Module	Circular pitch (p)	$m = p/\pi$
Outside diameter (D_o)	Pitch diameter and Module	$D_o = D + 2m$
Base circular diameter (D_b)	Pitch diameter and pressure angle	$D_b = D \cos \phi$

To find	*Having*	*Formula*
(*b*) *Helical Gear* :	(Refer Fig. 15.7)	
Normal module (m_n)	No. of teeth, pitch diameter (D) and helix angle (α)	$m_n = \frac{D}{N} \cos \alpha$
Normal module	Normal circular pitch (p_n)	$m_n = p_n/\pi$
Normal circular pitch (p_n)	Transverse circular pitch (p)	$p_n = p \cos \alpha$
Normal Pressure Angle ϕ_n (Refer Fig. 15.7)	Transverse pressure angle (ϕ) and helix angle	$\tan \phi_n = \tan \phi \times \cos \alpha$
Lead	Pitch circle radius and helix angle	$L = \pi \times D \cot \alpha$
Transverse teeth thickness (t)	Normal tooth thickness (t_n), and helix angle (Fig. 15.8)	$t = t_n/\cos \alpha$
(*c*) *Bevel Gears* :		
Pitch dia. of pinion (D_p)	No. of pinion teeth (n) and module	$D_p = n.m.$

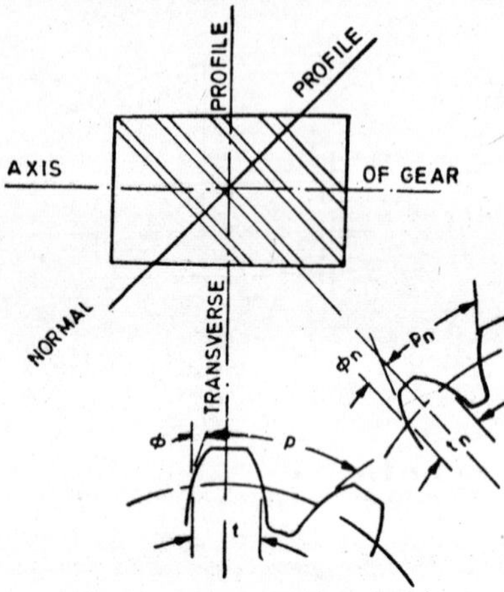

Fig. 15.7. Illustrations of normal and transverse profiles, pressure angles, tooth thickness and pitches of a helical gear.

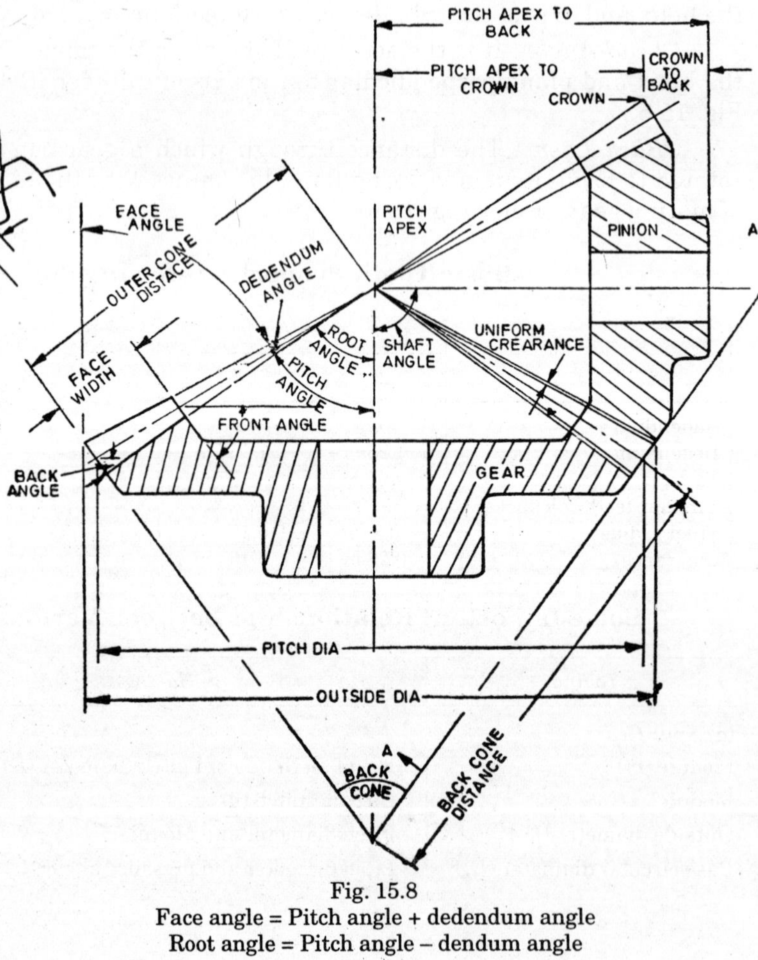

Fig. 15.8
Face angle = Pitch angle + dedendum angle
Root angle = Pitch angle – dendum angle

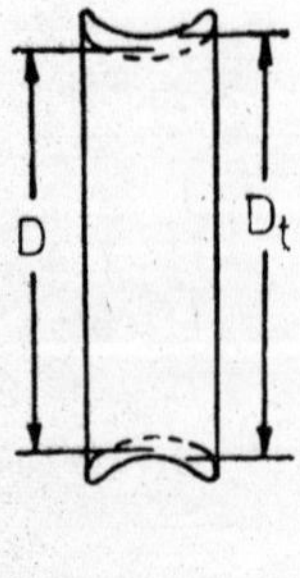

Fig. 15.9

To find	*Having*	*Formula*
Pitch dia. of gear (D_g)	No. of gear teeth (N) and module.	$D_p = N.m.$
Pitch angle of pinion (β_p)	No. of pinion teeth and No. of gear teeth	$\beta_p = \tan^{-1} n/N$
Pitch angle of gear (β_g)	Pitch angle of pinion	$\beta_g = (90 - \beta_p)°$
Outer cone distance (for pinion or gear)	Gear pitch diameter and pitch angle of gear.	$A_g = D_g/2 \sin \beta g$
Circular Pitch (p)	m	$\pi = m \times \pi$
Dedendum angle for pinion or gear	Dedendum of pinion or gear and outer cone distance	Ded. angle (p or g) = Ded. (p or g)/A_g
Outside Diameter of gear (D_o)	Pitch diameter of gear, addendum of gear and pitch angle of gear (g stands for gear, p stands for pinion)	$D_o = D + 2$ addendum of gear $\times \cos \beta_g$
Outside dia of pinion (d_o)	Pitch diameter of pinion (d), addendum of pinion and pitch angle of pinion	$d_o = d + 2$ addendum of pinion $\times \cos \beta_p$
(*d*) *Worm and Worm Gear* :		
Worm Lead (L)	No. of threads in worm (N_w) and axial pitch (p_x)	$L = N_w \times p_x$
Lead angle ψ	Lead (L) and pitch diameter (d)	$\tan \psi = L/\pi d$
Outside dia (d_o)	Pitch dia (d) of addendum (a),	$d_o = d + 2a$
Normal pitch (p_n) (*i.e.* in direction which is normal to the thread)	Axial pitch and lead angle	$p_n = p_x \times \cos \psi$
Worm Gear (Fig. 15.9)		
Pitch Diameter (D)	*No.* of teeth in gear (N_g) and axial pitch (of worm)	$D = (N_g \times p_x)/\pi$
Throat diameter (D_t)	Pitch diameter and addendum (a)	$D_t = D + 2 \times a$

15.4. Sources of Errors in Manufacturing Gears

The gears (gear teeth) are generally made by one of the following two methods :

(*i*) Reproducing method, in which the cutting tool is formed involve cutter, which forms the gear teeth profiles by reproducing the shape of the cutter itself. In this method, each tooth space is cut independently of the other tooth spaces.

(*ii*) Generating method, in which the cutting tool (hob) forms the profiles of several teeth simultaneously during constant relative motion of the tool and blank.

The various sources of errors in the gear made by reproducing method be due to (*i*) incorrect profile on the cutting tool, (*ii*) incorrect positioning the tool in relation to the work and (*iii*) incorrect indexing of the blank.

The sources of error when gears are made by generating method are : (*i*) Errors in the manufacture of the cutting tool (*ii*) errors in positioning the tool in relation to the work and (*iii*) errors in the relative motion of the tool and blank during the generating operation.

15.4.1. Gearing Operation Requirements. The gears can be classified into three categories according to the operation requirements as follows :

(*i*) High speed gears, (*ii*) High power gears, (*iii*) Precision gears.

The chief requirements, which an idle gearing must fulfil are : (*i*) The maintenance of a constant gearing ratio ; (*ii*) Provision of contact along the full length of the tooth, and (*iii*) Providing the proper backlash.

(The backlash or side play of a gear is clearance between teeth of meshing gears which provides for a slight rotation of one gear in reference to the mating gear.)

The important factor in case of high speed gears in a limitation of variations in the gearing ratio. When gears are operating at very high speeds then considerable acceleration appears even for small gearing ratio variations. These accelerations cause large dynamic loads which can damage the gear tooth.

High power gears are generally operated at low speeds but their teeth have to transmit large force. Therefore the gear teeth for this purpose should specify fullest possible contact along the tooth length.

The man requirement of the precision gears is the constant gearing ratio. These gears are generally used for counting and timing purposes, or in the measuring instruments and indexing devices, and usually have small face width and are of small module.

From what has been said above for the gearing operation requirements, it can be concluded that gears can be graded on the basis of the following accuracies :

(*i*) Kinematic accuracy of the gear. (*ii*) Smoothness of gear operation. (*iii*) Tooth bearing contact.

We shall now discuss about these accuracies in more details :

15.4.2. Kinematic Accuracy. This is represented by the permissible accumulated error in gear rotation per revolution of the gear.

The index of kinematic accuracy of a gear is the maximum kinematic error, which is defined as the maximum angular error in rotation of a gear, when it meshes in single-profile engagement with a precise (master) gear. This error is evaluated by drawing the kinematic error curve as shown in Fig. 15.10. This error is caused when the gear is being cut, due to errors in positioning the tool in relation to the blank, as well as to kinematic errors (errors in gear train) of the gear-cutting machine.

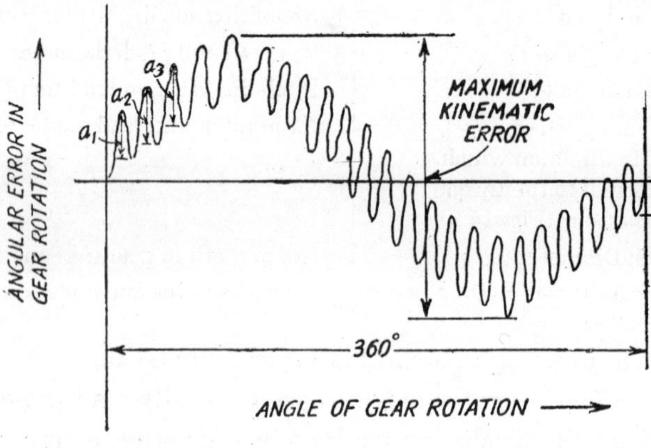

Fig. 15.10. Kinematic Error Curve.

The kinematic error may also be evaluated by the maximum accumulated error in the circular pitch. This is the maximum error between like profiles of any two teeth measured on a single gear.

Thus in Fig. 15.11 the maximum accumulated error in circular pitch = maximum plus error + maximum minus error. Another index of kinematic error may be the variations in the base tangent length. This constitutes the difference between the maximum base tangent and the minimum value on a single gear.

15.4.3. Smoothness of gear operation. Smoothness of gear operation is of vital importance to provide for noiseless engage-

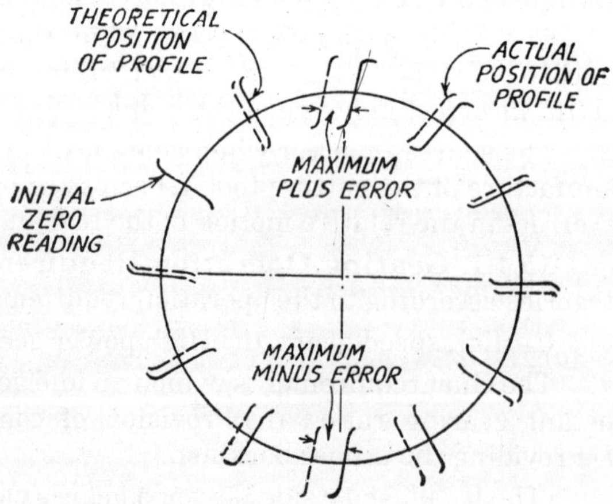

Fig. 15.11

ment and longer service life of gearing. It is represented by the value of the components of the full error in the angle or rotation of a gear, which are repeated many times per revolution of the gear. The index of smoothness of gear operation is the cyclic error which in the kinematic error curve in Fig. 15.12, is represented by Cyclic error = $(a_1 + a_2 + a_3 + \dots a_n)/n$.

Thus the cyclic error may be defined as the mean value of the range of variation in the kinematic error of the gear taken for all cycles during one revolution.

Smoothness of gear operation is also ensured by limiting the maximum permissible deviations in basic pitch and the involute profile tolerance.

15.4.4. Tooth bearing contact. The index of tooth bearing contact in gearing is the gear contact pattern, which is the part of the surface on a gear tooth profile showing traces of bearing contact with the teeth of the mating gear after they have been run together with the drive gear braked slightly.

Bearing contact for mating gears is represented by the relative size of the bearing contact pattern in per cent.

Along the tooth length it is denoted by the ratio of the traces of contact, subtracting gaps or skipped places large than the size of the module in mm, to the full length of the tooth.

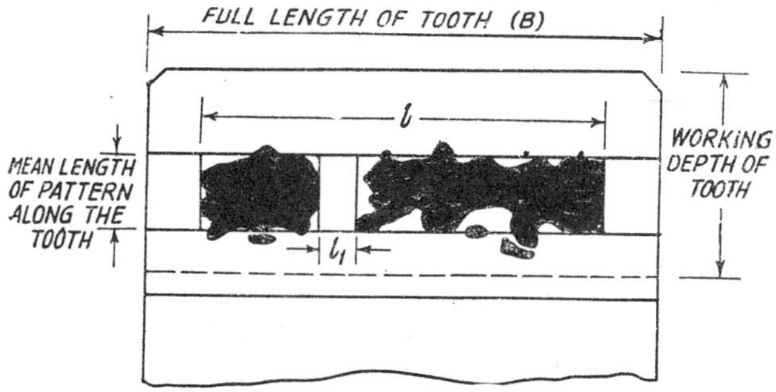

Fig. 15.12. Tooth bearing contact pattern.

Thus in Fig. 15.12, this is equal to $\frac{l - l_1}{B} \times 100\%$.

Along the depth of the tooth, the relative size of the bearing contact pattern

$$= \frac{\text{Mean height of the pattern along the full length}}{\text{Full length of tooth}} \times 100\%.$$

Gear errors. Various possible types of error on spur, helical, bevel and worm gears are described below :

(*i*) *Adjacent pitch error* — Actual pitch—design pitch.

(*ii*) *Cumulative pitch error* — Actual length between corresponding flanks of teeth not adjacent to each other—design length.

(*iii*) *Profile error* — The maximum distance of any point on the tooth profile form and normal to the design profile when the two coincide at the reference circle.

(*iv*) *The tooth to tooth composite error—single flanks* (Refer Fig. 15.13 *a*). — The range of difference between the displacement at the pitch circle of a gear and that of a master gear meshed with it at fixed centre when moved through a distance corresponding to one pitch with only the driving and driven flanks in contact.

(*v*) *The total composite errors—single flank* — The range of difference between the displacement at the pitch circle of a gear and that of a master gear meshed with it at fixed centre distance when moved through one revolution with only the driving and driven flanks in contact (Refer Fig 15.3 *a*).

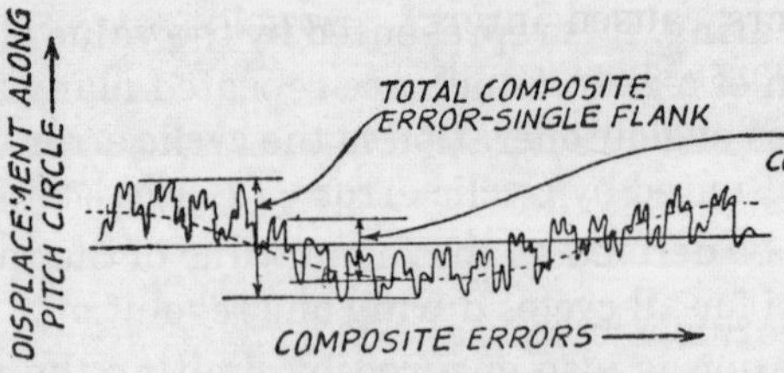

Fig. 15.13. (*a*) Typical single flank error chart.

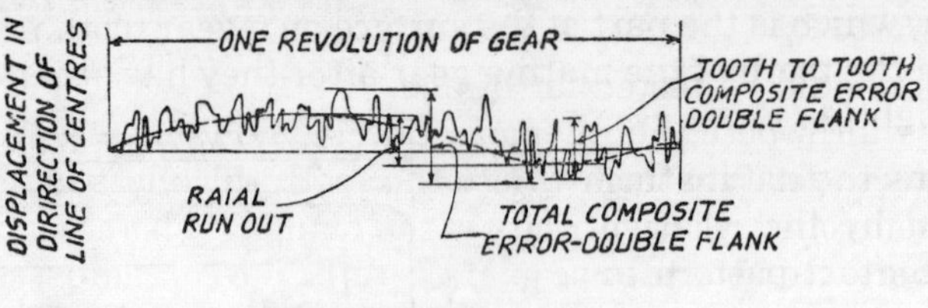

Fig. 15.13. (*b*) Composite Errors—Typical double flank error chart.

(*vi*) *The tooth to tooth composite error—double flank*	The range of variation in the minimum centre distance between a gear and a master gear when rotated through a distance corresponding to the pitch of the teeth (Refer Fig. 15.13 *a*).
(*vii*) *The total composite error—double flank*	The range of variation in the minimum centre distance between a gear and a master gear when the gear is rotated through one revolution (Refer Fig. 15.13 *b*).
(*viii*) *The tooth thickness error*	Actual tooth thickness measured along the surface of the reference cylinder—design tooth thickness.
(*ix*) *Cyclic error*	An error occurring during each revolution of the element under consideration.
(*x*) *Periodic error*	An error occurring at regular intervals not necessarily corresponding to one revolution of the component.
(*xi*) *Run out*	It is the total range of reading of a fixed indicator with the contact point applied to a surface rotated, without axial movement about a fixed axis.
(*xii*) *Radial run out*	It is the run-out measured along a perpendicular to the axis of rotation.
(*xiii*) *Eccentricity*	It is half the radial run-out.
(*xiv*) *Axial run-out (wobble)*	It is the run-out measured parallel to the axis of rotation, at a specified distance from the axis.
(*xv*) *Undulation*	A periodical departure of the actual tooth surface from the design surface (Refer Fig. 15.13 *b*).
(*xvi*) *Undulation height*	The normal distance between two surfaces that contain respectively the crests and the troughs of the tooth undulation (Refer Fig. 15.13 *c*).
(*xvii*) *Wavelength of an undulation*	The distance between two adjacent crests of an undulation (Refer Fig. 15.13 *c*).
(*xviii*) *Tooth alignment error*	The distance of any point on a tooth trace from the design tooth trace passing through a selected reference point on that tooth (Refer Fig. 15.13 *c*).

The presence of these errors caused interference in efficient operation of gears. These result in non-smooth and noisy operation which ultimately affect the working life.

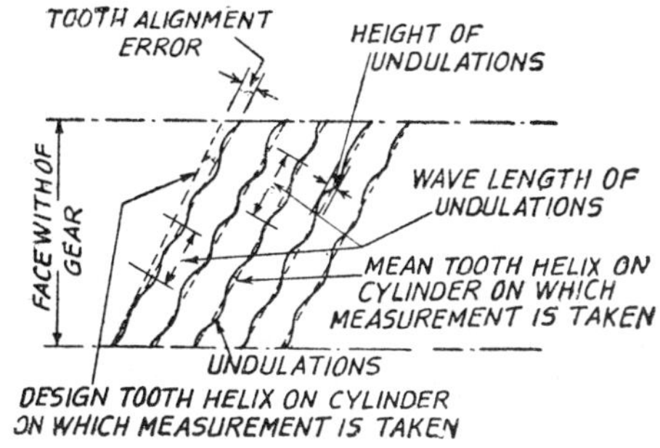

Fig. 15.13 (*c*)

15.5. Gear Measurement

For proper inspection of gear, it is essential to pay attention to the raw material, each process in the production cycle, machining the blanks, heat treatment, the cutting and surface finish of the teeth.

The gear blank should be tested for dimensional accuracy (face width, bore, hub, length, and outside diameter), and eccentricity. As outside diameter forms the datum from where the tooth thickness is measured, it forms an important item to be controlled. Concentricity of the blanks is also essential and the side faces should be true to the bore. On very precise gears details like tip radius, shape of root provided and surface finish are also measured.

Concentricity of teeth is an important item and should be checked to ensure that the set up and equipment is in good order. If teeth are not concentric then fluctuating velocity will be noticed on the pitch line while transmitting motion. This also leads to inaccuracy of parts, when being used for indexing purposes. Tooth concentricity can be checked by (*i*) mounting the gear between the bench centres, placing a standard roller in each tooth space and then using a dial indicator, (*ii*) using a projector in which case the teeth are brought against a stop and each image of tooth on screen should coincide with a line on the screen, (*iii*) using a gear testing fixture fitted with a spring loaded slide and dial indicator, in which case the spring exerts a constant pressure on the mating teeth and the movement of the dial indicator gives the measure of the eccentricity of teeth.

Good alignment of each tooth on a gear is essential, as otherwise the load will not be distributed evenly over its face. If teeth of a gear be machined poorly, it is quite probable that the load may be carried by one edge only introducing high bearing stresses. Tooth alignment can be checked by placing a standard roller in the tooth space and checking for parallelism off a surface plate. In the other method, the teeth on one gear ar lightly marked with prussian blue and mounted in a testing machine having a master gear. The contacts made on the mating gear give good idea of tooth alignment.

Hardness of gear tooth should be tested to ensure that heat treatment is proper and that the desired hardness due to provision of adequate thickness and grain size have been attained.

The method employed for measuring and testing of gears depends upon various factors, such as the precision of gears, method of manufacturing, equipment available etc.

The accuracy of any gear mainly depends upon the cutter accuracy and the setting of the machine. Thus for most of the gears, optical projection and rolling tests will suffice. But in manufacture of high precision gears, it is necessary to determine the accuracy of individual elements *e.g.*, tooth thickness, pitch of teeth and form of teeth etc.

Accuracy of Measurement. While the accuracy of measurement of gear depends upon the measuring equipment available, it must be emphasised that there are some in-built limitations in the gear itself, such as the inability of a gear to define its own axis of rotation. Thus if the reference circle of gear is eccentric, it would be reflected in pitch error. Similarly the errors in tooth surface finish such as undulation would zeopardise the validity of a single point measurement on a tooth flank.

The inspection of gear is mainly of two types (*a*) Analytical, and (*b*) Functional. By analytical inspection of gears we mean that all the individual elements of the gear teeth are checked. This method is slow and tedious and not of much use for industry. The discrete error values of pitch, tooth profile etc. can't give a true overall assessment of the accuracy of a gear. It is not easy to assess accurately how these elemental values combine in practice to give a prescribed performance under operational conditions. However this method is of great understanding of the subject to student. Nevertheless it may be stressed that all errors in pitch profile cause variations in the uniformity of rotary motion and the errors in tooth alignment or helix angle result in the concentration at small areas instead of being distributed uniformly.

The analytical inspection of the gears consists in determination of the following teeth elements in which the errors are caused due to manufacturing errors.

(A) Profile. (B) Spacing. (C) Pitch. (D) Runout or eccentricity or concentricity. (E) Thickness of tooth. (F) Lead. (G) Backlash.

The functional type of inspection consists of carrying out the running test of gear with another gear which is more accurate and is known as control gear or master gear, to determine composite vibration, noise level, or variation in action. If a pair of gears work together at the designed speed and under load with little noise, they are considered satisfactory for many purposes. If drive is noisy, then individual elements have to be measured. However master gear has to be measured on elemental basis only.

15.6. Rolling Tests

This is the most commonly used test under production conditions. This consumes much less time and gives quite accurate results. In rolling test, the gear to be tested is actually compared with a hardened and ground master gear. This test is generally performed on a most commonly used machine Parkson Gear Tester. This test reveals any errors in tooth form, pitch and concentricity of pitch line. When two gears are in mesh with each other, then any of the above errors will cause the variation of centre distance. This fact is utilised for testing the errors in gear by this machine.

It essentially consists of a base. Two carriages, one fixed and the other movable are mounted on the base. The position of the fixed carriage can be adjusted in order to accommodate a wide range of diameters. While in use, this fixed carriage is locked in one position. The movable carriage is spring loaded towards the fixed carriage. Two spindles are mounted in a parallel plane on each carriage and these are made to suit the bore of the gears.

The distance between the centre of two spindles is adjusted to be equal to the centre distance by slip gauges. A dial gauge is made to rest against the movable carriage and its reading is adjusted at zero at this time. The master gear is mounted on the spindle on fixed carriage and gear to be tested on the movable carriage. The gears when in mesh are then rotated by hand and the variations in the dial gauge readings are observed. If it falls outside the set limits, then gear is rejected. The variations might also be recorded by some electrical pick up in which the movement of carriage is first converted into electrical impulse which is magnified further and trace of variation obtained on a graph paper. The trace obtained

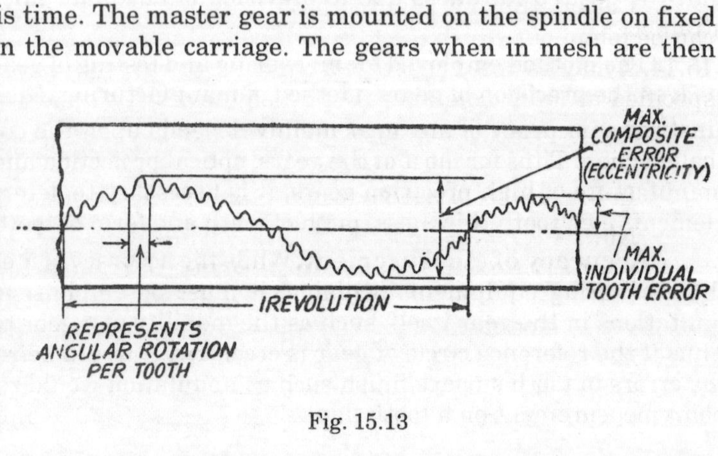

Fig. 15.13

will be depicting the compound errors *i.e.*, all errors like eccentricity and tooth form errors etc., which occur together and the trace will be as shown in Fig. 15.8.

The machine could also be used to carry out more complex tests by suitable modification in its operation, *e.g.*, by locking the movable carriage at the running centre distance of the gears, and by fixing the master gear, the backlash can be determined by setting a dial gauge at the pitch line of the production gear. It is also possible to check the gears for smooth running at this setting and this is very essential for gears. This is judged by the noise produced.

For these tests, if master gear is not available, then any two mating gears are mounted on the spindle and they are tested twice at relative angular positions of 180° to each other so that any compensating errors in one angular position in gears are also revealed.

15.7. Measurement of Individual Elements

15.7.1. Measurement of tooth thickness. The permissible error or the tolerance on thickness of tooth is the variation of actual thickness of tooth from its theoretical value. The tooth thickness is generally measured at pitch circle and is therefore, the pitch line thickness of tooth. It may be mentioned that the tooth thickness is defined as the length of an arc, which is difficult to measure directly. In most of the cases, it is sufficient to measure the chordal thickness *i.e.*, the chord joining the intersection of the tooth profile with the pitch circle. Also the difference between chordal tooth thickness and circular tooth thickness is very small for gear of small pitch. The thickness measurement is the most important measurement because most of the gears manufactured may not undergo checking of all other parameters, but thickness measurement is a must for all gears. There are various methods of measuring the gear tooth thickness.

(*i*) Measurement of tooth thickness by gear tooth vernier calliper. (*ii*) Constant chord method. (*iii*) Base tangent method. (*iv*) Measurement by dimension over pins.

The tooth thickness can be very conveniently measured by a gear tooth vernier. Since the gear tooth thickness varies from the tip of the base circle of the tooth, the instrument must be capable of measuring the tooth thickness at a specified position on the tooth. Further this is possible only when there is some arrangement to fix that position where the measurement is to be taken. The tooth thickness is generally measured at pitch circle and is, therefore, referred to as pitch-line thickness of tooth. The gear tooth vernier has two vernier scales and they are set for the width (w) of the tooth and the depth (d) from the top, at which w occurs.

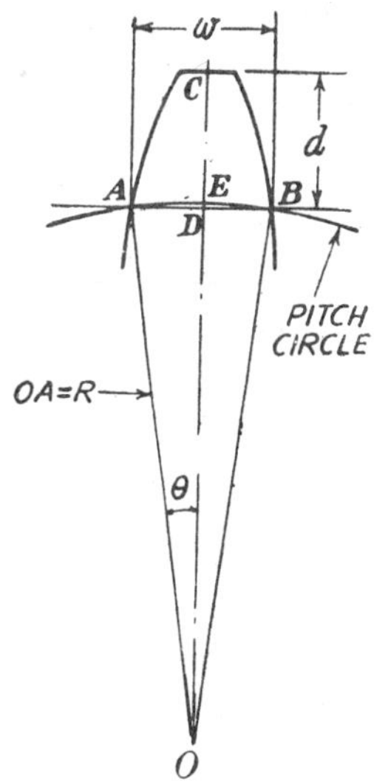

Fig. 15.14

Considering one gear tooth, the theoretical values of w and d can be found out which may be verified by the instrument. In Fig. 15.14, it may be noted that w is a chord ADB, but tooth thickness is specified as an arc distance AEB. Also the distance d adjusted on instrument is slightly greater than the addendum CE, w is therefore called chordal thickness and d is called the chordal addendum.

In Fig. 15.14, $w = AB = 2AD$

Now, $A\hat{O}D = \theta = 360°/4N$, where N is the number of teeth,

$w = 2AD = 2 \times AO \sin\theta = 2R \sin 360/4N$ (N = pitch circle radius)

$$\text{module } m = \frac{\text{P.C.D.}}{\text{No. of teeth}} = \frac{2R}{N}, \therefore \; R = \frac{N.m.}{2}$$

$$\therefore \quad w = 2\,\frac{Nm}{2}\sin\left(\frac{360}{4N}\right) = N.m.\sin\left(\frac{90}{N}\right) \qquad ...(1)$$

Also from Fig. 15.14, $d = OC - OD$

But $OC = OE + \text{addendum} = R + m = (Nm/2) + m$

and $OD = R\cos\theta = \frac{Nm}{2}\cos\left(\frac{90}{N}\right)$

$$\therefore \quad d = \frac{Nm}{2} + m - \frac{Nm}{2}\cos\left(\frac{90}{N}\right) = \frac{Nm}{2}\left[1 + \frac{2}{N} - \cos\left(\frac{90}{N}\right)\right] \qquad ...(2)$$

Any error in the outside diameter of the gear must be allowed for when measuring tooth thickness.

In the case of helical gears, the above expressions have to be modified to take into account the change in curvature along the pitch line. The virtual number of teeth Nv for helical gear $= N/\cos^3\alpha$ (α = helix angle)

Hence in Eqs. (1) and (2), N can be replaced by $N/\cos^3\alpha$ and m by m_n (normal module).

$$\therefore \quad w = \frac{Nm_n}{\cos^3\alpha}\sin\left(\frac{90}{N}\cos^3\alpha\right), \text{ and } d = \frac{Nm_n}{\cos^3\alpha}\left[1 + \frac{2\cos^3\alpha}{N} - \cos\left(\frac{90}{N}\cos^2\alpha\right)\right].$$

These formulae apply when backlash is ignored. On mating gears having equal tooth thickness and without addendum modifications, the circular tooth thickness equals half the circular pitch minus half the backlash.

Gear Tooth Calliper. (Refer Fig. 15.15). It is used to measure the thickness of gear teeth at the pitch line or chordal thickness of teeth and the distance from the top of a tooth to the chord. The thickness of a tooth at pitch line and the addendum is measured by an adjustable tongue, each of which is adjusted independently by adjusting screw on graduated bars. The effect of zero errors should be taken into consideration.

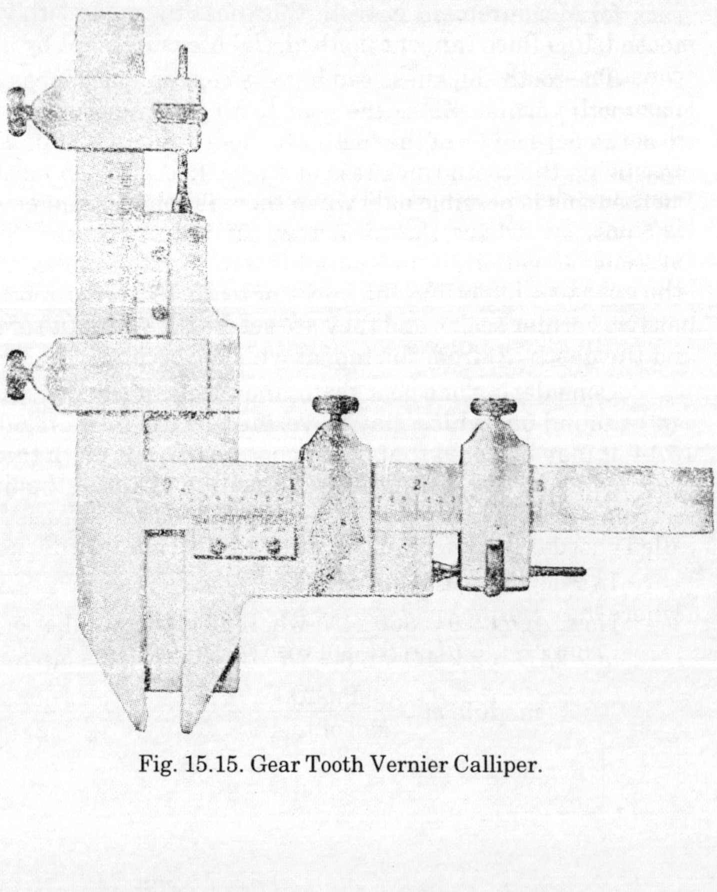

Fig. 15.15. Gear Tooth Vernier Calliper.

This method is simple and inexpensive. However it needs different setting for a variation in number of teeth for a given pitch and accuracy is limited by the least count of instrument. Since the wear during use is concentrated on the two jaws, the calliper has to be calibrated at regular intervals to maintain the accuracy of measurement.

15.7.2. Constant Chord Method. In the above method, it is seen that both the chordal thickness and chordal addendum are dependent upon the number of teeth. Hence for measuring a large number of gears for set, each having different number of teeth would involve separate calculations. Thus the procedure becomes laborious and time-consuming one.

The constant chord method does away with these difficulties. Constant chord of a gear is measured where the tooth flanks touch the flanks of the basic rack. The teeth of the rack are straight and inclined to their centre lines at the pressure angle as shown in Fig. 15.16.

Also the pitch line of the rack is tangential to the pitch circle of the gear and, by definition, the tooth thickness of the rack along this line is equal to the arc tooth thickness of the gear round its pitch circle. Now, since the gear tooth and rack space are in contact in the symmetrical position at the points of contact of the flanks, the chord is constant at this position irrespective of the gear of the system in mesh with the rack. This is the property utilised in the constant chord method of the gear measurement.

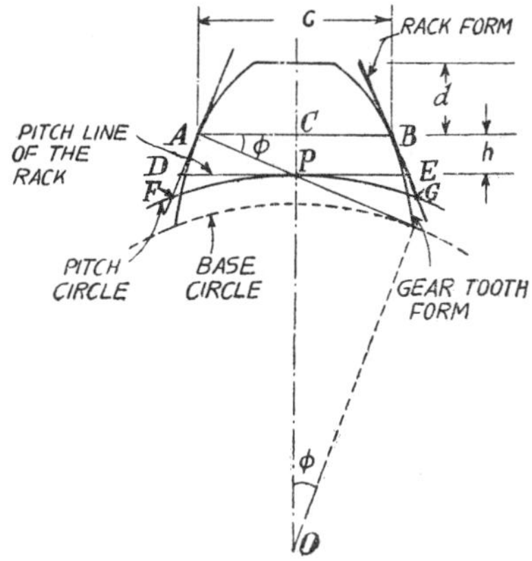

Fig. 15.16

The measurement of tooth thickness at constant chord simplified the problem for all number of teeth. If an involute tooth is considered symmetrically in close mesh with a basic rack form, then it will be observed that regardless of the number of teeth for a given size of tooth (same module), the contact always occurs at two fixed point A and B. AB is known as constant chord. The constant chord is defined as the chord joining those points, on opposite faces of the tooth, which make contact with the mating teeth when the centre line of the tooth lies on the line of the gear centres. The value of AB and its depth from the tip, where it occurs can be calculated mathematically and then verified by an instrument. The advantage of the constant chord method is that for all number of teeth (of same module) value of constant chord is same. In other words, the value of constant chord is constant for all gears of a meshing system. Secondly it readily lends itself to a form of comparator which is more sensitive than the gear tooth vernier.

In Fig. 15.16, $PD = PF = \text{arc } PF = 1/4 \times \text{circular pitch} = \frac{1}{4} \times \frac{\pi \times \text{P.C.D.}}{N} = 1/4 \times \pi \times m$

Since line AP is the line of action, *i.e.* it is tangential to the base circle, $\angle CAP = \phi$

$\therefore$ In right angled ΔAPD, $AP = PD \cos\phi = (\pi/4)m \cos\phi$

In triangle PAC, $AC = AP\cos\phi = (\pi/4)m\cos^2\phi$

$$c = \text{constant chord} = 2AC = (\pi/2)\, m \cos^2\phi \qquad ...(3)$$

where ϕ is the pressure angle (from Fig. 15.16)

For helical gear, constant chord $= (\pi/2)\, m_n \cos^2\phi_n$

where m_n = normal module *i.e.* module of cutter used and ϕ_n = normal pressure angle.

Now $\quad PC = AP\sin\phi = (\pi/4)\, m\cos\phi\sin\phi$

$$\therefore \quad d = \text{addendum} - PC = m - \frac{\pi}{4} m\cos\phi\sin\phi = m\left(1 - \frac{\pi}{4}\cos\phi\sin\phi\right) \qquad ...(4)$$

$$\left[\text{For helical gear, } d = m_n\left(1 - \frac{\pi}{4}\cos\phi_n\sin\phi_n\right)\right]$$

Also height of AB above pitch line $= PC = \frac{\pi m}{4} \sin \phi \cos \phi = \frac{\pi m}{8} \sin 2\phi$ ...(5)

15.7.3. Base Pitch. This is defined as the circular pitch of the teeth measured on the base circle. In Fig. 15.17, AB represents the portion of a gear base circle, CD and EF the sides of two teeth, FD being the base pitch. From the property of involute, if any line as GH is drawn to cut the involutes and tangential to the base circle, the $GH = FD$.

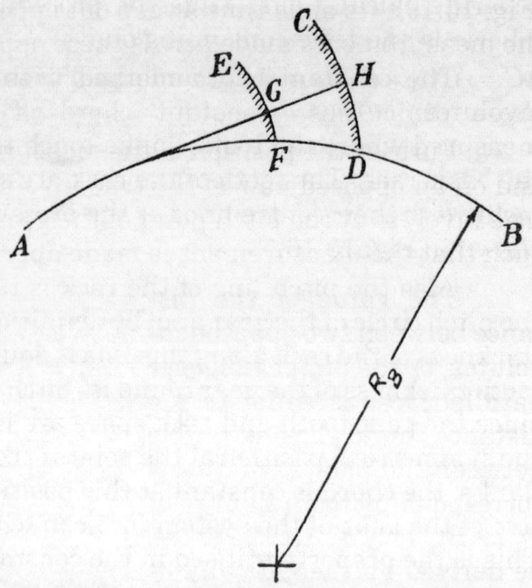

Fig. 15.17

Thus base pitch could also be defined as equal to the linear distance between a pair of involutes measured along a common generator.

Base circumference $= 2\pi R_B$

$\therefore$ Base pitch $= 2\pi R_B/N$

If ϕ is the pressure angle, then

$R_B = \text{P.C.R.} \times \cos \phi = (\text{P.C.D.}/2) \cos \phi$

$\therefore$ Base pitch $= (2\pi/N) \times (\text{P.C.D.}/2) \times \cos \phi$

$= \pi m \cos \phi$

This is the distance between tangents to the curved portions of any two adjacent teeth and can be measured either with a height gauge or on an enlarged projected image of the teeth. This principle is utilised in 'David Borwn' tangent comparator and it is the most commonly used method.

15.7.4. Base Pitch Measuring Instrument. This instrument has three tips. One is the fixed measuring tip, other one is the sensitive tip whose position can be adjusted by a screw and the further movement of it is transmitted through a leverage system to the dial indicator ; and the third tip is the supplementary adjustable stop which is meant for the stability of the instrument and its position can also be adjusted by a screw. The distance between the fixed and sensitive tip is set to be equivalent to the base pitch of the gear with the help of slip gauges. The properly set-up instrument is applied to the gear so that all the three tips contact the tooth profile. The reading on dial indicator is the error in the base pitch.

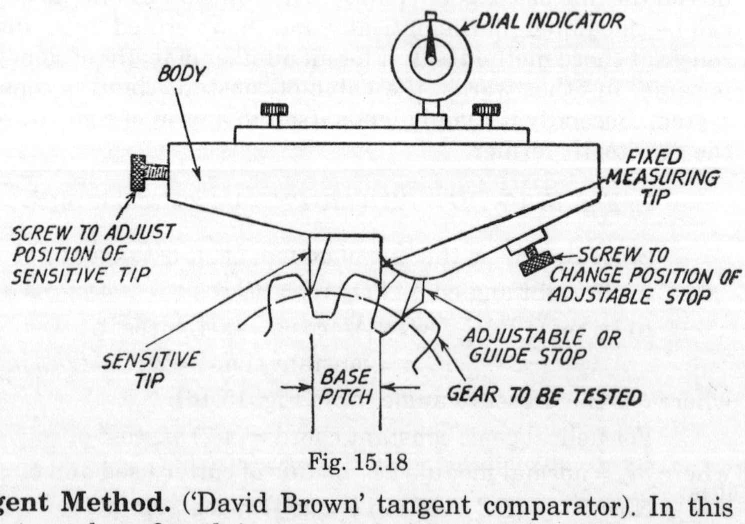

Fig. 15.18

15.7.5. The Base Tangent Method. ('David Brown' tangent comparator). In this method, the span of a convenient number of teeth is measured with the help of the tangent comparator. This uses a single vernier calliper and has, therefore, the following advantages over gear tooth vernier which used two vernier scales :

(*i*) the measurements do not depend on two vernier readings, each being function of the other.

(*ii*) the measurement is not made with an edge of the measuring jaw with the face.

Consider a straight generator (edge) ABC being rolled back and forth along a base circle (Fig. 15.19). Its ends thus sweep out opposed involutes $A_2\,AA_1$ and $C_2\,CC_1$ respectively. Thus the measurements made across these opposed involutes by span gauging will be constant (*i.e.* $AC = A_1C_1 = A_2C_2 = A_0C_0$) and equal to the arc length of the base circle between the origins of involutes.

Further the position of the measuring faces is unimportant as long as they are parallel and on an opposed pair of the true involutes. As the tooth form is most likely to conform to a true involute at the pitch point of the gear, it is always preferable to choose a number of teeth such that the measurement is made approximately at the pitch circle of the gear.

The value of the distance between two opposed involutes, or the dimension over parallel faces is equal to the distance round the base circle between the points where the corresponding tooth flanks cut *i.e.*, ABC in Fig. 15.19. It can be derived mathematically as follows :

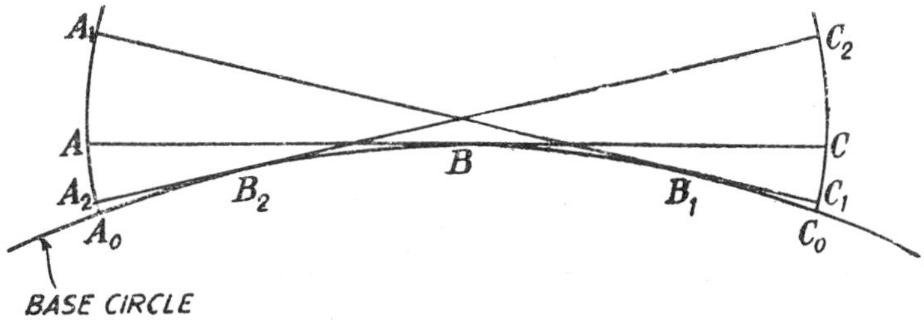

Fig. 15.19. Generation of pair of involutes by a common generator.

The angle between the points A and C on the pitch circle where the flanks of the opposed involute teeth of the gear cut this circle can be easily calculated.

Let us say that the gear has got 'N' number of teeth and AC on pitch circle corresponds to 'S' number of teeth. (Fig. 15.20) ; $\therefore$ Distance $AC = (S - 1/2)$ pitches

$\therefore$ Angle subtended by $AC = (S - 1/2) \times 2\pi/N$ radians.

Angles of arcs BE and BD

Involute function of pressure angle = $\delta = \tan\phi - \phi$

$$\therefore \text{ Angle of arc } BD = \left(S - \frac{1}{2}\right) \times \frac{2\pi}{N} + 2(\tan\phi - \phi)$$

$$\therefore \quad BD = \text{Angle of arc } BD \times R_b$$

$$= \left[\left(S - \frac{1}{2}\right) \times \frac{2\pi}{N} + 2(\tan\phi - \phi)\right] \times R_p \cos\phi \quad [\text{because } R_b = R_p \cos\phi]$$

$$= \frac{mN}{2}\cos\phi\left[\left(S - \frac{1}{2}\right)\frac{2\pi}{N} + 2(\tan\phi - \phi)\right] \left[\text{because } R_p = \frac{mN}{2}\right]$$

$$= Nm\cos\phi\left[\frac{\pi S}{N} - \frac{\pi}{2N} + \tan\phi - \phi\right]$$

As already defined, length of arc BD = distance between two opposed involutes and thus it is

$$= Nm\cos\phi\left[\tan\phi - \phi - \frac{\pi}{2N} + \frac{\pi S}{N}\right]$$

It may be noted that when backlash allowance is specified normal to the tooth flanks, this must be simply subtracted from this derived value.

Tables are also available which directly give this value for the given values of S, N and m.

This distance is first calculated and then set in the 'David Brown' tangent comparator (Fig. 15.21) with the help of slip gauges. The instrument essentially consists of a fixed anvil and a movable anvil. There is a micrometer on the moving anvil side and this has a very limited movement on either side of the setting. The distance is adjusted by setting the fixed anvil at desired place with the help of looking ring and setting tubes.

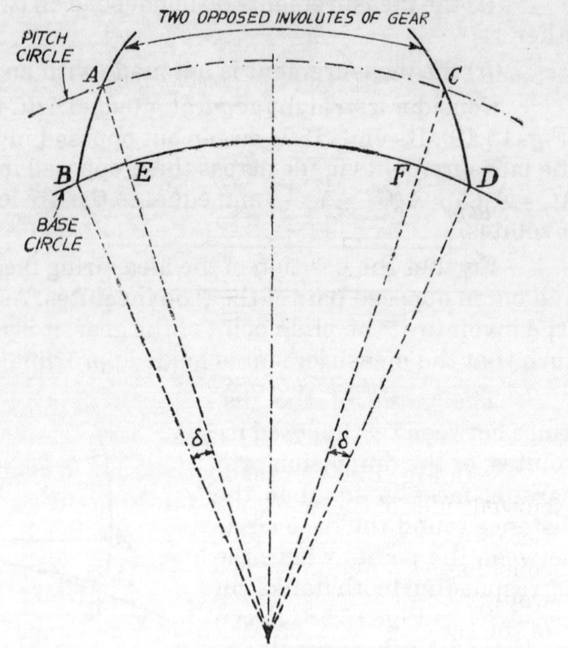

Fig. 15.20

15.7.6. Tangential Gear Tooth Calliper. It is utilised for measuring variations on the basic tooth profile from the outside diameter of spur and helical gears. The instrument consists of body, on the underside of which there are two slides having the tips acting like measuring contacts. The extended spindle of a dial indicator with the contact point A passes between the two tips along the vertical axis of symmetry of the instrument. The measuring tips are spread apart or brought together simultaneously and symmetrically in reference to the central axis by a screw which has a right-hand and a left-hand thread. The contact faces of the measuring tips are flat and arranged at angles of 14.5° or 20° with the central axis. The calliper is set up by means of a cylindrical master gauge of proper diameter based on the module of the gear being checked. After adjusting the tips by the screw, these are locked in position by locking nuts. The properly set up instrument is applied to the gear tooth and the dial indicator reading shows how much the position of the basic tooth profile deviates in reference to the outside diameter of the gear.

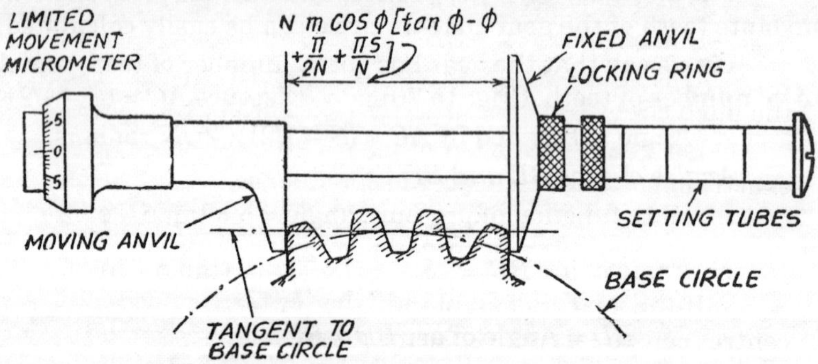

Fig. 15.21. 'David Brown' Base Tangent Comparator.

15.7.7. Test Plug Method for Checking Pitch Diameter and Tooth Spacing. Measurement over the rollers placed in the space between a pair of gear teeth gives a convenient method for checking tooth spacing and the pitch diameter. The special case of the roller with its centre on the pitch circle simplifies the problem. It is, therefore, considered desirable to find the diameter on the roller whose centre will lie on the pitch circle and to derive an expression for the distance over the rollers placed in opposite teeth spaces with the centres of rollers lying on the pitch circle.

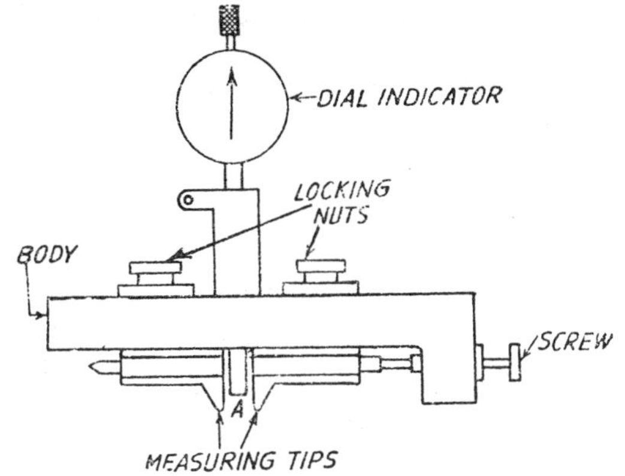

Fig. 15.22

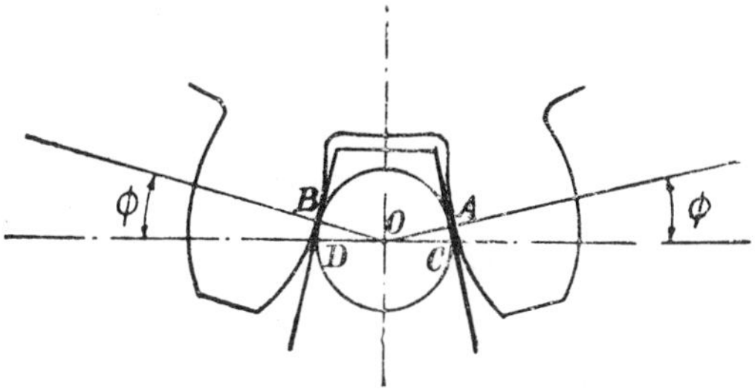

Fig. 15.23

In Fig. 15.23 a rack tooth is shown symmetrically in mesh with a gear tooth space, the curved sides of the gear tooth touching the straight rack sides at A and B on the line of action. Let us assume that the centre of the roller lies on the pitch point O. Now, if the rack tooth be considered as an empty space bounded by its outline, a circle with centre O and radius OB would fit and touch the rack space at A and B since OA and OB are perpendicular to the sides of the rack tooth. Thus the circle would touch the gear teeth at A and B.

In Δ OBD, OB is the radius of roller

$$OD = \text{Circular pitch}/2 = (\pi/4)m$$

$O\hat{B}D = 90°$, $B\hat{O}D = \phi$ = pressure angle, $\therefore$ $OB = OD \cos\phi = (\pi/4)\, m \cos\phi$

Dia. of roller $= 2 \times OB = 2 \times (\pi m/4) \cos\phi = (\pi/4)\, m \cos\phi$.

This is the diameter of a roller which will rest in tooth space and lie with its centre on the pitch circle. This value is constant for all gears of same pitch and pressure angle.

For gears with even number of teeth, a direct measurement by placing two rollers in exactly opposite tooth spaces is possible. In this case, the gauging diameter over the rollers

$$= \text{P.C.D.} + \text{Dia of roller} = mN + \frac{\pi}{2} m \cos\phi = m\left[N + \frac{\pi}{2}\cos\phi\right]$$

If the gear has an old number of teeth, a radial measurement with the gear between centres can be carried out, using a comparator with the gear. The accuracy of the spacing over any number of teeth may be checked by finding the angles subtended at the centre and comparing this with that obtained from a chordal check of the plugs.

As already indicated, precision gears and other gears are generally checked for tooth thickness by dimension over pins, as the dimensions over pins reading is mathematically related to the tooth thickness. This also verifies the correctness of profile and other elements of gear. Under this method two pins of equal diameter are placed in two opposite tooth spaces and compared with required value found using the equation, $M = \dfrac{m\,.\,N\,.\cos\phi}{\cos\phi_m} + d$

for gears having even number of teeth, and $= \dfrac{m\,.\,N\,.\cos\phi}{\cos\phi_m} \times \cos\dfrac{90}{N} + d$

for gears having odd number of teeth.

where M = Required size over the wires, m = Module, ϕ = Pressure angle, d = Diameter of wire $= 1.728 \times m$, ϕ_m = Pressure angle at centre of pin and is given by the relation

$$\text{inv. } \phi_m = \text{inv. } \phi + \frac{d}{mN \cos \phi} - \frac{\pi}{2N}, \text{ where inv. } \phi = \tan \phi - \phi \text{ (in radians)}.$$

Using the above equations, the size 'M' over wires can be calculated.

In case of helical gears the formulae used are as below :

$$M = \frac{m \,.\, N \,.\, \cos \phi_s}{\cos \alpha \times \cos \phi_{s1}} + d \text{ for even number of teeth} \quad ...(1)$$

Here $\tan \phi_s = \dfrac{\tan \phi_n}{\cos \alpha}$ where α is the helix angle ...(2)

and $\text{inv. } \phi_{s1} = \text{inv. } \phi + \dfrac{d}{m_N N \cos \phi_n} - \dfrac{\pi}{2N}$ ...(3)

The helical gears with odd number of teeth should not be measured with two wires, because in this case the correction factor cos (90/N) is not valid and can result in serious errors. In such cases the gear can be mounted on an arbour and a radial measurement made from the top of the wire to the axis of gear.

Size over wires/balls for helical gears

1. *When balls are used :*

Size over balls $M = \dfrac{mnN \,.\, \cos \phi_s}{\cos \phi_{1s}} + d$ for even number of teeth

$$\frac{m_N \,.\, N \,.\, \cos \phi_s}{\cos \phi_{1s}} \times \cos \frac{90}{N} + d \text{ for odd number of teeth}$$

Here ϕ_s is given by $\tan \phi_s = \tan \phi_n / \cos \alpha$

and angle ϕ_{1s} is given by $\text{inv. } \phi_{1s} = \text{inv. } \phi_s + \dfrac{d}{m_n \,.\, N \,.\, \cos \phi_n} - \dfrac{\pi}{2N}$

2. *When wires are used :*

(*a*) When No. of teeth are even, Size over wires $M = \dfrac{m_N \,.\, N \,.\, \cos \phi_s}{\cos \alpha \cos \phi_{1s}} + d$ for even number of teeth only

(Angles ϕ_s and ϕ_{ls} being found in same way as mentioned above).

(*b*) When No. of teeth are odd then Size over wires

$$M = y\sqrt{2\,[1 - \cos(\gamma - \theta)] + \frac{\sin^2(\gamma - \theta)}{\cos^2 \beta}} + d \quad ...(1)$$

where $y = \dfrac{m_N \,.\, N \,.\, \cos \phi_s}{2\cos \phi_{1s} \times \cos \phi}$ ...(2), *and* $\text{inv. } \phi_{1s} = \text{inv. } \phi_s + \dfrac{d}{m_N \,.\, N \,.\, \cos \alpha} - \dfrac{\pi}{2N}$ ...(3)

and angle $\gamma = 180\left(\dfrac{N-1}{N}\right)$ ...(4), *and* angle β is given by $\cos \beta = \cos \alpha \times \dfrac{m_N \,.\, N}{2y}$ ...(5)

and angle θ is $= \theta_1 + \Delta\, \theta_1$ ...(6), *where* $\tan \theta_1 = \dfrac{\sin y}{\cos \beta + \cos \gamma}$ ...(7)

and $\Delta\, \theta_1 = \dfrac{-\Delta h_1}{1 + \dfrac{\cos(y - \theta_1)}{\cot^2 \beta}}$ ...(8), *where* $\Delta h_1 = \theta_1 - \dfrac{\sin(y - \theta_1)}{\cot^2 \beta}$ ...(9)

Also $M \cong \dfrac{m_N \,.\, N \,.\, \cos \phi_s}{\cos \alpha \cos \phi_{1s}} \times \cos \dfrac{90}{N} + d$ for, odd number of teeth.

15.7.8. Checking of Profile of Involute Shape of Gear. Profile is the portion of tooth flank between the outside circle or start of tooth tip and the specified form circle of diameter approximately equal to pitch circle diameter minus twice the module.

Profile error is the deviation of the actual tooth from the theoretical profile in the designed reference plane of rotation. For testing profile, tip relief and any portion of the tooth surface below the active profile is not to be considered. The tolerance on the profile error is permitted as per the following table.

Table

Accuracy Class or Grade of Gear	*Profile Tolerance in Microns*
1	$2.0 + 0.06 \times k$
2	$2.5 + 0.10 \times k$
3	$3.0 + 0.16 \times k$
4	$4.0 + 0.25 \times h$
5	$5.0 + 0.40 \times k$
6	$6.3 + 0.63 \times k$
7	$8.0 + 1.0 \times k$
8	$10 + 1.6 \times k$
9	$16 + 2.5 \times k$
10	$25 + 4.0 \times k$
11	$40 + 6.3 \times k$
12	$63 + 10 \times k$

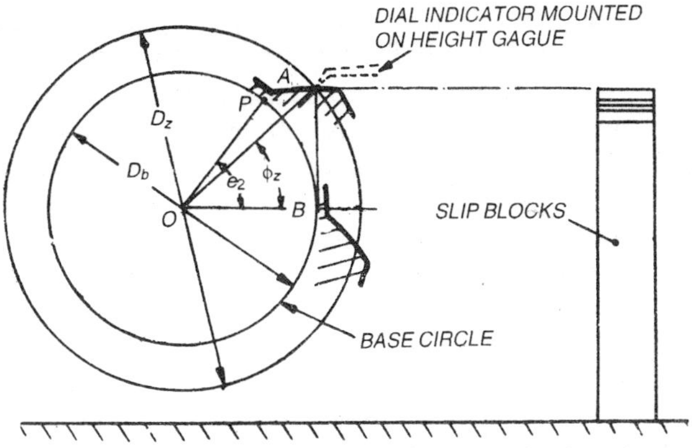

(P is the point corresponding to origin of involute)
Method of inspection of tooth profile using dividing head and slip gauge.
Fig. 15.24

Here k is a tolerance and is given by $k = m + 0.1\sqrt{D}$.

Where m is module, and D is the pitch circle diameter in mm.

(*a*) **Inspection of profile using dividing head and height gauge.**

This method has been derived from the property of the involute as shown in Fig. 15.24. The distance between points A and B measured on the generatix tangent to the base circle is equal to the arc from the tangent point B to the origin P of the involute on the base circle, *i.e.* arc $BP = AB = (D_b\,\pi/360) \times e_z$

S. No.	Dividing Head Reading	e_z *values*	*AB values* $= \dfrac{D_b \times \pi}{360} \times e_z$
1		e_z	
2		$e_z + 10'$	
3		$e_z + 20'$	
4		—	
5		—	
6		$e_z - 10'$	
7		$e_z - 20'$	
8		—	
9		—	
10		—	

When involute profile is rotated on dividing head by small roll angles, then the consecutive profile points A fall on to the vertical tangent line. It is possible to compute suitable roll angles for any diameter D_z from the corresponding pressure angle by equations :

$$D_z/D_b = \sec \phi_z,\ e_z = (180/\pi) \times \tan \phi_z$$

Initially for any known diameter ; say pitch diameter, the value of e_z and so AB is found. The highest gauge is set to zero at this height above the gear centre by means of slip gauges, then the corresponding position of gear profile is obtained by rotating the gear tooth towards indicating stylus until zero is obtained. Number of angles is increased or decreased in steps of 10′ or half degree as is convenient and for these values of e_z the values of AB are calculated, and the height gauge is set to this height by means of slips and then the dial reading over the tooth is compared. The deviation of dial reading gives the error of profile at these points.

This is a very time consuming method but best suited for calibration of master involute. It is therefore useful only for very precision components and involute master cams.

(*b*) **Gear involute measuring machine.** This machine is designed for checking the involute profiles of the spur and other gears. The machine is suitable for inspection of gear having module from 1 to 10 mm having maximum outside diameter upto 300 mm. The machine is provided with a measuring stylus. The kinematic design of the machine is such that when job is rotated the measuring stylus which is initially set at base circle radius of the gear by means of slip gauge, is also slided along the involute curve. The deviation of the tooth profile from the correct involute is indicated by a dial indicator of accuracy 0.001 mm connected by lever mechanism with the stylus. A master involute template is also provided with the machine for setting and calibration of the machine.

(*c*) **Checking of involute shape of gear.** As the involute curve is traced by the end *A* of a straight edge which rolls without slipping on a base circle diameter cylinder, any point *C* on the curve will correspond to the position *CE* of the straight edge, which, therefore, always remains tangential to the base circle. Conversely if the base circle cylinder were to roll on a fixed straight edge, any fixed point *e.g.*, *C* would move in an involute path, such as *CA*, as the cylinder rolls along the straight edge *CE*.

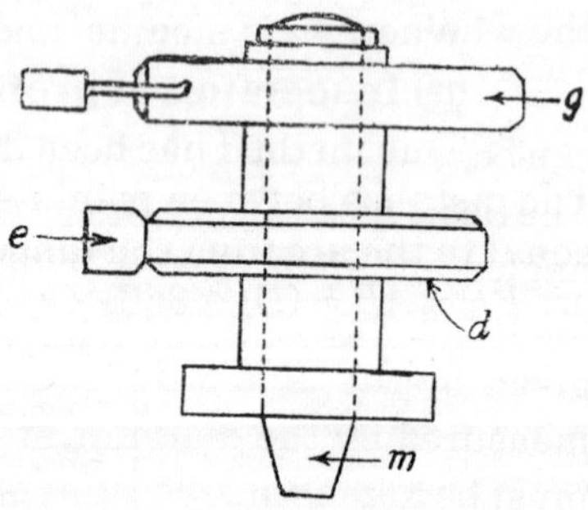

A straight edge rolled on the edge of a disc will be seen to be the equivalent of this arrangement, and provides the principle on which the involute tester operates. The gear to be tested is held on the mandrel *m*, which carries a ground disc *d* having exactly the same diameter as the base circle of the gear under test (Fig. 15.25). A straight edge *e* is mounted on a slide on the body of the instrument and in contact with base circle disc, so that as the straight edge moves along the slide, the base circle disc and gear are rotated without slip. A point on straight edge thus describes the true involute corresponding to the base circle, and if the top of an indicator of some kind is mounted exactly in the plane of the edge of the straight edge and in contact with the tooth flank it will register by its movements any departure of the tooth profile from the theoretical involute resulting either from errors or from deliberate modification of the profile. The indicator can also be replaced by the sensing element of a recorder so that permanent records of the gear teeth profiles can be made.

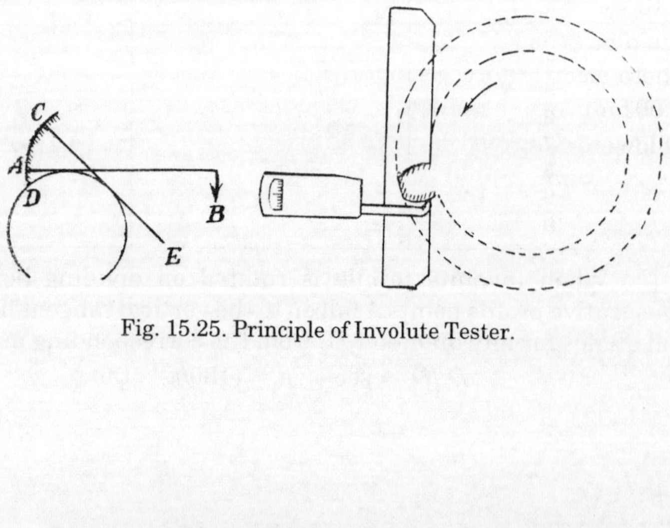

Fig. 15.25. Principle of Involute Tester.

(*d*) **Involute profile testing machine.** Such machines can be set to desired base circle radius values, thus doing away the necessity of having base disc for each gear of different dimensions. This facility is provided by a master base, disc or involute cam, built into the machine, and coupled to a linkage system which enables the base radius of the generated curve to be varied by adjustment of the linkage bar positions.

These machines generate the required involute shape, measure the departures from it that exist on the actual tooth flanks of the gear. The principle of operation is that of a base circle rolling without slip along a straight edge, or *vice versa*. A stylus detects the deviations in the metal condition of the tooth flank from the true involute in the form of a continuous trace on a strip chart recorder. A perfect profile will result in a straight trace parallel to the longitudinal axis of the chart paper.

15.7.9. Measurement of Gear Pitch. Gear pitch can be measured in the following ways :

(*i*) Cumulative circular pitch error over a span of teeth.

(*ii*) Adjacent pitch error or pitch variation.

(*iii*) Base pitch variation.

Here $L = K\pi m^2$, where K is the sector of pitches over which pitch error is to be checked.

The pitch error of gear can be easily determined by comparing the span length over a specified number of teeth *i.e.*, the cumulative error on a sector of predetermined pitches is measured.

The pitch variation is the difference between the longest and shortest circular pitch in the whole gear.

The measurement of cumulative error over a span of teeth and also the pitch variations can be conveniently measured using a dividing head and height gauge fitted with dial indicator. The basic method of measurement involves indexing the gear through single or multiple tooth angles and determining flank position circumferentially by means of a precision indicator mounted on a radially disposed slide. By means of the dividing head, rotation to the gear is given by the amount of theoretical angular pitch, the variation in the position of tooth is measured by the dial indicator. For larger gears, the angular accuracy of the dividing table must be higher in order that pitch error is determined accurately. Though dividing tables with ± 10 seconds of arc resolution are common, dividing tables are available which can be read to the nearest second of arc. Of course, these have to be calibrated using precision polygon and photo-electric autocollimator. The indicator unit should be capable of measuring reliably upto 0.001 mm over a range of 0.025 mm. For gears having larger cumulative pitch errors, indicators of lower sensitivity have to be used.

Tolerances on pitch errors

A. Tolerance over a sector of 'K' Pitches

Grade of Accuracy Class of gear	*Permissible Error in Microns*
1	$0.25\sqrt{L} + 0.6$
2	$0.4\sqrt{L} + 1.0$
3	$0.63\sqrt{L} + 1.6$
4	$1.0\sqrt{L} + 2.5$
5	$1.6\sqrt{L} + 4.0$
6	$2.5\sqrt{L} + 6.0$
7	$3.55\sqrt{L} + 8.0$
8	$5.0\sqrt{L} + 12$
9	$7.1\sqrt{L} + 17$
10	$10\sqrt{L} + 28$
11	$14\sqrt{L} + 33$
12	$20\sqrt{L} + 83$

B. Tolerance on Pitch Variations

Class	*Permissible Error in Microns*
1	$0.80 + 0.06\,F$
2	$1.25 + 0.1\,F$
3	$2.0 + 0.16\,F$
4	$3.2 + 0.25\,F$
5	$5.0 + 0.40\,F$
6	$8.0 + 0.63\,F$
7	$11 + 0.90\,F$
8	$16 + 1.25\,F$
9	$22 + 1.80\,F$
10	$32 + 2.50\,F$
11	$45 + 3.55\,F$
12	$63 + 5.0\,F$

The base pitch is the circular pitch of the teeth measured on the base circle, as we know that the base pitch = $p \times \cos \phi$. This is also the distance between the tangent to the curved positions of any two adjacent teeth. The base pitch can be very conveniently measured using ordinary base pitch measuring instruments which measure the straight distance between tangents to the adjacent teeth.

Base pitch can be accurately measured using tool makers or universal microscope.

Here $F = m + 0.25\sqrt{D}$ where D is the pitch circle diameter in mm.

Tooth-to-tooth pitch errors can be easily determined by use of two dial gauges by measuring the position of a suitable point on a tooth after the gear has been indexed through a suitable angle.

The gear is mounted in the centre with indexing arrangement. Two dial gauges are mounted as shown in Fig. 15.26. There is a spindle below the dial gauge having a small sphere at the end and touching the gear tooth at the pitch circle. The gear is then indexed through successive pitches to give a constant reading on dial gauge *A*. Any changes in the reading on dial gauge *B* indicate that pitch errors are present. The actual error can be determined by deducting the individual reading on dial *B* from the mean of the readings.

The pitch of the gear teeth can also be measured by measuring the distance from a point on one tooth to a suitable point on the next tooth.

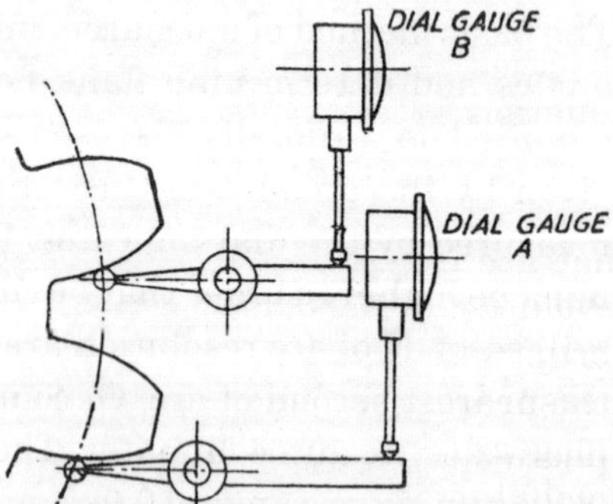

Fig. 15.26

Circular Pitch Measuring Machine. This instrument is used for checking the circular pitch of gear tooth. The two measuring contact tips are applied on the same sides of adjacent teeth of the gear. The left-hand tip is first set up to the required module by means of some suitable arrangement. The right hand tip is a two armed lever whose one contacts the gear tooth and the other one actuates the contact point of the dial indicator. Two guide points are also provided for the stability of the instrument.

The pitch variations can also be measured by the instrument shown in Fig. 15.27. It employs a fixed finger and stop for consistent positioning on successive pairs of teeth, and a movable finger whose movement can be sensed by a dial indicator.

It may be noted that readings obtained by above instrument will be affected by profile variations and runout of the gear.

In the case of helical gears, measurements may be made in the normal plane of the conjugate rack and divided by the cosine of helix angle for comparison with standard tolerances.

It may be mentioned that the above method of measuring pitch error can be applied to medium sized gears measuring from a few cms upto about a metre diameter. Small gears as used in watches and clock are inspected by optical projection in which the enlarged images of some of the teeth are compared against a master diameter at the screen of the projector. For thick gears convergent illumination is used to enable optical focussing at a selected section across the face-width.

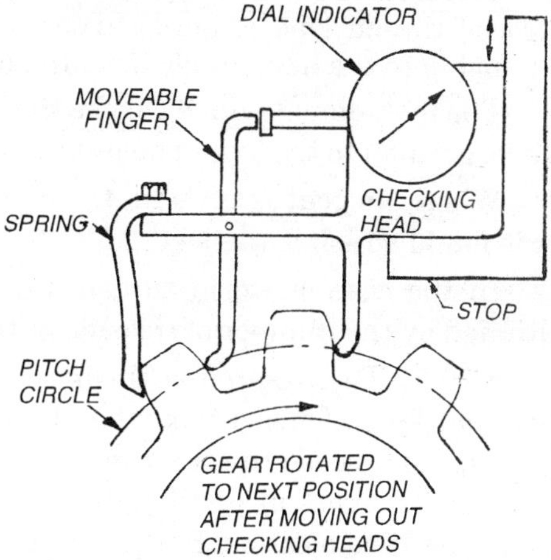

Fig. 15.27. Schematic arrangement of a pitch checking instrument.

Effect of Eccentricity of Pitch Error

It may be emphasized that gear pitch error and tooth eccentricity are inseparable because presence of one affects the other. Since eccentricity can take any position with reference to the tooth under consideration, the effective eccentricity for various positions of gear rotation at angle θ is expressed as equal to $e \sec \phi \sin (\phi + \theta)$, where e = eccentricity and ϕ = pressure angle.

Thus if a gear has been measured for tooth spacing error, the effect of its mounting eccentrically on pitch error can be taken care of by applying corrections progressively to the observed error values by calculating the correction applicable to each tooth from the above equation.

15.7.10. Runout. Runout means the eccentricity in the reference or pitch circle. Gears that are eccentric tend to have a vibration per revolution. A badly eccentric tooth may cause an abrupt gear failure. The runout in the gears is measured by employing gear eccentricity testers. The gear is held on a mandrel in the centres and the dial indicator of the tester possesses the special tip depending upon the module of gear being checked. The tip is inserted in between the tooth spaces. The gear is rotated tooth by tooth. The maximum variation is noted from the dial indicator reading and it gives the runout of the gear. The runout is twice the eccentricity. The adjoining table indicates the permissible runouts.

Class or Grade	*Permissible Runout in Microns*
1	$0.224\,F + 3.0$
2	$0.335\,F + 4.5$
3	$0.560\,F + 7.0$
4	$0.900\,F + 11$
5	$1.4\,F + 18$
6	$2.24\,F + 28$
7	$3.15\,F + 40$
8	$4.0\,F + 50$
9	$5.0\,F + 63$
10	$6.3\,F + 80$
11	$8.0 + 100$
12	$10.0\,F + 125$

Here $F = m + 0.25\sqrt{D}$, where D, is the pitch circle diameter.

15.7.11. Lead. It is the axial advance of the helix or the worm thread per turn. The control of thread lead is necessary to ensure adequate contact across the face width. The relationship to the helix angle has already been explained earlier.

The instrument which checks the lead consists of a probe being advanced along a tooth surface, parallel to the axis. The probe is a suitable dial indicator tip fixed in a suitable device.

When the gear is rotated, the displacement of the probe in one complete revolution of gear is found which is the lead.

In the case of worm thread, the axial pitch of the thread is first measured which multiplied by the number of threads in the worm gives the lead.

15.7.12. Backlash. Backlash in the gears is the play between the mating tooth surfaces. For the purposes of measurement and calculations, backlash is defined as the amount by which a tooth space exceeds the thickness on an engaging tooth. Backlash in the gear teeth results on account of errors in profile, pitch thickness of teeth etc. It is measured by mounting the gears in specified position. Backlash should be measured at the tightest point of the mesh. The pinion is held solidly against rotation and a rigidly mounted dial indicator is placed against the tooth at the extreme heel perpendicular to the surface. The backlash is determined by moving the gear back and forth. The backlash variation is measured by locating the points of maximum and minimum backlash in the pair and obtaining the difference. For precision gears the variation should not exceed 0.02 to 0.03 mm.

15.7.13. Lead Measurement. In order that tooth load be uniformly distributed across the face width of the gears, it is essential that lead per tooth of mating gears should be closely matched. Errors in the helix of either gear would result in non-uniform load concentration, resulting in noisy operation and damage. Irregularities in lead could occur due to either poor manufacture or from the presence of tooth undulations.

It is important to note that while the helix angle of pinion and gear is same but their lead is different depending on their diameters. Also the helix angle value increases from the roots to the tips of the teeth.

Lead can be measured either on a point-to-point basis, or by means of continuous generation using special purpose measuring machines. One type of generating machine incorporates a sine bar mechanism together with a means of converting a derived linear motion into a rotary motion.

15.8. Composite Method of Gear Checking

Composite testing of gears consists in measuring the variation in centre distance when a gear is rolled in tight mesh (double flank contact) with a specified or master gear. In composite gear checking two types of checkings are made :

(*a*) Total Composite Variation, (*b*) Tooth to Tooth Composite Variation.

Total composite variation is the centre distance variation in one complete revolution of the gear being inspected ; whereas tooth to tooth composite variation is the centre distance variation as the gear is rotated through any increment of $360°/N$. A uniform tooth to tooth variation shows profile variation whereas a sudden jump indicates the pitch variations.

Composite type of checking takes care of all the errors in the gears. It is specially very much suited for large gears as it also ensures control over the tooth spacing. The composite method of checking is very much suitable for checking worn gears.

Tolerance for Composite Errors. The following table gives the tolerance on total composite errors and tooth to tooth composite error.

Here factor $F = M + 0.25 \sqrt{D}$.

Class or Grade of Gear	*Total Composite Error in Microns*	*Tooth or Tooth Composite Errors in Microns*
1	4 + 0.32 *F*	2 + 0.16 *F*
2	6 + 0.30 *F*	3 + 0.224 *F*
3	10 + 0.08 *F*	4 + 0.32 *F*
4	16 + 1.25 *F*	6 + 0.45 *F*
5	25 + 2.0 *F*	9 + 0.56 *F*
6	40 + 3.2 *F*	12 + 0.90 *F*
7	56 + 4.5 *F*	16 + 1.25 *F*
8	71 + 5.6 *F*	22 + 1.8 *F*
9	90 + 7.1 *F*	28 + 2.24 *F*
10	112 + 9.0 *F*	36 + 2.8 *F*
11	140 + 11.2 *F*	45 + 3.55 *F*
12	180 + 14.0 *F*	56 + 4.50 *F*

Master Gears. Master gears are made with sufficient accuracy capable of being used as the basis for comparing the accuracy of other gears. These are mostly used in composite errors determination in which the master gears are rotated in close mesh (double flank) or in single contact with the gears under test. These can also be used for calibration of gear checking instruments used in shop-floor. Master gears are generally of two types *i.e.* Master gears type A used for checking precision gears of accuracy class upto 7 and type B master gears used for checking gears of grades from 8 to 12. Master gears are made from chromium-manganese tool steel or good quality gauge steel and are hardened to 62 H_{RC}. These are properly stabilised to relieve internal stresses. The master gears should preferably have lower module values because with coarse pitches the master gear would have either a very few teeth or else it will be quite big making it difficult to handle besides high-production cost.

15.8.1. Parkinson Gear Tester. The principle of this device is to mount a standard gear on a fixed vertical spindle and the gear to be tested on another similar spindle mounted on a sliding carriage, maintaining the gears in mesh by spring pressure. Movements of the sliding carriage as the gears are rotated are indicated by a dial indicator, and these variations are a measure of any irregularities in the gear under test; alternatively a recorder can be fitted, in the form of a waxed circular chart and records made of the gear variation in accuracy of mesh.

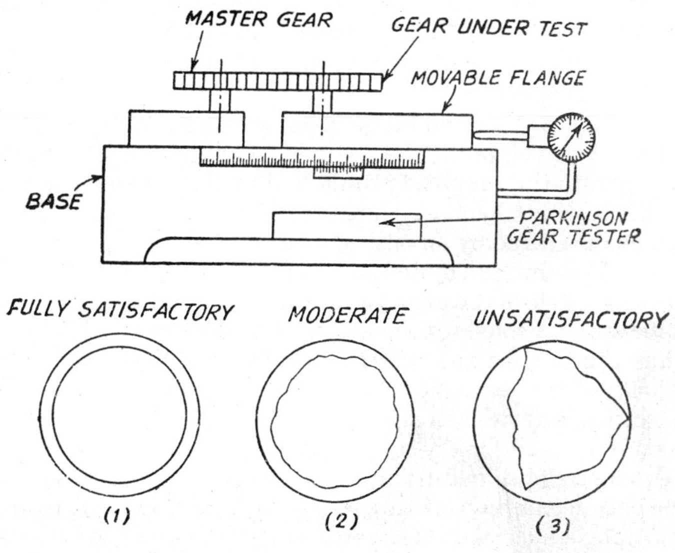

Fig. 15.28

Fig. 15.28 shows a gear tester for testing spur gears. (Testers are available for bevel, helical and worm gears also.)

The gears are mounted on the two mandrels, so that they are free to rotate without measurable clearance. The left spindle can be moved along the table and clamped in any desired position. The right mandrel slide is free to move, running on steel balls, against spring pressure and it has a limited movement. The two mandrels can be adjusted so that their axial distance is equal to the designed gear centre distance. The spring pressure can be regulated. There are also screws for limiting the movement of the sliding carriage. A scale is attached to one carriage and a vernier to the other ; this enables centre distances to be measured to within 0.025 mm. The dial indicator on the right contacts the right end of the sliding carriage and therefore indicates any radial variations of the gear under test as the gears are rotated.

When the waxed paper recorder is fitted, the chart makes a revolution for each one of the gears mounted on the sliding carriage. As the chart moves or rotates, the line traced records the movements of the floating carriage, a circle is drawn at the same time as the record. The figures shown in Fig. 15.28 are reproduction of a few typical charts with a reduced scale and the radial errors magnified about 50 times. The gear shown by No. 1 record is a fully satisfactory one, that at No. 2 is a moderate gear at No. 3 is an unsatisfactory one.

It may be noted that the method described above is dual flank method, *i.e.*, both tooth flanks come in contact which is seldom the case in actual practice. The chart records obtained by this test do not give a clear indication of true cumulative pitch error. This test is an expedient test for accepting or rejecting a gear but not for finding out detailed causes for rejection. It is used mainly to detect poor tooth form caused by worn or inaccurate cutting tool, and pitch circle eccentricity arising from inaccurate centring of the gear blank prior to tooth cutting, etc.

Technically more correct method of mesh testing is single-flank method in which, instead of measuring centre distance variation the angular variation is measured. The mesh tester is a complex system and more costly. The simplest machine of this type consists of two shafts each carrying a gear and a plain disc having diameter equal to the nominal pitch diameter of the gear. One shaft has a rotary joint between the gear and its associated pitch disc. An indicator is used to measure angular variation between the gear and disc on this shaft. In use, the two discs are brought into frictional contact so that one can drive the other without slip. This method is not popular because it requires the manufacture of two very accurate pitch discs for every gear pair of different size. Present day's single-flank mesh testers do not require different pitch discs. The two shafts carrying the gears are fitted with radial gratings having angular band of accurately spaced clear radial lines (one line for one minute of arc). When two such gratings (inclined at very small angle) rotate in close proximity, interference bands known as Moire fringes are formed moving in radial direction which generate electric pulses. These pulse trains are continuously phase-compared to provide a detailed chart record of gear transmission errors.

15.8.2. Modern Gear Measuring Machines. The surface shape of gear tooth can be defined by an equation that is linear though complex, and mechanisms can be constructed both to generate the required profile and to inspect the finished shape. In a gear testing (involute profile testing) machine, the gear is mounted on a spindle whose rotation (angular movement) can be measured by an optical grating.

A pusher touching the involute profile is mounted on a slide and moved tangentially to the base circle of the gear. According to the definition of the involute profile if the straight edge is moved at a constant linear velocity, the gear will be rotated at a constant angular velocity. Thus there is a fixed relationship between the linear velocity of the slide and the angular velocity of the gear, and by measurement and comparison of the instantaneous values of these velocities, any departure of the tooth profile from the true involute can be detected. A linear grating is used for the slide motion and a rotary grating for the gear movement. The pulse frequencies obtained from these grating systems are comparable to accuracy of the order of one part in a million. Since these two frequencies are not, in general related by a simple integral multiple, a special digital processing system is used for this purpose.

Pulses from rotary grating are fed directly to a phase comparator, and the train of pulses from linear grating are processed to convert them to the same frequency as those of the rotary grating and then fed to the phase comparator which derives the phase difference between the two inputs as an error signal which can be fed to an indicator or recorder.

15.8.3. Measurement of Gears using Gratings. As already mentioned, a gear could either be tested by checking involute profile, helix profile and pitch, or where faster speed of measurement is desired by method using interaction of all of above parameters, *i.e.*, composite test (meshing gear under test with a perfectly master gear). Mechanical methods of testing demand the highest standards of mechanical accuracy which some how is limited. However with gratings, it is possible to measure large absolute angular and linear displacements to great accuracy. A big advantage of the grating is its output in the form of digital pulses which can be conveniently manipulated by digital electronic techniques.

Gear transmission errors are measured conveniently using grating as shown in Fig. 15.29.

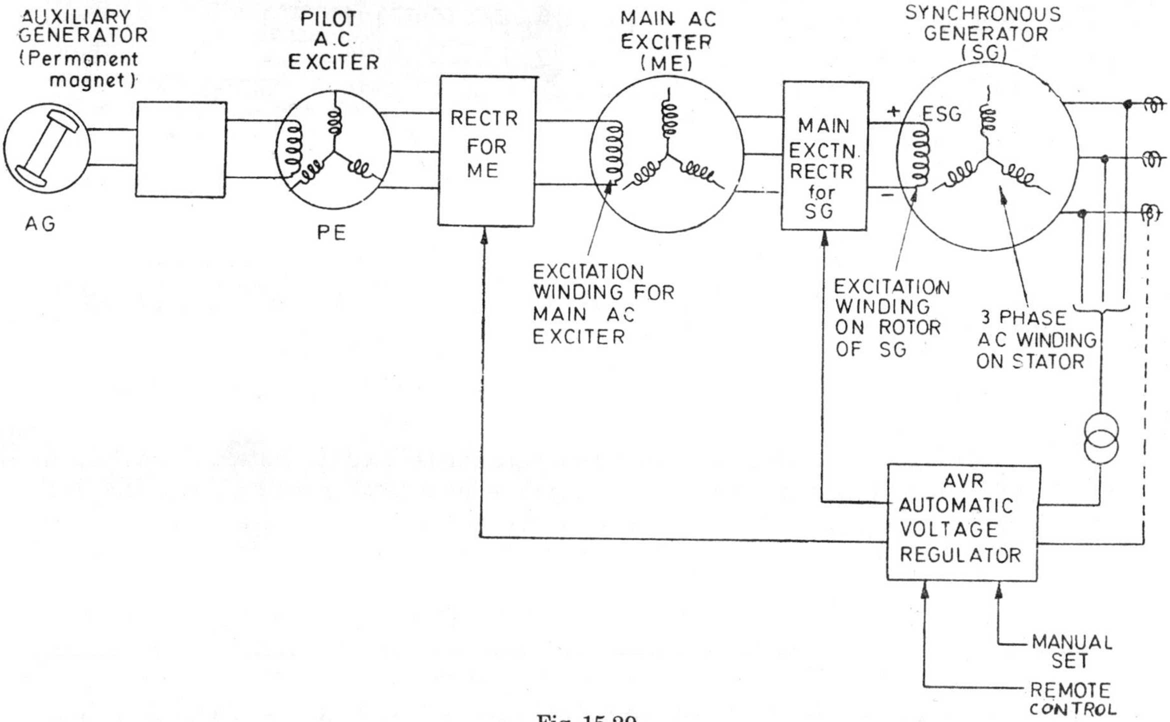

Fig. 15.29

The two gears (one to be tested and the master gear) are mounted on two shafts, each carrying a radial grating at its end. The output of the grating of driving gear is multiplied by T_1T_2

(where T_1 = No. of teeth on gear 1, and T_2 = No. of teeth on gear 2)

so that the resultant output is nominally identical to that obtained from the grating fitted with the driven gear. Difference between the two signals represents the transmission error of the gear pair, in the form of a phase error. In order to improve grating accuracy, four index gratings are used and common light source feeds them *via* fibre optics.

Fig. 15.30 shows the principle of pitch measurement using radial grating. The signal obtained from grating is multiplied by No. of teeth (T) on the gear being tested and divided by 21,600 (No. of lines provided on the circumference of the radial grating so as to obtained a signal having a frequency which gives exactly one pulse per tooth). A small displacement transducer is mounted on a reciprocating slide in contact with the profile of gear tooth space to measure pitch error. The motion of the slide, and the moment of time at which the transducer displacement is measured, are both governed by the reference pulses derived from the grating

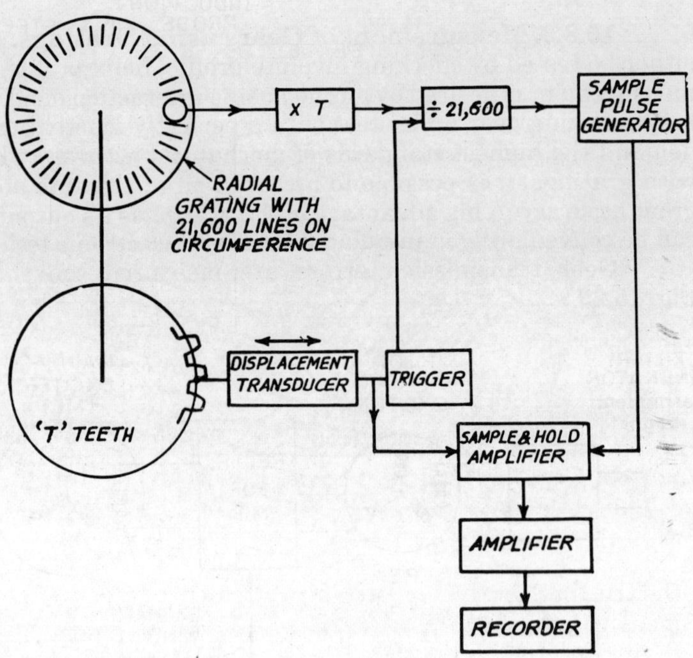

Fig. 15.30

when the gear is stationary. The displacement transducer is inserted into a tooth space and then gear is rotated slowly. The first tooth impinges on the stylus of the transducer and produces some output. At a pre-set value, the divide circuit is initiated, and this is the starting point of the train cf reference pulses. After this the transducer is withdrawn when the next tooth space is opposite the stylus, the transducer is again inserted automatically. The second tooth then deflects the stylus, and at a point determined by the second reference pulse, the intantaneous stylus deflection is measured and recorded. This provides pitch error of the second tooth, relative to the datum of first tooth. In this way pitch error of each tooth is measured. It provides direct readout of absolute indexing error and there is no need of summing up adjacent pitch errors.

For testing the gear involute profile, a radial grating is fitted concentrically on the gear shaft. A solid anvil, rigidly attached to the straight edge assembly is fitted with the tooth profile. A linear grating is installed to measure linear movement of the straight edge. As the gear is rotated, the straight edge and anvil move tangentially. The rotation of gear and movement of straight edge are thus sensed by gratings. For perfect gear tooth profile, there is a mathematical linear relationship between the two movements. The output of one of the gratings is multiplied by this linear constant and then two outputs are fed to a differential amplifier and the phase difference is fed to a recorder to indicate any error in the profile.

15.8.4. Automatic Gear Measuring Machine. The machine described below (Refer Fig. 15.31) is able to measure a gear more comprehensively in one automatic cycle. Gear to be tested an angle measuring system are mounted on the same shaft. For carrying out involute check, the base circle is set to an accuracy of 1 micron with the help of a digital counter by moving the setting carriage. The movement of setting carriage is measured by linear measuring system. The probe head is driven tangentially through the required length of area by the microprocessor with the linear measuring system 2 for that tangential axis having a resolution of 0.1 micron.

The horizontal carriage is driven by drive system 1 and its movement measured by linear measuring system 2. The vertical carriage with probe, mounted on horizontal carriage is driven by drive motor 1 *via* the steel tape operated pulley and it can move back and forth in axial direction of the gear. Provision to connect/disconnect drive to vertical column is available. The circular rotation of the component is controlled by DC drive motor 2 whose angular resolution gives a positioning accuracy of 0.5 second of arc. The microprocessor controls the relationship between the tangential movement and the angular rotation by manipulating drive motor 2 in such a way that the theoretical point of contact between the probe and the component follows

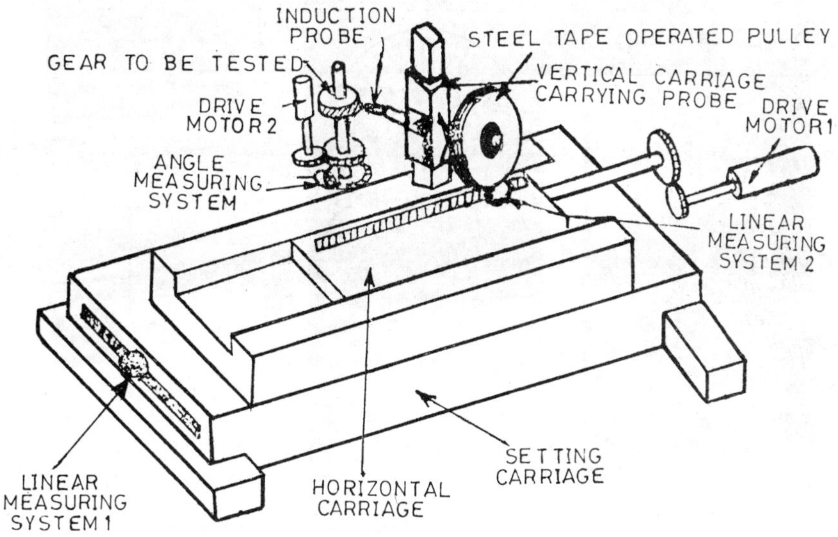

Fig. 15.31. Automatic gear measuring machine.

the true path of the involute. Variations from the perfect profile are recorded by the inductive measuring probe. Probe is made as short as possible and, at all times is kept as close as possible to the workpiece so that deflections and Abbe errors are eliminated.

For helix measurement, the vertical travel of the probe is controlled with a digital resolution of 0.1 micron. Helix angles between 0—90° can be checked. The microprocessor controls the drive 2 in such a manner that in accordance with the set helix angle, the contact point between probe and tooth flank traces a theoretically perfect helix angle and any deviations are picked up by the probe and amplified and recorded. Pitch can also be measured without changing probe. In fact same probe is used for measuring involute, lead and pitch.

15.8.5. Gear Testing Centre

For Involute, Helix Angle and Pitch. Checks tooth flank profiles and helix angles and determines the base circle diameter or pressure angle of internal and external gears with straight and helical teeth and gear hobs. In addition, single pitch error, composite pitch error and concentricity error tests can be performed.

Fully electronic and fully automatic control devices of the test cycle replace most of the so far necessary transmission components.

An electronic travel and angle measuring system with extremely high resolution capability allows the probe to follow the trace of a perfect involute of helix angle and the path pulses that are emitted are fed directly into the computer as comparison values between true and ideal dimension.

A software programmable computer controls the entire machine cycle and performs documentation and further processing of the measuring results.

Optional add-on equipment allows automatic gear loading and unloading. This equipment includes features such as automatic probe change, automatic driver, increased cycle speed, software controlled search for tooth gaps and a CNC controlled robot for feeding and removing the gears. The computer controls the automatic cycle of the machine and robot and retrieves, lists and processes all data. A multi-colour graphic plotter with automatic paper feed produces graphic printouts.

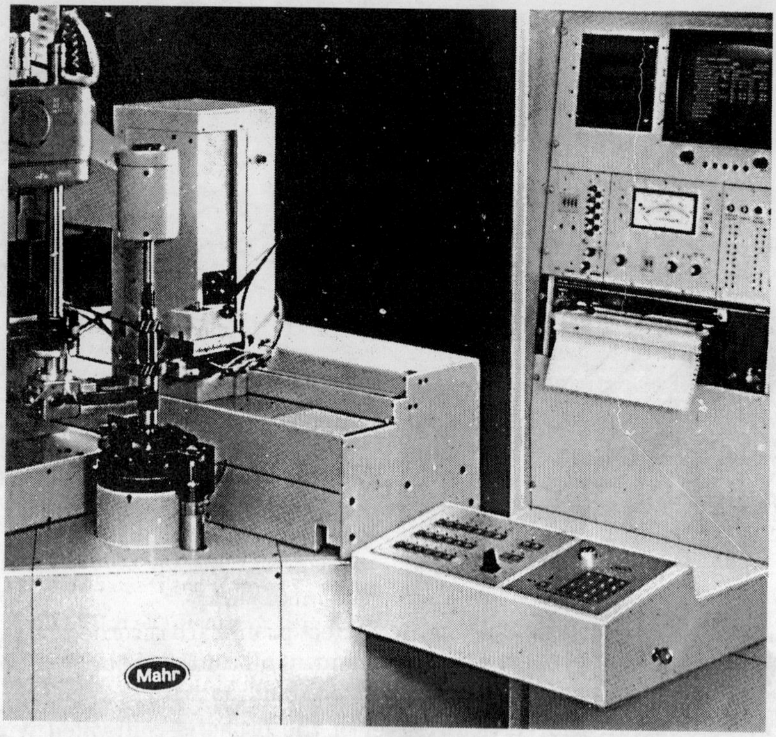

Fig. 15.32. Gear testing centre.

Base circle dia. range	0-550 mm
Max. outside dia. of gear	600 mm
Module range (DP)	0.2-20
Helix angle range	0-90°
Max. gear width	150 mm
Additional vertical probe travel	150 mm
Setting accuracy of base circle	1 μm
Adjustable error magnification of electronic printer	100x, 200x, 500x, 1000x, 2000x
Steady Center Attachment	
center distance	20-570 mm
throat depth	300 mm

WORKED EXAMPLES

Problem 15.1. Calculate the dimension of the maximum chord over four teeth when the gear under inspection has the following specifications :

No. of teeth = 32 ; module = 4, Pressure angle = 20°

Shift of the tool into the gear to provide backlash = 0.25 mm.

Solution. In this case the cutting tool is moving into the blank by 0.25 mm more, so the tooth thickness will get reduced. Hence, the correction needed on one side of tooth is

$$= 0.25 \tan \phi = 0.25 \tan 20° = 0.9010 \text{ mm}$$

and distance over 4 teeth for theoretical gear is

$$N.m. \cos\phi\left(\tan\phi - \phi - \frac{\pi}{2N} + \frac{\pi S}{N}\right)$$

$$= 32 \times 4 \times \cos 20\left(\tan 20 - \frac{\pi \times 20}{180} - \frac{\pi}{2 \times 32} + \frac{\pi \times 3}{32}\right)$$

$$= 42.42$$

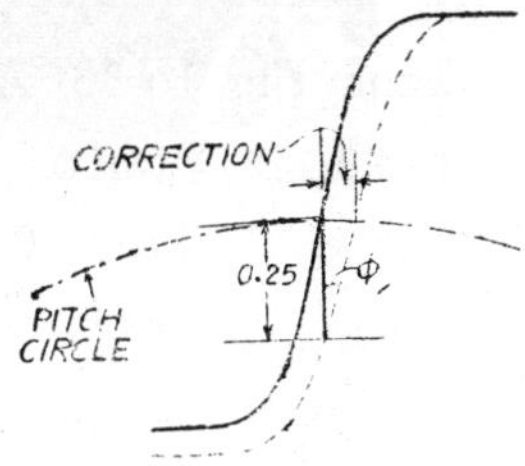

Fig. 15.33

and distance for the actual gear

$$= 42.42 - 2 \times 0.0190 = 42.382 \text{ mm.}$$

Problem 15.2. Calculate the dimension over pins in the following case. Also indicate the diameter of the pins to be used. N = 31, m = 3, ϕ = 20°.

Solution. Best in diameter = $\pi/2.m \cos\phi = \pi/2 \times 3 \times \cos 20° = 4.426$ mm

From Fig. 15.34, P.C.D. = $m.N = AC = 3 \times 31 = 93$ mm

and $\theta = \dfrac{360}{2 \times 31} = 5.8°$, $\therefore$ $\alpha = 5.8/2 = 2.9°$, and angle $CBA = 90°$

$CB = AC \cos\alpha = 93 \times \cos 2.9 = 92.87$

$\therefore$ Dimension over pin = 92.87 + 4.426 = 97.296 mm.

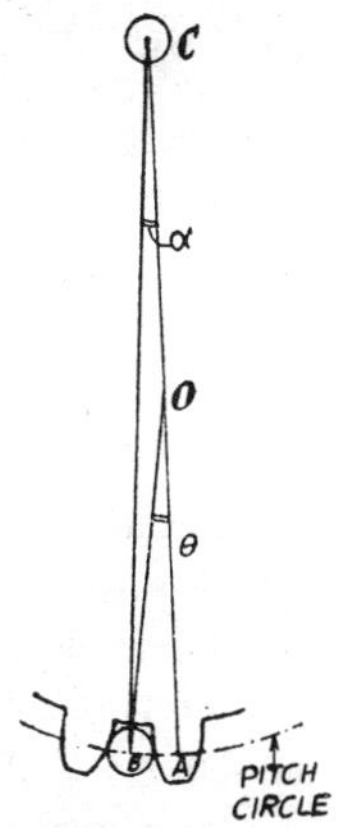

Fig. 15.34

Problem 15.3. For a 20° pressure angle gear having 36 teeth and 4 mm module, calculate :

(*a*) **Plug size (plug diameter)**

(*b*) **Distance over the plugs in opposite spaces**

(*c*) **Distance over the plugs spaced 12 teeth apart.**

Solution. (*a*) Plug size

$$= (\pi/2)\, m \cos\phi = \pi/2 \times 4 \times \cos \times 20° = (\pi/2) \times 4 \times 0.9397 = 5.905 \text{ mm.}$$

(*b*) Distance over the plugs in opposite spaces

$$= m\left[N + \frac{\pi}{2}\cos\phi\right]$$

$$= 4\left[36 + \frac{\pi}{2} \times 0.9397\right] = 155.815 \text{ mm.}$$

(*c*) Distance over the plugs spaced 12 teeth apart.

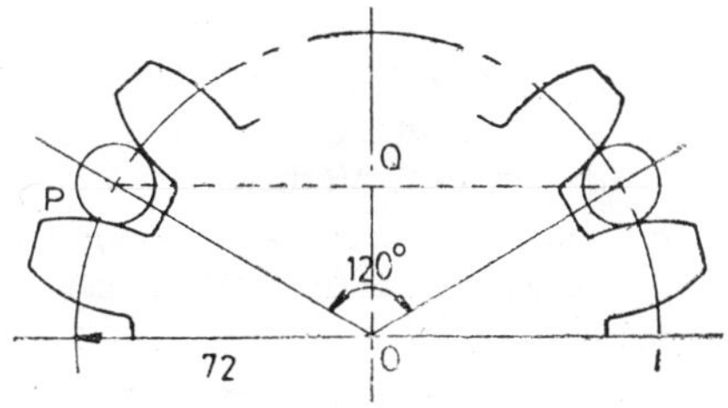

Fig. 15.35

Angle subtended by 12 teeth = 12 × 360/36 = 120°

In ΔOPQ (Fig. 15.35), $PQ = OP \sin 60°$

$$= 72 \times 0.866 = 62.352 \text{ mm.}$$

$\therefore$ Distance between the centre of the plugs = 2 × 62.352 = 124.704

Distance over the plugs = 124.704 + 5.905 = 130.609 mm.

Problem 15.4. Calculate the length of base tangent over two teeth for a helical gear having the following specifications : width of gear : 30 mm, No. of teeth : 10, Helix angle : 30°, pressure angle : 20°, module : 4.

Solution. The formula for parallel tooth (of spur gear) is :

Distance over S. No. of teeth in a gear

$$= N.m. \cos\phi\left(\tan\phi - \phi - \frac{\pi}{2N} + \frac{\pi S}{N}\right)$$

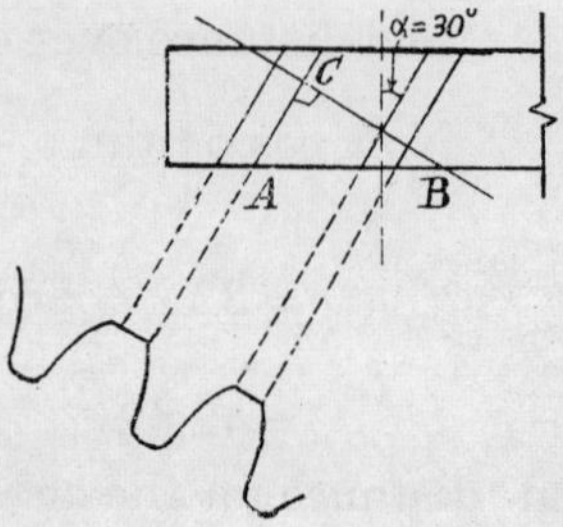

Fig. 15.36

From Fig. 15.36 : $AC = BA \cos \alpha$, $\therefore \quad m_{helical} = m_{spur} \cos \alpha$

$\tan \phi_{helical} = \tan \phi_{spur} . \cos \phi$, $m_{helical} = 4 \times \cos 30 = 3.464$ mm

$\tan \phi_{helical} = \tan 20 \times \cos 30 = 0.364 \times 0.866 = 0.3152$

$\therefore \quad \phi_{helical} = 17.5°$ or $\pi \times 17.5/180$ radians $= 0.3052$ radians.

Distance over two teeth of the helical gear

$$= 10 \times 3.464 \cos 17.5 \times \left(\tan 17.5 - 0.3052 - \frac{\pi}{2 \times 10} + \frac{\pi \times 2}{10} \right) = 15.554 \text{ mm.}$$

TYPICAL UNSOLVED QUESTIONS

1. What are the different elements of a spur gear which require inspection ? Name the instruments used in the inspection of these elements with their expected accuracies.

   Explain with the help of sketches the working principles of the instruments used in checking of profile and base pitch of the gear.
2. What are the elements of gears which are checked for accuracy ?

   Describe a gear tooth-vernier calliper and show how this is used for checking gears.
3. Describe a method for inspecting the involute profile of a spur gear tooth.

   Explain the principle of operation of a rolling gear tester. State the errors in a spur gear that can be detected by the rolling gear tester. Explain how the graphical recordings taken on such a tester may be interpreted.
4. Describe the gear tooth vernier caliper and explain how it is used to check gears. Calculate the settings for a straight spur gear having 50 teeth of module 3 mm.
5. What do you understand by "constant chord" caliper settings ? Calculate the chord length and its distance below the tooth tip for a gear of module 2.5 mm and 14.5° pressure angle.
6. A 40 teeth gear of module 2.5 mm is to be checked by measuring the distance over two pins inserted in the tooth spaces at opposite sides of the wheel. Calculate the diameter of the pin which will rest with its centre at the pitch point and determine the measurement over the pins.
7. Explain what is meant by constant chord caliper settings. Calculate the chord length and its distance below the tooth tip for a gear of module 3 mm and 20-degree pressure angle.
8. How will you go about checking the accuracy of a gear tooth vernier caliper ? What is the expected order of accuracy of this instrument.
9. Determine the base pitch of the base tangent of a gear of 20° pressure angle, having a module of 3 mm and 24 teeth.

16

Machine Tool Metrology

16.1. Introduction

The dimensions of a gauge, its surface finish and geometry are dependent upon the inherent quality and accuracy of the machine tool for its manufacture. Also in mass production the various components produced must be of such accuracy that they may be assembled on a non-selective basis. The accurate production of the component parts depends upon the accuracy of the machine tools.

The continuously increasing demand for highly accurate machined components has led to considerable research towards means by which the geometric accuracy of machines may be improved and maintained. It is very important, therefore, that under static conditions, the alignment accuracy of the machine tools is checked by some geometric tests. The tests which check the alignment accuracy of the various parts of a machine tool are called static rests.

A machine tool, at the same time must be able to display the desired behaviour and characteristics under normal operating conditions as operator is concerned with the actual working of machine tool. Therefore, some alignment tests are performed under dynamic loading conditions and these are referred to as dynamic tests. Thus it is obvious that the satisfactory working of a machine under static loading only is not sufficient but account should be taken of the vibration and deflection of various machine parts under actual dynamic loading also. In dynamic tests, the various members are actually loaded and alignment tests are carried out. Also various parts are excited at working frequency and behaviour of machine observed.

In other words it could be said that machine tools for the workshop must be able to produce work-pieces of given accuracy within prescribed limits consistently and without requiring artistic skill on the part of the operator and that the quality of workpieces depends upon :

(*i*) Stiffness and rigidity of the machine tool and its components parts.

(*ii*) The alignment of various machine parts in relation to one another. This is very important because the geometry of various shapes is based on the relative motion between various machine parts and hence on alignment of various parts.

(*iii*) The quality and accuracy of the control devices and the driving mechanism.

Stiffness and rigidity are a matter for designer and once tests on a prototype of a certain design have given satisfactory results, there is no need to test the machine of same design over and over again. The accuracy of manufacture, the precision of control devices and driving mechanisms such as lead screw of lathe, all should have relative alignment as accurate as possible. This depends upon the quality of manufacture and may vary from machine to machine. Therefore, each machine is subjected to an acceptance test concerning items (*ii*) and (*iii*).

The accuracy of machine tools (which cut metal by removing chips or swarfs) is tested by means of geometrical checks and practical tests.

Geometrical checks include checking of dimensions of forms and positions of components as well as checking of their displacement relative to one another. These comprise all the operations which affect the components of the machine and concern only sizes, forms, positions and relative movements which may affect the accuracy of the machine. It may be appreciated that the geometrical definitions are abstract and relate only to imaginary lines and surfaces. In practice, therefore, metrological definitions are followed which are concrete and take account of real lines and surfaces accessible to measurement. Metrological definitions cover in a single result all macro and micro-geometrical errors and allow a result to be reached covering all causes of errors without distinguishing them from one another.

Practical tests include the machining of test pieces appropriate to the fundamental purposes for which the machine has been designed and having predetermined limits and tolerances.

The results of practical tests and geometrical checks may be compared only in so far as these two kinds of tests have the same object ; in case both do not give the same result, the results obtained by carrying out practical tests are accepted as valid. In case where it is expensive or difficult to conduct both types of tests, the accuracy of a machine may be checked only by geometrical checks or only by practical tests.

The various geometrical checks generally made on machine tools are : (Before conducting these tests, it is essential that the machine is set up and principal horizontal and practical planes and axes are checked with spirit levels, etc.

(*a*) Straightness

(*i*) straightness of a line in two planes ;

(*ii*) straightness of slideways of machine tools ; (*iii*) straight line motion.

(*b*) Flatness

(*c*) Parallelism, equidistance and coincidence.

The parallelism includes (*i*) parallelism of lines and planes ; (*ii*) parallel motion.

(*d*) Rectilinear movements or squareness of straight lines and planes. Quality of the guiding and bearing surfaces of beds, uprights and base plates are also tested.

(*e*) Rotation

This includes (*i*) out of round, (*ii*) eccentricity, (*iii*) radical throw of an axis at a given point, (*iv*) out-of-true running (run-out), (*v*) periodical axial slip, (*vi*) camming.

Main spindle is the fundamental element of the machine and is tested for concentricity, axial slip, accuracy of axis and position, relative to other axes and surfaces.

(*f*) Movement of all the working components.

We shall now discuss the definitions and various tests to perform these geometrical checks.

Straightness. The definition of straightness of a line in two planes is given in Chapter 7. The various methods to check the straightness are :

(*i*) *Straight edge method.* This method is described in Chapter 7 and is limited to lengths below 1600 mm.

(*ii*) *Spirit level and optical methods.* These are also described in Chapter 7. The spirit level method enables checking the straightness only in the vertical plane and may be used for lengths below as well as above 1600 mm. The optical methods are used for lengths above 1800 mm only.

(*iii*) *Taut wire and microscope method.* In this method a steel wire of diameter 0.1 mm is stretched and arranged to be approximately parallel to the line to be checked. A microscope equipped with a horizontal micrometers displacement device is placed vertically on the surface of line to be tested and moved along the line to be tested. The deviation of the line to the taut wire in the horizontal plane of measurement can thus be noted down. This method should be avoided when the sag of the wire has to be taken into account as the sag is extremely difficult to determine with any great accuracy.

The conditions for the straightness of the slideways of machine tools are same as for a line. The methods indicated above are applicable for plane slideways : but, however for *V*-slideways the tests are performed on the surface of a cylinder or an intermediate piece (made to the shape of the slideways) placed on the *V*-slideways.

It may be noted that checking of straightness of slideways or beds is not as important as checking of straight line motion which takes into account all factors likely to effect motion. The expression "accuracy of straight line motion", refers to the trajectory of a point on a component of the machine when affecting a working or setting motion. The various possible cases of straight line motion are :

(*i*) motion of an axis on itself, when this axis remains within the two right angled planes which contain it at rest ;

(*ii*) motion of a plane surface in its own plane, when this surface remains in its own plane ;

(*iii*) motion of a component parallel to a straight line or a surface, when any point on this component remains at an equal distance from the line or from the surface ;

(*iv*) motion of a component perpendicular to a given plane, when each point of the component describes a trajectory perpendicular to the given plane.

The straight line motion can be tested for smaller travels with a straight edge and dial gauge and for longer travels with microscope and taut wire. In the method of checking with a straight edge and dial gauge, the dial gauge is fixed to the moving component of the machine whose straight line motion is to be tested and the feeler slides along the straight edge representing the reference line. In the method of checking with microscope and taut wire, the reference line is represented by a lightly stretched thin steel wire and the dial gauge is replaced by a microscope, the deviations being read directly from the scale of eyepiece by sighting the wire through the microscope.

The straightness of lathe slide displacement for lengths smaller than 1600 mm may be checked with mandrel and feeler gauge method and for the lengths longer than 1600 mm with taut wire and microscope method. In the former method, a mandrel is mounted between the two centres of the lathe and adjusted such that the dial gauge indications are equal at two ends. A dial gauge is mounted on the saddle is place of tool and its feeler brought into contact with the surface of the mandrel. The first set of readings is taken by traversing the dial gauge along the surface of mandrel, the second by rotating the mandrel through 180°. The similar two sets of readings are again noted after turning the mandrel end to end. The mean of the four sets of readings gives the actual error in straight line motion and this eliminates any possible error due to errors in mandrel.

Under some circumstances, the deviation of straight line motion in relation to a straight line can be permitted in one direction only. This should be clearly specified *e.g.*, as "trajectory concave only in the horizontal plane".

The tests for the remaining geometrical checks are covered in Chapter 7.

Dr. G. Schelesinger has for many years been concerned with the various alignment tests of machine tools for satisfactory working and much pioneering work has been done by him in devising and evaluating suitable tests. We will deal here latest techniques used in alignment testing of machines.

16.2. Machine Tools Tests

In most cases the production or machining of a given geometric surface is achieved through a combination of work-tool movements. In other words, the machined surface is generated, and the accuracy of the surface depends on the accuracy of the mating elements present in the machine tool.

The various tests applied to any machine tool could be grouped as below :

(*a*) Tests for the level of installation of machine in horizontal and vertical planes.

(*b*) Tests for flatness of machine bed and for straightness and parallelism of bed ways or bearing surfaces.

(*c*) Tests for perpendicularly of guideways to other guideways or bearing surfaces.

(*d*) Tests for the true running of the main spindle and its axial movements.

(*e*) Tests for parallelism of spindle axis to guideways or bearing surfaces.

(*f*) Tests for the line of movement of various members *e.g.*, saddle and table cross-slides etc., along their ways.

(*g*) Practical tests in which some test pieces are machined and their accuracy and finish is checked.

Following instruments are required for conducting these tests : dial gauges, tests mandrels, straight edges and squares, spirit levels, autocollimators, etc.

16.3. Alignment Tests on Lathe

16.3.1. Levelling of the Machine. Before the various tests on any machine tool are carried out, it is very essential that it should be installed in truly horizontal and vertical planes. In horizontal plane, both longitudinal and transverse directions are equally important. If, say, any long lathe bed is not installed truly horizontal the bed will undergo a deflection, thereby producing a simple bend and undesirable stresses will be introduced. If the bed is not installed truly horizontal in transverse direction, twist will be introduced. Thus the movement of the saddal can't be in a straight line and true geometric cylinder can't be generated.

For proper installation and maintenance of its accuracy, a special concrete foundation of considerable depth must be prepared. Also this must be insulated from the surrounding floor by introducing some form of damping.

The level of the machine bed in longitudinal and transverse directions is generally tested by a sensitive spirit level. The saddle is kept approximately in the centre of the bed support feet. The spirit level is then placed at *a-a* (Fig. 16.1), the ensure the level in the longitudinal direction. It is then traversed along the length of bed and readings at various places noted down. For test in transverse direction the level is placed on a bridge piece to span the front and rear guideways and then reading is noted. It is preferable to take two readings in longitudinal and transverse directions simultaneously so that the effect of adjustments in one direction may also be observed in the other. The readings in transverse direction reveal any

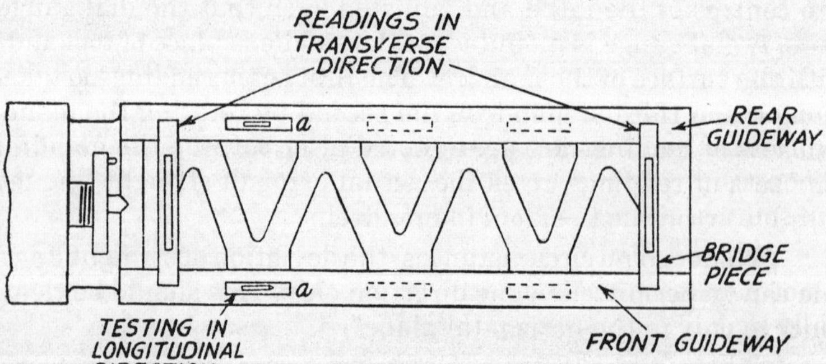

Fig. 16.1

twist or wind in the bed. It may be noted that the two guideways may be perfectly levelled in longitudinal direction, but might not be parallel to each other. This is revealed by the test in transverse direction.

The straightness of bed in longitudinal direction for the long beds can also be determined by other methods, *e.g.*, using straight edges, autocollimators or by taut wire method. But the test in transverse direction can be carried out only by spirit level.

It is desired that the front guideway should be convex only as the cutting forces and the weight of carriage act downward on it. If the front guideways are concave, then the effect will be comulative. The tendency of the carriage, under cutting forces is to lift upwards from the rear and this is prevented by a gib placed underneath the guideways. With the result, an upward force acts on the rear guideways ; which must, therefore, be made concave.

Transverse level may be in any direction, but no twist can be tolerated.

16.3.2. True Running of Locating Cylinder of Main Spindle. Locating cylinder is provided to locate the chuck or face plate. However locating surface can't be threaded one as threads get worn out soon and thus introducing play in face plate or chuck. Thus locating surface is cylindrical and this must run truly ; for only then the face plate etc., can run truly. The dial indicator is fixed to the carriage (or any other fixed member) and the feeler of the indicator touches the locating surface. The surface is then rotated on its axis and indictor should not show any movement of needle.

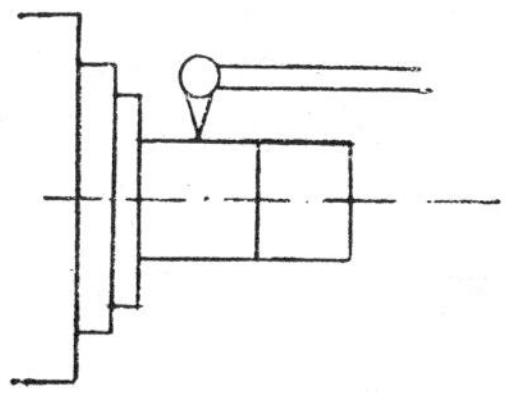

Fig. 16.2

16.3.3. Axial Slip of Main Spindle and True Running of Shoulder Face of Spindle Nose. Let us first distinguish between the axial play and the axial slip. Axial play means the indispensable freedom of spindle movement in axial direction to prevent it from seizing by heating. The spindle is supported between two bearings. Due to running of spindle, there will be a rise in temperature and thermal expansion of spindle would be there. If no axial play is allowed, it would try to bend. Thus there will be no adverse effect of axial play if the direction of cutting forces remains same. If the direction of cutting force changes, there would be some error introduced due to movement of spindle axially in either direction. Under such conditions, therefore, it is advisable to cut threads in one direction only.

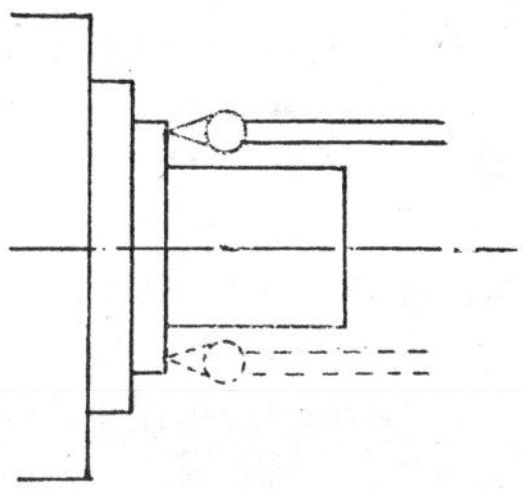

Fig. 16.3

Axial slip is defined as the axial spindle movement which follows the same pattern and is due to the manufacturing error. Actually this test is meant to check this error. To test this the feeler of the dial gauge rests on the face of the locating spindle shoulder and the dial gauge holder is clamped to the bed (Fig. 16.3). The locating cylinder is then rotated and the change in reading noted down. The readings are taken at two diametrically opposite points. The total error indicated by the movement of the pointer includes three main sources of errors.

(*i*) Axial slip due to error in bearings supporting the locating shoulder, *i.e.*, the bearings are not perpendicular to the axis of rotation and due to it a point on the shoulder will move axially in and out at diametrically opposite points.

(*ii*) Face of the locating shoulder not in a plane perpendicular to axis of rotation.

(*iii*) Irregularities of front face.

Due to axial slip, in screw cutting, the pitch will not be uniform due to periodic movement of the spindle. This, however, is not important while turning.

16.3.4. True Running of Headstock Centre. Headstock centre is live centre and the workpiece has to rotate with this centre. If it is not true with the axis of movement of the spindle, eccentricity will be caused while turning a work, as the job axis would not coincide with the axis of rotation of main spindle. For testing this error, the feeler of the dial indicator is pressed perpendicular to the taper surface of the centre (Fig. 16.4), and the spindle is rotated. The deviation indicated by the dial gauge gives the trueness of the centre.

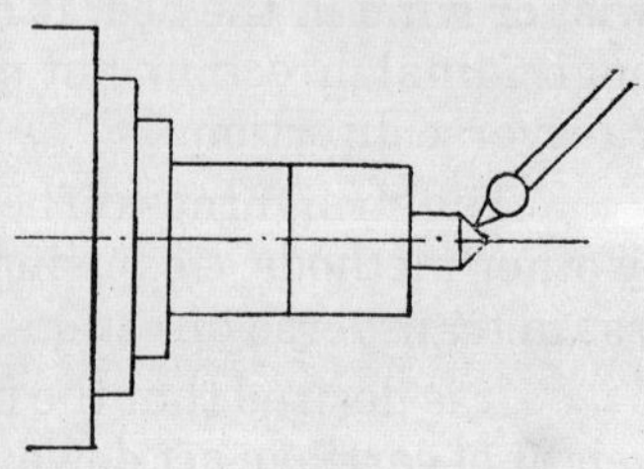

Fig. 16.4

16.3.5. Parallelism of the Main Spindle to Saddle Movement. This has to be checked in both vertical and horizontal planes. In this we require the use of mandrel. An important precaution in the use of mandrels and dial indicator is mentioned here. The mandrel must be so proportioned that its overhang does not produce appreciable sag, or else the sag must be calculated and accounted for. The rigidity indicator set up is also very important and must be carefully watched. Otherwise variations in readings are recorded by pointer may be solely due to deflection of the indicator mounting in different positions and it becomes very difficult to detect and isolate the spurious deflection from the true variations.

If axis of the spindle is not parallel to bed in horizontal direction, a tapered surface is produced.

Any deviation from parallelism of spindle axis from bed in vertical axis will produce a hyperboloid surface. For this test, a mandrel is fitted in the taper socket of the spindle. Mandrel has a concentric taper shank which is close fit to the spindle nose taper. The feeler of the dial indicator is pressed on the mandrel and the carriage is moved. The indication in horizontal plane is given by dial (*b*) and in vertical plane by dial (*a*) (Fig. 16.5). In vertical plane the mandrel should be rising towards the free end in order to counteract the weight of mandrel and job. But for counter-acting cutting forces, it should be lower towards free end. In horizontal plane, mandrel should be inclined in a direction opposite to the direction of tool pressure.

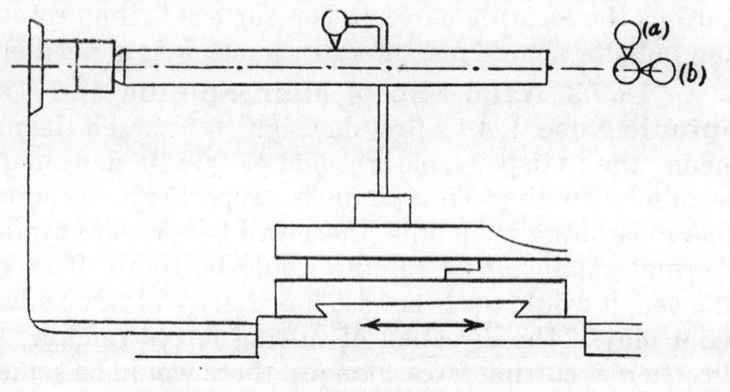

Fig. 16.5

16.3.6. True running of taper socket in main spindle. If the axis of tapered hole of the socket is not concentric with the main spindle axis, eccentric and tapered jobs will be produced. To test it, a mandrel is fitted into the tapered hole and readings at two extremes of the mandrel are taken by means of a dial indicator as shown in Fig. 16.6.

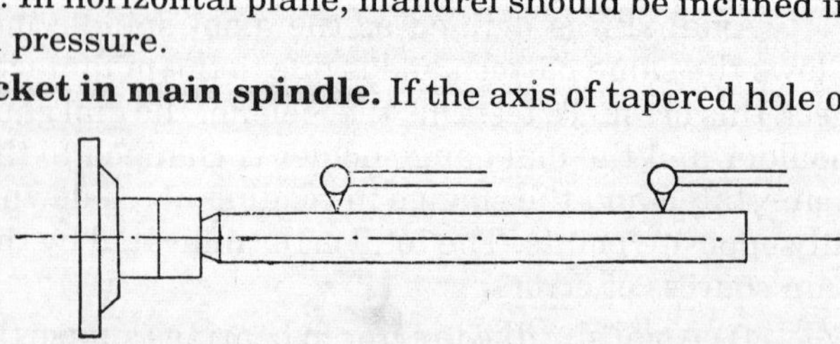

Fig. 16.6

16.3.7. Parallelism of tailstock guideways with the movement of carriage. Sometimes the job is held between head-stock and tail stock centre for turning. In that case the job axis must coincide with the tailstock centre. If the tailstock guideways are not parallel with the carriage movement there will be some offset of the tailstock centre and this results in taper turning.

To check the parallelism of tailstock guideways in both the planes *i.e.*, horizontal and vertical, a block is placed on the guideways as shown in Fig. 16.7 and the feeler of the indicator is touched on the horizontal and vertical surfaces of the block. The dial indicator is held in the carriage and carriage is moved. Any error is indicted by the pointer of dial indicator.

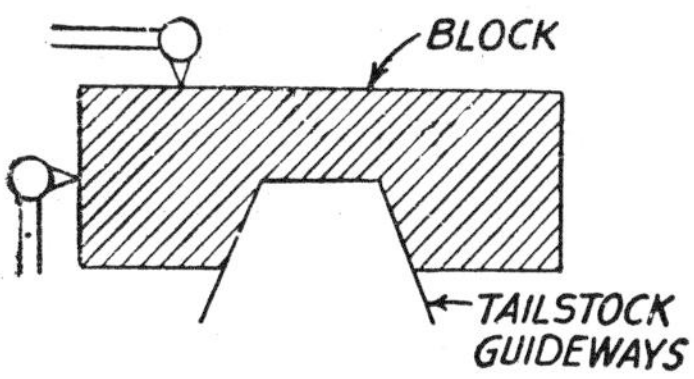

Fig. 16.7

16.3.8. Movement of upper slide parallel with main spindle in vertical plane. The dial indicator if fixed in the tool post. A mandrel is fitted in the spindle. The feeler of the dial gauge is pressed against the mandrel in vertical plane and the upper slide is moved longitudinally. This error is not tested in horizontal plane because there is swivelling arrangement for taper turning.

16.3.9. Parallelism of tailstock sleeve to saddle movement. If the tailstock sleeve is not parallel to the saddle movement, the height of dead centre would vary as varying lengths of sleeve are taken out. For the jobs held between two centres, it is necessary that the central axis of the dead centre be coaxial with the job axis in both the planes. If it is not so, the job

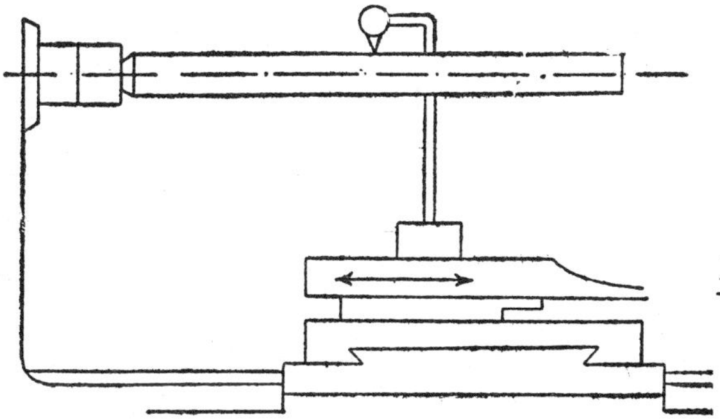

Fig. 16.8

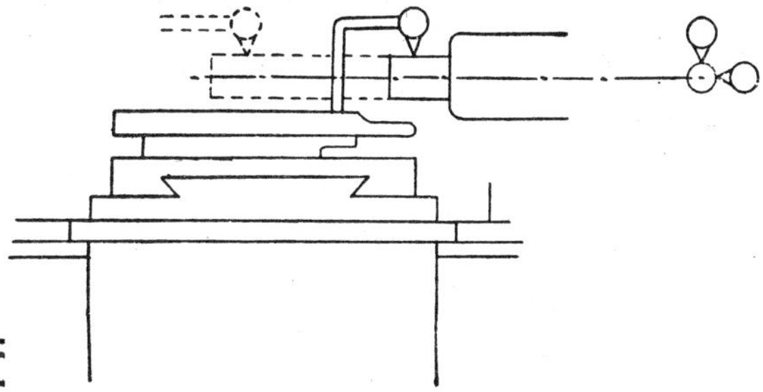

Fig. 16.9

may be tilted up or down or in sideways due to the support of the dead centre. The test is carried out by fixing the dial indicator on the tool post and pressing the plunger against the sleeves first in vertical and then in horizontal plane (Fig. 16.9). The carriage is moved along the full length of the sleeve and deviations as indicated by dial indicator are noted down. Tailstock sleeve should be rising towards the free end in vertical plane and should be inclined towards the tool pressure in horizontal plane.

16.3.10. Parallelism of tailstock sleeve taper socket to saddle movement. A mandrel is put in the sleeve socket. The dial gauge is fixed on the tool post and plunger is pressed against the mandrel and saddle is moved from one side to the other. This test is carried out in both the horizontal and vertical planes.

16.3.11. Alignment of both the centres in vertical plane. Besides testing the parallelism of the axes individually (main spindle axis and tailstock axis) it is necessary to check the relative position of the axes also. Both the axes may be parallel to carriage movement but they may not be coinciding. So when a job

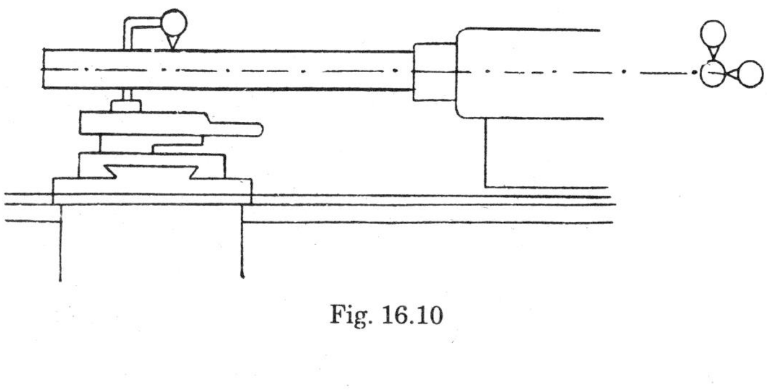

Fig. 16.10

is fitted between the centres, the axis of the job will not be parallel to the carriage movement. This test is to be carried out in vertical plane only. A mandrel is fitted between the two centres and dial gauge on the carriage. The feeler of the dial gauge is pressed against the mandrel in vertical plane as shown in Fig. 16.11 and the carriage is moved and the error noted down.

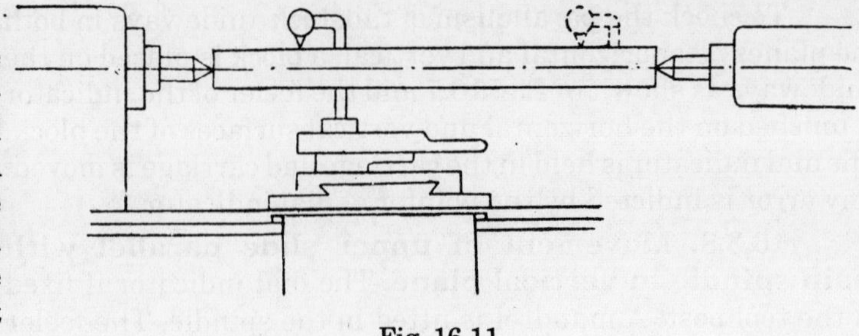

Fig. 16.11

16.3.12. Pitch accuracy of lead screw. The accuracy of the threads cut on any machine depends upon the accuracy of its lead screw. Thus it is very essential that pitch of the lead screw throughout its length be uniform.

Test for this is performed by fixing a positive stop on the lathe bed. Against the stop, the length bars and slip gauges can be located. An indicator is mounted on the carriage and first it makes contact against the calculated length of slip gauges. The initial loading of the dial gauge against the slip gauge is noted. The slip gauges are then removed and the carriage is connected to the lead screw and lead screw is disconnected from the gear train. An indexing arrangement is utilised for rotating the lead screw and lead screw is given some revolutions so that distance travelled by carriage is equal to the length of slip gauges. The reading of the dial indicator against the stop is noted down in this position. It it is same as before, there is no error, otherwise it can be recorded. In this method, care must be taken not to disturb the datum location when changing the gauges for testing different pitch lengths.

A suitable method for recording the progressive and periodic errors is by using a suitably divided scale, which is placed close to the line of centres. A microscope is rigidly mounted on the carriage in a convenient position to note the readings on the scale.

16.3.13. Alignment of lead screw bearings with respect to each other. The alignment of the bearings decides the position of the lead screw. Misalignment of lead screw *i.e.*, it not being parallel to the bed in vertical plane or horizontal plane can cause additional stresses due to bending, when carriage is moved. Due to it the lead screw might get damaged and the precision of the machine is reduced :

Alignment of lead screw bearing with split nut in both the planes is also essential.

16.3.14. Axial slip of lead screw. The thrust face and the collars of the lead screw (or the abuttment collar and the thrust bearing of the screw) must be exactly square to the screw axis, otherwise a cyclic endwise movement is set up which is of the same nature as the axial slip in the main spindle. Thus a periodic pitch error will be additional to any true periodic errors in the pitch of the screw.

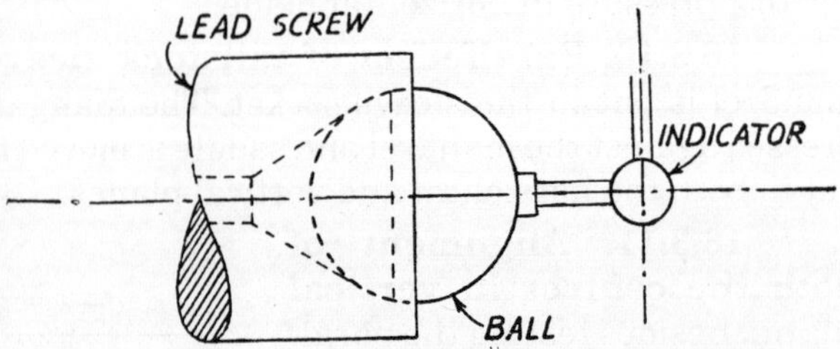

Fig. 16.12

For testing the axial slip in lead screw, a ball is fitted in the end of lead screw and the feeler of the dial gauge is pressed against the ball. The lead screw is rotated and deviation, if any, in any direction is noted down (Fig. 16.12).

16.3.15. Practical tests. These tests consist of the actual turning of some jobs on the machine. The job is made under prescribed condition of cutting speed, feed and depth of cut. The test piece is then measured for its geometry and surface finish and results compared with the standards as prescribed by the manufacturer. These tests are designed to reveal the combined effects of possible errors in alignment accuracy and the rigidity of machine.

The various jobs to be made as prescribed by M/s Hindustan Machine Tools are given below.

(*i*) Working accuracy machine of cylindrical turning (chucking). *S'* should be as small as possible.

Permissible error = 0.01 mm.

Swing over bed 'mm'	*d mm*	*l mm*
upto 300	50	10
over 300 and upto 500	80	15
Over 500	120	20

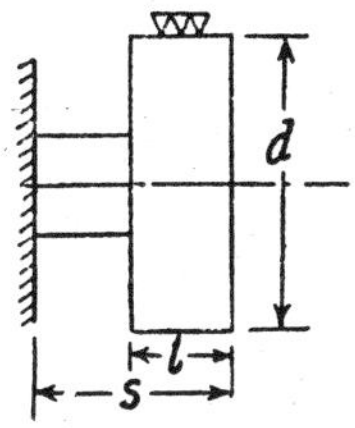

Fig. 16.13

(*ii*) Working accuracy of machine facing :

Permissible error = 0.02 mm over the diameter of test workpiece, which is taken as 300 mm for swing over bed of upto 500 mm, and 400 mm for swing over 500 mm.

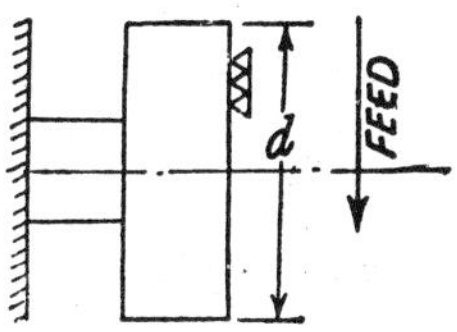

Fig. 16.14

(*iii*) Working accuracy of cylindrical turning, the job held between centres.

$D = L/8$ max.

L = half to full centre distance.

Permissible error is 0.02 mm/300 mm.

16.4. Alignment Tests on Milling Machine

Machine tools are very sensitive to impact or shock, even heavy cast ir on standards are not always solid and rigid enough to withstand stresses due to falling during transportation, and deformations may be set up. Although the machine is always carefully adjusted and aligned when on the test stand or in the assembly department of the manufacturer, it is well known from experience that erection in the workshop of the user is not always done with sufficient care and thus inaccuracies of the work may result from the faulty erection of the machine. So the machine should be carefully levelled up by means of a spirit level before starting with the actual trial tests.

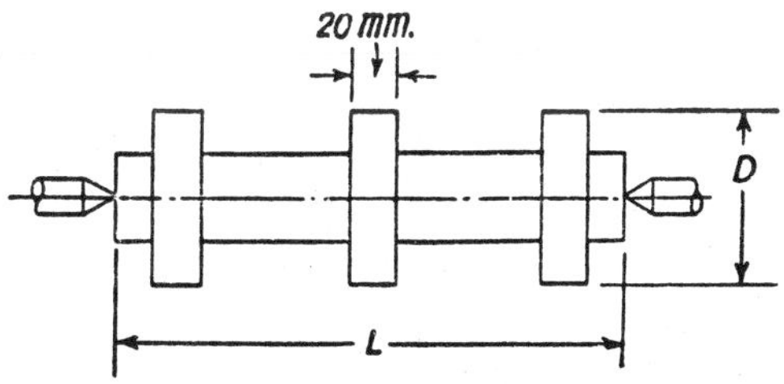

Fig. 16.15

Each trial measurement is based on the correct erection of the machine. No upright, base etc. can be made so rigid that it will be thoroughly free from deformation resulting from faulty erection.

Machine tools for the workshop must be able to produce workpieces of given accuracy within prescribed limits, consistently and without requiring artistic skill on the part of the operator.

For acceptance test of a machine, its alignment test is performed and to see its dynamic stability, which may be poor though alignment tests are right, certain specific jobs are prepared and their accuracy checked.

The relative alignment of all parts of machine and the accuracy of the control devices and driving mechanisms are measured under no load condition. The result of these measurements must lie within the prescribed limits given by the manufacturer depending upon the grade of the machine tool.

A specification for the alignment tests must comply with the following general requirements :

(1) The procedure for testing standard machine tools must not require more than 6 to 8 hrs of work provided all the tooling and measuring equipment are readily available.

(2) The permissible limits of accuracy of individual measurements must be wide enough to make economical manufacture possible while on the other hand the cumulative error of number of superimposed details should not be excessive.

The various tests performed on the milling machine are shown in (Fig. 16.16) and described below.

16.4.1. Cutter Spindle Axial Slip or Float. We have to distinguish between axial (or end) play and axial slip of the spindle.

End play means the indispensable freedom of a spindle moving in the axial direction to prevent it from seizing by heating. This end play is specially important on high speed machines and it should be within the prescribed limits.

Axial slip is defined as the axial spindle movement which may repeat positively with each revolution as a consequence of manufacturing errors. It is only this axial sliding movement which is to be tested, and the specified tolerance applies only to this movement.

When testing the axial slip of a spindle the feeler of the dial gauge rests on the face of the locating spindle shoulder and dial gauge holder is clamped to the table. The locating spindle shoulder is rotated and change in reading is noted. This is done at the two spots diametrically opposite to each other. The total error indicated by the movement of the pointer includes three main sources of errors.

(*i*) Axial slip due to error in bearing supporting the locating shoulder.

(*ii*) Face of the locating shoulder not in a plane perpendicular to axis of rotation.

(*iii*) Irregularities of front face.

Effects of this will be that in cutting spirals, the pitch will not be constant and we will get irregular pitch helix.

If the feeler touches at the same spot where the turning tool on the emery wheel has machined the spindle collar in the assembled machine, then the feeler will not show any deviation. Therefore axial slip must always be tested at two points 180° apart on the collar of the spindle.

16.4.1.1. Eccentricity of External Diameter. The feeler is placed on the cylindrical surface of the shoulder. The locating shoulder is rotated and any deviation in reading of dial gauge is noted.

It is due to eccentricity of the spindle in the hole in which it fits. Due to it, vibrations are produced and the cutter will float side ways and cut over, or under-size. Face mills may dig in when leading edges cease to cut.

16.4.2. True Running of Internal Taper. The table is set in its main position longitudinally and the mandrel 300 mm long is fixed in the spindle taper. A dial gauge is set on the machine table and feeler adjusted to touch the lower surface of the mandrel. The mandrel is then turned and the dial readings at two points are noted *i.e.*, one at the place nearest to spindle nose and other at about 300 mm from it. For shifting the position of dial gauge from *A* to B cross-slide of the machine is operated to bring the dial gauge at the bottom of the end of mandrel. There are can be two errors :

(*i*) Axis of the spindle and the axis of taper may not be parallel.

(*ii*) Eccentricity of the taper hole which, if present, should indicate same error at both the places.

The error in first case will give different readings at two places. Due to this error, cut will not be shared equally between teeth of cutters, and therefore vibrations and poor finish will result.

16.4.3. Table surface parallel with arbor rising towards overarm. In selecting the permissible errors of horizontal milling machines, care is taken to the fact that in the direction parallel to the cutter spindle, the work table extends towards the front face of the knee only, and never slopes down. While working, the table tends to incline downwards under the influence of the weight of work and cutting pressure, while the cutter arbor tends to deflect upwards. Great importance, therefore, has always been attached to the necessity of having the direction of table tolerance opposite to the deformation expected under cutting conditions. Parallelism between table face and the axis of the main spindle is checked as follows :

A dial gauge is set on the machine table. A mandrel 300 mm long is fitted in the spindle taper. The feeler of dial gauge is made to touch the lower surface of the mandrel. With mandrel in position (mean) the readings at the maximum travel of the table surface are observed. The stand of the dial gauge is moved and not the table itself remains stationary.

Effect of this error will be that the milled surface produced will not be square to the base and parallel cross ways.

16.4.4. Surface Parallel with Longitudinal Movement. For this test the dial gauge is fixed to the spindle. Feeler is directed upon the surface the machine table and latter moved longitudinally. The deviations from parallelism between the table surface and longitudinal motion are noted down. If the table is uneven, a straight edge may be placed on the surface. Due to this error the surface of the table will fluctuate up and down and cutter will not take equal cuts on the job which is clamped on the table and the milled surface will not be parallel to the base.

16.4.5. Traverse Movement Parallel with Spindle Axis. (*a*) in horizontal plane ; (*b*) in vertical plane. The table is set in its mean position and dial gauge fixed on the table. The table is moved crosswise and any deviation on reading of dial gauge is noted with feeler on one side of mandrel in horizontal plane and under the mandrel for error in vertical plane.

Due to this error, depth of cut will vary when cross slide is moved.

16.4.6. Central T-Slots Parallel with Longitudinal Movement. The T-slot, particularly the central one should be well machined on the internal vertical surface throughout its length because jigs and fixtures are located by T-slots. The general parallelism of the central slot with the longitudinal movement of the table is checked by using a bracket 150 mm long with a tennon which enters the T-slot. Against the upper surface of the bracket in vertical plane the feeler of the dial gauge is located. Having fixed the dial gauge to the spindle and adjusting its feeler to the surface of the bracket the table is moved longitudinal while the tennon block is held stationary by hand and deviations from parallelism are noted from dial gauge. During the process the tennon slides along the slot, thus eliminating the effects of local error.

Due to this error, the depth of cut will not remain constant as the job will be inclined according to inclination of T-slots with longitudinal movement and the axis of job held between tail stock and index head will not be perpendicular to cutter.

ALIGNMENT TESTS ON MILLING MACHINE

No.	*Test to be applied*	*Test Diagram*	*Gauges and Method*	*Permissible Error*
(1)	(2)	(3)	(4)	(5)
1*A* *B*	Cutter spindle Axial Slip or Float Eccentricity of External Diameter	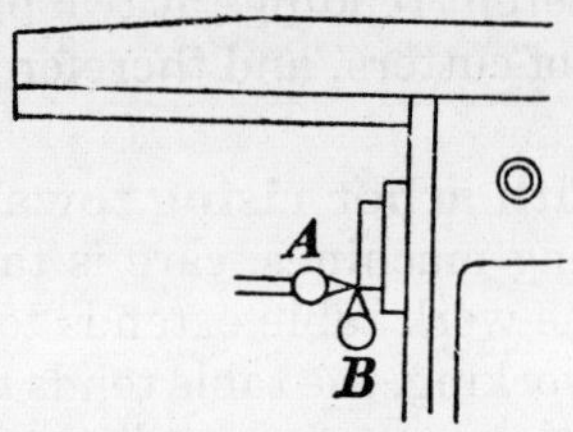	(*A*) Clock Indicator (*B*) Test two places at 180°	*A* 0.01 mm *B* 0.01 mm
2	Internal taper for true running	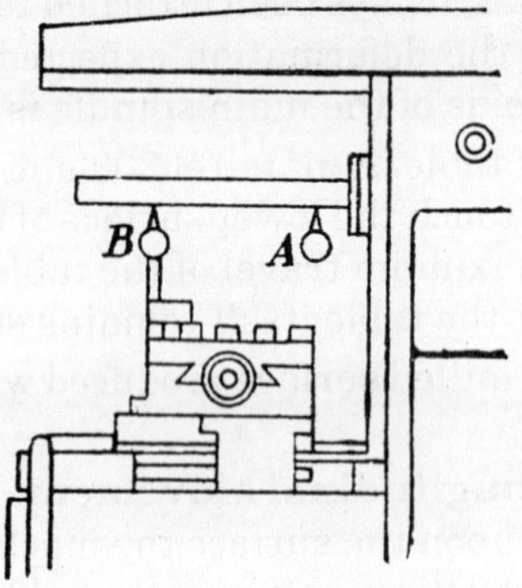	Clock Indicator slowly move mandrel for maximum eccentricity (*A*) Nearest to spindle nose (*B*) At a distance of 300 mm	 *A* 0.01 mm *B* 0.025 mm
3	Work Table Table surface parallel with arbor rising towards overarm	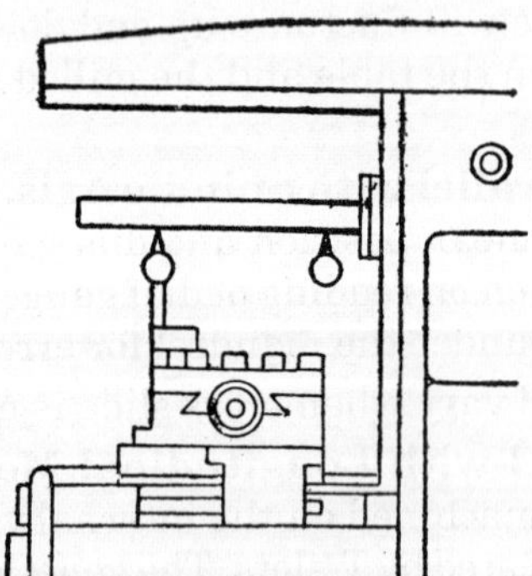	Stationary mandrel. Clock indicators under mandrel Test table surface for max. travel	0.025 mm per 300 mm
4	Surface parallel with its longitudinal movement.	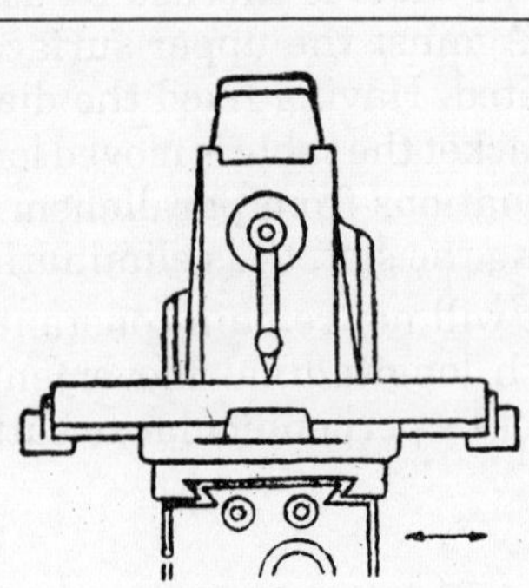	Clock Indicator in spindle Test table surface for max. travel	0.04 mm upto 600 mm movement 0.05 mm over 600 mm movement

5	Transverse movement parallelism with spindle axis in horizontal plane In vertical plane	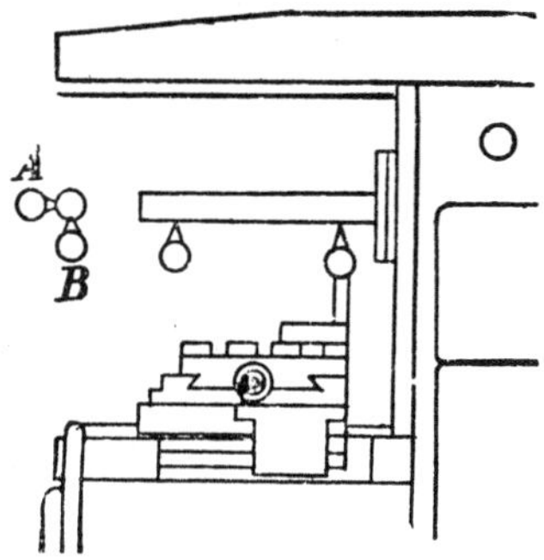	Stationary mandrel Clock indicator moved transverse with table. For *A* test on side of mandrel and for *B* test under mandrel	*A* 0.025 mm per 300 mm *B* 0.025 mm per 300 mm
6	Centre T-slot parallel with longitudinal table movement	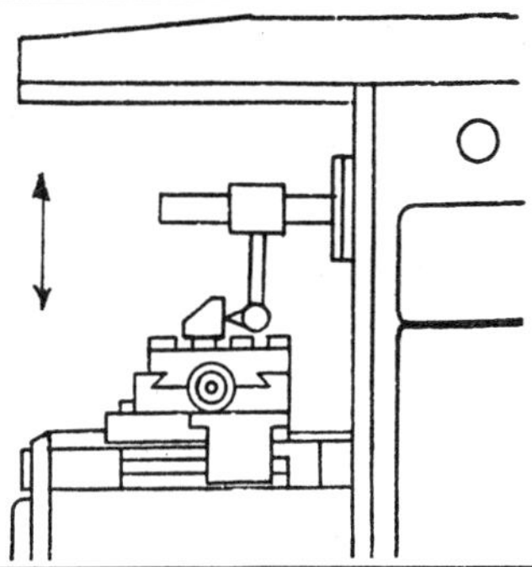	Clock indicator in spindle. Tenons in '*T*' slot	0.04 mm upto 600 mm movement 0.05 mm over 600 mm and upto 100 mm movement
7	Centre T-slot square with arbor.	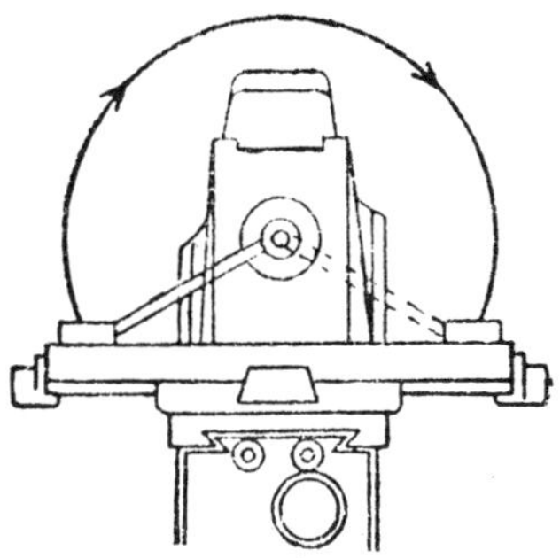	Clock in spindle. Tenon Block Turn over method	
8*A*	Column Column ways for knee square with table. Inclination front to rear	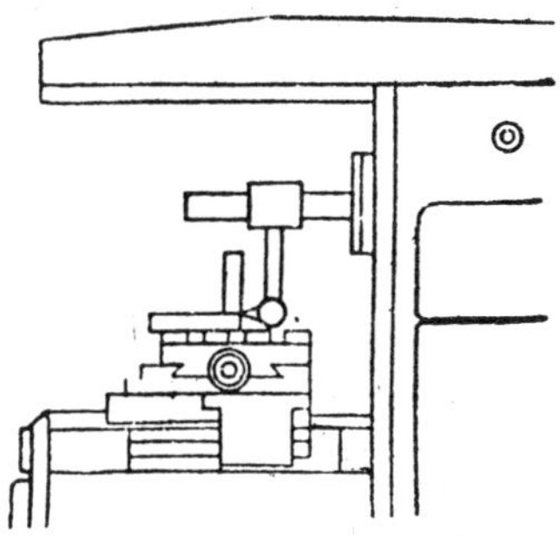	Clock in spindle. Square set vertically	A 0.025 mm per 300 mm
B	Side inclination			*B* 0.025 mm per 300 mm

9*A*	Overarm parallel with spindle In horizontal	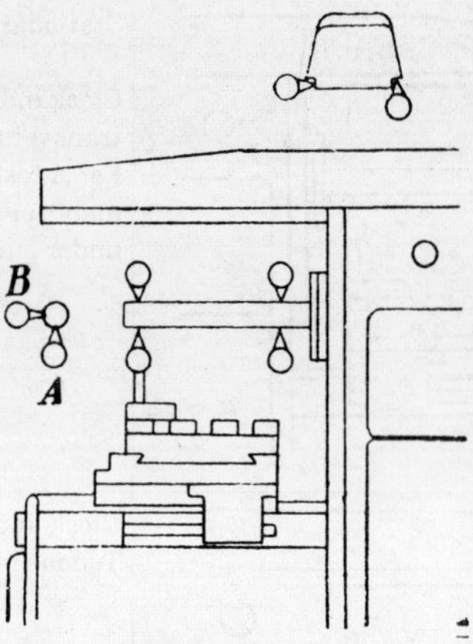	Clock in Spindle For *A* compare readings on under overarm and mandrel	*A* 0.025 mm per 300 mm
B	In vertical plane	Fig. 16.16	For *B* compare reading on side of everarm and mandrel	*B* 0.025 mm per 300 mm

16.4.7. Centre T-slot Square with the Arbor. If the central T-slot is not perpendicular to the arbor, the key way etc. cut on the machine will not be parallel to the axis of job.

For this test table is adjusted in the middle portion of its longitudinal movement and a tennon block 160 mm long inserted in the central T-slot. A dial gauge is fixed on the mandrel, the feeler being adjusted to touch the vertical face of the bracket. Observe the reading on the dial gauge when the bracket or tennon block is near one end of table. Then swing over the dial gauge and move the tennon block so that the corresponding reading can be taken near the other end of the table. Generally two tennon blocks are used.

16.4.8. Tests on Column. (*a*) Column ways of knee square with table, inclination front to rear (*b*) Side inclination.

Arrange table is its central position and fix a square with an arm about 300 mm long (Angle iron type bracket placed on the surface), on the table surface and attach a dial gauge to the spindle mandrel in such a way that the feeler rests on the arm (vertical face) of the surface near the bottom edge. Observe the reading of the dial gauge. Move the table upwards abut 300 mm and again observe the dial gauge. The difference in readings is a direct indication of the error of perpendicularity of the table surface (from guiding surface) and knee support or side guiding support. The above test is conducted for two positions of the square and in first position the dial gauge touches the square in front and in second position it faces the side of the square *i.e.* at 90° to the first position (for testing side inclination).

If column-ways for knee are not square with the table, as the table is fed upwards in facing operation or end milling, the surface produced will not be square with the table surface.

16.4.9. Overarm parallel with spindle

(*a*) in horizontal plane. (*b*) in vertical plane.

To check the parallelism of the overarm and the spindle, fix the dial gauge on the table and its feeler under the mandrel. Move the table crosswise and note any change in the reading. For error in the horizontal plane, repeat above readings so that feeler is under overarm and compare the two sets of readings. For vertical plane compare the readings on mandrel and side of overarm.

16.4.10. Alignment of the main spindle with bore of the bracket of the overarm. With the mandrel (parallel) in the bore of the overhanging bracket and gauge holder in the mandrel fitted to the spindle taper, the feeler is adjusted so that it touches the mandrel in the bore. The main spindle is turned slowly and reading of the dial gauge is noted at four points (opposite ends in horizontal and vertical planes *i.e.* 90° apart).

The difference between two 180° opposite readings and other two is twice the eccentricity of the mandrel in the vertical and horizontal directions respectively.

It the axis of the bearing of the supporting bracket is not co-axial with the spindle axis, the axis of the arbor which is held in spindle and supporting bracket will not be parallel with table surface and hence the cutter mounted on the arbor will take more cut on supporting bracket side if the bearing axis is somewhat lower than spindle axis and less cut if the bearing axis is above.

16.5. Alignment Tests on Pillar Type Drilling Machine

Before carrying out the alignment tests, the machine is properly levelled in accordance with the manufacturer's instructions. The various tests performed are :

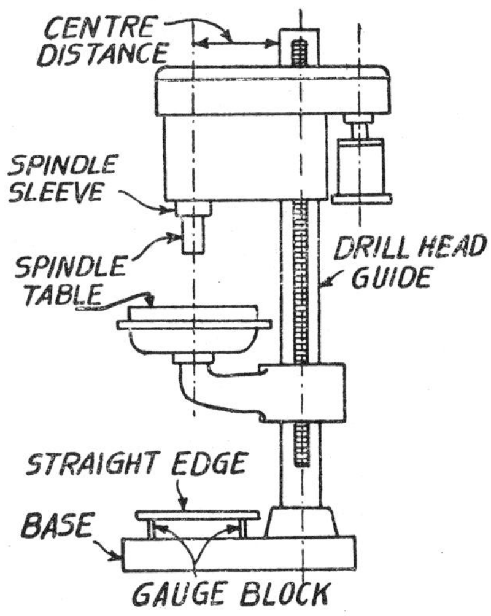

Fig. 16.17. Checking flatness of clamping surface of base.

1. *Flatness of clamping surface of base.* (Refer Fig. 6.17). The test is performed by placing a straight edge on two gauge blocks on the base plate in various positions and the error is noted down by inserting the feeler gauges. This error should not exceed 0.1/1000 mm clamping surface and the surface should be concave only.

2. *Flatness of clamping surface of table.* This test is performed in the same manner as test (1), but on the table. The permissible error is also same.

3. *Perpendicularity of drill head guide to the base plate.* The squareness (perpendicularity) of drill head guide to the base plate is tested :

(*a*) in a vertical plane passing through the axes of both spindle and column, and

(*b*) in a plane at 90° to the plane at (*a*).

The test is performed by placing the frame level (with graduations from 0.03 to 0.05 mm/m) on guide column and base plate and the error is noted by noting the difference between the readings of the two levels. This error should not exceed 0.25/1000 mm guide column for (*a*) and the guide column should be inclined at the upper end towards the front only, and 0.15/1000 mm for (*b*).

(Please note that Fig. 16.18 shows the similar test for the squareness of drill head guide with table, the only difference being that the frame level is to be placed on the base instead of table).

4. *Perpendicularity of drill head guide with table.* This test is performed exactly in the same way as (*b*). (Refer Fig. 16.18) and the permissible error is also same.

5. *Perpendicularity of spindle sleeve with base plate.* This test is performed in both the planes specified in test (3) and in the similar manner with the difference that the frame levels are to be placed on spindle sleeve and base plate. The error (*i.e.*, the difference between the readings of the two levels) should not exceed 0.25/1000 mm for plane (*a*) and the sleeve should be inclined toward column only ; and 0.15/1000 mm for plane (*b*).

6. *True running of spindle taper.* For this test, the test mandrel is placed in the tapered hole of spindle and a dial indicator is fixed on the table and its feeler made to scan the mandrel. The spindle is rotated slowly and readings of indicator noted down. The error should not exceed 0.03/100 mm for machines with taper upto Morse No. 2 and 0.04/300 mm for machines with taper larger than Morse No. 2.

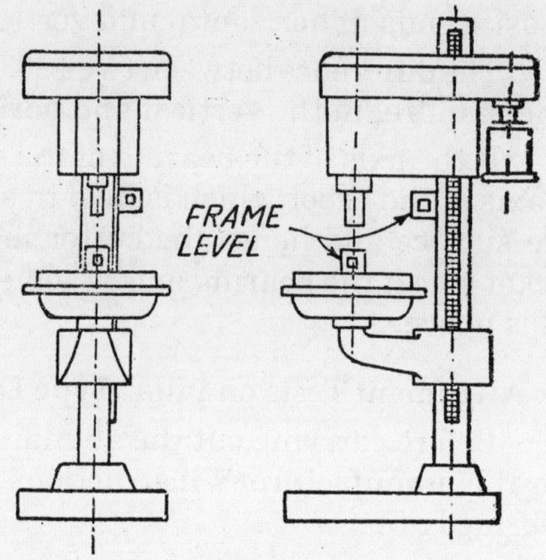

Fig. 16.18. Test for perpendicularity of drill head with table.

7. *Parallelism of the spindle axis with its vertical movement.* This test is performed into two planes (*A*) and (*B* at right angles to each other. The test mandrel is fitted in the tapered hole of the spindle and the dial indicator is fixed on the table with its feeler touching the mandrel. The spindle is adjusted in the middle position of its travel. The readings of the dial indicator are noted when the spindle is moved in upper and lower directions of the middle position with slow vertical feed mechanism.

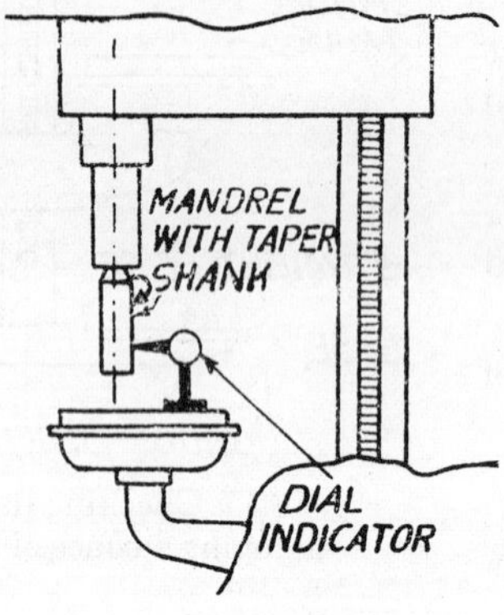

Fig. 16.19

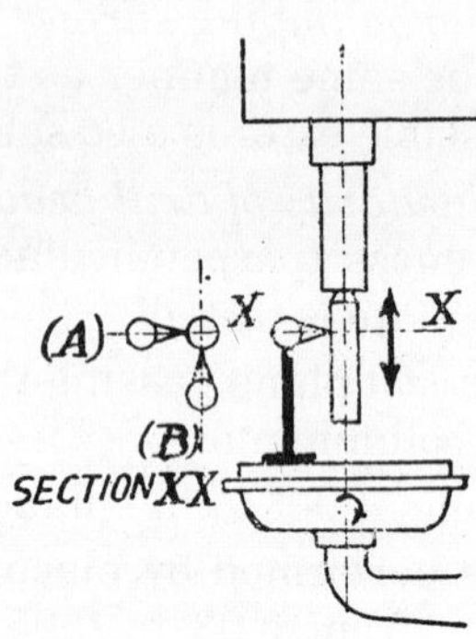

Fig. 16.20

The permissible errors are :

	for plane (*A*)	for plane (*B*)
For machines with taper upto Morse No. 2.	0.03/100 mm	0.03/100 mm
For machines with taper larger than Morse No. 2.	0.05/300 mm	0.05/300 mm

(Lower end of mandrel can be inclined towards column only).

8. *Squareness of clamping surface of table to its axis.* For performing this test, the dial indicator is mounted in the tapered hole of the spindle and its feeler is made to touch the surface of table (Refer Fig. 16.21). Table is slowly rotated and the readings of dial gauge noted down, which should not exceed 0.05/300 mm diameter.

9. *Squareness of spindle axis with table.* For this test a straight edge is placed in positions *AA′* and *BB′*. Work table is arranged in the middle position of its vertical travel. The dial indicator is mounted in the spindle tapered hole and its feeler made to touch the straight edge first say at *A* and reading noted down. The spindle is rotated by 180° so that the feeler touches at point *A′* and again reading is noted down. The difference of two readings gives the error in squareness of spindle axis with table. Similar readings are noted down by placing the straight edge in position *BB′*.

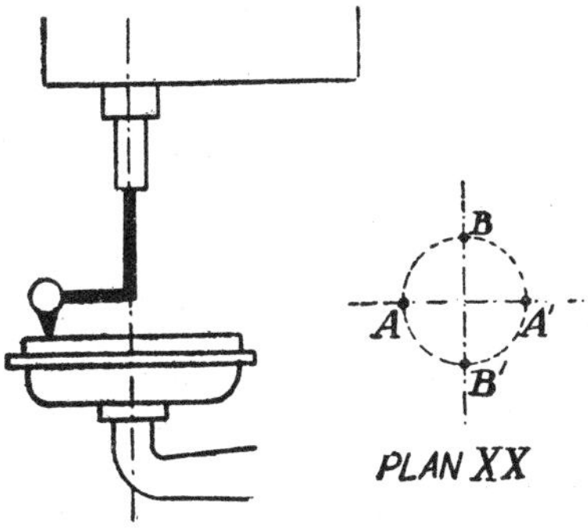

Fig. 16.21

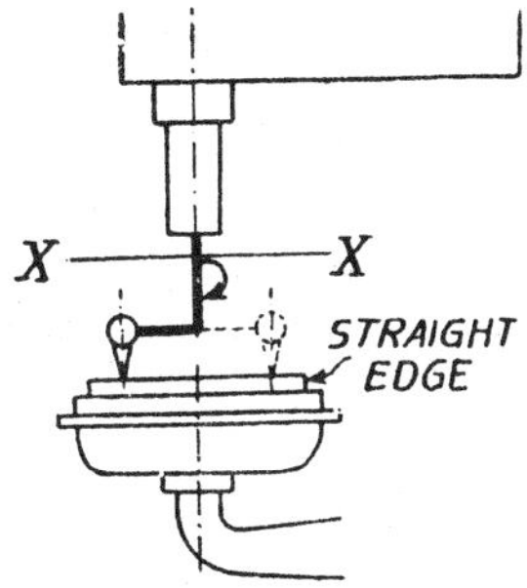

Fig. 16.22

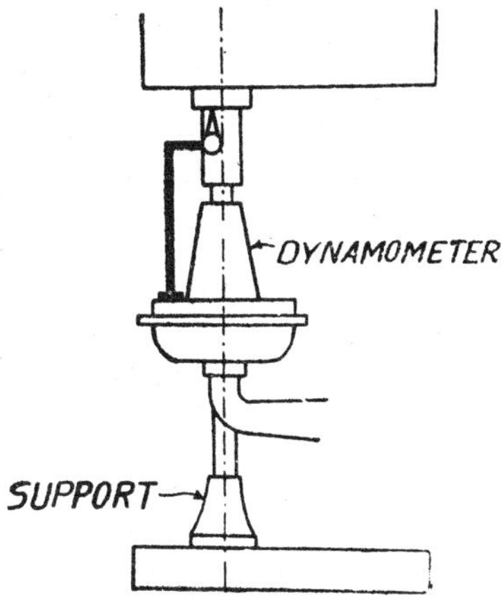

Fig. 16.23

The permissible errors are 0.08/300 mm with lower end of spindle inclining towards column only for set up AA' and 0.05/300 mm for set up BB'.

10. *Total deflection.* For this test, the drill head and table are arranged in their middle position. Dial indicator is mounted on table with its feeler touching the lower machined surface part and spindle stock as shown in Fig. 16.23.

Drill spindle is loaded with the dynamometer (load gauge) placed on table and the deflection of dial indicator noted down. The drill spindle pressure is set in accordance with the table on sideways.

Drilling Diameter in Open Hearth Steel	*kg*
6	100
10	200
16	350
20	550
25	750
32	950
40	1200

The permissible error is

	for centre distance upto and including
0.4 mm	200 mm
0.6 mm	200-300 mm
0.8 mm	300-400 mm
1.0 mm	400 and above mm

16.6. Acceptance Tests for Shapers

Following tests are conducted for shapers :

(*a*) Table top parallel to its movements by fitting dial gauge in tool box and keeping the ram stationary and moving the table.

(*b*) Ram movement parallel to table top by fitting dial gauge in tool box and moving the ram.

(*c*) Table support guide parallel to table movement by fitting dial gauge to front of table and traversing the table.

16.7. Acceptance Tests for Surface Grinders

Following tests are conducted for surface grinders :

(*a*) Table top parallel to its movement by fitting the dial gauge in the stationary spindle and dial feeler touching the table top and traversing the table.

(*b*) Slots parallel to table movement by fitting the dial gauge in the stationary spindle and dial feeler touching the slot and traversing the table.

(*c*) Spindle axis parallel to table top by fitting dial gauge on radial arm of the spindle and its feeler touching the square resting on table, and rotating the spindle by 180 degrees.

(*d*) Spindle axis square with the slots by fitting dial gauge on an arm fitted to spindle and rotating the spindle by 180 degrees.

(*e*) Vertical movement of spindle square with table top by fitting dial gauge in spindle and its feeler touching the square resting on table.

(*f*) Practical tests.

16.8. Mandrels

Mandrels are used in alignment tests to represent the axis and are of two types :

(*i*) Test mandrels with taper shanks. (*ii*) Mandrels between centres.

(*i*) In construction the *test mandrels with taper shanks* have a conical shank for inserting in the socket of the machine to be tested, and a cylindrical body which is used as a reference surface for the measurements. In alignment tests, the cylindrical surface of the mandrel represents the axis desired to be checked, either for out of true running or for position in relation to other elements of the machine tool. These are made of hardened and stabilised steel left either unplated or plated with hard chromium. The mandrels of such type having relatively slow, and steep tapers are shown in Fig. 16.24 (*a*) and (*b*) respectively.

The deflection of free ends of mandrels should be less than 0.65 μm for those having working length of 75 mm and than 5.8 μm for those having working length of 505 mm. These should be free from errors of roundness and the axes of the conical and cylindrical portions be straight and coincident. The cylindricity should be perfect over the whole of the measuring length.

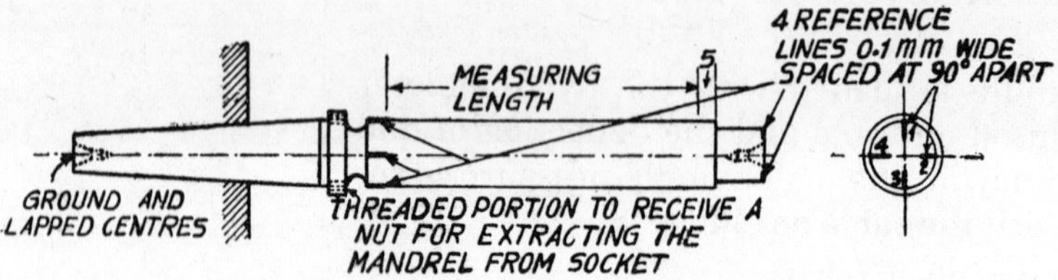

(*a*) Test mandrel with slow taper shank.

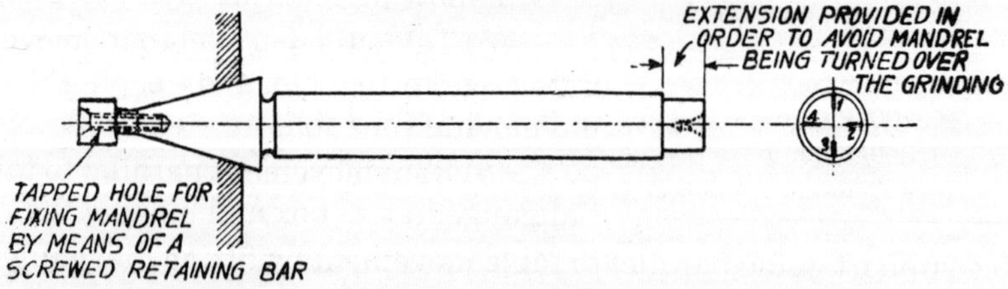

(*b*) Test mandrel with steep taper shank.

Fig. 16.24

Mandrels between centres are used to represent simply a straight line passing between two points. Its axis is made straight and exterior surface truly cylindrical having an accuracy of 3 microns over 300 mm length. It also has four reference marks located in two perpendicular axial planes and centres recessed for protection. Mandrels of large diameter are made tubular to reduce weight, but should be rigid enough.

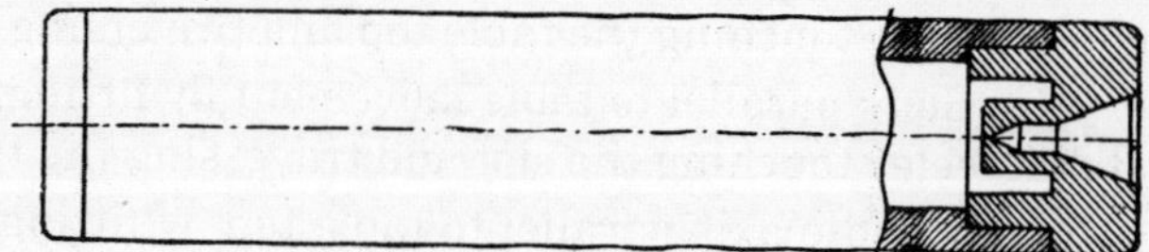

Fig. 16.25. Mandrel between centres.

TYPICAL EXAMINATION QUESTIONS

1. Distinguish between alignment test and performance test of a machine tool.
   Discuss with neat sketches how you would test any three of the following for an Engine Lathe :
   (*a*) Spindle centre run-out, (*b*) Spindle taper bore run-out, (*c*) Chuck run-out, (*d*) Cross slide alignment.
2. Distinguish between the "geometric" and "practical" tests on a machine tool.
3. Describe how you would perform the following alignment tests :
   (*i*) Squareness of horizontal milling machining centres tee-slot to spindle axis.
   (*ii*) Squareness of lathe cross-travel slide to spindle axis.
   (*iii*) Parallelism of axis of lathe centres and the saddle motion.
   (*iv*) Squareness of milling machine cross-slide ways to the vertical guides on the column.
   (*v*) Parallelism of quill movement of lathe tail-stock with machine bed guides.
   (*vi*) Parallelism of radial drill saddle movement and the table surface.
   Also state the tolerances normally permitted.
4. Describe two methods of testing a centre lathe for accuracy of facing. Why must the surface generated not be convex ?
5. What is the effect upon the work if the centre tee-slot of a horizontal milling machine is not square with the spindle axis ?
6. Describe with sketches the optical setup for checking the following :
   (*i*) Straightness of saddle-travel in horizontal plane for a centre lathe.
   (*ii*) Perpendicularity of spindle axis of the drilling machine to its table.
   (*iii*) Parallelism of transverse movement of a horizontal milling machine table with the spindle axis both in horizontal and vertical planes.
7. Explain what is understood by the term 'Alignment test' of a machine tool.
   Explain with neat sketch the method of checking the following in the acceptability test of drilling machine :
   (*a*) If the spindle is running true.
   (*b*) If the table and pillar are mutually perpendicular.
   (*c*) If the spindle perpendicular to the table.
8. What is the effect upon the work if a lathe tailstock centre line is parallel to, but slightly above, the headstock spindle axis ?
9. Describe in details, how would you check :
   (*i*) The Squareness to table of the spindle and the spindle feed of a radial drill.
   (*ii*) The spindle axis of a lathe is parallel to the bed.
   (*iii*) The cross-travel guides of a milling machine are square to the faces of the vertical guides.
   What will be the effect of each on the work and what is the permitted value of the tolerance in each case ?
10. With the aid of sketches describe how the following tests may be carried out on a centre lathe.
    (*i*) Straightness of a bed.
    (*ii*) Parallelism of the spindle axis with the guide ways.
    (*iii*) Parallelism of the axis of 'Centres' with the bed in the vertical plane.
11. In certain cases the permitted error in machine tool alignment is unilateral only. Why ? Give 3 examples giving reasons for unilateral tolerances.
12. What is the effect upon the work if the spindle of radial drill is not square to the table, being inclined towards the column at the bottom by 0.25 mm per metre ?
13. Describe two methods to test the alignment of the bore of the overarm support bracket and the spindle axis of a horizontal milling machine. What are the advantages of each ? What is the effect upon the work if it is not correctly aligned ?

17

Measuring Machines

17.1. Introduction

Measuring machines are generally used for measurement of length over the outer faces of a length bar or any other long member with ends that may be either rounded, or flat and parallel. The need of measuring machines is realized over vernier calipers, micrometer screw gauges etc., for linear measurement work due to measurement of considerably greater dimensions. The measuring machines are generally of more universal character and can undertake works of varied nature. A number of measuring machines of various designs are available, out of which some are discussed here.

17.2. Length Bar Measuring Machine

The general form of measuring machine called length bar measuring machine used to determine the length of a pin gauge consists of a long bed having vee and flat guide ways along the bed to allow suitable spacing of measuring head, anvil and work supports (Fig. 17.1). A fixed headstock and a movable tailstock each with a measuring face anvil are carried on the bed. A micrometer screw fixed on tailstock is used as a means of revealing difference in length between a setting master and workpiece. The machine can show the length of the work either by direct comparison with standard length bar or by the use of an accurate "built in" scale. In either case the tailstock has to be moved to such a position that the two anvils are just touching the end of the workpiece which is supported parallel to the bed of the machine. This is achieved by using a micrometer screw in tailstock and a sensitive indicator (dial gauge) in the headstock. Difference between the settings required for standard bar and workpiece is measured by the micrometer screw in the tailstock. The headstock anvil is carried in such a way as to show when contact is made with the workpiece. This part of the machine is so sensitive that the reading of the tailstock setting on any workpiece can be repeated to 1/40 (*i.e.* fortieth part of a mm). For correct measurement, the correct support in horizontal plane is very essential and for this the two support points should be spaced 0.577 L apart, L being the length of the workpiece. The supports used are of vee form and adjustable for height, and spacing laterally across the bed. This machine has two

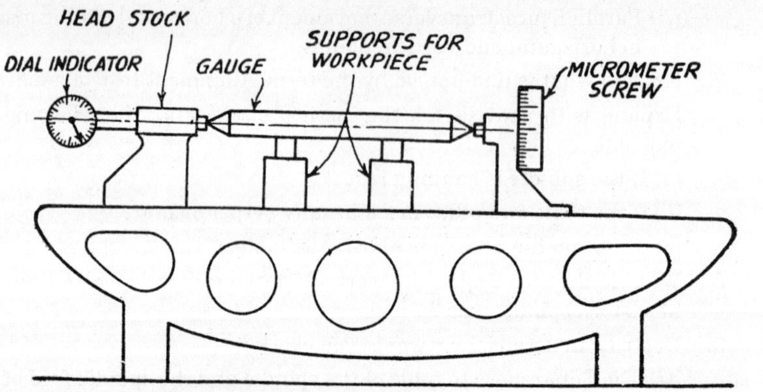

Fig. 17.1. Length-bar measuring machine.

special features which are essential in any set-up for fine measurement *i.e.*, the piece to be measured is positively located in relation to the measuring faces and the contact between measuring faces is proved mechanically without relying on any personal element.

When the length of any bar is to be measured, a length bar of appropriate grade whose dimensions are exactly known is placed over the vee supports which are then adjusted in height so that axis of the bar and axis of measurement are coincident. A reading on micrometer is then noted for some suitable position of the pointer of the dial indicator. The gauge is then replaced by the bar whose dimension is to be determined and a second micrometer reading is taken at the same reading of the dial indicator. The difference between two micrometer readings is then the difference between the length of the setting standard and the work. Its accuracy is made as that of micrometer and it is about $\frac{1}{400}$ *th*. In order to avoid undue movement of the screw the length of setting master should be close to the length of the workpiece. Before taking any readings, temperature stabilization should be attained for accurate results.

17.3. Newall Measuring Machine

This machine and Zeiss measuring machine and some others have the different features of the sensitive indicator on headstock side. The measurement of difference in length between the workpiece and standard bar length is done by the use of a micrometer screw. The sensitive indicator, in the Newall machine consists of a lever pivoted on a horizontal centre line just below the end of the anvil opposite to the measuring face as shown in Fig. 17.2. The lever is kept in contact with the end of the anvil member by a light spring pressure. The lever carries a sensitive spirit level at its upper end and the level has graduations, each graduation corresponding to the movement of about 0.0001 mm at the anvil. After the work has been placed between the anvils, the micrometer screw is adjusted until the bubble in the spirit level is brought to a fixed position. Again for other settings, the conditions are then kept same and setting can also be done accurately upto the said accuracy.

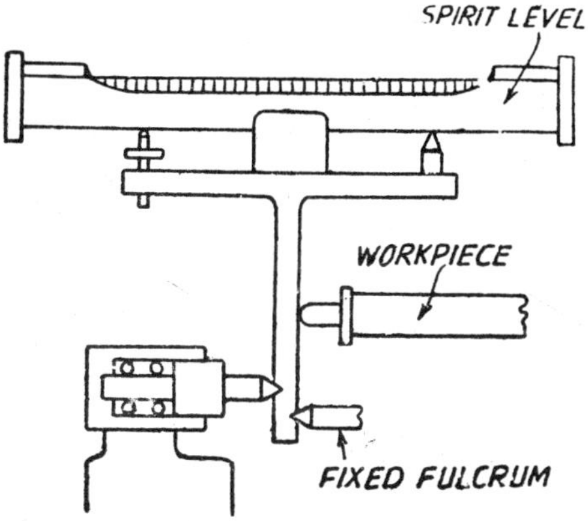

Fig. 17.2. Newall measuring machine.

In Zeiss measuring machine, indication of contact is provided by the optical system used in Zeiss optimeter (optical comparator) in which movement of the gauging anvil tilts a mirror and this causes movement of the image of the scale in the field of a microscope. By this arrangement, movements of anvil as small as $\frac{1}{4000}$ *th* mm can be detected. This machine includes a precision glass scale so that length of the work can be found without using a standard length bar. The scale is not placed in the same straight line as work, as this leads to a machine of bigger size. In order to avoid any error due to it or any tilt in the bed of the machine, a special optical system is used for reading the scale.

17.4. Universal Measuring Machine

These machines can be utilised to measure length and diameters of both plain and threaded work, tapers and the pitch of the screw threads to a high degree of accuracy. Some machines employ two and three co-ordinate measuring systems based on the use of line standards. The scales are generally enclosed within the machines and thus not liable to damage by physical contact. The basis of the accuracy of measurement of these machines is an accurately divided scale viewed through a microscope. The microscope is used to determine the movements of the measuring anvils of the machine. In order to satisfy the principle of alignment, the scale is positioned in line with the line of measurement.

17.4.1. The Societe Genevoise Universal Measuring Machine. It has two carriages which move at right angles on hand scraped flat and vee-shaped guiding surfaces. The upper carriage is capable of being lowered or raised and carries a tool holder at the front end. The tool holder is fitted with a sleeve of hardened steel which is finely ground to ensure the accurate centering of the various tools such as locating microscope, feller microscope, goniometer microscope, spotting tool and locating indicator. Both carriages carry a standard scale in box and the scale is viewed by a measuring microscope. An auxiliary scale is also attached to both carriages to register their approximate positions. There is arrangement to carry the cast iron table, glass table centre supports, rotary table and other accessories on the scraped bearing surfaces of the machine. It is possible to make measurements in all the three co-ordinates and also polar co-ordinates. Thus the curved surfaces such as aerofoils, turbine blades etc., can also be optically tested by use of a short-focus microscope carried in the special tool holder. The goniometer microscope comprises a viewing microscope fitted with two super imposed graticules each carrying diametral dotted lines. One graticule is fixed, and the other can be rotated and carries a circular glass scale divided in degrees. This enables the measurement of thread angles and flake angles etc.

17.4.2. The Matrix Internal diameter Measuring Machine. This machine is capable of measuring plain and threaded work, and also parallel or taper work. It can work in the range of 6 to 150 mm diameter and to a depth of 50 mm. The special feature of this machine is that it relies on end standard. The measurement of diameters is made by comparison with a box standard built up from matched slip gauges. This is also capable of measuring work which differs in size from the standard by several millimetres. These differences are measured by means of a micrometer used in conjunction with a fiducial indicator. The principle of its operation is rather unusual in the sense that it measures diameter by measuring two radii at exact 180° spacing. The gauge whose diameter is to be measured is supported on a circular table, which can be precisely rotated about a vertical axis. There is an arrangement by which the stylus can be exactly located at diameter end. For taking the measurements, the gauge is first approximately centered on the table with the help of concentric ring engraved on it. Further with the measuring stylus in contact, the whole table is traversed on a precision slide until a maximum reading is obtained. The table is then rotated by exactly 180° to a stop, and another reading is taken again after traversing to obtain the maximum reading. The readings on the gauge are compared with those obtained on a 'box' made up of two equal piles of slips on to which parallel end pieces are wrung and held with screw clamps.

For getting exact 180° rotation of the table, two limit stops are provided in the machine. For location of maximum diameter of workpiece, the table is provided the traverse motion by a knob and screw which transmits cross-wise movement to the table.

17.5. Optical Projection Comparator

Fundamentally projection comparator (projector) is a measuring instrument which projects an enlarged shadow of the part being measured on a screen, where it is compared to a master drawing. By this device, complicated shaped parts can be easily checked. In any projection system, there are four essential elements, *viz.*, source of light, collimating or condensing lens, projection lens and screen. The illumination required for any projector is a parallel beam of light, which can be achieved if the source of light is a point source and is placed at the principal focus of the collimating lens. The purpose of collimating lens is to render the beam of light from point source parallel. The projection lens consists of a system of lenses which form a real image on the screen of an object placed between it and collimator. The screen is generally translucent so that the image can be viewed from the opposite side.

17.5.1. Horizontal Projcctor. This was designed by N.P.L. and is the earliest form of projector. It consists of a base casting. A lens system and a sliding carriage carrying the object to be projected are mounted on the base. The carriage slides parallel to the optical axis and this enables the object to be focused. The disadvantage with this is that there is only one plane normal to the optical axis in which all sharply focused parts of the object must lie. The lamp housing and collimation lens are carried on two bars arranged to pivot about a vertical axis passing through the projection lens to enable the collimating beam to be inclined to the optical axis for projection of screw threads. Its magnification is of the order of $\times$ 50.

17.5.2. Vertical Projector. The horizontal comparator is not suitable for rapid inspection of small work and also it occupies more space. In vertical projector, the optical axis from the projection lens rises vertically and is reflected down again from a mirror on to a small horizontal screen. It has the advantage of occupying less floor space, the work is near the screen, and therefore the size of screen is also less. The horizontal screen enables user to stand and view the image conveniently. The profile drawings and templates can also be laid down conveniently without the necessity of fastening with tape etc. The illumination is obtained by a carbon arc in conjunction with a long focus collimator.

17.5.3. Cabinet Projector. The various advantages and conveniences of projection as a method of inspection have led to the development of a self-contained type of projector. It is now available in compact form and can be used anywhere. Some of the projectors available commercially are discussed here.

17.5.4. Bausch and Lomb Projector. The various parts are explained here in detail to understand the principle.

The optical system of this projector is shown in Fig. 17.3. This is discussed under two heads, illuminating system and projection system.

The illuminating system consists of a light source, a tungsten arc lamp. The glowing element in the lamp is a small cylinder of tungsten which is heated to incondescene by electron bombardment. It is enclosed in a ventilated lamphouse. The light from the lamp passes to a system of lenses called achromatic condenser. With single lens condensers, the screen

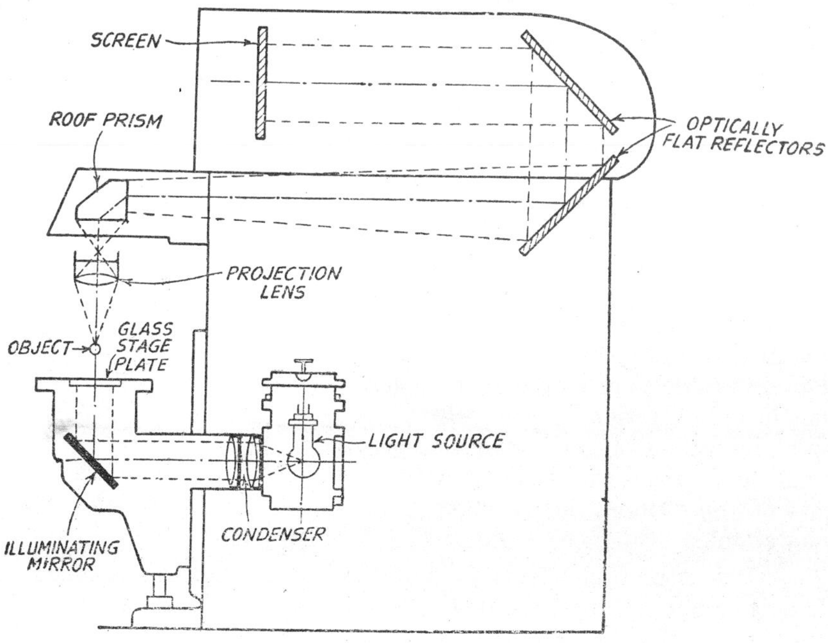

Fig. 17.3. Optical System of Bausch and Lomb Projector.

image consists of various colours. So to get rid of them, each lens assembly is composed of two kinds of glasses. The parallel beam of light from the condenser is then transmitted to the illumination mirror, which sends the beam vertically upward, through the glass stage plate in the worktable, past the object.

The projection system consists of the projection lens, roof prism, a pair of image reflectors and screen. The magnification of the projector can be adjusted by changing the projection lens assembly. The adjustments for changing the lens assembly are very easy and accurate. It is done by merely placing the assembly into the bracket. The light travels past the object to the projection lens from which it passes upward to the roof prism. The function of the roof prism is to direct the beam of light horizontally towards the back of the projector to assist in the projection of the image so that its aspects presented to the observer are correct. From the roof prism the light passes to two optically flat reflectors which change the direction of the beam of light and direct it horizontally to the screen.

When an observer looks the image, he will find the image as erect and its aspects same as that of the object, *i.e.*, movement of an object in any direction on worktable will cause a corresponding travel of the image. If the object is moved to right or back, a corresponding movement of image to the right or up will be observed.

17.5.5. Societe Genevoise Projector. This is a comprehensive measuring machine intended for the measurement of objects like gear cutting hobs and other complex tools. The work is held in the centres. For the spacing of every tooth, there is a headstock fitted with a dividing head. The longitudinal movement of the carriage is measured by a precision scale and microscope. The various types of indicating units can be set up for radial, longitudinal and transverse measurements, thus covering the measurement of any complicated shape. In this way in a gear hob cutter, the flute spacing, tooth relief concentricity etc., can be directly checked and the form and angle of teeth can be measured by projection.

17.5.6. Screw Thread Projection. In projection of a thread a part of the helix always crosses over in front of each part of the thread and interferes with the direct passage of the rays. This can be overcome by either inclining the axis of the thread at an angle equal to the helix angle so that the collimated beam gazes the thread without appreciable interference or by inclining the collimated beam to the helix angle by swivelling the lamp house about a vertical axis passing through the projection lens. The image formed on the screen is that of a plane cutting the thread normal to the helix and lying at an angle to the axis. In second method, the axis of the threads lies in the plane of the axis and hence defined profile is produced without loss of field and definition. The projection lens must be well designed, otherwise only small helix angle can be dealt with. In case of first system, the arrangement is very simple.

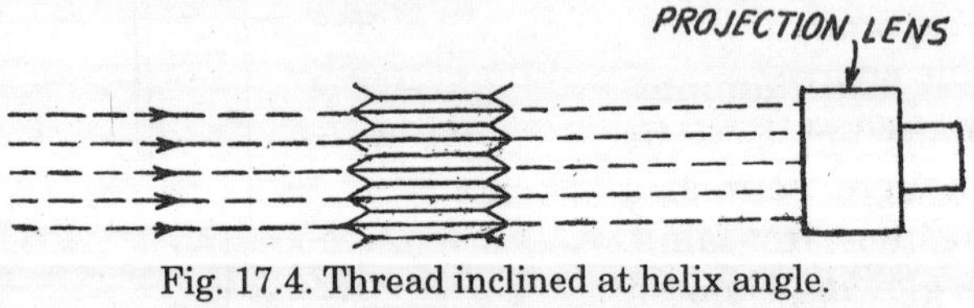

Fig. 17.4. Thread inclined at helix angle.

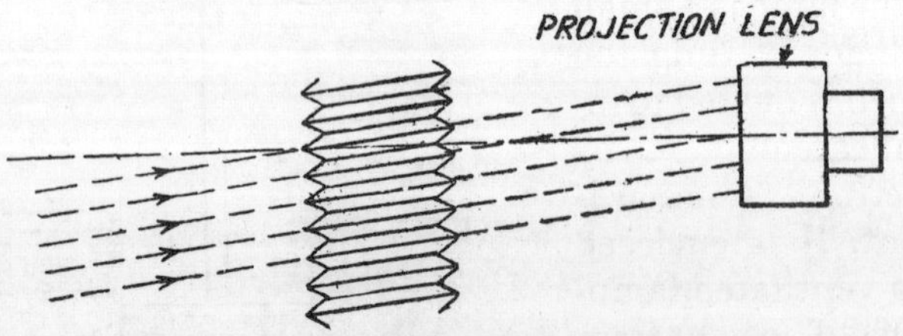

Fig. 17.5. Collimated beam inclined (Projection of Axial section).

It is found that difference in profile between the two methods is generally difficult to be detected and hence first method which is more convenient is used.

The advantages of thread inspection by projector are :

(1) It simultaneously compares all the elements of the thread with an accurate gauge.

(2) Right hand and left hand threads as well as wide range of diameters and pitches, can be measured.

(3) It eliminates the cost and uncertainty of gauge wear and is an inexpensive method.

(4) It is a rapid and more positive method.

17.5.7. Use of Standard Radius Chart and Standard Centreline Screen. A big advantage of optical contour projector is that it readily permits checking of dimensions from points and lines not located on the surface of the workpiece, like centres of holes, the intersections of two straight lines which are the expansions of rounded corners, the radius, centre of a fillet, etc. Fig. 17.6 shows the use of standard radius chart to locate the radius and centre of a fillet. Fig. 17.6 shows the use of a standard centre line screen to find the point of intersection of two sides where a corner has been rounded.

17.5.8. Checking of Profiles which can't be Projected. A special tracer unit on optical contour projector permits checking of internal or hidden contours which can't be projected on the contour screen. A ball tipped tracing stylus is moved over the part contour while an identical ball moves through an identical path at the end of a pantograph and the movement of other ball is projected on the screen where the enlarged desired internal contour is fixed up. The shape of the stylii is chosen properly depending on the kind of internal contour.

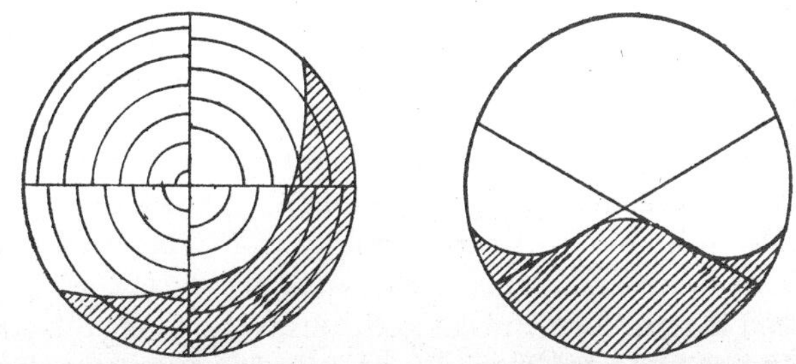

Standard Radius Chart Standard Centreline Screen

Fig. 17.6

17.5.9. Precaution in the Use of Optical Projectors. The workpiece should be so located on the table in the path of the projected light beam, and light and lens so focused that double shadow of false, silhouette edge and other abberations are avoided as far as possible. Dust, moisture, oil etc., should not be allowed to colled on such apparatus.

17.6. Engineering Microscopes

Engineering microscopes are optically assisted instruments for measuring the geometric dimensions and forms of small and medium sized technical parts. These present the enlarged view of the observed object area, either in its contours, or as a surface image. These are also used for referencing which can be accomplished by providing index lines on a transparent graticule inside the microscope tube, observed concurrently with the magnified image. The index lines guide in determining reference positions for specific surface elements. Index lines are also used for alignment and comparison with the observed part.

17.6.1. Tool Maker's Microscope. The tool maker's microscope is based on the principle of optics. The microscope consists of a heavy, hollow base ; accommodates the illuminating unit underneath ; and above this, on the top surface of the base, the work-table carriage is supported on balls and controlled by micrometer screws. Projecting up from the rear of the base is a column which carries the microscope unit and various interchangeable eyepieces. Light from the lamp at extreme right is collimated in the tube connecting the lamp to the centre of this instrument and is reflected as parallel beam by the prism at the end of the tube. On its way up this beam collects the image of the object to be inspected and this enters the microscope being turned by another prism into the inclined eyepiece. Here the inspection may be made by eye in the normal way or the projection attachment may be fitted which turns the beam by mirrors to bring up the image on the back of the transluscent screen. The magnification obtained depends upon the objective of the microscope but instruments are commonly supplied with facilities to give magnification from 10 *X* to 100 *X* on the projection screen. In order to adopt the apparatus to deal with a wide variety of work, various attachments may be fitted to the worktable. Microscope may be revolved on its mounting and normally has a recessed hole for accommodating a disc of glass upon which flat work may be rested and a clamp attachment is available for clamping. Linear movements of the table are controlled by micrometer screws having a movement upto accuracy of 0.0025 mm. The table slide is held against the ends of the micrometer spindle by light spring pressure so that movements greater than the micrometer travel may be controlled by interposing a slip gauge of the required dimension between the spindle and the point where it bears against this slide. This method is preferable even for movements within the range of this micrometer, as it is positive and less liable to error.

For the most effective manipulation, magnified image of the work is compared through the eyepiece (or projected) superimposed on a prepared background engraved on a glass disc in the eyepiece. There are several inter-changeable eyepiece units. The protractor unit is prepared with radial and cross-setting lines. This may be rotated by a knurled screw for setting any of the line with the work or image, and readings of the protractor may be made to 1 minute by means of a scale which divides each degree into 60 parts. The protractor which is illuminated is read by the eye through small hole in projecting eyepiece. Means are provided for setting the work table axis correct relative to protactor for zero reading. The thread template unit has selected thread forms arranged round the glass disc in the eyepiece and this may be rotated to bring any desired thread form into position for comparison with its magnified image of the work.

In order to view screws along the helix angle of the thread, the whole of the column unit with the underside illuminating arrangement attached may be swung at the helix angle of the thread. The pivot upon which this column swings is near the base. Most of the uses of this instrument will be on work where the shape of profile is projected from below, but surface shapes occur such as the edge of a recess, where surrounding metal prevents light from passing across the profile. These may be illuminated from above by a special arrangement which then allows the profile to be received in the normal manner, except that the intensity of light will be reduced from that when received direct. The chief applications of the tool room microscope are as follows :

1. The determination of relative position of various points on work.
2. Measurement of angle by using a protactor eyepiece.
3. Comparison of thread forms with master profiles engraved in the eyepiece, measurement of pitch and effective diameter.
4. Comparison of an enlarged, projected image with a tracing fixed to the projecting screen.

17.7. Optical Measuring Instruments

In these instruments, light beam is used as the amplifying lever. The light beam has the advantage of straightness and weightlessness. The optical measuring instruments include :

(*i*) Vertical optimeters. (*ii*) Horizontal optimeters. (*iii*) Tool maker's microscope.

These are used for highly precise measurements.

17.7.1. Vertical optimeter. It is mainly designed for external measurements of shafts, plates ; and can also be used for checking plug gauges and for varifying the slip gauges. The instruments are available in the range of measurement of 0 to 180 mm.

The general view of the instrument is shown in Fig. 17.7. It comprises the upright I mounted rigidly in a base 2. The base is made very heavy and wide so that it imparts ample stability to the instrument. The upright consists of a ground shaft on which a large pitch thread of special form is cut so that it could provide an easy grip for the bracket 3. The bracket 3, which is integral with the split clamp 4 can be easily moved up and down and adjusted and fixed in any position along the upright 1. The bracket 3 is clamped on the upright at any required position by the binding screw and raised or lowered by nut 10 mating with the special thread of the upright. The optimeter tube 5 which is of right-angular design is held by the split clamp 4 and is secured in position by the binding screw 6. The optimeter tube 5 can also be adjusted vartically in the clamp 4. The upper (working table) 8 is mounted on the optimeter table 7 and is under the measuring tip of optimeter tube. The distance between the measuring tip and the table is adjusted in two ways. The coarse adjustment is made by raising or lowering the bracket 3 and fine adjustment is made by raising or lowering the tables 7 and 8 by means of the adjusting nut 9. After setting up the instrument and before making measurement, all clamping screws of the brackets and tables must be firmly tightened.

1. Upright.
2. Base.
3. Bracket.
4. Split clamp.
5. Optimeter tube (Measuring head).
6. Binding screw.
7. Optimeter table.
8. Upper table.
9. Adjusting nut.
10. Nut.

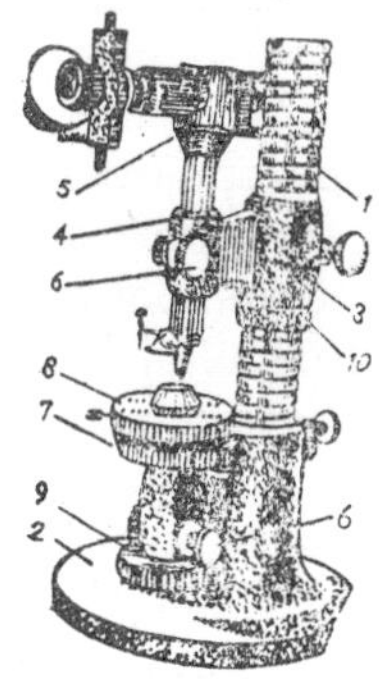

Fig. 17.7. Vertical Optimeter.

In the optimeter tube, measurements are made by the comparative method. The optimeter tube is of right-angle design. It consists of the measuring tip formed by a measuring spindle, at the bottom of which is secured a ball-shaped contact point. The optical system of the instrument is shown in Fig. 17.8. The movement of the measuring spindle (ball-shaped tip) is transmitted to a tilting mirror, at the top of which is placed an abjective. There is an eyepiece placed at right angles to the movement of tip. Through the eyepiece scale deviations are observed. There is a flat circular disc made of glass (reticule) on which scale is engraved. The disc is graduated on the side facing the triangular prism. The triangular prism is used to reflect the beam of light by 90°. The beam of light obtained from an electric lamp is directed by the hinged mirror to the scale on disc. The disc is covered with black photographic paper in which slots are provided. One such slot is opposite the scale. The reflected beam of light first passes through the triangular prism, then through the objective to the tilting mirror. For the initial position, only the position of the actual scale is seen through the eyepiece. Now if the mirror is tilted by the measuring spindle and is inclined in reference to the direction of the beam of light, beam will be reflected at a certain angle and returned to the eyepiece. The reflected image of the scale will be slightly displaced on the glass disc in reference to the index line. The amount of displacement can be easily read on scale of the disc.

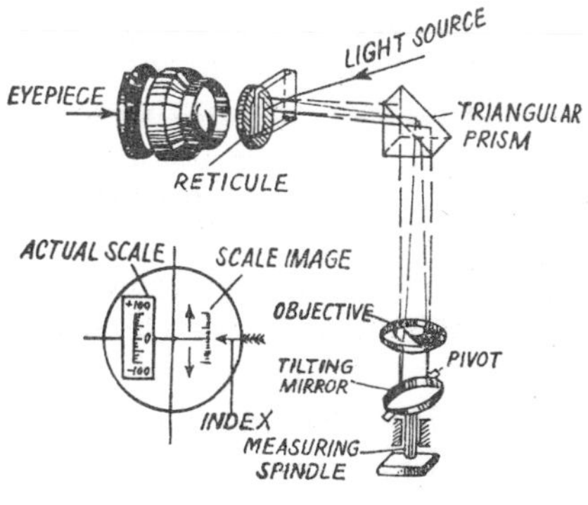

Fig. 17.8. Optical system of optimeter.

The reflected image of the scale is visible in the eyepiece because it is projected on the glass disc symmetrically on the other side of the optical axis in reference to the scale engraved on the disc. The slots in the paper covering the glass disc are arranged in such a way that the actual scale engraved on the disc is not visible and only the reflected scale is visible. The amplification of the instrument is equal to the

$$= \frac{\text{Scale spacing on disc to measure displacement of actual scale} \times \text{Magnification of eyepiece}}{\text{Scale division value of the optimeter}}$$

It's value is of the order of 1000. The scale has about 200 divisions and a range of measurement of ± 0.1 mm. As already pointed out, the instrument works on comparison basis, and therefore, the measuring tip is first set to the required dimension in relation to the table by means of slip gauges and then the component to be measured is placed on the table in place of slip gauges.

17.7.2. Horizontal Optimeter. The horizontal optimeter is in a way similar to the vertical optimeter, particularly the measuring head (tube) is the same for both the designs. It differs from the vertical type in the following features :

(*i*) The axis of the measuring spindle is horizontal.

(*ii*) The range of external measurement is from 0 to 500 mm.

(*iii*) The internal measurements in the range 13-180 mm is with the aid of auxiliary contact arms.

The design of the horizontal optimeter with set up for measuring internal surfaces is shown in Fig. 17.9.

The instrument consists of a wide base on which is supported a horizontal bar. The two brackets (right hand and left hand) move along this horizontal bar and can be adjusted in any position to a distance of 500 mm. For external measurements, the optimeter tube is secured in the right hand bracket, and the adjustable anvil which is having the same functions as the table in a vertical optimeter is supported on the left-hand bracket. The part to be checked is placed on the rotary table which can be adjusted in any direction. The rotary table itself rests on another self-aligning table which can be set in any position.

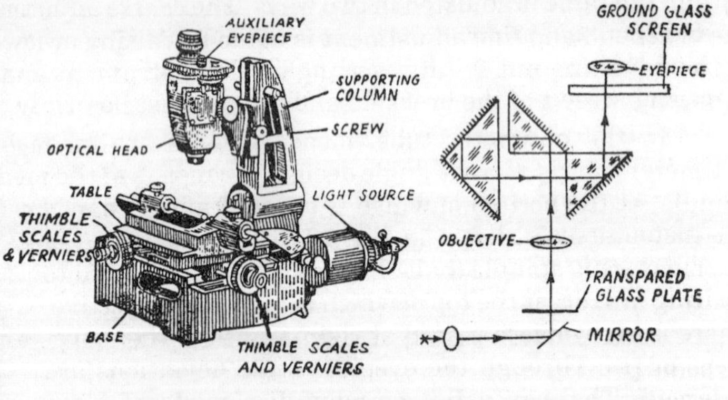

Fig. 17.9. Horizontal optimeter.

While using the instrument for internal measurements, two special contact arms are mounted on inner side of adjustable anvil and the optimeter tube. The contact arms carry the rocking links with the contact tips which come in contact with the internal part to be checked as shown in Fig. 17.9. The movement of the contact tips due to variation in the surface to be checked is transmitted to the contact point of the optimeter tube through the rocking links. The distance between the two contact tips is set up with a master ring gauge or with slip gauge.

17.7.3. Toolmaker's microscope. A general view of the small model toolmaker's microscope giving its design, and its optical system is shown in Fig. 17.10. This is designed for measurements on parts of complex forms *e.g.* profile of external threads, tools, templets and gauges. It can also be used for measuring centre-to-centre distances of holes in any planes, as well as the co-ordinates of the outline of a complex templet gauge, using the co-ordinate measuring system.

Basically, it consists of the optical head which can be adjusted vertically along the ways of a supporting column. The optical head can be clamped in any position by a screw. The working table on which the parts to be inspected are placed, is secured on a heavy hollow base. The table has a compound slide by means of which the measured part can have longitudinal and lateral movements. These movements are controlled by accurate micrometer screws having thimble scales and verniers. At the back of the base is arranged a ligh source which provides a horizontal beam of light which is reflected from a mirror by 90° upwards towards the table. The beam of light passes through a transparent glass plate on which flat parts to be checked are placed. A shadow image of the outline of contour of the part passes through the objective of the optical head and is projected by a system of three prisms to a ground-glass screen. Observations are made through the eyepiece of the optical head.

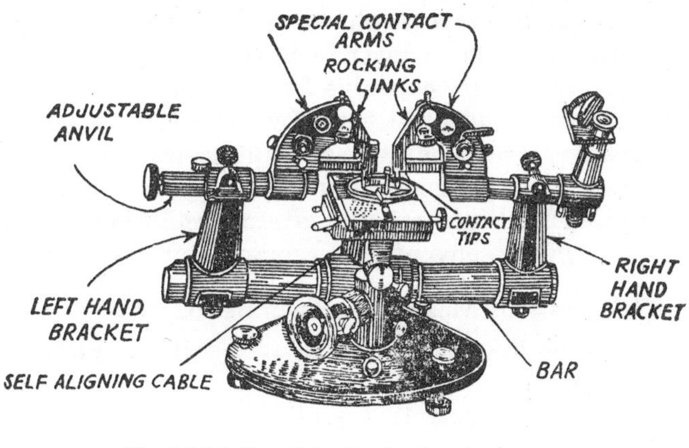

Fig. 17.10. Small toolmaker's microscope.

Cross-lines are engraved on the ground-glass screen which can be rotated through 360°, and the measurements are made by these cross-lines. The angle of rotation of screen can be read through an auxiliary eyepiece provided on the optical head. The eyepiece field of view contains an illuminated circular scale with a division value of 1 minute. The optical head tube is adjusted in height for focussing purposes, till a sharp image of the projected contour is seen on the ground glass screen.

Different types of graduated and engraved screens and corresponding eyepieces are used for measuring different elements *e.g.*, a revolving screen for measurement of standard threads will contain all the basic profiles of all standard threads in various pitch ranges and also various included angles of the threads.

Measurement of screw thread pitch. The image of the thread profile is set so that some point of the profile coincides with the cross-hairs as seen on the ground-glass screen. The reading on thimble of the longitudinal micrometer screw is noted down. Then the part is traversed by the micrometer screw until a corresponding point on the profile of the next thread coincides with the cross-hairs. The reading on thimble is again noted and the difference in two readings gives the actual pitch of the screw.

Measurement of angle of thread. It is determined by rotating the screen until a line on the screen coincides with one flank of the thread profile. The angle of screen rotation is noted and then the screen is further rotated till the same line coincides with the other flank of thread. The difference in two angular readings gives the actual angle of thread on the screw.

17.7.4. Atomic Force Microscope. An Atomic force microscope (AFM) is a very precise and multi functional device for measurement and visualisation of micro surface structure. The AFM uses sharp scanning of the surface of a sample. The sample is mounted on a piezo ceramic, which can be moved in three perpendicular directions. The sample can be scanned in horizontal (x and y) directions under the sharp tip. The tip is fixed at the free end of a cantilever on which a laser beam is focussed. The light from the laser is reflected on a split photo diode. By measuring the difference in signal, changes in the bending of the cantilever can be measured. In this way the topographical map of the sample can be produced with the vertical with resolution of 1 nm. A laser beam with a minimum spot size of less 0.06 μm for 50 × 0.55 numerical aperture objective for λ = 632 nm is focussed on the surface of the object and is moved to the required distance by auto tracking and autofocusing. The movements result in a 3-D profile corresponding to the surface profile. An atomic force microscope (AFM) can reveal a smple surface precisely up to nanometer size in three dimensions. Due to the sharper tip and very small loading force, the lateral resolution in AFM is largely improved when compared with that of the conventional profilometer.

17.8. Co-ordinate Measuring Machines (CMM)

Three dimensional measurements are essential for various components. CMMs are useful for this purpose. These machines have procise movements in $x - y - z$ coordinates which can be easily controlled and measured. Each slide in three directions is equipped with a precision linear measurement transducer which gives digital display and senses + ve/–ve direction. These are manufactured in both manual and computer-controlled models and come in a wide range of sizes to accommodate a variety of applications. The measuring head incorporates a probe tip, which can be of different kinds like taper tip, ball tip etc. Various type of CMMs are shown in Fig. 17.11. All these have very low measuring uncertainty, computer aided measuring runs, vibration free mechanical structure, and high rigidity. In addition all moving parts must be set very accurately, driven by fast motors, incorporate sensitive drive unit for fine adjustment of the axis, have rugged and precise probe system to facilitate exact dynamic probing. The cantilever type (CMM—refer Fig. 7.11) is easiest to load and unload, but is most susceptible to mechanical error because of sag or deflection in y-axis beam. Bridge type is more difficult to load but less sensitive to mechanical errors. Horizontal boring mill type is best suited for large heavy workpieces. Vertical bore mill type is highly accurate but usually slower to operate. A floating bridge type machine is also available in which the complete bridge can slide in y-direction on the slides. It has the compromises of both cantilever and bridge type, and is thus fast to operate, simple in alignment, and rugged construction affords consistent accuracy.

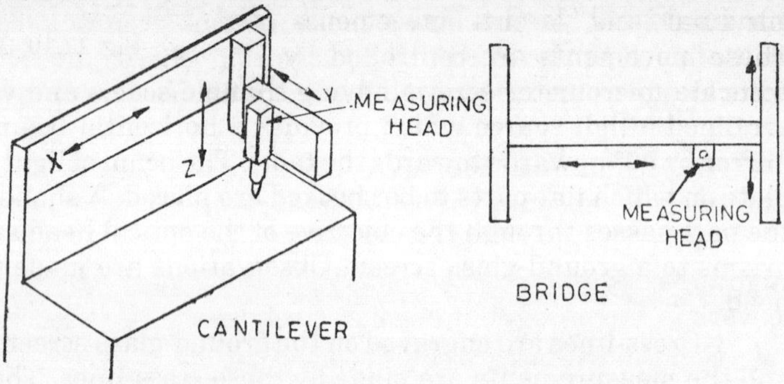

(Measuring head movement in plane perpendicular to paper)

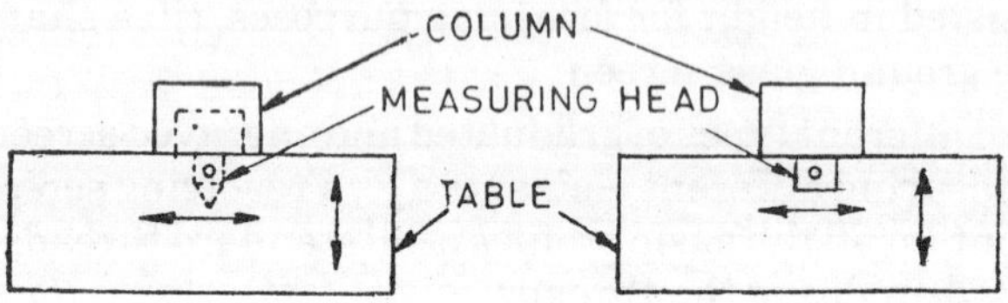

Horizontal Bore Mill Vertical Bore Mill

Fig. 17.11. Types of CMMs.

For measuring the distance between two holes, the workpiece is clamped to the worktable and aligned with the machine's three mutually perpendicular x, y and z measuring slides. The tapered-probe tip is then seated in first datum hole and the probe position digital readout is set to zero. The probe is then moved to successive holes, at each of which the digital readout represents the coordinate part-print hole location with respect to the datum hole. Machine is also equipped with automatic recording and data processing units which are essential when complex geometric and statistical analysis is to be carried out. In fact, in modern machines, automatic on-line processing of measurement data is possible when the part is still on the worktable.

In a special co-ordinate measuring machine, both linear (x and z axes) and rotary axes are incorporated. The machines can measure various features of parts whose shapes are objects of revolutions like cones, cylinders and hemispheres.

R-θ machines having motions of their measuring head in R, θ and ϕ direction are used for inspecting parts that are basically spherical.

As it is impossible to manufacture a mechanically perfect machine it is important to be able to analyse the geometry errors associated with each individual CMM and determine their effects on the machine's measurement accuracy. The result of such analyses can be used to compensate for these effects and thus provide a high degree of accuracy that could not otherwise be achieved.

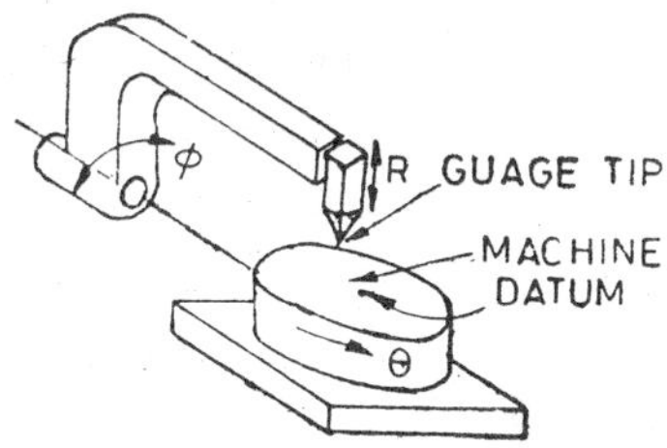

Fig. 17.12. Spherical co-ordinate (R-θ) Measuring Machine.

The prime advantage of co-ordinate measuring machine is the quicker inspection coupled with accurate measurements.

The co-ordinate measuring machine with mechanical gauge makes use of two-axis X and Y positioning tables to bring the work to the probe that engages the holes to the inspected.

Some machines are equipped with an optical comparator as well as travel dial indicator.

Present day co-ordinate measuring machines are three-axis digital read-out type and work up with an accuracy of 10 microns and resolution of 5 microns. These utilise a measuring element called *Inductosyn* data element which uses inductive coupling between conductors separated by a small air gap. As this element is not subjected to wear, it does not develop inaccuracy. It does not require reference standards or any other external device for its operation. The workpiece is aligned by a probe and by a switching adjustment on the worktable.

Many machines utilize Moire fringe concept for measurement.

Some coordinate measuring machines are available with accessories like optical viewing screen, (optical comparator), microscope attachment for the inspection of thin, soft, or delicate workpieces, and automatic print out. Some machines, in addition to measuring in three axes, are also designed to permit the checking of angularity, roundness, taper, and concentricity. Provision of rotary table makes such co-ordinate measuring machine more versatile because setting of a part need not be changed and all areas can be approached due to positioning of rotary table. The errors likely to occur in multiple set-ups are thus avoided.

Some co-ordinate measuring machines utilise electronic indicator probe (mounted on the end of the spindle) which can reach over and under the workpiece to check squareness in a single set up. Some machines are provided with linear air bearings on the horizontal slide motions to achieve finer slide position resolution.

17.8.1. Important Features of Co-ordinate Measuring Machines (CMM). In order to meet the requirement of faster machines with higher accuracies, the stiffness to weight ratio has to be high in order to reduce dynamic forces. To give maximum rigidity to machines without excessive weight, all the moving members, the bridge structure, Z-axis carriage, and Z-column are made of hollow box construction.

Principles of kinematic design are used in the three master guideways and probe location. Even whole machine with its massive granite worktable is supported on a three-point suspension.

A map of systematic errors in machine is build up and fed into the computer system so that error compensation is built up into the software.

All machines are provided with their own computers with interactive dialogue facility and friendly software.

Thermocouples are incorporated throughout the machine and interfaced with the computer to be used for compensation of temperature gradients and thus provide increased accuracy and repeatability.

With the advent of three-axis programming, computers enable CMM to measure three-dimensional object from variable datums.

The real benefit of today's CMM is its total flexibility and programmability, which makes it capable of handling virtually any measuring requirement within its physical size limit, thus rendering dedicated or specially designed gauging unnecessary.

Design improvements allied to a rapid growth in software for 3 and 4 axis movements enable CMMs to measure straight line relationships between basic features, *i.e.*, hole centre distances, etc. and also a variety of form measurements, such as turbine blades, cam profiles etc.

17.8.2. Possible Causes of errors in CMM. The table of CMM may not have perfect geometric form, or the table and probes may not be in perfect alignment. The probes may have a degree of runout, so it should be located at the same rotational position. The probes moving up and down in the *z*-axis may have some perpendicularity errors. There may be errors in optical readout of the digital system. It is, therefore, very essential that CMM should be calibrated with master plates before using the machine.

The magnitude of dimensional errors of a CMM is influenced by (*i*) Mechanics-associated with the straightness and perpendicularity of the guideways, (*ii*) Scale-division and adjustment of scales, (*iii*) Probe-form length and structure of probe configuration, (*iv*) Probe system calibration, repeatability, zero point setting and reversal error, (*v*) Electronics-interpolation errors due to digitization, (*vi*) Errors of data feeding by operator into compuer, (*vii*) Specimen-weight, clamping, surface finish, hardness, and due to (*viii*) Environment-(temperatures, oscillations, humidity, dirt).

While some conditions may be influenced by the operator/CMM user, other errors can be controlled and ascertained by the manufacturer and minimized by the measuring software. The length of the probe should be minimum and rigid in order to reduce deflection. The weight of the workpiece may change the geometry of the guideways and therefore, the workpiece must not exceed a maximum weight. Variation in temperature of CMM, specimen and measuring lab influence the uncertainty of measurement. Similarly, the smoke particle, a finger print, a dust particle and human hair may introduce uncertainty in measurement.

In addition to above, there can be several errors due to deviations in the guideway of CMM, but these are built in the machine and can not be influenced by the user. There are 21 components of such errors and are identified as translational errors, rotational errors and perpendicularity errors.

Translational errors result from errors in the scale division and errors in straightness perpendicular to the corresponding axis direction. Quite often the errors in scale division are termed as positional errors and are designated as *xpx*, *ypy* and *zpz*. Similarly translational errors due to straightness are termed as *xty*, *xtz*, *ytx*, *ytz*, *ztx* and *zty*. To clarify further it can be said as

Positioning Deviation (denoted by p)

xpx movement in x-axis and positional error in x-axis

ypy movement in y-axis and positional error in y-axis

zpz movement in z-axis and positional error in z-axis

Straightness Deviation (denoted by t)

xty movement in x-axis and straightness deviation in y-axis

xtz movement in x-axis and straightness deviation in z-axis

ytz movement in y-axis and straightness deviation in z-axis

ytx movement in y-axis and straightness deviation in x-axis

ztx movement in z-axis and straightness deviation in x-axis

zty movement in z-axis and straightness deviation in y-axis.

Rotational Deviation (denoted by r)

These errors are caused due to twisting deviations in guideways. Depending upon the behaviour of errors these can be split in roll error, pitch error and yaw error. When rolling in x- direction, pitch error is introduced in y-direction and yaw in z-direction.

yry movement in y-direction and roll in y-axis

yrx movement in y-direction and rotational deviation in x-axis (also known as pitch error)

yrz movement in y-direction and rotational deviation in z-axis (also known as yaw error)

xrx movement in x-direction and roll in x-axis

xry movement in x-direction and translational deviation in y-axis (also known as pitch error)

xrz movement in x-direction and translational deviation in z-axis (also known as yaw error)

zrz movement in z-direction and roll in z-axis

zrx movement in z-direction and translational deviation in x-axis

(also known as pitch error)

zry movement in z-direction and translational deviation in y-axis (also known as yaw error)

Perpendicularity Error (denoted by w)

This error is caused if the three axes are not orthogonal and deviation between these are termed as perpendicularity error.

xwy error in squareness between x and y axis

xwz error in squareness between x and z axis

ywz error in squareness between y and z axis

17.8.3. Error reduction and error compensation. In coordinate measuring machine which is not accurate but very repeatable, the inaccuracy can be compensated by error reduction and error compensation. In error reduction, the sources of problem are identified and physically eliminated. In error compensation, the errors are detected and the errors are mathematically eliminated, compensation being done by computers. Thus one by one, all the sources of errors are analysed and treating them as separate entity, all of them are eliminated one by one. In another approach, no attempt is made to separate the sources of errors, but it is assumed that the systems errors are part of an overall problem that can be dealt with as if it were one problem. This method deals with the error sources in combination and bases its algorithms on this assumption. It uses the smallest number of artifacts, to find the errors. Thus if one finds all the displacement errors existing on the perimeters of the measuring volume, then all errors inside the measuring volume are captured in combination. Since the recommended measurement error is usually no more than 10%, the error compensation should reduce the error to that extent. The data collected in small increments all around the measuring volume allow a grid of data points to be created that no longer has the bar errors.

17.8.4. Accuracy Specifications for Coordinate Measuring Machines. Two types of accuracies are defined in connection with coordinate measuring machines ; *viz geometrical accuracy* (determined by independent measurement because they make major contribution to overall accuracy of machine) and (*ii*) *total measuring accuracy* (determined by utilising the entire measuring machine system as applied to master gauges).

Geometrical accuracy concerns the straightness of axes, squareness of axes, and position accuracy. Total measuring accuracy concerns axial length measuring accuracy, and volumetric length measuring accuracy.

Straightness of axes. Straightness of axes is defined as deviation from a straight line in two orthogonal planes for each axis of movement, and thus following six measurement parameters need to be considered : Straightness of x-axis measured in y and z directions ; of y-axis in x and z directions ; of z-axis in x and y directions. Measurement is effected against a suitable straightness reference, *e.g.* laser beam and taking at least 10 readings at different points in each direction over full travel of each axis. Straightness is defined as the distance A (deviation bandwidth) between the two parallel lines containing the two graphs (Refer Fig. 17.14 *a*).

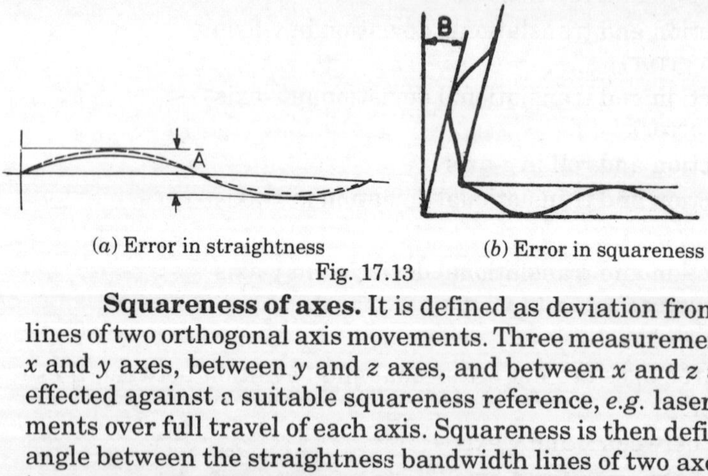

(*a*) Error in straightness (*b*) Error in squareness

Fig. 17.13

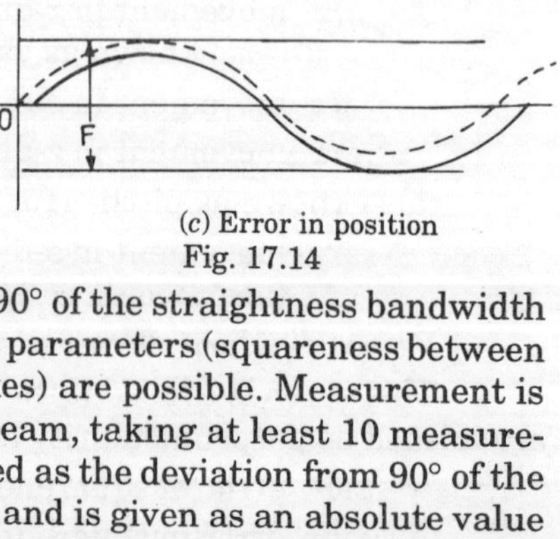

(*c*) Error in position

Fig. 17.14

Squareness of axes. It is defined as deviation from 90° of the straightness bandwidth lines of two orthogonal axis movements. Three measurement parameters (squareness between x and y axes, between y and z axes, and between x and z axes) are possible. Measurement is effected against a suitable squareness reference, *e.g.* laser beam, taking at least 10 measurements over full travel of each axis. Squareness is then defined as the deviation from 90° of the angle between the straightness bandwidth lines of two axes and is given as an absolute value in arc seconds (Refer Fig. 17.13 *b*).

Position accuracy. It is defined as difference between position readout of machine along an individual axis and value of a reference length measuring system. Following three measurement parameters are needed for position accuracy. Position accuracy of x axis, of y axis, and of z axis. Meaurement is effected along one measuring line for each machine axis located approximately at centre of measuring travel of remaining two axes. For this purpose, a suitable reference length measuring system, *e.g.* laser interferometer, is aligned to each machine axis within a permissible deviation of 1 arc minute (minimum 20 points measured over full travel of each axis). Fig. 17.14 shows a typical deviation record in which position accuracy F is defined as the distance between the two parallel lines containing the two graphs for the two directions.

Axial Length Measuring Accuracy. It is defined as difference between the reference length of gauges, aligned with a machine axis, and the corresponding measurement results from the machine. Three reference gauges are measured in each of the three axes x, y and z, with gauge length approximately 1/3, 1/2 and 3/4 of full travel of respective axis (upto a maximum of 1000 mm). Length measuring accuracy G is defined as the absolute value of the difference between the calibrated length of the gauge block and the actual measured value.

Volumetric Length Measuring Accuracy It is defined as difference between the reference length of gauges, freely oriented in space, and the corresponding measured results from the machine. Three reference gauges are measured, their lengths corresponding to approximately 1/3, 1/2 and 3/4 of the full travel of the longest axes (upto maximum of 1000 mm). Volumetric length measuring accuracy M is defined as the absolute value of the difference between the calibrated length of the gauge block and the actual measured values.

17.8.5. Calibration of Three-Coordinate Measuring Machine (CMM). The optical set up for the x-axis calibration is shown in Fig. 17.15 (a). The laser head is mounted on the tripod stand and its height is adjusted corresponding to the working table of CMM.

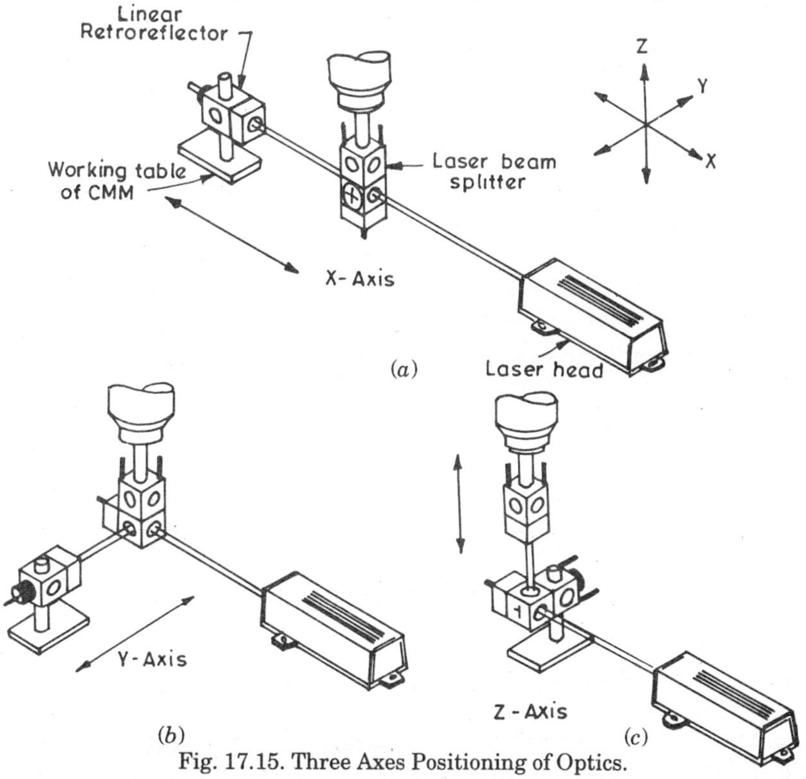

Fig. 17.15. Three Axes Positioning of Optics.

The interferometer contains a polarized beam splitter, which reflects the F1 component of the laser beam and let the F2 component pass through. The retroreflector is an optically ground and polished glass trihedral (three-surface) prism (sometimes called a "cube corner"). It reflects the laser beam back along a line parallel to the original beam, but offset from it by twice the distance at which the incoming beam is offset from the corner apex. For distance measurement, the F1 and F2 beams that leave the Laser Head are aimed at the Interferometer which splits F1 and F2 via polarizing beam splitter. Component F1 becomes the fixed-distance path, and F2 is sent to a target which reflects it back to the Interferometer. Relative motion between the Interferometer and the remote Retroreflector causes a Doppler shift in the returned frequency. Therefore, the Measurement Receiver Photo-detector in the Laser Head sees a frequency difference given by F1 – F2 ± ΔF2.

The F1 – F2 ± ΔF2 signal that is returned from the external Interferometer is compared in the Measurement Display Unit to the F1 – F2 Reference Signal. The difference, ΔF2, is related to the velocity and hence to the distance.

Laser head is adjusted for orientation and level so that the reflected beam from the Linear Retroreflector mounted on the working table of CMM gives maximum beam strength for the full range of its movement. The Laser Beam splitter fitted with Reference Retroreflector is mounted on the stationary bed of CMM between the Laser Head and Linear Retroreflector and properly aligned to obtain maximum beam strength.

The longitudinal micrometer microscope of CMM is set at zero and the Laser Display Unit is also set at zero. The CMM microscope is then set at the following points and the corresponding readings of Laser Display Unit are noted :

1 to 10 mm, every mm

10 to 200 mm, in steps of 10 mm

The accuracy of linear displacement measurements is affected by changes in air temperature, pressure and humidity. The refractivity of air and hence the wavelength of light propagating through it is dependent on three parameters. The automatic compensation for wavelength changes is provided by the material sensor attached to the machine.

The temperature is also noted at each reading to apply corrections for the thermal expansion of the standard steel scale with reference to the standard temperature of 20°C.

Similarly the readings for *y* and *z* axes are taken. The optical arrangements for *y* and *z* axes are shown in Fig. 17.15 (*b*) and 17.15 (*c*).

17.8.6. Performance of CMM. In evaluating the performance of a coordinate measuring machine, the following major aspects need consideration :

(*i*) Definition and measurements of "geometrical accuracies", such as positioning accuracy, straightness and squareness.

(*ii*) Master gauge measurement methods to define "total measuring accuracy" in terms of "axial length measuring accuracy, volumetric length measuring accuracy, and length measuring repeatability, *i.e.*, the coordinated measuring machine has to be tested as complete system. Measuring systems can be characterised by the combination of "mode of operation" and probe type. Modes include free floating manual, driven manual, and direct computer controlled. Probe types are passive, switching, proportional and nulling. The CMM is tested in each mode and with the probe that is commonly used.

(*iii*) Since environmental effects have great influence, explicit specifications on environmental conditions for the accuracy testing, including thermal parameters, vibrations and relative humidity are required.

It is usually difficult to establish a quantitative relationship between any particular environmental specification and the effect on machine's performance. Thus it is better to define what level of environmental influence is acceptable, and maintain those conditions.

The thermal effects dominate the environmental influences affecting a CMM. The sources of thermally induced errors include deviations of surrounding air temperature from 20°C, temperature gradients, radiant energy (*e.g.* sunlight), utility air temperature, and self-heating in machines with drive motors. Thermal effects may take the form of differential expansion between the workpiece and the machine scale system, drift between a workpiece origin and the machine scale system origin, and distortion of the machine structure leading to significant changes in the calibration and adjustment of the machine. The dominant effect of vibration is to degrade the repeatability of a machine. If the indicated relative motion between the machine table and the ram exceeds 50% of the working tolerance for repeatability, the vibration environment is deemed unacceptable.

It is important that suitable performance tests capable of testing the machine as a complete system are performed. It may be mentioned that use of parametric testing (straightness, squareness, angular motion) does not test the system performance and it is difficult to relate the results of these tests to expected performance. Performance test is carried out by measuring a mechanical artifact which provides some similarity between the machine testing and actual measurement of workpieces. Such testing must sample throughout the work zone. For performance test, linear displacement accuracy is checked by a step bar or a laser interferometer. These measurements are made along three orthogonal lines through the centre of the work zone to provide a thorough sampling of many combinations of x, y, and z errors that occur throughout the work zone of a machine.

Using the socketed ball bar provides a means of sweeping out the surface of a (nearly) perfect hemisphere with a physical object (ball). The CMM is used to measure the location of the centre of this ball at many locations on the hemisphere. The actual measurement data is compared to an ideal hemisphere simply by recording the range of the length of the ball bar computed from the data. The procedure calls for moving the socket defining the centre of the hemisphere to several locations in the work zone and repeating the meaurements. Three different lengths of the bar also are used. The performance is specified independently for the different lengths.

17.8.7. Applications of CMM. CMMs find applications in automobile, machine tool, electronics, space, and many other large companies. These machines are ideally suited for development of new products and construction of prototype because of their maximum accuracy, universatility and ease of operation.

Because of high speed of inspection, precision and reproducibility of coordinate measuring machines, these find application to check the dimensional accuracy of NC produced workpiece in various steps of production.

Regular inspection of workpieces by CMMs provides information on possible trends (due to tool wear, temperature factors and other influences) obtained through a statistical evaluation running simultaneously with production, allows efficient process inspection and control.

For safety components as for aircraft and space vehicles, 100% inspection is carried out and documented using CMM.

CMMs are best suited for the test and inspection of test equipment, gauges and tools.

CMMs can be used for determining dimensional accuracy of the bought in components, variation on same and thus the quality of the supplier.

These are ideal for determination of shape and position, maximum metal condition, linkage of results, etc., which other conventional machines can't do.

Because of verification facility, these can be assessed for their absolute accuracy, running in characteristics and variation. These make it possible to eliminate human error.

CMMs can also be used for sorting tasks to achieve optimum pairing of components within tolerance limits.

A coordinate measuring machine can replace several single purpose equipment with a low degree of utilisation like gear tester, gauge tester, length measuring machine, measuring microscope, etc.

To test a modified component, only a new component program is required whereas expensive modification of reference gauges is required in conventional machines.

CMMs are also best for ensuring economic viability of NC machines by reducing their downtime for inspection results. They also help in reducing reject costs, rework costs through measurement at the appropriate time with a suitable CMM.

17.8.8. Advantages of CMM. As the machined parts are becoming increasingly more complex with more features and tighter tolerances, inspection with surface plate and height gauges is becoming slow, inaccurate and costly. The various advantages of CMM are : increased inspection throughput, improved accuracy, minimisation of operator error, reduced operator skill requirements, reduced inspection fixturing and maintenance costs, uniform inspection quality, reduction of scrap and good part rejection, no need of separate go/no go gauges for each feature, reduction in calculating and recording time and errors, reduction in set-up time and fixturing costs through automatic compensation for misalignment, provision of a permanent record for process control and traceability of compliance to specifications, reduction in off-line analysis time, simplification of inspection procedures, possibility of reduction of total inspection time through use of statistical and data analysis techniques.

17.8.9. Computer Controlled Coordinate Measuring Machine. Fig. 17.16 shows the layout of a computer controlled CMM, a flexible measuring centre. The measurements as well as the inspection of parts for dimension, form, surface characteristics and position of geometrical elements are done at the same time in a fully coordinated manner (complete metrological description of a workpiece).

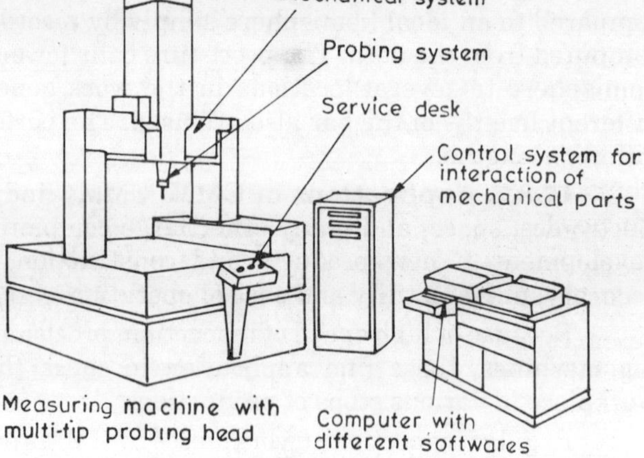

Fig. 17.16. System components of a computer controlled CMM.

Mechanical systems for the computer controlled CMMs can be subdivided into four basic types as shown in Fig. 17.17. The selection of a particular type depends on the application.

All these machines use probes (which may be trigger type or measuring type) Probe system is connected to the spindle in z-direction. It establishes the relation between the probing points in the measuring volume and the machine coordinate system.

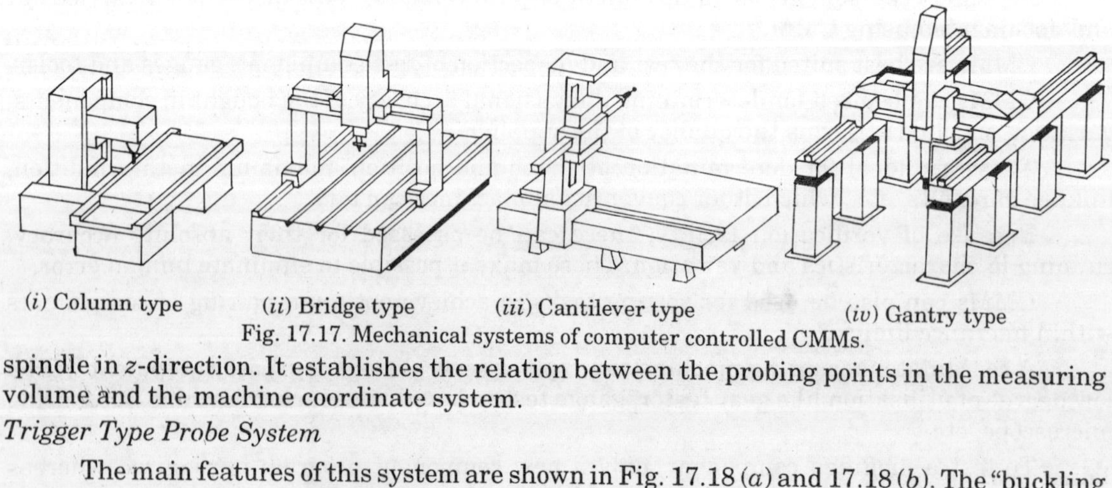

(*i*) Column type (*ii*) Bridge type (*iii*) Cantilever type (*iv*) Gantry type

Fig. 17.17. Mechanical systems of computer controlled CMMs.

Trigger Type Probe System

The main features of this system are shown in Fig. 17.18 (*a*) and 17.18 (*b*). The "buckling mechanism" is a three point bearing, the contacts of which are arranged at 120° around the circumference. These contacts act as electrical micro switches. When being touched in any

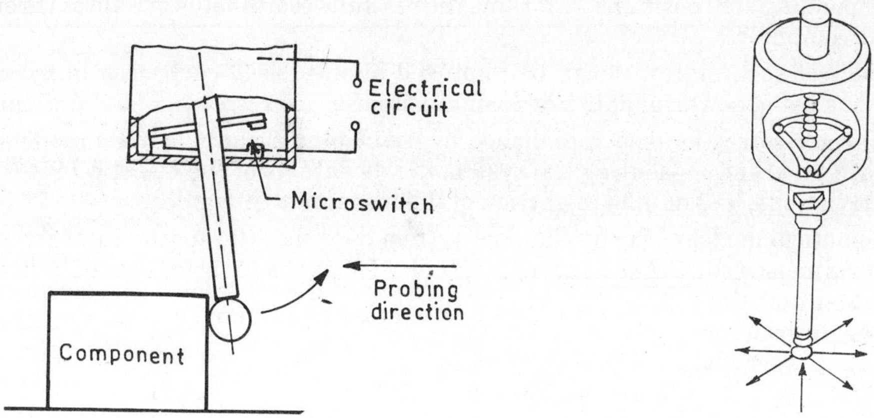

(a) Part section of Probe head. (b) Out line of Probe head.

Fig. 17.18. Trigger type Probe system.

probing direction one or more contacts is lifted off and the current is broken, thus generating a pulse. When the circuit is opened, the current co-ordinate positions are read and stored. After probing, a prestressed spring ensures the perfect zero position of the three point bearing. The probing force is determined by the prestressed force of the spring. With this probe system data acquisition is always dynamic and therefore the measuring time is shorter than in static principle.

Measuring Type Probe System

It is a small coordinate measuring machine in itself. The "buckling mechanism" of this system consists of parallel guideways. (Refer Fig. 17.19). At the moment of probing the spring parallelograms are deflected from their intitial position. Since the entire system is free from torsion, play and friction, a defined parallel displacement of probes as compared to their

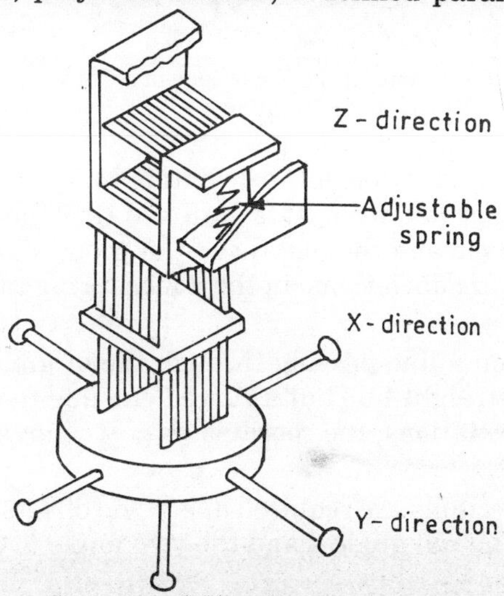

Fig. 17.19. Schematic of measuring probe head.

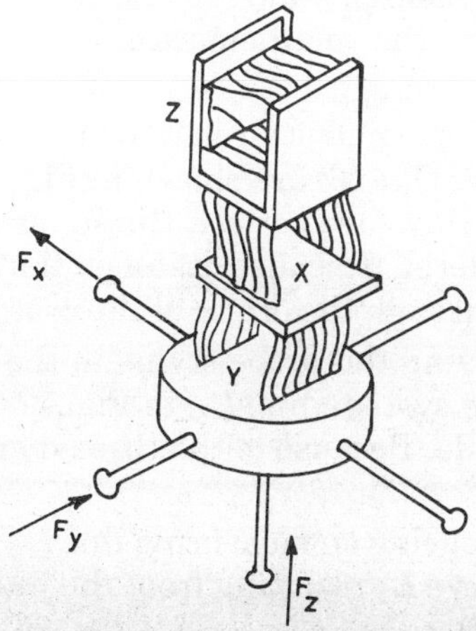

Fig. 17.20. Displacement of parallelogram when probing.

original arrangements can be measured. The displacement of parallelogram is shown in Fig. 17.20.

In static operation the electronic position regulating mechanism on probing of the specimen moves the slide of the probe axes till the inductive measuring systems of the probes are in their corresponding zero positions. After this the machine coordinates are automatically transferred to the computer.

The accuracy of such machines can be improved by in-process correction of the systematical deviations using powerful data processing systems.

Numerical error correction is accomplished by developing a mathematical model and defining a software program, measuring the systematic deviations of the CMM components and storing the deviations, and on-line correction of the measurement values.

The mathematical model of the mechanical system describes the spatial interaction of the systematic deviations of the CMM components. It also forms the basis for the computation of error propagation and final estimation of the overall accuracy concerning a specified measuring task.

If the components of the CMM are assumed as rigid bodies, the position deviations of a carriage can be described by three displacement deviations parallel to the axes (1), (2) and (3) and by three rotational deviations about the axes (4), (5) and (6) (Refer Fig. 17.21).

Similarly deviations (7)—(12) occur for *X*-carriage and (13) to (18) for *z*-carriage. Thus 18 path-dependent deviations and three squareness deviations (19), (20) and (21) are to be measured and to be treated in the mathematical model.

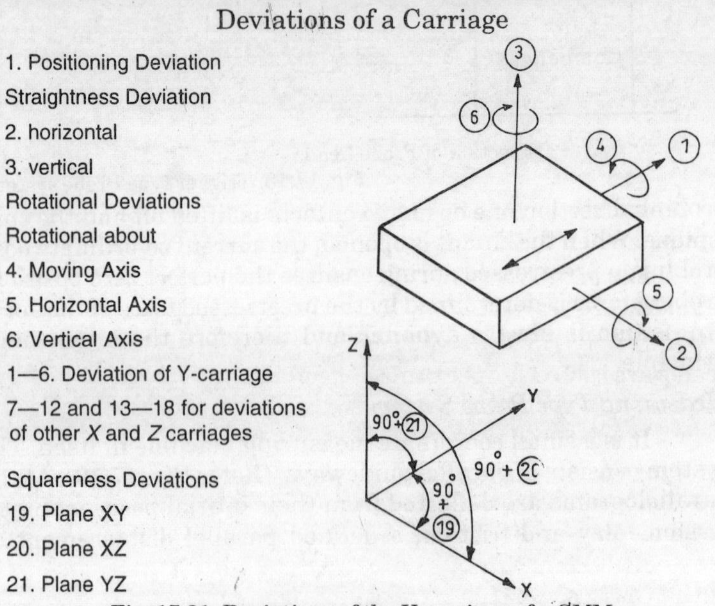

Fig. 17.21. Deviations of the Y-carriage of a CMM, squareness deviations.

The geometrical relationships of the spatial interaction of the deviation components in *y*-carriage (1) to (6) are shown in Fig. 17.19 in exhagrated form. It is assumed that the other two guideways are perfect, the squareness deviations and the lengths of the probe stylii are not considered. It is also assumed that the systematic deviations, in the *y*-measuring line are passing through the origin of the coordinate system.

Moving the probe stylus in the *y*-direction on a line passing through the origin of the coordinate system (line L_1) in Fig. 17.22) is not a straight line but a curved one due to errors in the guide. Because of the three displacement deviations, the coordinates *x, y, z* have to be corrected.

If moving on measuring line L_2, further corrections are required due to the offsets *x* and *z* from curve L_1 resulting from the pitch angle 5, the roll angle 4 and the yaw angle 6. (Refer Fig. 17.22)

Similarly the deviations of all three carriages and the squareness errors can be taken into account.

The treatment of all measured deviations is for (*a*) improvement of the accuracy by numerical error correction, and (*b*) computation of measurement uncertainty.

The effect of the error correction can be tested by means of calibrated step gauges.

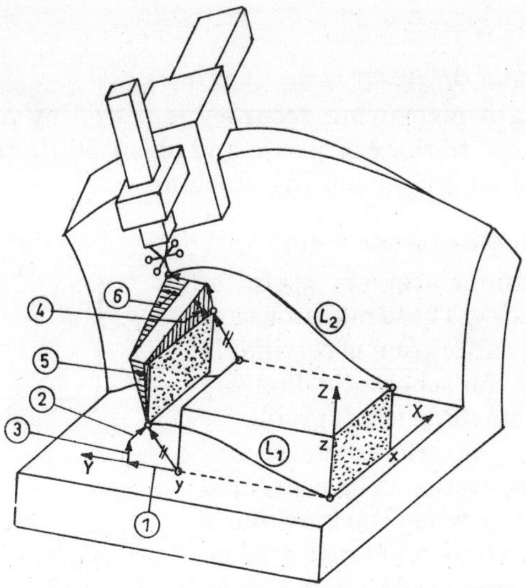

Fig. 17.22. Effects of systematic deviations of the Y-carriage.

In this way, on the basis of the mathematical model, a realistic estimation for the three dimensional length measurement uncertainty can be made.

For simple structured measuring instruments the relationship between accuracy and temperature is well known in contrary to CMMs with their high complexity and flexibility.

It is found that majority of the deviations can be corrected, if the temperatures of the scales are correctly taken into consideration. Thus one temperature sensor per scale should be the absolute minimum requirement for temperature measurements.

It has also been found that deformations of the CMM components (leading *e.g.* to straightness and squareness deviations) are caused by the changes of the spatial temperature gradients in the environment and by change of the spatial average temperature. The latter becomes obvious by the different thermal time constants of the CMM components, being lower for steel and higher for granite. Consequently one has to take into consideration the distortion of the CMM components caused by temperature influences, in order to guarantee the stated accuracy specifications even under poor environmental conditions.

The users of CMMs must always be sure that the measuring machine used complies with the specified uncertainty of measurement. This requirement can be met over a prolonged period of time only if the machines are re-verified at regular intervals. Several techniques are used to verify and improve the performance of CMMs.

The following accuracy/test items are carried out for CMM :

1. Measurement Accuracy

(*a*) Axial length measuring accuracy

(*b*) Volumetric length measuring accuracy

2. Axial Motion Accuracy

(*a*) Linear displacement accuracy

(*b*) Straightness

(*c*) Perpendicularity

(*d*) Pitch, yaw and roll

The axial length measuring accuracy is tested at the lowest position of the *z*-axis on the opposite side of the main axial guide of CMM. The length tested are approximately 1/10, 1/5, 2/5, 3/5, and 4/5 of the measuring range of each axis of CMM. The test is repeated five times for each measuring length and results plotted and permissible value of measuring accuracy is derived.

Volumetric Length Measuring Accuracy

The Volumetric length measuring accuracy is tested by measuring artefacts at two points on a spatial axis at 45° to the x or y axis and about 30° to the x-y plant. Measurments are made as in case of axial length measuring machine.

17.9. Three Dimensional Measuring Machine

3-D measuring machines are very useful in modern sophisticated industry. These machines are designed for 3-dimensional calibration of certifiable accuracy. Fig. 17.23 shows the schematic diagram of such a machine. Such a machine is adaptable for computer control. Laser interferometers are provided as scales. A cooling system is incorporated to reduce the temperature rise when the machine is in operation. The workpiece (gauge) is mounted on the table which moves to provide the x-measurement. y-motion is obtained by movement of the large carriage (carrying probe on z-slide) across the bridge. z slide mounted on y-carriage moves vertically up and down. Axes movement is controlled by stepping motors attached to lead-screws. The three carriages are mounted upon double—V ways, the x and y slides with roller bearings and the z-slide with plain ways.

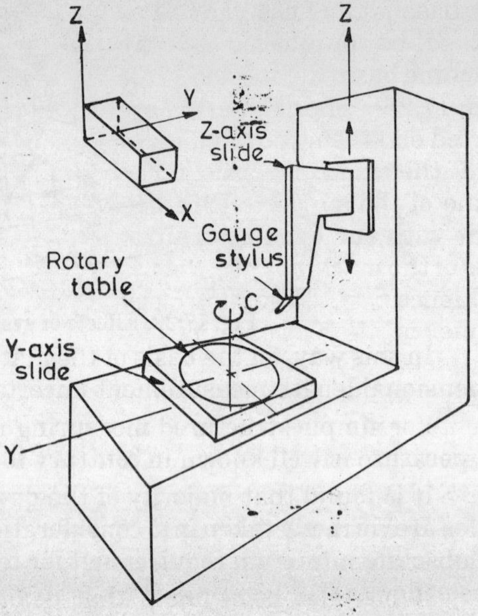

Fig. 17.23. 3-D measuring machine.

Fig. 17.23 shows a typical y-z measuring machine. The axi-symmetric part is centred upon the rotary table or the "c" axis. The rotary table is mounted on the horizontal (y) slide. The electronic gauge stylus is typically a ball-tipped, single axis, linear variable displacement transducer (LVDT) carried and positioned by the vertical (z) slide. The axis of the LVDT is typically mounted at a 45 degree angle with respect of the y and z axes. A correction is provided for the cosine error introduced when the direction of travel of LVDT is not normal to the part surface.

Displacement accuracy is achieved by laser interferometers operating in helium-shielded pathways. The interferometres are located in strict accordance with the Abbe principle, *i.e.*, the extension of the laser interferometer axis passes through the centre of the stylus ball at its null position (the centre of stylus ball being "functional point"). Refer Fig. 17.24.

On the y-axis slide, two laser interferometers, suitably separated, are provided. The difference in readings between these two lasers is used as a servo input to drive a piezoelectric crystal that supports one end of the y-axis table, thereby correcting the angular motion or pitch of the table.

Straightness accuracy is achieved by mounting straight edges parallel to each slide to measure and correct for slideway-straightness errors. For instance any error in the straightness of travel of z-slide will cause unwanted movement in the y direction. The LVDT gauge head that contacts the straight edge detects this movement and corrects it by zero shifting the y-slide. Similarly when nonstraightness of y-slide travel is detected, the z-axis is zero shifted in the proper direction to correct the travel.

However stiff a machine may be made, it deflects and distorts owing to the effects of changing and moving loads on the structure. The metrology system must therefore be made independent of the machine base, *i.e.*, the external forces upon the metrology system must be constant. The metrology base is thus designed so as not to be influenced by the machine base (Refer Figs. 17.24 and 17.25). The frame is supported on kinematic mounts inside the machine base. The plane of the supports is coincident with the bending neutral axis of the machine base, and its influence on the metrology frame is thereby minimised. The metrology frame houses the laser, laser pathways and remote interferometers and also supports the two straight edges.

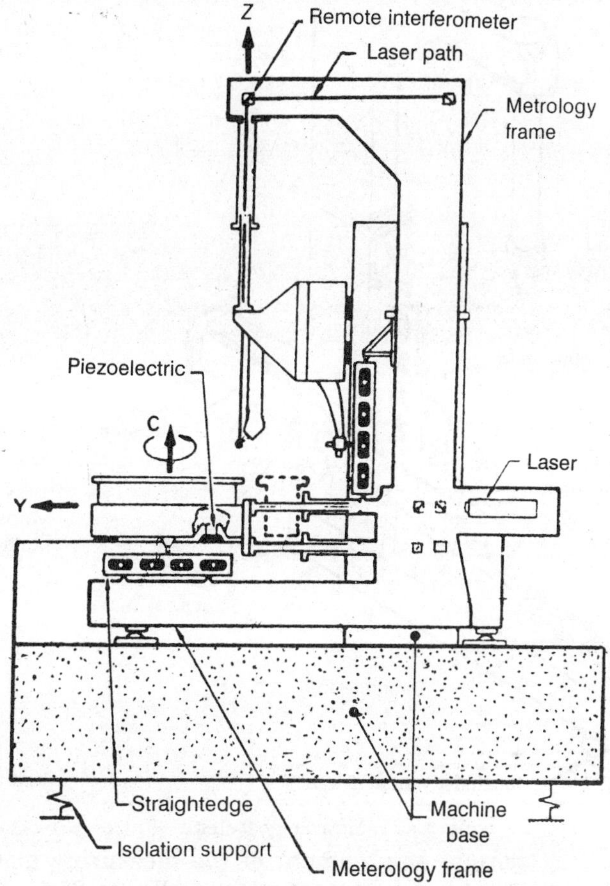

Fig. 17.24. Conceptual drawing of cross section of 3-D measuring machine.

Because of low coefficient of expansion of granite, it is chosen for building the machine base. The base is supported on three pneumatic isolators. The metrology frame is built of steel because temperature-controlled oil shower is included.

For stability of the laser (which depends on the stability of the medium in the pathways), helium at a pressure slightly above atmospheric pressure (maintained at constant value by a regulator) is provided in pathways. The effect of helium pressure change on the laser wavelength is taken into account.

y-axis and *z*-axis slides ride on and are guided by hydrostatic bearings. A portion of each bearing is evacuated and the evacuated section acts like a vacuum chuck to hold the bearing against the way (similar to preloading the bearing). The balance of the bearing surface provides lift. The *y*-slide support bearings are externally compensated to enhance the stiffness. The slide drive system (Fig. 17.26) can be considered as a rack and pinion drive wtihout gear teeth. The capstan is connected directly to the drive motor. The steel traction bar is squeezed between the capstan and the idler roller. One end of the traction bar is fastened to the slide with a spherical bearing. A coil spring supports the weight of the bar at the opposite end. Both the capstan and the idler are supported on hydrostatic bearings.

This type of drive system has minimum cost, minimum heat generation, maximum stiffness, minimum sliding friction, maximum linearity of displacement, no backlash, high reliability, compactness and minimum influence on slide straightness.

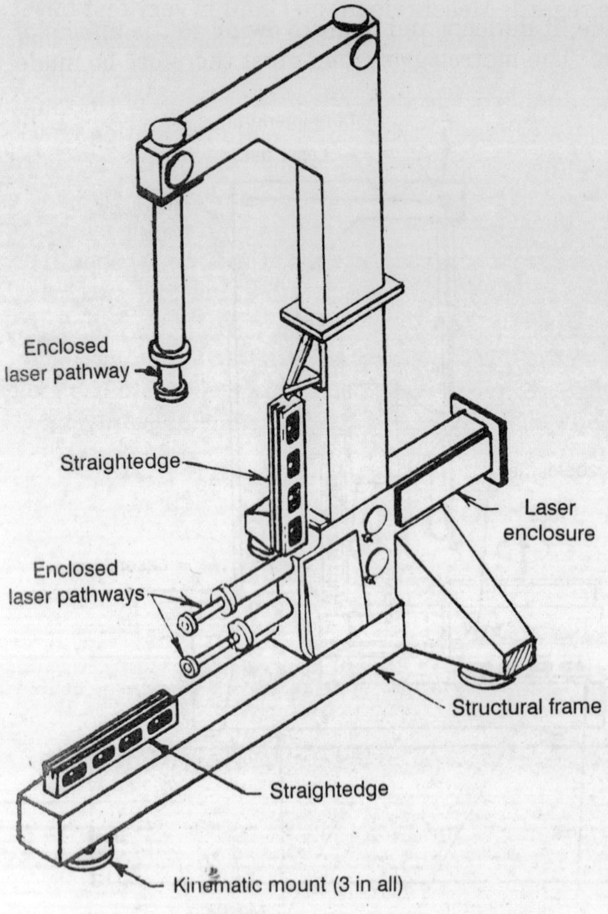

Fig. 17.25. Metrology system.

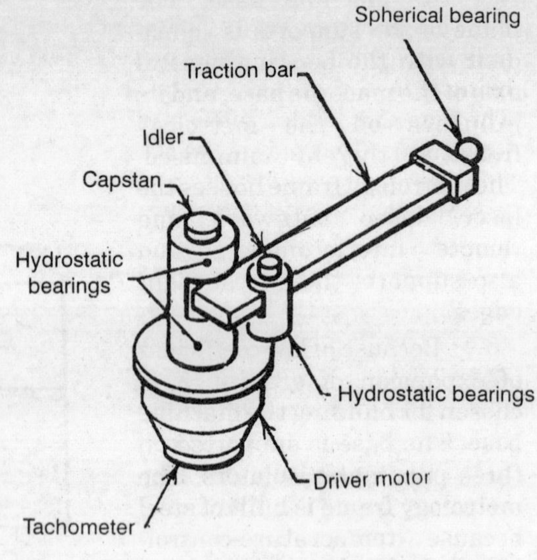

Fig. 17.26. Slide drive system.

The thermal environment of the measuring machine and the part is controlled by showering adequate quantity of oil controlled to 25°C. The shower is carefully sculptured to maintain machine temperature and to minimise splash. The primary advantages of liquid shower are its greater heat removal capability and the fact it is easily directed to the critical areas of the machine and workpiece surfaces. Liquids also have higher heat capacity than gases and accordingly it is possible to remove heat with corresponding lower temperature differences.

17.10. Universal Measuring Machine

With the introduction of concept of interchangeable manufacture parts, measurement of size and location of holes, contours and surfaces has attained more importance. Today the tolerances are specified in terms of fractions of a micron. Universal measuring machine, once thought to be a part of metrology laboratory, is becoming a part of the manufacturing process. The four essential components measured by universal measuring machine are : length, geometry, division of circle, and roundness.

17.10.1. Comparison of CMM and UMM. Coordinate Measuring Machine (CMM) is useful for rapid inspection of production parts. It does not inspect the geometric features of the part, while universal measuring machine (UMM) does it. CMM is not as accurate as UMM. If a part fails to qualify the inspection criteria, it is revealed by CMM but it can't reveal the reasons for same but UMM is useful in this regard. Measurement on CMM is very fast but it takes time on UMM. CMM is, therefore, very useful for quality control. For calibration of hole location gauges, UMM is virtually a primary standard. A hole-location gauge is used to inspect a CMM and is calibrated on UMM. UMM is capable of resolving the exact nature of the error of all part features, regardless of complexity. UMM discloses measurement information which in turn permits the improvement of the machines or methods by which the part is actually brought to finished form.

17.10.2. Inspection on UMM. It is very essential that the workpiece be *mounted* carefully on UMM. It must be flat and supported on accurate parallels. The flatness should be checked by traversing the machine axes and registering the indicator against the surface of workpiece. The clamping pressure should be directly over its supports.

For measuring the *location of hole* on a part, first the part needs to be clamped properly. For this purpose, the indicator probe is registered on the side of the workpiece and the long axis is traversed. The workpiece is held on one end and its side tapped at the opposite end to bring it into alignment with the axis. The part is then clamped and this process is repeated. The long axis is then moved until the spindle by eye appears to be centered over the workpiece longitudinally. This can be accomplished by extending the indicator linkage and noting if the indicator probe appears to be directly over each end when swung through a 180° arc. A "zero" is registered on one end of the workpiece at the lowest reading of the arc. Without moving the long axis, the lowest reading of the indicator is noted against the other end when swung through an arc. Leaving the indicator set against the latter end, the table is moved by fine-setting knob exactly half-way between this reading and zero on the meter. The above process is repeated to centre the spindle over the workpiece in the direction of short axis. Then the diameter can be measured using fine scale on the meter.

The *awkward shaped* workpieces can be located by using a rotary table to align reference holes or surfaces with the measuring axes. When the dimensions are specified w.r.t. centre hole in the part, then initial line-up is greatly facilitated by using the reference hole in the centre of the rotary table.

Microscope is used for measuring the location of small holes and for parts which might be damaged by physical contact. Parts like macro-circuit components, which have no reference surfaces against which an indicator probe-tip may register, can be measured by using microscope.

17.10.3. Mechanical Accuracy. The attainment of a "cubic concept" of accuracy in the Universal Measuring Machine requires the individual mastery of four mechanical arts-geometry, length, dividing the circle and roundness. These four mechanical arts are truly the "Foundations of Mechanical Accuracy".

Geometry—The foundation of all geometric accuracy and indeed of all dimensional measurements is the flat plane. Meehante cast iron (close grained) is a prime material component of most of the elements of geometry because it is readily available, can be easily cast to shape, relatively uninfluenced by relative-humidity and is inherently stable, provided certain precautions are taken. Proper composition of the iron improves machinability, wear qualities and the quality of the scraped surface. Stability of the iron depends mostly on slow, uniform cooling in the mould after casting. For maximum accuracy during machining, irregularities in the supporting surface are balanced out through the use of shims and most of the heat and stress of machining is removed in a first rough-cut. After that the workpiece is unclamped and permitted to relax. It is then supported evenly, clamped less tightly and finished with a light skin cut. Elastic deflection can be minimised through design and support. It is important to understand that geometric relationships are affected by the value of ambient temperature, provided it is steady. However, temperature differentials (which could occur due to proximity to radiators, doors, windows or drafts ; heat from handling ; changes in room temperature ; stratification, *i.e.*, non-uniform temperature from floor to ceiling) affect the accuracy. Localised heating/cooling of pieces subjects them to unpredictable distortion.

Standards of length. All the lengths measured should be traceable to an ultimate internationally acceptable standard. The accuracy of derived length-standards such as gauge blocks, precision scales, step gauges, lead screws, and laser interferometers depend in large measure on how truthfully they represent the ultimate length standard.

Dividing the circle. Accurate circle division is a requirement of all laboratories and machine shops. The authority for "angles" of angle gauges, rotary tables, or the angular relationships in mechanical parts and assemblies is derived from the self-proving principle of dividing the circle.

Roundness. Precision machine tools must achieve roundness of many mechanical parts and especially with their spindles. The authority for roundness is a precision spindle. Location, effective size and the fit between mating parts are all dependent on the ability to measure and define the conditions of roundness.

17.11. Numerical Controlled (NC) Co-ordinate Measuring Machines (CMM)

CMMs originally designed for high-precision measurement on relatively few workpieces in the metrology laboratory, are now being increasingly used both in the workshop and on the production floor because of the demand for accurate checking of dimensions, form and position of geometric elements of workpieces after machining. CMM (portable or bridge-type) with probe system (three dimensional probes suitable for dynamic measurements), probe changers, rotary table and workpiece handing devices becomes a multi function measuring system designed to maintain a high level of accuracy. With such an arrangement, a single measuring set up and hence a single workpiece clamping suffices to provide data on deviations in dimensions, form and position against three different set ups and three workpiece clampings in conventional methods.

Software, suitable for automatic calibration of the probe, determination of the workpiece coordinate system, calculation of geometric elements, automatic evaluation and recording of complete measurement is incorporated. Automatic compensation of workpiece misalignment is a common feature on many machines. Software for statistical evaluation of measured data, curve measurement, gear measurement and probing curved surfaces is also available. Fast feedback of measuring data significantly reduces idle times of the production equipment. After changing the probes and loading the workpieces, the automatic measuring sequence can be initiated by simply pressing a button.

The form or profile of cams, cutting dies, etc., is measured on coordinate measuring machine by a scanning technique which takes about fifty measurements and processes them every second.

In CMMs fitted with microprocessor based controls, adequate number of soft keys, *i.e.*, keys with variable functions, which hadle all the commands needed for man-machine dialogue are provided. The soft keys change their functions throughout the measuring sequence. At each stage their function is clearly defined on the display immediately adjacent to the keys. Various commands provided are ; full 3 D alignment, choice of three working planes, multi-point measurement of circle centres and diameters, probe diameter or radius compensation, datum translation or transfer, sub-datum facility to match drawing dimensioning, axis rotation. A repeat function simplifies inspection of similar features.

17.12. Electronic Inspection and Measuring Machines

To keep pace with fast production techniques utilising numerically controlled machines, for inspection purposes similar methods instead of traditional methods of measurements have been developed. Electronic inspection and measuring machines (usually three axis measuring machines) can be used in two modes, *viz.*, (*i*) point inspection, in which inspection head is moved to a predetermined point and any error is recorded from the true position ; and (*ii*) continuous inspection, which is used where inspection of whole surface is important such as aerofoils, turbine blades, etc. Such machines embody the new developments in hydrostatics and kinematics for their movement and use 'constraint' sensing. Their accuracy is of the order of $\pm$ 0.007 mm per axis over the full movement upto 1000 mm, and repeatability of $\pm$ 0.002 mm and resolution 0.001 mm.

The machine is designed for maximum stiffness in the measuring axis and the maximum stability is obtained by the one-piece fabricated torsion-box construction. The worktable is provided with a slow control for aligning the workpiece w.r.t. x- and y-axis. The worktable, the x-carriage and the probe column are supported on hydrostatic pads which are continuously fed with oil at a controlled pressure and temperature. Oil between the pads and the slideways gives a damped friction-free characteristic. An automatic control valve compensates for loads placed on the worktable. Because of the friction-free design accurate measurement can be made by 'constraint' sensing. Constraint sensing is a measuring technique in which a sensing probe is brought into contact with a selected feature of the workpiece and is allowed to settle naturally. It does not employ any sensitive indicator but the positioning accuracy is dependent on the freedom of movement of the carriage whose position is read by Moire fringe system of measurement.

17.13. Feeler Microscope

It is an attachment used on inspection machines. It contains a feeler at the bottom and the displacements of the feeler are optically magnified and viewed on a projection screen. The feeler is made to contact the surface whose dimensions are to be measured. A fiducial mark engraved on the screen provides an accurate reference position of the feeler. It can be used for several types of measurements. These are fitted with an automatic over-ride to prevent damage to the microscope through accidental mishandling. It, however, can't be used to measure irregular surfaces such as turbine blades, aerofoils etc.

17.14. Probes

A large variety of probes are available, depending on the application to transmit the motion from the surface being gauged to the measuring instruments. Simplest measuring probe has a short length of rod with a hemispherical end placed vertically in the ram of a measuring machine. For high precision, more elaborate probes are used. Proximity devices, which work on measurement of electrical capacitance between probe and workpiece, are used to eliminate any contact between probe and workpiece.

17.14.1. Optical Probe. The delicate components can be best inspected by an optical probe, which is a combination of an optical instrument with a photo-electric sensor. This can also be used for the location of inaccessible features.

17.15. Photo-electric Micrometer Microscope

It works on the principle of comparison of strained specimens. An unstrained sample, having two lines engraved on a polished surface at a separation of a fixed distance, is secured in a work holder, while a second sample with a known strain is also secured. The separation of the holders is then adjusted to permit line images from each sample to be superimposed *via* beam dividing prisms. The work holders are then displaced by means of a micrometer controlled stage to the nominal line pitch of the unstrained sample. It can be used to determine the difference in pitch to an accuracy of 0.5 μm.

17.16. Fluidic Numerical Control (NC) System

Such a system comprises a co-ordinate table and control unit operated by a combination of pneumatic and hydraulic power. It is used for two-axis point-to-point positioning. It uses cylinders for positioning and coarse measuring bar (for positioning in increment of 25 mm) and length bar for fine measurement unit with (0—25 mm range for positioning in 1/40th mm increments) for measurements. The control is accomplished by means of a low pressure pneumatic block paper tape reader, decoder, cycle timer and turbulance amplifier logic elements, which are coupled through high-gain pneumatic relays to conventional air-piloted hydraulic and pneumatic control valves which finally operate a hydraulic ram.

The program is punched on standard eight-hole paper tape. On command, the fluid cycle timer releases the table clamps, admits air to the table air bearing and releases the tape clamp. The paper tape advances on the information block, and the low-pressure signals from the tape reader are amplified and are used to set the fine and coarse measuring systems.

Such system fitted on machine can be used for accurate hole spacing, eliminating the use of jigs, thereby saving time. It has positioning accuracy of 0.05 mm.

17.17. Isometric Viewing of Surface Defects

An automatic scanning profiling system has been developed which utilises surface texture measuring instruments to collect surface topography data from areas of flat, and curved surfaces. The data is processed *via* a minicomputer for the measurement of wear volumes, and for the production of isometric plots, which permit the quantitative assessment of damage and help to distinguish between different types of damage such as wear and plastic deformation.

The curved surfaces to be assessed are measured on a rotary surface detector. The instrument "removes" the curvature and enables the surface profile to be plotted as a plane. Linear surface measurement is carried out on an instrument equipped with air bearing linear traversing table, which provides a datum accurate to better than 0.1 μm over its 200 mm stroke.

To provide multiple traverses at equal intervals across the surface being measured, the component is held in a stepper motor drive workholder. Two stepper motors are available to suit the part being measured, one providing linear motion in 1 micron steps over a range of 80 mm and the other continuous rotation in steps of 18 seconds of arc (5 milli-degrees).

Data collection and automatic control of the measuring instruments is performed by a minicomputer.

If the plotting option of the system is selected, a series of traverses by the stylus of rotary surface detector are traced by a graph plotter in isometric format, with hidden lines suppressed. The software is carefully designed for ease of use. For example, the computer automatically adjusts the effective horizontal and vertical magnifications of the plot, so that the data which occupies the full range of the surface profile instrument output will just fill the chart.

17.18. Laser Viewers for Production Profile Checks

The profiles of complex components like turbine blades can be checked by the use of optical sectioning techniques. It is based on use of laser and CCTV.

A section of the blade, around its edge, is delineated by two flat beams of laser light. This part of the edge is then viewed at a narrow angle by the TV camera, which also views, *via* the beam splitter, a graticule at the same focal distance (Refer Fig. 17.27). Both blade and graticule are displayed as magnified images on the monitor, the graticule position being adjustable so that its image can be superimposed on the profile image.

The graticule is effectively viewed at the same angle as the blade, so distortion due to viewing angle affects both blade and graticule equally. This means that the graticule images are direct 1 : 1 representations of the design profile.

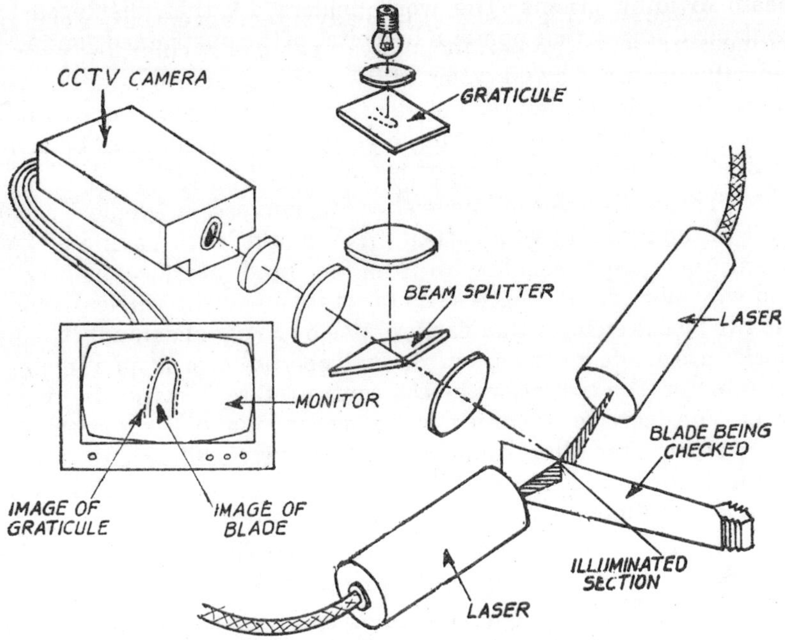

Fig. 17.27. Laser viewer using optical sectioning techniques.

17.19. The Image Shearing Microscope for Critical Dimension Measurement

Image shearing is the technique used for the precise measurement of lateral dimensions of objects under the microscope. It has several advantages over other methods of linear measurements. This instrument can be applied to the accurate measurement of integrated circuits, photomasks and wafers, without the need of time consuming calculations. It does not strain eyes. Precision settings of 0.01 micron can be easily achieved on 2 micron wide photomask lines.

In this instrument an optical system is used to produce two identical images of the object which can be superimposed or sheared across each other by means of a precisely calibrated mechanism. Measurements are taken when the two images are just in edge to edge contact, a condition that can often be set to better than one-tenth of the optical resolution of the microscope objective. The required "edge to edge" or 'just touch' setting can be made with great precision as the eye readily detects the desired condition of minimum intensity change across the images. For greater accuracy the images can be set so that they only just touch and are then sheared across one another until their opposite edges are touching (double shear) (Refer Fig. 17.28).

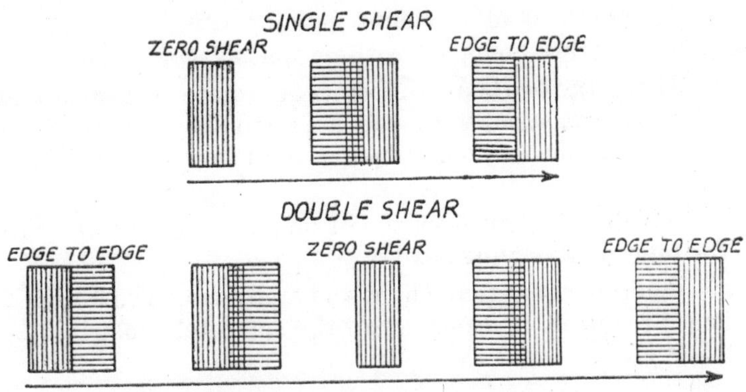

Fig. 17.28. Image shearing principle.

Image shearing has important advantages over the use of filar micrometers or graticules for measurements through the microscope. In particular, the precision of measurement is not radically impaired by vibration of the microscope or by small movements of the object. Also, many objects in a given field of view can be rapidly measured without the need to move each object to a reference point. And, a very high precision can be achieved in making the shear setting for the two contacting edges to virtually vanish.

The earlier type of image shearing microscope eyepiece consisted of an interferometer, built in the form of two identical prism blocks, each having a rhomboidal and a right-angle prism cemented together with a half-silvered interface. The two prism blocks are pivoted mid-way along their length so as to rotate about the vertical axis. A simple linkage between the two blocks ensures that each rotates by an equal amount in opposite directions. The prism blocks are made to move by manual operation of a micrometer screw. On operating the micrometer, the two prism blocks contra-rotate, moving the two images in relation to each other. The degree of shear is exactly proportional to the micrometer travel. Provision is made for inserting a red filter in one light path and green filter in the other light path to differentially colour the images. Differentially coloured images are useful in avoiding confusion when there are a large number of objects in the field of view.

In order to measure the width of any vertical or horizontal feature, the image shearing eyepiece is rotated to bring the shearing direction at right angles to the line to be measured.

Modified instruments use compact image shearing optical module which can be fitted directly between the top of a standard microscope and a binocular viewing head. Image forming light from microscope objective enters the base of the instrument and is reflected into the left-hand side arm by the 45 degree face of the prism.

Contained in this side arm is a K-prism which can be rotated by the operator about an axis in line with the entering beam. This allows the operator to rotate the microscope image as required.

The image forming light is now passed to the beam splitting cube. The 45 degree interface of this cube is specially coated to transmit light polarized in the horizontal plane and to reflect light polarized in the vertical plan.

Having been split into equal but oppositely polarized beams the light fractions travel through suitably orientated quarter wave plates to reflectors. The reflected beams pass back through their respective quarter wave plates to the cube interface. Double passage of the beams through the quarter wave plates turns their directions of polarization through 90 degrees.

The beam previously transmitted through the interface is now reflected by it and the beam previously reflected at the interface is now transmitted.

In zero shear adjustment the two oppositely polarized beams travel on from the interface exactly along the same path with no spatial separation.

If now the two beam reflectors are rotated together, taking the geometrical centre of the beam splitter as axis, the two beams will shear apart by an amount directly related to the degree of rotation. Any object images in the microscope field will now be split into two separate images.

The light beams from the beam splitting cube are reflected upwards by the exit prism to the binocular viewing head.

A birefringent component and analyser can be inserted directly above the module's exit prism if required, in order to tint the oppositely polarized beams in complementary colours, red and green.

The two beam reflectors are mounted on a carrier pivotted on an axis through the centre of the beam splitting cube. This carrier is rotated by means of a lever system coupled to a micrometer screw.

The calibrated control projecting from the side of the instrument operates the micrometer screw to control the amount of shear. A control knob on the other side of the instrument permits the operator to rotate the image through just over 180 degrees and a small control at the rear of the instrument permits the dichroic element for colouring the images to be inserted.

Direct dimensional readout is possible by coupling a multi-turn precision potentiometer to the shearing control and feeding the output of this potentiometer to a special electronic voltmeter fitted with a digital output display.

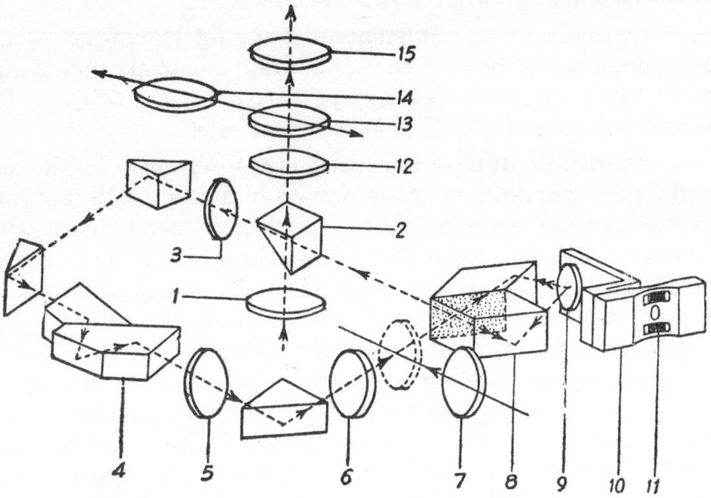

1. Window
2. Entry-exit prism
3. Lens field lens
4. "K" prism for image rotation
5. Quartz plate
6. Lens 2
7. Single image shutter (polar)
8. Rhomboid polarizing beam splitter
9. Image shearing mirror
10. Shearing flexure spring
11. Strain gauges
12. Lens 3
13. Dichroic filter
14. Compensator glass
15. Lens 4

Fig. 17.29. Optical layout of image shearing module.

A new image shearing module, developed with the aim of overcoming mechanical problems of wear in the micrometer screw thread, wear in the transfer linkage and mirror pivot system etc., is described below.

Fig. 17.29 shows the optical layout of a recently developed image shearing module to overcome above problems. The layout is similar in many respects to the earlier image shearing module, but the shearing components now consist of a double rhomboid polarizing prism and a tiltable mirror.

The image forming light entering the double rhomboid prism is divided into two equal intensity but oppositely polarised beams at the interface coating. The beam which is transmitted by this coating is reflected at the rear face of the prism and is then reflected off the tiltable mirror and front face of prism to be once again transmitted by the coating.

The beam which is reflected by the coating follows the same path as the other beam, but in the opposite direction, to be reflected once again at the coating. The two beams are then passed upto the binocular eyepiece by the combined entry-exit prism.

The tiltable mirror is fixed to a specially designed flexure spring on which are mounted four resistance strain gauges, connected together in the form of bridge network. This flexure acts both as a pivot for the mirror and as the transducer for converting the angular deflection of the mirror into a corresponding electrical signal. The output from the strain gauge bridge is fed directly into the electronic digital display unit.

17.20. Opto-electronic Dimensional Gauging

Opto-electronic dimensional gauging technique is a contactless method. It utilises photodiode arrays having high resolution due to arrangement of 1024 diodes at a pitch of 1/40 mm, having a narrow viewing aperture of about 1/20 mm. Photodiode arrays with 2048-element design are also available commercially.

Principle of Operation. This system works on the principle that each 'edge' of image produces a light-to-dark transition on individual cells. An image of the object to be gauged is produced on the array. All that is required is then to use the processing circuit with solid state memory which will remember where these transitions occur along the display, *i.e.*, image of the object. It is essential to ensure that the image presented to the array conforms exactly to the object to be gauged. For this purpose, the object is normally back-illuminated by a collimated light source obtained by passing the rays of a light source (tungsten lamp or quartz halogen lamp, or a laser diode) through a collimator. It may be noted that a photo diode array views a single line, rather than a two-dimensional image. The complete image is sensed by scanning the image line by line at predetermined incremental positions obtained by means of a precision slide or a stepping motor. It would be appreciated that a lot of raw data has to be handled from each scan of array and this is done by using dedicated logic circuits which 'latch' the edge information of any image falling on the array. An external clock is utilised to sequentially output the signal from each cell onto the subsequent processing system. The data generated on each scan of the array is processed and compressed by an intermediate circuit so that same can be usefully employed. In this way transitions occurring along the display are memorised in memory. This memory can be interrogated by an external device, and the information can then be further processed to extract size, length or positional information from the edge data. The stored data are updated each time a scan is completed, so that the image data can be enhanced or degraded by digital averaging and weighing techniques.

Schematic Configuration of the System. Fig. 17.30 shows the schematic configuration of a simple opto electronic inspection system.

The object to be gauged is back-illuminated by collimated light source. A lens assembly is incorporated to provide either magnification (*X* 2, *X* 3, or *X* 10) for dimesional inspections, and reductions (*X* 0.1 or *X* 0.05) for flaw detection. All the components involved in the system (shown in Fig. 17.30) are aligned in the factory so that they are on a common centre line. The optical path bettween the light source, object and lens systems is folded by using 45° prisms in order to reduce the space requirements and all the components are mounted on a precision slide to form an integral station of an inspection machine. As the depth of field (governed by the focal length of the lens) in case of dimensional inspection is of the order of 2.5 mm, the object has to be aligned very accurately.

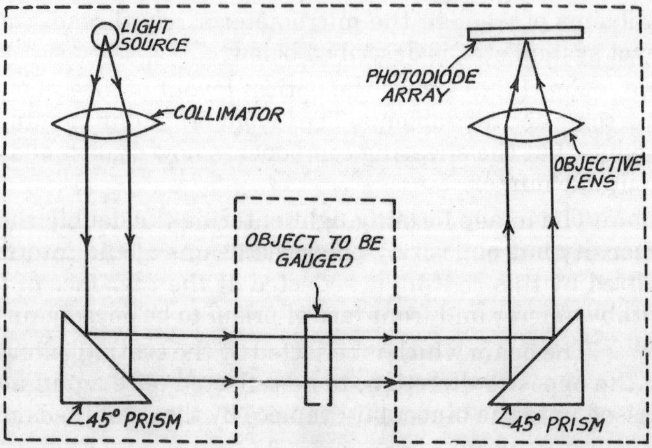

Fig. 17.30. Opto electronic inspection system.

Light Source. Selection of proper light source is an important aspect. Tungsten lamp is generally found to be quite suitable as it is less susceptible to edge diffraction than the laser.

Quartz halogen lamp is preferred for viewing surface defects because of the light intensity required and the need for front illumination. Laser diode is used for dynamic or vibratory measurements because of the stroboscopic effect due to pulsed light emitted from laser diode which can be used to freeze a moving object.

Some systems employ fibre optics to illuminate the object and to return the image to the photodiode array.

Example of Use. Fig. 17.31 shows a blanked component, in which the diameters of two holes and width of slot, and position of holes w.r.t. radiused edge of the slot have to be determined. For the sake of simplicity, it is assumed that camera viewing line has been located as the centre lines of the holes. It may also be noted that size of the important features lies within 25 mm, the measuring capability of photodiode array. The scan is started near the first feature and finished near the last feature of the object to be gauged as shown in Fig. 17.31. The latched edge information consists of a form of pulses as shown in Fig. 17.31, each pulse corresponding to light-to-dark transition of image of the object. The start and finish of scan correspond to diodes 0 and 1024 and knowing the positions of diodes at each pulse, any dimension can be easily calculated.

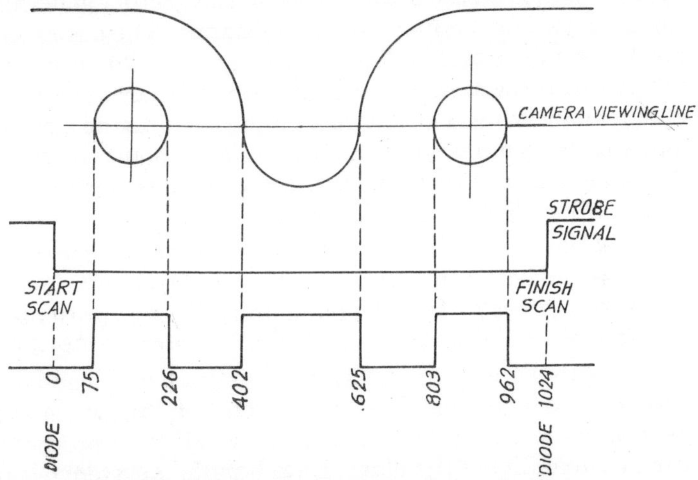

Fig. 17.31. Measurement with optoelectronic system.

Fig. 17.32 shows another application of detection of defects in the form of an object, by scanning the geometric form at various incremental positions, thus generating a snapshot of the form at predetermined indexed positions. The edge information may either be manipulated by the microprocessor to produce deviation values from a standard form or to plot the form automatically on a plotter.

It is also possible to detect flaws like inclusions, machining marks, cracks, indentations, etc. by front-illuminating the object, and by sensing the difference in reflectivity between the defect and the normal surface finish.

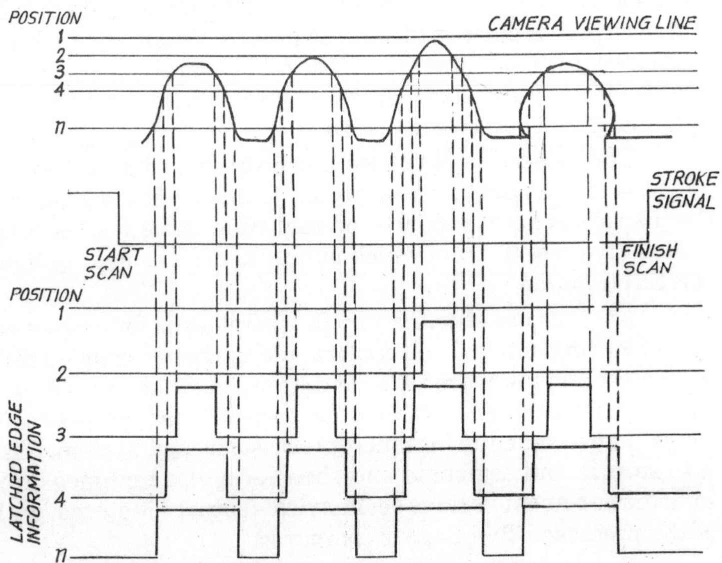

Fig. 17.32. Deflection of defects with opto electronic system.

17.21. Computers in Metrology

Like any other field, Metrology too is getting revolutionised very fast and it has also undergone enormous changes in recent years. The fast developments taking place in electronics, particularly microprocessors, has dramatically changed the nature and scope of metrology, too. The introduction of micro-processors, *i.e.* computer on chip has brought about a great revolution. These are cheap, versatile, fast and can be easily interfaced with metrology equipment. In such measurement systems where lot of data has to be gathered from several measuring instruments and data has to be processed, stored and displayed, the use of such device (microprocessor) is highly desirable. Either general purpose or special purpose software could be used depending on the application. The design of systems is such that operator need not have any programming knowledge. An alpha numeric display guides each step of its integrated inspection routines. Usually 'soft key' concept is used thereby eliminating the need of codes and complex keyboards. Programs to define axes (using non-linear data fitting library) and analysing circles (by accepting upto 20 points on the circumference of a circle) have been developed and are very useful and provide a powerful and comprehensive ability to cope with all variations of axes descriptions.

A big advantage of computer is its prcessing capability and presenting data in a form which would be of utmost interest to the metrologist. It clearly shows the trends of dimensions and summarises of dimensional details. Data is presented in graphical format. The features and faults, which otherwise could not be detected, are easily revealed.

With use of computer systems, it has become possible to prevent making defective parts rather than rejecting them at the end of the whole cycle. Vital information needs to be generated for the optimisation of the production process and the upgrading of the manufacturing capability of the plant. Data handling system organises all the inspection results into statistical form, analyses them in real-time, and performs full monitoring of the production line with diagnosis of the whole production process.

It is now not necessary to take parts to be inspected to measuring machine, but a robot arm having 4 degrees of freedom (3 rectangular linear motions plus wrist motion) is brought for inspection near the equipment while being machined. It incorporates various attachments and accessories such as electronic probes, automatic tool magazines, sensors and devices for parts recognition. The motion of the robot arm is controlled by a microprocessor which performs simultaneous multi-axis control of position, speed and acceleration and thus provides high dynamic performance of the arm.

The inspection data handling is organised at two levels ; one to perform the dimensional analysis of the part being inspected to discriminate good, bad and reworkable parts. Second, the inspection data is reduced on real-time statistical basis to derive trends, relationships and cumulative results from which potential dimensional problems can be spotted for immediate corrective action.

Another use of computers in metrology is for on-line non-contact dimensional measurement in which a television camera, using charge-coupled photodiode arrays, is *focused* on the workpiece and interfaced to a computer which can analyse shapes and sizes of the image. Laser illumination is commonly used for projecting shadows because the beam is well defined. The image is processed using microprocessor-based system. Vision systems break down images into sectors and segments and then recognise and measure their geometry. A vast array of information about various geometrical features is stored, with which the digitised information of the processed TV image is compared.

It uses solid state camera containing an array of photodetectors viewing a slit through a lens system. The photodetector signals are electronically sampled at very high rates to produce a video signal which is an electric analogue of the light intensity distribution across the scanned lines. The field of view and resolution of the camera are defined by the number of detectors within the array and the parameters of the lens system.

For moving objects, a scanner using a laser beam is used. It projects laser beam onto a multi-facet mirror to produce a fan beam. A parabolic mirror holds the beam at constant angle to the material surface across the entire beam span. When the light beam hits the material, any irregularity on the surface modulates the reflected light, which is detected by a video processor and microcomputer in an evaluation unit.

Stroboscope technology is used for freezing optically the motion of rotating or reciprocating parts. Remote inspection is possible by fibre optics.

17.21.1. Use of Computers in the Field of Metrology. In the field of metrology computer can be used for *(i)* controlling the function of a piece of measuring equipment (for which the movements of the measuring instrument/machine have to be motorised) and *(ii)* processing the output of the measurement process. Most manufaturers of measuring machines suitable for computer control offer a specific computer, and a range of general purpose software to cater for all the functions and capabilities of the machine. In these machines the measuring system is linked to the movement of motors in such a way that the movement will come to a halt in a precise predetermined position. The operation may be either fully automatic, (entirely unattended) which when once initiated, will run the machine till the task is complete or be semi-automatic. In automatic process, hundreds of readings may be taken, processed, analysed and hard copy results produced. Universal probes are used to carry out automatic inspection. In the semi-automatic operation, operator intervention is required at different stages and thus operation stops at different stages for operator to note and analyse and give command for further operation to continue.

The software for computer may be either written by the user, which is usually a difficult task, or be supplied by the manufacturer, or the computer be programmed to compile an operating program from the procedure used when a sample component is being measured manually. In the last technique, the procedure followed whilst measuring the sample will be automatically followed for subsequent components.

The application of using computer for processing the output of measuring instruments is unlimited. Nowadays micrometer screw gauges, vernier calipers, height gauges, etc. are available so that they can produce output in digitised form which can be fed into computers. A hand held computer is connected to these devices to store and process the inputs it collects from measuring instruments. It can also feed its contents into the large capacity central computer. The central computer can thus be fed with large quantities of data from different inspection locations in the plant and analyses them collectively. The output, after analysing, can be presented as numerical tabulations, or in a variety of graphical forms, or as three dimension represenation of complex profiles so that same surface is visible on CRT screen as would be viewed from all the corners.

Today, height gauges with built-in integral computer with keyboard and printer mounted on the height gauge itself, have been developed. The computer in such cases has limited printing and computing facilities required for all sorts of applications. The programs are permanently incorporated into the computer.

Similarly slip gauge box with bulit in computer and printer has been developed. The computer assists in selecting various combinations of gauges to develop any dimension. The required dimension has to be fed to computer and it displays the combination of gauges required. Not only this, it also illuminates the arrows above the actual block required to build the desired dimension. It also keeps in memory all the deviations from the precise size of all the blocks, prints the total deviation from the combination of blocks for a required size, stores and displays the information regarding calibration requirements.

The roundness measuring machine fitted with computer exploits fully the ability of a computer to carry out almost instantaneously, a large number of tedious calculations. Computer can easily define the circle if three points on its circumference are completely defined. Manually this job is painstaking and lot of time consuming. Similarly in gear measuring machine, the use of computer is made to do all tedious calculations and give final results in no time and very accurately.

17.22. Machine Vision

Machine vision or Computer Aided Inspection (CAI) is finding slowly its place for on line inspection, in line with the introduction of Computer Aided Machining (CAM). Machine vision is defined as a technique which allows a sensor to view a scene and derive a numerical or logical decision without further human intervention. Machine vision systems have the advantages of consistency, high speed and reliability.

The fundamental principle on which machine vision is based is that of digital image processing using a TV camera. The output of most TV cameras is a continuous video signal corresponding to a line-by-line scan of the image focused on the face of the camera tube. The signal is sampled at equally spaced intervals along each line, and a digital value assigned to the magnitude of the signal at those points. Line-by-line, a sampled image is constructed within the image processing computer as a 2-dimensional array. The sampling is analogous to laying a 2-D grid over the image of the scene and assigning numbers to the intensities within each element of the grid. The number of rows and columns comprising the grid are limited by both sensor and processor memory.

Similarly, the range of numbers used to describe the intensities are also dependent on these two factors. The number of picture elements in the grid (called pixels or pels) determines the spatial resolution of the stored image, and the range of numbers used to describe the brightness affects intensity resolution. By reducing spatial and intensity resolution, the amount of information held within the image decreases.

The design of vision systems requires a multi-disciplinary approach with a knowledge of sensors, optics, illumination, mechanical handling, as well as computer image processing techniques.

A machine vision system does not actually have to produce a recognisable image as part of its processing, as all the work is done on an electronic representation of the image. Vision system can be used for checking dimensions, checking overall shape conformity and checking surface finish.

With machine vision systems, it is quite possible to automatically measure dimensions to a fraction of a micron, using an image analyser and conventional microscope optics. The operator judgement about where to define an edge can be completely eliminated and highly reproducible measurements can be obtained.

In regard to checking overall shape conformity, it is done by checking the component against a master set of information and looking for discrepancies. The master information is a set of geometric measurements like area, perimeter, number of holes, amount of thinning of image possible before it breaks up, etc. Thus, the component need not be exactly aligned with the master as in conventional methods, and the handling requirements are also much simplified.

For checking surface finish by machine vision, the surface is treated as a kind of mirror (though it may not be perfect reflector). The vision system looks at an image of a light source reflected in the surface, and changes to that image signify a defective area. It is also possible to look for defects directly, if the defects can be relied upon to have a distinct contrast with the correct surface. The vision sensor can be mounted on a robot arm for inspecting complicated or large surfaces automatically.

The absolute resolution of the system is dependent upon the choice of optical lens system. Various types of image acquisition systems for different resolutions are : thermionic TV camera, solid state two-dimensional array cameras, linear array cameras. Thermionic TV tubes have the problem of drifting, especially during the first half hour or so after switching on, and under the influence of electromagnetic fields. However, they are best to distinguish small changes of tone than any solid state camera. Solid state sensors are preferred in general because these retain the spatial relationship between pixels (the smallest measurable unit, picture element) far more accurately. Solid state sensor systems can be calibrated for linearity of sensor and optics on a "one time" basis, allowing automatic correction of all subsequent measurements. The sensors in solid state cameras consist of an array of light-sensitive cells fabricated into a single silicon chip. These sensors are rugged, small and stable and thus usually fixed at the end of a robot arm. Solid state cameras are available with resolutions approaching those of CRT television camera devices (576 × 384 elements). In many cases, cameras embodying these devices give out digital data directly. Solid state sensors are also available as linear arrays. In such cases much higher resolutions (upto 4096 elements in one chip) are available. These are especially useful for applications requiring high precision as for dimensional checking.

Laser scanners can also be used. They comprise can arrangement of rotating mirrors to scan a laser beam spot rapidly over an object. The reflecting beam from the object is then directed on to a single photo-sensor, the output giving an indication of the surface reflectance at that point.

Where it is required to inspect a fast moving part, or to capture a transient event, it is possible to use an image sensor to freeze the motion. This is done by applying a short pulse of light and thereby storing the image as a coarse pattern on the sensor. The image may then be read, at a slower rate, from the sensor into the computer. For measuring length of a shaft very accurately, linear sensors are used at each end of the shaft with a known 'dead space' in between. If the component is such that it has several important dimensions to be checked in between also (*i.e.* not only the end dimensions of the component) then the vision system is integrated with a well-engineered stepping table. The specimen is then viewed in a number of steps which can be programmed to produce precisely abutting fields of view.

Image processing computer is usually a conventional mini- or micro-computer with the addition of a frame-grabber and frame store. The frame-grabber is a fast A/D converter which converts the analogue voltage levels of the camera into digital words needed by the computer. The frame store has memory to hold the entire image. The vision systems as such suffer from drift problem due to change in lighting. This is taken care of by compensating for varying ambient light, using a circuit which measures the peak white output of the sensor and adjusting the artificial light to maintain a constant 'peak white' output.

The various vision systems are binary vision, gray-scale vision and three dimensional vision. Binary vision technique involves location and shape analysis of flat objects. The object is backlighted to produce a high contrast image which is then converted to binary (black and white) image by thresholding. Several methods are then available to determine the position, orientation and shape of the object. The binary image can be thinned to produce a skeleton of the object which may then be analysed to identify links, nodes and end points. The position, size and number of these can be used to identify shape. Often, simple parameters such as area and perimeter can be used as simple shape descriptors.

By forming the convex hull of an object (equivalent to stretching a rubber band around it), the concavities or bays in the object may be isolated and labelled. A description of the number and size of these may be used as a crude description of shape.

To establish position and orientation of an object, the centroid may be easily computed and then the axis of least (or most) moment of inertia can define the orientation.

A series of concentric circles centred on the centroid of the object may be used to define shape. The intersections of the circles with the object's perimeter provide a number of (r, θ) pairs which can then be correlated with a reference set of (r, θ) pairs.

In *grey scale vision*, 2-dimensional digital signal processing techniques are used to enhance and then extract the relevant features from the image. Grey-scale operations include contrast enhancement, spatial filtering (to remove spot noise, etc.) edge enhancement (to highlight areas of rapid intensity change) and texture anlysis.

The third dimension in machine vision system can be measured be the structured light principle, by a process of triangulation. When a light pattern such as dots, bars or grids is projected onto an object, the pattern becomes deformed by the height contours of the object. A standard camera is used to sense these deformations is the projected pattern and a computer interperts this as height information. A single light strip is projected from a source inclined at an angle to a conveyor belt and spans its width. Objects passing underneath cause, by their height profiles, excursions in the strip when viewed from another direction. A range map is constructed from consecutive pictures of the deformed light line as the object moves along the conveyor and under the light pattern. The intensity in the picture corresponds to height information. And there is enough information held in the map for a surface plot to reconstitute the original shape.

Machine vision can carry out inspection tasks which are not practical for human to carry out at speed, like dimensional checking of hot metal.

17.22.1. Vision Systems. Machine vision, based on the processing of interpretation of electro-optical (television) images is being introduced into practice for a number of diverse programmable automation applications like inspection, material handling, and robot assembly. Vision machines are designed to recognise or identify workpieces, their stable states, and to determine their positions and orientations.

A vision system consists of a light source, an image sensor, an image digitiser, a feature extractor/data compactor, a system control computer, and output and peripherals. (Refer Fig. 17.33).

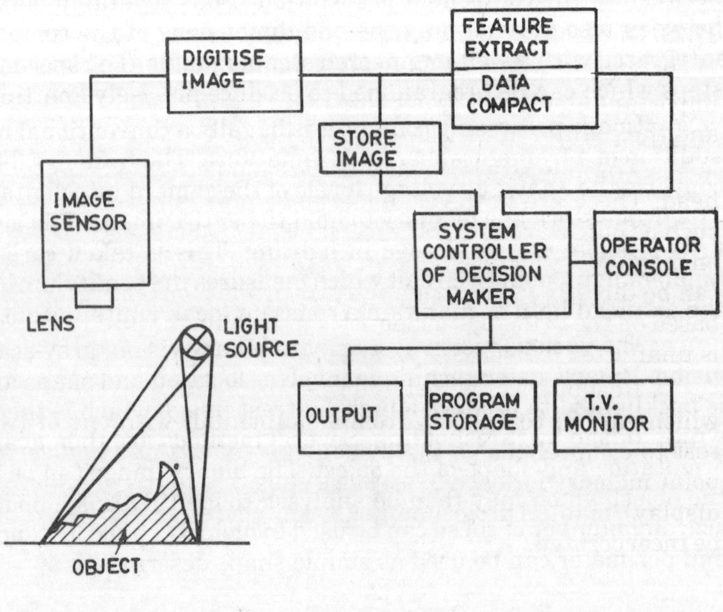

The image sensor is usually either a vidicon TV Camera or a solid-state, charge-coupled device camera, or charge-injected device camera. The image digitiser is usually a very fast, six-to-eight bit, analog-to-digital converter which stores the digitised image in the main memory.

Some systems use an analog computer and a computer-controlled threshold to convert the video information to a binary image rather than a 6 to 8 bit gray-scale image. In some systems, gray scale image (stored earlier) is converted to a binary scale image before any image processing takes place. The feature extractor/data compactor provides high speed processing of the input image data. Pattern recognition algorithms are employed to generate a simple feature data set. The system control computer makes the decisions about the part being inspected.

Two basic vision systems in use are : edge finders and correlators. Edge finders look for transitions form black to white, white to black, or from gray to non-gray level. When such a transition occurs, the edge finder system notes where it occurs (by line in the vertical direction, and by clock count in the horizontal direction). When a second transition occurs, it is also noted and the number of picture elements (pixels) between the transitions is noted. If a measurement is desired, then the system will have been calibrated (the size of pixel determine) and the physical distance between the two transitions can be computed. Correlation systems are template matching systems.

A complex image is stored in memory and the correlator searches the scene for a match to the image stored in memory. This method may be described by imagining a clear plastic template which has an image of the object printed on it. To locate the object, the template is moved until the printed image is aligned with the actual object. Using several templates, many objects can be identified and located. In a digital system the template is stored in memory as a two-dimensional matrix. This matrix is referred to as the reference. When a frame of video is loaded by the camera, the system overlays the reference on the upper left corner of the video and calculates the number of matches. This process, referred to as correlation, continues until the reference has been compared with the video throughout the field of view. The *x, y* coordinates where the best match occurs are then output. If there is no acceptable match anywhere in the field of view another reference may be used and the process repeated. This continues until the object is identified and located. Some vision systems employ light-striping systems. These use stripes of light projected onto an object with a laser or lamp. The distance of an object is determined by triangulation and the object's shape is determined from the distortion of the light stripes.

17.23. Digital Measurement Equipment

Digital indication is better by far when an exact quantitative value is desired. The important elements of any electronic digital redout system are the scale unit or transducer and the counter of digital readout unit.

The majority of linear digital readout systems comprise a glass scale and optical detector head. These are however, liable to be affected under workshop conditions by contaminants such as oil, swarf, and dirt. To overcome it some manufacturers have based their instruments on magnetic principles. These instruments also provide binary coded digit (BCD) output which can be directly fed to a computer or printer for further processing. In the case of instruments based on the principle of magnetic recording, the linear scale comprises a magnetic strip which is unaffected by workshop contaminants and can be of virtually and length.

Readout is provided by a range of digital display units and digital position readout units which vary in complexity from a simple unit with one or two axis display and thumb wheel rest to comprehensive three six display multi-function capability. Functions such as datum point memory for each axis, touch sensor (a touch to a workpiece starts counting or holds the display), manual programming by key operation and automatic programming by playback can be incorporated.

The advantages of such systems are : (*i*) Measuring element is free from errors caused by mechanical elements such as screws, gear etc. (*ii*) Learning time is short. Amateurs measure as quikly and as well, as professionals. Measurements takes half as long. (*iii*) More accurate measurements, the spread of measurements and the mean error are reduced to about 1/3. (*iv*) Excessive reading errors are eliminated. (*v*) The clear readability of digital readout is advantageous for persons with impaired vision. (*vi*) The display can be zeroed wherever desired. (*vii*) BCD output makes the instrument computer compatible.

17.24. Multiaxis Measuring Machines

Most of the modern measuring instruments and machines use digital type electronic instruments which do not use traditional machine divided and visually assessed scales. Instead they employ photo electric, magnetic or capacitance devices which, in conjunction with electronic counters, display the linear measurements to a high accuracy.

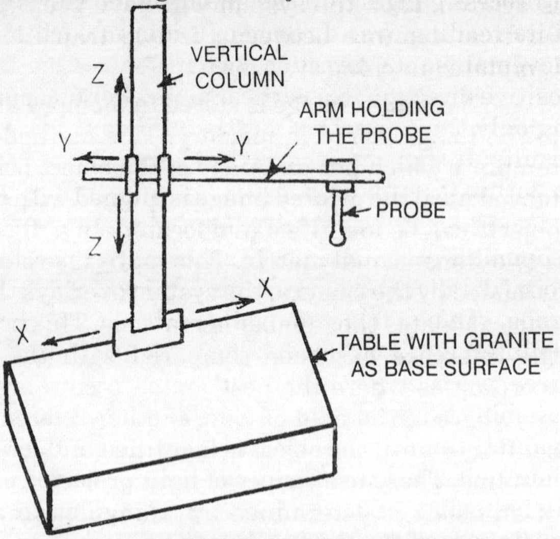

Fig. 17.34. Multiaxis measuring machine.

An electronic height gauge with a probe is a single axis machine because the measurements can only be made in a direction perpendicular to the surface on which the height gauge is mounted. If the height gauge is mounted against a linear scale so that its position can be measured in a horizontal direction, then we obtain two axis machine capable of measuring in horizontal and vertical directions. By incorporating a means of moving the height gauge against a scale in a horiozontal direction at right angles to the first horizontal movement, we can obtain three axis machine capable of measuring dimensions in any direction on a solid object. Fig. 17.34 shows in schematic form the *X, Y* and *Z* movements of positioning probe in such a machine. It would be noted that *X*-axis rail is parallel to the longest dimension of the base table and *Y*-axis is on the arm which runs along this rail. *Y*-axis is on the arm which runs along this rail. *Y*-dimension is obtained from the position of the probe along the arm. *Z* axis dimension is initially fixed by vertical column and then measured by the probe. For higher accuracy, end of arm could be supported to avoid deflection. The cantilever arm is thus substituted by a bridge. The weight of the moving parts in such an arrangement is overcome by the use of air bearings in which the members moving relative to each other are separated by a thin film of air which is pumped between them. Universal measuring machines based on this principle are nowadays available in a large variety and one can select the one appropriate to his requirements depending on range of measuring capability.

If the movements in *X, Y* and *Z* directions be controlled by electric motor, then such machines can operate automatically under the control of a prepared programme and used for inspection of a complex part automatically.

If a horizontal rotating part is included, then such machines become four-axis having additional movement in vertical axis of rotation. Similarly five axis machine can be designed by adding facility for rotation in a vertical plane. Usually such machines are special purpose machines to carry out specific tasks. In some of modern multi-axis measuring machines, a magazine is provided which contains a variety of measuring probes which can be automatically changed during the measuring cycle, thus enhancing the versatility of machine considerably.

The development of accurate electrical output signals capable of being fed to a computing device has resulted in the reas success of multi-axis, co-ordinate measuring machines. The computing devices not only assess the dimensions of the part to be measured but also provide a variety of analytical statistics. It is also possible to obtain hard copy of the desired dimensions from a printer.

The vital element in any measuring machine is the measuring probe which contacts the surface to be measured. For measurement of flat surfaces, a ball ended probe is very suitable. Rigid probes with conical ends are used for measuring hole positions. The contact probes used on measuring machines are available in two forms, *viz*. they may either only indicate a contact position or they may have a small measuring capability. The later type is very common for in-process gauging. In the analogue measuring mode, these probes have an accuracy of 0.5% of its reading over a range of 1 mm in each of five directions. These five directions are two movements in X direction, two in Y-direction in horizontal plane (at right angles) and one in positive direction in a vertical plane. For measuring machine the touch trigger probe (registering only the contact) is mostly used as it is capable of giving the most consistently accurate results. It triggers an electrical signal at the touch position. Probe indications are repeatable to within 0.1 micron (μm).

Fig. 17.35 shows in schematic form the constructional details of a touch trigger probe. The stylus is rigidly mounted in a carriage having three projecting rollers. This assembly is located by a spring against three pairs of balls mounted in the probe head. All six balls and the three rollers are linked to form an electrical circuit which is disturbed by any movement of the stylus relative to the head.

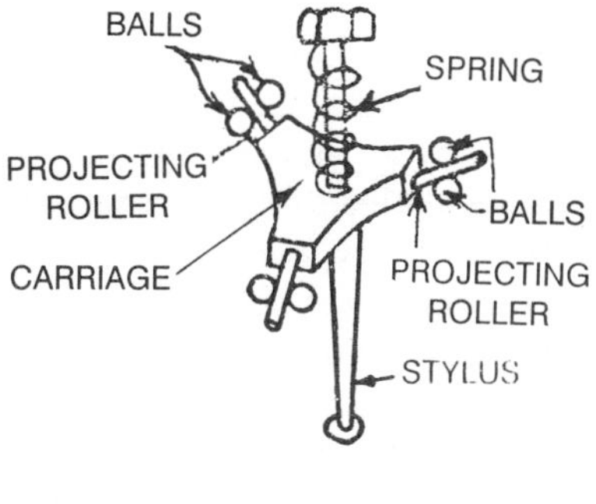

Fig. 17.35. Touch trigger probe.

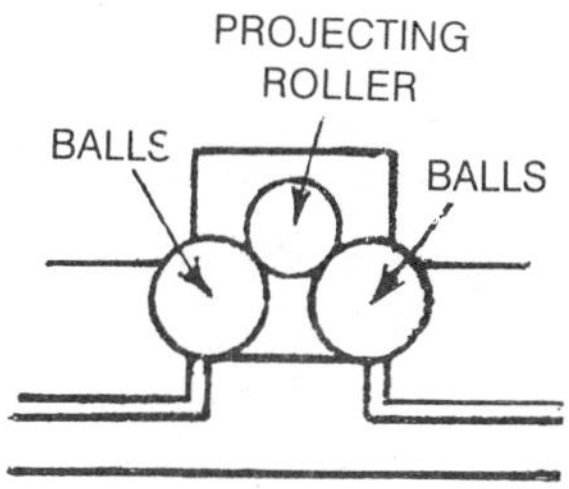

Fig. 17.36. Projecting roller and balls.

For measuring any dimension the probe is moved until the probe indicates that a touch has been made. It is then moved to the other face till a contact is obtained. The difference of two machine readings minus the diameter of the probe stylus thus gives the dimension to be measured. In actual use, the machine movements are motorised and the signal from the probe causes the machine reading at that point to be recorded automatically.

A large variety of styli are available to facilitate the use of probe in difficult positions. The versatility of the probes can be increased by the use of mountings which can hold more than one probe.

A motorised probe with special motor capable of moving the probe to a large number of predetermined positions with a repeatability of 0.5 μm has been developed. The movements of the probe can thus be computer controlled and programmed. In this way the limited movements which were to be made by machine alone can be carried out by the probe head.

17.25. Co-ordinate Measuring Robots

Nowadays more emphasis is being laid on on-line measurement of parts as they are being produced. With this philosophy it is possible to qualify the complete production process and not just the item in isolation. By in-line measuring techniques, the manufacture of defective components can be identified and prevented. Thus the rejection of completed assemblies, the waste of extermely costly machine and man-time as well as materials and other resources can be avoided.

Co-ordinate measuring robots to satisfy above philosophy have been developed for inspecting large components. A typical system consists of open modular type of structure. A horizontal robot arm, with its measuring head and probes, intrudes upon the production line space. Multi-arm configurations are used for inspecting wide range of different size products. Multi-arm units reduce inspection times considerably, especially if they are integrated with pick-up-and-place arms for parts handling, or if two or more arms operate simultaneously from different sides of the same piece on the production line.

The arm is fitted with a five-way measuring head for fast and repeatable insertion of the probes in five attitudes, allowing complete inspection of a component with a single set-up.

The robot arms are normally mounted on the side of a work surface fitted with guideways to allow longitudinal (*X*-axis) movement. More than one arm can be fitted on each side if required. The vertical (*Y*-axis) is counter-balanced by an adjustable pneumatic system which compensates for the weight of different probe configurations.

The axes are powered by d.c. motors through rack and pinion couplings, with the moving parts sliding on air bearings. Machine movements in all directions are monitored by high-resolution non contacting linear transducers.

This configuration allows resolution accuracies of between 2 and 10 microns-depending on the size of the component being measured at a maximum working speed of 500 mm/sec.

Accessories like full range of automatic electronic probes for on-the-fly measuring, etc. ultrasonic thickness detectors, etc. can be fitted to make the system more versatile. For inspecting multi-faced part indexable tables, to index the part and bring each of its sides to face the machine spindle for checking, can be used.

A powerful microprocessor supervises the system as well as processor and records the data. An easy-to-program controller employing a dedicated high-level interactive language is used so that persons not skilled in computing techniques can generate efficient measuring and data processing instructions for the machine. The changes in inspection cycles and production-line requirements can be easily made. Generally the system has a self teaching capability which enables a program to be generated automatically when the first item on a new batch of components is being inspected manually.

After collecting all the measurements needed to monitor he quality of a manufacturing process, the system produces the resulting data much faster than it can be interpreted by management personnel. The data is continuously organised in statistical form, thus making it possible to derive cumulative results, from which drift trends can be swifty spotted and corrected before any out-of-specification components are manufactured.

A supervisory computer graphic system, statistically analyses shop-floor inspection data, continuously monitors a process by recording and analysing the information produced, and presents the results graphically as histograms, frequency tables, bar charts, relation diagrams, trends, etc. The full-colour graphic presentation, compared to traditional alphanumeric display, is easier to digest and conveys more information.

17.26. 2-D and 3-D Measurements

Electronic height gauges with microprocessor based display are capable of performing 2-D measurement because of their ability to store, recall and manipulate data—speedly and accurately. With this, it is possible to carry out 2-D alignments and calculate the best fit through a number of holes.

Co-ordinate measuring machine (CMM) is used for 3-D measurements. Such a machine has a vee-flat guidance system which gives it a high degree of stability. *Y*-axis has an independent twin-beam, granite guideway, which effectively eliminates it as a stressed part of the bridge structure. The standard software packages provide a whole set of user friendly menus designed with the needs of the operator in mind. Do-and-learn facility which enables the user to write part programs is usually provided. These machines are however limited by the volume of the work area, the probe size and the complexity of the object shape as it is not always possible to bring the probe into contact with the desired position on the object.

Mapping technique is also used for 3-D measurements. Mapping technique is used in photogrammatric measurement system which is non-contact method for measuring of complex bodies. While determination of linear or planar quantities is relatively straight forward by the use of calibrated rules, grids, digitisers, projection equipment, electronic distance measurement (laser) techniques, the determination of three dimensional shape or position is more complex. A variety of techniques have evolved, often tailored to a particular object size or aimed at a particular measurement problem. However different situations call for flexible 3-D measurement tools which use the same equipment and standard methods of analysis for objects of vastly different dimensions, complexity, and purpose. Such systems should be capable of producing reliable 3-D data describing the object, obtain the required accuracy in a cost effective manner, make no assumptions concerning object size, allow rapid measurement by non-specialist operator, involve the minimum interference with the object, produce results which can be easily understood.

Photogrammatric measurement technique has been used in the production of topographic maps using aerial photographs. An overlapping pair of photographs form a stereo pair which can be used to create a 3-D model of the object of interest. Parallel measurements are made upon the photograph surface using high precision measurement instruments (stereo-comparators), while the operator views the 3-D model through a binocular viewing system. These measurements are used to compute accurate 3-D coordinates which describe the object and so the output is similar to that of a conventional 3-axis measurement machine.

The availability of low-cost micro computers with graphics capability and sufficient computing power has allowed photogrammetry to develop as an industrial measurement tool using many different kinds of imaging system (X-ray devices, microscopes, conventional cameras etc.).

A variety of cameras can be used to provide input data into the system. Optical microscope work uses 35 mm or glass plate cameras and aerial photographs can be 70 mm or 23 cm format. In each case the operator creates a 3-D model and then digitises selected features using a floating mask tracked by rotary or linear encoders. The encoder output is passed over a serial line to the micro-computer which computes a 3-D ordinate and can display an orthogonal or perspective elevation of the object on a graphics screen.

The operator can quickly build up a 3-D map of the object and can change scale, rotate the image, edit or change the projection as required. All data is stored as feature coded data records which can be accessed or further manipulated.

The basic system consists of a stereocomparator, 16 bit micro-computer, plotter and printer. A standard set of analysis and display routines is employed which have been written is a modern high level structured language.

The features of such a system are : non-contact measurement, ability to measure objects from microns to km in size, perform in situ measurement, objects may be in the range of cm to m, used to measure areas of difficult access or hostile environment, to perform dynamic analysis, allow rapid measurement by non-specialised operators, present results in graphical and digital form. Time on site is dictated by time required for photography and not the analysis.

The photographs provide a wealth of dimensional information from which either key elements can be extracted to provide parameters of the object (key distances, orientations, etc.) and enough data can be extracted to describe the complete surface shape. Moving parts can be frozen at the instant of photography using shutter speeds.

Such a photometric instrument has been used for measuring sub-micron line widths in the production of wafers and masks. It is also used for similar measurements in manufacture of multi layer semi-conductors, and the measurement of profiles of glass fibres used in the communications industry.

High accuracy and reproducibility is achieved by a variable slit aperture, that can be adjusted to exactly match the size, shape and surface characteristics of the structure to be measured.

The measuring process is controlled by a built-in microprocessor interfaced to a micro-computer. The brightness of the microscope illumination is also computer controlled. The pilot lamp setting is also automated, as well as the gain and high voltage control for the signal height adjustment. All the controls for the system can be operated via function keys on the control panel.

The micro-processor automatically controls the movement and adjustement of the measuring slit as well as the selection of the optimum measuring speed and the scan process.

It is possible for the operator to view both the adjustments and measuring process and also to videotape the entire sequence.

17.27. Digital Height Gauge with Statistical Production Control

Today's modern production techniques reflect a growing need for statistical Process Control (SPC). Systems capable of predicting or anticipating "out of tolerance" are thus of great value. The improvement in quality, the reduction of scrap and the increase of efficiency in production capacity are all closely related. The digital height gauge with capability to provide SPC is best suited in this direction. Such a gauge has following features :

— Storage of upto 400 dimensions
— data retention on failure of power
— automatic entry of data as component is measured
— entry of data via the custom touch key-pad
— calculation of capability indices for quick and easy assessment of machine/plant or process capability
— easy identification of component features out of tolerance
— calculation of mean, standard deviation and range
— assessment of 2-sigma, 3-signal, and 4-sigma confidence intervals
— comprehensive print-out of statistical data including histograms
— each sequence of measurement is recorded in numerical order and can be recalled subsequently and printed out in the format required.

17.28. Air Gauging with Electronic Sensors

Air gauging systems operate on either low or high air pressure. While low pressure systems have greater sensitivity, quicker response time and minimal distortion when measuring flexible components, the high pressure systems are self cleaning type and have a large measuring range. Basically air gauging comprises air jet gauges—such as ring or plug, and air operated liquid columns for multi-dimensional measurement.

Now-a-days electronic flowmeters are used in place of air operated liquid columns. They have the advantage of measuring flow of air with the added benefit of electronic display. Such instruments can easily have 2 or 3-range selection to give an extra magnification factor. Tolerance limit lights can be incorporated to indicate whether parts are inside or outside manufacturing tolerances. Response time increases many folds.

The versatility of air gauging is enhanced by the wide range of measuring tools like 2 and 3 jet non-contact air plug gauges, setting rings and air jet gauges.

17.29. Multi-Gauging Systems

Multi-gauging systems are used to measure a number of dimensions simultaneously. Part to be gauged are compared with a setting master which simulates the component. The feature gauged could be external/internal diameters, lengths, straightness, squareness ovality, run-out of faces, etc. The measuring head gauging fixture is specifically designed to suit the component to be measured and may be completely special or it may be built using a series of modular elements. It contains the means for sensing the dimensional difference between the components and the master which may take the form of mechanical or electronic probes or air jets connected to the means for amplifying the difference.

The amplifying and display of readings may take the form of dial gauges or some form of electronic or air/electronic system.

Displays may be analogue, digital or graphic and may be augmented by out of tolerance indication.

Using electronic differential methods, the relationship between different features can be related to a common datum.

The choice of system depends on number of factors like initial cost, dimensional tolerances of the features to be measured, complexity of the component, complexity of the features to be measured, number of features to be measured, speed of measurement required, skill or otherwise of the user.

The reasons affecting the choice of displays are given below :

(*i*) Pointer displays—best where a rapid check run-out or concentricity is required.

(*ii*) Columns—natural choice where a considerable number of dimensions are involved.

These are fastest and most convenient form of displaying the readings for every dimension.

(*iii*) Digital—provides high accuracy over a longer measuring range and best for situations where it is required to work in drawing dimensions. Can be viewed without strain over longer distances.

(*iv*) Graphic on VDU—the most sophisticated display essential when statistical process control systems are employed.

Systems capable of dealing with very large number of interrelated dimensions can be developed.

Automatic inspection machines incorporate both automated loading hopper, magazine, pick and place robot and—automated segregation of inspected components. Automatic inspection is essential where the complexity of the component is such that manual methods can not achieve the desired levels of accuracy.

The results of inspection could be fed to electronic computer based system which may also control the machine operation. The use of such computer based processing also allows the results obtained to provide a wide range of control facilities including feedback for control of the manufacturing process.

Coordinate measuring machines accommodate multi-dimensional inspection by using a single point contact to take successive measurements over the component profile. The contact movement and processing of the dimensional information is under computer control, which can also provide similar facilities to those offered by multi-point gauging.

17.30. Universal Measuring Machine (Fig. 17.37)

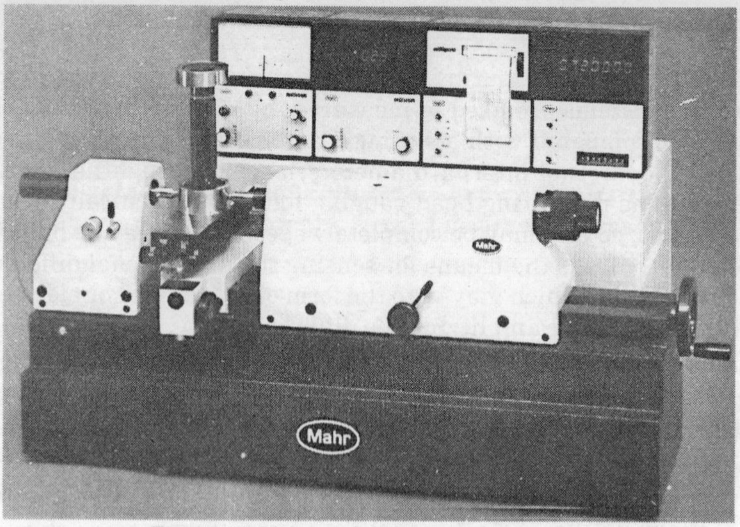

Fig. 17.37. Universal measuring machine.

Measuring range	mm	0 – 1000
Max. deviation	µm	0, 7
Repeatability	µm	0, 1
Incremental measuring system		
Resolution	µm	0, 1
Length of scale	mm	200
Measuring face dia.	mm	7, 5
Adjustable measuring pressure	N	0–10
Height of measuring spindles	mm	110

Application

For precision measurements of outside and inside diameters, lengths and threads. Ideal for frequently changing dimensions within a 200 mm range since no setting gauges are required. Therefore, this machine is excellent for checking plug gauges, ring gauges, snap gauges, thread gauges and gauge blocks in the inspection room, and also for checking precision parts in the production line.

Design Features

Heavy duty, sturdy machine base, made of high density, black granite, which carries the measuring head, setting carriage and if required support tables. A hand wheel adjusts the setting carriage which accommodates 200 mm incremental measuring system, connected to a digital comparator. The measuring anvil in the left hand head is mounted on a parallel having assembly virtually friction-free and without play. The measuring pressure is adjustable. Anvil retraction allows a no-contact mounting of the part. Both measuring faces are **carbide tipped**.

Test Procedures

When determining the **deviation** of the component under test, the measurement displayed is the deviation from the nominal value. This nominal dimension of the part is preset on the digital indicator, which is connected to the incremental measuring system. Readings are taken from the indicator on the left hand measuring head.

When checking the **actual dimension,** the latter is displayed. For this operation, the indicator on the left hand measuring head serves as zero indicator. The actual dimension is read from the digital display on the right hand head.

For **processing** of the measuring results of deviation and actual dimension printers or computers may be connected.

Advantages

Versatility. A large selection of accessories and attachments for many different kinds of inside and outside measurements on gauges and workpieces are available.

Direct readings within 200 mm range. No setting gauges required within that range.

Digital readings with resolution of 0.1 μm on indicating instrument which is connected to increamental measuring system.

Optimum sensitivity and accuracy guaranteed through a smooth and even movement of the anvil which is mounted friction-free and without play on a parallel spring assembly. The latter is maintenance-free thereby providing consistent accuracy even in continuous operation.

Adjustable measuring pressure to match size and shape of parts. Even thin-walled, small components can be checked without deforming them.

Constant measuring pressure is generated by means of weights, thereby eliminating any interference from the operator's personal feel.

Maximum wear resistance. A lever retracts the anvil thereby allowing no-contact positioning of the component which together with carbide tipped measuring faces reduces wear to a minimum.

17.31. Workshop Measuring Instrument for Dynamically Balanced Components (Fig. 17.38)

Application

For testing dynamically balanced components. Measures lengths, outside and inside diameters, lengths and distances of registers and steps, widths and depths of grooves. Also for testing concentricity, diameter and pitch of threads and diameter of gears and splines.

Design Features

Sturdy base, made of hard granite with two tailstocks which are mounted in a T-slot and designed to hold the parts during the testing operation. In front of this assembly two measuring carriages are moving in x- and y-axis. The carriages, mounted on high precision ball bearings, move on chromium plated steel guideways, which are ground and precision lapped. Both carriages are equipped with sealed incremental measuring systems. Adjustment knobs with fine setting are provided for shifting the measuring carriages. They can be locked in place at any given point. The compact control–and indicating unit is furnished with a computer and integrated printer.

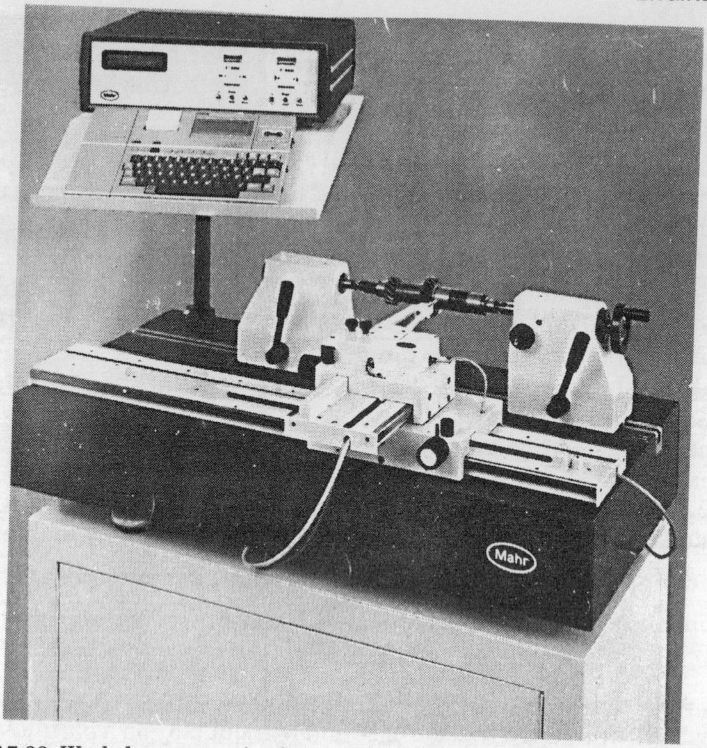

Fig. 17.38. Workshop measuring instrument for dynamically balanced components.

Measuring Range	
x-axis	500 mm
y-axis	150 mm
Outside dia.	0–300 mm
Outside dia. with measuring jaw	0–100 mm
Inside dia. with inside measuring attachment	15–90 mm
Resolution	
x-axis	5 μm
y-axis	1 μm
Max. deviation	
x-axis	10 μm
y-axis	5 μm
Pressure at data retrieval	3 N
Ball anvil dia. (carbide tipped)	8 mm

17.32. Universal Measuring Instrument (Fig. 17.39)

Measures inside-and outside diameters, registers, recesses, small collars and grooves. Also eccentricity and concentricity of various outside and inside diameters in regard to each other and right angle of outside diameters to the face of the work.

Design Features

Rugged box type housing with hardened mounting table. Three holders, one moveable, two fixed, for mounting interchangeable measuring arms. Moveable measuring arm holder is mounted virtually without play and friction-free in high precision ball bush-guide and retractable by easy to reach lever. Measuring pressure adjustable. One fixed measuring arm holder may be shifted for coarse setting, the other serves for centering the work and is arranged rectangularly to the measuring axis. Measuring arms for all sorts of measuring problems available. Mounting table adjustable in height and for aligning work inclinable. Post with support arm for testing concentricity by a dial comparator can be attached.

Fig. 17.39. Universal measuring instrument.

Advantages

Versatility. A large variety of accessories which can be mounted and exchanged in a minimum of time permit inside and outside measurements on smooth and threaded parts.

Super sensitive and maximum accuracy, by high precision ball bush-guide of the moveable arm holder virtually without play and friction-free.

Adjustable measuring pressure for adapting the instrument to size and shape of work. Small and thin-walled work pieces can be measured without deformation.

Direct indication of measuring results. Actual dimension of the part can be determined with absolute accuracy.

Mounting table inclineable for aligning the work.

17.33. 3-D Optical Probe

The development of universal 3-D optical probe would enable conversion of touch-probe system to a precision non-contact vision system.

The optical system of such a probe is based on miniature charge coupled device (CCD) technology, optical prism design and lens technology. This provides a very high definition image on the VDU screen. The video measuring optical design is based on a fixed focal length lens system. As power drive zoom lens traverse system is incorporated. The zoom lens brings the system into the micro measurement field with magnifications ranging upto 140 X. An LED panel on the optical unit gives a digital readout of the actual zoom magnification.

Automatic compensation for the Z-axis for zoom position ensures that the screen picture and vertical foucs system are correctly sealed. Irrespective of up or down zoom movement the object on the VDC screen always remains in foucs.

Lighting of the object is by a fibre optic cold ring light system which surrounds the imaging lens.

17.34. Automatic Shaft Inspection Machine

This machine works on optical measuring method and can measure shaft-like components, analyse the results and feed back correction data to the production process.

Measurement of cylindrical components is by analysis of a projected 2-D image. Component could be fed either manually or by robot.

Operation is by a simple sealed Keypad with self illuminating key prompts on a visual display unit (VDU) for the operator. There is in addition a fold away alpha-numeric key board with key holder only access. The gauging sequence and component control data is pre-programmed via the key board. Alternatively a completely new component can simply be placed in the machine, be scanned by it, and a picture of it displayed on the VDU Measurements can then be taken directly by the operator who simply indicates via the key board or a computer 'mouse' what features are of interest. In the course of doing this the machine dimensions the components and also compiles a measuring routine for future use.

Programs are stored against component type or part number. In the gauging mode, the machine is speedily reset to accept different components simple by selecting the appropriate identity from a displayed directory using the inbuilt keypad. Measurement proceeds automatically and on completion of a cycle the values obtained are displayed with reference to permitted deviation from nominal sizes. The operator can select numeric or graphic display. In both the cases, the out-of tolerance features are high lighted. The measurement data can be stored as long as desired.

Stored data can be analysed in several ways. The individual measurement chart gives a trend plot of the last 100 component measurements with reference to control limits. An individual measurement report provides a numeric listing of mesurements with indicated X and R values for each sub-group sample of components measured. These are supplemented by a control chart giving a plot of X-bar and R for the last 25 sub-groups with respect to process control limits.

17.35. Sub-micron Accuracy from non-contact Gauging System

The non-contact gauging system consists of capacitive sensor arrays. It consists of a sensing head and a microprocessor controller/computer interface. Each sensing head is tailored to the application and is diametrically opposite in shape to the component being measured. Embedded in the head are the non-contacting capacitive sensors which are used to measure the gaps between the two surfaces and therefore determine whether the components conform to the required standard in terms of both dimension and shape.

Such a system has been developed for production rate gauging of geometrically complex components. Each individual sensor can provide sub-micron resolution plus accuracy of 0.1% of the useable range. The system can be readily interfaced with a microcomputer to enable information and data to be used for reports, quality and production control records-including statistical process control (SPC).

17.36. Process Control Robot

The process control robot is an automated flexible gauge built like a machine tool. It is designed to work adjacent to or even in the middle of machine tools. It has the capability of carrying out automatic inspection routines on parts uninterruptedly.

It is designed as a horizontal arm (single or dual) measuring machine and can be incorporated in any production line. It is also available with an integral rotary or flat table enabling ideal access for pallet or gantry loading.

The high precision mechanical slides have pre-loaded roller packs as bearings, and rack and pinion drive systems. These enable acceleration and speeds to be attained which are far beyond the reach of normal coordinate measuring machines (CMM)—0.3 g acceleration and 20 m/sec velocity. In this way inspection cycles become four to five times faster than conventional CMM.

The robots are controlled by Computer Numerical Control (CNC) path control system with integrated linear and circular interpolation. This ensures precise vectorial movements for contour scanning. Circular interpolation increases the measuring speed substantially for parts with circular, spherical or cylindrical geometry. Built in diagnostic system is incorporated enabling verification of all input/output signals.

17.37. Nanotechnology

The term Nanotechnology relates to the manufacture and measurement of components to dimensions and tolerances of nanometers (10^{-9} m) and of fraction of nanometers. Such technology finds application for production of X-ray optical components, production of surface finish of precision bearings, semiconductor substrates and laser gyroscope mirrors.

X-ray interferometers are used to provide traceable measurements down to 10 picometers and scanning tunneling microscopes for measuring surface topography with sub-manometer resolution parallel to the surface.

Nanotechnology promises to make major contributions to mechanical engineering, electronics, optics and materials technology. In the materials field, the reduction of grain and fibre sizes in the starting materials for ceramics, sintered metals and composite down to the nanometric dimensions significantly enhances the mechanical, thermal electrical and magnetic properties.

In machining there is substantial benefit in exploiting technique of nanotechnology to allow ductile machining of brittle materials such as ceramics, glasses, and silicon. If the depth of cut can be controlled to 10 nm or better, the need for polishing and lapping may be obviated.

17.38. Microprocessors in Metrology

The microprocessor is becoming a constantly increasing part of our environment. The microprocessor provides a cheap and compact means of manipulating digital data. It replaces a quantity of logic circuits in all applications requiring logical decisions, calculations, etc. It can solve data management applications and the processing of results speedily and effectively. The use of microprocessors in gauges has become generally accepted. Microprocessor in a gauge can mix signals from displacement transducers, compare the measurements with nominal value and display the results on a visual display unit (VDU) screen. The microprocessor based gauge is thus ideal for measuring and data processing instrument used in quality control.

While in analog measurement the total measuring range may be divided into 100 graduations, with digital electronics it is possible to break the same into 5000 parts, providing much greater precision for the same measuring range.

In the manipulation of data a processor can remove the tedium, preclude the risk of error, and provide results rapidly, more reliably and with constant interpretation. Microcomputers can handle very complex signal combinations with no reduction in accuracy.

The microprocessor also makes it possible to operate in realtime within a machining process, providing in-process dimensional gauging with adaptive feedback to the machining algorithms to achieve a consistent accurately dimensioned components.

The application of micro-processor to metrology field has resulted in an extensive range of digital measuring instruments combining accuracy and ease of use.

The processor improves set up procedures and de-skills operations in general. For example, hours need not be wasted in setting up for spindle error measurement to minimise misalignment of the measuring system. A microprocessor can log errors at high speed on-the-fly, and then remove set-up errors using mathematical techniques ; such as least squares fit. Automatic correction of systematic errors can be performed within a measuring system in real-time, with interpreting between fewer reference points, to increase the measurement accuracy by at least a factor of ten. The microprocessor also makes it possible to operate in real-time within a machining process, providing in-process dimensional gauging with adaptive feedback to the machining algorithms to achieve consistent accurately dimensioned components.

While analog computing systems are expensive and a one-of-a kind device for a particular application, microprocessor based systems are less expensive and perform the same function developed in a software only and thus hardware remains same.

A process can also synchronise continuous movement, in realtime, of multiple axes of a machine to generate mathematically correct 2- or 3-dimensional standards—as in gear measurement.

Use of microprocessor permits a very high degree of flexibility in comparison with conventional methods. The same instrument can be used for very varied tasks simply by modifying the software.

These have low initial cost, low operating cost, and are accurate across the full range of scale (pneumatic gauges tend to lose precision near the ends of the scale).

Microprocessors can reduce the costs of multi-dimesional gauges. It is quite easy to design amplification for different scale ranges and to transfer the gauging output signal to some other more sophisticated statistical or record gathering and reporting system.

Different types of readout on the screen are possible. One type is a series of bars resembling the readout from a bank of column gauges. Alternatively, a graphics display can depict the workpiece and indicate the amounts of error at gauged points on the surface of the workpiece.

Memory is an important facility offered by the microprocessor, which allows the data to be stored and recalled. Thus a sequence of measurements can be subjected to further processing. While a mechanical gauge (GO and NO GO type) can only indicate whether or not a workpiece is acceptable ; a microprocessor based gauge can serve as an important information system which can be used for data collection, recording and printout of data, networking, closed-loop tool compensation, production control, statistical process control (SPC), etc.

Hardware and Software involved

The hardware of microprocessor based dimensional control instrument broadly consists of :

— a processing system composed of a microprocessor, with a read-only memory containing the operating system, and a memory containing the application program and variables.

— an interface with the operator composed of a keyboard, screen display signal lamps, etc.

— an interface to measure the lengths, etc.

— Interface with different peripherals such as a printer, another computer, mass memory etc.

Softwares are of two types :

— Basic software (operating software). It allows access to different resources of the system by means of simple and well defined orders. These orders constitute the part accessible to the programmer who can also work in high level-languages.

17.39. Fiber Optic Probes for In-process Control

Fiber optic transducers are based on following principles : (*i*) intensity-modulated, (*ii*) interferometric, (*iii*) polarised based, (*iv*) wavelength based.

Intensity modulated devices have the advantage of ease of installation, low cost and reliability. These devices may be based on either reflective or transmissive category. Both of these methods can be used in either the on-off mode or the proportional mode. The latter method provides a relative output signal which can be interpreted as a measurement of some physical property (position displacement or a dimension).

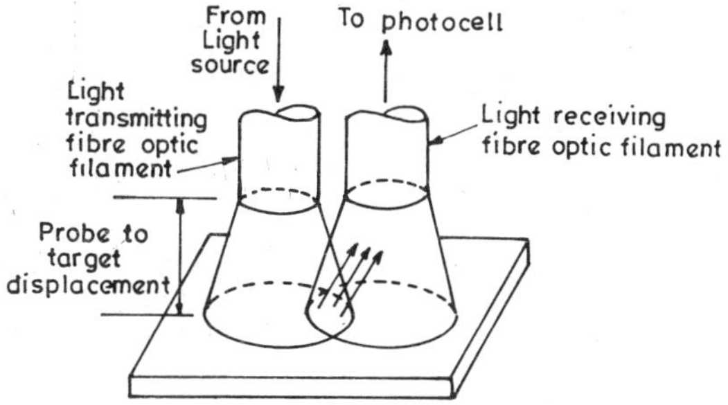

Fig. 17.40. Displacement sensing mechanism of adjacent fiber optic elements.

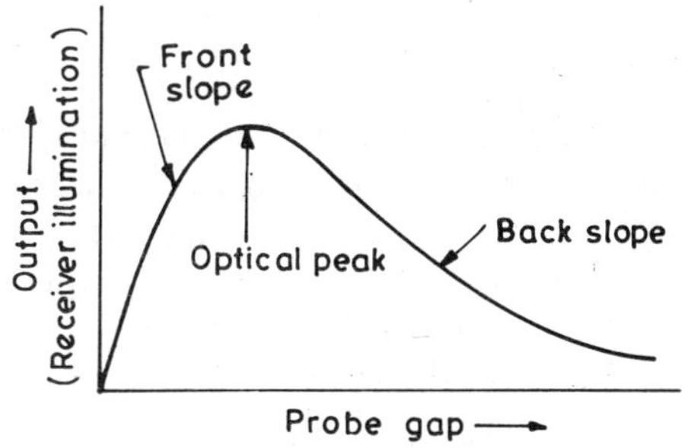

Fig. 17.41. Typical fiber optic sensor calibration curve.

Fiber optic level displacement sensor utilises adjacent pairs of light-transmitting and light-receiving fibers. Operating mechanism is based on the interaction between the source fibers and the field of view of the receiving, or detector fibres. (Refer Fig. 17.40).

When the gap between fibre optic element and target surface is nearly zero, then the light coming from the light transmitting fibre optic filament is reflected directly back into it, providing very little light to the receiving fibre and thus producing zero output signal. As the distance between probe and target increases, more reflected light will be captured by the receiving fibre. This increase will continue until the entire face of the receiving fibre is illuminated with reflected light. Further increases in distance will cause the diverging field of reflected light to exceed the field of view of the receiving fibre, reducing the output signal. The relationship between probe gap and output is shown in Fig. 17.41. The three key segments are shown in Fig. 17.41. At optical peak, the receiver fiber captures the maximum reflected energy relative to the surface reflectance and optical path attenuation. The front slope response characteristic is a function of the fiber diameter, the relative position of the transmit and receive fibers, and the numerical aperture of the fibers. The back slope response characteristics is primarily a field intensity inverse square law function.

The responses for three different fiber relationships or arrays are shown in Fig. 17.42. Curve (1) is for random or thoroughly inter mixed transmitting and receiving fibers and it gives the most sensitive front slope and the smallest sensing range. Curve (2) is for concentric transmit inside fiber grouping and it provides an intermediate response Curve (3) for semi-circular fiber grouping gives the least sensitive front slope and greatest range.

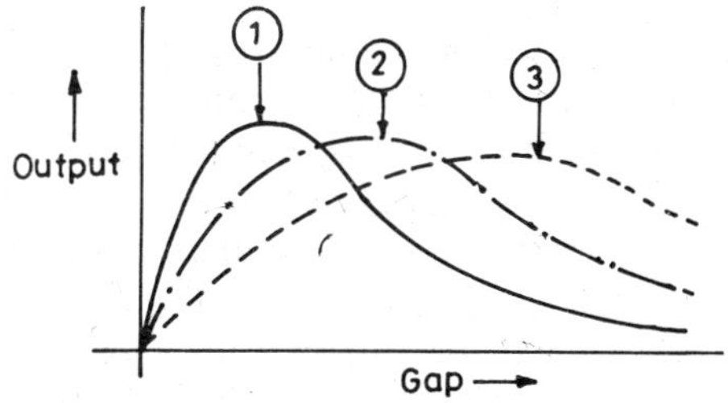

1. Random grouping array
2. Concentric transmit inside grouping array
3. Hemispherical grouping array

Fig. 17.42

It is also seen by experiments that when source intensity is same and surface reflectance is varied, then output changes slightly (following nearly same pattern). The form of the response curve is not affected by changes in surface colour. For changes in source intensity, the magnitude of received light at the optical peak remains unchanged regardless of the reflectance, thus creating a sensing system having repeatable calibration characteristics.

If the geometry of the reflecting surface is changed from a specular to a diffuse nature, a slight change occurs in the back slope of the curve. The front slope range of random configuration is thus found to be basically independent of the bundle diameter. However, the back slop range is proportional to the overall diameter of bundle. For hemisphere configuration, the front slope is directly proportional to the bundle diameter.

The output signal is modulated with change in target material and the calibration curve is altered with change in the media between fiber and target.

When a focusing lens system is placed near the sensing tip of the fibre optic probe, then two peaks are seen with a very sharp null point midway between the peak. The operating gap or stand off distance with lens system is approximately 100 times greater than without the lens system. It is also possible to design fiber optic lever displacement system so that it is not sensitive to reflectivity of target surface.

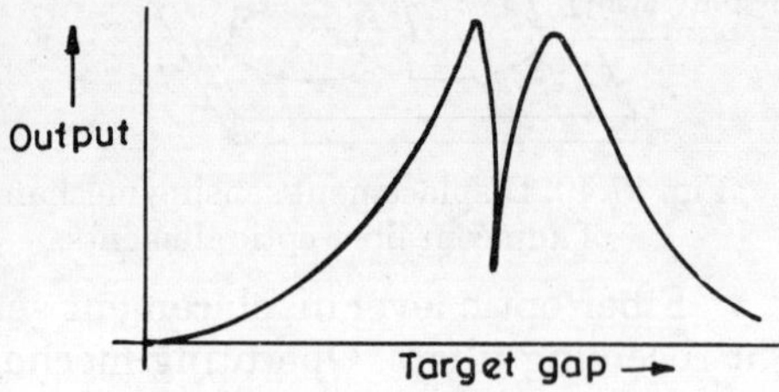

Fig. 17.43. Characteristics of fibre optic probe with lens system.

The fiber optic lever displacement transducer offers excellent opportunities to be used for several unique applications because of its noncontacting nature, high-frequency response, small size coupled with inherent simplicity, operational practicality, zero hysterisis operation, etc.

17.40. Inspection of Printed Circuit Boards (PCB)

Printed circuit boards are the backbone of any electronic equipment. The interconnection conductors are fabricated on PCBs and components are mounted and soldered to get a PCB assembly.

A careful selection of the fabrication process and a good evaluation procedure is essential for reliable PCB. Automatic routing of lines and photoplating of the entire design is done using computer aided design techniques. Usually multilayer PCBs are used because of higher interconnection density and because the design calls for a thermal management and more ground and power lane area. A multilayer is fabricated using thin laminates of 0.2 mm to 0.8 mm size first by printing and etching inner layers. The inner layers and films are prepunched before these operations. Blackening of copper conductor is done and then final lay up of the inner layers, prepregs (semicured epoxy filled glass used for bonding and separation of layers) and outer layers are done on smooth press plates. The whole 'Book' is pressed as per the cycle prescribed for preheat, cure and cooling on heated press machine. After the boding, the multilayer laminates are drilled and processed.

Laminates are inspected for surface uniformity and appearance standards, silk screen adherence tests, punchability and machinability, peel strength, insulation resistance and solder resistance using ASTM test methods or IPC test methods.

After the boards are inspected for any major defect, they go for cleaning and pattern plating. A properly cleaned surface ensures good adhesion of plating. The cleaned boards are plated copper to get the minimum 25 μm thickness in acid copper plating bath. The boards are then tin plated. The thickness built in both operations are monitored.

The plated boards are stripped of the photo resist in 1—2% caustic solution and then inspected. The etching is done on a good spray etching machine with conveyor control, temperature control and spray control. Uniform spray and neutralisation after that are important to get a board free from undercut or extra copper.

Etched boards are inspected and then taken for stripping of tin. The boards are thoroughly cleaned and a black oxide coating is given. Stencils are prepared for solder masking each side using a five star film by exposing and developing in hydrogen peroxide. The stencil is fixed on screen mesh and then aligned. UV ink or a thermal ink is used for solder masking. After curing one side the other side is printed. Exposing, developing and then UV curing will complete the solder masking process. This is followed by edge gold plating and cutting. Routing machines are used for profiling a PCB. After a final cleaning and a thorough inspection using many inspection equipments, the boards will be ready for shipment.

The evaluation of PCB is standardised by the customer. The incoming inspection of commercial boards are normally followed using a checklist, which lists number of *A* and *B* defects. *A* defects include critical and major defects that affect the function of the board and likely to create hazardous condition for individual.

B defects are minor ones which may not impair function but may degrade the board.

Acceptable quality level is 1% for *A* defects and 2% for *B* defects.

Commonly used instruments for inspection of PCBs are :

1. Standard caliper with a resolution of 0.1 mm.
2. A large caliper with a span of at least equal to the largest board dimension and a resolution of 0.1 mm.
3. A micrometer with a resolution of 0.01 mm and a throat of app. 50 mm.
4. A 10X magnifier with graticule having a resolution of 0.1 mm.
5. A set of cylindrical plug gauges 0.5 to 1.5 mm in steps of 0.05 or 0.02 mm and 1.5 to 6.3 mm in steps of 0.1 mm.
6. A set of feeler gauges in steps of 0.1 mm.
7. A 2X magnifier with built in illumination.
8. A stereo microscope or microscope with magnification upto 100X.
9. A magnifier with a special optical system allowing a quick inspection of the entire hole wall.
10. Meg Ohm meter for measuring insulation resistance.
11. A light table with tilted work plane.
12. A four terminal measuring instrument for measuring copper thickness.
13. Equipment for preparing microsections.

The inspector carries out initial inspection, base material inspection, main warp direction (if specified), thickness of board with a vernier at four different places.

The edge connector boards must be strictly checked for the thickness. Warp and twist must be within the limits otherwise unsatisfactory results will be achieved in mass soldering, contact mating problems in plug in boards, and overstressing of components after soldering.

Warp and twist are inspected using the feeler gauges and rulers.

Some of the defects in the base materials are :

(*i*) **Measling.** It is caused by small cavities where the fibres of the glass cloth intersect.

(*ii*) **Crazing.** Merging of measle spots is called crazing. It is rated as A—defect because electrolyte can penetrate inside crazing.

(*iii*) **Blistering.** It refers to local swelling and separation (delamination) of the base material, either between some of the glass cloth layers or between the base material and the copper foil. The copper pad must not be lifted in the blister area for hand soldering. For mass soldering it is classified as A defect.

(*iv*) **Delamination.** It refers of the widespread separation between the individual glass cloth layers of the base material. It is also referred as A—defect and is because of epoxy starvation in the glass cloth.

(*v*) **Haloing.** It is a light area at machined areas of the board (as a light ring around a hole).

(*vi*) **Weave exposure.** It refers to uncovered glass weave in the surface of board. This defect is classed as 'A' defect because laminate can absorb chemicals used for processing.

(*vii*) **Incomplete curing : (A defect).** It can be checked by H pencil by applying movements of 1 cm at 45 deg. Penetration of pencil leads to rejection.

(*viii*) **Surface appearance.** Dents and pits, scratches, undesired metal particles etching faults, contaminated surface, salt residues on the board.

(*ix*) **Insulation resistance.** For 100 mm conductor length with spacing 0.3 mm, the resistance should be at least 500 Mohms.

(*x*) **Pattern Inspection** is done by laying a positive and then a negative film on top of the solder and component sides, and viewing the board to see possible short circuits between conductors and solder pads when using the positive film and missing pads, conductors or parts of them when using negative of the master.

A projector is also used, and the pattern projected at a suitable magnification (10 X) on a Matte glass screen.

(*xi*) **Short circuits** are cause for rejection.

(*xii*) **Metallic growth.** This can occur due to storage at temperature above 50 deg. C and humidity > 80% but solder masking and reflowing reduces this occurrence. Pure tin is more susceptible but addition of even small percentage of lead prevents this occurrence.

(*xiii*) Broken conductors : (cause for rejection).

(*xiv*) Displacement of edge of pattern :

(*xv*) **Edge definition.** This is defined as the unevenness of the track.

(*xvi*) **Nodules.** Outgrowth along the pattern edges are usually caused by overplating or by too large a current density during plating, poor hole drilling or contaminated plating bath. Large nodules are classed as A defects.

(*xvii*) Projection and indentation along rim of solder pad and along conductor edges.

(*xviii*) Pin holes and voids in the pattern.

(*xix*) Broken annular ring (eccentrically drilled solder pads).

(*xx*) Overhangs are edges of conductor length that are set loose due to overetching or starting with a higher thickness of base copper.

(*xxi*) **Pattern adhesion defect (classed as A defect).** This may be due to more undercut. It can be checked by peel test.

(*xxii*) Inspection of drilled holes in plates. Rough and gauged holes should not be accepted. Voids inside hole if greater than 2% of holes, and 5% of area in any hole and any void at pad to barrel transition are causes for rejection. Cracks are also very dangerous and this is indicated in resistance measurement.

(*xxiii*) **Protruding glass fibres.** A drill bit can catch a glass fibre and tear it out along part of the hole circumference resulting in a glass fibre protruding into the hole. This fibre receives excessive deposit during plating leaving less plating in adjacent area. This is classed as A defect.

(*xxiv*) Levelling effect is the thermal craking which shows up deep pits in glass fibre weaving. This is normally visible only in microsectioining.

SOLVED PROBLEMS

Problem 17.1. *What means are incorporated in engineering microscope to make them suitable for metrology functions ?*

Solution. (*a*) The primary function of microscope is to magnify and present the enlarged view of the object area, either in its contours, or as a surface image.

(*b*) Referencing or aiming (achieved by incorporating index lines on a transparent graticule inside the microscope tube). With its aid the magnified image of the object can be compared.

(*c*) Staging-means for holding the object and for moving it linearly and/or rotationally on a calibrated scale.

(*d*) Illumination.

Problem 17.2. *What are the two basic*

(*a*) *Types of illumination sources used in microscopes.*

(*b*) *Which of these two system produces a sharper definition of contour.*

Solution. (*a*) These are

(*i*) The transmitted-light system with substage source, resulting in a silhouette image.

(*ii*) The incident-light system, producing an image by light reflection.

(*b*) Silhouette image provides a sharper definition of contour.

Problem 17.3. *What do you understand by referencing errors which could arise in measurement of cylindrical objects with highly reflective surfaces, using microscope.*

Solution. When light source aperture is large, stray rays may glance from the object surface above its cross-sectional plane, thus transposing the observable silhouette line closer to the axis of the object. This may result in object being observed smaller by as much as 0.05 mm.

The effect of this observation can be reduced by adjusting diaphragm provided in front of the light source.

Problem 17.4. *What are the various methods of referencing in microscopes.*

Solution. (*i*) Referencing by optically indicating mechanical contact

(*ii*) Referencing by grazing incidence light beam

(*iii*) Referencing by double-image sighting.

Problem 17.5. *What are the important measuring characteristics with optical projectors.*

Solution. Optical projectors have following favourable characteristics.

(*a*) Since no physical contacts are made, no distortion of probe or workpiece is possible.

(*b*) Projected image can be observed by many inspectors simultaneously, permitting analysis by a group.

(*c*) Several dimensions and form characteristics can be compared and evaluated in a single setting.

(*d*) Complicated parts requiring large number of dimensions to be checked can be inspected quickly by comparing with standard profiles.

(*e*) Direct measurement of various length dimensions is possible by measuring image with a glass scale with etched graduations.

(*f*) Angles can be measured on image on screen by using regular drafting protractors.

(*g*) No wear out problems and as such no regular checking/calibration is required.

(*h*) The use of work holding and locating fixtures enhance the usage capabilities further.

(*i*) The tracing of optically obstructed surface elements, and the application of optical cross sectioning, provide means for accurate measurement of difficult dimensions.

(*j*) Surface characteristics by light reflection is possible.

Problem 17.6. *What are the potential applications of optical projectors for precision measurements and inspection ?*

Solution. Optical projectors are adaptable for a wide variety of applications. Some of these are :

(*a*) Producing image of object on screen by projection (shadow silhouette) or reflection (mirror image of surface) or combination of both, and comparing it with standard profiles engraved.

(*b*) Inspection of surface properties, general contours and limit gauging.

(*c*) Inspection by comparison with master charts (screen charts or overlay charts).

(*d*) Inspection by direct measurements on screen image using graduated rules/glass scales for linear measurements, tool maker protractors for angular measurements, transparent template for radius, etc.

(*e*) Inspection with measuring devices built into the optical projector (co-ordinate tables fitted with micrometer heads ; screen projectors for angular measurement).

(*f*) Inspection with the aid of fixturing and special attachments (helix angle adjustment of table to project thread forms normal to contour plane ; optical sectioning with special illumination).

Problem 17.7. *In mechanical gauges, the repeat accuracy is influenced by gauging force and associated deflection which is absent in case of optical measurements. On what factors the repeat accuracy depends in optical measurements ?*

Solution. The attainable repeat accuracy while measuring on optical projectors depends on the (*i*) precision of point (to establish reference positions, accomplished by bringing into coincidence one of the screen hairlines with the selected element of image) and (*ii*) accuracy of the distance measurement on screen.

Problem 17.8. *The radius of a concave surface on a component was measured by fastening it an a milling machine and taking three readings at different locations on longitudinal and cross slides as under fitted with micrometer heads.*

	P_1	P_2	P_3
On longitudinal slide	*0.25*	*44.92*	*92.52 mm*
On cross slide	*0.35*	*3.78*	*16.17 mm.*

Determine the radius. Also explain some other method of determining the same.

Solution. The equation of a circle on rectangular co-ordinates is

$$x^2 + y^2 + 2ax + 2by + c = 0$$

where x and y are the co-ordinates of any point on the circle w.r.t. X and Y axes and a, b and c are constants.

The radius of the circle is then $\sqrt{a^2 + b^2 - c}$

It will be a appreciated that the work of calculation can be much simplified if one point say P_1 be converted to (0, 0) origin by suitably shifting the axes. In that case c will be zero and only two equations will have to be solved.

The new co-ordinates of P_1, P_2 and P_3 will then be and

$$P_1(0, 0), P_2(45.92 - 0.25, 3.78 - 0.35)\ P_3(92.52 - 0.25, 16.17 - 0.35)$$

or $$P_1(0, 0), P_2(45.67, 3.43) \text{ and } P_3(22.27, 15.82)$$

The two equations then become

$$x_1^2 + y_1^2 + 2ax_1 + 2by_1 = 0 \quad ...(i)$$

or $$x_2^2 + y_2^2 + 2ax_2 + 2by_2 = 0 \quad ...(ii)$$

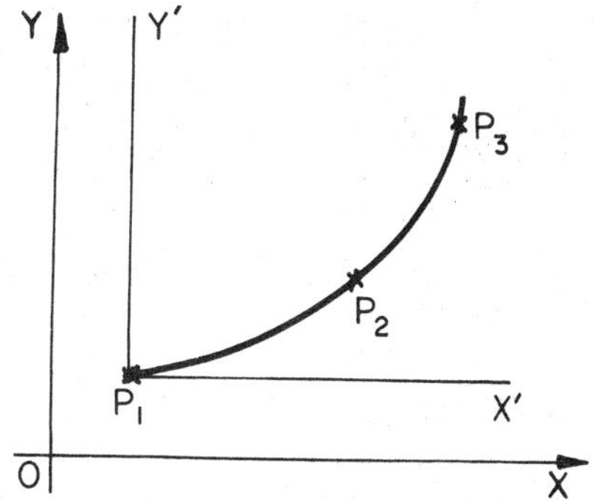

Fig. 17.44

Multiply (*i*) by y_2 and (*ii*) by y_1

$\therefore$ $$x_1^2y_2 + y_1^2y_2 + 2ax_1y_2 + 2by_1y_2 = 0$$

and $$x_2^2y_1 + y_2^2y_1 + 2ax_1y_1 + 2by_1y_2 = 0$$

Subtracting these two equations,

$$2a(x_1y_2 - x_2y_1) = x_2^2y_1 + y_2^2y_1 + x_1^2y_2 - y_1^2y_2$$
$$= y_1(x_2^2 + y_2^2) - y_2(x_1^2 + y_1^2)$$

Similarly by multiplying (*i*) by x_2 and (*ii*) by x_1 and substracting

$$2b(x_1y_2 + x_2y_1) + x_2(x_1^2 + y_1^2) - x_1(x_2^2 + y_2^2)$$

Solving these $R = \sqrt{a^2 + b^2} = \dfrac{\sqrt{(x_1^2 + y_1^2)(x_2^2 + y_2^2)[(x_1 - x_2)^2 + (y_1 - y_2)^2]}}{2(x_1y_2 - x_2y_1)}$

Substituting the values, we have

$$R = \frac{\sqrt{\{(45.67)^2 + (3.43)^2\}\{(92.27)^2 + (15.82)^2\}[(45.67 - 92.27)^2 + (3.43 - 15.80)^2]}}{2(45.67 \times 15.82 - 92.27 \times 3.43)}$$

$$= \frac{\sqrt{(215.0 + 11.75)(8500 + 250)[(46.60)^2 + (-12.39)^2]}}{2(740 - 317)} = \frac{\sqrt{(2191.75)(8750)(2170 + 153)}}{2(423)}$$

$$= \frac{\sqrt{(2191.75)(8750)(2323)}}{846} = \frac{1000 \times \sqrt{45000}}{846} = \frac{1000 \times 100 \times 2.12}{846} = \mathbf{251\ mm.}$$

Another method of determining radius of a circle passing through three points is by finding intersection of the normal bisecting two chords of the radius.

If $P_1(x_1, y_1)$, $P_2(x_2, y_2)$ and $P_3(x_3, y_3)$ are three points, then centre of line between P_1 and P_2 is $\left(\dfrac{x_1 + x_2}{2}, \dfrac{y_1 + y_2}{2}\right)$ and slope of bisecting normal $= -\dfrac{x_1 - x_2}{y_1 - y_2}$.

Problem 17.9. *Write brief note about (i) Coordinate measuring machine equipped with a laser probe and (ii) Virtual measuring system.*

Ans. (*i*) A CMM equipped with a laser probe can convert a part of physical model into a digitised file. Such a file can be compared with other files and can be manipulated by designers to improve quality. Manufacturers can verify that each finished part measures exactly as designed.

(*ii*) Virtual measuring system uses a microscope system to examine an electronic replica of the surface texture of part. Such a system is non-contact 3-D surface measurement system and provides image of the surface. The images are processed on a PC using vertical scanning interferometry and vision analysis software to produce 2-D profiles, 3-D plots and contour plots. It generates statistics for average roughness, average profile height, reduced peak height, core roughness depth, reduced valley depth and a number of other parameters. It also determines the depth, spacing and angle of groove in a honed surface. Optical probe for examination of a cylinder bore can be rotated 360 degrees and moved vertically along the cylinder wall.

TYPICAL EXAMINATION QUESTIONS

1. Sketch the optical path of a profile projector and discuss use of such projectors.
2. What is toolmaker's microscope and give more typical application of its use.
3. Describe with sketches the working principles and the applications of the following :
   (*a*) Toolmaker's microscope. (*b*) Angle Dekkor. (*c*) Solex Pneumatic Gauge. (*d*) Sine Table. (*e*) Autocollimator. (*f*) Universal Measuring machine. (*g*) Projection Comparator. (*h*) Optical Dividing Head. (*i*) Gear Pitch Testing Machine. (*j*) Gear Tester.
4. Describe, with the help of a sketch, the working principle of any instrument used for the inspection of the profile and size of a cam. How will you check it on this instrument ?

18

Quality Control and Quality Assurance

18.1. What is Quality ?

Everybody now-a-days talks of quality, thinks about it and insists for it while purchasing any article. To buy any thing we all look for good quality at a reasonable price. Each of us may have a different notion about the quality of the same product. We want as many of the qualities in a product as possible. Quality product or quality service should be provided at an affordable price.

Question is how we can define quality ? In very simple words we may say that quality means "fitness for use". So when we talk of quality we mean that the product should be fit for use and it is the user who judges and decides what qualities he wants in the product. Thus quality is customer's determination and the quality products are those which totally satisfy the needs and expectations of customer in every respect on continuous basis.

All human institutions, such as factories, schools, hospitals, temples, governments, etc. exist to provide products or services to human beings. What is *essential* in these products or services is that *they should be fit for use* and provide excellence in service. Such products should be well designed with functional perfection and be produced right in the first instance. If a hospital is well equipped but not serving up to your expectations then the service is not good or the quality of service is poor. If a factory produces the cars, which give lot of trouble and do not run properly then the car is not fit for use in the real sense or you say the quality of the car is poor. So, whether it is a product or a service it should be fit for use or should be of good quality.

According to ISO : 8402, quality is defined as the totality of features and characteristics of a product or service that bear on its ability to satisfy stated or implied needs. There has been a wrong notion in minds of a few that quality can be achieved at higher cost. On the other hand experience has shown that it is poor quality that implies waste of material, effort of labour, equipment utilisation and thus results in higher cost. One must remember that quality has to be created.

18.2. Quality Characteristics

Now, let us examine a little more in detail as to what quality is. The quality of a product comprises several engineering and manufacturing characteristics which go to make the product meet the ultimate performance expectations of the designer/customer.

What is a characteristic ?

A dimension is a characteristic.

A chemical property is a characteristic.

A sensory property is a characteristic.

All these characteristics contribute towards the quality of the product.

So we can say that appearance of a product is a quality characteristic of the product. Similarly performance, length of life, dependability, reliability, durability, maintainability, the smell, taste, feel, sound, etc. are all characteristics which go to build the quality of the product. Each of these characteristics is like a building block that goes into the construction of, say, an arch, which may be considered to be quality.

Now, for any product to be successful, apart from quality the price also should be reasonable from the buyer's point of view. Today, quality is becoming significant factor in the business strategy planning process.

18.3. Facets of Quality

The quality characteristics can be grouped into 4 categories.

(*i*) Quality of design. (*ii*) Quality of conformance. (*iii*) Availability. (*iv*) Customer service.

(*i*) *Quality of Design*. The design must meet all the requirements of the customer. The designer has to take into account the question of appearance, reliability, maintainability of the product.

The designer should aim at a design which fulfils all the qualities or almost all qualities desired by the customer. He should also aim to make a product which would become competitive.

The designer should be fully aware as to what the requirements of the customer are, what characteristics are to be put in to make the product acceptable to the customer, what are the service conditions, etc. He should be able to translate all these into clear specifications and drawings. There should be no ambiguity in the information he supplies to shop, etc.

(*ii*) *Quality of Conformance.* The degree or extent to which a product actually conforms to in accordance with the specification and drawings is called the "quality of conformance."

(*iii*) *Availability.* This includes such characteristics as reliability, maintainability and length of life. All these indicate as to what extent the product will be available for use.

Reliability of a product may be taken as the extent to which it can give continuous service. If a product breaks down frequently it is not reliable as you are not getting its service continuously as you want.

A product should also be maintainable. It should lend itself to easy maintenance within the least possible time.

If a product is such that to take off an external assembly, one has to dismantle the whole machine then this is not an easily maintainable product.

If a machine was supposed to have lasted for 20 years, but it had to be scrapped in 10 years, the life of the machine is less and its availability has come down very much.

(*iv*) *Customer Service.* This indicates the extent to which the customer can get assistance from the manufacture or to keep the product in continuous use. This service is an after-sale function. Replacement of defective components, conveying sufficient technical knowhow to customer for maintenance through maintenance manuals, leaflets, etc., deputing experts to customers place for commissioning and attending to breakdowns, training the customers staff for proper maintenance of the product, etc. are some of the services given by the manufacturer.

18.4. Quality—A Business Strategy

Quality results in creation of goodwill and offers a competitive edge. Today's strategic business plans are based on maximisation of profit, adjusting to market changing needs in shortest time, reduction in cycle time, and lastly optimum utilisation of resources. These strategies need to be well built into the business plan and can be achieved through proper quality management programmes.

In the present age of liberalisation and globalisation, there is no choice but to produce quality goods and quality services in order to survive. Market compulsions driven by customer needs and the complexity of products force the manufacturers to adopt total quality management approach. Newer quality management practices have to be adopted to achieve substantial reduction in costs to maintain quality.

18.5. Quality Control

Generally, inspection, which is the process of sorting good from bad, is considered as a form of quality control. But Quality Control and Inspection are different.

Conceptually, Quality Control aims at prevention of defects at the very source, relies on effective feedback system and corrective action procedure. Following examples illustrate the point :

Acceptance of material after subjecting them to inspection at inward goods store is inspection concept. *QC* concept is to take early step to ensure that the material to be supplied is going to be manufactured to the plant's requirements so as to ultimately eliminate the inward goods inspection.

Discovering that the components produced are not as per the required dimensional specifications due to wrong tool or improper setting after the production is over, is inspection concept. To prevent the issue of a wrong tool and to have set up approval procedure prior to starting continuous production and keeping a simple check during production, are *QC* concepts.

To reprimand or to inform an operator about his poor performance at the end of the shift it essentially an inspection approach. On the other hand to alert the operator against poor performance prior to his commencing the work is an effective prevention oriented activity.

Quality control uses inspection as a valuable tool. In quality control programme inspection data are effectively used to take prompt corrective actions to arrest the incidence of defects in the job on hand and to plan the prevention of occurrence of the same or similar defects in future.

Technically, a well organised Quality programme concerns itself with studies aimed at defect prevention—no matter where the defect is originated—design, purchase, sales, despatch, etc. Process capability analysis is used to ensure that the facilities available can meet the required design standards and customer quality requirements, initiating preventive and corrective actions.

Organisationally, it has to be recognised that in achieving the required quality, every functional group in the plant—design, purchases, manufacturing, despatch, sales, etc.—has its own unique contribution to make. Therefore each functional group has to have within its own sphere a charter of quality programme suitably structured to discharge well all the responsibilities it has towards product quality. Thus, it becomes obvious that quality programme has to be a plant-wide activity unlike inspection which is essentially a localised activity confined mostly to the manufacturing shops.

Further, organisationally, it has to be recognised that the best principles of control are 'self-control'—control of quality by operator instead of 'controlling him for quality'—consisting of building a system where control is delegated down to worker level, where the action takes place and such a principle of the quality programme must be built in each functional group of the plant.

18.5.1. Quality Control Defined. Quality control may be defined as under by various management experts.

(*i*) Alford and Beatty define *QC* as "In the broad sense, quality control is the mechanism by which products are made to measure up to specifications determined from customers, demands and transformed into sales engineering and manufacturing requirements, it is concerned with making things right rather than discovering and rejecting those made wrong. Quality control may be defined as that industrial management technique by means of which products of uniform acceptable quality are manufactured".

(*ii*) Bethel, Alwater and Stackman define *QC* as "Systematic control of those variables encountered in a manufacturing process which affect the excellence of the product. Such variables result from the application of materials, men, machines and manufacturing conditions only when these variables are regulated to the extent that they do not detract unnecessarily from the excellent of the manufacturing process as reflected in the quality of the finished product, can the control of quality be said to exist."

According to Deming, quality should be aimed at the needs of the customer, present and future.

Juran defines Quality as

(*a*) The degree to which a specific product satisfies the wants of a customer.

(*b*) The degree to which a specific product conforms to the design or specification.

According to Broh, quality is the degree of excellence at an acceptable price and control of variability at an acceptable cost.

The European organisation for Quality Control defines that "Quality of a product is the degree to which it meets the requirements of customer".

While some of the attribute which contribute to the product quality are measurable, some like appearance, quality are not measurable. The advantages that a plant derives by pursuing the activities for bringing about product quality are—

1. Retaining existing customers and gaining new ones 2. Enhanced prestige and reputation :

18.5.2. Objectives of Quality Control. The main objectives of a quality control programme in an industry are :

(*i*) Evaluation of quality standards of incoming material, product in actual manufacture, and of outgoing product.

(*ii*) Judging the conformity of the process of the established standard and taking suitable action when deviations are noted.

(*iii*) Evaluation of optimum quality obtainable under given conditions.

(*iv*) To improve quality and productivity by process control and experimentation.

(*v*) Developing procedures for good vender-vendee relations.

(*vi*) Developing quality consciousness both within and outside the organisation.

(*vii*) Deploy trained and qualified manpower, and motivate them.

(*viii*) Make use of automated processes which have repeatability and thus high consistency and uniform quality of products.

18.6. Areas of Application of Quality Control

The broad areas of application of quality control are incoming material control, process control and product control.

The problem of incoming material control may be defined as procuring and stocking materials at economic levels best suited for producing the established quality standards. From the quality control point of view, procuring the right quality of materials assumes great importance. In any manufacture, it is essential that acceptable quality standards have to be evaluated and maintained in all the stages of manufacture ; and this may not be met economically, if the incoming material, quality of raw materials are purchased or sub-contracted from outsiders. Incoming material may be defined to include materials, parts, assemblies, etc., moving from one department to another, and shipping of the finished product to the consumer. The material may be in the solid, liquid, or gaseous form. Examples would be steel rods, flats, castings, machined parts, wires, sheets, powders, acids, oils, cotton, ore, electronic equipment, lamps, glass bottles, packing materials etc.

The need for proper incoming material control from quality control point of view is fairly obvious. Poor quality of the materials accepted will not only affect the quality, but also production, through increased rejections, downtime on machines, idle time for men, and/or longer time for processing. Consider the case of a factory producing sledge hammers from flats purchased from an outside firm, which had to reject a large number of almost finished hammers due to the poor quality of the flats like cracks, blow holes, and cavities, which were not apparent at the time of the starting of the operations. In addition to the rejections, the factory had to pay the operator extra production bonus on these items, since it was not the fault of the operators. Similar situations may arise in machining operations, castings purchased from outside suppliers or from the foundry department in the same factory. Component parts purchased may not fit properly on the assemblies. Again, there have been situations where wrong quality materials, sometimes at higher cost, have been purchased by firms, due to lack of understanding as to what should be purchased. Ultimately, one may also have to face the problem of customer complaints in actual field service due to lack of good incoming control procedures. From these considerations the object of incoming material control programme may be summarised in the statement of a quality control manager of a firm : "We do not consider a problem cleared until the suppliers, process has been altered to our satisfaction, quite apart from delivery of corrected components. For this reason liaison is maintained not only with the supplier, but also with several departments in the factory for any latent defects being shown up while processing".

In regard to the incoming material problem a few basic facts have to be kept in mind. First, this is a problem of dealing with outside suppliers, who are generally very large in number. Secondly, it may not always be possible for all items required to select vendors at will. Hence the approach involves much more than mere prompt payment and usual business courtesies. In addition, the following points have to be recognised :

(*i*) A basic quality control approach to qualify problems is that quality is built into the product while manufacture is continuing. This principle should also be recognised in incoming material control procedures and its influence on vendor-vendee relations with respect to quality and selection of vendors. The procedures should aim at establishing good relations between vendor and vendee by suitable quality evaluation of material based on facts and through good communication, assisting the supplier in improving the quality, if necessary.

(*ii*) Material specifications should be prepared before placing orders. Obviously, the supplier can not be held responsible for the quality of material supplied, if the requirements are not clear.

(*iii*) It would be highly desirable if one could do without incoming inspection of materials. (After all, this would be duplicate inspection as somebody else is expected to have checked the quality before supplying). Incoming material control procedures ultimately should aim at this objective and at least in many situations reduce the inspection to a minimum providing a definite quality assurance.

From the foregoing discussion, it may be said that a successful incoming material control programme is characterised by a good vendor-vendee relation ; this may be achieved by documented quality specifications, quality evaluation, communication, mutual discussion and assistance, and vendor rating. Some of the popular techniques are—vendor selection procedures, sampling plans for quality evaluation, and vendor rating plans.

Process control refers to the procedures adopted to evaluate, maintain and improve quality standards in several stages of manufacture ; it takes into consideration the economic aspects of the situation, *viz.* the quality/cost ratio. Since the control exercised at several stages in the manufacture of a product helps in building the desired quality in the finished product, process control procedures and techniques may be looked upon as ensuring quality economically.

A process may be considered satisfactory as long as it continues to produce items within specifications. It is thus necessary to watch the process continuously to see that it behaves in the way we want it to behave. Process control techniques aim in evaluating the process standards in terms of dimensions, scrap, rejections, rework, etc. by studying the process behaviour and helping in maintaining the standards wherever they are found suitable. When they are not suitable, corrective action is initiated to bring the process to the desired standards. This approach assists in building quality into the product, and it is only then that control is exercised over the process.

The implications of process control may be appreciated by analysing the problem in some detail. Naturally, this should start with the objectives of process control, which may be stated as under :

(*i*) The knowledge that the process is in a state of control. This is essential to see that items produced conform to the requirements as the manufacturing is continuing. This means that the process should be controlled at the right level and variability so that the stipulated specifications are met.

(*ii*) Whenever deviations occur from the standard established, it must be found out as quickly as possible, so that corrective action could be instituted.

(*iii*) The ultimate object is not merely to detect trouble, but also to find it. This means building up an efficient information system.

(*iv*) An efficient system of process control would be one where the time lag between detection of trouble and taking action is the least.

In devising process control procedures, the above objectives have to be borne in mind. The various steps involved are briefly discussed below :

Planning : The determination of quality characteristics of control, inspection, equipment and procedures, development of criteria for judging the process of control, making and recording measurements and observations that would answer questions implied in the objectives.

Analysis : Making use of the information complied for initial study and also on a continuing basis.

Process Engineering : The study made to eliminate the causes of poor quality or to initiate corrective action as and when required, based on the analysis. This would involve the association of process behaviour patterns obtained from measurements to the technical and operating conditions of the process.

Process Capability : Determination of the 'best' the process can do under given conditions.

Product control refers to the schemes for evaluating the quality of the out-going product including finished products in stores and warehouses, analysis of information and data obtained from the field where the product is in actual use, and feeding back to the production processes, as an endeavour for constantly attempting to improve upon product quality. Here again the object would be to build on information system regarding the final product, which may come from quality audit studies, complaints, analysis, self life studies, reliability studies, opinion surveys etc.

The approach to the solution of quality problems involves three aspects :

(*i*) **Engineering.** The creation and development of a product is basically engineering ; the development of quality evaluation through improved inspection procedures is also engineering ; again, the knowledge of the causes of defects and sub-standard products and their rectification is engineering.

(*ii*) **Statistical.** The concept of the behaviour of a process, which as brought in the idea of 'prevention' and 'control' is statistical ; building an information system to satisfy the concepts of 'prevention' and 'control' and improving upon product quality, requires statistical thinking.

(*iii*) **Managerial.** The competent use of the engineering and statistical technology is managerial ; the creation of a climate for quality consciousness in the organisation depends upon the policies and practices of the management ; the effective co-ordination of the quality control function with those of others is managerial.

The reputation and survival of a company will depend upon its ability to manufacture goods of acceptable and uniform quality products at a given price, and do continue to maintain the established standard of quality. This is actually what creates better sales for the products. The quality of the product, thus, should always be thought of in terms of its cost or price, maximising the ratio, taking into-consideration the consumers' requirements. Maximization of the ratio can be done in several ways, *viz.*, better quality at the same cost, or if the quality is satisfactory to reduce the cost per unit for the same quality or improve the quality, at the same time reducing the cost.

The above point brings to the fact that the quality control function in an organisation is a continuous, not an *ad-hoc* one. It also helps in specifying the objectives of a quality control programme. First, the management must have the assurance that the quality specified has been shipped to the customers this may be termed as quality assurance and secondly, this specified quality has been achieved at a minimum cost.

Lastly, quality activities are continuous and perpetual processes which must go on all the time, starting from start to finish.

18.7. Quality Control Concepts

18.7.1. Process. A process is defined as : "A set of interrelated work activities that are characterized by a set of specific inputs and value-added tasks that produce a set of specific output".

A process can be contained within a functional organization, or it can span several functional organizations.

18.7.2. Charts. The charts are the pictorial tools for studying the processes and situations and for planning improvement. The use of charts would result in standardisation, simplification, variety reduction, and optimisation of process average.

18.7.3. Quality Control. It is defined as : "Assurance of conformity of the manufactured products or the processes to established specifications by systematic observations, inspections or tests and by regulation of materials, workmanship or other influences.

18.7.4. Quality Assurance. It is defined as : "All those planned and systematic actions necessary to provide adequate confidence that a system or component will perform satisfactorily in service".

Difference between Quality Control and Quality Assurance. Quality control is related to actual inspection or test in order to determine whether a particular component or process is acceptable or not : whereas Quality Assurance encompasses numerous activities such as planning the required actions, establishing acceptance criteria, assigning qualified personnel for the process as well as for the inspection etc.

18.7.5. Action Plan to Achieve Quality. Fig. 18.1 shows a popular PDCA cycle to improve quality of any process.

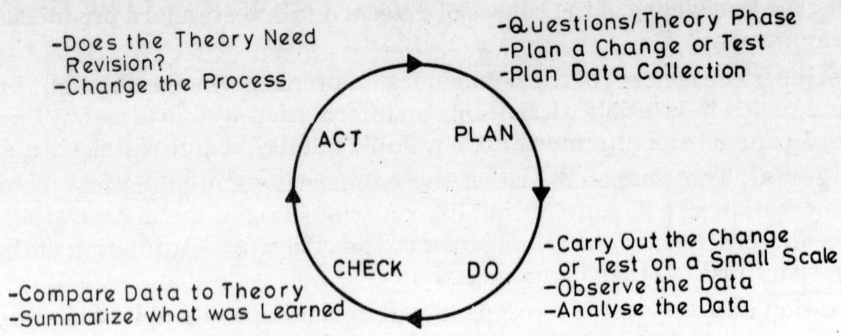

Fig. 18.1. PDCA Cycle (Plan, DO, Check, Act Cycle).

Any process comprises of five elements :

Material, Equipment, People, Method, and Environment. Any variation in any of these 5 elements will have effect on quality. This relationship is shown by a fishbone chart as shown in Fig. 18.2.

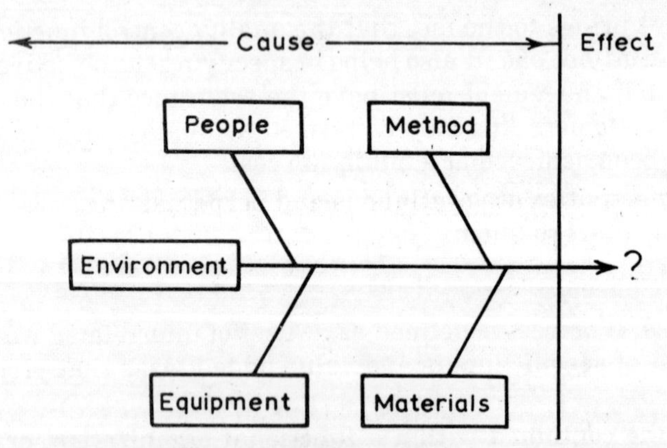

Fig. 18.2. Fishbone Chart.

The various potential causes for all the five elements on fishbone diagram can be identified by following a brainstorming session.

The steps to be followed in this technique are :

(*i*) Give attendees time to individually formulate a list of potential causes (may be even prior to the meeting).

(*ii*) Go around the group and have each person say one of their ideas, continuing this way until everyone's list is complete. It is important that no evaluation should take place at this stage. Every one should be allowed to express freely.

(*iii*) As new ideas come up, add bones to the fish. Don't take a lot of time trying to decide whether the cause is people, materials, methods, etc. It is only important that the idea gets up there somewhere.

(*iv*) Review the final product, clarifying and combining ideas. Make an attempt to prioritize the causes by having everyone select their top three and tabulating the results. The task now is to support the prioritization with data.

Fig. 18.3 shows the cause and effect diagram for failure of a seal and Fig. 18.4 shows the flow diagram for the same.

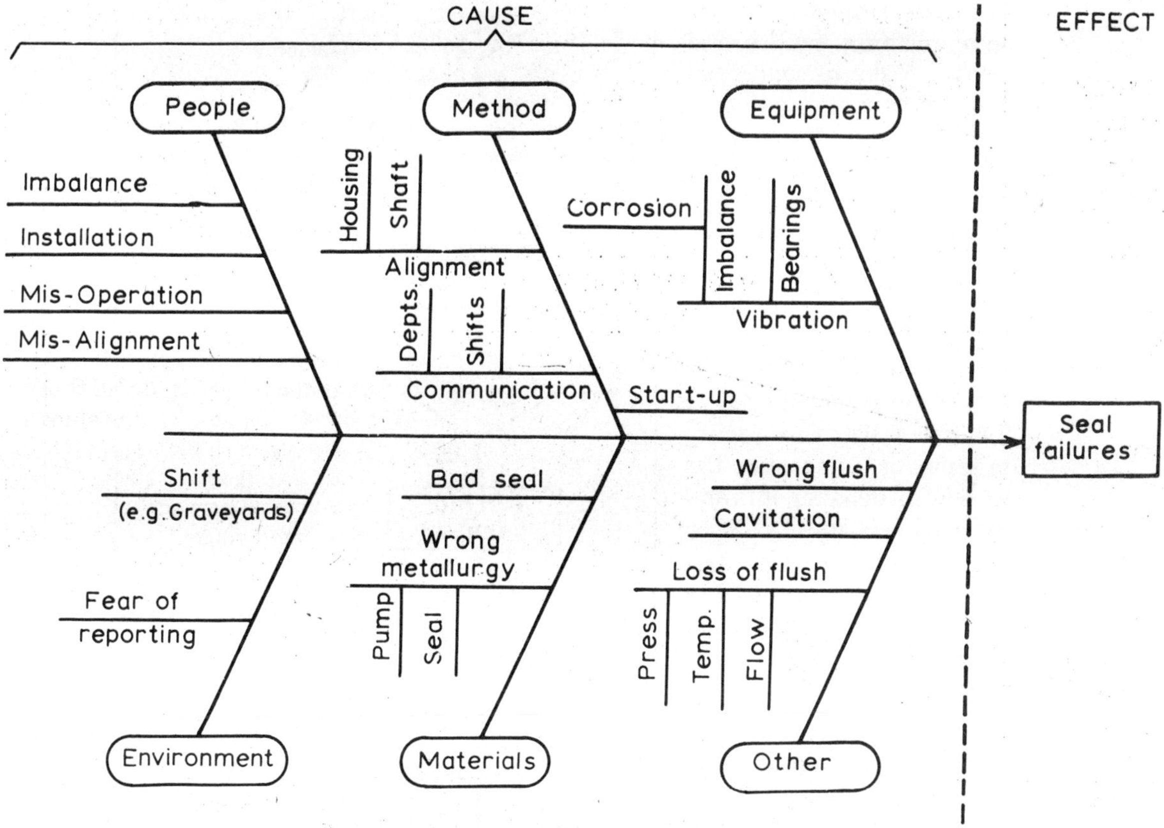

Fig. 18.3. Seal failure cause and effect diagram.

18.7.6. Factors Affecting Quality. The following nine M's directly affect the quality of products and services, and thus these must be thoroughly recognised and dealt with :

(*i*) Material
(*ii*) Manpower
(*iii*) Money
(*iv*) Market for products, services
(*v*) Motivation of employees
(*vi*) Modern information approaches
(*vii*) Management
(*viii*) Machines used
(*ix*) Mounting product needs.

18.7.7. Organising Quality. The quality control function can only be performed effectively and economically if it is organised and managed properly. A properly organised quality control organisation offers the advantages of simplified communication, reduced conflicts in responsibilities and activities, reduced costs, satisfied customers, etc. A well-organised quality control organisation becomes instrumental in producing an item of acceptable quality at competitive price.

Organising for quality involves the following tables :

- Identifying quality related tasks.
- Assigning responsibility for accomplishing quality and related tasks to concerned departments or external agencies.
- Breaking down the overall work into elements, for each department.
- Clearly defining the authority and responsibility for each job.
- Clearly defining the inter-relationship among jobs.
- To accomplish the quality mission by composing the work of all departments and external agencies.

18.7.8. Planning an Organisation. The important guidelines for planning an organisation that would function properly are :

- Organisation levels should be minimum.
- Every person should have clearly defined responsibility and he must be accountable (answerable for his duty).
- The responsibility and authority should be close to the point of action and should be clear throughout the organisational structure.
- A person should receive orders from only one superior authority to avoid confusion and complexity.
- Each component must accomplish its own objective and that of the organisation.
- There should be no overlap between the functional and managing duties.

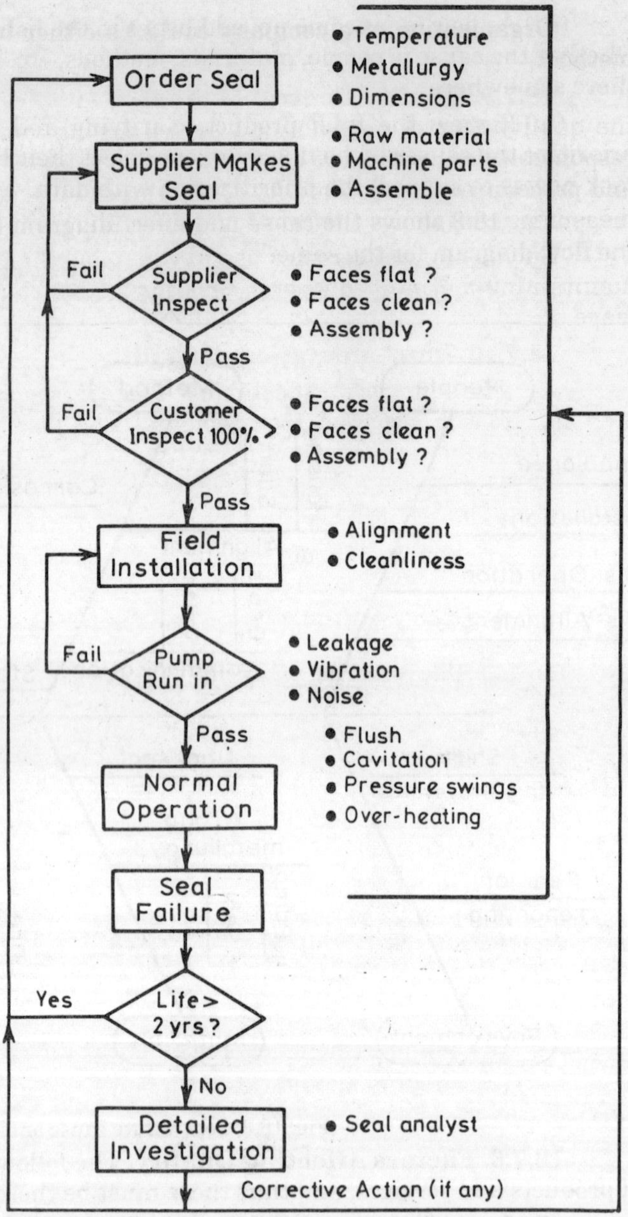

Fig. 18.4. Flow Diagram for Analysing Mechanical seal failure.

An organisation chart should be designed carefully ensuring that it identifies authority and fundamental relationships, assigns responsibilities, identifies strong or weak control, implifies the management function, and improves the communication channels.

Organisation of a company could be either by function, or by project, or by matrix method.

18.7.9. Quality Control Manual. Quality control manual serves as the backbone of the quality control program and serves a reference document. It contains information on responsibilities ; statistical methodology ; personnel ; organisational charts ; quality policies and procedures ; vendor quality control procedure ; quality costs and inspection procedures ; measuring equipment ; defect prevention ; marketing.

Such a manual becomes useful when making quality related decisions. It also helps in maintaining continuity of operation of the quality control organisation even if trained persons leave.

18.7.10. Implementation of Quality

— The supplier must establish a *quality system* to demonstrate that products conform to the specified requirements. The documented quality system should include quality management objectives, policies, organisation and procedures.

— The supplier should delegate the responsibility and authority to appropriate personnel to identify quality problems and provide effective solutions.

— The supplier should appoint a management representative (preferably independent of other functions) who shall have the necessary authority and the responsibility for ensuring that the requirements of the standard are implemented and maintained.

— The quality system should be periodically and systematically reviewed to ensure its continued effectiveness.

— The supplier should establish a procedure for conducting a sufficiently extensive and timely review of the specified requirements.

— The supplier should develop and maintain clear and complete documented instructions that prescribe the communication to meet the specified requirements and the performance of work in design, development, manufacture and installation, which would be adversely affected by lack of such instructions.

— The supplier should develop and maintain records to demonstrate achievement of the required quality and the effective operation of the quality system.

— The supplier should establish and maintain documented procedures for corrective action.

— The supplier should establish and maintain control of design functions.

— The supplier should establish and maintain control of all documentation.

— The supplier should provide control, calibrate and maintain inspection, measuring and test equipment suitable to demonstrate the conformity of equipment and materials to the specified requirements.

— All purchased equipment, materials and services should conform to the specified requirements.

— The manufacturing control is absolutely essential by the supplier. Satisfactory work instructions on following should be made clear ; identification of material, details of operations to be performed, details of tools and test equipment to be employed, checks/tests to be done and calibration procedures, inspection procedure details, environmental conditions to be maintained during operation and inspection, criterion for selection/rejection, sampling techniques etc.

— The supplier should establish and maintain control of all special processes (like welding, casting, forging, forming heat treatment etc.), which can be done by assessing temperature and humidity cycling, vibration, radiograph, magnetic particle inspection, penetrant inspection, ultrasonic inspection ; spectrographical analysis, etc.

— The supplier should perform all tests on the finished product in support of evidence of full conformance to specified requirements.

— The sampling procedures used should be in accordance with the specified requirements and agreeable to purchaser.

— The supplier should establish and maintain procedures for controlling materials that does not conform to the specified requirements.

— The supplier should establish and maintain a system for identifying the inspection status of material during all stages of manufacture. Some suitable method be adopted to identify the inspected and non-inspected parts.

— The supplier should establish and maintain a system to control packing, preservation and marking process to ensure conformity to specified requirements.

— It should be ensured that adequate personnel who have to handle contracts, designs, manufacturing, installation quality, erection functions are properly trained to perform their duty effectively.

18.7.11. Quality Assurance System. The objective of a quality assurance system is to maintain the necessary standard of quality. The important components of the quality assurance system are :

— Developing personnel with adequate understanding and training.

— Monitoring quality assurance programme of suppliers.

— Checking the accuracy of test/measuring equipment.

— Evaluating and controlling the quality of product in the field.

— Establish a feedback link about quality related information to the management.

— Evaluating, planning and controlling product quality.

— Conducting special quality studies and managing the entire quality system.

— According for the quality and reliability needs during the product design and development phase.

18.8. Quality Policy

Quality policy refers to the basic principles which are to guide the actions of the company. Regarding Quality functions to be carried out in the company, it is essential to enunciate the company's Policy on the following basic and important questions :

(*a*) The standard of outgoing Quality—whether the Policy is to strive for Quality leadership or price leadership or no leadership at all ;

(*b*) The pattern of customer leadership including guarantees of the product, the extent of rigidity or flexibility in settling customer's claim for defectives ;

(*c*) The extent of leadership in meeting customer Quality needs ; and

(*d*) The pattern of vendor relations.

18.8.1. Documenting the Quality Policy. It is important to document the Policy. Putting the Policy in writing has the following advantages :

(*a*) Compels those responsible for policy formulation to think on the problem to a depth never before achieved ;

(*b*) Renders the communication authoritative and uniform ; minimises misinterpretation ;

(*c*) Provides a basis for management by agreed objective rather than by crisis or opportunism ;

(*d*) Allows for auditing the performance against the Policy.

The experience in this regard in several establishments in different parts of the country shows that the absence of a clear Quality Policy is the main limitation to a company's progress in solving major Quality problems. It is generally observed that middle management, in general, is aware of the need for a clear Policy ; but does not take initiative because of a feeling : the 'setting Policy', is a Top Management responsibility. It is a mistake for Middle management to sit back on any such theory. Instead, Middle Management should draft its proposals for Quality Policies and submit these to Top Management for approval.

Quality Target is a specific attainable aim, result or goal.

Example : To hold scrap losses to 3.5% during 1996.

It should be stated in a way that serves as a basis for action. For such targets to be clear, they must be stated quantitatively. Objectives expressed quantitatively helps to :

— Unify the thinking of managerial personnel ;

— Stimulate action ;

— Run the company on a planned basis rather than on a crisis-to-crisis basis ;

— Compare the performance against objective.

Development of a positive programme to accomplish these above is not an easy task. The main difficulty is to reach agreement on company objectives. This requires co-ordination of the views of executives. For this purpose, organisational machinery must be used for securing the proposals and views of the executives ; estimating the relative importance and potential gains of the various proposals and agreeing on the targets to be achieved.

If such a co-ordinated approach is not evolved in the company, each executive is forced to work at what he regards as important, based on his departmental and personal aspirations.

Based on these ideas, one company has accepted the following quality objective :

"To achieve and sustain a reputation for quality at competitive prices in the national and international markets for our entire product range"

To fulfill the above Objective, company has outlined the following Quality Policy :

(*i*) Product design to fully meet customer Quality requirements including functional, aesthetic and life characteristics, with adequate attention to economy of manufacture.

(*ii*) Strict adherence to specifications during manufacture, despatch, erection and commissioning with emphasis on defect prevention.

(*iii*) Use of scientific selection, development, evaluation, surveillance and technical assistance to our Suppliers and Sub-contractors.

(*iv*) Planned and systematic programme for acquainting our Customers adequately on proper usage of our products.

(*v*) Continuous product review and improvement based on market survey and feed back of field performance.

18.9. Quality Planning

Vendor Quality problem is immense. Purchased goods contribute not only to company's cost but also to the company's Quality. This contribution is decisive regarding final product Quality—for better or for worse. New plans of vendor relations need to be evolved having "Vendor development" as its focal point and "interdependence" on a broad front-economic ; managerial and technological, as a key feature.

The ultimate objective in vendor Quality relations is production of material which so adequately conforms to the buyer's requirements that there is no need for extensive acceptance or corrective procedures by the buyer.

The activities to be carried out to achieve this objective are as follows :

(*a*) Thinking through and publishing the company's policy of vendor Quality relations ;

(*b*) Using multiple vendors for major procurement items ;

(*c*) Developing methods for identifying qualified vendors, and for eliminating those who are unable to meet Quality requirements ;

(*d*) Communicating essential and helpful information-designs, specifications, standards, standard practice, etc.

(*e*) Communicating engineering changes promptly ;

(*f*) Developing methods of detecting deviations promptly, especially through pre-production or trial runs ;

(*g*) Providing for use of vendor Quality data in lieu of incoming Inspection ;

(*h*) Developing methods of giving helpful assistance to vendors on their Quality problems ;

(*i*) Reviewing the performance of vendors through vendor rating or other plans and following up on the chronically poor vendors.

In order to understand the mutual vendor-vendee responsibilities, a statement of *VENDOR QUALITY POLICY* is essential. Such a policy should reflect :

— The responsibilities of a company to its suppliers ;

— Suppliers' responsibilities to the company.

— The concrete measures to be adopted for fulfilling the mutual responsibilities.

The evaluation of vendors as part of vendor selection process can be performed in the following ways :

(*a*) General reputation of vendors

(*b*) Data from other buyers

(*c*) Data bank

(*d*) Vendor surveys.

General reputation of any but large well-known companies is a vague, imprecise basis for vendor evaluation. The difficulty with respect to data from other buyers lies in the reluctance of the companies to part with the information in case it affects the competition. The concept of data bank for vendor evaluation is new. It consists in pooling vendor performance and survey data on Quality from several buyers and a summary made available to users of the data bank. The idea of bank can be tried to start with in an organisation having a number of sister concerns. It is worth to bring out periodically a hand-book containing data on :

"Quality performance and survey on all major items purchased from different suppliers" ;

"Shop rejection rate as well as field failure rate" ; and

"Cost of remedying the failure for specific purchased components" ;

Based on the information pooled from all the sister units. Such information is valuable in negotiating new or renewal of contracts with vendors.

A VENDOR QUALITY SURVEY is an evaluation of a vendor's ability to meet Quality requirements. The results of the survey are used in the vendor selection decision or, if the vendor has already been chosen, the survey alerts the purchaser to areas where the vendor may require assistance in meeting requirements. The survey evaluates the vendor on Quality policies and practices, facilities, procedures and appraisal of personnel.

The ingredients of initial planning include :

(*a*) Elaboration on the meaning of specification, including the setting of standards for sensory Qualities ;

(*b*) Defining key words and phrases to establish a common language ;

(*c*) Establishment of seriousness, classification of defects ;

(*d*) Standardising the methods of measurement and test ;

(*e*) Establishing sampling plans, acceptance levels and other criteria for action ;

(*f*) Establishing systems of lot identification and preserving the order of manufacture ;

(*g*) Packing instructions.

The concept adopted in formulating the Quality plan with vendors is one of '*INTERDEPENDENCE*'. It means that buyer and vendor are equally concerned with prevention of defects or at least detecting them as early as possible. It is desirable to incorporate the Quality plans into the purchase contracts.

Aim of routine vendor control is to ensure that vendors on regular orders are kept continuously alert in maintaining the Quality of their supplies at specified standards. For this purpose, the following methods are adopted :

Vendor rating

Technical assistance

Vendor training

Vendor certification, and

Vendor incentive.

Vendor rating is a method of comparing the relative merits of different vendors on the basis of the Quality of supplies, price and serviceability reflected by the promptness in adhering to delivery dates and stipulated quantities. Vendor rating scheme helps suppliers to know their position in relation to their competitors and thus acts as an inducement to improve their own business-status.

Technical assistance to vendors is a special facility extended to very deserving cases wherein the plant assists the vendor in overcoming some of the production difficulties affecting product conformance to set standards. Giving gauges and instruments on loan also comes under technical assistance.

Vendor training consists in conducting courses on Quality Management Concepts, Quality Control techniques and procedures and specific methods for the personnel drawn from the vendor's plant. Such training will also aim to acquaint then well with the Quality requirements of the products supplied and to ensure that similar Inspection practices are adopted in the supplier's plant.

18.10. Quality Improvement and Quality Motivation

It is a common experience that management controllable defects greatly outnumber operator controllable defects. According to Dr. J.M. Juran, about 8% of the defects are management controllable and this does not vary from industry to industry, but varies greatly among operations.

Lack of awareness on the part of many industrial managers regarding the fact that the bulk of defects are management controllable compels them to believe that most defects are the result of operator carelessness, indifference and even sabotage. This leads to embarking on operator motivational schemes.

18.10.1. Dominant System. In all processes the input materials are subjected to wide assortment of operational "Systems" (machines, tools, operators etc.) in order to achieve the desired end result. These operational systems are subject to variation, and it is regulation of these variables which is the basic problem of process control. Although the process systems are numerous, it is common to find that one of them "dominates" *i.e.*, it is more important than all the rest put together. It is most helpful to identify the dominant system.

Four of the most useful/usual dominant systems are :

(*a*) Set-up dominant System : The process has been engineered to so high a degree of reproducibility that, if properly set up in the first time, it will run the entire lot off successfully. Quality in such cases depends on the validity of the original set up of the machine.

(*b*) Machine Dominant System : The process keeps changing during the fabrication of the lot, requiring periodic check and readjustment.

(*c*) Operator Dominant System : The operation has not been fully engineered and the unengineered residue, which depends on the skill and attention of the operator, is the major cause of defectives.

(*d*) Component Dominant System : The input materials or components are the main variables affecting the Quality of the finished product.

The dominant system is not the sole source of defects for the process in question, but it is the main source ; *e.g.* a printing operation is set up dominated as to defect such as spelling or colour of ink. It is operator dominated as to the finger prints on the sheet.

If the operator controllable defects predominate, the Quality Improvement Programme must be designed to influence a large number of people to *CHANGE THEIR ATTITUDE TOWARDS QUALITY*. If the management controllable defects predominate, then the programme must be designed to *DIAGNOSE THE CAUSE FOR THE DEFECTS AND TO DEVELOP REMEDIES*.

18.10.2. Quality Improvement. *Sporadic Troubles—Trouble Shooting*. All continuing performances are subject to variation. Some of these variations are non-significant and are ignored or deferred as to corrective action. Other variations are so significant that they trigger the alarm signals of the control systems. Significance may be either statistical or economic or both. For a discussion of statistical significance, statistical tests are applied. Economic significance is influential in establishing the order of priority of corrective actions to be taken. These significant variations are a form of sporadic departure from standard which call for quick diagnosis of the cause and removal of the cause to restore status quo. The sequence of events used to *RESTORE THE STATUS QUO* is known as trouble shooting or fire-fighting.

Chronic Trouble : Break Through

There are cases in which the same problem comes up over and over again, suggesting that the cause is not a reported error but a basic deficiency in the design, process system etc. Such problems are usually major and cause only minor irritations (as its continuous nature leads all concerned to accept it as unavoidable). The solution calls for extensive and complex data collection over long periods, analysis of data by technical and SQC personnel using SQC techniques. The solution when implemented *CHANGES THE STATUS QUO* and a new level of performance that is superior to the previous level is achieved.

Thus the two essential criteria for "Break Through" are ;

(*a*) The new (superior) level of performance has never previously been attained. *IT SETS A NEW RECORD*.

(*b*) The change is the result of human determination to set a new record. It is not the result of luck.

MAN, NOT CHANCE, IS THE MASTER.

18.11. Guidelines for Development of Q.C. Plan

There is a need for preplanned, systematic and integrated approach for total quality management in the manufacture of all products. This can be achieved through development of quality control plan whose objectives are :

(*a*) To establish a defect-free manufacturing process.

(*b*) To control and monitor the process according to the design requirements and to obtain on line and final product inspection data.

The steps involved in developing a Quality Control Plan have been shown by means of flow diagram in Fig. 18.5 and are summarised as below :

(*a*) Before preparation of the plan, the current Quality at every stage of manufacture of the product must be assessed.

(*b*) After assessment of the current rate of rejection, rework and non-conformance of each state, it is possible to judge whether the current Quality is satisfactory or not.

(*c*) If it is found not satisfactory, the Quality data are plotted on a graph in the sequence of production and then it is seen whether the process is under statistical control or not.

(*d*) If the process is not under statistical control, the process parameters and the associated technological features must be regulated to the extent that the process is brought under the influence of chance causes only.

(*e*) Once the process is under statistical control, the process capability of that stage of manufacturing is found out and compared with the design requirements and specification tolerances.

(*f*) If the process capability is unsatisfactory, then the process is studied in depth and by a Break through Operation, the process is made capable.

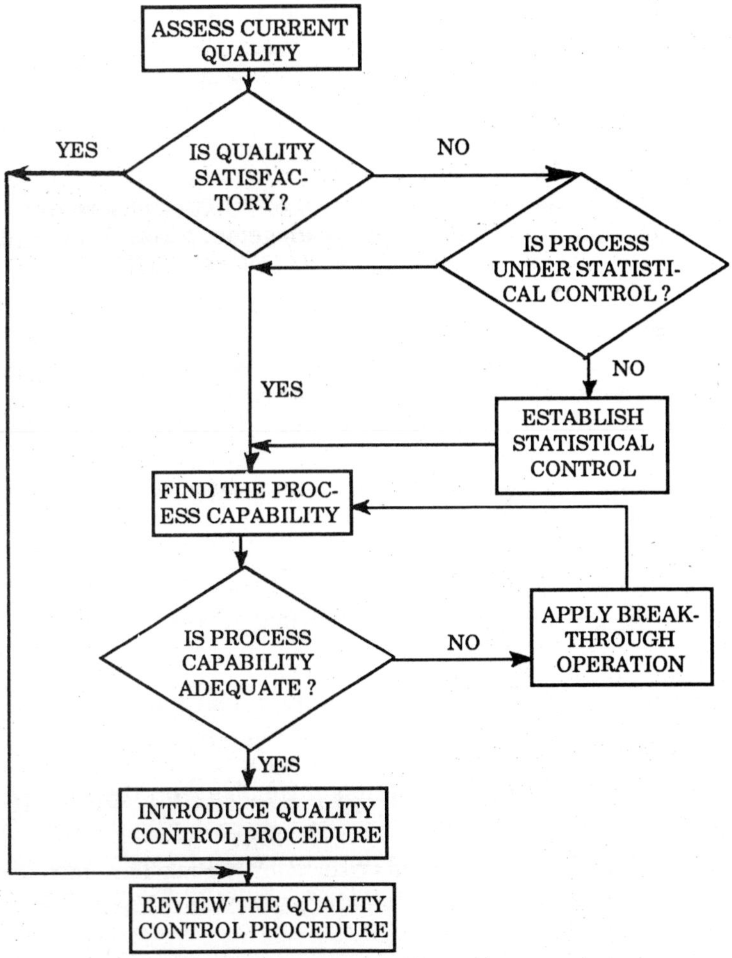

Flow diagram for development of a quality control plan
Fig. 18.5

(*g*) A scientific and robust QC procedure is then introduced to control the improved process.

(*h*) After achieving a satisfactory Quality status, review of the QC procedure is carried out periodically so that the process remains all the time in a state of Quality Control.

18.12. The Fourteen Steps to Quality Improvement are :

Step one	:	Management Commitment
Step two	:	Quality Improvement Team
Step three	:	Quality Measurement
Step four	:	Cost of Quality Evaluation
Step five	:	Quality Awareness
Step six	:	Corrective Action
Step seven	:	Establish and Adhoc Committee for Quality Circle Movement
Step eight	:	Supervisor Training
Step nine	:	Quality Circles Day
Step ten	:	Goal Setting
Step eleven	:	Error Cause Removal
Step twelve	:	Recognition
Step thirteen	:	Quality Councils
Step fourteen	:	Do it over Again

18.13. Quality Motivation

Motivation is an internal process that causes people to work towards goals that they feel will satisfy their needs. Motivation is clearly a matter of being attracted, not pushed, towards a goal or a set of goals.

There are various factors of causes resulting in some form of Quality failure. The most important of these being lack of involvement of concerned men in Quality. This results in the errors by the operators and other men who contribute to Quality. Several studies carried out in Indian Industries on the root causes for Quality deficiencies have revealed that more than 50% of these deficiencies are traceable directly or indirectly to human errors at various levels of organisation.

Experts in the field of Quality Control have classified three errors and their root causes. The remedial measures depend upon the type of errors and so is motivational programme to minimise them. They are :

(*a*) Errors due to inadvertence

(*b*) Errors due to lack of techniques

(*c*) Wilful errors.

(*a*) **Errors due to inadvertence.** This type is primarily due to lack of knowledge of the product, lack of information on the consequence of certain Quality deficiency, working in isolation from the user departments of the product, lack of systematic feed back of information on the Quality failures, adoption of collaborators' design without analysing various product specifications etc.

To motivate the operators to commit lesser and lesser inadvertent mistakes, the essential step to be taken is to create Quality awareness through the entire cross section of organisation.

Inadvertent failures are not due to a specific reason, but are because the operators or the inspectors at that time were not aware that they were making the errors.

(*b*) **Errors due to lack of technique.** There are situations in which, for specific defects, there are consistent differences in error levels among the operators despite the fact that the specifications, materials, processes etc. are outwardly the same for all. These are due to lack of knowledge, skill, technique etc., and how to avoid errors.

Technique errors can be controlled by investigating the performance of "better" operators and comparing them with the "poor" or defect prone operators.

(*c*) **Wilful Errors.** These are not because of deliberate planning by the people but because of various factors that operate in the manufacturing plants. Wilful errors are distinguished by the fact that

(*i*) The operator knows how to avoid the error

(*ii*) Despite this knowledge, he deliberately makes the errors *i.e.*, fails to meet the Quality requirement.

Wilful errors can broadly be classified into two categories :

(*i*) Management initiated

(*ii*) Operator initiated

In many situations the management emphasis changes from time to time *e.g.* Quality, Cost, Delivery schedule, Safety, etc. or due to conflicting Quality specifications, such as official and unofficial specifications. Shop floor specifications are sometimes too unrealistic and cannot be achieved on many situations, yet the products are acceptable on "consideration" as and when the need arises. The operators thus lose interest in attention towards meeting specifications and achieving Quality.

Managers like-wise sometimes commit wilful errors when it is believed that Quality standards are not that important.

18.13.1. Improving Motivation. Commonly a campaign is designed to secure new action, alertness or awareness from the operators. This campaign should be distinguished from other campaigns, which are common to the organisation, usually launched for various other purposes.

The first step of such campaign is through usual media *e.g.* mass meetings supported by trade unions, managers etc., Quality posters, Exhibition of scraps or other quality deficiencies and so on. Secondly, the work force need to be convinced about the need for such attention on Quality. An important aspect of making the motivational programmes successful and to ensure that operators take interest in so called management activities is *TRAINING*. The operator controllable problems can be solved through the *PARTICIPATIVE APPROACH INVOLVING QUALITY CIRCLES*. Quality Circles are small group of employees doing similar or related work who meet regularly to identify, analyse and solve product Quality and production problems to improve general operations. Workers are given training in statistical 'techniques designed to help them correctly evaluate information.

The ideal circle size is seven to ten with members meeting once a week. Though QC circles are not a panacea for problems, they do offer some constructive possibilities. The answer will have a significant impact on the character of all our lives in the years to come.

The basic objectives of the QC circles are as follows :

To encourage firstline supervisors to educate themselves and to develop leadership and Supervisory qualities. Quality Circles provide for the progressive development of each individual's native intelligence and shrewdness and allow him to gain recognition and prestige through the success achieved by the group. The big driving force behind the Quality Circles is management.

A new slogan "Zero Defect" has been introduced at many places. This programme is partly motivational where the objective was primarily to reduce the defect level from the present. The basic contents of this programme are :

— A motivational package aimed at operator controllable defects through meetings, posters, pledge cards, etc.,

— A prevention package aimed at reducing management controllable defects through cause and effect analysis.

— Analysis and appropriate actions, procedures.

— Provision for prompt feed back to workers.

It may be emphasized that any motivational programme will be successful provided proper recognition of sources of Quality failures is possible and the programmes are directed towards solving these problems or bottleneck areas.

18.14. Quality Losses and Avoidance

The total of all quality losses is termed as 'gold-in-the-mine'. Individual major losses leading to gold-in-the-mine can be classified as follows :

Tangible ;
Materials scrapped
Labour and overhead on product scrapped
Labour, materials and overhead necessary to effect repairs on salvageable product
Extra operations added because of presence of defects
Excess inspection costs
Discount on 'second grade quality' product
Charges on quality guarantees-free replacements
Downgrading of product
Returned materials, repair and processing
Intangible ;
Delays and stoppages caused by defectives
Customer good-will
Loss in morale due to friction between departments.

An assessment of the total quality lossess is a prerequisite of utmost importance before launching a programme for improving quality and at the same time reducing costs.

The quality control activities do cost something. However the small investment in preventive measures results in substantial savings and avoids many 'costs' which, otherwise, are later incurred in an attempt to ensure right quality to the consumer. It may be appropriate to remember that 'a stitch in time saves nine'.

A study on Quality Costs in Engineering Industries has shown that by increasing the prevention costs, the profit was stepped up from 71/2% to 12%. The break up of Quality Costs before and after increasing prevention costs is shown in Fig. 18.6 and 18.7.

This implies that the profit of engineering industry could be increased by over 50%.

Quality costs can be estimated for identifying the priority areas that need attention by suitably classifying and breaking up the costs from the existing records. However, there might be situations demanding a little extra effort.

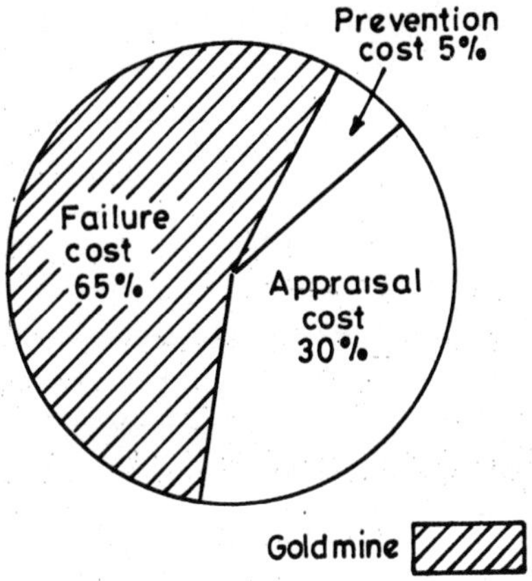

Profit—7.5%
Quality Cost Distribution in a typical Engineering Industry
Fig. 18.6

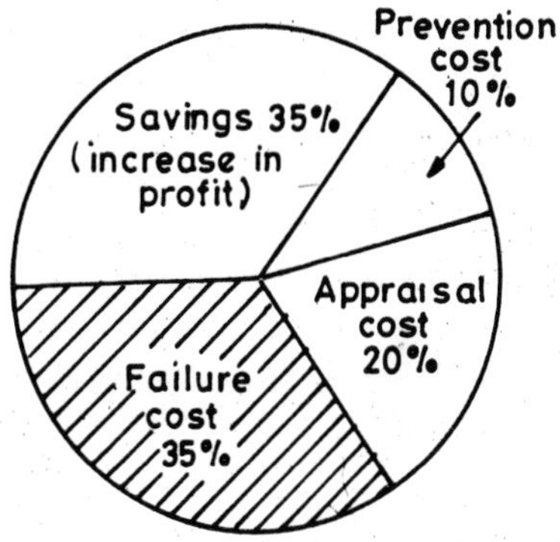

Profit—12%
Result of Doubling the Prevention cost
Fig. 18.7

This will involve defining the elements of activities and estimating the time spent on these as a proportion of the total activity for which costs are available from existing records, or it can be fairly approximated after discussing the relevant details with the Accounts and allied departments. Ultimately, the Accounts department should take up the analysis of quality costs and their reporting for periodic review as a routine function.

The quality costs can be related to net sales billed or production in terms of standard weighted units or direct labour or cost of manufacture depending on their sensitivity to changes in production schedule or prices or seasonal fluctuations.

The next step requires the evaluation of investment in the preventive measures and gains expected for a rational decision on the course of action to be taken. This requires thorough examination for identifying 'avoidable costs'—the costs that will disappear if all defects were to disappear. Many quality costs can be avoided at little expense such as careful marking or setting, first off or patrol inspection, revision of tolerances where economical, proper communication, etc.

There are other quality losses which can be avoided at relatively more cost but still offer a profitable proposition such as change in technology, new tools and jigs, etc. Still there are a few others the avoidance of which requires more investment that the returns they promise and therefore need to be ignored.

Thus an objective analysis can help in laying guidelines for planning a strategy for cost reduction and control.

18.15. Quality Economics

The objective of quality programme in a company can be set as achieving the required quality at economic cost through minimising quality losses. This is a complex task. It calls for an integrated approach to quality, touching basic and fundamental issues like management policy objectives, organisation and manual on quality, etc. These topics are discussed hereafter.

18.15.1. Elements of Quality Costs

Item	*Definition*
1.0 Cost of Failures	**Losses due to non conforming parts and assemblies.**
1.1 Scrap	Material, Labour and burden of non-usable parts.
1.2 Repairs—inside	Labour of making good defective parts and assemblies.
1.3 Repairs—outside	Material and Labour of fixing defective items returned by customers.
1.4 Service	Administrative cost of service to customer who receives defective items.
1.5 Others	Delays, stoppages and loss of customer goodwill caused by defectives.
2.0 Cost of Appraisal	**Cost of evaluating quality and of identifying and segregating non-conforming parts and assemblies.**
2.1 In-coming Inspection	Labour in sampling and sorting of purchased parts.
2.2 100 per cent Inspection	'Direct' labour of 100% inspection and 100% tests to separate 'bad' from 'good' items.
2.3 Spot-check Inspection	Indirect labour of roving and spot-check inspections to detect defective lots and to prevent production of defects.
2.4 Outgoing Quality Rating	Labour and indirect costs of sampling, testing and rating, outgoing product.
3.0 Cost of Prevention	**Engineering, Technical and Supervisory costs of preventing recurring defects.**
3.1 Quality Control Engineering	Cost of investigation, analysis and correction of causes of defects by quality control department.
3.2 Engineering	Cost of investigation, analysis and corrective measures by engineering department.

18.15.2. Quality Costs Assessment—Their Constituents. An essential element in operating a total quality control programme is the identification, analysis, and control of quality costs for the business. The specifically typical items that make up these operating quality costs are listed below :

I. Prevention :

1. *Quality planning and process control planning.*

(*a*) Quality Planning : The costs associated with the time which the personnel in the quality control function spend in planning the systems of control of materials, processes and products through formal methods, procedures and instructions.

(*b*) Process Quality Control : The costs associated with the time spent by the personnel of the quality control in studying and analysing manufacturing processes, for the purpose of establishing a means of control as well as improving existing process capability, initiating and maintaining control over operating processes.

2. *Quality planning by functions other than quality control.* This includes costs associated with the time spent in quality planning work such as pre-production analysis, reliability studies, writing instructions or operating procedures for test, inspection and process control by personnel who do not work for the Quality Control Manager.

3. *Design and development of quality measurement and control equipment.* This represents the costs associated with the time which the personnel spend in designing quality measurement and control device. It does not include the cost of equipment or depreciation.

4. *Quality training.* This represents the costs of developing and operating formal quality training programmes throughout the company operations, designed to train personnel in the understanding and use of quality control techniques. It does not include training costs of instructing operations to achieve normal quality proficiency.

5. *Other preventive expenses.* These include miscellaneous items of expenditure not included above.

II. Appraisal :

1. *Test and inspection of purchased materials.* The costs associated with the time which the inspection and testing personnel spend on evaluating the quality of purchased materials and any applicable costs of supervisory and clerical personnel. Also this may include the costs of inspectors travelling to vendors to evaluate purchased materials.

2. *Laboratory Acceptance Testing.* The costs of all tests provided by the laboratory or testing units to evaluate the quality of purchased materials.

3. *Inspection and Testing.* The costs associated with the time that inspection personnel spend evaluating the quality of the product in the shop and applicable costs of supervisory and clerical personnel. It does not include the cost of inspection and testing of purchased material in (1).

4. *Checking Labour.* The costs associated with the time which the shop-operators spend on checking quality of their own work as required by the quality plan, checking product or process for quality conformance at planned points in manufacturing, sorting lots which are rejected for not meeting quality requirements and other in-process evaluations of product quality.

5. *Set up for Inspection and Test.* The costs associated with the time which the personnel spend in setting up product and associated equipment to permit functional testing.

6. *Test and Inspection Material.* The cost of power and material for testing major apparatus such as steam or oil, electricity and materials consumed in destructive testing.

7. *Quality Audits.* The costs associated with the time spent on performing routine quality audits on the in-process or on the finished product.

8. *Outside Endorsement.* These represent the underwriter's laboratory fees, (ISI certification, Lloyds Testing), insurance inspection costs, etc.

9. *Maintenance and calibration of inspection and test equipment.* The costs associated with the time spent by the maintenance personnel in calibrating and maintaining test and inspection equipment.

10. *Review and test of inspection data.*

The cost associated with the time of product engineers who review test and inspection data prior to release of the product for shipment.

11. *Field Testing.* The costs incurred by the company in field testing of the product at the customer's site prior to final release. These costs might include travelling costs, staying expenses, etc.

III. Failure :

A. Internal

1. *Scrap.* The losses incurred in the course of obtaining the required level of quality. It should not include materials scrapped for other reasons such as obsolescence and product design changes resulting from further evaluation of customer needs. Scrap might be further. Sub-divided, *e.g.* between fault of own manufacture of fault of vendor.

2. *Rework and repair.* The extra payments made to operators in the course of obtaining the required level of quality may be further sub-divided as between fault of own manufacture and fault of vendor.

3. *Material procurement costs.* The additional costs incurred by materials procurement personnel in handling both rejects and complaints on purchased material. Such costs may include getting disposition from vendors for rejected materials, making certain that the vendors understand quality requirements for either rejects or complaints, etc.

4. *Trouble Shooting.* The costs associated with the time spent by production engineers on solving problems involving quality, *e.g.*, if a product, component or material does not conform to quality specifications, a production engineer may be requested to review the feasibility of product specification changes. It does not include engineering development work which may be performed on the factory floor.

5. *Down-grading.* In some industries, such as textiles or ceramics, final products are rarely scrapped. Rather they are "down-graded" and sold as seconds at a price discount. The loss associated with the down-grading represents the cost.

B. External

1. Expenditure for the adjustment of customer complaints.

2. Product or customer service. Expenses directly attributable to correcting imperfections on special testing not included in complaints. It does not include installation, service or maintenance contracts.

18.16. Quality Assurance

Recognising that inspection should primarily be a service function to the manufacturing, modern companies have reorganised their quality functions in tune with basic philosophy that "the task performer should be responsible for Quality as well". Every department contributes to the end product quality. Building quality into the product needs a total participation from all the departments, starting from pre-contract stage, through Design, Purchase, Production Engineering, Technology, Manufacturing, Testing and Despatch right upto the product commissioning and subsequent services. To coordinate the activities of various departments Quality Assurance groups are created at plant level.

The primary objective in creation of QA is to provide the necessary confidence to the customer as well as to the top management that all concerned are carrying out their activities effectively and that the product quality is achieved to the customers' satisfaction at an economic cost.

18.16.1. Responsibilities of Quality Assurance. While emphasizing that participation from all the departments is necessary for achieving product quality, certain responsibilities are listed out for the quality departments, namely, Quality Assurance and Quality Control. Responsibilities of Quality Assurance include :

Plan, develop and establish Quality policies and procedures required by the customer.

Issue and maintain the manual containing policies of the general directive or definitive nature, up-to-date descriptions of the organisation, instructions and methods of implementation of these policies in a written form :

Assure that only products that conform to applicable specifications are despatched to the customer and that necessary corrective actions are taken when deviations occur.

Constantly evaluate the effectiveness of the Quality programmes and develop such indices of performances which reflect the status of quality in the plant and outside.

Carry out special studies and investigations for solving quality problems.

Act as liaison between the company, customer, Quality Control, third party inspection agencies or supplier representatives on matters pertaining to the Quality of the product or the company Quality Assurance programmes.

Coordinate actions on customers Quality complaints to assure the earliest possible satisfaction of such complaints.

Perform systems audits, outgoing product audits and inplant process audits as a means of evaluating the effectiveness of the established QA systems.

Plan and coordinate vendor quality surveys and constantly evaluate results of vendor performance.

Organise for training and education programmes on quality and related matters for the plant, customers' and suppliers' personnel.

Liaise with outside inspection, consultants and corporate management on quality matters.

18.16.2. Responsibilities of Quality Control. Perform acceptance and process control inspections as required by Quality Control procedures. Also assure that necessary corrective actions are taken when deviations occur.

Provide assistance to Manufacturing to assure that quality is maintained during manufacturing processes.

Attach to raw material purchase orders the necessary material specifications and any specific requirements to assure that the purchased material is capable of meeting any special customer requirements.

Perform and document inspection, tests and verification of purchased material as specified in procedures.

Request and maintain documentation on corrective action requests.

Perform in-process tests and verification as specified in Quality Control Procedures.

Perform final inspection, tests and verification as required by Quality Control Procedures.

Maintain the calibration of all equipment and tools in accordance with the methods and procedures established and monitored as in Quality Manual.

Ensure that all equipment and tools are calibrated when required and make necessary arrangements to bring tools back into calibration before releasing for use.

Maintain required records indicating traceability of standards and calibrations performed.

18.16.3. Quality Assurance Manual. Purpose : A Quality Assurance Manual is a document containing policies of the company and procedures affecting product quality. It includes the following :

— REFERENCE for the policies, procedures and the reasons behind them. In a way it is a proof that policies and procedures received due thought. These help the executors of the plans.

— TEXT BOOK for training quality assurance and quality control personnel and also allied production and engineering personnel.

— PRECEDENT for guiding future decisions since the manual codifies past practices and agreements.

— CONTINUITY of operations despite employee turnover. Lack of it might lead to uncontrolled and undesired changes.

— BASE for auditing the policies and procedures in current use.

Some companies have insisted on introduction of the manual by vendors. Some vendors have been benefitted. Others ended with increased costs and unnecessary details. It is therefore very essential to bring out o 's manual carefully that will serve the purpose it is intended, commensurate with its returns direct and indirect.

Evolution of the Manual : The Manual is synonymous with the code of law for disciplined activity that spells efficiency. Quality Manual is a necessity with the coming up of age of complex industry faced with multiple quality requirements and problems. The size of the company and the extent and complexity of quality practices determine the contents of the manual. Manual may be evolved to meet the need of various departments like design, personnel, etc. Wherever possible a composite manual may be evolved.

Development of Manual : A beginning may be made with recording existing practices that have stood the test of time and adding rational, useful and necessary procedures. This may be done in the following stages :

— Recording of ideas.

— Review by concerned personnel.

— Preparation of formal document by a composite team.

— Approval by all concerned departments and authority.

— Issuance by a centrally authorised departments like Quality Assurance.

— Usage or application by concerned departments.

— Updating.

The spectrum of procedure may cover the following aspects :

— Title and identifying number.

— Effective period.

— Sections replaced in the current manual, if any.

— Table of contents.

— Purpose and or objective.

— Scope.

— Summary in flow chart, wherever convenient.

— Detailed procedures, forms, etc.

— Glossary of terms.

— Authorising signature.

The manual should be operative to avoid its becoming an obsolete document. The manual should be suitably printed and circulation documented to ensure that all concerned have seen issue.

Quality Assurance engineers might collect necessary information and prepare a draft manual. Line department and quality engineers might review and modify as also provide necessary coordination. General Manager might authorise its approval and Quality Department should publish and maintain the manual.

Contents of the Manual : The following may be used as guidelines for compiling a manual to suit one's own needs. The various sections in a manual include : Information, about manual, administration of quality functions, quality and inspection planning, design, vendor relations, process control, finished goods control, field performance, general test methods and measurements, specific products, specific processes, personnel, obligatory laws, consumer requirements, control and cost of quality functions, etc.

While documenting, policies and procedures may be given the importance they deserve in each situation. It should confine to statement of task to be accomplished and should provide guidelines for accomplishing it without getting involved in unnecessary details.

Audit of Actual Practice versus Manual : Supervisors have the responsibility for implementing the procedures stipulated in manuals. There might be forces that impel deviations and gradually form new practices, some of which might be undesirable. Supervisors do not have the time to assess the extent to which manual is being adopted. It, therefore, becomes necessary of enforce periodic audit of practices in vogue versus manual stipulations by an independent body. The reasons for deviations like lack of communication, training or facilities must be looked into and situations remedied.

18.17. Attributes of Quality Control Manager

In order to discharge his duties the quality control manager should imbibe following attributes :

— Broad knowledge of engineering, production, research, competance in statistical analysis, finance and marketing functions.

— Ability to listen and to motivate ; to plan, organise ; delegate, and sell effectively ; to think and utilise time effectively ; to communicate effectively.

— Creativeness and fairness ; honesty and compassion, performance mindedness ; impartiality and decision making ability ; skill in negotiating ; patience and persistence ; enthusiasing etc.

18.18. Quality Auditing

Quality audits are performed to find whether the end product satisfies the desired quality specifications ; proper functioning of all equipment and machinery ; operators are carrying out their duties as per specified quality plans.

Special care must be given for auditing of tool and gauge, testing ; quality plans, specifications and drawings.

Audits should be carried by unbiased persons and without prior announcements and cover each and every person. Audits should be made by check lists and its results recorded and made available to all concerned people. The aim of audit should not be to penalise but to take measures to correct the deficiencies found during the audits.

18.19. Basic Concepts of Quality Assurance Programmes

Nowadays, there is a growing need for the recognition to provide real value to the customers, which in turn requires improvement in the quality to services. The improvement in the quality is more beneficial to the industry that it is to the customers ; because the improved quality results in substantial reduction in the cost of the process and the services. Thus Quality Assurance is a mutually advantageous entity.

According to the concept of Quality management, attempts should be made toward working on improving the system, and working together to understand and measure our work processes for continuous improvement.

Guidelines for Application of Quality Management

1. Create constancy of purpose toward improvement of product and service, with the aim to become competitive and to stay in business.

2. Adopt the new philosophy. We are in a new economic age ; we must awaken to the challenge, must learn the responsibilities, and take leadership for change.

3. Cease dependence on inspection to achieve quality. Eliminate the need for inspection on a mass basis by building quality into the product in the first time. (Learn how to "do it right the first time !").

4. "Continual Improvement" is essential part of everyone's job. Everyone should think in the terms ; "*We did the right thing the first time and saved rework or other unnecessary time*".

5. The practice of awarding business on the basis of price tag may not prove economical

ultimately. Instead the efforts should be made to minimise total cost.

6. Improve constantly and forever the system of production and service, to improve quality and productivity, and thus constantly decrease costs.

7. Institute training on the job.

8. Institute leadership. The aim of leadership should be to help people, machines and system to do a better job. Leadership of management is in need of overhaul, as well as leadership of production workers.

9. Derive out fear, so that everyone may work effectively.

10. Break down barriers between departments. People in research, design, sales and production must work as a team, to foresee problems of production and in use that may be encountered with the product or service.

11. Eliminate slogans, exhortations, and targets for the work force asking for zero defects and new levels of productivity.

(Ask yourself ; "did we improve the process ?")

12. Eliminate work standards (quotas) on the production floor. Substitute leadership. Eliminate management by numbers, numerical goals. Substitute leadership.

13. Remove barriers that deprive the worker of his right to pride of workmanship. The responsibility of supervisors must be changed from sheer numbers to quality. Remove barriers that deprive people in management and in engineering of their right to pride of workmanship.

14. Institute a vigorous program of education and self-improvement.

15. Put everybody in the organisation to work to accomplish the transformation, which is possible only with cooperation of each and every individual.

16. Continual Improvement Process calls for the effective Team Work. Every employee has a role and responsibilities to ensure continual improvement in quality, productivity and profitability. The Quality Assurance Department plays a vital role in creating the *team spirit* and will be the driving force in keeping the team together and keep it working together to continually improve the process. The team concept is a departure from the traditional management style, according to which assignments were given to individuals and the success of these assignments represented the way to get ahead. Today, this concept is changing in favour of success of the assignments made to the teams.

18.20. Process of Introducing Quality Program in an Existing Organisation

Fig. 18.8 depicts the process by which the quality program can be introduced in the existing organization.

The Quality Assurance (QA) department should function independently ; but it must interface with every department wherever improvement is necessary. The interface of QA department with the other departments is depicted in the Fig. 18.9.

The QA department must have three sections :

1. Quality Engineering,

2. Inspection and Testing, and

3. Auditing and Monitoring.

Training functions are performed under Quality Engineering. The functions of each of these sections are described in Fig. 18.10.

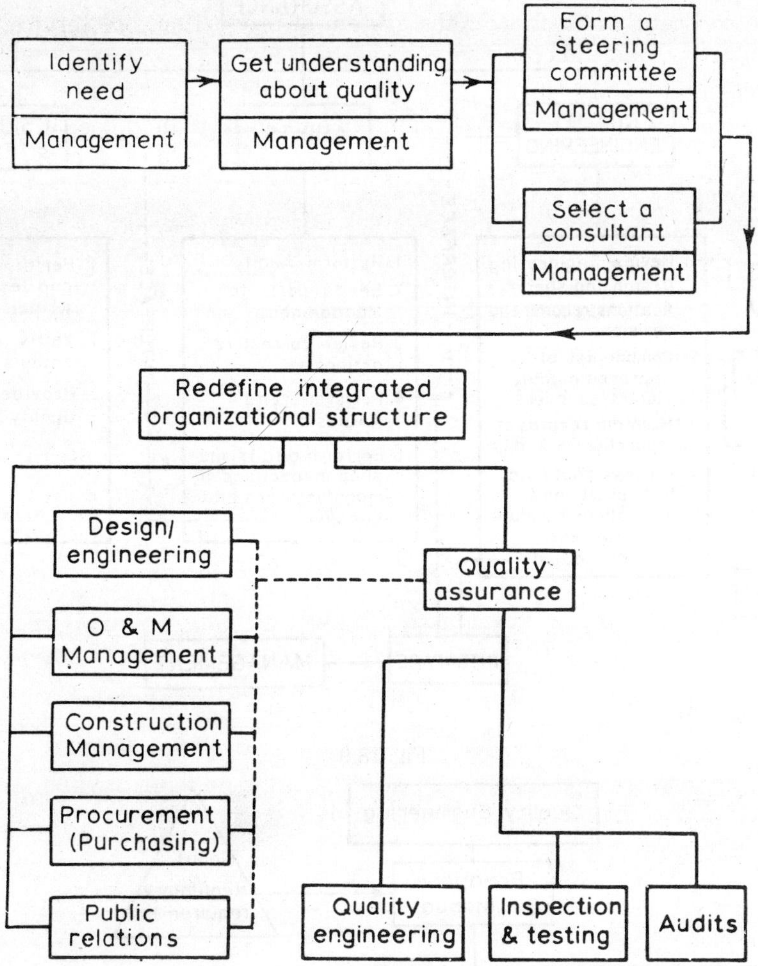

Fig. 18.8

The first task of the Quality Engineering section is to prepare the QA Manuals.

The QA Manual must identify the structures, systems, components and activities which are within the scope of the application of the QA requirements.

An essential requirement of the Quality Assurance program is the performance of all the quality related tasks in accordance with the written procedures. Hence, it will be required of all the departments that the written procedures to prepared for all the activities affecting (which are identified in the QA Manual) and submitted to the QA department for review. Review of the procedures by the Quality Engineering ensures that the essential quality elements are incorporated in the procedures.

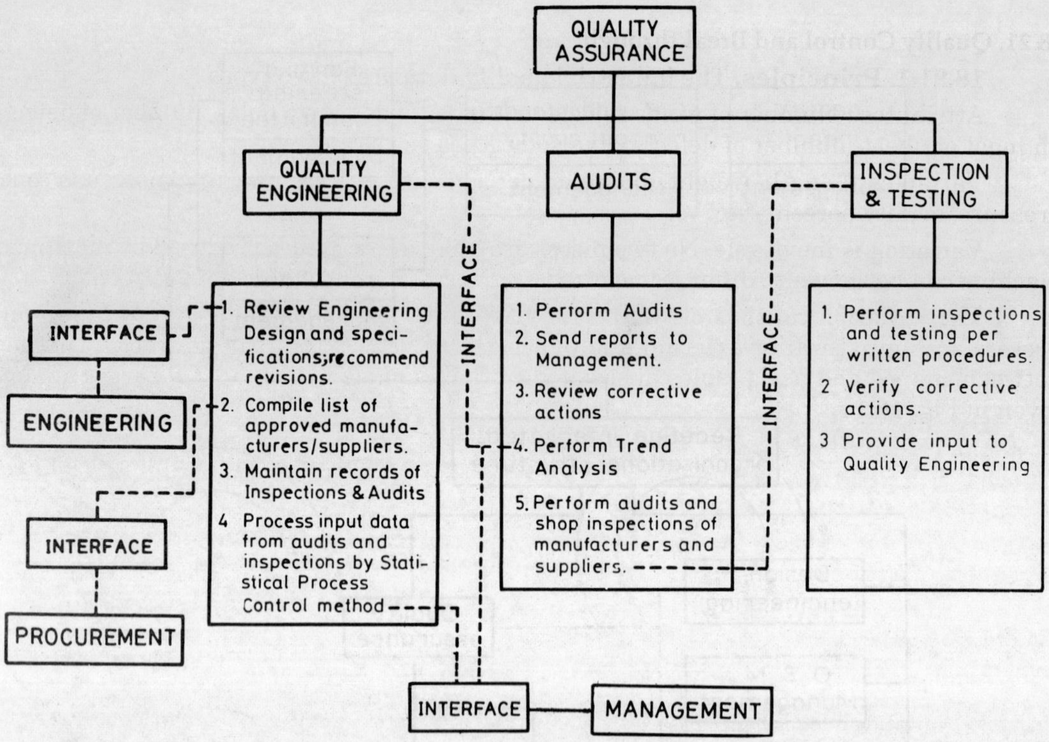

Fig. 18.9

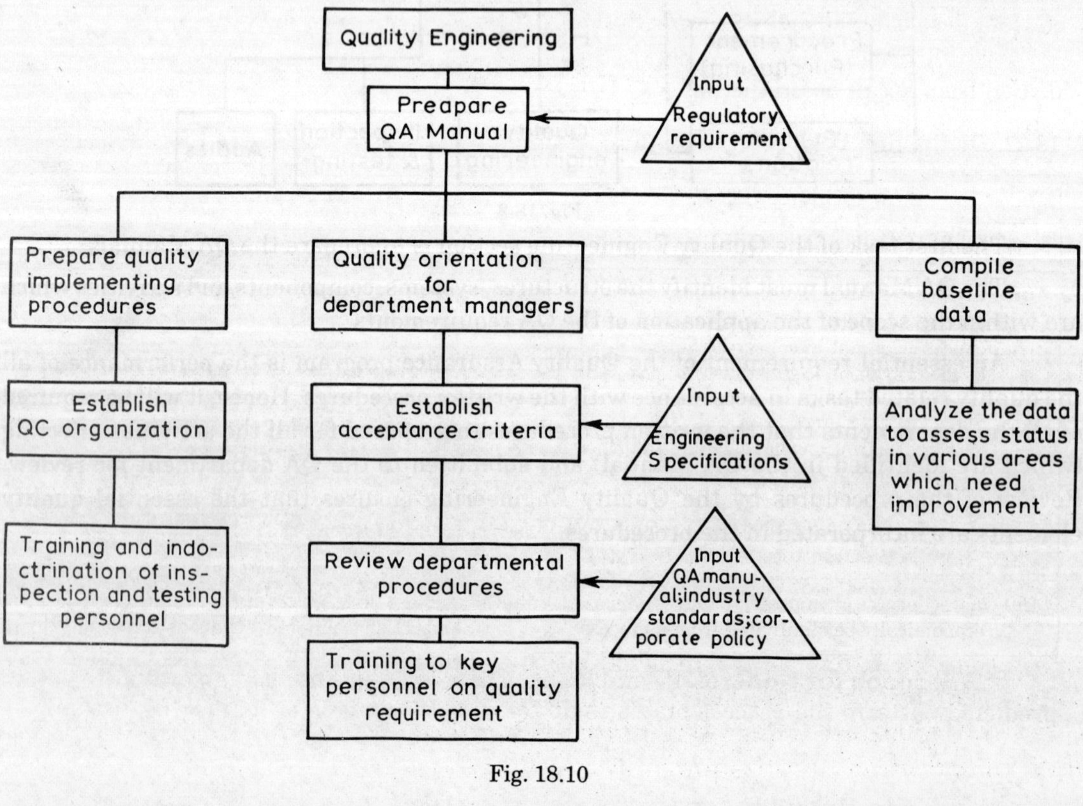

Fig. 18.10

18.21. Quality Control and Breakthrough

18.21.1. Principles. The industrial quality data are of two types :

Attributes : Number of welds rejected, number of defective holes, number of defects on an ingot or sheet, number of defects of various types on casting, etc.

Variables : Length, weight, diameter, percent carbon, hardness, thickness, ash content, pressure, acidity percent, etc.

Variation is inevitable. No two pieces are alike. Even pieces produced by most modern machine can be shown to differ if inspected by a corresponding precise instrument.

The variation consists of two parts—natural or inherent due to chance causes in the process and unnatural or external due to assignable causes that can be traced. Distinction between chance and assignable causes of variation is made in Art. 18.22. Target analogy is given in Fig. 18.11.

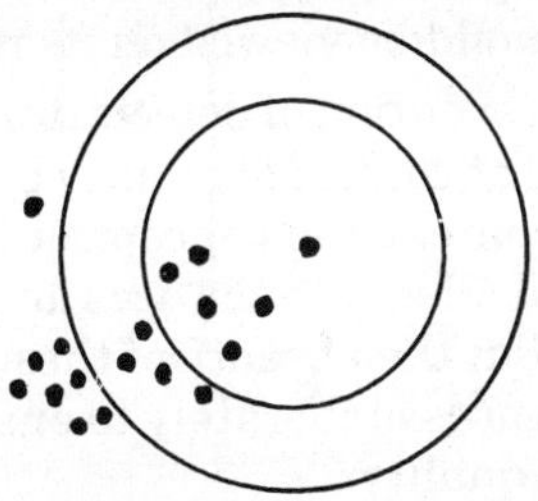

Variation due to chance plus assignable causes

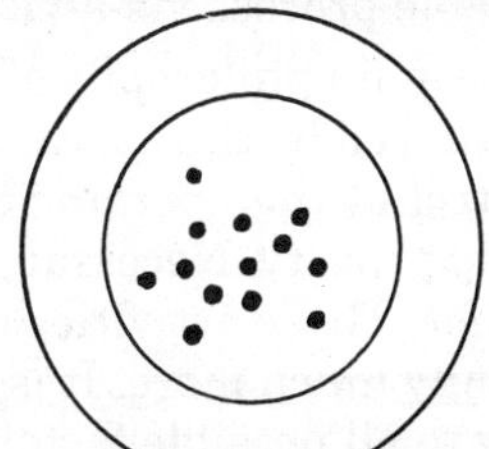

Variation due to chance cause only

Fig. 18.11

Variation resulting from natural causes alone reveals certain pattern or follows certain statistical law—such as posson, normal, etc. Reference may be made to the pattern of variation in thickness of steel sheets, number of machines under breakdown, etc.

From known pattern or population distribution, the pattern of distribution of samples drawn from it can be predicted.

18.22. Distinction Between Chance and Assignable Causes of Variation

	Chance Causes	Assignable Causes
(*i*)	Consist of many individual causes.	Consists of one or just a few individual causes.
(*ii*)	Any one chance causes results in only a minute amount of variation. (However, many of the chance causes act simultaneously so that the total amount of chance variation is substantial).	Any one assignable cause can result in a large amount of variation.
(*iii*)	Some typical chance causes of variation are : (*a*) Slight variations in raw material (though within the specifications) (*b*) Slight vibration of machine (*c*) Lack of human perfection in reading instruments and setting controls.	Some typical assignable causes of variation are : Batch of defective raw material Faulty set up Untrained operator
(*iv*)	As a practical matter, chance variation cannot economically be eliminated from a process.	The presence of assignable variation can be detected and action to eliminate the causes is usually justified.

Approach for Control. If samples from process or incoming lots continue to follow this predicted pattern, the process may be said to be under control, or lot may be presumed to be of quality conforming to process standard. Alternatively, we may infer that the process has undergone change, or that the lot does not conform to attainable standard of quality. Attempts are made to retain this change if it is desirable (*i.e.* revise standard also) or to avoid it, if it is undesirable. Of course there is a risk of taking wrong decisions which may be restricted to desired degree.

Results of samples may be graphically presented in the form of a chart with Control limits based on process capability (dealt with in next chapters) to detect any undesirable change in early stages (through any shift in the expected pattern) to enable preventive measures during production. This avoids wastage of resources like material, efforts of operators and technicians, machine, etc. It is important that the order of production is recognised and all allied information about production such as product size, delivery time, operator, machine, etc. are maintained on the graph.

A process under statistical control, need not yield products informing to specifications. An incapable process vis-a-vis its specifications should be viewed on its merit.

The main philosophy of Quality Control is prevention of defects during production. This is best achieved by the operator himself. This aims at control of quality by the operator rather than control of the operator for quality. In this context the concept of Self Quality Control attains top priority. Necessary education and allied adequate services and facilities have to be provided for. These activities will cost something, but the absence of these will ultimately cost the company much more. It is essential to keep records elaborately to enable its classification according to all possible factors that influence the quality.

Approach for Breakthrough. Contrary to control aimed at maintaining the attainable process standards, "Breakthrough" aims at improving existing standards of performance. A reference may be made to Figs. 18.12 and 18.13.

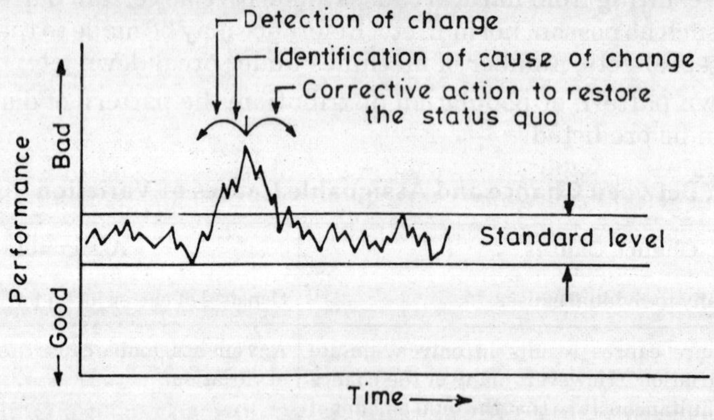

Fig. 18.12

As a first step, enumeration of all causes or sources having impact on the totality of outcome, namely, cost, production, productivity and quality including reliability it made. This is done through a 'brain storming' session involving all concerned personnel from design, production, inspection, etc. from all levels including operators. This is conveniently presented in the form of 'Cause and Effect Diagram'. This is also known as 'Ishikawa Diagram' after the name of its introducer or 'Fish-Bone Diagram' by virtue of its appearance.

Figs. 18.12 and 18.13 show typical examples of Irshikawa Diagram of 'Welding output Quality' and 'Shift' in castings. From a study of the diagrams it is possible to know the diversity of factors influencing the quality characteristic of interest. Accordingly appropriate planning for further study for a break-through approach can be organised.

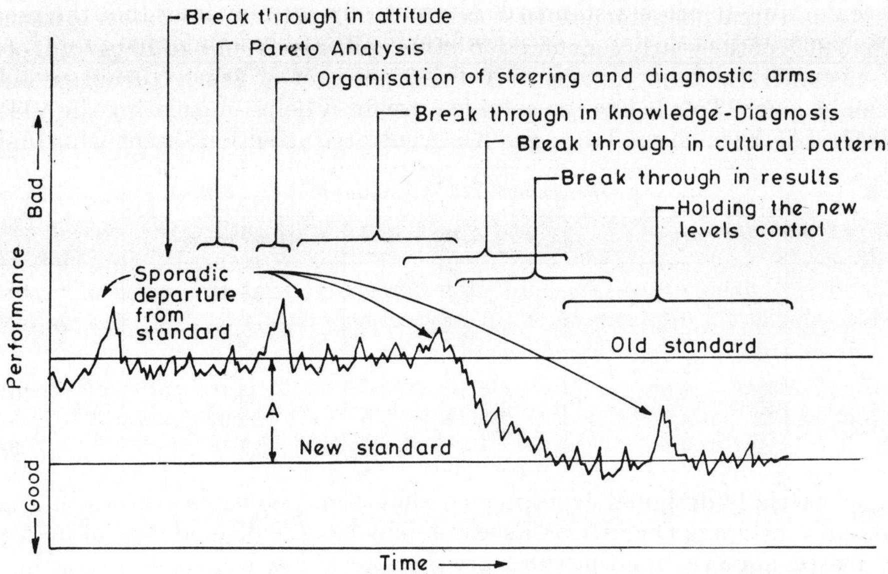

Fig. 18.13

A—The difference between the old standard and the new standard is regarded as a chronic ailment which can economically be curved.

Next step involves classifying all the quality faults observed hitherto according to the causes and sources and identifying the vital few (Pareto Principle) among these for incorporating preventive measures that will ensure high ratio of gains accrued to input investment in various forms.

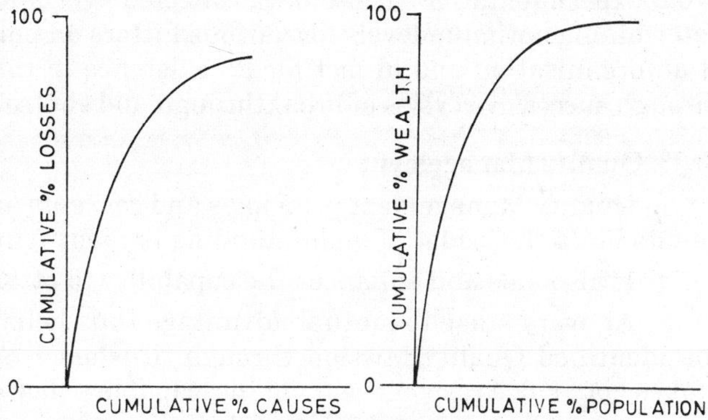

Fig. 18.14. Analogy between distribution of wealth and losses.

Pareto type curves originally intended to exhibit unequal distribution of wealth can be applied to quality losses in any form. It is experienced that only a few of the many causes or sources account for major portion of the losses in the form of rejections, rework, scrap and even capital locked up in inventory of raw material of semi-finished product or finished products. The curve showing cumulative per cent of causes or sources or items versus cumulative losses on account of defects or capital blocked is shown in Fig. 18.14.

Consider monthwise data or rejection of stators in winding section. Summary is presented in Table 18.1. It is seen that from among seven different causes of rejections 'HVF' alone accounts for about 70% rejections every month. The rejections due to 'HVF' and 'UBC' account for more than 80% of the total rejections each month. Thus rejections recur in such a way that a few (a small percentage of) causes account for a large percentage of rejections of stators.

Table 18.1 Illustrative example of Pareto Analysis of Rejections of stators.

This is a universal phenomenon. In stores, 15 to 20% of the total number of items account for about 85 to 90% of the yearly consumption value and only about 8 to 10 percent of the total number of customers account for about 70 to 80 per cent of yearly turnover. Many such examples can be given. This is a very useful feature for it helps in isolating the VITAL FEW from TRIVIAL MANY which need to be given concentrated attention for ensuring best results.

Month	*Percentage of Rejections due to*			*% Rejection Total*
	HVF	*UBC*	*Other causes*	
FEB	7.1	1.6	2.1	10.8
MARCH	5.2	1.8	1.6	8.6
APRIL	7.2	1.3	2.8	11.3
MAY	9.1	1.7	2.3	13.1
JUNE	4.3	1.5	4.7	10.5
JULY	4.8	1.9	2.0	8.7
AUG	4.1	1.2	1.1	6.4

Besides specially designed experiments, simple and multiple regression studies and allied techniques including Operations Research may have to be made use of in locating the factors that matter and in estimating the degree by which these need to be altered for effecting desired improvements. It may also be necessary to conduct pilot confirmatory runs before making formal process changes that should be incorporated in the process standard to make it upto date. In many cases it is possible to maintain process details in such a manner as might avoid experimentation. These data collected over a long period relatively form the basis for determining optimum levels for various factors on plant scale itself. For growth and stability of an organisation and in fact for its existence in this, world, it is necessary that it passes through successive cycles of breakthrough and control.

18.23. Quality Management

Quality Management principles and concepts are comparatively easy to grasp and they help to build the bridge of understanding in identifying the common Quality Mission.

It also aids and enhances the capability to attain the "Fitness For Use".

At every stage to mutual advantage thus helping in short and long term fulfilment of the identified Quality Mission through "*Customer Satisfaction*". Properly understood, this makes both sides responsive to an inescapable policy of "*Team Work*".

Any Scientific Vendor Development Programme encompasses a Vendor Training programme for providing a platform for opening up a mutual dialogue for effective communication and mutual education through a common language and understanding, first at Upper Management level and thereafter at other levels. Managerial, Technical, Financial and other assistance flow from that understanding. One must strive for it to achieve continuously. Maturity of relationship comes only in time.

18.23.1. Premises of Quality Management

Dynamic character of Quality—the urge for it.

"Quality is never an accident.

It is the result of intelligent efforts.

There must be the will to produce a superior thing".

System approach to Quality :

(*a*) Coordinates and integrates efforts of various groups in the organisation in response to environmental input output needs.

(*b*) Translate these efforts into a well coordinated plan of action.

(*c*) Spells out clearly the responsibilities and functions of each group as well as interfaces.

Quality Assurance comprises of :

All those planned and systematic actions necessary to provide adequate confidence that a system, structure or component will perform satisfactorily in service.

Quality Assurance includes Quality Control, Quality Control Comprises of those QA activities related to physical characteristics of a material, structure, component or system.

Which provide means to control Quality of those to predetermined requirements.

Here the emphasis is for on line appraisal of status for data base, for feed back, for *QC* by defect, prevention. Inspection in include in QC only as a means for appraisal in a postmortem situation, but is useful when designed for providing the required data base. We are speaking of the required. "*Quality of Conformance*" to specification, while QA takes up the planning for Quality.

— Quality of design to meet customer needs, performance, ability, safety, aesthetics, life characteristics, etc.

Some amount of inspection is essential. Preplanning and on line appraisal assure Quality data generation and Quality Control. By itself it can merely segregate bad from the good, but cannot create Quality. Building Quality into the product calls for defect prevention through Quality Planning and Quality Control as part of systematical Quality Assurance measures.

18.23.2. Principles of Scientific Quality Control

Good Quality depends on Good Process.

Proposition 1 : No Industrial process operates more than a short time in a state of adequate adjustment.

Proposition 2 : Knowing when and where differences and changes exist, or occur in a process are important bits of information.

By seeking clear specification, classification of Quality characteristics as to their seriousness and if possible demonstration about the use of the product, a vendor will be able to get his manufacturing facility in a state of readiness for production after mastering the financial and technical resources, including trained manpower.

But he will still need to know "Process Capability" by application of "Statistics" besides all those techno-managerial inputs. This is done by application of SQC.

(*a*) Quantifying the process capability *i.e.*, evaluation of inherent variation in the process.

(*b*) Comparing the process capability with the Quality requirements.

Both these are control information based on propositions 1 and 2 mentioned earlier.

There is a paramount necessity to understand the relationship between process variables and product results through specific scientific data collection and analysis.

There are Quality Control tools available to measure process capability and Quality specialists shall be familiar with these tools. It is usually feasible to secure early assurance that the process will be adequate.

Quality Management for both successful control and breakthrough approaches require use of three different disciplines :

1. Management
2. Technology
3. Statistics

After understanding this, following steps can be taken :

(*a*) Launch a new programme at vendor's premises with assurance that process is adequate.

(*b*) Establish sound plan for process controls.

(*c*) Train for supervision and work leading to a status of self-control.

(*d*) Improve current yield or productivity with focus on ultimate competitiveness.

18.23.3. Organisation for Quality. A suitable organisation should be setup on abroad premise of trust, to achieve the required standard of Quality. The Quality Management programmes should be based on the fundamental principle that the "Task Performer" shall own the responsibility to execute his tasks without defects and necessary resources and services will be made available to the "Task Performer" to fulfil this responsibility. Quality Assurance and Quality Control Departments render necessary services to different functional divisions and ensure that the total Quality Management is carried out effectively. The Quality Assurance Department, in particular, will endeavor to secure the customer's confidence and provide assurance regarding performance by the organisation and its vendors, besides ensuring internally an assurance to the Unit and Corporate Management regarding the maintenance of the required Quality status all through, right from input through process to output at customer's end.

Thus can be easily extended to vendor's premises and Quality Management at vendor's end demands a similar planned and systematic approach to Quality Assurance at every stage, enabling its every "Task Performer" to execute his tasks without defect with creation of evidence for his own and customer satisfaction.

Management, Technology and *SQC* are effective aids for Total Quality Assurance for both vendors and customers alike.

Basically all three above disciplines only enable a "Task Performer" to know :

(*a*) The means for knowing what he should be doing – *e.g.* Specification.

(*b*) The means for knowing what he is actually doing – *i.e.* measuring devices, instruments and observed/test data.

(*c*) The means for regulating the process, so that he can aim and *Do Right at the First Instance in a Status of Self-Control*.

Retaining this capability, a coordinated and integrated organisation frame can be suitably devised for various groups of "Task Performer" for different functionaries (including QA and QC) as a system and sub systems under it with defined interfaces. Health of each module of sub-systems and interfaces determines the system health and the effectiveness of contribution to the integrated company Objective of Quality.

18.23.4. Pitfalls of Breakthrough in Quality

(*a*) Pre-occupation with controls

(*b*) No alarm signal

(*c*) No standard approach

(*d*) Vague responsibility

(*e*) Fear of the amount of work required.

18.23.5. Guidelines for Introducing Change-Seven Steps

(*a*) Study existing cultural pattern

(*b*) Identify those aspects which need change

(*c*) Secure active participation of others involved (within and without). This may involve customer participation as well as his active assistance.

(*d*) Start on a small scale and use results to broaden application

(*e*) Make use of available tools

(*f*) Provide sufficient time for mental changes to take place

(*g*) Avoid surprises.

18.23.6. Quality Improvement Projects. *All quality improvements are made project by project and in no other way.*

Studies are required, by application of simple to sophisticated tools of statistical quality control and operation research.

The training and self-development programme should also be carried out as planned projects for real Quality break-throughs.

18.23.7. Advance Quality Plan, Surveillance and Quality Audit. For assurance of self and customers, some amount of pre-planning always helps each Task Performer and makes coordination interaction far more easier. Advance Quality Plan, as part of contractual obligation, with specific check lists and minimum agreed customer "hold-points" serves this purpose to create evidences for Quality operation and confidence.

Similarly a vendor must mentally prepare and subject himself to Quality Surveillance from his customer, who wants to see evidence of planning and defect prevention as part of on-line-control.

Further, as periodic re-look at the system effectiveness to ensure that all is well on the Quality front, in-house Audits may be organised. The vendor can accept and welcome the Customer Quality Audit from a position of strength and confidence, preparing himself for further improvements.

18.24. Features of Japanese Quality Management

It is desirable to study the salient features of Japanese Quality Management for drawing some lessons from it.

First of all Japan is an Asian Country like ours, who has been successful of late in technical capability and Quality in particular. Rather, it is the undoubtable leader in Quality in the whole world today.

Secondly, although its technological base is largely from Europe or America, it has a national talent to absorb by adopting some Scientific Management Principles, globally practised more successfully even compared to their soil of origin.

Thirdly, though Japan had a poor reputation for Quality even as late as pre-second World War days, within a short span, it has achieved a unique status in Quality.

It has a robust private industry with strong national backing. The example of large number of tiny, small, medium sector industries living with the large and giant units and still be able to meet all Quality needs because of nationwide Quality consciousness and commitment is spectacular and a lesson for manufacturers and vendors in our country.

Therefore, it is essential to study JAPAN as a model of Quality Management from a closer angle.

The Salient Features of the Japanese Quality Management are :

(*a*) **Company-wide quality control :**

(*i*) Company-wide participation, from top management to all levels of employees.

(*ii*) Participation

Not only by departments of technology, design, research and manufacturing but also by sales, materials, clerical and other management departments like planning, finance, personnel, business development, etc.

(*iii*) Q.C. concepts used

Not only in incoming material control, production process, new product design control, but also in business analysis for new company policy, for checking if top management policy is being implemented, for control of personnel and labour, for solving sales problem, etc.

(*b*) ***Quality education and training :***

(*i*) Company-wide Quality Control, requiring participation by all in Q.C. activities, call for education from top management to Bottom level worker.

(*ii*) Such education and training enables each employee to display his capabilities fully.

(*iii*) Eventually infinite possibilities emerge, both individual and company gaining benefit and profit.

(*iv*) Organisation of large number of long term seminars.

(*v*) For each company, such initial training requires more than three years to educate all executives involved.

(*c*) **Application of S.Q.C. methods :**

(*i*) Through education and training in SQC methods, all employees can apply SQC in practice.

(*ii*) Application of SQC is one of the pillars of Japanese QC.

(*iii*) Simple SQC methods very widely employed.

(*iv*) Extremely high level SQC methods developed to suit different problems.

(*d*) ***Quality control audit :***

(*i*) QC Audit carried out by Top Management of Company.

(*ii*) Each department Head carries out internal audit in his department.

(*iii*) QC Audit for affiliated companies and sub-suppliers.

(*e*) ***Q.C. circle activities :***

(*i*) A QC Circle is a *permanent* group with a leader and a few Circle members, usually 7-10 persons per group.

(*ii*) At present, in Japan, over one million QC Circles exist.

(*iii*) QC Circle leader and some members trained in SQC methods and application.

(*iv*) Each circle voluntarily performs QC activities involving in self-development Quality Control and improvements.

(*f*) **Nation-wide QC promotion activities :**

(*i*) In November every year, various events organised to observe "Quality Month".

(*ii*) QC Circle conferences-about 100 per year all over Japan.

(*iii*) QC Annual Conferences.

18.25. Quality Circles-Understanding and Practicing

18.25.1. Birth of Quality Circles

Quality Control Circle concept, which had its birth in Japan 25 years ago, has made its impact the World Over. In order to go into the values encompassed in the Quality Circles, it is necessary to briefly recapitulate as to *WHY* and *HOW* quality Circles come into being.

Today, it is well known that the Quality Circles came into being in Japan after the World War-II. It was necessary for Japan to put their shattered economy on its rails. To do so, they had to wipe out their poor image in Quality. It was a question of survival of a nation. Therefore, with the help of American Quality Management experts. Dr. Deming and Dr. Juran, the

Japanese managers learnt the Quality Control Techniques, and broadly all the aspects of Quality Management. It was at this point that Dr. K. Ishikawas of MUSHASHI Institute of Technology. Tokyo, added a new dimension to this customer effort by involving task performers at the grass roots level also to work towards the improvement of the quality. He was a visionary and he envisaged the advantages of small group working and asked the workers through his writings to form small voluntary groups and to undergo training in Quality related subjects to solve themselves their work related problems. He christened those small groups as Quality Control Circles.

According to the formal definition given by the Union of Japanese Scientists and Engineers (JUSE) "Quality Circle is a small group formed to perform voluntarily Quality Control activities leading to self development within the work place/workshop utilising Quality Control techniques with all members participation.

The focus naturally is on Quality and the QC Circles function as an integral part of Company Wide Quality Control. Training is an important component of the Quality Circle.

18.25.2. How Much of the Quality Problems could be Solved by Quality Circles. The mere introduction of Quality Circles does not mean that we can find solutions to all Quality problems. The Japanese made a top to bottom approach first and then bottom-up. As Dr. Juran has pointed out, actions for the improvement of Quality and solving Quality problems lie in the domain of the management to the tune of 80-85%. Dr. Juran says that upto 1981 the Japanese QC Circles had solved 15 million problems and out of these only one third related to Quality and the rest related to productivity, cost reduction, safety etc. To put in his words, "the QC Circles have at best accounted for 10% of the overall Japanese revolution in Quality". For the major problems of Quality it is in management cadres who have to take the bull by the horn. One example in this context is the lead taken by the Chief Executives of the Japanese Companies in heading the Quality Audit teams for their Annual Quality Audit.

18.25.3. The Impact of Quality Circles on other Aspects of the Organizational Activities. The impact of Quality Circles on the organization has to be examined on its totality to understand its values. One is from the stand point of the workmen, who would be directly involved in them ans second is from the stand point of the managers, who are expected to support them. In this context, according to Dr. Juran. "Many perhaps most of the projects of QC "Circles are now oriented to productivity and cost. From the employee view point, such studies help to improve the company's health, in turn, healthier the company more secure the employees since most Japanese join a company for life".

"From the company's view point, the time spent on such projects is cost effective and the efforts to carry out projects yield a by-product in the form of constructive team work between workers and managers. Some managers feel that this by-product is more important than the subject matter of the projects"

It transpires from Dr. Juran's observation that though QC Circles function as an integral part of company wide quality circle (CWQC) or total quality, the by-products or the spin off benefits of quality Circles in enhancing the organisation's group culture, team work, harmonious human relations etc. should not be lost sight of.

18.25.4. The Right Approach to Installing Quality Circles. In most cases the reasons for failure of Quality Circles could be found in the very first stage of the installation of Quality Circles. Here again the value systems come into focus.

Dr. Ishikawa in one of his papers on CWQC has listed 14 differences in the western and Japanese approach to management. A few of these are very relevant to us also.

Japanese Quality Circles are still going strong because they are built on sound foundations and principles.

18.25.5. Quality Circles in Japan. Some of the values which the Japanese cherish religiously uphold and meticulously practice, explain for their value based approach to Quality Circle, anchoring it into their cultural moorings.

(*a*) Respect for the human individual is very evident in all walks of their lives. For instance, bowing three times in solution to one another, irrespective of one's status in life is demonstrative of this principle.

(*b*) **WAAH :** Buddhism has taught them what is called the WAAH culture, which means "being at peace" : with oneself. When one does a good deed, a good job treating work itself as an activity of producing something of value to other, there is immense satisfaction—joy the of doing a thing right the first time. Could there be a better way of being at peace then working as a small team to do a good job ?

(*c*) **KAIZEN :** This is another nature of the Japanese which had brought about their economic miracle through improvements. Quality year after KAIZEN means "changing for the better". When Dr. Juran taught them to have annual Quality improvement plans as one of the three mantras for Quality improvement, the Japanese have put it into practice in letter and spirit.

(*d*) **ICHIGAI :** This is yet another term and a unique Quality characteristics of the Japanese. This simply means "being Worthwhile". Every Japanese wants to be worth while to his country, company and society. He wants to be worthy in the estimate of the others. Nothing can be more insulting to the self respect of the Japanese to than be called a worthless person. So, every Japanese generally, gives of his best ; wherever he goes, no matter whether his job is a president or a janitor. The Quality Circles give the workmen opportunity to demonstrate their self-worthiness in ample measure.

(*e*) **SEIRI, SEITON, SEIKSTA :** These three S's extoll the virtues of tidiness, orderliness and cleanliness at all times and at every place. These account for the clinically clean shop floors, road and also in their personal habits.

(*f*) **SEMPAI-KOHAI :** These two words explain in a nutshell the senior-junior relationship and accounts for the harmonious relationship between workers, supervisors and managers. The senior looks upon his junior like a younger brother and develops him. The senior earns his respect from the junior through his role as a teacher and guide. The senior does not view his junior as a threat to his future when he develops him. It is this spirit that pervades the working of Quality Circles also, where the Circles get full support of the manager.

(*g*) **TATEMAE-HONNE :** The Japanese always believe in consensus decision making and consensus forms the basic principle of respect for the human individual. However a leader does not absolve himself of his responsibility to lead, the team. The world TATEMAE-HONNE means "leader always in front". While taking the responsibility to lead, the leader gives the credit to the team for its success. This is yet another principle on Quality Circles it may seem that, though the Quality Circle concept was started to focus mainly on Quality first, the concept itself is spun around sound value systems without which the concept itself would have fallen on the way side like many other concepts.

The value based approach even in business could be seen in the company's policies and objectives laid down by them. For example, one of the Japanese industrial grants has listed seven spiritual objectives as a part of their company's basic business principles, beliefs and values.

SOLVED PROBLEMS

Q. 18.1. Define the terms : quality, quality assurance, and quality control.

Ans. Quality is the totality of features and characteristics of a product or service that bear on its ability to satisfy a given need.

The quality can be assured, controlled and improved only if it can be evaluated. It is thus essential that those characteristics and features, which have bearing upon fitness for purpose of a product or service, be identified. The ability to satisfy a given need includes economics as well as availability, maintainability, reliability, design and other characteristics.

Quality Assurance concerns with all activities and functions to achieve quality. This objective can be met by the application of disciplines and techniques defined in standards. Quality control concerns the operational techniques and activities that sustain the product or service quality to specified requirements.

Q. 18.2. What is the role of standards for quality ?

Ans. Standards promote consistent quality and economic production. They provide for quality of goods and services by defining those features and characteristics that govern their ability to satisfy given needs.

Standard specifications for materials, components and processes form the backbone of any quality assurance programme.

Standards for sampling, gauging and measurement lay down the methods of test and calibration which are basic tools for assessing quality.

Q. 18.3. What are the roles of management and suppliers to ensure quality of the product ?

Ans. Management has to define the objective clearly based on the need of customer and market. Management has to be clear about adoption of total quality management, *i.e.*, assure quality at all stages, right from design, planning, procurement, manufacturing to assembly ; testing and packing, etc. All persons have to be pursuaded to have the correct altitude of mind towards quality. It must be remembered that prevention is usually less costly than cure. It is essential to clearly define the purpose and aims to be achieved. The objectives have to be clearly communicated. A programme for quality assurance must be established to provide basis upon which periodic reviews can be taken and evaluations made based on data feedback. Management has to ensure that each person in the organisation is committed to quality ; he clearly understands that quality assurance is important to his future, he understands the procedures and is convinced that he can assist in achievement of adequate quality.

To ensure quality should not be the function of quality department but each department should contribute to quality.

Role of supplier. The supplier is responsible for providing goods or services in accordance with specification on schedule at the agreed price. A precise specification for this purpose is essential. The inspection of end product alone is not sufficient for quality but his quality assurance programme must be evaluated. It must be appropriate, practical, up-to-date, correspond to reality and be fully implemented. The quality should not be adversely affected by chance variations, like changes in key personnel, materials obtainable, etc. The supplier's quality management system has to be thoroughly assessed.

Q. 18.4. How the quality of products can be improved ?

Ans. Quality may be defined as providing fitness for purpose with value for money. A quality product is one which full conforms to the agreed requirements of the customer. A customer is concerned with

— the specifications (his expectations)
— conformance to specification (what he receives)
— reliability (how well it performs)
— price (how much does he pay during the life time use of the product)
— delivery (how long he waits)

Quality products can be obtained only when the top management dedicates itself to achieving improved quality and persuades its entire work forces to accept that "either they deliver the goods defect free" ; and to further accept that "only a quality business can deliver a quality product". Quality is a matter of personal commitment of the top management and the total organisation, leading on to good team performance through effective leadership.

Q. 18.5. Write a short note on quality control.

Ans. Quality control is an important sub-set of overall quality assurance. Control means measuring and applying corrective action. It must be remembered that quality can not be inspected in, it can only be designed and built into the product. Prevention of defects of any type is the essential feature of modern quality control. The sequence of actions to control quality can be likened to a closedloop, error-feedback servo-system. Maximum efficiency of the control of quality can be achieved when the speed of response of this servo is faster than the rate of change of those processes/parameters which cause errors. Thus, the more closely the control loop is applied to the point of manufacture the greater the efficiency. Quality control must be planned into manufacturing systems at the time of designing those systems.

Q. 18.6. Pick up the correct choice from three alternatives provided for each question.

(*a*) ...... is a management function, whereby control of the quality of manufactured products and raw materials is exercised to prevent the manufacture of defective parts.
(Reliability, Engineering design, Quality control)

(*b*) ...... is the probability that an item will perform a stated function satisfactorily for a stated time period under specified conditions. (Reliability, Quality, Availability)

(*c*) ...... is the probability that the item is functioning normally at any point in time when operated subject to specified conditions. (Reliability, Availability, Quality assurance)

(*d*) ...... is the search for the solution that will generate the maximum benefit.
(Redundancy, Optimisation acceptable quality level)

(*e*) ...... is the existence of more than one means of performing a stated function.
(Optimisation, Redundancy, Reliability)

(*f*) ...... is the termination of a unit that carries out a specified function.
(Hazard rate, Failure, Useful life)

(*g*) ...... is the length of time a unit functions with a failure rate considered to be acceptable. (Useful life, Hazard rate, Acceptable, Quality level)

[**Ans.** (*a*) quality control, (*b*) reliability, (*c*) availability, (*d*) optimisation, (*e*) redundancy, (*f*) failure, (*g*) useful life]

Q. 18.7. Match parts A and B regarding quality.

A	*B*
1. Totality of features and characteristics of a product or service that bear on its ability to satisfy a given need.	(*a*) Quality assurance
2. The document that prescribes in detail the requirements with which the product or service has to comply.	(*b*) Design
3. An activity giving solutions for repetitive application, to problems essentially in the spheres of science, technology and economics.	(*c*) Quality control
4. The process of task recognition and problem-solving with the object of fulfilling needs by the creation of products and services.	(*d*) Reliability
5. All activities and functions concerned with the attainment of quality.	(*e*) Inspection by attributes
6. The ability of an item to perform a required function under stated conditions for a stated period of time.	(*f*) Standardisation
7. A procedure in which certain characteristics of an item are assessed without measurement and classified as conforming or not conforming to specified requirements.	(*g*) Specification
8. The operational techniques and activities that sustain the produced or service quality to specified requirements.	(*h*) Quality

ANSWER

1. (*h*)	**2.** (*g*)	**3.** (*f*)	**4.** (*b*)	**5.** (*a*)
6. (*d*)	**7.** (*e*)	**8.** (*c*).		

Q. 18.8. Quality can be defined as conformance to specifications, fitness for use, and customer satisfaction at competitive cost. Define the terms used in this definition.

Ans. The customer finds a key place in quality. His requirements have to be clearly spelled through specification. The quality of conformance to specifications and workmanship criteria during manufacturing is important. It is not enough to ensure quality of design, but to make sure quality of conformance, *i.e.* the product will perform as desired.

Products or services provided to human beings should respond to the overall needs of the user in price, delivery and fitness for use.

Fitness for use is determined by those features of the product which the user can recognise as beneficial to him. Fitness of use is judged by user and not manufacturer.

Product or service quality refers to the producer's ability to satisfy user at a cost which he can afford and still be able to realise a profit. An organisation exists because it has to meet the needs of consumer but it has also to provide services at a competitive cost and still make a profit. One must remember the rule that there will always be a way of improving quality and reliability at reduced cost.

Q. 18.9. What do you understand by company-wide quality management and what are its advantages ?

Ans. Quality conscious companies prefer to adopt total quality management or total quality control which is also referred to as company wide quality management. Involvement of all employees in such a programme is essential. Quality at competitive cost, keeping delivery commitment and safety aspects is key essence of total quality control. Number of defectives reduce considerably with adoption of such systems. Quality becomes uniform, reliability increases, confidence in product and customer's trust improves and number of complaints get lowered. Human relationship within company improves and flow of information becomes smooth.

Q. 18.10. Provide a single word for the following sentences associated with quality expected :

1. A measure of the degree to which device conforms to applicable specification and workmanship standards.
2. The probability that a system will operate at any specified instant of time, when used under stated conditions.
3. The ability of an item to perform a required function under stated conditions for a stated period of time.
4. A gradual impairment in ability to perform.
5. The operation of an equipment (complex item) prior to use to detect and replace parts that are defective or expected to fail and to correct errors in fabrication or assembly.
6. The termination of the ability of an item to perform its required function.
7. Actions necessary for retaining a device in or restoring it to a specified condition.
8. The existence of more than one means of performing a given function.
9. The probability of rendering the wrong decision based on pessimistic data or analysis.
10. The process of detecting and eliminating a failure with the aim of restoring the system to its operational effectiveness.
11. A science based on all the known methods of quality control and quality engineering to assure the manufacture of a product of acceptable quality standards.
12. The operation of a device under stress prior to its ultimate application intended to stabilise its characteristics and to identify early failures.
13. Sudden and complete failure.
14. Total time during which a system is not an acceptable operating condition or can't be operated without some repair or maintenance activity.
15. A failure that occurs as a result of deterioration process or mechanical wear and whose probability of occurrence increases with time.
16. The probability that a failed system will be restored to operable condition within a specified active repair time.
17. Tests performed on all equipment prior to shipment to a customer in order to ensure a satisfactory product.
18. Standardisation in order to create national, regional, and international product compatibility and transparency.

19. A deductive type of failure analysis that focusses on one particular undesired event at a time, and then provides a method for determining the possible causes of that event.

20. The extent to which the product complies with the specifications, standards, and workmanship criteria.

21. A graphical display of a quality characteristic that has been measured or computed from a sample versus the sample number of time.

22. Judging whether an article/device/ a lot is defective or non-defective by comparing the result of a test carried out by some means or with a quality criteria.

23. Procedure by which an authoritative body recognises the competence of a group to conduct registration activities, tests, or other conformity assessment activities.

24. Conformance assessment mechanism where third party conducts an assessment to determine compliance to ISO 9000 standards.

25. Pursuing the thrust areas of maximisation of profit, fast response to market needs, cycle time reduction and optimum utilisation of resources.

Answers

1. Quality **2.** Availability **3.** Reliability **4.** Degradation
5. Debugging **6.** Failure **7.** Maintenance **8.** Redundancy
9. Risk **10.** Repair **11.** Quality assurance
12. Burn-in **13.** Catastrophic failure **14.** Downtime
15. Wearout failure **16.** Repairability **17.** Acceptable test
18. Harmonisation **19.** Fault tree analysis
20. Conformance **21.** Control chart **22.** Inspection **23.** Accreditation
24. Quality audit **25.** Strategic business plan.

19

Statistical Quality Control

19.1. Introduction

It has been the aim of engineers to strive at higher and higher accuracy, but absolute accuracy is unattainable. Every process and its output are subject to variation. No two things can be made exactly alike. In mass-production, therefore, no attempts are made to attain absolute perfection but to control the degree of variability so that parts lie within the prescribed tolerances. Statistics acts as a practical tool in this direction. Statistics refers to the collection, organisation, analysis, interpretation, and presentation of data. Statistical quality control (*SQC*) involves analysis of the characteristics of a production output by inference from sampling the output. The analysis, in turn, is used to make the necessary corrections to ensure that no defectives are produced and process remains in control. There are two types of causes responsible for change of dimension on any component manufactured.

1. *Inherent or random (chance) sources of variation* :

This is caused by a large number of chance factors and depends upon the rigidity of machine tool and its process capability. Some examples are human variation in setting control dials, slight variation in machine, slight variation in raw material. This natural variability of the process can't be controlled and it occurs even under the best of methods and circumstances.

Though variation within stable pattern is inevitable due to inherent/chance causes, but as long as process operates within this established pattern, it is said to be in statistical control.

2. *Assignable causes of variation* :

Variations in size due to these causes can be checked as they are within the reach of man. Variations of this type are caused by human factors, tool wear, machine wear, or tool getting loose, or due to change of initial setting, differences in raw materials, environments, etc. It is the function of a system of process control to distinguish between the two types of variability causing the variation of size. The assignable faults, if present, must be detected and corrected before defective pieces are produced. The statistical quality control techniques attempt to locate the defect while the work is in process of production.

In statistical quality control techniques, the actual physical measurements are first secured by means of inspection. The degree of variation among the measurements is analysed by statistical methods in terms of a standard value of variation among acceptable parts. By such comparisons, quality control is made to predict the significance of the variation analysed. Variation in product quality is constantly studied within batches, on processing equipment, between different lots of the same article on critical quality characteristics and standards. This type of prediction can then be used to exert a positive influence over the control of product quality by studying the assignable causes. The quality control is thus directly associated with inspection, and for quality control, controls are maintained. In order that things may be economical and quick, 100% inspection is not desirable. Rather 100% inspection leads to fatigue of human brain and faulty parts are likely to be accepted. Control charts tell immediately whether the product is satisfactory or machine should be shut down and cause searched for.

The modern quality control is defined as the combination of all the devices and techniques that are used to control product quality at the most economical cost which yield adequate customer satisfaction ; whereas the statistical quality control is that part of the means for establishing and achieving a quality specification, which requires use of the tools of statistics. The aim of the statistical methods of quality control is to provide a rapid and objective way of deciding when measurable quality variations should be investigated and when they should be dismissed as chance variations which probably would not be profitable to investigate. The statistical methods help to increase the value of the ratio "$\left(\dfrac{Quality}{cost}\right)$" *i.e.* offer better quality at the same cost, or by offering the same quality at a lower cost, or by both raising the quality and lowering the cost. In addition these methods also offer to achieve the following :

(*i*) Better quality level.

(*ii*) Better uniformity of quality.

(*iii*) Better utilisation of raw materials.

(*iv*) Most efficient utilisation of equipment.

(*v*) Less scrap and rework.

(*vi*) Better inspection.

(*vii*) Improved producer-consumer relations.

(*viii*) Better specifications.

Before we proceed to study the theory and the methods of quality control, it is better to understand a few terms.

Attributes and Variables. Those measurements in which a quality characteristics is present or absent, or falls into a limited number of discrete categories, are referred to attribute measurements *e.g.*, methods in which item examined is considered satisfactory or unsatisfactory. *GO* and *NO GO* gauges provide measurement by attributes. Measurements by attributes are generally less expensive, but many more measurements have to be taken in such cases as there is no way of telling to what extent an item fails to meet the desired quality level when it is rejected on an attribute basis.

Measurements by variables involve the use of a scale which can be theoretically sub-divided infinitely and the results of the quality can be expressed quantitatively, *i.e.*, in numbers. The measures of length, size, weight, surface finish etc. are the examples of measurements by variables.

Variability. The two things are said to be alike when the differences among them are so small that we do not care about them. But generally two things considered to be identical vary from each other when sufficiently precise measuring equipment is available. In other words, the things can be considered alike, if the existing variability is unimportant from a practicable point of view.

19.2. Data Presentation

In all the industries, volumes of data are collected every day on the various aspects. These data are collected to serve as the base material for making useful and meaningful decisions. The inspectors or supervisors collect data on the various aspects of quality and productivity. The main objective of this exercise is to assure that quality which the customer wants at a cost which he can afford to pay and to deliver at the right time. Another important objective is to continuously assess the current status for further improvement. How could these objectives be achieved ? Reliable and required data should be collected with the desired accuracy. These data should be analysed using the right statistical tools for drawing meaningful conclusions.

The summary for results should be presented in suitable tables, graphs, and diagrams.

Data Bank

Prior to collection of data one should have a definite and clear cut idea of the nature of problem on hand and the basic objective of the study. The particular aspects to be studied and the desired accuracy must be clearly specified. Every type of difficulty that is likely to arise in the course of investigation should be foreseen and suitable course of action must be preplanned. It is always better to first arrange a pilot survey and ascertain the loopholes. Data sheets should thereafter be modified. No data sheet can, however, be made perfect. Truthful recording of various factors is a necessary pre-requisite. Many a time it is noticed that conclusions are drawn prior to or in course of data collection. In such cases data are merely collected to confirm the predrawn conclusions. Such dogma must be avoided. Results of statistical investigation are likely to be affected by different types of errors. It is essential, therefore, that data should be cross-examined for genuineness during collection and also before statistical analysis is undertaken. Also, after the analysis is over, final results should be carefully checked against the results of other comparable investigations.

Scrutiny of data

Important decisions are always taken on the basis of the collected data and if it is erroneous, it is very likely that a wrong decision would be favoured which defeats the purpose of data collection. The errors may creep in because of sampling and testing error, errors in the measuring devices and so on.

Compilation of data : Raw data should first be compiled for suitable presentations. A statistical table is a convenient way of presenting numerical information so that essential comparisons are brought by suitable placement of relevant figures near one another. It is two way arrangement of data in rows and columns against classified headings.

Every table has broadly five parts :

(*a*) **Title of the table :** This comes at the top of the table. It should be brief, explicit and self-explanatory.

(*b*) **Column caption :** Since it is easier for the eye to compare entries in a row, usually the more important way of classification is arranged columnwise.

(*c*) **Row Caption :** Relatively less important items are covered here.

(*d*) **Body :** The set of entries in the appropriate cells of the table together with totals, sub-totals etc., form the body of the table.

(*e*) **Foot note :** The source of information is generally cited at the bottom right hand corner of the table.

Methods of Data Presentation

(*i*) **Pie Diagram.** A pie diagram is a circle divided into different sectors, with angles at centre proportional to different components of a total. It is usual to start from a horizontal radius to the right and proceed in the anti-clockwise direction giving the quantities in a decreasing order of magnitude except the miscellaneous item which is given last.

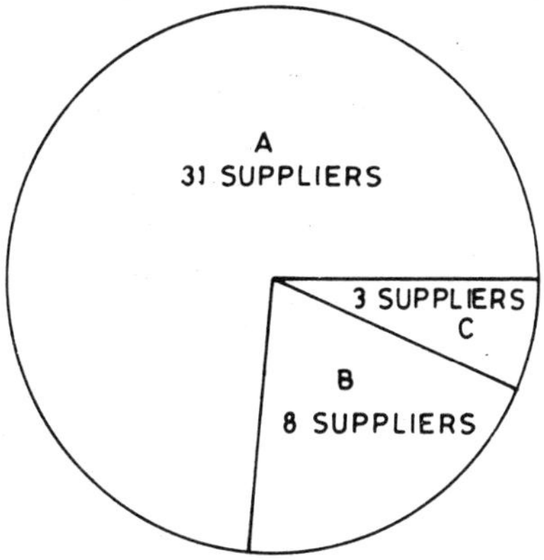

Supplier	*Rejections*
A	00.00%
B	19.50%
C	80.50%

Fig. 19.1. Classification of suppliers with regard to quantity of rejections.

Sectors can also be coloured or marked to emphasise divisions.

Supplierwise lot acceptance of castings in specified grade, other grades and rejected or retest bars ; breakdown of the production cost into labour, material and overheads ; apportionment of quality cost into appraisal, failure and preventive costs ; Percentage contribution of various causes of defectives etc., can well be represented by Pie-Diagram.

For example, supplierwise and causewise analysis for rejections of a particular product is given in Figs. 19.1 and 19.2 respectively.

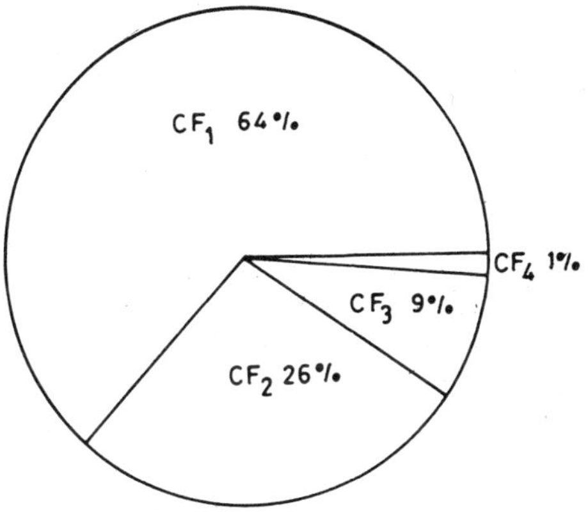

CF_1 Visual/*RT* / *UT* Failure
CF_2 Dimentional Deficiency
CF_3 Non Availability of *TC*
CF_4 Lab Result and *TC* Disagree

Fig. 19.2. Causewise rejection analysis.

(*ii*) **Bar diagram.** It consists of vertical or sometimes horizontal bars of width and of height proportional to the magnitudes of the variable for the different items. They all stand on the same bar line and are separated from each other by equal intervals. A scale parallel to the bars is provided with graduations to read the magnitudes easily. Below each bar, the period or details to what it stands for, is written. Causes of scrap or rework, reasons for non-utilisation of machine, etc., are represented along the X axis, and Y axis gives the magnitudes.

If the data are classified by two characteristics, bars are drawn with heights adjusted according to one characteristic and each bar may be divided into sections according to other classification. The details of quantity supplied and the number rejected for 7 suppliers are presented in the chart in Fig. 19.3.

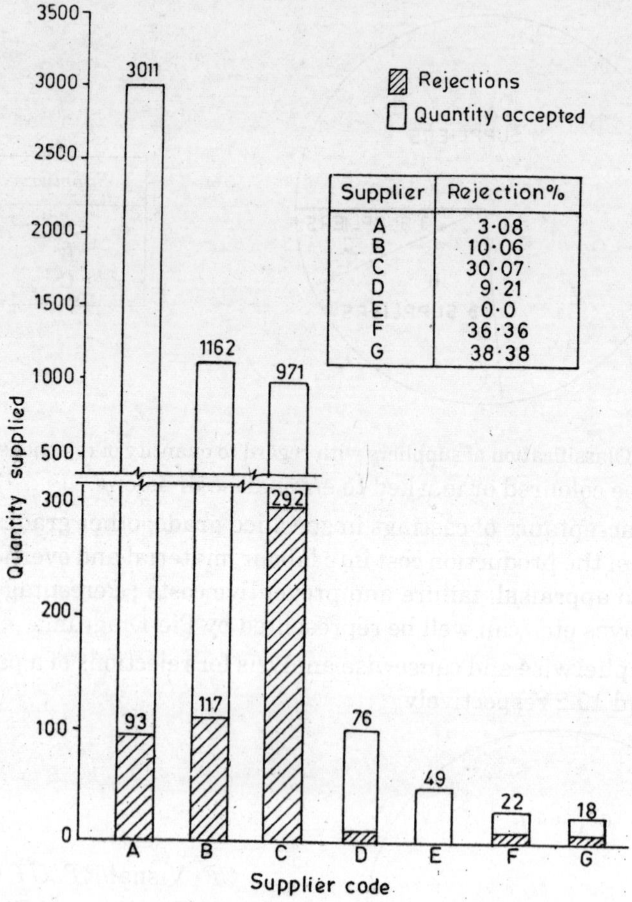

Supplier	Rejection%
A	3·08
B	10·06
C	30·07
D	9·21
E	0·0
F	36·36
G	38·38

Fig. 19.3. Supplierwise quantity supplied and rejections detected.

(*iii*) **Histogram.** It is one of the diagrams especially suited to represent frequency distribution. The X-axis is divided into class intervals and on them the vertical bars are erected with heights proportional to the class frequency. Histogram is thus represented by a system of continuous rectangles in which the area of a rectangle represents the frequency. Outgoing quality evaluation of component dimensions, pattern of variation of chemical and physical properties of castings can well be brought under the purview of this.

The histogram showing the distribution of *Green compression strength* in Kg/cm^2 is given in Fig. 19.4.

(*iv*) **Rectilinear Chart.** This is very useful to gauge the increase or decrease of average number of exposures per casting,

— sales to inventory ratio,

— production, machine utilisation, etc.

Usually, *X*-axis represents month, date and shift, machine number etc. The production figures, utilisation particulars of scrap or rework percentage, etc. are plotted as dots on the *X*-axis. They are joined together by straight lines to form a line chart. Monthly production, reworks, utilisation of different work centres, group efficiency, etc., can be plotted on such graphs.

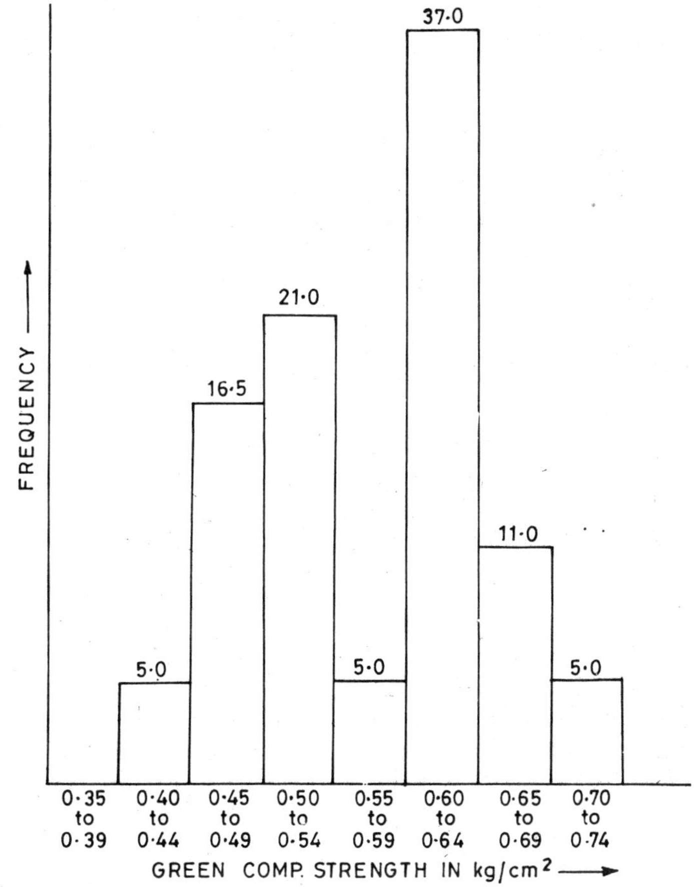

Max : 0.70 kg/cm^2, Min : 0.40 kg/cm^2, Range : 0.30 kg/cm^2.

Fig. 19.4. Histogram showing the green compression strength in kg/cm^2.

When two characteristics are compared, for example, yearly purchases and consumption of stores, etc., two bars are drawn against the period and are coloured or marked differently. For example, the percentage of defectives of a particular product is given in Fig. 19.5.

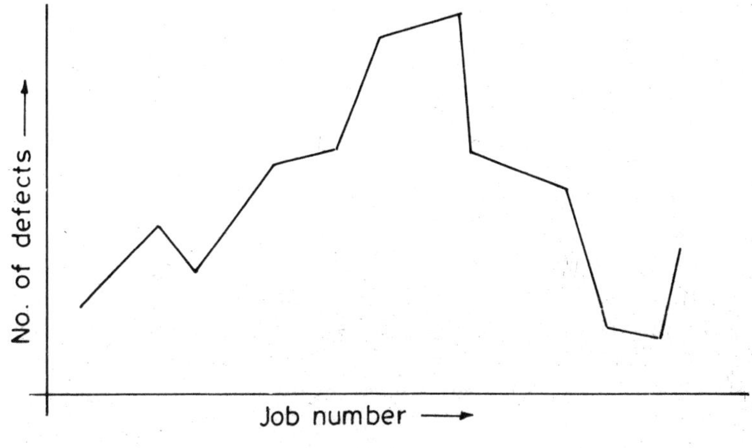

Fig. 19.5

19.3. Some Statistical Measures

The various useful statistical measures are average (arithmetic mean), range, standard deviation etc. Arithmetic mean represents the measure of central tendency. If $x_1, x_2, x_3 \ldots x_n$ are the values of n items observed, then arithmetic mean.

$$\bar{x} = \frac{x_1 + x_2 + x_3 + \ldots + x_n}{n}$$

Though arithmetic mean is rigidly defined as it is based on all observations of a series thus being least affected by fluctuations of sampling, it has the following drawbacks : *(i)* It gives greater importance to bigger items of a series and lesser to smaller ones ; *(ii)* It may give a wrong idea about actual series as two series (set of observations) entirely different from each other may have same value of arithmetic mean. Thus to have complete picture of set of observations, either range or standard deviation must also be specified.

Other term commonly used is median which is the middle value of distribution, half the values being below it and half above it. The mode is the value occurring most frequently in a distribution. The mid-range is the average of the smallest and the largest readings.

Range (R) which is the measure of dispersion, is the difference between the largest and the smallest observed values.

Standard deviation (σ) is the most useful measure of dispersion of a set of observations. σ is equal to root mean square deviation of the observed numbers from their arithmetic means $\bar{x}$, *i.e.*,

$$\sigma = \sqrt{\frac{(x_1 - \bar{x})^2 + (x_2 - \bar{x})^2 + \ldots (x_n - \bar{x})^2}{n}}$$

The standard deviation is the most significant characteristics of a distribution because it provides an arbitrary measure of the probability that a certain number of measurements will fall within a particular range of values. Variance (σ^2) is defined as the square of the standard deviation.

Since usually the data present is huge, for finding the mean, range and standard deviation of any number of observations, the various dimensions are arranged at random in sub-groups or it may be observed that how many times a dimension occurs. Thus we have a tabulation of dimension and frequency ; then mean size $\bar{x}$

$$= \frac{\Sigma (fx)}{\Sigma f}$$

x = size of any article or any dimension and f = frequency

and standard deviation $= \sqrt{\dfrac{\Sigma (x - \bar{x})^2}{\Sigma f}}$

In case of mass production, at random four or five articles are chosen depending upon the size of sub-group and their mean, range and standard deviation found.

19.4. Summarisation and Pictorial Presentation of Data

In inspection of any item, many observations are required to be taken and thus it results in the accumulation of so many numbers. Obviously it is very difficult for human being to conclude anything or draw any inference from these numbers as such. It is, therefore, necessary to condense data into a few numbers that we consider to be the important aspects of the original numbers. The condensed numbers which are representatives of original data are arithmetic mean, median, mode, range and standard deviation.

The most commonly used measure of central trend is the arithmetic mean, which consists of the sum of numbers divided by the number of items. The other measures of central trend are median and mode.

The median is that number which is exceeded by as many numbers as fall below it. It is most conveniently determined by first arranging the numbers in sequence of magnitude and then determining the middle number. Its chief advantage is that it is generally easy to determine and it is unaffected by extreme items.

The mode is that value which occurs most frequently in a series of numbers. In case, no two numbers are the same then mode may be determined by arranging the numbers into class intervals and mode will be that class interval in which the greatest number of observations fall.

Measures of dispersion are generally expressed by range and standard deviation and their aim is to evaluate the extent to which observations are scattered around the mean. Range is the difference of maximum and minimum readings in a set of observations. Standard deviation (σ) is defined as the root mean square of the differences between the observed values and the mean.

Pictorial Presentation of Data. There are many different ways of presenting data pictorially, but only a few most commonly used ways in the quality control work are described here.

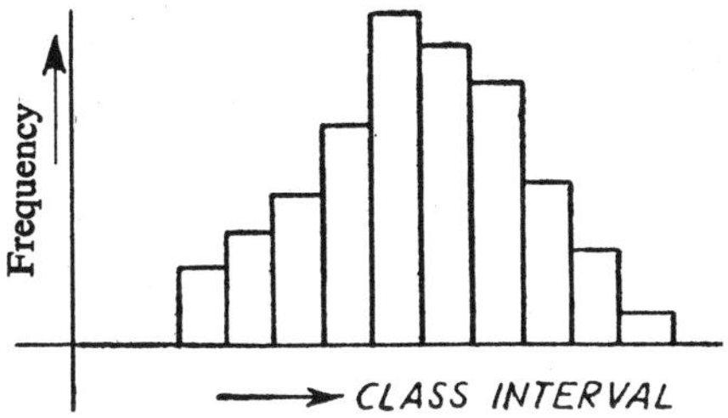

Fig. 19.6. Frequency in Histogram.

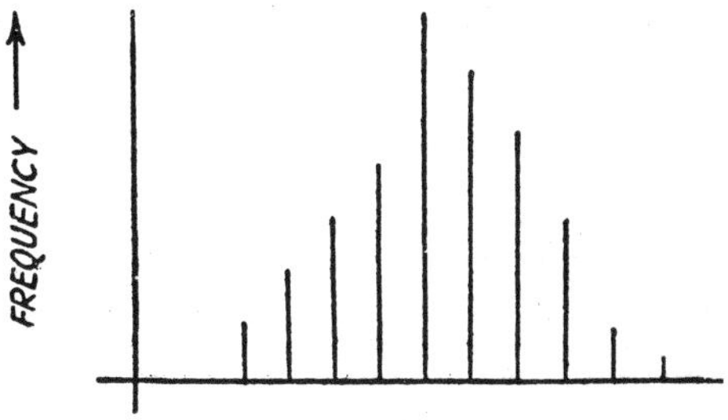

Fig. 19.7. Frequency Bar Chart.

(*i*) *Frequency Histogram.* In some circumstances, this is the best way of graphically representing the frequency distribution. In this graph the sides of the columns represent the upper and lower cell boundaries and their heights (and areas) are proportional to the frequencies within the cells.

(*ii*) *Frequency Bar Chart.* In this, bars are centered on the midpoints of the cells ; the heights of the bars being proportional to the frequencies in the respective cells.

(*iii*) *Frequency Polygon.* It consists of a series of straight lines joining small circles which are plotted at cell mid-points with a height proportional to cell frequency. The frequency polygon can generally be used to better advantage if two or more distributions are to be plotted on the same diagram. The frequency polygon implies that all the values in an interval are concentrated at the midpoint, whereas in the histogram they are spread uniformly over the interval.

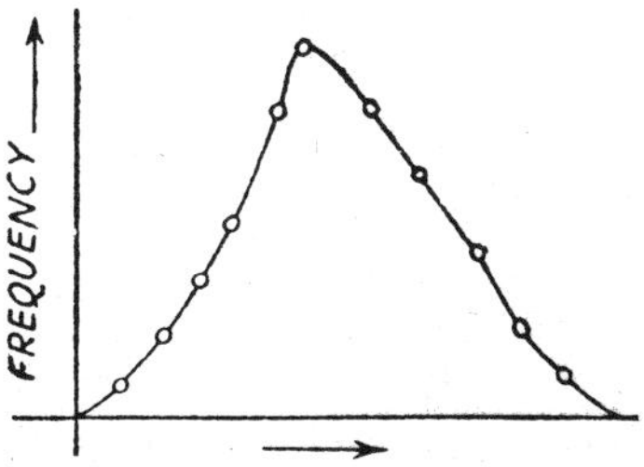

Fig. 19.8. Frequency Polygon.

(*iv*) *Ogive (Cumulative Frequency Curve).* The chief advantage of the ogive is the ease with which cumulative percentages can be read.

(*v*) *Graph Papers.* The various types of graph papers generally used in statistical work are :

(*a*) *Semi-logarithmic paper.* On this paper, one of the scale consists of equal sub-divisions and the other scale is logarithmic. The advantage of such a paper is that it permits greater precision when reading smaller values on the log scale. Also the logarithmic relations can be plotted as straight lines on such a paper.

(*b*) *Log-log paper.* This type has logarithmic scales in both directions. In addition to permitting greater precision when reading smaller values, this paper eliminates the convergence of lines (straight or curved) having a common origin on equal interval paper.

(*c*) *Probability paper.* It provides a way of quickly checking the normality of a distribution.

19.5. Variations in Quality

Variation in quality characteristic is inevitable, however well the process may be controlled and tests carried out. Thus, for example, measurements taken on some components are bound to vary from piece to piece even though the process may be well under control.

This is a common shop experience. This variation is attributable to two types of causes.

—Natural or Chance Causes and —Assignable Causes.

Variations due to natural or chance causes are inherent in a process. It is due to multitude of causes which are difficult to identify and uneconomical to eliminate. Further, variations due to chance causes follow statistical laws. Assignable causes of variation are generally due to few individual causes which can be identified and eliminated. These two types of causes can be distinguished as follows :

Sl. No.	*Chance Causes*	*Assignable Causes*
(*i*)	Consists of many individual causes	Consists of a few individual causes
(*ii*)	Any one chance cause results in only a very minute amount of variation.	Any one assignable cause can result in a large amount of variation.
(*iii*)	Some typical chance causes of variation are small ; variation in raw material ; small vibration of a machine ; lack of perfection in reading instruments.	Some typical assignable causes of variations are : A batch of defective raw material ; Faulty set-up ; New operator.
(*iv*)	As a practical matter chance variation cannot economically be eliminated from a process.	The presence of assignable variation can be detected and action to eliminate the causes is usually economically justified.
(*v*)	Follow statistical laws of variation.	Do not follow any statistical law.

A process is said to be in a state of statistical control if the variations in the quality characteristics of a product follow the appropriate statistical law of variation. Thus, the objectives of controlling a process can be stated as :

(*a*) to restrict the causes of variation in the quality characteristic to chance causes ;

(*b*) to detect and eliminate the assignable causes of variation.

Analysis of pattern of variation exhibited by data on quality characteristics gives a clue to the behaviour of the process and helps in bringing the process under control.

Frequency Distribution—A Simple Method for Analysis of Data

A first step in examining a set of data is to summarise them in a form such that the main features *viz.*, the extent of variation and the pattern of the observed readings in that range are available in necessary details. This is accomplished by forming a frequency distribution of the data. An understanding of frequency distribution will provide information on ways to measure variability in products, processes and other operations and permits one to develop the ability of recognising and explaining types and significance of variations.

An example of a frequency distribution is given for the outside diameter measurements on yoke for 15/40 stop valve.

The range of observed variation is from 47.70 mm to 48.02 mm. This range has been divided into class intervals each of width .03 mm.

Frequency Distribution of O.D. of Yoke for 15/40 Stop Valve

O.D. (mm) Range (Class Interval)	*No. of Observations (Frequency)*
47.70—47.73	1
47.73—47.76	2
47.76—47.79	6
47.79—47.82	13
47.82—47.85	18
47.85—47.88	21
47.88—47.91	16
47.91—47.94	12
47.94—47.97	6
47.97—48.00	3
48.00—48.03	2
Total	100

It is evident from the table that :

(*a*) The observations cluster around a central value of 47.87 mm.

(*b*) The values are spread roughly symmetrically around the central value.

(*c*) Small deviations from central value are more frequent than large ones.

The histogram and the frequency curve of the above frequency distribution are given in Fig. 19.9 and Fig. 19.10.

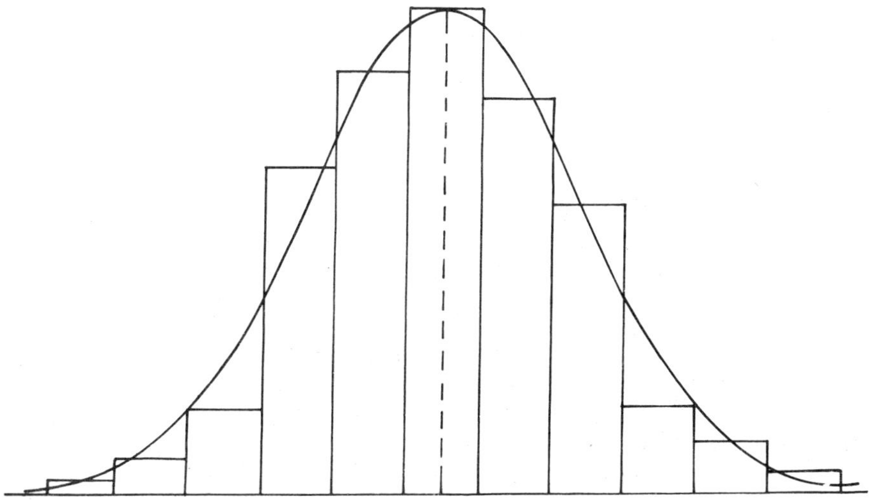

Fig. 19.9. Histogram for data on O.D. measurement on yoke for 15/40 stop valve.

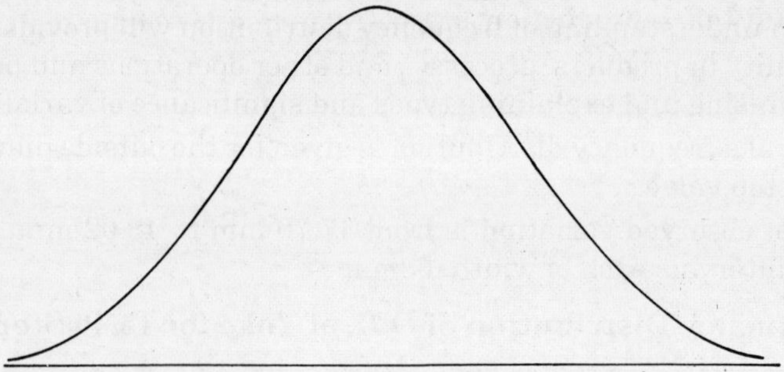

Fig. 19.10. Frequency curve for data on O.D. measurements on yoke for 15/40 stop valve.

19.6. Frequency Curve

If the number of observations were very much larger and if the groups were made much narrower, then the histogram approaches a smooth curve for the frequency distribution of observations on outside diameter measurement on 15/40 stop valve.

The frequency curve gives the spread of the whole population and the area covered by the curve between any two points on the range is proportional to the frequency in that range. When the curve is so expressed that the total area under the curve is unity, it is referred to as Probability Distribution. In such a case the probability or chance that any observation will lie between two given points is the total area under the curve enclosed by two vertical lines at those points.

19.7. Shape of Frequency Distribution

If the shape of the graphical representation is close to the bell shaped normal distribution pattern given in figure 19.10 (stricter statistical tests are available to confirm this), then the variation in the recorded measurements can be considered to be due to chance causes and the process from which the data were obtained can be considered to be in a state of control. In such cases 99.73% of the values are expected to lie within $\mu \pm 3\sigma$. Some of the typical shapes of Industrial Frequency Distribution are given in Figs. 19.11 to 19.14.

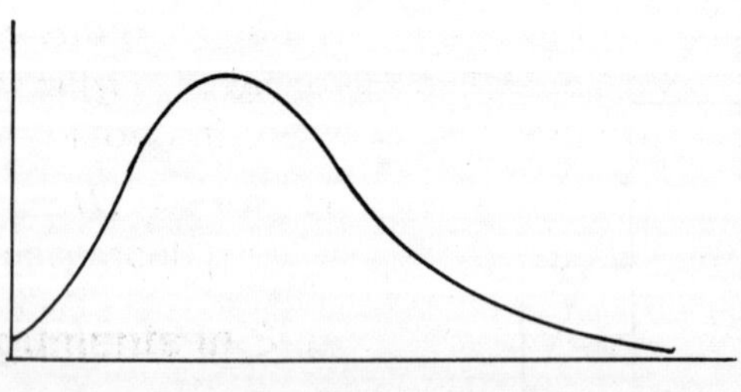

Fig. 19.11. Shows the skew frequency distribution.

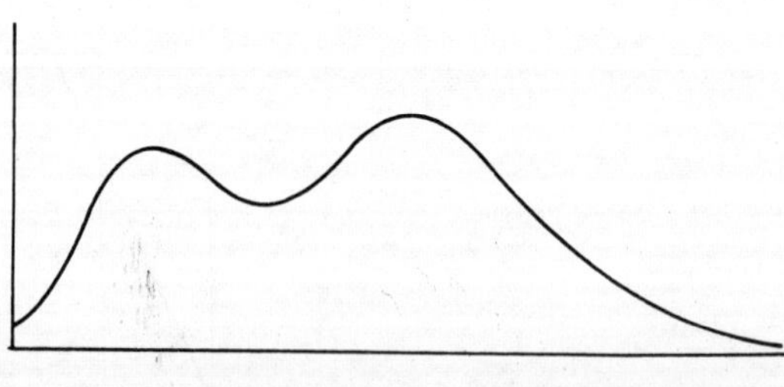

Fig. 19.12. Shows the 'Bimodal' distribution.

Practical Implications of Frequency Distributions :

(*a*) Frequency distribution patterns reveal the relative amount of variation in a lot.

(*b*) Frequency distributions help us to determine whether or not the product is within tolerances.

(*c*) Frequency distributions are particularly useful in evaluation of the quality of incoming lots.

(*d*) Capability of a production process can also be evaluated by means of frequency distributions.

This is often the result of some dominant factors.

It indicates batches of similar product obtained from two different sources-different vendors, machines, shifts, operators.

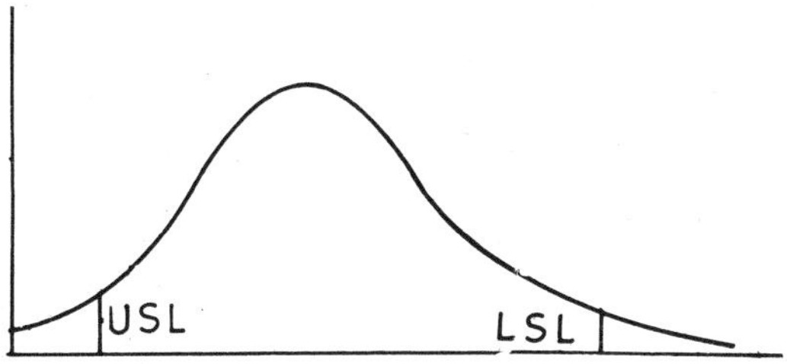

Fig. 19.13
LSL—Lower specification Limit.

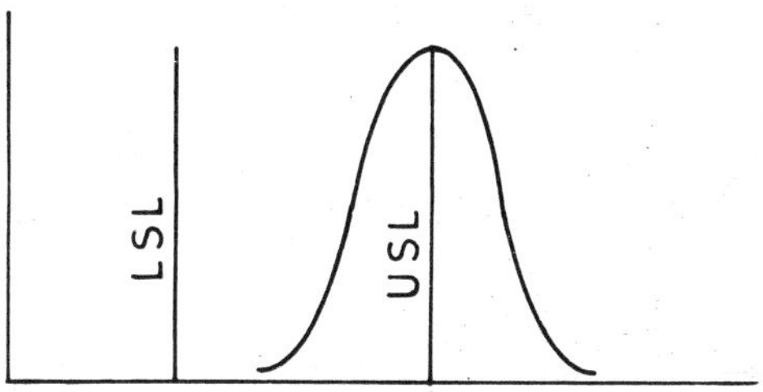

Fig. 19.14
USL—Upper specification Limit.

Fig. 19.13 shows the normal distribution when products having variation wider than that of the required specification are subject to 100% inspection. The accepted products exhibit variation of the type shown in figure 19.14.

Fig. 19.14 shows that variability of product is smaller compared to the required specification and centering of the process close to the upper limit leading to over-sized items.

DIFFERENT DISTRIBUTIONS USED IN SQC

Type	*Range*	*Mean*	*Variance*	*Probability Function*
BINOMIAL Two mutually excluding events of probabilities ; $P(A) = p$ $P(B) = 1 - p$	0, 1, 2, ... n 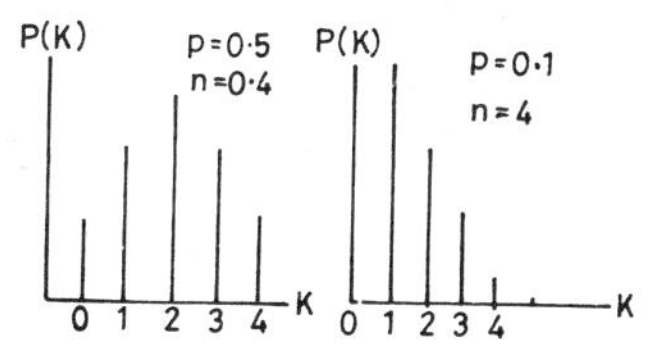 	np	$np(1-p)$	$P(K)$ = Probability of the event A occuring K times in n trials
POISSON Occurrence of isolated events in a continum of time	0, 1, 2, ... ∞ 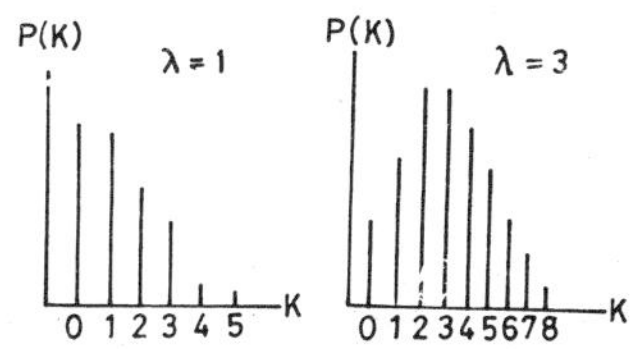 	λ	λ	$P(K)$ = Probability of K events occuring per unit time.

Type	Range	Mean	Variance	Probability Function
RECTANGULAR	0 1 2 ... n 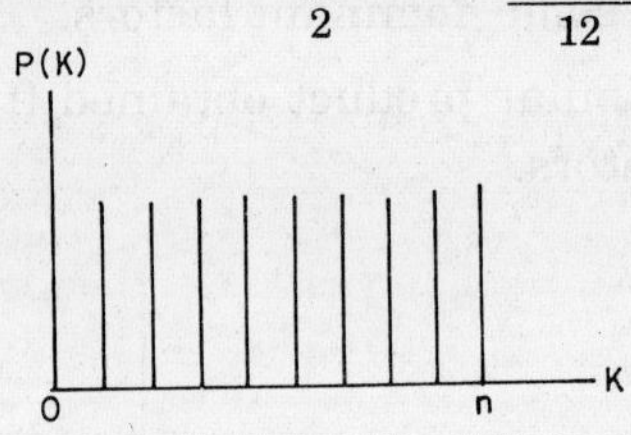	$\frac{n}{2}$	$\frac{n^2+2n}{12}$	$P(K) = \left(\frac{1}{n}+1\right)$
GEOMETRIC	0 1 2 ... ∞ 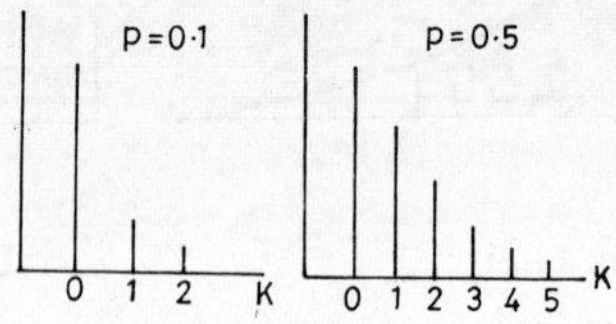	$\frac{p}{1-p}$	$\frac{p}{(1-p)^2}$	$P(K) = p^k (1-p)$
NOTE : The Poisson distribution can always be used as an approximation to the Binomial distribution when p in the Binomial is fairly small and n is fairly large. The nearer p is to zero and the larger n, the better is the approximation.				

Table 19.2. Continuous Distributions

Type	Range	Mean	Variance	Probability Function
NORMAL The average is also the most probable value and there is an equal probability of any value being below or above average	$-\infty$ to ∞ 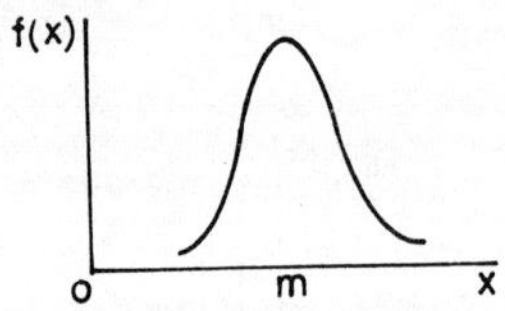	m	σ^2	$= \frac{1}{\sigma\sqrt{2}\,\pi} e^{\frac{-(x-m)^2}{2\sigma^2}}$
UNIFORM The probability for all values of the variable x is the same.	0 to a 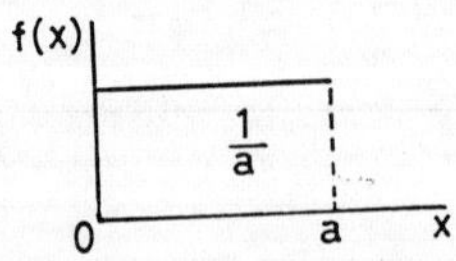	a	$\frac{a^2}{12}$	$P(K) = \frac{1}{a}$
EXPONENTIAL	0 to ∞ 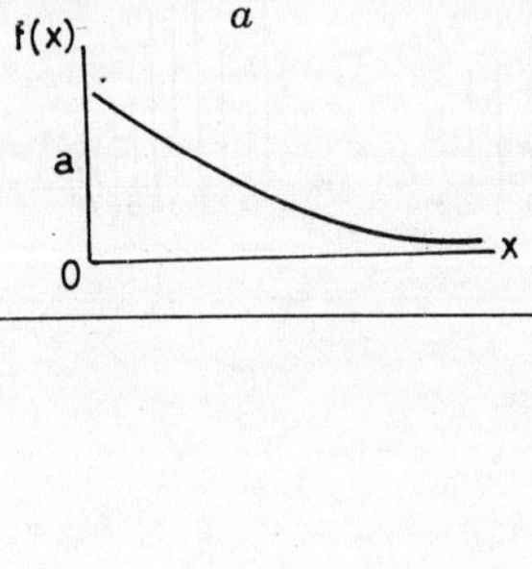	$\frac{1}{a}$	$\frac{1}{a^2}$	$P(x) = ae^{-ax}$

Type	Range	Mean	Variance	Probability Function
GAMMA	0 to ∞	$\frac{a+1}{b}$ 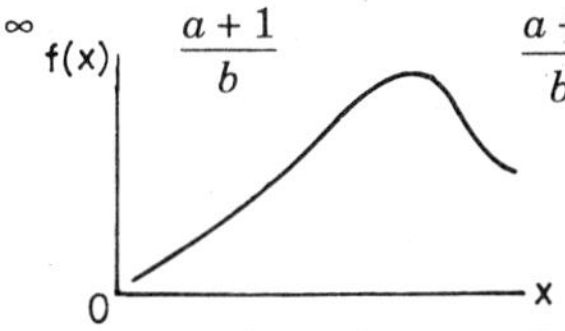	$\frac{a+1}{b^2}$	$P(x) = \frac{b^{a+1}\, x^a\, e^{-bx}}{a\,!}$
BETA	0 to 1	$\frac{a+1}{a+b+2}$ 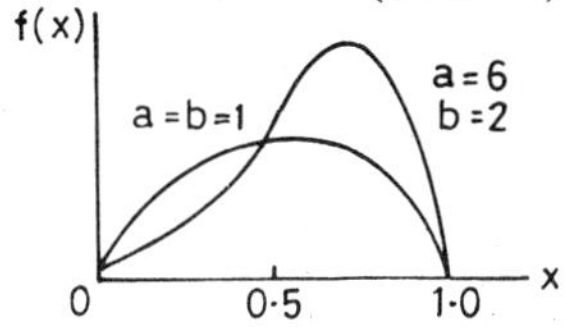	$\frac{(a+1)(b+1)}{(a+b+2)^2\,(a+b+3)}$	$P(x) = \frac{(a+b+1)\,!}{a\,!\,b\,!}\, x^a\,(1-x)^b$
NOTE : The Binomial distribution can be used as an approximation to the Normal distribution when *np* exceeds a value approximately equal to 5.				

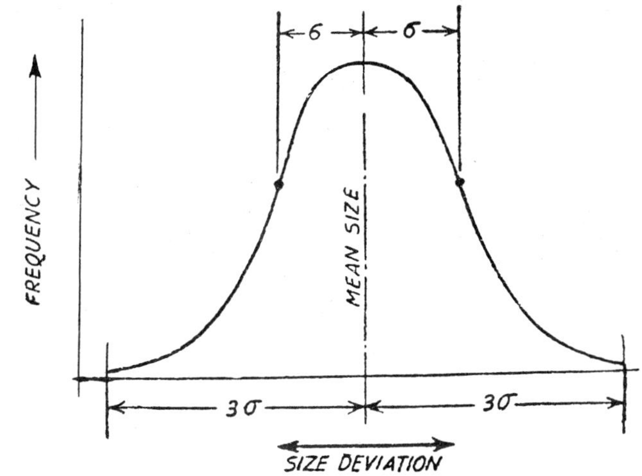

Fig. 19.15

σ = standard deviation of sub-groups = $\frac{\sigma_x}{\sqrt{n}}$

σ_x is the standard deviation of individual items,
and n = sub-group size

Normal Curve. It is the most useful curve used in quality control techniques and looks as shown in Fig. 19.15. In effect this is the curve which shows the frequency of variation of a number of observations for the various values around the arithmetic mean. It may be noted that it is a bell shaped and symmetrical curve. The other characteristics of this curve are : It represents a population of infinite size ; it is asymptotic about X-axis, about 99.74% area being covered within $\bar{X} \pm 3\sigma$ range. It is described fully by the terms $\bar{X}$ and σ. For normal distribution, 68.26% of the observations will fall within ± 1σ from the mean, and 95.46% within ± 2σ from the mean.

Standard Deviation of the Average Size of Sub-groups. If a large number of parts be split at random into groups, then the grand average of the average of sub-groups will be same as the average of universe (all individual items).

It is observed that the averages of sub-groups are more closely spread around the mean size than the sizes of individual items *i.e.* standard deviation of the sub-groups is less than the standard deviation of individual items.

$$\sigma = \text{standard deviation of sub-groups} = \sigma_x/\sqrt{n}$$

σ_x is the standard deviation of individual items, and n = sub-group size.

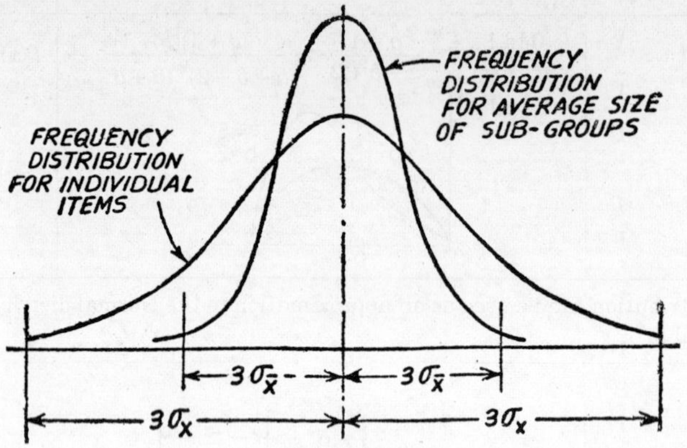

Fig. 19.16

Example 19.1. *The values of arithmetic mean and standard deviation of 25 number observations are 32.6 and 0.7 respectively. These observations were made in two sets of 10 and 15 ; and the values observed in first set were 32.2, 32.8, 32.3, 32.4, 32.5, 32.6, 32.3, 32.7, 32.4, 32.8. Determine the mean and standard deviation of both the series.*

Solution. For the first series.

$$\text{Mean } (\bar{x})_1 = \frac{32.2 + 32.8 + 32.3 + 32.4 + 32.5 + 32.6 + 32.3 + 32.7 + 32.4 + 32.8}{10}$$

$$= \frac{325.0}{10} = 32.50. \ \textbf{Ans.}$$

Standard deviation

$$\sigma_1 = \sqrt{(-0.3)^2 + (0.3)^2 + (-0.2)^2 + (-0.1)^2 + (0)^2 + (0.1)^2 + (-0.2)^2 + (0.2)^2 + (-0.1)^2 + (0.3)^2}$$

$$= \sqrt{0.09 + 0.09 + 0.04 + 0.01 + 0 + 0.01 + 0.04 + 0.04 + 0.01 + 0.09}$$

$$= \sqrt{0.42} = 0.65 \ \textbf{Ans.}$$

Now $$\bar{x} = \frac{n_1 \times (\bar{x}_1) + n_2 \times (\bar{x}_2)}{n_1 + n_2}$$

where $(\bar{x})_1$, $(\bar{x})_2$, $\bar{x}$ are the arithmetic means of items n_1 and n_2, and $n_1 + n_2$ respectively.

$$\therefore \quad 32.6 = \frac{32.5 \times 10 + (\bar{x})_2 \times 15}{25};$$

or $$(\bar{x})_2 = \frac{815 - 325}{15} = \frac{490}{15} = 32.6 \ \textbf{Ans.}$$

Now $$\sigma^2 = \frac{n_1\sigma_1^2 + n_2\,\sigma_2^2}{n_1 + n_2} + \frac{n_1 n_2}{(n_1 + n_2)^2}\{(\bar{x})_1 - (\bar{x})_2\}^2;$$

where σ = standard deviation of both the sets

σ_1 and σ_2 = standard deviation of set 1 and 2 of n_1 and n_2 items respectively.

$$\therefore \quad (0.7)^2 = \frac{10 \times (0.65)^2 + 15 \times \sigma_2^2}{25} + \frac{15 \times 10}{25 \times 25}(32.5 - 32.6)^2$$

or $$0.49 \times 25 = 10 \times 0.42 + 15 \times \sigma_2^2 + 0.06$$

or $$12.25 = 4.2 + 15\sigma_2^2 + 0.06$$

or $$\sigma_2^2 = \sqrt{\frac{8.05 - 0.06}{15}} = \sqrt{\frac{7.99}{15}}$$

$$= \sqrt{0.5333} = 0.73. \quad \textbf{Ans.}$$

Example 19.2. *The specification limits on the gross weight of an ink bottle are 110 ± 3 gm. It was found that weight of an empty bottle has a mean of 53.9 gm and standard deviation of 0.7 gm. Empty bottles are fed into the filling machine in random order. Find out what is the maximum allowable standard deviation at the filling stage (i.e., the standard deviation of the weight of ink filled in a bottle) so that the final product meets the specification limits on the gross weight.*

Solution. Since the final specification limits are 110 ± 3 gms ; obviously to meet these specification limits, $\pm 3\sigma'$ should be equal to ± 3 gm,

i.e. σ' = Standard deviation of universe = 1 gm.

If σ_i = Standard deviation of ink bottle = 1 gm.

σ_b = Standard deviation of empty bottle = 0.7 gm.

σ_f = Standard deviation of filling stage

and $\overline{X}_i$ = Mean of ink bottle = 110 gm.,

$\overline{X}_b$ = Mean of empty bottle = 53.9 gm.

and $\overline{X}_f$ = Mean of ink to be filled

$= \overline{X}_i - \overline{X}_b = 112 - 53.9 = 56.1$ gm

then $\sigma_i = \sqrt{\sigma_b^2 + \sigma_f^2}$ or $1^2 = (0.7)^2 + \sigma_f^2$

or $1 = 0.49 + \sigma_f^2$ or $\sigma_f = \sqrt{0.51} = 0.71$.

$\therefore$ Standard deviation at filling stage should not be greater than 0.71 gm.

19.8. Some Simple Statistical Tools

In this chapter, we shall discuss four separate but related techniques that constitute the most common working statistical tools in quality control. These are :

1. *The Shewhart control charts of measurable quality characteristics.* These are also described as charts for variables or as charts for $\overline{X}$ and R (average and range) and charts for $\overline{X}$ and σ (average and standard deviation). These charts help in trouble shooting of the manufacturing process by providing information on three matters all of which need to be known as a basis for action.

(*a*) Basic variability of the quality characteristics.

(*b*) Consistency of performance.

(*c*) Average level of the quality characteristics

No production process can produce identical objects. Some variability is unavoidable which depends upon the various characteristics of the production process, such as the machines, the materials, the operators. When both upper and lower values (*e.g.*, dimensional tolerances) are specified and control chart shows that the process variability is great, then it is necessary to make change in the process or to take the trouble of sorting good product from bad. Sometimes, it may be observed that tolerances are tighter than necessary for the functioning of the product and they can be widened instead of choosing a very precise and costly process.

The control limits on the charts are so placed as to disclose the presence or absence of the assignable causes. The control charts tell when to leave process alone and when to take action to correct trouble. The elimination of assignable causes of erratic fluctuation is described as bringing a process under control and is responsible for many of the cost savings resulting from statistical quality control.

In some instances the process variability is such that the natural tolerance range is narrower than the specified tolerance range, still the products are unsatisfactory because the average level of the quality characteristic is too high or too low. This will also be disclosed by control charts and correction of the average level can be taken by changing machine setting or taking some suitable action.

Once the control chart shows that a process is brought under control at a satisfactory level and with satisfactory limits of variability, one may feel confident that the product meets specifications. This suggests the possibility of basing acceptance procedures on the control charts and thereby result in substantial savings in costs related to inspection.

2. *The Shewhart control chart for fraction defective*. This is also described as p-chart in technical language. In most of the routine inspection work (either 100% inspection or sampling inspection), the products are classified as either accepted or rejected. In such inspection it is a common practice to make a record of the number of items rejected. The ratio of the number of items rejected to the number of items inspected is the fraction defective which is commonly expressed as a decimal fraction. The control limits of the chart which easily detect the presence of assignable and chance causes are set by simple statistical calculations. The p-chart is somewhat less sensitive than $\overline{X}$ and R charts and it does not have a great diagnostic value, but it serves to point out those situations needing diagnosis of trouble by the control chart for variables.

Nevertheless, it is an extremely useful aid to production supervision in giving information as to when and where to exert pressure for quality improvement. It results in the reduction of the average fraction defective and in the improvement of inspection practices by disclosing erratic fluctuations in the quality of inspection. The p-chart serves as a very good tool in dealing with outside vendors by pointing out whether the vendors differ in quality level or in the variability of quality level.

3. *The Shewhart control chart for defects per unit.* In technical language of the subject, it is described as c-chart. This type of chart applies to a limited number of manufacturing situations involving quality, *e.g.*, inspection of a fairly complex assembled unit in which there are a great many opportunities for occurrences of defects of various types, and the total number of defects of all types found by inspection are recorded for each unit. Other situation can be one in which, for example, it is desired to count the number of imperfections in a bolt of cloth, or weak spots in the insulation of *PVC* insulation on electrical conductor, or number of defects of such type as blemishes on a painted area.

In c-chart also, the control limits are set in a way to detect the presence or absence of assignable causes of variations ; thereby telling when to take action on the process and when not to do so. It has been observed that erratic variation in inspection standards and inspection practices exist in this type of inspection and thus the control charts for defects per unit generally prove helpful in standardising inspection methods.

4. *Acceptance Sampling.* Acceptance inspection (generally done by sampling techniques) is a necessary part of manufacturing and is applied at various stages in a factory, *e.g.*, it can be applied to incoming materials, to partially finished products at various intermediate stages of the manufacturing process, to final product, by the purchaser of manufactured products.

It may be emphasised here that the best protection against the acceptance of defective product is in having the product made right in the first place and that the good sampling acceptance procedures often contribute to this objective through more effective pressure for quality improvement than can be exerted with 100% inspection.

The modern sampling acceptance procedures are based on the laws of probability and though much superior but some defectives are likely to be passed by it if a part of the stream of product submitted for acceptance is defective. It attempts to evaluate the risk assumed with alternative sampling procedures and to make a decision as to the degree of protection needed at any instance. It is then possible to choose a sampling acceptance scheme that gives a desired degree of protection with due consideration for the various costs involved.

19.9. Theory of Quality Control

Whenever the variability in any article is due to chance factors, the pattern of variability follows the laws of probability. References about universe (entire population) are derived by taking samples from it and carrying out statistical analysis of the sample. In other words, the behaviour of the process is estimated on the basis of the behaviour of samples taken from the process. It is important that samples are taken judiciously and at appropriate intervals. If sample size is great, then its statistics more closely reflects the parameters of universe. Generally the distribution of dimensions with inherent sources of variation follows normal distribution curve. In actual practice, most of the inferences are derived from samples, and as per the central-limit theorem of probability, under very general conditions, the mean of samples follows a normal distribution regardless of the distribution of the population. In other words, no matter what is the shape of the distribution of a universe, the distribution of average values ($\bar{x}$) of samples of size 4 or more drawn from that universe will tend toward a normal distribution. This very powerful relationship permits probability statements to be made about a process whose actual population distribution is not known.

The characteristics of the process variability are shown by plotting a graph between the work size (dimension) and frequency (number of times each work size occurs). The total spread of the curve thus represents the limit between which work will normally be produced. It may be observed that the curve is asymptotic near the ends of horizontal axis and maximum area of the curve lies within $\pm 3\sigma$ limits, where σ is the standard deviation of the process. Standard deviation is the distance from the vertical axis of the normal curve to the point where its curvature changes from convex to concave shape. The law of the normal distribution curve is,

$$y = \frac{1}{\sigma \sqrt{(2\pi)}} e^{\frac{-(x-\bar{x})^2}{2\sigma^2}}$$

where y is the frequency with which a size occurs

x is the size of the individual part

$\bar{x}$ is the mean size

and σ is the standard deviation.

The areas of the normal curve enclosed by the base line, the ordinate at the mean, the curve, and the ordinate at various sigmas from the mean are given below in Table 19.3.

Table 19.3

σ	*0.0*	*0.1*	*0.2*	*0.3*	*0.4*	*0.5*	*0.6*	*0.7*	*0.8*	*0.9*
0.1	0.00000	0.03983	0.07926	0.11791	0.15542	0.19146	0.22575	0.26804	0.28814	0.31594
1.0	0.34134	0.36433	0.38493	0.40320	0.41924	0.43379	0.44520	0.45543	0.46407	0.47128
2.0	0.47725	0.48214	0.48610	0.48928	0.49180	0.49379	0.49534	0.49653	0.49744	0.49813
3.0	0.49865	0.49903	0.49931	0.49952	0.49966	0.49977	0.49984	0.49989	0.49993	0.49995

It may be noted that the above table gives the areas of the normal curve in one direction from the mean and areas in the other direction will also be the same. It may be also noted that about 68.27% area lies within $\pm 1\sigma$, 95.45% within $\pm 2\sigma$ and 99.73% within $\pm 3\sigma$ limits. Other important thing for the normal curve is that the median, mode and mean, all coincide.

It may so happen sometimes that the cause system involved will produce a lopsided distribution. In such a case, the area under the curve for any given number of standard deviations from the mean will differ for the sides of the curve and also from the normal curve. The deviation of such a curve from the normal curve is known as skewness and it may be ignored, if of small order. However, one cannot afford to ignore large amounts of skewness.

Most important property of this curve is that 99.7% of area under the curve, which represents on some scale nearly all the number of articles considered, lies within $\pm 3\sigma$ limits from the mean. It can, therefore, be assumed that as long as machine is in control, (*i.e.*, no assignable causes are present) all the articles produced will lie within $\pm 3\sigma$ limits from the mean size. However, whenever an unnatural or assignable cause of variability influences the production process, an additional variation is superimposed on the chance variation and the probable $\pm 3\sigma$ limits of variability are exceeded. Thus if any article does not confine itself within $\pm 3\sigma$ limits, it is indication of the assignable cause being present and machine must be stopped and cause hunted for. One thing to note is that however 0.3% probability is there that the product may lie outside the control limits, though machine is in control. Thus 0.3% chances are there of getting wrong information. This is also known as type I error. Thus the principle benefit of quality control is to point out when and where corrective action is required and to alert the management to locate and eliminate the assignable causes of variation by hunting for it.

The various advantages of quality control may be summarised as follows : to detect the cause of excessive variability and forecasting the trouble before rejections, thus reducing the amount of spoiled work ; also output is increased and wasted machine and men hours are reduced. It also insures rapid and efficient inspection at a minimum of cost and points out when to use 100% inspection for screening. It is also possible to maintain complete records of daily quality and data required for beneficial changes in processing and specifications are also furnished. Finally it results in better customer relations through a general improvement in product quality.

Quality control technique falls broadly into two categories known as method of attributes and method of variables. In the method of attributes, an effort is made to determine whether the given parts are satisfactory or defective or whether it lies within the specified range or not. This type of information can best be had by 'GO' and 'NO GO' type of gauges. So this method simply notes an intrinsic property of a quality characteristic in the part examined. In the method of variables, the size of the specimen is actually measured and an attempt made to analyse and control the results of a process.

Example 19.3. (*a*) *What do you understand by Camp-Meidell Inequality, and Tchebycheff's Inequality and when is each applied ?*

(*b*) *What per cent of cases will lie outside $\overline{X} \pm 3\sigma$ limits in a frequency distribution :*

(*i*) *if it is normal distribution,*

(*ii*) *if it satisfies conditions of Camp-Meidell Inequality,*

(*iii*) *if nothing is known about the distribution.*

Solution. (*a*) Camp-Meidell Inequality is applied to the distribution which has only one mode whose value is equal to arithmetic mean, and the frequencies decline continuously on both sides of the mode. If so, then more than $\left[1 - \left(\frac{1}{2.25t^2}\right)\right]$ items of the distribution will lie within the closed range of $\overline{X} \pm t\sigma$.

Tchebycheff's Inequality is applicable when nothing is known about the pattern of variation ; and according to it more than $\left(1 - \frac{1}{t^2}\right)$ items of any set of finite numbers will lie within the closed range of $\overline{X} \pm t\,\sigma$.

(*b*) (*i*) In case of normal distribution percentage of total area within the limits of $\overline{X} \pm 3\sigma$ = 99.73 ; therefore, percentage of items (cases) outside the limits will be 0.27%. **Ans.**

(*ii*) When the distribution satisfies the conditions of the Camp-Meidell Inequality, then more than $1 - \frac{1}{2.25\,t^2}$ of any distribution will fall within the $\overline{X} \pm t\sigma$ limits, or $\frac{1}{2.25 \times (3)^2}$ will fall outside $\overline{X} \pm 3\sigma$ limits.

∴ Cases outside the specified limits will be

$$= \frac{1}{2.25 \times 3^2} = \frac{1}{20.25} = 0.05 = 5\%. \quad \textbf{Ans.}$$

(*iii*) When nothing is known, then Tchebycheff's Inequality will apply and $\frac{1}{t^2}$ of any set will fall outside the closed range of $\overline{X} \pm t\sigma$.

∴ Cases outside the specified limits will be

$$= \frac{1}{t^2} \times 100\% = \frac{1}{9} \times 100 = 11.1\% \quad \textbf{Ans.}$$

Example 19.4. *The mean value and standard deviation of set of observations have been found to be 100 and 10 respectively. Determine what percentage of observations will lie between 84 and 116, (i) above 82, and (ii) below 79 ?*

Solution. Table 19.3 gives the area of the normal curve enclosed by the base line, the ordinate at mean, the curve, and the ordinate at various sigmas from the mean.

In this case $\sigma = 10$

and $\overline{X} = 100.$

(*i*) For observations between 86 and 100, *i.e.* corresponding to 6σ.

Area = 0.4452.

Similarly between 100 and 116, Area = 0.4452

Total area = 0.8904

∴ Percentage of observations between 86 and 116 = 89.04%

(*ii*) Area between mean and 82, *i.e.* corresponding to $1.8\sigma = 0.46407$

and area above mean to other extreme = 0.5000.

$\therefore$ Total = 0.96407.

$\therefore$ Percentage of observations above 82 = 96.4%

(*iii*) Area between 100 and 79 is = corresponding $2.1\sigma = 0.48214$.

$\therefore$ Area below 79 = 0.5 – 0.48214 = 0.01786.

$\therefore$ Percentage of observations below 79 = 1.79%.

19.10. Process Capability

The process capability (or machine accuracy study) of a machine or manufacturing process can be defined as the minimum tolerance to which machine can possibly be expected to work and produce no defectives under the specified conditions. The knowledge of equipment capability can be used in assigning work to machines. The process capability studies lead to improved equipment maintenance, improved operator training and reduced costs.

To start with, the idea of process capability of a machine can be had by plotting the measurements of 50 to 100 consecutive pieces of product in the exact order in which they are made. Such a chart of the individual measurements is a convenient tool for comparing ability of a process with a specification requirement. The important points to be kept in mind while making such a capability study are : parts should be kept strictly in the order of manufacture ; measurements should be taken with a gauge capable of dividing tolerance into tenths ; a log of changes in tool settings, material lots, operators, machine speeds, temperature or other variables should be maintained. Then, the various measurements may be plotted on a chart and the significant patterns and associations will be quite visible on such charts. Fig. 19.17 shows some patterns typical of those found in these charts of individual measurements. The charts easily reveal tool wear, improper setting of tool, etc. It is not unusual to find that a machine which allegedly "can't hold that tolerance" is actually adequate but is not being operated in the best possible way due to various causes of variability in the process.

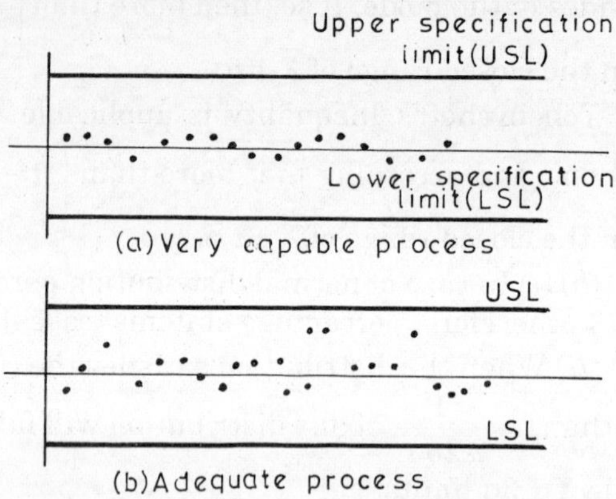

Charts of individual measurements in the study of capability
Fig. 19.17

Statistical Determination of Capability. Every machine has particular characteristics with respect to the inherent variation due to chance cause systems. The capability study establishes for each process those variations which are then interpreted as the minimum attainable tolerances in that specific process.

A machine's capability study can be conducted by (*i*) selecting a part to be produced that is representative of the type of operation performed by the machine, (*ii*) inspecting each part (total of at least 100) as it is produced and noting its characteristic with a measuring equipment capable of measuring beyond the accuracy of the machine being studied, (*iii*) tabulating the results as the frequency (No.) of components for various values of deviation from nominal value.

If $f =$ frequency of occurrence of each measurement.

$X =$ deviation of the dimension from the nominal dimension.

$n =$ Total number of measurements ($n \geq 100$)

then $\bar{X} = \dfrac{\Sigma fX}{n}$ and $\sigma = \sqrt{\dfrac{\Sigma fX^2}{n} - \bar{X}^2}$

Then, the machine accuracy is 6σ.

The values $\bar{X} \pm 3\sigma$ are called the natural limits of the process. With $n \geq 100$, it is highly likely that the natural limits so formed will contain in excess of 99% of all products manufactured. It is very important to note that process capability study must be made under highly controlled experimental conditions in which every effort is made to eliminate all sources of variation except machine variation.

For small samples, process tolerance must be stated with care. For a random sample of size n, where the population or universe is normal,

$$\bar{X} = \frac{\sum_{i=1}^{n} X_i}{n}$$

and $$S = \sqrt{\frac{\Sigma X_i^{\,2} - n\bar{X}^2}{n-1}}$$

where $X_i =$ value of i_{th} measurement.

(It may be noted that the sample standard deviation of a set of observations tends to be slightly less than that for population or distribution from which it is drawn. If the sample standard deviation is computed to serve as an estimate of the population standard deviation, this bias may be automatically corrected by dividing the sum of the squared deviations by $n - 1$ instead of by n. The letter S is used instead of the Greek letter σ to indicate clearly that the measure is obtained from sample data. Thus S is the estimator of σ).

Natural limits are given by $\bar{X} \pm KS$, where values of factor K are tabulated in table 19.6.

What it is desired to estimate the mean and standard deviation of the process from a small size sample then point estimators (which are good estimations in a statistical sense given below) are used.

Population of Universe	**Estimator from sample**
Mean	$\bar{\mathbf{X}}$
Standard Deviation	**S**

The question of the accuracy of the estimators is dealt with by forming confidence intervals in which it is assumed that the random variable (quality characteristic) X has a normal probability distribution and the sample is a random sample. Confidence limits on the true mean are :

$$\bar{X} \pm t_{\alpha/2,\, n-1} \frac{S}{\sqrt{n}}$$

where $1 - \alpha$ is called the confidence level and it represents the probability that the procedure used will yield an interval that contains the population mean quality characteristic. Usual values of confidence level used are 90%, 95% and 99%.

$t_{\alpha/2,\, n-1}$ is read from the t-distribution table 19.4 with $n - 1$ degrees of freedom.

19.11. Determining Confidence Limits for the Population Mean

Since only a sample of data is used in statistical means, the decisions on estimates always involve the possibility of error. A good estimate of the population mean μ can be had by the sample mean, $\bar{X}$. However it can't be said with certainty that population mean lies at this particular point, but certain limits can be specified within which the true value for the population mean will lie. This interval estimate is often specified together with a measure of certain confidence one has in interval estimate. The procedure of computing it is described below. It may be noted that the procedures are for normal distributions but they will give satisfactory results in cases where a distribution only approximates the normal as long as the sample size is large.

Case I. When population standard deviation σ is unknown. In this case, the limits between which the population mean μ will lie can be computed from the sample standard deviation by the formula :

$$\text{Limits between which } \mu \text{ lies} = \bar{X} \pm t \frac{S}{\sqrt{n}}$$

where $\bar{X}$ = Mean for the sample, S = Sample standard deviation and n = No. of items in the sample.

Value of t can be obtained from table 19.4 for $n - 1$ degrees of freedom and the desired amount of confidence in determining the confidence limits between which true value lies.

Table 19.4. Values of t

Degrees of Freedom n–1	*Two Tailed Probabilities*										
	0.90	*0.80*	*0.70*	*0.60*	*0.50*	*0.25*	*0.10*	*0.05*	*0.025*	*0.010*	*0.005*
1	0.158	0.325	0.510	0.727	1.000	2.414	6.314	12.71	25.45	63.66	127.3
2	0.142	0.289	0.445	0.617	0.817	1.604	2.920	4.303	6.205	9.925	14.09
3	0.137	0.277	0.424	0.584	0.765	1.423	2.353	3.183	4.177	5.841	7.453
4	0.134	0.271	0.414	0.569	0.741	0.344	2.132	2.776	3.495	4.604	5.598
5	0.132	0.267	0.408	0.559	0.727	1.301	2.015	2.571	3.163	4.032	4.773
6	0.131	0.265	0.406	0.553	0.718	1.273	1.943	2.447	3.969	3.707	4.317
7	0.130	0.263	0.404	0.549	0.711	1.254	1.895	2.365	2.841	3.500	4.029
8	0.130	0.262	0.399	0.546	0.706	1.240	1.860	2.306	2.752	3.335	3.833
9	0.129	0.261	0.398	0.543	0.703	1.230	1.833	2.262	2.685	3.250	3.690
10	0.129	0.260	0.397	0.542	0.700	1.221	1.813	2.228	2.634	3.169	3.589
11	0.129	0.260	0.396	0.540	0.697	1.215	1.796	2.201	2.593	3.106	3.500
12	0.128	0.259	0.395	0.539	0.695	1.209	1.782	2.179	2.560	3.055	3.428
13	0.128	0.259	0.394	0.538	0.694	1.204	1.771	2.160	2.533	3.012	3.373

14	0.128	0.258	0.393	0.537	0.692	1.200	1.761	2.145	2.510	2.977	3.326
15	0.128	0.258	0.393	0.536	0.691	1.197	1.753	2.132	2.490	2.947	3.286
20	0.127	0.257	0.391	0.533	0.687	1.185	1.725	2.086	2.423	2.845	3.153
25	0.127	0.256	0.390	0.531	0.684	1.178	1.708	2.060	2.385	2.787	3.078
30	0.127	0.256	0.389	0.530	0.683	1.173	1.697	2.042	2.360	2.750	3.030
40	0.126	0.254	0.388	0.529	0.681	1.167	1.684	2.021	2.329	2.705	2.971
60	0.126	0.254	0.387	0.527	0.679	1.162	1.671	2.000	2.229	2.660	2.915
120	0.126	0.254	0.386	0.526	0.677	1.156	1.658	1.980	2.270	2.617	2.860
α	0.126	0.253	0.385	0.524	0.674	1.150	1.645	1.960	2.241	2.576	2.807

It may be noted that the value in the table heading show the probability of the true value being outside one or the other of the computed limits. The probability of it being beyond a specified one of the limits is one-half the value in the table heading. Thus if a probability of 0.05 is used, the probability is 0.05 or 1 out of 20 that the real mean is outside the limits and 0.95 or 19 out of 20 that it is inside the limits. The probability that it is greater than the upper limit or smaller than the lower limit is one-half the value in the table heading or 0.025.

Example 19.5. *A factory receives its raw material in containers, and it is found by testing 30 containers that on the average about 159 parts are produced per container and that the standard deviation comes out to 15. Determine the reliability of this average figure for 95% confidence limits.*

Solution. Referring to table 19.4, for 95% confidence limit, *i.e.* 0.05 probability, and 30 – 1, *i.e.* 29 degrees of freedom, value of $t = 2.042$.

∴ Reliability of the given average figure.

$$= 159 \pm 2.042 \frac{15}{\sqrt{30}} = 159 \pm 5.57 \text{ parts.} \quad \textbf{Ans.}$$

Case II. When the range is known

When the number of observations in a sample is small, then a measure of dispersion can be obtained by a sample procedure using the sample range R. However, it may be noted that the range is not as efficient as the standard deviation as the measure of the dispersion.

The limits in this case are $\overline{X} \pm t_R \,.\, R$

Values of t_R can be obtained from table 19.5. Like previous case, in this case also the probability given is for the mean to lie beyond one limit or the other, or a probability of $1 - P$ to lie between them.

Table 19.5. Values of t_R

Sample Size	*Probability*			
	0.10	*0.05*	*0.02*	*0.01*
2	3.16	6.35	15.9	31.8
3	0.89	1.30	2.11	3.01
4	0.53	0.72	1.02	1.32
5	0.39	0.51	0.69	0.84
6	0.31	0.40	0.52	0.63
7	0.26	0.33	0.43	0.51
8	0.23	0.29	0.37	0.43
9	0.21	0.26	0.32	0.37

10	0.19	0.23	0.29	0.33
11	0.17	0.21	0.26	0.30
12	0.16	0.19	0.24	0.28
13	0.15	0.18	0.22	0.26
14	0.14	0.17	0.21	0.24
15	0.13	0.16	0.20	0.22
16	0.12	0.15	0.19	0.21
17	0.12	0.14	0.18	0.20
18	0.11	0.14	0.17	0.19
19	0.10	0.13	0.16	0.18
20	0.10	0.13	0.15	0.18

19.12. Confidence Limits for Standard Deviation

The best estimate for the population standard deviation, σ can be had by the sample standard deviation, S calculated from the equation

$$S = \sqrt{\frac{\Sigma(x - \bar{x})^2}{n - 1}}$$

Again for this also, the limits between which the true value for the population standard deviation will lie can be calculated for any desired degree of confidence, by the formula

$$\text{Upper limit for } \sigma = \sqrt{FS^2} \quad \text{and} \quad \text{Lower limit for } \sigma = \sqrt{\frac{S^2}{F}}\,.$$

For values of F also standard tables are available which have, however, not been reproduced here.

19.13. Tolerance Limits for Items in a Population

Many times, one is confronted with the problem of determining the limits between which will lie the measurements for all or same specified proportion of the measurements for individual items in a population. A simple statistical method to determine the limits in such a case with any desired degree of confidence is outlined below :

Limits $= \bar{X} \pm KS$

where $\bar{X}$ = mean

S = standard deviation of sample and value of K can be had from Table 19.6.

For any sample size or number of observations the value of K depends on the level of confidence desired and on the proportion of the population that must lie between the limits. It may be noted that results will be correct only when the distribution is normal.

Example 19.6. *50 items produced from a machine were inspected and it was found that mean diameter is equal to 12.258 mm with the standard deviation S equal to 0.03 mm. Determine the limits between which virtually all the items produced will measure. Assume 95% confidence that 99% of the items will lie within the required limits.*

Solution. From Table 19.6 (for determining the value of K), for $n = 50$ and 6th column, $K = 3.126$.

∴ Limits are 12.258 ± 3.126 (0.03) = 12.258 ± 0.09378 mm **Ans.**

19.14. Control Charts

Control charts are used to separate out the assignable cause of quality variation. They are used to distinguish between the inherent random variability of a process and the variability attributable to assignable causes. They accomplish this through an appropriate choice of control limits calculated from the laws of probability.

Random samples of work in process are taken at random and inspected. Data collected are then presented graphically in chart form and chart method is an essential part of quality control systems. The control limits on the control charts are then derived from simple calculations. The control limits are directly related to $\pm 3\sigma$ limits and become the basis to judge the significance of variations in process. Sample size chosen must be such that it protects quality at a minimum cost.

If a machine be set to produce work to given dimension, then two changes in product quality can occur.

(1) The mean size at which the machine is set may shift due to tool wear or setting might have moved.

Table 19.6. Values of K for Normal Distribution

n	*90% confidence that percentage of population between limits is*			*95% confidence that percentage of population between limits is*			*99% confidence that percentage of population between limits is*		
	90%	*95%*	*99%*	*90%*	*95%*	*99%*	*90%*	*95%*	*99%*
2	15.98	18.80	24.17	32.02	37.67	48.43	160.2	188.5	242.3
3	5.847	6.919	8.974	8.386	9.916	12.86	18.93	22.40	29.06
4	4.166	4.943	6.440	5.369	6.370	8.299	9.398	11.15	14.53
5	3.494	4.152	5.423	4.275	5.079	6.634	6.612	7.855	10.26
6	3.131	3.723	4.870	3.712	4.414	5.775	5.837	6.345	8.301
7	2.902	3.452	4.521	3.369	4.007	5.248	4.613	5.488	7.187
8	2.743	3.264	4.278	3.136	3.732	4.891	4.147	4.936	6.468
9	2.626	3.125	4.098	2.968	3.532	4.631	3.822	4.550	5.966
10	2.535	3.018	3.459	2.839	3.379	4.433	3.582	4.265	5.594
11	2.463	2.933	3.849	2.737	3.259	4.277	3.397	4.045	5.308
12	2.404	2.863	3.758	2.655	3.162	4.150	3.250	3.870	5.079
13	2.355	2.805	3.682	2.587	3.081	4.044	3.130	3.727	4.893
14	2.314	2.756	3.618	2.529	3.012	3.955	3.029	3.608	5.737
15	2.278	2.713	3.562	2.480	2.954	3.878	2.945	3.507	4.608
16	2.246	2.676	3.514	2.437	2.903	3.812	2.872	3.421	4.492
17	2.219	2.643	3.417	2.400	2.858	3.754	2.808	3.315	4.393
18	2.194	2.614	3.433	2.366	2.819	3.702	2.753	3.279	4.307
19	2.172	2.588	3.399	2.337	2.784	3.656	2.703	3.221	4.230
20	2.152	2.564	3.368	2.310	2.752	3.615	2.659	3.168	4.168
21	2.135	2.543	3.340	2.286	2.723	3.577	2.620	3.121	4.100
22	2.118	2.524	3.315	2.264	2.697	3.543	2.584	3.078	4.044
23	2.103	2.506	3.292	2.244	2.673	3.512	2.551	3.040	3.993
24	2.089	2.489	3.270	2.225	2.651	3.483	3.522	3.004	3.947
25	2.077	2.474	3.251	2.208	2.631	3.457	2.494	2.972	3.904

26	2.065	2.460	3.232	2.193	2.612	3.432	2.469	2.941	3.865
27	2.054	2.447	3.215	2.178	2.595	3.409	2.446	2.914	3.828
28	2.044	2.435	3.199	2.164	2.579	3.388	2.424	2.888	3.794
29	2.034	2.424	3.184	2.152	2.554	3.368	2.403	2.864	3.763
30	2.025	2.413	3.170	2.140	2.549	3.350	2.385	2.841	3.733
35	1.988	2.368	3.112	2.090	2.490	3.272	2.306	2.748	3.611
40	1.959	2.334	3.066	2.052	2.445	3.213	2.247	2.677	3.518
50	1.916	2.284	3.001	1.996	2.379	3.126	2.162	2.576	3.385
60	1.887	2.248	2.955	1.958	2.333	3.066	2.103	2.506	3.293
80	1.848	2.020	2.894	1.907	2.272	2.986	2.026	2.414	3.173
100	1.822	2.172	2.854	1.874	2.233	2.934	1.977	2.355	3.096
200	1.764	2.102	2.762	1.798	2.143	2.816	1.865	2.222	2.921
500	1.717	2.046	2.689	1.737	2.070	2.721	1.777	2.117	2.783
1000	1.695	2.019	2.654	1.709	2.036	2.676	1.736	2.068	2.728
α	1.645	1.960	2.576	1.645	1.960	2.576	1.645	1.960	2.876

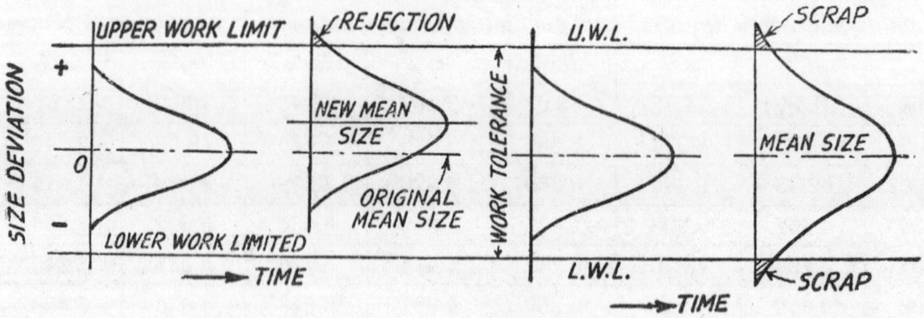

Effects of mean size drifting. Effect of change in process variability.

Fig. 19.18

(2) The process variability may change, *i.e.* the range in which the size were produced may increase. This is due to some fundamental mistake in machine.

The effects of these two changes are shown in Fig. 19.18 and can be best studied by control chart for average ($\bar{X}$-chart) and control chart for range (range-chart). If these defects occur, then defective work will be produced which will require 100% inspection. Thus these two control charts are very essential. It may be noted that $\bar{X}$-chart indicates the centering of the process and R-chart indicates the variability about the aim of the process.

19.14.1. Pre-Control (PC) Technique for Checking Defects. This technique can detect process conditions and changes that may cause defects. It can be useful for the duration when control charts are being established. This technique is based on the assumption that the parameters of the population are identical with those of the sample and that the distribution is normal so that ± 3σ limits are the natural tolerance limits. Pre-control assumes, that the natural tolerance of the process is the same as the part tolerance and that the process is precisely centered about the average. Thus any shift in the process would produce defective work. For such conditions, 86% of the parts will fall within 1/2 part tolerance, provided that only common random variations are present. Half part tolerance limits, are designated as pre-control (PC) lines, each 1/4 of the way in from each specification part tolerance limit (Refer

Fig. 19.19). Under normal circumstances, 7% parts will occur outside a PC line. The chances for two parts to fall outside PC line are 0.07 × 0.07 = 0.0049 or ≈ 1/196. Thus if two parts in a row fall outside the same PC line, the chances are 195/196 that the process has shift and there is need to recenter it by making proper adjustment.

Similarly if one piece falls beyond one PC line and the next piece falls beyond the other PC line, then it is sufficient evidence for concluding that process variation or distribution has widened and there is need for appropriate correction.

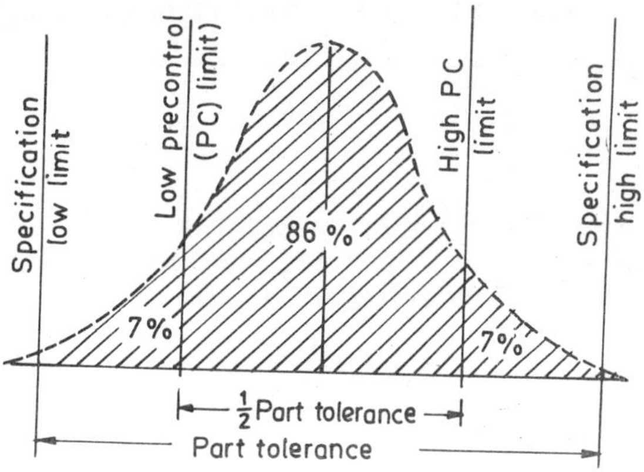

Fig. 19.19

19.15. Control Chart for Variables

$\overline{X}$-chart and $\overline{R}$-chart provide maximum utilisation of information available from data and provide detailed information on process average and variation for control of individual dimensions.

In any process, when the variations in successive observations of the quality characteristic involved are due solely to a constant system of causes, that process is said to be in a state of statistical control. It is thus, obvious that in every process, causes of quality variation are at work. Individually, their effect on quality variation is negligible and collectively they make up what is known as inherent variability or process capability. No efforts should be made to eliminate these causes or even to identify them. Depending upon the rigidity and design of the machine, these variations are controlled.

The control charts help to easily identify the assignable causes of quality variation and these should be controlled or eliminated economically, whenever detected.

In gathering data for the preparation of control charts the following points should be remembered :

(*i*) We should be sure that we are measuring the thing we want to measure.

(*ii*) Measuring equipment and computing equipment should be operating properly and should be operated properly.

(*iii*) Operators should be properly instructed and free from bias.

(*iv*) Transcription errors should be avoided by adequate checking.

(*v*) At least 40 to 100 observations should be taken. (It depends upon size of sub-group also). The important thing to be noted in the control charts is the grouping of data, the idea behind which is explained here. It has already been pointed out that the primary purpose of a control chart is to reveal the presence of assignable causes of variation.

It is, therefore, necessary that the control chart limits be based on only the variability that is inherent in the process. If the standard deviation is computed on the basis of individual measurements, it will include the determination of all the assignable causes as well as chance causes of variation. As a result, the estimate of variability will be too large, if the process is actually not in control. The spread of the same process by considering individual items and by taking sub-groups is shown in Fig. 19.16.

Sub-group Size. It has been observed that for group size of four or more the distribution of group averages is normal without regard to the shape of the parent distribution. In other words, the three sigma limits for averages are valid when group size is four or more, though the parent population is definitely non-normal. Using bigger group size (more than four) is not also very desirable as otherwise the points will have to be plotted after long intervals whereas with small group, the points are plotted more frequently and indications of lack of control thereby caught more quickly. Another advantage of small group size is that it is more likely to contain inherent variability than a larger group. Further, when the individual items are approximately distributed and it is desired to check them, the control limits for individuals can be very easily computed, since they are exactly twice as far from the central line as the control limits for averages of fours :

$$\left(\text{By relation } \sigma_{\bar{x}} = \frac{\sigma'}{\sqrt{n}}\right)$$

where $\sigma_{\bar{x}}$ is the standard deviation of group averages.

σ' is the standard deviation of parent population and n is group size.

The characteristics of various sample sizes are summarised below.

Sample size of 2. The estimates of limits with $n = 2$ are likely to be erratic and therefore it should not be used unless there is a good engineering reason for it.

Sample size of 3. Though it can be used satisfactorily, but its use makes calculations tedious as recurring decimals appear, frequently.

Sample size of 4. It is the best sample size. It is stable and yet sensitive.

Sample size of 5. It is also very satisfactory and best to use, as calculations become very easy, particularly if measurements are in decimals. However relatively more measurements have to be taken to get a minimum number of samples.

Sample size of 6. It is generally not recommended unless engineering or sub-grouping reasons call for it. Two samples of 3 are generally considered to be better than a sample of 6.

Sample size of more than 6. It is not recommended. If n is more than 10, then range cannot be taken as efficient measure of dispersion.

Relation between sample size n, population standard deviation σ', and error E is given by

$$n = \left(\frac{t\sigma'}{E}\right)^2$$

where t is the number corresponding to the confidence required, and t is the multiple of σ' such that the given confidence, *i.e.* the area of the normal curve lies within $\overline{X} \pm t\sigma'$.

Example 19.7. *The population standard deviation of a process has been determined to be 0.12 ; and it is required to determine the sample size to assure with 95% confidence that the average quality will lie within ± 0.10 of the mean of the determination. What will be the sample size if confidence level desired is 99.7%.*

Solution. 95% of the area of the normal curve will be within $\overline{X} \pm 2\sigma'$

$$\therefore \quad t = 2$$

$$\therefore \quad n = \left(\frac{2 \times 0.12}{0.10}\right)^2 = 5.76, \text{ say } 6. \quad \textbf{Ans.}$$

For 99.7% confidence level, $t = 3$

$$\therefore \quad n = \left(\frac{3 \times 0.12}{0.10}\right)^2 = 12.96 = 13. \quad \textbf{Ans.}$$

19.15.1. Calculation of 3-Sigma Limits on Control Charts for $\overline{X}$. The control limits for $\overline{X}$ chart can be derived either from standard deviation or from range. It has been observed for a given process, that the general pattern of the points for standard deviation is the same as the pattern of points for ranges. Thus in view of the large amount of labour involved in computing the standard deviations, it is generally preferable to use ranges. Before starting a chart, the given data should be accumulated in 25 groups. After grouping the data, the averages and ranges of the sub-groups are calculated and from that the grand average of sub-groups $\overline{\overline{X}}$ and average of ranges of sub-groups ($\overline{R}$) is determined. Now, since there exists a relationship between the population standard deviation (σ') and the average of ranges ($\overline{R}$), which of course depends upon the size of the sub- groups ; σ' can be easily estimated. The factor for determining σ' from $\overline{R}$ is d_2 and is tabulated in table 19.7 for various sizes of sub-groups.

Also $\quad \sigma_{\overline{X}} = \dfrac{\sigma'}{\sqrt{n}}$

$\therefore$ 3-sigma limits for $\overline{X}$-chart will be $3\sigma_{\overline{X}} = \dfrac{3\sigma'}{\sqrt{n}} = \dfrac{3\overline{R}}{\sqrt{n} \times d_2}$

$\therefore$ Upper control limit $\quad UCL_{\overline{X}} = \overline{\overline{X}} + \dfrac{3\overline{R}}{\sqrt{n}\, d_2}$

and lower control limit $\quad LCL_{\overline{X}} = \overline{\overline{X}} - \dfrac{3\overline{R}}{\sqrt{n} \times d_2}$

In order to shorten the calculation of control limits $\overline{R}$, the factor $\dfrac{3}{\sqrt{n}\, d_2}$, the multiplier of $\overline{R}$ has been computed for each value of n and tabulated in Table 19.7. The factor is designated A_2. The three-sigma control limits on $\overline{X}$-chart then become

$$UCL_{\overline{X}} = \overline{\overline{X}} + A_2 R \quad \text{and} \quad UCL_{\overline{X}} = \overline{\overline{X}} - A_2 R$$

If control limits are calculated from σ rather than from R then σ' will be equal to $\dfrac{\sigma'}{C_2}$ (value of C_2 is tabulated in Table 19.7 and control limits become)

$$UCL_{\overline{X}} = \overline{\overline{X}} + \frac{3\overline{\sigma}}{C_2 \sqrt{n}} \quad \text{and} \quad LCL_{\overline{X}} = \overline{\overline{X}} - \frac{3\overline{\sigma}}{C_2 \sqrt{n}}$$

Here also, in order to shorten the calculations of control limits from σ, the factor $\dfrac{3}{C_2 \sqrt{n}}$ has been computed for various values of n and tabulated in Table 19.7. This factor is designated by A_1 and thus 3-sigma control limits by using this factor become

$$UCL_{\overline{X}} = \overline{\overline{X}} + A_1 \overline{\sigma} \quad \text{and} \quad UCL_{\overline{X}} = \overline{\overline{X}} - A_1 \overline{\sigma}$$

It is important to understand that control limits on control charts and tolerance limits are different and should not be confused.

After the control charts have been established then all points will fall within upper and lower control limits as long as process is in control. The lack of control is indicated by the following four conditions. Occurrence of any one of these conditions is indication of presence of assignable cause which has to be investigated and rectified.

(*i*) A point falling outside the control limit line.

(*ii*) A run of more than 8 points on the same side of the control line.

(*iii*) A trend with at least 7 points.

(*iv*) 13 out of 15 points falling on the same side of the control line.

19.15.2. Control Chart for Standard Deviation. It has already been emphasised that these two charts tell practically the same story and hence only range chart may be used. It has also been observed in many industries that R rarely goes out of control though $\overline{X}$ does so frequently. In such cases maintaining R-chart adds to cost without giving any useful information.

Table 19.7

No. of observations sub-group ; n	*Factor for estimating σ′ from $\overline{R}$* $d_2 = \frac{\overline{R}}{\sigma'}$	*Factor for estimating σ′ from $\overline{\sigma}$* $C_2 = \frac{\overline{\sigma'}}{\sigma}$	*Factor for determining 3-sigma limits for $\overline{X}$ from $\overline{R}$* A_2	*Factor for determining 3-sigma limits for $\overline{X}$ chart from $\overline{\sigma}$* A_1
2	1.128	0.5642	1.88	3.76
3	1.693	0.7236	1.02	2.39
4	2.059	0.7979	0.73	1.88
5	2.326	0.8407	0.58	1.60
6	2.534	0.8686	0.48	1.41
7	2.704	0.8882	0.42	1.28
8	2.847	0.9027	0.37	1.17
9	2.970	0.9139	0.34	1.09
10	3.078	0.9227	0.31	1.03
11	3.137	0.9300	0.29	0.97
12	3.258	0.9410	0.27	0.93
13	3.336	0.9027	0.25	0.88
14	3.407	0.9139	0.24	0.85
15	3.472	0.9227	0.22	0.82
16	3.523	0.9532	0.21	0.79
17	3.588	0.9551	0.20	0.76
18	3.640	0.9576	0.19	0.74
19	3.689	0.9599	0.19	0.72
20	3.735	0.9619	0.18	0.70
21	3.778	0.9638	0.173	0.68
22	3.819	0.9655	0.169	0.66
23	3.858	0.9670	0.162	0.65
24	3.895	0.9684	0.157	0.63
25	3.931	0.9696	0.153	0.62
30	4.086	0.9748	0.134	0.56
35	4.213	0.9784	0.120	0.52
40	4.322	0.9811	0.110	0.48
45	4.415	0.9832	0.101	0.45
50	4.498	0.9849	0.095	0.43

55	4.572	0.9863	0.089	0.41
60	4.669	0.9874	0.084	0.39
65	4.699	0.9884	0.079	0.38
70	4.755	0.9892	0.075	0.36
75	4.806	0.9900	0.071	0.35
80	4.851	0.9906	0.069	0.34
85	4.898	0.9912	0.066	0.33
90	4.939	0.9916	0.064	0.32
95	4.978	0.9921	0.062	0.31
100	5.015	0.9925	0.060	0.30

In σ-control chart, the central line on the control chart is set at $\overline{\sigma}$ and the limits are set at $\overline{\sigma} \pm 3\sigma_\sigma$. The approximate value of σ_σ for a normal universe is

$$\sigma_\sigma = \frac{\sigma'}{\sqrt{2n}}$$

Modern statistical theory gives the exact value as

$$\sigma_\sigma = [2(n-1) - 2nC_2{}^2]^{1/2} \frac{\sigma'}{\sqrt{2n}}$$

When n is large, the difference between above two equations becomes negligible. In order to simplify the calculation of 3-sigma limits on control chart for standard deviation, factors B_3 and B_4 are tabulated in Table 19.8 for various sub-group sizes.

Thus 3-sigma limits on a σ-chart from an observed σ_σ are

$$UCL_\sigma = \overline{\sigma} + 3\sigma_\sigma = B_4\overline{\sigma}$$

$$LCL_\sigma = \overline{\sigma} + 3\sigma_\sigma = B_3\overline{\sigma}$$

In Table 19.8, the values of constants D_3 and D_4 factors for determining 3-sigma limits for range charts are also given.

19.15.3. Control Charts for Range. In order to check the process variability, the range chart is required. For finding the upper and lower control limits for R-chart, constants, D_3 and D_4 are tabulated corresponding to sample size.

$$U.C.L. \text{ for } \overline{R} \text{ chart} = \overline{R} \times D_4 \quad \text{and} \quad L.C.L. \text{ for } \overline{R} \text{ chart} = \overline{R} \times D_3.$$

An important relationship to be noted is that the ratio of average range $\overline{R}$-from samples of size n from a normal universe to the estimated standard deviation of the universe follows a predictable pattern as a function of sample size.

In range chart actually there is no need of lower control limit and sometimes it is not used. In a range-chart, if any plot lies below lower control limit, it means that for some reason the process variability has decreased and process has improved. The cause for it should be found out and with it, it may be possible to incorporate the improvement in future to improve the overall quality.

If any plot goes beyond the *U.C.L.*, then the machine must be immediately stopped. Statistically range should be within the control limits and its going out of *U.C.L.*, means some assignable cause is present, which must be looked for.

Both these charts were the charts for quality control by variable, as actual measurement had to be taken. It has the advantage of giving the information of machine getting out of order at the same instant, thereby reducing the amount of scrap. But it has some disadvantages also.

Table 19.8

Number of observations in sub-group n	*Factors for σ-chart*		*Factors for R-chart*	
	Lower Control Limit B_3	*Upper Control Limit B_4*	*Lower Control Limit D_3*	*Upper Control Limit D_4*
2	0	3.27	0	3.27
3	0	2.57	0	2.57
4	0	2.27	0	2.28
5	0	2.09	0	2.11
6	0.03	1.97	0	2.00
7	0.12	1.88	0.08	1.92
8	0.19	1.81	0.14	1.87
9	0.24	1.76	0.18	1.82
10	0.28	1.72	0.22	1.78
11	0.32	1.68	0.26	1.74
12	0.35	1.65	0.28	1.72
13	0.38	1.62	0.31	1.69
14	0.41	1.59	0.33	1.67
15	0.43	1.57	0.35	1.65
16	0.45	1.55	0.36	1.64
17	0.47	1.53	0.38	1.62
18	0.48	1.52	0.39	1.61
19	0.50	1.50	0.40	1.60
20	0.51	1.49	0.41	1.59

(*i*) Various charts are required for various dimensions in one part. Thus, it must be maintained for important characteristics and dimensions only.

(*ii*) A large number of actual measurements have to be made and 'GO' and 'NO GO' gauges can't work.

(*iii*) Some characteristics are such which can't be expressed in terms of numbers, *e.g.*, the surface equality of glass. It is either good or bad. For this the charts which work on attributes are required. These are also known as control charts for fraction defectives.

19.15.4. Important points to be observed in drawing the quality control chart. As the quality control charts form important documents and have to be examined by senior personnel, these should be drawn in convenient manner and following points considered :

(*i*) The size of the chart should be handy one.

(*ii*) The chart should be as neat as possible.

(*iii*) There should be provision for accommodating additional future points on the charts.

(*iv*) The chart should be labelled properly and sufficiently.

(*v*) When it becomes necessary to draw revised control chart limits on the basis of additional data, they should not be drawn over the original limits but after these.

(*vi*) Plotted points should be quite legible.

(*vii*) The average lines are drawn by solid and control limits by dotted lines.

(*viii*) There should be minimum number of background lines on the graph paper selected for drawing the control charts. Such charts are often plotted on rectangular cross-section paper.

Sometimes profile paper may also be used. The vertical scale is used for the statistical measures $\overline{X}$ and R. The horizontal scale is used for sub-group number. Dates, hours, or lot numbers may also be indicated on horizontal scale.

19.15.5. Necessary Steps in Connection with Using of Control Charts. Here an attempt is made to visualise the decision and calculations that must be made and the actions that must be taken at various stages of the control charts. The various stages in the use of control charts that need consideration are :

(*i*) Decisions preparatory to the control charts.

(*ii*) Starting the control charts.

(*iii*) Determining the trial control limits

(*iv*) Drawing preliminary conclusions from the charts.

(*v*) Continuing to use the charts.

We shall discuss these now one by one.

(1) Decisions preparatory to the control charts. The various decisions have to be taken prior to starting the work on control charts. The decisions are in respect of the objectives of charts, the choice of variables, the basis of sub-grouping the size and frequency of sub-groups, setting up the forms for recording the data and the method of measurement.

Objectives of control charts. The various objectives for starting the control charts can be :

(*a*) To determine whether a process can meet the specifications or to collect further information for establishing or changing the specifications.

(*b*) To get information for changing the production procedures.

(*c*) To get information for changing the inspection techniques or acceptance procedures.

(*d*) Determination, as to when to hunt causes of variation and take action to correct them and when to leave the process alone.

(*e*) To take decision on acceptance or rejection of manufactured or purchased product.

Choice of variables. The variable chosen must be one which can be measured and expressed in numbers. As the control charts are very costly techniques, only very important variables must be considered.

Basis of sub-grouping. The success of these techniques depends upon the division of observations into rational sub-groups, *i.e.*, the discrimination used in the selection of the sub-groups. The sub-groups should be selected in a way that gives the maximum opportunity for variation from one sub-group to another *e.g.*, in production process when the aim is to detect the shifts in process average, one sub-group should consist of items produced as nearly as possible at one time, the next group of items produced at a single later time and so forth.

If the aim of control chart is to provide a basis for acceptance then method of sub-grouping will be different and in this case the items in a sub-group should be selected at random from the production in some given period of time so that the sub-group is representative of the complete production over a given period of time.

Size and frequency of sub-groups. In the control charts, the sub-groups should be selected in such a way which gives minimum opportunity for variation within a sub-group and for this purpose the sub-group size should be as small as possible. Usually size of 4 or 5 is chosen which is better than size of 2 or 3 on statistical grounds, *i.e.*, the distribution of $\overline{X}$ is nearly normal though universe is non-normal. The use of sub-group size as 5 also makes the calculations easier. When cost of measurement is high, sub-group of size 2 or 3 may also be used. The sub-group size 10 to 20 is used when it is desired to make the control chart sensitive to small variations in the process average ; because larger is sub-group size, the narrower the control limits and easier it is to detect small variation. This is, of course, true when sub-groups are selected such that variations in process average occur between and not within sub-group. When sub-group size is large, it is preferable to use σ charts and not R-chart. The frequency of taking a sub-group may be expressed either in terms of time or as a proportion of the items produced.

Forms for recording the data. In one type of form the successive measurements in each sub-group are recorded below the other and this form avoids the necessity of mental arithmetic in figuring averages. In another type, each line contains all the measurements in one sub-group. This type avoids the necessity of copying the $\overline{X}$ and R values to compute the grand average and generally allows room for more sub-groups on a page.

In all the forms, space must be provided for indicating the item measured, the unit of measurement and other relevant information. Space for putting remarks like production changes (*i.e.* changing of machine setting, sharpening of tool, changed operator, inspection changes and any other change in existing conditions) should be made so that the quality control inspector may get clues to the causes of any out of control points.

(2) Starting the Control Charts. The various steps in starting a control chart are (*i*) making and recording the measurements of readings and other relevant data, (*ii*) calculation of $\overline{X}$ and R for each sub-group, (*iii*) plotting the $\overline{X}$ and R charts.

(3) Determining the Trial Control Limits. In determining the trial control limits, the determination of minimum number of such groups required before control limits are calculated plays an important role. It is actually a compromise between a desire to obtain the guidance given by averages and control limits as soon as possible after the start of collecting data and a desire that the guidance be as reliable as possible. If the number of sub-groups are less, it will soon provide the basis for action but will be less reliable. On statistical grounds it is desirable that at least 25 sub-groups be taken for determining the control limits. It may be noted that the first two sub-groups may not be representative of what is measured later as the mere act of taking and recording measurements is sometimes responsible for a change in the pattern of variation. Thus, if possible, about 50 sub-groups should be considered. However, if sub-groups are obtained slowly, control charts may be started with less number of sub-groups with subsequent modification of limits as more sub-groups are obtained.

After forming the sub-groups, the next step is to calculate the trial control limits or plot the central line and the control limits on the charts.

(4) Drawing Preliminary Conclusions from the Charts. It has already been emphasised that the quality control charts act as highly effective means for detecting the assignable causes of variation in product quality. In order to detect the actual cause of trouble, it is advisable and standard practice to break the process into four general sources of product quality variation, *i.e.*, raw materials, machines, men and methods and to maintain control charts for different sources of raw materials, different machines, different operations of shifts, and different methods of processing. Further the length of time required for an obvious shift in quality level may give an important clue to the nature of the assignable cause involved.

Indication of control or lack of control. When any point falls outside the control limits on either $\overline{X}$ or R-charts, it is indication of lack of control and presence of assignable causes. It can be said with confidence that the constant cause system is no more present ; as a constant cause system will seldom be responsible for points falling outside control limits (0.3% probability).

On the other hand, when all the points fall inside the control limits, we can't say with the same confidence that no assignable causes of variation are present. Actually the statement that process is in control only means that for practical purposes, it tends to act as if no assignable causes of variation are present.

In such cases, lack of control may be indicated by runs of seven or more points in succession on the same side of the central line (Theory of Trend).

It has also been observed that even in the best manufacturing process, occasional errors constitute assignable causes of variation but they may not constitute a basis for action. Therefore, sometimes if only one out of 35 or 2 out of 100 points fall outside the control limits, the process is considered to be under the control.

When a process is under control, it becomes possible to judge as to what is necessary to permit the manufacture of product that meets the specifications for the desired quality characteristic. By control charts one can find the estimate of the centering and the dispersion of the process. But this information of $\overline{X}$ and σ' can be relied upon only when large number of sub-groups are considered. For all practical purposes, the spread of the distribution of any process is approximately $6\sigma'$. Knowing this and the specification limits prescribed, the necessary action can be taken.

Relationship between control limits and specification limits. There can be either two specification limits, maximum or upper limit, U and a minimum or lower limit L, or only one specification limit, either U or L. We shall discuss the relationship between control limits and specification limits with assumption that the specification limits are neither tight nor loose and they have to be met in order to perform the required function, and that the process in under control.

A very important point to be remembered is that the specification limits normally relate to the individual test and the control chart limits to averages. Therefore, control chart limits can't be directly related to the specification limits. But when sufficient number of sub-groups are being considered, it is possible to estimate the 3-sigma limits for individual value, *i.e.* $3\sigma'$ limits by relation $\sigma' = \frac{\overline{R}}{d_2}$. The situation becomes satisfactory and it is desirable also that the width of the control limits band for individuals should not exceed 3/4th of the width of the specification band.

However, three possible cases can be considered :

(*i*) The spread of the process ($6\sigma'$) is appreciably less than the difference between specification limits, $(U - L)$

(*ii*) $6\sigma' = (U - L)$

(*iii*) $6\sigma' > (U - L)$.

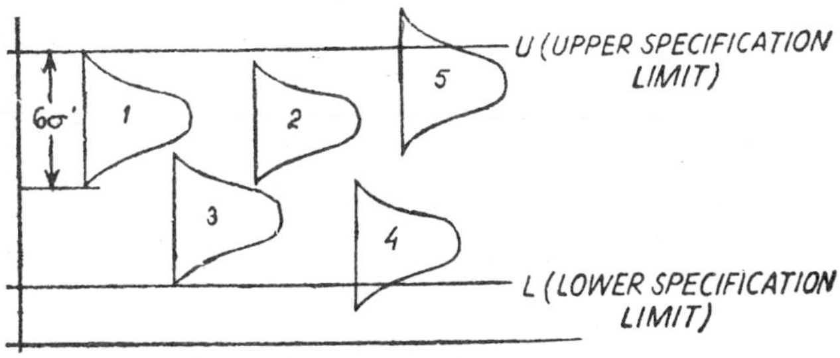

Fig. 19.20

The first situation is explained in Fig. 19.20. It may be observed that with any of the position 1, 2, or 3, practically all the products manufactured will meet specifications as long as process remains under control (whose indication can be had by R-chart). Thus here the machine setting need not be disturbed between positions 1 and 3, thereby avoiding the cost of frequent machine set-ups and of delays due to hunting for assignable causes of variation. In such cases the following can be done :

(*a*) To replace 100% inspection by acceptance, based on $\overline{X}$ and R-charts.

(*b*) If advantage is gained by tightening the specification limits, that may be allowed.

(*c*) Discontinue the use of control chart or at least increase the time interval between control chart inspections.

With the process in position 4, some product will fall below the lower specification limit and in position 5, some product will fall above the upper specification limit. In either case, the obvious action is to change the centering of the process bringing it closer to position 2 or 3 respectively. Once this is done, consideration may be given to action enumerated above.

The second case, in which the process is exactly centred between the specification limits, three cases are possible (Fig. 19.21). In position 1, all products will be good, but in positions 2 and 3, products will fall outside the limits. In such cases, all possible steps have to be taken to maintain the centering of the process by continuously using control charts for $\overline{X}$ and R with sub-groups at frequent intervals and immediate attention to points out of control.

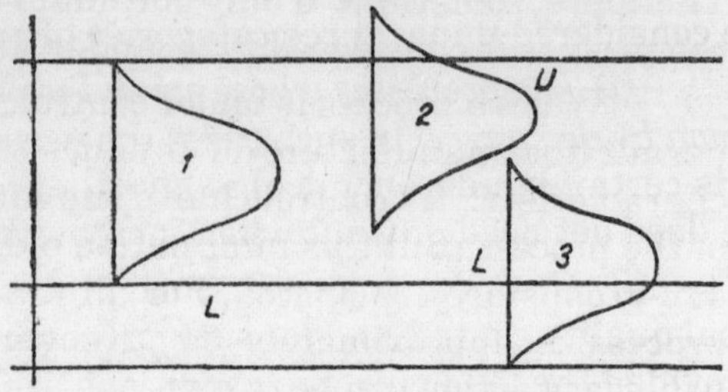

Fig. 19.21

In this case better solution would be to use a process with less process variation or to relax the tolerances, if possible.

In third case where specification limits are so tight that even when the process is in control and perfectly centred, some nonconforming product will be formed. This situation definitely calls for fundamental change in process, revision of tolerances, strict control on centering of process and to use 100% inspection.

In certain cases only one specification limit is specified as might be true of a specified minimum tensile strength or minimum weight of the contents of a container etc. In such cases also, the various possibilities can be considered by examining the portion of $\overline{X} - 3\sigma'$ w.r.t. L. The three possible situations are :

(*a*) The low value of the process distribution $X' - 3\sigma'$ is appreciably above L. (There is plenty of margin of safety).

(*b*) The low value of the process distribution $\overline{X} - 3\sigma'$ is approximately at L. (Specification is just barely met as long as process stays in control).

(*c*) It is appreciably below L. (In this case some non-conforming product is inevitable till some change is made in the process).

All these possibilities are shown in Figs. 19.22 (*a*), (*b*) and (*c*).

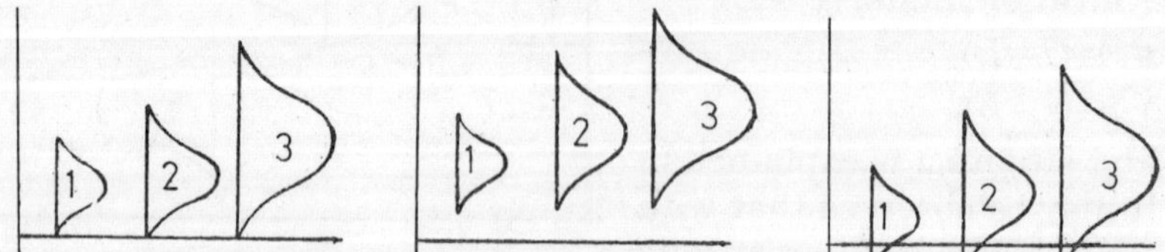

Fig. 19.22

In Figs. 19.22 (*a*) and (*c*), (*b*) three curves, 1, 2 and 3 with different dispersions have been considered in order to emphasise the inter relationship of process average, process dispersion and the specification minimum limit. In first case where all the low limits are at same value, distribution 2 with a greater dispersion must have a greater process average than 1 for the low points to be at same value ; similarly 3 must have a greater process average than 2. It is evident that greater the dispersion, higher must be the average for the entire distribution to fall above the specification minimum limit. On the other hand, the less the dispersion the more important it is that the process average does not go out of control *e.g.*, in the last case shown in Fig. 19.22 (*c*) if the process average of all the three distributions 1, 2 and 3 be shifted an equal amount below their position in second case, then the proportion of bad product in distribution 1 will be much greater than that in 3, because the area of the distribution below the specification limit will be more for distribution 1.

(5) Continuing to use the charts. The study of continuance of the control charts can be considered under three categories of processes.

(*i*) *The processes whose spread is very less in comparison to the specification limits laid down by designers.* In such cases the use of control charts will not serve any useful purpose as it is certainly uneconomical to give the consumer a degree of product quality uniformity that he does not need or want when increased uniformity raises production costs.

(*ii*) *Unstable processes.* Such processes are normally very difficult to bring under statistical quality control and for such processes it will be neither necessary nor economical to do so provided the following two conditions are met :

(*a*) The inherent variability of the process must remain essentially the same regardless of the process level. In other words, the ranges must show statistical control even though the averages do not.

(*b*) The distance between the specification limits must be materially greater than six times the inherent variability of the process (the standard deviation of the individuals).

(*iii*) *The process which is under statistical control and capable of producing the articles within the specified tolerance limits.* To start with the control charts on any new machine, the trial control limits are calculated only in order to determine if the past operations were in control. These limits could be continued with as long as no point falls outside. In case any point falls outside these trial limits, then it may be necessary to revise the control limits. However, if no point falls outside and the process seems to be in control with both average and dispersion satisfactory from the stand point of the specification limits, the trial control limits should be reviewed from time to time ; may be once a week or month or after 50 to 100 sub-groups.

Once a process is brought into control with a satisfactory average and dispersion, the control charts further help to :

(*i*) Take action to remove the assignable cause of variation (indicated by out of control points).

(*ii*) Take action to establish both the process average and the process dispersion. The use of control charts also helps in taking action on the specification as these very easily indicate the capability of the manufacturing process ; thus advising either to change the specification limits or use other process.

19.15.6. Use of the Variable Control Charts for Reducing the Volume of Inspection. It is well-known fact for reducing the cost of inspection and for obtaining the satisfactory product that is may be made right in the first place. When the control charts are maintained for any process then the control charts give evidence that a process is in statistical control and thus provide a rational basis for the acceptance of the process and the product.

19.15.7. Some Important Points in the Application of Quality Control Charts. Some of the principle points are discussed here in connection with the application and interpretation of the quality control charts, which are otherwise likely to be misunderstood.

(*i*) In any manufacturing process there are many variables resulting from the changes in the raw materials, methods, men and machines. Therefore, it is impracticable to produce two items exactly identical. The aim of the quality control charts is to make the pieces which are essentially alike and ensure a satisfactory quality level, irrespective of the number of variables in the process. The quality control charts are applicable when too many changes are made in the process from time to time and by these methods the products within the capability of the process can be assured.

(*ii*) A sample may be however small, it represents the production process and enables to take the action when the process is not in control ; of course ; it may not properly represent the whole of the lot.

(*iii*) The method of testing need not be special or very precise one ; all that desired is that it should be able to form the basis of acceptance or rejection of the material tested. In other words, the amount of possible error in the test results should be small in comparison to the scale over which test results will normally vary. Further, as the function of the quality control chart is to measure the degree and kind of control existing in a repetitive process, this tool can be safely applied at any time to repetitive process designed to produce successive items all essentially alike.

(*iv*) A process may continue for any amount of time, the inherent variability in the type of product produced by it does not change significantly till fundamental changes occur in the production process. If in any process which has continued for a long time and significant shift in quality level or inherent variability is observed then either the cause of shift should be apparent (due to change in raw material, method, operator or machine), if not, the manufacturer should search diligently for the causes until it is clear that further search would probably not be profitable. Thus whenever any point falls outside control limits it should be regarded as danger signal and efforts made to find the cause for it. Whenever a slight or large shift occurs in the process it results in falling of a point outside the control limits and it is very rare that point may fall outside without any assignable cause.

(*v*) It may be appreciated that each plotted point on the control charts tells a most eloquent story in relation to its control limits *e.g.*, if a trend of points shows that control limits should be changed and after that a point falls outside the new limits, it is indication of the condition that the control is probably not being maintained at the new level. Thus the significance of any point on a control chart depends upon its position in relation to the control limits that apply to it.

(*vi*) In cases where the specification of a product is concerned with only the minimum limit, the falling of a point above the upper control limit for averages must be considered as a danger signal as it indicates the unstable process.

(*vii*) In the case of $\overline{X}$-chart, the group averages that go above the upper control limit may possibly be the result of a gradually rising quality level ; but in R-chart, a range that goes above its upper control limit can indicate only a lack of control.

(*viii*) A controlled product at a lower quality level is generally considered to be preferable to an uncontrolled product at higher quality level. To understand it better let us compare two sources from where products are taken. Say, in first source although the quality level is high, but the various products are scattered about the central line at longer distances. Thus in this case there is no telling what the rest of the lot is like and there is no basis for predicting the future. In another case, say the quality level is low but the various products are very closely placed around the central line, then it has a high degree of assurance that the quality average and variability of the untested material are essentially the same as for the tested material.

19.15.8. Use of Control Charts when known Trend Exists. In many processes, the various trends exist which cause a gradual shift in quality level *e.g.*, the tool wear in machining process gradually keeps on resulting in the increase of dimensions of articles produced. It is also obvious that it is impossible or impracticable to control such a trend and suitable procedure has to be adopted to deal with such problem. In order to take care of the known trend, it is apparent that the specification limits must be wide enough to allow a marked shift in process level if defective items are to be avoided.

In such cases, first the nature of the trend (*i.e.* whether straight line or curvilinear) has to be determined by examining several runs and for this purpose it may be necessary to inspect more number of items during testing period. It is also essential to determine the rate at which the level shifts.

A particular trend in a process can be determined by the theory of runs (*i.e.* movement of successive points in the same direction). Runs frequently give advance warning of a departure from control standards. A process can be suspected to be out of control or for shift in its process average when either all the points out of seven successive points observed are on the same side of the central line, or 10 out of 11, or 12 out of 14, or 14 out of 17, or 16 out of 20 successive points observed are on the same side of the central line.

Run-sum test. It is applied on $\overline{X}$-chart to judge whether the process is in control. The test is very simple and can be described as follows :

Divide the control chart into zones, one standard deviation wide, measuring in either direction from the centre line of the chart (usually the process average). As successive observations are plotted, assign each a score according to the zone in which it falls. If the point falls above the central line and no higher than the limit of the first zone above the central line, give it a score of + 0 ; if the point falls in the second zone above the central line, but no higher than the upper limit of the second zone, give it a weight of + 1 ; and so forth. Assign negative scores to points occurring below the central line according to this scheme. If the point is on the centre line, give it a score of zero with the same sign as of the score immediately preceding. If the first observation from a process falls on the centre line, assign it a score of + 0.

Accumulate the assigned score, but begin the accumulation a new whenever there is a change in the sign of the scores or whenever there is a known change in the process conditions. Call the accumulated scores the run sum, S, and when S becomes greater in absolute value than 4, conclude that the process average is shifting.

Types of errors. When $\pm\, 3\sigma$ limits are used on the control chart, even if the process is in full control, there is a probability of 0.27% that a point on control chart may fall outside the control limits. This is known as type I error. Similarly if the process goes out of control, there is probability greater than zero that a point will fall within control limits. This is known as type II error and can be minimised by increasing the sub-group size. It may be noted that these two errors are inter-related and attempts to decrease one will result in an increase of other.

Example 19.8. *Control charts for $\overline{X}$ and R are maintained for control of an important dimension of a component. The sub-group size is settled as 5. The values of $\overline{X}$ and R are computed for each sub-group and the values of $\Sigma\overline{X}$ and ΣR are computed for each sub-group and the value of $\Sigma\overline{X}$ and ΣR after 25 sub-groups are found, to be 614.8 and 120.0 respectively. Compute the values of 3-sigma limits for the $\overline{X}$ chart. Assume $\overline{R} = 2.326\sigma'$.*

Solution. $\Sigma\overline{X} = 614.8$, $\Sigma R = 120$,

no. of sub-groups = 25, sample size $n = 5$

$$\overline{\overline{X}} = \frac{614.8}{25} = 24.592$$

$$\overline{R} = \frac{120}{25} = 4.8$$

Since $\overline{R} = 2.326\sigma'$,

$$\therefore \quad \sigma' = \frac{\overline{R}}{2.326} = \frac{4.8}{2.326} = 2.06$$

$$\therefore \qquad 3\sigma_{\overline{X}} = \frac{3\sigma'}{\sqrt{n}}$$

$$= 3 \times \frac{2.06}{\sqrt{5}} = 2.774$$

$$UCL_{\overline{X}} = \overline{\overline{X}} + 3\sigma_{\overline{X}}$$

$$= 24.592 + 2.774$$

$$= 27.336$$

and

$$LCL_{\overline{X}} = \overline{\overline{X}} - 3\sigma_{\overline{X}}$$

$$= 24.592 - 2.774$$

$$= 21.818.$$

Example 19.9. *The following table gives the ΣX and range values of 25 samples of 5 items each. The data are taken from a production line with a view to setting up control charts.*

(*a*) *Determine the upper and lower control limits of* $\overline{X}$ *and R charts.*

(*b*) *Explain if the process is in control.*

(*c*) *Examine if process can be kept in control.*

(*d*) *The specification limits given for individual items are 400 ± 30. Compare the process capability with the specification.*

(*e*) *What will be the upper and lower control limits if the process is set at nominal size ?*

Sub-group No.	ΣX	R
1	2603	30
2	1922	25
3	1966	27
4	1930	16
5	2010	21
6	1956	24
7	2000	23
8	1979	14
9	1982	15
10	1967	31
11	2021	38
12	1941	19
13	1949	29
14	1933	21
15	1980	13
16	1994	24
17	1990	20
18	1961	18
19	1990	28
20	1951	33
21	1958	13
22	1965	36
23	1988	5
24	1982	8
25	1949	7

Solution. Sub-group size = 5.

Sub-group No.	ΣX	$\bar{x}$	R
1	2003	400.6	30
2	1922	384.4	25
3	1966	393.2	27
4	1930	386.0	16
5	2010	402.0	21
6	1956	391.2	24
7	2000	400.0	23
8	1979	395.8	14
9	1982	396.4	15
10	1967	393.5	31
11	2021	404.2	38
12	1941	388.2	19
13	1949	389.8	29
14	1933	396.6	21
15	1980	396.0	13
16	1994	398.8	24
17	1990	398.0	20
18	1981	392.2	18
19	1990	398.0	28
20	1951	390.2	33
21	1958	391.6	13
22	1965	393.0	36
23	1988	397.6	5
24	1982	396.4	8
25	1949	398.8	7
		$\Sigma \bar{X} = 9863.4$	$\Sigma R = 538$ $\bar{R} = \frac{538}{25} = 21.52$
		$\bar{\bar{X}} = \frac{9863.4}{25} = 394.536$	

Solution. Sub-group size = 5.

(*a*) for $n = 5, A_2 = 0.58$ (from Table 19.7)

$$UCL_{\bar{X}} = \bar{\bar{X}} + A_2\bar{R} = 394.536 + 0.58 \times 21.52$$
$$= 394.536 + 12.460 = 406.996$$
$$LCL_{\bar{X}} = 394.536 - 12.460 = 382.076$$

(*b*) From Table 19.8, we compare all values of $\bar{X}$ and R with the UCL and LCL of both $\bar{X}$ and R.

$$\text{for } n = 5, D_3 = 0, D_4 = 2.11$$
$$UCL_R = D_4\bar{R} = 2.11 \times 21.52 = 54.4$$
$$LCL_R = 0 \times 21.52 = 0.$$

As both the charts are in control.

$\therefore$ Process is in control.

(*c*) $n = 5, A_2 = 0.58 = \dfrac{3}{d_2 \sqrt{n}}$ and $d_2 = \dfrac{3}{0.58 \sqrt{5}} = \dfrac{3}{0.58 \times 2.236} = 2.236$

$$\sigma' = \frac{\overline{R}}{d_2} = \frac{21.52}{2.232} = 9.23$$

$$3\sigma' = 3 \times 9.23 = 27.69, 6\sigma' = 55.38.$$

$$USL - LSL = 30 + 30 = 60$$

As $6\sigma'$ < USL-LSL, the process can be easily kept in control by adjusting the centre or $\overline{X}$

(*d*)
$$UCL = 394.536 + 27.69 = 427.226$$
$$LCL = 394.536 - 27.69 = 366.849$$
$$USL = 400 + 30 = 430$$
$$LSL = 400 - 30 = 370$$

(*e*) Let us centre the process at nominal size so that area above and below control limits is same.

$$UCL = 400 + 27.99 = 427.69$$
$$LCL = 400 - 27.69 = 372.31$$
$$USL = 430, LSL = 370$$

Ans. 427.69, 372.31.

Example 19.10. *Control charts for $\overline{X}$ and σ are maintained on a process. The sub-group size is 10. The values σ are computed for each subgroup. After 18 subgroups $\Sigma\sigma = 8.24$ and $\Sigma\overline{X} = 5958$. Compute the value of 3-sigma limits for the $\overline{X}$ and σ charts on the assumption that the process is in statistical control.*

For $n = 10$, take $C_2 = 0.9227$.

Solution. (Lot size) is quite large, $\dfrac{\Sigma\overline{X}}{N}$ approaches $\overline{\overline{X}}$ or $\overline{X}'$.

$$N = n \times 18 = 10 \times 18 = 180$$

$$\overline{\overline{X}} = \frac{5958}{180} \; 33.1$$

$$\overline{\sigma} = \frac{8.24}{18} = 0.458$$

$$\sigma' = \frac{\overline{\sigma}}{c_2}, \sigma' = \frac{0.458}{0.9227} = 0.496$$

$$\sigma_{\overline{X}} = \frac{0.496}{\sqrt{10}} = 0.157$$

Since the value of $n = 10$.

$$\therefore \quad \sigma_\sigma = 2[(n-1) - 2nc_2{}^2]^{1/2} \frac{\sigma'}{\sqrt{2n}}$$

$$= 2[(10-1) - 2 \times 10(0.9227)^2]^{1/2} \frac{0.496}{\sqrt{20}} = 0.111$$

3-sigma-control limits for $\overline{X}$ chart are $\overline{\overline{X}} \pm 3\sigma_{\overline{X}}$

$$UCL_{\overline{X}} = 33.1 - 3(0.157) = 33.571$$
$$LCL_{\overline{X}} = 33.1 - 3(0.157) = 32.620$$

3-sigma control limits on σ chart are $\overline{\sigma} \pm 3\sigma_\sigma$

$$UCL_\sigma = 0.458 + 0.333 + 0.781$$

$$LCL_\sigma = 0.458 - 0.333 = 0.125.$$

Example 19.11. *The values of $\Sigma\overline{X}$, ΣR for a process for 25 subgroups of 5 sample are 3569.2 and 196 respectively. Specification limits for process are 171 ± 11.*

(a) Determine the control limits for $\overline{X}$ and R charts.

(b) Sub-groups 1 and 3 have R value of 25 and 20. As these are very much out, may be discarded. Determine new control limits now.

(c) The scrap is very costly and as such it should be avoided. How the process should be changed for achieving this and what will be new control limits ?

Solution. (a) $\Sigma\overline{X} = 3569.2$, $\Sigma R = 196$, $n = 5$.

$$N = 20$$

$$\overline{\overline{X}} = \frac{3569.2}{20} = 178.46, \overline{R} = \frac{196}{20} = 9.8$$

$$UCL_{\overline{X}} = 178.46 + A_2\overline{R} = 178.46 + 9.8 \times 0.58$$

$$= 178.46 + 5.69 = 184.15$$

$$LCL_{\overline{X}} = 178.46 - 5.69 = 172.77$$

(b) Control limits of R-chart :

$$UCL_R = D_4\overline{R} = 2.11 \times 9.8 = 20.7,$$

$$LCL_R = 0 \times 9.8 = 0$$

It will be seen that sub-groups 1 and 3 are out of control. Eliminating them, ΣR for 18 groups = 196 – 45 = 151.

$$\therefore \qquad \overline{R} = \frac{151}{18} = 8.4$$

then $$UCL_R = 2.11 \times 8.4 = 17.724, LCL_R = 0.$$

All values lie within control limits, so R chart is in control at a centering of 8.4.

Now $\overline{X}$-chart limits will be :

$$UCL_{\overline{X}} = 178.46 + 8.4 \times 0.58 = 178.46 + 4.875$$

$$= 183.335$$

$$LCL_{\overline{X}} = 178.48 - 4.875 = 173.605.$$

(c) Specifications limits = 171 ± 11, LSL = 160, USL = 182

$$\sigma' = \frac{R'}{d_2} = \frac{8.4}{2.326} = 3.61, 3\sigma' = 3 \times 3.61 = 10.83$$

As scrap is costly, we keep $LSL = LCL$ and adjust centering accordingly.

$$\overline{X} = LCL = + 3\sigma' = 160 + 10.83 = 170.83.$$

(d) $$UCL_{\overline{X}} = 170.83 + 4.875 = 175.705.$$

$$LCL_{\overline{X}} = 170.83 - 4.875 = 165.955.$$

$$\therefore \qquad R = 8.4, UCL_R = 17.124, LCL_R = 0.$$

Example 19.12. *Control charts for $\overline{X}$ and σ are maintained on the resistance in ohms of a certain rheostat coil based on a subgroup size of five. After 30 subgroups, $\Sigma\overline{X}$ = 58,395 and Σσ = 1,356.*

(a) Determine the central lines and 3-sigma control limits for this process.

(*b*) *Estimate the value of σ′ assuming that the process is operating in statistical control.*

(*c*) *Assuming that the distribution generated by the process is approximately normal, what proportion of the rheostat coils meets specifications of 2,000 ± 150 ohms ?*

Solution. For a subgroup size of 5,

$$c_2 = 0.8407$$

Values of $A_1 = 1.60, B_3 = 0, B_4 = 2.09$

(*a*) $$\overline{\overline{X}} = \frac{\Sigma \overline{X}}{30} = \frac{58,395}{30} = 1,946.5$$

$$\overline{\sigma} = \frac{\Sigma\sigma}{30} = \frac{1,356}{30} = 45.2$$

$$UCL_{\overline{X}} = \overline{\overline{X}} + A_1\overline{\sigma} = 1,946.5 + (1.60 \times 45.2) = 2,018.82$$

$$UCL_{\overline{X}} = \overline{\overline{X}} - A_1\overline{\sigma} = 1,946.5 - (1.60 \times 45.2) = 1.874.18$$

$$UCL_{\sigma} = B_{4\overline{\sigma}} = 2.09 \times 45.2 = 94.47$$

$$UCL_{\sigma} = B_{3\overline{\sigma}} = 0 \times 45.2 = 0.$$

(*b*) Since $c_2 = 0.8407$ $\sigma' = \dfrac{\overline{\sigma}}{c_2} = \dfrac{45.2}{0.8407} = 53.76$

For upper specification limit of 2,150,

$$\sigma = \frac{2,150 - 1,946.5}{53.76} = 3.79$$

From Table 19.3. when σ = 3.79, the area of normal curve between mean value of *UCL* = 0.49993.

For lower specification limit of 1,850, $\sigma = \dfrac{1,850 - 1,946.5}{53.76} = 3.79$

From Table 19.3, when σ = – 1.795, the area of the normal curve between mean name of *LCL* = 0.46407.

Therefore, area of curve between the upper and lower specification limits = 0.49993 + 0.46407 = 0.964.

Hence about 96.4% of the rheostat coils will meet the specification limits of 2,000 ± 150 ohms.

Example 19.13. $\overline{X}$ *and R charts have been maintained on a certain quality characteristic. All points have been found to be within control limits on both charts. A sudden change in the process occurs that increases* $\overline{X}$ *by 1.5 σ′ but does not change σ′. Assuming that the quality characteristic is normally distributed both before and after the change and that the control limits are based on observations made before the shift in process centering, answer the following :*

(*a*) *If the subgroup size is 3, approximately what percentage of points would you expect to fall outside control limits on the* $\overline{X}$ *Chart because of change in* $\overline{X}$ *? What percentage on the R chart ?*

(*b*) *Answer the same question assuming a sub-group size of 5, and 8.*

(*c*) *Relate the answers to the above questions to Type II error and the effect of subgroup size on the detection of shifts in* $\overline{X}$ *on the* $\overline{X}$ *chart.*

Solution. Fig. 19.23 shows that $\overline{X}$ chart before and after the change in $\overline{X}$ with the control limits based on $\overline{X}$ only.

When the process shifts to $\overline{\overline{X}} + 1.5\sigma'$ and control limits remaining the same *i.e.* $\overline{\overline{X}} \pm 3\ \sigma_{\overline{X}}$, area of the curve between $UCL_{\overline{X}}$ and $LCL_{\overline{X}}$ can be calculated with the help of Table 19.3. For using Table 19.3, we first find σ.

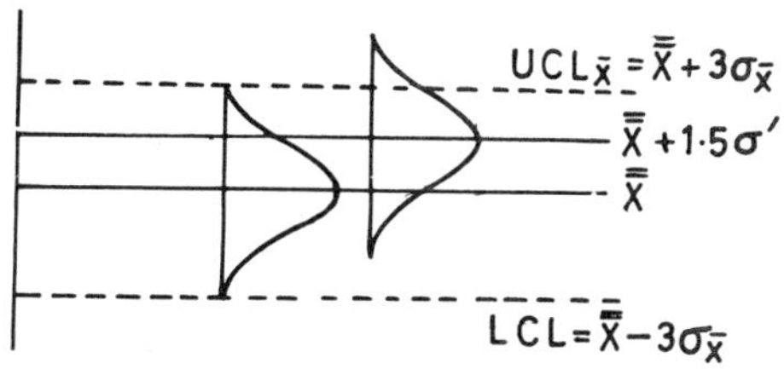

Fig. 19.23. $\overline{X}$-chart before and after the shift in $\overline{\overline{X}}$.

When $\quad UCL_{\overline{X}} = \overline{\overline{X}} + 3\sigma_{\overline{X}}$,

$$\sigma = \frac{(\overline{\overline{X}} + 3\sigma_{\overline{X}}) - (\overline{\overline{X}}' + 1.15\sigma')}{\sigma_{\overline{X}}} = \frac{3\sigma_{\overline{X}} - 1.5\sqrt{n}\ \sigma_{\overline{X}}}{\sigma_{\overline{X}}}$$

$$= 3 - 1.5\sqrt{n} \qquad \left(\text{Since } \sigma_{\overline{X}} = \frac{\sigma'}{\sqrt{n}}\right)$$

Similarly when $LCL_{\overline{X}} = \overline{\overline{X}} - 3\sigma_{\overline{X}}$

$$\sigma = -3 - 1.5\sqrt{n}$$

(*a*) When $\quad n = 3$

Value of σ corresponding to $UCL_{\overline{X}} = 3 - 1.5\sqrt{3} = 0.402$

Value of σ corresponding to $LCX_{\overline{X}} = -3 - 1.5\sqrt{3} = -5.598$

From Table 19.3, the area of curve upto $LCL_{\overline{X}}$ is negligible and area of curve upto $UCL_{\overline{X}}$ is 0.6560.

Hence the area of curve beyond the control limits is 1.0000 – 0.6560 = 0.3440 or about 34.4% of the points will fall outside the control limits.

(*b*) When $\quad n = 5$

Following the above procedure and taking the value of *n* as 5, it can be found that 63.8% of the points will fall outside the control limits.

When $\quad n = 8$

Percentage of points falling outside the control limits is 89.3%.

It may be noted that in all the above cases, the quality characteristic is normally distributed and σ′ and hence $\overline{R}$ remains the same. Therefore the number of points falling beyond the control limits or *R* chart is same as before the shift.

(*c*) If we conclude that universe has not changed, when it really has changed, this conclusion is described as Type II error.

In all the above cases the universe has changed. In (*a*) about 34.4% of the points outside the control limit or 65.6% of the points fall within the control limits. When the points fall within the control limit we will be concluding that universe has not changed even when it has really changed. Hence the probability of Type II error 65.6%. Similarly in (*b*) this probability is 36.2% and 10.7% respectively.

From above it can be concluded that an increase in sub-group size reduces the frequency of Type II error.

Example 19.14. *A machine is working to a specification of 12.58 ± 0.05 mm. A study of 50 consecutive pieces shows the following measurements put into 10 groups of 5 each :*

1	2	3	4	5	6	7	8	9	10
12.62	12.63	12.62	12.61	15.59	12.57	12.57	12.58	12.61	12.76
12.60	12.56	12.56	12.66	12.58	12.63	12.56	12.57	12.60	12.59
12.62	12.60	12.57	12.62	12.57	12.60	12.61	12.60	12.62	12.62
12.61	12.59	12.58	12.61	12.59	12.60	12.59	12.60	12.60	12.58
12.65	12.60	12.63	12.60	12.56	12.59	12.59	12.61	12.65	12.54

(i) Determine the process capability.

(ii) Determine the 3 sigma limits for $\overline{X}$-chart.

(iii) Does it appear that the machine is capable of meeting the specification requirements ?

(iv) Calculate the per cent defective, if any.

(v) Suggest possible ways by which the per cent defective can be reduced.

Assume :

(i) normal distribution and (ii) d_2 for-group size 5 is 2.326.

Solution. Values of $\overline{X}$ and R for the 10 groups from the given data are :

Group	$\overline{X}$	*R*	*Group*	$\overline{X}$	*R*
1.	$\frac{63.10}{5} = 12.62$	0.05	2.	$\frac{62.98}{5} = 12.60$	0.07
3.	$\frac{62.96}{5} = 12.59$	0.07	4.	$\frac{63.10}{5} = 12.62$	0.06
5.	$\frac{62.89}{5} = 12.50$	0.03	6.	$\frac{62.99}{5} = 12.60$	0.06
7.	$\frac{62.92}{5} = 12.58$	0.05	8.	$\frac{62.96}{5} = 12.59$	0.04
9.	$\frac{63.08}{5}$ 12.62	0.05	10.	$\frac{62.99}{5} = 12.60$	0.06

$$\therefore \quad \overline{\overline{X}} = \frac{126.02}{10} = 12.602 \text{ and } \overline{R} = \frac{0.54}{10} = 0.054$$

Now σ' = population standard deviation

$$= \frac{\overline{R}}{d_2} = \frac{0.054}{2.326} = 0.0232$$

(i) Process capability is the minimum spread of the measurement variation which includes 99.73% of the measurements from the given process, *i.e.* $6\sigma'$.

$\therefore$ Process capability = $6 \times 0.0232 = 0.1392$ mm.

(ii) 3σ limits for $\overline{X}$-chart,

$$3\sigma_{\overline{X}} = \frac{3\sigma'}{\sqrt{n}} = 3 \times \frac{0.0232}{\sqrt{5}} = \frac{3 \times 0.0232}{2.236} = 0.0311.$$

$$\therefore \qquad UCL_{\overline{X}} = \overline{\overline{X}} + 3\sigma_{\overline{X}} = 12.602 + 0.0311 = 12.6331 \text{ mm.}$$

$$\text{or} \qquad LCL_{\overline{X}} = \overline{\overline{X}} - 3\sigma_{\overline{X}} = 12.602 - 0.0311 = 12.5609 \text{ mm.}$$

(iii) It will be noted that first point of $\overline{X}$ lies outside the upper control limit, indicating thereby that process is not in control. However, assuming that process in future will be in control and therefore, eliminating this point

$$\overline{X} = \frac{113.38}{9} = 12.598 \text{ and } \overline{R} = \frac{0.49}{9} = 0.0544$$

$$\therefore \qquad \sigma' = \frac{\overline{R}}{d_2} = \frac{0.0554}{2.326} = 0.0233.$$

Now process is centred at 12.598 with $3\sigma'$ spread of 0.0699 mm.

$\therefore$ Maximum and minimum points can fall upto

$$12.598 + 0.0699 = 12.6679$$

and

$$12.598 - 0.0699 = 12.5281$$

whereas specification limits are 12.58 ± 0.05, *i.e.* 12.63 mm and 12.53 mm.

Obviously the process capability being more than tolerance value, rejects will be there.

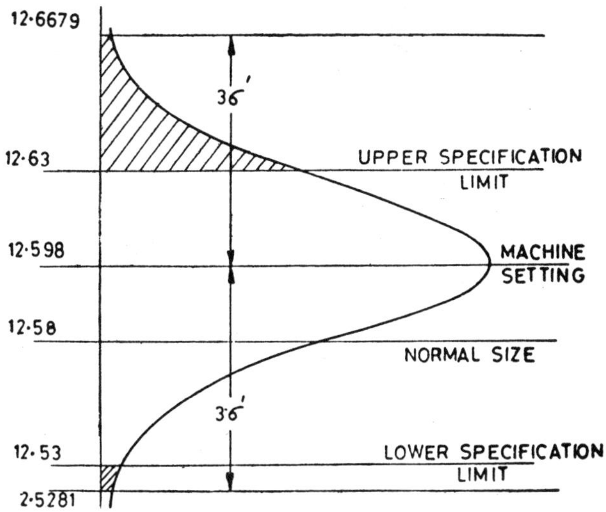

Fig. 19.24. Percent defective likely to occur shown by shaded area.

(*iv*) Per cent defective will correspond to area between the lines for dimensions 12.6679 and 12.63 on upper side, and 12.5281 and 12.53 on lower side (refer Fig. 19.24).

Area between dimensions 12.6679 and 12.598 is 49.86% and area between 12.63 and 12.598 corresponds to $\frac{12.63 - 12.598}{\sigma'}$

$$= \frac{0.032}{0.0232} = 1.37\sigma \text{ limits} = 41.924\% \text{ (From table 19.3)}$$

$\therefore$ Percentage defective in shaded area = 49.86 – 41.924 = 7.936

The percentage defective on lower side is negligible and may be ignored,

$\therefore$ Per cent defective = 8% (Approx.)

(*v*) Actually the process is not capable of meeting the specification limits. However, 0.4% defective can be reduced further by changing the machine setting to 12.58 mm.

Example 19.15. On the use of Control Charts :

It was desired to manufacture above 10,000 bushings having diameter 25.000 ± 0.05 mm on a four-spindle automatic screw machine. On starting the work it was found that about 10% of the bushings were outside the limits. Tool setter feels that is the best that could be done, and that if tolerances are not made wider then sorting will have to resorted to. The foreman found from his records that same machine was used earlier for this tolerance without any trouble. It was decided to get the facts by means of a chart of individual measurements.

Diameters of about 100 bushings manufactured were measured in sequence and the dimensions plotted. At first glance chart showed that about 10% dimensions are lying outside ; however, on closer examination, it was found that bushings coming from spindle No. 2 were about 0.0375 mm larger than those from spindles 4 and 1, the bushings from spindle 3 were usually the smallest in each group of four. Getting this clue, the points were regrouped according to the source, *i.e.* spindlewise. It was then found out that an adjustment at spindle 2 was necessary to lower the average diameter of bushings made on it. Further it was also found that an adjustment at spindle 3 be made to raise the average diameter of its bushings.

Example 19.16. *The dimension of a certain manufactured part is specified as 2.050 ± 0.020. If the dimension falls below L, the part must be scrapped, if above U, rework is required. Control charts for* $\overline{X}$ *and R are maintained on this dimension. The subgroup size is 4. After 20 sub-groups,* $\Sigma\overline{X} = 41.340$ *and* $\Sigma R = 0.320$. *If the process is in statistical control and normally distributed, what can you conclude regarding its ability to meet specifications ? Can you make any suggestion for improvement ?*

Solution :

$$\Sigma R = 0.320$$

$$\overline{R} = \frac{0.320}{20} = 0.016$$

$$\sigma' = \frac{\overline{R}}{d_2} = \frac{0.016}{2.059} = 0.0078$$

$$\Sigma\overline{X} = 41.340$$

$$\overline{\overline{X}} = \frac{41.340}{20} = 2.067$$

$$6\sigma' = 0.0078 \times 6 = 0.0468$$

$$USL - LSL = 0.040$$

It is evident that the natural tolerance ($6\sigma'$) of this process is considerably greater than the specified tolerance, and therefore, some product is likely to fall outside the specification limits. Hence the process is not able to meet the specifications.

$$USL = 2.050 + 0.020 = 2.070$$

$$\sigma \text{ corresponding to } USL = \frac{2.070 - 2.067}{0.0078} = 0.3846$$

From Table 19.3, area of normal curve upto $USL = 0.6498$

Area of the curve above $USL = 1 - 0.6498 = 0.3502$

Hence, about 35% of the product will require rework.

Also since LSL is considerably lower than $\overline{\overline{X}} - 3\sigma'$, the product below LSL is practically 0%. Hence the scrap is 0%.

The situation can be improved by shifting the centering of the process. In order to keep the scrap practically to 0%, let the centering be shifted to a value $\overline{\overline{X}}$ so that $\overline{\overline{X}} - 3\sigma' = LSL$.

With this new value of centering, σ corresponding to USL

(Hence $\overline{X} - 0.0234 = 2.030$, $\overline{X} = 2.030 + 0.0234 = 2.0534$)

$$= \frac{2.070 - 2.0534}{0.0078} = 2.128$$

Area of the curve upto USL = 0.9834

Area of the curve above USL = 1 − 0.9834 = 0.0166

Hence, the % of rework = 1.66%.

From the above results, it is evident that a shifting in process centering will result in considerable improvement.

Example 19.17. (*a*) *Control chart for $\overline{X}$ is to be prepared for a certain dimension of component. The sub-group size is 4. After 20 sub-groups it is found that $\Sigma\overline{X} = 825.60$ mm and $\Sigma R = 5.60$ mm.*

Compute the central line and the control limits for $\overline{X}$-chart.

d_2 for sub-group size 4 is equal to 2.059.

(*b*) *If the specified dimension is 41.0 ± 0.40 mm and the above process is in control and is normally distributed, can it meet the specification requirement ? If not, determine the percentage of rejections.*

Solution. (*a*) $\Sigma\overline{X} = 825.60$

$$\therefore \quad \overline{\overline{X}} = \frac{\Sigma\overline{X}}{20} = \frac{825.60}{20} = 41.28$$

∴ Central line = 41.28 mm

$$\overline{R} = \frac{\Sigma R}{20} = \frac{5.60}{20} = 0.28 \text{ mm}$$

$$\sigma' = \text{population standard deviation} = \frac{R}{d_2} = \frac{0.28}{2.059} = 0.1115$$

$$3\sigma_{\overline{x}} = \frac{3\sigma'}{\sqrt{n}} = \frac{3 \times 0.1115}{\sqrt{4}} = \frac{0.3345}{2} = 0.1673$$

$$\therefore \quad UCL_{\overline{x}} = 41.28 + 0.1673 = 41.4473$$

and $$LCL_{\overline{x}} = 41.28 - 0.1673 = 41.1127.$$

∴ Control limits on $\overline{X}$-chart are 41.4473 and 41.1127 mm.

(*b*) Specified dimension = 41.0 ± 0.40 mm

is between = 41.4 and 40.6 mm.

Central line = 41.28

and $3\sigma' = 3 \times 0.1115$

$= 0.3345$

∴ For this process, dimensions can vary from 41.28 – 0.3345, *i.e.* 40.9455 to 41.28 – 0.3345 = 41.6145 mm.

Obviously the process variability (0.669 mm) being smaller than the specification limits (0.8 mm), it can meet the specification limits.

But since centring is not properly done, rejects will be expected corresponding to the shaded area shown in Fig. 19.25.

Area between limits 41.6145 and 41.28 = Corresponding to 3σ′ = 0.49865.

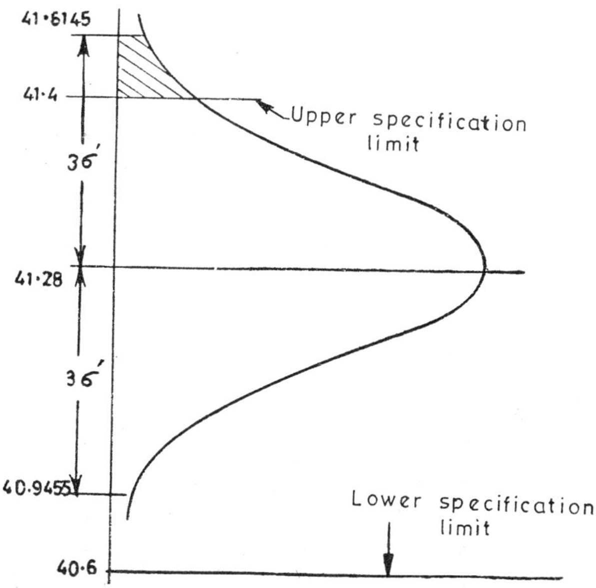

Fig. 19.25

Area between lines 41.4 and 41.28, *i.e.* corresponding to

$$= \left(\frac{41.4 - 41.28}{0.1115} = \frac{0.12}{0.1115} = 0.108 \right) \sigma'$$

$$= 0.35973$$

$\therefore$ Area between lines 41.6145 and 41.4

$$= 0.49865 - 0.35973 = 0.13892$$

$\therefore$ Number of rejects will be 13.89%. **Ans.**

19.16. Fraction Defective and the Binomial Distribution

It has already been stressed that in industry it is not always possible to measure a quality, and in such situations quality could be said to be either satisfactory or not satisfactory. In certain cases, although it is possible to measure the quality but there is not much of time, equipment and personnel to make measurements and the products are classified as satisfactory or not by *'GO'* and *'NO GO'* gauges. Such inspection is referred to as the 'attributes inspection' (Inspection by measurements is called the 'variables inspection'). In the factory language the non-satisfactory products are referred to as 'defectives' and satisfactory as 'non-defectives'. In such type of inspection, there are only two possibilities, *i.e.* the product can be either defective or non-defective. In other words, these two classes are mutually exclusive and exhaustive. There is another important term in this connection, known as fraction defective or proportion defective. If the whole aggregate of production has a certain fraction defective (p), say 0.02, we expect 2 out of 100 to be defectives and the other 98 to be non-defectives. In this definition, it is assumed that the production runs over a large number of days and that the aggregate of production is large ; day's production is a part of this aggregate or universe. If in a day about 10,000 units are manufactured, it is expected that nearly 2% or 200 units may be expected to be defectives. However, this does not mean that everyday exactly 200 units will be defective, rather some days we may have 201 defectives, some days 199 and so on. But on one day if we have 500 defectives, it is the alarming condition and indicates that some deterioration has occurred and we should look for assignable causes. We will now explain by statistical theory as upto what limit we could expect the fluctuations in excess or deficit of 200 units as due to chance causes.

Starting from a very simple problem, say we toss a coin. Obviously enough, either head or tail should turn up and not both in a single tossing. If our coin is not biased, we can anticipate head or tail to turn up with equal chances. In other words, in single tossing the probability of head or tail is only half. If the coin is tossed twice, then the various possibilities can be : heads in both occasions (*HH*) : first head and then tail (*HT*) ; first tail and the head (*TH*) ; or tails in both the instances (*TT*). Further, as the second toss is not influenced in any way by the result of first, these tosses are said to be independent. Now the probability of occurrence of (*HH*) is $\frac{1}{2} \times \frac{1}{2}$. Similarly the probability of occurrence (*HT*) is probability of *H* in first toss × probability of occurrence of *T* in second toss $= \frac{1}{2} \times \frac{1}{2} = \frac{1}{4}$. Similarly probability of (*TH*) and (*TT*) are also $\frac{1}{4}$ and $\frac{1}{4}$ respectively.

In the theory of defectives ; the order of occurrence has no importance, *i.e.*, *TH* and *HT* are one and the same thing. If so, probability of *HH* $\frac{1}{4}$ probability of *TH* or (*TH*) $= \frac{1}{4} + \frac{1}{4} = \frac{1}{2}$ probability of $TT = \frac{1}{4}$.

If two coins are tossed at the same time, then the chances of various possibilities (*HH, HT* or *TH*) remain the same. If three coins are tossed at the same time, then the various possibilities have the following chances :

All heads (*HHH*) Probability $= \frac{1}{2} \times \frac{1}{2} \times \frac{1}{2} = \frac{1}{8}$

Two heads and one tail

(*HHT, HTH, THH*) Probability $= \frac{1}{8} + \frac{1}{8} + \frac{1}{8} = \frac{3}{8}$

Two tails, and one head

(*TTH, THT, HTT*) Probability $= \frac{1}{8} + \frac{1}{8} + \frac{1}{8} = \frac{3}{8}$

All tails (*TTT*) Probability $= \frac{1}{8}$

These various probabilities could be expressed in terms of Binomial expansion, according to which

$$\left(\frac{1}{2}+\frac{1}{2}\right)^3 = \left(\frac{1}{2}\right)^3 + {}^3C_1\left(\frac{1}{2}\right)^2\left(\frac{1}{2}\right) + {}^3C_2\left(\frac{1}{2}\right)\left(\frac{1}{2}\right)^2 + {}^3C_3\left(\frac{1}{2}\right)^3$$

$$= \frac{1}{8} + 3 \times \frac{1}{8} + 3 \times \frac{1}{8} + \frac{1}{8}$$

It may be noted that the total sum of all these probabilities = 1.

In general if we toss n coins at the same time (the probability of head or tail in single toss being $\left(\frac{1}{2}\right)$, the probabilities could be expressed as the Binomial expansion of $\left(\frac{1}{2}+\frac{1}{2}\right)^n$ which is equal be

$$\left(\frac{1}{2}\right)^n + {}^nC_1\left(\frac{1}{2}\right)^{n-1} \times \left(\frac{1}{2}\right) + {}^nC_2\left(\frac{1}{2}\right)^{n-2} \times \left(\frac{1}{2}\right)^2 + \ldots\ldots + {}^nC_n\left(\frac{1}{2}\right)^n$$

There are cases, when the probability of toss need not be $\frac{1}{2}$ and $\frac{1}{2}$ *e.g.*, in throwing a dice, the chance of number '1' turning up is $\frac{1}{6}$ and similarly $\frac{1}{6}$ for any of the other numbers because there are six faces on the dice and any one of them is equally likely to turn up in a throw. It may be noted here that we are considering only such cases in which we are interested in finding the probability of an event either to occur or not to occur and in that case the sum of both the probabilities is 'one' *e.g.*, in the case of dice also, probability of No. 1 turning up is $\frac{1}{6}$ and the probability of its not turning up is $\frac{5}{6}$.

Coming back to our problem of fraction defectives, we know that inspection is carried out by checking random sales from the total aggregate of production. If the standard proportion defective for the aggregate is p, the chance that any one product chosen from aggregate (in a random selection) would be defective is p, and the chance that it would be non-defective is $(1 - p)$ say q. We now find that all the necessary conditions (listed below) of the binomial distribution are satisfactory by this case also, provided it is ensured that the sample chosen from the aggregate is by random selection and the probability for a defective remains as p from trial to trial.

(*i*) A product can be either defective or non-defective, it can't be both (classifications are mutually exclusive).

(*ii*) Further it has to be either defective or non-defective and nothing else. Or in other words $p + q = 1$. (Classifications are exhaustive).

Say we take, random sample of n units from a day's production, then the probabilities of various possibilities in a sample of n units are as follows :

All n defectives	Probability = p^n
$(n - 1)$ defectives and 1 non-defective	Probability = ${}^nC_1\, p^{n-1}$
$(n - 2)$ defectives and 2 non-defectives	Probability = ${}^nC_2\, p^{n-2}\, q^2$
$(n - 3)$ defectives and r non-defectives	Probability = ${}^nC_r\, p^{n-r}\, q^r$
All n non-defectives	Probability = ${}^nC_n\, q^n = q^n$

These various probabilities are the terms of the binomial expansion of $(p + q)^n$, where $p + q$ is always positive and equal to unity. The binomial distribution is a frequency distribution where the variate (x = number of defectives) takes the values $n - 1, n - 2, \ldots 2, 1, 0$ and the frequencies are the corresponding probabilities of occurrence (of x) $p^n, {}^nC_1p^{n-1}q, {}^nC_2p^{n-2}q^2, \ldots, {}^nC_rp^{n-r}q^r, \ldots {}^nC_{n-1}\, pq^{n-1}, {}^nC_nq^n$.

Mathematically by adding up these it can be proved that the chance of getting number of defectives (x) outside the range $np \pm \sqrt{3npq}$ is very small and that if we do get any number of defectives outside these limits in any sample, we can suspect that sample to be not from our universe. Further if n is very large, then the binomial distribution tends to the normal curve pattern of variation and in that case $\sigma = \sqrt{npq}$ and mean = np. So we can use np as our central line, $np + 3\sqrt{npq}$ as U.C.L. and $np - 3\sqrt{npq}$ as L.C.L. and can evolve a control chart analysis. This is known as np-chart and is used in case where n is the same for all the samples. But, if the sample n varies then instead of considering the actual number of defectives we can consider the proportion defective $\left(\frac{p}{n}\right)$ in the sample. Then in that case the central line will be at p (of the universe) and U.C.L. $= p + 3\sqrt{\frac{pq}{n}}$ and L.C.L. $= p - 3\sqrt{\frac{pq}{n}}$. This is known as p-chart.

In quality control work, to differentiate the sample proportion defective and the aggregate (or universe proportion defective) the symbol p' is used to denote the universe proportion defective and $\bar{p}$ to denote the sample proportion defective.

$$p = \frac{\text{Number of defectives in rational sub-groups}}{\text{Number of units in the rational sub-groups}}$$

if p' of universe is not known, we estimate p' by $\bar{p}$, where

$$\bar{p} = \frac{\text{Total number of defectives in all the rational subgroups}}{\text{Total number of units in all the rational subgroups}}$$

When the standard of the universe is established by control and is estimated, it can be confidently said that the proportion defectives will lie within such and such limits. The consumer and producer know what is being bought and what is being sold. Both are confident of the quality and thus the confidence in trade is increased.

The p-chart, though not as useful as $\overline{X}$ and R-charts for diagnosis purposes, has definitely been appreciated by many modern manufactures as a device to establish better relationship between vendor and purchaser. Further the use of p-chart is very simple and economical. In most of these cases, mere examination of inspection records gives valuable information to the management. Even when rational sub-group size is not constant from day to day, the p-chart with variable limits helps to analyse the process standard scientifically.

19.17. Control Charts for Fraction Defectives

These charts are desirable at that place where completeness of the product is important and not the dimensional accuracy, *e.g.*, completeness of die-casting or some moulding etc. This is also useful where the process control is not possible by controlling a few dimensions. Rather we are interested in finding the number of defects in a part. If a customer is checking his parts by a sampling system and can accept 2.5% defectives then it is desirable to ensure that on average the process is not producing in excess of 2.5% defectives.

In above instances, a control chart for the fraction of a sample which is defective, is desirable. Although the chance factor in sampling varies, it can be confined by means of mathematical calculations. The control limits must be set at such a value for a given process average per cent defective, that if any plot goes outside the limits ; then that will be due to assignable causes. Variations within control limits are due to inherent sources of variations.

According to statistics, limits of control chart are kept as

$$\overline{p} \pm 3\sqrt{\frac{\overline{p}\,(1-\overline{p})}{n}} \text{ from tolerable \% age defectives.}$$

$\overline{p}$ = Mean fraction defective, n = sample size.

Usually lower control limit becomes negative and is not applicable. In such case it should be treated as zero.

In case sample size is not constant, then $n.p.$ chart is used and the control limits then become $n.\overline{p} \pm 3\sqrt{n\overline{p}\,(1-\overline{p})}$.

Sample size chosen should be such as to contain at least one defective. For example, with a process average of 1%, the smallest sample size chosen should be $\frac{100}{2} = 50$. To obtain more accurate estimates and to secure greater protection, the size of sample must be large one. But considering the inspection expenses, it should be small one. Thus a balance must be striked somewhere so as not to sacrifice any important thing.

When the process average per cent defective is unknown, it can be found out by taking a large number of samples when the process is running normally and finding the mean value of per cent defectives in the samples.

By maintaining p-charts, work gradually improves in quality as more and more assignable causes are discovered.

Example 19.18. (*a*) *How do you compare p-chart with $\overline{X}$ and R charts.* (*b*) *A manufacturer purchases small bolts in cartons that usually contains several thousand bolts. Each shipment consists of a number of cartons. As part of the acceptance procedure for these bolts, 400 bolts are selected at random from each carton and are subjected to usual inspection for certain defects. In a shipment of 10 carbons, the respective percentages of defectives in the samples from each carton are 0, 0, 0.5, 0.75, 0, 2.0, 0.25, 0, 0.25 and 1.25. Does this shipment of bolts appear to exhibit statistical control with respect to the quality characteristics examined in this inspection ?*

Solution. (*a*) The various comparisons between two types of charts are given below :

(*i*) *p*-chart is an attribute control chart, *i.e.*, for quality characteristics which can be classified as either conforming or non-conforming to specification whereas $\overline{X}$ and R charts are for quality characteristics that can be measured and expressed in numbers.

(*ii*) Cost of maintaining *p*-charts is less than the corresponding $\overline{X}$ and R charts, because that data for *p*-chart can be collected by using limit gauges and is generally available whereas the data for $\overline{X}$ and R charts have to be collected specifically for this purpose and measuring instruments employed are generally costly.

(*iii*) *p*-charts can be applied to any number of quality characteristics observed on one article but separate $\overline{X}$ and R charts are required for each measured quality characteristic.

(*iv*) *p*-chart is best suited in case where inspection is carried out with a view to classifying an article as accepted or rejected. $\overline{X}$ and R charts are best suited for critical dimensions.

(*v*) *p*-chart, though discloses the causes of variations, but is not as sensitive as the $\overline{X}$ and R charts. For actual diagnosis of causes of troubles, $\overline{X}$ and R charts are best, but *p*-chart can be used effectively in the improvement of quality.

(*vi*) The sample size is generally larger for *p*-chart than $\overline{X}$ and R charts. For *p*-charts, the sample should be large enough so that at least nine times out of ten or more defectives will be found. The variation of sample size influences the chart control limits much more in $\overline{X}$ and R charts than in *p*-chart.

(*vii*) *p*-chart is superior to $\overline{X}$ and R charts because it is more sensitive in detecting a shift in the process.

(*b*) As the %age of defectives in 10 cartons are 0, 0, 0.5, 0.75, 0, 2.0, 0.25, 0, 0.25 and 1.25.

$$\therefore \quad p\% = \frac{0+0+0.5+0.75+0+2.0+0.25+0+0.25+1.25}{10} = \frac{5}{10} = 0.5$$

$$\therefore \quad \bar{p} = 0.005$$

$$3\sigma_p = 3\sqrt{\frac{\bar{p}\,(1-\bar{p})}{n}} = 3\sqrt{\frac{0.005\ (0.995)}{400}} = 0.01058$$

$\therefore$ U.C.L. = 0.005 + 0.01058 = 0.01558 and L.C.L. = 0.005 – 0.01058 = 0

($\because$ It comes out to be a negative value which has no meaning).

On plotting the control chart, it will be observed that one point (2.0%) or 0.02 is outside the upper control limit. Therefore, it is concluded that this shipment does not exhibit statistical control with respect to the quality characteristics examined in the inspection.

Example 19.19. *In a manufacturing process, the number of defectives found in the inspection of 20 lots of 100 samples is given below :*

Lot No.	*No. of defectives*	*Fraction defectives*	*Lot No.*	*No. of defectives*	*Fraction defectives*
1	6	0.06	11	4	0.04
2	11	0.11	12	6	0.06
3	15	0.15	13	7	0.07
4	9	0.09	14	8	0.08
5	6	0.06	15	2	0.02
6	4	0.04	16	4	0.04
7	5	0.05	17	5	0.05
8	3	0.03	18	3	0.03
9	4	0.04	19	6	0.06
10	5	0.05	20	7	0.07

(*a*) *Determine the control limits of p-chart and state whether the process is in control.*

(*b*) *What will be new value of mean fraction defective if some abvious point outside control can be analysed, remedied and eliminated. What will be corresponding upper and lower limits and examine whether the process is still in control or not.*

(*c*) *Determine the sample size when a quality limit not worse than 8% is desirable and a 10% bad product will not be permitted more than three times in a thousand.*

Solution.

(*a*) Mean fraction defective $\bar{p} = \dfrac{\text{Total No. of defectives}}{\text{Total No. of items inspected}}$

Total No. of defectives

$$= 6 + 11 + 15 + 9 + 6 + 4 + 5 + 3 + 4 + 5 + 4 + 6 + 7 + 8 + 2 + 4 + 5 + 3 + 6 + 7$$
$$= 120$$

$$\therefore \quad \bar{p} = \frac{120}{20 \times 100} = 0.06$$

$$\therefore \quad UCL_p = \bar{p} + 3\sqrt{\frac{\bar{p}(1-\bar{p})}{n}} = 0.06 + 3\sqrt{\frac{0.06(1-0.06)}{100}} = 0.06 + 3\sqrt{\frac{0.06 \times 0.94}{100}}$$
$$= 0.06 + 3 \times 0.02365$$
$$= 0.06 + 0.07095 = 0.13095 \cong 0.131$$

and
$$LCL_P = \bar{p} - 3\sqrt{\frac{\bar{p}(1-\bar{p})}{n}} = 0.06 - 0.07095 = -0.01095$$

As the fraction defective can't be negative. LCL_P is taken as zero.

Looking at the values of fraction defectives in the table, it will be noted that only third lot having value of 0.15 is above the UCL_p and all others are well below it. The reasons for only this point to be out could be analysed and eliminated, and then it can be said that the process is in control.

(*b*) The revised values of $\bar{p}$ after eliminating third lot will be $\dfrac{120-25}{100 \times 19} = \dfrac{195}{1900} = 0.056$

The new control limits will be $0.056 \pm 3\sqrt{\dfrac{0.056 \times 0.944}{100}} = 0.056 \pm 0.069 = 0.125$ and 0

Now all the points are within the control limits. The revised quality level is 0.056.

Since probability that a defective whose more than 8% defective quality will not be permitted is more than 3 times in a thousand (0.3%). *i.e.*, corresponding to 3σ limits.

$$\therefore \quad \bar{p} + 3\sigma_p = 0.08$$

or
$$0.056 + 3\sqrt{\frac{0.056(1-0.056)}{n}} = 0.08$$

or
$$\sqrt{\frac{0.056 \times 0.944}{n}} = \frac{0.08 - 0.056}{3} = \frac{0.024}{3} = 0.008$$

or
$$n = \frac{0.056 \times 0.944}{0.008 \times 0.008} = 826.$$

19.18. The C-Chart

We have seen that $\bar{X}$ and R charts can be used when quality can be measured in numbers, the *p*-chart can be tried to attain control when the quality is classified by defectives and non-defectives. But there are situations in which quality is given as number of defects per unit, *e.g.*, the number of surface defects in a painted area or photographic film, the number of seeds in a glass bottle etc. The argument could be extended even to those manufactured parts which have large number of dimensions to be measured and part can't be just rejected owing to only one dimension being out of order ; under such condition the out of order dimensions can be expressed as number of defects over the number of dimensions measured.

In industry, no product can be absolutely perfect and flaws are bound to occur, though rarely. Our aim is to control and keep even these rare occurrences at a minimum possible. The *c*-chart techniques help us to keep these defects per unit at the lowest level. In the language of probability, it could be said that the opportunities for the occurrence of a defect are infinite : but the probability of occurrence at some chosen spot or time is very small (infinitesimal) and as a consequences a small (finite) number of defects actually occur. The methods to study such situations have been evolved by statistics, according to which the fluctuation from a standard (or a norm) are described what is known as a Poisson distribution if the following conditions are satisfied.

(*a*) There should be infinite opportunities for occurrence, with a low (but steady) probability of occurrence in an individual case, and a few occurrences do actually take place.

(*b*) The samples are chosen by random selection and are independent of one another.

(*c*) The choice of unit area (area of opportunities) should be the same from sample to sample.

In most of the cases, the above conditions are satisfied and where the "area of opportunity" varies, the use of c/n-chart can also be made.

According to the Poisson Law of Distribution, (which states that if m is the "standard" defects per unit, the sampling fluctuations can take a value x in the long run with the proportional frequency equal to $f(x) = e^{-m} \frac{m^x}{x!}$ (for x = 1, 2, ... α) the defects per unit have to be integers in actual counting of defects in an experiment, and that a large value of x can occur but has a low probability.

The various properties of the Poisson (frequency) distribution are summed up below :

(*i*) The mean of the distribution is equal to the "standard" defects per unit (m).

(*ii*) The standard deviation of the distribution = $\sqrt{\text{mean}} = \sqrt{m}$.

(*iii*) For large values of m, the Poisson distribution approaches the normal curve pattern. (For normal work if $m \geq 3$, normal approximation for the Poisson distribution could be used but it is more satisfactory when $m \geq 6$).

(*iv*) Poisson distribution is completely specified by one parameter, *i.e.*, m.

When the 3 conditions (*a*, *b* and *c*) given above are satisfied, the defects per unit (*c*) fluctuate in the Poisson Law of Pattern. Now if m is large. Poisson Law of Pattern approaches normal curve pattern of variation and $\pm 3\sqrt{m}$ would contain most of the possible fluctuations unless an assignable cause affects our standard. If the sample size (area of opportunity) is kept constant, the values of m and $3\sqrt{m}$ can be used in the construction of *c*-chart. In actual practice the value of m is not known, which can be estimated by $\bar{c}$, where

$$\bar{c} = \frac{\text{Number of defects in all units inspected}}{\text{Total number of units inspected}}$$

So in *c*-chart, Central line $= \bar{c}$

U.C.L. $= \bar{c} + 3\sqrt{\bar{c}}$

L.C.L. $= \bar{c} - 3\sqrt{\bar{c}}$

For c-chart the sample size may be set at one single unit of product *e.g.*, 50 meters of rope, a dozen steel containers, a pair of chappals etc. In other words, the sample size should be a fixed number of units of product. It may be noted that c-chart controls variation in number of defects per sample.

Where sub-group consists of n units, U-chart is used where $U = \frac{c}{n}$ and $\overline{U} = \frac{\Sigma c}{\Sigma n}$

Central line $= \overline{U}$

and $\text{L.C.L.} = \overline{U} + \frac{3\sqrt{\overline{U}}}{\sqrt{n}}$

and $\text{L.C.L.} = \overline{U} - \frac{3\sqrt{\overline{U}}}{\sqrt{n}}$.

Example 19.20. *The following table shows the number of defects observed on 25 similar big castings :*

Casting No.	*No. of defects observed*	*Casting No.*	*No. of defects observed*
1	7	14	24
2	14	15	14
3	14	16	9
4	18	17	9
5	8	18	11
6	14	19	10
7	8	20	8
8	11	21	9
9	20	22	11
10	12	23	7
11	22	24	26
12	15	25	8
13	8		

Find $\bar{c}$ and compute the trial control limits. What value of c' can be suggested for subsequent period ?

Solution. Total No. of defects observed on 25 castings (added all the defects) = 328

$\therefore \quad \bar{c} = \frac{328}{25} = 13.12$

Now $\quad 3\sqrt{\bar{c}} = 3\sqrt{13.12} = 3 \times 3.62 = 10.86$

$\therefore \quad \text{U.C.L.} = \bar{c} + 3\sqrt{\bar{c}} = 13.12 + 10.58 = 23.98$

and $\quad \text{L.C.L.} = 13.12 - 10.89 = 2.26$

Observing the data it is seen that only casting No. 24 lies outside the U.C.L. Eliminating it, the suggested value of $c' = \frac{302}{24} = 12.58$.

Example 19.21. *Six hundred bicycle mudguards were observed for point defects-air bubble, scratch, foreign particle, etc. In all 1200 defects were found in 600 mudguards. What type of control chart be used here ? If sample size 10, what will be the limits.*

Solution. For this application, since defects are observed in a sample of 10, U-chart will be used.

$$U = \frac{1200}{600} = 2$$

$$\text{U.C.L.} = \overline{U} + 3\sigma_U = \overline{U} + 3\frac{\sqrt{U}}{\sqrt{n}} = 2 + 3\frac{\sqrt{2}}{\sqrt{10}} = 2 + \frac{3}{10}\sqrt{20}$$

$$= 2 + 0.3 \times 4.47 = 2 + 1.441 = 3.441$$

and $$\text{L.C.L.} = \overline{U} - \frac{3\sqrt{U}}{\sqrt{n}} = 2 - 1.441 = 0.559.$$

Example 19.22. (*a*) *What is the difference between a defect and defective ? (b) The following tabulation gives defects observed in an aircraft sub-assembly operation and shows the number of units produced on each day during a week. Prepare an U-chart for this. Assume U′ = 4.2.*

	Mon	*Tue*	*Wed*	*Thurs*	*Fri*	*Sat*
n	5	5	5	5	4	5
c	19	25	15	29	36	27

Solution. (*a*) A defective is an article that in some way fails to conform to one or more given specifications ; whereas each instance of article's lack of conformity to specifications is a defect. Obviously a defective can contain one or more defects.

(*b*) The value of U', *i.e.*, control line on the control chart = 4.2. Control chart limits are computed as follows :

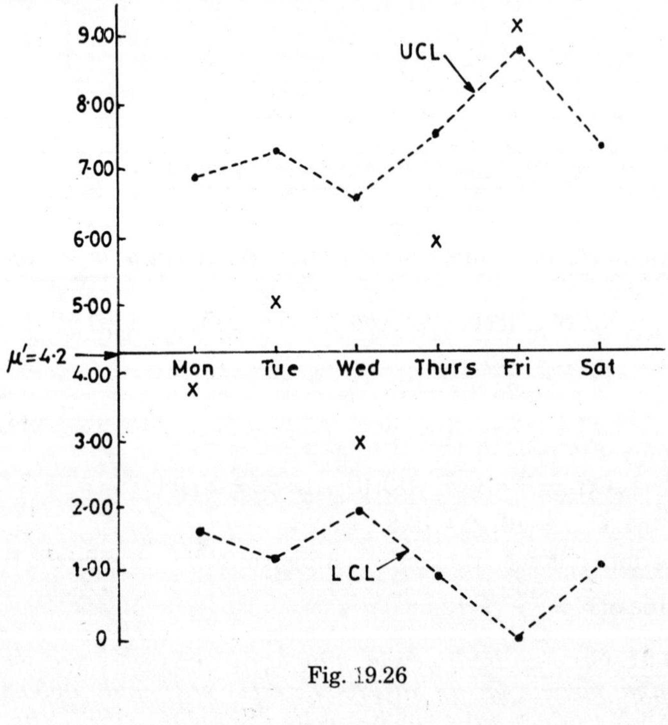

Fig. 19.26

Day	No. of defects	Unit produced n	Defects per unit $U = \frac{c}{n}$	$3\sigma = \frac{3\sqrt{U}}{\sqrt{n}}$	U.C.L. = $U' + 3\sigma$	L.C.L. = $U' - 3\sigma$
Mon	19	5	3.8	2.61	6.81	1.59
Tue	25	5	5.0	3.00	7.20	1.20
Wed	15	5	3.0	2.32	6.52	1.88
Thurs	29	5	5.8	3.23	7.43	0.97
Fri	36	4	9.0	4.50	8.70	0.00
Sat	27	5	5.4	2.11	7.31	1.09

Example 19.23. *How can you adapt c-chart for quality rating ? (Demerit Control Chart).*

Solution. For some products, all types of defects are not of equal importance ; and to plot the sum of all defects, regardless of type, on one control chart is misleading and uneconomical. In such cases where some defects are more serious than others, it is advantageous to weigh defects according to some scale that measures their seriousness.

In such cases, the defects are classified into certain groups. One such classification is as follows :

Class A Defects—Critical (very serious).

Will render product totally unfit for service ; useless or unsaleable.

Liable to cause personal injury or damage.

Class B Defects—Major (serious).

Will cause considerable consumer dissatisfaction, will shorten the life of the product, lessen its efficiency, or require exclusive servicing.

Class C Defects—Moderately serious.

Will have the same effects as those of class B, but to a lesser degree. It may include major defects of appearance, finish or workmanship.

Class D Defects—Minor (Not serious).

These will not cause operating failure of unit in service. These include defects of appearance, finish or workmanship.

Each class of defect is given some weightage which will ordinarily be somewhat arbitrary, but is supposed to reflect the economic consequences of failure to correct each type of defect. The standard deviation is given by $\sqrt{\Sigma_1^r x_i^2 c_i}$, where x_i is the weight assigned and c_i is the expected number of defects per sample and r is the number of classes of defects. With some product, it may be required to have a single index of its product quality. Such a quality index may be obtained from use of demerit-per-unit data as follows :

$$\text{Quality Index} = \frac{\text{Observed demerits per unit}}{\text{Expected demerits per unit}}.$$

Example 19.24. *Four types of defects are classified in a telephone assembly industry. The weightage to each defect is shown below. A sample of 25 units is inspected and number of defects of each class observed is tabulated below. Determine the value of U.C.L. and L.C.L. for demerit control chart.*

Class of defect	*Weightage* x_i	*No. of defects* c_i
1	0.75	5
2	0.60	15
3	0.20	75
4	0.05	50

Solution. The control limits are calculated as under :

Std. deviation $= \sqrt{\Sigma_i^{\ 4}\, x_i^{\ 2}\, c_i} = \sqrt{0.455} = 0.675$

Central line should be placed at $\Sigma_1^{\ 4}\, x_i\, c_i = 1.21$

$\therefore$ U.C.L. = 1.21 + 3 × 0.675 = 1.21 + 2.025 = 3.235

and L.C.L. = 1.21 – 3 × 0.675 = 0. **Ans.**

Class of defect	*Weightage* x_i	*No. of defects* C_i	*Defects* c_i	*Demerits per sample* $x_i\, c_i$	*Variance* $x_i^{\ 2}\, c_i$
1	0.75	5	0.2	0.15	0.11
2	0.60	15	0.6	0.36	0.22
3	0.20	75	3.0	0.60	0.12
4	0.05	50	2.0	0.10	0.005
				1.21	0.455

19.19. Sampling Inspection

Whenever a large number of parts are to be inspected, 100% inspection is not desirable. Firstly it is very expensive and secondly, the job will be very much monotonous for an inspector and due to fatigue of brain, he is likely to pass the defective parts also. Quality assessment of large batches of parts can best be carried out by statistical sampling techniques. In this technique the whole lot of parts to be inspected is divided into a large number of sub-groups. The size of sub-group is determined carefully based on statistical theory of probability etc.

A random sample of few items, whose size is again dependent upon the quality protection and determined statistically, is selected. If it contains less than a certain number of defectives, the lot is accepted. The number is called acceptance number. If a sample contains more than a certain number (rejection number), then complete lot is rejected and 100% inspection made for that.

Statistical sampling technique is based on the knowledge of the probability of acceptance or rejection of a lot containing a given percentage of defective items. It is also assumed that a sample taken is representative of the lot.

19.20. Theory of Probability

If an event is sure to occur it has a probability of one. An event that is certain not to occur has a probability of zero. All other probabilities fall between these extremes of zero and one.

If in a system, an event has the probability of occurrence $= p_1$ and some other event has probability of occurrence $= p_2$, then probability of either of two (p) is the sum or two probabilities *i.e.* $p = p_1 + p_2$. This is called the theorem of addition.

If the probability of an event in a system is p_1 and of some other event in other system is p_2, then the probability (p) of both to occur simultaneously is product of two, *i.e.*, $p = p_1 \times p_2$. This is called the theorem of multiplication.

The probability of occurrence of an event can be calculated by Binomial system or by Poisson Probability method. In the sampling inspection the following notations are used :

N = lot size, n = sample size,

c = acceptance number, p = %age defective.

The probability of zero defectives in sample $= e^{-np}$

Probability of 1 defective $= npe^{-np}$

Probability of 2 defectives $= \dfrac{(np)^2 e^{-np}}{2!}$

Probability of 3 defectives $= \dfrac{(np)^3 e^{-np}}{3!}$ and so on ;

But for finding the probability of occurrence of any number of defectives corresponding to np and acceptance number is directly tabulated. (Refer Table 19.9).

Example 19.25. (*a*) *What are the rules of combining probabilities ?*

(*b*) *What do you understand by permutation ?*

(*c*) *What do you understand by combination ?*

Solution. (*a*) 1. The probability of an event A, denoted by $P[A]$, is a number between 0 and 1 inclusive, an impossible event having probability 0 and one that is certain to occur having probability 1.

2. The probability that *at least one of* two events A and B occurs is given by

$$P[A + B] = P[A] + P[B] - P[AB] \qquad ...(1)$$

3. The probability of the simultaneous occurrence of both A and B is given by

$$P[AB] = P[A]P[B/A] = P[B]P[A/B] \qquad ...(2)$$

If the two events A and B are *mutually exclusive* (meaning that it is impossible for both to occur together), $P[AB] = 0$, then Eq. (1) becomes

$$P[A + B] = P[A] + P[B] \qquad ...(3)$$

This is called the *additional theorem* for probabilities and is the reason for the use of + sign to denote "and/or" which when A and B are mutually exclusive, reduces to strict alternate "either-or"

Since the events A and $\bar{A}$ are *mutually exclusive* and one of these must occur therefore,

$$P[A] + P[\bar{A}] = 1 \qquad ...(4)$$

The events A and B are said to be *independent* if the probability of occurrence of A is the same whether B occurs or not, that is, if,

$$P[A/B] = P[A/\bar{B}]$$

Now $\qquad P[A] = P[AB] + P[A\bar{B}]$

$$= P[B]P[A/B] + P[\bar{B}]P[\bar{A}/\bar{B}] \text{ from Eq. (1)}$$

If $\qquad P[A/B] = P[A/\bar{B}]$, this becomes

$$P[A] = P[A/B]\,(P[B] + P[B]) = P[A/B]$$

and therefore, Eq. (2) can be written for independent events,

$$P[AB] = P[A]P[B] \qquad ...(5)$$

This is called *Multiplication theorem* for probabilities and is the reason for using AB to mean "A and B". These theorems can be generalised to any number of events.

(*b*) *A permutation* of a finite number of distinguishable objects is any of their arrangements in a definite order. For example with n objects, by considering that the first place in the ordered arrangement can be filled in n ways, since any one of the n objects can be chosen. The next place can be filled in $n - 1$ ways, since only $n - 1$ objects are left to be picked. Similarly the third place can be filled in $n - 2$ ways, and so on until the last place can be filled in only one way, with the last object left. The total number of ways is therefore,

$$n(n - 1)(n - 2)(n - 3) \;......\; (3)(2)(1)$$

and this number is denoted by $n\,!$ or $\angle n$ (read as "n factorial"). If the problem is to select r objects out of n distinguishable objects, and arranging them in order, then

$$^nP_r = n(n-1) \ldots\ldots (n-r+1)$$
$$= n(n-1) \ldots\ldots (n-r+1)\frac{\angle n-r}{\angle n-r} = \frac{\angle n}{\angle n-r}.$$

(c) A *combination* is the number of ways in which a certain number of articles are selected from a collection, *note caring in what order they are arranged*. The number of ways of picking out r objects from n distinguishable objects, called *the number of combinations of n things taken r at a time*, will be denoted by ^nC_r and is given by

$$^nC_r = \frac{n!}{r!\,(n-r)!}.$$

Example 19.26. *In an acceptance plan 50 articles are inspected out of a lot of 1000 articles. The lot is accepted, if sample contains no defectives, otherwise it is rejected. If the lot submitted contains 1% defectives, then what is the probability that it will be accepted ?*

Solution. Since lot is 1% defective.

$$p' = 0.01$$

Sample size $n = 50$

$\therefore$ $np' = 50 \times 0.01 = 0.50$

From Table 19.9, under the column headed $c = 0$ (*i.e.* no. of defectives acceptable), the probability of acceptance of lot = 0.607.

Thus 60.7% lots will be accepted and rest rejected.

Example 19.27. *In the above example, if procedure is changed as ; If the sample contains 0 defective, lot is accepted ; if it contains 1 defective, a second sample is required ; and if it contains 2 or more defectives then lot is rejected. Determine, all the probabilities.*

Solution. As already determined, probability of acceptance with zero defective = 60.7%.

Probability of zero or one defective in sample from Table 19.9, corresponding to np' = 0.50 and $c = 1$ is 0.910, *i.e.* 91.0%.

$\therefore$ Probability of only one defective, or probability of taking second sample is 91.0 – 60.7, *i.e.*, 30.3%.

Probability of getting 2 or more defectives

= 1 – Probability of getting zero or 1 more defective

= 1 – 0.91 = 0.09 = 9%.

$\therefore$ Probability of rejecting lot = 9%.

Example 19.28. *3 defectives are found in a lot of 10 articles. If a sample of 2 articles be inspected, then what are the various probabilities ?*

Solution. Let of 10 contains 3 defective and 7 non-defective pieces.

Now there are $^{10}C_2$ ways of taking random sample of 2 out of lot of 10.

$$^{10}C_2 = \frac{10!}{2!\,8!} = \frac{10 \times 9 \times 8!}{2 \times 8!} = 45.$$

Probability that the random sample contains 0 defective is possible when both pieces are taken from 7 non-defectives and none from 3 defectives, for which probability

$$= \frac{^7C_2\,^3C_0}{^{10}C_2} = \frac{7!}{2!\,5!} \times \frac{1}{45} = \frac{7 \times 6 \times 5!}{2 \times 5!} \times \frac{1}{45} = \frac{21}{45} = 0.467.$$

Similarly probability that random sample contains 1 defective

$$= \frac{^7C_1\,^3C_1}{^{10}C_2} = \frac{7!}{6!} \times \frac{3!}{2!} \times \frac{1}{45} = \frac{7 \times 6!}{6!} \times \frac{3 \times 2!}{2!} \times \frac{1}{45} = \frac{21}{45} = 0.467.$$

Probability that sample contains 2 defectives $\frac{{}^7C_0\,{}^3C_2}{{}^{10}C_2} = \frac{7\,!}{7\,!} \times \frac{3\,!}{2\,!} \times \frac{1}{45} = \frac{3}{45} = 0.067.$

Probability that sample contains at least one defective = 0.467 + 0.067 = 0.534.

19.21. Basic Principle of Lot Sampling

Sampling inspection techniques are used to estimate the lot quality for lot acceptance purposes. As already pointed out, 100% inspection (or screening) is not at all economical and in some cases it is impossible ; whereas the sampling inspection is generally more practical, quick, economical, and provides incentive to the producer to improve his quality as the rejection of a sample means rejection of the whole lot.

Table 19.9. Summation of Terms of Poisson's Exponential Bionomial Limit 1,000 × probability of c or less occurrences of event that has average number of occurrences equal to e′ or np′

$c \rightarrow$ c or np' ↓	0	1	2	3	4	5	6	7	8	9
0.02	980	1000								
0.04	961	999	1000							
0.06	942	998	1000							
0.08	923	997	1000							
0.10	905	995	1000							
0.15	861	990	999	1000						
0.20	819	982	999	1000						
0.25	779	974	998	1000						
0.30	741	963	996	1000						
0.35	705	951	994	1000						
0.40	670	938	992	999	1000					
0.45	638	925	989	999	1000					
0.50	607	910	986	998	1000					
0.55	577	894	982	998	1000					
0.60	549	878	977	997	1000					
0.65	522	861	972	996	999	1000				
0.70	497	844	966	994	999	1000				
0.75	472	827	959	993	999	1000				
0.80	449	809	953	991	999	1000				
0.85	424	791	945	989	998	1000				
0.90	407	772	937	987	998	1000				
0.95	387	754	929	984	997	1000				
1.00	368	736	920	981	996	999	1000			

c → c or np′ ↓	0	1	2	3	4	5	6	7	8	9
1.1	333	699	900	974	995	999	1000			
1.2	301	663	879	966	992	998	1000			
1.3	273	627	857	957	989	998	1,000			
1.4	247	592	833	946	986	997	999	1,000		
1.5	223	558	809	934	981	996	999	1,000		
1.6	203	525	783	921	976	994	999	1,000		
1.7	183	493	757	907	970	992	998	1,000		
1.8	165	463	731	891	964	990	997	999	1,000	
1.9	150	434	704	875	956	987	997	999	1,000	
2.0	135	406	677	857	947	983	995	999	1,000	
2.2	111	335	623	819	928	975	993	998	1,000	
2.4	091	308	570	779	904	964	988	997	999	1,000
2.6	074	267	518	736	877	951	983	995	999	1,000
2.8	061	231	469	692	948	935	976	992	998	999
3.0	050	199	423	647	815	916	966	988	996	999
3.2	041	171	380	603	781	895	955	983	994	998
3.4	033	147	340	558	744	871	944	977	992	997
3.6	027	126	303	515	706	844	927	969	988	996
3.8	022	107	269	473	668	816	909	960	984	994
4.0	018	092	238	433	629	785	889	949	979	992
4.2	015	078	210	395	590	753	867	936	972	989
4.4	012	066	185	359	551	720	844	921	964	985
4.6	010	056	163	326	513	686	818	905	955	980
4.8	008	048	143	294	476	651	791	887	944	975
5.0	007	040	125	265	440	616	762	867	932	968
5.2	006	034	109	238	406	581	732	845	918	960
5.4	005	029	095	213	373	546	702	822	903	951
5.6	004	024	082	191	342	512	670	797	886	941
5.8	003	021	072	170	313	478	638	771	867	929
6.0	002	017	062	151	285	446	606	744	847	916
C →	10	11	12	13	14	15	16			
2.8	1,000									
3.0	1,000									
3.2	1,000									
3.4	999	1,000								
3.6	999	1,000								
3.8	998	999	1,000							
4.0	997	999	1,000							
4.2	996	999	1,000							

$c \rightarrow$ c or $np' \downarrow$	0	1	2	3	4	5	6	7	8	9
4.4	994	998	999	1,000						
4.6	992	997	999	1,000						
4.8	990	996	999	1,000						
5.0	986	995	998	999	1,000					
5.2	982	993	997	999	1,000					
5.4	977	990	996	999	1,000					
5.6	972	988	995	998	999	1,000				
5.8	965	984	993	997	999	1,000				
6.0	957	980	991	996	999	999	1,000			
6.2	002	015	054	134	259	414	574	756	826	902
6.4	002	012	046	119	235	384	542	687	803	886
6.6	001	010	040	105	213	355	511	658	780	869
6.8	001	009	034	093	192	327	480	623	755	850
7.0	001	007	030	082	173	103	450	599	729	830
7.2	001	006	025	072	156	276	420	569	703	810
7.4	001	005	022	063	140	253	392	539	679	788
7.6	001	004	019	055	125	231	365	510	648	765
7.8	000	004	016	048	112	210	338	481	620	741
8.0	000	003	014	042	100	191	313	453	593	717
8.5	000	002	009	030	074	150	256	386	523	653
9.0	000	001	006	021	055	116	207	324	456	587
9.5	000	001	004	015	040	089	165	269	392	522
10.0	000	000	003	010	029	067	130	220	333	458

$c \rightarrow$	10	11	12	13	14	15	16	17	18	19	20	21	22
6.2	949	975	989	995	998	999	1,000						
6.4	939	969	986	994	997	999	1,000						
6.6	927	963	982	992	997	999	999	1,000					
6.8	915	955	978	990	996	998	999	1,000					
7.0	901	947	974	987	994	998	999	1,000					
7.2	887	937	967	984	993	997	999	999	1,000				
7.4	871	926	961	980	991	996	998	999	1,000				
7.6	854	915	954	976	989	995	998	999	1,000				
7.8	835	902	945	971	986	993	997	999	1,000				
8.0	816	888	936	966	983	992	996	998	999	1,000			
8.5	763	849	909	949	973	989	993	997	999	999	1,000		
9.0	706	803	876	926	959	978	989	995	998	999	1,000		
9.5	645	752	836	838	940	967	982	991	996	999	999	1,000	
10.0	583	697	792	864	917	951	973	986	993	997	998	999	1,000

$c \rightarrow$ c or $np' \downarrow$	0	1	2	3	4	5	6	7	8	9
10.5	000	000	002	007	021	050	102	179	279	397
11.0	000	000	001	005	015	038	079	143	232	341
11.5	000	000	001	003	011	028	060	114	191	289
12.0	000	000	001	002	008	020	046	090	155	242
12.5	000	000	000	002	005	015	035	070	125	201
13.0	000	000	000	001	004	011	026	054	100	166
13.5	000	000	000	001	003	006	019	041	079	135
14.0	000	000	000	000	002	006	014	032	062	109
14.5	000	000	000	000	001	004	010	024	048	088
15.0	000	000	000	000	001	003	008	018	037	070
$c \rightarrow$ $c \downarrow$	10	11	12	13	14	15	16	17	18	19
10.5	521	639	742	825	888	932	960	978	988	994
11.0	460	579	689	781	854	907	944	968	982	991
11.5	402	520	633	733	815	878	924	954	974	986
12.0	347	462	576	682	772	844	899	937	963	979
12.5	296	406	519	628	725	806	869	916	948	969
13.0	252	353	463	573	675	764	835	890	930	957
13.5	211	304	409	518	623	718	798	861	908	942
14.0	176	260	358	464	570	669	756	827	883	923
14.5	145	220	311	413	518	619	711	790	853	901
15.0	118	185	268	363	466	568	664	749	819	875
$c \rightarrow$ $c \downarrow$	20	21	22	23	24	25	26	27	28	29
10.5	997	999	999	1,000						
11.0	995	998	999	1,000						
11.5	992	996	998	999	1,000					
12.0	988	994	997	999	999	1,000				
12.5	983	991	995	998	999	999	1,000			
13.0	975	986	992	996	999	999	1,000	1,000		
13.5	965	980	989	994	997	998	999	1,000		
14.0	952	971	983	991	995	997	999	999	1,000	
14.5	936	960	976	986	992	996	999	998	999	1,000
15.0	916	947	967	981	989	994	997	997	999	1,000

$c \rightarrow$ c or $np' \downarrow$	0	1	2	3	4	5	6	7	8	9
16	000	000	004	010	022	043	077	127	193	275
17	000	000	002	005	013	026	049	085	135	201
18	000	000	001	003	007	015	030	055	092	143
19	000	000	001	002	004	009	018	035	061	098
20	000	000	000	001	002	005	001	021	039	066
21	000	000	000	000	001	003	006	013	025	043
22	000	000	000	000	000	002	004	008	015	028
23	000	000	000	000	000	000	002	004	009	017
24	000	000	000	000	000	000	001	003	005	011
25	000	000	000	000	000	000	001	001	003	006
$c \rightarrow$ $c \downarrow$	14	15	16	17	18	19	20	21	22	23
16	368	467	566	659	742	812	868	911	942	963
17	281	371	468	664	655	736	805	861	905	937
18	208	287	375	469	562	651	731	799	855	899
19	150	215	292	378	469	561	647	725	793	849
20	105	157	221	297	381	470	559	644	721	787
21	072	011	163	227	302	384	471	558	640	716
22	048	077	117	169	232	306	387	472	556	637
23	031	052	082	123	175	238	310	389	472	555
24	020	034	056	087	128	180	243	314	392	473
25	012	022	038	060	092	134	185	247	318	394
$c \rightarrow$ $c \downarrow$	24	25	26	27	28	29	30	31	32	33
16	978	987	993	996	998	999	999	1,000		
17	959	975	985	991	995	997	999	999	1,000	
18	932	955	972	983	990	994	997	998	999	1,000
19	895	927	951	969	980	988	993	996	998	999
20	843	888	922	948	966	978	987	992	995	997
21	782	838	883	917	944	963	976	985	991	994
22	712	777	832	877	913	940	959	973	983	989
23	635	708	772	827	873	908	936	956	971	981
24	554	632	704	768	823	868	904	932	953	969
25	472	553	629	700	763	818	863	900	929	950
$c \rightarrow$ $c \downarrow$	34	35	36	37	38	39	40	41	42	43
19	999	1,000								
20	999	999	1,000							
21	997	998	999	999	1,000					
22	994	995	998	999	999	1,000				
23	988	993	996	997	999	999	1,000			
24	979	987	992	995	997	998	999	999	1,000	
25	966	977	985	991	994	997	998	999	999	1,000

According to the method of inspection, the quality (*i.e.*, the characteristic of interest) of a product or material can be broadly classified as (*i*) attributes, (*ii*) count of defects, (*iii*) variables. In the method of classification by attributes, each item inspected is classified as satisfying one particular characteristic or not, *e.g.*, in inspection by gauges-either a component is satisfactory or not. In inspection by attributes, the object is to limit the number of items having one or more defects. In second method (count of defects), and number of defects on each inspected item are counted and it is carried out when the object is to limit the average number of defects in the items. In the method of classification by variables, the characteristic is measured on a continuous scale. For these three methods of quality classification, the lot quality is described respectively in terms of (*i*) percentage defective, (*ii*) number of defects per item or for 100 items, (*iii*) arithmetic mean or standard deviation or both, or coefficient of variation which is the ratio of the standard deviation to the average.

19.21.1. Whether to go for attributes inspection, or inspection by variables. Whenever there is a choice to decide which of the two methods of inspection should be followed, the following considerations may be looked for :

	Consideration	*Attributes Inspection*	*Inspection by variables*
(*i*)	Cost of inspection	Since it involves use of GO or *NO GO* gauges, inspection cost is low.	Measurement of any characteristic by measuring instruments involves more time and labour and thus cost. It also involves more record keeping and calculations for computation of mean, standard deviation etc.
(*ii*)	Items to be inspected for the same degree of efficiency in drawing inference about the lot.	More	Less. This method is, therefore, suited in case were destructive or costly testing is involved.
(*iii*)	Method of estimating the sample quality from the item quality.	Less complicated	More complicated.
(*iv*)	Information furnished	It does not give much information	It gives much more information about the quality of item lot.
(*v*)	Assumptions	No. Therefore, more applications	Characteristics, under consideration should have normal distribution.
(*vi*)	Limitations	No. Many characteristics can be considered at a time to classify an item as defective or non-defective.	Only one characteristics can be considered at a time.
(*vii*)	General	Inspection is slightly subjective, *i.e.* items of border-line quality may be classified as defective or non-defective depending on the sense of the operator.	Inspection is more objective. As it involves taking actual measurements, it minimises the possibilities of inspection bias and error.

19.21.2. Principles of Sample Selection. It has been realised in most of the industrial and trade applications that 100% inspection (checking) and testing of components is not feasible and the sampling procedures have become indispensable tools because of their economy, reliability and practicability. However, the reliability of the conclusions drawn (*e.g.*, estimating the quality of a lot or ascertaining its conformity to the requirements of a specification) on the basis of the sample depends on its representativeness and the method of its selection. The method of sampling adopted should be such that it ensures a truly random and representative sample leading to sound and satisfactory estimation of lot quality. Bureau of Indian Standards has prepared a specification (IS : 4905-1968 Methods for Random Sampling) to achieve this objective.

It may be understood that in the lot sampling, the inference about the lot is drawn on the basis of information given by a sample, and there is every possibility that the sample may not be true representative of the lot because out of any lot many samples can be made and all the samples cannot be identical. Thus there are chances that a bad lot may be accepted or good lot may be rejected ; the error on this account is known as the sampling error. Similarly when lot sampling is being carried out to estimate the lot quality then due to sampling error the sample estimate may be different from the actual value of the lot quality.

The sampling error depends upon :

(*i*) the degree of homogeneity of the lot, (*ii*) the size of the sample, (*iii*) method of sample selection.

It may be stressed here that the method of sampling adopted should be such that it avoids bias of any form (*i.e.*, giving preference to certain positions in the lot, intentionally choosing either defectives or non-defectives only, unplanned sampling). Method of random sampling (in which the chance for inclusion of any items in the sample is predetermined and is independent of the quality of the item and is also independent of the quality of other items selected for the sample) avoids all types of biases.

For random selection, if possible, the lot should be thoroughly mixed, sample units selected blindly from all parts of the lot, the various parts so arranged that sampler has access to all, the position of selection of sample unit in each part being changed.

The method of **simple random sampling** can be followed if the lot consists of a number of items such that each item is easily identifiable and, apart from the lot size, all other information about the composition of the lot is available. According to this method, once an item has been drawn, it will not be put back into the lot till complete sampling is over. IS : 4950-1968 gives a table of random numbers containing 3000 digits (from 0 to 9), arranged randomly in 15 pages. For convenience of reading, the digits have been grouped into sets of 5 and arranged in 40 rows and 50 columns at each page. All the items in a lot are numbered at random and then the item numbers to be selected from the lot to form a random sample are found out from this table. The random numbers are selected by choosing any page and starting at random from any place vertically downwards. The digits greater than the lot size may be ignored.

19.21.3. Methods of Random Sampling. Other methods of sampling include :

(*i*) Systematic sampling.

(*ii*) Stratified sampling.

(*iii*) Cluster sampling.

(*iv*) Sampling in stages.

Method of **systematic sampling** is applicable to cases where the items are presented in an orderly manner and the sample units are chosen at regular intervals. Here it is assumed that there is no deliberate attempt to manipulate the sequence of the items in the lot in any desired manner while the lot is presented for inspection.

In the method of systematic sampling, first a single sample item is selected from the lot of N items at random and then the items are selected at regular predetermined intervals (of N/n) to make up the desired sample of size n.

The method of **stratified sampling** is possible where the lot can be sub-divided into a certain number or more homogeneous groups or strata and then each stratum sampled separately. As this method draws items from each stratum of the lot, the sample drawn by this method is considered to be more representative of the lot.

The division of a lot into the strata may be on the basis of the homogeneity of the items within a lot, or convenience of sampling, or such other considerations which would make the item within each stratum as much alike as possible. The number of items selected from each stratum are proportional to the size of the stratum, ensuring that at least two items are drawn from each stratum.

When the lot submitted for inspection consists of certain groups of clusters of items, (*e.g.*, when a lot consists of items packed in cartons and it is either impracticable or costly to repack the cartons opened for selecting sample items) it is advisable to go in for **cluster sampling method.** In this method, first a few clusters are selected and then all the items are inspected in the selected clusters. The results of this method will be satisfactory only if the items within a cluster are quite heterogeneous.

In cases where the lot is in the form of packages and it is impossible to take sample from each package after opening it, then first some packages are selected at random and then from each package a random sample of items proportional to the size of the package is chosen. This method is known as sampling in stages. In this method, in the first stage, a desired number of primary units are selected at random and in the second stage, the required number of items are chosen at random from the selected primary units.

19.21.4. Planning sample selection. Proper planning of the sample selection is very important to satisfy the assumptions made in sample theory and to achieve the maximum possible sampling efficiency. For this purpose it is desirable that the maximum possible information be gathered about the lot, the nature of items and the quality characteristics etc. Some of the important points requiring due considerations are discussed below :

(*i*) We should be clear about the purpose for which the sample has to be selected, *i.e.* whether for assessing the quality of the lot, or for lot acceptance. It may be noted that though the method of sampling will be same for both the purposes, the size of the sample will be different in two cases.

(*ii*) The next important step is the formation of lots, for which due consideration has to be given to the nature of the lot, *i.e.* stationary lot or moving lot, item lot or bulk lot, homogeneity of lot, formation of sub-lots etc.

It may be emphasised that for inspection purposes, a lot is viewed as an aggregate of items which may be in the form of a batch, a group, a continuous stream or a bulk of products of raw materials.

A stationary lot is one in which all the items are simultaneously available for inspection. Its advantages are that the planning is easier, the various items can be easily identified and it is possible to have re-sampling and sequential sampling. However, to arrange all the items of the lot such that each item is accessible to the inspector to avoid bias is a difficult problem. In the case of moving lot, the various items flow past the point of inspection and as designed, a few items are selected at a time from the moving batch. Sampling from moving lot is very convenient in the case of bulk materials like coal, grains, etc.

An item-lot consists of an aggregate of discrete items, or an aggregate of items of continuous or bulk materials which have been rendered discrete in the form of specimens or increments. A bulk lot is presented in the form of continuous or bulk material. As stated above if the continuous or bulk material is presented in the form of increments, (*e.g.*, continuous length of cloth presented in the form of pieces of 1 metre, or oil tinned into containers of 1 kg capacity) then it will be treated as item lot.

A homogeneous lot is one which comes out from a process under statistical control. In order to have high efficiency of sampling it is desirable to maintain high degree of homogeneity of a lot, and for this purpose the lot should be confined to the items or products originating from essentially similar conditions. Under the conditions where a lot is formed by combining various sub-lots coming from different sources then, as far possible, the identity of the sub-lots should be preserved so that efficient stratification can be achieved.

When the lot size is very big, then for considerations of economy in inspection or efficiency of sampling, it may be considered desirable to divide the lot into sub-lots. Even if the sub-lot is very bulky then the division into sub-lots may be carried on two or more stages as well. An example of such a case will be to find the quality of coal coming from a mine. Say the coal is coming in the truck-load by the rails. First it may be decided to sample the rails, *i.e.* sampling be done once in a week. After a particular train is selected, then a small number of wagons are selected and further from each wagon equal quality samples are taken and inference drawn. It may be stressed here that inference will finally be made about the lot as a whole and not about each sub-lot. Such a method of sampling is also known as sampling in stages.

(*iii*) In the formation of lot, the decision about the size of the lot is also very important item. As a general rule the lot size should be as large as possible keeping two things in view, *viz.*,

(*a*) a reasonable degree of homogeneity should be maintained, *i.e.*, the items in the lot should not differ widely in quality, and

(*b*) we should be able to arrange the various items of the lot such that the inspector will have easy access to all the items for selection of a sample. This calls for proper storage and handling facilities.

The advantage in having a bigger lot size is that, for any given degree of efficiency of sampling and for the same degree of homogeneity in the lot, the sample size will not increase as rapidly as the lot size and will not increase after a certain size for a lot. In other words, having a bigger lot size means the inspection cost is low.

On the other hand, if a large lot is rejected, then 100% inspection means cost of inspection will increase enormously and therefore a suitable compromise is essential in deciding the lot size.

As it is very important condition in the formation of a lot that the items must be of a single type, grade, class, size etc., produced under relatively uniform conditions of a manufacture ; it is proposed that the items for a lot be taken from a single batch of raw material or from component parts obtained from a single source, or manufactured by a single production method, or obtained from a single production line with the same dies and fixtures, or manufactured during single production shift, or produced from one setting of the machine and so on.

(*iv*) *Determination of sample size* : It may be again stressed that the method of determining the sample size for the purpose of assessing the lot quality is different than the method of determining the sample size for the purpose of lot acceptance. Both these methods are discussed in details later.

Here we are broadly outlining the factors which require due consideration in deciding the size of the sample :

(*a*) We should be clear about the extent of error (risk element) that we can tolerate due to sampling.

(*b*) Degree of homogeneity in the lot.

(*c*) *Lot size* : As already stated a smaller lot size means that the sample size will be bigger in percentage (relatively), whereas the percentage of items in a sample for a bigger lot will be comparatively very less.

(*d*) The inspection cost should be minimum.

(*v*) *Determination of item size or sample-unit size* : This problem arises in case of bulk or continuous materials. (It is obvious that items in the sample can be either in the discrete form, *i.e.* single integral unit of product which can't be further sub-divided, or in the continuous or bulk materials form). Under such situation, item is defined as a specimen of specified length, or as a specimen of a specified area, or as a relatively small portion of specified weight or volume of the bulk or the increment.

The determination of size of the item in such cases is very important factor in sampling techniques.

In the case of bulk-lots, usually larger the size (area, volume or weight) of the item, greater would be sampling efficiency, but beyond a certain size for the item the rate of gain in the sampling efficiency may be negligible. Besides from practical considerations, the large size of the increment for bulk products like iron ore, coal etc., may also pose problems like limitations of manual sampling, introduction or errors due to crushing, etc.

However, if sampling efficiency is the only criterion for determining the item size, then that size, beyond which the rate of gain in efficiency would be very small, may be taken as the appropriate item size.

For determining the item size following this principle, statistically designed experiments should be conducted covering different sizes for the item with respect to the item quality characteristics of interest. Corresponding to each size considered, from the values of the item quality, a measure of variability in item quality should be calculated. These values plotted against different sizes for the item would give a curve. In general, this curve would slope downwards in the direction of increase in the item size and would be more or less flat after a certain point. The size corresponding to this point would be an appropriate item size.

In an investigation conducted on coal of sizes 38 to 60 mm, seven series of 35 increments (to be considered as items) each were collected. The increments in each series were of approximately equal size which were analysed for ash content, individually. The average size of the increments in each series and the corresponding measure of variability (co-efficient of variation, v) of ash content were as follows :

Series	*Average size of increments (weight)*	*Co-efficient of variation v of ash content*
1	94 g	61.1
2	345 g	50.7
3	978 g	41
4	1.7 kg	33.5
5	2.4 kg	29
6	3.6 kg	23.4
7	3.9 kg	22

The co-efficient of variation of ash content between the increments has been plotted against the average size of the increments in Fig. 19.27.

On the basis of this investigation about 3.5 kg could be taken as a suitable size of the increment as increments of larger size do not decrease the co-efficient of variation appreciably.

(vi) General conditions for gross sample, i.e. mixture of all items (increments) selected from single bulk-lot or a sub-lot.

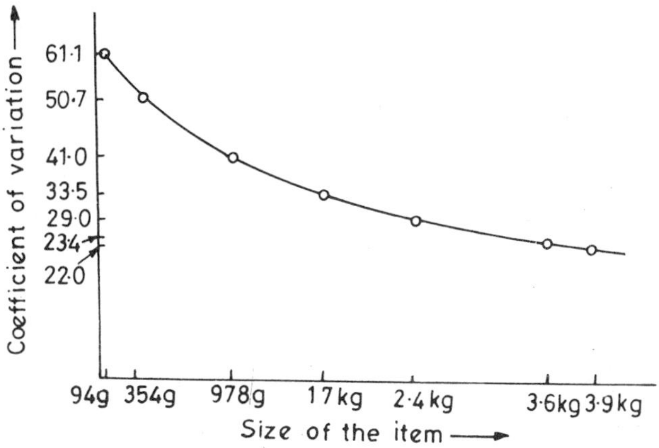

Fig. 19.27. Determination of Item Size.

In order to gather maximum information regarding item quality in case of bulk material, it is advisable to keep each increment in a separate container so that its identity is maintained. However, in cases, where the inspection cost of bulk material is high ; the increments selected in accordance with the principles of the sample selection may be mixed up thoroughly to form a gross sample (it may be noted that the samples being of smaller size, compared to the whole bulk, can be easily mixed up thoroughly so that a small sample taken out of the gross sample will be representative of the full sample ; this process is sometimes known as *reduction*) and inspection done on it. This method provides an estimate of the average quality as efficiently as the one based on inspection of each individual increment within the gross sample. However, it may be appreciated that a single gross sample is not sufficient to give any information about the variability in the lot, or to provide a measure of the sampling error involved and of the reliability of the assessment made of the average quality level.

19.22. Lot Sampling for Estimation of Lot Quality

The method of sample selection and the sampling theory used for estimating the lot quality by lot sampling is same as for the lot acceptance.

The method of determining the sample size for estimating the average quality of a lot or process is given in Art. 19.22.1. Once the sample size has been determined, then the lot fraction defective (in case of inspection by attributes) can be considered same as the sample fraction defective which is

$$= \frac{\text{No. of defectives in sample}}{\text{Sample size}}$$

Similarly in the case of inspection by count of defects and inspection by variables, the average count of defects per item in the sample $\left(\frac{\text{Total no. of all the defects found in sample}}{\text{Sample size}}\right)$ and the sample average $\left(\frac{\text{Sum of all the observations}}{\text{Sample size}}\right)$ give a good estimate of the average count of defects per item in the lot and the lot average respectively.

Though the above method of estimating the lot quality on the basis of sample quality gives sufficiently accurate results, but to account for the sampling errors, it is often desirable to give an indication of the reliability of the estimate ; and this is usually done by specifying two limits, upper and lower limits (the difference between them being known as confidence interval). The significance of these limits is that if different samples are taken from the same lot, then their averages will never exceed these limits or will be within these limits. As the sampling procedure is based on the theory of probability and 100% guarantee can never be given, this confidence interval can also be specified with only certain guarantee usually known as confidence co-efficient. Thus saying that confidence coefficient of 99% has been used in determining the upper and lower limits based on the results of a certain sample, would mean that on an average 99% of these intervals will contain the true value of lot quality.

For most of the industrial applications, a confidence coefficient of 95% is considered to be sufficient. It may be noted that the confidence interval is not dependent on lot size if sample size is small compared to lot size.

The idea of the lot dispersion, *i.e.* the lot standard deviation can also be had by determining the sample standard deviation

$$\left(\sqrt{\frac{\Sigma (x - \bar{x})^2}{n - 1}}\right)$$

If the sample size is more than 20, then sample standard deviation itself gives a good estimate of the lot standard deviation. However, for sample size less than 20, the sample standard deviation has to be multiplied by a factor C given below :

Table 19.10

Sample Size	*Factor C*	*Sample Size*	*Factor C*
2	1.2535	12	1.0235
3	1.1281	13	1.0208
4	1.0851	14	1.0191
5	1.0639	15	1.0181
6	1.0514	16	1.0165
7	1.0425	17	1.0155
8	1.0363	18	1.0148
9	1.0313	19	1.0144
10	1.0283	20	1.0133
11	1.0250		

In case it is felt that calculation of sample standard deviation involves laborious calculation then the value of lot standard deviation can be estimated by determining the sample range (provided sample size is less than 20) also, but this method does not give as accurate results as the former one.

When sample size is large, it may be considered into 5, 8 or 10 sub-groups and then the average of the range of each sub-group found out.

The lot standard deviation is then found by multiplying the sample range R in case of sample (or average of range of sub-groups $\bar{R}$ in case of large sample) by factor D, given in the following table 19.11.

Table 19.11

Sample Size n	*Factor D*	*Sample Size n*	*Factor D*
2	0.8862	12	0.3069
3	0.5908	13	0.2998
4	0.4857	14	0.2935
5	0.4299	15	0.2880
6	0.3946	16	0.2831
7	0.3698	17	0.2787
8	0.3512	18	0.2747
9	0.3367	19	0.2711
10	0.3249	20	0.2677
11	0.3152		

Whatever has been talked about the confidence limits and confidence coefficient for estimating the lot quality from sample quality holds for estimating the lot dispersion from sample dispersion or sample range.

19.22.1. Interval estimates of lot quality. (*a*) *Confindence Interval for lot proportion defective.* The values of the upper and lower confidence limits can be had from Table 3 of **IS** : 1548-1969 corresponding to the sample size and the number of defectives found in the sample.

If sample size is > 30, then the formula for confidence limits for the lot proportion of defectives given by India Standard is

$$\frac{1}{n}\left\{d + 2\sqrt{\frac{d(n-d)}{n}}\right\} \text{ and } \frac{1}{n}\left\{d - 2\sqrt{\frac{d(n-d)}{n}}\right\}$$

where n = sample size

and d = number of defectives found in the sample.

(*b*) Confidence interval for the average count of defects per item in the lot.

The values of upper and lower confidence limits in this case are determined by dividing the values of constants, obtained from table 4 of IS : 1548-1969 corresponding to sample size and the total number of defects found in sample, by the sample size.

In case number of defects in a sample is > 30, Indian Standard recommends the formula

$$\left\{c \pm 2\sqrt{\frac{c}{n}}\right\}, \text{ where } c = \frac{\text{Total no. of defects in sample}}{\text{Sample size } (n)}$$

for determining these limits.

(*c*) Confidence interval for lot average using sample standard deviation (S) or sample range (R) or average of ranges of sub-groups in a sample ($\bar{R}$).

The formula given by IS : 1548-1969 for the upper and lower confidence limits in this case is

$$\bar{x} + h' S \text{ and } \bar{x} - h' S$$

(value of h' corresponding to given sample size)

n can be had from Table 5 of IS : 1548-1969

or
$$\bar{x} + hR \text{ and } \bar{x} - hR$$
$$\bar{x} + h\bar{R} \text{ and } \bar{x} - h\bar{R}$$

(value of h being obtained from Table 19.8 of IS : 1548-1969).

(*d*) Interval estimate of lot dispersion, *i.e.*, confidence interval for the lot standard deviation.

The formulae for upper and lower confidence limits given by IS : 1548-1969 are :

$$f_1 S \text{ and } f_2 S$$

or

$$f_3 R \text{ and } f_4 R$$

The values of f_1, f_2, f_3 and f_4 corresponding to given sample size n can be had from Tables 7 and 8 of IS : 1548-1969 respectively.

19.22.2. Methods for determination of sample size to estimate the average quality of a lot or process. In statistical quality control it is very desirable to devise methods by which the average quality of a lot of material or a process be estimated without going in for 100% inspection, but testing only a sample. The most important thing is that sample size chosen should be representative of the lot and enable the estimation of the average quality with a specified degree of accuracy, *i.e.*, with a specified limit of error and probability level. (It may be stressed here that sample size referred to here is different from sample size used in sampling inspection techniques for the purpose of accepting or rejecting a lot).

Two cases arise, *viz*., some empirical knowledge about the variability in the lot or process (*i*) is available from the analysis of past data or from a study of process, (*ii*) is not available ?

The empirical knowledge is said to be available if an estimate of the percent defective is known in case of attributes, or if an estimate of standard deviation or an estimate of range together with the shape of distribution of the values is known in case of variables.

(*i*) When empirical knowledge is available :

(*a*) *Sampling for attributes*. When the characteristics under consideration of of attribute type, the error limit can be specified as the maximum difference in the true value of per cent defective and its estimate that can be tolerated.

For estimating the per cent defective in the lot or process, some prior knowledge of the per cent defective is desirable for determination of the sample size. The sample size (n) in such cases is given by the formula :

$$n = \frac{u^2 p (100 - p)}{d^2}$$

where u = factor corresponding to the selected probability level and its values are given in table 19.12.

Table 19.12

Probability level per cent	u
99	2.58
98	2.33
95	1.96
90	1.64

p = estimate of the per cent defective from past data, and

d = limit of error expressed as the difference in the true per cent defective and its estimate.

(*d*) *Sampling for variables*. When the characteristic is of a variable type, the error limit can be specified either (1) in terms of the units in which the characteristic is being measured, or (II) in terms of the percentage of the true mean value of the characteristic.

(Method I is to be chosen where the standard deviation of the characteristic is constant over the range of possible values of the average and Method II is to be chosen where the coefficient of variation of the characteristic is constant over the range of possible values of the average).

For method I, *i.e.*, when the standard deviation is known and is constant, the sample size (n) is given by the formula

$$n = (uS/E)^2$$

where u = factor corresponding to the selected probability level (values given in the table 19.12)

S = known standard deviation

E = error limit expressed in the same units of measurement as the standard deviation.

For method II, S is to be replaced by coefficient of variation in per cent and E by the limit of error expressed in per cent, in the above formula.

The value of the standard deviation, if not known directly, can be estimated if some knowledge of range and the shape of the distribution is available from the past behaviour of the process.

In that case,

$$\text{Standard deviation} = \frac{\text{Range}}{3.5} \text{ for rectangular distribution}$$

$$= \frac{\text{Range}}{4.2} \text{ for right angle distribution}$$

$$= \frac{\text{Range}}{4.9} \text{ for isosceles triangle distribution}$$

$$= \frac{\text{Range}}{6.0} \text{ for normal distribution.}$$

In case the information is available in groups, the standard deviation = $D \times \overline{R}$.
where $\overline{R}$ = average of the range of various groups and value of constant D can be had from the Table 19.13.

Table 19.13

Group Size	*D*
4	0.4857
5	0.4299
6	0.3946
7	0.3698
8	0.3512
9	0.3367
10	0.3249

(*ii*) When empirical knowledge is not available.

(*a*) *Sampling for attributes.* An initial sample of fairly large size may be selected from the lot and inspected. If n_1 is the initial sample size and m_1 is the number of defectives in the initial sample, then the estimate of the per cent defective (p_1) is given by

$$p_1 = \frac{100 \times m_1}{n_1}$$

The sample size necessary to estimate the per cent defective with given error limits is then given by

$$n = \frac{u^2 \, p_1 \, (100 - p_1)}{d^2}$$

u and d have same meaning as for the method when empirical knowledge is available. If the estimate of the sample size (n) is greater then n_1, then $n - n_1$ additional items may be selected and inspected.

(*b*) *Sampling for variables.* From the lot or process, an initial sample of a small size, say n_1, may be selected and tested for the characteristic. From the test results so obtained, an estimate of the variability in the lot may be ascertained as follows :

$$S^2 = \frac{1}{n_1 - 1}\left[(x_1{}^2 + x_2{}^2 + \ldots \, x_{n1}{}^2) - \frac{(x_1 + x_2 + x_{n1})^2}{n} \right]$$

where x_1, x_2, ... x_{n1} denote the test results.

The sample size necessary to obtain an estimate of the average quality with the given limit of error is then given by

$$n = \frac{t^2 \, S_1{}^2}{E^2}$$

where t = factor corresponding to the initial sample size (n_1) and the probability level (the value of t can be had from Table 3 of IS : 5002-1969)

S_1 = estimate of the standard deviation obtained from the initial sample, and

E = limit of error expressed in the same units as the standard deviation.

If the estimate of the sample size (n) so obtained is greater than the initial sample size, then $n - n_1$ additional items may be selected from the lot or process and tested.

In the above formulae, while deciding about the limit of error the cost of inspection may also be borne in mind, which is equal to the overhead cost plus the product of the cost of sampling and testing a single item, and the sample size. In most cases, the cost of inspection and the limit of error may be suitably balanced so as to keep both within reasonable limits.

19.23. Lot Sampling for Lot Acceptance

Having realised that 100% inspection is not feasible, it was considered desirable to go in for sampling inspection in which the decision about the acceptance or rejection of a complete lot is made on the basis of evidence obtained from a small sample. Due to sampling error there are chances that good lot is likely to be rejected and producer suffers in that case ; also chances are there that bad lots may be accepted and thus the consumer suffers in that case. Also in any case the accepted lots will contain some defectives and they can't be eliminated altogether. The rejected lots are 100% inspected and defective parts are replaced by non-defectives. The 100% inspection of lots definitely exerts a positive influence on the producer to improve the quality of his products.

It is thus obvious that both the producer and the consumer have to take certain amount of risk (estipulated above) and then the sampling plans can be designed to ensure that neither the producer nor the consumer are going to be affected to an extent more than specified.

If producer is maintaining the control charts for his process, he knows the average quality of his products, and he can't produce goods better than this quality. In case it is acceptable to the consumer then the producer desires that all his lots upto this quality should be accepted by the sampling inspection plan. This limit in the control terminology is called acceptable quality level (*AQL*) and all goods of this quality and better than this are considered as good.

On the other hand consumer, depending, on his requirements, can specify a quality (bad quality, in control terminology known as lot tolerance per cent defective or *LTPD*) beyond which he does not want to accept any lot. The goods of quality *LTPD* or worse than this are considered as bad.

As already indicated, both the producer and the consumer are made to take some risks due to sampling errors. The producer's risk is usually taken as 5% [*i.e.* 5% of the lots of quality (*AQL*) are likely to be rejected] and the customer's risk is taken as 10% (*i.e.* 10% of lots of quality worse than *LTPD* are likely to be accepted).

19.23.1. Operating Characteristic Curve. (O.C. Curve) (Fig. 19.28). The characteristics of the system (sampling plan) is best shown by plotting a graph between probability of acceptance against percentage of defectives. Such a curve is called the O.C. Curve. It compares the performance of sampling plans over a range of possible quality levels of submitted product. The two important points of the curve are :

The sampling plan chosen should be such as to pass through these two points. The interest of producer is that most of those components which are less defective should be accepted in full. The consumer's interest lies in accepting the more defective parts to a minimum extent. Thus saying producer's risk is represented by $p_{0.95}$ = 0.25% means producer wants if a lot contains 0.25% defectives, then 95% of the lots must be accepted. Similarly saying consumer's risk is $p_{0.10}$ = 2.5% means that consumer does not want a worse quality containing more than 2.5% defectives and he would at the most accept 10% of lots containing 2.5% defectives, 2.5% is also called lot tolerance per cent defective (*LTPD*), *i.e.* the worst lot acceptable.

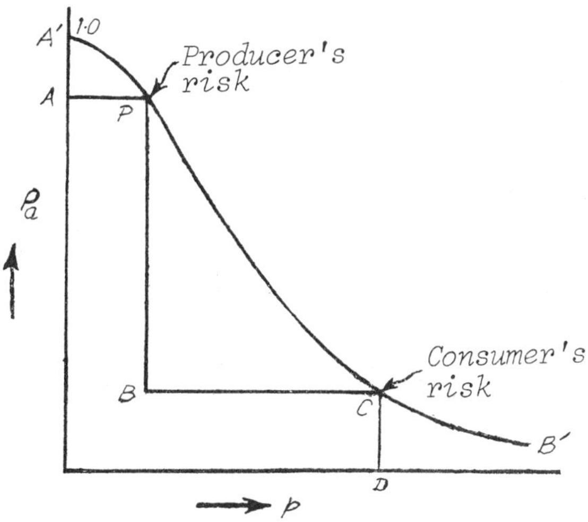

Fig. 19.28
(*i*) Producer's risk and (*ii*) Customer's risk.

Theoretically the O.C. curve should be as *APBCD*. But this is not possible and curve is usually as shown by *A′PCB′*. Here, if a large sample and acceptance number are used, the plan becomes more discriminating *i.e.* a greater number of good lots are accepted and a greater number of bad lots are rejected. At the same time it is desired to have economical plan also in which the total number of articles inspected will be minimum and which will give the required degree of protection to both producer and consumer.

Type A, O.C. Curve. This gives the probabilities of acceptance for various fraction defectives as a function of the lot quality of finite lots. The computation for this curve is based on hypergeometric probabilities ; the binomial or Poisson distributions often give satisfactory approximations. The curve is definitely not a continuous one ; *e.g.*, a lot of 200 items may be 0.5% or 1.0% defective but not 0.8% ; but in practice it is customary to draw it as continuous one.

Type B, O.C. Curve. This gives the probabilities of acceptance of a lot as a function of product quality. The computation for such a curve is done as if the lot size is infinite. For type B curve the binomial is exact and the Poisson often gives a satisfactory approximation. Such a curve is correctly viewed as continuous. When the sample size n is not more than 1/10th of the lot size N, Type A and Type B curves may be considered as identical for most practical purposes. It may be noted that for some conditions type B curve always falls above type A curve, meaning thereby that the use of type B curve tends go give a figure for Consumer's Risk that is somewhat too high.

Characteristics of O.C. Curve. Following are a few characteristics associated with every O.C. curve :

(*a*) With the fixed value of acceptance number c and N, the larger the value of n, the steeper the O.C. curve and thus it gives better discrimination between good and bad lots.

(*b*) With the fixed value of acceptance number and sample size, the O.C. curves for different values of N are not appreciably different and thus indicating that it is the absolute size of the sample that is more important rather than its relative size compared to the size of the lot.

(*c*) With the fixed value of N, if a number of O.C. curves all of which necessarily satisfy only one set of co-ordinates, then the O.C. curve with larger n and c gives a better discrimination between good and bad lots.

The most economical plans for *LTPD* and consumer's risks are published by Messrs Dodge and Roming, popularly known as Dodge and Roming tables. Sampling Inspection Tables for inspection by attributes and by count of defects, and for inspection by variables for per cent defective have been prepared by Bureau of Indian Standard and we shall be discussing the practice recommended by *BIS* here.

In acceptance sampling (Dodge and Roming, tables as well as *BIS* tables), the rejected lots are 100% inspected. The average outgoing quality therefore first increases with increasing %age defectives in incoming products. As it increases too much, the rejected parts are replaced by good parts and outgoing quality again improves. Thus average outgoing quality level (*AOQL*) attains maximum value and then falls down.

As operating characteristic curve effectively summarises the nature of quality protection obtained by operating a particular plan on a number of lots (whose quality may vary within any limits), many of the published plans furnish the O.C. curves of the plans for selection purposes to enable choice from the point of view of quality protection.

Many a time, in addition to the producer's risk and consumer's risk being used as a measure of protection offered by a plan, the lot quality which has a 50% chance of being accepted or rejected by the plan (normally referred to as indifference quality) is also sometimes considered.

The O.C. curve does not give one useful information *i.e.*, the average outgoing quality limit (*AOQL*). The consumer is very much concerned with (*AOQL*) value as it ensures him that the submitted lots may be of any quality but he will not get quality worse than *AOQL* value.

It may be noted that if the incoming quality is very poor, then many of the lots will be rejected and they have to be 100% inspected and defective items are to be rectified. As a result of this after certain value of the incoming quality the outgoing quality will start improving as the rejected lots do not contain any defective. This is shown in Fig. 19.29.

It may be stressed here that the adoption of a certain particular sampling inspection plan will ensure its *AOQL* value only in the long run and it does not guarantee this value for short periods.

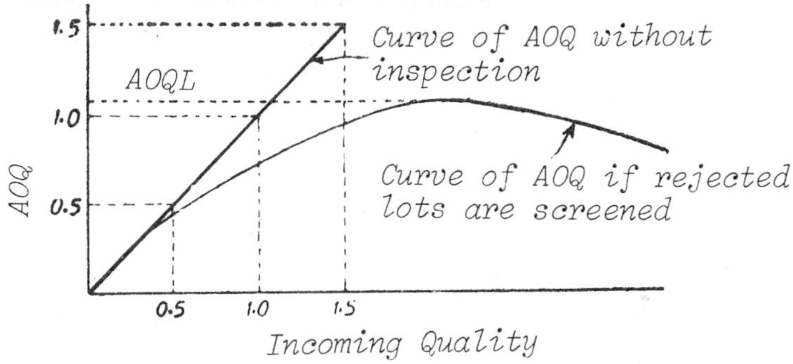

Fig. 19.29

19.23.2. Definitions. We shall now define the various terms commonly used in connection with the acceptance sampling.

(*i*) *AQL*. Acceptable quality level ; this is the fraction defective that can be tolerated without serious effect upon further processing operation or customer reaction. In other words *AQL* is the maximum per cent defective that, for the purpose of sampling inspection, can be considered satisfactory as a process average. Probably the first step in deciding upon a satisfactory sampling plan will be to decide upon a quality level that is acceptable. The level (*AQL*) is commonly designated by the symbol P_1.

(*ii*) *RQL*. Rejectable quality level ; it is also commonly known as lot tolerance per cent defective (*LTPD*) and designated by either p_1 or p_2. The only way to get real protection against unsatisfactory material is to use a sampling plan that will reject most of the lots offered that would seriously interfere with further processing operations or cause too much unfavourable customer reaction. *LTPD* represents the percentage defectives in a lot that can be tolerated in only a specified proportion of lots.

(*iii*) *Producer's risk* (α). This is the risk which the producer runs of having lots of quality p_1 (specified *AQL*) rejected. Saying $\alpha = 0.05$ means that in the long run about one lot in 20 will be rejected, provided the lots are coming from a process controlled at quality level p_1. This risk should be kept as low as possible. Reducing this risk too much would require an increased sampling that would generally be uneconomical. Of course, the producer can decrease his risk by producing material at a better quality level than p_1 depending on the other economical considerations involved.

(*iv*) *Consumer's risk* (β). This is the risk which the consumer runs of accepting lots of quality p_2 (specified *LTPD*). Saying $\beta = 0.01$ means that in the long run, only about one lot in ten of quality p_2 will be accepted. The risk of accepting lots of quality worse than the *LTPD* will be smaller than the designated consumer's risk.

(*v*) *N*. Number of pieces in given lot (lot size).

(*vi*) *n*. Number of pieces in sample (sample size).

(*vii*) *Defect*. A failure to meet a specified quality standard.

(*viii*) *Defective*. An item that contains one or more defects.

(*ix*) *Group size*. A number of successive observations that are considered together for control chart purposes.

(*x*) *Acceptance number* (*c*). The maximum allowable number of defective pieces in a sample of size *n*, for acceptance of the lot.

(*xi*) *Rejection number* (*r*). The minimum number of defectives in the sample for the rejection of lot.

(*xii*) *Fraction defective (p)*. For a sample, this is equal to the ratio of the number of defectives (m) to the number of items in a sample (n) ; and for a lot this is the ratio of the number of defectives (M) to the number of item in lot (N).

(*xiii*) *Process average*. It represents the average per cent defective of the products submitted by the producer for original inspection.

(*xiv*) Probability of acceptance (P_a).

(*xv*) Probability of rejection ($1 - P_a$).

(*xvi*) Operating characteristic (O.C.) curve. It, at a glance, shows the percentage of lots likely to be accepted for varying percentage of defectives.

(*xvii*) $P_{0.95}$, $P_{0.50}$ etc. Fraction defective having a probability of acceptance 0.95, 0.50 etc. under any given acceptance criteria.

(*xviii*) *Single Sampling Plan*. The procedure in acceptance sampling is to consider each submitted lot of product separately and to base the decision on acceptance or rejection of the lot on the evidence of one or more samples chosen at random from the lot. In single sampling plan only one sample is taken out of lot and the decision to accept or reject the lot is based on this sample.

Any systematic plan for simple sampling requires that three numbers be specified.

(*a*) N. Number of articles in the complete lot from which the sample is to be drawn.

(*b*) n. Sample size *i.e.*, the number of articles in the random sample drawn from the lot.

(*c*) *c. acceptance number*. This is the maximum allowable number of defectives in a sample and more than c defectives will cause the rejection of the lot.

Thus saying a plan $N = 100$, $n = 5$, $c = 0$ would mean "Take a random sample of 5 out of a lot of 100. If the sample contains more than zero (0) defectives, reject the lot, otherwise, accept the lot".

(*xix*) *Double Sampling Plan*. If a lot is very good or very bad, a small sample can detect it. If it be somewhere in between, then second sample can give still better idea. Thus total inspection is reduced in double sampling plan.

In double sampling plans, the attributes are usually specified by five constants. *viz.*,

N = lot size.

n_1 = the first sample size.

c_1 = acceptance number for the first sample.

n_2 = the second sample size.

c_2 = maximum number of defectives allowed in the combined first and second samples.

(*xx*) *Average Outgoing Quality (AOQ)*. It represents the average quality (average percentage defectives in the outgoing products (after inspection) including all accepted and all rejected lots which have been 100% inspected and defectives replaced by non-defectives.

In the case of single sampling plan, for lots of quality p, if the probability of acceptance is P_a, then $AOQ = (p)(P_a)$, if defectives found in samples are not to be replaced. This may be proved as follows :

Let the total lots each containing N items submitted be $X + Y$.

Let the lots accepted be X.

Defectives in X lots are pNX.

Since the rejected lots are fully screened and therefore are free from defectives. Therefore Y lots will not contain any defective.

Therefore total defectives in lots are pNX.

$$\therefore \quad AOQ = \frac{pNX}{N(X+Y)} = p\left(\frac{X}{X+Y}\right)$$

$$= p\,P_a$$

Also $\quad \%AOQ = (100p)\,(P_a)$

Similarly when the defectives found in the samples are to be replaced by non-defectives,

$$\%AOQ = \frac{N-n}{N}\,(100p)\,(P_a)$$

(*xxi*) *Average Outgoing Quality Limit (AOQL).* It represents the maximum average per cent defective in the outgoing products.

(*xxii*) *Important points about Sampling Inspection.* In judging various acceptance sampling plans it is desirable to compare their performance over a range of possible quality levels of submitted product. An excellent picture of this performance is given by the O.C. curve (which shows the ability of the plan to distinguish between good and bad lots). As already explained the O.C. curve shows the probability P_a (for a given fraction defective p, in a submitted lot) that such a lot will be accepted by the given sampling plan.

In connection with various sampling plans *i.e.* different values of N, n and c the following comments may be offered as are observed by drawing their O.C. curves.

(*i*) Sampling acceptance plans with same per cent samples give very different quality protection. Thus it is very wrong notion that the sample size should be in proportion to the lot size.

(*ii*) Fixed sample size tends towards constant quality protection. In other words, the size of the lot has not much influence on the shape of O.C. curve.

(*iii*) In sampling plans developed without benefit of statistical analysis c is often specified as zero under the illusion that if sample is perfect, the lot will be perfect. Actually the acceptance number need not be zero.

(*iv*) The sampling plan chosen should be such which satisfies both producer and the consumer and requires total minimum inspection.

Example 19.29. *A random sample of 20 is to be taken out of a lot of 150 pieces, 15 of which are defective. Determine the probability that the sample will contain 2 or more defectives.*

Solution. $\quad p = \text{fraction defective} = \dfrac{15}{150} = 0.1$

$$\therefore \quad np = 20 \times 0.1 = 2,$$

from table 19.9, corresponding to $np = 2$,

Probability of 1 or less defective

$$P_{1\,or\,less} = 0.406$$

$$P_{2\,or\,less} = 0.677$$

$$\therefore \quad P_2 = 0.677 - 0.406 = 0.273$$

Similarly $\quad P_3 = P_{3\,or\,less} - P_{2\,or\,less}$

$$= 0.857 - 0.677 = 0.180$$

$$P_4 = 0.947 - 0.857 = 0.090$$

$$P_5 = 0.983 - 0.947 = 0.036$$

$$P_6 = 0.995 - 0.983 = 0.012$$

$$P_7 = 0.999 - 0.995 = 0.004$$

$$P_8 = 1.000 - 0.999 = 0.001$$

$$P_{2 \text{ or more}} = 0.180 + 0.09 + 0.036 + 0.012 + 0.004 + 0.001 = 0.596.$$

Example 19.30. (*a*) *For the single sampling plan N = 1000, n = 100, c = 2 ; find the values of* $p_{0.95}$, $p_{0.50}$ *and* $p_{0.10}$

(*b*) *Prepare an AOQ curve for the above single sampling acceptance rectification plane and hence determine the value of AOQL.*

Solution. (*a*) $p_{0.95}$ means the % of defective in the lot corresponding to 95% probability of acceptance.

Referring to Table 19.9 (summation of terms of Poisson's exponential Binomial limits).

Corresponding to $c = 2$, and $P_a = 95\%$

and $np' = 0.819$ (by interpolation)

$\therefore$ $p_{0.95} = 0.00819$ or 0.819%

Similarly for $c = 2$ and $P_a = 50\%$

$np' = 2.6735$ (by interpolation)

and for $c = 2$ and $P_a = 10\%$

$np' = 5.33$ (by interpolation)

and $p_{0.10} = 0.0533$ or 5.33%

(*b*) *AOQ* for $p_{0.95} = P_a \times p_{0.95} = 0.95 \times 0.819 = 0.727$

Similarly *AOQ* value for other defectives

$= P_a \times p_{0.50} = 0.50 \times 2.6735 = 1.3368$

and $P_a \times p_{0.10} = 0.10 \times 5.33 = 0.533$

Since only 3 values of *p* have been plotted as shown in Fig. 19.30 according to it maximum value of *AOQ* = 1.337

$\therefore$ *AOQL* = 1.337%. **Ans.**

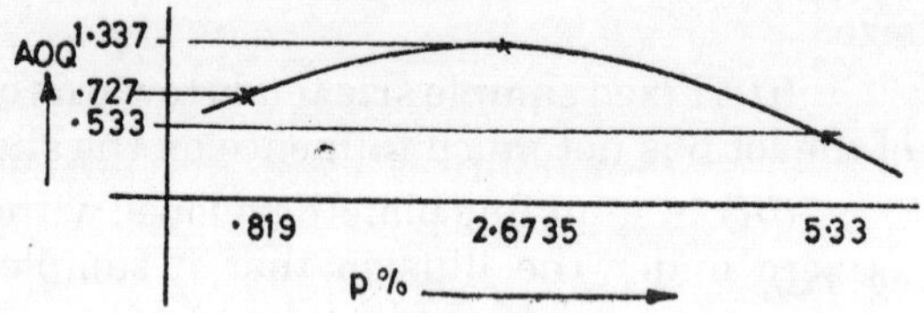

Fig. 19.30

Example 19.31. *A lot containing 1000 items out of which 10 are defective is to be inspected. A sample of 36 items is taken and if it contains 0 defective, the lot is accepted. If it contains 1, 2 or 3 defectives in first sample then a second sample of 59 items is taken and if total number of defectives in 1st and 2nd samples is less than 3 then lot is accepted. Calculate the probability of acceptance of the lot.*

Solution. Probability of acceptance of lot

= Probability of zero defective in first sample

+ probability of 1 defective in first sample followed by 0, 1 or 2 defectives in second sample

+ Probability of 2 defectives in first sample followed by 0 or 1 defective in second sample

+ Probability of 3 defectives in first sample followed by 0 defectives in second sample

$$np' \text{ for first sample} = 36 \times \frac{10}{1000} = 0.36$$

From Table 19.9, for $np' = 0.36$ and $c = 0$,

i.e. $p_0 = 0.698$

$P_{1 \text{ or less}} = 0.948$

$\therefore$ $P_1 = 0.948 - 0.698 = 0.250$

$\therefore \quad P_{2\,or\,less} = 0.994$

$\therefore \quad P_2 = 0.994 - 0.948 = 0.046$

$P_{3\,or\,less} = 1.000$

$P_3 = 1.000 - 0.994 = 0.006$

If there is a defective in first sample, then for second sample,

$$np' = 59 \times \frac{9}{(1000-36)} = \frac{59 \times 9}{964} = 0.55$$

$\therefore \quad$ for $np' = 0.55$, $P_{2\,or\,less} = 0.982$

If there are 2 defectives in the first sample,

$$np' = 59 \times \frac{8}{964} = 0.49$$

$\therefore \quad P_{1\,or\,less} = 0.913$

If there are 3 defectives in first sample, then

$$np' = 59 \times \frac{7}{964} = 0.42$$

$\therefore \quad P_0 = 0.648$

Now $(P_0)_1$ *i.e.* probability of 0 in first sample = 0.698

$$(P_1)_1 \times (P_{2\,or\,less})_2 = 0.250 \times 0.982 = 0.246$$

$$(P_2)_1 \times (P_{1\,or\,less})_2 = 0.046 \times 0.913 = 0.042$$

$$(P_3)_1 \times (P_0)_2 = 0.006 \times 0.648 = 0.004$$

$\therefore$ Probability of acceptance = 0.698 + 0.246 + 0.042 + 0.004 = 0.990

Example 19.32. *In a double sampling plan $N = 10{,}000$, $n_1 = 100$, $c_1 = 0$, $n_2 = 150$, $c_2 = 2$. Determine the probability that a lot containing 1% defective will be rejected.*

Solution. For first sample $n_1 p' = 100 \times 0.01 = 1$

From Table 19.9, Probability of 0 defective = 0.368

Probability of 1 or less defective = 0.736

$\therefore \quad P_1 = 0.736 - 0.368 = 0.368$

$P_2 = P_{2\,or\,less} - P_{1\,or\,less} = 0.920 - 0.736 = 0.184$

Since the sample from a lot of 10,000 does not affect the quality to that extent, for second sample also

$$n_{2\,p'} = 100 \times 0.01 = 1$$

$\therefore \quad P_{1\,or\,less} = 0.736$

and $\quad P_0 = 0.368$

Probability of acceptance of lot

$$= (P_0)_1 + (P_1)_1 \times (P_{1\,or\,less})_2 + (P_2)_1 \times (P_0)_2$$

$$= 0.368 + (0.368) \times (0.736) + (0.184) \times 0.368$$

$$= 0.368 + 0.271 + 0.0677 = 0.7067$$

$\therefore$ Probability of rejection of lot $= 1 - P_a = 1 - 0.7067 = 0.2933$. **Ans.**

Example 19.33. *A single sampling plan is given as $n = 100$ and $c = 3$. The lot size is large in comparison with sample size. Compute the approximate probability of acceptance of lots with 1% defective (use Poisson).*

(a) Determine the AOQ value for the above lots.

(b) What will be the average inspection in per cent ? (Assume acceptance, rectification plan).

Solution. (*a*) $np' = 100 \times 0.01 = 1$

Using table 19.9, Probability of 3 or less defectives = 0.981

∴ Probability of acceptance of lot = 98.1%

(*b*) $AOQ = p \times P_a = 1\% \times 98.1\% = 98.1\%$. **Ans.**

(*c*) Total Inspection of say 100 lots = 100 articles in 98.1 lots + N in 1.9 lots

$= 100 \times 98.1 + 1.9\,N = 9810 + 1.9\,N$

$$\therefore \text{ Total \% age inspection} = \frac{9810 + 1.9\,N}{100\,N} \times 100 = 1.9 + \frac{9810}{N}\ \%$$

Example 19.34. *Design a single-sampling inspection plan by attribute which will meet or nearly meet the following requirements :*

$\alpha = 0.05$ $\quad p_1 = 0.01$

$\beta = 0.10$ $\quad p_2 = 0.05$

Solution. Producer's risk $\alpha = 0.05$ *i.e.* for quality of $p_1 = 0.01$, *i.e.* 1% of defectives, he expects 95% probability of acceptance.

Trying with acceptance number $c = 0$

Referring to table 19.9 corresponding to $c = 0$ and $P_a = 0.95$,

$np' = 0.05$ (by interpolation)

∴ $n \times 0.01 = 0.05$

and $n = 5$

Now seeing if it satisfies the customer's risk also ; for which $np' = 5 \times 0.05 = 0.25$ and corresponding to $c = 0$; $P_a = 0.779 = 77.9\%$ but actually it should have been $\beta = 0.10 = 10\%$ only.

Trying with other acceptance numbers, results are given in tabular form as below :

c acceptance number	*np_1 No. of items × 0.01 corresponding to $P_a = 0.95$ (from table 19.9)*	*n No. of items*	*np_1 No. of items × 0.05*	*$(P_a)_0$ Probability of acceptance or consumer's risk (from table 19.9)*
0	0.05	5	0.25	0.779
1	0.35	35	1.75	0.478
2	0.82	82	4.10	0.224
3	1.364	136	6.80	0.093

$c = 3$ seems to be giving very close results and satisfying both the producer's and consumer's risk nearly

∴ $n = 136$ and $c = 3$ is the plan. **Ans.**

Example 19.35. *In a double sampling 2% AOQL acceptance rectification plan :*

$n_1 = 32,$ $\quad c_1 = 0$ $\quad n = 1000$

$n_2 = 38,$ $\quad c_2 = 2$

Determine (i) the probability of acceptance of a 2% defective lot, (ii) the average total inspection.

Solution. (*i*) For first sample

$n_1p_1' = 32 \times 0.02 = 0.94$

∴ $P_0 = 0.527$ (by interpolation from Table 19.9)

$P_{1\ or\ less} = 0.864$

∴ $P_1 = 0.864 - 0.527 = 0.337$

$$P_2 = P_{2\,or\,less} - P_{1\,or\,less} = 0.973 - 0.864 = 0.109.$$

For second sample, if first sample contains no defective, then

$$n_2p_2' = 38 \times \frac{20}{1000 - 32} = \frac{760}{964} = 0.79$$

$\therefore \quad P_{2\,or\,less}$ (for $n_2p_2' = 0.79$) $= 0.954$

if first sample contains 1 defective, then

$$n_2p_2' = 38 \times \frac{19}{964} = 0.75$$

$$P_{1\,or\,less} = 0.827$$

if first sample contains 2 defectives, then

$$n_2p_2' = 38 \times \frac{18}{964} = 0.71$$

$\therefore \quad P_0 = 0.492$

$\therefore$ Probability of acceptance

$$= (P)_1 + (P_1)_1 \times (P_{1\,or\,less})_2 + (P_2)_1 \times (P_0)_2$$
$$= 0.527 + (0.337)(0.827) + (0.109)(0.492)$$
$$= 0.527 + 0.279 + 0.0537 = 0.8597 = 85.97\%.$$ **Ans.**

(*ii*) Average total inspection :

Say there are 100 lots

Probability of acceptance in the first sample = 52.7%

$\therefore$ for 52.7 lots, total inspection = $52.7 \times 32 = 1686.4$ 1686

Probability of acceptance in second sample = 0.3327 *i.e.* 33.27%

$\therefore$ for 33.27 lots, total inspection = $(32 + 38) \times 33.27$

$$= 70 \times 33.27 = 2328.9 \cong 2329$$

For rejected lots 14.03%, 100% inspection is required

$\therefore$ Inspection = $14.03 \times N = 14.03 \times 1000 = 14030$

$\therefore$ Total inspection = $14030 + 1686 + 2329 = 18045$

$\therefore$ average inspection = $\dfrac{18045}{1000 \times 100} \times 100\% = 18.045\%$. **Ans.**

19.24. Sampling Plans for Inspection by Attributes and by Count of Defects

In industry, very often we come across situations when the quality of a lot of manufactured products has to be evaluated with a view to determining its acceptability or not. This is achieved by carrying out sampling inspection according to certain plans. The sampling plans discussed in this article can be used in those situations where either the items could be classified as defectives or non-defectives (inspection by attributes), or the number of defects of an item could be counted (inspection by count of defects).

A necessary condition for using any sampling plan is that the lot quality should be specified in terms of fraction (or percentage) defective in case of inspection by attributes and in terms of number of defects per item (or per 100 items) in case of inspection by count of defects. It may be mentioned that if the lot quality is uniform and lot comprises of homogeneous items, then a small sample can be taken to represent the lot and the inspection cost will be low. The requirement of good sampling plans is that these should be sufficiently flexible to meet a wide variety of quality requirements and administrative and technical conditions.

The various types of sampling are :

(a) AQL Plans

(i) Single sampling plans. (ii) Double sampling plans.

(iii) Multiple sampling plans.

(b) One's own Plans.

(i) Single sampling plans for any desired values of AQL and LTPD

(ii) Sequential plans.

The selection of any particular plan depends upon the following factors :

(a) Cost of inspection.

(b) Protection desired by the producer and the consumer.

As regards the quality protection, all the plans are capable of giving the same equally, but it is mainly due to other factors that one has to choose one particular plan. The other factors are ; (i) the sampling inspection cost *i.e.*, the size of the sample. If inspection cost depends only on the sample size then the single, double, multiple and sequential plans stand in the decreasing order of cost ; (ii) amount of information provided on the quality of each lot (in other words on the process producing these lots). If the process average (*i.e.*, average per cent defective of the products) is to be estimated then only first sample is considered, and as single sampling plan gives biggest first sample, single sampling plan will be selected in such a case ; (iii) the administrative cost involved in using these plans *e.g.*, sequential plans require more computation and recording whereas single plans are cheapest in this respect. However, the double, multiple and sequential sampling plans have the psychological advantage in the sense that the lots are given 'more than one chance' for acceptance.

It may once again be stressed that the protection provided by any particular sampling plan to the producer and the consumer can be judged at a glance by its operating characteristic curve ; and steeper is the O.C. curve, (which is closely related to the amount of inspection) better it discriminates between good and bad lots.

It is obvious that for choosing any sampling plan, the choice of *AQL* (acceptable quality level) is very important. The decision about *AQL* value is arrived at by striking of compromise between the quality desired and the quality attainable *e.g.*, if *AQL* is specified more than the quality which a production process (from which lots are coming) is capable of giving then the amount of rejection will be very high. On the other hand if the *AQL* is not exacting enough then an excessive amount of inferior product are likely to be accepted. The appropriate value of the *AQL* may be arrived at on the basis of the past performance of the supplier of the product ; or else it may be stipulated in accordance with the agreement between the parties concerned.

Indian Standard Specification No. IS : 2500 (Part I)-1973 [Sampling Inspection Tables for Inspection by attributes and by Count of Defects], while preparing the Sampling Tables has taken into consideration the *inspection level* also. By inspection level is meant the relative amount of inspection one is required to do. Obviously if the inspection level is high, then there will be less risk of accepting lots of quality worse than the chosen *AQL*, and *vice-versa*. In actual practice one has to strike a compromise between a large sample which gives a reliable estimate of the lot quality (but inspection cost is high) and a small sample which reduces the inspection cost.

The Indian Standard Specification referred above specifies five inspection levels ; namely I, II, III, IV and V. Inspection levels I and II are applicable in the selection of 'special small sample inspection plans'. These are intended to be used in cases where relatively small sample sizes are appropriate due to consistent supply of good material or as agreed to between the producer and the consumer. Specification recommends inspection level IV to be most suitable for majority of cases and under normal conditions of acceptance inspection, it gives a reasonable compromise between high inspection costs and the risks involved. Under the cases where the cost of inspection does not matter much but acceptance of defective item is considered as a serious matter then inspection level V may be used and in the reverse case level III.

For using the sampling table given in this specification, first the 'sample size code letter' has to be chosen from Table 1 of IS : 2500 (Part I)-1973 corresponding to the lot size and the inspection level chosen. Then from relevant tables, the sample plan can be chosen corresponding to this sample size code letter and the *AQL* value. For *single sampling plan*, from Table 19.4 of IS : 2500 (Part I) 1973, one gets the sample size (n) and the acceptance number (a). Thus if out of this sample inspected the number of defectives are more than the acceptance number, the lot is rejected otherwise accepted. For *double sampling plan,* from Table 19.4 of IS : 2500 (Part I)-1973, on the basis of the *AQL* value chosen and the sample size code latter selected, one gets the values of two samples of the same size (n_1) and acceptance and rejection numbers (a_1 and r_1 respectively) corresponding to the first sample size and the acceptance and rejection numbers a_2 and r_2) corresponding to the cumulative sample size. Thus first a sample of n_1 items is taken and if the number of defectives (defects) is less than or equal to a_1 the lot is accepted. If the number of defectives (defects) is equal to or more than r_1 then the lot is rejected ; and if it is in between a_1 and r_1, then a second sample of n_1 items is taken. The lot is accepted if the number of defectives (defects) in the combined sample of $2n_1$ items is less than or equal to a_2 and rejected if it is more than or equal to r_2.

For *multiple sample plans,* from table 4 of IS : 2500 (Part I)-1973, corresponding to the chosen value of *AQL* and the sample size code letter chosen from table 1, one finds the value of seven samples of the same size (n_1) and the relevant acceptance and rejection numbers corresponding to the given stages. Thus first sample of size (n_1) is selected in a random order and examined for defectives (defects). If this sample is found to contain the defectives (defects) less than or equal to the acceptance number given corresponding to the first sample then the lot is accepted. It will be rejected if the number of defectives (defects) is equal to or more than the rejection number given against first sample. If, however, the number of defectives (defects) in the first sample lies in between the acceptance and rejection numbers given for the first sample, then a second sample of size n_1 is selected and examined for defectives (defects). The number of defectives (defects) in the combined sample (*i.e.*, the first and the second sample taken together) is then compared against the acceptance and rejection number corresponding to the second stage of sampling for taking decision with regard to the acceptance or rejection of the lot. If no decision is reached, then a third sample and so on are drawn upto the seventh stage till a decision finally reached so as to accept or reject the lot.

19.24.1. Normal reduced and tightened inspection. When the quality of the submitted lots shows significant shift, it is desirable to make appropriate changes in the sampling plans. If the quality deteriorates, it is necessary to tighten the inspection ; if the quality improves, it may be desirable to relax the inspection.

Normal Inspection. Inspection under a sampling plan that is in force for a particular product by producer is called 'normal inspection'. It may be continued as long as the quality of the product submitted is better than or equal to the chosen *AQL*. The consistency in maintaining the level of quality by the producer can be ascertained either from a continuous record of inspection data which can be used to estimate the process average of the producer or from a knowledge of the production of the lots that are not accepted. In case the quality becomes consistently better than the stipulated *AQL*, 'reduced inspection' may be undertaken. If however, the quality becomes consistently worse than the chosen *AQL*, 'tightened inspection' is to be restored to.

Tightened Inspection. Inspection shall be tightened either by raising the inspection level, *i.e.*, by selecting a sample size code letter higher than the one adopted for normal inspection or by employing a smaller *AQL*. Since the former approach leads to an increased amount of inspection, tightening is done by using a sampling plan with an *AQL* smaller than that used previously. When tightening of inspection is done, the smaller *AQL* shall be chosen always with respect to the actual sample size code letter used for normal inspection and not with respect to the intended letter which may not have any sampling plan.

The following criteria is applied for changing from normal to tightened inspection and *vice-versa.*

(*a*) If 2 out of 5 (or less) consecutive lots have been rejected while on normal inspection, change over to tightened inspection.

(*b*) If, while on tightened inspection, 5 consecutive lots have been accepted, change over to normal inspection.

Reduced Inspection. If the quality of the submitted lots is considerably better than the *AQL* chosen, reduced inspection may be restored to either by selecting lower sample size code letter than the one used for normal inspection or by relation to the *AQL*. Since the former approach leads to economies in inspection it is preferable unless there is an agreement to the contrary, to reduce inspection by changing over to a plan with a lower sample size code letter than the one adopted for normal inspection.

The following criteria may be applied for changing over from normal to reduced inspection and *vice-versa.*

(*a*) If none out of 10 consecutive lots has been rejected while on normal inspection ; change over to reduced inspection.

(*b*) If a lot is rejected, and if at the same time the rejected lot is preceded by less than 10 lots accepted on reduced inspection, change over to normal inspection.

Table 19.14. Formulae for Constructing Single Sampling AQL, LTPD plans

LTPD/AQL	*Sample size × AQL*	*Acceptance Number*	*Sample Size × AOQL*	*LTPD/AOQL*
(1)	(2)	(3)	(4)	(5)
44.890	0.052	0	0.368	12.514
10.946	0.355	1	0.840	9.262
6.509	0.818	2	1.371	7.764
4.890	1.366	3	1.942	6.879
4.057	1.970	4	2.544	6.285
3.549	2.613	5	3.168	5.855
3.206	3.286	6	3.812	5.526
2.957	3.981	7	4.472	5.264
2.768	4.695	8	5.146	5.051
2.618	5.426	9	5.831	4.872
2.497	6.169	10	6.528	4.720
2.397	6.924	11	7.233	4.589
2.312	7.690	12	7.948	4.475
2.240	8.464	13	8.670	4.373

(1)	*(2)*	*(3)*	*(4)*	*(5)*
2.177	9.246	14	9.398	4.283
2.122	10.035	15	10.134	4.204
2.073	10.831	16	10.875	4.129
2.029	11.633	17	11.622	4.061
1.990	12.442	18	12.374	4.000
1.954	13.254	19	13.131	3.945
1.922	14.072	20	12.892	3.894
1.892	14.894	21	14.657	3.848
1.865	15.719	22	15.427	3.799
1.840	16.548	23	16.200	3.759
1.816	17.382	24	16.976	3.723
1.795	18.218	25	17.756	3.683
1.775	19.058	26	18.540	3.652
1.757	19.900	27	19.326	3.617
1.739	20.746	28	20.115	3.589
1.723	21.594	29	20.907	3.559
1.707	22.444	30	21.702	3.530
1.692	23.298	31	22.499	3.507
1.679	24.152	32	23.298	3.481
1.665	25.010	33	24.100	3.456
1.653	25.870	34	24.904	3.443
1.641	26.731	35	25.711	3.411
1.630	27.594	36	26.519	3.394
1.619	28.460	37	27.330	3.374
1.609	29.327	38	28.142	3.354
1.559	30.196	39	28.956	3.336
1.590	31.066	40	29.773	3.318
1.581	31.938	41	30.590	3.302
1.572	32.812	42	31.410	3.286
1.564	33.686	43	32.231	3.270
1.556	34.563	44	33.054	2.255
1.548	35.441	45	33.878	3.241
1.541	36.320	46	34.704	3.224
1.534	37.200	47	35.532	3.211
1.527	38.082	48	36.360	3.199
1.521	38.965	49	37.191	3.186

19.24.2. Construction of ones own sampling plan. (Single sampling AQL, LTPD plans for inspection by attributes and by count of defects) :

Single sampling plans for inspection by attributes and by count of defects corresponding to any specified combination of AQL and LTPD values can be constructed by making use of Table 19.14. The plans so obtained are virtually independent of the lot size. They assume the producer's risk to be 5% and the consumer's risk to be 10%. Since the acceptance of a bad lot is more serious when the lot is large than when it is small while selecting a sampling plan for inspection of a large lot such as AQL and LTPD combination should be preferred which lead to large sample sizes.

All values of AQL, LTPD and AOQL are to be expressed as fraction defective or number of defects per item in Table 19.14.

Example 19.36. *Suppose lots containing 1000 boiler rivets are submitted for inspection for shank diameter. It is desired to construct a single sampling plan with AQL of 1 per cent and LTPD value of 5 per cent. (The producer's risk and consumer's risk are taken to be 5% and 10% respectively).*

Solution. From values of AQL and LTPD calculate $\frac{\text{LTPD}}{\text{AQL}} = \frac{0.05}{0.01} = 6$. Referring to Table 19.14 above, the nearest tabulated value of LTPD/AQL is obtained as 4.890 and the corresponding value of sample size × AQL and acceptance number are 1.366 and 3 respectively. So dividing the value of $n \times$ AQL by the AQL value, the sample size is obtained as $\frac{1.366}{0.01} = 137$ approximately. The sampling plan would then work as follows :

From each lot of 1000 rivets pick out a sample of 137 rivets at random. Inspect the shank diameter of the rivets in the sample with the help of prescribed gauges and note down the number of defective rivets. The lot shall be accepted if the number of defective rivets in the sample is 3 or less otherwise the lot shall be rejected.

If the sampling plan is operated on the acceptance-rectification basis, then the resulting AOQL of the plan may also be obtained with the help of above Table. The value of $n \times$ AOQL corresponding to the figure of 4.890 of LTPD/AQL is read as 1.942. Dividing this figure by the sample size n which is 137, the AOQL is obtained as 1.4%.

Note. The above table can also be used for construction of ones own plan when the stipulations are made in terms of AOQL and LTPD in the same fashion as above entering the table through column 5 instead of column 1.

19.24.3. Sequential Analysis. It has already been stressed that in the case of double sampling for attribute inspection, the average amount of sampling required is less for the same protection, if the lot quality is not appreciably different. Theoretically the amount of sampling (in other words inspection cost) could be reduced further by designing triple or quadruple sampling plans, but in actual practice it becomes very difficult to construct and administer such plans. However sequential sampling plans offer both of these advantages. Due to least inspection cost, these are best suited in cases where inspection is of destructive type or samples are difficult to obtain.

Two types of sequential sampling plans are available for inspection by attributes and by count of defects, namely, (*a*) item by item sequential plans, and (*b*) group-by-group sequential plans.

In sequential analysis (which is the ultimate in multiple sampling), the number of items to be inspected is not predetermined. The sample items are inspected one at a time in the case of item-by-item sequential plans (or in small groups in the case of group-by-group sequential sampling plans). After inspecting it, a decision is made using appropriate criteria based on the inspection data available upto that stage, either

(*i*) to accept the lot

or (*ii*) to reject the lot

or (*iii*) to continue sampling.

The procedure is continued until the lot is either accepted or rejected.

For the operation of this type of sampling plan either the graphical procedure shown in Fig. 19.31, or a tabular procedure is followed. The formulae for the acceptance and rejection criteria are given below.

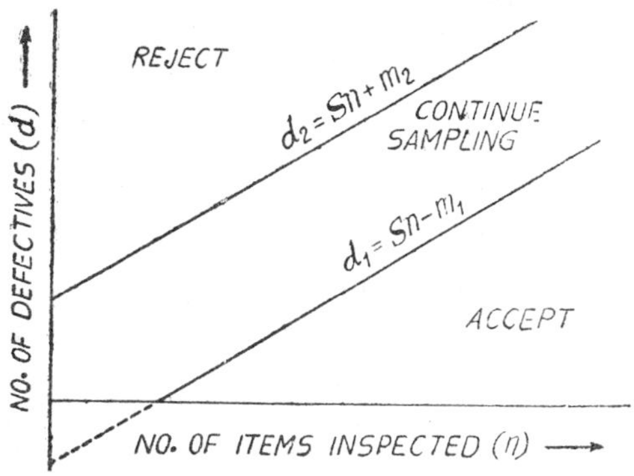

Fig. 19.31. Sequential analysis plan for attribute inspection.

For designing a sequential analysis plan for inspection by attributes, first we should decide about the values of the following parameters which should be based on the practical and economic aspects of the product involved :

(*i*) Good quality (AQL) $= p_1$ (expressed as fraction defective)

(*ii*) Bad quality (LTPD) $= p_2$ (—do—)

(*iii*) Producer's risk $= \alpha$ (expressed as a fraction)

(*iv*) Consumer's risk $= \beta$ (—do—)

Then the values of m_1, m_2 and S in the equation of acceptance and rejection lines (*i.e.* $d_1 = Sn - m_1$, and $d_2 = Sn + m_2$ respectively) are computed as follows :

$$m_1 = \frac{B}{g_1 + g_2}\ ;\ m_2 = \frac{A}{g_1 + g_2} \text{ and } S = \frac{g_2}{g_1 + g_2}$$

where $g_1 = \log p_2 - \log p_1$ and $g_2 = \log(1 - p_1) - \log(1 - p_2)$

$A = \log(1 - \beta) - \log \alpha$ and $B = \log(1 - \alpha) - \log \beta$.

In preparing a sampling table, d_2 is taken to be the next higher whole number to the computed value, and d_1 is taken to be the next smaller.

If d_2 is equalled or exceeded after n items have been inspected the lot is rejected.

If d_1 is equalled or a value less than d_1 is obtained after n items have been inspected, the lot is accepted.

If a value between d_1 and d_2 is obtained after inspecting n items then inspection is continued. For facilitating the use of this plan, the values of d_1 and d_2 for different values of n can be calculated before hand and kept ready either in the form of a graph or a table.

If it is not convenient to take samples one by one (*i.e.* item-by-item sequential sampling), then sample may be taken in groups (group-by-group sequential plans) of say 10, 20, 50 or any other convenient number at a time and the decision based upon the total inspected. (In many instances it is found more practical to inspect a sequence of groups of items rather than a sequence of items themselves).

The above formulae for computing the acceptance and rejection criteria for item-by-item sequential sampling plans for inspection *by count of defects* get modified as below :

Let $c_1 =$ AQL expressed as number of defects per item

$c_2 =$ LTPD expressed as number of defects per item

$\alpha =$ Producer's risk

$\beta =$ Consumer's risk

Then $A = \log(1-\beta) - \log\alpha$ and $B = \log(1-\alpha) - \log\beta$

$$g = \log c_2 - \log c_1$$

$$m_1 = \frac{B}{g} \quad \text{and} \quad m_2 = \frac{A}{g}$$

$$S = \frac{c_2 - c_1}{g}$$

Then acceptance limit $d_1 = Sn - m_1$

and rejection limit $d_2 = Sn + m_2$

The average amount of sampling required at quality levels p_1 and p_2 will respectively be

$$n_{p1} = \frac{(1-\alpha)\,m_1 - \alpha m_2}{S - p_1} \quad \text{and} \quad n_{p2} = \frac{(1-\beta)\,m_2 - \beta m_1}{p_2 - S}$$

and the maximum amount of sampling (on the average) required to make a decision will generally occur when the quality of lots submitted lies between p_1 and p_2.

It may sometimes so happen in certain border line cases that the sample size required to reach a decision could reach infinity. (Though in practice sample size for taking a decision will never exceed 2-3 times n_{p1} or n_{p2}).

For practical reasons it is desirable to stop at a point and decide arbitrarily in one of the following ways :

(*i*) Reject the lot, if not accepted so far (In this case producer's risk is high).

(*ii*) Accept the lot, if not rejected so far (In this case, the consumer's risk is increased).

(*iii*) Take decision to accept on the basis that number of defectives are closer to d_1 or d_2.

Smallest sample number at which lot may be accepted (n_1) or rejected (n_2) can be determined by the formula

$$n_1 \rightarrow \text{next larger integer to } \frac{m_1}{S}, n_2 \rightarrow \text{next larger integer to } \frac{m_2}{1-S}.$$

If a comparison was to be made between the amount of sampling required for the three types of sampling, the curves for these will look like as shown in Fig. 19.32, provided other things remain same, *i.e.* OC characteristic is same. It may be noted that sequential sampling

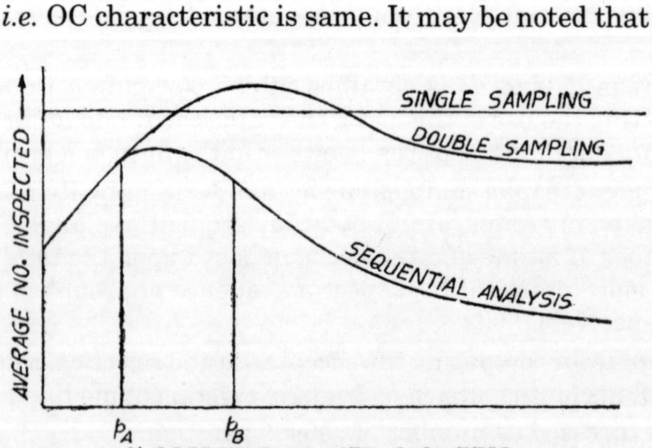

Fig. 19.32

plan has marked advantage over single sampling plan regardless of the quality of lot submitted. The sequential sampling plan is better than double sampling plan only if lot contains more than $P_A\%$ of defectives and after $P_B\%$ it is definitely advantageous.

Example 19.37. (*a*) *What is the difference between sequential sampling plan and the multiple sampling plan ?*

(*b*) *Prepare a unit sequential sampling plan to meet the following specifications :*

$$\alpha = 0.05 \qquad \beta = 0.10$$
$$p_1 = 0.02 \qquad p_2 = 0.05$$

Determine the probability of acceptance of a lot with 0.03 fraction defective.

Solution. (*a*) In multiple sampling plans, three or more samples of a stated size are inspected and decision on acceptance or rejection is reached after inspecting the stated number of samples.

In sequential sampling plan, decision is possible after each item has been inspected and there is no specified limit on the total number of units to be inspected.

(*b*) Symbol α represents $1 - P_a$ for a lot of quality p_1, symbol β is P_a for a lot of a quality p_2.

The sequential sampling plan is fully defined by the equation of the rejection line, $d_2 = Sn + m_2$ and the acceptance line, $d_1 = Sn - m_1$ (Refer Fig. 19.31). To compute S, the slope of these lines, and m_1 and m_2, the intercepts, certain auxiliary symbols, g_1, g_2, A and B are used.

$$g_1 = \log \frac{p_2}{p_1} = \log \frac{0.05}{0.02} = \log 2.5 = 0.3979$$

$$g_2 = \log \frac{1 - p_1}{1 - p_2} = \log \frac{0.98}{0.95} = \log 1.031 = 0.0132$$

$$A = \log \frac{1 - \beta}{\alpha} = \log \frac{0.90}{0.05} = \log 18 = 1.2553$$

$$B = \log \frac{1 - \alpha}{B} = \log \frac{0.95}{0.10} = \log 9.5 = 0.9777$$

$$m_1 = \frac{B}{g_1 + g_2} = \frac{0.9777}{0.3979 + 0.0132} = \frac{0.9777}{0.4111} = 2.38$$

$$m_2 = \frac{A}{g_1 + g_2} = \frac{1.2553}{0.4111} = 3.05$$

$$S = \frac{g_2}{g_1 + g_2} = \frac{0.0132}{0.0132 + 0.3979}$$

$$= \frac{0.0132}{0.4111} = \frac{132}{4111} = 0.032$$

$$\therefore \qquad d_2 = 0.032 \times n + 3.05 \text{ and } d_1 = 0.032 \times n - 2.38.$$

With the help of these equations, the acceptance and rejection lines can be converted into an item-by-item table of acceptance and rejection numbers.

It will be noted that the computed values of d_1 and d_2 are generally not whole numbers. Usually the rejection number is taken as the next whole number above d_2, and the acceptance number is taken as the next whole number below d_1.

19.25. Dodge-Romig Tables

Dodge and Romig have developed an extensive and valuable set of ready-made tables, the plans of which have been designed to minimise the total inspection, including screening. Here four separate groups of plans are described in brief :

(*i*) Single-sampling lot tolerance tables
(*ii*) Double-sampling lot-tolerance tables
(*iii*) Single-sampling AOQL tables
(*iv*) Double-sampling AOQL tables.

Of these first two groups of plans are tabulated according to lot tolerance per cent defective (LTPD). (LTPD) being 0.5%, 1.0%, 2.0%, 3.0%, 4.0%, 5%, 7.0% and 10.0%) with a consumer's risk of 0.10 and the last two groups of plans are tabulated according to the average outgoing quality which these assure (AOQL being 0.1%, 0.25%, 0.5%, 0.75%, 1.0%, 1.5%, 2.0%, 2.5%, 3.0%, 4.0%, 5.0%, 7.0% and 10.0%).

Of these four kinds of plans, the double sampling AOQL type plans have found the widest acceptance. The first two plans give considerable assurance that individual sub-standard lots will be rejected. These plans are particularly applicable in cases where lots retain their identity after inspection, as when a lot is shipped as inspected to a consumer in a specific transaction. On the other hand the last two, *i.e.*, AOQL plans stress the limit on poor quality in the long run but do not maintain uniform assurance that individual low quality lots will be rejected. These (AOQL) plans are particularly applicable when lots are merged in a common supply whose average performance is of interest.

The various plans given in table 19.15 (Dodge-Romig table for a single sampling plan. LTPD = 3%) afford the same quality protection as measured by the tolerance. The procedure of using this table is : For each lot, a sample of n items must be inspected. If it contains less than or equal to c defectives, lot is accepted otherwise it is rejected. In case of rejection, the complete lot must be inspected.

In case the average per cent defective to certain lots in the past is known and no change in the process is foreseen, then for having least total amount of inspection, the plan must be chosen corresponding to column having the same process average. For example, if the item to be inspected has averaged 0.7% defective in the past and no change is foreseen, the least total amount of inspection can be anticipated if a plan is chosen from the column headed 0.61-0.90.

Table 19.16 shows a typical double sampling AQOL plan, in which case also a whole series of sampling plans is provided to suit varying needs as to lot size and past process average fraction defective. These plans differ considerably from each other as to lot tolerance but have the same AOQL. The procedure for this table is : Collect and inspect at random first sample of n_1 items. If it does not contain more than c_1, defective, accept the entire lot. If defectives are more than c_1, take a second random sample of n_2 items and inspect. If the total number of defectives, found in the first and second samples together is not more than the acceptance number given for the second sample (c_2), accept the entire lot, otherwise reject it or inspect fully.

Table 19.15. Typical Dodge-Romig Table for a Sample Sampling Plan Simple Sampling (LTP = 3%)

Process Average %	*0-0.03*			*0.04-0.30*			*0.31-0.60*			*0.61-0.90*			*0.91-1.20*			*1.21-1.50*		
Lot size	*n*	*c*	*AOQL %*	*n*	*c*	*AOQL %*	*n*	*c*	*AOQL %*	*n*	*c*	*AOQL %*	*n*	*c*	*AOQL %*	*n*	*c*	*AOQL %*
1-40	All	0	0	All	0	0	All	0	0.08	All	0	0	All	0	0	All	0	0
41-55	40	0	0.18	40	0	0.18	40	0	0.18	40	0	0.18	40	0	0.18	40	0	0.18
56-100	55	0	0.30	55	0	0.30	55	0	0.30	55	0	0.30	55	0	0.30	55	0	0.30
101-200	65	0	0.38	65	0	0.38	65	0	0.38	65	0	0.38	65	0	0.38	65	0	0.38
201-300	70	0	0.40	70	0	0.40	70	0	0.40	110	1	0.48	110	1	0.48	110	1	0.48
301-400	70	0	0.43	70	0	0.43	115	1	0.52	115	1	0.52	115	1	0.52	115	2	0.54
401-500	70	0	0.43	70	0	0.45	120	1	0.53	120	1	0.53	160	2	0.58	160	2	0.58
501-600	75	75	0.43	75	0	0.45	120	1	0.56	160	2	0.63	160	2	0.63	200	3	0.65
6001-800	75	75	0.47	125	1	0.57	125	1	0.57	165	2	0.66	205	3	0.71	240	4	0.74
801-1,000	75	75	0.48	125	1	0.59	170	2	0.67	210	3	0.73	250	4	0.76	290	5	0.78
1,001-2,000	75	75	0.63	130	1	0.60	175	2	0.72	260	4	0.85	300	5	0.90	380	7	0.95
2,001-3,000	75	75	0.63	130	1	0.62	220	3	0.82	300	5	0.95	385	7	1.0	460	9	1.2
3,001-4,000	130	130	0.63	175	2	0.75	220	3	0.84	305	5	0.96	425	8	1.0	540	11	1.2
4,001—5,000	130	130	0.64	175	2	0.76	260	4	0.91	345	6	1.0	465	9	1.1	620	13	1.2
5,001-7,000	130	130	0.64	175	2	0.76	265	4	0.92	390	7	1.1	505	10	1.2	700	15	1.3
7,001-10,000	130	130	0.64	175	2	0.77	265	4	0.93	390	7	1.1	550	11	1.2	775	17	1.4
10,001-20,000	130	130	0.64	175	2	0.78	305	5	1.0	430	8	1.2	630	13	1.3	900	20	1.5
20,001-50,000	130	130	0.65	225	3	0.86	350	6	1.1	520	10	1.2	750	16	1.4	1,090	25	1.6
50,001-100,000	130	130	0.65	265	2	0.96	390	7	1.1	590	12	1.3	830	18	1.5	1,215	28	1.6

n = sample size, c = acceptance number (allowable defects in a sample), $AOQL$ = average outgoing quality limit.

19.26. Dodge Plan for Continuous Sampling

One of the plans for use where articles are manufactured continuously, as on a conveyor, is shown below.

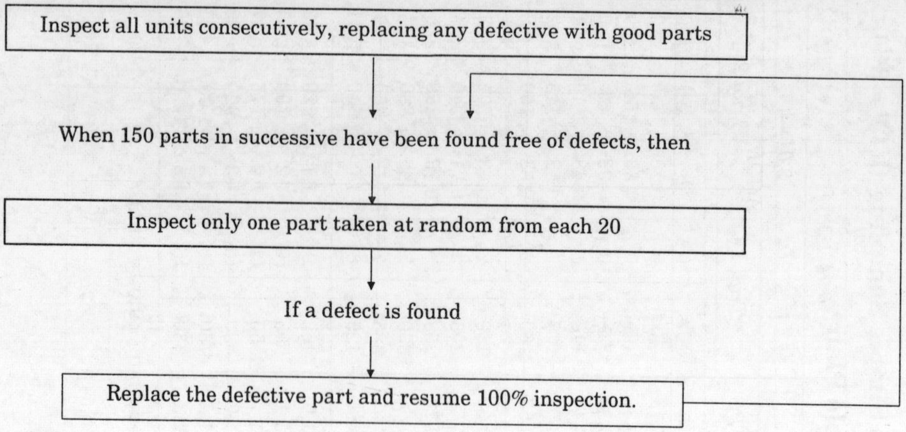

19.27. Philips Standard

Philips Standard is applicable where screening inspection by the consumer is not possible. In fact screening does not exert such a pressure for quality improvement as the out right rejection of one or more of the entire lots. This standard uses $P_{0.50}$ the point of control or indifference quality, as the basis for referencing all its sampling plans. Some notable features of Philips Standard are as follows :

1. All plans are classified by a particular point on the *OC* curve, the product quality for which the probability of acceptance is 0.50.

2. Producer and consumer, in effect, share equally in the risk of a "wrong" decision by any sampling plan included in the system. Thus if 2.5%, plan is used, product exactly 2.5% defective is as likely to be accepted as it is to be rejected.

3. The relative discrimination between good and bad product is approximately constant at the indifference quality or point of control. This is because, all the sampling plans in the system have the same ratio of the slope of the *OC* curve at $p_{0.50}$ to the value of the quality standard $p_{0.50}$.

4. A consequence of item 3 is that the producer's risk of rejection of product at some stated amount moderately better than the quality standard is roughly comparable to the consumer's risk of acceptance of product at the same amount worse than the standard.

5. Another consequence of item 3 is that the sample size is relatively large for plans designed for maximum consumer protection and decreases as the quality standard is relaxed.

6. Single sampling is stipulated whenever the lot is 1,000 or less. Double sampling is stipulated whenever the lot is greater than 1,000.

Philips Standard Sampling Table. Table 19.17 represents the Philips Standard.

Table 19.17
Typical Dodge-Roming Table for a Double Sample Plan

Process Average	*0-0.20*					*0.21—2*					*2.01—4*				
Lot size	*Trial 1*		*Trial 2*		*LTPD %*	*Trial 1*		*Trial 2*		*LTPD %*	*Trial 1*		*Trial 2*		*LTPD %*
	u_1	c_1	c_2	c_2		n_1	c_1	n_2	n_2		n_1	c_1	n_2	c_1	
1-3	All	0	—	—	—	All	0	—	—	—	All	0	—	—	—
4-15	3	0	—	—	50.0	3	0	—	—	50.0	3	0	—	—	50.0
16-50	5	0	3	1	53.5	5	0	3	1	53.5	5	0	3	1	53.5
51-100	5	0	3	1	55.0	0	0	8	2	43.0	6	0	8	2	43.0
101-200	5	0	4	1	52.0	7	0	7	2	42.0	7	0	12	3	38.0
201-300	7	0	7	2	42.5	7	0	7	2	42.5	7	0	13	3	37.0
301-400	7	0	7	2	42.5	7	0	7	2	42.5	8	0	17	4	35.0
401-500	7	0	8	2	40.0	7	0	8	2	40.0	8	0	18	4	34.0
501-600	7	0	8	2	40.0	8	0	13	3	35.0	8	0	18	4	34.0
601-800	7	0	8	2	40.0	8	0	13	3	35.0	8	0	18	4	34.5
801-1,000	7	0	8	2	40.5	8	0	13	3	35.0	9	0	18	4	33.0
1,001-2,000	7	0	8	2	40.5	8	0	14	3	34.0	9	0	23	5	31.0
2,001-3,000	7	0	8	2	41.0	8	0	14	3	34.0	9	0	24	5	30.0
3,001-4,000	7	0	8	2	41.0	8	0	14	3	34.5	9	0	24	5	30.5
4,001-5,000	7	0	8	2	41.0	8	0	14	3	35.0	10	0	29	6	29.5
5,001-7,000	7	0	8	2	41.0	8	0	18	4	32.5	16	1	29	7	28.5
7,001-10,000	7	0	8	2	41.0	9	0	18	4	32.5	17	1	38	8	26.0
10,001-20,000	7	0	8	2	41.0	9	0	18	4	32.5	17	1	38	8	26.0
20,001-50,000	7	0	8	2	41.0	9	0	18	4	32.5	18	1	42	9	25.5
50,001-100,000	7	0	8	2	33.5	9	0	25	5	30.0	18	1	52	10	24.5

(Contd...)

0—0.20					0.21—2					2.01—4				
Trial 1		Trial 2		LTPD %	Trial 1		Trial 2		LTPD %	Trial 1		Trial 2		LTPD %
u_1	c_1	n_1	n_2		n_1	c_1	n_2	n_2		n_1	c_1	n_2	c_1	
All	0	—	—	—	All	0	—	—	—	All	0	—	—	—
3	0	—	—	50.0	3	0	—	—	50.0	3	0	—	—	50.0
6	0	6	2	48.0	6	0	6	2	48.0	6	0	6	2	48.0
7	0	11	3	38.5	7	0	11	3	38.5	7	0	16	4	36.5
8	0	16	4	35.5	13	1	20	6	33.5	14	1	24	7	32.0
8	0	17	4	35.0	14	1	26	7	31.5	19	2	29	9	31.0
8	0	22	5	34.0	15	1	30	8	31.0	21	2	44	12	29.0
15	1	23	6	30.5	16	1	37	9	28.5	22	2	50	13	27.0
16	1	28	7	28.4	22	2	38	10	27.5	28	3	52	14	26.0
16	1	28	7	29.0	22	2	43	11	27.0	29	3	56	15	25.5
16	1	34	8	28.0	24	2	56	13	25.5	36	4	69	18	24.5
17	2	38	9	27.5	24	2	61	14	25.0	45	5	95	23	23.0
17	2	38	10	26.0	33	3	72	16	23.0	50	6	115	27	22.0
24	2	46	11	25.0	41	4	99	21	21.5	70	8	150	34	20.0
26	2	54	12	23.5	44	4	111	22	20.0	80	9	195	41	19.5
27	2	63	13	22.5	50	5	120	24	19.5	90	10	240	47	18.0
27	2	68	14	22.0	60	6	145	28	18.5	110	12	265	53	17.5
28	2	77	15	22.0	70	7	205	32	18.0	125	14	320	62	17.0
28	2	87	17	21.0	80	8	245	39	17.5	140	16	355	69	16.6
36	4	99	20	21.0	85	8		44	17.0	150	17	390	77	16.6

19.28. ABC Standard

ABC standard is based on AQL ("Acceptable Quality Level") which is a point on the *OC* curve at which a batch is very likely to be accepted. The probability of a lot or batch containing exactly the AQL percentage of defectives being accepted is often set at 0.95 but other values near this are sometimes used.

Many plans based on AQL have been developed and revised from time to time. The latest development was in the year 1963, when United States adopted a plan known as MIL-STD-105D. The international designation of MIL-STD-105D is ABC-STD-105.

ABC Standard Tables have been developed which give sample size and acceptance and rejection numbers. To enter any one of these tables AQL and sample size code letter must be known. To determine which table should be used, it is necessary to know whether single, double or multiple sampling is to be used and to know whether normal, tightened or reduced inspection is to be used.

The AQL values in the standard may be interpreted either as defects per hundred units or as per cent defective depending whether acceptance criteria are to be based on the number of defects or the number of defectives observed in a sample. However, AQL values above 100 are interpreted as applying to defects per hundred units.

All the AQL values are multiples of the numbers 1, 1.5, 2.5, 4.0 and 6.5. These numbers are roughly in a geometrical progression and correspond to systems of "preferred numbers" in common use for other industrial purposes.

There is a table which relates the lot or batch size and sample size code letter that determines the sample size. It assumes three general inspection levels which are ones used in most cases. The Standard states : "Unless otherwise specified, Inspection Level II will be used. However, Inspection Level I may be specified when less discrimination is needed, or Level III may be specified for greater discrimination".

It also assumes four special levels S-1 to S-4 which are to be used for the special case where relatively small sample sizes are necessary and larger sampling risks can or must be tolerated.

In ABC standard, AQL is defined as the maximum per cent defective (or the maximum number of defects per hundred units) that, for purpose of sampling inspection, can be considered satisfactory as a process average.

In all AQL systems, the acceptance criteria under normal inspection have been chosen to protect the producer against rejection of lots meeting the quality standard. However, in most AQL systems the producer's risk that such lots will be rejected is not the same for all sampling plans. It depends on the sample size-'smaller the sample size, more the risk the producer takes that a lot will be rejected when it contains defectives exactly-equal to AQL% with a smaller sample size, the consumer's risk of accepting a lot much worse than the AQL is also much greater.

All the military AQL systems have used a somewhat empirical relationship between lot size and sample size. Unless a sample is a substantial fraction of its lot, the OC curve of the sampling plan is practically independent of lot size, depending almost entirely on sample size and acceptance number. Although the absolute sample size increases with lot size in the military AQL system, the relative sample decreases.

In ABC standard, the sample size code letter depends only on the lot size and inspection level and is not influenced by the AQL. The military standards have not provided an automatic decrease in sample size with a relation of the quality standard, as in the case of Philips Standard sampling system. In ABC standard, the sample size is further determined by the AQL and the sample size code letter.

Table 19.18

Philips Standard Sampling System

Point of control	*Lot size*	*0.25%*			*0.5%*			*1%*			*2%*			*3%*			*5%*			*7%*			*10%*		
		n	*c*		*n*	*c*		*n*	*c*		*n*	*c*		*n*	*c*		*n*	*c*		*n*	*c*		*n*	*c*	
	20-50	All	—		All	—		All	—		30	0		20	0		13	0		10	0		7	0	
	51-100	All	—		All	—		60	0		30	0		20	0		13	0		10	0		7	0	
Single Sampling	101-200	All	—		100	0		60	0		35	0		55	1		35	1		25	1		17	1	
	201-500	175	0		100	0		135	1		75	1		55	1		35	1		30	2		25	2	
	501-1,000	225	0		225	1		110	1		85	1		85	2		45	2		35	3		35	3	
		n_1	c_1	c_2	n_1	c_1	c_1	n_1	c_1	c_2	n_1	c_1	c_2	n_1	c_1	c_2	n_1	c_1	c_2	n_1	c_1	c_2	n_1	c_1	c_2
Double Sampling	1,001-2,000	300	0	1	150	0	1	110	0	2	55	0	2	45	0	3	25	0	3	30	1	5	22	1	5
	2,001-5,000	425	0	2	200	0	2	135	0	3	70	0	3	70	1	5	45	1	5	55	2	10	40	2	10
	5,001-10,000	525	0	3	260	0	3	220	1	5	110	1	5	125	2	10	75	2	10	75	3	15	45	3	15
	10,001—20,000	875	1	5	440	1	5	380	2	10	190	2	10	180	3	15	110	3	15	100	4	20	70	4	20
	20,001—50,000	1,500	2	12	750	10	10	540	3	15	270	3	15	24	4	20	140	4	20	120	5	25	85	5	25
	50,001 and over	2,200	3	13	1,100	15	15	700	5	20	350	4	25	290	5	25	175	5	25	145	6	30	105	6	30

Criteria for Shifting to Tightened Inspection and Requalification for normal Inspection.

ABC standard states : "When normal inspection is in effect, tightened inspection shall be instituted when 2 out of 5 consecutive lots or batches have been rejected on original inspection (*i.e.* ignoring resubmitted lots or batches for this procedure".

The statement regarding requalification for normal inspection is as follows : "when tightened inspection is in effect, normal inspection shall be instituted when five consecutive lots or batches have been considered acceptable on original impaction".

Criteria for Qualification and loss of Qualification for Reduced Inspection. The ABC standard states the following conditions for a shift from normal to reduced inspection :

(*a*) The preceding 10 lots or batches (or more as indicated in relevant table have been on normal inspection and none has been rejected on original inspection ; and

(*b*) The total number of defective (or defects) in the samples from the preceding 10 lots or batches (or such other number as was used for condition (*a*) above is equal to or less than the applicable number given in the relevant table. If double or multiple sampling is in use, all samples inspected should be included, not "first" samples only ; and

(*c*) Production is at a steady rate ; and

(*d*) Reduced inspection is considered desirable by the responsible authority.

Normal inspection must be reinstated whenever "a lot or batch is rejected" or "production becomes irregular or delayed, or other conditions warrant that normal inspection shall be instituted.

19.29. Sampling Plans for Inspection by Variables for percent Defective

Here we shall be dealing only with the single sampling plans, and not double and multiple sampling plans which are found to be rather complicated for practical use. For these plans it is essential condition that the lot quality be known in terms of per cent defective. The various types of single sampling variable plans are :

I. AQL plans

(*i*) *Variability known plans, i.e.* when variability (σ^2) is known from past experience, control chart data etc. These plans are most economical to operate as sample is minimum for these and calculation also least. However, the past knowledge of variability is essential.

(*ii*) *Variability unknown plans.*

(*a*) *Standard deviation method, i.e.* variability in the lot is not known and is estimated from the sample standard deviation (s). These require cumbersome calculations.

(*b*) *Range Method.* In this method the variability is estimated from the sample range (R). It requires biggest sample but the calculations are not as cumbersome as for standard deviation method.

All these plans can be considered separately for the following two categories :

(*1*) Only one sided specification limit is given.

(*2*) Two-sided specification limits are given.

II. One's own plans. Sampling plans of one's own choice can be made depending on the values of AQL and LTPD desired. The various types of plans possible are same as given under AQL plans.

For AQL plans which are given in Indian Standard Specification IS : 2500 (Part II)-1965, first the suitable value of AQL has to be chosen depending on the quality desired and the quality attainable. Then corresponding to the inspection level desired (*i.e.* either more inspection but lesser risk of accepting lots of quality worse than the chosen AQL or *vice versa*), the sample size code letter can be chosen.

Lot acceptability. The acceptability of a lot of materials submitted for inspection shall be determined by using one of the sampling plans associated with a specified value of AQL. Depending upon the specification of one-sided or two-sided limits, the acceptability criteria for the three types of sampling plans would be as given below :

(*ii*) **Variability known method**

(1) *For one-sided specification limits.* On this basis of the AQL and the sample size code letter chosen, the value of the sample size (n) and a suitable factor (k'') shall be obtained from Table 2 of IS : 2500 (Part II)—1965. The mean of the n test results shall then be calculated and the lot shall be declared as acceptable if :

(*a*) the value of the expression $(\bar{x} + k''\,\sigma) \leq U$, when the upper specification limit (U) is given ; or

(*b*) the value of the expression $(\bar{x} - k''\,\sigma) \geq L$, *when the lower specification limit (L) is given.*

(2) *For two-sided specification limits.* The lot shall be declared as acceptable if :

(*a*) The value of the expression $\sigma/(U - L) \leq$ the maximum value specified below for the chosen AQL.

AQL%	0.10	0.15	0.25	0.40	0.65	1.00	1.50	2.50	4.00	6.50	10.0
Upper Limit of (U–L)	0.152	0.158	0.165	0.175	0.184	0.194	0.206	0.223	0.243	0.270	0.304

(*b*) the value of the expression $(\bar{x} + k''\,\sigma) \geq U$, and

(*c*) the value of the expression $(\bar{x} - k''\,\sigma) \geq L$.

(*ii*)**Variability unknown** (Standard deviation method).

(1) *For one-sided specification limits.* On the basis of the AQL and the sample size code letter chosen, the value of the sample size (n) and a suitable factor (k') shall be obtained from Table 3 of IS : 2500 (Part II)—1965. The mean ($\bar{x}$) and the standard deviation (s) shall then be calculated from the test results and the lot shall be declared as acceptable if :

(*a*) the value of the expression $(\bar{x} + k's) \leq U$, when the upper specification limit (U) is given ; or

(*b*) the value of the expression $(\bar{x} - k's) \geq L$, when the lower specification limit (L) is given.

(2) *For two-sided specification limits.* The lot shall be declared as acceptable if

(*a*) the value of the expressions (U-L) $\leq$ the maximum value specified in Table 5 of IS : 2500 (Part II)—1965 for the particular combination of the chosen AQL and the sample size code letter.

(*b*) the value of expression $(\bar{x} + k's) \leq U$, and

(*c*) the value of the expression $(\bar{x} + k's) \geq L$.

(*iii*) **Variability unknown** (Range method).

(1) *For one-sided specification limits.* On the basis of the AQL and the sample size code letter chosen, the value of the sample size (n) and a suitable factor (k) shall be chosen from Table 4 of the IS : 2500 (Part II)—1965. The mean ($\bar{x}$) and the range (R) if the sample size is less than 10, or the mean range (R) if the sample size is 10 or more, shall then be calculated from the test results. The lot shall be declared acceptable if :

(*a*) the value of the expression $(\bar{x} - kR)$ or $(\bar{x} - \overline{kR}) \leq U$, when the upper specification limit (U) is given ; or

(b) the value of the expression $(\bar{x} + kR)$ or $(\bar{x} - k\overline{R}) \geq L$, when the lower specification limit (L) is given.

(2) *For two-sided specification limits.* The lot shall be declared as acceptable if :

(a) the value of expression $(U\text{-}L) \leq$ the maximum value specified in Table 7 of IS : 2500 (Part II)-1965 for the particular combination of the chosen AQL and the sample size code letter ;

(b) the value of the expression $(\bar{x} + kR)$ or $(\bar{x} + k\overline{R}) \leq U$, and

(c) the value of the expression $(\bar{x} - kR)$ or $(\bar{x} - k\overline{R}) \geq U$.

19.30. Reliability and Life Testing

Reliability may be explicitly defined as the probability that a system will perform satisfactory for at least a given period of time when used under stated conditions. The *reliability function* is thus same probability expressed as a function of the time period.

Failure is the inability of an equipment to perform its required function. Toe determine the life of a manufactured product for purposes of acceptance inspection, elaborate test apparatus is required to determine the exact moment at which the performance ceases to be satisfactory (or life ends or failure occurs). The definition of satisfactory performance is defined by the specification which depends on the use to which the product is to be put, *e.g.*, a certain product according to one specification may have short life, but long life under another set. In most of the cases, the failure of a component is viewed as terminating its life. In the case of complex manufactured devices where the device can be restored to its original satisfactory performance by replacing one or more components on occurrence of its failures the time expected to elapse between successive failure is important.

A picture of the failure pattern of certain types of electronic components as well as certain kinds of complex systems is depicted in Fig. 19.33. Initially high failure rates may be expected (premature failures), but on replacement of all the prematurely failing items the failure rate reaches a lower value and then remains fairly constant for very long time (known as normal operating period). During normal operating period the average failure rate remains essentially constant. Most of the acceptance sampling plans, in which case life is the quality characteristic specified by the designer, are based on the assumption that the failure rate is constant throughout life ; *i.e.* flat portion of the curve in Fig. 19.33. Obviously, therefore, acceptance sampling can't be started unless all early premature failures have occurred. Since generally, the uniform failure rate continues for a longer time than the intended period of use, the final wear out period can be disregarded in the design and interpretation of acceptance sampling plans. Two related quantities used in computing OC curves are failure rate (λ') and arithmetic means or mean life (θ'). λ' is the probability of a failure in a stated unit of time ; unit of time may be any time (one hour is the common practice ; in some instances instead of specifying in hours, it may be

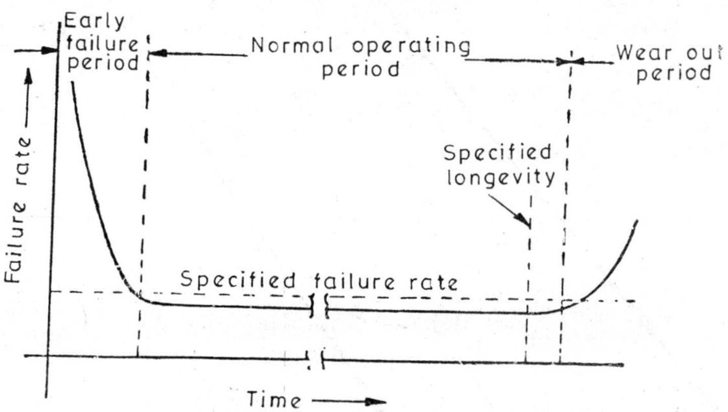

Fig. 19.33. Failure rates vs. time under constant environmental operating conditions.

specified as number of operating cycles in case of switching devices). In the case of complex devices, θ' is interpreted as the mean time between failures (MTBF). θ' is also expressed in the same units as used.

For λ'. Considering an acceptance sampling plan based on a stipulated maximum number of test hours, which states : Select 10 items at random from a lot and start testing them. Whenever an item fails, replace it with another item selected at random from the lot. If the test continues for more than 500 hours with not more than 1 failure, accept the lot. If two failures occur before the 500 hours of testing, reject the lot and terminate the test. Obviously, acceptance under this plan requires $10 \times 500 = 5000$ item hours of test with an acceptance number of 1. If it be assumed that the probability of a failure is the same for every item hour, the *OC* curve can be calculated in the same way as if it were an ordinary single sampling attributes plant with $n = 5000$ and $c = 1$. Here the failure rate λ' takes the place of the fraction defective p', For life testing plans, the OC curve is plotted in terms of mean life ($\theta' = 1/\lambda'$) rather than in terms of failure rate (λ' corresponding to p' or ordinary OC curve).

Thus for various values of λ', $\left(\theta' = \frac{1}{\lambda'}\right)$, the probability of acceptance can be calculated as described below :

Failure rate per hour, λ'	*Mean life hours* $\theta' = \frac{1}{\lambda'}$	*Expected average number of failures in 5000 test hours (5000 λ)*	*Probability of acceptance (Probability of one or less failure ; Read from Table 19.9)*
0.00001	100000	0.05	0.9958
0.00002	50000	0.10	0.995
0.00005	20000	0.25	0.974
0.00006	16667	0.30	0.963
0.00008	12500	0.40	0.938
0.00010	10000	0.50	0.910
0.00020	5000	1.00	0.736
0.00030	3333	1.50	0.558
0.00040	2500	2.00	0.406
0.00050	2000	2.50	0.293
0.00060	1667	3.00	0.199
0.00080	1250	4.00	0.092
0.00100	1000	5.50	0.040

It may be noted that the probability of acceptance for a lot that has mean life of 10,000 hours in 0.910 ; or in other words the producer's risk is 0.09 or 9% of rejecting a lot that has a mean life of 10,000 hours. Similarly the consumer's risk of accepting a lot that has a mean life of 1,250 hours is 9.2%.

The mean life at various stages in a life test can be estimated by the formula

$$\text{Estimated } \lambda' = \frac{\text{Number of failures observed}}{\text{Number of item hours of test}} \text{ and estimated } \theta' = \frac{1}{\lambda'}$$

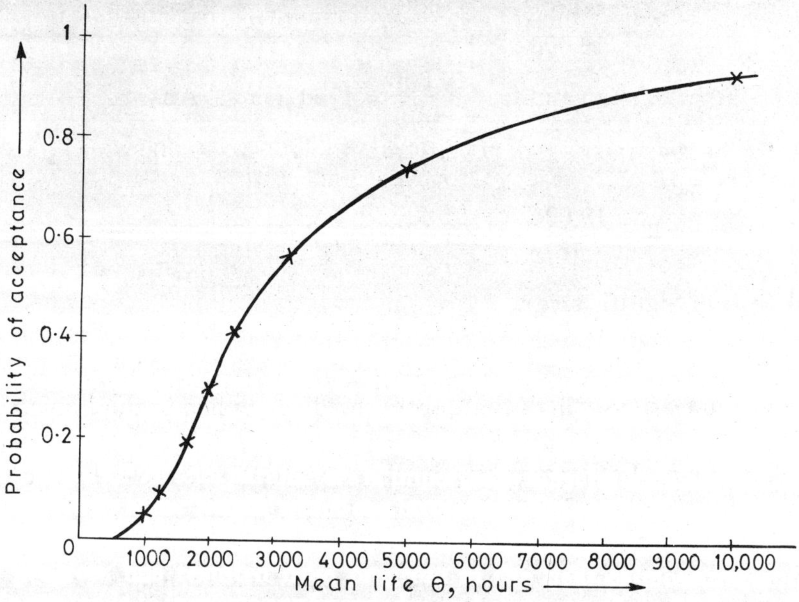

Fig. 19.34. OC curve for a test of 5000 item hours with acceptance number of 1 (based on assumption of a constant failure rate).

Further if λ' represents a constant probability of failure per hour, then the probability that any one trial will not result in a failure is $1 - \lambda'$; and if trials are to be continued until a failure occurs then the theorem of conditional probabilities can be applied to determine the probability of survival, *i.e.*, of no failure and thus the probability of an item surviving for H hours $= (1 - \lambda')^H$. It may be noted that reliability is also nothing but the probability of no failure throughout some specific period of time.

Reliability $= (1 - \lambda')^H$; and if the value of λ' is very small and H is very large, as is generally the case, then $(1 - \lambda')^H$ can be written $= e^{-H\lambda'}$

The expression $e^{-H\lambda'}$ is usually referred to as exponential reliability function.

If H' represents the chosen unit time period in hours and P_f' is the probability of failure during period H' the exponential reliability function yields the following formula for mean life in hours :

$$\theta' = \frac{H' \log e}{\log (1 - P_f')} .$$

Example 19.38. *An acceptance sampling plan for life testing requires that a sample of 19 items be tested with replacement for 1000 hours. If not more than 7 failures occur, the lot is Compute the mean life for which—*

(i) the producer's risk of a lot rejection is 0.05,

(ii) the consumer's risk of a lot acceptance is 0.10.

Solution. Acceptance plan requires 19 × 1000 items hours = 19,000 items hours. Acceptance number = 7.

(*i*) For producer's risk of a lot rejection of 0.05, probability of acceptance, or probability of 7 or less failures should be 0.95. Referring to Table 19.9, corresponding to $P_a = 0.95$, 19,000 λ' should be equal to 4.0 (λ' = failure rate per hour)

$$\therefore \qquad \lambda' = \frac{4.0}{19,000}$$

and $$\theta' = \text{mean life} = \frac{19,000}{4.0} = 4750 \text{ hours.} \quad \textbf{Ans.}$$

(*ii*) Similarly for consumer's risk of 0.10 from table 19.9, 19,000 λ' or $P_a = 0.10$ for 7 or less failures should be = 16

or $$\theta' = \frac{19,000}{16} = 1187.5 \text{ hours.} \quad \textbf{Ans.}$$

19.31. Statistical Tests of Significance

Statistical tests of significance are very important tools in the hands of industrialists for taking decisions.

The various applications where these tests could be applied in the industry are of two kinds.

(*a*) (*i*) To determine whether a given sample is the representative of a specific lot.

(*ii*) To determine whether the mean sample value of a given lot differs significantly from given specified value.

(*iii*) To determine whether the mean values of two given specified lots differ significantly from each other.

(*iv*) To find out if a new chemical is more effective in treating water than a standard chemical being used.

(*v*) To determine whether the new method of preserving milk is better than the existing one in dealing with the bacteria.

In all these cases the mean values are to be obtained and examined if they are the same or different.

(*b*) (*i*) To find whether one draughtsman is more consistent than another in preparing the drawings.

(*ii*) To determine if the coal from a new source has resulted in a change in the variability of steam output.

(*iii*) To find whether the given temperature and humidity conditions over the rollers affect the uniformity of the quality of paper.

In all these cases the values of the variance are to be obtained and examined if they are the same or different. The most commonly used tests of significance are :

(*i*) **t-test.** It is used for testing the significance of the difference between the mean of the lot from which the sample has been taken and a specified value. It is also used for testing the significance of the difference between the means of two lots having the same variance.

(*ii*) **F-test.** It is used for testing the quality of the variances of two independent lots.

(*iii*) **Chi-squared (χ^2) test.** It is used for testing the equality of the variance of a given lot and a specified value ; *i.e.* whether the variance of the lot from which the sample has been drawn differs from a specified value.

It could also be used for testing whether the frequencies observed from certain categories differ significantly from the expected theoretical frequencies.

It is important to note that all these tests are based on certain assumptions (given later) and, therefore, these should not be used in discriminately, and the feasibility of application of these tests should be ascertained.

The procedure commonly adopted for these tests is to first set up a null hypothesis regarding the equivalence (or no difference) between the values of the parameters of the lots from which the samples have been drawn. (In certain cases, the null hypothesis may simply refer to the equivalence of the lot parameter to some specified value). The appropriate test of significance is then applied on the basis of the statistics computed from the sample and determining whether the observed difference between the statistics could have arisen just due to chance. If the probability of obtaining the observed difference due to chance alone is quite small, it is concluded that the observed difference is statistically significant. The null hypothesis is then rejected and the values of the lot parameters are taken to be significantly different from each other. In industrial applications the levels of probability at which the significance is judged are normally taken to be 5 to 1 per cent. If the observed probability is less than 5 but more than 1 per cent, it is used to be significant whereas if it is less than 1 per cent, it is said to be highly significant. But 5 and 1 per cent levels of significance are the conventional levels. The particular level to be adopted in practice depends upon the nature of the problem and the seriousness of the decision.

There are two kinds of risks involved in accepting the conclusions based on the tests of significance, namely (*a*) deciding that a significant difference exists when there is no real difference and (*b*) deciding that no difference exists where there is a real difference. It is of course not possible to minimise both the risks at the same time. Hence, one of the risk (usually of the first kind) is controlled by assigning to it a chosen level of probability and then the other risk is minimised.

We are defining below three terms commonly used in the statistical tests of significance.

(*i*) *Null Hypothesis.* The hypothesis (or assumption) of the equivalence (or no difference) between the effects of method(s) so that the sample(s) emanates from the same lot.

(*ii*) *Level of significance.* The probability (or risk) of rejecting the null hypothesis when it is true. Conventionally it is taken to be 5 or 1 per cent.

(*iii*) *Degree of freedom.* The number of independent component values which are necessary to determine a statistic.

19.31.1. t-tests. *Assumptions for t-test.* Before applying *t*-test, it is important to ensure that the following requirements are met with

(*i*) Observations follow normal distribution.

(*ii*) Samples are taken at random.

(*iii*) In case of two lots, their variance are not significantly different.

(*a*) *t*-test to test the difference between the mean of the lot and some specified value :

In case the difference of mean (of the lot) is to be tested against the specified value without giving any importance to the sign of the difference (*i.e.* positive or negative) then the test is said to be two-sided. However, if the sign is of importance *i.e.* the variation is allowed in one direction only then the test is said to be one-sided.

For this test, a sample of size n is taken from the lot and its mean and standard deviation calculated first. Then the t-value is found by the formula

$$t = \frac{\bar{x} - \mu}{s/\sqrt{n}}$$

where $\bar{x}$ = mean of the sample = $\frac{\Sigma x}{n}$

s = Standard deviation of the sample = $\sqrt{\frac{\Sigma (x - \bar{x})^2}{n - 1}}$

μ = Specified value.

(Please note that this t-value has n-1 degrees of freedom).

Table 19.19. Critical Values of t-distribution

Degree of freedom (n – 1)	*Significant Levels*			
	One-sided Test		*Two-sided Test*	
	0.05	*0.01*	*0.05*	*0.01*
1	6.314	31.821	12.706	63.657
2	2.920	6.965	4.303	9.925
3	2.353	4.541	3.182	5.481
4	2.132	3.747	2.776	4.604
5	2.015	3.365	2.571	4.032
6	1.943	3.143	2.447	3.707
7	1.895	2.998	2.365	3.499
8	1.860	2.896	2.306	3.355
9	1.833	2.821	2.262	3.250
10	1.812	2.764	2.228	3.169
11	1.796	2.718	2.201	3.106
12	1.782	2.681	2.179	3.055
13	1.771	2.650	2.160	3.012
14	1.761	2.624	2.145	2.977
15	1.753	2.602	2.131	2.947
16	1.746	2.583	2.120	2.921
17	1.740	2.567	2.110	2.898
18	1.734	2.552	2.101	2.878
19	1.729	2.539	2.093	2.861
20	1.725	2.528	2.086	2.845
21	1.721	2.518	2.080	2.831
22	1.717	2.508	2.074	2.819
23	1.714	2.500	2.069	2.807
24	1.711	2.492	2.064	2.797
25	1.708	2.485	2.060	2.787
26	1.706	2.479	2.056	2.779
27	1.703	2.473	2.052	2.771
28	1.701	2.467	2.048	2.763
29	1.699	2.462	2.045	2.756
30	1.697	2.457	2.022	2.750
α	1.645	2.326	1.960	2.576

If this t-value is greater than or equal to the critical values of t given in table (Table 19.18, specifies the critical values of t at 5 and 1 per cent levels of significance for one-sided and two-sided test and for the different degrees of freedom and is reproduced from Appendix B of IS : 6200-1971), then the difference is taken to be significant and the null hypothesis of no difference is rejected. If, however, this t-value is less than the critical value of t then it is assumed that the lot mean does not differ significantly from the given specified value. Obviously in the Table 19.19, the critical value has to be chosen for either one-sided test or two-sided test (as the case may be) and for the given degree of freedom $(n - 1)$. This test will be made clear by example 19.39.

Example 19.39. *On an automatic machine producing cylindrical plugs, a sample of 10 pieces was taken and the following readings were observed.*

25.1, 25.3, 24.9, 25.4, 24.2, 26.1, 25.9, 25.7, 24.5, 24.8.

The machine was set to produce cylinder plugs of diameter 25.0 mm.

Find out if average diameter of the pins differs significantly from the specified diameter of 25.0 mm.

For 9 degrees of freedom, for two-sided test, critical values of t are 2.62 and 3.250 respectively for 5% and 1% levels of significance.

Solution. (*a*) Mean of the 10 pieces

$$\bar{x} = \frac{\Sigma x}{n} = \frac{25.1 + 25.3 + 24.9 + 25.4 + 24.2 + 26.1 + 25.9 + 25.7 + 24.5 + 24.8}{10}$$

$$= \frac{251.9}{10} = 25.19 \text{ mm.}$$

$$\text{Standard deviation } s = \frac{\sqrt{\Sigma x^2 - \dfrac{(\Sigma x)^2}{n}}}{n - 1}$$

$$= \sqrt{\frac{630.01 + \ldots\ldots + 615.14 - (251.9)^2/10}{(10 - 1)}}$$

$$= \sqrt{\frac{6358.61 - 6350.4}{9}} = \frac{1}{3}\sqrt{8.21} = \frac{1}{3} \times 2.17 = 0.72$$

Null hypothesis is that the mean diameter of the pins manufactured by this machine is equal to 25.0 mm.

For applying t-test, first calculate the value of t by the formula

$$t = \frac{\bar{x} - \mu}{\dfrac{s'}{\sqrt{n}}} = \frac{25.19 - 25.0}{\dfrac{0.72}{\sqrt{10}}} = \frac{0.19}{0.72} \times 3.16 = 0.83.$$

Since the calculated value of t is less than 2.262 (the 5% point of the t-distribution, two sided, with 9 degrees of freedom), the validity of the null hypothesis is not doubted and the lot mean diameter is taken to be in agreement with the specified diameter of 25.0 mm.

(*b*) t-test for testing the difference between the means of two independent lots.

In this test, the aim is to find out whether the means of two independent lots differ significantly from each other.

For this, first two samples of sizes n and n' are taken from the two lots and their means ($\bar{x}$ and $\bar{x}'$) calculated. Also the value of the standard deviation of the pooled samples s' is calculated.

where
$$s' = \sqrt{\frac{\Sigma (x - \bar{x})^2 + \Sigma(x' - \bar{x}')^2}{n + n' - 2}}$$

Then the value of t is found by the relation

$$t = \frac{\bar{x} - \bar{x}'}{s' \sqrt{\frac{1}{n} + \frac{1}{n'}}}$$

(Please note that in this case the degrees freedom)

$$= (n + n' - 2)$$

In case this value is greater than the critical value of t given in Table 19.18 then the two means are supposed to be significantly different, otherwise not.

This test will become clear by the example given below.

Example 19.40. *Two automatic machines are set to produce cylindrical plugs of 25.0 mm diameter. A sample of 10 pieces is taken from 1st machine and a sample of 8 pieces is taken from the 2nd and the dimensions noted are given below.*

It is required to examine whether the pieces manufactured by 1st machine are bigger in size than those manufactured by 2nd machine.

Diameters of pieces taken from 1st machine :

25.1, 25.3, 24.9, 25.4, 24.2, 26.1, 25.9, 25.7, 24.5, 24.8.

Diameters of pieces taken from 2nd machine :

25.2, 25.0, 24.9, 24.6, 24.5, 24.3, 24.8, 25.1.

Critical value of t for 16 degrees of freedom for one-sided test = 1.746 and 2.583 respectively for 5% and 1% significance levels.

Solution. The value of the standard deviation of the pooled samples.

$$s' = \sqrt{\frac{\Sigma (x - \bar{x})^2 + \Sigma(x' - \bar{x})^2}{n + n' - 2}}$$

$$\bar{x} = \frac{\Sigma x}{n} = \frac{251.9}{10} = 25.19$$

and

$$\bar{x}' = \frac{\Sigma x'}{n'}$$

$$= \frac{25.2 + 25.0 + 24.9 + 24.6 + 24.5 + 24.3 + 24.8 + 25.1}{8}$$

$$= \frac{198.4}{8} = 24.8$$

$$\Sigma(x - \bar{x})^2 = (-0.09)^2 + (0.11)^2 + (-0.29)^2 + (0.21)^2 + (-0.99)^2 + (0.71)^2 + (0.51)^2 + (-0.69)^2 + (-0.39)^2$$

$$= 0.0081 + 0.0121 + 0.0841 + 0.0441 + 0.9801 + 0.8281 + 0.5041 + 0.2601 + 0.4761 + 0.15256$$

$$= 3.3495$$

$$\Sigma(x' - \bar{x}')^2 = (0.28)^2 + (0.08)^2 + (-0.02)^2 + (0.32)^2 + (0.42)^2 + (0.38)^2 + (-0.12)^2 + (0.81)^2$$

$$= 0.0784 + 0.0064 + 0.0004 + 0.1024 + 0.1764 + 0.1444 + 0.0144 + 0.0324$$

$$= 0.5552.$$

$$\therefore \quad s' = \sqrt{\frac{3.3495 + 0.5552}{10 + 8 - 2}} = \sqrt{\frac{3.9037}{16}} = \frac{1.976}{4} = 0.494$$

Now $$t = \frac{\bar{x} - \bar{x}'}{s'\sqrt{\frac{1}{n} + \frac{1}{n'}}} = \frac{25.19 - 24.92}{0.494\sqrt{\frac{1}{10} + \frac{1}{8}}}$$

$$= \frac{0.27 \times \sqrt{80}}{0.494 \times \sqrt{18}} = \frac{0.27}{0.494} \times \frac{\sqrt{40}}{\sqrt{9}} = \frac{0.27 \times 6.324}{0.494 \times 3} = \frac{1.70748}{1.482} = 1.245.$$

Since the value is less than 2.583.

∴ The pieces manufactured by first machine are not bigger than those manufactured by 2nd machine.

19.31.2. Paired t-Test. This test is applied to determine whether or not there is a significant difference between the means of two related lots. Such a situation arises when two samples are so given that the observations of one sample correspond to the observations of the other.

For this test, the differences between observations in each pair are found out ; (the null hypothesis for such a case is that the mean of the difference is zero) ; and the t-test applied on these differences as if testing of the mean of these against zero.

Thus in case of two machines set to produce cylindrical plugs at a specified diameter, if the observations of two lots are quite close to each other, then this test is applied as under :

S. No.	*Observations of diameters on two lots*		
	I	*II*	*Difference I-II*
1	d_1	d_1'	$d_1 - d_1'$
2	d_2	d_2'	$d_2 - d_2'$
3	d_3	d_3'	$d_3 - d_3'$
:	:	:	:
:	:	:	:
n	d_n	d_n'	$d_n - d_n'$
			$\Sigma d = (d_1 - d_1') - (d_2 - d_2') + \ldots\ldots (d_n - d_n')$

$$\bar{d} = \frac{\Sigma d}{n}$$

$$s_d = \sqrt{\frac{1}{n-1}\left[\Sigma d^2 - \frac{(\Sigma d)^2}{n}\right]}$$

To examine if the mean difference is significantly different from zero, t is computed as

$$t = \frac{|\bar{d} + 0|}{\frac{s_d}{\sqrt{n}}}$$

If this value is less than the critical value of t found from Table 19.18 for critical values of t-distribution for 5% significance level from two-sided or one-sided as the case may be with $n - 1$ degrees of freedom, then it is presumed that the means of the two lots are same or in other words, the null hypothesis that the average of the two lots does not differ with each other is not questioned.

Note. It may be noted that the t-test is applied when the sample size is small and the information about the standard deviation is not available and, therefore, inferences are made about the lot standard deviation by using the sample standard deviation as an estimate. It may be recalled that the lot standard deviation (σ) is the square root of the mean the squares of deviations of the observations from their mean, whereas the sample standard deviation(s) is the square root of the quotient obtained by dividing the sum of squares of deviations of the observations from their mean by one less than the number of observations in the sample. If the number of observations in a sample is very large, *i.e.* more than 30 and the population standard deviation is known from previous experience or calculated from very large number of the observation, then *t*-test approaches normal test and actually under that eventuality the normal test may be applied and the critical values to infer regarding the rejection of null hypothesis are to be taken from table 19.19 (for critical values of *t*-distribution) corresponding to α (infinity).

For normal test the test statistic U is given by

$$U = \frac{\bar{x} - \mu}{\sigma/\sqrt{n}} \text{ for one sample}$$

and

$$U = \frac{\bar{x}' - x'}{\sqrt{\frac{\sigma^2}{n} + \frac{\sigma'^2}{n'}}} \text{ for two sample case.}$$

19.31.3. F-test. It is used to test the equality of two variances. (Variance is defined as the square of the sample standard deviation).

The statistic F is defined as $F = \dfrac{s_1^2}{s_2^2}$

Always s_1^2 should be bigger than s_2^2

where $$s_1^2 = \frac{\Sigma (x - \bar{x})^2}{n_1 - 1} \quad \text{and} \quad s_2^2 = \frac{\Sigma (x - \bar{x}')^2}{n_2 - 1}$$

$\bar{x}$ and $\bar{x}'$, m and n_2 are the means and the sample sizes of two samples.

As F involves comparison of two samples, the degrees of freedom $(n_1 - 1)$ and $(n_2 - 1)$ are associated with it ; and the critical values of the F-distribution for 0.05 and 0.01 significance levels, for one-sided and two-sided tests and for degrees of freedom $(n_1 - 1)$ and $(n_2 - 1)$ are given in Appendix C of IS : 6200—1971.

Note that $(n_1 - 1)$ refers to the degrees of freedom for the larger mean square placed in the numerator.

The significance of F-value as calculated from above equation is compared with the critical values in this table. Again, as before if the calculated value is greater than this critical value, then the null hypothesis regarding the qualities of variances is rejected.

The assumptions for F-test are as for t-test.

Example 19.41. *In example 19.40, find out whether any one sample gives more consistent results.*

Critical value of F for 0.05 significance level for two-sided test for 9 and 7 degrees of freedom is 4.82.

Solution. $$F = \frac{s_1^2}{s_2^2}$$

$$s_2^2 = \frac{\Sigma (x - \bar{x})^2}{n - 1} = \frac{3.3495}{10 - 1} = \frac{3.3495}{9} = 0.388$$

$$s_2^2 = \frac{\sigma (x - \bar{x}')^2}{n_2 - 1} = \frac{0.5552}{8 - 1} = \frac{0.5552}{7} = 0.0793$$

$$\therefore \qquad F = \frac{s_1^2}{s_2^2} = \frac{0.388}{0.0793} = 4.86.$$

This value is greater than the critical value and thus it could be said that s_2^2 is significantly smaller than s_1^2, and it is concluded that the sample 2 gives better reproducible results as compared to sample 1.

19.31.4. x²-Test. (*i*) For testing of a variance of the lot from which a sample is drawn and a specified value :

Let sample variance $= s^2$

and specified or lot variance $= \sigma^2$

Then
$$x^2 = \frac{(n-2)s^2}{\sigma^2}$$

(Note, it has got $n - 1$ degrees of freedom, n being the sample size.)

If this value is greater than or equal to the tabulated value for the chosen level of significance given in table 19.19 (or critical values of x^2-distribution) then the null hypothesis regarding the quality of the lot variance to the specified value is rejected.

Assumptions for this test are same as in t-test.

(*ii*) Testing goodness of fit, *i.e.* to test whether the observed frequencies differ significantly from those expected on the basis of certain hypothesis when the observations in a sample fall into certain specified categories or classes.

Table 19.19. Critical Values of x²-Distribution

Degrees of Freedom	*Significance Level*		*Degrees of Freedom*	*Significance Level*	
	0.05	*0.01*		*0.05*	*0.01*
1	3.84	6.64	21	32.67	38.93
2	5.99	9.21	22	33.92	40.29
3	7.82	11.34	23	35.17	41.64
4	9.49	13.28	24	36.42	42.98
5	11.07	15.09	25	37.65	44.31
6	12.59	16.81	26	38.89	45.64
7	14.07	18.48	27	40.11	46.96
8	15.51	20.09	28	41.34	45.28
9	16.92	21.67	29	42.56	49.59
10	18.31	23.21	30	43.17	50.89
11	19.68	24.73	40	55.75	63.69
12	21.03	26.22	50	67.50	76.15
13	22.36	27.09	60	79.08	88.38
14	23.69	29.14	70	90.53	100.42
15	25.00	30.58	80	101.88	112.33
16	26.30	32.00	90	113.14	124.12
17	27.59	33.41	100	124.34	135.81
18	28.87	34.81			
19	30.14	36.19			
20	31.41	37.57			

If $f_1, f_2 ..., f_n$ represent the observed frequencies and $f_1', f_2', ..., f_n'$ represent the expected frequencies then

$$x^2 = \sum_{i=1}^{n} \frac{(f_i - f_i')^2}{f_i'}$$

As before, this value is to be compared with the critical values given in Table 19.19 for the given level of significance (generally 5%) and the degrees of freedom.

If this test is used to find out whether theoretical distribution like Binomial, Poisson, Normal etc., fits the given observations satisfactorily or not, then the degrees of freedom = $n - 1 - p$, where p = number of parameters estimated from the sample.

Further in applying this test, if it is found that the frequency in any class is less than 5, then that frequency may be pooled with that of adjacent class so that the frequency becomes 5 or more.

Example 19.42. *In an automatic machine, sample of 500 items is taken from each day's production and the number of defective items calculated is given below :*

Day	*No. of defectives*
Monday	*6*
Tuesday	*8*
Wednesday	*5*
Thursday	*7*
Friday	*6*
Saturday	*9*
Sunday	*8*
Total	*49*

It is required to test the hypothesis that the proportion of defective is constant from day-to-day.

Given the critical value of x^2 for 5% level of the significance for 6 degrees of freedom = 12.59.

Solution. According to the null hypothesis the expected number of defectives for each of the 7 days = $\frac{49}{7} = 7$.

$$\therefore \quad x^2 = \frac{(6-7)^2}{7} + \frac{(8-7)^2}{7} + \frac{(5-7)^2}{7} + \frac{(7-7)^2}{7} + \frac{(6-7)^2}{7} + \frac{(9-7)^2}{7} + \frac{(8-7)^2}{7}$$

$$= \frac{1}{7} + \frac{1}{7} + \frac{4}{7} + 0 - \frac{1}{7} + \frac{2}{7} + \frac{1}{7} = \frac{10}{7} = 1.43.$$

As this value is very much less than 12.59, therefore, the number of defectives observed each day do not vary from day-to-day.

19.32. Regression and Correlation

Many times we are interested in determining the relationship between the results of one test and the results of some other test conducted on the same specimen. An excellent example of such a case would be to find out the relationship between the tensile strength and the hardness test (BHN—Brinell hardness number) of the steel. Obviously the tensile strength is a destructive test and if we can find that tensile strength follows some trend with variation of BHN then the latter will be a very good substitute for destructive test.

There are two important things in predicting the adoption of one test for other test.

(*i*) To determine the trend line (regression line).

(*ii*) To find out the closeness of the relationship (correlation) of the two methods of testing.

It may be appreciated that simply determining the trend of variation between the observations of two test results is not that important as the colseness of the relationship between the two. *i.e.* with what guarantee we can say that one test substitutes for the other test. We shall study the statistical methods to determine both trend line and the correlation.

For determining the trend line we generally adopt the *method of least squares*. For this purpose, the independent variable (cause, of thing from which the prediction is being made) is plotted along the X-axis ; and the dependent variable (effect, or thing being predicted) is plotted along the Y-axis. For example, in the example cited above, BHN is plotted along X-axis and tensile strength along Y-axis. The plot of these two observations obtained by making tests on different specimens will be a sort of scatter diagram. The trend line is to be drawn in between these plotted points in such a way that—

(*a*) the algebraic sum of the vertical deviations from the trend line equals zero. (This will be true for any line, other than vertical, which passes through the centre of gravity of the plotted points but such a line can't give the trend between two sets of parameters. Therefore, second condition should come in ;

(*b*) the sum of the squares of the vertical deviations must be less than the sum of the squares from any other straight line. (Obviously this condition limits the line in one position which most closely shows the trend between two parameters).

If the test results of two tests for n observations are $x_1, x_2, x_3 \ldots\ldots, x_n$

and $y_1, y_2, y_3 \ldots\ldots y_n$

respectively, then for first condition $\Sigma y = an = + b\Sigma x$ ...(*i*)

and for second condition, $\Sigma xy = a\ \Sigma x^2 + b\Sigma x^2$ ...(*ii*)

By solving these two equations, the values of constants a and b can be determined and then the trend line is represented by the equation

$$y_c = a + bx \qquad \ldots(iii)$$

where y_c gives the computed or most likely values of y for given values of x. It may be cautioned here that this equation can't be used in the reverse way, *i.e.* for estimating the value of x for given value of y. For that purpose the values of a and b should be determined separately taking y as cause and x as effect, *i.e.* x and y may be interchanged and above process repeated. Further as nothing can be said about the behaviour of trend for extreme values, it is not good to make any extrapolations.

Solving of equations (*i*) and (*ii*) can be avoided by using the formulae for a and b as

$$a = \frac{\Sigma x^2 \Sigma y - \Sigma x \Sigma xy}{n\Sigma x^2 - (\Sigma x)^2} \qquad \ldots(iv)$$

and

$$b = \frac{n\ \Sigma xy - \Sigma x \Sigma y}{n\Sigma x^2 - (\Sigma x)^2}$$

Further as $a = \bar{y} - b\bar{x}$ ...(*v*)

$\therefore$ $b = \dfrac{\bar{y} - a}{\bar{x}}$ ...(*vi*)

It suggests that only value of a may be calculated from equation (*iv*) and then the value of b can be found out from equation (*vi*).

Standard error of the estimate : Since the estimate value of y (*i.e.* y_c) is only an estimate ; it is important to know the extent to which the actual values may be expected to scatter around the estimate. This is given by

$$\sigma_{ys} = \sqrt{\frac{\Sigma y^2 - (a\Sigma y + b\Sigma xy)}{n}}$$

Thus for a computed value of y, observed values may be expected to lie between $y_c \pm 3\sigma_{ys}$ since variance (standard deviation squared) is an additive quantity, the total variance of the y values about the mean of y is equal to the sum of the variance of the computed values of y about the mean of y, plus the variance of the observed values about the trend line.

Correlation. It has already been pointed out that once it has been determined that there exists a cause and effect relationship between two variables, then the correlation provides a method for evaluating the extent to which variations in one variable are associated with variation in another variable. In other words, it permits appraisal of the closeness of the cause and effect relationship.

The extent or degree of correlation between two variables is determined by computation of correlation co-efficient (r). When variations in the dependent variable are due to the independent variables alone (*e.g.*, distance of a falling body is proportional to the square of time lapsed), such variables are said to have perfect correlation which is expressed by $r = 1.00$. On the contrary when there is no correlation between the two variables, the correlation is said to be zero or neutral and expressed by $r = 0$. When r lies between 0 and 0.25, the correlation is considered to be of doubtful value ; when r is between 0.25 and 0.50, the correlation is considered to be fair : it is good when r lies between 0.50, and 0.75 and best (or excellent) when r is between 0.75 and 1.00. Further correlation is positive when both variables increase simultaneously and negative if one increases as other decreases.

The correlation co-efficient (r) can be calculated by any of the three formulae given below. However, it is important that for computing the value of r, at least a sample of $n = 30$ pairs of values should be obtained,

(*i*) $$r = \sqrt{\frac{(a\Sigma y + b\Sigma xy) - \bar{y}\Sigma y}{\Sigma y^2 - \bar{y}\Sigma y}}$$

the sign of r is same as of co-efficient b.

(*ii*) $$r = \sqrt{1 - \left(\frac{\sigma y_s}{\sigma y}\right)}$$

where $\sigma_s^2 = \dfrac{\Sigma y^2 - (a\Sigma y + b\Sigma xy)}{n}$ and $\sigma_y^2 = \dfrac{\Sigma y^2}{n} - \dfrac{(\Sigma y)^2}{n}$

the sign of r is same as of co-efficient b.

(*iii*) $$r = \frac{n\,\Sigma xy - (\Sigma x)(\Sigma y)}{\sqrt{[n\Sigma x^2 - (\Sigma x)^2][n\Sigma y^2 - (\Sigma y)^2]}}$$

the sign of r is obtained automatically by calculating this equation.

19.33. Statistical Analysers

In normal practice, the statistical quality control techniques are not used directly on the shop floor. The inspector notes various readings on shop floor by gauges and with the help of these readings, the curves are plotted separately and SQC techniques used to predict about process behaviour. Thus measurement and SQC aspects were performed at different times and different places. In such a system there could have been chances of errors due to the manual entry of arrays of measured values into the computer system to predict results by SQC procedures. However, with the introduction of microcomputer (chips), the cheap computing power has become available right at the measuring point. The conventional micrometer gauge has to be replaced by an electronic digital caliper which is linked *via* an interface to small table-top microcomputer. Usually an LCD readout and desired parameters are built in with this computer system. This type of inspection system does away with tedious manual record keeping for SQC. Manual charting is a cumbersome and slow process. In multidimensional checks, the problem of maintaining several charts is tedious one. The inspection productivity is greatly improved by this system. The operator picks up a part, measures it and then presses a button to register the result in the microcomputer. Thus inspection of a sample batch can be done in one continuous motion.

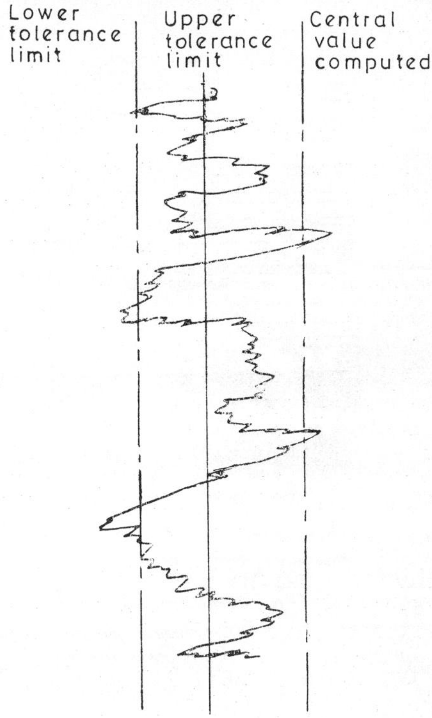

Fig. 19.35. Graphical display of measured results.

The digital display provides a readout precise to the last digit, and these values can now be stored for further processing as required for statistical quality control.

It must be understood clearly that the processes are variable and as such the data derived from measuring the processes also vary. The correct conclusions cannot be arrived at by examining individual data but only by using statistical methods which base their findings on accumulation of data.

The table-top microcomputer can be programmed for a number of SQC functions. Before starting checks, the inspector keys in the data relevant to the part number and the dimensional feature being checked (nominal dimension, upper and lower tolerance limits etc.). The inspection of the sample batch then starts, the measured results being automatically transferred to the microcomputer as explained earlier. The microcomputer system can print every measured result with a special indication whenever the result is out-of-tolerance. The results of measurements of a batch of parts, can be displayed graphically within two lines to represent the upper and lower tolerance limits. Fig. 19.35 shows such a graphical display. It would be appreciated that one can at a glance have the idea of complete process and is much easier than scanning through columns of the figures. The computer can also work out central value, scrap and rework percentages, if the rejection limits etc. are specified.

It is possible to draw a frequency histogram as shown in Fig. 19.36. Each bar's length indicates the number of parts accepted.

In the tolerance class, it is also possible to generate an exception report indicating the out-of-tolerance values for defectives, alongwith the identification like scrap, rework. The standard deviation value can be calculated and displayed. With statistical method information fed in computer, it can say whether lot is acceptable or not.

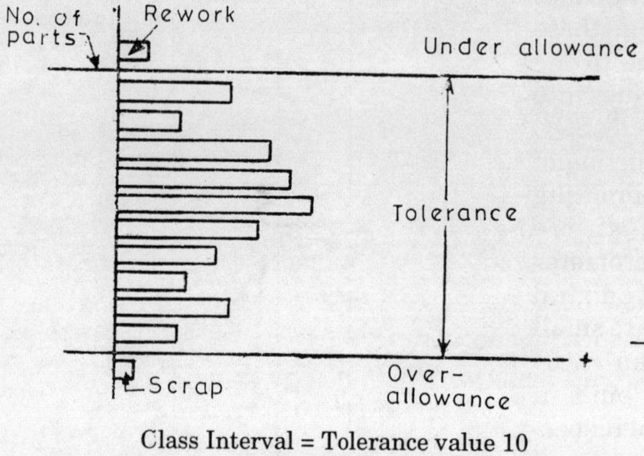

Class Interval = Tolerance value 10
Fig. 19.36

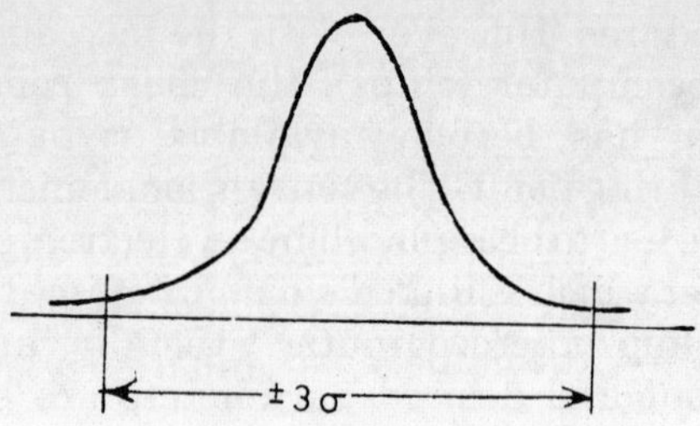

Fig. 19.37

Thus statistical analysis involving complex mathematics can be presented in an easily understood graphical form.

While the histogram shown in Fig. 19.36 goes some way to providing a summary of measured results, the general pattern can be difficult to assimilate.

By data reduction procedure it is possible summarise the key aspects of the distribution in one or two characteristics. The first step in data reduction is to convert the histogram into an analogue graphical form (frequency polygon) by incorporating suitable software in the computer system. The general form of this curve is symmetrical and bell shaped (Refer Fig. 19.37). The highest point of the curve corresponds to the maximum value of the frequency distribution (mode, or arithmetic mean of the entire distribution). The value of sigma can be graphically identified as the point where the curve changes direction, or it can better be calculated. The six times spread is important because statistically 99.73% of all measured values will occur between the three sigma limits. Most SQC software is, therefore, designed to display how the six sigma spread of a frequency distribution curve fits within the given tolerance limits of an inspected dimension. A six sigma value which is lower than the tolerance band indicates a process which is well under control. Additionally, the graphical depiction can provide the inspector with a visual guide for process control.

The displacement mode of the frequency distribution curve from the centre of the tolerance field can be easily calculated and this deviation from the mean is displayed so that machine operator can reset the tool to get the mode of the frequency distribution curve to coincide with the nominal dimension.

Microcomputer system is also able to convert a set of measured values into meaningful graphics (X and R-charts) which quickly show if the process needs correction. The operator thus gets on-line information about any change in setting, tool wear, vibration, uneven material characteristics etc.

Statistical process control techniques enable an engineer to assess whether or not variations in a process are due to chance.

The various facilities provided by microprocessor based statistical analysers can be summed up as follows :

— capability to handle metric and british system units.

— a measuring mode of either the actual, maximum, minimum, total indicator reading or nominal.
— total number of workpieces measured.
— both minimum and maximum measured value within the sample.
— sequence number and measure of each workpiece with out-of-tolerance values indicated by an asterisk.
— the difference between the minimum and maximum measured values.
— a complete pictorial histogram of the measured sample with optimum display arrangements.
— average of all measured values.
— standard deviation of all measured values.
— the value of 3-standard deviations which will include 99.7% of all pieces within the lot.
— a true statistical estimate of the per cent under and over tolerance in the lot being sampled.
— a total per cent out-of-tolerance based on the normal distribution computations.
— the fit factor of the sample data which is a measure of normality.
— a discretionary analysis with or without any stray pieces in the sample.

19.34. Microprocessors in SQC

SQC techniques involve collection, organisation, analysis, presentation and interpretation of data in a timely fashion. It thus calls for lot of record-keeping and performing calculations.

With the advent of microprocessors, the task of operator is considerably simplified. Smart microprocessor-based gauges and related software virtually eliminate the need for tedious record-keeping and accurate calculations. In addition to gathering and processing the data, these devices also generate a variety of graphic statistical tools, including histograms and control charts.

Data analyzers are typically microprocessor-based data loggers that include data-processing capability for automatically calculating the various statistical characteristics of sample data and using them to perform basic analysis routines.

Analyzers can generally be used with a variety of gauges, provided they have an appropriate electronic output and are plug-compatible. Most of these systems are now available from many of the traditional gauge makers, which have all moved from the age of dial indicators to the age of digital readouts and smart microprocessor-based gauges.

SQC software is generally incorporated in coordinate measuring machines (CMMs), which include SQC routines in the software.

Typical statistical analysers (multistatistical analyzer), have capability of carrying out automatic $\overline{X}$ and R analysis of multidimensional workpieces. These analysers accept upto 16 inputs simultaneously and print out setup and sample data, out-of-tolerance conditions, $\overline{X}$ and range charts plus a histogram for each dimension monitored. The measuring routines can be quickly changed, *via* simple coded input buttons to suit the application.

In-process analysers for monitoring individual process measurements have been developed. These are used with gaugeheads and include built-in clock, which keep track of times and dates for later documentation on a printout and prompt the operator through an audible signal to take sample measurements at set intervals.

After the sample size has been entered into the system, an AUTO setup mode automatically calculates the various control limits for $\overline{X}$ and R charts from subgroup measurements taken during a prerun. Subgroup size can be programmed from 2 to 20 samples. Alternatively, control limits can be entered manually. In the run mode, this analyzer either takes the manually entered control limits using data gathered during the prerun.

This analyser automatically spots trends and flashes appropriate signals before out-of-tolerance limits are reached. Process-control message as well as gauge-status information appear on the digital display to simplify decision-making for the operator. In addition, an optional printer produces hard-copy printouts of data, including histograms and $\overline{X}/R$ charts.

Computer Aided Dimension Analysis and Reporting System (CADAR). It is basically an electronic digital micrometer, which collects data using the Moire fringe transducer system and transfers the data from the output terminal on the measuring head *via* cable to the input/output interface board. The data is recorded and processed within the central processor and data in desired form is presented on a visual display unit and printer. The memory and facilities for data manipulation are contained within the central processor. Due to use of electronic digital micrometer, the measurement is done very fast and accurately. There is a great consistancy of measurement between different operators and reading errors are greatly minimised.

It can provide comparisons for checking variance from a standard, display readings in digits, tolerances for checking variance from pre-set upper and lower limits, maximum/minimum indications.

It can also provide a calculation of statistical mean, standard deviation etc. of the number of readings noted. The graphic displays give a control plot, distribution plots (curve or histogram).

Statistical analysers have also been designed to accept input from standard electronic probes from a group of different fixtures and provide immediate digital or printed readout of each measurement. After the last part in a batch is inspected the system produces a printout of certain critical parameters, including number of parts inspected, classifications in 20 classes, mean, and standard deviation. Parameters are programmed and modified from the keyboard, and the program can be recorded on an optional data cassette for storage and later reuse.

Computer-Aided Dimension Analysis and Reporting. The information obtained from a micrometer, could be fed to a microprocessor, and can be of great help and act as quality control tool. Analysis of a sequence of measurements held in a storage medium can indicate an adverse trend in the manufacturing process. Such systems acquire data, store it, analyse and transmit to VDU (visual display unit). Data is acquired from an electronic micrometer which operates on principle of Moire fringes. A keyboard is also provided to enable user to enter data manually from other sources such as dial gauges, verniers, etc. The data collect terminal collects data from batches of components. The data is collected feature by feature and subsequently processed by statistical techniques. It provides mean, range and standard deviation values for each feature of the component inspected. Routine assessment of a machine tool by examining batches of components enables the process engineer to determine the overall stability and capability of the machine. It enables user to identify untoward trends of process variability.

Computer offers a means to automate the evaluation and presentation of statistical data and a major benefit of this automation is the speed with which the required information becomes available to those controlling the manufacturing process. There is no need of manually plotting of QC chart which is time consuming, and prone to error.

Computer-aided gauging system, as self-contained computer-based gauging system designed specifically for dimensional measurement and data accumulation has been developed. Using a computer, the system works with electronic or air-electronic gauging equipment to provide automatic metering, digital and bargraph display of gauged dimensions, histograms and statistical representation of sizes, accept and reject counting, and interactive diagnostic routines. Menu prompting makes the system user-friendly, and the results can be displayed on a color CRT or associated printer.

In addition, the system can be used to feed correction information to a machine tool for in-process control. Each cell or group of machines will eventually be monitored by a suitable automatic gauge with such feedback capability. This will provide substantial help in the reduction of scrap and a step forward toward the automated plant.

An interesting example of the application of statistical method for machine-tool control is automatic size control through tool compensation by running $\overline{X}$ estimates to determine the requirement for tool compensation. This technique prevents excessive adjustments based on random errors.

Statistical-analysis routines can be implemented on any of a variety of computers ranging from special-purpose minis to corporate host computers.

Software has been developed to provide the ability to write inspection programs, which can include data collection with any of the associated gauges accompanied by accept/reject decisions on each feature measured and followed by statistical evaluation of the entire lot. A MIL-STD sampling plan automatically determines the sample sizes and suggests acceptance or rejection of the lot based on overall results. A "defect map" shows the frequency with which each feature falls above or below its tolerance zone. In addition, mean, σ, and histograms can be printed for any measured characteristic.

Comprehensive quality-assurance data-management (QADM) systems have been developed. These include control-chart analysis, descriptive statistics, and true-position analysis. Control-chart analysis provides a comprehensive set of Shewart control-chart techniques including $\overline{X}$, R, and σ to determine capabilities of machines, methods, and processes. Control-chart limits are generated automatically either from current data or from user-su plied standards.

Descriptive statistics provides general data-analysis calculations, such as mean, standard deviation, skewedness (a measure of how nonsymmetric the distribution is), kurtosis (a measure of peakedness distribution) and many other tests for normality. Frequency tables. histograms, probability, and normal inference plots summarize the data.

Computer-Aided Inspection and Reporting system (CAIR), for uniform collection, storage, and analysis of quality-related information have also been developed. The analysis programs include the following :

(*a*) Histograms that include several algorithms that automatically test the data set for its distribution characteristics, including skewness and kurtosis.

(*b*) Sequential plots to analyze trends. One of the main practical applications of this routine is to determine manufacturing-process capability.

(*c*) Feature analysis, which determines how a part positional data compares with tolerance boundaries (best-fit analysis).

(*d*) Elementary statistics, which compute mean, standard deviation and other statistical variables for each dimension.

(*e*) $\overline{X}$ and R control-chart plots for process monitoring and for machine-tool capability studies.

SOLVED PROBLEMS

Q. 19.1. Match the Parts A and B

	A	*B*
1.	characteristics evaluated on the basis of qualitative data such as good or bad, pass or fail, go or no go, accept or reject.	(*a*) Binomial distribution
2.	the value in a frequency distribution representing the most frequently occuring class.	(*b*) statistical control
3.	a large source of items from which sample is taken.	(*c*) regression analysis
4.	characteristics evaluated on the basis of actual measured values (such as dimensions)	(*d*) mode
5.	the measured inherent reproducibility of the product turned out by a process.	(*e*) poisson distribution
6.	a probability distribution that applies when random samples are identified in terms of two classifications such as conforming or nonconforming.	(*f*) attributes
7.	a measure of dispersion calculated as the sum of the squares of the deviation from the arithmetic mean divided by the number of observations.	(*g*) population
8.	a simple approximation of any term in a binomial expression, which reduces the necessary calculations.	(*h*) statistical control.
9.	chance variation (natural variation) inherent (always present) in a process.	(*i*) process capability.
10.	the collection, organisation analysis, interpretation, and presentation of data.	(*j*) variance.
11.	the condition describing a process from which all special causes of variations have been eliminated and only common causes remain.	(*k*) common cause.
12.	the study of the relationship of two or more variables.	(*l*) variables.

Answers

1. *f*	**2.** *d*	**3.** *g*	**4.** *l*	**5.** *i*
6. *a*	**7.** *j*	**8.** *e*	**9.** *k*	**10.** *h*
11. *b*	**12.** *c*.			

Q. 19.2. What do you understand by frequency distribution ? What are the common probability distributions ? What are their forms, applications, etc. in SQC ?

Ans. Every process follows an inherent pattern of distribution or process capability which must be identified before attempting process control through SQC. The frequency distribution is a graphical representation of frequency of occurrence versus the quality characteristics. It may be noted that raw data tabulated initially in random order does not provide much useful information, but when arranged in order of size or grouping in adjacent cells (to produce graphic representation of frequency distribution) can provide insights into the process.

One normally comes across following five types of probability distributions :

(*i*) **Normal distribution.** (Bell shaped curve) This is the most common distribution found in natural occurrences, as well as in manufacturing. It is symmetric curve, high in the middle and tapering off smoothly toward zero at both sides. It results from chance occurrence. (Fig. 19.38)

(*ii*) **Exponential Distribution.** Fig. 19.39 shows it graphically. It is represented as $y = \frac{1}{x} e^{-x/\bar{x}}$.

It is applicable when more observations are likely to occur below average than above.

(*iii*) **Weibull Distribution.** It is applicable in describing a wide variety of patterns of variation, including departures from the normal and exponential. (Fig. 19.40)

$$\beta = \text{Shape parameter.}$$

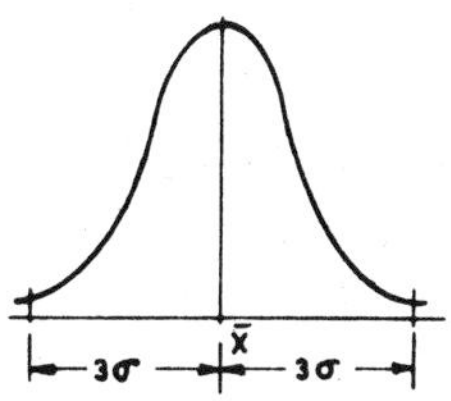

Fig. 19.38

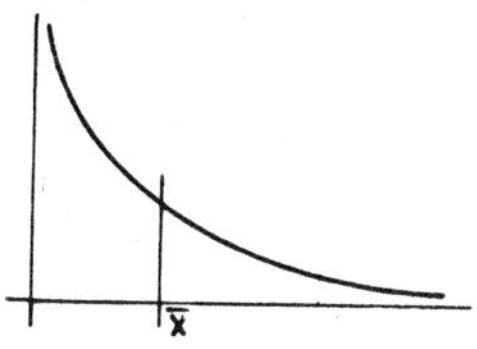

Fig. 19.39

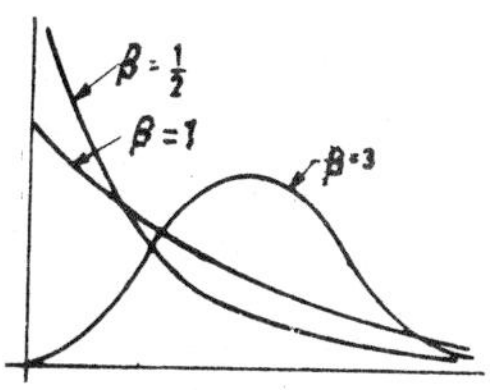

Fig. 19.40

(*iv*) **Poison distribution.** (Fig. 19.41) It is particularly applicable in establishing probabilities associated with sampling procedures. It is same as binomial but particularly

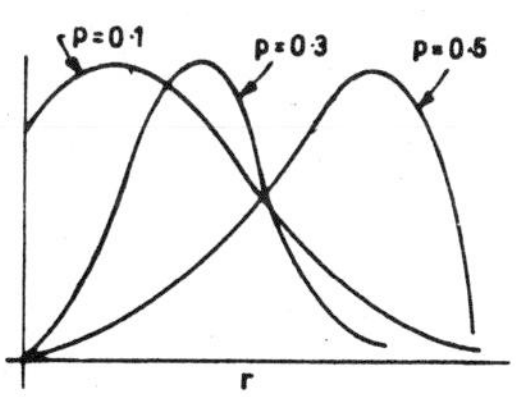

Fig. 19.41

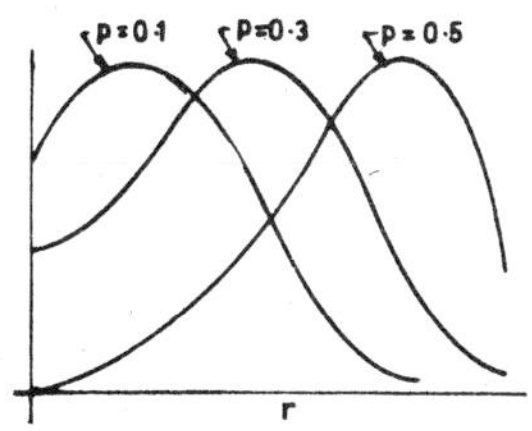

Fig. 19.42

applicable when there are many opportunities for occurrence of an event but a low probability (less than 0.1) on each trial.

p = number of trials, r = number of occurrences.

(*v*) **Binomial Distribution.** (Fig. 19.42) It is applicable in defining the probability of r occurrences in n trials of an event that has a probability of occurrences of p on each trial.

Q. 19.3. Explain clearly the differences between the random or chance causes of variation and the assignable causes of variation.

Ans. In any manufacturing process, there can be several types of variations caused by many factors. Some of these factors involve *chance* occurrences, that produce random variations inherent to the process. The random causes consist of many individual causes which are prevalent together at a time. While one random cause results in a minute amount of variation, many random causes act together to yield a substantial total. It has been established that variation due to chance causes follow normal distribution curve whose shape is dependent on the quality of the machine tool.

The assignable cause of variation may be either one or just a few individual causes. Some of the examples are operator blunder, faulty setup, defective raw material. Since the assignable causes results in a large amount variation, it can be easily detected.

Random causes being inherent part of the process can't be economically eliminated from a process. But the assignable variation can be easily detected and action taken to eliminate the cause.

As long as a process is operated under the influence of chance causes, it produces variation within a prescribed pattern well known and the process is said to be in statistical control. However with presence of assignable cause, the process is no more in control till this cause is eliminated.

If a process is operating only under the influence of chance causes, it will produce components within the prescribed control limits. As long as products fall within these limits, no adjustment is necessary. Any observation outside these limits can be due to assignable cause which needs to be detected and corrected.

With only random variation, the process is sufficiently stable to use sampling procedures to predict the quality of total production but with the presence of assignable variation, the process is not sufficiently stable to use sampling procedures for prediction.

Q. 19.4. (*a*) **Define control charts and how many types of control charts are in use.**

(*b*) **Define control limits and write the formulae for central line and upper and lower control limits of various types of charts.**

Ans. (*a*) **Control Charts.** To avoid defects, the process should be uniform *i.e.*, the variability in the product should be due to numerous small chance causes of variation rather than a few assignable causes of variation. Controlling a process means that cause of chance variation be restricted and assignable causes be eliminated. Control charts help to detect the presence of assignable causes of variation. Control charts needed for attribute and measurable type of characteristics are listed below :

Characteristic	Type of Charts
Attribute	1. C-Chart or number of defects per unit chart
	2. *np*-chart or number defective chart
	3. *p*-chart or fraction defective chart
Measurable	$\overline{X}$-R Chart-Average and Range Chart

(*b*) **Control limits.** A control chart consists of three horizontal lines *viz.*,

(*i*) A central line (*CL*) to indicate the desired standard or the level of the process :

(*ii*) Upper control limit (UCL) and

(*iii*) Lower control limit (LCL)

The central line, upper control limit and lower control limit for the various types of control charts are given below :

Central Line and Control Limits

Types of chart	*Central Line*	*UCL*	*LCL*
$\overline{X}$-*R* Chart	$\overline{X}$ or desired level	$\overline{X} + A_2\overline{R}$	$\overline{X} - A_2\overline{R}$
	$\overline{R}$	$D_4\overline{R}$	$D_3\overline{R}$
C-Chart	$\overline{C}$	$\overline{C} + 3\sqrt{C}$	$\overline{C} - 3\sqrt{C}$
np-Chart	np	$np + 3\sqrt{np(1-p)}$	$np - 3\sqrt{np(1-p)}$
p-Chart	$\overline{p}$	$\overline{P} + 3\sqrt{\frac{\overline{p}(1-\overline{p})}{n_i}}$	$P = 3\sqrt{\frac{\overline{p}(1-\overline{p})}{n_i}}$

$\overline{X}$ = Average of the characteristic of the sub-group (sample)

R = Average of the ranges of the characteristic of the sub-groups

C = Average number of the defects

np = Average number of defectives per sample of size N

$\overline{p}$ = Average proportion of defectives

n_i = Sample sizes.

Q. 19.5. Write short notes on (*a*) **process capability** (*b*) **process capability vs. tolerance.**

Ans. (*a*) *Process capability* refers to the minimum variation that has to be tolerated on any process under the existing situation. Thus process capability is quality-performance capability of the process with given factors under normal, in-control conditions. Without an action of fundamental importance like reconditioning the machine or replacement of the machine or improving operator's skill etc., the process capability cannot be improved. Two significant elements in this concept of process capability are :

(*i*) *Process factors* : The first consideration necessary to the concept of process capability is that a process is made up of a number of distinct factors. These factors include raw material, machine or equipment, the operator's skill, measuring devices and the inspector's skill. A change in one or more of these factors may change the process capability. Hence a process capability to be meaningful must be stated with respect to a given set of specifically listed process factors.

(*ii*) *Process conditions* : The second element implies that for process capability study to be meaningful, the process being analysed should be one that has measurements normally distributed and in a state of control.

(*b*) *Process capability Vs. tolerance.* When $\overline{X}$-*R* chart shows that the process is in control, it means that variation in the dimension subjected to control is within the limits of the natural variation. But it is not possible to say before hand whether the dimensions turned out in a state of statistical control meet the design specifications or not. The answer to this question depends on the nature of relationship that exists between the process capability and design specification. Thus it is not enough to bring the process within control ; it is also necessary that the process average and the process capability should be such as to permit the production of items almost all of which comply with the specifications.

Q. 19.6. What is the task of Statistical Quality Control (SQC) and what are the statistical quality tools available ?

Ans. The task of SQC is to distinguish the out-of-control variation of a process from the random causes always present. This is possible because the variations due to assignable causes, are large and persistent enough to alter the accustomed statistical pattern, or distribution, of the quality being examined.

Various statistical tools, for SQC available are :

(*i*) **Frequency distribution.** It provides a tabulation of the number of times a given quality characteristic occurs within the sample of product being checked. It may be used to show at a glance the average quality, the spread of the quality, and the comparison of the quality with specific requirements. It is used in the analysis of the quality of a given process/product.

(*ii*) **Control Chart.** It is a graphic method of evaluating whether a process is or not in a state of statistical control. If the prescribed limit on the control chart is exceeded, process need to be investigated to look for assignable cause of variation. It may be used to maintain control over a process after the frequency distribution has demonstrated that the process is in control.

(*iii*) **Sampling Tables.** These are specific set of procedures that usually consist of acceptance sampling plans in which lot sizes, sample sizes, and acceptance criteria, or the amount of 100% inspection is related. It is useful to assure the quality of material produced/procured.

(*iv*) **Special Methods.** (analysis of tolerances, correlation, analysis of variance). These are used for special analysis of engineering designs or process troubles. It is important to note that statistical tools provide an indication only of change. These can provide a signal to alert someone that a change in normal pattern has occurred but the possible cause has to be investigated based on experience. Various statistical tools may be used separately or in combination.

Q. 19.7. Write short notes on :

(*a*) **Acceptance sampling plans**

(*b*) **Sampling plans.**

Ans. (*a*) **Acceptance sampling plans.** The function of inspection in industry is to determine conformity to specification. There may be inspection of incoming materials, inspection of processes at several stages, and final inspection of finished product either by the producer or consumer or both. The usual problem is how much of inspection ? Will it always be possible to examine each and every unit of product ? These questions lead to the possibility of sampling inspection.

In some cases complete inspection of all the units in a lot may not be possible due to urgent demand from the production line.

Insistence upon complete inspection may result in hold-ups of the production line for want of material and the consequent cost of the production delays may be far greater than the cost of acceptance of defective product.

The evaluation of quality may sometimes be time consuming or costly and appraisal cost may be more than the failure cost arising out of passing defective unit.

Again inspected, the inspection fatigue impairs the inspector's accuracy. Due to this fatigue, screening will rarely, in practice, eliminate all defective items.

Also, there are situations where 100 per cent inspection cannot be employed *e.g.* life tests and other destructive tests.

Thus, in practice, there exists many situations where the quality of a product has to be evaluated by inspecting a few but not all the items. Such an inspection is known as sampling inspection and its principal uses in industry are :

(*i*) Determination of the quality and acceptability of incoming raw materials, component parts, etc.

(*ii*) Decisions as to the quality and acceptability of semifinished product for further processing as it passes from section to section within the factory.

(*iii*) Determination of quality of outgoing products.

(*iv*) Improvement and control of quality.

(*b*) **Sampling plans.** A sampling plan gives the sampling procedure to be followed for acceptance or rejection of product submitted for acceptance. Plans where item *quality* is defined by *attributes* and lot *quality* by fraction defective are called *attribute* sampling plans (Dodge-Romig and IS 2500 Part-I). Plans where items quality is an actual measurement and lot quality tolerances (mean dimension and dispersion for the whole lot) are called variable sampling plans (IS 2500 Part-II).

Q. 19.8. What do you understand by statistical quality control ?

Ans. The fundamental purpose of SQC is to guard against the manufacture of unacceptable products by utilising the process of inspection, analysis of data and action. SQC provides a scientific approach to collection of data, analysis of data so compiled and to draw inference therefrom in order to install suitable remedial measures and procedures.

There is no such thing as constancy in nature ; all things vary. All operations whether done by hand or machine are also subject to variability. This variability may be caused by inherent variability of the process (known as chance or random variability) or due to assignable (non-random) causes. SQC techniques provide criteria to discriminate between random and non-random variability. Once the statistical techniques have disclosed the existence of assignable causes, the determination of the assignable causes for such non-random variations requires engineering judgement, experience and at times experimentation. The decision to remove an assignable cause, when discovered, in neither a statistical nor an engineering decision, but one of the economics of management.

Q. 19.9. What is the difference between traditional inspection and statistical quality control (SQC) ?

Ans. Traditional inspection techniques can catch defects only after they have occurred. No amount of inspection can prevent the manufacture of scrap.

Defect-free output requires that the manufacturing process be controlled, which is achieved by statistical quality control. SQC studies prewarn problems in on going processes in a timely fashion in order to keep them in satisfactory control. Thus statistical quality control aims at defect *prevention* instead of defect detection.

Q. 19.10. What are the underlying principles of SQC ?

Ans. The underlying principles of SQC are set standards of quality ; establish the capability of the process ; monitor the process to detect any significant changes that indicate that the process is going out of control, and make corrections to keep the process in control.

Q. 19.11. Draw a comparison in tabular form between random and assignable causes of variation.

Ans.

Random (chance) causes	*Assignable causes*
Consists of many individual causes.	Consists of one or just few individual causes.
Any one random cause results in a minute amount of variation (but many random causes act together to yield a substantial total).	Any one assignable cause can result in a large amount of variation.
Examples are human variation in setting control dials, slight variation in machines and in raw material.	Examples are operator blunder, a faulty setup, or defective raw materials.
Random variation cannot economically be eliminated from a process.	Assignable variation can be detected ; action to eliminate the causes is usually economically justified.
When only random variation is present, the process is operating at its best ; if defectives are still being produced, a basic process change must be made or the specifications revised to reduce the defectives.	If assignable variation is present, the process is not operating at its best.
An observation within the control limits of random variation means that the process should not be adjusted.	An observation beyond control limits usually means that the process should be investigated and corrected.
With only random variation, the process is sufficiently stable to use sampling procedures to predict the quality of total production or make process optimization studies.	With assignable variation present, the process is not sufficiently stable to use sampling procedures for prediction.

Q. 19.12. Explain the difference between statistics, SQC, and sample inspection ?

Ans. *Statistics* refers to the collection, organization, analysis, interpretation, and presentation of data. SQC involves analysis of the characteristics of a production output by inference from sampling the output. The analysis, in turn, is used to make the necessary corrections that will keep the process under control. SQC attempts to detect trends that will ultimately lead to the production of rejects. The objective is to identify the causes for such trends and then make the necessary corrections before any rejects are produced.

In sample inspection, decision regarding the acceptance or rejection of the lot is based on the characteristics of the samples.

Q. 19.13. Explain what you mean by variation.

Ans. It is a fact that every process and, its output are subject to variation. No matter how good or precise a manufacturing process may be, no two items coming from it will be perfectly identical in every way. They may look alike, pass all the same criteria for a satisfactory product, and even function identically, but very close inspection will reveal minute measurable differences over some finite span of time.

Q. 19.14. When a process can be said to be in statistical control ?

Ans. In a typical manufacturing process, the variations may result from causes like-human factors, tool wear, machine wear, differences in raw material, environmental factors, etc. Some of these factors involve chance occurrences that produce random variations inherent to the process. In fact, some stable "system of chance causes" is inherent in any particular scheme of production and inspection. Variation within this stable pattern is inevitable, as long as a process operates within this established pattern, it is said to be in statistical control.

Q. 19.15. What is the function of SQC ?

Ans. Any variation outside the stable pattern is an indication that the process is no longer in statistical control and requires attention. This task of SQC is to distinguish the out-of-control variations of a process from the random "noise" that is always present. That distinction can be made only if the out-of-control variations are large enough or persistent enough to alter the accustomed statistical pattern, or distribution, of the quantity being examined.

The out-of-control variations are said to be due to "assignable" causes, in distinction from random causes. Ultimately, the task of SQC is to identify these assignable causes so that they can be dealt with in a timely fashion.

Q. 19.16. Explain the four main statistical tools used in SQC techniques.

Ans. Statistical tools can provide an indication only of change. A signal to alert that a change in the normal pattern of behaviour has occurred and that possible causes should be investigated. SQC does not provide an absolute measure of process parameters, nor does it necessarily provide a clear-cut indication of exactly what needs to be done.

The following statistical tools may be used separately or in combination.

(*i*) **Frequency Distribution.** It is a tabulation or tally of the number of times a given quality characteristic occurs within the sample of product being checked. As a picture of the quality of the sample, it may be used to show at a glance the average quality, the spread of the quality, and the comparison of the quality with specific requirements. This tool is used in the analysis of the quality of a given process or product.

(*ii*) **Control Chart.** It is a graphic method of evaluating whether a process is or is not in a state of statistical control. When the curve of the chart approaches or exceeds the limits, some change in the process may be used to maintain control over a process after the frequency distribution has demonstrated that the process is "in control".

(*iii*) **Sampling Tables.** It is a specific set of procedures that usually consist of acceptance sampling plans in which lot sizes, sample size, and acceptance criteria, or the amount of 100% inspection is related. This tool is used when assurance is desired for the quality of material either produced or received.

(*iv*) **Special Methods.** Techniques like analysis of tolerances, correlation, and the analysis of variance, are used for special analyses of engineering designs or process troubles.

Q. 19.17. The inherent pattern of behaviour of any process (or the process capability) can be described in terms of a frequency distribution. What are the various ways of showing frequency distribution ?

Ans. (*i*) Frequency distribution can be shown in simplest way in the form of a vertical bar chart, or histogram. Each bar width, or cell, of the histogram represents the interval of values chosen to represent the frequency distribution of measured values. The height of the bars is proportional to the number of occurrences for a particular cell.

(*ii*) Frequency distribution can also be shown as a simple plot of frequency of occurrence on one axis against the quality characteristics being measured on a coordinate axis. Such a frequency distribution curve matches the envelope of a histogram for the same data.

Q. 19.18. Show the characteristics of a normal distribution on a diagram.

Ans. (Refer Fig. 19.42). In the case of a normal distribution, 99.73% of population lies within the spread of $\pm 3\sigma$ limits.

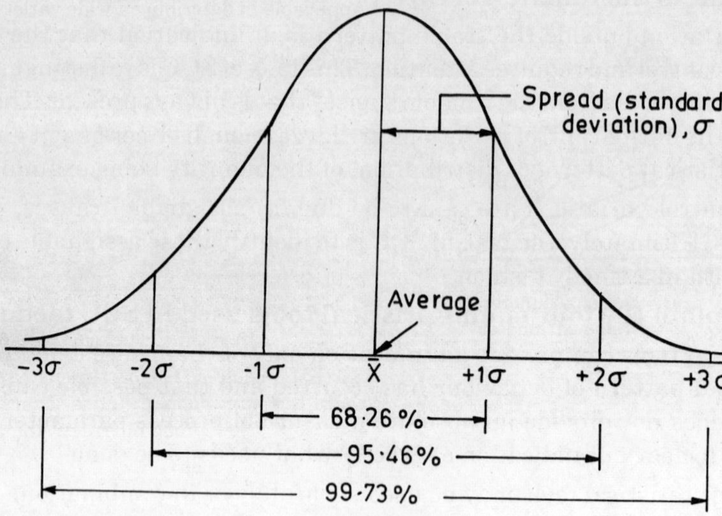

Fig. 19.42

Q. 19.19. What are the various types of frequency distributions normally encountered ? Discuss their characteristics and applications.

Ans. The most common distribution found in natural occurrences as well as in manufacturing is the bell-shaped curve. It is symmetric ; high in the middle and tapering off smoothly toward zero at both sides. This "normal-distribution" curve results from chance occurrences.

Other types of distributions are commonly associated with manufacturing processes ; exponential, applicable when more observations are likely to occur below the average than above ; Weibull, applicable for a wide variety of patterns of variation ; and Poisson, particularly applicable in establishing probabilities associated with sampling procedures.

Table below shows the form of application of various distributions.

Common Probability Distributions

Distribution	*Form*	*Applications*
Normal	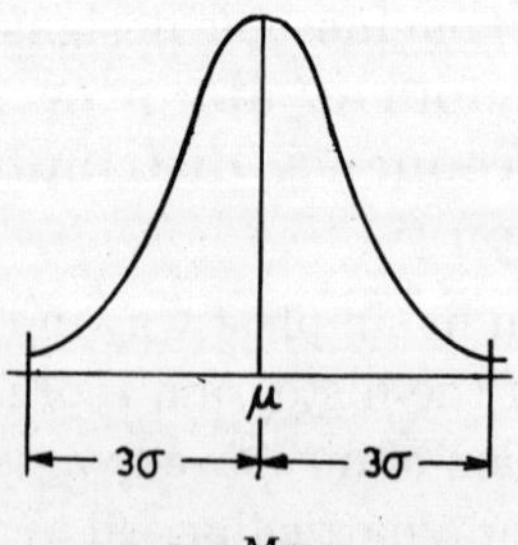 μ = Mean σ = Standard distribution	Applicable when there is a concentration of observations about the average and it is equally likely that observations will occur above and below the average. Variation in observations is usually the result of many small causes.
Exponential	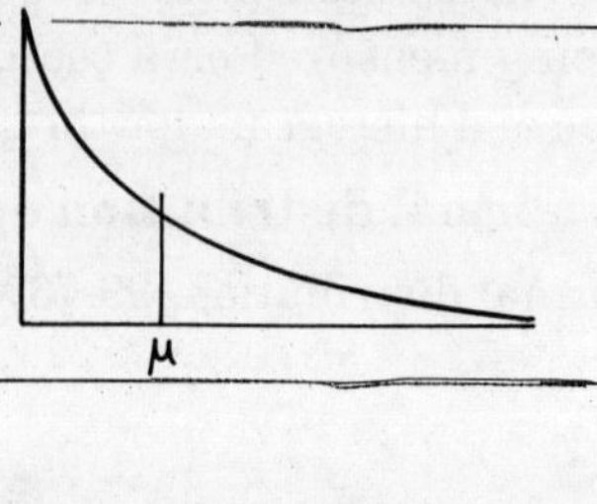	Applicable when it is likely that more observations will occur below the average than above.

Weibull	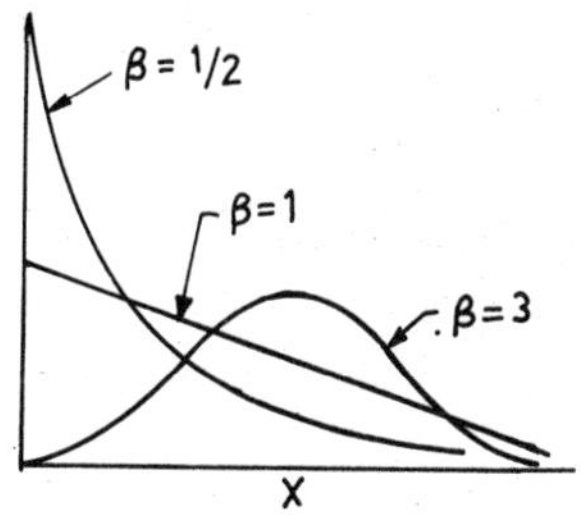 α = Scale parameter β = Shape parameter	Applicable in describing a wide variety of patterns of variations, including departures from the natural the exponential.
Poisson	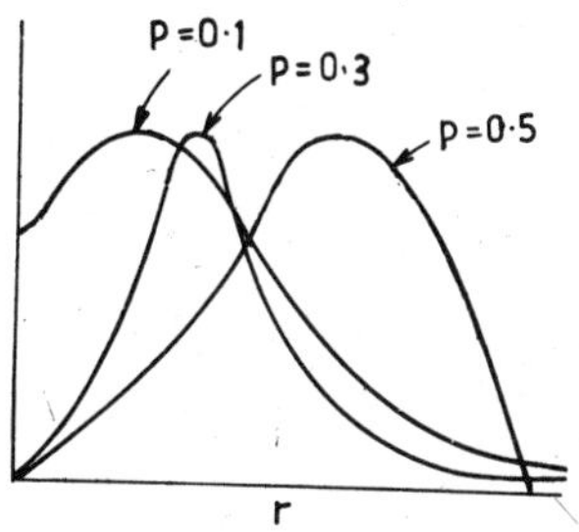 *r* = Number of occurrences *p* = Probability of occurrence	Same as binomial but particularly applicable when there are many opportunities for occurrence of an event but a low probability (less than 0.10) on each trial).
Binomial	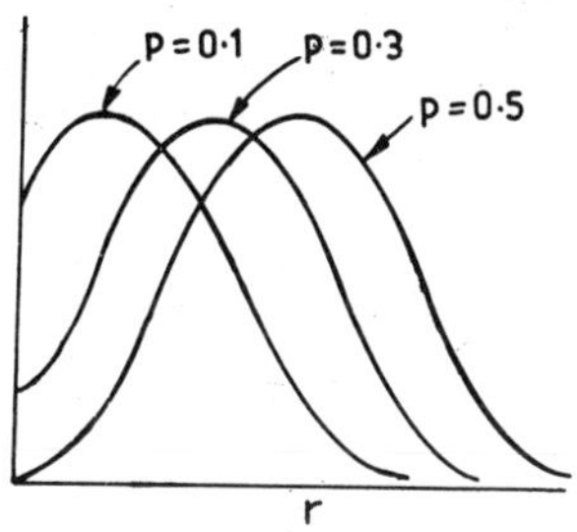 *r* = Number of occurrences *p* = Probability of occurrence	Applicable in defining the probability of *r* occurrences in *n* trials of an event that has a probability of occurrence of *p* on each trial.

Q. 19.20. What are the two important characteristics of any distribution ?

Ans. The two most useful characteristics of distribution are its central tendency, or its most representative value, and the spread, or dispersion, of the data.

Central tendency is commonly indicated as the mean, the median, the mode, or the midrange of the distribution. The mean—more properly called the arithmetic mean—is simply the average value and is calculated by adding up all the numerical values represented in the distribution and dividing the sum by the number of such values, or the total frequency.

The median is the middle value of the distribution ; half the values are below it ; half are above. The mode is the value occurring most frequently in a distribution. The midrange is the average of the smallest and the largest readings.

In a perfectly symmetrical distribution, such as the normal distribution, these four measures are all equal. The arithmetic mean, usually represented by a bar over the symbol for the variable being measured (such as $\overline{X}$), is the most commonly used measure for central tendency.

The most common measure of spread is the root mean square (RMS) deviation, referred to as the standard deviation (σ). The standard deviation is calculated as the square root of the sum of the squares of deviations divided by the number of measurements.

$$\sigma = \sqrt{\frac{\Sigma (X - \overline{X})^2}{n}}$$

The standard deviation is one of the most significant characteristics of a distribution because it provides an arbitrary measure of the probability that a certain number of measurements will fall within a particular range of values.

Q. 19.21. SQC uses sample measurements and statistical analysis of the sample measurements to estimate corresponding characteristics, or parameters for the universe. Explain how it works.

Ans. (*i*) The behaviour of the process is estimated on the basis of the behaviour of samples taken from the process. Taking judicious samples at appropriate intervals enables the process to be kept under statistical control.

There are some very convenient relationships between the statistics obtained from a sample and the parameters of the universe from which those samples are taken. The larger the sample, the closer it is to the universe ; therefore, its statistics more closely reflects the parameters of the population.

One significant relationship is the central-limit theorem of probability, which states that, under very general conditions, the sample means means will follow a normal distribution regardless of the distribution of the underlying population. In other words, no matter what is the shape of the distribution of a universe, the distribution of average values ($\overline{X}$) of samples of size n drawn from that universe will tend toward a normal distribution as the sample size n grows without bound.

It has been observed that, even if the sample size is only 4 and the universe is far from normal, the distribution of the averages of samples will be very close to normal. This very powerful relationship permits probability statements to be made about a process when the actual population distribution isn't known.

Another important relationship among statistical measures is that the ratio of the average range ($\overline{R}$)—from samples of size n from a normal universe-to the estimated standard deviation of the universe follows a predictable pattern as a function of sample size.

Q. 19.22. The Schewart control charts are significant and widely used tools of SQC. Explain them.

Ans. The power of control chart lies in its ability to separate out the assignable cause of quality variation. The purpose of setting up control charts is to distinguish between the inherent random variability of a process and the variability attributed to assignable causes. They accomplish this through an appropriate choice of control limits calculated from the laws of probability.

Table below shows comparison of 3 types of control charts

Statistical measure plotted	*Average $\overline{X}$ and range R*	*% defective (p) Number of defectives (np)*	*Defects per unit (u) Number of defects (c)*
Types of data required	Variables data (measured values of a characteristic)	Attribute data (number of defective units of product)	Attribute data (number of defects per unit of product)
General field of application.	Control of individual characteristics	Control of overall fraction defective of a process.	Control of overall number of defects per unit.
Significant advantages	1. Provides maximum utilisation of information available from data.	1. Data required are often already available from inspection records.	Same advantages as *p* chart but also provides a measure of degree of defectiveness.
	2. Provides detailed information on process average and variation for control of individual dimensions.	2. Easily understood by all personnel.	
		3. Provides an overall picture of quality	
Significant disadvantages.	1. Not understood unless training is provided. Can cause confusion between control limits and tolerance limits.	1. Does not provide detailed information for control of individual characteristics.	Does not provide detailed information for control of individual characteristics.
	2. Cannot be used with go/no-go type of data.	2. Does not recognize different degrees of defectiveness in units of product.	

Q. 23. The various types of control charts depend on the characteristics being tested. Explain them.

Ans. The $\overline{X}$ chart plots the average of the measurements in the sample. This tends to indicate the aim or centering of the process.

The *R* chart plots the range of the measurement in the sample, indicating the variability about the aim of the process.

The *p* chart plots the per cent defective in the sample.

The *c* chart plots the number of defects in the sample.

The $\overline{X}$ and *R* chart are used to track variables (characteristics evaluated on the basis of actual measured values, such as dimensions). These are usually plotted as a pair with the $\overline{X}$ chart above the *R* chart. The *p* and *c* charts are used for attributes, characteristics that are evaluated on the basis of qualitative data, such as go or no-go.

Q. 24. What is a control chart and how it is made ?

Ans. A control chart plots a series of sample measurements taken at regular intervals and includes a line representing the grand average, or average of sample averages, and a set of upper and lower control limits. Typically, successive samples are plotted along the horizontal axis, and sample values are plotted on the vertical axis.

In order to set the control limits, it is first necessary to establish the capability of the process under consideration. Typically, the process capability is calculated as ± 3 standard deviations using the average of the sample averages in the standard-deviation formula. To ensure that the sample standard deviation reflects a conservative standard deviation for the population, the denominator in the formula is modified by replacing n with $n - 1$.

$$s = \sqrt{\frac{\Sigma (X - \overline{X})^2}{n - 1}}$$

Ideally, the process capability should equal 75% of the product tolerance to allow some shift of the central tendency as production proceeds. The upper and lower control limits (UCL and LCL) are then set at $\pm\, 3\sigma$ from the mean. A mean value is also computed for all the measurements and is used as a horizontal central line on a control chart.

To establish the control chart, samples are taken periodically from the process, and the required measurement is made on each sample item and recorded, along with identification of the source, time of operation, and any other relevant information that could shed light on problems that may arise. Usually, the samples are taken in subgroups of four or five, with the items in each subgroup closely spaced in time so that conditions are relatively constant between items. To arrive at valid control limits, 20-25 subgroups at more widely spaced intervals should be collected.

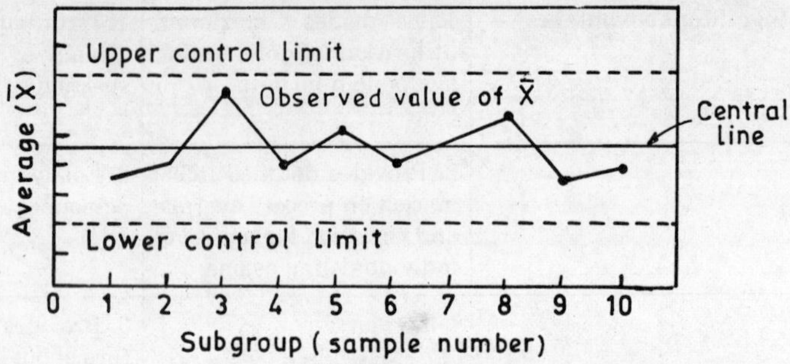

Fig. 19.43

Q. 25. How the lack of control is indicated on control charts ?

Ans. Once the control charts are established, lack of control is indicated by the following conditions :

— A point falling outside the control-limit lines.

— A run of more than 8 points on the same side of the central line.

— A trend with at least 7 points.

— 13 out of 15 points falling on the same side of the central line.

All of these conditions are an indication that some assignable cause is affecting the process, and an investigation should be conduced to discover and remove the cause.

Q. 26. Given below are some commonly used terms in Quality Control in part B and their description in part A. Match them :

A	B
1. A measure of the spread of the process output.	(*a*) Sub-group
2. The condition describing a process from which all causes of variation have been eliminated and only common causes remain evidenced on a control chart by the absence of points beyond the control limits and by the absence of nonrandom patterns or trends within the control limits.	(*b*) Binomial distribution
3. The collection organisation analysis interpretation and presentation of data.	(*c*) Cell

4. A sample of collection of values chosen from the process output as representative of a particular period or point in time and analyzed together.	(*d*) Process capability
5. Characteristics evaluated on the basis of actual measured values such as dimensions.	(*e*) Control chart
6. A measure of dispersion calculated as the sum of the squares of the deviation from the arithmetic mean divided by the number of observation ; same as the square of the standard deviation ; designated as s^2.	(*f*) Standard deviation
7. Characteristics that are evaluated on the basis of qualitative data such as conforming or nonconforming pass or fail go or no-go present or absent.	(*g*) Variance
8. A probability distribution that is applied when random samples are identified in terms of two classification such as 'accept' and 'reject'.	(*h*) Common cause
9. The measured inherent reproductibility of the product turned out by a process.	(*i*) Statistics
10. The interval of values chosen to represent the frequency distribution of measured values ; the width of each bar in a histogram.	(*j*) attributes
11. Chance variation that is always present in a process the natural variation inherent in the process itself.	(*k*) Statistical control
12. A graphic representation of variability of a process showing plotted values of sample measurements from its output with respect to time. It includes a central line representing the average value standard value of the measurements and one or more control limits.	(*l*) Variables.

Answer to Question 19.26

1. (*f*)	**2.** (*k*)	**3.** (*i*)	**4.** (*a*)	**5.** (*l*)
6. (*g*)	**7.** (*j*)	**8.** (*b*)	**9.** (*d*)	**10.** (*c*)
11. (*h*)	**12.** (*e*).			

Q. 27. Given below are some of the terms commonly used in quality control in Part B and their description in Part A. Match them :

A	*B*
1. A line on a control chart representing the maximum extent of variation in the measurements being plotted that could be reasonably expected to occur if only common causes of variation were present.	(*a*) Normal distribution

2. A probability distribution in which the characteristics being measured can take on only certain specific values.	(*b*) Random samples
3. A way of describing the output of a common-cause system of variation in which the behaviour of individual values is not predictable but the outcome as a group form a pattern that can be described in terms of locations spread and shape.	(*c*) Poisson distribution
4. A tabulation of data arranged according to size.	(*d*) Distribution
5. A vertical bar chart of a frequency distribution.	(*e*) range
6. A symmetrical bell-shaped frequency distribution for variables ; commonly used to establish control-chart limits.	(*f*) Control limit
7. A simple, approximation of any term in a binomial expression, which reduces the necessary calculations.	(*g*) regression analysis
8. A large source of items from which the sample is taken.	(*h*) histogram
9. Samples chosen in such a way that any member of the process output is equally likely to be selected without regard for any ordering property such as sequence of production that might be present.	(*i*) rational subgroups
10. The difference between the highest and lowest measurement value in a subgroup.	(*j*) Discrete distribution
11. Subgroups chosen so that the variation represented within each subgroup is as small as feasible for the process and any changes in the process performance will appear as differences between subgroups.	(*k*) Population
12. The study of the relationship of two or more variables.	(*l*) Frequency distribution

Answers to Questions 9.27

1. (*f*)	**2.** (*j*)	**3.** (*d*)	**4.** (*l*)	**5.** (*h*)
6. (*a*)	**7.** (*c*)	**8.** (*k*)	**9.** (*b*)	**10.** (*l*)
11. (*i*)	**12.** (*g*).			

Unsolved Problems

Q. 1. Define quality. Explain how will you organise for effective quality management.

Q. 2. Discuss the various types of quality costs. Explain how will you control the cost of quality ?

Q. 3. Explain what do you understand by Management of Product Quality. Discuss the concept of Total Quality Management.

Q. 4. Explain the difference between control charts for variable and control charts for Attributes. Draw a p chart and comment on it. 25 samples of 100 items were inspected, they reveal the following.

Sample No. 1, 2, 3, 4, 5, 6, 7, 8, 9, 10, 11, 12, 13, 14

Number of defectives 14, 22, 25, 15, 20, 14, 12, 24, 10, 17, 35, 36, 16, 23

Sample No. 15, 16, 17, 18, 19, 20, 21, 22, 23, 24, 25, Total number of defectives

Number of defectives 14, 6, 7, 33, 17, 34, 11, 16, 25, 36, 18, 500

Q. 5. What is the difference between $\bar{c}$ chart and $\overline{np}$ chart. Draw a $\bar{c}$ chart and comment on it. Inspection of 20 compressors is given below :

Compressor No. 1, 2, 3, 4, 5, 6, 7, 8, 9, 10

Number of defects 2, 5, 7, 3, 5, 4, 1, 6, 3, 2

Compressor No. 11, 12, 13, 14, 15, 16, 17, 18, 19, 20

Number of defects 4, 4, 3, 7, 12, 2, 3, 1, 4, 2

Q. 6. Draw control charts for Mean $\overline{X}$ and Range R from the following data relating to 20 samples of size 5 each and comment whether it is in Statistical Quality Control or not.

S.No., 1, 2, 3, 4, 5, 6, 7, 8, 9, 10, 11, 12, 13, 14, 15, 16, 17, 18, 19, 20

$\overline{X}$, 38, 33, 24, 36, 27, 30, 31, 27, 24, 29, 32, 22, 21, 28, 28, 24, 30, 25, 37, 31

R, 15, 1, 22, 24, 28, 33, 21, 20, 29, 18, 30, 11, 28, 21, 15, 18, 19, 33, 17, 17

For a sample of size 5, $D_3 = 0.000$, $D_4 = 2.115$, $A_2 = 0.577$.

Q. 7. Discuss the procedure of (1) Single sampling plan (2) Double Sampling Plan (3) Item by item Sampling Plan for Attribute. Explain the OC curve and Producers Risk and Consumer's Risk.

Q. 8. Find a single-sample fraction defective sampling inspection plan that will come close to having $p_1 = 0.03$ $p_2 = 0.10$,

$$\alpha = 0.05 \text{ and } \beta = 0.10.$$

Q. 9. Develop a Double Sampling Plan that will come close to having $p_1 = 0.03$, $p_2 = 0.10$, $\alpha = 0.5$ and $\beta = 0.10$.

Q. 10. Write an essay on the role of statistical quality control in modern production technology.

Q. 11. How the process charts check the working of machines and put-down the number of rejected parts ?

Q. 12. (*a*) Explain the first and second type of errors :

(*b*) Where it is desirable to use following charts ?

(*i*) Average Chart ($\overline{X}$-charts) ?

(*ii*) Defective Charts (p-chart)

(*iii*) Range Charts (R-charts).

Q. 13. (*a*) What do you understand by operating characteristic curve ?

(*b*) Hot it can be said that sampling inspection gives results than 100% inspection ?

(*c*) Describe various sampling techniques.

Q. 14. Control charts for $\overline{X}$ and R are maintained on the tensile strength in kilograms of a certain yarn. The subgroup size is 5. The values of $\overline{X}$ and R are computed for each subgroup. After 25 subgroups $\Sigma \overline{X} = 2005$, $\Sigma R = 515$. Compute the values 3 sigma limits for the $\overline{X}$ and R-charts, and estimate the value of σ on the assumption that the process is in statistical control.

Q. 15. What do you understand by the term "Process is in control" ? Control charts for $\overline{X}$ and σ are maintained on the weight in ounces of the contents of ascertain container. The subgroup size is 10. After 20 subgroups $\Sigma\overline{X} = 6916$ and $\Sigma\sigma = 8.64$, compute

(*a*) 95% confidential limits on $\overline{X}$ chart.

(*b*) 3 sigma limits for $\overline{X}$ and σ charts.

(*c*) σ

Assume that the process is in Statistical Quality Control.

State any other assumption you would like to make.

Q. 16. (*a*) Write a short note on the process capability.

(*b*) What is meant by the "natural tolerances of the process"?

(*c*) Control chart for $\overline{X}$ and *R* are to be established at a work station at which a hole of a certain diameter being drilled in a part. The analyst begins by estimating the magnitude μ and *R*. He does this by takings 25 samples of size 5, measuring the diameter of the holes contained in each sample, and calculating the sample average and ranges. These averages and ranges are given in the following table :

Sample Number	*Sample average mm*	*Sample range mm*
1	35.492	0.05
2	35.546	0.04
3	35.488	0.09
4	35.464	0.08
5	35.516	0.15
6	35.496	0.06
7	35.548	0.07
8	35.544	0.10
9	35.506	0.03
10	35.470	0.05
11	35.534	0.08
12	35.526	0.12
13	35.468	0.20
14	35.480	0.04
15	35.488	0.08
16	35.506	0.07
17	35.534	0.09
18	35.470	0.10
19	35.608	0.11
20	35.540	0.23
21	35.506	0.05
22	35.498	0.11
23	35.476	0.06
24	35.520	0.10
25	35.514	0.06
Total	887.738	2.22

(*i*) Compute the values of central line and the 3 sigma limits for the control charts.

(*ii*) What are the natural tolerances of this process if the individual items are normally distributed ?

(*iii*) The specifications for the dimension under consideration are 35.500 ± 0.076. If the individual items are normally distributed, what percentage defective output can be firm expect if the process remains in control at present level ?

Q. 17. $\bar{X}$ and R charts have been maintained on a process with a subgroup of size 4. Past data indicate an $\bar{X}$ of 20 and a σ of 2.0. The following 3. sigma limits are used $UCL_{\bar{x}} = 23.0$, $LCL_{\bar{x}} = 17.0$, $UCL_R = 9.4$ and $LCL_R = 0$.

A suggestion is made to increase the sample size from 4 to 5 but to maintain the same limits as before on the $\bar{X}$ and R charts. Discuss the consequences that would be expected. If the suggestion is adopted, would points be more or less likely than before to fall outside the control limits on the $\bar{X}$ chart ? On the R-chart ? Explain your answers.

Q. 18. Control charts for $\bar{X}$ are maintained on the tensile strength of a particular yarn, the subgroup size is 5. After 25 subgroups $\Sigma\bar{X} = 626.8$ and $\Sigma R = 120$. If there is a single specification limit L = 16, make the necessary calculations, based on the control chart, to determine whether or not this product would meet its specification. If not, approximately what percentage is expected to be below the lower specification limit ? Assume that the normal distribution is applicable.

Q. 19. Control chart for $\bar{X}$ and R are maintained on the outside diameter of the bearing. The subgroup size is 4. The values of $\bar{X}$ and R are computed for each subgroup. After 20 subgroups $\Sigma\bar{X} = 43.275$, $\Sigma R = 0.286$. The dimension specification for the product is 2.150 ± 0.020. If the dimension falls above the specification limit, rework is required ; if below the specification limit, the bearing must be rejected. If the process is in statistical quality control and normally distributed, what can you conclude regarding is ability to meet the specification ? Can you make any suggestion ?

Q. 20. Over the past eight or nine months an assembly line has been producing 400 units of a product per day. Final inspection records show that on average 20 units a day are defective. A survey of the daily reports shows that on an occasional day as few as one or two are defective. (*a*) Do these records indicate statistical control of quality ? (*b*) What help might be obtained if any, from these records in improving quality performance at this line ?

Q. 21. A company purchases small steel balls in cartons that usually contain several thousand balls. Each shipment consists of a number of cartons. As part of the acceptance procedure for these balls, 400 balls are selected at random from each carton and are subjected to visual inspection for certain defects. In a shipment of 10 cartons, the respective percentages of defectives in the sample from each carton are 0.5, 0, 0.75, 2, 0, 0, 0.25, 0, 0.25 and 1.25. Does this shipment of balls appear to exhibit statistical control with respect to the quality characteristic examined in this inspection ?

Q. 22. A double sampling plan is as follows : (*a*) Select a sample of 2 from a lot of 20. If both articles inspected are good, accept the lot. If both are defective, reject the lot. If 1 is good and 1 defective, take a second sample of one article. (*b*) If the article in the second sample is good, accept the lot. If it is defective, reject the lot.

If a lot 20% defective is submitted what is the probability of acceptance ?

Q. 23. In a single sampling plan, N = 10,000 and $c = 1$. Find the sample size when the probability of acceptance of 0.5% defective lot is 0.025. Compute the average outgoing quality.

Q. 24. In a double sampling plan,

$$N = 5{,}000\ n_1 = 100,\ c_1 = 0,\ n_1 + n_2 = 200 \text{ and } c_2 = 1$$

(*a*) Compute the probability of acceptance of a 1% defective lot.

(*b*) Assume that a lot rejected by the proposed sampling plan will be 100% inspected and defectives found in accepted samples not to be replaced. What will be the AOQ if submitted lots are 1% defective.

(*c*) Consider both the inspection of samples and inspection of rejected lots, what will be the average number of articles inspected per lot if the submitted product is 1% defective ?

(*d*) Compute the average size.

Q. 25. A single sampling acceptance criteria for any desired $P_{0.50}$ is given by approximate relation :

$$n = \frac{c + 0.67}{p_{0.50}}, \text{ in which}$$

n = sampling size

c = acceptance number

$p_{0.50}$ = fraction defective for which the probability of acceptance is 0.5.

A vendor selects $p_{0.50} = 0.025$. The vendor has 3,000 lots, of which 1,000 are of 2% quality ; 1,000 of 2.5% quality ; and the rest of 3% quality. The vendor chooses the acceptance number 0. Find how many lots are likely to pass vendor's inspection ?

A suggestion is made to choose the acceptance number as 1. If this suggestion is adopted, how many lots will now be passed by vendor's inspection ?

Q. 26. A multiple sampling plan is as follows :

Sample No.	*Individual sample size*	*Combined sample size*	*Acceptance number*	*Rejection Number*
1.	20	20	'X'	2
2.	20	40	0	3
3.	20	60	1	3
4.	20	80	2	4
5.	20	100	2	4
6.	20	120	2	4
7.	20	140	3	4

X = Acceptance not allowed on first sample.

If 1,000 lots all of which are 2% defective are submitted to this sampling plan, find the value of average outgoing quality, assuming that only rejected lots are to be screened and defectives found are to be replaced.

Q. 27. Design a single sampling plan for which the 'Consumer's Risk' of acceptance of 2% defective product is 0.10 and the 'Producer's Risk' of rejection of 0.6% defective product is 0.05.

Determine the values of $P_{0.50}$ and 'indifferent quality' for this sampling plan.

Q. 28. Define the terms (*i*) Measurement process uncertainty (*ii*) Capability (*iii*) Production process capability indices C_p and C_{pk}.

Ans. (*i*) **Measurement Process Uncertainty.** It is defined as a combination of measurement bias, repeatability and reproducibility, attributed to a measuring procedure. For the purposes of these procedures, measurement uncertainty (U) can be defined as the half interval (or +/– interval) about the true value of measurand within which atleast 99.73% of the results of measurement will fall. This is also referred to as a 99.73% confidence level or the six sigma uncertainty.

(*ii*) **Capability.** It is defined as the ratio of the uncertainty of a measuring process to the range of interest of the measuring process.

(*iii*) Production process capability indices C_p and C_{pk} are defined as under.

$$C_p = \frac{\text{Tolerance}}{6\,\sigma_p} = \frac{\text{Tolerance}}{6\sqrt{\sigma_{ma}^2 + \sigma_{me}^2}}$$

where σ_p = standard deviation of the production process.

σ_{ma} = standard deviation of the manufacturing process.

σ_{me} = standard deviation of the measurement process.

$$C_{pk} = C_p(1-k) = \left(\frac{\text{Tolerance}}{6\,\sigma_p}\right)\left(1 - \frac{b_p}{1/2\ \text{Tolerance}}\right)$$

and

$$k = \frac{b_p}{1/2\ \text{Tolerance}} = \frac{b_{ma} + b_{me}}{1/2\ \text{Tolerance}}$$

where b_p = bias of the production process.

b_{ma} = bias of the manufacturing process.

b_{me} = bias of the measurement process.

$$C_{pk} = \left(\frac{\text{Tolerance}}{6\sqrt{\sigma_{ma}^2 + \sigma_{me}^2}}\right)\left(1 - \frac{b_{ma} + b_{me}}{1/2\ \text{Tolerance}}\right).$$

Q. 29. Define process control and what is its purpose ?

Ans. Process control is defined as technique to provide feed- back to the manufacturing process to keep it targeted and reduce scrap. Its purpose is to make sure that only good parts are shipped and to provide feedback to the manufacturing process to keep it targeted and reduce scrap.

Objective Type Questions

Choose the correct answer

1. Which is true for Dodge-Roming sampling scheme ?
   (*a*) it is designed for minimum total inspection.
   (*b*) it is indexed on AOQL and LTPD.
   (*c*) it provides for only single sampling plans.
   (*d*) all of above.
2. The maximum percent defective that, for the purpose of sampling inspection, can be considered as a process average.
   (*a*) AQL (*b*) AOQL (*c*) LTPD (*d*) ABC.

3. Rule of thumb sampling plan is generally based on

(*a*) 1% sampling size (*b*) 2% sampling size

(*c*) 5% sampling size (*d*) 10% sampling size.

4. Number of inspection levels used in ABC standard

(*a*) 1 (*b*) 2 (*c*) 3 (*d*) 5.

5. Different inspection levels are provided in ABC standard

(*a*) to balance the cost of inspection against the amount of protection needed.

(*b*) to balance two risks.

(*c*) to enable achieving derived AOQL.

(*d*) to minimise average total inspection.

6. The ratio of inspection in 3 general inspection levels of military standards is

(*a*) 0.4 : 1.0 : 1.6 (*b*) 0.5 : 1.0 : 1.5

(*c*) 0.2 : 1.0 : 1.8 (*d*) 0.3 : 1.0 : 1.2.

7. For which type of distribution, the value of mean, mode and median coincide ?

(*a*) normal (*b*) binomial

(*c*) Poisson (*d*) hyper geometric.

8. Which in correct relationship when distribution is only moderately skewed ?

(*a*) Mean – Mode = 3 (Mean – Median) (*b*) Mean + Mode = 3 (Mean + Median)

(*c*) Mode – Mean = (Mean + Median)/2 (*d*) Mean – Mode = Mean – Median.

9. Which of following is not measure of dispersion ?

(*a*) Range (*b*) Standard deviation

(*c*) Mean deviation and quartile deviation (*d*) Median and mode.

10. Which of the following is near the parametric value of the population ?

(*a*) arithmatic mean (*b*) mode

(*c*) average of sample averages (*d*) standard deviation.

11. The value of the variate which occurs with the greatest frequency

(*a*) median (*b*) mode (*c*) average (*d*) standard deviation.

12. If producer's risk is α, probability of acceptance of AQL is

(*a*) α (*b*) $1 - \alpha$ (*c*) $1 + \alpha$ (*d*) $(1 - \alpha)/2$.

13. Type I and Type II errors in acceptable sampling refer to

(*a*) rejection of good lot ; and acceptance of bad lot.

(*b*) acceptance of good lot ; and rejection of bad lot.

(*c*) rejection of both good and bad lots respectively.

(*d*) acceptance of both good and bad lots respectively.

14. Which two points are specified in choosing a sampling plan ?

(*a*) α, β (*b*) $1 - \alpha, 1 - \beta$

(*c*) AQL and LTPD (*d*) AQL, $1 - \alpha$ and LTPD, β.

15. Which is false statement regarding operating characteristics (OC) curve.

(*a*) It shows ability to distinguish between good and bad lots.

(*b*) No sampling plans can give complete protection against acceptance of defective.

(*c*) Larger the sample size, steeper the slope of OC curve.

(*d*) Acceptance number is zero for ideal sampling plan.

16. Statement of number of sample points before specific change is detected

(*a*) ARL (average run longer) (*b*) Type I error

(*c*) $\pm 3\ \sigma$ limits (*d*) normal distribution.

State whether following are true or false

1. Quality of a product in a factory can be improved if variability is decreased.
2. Control charts enable to detect assignable causes of variation.
3. Shewhart control charts are used only for variables.
4. Variables are measurable quality characteristics and attributes are those which could be classified as conforming or non conforming.
5. Arranging raw data as frequency distribution reveals the central tendency and the quantum of variation.
6. A histogram enables one to examine shape, central tendency, scatter and capability of process to meet specification requirements.
7. A cumulative frequency distribution is called ogive.
8. For 'less than' ogive, the cumulative frequency diminishes and increases for 'greater than' ogive.
9. A big advantage of frequency polygon over histogram is that frequency polygons of several distributions can be plotted on the same axis, enabling their comparisons.
10. A normal distribution can be described completely in terms of mean and standard deviation.
11. When only chance causes are present in a process, it is in control. When assignable cause of variation is observed, process is out of control.
12. As sample size is increased, cost of inspection increases, control chart becomes sensitive to detection in variation of process average, range chart does not give a good estimate.
13. Narrowing the control limits on control charts will increase probability of type I and II errors.
14. Keeping control limits of $\pm 3\sigma$ away from mean will strike a rational balance between type I and type II errors.
15. Variations which are likely to occur in random manner can be improved only by changing machine tool/process.
16. Assignable variation occurs on account of external causes and these can be detected by control charts.
17. Binomial distribution is useful in case of two mutually exclusive events and when sample size is less than 20.
18. Poisson distribution is useful for discrete, independent events which occur at random and infrequently.
19. Control chart of sample mean is used instead of individual variable values because former tends towards normal distribution.
20. A sampling plan defines the size of sample and the allowable number of defectives.
21. Scientific acceptance sampling provides the estimate of the lot quality.
22. Rule of thumb sampling plan envisages 10% sample size.
23. Tables of ABC standard and military standard are identical.
24. Operating characteristics curve shows the probability of rejection of various incoming percentage defectives.
25. Taguchi method of process capability study uses a truncated experimental design to determine which process inputs have the greatest effect on process variability and setting the greatest influence variables at such levels as to minimise their effect on process variability.
26. $\overline{X}$ chart tracks the accuracy of the process and R chart tracks the precision.
27. Six sigma philosophies are related to statistical process control, stochastic control and engineering process control.
28. Lean philosophy is a technique as well as a continuous effort that is used to accelerate and minimise the cost of any process by eliminating the waste in either manufacturing or service.
29. Lean philosophy is about identifying and removing inefficiencies like non-value-added cost or unneeded wait time within the process caused by defects, excess production, etc.

30. Lean philosophy recognises seven types of wastes *viz*, over production, delay and wait time, transportation, processing and complexity, excess inventory, wasted motions/unutilised talent and defects.

Q. Fill in the blank selecting right choice from given two options.

1. Number of times a variable occurs is called .... (frequency, standard deviation)
2. A histogram of the given data is indicative of centre and ...... of the data. (variation, frequency)
3. A graphical representation of a frequency table is show by .... (ogive, histogram)
4. A graphical representation of a cumulative frequency table is shown by ...... (ogive, histogram)
5. Statistical process control ensures output to be within .... (process variability, specification)
6. In a set of data, the middle observation is .... (mode, median)
7. ...... shows the probability of accepting batches of differing incoming %age of defectives. (OC curve, Ogive)
8. The probability of observing at least one defective in a random sample size of 10 drawn from a population producing 5% defective unity on average is .... (0.95^{10}, $1 - 0.95^{10}$).
9. If subgroup size is 4, value of $\sigma_{\bar{x}}$ will be .... ($\sigma/2$, 2σ)
10. In control chart theory the distribution of number of defects per unit follows very closely ...... distribution. (binomial, poisson)
11. ...... limits are marked on $\overline{X}$–chart. (specification, control)
12. An attribute sampling plan is specified by ...... (n, C_i ; α, β)
13. Steeper the OC curve, more is protection to .... (consumer, both consume and producer)
14. Dodge-Roming sampling schemes aim at minimising .... (AOQL, total inspection)
15. The probability of accepting material produced at an AQL is .... {$(1 - \alpha)$, $(1 - \beta)$}
16. Dodge-Roming sampling plans provide sampling schemes indexed on .... (AOQL and LTPT, $1 - \alpha$ and $1 - \beta$)
17. Single, double and multiple sampling plans are provided in .... (Dodge & Roming plans, MIL–STD plans)
18. Normal, reduced and tightened inspection provision exists under .... (ABC standard, Dodge and Roming)
19. The minimum sample size in $\overline{X}$-charts and R-charts is .... (3, 4)
20. $\overline{X}$-charts control the variability ...... sub groups. (within, between)
21. Narrowing the control limits on process control chart will increase ...... error. (type I, type II)
22. All manufacturing processes display some level of .... (uniformity, variability)
23. The capability of manufacturing process is determined by measuring the ...... of the process. (input, output)
24. Variation in material properties, vibration and chatter, wear of sliding components, operator errors etc. are causes of .... (inherent variability, assignable causes)
25. Process capability ratio is defined as ratio of tolerance spread (USL–LSL) and .... ($3\sigma'$, $6\sigma'$)
26. Value of process capability of ≥ ...... is considered good. (0.87, 1.33)

Answers

Objective Type Questions

1. (*d*)	2. (*a*)	3. (*d*)	4. (*c*)	5. (*a*)	6. (*a*)
7. (*a*)	8. (*a*)	9. (*d*)	10. (*c*)	11. (*b*)	12. (*b*)
13. (*a*)	14. (*d*)	15. (*d*)	16. (*a*)		

True (T) or False (F)

1. T	2. T	3. F	4. T	5. T	6. T
7. T	8. F	9. T	10. T	11. T	12. T
13. F	14. T	15. T	16. T	17. T	18. T
19. T	20. T	21. F	22. T	23. T	24. F
25. T	26. T	27. T	28. T	29. T	30. T

Fill in the blanks

1. frequency
2. variation
3. histogram
4. ogive
5. specification
6. median
7. OC curve
8. $1 - 0.95^{10}$
9. $\sigma/2$
10. poisson
11. control
12. n, c_i
13. both consumer and customer
14. total inspection
15. $(1 - \alpha)$
16. AOQL and LTPT
17. MIL-STD
18. ABC standard
19. 4
20. between
21. type I
22. Variability
23. Output
24. inherent variability
25. $6\sigma'$
26. 1.33.

20

Total Quality Management (TQM)

20.1. Introduction

Today, total quality management (TQM) has been well-accepted by managers.

Earlier the emphasis of quality management was based on the use of mathematical and statistical tools for improving processes in mass-production settings. In the case of Shewhart's early work, statistical methods such as control charts and probability sampling theory were used to sample for quality variation in production processes. These variations from acceptable tolerance limits were then investigated and their causes subsequently eliminated, thus resulting in more consistent final products.

Besides control charts, the approach of plan-do-check-action (PDCA) cycle provided excellent approach to problem-solving among workers. Emphasis was then laid an ways to institutionalise programs that would yield continuous quality improvement. Quality management was extended beyond statistical tools to include quality of design and quality of conformance. Entire managerial hierarchies were then equipped with how to manage for quality. Quality control (QC) circles were introduced to involve employees, and quality began to be included in business plans. Quality management was starting to be described as an integrated process of habitual improvement. Quality management was extended beyond the boundaries of manufacturing to the management of work in general.

Today it is important to adopt process control through statistical quality control and statistical process control and employee participation through quality circles. TQM looks for involvement of all departments of the organisation, procurement, service, sales, manufacturing personnel etc. Systems in all departments are required to work in a concerted and cohesive manner. Top management should be commited to Quality. Quality Policy has to clearly spell out the objectives in this direction.

There is need for integrating quality control at all stages of manufacture, from design to production to the shipment stage. The quality chain concept is concerned with assurance extended to every part of organisation.

In the new scenario of domestic competition and globalisation, organisations have no option but to remain quality conscious and cost competitive. Quality is essential for successful innovation. Today customer demands quality in products in services and in life. Competition and customer expectations demand constant efforts to improve, even on success.

Quality is the result of untiring and intelligent effort. A will is needed to produce a qualtiy product for which hard work and devotion with involvement of all is necessary.

Cleanliness of the work place, team work and education are essential ingredients. Effort should be to make product right for first time and all times.

20.2. Evolution of TQM

If we look in fast at various stages in quality management we find clearly the four phases in this process.

(*i*) **Quality Control Phase** which focussed on inspector to detect defect after production. The inspector was responsible for weeding out the defectives at the end of the production cycle. No corrective action could took place before waste was created,

(*ii*) **Quality Assurance (QA) Phase** saw introduction of statistical methods to monitor trends and to ensure that only such products which fall within the specified tolerance limits are produced. The QA movement recognised the fact that responsibility for quality cannot be assigned to inspectors alone but all concerned departments have to contribute. Entire manufacturing process must be committed to meeting the quality needs of the design. QA focuses on procedure compliance and product conformity to specifications through production and operation management. QA relies on a comprehensive system of planning, documentation, statistical process control and product certification.

(*iii*) **Total Quality Control(TQC) Phase** emphasised on being pro-active and building stronger links with customers and suppliers and introducing a philosophy of continuous improvement. The focus clearly shifted from inspection, to managing *via* systems and procedures and to a pro-active role of problem solving *via* continuous improvement programmes. TQC programmes attempt to expand the QA philosophy beyond manufacturing operations into other areas of organisation. The purpose of problem analysis is to develop the long term solutions rather than respond to short term variations.

Utilisation of computer integrated manufacuring system and just-in-time (JIT) operations management methods, also facilitate this control.

(*iv*) **Total Quality Management (TQM) Phase** emphasises that the responsibility for quality does not stop with the quality department but extends to every single individual in the organisation. The goal is to be transmitted among various groups, departments and levels and making them work as team.

Concepts of productive maintenance, or fool proofing, quality circles and company-wide-quality control are accepted worldwide today. The customer satisfaction is the key issue in ensuring quality into products.

ISO 9001 helps in assessing company's quality requirements, translates quality concepts into achievable targets, provides an opportunity to create and improve quality system, and spreads the message of quality across the organisation. TQM is achieved through Statistical Process Control together with worker motivaton through job enrichment, job enlargement and job rotation plus keeping sight of financial goals through cost control.

The model of TQM is based on the use of systems and procedures for controlling quality. Quality systems entail having the organizational structure, responsibilities, procedures, processes and resources for implementing QM, such that there is a guiding framework to ensure that every time a process is performed the same information, methods, skills and controls are used and practised in a consistent manner. Such systems usually require companies to document all their procedures, work instructions, specifications and methods for all functions and aspects of the organization, thus providing employees with a referecne system to assess their work and work improvements.

ISO 9001 is established to provide customers with an assurance that the quality of products and/or services provided by a supplier meet their requirements. However, quality standards have been in existence in one form or another for some time ; the early standards were provided by major purchasers to their suppliers, but these early standards were mainly inspection-based activities, where purchasers develop their own systems of assessment, which "involved visiting the supplier to examine the degree to which their operating procedures and systems followed the requirements of the (purchasers') standard". Today, the ISO 9001, quality standards is recognized as national standard by many countries, and third-party certification and registration facilities are also available internationally, thus providing companies world-wide with an infrastructure for systematic quality assurance and control. A quality system is widely understood to be a fundamental pillar in a company's holistic approach to QM and can help in ensuring that any improvements made are held in place.

ISO 9001 quality system offers advantages like internal operating efficiency, increased profitability and improved marketing. A system based on the ISO 9001 is recognized to provide only the foundation to TQM, but provides no assurance that there is a commitment to continuous and company-wide improvement.

Despite its origins in statistical and engineering backgrounds, TQM today contains a large element of human relations emphasis. TQM model provides focus on the qualitative aspects such as greater customer orientation, employee involvement, teamworking and the generally better management of employees within the company.

With the increasing focus on creating and internalizing a customer-oriented work culture, this approach of TQM places much significance on education, training and communication. To ensure that employees understand the concepts of TQM and their roles in the quality chain, training and education in quality principles are usually provided for all staff. Education and tranining are also supplemented with an increase in communication from management to encourage employees to 'do it right the first time'. The use of posters and slogans on the shop-floor stressing the importance of quality improvement and customer satisfaction is also very much the norm for this approach to TQM.

TQM approach calls for the managment of the organizational culture that is geared towards continuous improvement. There is need for a major transformation of the company, which is achieved not by changes in the production processes but rather through changing people's mindset with a shift in the responsibility of producing quality work from a functional department towards the individual employee. The TQM advocates also call for the involvement of all employees in decision-making process of the organization for example through their involvement in QC circles, cross-functional project teams and so on.

20.3. What is TQM ?

TQM is an effective management tool for ensuring continuous customer satisfaction, leading to attainment of organisational growth. It is a systematic way to ensure that activities happen the way they are planned, and thus pushing the total organisation to higher standards of performance. Quality in all spheres of activities must start with "quality consciousness" within an organisation.

TQM is not conformation to specifications alone, but it demands continuous improvement with a positive attitude to work with commitment of all.

TQM is a management philosphy to a journey to organisatioanal excellence and in sastisfying customer needs in totality on a continuing basis through involvement of each and every employee in the organisation making continuous improvement.

20.4. Principles of Total Quality Managment

Management Commitment and Vision. Top down commitment and bottom up initiative. Impart suitable training and support the effort. Future vision of an organisation leads to directions, values and goals which lift horizons and raise expectations.

Customer Orientation. Meet the needs of customers through regular feed back.

Quality Streategy and values. This emanates from top. Defect prevention strategy leads to strategic goals and plans of action. Translate the organisational strategy into thrust areas.

Quality Culture. Dedication to continuously strive for improvement in all procedures, activities and services. Constantly work for eliminating all waste. Every employee to strive for professional excellence.

Employee Emprovement and Participation. Bring about work ethics and attitudinal changes to harness the potential, commitment and involvement of people.

Team Work. It envisages breaking barriers between departments (reducing layers and encouraging cross functional teams.

Communication. Two way, clear, effective, consistent and forceful. Essential for open and active involvement of all in company's vision and values.

Continuous Improvement. Dynamic concept of improvement by raising "bench mark" in all areas to achieve organisational excellence.

Measurement and Audit to ensure excellence.

20.5. Elements of TQM System

TQM system should apply to and interact with all the activities. The International Standards Organisation (ISO) standard 9001 sets out the methods by which a management system, incorporating all the activities associated with quality, can be implemented in an organisation to ensure that all the specified performance requirements and needs of the customers are fully met and well documented. The importance is to have a formal system for documentation and information. Design of the system will not be one time effort, but will undergo continuous change depending upon organisation and customer needs. Fig. 20.1 shows various elements of TQM system.

TQM is increasingly being seen as a new management paradigm. TQM provides new way of managing to improve effectiveness, flexibility and competitiveness of a business as a whole to meet the requirements of customers.

TQM is essentially a way of planning, organising and understanding each activity of the organisation. For an organisation to be truly effective, each part of it must work together towards the same goals, recognising that each person and each activity affects and in turn is affected by others. TQM is also a way of ridding people's lives of wasted efforts by bringing everyone into the processes of improvement, so that results are achieved in less time. The methods and techniques used in TQM can be applied throughout any organisation.

TQM is today seen as a compnay-wide effort that emphasizes three core principles *viz* customer orientation, process orientation, and continuous impovement. To implement these principles, there are essentially some core areas that organizations need to follow.

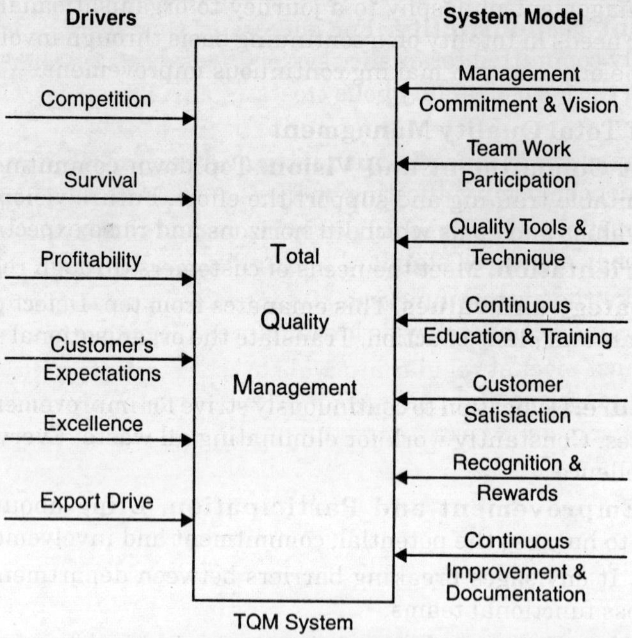

Fig. 20.1. TQM system.

20.6. Guidelines for TQM Implementation

- *Management leadership.* This calls for the leadership and involvement of senior management in creating and sustaining a customer orientation and clear and visible quality values for the company.
- *Information and analysis.* To drive TQM, there is a need for adequate company data, information and the analysis system to support the improvement process. Measurement systems, both in terms of the traditional measurement of defects or variations, and the monitoring of Cost of Quality need to be implemented.
- *Strategic quality planning.* All the planning process needs to start at the top, and all the key quality requirements need to be integrated into overall business planning. It also specifies the need for company's short-term and longer-term plans and how quality and operational performance are to be deployed to work units.
- *Human resource development and management.* The criteria call for establishing an environment conductive for building quality excellence. It looks for human resource planning and management, employee involvement, education and training, and employee performance and recognition.
- *Management of process quality.* There is a need for good process management in any TQM company and these may include R&D, design, management of process quality for all work units and suppliers, systematic quality improvement and quality assessment.
- *Quality and operational results.* These criteria call for monitoring the quality levels and improvement trends in quality, company operational performance and supplier quality. Indicators and benchmarks relating to competitors are also looked for.
- *Customer focus and satisfaction.* The company's relationship with customers and its knowledge of customer requirements and market are looked for.

20.7. Quality Management and Quality System Requirements

1. Management responsibility. For success of this system, it is essential that management be totally committed for quality system. It is expected that direction be given by the management in the form of quality policy to its organizational associates to get desired quality.

Further the management must fix **authority and responsibility** of every employee to detect and control nonconformity affecting quality.

Management must provide all **resources** in the form of adequate material, machinery, men, methods and money of desired standard to result in quality work.

In order to make the system functional, management must appoint an experienced person of desired talent and competency as **Management Representative.**

Management must meet at regular intervals to take stock of quality system with help of Management Representative and heads of all departments. Such **management review meetings** enable management to be current with the status of quality system and take timely corrective actions.

2. Quality system. It is essential that a quality system be established in a documented manner (*i.e.* proper records on registers). A **quality manual** providing clause by clause interpretation in the supplier's own context has to be prepared. **Work instructions and checklists** for ecery procedure affecting quality must be drafted suitably. **Qulaity plans** specifing all the resources, their schedule of acquisition to match with desired activity schedule also need to be prepared.

3. Contract review. The supplier has to make contract review in a documented manner to cover the **contract** (specifing the requirements of the customer as per order) and the **Capability Assesment** (Based on quality planning, the supplier, before entering into the contract must assess his own capabilities to meet the ordered requirements.)

4. Design control. It envisages that the supplier shall also have design control of its product in a documented manner. The supplier should identify qualified personnel with adequate resources for developing the design, and also identify design input requirements precisely to get design output to the satisfaction of customer requirements.

5. Document and data control. A control system of documents and data in a documented manner is essential. Documents pertinent to perform any organizational activity must be available at desired location. Any document and data has to be reviewed and approved by authoized personnel prior to issue.

6. Purchasing. A great emphasis is laid on doing all activities in a documented manner. For example, evaluation, before selecting, of subcontractor's ability to meet the supplier's requirements must be documented. Similarly it is essential to defne controls that the supplier shall have on the subcontractors to get quality goods and services from the subcontractors. The purchase documents must be properly maintained and reviewed to clearly reflect the suppliers' requirements.

7. Control of cutomer supplied products. In cases where the customer supplies any product to be incorporated in the related activity, the supplier must establish and maintain documented procedures for the control of verification, storage and maintenance of such customer supplied products.

8. Product identification and traceability. Wherever possible and required, the supplier must have procedures of identification and traceability of products used to fix responsibility of non-confromity so that, in future, that non-conformity can be avoided.

9. Process control. The supplier must identify, plan and incorporate detailed procedures defining the manner of production, installation and servicing the processes which directly affect the quality. It must also ensure compliance with reference standards/codes, and approval of process and equipments.

10. Inspection and testing. ISO 9001 requires such inspection and testing at three stages as under :

Receiving inspection and Testing

It shall be ensured that no product is used without inspection and testing.

Wherever the subcontractor is liable to have his own inspection and testing, the documents of such inspection and testing must be provided.

In Process inpsection and testing

Inspection and testing should be done at identified stages of production as laid down in quality plan.

Final inspection and testing

All final inspection and testing to complete the evidence of conforming with specified requirements, must be done in accordance with quality plan.

No product shall be dispatched until all activities related to inspection and testing as specified in quality plan are satisfactorily completed and associated data and documentation are available and authorized.

11. Control of inspection measuring and test equipment. The supplier must maintain documented procedures to control, calibrate and maintain inspection, measuring and test equipment to ensure the conformance of product to the specified requirements.

12. Inpsection and test status. The inspection status of product and material to be used in production should be identified by suitable means so that neither uninspected nor rejected products are put to use in production.

13. Control of non-confirming products. The supplier has to ensure that no nonconforming product is put to use or installed by establishing documented procedures for such avoidance of use. The responsibility in regard to disposition of such products must be fixed.

14. Corrective and preventive action. The supplier msut establish and maintain documented procedures for implementing corrective actions to eliminate causes of actual non-conformities and preventive actions to eliminate cause of potential non-conformities.

15. Handling, Storage, Packaging, Preservation and Delivery

The supplier must establish and maintain documented procedures for handling, storage, package, preservation and delivery of product so that damage of deterioration to products is prevented during handling.

16. Control of quality records. Quality records are an output of quality documents and plans related to quality system.

The supplier must establish and maintain documented procedures for .

(*a*) Storing quality records to avoid damages or deterioration, and

(*b*) Identification, collection, indexing, access, filing, storage, maintenance and disposition of quality records.

17. Internal quality audits. Supplier must (*a*) plan and implement internal audit by independent and trained personnel to find out whether quality activities and related results comply with planned arrangements ; (*b*) schedule the internal audit and record the findings of such audit and bring them to the attention of the personnel having responsibility for the area of audit for taking corrective actions in time.

18. Training. Identify training needs and provide for the training of all personnel performing activities affecting quality.

19. Servicing. Supplier must develop adequate documentation to ensure after sales quality of product to ensure utmost customer satisfaction.

20. Statistical techniques. Supplier must identify the need for statistical techniques necesary for quality and implement and control the application of such statistical techniques.

In ISO-9001 : 2000, all the above 20 requirements are considered in following 5 heads.

(*i*) Quality managment system

(*ii*) Management responsibility

(*iii*) Resource management

(*iv*) Product realisation

(*v*) Measurement, analysis, and improvement.

20.8. Tools and Techniques for Total Quality Management

Total Quality Management envisages to continually improve the system in terms of design, material, maintenance, machine capability, training, supervision to create consistancy and for this purpose it has a set of guiding principles. It employs quantitative methods to continuously improve an organisation's process by development of quality culture, and the empowerment of employees.

Various tools and techniques for TQM are as follows :

(*i*) **Workplace utitilization**

For quality and productivity improvement, work place utilization by proper arrangement, orderliness, cleanliness, clean up and discipline are important.

(*a*) Sort out unnecessary items. Unwanted things be removed to avoid confusion and obstacle to working.

(*b*) Orderliness implies keeping everything in a predetermined place so that searching is easier.

(*c*) Cleanliness implies keeping work area very clean with no oil spots, burrs, dirt or dust. A pleasant working environment results in quality products.

(*d*) After work is done, machines and workplace be cleaned. Have routine checks for the machines like lubrication checking, air pressure etc.

(*e*) Discipline implies maintaining good habits, following work instructions/standards, safety precautions etc.

(*ii*) **Total Quality Control.** The latest concept today is that responsibility for quality control rests with all departments.

Total quality control looks for quality at source, variability reduction by systematic problem solving methods and procedures, doing things right first time and inching towards Zero variability.

Total Quality Control focuses on

- Waste Elimination
- Houses Keeping and Visual Control
- Demand and Pull Production

- Quick Changeover
- Zero Defect Production
- Total Productive Maintenance
- JIT Purchasing and Packaging
- JIT Dealers and Distributors

(*iii*) **Total Employee Involvement.** To achieve quality, every one in the organisation from top to bottom must cooperate, work as a team and develop feeling of owning. Aim is to have co-operation, mutual enlightment, spontaneous manifestation of individual capabilities, and drive towards the common goal of productivity and profitability.

Group activities and schemes like cross functional task forces, zero defect groups, suggestion scheme, quality circles, brain storming sessions, awareness programmes and training help to achieve desired results.

(*iv*) **Problem Solving Process.** Problems are best solved by following cycle: Plan–Do-Check-Action.

(*a*) **Plan.** A course of action is planned keeping is mind the customers' requirements and service conditions. The problem should be first defined by identifying its importance, understanding background. Expressing qualitatively the undesirable result of poor performance and estimating the possible improvement, defining theme and target, setting up a team to undertake the task efficiently, preparing budget estimate, and setting a time table to do it. Next setp is to observe, *i.e.*, recognise the features of the problem by investigating time, place, type and symptom to discover the features of the problem, considering all points of view to discover variations, and collecting necessary information and present properly. Next analyse the problem by setting up hypothesis.

Using cause and effect diagram, identifying main cause, obtaining more data to reinforce the hypothesis, and integrating all information to arrive at a conclusion.

(*b*) **Do.** Work must be executed according to the plan.

(*c*) **Check.** Verify results of work with original plan and whether it meets expectations of customer. Compare data before and after the action was taken. Convert the effects into monetary terms and compare results with set targets.

(*d*) **Action.** Initiate necessary corrective action and feed forward results to start the next plan.

(*v*) **Quality Circles.** In quality circles methodology, task performers at the grass root levels are motivated to work towards the improvement of quality. Workmen follow quality control technique in the shop floors by forming small groups and solve the daily problems of the organisation.

(*vi*) **Use of Statistical Tools.** Statistics is concerned with the collection, analysis, interpretation, and presentation of data. Statisticasl process control techniques are used for acheiving quality.

Base statistical tools are : graphs/run charts (line charts, bar charts, piecharts), check sheets, cause and effect diagrams, pareto diagrams (isolating the vital few from the trivial many), histograms, scatter diagrams, stratification of information (breaking down whole into smaller related sub-groups).

Statistical Process Control

Every process has some inherent variation which is random and predictable. It can be reduced only with a process change. Any variation outside that expected for a process due to bias or some special cause can be detected and eliminated by statistical process control techniques.

Statistical process control enables a production operation to consistently meet demanding specifications. Control charts are used as diagnostic tools in control of production processes. Control chart is a graphic representation of measured actual process performance relative to computed control limits.

Control charts help to understand the inherent capability of process, bring the processes under control by eliminating special causes of variation, reduce tampering with processes that are under statistical control, and monitor the effects of process changes aimed at improvement. For statistical control of variables three control charts are normally used, namely (*a*) mean chart, (*b*) range chart, and (*c*) standard deviation charts.

(*vii*) **Quality Function Deployment.** Quality function deployment tool transforms customer requirements into product parameters. It is basically a matrix indicating "what is required" on the vertical axis and "how to achieve it" on the horizontal axis. It helps in identifying a specific product parameter which could satisfy a specific customer requirement. It also helps in identifying inter-relationship between parameters during making changes.

(*viii*) **Taguchi Approach for Quality.** Taguchi's approach to quality have been combining engineering and statistical methods to achieve rapid improvement in cost and quality by optimising the product design and manufacturing processes.It integrates product design and manufacturing process.

Taguchi's systematic approach to quality engineering is based on integrating (1) system design (initial design stage to establish materials, parts and initial product parameters) (2) parameter design, (testing various combinations of materials to maximise the production and lowering the costs ; (3) tolerance design, (buying better grade materials).

Parameter-design is the key step as it offers concept of loss and uncontrollable factors. Losses occur when a product's functional characteristcs deviate from desired parameters or values. There should be a minimum of variation in its function characteristics. Taguchi emphasis that instead of finding and eliminating causes for variation, it is better to remove or reduce the impact of causes. Taguchi's parameter design focuses on finding a product's parameter values and operating levels, which are least sensitive to change in environmental conditions and controllable factors. This is the key to getting high quality and low cost products.

(*ix*) **Failure Mode and Effect Analysis (FMEA) and Fault Tree Analysis (FTA).** FMEA and FTA are tools used for reviewing the design of any new product w.r.t. customer requirements or specified product performance before it is sent for production.

FMEA is basically used for analysing the effect of failure of individual components on the system as a whole. It looks at the product and its elements from the points of how and in what manners it can fail.

Fault Tree Analysis starts from system or product failure of major assembly failure to sub assemblies failure to component failure. A failure mode for the system is imagined and the root cause is analysed upto individual components.

(*x*) **Eliminating Human Error by Fool Proofing.** Human beings are liable to commit error unknowingly. Constantly aim at the total elimination of the possibility of human error to acheive zero defect.

Fool proofing system is a means to create devices that can discover defectives/disorders. Some examples of fool proofing are :

If a disorder is found in the goods, the device does not allow the machine to start processing.

If there is a miss step the device does not allow the machine to start processing.

If there is a miss step in work process or in motion it is automatically adjusted, and the device will alllow processing to proceed.

The disorder that has occurred in preceding process is examined at the next process, and device will stop defectives.

If a certain operation is forgotten or skipped the device does not allow the next operation to begin.

(*xi*) **Continuous Improvements.** Continuous improvement is possible by small improvements on a continuous basis involving all the people in the ogranisaion. Very large number of small improvements are more effective in an organisatonal environment than a few improvement of large value. Do not believe what is practiced now is the best. There could be a better method of doing things.

(*xii*) **Improving the Process on basis of Capability study**

Process capability index

C_p = Process capability potential or Process capability variation.

C_{pk} = Process capability index or Measure of process capability.

$$C_p = \frac{\text{Specification width}}{\text{Process width}} = \frac{USL - LSL}{6\sigma}$$

and $C_{pk} = (1 - K)\, C_p$ where K = correction factor

$$= \frac{(D - \overline{X})}{1/2 \times (USL - LSL)}$$

where D = designed mean value

$\overline{X}$ = observed mean value

K = correction factor and it has always + ve value.

When the process average $\overline{X}$ and design centre D or target value, coincidè K is reduced to zero, making C_p and C_{pk} equal.

If, however, the process average is skewed towards one end or the other of a specification limit, away from design centre D, the value of K increases, causing a decrease in C_{pk}, relative to C_p.

Routine data collection and C_p, C_{pk} reports are meaningless, if process is not under "statistical control".

(*xiii*) **Quality Improvement Techniques**

(*a*) **Affinity diagram.** This diagram gathers large amount of data and organises it into grouping based on natural relationship between each item, so that these data can be analysed systematically.

(*b*) **Inter relationship diagram** shows relationships between inter-related factors.It is used to realte complex cause and effect relationships. This diagram explores and displays inter related factors involved in complex problems.

(*c*) **Tree diagram.** This diagram systematically maps out the full range of tasks/methods needed to achieve the goal. This diagram helps planning effectively so that objectives are achieved. It does this by showing the sequencing and inter relationships so that the resulting actions are error free.

(*d*) **Matrix data analysis.** It requires complex mathematical skills and relies on systematic organisation and presentation of data so that relations can be examined. It shows the strength of the relationship between variables which have been statistically determined.

(*e*) **Process Decision Program Chart (PDPC).** It helps selection of best process to obtain the desired results. It not only evaluates existing process but also looks at alternatives through newly gained knowledge. PDPC maps out every conceivable event that may occur when moving form a problem statement to the possible solution.

(*f*) **Arrow diagram.** This diagram is used to plan the most appropriate schedule for any task and to control it effectively during the progress. It helps to establish most suitable plan for a project and also controls its progress.

20.9. World Class Quality at Competitive Cost

A few guidelines are given below to achieve this aim.

The main aim of any industry is to enhance productivity and competitiveness. Costs have to be kept down using defect free materials and keeping inventory costs down by using JIT delivery methods. Better product quality demands compatible design and manufacturing systems.

Information can change operations. Collection of reliable desired information is essential at every stage to improve the situation. Test and inspection data should be used as a prime source of informations. Information should be communicated in a readily understandable form so that causes of problem can be identified. Further the information should be timely, relevent, accurate and scannable. Reports should generate action to save time and effort.

Adopt a multifunctional approach for corrective action by organising timely meetings with all concerned. Make it sure that corrective actions happen. Provide the data collected to all concerned so that future problems are prevented. The quality system must be dynamic and be received regularly based on the results of inspection, testing, and customer reaction. Use statistical methods to control processes or achieve permanent improvement.

Use statistical methods to aid in the continual search for process or product improvements. Establish a yield specification and measure performance against it so that proper attention is devoted by lower level management to satisty upper level management demands. The yield specification must set new bench-marks once old ones are acheived. Establish a reliability specificaton, *i.e.*, long-duration trouble-free performance. Top management must state the objects and ask the right question relating to it at all levels. The cost of acheiving conforming products that satisfy customer demands must be known. A strategic plan for quality both long term and short term must be established. Such goals could be : tenfold improvement in 5 years. Use high technology to improve product and systems. While products may be improved beyond customer expectation, economics must be kept in mind. Vendors can be a great asset ; use them accordinlgy. Stay close to your customers and know your competition. Seriousness about quality improvement must be reflected in actions and words with daily reminders. Also provide an environment in which people can be self-motivating.

Quality system must be adjusted as product quality improves. Adjust personnel, material, or specification as changes occur.

SOLVED PROBLEMS

Q. 1. What is need of quality and total quality management ?

Ans. Over a period of time, man has realised that it is advisable and more economical to buy quality products. In the present day when the plants have become complex and large, nobody can afford to have unreliable equipment since outage of one small portion of plant could mean shut down of complete costly plant. To achieve the present day quality of products, industry has witnessed several phases. There was time when customers were expected to, and wise customers were prepared to pay extra for quality. However in the competitive business climate of today, quality is no longer anyone's option but it is a positive requirement without which no organisation can survive. Today we are in the midst of quality revolution in industry and every body is enjoying the fruits of quality products.

Ordinarily, quality is referred to as customer satisfaction and fitness for use, meeting all the expectations of the buyer of goods or services. As per ISO 8402, quality is defined as "The totality of features and characteristics of a product or service that bear on its ability to staisfy stated or implied needs". In TQM, the quality is considered as customer delight and not just satisfaction. All this is possible when an organisation plans for "Above Customer Expectations" (ABC), for which the quality has to be integrated or embedded in the culture of the organisation. TQM is today a well established science and organisations serious about it can strictly follow ISO 9001 of International Organisation for Standardisation (ISO) of International Quality Standards or its equivalent IS-14001 published by Bureau of Indian Standards.

Q. 2. How development in quality history taken place ?

Ans. Developments in quality circles in industry have taken place in various phases as follows :

(*i*) Quality control [statistical quality control (SQC) phase]. The industry took necessary care to ensure production of quality components to the satisfaction of customer. Various important tools like quality control charts, sampling plans and other statistical techniques were adopted.

(*ii*) Quality Assurance Phase. While SQC phase depended on meeting quality by sorting out rejects, QA recognised that inspection alone is not the answer to achieve quality needs of the design. Thus focus shifted to procedure compliance and product conformity to specification through production and operations management, using statistical process control as a tracking tool.

(*iii*) Total Quality Control Phase. It attempted to expand the QA philosophy from manufacturing and production shops to other areas of the organisation with a view to create awareness of quality and participation of total staff in it. Focus got shifted from direct cost reduction and preoccupation with efficiency to the pursuit of quality through the elimination of waste and non-value added procedures and assuring continuous improvement through the refinement and expansion of quality control systems and procedures. Use of computer and modern management techniques like "just-in-time operation" have greatly helped in achieving total quality control phase.

(*iv*) Total Quality management. In the case of TQM the quality control is embedded within and is driven by the culture of the organisation. TQM has emerged as the most integrated form of organisation development.

The ISO 9001 standard on quality system provide a sound base for TQM. Fig. 20.2 shows the essential features related with success of TQM. It would be seen that commitment of top mangagement is essential. TQM must be management led, company-wide in implementation and dedicated to continuous improvement. Total involvement of each employee is essential. TQM provides an opportunity to all employees to directly participate in their work and take decisions concerning their job. People participate voluntarily in quality circles and quality improvement teams. The job satisfaction and employee satisfaction are the important outcomes of TQM implementation. The aim of TQM is that everybody connected in one way or other with the organisation, must be able to feel comfortable and realise their potential. In broader sense, TQM attempts to create a work place where humanity is respected, thus it lets the unlimited potential of human beings blossom. The third leg of triangle in Fig. 20.1 refers to quality control and standards. However, the potentials of quality tools and techniques are exploited by building an organisational culture so that quality improvements, quality development, and quality maintenance to meet full customer satisfaction at economical levels are embedded into the work and activities of the organisation.

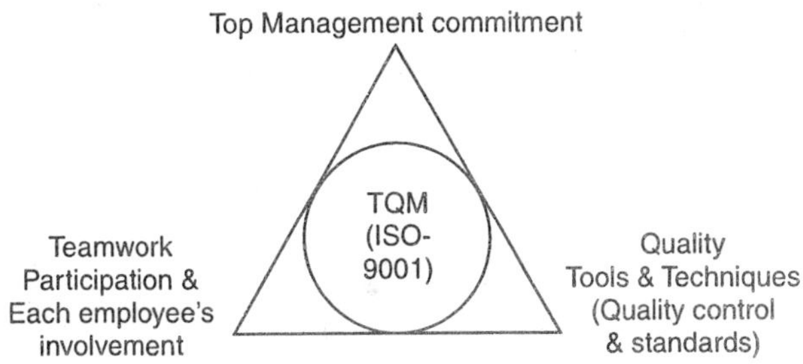

Fig. 20.2. Essential features of TQM.

Q. 3. Write a short note on Total Quality Management (TQM).

Ans. The production of an item in an organisation involves a series of processes. In order to produce a quality product, every process from the beginning to the end has to be correct.

The function of TQM is to improve the quality of work of all the people at all the functional areas of the organisation. Implementing TQM calls for a complete change in the behaviour of people working in the organisation. Thus TQM can be defined as an all encompassing dynamic process in an organisation to promote never ending improvement in the effectiveness and efficiency element of a business.

TQM uses the fundamental ideas of group activity, participation of all, application of statistical and other *QC* tools, self-development and creativity. It imbibes the philosophy that there is always a better way of doing things. The entire work force in the organisation actively pursues to achieve the specific corporate goal in regard to customer satisfaction for quality, price and delivery. The three dimensions of TQM are excellence, universality and perpetuality.

TQM is a dynamic process that promotes unending improvement in the effectiveness and efficiency of all elements of a business. Quality finds first place in the entire business thinking and action. The total work force in organisation pursues the specific corporate goal of making quality as customer oriented event instead of product oriented event. All efforts are made to achieve customer satisfaction with regards to quality, price and delivery, use techniques like statistical process control, delegation of responsibility to employees who have thorough understanding of customer needs and are committed to enforcement of total quality. The three dimensions of TQM are excellence, universality, and perpetuality. TQM is not confined only to products, process, customers, employees but to the entire gamut of organisational activities including each and everyone with total support and commitment of the management. Thus total participation and teamwork, judicious choice of tools to analyse problems, and company wide approach are essential for success of TQM. The characteristics of TQM are that the effort should be in totality (all encompassing) ; well thought documentation properly disseminated to all persons involved ; continuous improvements of operations, products and services ; solid foundation of organisational structure and systems to manage the broad scope of quality activities.

Q. 4. What are the objectives of TQM ?

Ans. Important objectives of TQM are :

— It is important to understand that quality is conformance to requirements and not goodness. It must have strong customer focus.

— TQM integrates people, machines and information to provide customer quality satisfaction at affordable cost.

— The system must work on the basis of prevention and not appraisal.

— TQM calls for organisation of quality activities in conformance with stated quality policies and objectives.

— Strong quality mindedness and company wide quality motivation, involvement, commitment, and training/retraining and orientation.

— To set aim at zero detect and not approaching to perfection.

— Continuous tracking of entire system using feed forward and feedback information with analysis of results and comparison with set targets.

— Structure the system to provide linkage between various processes and quality related activities so that audit trail is provided for.

— The end aim should be to judge the price of non-conformance to requirements instead of quality indices.

Q. 5. What do you understand by total quality culture (TQC) ?

Ans. The creation of TQC means that the values, attitudes and competencies of people can be influenced permanently through appropriate training and development efforts.

Establishing standards, testing and certifying compliance to established standard (ISO-9001) is the first step in the journy towards achieving total quality.

Total quality is management science. Standards and standardisation, quality system certification and inspection, measurement systems, testing laboratories, their accreditation and calibration services, production and supply of standard reference materials, etc., are all important building blocks.

Q. 6. Write preachings of TQM by some renowned personalities.

Ans. The present state of TQM is due to various renowned personalities in Japan and U.S.A. who gave certain cardinal principles on the basis of their experience and success in establishing TQM systems. Some of these are :

(*i*) **Juran.** Clearly spell out the need, set goals for improvements, organise to achieve goals, train/retain and orient personnel, adopt problem solving approach, report progress, give recognition, ensure no communication gaps, keep score, and maintain momentum by methods like annual customer satisfaction measurement survey, annual management behaviour survey, cost of quality calculation, benchmarking.

(*ii*) **Deming.** Cream constancy of purpose for continual improvement for long range needs, adopt new philosophy, avoid mass inspection, minimise total cost and not merely initial cost, constantly improve every process, organise training on job, adopt and institute leadership, effective two way communication and driving out fear, introduce spirit of teamwork by breaking drown barriers, no slogans/posters/exhortations, eliminate arbitrary numerical targets, provide right to pride of workmanship, encourage education and self improvement programmes, top management's commitment to ever-improving quality and productivity.

(*iii*) **Taguchi.** Identify the main function, noise factors, quality characteristic and control factors ; define data analysis procedure in the form of matrix ; determine optimum level and predict performance ; verification and future actions.

(*iv*) **Crosby.** Management stand on quality, team for quality improvement, objective evaluation and corrective action of quality, ingredients in cost of quality, creating quality awareness at all levels, provisions for corrective action, plan for zero defect programme, quality training, realisation among employees about change, establish improvement goals, free communication about error-cause removal, appreciation, recognition and rewards, quality councils, emphasis that quality is an unending programme.

Q. 7. Why TQM is needed by industry ?

Ans. It is generally said that we are in the midst of a "Quality Revolution". Competitiveness could be ensured by quality, price and delivery.

The price could be a strategic decision which once taken can be implemented immediately. The delivery could be viewed as one which may require harnessing of resources depending on the quantum and lead-time. Quality is one aspect which has to be first understood correctly, planned well ahead and articulated in order that it is sustained and improved. Quality is not just built by a few but is the net resultant of everybody's action. The implementation of quality reduces cost, reduces hassels and improves delivery performance. Quality thus has developed as the most important competitive tool and is widely appreciated. The route to achieve this competitive edgc is "Total Quality Management" (TQM). There are people who understand quality to mean as conformance to requirements, reduction of variability, availability on demand, doing it right the first time, having good aesthetic appearance, reliability etc. One has to have clarity in understanding what we mean by the term quality. It is necessary to have a clear understanding and it interestingly could be correlated to the hierarchy in an organisation. The employees of organisation could be broadly classified as follows :

— Task performer — Supervision

— Middle Management — Top Management

The way these people should view quality, their role towards ensuring this are given below :

To first group of people, quality means conformance to specifications. They are given clear task guide line or work instructions based on which they execute the job. Their role is action oriented. Supervision has an important role of ensuring routine operation optimisation. They coordinate the various resource inputs for this and based on this optimisation only, the quality cost can be got to a minimum. To them quality means doing it right the first time. Towards this, they may be employing certain tactics which will yield them the optimisation.

The role of the middle management in an organisation is to integrate all functions and to assure the quality of the products and services to the customers. Towards this, they should evolve proper strategies.

In a true competitive world, just satisfying the customer is not good enough. There are many organisations who go beyond this. To them quality means customer delight. This is best achieved by concentrating on quality of all functions beyond just product quality and giving emphasis on human aspects so that the involvement and commitment of all employees is of a high order.

Assuring product quality involves most of the functions in an organisation. Networking these activities to result in customer satisfaction must be well planned, resources coordinated and controlled (*i.e.*, well managed). Managing for quality over the decades has undergone significant strategic stages-inspection, quality control and quality assurance. Appreciating the need for assurance of quality in sourcing decisions, scope for reduction in quality cost by a systematic management of quality, the International Organisation for Standardisation brought out a series of quality management standards called ISO 9000. Today, an organisation could reorient its quality management system as per ISO 9000 and obtain a system certification from a certifying body evidencing to the market, its capabilities of sustaining quality of its products/services. The intrinsic quality of the product/service will depend upon not only the systems standard but also on the product/process standards. When these two complimentary standards are adhered to, one could assure quality. The benefits that could be realised by organisations introducing ISO 9001 system are—arouses abundant awareness, building block for betterment, catalyst for cultural change, documentation for detailed do, ensures effectiveness and efficiency, and fusion of field feedback.

Q. 8. Write short note on TQM culture.

Ans. The TQM model provides a distinctly different way of looking at the management style. TQM is management led, company-wide in implementation, dedicated to continuous improvement, and the responsibility of every employee. Although TQM got evolved from quality control but there is a basic shift in management philosophy and is a complete way of managing an organisation. It tries to build an organisational culture where quality improvements are embedded into the work and activities. TQM cultures ensure that things are done right first time.

The traditional approach where in all blue collar employees were normally workers, and senior white collar employees where managers can't hold good for TQM culture. For success of TQM it is essential that the concept and role of workers and managers as considered earlier be done with and both are considered as team members and the facilitators to achieve desired goal. TQM involves everyone including workers, clerical and administrative staff, managers, management etc. Everyone is equal and works as a part of a team.

The workers participation need not be legislated but each employee is made to join voluntarily to look at problems. Instead of just participating, he gets fully involved to find ways and means to eliminate errors and prevent waste. In the traditional approach the participation of worker was considered with view to bargain but the concept of TQM is involvement *i.e.*, participation plus commitment and pride. In mere workers' participation, both unions and employees are automatically taken care of in TQM culture and one has only to address organisational needs, especially those of meeting customer requirements. In TQM culture emphasis is laid on recognisation of employee rights and needs. Work is redesigned to benefit people. Stress is on participative decision making and problem solving and integrated approach to change. In TQM culture all work purposefully, with pride, towards a common goal—total quality. TQM approach provides an answer to transform not only employee employer relationship but also focusses attention on economic development. Family oriented relationship exists where everyone contributes his best. Workers are empowered and take decision to do right. The entire culture is integrated and cooperative.

The values of TQM culture are :

—transparent thinking —openness —trust

—cooperation and integrated approach —discipline —respect

—voluntary working with commitment —family like working and team approach
—customer requirement utmost —participation with commitment and pride.

The results of adoption of TQM are :

—Quality improvement, quality development and quality maintenance.
—Full customer satisfaction at economic levels —Reduced costs
—Improved certainty in operations —Improved image of organisation
—Satisfied employees with sense of involvement
—Job satisfaction and employees satisfaction
—Potential of human beings blossoms —Respect for humanity
—All round prosperity.

Q. 9. What are the strategies and elements in TQM Implementation ?

Ans. For implementing quality improvement project, it is necessary to develop implementation process in line with the quality improvement process comprising five major parts. (Fig. 20.3)

1. Planning for TQM implementation
2. Organisation for TQM implementation
3. Execute the implementation plan
4. Evaluating and holding the gains
5. Recycling to assess new customer needs.

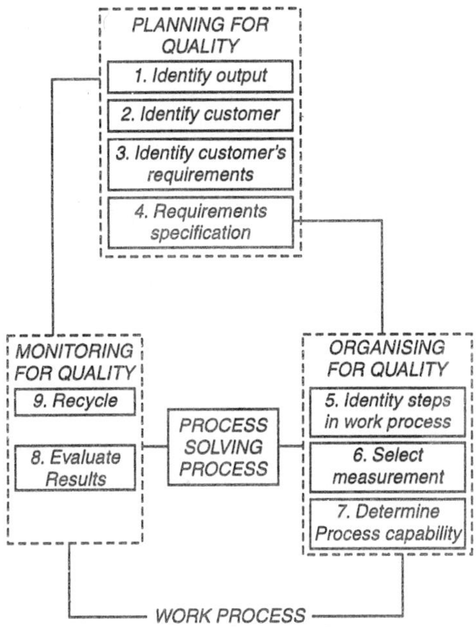

Fig. 20.3. The quality improvement process.

Planning the elements of the implementation strategy

TQM strategy is based on six key enablers in fishbone framework.

— Leadership should be established through quality as a means to an end and not an end in itself.

— Work for constancy in the enablers, viewing their effectiveness through the eyes of the customer, internal and external.

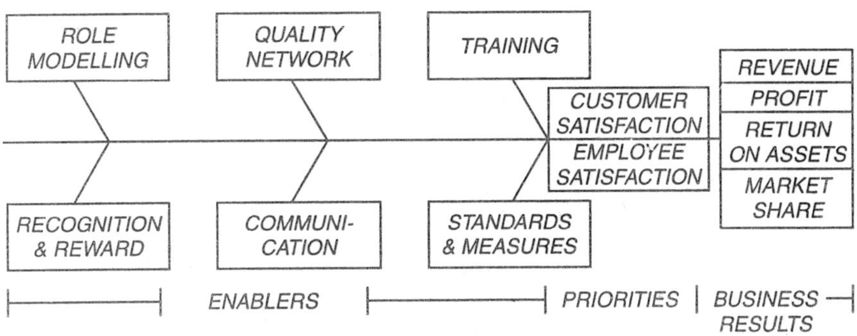

Fig. 20.4. Model for TQM implementation.

Training

Training programme for cost of quality, quality improvement process, problem solving process and interactive skills for all employees is essential.

— The course design should have a top down cascaded approach with each boss going through the programme twice-once as a participant and once as a family group manager.

— At the end of the training programme, each family group should identify critical bottleneck areas for improving consistency in results as quality improvement projects.

Role modelling by senior management

Adopt the principle of learn-use-teach-inspect. Senior management behaviour should demonstrate quality in their daily work.

Recognition and Rewards

Develop recognition and reward scheme incorporating :

— A 100-point rating sheet to evaluate process disciplines, innovative, team working and business results.

— Three-tier grading of quality improvement efforts for recognition, through evaluation panels.

Communication

A quarterly newsletter featuring significant strides in leadership through quality.

Corporate communications carrying quality and customer satisfaction as underlying themes—whether in print, talk or video.

Standards and Measures

— Training : 100% of all employees who have completed 6 month's stay in company should be trained on leadership through quality.

— Employee involvement : 100% of employee participation in business priority linked quality projects.

80% of all customer Satisfaction linked projects to be completed annually.

50% of all business priority linked projects to be completed annually.

Recognition of business excellence

— Work process standards and improved by at least 30% annually and recognized.

Senior Management Process Review

— These are structured meetings scheduled for review round the year to evaluate process, efficacy and business results.

TQM Effectiveness

— Annual surveys on customer satisfaction and management behaviour.

Organising for TQM Implementaion

TQM should be a blend of charisma and process disciplines to be effective.

Efforts are required for continuous harnessing of 3 forces : vision, sensitivity and technology.

Vision. The direction and pace of change required becomes more clear and evident when vision of a company is clearly spelled out.

Sensitivity. The four "antennae" used for guiding the quality process can be derived out of :

— Annual Customer Satisfaction Measurement Survey.

— Annual Management Behaviour Survey.

— Cost of Quality Calculations (these documents reflect the financial consequence of rework and waste in operations).

— Benchmarking—to decide upon the standards of excellence.

Technology. The Corporate Quality Movement is based on a conscious development in technology for :

1. Competency assessment for TQM and deployment.
2. Organisational change.
3. Accelerating the change process.

Q. 10. More and more organisations are beginning to implement Total Quality Management (TQM) to sustain their competitive edge in industry. What steps need to be taken to create a TQM culture ?

Ans. Implementing TQM calls for a complete change in the behaviour of people working in the organisation.

Considering behavioural change, two inter-related issues have to be addressed—attitude and value system.

Behaviour is what is visible outside a person. Coming to office poorly dressed, not attending Quality Circle meetings, not wearing safety goggles, and the like.

Attitude is what causes that behaviour. An attitude is a tendency or a predisposition to behave in certain ways in particular situations.

Q. 11. Explain the following terms, clearly bringing out difference between them :

(*i*) Quality control (QC)

(*ii*) Quality Assurance (QA)

(*iii*) Total Quality Control (TQC)

(*iv*) Total Quality Management (TQM).

Ans. (*i*) Quality Control is concerned basically with the detection of defects in a product being manufactured. The detection of faults may take place either by inspection after production, or by inspection during production/on-line inspection. Quality Control is also concerned with the sorting out rejects and with the adherence to quality standards. Most of the techniques are based on statistical methods.

(*ii*) Quality Assurance (QA) goes one step further. While inspection is essential to check quality, it is not the only answer to achieve quality. QA recognises that the entire manufacturing process must be committed to meeting the quality needs of the design. Thus quality control must apply at all the stages, *viz.* procurement of raw materials, design, manufacturing, assembly, erection, commissioning, and maintenance phase of the equipment. Quality Assurance focuses on procedure compliance and product conformity to specifications through production as operating management, often using Statistical Process Control (SPC) as a "tracking" tool.

(*iii*) Total Quality Control (TQC) expands the scope of quality beyond manufacturing to other aspects of organisation. Thus while TQC makes best use of all the tools used in QA, but it goes further to develop long term solutions for quality by eliminating waste and non-value added procedures. TQC aims at assuring continuous improvement through the refinement and expansion of quality control systems and procedures. It utilises computer systems and management methods like just-in-time operations. TQC offers company-wide movement and is largely limited to the manufacturing department. But for overall quality, it is essential that the parties outside factory, *viz.* suppliers of other equipment/products/materials, distributors and customers also participate in the "chain of quality".

(*iv*) Total Quality Management (TQM) goes still one step further and is responsible for the quality revolution in the industry. It is concerned with the integration of all the efforts in the organization towards quality improvement, quality development and quality maintenance to meet full customer satisfaction at economical levels. The major difference between TQM and TQC is that while the systematic analysis, preplanning, blueprinting of operation in TQC is dictated by external controls through procedural compliance and enhancement, but same is governed by a process of habitual improvement, and control is embedded within and is driven by the culture of the organisation in case of TQM. The management of organisation provides leadership and support. Management commitment is essential for success of TQM. Other important requirement is teamwork participation. In TQM, a customer orientation achieved through continuous quality improvement becomes a valid "taken for granted" atmosphere shared by everyone, unquestioned and habitually enacted. TQM got evolved from quality control but it represents a basic shift in management philosophy and is a complete way of managing an organisation. Quality control and standards, of course, remain an important component of TQM. It tries to build an organisational culture where quality improvements are embedded into the work and activities, and workers feel full participation in achieving desired quality.

Q. 12. What is the aim and goal of quality system ?

Ans. The goal of the quality system requirements is the development of fundamental quality systems that provide for the continuous improvements, emphasizing defect preventing and the reduction of variation and waste in the supply chain.

Q. 13. Identify opportunities for quality and productivity improvements.

Ans. Some examples are :

Unscheduled machine downtime.

Machine set up, die change and changeover times

Excessive cycle time

Scrap, rework, and repair

Non value added use of floor space

Excessive variation

Less than 100% first run capability

Process averages not centered on target values (bilateral specifiction)

Testing requirements not justified by accumulatd results.

Waste of labor and materials

Excessive cost of non quality

Difficult assembly or installation of the product

Excessive handling and storage

New target values to optimize customer process

Marginal measurements system capability

Customer dissatisfaction

Complaints, repairs, returns, misshipments, incomplete order, customer plant concerns, warranty etc.

Q. 14. List various techniques for continuous improvement is quality and productivity improvement.

Ans. Some examples are :

Capability indices

Control charts (variables, attributes)

Cumulative sum charts

Design of experiment

Evolutionary operation of process

Theory of constraints

Overall equipment effectiveness

Cost of quality

Parts per million analysis

Value analysis

Problem solving

Benchmarking

Analysis of motion/ergonomics

Mistake proofing

Plant layout should minimize material travel and handling, facilitate synchronous material flow and maximize value added use of floor space. Method should be developed for evaluating the effectiveness of existing operations and processes considering the following factors : overall work plan, appropriate automation, ergonomics and human factors, operator and line balance, storage and buffer inventory levels, value added labour content.

Q. 15. Define TQM to cover all important aspects of it.

Ans. TQM is a common method to improve the whole organisation step-wise, in a structured, and systematic manner, on the basis of hard work, discipline, intensive training, and consistent implementation of techniques and resources. For its success, everyone in the organisation must master the quality attitude, mentality and skills. It must be a unending continuous process, *i.e.*, it be cyclic, iterative, and never ending activity. It is a disciplined method of defining a problem, collecting all relevant information, observing, analysing to find root cause, taking action, checking effectiveness of action, instituting the solution with aim to modify further based on future developments, evaluate the process, involve everyone in the process, collect feedback and take corrective actions.

Q. 16. State the important Kaizen rules.

Ans. — Work with as per guideline

— Take problems as opportunities for improvement

— Collect relevant information

— Consider all important facts

— Work as per plan

— Avoid waste

— Maintain order and neatness

— Keep appointments

— Follow PDCA cycle continually.

Q. 17. Six sigma technique adopts a specific time sequence. What do you know about it ?

Ans. The five steps as per this are

(*i*) Define the project clearly and the expected result must be feasible.

(*ii*) Ensure a clear measure for the problem and define the starting value

(*iii*) Analyse using statistics and depict full and clear picture of the problem

(*iv*) Improve by developing a break-through solution and implement it into the organisation.

(*v*) Control the process by embedding the solution in the standard management system so that the improvement cannnot slip away.

Q. 18. What are the eight principles on which continuous organisational improvement is possibe as per ISO 9001 standard ?

Ans. (*i*) Customer focuss

(*ii*) Leadership

(*iii*) Involvement of people

(*iv*) Process approach

(*v*) Systems approach to management

(*vi*) Continuous improvement

(*vii*) Decision making based on factual data and information

(*viii*) Supplier partnership.

Q. 19. What points were given by Deming to improve quality ?

Ans. Deming gave following points to improve quality.

(*i*) Create consistency of purpose towards improvement of product and service.

(*ii*) Adopt new philosophies.

(*iii*) Cease dependence on mass inspection to acheive quality.

(*iv*) End the practice of awarding business on the basis of price tag.

(*v*) Improve constantly and forever the system of production and service.

(*vi*) Institute modern methods of training and education on job including management.

(*vii*) Institute leadership.

(*viii*) Drive out fear.

(*ix*) Break down barriers between departments.

(*x*) Eliminate slogans, exhortions and targets for the work force asking for zero defect and new level of productivity.

(*xi*) Eliminate work standards that prescribe numerical quotas for both the work force, and for managers.

(*xii*) Remove barriers that impede hourly paid workers and managers from enjoying pride of workmanship.

(*xiii*) Institute a vigorous programme of education and retraining.

Q. 20. What points were given by Juran for TQM ?

Ans. Juran developed TQM message around the following steps :

(*i*) Create awareness of the need and opportunity for quality improvement.

(*ii*) Set goals for continuous improvement.

(*iii*) Build an organisation to achieve goals by estabilishing a quality council, identify problems, select a project, and choosing facilitators.

(*iv*) Give every one training.

(*v*) Carry out projects to solve problems.

(*vi*) Report progress.

(*vii*) Show recognition.

(*viii*) Communicate the result.

(*ix*) Keep a record of successes.

(*x*) Incorporate annual improvements into the company's regular systems and processes and thereby maintain momentum.

Q. 21. What points were given by Crosby to improve quality ?

Ans. For quality improvements, Crosby gives fourteen steps of quality improvement programme :

(*i*) Management commitment.

(*ii*) Quality improvement teem.

(*iii*) Quality measurement

(*iv*) Cost of quality evaluation.

(*v*) Quality awarness.

(*vi*) Corrective action.

(*vii*) Establish an adhoc committee for zero defect.

(*viii*) Supervisor training.

(*ix*) Zero defects day.

(*x*) Goal setting.

(*xi*) Removal of error causes.

(*xii*) Recognition.

(*xiii*) Quality councils.

(*xiv*) Repeat the whole process agian.

Q. 22. What is the philosophy of six sigma ?

Ans. To identify what really matters to customer and to delivery at such a low level of error and defect that perfection close to zero defects is achieved in practice.

Q. 23. Explain the intent of using words "shall", "will", "should" and "may" in any contractual document.

Ans. "Shall" is used to express a provision that is binding between two or more parties ; "will" is used to express a declaration or purpose or intent by one party ; "should" is used to express a recommendation among other possibilities ; and "may" is used to indicate a course of action permissible within the limits of the quality olan.

Q. 24. Explain the difference between quality system and quality plan.

Ans. Quality system is organizational structure, procedures, processes and resources needed to implement quality management. It is as comprehensive as needed to meet the quality objectives.

Quality plan is document setting out the specific quality practices, resources and sequence of activities relevant to a particular product, project or contract.

Q. 25. Arrange the following stages in evolution of quality control in ascending order

(*a*) **TQM,** (*b*) **100% inspection,**

(*c*) **Foreman vertification,** (*d*) **statistical quality control,**

(*e*) **Operator inspection** (*f*) **Statistical process control,**

(*g*) **Quality asurance.**

Ans. The ascending order is (*e*), (*c*), (*b*), (*d*), (*f*), (*g*), (*a*).

Q. 26. Explain difference between quality assurance and total quality management.

Ans. Quality assurance concerns all activities needed to provide adequate confidence that a product or service will fulfil requirements for quality. Total quality management means a management approach centred on quality, based on participation of all members and aiming at long term success through customer satisfaction.

Q. 27. What is the basic concept of the ISO 9001 ?

Ans. ISO 9001 standard is practical standard for quality management systems and it can be applied by all types of businesses : manufacturing, distribution and service. It has found wide application as a basis for agreement between buyers and suppliers.

The standard is based on two main foundations :

— Management responsibility and commitment to quality which should be expressed in a formal policy statement and implemented through appopriate measures ;

— A set of requirements that deal with each aspect of the company activity and organization that affects quality.

An important condition for conformance to the standard is appropriate documentation of the quality policy and procedure which, put together, form the Quality Manual of the company.

The models prescribed for the quality systems include requirements to the different aspects/activities affecting quality. 20 such aspects for quality system requirements are :

Management responsibility
Quality system
Contract review
Design control
Document and data control
Purchasing
Control of customer-supplied product
Product identification and traceability
Procees control
Inspection and testing

Control of inspection, measuring and test equipment
Inspection and test status
Control of nonconforming product
Corrective and preventive action
Handling, storage, packaging, preservation and delivery
Control of quality records
Internal quality audits
Training.
Servicing
Statistical techniques.

Q. 28. Stated below are few terms commonly used in connection with quality management. Match the parts A and B.

A	B
1. Quality assurance	(*a*) The totality of features and characteristics of a product or service that bear on its abilitiy to satisfy stated or implied needs.
2. Quality plan	(*b*) Conceptual model of interacting activities that influence the quality of a product or services in the various stages ranging from the idenfication of needs to the assessment of whether these needs have been satisfied.
3. Quality	(*c*) The overall quality intentions and direction of an organization as regards quality as formally expressed by management.
4. Quality audit	(*d*) Those aspects of the overall management function that determines and implements the quality policy.
5. Quality survellience	(*e*) All those planned and systematic actions necessary to provide adequate confidence that a product or service will satisfy given requirement for quality.
6. Quality loop	(*f*) The operational techniques and activities that are used to fulfil requirements for quality.
7. Quality policy	(*g*) The organizatonal structure, responsibilities, procedures, processes and resources for implementing quality.
8. Quality control	(*h*) A document setting out the specific quality practices, resources and sequence of activities, relevant to a particular product service, contract or project.
9. Quality system	(*i*) A sysematic and independent examination to determine whether quality activities and related results comply with planned arrangement and whether these arrangements are implemented effectively and are suitable to achieve effectiveness.
10. Quality management	(*j*) The continuing monitoring and verification of the status of procedures, methods, conditions, processes, products and services, and analysis of records in relation to stated reference to ensure that specified

Ans.
1—(*e*), 2—(*h*), 3—(*a*), 4—(*i*), 5—(*j*), 6—(*b*), 7—(*c*), 8—(*f*), 9—(*g*), 10—(*d*)

Q. 29. Match the following views about quality and the quality gurus who enunciated them.

A	B
1. Create a positive climate for improvement and have knowledgeable workers than rigid systems	(*a*) Philip B.Crosby
2. Quality happens only through projects, "quality improvement projects"	(*b*) Shigeo Shingo
3. Major objective of total quality is zero defect and organizations have to be serious about prevention.	(*c*) J.M. Juran
4. Originator of quality circles and fishbone diagram	(*d*) Bill Convay
5. Quality has to come from new management way of thinking, and wide utilisation of statistical tools.	(*e*) W.F. Deming
6. Complete elimination of human error by following checklist and acheive zero quality control	(*t*) Kaoru Ishikawa

Ans. 1—(*e*), 2—(*c*), 3—(*a*), 4—(*f*), 5—(*d*), 6—(*b*).

Q. 30. Different experts defined quality in different ways. Match parts A and B in this regard.

1. A Feigenboum	(*a*) The totality of features and characteristics of a product or services that can bear on its ability to satisfy stated or implied needs.
2. Taguchi	(*b*) Quality is fitness for use.
3. Juran	(*c*) Quality is conformance to requirements or specifications.
4. A.V. Feigerabaum	(*d*) Quality is what the customer says it is.
5. Crosby	(*e*) Quality is an effective system of integrating quality improvement efforts of various groups of the organization so as to provide products/ service at level which allow customer satisfaction.
6. ISO-9001	(*f*) Quality is the loss that a product costs to society after being shipped other than any losses by its intrinsic functions.

Ans. 1—(*d*). 2—(*f*). 3—(*b*). 4—(*e*). 5—(*c*). 6—(*a*).

Q. 31. (*a*) **All designs are based on needs which are usually translated into features with specified criteria. What are these ?**

(*b*) **What abilities a product should have for satifactory performance ?**

Ans. (*a*) Needs are translated into characteristics with criteria like safety, availability, reliability, usability, manitainability, economics, environment, reusability, etc.

(*b*) The quality requirements for any product to meet its purpose are : durability, serviceability, reliability, safety, affordability, suitability, applicability, ease of operation and maintenance, etc.

21

International Standard on Quality Management System (ISO 9001-2000)

21.1. Introduction

With globalisation and opening up of Indian economy, industry has to complete in international markets. The traditional practices of quality control in checking the products for compliance can not enable industries to compete. Today, instead of sorting out good and bad parts, every effort is needed to produce acceptable goods in first instance so that non-conforming material is not produced. Improvements have to be made at all levels in the organisation and these must be visible.

All quality conscious industries desirous of standing on firm foot in international market get themselves accredited to ISO 9001 standard on 'Quality Management Systems'.

Earlier International Organisation for Standardisation (ISO) had three standards ISO-9001, ISO-9002, and ISO-9003 providing models for quality assurance in (*i*) design, development, production, installation and servicing, (*ii*) production, installation and servicing (*iii*) final inspection and testing respectively. In addition ISO-9004 covered quality management and quality system guidelines. It prescribed 20 elements for this purpose.

ISO 9001-2000 was introduced in Dec., 2000- and thus three standards ISO 9001–1994 ; ISO 9002-1994 and ISO 9003-1994 are now obsolete.

ISO 9001 : 2000 is applicable to all types of industries and organisations, irrespective of their size, nature and functions. Both product and service oriented organisations can make use of this standard to achieve quality. ISO-9001 : 2000 certified organisation enjoys world wide recognitions. Such organisations incorporate proper system of quality in their functioning and produce products and services of desired quality with consistency acceptable to customer.

ISO 9001-2000 prescribes internationally recognised set of requirements to improve customer satisfaction, eliminate poor quality and thus improve process efficiency and effectiveness. These standards are user friendly, generic in nature, and follow a logical and easily understood format.

Adoption of ISO 9001 : 2000 compliance ensures that company has complete compliance towards total quality management and has documented quality systems in built. Its parts are of best quality, most competitive even in global market and thus can look for increased business, guarantee customer satisfaction and increased customer demand.

21.2. Principles upon which ISO-9001 : 2000 is based

(*i*) It has total focus on customer. It treats customer as supreme and all actions are to ensure quality and low cost to consumer. It works on estabilishing customer confidence on the product and services. Customer requirements get top most priority.

(*ii*) It involves all the employees, management, customer and all people concerned in the process. Every person understands his responsibility and the contribution expected of him. Accountability of each person in the entire process is defined and fixed.

(*iii*) It is leadership based. People have better trust and faith in each other due to high transparency. Confidence level of management is high due to lesser problems at working level. Fire fighting on day to day basis in replaced by time utilised for proper planning.

(*iv*) It utilises process approach. Entire working is based on achieving desired end results in optimum manner.

(*v*) Management need not waste its time in trivial matters. Entire functioning is so systematically organised that it is system that works and not individuals. Each person contributes to make system stronger. Management thus concentrates in improving system.

(*vi*) System results in continual improvement in system. System is never stagnant but looks for more and more refinements so as to result in quality products and at lesser cost.

(*vii*) Decision making is based on factual approach.

(*viii*) It results in mutually beneficial supplier relationship.

21.3. Steps to be Taken for Certification

1. Documentation. Entire ISO-9001 process is based on preparing and following as per documentation Preparing Policy and Procedure Manuals is most important phase.

2. Implementation. ISO 9001 : 2000 is basically an instructions manual for the business. Implementing ISO means putting it into practice within the organisation. It typically takes between three to nine months, and is the second phase of the process.

3. Inspection audit. When all documents are ready and these are being implemented throughout the organization, call the Auditing Organization, known as a Registrar. They will inspect the documents, do an on-site inspection to verify that these are being implemented properly, and make a report to the Registrar for Registration after interviewing entire team and satisfying themselves fully.

4. Registration. When audit is cleared, the company is registered as an ISO-Certified organization and published in the Registered Company Directory.

21.4. Implementing an ISO 9001 : 2000

1. **Take timely decision.** Commit to implement (QMS). Make a senior person of organisation responsible for implementing system. This person will be in direct communication with the CEO on the progress of the QMS implementation. This person must believe in team working and be acceptable to entire team.

2. Form an implementation team and select a QMS consultant.

Select up to eight personnel from cross section of organisation representing all departments and levels of employees. Having a good representation of employees on this team will help ensure that the decisions made are appropriate and are implemented.

3. Conduct an assessment of gap analysis and report to senior management regarding adequacy of existing procedures and modifications needed.

4. Identify the processes. Prepare flow chart of processes, describe process inputs and outputs. Prepare sequence and interaction of these processes, needed for the quality management system and their application throughout the organization.

5. Determine criteria and methods to ensure that operation and control of these processes are effective.

6. Create a Quality Policy. Communicate the policy across the organization. Set agreed objectives and targets, and create an action plan to achieve them.

7. Develop leaders and the implementation team who can learn, understand and explain the policy, criteria and methods so as to sustain employee awareness.

8. Define organizational structure. Define and document responsibilities of the QMS. Keep this list or matrix up to date with job titles as a formal document.

9. Involve all employees in developing and improving their system.

10. Decide on the procedure for documentation structure.

11. Flowchart key processes showing all interfaces.

12. Create all forms in line with agreed coding procedure.

13. Correlate all forms to flow charts and review them for accuracy.

14. Develop and use flowcharts as documented procedures.

15. Review, reconcile, approve and issue to all concerned.

16. Prepare new procedures and train staff to implement them.

17. Issue new procedures and the environmental management program. Describe the whole system in QMS manual.

18. Launch the system and respond quickly to revision requests for continuous improvement.

19. Start auditing to continually improve the integrated system.

20. Hold review meetings every month. Conduct pre-assessment at least four months before any registered systems auditor and ensure all corrective actions are up to date.

21.5. Systemic Requirements of ISO 9001 : 2000

(*i*) **Establish Quality System**

- Develop, implement and improve quality management system
- Identify the processes that make up quality system.
- Describe quality management processes.
- Use quality system processes
- Manage, monitor and improve process performance.

(*ii*) **Document quality system**

- Develop documents to implement quality system and reflect working of organization.
- **Prepare quality system manual** by documenting procedures, describing how processes interact, and defining the scope of quality system.
- **Control quality system documents**
- Approve documents before distributing
- Provide the correct version of documents at points of use.
- Review and re-approve documents whenever these are updated.
- Specify the current revision status of all documents.

- Monitor documents that come from external sources.
- Prevent the accidental use of obsolete documents.
- Preserve the usability of quality documents.
- **Maintain quality system records**

Use records to prove compliance of requirements. Develop a procedure to control records, and ensure that records are useable.

21.6. Management Requirements of ISO 9001 : 2000

(*i*) **Support quality**

Promote the importance of quality by promoting the need to meet (*a*) customer requirements, (*b*) regulatory requirements and (*c*) statutory requirements.

Develop a quality management system by supporting the development of a quality system ; Formulating organization's quality policy ; Setting organization's quality objectives, and providing quality resources.

Implement quality management system by providing resources to implement quality system, and encouraging personnel to meet quality system requirements.

Improve quality management system by performing quality management reviews and providing resources to improve the quality system.

(*ii*) **Satisfy Your customers**

Identify and meet customer requirement and enhance customer satisfaction

(*iii*) **Establish a quality policy**

Define organization's quality policy by ensuring that (*a*) It serves the purpose of organization, (*b*) it emphasizes the need to meet requirements, (*c*) it facilitates the development of quality objectives, (*d*) and it makes a commitment to continuous improvement.

Manage organization's quality policy by communicating the policy to organization and review the policy to ensure that it is still suitable.

(*iv*) **Carry out quality planning**

Formulate quality objectives and ensure that objectives are set for functional areas and are set at organizational levels, that these facilitate product realization, support the quality policy, and that these are measurable.

Plan quality management system by planning the development, implementations, improvement and modifications of quality management system.

(*vi*) **Control quality system**

Define, clarify and communicate responsibilities and authorities

Appoint management representative to oversee quality management system, report on the status of quality management system, and support the improvement of quality management system.

Support internal communications to ensure that internal communication processes are established and that communication occurs throughout the organization.

(*vi*) **Perform management reviews**

Review quality management system (Evaluate the performance of quality system and it should be improved on continuous basis).

Examine management review inputs

Examine audit results, product conformity data, opportunities to improve feedback from customer, process performance information, corrective and preventive actions, changes that might affect system, and previous quality management reviews.

Generate management review outputs. (Generate actions to improve quality system, products and generate actions to address resource needs.)

21.7. Resource Requirements

(*i*) **Provide Quality Resources**

Identify resources needed to support the quality system and provide resources needed to support the quality system and to improve customer satisfaction.

(*ii*) **Provide Quality Personnel**

Support competence. Define acceptable levels of competence. Identify training and awareness needs.

Deliver training and awareness programms. Evaluate effectiveness of training and awareness. Maintain a record of competence.

(*iii*) **Provide quality infrastructure**

Identify infrastructure needs (Identify building needs, workspace needs, hardware needs, software needs, utility needs, equipments needs, and support service needs)

Provide needed infrastructure, *i.e.*, buildings, workspaces, hardware, software, utilities, equipment, and support services and maintain them properly.

(*iv*) **Provide quality environment**

Identify needed work environment and manage needed work environment.

21.8. Realization Requirements

(*i*) **Control Realization Planning**

Plan product realization processes by defining product quality objectives and requirements; identify product realization needs and requirements.

Development product realization processes, *i.e.*, Develop product realization documents, and record keeping systems and develop methods to control quality during product realization.

(*ii*) **Control customer process**

Identify customer's product requirements (Identify the requirements (*i*) that customers want to be met (*ii*) that are dictated by the product's use, (*iii*) that are imposed by external agencies, and that the organization wishes to meet.)

Review customer's product requirements (Review requirements before accepting orders from customers, maintain a record of product requirement reviews, and control changes in product requirements.)

Communicate with customers (Develop a process to control communications with customers and implement customer communications process.

(*iii*) **Control product development**

Plan design and development

Define product design and development stages.

Clarify design and development responsibilities and authorities.

Manage interactions between design and development groups.

Update design and development plans as changes occur.

Define design and development inputs

Specify product design and development inputs.

Record and review product design and development input definitions.

Generate design and development outputs

Create product design and development outputs.

Approve design and development outputs prior to release.

Use design and development outputs to control product quality.

Carry out design and development reviews

Perform and record product design and development reviews.

Perform design and development verifications

Carry out and record product design and development verifications.

Conduct design and development validations

Perform and record product design and development validations.

Manage design and development changes

Identify, record, review, verify and validate changes in product design and development and approve changes before they are implemented.

(*iv*) **Control Purchasing Function**

Control purchasing process

Ensure that purchased products meet requirements and that suppliers meet requirements.

Document product purchases

Describe the products being purchased and specify the requirements that must be met.

Verify purchased products

Verify purchased products both at production shop premises and at suppliers' premises (when required).

(*v*) **Control operational activities**

Control production and service provisions including processes, information, instructions, equipment, measurements. and activities.

Validate production and service provision

Prove that special processes, process personnel and process equipment can produce planned outputs.

Identify and track products

Establish and maintain the identity of products (when appropriate).

Identify the status of products (when appropriate).

Record the identity of products (when required).

Protect, identify, verify, and safeguard property supplied by customers.

Preserve the products and components during internal processing, and during final delivery.

(*vi*) **Control Monitoring Devices**

Identify monitoring and measuring needs

Select and calibrate monitoring and measuring devices

Protect monitoring and measuring devices from unauthorized adjustment and from damage or deterioration.

Validate monitoring and measuring software and revalidate monitoring and measuring software when necessary.

Use monitoring and measuring devices to ensure that the products meet requirements.

20.9. Remedial Requirements

(*i*) **Perform Remedial Processes**

Planning of remedial processes is useful to assure conformity, and to improve the system.

Implement remedial processes to demonstrate conformance and to improve quality management system.

(*ii*) **Monitor and Measure Quality**

Monitor and measure customer satisfaction by indentifying ways to monitor and measure customer satisfaction. Use customer satisfaction information.

Plan and perform regular internal audits

Set up an internal audit program and develop an internal audit procedure. Plan internal audit projects and perform regular internal audits.

Solve problems discovered during audits and verify that problems have been solved.

Monitor and measure quality processes by using suitable methods to monitor and measure processes and taking action when processes fail to achieve planned results.

Monitor and measure product characteristics

Verify that product characteristics are being met and keep a record of product monitoring and measuring activities.

(*iii*) **Control Nonconforming Products**

Develop a procedure to control nonconforming products by defining how nonconforming products can be identified and handled.

Identify and control your nonconforming products

Eliminate or correct product nonconformities, prevent the delivery or use of nonconforming products, and avoid the inappropriate use of nonconforming products.

Re-verify nonconforming products that were corrected and prove that corrected products now meet requirements.

Control nonconforming products after delivery or use

Maintain records of nonconforming products by describing products nonconformities and the actions taken to deal with nonconformities.

(*iv*) **Analyze quality information**

Define quality management information needs to evaluate and improve quality system.

Collect quality management system data by monitoring and measuring the suitability and effectiveness of quality system.

Provide quality management information about customers, suppliers, products and processes.

(*v*) **Make Quality Improvements**

Improve quality management system

Use audits to generate improvements.

Use quality data, quality policy, and quality objectives to generate improvements.

Use management reviews to generate improvements.

Use corrective and preventive actions to generate improvements.

Correct actual nonconformities

Review nonconformities and figure out what causes nonconformities.

Evaluate whether corrective actions need to be taken and develop corrective actions to prevent recurrence. Take corrective actions when necessary and record the results achieved by corrective actions.

Examine the effectiveness of corrective actions.

Prevent potential nonconformities

Detect potential nonconformities and identity their causes

Study the effects of potential nonconformities and evaluate whether there is need to take preventive action.

Develop preventive actions to eliminate causes and take preventive actions when necessary.

Record the results achieved by preventive actions and the effectiveness of preventive actions.

21.9. Quality System Documentation

It is developed at four levels as follows.

1. **Quality manual.** An operating document containing company's policies and procedures. It describes the quality management system in accordance to quality policy.

2. **Quality procedures.** This is a confidential document for internal use of organization. It describes the activities of individual functional units. The work force is instructed in broad terms, how policies and objectives are expressed in the quality system manual.

3. **Work instructions.** These describe in detail how a specific activity is to be performed.

4. **Quality records.** These demonstrate the performance of activity and how the products are developed and produced in accordance with the specified requirements as per quality system.

21.10. Preparation of Quality Manual

The means and procedures required for quality assurance are described in the quality manual (QM) which is a document setting out the quality policy, quality system and quality-relevant practices of an organisation. It serves the current documentation of all quality-relevant measures. Its main purpose is to furnish an appropriate description of the quality assurance system. It may be related tc the overall scope of tasks of an organisation or to partial tasks. The QM establishes rules for the general acceptance and maintaining of the system. It needs to be constantly updated and used in everyday work.

A quality manual must contain at least

- a quality policy statement
- the structure of the laboratory (organisational charts)
- the operational and functional activities pertaining to quality so that each person concerned will know the extent and the limits of his responsibility
- general quality assurance procedures.
- reference to quality arrurance procedures specific for each test, where appropriate.
- reference to proficiency testing, use of reference material etc., where appropriate
- satisfactory arrangements for feedback and corrective action whenver testing discrepancies are detected during tests
- procedure for dealing with complaints

The quality system of an enterprise or an organisation is influenced by the organisation's objectives, the kind of scientific and technical services and the organisation's specific working method.

The QM must be clearly structured. A general part should contain information on the organisation, the staff, the use of maintenance of the documents, including the manual. The task-specific parts provide information on the quality assurance elements and references to devices, procedural provisions and working instructions, documents also applicable, publication, standards, directives and the like.

For ease of handling and maintenance, a QM should be presented in a loose-leaf binder ; depending on the total number of pages of the QM, page identification should be per chapter or section.

Documented quality system procedures

The basic documentation for the overall planning and administration of activities with regard to quality should cover all the applicable elements. They should describe the responsibilities, authorities and interrelationships of the personnel who manage, perform, verify or review work affecting quality. It is advisable to prepare each documented procedure in the same structure and format so that the users get familiar with the consistent approach.

A quality manual should contain procedures for overall planning and administration of activities affecting quality.

Quality manuals enable communicating of organization's quality policy, procedures and requirements ; describing and implementing an effective quality system ; providing the documented bases for auditing the quality system ; providing continuity of the quality system and its requirements during changing circumstances ; training personnel in the quality system.

A quality manual should normally contain the following :

(*a*) title, scope and field of application ; organisation to which it applies.

(*b*) table of contents ; showing the titles of the sections. Numbering should be clear and logical.

(*c*) introductory pages providing minimum information about organisation and the quality manual.

(*d*) the quality policy and objectives of the organization ;

(*e*) a description of the organizational structure, responsibilities and authorities ;

(*f*) a description of the elements of the quality system,

(*g*) a definitions section,

(*h*) a guide to quality manual,

(*i*) an appendix for supportive data.

21.11. Preparation of Quality Plans

Quality plans provide a mechanism to tie specific requirements of the product/project to existing generic quality system procedures with some additional documented procedures. Thus the quality plan is supplemental to the company's generic quality system documentation.

A quality plan may be used to monitor and assess adherence to the requirements for quality. A quality plan may also be useful where a documented quality system does not exist.

The quality plan shows how the required activities are to be carried out.

The format and level of detail should satisfy customer requirement, as also the supplier's method of operation and the complexity of the activities. Quality plan may be separate for design, purchasing, production or inspection and test.

A quality plan may be submitted to the supplier for customer by the supplier for review and acceptance. The quality plan should be reviewed for adequacy and formally approved by representatives from all affected functions within the supplier's organization.

The supplier should revise the plan/to reflect changes made to the product/project changes in the manner in which the product or the service is provided, or changes in quality assurance practices.

The scope of the quality plan should include the product or project, the scope of the contract, the product/project quality objectives (in measurable terms if possible), exclusions and the conditions of validity.

It is important to identify individuals within the supplier's organization who are responsible for ensuring that the activities required by the specified quality system or contract are planned, implemented and controlled and their progress monitored ; communicating requirements peculiar to the specific product/project to all affected departments, subcontractors and customers, and resolving problems that arise at the interfaces between such groups ; reviewing the results of any audits conducted ; controlling corrective actions.

The quality plan should indicate when, how and by whom the requirements specified for the product/project are to be reviewed ; and how the results of review are to be recorded and how conflicts or ambiguities in requirements are to be resolved.

The quality plan should also indicate when, how and by whom design process is to be carried out, controlled and documented, and reviewed and also the extent of involvement of customer.

The plan should also indicate important products that are to be purchased ; methods to be used to evaluate, select and control subcontractors ;

The plan should further indicate : how products provided by the customer (such as material, tooling, test equipment, software, data or services) are identified and controlled ;

The quality plan should indicate how the production, installation and servicing processes will be controlled to ensure that specified requirements are met.

The quality plan should indicate relevant inspection and test plan and where inspections or tests are required to be witnessed or performed by customer or by regulatory authorities :

The quality plan should also indicate the control system to be used for inspection, measuring and test equipment.

The quality plan should indicate how nonconforming products are identified and controlled to prevent misuse until proper disposal.

The quality plan should indicate the preventive and corrective actions and follow-up activities.

The plan should indicate how the specified requirements for handling, storage, packaging and delivery are to be met ;

The plan should indicate how records specific to the product/project are to be controlled.

The plan should indicate the nature and extent of quality audits to be undertaken and how the results are to be used to correct and prevent recurrence of nonconformities which affect the product/project.

21.12. Internal Audit

Internal audit of the quality system can be done only by a person who has passed the examination conducted by the Bureau of Indian Standards (BIS) and successfully completed the training. A company can avail services of audit only from the accredited persons or bodies.

Internal quality audits are carried out (*a*) to find out whether all quality related functionaries comply with the laid down procedure ; and (*b*) to determine whether the implemented systems as such are effective.

An internal audit of quality system should be conducted to check that following requirements are met.

1. Management Policy and organization

— Is there clear written policy on quality and whether same is publicised.

— Is there a clearly defined organization for executing this and with responsibilities well defined.

— Are there defined procedures ?

— Are all employees clear of their role to achieve quality ?

— Are the responsibilities defined ?

— Is there assigned authority ?

— Whether top management is convinced ?

— Is there a training programme ?

2. The Quality System ISO-9001

— Does the quality conform to the requirement of relevant standards ?

— Whether quality plan and documents preferred ?

— Are there written procedures ?

— Are there relevant controls ?

— Is there proper quality audit system ?

3. Contract Review

— Is contract formally reviewed with customer ?

— Are promises being met ?

— Are capacity and equipment capable of meeting the contract requirement ?

4. Design and Design Change Control

— Is there a control system for design control ?

— Is the design manufacturing interface controlled ?

— Is there approved system for changes ?

— Is there traceability ?

— Is there a system to obtain feedback from customer ?

— Are the products conforming to the requirement ?

5. Documentation Control

— Is the quality manual issued and pages updated ?

— Is every employee given enough copies of relevant documentation.

— Is all documentation available where needed ?

6. Procurement

— Is there a supplier assessment system ?

— Is there an established system to identify reliable vendors ?

— Is a supplier register maintained ?

— Are suppliers expected to conform to ISO-9001 ?

— Is there traceability and identification ?

— Are management procedures documented ?

7. Production

— Are procedures, drawings, diagrams, instructions, reference model, documented ?

— Are appropriate instructions in the relevent places, flow chart at the shop etc. ?

— Is there a system for pass, indoubt or fail ?

8. Inspection and Test

— Is there a procedure for inspection of incoming material ?

— Is non-conformity controlled ?

— Is there traceability ?

— Is there inprocess inspection ?

— Is non-conformity identified, segregated and disposed off ?

— Is there final inspection ?

— Whether inspection staff have authority to release report ?

— Are rework items inspected ?

9. Inspection Measurement and Test Equipments

— Is there an inventory of all equipments ?

— Is there a system of calibration ?

— Is there a planned equipment maintenance ?

— Is equipment labelled for frequency of check and calibration ?

— Are non-conforming pieces of equipment controlled ?

10. Handling, Storage, Packaging and Delivery

— Are all stores, production, finished products etc. clearly segregated and controlled ?

— Are all the items marked in accordance with the status ?

— Can handling methods prevent damage to the product ?
— Is there a feed back mechanism ?

11. Audits

— Are there internal audits ?
— Are there optional external audits from consultants ?
— Is the audit covered by plan and procedures ?
— Do corrective actions arise from the audits ?
— Does management review the audits results ?

12. Training

— Is there a regular training programme ?
— Whether training schedule exists for all employees ?
— Are all training courses fully documented ?
— Is the responsibility clearly defined ?
— Do they take aid of magazines, books, video cassettes ?

13. House Keeping Applicable to TQM

— What is the state of front yard ?
— How does reception/lobby look ?
— Are plant and equipment properly maintained.
— Whether walk ways are clear without obstructions ?
— Is work place clean ?
— How do the toilets look ?
— Whether all safety practices are followed ?
— Are the walls, floors, roofs clean ?
— Is there scrap lying around ?
— Whether operators use personal protective equipment ?
— Is the canteen clean and pleasant ?

SOLVED PROBLEMS

Q. 1. Write a brief note on summary of elements of ISO 9001 standard.

Ans. Elements of ISO 9001 standard are

(1) *Management Responsibility*. Requires that quality policy be defined, documented, and communicated throughout the organization ; that responsibility regarding quality be clearly defined ; that in-house resources are available for verification activities ; that a management representative be appointed to ensure that quality system requirements are being met ; and that the management representative leads a management review periodically to ensure the continuing suitability and effectiveness of the quality system.

(2) *Quality System*. Requires a quality system that meets the criteria of the applicable ISO 9000 series standard and it is established and maintained (documented as a quality system manual and implemented) as a means of ensuring that production conforms to requirements.

(3) *Contract review*. Requires review of contracts to ensure that requirements are adequately defined and to ensure that capability exists to meet the requirements.

(4) *Design control*. Requires procedures for controlling and verifying product design to ensure that specified requirements are being met and to include procedures for design/developments planning, design input/output, design verification and design changes.

(5) *Document control*. Requires establishing and maintaining procedures for controlling documentation through approval, issue, change, and modification.

(6) *Purchasing*. Requires that purchased products conform to specified requirements ; ensured through sub-contractor assessments, clear and accurate purchasing data and verification of purchased product.

(7) *Purchaser-Supplied Product*. Requires procedures for verification, storage and maintenance of purchaser supplied product.

(8) *Product Identification and Traceability*. Requires procedures for identifying during all stages of production, delivery and installation, and individual product or batch-unique identification as needed.

(9) *Process Control*. Requires procedures to ensure that production and installation processes are carried out under controlled conditions, which include documentation, monitoring and control of suitable process and product characteristics, use of approved equipment and criteria for workshop.

(10) *Inspection and Testing*. Requires that procedures for inspection and test are followed for incoming material, during processing, etc. and final stations be in place as documented in quality plan ; must include maintenance of records and disposition of product.

(11) *Inspection, Measuring and Test Equipment*. Requires procedures for selection, control, calibration and maintenance for measuring test equipment.

(12) *Inspection and Test Status*. Requires that markings, stamps or labels be affixed to product throughout production and installation to show conformance or non-conformance to tests and inspections.

(13) *Control of Nonconforming Product*. Requires control of non-conforming product to ensure it is not inadsvertently used, includes identification, segregation and evaluation.

(14) *Corrective Action*. Requires procedures for investigating causes of nonconformance, taking action to rectify them and creating controls to prevent future occurrence.

(15) *Handling, Storage, Packing and Delivery*. Requires procedures for handling, storage, packaging and delivery of product.

(16) *Quality Records*. Requires procedures for identification, collection, indexing, filling and storage of quality records.

(17) *Internal Quality Audits*. Requires a system of internal audits to verify whether quality activities comply with requirements and to determine the effectiveness of the quality system.

(18) *Training*. Require procedures for identifying training needs and providing training for all personnel to meet those needs.

(19) *Servicing*. Requires procedures for performing servicing as required by contract.

(20) *Statistical Techniques*. Requires procedures for identifying the use of statistical techniques in process, product and service.

Q. 2. What are the advantages of adopting ISO 9001 ?

Ans. They set the basic rules for quality systems from concept to implementation, irrespective of type of product or service.

Motivates exporters

Pressure on work force to rectify and meet customer-commitment

Independent audit and management reviews provide better communication

Sets a base line

Establishes reasonable standard for Government procurement

Reduces time consuming audits by customers and regulators

Focuses training and professional development

Improves the company's quality image

Improves efficiency : Reduces scrap and rework

Gives marketing advantage

Gives a recognizable logo that can be added to sales literature

Technical barrier will become obsolete by a common standard of ISO

Inconsistencies in audit will disappear

Q. 3. What are disadvantages of ISO 9001 ?

Ans. The implementation of a formal quality system is very demanding of resources. It is time consuming and may involve considerable secretarial expense. Assessment and registration are costly. A lot of team work is called for, unless carefully planned, the system can become non-cost effective and burdensome. If a company has made some progress on total quality, the effort will reduce, once the system is introduced, the work will reduce dramatically.

By exposing people's actions and cherished practices as not conductive to quality the process can become traumatic.

The need to change attitudes and accept new working practices may strain the management capability of the company beyond its ability to cope.

Registration to ISO 9001 may not completely eliminate second party assessment. It may be used as a guide for selection of firms for looking at more closely for meeting the specific requirements of the customer.

Q. 4. Write short note on preparation of documentation for ISO 9001.

Ans. Every company has processes whereby it manufactures products, provides service to customers, stores materials or products, disposes non-conforming products and so on. Many of these activities are perhaps already documented to some extent in the form of policies, operational procedures, standard operating procedures, flow diagrams, Pie charts, hand written notes, inspection proceeds memoranda, letters and manuals. The first step for those companies that have better or no experience in developing a quality assurance system is analyzing the previously maintained documents to learn what is being done at present. The documents can be stored into stacks according to how the information applies to the particular standard with which it is required to comply. Each document should then be assessed for :

Completeness, consistency, redundancy with other documents, clarity and appropriateness and necessary and this work can be done by internal teams.

The next step is to develop additional documentation that is needed to make the quality assurance procedure complete. What was developed, found or used to satisfy the requirements in the proceeding paragraph should be written down or stored in a computer memory as a documented quality assurance procedure.

The next task is evaluating the quality assurance system as documented earlier. This is done through the adequacy audit. This is an office exercise which determines the extent to which the documented Q.A.P. adequately meets the relevant ISO 9001 standard and if it provides objective evidence that the system has been designed to do so. It is preferred that the adequacy audits be done by an independent assessor.

The three key assessor attributes as :

1. Knowledge of the ISO standard.
2. Assessing experience.
3. Objectivity.

Based on the adequacy audit, corrections regarding procedural requirements to meet the standard are made and once for all the procedures have to be firmed up and documented.

This may be called by any name, perhaps, quality assurance, procedure manual.

The next step is the compliance with the requirement of the Q.A.P. as documented.

Are the practices intended to be followed in the Q.A.P. manual being actually followed ? Is there any evidence to confirm that what is being done is as intended.

This calls for training of all the employees to follow the Q.A.P. Then an internal compliance audit is conducted. This is the audit which seeks to establish the extent to which the documented procedures are implemented and observed by the work force.

Once, the compliance audits result are found to be positive, the firm is in a position to apply for registration. When the concerned third party comes to the firm for auditing for the purpose of registration, they may find discrepancies during the audit. The discrepancies are of two types :

1. Major discrepancies—Total absence or break down of a mandatory procedure.
2. Minor discrepancies—Lapses to comply with the mandatory procedure in practice.

Number of identical similar discrepancies can be an indication of a major discrepancy—registration will be denied in case of a major discrepancy.

Thus a company intending to register under ISO 9000 series has to establish and maintain a documented quality management system which establishes that the product/service conforms to specified requirements.

Various documents required for this purpose include : (*i*) Organisational plan, (*ii*) System documentation and (*iii*) Quality manual.

Organisation chart should define the structure of organisation, clearly listing the responsibilities of various personnel who control the quality. The structure may be in three or four levels, namely top management, departmental heads, middle management, and supervisors.

Quality manuals serve as permanent reference for implementation and maintenance of quality management system in a company. It communicates the quality policy and objectives of the company to be used by staff and for information of customers. It clearly lays down the responsibility and authorities in respect of quality assurance.

Q. 5. What steps are to be followed for installation of ISO 9001 quality system ?

Ans. Preparation Steps

— Prepare to create a quality organization.

— Management of commitment to ISO 9001 and express Quality Policy.

— Set up an implementation committee and appoint a coordinator.

— Start ISO awareness programme.

— Conduct status review and prepare action plan.

— Develop organizational structure for quality management.

— Define the responsibilities and authorities of the personnel performing quality related activities.

— Develop quality system documentation.

— Prepare the resource system in line with the standard.

II. Implementation Steps

— Implement the documented quality system.

— Establish internal quality audit.

— Monitor and stabilize the quality system.

— Conduct pre-registration internal audit.

III. Registration and Certification Steps

— Apply for registration.

— Adequacy audit by certification body.

— Compliance audit by certification body.

— Certification.

Q. 6. How the assessment and certification/registration of quality systems complied with ?

Ans. The third-party certification by registration bodies carrying out assessment and giving a certificate to the company is accepted by all buyers.

The method for credibility of certifiers/registrars and of the assessment of their competence as a basis for their recognition needs to be established by accreditation bodies.

Many developed countries have established accreditation bodies to control the competence and impartiality of quality system certification/registration bodies. This ensures a reasonable degree of recognition at the national level.

To achieve recognition at the international level, some of the accreditation bodies (and in their absence the certification/registration bodies) have concluded mutual recognition agreements with similar bodies in other countries. Again accreditation/certification/registration bodies are faced with the prospect of multiple assessments (by their peers) and limited recognition. The optimum solution to the problem of recognition at the international level lies in an *international system of recognition*.

Q. 7. What are the metrological requirements of ISO 9001 standard ?

Ans. Three out of the twenty sub-clauses of ISO 9001 stipulate requirements for the quality system that deal with inspection and testing.

The sub-clauses on inspecting and testing and test status deal with inspection and testing at different stages (receiving, in-process and final), with inspection and test records and the status of tested products (identification of conforming and non-conforming products).

Sub-clause devoted to the inspection, measuring and testing equipment contains 9 requirements concerning measuring and testing equipment.

(*a*) *determine the measurements to be made and the accuracy required, and select the appropriate inspection, measuring and test equipment ;*

(*b*) *identify, all inspection, measuring and test equipment that can affect product quality, and calibrate and adjust them at prescribed intervals, or prior to use, against certified equipment having a known valid relationship to internationally or nationally recognized standards. Where no such standards exist, the basis used for calibration shall be documented ;*

(*c*) *define the process employed for the calibration of inspection, measuring and test equipment, including details of equipment type, unique identification, location, frequency of checks, check method, acceptance criteria and action to be taken when results are unsatisfactory ;*

(*d*) *identify inspection, measuring and test equipment with a indicator or approved identification record to show the calibration status ;*

(*e*) *maintain calibration records for inspection, measuring and test equipment ;*

(*f*) *assess and document the validity of previous inspection and test results when inspection, measuring and test equipment is found to be out of calibration ;*

(*g*) *ensure that the environmental conditions are suitable for the calibrations, inspections, measurements and tests being carried out ;*

(*h*) *ensure that the handling, preservation and storage of inspection, measuring and test equipment is such that the accuracy and fitness for use is maintained. ;*

(*i*) *safeguard inspection, measuring and test facilities, including both test hardware and test software, from adjustments which would invalidate the calibrations setting.*

Careful consideration shows that these laconic, concise requirements contain all the necessary and sufficient conditions to ensure adequate measurement results. Requirement (*a*) deals with the identification of those measurements that should be carried out on the product and the selection of suitable equipment. This action, which can be called *design of measurement and test operations,* is normally undertaken before production starts.

The remaining requirements are intended to ensure the accuracy of inspection, measuring and test equipment and the validity of the results of measurements and tests. This is based on the following foundations:

— traceability (requirement *b*) through certified standard equipment ;

— adequate calibration procedures (requirement *c*) which should be documented and should include equipment types and locations, frequencies of checks, check method, and acceptance criteria ;

— suitable environment (requirement *g*) for calibrations as well as for measurements and tests ;

— adequate identification and records of measuring and test equipment (requirements *d* and *e*) ;

— careful handling of the equipment and safeguards against the inadvertent upsetting of their adjustment (requirements *h* and *i*).

22

Non-Destructive Testing of Metals and Alloys

22.1. Introduction

As the name itself implies, non-destructive testing means testing of materials without destroying them. Its purpose is to detect and evaluate any defects or predict the strength and serviceability of the parts. Flaws may be inherent within a material or they may arise during any stage of building the complete machine. The non-destructive techniques have become important tools for the industries to initiate preventive maintenance programmes. With the help of these techniques, it has become possible to accurately predict machine life by knowing the true conditions of its internal parts. It is possible to test the materials in their raw form as supplied to the equipment builder, test during fabrication, test after final shaping and assembling and finally the regular checks during machine's operating life.

The purpose of testing metals is to ensure that the recommended requirements have been observed and that the components are fit for the service intended to be performed. Non-destructive tests are becoming popular because these can be carried out in situ without damaging the parts in use. Destructive tests are carried out occasionally as demanded by quality control procedures. Parts in actual use may be fabricated by casting, forging, welding, etc., or be in the form of plates, tubes, etc. The typical defects in castings are porosity, internal shrinkage, tears, cracks, scabs, etc. Defects in welds are porosity cracking, lack of fashion, lack of penetration, over penetration, slag inclusion, undercut, etc. Defects in forgings could be central cavities, cracks, etc. and those in tubes could be laminations, draw marks, defective longitudinal welds, etc. In operating plant, the problem may be reduced to studies of crack formation due to fatigue or creep processes, and changes in section due to wear-and/or erosion. Thus the detection of some discontinuity in the structure (which may always appear on the surface and even if it does it may not be visible to naked eye) is an essential requirement.

Visual Inspection. Initial visual inspection of the geometry of the component and the type and nature of the defect likely to be present is the first stage of any inspection. The inaccessible surface can be inspected by introscopes (illuminated optical device using a system of lenses and mirrors). Miniature conventional cameras have been developed to examine internal condition of long parts. The photographs can be taken by remote control. Miniature television cameras have also been developed. These transmit a continuous record of the inner surface of the part, to be viewed on a monitoring screen by the operator.

Non-destructive testing of metals and alloys can be carried out by (1) Magnetic particle inspection, (2) X-ray radiography, (3) Gamma radiography, (4) Ultrasonic testing, (5) Electrical method, (6) Damping test, (7) Non-magnetic methods of non-destructive testing, (8) Optical holography method, and (9) Hardness testing.

22.2. Magnetic Particle Inspection

In order that minimum material be used for high strength, it is essential that minute irregularities and discontinuities be discovered in the material.

Magnetic-particle inspection is a non-destructive method for detecting flaws or defects in magnetizable materials, by which a visible indication of the defects is produced on the surface of the part being inspected. This method will locate cracks and surface defects in a wide range of products. The name "Magnaflux" is commonly associated with this process.

Fields of Application. (1) In the inspection of parts which are cyclically stressed in service, the danger of fatigue must be considered. Fatigue cracks start at points of local high tensile strength. Such conditions are frequently due to the design of the part causing severe inequalities in stress distribution. This method is, therefore, frequently applied at various stages in the processing of a part to locate defective parts as soon as they become defective thus saving the cost of subsequent machining and finishing process. Grinding cracks and invisible surface cracks which may be inherent in material or added during processing, lower the dynamic strength of materials, and can be detected by this method.

(2) To locate fatigue cracks in parts already in service.

Method of Detecting the Defects. This method is relatively easy and employs simple technique. Materials (magnetic) of any shape, size, composition can be tested. There are two steps involved : (*a*) Magnetization of the material, (*b*) Application of finely divided magnetic particles (either dry or suspended in a liquid).

Inhomogeneties such as blowholes, cracks and inclusions in a magnetic material produce distortion in an induced magnetic field. The path of the magnetic flux is distorted because the inhomogeneties have different magnetic properties than the surrounding material. All the magnetic methods of non-destructive testing employ some means by which distortion called leakage flux, can be detected.

If the flaw is a surface flaw or lies sufficiently close to the surface, the lines of force set up a pair of poles at the surface, which act like a small horse-shoe magnet to attract and hold some of the magnetic powder. These powder particles form a visible indication of the location and extent of the defect. The surface to be tested must be clean, dry and free of rust.

For maximum sensitivity, the specimen should be magnetized with the direction of magnetic field perpendicular to the defect. Fig. 22.1 shows how a circular field will locate a defect lying parallel to the longitudinal axis of the specimen but will not locate a transverse flaw.

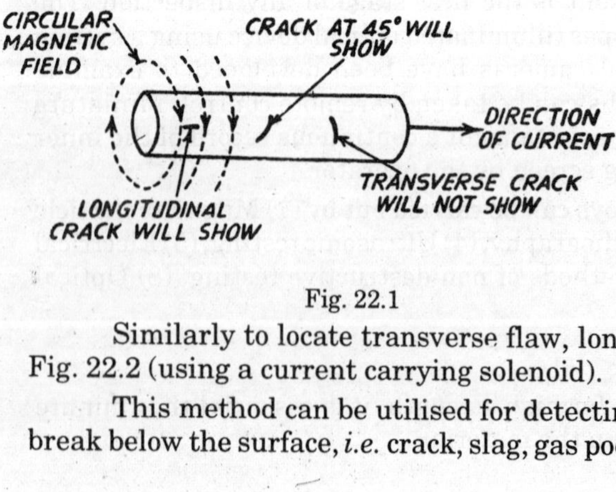

Fig. 22.1

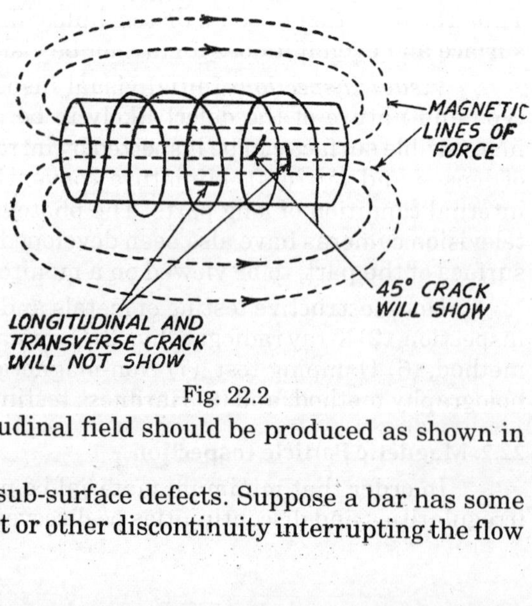

Fig. 22.2

Similarly to locate transverse flaw, longitudinal field should be produced as shown in Fig. 22.2 (using a current carrying solenoid).

This method can be utilised for detecting sub-surface defects. Suppose a bar has some break below the surface, *i.e.* crack, slag, gas pocket or other discontinuity interrupting the flow of the magnetic flux (Fig. 22.3). The flux lines try to get around the increased reluctance caused by the defect, and some flux lines are crowded out on the surface. Again local poles are set up where the leakage field leaves and re-enters the surface of the part. Indications of the leakage can be obtained by applying suitable magnetic powder.

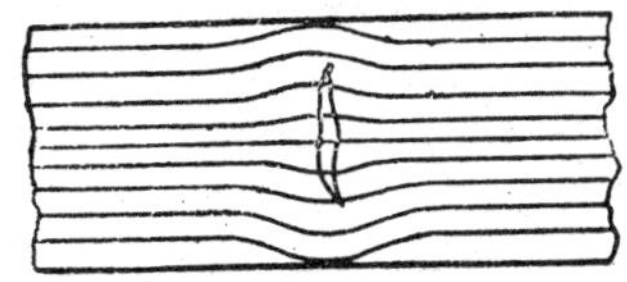

Fig. 22.3. Flux distribution sub-surface defect.

The magnetic powder can be used either in the form of a powder (dry technique) or in a liquid suspension (wet-technique). The material used for the dry technique consists of finely divided ferromagnetic particles having high permeability and low retentivity. The particles may be coloured to give maximum contrast on the specimen. In the wet technique finely divided particles of red or black iron oxide are suspended in a light petroleum distillate or in water.

By this method the following defects can be detected :

(1) Quench cracks,
(2) Fatigue cracks,
(3) Thermal cracks in welds,
(4) Surface defects in castings,
(5) Forging cracks etc.

22.3. Magnetic Crack Detection Method

Magnetic crack detection method is valuable for the detection of crack present at the surface of a component made from a magnetic material. This method is based on the fact that the faces of a crack tend to form north and south poles if a magnetic flux is established in the component. The flux is induced by the use of a permanent magnet or an electromagnet, by causing current flow through the component or by wrapping coils around it and then passing a current through the coils. If the system is so arranged that the crack interrupts the flux lines, then it may be revealed by the application of either dry iron powder or iron powder in suspension is a liquid. The particles of iron are attracted to the pole formed at the crack which is then delineated as a black line.

Subsurface defects are indicated by the use of heavy d.c. currents which produce a flux below the surface of the material.

22.4. Penetrant Inspection Method

Penetrant inspection method is suitable only for the detection of cracks open at the surface but can be used on virtually all the metals. The machined surface to be tested is cleaned and degreased. The liquid penetrant is applied at surface and permitted to remain for several hours, during which the liquid penetrant into any defects open at the surface. The excess penetrant that remains is then removed and an absorbent powder (developer) is sprayed uniformly on the surface. The developer acts as a blotter and draws out a portion of the penetrant which had previously seeped in the opening. As the penetrant is drawn out it diffuses into the developer forming an indication much wider than the surface opening. The penetrant is usually coloured with some dye and the developer is white. Thus the crack easily shows up on the surface of the developer.

The method is much simplified if fluorescent magnetic penetrant is used and viewed in ultraviolet light.

22.5. X-ray Examination

X-ray Radiography. X-ray examination is a non-destructive method of detecting the presence and location of internal cavities or other discontinuities in materials. X-ray examination of materials may be carried out by processes known as radiography, photofluoroscopy and fluoroscopy. All are dependent upon the differential penetration of matter and differ only in the manner in which the emergent radiation is detected or recorded. The absorption of X-ray follows the law,

$$I = I_o e^{\mu x}$$

where I_o = Intensity of incident X-radiation.
I = Intensity of emergent X-radiation.
x = Thickness of absorbing material.
μ = Linear absorption coefficient.
e = Napierian constant.

The range of X-ray tube voltage ordinarily used for radiography of metals is from about 5 to 2,000 kV.

If a beam of X-radiation is directed normal to the surface of a mass gradually tapering cross-section, such as a steel wedge, a high percentage of the incident beam will penetrate at pointed edge and emerge on the opposite side. As the wedge section increases in thickness, less X-radiation will be transmitted until a point is reached at which practically all the incident X-ray beam is absorbed. If the specimen is of uniform thickness, entire emergent or transmitted beam will be of equal intensity. However, if the specimen contains a cavity, groove or similar change in cross-section, with a different X-ray absorption characteristic, a variation in X-ray transmission at these points will result. It is this variance of the emergent X-ray beam which makes it possible to study the internal character of matter. As X-rays are invisible the effect of differential absorption caused by discontinuities in matter must be shown by some means of measuring or recording emergent radiation. Photosensitive film is one of the most reliable methods. X-rays have the same effect on a photosensitive emulsion as does visible light. The greater the intensity of the X-ray beam, the greater becomes the resulting density on the film. The image of a cavity inside the specimen will appear darker than the surrounding area because of the greater X-ray transmission at this point. The position or depth from the surface of the cavity or discontinuity has little or no effect upon the appearance of the resultant X-ray image.

22.6. Radiography

When we talk about radiographic inspection we include X-ray, gamma ray and variations of these methods. Basically, method involves passing rays through materials to be tested. Rays impinge on a film or screen, reveal and record the internal structure of test object. Thus radiography records the transmitted X-ray radiation on a film to produce an image or shadow picture termed as "radiograph". Applications of radiography in industry fall into two basic categories : (1) Foundry products ; (2) Welded products.

Defects that occur in castings and which can be detected by radiographic examination are : shrink, hot tears, porosity, lack of fusion around chaplets, inclusion of sand and tramp metal, coarse dendritic structure and cracks. Shrink resulting from certain foundry practice faults is characterised by its location and appearance. As the radiographic image gives a true projection of this condition, the foundaryman can use this information in making remedial changes. Under certain circumstances shrink produces cracklike void commonly called "hot tears". These flaws generally occur at the junction of heavy and thin section of the casting and can easily be identified from the X-ray film.

Gases not liberated from the mould create various forms of porosity in the casting and this will be revealed by radiographic inspection. Similarly other defects are also inspected. In all these cases the sensitivity limit is about 2 per cent. (Radiographic sensitivity is a measure in percentage of the minimum thickness change evident in a radiographic image. Standard practice requires that radiographs show evidence of a minimum sensitivity of 2 per cent. Thus if a 25 mm thick plate of steel is radiographed, a section change in the form of an increase or decrease in thickness equal to 0.5 mm or 2 per cent must be visible on the film).

Radiography has been extensively used to examine welds of pressure vessels, storage containers, aircraft landing gears and, propellers. When faulty welding is located by a radiography the weld metal is chipped out and the seam rewelded. X-ray examination is then made to determine the quality of repair. Most common flaws detected are : gas porosity, trapped slag, lack of fusion or incomplete penetration, cracks etc. Radiography is not reliable for the detection of extremely fine flaws originating at the weld or plate surface.

22.7. Fluoroscopy

Fluoroscopy makes use of certain chemicals such as zinc sulphide, barium platinocyanide and calcium tungstate which fluoresce when acted on by X-ray radiation. These materials are distributed on card-board or similar material to make a screen and absorb part of the X-ray radiation, transforming it into a visible long wavelength of light. The image of a cavity in any object examined by this method appears as a light area. The advantages of fluoroscopy are :

(1) The image is directly visible.

(2) The object can be viewed while in motion.

(3) Inspection is rapid.

(4) Cost of producing an image is very low.

The chief limitations are :

(1) Use is limited to low-absorption materials.

(2) Sensitivity of inspection is lower than with radiographs (5% under ideal conditions).

(3) There is no permanent record.

(4) X-ray protection requirements limit its applications.

22.8. Photofluoroscopy

It combines some of the advantages of both radiography and fluoroscopy. In this case a permanent record is made and film cost is considerably reduced. But this method has its limitations of fluoroscopy regarding sensitivity and inspection of flow absorption materials.

22.9. Gamma Radiography

Principle of detecting defects is the same as X-ray radiography.

Gamma rays are emitted during the disintegration of radioactive material and like X-rays are electromagnetic radiation. Gamma-rays are of shorter wavelength and consequently are more penetrating. The chief merits of Gamma sources are (1) small size, (2) high penetration, (3) relatively low cost compared with X-ray units, (4) independence from supply of electricity and water. The major disadvantages of such sources are :

(1) They are of low intensity requiring long explosure.

(2) Some sources have a short half-life which necessitates decay correction or replacement.

(3) Since they cannot be 'shut off' protection is always necessary.

22.10. Ultrasonic Inspection

Introduction. Ultrasonic is the general subject of sound whose frequency is above the upper pitch limit of the human ear. Ultrasonic frequencies are generally considered to range from 15 kilocycles upwards.

In the metals field, ultrasonic testing and inspection is one of the most useful non-destructive methods. Railroads, water or oil pipes, boilers, shafts, die blocks, aircraft parts of forged materials, extruded and forged aluminium parts, etc., are tested for cracks, segregations, inclusions or other internal discontinuities by this method. Certain types of welds are also inspected.

Ultrasonic Wave Production. There are a number of ways in which ultrasonic waves may be produced. However, the method of producing ultrasonic waves for non-destructive testing is to utilize the piezoelectric effect. When mechanical pressure or tension is applied to certain crystals, electric charges develop in the crystal face. The magnitude of the electric charge produced is directly proportional to the mechanical pressure or tension applied. The sign of the electric charge changes when the mechanical force changes from compression to tension. This phenomenon is known as the piezoelectric effect. This effect has been observed in a number of crystals like quartz, tourmaline and rochelle salts. There is also a reverse piezoelectric effect in which the crystal changes dimensions if subjected to a suitable field. If a quartz plate is put unto an a.c. field the quartz will be compressed in one-half of the cycle and expanded in the other half. The crystal will contract and expand with the same frequency as the applied voltage. The amplitude of vibration is maximum when the frequency of the electric field is the resonant frequency of the crystal.

Any device which converts one type of energy into another type is known as a 'Transducer'. The term transducer is used here to mean a device to convert electric energy to mechanical energy and *vice versa.* Transducers used for this purpose are barium titanate, quartz, lithium sulphate, etc.

22.10.1. Inspection of Flaw in Materials. *Pulse Echo Technique.* In this technique a pulsed ultrasonic beam is sent through the couplant into the specimen. At the opposite face, the beam is reflected and the echo picked up by a transducer as shown in Fig. 22.4.

A discontinuity or flaw in the specimen will also send back an echo. The time intervals that elapse between initial pulse and the arrival of the echoes are measured with a C.R.O. In the echo pattern, a flaw can be recognised by the relative position and amplitude of its echo. There may also be a number of multiple reflections displayed in the echo pattern. The

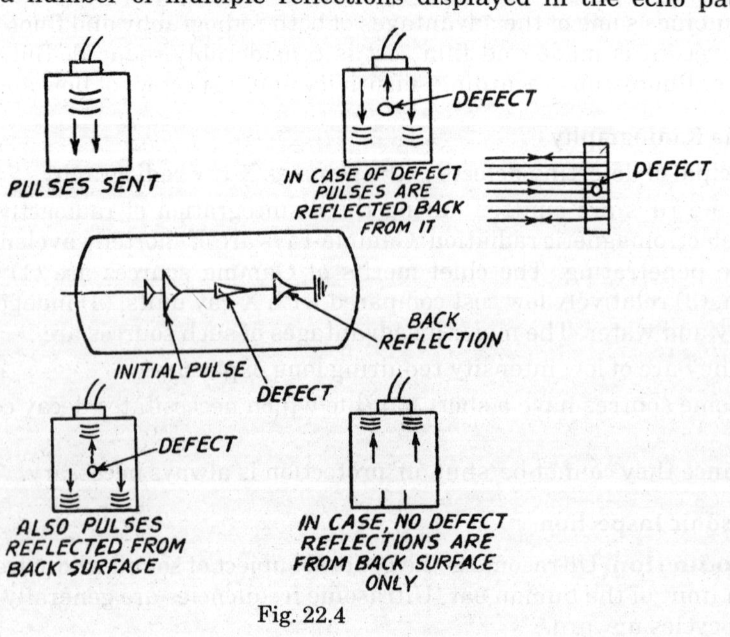

Fig. 22.4

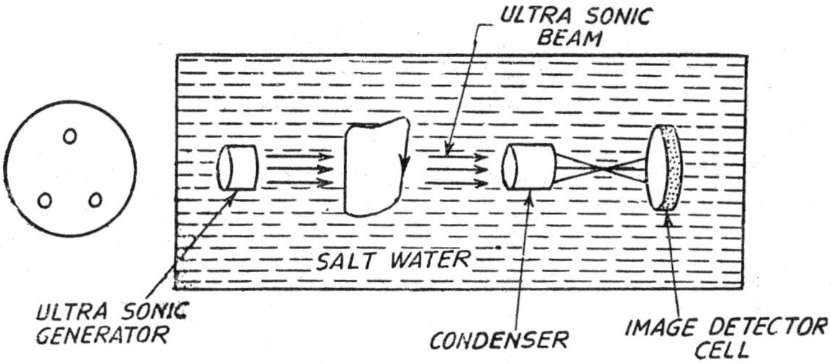

Fig. 22.5

representation of this technique depends on the duration of the ultrasonic pulses. The shorter the duration of the pulse, the thinner the specimen that can be successfully tested.

Large pipes upto 30 m diameter can also be tested by this method.

22.10.2. Accoustic Image method. An ultrasonic wave is passed through the specimen and focussed on the image forming cell. The image cell consists of flake shaped aluminium particles 0.05 to 0.6 micron in size suspended in a light viscous liquid. In the absence of any ultrasonic energy, the particles assume random positions as a result of thermal agitation.

The particles are oriented by the ultrasonic beam. The particles which are oriented to the incident beam reflect considerable light and appear bright. Particles randomly oriented reflect negligible light and appear as a dark background. The degree of orientation depends on the intensity of the ultrasonic beam at a given point in the image. Thus the variation of intensity in the reflected light pattern indicates the amount of ultrasonic energy transmitted through the specimen. Flaws appear as dark spots in the images. Such cell is limited in its sensitivity and is slow because the alignment of the particles may require several seconds. Also the particles must be reagitated after each image is formed. It is not possible to locate the position of the defect.

22.11. Electrical Method

A number of electrical methods have been employed in non-destructive inspection and testing of machinery and wide variety of metallic materials and parts of dimensional inaccuracies and physical defects. The method consists of measuring the electrical resistance of the material and then to note the variation in the electrical resistance. This variation is correlated to the physical defect.

A crack detector operates on the principle that if a crack occurs anywhere within a piece of metal, it interferes with the flow of electric current through the metal, therefore, increasing its overall resistance. This holds true regardless of the shape of the piece. Operation of the instrument consists of accurately measuring the electrical resistance of some critical machine part between two definitely established contact points, usually chosen at the extreme opposite ends and of repeating the measurements at regular intervals. When successive measurements show that the resistance is increasing at a progressively high rate, a fatigue fracture is beginning to propagate and the part should be removed from service. Fig. 22.6 shows a typical resistance

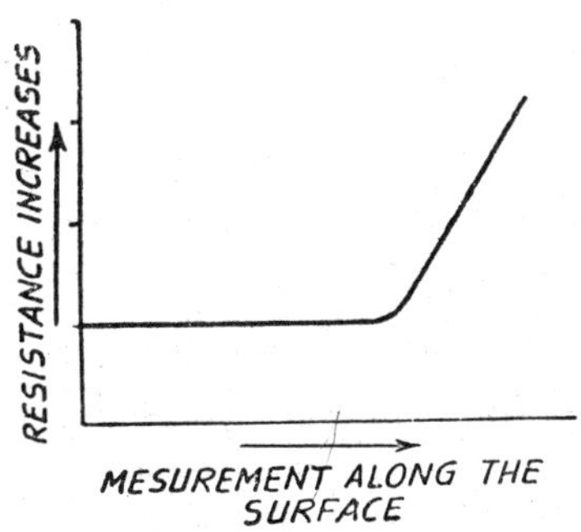

Fig. 22.6

curve showing fracture in one of the main crank pins of steam locomotive. This method of detecting fatigue fracture is particularly applicable to parts assembled in the hydraulic press, railway-car axles, locomotive driving axles and crankpins pressed into wheel centres, because testing may be carried out without dismantling the assembly.

22.12. Electrical/Magnetic Methods

Electrical and/or magnetic tests are used for detecting differences in composition, condition, dimension and the presence of defects. These methods are based on the characteristics (relationships) between electrical magnetic properties and the constitution of alloys.

An instrument based upon a probe containing two coils in tandem has been designed to detect flaws and the effects of corrosion and/or erosion in non-magnetic tube. The probe is passed down the bore of the tube to be examined. The primary coil is excited by an alternating current which produces a magnetic flux and induces a current in the secondary (search) coil. When the probe is placed inside a tube an eddy current is induced in the tube and modifies the current in the secondary coil. The coils form a part of a bridge and the electrical system is arranged to permit adjustment so that the output of the coils may be balanced when the probe is placed within a round tube of the correct dimension and material. Differences in tube properties which are influenced by composition and dimension—such as incorrect material or loss of material by thinning or pitting upset the balance of the bridge circuit, producing a voltage which can be made to operate a pen recorder.

22.13. Damping Test

When a solid specimen vibrates, its free oscillations decay when isolated from their environment. Some of the energy is always converted into heat. The various mechanisms by which this transfer of energy occurs are collectively termed 'Internal Friction.' Measurement of damping can give information on the origin of defects such as the formation of quenching cracks. For example, an increase in damping was found in a steel specimen which had been quenched. With cylindrical pieces it is possible to determine the position of a crack in the piece.

The method of determining the damping factor consists of finding the time for amplitude to decrease by a specific fraction, the ratio between the amplitudes of successive free oscillations is constant. The natural log of this ratio delta, called the logarithmic decrement is taken as a measure of the internal friction. (damping factor $Q = \pi^8$). Fig. 22.7 shows how the damping factor varies in the case of a sound specimen and for a cracked specimen. Using torsional vibrations, sensitivity with which flaws can be detected is least when they are situated in the centre of the bar. High-strain damping is sensitive to the presence of cracks and this fact has been applied successfully in the routine detention of season cracks in brass cartridge cases.

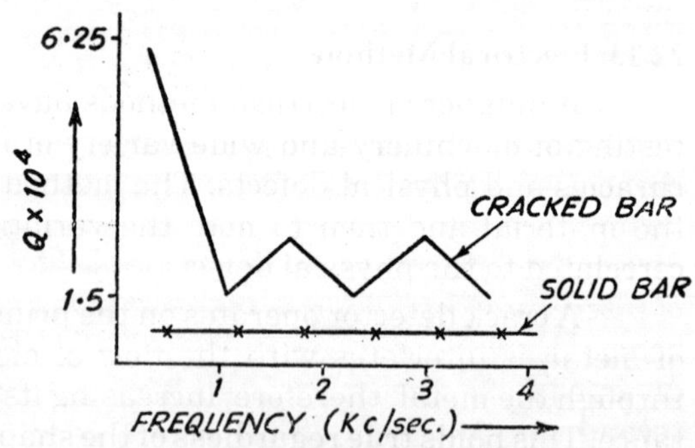

Fig. 22.7

22.14. Non-magnetic Methods of Non-destructive Testing

Several effective non-magnetic methods for inspecting both magnetisable and non-magnetisable materials have been recently developed.

(*a*) Zyglo, fluorescent method is most suitable for detecting surface cracks, minute holes and season cracks in metals like ceramics, glass and plastics. The part is first dipped into a highly fluorescent, liquid penetrant. By dipping, the penetrant settles in surface discontinuities and is examined under black light. With all other light subdued, near ultraviolet activates the zyglo penetrant and the mere scratches scarcely noticeable turn out to be deep cracks in the materials.

This method is highly sensitive, particularly in the location of underlying defects. The scratch etc. can be very easily focused, thus increasing the speed of testing.

(*b*) Statiflux is an electrified particle method of inspection for minute surface discontinuities in non-conducting materials. The non-visible discontinuities in the materials are rapidly located by a large visible indication marking each defect and its extent. The part to be tested is dipped or coated with a hot liquid penetrant conductor which enters even very fine discontinuities. The surface is then dried and a cloud of electrostatically charged particles is blown. The particles are held electrostatically at the defect. A visible powder indication is thus built up around the defect.

22.15. Optical Holography Method

Holography is basically a two-step process for creating a three-dimensional image of a diffusely reflecting object having some arbitrary shape. In the first step both the amplitude and phase of any type of coherent wavemotion emanating from the object are recorded by encoding this information in a suitable medium. This encoding is called a hologram. Subsequently the wave motion is reconstructed for the hologram by a coherent beam in a process that results in the generation of an image having the true shape of the object. This regenerated image can be used as a kind of three-dimensional template against which any deviation in the shape or dimension of the object can be observed and measured, thus proving this technique useful for non-destructive testing.

In principle, holography can be performed with (*a*) any wave radiation encompassed in the entire electromagnetic spectrum ; (*b*) any particulate radiation, such as neutrons and electrons that possess wave-equivalent properties ; and (*c*) non electromagnetic wave radiation, such as sound waves. The two methods currently available for practical non-destructive inspection are optical holography, visible light wave, which is discussed here ; and acoustical holography using ultrasonic waves.

Holographic Recording. When visible light waves are employed in holography, the hologram is recorded using an optical system called a holocamera. (Ref. Fig. 22.8). A monochromatic laser beam of phase coherent light is divided into two beams by a variable beam splitter. One beam, the object beam, is expanded by a spatial filter into a divergent beam directed to illuminate the object uniformly. A portion of the laser light reflected from the object is intercepted by a high resolution photographic plate, as shown in Fig. 22.8 (*a*). The second beam, the reference beam, originating from the beam splitter is directed by a mirror, and diverges from a second spatial filter being further directed onto the photographic plate by a second mirror. With either object beam or reference beam absent, a uniformly exposed photographic plate will result. However, with both coherent beams falling on the plate simultaneously, an interference pattern is generated as a result of the coherent interaction of the two beams, and this pattern is recorded by the photographic emulsion.

The fixed wavelength of the light used and the varying magnitude of the offset angle (the angle between the object beam and reference beam at any point on the photographic plate) determine the orientation and spacing of the fringe comprising the recorded interference pattern. The angle between the reference beam and the photographic plate, in combination with the offset angle and wavelength of the light, influence the resolution requirements of the recording medium. These requirement generally are of the order of 1,000 to 3,000 lines per millimeter. The photographic emulsion must have sufficient resolution and contrast to clearly

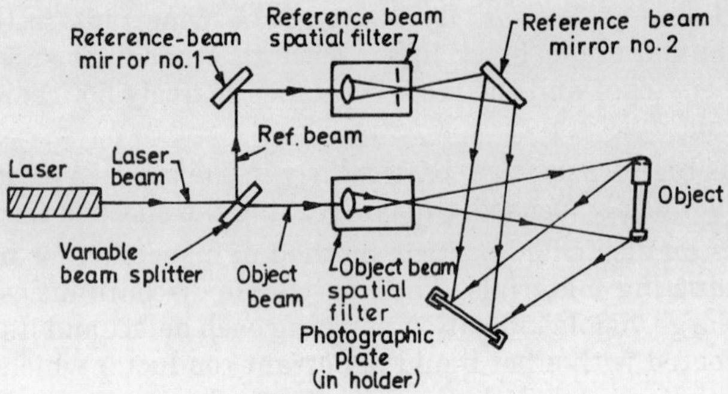

(*a*) Halographic recording

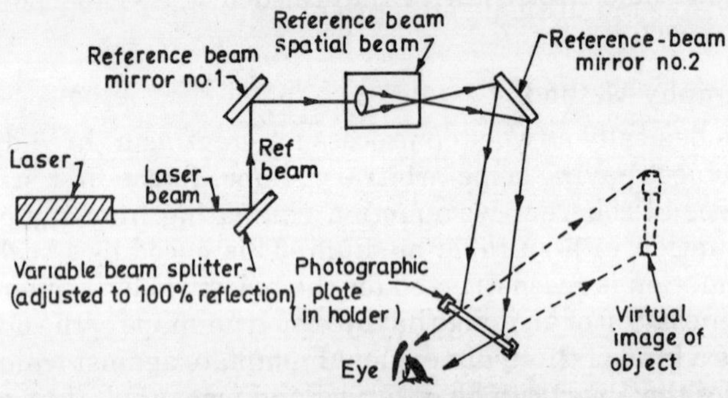

(*b*) Halographic reconstructing

Fig. 22.8. Schematic diagrams of the basic optical system. (*a*) Halocamera used to record hologram ; (*b*) Optical system for reconstructing a virtual image of the object.

record these interference fringes. Ordinary photographic emulsions are unsatisfactory, because they can resolve only about 200 lines per millimeter. For higher resolution, exceptionally fine-grained emulsions are required.

The actual difference between the lengths of the paths of the reference beam and the object beam must be less than the temporal coherence length of the laser beam, and should be as close to zero as possible to obtain fringes of the highest contrast. The holocamera must be isolated from vibration to the extent that there is less than one-fourth to one-eighth variation in wavelength in the object-beam and reference-beam paths during an exposure. An apparent variation in wavelengths implies a time-varying difference between the path lengths. This can be caused by relative motion of various elements in the object and reference portions of the holocamera.

The contrast of the interference fringes is directly related to the amplitude of the reflected light from the object and the amplitude of the reference beam. The fringe separation is inversely proportional to the sine of one-half the offset angle. The complex interference recorded on the high-resolution photographic plate contains all the information about the object.

Holographic Reconstruction. In the reconstruction process, (Ref. Fig. 22.8 (*b*)), the hologram is used as a diffraction grating. When it is illuminated with the reference beam only, three beams emerge—a zero-order, or undeflected beam, and two first-order diffracted beams. The diffracted beams produce the real and virtual images of the object to complete the holographic process. The real image is pseudoscopic, or depth inverted in appearance. Hence the virtual image (also referred to as the true, primary or nonpseudoscopic image) is the one that is of primary interest in practical applications of holography. In Fig. 22.8 (*b*) only first-order diffracted beam that yields the virtual image has been shown, the other two beams having been omitted from the figure for the sake of clarity. If the original object is three-dimensional, the virtual image is a three-dimensional replica of the object, possessing both parallax and depth of focus.

The light intensity in the reconstructed image depends on the efficiency of the hologram. For a typical hologram, the intensity is less than 3% of the incident light. The light in the image may be utilized just as one would utilize the light from an illuminated object viewed through a window. Pictures can be made photographically or electronically permitting interpretation.

22.16. Hardness Test

Hardness tests form a part of inspection to determine the suitability and uniformity of materials. These are also used widely as they are non-destructive tests and with hardness value, idea of the tensile and ultimate strength of materials can be had by certain conversion factors.

Hardness is a property of material which is defined as the resistance to penetration, abrasion, scratch cutting etc. The hardness is a relative concept. Various methods of hardness testing for various criteria are given below.

22.16.1. Brinell Hardness Test. This determines the resistance of a material to penetration. Hardness is determined by measuring the impression made by steel ball forced into a specimen under a predetermined load. It has got only one scale of hardness and Brinell Hardness number is defined as the average axial stress over the surface of the indentation produced by a steel ball, assuming the indented surface is spherical. The advantage of Brinell hardness test is that it has got direct relations with maximum tensile strength of the material. The diameter of the spherical impression is measured by a high magnification microscope and the spherical area calculated from the diameter of impression and diameter of ball. The surface to be tested must be smooth and free from surface defects. Brinell test makes a large impression, so it can be applied effectively to the heterogeneous surface. It has the disadvantage that part tested is ruined and a big piece is required because specimens must be thick enough to prevent the anvil support from influencing the penetration of the ball. This cannot be used for very hard materials testing as the ball may itself get deformed.

22.16.2. Rockwell Hardness Test. This test consists of measuring the difference in depth of penetration of steel ball or diamond cone which is forced into the metal first by a minor load and then by a major load. The advantage of the differential load is that errors resulting from mechanical defects in the testing system are eliminated. Also errors due to the slight surface imperfections which are usually hard and the different conditions of contact which exist between the penetrator and the surface of specimen are eliminated.

22.16.3. Vickers Pyramid Hardness Test. This is used in the precision hardness testing of thin sheets and hardness surfaces.

It uses diamond indentor which undergoes very little deformation when in contact with very hard materials. The indicator is loaded through a leverage by dead weights for a predetermined time, usually 30 sec. The load is applied very gradually and acts for pre-determined time and released automatically through and oil dash arrangement and cam mechanism. The measurement is made by means of a swinging microscope mounted on the side of the machine. The penetrator used is a square based diamond pyramid having the angle of pyramidal point equal to 136°. The Vickers hardness number is determined by measuring the diagonal of a square surface impression. Its advantage is that the impression is so fine that it cannot be seen by naked eye. Therefore, test can be easily performed on the finished product.

22.16.4. Shore Scleroscope Hardness Test. This test is used for finding the hardness of very soft materials like rubber and those heavy parts which can't be taken below the hardness testing machine *e.g.*, lathe bed. This records the rebound rather indentation. The instrument used is very cheap and can be operated very easily.

In it, a diamond tipped hammer rebounds from the specimen, which is first dropped from a certain height. The height of rebound is a measure of hardness. If the material is soft then more energy will be absorbed in forming the indentation and rebound will be less and *vice versa.* The hammer used is of cylindrical shape and is guided in its fall by a tube. The tube is kept in vertical position.

22.16.5. Mohr's Scale of Hardness. The test is used to determine the scratch hardness of a material. This is of particular use in case of very thin specimens *e.g.* blades. Some standard minerals of very high hardness are used to produce permanent scratches on another material. The indentation made is measured by high magnifying microscope and hardness found.

22.16.6. File Hardness Test. This is the simplest test to determine the abrasive hardness of a metal. The test piece is subject to the cutting action of a file and it is observed whether a visible scratch is produced or not. This is very crude test and must more depends upon the skill of the operator.

TYPICAL EXAMINATION QUESTIONS

1. State whether following statements are true (T) or false (F) :

(*a*) In proof testing, equipment is subjected to load equal to or greater than the designed value to check that it remains intact.

(*b*) In destructive testing, little or no specimen preparation is required.

(*c*) In nondestructive testing, the test equipment is often portable.

(*d*) In nondestructive testing the properties are measured indirectly and results are often qualitative or comparative.

(*e*) In destructive testing different observers may interpret the results differently and results often require interpretation by skilled operators.

(*f*) In liquid penetrant inspection, a liquid penetrant is drawn into surface by capillary action and subsequently revealed by developer material.

(*g*) Liquid penetrant test can reveal both surface and subsurface flaws and inclusions.

(*h*) In magnetic particle inspection, if current flows in circular path around object, magnetic field will be longitudinal and thus longitudinal flaw will not show.

(*i*) A series of inspections utilising different forms of magnetisation may be required to fully inspect a part in magnetic particle inspection.

(*j*) In magnetic particle inspection technique, alternating-current methods are more sensitive to surface flaws, whereas direct-current methods are more capable of detecting subsurface defects.

(*k*) Air being good transmitter of ultrasonic waves, it is used as acoustic coupling medium between transducer and object to be inspected.

(*l*) In radiography testing, complex shapes can pose problems in setting exposure condition and obtaining proper orientation of source, specimen and film.

(*m*) In eddy current testing method, electrical conducting test material when brought near a.c. coil generating alternating magnetic field, causes eddy currents at surface and interact with main field, altering the impedence of coil. Defects affect the magnitude and direction of induced eddy current and can be detected by instruments.

(*n*) Only growing flaws can be detected and not the mere presence of defects in acoustic emission testing method.

(*o*) Thermal methods can be used to evaluate the soundness of engineering materials and components.

(*p*) Holographic interferometry can be used to detect minute changes in the shape of an object under stress.

(*q*) If a defect is sufficiently small, it is possible for it to remain dormant throughout the useful lifetime of a product and thus such pieces need not be rejected but can be accepted.

2. Choose correct word in blank from 2 choices given against each.

(*a*) ...... testing provides a direct and reliable measurement of how a material will respond to service conditions. (destructive, nondestructive)

(*b*) In nondestructive testing, ... specimen preparation is required. (high, little)

(*c*) In nondestructive testing, the properties are measured ...... (directly, indirectly)

(*d*) The simplest and most widely used nondestructive testing method is ...... inspection. (visual, hardness)

(*e*) A liquid penetrant is drawn into surface flaws by ..... action. (viscous, capillary)

(*d*) The surface for liquid penetrant inspection penetrant inspection must be .... (porous , non porous)

(*e*) Liquid penetrant inspection can test flaws that are ...... surface. (open to, inside)

(*f*) In magnetic particle inspection method, a ...... magnetic filed exists in the vicility of flaws and defects. (distorted, uniform)

Answers

1. (*a*) T (*b*) F (*c*) T (*d*) T (*e*) F (*f*) T
(*g*) F (*h*) T (*i*) T (*j*) T (*k*) F (*l*) T
(*m*) T (*n*) T (*o*) T (*p*) T (*q*) T.

2. (*a*) destructive (*b*) little (*c*) indirectly
(*d*) visual (*e*) capillary (*f*) non-porous
(*g*) open to (*h*) distorted

23

Machine Vision Systems

23.1. Introduction

Machine vision (also sometimes called computer vision or intelligent vision), can be defined as a means of simulating the image recognition and analysis capabilities of the human eye/brain system with electronic and electromechanical techniques. In human vision system, eyes sense the image and brain analyses the information and takes action on the basis of analysis. Similarly in machine vision system, visual sensing of information is done after forming an image of the object and information is analysed to make a useful decision about its content. Machine vision system thus needs both visual-sensing and interpretive capabilities.

Image sensing device may consist of vidicon camera or a charge-coupled device (CCD) image sensor. It receives light through a lens system and converts this light into electrical signals. Micro-computers are used to refine and analyse these electrical signals to provide an interpretation of the object that generated these signals.

Machine vision systems are now-a-days used to provide accurate and inexpensive 100% inspection of workpieces. It is ideally suited for in-process gauging and provides timely corrective actions to control the process. Machine vision systems are used for functions like gauging of dimensions, identification of shapes, measurement of distances, determining orientation of parts, quantifying motion, detecting surface shading etc. These systems offer excellent consistency and reliability of results. The resolution is limited by pixel array size and discrimination is limited by high-contrast images. Accuracy remains consistent in high production volume. While human vision also possesses such capablities, but is best suited for low or moderate production volume. On the other hand machine vision systems are best suited for high production volume because of high cost and consistency of results at higher production. These systems function reliably without fatigue. A big advantage of machine vision systems is that they can be designed to operate for wavelength in ultraviolet and infrared ranges to which human eye is insensitive.

Machine vision system is better suited for inspecting the masks used in the production of microelectronic devices. It is a noncontact measuring technique and thus suited for fragile workpieces and which would distort under measuring force and would get contaminated by touch. Standoff distance upto one metre is possible. Accuracy of $\pm$ 3μ and better is achieved.

23.2. What is a Machine Vision System ?

Machine vision system may be defined as means of electro-optically simulating the image recognition capability of the human eye/brain system.

It is a technique by which visual sensory inputs can be obtained and processed in a manner, such that meaningful information about the image is generated. The availability of microprocessors coupled with advanced pattern recognition software makes machine vision an industrial reality. Machine vision systems perform the functions of image sensing, image analysis and image interpretation. These systems have ability to automatically acquire data about an object, measure image features, recognise objects and make appropriate decisions.

23.3. Principle of Working

The machine vision system involves following four basic steps :

(*i*) Image formation

(*ii*) Processing of image in a form suitable for analysis by computer

(*iii*) Defining and analysing the characteristics of image

(*iv*) Interpretation of image and decision making.

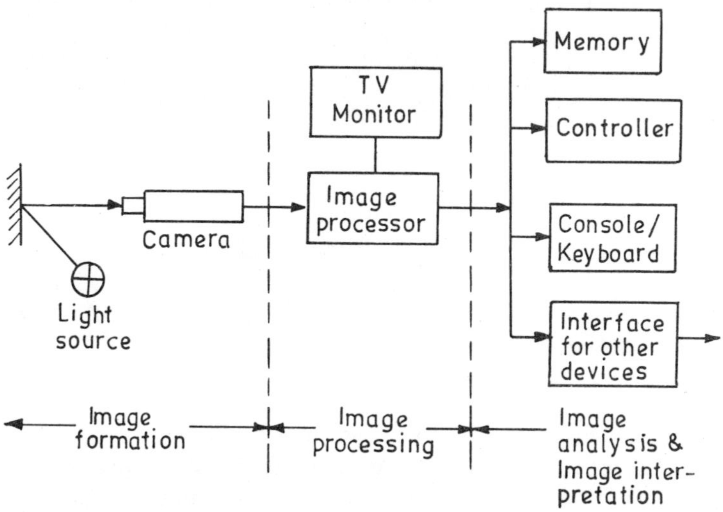

Fig. 23.1. Overview of machine vision process.

We will now discuss these four steps in more details.

(*i*) *Image formation.* For formation of image suitable light source is required. It may consist of incandescent light, fluorescent tube, fiber-optic bundle, arc lamp, or strobe light. Laser beam is used for triangulation system for measuring distance. Polarised or ultraviolet light is used to reduce glare or increase contrast. It is important that light source is placed correctly since it influences the contrast of the image. Selection of proper illumination technique, (*viz.*, back lighting, front lighting-diffused or directed bright field, or directed dark field, or polarised, structured light) is important. Back lighting is suited when a simple silhouette image is required to obtain maximum image contrast. (Refer Fig. 23.2).

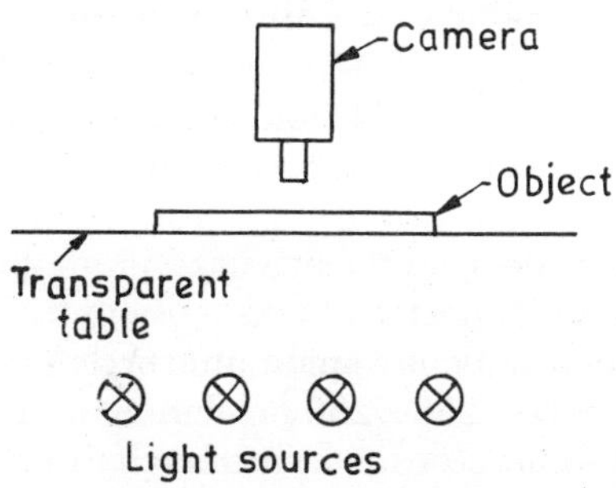

Fig. 23.2. Back lighting.

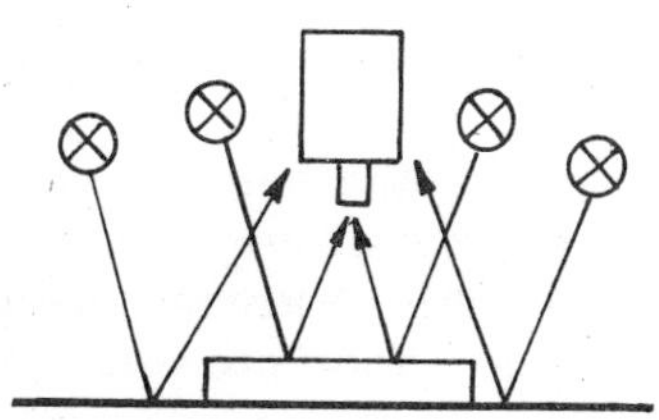

Fig. 23.3. Diffused front lighting.

Frontlighting is used when certain key features on the surface of the object are to be inspected.

If a three-dimensional feature is being inspected, side lighting or structured lighting may be required.

The proper orientation and fixturing of part also deserve full attention.

An image sensor like vidicon camera, CCD or CID camera is used to generate the electronic signal representing the image. The image sensor collects light from the scene through a lens and using a photosensitive target, converts it into electronic signal. Most image sensors generate signals representing two-dimensional arrays (scans of the entire image).

Vidicon Camera used in closed-circuit television systems can be used for machine vision systems. In it, an image is formed by focussing the incoming light through a series of lenses onto the photoconductive face plate of the vidicon tube. An electron beam within the tube scans the photo conductive surface and produces an analog output voltage proportional to the variations in light intensity for each scan line of the original scene.

It provides a great deal of information of a scene at very fast speeds. However they tend to distort the image due to their construction and are subject to image burn-in on the photo conductive surface. These are also susceptible to damage by shock and vibration.

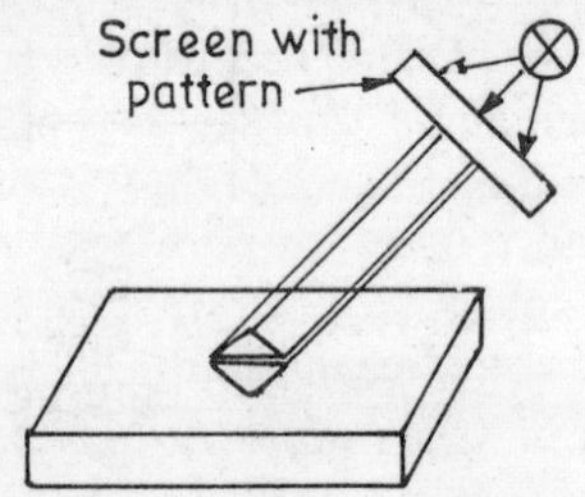

Fig. 23.4. Structured light.

Solid State Cameras. These are commonly used in machine vision systems. These employ charge coupled device (CCD) or charge injected device (CID) image sensors. They contain matrix or linear array of small, accurately spaced photo sensitive elements fabricated on silicon chips using integrated circuit technology. Each detector converts the light falling on it, through the camera lens, into analog electrical signal corresponding to light intensity. The entire image is thus broken down into an array of individual picture elements (pixels).

Typical matrix array solid state cameras may have 256 × 256 detector elements per array. Solid-state cameras are smaller, rugged and their sensors do not wear out with use. They exhibit less image distortion because of accurate placement of the photodetectors. CCD and CID differ primarily in how the voltages are extracted from the sensors.

(*ii*) *Image Processing.* The series of voltage levels available on detectors representing light intensities over the area of the image need processing for presentation to the microcomputer in a format suitable for analysis. A camera may typically form an image 30 times per sec *i.e.* at 33 m sec intervals. At each time interval the entire image has to be captured and frozen for processing by an image processor. An analog to digital converter is used to convert analog voltage of each detector into digital value.

If voltage level for each pixel is given either 0 or 1 value depending on some threshold value, it is called Binary System. On the other hand gray scale system assigns upto 256 different values depending on intensity to each pixel. Thus in addition to black and white, many different shades of gray can be distinguished. This thus permits comparison of objects on the basis of surface characteristics like texture, colour, orientation, etc., all of which produce subtle variations in light intensity distributions. Gray scale systems are used in applications requiring higher degree of image refinement. For simple inspection tasks, silhouette images are adequate and binary system may serve the purpose. It may be appreciated that gray-scale system requires huge storage processing capability because a 256 × 256 pixel image array with upto 256 different pixel values will require over 65000-8 bit storage locations for analysis, at a speed of 30 images per second. The data processing requirements can thus be visualised. It is, therefore, essential that some means be used to reduce the amount of data to be processed. Various techniques in this direction are :

(*a*) *Windowing.* This technique is used to concentrate the processing in the desired area of interest and ignoring other non-interested part of image. An electronic mask is created around a small area of an image to be studied.

Thus only the pixels that are not blocked out will be analysed by the computer.

(*b*) *Image Restoration.* This involves preparation of an image in more suitable form during the pre-processing stage by removing the degradation suffered. The image may be degraded (blurring of lines/boundaries ; poor contrast between image regions, presence of background noise, etc.) due to motion of camera/object during image formation, poor illumination/poor placement, variation in sensor response, poor contrast on surface, etc.).

The quality may be improved, (*i*) by improving the contrast by constant brightness addition, (*ii*) by increasing the relative contrast between high and low intensity elements by making light pixels lighter and dark pixels darker (contrast stretching) or (*iii*) by fourier domain processing.

Other techniques to reduce processing are edge detection and run length encoding. In former technique, the edges are clearly found and defined and rather than storing the entire image, only the edges are stored. In run-length encoding, each line of the image is scanned, and transition points from black to white or *vice versa* are noted, along with the number of pixels between transitions. These data are then stored instead of the original image, and serve as the starting point for image analysis.

(*iii*) *Image Analysis.* Digital image of the object formed is analysed in the central processing unit of the system to draw conclusions and make decisions. Analysis is done by describing and measuring the properties of several image features which may belong to either regions of the image or the image as a whole. Process of image interpretation starts with analysis of simple features and then more complicated features are added to define it completely. Analysis is carried for describing the position of the object, its geometric configuration, distribution of light intensity over its visible surface, etc.

Three important tasks performed by machine vision systems are measuring the distance of an object from a vision system camera, determining object orientation, and defining object position.

The distance of an object from a vision system camera can be determined by stadimetry (direct imaging technique, in which distance is judged by the apparent size of an object in the field of view of camera after accurate focussing), or by triangulation technique, or by stereo vision (binocular vision technique using the principle of parallax).

The object orientation can be determined by the methods of equivalent ellipse (by calculating an ellipse of same area as the image of object in two-dimensional plane, and orientation of object being defined by the major axis of the ellipse), the connecting of three points (defining orientation by measuring the apparent relative position of three points of image), light intensity distribution (determining orientation based on relative light intensity), structured light method (in which the workpiece is illuminated by the structured light and the three dimensional shape and the orientation of the part are determined by the way in which the pattern is distorted by the part).

Image can be interpreted by analysis of the fundamental geometric properties of two-dimensional images. Usually parts tend to have distinct shapes that can be recognized on the basis of elementary features. For complex three-dimensional objects, additional geometric properties need to be determined, including descriptions of various image segments (process being known as feature extraction). In this method the boundary locations are determined and the image is segmented into distinct regions and their geometric properties determined. Then these image regions are organised in a structure describing their relationship.

An image can also be interpreted on the basis of diference in intensity of light in different regions. Analysis of subtle changes in shadings over the image can add a great deal of information about the three-dimensional nature of the object.

(iv) Image Interpretation. Image interpretation involves identification of an object based on recognition of its image. Various conclusions are drawn by comparing the results of the analysis with a prestored set of standard criteria.

In a binary system, the image is segmented or windowed on the basis of clustering of white and black pixels. Then all groups of black pixels within each segment (called blocks) and groups of white pixels (called holes) are counted and total quantity is compared with expected numbers to determine how closely the real image matches the standard image.

Statistical approach can be utilised to interpret image by identification of a part on the basis of a known outline. The extent of analysis required for part recognition depends on both the complexity of the image and the goal of the analysis.

The complex images can be interpreted by use of gray-scale interpretation technique and by the use of various algorithms.

The most commonly used methods of interpreting images are feature weighing (several image features and measured to interpret an image, a simple factor weighing method being used to consider the relative contribution of each feature to be analysed) and template matching (in which a mask is electronically generated to match a standard image of an object). In actual practice, several known parts are presented to the machine for analysis. The part features are stored and updated as each part is presented, until the machine is familiar with the part. Then the actual parts are studied by comparison with this stored model of a standard part.

Similarly mathematical models of the expected images are created. For complex shapes, the machine is taught by allowing it to analyse a simple part. Standard image-processing software is available for calculating basic image features and computing with models.

23.4. How Machine Vision System Functions

Fig. 23.5 shows a general block diagram of a vision system on next page.

Lighting and presentation of object to be evaluated is very important task in implementing a vision system. It has great impact on system repeatability, reliability, and accuracy. Lighting source and projection should be chosen such that it accentuates the key features of the object, and gives sharp contrast and detail of the image. The specular reflections by small angle lighting and other techniques which provide diffused reflection should be avoided.

Image sensor usually comprises of a TV camera, which may be vidicon TV camera which has greater resolution and low in cost, or it may be solid state camera (charge coupled device CCD or charge injection device CID). The solid state cameras have greater geometric accuracies, no image lag, and longer life.

Image digitiser is usually a six to eight bit analog to digital A/D converter which is designed to keep up with the flow of video information from camera and store the digitised image in memory. For simple processing, analog comparator and a computer controller threshold to convert the video information to a binary image is used. The binary images (having only two values for each pixel) are much simpler and facilitate high speed processing. However gray scale images contain a great deal more picture information and must be used for complex images with subtle variation of gray level across the image.

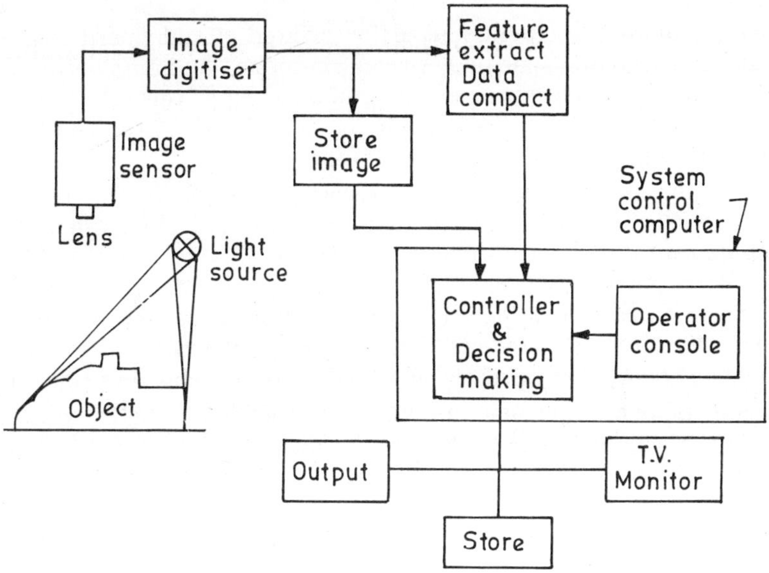

Fig. 23.5

Feature extractor/data compactor employs a high speed array processor to provide very high speed processing of the input image data. To generate a relatively simple feature data set, pattern recognition algorithms need to be implemented.

System control computer communicates with the operator and makes decisions about the part being inspected. These decisions are usually based on some simple operations applied to the feature data set representing the original image.

The output and peripheral devices operate under the control of the system control computer. The output enables the vision system to either control a process or provide cation and orientation information two a robot, etc.

23.5. Fields of Machine Vision Systems

The machine vision could be used for :

(*a*) **Inspection.** The ability of an automated vision system to recognise well-defined patterns and determine if these patterns match those stored in the system makes machine vision ideal for inspection of raw materials, parts, assemblies, etc.

(*b*) **Part identification.** Machine vision systems due to their ability of part recognition provide positive identification of an object for decision making purposes. Part identification applications of machine vision systems range from the familiar optical character recognition process to pårt sorting and bin picking operations.

(*c*) **Guidance and Control.** Machine vision systems are being used to provide sensory feedback for real-time guidance and control applications, ranging from visual servoing of industrial robots and weld seam tracking to calculations of geometric offsets for part processing and assembly operations.

23.6. Machine Vision System Applications

Machine vision systems can be used to replace human vision for applications like welding, machining to ensure that correct relationship is maintained between tool and workpiece, assembly of parts to analyse the position of parts so that other parts can be correctly aligned for insertion or some other form of mating.

Machine vision systems are frequently used for printed circuit board inspection to ensure minimum conductor width and spacing between conductors and many other features. These are used for weld seam tracking, robot guidance and control, inspection of microelectronic devices and tooling, on-line inspection in machining operation, on-line inspection of assemblies, monitoring high speed packaging equipment, etc.

The ability of an automated vision system to recognise well-defined patterns and to determine if these match with those stored in the memory make it ideal for the inspection of parts, assemblies (both quantitatively *i.e.* dimensions, number of parts, number of holes etc.) and qualitatively, *i.e.* defects, missing parts, misaligned parts, etc.). Machine vision systems in place of human inspectors are able to perform 100% on-line inspection of parts at high speeds, particularly in unpleasant environments. Without tiring or boredom, these can operate continuously with consistent results over prolonged periods of time, achieving 100% accuracy.

Another important application of machine vision system is for the recognition of an object from its image. (Part identification). These systems are designed to have strong geometric feature interpretation capabilities and are used in conjunction with part-handling equipment like robots.

23.7. Developmental Programmes

Advanced vision systems are being developed to overcome limitations like limited 3-D interpretation ; limited interpretation of surfaces ; limited capability to examine a part if it is not properly oriented or obstructed by other overlapping parts ; etc.

Future development areas in machine vision system are :

— improved camera resolution
— ability to sense colour
— range sensing capability
— ability to detect overlap
— improved gray scale algorithms
— standardised software algorithms
— robot wrist-mounted vision systems
— motion-sensing capability
— parallel processing of whole image
— optical computing
— vision systems in servo mode.

23.8. Gray Scale Vs. Binary Imaging

The machine vision inspection systems involve conversion of continuous tone (or gray scale) image as seen by a black and white television camera into a binary image consisting of points which are all black or all white. It calls for use of considerable skill in optics and lighting to ensure that the "threshold" image looks like a reasonable facsimile of the part. Once an image is converted to binary even with good lighting and optics, much information is lost. Thus gray scale image processing techniques make machine vision more flexible and easier to use.

Sophisticated gray scale image processing combining edge detection and correlation provides a powerful technique to deal with varying parts position and varying lighting.

23.9. Gray Scale Image Processing Technique

There are four steps involved in this inspection process :

(*i*) **Find the part.** This involves finding the position of the part in the field of view and how it is oriented.

(*ii*) **Recognizing the part.** To establish above two steps, use is made of a number of graphic jigs and gauges, which can be displayed on the programmer's screen. These are manipulated using a light pen and placing them over jig points and features to be inspected on images of parts. These jigs and gauges have mechanical counterparts which simplify understanding how they work.

If during inspection, the orientation is precisely the same from part to part, the part may by electronically jigged using positioning pins. These pins operate like the positioning pins a tool maker uses in designing hard mechanical jigging. To find and recognise a part, the operator positions three pins over the gray scale image on the programmer screen by using a light pen. These pins serve to locate the part. If during inspection, parts move through translation or through rotation, the coordinate system of the part is translated or rotated with the location of positioning pins.

Operation of the positioning pins relies on feature extraction techniques. Each pin has an associated search area, (*i.e.* how much the pin, and therefore the part is expected to move). A normal size for the search area is 32×24 pixels, *i.e.,* 10% of a 320×240 field of view. Since the search area size depends on how much the part position can vary, the search area can be made larger or smaller by the use. The larger the search area, long it takes to find the part.

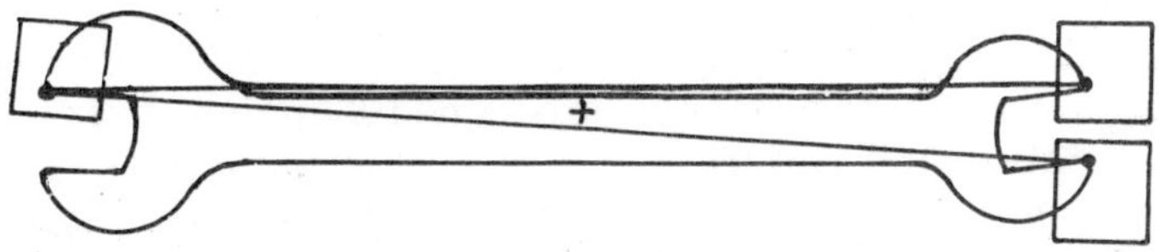

Fig. 23.6. Finding an object using positioning pins.

The "patch" of the picture inside each of the three search areas is edge detected using the Sobel edge detector (Refer Fig. 23.7). The central 7×7 portion of the Sobel pattern is stored away as a template for each positioning pin while the object is being trained. The distance between the centres of the positioning pins and the angles between them is stored away. These distances and angles can be thought of as describing a triangle. The apexes of the triangle are the centres of the positioning pins.

For *pixel E* in Fig. 23.7, the Sobel value is

$$\text{Sobel} = \sqrt{\text{Grad } X^2 + \text{Grad } Y^2}$$

where grad $X = (A + 2B + C) - (G + 2H + I)$

and. grad $Y = (C + 2F + I) - (A + 2D + G)$

A	B	C
D	E	F
G	H	I

Fig. 23.7. Edge detection.

When inspecting parts, the edge template is correlated or matched over the Sobel edge-detected search area for each pin. When correlation is complete, correlation peaks determine the location of best match between the template and the patch.

Next, a correlation operation is carried out to best-fit the triangle to the peaks of correlation in three patches. The result is the *X, Y* and angular offset corresponding to the shift of the part. Once the offsets are known, all subsequent measurements can be done on a rotated or translated coordinate system.

Operating with edges provides the ability to deal with lighting changes, both in intensity and direction. This technique is quite tolerant of variations in part-to-part glare and shadows.

Positioning pins can also be used for sorting of parts. By choosing three features which appear in one object and not in another, reliable sorting can take place.

(*iii*) **Inspecting the parts and measuring them.** Once the part has been located and identified using positioning pins, inspection and measurement take place. Inspection is carried out using graphic gauges. All gauges work over a broad range of lighting, and over a broad range of part motion. Three types of gauges used are : Calipers, defect finder, and font reader. Calipers are used to make measurements and to report if a measurement is within tolerance or not. The caliper is positioned over the features, to be measured. The blade is then snapped onto the appropriate edges. The caliper memorises the edge pattern in the neighbourhood of the blades and stores this pattern. While inspecting, the templates corresponding to each of the caliper blades are correlated with the edges of the image along the length of the caliper. The distance between the correlation peaks is the distance measured.

The defects can be found by determining consistent the amount of 'edginess' in a given search area is. Edginess is defined as the total number of edge pixels in the area, where an edge pixel is one whose Sobel value is greater than 16.

Fonts of characters are read with seven segment displays.

(*iv*) **Reporting the results.** The test results of machine vision inspection are reported to a host computer.

23.10. Mathematical Morphology for Shape Analysis

Mathematical morphology is a method of image processing and analysis based on set theoretical descriptions of image transformations and measurements. These transformations employe the concept of a structuring element, a predefined shape which is employed as a probe to test out the spatial nature of a pattern. Because all discriminations in such systems are made with respect to known geometric structures, their behaviour is similar to that of an optical comparator where a known good shape is compared to an unknown shape. Such systems function as non-contact gauging systems, analysing and measuring objects by performing morphological transformations on their video images.

23.11. Circular Scanning

Circular scanning refers to a solid state device that is configured as a light sensitive discretely segmented annular ring. Fig. 23.8 shows the difference between matrix scanner, linear array (linear scanner) and circular scanner. Circular scanning devices are used for angular position measurement, line following (seam or line tracking), surface inspection of bore and cylinder outside diameter etc.

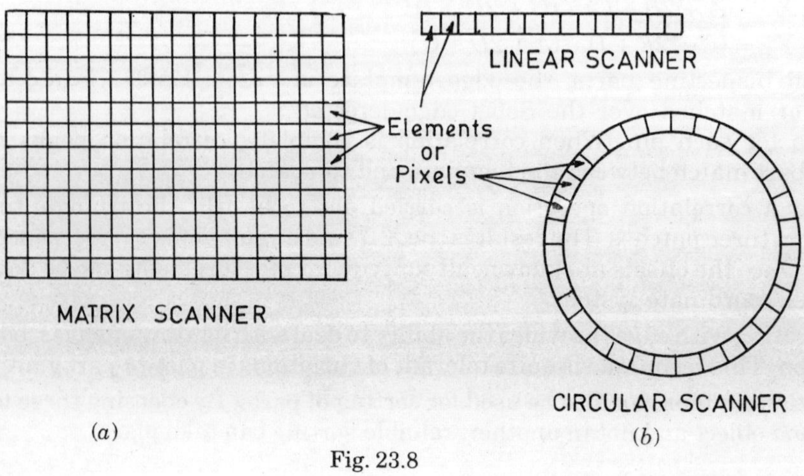

(*a*) (*b*)

Fig. 23.8

Circular scanners are self-scanned photodiode arrays which have their geometry, (like number of elements, aperture width, angular resolution, element spacing etc.) defined. These devices consist of monolithic arrays of silicon photodiodes. Each discrete photodiode site will output a signal that is directly proportional in amplitude to the illumination incident upon it. The output is a serial analog data stream whose rate is determined by the input clock frequency. The total time required to read out the number of elements is a function of the input clock and the number of elements in the array. The requirements of light intensity are a function of the circular scan time and the physical size of each photodiode.

If a cross hair target is imaged onto a circular array, the output would be as shown in Fig. 23.9 (*b*). This geometry, coupled with the appropriate optical imaging system, is capable of making direct angular measurements, defining coordinates for tracking or line following, sensing discontinuities in cylindrical surfaces, etc.

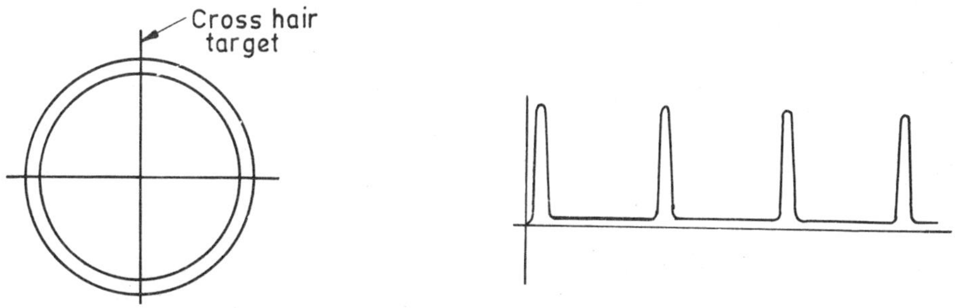

(*a*) (*b*)

Fig. 23.9

The various applications of circular scanning could be for detecting presence/absence of a feature, tracking (obtaining the coordinate data necessary to position the sensing system along defined patterns), displacement (calibration, verification, and alignment of an object moving in an angular motion about a pivotal centre), cylindrical bore or other defect inspection, location recognition, etc.

23.12. Photogrammetry

It is a dimensional measurement technique achieved through the use of photographic images. By precise two-dimensional optical measurement of a set of photographs, three-dimensional coordinates of points of interest can be extracted. It is the basis of all aerial surveying.

23.13. Industrial Computed Tomography (CT) for Evaluation of Internal Structures of Large, Dense Parts

Computed tomography (CT), popularly called the "CAT scan" is replacing X-ray technology and ultrasonic technology since it provides accurate and highly flexible images of the internal structures of composite and metal industrial components. CT is an advanced form of X-rays technology. While X-ray technology is used to detect internal flaws like cracks (which causes big density change), inclusions, porosity, or voids in castings/forgings ; CT is capable of detecting density changes of 1 to 2 percent (as may be noticed in porosity of a casting). CT is a filmless, digital imaging technique that can produce fast, digital radiography (DR) images of complete parts and show, more detailed computed tomography (CT) images. Thus CT generates a computer image of a thin, cross-sectional slice of an object. The CT image is a 2-D slice through a 3-D object. X-ray source emits photons in a thin, fan shaped beam, which pass through the object and are picked up by the detector array. The X-ray source, is a single examined, and the detector array opposite the X-ray beam, the section of the part being plane.

For DR the part is moved straight up through the X-ray beam and the detector collects data a line at a time. The lines are "stakked" to form the picture that resembles a regular X-ray. For CT the part is held at the height desired and rotated in the X-ray beam. Line of data are recorded for hundreds or thousands of angular positions. The computer composes the data into a single image of a slice through the part at the plane of the X-ray. Thus CT depends upon a specialised computer system to calculate and show the relative densities of the internal structures, within the section being examined, based on the relative attenuation of the X-ray beam passing through all the points within the plane. In other words, the number and intensity of the photons that pass through the object as it is rotated during the inspection, is computer-combined to produce a 2-D tomogram or cross-section, including dimensions, densities, and the exact locations of materials and any material flaws. Since the technique is sensitive to relative density differences between areas of under 1%, it is capable of detecting even minor internal-structural irregularities.

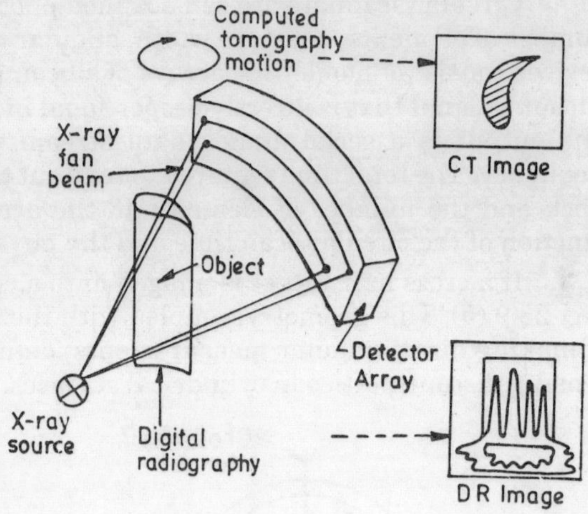

Fig. 23.10

The computer can combine series of CT slices to build up solid models of the object. These models may then be once again "sliced" by the computer to totally different ways to provide alternate images of internal structures. The internal structures inaccessible to a co-ordinate measuring machine or ultrasonic sensors may be dimensionally inspected by extracting the information from the computer-integrated solid model.

For wider range of denser materials 420 Ke V X-ray source is used. The linear detector array consists of 1280 channel, employing a gas-ionisation process, and can scan the field of view of 300 mm. Scan times for a single slice through a part may vary from 30 sec to 2 minutes. System can take 300 slices at 60 sec/slice. This way full, solid model of the part is captured. The parts can be inspected in position and need not be cut for inspection.

23.14. Grazing Incident Technique to Evaluate Surface Flatness

Laser-based system using grazing incidence interferometry to view part surface topography has been developed. It is a menu-driven, microprocessor based system which can analyse over 65,000 data points and computer flatness in less than 20 seconds and provide Total Indicated Reading (TIR), Local Slope (LS), and slice profiles. It measures the entire part surface at one time and produces two and three dimensional profiles with sub-micron accuracy and resolution. Test results can be viewed on the video screen and hardcopied on a colour printer. It correctly evaluates flatness by looking at the entire surface simultaneously and does not rely on several traces across a surface to determine indirectly overall flatness.

In this instrument, a laser beam inside the interferometer is expanded into 200 mm column of light. At the hypotenuse of the system's precision prism, the light column is divided into two beams : the reference and test beam.

The test beam exits the prism and strikes the irregular surface of the part at a grazing angle. The test beam reflects off the part surface, and re-enters the prism to recombine with the plane reference beam. When these beams recombine, they interfere with one another to create an interference pattern or a series of light and dark bands. The interference pattern represents the shape of the part of surface.

The measurement performance can be improved by striking the test beam on the part surface at a grazing angle. When normal incidence light strikes a rough surface, most or all energy is lost due to scatter. It would be appreciated that sunlight, when it strikes the water's surface at sunset (in comparison to at high noon) at a grazing angle, the surface easily reflects the sun's rays and appears to have mirrored finish. The gazing incidence design turns non-reflective, rough surfaces on ceramics, plastics, metals and composite into polished mirror surface to facilitate measurement.

The interference pattern is automatically interpreted with phase measuring interferometric (PMI) system. PMI system can capture over two million data samples, interpret them and compiles the data into a three dimensional image of the test surface.

A high contrast video system captures images of the interference pattern and divides the test surface into 65,536 picture elements or pixels. The pixel map is superimposed over the interference pattern and an intensity profile for the pattern is recorded into the system microprocessor. The interference pattern is shifted under computer control. At each pixel the intensity variation, which appears as a sine wave, is recorded by the microprocessor and used to determine its relative phase. The phase differences throughout the data matrix are used to define the surface topography.

23.15. Visual Inspection

It is a non-destructive technique used to detect and examine a variety of surface flaws and surface cracks. The visual inspection is aided by borescope (which illuminates and observes internal, closed or inaccessible areas), interference microscope (which measures the depth of scratches in the finish of finely polished/lapped surfaces), image sensors (which develop permanent visual records of remote cbject) ; magnifying system to evaluate surface finish, shape and microstructure) and dye and fluorescent penetrants (which enchance the observation of surface cracks).

Borescope. It consists of a long tube (rigid or flexible) available in a wide variety of lengths and diameters. A light source fitted at observer end sends light signal at other end where the surface in difficult to-reach areas is to be examined. The optical connection between the viewing end and an objective lens at the tip of borescope is possible by,

(*i*) using a series of achromatic relay lenses in a rigid optical tube.

(*ii*) using a bundle of optical fibers (rigid/flexible tube) that conduct light from external source.

(*iii*) using wiring to carry the image signal from a charge-coupled device (CCD) imaging sensor installed at the distal tip of a flexible tube.

The distal tip with combination of prisms and mirrors can be designed to define the direction and field of view and thus used for a variety of applications. Usually the surface to be inspected is illuminated by a fibre optic light guide.

The tube diameter of rigid borescope may range from 0.9 to 70 mm and length from 0.15 to 30 metres. Its use is limited to applications with a straight line path between the observes and surface to be observed. The field of view can range from 10 to 90°. It is possible to have orbital scan which involves the rotation of the optical shaft for scanning purpose. The illumination system incorporates features like orbital scan, different viewing head, and adjustable focussing of the objective lens. Some variants available are miniborescope (using a single image-relaying rod or quartz fibre in the optical tube in place of conventional relay lenses), hybrid borescopes (using rod lenses combined with convex lenses to relay the image), extendable borescopes (enabling use of extension tubes, or rigid chamberscope (having zoom facility and used for more rapid inspection of large chambers).

Fig. 23.11. Rigid Borescope.

Flexible Fiberscopes. These are available in 1.4 to 13 mm diameter and upto 12 m length (Fig. 23.12). For distances more than 15 m, videoscopes with CCD probes are used for electronic transmission of colour or black and white images to a video monitor. A CCD chip (consisting of thousands of light-sensitive picture elements, known as pixels arranged in a pattern of rows and columns) is placed at the distal end. The objective lens focusses the image of object on the surface of the CCD chip. The light at CCD chip is converted to electrons that are stored in each pixel. A voltage proportional to the number of electrons at each pixel is determined electronically for each pixel site. This voltage is then amplified, filtered, and sent to the input of a video monitor. Generally videoscopes produce higher resolution than fiberscope and have longer working length. The display reduces eye fatigue and enables point-to-point measurement on the viewing screen using reticles. The electronic form of the image signal allows digital image enhancement and the potential for integration with automatic inspection systems.

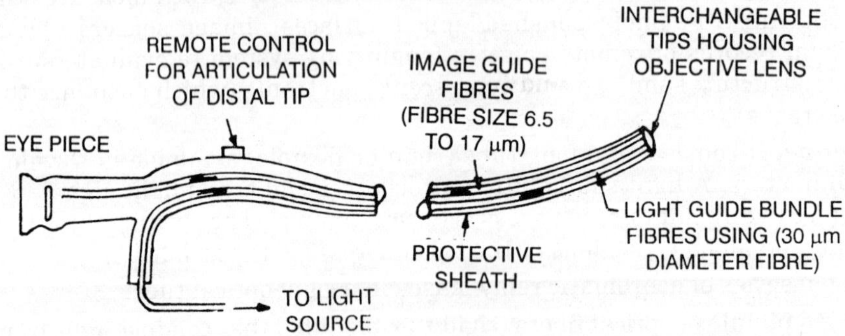

Fig. 23.12. Flexible fibrescope.

23.16. Laser Inspection

Metrology lasers are low-power instruments. Most are helium-neon continuous-wave output lasers that emit visible or infrared light. He-Ne lasers produce light at a wavelength of 6328 Å (0.6 μm) that is in phase, coherent, and a thousand times more intense than any other monochromatic source.

Laser inspection systems enable measurement of a part as it is produced, thus permitting 100% quality. Laser systems have wide dynamic range, low optical cross talk, and high contrast.

Lasers find applications in dimensional measurements and surface inspection because of the properties of laser light (bright, unidirectional, collimated beam, with a high degree of temporal and spatial coherence). These are useful where precision, accuracy, rapid non-contact gauging of soft, delicate, hot or moving parts is called for.

Various technique for dimensional measurements are :

(*i*) **Scanning Laser Gauge.** Fig. 23.13 shows a schematic diagram of a scanning laser gauge. It basically utilises a transmitter, receiver and processor electronics. A thin band of scanning laser light is made to pass through a linear scanner lens to render it parallel beam. The object placed in a parallel beam, casts a time-dependent shadow. Signals from the light entering the photo cell (receiver) are processed by a microprocessor to provide display of the dimension represented by the time difference between the shadow edges. It can provide results to an accuracy of ± 0.25 μm for 10-50 mm diameter objects. It can be used for objects 0.05 mm to 450 mm diameter and offers repeatability of 0.1 μm.

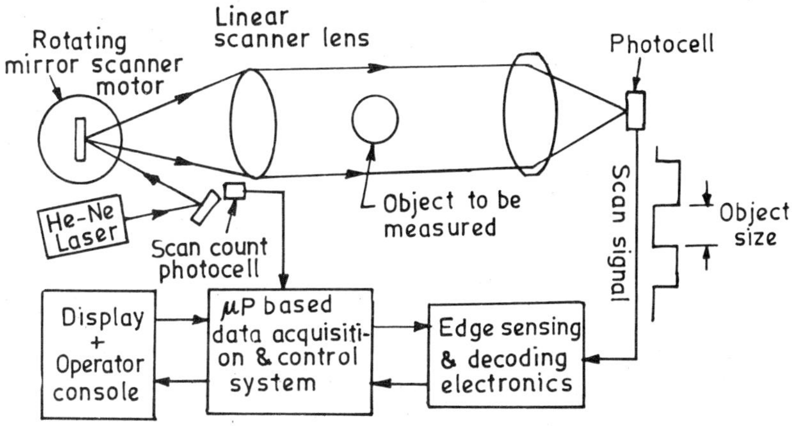

Fig. 23.13. Schematic of scanning laser gauge.

(*ii*) **Photo Diode Array Imaging.** In this method, shadow of stationary part is projected on a solid-state diode array image sensor. The system comprises of laser source, imaging optics, photodiode array, and signal processor and display unit. For large parts, two arrays, one for each edge are used. Accuracies as high as ± 0.05 μm have been achieved.

(*iii*) **Diffraction Pattern Technique.** These are used to measure small gaps and small diameter parts. In this method, a parallel coherent laser beam is diffracted by a small part, and the resultant pattern is focussed by a lens on a linear diode array. Since diffraction is not suitable for diameters larger than a few millimeters, its use is restricted to small wires, etc. The measurement accuracy is more for smaller parts. The distance between the alternating light and dark bands in the diffraction pattern is a direct function of the wire diameter, wavelength of laser beam, and the focal length of the lens.

(*iv*) **Laser Triangulation Sensors.** Fig. 23.14 shows the principle of laser triangular sensor. A finely focussed laser spot of light is directed at the part surface. A lens in the sensor images this bright spot onto a digital solid-state, position-sensitive photo detector. The measuring range is very small. The stand off distance is calculated and fixed carefully. The location of the image spot is directly related to the stand off distance from the sensor to the object surface. A change in the standoff distance results in a lateral shift of the spot along the sensor array. This technique provides quick measurement of deviations due to changes in surface and thus it can perform automatic

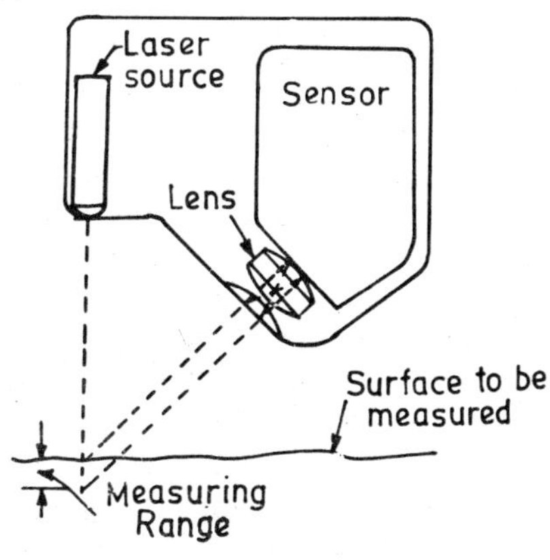

Fig. 23.14

calculations on shell metal stampings for gap and flushness, hole diameters, and edge locations very quickly. It is possible to measure part thickness or the inside diameter of bores by using two sensors. The accuracy of measurement and range of measurement are directly related.

(*v*) **Two-Frequency Laser Interferometer.** Fig. 23.15 shows schematic arrangement comprising of two-frequency laser head, beam directing and splitting optics, measurement optics, receivers, wavelength compensators, and electronics. It is ideally suited for measuring linear positioning, straightness in two planes, pitch and yaw. The two-frequency laser head provides, one frequency with a *P* polarisation (measuring beam) and another frequency with an *S* polarisation (reference beam). The laser beam is projected to a remote interferometer and is split at the polarising beam splitter into its two separate frequencies. The measuring beam is directed through the interferometer to reflect off a target mirror or retroreflector attached to the object to be measured.

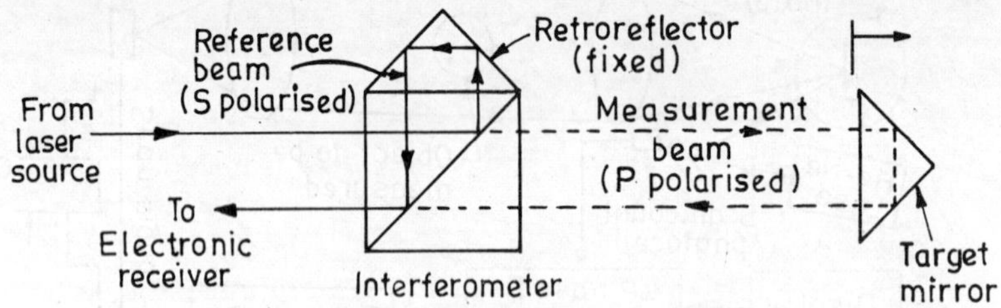

Fig. 23.15

The reference beam is reflected from fixed retroreflector. The measurement beam on its return path recombines with the reference beam and is directed to the electronic receiver. A frequency shift is induced into the returning measurement beam by Doppler effect due to movement of the target mirror. Because of their orthogonal polarisation, the frequencies do not interfere to form fringes until the beam reaches the receiver which monitors the frequency shift and provide output of the displacement.

The advantage of this technique lies in independence of system on the beam intensity and thus it provides high sensitivity, being independent of noise due to air turbulence, electrical noise, and light noise. Laser head can be mounted any where. Further due to use of single photodetector per measurement axis, sensitivity to optical alignement is less. The linear resolution of this method is 1 nm, straightness resolution is 40 nm and angular resolution is 0.03 arc seconds.

(*vi*) **Laser Scanning Gauge.** Fig. 23.16 shows a laser scanning gauge used for non-contact measurement of large objects to an accuracy of about 0.01 mm using the low divergence of a laser beam. A rotating mirror at the focus of the first lens produces a beam parallel to the optic axis. The time between the leading edge of the beam being extinguished by the object and when it emerges from the other side is a measure of the cross-section which is a function of the tangent of the angle subtended by the object at the focus of the first lens. This technique is useful to measure the roundness and diameter of hot steel bars under conditions of vibration to an accuracy of 0.025 mm over diameters of 5-25 mm.

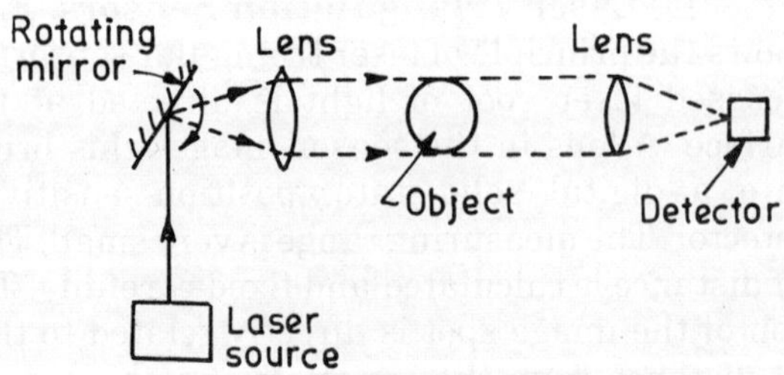

Fig. 23.16

(*vii*) **Gauging Wide Diameter from the Diffraction Pattern formed in a Laser Beam.** Fig. 23.17 shows a method of measuring the diameter of thin wire using the interference

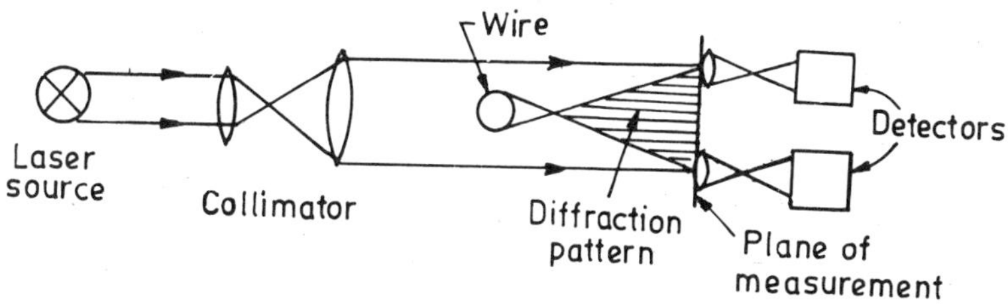

Fig. 23.17

fringes resulting from diffraction of the light by the wire in the laser beam. The detection of a null magnified to fill the dectector's aperture indicates that the wire is outside its dimensional tolerance. The system uses the changes in fringes formed by diffraction with wire diameter which cause a variation in the output from the photodetector and can be used as part of a control system or warning indicator.

A measure of the diameter can be obtained by moving the photodetector until the output is restored to its original value. Changes in wire diameter as small as 0.2% over wire diameters from 0.005 to 0.2 mm can be measured. Fig. 23.18 shows the method of length measurement by fringe counting. The laser output, which may be incoherent, illuminates three slits at a time in the first plane which form interference fringes. The movement of the interference fringes is determined by a detector on the other side of a short length of conventional Moire grating. The total number of slits in the first plane is governed by the length over which measurement is required. The spacing between the slits and distance of the slit to the plane of the grating depend on the wavelength of the light used. Using He-Ne laser at 0.63 μm, a fringe spacing of 1 μm is obtained at 1.4 μm from the slits if a slit separation of 1 μm is used.

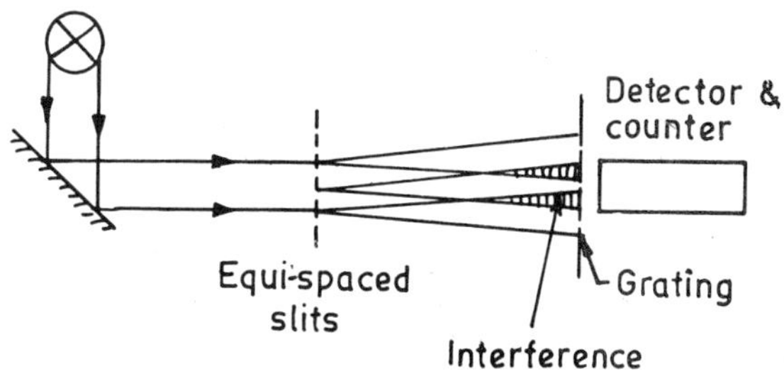

Fig. 23.18

This method is capable of accurate measurements over relatviely short distances of the order of 100 mm independent of variation of refractive index and turbulence in air path.

23.17. Robotic Gauging and Inspection Systems

Robots with laser probes and machine vision systems are finding more and more applications in flexible inspection. In flexible inspection, sensors are moved about the part being inspected along a programmed path trajectory. Robots with articulated arms can reach inside of parts with cavities, such as car bodies. Robots can be programmed to measure reference datum points and planes on a part. Algorithms, using these reference point measurements, can calibrate the part co-ordinate system, which allows the CAD/CAM database to be integrated into the robotic measuring system. This integration eliminates the need for precision part fixturing, allows real-time adaptive robot path trajectory control, and provide off-line programming of robot path trajectories. Control of robotic inspection systems by powerful minicomputers, allows quick down loading of path trajectory programs to recalibrate a current part co-ordinate system, or to reprogram robots to inspect a new part.

Robots equipped with optical and vision sensors provide the maximum flexibility to carry out a wide range of inspection tasks. While optical sensors provide accurate, non-contact measurement of distances and ability to accurately measure co-ordinates on plane and curved surface that would require structured lighting for vision measurements, the vision systems allow the inspection of part features, having a variety of geometrical shapes.

Inspection robots are presently used for moderate through-put rates. Robot positioning accuracy is the main source of error in robots inspection systems.

Selection of Robot Based Inspection System

Robot size. The robot size is determined by the inspection volume required by the application. The revolute co-ordinate geometry design or gantry design of robot is widely used for inspection applications. The gantry design, provides high repeatable positioning accuracy due to its geometric design. Robot positioning accuracy is determined by its ability to match its actual position in 3-D space to the command position called for by the program's position instruction. The error between actual and command positions may be caused by servo component design, structural natural frequencies, bearing friction, gear backlash and load torque. As the robot approaches its command position, the dynamic errors produce an effect called "setting time".

Sensors. Either vision systems or optical sensors are used. In vision systems, binary images allow the use of simple algorithms for image processing and introduce least delays into the process inspection cycle. With good silhouette lighting and binary image processing, it is possible to obtain reliable measurements to a one pixel accuracy. Since this accuracy is limited by robot positioning accuracy, the pixel calibrations is set equal to 2.3 times, the standard deviation of the robot positioning error distribution.

Optical sensors employ a light beam to illuminate the measurement location and use reflected light and triangulation techniques to measure distance. In order to permit measurements on surfaces with poor reflectivity light frequencies in the near infrared range are used. Detectors used to sense light beam deflection, caused by the trigonometric relationships, are of either the line scan matrix type or a semiconductor film employing analog circuitry. For latter type close-loop control of light beam intensity is necessary to maintain constant reflected light intensity. The measurement accuracy of optical probes is high and these have high speed measurement capabilities.

Computer control systems. A robot inspection system uses 3 or 4 small computers to perform difference distributed processing functions. These computers are interconnected by a data communication and control network, thus permitting a host computer to co-ordinate and control the robot and sensors.

System software. It must provide the following functions :

— initiate and co-ordinate the process inspection cycle.

— read in and store sensor measurement data during inspection cycle execution.

— perform data processing.

— output process data results in some acceptable report format.

— provide a limited operator interface to handle calibration and diagnostic requirements.

— output warning and diagnostic messages, to indicate system malfunction and type of errors.

23.18. Computers in Metrology

Computers can be advantageously applied in the field of Engineering Metrology for tasks like processing of acquired data, control of calibration equipment, in coordinate measuring machine, etc. Computers are capable of automating and streamlining most of the fields of metrology. It is for the metrology engineers to visualise applications of computers and exploit their power to yield better results in both automatic data acquisition and the control of the measuring process. However it may be mentioned that all problems may not get solved with use of computer alone as there are some disadvantages and limitations of computers which must be clearly understood and appreciated. The design and validation of software for any application is one such area.

Computers find extensive applications in the field of roundness measurement, form and surface texture measurement. These employ proven computational methods to process measurement data and present the final result in a way which is eaily understood by the operator.

Advantages and Limitations

The advantages of using computer for processing of acquired data and control are as under :

— A particular measurement sequence is strictly adhered to since computer accepts the information in a sequential manner and also provides necessary guidance to operator in this regard.

— Inspection time is reduced considerably.

— The calculation of final result is available immediately on completion of the last measurement.

— Rejected readings can be repeated straight away, before the set up is disturbed.

— Scope for copying and calculation errors is virtually eliminated.

— The checking of the result is made much more simple.

— Time for calibration is reduced considerably.

— New Operators can be trained quickly and they need not be highly qualified.

The limitations on use of computers for this application could be :

— Computer adheres to a given criteria rigorously and thus all the qualifying requirements and ability of operator in accepting/rejecting a reading need to be told to computer clearly without any ambiguity.

— Sometimes a number may be entered incorrectly due to transposing error or key-bounce.

— Strict control is needed over the use, amendment and copying of programme tapes to ensure that unauthorised modifications are not made.

— Checking procedure to ensure correct loading of program from tape needs to be followed.

— Measurement process gets remote from the operator.

— It is difficult to locate the source of problem by normal operator.

— While the effects of drifts, environment influences are hidden or not noticed ; but operator may not get that confidence.

— While human eye and memory are extremely good at detecting drifts and averaging high frequency noise on signals, careful programming has to be undertaken to give a computer a similar facility.

Steps in producing software for engineering metrology

— The precise and detailed definition of geometric form.

— The specification of the measurement procedure.

— The mathematical modelling of the measurement.

— The design of the mathematical algorithm.

— The implementation, validation and dissemination of the software.

23.19. Software Control Hierarchy

Fig. 23.19 shows the software control hierarchy to be adopted if computer measurement and control is to be used for integrating of measuring machine into a full flexible manufacturing system. The different levels of system control for such application can be :

Computer operating software. The executive program which controls the operation of the rest of the computer equipment.

Communications Software. It enables the computer to establish communication with various measuring machines, other computers, remote terminals, etc.

Probe control software. It helps in interpreting the switching signals from the probe.

Machine Control software. It contains necessary instructions to control the machine correctly under control of computer.

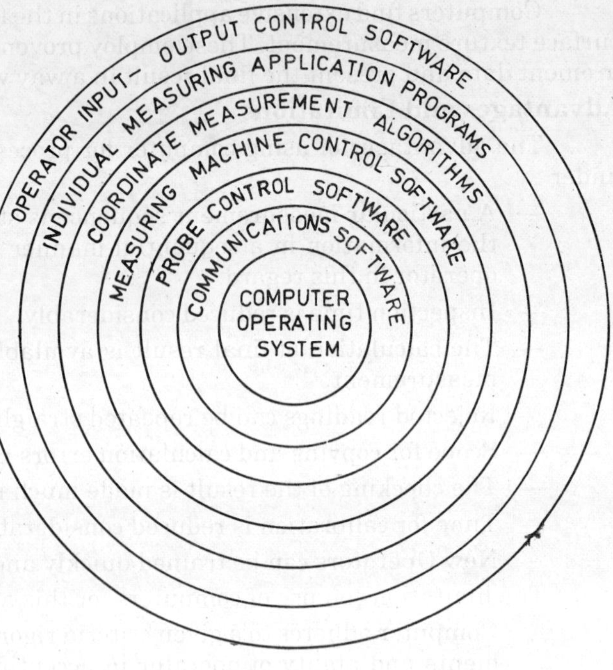

Fig. 23.19

Measurement algorithms. It contains a library of standard measurement formulae which are required and can be selected for use in conducting measurements on specific geometrical features.

Individual Measuring Application Software. It is a program of computer instructions defining a specific inspection routine and measurement sequence.

Operator Input/Output Software. It enables operators to construct the application software and produce calibration certificates.

23.20. Automatic Inspection Machines

A self contained inspection system requires many facilities. It must employ reliable and practical sensors. For in-process control, high accuracy on-line inspection systems with control feedback to the machine control are required. The measurement parameters that encompass any measuring tool are :

— Component dimensional value to the desired accuracy and repeatability

— Size

— Form errors-Out of roundness, lobing

— Surface roughness

— Tool wear

— Tool presence

— Vibration
— Cutting forces
— Collant and oil condition
— Flexing of the machine structure.

The data entered into machine must be properly validated by verification checks.

The measurement values need to be converted to machine coordinates. The machine needs facilities to check presence/absence of metal and ensuring that it is placed at desired time so as not to cause damage to measuring probes. The machine then instructs for complete measurement procedure to be followed and results outputted.

Data created by inspection part programmes is stored in operator assigned file names for use by subsequent data processing program.

Since printing is a slow process, the through-put of inspection machine can be increased by temporarily storing the information to be printed on a disk file and then outputting from file to printer.

23.21. Expert Knowledge Based Real Time Inspection System

An expert system provides the computer with the art of decision making, which is gained via human experts in the particular field of application. The primary objective of "Expert knowledge Base Inspection" is the ability to analyse the data/information originated by inspecting a particular item under inspection. The information signals available must be understood and correlated by assigning appropriate interpretations of the signals. The interpretation of the information to extract the contents may be classified as the knowledge acquisition stage. Quantifying the measurement of the magnitude of information gives a measurement of a parameter. The acceptability criteria is defined by the threshold window that meets the upper and lower boundary conditions. To define an end limit of a particular parameter is debatable and its quality is arbitrarily fixed and often subjective with experts tending to employ heuristic rules or rule of thumb. Any engineering application certainly obeys laws of physics and certain amount of heuristics, as every skilled individual could interpret the information in a different way. It is this area of human expertise that is attempted to be automated. The streamlining of the expertise management system would eventually enable full automation of the inspection system.

Knowledge base expert systems employ human knowledge to solve problems that require human intelligence. Knowledge systems represent and apply knowledge electronically.

A knowledge based expert system can be classified as a computer system that has within its software operating system, a specialised software package that enables knowledge to the encoded and stored in its data base. The stored information may be retieved by the operator or the running program. The information could be modified and stored after processing. Knowledge base system consists of :

(*i*) knowledge base

(*ii*) inference engine

(*iii*) a set of rules.

Q. 23.1. List of different methods of dimensional measurements using laser.

Ans. Laser techniques are used for measurement of dimensions in the following ways :

(*i*) **Scanning laser gauges.** In this technique, they can measure within a range of 0.05 to 450 mm with repeatability of 0.1 μm for smaller diameters. These are used for inspection on centerless grinding, precision machines, turbine blades etc.

(*ii*) **Photodiode array imaging.** In this technique they have accuracy of $\pm$ 0.05 μm and are used for part edge location and part dimensioning.

(*iii*) **Diffraction pattern systems.** This technique is used for small diameter parts and is more accurate for more small parts. It is thus used for measurement of small gaps and small diameters (one or two mm) of thin wire, needles, fibre optics, etc.

(*iv*) **Laser triangulation sensors.** These are used for measurement of gap and flushness on sheet metal stampings, hole diameters, and edge locations. These are best suited for electronics industry. The accuracy depends on stand-off and range, being less accurate at larger ranges.

(*v*) **Interferometers.** These are used for calibration of length measuring standards like glass scale and machine tools, coordinate measuring machines. The linear resolution of two frequency displacement interferometer is 1 nanometer.

(*vi*) **Holography.** Laser holography is used to detect faults in high pressure vessels, gas pipelines, brake drums etc. It is capable of measuring deviations as small as 0.1 μm in surface shape.

Q. 23.2. What do you understand by scanning laser gauges ?

Ans. A laser beam can be deflected along a line not necessarily straight-across the object under investigation one or more times. Scanning systems for 3-D inspection or dimensional measurement project a pattern of "structured light" from a laser source onto an object. An imaging system analyses the deformation of the projected pattern at the object. That deformation reflects the object's shape and critical dimensional parameters in both the image plane and in depth. Each time the beam hits an edge or similar feature within the scanned field, it changes the levels of scattered or transmitted light.

Laser scanning can also configure a system to make the spatial and depth resolutions independent.

Q. 23.3. What do you understand by Video Metrology or vision methodology as applied to industrial inspection ?

Ans. Vision methodology is analogous to replacing the human inspector who uses his eyes. For vision, brain for storage is taught an inspection technique with a camera, video memory and microcomputer. It uses low light level camera based with X-ray intensifier fluoroscopic device and visible light and laser non-contact metrology systems to detect and analyse surface. The camera image is temporarily stored in a solid state memory device. Computer logic reviews subsequent stored image data and makes comparison of picture contrast differences.

Q. 23.4. What are the three distinct groups of categorisation of image cameras ?

Ans. Line scan, solid state matrix, and analog.

Q. 23.5. What are the advantages and disadvantages of analog image sensors ?

Ans. The advantages are : resolution, low lighting, contrast sensitivity, capability to preprocess cost. The disadvantages are : poor linearity image drift, and image burn.

Q. 23.6. Computers can work only with digital information. Explain how a picture captured by camera is digitised ?

Ans. Picture is disected by rows and columns into five segments, each segment having a known size and contrast value.

Current hardware available to transfer analog video signals to digital matrix are available in 128 × 128, 256 × 256, 512 × 512, 1024 × 1024. Further it can distinguish between many shades of gray and available hardware shades are 16, 64 and 256 contrast difference from white to black.

Increased point definition improves as the resolution increases.

Q. 23.7. Define gray scale analysis.

Ans. In this technique, discrete areas or windows are formed around only the portions of the image to be inspected. For determining if brackets are present, high intensity lighting is positioned so that a bracket, when the bracket is missing no shadow will be cast. When the bracket is present, a large number of darker pixel can be observed in the window due to the cast shadow than when a bracket is missing—A contrast threshold between the dark and light pixel value area can be set. This type of discrete area analysis is a powerful tool can be used for inspection of absence, correct part assembly, orientation, part integrity, etc.

Q. 23.8. Write short note on automated video system.

Ans. Video systems measure without contact and this eliminates all operator influences. These make off-line, dimensional measurements of a part using charge-coupled device (CCD) video cameral with a fixed or zoom lens. A motorised drive moves the part and a PC loaded with software displays an image of the part being measured. It also enables pre-programmed points to be located and measured. Measurements made are displayed and can be downloaded directly to a statistical-process-control software program for further analyses. These can measure 10-100 times faster than using a manual microscope or optical comparator. However gauge repeatability and reproducibility studies should be carried out in order to confirm the accuracy of such systems.

24

Reliability

24.1. System Reliability

The reliability factor or the probability of failure of a system can be determined by breaking it into sub-systems and elements whose individual reliability factors can be estimated. Depending on the manner in which the sub-systems and elements are connected to constitute the system, the combinatorial rules of probability can be applied to obtain system reliability.

Thus once the sub-systems and elements which constitute the system are determined and their individual reliability factors estimated (identified), these are represented by a block diagram to show how these are connected. Next the condition for the successful operation of the system is determined, *i.e.* whether all the units have to be operative or it is sufficient for particular units to function for successful operation of the system. Finally the combinational rules of probability theory are applied to arrive at the system-reliability factor.

24.2. Probability and its Application to Series and Parallel Reliability (Redundancy)

Philosophy of probability. It is very important to understand in connection with philosophy of probability that for any given cause, we can know only the event which is most likely to occur, but we can never have complete and absolute certainty of the result until it occurs. With the mass of previous data, the probability can be predicted with a high degree of certainty but probability never means that in a few trials the events will occur exactly according to probability. All probability indicates is that as the total number of trials is increased indefinitely the number of events would occur closer to the predicted probability.

When an event can occur in different ways, each of which is mutually exclusive, then the probability of its occurrence in any one of many possible ways is the sum of the probability of occurrences of several different possible ways. (Term-"mutually exclusive" means that if there are two occurrences X and Y, then Y cannot occur if X occurs).

If an event is made up of a number of separate and independent subevents which must occur simultaneously or in series, the probability of occurrence is the product of the probabilities that each will happen.

Permutation of a number of things involves the number of different orders of arrangement of these things which can be made. Permutation or number of possible arrangements of n things $= n \times (n-1) \times (n-2)$ ...... $3 \times 2 \times 1$, or $n\,!$, ${}^nP1$ (n-factorial). In general the number of permutations of n things with r number of different things appearing in each permutation is

$$ {}^nP_r = \frac{n\,!}{(n-r)\,!} $$

It may be emphasised that the order of arrangement is important in the permutations. However if we are not interested in order, then combinations are formed. As long as each specific arrangement has a number of different things involved, it is called a combination, regardless of their order. Total number of combinations

$$= \frac{\text{Total number of permutations}}{\text{Number of permutations of the number of things appearing in each permutation}}$$

or $\quad {}^nC_r = \dfrac{{}^nP_r}{r\,!}$, where ${}^nC_r$ = Number of combinations of n events taken r at a time.

100-Hour Reliability

While probability is concerned with occurrence or non occurrence of events without reference to time, the reliability is expressed in terms of probability and as a function of time. 100 hours is the arbitrarily selected time. Thus if a series of tests are conducted on number of same equipment for 100 hours, and say, 92% of them survive the test, then reliability would be indicated as 0.92 with reference to 100-hour reliability.

Series Reliability

If a number of elements of a system are connected such that the failure of any one element would cause complete failure, then such elements are stated to be functionally in series and survival of each is essential for system to survive.

Reliability of such equipment is determined by multiplying the individual reliability of each element together.

Parallel Reliability (Redundancy)

This is justified where the increase in cost due to additional components will be compensated for by the increased reliability obtained. In this system probability of failure or probability of unreliability (1-reliability) of overall system is the product of unreliability of each component.

Series Parallel Reliability

This would be clear from following example :

An equipment consists of 100 parts, of which 20 parts are connected functionally in series, which in turn are connected in series to the parallel branches of 60 and 20 parts, the individual parts in 60 and 20 being connected in series. The reliability of each part in the three groups is 0.95, 0.96 and 0.93 respectively.

Since reliability of each part in first group of 20 parts is 0.95, *i.e.*, $P_a = 0.95$.

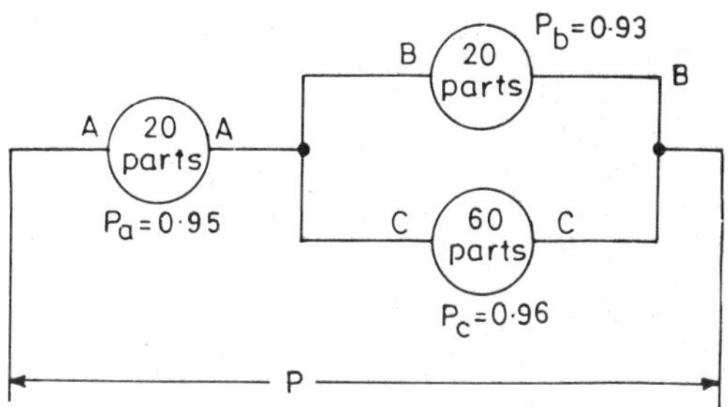

Fig. 24.1

$\therefore$ Overall reliability of all 20 parts connected in series $P_{AA} = {P_a}^{20} = 0.95^{20} = 0.358$.

Similarly $\quad P_{BB} = 0.93^{20} = 0.240$ and $P_{CC} = 0.96^{60} = 0.0863$.

Unreliability of groups BB and CC will be $(1 - P_{BC}) = (1 - P_{BB})\,(1 - P_{CC})$

$$= (1 - 0.240)\,(1 - 0.0863) = 0.760 \times 0.914 = 0.695$$

or $\quad$ reliability of branch BB and CC is $\;P_{BC} = 1 - 0.695 = 0.305$.

Overall reliability of equipment $\qquad P = P_{AA} \times P_{BC} = 0.358 \times 0.305 = \mathbf{0.109}$.

24.3. Reliability Analysis

Reliability analysis, which is based on statistical concepts, may be used (*i*) to predict the probable life of a part or assembly in operation, (*ii*) to evaluate the life test data for different items, and (*iii*) to control mass-produced components or assemblies in order to guarantee a preset service life.

Let N = the total number of devices on test with a specified operating cycle. If all the devices are functionally tested for a specified time of operation and if N_s devices operate successfully, and N_f devices fail, then the reliability (P) and probability of failure or unreliability of devices for the specified conditions, becomes $P = N_s/N$ and $U = N_f/N$.

It will be noted that $P + U = 1$. The number of failures (N_f) will be observed to increase progressively if the testing is continued, which implies that reliability (P) is a function of time of operation.

The instantaneous rate (λ) of failure of the devices is known as the 'failure rate'. If the failure rate (λ) curve of the above-stated group of devices is plotted against time, a typical curve will be observed to include three noticeable features in the following time sequence :

(*i*) An initial period (α) of high failure rates, often called "infant mortality" or "break in" period ; these failures are attributed to inherently weak devices resulting from poor quality of materials and the defects creeping in from the manufacturing process. Very rigorous quality control of the materials from the early stages of procurement and during the manufacturing process will enable the minimisation of such failures ; this weeding out (for certain time after manufacture) of the initial defectives is known as "burning in" of the device, after which delivery is made to the customer ; (*ii*) A period (β) of chance (or random) failures which is the useful service life period ; this period is very important for this decides the operating life of the device, and during this period, the failure rate (λ) is constant ; and (*iii*) A wear out period (θ) which is characterised by a sharp increase in the failure rate owing to the effects of aging ; wear out can be delayed by proper material selection and design. Preventive maintenance schedules, involving repairs or replacements prior to wear-out, are required to be adopted to preclude the use of the device in this high-failure-rate period.

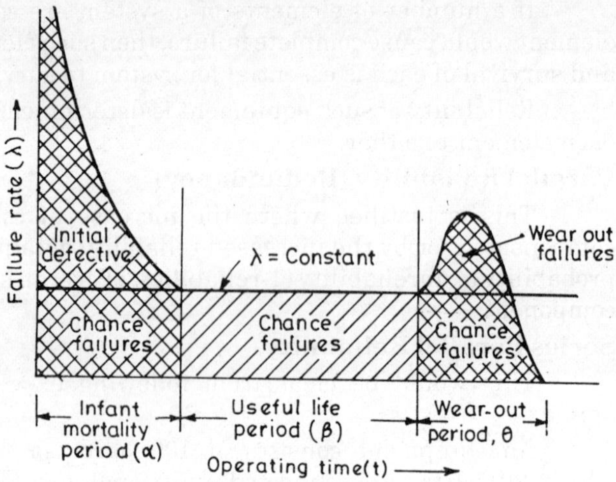

Fig. 24.2

Fig. 24.2 indicates a very typical failure rate-life characteristic curve based upon empirical test observation for many types of devices.

In the second part (*i.e.* useful period) of the above, the random nature of the breakdown produces a constant rate of failure per fixed number of the survivors so that

$$dN_f/dt = -\lambda N_s,$$

where λ = Failure rate per unit number of survivors ; its units are as follows : λ = % per 1000 hr = 1 failure per 10^5 hr = 10×10^{-6} failures per hr per part ; 1 bit = 1 failure per 10^8 hrs. By differentiation of the expressions of U and P, and rearranging the results obtained,

$$\lambda = -\left(\frac{1}{P}\right)\left(\frac{dP}{dt}\right), \text{ or } P = \exp.(-\lambda t) = e^{-\lambda t}$$

which is the basic expression for reliability of a device during its useful life. The reciprocal of failure rate is called the Mean Time Between Failures (M.T.B.F.) which is *not* wear-out life, but is usually larger than this life period.

24.4. Reliability Analysis of Systems

Reliability of a product is determined by design. Thus it's analysis during design state is essential. Since no hard data are available for a quantitative reliability assessment it is guided largely by experience. Some of the procedures used to stimulate reliability improvements are discussed below :

(*i*) *Some basic rules to improve reliability.* It is difficult to tabulate all the principles of durability and reliability. In this direction, experience plays an important part. Some of the observations are :

(*a*) Trace element principle. Properties like creep strength, corrosion resistance, elasticity, hardness, etc. can be altered by the judicious use of trace element techniques (in which extremely small amounts of impurities are added and these have profound effects on the service characteristics of materials).

(*b*) *Alloying and mixing technique.* Many new substances having extraordinary properties have been obtained by alloying elements and binders to materials. Some examples are ceramic-metal combinations, plastic-glass mixtures, blends.

(*c*) *Chemical saturation principle.* Corrosion can be reduced by nullifying chemical activity by attaching pre-saturated bonds in a material.

(*d*) Quality control principle. The testing of samples to obtain a statistical estimate of the expected performance of a production lot.

(*e*) Prestressing principle. By applying stress to an unloaded structural member, it can be made to have little or no stress when loaded.

(*f*) Hard and soft principle. The wearing out component is made of soft material and mating part of hard material.

(*g*) Large and small principle. The effects of destructive forces rise in a nonlinear relation to overall dimensions.

(*h*) *Excess material principle.* Strength and durability are not always gained by heavy construction.

(*i*) *Underrating principle.* Electronic components if operated at less than rated voltage, show a life expectancy several times that obtained at nominal voltage.

(*ii*) *Failure mode analysis.* It is a procedure by which potential failure modes, in a system are identified and analysed. The main purpose of this technique is to identify and eliminate failure modes, early in the design cycle where they are most economically dealt with. It starts with the selection of a subsystem or component and then identifies and documents all potential failure modes, failure mechanism, failure cause, identification of effects of higher level systems, and assign criticality rating.

(*iii*) *Fault-tree analysis.* It is a technique used for system safety and reliability analysis. The analysis proceeds from a designated "top event" to basic failure causes called "primary events". A fault tree is a model that graphically portrays the combination of events leading up to the undesirable top event.

(*iv*) *Reliability design guidelines*

(*a*) Simplicity. Use of a single part to perform multiple functions helps to reduce the number of failure modes and thus results in high reliability.

(*b*) Use of proven components. Use of proven components minimises analysis and testing to verify reliability.

(*c*) Use of probabilistic design approach. A reliable system can be obtained from less reliable components by using concept of redundancy.

In case of severe local environment (like shock and vibration, heat, corrosion etc.), environment should be modified to achieve high reliability.

Identification and elimination of critical failure mode through fault tree analysis helps improve reliability.

Incorporation of automatic sensing and switching devices, and self diagnostic features helps in obtaining high reliability.

Other features include detection of impending failure, preventive maintenance, tolerance evaluation, and human engineering.

(*d*) *Probabilistic approach to design.* Reliability is basically a design parameter and can be quantified and designed by probabilistic approach. The reliability of any system is a function of the reliabilities of its components. A designer can increase reliability by increasing mean strength, decreasing strength variation (improving and controlling the process), decrease average stress, and decrease stress variations.

Reliability Testing

Reliability can be assessed through product testing. During testing, the actual site conditions must be simulated so that test results are close to field condition.

The reliability test data can be used to estimate the reliability.

— set a confidence limit on reliability

— test the hypothesis that a reliability goal has been met

Various approaches are based on :

(*a*) exponential distribution

(*b*) Weibull distribution

(*c*) Binomial distribution for success-failure testing.

24.5. System Reliability for Series and Parallel Configuration

If a system consists of '*n*' identical independent items connected in series, the system reliability

$$P(S) = P(X_1) \times P(X_2) \times \ldots\ldots \times P(X_n)$$

If each component exhibits a constant hazard rate, then reliability factor for each item is $e^{-\lambda_i t}$

i.e.
$$P(S) = e^{-\lambda_1 t} \times e^{-\lambda_2 t} \times \ldots \times e^{-\lambda_n t}$$

$$= \exp.\ [-(\lambda_1 + \lambda_2 + \ldots + \lambda_n)t] = \exp.\left(-\sum_{i=1}^{n} \lambda_i t\right)$$

If various items are governed by the rule for linearly increasing hazard, then reliability of *i*-th item is exp. $(-k_i t^2/2)$

$$\therefore \quad P(S) = e^{-K_1 t^2/2} \times e^{-K_2 t^2/2} \times \ldots\ldots\ e^{-K_n t^2/2}$$

$$= \exp.\ [-t^2 (K_1 + K_2 + \ldots K_n)/2] = \exp.\left[-\sum_{i=1}^{n} K_i t^2/2\right]$$

If *p* out of *n* components of system follow constant hazard rate and balance are governed by linearly increasing hazard rule, then

$$P(S) = \exp.\left(-\sum_{i=1}^{P} \lambda_i t\right) \times \exp.\left(-\sum_{i=p+1}^{n} K_i t^2/2\right)$$

For items connected in parallel, and governed by constant hazard rule

$$P(S) = 1 - (1 - e^{\lambda_1 t}) \times (1 - e^{-\lambda_2 t}) \times \ldots \times (1 - e^{-\lambda_n t})$$

If all items be identical and independent,

then $$P(S) = 1 - (1 - e^{\lambda_1 t})^n.$$

24.5.1. Reliability of system in which r out of n components should function properly. Some examples are :

(*a*) A six cylinder automobile engine in which the successful operation of at least five cylinders is enough for satisfactory performance of the automobile.

(*b*) A piece of stranded wire with 8 strands in which at least 6 strands are necessary to pass the required current.

(*c*) A shaft lift operated by four cables out of which at least two are necessary for safe operation.

Let us discuss the reliability of an r-out-of-n system when the n components are independent and identical. First we shall obtain an expression for the probability of *exactly r successes* out of n identical independent trials. This can be obtained by considering the several ways of achieving r successes in n trials. One way is to have r consecutive successes followed by $n - r$ consecutive failures. Since each success and failure is independent, the probability of the assumed sequence is obviously.

$$p^r(1-p)^{n-r},$$

where p is the probability of one success. But this is only one method of getting r successes. There are many other ways of doing this ; in fact the number of ways in which r successes can be obtained in n trials is equal to the number of ways in which a group of r things can be selected out of a group of n given things. Where the order of selection is not important we have

$$\frac{n \times (n-1) \times (n-2) \times \ldots \times (n-r+1)}{r \times (r-1) \times \ldots \times 1} = \frac{n\,!}{(n-r)\,!\,r\,!} = \binom{n}{r}.$$

Hence, the total probability of getting r successes out of n trials is

$$\frac{n\,!}{(n-r)\,!\,r\,!} p^r(1-p)^{n-r} = \binom{n}{r} p^r(1-p)^{n-r}$$

This is known as the Bernoulli trial or the binomial experiment.

Next we shall try to obtain an expression for the probability of at least r successes out of n identical independent trials. This is easily obtained as follows :

The probability of exactly r successes $= \binom{n}{r} p^r(1-p)^{n-r}$,

The probability of exactly $r + 1$ successes $= \binom{n}{r+1} p^{r+1}(1-p)^{n-r+1}$,

....

the probability of exactly n successes $= p^n$.

Hence, the probability of getting at least r successes is the sum of the above probabilities, that is,

$$\binom{n}{r} p^r (1-p)^{n-r} + \binom{n}{r+1} p^{n+1} (1-p)^{n-r+1} + \ldots + p^n = \sum_{k=1}^{n} \binom{n}{K} p^K (1-p)^{n-K}.$$

Q. 24.1. Define the word 'reliability'. What are the four important factors in the assignment of a reliability factor.

Ans. Reliability is the probability of a component to behave adequately in the manner expected of it for the period intended under the given operating conditions.

The four factors important in the determination of a reliability factor are :

(*i*) the reliability of a component is expressed as a probability

(*ii*) the component/device must give adequate performance. It describes in unambiguous terms what is expected of device or system.

(*iii*) duration of adequate performance is specified. It represents a measure of the period for which the performance is satisfactory.

(*iv*) the environment of operating conditions (with regard to temperature, humidity, vibration, shock, etc.) under which the component has to perform are specified.

Q. 24.2. Define the terms (*i*) failure density, (*ii*) failure rate, and (*iii*) reliability. How these parameters will vary with time in a typical failure-data analysis ?

Ans. (*i*) Failure density is the ratio of the number of failures, during a given unit interval of time to the total number of tests at the very beginning of the test. Failure density is also called ratio part-failure rate.

(*ii*) Failure rate is the ratio of the number of failures during a particular unit interval to the average population during that interval. Failure rate is also known as the hazard rate.

(*iii*) Reliability is the ratio of the survivors at any given time to the total initial population. It is the probability of a device functioning satisfactorily for a given period under stipulated operating conditions. Reliability factor is also called the probability of survival.

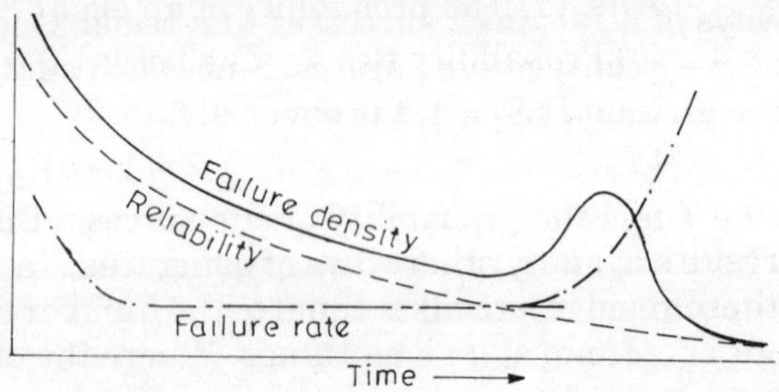

Fig. 24.3

Fig. 24.3 shows the behavior of these three parameters verses time.

Q. 24.3. Explain the difference between following three types of failure.

(*i*) **Instant failures ;** (*ii*) **service failures ;** (*iii*) **Wear-out failures.**

Ans. (*i*) Intsant or early failure is the short initial period on the curve plotted between failure rate or hazard versus time (Refer Fig. 24.4). This period is the break-in period when components fail due to defects in manufacture, being inherently weak because of weak parts, bad assembly, poor insulation, etc.

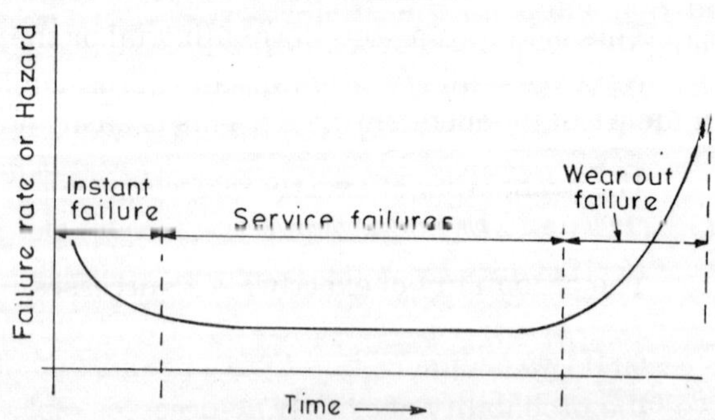

Fig. 24.4

(*ii*) Failure in second zone in Fig. 24.4 is termed service failure and during this period the failure or hazard rate is constant. In this constant-hazard period, the parts, do not deteriorate with time. Random failures, take place at constant rate.

(*iii*) The failures, in the third zone in Fig. 24.4 are the wear-out failures. The incidence of failure in this zone is high since most of the components have exceeded their service life and have deteriorated.

Q. 24.4. Explain the steps involved in determining the system reliability.

Ans. The entire system is first broken down into elements and sub-systems whose individual reliability factors can be estimated or identified. Next, the block diagram is drawn to represent the way in which the elements are connected in the logical manner. The condition for the successful operation of the system is then determined, that is, it may be decided as to how the units should function. In other words it is established whether all units have to function necessarily for system to operate, or system can operate with some units failing, or combinations of such conditions in various sub-systems. Finally, the combinational rules of probability theory, *i.e.* addition rule, multiplication rule, and their combinations are applied to arrive at the system reliability factor.

In case where components are connected in series, the overall reliability is product of reliability of each component and thus the reliability of a series system is always worse than the poorest component in the system. On the other hand the parallel greatly increases system reliability. If R_1, be reliability of one component, R_2, R_3 and R_n of other components, then reliability of n components in parallel is

$$1 - [(1 - R_1)(1 - R_2)(1 - R_3) \ldots (1 - R_n)]$$

Q. 24.5. Write down the important probability rules.

Ans. (*a*) The probability of an event E occurring is denoted by $P(E)$.

(*b*) Probability has a value between zero and one, *i.e.*, $0 \le P(E) \le 1$.

(*c*) If $P(E) = 1$, the event definitely occurs, *i.e.*, in almost every case.

(*d*) If $P(E) = 0$, the event never occurs, *i.e.*, very rarely does the event occur.

(*e*) *Addition rule* : The probability of occurrence in a certain operation of any one of the results $E_1, E_2 \ldots, E_n$ (it being immaterial which one), is equal to the sum of the probabilities of these results, provided that every pair of them is *mutually incompatible*. Thus,

$$P(E_1 \text{ or } \ldots \text{ or } E_n) = P(E_1) + P(E_2) + \ldots\ldots + P(E_n).$$

(*f*) The sum of the probabilities of mutually exclusive but collectively exhaustive events in a system is equal to one. If $E_1, E_2 \ldots, E_n$ are mutually exclusive and collectively exhaustive, then

$$P(E_1) + P(E_2) + \ldots + P(E_n) = 1.$$

(*g*) *Multiplication rule* : The probability of the simultaneous occurrence of two events A and B is equal to the product of the probability of the first event A and the conditional probability of the second event B computed for the assumption that the first event has occurred. If E_1 and E_2 are two events, then

$$P(E_1 \text{ and } E_2) = P(E_1) \times P(E_2/E_1).$$

If $E_1, E_2, \ldots, E_n$ are n events, then

$$P(E_1 \text{ and } E_2 \text{ and } \ldots \text{ and } E_n) = P(E_1) \times P(E_2/E_1) \times \ldots \times P(E_n/E_1 \text{ and } E_2 \text{ and } \ldots \text{ and } E_{n-1}).$$

(*h*) If events $E_1, E_2, \ldots E_n$ are independent events, then

$$P(E_1 \text{ and } E_2 \text{ and } \ldots \text{ and } E_n) = P(E_1) \times P(E_2) \times P(E_3) \times \ldots \times P(E_n).$$

(*i*) If two events E_1 and E_2 are not *mutually exclusive,* then the probability of occurrence of E_1 or E_2 is

$$P(E_1 \text{ or } E_2) = P(E_1) + P(E_2) - P(E_1) \times P(E_2/E_1).$$

Q. 24.6. (*a*) **What is the condition for successful operation of a system comprising of n units connected in series ?**

(*b*) **What is the probability of system to function satisfactorily when** (*i*) **each unit is independent of other,** (*ii*) **units are not independent of one another ?**

(*c*) **A system comprises of n identical units in series, each having a high reliability factor p. If all units function independently, what is the probability of system to function satisfactorily.**

(*d*) **Ten identical components are connected in series in a system. If system reliability is required as 0.95, what should be reliability of each component ?**

Ans. (*a*) For successful operation of system with all units in series, it is essential that all units function satisfactorily. Thus the reliability of a series system is always worse than the poorest component in the system.

(*b*) If the successful operation of *n* units connected in series be represented by X_1, X_2, X_3 ... X_n and their respective probabilities by $P(X_1), P(X_2), P(X_3), \ldots\ldots P(X_n)$, then system reliability (when (*i*) each unit is independent) of the successful operation of the remaining units, is

$$P(S) = P(X_1) \times P(X_2) \times P(X_3) \times \ldots \times P(X_n)$$

and when (*i*) individual units are not independent of one another, *i.e.* the successful operation of unit might affect the successful operation of other units, and so on, then

$$P(S) = P(X_1 \text{ and } X_2 \text{ and } X_3 \text{ and } \ldots \text{ and } X_n)$$

$$= P(X_1) \times P(X_2/X_1) \times P(X_3/X_1 \text{ and } X_2) \times \ldots P(X_n / X_1 \text{ and } X_2 \ldots \text{ and } X_{n-1})$$

Here $P(X_2/X_1)$ means the probability of the successful operation of unit 2 under the condition that unit 1 operates successfully. Similarly $X(X_n/X_1$ and X_2 and ... and $X_{n-1})$ represents the probability of the successful operation of unit *n* under the condition that all the other remaining units 1, 2, ..., $n - 1$ are working successfully.

Reliability factor of 1 unit is *p*

$\therefore$ Probability of failure of each unit of $= 1 - p$ and $p = 1 - q$

System reliability $P(S) = p \times p \times p \times \ldots \times p = p^n = (1 - q)^n$

If *p* is very high, *q* is very small,

$\therefore \quad P(S) = 1 - nq$

(*d*) $\quad P(S) = p^{10}$, or $p^{10} = 0.95$, or $\log p^{10} = \log 0.95$

and $\quad 10 \log p = \log 0.95$ and $p = 0.9949$

$\therefore$ Reliability of each component should be 0.9949 so that system reliability is 0.95.

Q. 24.7. (*a*) **Find an expression to determine the probability of failure of a system comprising of n components connected in parallel.**

(*b*) **Determine the reliability of a system consisting of four identical units connected in parallel, each unit having reliability factor of 0.9. Assume that unit failures are independent of one another.**

(*c*) **System reliability of a system consisting of 10 identical components connected in parallel is desired to be 0.95. Determine the reliability of each component.**

Ans. (*a*) Let $X_1, X_2, \ldots\ldots X_n$ represent the successful operation of units 1, 2, ... *n* and $\overline{X}_1$, $\overline{X}_2, \ldots, \overline{X}_n$ represent the unsuccessful operation (failure) of *n* units. Let $P(X_1)$ be probability of successful operation of unit 1 and $P(\overline{X}_1)$ be the probability of its failure, then $P(\overline{X}_1) = 1 - P(X_1)$

Probability of failure of the system $P(\overline{S}) = P(\overline{X}_1 \text{ and } \overline{X}_2 \ldots \text{ and } \overline{X}_n)$

$$= P(\overline{X}_1) \times P(\overline{X}_2/\overline{X}_1) \times P(\overline{X}_3/\overline{X}_1 \text{ and } \overline{X}_2) \times P(\overline{X}_n/\overline{X}_1 \text{ and } \overline{X}_2 \text{ and } \ldots \text{ and } \overline{X}_{n-1})$$

where $P(\overline{X}_3/\overline{X}_1$ and $\overline{X}_2)$ represents the probability of failure of unit 3 under the condition that units 1 and 2 have failed, and so on.

Now $\quad P(\overline{S}) = P(\overline{X}_1) \times P(\overline{X}_2) \times \ldots \times P(\overline{X}_n)$

and $\quad P(S) = 1 - P(\overline{S}) = 1 - [1 - P(X_1)] \times [1 - P(\overline{X}_2)] \times \ldots\ldots \times [1 - P(X_n)]$

If all *n* components be identical and unit failures be independent of one another,

then $\quad P(S) = 1 - P(\overline{S}) = 1 - [1 - P(X)]^n$

(*b*) $P(S) = 1 - (1 - 0.90)^4 = 1 - 0.1^4 = 1 - 0.0001 = 0.9999$

(*c*) $P(S) = 1 - [1 - P(X)]^{10}$

or $0.95 = 1 - [1 - P(X)]^{10}$ or $[1 - P(X)]^{10} = 0.05$

and $1 - P(X) = 0.7419$ and $P(X) = 0.2581$

Thus each unit could have poor reliability of 0.2581 so as to contribute a high reliability factor of 0.95.

Q. 24.8. Determine system reliability for following systems :

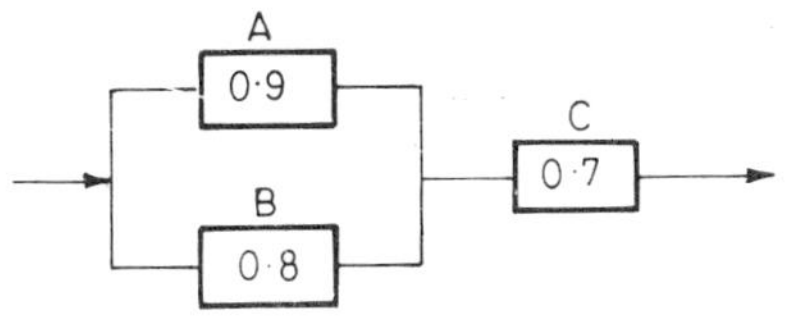

Fig. 24.5 (*a*)

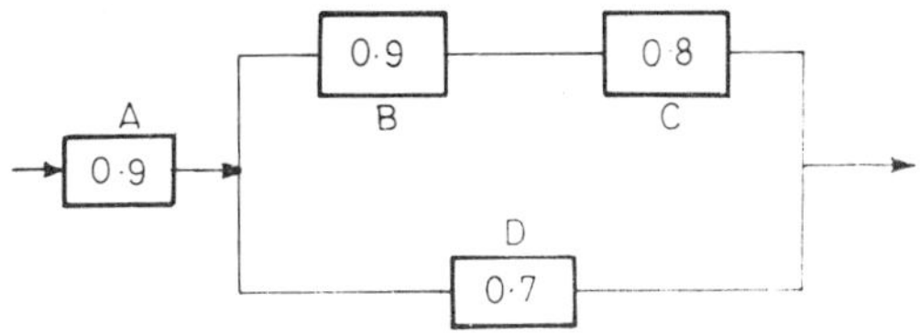

Fig. 24.5 (*b*)

Ans. (*a*) It is best to tackle cases of complex systems by systematically replacing each parallel path by an equivalent single path, thus reducing the given system to various elements connected in series.

As per probability rules, the probability of two elements in parallel is

$P(A \text{ or } B) = P(A) + P(B) - P(A \text{ and } B) = P(A) + P(B) - P(A) \times P(B/A)$

Here $P(B/A)$ means probability of successful operation of B when A is working. If elements A and B are independent, then $P(B/A) = P(B)$

$\therefore \quad P(A \text{ or } B) = P(A) + P(B) - P(A) \times P(B)$

$= 0.9 + 0.8 - 0.9 \times 0.8 = 1.7 - 0.72 = 0.98$

After converting two items in parallel to a single item, the resultant circuit will be as shown in Fig. 24.6.

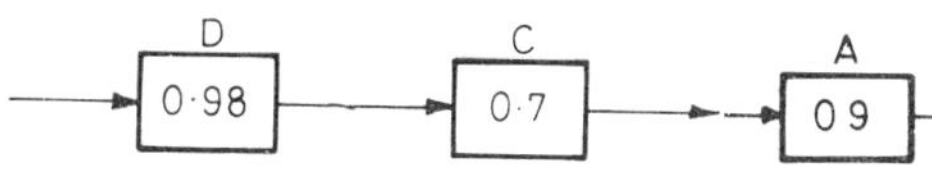

Fig. 24.6

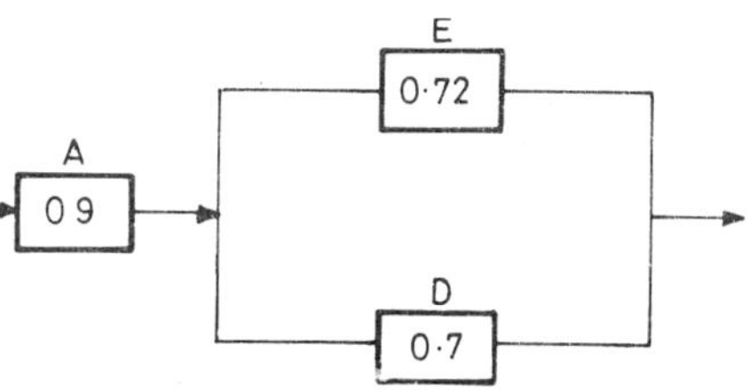

Fig. 24.7

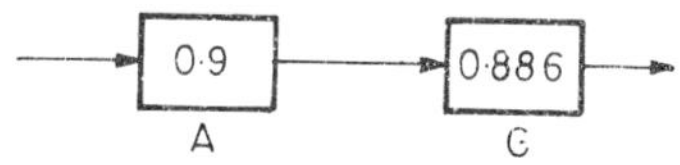

Fig. 24.8

$\therefore$ Reliability of system $= P(D \text{ and } C) = P(D) \times P(C/D) = P(D) \times P(C)$

$= 0.98 \times 0.7 = 0.686$

Considering elements B and C,

$P(B \text{ and } C) = P(B) \times P(C/B) = P(B) \times P(C)$

$= 0.9 \times 0.8 = 0.72$

The system then gets modified as shown in Fig. 24.7.

For parallel elements F and D,

$P(F \text{ or } D) = P(F) + P(D) - P(F \text{ and } D)$

$= P(F) + P(D) - P(F) \times P(D)$

$= 0.72 + 0.7 - (0.72 \times 0.7) = 0.886$

The system now gets modified as shown in Fig. 24.8 and

$$P(S) = P(A) \times P(G/A) = P(A) \times P(G) = 0.9 \times 0.886$$
$$= 0.7974 \cong 0.8.$$

24.6. Optimum Reliability

One method to increase the reliability of a system is to increase the redundancy in the system, *i.e.* to provide for components in parallel, such that if one fails the other keeps the system operational. However incorporation of redundancy is at a cost. Soon a point is realised where additional investments in reliability increasing redundancy do not justify the corresponding increases in cost.

The following example will illustrate how a trade-off between the additional reliability and additional cost of adding redundancy can be arrived at.

Let us consider power supply arrangement for an artificial satellite comprising of 36000 solar cells. These cells are arranged in 600 branches, each branch consisting of 60 cells connected in series. All branches are connected in parallel. If any one cell fails in the branch, the complete branch is not able to contribute to the system. The reliability of each cell is 0.999. The reliability of complete string of 60 cells will now be $(0.999)^{60} = 0.9462$. In order to obtain the minimum desired power, all the 600 branches have to contribute. To take care of this requirement, some redundant branches have to be added in the system. The quantum of redundant branches should be optimum because addition of branches not only increases cost but adds to weight also which is not desirable for satellite. Let us assume n redundant branches are added so that total number of branches will now be $600 + n$. We can determine the probability of at least 600 branches operational out of $600 + n$, with each branch having reliability of 0.9462. This is obtained by adding the probability of 600, 601, 602, ... $600 + n$ branches operating which is equal to

$$\sum_{x=600}^{600+n} \frac{(600+n)\,!}{x\,!\,(600+n-x)\,!} \times (0.9462) \times (1-0.9462)^{600+n-x}$$

Let us plot this reliability versus value of n and curve obtained will be as shown in Fig. 24.9. From this figure it will be realised that reliability increases slowly from 20 and reaches maximum limit near 50. Thus provision of more than 50 branches does not help in improving the reliability and is the optimum level.

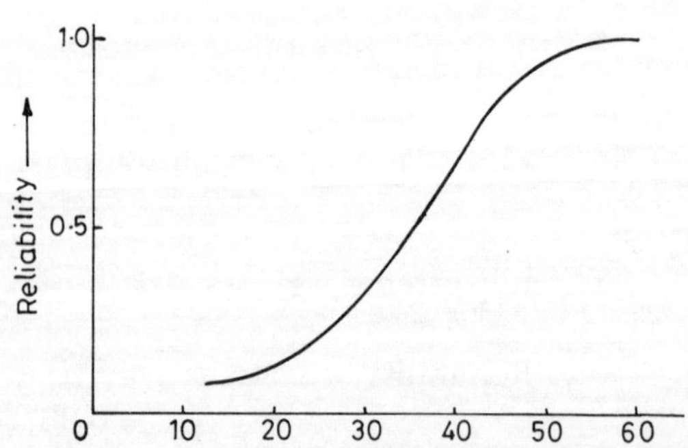

Fig. 24.9

Designing for Optimum Reliability

Mathematical Treatment. Reliable systems can be produced only when the reliability is considered at the design stage itself. At design stage, the system should be examined carefully from different angles. The weak areas of the system from the reliability point of view are identified by performing reliability analysis. In addition, the weakness in the system is overcome by introducing redundancy or using more reliable system components or modifying the design.

Failure data analysis. If the failure data on a component or a device is available, it can be analysed. In reliability work, the exponential distribution is frequently used.

For exponential distribution, the mean of exponential distribution

$$\mu = \frac{1}{n} \sum_{i=1}^{n} t_i$$

where n = number of parts tested

t_i = failure time of i part

Variance of the mean of exponential distribution = μ^2/n.

Example 24.9. *Failure times of 10 identical electric bulbs are 2000, 1800, 2200, 1900, 1700, 2500, 2300, 2100, 2400, hours respectively. Estimate the value of the mean life of bulbs.*

Solution.

$$\mu = \frac{1}{10}(2000 + 1800 + 2200 + 2100 + 1900 + 1700 + 2500 + 2300 + 2100 + 2400)$$

$$= \frac{1}{10} \times 21000 = 2100 \text{ hours}$$

Thus mean life = 2100 hours.

24.7. Reliability Allocation

Reliability allocation is the process of assigning reliability requirements to system parts or components to achieve the overall specified reliability of a system.

A system could be considered as composed of subsystem forming a series configuration. (If one subsystem fails, the entire system will fail). The failure rates of subsystems are supposed to be constant and subsystem failures independent. Further subsystem mission times are equal to the system mission time. The method will be clear from following example.

Example 24.10. *A series system is composed of 4 independent subsystems. The estimated failure rates of these are 0.0001, 0.0002, 0.0003, and 0.0004 failures/hour. The specified failure rate of the series system is $\lambda = 0.002$ failures/hr.* Calculate the value of failure rate to be allocated to each subsystem.

Solution. Constant failure rate of the series system

$$\lambda_{cs} = \sum_{i=1}^{n} \lambda_i$$

$$= 0.0001 + 0.0002 + 0.0003 + 0.0004 = 0.001 \text{ failures/hr.}$$

The subsystems' weights are

$$\beta_1 = \frac{\lambda_1}{\lambda_{cs}} = \frac{0.0001}{0.001} = 0.1 \qquad \beta_2 = \frac{\lambda_2}{\lambda_{cs}} = \frac{0.0002}{0.001} = 0.2$$

$$\beta_3 = \frac{\lambda_3}{\lambda_{cs}} = \frac{0.0003}{0.001} = 0.3 \quad \text{and} \quad \beta_4 = \frac{\lambda_4}{\lambda_{cs}} = \frac{0.0004}{0.001} = 0.4$$

The allocated failure rate for subsystem 1 is $\beta_1 \times \lambda = 0.1 \times 0.0002$ failure/hr. Similarly allocated failures rates of subsystems 2, 3 and 4 are 0.00004, 0.00006 and 0.00008 failure/hr.

24.8. Repairable System Allocation Method

This method is used to allocate failure rate for systems composed of subsystems in series. If system is composed of n independent subsystems in series, and all subsystems are identical ; and if each subsystem is assigned to one repairman, *i.e.*, $\mu/\lambda >> 1$ (λ = subsystem constant failure rate and μ is subsystem constant repair rate) :

Then series system steady state availability $A_{ss} = \sum_{i=1}^{n} A_{si}$, n = no. of subsystems

and A_{si} = steady state availability of ith subsystem

$$A_{si} = \frac{\mu_i}{\mu_i + \lambda_i}$$

(μ_i = subsystem constant repair rate of ith subsystem,

and λ_i = corresponding failure rate)

For identical subsystems,

$$A_{ss} = A_s^{\,n} = \left(\frac{\mu}{\mu + \lambda}\right)^n = \left(1 + \frac{1}{\theta}\right)^n, \text{ where } \theta = \frac{\lambda}{\mu}$$

The value to be allocated to each subsystem $\lambda = \theta \cdot \mu$.

Example 24.11. *A series system is composed of seven statically independent subsystems ; all being identical. Each subsystem is assigned to one repairman. The specified system availability and repair rate are 0.98 and 0.25 repairs/hr*, respectively. Calculate the value of failure rate to be allocated to each subsystem.

Solution. $A_s = (0.98)^{1/7} = 0.99712$

$$A_s = \frac{\mu}{\mu + \lambda} = \frac{1}{1 + \frac{\lambda}{\mu}} = \frac{1}{1 + \theta}$$

or $$\theta = \left(\frac{1}{A_s} - 1\right) = \left(\frac{1}{0.99712} - 1\right) = 0.002888$$

$\therefore$ Specified repair rate $\lambda = 0.002888 \times 0.25 = 0.000722$ failures/hr.

24.9. Reliability Analysis Techniques

Fault free technique is commonly used to evaluate reliability of a system. This analysis begins by identifying the top event, known as undesirable event of the system. This undesirable event may be caused by several events which can be described to be connected by logic gates (like AND, OR gates). Thus to solve a problem first of all, system definition is established and a fault tree constructed, which is evaluated qualitatively. Basic data such as component's failure rate, repair rate, and failure occurrence probability failure rate, repair rate, and failure occurrence probability are collected. The fault tree is evaluated quantitatively and corrective measures recommended. Fault tree symbols used are shown in Fig. 24.10.

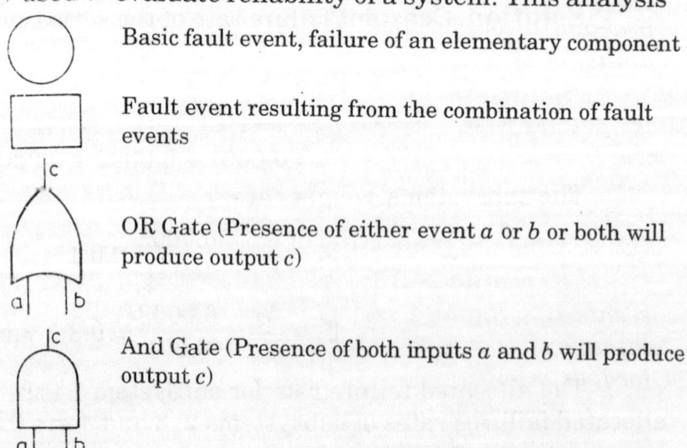

Fig. 24.10

24.10. Probability Evaluation

Case 1. When logic gate input fault events are non-repairable.

For OR gate, E_0 = Output event and E_1 to E_n are input events.

Thus $$E_0 = E_1 + E_2 + \ldots\ldots E_n$$

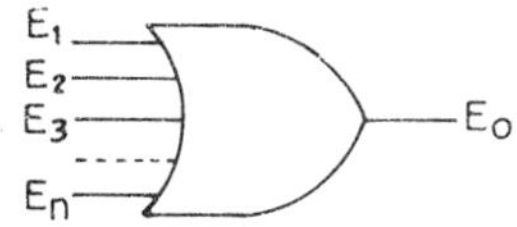

Fig. 24.11

and $$P(E_0) = 1 - \sum_{i=1}^{n} [1 - P(E_i)]$$

where $P(E_0)$ = Probability of occurrence of output event E_0

and $P(E_i)$ = Probability of occurrence of input event E_i

For AND gate, $$P(E_0) = \sum_{i=1}^{n} P(E_i)$$

Case 2. When logic input fault events are repairable.

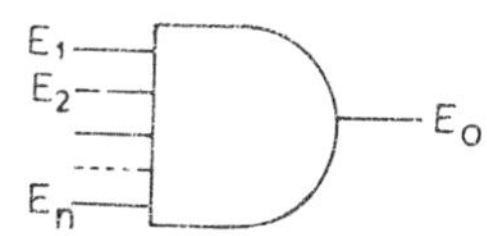

Fig. 24.12

In this case for OR gate $$u_0 = 1 - \sum_{i=1}^{n} (1 - u_i)$$

u_0 = Unavailability (probablity of occurrence) associated with output event

u_i = Steady-state unavailability of ith independent input fault event, $u_i = \dfrac{\lambda_i}{\lambda_i + \mu_i}$

λ_i = Constant failure rate of ith component (fault event occurence rate)

and u_i = Constant repair rate of the ith component.

For AND gate $$u_0 = \sum_{i=1}^{n} u_i.$$

Example 24.12. *A three-independent-input OR gate is shown in Fig. 20.13. Each input fault event occurrence probability is specified. Calculate the output event probability of occurrence.*

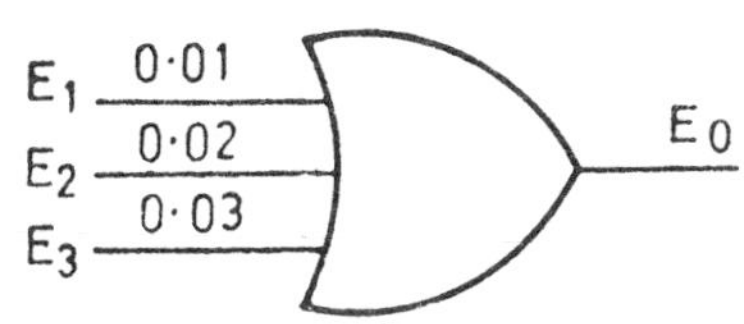

Fig. 24.13

Solution.

$$P(E_0) = 1 - [1 - P(E_1)]\ [1 - P(E_2)]\ [1 - P(E_3)]$$
$$= 1 - (1 - 0.01)\ (1 - 0.02)\ (1 - 0.03)$$
$$= 1 - 0.99 \times 0.98 \times 0.97 = 1 - 0.941094 = 0.058906.$$

Thus the probability of output event occurrence = **0.0589.**

Example 24.13. *Fig. 24.14 shows a four-independent-input AND gate. For input fault events 1, 2, 3 and 4, constant occurrence rates are 0.001, 0.002, 0.003, 0.004 failures/hr. The input fault events constant repair rates are 0.003, 0.004, 0.005 and 0.006 repairs/hr respectively for four inputs.*

Determine the—

(i) steady state unavailability of each input fault event, and

(ii) the probability of occurrence of the AND gate output event.

Solution. Unavailability of input 1

$$u_1 = \frac{\lambda_1}{\lambda_1 + \mu_1} = \frac{0.001}{0.001 + 0.003}$$

$$= \frac{0.001}{0.004} = 0.25$$

$$u_2 = \frac{0.002}{0.002 + 0.004} = 0.33$$

$$u_3 = \frac{0.003}{0.003 + 0.005} = 0.375$$

$$u_4 = \frac{0.004}{0.004 + 0.006} = 0.4$$

Probability of output event to occur

$$= u_1 \, u_2 \, u_3 \, u_4$$

$$= 0.25 \times 0.33 \times 0.375 \times 0.4$$

$$= \mathbf{0.0125}.$$

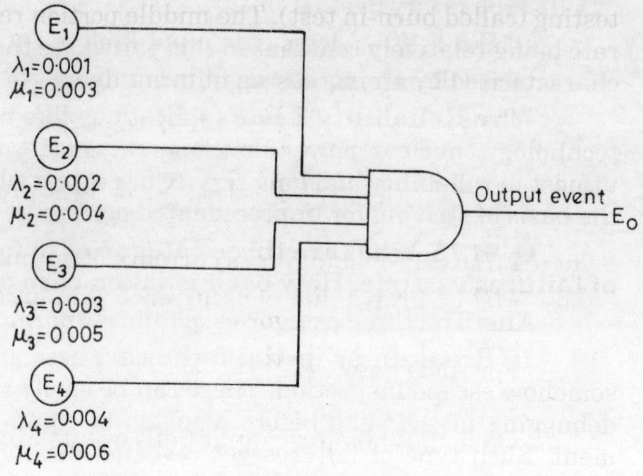

Fig. 24.14

Q. 24.14. What do you understand by the term reliability and why it is required ?

Ans. Reliability is concerned with failure-free performance of the system and also duration of the failure-free performance. It is established at the design stage.

Reliability is defined as the probability that a given system will perform its intended function satisfactorily for its intended life under specified operating condition. It is related to the probability of successful performance for any system.

Reliability is the property of an item or machinery enabling it to accomplish its functions during a given life period or given operating time, retaining its operating indices within the given limits. The reliability of machines depends upon their trouble free dependability, durability, maintainability and retainability.

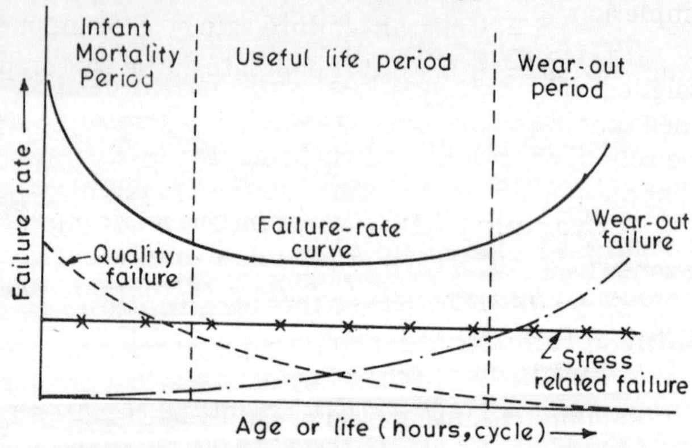

Fig. 24.15

One measure that is frequently used as an indirect indicator of system reliability is called the "mean time to failure", which is the expected or mean value of the time-to-failure. If we are interested in studying the reliability of a large population of items, then our interest lies in replacement and maintenance. For this purpose we need to know the rate at which the items in population, which have survived at any point in time will fail. This is called the failure rate or hazard rate. The failure rate for most components follows the life-characteristic curve (bath-tub curve). Three types of failures (*viz.* quality failures, stress related failures, and wear out failures) have been considered. The sum total of these failures gives the overall failure rate.

This curve has three distinct periods. Infant mortality is due to the early failure of substandard products which could be due to latent material defects, poor assembly methods, and poor quality control. These failures can be eliminated by a short period of in-plant product testing (called burn-in test). The middle portion represents the design failure rate, the failure rate being relatively constant in this period. As the products age, they reach a wear-out phase characterised by an increasing failure rate.

Why Reliability ? For applications like outer space exploration, high-speed missile technology, nuclear power development, modern testing and calibration devices, etc., the utmost in reliability and long service are essential. Thus the design of such items has to be on the basis of striving for unprecedented performance.

Q. 24.15. What are three categories of failures ? Draw a graph showing number of failures vs. time. How each is taken care of in design ?

Ans. The three categories of failures normally encountered are :

(*i*) **Break-in or initial failures.** These arise because of manufacturing faults which somehow escape inspection. These can be easily taken care of by carrying out burn-in test or debugging in the shop before dispatch of equipment. Such type of failures get revealed during burn-in test or debugging and can be replaced with healthy parts.

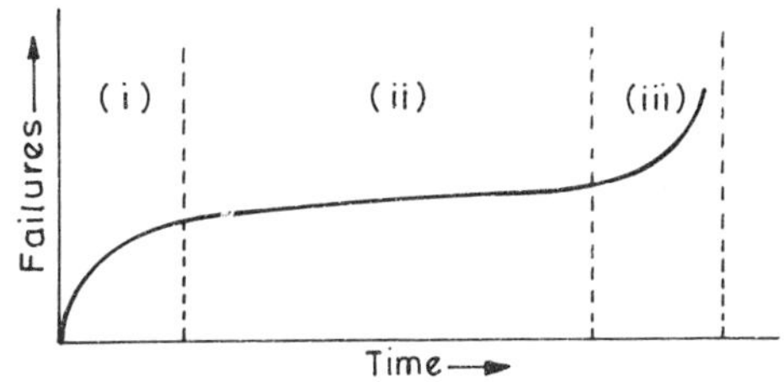

Fig. 24.16

(*ii*) **Wear-out failures.** The occur after the product has put in its design life. These are taken care of by following preventive maintenance procedures so that such components are replaced before they actually fail when their life is nearing completion.

(*iii*) **Chance failures.** These refer to premature failure to which no cause can be readily assigned. It is a cause of real worry for the designer. The designer can control it only in the sense that he can regulate its rate of occurrence to some extent by determining its reliability. The reliability of a product is measured by the rate of occurrence of the chance failures. The reliability for time '*t*' is defined as the probability of the survival of a system through time '*t*'.

Fig. 24.16 shows how number of failures increase with time, the three types being shown by zones (*i*), (*ii*) and (*iii*).

Q. 24.16. Failure is the termination of the ability of an item to perform its required function. Match parts A and B in regard to types of failures.

A	*B*
1. Failures resulting from deviation in characteristics/parameters beyond the specified limits : but not causing complete lack of required function.	(*a*) Misuse failure
2. Failures resulting from deviations in characteristics beyond the specified limits, causing complete lack of required function.	(*b*) Sudden failure
3. Failures attributable to the application of stress beyond the stated capabilities.	(*c*) Complete failure
4. Failures attributable to weakness inherent in the product when subjected to stresses within the stated capabilities.	(*d*) Catastrophic failure
5. Failures that could not be anticipated by prior examination.	(*e*) Partial failure

6.	Failures that could be anticipated by prior examination.	(*f*) Degradation failure
7.	Failures that are both sudden and complete.	(*g*) Inherent weakness failure
8.	Failures that are both gradual and partial.	(*h*) Gradual failure

Answers

1. (*e*)	2. (*c*)	3. (*a*)	4. (*g*)	5. (*b*)
6. (*h*)	7. (*d*)	8. (*f*).		

Q. 24.17. Define the term system effectiveness.

Ans. The term often used to describe the overall capability of a system to perform its intended function is "system effectiveness" which is defined as the probability that the system can successfully meet an operational demand within a given time when operating under specified condition. Effectiveness is influenced by the way the system is designed, manufactured, used and maintained, and thus is a function of all life-cycle activities.

The effectiveness of a system is a function of several attributes, such as design adequacy, performance measures, safety, reliability, quality, and maintainability. (Refer Fig. 24.17).

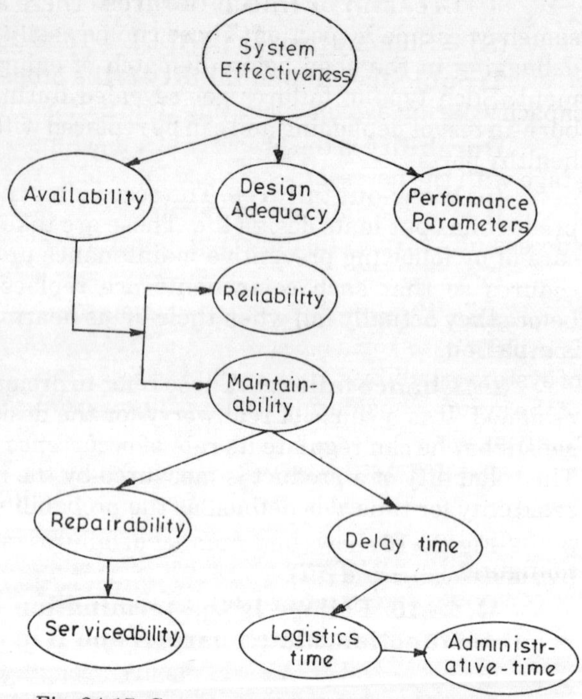

Fig. 24.17. Components of system effectiveness.

Q. 24.18. Define the terms maintainability, serviceability, availability and durability.

Ans. Maintainability. It is the suitability of a machine to the prevention, detection and elimination of breakdowns and trouble by carrying out proper servicing and repairs.

It is also defined as the probability that a failed system will be restored to a satisfactory operating condition within a specified interval of down time. The ease of fault detection, isolation, and repair are all influenced by system design and are the main factors contributing to maintainability. The timely supply of spare parts, the supporting repair organisation and the preventive maintenance practice, also contribute to maintainability. Good maintainability may somewhat offset low reliability. Thus if the desired reliability can not be met because of performance constraints, improvements in system maintainability can help overcome the problem.

Serviceability. It affects maintainability. It is a characteristic of design deserving attention at design stage. It is defined as the degree of ease with which a system can be repaired. It specifically considers the fault detection, isolation, and repair. Repairability considers only the actual repair time and is defined as the probability that a failed system will be restored to operation within a specified interval of active repair time. Access covers, plug-in-modules, standardised components and other features to allow easy removal and replacement of failed components, improve the repairability and serviceability.

Availability. It is the function of both reliability and maintainability.

$$\text{Inherent availability} = \frac{\text{MTBF}}{\text{MTBF} + \text{MTTR}}$$

Achieved availability is used for development and initial production of a system and is given by

$$\frac{\text{Operating time (OT)}}{\text{OT + Total preventive maintenance time (TPM) + Total corrective maintenance time (TCM)}}$$

Operational availability covers all segments of time and is given by

$$\frac{\text{OT + ST}}{\text{OT + ST + TPM + TCM + TALDL}}$$

(ST = Standby time and TALDL = Total administrative and logistics delay time).

Trouble-free dependability is the property enabling a machine to retain its working capacity during a given operating time without mechanical stoppages.

Durability is the property of a machine to retain its working capacity upto its limiting stage with the necessary stoppage for repair and servicing.

It is a special case of reliability. It is defined as the probability that an item will successfully survive its projected life, or rebuild point without a durability failure.

Q. 24.19. What do you understand by reliability analysis of systems ?

Ans. A system reliability model is used as the basis of reliability analysis. The analysis is usually based on a block diagram that represents system success according to the definition of system reliability. The reliability block diagram is constructed from an engineering analysis of the system considering modes of failure and understanding the grouping of similar parts for analysis purposes.

The reliability models are considered for both static and dynamic conditions. A system could comprise of series configuration, parallel configuration, or combination of both. In a series configuration, all subsystems must survive for the system to operate successfully. In a parallel configuration, all subsystems must fail before the system is considered to be in a failed state.

Q. 24.20. On what factors the reliability of any system is dependent ?

Ans. The reliability of any system is governed by :

(*a*) its simplicity or complexity. Complex system being made of a number of series components, has less overall reliability.

(*b*) time for which it is called upon to perform.

(*c*) the environment in which it operates.

(*d*) the state of the design art for the system.

(*e*) quality of design.

Q. 24.21. Explain the trade off between cost and reliability.

Ans. Fig. 24.18, shows how the initial cost and cost of operation and maintenance of system increase as the system is designed for higher and higher reliability. The initial cost obviously increases in proportion to the quantum of reliability desired. However, cost of operation and maintenance is high for less reliable system and it decreases for reliable systems. Thus variation of total cost shows that there is an economic cost of risk and risk reduction.

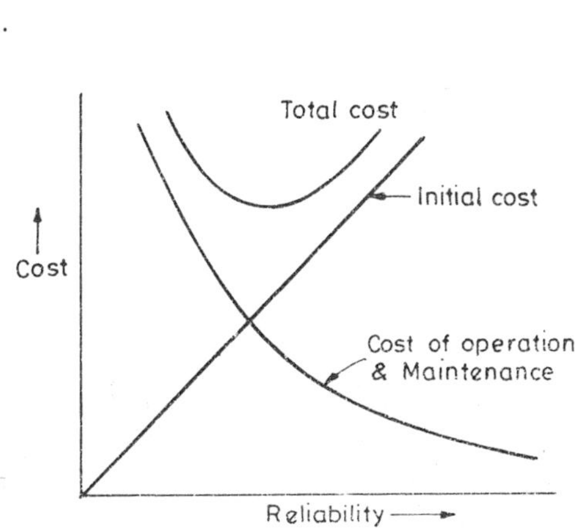

Fig. 24.18

24.11. Fault Diagnostics and Trouble Shooting

Fault diagnostics aids are very important now-a-days so that in the event of non-operation of any equipment the faulty component is detected as quickly as possible and same is replaced to make the equipment functional again.

For fault detection, its presence has to be established and all symptoms noted accurately. This is possible by performing a functional test *i.e.* checking the system's actual performance against its specification. This way a full set of fault symptoms can be obtained.

Fault location or trouble shooting is the next task which concerns the cause of the fault.

Whenever any component in any system fails, it results in a set of trouble shooting symptoms. These symptoms are often unique to the type of fault. By interpreting the symptoms it is possible to pin-point the faulty component.

To simplify the task, it is advisable to divide the system into several functioning blocks. Thus first the block is located and then with some logical approach the actual faulty component in it is located.

Manually this job is done by adopting sequential trouble shooting method. Usually various systems are not connected in series blocks but may have parallel branches and possibly feedback loops. These make the trouble shooting very complex. The non-sequential fault location method based on automatic testing with computer-aided fault analysis is preferred in such cases. The faulty system, linked to computer-controlled test gear, would have its whole system checked, and results would be matched with those resulting from typical fault conditions held in computer memory.

To achieve rapid trouble shooting and repair (for minimum mean time to repair), the technician needs to be provided with aids to back up his trouble shooting skill. These aids are :

(*a*) maintenance manuals and trouble shooting guides.

(*b*) test instruments.

(*c*) special tools and tackles.

24.12. System Maintenance

Any system consists of number of component parts connected together to make a regular and complete whole. In order that the overall system functions, it is essential that each component and the subsystem work correctly.

The purpose of maintenance is to achieve a satisfactory level of system availability at a reasonable cost and maximum efficiency.

$$\left[\text{Availability} = \frac{\text{Mean Time Between Failures (MTBF)}}{\text{MTBF + Mean Time to Repair (MTTR)}}\right]$$

High availability can be achieved only when the MTTR value in low, *i.e.*, the system can be maintained relatively easily. MTTR is the average of the times taken to repair any fault in the system. The designer can achieve low value of MTTR by paying close attention to the accessibility of components, building in fault display panels, and by providing internal test facilities.

Maintainability is defined as the probability that a system that has failed will be restored to a full working condition within a given time period. The measures of maintainability are the MTTR, and the μ (repair rate), which are related as $\mu = \frac{1}{\text{MTTR}}$ and maintainability = $1 - e^{-\mu t} = 1 - e^{-t/\text{MTTR}}$, where t = time allowed for the maintenance action. The maintenance policy adopted for a particular system is dependent on several factors like type of system, its location and operating and environmental condition, the required levels of reliability and availability, the overall competence of a skilled maintenance staff, and the provision of spares. For certain types of systems the maintenance policy may include in its program details of recalibration (checking amount of drift from the specified figure and to correct for any partial failure that may have cause the system's performance to be outside the tolerance limits) and preventive actions.

Two types of maintenance are in vogue :

(*i*) **Preventive maintenance.** In this system, the components of a system that are nearing the end of their life (wearing out) are replaced before the component actually fails. Failures of components entering the wear-out period in some cases are not random and can be predicted (as in the case of components with moving parts, filament lamps etc.). However in many cases, accurate prediction of a component entering a wear-out period is difficult. A disadvantage of this system is that the disturbance created during any preventive maintenance action may itself cause failures.

(*ii*) **Corrective maintenance.** When the failures of the component parts are entirely random (as in case of electronic systems), the parts are replaced on their failure. The service checks for such systems may therefore not be desirable as these may lead to further faults. Thus making routine checks on equipment where faults are entirely random can only result in degraded reliability. Corrective maintenance is concerned with the detection, location and repair of faults as they occur. This requires good understanding of system trouble shooting methods. The three distinct phases in the corrective maintenance task are :

(*a*) **Fault detection.** It is essential to establish the presence of fault and all symptoms in this regard need to be accurately noted. For this purpose, a functional test, checking the system's actual performance against its specification, must be made.

(*b*) **Fault location (trouble shooting).** The purpose here is to narrow down the search for the cause of the fault, first to one block or subsystem within a system, and finally to one component within that block.

(*c*) **Fault rectification.** The faulty component has to be repaired or replaced, as the need be.

24.13. Trouble Shooting Systems

When a component fails in a system, a certain set of trouble-shooting symptoms result. These symptoms are often unique to the fault, like high vibration, high temperature, or change in output signal, etc. The faulty component can often be pinpointed by interpreting the symptoms. The task of finding a single faulty component among several thousand may be difficult because of the size and complexity of such a system. However the problem can be tackled by considering the system in block diagram form. For instance in electronic circuits, the system can be divided up into several functioning circuit blocks and, by measurement, the portion or block that has failed can be located, and then detailed measurements can be made on that block to find the actual faulty component. It would be appreciated that the block diagram is an essential aid to system fault location and also to assist in helping the understanding of the operation of complex systems. The use of block diagram and a sort of logical approach makes fault finding very simple.

For electronic circuits, a powerful method of trouble shooting is *nonsequential fault location*. This uses automatic testing, for example, based on the theoretical analysis of the transfer characteristics (response of output to input) of the system. This method is better suited to computer aided fault analysis rather than the individual service technician. The faulty system, linked to computer controlled test gear, would have its whole checked, and the results would be matched with those resulting from typical fault conditions held in the computer's memory.

The technician, to locate a fault in a system, has to select one or a mixture of sequential trouble shooting methods. The sequential methods of system trouble shooting could either be non-systematic (like random checks ; these are rarely used), or systematic. The systematic methods (based on a set of rules) could be either based on reliability data (not often used), or based on the functional structure of the system (in which case three systematic trouble shooting methods commonly used are : (*i*) input to output ; (*ii*) output to input ; and (*iii*) half-split. The input to output, and output to input methods are straight forward. A suitable input signal is injected into the first block and then measurements are made sequentially at the output of each block in turn, either from input to the output or *vice versa,* until the faulty block is found. The half-split method is useful when the system is made up of a large number of the blocks in series. In this method, the entire system is splitted into two sections. Each half is tested to find which is faulty one. The faulty section is again splitted further to locate the faulty half.

In series connected circuit, if there are n number of blocks, then the mean number of measurements required to locate faulty block would be

$$\frac{1}{2n}(n-1)(n+2)$$

(in the case of input to output, and output to input method)

and $3.32 \log_{10} n$ (in the case of half-split method).

From above it is obvious that number of measurements are much less in the case of half-split method. Thus half-split method is very powerful for trouble shooting a long chain of series circuits. It may be mentioned that many times, visual inspection of mechanical conditions, broken wires, cold solder joints, solder bridges, damaged printed circuit board traces, or burnt and damaged components also provide very useful information. However actual measurements and tests are essential to locate the faulty component.

The systems in actual practice are more complex and not simply series-connected blocks. There may be divergence (an output from one block feeding two or more units), convergence (two or more input lines feeding a circuit block), or feedback (to modify the characteristics of the system). In the case of divergent systems, each output is checked in turn and the search is continued for a faulty block in the areas that is common to the incorrect outputs. In the case of convergence arrangement, first all inputs are checked, one by one at the point of convergence. If any one is incorrect, then the fault must lie in the block providing it. If all inputs are correct, then the fault can only be in the preceding block. Systems with feedback loops which connect the output of some block with the input of an earlier block *via* some network, present one of the more difficult problems in trouble shooting. In such cases, first of all the type of feedback being used in the system and its purpose must be understood. With modifying feedback it may be possible to break the feedback loop, thus enabling each block to be tested separately without a fault signal being fed around the loop. The feedback is best disconnected at the input block end. If sustaining type feedback is disconnected from the input, it may be possible to inject a suitable signal in its place and then check circuit block outputs, including of course the feedback element. In such systems, the knowledge of the system, understanding of the operation, and a logical approach are essential for trouble shooting.

24.14. Trouble Shooting Aids

For rapid trouble shooting and repair in order to have low down-time, the technician must have the maintenance manuals and trouble shooting guides, test instruments and specialised tools readily available. Of course, to obtain high values of maintainability, during design stage it should be ensured that the components are easily accessible, provision for fault display made, and built in test circuits provided.

A good maintenance manual should provide description of the system with an explanation of its use, performance specification, theory of operation starting with block diagram and elaborating to circuit diagrams, maintenance procedures, spare parts list and mechanical layout. The maintenance procedures could be preventive (replacement of parts subjected to wear, recalibration and lubrication) and corrective. The corrective maintenance methods should include methods for dismantling along with safety procedures, list of test instruments and special tools, test instructions, and trouble shooting guides and suggested trouble shooting procedures.

A maintenance manual should not include irrelevant material. The written instructions must be clear and concise so that ambiguous meanings are not implied. The diagrams should be arranged so that the technician can quickly and easily understand its meaning. This can be achieved by using correct symbols, good relative arrangement of symbols, and careful routing of inter connections. Where there is a clear sequence from cause to effect, this should be drawn from left to right. Instructions on dismantling and safety should be given. Safety instructions are to avoid accidents and for protection of sensitive components.

Trouble shooting guidance can be provided in several forms. Tables can be provided showing typical symptoms for various fault conditions, together with the most probable causes and/or suggested courses of action to narrow down the area of the fault. Trouble shooting guide can also be in the form of questions. By obtaining an answer of yes or no to a question at each stage, mostly by measurement or observation, technician performing the trouble shooting is guided to the faulty part. Such guides are very helpful in speeding up the location and repair of faults.

Test instruments are essential to speed up the trouble shooting process, if the performance and limitations (accuracy resolution, loading effect, and bandwidth etc.) of the type of instrument used are fully appreciated.

24.15. Maintainability

Design engineers should attach high priority to maintainability aspect. For successful and long trouble free operation of any system, this aspect deserves serious attention.

A system designed for maintainability should incorporate all safety features and be equipped with assembly and access techniques that offer minimum interference to operator, and be assembled by an organised team, adequately identified with labels, and permit expedient replacement of equipment. It must be good practical design which is both accessible for maintenance and flexible enough to incorporate future changes.

From maintenance point of view, following points, which are likely to be overlooked, should be got incorporated in the design of, for instance, control systems.

- Provision of shut off valves and plugged test-tees on all direct measurement gauges, and shut off valves and test-tees/quick connect on transmitted signal line where calibration/checking may be required.
- Provision of siphons for liquid seals to avoid direct contact of hot liquid with transmitter/gauge.
- Proper valving for draft gauge.
- Provision of electrical isolating switches for power disconnects.
- Protection of open terminals, may be by insulated sheet material mounted over group of terminal blocks, or totally enclosed metal housing in dusty environment.

— Convenience outlets with separately fused circuit and internal illumination.
— Segregation of hardware for one sub-system into a nice, neat group for easy access.
— Painting white the interior surfaces of control panels/desks.
— Provision of a service loop (extra length of connection wires) for hardware mounted in an awkward working area to enable mounting to more accessible position for maintenance.
— Provision of shelf or tray mounting for quick replacement.
— Provision of manual control facility for critical control loops.
— Proper car for neatness of wiring and tubing in rear of panel and for easy accessibility to complete hardware provided.
— Check for compatibility of valve size, the lug, and the switch terminal.
— Only those contacts actually used be wired, making provision of 10-20% spare terminals for future requirements.
— Clearly define the plans for spare parts and the level of repair to be done at site.
— Provision of typed or engraved legends for identification of equipment both in front and rear of panel/desk.
— Labelling of each disconnect with information of loop number serviced by it.
— Labelling of each terminal block consistent with the as-built wiring documentation and consecutively numbering of each terminal point.
— Cross location making at both ends of wires (for smaller systems) which permits maintenance personnel to trace through a circuit without use of drawings.
— Provision of good communication facilities which is essential for good maintenance.
— Thorough checking to ensure the compatibility of the elements of the overall design and to insure that there are no interferences that will cause maintenance problems.
— Provision of adequate doorways into the control room and doors be wide enough to permit entry of test panels for routine maintenance.
— Provision of ramps (in place of steps) where raised floors are provided for control room.
— Provision of adequate space around panels/racks/desks.
— Provision of good filtration of particulate matter and oxidants in control room to remove any substances the might cause oxidation, coating, or corrosion of electrical contacts.
— Provision of different power receptacles for instruments in comparison to regular power receptacles to plugging into wrong receptacle.
— Provision of separate power receptacles of panels for lighting and connection of test equipment.
— Provision of separate good solid ground (obtained by having copper rods to the ground) for low signal level solid state hardware.

24.16. Reliability

Reliability predictions refer to the future behaviour of items and this can be expressed as probabilities. It is important to remember that reliable systems are not obtained by chance but only by careful outline. Reliability has to be designed in and defects tested out. The reliability concept starts with picking good parts and using them right.

The reliability of an item is the probability that it will perform a required function under a specified condition for a stated period of time.

The complement of the reliability is the failure probability. This is the probability that the lifetime of an item is shorter than or equal to a given time.

Wear-out. Failures (as in the case of light-bulbs) occur as a result of deterioration process or mechanical wear. The lifetimes of items which have failed by wear outs correspond to the normal distribution. For such items, there is a period of time in which they do not fail and in which the reliability remains unity. If these items have crossed initial safe period, failure rates start increasing at a faster rate. The items are replaced before entering the region with decreasing reliability, then an operation without breakdowns is possible. Thus preventive maintenance is very useful for such failures.

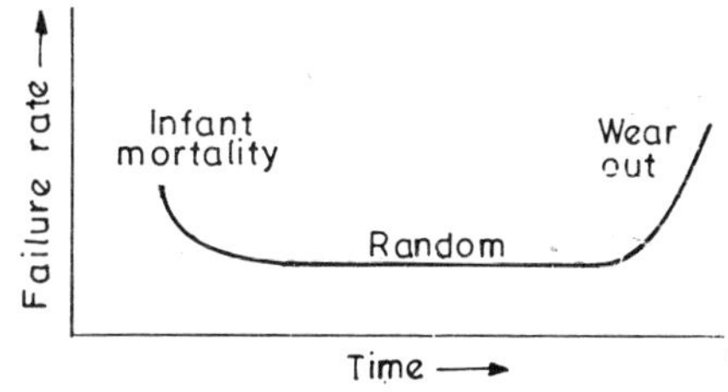

Fig. 24.19

For electronic devices, the failures are usually random ones, which are predictable only in a probabilistic or statistical sense. For such cases, a preventive maintenance does not improve the reliability, but may deteriorate the performance. Such failures are best described by bath-tub failure rate curve. Three distinct phases can be observed. The infant mortality is concerned with the early, life failure of a part and usually associated with material or manufacturing defects. The failure rate then decreases rapidly and stabilises when weak units have been replaced (this period may be some hundred hours for discrete semiconductors and some months for integrated circuits). The useful portion of life is the region with the time independent failure rate caused by random events. The low failure rate depends upon operational and environmental stresses. At the end-of useful life (of the order of 20 years), the wearout phase associated with rapid failure rates starts. The parts must be replaced before this stage.

Increasing Reliability by Adopting Concept of Redundancy

The process plants nowadays have greatly increased in size and complexity. For safe, efficient and reliable operation of any process, it is essential that a control and instrumentation system should be highly reliable and have very high integrity. It is only through the reliable control and instrumentation system that the user's faith can be bolstered and system kept in operation without problems. One of the ways to increase reliability of each systems is to adopt labour intensive periodic testing. However, such test arrangements increase the cost of the system and degrade the plant safety. The possibility of failures does exist between the time of finishing test and use of equipment. The present day emphases to make the system reliable is to adopt concepts of self-testing, fault tolerance and fail safe features etc.

Single Path Systems. If only single device is used for control purposes then, its failure would lead to complete breakdown. In such systems, it is absolutely essential that all the devices used should be of best quality and highly reliable. The system needs to be designed such that any failure would lead the plant to safe condition (Fail safe system). A single path arrangement is the simplest for alarm system and those controls which would cause little damage in the event of malfunction.

Test Arrangements. Where a high degree of security is required, particularly when a malfunction could result in a potentially dangerous situation, some means of testing the apparatus at suitable intervals should be incorporated. Such tests are usually carried out with the protected plant on-line and must therefore be isolated from the apparatus for the duration of the test. For this period the plant may carry an additional risk in that it has lost some of its protection.

Such a test could be done with automatic or manual switchover of apparatus outputs and isolation of transducers from the plant. After the test, the apparatus is restored to its normal function. While testing facility enables operator to have psychological satisfaction, it must be remembered that such testing shows merely that the apparatus was in a workable condition only at the time of the last test. A fault may well have developed within the next few seconds, or even as a result of resetting the apparatus back to normal working, which would not be detected until the next test, perhaps a long period later. Further, in some apparatus it is difficult to provide a sufficiently comprehensive test facility without at the same time increasing substantially the overall failure risk of the apparatus as a direct result of the additional items needed. Also, an apparatus provided with a test facility may have a higher capital cost, a higher maintenance cost because of the additional components it contains, together with an extra layout cost both for repair and for the testing itself.

Multi-Path (Redundant) Systems. Such systems are essential where failure to control the plant concerned is expensive or hazardous and such control must remain operative for long periods. The idea that a 'worst component' should be duplicated in direct parallel (the 'weak-link' theory is tenable provided that the working life of the plant to be controlled is very much less than 10% and preferably not greater than 1% of the m.t.f. of that component type, depending on the repercussions that the loss of control of that particular item of plant may have. The same argument is applicable to the direct parallelism of two complete paths. The basic objection to both these arrangements is that will be no indication of the loss of one component or path until the second also fails.

This can be achieved in one of the following ways :

(*a*) The two paths be separated sufficiently immediately prior to the combined control output to the plant to indicate that both paths are operative separately.

(*b*) The two paths are separated completely including separate sensors and that the first path is regarded as the duty path and is so arranged that the second path is 'standby path' to which automatic changeover would occur should the 'duty path' fail.

This can be implemented by ensuring the means by which the integrity of either path may be identified, since in either the digital or the analogue cases it must be possible to distinguish a fault from required action states or values, respectively. In practice, there are three basic means available, some of which may be used conjointly, by which a faulty path can be revealed.

(*i*) By using one or more additional paths arranged in parallel with both main paths, with further indication(s) as in (*a*) above.

(*ii*) The failure mode could be arranged so that the path would revert preferentially to a state corresponding to a plant abnormality.

(*iii*) By using a 'validation system'.

2. **Output of 3 Systems.** This arrangement is probably the best. The advantage of a third path lies its ability, should any two paths disagree, to provide the equivalent of a casting vote in favour of the one of the first two with which it agrees. Any path at any time may be cast in the role of arbitrator to the other two ; were this not so, the two paths which disagree would cancel and the third path would, in effect, cause the apparatus to become single path.

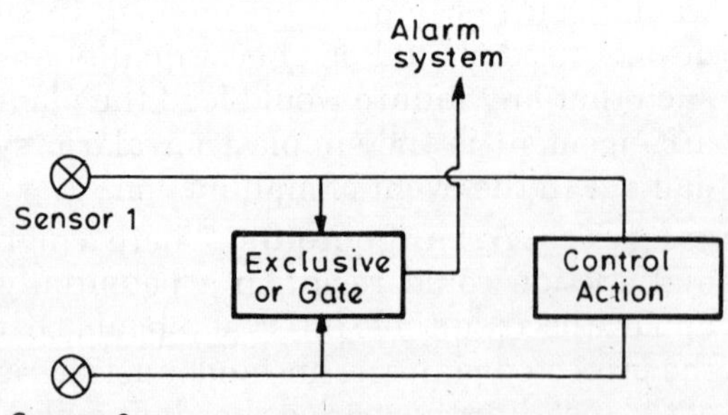

Fig. 24.20. Simplified block diagram of validation system.

Validation System. In validation system the two paths are arranged to have identical properties in respect of sensing the abnormal plant state, but their breakdown properties are, as far as possible, decorrelated by making them entirely separate with independent power supplies etc.

Both paths remain in separate operation continually and a continuous comparison is made between their outputs immediately prior to the combined control output. Such a pair, if they agree, cause whatever may be required action. Their disagreement beyond predetermined limits raises a discrepancy alarm (Fig. 24.20). If needed it can provide the same action, a different action, or no action other than an alarm depending on the application. Additionally, use may be made of natural laws against which the indication of each path may be validated.

Comparison of Various Systems. The relative reliabilities of various systems can be evaluated very briefly in simple approximate terms, in which various arrangements of identical independent paths each having a defined failure mode are assumed.

Let each path have a failure probability of q in N time periods, each of M duration, and each of which results in either an 'action' or a "no action" decision (q is very much less than 1).

Single-path systems. Whatever failure mode is chosen, an apparatus breakdown will result either in 'missing actions' (failure to protect the plant) or 'unwanted action' (false tripping, unwanted plant shutdown). Confidence has often to be supported by testing with attendant disadvantages and hazards. The system failure probability is $Q = q$. A fault can only be revealed by testing or by anomalous plant behaviour.

2 Out of 3 Systems. In such a system it is assumed that the same failure mode as for the single-path system will apply. Any two of the three paths will thus produce the same overall result provided that their decisions coincide within the period of N. Because there are three possible parts (A/B, B/C, C/A), the indications of which could cause the appropriate failure mode action, the system failure probability, $Q = 3q^2$. By choice of appropriate failure mode, any path fault may be made self-declaring in period N.

2 Out of 2 Validated Pair Systems. Since there is only one pair of paths available for failure as opposed to three pairs for the 2 out of 3 system, $Q = 2q^2$. By choice of appropriate failure mode any path fault may be made self-declaring in period N.

The increasing size and hazards of process plant demand more reliable instrumentation, protection and control systems. Improvements in the reliability of individual components may be offset by the increasing complexity of single-path systems, which cannot be fault tolerant and need functional testing. Possible solutions include partial or total duplication of paths, and various forms of multiple-path system such as the familiar 2 out of 3 scheme.

Many plant monitoring systems can no longer be considered in isolation ; they must be considered as an integral part of the whole process operation including the optimisation of all the failure modes. The growth in the number and size of process plants demands greater consideration of the environmental, economic and human risk factors.

Sensing of plant conditions requires apparatus, the cost of which is usually small in relation to that of the plant which it protects. Because of the increasing cost of ownership of systems which are not self-testing, it becomes even more essential for improved reliabilities to be achieved by the use of multipath systems rather than by single-path methods.

The 'validated pair', incorporating deliberate choice of the failure mode and compliance with selected natural laws, appears to offer a useful solution. The costs and hazards of test arrangements, together with growing plant security requirements, emphasise the need for evolving improved systems, in which fault tolerance and self-proclamation of faults are inherent features. Such systems can be made :

— *Fault-tolerant.* Under conditions of fault they will provide useful service and thus have survivability.

— *Non-lieing.* Under fault conditions they cannot give indications or actions which are in error by more than acceptable tolerance without also indicating that the apparatus is lieing.
— *Self-validating.* Any fault is declared and alarmed as it occurs with no routine testing being required, and with the apparatus remaining in useful service.
— *Fail-Safe.* Failure in the apparatus cannot cause hazard to the plant.
— *Intelligent.* Indications or actions delivered are self consistent and also consistent with natural laws.
— *Economics.* The fewer paths required in validation systems for the same coverage as their alternatives mean lower cost.

Q. 24.22. Define the terms : Integrity, Reliability and Maintainability.

Ans. Integrity. Integrity is defined as the assurance provided by the system that all functions and tasks assigned to it are performed correctly, unless system clearly indicated about its incapacity for the same. Any malfunction of any part of the system should be diagnosed by the system through hardware or software and reported to the outside world. For high integrity the messages of malfunction must be consistent and correct, and there should be minimum time lag between occurrence and reporting of the malfunction.

Reliability. The reliability of a system is enhanced by equipping it with redundant parts. On occurrence of any malfunction, the correct operation of the redundant part is essential for reliability of the system.

Maintainability. Maintainability of a system depends upon as to

(*i*) what level is filed maintenance permitted (*i.e.*, upto unit, or board or component level).
(*ii*) whether the system or its subsystems have to be powered down for replacement of failed items and whether on-line repair is possible.
(*iii*) whether the integrity mechanisms of system give sufficient information for failure search and repair.
(*iv*) the type of tools required for maintenance.

Q. 24.23. Define maintainability.

Ans. Maintainability is the probability that, when maintenance action is initiated under stated conditions, a failed system will be restored to operable condition within a specified time. System maintainability can be improved by providing accessible test points, built-in test equipment, built-in diagnostic aids, training the operating personnel, and providing spare parts and equipment of incorporating repairs.

Q. 24.24. From consideration of maintainability, what important points should be taken care of by designer ?

Ans. The following is the check list for a designer to ensure high maintainability :

— Use modular or unit packaging.
— Use standardised parts.
— Use self lubricating principles.
— Use sealed and lubricated components.
— Provide built-in testing and calibration feature.
— Employ self-adjusting mechanisms.
— Use gear-driven accessories in preference to belt and pulleys.
— Simplify operator and maintenance functions and use simple design.

— Design for quick and positive recognition of marginal performance/malfunction.
— Design for quick position recognition of the replaceable defective components.
— Design to minimise the types and number of tools and test equipment required for maintenance.
— Design for optimum accessibility of all components/system/equipment.
— Design for maximum safety and protection of personnel.

Q. 24.25. What do you understand by "Logistics" ?

Ans. Logistics is a design discipline to reduce support costs by concentrating on the design of all support systems. Support system refers to all the activities required to keep a system operational for life cycle. The purpose of logistics is to assure that the system can be economically supported when placed in its operational environment. This includes the coordination of different design efforts, such as support-equipment design, technical publications, maintenance procedures, provision of spares, stocking of spares, field services, training of personnel, facility engineering, etc.

Q. 24.26. What do you understand by anthropometric data ?

Ans. Anthropometric data concerns the :

(*i*) maximum or comfortable spans for reaching controls and reading instruments.

(*ii*) limitations of muscular efforts that can be applied to controls in various positions, of various shapes, for varying periods of time.

(*iii*) mental reactions to signals (audio and visual) coming from a system.

(*iv*) interactions of control movements and the resultant effect (compatibility), etc.

Q. 24.27. (*a*) **What are the general conditions for which Preventive Maintenance is appropriate for a single machine ?**

(*b*) **What kinds of costs are associated with machine breakdown ? Discuss the general methods by which the reliability of production system can be maintained.**

Ans. (*a*) If the break down of a machine tool results in complete shut down of a plant it is essential that such machine be kept in working order all the time. The most effective way is to adopt preventive maintenance techniques.

Other conditions for carrying out preventive maintenance are :

— If cost of preventive maintenance is less than the average cost of repair after breakdown.
— If machine break down results in high idle-labour costs, then preventive maintenance can be carried out whenever machine is idle.
— Machine tool is critical and bottleneck in the plant.

(*b*) Costs associated with machine break down are as follows :

— Loss due to breakage of part
— Loss due to loss in production as a result of down time of machine
— Cost of repair
— Cost of maintenance personnel
— Penalties due to delayed deliveries
— Reliability of production system can be improved by
— Using improved component design
— Using simple and standardised parts
— Providing testing facilities to ensure healthiness
— Provide adequate redundancy whenever necessary

— Ensure system load below design value
— Take quality aspects into account at all stages
— Perform periodic and preventive maintenance.

Q. 24.28. What is the difference between discrete and continuous random variables ?

Ans. A discrete random variable is one that can have one of the two states, *i.e.* yes or no ; Go or No Go, or only a discrete number of values. In tossing a coin there are only two discrete states. In throw of a dice, there can be six discrete states.

A continuous random variable is one that can have an infinite number of values. For example, length of a specimen may vary between 2 and 3 cm. Since any number of values of length are possible in the range of 2 to 3 cms, it is a case of continuous random variable.

Q. 24.29. (*a*) **Under what conditions is Binomial distribution applicable ?**

(*b*) **What is the probability of *x* occurrences if *n* trials of an event take place, if the constant probability of occurrence is *p*.**

(*c*) **A process has past records to show its capability of producing 10% defective parts. A random sample of 10 is taken. What is the probability that (*i*) 1 part will be defective, (*ii*) 2 parts will be defective, (*iii*) one or less parts will be defective, and (*iv*) three or more parts will be defective in the sample.**

Ans. (*a*) Binomial distribution (basic discrete sampling distribution) is applicable where probability of certain occurrences in a number of trials of an event is sought and the event has a constant probability of occurrence, *i.e.*, population being sampled is infinite in size and the replacement of sampled unit takes place.

(*b*) Desired probability $P(x) = \dfrac{n\,!}{x\,!\,(n-x)\,!}\,p^x(1-p)^{n-x}$

(*c*) Given $n = 10$, $p = 0.1$,

(*i*) Probability of one part being defective is

$$P(1) = \frac{10\,!}{1!(10-1)\,!}(0.10)^1\,(1-0.1)^9 = 10(0.1)\,(0.3875) = 38.75\%$$

(*ii*) Probability of 2 parts being defective

$$P(2) = \frac{10\,!}{2\,!\,(10-2)\,!}(0.10)^2\,(1-0.1)^8 = 19.35\%$$

(*iii*) Probability of 1 or less parts being defective is $P(1 \text{ or less}) = P(0) + P(1)$

$$= \frac{10\,!}{0\,!(10-0)\,!}(0.1)^0\,(0.9)^{10} + 0.3875$$

$$= 0.3485 + 0.3875 = 0.736 = 73.6\%$$

(*iv*) Probability of 3 or more defective

$$= 1 - P(2 \text{ or less}) = 1 - (P_0 + P_1 + P_2)$$

$$= 1 - (0.3485 + 0.3875 + 0.1935) = 1 - 0.9295 = 7.05\%.$$

Q. 24.30. (*a*) **Under what conditions the Poisson distribution is applicable ? What is the probability of exactly *x* occurrences of an event of probability *p* in a sample of size *n* ; what is the Poisson density function, and the cumulative distribution function ?**

(*b*) **In a wire factory the average number of faults per year per 100 km of wire is 0.5. In a piece 10 km long, what will be the probabilities of 0, 1, 2, 3, 4, 5 faults occurring in a period of 20 years.**

Solution. (*a*) Binomial distribution tends to approach Poisson distribution when the fraction defective is small and number of trials are infinite. The Poisson distribution is most useful in determining probabilities for defects/failures per unit of time, area, or amount, *i.e.*, when failure rate is small and test times are relatively large.

The probability of x occurrences of an event of probability p in a sample of size n is

$$P(x) = \frac{(\mu)^x e^{-\mu}}{x\,!}$$

where μ = mean = np or λt in case of reliability

t = time, λ = constant failure or arrival rate.

Occurrence of each event is represented as a point on a time scale. In reliability engineering each event represents a failure. The Poisson density function is defined as

$$f(n) = \frac{(\lambda t)^n \exp.(-\lambda t)}{n\,!}, \; n = 0, 1, 2$$

where t is the time and λ is the constant failure or arrival rate.

The cumulative distribution function F is given by

$$F(n) = \sum_{i=0}^{n} \frac{(\lambda t)^i \exp.(-\lambda t)}{i\,!}.$$

(*b*) Assuming the average failure rate data to be valid for the 10 km wire and for the two periods being considered, the expected failure rate λ is,

$$\lambda = \frac{0.5 \times 10}{100} = 0.05 \text{ faults per year}$$

For a 20 year period,

$$\mu = \lambda t = 0.05 \times 20 = 1.0$$

and

$$P(x) = \frac{(1.0)^x e^{-1.0}}{x\,!} \text{ for } n = 0, 1, 2, 3, 4, 5$$

This failure density function is shown in Fig. 24.21 (*a*) and the cumulative failure distribution is shown in Fig. 24.21 (*b*).

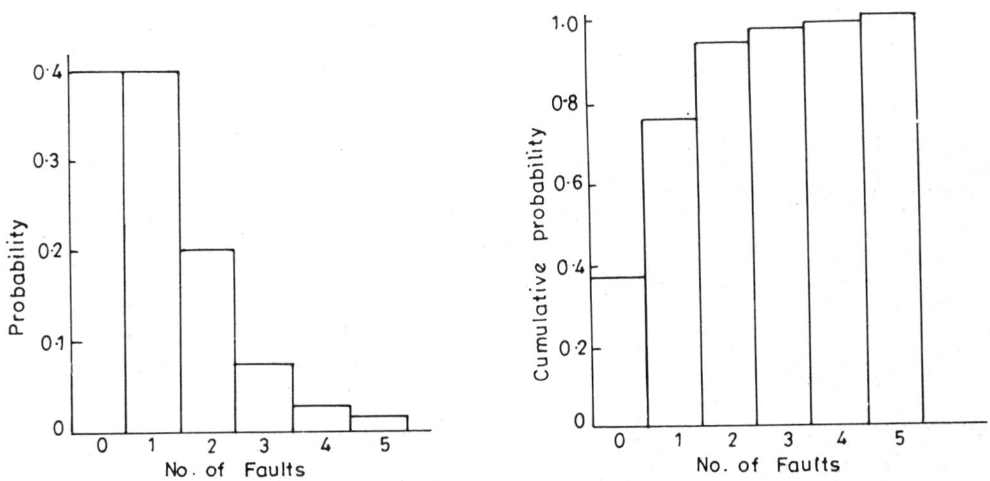

(*a*) Failure density function. (*b*) Cumulative distribution function.

Fig. 24.21

Q. 24.31. Write down equation for probability density function and cumulative distribution for an exponential distribution.

Ans. Probability density function $f(t) = \lambda e^{-\lambda t}$

where λ = constant failure rate and t = time

Cumulative distribution $\quad F(t) = \int_0^t \lambda e^{-\lambda t}\, dt = 1 - e^{\lambda t}$

Reliability $\quad R(t) = e^{-\lambda t}$.

The failure density function $f(t)$, cumulative failure distribution $F(t)$ and hazard rate $\lambda(t)$ are shown in Fig. 24.22 (*a*), (*b*), (*c*) respectively.

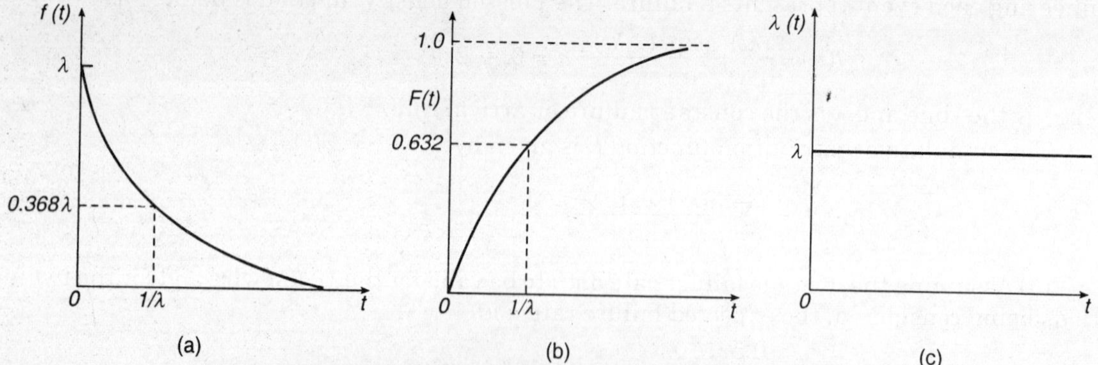

Fig. 24.22. Exponential reliability functions (*a*) Failure density function (*b*) Cumulative failure distribution (*c*) Hazard rate.

Q. 24.32. Failure rate of component is 0.0007 failure/hr. Compute the reliability of this component for a 300 hr. operation.

Ans. $\quad t = 300$ hr., $\quad \lambda = 0.0007$

$$R(t) = R(300) = e^{-(0.0007)(300)} = 0.8105$$

The reliability of the component for the specified mission time is 81%.

Q. 24.33. A component is to be selected from a bunch of components whose life is known to be exponentially distributed with a mean μ = 1,000 hr. What is the probability of this sample component not exceeding 1,000 hours ?

Ans. Exponential distribution function is

$$f(x) = \frac{1}{a} e^{-x/a} \qquad 0 \le x \le \infty$$

Probability of occurrence of an event exceeding the mean value is only $1 - 0.632 = 0.368$. Exponential function over the range $x = 0$ to 1,000

or $$P(x \le 1{,}000) = \int_0^{1{,}000} f(x)dx = \int_0^{1{,}000} \frac{1}{1{,}000} e^{-x/1{,}000}\, dx$$

$$= \left| e^{-x/1{,}000} \right|_0^{1{,}000} = 1 - e^{-1} = 0.632.$$

Q. 24.34. A company manufactured 2,000 components having an average life of 1,000 hours with a standard deviation of 200 hours. How many items might be expected to fail in the first 700 hours ?

Ans. The probability density function for this problem is shown in Fig. 24.23 in which $\mu = 1{,}000$ and $\sigma = 200$. The required probability is shown by the hatched area in figure.

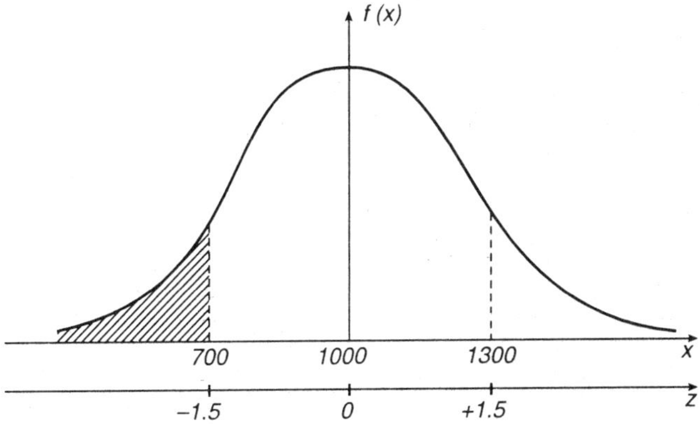

Fig. 24.23. Required area of normal density function.

$$Z = \frac{700 - 1000}{200} = -1.5$$

From the statistical table of normal distribution, the required area is

$$= 0.5 - 0.4332 = 0.0668$$

Expected number of failures = 2,000 × 0.0668 = 134 items.

Q. 24.35. When Weibull distribution is used ?

Ans. The Weibull distribution is widely used for reliability analysis of mechanical components.

Weibull distribution is a three parameters distribution whose probability density function is as :

$$f(t) = \frac{\theta}{\mu^{\theta}} t^{\theta - 1} . \exp.\left[-\left(\frac{t}{\mu}\right)^{\theta}\right] \qquad \text{for } \theta > 0, \mu > 0, t \geq 0$$

where μ is the scale parameter, θ is the shape parameter, t is the time and cumulative distribution function is as :

$$F(t) = 1 - \exp.\left[-\left(\frac{t}{\mu}\right)^{\theta}\right]$$

Therefore, reliability function is :

$$R(t) = \exp.\left[-\left(\frac{t}{\mu}\right)^{\theta}\right].$$

Q. 24.36. A spring is found to have average life of 1000 hours. What is its reliability for 250 hour operation if shape parameter on Weibull distribution is 2 ?

Ans. Reliability of spring is given by

$$R(t) = \exp.\left[-\left(\frac{t}{\mu}\right)^{\theta}\right]$$

Putting the values of μ = 1000 hr., θ = 2, and t = 250 hr.

$$R(250) = \exp.\left[-\left(\frac{250}{1000}\right)^{2}\right] = \exp.\left[-\frac{1}{16}\right] = 0.939.$$

25

Evaluating Uncertainty in Measurement

25.1. Introduction

Every measurement has an uncertainty associated with it. Quality of any measurement can be judged by evaluation and expression of uncertainty in it. Without knowing the measure of uncertainty in any measurement, the credibility of results of measurement cannot be established.

Uncertainty of measurement can thus be defined as the range of values that could reasonably be attributed to the measured quantity of a parameter, associated with the result of a measurement. In other words, when uncertainty in any measurement is evaluated and reported in a specified way, it indicates the level of confidence that the value actually lies within the range defined by the uncertainty interval.

An approach to calculation of uncertainty, its evaluation and expression has been prescribed by "Guide to the Expression Uncertainty in Measurement (GUM)" published by ISO Technical Advisory Group on Metrology.

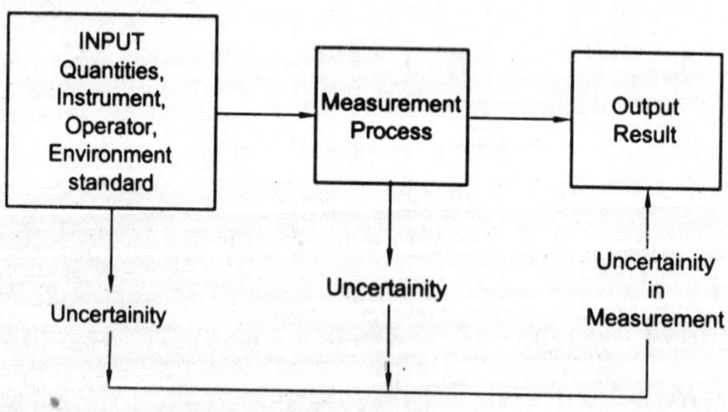

Fig. 25.1. Output result accompanied by statement of uncertainty.

25.2. Sources of Uncertainty in Measurements

Every measurement is accompanied with a variety of sources of uncertainty. Some sources of uncertainty are listed below :

(*a*) Incomplete definition of measurand.

(*b*) Imperfect realization of the definition of the measurand.

(*c*) Non-representative sampling—the sample measured may not represent the defined measurand.

(*d*) Configuration of the experimental set-up including uncertainty of standard, instrument, material, etc.

(*e*) Inadequate knowledge of the effects of environmental conditions on the measurement or imperfect measurement of environmental conditions, *e.g.* in case of dimensional measurement the uncertainty of measurement in Laboratory temperature or its variation will cause uncertainty in measured value of length due to thermal expansion effects. Further temperature measured for corrections, itself has certain uncertainties.

(*f*) Personnel efforts including personal bias in reading analog instruments.

(*g*) Calibration method/procedure.

(*h*) Finite instrument resolution or discrimination threshold.

(*i*) Inexact values of measurement standards and reference materials. (*e.g.* uncertainty associated with standard given in calibration certificate).

(*j*) Inexact values of constants and other parameters obtained from external sources and used in the data-reduction algorithm *e.g.* uncertainty in value of linear thermal expansion coefficient value, in case of dimensional measurements.

(*k*) Approximations and assumptions incorporated in the measurement method and procedure.

(*l*) Environmental conditions under which the calibration is carried out, *e.g.* (temperature, vibration, dust, humidity, etc.).

(*m*) Varations in repeated observations of measurand under apparently identical conditions.

Other sources of uncertainty are :

Master Gauge Calibration ;

Long Term Reproducibility ;

Thermal Expansion [Thermometer calibration, Coefficient of thermal expansion, and Thermal gradient (internal, gauge to gauge, scale)] ;

Elastic Deformation (Probe contact deformation, and Compression of artifacts under their own weight) ;

Scale Calibration (Uncertainty of artifact standards, Linearity, Fit routine, Scale thermal expansion, and Index of refraction correction) ;

Instrument Geometry [Abbe offset, Instrument geometry errors, Scale and gauge alignment (cosine errors, obliquity), Gage support geometry (anvil flatness, block flatness, surface finish)] ;

Artifact Effects (Flatness, Parallelism, Roundness, Phase corrections on reflection) ;

Effect of Temperature in Uncertainty (Length based measurements are a function of : (*i*) the length of the object being measured, (*ii*) the temperature of the object at the time of measurement, (*iii*) the coefficient of linear thermal expansion of the material of the object, (*iv*) the uncertainties in each of these quantities).

25.3 Method of Evaluation of Uncertainty

Uncertainty in a measurement is evaluated by :

(*i*) Identifying all the quantities that are involved in or influence the measurement. Thus a measurand Y may be function of input quantities $X_1, X_2 \ldots\ldots X_n$, *i.e.* $Y = f(X_1, X_2 \ldots\ldots X_n)$. Hence function f includes every quantiy that contributes a major component of uncertainty to measurement results.

(*ii*) Writing a model/relationship formula between input estimates and output estimates.

(*iii*) Evaluating uncertainty in measurement associated with each of the input estimates.

(*iv*) Combining all these to get the uncertainty associated with output estimate.

(*v*) Statement of uncertainty is given by expanding the combined uncertainty by a coverage factor.

25.3.1. Evaluating Uncertainty of Measurement

Two approaches used for evaluating the uncertainty of measurements are :

(*i*) Conventional approach based on (*a*) Random errors and (*b*) Systematic errors.

Random errors follow Gaussion / normal distribution approach and is determined by statistical methods ; whereas a systematic error follows Rectangular, Triangular or U-shaped distribution approach and is determined by other means.

(*ii*) Modern Approach based on (*a*) Type A method and (*b*) Type B method.

Modern approach is accepted as an international procedure. It is problem specific and defines the method explicitly. It is useful even for those problems where limited observations are possible.

Experimental methods/set up is the basis of determining the uncertainty in modern approach but the principal of evaluating the uncertainty is nearly the same as followed in conventional approach.

25.3.2. Type A Evaluation of uncertainty

This method is used when a large number of independent observations are taken for an input quantity under same conditions of measurement. Uncertainty is estimated by statistical analysis of a series of observations scattered around some mean value. If for a quantity Q there are independent observation $q_1, q_2, q_3, \ldots\ldots q_n$, then arithmetic mean $\overline{q} = \sum q_j / n$

The experimental variance $s^2(q)$ of values q_j is given as

$$s^2(q) = \frac{1}{n-1}\sum_{j=1}^{n}\left(q_j - \overline{q}\right)^2$$

Its positive square root is called the **Experimental Standard Deviation**. The best estimate of the variance of the arithmetic mean $\overline{q}$ is the experimental variance of the mean given by

$$s^2(\overline{q}) = \frac{s^2\,(q)}{n}$$

Its positive square root is called the **Experimental Standard Deviation of the Mean**. The standard uncertainty $u(\overline{q})$ associated with the input estimate $\overline{q}$ is the experimental standard deviation of the mean, *i.e.*,

$$u(\overline{q}) = s\,(\overline{q})$$

25.3.3. Type B Evaluation of Standard Uncertainty

In type B method of evaluation, the standard uncertainty is evaluated by the judgement of the available information on possible variability of input variables.

For instance, in this method one could make use of manufacturer's specifications, *e.g.*, for material/measuring instrument previous measurements data, data provided in calibration and other certificates *e.g.*, those for standards used, experience with or general knowledge of the behaviour and properties of relevance materials and instruments, uncertainties assigned to reference data taken from handbooks *e.g.* value of thermal expansion coeff. etc.

It is obvious that for making **Type B** evaluations critical thinking, intellectual honesty and professional skill are required. The value of uncertainty may be taken from a resultant value of a previous measurement, a reference value from the literature, using the value of uncertainty associated with a standard for which a calibration certificate is available. The standard uncertainty $u(x_i)$ associated with x_i is to be adopted where it is given. If data of this kind are not available, the uncertainty could be evaluated on the basis of experience.

The calibration certificate may give the expanded uncertainty in the certificate along with the coverage factor, *i.e.* standard uncertainty $u(xi)$ = Expanded uncertainty U in certificate / coverage factor.

The coverage factor may be specified either as one numerical value or it may be stated at a multiple of sigma. The standard uncertainty will be obtained by dividing the given value by the multiple. For example, if a calibration certificate states the uncertainty at 3σ, the standard uncertainty is then simply, quoted uncertainty/3. Further the quoted uncertainty may define an interval having a level of confidence, then students t table has to be used to find the factor by which this value should be divided to get u.

Type B evaluation where limits of variability of an input quantity are known and a probability distribution assumed for the quantity, X_i.

(*a*) **Rectangular distribution** is used when the limits can be determined and the value of the measurand is equally likely to lie anywhere within the limits.

If temperature of a room is likely to vary between 24° and 25°C, the temperature is equally likely to be anywhere within these limits. The distribution is represented graphically in Fig. 25.2. Here μ is the expected value, *i.e.* 24.5°C, value of a semirange of limits is 0.5°C. This is a uniform rectangular distribution and the Variance (u^2) and standard uncertainty u due to variation in temperature is given by $u^2(x_i) = \frac{1}{3}a^2$, *i.e.* $u(x_i) = a / \sqrt{3}$.

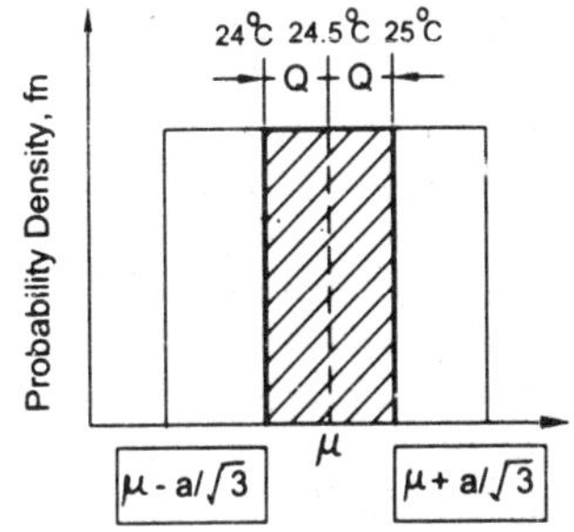

Fig. 25.2

(*b*) **Triangular distribution** is used when it is known that most of the values are more likely to be near the centre of the distribution, rather than at the extreme limits. The distribution is represented graphically in Fig. 25.3. Here μ is the expected value, the Variance u^2 and standard uncertainty u are given by

$$u^2(x_i) = \frac{1}{6}\, a^2, \text{ and } u^2(x_i) = a/\sqrt{6}$$

(*c*) **Trapezoidal distribution is** used when values are likely to be near the mid-point than those near the boundary. Trapezoidal distribution has equal sloping sides (an isosceles trapezoid), a base of width = $2a$, and a top of width $2b$, with $b / a = \beta$, where $0 \le \beta \le 1$. Variance associated with expectation is $u^2(x_i) = a^2(1 + \beta^2)/6$.

(*d*) **U-probability distribution** is used in the cases where the value at extreme limits is most likely to occur but the value at mean is least likely, *e.g.* Arcsin distribution. In this case the standard uncertainty is : $u = a / \sqrt{2}$.

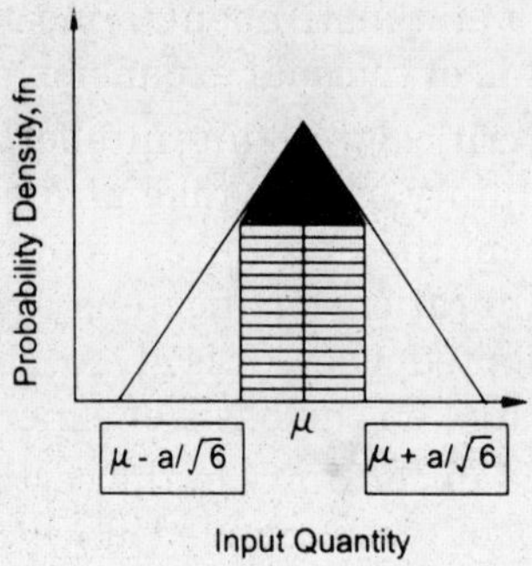

Fig. 25.3

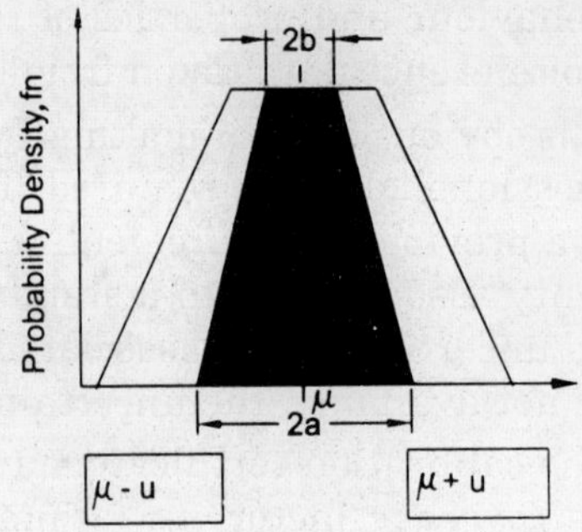

Fig. 25.4

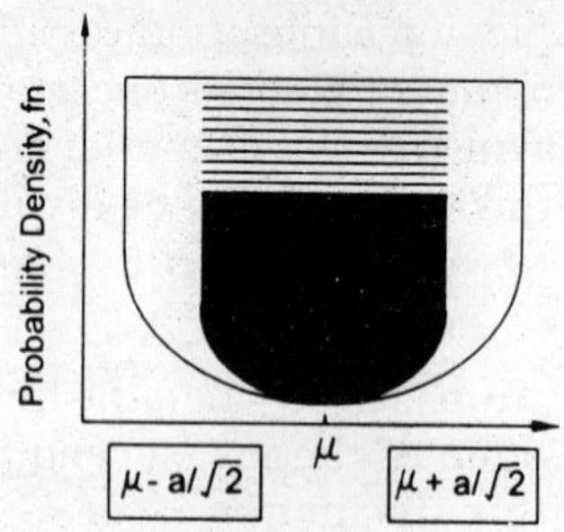

Fig. 25.5

Standard Uncertainty of Output Estimates

The uncertainty in each input quantity $u(x_i)$ contributes to the uncertainty of the output quantity $u(y_i)$, $u(y_i) = c_i u(x_i)$; where, c_i is known as sensitivity coefficient. It is calculated as $c_i = \partial f / \partial x_i$.

The sensitivity coefficient describes to what extent the output extimate y is influenced by variation of the input estimate x_i. In many cases this can be replaced using a more practical approach. A list of contributions to the uncertainty budget can be prepared and the sensitivity coefficient can be determined by varying the quantity over its range and measuring its influence on the measurand.

Let us work out values of c_i and u (y_i) for the *case of length measurement where the temperature dependance of length is given by linear thermal expansion coeff.* α *(for steel* α = *11.5). If the measurements were being carried for a nominal length (artifact of steel) and the contribution to this from the variation of temperature variation of ±0.5°C is to be calculated, the standard uncertainty* **u(x$_i$) = u(t) = 0.29°C under rectangular distribution.**

c_i sensitivity coefficient is equal to change in length of L = 100 mm per °C

$$c_i = \alpha.L \text{ m per °C} = 11.5 \times 10^{-6} \times 0.1 \text{ m per°C} = 1.15\ \mu\text{m per °C}$$

Thus, $u(y_i) = c_i u(x_i)$, becomes $u(l) = 1.15\ \mu m\ °C^{-1} \times 0.29°C = 33\ \mu m$

Combined standard uncertainty for uncorrelated input quantities

In this case all the input quantities are assumed to be **independent** and the combined stadard uncertainty $u_c(y)$, the positive square root of the combined variance $u_c^2(y)$ is treated below

$$u_c^2(y) = \sum u_i^2\,(y)\text{; and combined uncertainty } u_c(y) = \sqrt{\sum u_i^2\,(y)} = \sqrt{\sum (c_i)^2\,u^2\,(x_i)}$$

where, c_i is the sensitivity coefficient.

Let us consider the two simple cases :

(*a*) If function f is a sum or difference of the input quantities X_i,

$$f(X_1, X_2, \ldots X_i \ldots X_n) = \sum p_i X_i$$

the output estimate is given by the corresponding sum or difference of the inputs estimates $y = \sum p_i x_i$

where the sensitivity coefficients equal p_i and the combined uncertainty is given by

$$u_c^2(y) = \sum_{i=1}^{N} (p_i)^2 u^2(x_i)$$

Each $u(x_i)$ is a standard uncertainty evaluated as Type A or Type B.

(*b*) If function f is a product or quotient of the input quantities X_i,

$$f(X_1, X_2, X_3, \ldots\ldots X_n) = c.\prod_{i=1}^{N} X_i^{p_i}$$

the output estimate is given by the corresponding product or quotient of the inputs estimates $y = c \prod x_i^{pi}$

The sensitivity coefficients equal $p_i\,y/x_i$ and the relative combined uncertainty is given by $[u_c(y)/y]^2 = \sum_{i=1}^{N} [p_i\,u\,(x_i)/x_i]^2$

The combined standard uncertainty $u_c(y)$ is an estimated standard deviation and characterises the dispersion of the values that could reasonably be attributed to the measurand y.

25.3.4. Expanded Uncertainty

For calibration labs it is a requirement that the labs will state an *expanded uncertainty of measurement U*. The expanded uncertainty U is obtained by multiplying the combined standard uncertainty $u_c(y)$ by a coverage factor k.

$$U = k\,u_c(y)$$

The coverage factor may be selected or calculated for a stated level of confidence. Assuming that the combined uncertainty has a normal distribution the coverage factor is taken for a confidence level from student t table for effective degrees of freedom. Assumption of normal distribution is justified by invoking the Central Limit Theorem, which states that if many distributions are combined, irrespective of their own shape, the combined distribution will approximate a normal distribution. Hence the combined uncertainty will tend towards a normal distribution, as more and more components are included.

Values of k can be found for any desired level of confidence provided we have the number of degrees of freedom.

However if one or two components of uncertainty are dominant, normal distribution cannot be assumed and the combined uncertainty follows the distribution of the dominant component.

25.3.5. Statement of Uncertainty in Measurement

In calibration certificates the complete result consisting of the estimate y and associated expanded uncertainty U is given in form (y ± U). To this a note is added which has following centent

The reported expanded uncertainty in measurement is stated as the standard uncertainty in measurement multiplied by the coverage factor $k = xx$, which for a normal distribution with, $\nu_{eff} = yy$, effective degrees of freedom corresponds to a coverage probability of approximately 95%.

25.3.6. Uncertainty Budget

Uncertainty budget includes a list of all sources of uncertainty together with the associated standard uncertainties of measurement and methods of evaluating them.

25.3.7. Degrees of Freedom

Degrees of freedom is a term used to indicate confidence in the quality of the estimate of a particular input quantity to the uncertainty budget. The reason for determining the number of degrees of freedom associated with an uncertainty component is to allow the correct selection of the value of k factor. It is a measure of reliability of a component. A high no. of degrees of freedom is associated with a large number of measurements or a value with low dispersion.

Every component of uncertainty can have an appropriate no. of degrees of freedom ν assigned to it.

For the mean of n number of observations the uncertainty evaluated by Type A method, $\nu_i = n - 1$.

For Type B evaluation, when estimated upper and lower limits for a distribution are known, such that we have complete confidence in the values of these limits, then $\nu_i \to \propto$.

It is better to try to determine the limits more definitely, particulary if the uncertainty is a major one.

But if the limits themselves have some uncertainty, then a lesser number of degrees of freedom must be assigned. The ISO guide to the expression of uncertainty in measurement (GUM) gives a formula that is applicable to all distributions. It is given as :

$$\nu_i \approx 1/2[\Delta u(x_i) / u(x_i)]^{-2}$$

where $\Delta u(x_i)/u(x_i)$ is the relative uncertainty in the uncertainty, if we take the reliability *i.e.* $\Delta u(x_i) / u(x_i) = 0.1$ or 10% then $\nu \approx 50$, and if it is 5% then $\nu \approx 200$. Similarly $\nu = 2$ for relative uncertainty in uncertainty = 50%.

Once the combined uncertainty has been obtained, the degrees of freedom ν for each component must be combined to get effective degree of freedom, ν_{eff}, obtained by Welch-Satterthwaite formula given as :

$$\nu_{eff} = \frac{\{u_c^4(y)\}}{\sum_{i=1}^{N} [\{u_i(y)\} / \nu_i]^4}$$

where the $u_i(y)$ [i = 1, 2, ... N] is the contribution to the standard uncertainty associated with the output estimate y resulting from the standard uncertainty associated with the input estimate x_i, ν_i is the degree of freedom for each standard uncertainty component.

If calculated value of ν_{eff} is not an integer, it is truncated to the next lower integer and the corresponding **coverage factor k** value is obtained from the table of student's '*t*'-distribution.

25.3.8. Definitions

Uncertainty. An estimate characterizing the range of values within which the true value of the measurand lies. It has three components namely, the *standard uncertainty*, the *combined uncertainty* and the *expanded uncertainty*.

The standard uncertainty of measurement U(*xi*) in calibration is evaluated by two different methods denoted by Type-A or Type-B.

The Type-A method of evaluation of uncertainty utilizes the rules of basic statistics and quantitatively is equal to the mean of the standard deviation of a set of measured values.

The Type-B method of evaluation is not obtained from the repeated observations, but evaluated by scientific judgement based on available information or by estimate on the basis of a single value or taking data from the calibration certificate and stated in such a way that it corresponds to a standard deviation.

Combined Standard Uncertainty (u_c) can be evaluated by combining the standard uncertainties of the input estimates, $u(x_1)$, $u(x_2)$ ...... $u(x_n)$ evaluated by either by Type-A or Type-B method. According to the law of propagation of uncertainty, the combined standard uncertainty can be expressed by the relation :

$$u_c = (u_A^2 + u_B^2)^{1/2}$$

Expanded Uncertainty U, defines an interval about the result of a measurement, that may be expected to encompass a large fraction of distribution of values that could reasonably be attributed to measurand. It is obtained by multiplying the combined standard uncertainty by a coverage factor *k*-2 for 95% confidence level. So $U = k.u(y)$. The result of measurement is then expressed conveniently as $Y = y \pm U$.

The coverage factor *k* is evaluated by first calculating the effective degree of freedom ν_{eff}, and then from student's-*t* table, *k* is found to the value corresponding to calculated value of ν_{eff}

Q. 25.1. Determine the value of uncertainty of measurement in calibration of a digital thermometer with RTD probe (PT-100) in dry block calibrator at a temperature of 300°C by comparison with a calibrated standard platinum resistance thermometer as a reference standard. The temperature indicated by the reference standard (SPRT) is determined by a measurement of its resistance with A.C. Thermometry Bridge.

Ans. The temperature t_x to be assigned as the temperature of the digital thermometer is given by $E_x = t_x - t_i$.

where E_x is the error in the indicated temperature and t_i is the indicated value by the digital thermometer and mathematical model used for evaluation of uncertainty is as under :

$$t_x = t_s + \delta t_b + \delta t_r + \delta t_s + \delta t_i + \delta t_u + \delta t_c + \delta t_z + \delta t_h + \delta t_R + \delta t_{Ro} + \delta t_x$$

- t_s – Temperature of the reference thermometer (SPRT) derived with the A.C. thermometer bridge measurements
- δt_b – Temperature correction due to bridge measurement
- δt_r – Temperatrue correction due to standard resister used with the bridge
- δt_s – Temperature correction due to short term stability of the block calibrator
- δt_i – Temperature correction due to resolution of the instrument under calibration
- δt_u – Temperature correction due to the uniformity of the dry block calibrator
- δt_c – Temperature correction due to conversion of resistance to temperature of the digital thermometer

δt_z – Temperature correction due to stability of ice point

δt_h – Temperature correction due to hysteresis in the increasing and decreasing cycle of temperature

δt_R – Repeatability of DUC at the temperature of 300°C (measurement scatter)

δt_{Ro} – Repeatability of the DUC at 0°C before and after calibration at the maximum of the calibration range (measurement scatter)

δt_x – The temperature correction due to stem conduction, It has been observed that immersion of 15 cm of standard as well as of under test is sufficient to neglect effect of stem conduction and hence δt_x can be taken as zero.

Uncertainty Equation :

As none of the input quantities are considered to be correlated, the combined standard uncertainty is given by :

$$u_c^2(y) = \sum_{i=1}^{n} (c_i)^2 u^2(x_i)$$

where $u_c(y)$ = Combined uncertainty of measurement

$u(x_i)$ = Components of uncertainty

$c_i = \partial f / \partial x_i$, Sensitivity coefficients

n = No. of factor contributing to uncertainty.

Type B Uncertainty Components

1. **Uncertainty due to SPRT at (t_s).** The calibration certificate of the SPRT used as a reference standard states for the measured temperature value 299.999°C, the expanded uncertainty of measurement $U = \pm$ 3 m°C (0.003°C) at 95% confidence level with coverage factor 'k' = 2. So standard uncertainty at Normal Probability Distribution is $u(t_s)$ = 0.003°C/2 = 0.0015°C, Sensitivity coefficient $c_1 = 1$, Degree of freedom $\nu_1 = \infty$.

2. **Uncertainty due to A.C. Thermometry Bridge (δt_b).** The uncertainty of A.C. thermometry Bridge used for measurement of resistance of SPRT is given in the specifications of the bridge as 1 ppm of f.s. corresponding to resistance ratio of *Rt*/*Rs*, which in terms of temperature is stated to be 1 m°C (0.001°C). Assuming Rectangular Probability Distribution, the standard uncertainty $u(\delta t_b) = 0.001/\sqrt{3}$ = 0.000577°C, Sensitivity coefficient $c_2 = 1$, Degree of freedom $\nu_2 = \infty$.

3. **Uncertainty due to internal reference register (δt_r).** Uncertainty due to internal reference register *Rs* (100 ohm) mounted in a temperature controlled oven within ± 1°C and temperature coefficient is quoted as ≤ ± 1 ppm/°C, which correspond to 0.001/°C in terms of temperature as quoted in the specification oa A.C. bridge. Assuming Rectangular Probability Distribution, the standard uncertainty $u(\delta t_r) = 0.001/\sqrt{3}$ = 0.000577°C, Sensitivity coefficient $c_3 = 1$, Degree of freedom $\nu_3 = \infty$.

4. **Uncertainty due to the stability of block calibrator temperature (δt_s).** Uncertainty due to the stability of temperature of block calibrator as stated in the stability test certificate is ± 40 m°C (0.040°C) at a temperature of 300°C. Based on our experience and study of the block calibrator, the control cycle of the calibrator over a short period of time of 5 minutes, when the equilibrium has reached (as the time taken for completing set of 10 readings does not exceed this time), is ± 010 m°C (0.010°C). Assuming Rectangular Probability

Distribution, the standard uncertainty $u(\delta t_s) = 0.010/\sqrt{3} = 0.00577$°C, Sensitivity coefficient $c_4 = 1$, Degree of freedom $\nu_4 = \propto$.

5. **Uncertainty due to resolution of the DUC (δt_i).** The resolution of the digital thermometer under calibration is 1 m°C (0.001°C). This gives a temperature resolution limit of ± 0.5 m°C. Assuming Rectangular Probability Distribution, the standard uncertainty $u(\delta t_i) = 0.0005/\sqrt{3} = 0.000289$°C, Sensitivity coefficient $c_5 = 1$, Degree of freedom $\nu_5 = \propto$.

6. **Uncertainty due to the uniformity of the block calibrator (δt_u).** The overall uniformity of the block calibrator as mentioned in its technical specifications in the range 50–700°C in the insert of the block is stated as ± 30 m°C (0.030°C). So assuming Rectangular Probability Distribution, the standard uncertainty $u(\delta t_u) = 0.030/\sqrt{3} = 0.017321$°C, Sensitivity coefficient $c_6 = 1$, Degree of freedom $\nu_6 = \propto$.

7. **Uncertainty in the conversion from resistance to temperature of DUC (δt_c).** Uncertainty in the conversion from resistance to temperature of digital thermometer is stated as 15 m°C (0.015°C) at a reading of 300°C in the specification of digital thermometer under calibration. So assuming Rectangular Probability Distribution, the standard uncertainty $u(\delta t_c) = 0.015/\sqrt{3} = 0.00866$°C, Sensitivity coefficient $c_7 = 1$, Degree of freedom $v_7 = \propto$,

8. **Uncertainty due to stability of ice point bath (δt_z).** The stability of ice point depending upon the study and experimental data is ± 5 m°C (0.005°C), so assuming Rectangular Probability Distribution, the standard uncertainty $u(\delta t_z) = 0.005/\sqrt{3} = 0.0028867$°C, Sensitivity coefficient $c_8 = 1$,Degree of freedom $\nu_8 = \propto$

9. **Uncertainty due to the hysteresis effect in digital thermometer (δt_h).** Uncertainty due to the hysteresis effect in digital thermometer under calibration can be evaluated from the readings of device under calibration in the increasing and decreasing cycle of temperature measurement. This has been estimated to be within ± 10 m°C (0.010°C). So assuming Rectangular Probability Distribution, the standard uncertainty $u(\delta t_h) = 0.010/\sqrt{3} = 0.00577$°C, Sensitivity coefficient $c_9 = 1$, Degree of freedom $\nu_9 = \propto$.

Type 'A' Uncertainty Components

1. **Uncertainty due to Measurement Scatter of DUC at 300°C (δt_R) :**

S. No.	*DUC reading (q_i) in °C*	*($q_i - q_{ave}$) in °C*	*$(q_i - q_{ave})^2$ in °C*
1.	300.161	$-1.0000\,E-04$	$1.0000\,E-8$
2.	300.161	$-1.0000\,E-04$	$1.0000\,E-8$
3.	300.160	$-1.1000\,E-03$	$1.2100\,E-6$
4.	300.159	$-2.1000\,E-03$	$1.0000\,E-6$
5.	300.158	$-3.1000\,E-03$	$9.6100\,E-6$
6.	300.159	$-2.1000\,E-03$	$4.4100\,E-6$
7.	300.163	$1.9000\,E-03$	$3.6100\,E-6$
8.	300.164	$2.9000\,E-03$	$8.4100\,E-6$
9.	300.164	$2.9000\,E-03$	$8.4100\,E-6$
10.	300.162	$9.0000\,E-04$	$8.1000\,E-7$
	Sum = 3001.611		$4.0900\,E-05$

$q_{ave} = 300.1611$ and $n = 10$

$S^2 = 4.5444\,E - 06°C$

$S = 2.1318\,E - 03°C$

Considering normal distribution and sensitivity coefficient, $c_{10} = 1$

$$u(\delta t_R) = S/\sqrt{10} = 2.1318\,E - 04$$

Degree of freedom $\nu_{10} = n - 1 = 10 - 1 = 9$.

2. **Uncertainty due to the repeatability of the DUC at 0°C before and after calibration at the maximum of the calibration range (δt_{Ro}) :**

S. No.	*DUC reading (q_i) in °C*	*($q_i - q_{ave}$) in °C*	*($q_i - q_{ave})^2$ in °C*
Ice point measurement before calibration			
1.	0.062	– 2.0000 E – 04	4.0000 E – 8
2.	0.063	8.0000 E – 04	6.4000 E – 7
3.	0.068	5.8000 E – 03	3.3640 E – 5
4.	0.058	– 4.2000 E – 03	1.7640 E – 5
5.	0.061	– 1.2000 E – 03	1.4400 E – 6
Ice point measurement after calibration			
6.	0.064	1.8000 E – 03	3.2400 E – 6
7.	0.058	– 4.2000 E – 03	1.7640 E – 5
8.	0.062	– 2.0000 E – 04	4.0000 E – 8
9.	0.065	2.8000 E – 03	7.8400 E – 6
10.	0.061	– 1.2000 E – 04	1.4400 E – 6

Sum = 0.622 8.3600 E – 05

$q_{ave} = 0.0622$ and $n = 10$

$S^2 = 9.2889\,E - 06°C$

$S = 3.0478\,E + 03°C$

Considering normal distribution and sensitivity coefficient, $c_{11} = 1$

$$u(\delta t_{Ro}) = S/\sqrt{10} = 9.63798\,E - 04$$

Degree of freedom $\nu_{10} = n - 1 = 10 - 1 = 9$.

Combined standard uncertainty, $u_c(t_x)$:

$$u_c^2(t_x) = c_1^2\,u^2(t_s) + c_2^2\,u^2(\delta t_b) + c_3^2\,u^2(\delta t_r) + c_4^2\,u^2(\delta t_s) + c_5^2\,u^2(\delta t_i) + c_6^2\,u^2(\delta t_u) + c_7^2\,u^2(\delta t_c) + c_8^2\,u^2(\delta t_z) + c_9^2\,u^2(\delta t_h) + c_{10}^2\,u^2(\delta t_R) + c_{11}^2\,u^2(\delta t_{Ro})$$

$$= (0.0015)^2 + (0.000577)^2 + (0.000577)^2 + (0.00577)^2 + (0.000289)^2 + (0.017321)^2 + (0.00886)^2 + (0.0028867)^2 + (0.00577)^2 + (0.0002132)^2 + (0.0009638)^2$$

$$= 2.25E - 06 + 3.33E - 07 + 3.33E - 07 + 3.33E - 05 + 8.33E - 08 + 3.00E - 04 + 7.50E - 05 + 8.33E - 06 + 3.33E - 05 + 4.55E - 08 + 9.29E - 07$$

$$= 4.54E - 04$$

$$u_c(t_x) = (4.54E - 04)^{1/2} = 0.021°C$$

Effective Degree of Freedom. The effective degree of freedom of combined standard uncertainty is given by the following equation :

$$\nu_{eff} = \frac{u_c^4\,(t_x)}{\sum_{i=1}^{n} u_i^4 / \nu_i}$$

$$= \frac{(0.021)^4}{\begin{array}{l}(0.0015)^4/\infty + (0.000577)^4/\infty + (0.000577)^4/\infty + (0.00577)^4/\infty \\ + (0.000289)^4/\infty + (0.017321)^4/\infty + (0.00866)^4/\infty + (0.0028867)^4/\infty \\ + (0.00577)^4/\infty + (0.0002132)^4/9 + (0.0009638)^4/9\end{array}}$$

$$= 1.94481E-07 / 2.29565E-16 + 9.58749E-14$$

$$= 1.94481E-07 / 9.61045E-14 = 2023640.5 \cong \infty$$

Expanded Uncertainty, *U*. For degree of freedom, $\nu_{eff} \cong \infty$ at approximately 95.45% level of confidence, the '*t*' factor from student's '*t*' table is 2. So the coverage factor '*k*' = 2.

The expanded uncertainty *U* is :

$$U = k\,u_c(y) = k\,u_c(t_x) = 2 \times 0.021 = 0.042°\text{C}$$

Results. The temperature t_x to be assigned to the digital thermometer indicating a temperature of 300.161°C is 299.999°C ± 0.042°C.

The reported expanded uncertainty of measurement is stated as the standard uncertainty of measurement multiplied by the coverage factor '*k*' = 2 based on students' *t* – distribution for effective degree of freedom $\nu_{eff} = \infty$ estimated at a confidence level of approximately 95%.

Uncertainty Budget

Qty	*Estimated Value x_i °C*	*Limits Dx_i °C*	*Standard Uncertainty $u(x_i)$ / °C*	*Probability Type-A & B Divisor*	*Degree of freedom*	*Sensitivity Coefficient C_1*	*Uncertainty Contribution $u_i(y)$ / °C*	*$u_i^2(y)$ (°C)^2*
t_s	299.999	0.003	0.0015	Normal Type B, 2	∞	1	0.0015	2.25*E*-06
δt_b	—	0.001	0.000577	Rectangular Type B, √3	∞	1	0.000577	3.33*E*-07
δt_r	—	0.001	0.000577	Rectangular Type B, √3	∞	1	0.000577	3.33*E*-07
δt_s	—	0.010	0.00577	Rectangular Type B, √3	∞	1	0.00577	3.33*E*-05
δt_i	—	0.0005	0.000289	Rectangular Type B, √3	∞	1	0.000289	8.33*E*-08
δt_u	—	0.030	0.017321	Rectangular Type B, √3	∞	1	0.017321	3.00*E*-04
δt_c	—	0.015	0.00866	Rectangular Type B, √3	∞	1	0.00866	7.50*E*-05

δt_z	—	0.005	0.0028867	Rectangular Type B, √3	∞	1	0.0028867	8.33*E*-06
δt_h	—	0.010	0.00577	Rectangular Type B, √3	∞	1	0.00577	3.33*E*-05
δt_R	—	—	0.0002132	Normal Type A, √10	9	1	0.0002132	4.55*E*-08
δt_{Ro}	—	—	0.0009638	Normal Type A, √10	9	1	0.0009638	9.29*E*-07
Combined standard uncertainty, $u_c\ (t_x)$			—	—	—	—	√4.54E-04 = 0.021°C	
Expanded Uncertainty, *U*						'*k*' = 2	0.021 × 2 = 0.042°C	
$t_x = t_s$ = 299.999							± 0.042°C	

25.4. Competence of Testing and Calibration Laboratories

The confidence that a testing or calibration laboratory consistently obtains reliable results is of utmost importance to users of laboratory services.

The testing laboratory practices need to be harmonised and fall in line with International practices so that such labs get accredited to be accepted as world class labs.

ISO/IEC 17025 is the new standard for the competence of testing and calibration laboratories. It was published in Dec. 1999 and aligned with 1994 edition of ISO. Such laboratories need to implement quality procedures for montoring the validity of test and calibration performed.

Measurement assurance is the application of broad quality control principles to measurement and calibration. Monitoring the validity of test and calibration performed by a lab on a continuous basis is essential for its survival.

Laboratories need to adopt the measurement assurance practices which can be implemented within the laboratory.

International standard ISO 17025 stipulates the managerial as well as the technical requirements that are required to be complied by the calibration and testing lab in order to demonstrate their technical competence.

Standard ISO 17025 has several aspects for harmonization and better interpretation of clauses. This standard gives more emphasis on some management system requirements. It gives minimum required contents of quality policy. It incorporates specific requirements for control, review and approval, issue and amendment of documents.

As per standard ISO 17025, it is the responsibility of the lab to ensure that clients' needs are identified and understood, and that the calibration/test method selected should be capable of meeting these needs. Thus if a client has proposed a method ; it is the duty of the lab to inform client that the method proposed is inappropriate or out of date. This standard calls for specific procedures for cause analysis, selection and implementation of corrective action, subsequent monitoring and follow-up audits. This standard also provides for identification of needed improvements and potential non-conformances and highlights the need to look out for potential problems and opportunities for improvement, *i.e.,* a more proactive approach rather than waiting for non-conformances to occur. This standard also specifically calls for co-operation with clients and gives guidance on lab client relationship in the form of notes.

This standard permits the inclusion of opinions and interpretations in reports and certificates. Opinions and interpretations require clear identification in reports and their basis must be documented. The inclusion of opinions and interpretations provides the opportunity for laboratories to offer a more comprehensive, accredited service thus enhancing the value of their accreditation. Only very competent persons should be assigned this task, who not only have the required knowledge but also know limitations of their knowledge.

International standard ISO/IEC 17025 will help to level the "playing field" for the operation of laboratories and thus facilitate worldwide acceptance of certificates from the laboratories accredited to this standard, and improve the standing of international mutual recognition agreements.

An important condition stipulated in ISO-17025 is that labs have to assure the quality of test and calibration results. The foremost requirement of assuring the quality of calibration is the metrological confirmation of measuring equipment and controlling the measurement process. Further the validity of test and calibration needs to be monitored for which suitable quality control methods appropriate for the type and volume of work in lab need to be followed.

Measurement assurance practices could include maintaining control charts for reference standards, intermediate check of instruments/standards, retesting of items using same or different methods, safeguarding of any software or hardware adjustments, and correlation of results.

25.4.1. Apex Level Calibration and NPL

National Physical Laboratory, New Delhi is the custodian of Indian national standards of physical measurements and represents the country in the Meter Convention (Treaty originally signed in 1875). It has the responsibility of maintaining and continuously upgrading by R and D national standards as per international trends and practices. NPL also provides apex level calibration to the industry and other users. It ensures equivalence of national standards with the international system. For this purpose it has signed the Mutual Recognition Arrangement (MRA) to remove technical barriers to trade as an important requirement of the WTO agreement. For this purpose there is an intense activity of participating in international key comparisons organized by International Bureau of Weights and Measures (BIPM)/CIPM and Regional Metrology Organization (RMO), and Asia Pacific Metrology Programme (APMP).

25.5. International Traceability

As markets open up and trade takes place at global level, all products need to be manufatured and tested for international acceptance. The method of testing and calibration need to have international traceability. National standards need to be traced to international standards. Obviously National Metrology Institutes have to play an important role in maintaining the measurement traceability to international systems. The precision measurement is related to the uncertainty in the measurement and traceability to SI units. Modern instrumentation requires various measurements with a reproducibility that exceeds the uncertainty of the realisation of the SI units. For this purpose metrologists have developed artifacts and experiments that can generate various parameters that are stable and reproducible to a level approaching 0.001 parts in 10^6.

National Physical Laboratory, India is playing a dominant role for the realisation of the standards of physical, chemical and electrical measurements with highest levels of accuracy and traceability to international measurement system. Considerable efforts are needed in building the confidence that measurements made in one location in the world are compatible or equivalent to those made in other locations on the same or related products. For this, the measurements have to be applicable, comparable and traceable to both national and international standards.

Following techniques are involved in validating these efforts :

(*i*) Mutual Recognition Arrangement

(*ii*) Accreditation Procedures

(*iii*) Audit Processes

(*iv*) Proficiency Testing

(*v*) Measurement Assurance Programmes

(*vi*) Intercomparison and Calibration

(*vii*) Achieving Traceability and Timeliness in Traceability.

Traceability means to make the measurements traceable to SI units. Hierarchical chain follows following orders : SI units, primary standards, secondary standards, reference standards, working standards, and measuring standards. The definition and realisation of SI units are given by CIPM. The recommendations of CIPM give standards (primary) of highest metrological quality. As per latest norms, most of these standards based on quantum phenomena need to be intercompared before establishing these as primary standards. The intercomparison of primary standard establishes a relationship between primary standards used by various National Metrology Institutes. (NMI). This is known as equivalence of standards. Thus the intercomparison establishes traceability among standards of same quality (horizontal traceability). The vertical traceability is achieved through calibration. Both vertical and horizontal traceability, *i.e.* calibration and intercomparison are ways to establish traceability and to ensure reliable measurement of parts to be compatible worldwide.

The primary standards maintained at various NMIs have to be equivalent, there being no hierarchical relationship between them. The measurement capabilities of various NMIs are demonstrated to be equivalent within a certain range of uncertainty. Equivalence is declared by signing Mutual Recognition Arrangement (MRA) and declaring measurement capabilities on the basis of quality systems and intercomparisons/key comparisons among NMIs.

The assurance of laboratories technical competence to perform measurements by interlaboratory comparison is achieved by proficiency testing (PT). Participation in PT helps in building confidence among staff and provides independent and objective evidence to the measurement capability and quality assurance to participating labs.

Timeliness in traceability, *i.e.* having calibration at smaller periods (3 to 6 months) to take care of drifts helps to guarantee maximum reliability in measurement.

25.6. Mass Metrology

Metrology is concerned with units of measurements, standards of measurements, measuring instruments and measuring procedures. Measuring is the experimental procedure whereby a special value of a physical or measured quantity or a multiple or fraction of such unit is determined. The measured value is the special value of the measured quantity to be determined and is expressed as a numerical value and a unit. Processing the data to the measuring results is also part of measuring. Calibration is the determination of the relationship between output quantity and intpu quantity. The indication error of the scale or dial of the measuring instrument is determined by calibration. On the other hand adjusting a measuring instrument means setting the device so that the display deviates as less as possible from the correct value or the errors are kept within prescribed limits.

Unit of Mass. According to the International System of Units (SI) the unit of mass is kilogram and is defined as the mass of the International Prototype Kilogram kept under the custody of the International Bureau of Weights and Measures (BIPM) at Paris in France. The Interantional Protctype kilogram is a right circular cylinder whose height is 39 mm, and is equal to its diameter. It is made of a single-phase alloy of platinum-iridium with 10% iridium by weight. This is only weight in the world whose mass is taken exactly one kilogram and is called "international kilogram".

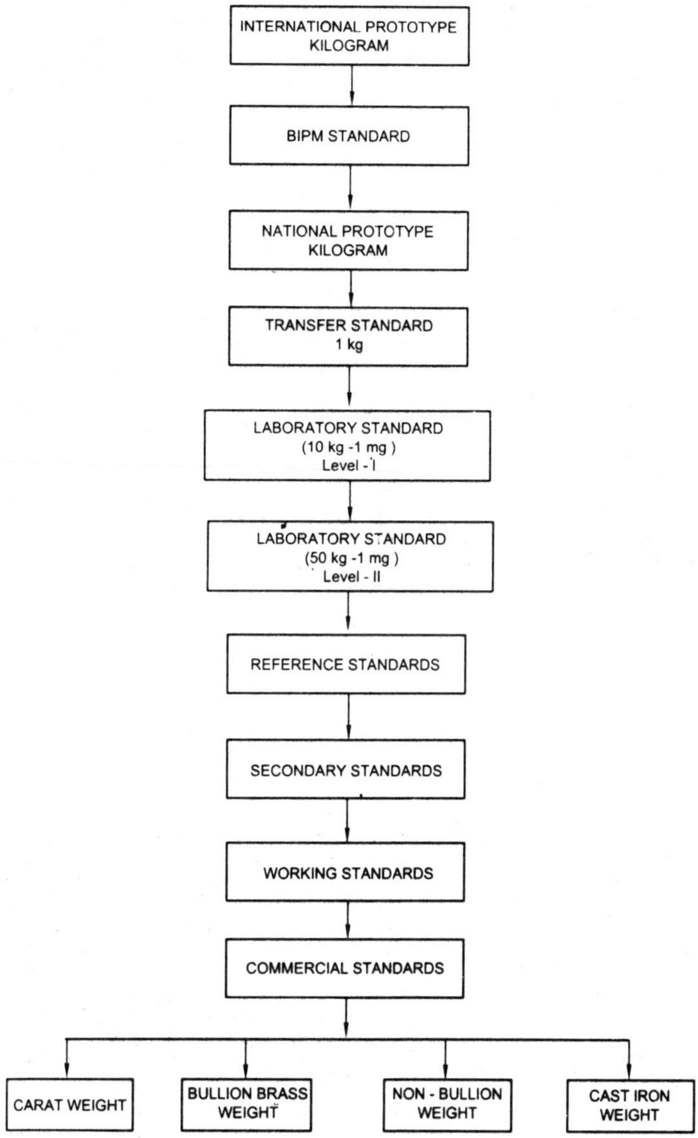

Fig. 25.6. Hierarchy chart of mass standards.

Standards of Mass. Fig. 25.6 shows the hierarchy chart of standards for mass. The grouping starts with the International Prototype kilogram and ends with the working standards in national and industrial metrology.

Next to international kilogram, BIPM has other 65 such cylinders whose mass values have been derived from the international kilogram. Six of them namely k_1, k_7, k_8, k_{32}, k_{43} and k_{47} are called the copies of the international kilogram and are preserved under the same conditions as the international kilogram. The remaining cylinders are serving as the national prototypes and transfer standards of BIPM. Every country, that is the member of the 'Convention of meter', is entitled to have one copy of the international prototype kilogram, which serves as its national prototype kilogram. This prototype is maintained as primary standard of mass at the national metrological laboratory of that country. The copy No. 57 of the International Prototype Kilogram serves as the National Primary Standard of Mass of our country and is kept under the custody of the National Physical Laboratory (NPL). National Physical Laboratory, under the subordinate legislation of the standards of Weights and Measures Act of 1976, is responsible for the realization, establishment, custody, maintenance, determination, reproduction and updating the national standards of weights and measures. The national prototype kilogram is periodically calibrated against the International Prototype Kilogram by BIPM at least once in a period of 10 years. In addition to primary standard a number of transfer and laboratory standards are maintained by the NPL. The Department of Legal Metrology enforcing the laws of weights and measure Act maintains a series of standards at different levels. This ensures the traceability of mass measurement carried out in the country at any level to the national standards and hence the international kilogram.

Mass Measuring Instruments. For high accuracy mass determination, analytical balances and comparator balances are used. An analytical balance is characterized by a small-scale interval (d) compared to the maximum capacity (Max.), the ratio d/Max generally being smaller than 10^{-5} and the maximum capacity usually not exceeding 10 kg. Analytical balances are grouped into weighing instrument of special accuracy (class I) and of high accuracy (class II). For weighing instruments of special accuracy, the following classification is generally used :

macro balance	Max. > 100 g	$d < 10^{-5}$ Max.	(*e.g.* d = 0.1 mg)
half-micro balance	Max. ≤ 100 g	$d < 10^{-5}$ Max.	(*e.g.* d = 0.01 mg)
micro balance	Max. ≤ 50 g	$d < 10^{-5}$ Max.	(*e.g.* d = 0.001 mg)
ultra micro balanced	Max. ≤ 5 g	$d < 10^{-6}$ Max.	(*e.g.* d = 0.0001 mg)

For weighing instrument with resolutions higher than these (*i.e.* $d \leq 10^{-7}$ Max), the term comparator balance or mass comparator is used. The mass comparator allows only differential weighing whereas analytical balances can be used both for differential weighing as well as for proportional weighing. Differential weighing stands for mass comparison using mass standards or other suitable objects, whereas proportional weighing is the simple mass comparison without having recourse of reference standards.

Selection of Weighing Instrument. For a given weighing task selection of suitable weighing instrument depends on the maximum load to be weighed and the resolution (also known as readability) required. The weighing instruments are often referred to number of decimal places in grams that the instrument is capable to indicate in its scale. A six-place balance will resolve up to sixth decimal place of a gram *i.e.* to a microgram. Sometimes not only the resolution but the maximum capacity is also specified for example a 200 g five place balance or a 5 g six place balance.

In legal metrology, as per OIML (International Organization on Legal Metrology) International Recommendation No. R 76-I, all non-automatic weighing instruments are classified in four accuracy classes. The maximum permissible error (MPE) of each class of weighing instrument is based on its verification scale interval (e) and is a function of the applied load on the instrument. The manufacturer for a particular weighing instrument specifies the value of e, if it is not given it is assumed to be equal to the scale interval (d).

The correctness of the instrument is defined through the value of the applied load *i.e.*, mass of the weight used. This means certain accuracy is required for the standard weight used. All the weights used in trade, commerce, industry and other applications (up to 50 kg) are classified into seven accuracy classes as per OIMI International Recommendation No. R-111 and higher capacity weights into four other classes as per OIML R-47. Appropriate accuracy class weights are used depending on the required accuracy in measurement results.

The weighing instrument also contributes to the testing uncertainty and must therefore fulfill certain requirements with regard to its repeatability. Concerning the measurement uncertainty in the non-automatic weighing instrument, OIML R-76-1 states three important requirements as follows :

(*i*) **Permissible differences between results.** Regardless of what variation of results is permitted, the error of any single weighing result shall by itself not exceed the maximum permissible error for the given load.

(*ii*) **Repeatability.** The difference between the results of several weighing of the same load should not be greater than the absolute value of the maximum permissible error of the instrument for that load.

(*iii*) **Weights.** The standard weight used for the verification of an instrument should not have an error greater than 1/3 of the maximum permissible error of the instrument for the applied load.

These three single demands are enough for correctly judging the performance of a weighing instrument, generally with 95% confidence level.

Adjustment of an Electronic Balance. Mechanically balances are normally not adjusted because they indicate weighing differences in mass units using their sensitivity ditermined separately by independent measurement. But in the case of electromagnetic or electronic balance the indication is adjusted in mass units. Now-a-days, only electronic balances are commonly used.

Normally comparisons of two weights are carried out in air using suitable balance. Thus a balance serves to compare not only the weight forces but also buoyancy forces. For electromagnetic force compensation balances, the indication m_w, in units of mass, is proportional to the electromagnetic force F which compensates the weight and buoyancy forces such that :

$$m_w = c_j F = c_j mg(1 - \rho_a/\rho)$$

where c_i = adjustment constant

m = mass of the test weight

g = local acceleration due to gravity

ρ_a = density of air during the weighing

ρ = density of the test weight.

Measuring Procedure. The mass of a body is determined by weighing *i.e.* the comparison of the mass of the object to be weighed with mass of a reference weight of known mass on a balance after allowing all necessary corrections, particularly, the buoyancy correction. In order to obtain relative uncertainties < 1×10^{-5}, differential weighing is indispensable.

For differential weighing, following two different weighing principles are normally used :

Transposition (Gauss) Weighing. This principle can be used only with equal-arm balances. The test weight and the reference weight are interchanged on the pans at least once and the results of the two weighing are averaged.

Substitution (Borda) Weighing. This principle can be used with all weighing instruments. The test weight and the reference weight are weighed one after another on the same pan. In the case of equal arm balances the second pan is loaded with a fixed auxiliary weight.

For modern analytical and comparator balance, the substitution principle is generally used because, unlike the Gaussian principle, it allows a simple balance design and offers more ease of handling and, thus, less time of measurement. It also makes it relatively simple to automate the weighing sequence and, thus, to increase the accuracy of measurement.

25.7. Coordinate Measuring Machine and Uncertainty in Measurements

Coordinate measuring machines (CMM) can be used for measuring any type of object both during setting up process as well as during actual production. In fact the designers have been able to specify close tolerances ; and success in NC machines and flexible manufacturing systems has been possible because of presence of reliable CMMs capable of giving accurate measurement in short span of time. CMMs have capability to position the measuring conical probe attachment anywhere in three axes *x, y* and *z* (spatial coordiate system). Machine basically has a quill capable of moving alongwith measurement probe in three axes similar to NC machines. The displacements along the axes can be read by length measuring system which transfer readings to electronic control cabinet.

Measuring spacing between two bores. The conical probe of CMM is capable of self-centering itself with reference to bore and thus sets itself at the centre of bore. With this as reference the distance between two bores can be assessed very conveniently. Even if part is not aligned exactly along the axis of the machine. The computer program takes care of cosine errors in measurements and provides exact distance between any two holes in a workpiece. This software alignment facility in CMM increases its versality considerabily.

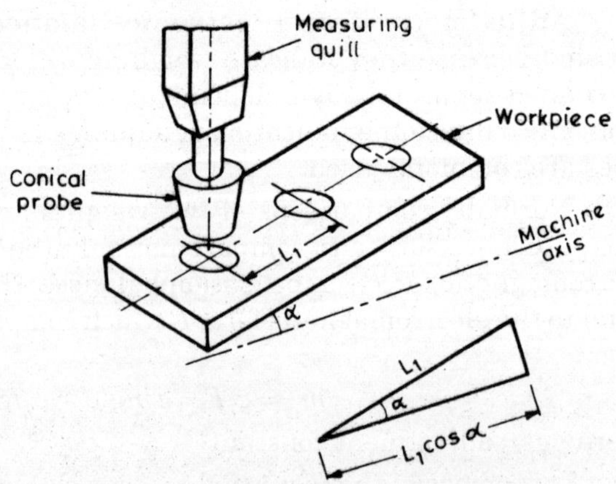

Fig. 25.7. Measurement of bore distances and alignment error.

Measurning diameter of a bore. For most of the measurements, CMMs use spherical probe attachment. Ball tip probe has an advantage that its mid-point is always equidistant from any point on the sphere surface and thus the displayed coordinates of tip to computer refer to its mid-point. For mesuring diameter of a hole, the spherical probe contacts any three points of the surface of the hole and computer program with help of this information of three points determines the circle diameter and its mid-point coordinates. The ball tips are made of ruby (highly homogeneous and hard).

Probing in three dimensions. The measuring quill has a large number of measuring tips with spheres (probes) aligned in different directions so that any feature on the surface of a workpiece can be accessed with one of the probes. The ball tip diameters and their mid-point distances on all probes are measured on same CMM by probing a high precision calibration sphere of known diameter with each of the mounted probes. The computer calculates and stores the above probe data for different probes. In this way an imaginary probing sphere of zero dimension is created which can access any point.

A universal 3-D measuring probe, permitting deflection of the probe in all probing directions and making contact with a defined measuring force, ensuring repeatable and constant, zero position in all axes and confirming the same by electrical probing signal, is considered the most basic element of CMM.

Dynamic Probes. Dynamic or switching probe is the simplest probe head working on a simple mechanical principle. Basically it comprises three spheres resting on 3 sets of two cylinders each. Thus brobe has flexibility to move and still it arrests the probe in an exact

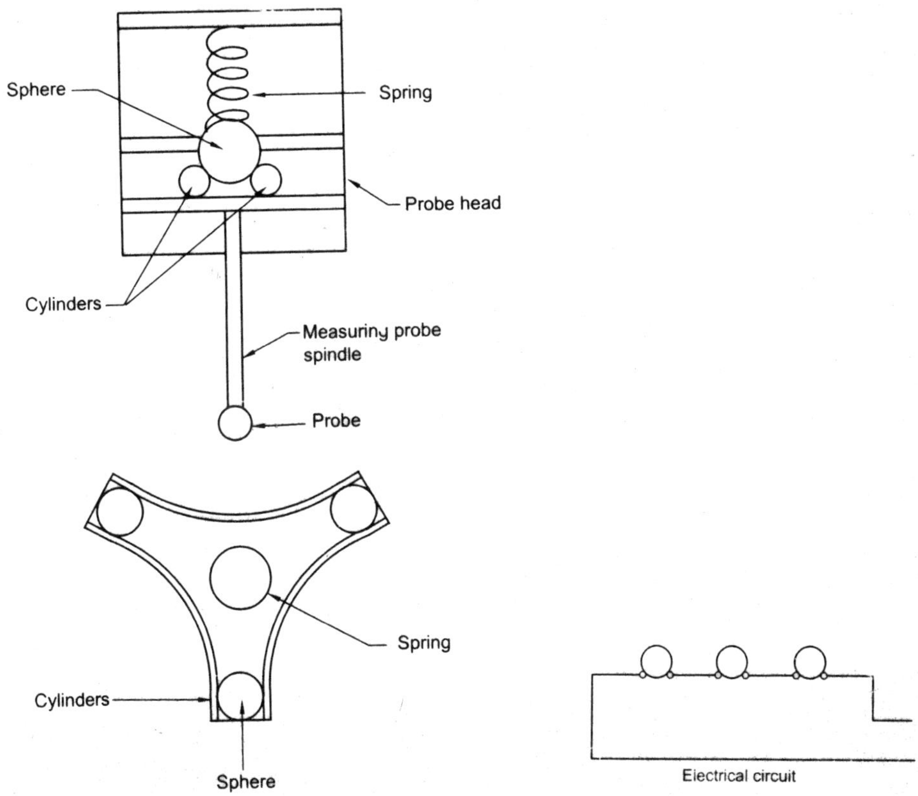

Fig. 25.8. Dynamic probe ideally suited for measuring individual points.

volumetric flexible position. Three spheres are located at the end of an arm and 120° radially away from each probe. Probe is supported from this arm and arm itself is supported against a spring and three spheres rest on three sets of cylinders each located on a fixed plate. Entire system thus acts as prestressed buckling machanism. When probe head makes contact on the surface to be measured, it is deflected and then returns to its spatially neutral position defined by three bearing points which also provide electrical contact connected in series. When probe is touching a surface, the electrical circuit is disconnected as one probing point will be open. However when probe is just touching the surface electrical circuit is through and as soon as it occurs, the reading ia frozen in memory. Such an electromechanical principle is good but not accurate due to prestressing and probing direction. Piezoelectric sensors display identical sensitivity from all probing directions and generate the pulse before the mechanical contact opens.

Probe heads are so designed that any set of measuring probe combinations can be mounted on it by removing one adapter plate and fixing another one with desired probes. Fixing of adapter plate on measuring head by magnet ensures that the probe is always mounted in an identical spatial position by three point bearing with an accuracy of fractions of a micron.

All probe combinations to be used are pre-calibrated and their calibration data is stored in memory and can be called for CMM to become operational with desired probe.

Probe Heads Capable of measuring in 3-axes. Dynamic probe described earlier is suitable for measuring in vertical axis. The versatility of CMM can be increased considerably by adopting probe heads for measuring in three axes. Measurement thus becomes static and free from any distorting forces resulting in increase of accuracy and permitting length measurement with CMM axes being completely immobile.

Such a probe basically consists of three superimposed, orthogonal spring parallelograms. The deflection of parallelogram is sensed by an LVDT. A motorised stop machanism allows the parallelogram to be clamped in its centre position. Each parallelogram is also mechanically connected to a solenoid which moves in a split-ring magnet and generates a defined measuring force due to flow of current in solenoid.

When probe contacts the workpiece, its deflection energises the drive of the measuring component and the output of the inductive transducer controls the positioning. This results in the drive settling in zero position (centre position of the parallelogram).

In case of other two axes also, the clamping force holds the probe head in its spatial zero point. The positioning control is independent to act independently in two or three axes simultaneously. The readings are noted at CMM.

Even if probe head is not in zero position during probing, the deflection in probe head axis is copied in length measuring system of the corresponding CMM axis. The probe head output and the CMM length measurement values are added in all three measuring axes. *i.e.* Resulting measurment value = CMM coordinates + probe head output.

The probe capable of measuring in three axes in conjunction with electronic control enables continuous probing of a surface. The time—or distance—dependent data is transferred continuously from single point probe to the computer while the surface is being scanned by the probe.

The probe head is continuously repositioned by the drive system smoothly in the probing axis so that probe head remains within measuring range all the time.

Optical non-contact probing is used for probing soft and delicate parts. Optical probes have high speed of operation. For measurement of edges and for determination of a geometrical shape, specially designed procedures are used. However versatility of tactile sensors with ball tips is much superior. It is also possible to combine optical and touch probing to get advantages of both systems. Touch probing is best suited for measuring vertical contours, lateral bores, sloped edges, optically in accessible features, etc. and optical probing is used for all other features to save time. Optical probes measure the topography of parts of any shape by using triangulation method.

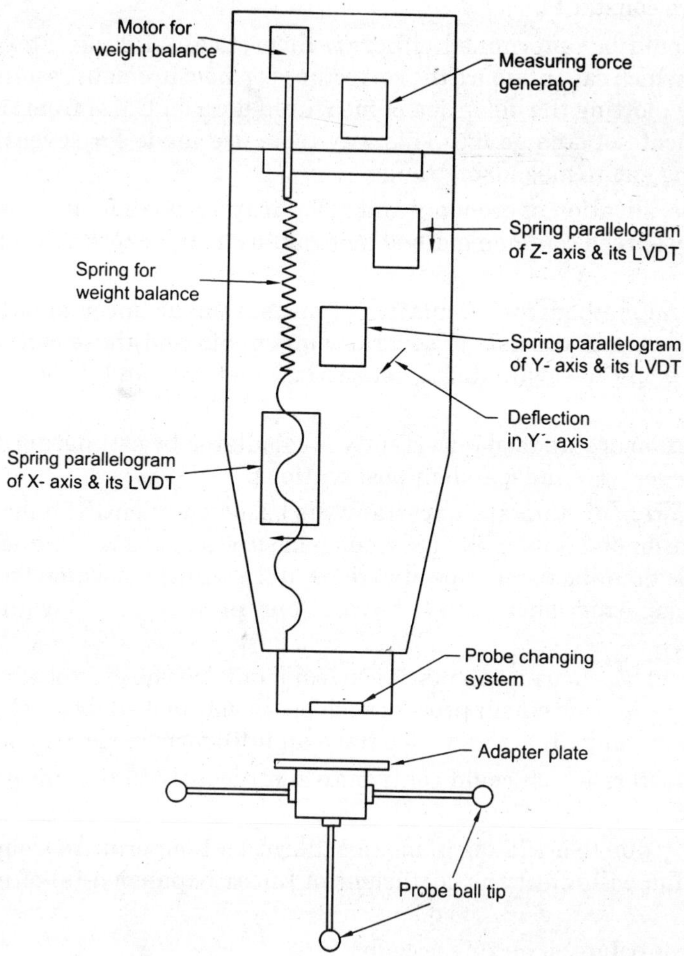

Fig. 25.9. Schematic of measuring probe head.

In order to enable probing an object from all sides, a computer controlled articulating probe holder is used as it can assume any spatial angle with a high resolution. Thus articulating probe holder can orient the measuring probe in any required position.

25.8. Length Measurement Uncertainty of CMM

The suitability of CMM can be estabilished by specifying its length measurement uncertainty *i.e.* the range in which the measured value lies. For CMM, uncertainty estimation is essential part as CMM is used to inspect components which are to serve as standards. Quality in today's era is an essential requirement and it can be only guaranteed through proper measuring techniques.

CMMs are normally calibrated using precise physical standards of known length. Step gauge can also be used for this purpose.

Uncertainty in length measurement is the uncertainty with which a CMM can determine the distance between two ponts located on parallel surfaces. It can be specified as

$$u = c_1 + kL \leq c_2$$

u = length measurement uncertainty

c_1, c_2 and k are constants.

Uncertainty could be determined either (*i*) theoretically taking into account all the contributing factors, which can influence the correctness of measurement results, or (*ii*) graphically, *i.e.*, by actually plotting the deviation of measured data on CMM from true value given by calibration certificate of step gauge. This way plots are made for several readings at a number of steps along entire measuring range.

Because of liberalisation of economy and globalisation of trade, it is essential that we adopt internationally agreed common method for evaluation and expression of uncertainty of measurement.

As per universal method, for calculation of measurement uncertainty, uncertainty of the result of a measurement consists of several components and these can be grouped into two categories, size Type A—evaluated by statistical methods and Type B—evaluated by other means.

Type A evaluation of standard uncertainty is calculated by estimating standard deviation of the mean of a series of independent observations.

Type 'B' evaluation of standard uncertainty is based on scientific judgement using all the relevant information available, like previous measurement data, data available in calibration/other reports, manufacturer's specifications, uncertainties assigned to reference data taken from handbooks, experience with behaviour and property of relevant materials and instruments.

The reliability of the measurements depends not only on the metrological confirmation of CMM but also on the measurement process such as the method adopted, skill of personnel, environment condition, and all factors which have an influence on the measurement.

The possible factors which could contribute significantly to the uncertainty of measurement in CMM are

(*i*) Uncertainty due to deviation of mean calibration temperature from standard temperature of 20°C. This will affect the coefficient of linear expansion (s) of material of step gauge and CMM.

Uncertainty contribution on this account

$$a_1 = L \times \Delta t \times (\alpha_s - \alpha_c)$$

Since, this uncertainty contribution a_1 follows rectangular probability distribution, standard uncertainty due to a_1 is given by

$$u_1 = a_1/\sqrt{3}$$

(*ii*) Uncertainty quoted in calibration certificate of step gauge

$$a_2 = c_1 + kL$$

Since, this uncertainty follows a normal probability distribution, assuming confidence level of 95%,

$$u_2 = a_2/2$$

(*iii*) Uncertainty contribution due to variation of temperature during calibration for the entire range, *ve* . a_3 = Range × Δt × α_s

This being rectangular probability distribution, $u_3 = a_3/\sqrt{3}$

(*iv*) Uncertainty contribution due to resolution of CMM, *i.e.*, . a_4 = resolution/2

Again being rectangular probability distribution, $u_4 = a_4/\sqrt{3}$

(*v*) Uncertainty contribution due to flatness error of measuring faces of step gauge (a_5)

$$u_5 = a_5/\sqrt{3}$$

(*vi*) Uncertainty contribution due to parallelism of measuring faces of step gauge (a_6)

$$u_6 = a_6/\sqrt{3}$$

Combined standard uncertainty of measurement result $U_c = \sqrt{u_1^2 + u_2^2 + \ldots\ldots u_6^2}$

Expanded uncertainty is defined as the measure of uncertainty which defines an interval about the measurement result within which the value of measurand can be said to lie confidently.

This expanded uncertainty U is determined by multiplying value of U_c with a coverage factor k, the value of which is chosen on the basis of desired level of confidence, for instance for normal probability distribution and for level of confidence of 95%, value of $k = 2$.

SOLVED QUESTIONS

Q. 25.1. What are the requirements for guideways of measuring axes of CMMs ? What is the material used for guideways ?

Ans. (*i*) Least friction as can be achieved with air bearings, recirculating ball bearing cages, anti-friction bearings, etc.

(*ii*) Optimum straightness

(*iii*) No short-term guide way deviation.

Guideway components are usually made from granite with long-term stability. Guideways for large CMMs are made from steel.

Q. 25.2. Why granite is preferred for quideway materials in CMM ? What are its limitations ?

Ans. Granite has long-term stability. Any damage does not result in unevenness of the surface. Further granite is non-corrodible and light in weight (comparable to aluminium).

Granite suffers from poor thermal conductivity. Temperature differences can lead to bending and deformations.

Q. 25.3. What is the material used for table of CMM, scales, and quills/cross beams ?

Ans. Machine bed of CMM is often made from granite. To avoid distortion due to thermal influences, it is thermally insulated. However temperature is measured at various points and any residual bend due to thermal influences is calculated and corrections made in display.

Scales are made from cerodur—a glass ceramic with nearly zero thermal expansion coefficient. Quills/crossbeams and other long, slim guideway components are made from extruded, diamond machined wrought aluminium alloy. These can be anodized, are non-ageing and thermally stable.

Q. 25.4. Whether Abbe principle is followed in CMM ? How extreme accuracy is guaranteed in CMM ?

Ans. Abbe principle (*i.e.* locating object to be measured and the measuring scale on same axis) is not possible in CMM. For achieving high accuracy, guideways are constrained to move in one axis, other five degrees of freedom being constrained. These are manufactured, reworked with sufficient precision. Software corrected computer aided accuracy is incorporated, *i.e.*, systematic portions of all deviations are measured at specific distances over the entire range and stored in the computer. During actual measurement of any object, each measured point is software corrected for all parameters, *viz*., three squareness deviations between the axes, six individual deviations per axis, etc.

Q. 25.5. What are the functions of an electronic control for computer numerical control (CNC) ? What are the characteristics of drivers provided ?

Ans. Both straight line and continuous path controls are adopted. Necessary controls for drives, probe head, probe changing system, measuring systems, etc. are provided. Subcontroller is provided for material flow systems. A computer system for comprehensive computation based on data provided by pickups is incorporated to display measurement results on dimensions, form and position of the probed features. Drives (d.c. torque motors) can operate over extremely wide range of speeds ($1 : 10^5$) and with very high efficiency so that heat generation is minimal. Power transmission is based on tape drive so that guideway properties are not impaired. Spindle drives incorporate universal suspension joints.

Q. 25.6. The axis of rotation and angle measuring system of a rotary table must be of the utmost precision. What is the conventional angle measuring resolution of CMMs ?

Ans. 0.5″.

Q. 25.7. How is measuring uncertainty of result in CMM expressed ?

Ans. Measuring uncertainty for CMM is specified in terms of value (*i*) parallel to axis (u_1) and (*ii*) volumetric value (u_3).

This data consists of a constant and a length-dependent value. It could thus be expressed as :

$$u_1 = (1.0 + L/500)\ \mu m$$
$$u_3 = (1.6 + L/400)\ \mu m$$

L is measured length in mm.

The environmental requirements for vibrations and temperature during measuring conditions have to be met.

Q. 25.8. How CMM is monitored regularly ?

Ans. Each CMM is provided with one calibrated test standard made of invar, which comprises of nine spheres and a ring gauge (Refer Fig. 25.9). It permits virtually automatic monitoring of CMM in very little time.

Test standard is measured in two vertical setups displaced to each other. Thus 72 measurement lines, including the space diagonals get checked. The test standard can be checked by means of a test and calibration procedure incorporating measurement in 2 setups rotated through 90° and gauge block comparison.

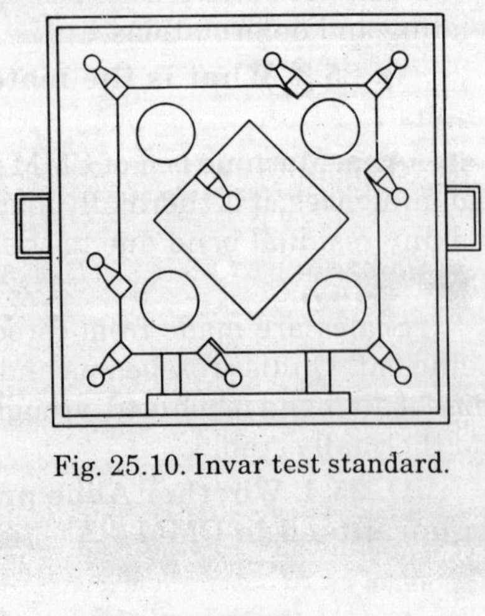

Fig. 25.10. Invar test standard.

Q. 25.9. What are the broad functions performed by a computer provided with CMM ?

Ans. Computer incorporates necessary software to display results of measurement on dimensions, form and position of probed features, collecting data and analysing them as obtained from readings of probes. It performs several important functions like

(*i*) Performing corrections on each probing point (to reduce and eliminate all systematic influences) with regard to guideway deviations, squareness deviations, bend influences, thermal influences, probe tip radius and mid-point probe bend, brobe head deflection, etc.

(*ii*) Transforming the machine coordinate data in volumetric coordinate system (described by six parameters—3 angles and spatial zero point defined by 3-coordinates) to workpiece coordinate system. The workpiece may have several coordinate systems.

(*iii*) Computing of substitute elements wherever feasible—In general the probing points must be as large as desirable to cover the geometric element completely. However in some cases, shape may differ minutely from ideal substitute elements within permitted tolerances, then number of probing points can be reduced and properties of substituted geometry computed. Different best fit methods can be used by computer to obtain consistent results even with only a few probing points.

(*iv*) *Evaluation and Control.* The dimensional specifications in the technical drawings often make it necessary for the measured elements to be linked mathematically so that further related elements are obtained. Very often measurement results have to be converted from one type of coordinate system to another.

Large data quantities need to be processed for complex shapes. Complex shapes are given numerical description with patches, and also using tangential bordering lines between all adjacent patches.

26

Metrology Laboratories

26.1. Metrology Laboratory Calibration Programme Management

Metrology is backbone of quality system and plays a vital role for industry to keep pace with fast developing technology and high quality standards.

Measurements are made at several points in the production process. For the measurement to be meaningful accuracy or uncertainty statement should be specified. Accuracy of measurements determines that production meets the goals of quality, safety, reliability and productivity. The process of quantifying the measurement uncertainty of an instrument which are referenced to standards of measurement that stem from a common source is termed calibration. Accurate measurements play a crucial role at each stage in the developement and production of quality products. The effectiveness of the quality control steps depends directly on the accuracy and confidence with which the test and measuring instrument can yield test results. Systematic periodic checking, *i.e.*, the calibration of instruments is thus essential for development and maintenance of market competitiveness. Customer would like to be assured of dependable calibration for guaranteeing the product quality. The aim of calibration programme management is to assure that instruments are maintained within accuracy limits sufficient for the intended task.

Calibration involves a set of operations, which under specified conditions, establish the relationship between values indicated by a measuring instrument or measuring system, or values represented by a material measure or a reference material and the corresponding values of a quantity realised by a reference standard. In its simplest form it is the comparison of two measurement devices or systems, one of known uncertainty (standard) and the other of unknown uncertainty and to estimate the correct value of the unknown and determine its uncertainty. In order that there is a common reference for the measurements taken, the reference standards used should have traceability to national/international standards.

If calibration is done too often it is waste of time and money. If delayed too much, it is waste of product. The optimum interval is to be determined, though difficult.

The calibration should be done in the laboratory maintained at proper environmental conditions. Suitable standards should be used as reference. Procedures professionally accepted should be performed by trained personnels. Factors for environmental control are temperature, relative humidity, cleanliness, vibration, electromagnetic interference and voltage regulation of electric power.

Where onsite calibration is desired, transportable, rugged, environmentally compensated and easily operated systems are needed. In-situ calibration is performed for sophisticated and delicate instruments. Suitable artifacts need to be developed for this purpose.

The calibration standards should have necessary range and accuracy, stability, precision and sensitivity or resolution, capabilities to match the defined requirements of instruments. The standard equipment should have valid and traceable calibration. The calibration procedure should include testing conditions, sequential test steps, and the details regarding the data collection. Once the calibration is completed the instrument is labelled to identify it from instruments which are not calibrated. Access to adjustable controls/devices on instruments whose setting affects the performance are sealed in order to prevent tampering by unauthorised personnels.

Some of the basic elements of the calibration programme management based on the above aspects of calibration and the guidelines of ISO are given below.

(*i*) Identifying the calibration requirement of instruments.

(*ii*) Selection of standards and measuring devices with consideration for accuracy requirement.

(*iii*) Environmental controls.

(*iv*) Establishing calibration traceability.

(*v*) Calibration procedures.

(*vi*) Personnel requirement and training.

(*vii*) Calibration interval and calibration labelling.

(*viii*) Sealing for integrety and calibration records.

The calibration may be performed either inhouse or by the outside agency as per the schedule. The calibrated instruments are suitably labelled giving the date of calibration, calibration due date and limitations for use if any and the calibration agency. The calibration records pertaining to the instruments are maintained for sufficient period of time to satisfy regulatory and contractual requirement and to allow for any interval adjustment.

26.2. Quality Management System for Metrology Laboratories

Laboratories which design and deliver calibration and testing services should adopt the ISO : 9001 standard as their quality system. Further for customer satisfaction that the quality specifies and assures of the technical capabilities and proficiencies of the laboratory, ISO/IEC Guide 25 should be adopted.

Guide 25 purports to contain all of the systems elements addressed in ISO 9001, expressed in language more specifically directed at laboratory. In addition, there are a number of technical requirements which must be applied to ensure reliable outcomes for each "product"-test reports. Guide 25 also pays particular attention to the requirements of test methods being met and is much more explicit in its requirements for laboratories on specific issues such as equipment calibration and the like.

The various features of **ISO/IEC guide 25** are discussed below.

Management Responsibility. It lays stress on organisational requirements viz. supervision, protection of clients rights, confidence in its independence of judgment etc. Managerial staff must have the authority and resources to discharge all duties. The quality manager is responsible for the quality system and its implementation and a Technical manager is responsible for all technical issues. Review at least once a year is mandatory.

Quality system requirements. Explicit guidance regarding the procedures to be included in the manual for quality system implementation in laboratories are identified.

Contract review. Requires review of facilities and resources for undertaking new work.

Design control. Requires test/calibration methods to be in line with national/international standards. In case of exceptions, methods adopted should be with mutual consent of client.

Document and data control. Requires procedure for control and maintenance of data and documents. In addition it specifies clearly the procedures to be included as part of quality system.

Purchasing. Specifies procedures for evaluation of sub-contractors and verification of purchased products. Additionally also stipulates that sub-contracting work of testing/calibration is placed with supplier meeting requirements of this Guide and same is intimated to client in writing.

Control of customer supplied product. Procedures for verification, storage and maintenance of customer supplied product are dealt meticulously and elaborately.

Product identification and traceability. Requirements lay stress on unique identification and traceability of resources involved.

Process control. The requirements of resources and facilities viz. personnel, equipment, methods, materials and environment are detailed out.

Inspection and testing. Requires procedure for inspection of support services and supplies. Quality of sub-contracting of calibration/testing is ensured by requirement of assigning work on laboratories meeting the criterion of this Guide. In addition it elaborates requirements of checks viz. QC schemes, proficiency testing etc.

Control of inspection, measuring and test equipment. Requires procedure for selection, calibration, traceability, identification, handling etc. of test and measuring equipment.

Inspection and test status. Status identification by records and ensures traceability.

Control of non conforming product. Permits amendments to reports/certificates.

Corrective and preventive action. Prescribe procedure for corrective actions and handling of complaints. Requirement of audit in areas where non compliance is reported and/or suspected is also addressed.

Handling, storage, packaging and delivery. More specific w.r.t. handling and delivery of product (calibration/test report).

Control of quality records. Specifies minimum requirements of contents of records and maintenance of confidentiality.

Internal Quality Audits. Periodic and planned internal quality audits required.

Training. Requires ensuring of training for keeping personnel uptodate.

Statistical techniques. Requires use of such techniques for quality control checks, uncertainty determination and sampling.

26.3. Role of Calibration Services

For global trade it is essential that dissemination of units takes place successfully. An essential prerequisite for it is a successful, reliable realisation and maintenance of units to be ensured by the National Metrology Institutes.

For progress of industry, products of acceptable quality need to be manufactured. For this strengthening of metrological intrastructure through **calibration service** is prerequisite for international acceptance of products.

The main objectives of the calibration services are :

- to create an **infrastructure of distributed metrological competence** for tracing back measurement results to national standards. This calls for technical completence of higher order.
- to accredit calibration laboratories and to authorize them to issue calibration certificates as document of traceability to national standards.
- traceability has to be assumed on a permanent basis by interlaboratory comparisons of standards between National Metrology Institutes.
- to ensure that the accreditation criteria are met in the long term.
- traceability of measurement and test equipment to national measurement standards plays a decisive role.
- to meet the requirements of the relevent ISO/IEC Guides and additional interpreting or application documents for the calibration laboratories and the accreditation body.
- to **promote international recognition of the calibration certificates** of the calibration service for avoidance of duplication of calibration work by mutual recognition of certificates across border.

26.4. Laboratory Accreditation

For free and smooth international trade it is necessary that there are arrangements for giving recognition to laboratories for trade and regulatory purpose and the development of mutual recognition agreements between national laboratory accreditation bodies as the mechanism for facilitating international acceptance of test data. Laboratories should be competent and reliable for the purposes of providing test and calibration data for use in control of manufacturing processes, quality management, purchasing and enforcement of regulations.

Laboratories are established to perform tests and make measurements and, generally, laboratory workers have been trained to follow sound scientific practices and management principles. Focus is needed both on the technical issues as well as quality management.

With trade becoming more demanding in its technical requirements with test data often a critical and integral part of it, accreditation of laboratories has become necessary.

Accreditation of laboratories helps in ensuring that testing carried out in country of origin is accepted by customers and authorities in the country of import. It is necessary that users should have *confidence* in the *competence* of laboratories providing test data, irrespective of where the laboratories were located. Accreditation offers mechanism which is nondiscriminatory and transparent and recognises laboratories on the basis of competence.

Further the accreditation bodies should demonstrate their competence in much the same way as laboratories. Competent national laboratory accreditation bodies should achieve international recognition through a process of *peer review* by other national bodies leading to the conclusion of *mutual recognition agreements* between the bodies.

For this purpose each country should have a single comprehensive laboratory accreditation body, or multiple bodies with more focused mandates, who would accredit laboratories using internationally agreed criteria and themselves adopting internationally agreed criteria and best practices.

There is a reasonable level of harmonisation of practice between various countries through the application of Guides 58 and 25. ISO/IEC Guide 25 deals with technical and quality management system requirements for laboratories, and ISO/IEC Guide 58 deals with the similar requirements for laboratory accreditation bodies.

26.5. Third-party Accreditation of Testing and Calibration Laboratories

A network of accreditation bodies coordinated through mutual recognition arrangements (MRAs) represents the optimum solution acceptable to all. The following conditions should be fulfilled to ensure recognition throughout such a network :

(*a*) the requirements for the competence of laboratories should be harmonized;

(*b*) the rules for the operation of the accreditation system, *i.e.* the accreditation procedures and practices should be harmonized;

(*c*) the organization and structure of the accreditation body should conform to harmonized rules accepted by all accreditation bodies in the system.

In addition to the above conditions a number of measures that promote competence and mutual confidence should be practised such as proficiency testing and other inter-laboratory comparisons at the level of accredited laboratories and exchange of staff and joint accreditations at level of accreditation bodies.

26.6. International Organisation of Legal Metrology (OIML)

OIML was created in 1955 with aim of establishing and maintaining credibility and confidence in measurement and conformity assessment and verification on a global level. Today with international competition and globalisation of trade, international cooperation is must and role of OIML is very relevant to achieve this objective and also to contribute to quality and competitivness and benefit the trade.

OIML also plays important role to harmonise requirements for measurements and measuring instruments and the method of their verification, meeting the increasing requirements to credibility and confidence in conformity assessment and conformity verification, to disseminate metrology and to contribute to the future of metrology, impart training in metrology and maintain leadership in the international development and harmonisation of legal metrology activities.

For international trade we need to have common and well defined units of measurements. The Meter Convention established in 1875 is the most important advance in the history of metrology in settling the question of international harmonisation of the units of measurements and the measurements standards. Today the globalisation of metrology is just as important as trade. We need more than units and measurements standards to succeed in metrology. We need credibility and confidence in measurement results. The demand for credibility and confidence in measurement results assumes greater significance in global trade.

To achieve confidence, credible conformity assessment and conformity verification certificates are required. This is possible when well defined and realistic requirements to measurements and instruments are achieved, and there are structures for an internationally harmonised implementation of such requirements.

OIML is responsible for the development of the *International Recommendations* by offering the well defined, up to date and realistic requirements to measurements and instruments covering scope and application, metrological requirements, technical requirements, methods for testing and verifying conformity to the requirements, format for reporting test results and terminology. According to the GATT/WTO Code of Standards, the

OIML International Recommendations are considered as *international standards*. They focus on metrological performance and testing procedures and are intended to be compatible with other relevant international specifications. They are drafted by the *OIML Technical Committees* and *Subcommittees* and are approved by the *International Committee of Legal Metrology*.

OIML also issues *certificate* for a type of instrument that complies with all requirements of the relevant *OIML Recommendation.* Certification is done for patterns, models or types of instruments and does not apply to individual instruments.

OIML also disseminates knowledge and information through seminars with a view to achieve goals like-to disseminate metrological information, to promote cooperation between Member States, and to define sound metrological basis for the development of future Recommendations. These seminars encourage technical discussions between representatives of national legal metrology services, manufacturers, standardisation bodies and other interested parties.

The *OIML Development Council* provides assistance in dependable measurements by focusing on basic legal metrology aspects, general operation of metrology services, staff training, laboraories and equipment.

The *Technical Committees (TC) and Subcommittees (SC)* of OIML develop the *International Recommendations* and *Documents.*

The technical work is carried out in 18 *Technical Committees*, with their *Subcommittees*

TC 1 Terminology
TC 2 Units of measurement
TC 3 Metrological control
TC 4 Measurement standards and calibration of verification devices
TC 5 Electronic instruments
TC 6 Prepackaged products
TC 7 Measurement instruments for length and associated quantities
TC 8 Measurement of quantities of fluids
TC 9 Instruments for measuring mass and density
TC 10 Instruments for measuring pressure, force and associated quantities
TC 11 Instruments for measuring temperature and associated quantities
TC 12 Instruments for measuring electrical quantities
TC 13 Measuring instruments for acoustics and vibration
TC 14 Measuring instruments used for optics
TC 15 Measuring instruments for ionising radiations
TC 16 Instruments for measuring pollutants
TC 17 Instruments for physico-chemical measurements
TC 19 Medical measuring instruments

26.7. Core Values and Concepts for Industries

For any industry to be successful, it must deliver ever improving value to customer and take steps for improvement of overall company performance and capabilities.

A set of following core values and concepts will help considerably for integrating customer and company performance requirements.

(*i*) **Commitment to quality.** For producing quality product, quality assurance and quality management should be put in actual practice through system approach and system standardication.

(*ii*) **Customer satisfaction.** All product and service characteristics that contribute value to customers and lead to customer satisfaction and preference must be a key focus of a company's management system. Quality should be judged by customers.

(*iii*) **Commitment of leadership.** Senior management should set directions and create a customer orientation, clear and visible values, and high expectations. Reinforcement of the values and expectations requires personal commitment and involvement. Pursuit of market leadership requires a strong future orientation and a willingness to make long-term commitments to all stakeholders.

(*iv*) **Continuous improvement and learning.** Achieving the highest levels of performance requires a well-executed approach to continuous improvement and learning and adopting latest practices and techniques.

(*v*) **Employee participation and development.** A company's success in improving performance depends increasingly on the skills and motivation of its work force. Employees must be imparted training at regular intervals.

(*vi*) **Fast response and preparedness to change.** Success in competitive markets increasingly demands ever-shorter cycles for new or improved product and service introduction. Also, faster and more flexible response to customer is now a more critical requirement.

(*vii*) **Design quality and prevention.** Business managements should place strong emphasis on design quality—problem and waste prevention achieved through building quality into products and services and into production and delivery processes.

(*viii*) **Managing quality.** Pay proper attention to principle of metrology and their application in workplace ensure technical competence, calibration hierarchy, traceability etc. for this purpose.

(*ix*) **Management by fact.** A modern business management system needs to be built upon a framework of measurement information, data, and analysis. Companies should seek to build internal and external partnerships to better accomplish their overall goals.

(*x*) **Results orientation.** A company's performance system needs to focus on results. Results of company should be guided by and balanced by the interests of all stakeholders—customers, employees, stockholders, suppliers and partners, the public, and the community.

(*xi*) **Corporate responsibility.** A company's management should stress corporate responsibility *i.e.*, basic expectations of the company—business ethics and protection of public health, safety, and the environment.

(*xii*) **Compliance with ISO 9001.** Buyers are putting demands on sellers for certification of compliance with ISO 9001, and certification that products or processes are environmentally friendly.

26.8. Automated Measurement Techniques

For measuring geometric dimensions by automatic methods (which is necessary for calibration), *i.e.*, involving no human interventions, various steps involved are :

(*i*) Staging, *i.e.*, positioning the instrument or workpiece in relation to one another to estable the mutual reference position for subsequent measurements.

(*ii*) Movement over a predetermined distance to a point of coincidence where the significant surface element of the object coincides with the fixed target.

(*iii*) Indication of amount of deviation of the measured objects from nominal value.

(*iv*) Appraisal of measurement result.

(*v*) Decision to accept or reject or segregate.

Usually automatic multidimensional ganging devices are employed to measure all features in one go. The role of coordinated measuring machine has increased considerably. The geometric and dimensional conditions of particular features are conveniently assessed by optical projection methods having facilities for automatically controlled displacement movements of the optical projector stage.

The gauging systems selected must satisfy the required degree of discriminating sensitivity and operating speed and also incorporate requirements of space and access conditions of the workpiece. These machines are fitted with mechanised workhandling and transport system positioning devices, and means for segregation and disposal of parts.

Depending on complexity of workpiece, it may be required to be inspected in different places and/or orientation and machine should be capable of imparting necessary linear, rotary, indexing movements to workpiece, as desired. These machines are also equipped with special devices to evaluate the combined effect of signals produced by a succession of measurements.

In case of any malfunction various safety features incorporated act. Necessary indications and displays are provided for intervention of operator. Reject parts can be classified into groups as desired.

The proper orientation of workpieces is checked by limit switches or by sensors using infrared light beams.

Some machines are fitted with instruments and devices to assist in size control of the process by signalling of drift before the limit of toleranced size is reached by alerting the operator. Most of machines incorporate automatic calibration system. A minimum and maximum tolerance gauge master is installed in each gauge station, and all gauge circuits are calibrated to their proper tolerance range by simply pushing the calibration button on the machine.

Coordinate measuring machines have facilities to sense the axis alignment automatically. If the workpiece is offset by tilt in one direction, it produces three dimensional misalignment. The interference with the measurement results due to misalignment can be eliminated by applying the calculation of an appropriate type of automatic axis alignment program. It is also possible to convert cartesian coordinatesinto polar coordinates. Programmable calculators are used for data interpretation.

A private prints out all the results of measurements. Optical inspection with automated functions enhance the capability of such machines.

26.9. Geometric Dimensioning and Tolerancing

The plus and minus system of dimensioning and tolerance of parts is insufficient to convey the design intent in totality. Geometric dimensioning and tolerancing only provides complete language of design. This system defines the geometric characteristics of part size, shape, orientation and location which are essential to ensure that parts will mate with other mating parts in assembly process.

For any part, the exact configuration and location of tolerance zone and the origin of measurement need to be defined.

Any feature needs following to be defined : a desired size, tolerance on size, desired location, and tolerance on location. Geometric dimensioning and tolerancing as developed by American National Standards Institute (vide standard ANSI Y 14.5) provides a systems of symbols to define part shapes, feature form, orientation, runout, profile and location.

Refer to Fig. 26.1 in which basic dimensions of hole in workpiece have been specified.

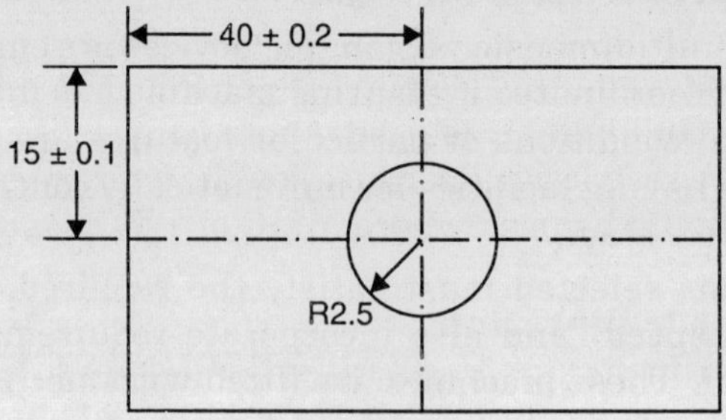

Fig. 26.1

Here it is not clear whether reference point is axis of hole, or edges of workpiece. So datum features need to be provided. Further hole size tolerance is not given. Complete information is available in Fig. 26.2.

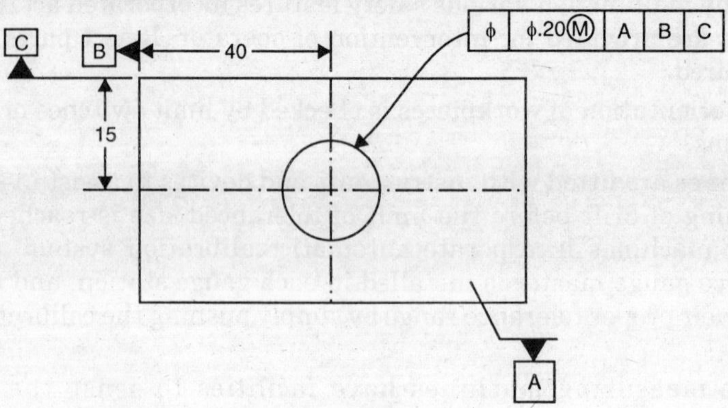

Fig. 26.2

In Fig. 26.2, the information blocks read from left to right are *A*, *B*, and *C*. So primary datum feature is A, the secondary *B* and tertiary *C*.

The feature control frame for hole can be read as : — The axis ϕ may be out of position ⊕ a diameter of 0.20 if the hole is produced at a diameter of 2.5 holding perpendicularity to datum *A*, and location to datums *B* and *C*. Further since hole uses the maximum material condition symbol, it allows the geometric tolerance of location to grow as the hole grows.

Thus if hole is made at a diameter of 2.51, then 0.21 would be allowed and so on.

SOLVED PROBLEMS

Q. 1. Describe the interpretation of following three conditions :

(*i*) **Maximum material condition (MMC)**

(*ii*) **Least material condition (LMC)**

(*iii*) **Regardless of feature size condition (RFS).**

Ans. (*i*) In MMC, the feature contains the most material or is the heaviest feature. It is the smallest hole or the largest shaft. It allows more parts to pass the inspection based on functional requirements. It is used when mating is most important.

(*ii*) In LMC ; the feature contains the least material, *i.e.* largest hole or smallest shaft. It is used for features when the preservation of material is of great importance and cost is a significant factor. It is used when wall thickness is thought to be endangered and the holes stand a chance of approaching a breakdown condition or when enough material for machining is required in casting.

(*iii*) RFS concept is used when the geometric tolerance is to remain the same irrespective of limits of size of shaft and hole. It is used where balance is important as on rotating parts. It creates tighter, more restrictive tolerances on the average produced part. Since part with RFS will be costly it should be used where MMC or LMC concepts may endanger part function.

Q. 2. Define the term 'calibration'.

Ans. The process of quantifying the measurement uncertainty of an instrument which are referenced to standards of measurement that stem from a common source is termed calibration.

Q. 3. What responsibilities are assigned to metrology group and calibration laboratories is modern engineering industry ?

Ans. Metrology group in an engineering industry is charged with the following responsibilities :

- To ensure close tolerances, conformance to specifications, compatibility of measurements with the help of sophisticated metrological and test equipment.
- To plan and organize a precision measurement and metrology laboratory with all latest metrology equipment.
- To ensure uniformity of measurement methods and correctness of measuring instruments.
- To assist in drawing adequate purchase and acceptance norms, instruments and testing devices.
- To coordinate the adjustment, repair and maintenance of instruments.
- To check the conformance of bought out or manufactured tools, gauge and measuring devices.
- To register, calibrate and re-calibrate periodically all such instruments during their useful life.
- To associate with quanlity imporvement programs in scientific and technical activities for extending assistance to different working groups engaged in design, production, purchase, maintenance, erection and commission and sales.
- To maintain the basic standards after getting them calibrated from approved calibration agencies as per national measurement standards for traceability of measurements to international standards.
- To undertake operative measurement assignments in different work areas.
- To organise training/retraining programmes on latest trends.
- To arbitrate the measurement problems when conflicting readings are reported from two or more task performers.
- To disseminate the knowledge of metrology to the operating staff and create awareness about the importance of metrology for quality improvement programs amongst all the employees of the organization.
- To coordinate with external agencies for accreditation of metrology systems and facilities of the plant.
- To develop measurement instruments and techniques to keep abreast with the latest researches in the field.

OBJECTIVE TYPE QUESTIONS

1. Who made the following classic statement?

   "When you can measure what you are speaking about and express it in numbers, you know something about it, and when you can't express it in numbers, your knowledge, is of a meagre and unsatisfactory kind. It may be the beginning to knowledge, but you have scarcely in your thought advanced to the stage of science."

   (*a*) Arnold Young (*b*) Taylerson
   (*c*) Johanson (*d*) Lord Kelvin
   (*e*) Taylor.

2. The ease with which observations can be made accurately is referred to as

   (*a*) readability (*b*) sensitivity
   (*c*) accuracy (*d*) precision
   (*e*) repeatability

3. Accuracy of measuring equipment is

   (*a*) the closeness with which a measurement can be read directly from a measuring instrument
   (*b*) a measure of how close the reading is to the true size
   (*c*) the difference between measured value and actual value
   (*d*) the smallest change in the measurand that can be measured
   (*e*) the capability to indicate the same reading again and again for a given measurand

4. Which of the following errors are generally distributed in accordance with the Gaussian distribution

   (*a*) controllable errors (*b*) calibration errors
   (*c*) avoidable errors (*d*) random errors
   (*e*) errors due to ambient conditions.

5. Which of following is true

   (*a*) Systematic errors are caused due to system interaction errors
   (*b*) gross errors are know as compensating errors
   (*c*) Systematic errors follow probability laws
   (*d*) (a) and (c) above

6. Which of following are know as cumulative errors

   (*a*) systematic errors (*b*) random errors
   (*c*) gross errors (*d*) none of above

7. Inadvertent errors are caused due to

   (*a*) less information about product (*b*) less knowledge about product
   (*c*) improper timely feedback (*d*) all of above

8. Which of following causes are responsible for systematic errors
   (*a*) zero offset (*b*) change in sensitivity
   (*c*) non linearity (*d*) all of above
9. Fig.1 shows the dimensions obtained on a component by a certain instrument. This instrument is
   (*a*) precise but not accurate
   (*b*) accurate but not precise
   (*c*) both accurate and precise
   (*d*) neither precise nor accurate
   (*e*) sensitive.

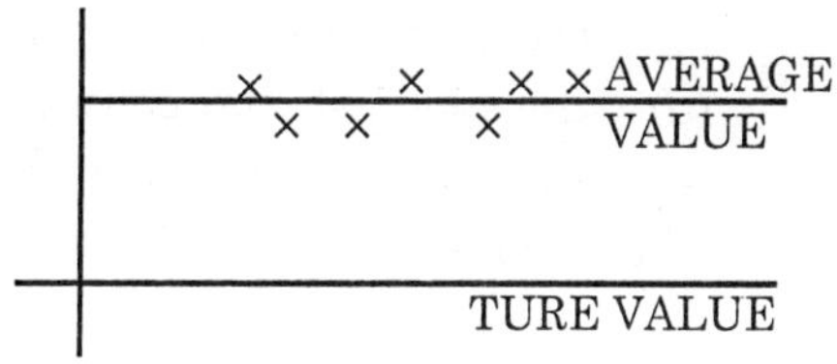

Fig. 1

10. Tolerances are specified
    (*a*) to obtain desired fits
    (*b*) because it is not possible to manufacture a size exactly
    (*c*) to obtain high accuracy (*d*) to have proper allowance
    (*e*) to have proper inspection
11. Which of the following is the most important characteristic of a measuring instrument in general
    (*a*) precision (*b*) accuracy
    (*c*) repeatability (*d*) sensitivity
    (*e*) readability
12. Sensitivity and range of measuring instrument have
    (*a*) direct relationship (*b*) linear relationship
    (*c*) inverse relationship (*d*) unpredictable relationship
    (*e*) no relationship
13. Systematic errors are
    (*a*) randomly distributed (*b*) regular repetitive in nature
    (*c*) distributed on both +ve and –ve sides of mean value
    (*d*) unknown errors (*e*) of unpredictable nature.
14. The foundation for all dimensional measurements is the
    (*a*) datum plane (*b*) accuracy
    (*c*) datum line (*d*) geometry of work piece
15. Granite surface plates are free from burrs or protrusions because of
    (*a*) lower coefficient of expansion (*b*) light weight
    (*c*) very fine grain structure (*d*) lapping operation
16. The perferred instrument for measuring holes, grooves, and recesses is
    (*a*) mechenical comparator (*b*) vernier calliper
    (*c*) dial gauge (*d*) depth gauge
17. Accurate measurement of 'thread plug gauges', such as outside dimeter, pitch diameter, and root diameter are measured with the help of
    (*a*) vernier caliper (*b*) calipers
    (*c*) slip gauges (*d*) Floating carriage micrometer

18. Precision of measuring equipment is
   (*a*) the closeness with which a measurement can be read directly from a Measuring instrument
   (*b*) a measure of how close the reading is to the true size
   (*c*) the difference between measured value and actual value
   (*d*) the smallest change in measurand that can be measured
   (*e*) the capability to indicate the same reading again and again for a given measurand
19. The maximum amount by which the result differs from the true value is called
   (*a*) correction (*b*) discrepancy
   (*c*) error (*d*) accuracy
   (*e*) uncertainty
20. Response is defined as the measure of a system's fidelity to purpose. The purpose of measuring instruments may be considered for the following cases.
   (*a*) amplitude response (*b*) frequency response
   (*c*) phase response (*d*) delay or rise time
   (*e*) all of the above
21. Who stated that maximum accuracy in linear measurements may be obtained only when the measurement scale or the standard is in line with the axis of part being measured
   (*a*) Johnson (*b*) Abbe
   (*c*) Taylor (*d*) Moore
22. Simple steel rule is very convenient for linear measurements but it has limitations on account of
   (*a*) accuracy (*b*) resolution
   (*c*) mechanism to ensure ease of measurement
   (*d*) all of above
23. Surface plates provide datum plane for measurements and these are usually fabricated in sets of
   (*a*) two (*b*) three
   (*c*) four (*d*) there is no such criteria
24. Which is most commonly used instrument for measuring holes, grooves and recesses
   (*a*) vernier caliper (*b*) hermaphrodite caliper
   (*c*) depth gauge (*d*) V-block
25. In order to prevent rapid wear on micrometer screw gauge surfaces, the measuring faces are tipped with
   (*a*) tungsten carbide (*b*) chromium
   (*c*) ceramic (*d*) silicon carbide
26. If two components are manufactured at different plants and assembled else where, which characteristic of measuring instrument becomes more important
   (*a*) precision (*b*) accuracy
   (*c*) sensitivity (*d*) repeatability
27. Pick of false statement
   (*a*) use of more sensitive instrument than desired for given application is not advantageous.

(*b*) As the requirement of accuracy increases, the cost of production increases exponentially.
(*c*) Accuracy of measurement is important for manufacture of quality product
(*d*) Temperature variations have no influence on accuracy of measuring instruments

28. The process of validation of the measurement to ascertain whether the given quality conforms to national standard of measurement is known as
(*a*) calibration (*b*) resolution
(*c*) traceability (*d*) repeatability

29. Which of the following can be used to scribe lines parallel to the edges of a part
(*a*) vernier calipers (*b*) screw gauge
(*c*) divider (*d*) hermaphrodite caliper
(*e*) combination set

30. Pick up false statement
(*a*) The square head alongwith graduated rule can be used as height/depth gauge
(*b*) Centre head can be used for measurement of diameter.
(*c*) protector head comprise a rotatable turret (having angular scale graduated in degrees) within a stock
(*d*) steel rule can be positioned diametrically across the jobs to measure the diameter

31. Pick up correct statement
(*a*) calipers can be used to physically duplicate the separation between the reference point and the measured point of any dimension.
(*b*) calipers are available only as outside callipers
(*c*) the use of inside and outside callipers does not need high skill of operators
(*d*) A divider essentially consists of one divider leg and one caliper leg.

32. Which of the following can't fall under the head–controllable errors
(*a*) calibration errors (*b*) environmental errors
(*c*) avoidable errors (*d*) random errors
(*e*) non-similarity of conditions while calibrating and measuring

33. Which of the following errors are regularly repetitive in nature
(*a*) systematic errors (*b*) random errors
(*c*) illegitimate errors (*d*) controllable errors
(*e*) avoidable errors

34. Match the two parts below

1.	Degree of closeness of the measured value with its true value	A. Error
2.	Ability of instrument to detect small differences in the quantity	B. Accuracy
3.	To compare an unknown with a standard through a calibrated system	C. Sensitivity
4	Difference of true value and measured value	D. Indirect

(*a*) 1—B, 2—C, 3—A, 4—D (*b*) 1—B, 2—A, 3—C, 4—D
(*c*) 1—B, 2—C, 3—D, 4—A (*d*) 1—C, 2—B, 3—A, 4—D

35. Phenomenon of wringing in slip gauges occurs due to
(*a*) molecular attraction (*b*) atmospheric pressure
(*c*) both (a) and (b) (*d*) none of above

36. One Angstrom is equal to
(*a*) 10^{-6} m (*b*) 10^{-8} m
(*c*) 10^{-9} m (*d*) 10^{-10} m

37. Theoretical size common to both the parts of a mating pair is called
(*a*) nominal size (*b*) normal size
(*c*) base size (*d*) actual size

38. Which of the following errors are inevitable in the measuring system and it would be vainful exercise to avoid them
(*a*) systematic errors (*b*) random errors
(*c*) calibration errors (*d*) environmental errors
(*e*) deformation errors.

39. The misalignment error (D′ – D) in the case shown in Fig. 2 in equal to
(*a*) $D \sec\theta + d\tan\theta$
(*b*) $D \sec\theta + \frac{d}{2}\tan\theta$
(*c*) $D(\sec\theta + 1) + d\tan\theta$
(*d*) $D(\sec\theta - 1) + d\tan\theta$
(*e*) $D(\sec\theta - 1) + \frac{d}{2}\tan\theta$.

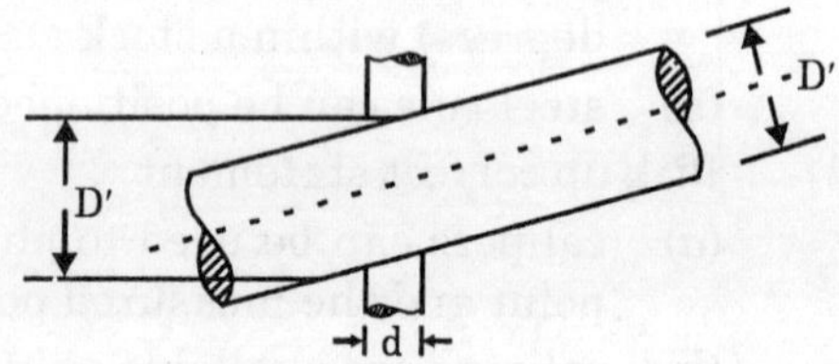

Fig. 2

40. If variation in manufacturing an instrument be V_1 and in measuring process it be V_2, then variation in observing the reading will be
(*a*) $V_1 + V_2$ (*b*) $V_1 - V_2$
(*c*) $V_1 \times V_2$ (*d*) $V_1 \div V_2$

41. Least count of dial indicator
(*a*) 0.01 mm (*b*) 0.05 mm
(*c*) 0.1 mm (*d*) 0.2 mm

42. The hardness of gauging surface of screw ring gauge is generally measured by
(*a*) diamond pyramid machine (*b*) scratch test
(*c*) scleroscope (*d*) charpy machine

43. What is the accuracy of measurement of small gap gauges using slip gauges
(*a*) ±0.001 mm (*b*) ±0.002 mm
(*c*) ±0.005 mm (*d*) ±0.01 mm

44. Which is most convenient method to measure the cylindrical ring ganges for their taper and ovality
(*a*) slip gauge (*b*) micrometer
(*c*) dial indicator (*d*) pneumatic comparator

45. Which of the following instruments is most accurate
(*a*) vertical caliper (*b*) manaometric screw gauge
(*c*) optical projector (*d*) mechanical comparator
(*e*) slip gauges

46. Which of the following refers to parasitic error
   (*a*) Error, often gross, which results from incorrect execution of measurement
   (*b*) algebraic difference between the results of measurement and the value of comparison
   (*c*) error which varies in an unpredictable manner in absolute value and in sign when a large number of measurements of the same value of a quantity are made under practically identical conditions.
   (*d*) disagreement between the result of measurement and the value of the quantity measured.
   (*e*) error which during several measurements, under the same conditinos of the same value of a certain quantity, remains constant in absolute value and sign or varies in accordance with a specified law when the conditions change.

47. Match the 2 parts below relating to characterisitic of measuring instruments

1. The ratio of the change of instrument indication to the change of quantity being measured	A. Range
2. Successive readings of measured quantity are same every time	B. Sensitivity
3. Difference between the measured value and true value	C. Consistency
4. Difference between the lower and higher values an instrument in capable of measuring	D. Error

   (*a*) 1—D, 2—A, 3—C, 4—B (*b*) 1—B, 2—C, 3—D, 4—A
   (*c*) 1—C, 2—B, 3—D, 4—A (*d*) 1—B, 2—C, 3—A, 4—D

48. Pick up wrong statement from following
   (*a*) If error of measurement is low, its accuracy will be more
   (*b*) Higher accuracy instruments tend to be more sensitive
   (*c*) A highly sensitive instrument may not be consistent
   (*d*) If range of instrument increases its sensitivity also increases

49. Which of the following characterises the dispersion of the results obtained in a series of measurements of the same value of a quantity measured
   (*a*) absolute error (*b*) relative error
   (*c*) root mean square deviation (*d*) uncertainty of measurement
   (*e*) variation of indication

50. A surface gauge is used for
   (*a*) levelling the surface plate (*b*) checking the surface finish
   (*c*) laying out the work accurately (*d*) finding the depth of the surface
   (*e*) finding flatness of surfaces.

51. The centre head attachment on a combination set is used to
   (*a*) measure circular workpieces
   (*b*) measure height, depth and distance between centres
   (*c*) locate the centre of a bar stock or a circular job
   (*d*) all of above

52. A scribing tool comprising one divider leg and one calliper leg is known by the name
   (*a*) calliper (*b*) transfer calliper
   (*c*) hermaphrodite calliper (*d*) firm joint calliper

53. Which is not correct about ratchet stop
(*a*) It enables optimum contact force and avoids distortion of workpiece
(*b*) avoids stress on the instrument
(*c*) accuracy of instrument becomes better due to minimum dependence on skill of user
(*d*) It allows ease of signal processing

54. Maximum deviation from flatness for big surface plates weighing around 1000 kg for grade 0, I, and II is of the order of
(*a*) 8. 16, and 33 microns (*b*) 12, 20, 42 microns
(*c*) 16, 33, 48 micorns (*d*) 32, 49, 67 mincrons

55. V-blockes are extensively used for inspection of jobs with a
(*a*) circular cross section (*b*) elliptical cross section
(*c*) square cross section (*d*) V-cross section

56. Major parts in a combination set are
(*a*) graduated rule, caliper and protractor head
(*b*) centre head, square head, graduated rule and graduated protractor head
(*c*) square head and protractor head
(*d*) graduated rule and square head

57. Parasitic error is caused due to
(*a*) improper use of measuring instrument
(*b*) wrong design of instrument (*c*) changes in ambient conditions
(*d*) errors in computation (*e*) deflection of stylus

58. Examine following statements and state which is false
(*a*) Random errors can be minimised by taking number of readings but systematic errors can't be eliminated by repeated measurements
(*b*) Calibration helps to reduce random errors but not calibration errors
(*c*) lesser the random errors, more is repeatability and high precision
(*d*) lesser the systematic errors, higher is accuracy

59. Errors due to parallax effect, misalignment, zero setting etc fall under the head of
(*a*) controllable errors (*b*) random errors
(*c*) calibration errors (*d*) avoidable errors

60. Which of following are controllable errors
(*a*) systematic (*b*) random
(*c*) traceability (*d*) which can be analyzed statistically

61. Calibration of instruments is necessary to
(*a*) comply with standards
(*b*) detect and ensure maintainability of accuracy
(*c*) satisfy customer needs (*d*) comply with measurement procedures

62. If a quantity Q is dependent on three other quantities q_1, q_2 and q_3 related such that $Q = K(q_1)^{n1} \times (q_2)^{n2} \times (q_3)^{n3}$

then overall error $\dfrac{\delta Q}{Q} =$

(a) $n_1\left(\frac{\delta q_1}{q_1}\right)+n_2\left(\frac{\delta q_2}{q_2}\right)+n_3\left(\frac{\delta q_3}{q_3}\right)$ (b) $\frac{1}{n_1}\frac{\delta q_1}{q_1}+\frac{1}{n_2}\frac{\delta q_2}{q_2}+\frac{1}{n_3}\frac{\delta q_3}{q_3}$

(c) $\frac{\delta q_1}{q_1}+\frac{\delta q_2}{q_2}+\frac{\delta q_3}{q_3}$ (d) $\left(\frac{\delta q_1}{q_1}\right)^{n_1}+\left(\frac{\delta q_2}{q_2}\right)^{n_2}+\left(\frac{\delta q_3}{q_3}\right)^{n_3}$

(e) $n_1.n_2.n_3.\frac{\delta q_1}{q_1}.\frac{\delta q_2}{q_2}.\frac{\delta q_3}{q_3}$.

63. A feeler gauge is used to check
(a) radius (b) screw pitch
(c) surface roughness (d) unsymmetrical shape
(e) thickness of clearance

64. To eliminate the personal element of feel during setting a measuring instrument is achieved by
(a) zero setting device (b) auxiliary device
(c) measuring standard (d) fiducial indicator

65. The maximum amount by which the result differs from the true value is
(a) error (b) discrepancy
(c) correction (d) accuarcy

66. Which error does not fall under the head of controllable errors?
(a) environmental errors (b) random errors
(c) avoidable errors (d) calibration errors

67. Which of following errors are regularly repetitive in nature
(a) random errors (b) contollable errors
(c) illegitimate errors (d) sytematic errors

68. Which of following errors are generally distributed in accordance with the Gaussion distribution
(a) random (b) controllable
(c) calibration (d) avoidable

69. Measuring mechanism whose mobile component attains its equilibrium position without oscillations around new position is called
(a) damped measuring mechanism (b) aperiodic measuring mechanism
(c) stable measuring mechanism (d) precise measuring mechanism
(e) analogue measuring mechanism

70. Measuring instrument which conforms to all the specified statutory provisions is called
(a) ordinary measuring instrument
(b) measuring instrument acceptable for verification
(c) auxiliary measuring instrument (d) legal measuring instrument
(e) statutory measuring instrument

71. Measuring instrument intended to define or present physically, conserve or reproduce the unit of measurement of a quantity (or a multiple or sub-multiple of that unit) and to transfer it to other measuring instruments by comparison is known as

(*a*) legal measuring standard (*b*) secondary standard
(*c*) working standard (*d*) primary standard
(*e*) standard

72. A workpiece is usually required to be held in a vertical position for laying out. For this purpose, it is clamped to
(*a*) surface plate (*b*) an angle plate
(*c*) a V-block (*d*) a machine bed
(*e*) enginee's square

73. The phenomenon shown by a measuring instrument which gives different indications in a series of measurements of the same value of the quantity measured is called
(*a*) repeatability of measuring instrument
(*b*) error of repeatability (*c*) dispersion of indications
(*d*) error of trueness (*e*) discrimination of measuring instrument

74. The quality of a measuring instrument which characterises the ability to respond to small changes of the quantity measured is called
(*a*) discrimination of a measuring instrument
(*b*) response of a measuring instrument
(*c*) accuracy
(*d*) precision (*e*) repeatability

75. Instrument which is designed to eliminate the personal element of feel when setting a measuring instrument is called
(*a*) fiducial indicator (*b*) zero setting device
(*c*) auxiliary measuring instrument (*d*) measuring standard
(*e*) indicating element

76. The thickness of light gauge sheet steel can be best checked with a
(*a*) finely divided steel scale (*b*) depth gauge
(*c*) hermaphrodite caliper (*d*) micrometer
(*e*) thickness measuring machine fitted with dial gauge

77. Which of the following gives an idea about the ability of the equipment to detect small variation in the input signal (quantity being measured)
(*a*) readability (*b*) accuracy
(*c*) sensitivity (*d*) precision
(*e*) repeatability

78. If attempts are made to make an instrument very sensitive, which of the following qualities is likely to be impaired
(*a*) precision (*b*) accuracy
(*c*) readability (*d*) rangeability
(*e*) all of the above

79. Optical flats are made of
(*a*) quartz (*b*) glass
(*c*) plastic (*d*) steel
(*e*) silicon

80. The axis of measurement and the scale or other dimensional reference should coincide. This principle is called the
(*a*) principle of kinematic design (*b*) principle of alignment
(*c*) principle of linear measuring instruments
(*d*) principle of collinearity
(*e*) principle of location and movement.

81. Pick out the wrong statement about flexible strips
(*a*) These are used in instruments where small movements are required between component parts
(*b*) no force or torque is required to displace a member located on flexible strip from its mid position
(*c*) it has no friction or backlash (*d*) it is not subjected to wear
(*e*) it has negligible hysterisis

82. The least count of a metric vernier caliper having 25 divisions on vernier scale, matching with 24 divisions of main scale (1 main scale division = 0.5 mm) is
(*a*) 0.05 mm (*b*) 0.01 mm
(*c*) 0.02 mm (*d*) 0.001 mm
(*e*) 0.005 mm

83. A scale in which the distance between graduations is proportional to the value of that graduation is called
(*a*) regular scale (*b*) equidistant scale
(*c*) linear scale (*d*) line scale
(*e*) continuous scale

84. A scale whose graduation marks are in a discontinuous manner and are composed of aligned numbers indicating directly the numerical value of the quantity measured is called
(*a*) linear scale (*b*) base of a line scale
(*c*) equidistant scale (*d*) regular scale
(*e*) digital scale

85. Element of the indicating device carrying the scale is called
(*a*) dial (*b*) housing
(*c*) transducer (*d*) index
(*e*) frame

86. The thread micrometer measures
(*a*) the major diameter of the thread (*b*) the minor diameter of the thread
(*c*) the effective diameter of the thread
(*d*) the root diameter of the thread (*e*) all the diameters of the thread

87. V-block is used in the workshop to check
(*a*) roundness of a cylindrical work (*b*) surface roughness
(*c*) dimensions of oval job (*d*) taper on a job
(*e*) none of the above

88. Repeatability of measuring equipment is
(*a*) the closeness with which a measurement can be read directly from a measuring instrument
(*b*) a measure of how close the reading is to the true size
(*c*) difference between measured value and actual value
(*d*) the smallest change in measurand that can be measured
(*e*) the capability to indicate the same reading again and again for a given measurand

89. The purpose of ratchet screw in micrometer screw gauge is
(*a*) to lock a dimension (*b*) to impart blow motion
(*c*) to maintain sufficient and uniform measuring pressure
(*d*) to take care to wear of screw threads
(*e*) to allow zero adjustment

90. The purpose of adjusting nut in a micrometer screw gauge is to
(*a*) take care of zero error (*b*) impart body motion
(*c*) compensate for wear between screw and nut
(*d*) take care of backlash (*e*) ensure uniform measuring pressure

91. The taper of internal dovetail can be measured with the help of
(*a*) sine bar (*b*) combination set
(*c*) balls of standard dimensions and slip gauges
(*d*) clinometer (*e*) dial gauge

92. External taper can be accurately measured with the help of
(*a*) sine bar and slip gauges (*b*) dividing head
(*c*) precision balls and height gauge (*d*) combination set
(*e*) clinometer

93. Stick micrometers are designed for measuring
(*a*) bore of cylinders (*b*) longer external lengths
(*c*) cylindricity (*d*) longer internal lengths
(*e*) diameters which can't be easily accessed

94. Differential screw micrometer is used
(*a*) to give direct indication of difference between two readings
(*b*) to measure gap between grooves
(*c*) for digital readout (*d*) to measure diameter of inaccessible holes
(*e*) for very high degree of accuracy

95. A sine bar is specified by
(*a*) its total length (*b*) the centre distance between the two rollers
(*c*) the size of the rollers
(*d*) the distance between rollers and upper surface
(*e*) weight of sine bar

96. The diameter of very large bores can be measured accurately by
(*a*) flexible graduated tape (*b*) cylindrical gauge
(*c*) keilpart gauge (*d*) four balls method
(*e*) swinging a pin gauge in the bore

97. The number of slip gauges in a set are
(*a*) 87 (*b*) 45
(*c*) 103 (*d*) 31
(*e*) all of the above sets are available

98. Profile of a gear tooth can be checked by
(*a*) sine bar (*b*) bench micrometer
(*c*) optical pyrometer (*d*) optical projector
(*e*) slip gauges

99. Gear tooth caliper is used to find the chordal thickness of the following type of gear tooth
(*a*) spur gears (*b*) helical gears
(*c*) worm gears (*d*) bevel gears
(*e*) any type of gear

100. Fig. 3 shows a case of error in relative location of surfaces. This case is for
(*a*) misalignment
(*b*) axial runout
(*c*) radial runout
(*d*) non parallelism of axes
(*e*) squareness error.

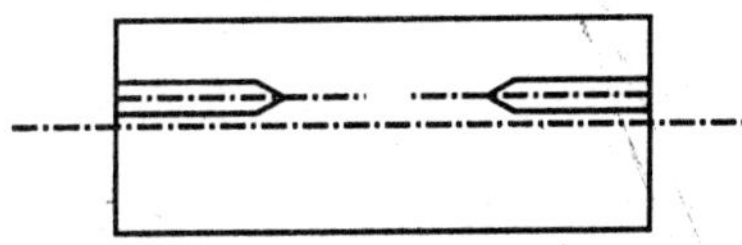
Fig. 3

101. The M-and E-system in metrology are related with measurement of
(*a*) gears (*b*) screw threads
(*c*) flatness (*d*) angularity
(*e*) surface finish

102. All the thread characteristics can be measured precisely with
(*a*) screw pitch gauge (*b*) micrometer with V anvil
(*c*) tool room microscope (*d*) thread gauge
(*e*) thread measuring machine

103. The advantage of vernier caliper over micrometer is that it
(*a*) is easier and quicker to use (*b*) is more accurate
(*c*) can be used to make both inside and outside measurements over a range of sizes
(*d*) all of the above (*e*) none of above

104. Standards to be used for reference purposes in laboratories and workshops are referred to as
(*a*) primary standards (*b*) secondary standards
(*c*) tertiary standards (*d*) working standards
(*e*) none of the above

105. The cross section of International Prototype Metre is shown in Fig. 4.
(*a*) Figure A
(*b*) Figure B
(*c*) Figure C
(*d*) Figure D
(*e*) Figure E

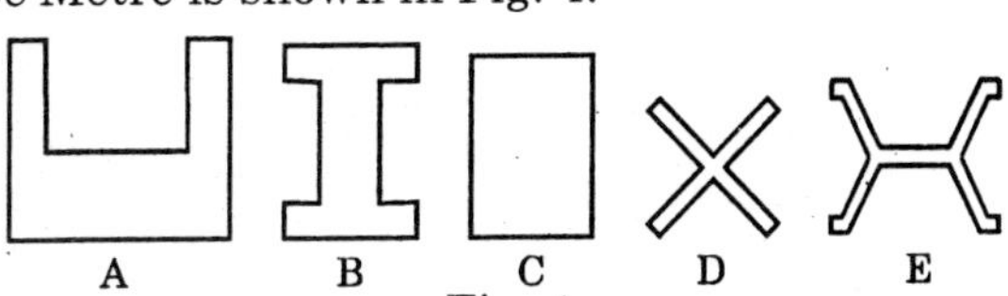

Fig. 4

106. The combination set can be used to
(*a*) check angular surfaces
(*b*) draw circles and arcs
(*c*) scribe lines
(*d*) all of the above
(*e*) none of the above

107. In layout work, a pencil should not be used to draw lines on metal because
(*a*) it will wipe off easily
(*b*) the line will be too wide for accurate work
(*c*) the lines will smudge and be difficult to see
(*d*) the lines do remain on metal even after good rubbing
(*e*) all of the above

108. Surface plate is usually made of grey cast iron because it provides
(*a*) non wearing plate
(*b*) very hard plate
(*c*) easy to cast plate
(*d*) lubrication due to graphite flakes
(*e*) stable plate

109. The term traceability in Engineering Metrology is concerned with
(*a*) measuring machines
(*b*) optical instruments
(*c*) pneumatic comparator
(*d*) standards
(*e*) limits and fits

110. In selective assembly
(*a*) parts in an assembly can be replaced by a similar part without any further alteration
(*b*) parts are produced on hole basis system
(*c*) all the parts are always interchangeable
(*d*) the size of one of the components is measured accurately and then mating component is made to match with this
(*e*) the parts of any one type are classified into several groups according to size.

111. Constant measuring pressure in micrometer screw gauges is ensured by
(*a*) locknut
(*b*) barrel and thimble
(*c*) spindle
(*d*) spanner
(*e*) ratchet

112. The advantage of wavelength standard is that it would be possible to repeat measurement to a very high degree of accuracy. The wavelength standards can be reproduced to an accuracy of about one part in
(*a*) 1000
(*b*) 10^5
(*c*) 10^7
(*d*) 10^8
(*e*) 10^9

113. The basic unit in angular measurement is
(*a*) degree
(*b*) minute
(*c*) second
(*d*) right angle
(*e*) 360°

114. Optical gauge works on the principle of
(*a*) refraction
(*b*) reflection
(*c*) dispersion
(*d*) polarisation
(*e*) interference of light rays

115. The micrometer show in Fig. 5 is called
(*a*) bench micrometer
(*b*) spline micrometer
(*c*) disc type micrometer
(*d*) tube micrometer
(*e*) sheet metal micrometer

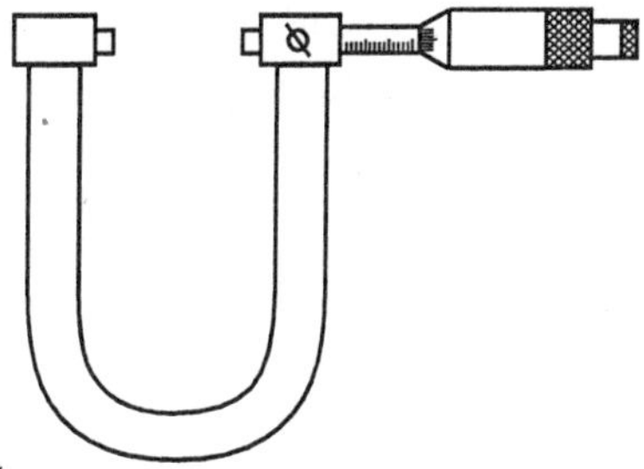
Fig. 5

116. Fig. 6 below shows an outside micro meter
Sleeve, thimble, ratchet and spindle are shown by letters
(*a*) B, C, A and D respectively
(*b*) C, B, D and A respectively
(*c*) B, C, D and A respectively
(*d*) C, D, B, A respectively

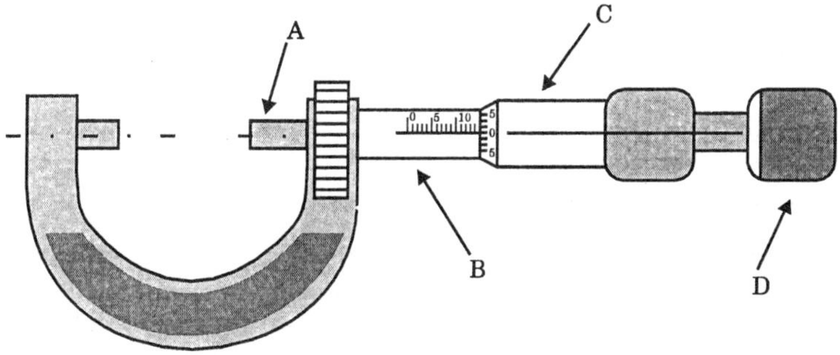

Fig. 6

117. The micrometer shown in Fig. 7 is called
(*a*) tube micrometer
(*b*) sheet metal micrometer
(*c*) height micrometer
(*d*) disc type micrometer
(*e*) depth micrometer

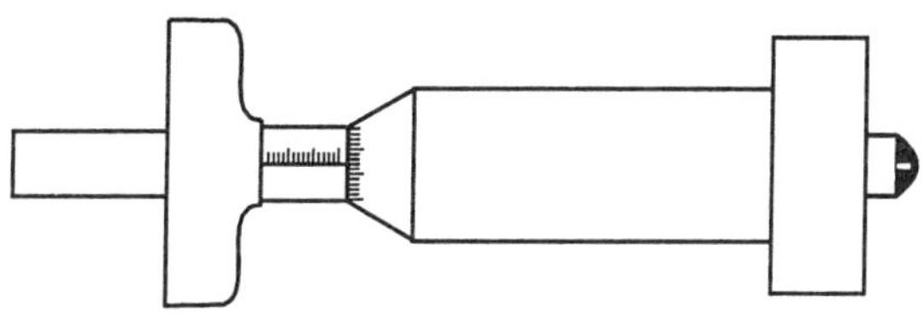
Fig. 7

118. Pick up false statement about micrometer screw gauge
(*a*) It can be used both for line and end measurements
(*b*) It is more precise than vernier
(*c*) It has no parallax error
(*d*) compared to vernier, measureing range is lower
119. Wringing w.r.t slip gauges implies
(*a*) causing slip gauges to adhere by molecular attraction
(*b*) applying little oil for protection and removing oil at assembly
(*c*) applying wear gauges at ends
(*d*) twisting two gauges under pressure
120. Least count of vernier caliper is
(*a*) 1 main scale division (MSD) — 1 vernier scale division (VSD)
(*b*) MSD + VSD (*c*) MSD ÷ VSD
(*d*) MSD × VSD

121. Pick up correct statement
   (*a*) In forward vernier, IOVSD = 9 MSD
   (*b*) In backward vernier, IOVSD = 11 MSD
   (*c*) Both (*a*) and (*b*) are correct (*d*) Neither (*a*), nor (*b*) is correct
122. Millimeter scale in a micrometer is marked on
   (*a*) barrel (*b*) thimble
   (*c*) spindle (*d*) anvil
   (*e*) ratchet
123. Match the two parts below relating to parts of micrometer

1. Rigidly fixed part on left side of frame	A. Anvil
2. Fixed cylindrical surface on right side of frame having engraved graduation marks above and below reference line	B. Barrel
3. A tubular cover fastened and integrated with a screwed spindle	C. Ratchet
4. A small extension at end of thimble to control pressure applied on workpiece, maintaining standards conditions of measurement	D. Thimble

Code for answers
   (*a*) 1 — A, 2 — B, 3 — D, 4—C (*b*) 1 — A, 2 — B, 3 — C, 4 — D
   (*c*) 1 — B, 2 — A, 3 — D, 4 — C (*d*) 1 — B, 2 — A, 3 — C, 4 — D
124. The least count of micrometer 0.01 mm and pitch of spindle thread in micrometer screw gauge is
   (*a*) 0.1 mm (*b*) 0.5 mm
   (*c*) 1.0 mm (*d*) 2.0 mm
125. Circular scale of a micrometer is marked on
   (*a*) anvil (*b*) barrel
   (*c*) ratchet (*d*) thimble
   (*e*) spindle
126. A feeler gauge is used to check
   (*a*) flatness and straightness (*b*) unsymmetrical shape
   (*c*) thickness of clearance (*d*) surface roughness
127. A surface gauge is used to
   (*a*) level the surface plate (*b*) determine depth
   (*c*) laying out the work accurately (*d*) check surface finish
128. Which of following is most accurate instruments
   (*a*) micrometer screw gauge (*b*) electronic comparator
   (*c*) optical protector (*d*) universal testing machine
129. According to well accepted practice, slip gauges which have been handled for a few minutes should be left for sometime. The reason for this could be
   (*a*) the moisture transferred from hand would evaporate

(b) the structure of gauges, if disturbed would stabilise
(c) gauges attain the room temperature again
(d) with continuous use, gauges may wear quickly
(e) there is no such practice

130. Which of the following is not the essential requirement for accuracy of measurement with a sine bar
(a) flatness of upper surface
(b) equality of size and roundness of rollers
(c) exact distance between top and bottom surfaces
(d) parallelism between top and bottom surfaces
(e) parallelism of rollers to upper surface and equality of axis distance from surface.

131. Match the 2 parts below relating to Linear measurements

1. Measurement made between two flat parallel surfaces	A. Built in datum
2. It has parallax error	B. End Measurement
3. Ease of alignment with axis of measurement in end standards	C. Length standard
4. Airy points	D. Line standard

(a) 1—D, 2—B, 3—A, 4—C
(b) 1—B, 2—D, 3—A, 4—C
(c) 1—B, 2—D, 3—C, 4—A
(d) 1—A, 2—B, 3—C, 4—D

132. Modern metre is defined by length of the path travelled by light in vacuum. The reproducibility is of the order of
(a) 1 part in 10^{11}
(b) 2 parts in 10^{11}
(c) 3 parts in 10^{11}
(d) 4 parts in 10^{11}

133. Sub-division of end standard can be carried out using
(a) airy supports
(b) Brookes level comparator
(c) wringing of slip gauges
(d) sub-division method

134. The following type of gauge has gauging sections combined on one end
(a) combination gauge
(b) limit gauge
(c) Go and No Go gauge
(d) fixed gauge
(e) progressive gauge

135. The vernier reading should not be taken at its face value before an actual check has been taken for
(a) zero error
(b) its calibration
(c) flatness of measuring jaws
(d) temperature equalisation
(e) all of the above

136. Which comparator utilises the principle of a button spinning on a loop of string
(a) Sigma comparator
(b) Aramson comparator
(c) Optical comparator
(d) Zeiss interferometer
(e) Eden-Rolt comparator

137. Gear tooth vernier is used to measure
(a) circular pitch
(b) depth of tooth
(c) tooth thickness
(d) addendum and dedendum
(e) pitch line thickness of tooth

138. Error of measuring equipment
   (*a*) the closeness with which a measurement can be read directly from a measurand
   (*b*) a measure of how close the reading is to the true size
   (*c*) the difference between measured value and actual value
   (*d*) the smallest change in measurand that can be measured
   (*e*) the capability to indicate the same reading again and again for a given measurand.

139. Which of the following is not provided on combination set
   (*a*) centre head (*b*) protractor head
   (*c*) vernier scale (*d*) spirit level
   (*e*) squaring head

140. Universal surface gauge is used
   (*a*) for flatness testing (*b*) for layout work and inspection
   (*c*) for measuring profile of complex surface
   (*d*) for measuring surface roughness
   (*e*) for measuring concavity of surface

141. Accuracy is
   (*a*) the repeatability of a measuring process
   (*b*) error of judgment in recording an observation
   (*c*) the ability of instrument to reproduce same reading under identical situations
   (*d*) agreement of the result of a measurement with the true value of the measured quantity
   (*e*) the least resolution of an instrument

142. Random errors follow the following distribution
   (*a*) claussian (*b*) hyperbolic
   (*c*) parabolic (*d*) t-student
   (*e*) β-distribution

143. Which one of the following is the least accurate measuring device
   (*a*) air gauge (*b*) micrometer screw gauge
   (*c*) optical projector (*d*) vernier micrometer
   (*e*) steel scale

144. Pick up the wrong statement
   (*a*) Accuracy of an instrument is closeness to the true dimension
   (*b*) Precision represents the degree of repetitiveness
   (*c*) Sensitivity refers to minimum change in value that the instrument can reliably indicate
   (*d*) As the sensitivity of an instrument increases, its range of measurement also increases.
   (*e*) If an instrument is not precise it will give different results for same dimension when measured again and again

145. Fiducial indicators contain
   (*a*) calibrated scale (*b*) a single index mark
   (*c*) micrometer screw movement (*d*) optical head
   (*e*) interferometric devices

146. A comparator for its working depends on
(a) accurately calibrated scale
(b) comparison with standard such as slip gauges
(c) accurate micrometer gauge (d) optical devices
(e) determining zero error of scale correctly

147. Precision is
(a) the repeatability of a measuring process
(b) agreement of the result of a measurement with the true value of the measured quantity
(c) the ability of a measuring device to detect small differences in a quantity being measured
(d) the ability of an instrument to reproduce same reading under identical conditions
(e) error of judgment in reading an observation

148. The thickness of oil film at the surface of slip gauges is of the order
(a) 0.005 micron (b) 0.1 micron
(c) 1 microns (d) 10 micron
(e) 100 microns

149. Pick up the wrong statement. Surface plates are usually made of granite because of following advantages
(a) Because of long period of time for relaxing, it is free from build in residual stresses. There is no corrosion effect also
(b) there is less tendency for it to warp
(c) if a tool or workpiece drops accidentally over its surface, residual stresses are not induced
(d) it simply powders somewhat at the point of impact by falling object
(e) it enables the phenomenon of wringing flat surface over it.

150. Pick up the wrong statement
(a) Error is the disagreement between the result of measurement and actual value.
(b) Random errors are regularly repetitive in nature and result from improper conditions or procedures that are consistent in action
(c) Parasitic error results from incorrect execution of measurement
(d) Uncertainty of measurement represents the dispersion of the result of measurement defined by the limits of the error
(e) Absolute error is the algebraic difference between the result of measurement and the value of comparision

151. Fig. 8 shows principle of

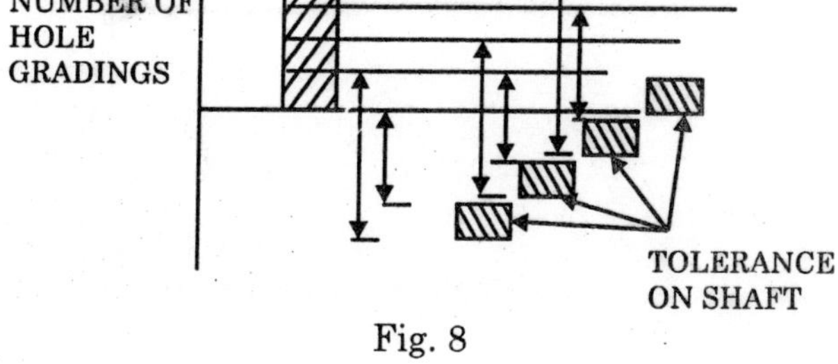

Fig. 8

(a) traceability (b) interchangeability
(c) matched fits (d) selective assembly
(e) limits, fits and tolerances

152. As per IS specification, Fig. 9 shows vernier caliper of type

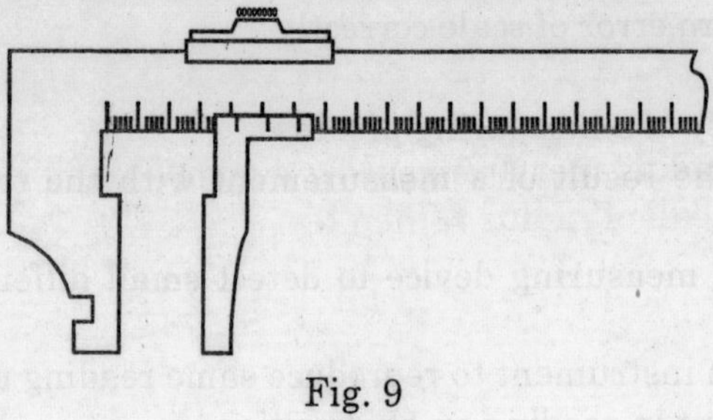

Fig. 9

(a) A (b) B
(c) C (d) AA
(e) BB.

153. Which one of the following instruments is the most accurate
(a) steel scale (b) micrometer screw gauge
(c) vernier caliper (d) vernier dial gauge
(e) optical projector

154. In Fig. 10 which is the best method of wringing slip gauge
(a) A
(b) B
(c) C
(d) D
(e) E

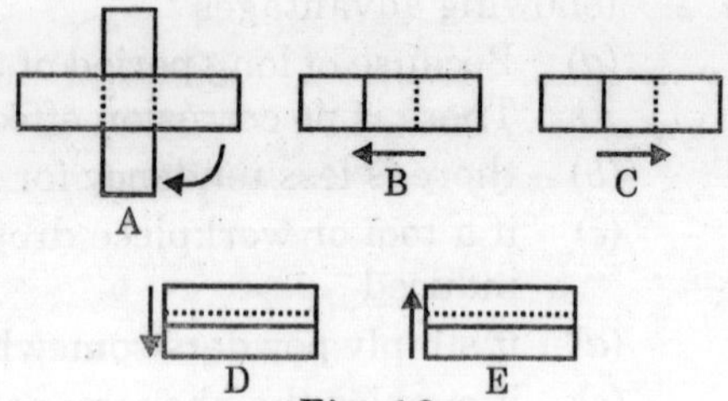

Fig. 10

155. In the case of high precision surface plates of diameters upto 200 mm, the working surface should lie between two parallel planes, whose maximum distance apart is
(a) 0.005 mm (b) 0.0005 mm
(c) 0.05 mm (d) 0.5 mm
(e) 0.001 mm

156. A plug of diameter d rests in an angle as shown in Fig. 11. An equation giving distance L in terms of d, H and α would

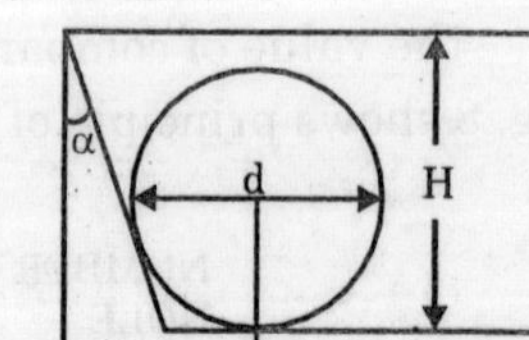

Fig. 11

(a) $L = H \tan\frac{\alpha}{2} + \frac{d}{2}\tan(45 - \alpha)$

(b) $L = H \cot\frac{\alpha}{2} + \frac{d}{2}\tan\left(45 - \frac{\alpha}{2}\right)$

(c) $L = H \tan\alpha + \frac{d}{2}\tan\alpha$

(d) $L = H \tan\alpha + \frac{d}{2}\tan\left(45 + \frac{\alpha}{2}\right)$

(e) $L = H \tan\alpha + \frac{d}{2}\tan\left(45 - \frac{\alpha}{2}\right)$.

157. Fig. 12 shows the measurement of

(a) roundness
(b) radius of curvature
(c) cylindricity
(d) flatness
(e) surface texture

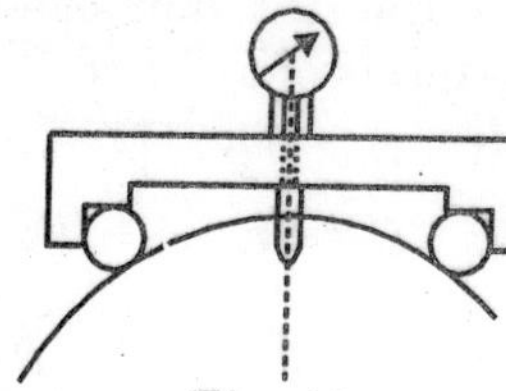

Fig. 12

158. Fig. 13 shows the radius measurement by dynamic method for concave surface. If T is the time of SHM of ball of radius r, then R =

(a) $r+\frac{Tg^2}{6\pi^2}$
(b) $r+\frac{Tg}{6\pi^2}$
(c) $\frac{r}{2}+\frac{T^2g}{3\pi^2}$
(d) $2r+\frac{Tg}{3\pi^2}$
(e) $r+\frac{Tg^2}{6\pi^2}$.

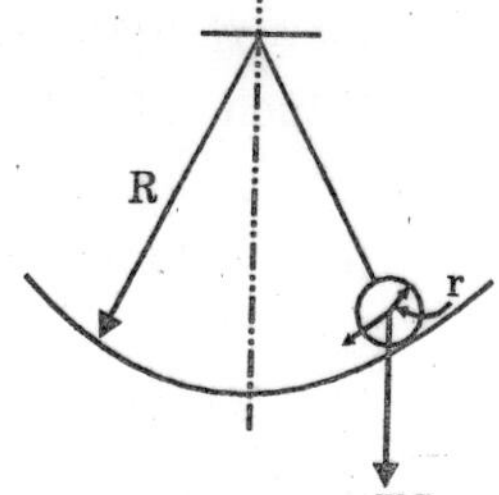

Fig. 13

159. The radius of a pulley block is measured as shown in Fig. 14

(a) $R=\frac{l-d}{8d}$
(b) $R=\frac{(l-d)^2}{4d}$
(c) $R=\frac{(l-d)^2}{8d}$
(d) $R=\frac{l^2}{8d}$
(e) $R=\frac{(l-d)^2}{2d}$

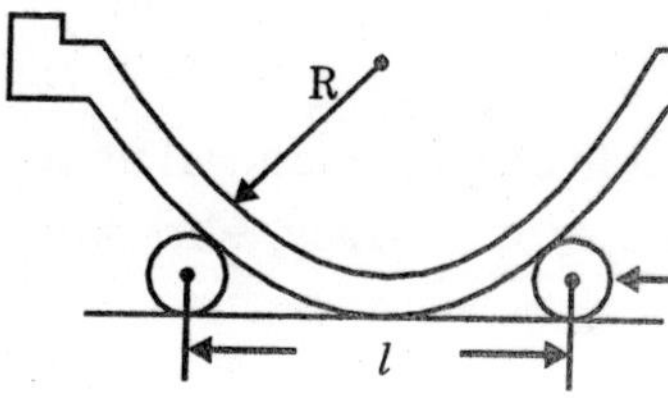

Fig. 14

160. The radius of concave surface can be easily determined by a depth micrometer as shown in Fig. 15. Radius of curvature R =

(a) $\frac{d^2}{8h}+\frac{h}{2}$
(b) $\frac{d^2}{8h}-\frac{h}{2}$
(c) $\frac{d^2}{4h}+h$
(d) $\frac{d^2}{8h}+h$
(e) $\frac{d^2}{8h}+2h$.

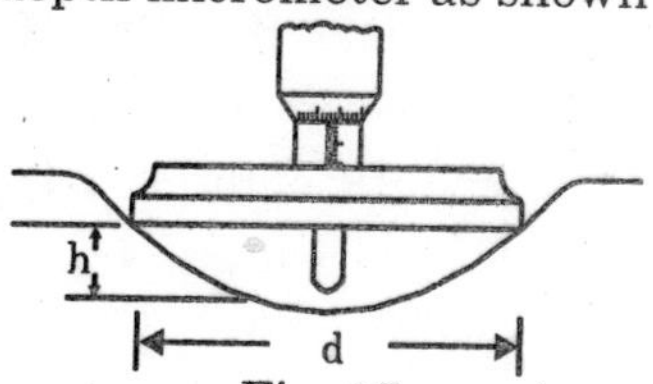

Fig. 15

161. For the situation shown in Fig. 16 the expression for H in terms of r, B and D would be

(a) $H = \sqrt{R^2 + r^2}$

(b) $H = (R + r) + (D + r)$

(c) $H = (R + r) + \sqrt{R^2 - D^2}$

(d) $H = (R + r) + \sqrt{r^2 + R^2 - D^2}$

(e) $H = (R + r) + \sqrt{2D(R + r) - D^2}$.

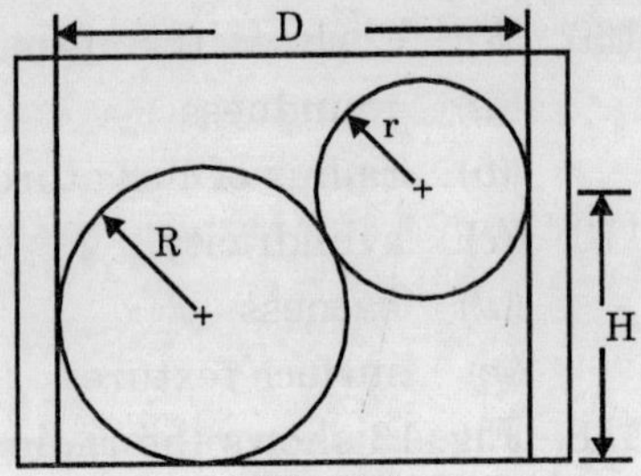

Fig. 16

162. Which of the following is not the accessory of slip gauges

(a) scribing and centre points (b) measuring jaws

(c) holder (d) calipers

(e) base

163. The lateral faces of slip gauges are at right anagles correct to within

(a) ± 1 degree (b) ± 30 minutes

(c) ± 10 minutes (d) ± 1 minute

(e) ± 10 seconds.

164. Protector gauge blocks in slip gauges are

(a) used as reference blocks (b) mounted in the centre of pile

(c) never touched (d) wrong on the end of combinations

(e) made of same material as the slip gauges

165. According to accuracy, slip gauges are classed under following number of accuracy classes

(a) two (b) three

(c) five (d) seven

(e) ten

166. A protector slip gauge is provided to

(a) protect slip gauges when not in use (b) take up all the wear when in use

(c) clean the slip gauges (d) facilitate wringing of slip gauges

(e) assemble the slip gauges properly

167. The ratio of total surface area A and the volume V of a cylinder of diameter d and length l is

(a) $\frac{A}{V} = \frac{6d}{l}$ (b) $\frac{A}{V} = \frac{4l}{d}$

(c) $\frac{A}{V} = \frac{l + d}{dl}$ (d) $\frac{A}{V} = \frac{6d + 4l}{l}$

(e) $\frac{A}{V} = \frac{2d + 4l}{dl}$

168. Tolerances are specified

(a) to obtain high accuracy (b) to have proper allowance

(c) to obtain desired fits

(d) because exact size components can't be manufactured.

169. Which is most important characteristic of a measuring instrument for general applications
(*a*) accuracy (*b*) sensitivity
(*c*) repeatability (*d*) compact and sturdy

170. Which relationship generally holds for sensitivity and range of measuring instrument
(*a*) direct (*b*) normal
(*c*) linear (*d*) inverse

171. Parasitic errors is caused due to
(*a*) errors in computation/incorrect execution of measurement
(*b*) improper use of instrument (*c*) changes in ambient condition
(*d*) wrong design of instrument

172. Which errors are inevitable in instruments
(*a*) systematic (*b*) calibration
(*c*) random (*d*) parasitic

173. Which of the Fig. 17 below gives the correct disposition of manufacturing tolerances for screw thread limit gauges for both Go and Not Go gauges in case of ring caliper gauges

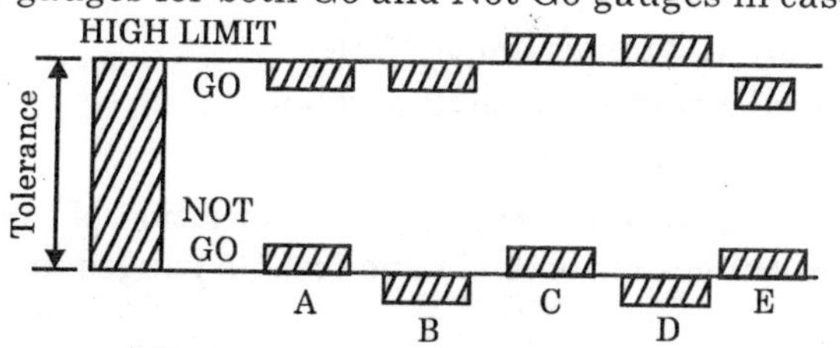

Fig. 17

(*a*) A (*b*) B
(*c*) C (*d*) D
(*e*) E

174. In hole basis system, basic size is equal to the
(*a*) maximum hole size (*b*) minimum hole size
(*c*) average hole size (*d*) none of above

175. The two slip gauges in precision measurement are joined by
(*a*) assembling (*b*) sliding
(*c*) adhesion (*d*) wringing
(*e*) slipping

176. Plug gauges are used to
(*a*) measure the diameter of the workpieces
(*b*) measure the diameter of the holes in the workpieces
(*c*) check the diameter of the holes in the wokpieces
(*d*) check the length of the holes in the workpieces
(*e*) check the outside diameter of workpieces

177. Pick out the wrong statement about Taylor's principle of gauging
(*a*) Go gauges should be full form gauges
(*b*) Go gauges should check all the related dimensions simultaneously
(*c*) It is sufficient to use Go gauges on the width and length of the component

(*d*) Not Go gauges should check only one dimension at a time

(*e*) It is useless to have the Not go gauges of the full form

178. Which is not true for selective assembly?

(*a*) suitable for industries like automobiles, ball and roller bearing units, etc.

(*b*) cost reduces

(*c*) wastages is reduced

(*d*) tolerances derived are very wide

179. In a limit system, a series of tolerances are arranged to suit

(*a*) a specific range of size (*b*) cost

(*c*) desired function (*d*) allowances, desired

180. What does ITOI represent

(*a*) tolerance grade (*b*) tolerance class

(*c*) tolerance factor (*d*) none of above

181. 'Go limit' refers to

(*a*) upper limit of shaft and lower limit of hole

(*b*) lower limit of shaft and upper limit of hole

(*c*) minimum material condition of shaft and hole

(*d*) middle limits of hole and shaft

182. In Fig. 18 which of the below gives the correct disposition of manufacturing tolerances for screw thread limit gauges for both Go and Not Go gauges in case of plug gauges

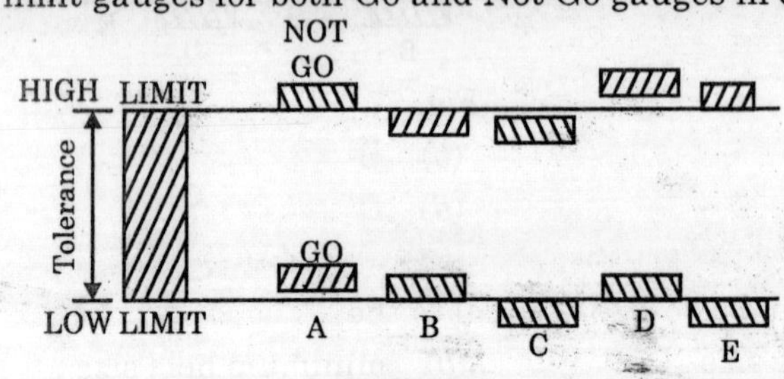

Fig. 18

(*a*) A (*b*) B

(*c*) C (*d*) D

(*e*) E.

183. The consideration for deciding the limits of a component are

(*a*) functional requirement

(*b*) interchangeability for ease of replacement

(*c*) minimum cost and time of producing it

(*d*) all of above

184. To achieve full interchangeability in mass production, the manufacturing process selected should have ________ manufacturing tolerance

(*a*) equal to (*b*) grater than

(*c*) less than (*d*) less than or equal to

185. Which are associated with interchangeable production.

(*a*) standardized items be used

(*b*) components follow normal distribution.
(*c*) machine having better process capability compared to tolerances desired be used
(*d*) all of above

186. Which is not correct about selective assembly
(*a*) automatic gauge is used as it enables segregation of parts in desired number of groups
(*b*) Number of groups equal to process capability ÷ tolerance desired
(*c*) high quality at low cost is enabled
(*d*) all are correct

187. The term "Allowance" in limits and fits is usually referred to
(*a*) minimum clearance between shaft and hole
(*b*) maximum clearance between shaft and hole
(*c*) difference of tolerance of hole and shaft
(*d*) difference between maximum size and minimum size of the hole
(*e*) difference between maximum size and minimum size of the shaft

188. Difference of mean size of hole and shaft is called
(*a*) mean tolerance (*b*) mean clearance
(*c*) mean allowance (*d*) mean fit

189. Which of following is not associated with quality control charts
(*a*) binomial distribution (*b*) normal distribution
(*c*) random sampling (*d*) all of above

190. In mass production
(*a*) bilateral tolerance are preferred
(*b*) basic size is taken half way between upper and lower limits
(*c*) If unilateral tollerance is specified, the basic size should be changed to make it bilateral
(*d*) all above are true for mass production

191. Interchangeable production results in
(*a*) higher output (*b*) lesser assembly time
(*c*) enabling mass production (*d*) all are correct

192. Which of the following is the correct way of designating fit
(*a*) H_8/g_7 (*b*) g_7/H_8
(*c*) 50 H_8/g_7 (*d*) H_8/g_7–50
(*e*) $50H_8/50h_7$

193. Fig. 19 shows the typical representation of different types of fundamental deviations for holes and shafts
What do A_h, B_h, C_s, D_s represent
(*a*) A_h – lower deviation for hole, B_h – upper deviation for hole, C_s – upper deviation for shaft, D_s – lower deviation for shaft
(*b*) A_h – upper deviation (UD) for hole, B_h– lower deviation (LD) for hole, C_S – UD for shaft, D_S – LD for shaft.
(*c*) A_h – LD for shaft, B_h – UD for hole, C_S – LD for shaft, C_d = UD for shaft.
(*d*) none of above

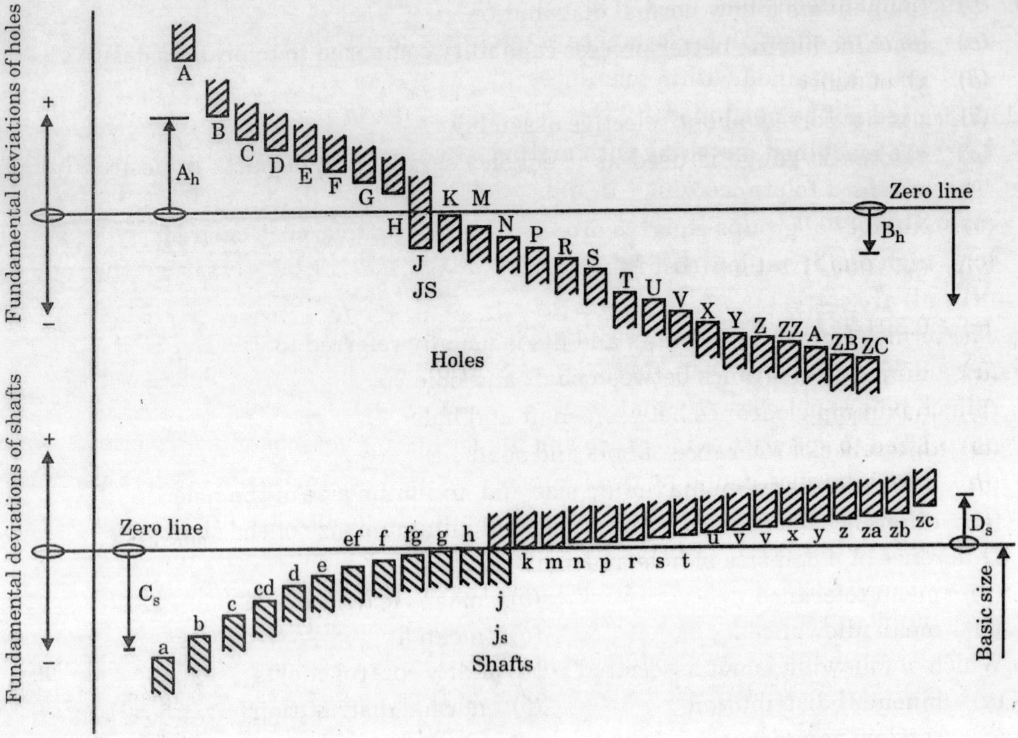

Fig. 19

194. Drilled holes, and honed holes could be designated by following grades respectively
   (*a*) H_5, H_{11} (*b*) H_6, H_{10}
   (*c*) H_8, H_6 (*d*) H_{10}, P_5
   (*e*) H_{11}, H_8

195. The number of tolerances and the basic deviations specified are
   (*a*) 18, 28 (*b*) 16, 21
   (*c*) 18, 22 (*d*) 20, 28

196. It limits ES and EI for a 25 mm hole be +0.025 and +0.002 mm, then the maximum and minimum hole size will be
   (*a*) 25.025 mm and 25.002 mm (*b*) 25.025 mm and 25.002 mm
   (*c*) 24.998 mm and 25.002 mm (*d*) 24.990 mm and 24.975 mm

197. Which type of fit may sometime provide clearance and sometime interference fit
   (*a*) interference (*b*) clearance
   (*c*) transition (*d*) none of above

198. A basic hole has zero ______
   (*a*) lower deviation (*b*) upper deviation
   (*c*) middle deviation (*d*) nominal deviation

199. It the dimension of a component is finished within the prescribed limits of size, then it is called
   (*a*) basic size (*b*) nominal size
   (*c*) actual size (*d*) acceptable size

200. Functional dimensions
   (a) have no effect on performance of quality of component
   (b) are maintained within tolerances
   (c) need not be machined
   (d) are machined matching with mating machined part

201. The standard tolerance unit I in the case of limit and fits for sizes above 500 mm and upto 3150 mm is
   (a) $0.45(\sqrt[3]{D})$ + 0.001 D   (b) $0.52(\sqrt[3]{D})$ + 0.001 D
   (c) $0.30(\sqrt[3]{D})$ + 0.042 D   (d) $0.005(\sqrt[3]{D})$
   (e) 0.004 D + 2.1 (D is in mm)

202. Which pair is not correctly matched
   (a) Used to specify the tolerance and fits for mating components—Tolerance symbols
   (b) The tolerance bound by two limits of size of the components—Tolerance zone
   (c) A set of tolerances varying according to the basic size for uniform level of accuracy within the grade—International tolerance grade (IT)
   (d) Designated by the letters representing the fundamental deviation followed by the number representing the standard tolerance grade—Fundamental tolerance.

203. Sensitivity of measuring equipment is
   (a) the closeness with which a measurement can be read directly from a measuring instrument
   (b) a measure of how close the reading is to the true size
   (c) the difference between measured value and the actual value
   (d) the smallest change in measurand that can be measured
   (e) the capability to indicate the same reading again and again for a given measurand

204. Pick up wrong statement about fundamental deviation
   (a) Fundamental deviation is the maximum difference between the size of a component and its basic size.
   (b) Fundamental deviation is identical to the upper deviation for shafts and lower deviation for holes
   (c) It is the closest deviation to the basic size and it is zero for H hole and h shaft
   (d) Fundamental deviation and fundamental tolerance indicate the type and quality of fit, respectively.

205. Newal system of limits and fits is the oldest system working on hole basis system. The grades of holes and shafts specified respectively are
   (a) 2, 6   (b) 1, 8
   (c) 4, 12   (d) 6, 20
   (e) 8, 26

206. Which pair is not matched rightly
   (a) The minimum difference between the size of a component and its basic size—Deviation.
   (b) The algebric difference between the actual size and the basic size—Actual deviation.
   (c) The algebraic difference between the maximum limit of size and basic size—Upper deviation

(d) The algebraic difference between the minimum limit of size and basic size—Lower deviation.

207. ISA tolerance system consists of following numbers of qualities of tolerance, and grades of fit respectively
(a) 6, 15 (b) 8, 20
(c) 12, 21 (d) 16, 21
(e) 21, 26

208. Expressing a dimension as $25.3^{\pm 0.05}$ mm is the case of
(a) unilateral tolerance (b) bilateral tolerance
(c) limiting dimensions (d) all of the above
(e) none of the above

209. Surface roughness on a drawing is represented by
(a) triangles (b) circles
(c) squares (d) rectangles
(e) none of above

210. Expressing a dimension as $\frac{32.5}{32.4}$ mm is the case of
(a) unilateral tolerance (b) bilateral tolerance
(c) limiting dimension (d) all the above
(e) none of the above

211. A bore of 14.67 mm in a workpiece can be measured by
(a) steel rule (b) vernier caliper
(c) pneumatic gauge (d) micrometer
(e) plug gauge

212. In Fig. 20, size A refers to
(a) lower deviation
(b) minimum interference
(c) minimum clearance
(d) maximum clearance
(e) maximum transition.

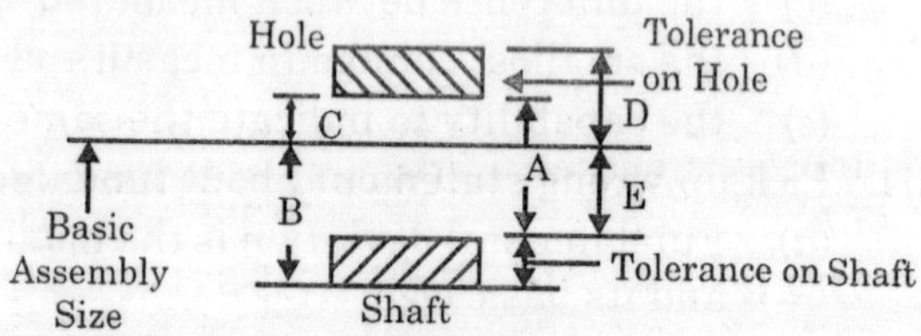

Fig. 20

213. In Fig.20 sizes B and C refer to
(a) lower deviation (b) upper deviation
(c) maximum clearance (d) minimum clearance
(e) minimum allowance

214. In Fig. 20, sizes D and E refer to
(a) lower deviation (b) upper deviation
(c) maximum clearance (d) minimum clearance
(e) minimum allowance

215. Match the following in regard to symbols used to represent deviation

1. upper deviation for hole	A. ei
2. lower deviation for hole	B. es
3. upper deviation for shaft	C. EI
4. lower derivation for shaft	D. ES

(a) 1—D, 2—C, 3—B, 4—A (b) 1—A, 2—B, 3—C, 4—D
(c) 1—D, 2—C, 3—A, 4—B (d) 1—C, 2—D, 3—A, 4—B

216. Fig. 21 shows upper and lower deviations for shafts and holes
Upper and lower deviations for shaft es and *ei* and corresponding deviations for hole ES and EI are represented by
(a) B and D, A and C, F and G, E and H respectively
(b) A and C, B and D, G and H, E and G respectively
(c) A and B, C and D, E and F, G and H respectively
(d) A and C, B and D, E and H, G and F respectively

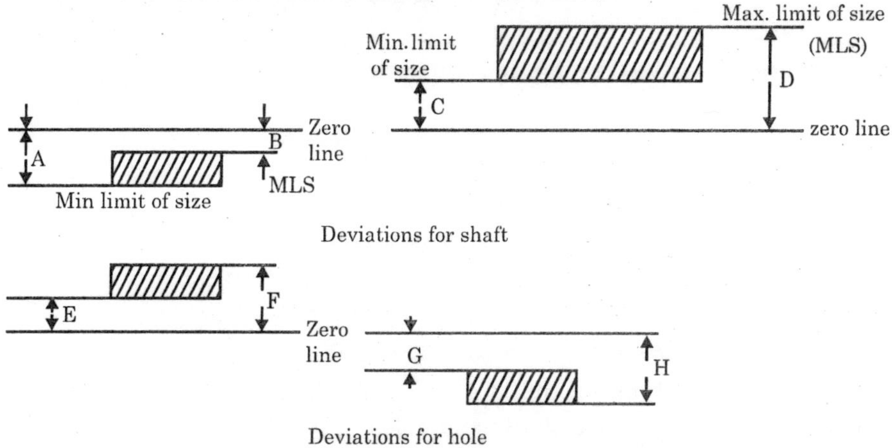

Fig. 21

217. The diameter of finish turned shaft can best be checked with a
(a) combination set (b) slip gauge
(c) height gauge
(d) micrometer screw gauge (e) dial indicator

218. One micron is equal to
(a) 1 mm (b) 0.1 mm
(c) 0.01 mm (d) 0.001 mm
(e) 0.0001 mm

219. In Fig. 22, which of the combination of hole and shaft represents hole basis system
(a) A (b) B
(c) C (d) D
(e) none of the above

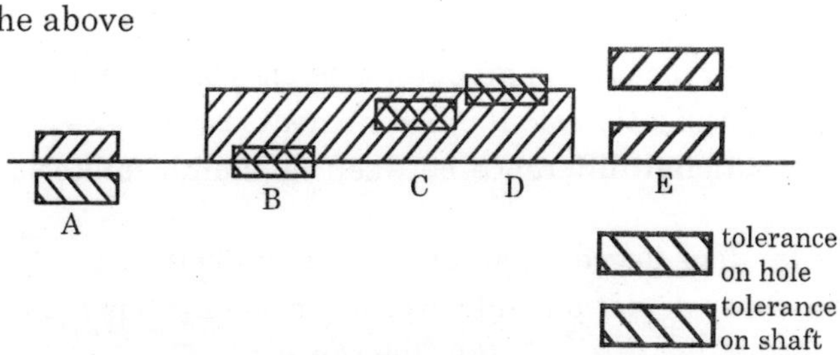

Fig. 22

220. In Fig. 23 which of the following combination represents interference fit
(*a*) A and E (*b*) B and D
(*c*) B, C and D (*d*) C only
(*e*) E only

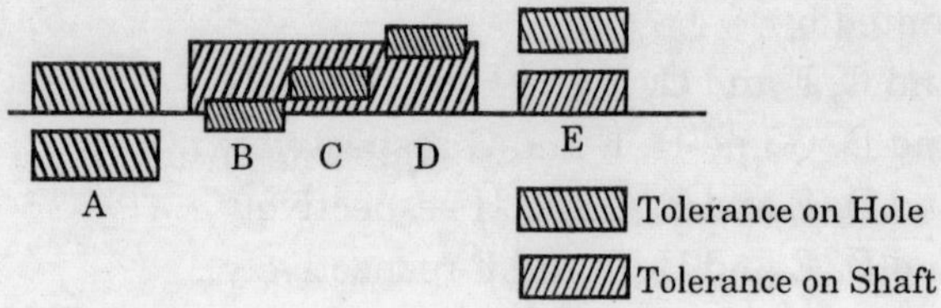

Fig. 23

221. In the designation 50 H8/d7, match the two parts below in respect of meaning of various symbols used
1. 50 A. Toreance grade
2. H B. Type of fit
3. d C. Basic size
4. 8 and 7 D. Nature of fit for hole
(*a*) 1—D, 2—C, 3—A, 4—B (*b*) 1—C, 2—D, 3—B, 4—A
(*c*) 1—C, 2—D, 3—A, 4—B (*d*) 1—D, 2—C, 3—B, 4—A

222. Numerous fits can be selected using a suitable combination of
(*a*) fundamental deviations on shafts (*b*) fundamental deviations on hole
(*c*) fundamental tolerances on shafts and holes
(*d*) all of above

223. Wear allowance is provided on
(*a*) 'Go' gauge (*b*) 'No Go' gauge
(*c*) both 'Go' and 'No Go' gauges (*d*) Neither 'Go' nor 'No Go' gauge

224. Accurate centering of work mounted in an independent chuck can be determined by using a
(*a*) centre gauge (*b*) height gauge
(*c*) dial indicator (*d*) surface gauge
(*e*) micrometer

225. In limits and fits system, basic shaft system is one whose
(*a*) lower deviation is zero (*b*) upper deviation is zero
(*c*) minimum clearance is zero (*d*) maximum clearance is zero
(*e*) standard tolerance is zero

226. Pick wrong pair
(*a*) Shaft – refers to specification of all outer elements of the part
(*b*) Hole – refers to specification of all inner elements regardless of shape
(*c*) Deviation – Intentional difference between maximum material limits of mating parts
(*d*) Tolerance grade – The degree of accuracy of manufacturing

227. Difference between upper deviation and lower deviation is known as
(*a*) allowance (*b*) tolerance
(*c*) tolerance grade (*d*) fundamental deviation

228. In case of interference fit, difference between maximum size of hole and minimum size of shaft before assembly is called
   (*a*) maximum interference (*b*) minimum clearance
   (*c*) maximum clearance (*d*) minimum interference
229. Hole 'A' and shoft 'a' have the largest
   (*a*) fundamental deviation (*b*) tolerance zone
   (*c*) tolerance class (*d*) upper deviation
230. Fundamental tolerance unit is function of
   (*a*) diameter of part (*b*) tolerance grade
   (*c*) fundamental deviation (*d*) tolerance zero
231. H_{11}/a_{11} will result in
   (*a*) clearance fit (*b*) transition fit
   (*c*) force fit (*d*) interference fit
232. R7/h6 will result in
   (*a*) clearance fit (*b*) transition fit
   (*c*) force fit (*d*) interference fit
233. H8/j7 will result in
   (*a*) clearance (*b*) transition fit
   (*c*) force fit (*d*) interference fit
234. Equivalent shaft basis fit for hole basis H8-C9 clearance fit will be
   (*a*) C9-h8 (*b*) H8-c7
   (*c*) J8-h7 (*d*) G9-h8
235. Tolerance on reference gauges is ______% of work tolerance, and on inspection gauges it is ______% of work tolerance
   (*a*) 5%, 10% respectively (*b*) 10%, 5% respectively
   (*c*) 10%, 10% respectively (*d*) 5%, 5% respectively

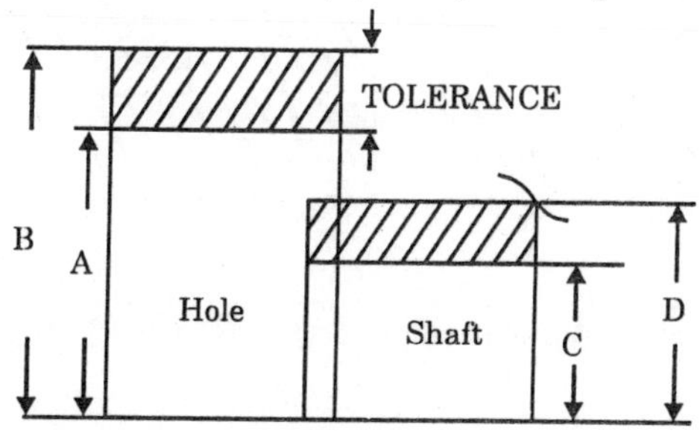

Fig. 24

236. In above Fig. 24 maximum metal limit and least metal limit are expressed by
   (*a*) A and D, and B and C
   (*b*) A and B, and C and D
   (*c*) A and C, B and D
   (*d*) A and B only
237. In Fig. 25 below, minimum clearance, maximum clearance, minimum interference, and maximum interference are represented by

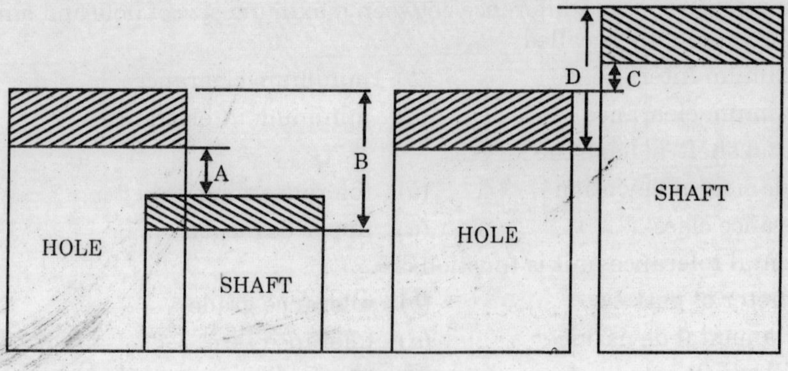

Fig. 25

(*a*) A, C, B and D respectively
(*b*) C, D, A and B respectively
(*c*) A, B, C and D respectively
(*d*) B, A, D and C respectively

238. Which type of fit is represented by H7/g7, H7/s6, H7/k6 respectively
(*a*) clearance (C), interference (I), and transition (T)
(*b*) T, I and C
(*c*) I, T, and C (*d*) C, T, and I

239. What type of fit is represented in figure 26
(*a*) clearance fit (*b*) transition fit
(*c*) interference fit (*d*) all of above

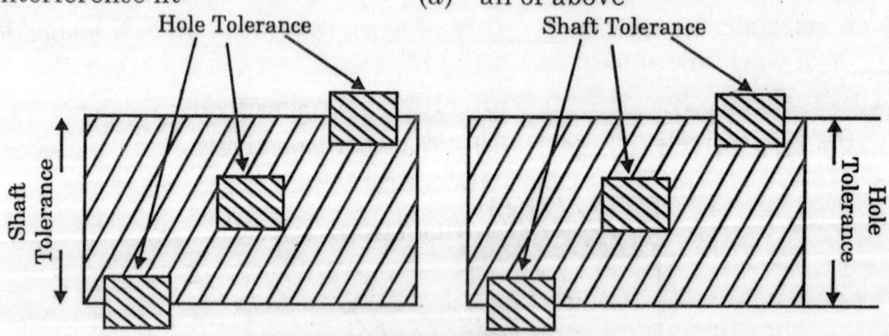

Fig. 26

240. Standard tolerance unit (*i*) for tolerance grades 1T6 and 1T16 are
(*a*) 10 *i* and 1000 *i*
(*b*) 16 *i* and 640 *i*
(*c*) 25 *i* and 400 *i*
(*d*) 40 *i* and 250 *i*

241. Which is GO plug gauge and which is NO GO plug gauge in Fig. 27
(*a*) A and B (*b*) B and A
(*c*) both A and B (*d*) neither A, or B

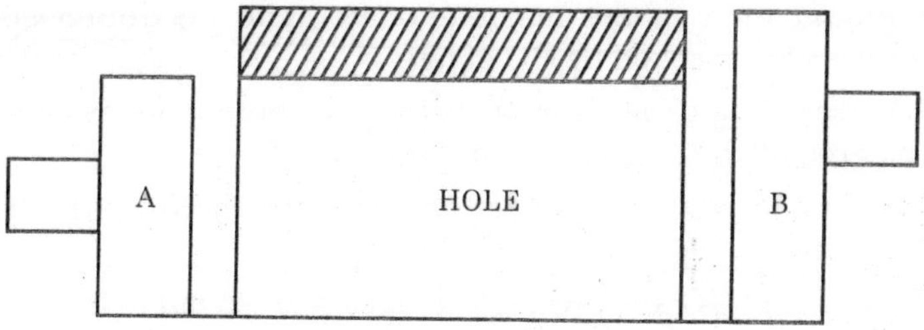

Fig. 27

242. Pick up correct statement with regard to designing gauges
   (*a*) The form of GO gauges should be a replica of the form of the mating part.
   (*b*) GO gauges enable serveral related dimensions to be checked simultaneously and these must always be put into conditions of maximum impassibility.
   (*c*) NOT GO gauges check a single element of feature at a time and must always be put into conditions of maximum passability
   (*d*) All of above statements are correct

243. Which statement is not correct about the material used to manufacture gauges
   (*a*) The material used in the manufacture of gauges should be hard and wear resistant for a prolonged life.
   (*b*) It should be capable of maintaining dimensional stability and form and be corrosion resistant.
   (*c*) It should not be easily machinable, in order to maintain the required degree of accuracy.
   (*d*) It should have low coefficient of expansion in order to avoid temperature effect

244. Pick up false statement about materials for gauges
   (*a*) High-carbon steel is the most suitable and inexpensive material used for manufacturing gauges.
   (*b*) Chromium-plated gauges have hard surface and are very popular and extensively used for gauging.
   (*c*) Glass gauges are not very popular because they have very high coefficient of thermal expansion.
   (*d*) Elinvar (42% nickel alloy) is more stable than Invar (36% nickel alloy) and it has low coefficient of expansion.

245. Which of the following is not the angle measuring device
   (*a*) angle plate (*b*) sine bar
   (*c*) bevel protector (*d*) angle gauge
   (*e*) combination square

246. To check the diameter of a twist drill with a micrometer, the measurement must be taken across the
   (*a*) margins of the drill (*b*) flutes of the drill
   (*c*) cutting edges of the drill (*d*) lips of the drill
   (*e*) web of the drill

247. Pick out the wrong statement about gauges for internal threads
   (*a*) The GO screw plug gauge is made to the minimum metal limit and of full form and checks the virtual effective diameter. Any error in the pitch or flank of screw affects the effective diameter.

(b) The minimum limit of the major diameter and maximum size of the effective diameter are also checked by GO gauge.

(c) NOT GO screw plug gauge is made to the maximum effective diameter of the screw thread cleared at the root and crest

(d) The plain go and NOT GO gauges are used for checking the limits of the size of minor diameter

(e) The dimensions of GO and NOT GO gauges correspond to maximum and minimum minor diameters

248. Which is false statement relating to Taylor's principle for gauge.

(a) The NOT GO side of gauge should not enter or pass.

(b) There is no need to provide an allowance for wear in case of NOT GO gauge

(c) The GO side of the gauge should enter the hole or just pass over the shaft under the weight of the gauge without application of undue force.

(d) The GO gauges do not suffer wear on the measuring surface and this no wear allowance is provided for GO gauges.

249. Tolerances of NO GO gauges, tolerance for GO gauges and wear allowance on GO gauges are represented in Fig. 28 by

(a) A, B and C respectively (b) B, A and C respectively

(c) C, C, and A respectively (d) C, C and B respectively

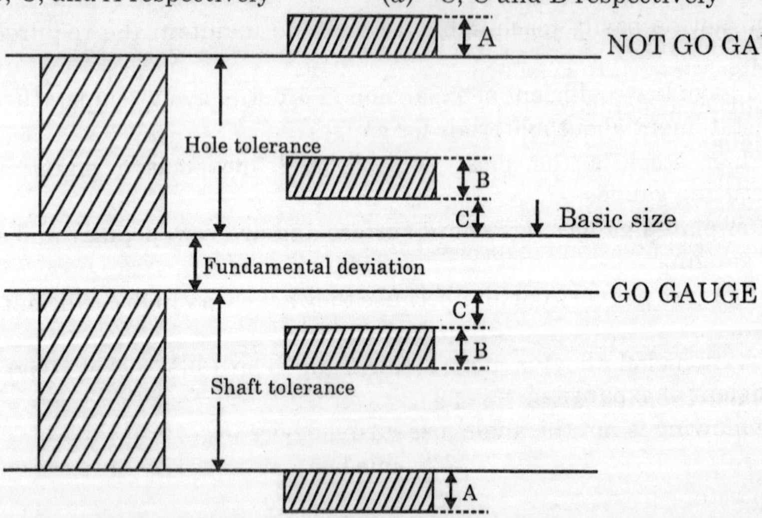

Fig. 28 shows disposition of tolerances and allowances on gauges

250. Expressing a dimension as $18.3^{+0.00}_{-0.02}$ mm is the case of

(a) unilateral tolerance (b) bilateral tolerance

(c) limiting dimensions (d) all of the above

(e) none of the above

251. In instrumentation terminology, a correction is

(a) an error

(b) the revision applied to the indicated value so that the final result obtained improves the worth of the result

(*c*) reading error
(*d*) range of error–degree of correctness
(*e*) lowest value of input which does not indicate the result

252. Many external comparators have anvils or work tables which are grooved. The purpose of this is to
(*a*) facilitate supporting of work
(*b*) provide three point support to the work
(*c*) Not to pass on inaccuracy of surface to the measurement
(*d*) trap any dirt on the table so that it does not interfere with the measurement
(*e*) avoid sticking of standards on the table.

253. Johansson microcator is a type of
(*a*) mechanical comparator
(*b*) electrical comparator
(*c*) pneumatic comparator
(*d*) optical comparator

254. Which is not correct about mechanical comparators
(*a*) These are compact and easy to handle
(*b*) parallax error is possible
(*c*) Due to low inertia, these are sensitive to vibrations
(*d*) Sigma comparator is mechanical comparator

255. Straight edge are used to measure
(*a*) straight length of parts (*b*) flatness
(*c*) parallelism (*d*) perpendicularity
(*e*) circularity

256. IS : 919 on limits and fits specifies following numbers of grades of fundamental tolerances, and fundamental deviations respectively
(*a*) 25, 18 (*b*) 25, 16
(*c*) 18, 22 (*d*) 20, 28
(*e*) 18, 25

257. For general use the measuring tip of a comparator should be
(*a*) flat (*b*) spherical
(*c*) conical (*d*) concave
(*e*) grooved

258. Which instrument is shown in Fig. 29 below
(*a*) Johansson Mikrokator
(*b*) Sigma Mechanical Comparator
(*c*) Dial indicator
(*d*) mechanical comparator

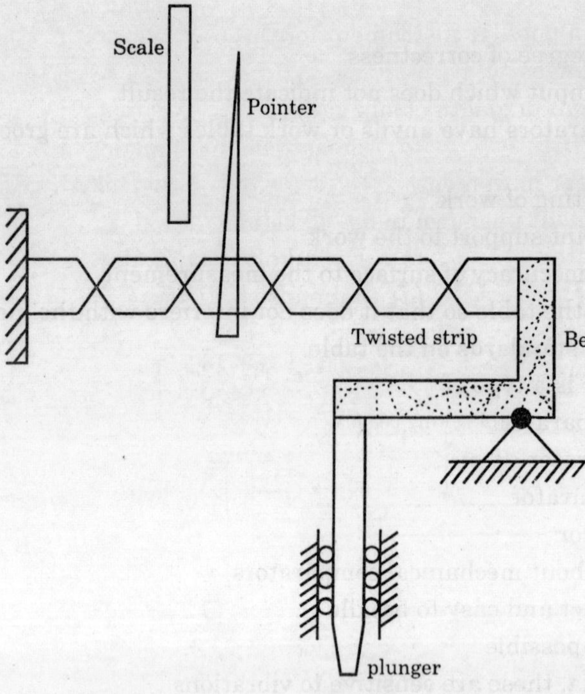

Fig. 29

259. Which comparator employs Abramson movement
   (*a*) Johansson Mikrokator (*b*) clinometer
   (*c*) Sigma comparator (*d*) zeiss ultra-optimeter

260. In which comparator, magnification is obtained in two stage
   (*a*) Johansson Mikrokator (*b*) clinometer
   (*c*) Sigma comparator (*d*) zeiss ultra-optimeter

261. Dial indicator usually has—contact point
   (*a*) flat (*b*) button
   (*c*) spherical (*d*) conical

262. Basic shaft and basic hole are those whose upper deviation and lower deviation respectively are
   (*a*) + ve, – ve (*b*) – ve, + ve
   (*c*) minimum, minimum (*d*) minimum, maximum
   (*e*) zero, zero

263. The standard tolerance unit is equal to
   (*a*) $0.45\ (\sqrt[3]{D})\ + 0.001\ D$ (*b*) $0.45\ (\sqrt{D}) + 0.001\ D$
   (*c*) $0.45\ (\sqrt[3]{D})\ + 0.01\ D$ (*d*) $0.45\ (\sqrt[4]{D})\ + 0.01\ D$
   (*e*) $0.45\ (\sqrt[3]{D})$

   where D = geometric mean of the lower and upper diameters of a particular diameter step.

264. Eden-Rolt comparator is a popular instrument for the
(*a*) calibration of slip gauges
(*b*) absolute measurement of length of slip gauges
(*c*) measurement of flatness (*d*) measurement of angles
(*e*) measurement of linear movement

265. Which measuring instrument in shown in fig. 30 below
(*a*) Mechanical comparator (*b*) Eden-Rolt Comparator
(*c*) Sigma comparator (*d*) Johannson Microcator

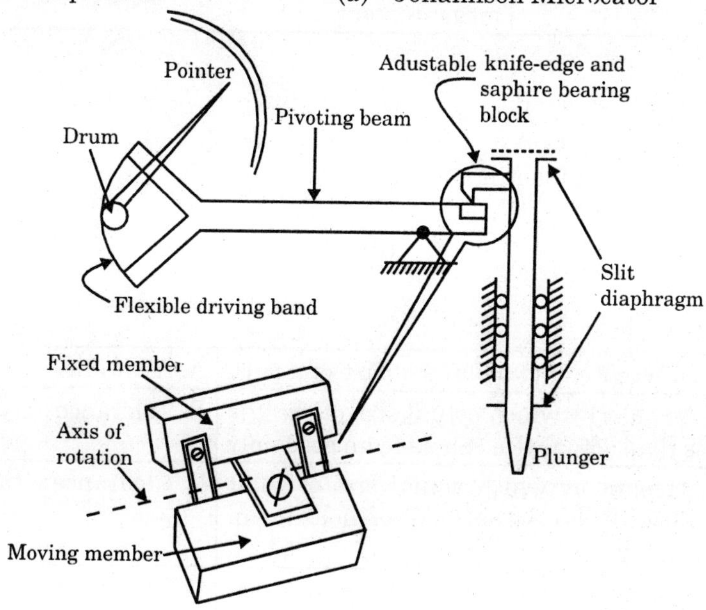

Fig. 30

266. It is desirable to handle the slip gauges with a cloth or chamois leather in order to
(*a*) avoid injury to hands (*b*) protect the surfaces of slip gauges
(*c*) insulate them from the heat of the hand
(*d*) ensure that the varnish applied on gauge does not come out
(*e*) none of the above

267. Slip gauge marked with letter 'P' is
(*a*) protected gauge (*b*) poorest gauge
(*c*) wear gauge (*d*) preferred gauge

268. Pick up wrong statement w.r.t. verniers calipers versus micro-meter screw gauges.
(*a*) Micrometer is more reliable than verier scale
(*b*) Range of vernier is less compared to micrometer
(*c*) Principle of alignment is followed in micrometer but not in vernier
(*d*) Micrometers have better least count than vernier

269. Which is best material for manufacture of surface plates
(*a*) cast iron (*b*) mild steel
(*c*) alloy steel (*d*) tungsten carbide

270. Match the two parts below pertaining to various gauges and their applications

Gauge	Application
1. Universal surface gauge	A. To secure variable heights above surface plate
2. Telescopic gauge	B. To measure the clearance
3. Planer gauge	C. To measure internal diameter of holes slots, grooves etc.
4. Feeler gauge	D. More versatile instrument used in non-precision measurement

Code:

(*a*) 1—D, 2—C, 3—A, 4—B (*b*) 1—C, 2—D, 3—A, 4—B
(*c*) 1—D, 2—C, 3—B, 4—A (*d*) 1—A, 2—B, 3—C, 4—D

271. For grade IT 7, value of tolerance is equal to

(*a*) 7 i (*b*) 10 i
(*c*) 16 i (*d*) 25 i
(*e*) 40 i.

272. Match the 2 parts below

1. Difference of two limits of size of a part	A. Allowance
2. Amount by which actual size of shaft is less than actual size of hole in an assembly	B. Tolerance
3. Amount by which actual size of shaft is larger than actual size of hole in an assembly	C. Clearance
4. A planned deviation between an exact dimension and a nominal dimension	D. Interference

(*a*) 1—B, 2—C, 3—D, 4—A (*b*) 1—C, 2—B, 3—D, 4—A
(*c*) 1—B, 2—C, 3—A, 4—D (*d*) 1—A, 2—C, 3—D, 4—B

273. Match the 2 parts below

1. ______ is provided to accommodate inherent inaccuracies of the manufacturing process	A. Limits
2. The relationship between the two mating parts before assembly	B. Allowance
3. Intentional difference between the maximum material limits	C. Fit
4. Every component can be manufactured to lie between two ______	D. Tolerance

(*a*) 1—D, 2—C, 3—B, 4—A (*b*) 1—A, 2—B, 3—C, 4—D
(*c*) 1—D, 2—C, 3—A, 4—B (*d*) 1—C, 2—D, 3—B, 4—A

274. Examine the 2 statements below

S–A: Go gauges are designed to check maximum metal limit and No Go gauges are designed for maximum possibility

S–B: Maximum metal limits corresponds to the lower limit of a hole and higher limit of the shaft and lower metal limit corresponds to higher limit of a hole and lower limit of the shaft

(*a*) S-A is true and S-B is false (*b*) S-A is false and S-B is true
(*c*) Both are true (*d*) Both are false

275. For defining length the standard generally followed is
(*a*) bar standard (*b*) end standard
(*c*) light wave standard (*d*) any of the above
(*e*) none of the above

276. Planer gauge is used for
(*a*) testing flatness of surfaces
(*b*) adding to utility of measurements on surface plate
(*c*) angular measurement (*d*) testing radius of corners
(*e*) testing thickness of small gaps

277. Which is not correct about LVDT
(*a*) Ideally suited for dynamic measurements
(*b*) Highly sensitive with good magnification
(*c*) Insensitive to temperature changes and is economical also
(*d*) Can't be overloaded mechanically as there is no elastic member

278. Which is not correct about requirements of comparators
(*a*) Provide high accuracy, precision, resolution and amplification
(*b*) Be suitable for rough and tough environment and be versatile also
(*c*) Scale should be linear and provision for compensation for temperature variation be incorporated.
(*d*) all of above are correct

279. A comparator works on
(*a*) interchange method (*b*) direct measurement
(*c*) displacement method (*d*) magnification method

280. IS specifications specify vernier calipers as type A, B and C. This classification is based on
(*a*) accuracy (*b*) least count
(*c*) range
(*d*) internal or external measurement and for marking purpose
(*e*) type of graduations

281. Which is not the feature of pneumatic comparator
(*a*) non–contact inspection with high range of amplification
(*b*) minimum gauging force
(*c*) suitable for attribute inspection only
(*d*) suited for varied inspections like length, circularity cylindricity, etc

282. What is possible with solex pneumatic comparator
(*a*) High amplification (upto 50,000) (*b*) high degree of resolution
(*c*) sensing variation in dimension upto micron level
(*d*) All of above

283. Which is best comparator for inspecting multiple dimensions of a part in a single setting
(*a*) electronic comparator (*b*) optical comparator
(*c*) pneumatic comparator (*d*) ultra-optimeter

284. Pick up false statement about comparators
(*a*) A comparator gives dimensional differences in relation to master setting
(*b*) A comparator works on principle of relative measurement
(*c*) Mechanical comparators are able to achieve exceptionally high magnification
(*d*) optical comparators have very high precision because of lesser moving parts

285. Sigma electronic comparator is extensively used because it has
(*a*) high accuracy and repeatability (*b*) setting tolerances on line
(*c*) ease of integration with computer aided devices
(*d*) no moving part at all

286. Which type of comparator is best suited for dynamic measurement due to its high speed of response
(*a*) pneumatic comparator (*b*) optical comparator
(*c*) electronic comparator (*d*) electro-mechanical comparator

287. The cross-section of straight edges upto 180 mm length is
(*a*) rectangular (*b*) circular
(*c*) I-section (*d*) elliptical
(*e*) L-shape

288. Optical micrometer is used to
(*a*) measure small linear displacements (*b*) measure surface profiles
(*c*) measure surface roughness
(*d*) set very small displacement by rotating the glass block through relatively large angles
(*e*) check parallelism

289. Airy points of support are
(*a*) 0.577 L apart (*b*) 0.554 L apart
(*c*) 0.5 L apart (*d*) 0.58 L apart
(*e*) 0.612 L apart
where L = length of bar

290. The maximum number of faces in precision polygons can be
(*a*) 6 (*b*) 8
(*c*) 12 (*d*) 20
(*e*) 72.

291. In precision polygon, a central hole and small holes are drilled through the thickness
(*a*) for mounting purposes (*b*) to achieve high accuracy
(*c*) for ease of manufacture (*d*) to make them light
(*e*) for decoration

292. Precision polygons are calibrated from first principles using
(*a*) one autocollimator (*b*) two autocollimators
(*c*) three autocollimators (*d*) two precision spirit levels
(*e*) angle gauges.

293. The fact that how closely the instrument reading follows the measured variables is termed as
(*a*) fidelity (*b*) accuracy
(*c*) threshold sensitivity (*d*) precision
(*e*) hysteresis

294. Slip gauge are usually calibrated using
(*a*) collimator (*b*) interferometer
(*c*) universal measuring machine (*d*) optical projector

295. In calibration of slip gauges using interferometer,
(*a*) the end of slip gauge is seen one after olter
(*b*) both ends of slip gauge are checked simultaneously
(*c*) any one of above
(*d*) none of above

296. The accuracy in calibration of slip gauges with interferometer is of the order of
(*a*) 0.02 mm (*b*) 0.002 mm
(*c*) 0.0002 mm (*d*) 0.00002 mm

297. Zeiss interferometer used for measurement of slip gauges uses
(*a*) one wavelength of cadmium light
(*b*) two wavelengths of cadmium light
(*c*) three wavelengths of cadmium light
(*d*) four wavelengths of cadmium light

298. Which of the following methods is not used for testing straightness
(*a*) spirit level method (*b*) autocollimator
(*c*) interference method (*d*) beam comparator
(*e*) laser beam

299. The standard magnification used to inspect profile gauges, using optical projector is
(*a*) 10 (*b*) 25
(*c*) 150 (*d*) 100

300. The height of a component can be measured by
(*a*) depth micrometer (*b*) bench micrometer
(*c*) comparators (*d*) any one of above

299. Taper ring gauges of size below diameter of 20 mm are best measured by
(*a*) inserting precision balls of different sizes in the gauge
(*b*) microscope
(*c*) optical projector (*d*) comparator

300. Optical square is
(*a*) Engineer's square having stock and blade set at 90°
(*b*) a constant deviation prism having the angle of deviation betwen the incident ray and reflected ray, equal to 90°
(*c*) a constant deviation prism having the angle of deviation between the incident ray and reflected ray, equal to 45°
(*d*) used to produce interfence fringes (*e*) used for angular measurement

301. In a sine bar the standard length is measured from
(*a*) edge to edge
(*b*) between inner circumference of two rollers
(*c*) between outer circumference of two rollers
(*d*) between the centres of two rollers
(*e*) none of the above

302. Electronic level instrument is a replacement for
(*a*) vernier depth gauge
(*b*) microscope
(*c*) auto-collimator
(*d*) angle dekkor
(*e*) spirit level

303. Match the 2 parts below regarding maximum permissible errors in various measuring devices

Measuring Device	Permissible Error
1. Wear of centres	A. 0.002 mm
2. Straight movement of measuring table in horizontal plane	B. 0.06 mm
3. Class '1' micrometer	C. 0.01 mm
4. Perpendicular motion of ocular microscope column w.r.t. work stage over 100 mm movement	D. 0.004 mm

(*a*) 1—C, 2—A, 3—D, 4—B
(*b*) 1—A, 2—C, 3—B, 4—D
(*c*) 1—C, 2—A, 3—B, 4—D
(*d*) 1—C, 2—A, 3—B, 4—D

304. Least count of clinometer used to check reading of column rotation for setting of helix angle in universal micrometer is
(*a*) 0.5′
(*b*) 1′
(*c*) 2′
(*d*) 2.5′

305. Pick up correct statement:
Optical measuring instruments require
(*a*) no preventive maintenance
(*b*) periodical calibration
(*c*) no care
(*d*) daily calibration before each use

306. Polygons in metrology are concerned with
(*a*) method of circular dividing
(*b*) testing of parallelism
(*c*) testing of circularity
(*d*) interferrometry measurements
(*e*) linear measurement

307. A sine bar should not be used for measurement of angles greater than
(*a*) 15°
(*b*) 20°
(*c*) 30°
(*d*) 45°

308. With longer sine bar, the measurement accuracy is
(*a*) better
(*b*) less
(*c*) unaffected
(*d*) accuracy depends on angle to be measured

309. Angle gauges are available in sets of
(*a*) 6, 11 and 16
(*b*) 8, 14 and 30
(*c*) 4, 8, and 12
(*d*) 12, 22, and 32

310. A set of 16 angle gauges can form angles between 0° of 90° in intervals of
(*a*) 1″ (*b*) 5″
(*c*) 1′ (*d*) 1°

311. The long straight edges are supported at two points for minimum deflection at centre. The distance between supports compared to the length of straight edge should be
(*a*) 0.5 (*b*) 0.554
(*c*) 0.577 (*d*) 0.6
(*e*) 0.677

312. Bevel protractor is used for
(*a*) angular measurements
(*b*) linear measurements
(*c*) height measurements
(*d*) flatness measurement
(*e*) parallelism measurement

313. Clinometer is related with
(*a*) Engineer's parallels
(*b*) angle gauges
(*c*) spirit level
(*d*) bevel protractor
(*e*) tolerance measurement

314. Which measuring instrument is shown in Fig. 31 below
(*a*) sine bar (*b*) angle dekkor
(*c*) sine centre (*d*) clinometer

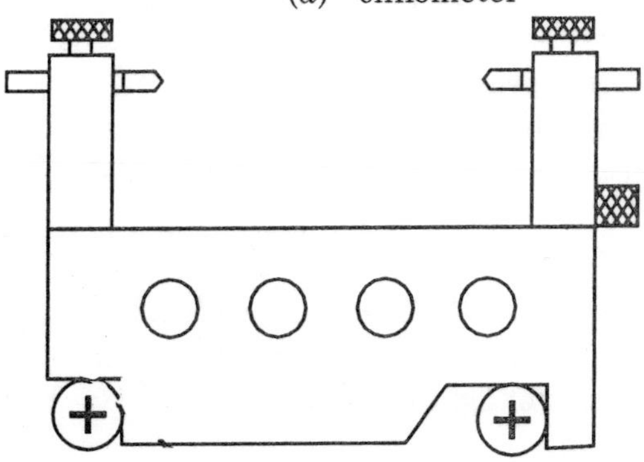

Fig. 31

315. The set up shown in Fig. 32 below for measurement of included angle of taper plug gauge uses following equipments
(*a*) sine centre
(*b*) sine bars
(*c*) surface plate to provide datum surface
(*d*) all of above

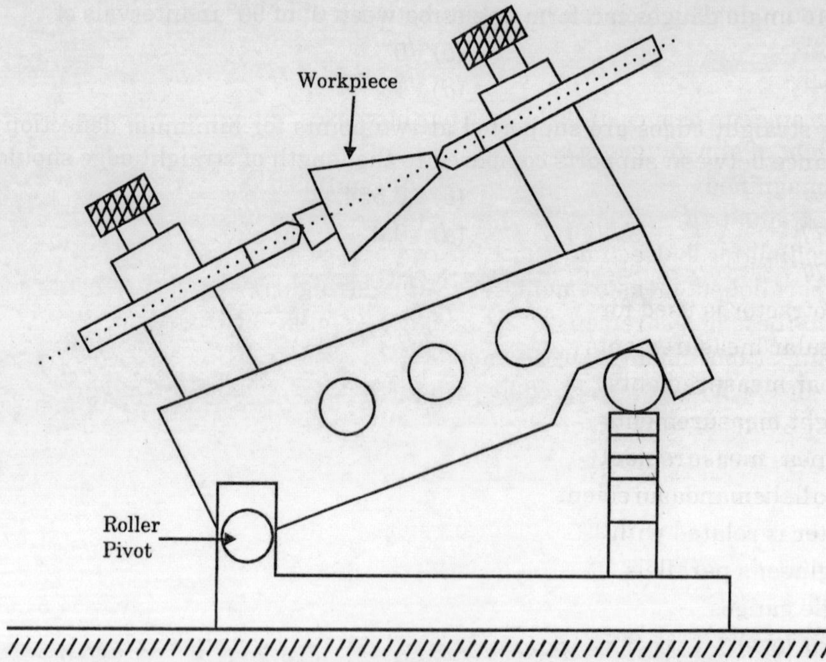

Fig. 32

316. Pick up the correct statement. A comparator
(*a*) needs to be calibrated (*b*) need not be calibrated
(*c*) contains a calibrated scale
(*d*) is highly accurate over its complete measuring range
(*e*) is best suited for measurement of absolute dimensions

317. Sensitivity of spirit level depends upon
(*a*) radius of curvature of bubble tube (R)
(*b*) base length of level (L) (*c*) both (R) and (L)
(*d*) none of above

318. Which of the following is the essential condition for interferometry measurement
(*a*) an air gap (wedge) of varying thickness must exist between the two surfaces
(*b*) an optical flat is required (*c*) the work surface must be reflective
(*d*) monochromatic source of light is required
(*e*) all of the above

319. Angle dekkor (a comparator to measure the change in angular position of the reflector in two planes) is a small variation of
(*a*) clinometer (*b*) sine bar
(*c*) angle gauges (*d*) autocollimator

320. Which principles govern the application of optical instruments in metrology
(*a*) magnification, accuracy, alignment, and interferometry
(*b*) laser source, objective lens, plane reflector, and prism
(*c*) laser collimator, alignment and monitoring, collimator
(*d*) angle comparisons, verification, high resolution, maximum measuring range.

321. Maximum angle which can be measured using sine bars is
(*a*) 15° (*b*) 30°
(*c*) 45° (*d*) 60°

322. As the angle to be measured by sine bar increases, the error in measurement due to error in height of slip gauges gets
(*a*) magnified (*b*) nullified
(*c*) attenuated (*d*) remains unaffected

323. Autocollimator is used for
(*a*) parallelism measurement (*b*) straightness measurement
(*c*) flatness measurement (*d*) angular measurement
(*e*) linear movement measurement

324. Angle of V-groove is measured by using
(*a*) two pins of same size
(*b*) two discs of same size
(*c*) one pin and one disc
(*d*) two pins or two discs of different size

325. The angle of V-groove of threads can be measured by two discs of different diameters, touching the sloping sides of thread and the difference in heights of discs is measured by using
(*a*) slip gauges (*b*) height gauge
(*c*) dial indicator (*d*) comparator

326. The internal taper in a blind hole is measured using
(*a*) one pin (*b*) single ball
(*c*) two pins (*d*) two balls

327. In measurement of internal taper, after placing the two balls, gap between balls is filled with combination of.
(*a*) slip gauges (*b*) plug gauges
(*c*) pins (*d*) discs

328. The correct relationship for pneumatic comparator in Fig. 33 is represented by
(*a*) Curve A (*b*) Curve B
(*c*) Curve C (*d*) Curve D
(*e*) none of the above

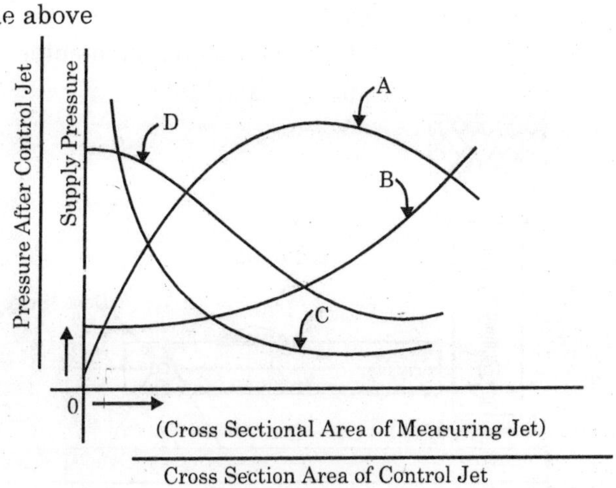

Fig. 33

329. Pick up false statement about angle dekkor
  (*a*) It is a type of autocollimator instrument
  (*b*) It is highly sensitive instrument
  (*c*) It is externally useful for a wide variety of short range angular measurements
  (*d*) The whole optical system is enclosed in a tube which is mounted on an adjustable table.

330. Fig. 34 shows
  (*a*) pneumatic comparator
  (*b*) flatness tester
  (*c*) fluid gauge or hydraulic comparator
  (*d*) level gauge
  (*e*) manometer

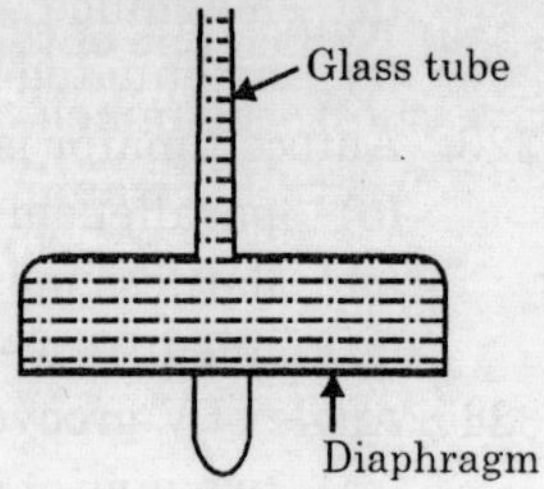

Fig. 34

331. Which instrument is shown in Fig. 35 below
  (*a*) rotary table (*b*) vernier clinometer
  (*c*) sine table (*d*) angle dekkor

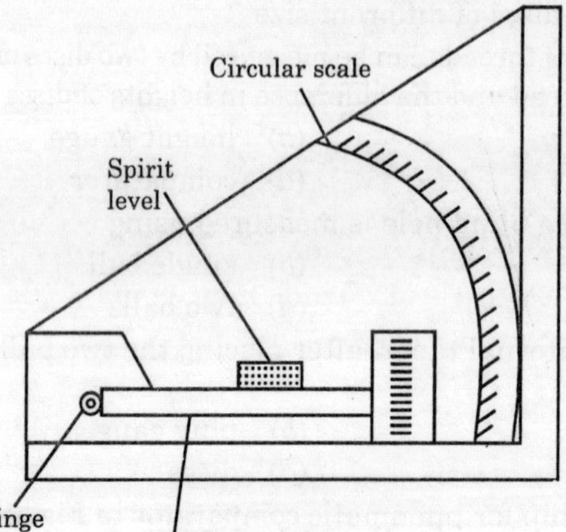

Fig. 35

332. Which instrument is shown in Fig. 36 below
  (*a*) Vernier clinometer (*b*) micrometer clinometer
  (*c*) sine table (*d*) rotary table

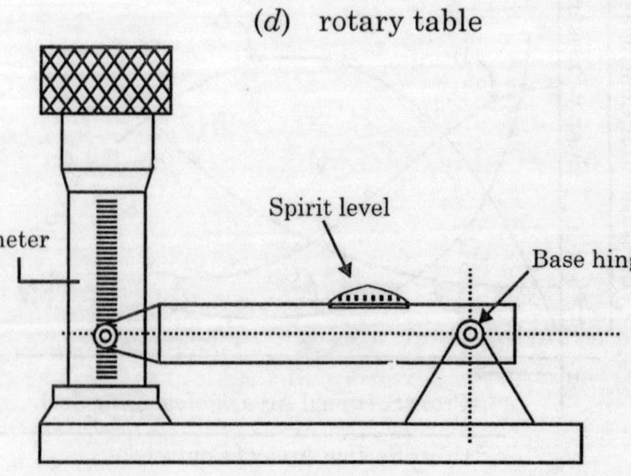

Fig. 36

333. If the fringe pattern on an optical flat is as shown in Fig. 37, the surface under flat would be
(a) Perfectly flat (b) Cylindrial
(c) Spherical (d) Ridge or Valley
(e) Convex

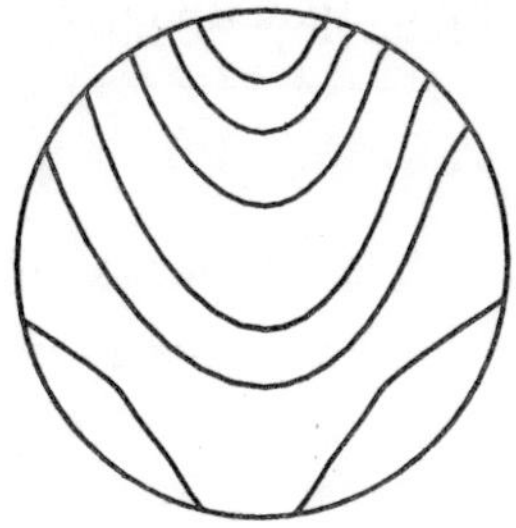

Fig. 37

334. Which one of the following is an angle measuring device
(a) trammel
(b) hermaphrodite caliper
(c) divider (d) angle iron
(e) sine bar

335. Filar microscopes
(a) have no reticles (b) have fixed reticles
(c) have moving reticles
(d) can be moved relative to the work by means of a fine-pitch lead screw
(e) use a scale on the side of the optical tube to give a measure of the focusing position

336. A collimator is simply a
(a) source of a bundle of parallel light rays
(b) source of point light
(c) sort of alignment telescope (d) standard for flatness
(e) device used in interofermetric measurements

337. If x be half the included angle of thread and p its pitch, then best size wire's diameter for measurement of effective diameter of thread is
(a) p/2 sec x (b) p sec x/2
(c) p cos x/2 (d) p/2 cos x/2
(e) p/2 sec x/2

338. Angle Dekkor is another type of
(a) auto-collimator (b) optical square
(c) clinometer (d) angle guage
(e) electronic level

339. The angles of angle guages in the degrees series are
(a) 1, 3, 9, 25, 42 (b) 1, 3, 9, 27, 81
(c) 1, 3, 9, 27, 41 (d) 1, 5, 10, 25, 45
(e) 2, 3, 5, 27, 41

340. Wickman gauge of used for inspection of
(a) holes (b) shafts
(c) gears (d) tapers
(e) screw threads

341. Which one of the following is not an angle measuring device
(a) bevel protector (b) sine bar
(c) combination square (d) angle iron
(e) angle gauge blocks

342. Fig. 38 shows a

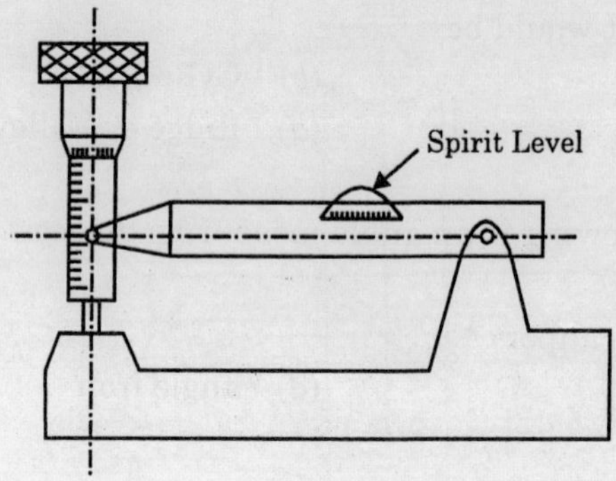

Fig. 38

(*a*) flatness tester (*b*) micrometer clinometer
(*c*) sine centre (*d*) pendulum clinometer
(*e*) taper tester

343. The reflector combined with autocollimator can be used for checking
(*a*) alignement (*b*) parallelism
(*c*) circularity (*d*) distance between two far off points
(*e*) surface finish

344. As per laser Doppler principle, light when coming in contact with a moving object experiences change in its frequency depending upon
(*a*) velocity of moving object (*b*) direction of moving object
(*c*) both velocity of direction of moving object
(*d*) none of above is correct

345. Some substances generate voltage when they are subjected to mechanical forces or stresses along specific planes. Such substances are known as
(*a*) thermo-electric (*b*) magneto-electric
(*c*) piezo-electric (*d*) photo-electric
(*e*) radio-active

346. Tomlinson recorder is associated with measurement of
(*a*) surface flaws (*b*) surface perpendicularity
(*c*) surface finish (*d*) surface curvature
(*e*) surface flatness

347. Which of the following methods is not concerned with surface finish measurement
(*a*) spectrophotometry method (*b*) ultrasonic method
(*c*) field emission method (*d*) critical angle of attack method
(*e*) Talysurf principle

348. A spirit level is required to have 2 mm of bubble movement for 2 second inclination. The radius of tube should be around
(*a*) 2 m (*b*) 20 m
(*c*) 80 m (*d*) 120 m
(*e*) 200 m

349. Fig. 39 shows the following type of interferometer

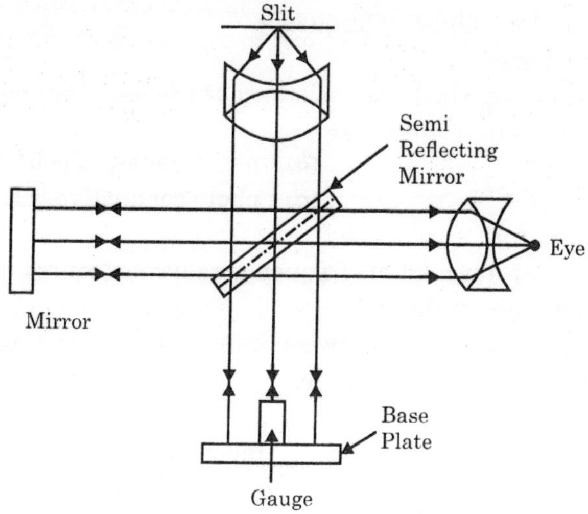

Fig. 39

(*a*) NPL gauge interferometer (*b*) Michelson interferometer
(*c*) Twyman-Green modified Michelson interferometer
(*d*) Farbry-Perot interferometer (*e*) Kosters gauge interferometer

350. Pick up false statement about gratings
(*a*) Gratings are scales with continuously repeating pattern of lines or groves that are closely spaced (line spacing may be of the order of 50–1000 per millimetre).
(*b*) Ronchi rulings consist of strips that are alternatively opaque and transmitting, with a spacing of 300–1000 per millimetre.
(*c*) Phase gratings consist of triangularly shaped, contiguous grooves similar to spectroscopic diffraction gratings.
(*d*) Reticles are special type of gratings.

351. The only natural material producing a spectral line (6440° A red) almost completely symmetrical is
(*a*) Cadmium (*b*) Mercury
(*c*) Krypton (*d*) Helium
(*e*) Neon

352. For which type of surface, the fringes are seen to move towards the centre
(*a*) concave (*b*) convex
(*c*) tapered (*d*) flat

353. Which instrument measures roughness directly
(*a*) micro interferometer (*b*) profilometer
(*c*) wallace surface dynamometer (*d*) all of above

354. Which of following is responsible for first order surface irregularity.
(*a*) lack of sraightness (*b*) lack of flatness
(*c*) lack of rigidity (*d*) vibrations

355. Which type of fringe is formed when there is path difference of $\lambda/2$ in interference pattern

(a) dark fringe (b) bright fringe
(c) dark or bright depending on frequency
(d) overlapping frings

356. A photo-electric device in which the resistance of the metal changes directly proportional to the light striking on it, is known as
(a) photo-cell (b) photo-emission cell
(c) photo-voltaic cell (d) photo conductive cell
(e) none of the above

357. Why laser source must have double frequency radiation source
(a) for clarity and high brightness
(b) for detecting the phase shift between interfering measuring and reference beams.
(c) for formation of interference fringes
(d) all of above

358. Laser transducers can measure upto ___ independent axes of displacement with one transducer head
(a) two (b) three
(c) four (d) six

359. Universal surface gauge is used for
(a) checking straightness (b) checking flatness
(c) checking parallelism (d) layout work and inspection
(e) checking the surface finish.

360. Laser has sine wave and its wavelength is constant for a given
(a) colour (b) pressure and temperature
(c) set up (d) under all conditions

361. A laser transduer measures
(a) absolute measurement of length
(b) the relative change of position between interferometer and retro-reflector
(c) both of above
(d) none of above

362. Lasers are employed in the bandwidth of
(a) 4-5 μm (b) 20-40 μm
(c) 50-100 μm (d) 100-200 μm

363. Laser light has advantage that it is
(a) monochromatic (b) coherent
(c) naturally collimated (d) all of above

364. Doppler effect is observed in instruments based on
(a) probe type sensors (b) laser principles
(c) electro magnetic phenomena (d) pneumatic comparison principles

365. Pick up wrong statement
(a) When two light waves interact with each other, the wave effect leads to interference of light
(b) When light from a monochromatic light source falls on an optical flat oriented at a very small angle with respect to a flat reflecting surface, a band of alternate light and dark patches (fringes) are produced.

(c) An optical square is a pentagonal prism used for turning the line of sight by 90° from its original path regardless of the angle at which the incident beam strikes the face of the prism.

(d) Tool maker's microscope is a multifunctional device used for interferometric measurements.

366. Pick up false statement

(a) In scale, rulings are spaced relatively far apart but in 'grating' rulings are more closely spaced.

(b) Gratings cannot be either generated or read manually and these require special read-out systems, usually photoelectric.

(c) Laser-based interferometers are increasingly becoming popular in metrology applications because of lower cost.

(d) NPL flatness interferometer is used for checking flatness between gauge surfaces and the Pitter-NPL gauge interferometer is used for determining the actual lengths of slip gauges.

367. Which of the following is not the correct method of specifying numerical value of surface roughness

(a) Centre-line average (CLA) value (b) mean-line and envelop line systems

(c) r.m.s. value (d) peak-to-valley height

(e) roughness characteristic like depth of smoothness, mean depth, etc.

368. Which pair is not correctly matched

(a) Lay–The direction of predominant surface pattern produced based on the machining operation used to produce it.

(b) Form–Consisting of surface irregularities resulting from the various machining processes.

(c) Roughness Width– The distance parallel to the nominal surface between successive peaks or ridges which constitute the predominate pattern of the roughness.

(d) Waviness–The widely spaced component of the surface texture, as a result of workpiece or tool deflection during machining, vibrations or tool runout.

369. Fig. 40 shows the roughness characteristics of a surface. Distances A and B represent

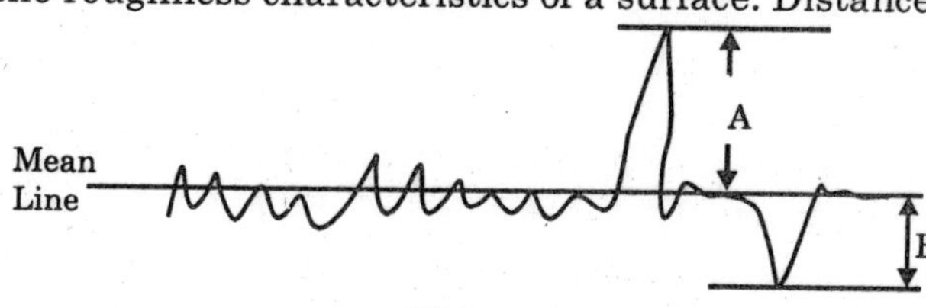

Fig. 40

(a) CLA values (b) peak to valley height

(c) depth of smoothness and mean depth respectively

(d) mean depth and depth of smoothness respectively

(e) fullness grade and emptiness grade respectively

370. Effective profile in definition of surface texture is

(a) real contour of a surface

(b) imaginary surface having same roughness as specimen

(*c*) obtained after removing repetitive irregularities

(*d*) imaginary profile to enable measurement of roughness.

371. Roughness is referred to as _________ texture, and waviness is called _________ texture.

(*a*) primary, secondary (*b*) secondary, primary

(*c*) error of form, error of geometry (*d*) error of geometry, error of form

372. Surface texture to a large extent is dependent on

(*a*) manufacturing method (*b*) process capability of machine

(*c*) wear and tear of machine tool

(*d*) material composition and skill of operator

373. Which is the prevalent standard for measuring surface roughness

(*a*) Ten-point height average value (*b*) RMS value

(*c*) Peck to valley height value (*d*) Ra (centre line average value)

374. According to Taylor's Principle, No Go gauge checks

(*a*) only one feature at a time (*b*) only important dimensions at a time

(*c*) all the dimensions at a time

(*d*) only the related dimensions at a time

(*e*) as many dimensions as possible at a time.

375. The primary texture or roughness or micro-errors on surface results due to

(*a*) normal action of the tool in production process

(*b*) vibrations and non-uniformity of cutting process

(*c*) flaws in material (*d*) dominant direction of tool marks (lay)

(*e*) all of the above

376. In style instrument used for surface finish, if f is the frequency of stylus movement, V traverse speed, then surface wavelength is equal to

(*a*) f/V (*b*) fV

(*c*) V/f (*d*) none of above

377. The peaks and valleys of surface irregularities are called

(*a*) roughness (*b*) texture

(*c*) lay (*d*) asperities

378. The degrees of surface roughness can be assessed by following numerical value.

(*a*) 10-point height average (*b*) RMS value

(*c*) centre line average height (*d*) all of above

379. Which instrument enable a quick-fix means of conducting initial investigation of a surface quality

(*a*) Tomlinson surface meter (*b*) Taylor-Hobson Talysurf

(*c*) both of above (*d*) none of above

380. The secondary texture or waviness, or macro-error on surfaces results due to

(*a*) normal action of the tool in production process

(*b*) vibrations and non-uniformity of cutting process

(*c*) flaws in material (*d*) dominant direction of tool marks (lay)

(*e*) all the above

381. The function of a transducer is
   (*a*) to amplify input signal
   (*b*) to modify the input signal
   (*c*) to convert the primary signal into a more useful quantity usually an electrical impulse
   (*d*) to codify the input signal
   (*e*) to decodify the input signal

382. Which method is used to test the straightness
   (*a*) wedge method
   (*b*) interference method
   (*c*) dial indicator
   (*d*) interferometer

383. Allowance represents
   (*a*) minimum interference for interference fits
   (*b*) minimum clearance (and positive) in clearance fits
   (*c*) maximum clearance in clearance fits
   (*d*) maximum clearance (and negative) in clearance fit.

384. Match the 2 parts regarding surface roughness

1.	real contour	A.	effective profile
2.	primary texture	B.	irregularities at one place
3.	secondary texture	C.	waviness
4.	flaw	D.	roughness

   (*a*) 1—A, 2—D, 3—C, 4—B
   (*b*) 1—A, 2—C, 3—D, 4—B
   (*c*) 1—A, 2—B, 3—D, 4—C
   (*d*) 1—D, 2—A, 3—B, 4—C

385. Pick up the correct statement in connection with surface finish.
   (*a*) Pitch of secondary texture is same as pitch of primary texture
   (*b*) Pitch of secondary texture is more than pitch of primary texture
   (*c*) Pitch of secondary texture is less than pitch of primary texture
   (*d*) Pitch of primary and secondary texture can't have any relationship
   (*e*) Pitch of primary and secondary texture is related mathematically

386. Pick up incorrect statement from following:
   (*a*) The direction of the predominant surface pattern, ordinarily determined by the production process used for manufacturing the components, is referred to as lay
   (*b*) Irregularities that occur in isolation or infrequently because of specific causes such as scratches, cracks, and blemishes are called surface texture
   (*c*) The average height from a mean line of all ordinates of the surface, regardless of sign, is the central line average value.
   (*d*) The datum created by a skid or shoe is the locus of its centre of curvature as it slides along the workpiece surface.

387. Match the two parts below pertaining to surface roughness

1.	Irregularities of small wavelength	A.	Effective profile
2.	Irregularities of large wavelength	B.	Primary texture
3.	Real contour of a surface obtained by using instrument	C.	CLA value
4.	Average values of the ordinates from the mean line, regardless of the arithmetic signs of the ordinates	D.	Secondary texture

Code,

(*a*) 1—B, 2—D, 3—C, 4—A (*b*) 1—B, 2—D, 3—A, 4—C

(*c*) 1—D, 2—B, 3—C, 4—A (*d*) 1—B, 2—A, 3—C, 4—D

388. Surface roughness is usually measured by average roughness (Ra). Other name given to Ra is

(*a*) Arithmetic average (AA) (*b*) Centre line average (CLA)

(*c*) Arithmetic mean deviation of the profile

(*d*) all of above

389. As per 1S:3073 which characteristics of surface texture are indicated on drawing

(*a*) Roughness value (Ra) (*b*) Sampling length or cut off length

(*c*) machining method, machining allowance, and direction of lay

(*d*) all of above

390. ISO recommends the use of grade for specifying the surface texture. How many number of grades are specified and what is the range of roughness value for them.

(*a*) 12, 0.025 μm to 50 μm (*b*) 8, 0.02 μm to 25 μm

(*c*) 6, 0.01 μm to 10 μm (*d*) 4, 0.01 μm to 5 μm

391. If graduations on beam of a vernier gauge are marked at every 1/2 mm and 10 divisions on vernier scale are on a distance of 9.5 mm, then least count is

(*a*) 0.1 mm (*b*) 0.05 mm

(*c*) 0.01 mm (*d*) 0.02 mm

(*e*) 0.005 mm

392. Scale sensitivity is defined as

(*a*) Ratio of a change in scale reading to corresponding change in pointer deflection

(*b*) Least reading of scale/range of scale

(*c*) Least reading of scale/unit measurable quantity

(*d*) Least count of scale/range of scale (*e*) none of the above

393. A three-lobed part if checked on 60° V-block would provide following magnification of the radial out of round characteristics

(*a*) 1 time (*b*) 2 times

(*c*) 3 times (*d*) 4 times

(*e*) 5 times

394. A five-lobed part, if gauged in the 60° V-block would produce the following magnification of the radial out-of-round characteristics

(*a*) 0 times (*b*) 1 times

(*c*) 2 times (*d*) 3 times

(*e*) 5 times

395. A beam comparator is used to measure

(*a*) straightness (*b*) flatness

(*c*) height and depth (*d*) circularity

396. A circle where the sum of areas inside this circle are equal to the sum of areas outside the circle is called

(*a*) minimum zone circle (*b*) minimum air cirumscribed circle

(*c*) maximum inscribed circle (*d*) least square reference circle

397. Concentrity is ________ times the eccentricity

(*a*) one (*b*) two

(*c*) three (*d*) two and half

398. Cylindricity of a component is described by _______ zone method of analysis

(*a*) minimum (*b*) maximum

(*c*) average (*d*) zero

399. Diametrical gauging for out-of roundness measurement is not sufficient because

(*a*) it can't sense even lobed parts

(*b*) radial changes always occur which can't be sensed by it

(*c*) it is difficult to find true centre (*d*) its readings can't be fed to computer

(*e*) it provides measurement at 2 points only.

400. A plug guage is used for measuring

(*a*) cylinders (*b*) cylindrical bores

(*c*) spherical holes (*d*) screw threads

(*e*) angles

401. For grade '1' micrometer screw gauge, the total error at 20°C should not exceed

(*a*) $\left(4+\frac{L}{100}\right)\mu m$ (*b*) $\left(10+\frac{L}{50}\right)\mu m$

(*c*) $\left(5+\frac{L}{50}\right)\mu m$ (*d*) $\left(4+\frac{L}{200}\right)\mu m$

(*e*) $\left(10+\frac{L}{20}\right)\mu m$

Where L = upper limit of the measuring range in mm.

402. Pick out the wrong statement about measuring out-of-roundness by V-block gauging

(*a*) 60 degree V-block provides triple magnification of out-of-round chracteristics of 3 lobed part

(*b*) A-five lobed part may be gauged in the 60° V-block and produce a zero reading, registering wrongly that the part is round

(*c*) for checking a five lobed part, 108° V is required

(*d*) The use of series of V blocks would not necessarily detect all types of out-of-roundness

(*e*) With a V-block, ovality and other even-numbered lobing are magnified

403. A master gauge is

(*a*) a new gauge (*b*) an international reference standard

(*c*) a standard gauge for checking accuracy of gauges used on shop floors

(*d*) a guage used by experienced technicians

(*e*) the most accurate gauge.

404. Gear tooth vernier is used to measure

(*a*) gear tooth profile (*b*) gear tooth thickness

(*c*) pitch line thickness of gear tooth

(*d*) module (*e*) addendum and dedendum

405. Moire fringes are observed when
(*a*) an optical flat is placed over smooth surface
(*b*) a microscope is used to observe surface texture
(*c*) index grating is moved over scale grating
(*d*) white light is diffused through a prism
(*e*) two monochromatic light beams have a phase difference of half the wave-length

406. The surface roughness on a drawing is represented by
(*a*) circles (*b*) squares
(*c*) zig-zag lines (*d*) triangles
(*e*) curves

407. Inside and outside diameters of a thin tube are to be measured. Only one of the following tools is to be selected. Which one would you choose.
(*a*) inside caliper (*b*) outside caliper
(*c*) trammel (*d*) odd-leg caliper
(*e*) vernier caliper

408. Fig.41 shows the principle of
(*a*) microscope (*b*) telescope
(*c*) optical clinometer (*d*) autocollimator
(*e*) projector

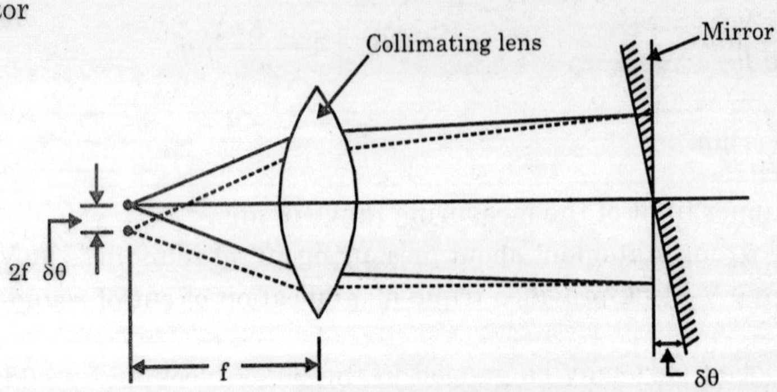

Fig. 41

409. Electronic level contains
(*a*) a pendulum (*b*) spirit level
(*c*) micrometer (*d*) clinometer
(*e*) sine bar

410. Keilpart gauge is used to measure
(*a*) surface roughness (*b*) taper
(*c*) inside diameter of a tool (*d*) gear eccentricity
(*e*) effective diameter of a screw thread

411. In interferometric methods, the path difference between one bright band and the next is varied by
(*a*) half wave length (*b*) two half wave lengths
(*c*) one quarter wavelength (*d*) two wavelengths
(*e*) none of the above

412. The 'best size wire' for measuring the effective diameter of threads is of diameter

(a) $\frac{p\sec\theta}{2}$ (p = pitch of thread, θ = semi-angle of thread)

(b) $\frac{p\cos\theta}{2}$

(c) $p \sec\theta$ (d) $\frac{p\sec\theta}{4}$

(e) $2p \sec\theta$

413. On a triple thread screw

(a) lead = pitch (b) lead = 3 pitch

(c) lead = $\frac{1}{3}$pitch (d) lead = 9 pitch

(e) 9 lead = pitch

414. Gratings are used in connection with

(a) flatness measurement (b) roundness measurement

(c) surface texture measurement (d) convexity/concavity measurement

(e) linear displacement measurements

415. The method of fractional coincidences in interferometry techniques is used for

(a) Measurement of end gauges (b) flatness of surface

(c) linear displacement measurements

(d) convexity/concavity of surfaces (e) surface roughness measurement

416. An optical guage works on the principle of

(a) reflection of light (b) polarisation of light rays

(c) interference of light rays (d) refraction of light rays

(e) dispersion of light rays

417. The light intensity of fringes with laser compared to micro-chromatic light is

(a) very high (b) low

(c) same

(d) proportional to degree of flatness of optical flat

418. Laser inter terometry has major limitation because

(a) it generates only a single wavelength

(b) it generates two wavelengths (c) it generates wavelengths and varying length

(d) it generates straight wave length

419. An optical flat can enable measurement of height difference in the range of

(a) 0.0001 to 0.001 mm (b) 0.01 to 0.1 mm

(c) 0.1 to 1 mm (d) 1 to 10 mm

420. Match the 2 parts below relating to devices used for optical measurements

1. Disk made of quartz with its surface ground and lapped to a high degree of flatness — A. scale

2. a flat surface having rulings on it and these are spaced relatively far apart — B. reticles

3.	a flat surface having rulings move closely spaced	C.	optical disc
4.	provides a reference in the form of cross-wires for taking measurement	D.	gratings

(*a*) 1—A, 2—C, 3—D, 4—B (*b*) 1—C, 2—A, 3—D, 4—B
(*c*) 1—C, 2—A, 3—B, 4—D (*d*) 1—C, 2—B, 3—D, 4—A

421. Which is the application of optics in metrology
(*a*) comparators and alignment of parts
(*b*) absolute standard of length (*c*) interferometers
(*d*) all of above

422. What is not associated with fringes (band of alternate light and dark patches)
(*a*) monochromatic light source (*b*) optical flat
(*c*) flat reflecting surface (*d*) compressed air

423. Which is not associated with optical measurement
(*a*) lenses and mirrors (*b*) viewing screen, eyepiece
(*c*) orifice (*d*) light source

424. Which is not true for optical flat
(*a*) it is a disk of high-quality glass or quartz
(*b*) it reflects incident ray by 90°
(*c*) its surface is ground and lapped to a high degree of flatness
(*d*) it is popular for measurements associated with interferometry technique.

425. Which is not correct with reference to interferometers
(*a*) It is used for large linear measurement
(*b*) it is used for measuring flatness errors and also accuracy of slip gauge
(*c*) it works on same principle as of optical flat
(*d*) it has adequate arrangements to control the lay and orientation of fringes

426. Pentaprismn turns the line of sight of light beam by ______.
(*a*) 45° (*b*) 60°
(*c*) 90° (*d*) 180°

427. When two light waves interact with each other, the wave effect leads to a phenomenon called ______.
(*a*) dispersion (*b*) refraction
(*c*) deflection (*d*) interference

428. With which isotope the wavelength is easily reproducible for measurement of length
(*a*) kypton-86 (*b*) mercury 198
(*c*) cadmium (*d*) sodium

429. A big advantage of optical square (penta prism) is that
(*a*) irrespective of angle of incident beam on face of prism, beam is turned through 90°
(*b*) Accuracy is not affected by errors of mounting of prism
(*c*) Accuracy is not affected by not being able to maintain base parallel to datum
(*d*) All of above are true

430. Involute function of a gear is defined as
(*a*) (cos α – α) radians (*b*) (sin α – α) radians
(*c*) (tan α – α) radians (*d*) (cot – α) radians
(*e*) (sec – α) radians

431. The pitch circle radius r_p and base circle radius r_b of a gear are related by the following relationship (α = pressure angle)
(*a*) $r_b = r_p \cos \alpha$ (*b*) $r_b = \dfrac{r_p}{\cos \alpha}$
(*c*) $r_b = r_p \sin \alpha$ (*d*) $r_b = r_p \tan \alpha$
(*e*) $r_b = r_p (\tan \alpha - \alpha)$

432. The diameter of a large bore is measured by rocking a pin guage in it as shown in Fig. 42. If L be length of pin, and 2*l* the swing, then diameter D is nearly equal to

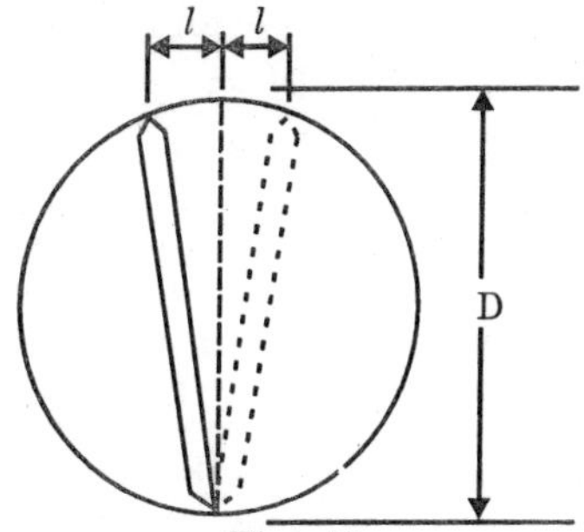

Fig. 42

(*a*) $L = \dfrac{l^2}{2L}$ (*b*) $L + \dfrac{l^2}{L}$
(*c*) $L + \dfrac{l^2}{4L}$ (*d*) $L + \dfrac{l^2}{8L}$
(*e*) $L + \dfrac{l^2}{3L}$

433. Pick up false statement
(*a*) Tool maker's microscope is used to measure thickness
(*b*) Parkirson's gear tester is used to measure variation in centre distance
(*c*) Analytical checking of gears includes checking of tooth profile
(*d*) David Brown Tangent micrometer is used to measure base pitch of gear

434. Functional inspection of running test of gear is not used to determine
(*a*) lead of gear (*b*) composite vibration
(*c*) noise level (*d*) variation in action

435. A sine bar is used to measure
(*a*) surface roughness (*b*) gear profiles
(*c*) internal tapers (*d*) external tapers
(*e*) all of the above

436. The normal chordal tooth thickness of a gear is the shortest distance from the
(*a*) tooth crest to the point of the chord
(*b*) tooth crest to the mid point of the chord
(*c*) mid point of the constant chord to the tip of the tooth
(*d*) from the tooth crest to any point on the chord
(*e*) mid point of the constant chord to tip of the tooth

437. Addendum of a gear is equal to
(*a*) Pitch p (*b*) 0.3 p
(*c*) 0.3183 p (*d*) 0.3683 p
(*e*) 0.6866 p.

438. Optical flats are used in conjunction with
(*a*) angular measurements (*b*) surface flatness
(*c*) surface parallelism (*d*) interferrometric measurement
(*e*) surface profiles

439. In the phenomenon of optical interference, the two beams of light, both originating from the same source, can cause interference patterns when the path difference between them amounts to
(*a*) 1 micron (*b*) 1 mm
(*c*) one wavelength of the light being used
(*d*) an odd half-wavelength of the light being used
(*e*) an even half-wavelength of the light being used

440. The NPL gauge interferometer is designed and used for
(*a*) absolute measurement of length of slip gauges
(*b*) judging flatness of surface
(*c*) comparing height with a standard reference
(*d*) measurement of fringe displacement
(*e*) parallelism of two ends of slip gauge

441. Which method gives accurate results when effective diameter is measured without considering the thread angle.
(*a*) 2-wire method (*b*) 3-wire method
(*c*) best wire size (*d*) bench micrometer

442. Wear allowance is provided on
(*a*) Go gauge (*b*) No Go gauge
(*c*) both Go and No Go guages (*d*) when both are combined in one guage
(*e*) neither Go nor No Go Gauge

443. Which of the following gauge is of full form
(*a*) Go gauge (*b*) No Go gauge
(*c*) snap gauge (*d*) thread gauge
(*e*) double ended gauge.

444. Taylor's principle is concerned with
(*a*) pneumatic comparators (*b*) inferometric measurements
(*c*) gauging measurements (*d*) angular measurements
(*e*) roundness measurements

445. Accuracy of setting a sine bar
(*a*) is poor for smaller angles
(*b*) is maximum when angle of measurement is 45°
(*c*) decreases appreciably with steep angle
(*d*) is depended on accuracy of outer dimensions of sine bar
(*e*) is function of size of rollers of sine bar

446. The radius of curvature of tube in precise spirit level is of the order of
(*a*) 10 cm (*b*) 50 cm
(*c*) 1 m (*d*) 10 m
(*e*) 50 m.

447. Profile of a gear tooth is to be checked. Which one of the following device would you choose
(a) optical pyrometer (b) bench micrometer
(c) sine bar (d) telescopic gauge
(e) optical projector

448. Clinometer is an instrument concerned with
(a) temperature measurement (b) flatness measurement
(c) linear measurement (d) roundness measurement
(e) angular measurement

449. Auto-collimator is used for measurement of
(a) small angular differences (b) flatness
(c) linear surfaces (d) concavity
(e) surfaces roughness

450. Optical square is used in connection with
(a) alignment tests involving two surfaces at right angles
(b) interferometry measurements (c) angular measurements
(d) circular division (e) measurement of flatness of surfaces

451. According to Indian standards, the slip gauges according to accuracy are graded into following categories
(a) 2 (b) 3
(c) 4 (d) 5
(e) 6.

452. In which of the following inspection system, the inspection keeps pace with the production
(a) sampling inspection (b) cent percent inspection
(c) cage inspection (d) centralised inspection
(e) floor or partrol inspection

453. In inspection by attributes
(a) good quantities are separated from bad
(b) quality under consideration is measured on some scale expressing it quantitatively
(c) variations due to assignable factors are determined
(d) variations due to change factors are determined
(e) theory of probability is applicable

454. The square of the standard deviation is also called
(a) skewness (b) variance
(c) medium (d) mode
(e) range

455. In an interferometer experiment, following pattern of fringes (In Fig. 43) was observed. The surface being tested is
(a) flat (b) convex
(c) concave
(d) ridge/valley in the middle
(e) smooth cylindrical surface

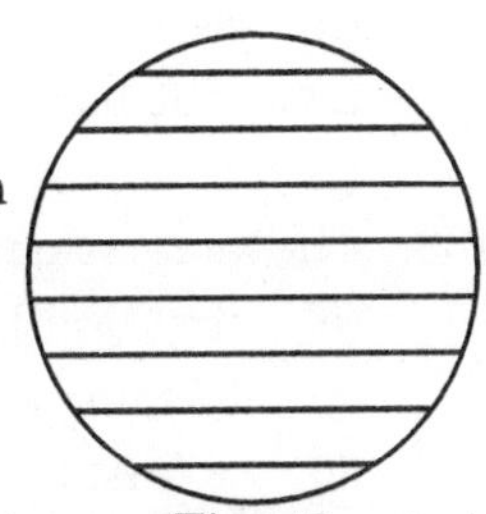
Fig. 43

456. If the interference bands with optical flat are as shown in Fig. 44, the surface would be
(a) convex
(b) concave
(c) ridge
(d) progressively convex
(e) uneven

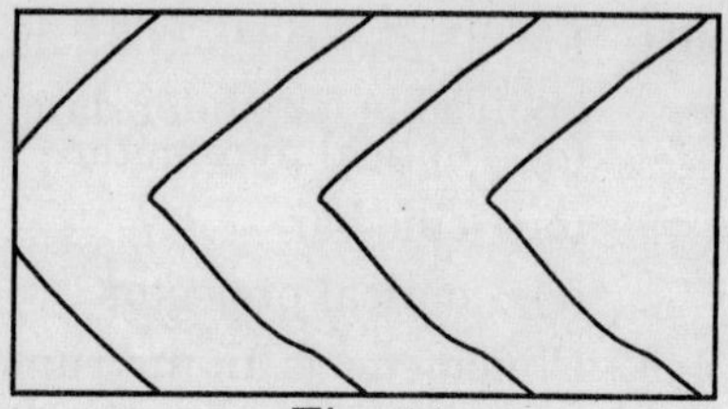

Fig. 44

457. A scale in which the distance between graduation is proportional to the value of that graduation is called
(a) equidistant scale (b) regular scale
(c) linear scale (d) scale with a constant value of graduation
(e) non-linear scale

458. Coefficient of variation in term of standard deviation σ is defined as
(a) $\sigma^2 \times 100$ (b) $\sqrt{\sigma} \times 100$
(c) $\frac{\sigma}{\overline{\overline{X}}} \times 100$ (d) $\frac{\sigma}{R} \times 100$
(e) $\pm 3\sigma \times 100$

459. Profilometer is an instrument used to measure
(a) gear involute (b) thread profile
(c) taper (d) surface roughness
(e) surface flatness

460. Variance is defined as
(a) $\frac{\sigma}{2}$ (b) $\pm 3\sigma$
(c) $\sqrt{\sigma}$ (d) σ^2
(e) $\frac{\sigma}{\overline{\overline{X}}}$.

461. The value of a set of data at which the greatest number of cases is concentrated is called
(a) mean (b) median
(c) range (d) standard deviation
(e) mode

462. If the correctness of profile of a thread is to be checked, which instrument would you choose
(a) bench micrometer (b) screw pitch guage
(c) sine bar (d) telescopic gauge
(e) optical projector

463. The more sensitive instrument
(a) first oscillates more (b) oscillates more slowly
(c) has no oscillations (d) is never stable
(e) is unstable

464. Fig. 45 shows the time displacement relation for damped motion. Which of the curves is applicable for under damped system

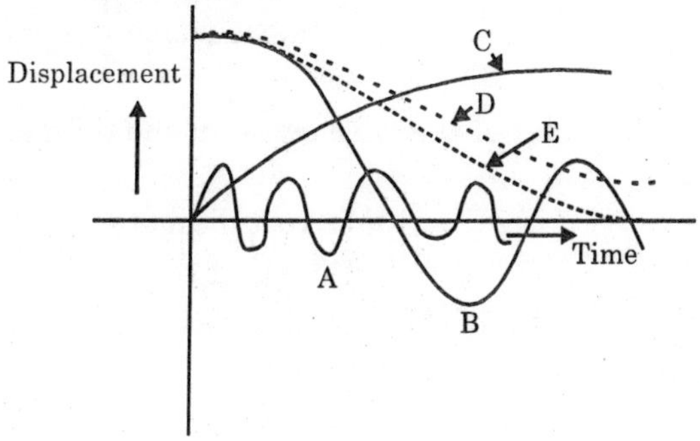

Fig. 45

(*a*) A (*b*) B
(*c*) C (*d*) D
(*e*) E.

465. Frequency of oscillation of an instrument indicating device is a function of
(*a*) mass (*b*) damping
(*c*) sensitivity (*d*) both damping and sensitivity
(*e*) all of above, i.e, mass, damping and sensitivity

466. The lost motion of the spindle in micrometer screw gauge when the rotation of the thimble is changed in direction is referred to as
(*a*) wear (*b*) least count
(*c*) total error (*d*) backlash
(*e*) accuracy

467. Stick micrometers are used for measuring
(*a*) depth of holes (*b*) longer internal lengths
(*c*) longer external lengths (*d*) height of parts
(*e*) very small diameters

468. All the working surfaces and the cylindrical surfaces of the rollers of sine bar have a surface finish of the order of
(*a*) 0.2 micron (*b*) 0.5 micron
(*c*) 1 micron (*d*) 5 micron
(*e*) 10 micron

469. The effect of errors in spacing of the rollers, or height of slip gauge combination is a function of the
(*a*) sin θ (*b*) cos θ
(*c*) tan θ (*d*) cosec θ
(*e*) sec θ

470. Sine centre is used for measurement of
(*a*) included internal angle between two faces
(*b*) height of projected parts
(*c*) semi-angle of taper of a job (*d*) flatness
(*e*) any one of the above

471. Clinometer is used for

(a) angular measurement (b) linear measurement
(c) bore measurement (d) measurement of environmental conditions
(e) level of flat surfaces

472. In wire methods, the diameter of the wire selected should make contact with the screw along the

(a) outer diameter (b) root diameter
(c) pitch cylinder (d) axis of the screw

473. In a two-wire method, diameter of the best size wire is proportional to

(a) included angle of thread (b) $\theta/2$
(c) sec $\theta/2$ (d) cot $\theta/2$

474. The pitch error occurring due to incorrect tool work velocity ratio which is constant is referred to as a

(a) cyclic error (b) periodic pitch error
(c) progressive error (d) non-progressive error

475. Which thread gauge is used to inspect external thread form and internal thread respectively

(a) plug screw guage and ring screw gauge
(b) ring screw gauge and plug screw gauge
(c) external screw gauge and internal screw gauge
(d) none of these

476. A NOT GO screw gauge will primarily check for

(a) outer diameter and nothing else (b) inside diameter and nothing else
(c) effective diameter and nothing else
(d) none of these

477. The change in pitch diameter of a screw thread (δd) and the pitch error (δp) are approximately related as under

(a) $\delta d = \delta p$ (b) $\delta d = \delta p/2$
(c) $\delta d = 2\delta p$ (d) $\delta d = 1.5\ \delta p$
(e) $\delta d = 3\delta p$.

478. Pick up the correct statement. The effect of pitch error and angle error is to

(a) increase the simple effective diameter of a bolt and increase that of a nut
(b) decrease the simple effective diamter of a bolt and decrease that of a nut
(c) increase the simple effective diameters of both bolt and nut
(d) decrease the simple effective diameters of both bolt and nut
(e) have no effect on simple effective diameters of bolt and nut

479. Which is not correct about two and three wire method to measure effective diameter of threads

(a) Deformation of wire and thread will be negligible to warrant some type of correction
(b) The wires are accurately finished and hardened steel cylinder and these should be provided with suitable means of suspension.
(c) The wires should be free to assume their positions in the thread grooves without any restraint
(d) As far as possible, measure the effective diameter by means of wires of best size.

480. If θ is the included angle of a screw thread then change in simple effective diameter due to change in angle error is proportional to
    (*a*) tan θ (*b*) sin θ
    (*c*) cos θ (*d*) sec θ
    (*e*) cosec θ
481. Pitch error result in ______ of effective diameter of nut and in ________ effective diameter of bolt
    (*a*) increase, decrease (*b*) decrease, increase
    (*c*) increase, increase (*d*) decrease, decrease
482. Major diameter of an external thread can be measured by
    (*a*) thread micrometer (*b*) one wire method
    (*c*) bench micrometer (*d*) any one of above
483. Snap gauge for Go dimension corresponds to
    (*a*) maximum metal condition (*b*) minimum metal condition
    (*c*) average metal condition (*d*) none of above
484. The best size wire for ISO metric thread for measuring pitch diameters of screw thread in terms of its pitch p is
    (*a*) 0.5 p (*b*) 0.6 p
    (*c*) 0.75 p (*d*) 0.5773 p
    (*e*) 0.4227 p
485. In case of threads, crest diameter is the ________ diameter for nut, and ________ diameter for bolt thread
    (*a*) minor, major (*b*) major, minor
    (*c*) major, major (*d*) minor, minor
486. A tolerance class for screw threads for bolts and nuts is specified by
    (*a*) tolerance grade (*b*) tolerance position
    (*c*) both (a) and (b) (*d*) none of above
487. A profilometer is a compact device that can be used for direct measurement of
    (*a*) elements of gears (*b*) thread dimensions
    (*c*) complex profiles (*d*) surface texture
488. Which instrument assesses surface irregularities through frictional properties and the average slope of the irregularities.
    (*a*) pneumatic comparator (*b*) Tomlinson
    (*c*) Mercin (*d*) light interference microscope
489. For fine irregular surfaces, what is the cut off length used
    (*a*) 0.1 mm (*b*) 0.8 mm
    (*c*) 1.2 mm (*d*) 2.5 mm
490. Which instrument can be used to measure base pitch of a gear
    (*a*) Gear tooth vernier caliper (*b*) David Brown tangent caliper
    (*c*) David Brown tangent comparator
    (*d*) David Brown tangent micrometer

491. Which parameter is measured using the Pitter—NPL gauge interferometer
(*a*) length (*b*) angle
(*c*) straightness (*d*) flatness

492. Analytical checking of géars includes checking of
(*a*) tooth profile (*b*) vibrations
(*c*) noise (*d*) smooth running

493. The relationship between chordal thickness t_c and chordal height (h_c) for a gear is
(*a*) h_c = module (m) + $\frac{t_c^2}{4(PCD)}$ (*b*) $h_c = m + t_c^2/PCD$
(*c*) $h_c = m + \frac{4t_c^2}{2(PCD)}$ (*d*) $h_c = m + \frac{t_c^2}{2(PCD)}$

494. If a gear has N no. of teeth m-module, ϕ-pressure angle and S– the span of measurement, then theoretical span width for tooth span micrometer will be
(*a*) $Nm \cos\phi\left(\frac{\pi S}{N} - \frac{S}{2N} + \tan\phi - \phi\right)$ (*b*) $Nm \cos\phi\left(\frac{\pi S}{N}\right)$
(*c*) $Nm \cos\phi\left(\frac{\pi S}{N} + \tan\phi\right)$ (*d*) $Nm \cos\phi\left(\frac{\pi S}{N} + \tan\phi - \phi\right)$

495. A gear calliper has two vernier scales
(*a*) one in vertical direction and one in horizontal
(*b*) both in horizontal direction
(*c*) both in vertical direction (*d*) both inclinable at any angle

496. If gear is used for high degree of positional accuracy for a machine slide, then which error in gear can't be tolerated
(*a*) pitch error (*b*) tooth profile error
(*c*) run out error (*d*) assembly error

497. Gear tooth caliper is used to measure following elements of gear
(*a*) pitch and backlash (*b*) lead and tooth thickness
(*c*) chordal thickness and chordal height
(*d*) base circle diameter and PCD

498. Base tangent length measured by tooth span micrometer for gear testing is function of
(*a*) no. of teeth on the gear (*b*) no. of teeth spanned
(*c*) both (a) and (b) (*d*) involute function and pressure angle

499. Parkinson gear tester measures the composite error which reflects errors due to
(*a*) run out (*b*) tooth-to tooth spacing
(*c*) profile variation (*d*) all of above

500. Parkinson's gear tester is used to measure
(*a*) tooth thickness (*b*) tooth profile
(*c*) variation in centre distance (*d*) all of above

501. Universal measuring machine is a combination of optical measuring machine and
(*a*) tool maker's microscope (*b*) dividing heat
(*c*) clinometer (*d*) parkinson's gear tester

502. To enable a workpiece to be held in a vertical portion for laying out, it is clamped to
(*a*) surface plate (*b*) V-block
(*c*) machine bed (*d*) angle plate

503. For judging the quality of a measuring instrument, which of following characteristises its ability to respond to small changes in the quantity to be measured
(*a*) accuracy (*b*) precision
(*c*) response (*d*) discrimination

504. Gear tooth thickness can be measured by
(*a*) gear tooth verniers (*b*) constant chord method
(*c*) base tangent method (*d*) all of above

505. According to Indian standard specification, spur gears are classified into ________ quality accuracy grades from high precision to coarse quality, low speed gears
(*a*) 5 (*b*) 7
(*c*) 10 (*d*) 12

506. High precision gears, finished by grinding can be operated at high speed of _______ m/s.
(*a*) 0.8 (*b*) 5
(*c*) 8 (*d*) 15

507. Which errors are possible in spur gears
(*a*) gear blank run-out errors
(*b*) gear tooth profile errors, tooth spacing and tooth thickness, alignment, etc.
(*c*) combined errors of gear in assembly
(*d*) all of above

508. The portable base pitch measuring instrument has one dial indicator and ____ number of tips
(*a*) one (*b*) two
(*c*) three (*d*) four

509. Involute profit can be checked using involute measuring machine for
(*a*) small-sized and thin gears (*b*) large-sized gears
(*c*) both (a) and (b) (*d*) there is no such criteria

510. The bright areas and dark areas in fringes occur when the path difference of two light ray from the same source amounts to an ________ and an ________ number of half wavelengths
(*a*) odd, even (*b*) even, odd
(*c*) equal, unequal (*d*) unequal, equal

511. Which element enables a microscope to be used as a measuring instrument
(*a*) laser light beam (*b*) optical flat
(*c*) reticle (*d*) pentaprism

512. Which is related with measurements using light waves
(*a*) lapped surfaces (*b*) optical flats
(*c*) comparators (*d*) spirit level

513. The advantage of using krypton-86 isotope in measurements is that it
(*a*) is less harmful to eyes (*b*) has attractive colour
(*c*) wavelength is constant (*d*) is easily reproducible wavelength

514. A big advantage of pentaprism is that it
 (*a*) has highly flat lapped surface (*b*) can be easily installed
 (*c*) has capability to split light into multiple rays
 (*d*) is insensitive to mounting arrangement

515. Why light has become a standard for measurement of length
 (*a*) wavelength of light is known and remains constant
 (*b*) it enables measurement of length accurately and precisely
 (*c*) it is reproducible
 (*d*) it enables formation of interference fringes

516. A tool maker's microscope finds wide application in metrology laboratories to measure
 (*a*) accuracy of length standard
 (*b*) counting the fringes in measurement of length
 (*c*) profiles of machines
 (*d*) shape, size, angle and profile of small components

517. Which of following is not the application of optics
 (*a*) absolute standard of length (*b*) alignment
 (*c*) interferometers (*d*) biological degradation

518. Geometric features of different components can be inspected on
 (*a*) Universal measuring machine (*b*) coordinate measuring machine
 (*c*) machine vision systems (*d*) laser interferometers

519. Universal measuring machine is a combination of tool maker's microscope and
 (*a*) laser interferometer (*b*) optical measuring machine
 (*c*) profilometer (*d*) surface dynamometer

520. Distance between maximum inscribing circle and minimum circumscribing circle measured radially is
 (*a*) error of circularity (*b*) axial run out
 (*c*) error of concentricity (*d*) radial run out

521. Which dimensions can be measured using image shearing microscope
 (*a*) linear dimensions (*b*) angular dimensions
 (*c*) flatness (*d*) parallelism

522. Tool maker's microscope can measure following components
 (*a*) gauges (*b*) screw threads and gears
 (*c*) dies and fixtures (*d*) all of the above

523. What is the application of CMM
 (*a*) small batch production (*b*) high value products
 (*c*) components of varied features and complex geometry
 (*d*) all of above

524. For which situation , CMM is good choice
 (*a*) integrated automated inspection system
 (*b*) interfaced with CNC machine (*c*) reverse engineering
 (*d*) all of above

525. Which is the typical subroutine used in CMM
(*a*) circle and cylinder (*b*) perpendicularity between two planes
(*c*) parallelism between two planes (*d*) all of above

526. Match the two parts below relating to types of probes used in CMM

1.	Using electronic technology like LVDT to ensure optimum contact pressure between probe and workpiece	A. Touch trigger probe
2.	using difference in contact resistance to indicate defection of probe	B. Soft probe
3.	Probes using light beam stylii	C. Hard probes
4.	Stylus integral with probes (which may be conical, pointed or ball end)	D. Non-contact probe

(*a*) 1—B, 2—A, 3—D, 4—C (*b*) 1—A, 2—B, 3—D, 4—C
(*c*) 1—B, 2—A, 3—C, 4—D (*d*) 1—A, 2—B, 3—C, 4—D

527. CMMs may assume following mode of operation
(*a*) manual (*b*) Semi-automatic
(*c*) computer controlled (*d*) any one of above

528. The software associated with CMM has following component
(*a*) move commands (*b*) measurement commands
(*c*) formatting commands (*d*) all of above

529. Which parameter is of interest while finding the minimum separation of a pair of parallel planes to contain all the points on the surface
(*a*) flatness (*b*) surface roughness
(*c*) parallelism (*d*) straightness

530. The cylindricity of a component is usually described by
(*a*) minimum zone method (*b*) maximum zone method
(*c*) average zone method (*d*) cylindrical zone method

531. Which of following is contact type of instrument
(*a*) stroboscope (*b*) pyrometer
(*c*) micrometer (*d*) infrared based instrument

532. Which machine is capable of measuring all three orthogonal axes
(*a*) profilometer (*b*) laser based transducers
(*c*) coordinate measuring machine (*d*) Talysurf

533. Which structure provides high rigidity and accuracy to CMM
(*a*) gantry (*b*) bridge
(*c*) column (*d*) cantilever and horizontal arm

534. Which measuring system enables reverse engineering
(*a*) CMM (*b*) vision system
(*c*) laser interferometry
(*d*) remote controlled electronic probes and sensors

535. Image generation and digitisation is a stage used in
(*a*) flexible inspection system (*b*) CMM
(*c*) Machine vision inspection system
(*d*) On-line inspection in CIM

536. Pick up wrong statement
(*a*) Any cost associated with correcting failure or waste comes under quality cost.
(*b*) Cause and effect diagram also known as "Fishbone diagram" or Ishikawa diagram is used to identify possible causes of solution.
(*c*) Brainstorming, quality circles, and Benchmarking process are methods for generating solution to improve quality.
(*d*) In deming circle, the four phases in PDSA are: Plan, Do, Study and Act

537. Pick up incorrect statement
(*a*) Lower limit in control charts is three sigma away from upper limit
(*b*) Pareto's law is also known as 20-80 law
(*c*) Pareto's law helps to locate bottlenecks
(*d*) All are incorrect

538. Which of following is not concerned with Quality of Conformance
(*a*) how well the manufactured product conforms to the quality of design.
(*b*) translating user-based characteristics into identifiable product attributes.
(*c*) extent to which a firm and its suppliers can create products with a predictable degree of dependability.
(*d*) organizing the manufacturing process to ensure that product quality stringently adheres to specifications.

539. Consider the following statements regarding Quality of performance (QOP) and state which is not concered with QOP
(*a*) how well the manufactured product gives its performance
(*b*) focusing on ascertaining how quality characteristics is determined in quality of design.
(*c*) tightness of specifications for the manufacture of the product
(*d*) profit incurred

540. What is the correct sequence for PDCA cycle?
A. Monitor and measure the processes and the products and evaluate their progress according the objectives of the plan.
B. Implement improvement processes
C. Establish the objectives and processes necessary to deliver
D. Take actions to continually improve processes performance
(*a*) A-D-B-C (*b*) C-D-B-A
(*c*) C-B-A-D (*d*) D-B-A-C

541. When using the possion as approximation to the binomial, the following conditions apply for the best approximation
(*a*) larger sample size and larger fraction defective
(*b*) larger sample size and smaller fraction defective

(*c*) smaller sample size and larger fraction defective
(*d*) smaller sample size and smaller fraction defective

542. The probability of observing at least one defective in a random sample of size seven drawn from a population that has been producing, on the average five percent defective units is

(*a*) $(0.05)^7$ (*b*) $(0.95)^7$
(*c*) $1 - (0.05)^7$ (*d*) $1-(0.95)^7$

543. A histogram of the given data spotlights

(*a*) center and variation of the data (*b*) center and standard deviation of the data
(*c*) reasons for variations (*d*) assignable causes

544. In order to set up a shewhart control chart initially for a repetitive industrial process, information on the following process characteristic is sought.

(*a*) process capability (*b*) process performance and process tolerance
(*c*) process reliability (*d*) process conformance

545. In control-chart theory, the distribution of the number of defects per unit follows very closely the

(*a*) normal distribution (*b*) binomial distribution
(*c*) chi-square distribution (*d*) poisson distribution

546. The steeper the OC curve, the

(*a*) less protection for both producer and consumer
(*b*) more protection for both producer and consumer
(*c*) lower the AQL (*d*) smaller the sample size

547. The AQL for a given smapling plan is 0.5 percent. This means that

(*a*) the producer takes a small risk of rejecting product which is 0.5 percent defective or better
(*b*) all accepted lots are 0.5 percent defective or better
(*c*) the average quality limit of the plan is 0.5 percent
(*d*) the average quality level of the plan is 0.5 percent
(*e*) all lots are 0.5 percent defective or better

548. The probability of accepting material produced at an acceptable quality level is definfed as

(*a*) α and β (*b*) AQL
(*c*) $1 - \alpha$ (*d*) $1 - \beta$

549. Dodge-Roming sampling schemes aim at minimising

(*a*) the total inspection (*b*) the acceptance number
(*c*) AQL (*d*) AOQL

550. Dodge-Roming sampling plans provide sampling schemes indexed on

(*a*) AQL and LTPD (*b*) AOQL and LTPD
(*c*) AOQ and LTPD (*d*) LQ and LTPD

551. Reliability prediction and measurement are primarily useful in

(*a*) evaluating feasibility and overall reliability
(*b*) establishing reliability goals (*c*) defining problem areas
(*d*) all of the above

552. Reliability testing of parts is performed to yield information on
(*a*) application suitability (*b*) environmental capability
(*c*) measurement of life characteristics
(*d*) all of the above.

553. Quality costs are best classified as
(*a*) cost of inspection and test, cost of quality engineering, cost of quality administration, and cost of quality equipment
(*b*) direct, indirect, and overhead
(*c*) cost of prevention, cost of appraisal, and cost of failure
(*d*) cost of training and education

554. TQM is about
(*a*) moving towards perfect quality and involving everyone in quality
(*b*) trying to avoid customer complaints
(*c*) getting the right quality level
(*d*) management being concerned about quality

555. Burning in operation
(*a*) demages products (*b*) increases costs
(*c*) prevents early failure (*d*) ensures consistent quality

556. Match List-I with List-II and select the correct answer using the codes given below the lists:

	List-I		List-II
A.	Pareto chart	1.	Control
B.	Histogram	2.	Analyse
C.	Control chart	3.	Analyse and improve
D.	Scatter diagram	4.	Measure, analyse

Codes:

	A	B	C	D
(*a*)	2	4	1	3
(*b*)	1	2	3	4
(*c*)	3	2	4	2
(*d*)	4	3	2	1

557. Match List-I and List-II and select the correct answer using the codes given below the lists:

	List-I		List-II
1.	X- Chart	A.	Precision
2.	R-Chart	B.	Number of defects
3.	P-Chart	C.	Fractional defects
4.	C-Chart	D.	Accuracy

Codes:

	A	B	C	D
(*a*)	2	3	1	4
(*b*)	4	3	1	2
(*c*)	2	1	3	4
(*d*)	4	1	3	2

558. Normal distribution tables give
   - (a) the value of the mean
   - (b) the area under the curve
   - (c) the height of the curve
   - (d) the value of the standard deviation

559. Quality can be defined as the totality of feature and characteristics of a product or service that bears on its ability to satisfy
   - (a) given needs
   - (b) factory needs
   - (c) stated or implied needs
   - (d) market needs

560. Pick up wrong statement
   - (a) Total Quality management is a strategic approach to ensuring quality in an organization
   - (b) Quality function deployment (QFD) follows voice of the customer
   - (c) The main components of quality management are: Quality planning, quality assurance, quality control, and quality improvement
   - (d) Attributes in a process are measurelable and variables are countable

561. Which pair is not correctly matched
   - (a) A statistical test to determine cause and effect relationship between process variable and output—DOE
   - (b) Philosophy of identifying and eliminating non-essential and non-value adding activities—Lean
   - (c) Change in management methodology to manage and improve business processes to limit process variations to 3.4 defects per million opportunities—Six Sigma
   - (d) A measure to quantify the amount of variation or dispersion of a set of data value—Range

562. Optimal quality is reached when the
   - (a) Stakeholder accepts the project deliverable
   - (b) revenue from improvements equals the costs of conformance
   - (c) revenue from improvement equals the incremental costs to achieve the quality
   - (d) revenue from corrective actions equals the cost of the improvement

563. Which analysis tool allows to determine the cause-and-effect of the production faults?
   - (a) A flowchart
   - (b) Pareto diagram
   - (c) Ishikawa diagram
   - (d) control chart

564. Which strategy allows you to design experiments to look for ways to improve quality.
   - (a) Allow the project manager to move the relationship of activities to complete the project work with the best resources available.
   - (b) Allow the project manager to experiment with the project design to determine what variables are causing the flaws.
   - (c) Allow the project manager to experiment with variables to attempt to improve quality.
   - (d) Allow the project manager to experiment with the project design document to become more productive and to provide higher quality.

565. Which chart shows the distribution of problems and their frequencies.
   - (a) A control chart
   - (b) An Ishikawa chart
   - (c) A pareto diagram
   - (d) A flowchart

566. Which is wrong statement
   (*a*) The Cause-and-Effect (C and E) diagrams or fishbone diagrams are used to show how different factors relate together and might be tied to potential problems.
   (*b*) Pareto law also known as 80/20 rule, states that 80% of the problems come from 20% of the causes
   (*c*) The costs of items that do not conform to quality are known as cost of poor planning.
   (*d*) A scatter diagram is a powerful tool for spotting trends in data.

567. Reduced quality on a project would likely lead to
   (*a*) rework and increased cost risk (*b*) absenteeism and increased cost
   (*c*) increased inspections and decreased cost
   (*d*) intervention by customer

568. Which of following best describes the function of quality assurance
   (*a*) overall project performance (*b*) scope verification
   (*c*) project deliverable (*d*) project stakeholders

569. Which of following tool easily checks if results are out of control?
   (*a*) Pareto chart (*b*) Control chart
   (*c*) Ishikawa diagram (*d*) benchmarking

570. Which is the best tool to look for causes that have the highest number of defects associated with them.
   (*a*) Pareto chart (*b*) Control chart
   (*c*) Benchmarking (*d*) Cause and effect diagram

571. Pick up wrong statement
   (*a*) Internal failure cost is not attributed to failure that results in rework
   (*b*) Quality planning is the process of creating a plan to meet the requirement of quality.
   (*c*) Quality is planned into a project and inspections ensure that it exists.
   (*d*) Kaizen technologies are small changes to processes and products on a steady, continuous basis to save costs and improve quality.

572. Perform Quality Control is the process that uses.
   (*a*) scatter diagram and flow charts (*b*) run diagram and control charts
   (*c*) inspection (*d*) all of above

573. Pick up wrong statement
   (*a*) In inspection, the product of the project is compared by the customer with best benchmarked targets
   (*b*) Run charts show trends and variances over time and are used in quality management
   (*c*) Benchmarking is a tool of quality management for both the Plan Quality and Perform Quality Assurance processes.
   (*d*) The quality policy is a document used by the performing organization that describes their attitude regarding quality.

574. What are Kaizen technologies?
   (*a*) Small improvements for optimum results
   (*b*) Small improvements for all projects

(*c*) Small process and product improvements are carried out on a continuous basis.
(*d*) Small process improvements that are made to shorten the project duration.

575. Which of the following is most true about quality?
(*a*) It will cost more money to build quality into the product
(*b*) It will cost less money to build quality into the product process.
(*c*) Quality is inspection-driven. (*d*) Quality is prevention-driven

576. Which of the following can be described as a business philosophy to find methods that will continuously improve products, services, and business practices?
(*a*) TQM (*b*) ASQ
(*c*) QA (*d*) QC

577. Pick up wrong statement
(*a*) Quality is the sum of the characteristic of a product that allow it to meet the demands or expectations of the customer.
(*b*) Grade is a rank for entities having the same functional use but different technical characteristics.
(*c*) Low quality is always a problem, but low grade may not be.
(*d*) Level of quality and cost of quality vary in direct linear proportion.

578. With reference to quality circles, consider the following statements:
1. It is a group of workers who voluntarily meet to discuss ways of improving products or process.
2. The working staffs involved in quality circles meet frequently either at someone's home or at the plant before shift begins.

Which of the above statements is/are true?
(*a*) 1 and 2 (*b*) 1 only
(*c*) 2 only (*d*) none

579. Which of the following statements correctly define 'Quality'.
(*a*) Meeting customers, both internal external expectation
(*b*) Degree of excellence at an acceptable price and control of variability at an acceptable cost.
(*c*) Meeting product specifications within specified tolerance.
(*d*) All of above

580. Robustness of a product or process implies.
(*a*) Intended performance of a product or service
(*b*) Measure of life of product before it is discarded
(*c*) Ability to perform even in face of uncontrollable outside influences or adverse condition
(*d*) Mean time to failure or between the failure

581. Select wrong statement
(*a*) In SPC there is no concept of special causes and common causes.
(*b*) While common causes are normal special causes are typically considered unusual and preventable.
(*c*) Tolerances deal with the limits set for the product acceptance.
(*d*) Control limits focus on whether the process itself is acceptable

582. A quality control programme is considered to be
(*a*) a collection of quality control procedures and guidelines
(*b*) a step-by-step list of all quality control checkpoints
(*c*) a summary of company quality control policies
(*d*) a system of activities to provide quality of products and service.

583. A fully developed position description for a quality engineer must contain clarification of
(*a*) responsibility (*b*) accountability
(*c*) authority (*d*) All of above

584. Match the two parts below

1.	Pareto chart	A.	Step in process and their relationships
2.	Control chart	B.	Deciding when to investigate the process
3.	Fishbone diagram	C.	The vital few from many trival
4.	Process flow chart	D.	To brainstorm the causes of an effect

Codes
(*a*) 1–C, 2–B, 3–D, 4–A (*b*) 1–B, 2–C, 3–D, 4–A
(*c*) 1–B, 2–C, 3–A, 4–D (*d*) 1–C, 2–B, 3–A, 4–D

585. Pick up false statement
(*a*) Cause-effect diagram is used in problem analysis
(*b*) p-chart is used to monitor attribute
(*c*) Source of problem can be found by using control chart alone
(*d*) SPC helps in identifying assignable causes disturbing the process

586. Pick up correct statement
(*a*) A control chart displays process capability
(*b*) Inspection assures that quality problems are resolved
(*c*) Type-1 error is associated with consumer's risk
(*d*) In case of C-chart, the central limit, and upper/lower control limits are $\bar{C}, \bar{C} \pm 3\sqrt{\bar{C}}$

587. Pick up right choice
(*a*) The customer and other stakeholders are responsible for the quality of the product.
(*b*) The Quality team is responsible for the quality of the project deliverables.
(*c*) Control charts illustrate how a process moves through a system and how the components are interrelated.
(*d*) Pareto diagram is concerned with overall project quality performance.

588. Match the two parts about quality management processes

A.	Quality planning	P.	Identifying, which quality standards are relevant to the project and determining how to satisfy them
B.	Quality assurance	Q.	Evaluating overall project performance on a regular basis to provide confidence that the project will satisfy the relevant quality standards
C.	Quality control	R.	Monitoring specific project results to determine if they comply with relevant quality standards and identifying ways to eliminate causes of unsatisfactory performance

Code.

(*a*) A–P, B–Q, C–R (*b*) A–Q, B–P, C–R

(*c*) A–R, B–Q, C–P (*d*) none of above

589. Who is responsible for the quality of the project?

(*a*) Project CEO (*b*) The project team

(*c*) Stakeholders (*d*) Customers

590. Pick up wrong statement

(*a*) The two elements of quality as fitness for use are quality of design and quality of conformance

(*b*) The quality hierarchy progresses from inspection, through quality control, quality assurance to total quality management.

(*c*) Statistical process control is a method of controlling processes and remove defects

(*d*) The simple objective of TQM is to 'Do the right things, right the first time, every time'.

591. Pick up incorrect statement

(*a*) ISO 9000 is an international standard, not a quality system, but it helps organisations to follow their own quality procedures.

(*b*) The cost of avoiding mistakes is always much more than the cost of correcting them.

(*c*) Success in implementation of quality requires the participation of all memebers of the team and the Commitment of management.

(*d*) The Cost of Poor Quality (COPQ), is derived from the non value adding activities of waste in a process

592. Which of the following not correct?

(*a*) Total quality management is a business philosophy to find methods to continuously improve.

(*b*) A safety measure is not an attribute of the cost of nonconformance, but rather a cost of adhering to quality.

(*c*) Training to meet the quality expectations are not attributed to the cost of quality

(*d*) When project work results are faulty and must be done again, it is attributed to the cost of nonconformance to quality.

593. Pick up wrong statement

(*a*) Design of experiments uses experiments and "what-if" scenarios to determine what variables are affecting quality.

(*b*) Checklists are simple and effective quality management tools to ensure that the project team is completing the required work

(*c*) Pareto diagram maps out the causes of defects and illustrates their frequency.

(*d*) Flowcharts show how a process moves through a system and how the components are related

594. Normally distributed data can be represented by a bell shaped normal curve. Which of following is not true for % age of data

(*a*) % data between $\mu \pm 1\sigma = 68.25$ (*b*) % data between $\mu \pm 2\sigma = 95.46$

(*c*) % data between $\mu \pm 3\sigma = 99.73$ (*d*) % data between $\mu \pm 6\sigma = 100$

595. From an operations perspective, quality concerns.
   (*a*) determining customers' needs and embodying these in the design specifications of services and products
   (*b*) consistently meeting services and product design specifications
   (*c*) defining quality tools properly (*d*) development of cause and effect diagrams

596. The first step in the quality conformance procedure is to.
   (*a*) Identify causes of below-standard quality
   (*b*) Define the quality characteristics
   (*c*) Measure and control the quality characteristics
   (*d*) Reducing the cost of doing business by eliminating errors

597. Which of the following is not an example of a quality tool or technique
   (*a*) Network analysis (*b*) Six sigma practice
   (*c*) Cause and effect diagram (*d*) Process mapping

598. Which statement is true for sampling inspection
   (*a*) No risk is involved
   (*b*) Characteristics of entire lot can be completely specified
   (*c*) less fatigue results in less mistakes (*d*) all are true

599. Which sampling method is used when simple random sampling method becomes inconvenient to use in large lots
   (*a*) stratified sampling (*b*) double sampling plan
   (*c*) systematic sampling (*d*) cluster sampling

600. Which sampling plan involves maximum data handling analysis
   (*a*) sequential (*b*) single
   (*c*) double (*d*) cluster

601. In order to implement total quality management (TQM) in a business the following culture change is necessary
   (*a*) Adopt statistical quality control techniques
   (*b*) Improvement is the responsibility of the quality department
   (*c*) Mistakes should be criticized and corrected
   (*d*) Suppliers and customers need to be part of the improvement process

602. Which of following is not correct?
   (*a*) Two basic techniques of SQC are: SPC (Statistical Process Control), and SAS (Scientific Acceptance Sampling)
   (*b*) Statistical Process Control (SPC) techniques test for assignable, but not random variations in the system.
   (*c*) A control chart is the evaluation tool in statistical process control.
   (*d*) There are two control charts for attribute: an $\overline{X}$ bar chart for sample means and a sample range chart R.

603. Match the 2 parts below in connection with control charts

Basis of Chart	**Type of Chart**
1. On basis of number of defects/limit	A. C chart
2. On basis of size of variable	B. P chart
3. On basis of defective units in a subgroup	C. R chart
4. dispersion of measured data	D. $\overline{X}$ chart

Use code for answer

(*a*) 1–C, 2–A, 3–B, 4–D (*b*) 1–A, 2–C, 3–D, 4–B

(*c*) 1–C, 2–A, 3–D, 4–B (*d*) 1–A, 2–C, 3–B, 4–D

604. Which chart pattern shows that assignable cause is present

(*a*) trend (*b*) shift

(*c*) extreme variation (*d*) all of above

605. What is the name given to ratio of specification range and process capability

(*a*) tolerance pointers (*b*) capability ratio

(*c*) process index (*d*) production sensitivity

606. Indifferent quality range is the range between LTPD and

(*a*) AQL (*b*) consumer's risk

(*c*) producer's risk (*d*) none of above

607. Pick up wrong statement

(*a*) In sampling operating curve (OC), if we increase the acceptance number and keeping constant sample size then, probability of acceptance increases, consumer risk increases, and producer risk decreases

(*b*) In sampling OC curve if we keep constant acceptance number and increasing the sample size then, Consumer risk decreases and Producer risk increases

(*c*) Tripple sampling plan is considered best option

(*d*) In acceptance sampling, when there is finite-probability that the lot may be accepted even if the quality is not really good is called consumer's risk

608. The UCL and LCL for $\bar{X}$ chart, using average value of range $\bar{R}$ for sample size n will be

(*a*) $\bar{X} \pm \dfrac{\bar{R}}{d_2\sqrt{n}}$ (*b*) $\bar{X} \pm \dfrac{3\bar{R}}{d_2\sqrt{n}}$

(*c*) $\bar{X} \pm \dfrac{3\bar{R}}{d_2 n}$ (*d*) $\bar{X} \pm \dfrac{3\bar{R}}{\sqrt{d_2 n}}$

609. Population standard deviation and process capability for a process with mean $\bar{X}$ and average $\bar{R}$ will be

(*a*) $\dfrac{\bar{R}}{d_2}$ and $\dfrac{6\bar{R}}{d_2}$ respectively (*b*) $\dfrac{\bar{R}}{\sqrt{d_2}}$ and $\dfrac{3\bar{R}}{\sqrt{d_2}}$

(*c*) $\sqrt{\dfrac{\bar{R}}{d_2}} + 6\sqrt{\dfrac{\bar{R}}{d_2}}$ (*d*) $\dfrac{\bar{R}}{d_2}$ and $\dfrac{3\bar{R}}{d_2}$

610. UCL for R chart, taking centerline = $\bar{R}$ will be

(*a*) $\bar{R} + \dfrac{3d_3\bar{R}}{d_2}$ (*b*) $\bar{R} + \dfrac{3\bar{R}}{d_2}$

(*c*) $\bar{R} + \dfrac{d_3\bar{R}}{d_2}$ (*d*) $\bar{R} + \dfrac{3\bar{R}d_3}{\sqrt{d_2}}$

611. UCL and LCL for p chart with sample size n and centre line = p will be

(*a*) $p \pm 3\sqrt{\frac{pq}{n}}$ (*b*) $p \pm \frac{3pq}{\sqrt{n}}$

(*c*) $p \pm \frac{3\sqrt{pq}}{n}$ (*d*) $p \pm \frac{3\sqrt{p(1-q)}}{\sqrt{n}}$

612. In the term TQM, Q refers to

(*a*) quality (*b*) product quality

(*c*) process quality (*d*) organisational excellence

613. Process capability ratio is given by

(USL – upper spec limit, LSL – lower spec limit, σ = process standard deviation), (UCL – upper control limit, LCL – lower control limit)

(*a*) $\frac{USL - LSL}{6\sigma}$ (*b*) $\frac{UCL - LCL}{6\sigma}$

(*c*) $\frac{6 \times \sigma}{USL - LSL}$ (*d*) $\frac{6\sigma}{UCL - LCL}$

614. Which pair is not correctly matched

(*a*) ABC control – based on usage value of items

(*b*) XYZ control – based on variance of demand

(*c*) SDE control – based on availability of items

(*d*) HML control – based of defectives in lots.

615. The control chart is a process

(*a*) planning tool (*b*) process monitoring tool

(*c*) control tool (*d*) all are true

616. Which pair is not correctly matched

(*a*) X-chart – Normal distribution (*b*) P-chart – Binomial distribution

(*c*) Quality control chart – β-distribution

(*d*) C-Chart – Poisson's distribution

617. Which of following is popular sampling plan

(*a*) simple random sampling and constant interval sampling plans

(*b*) stratified sampling (*c*) cluster sampling

(*d*) all of above

618. The sampling plan (SP) which offers the required degree of protection to the producer for the least total amount of inspection for a batch is called

(*a*) optimum SP (*b*) economic SP

(*c*) most popular SP (*d*) best SP

619. The population is divided into well defined groups in

(*a*) cluster sampling (*b*) stratified sampling

(*c*) both (*a*) and (*b*) above (*d*) sequential sampling

620. The success of quality control depends on

(*a*) appropriate sampling method (*b*) selecting suitable sample size

(*c*) both (c) and (b) above (*d*) none of above

621. Pick up incorrect statement
   (*a*) Statistical process control (SPC) methods are used to detect the presence of assignable causes of variation and inspection identifies whether processes need corrective action.
   (*b*) Statistical process control charts are useful for measuring the current quality generated by the process and for detecting whether the process is producing desired quality.
   (*c*) Crucial decisions in the design of control-charts are sample size and control limits.
   (*d*) The use of p-charts requires a small sample size than does the use of $\overline{X}$ -charts.

622. Pick up the incorrect statement.
   (*a*) The central line of a control chart is decided by management.
   (*b*) The spread in control limits affects the chances of searching for assignable causes when none exists.
   (*c*) A process can be in statistical control but still may not be producing all of its output within design specifications.
   (*d*) The process capability ratio and the process capability index are quantitative measures used to assess the capability of a process.

623. Pick up wrong statement
   (*a*) Acceptance sampling is concerned with the decision to accept or reject a batch of goods and its design includes parameters AQL, α, LTPD, β, sample size, and the acceptable number of defective items in the sample.
   (*b*) OC curve depicts the chances of rejecting poor quality material, and accepting good quality materials.
   (*c*) Best acceptance sampling plan can be selected by using an operating characteristic curve.
   (*d*) When the lot proportion defective is either very low or very high, double sampling and sequential sampling plans have an advantage over single-sampling plans.

624. Who defines quality as "The totality of features of characteristics of a product/service that bears on its ability to satisfy stated or implied needs".
   (*a*) Juran (*b*) Crossby
   (*c*) Feigenbaum (*d*) ISO 8402

625. What is the aim of fool proofing technique used for total quality management
   (*a*) to specify time schedule (*b*) to ensure just in time criteria
   (*c*) to achive zero defects (*d*) to deploy six sigma technique

626. Select wrong statement
   (*a*) An operating characteristic curve shows how well an acceptance plan discriminates between good and bad lots.
   (*b*) An acceptance plan is mutually agreed to by a supplier and a purchaser, regarding whether to accept a lot of goods based on sampling and testing from the lot or batch.
   (*c*) Three Scientific Acceptance Sampling (SAS) plans are single, double and sequential sampling.
   (*d*) Lot tolerance per cent defective (LTPD) is the level of quality considered acceptable.

627. Select the right statement
   (*a*) Inspection enables defects from occurring.

(*b*) Prevention helps in catching the errors that have occurred
(*c*) Attribute sampling is binary
(*d*) Variable sampling does not exist

628. Which of following represents attribute gauge and variable type gauge respectively
(*a*) dial gauge, plug gauge (*b*) plug gauge, dial gauge
(*c*) slip gauge, ring gauge (*d*) micrometer, angle gauge

629. A system comprising a number of analytical measurement and tools used to assess the quality of products is called
(*a*) Quality control (*b*) Quality assurance
(*c*) Quality management (*d*) SQC

630. A system pertaining to the overall management plan to guarantee the integrity of data is called
(*a*) Quality control (*b*) Quality assurance
(*c*) Quality management (*d*) SQC

631. A system using statistical techniques to provide a reliable and economical method for ensuring quality control is called
(*a*) Quality control (*b*) Quality assurance
(*c*) Quality management (*d*) SQC

632. Pick up incorrect statement
(*a*) The OC curves show how the probability of accepting a lot decreases with an increasing proportion defective in the lot
(*b*) When rectified inspection is used to replace defective items with good items, the material leaving the inspection area will have a lower quality level.
(*c*) Process capability is a measure of the natural behavior of the process after special causes of variation are eliminated.
(*d*) Process capability is usually expressed via indices and ratios relating the normal variation of the process to the engineering specifications.

633. What does the standard deviation tell about a data set?
(*a*) How diverse the population is (*b*) Provides variance analyses
(*c*) The specification limits of the population
(*d*) The range of data points within the population

634. Quality management theory is characterized by which of the following statements:
(*a*) Inspection is the most important element for ensuring quality
(*b*) Planning and prevention for quality must be emphasized over inspection
(*c*) Contingency planning, part of risk management, is a critical element of quality assurance
(*d*) Quality planning quantifies efforts to exceed customer expectations.

635. Which of the following is not emphasized in quality management
(*a*) Customer satisfaction (*b*) Team responsibility
(*c*) Phases within process (*d*) Prevention over inspection

636. Quality audits are an important part of quality management because they
(*a*) allow for quantification of the risk
(*b*) randomly audit product results to check that quality standards are met

(*c*) check to see if the quality process is being followed
(*d*) do not allow team members time to cover up defects.

637. Which is not the concept of TQM
(*a*) Customer focus and continuous improvement
(*b*) robust design of experiments
(*c*) employee empowerment and use of quality tools
(*d*) product design to meet customer expectation, and building quality into the process

638. Match the two parts below regarding quality control tools

1.	Identifying potential causes for a particular quality related problem	A.	Scatter diagram
2.	To asses the degree of correction between two variables	B.	Checklist
3.	To separate the most significant problems from the relatively less significant ones.	C.	Fish-bone diagram
4.	A list of common defects and their frequency of occurrence	D.	Pareto analysis

Code
(*a*) 1–A, 2–C, 3–D, 4–B (*b*) 1–C, 2–A, 3–B, 4–D
(*c*) 1–A, 2–B, 3–C, 4–D (*d*) 1–C, 2–A, 3–D, 4–B

639. Pick up incorrect statement
(*a*) Benchmarking takes results from previous projects and thus is not suitable to help measure quality on the project.
(*b*) Audits are tools of Perform Quality Assurance program to ensure that the process is being followed.
(*c*) Pareto charts rank defects from greatest to least.
(*d*) Total Quality Management stresses, among other things, that everyone contributes to the quality of the product and process.

640. The goal of six sigma is to _________ the standard derivation of the process.
(*a*) reduce (*b*) increase
(*c*) make zero (*d*) none of above

641. The aim of six-sigma is to see that six standard deviation can ________ within the tolerance limits.
(*a*) exceed (*b*) fit
(*c*) lie much below (*d*) none of above

642. Six sigma system has developed a green belt programme meant for
(*a*) Project team members (*b*) quality circles
(*c*) environmental group (*d*) focussing on customer needs

643. Six sigma system intends to ensure that six standard deviation of process can fit within
(*a*) control limits (*b*) range of observed values
(*c*) process capability of machine (*d*) tolerance limits of product

644. Pick up wrong statement about six sigma
   (*a*) Six-Sigma is a popular philosophy of quality management that focuses on achieving very high levels of quality by controlling the process and reducing defects.
   (*b*) The basic tool of six-sigma process is DMAIC. (Define, Measure, Analyse, Improve, Control)
   (*c*) Six-Sigma can be used for improvements to existing processes, and to develop and implement major changes in process.
   (*d*) Six-sigma is a technique of examining controlled changes of input factors and the observation of resulting changes in outputs.

645. Match the two parts below relating to TQM techniques

1.	Define, measure, analyse, improve and control (DMAIC)	A.	Quality circles
2.	Team building and employee involvement	B.	Quality Function Deployment (QFD)
3.	House of quality	C.	Poka Yoka
4.	Mistake proofing	D.	Six Sigma

   Codes for answer
   (*a*) 1–D, 2–A, 3–B, 4–C (*b*) 1–A, 2–D, 3–C, 4–B
   (*c*) 1–D, 2–A, 3–C, 4–B (*d*) 1–A, 2–B, 3–C, 4–D

646. Concept of 80–20 rule is used in
   (*a*) fish bone diagram (*b*) scatter diagram
   (*c*) Pareto analysis (*d*) Quality function deployment (QFD)

647. Which tool is applicable to ensure fool proof system to ensure no defectives
   (*a*) Poka Yoka (*b*) Six sigma
   (*c*) Fish bone diagram (*d*) Pareto chart

648. Pick up false statement
   (*a*) Six sigma is represented by normal distribution curve
   (*b*) Goal of six sigma is to reduce number of defects to 2.4 parts/billion
   (*c*) Poka yoka is helpful tool to eliminate human errors and other mistakes
   (*d*) Fault tree analysis method is used to determine reliability of product

649. Quality assurance deals with activities which
   (*a*) prove that products and services meet the required quality standard
   (*b*) aim at customer satisfaction
   (*c*) enable control of quality by inspection
   (*d*) all of above

650. Waterfall method is observed in
   (*a*) flow chart (*b*) brainstorming
   (*c*) fish bone diagram (*d*) product design and development

651. Quality of performance is the sum of quality of design and
   (*a*) quality of conformance (*b*) quality of marketing
   (*c*) quality of manufacturing (*d*) quality of company

652. The most economic sampling plan is one whose acceptance number results in
(*a*) minimum average total inspection per batch
(*b*) achieving close to LTPD
(*c*) satisfying consumer's risk (*d*) satisfying producer's risk

653. If n be sample size in a batch of N items and α be producer's risk, then total inspection per batch in the long run will be
(*a*) n + nR (*b*) n + (N – n)R
(*c*) n + (1 – n)R (*d*) nR + (N – n)

654. A qualitative variable that can take one of only two values is called
(*a*) digital (*b*) variable
(*c*) attribute (*d*) limit

655. The primary focus in TQM is
(*a*) to identify and meet customer needs and preference
(*b*) to do work right first time (*c*) to adopt zero defects philosophy
(*d*) to user in quality revolution by quality circles

656. Match the two parts below in relation to quality control

1.	Relative precision index. (work tolerance/average value of ranges) of sub-groups is a useful indicator to determine whether	A.	range
2.	A convenient substitute for the variability of the process	B.	Quality function deployment (QFD)
3.	Techniques that promote team building and employee involvement	C.	process can be allowed to drift before taking corrective action
4.	Matrix looking like house of quality	D.	Quality circle

Codes for answer
(*a*) 1–C, 2–A, 3–D, 4–B (*b*) 1–C, 2–A, 3–B, 4–D
(*c*) 1–A, 2–B, 3–C, 4–D (*d*) 1–D, 2–C, 3–B, 4–A

657. Poka-Yoka, a Japanese technique prevents occurrence of mistakes/errors during production. It involves
(*a*) study entire manufacturing process and identify critical areas
(*b*) evolving fool proofing mechanisms
(*c*) principles of elimination, prevention, replacement, facilitation, detection, and mitigation
(*d*) all of above

658. Which of following statement are correct
(*a*) The most usual way of modelling continuous improvement is Plan-Do-Check Act cycle
(*b*) QFD is a system of designing a product or service, based on customer requirements, with the participation of members of all functions of the organization.

(c) While ISO 9000 incurs appraisal costs and prevention costs, TQM involves investment in prevention through improvement activities.
(d) All of above are correct

659. Which is not correct about the main principles involved in TQM?
(a) Everyone in the organizaton is involved in improving on a continuous basis the process under his or her control and each individual takes responsibility for the quality assurance
(b) teamwork is practised and every individual's skills and abilities are fully utilized
(c) supplier and customers are integrated into the improvement process and participation and development of employees in the business are encouraged;
(d) Training and education are given low priority

660. Pick up false statement
(a) Quality function deployment (QFD) uses a chart called the "house of quality" to encourage interfunctional planning and communication
(b) Continuous improvement involves identifying benchmarks of excellent practice and instilling a sense of ownership among employees.
(c) Total Quality Management occurs when an entire organizational culture becomes focused on quality and customer satisfaction
(d) Top management commitment and leadership is immaterial to attain and sustain quality improvement but benchmarking and focusing on customers are important for quality.

661. Which pair is not correctly matched
(a) ISO 9000-series – Improve product quality and services in industries/organization
(b) ISO:14000-series – Environmental management
(c) ISO 14040 series – Life cycle assessment
(d) ISO:18000-series – Emission standard.

662. The objective of ISO:9000 series on quality management
(a) customer satisfaction (b) Employer satisfaction
(c) setting up quality circles (d) Environment issues

663. The main objectives of ISO standard are
(a) Achieve, maintain and continuously improve product quality
(b) improve quality of operations to continuously meet customer's and investor's needs
(c) provide confidence to integral management and other employees that quality requirements are being fulfilled.
(d) All of the above

664. Which of following is not associated with machine vision system?
(a) image generation and digitization
(b) image processing, analysis and interpretation
(c) giving direction to instruction station for action
(d) actualizing scanning inspection microscope

665. Modern vision systems use following type of camera
(a) Vidicon or analog camera (b) solid state camera
(c) digital camera (d) charge-coupled-device (CCD) camera

666. Match the two parts below relating to image processing and analysis in machine vision system

1.	Popular technique for image processing	A. edge detection
2.	Process of converting the pixel values into either of the 2 values	B. feature extraction
3.	Process of distinguishing the image of the object from its surroundings	C. thresholding
4.	Process of identifying the work part by comparing with the templates of known objects	D. segmentation

Codes:

(*a*) 1 – C, 2 – B, 3 – A, 4 –B
(*b*) 1 – D, 2 – C, 3 – A, 4 –B
(*c*) 1 – D, 2 – C, 3 – B, 4 –A
(*d*) 1 – A, 2 – B, 3 – D, 4 –C

667. Pick up false statement about machine vision systems.
(*a*) It enables geometric tests of alignment accuracy of machine tools
(*b*) It enables identification and orientation of a work part within the field of vision
(*c*) It facilitates automatic inspection
(*d*) It enables identification of surface defects and verification of holes.

668. In machine vision systems, template matching is used for
(*a*) image digitization (*b*) image processing
(*c*) image interpretation (*d*) generation of actuation signals.

669. Coordinate measuring machine (CMM) enables the location of point coordinates in
(*a*) 0–D space (*b*) 1–D space
(*c*) 2–D space (*d*) 3–D space

670. In template matching and pattern recognition process in machine vision systems, if two objects have closely resembling features, the correct identification is done by
(*a*) feature weighting (*b*) feature highlighting
(*c*) feature indexing (*d*) feature elimination

671. In machine vision systems, a discrete addressable picture element is called
(*a*) pixel (*b*) pixar
(*c*) vision perimeter (*d*) buffer

672. In machine vision system, 3D image is captured using
(*a*) one camera (*b*) at least two cameras
(*c*) electron microscope (*d*) optical microscope

673. The most popular technique for image processing in machine vision system is
(*a*) ADC (*b*) segmentation
(*c*) thresholding (*d*) pattern recognition

674. Nanometrology–*i.e.*, techniques for measuring at nanoscale are based on physical phenomena observed on
(*a*) particle interactions (*b*) forces at nanoscale
(*c*) atomic force microscopy and X-ray diffraction
(*d*) All of the above

675. Match the two parts below concerning nanometrology

1. This measurement technique consists of a silicon cantilever with a sharp tip (probe)	A. Scanning Electro Microscope

with a radius of curvature of a few nm. The forces at atomic level between tip and specimen deflect the tip and it is sensed by a laser spot and recorded by an array of photodiodes.

2. In this instrument a conducting tip is brought near the surface, and voltage difference applied between tip and surface allows electrons to tunnel through the vacuum between them	B. Atomic force microscope
3. In this instrument, the surface is scanned with a high energy electron beam. Interactions (back transmission of electron, secondary electrons scattering) between them are detected	C. X-ray diffraction
4. In this method, a crystalline structure causes X-rays to diverge. Angle of diffractions is measured.	D. Scanning Tunnelling Microscope

Codes:

(*a*) 1 – B, 2 – C, 3 – A, 4 – D
(*b*) 1 – D, 2 – A, 3 – B, 4 – C
(*c*) 1 – B, 2 – D, 3 – A, 4 – C
(*d*) 1 – D, 2 – B, 3 – C, 4 – A

676. Match the two parts below pertaining to microscopes used in nanometrology

Microscope	**Application**
1. Scanning tunnelling microscope	A. Most versatile microscope with magnification ranging from 5x × to 10^6x.
2. Scanning electron microscope	B. Allows user to switch between image mode to diffraction mode
3. Atomic force microscope	C. Provides 3-D atomic scals images and used for plotting, surface topography of nanomaterials
4. Transmission electron microscope	D. Having best resolution upto atomic level

Codes:

(*a*) 1 – A, 2 – C, 3 – B, 4 – D (*b*) 1 – C, 2 – A, 3 – B, 4 – D
(*c*) 1 – A, 2 – C, 3 – D, 4 – B (*d*) 1 – C, 2 – A, 3 – D, 4 – B

677. Which of following as a not a 2-D nanostructure?
(*a*) nano coating (*b*) nanofilm
(*c*) nanosheet (*d*) nanotube

678. In which technique, primary electrons are focussed into a small diameter electron probe?
(*a*) SEM (*b*) TEM
(*c*) STEM (*d*) XRD

679. Which measurement technique offers two modes of usage (viz, imaging mode, and differ action mode)?

(*a*) SEM (*b*) TEM
(*c*) STEM (*d*) XRD

680. In which technique, the atoms of the material are excited with ionized radiation and the electrons subsequently emitted by the excited atoms, are detected and analysed?
(*a*) X-ray reflectometry (XRR) (*b*) spectral ellipsometry (SE)
(*c*) X-ray photo electron spectroscopy (XPS)
(*d*) Transmission electron microscope (TEM)

681. Which of following technique has very high accuracy because of atomic resolution?
(*a*) XRR (*b*) SE
(*c*) XPS (*d*) TEM

682. Which inspection system acquires image data of an object, processes and interprets data by a computer program?
(*a*) Flexible impection system (*b*) On-line process in CIM environment
(*c*) Machine vision system (*d*) microscopy

683. The size of atoms is approximately
(*a*) 0.2 nm (*b*) 0.02 nm
(*c*) 2 nm (*d*) 20 nm

684. A nanocomputer uses
(*a*) Y-junction carbon nanotubes (*b*) bioactive nanoparticles
(*c*) nanocrystals (*d*) silicon nanofilaments

685. Which is not a device used for measurement of nanosize materials?
(*a*) microscopy (*b*) X-ray diffraction
(*c*) nanolithography (*d*) atomic force microscope

686. Which microscope works on the principle of quantum tunnelling?
(*a*) AFM (*b*) XRD
(*c*) SEM (*d*) STM

687. Which microscope provides 3-D atomic scale images of the surface being tested?
(*a*) STM (*b*) SEM
(*c*) TEM (*d*) quantum microscope

688. Pick of false statement about nanotechnology?
(*a*) Composites made from nano-size particles are much stronger compared to existing materials-science models.
(*b*) Metals with grain size of around 10 nm are about 7 times harder and tougher than their ordinary counter parts with grain sizes in range of 100 nm.
(*c*) Quantum effects can begin to dominate the behaviour of matter at the nanoscale, affecting the optical/electrical/magnetic behaviour of materials.
(*d*) Nanoparticles have a relatively smaller surface area compared to same mass of material produced in a larger form.

689. The best device for plotting surface topography of nanomaterials is
(*a*) optical microscope (*b*) electron microscope
(*c*) scanning tunnelling microscope (*d*) nanosensors

690. Which of following in included under nanomaterials?
(*a*) bulk powders in form of dispersions of nanoparticles
(*b*) bundles of nanowires
(*c*) nanotubes and multi-nanolayers (*d*) all of the above

691. The basic component of an atomic force microscope is
(*a*) piezoelectric transducer (*b*) field emission gun
(*c*) gonimeter (*d*) electron gun

692. In an XRD system, which component enables to control the geometric relationship between the incident beam, sample, and detector
(*a*) gonimeter (*b*) electron gun
(*c*) laser microscope (*d*) interlaced scanner

693. Which choice is correct about nanotechnology?
Nanotechnology is manipulation of matter on
(*a*) atomic scale (*b*) supramolecular scale
(*c*) molecular scale (*d*) all of the above

694. Nanoparticles processing nanodimensions in all the three principal directions are called
(*a*) quantum dots (*b*) nanotubes
(*c*) nanowires (*d*) thin film layers

695. While properties of nanoparticles with size of 50 nm is similar to those of bulk materials, but in the range of 10-50 nm size, properties vary _____.
(*a*) linearly (*b*) remain same
(*c*) parabolically (*d*) erratically

696. Topography i.e., size and shape/surface features composition/crystellographic/information of nanoparticles can be measured using.
(*a*) atomic force microscopy (*b*) X-ray diffraction techniques
(*c*) scanning electron microscopy (*d*) transmission electron microscopy

697. Very high resolution of the order of 0.1 nm is obtained in
(*a*) AFM (*b*) SEM
(*c*) TEM (*d*) AFM and TEM

698. Nanotechnology involves understanding and control of matter at dimensions between approximately 1 and 10 nm.
Which is true about nanoscale science. It involves
(*a*) imaging (*b*) measuring
(*c*) modelling and manipulating of matter at length scale of 1-100 nm.
(*d*) all of the above

699. The resolution of optical microscope can be improved by
(*a*) glazing the surface
(*b*) coating the surface with thin polished metal
(*c*) increasing intensity of incident light
(*d*) decreasing the wavelength of light

700. Which incident technique enables 3-D atomic scale images of the surface?
(*a*) STM (*b*) SEM
(*c*) TEM (*d*) all of the above

701. Which law provides relationship between the diffraction pattern and the structure of material
(*a*) Rutherford law (*b*) Bragg law
(*c*) Ludlow law (*d*) Louis law

702. In which microscope, image is not formed by focussing light or electrons on to a surface
(*a*) AFM (*b*) Nanomer

(c) Graphene (d) SEM

703. Traceability at microscale is achieved using artefacts like
(a) scales and straight edges (b) step gauges
(c) laser interferometers (d) all of the above

704. At nanoscale, the following surface is used as calibration artefact for achieving traceability
(a) Mica (b) Silicon
(c) Crystalline highly oriented pyrolytic graphite
(d) all of the above

705. Zero dimensional (0-D) nanostructure is one in which
(a) one dimension is outside the nm rangle
(b) three dimensions of nanostructure are outside the nm range
(c) all dimension of nanostructure are in the nm range
(d) none of the above

706. Which of following is not a measurement tool for nanomaterials/nanoparticles?
(a) AFM, SEM, interferometery
(b) opto-acoustic tools
(c) gas adsorption method
(d) motion caused by thermal drift

Answers

1. (d)	2. (a)	3. (b)	4. (d)	5. (a)
6. (a)	7. (d)	8. (d)	9. (a)	10. (b)
11. (c)	12. (c)	13. (b)	14. (a)	15. (c)
16. (d)	17. (d)	18. (a)	19. (d)	20. (e)
21. (b)	22. (d)	23. (b)	24. (c)	25. (a)
26. (b)	27. (d)	28. (c)	29. (d)	30. (d)
31. (a)	32. (d)	33. (a)	34. (c)	35. (c)
36. (b)	37. (c)	38. (b)	39. (d)	40. (a)
41. (a)	42. (b)	43. (a)	44. (d)	45. (e)
46. (a)	47. (b)	48. (d)	49. (c)	50. (c)
51. (d)	52. (c)	53. (d)	54. (a)	55. (a)
56. (b)	57. (d)	58. (b)	59. (d)	60. (a)
61. (b)	62. (a)	63. (e)	64. (d)	65. (d)
66. (b)	67. (d)	68. (d)	69. (b)	70. (d)
71. (e)	72. (b)	73. (c)	74. (a)	75. (a)
76. (d)	77. (c)	78. (a)	79. (a)	80. (b)
81. (b)	82. (c)	83. (c)	84. (e)	85. (a)
86. (c)	87. (a)	88. (e)	89. (c)	90. (c)
91. (c)	92. (a)	93. (d)	94. (c)	95. (b)
96. (e)	97. (e)	98. (d)	99. (a)	100. (c)
101. (e)	102. (c)	103. (c)	104. (c)	105. (e)

106. (*a*)	107. (*a*)	108. (*d*)	109. (*d*)	110. (*e*)
111. (*e*)	112. (*e*)	113. (*d*)	114. (*e*)	115. (*e*)
116. (*c*)	117. (*e*)	118. (*a*)	119. (*a*)	120. (*a*)
121. (*c*)	122. (*a*)	123. (*a*)	124. (*b*)	125. (*d*)
126. (*c*)	127. (*c*)	128. (*c*)	129. (*c*)	130. (*d*)
131. (*b*)	132. (*c*)	133. (*b*)	134. (*e*)	135. (*a*)
136. (*b*)	137. (*e*)	138. (*c*)	139. (*c*)	140. (*b*)
141. (*d*)	142. (*a*)	143. (*e*)	144. (*d*)	145. (*b*)
146. (*b*)	147. (*a*)	148. (*a*)	149. (*e*)	150. (*b*)
151. (*d*)	152. (*b*)	153. (*e*)	154. (*a*)	155. (*b*)
156. (*e*)	157. (*a*)	158. (*c*)	159. (*c*)	160. (*a*)
161. (*e*)	162. (*d*)	163. (*a*)	164. (*d*)	165. (*c*)
166. (*b*)	167. (*e*)	168. (*d*)	169. (*c*)	170. (*d*)
171. (*a*)	172. (*c*)	173. (*b*)	174. (*b*)	175. (*d*)
176. (*c*)	177. (*c*)	178. (*d*)	179. (*a*)	180. (*a*)
181. (*a*)	182. (*a*)	183. (*d*)	184. (*d*)	185. (*d*)
186. (*d*)	187. (*a*)	188. (*b*)	189. (*b*)	190. (*d*)
191. (*d*)	192. (*c*)	193. (*a*)	194. (*d*)	195. (*a*)
196. (*a*)	197. (*c*)	198. (*a*)	199. (*d*)	200. (*a*)
201. (*e*)	202. (*d*)	203. (*d*)	204. (*a*)	205. (*a*)
206. (*a*)	207. (*d*)	208. (*b*)	209. (*a*)	210. (*c*)
211. (*d*)	212. (*c*)	213. (*a*)	214. (*b*)	215. (*c*)
216. (*a*)	217. (*d*)	218. (*d*)	219. (*a*)	220. (*e*)
221. (*b*)	222. (*d*)	223. (*a*)	224. (*c*)	225. (*b*)
226. (*c*)	227. (*b*)	228. (*b*)	229. (*a*)	230. (*a*)
231. (*a*)	232. (*d*)	233. (*b*)	234. (*a*)	235. (*b*)
236. (*a*)	237. (*c*)	238. (*a*)	239. (*b*)	240. (*b*)
241. (*a*)	242. (*d*)	243. (*c*)	244. (*c*)	245. (*a*)
246. (*a*)	247. (*e*)	248. (*d*)	249. (*a*)	250. (*a*)
251. (*b*)	252. (*d*)	253. (*a*)	254. (*c*)	255. (*b*)
256. (*d*)	257. (*b*)	258. (*a*)	259. (*a*)	260. (*c*)
261. (*c*)	262. (*d*)	263. (*a*)	264. (*a*)	265. (*c*)
266. (*c*)	267. (*c*)	268. (*b*)	269. (*a*)	270. (*a*)
271. (*c*)	272. (*a*)	273. (*a*)	274. (*b*)	275. (*d*)
276. (*b*)	277. (*a*)	278. (*d*)	279. (*a*)	280. (*d*)
281. (*c*)	282. (*d*)	283. (*c*)	284. (*c*)	285. (*d*)
286. (*c*)	287. (*a*)	288. (*d*)	289. (*a*)	290. (*e*)
291. (*a*)	292. (*b*)	293. (*a*)	294. (*b*)	295. (*b*)
296. (*d*)	297. (*c*)	298. (*c*)	299. (*c*)	300. (*d*)
301. (*a*)	302. (*e*)	303. (*a*)	304. (*b*)	305. (*b*)
306. (*a*)	307. (*d*)	308. (*b*)	309. (*a*)	310. (*a*)

311. (*b*)	312. (*a*)	313. (*c*)	314. (*c*)	315. (*d*)
316. (*a*)	317. (*c*)	318. (*d*)	319. (*d*)	320. (*a*)
321. (*c*)	322. (*a*)	323. (*d*)	324. (*d*)	325. (*c*)
326. (*d*)	327. (*a*)	328. (*d*)	329. (*b*)	330. (*c*)
331. (*b*)	332. (*b*)	333. (*d*)	334. (*e*)	335. (*c*)
336. (*a*)	337. (*a*)	338. (*a*)	339. (*c*)	340. (*e*)
341. (*d*)	342. (*b*)	343. (*a*)	344. (*c*)	345. (*c*)
346. (*c*)	347. (*a*)	348. (*e*)	349. (*e*)	350. (*d*)
351. (*a*)	352. (*a*)	353. (*b*)	354. (*b*)	355. (*a*)
356. (*d*)	357. (*b*)	358. (*d*)	359. (*d*)	360. (*d*)
361. (*b*)	362. (*a*)	363. (*d*)	364. (*b*)	365. (*d*)
366. (*c*)	367. (*b*)	368. (*b*)	369. (*b*)	370. (*a*)
371. (*a*)	372. (*a*)	373. (*d*)	374. (*a*)	375. (*a*)
376. (*c*)	377. (*d*)	378. (*d*)	379. (*c*)	380. (*b*)
381. (*c*)	382. (*a*)	383. (*b*)	384. (*a*)	385. (*c*)
386. (*b*)	387. (*b*)	388. (*d*)	389. (*d*)	390. (*a*)
391. (*b*)	392. (*a*)	393. (*c*)	394. (*a*)	395. (*b*)
396. (*d*)	397. (*b*)	398. (*a*)	399. (*b*)	400. (*b*)
401. (*a*)	402. (*e*)	403. (*c*)	404. (*c*)	405. (*c*)
406. (*d*)	407. (*d*)	408. (*d*)	409. (*a*)	410. (*c*)
411. (*b*)	412. (*a*)	413. (*b*)	414. (*e*)	415. (*a*)
416. (*c*)	417. (*a*)	418. (*a*)	419. (*b*)	420. (*b*)
421. (*d*)	422. (*d*)	423. (*c*)	424. (*b*)	425. (*a*)
426. (*c*)	427. (*d*)	428. (*a*)	429. (*d*)	430. (*c*)
431. (*a*)	432. (*a*)	433. (*a*)	434. (*a*)	435. (*d*)
436. (*b*)	437. (*c*)	438. (*d*)	439. (*d*)	440. (*a*)
441. (*c*)	442. (*a*)	443. (*a*)	444. (*c*)	445. (*c*)
446. (*c*)	447. (*e*)	448. (*e*)	449. (*a*)	450. (*a*)
451. (*b*)	452. (*c*)	453. (*a*)	454. (*b*)	455. (*e*)
456. (*a*)	457. (*c*)	458. (*c*)	459. (*d*)	460. (*d*)
461. (*e*)	462. (*e*)	463. (*a*)	464. (*b*)	465. (*d*)
466. (*d*)	467. (*b*)	468. (*a*)	469. (*c*)	470. (*c*)
471. (*a*)	472. (*b*)	473. (*c*)	474. (*c*)	475. (*b*)
476. (*c*)	477. (*c*)	478. (*a*)	479. (*a*)	480. (*e*)
481. (*b*)	482. (*c*)	483. (*a*)	484. (*d*)	485. (*a*)
486. (*c*)	487. (*d*)	488. (*c*)	489. (*b*)	490. (*c*)
491. (*a*)	492. (*a*)	493. (*a*)	494. (*a*)	495. (*a*)
496. (*a*)	497. (*c*)	498. (*c*)	499. (*d*)	500. (*c*)
501. (*a*)	502. (*d*)	503. (*d*)	504. (*d*)	505. (*c*)
506. (*d*)	507. (*d*)	508. (*c*)	509. (*c*)	510. (*b*)
511. (*c*)	512. (*b*)	513. (*d*)	514. (*d*)	515. (*a*)

516. (*d*)	517. (*d*)	518. (*a*)	519. (*b*)	520. (*a*)
521. (*a*)	522. (*d*)	523. (*d*)	524. (*d*)	525. (*d*)
526. (*a*)	527. (*d*)	528. (*d*)	529. (*a*)	530. (*a*)
531. (*c*)	532. (*c*)	533. (*c*)	534. (*a*)	535. (*c*)
536. (*b*)	537. (*d*)	538. (*d*)	539. (*c*)	540. (*c*)
541. (*b*)	542. (*d*)	543. (*a*)	544. (*a*)	545. (*d*)
546. (*a*)	547. (*a*)	548. (*c*)	549. (*a*)	550. (*b*)
551. (*d*)	552. (*d*)	553. (*c*)	554. (*a*)	555. (*c*)
556. (*a*)	557. (*d*)	558. (*b*)	559. (*c*)	560. (*c*)
561. (*d*)	562. (*c*)	563. (*c*)	564. (*c*)	565. (*c*)
566. (*c*)	567. (*a*)	568. (*a*)	569. (*b*)	570. (*a*)
571. (*a*)	572. (*d*)	573. (*a*)	574. (*c*)	575. (*d*)
576. (*a*)	577. (*d*)	578. (*a*)	579. (*d*)	580. (*c*)
581. (*a*)	582. (*d*)	583. (*d*)	584. (*d*)	585. (*c*)
586. (*b*)	587. (*b*)	588. (*a*)	589. (*b*)	590. (*c*)
591. (*b*)	592. (*c*)	593. (*c*)	594. (*d*)	595. (*b*)
596. (*b*)	597. (*a*)	598. (*c*)	599. (*a*)	600. (*a*)
601. (*d*)	602. (*d*)	603. (*a*)	604. (*b*)	605. (*b*)
606. (*a*)	607. (*c*)	608. (*d*)	609. (*a*)	610. (*a*)
611. (*a*)	612. (*d*)	613. (*a*)	614. (*b*)	615. (*d*)
616. (*c*)	617. (*d*)	618. (*b*)	619. (*c*)	620. (*c*)
621. (*d*)	622. (*a*)	623. (*b*)	624. (*d*)	625. (*c*)
626. (*d*)	627. (*c*)	628. (*b*)	629. (*a*)	630. (*b*)
631. (*d*)	632. (*b*)	633. (*a*)	634. (*b*)	635. (*c*)
636. (*c*)	637. (*b*)	638. (*d*)	639. (*a*)	640. (*a*)
641. (*b*)	642. (*a*)	643. (*d*)	644. (*d*)	645. (*a*)
646. (*c*)	647. (*a*)	648. (*b*)	649. (*a*)	650. (*d*)
651. (*a*)	652. (*a*)	653. (*b*)	654. (*c*)	655. (*a*)
656. (*a*)	657. (*d*)	658. (*d*)	659. (*d*)	660. (*d*)
661. (*d*)	662. (*a*)	663. (*d*)	664. (*d*)	665. (*d*)
666. (*b*)	667. (a)	668. (*c*)	669. (*d*)	670. (*a*)
671. (*a*)	672. (*b*)	673. (*b*)	674. (*d*)	675. (*c*)
676. (*d*)	677. (*d*)	678. (*a*)	679. (*b*)	680. (*c*)
681. (*b*)	682. (*c*)	683. (*a*)	684. (*a*)	685. (*c*)
686. (*d*)	687. (*a*)	688. (*d*)	689. (*c*)	690. (*d*)
691. (*a*)	692. (*a*)	693. (*d*)	694. (*a*)	695. (*a*)
696. (*c*)	697. (*d*)	698. (*a*)	699. (*d*)	700. (*a*)
701. (*b*)	702. (*a*)	702. (*a*)	703. (*d*)	704. (*d*)
705. (*c*)	706. (*d*)			

EXPLANATION

58. (*b*) Calibration has no effect on random errors but it helps to reduce systematic errors.

206. (*a*) The correct answer is fundamental deviation.

537. (*b*) Cause and effect diagram enables to focus on the root causes of the problem rather than on the history of the problem.

688. (*d*) Larger surface area of nano-particles makes them more chemically reactive and stronger.

Solved Questions & Answers

1. Give the following Answers.
   (*a*) What is the relationship between tolerances and the size of the products?
   (*b*) For tolerance grades IT6 to IT16, how much each grade increases from the previous one?
   (*c*) The tolerance zone is governed by two limits. What are these limits?
   (*d*) The position of tolerance zone from zero line is governed by ______?
   (*e*) How many classes of basic deviations for holes and shafts are defined by the ISO?

2. The hole $40^{+0.020}_{+0.000}$ and shaft $40^{+0.010}_{-0.010}$ when assembled will result in
   (*a*) clearance fit (*b*) interference fit
   (*c*) transition fit (*d*) drive fit

3. A shaft of diameter $20^{+0.05}_{-0.10}$ mm and a hole of diameter $20^{+0.20}_{+0.10}$ mm when assembled would yield
   (*a*) Transition fit (*b*) Interference fit
   (*c*) Clearance fit (*d*) None of these

4. A bush was turned after mounting the same on a mandrel. The mandrel diameter in millimeters is $40^{+0.00}_{-0.05}$ and bore diameter of bush is $40^{+0.06}_{-0.010}$. The maximum eccentricity of the bush in mm will be
   (*a*) 0.01 (*b*) 0.055
   (*c*) 0.1 (*d*) 0.11

5. In an engineering drawing one finds the designation of 20G7f8, the position of tolerance of the hole is indicated by
   (*a*) Letter G (*b*) Letter f
   (*c*) Number 7 (*d*) Number 8

6. The fit on a hole-shaft pair system is specified as H7-h6, the type of fit is
   (*a*) Clearance fit (*b*) Running fit (Sliding fit)
   (*c*) Push fit (transition fit) (*d*) Force fit (Interference fit)

7. Two shafts A and B have their diameters specified as 100 ± 0.1 mm and 0.1 ± 0.0001 mm respectively. Which of the following statements is/are true
   (*a*) Tolerance in the dimension is greater in shaft A
   (*b*) The relative error in the dimension is greater in shaft A
   (*c*) Tolerance in the dimension is greater in shaft B
   (*d*) The relative error in the dimension is same for shaft A and shaft B

8. 3 blocks B_1, B_2 and B_3 are to be inserted in a channel of width S, maintaining a minimum gap of width T = 0.125 mm, as shown in figure. For P = 18.75 ± 0.08, Q = 25.00 ± 0.12, R = 28.125 ± 0.1 and S = 72.35 + X, the value of tolerance X is?

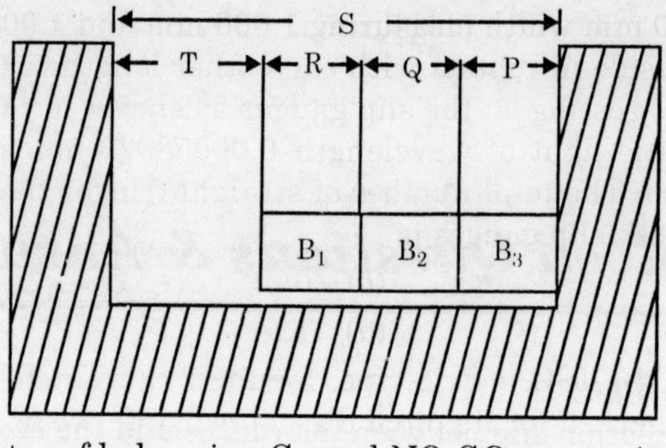

9. Checking the diameter of hole using Go and NO-GO gauges in an example of inspection by ____________ (Variable/attributes).
10. Slip gauges are calibrated by outside micrometer (T/F)
11. The dimensional limits on a shaft of 25h7 are
    (*a*) 25.000, 25.021 mm (*b*) 25.000, 24,979 mm
    (*c*) 25.000, 25.007 mm (*d*) 25.000, 24.993 mm
12. In an interchangeable assembly, shafts of size $25.000^{-0.040}_{-0.0100}$ mm mate with holes of size $25.000^{+0.020}_{-0.000}$ mm. The maximum possible clearance in the assembly will be
    (*a*) 10 microns (*b*) 20 microns
    (*c*) 30 microns (*d*) 60 microns
13. GO and NO-GO plug gauges are to be designed for a hole $20.000^{-0.050}_{-0.010}$ mm. Gauge tolerances can be taken as 10% of the hole tolerance. Following ISO system of gauge design, sizes of GO and NO-GO gauge will be respectively
    (*a*) 20.010 mm and 20.050 mm (*b*) 20.014 mm and 20.046
    (*c*) 20.006 mm and 20.054 mm (*d*) 20.014 mm and 20.054
14. A part shown in the figure is machined to the sizes given below

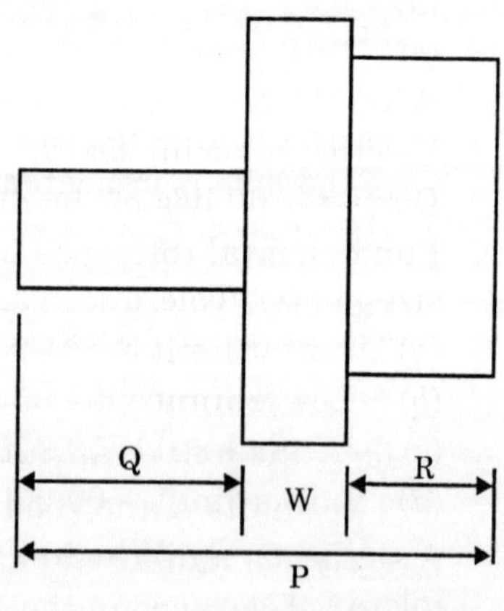

    $P = 35.00 \pm 0.08$ mm
    $Q = 12.00 \pm 0.02$ mm
    $R = 13.00^{+0.04}_{-0.02}$
    With 100% confidence, the resultant dimension W will have the specification
    (*a*) 9.99 ± 0.03 mm
    (*b*) 9.99 ± 0.13 mm
    (*c*) 10.00 ± 0.03 mm
    (*d*) 10.00 ± 0.13 mm
15. Allowance in limits and fits refers to
    (*a*) maximum clearance between shaft and hole
    (*b*) minimum clearance between shaft and hole
    (*c*) difference between maximum and minimum size of hole
    (*d*) difference between maximum and minimum size of shaft

16. Two slip gauges of 10 mm width measuring 1.000 mm and 1.002 mm are kept side by side in contact with each other lengthwise. An optical flat is kept resting on the slip gauges as shown in the figure. Mono-chromatic light of wavelength 0.00058928 mm is used in the inspection. The total number of straight fringes that can be observed on both slip gauges is

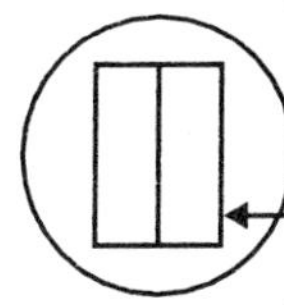

(*a*) 2 (*b*) 6
(*c*) 8 (*d*) 13

17. A threaded nut of M 16, ISO metric type, having 2 mm pitch with a pitch diameter of 14.701 mm is to be checked for its pitch diameter using two or three number of balls or rollers of the following sizes
(*a*) Rollers of 2 mm φ (*b*) Rollers of 1.155 mm φ
(*c*) Balls of 2 mm φ (*d*) Balls of 1.155 mm φ

18. For a hole and shaft assembly of basic size of 30 mm, tolerances on hole and shaft are $^{+0.021}_{-0.000}$ mm and $^{-0.040}_{-0.075}$ mm respectively. Determines the
(*a*) value of tolerances on hole and shaft
(*b*) maximum and minimum clearances
(*c*) allowance
(*d*) maximum metal limit of hole and shaft, and
(*e*) type of fit

19. Determine the high and low limits of a bush to obtain maximum clearance of 0.10 mm, and minimum clearance of 0.02 mm w.r.t. a shaft of size 40 ± 0.02 mm size.

20. A hole is specified as $40^{0.050}_{0.000}$ mm. The mating shaft has a clearance fit with minimum clearance of 0.01mm. The tolerance on the shaft is 0.04 mm. The maximum clearance in mm between the hole and the shaft is
(*a*) 0.04 (*b*) 0.05
(*c*) 0.10 (*d*) 0.11

21. What are the upper and lower limits of the shaft represented by 60 f_8?
Use the following Data:
Diameter 60 lies in the diameter step of 50-80 mm.
Fundamental tolerance unit, *i*, in μm = 0.45 $D^{1/3}$+0.001D, where D is the representative size in mm; Tolerance value for IT8 = 25i. Fundamental deviation for f shaft = –5.5 $D^{0.41}$
(*a*) Lower limit = 59.925 mm, Upper Limit = 59.970 mm
(*b*) Lower limit = 59.954 mm, Upper Limit = 60.000 mm
(*c*) Lower limit = 59.970 mm, Upper Limit = 60.016 mm
(*d*) Lower limit = 60.000 mm Upper Limit = 60.046 mm

22. A shaft has a dimension, $\phi 35^{-0.009}_{-0.025}$. The respective values of fundamental deviation and tolerance are
(*a*) –0.025, ± 0.008 (*b*) –0.025, ± 0.016
(*c*) –0.009, ± 0.008 (*d*) –0.009, ± 0.016

23. In order to have interference fit, it is essential that the lower limit of the shaft should be
(*a*) greater than the upper limit of the hole
(*b*) lesser than the upper limit of the hole

(*c*) greater than the lower limit of the hole
(*d*) lesser than the lower limit of the hole

24. A ring gauge is used to measure
(*a*) outside diameter but not roundness (*b*) roundness but not outside diameter
(*c*) both outside diameter and roundness (*d*) only external threads

25. A displacement sensor (a dial indicator) measures the lateral displalcement of a mandrel mounted on the taper hole inside a drill spindle. The mandrel axis is an extension of the drill spindle taper hole axis is an extension of the drill spindle taper hole axis and the protruding portion of the mandrel surface is perfectly cylindrical. Measurements are taken with the sensor placed at two positions *P* and *Q* as shown in the figure. The readings are recorded as R_x = maximum deflection minus minimum deflection, corresponding to sensor position at X, over one rotation.

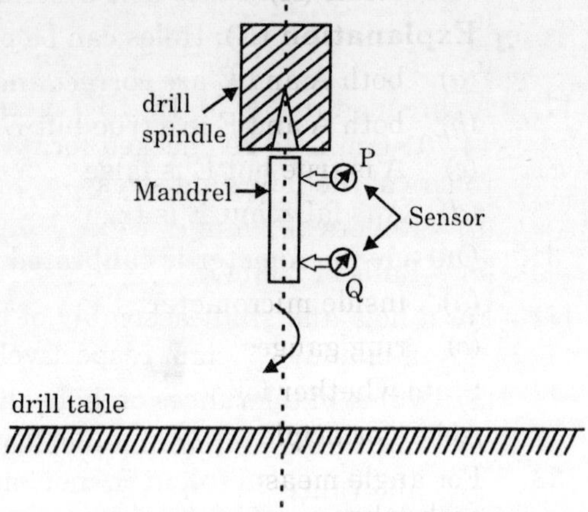

If $R_P = R_Q > 0$, which one the following would be consistent with the observation?
(*a*) The drill spindle rotational axis is coincident with the drill spindle taper hole axis.
(*b*) The drill spindle rotational axis intersects the drill spindle taper hole axis at point P.
(*c*) The driil spindle rotational axis is parallel to the drill spindle taper hole axis
(*d*) The drill spindle rotational axis intersects the drill spindle taper hole axis at point Q

26. Match the following:

Feature to be inspected	Instrument
P. Pitch and angle error of screw thread	1. Auto Collimator
Q. Flatness error of a surface	2. Optical interferometer
R. Alignment error of a machine slideway	3. Dividing Head and Dial Gauge
S. Profile of a cam	4. Spirit Leve!
	5. Sine bar
	6. Tool maker's Microscope

(*a*) P–6, Q–2, R–4, S–6 (*b*) P–5, Q–2, R–1, S–6
(*c*) P–6, Q–4, R–1, S–3 (*d*) P–1, Q–4, R–4, S–2

27. A hole is of dimension $\phi 9^{+0.015}_{+0.000}$ mm. The corresponding shaft is of dimension $\phi 9^{+0.010}_{+0.001}$ mm. The resulting assembly has
(*a*) loose running fit (*b*) close running fit
(*c*) transition fit (*d*) interference fit

28. In an interchangeable assembly, shafts of size $25.000^{+0.040}_{-0.010}$ mm mate with holes of size $25.000^{+0.030}_{+0.020}$ mm. The maximum interference (in microns) in the assembly is
(*a*) 40 (*b*) 30
(*c*) 20 (*d*) 10

29. Outside diameter of bush is turned using a mandrel. The mandrel diameter is maintained as $30.000^{-0.000}_{-0.050}$ mm and the bore diameter is $30^{+0.050}_{-0.000}$ mm. The maximum value in mm of the eccentricity on the bush due to the locating mandrel

(*a*) 0.050 (*b*) 0.100

(*c*) 0.020 (*d*) 0.025

30. **Assertion (A):** Basic hole system is advocated by Indian Standards for fit design.

Explanation (E): Holes can be produced to any size

(*a*) both A and E are correct and E explains A

(*b*) both A and E are true but E does not explains A

(*c*) A is true but E is false

(*d*) A is false but E is true

31. Outside micrometer is calibrated with the help of

(*a*) inside micrometer (*b*) depth micrometer

(*c*) ring gauges (*d*) slip gauges

32. State whether following statement is true or false.

Abbe's principle of alignment states that the scale axis must coincide with workpiece axis.

33. For angle measurement in metrology, the following pair can be used in conjunction with each other

(*a*) sine bar and vernier calipers (*b*) bevel protractor and slip gauges

(*c*) slip gauges and sine bar (*d*) sine bar and bevel protractor

34. In mass manufacturing, two holes in a plate type of component are ideally located with the help of

(*a*) one solid pin in the larger hole and one diamond pin in the smaller hole

(*b*) two solid pins in either holes

(*c*) two diamond pins in either holes

(*d*) one solid pin in the smaller hole and one diamond pin in the larger hole

35. The skid in a stylus type roughness measuring instrument is used to provide ... A.... and define ..... B ..... geometrical profile of the surface.

A. support/datum B. micro/macro

36. A taper hole is inspected using a CMM, with a proble of 2 mm diameter. At a height, Z = 10 mm from the bottom, 5 points are touched and a diameter of circle (not compensated for probe size) is obtained as 20 mm. Similarly, a 40 mm diameter is obtained at a height Z = 40 mm. The smaller diameter (in mm) of hole at Z = 0 is

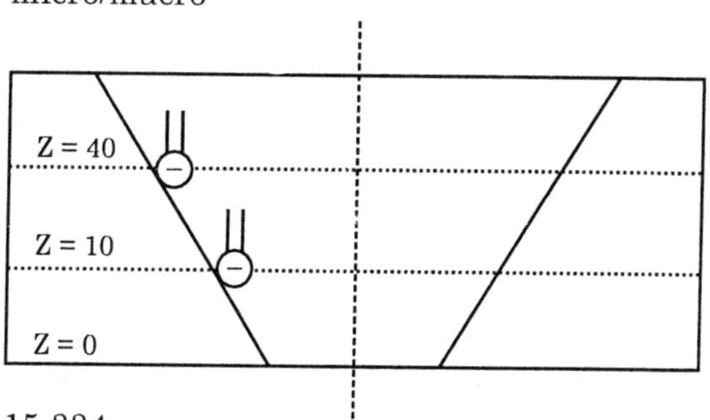

(*a*) 13.334 (*b*) 15.334

(*c*) 15.442 (*d*) 15.542

37. **Statement (I):** In sugarcane crushing rollers, the fit between the cast roll and the forged steel shaft is of interference type

Statement (II): This helps in removing the roll from the shaft whenever not needed.

(*a*) Both Statement (I) and Statement (II) are individually true and Statement (II) is the correct explanation of Statement (I).
(*b*) Both Statement (I) and Statement (II) are individually true but Statement (II) is not the correct explanation of Statement (I).
(*c*) Statement (I) is true but Statement (II) is false.
(*d*) Statement (I) is false but Statement (II) is true.

38. **Assertion (A):** In interference fit, the outer diameter of the shaft is greater than the inner diameter of the hole.
Reason (R): The amount of clearance obtained from the assembly of hole and shaft resulting in interference fit is called positive clearance.
(*a*) Both A and R are true and R is the correct explanation of A
(*b*) both A and R are true but R is not a correct explanation of A
(*c*) A is true but R is false (*d*) A is false but R is true

39. Consider the following statements
In case of assembly of mating parts
1. the difference between hole size and shaft size is called allowance
2. in transition fit, small positive or negative clearance between the shaft and hole member is employable
Which of the above statement(s) is/are correct?
(*a*) 1 only (*b*) both 1 and 2
(*c*) 2 only (*d*) Neither 1 nor 2

40. Which one of the following is clearance fit?
(*a*) $\Phi H50^{+0.015}_{+0.005}\ h50^{-0.010}_{+0.000}$ (*b*) $\Phi H50^{+0.010}_{+0.000}\ h50^{+0.025}_{+0.015}$
(*c*) $\Phi H50^{-0.015}_{+0.000}\ h50^{+0.025}_{+0.005}$ (*d*) $\Phi H50^{-0.010}_{-0.000}\ h50^{+0.030}_{+0.005}$

41. **Assertion (A):** In centralized inspection, material handling is less
Reason (R): Less number of gauges and instruments are required as inspection is carried out in one location
(*a*) both A and R are true and R is the correct explanation of A
(*b*) both A and R are true but R is not a correct explanation of A
(*c*) A is true but R is false (*d*) A is false but R is true

42. Which of the following in an interference fit?
(*a*) Push fit (*b*) Running fit
(*c*) Sliding fit (*d*) Shrink fit

43. In the tolerance specification 25 D6, the letter D represents
(*a*) grade of tolerance (*b*) upper deviation
(*c*) lower deviation (*d*) type of fit

44. Match List-I (Measuring Device) with List-II (Parameter Measured) and select the correct answer using the codes given below the lists:
List-I
A. Diffraction grating B. Optical flat
C. Auto collimators D. Laser scan micrometer
List-II
1. Small angular deviations on long flat surfaces

2. Online measurement of moving parts
3. Measurement of gear pitch
4. Joining hydraulic piston rods for agricultural machinery
5. Measurement of very small displacements

Codes:

	A	B	C	D		A	B	C	D
(*a*)	5	4	2	1	(*b*)	3	5	1	2
(*c*)	3	5	4	1	(*d*)	5	4	1	2

45. What is the dominant direction of the tool marks or scratches in a surface texture having a directional quality, called?
(*a*) primary texture (*b*) Secondary texture
(*c*) Lay (*d*) Flaw

46. The M and E-system in metrology are related to measurement of
(*a*) screw threads (*b*) flatness
(*c*) angularity (*d*) surface finish

47. What symbol is used to indicate surface roughness?
(*a*) = (*b*) $\sqrt{\ }$
(*c*) $^{0.1}\sqrt{\ }$ (*d*) Δ

48. Match **List-I** (Symbols for direction of lay) with **List-II** (Surface texture) and select the correct answer using the code given below the lists:

List-I		**List-II**
A. =	1.	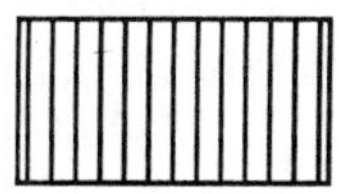
B. ×	2.	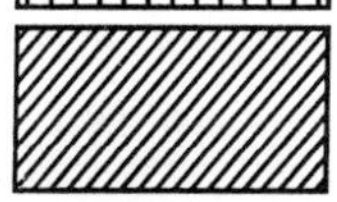
C. |	3.	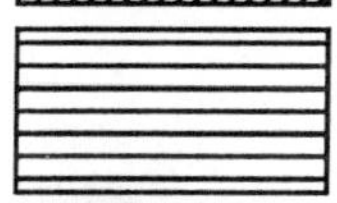
D. **R**	4.	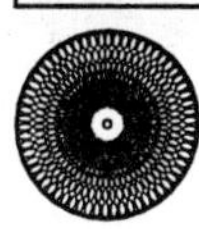

Codes:

	A	B	C	D		A	B	C	D
(*a*)	4	2	1	3	(*b*)	3	2	1	4
(*c*)	4	1	2	3	(*d*)	3	1	2	4

49. Interference fit joints are provided for
(*a*) Assembling bush bearing in housing (*b*) Mounting heavy duty gears on shafts
(*c*) Mounting pulley on shafts (*d*) Assembly of flywheels on shaft

50. A fit on a hole-shaft system is specified as H7-s6. The type of fit is
(*a*) Clearance fit (*b*) Running (sliding) fit
(*c*) Transition fit (*d*) Interference fit

51. Clearance in a fit is the difference between
(*a*) Maximum hole size and minimum shaft size

(*b*) Minimum hole size and maximum shaft sıze
(*c*) Maximum hole size and max¹mum shaft size
(*d*) Minimum hole size and minimum shaft size

52. **Statement (I):** In interference fit, the outer diameter of the inner cylinder will be more than the inner diameter of the hollow outer cylinder.
Statement (II): These fits are recommended for two parts frequently dismantled and assembled.
(*a*) Both Statement (I) and Statement (II) are individually true and Statement (II) is the correct explanation of Statement (I)
(*b*) Both Statement (I) and Statement (II) are individually true but Statement (II) is NOT the correct explanation of Statement (I)
(*c*) Statement (I) is true but Statement (II) is false
(*d*) Statement (I) is false but Statement (II) is true

53. A hole and a shaft have a basic size of 25 mm, and are to have a clearance fit with a maximum clearance of 0.02 mm and a minimum clearance of 0.01 mm. The hole tolerance is to be 1.5 times the shaft tolerance. The limits of both hole and shaft using hole basis system will be
(*a*) low limit of hole = 25 mm, high limit of hole = 25.006 mm, upper limit of shaft = 24.99 mm and low limit of shaft = 24.986 mm
(*b*) low limit of hole = 25 mm, high limit of hole = 25.026 mm, upper limit of shaft = 24.8 mm and low limit of shaft = 24.76 mm
(*c*) low limit of hole = 24 mm, high limit of hole = 25.026 mm, upper limit of shaft = 25 mm and low limit of shaft = 24.99 mm
(*d*) low limit of hole = 25.006 mm, high limit of hole = 25 mm, upper limit of shaft = 24.99 mm and low limit of shaft = 25 mm

54. The maximum interference in mm after assembly between a bush of size $30.00^{+0.06}_{+0.03}$ and a shaft of size $30.00^{+0.04}_{-0.02}$ is
(*a*) 0.07 (*b*) 0.05
(*c*) 0.02 (*d*) 0.01

55. The dia of a hole is given as $50^{+0.15}_{+0.00}$ mm. The upper limit on the dimensions in mm, of the shaft for achieving maximum interference of 50 microns is ----------------.

56. A 30 mm hole shaft assembly results in minimum and maximum clearances of 0.03 mm and 0.30 mm respectively. The hole has a unilateral tolerance with zero fundamental deviation. If the tolerance on the shaft is 0.08 mm, the maximum size of the hole is
(*a*) 230.11 (*b*) 30.16
(*c*) 30.19 (*d*) 30.27

57. A hole is specified as $\phi 30 \pm 0.04$ mm with $-\phi 0.01$ mm. The virtual diameter of the hole (i.e., the maximum diameter of the pin that can enter the hole) is ----------------.

58 A Shaft of diameter $20^{+0.05}_{-0.15}$ mm and a hole of diamter $20^{+0.20}_{+0.10}$ mm when assembled would yield
(*a*) Transition fit (*b*) Interference fit
(*c*) Clearance fit (*d*) None of these

59. Abbe's principle of alignment is used to be followed in
(*a*) vernier calipers (*b*) depth vernier
(*c*) internal caliper micrometer (*d*) height vernier

60. To obtain dimension of 61.18 mm using slip gauges, the most appropriate combination is

(*a*) 1.18 + 50.00 + 10.00 (*b*) 0.08 + 1.10 + 60.00

(*c*) 1.08 + 0.10 + 50.00 + 10.00 (*d*) 1.08 + 1.10 + 50.00 + 9.00

61. The combination of slip gauges to obtain a dimension of 10.35 mm will be

(*a*) 10.00 + 0.30 + 0.05 (*b*) 8.00 + 1.30 + 1.05

(*c*) 10.00 + 0.35 (*d*) 5.00 + 4.00 + 1.00 + 0.35

62. In an interference microscope, a groove produces a band distortion of 4 band spacing. If the wave length of the monochromatic light source is 0.5 microns, the groove depth is ... mm

63. A straight edge is inspected on a coordinate measuring machine. The readings obtained at intervals of 100 mm along the top edge are 100.1, 100.5 and 100.4 mm. The straightness error, making the end points as reference is .... mm.

64. **Assertion (A):** End standards are used for precision component inspection.

Reason (R): In end standards line of measurement coincides with line of indication.

(*a*) Both A and R are correct and R explains A

(*b*) Both A and R are true but R not explains A

(*c*) A is true but R is false

(*d*) A is false but R is true

65. Match the two parts below

Instrument

P. Dial indicator Q. Pneumatic gauge

R. GO/NO GO gauge

Principle of inspection

1. Non contact 2. Limit of size

3. Comparator

	P	Q	R
(*a*)	2	3	1
(*b*)	3	1	2
(*c*)	1	2	3
(*d*)	2	1	3

66. Which one of the instrument is a comparator

(*a*) Tool makers microscope (*b*) GO/NOGO gauge

(*c*) Optical interferometer (*d*) Dial gauge

67. The geometric tolerance that does not need a datum for its specification is

(*a*) Concentricity (*b*) Run out

(*c*) Perpendicularity (*d*) Flatness

68. The surface finish of the component produced by a grinding process was measured. Based on roughness parameter obtained, which one of the following is true.

(*a*) $R_a < R_s < R_z < R_t$ (*b*) $R_s < R_a < R_z < R_t$

(*c*) $R_z < R_a < R_s < R_t$ (*d*) $R_a < R_z < R_a < R_t$

69. In a three wire inspection of external thread of M16 × 2, the measurement over the wires was found to be 16.455 mm

(A) The best wire size is

(*a*) 1.155 mm (*b*) 2 mm

(c) 2.309 mm (d) 4 mm

(B) The effective pitch diameter of the thread is

(a) 14.723 (b) 16.705

(c) 17.122 (d) 17.455

70. In an interchangeable assembly, a shaft of $20.00^{+0.015}_{-0.005}$ mm diameter and a hole of $20.00^{+0.025}_{-0.010}$ mm diameter form a mating pair. In the worst assembly conditions, clearance between them will be

(a) 40 microns (b) 30 microns

(c) 25 microns (d) 15 microns

71. A cylindrical pin of diameter $1.996^{+0.0015}_{-0.0015}$ mm is assembled into a hole of diameter $2.000^{+0.0015}_{-0.0015}$ mm. The allowance (in mm) provided for this assembly is

(a) 0.001 (b) 0.015

(c) 0.025 (d) 0.035

72. A keyway of 5 mm depth is to be milled in a shaft of diameter 60 ± 0.1 by positioning the shaft in a V-block of angle 120 deg and setting the tool with reference to the intersection of the face of the V-block. The error due to shaft tolerance which would be transferred on to the component is

(a) 0.09 mm (b) 0.119 mm

(c) 0.22 mm (d) 0.3 mm

73. Tolerance on the dimension x in the two component assembly shown below is (all dimensions are in mm)

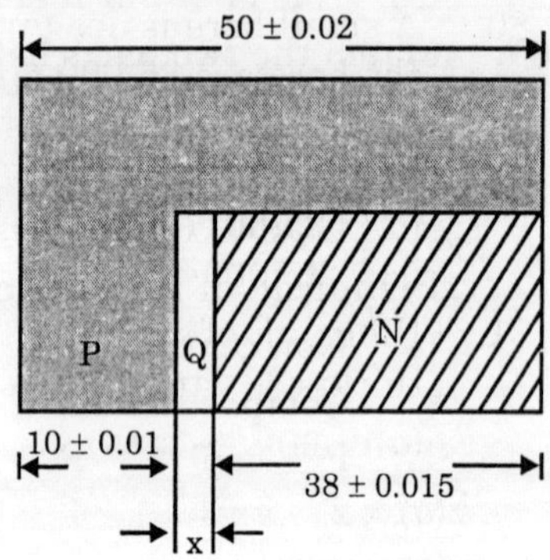

(a) ±0.025 (b) ±0.030

(c) ±0.040 (d) ±0.045

74. An autocollimator is used to

(a) Measure small angular displacements on flat surfaces

(b) Compare known and unknown dimensions

(c) Measure the flatness error (d) Measure roundness error between centers

75. A sine bar has a length of 250 mm, each roller has a diameter of 20 mm, during taper angle measurement of a component, the height from the surfaces plate to the centre of a roller is 100 mm. The calculated taper angle (in degrees) is

(a) 21.1 (b) 22.8

(c) 23.6 (d) 68.9

76. An experimental setup is planned to determine the taper of workpiece as shown in the figure. If the two precision rollers have radii 8 mm and 5 mm and the total thickness of slip gauges inserted between the rollers is 15.54 mm, the taper angle "θ" is

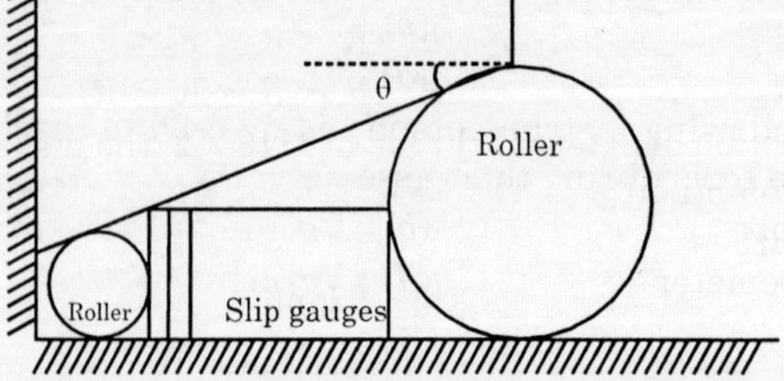

(*a*) 6 deg (*b*) 10 deg
(*c*) 11 deg (*d*) 12 deg

77. Observation of a slip gauge on flatness interferometer produced fringe counts numbering 10 and 14 for two readings. The second reading is taken by rotating the setup by 180°. Assume that both faces of the slip gauge are flat and the wavelength of the radiation is 0.5086 μm. The parallelism error (in μm) between the two faces of the slip gauge is
(*a*) 0.2543 (*b*) 1.172
(*c*) 0.5086 (*d*) 0.800

78. In an interchangeable assembly, shafts of size $25.000^{+0.040}_{-0.010}$ mm mate with holes of size $25.000^{+0.030}_{+0.020}$ mm. The maximum interference (in microns) in the assembly is
(*a*) 40 (*b*) 30
(*c*) 20 (*d*) 10

79. The following data are given for calculating limits of dimension and tolerances for a hole: Tolerance unit i (microns) = 0.45 ($D^{1/3}$) + 0.001D. The unit of D is mm. Diameter step is 18-30. If the fundamental deviation for hole H is zero and IT8 = 26i, the maximum and minimum limits of dimension for a $25H_8$ hole (in mm) are
(*a*) 24.984 and 24.967 (*b*) 25.017 and 24.987
(*c*) 25.033 and 25.00 (*d*) 25.00, 24.967

80. A small bore is designated 25H7. The lower (minimum) and upper (maximum) limits of the bore are 25.000 mm and 25.021 respectively. When the bore is designated as 25H8, then the upper limit is 25.033 mm. When the bore is designated as 25H6, then the upper limit of the bore in mm is
(*a*) 25.001 (*b*) 25.005
(*c*) 25.009 (*d*) 25.013

81. Diameter of a hole after plating needs to be controlled between $30^{+0.050}_{+0.010}$ mm. If the plating thickness varies between 10-15 microns, diameter of the hole before plating should be
(*a*) $30^{+0.07}_{+0.030}$ mm (*b*) $30^{+0.065}_{+0.020}$ mm
(*c*) $30^{+0.08}_{+0.030}$ mm (*d*) $30^{+0.070}_{+0.040}$ mm

82. For the given assembly : 25 H7/g8, match Group A with Group B

Group-A		Group-B	
P	H	I	Shaft Type
Q	IT8	II	Hole Type
R	IT7	III	Hole Tolerance Grade
S	g	IV	Shaft Tolerance Grade

(*a*) P-I, Q-III, R-IV, S-II (*b*) P-I, Q-IV, R-III, S-II
(*c*) P-II, Q-III, R-IV, S-I (*d*) P-II, Q-IV, R-III, S-I

83. The flatness of a machine bed can be measured using
(*a*) Vernier calipers (*b*) Auto collimator
(*c*) Height gauge (*d*) Tool maker's microscope

84. Which one of the following instruments is widely used to check and calibrate geometric features of machine tools during their assembly?
(*a*) Ultrasonic probe (*b*) Coordinate Measuring Machine (CMM)
(*c*) Laser interferometer (*d*) Vernier Calipers

85. Which one of the following is modeled based on adaptation capabilities of biological systems?
(*a*) Relational database (*b*) Fuzzy system
(*c*) Simulated annealing algorithm (*d*) Genetic algorithm
86. Which one of the following statements is TRUE?
(*a*) The 'GO' gage controls the upper limit of a hole
(*b*) The 'NO GO' gage controls the lower limit of a shaft
(*c*) The 'GO' gage controls the lower limit of a hole
(*d*) The 'NO GO' gage control the lower limit of a hole
87. Cylindrical pins of $25^{+0.020}_{+0.010}$ mm diameter are electroplated in a shop. Thickness of the plating is $30^{\pm 2.0}$ micron. Neglecting gage tolerance, the size of the GO gage in mm to inspect the plated component is
(*a*) 25.042 (*b*) 25.052
(*c*) 25.074 (*d*) 25.084
88. Holes of diameter $25.0^{+0.040}_{\pm 0.020}$ mm are assembled interchangeably with the pins of diameter $25.0^{+0.005}_{-0.008}$ mm. The minimum clearance in the assembly will be
(*a*) 0.048 mm (*b*) 0.015 mm
(*c*) 0.005 mm (*d*) 0.008 mm
89. A GO-NO GO plug gauge is to be designed for measuring a hole of nominal diameter 25 mm with a hole tolerance of ± 0.015 mm. Considering 10% of work tolerance to be the gauge tolerance and no wear condition, the dimension (in mm) of the GO plug gauge as per the unilateral tolerance system is
(*a*) $24.985^{+0.003}_{-0.003}$ (*b*) $25.015^{+0.000}_{-0.006}$
(*c*) $24.985^{+0.03}_{-0.03}$ (*d*) $24.985^{+0.000}_{-0.003}$
90. The diameter of a recessed ring was measured by using two spherical balls of diameter d_1 = 60 mm and d_2 = 40 mm as shown in the figure.

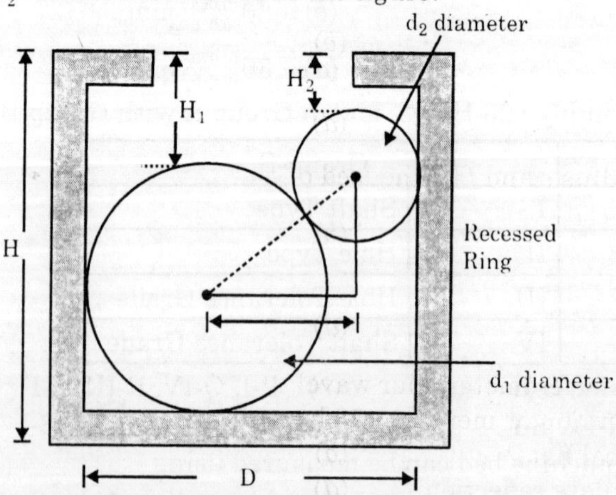

The distance H_1 = 35.55 mm and H_2 = 20.55 mm.
The diameter (D, in mm) of the ring gauge is ______
91. Match the following:
P. Feeler gauge Q. Fillet gauge
R. Snap gauge S. Cylindrical plug gauge
(*i*) Radius of an object (*ii*) Diameter within limits by comparison
(*iii*) Clearance or gap between components

(*iv*) Inside diameter of straight hole

(*a*) P-(*iii*), Q-(*i*), R-(*ii*), S-(*iv*) (*b*) P-(*iii*), Q-(*ii*), R-(*i*), S-(*iv*)
(*c*) P-(*iv*), Q-(*ii*), R-(*i*), S-(*iii*) (*d*) P-(*iv*), Q-(*i*), R-(*ii*), S-(*iii*)

92. Two optically flat plates of glass are kept at a small angle θ as shown in the figure Monochromatic light is incident vertically.
If the wavelength of light used to get a fringe spacing of 1 mm is 450 nm, the wavelength of light (in nm) to get a fringe spacing of 1.5 mm is

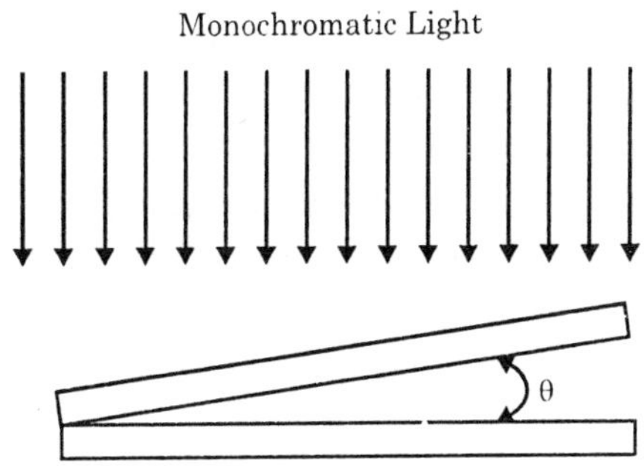

93. In the assembly shown below, the part dimensions are:
$L_1 = 22.0^{\pm.01}$
$L_2 = L_3 = 10.0$ mm $^{\pm 0.005}$ mm
Assuming the normal distribution of part dimensions, the dimension L_4 in mm for assembly condition would be:
(*a*) $2.0^{\pm 0.008}$ (*b*) $2.0^{\pm 0.012}$
(*c*) $2.0^{\pm 0.016}$ (*d*) $2.0^{\pm 0.020}$

94. A cylindrical pin of $25^{+0.020}_{+0.010}$ mm diameter is electroplated, plating thickness is $2.0^{\pm 0.005}$ mm. Neglecting the gauge tolerance, the diameter (in mm, up to 3 decimal points accuracy) of the GO ring gauge to inspect the plated pin is ________.

95. Determine the effective diameter of a threaded plug of 24 mm diameter with 3 mm pitch and included angle of 60° by using a floating carriage micrometer. Micrometer readings were taken over standard plug of 22.002 mm diameter and over the threaded plug with two best size wires. These readings were 12.9332 mm and 12.1135 mm respectively.

96. With a sharp tool without nose radius, it maximum height of uneveness be R_{max}, then surface roughness value is equal to

(*a*) $\frac{R_{max}}{2}$ (*b*) R_{max}

(*c*) $\frac{R_{max}}{4}$ (*d*) $\frac{R_{max}}{3}$

97. If tool has nose radius *r* and *f* be the feed of tool, then roughness value is proportional to

(*a*) $\frac{f}{r}$ (*b*) $\frac{f^2}{r}$

(*c*) $\left(\frac{f}{r}\right)^2$ (*d*) $\frac{f}{r^2}$

98. In NPL flatness interferometer, four wavelengths of monochromatic light sources can be obtained from cadmium or mercury 198 lamp. Different wavelengths are obtained by
(*a*) rotating optical flat (*b*) adjusting position of collimating lens
(*c*) tilting glass plate reflection (*d*) using different filters

99. Interferometer is used as absolute measuring instrument by measuring fractions obtained with three or four different wavelengths. Determine the fractions obtained for a 2.90 mm slip gauge using cadmium lamps with red, green and blue wavelengths of 0.643850537 μm, 0.50858483 μm, and 0.47999360 μm respectively.

100. The minimum length of slip gauge in any set of slip gauges is
(*a*) 1.0005 mm (*b*) 1.001 mm
(*c*) 1.01 mm (*d*) 10.001 mm

Hints and Solutions

1. (*a*) parabolic
   (*b*) about 60%
   (*c*) size of the components, and its position related to the basic size
   (*d*) fundamental deviation
   (*e*) 28
2. (*c*) To determine the type of fit, let us find maximum and minimum clearance
   Max. clearance = High limit of hole – low limit of shaft
   = 40.01 – 40.00 = 0.01 mm
   Min . clearance = low limit of hole – high limit of shaft
   = 39.99 – 40.02 = – 0.03
   Since one difference is negative, it is transition fit

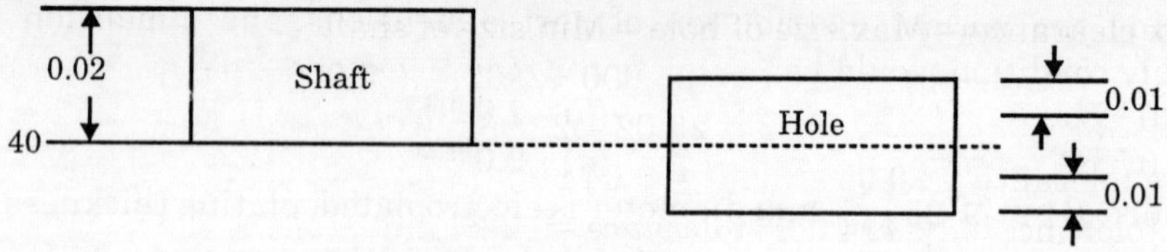

3. (*c*) Here max. clearance = 20.20 – 19.85 = 0.35
   Min. clearance = 20.10 – 20.05 = 0.15 mm
   Both being +ve, it is clearance fit
4. (*b*) Eccentricity = max. radial clearance

$$= \frac{40.06 - 39.95}{2} = 0.055 \text{ mm}$$

5. (*a*) Position of tolerance of hole is specified by capital letter i.e., 'G' here.
6. (*a*) In fit H_7h_6 fundamental deviation of hole of shaft is zero and tolerances are above zero line for hole and below zero line for shaft. The maximum clearance is positive and thus it is clearance fit

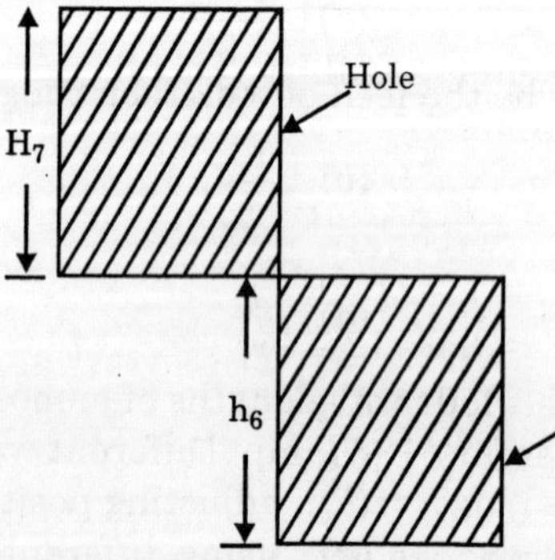

7. (*a*) Tolerance on shaft A = 100.1 – 99.9 = 0.2 mm
   and on shaft B = 0.1001 – 0.0999 = 0.002 mm
   ∴ Tolerance in shaft A is higher.
8 (*a*) Here value of T = 0.125 mm is to be assured, i.e., considering upper tolerances of B_1, B_2 and B_3, T = 0.125 mm should be possible, i.e.,
   18.75 + 0.08 + 25.00 + 0.12 + 28.125 + 0.1 + 0.125 = 72.35 + X

or $18.83 + 25.12 + 28.225 + 0.125 = 72.35 + X$

$$72.300 = 72.35 + X$$

$$X = -0.05 \text{ mm}$$

9. ( ) Attributes
10. (*a*) False
11. (*b*) Size 25 lies between 18 and 30

$\therefore \quad D = \sqrt{18 \times 30} = 23.24$ mm

Tolerance unit $i = 0.45\sqrt[3]{23.24} + 0.001(23.24) = 1.31$ micron

for quality 7, tolerance grade

$$= 16\,i = 16 \times 1.31 = 20.96\ \mu$$

$\therefore$ Shaft tolerance $= 25 - (20.96 \times 10^{-3})$ mm

$$= 25 - 0.021 = 24.979 \text{ mm}$$

12. (*d*) Max clearance = Max.size of hole – Min.size of shaft

$$= (25.000 + 0.02) - (25.000 - 0.04)$$

$$= 25.02 - 29.96 = 0.060 \text{ mm} = 60\ \mu.$$

13. (*b*) Hole tolerance = 20.05 – 20.01 = 0.04 mm

Gauge tolerance = 10% of hole tolerance = 0.004 mm

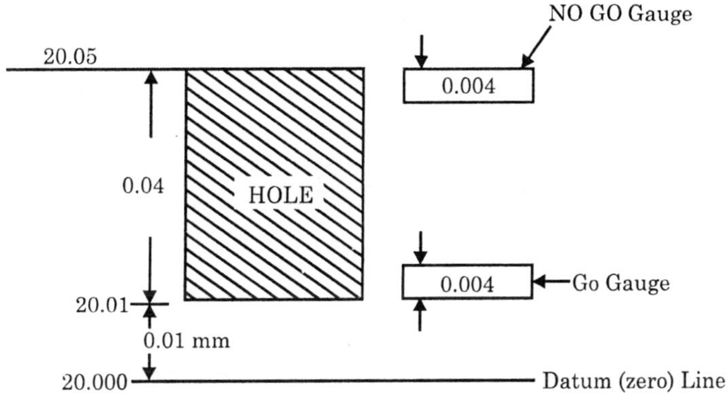

Thus sizes of GO and NO GO gauges will be

20.000 + 0.01 + 0.004 = 20.014 mm

and 20.05 – 0.004 = 20.046 mm

14. (*a*) Basic dimension of W

$$= P - Q - R$$

$$= 35 - 12 - 13$$

$$= 10 \text{ mm}$$

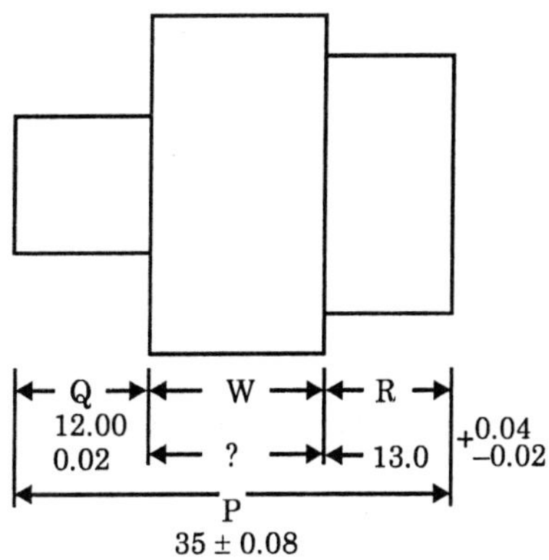

If tolerances of Q and R on +ve side are considered, total tolerance = 0.04 + 0.02 = 0.06

Thus W can have tolerance of 0.08 – 0.06 = 0.02 mm

Similarly, on negative side, tolerance of W can be 0.08 – (0.02 + 0.02) = 0.04 mm

This tolerance can be ± 0.03, if base size is taken as 10 – 0.01 mm = 9.99 mm

Thus dimension of W = 9.99 ± 0.03 mm

15. (*b*)

16. (*c*) Difference in height $= \frac{\lambda}{2} N$

or $$1.002 - 1.000 = \frac{0.00058928}{2} \times N$$

and $$N = \frac{0.002 \times 2}{0.0006} = 8$$

17. (*b*) Pitch diameter of screw threads is measured by wires (or rollers) of best size. Diameter of such roller

$$= \frac{P}{2} \sec\frac{\theta}{2} = \frac{2}{2} \sec\left(\frac{60}{2}\right) = 1.155 \text{ mm}$$

18. (*a*) (*i*) Tolerance on hole = 0.021 – 0.000 = 0.021 mm
and tolerance on shaft = 0.075 – 0.040 = 0.035 mm

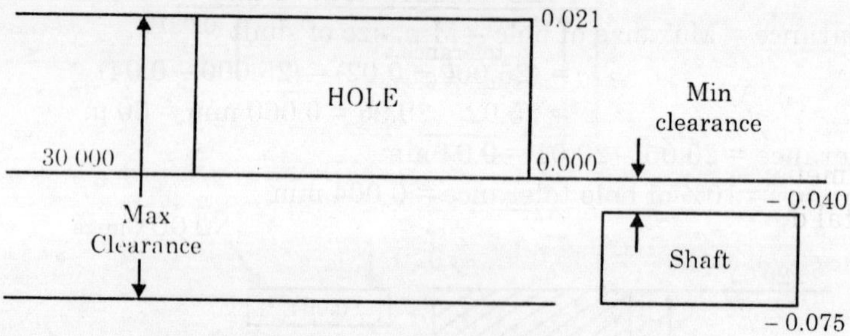

(*ii*) Max. clearance = 30.021 – (30.000 – 0.075) = 0.096 mm
Min. clearance = 30.000 – (30.000 – 0.04) = 0.04 mm
(*iii*) Allowance = Lower limit of hole – higher limit of shaft
= 30.000 – (30 – 0.040) = 0.040 mm
(*iv*) Max. metal limit of hole = 30.000 mm
and max. metal limit of shaft
= 30.000 – 0.040 = 29.960 mm
(*v*) Type of fit – Since max and min size of hole is bigger than the biggest size of shaft, it is clearance fit.

19. (*a*) Refer fig. showing size of shaft and location of bush to obtain desired max. and min. clearance
Max limit of bush = (40.00 – 0.02) + 0.1 = 40.08 mm
and Min limit of bush
= 40.00 + 0.02 + 0.02
= 40.04 mm

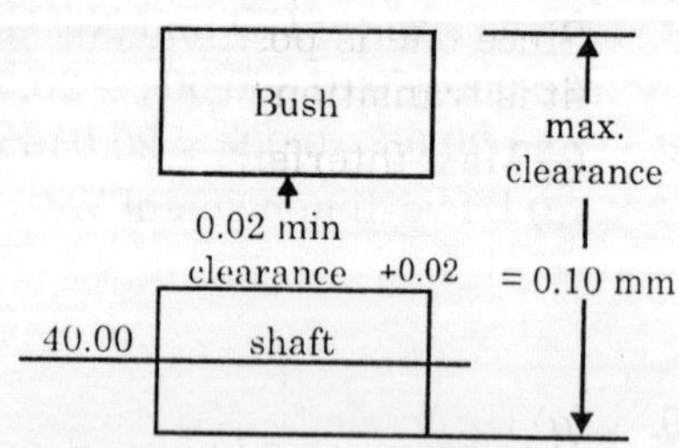

20. (*c*) Max. clearance = 0.050 + 0.01 + 0.04 = 0.10 mm

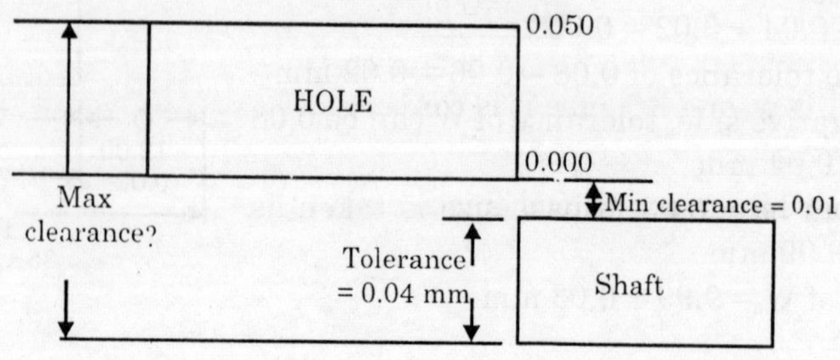

21. (*a*) Shaft is 60 f 8

Dia for calculation of tolerance unit D = $\sqrt{50 \times 80}$ = 63.246 mm

$$i = 0.45\ \sqrt[3]{D} + 0.001\,\text{D} = 1.859\ \mu\text{m} = 0.00186\ \text{mm}$$

IT8 = 25i = 0.04646 mm

Fundamental deviation of shaft f = – 5.5 $D^{0.41}$ = – 0.0312 mm

Upper limit of shaft

= 60.000 – 0.03012 ≈ 59.97 mm

Lower limit of shaft

= 59.97 – 0.0464 ≈ 59.924 mm

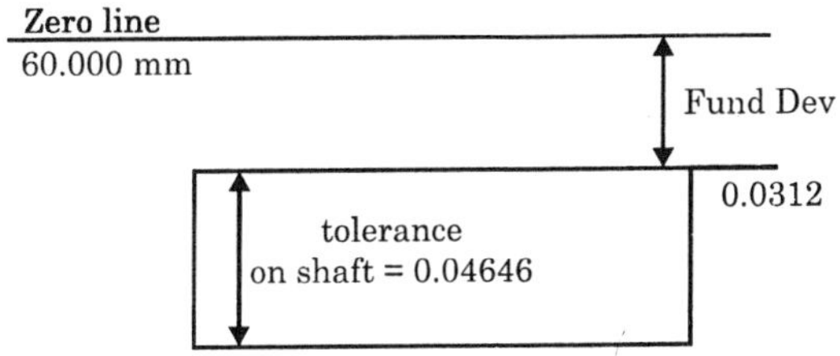

22. (*d*) Shaft dimension is $35^{-0.009}_{-0.025}$

Fundamental deviation = –0.009

and Tolerance of shaft = 35.000(–0.025) – 35.000(–0.009) = 0.016 mm

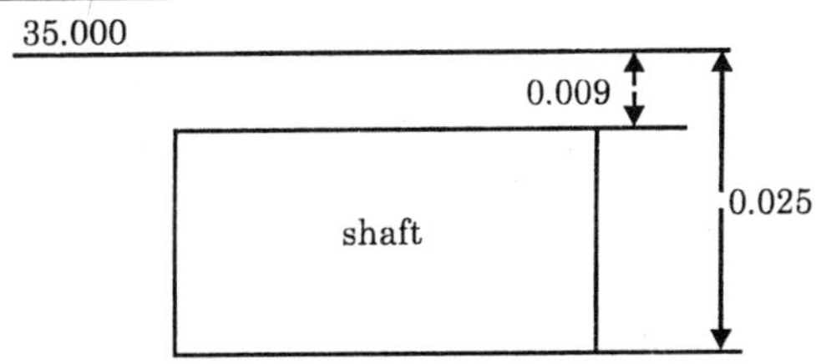

23. (*a*)

24. (*c*)

25. (*c*)

26. (*b*)

27. (*c*) Max. size of hole – min. size of shaft = 0.015 – 0.001 = 0.014 mm

Min hole size – max shaft size

= 0.000 – 0.01

= – 0.01 mm

Since one is positive and one is negative, the fit is transition fit

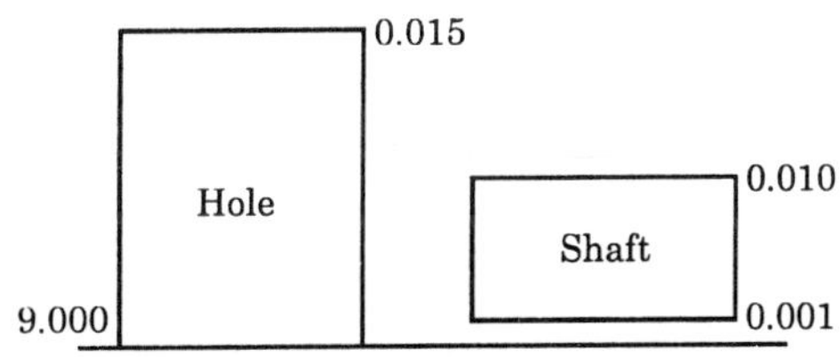

28. (*c*) Max. interference

= Min. size of hole – max. size of shaft

= 25.020 – 25.040

= –0.020 mm = 20 μm

29. (*a*) Eccentricity = Max. radial clearance

$$= \frac{30.05 - 29.95}{2} = 0.05\ \text{mm}$$

30. (*c*) Explanation E is wrong because it is convenient to produce shafts of any size

31. (*c*)

32. (*a*) Scale axis must coincide with workpiece axis

33. (*c*)

34. (*a*)

35. A. datum, B. macro

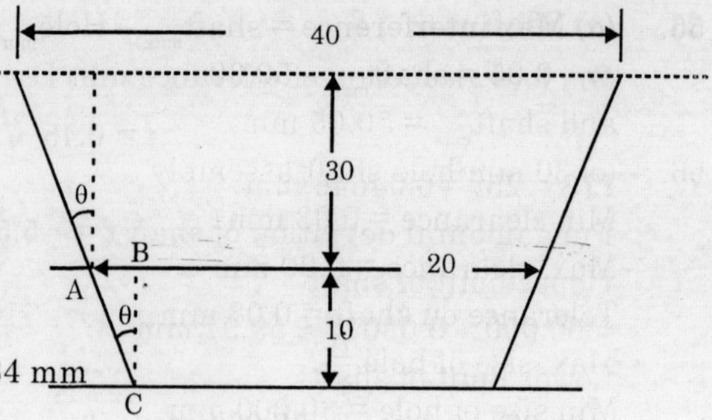

36. (*a*) $\tan\theta = \dfrac{\frac{(40-20)}{2}}{30} = \dfrac{1}{3}$

Also $\tan\theta = \dfrac{AB}{BC} = \dfrac{1}{3}$

and AB, (x) = 3.33 mm

∴ dia. at C = 20 – 2x = 20 – 6.66 = 13.334 mm

37. (*c*)

38. (*c*)

39. (*c*)

40. (*a*) For clearance fit, upper size of shaft should be lower than the lower limit of hole. This is seen in option (*a*).

41. (*b*)

42. (*d*)

43. (*a*) D 6 represents grade of tolerance.

44. (*b*)

45. (*c*)

46. (*d*)

47. (*c*)

48. (*b*)

49. (*a*)

50. (*d*)

51. (*b*)

52. (*c*)

53. (*a*) Max clearance = upper size of hole – lower size of shaft = 0.02 mm

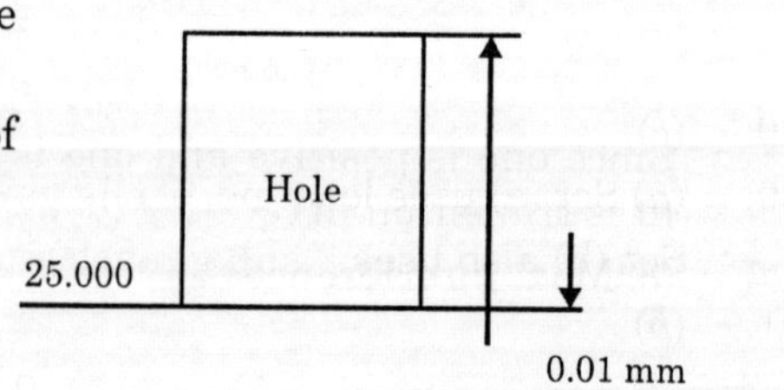

Min clearance = low size of hole – upper size of shaft = 0.01 mm

Tolerance of hole + 0.01 + Tolerance of shaft (T_s) = 0.02 mm

$1.5 \times T_S + 0.01 + T_S = 0.02$ mm

$2.5\,T_S = 0.01$ and $T_S = \dfrac{0.01}{2.5} = 0.004$ mm

∴ Tolerance of hole = 0.006 mm

∴ High size of hole = 25.000 + 0.006 = 25.006

upper size of shaft = 25.000 – 0.01 = 24.99 mm

low limit of shaft = 24.99 – 0.004 = 29.986 mm

Thus (*a*) is right choice

54. (*d*) Max. interference = Max. size of shaft – min. size of hole = + 0.04 – 0.03 = 0.01 mm

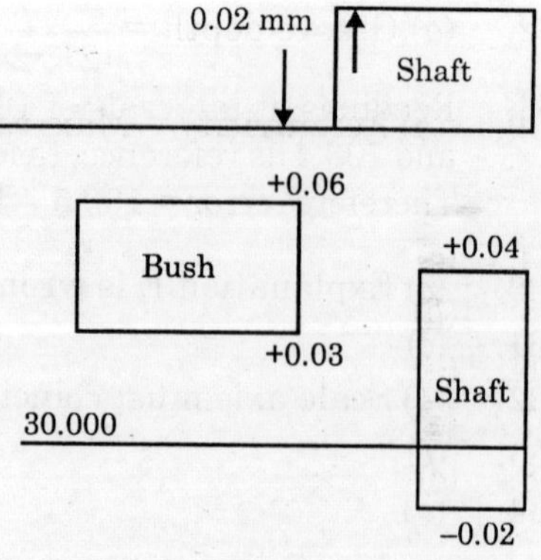

55. (*a*) Max interference = $shaft_{max}$ – $Hole_{min}$
or 0.05 = $shaft_{max}$ – 50.00
and $shaft_{max}$ = 50.05 mm

56. (*c*) 30 mm hole shaft assembly
Min clearance = 0.03 mm
Max. clearance = 0.30 mm
Tolerance on shaft = 0.08 mm
Max. size of hole = ?
Min size of hole = 30.000 mm
Max. clearance = Tol. of hole + 0.03 + 0.08
or 0.30 = Tol. of hole + 0.11
and Tol. of hole = 0.30 – 0.11 = 0.19 mm
∴ upper size of hole = 30.19 mm

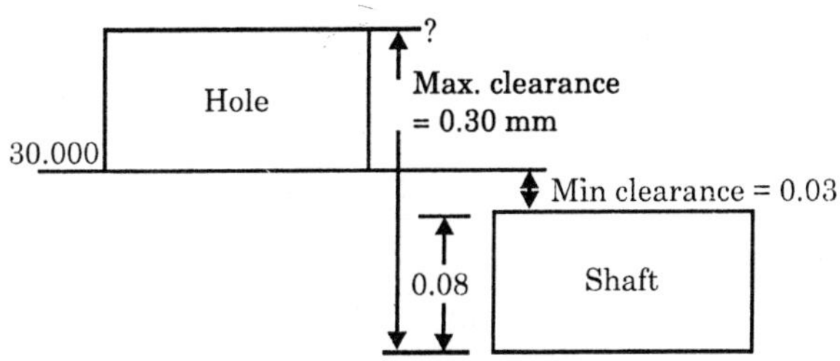

57. Virtual diameter of hole
i.e., max dia of pin that
car enter hole = 30 – 0.04 – 0.01
= 29.95 mm

58. (*c*) Shaft of $20^{+0.05}_{-0.15}$ and hole of $20^{+0.20}_{+0.10}$ will result in clearance fit because max size of shaft is less than min. size of hole.

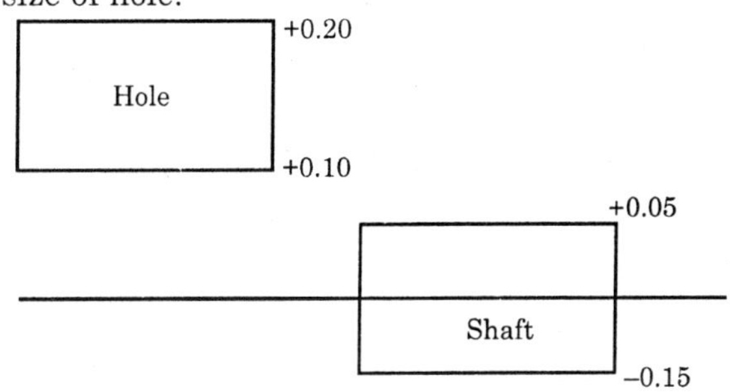

59. (*c*)

60. (*a*) Set (*a*) uses least no. of slip gauges of appropriate size to obtain dimension of 61.18 mm
Set (b) also uses 3 slip gauges and gauge of 0.08 mm which is not available

61. (*b*)

62. (*a*) Groove depth = $\frac{N}{2} \times \frac{\lambda}{2} = \frac{4}{2} \times \frac{0.5}{2}$ = 0.5 μm = 0.0005 mm

63. Readings at intervals of 100 mm are 100.1, 100.5 and 100.4 mm. Considering points 100.1 and 100.4 as reference, reading at mid point for error free edge should have been 100.25 mm. Therefore, error = 100.5 – 100.25 = 0.25 mm

64. (*c*)

65. (*b*)

66. (*d*)

67. (*d*)

68. (*c*)

69. (A) (a) Best wire size = $\frac{P}{2}\sec\theta = \frac{3\sec 60°}{2}\frac{}{2} = 1.155$ mm

(B) (a) Effective pitch diameter = $M - \left(d + \frac{P}{2}\tan\theta\right) = 16.455 - \left[1.155 + \frac{2}{2}\tan\frac{60}{2}\right]$

= 14.723 mm

70. (*b*) In worst assembly condition, the clearance will be maximum, *i.e.*, difference between hole with minimum metal and shaft with minimum metal
i.e., 0.025 + 0.005 = 0.030 mm or 30 microns

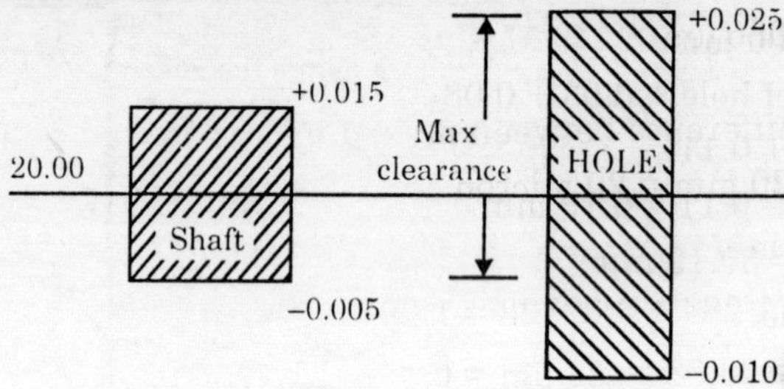

71. (*a*) Allowance = Min. hole dia – Max. shaft . dia
= (2.000 – 0.0015) – (1.996 + 0.0015)
= 0.004 – 0.003 = 0.001 mm

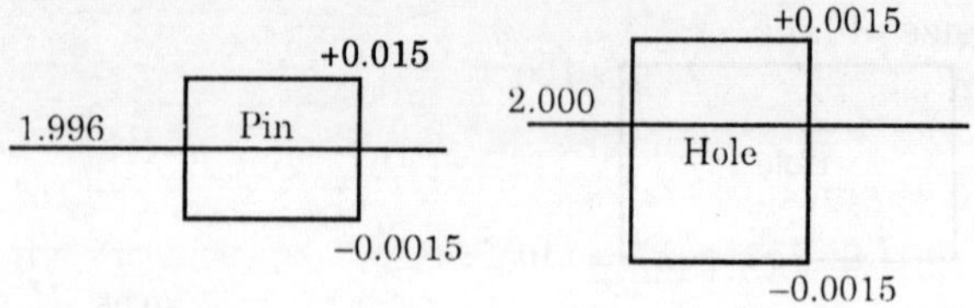

72. (*c*) $OC_{min} = \frac{OA}{\sin 60°} = \frac{60 - 0.1}{\sin 60°} = 34.58$

and $OC_{max} = \frac{60 + 0.1}{\sin 60} = 34.69$

∴ Error in depth due to tolerance of ± 0.1 mm on shaft
= (34.69 – 34.58) × 2 = 0.22 mm

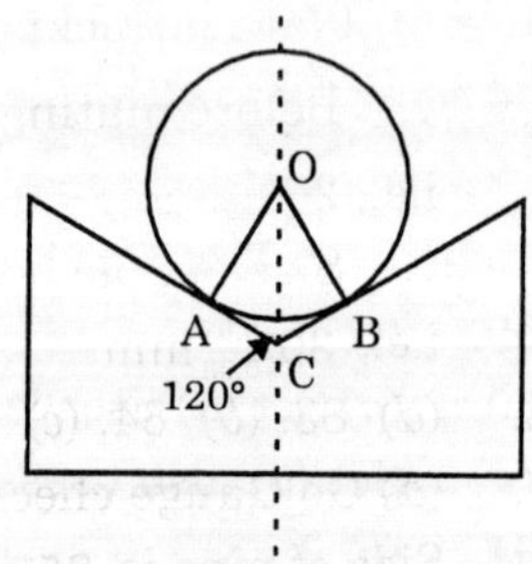

73. (*d*) Tolerance on dimension *x* is

$x_{max} - x_{min}$
= [50.02 – (37.985 + 9.99)] – [49.98 – (38.015 + 10.0)]
= 2.045 – 1.955 = 009 mm = ± 0.045

74. (*c*)

75. (*a*) $\sin\theta = \frac{AB}{OA} = \frac{100 - 10}{250} = \frac{90}{250}$

and θ = 21.1°

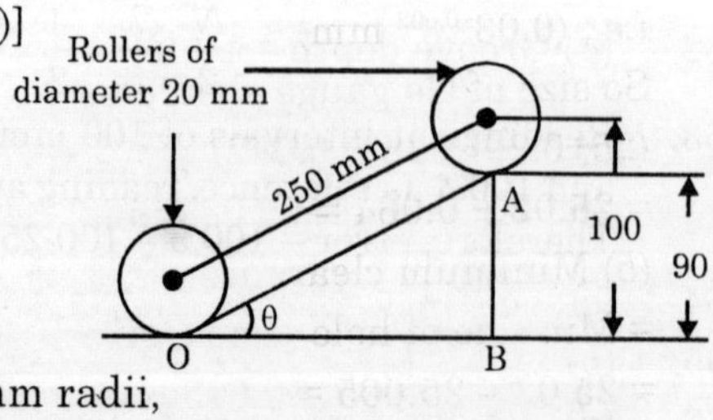

76. (*d*) O_1 and O_2 are centres of rollers of 5 mm and 8 mm radii,
θ = taper angle of workpiece
Draw O_1 B parallel to surface of workpiece and O_1 A parallel to surface plate
Angle $AO_1B = \theta$

and A O_1 O_2 = θ/2

$$\tan \theta/2 = \frac{AO_2}{O_1O_2} = \frac{8-5}{5+8+15.54} = \frac{3}{28.54}$$

and θ/2 = 6° and θ = 12°

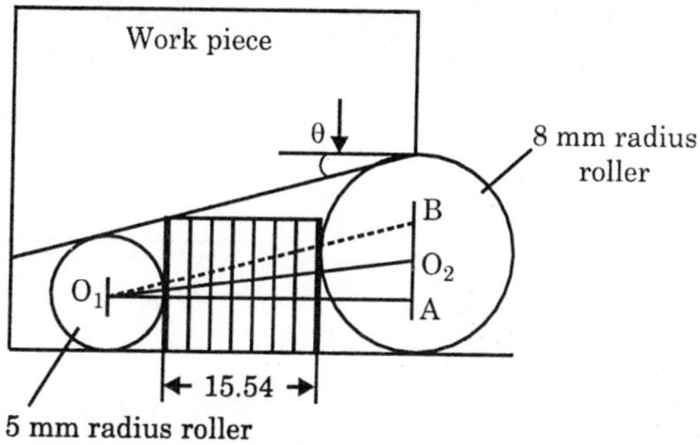

77. (*c*) Error in parallelism

$$= \frac{\text{Difference in no. of fringes}}{2} \times \frac{\lambda}{2}$$

$$= \frac{14-10}{2} \times \frac{0.5086}{2} = 0.5086\ \mu m$$

78. (*c*) Max interference = difference between max. dia. of shaft and min size of hole
= 25.040 – 25.020 = 0.020 mm = 20 micron

79. (*c*) D = $\sqrt{18 \times 30}$ = 23.23 mm

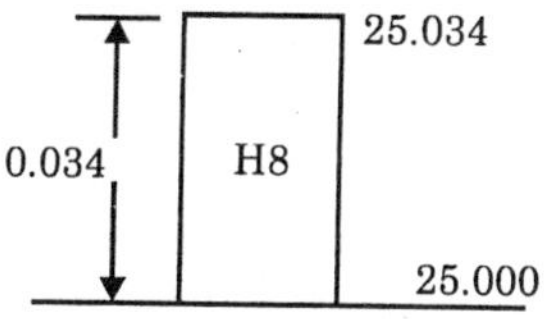

i = 0.45 $D^{1/3}$ + 0.001 D = 1.284 + 0.02323 = 1.307 micron

IT8 = 26 i = 26 × 1.307 = 33.98 ≃ 34 μm = 0.034 mm

∴ Min and max size of hole 25 H8 are 25.000 and 25.034 mm

80. (*d*) For 25H7, tolerance = 21 μm

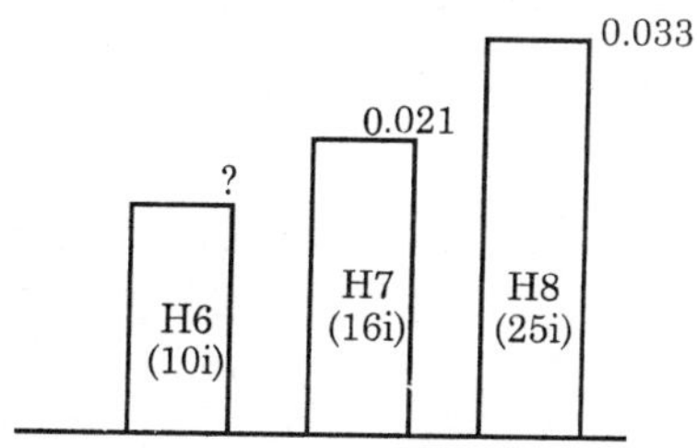

H7 is IT7 = 16 i

$$\therefore \quad i = \frac{21}{16} = 1.3125$$

For H6, IT6 = 10 i = 13 μm

∴ upper limit for 25H 6 = 25.13 mm

81. (*a*) Plating thickness varies between 10 to 15 microns

∴ Before plating, diameter of hole should be higher by twice the plating thickness i.e., 20 μm and 30 μm

∴ Size of hole be $30^{+0.050+0.020}_{+0.010+0.020}$

i.e., $30^{+0.07}_{+0.03}$ mm

82. (*d*) 83. (*b*) 84. (*c*) 85. (*d*) 86. (*c*)

87. (*d*) Go Gauge checks max. metal limit i.e., upper dimension of pins.
Size of pins is $25^{+0.020}_{-0.010}$ and thickness of plating is $30^{\pm 2.0}$ μm.
i.e., ($0.03^{\pm 0.002}$ mm)

So size of Go gauge should be
(25+0.20) + (2 × 0.032)
= 25.02 + 0.064 = 25.084 mm

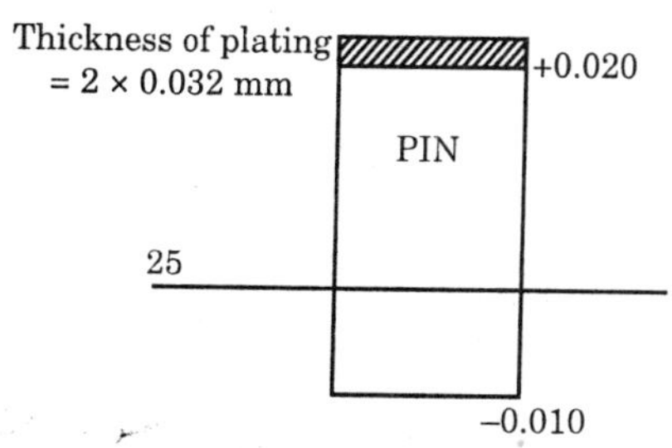

88. (*b*) Minimum clearance
= Min size of hole – max. size of shaft
= 25.02 – 25.005 = 0.015 mm

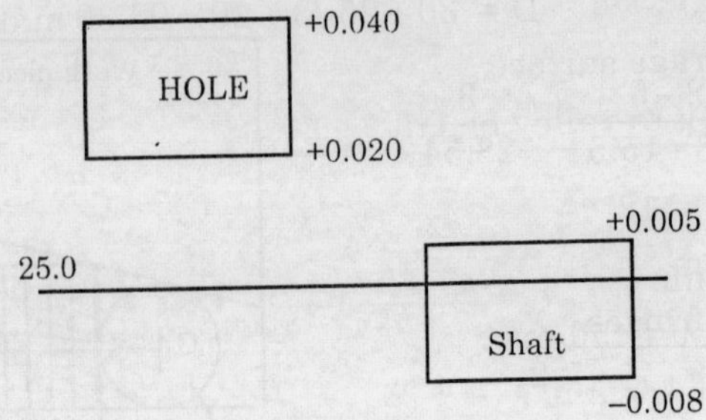

89. (*d*) Hole tolerance = ±0.015 mm

Gauge tolerance = 10% of tolerance of hole = 0.003 mm

Size of Go gauge (corresponding to min. size of hole, i.e., max metal limit)

$= (25 - 0.015)^{+0000}_{-0.015} = 24.985^{+0.000}_{-0.015}$ mm

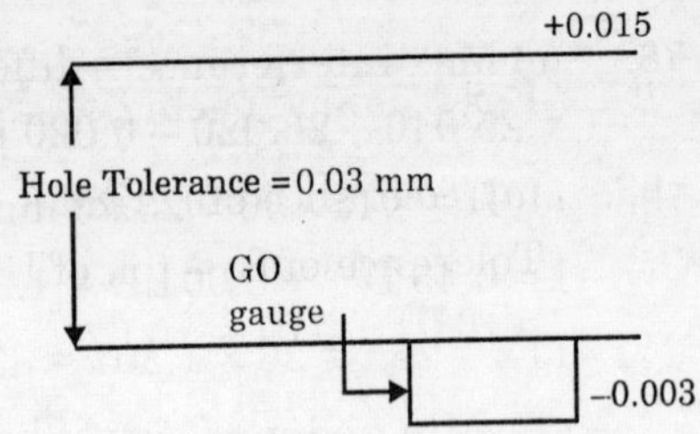

90. $d_1 = 60$ mm

$d_2 = 40$ mm

$H_1 = 35.55$ mm

$H_2 = 20.55$ mm

$$D = \frac{d_1}{2} + AB + \frac{d_2}{2}$$

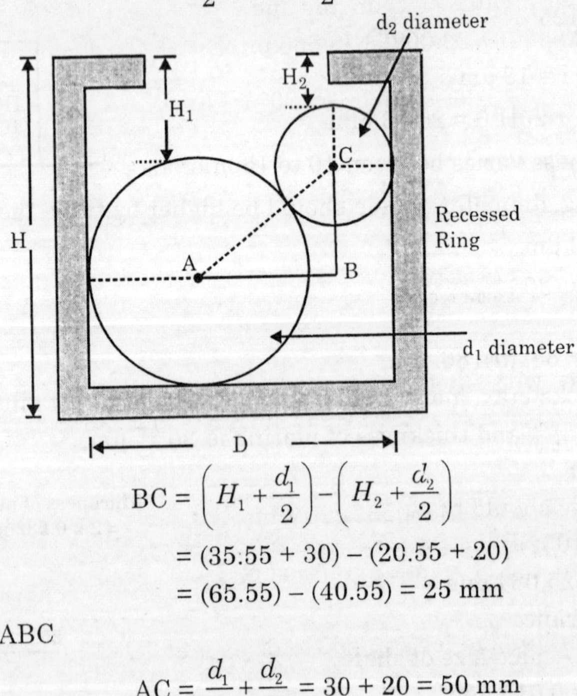

$$BC = \left(H_1 + \frac{d_1}{2}\right) - \left(H_2 + \frac{d_2}{2}\right)$$

$$= (35.55 + 30) - (20.55 + 20)$$

$$= (65.55) - (40.55) = 25 \text{ mm}$$

In right angle ΔABC

$$AC = \frac{d_1}{2} + \frac{d_2}{2} = 30 + 20 = 50 \text{ mm}$$

$$AB = \sqrt{AC^2 - BC^2}$$

$$= \sqrt{50^2 - 25^2} = 25\sqrt{3} \text{ mm}$$

$$D = 30 + 25\sqrt{3} + 20 = 93.30 \text{ mm}$$

90. (*a*) Central line average surface

roughness $R_a = \dfrac{\text{Peak to valley height}}{4} = \dfrac{20}{4} = 5\ \mu\text{m}$

91. (*a*)

92. Fringe spacing = 1 mm

Wavelength of light $\lambda = 450$ nm

For fringe spacing of 1.5 mm, $\lambda' = ?$

For 1.0 mm spacing, $1.0 = n \times \dfrac{\lambda}{2} = 1 \times \dfrac{450}{2}$

For 1.5 mm spacing, $\lambda' = \dfrac{1.5 \times 450}{2} = 675$ nm

93. (*d*) Tolerance of L_2 and $L_3 = 2 \times \pm 0.005 = \pm 0.01$ mm

Tolerance of L_4 = tol. of L_2 and L_3 + Tol. of L_4

$= \pm 0.01 \pm 0.01 = \pm 0.02$

$\therefore$ Dimension of $L_4 = (22 - 2 \times 10)\ \pm 0.02 = 2^{\pm 0.02}$

94. Go ring gauge measures max. metal limit, i.e., upper size of shaft after plating. Upper limit of shaft after plating

$= 25.02 + 2.005 + 2.005 = 29.030$ mm

95. Effective diameter = value of diameter under wire + pitch value

$= \{(12.1135 - 12.9332) + 22.002\} + (0.866 \text{ pitch} - \text{best size wire diameter})$

$= -0.8197 + 22.002 + (0.866 \times 3 \times \dfrac{3}{2} \sec 30°)$

$= 21.1823 + 2.598 \times 1.732$

$= 21.1823 + 0.8659$

$= 22.0482$ mm

96. (*c*)

97. (*b*)

98. (*d*)

99. Total no. of fringes with fraction formed $= \dfrac{\text{length of surface}}{\dfrac{\lambda}{2}}$

For red light, this is $= \dfrac{2.9 \times 2}{0.063850537 \times 10^{-3} \text{ mm}} = 9080.3042$

Fraction R = 0.3042

Similarly for green and blue lights fraction G and fraction B

= 0.1838 and 0.4948

These fractions are unique for slip gauge of 2.9 mm length and can not be obtained for any other length.

100. (*a*)

APPENDIX-I

Rules Relating to the Circle

To Find Circumference :

Multiply diameter by 3.1416

Or divide diameter by 0.3183

To Find Diameter :

Multiply circumference by 0.3183

Or divide circumference by 3.1416

To Find Radius :

Multiply circumference by 0.15915

Or divide circumference by 6.28318

To Find Side of an Inscribed Square :

Multiply diameter by 0.7071

Or multiply circumference by 0.2821

Or divide circumference by 4.4428

To Find Side of an Equal Square :

Multiply diameter by 0.8862

Or divide diameter by 1.1284

Or multiply circumference by 0.2821

Or divide circumference by 3.545

Square

A side multiplied by
1.4142 equals diameter of its circumscribing circle.

A side multiplied by
4.443 equals circumference of its circumference circle

A side multiplied by
1.128 equals diameter of an equal circle

A side multiplied by
3.547 equals circumference of an equal circle.

Square inches multiplied by
1.273 equals square inches of an equal circle.

To Find the Area of a Circle :

Multiply circumference by one-quarter of the diameter.

Or multiply the square of diameter by 0.7845

Or multiply the square of circumference by 0.7958

Or multiply the square of 1/2 diameter by 3.1416

To Find the Surface of a Sphere of Globe :

Multiply the diameter by the circumference

Or multiply the square of diameter by 3.1416

Or multiply four times the square of radius by 3.1416

Tapers and Angles

Taper per Foot	*Included Angle*			*Angle With Centre Line*			*Taper per inch*	*Taper per inch from Centre Line*
	Deg.	*Min.*	*Sec.*	*Deg.*	*Min.*	*Sec.*		
1/8	0	35	47	0	17	54	0.010416	0.005208
3/16	0	53	44	0	26	52	0.015625	0.007812
1/4	1	11	38	0	35	49	0.020833	0.010416
5/16	1	29	31	0	44	46	0.026042	0.013021
2/8	1	47	25	0	53	42	0.031250	0.015625
7/16	2	5	18	1	2	39	0.036458	0.018229
1/2	2	23	12	1	11	36	0.041667	0.020833
9/16	2	41	7	1	20	34	0.046875	0.023438
5/8	2	59	3	1	29	31	0.052084	0.026042
11/16	3	16	56	1	38	28	0.057292	0.028646
3/4	3	34	48	1	47	24	0.062500	0.031250
13/16	3	52	42	1	56	21	0.067708	0.033854
7/8	4	10	32	2	5	16	0.072917	0.036456
15/16	4	28	26	2	14	13	0.078125	0.039063
1	4	46	19	2	23	10	0.083330	0.041667
11/4	5	57	45	2	58	53	0.104166	0.052084
11/2	7	9	10	3	34	35	0.125000	0.062500
13/4	8	20	28	4	10	14	0.145833	0.072917
2	9	31	37	4	45	49	0.166666	0.083332
21/2	11	53	38	5	56	49	0.208333	0.104166
3	14	15	0	7	7	30	0.250000	0.125000
31/2	16	35	39	8	17	49	0.291666	0.145833
4	18	55	31	9	27	44	0.333333	0.166666
41/2	21	14	20	10	37	10	0.375000	0.187500
5	23	32	12	11	46	6	0.416666	0.208333
6	28	4	20	14	2	10	0.500000	0.250000

Decimal Equivalents of Millimeters

mm.	*Inches*	*mm.*	*Inches*	*mm.*	*Inches*	*mm.*	*Inches*	*mm.*	*Inches*
0.01	0.00039	0.41	0.01614	0.81	0.03189	21	0.82677	61	2.40157
0.02	0.00079	0.42	0.01664	0.82	0.03228	22	0.86614	62	2.44094
0.03	0.00118	0.43	0.01693	0.83	0.03268	23	0.90551	63	2.48031
0.04	0.00157	0.44	0.01732	0.84	0.03307	24	0.94488	64	2.51968
0.05	0.00197	0.45	0.01772	0.85	0.03346	25	0.98425	65	2.55905
0.06	0.00236	0.46	0.01811	0.86	0.03386	26	1.02362	66	2.59842
0.07	0.00276	0.47	0.01850	0.87	0.03425	27	1.06299	67	2.63779
0.08	0.00315	0.48	0.01890	0.88	0.03465	28	1.10236	68	2.67716
0.09	0.00354	0.49	0.01929	0.89	0.03504	29	1.14173	69	2.71653
0.10	0.00394	0.50	0.01969	0.90	0.03543	30	1.18110	70	2.75590
0.11	0.00433	0.51	0.02008	0.91	0.03583	31	1.22047	71	2.79527
0.12	0.00472	0.52	0.02047	0.92	0.03622	32	1.25984	72	2.83464
0.13	0.00512	0.53	0.02087	0.93	0.03661	33	1.29924	73	2.87401
0.14	0.00551	0.54	0.02126	0.94	0.03701	34	1.33858	74	2.91338
0.15	0.00591	0.55	0.02165	0.95	0.03740	35	1.37795	75	2.95275
0.16	0.00630	0.56	0.02205	0.96	0.03780	36	1.41732	76	2.99212

(Contd.)

mm.	*Inches*	*mm.*	*Inches*	*mm.*	*Inches*	*mm.*	*Inches*	*mm.*	*Inches*
0.17	0.00669	0.57	0.02244	0.97	0.03819	37	1.45669	77	3.03149
0.18	0.00709	0.58	0.02283	0.98	0.03858	38	1.49606	78	3.07086
0.19	0.00748	0.59	0.02323	0.99	0.03898	39	1.53543	79	3.11023
0.20	0.00787	0.60	0.02362	1.00	0.03937	40	1.57480	80	3.14960
0.21	0.00827	0.61	0.02402	1	0.03937	41	1.61417	81	3.18897
0.22	0.00866	0.62	0.02441	2	0.07874	42	1.65254	82	3.22834
0.23	0.00906	0.63	0.02480	3	0.11811	43	1.69291	83	3.26771
0.24	0.00945	0.64	0.02520	4	0.15748	44	1.73228	84	3.30708
0.25	0.00984	0.65	0.02559	5	0.19685	45	1.77165	85	3.34643
0.26	0.01024	0.66	0.02598	6	0.23622	46	1.81102	86	3.38382
0.27	0.01063	0.67	0.02638	7	0.27559	47	11.85039	87	3.42519
0.28	0.01102	0.68	0.02677	8	0.31496	48	1.88976	88	3.46456
0.29	0.01142	0.69	0.02717	9	0.35433	49	1.92913	89	3.50393
0.30	0.01181	0.70	0.02756	10	0.39370	50	1.96850	90	3.54330
0.31	0.01220	0.71	0.02795	11	0.43307	51	2.00787	91	3.58267
0.32	0.01260	0.72	0.02835	12	0.47244	52	2.04724	92	3.62204
0.33	0.01299	0.73	0.02874	13	0.51181	53	2.08661	93	3.66141
0.34	0.01339	0.74	0.02913	14	0.55118	54	2.12598	94	3.70078
0.35	0.01378	0.75	0.02953	15	0.59055	55	2.16539	95	3.74015
0.36	0.01417	0.76	0.02992	16	0.62992	56	2.20472	96	3.77952
0.37	0.01457	0.77	0.03032	17	0.66929	57	2.24409	97	3.81889
0.38	0.01496	0.78	0.03071	18	0.70866	58	2.28346	98	3.85826
0.39	0.01535	0.79	0.03110	19	0.74803	59	2.32283	99	3.89763
0.40	0.01575	0.80	0.03150	20	0.78740	60	2.36220	100	3.93700

Decimal Equivalents of 8ths, 16ths, 32nds, 64ths

8ths	*32nds*	*64ths*	*64ths*
1/8 = 0.123	1/32 = 0.03125	1/64 = 0.015625	11/64 = 0.515625
1/4 = 0.250	1/32 = 0.09375	1/64 = 0.046875	25/64 = 0.546875
3/8 = 0.375	5/32 = 0.15625	5/64 = 0.078125	37/64 = 0.578125
1/2 = 0.500	7/32 = 0.21875	7/64 = 0.109375	39/64 = 0.609375
5/8 = 0.625	9/32 = 0.28125	9/64 = 0.140625	41/64 = 0.640625
3/4 = 0.750	11/32 = 0.34375	11/64 = 0.171875	43/64 = 0.671875
7/8 = 0.875	13/32 = 0.40625	13/64 = 0.203125	45/64 = 0.703125
16ths	15/32 = 0.46875	15/64 = 0.234375	47/64 = 0.734375
1/16 = 0.0625	17/32 = 0.53125	17/64 = 0.265625	49/64 = 0.765625
3/16 = 0.1875	19/32 = 0.59375	19/64 = 0.296875	51/64 = 0.796875
5/16 = 0.3125	21/32 = 0.65625	21/64 = 0.328125	53/64 = 0.828125
7/16 = 0.4375	23/32 = 0.71875	23/64 = 0.359375	55/64 = 0.850375
9/16 = 0.5625	25/32 = 0.78125	25/64 = 0.390625	57/64 = 0.890625
11/16 = 0.6875	27/32 = 0.84375	27/64 = 0.421875	59/64 = 0.921875
13/16 = 0.8125	29/32 = 0.90625	29/64 = 0.453125	61/64 = 0.953125
15/16 = 0.9375	31/32 = 0.96875	31/64 = 0.484375	63/64 = 0.984375

Decimal Equivalents of 7ths, 14ths, and 28ths

7th	*14th*	*28th*	*Decimal*	*7th*	*14th*	*28th*	*Decimal*
1		1	0.035714			15	0.535714
	1		0.071420	4			0.571429
		3	0.107143			17	0.607143
1	3		0.142857		9		0.642867
		5	0.178571			19	0.678571
			0.214286	5			0.714286
		7	0.25			21	0.75
			0.285714		11		0.785714
2		9	0.321429			23	0.821429
	5		0.337143	6			0.857143
		11	0.392857			25	0.892857
			0.428571		13		0.928571
3		13	0.464286			27	0.964286
	7		0.5				

Decimal Equivalents of 6ths, 12ths, and 24ths

6th	*12th*	*24th*	*Decimal*	*6th*	*12th*	*24th*	*Decimal*
	1	1	0.041667	3			0.5
			0.083333			13	0.541666
1		3	0.125		7		0.583333
			0.166666			15	0.625
		5	0.208333	4			0.666666
	3		0.25			17	0.708333
		7	0.291666		9		0.75
2			0.333333			19	0.791666
		9	0.375				0.833333
	5		0.416666			21	0.875
		11	0.458333		11		0.916666
						23	0.958333

Miscellaneous Measurements

Measures of Length

1 mile = 1760 yards = 5280 feet.
1 yard = 3 feet = 36 inches.
1 foot = 12 inches.

The following measures of length are also used occasionally :

1 mil = 0.001 inch, 1 fathom = 2 yards = 6 feet
1 rod = 5.5 yards = 16.5 feet, 1 hand = 4 inches, 1 span = 9 inches.

Surveyor's Measure

1 mile = 8 furlong = 80 chains
1 furlong = 10 chains = 220 yards
1 chain = 4 rods = 22 yards = 66 feet = 100 links.
1 link = 7.92 inches

Nautical measure

1 league = 3 nautical miles.

1 nautical mile = 6080.26 feet = 1.1516 statute mile.

One degree at the equator = 60 nautical miles = 69.096 statutemiles

360 degree = 21,600 nautical miles = 24,874.5 statue miles

= circumference of earth at the equator

Square Measure

1 square mile = 640 acres = 6400 square chains.

1 acre = 10 square chains = 4840 square yards = 43,560 square feet

1 square chain = 16 square rods = 484 square yards = 4356 square feet.

1 square rod = 30.25 square yards = 272.25 square feet = 625 square links.

1 square yard = 9 square feet.

1 square feet = 144 square inches.

An acre is equal to square, the side of which is 208.7 feet.

Metric Conversion Table

Millimetres	... =	0.03937	x Inches
Millimetres	... x	25.400	= Inches
Metres	... =	3.2809	x Feet
Metres	... x	0.3048	= Feet
Kilometres	... =	0.621377	x Miles
Kilometres	...x	1.6093	= Miles
Square centimetres	... =	0.15500	x Square inches
Square centimetres	... x	6.4515	= Square inches
Square metres	... =	10.76410	x Square feet
Square metres	... x	0.09290	= Square feet
Square kilometres	... =	247.1098	x Acres
Square kilometres	... x	0.00405	= Acres
Hectares	... =	2.471	x Acres
Hectares	... x	0.4047	= Acres
Cubic centimetres	... =	0.061025	x Cubic inches
Cubic centimetres	... x	16.3867	= Cubic inches
Cubic metres	... =	35.3156	x Cubic inches
Cubic metres	... x	0.02832	= Cubic feet
Cubic metres	... =	1.308	x Cubic yards
Cubic metres	... x	0.765	= Cubic yards
Litres	... =	61.023	x Cubic inches
LItres	... x	0.01689	= Cubic inches
Litres	... =	0.026418	x U.S. gallon
Litres	... x	3.7854	= U.S. gallon
Grams	... =	15.4324	x Grains
Grams	... x	0.0648	= Grains
Grams	... =	0.03527	x Ounces, avoirdupois
Grams	... x	28.3495	= Ounces, avoirdupois
Kilograms	... =	2.2046	x Pounds

Kilograms	... x	0.4536	= Pounds
Kilograms per square centimetres	... =	14.2234	x Pounds per square inch
Kilograms per cubic centimetres	... x	0.0703	= Pounds per square inch
Kilograms per cubic metre	... =	0.06243	x Pounds per cubic foot
Kilograms per cubic metre	... x	16.01890	= Pounds per cubic foot
Metric tonnes (1,000 kilograms)	... =	1.1023	x Tons (2,000 pounds)
Metric tonnes	... x	0.9072	= Tons (2,000 pounds)
Kilowatts	... =	1.3405	x Horse power
Kilowatts	... x	0.746	= Horse power
Calories	... x	3.9683	x B.T. units
Calories	... x	0.2520	= B.T. units

THE METRIC SYSTEM OF MEASUREMENT

Measures of Length

1 Millimetre (mm)	=	... 0.03937079 inch, or about 1/25 inch
10 Millimetre = 1 Centimetre (cm).	=	... 0.39377079 inch
10 Centimetres = 1 Decimetre (dm)	=	... 3.937079 inches
10 Decimetres = 1 Metre (M.)	=	... 39.37079 inches, 3.208992 feet, or 1.09361 yards
10 Metres = 1 Decametre (Dm.)	=	... 32.808922 feet
10 Decametres = 1 Hectometre (Hm.)	=	... 19.927817 rods
10 Hectometres = Kilometres (Km.)	=	... 1093.61 yards, 0.6213824 miles
10 Kilometre = 1 Myriametre (Mm.)	=	... 6.213824 miles
1 Inch = 2.54 cm, 1 foot = 0.3048 m	=	1 yard = 0.9144 m.

1 rod = 0.5029 Dm., 1 mile = 1.6093 km.

Measures of Weight

Gram (g.) = 15.4324874 gr. Troy, or 0.3214 oz. Troy or = ... 0.03527398 oz. avoidupois

10 Grams = 1 Decagram (Dg.)	=	... 0.3527398 oz. avoidupois
10 Decagrams = 1 Hectogram (Hg.)	=	... 3.527398 oz. avoidupois
10 Hectrograms = 1 Kilogram (Kg.)	=	... 2.2046125 lb.

1000 Kilograms = 1 Ton (T) = 2204.62125 lb or 1.1023 tons or 2000 lb. or 0.9842 ton of 2240 lb., or 19.68 cwt.

1 grain = 0.0648 g., 1 oz. avoidupois = 28.35 g., 1lb. = 0.4536 kg.,

1 ton (2000 lb.) = 0.9072 T.,

1 ton (2240 lb.) = 1.016 T., or 1016 kg.

Measures of Capacity

1 Litre (L.) = 1 cubic decimetre = 61.0270515 cubit in., or 0.03531 cu. ft., or 1.0567 liquids qts., or 0.908 dry qt., or 0.26417 Amer. gal

10 Litres = 1 Decalitre (Dl.) = 2.6417 gal., or 1.135 pk.

10 Decalitres = 1 Hectolitre (Hl.) = 2.8375 bu.

1 cu. foot = 28.317 Litres, 1 gallon (American) = 3.785 Litres,

1 gallon (British) = 4.543 Litres.

APPENDIX-II

INDIAN STANDARD SPECIFICATIONS ON ENGINEERING METROLOGY

IS : 919—1963 : *Recommendations for limits and fits for engineering*. It covers recommendations relating to limits of size of parts of components and to fits constituted by their assembly.

IS : 2092—1962 : *Dial Gauges*. It relates to two types of dial gauges for general engineering purpose manufactured in two grades with total plunger movement or lift of 3.5 and 10 mm. It also gives the requirements of dimensions and permissible error for dial gauges, their testing and notes on the care in use and their selection. Recommended requirements of high precision dial gauges have also been covered.

IS : 2103—1972 : *Engineer's Squares*. It specifies dimensions, manufacture, workmanship and finish, and accuracy requirements of fixed blade engineer's squares made in three grades of accuracy and of sizes upto 100 mm suitable for workshop and inspection.

IS : 2186—1976 : *Dimensions for External Interference Threads*. It gives the dimensions for external interference of threads in the diameter range 1 to 150 mm.

IS : 2220—1962 : *Steel straight Edges*. It lays down material, accuracy and other requirements, such as dimensions, workmanship and finish of two grades of steel straight edges of sizes 100 to 2000 mm. for use on surface plates and in testing machining tools. The procedure or testing the straightness of the straight edges is also given.

IS : 2251—1965 : *Plug and Ring Gauges for Self-holding Tapers*. It prescribes the requirements regarding material, hardness, dimensions, workmanship and finish of plug and ring gauges for the verification of internal and external self-holding tapers conforming to IS : 1715—1963. The basis of design and recommendations for use of plug and ring gauges are also given.

IS : 2285—1963 : *Cast Iron Surface Plates*. It specifies requirements regarding material, dimensions, accuracy, designations, workmanship and finish of two grades of rectangular and square surface plates made of cast iron, used for inspection and marking purposes.

IS : 2334—1975 : Gauging practice for ISO metric screw threads.

IS : 2554—1963 : *Cast Iron Angle Plates*. It specifies the material and dimensions of two grades of cast angle plates and gives recommendations for box angle plates.

IS : 2909—1964 : *Guide for the Selection of Fits*. It gives recommendations for the selection of fits in engineering problems concerned with the mating of a shaft and a hole. The recommendations given in this specification are also applicable to non-cylindrical parts.

IS : 2921—1964 : *Vernier Height Gauges*. It is intended primarily to indicate standards of accuracy relating to vernier height gauges. The measuring range is up to a metre. The graduations on the beam shall be one of the types given. Diagrams have been included for purpose of illustration only.

IS : 2949—1964 : *V-block*. It specifies the sizes and accuracies of two grades of V-blocks which are applicable to either single of V-blocks or matched pairs of both grades. It recommends only 90° V-blocks.

IS : 2966—1964 : *Internal Micrometer (including stick micrometers)*. It deals with the internal micrometers of a measuring range upto 1000 mm, comprising a measuring head, extension rods with or without spacing collars and in smaller sizes, a handle. It also deals with stick micrometers which are used for measurement of longer internal lengths.

IS : 2967—1964 : *External Micrometers*. It deals with external micrometers of range of 25 mm and unit of graduation of 0.01 mm, used in general engineering practice.

IS : 2984—1966 : *Slip Gauges*. It specifies dimensions and accuracies of slip gauges of three grades (0, 1 and 2) and also their flatness parallelism and hardness. It also gives notes on the care in use of slip gauges and recommended set for general use.

IS : 3073—1967 : *Assessment of Surface Roughness*. It relates to terminology and parameters to be employed in statements relating to surface roughness, preferred values of surface roughness for grading of surfaces, standard sampling lengths to be used in graphical procedure and instrument construction, instruments and methods to be employed in the quantitative assessment of surface roughness and information to be given in statements relating to surface roughness.

IS : 3139—1966 : *Dimensions for Screw Threads for Bolts and Nuts (diameter range M 42 to M 150)*. It deals with basic and design profile as well as dimensions, allowances, tolerance had designation of screw threads for bolts and nuts of normal diameter range 42 to 140 mm. It also specifies tolerances for one tolerance class. 6g for bolt threads and 6H for nut threads.

IS : 3179—1965 : *Feeler Gauges (0.03 mm. to 1 mm.)*. It applies to feeler gauges, comprising a series of blades of graded thickness, assembled in a protective sheath, from 0.03 to 1 mm.

IS : 3455—1966 : *Tolerances for Plain Limit Gauges*. It covers simple forms of limit gauges, such as cylindrical plug, ring gauges and snap gauges for gauging dimensions from 1 to 500 mm. It also prescribes allowances for wear, and requirements for hardness and surface roughness.

IS : 3477—1973 : *Snap Gauges*. It prescribe material, shapes and dimensions and other requirements for snap gauges.

IS : 3484—1966 : *Plain Plug Gauges.* It prescribes material, dimension and general requirements of plain plug gauges. This standard has been revised and replaced by IS : 6137—1971 and IS : 6138—1971 which deal with gauging members for plain gauges 'Go member' (size range 1 to 30 mm) and 'Gauging members' for plain plug gauges 'No Go member' (diameter range 1 to 30 mm) respectively.

IS : 3485—1972 : *Plain Ring Gauges.* It prescribes material dimensions and general requirements and method of designing plain ring gauges.

IS : 3510—1966 : *Toolmaker's Flats and High Precision Surface Plates.* It prescribes material, general design and dimension and accuracy requirements.

IS : 3512—1966 : *Toolmaker's Straight, Edges.* It prescribes the material, dimensions, accuracy and general requirements.

IS : 3651—1976 : *Vernier Calipers.* It prescribes requirements in respect to material, hardness, dimensions, graduations, accuracy and designation.

IS : 4019—1967 : *Firm-Joint Dividers.* It specifies material for legs, screw, nut and washer ; shape and dimensions, heat treatment and hardness, construction and workmanship, finish, functional test and joint bend test for firm-joint dividers of 75, 100, 150, 200 and 300 mm nominal sizes. It also gives the detailed consideration to those parts which directly affect the quality, functioning and life of the complete instrument.

IS : 4052—1967 : *Spring Calipers.* It covers six sizes of spring calipers, *viz.*, 75, 100, 150, 200, 250 and 300 mm. It specifies material for legs, spring another components, shape and dimensions, heat treatment and hardness, construction and workmanship, finish functional test, marking and packing etc.

IS : 4083—1967 : *Spring Dividers.* It gives detailed consideration to those parts which effect the quality, functioning and life of the complete instrument. It also specifies material for legs, springs and other components, shapes and dimensions, heat treatment and hardness ; construction and workmanship, finish functional and joint tests, marking and packing particulars for 75, 100, 150, 200, 250 and 300 mm spring dividers.

IS : 4189—1967 : *Firm-joint Inside and Outside Calipers.* It prescribes requirements for firm-joint inside and outside calipers of size 100, 150, 200 and 300 mm. It also specifies material, hardness, shape and dimensions, designation, construction and workmanship, finish, functional test, marking and packing.

IS : 4210—1967 : *Strip Feeler Gauges for Electrical Purposes.* This standard is applicable to strip feeler gauges used for measuring air-gaps in electrical machinery and comprising a series of gauging blades of graded thickness form 0.1 to 1.0 mm and 1.0 to 20 mm. It also specifies the material dimensions and tolerances and general requirements.

IS : 4211—1967 : *Thread Pitch Gauges for ISO Metric Screw Threads (pitchgerane 0.25 to 6.0 mm).* It specifies the requirements for thread-pitch gauges consisting of an assembly of 24 blades which help to identify the thread pitch of ISO metric screw threads in the pitch range 0.25 to 6.00 mm, and a blade having ISO profile 60° (included angle) assembled in a protective sheath.

IS : 4213—1967 : *Vernier Depth Gauges*. It prescribes requirements of vernier depth gauges of short depth of 200, 300 and 400 mm, reading to 0.05 mm. It also specifies the procedure for testing of reading accuracy, flatness of measuring face and thickness of graduation lines.

IS : 4218 : ISO metric screw threads.

IS : 4218 (Part I)–1967 : *Basic and Design Profiles*. It specifies basic and design profiles per ISO metric screw threads.

IS : 4218 (Part II)–1967 : *Pitch Diameter Combinations*. It specifies a series of diameter and pitch combinations for ISO metric screw thread in the diameter range 1 to 300 mm.

IS : 4218 (Part III)–1967 : *Basic Dimensions for Design Profiles*. It specifies basic dimensions for design profiles for ISO metric screw threads in the diameter range 1 to 300 mm.

IS : 4218 (Part IV)–1967 : *Tolerancing System*. It specifies tolerances for ISO metric screw threads for the diameter range 1 to 300 mm. The tolerance values have been tabulated for the normal length of engagement only.

IS : 4218 (Part V)–1967 : *Tolerances*. It tabulates tolerance for ISO metric screw threads for the tolerance class covered in Part IV of the standard, in the diameter range 1 to 300 mm. The tolerances have been arrived at based on the tolerancing system specified in Part IV of the standards.

IS : 4218 (Part VI)–1967 : *Limits of Sizes for Commercial Bolts and Nuts (Diameter range 1 to 39 mm)*. It specifies limits of sizes for ISO metric screw threads in the diameter range 1 to 39 mm for commercial bolts and nuts.

IS : 4239—1970 : *Mechanical Bevel Protractors*. It prescribes requirements for mechanical and optical bevel protractors for general engineering measurement purposes.

IS : 4241—1967 : *Engineer's Parallels*. It covers three types of engineer's parallel, *viz*., solid type, box type and adjustable type. There are two grades in solid type, namely, Grade A for reference purposes and Grade B for workshop purposes. It also specifies the material hardness and construction requirements.

IS : 4290—1968 : *Roughness Comparison Specimens*. It prescribes requirements for roughness comparison specimen of turned, ground, bored, milled, shaped and planed surfaces which are intended for tactile and visual comparison with workshop surfaces of similar lay, produced by similar machining methods.

IS : 4349—1967 : *Precision rollers*. It specifies four types of precision cylindrical rollers ranging from 1 to 120 mm used in precision engineering measurements. It also specifies dimensions, finish, surface roughness designation etc.

IS : 4440—1967 : *Slip Gauge Accessories*. It specifies material, design dimensions, accuracy and finish, and gives the nomenclature for measuring jaws, scribing and centre points and holder and base.

IS : 4492—1968 : *Welded V-blocks*. It deals with the requirements of welded V-blocks for cylindrical workpieces in the diameter range of 200 to 300 mm.

IS : 4537—1968 : *Test Chart for Capstan and Turret Lathes*. It provides a convenient basis for the preparation of a performa for preparing test certificate. It also prescribes limits of accuracy and procedures for carrying out the tests.

IS : 4563—1968 : *Block Squares*. It specifies requirements regarding material, dimensions, tolerances, design features, and finish for engineer's precision blocks, squares of the solid and hollow types.

IS : 4583—1968 : *Plate Snap Gauges*. It specifies shapes and overall dimensions of plate snap gauges in the size range over 2 to 250 mm of both double-ended and single-ended progressive type. It also specifies for the double-ended type plate snap gauges in the size range over 2 to 100 mm and single-ended progressive type plate snap gauges in the size range over 100 to 250 mm.

IS : 4960—1968 : *Universal and Elongated Type V-blocks*. It prescribes requirements regarding material, dimensions, hardness, finish etc., universal type of V-blocks of size 100, 160, 200, 300 mm (workpiece size range from 8 to 300 mm) and elongated type V-blocks of sizes 40, 50, 63, and 100 mm (workpiece size range from 5 to 100 mm). Also applies to either single V-blocks or machined pairs of sane accuracy.

IS : 5049—1969 : *Gauge for Wire Diameters*. It gives details of gauges to check the diameter or wires in range 0.125 to 8 mm as per IS : 280. The gauges is meant to determine conformance or otherwise to the size of wires. Does not indicate the actual diameter.

IS : 5268—1969 : *Camel Back Type Straight Edge*. It lays down material, accuracy and other requirements for two grades of camel back type straight edges of sizes 300 to 5000 mm for use on surface plates, testing machine tools etc.

IS : 5273—1969 : *Radius Gauges*. It prescribes requirements for radius gauges comprising a series of graded gauging blades for checking concave and convex radii in the range 0.6 to 25 mm assembled in a protective sheath.

IS : 5359—1969 : *Sine Bars*. It gives the essential requirements of sine bars of three size, for example, 100, 200 and 300 mm (these being the distances between the axis of the rollers).

IS : 5388—1969 : *Handles for Plug Gauges, Plain and Threaded*. It specifies handles for plug gauges, plain and threaded in the diameter range 1 to 300 mm.

IS : 5420 (Part I)—1962 : *Principles and Applications*. It explains the various concepts underlying the precision of test methods. It gives procedures to calculate repeatability and reproducibility of test methods. It also lays down the procedures for determining acceptability of the test results and for comparing the test methods on the basis of their precision. Finally, it recommends the procedures for writing the specification limits taking into account the precision of the test methods.

IS . 5440—1969 : *Optical Flats*. It gives the requirements for optical flats which are used for determination of flatness and parallelism of precision gauges.

IS : 5812—1970 : *Optical Bevel Protractors.* It prescribes requirements of optical bevel protractors for general engineering purposes.

IS : 5939—1970 : *Since Angle Sine Tables.* It prescribes the requirements of sine angle sine tables used in engineering metrology.

IS : 5943—1970 : *Compound Angle Sine Tables.* It specifies requirements of compound angle sine tables used in engineering metrology.

IS : 5979—1970 : *Sine Centres.* It specifies the requirements for sine centres with distance of roller 200 and 300 mm.

IS : 5980—1970 : *Bench Centres.* It specifies requirements for bench centres for checking run-out, concentricity, etc.

IS : 6231—1971 : *Prismatic Angle Gauges.* It prescribes requirements of prismatic angle gauges for building any angle from 0 to 360° in steps of 6 seconds.

IS : 6232—1971 : Cast iron box angle plates.

IS : 397—1952 : *Methods for Statistical Quality Control During Production by the Use of Control Chart (Tentative with Amendment No. 1).* It outlines the control chart method of controlling quality during production. The principles and procedures have been illustrated with practical examples to show how they can be applied to specific cases. Guidance has been provided on choice of quality characteristic, collection of data, selection of statistical measure and types of control charts. Simplified procedures are given to help in establishing and operating the control charts. Computation involves required knowledge of simple arithmetic.

IS : 1548—1960 : *Manual on Basic Principles of Lot Sampling.* It explains the various statistical concepts underlying the sampling inspection schemes. Lays down some of the basic principles of sampling lot by lot. The manual deals with sampling inspection techniques as a means to solve two types of problems, namely (*a*) estimation of lot quality ; and (*b*) lot acceptance starting from classification and specification of quality ; the manual elaborates on the merits and demerits of sampling inspection as compared to hundred per cent inspection. Some broad guide-lines on the principles and planning of sampling inspection are also given. At the end, tables and illustrative examples are provided for a variety of solutions encountered in the point and interval estimates for parameters of lot quality.

IS : 2500—1974 : *Sampling Inspection Tables.*

IS : 2500 (Part I)—1974 : *Inspection by Attribute and by Count of Defects (With Amendment No. 1).* It provides tables for sampling inspection plans for lot by lot inspection when the inspection is done by attributes or by count of defects. Broad details relating to the preliminaries of the acceptable sampling plans are also given. Ready-made tables for different types of sampling plans such as single, double, multiple, classified in terms of acceptance quality level (AQL) are expected to meet most of the practical situations. Rules for changing from normal inspection to reduced or tightened inspection and *vice versa* are also clearly indicated. Besides, formulae are given for the construction of one's own sequential sampling plans. The illustrative examples at end, as also the operating characteristic (OC) curves, can be of immense help in understanding the operating of the plans together with the protection afforded by them.

IS : 2500 (Part II)–1965 : *Inspection by Variable for Per cent Defective.* It provides tables for single sampling plans for lot by lot inspection when the inspection is done by variables. After listing the various considerations for the choice between the attributes and variable inspection, the standard gives some broad details relating to the preliminaries in the selection of variable inspection plans. Readymade tables for single sampling plans classified in terms of the acceptable quality level (AQL) are also provided for the three situations, *viz.*, variability known, variability unknown (standard deviation method) and variability unknown (range method). The application of the plans to one sided and two sided specification limits are dealt with separately. Rules for changing from normal to reduced or tightened inspections and *vice versa* are also clearly indicated. The standard also gives formulae and necessary tables for the construction on one's own sampling variable plans. The illustrative example given at the end, as also the operating characteristic (OC) curves, can be of immense help in understanding the operation of the plans along with the risk associated with them.

IS : 4905—1968 : *Methods for Random Sampling.* It lays down general procedure for the selection items from lot on an objective basis using different random sampling techniques, such as simple random sampling, stratified sampling, systematic sampling, cluster sampling and two stage sampling. It also prescribes the methods for calculation of sample estimates like mean and proportion of defectives from the sample data. The method of random selection as well as calculation of the estimates are illustrated with plenty of examples. A total of 30,000 random digits are also appended for practical use.

IS : 5002—1969 : *Methods for Determining the Sample Size to Estimate the Average Quality of a Lot or Process.* It provides methods for calculating the sample size of estimate the average quality of a lot or process with a specified limit of error and probability level when the quality characteristic is of attribute as well as variable type. Methods are given to make use of available empirical knowledge of past data as also the case of nonavailability of such an information is considered. Method for arriving at the sample size in case of two stage sampling is also included.

APPENDIX-III

REFERENCES FOR FURTHER READING

Title of book	*Author*	*Publishers*
Handbook of Industrial Metrology	A.S.T.M.E.	Prentice Hall
Engineering Inspection	A.C. Parkinson	Pitman
Engineering Metrology	K.J. Hume	Macdonald
Practical Metrology	K.J. Hume and G.H. Sharp	—do—
Handbook No. 2 British Standards for Workshop Practice		British Standards Institution
Engineering Precision Measurements	A.W. Judge	Chapman and Hall
Practical Engineering Metrology	K.W.B. Sharp	Pitman
Metrology and Gauging	S.A.J. Parsons	Macdonald
Fundamentals of Mechanical Inspection	R. Jenkins	McGraw Hill
Precision Workshop Methods	H.J. Davis	Arnold
Engineering Dimensional Metrology	L. Miller	Arnold
Fundamentals of Dimensional Metrology	T. Busch	Wilkie Bros.
Engineering Inspection, Measurement and Testing	H.C. Town and R. Colebourne	Odhams
Inspection and Gauging	C.W. Kennedy	Machinery Pub. Co.
Symposium on Dimensional Metrology	N.P.L.	H.M.S.O. 1953

Index

D

T